(continued on back endsheets)

College Algebra and Calculus

An Applied Approach

RON LARSON

The Pennsylvania State University
The Behrend College

ANNE V. HODGKINS

Phoenix College

with the assistance of

DAVID C. FALVO

The Pennsylvania State University
The Behrend College

BROOKS/COLE
CENGAGE Learning™

Australia • Brazil • Japan • Korea • Mexico • Singapore • Spain • United Kingdom • United States

BROOKS/COLE
CENGAGE Learning™

College Algebra and Calculus: An Applied Approach
Ron Larson and Anne V. Hodgkins

VP/Editor-in-Chief: Michelle Julet

Publisher: Richard Stratton

Senior Sponsoring Editor: Cathy Cantin

Associate Editor: Jeannine Lawless

Editorial Assistant: Amy Haines

Associate Media Editor: Lynh Pham

Senior Content Manager: Maren Kunert

Marketing Specialist: Ashley Pickering

Marketing Assistant: Angela Kim

Marketing Communications Manager:
    Mary Anne Payumo

Associate Content Project Manager, Editorial Production:
    Jill Clark

Art and Design Manager: Jill Haber

Senior Manufacturing Buyer: Diane Gibbons

Senior Rights Acquisition Account Manager: Katie Huha

Text Designer: Jean Hammond

Art Editor: Larson Texts, Inc.

Senior Photo Editor: Jennifer Meyer Dare

Illustrator: Larson Texts, Inc.

Cover Designer: Sarah Bishins

Cover Image: © Vlad Turchenko, istockphoto

Compositor: Larson Texts, Inc.

For product information and technology assistance, contact us at **Cengage Learning Customer & Sales Support, 1-800-354-9706**

For permission to use material from this text or product, submit all requests online at **www.cengage.com/permissions.** Further permissions questions can be emailed to **permissionrequest@cengage.com**

Library of Congress Control Number: 2008929478

ISBN-13: 978-0-547-16705-3

ISBN-10: 0-547-16705-9

Brooks/Cole
10 Davis Drive
Belmont, CA 94002
USA

Cengage learning is a leading provider of customized learning solutions with office locations around the globe, including Singapore, the United Kingdom, Australia, Mexico, Brazil, and Japan. Locate your local office at: **international.cengage.com/region**

Cengage learning products are represented in Canada by Nelson Education, Ltd.

For your course and learning solutions, visit **www.cengage.com**

Purchase any of our products at your local college store or at our preferred online store **www.ichapters.com**

Printed in Canada
1 2 3 4 5 6 7 12 11 10 09 08

Contents

*Available online at the text's companion website.

A Word from the Authors

Welcome to the first edition of *College Algebra and Calculus: An Applied Approach*! This textbook completes the publication of a whole series of textbooks tailored to the needs of college algebra and applied calculus students majoring in business, life science, and social science courses.

> *College Algebra with Applications for Business and the Life Sciences*
> *Calculus: An Applied Approach,* Eighth Edition
> *Brief Calculus: An Applied Approach,* Eighth Edition
> *Applied Calculus for the Life and Social Sciences*
> **College Algebra and Calculus: An Applied Approach**

Many students take college algebra as a prerequisite for applied calculus. We wrote all of these books using the same design, writing style, and pedagogical features, with the goal of providing these students with a level of familiarity that encourages confidence and a smooth transition between the courses. Additionally, by combining the college algebra and applied calculus material into one textbook, we have given students one comprehensive resource for both courses.

We're excited about this new textbook because it acknowledges where students are when they enter the course—and where they should be when they complete it. We review the basic algebra that students have studied previously (in Chapter 0 and in the exercises, notes, study tips and algebra review notes throughout the text), *and* present solid college algebra and applied calculus courses that balance understanding of concepts with the development of strong problem-solving skills.

In addition, emphasis was placed on providing an abundance of real-world problems throughout the textbook to motivate students' interest and understanding. Applications were taken from news sources, current events, government data, and industry trends to illustrate concepts and show the relevance of the math.

We hope you and your students enjoy *College Algebra and Calculus: An Applied Approach*. We are excited about this new textbook program because it helps students learn the math in the ways we have found most effective for our students — **by practicing their problem-solving skills and reinforcing their understanding in the context of actual problems they may encounter in their lives and careers.**

Please do tell us what you think. Over the years, we have received many useful comments from both instructors and students, and we value these comments very much.

Ron Larson

Ron Larson

Anne Hodgkins

Anne V. Hodgkins

Acknowledgments

Thank you to the many instructors who reviewed *College Algebra with Applications for Business and the Life Sciences, Calculus: An Applied Approach* Eighth Edition, and *Brief Calculus: An Applied Approach* Eighth Edition, and encouraged us to try something new. Without their help, and the many suggestions we've received throughout the previous editions of *Calculus: An Applied Approach*, this book would not have been possible. Our thanks also to Robert Hostetler, The Behrend College, The Pennsylvania State University, and Bruce Edwards, University of Florida, for their significant contributions to previous editions of this text.

Reviewers of *College Algebra with Applications for Business and the Life Sciences*

Michael Brook, *University of Delaware*
Tim Chappell, *Metropolitan Community College—Penn Valley*
Warrene Ferry, *Jones County Junior College*
David Frank, *University of Minnesota*
Michael Frantz, *University of La Verne*
Linda Herndon, *OSB, Benedictine College*
Ruth E. Hoffman, *Toccoa Falls College*
Eileen Lee, *Framingham State College*
Shahrokh Parvini, *San Diego Mesa College*
Jim Rutherfoord, *Chattahoochee Technical College*

Reviewers of the Eighth Edition of *Calculus: An Applied Approach*

Lateef Adelani, *Harris-Stowe State University, Saint Louis;* Frederick Adkins, *Indiana University of Pennsylvania;* Polly Amstutz, *University of Nebraska at Kearney;* Judy Barclay, *Cuesta College;* Jean Michelle Benedict, *Augusta State University;* Ben Brink, *Wharton County Junior College;* Jimmy Chang, *St. Petersburg College;* Derron Coles, *Oregon State University;* David French, *Tidewater Community College;* Randy Gallaher, *Lewis & Clark Community College;* Perry Gillespie, *Fayetteville State University;* Walter J. Gleason, *Bridgewater State College;* Larry Hoehn, *Austin Peay State University;* Raja Khoury, *Collin County Community College;* Ivan Loy, *Front Range Community College;* Lewis D. Ludwig, *Denison University;* Augustine Maison, *Eastern Kentucky University;* John Nardo, *Oglethorpe University;* Darla Ottman, *Elizabethtown Community & Technical College;* William Parzynski, *Montclair State University;* Laurie Poe, *Santa Clara University;* Adelaida Quesada, *Miami Dade College – Kendall;* Brooke P. Quinlan, *Hillsborough Community College;* David Ray, *University of Tennessee at Martin;* Carol Rychly, *Augusta State University;* Mike Shirazi, *Germanna Community College;* Rick Simon, *University of La Verne;* Marvin Stick, *University of Massachusetts – Lowell;* Devki Talwar, *Indiana University of Pennsylvania;* Linda Taylor, *Northern Virginia Community College;* Stephen Tillman, *Wilkes University;* Jay Wiestling, *Palomar College;* John Williams, *St. Petersburg College;* Ted Williamson, *Montclair State University*

How to get the most out of your textbook . . .

Establish a Solid Foundation

CHAPTER OPENERS

Each opener has an applied example of a core topic from the chapter. The section outline provides a comprehensive overview of the material being presented.

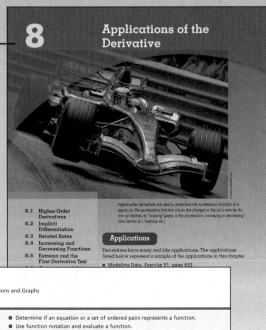

8 **Applications of the Derivative**

8.1 Higher-Order Derivatives
8.2 Implicit Differentiation
8.3 Related Rates
8.4 Increasing and Decreasing Functions
8.5 Extrema and the First-Derivative Test

Higher-order derivatives are used to determine the acceleration function of a sports car. The acceleration function shows the changes in the car's velocity. As the car reaches its "cruising" speed, is the acceleration increasing or decreasing? (See Section 8.1, Exercise 45.)

Applications

Derivatives have many real-life applications. The applications listed below represent a sample of the applications in this chapter.
■ Modeling Data, Exercise 51, page 633

SECTION OBJECTIVES

A bulleted list of learning objectives enables you to preview what will be presented in the upcoming section.

194   CHAPTER 2   Functions and Graphs

Section 2.4

Functions

■ Determine if an equation or a set of ordered pairs represents a function.
■ Use function notation and evaluate a function.
■ Find the domain of a function.
■ Write a function that relates quantities in an application problem.

DEFINITIONS AND THEOREMS

All definitions and theorems are highlighted for emphasis and easy recognition.

Definition of a Function

A **function** f from a set A to a set B is a rule of correspondence that assigns to each element x in the set A exactly one element y in the set B. The set A is the **domain** (or set of inputs) of the function f, and the set B contains the **range** (or set of outputs).

Vertical Line Test for Functions

A set of points in a coordinate plane is the graph of y as a function of x if and only if no vertical line intersects the graph at more than one point.

Example 7 **The Path of a Baseball** ®

A baseball is hit 3 feet above home plate at a velocity of 100 feet per second and an angle of 45°. The path of the baseball is given by the function

$$y = -0.0032x^2 + x + 3$$

where y and x are measured in feet. Will the baseball clear a 10-foot fence located 300 feet from home plate?

SOLUTION When $x = 300$, the height of the baseball is given by

$$y = -0.0032(300)^2 + 300 + 3 = 15 \text{ feet.}$$

The ball will clear the fence, as shown in Figure 2.42.

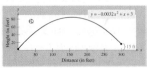

FIGURE 2.42

Notice that in Figure 2.42, the baseball is not at the point $(0, 0)$ before it is hit. This is because the original problem states that the baseball was hit 3 feet above the ground.

EXAMPLES

There is a wide variety of relevant examples in the text, each titled for easy reference. Many of the solutions are presented graphically, analytically, and/or numerically to provide further insight into mathematical concepts. Examples based on a real-life situation are identified with an icon ®.

Tools to Help You Learn and Review

CONCEPT CHECK

These noncomputational questions appear at the end of each section and are designed to check your understanding of the key concepts.

(CONCEPT CHECK)

1. Determine whether the following statement is true or false. Explain your reasoning.

   The points (3, 4) and (−4, 3) both lie on the same circle whose center is the origin.

2. Explain how to find the x- and y-intercepts of the graph of an equation.

3. For every point (x, y) on a graph, the point $(-x, y)$ is also on the graph. What type of symmetry must the graph have? Explain.

4. Is the point (0, 0) on the circle whose equation in standard form is $(x - 0)^2 + (y - 0)^2 = 4$? Explain.

✓ CHECKPOINT 4

Evaluate the function in Example 4 when $x = -3$ and 3. ▧

CHECKPOINT

After each example, a similar problem is presented to allow for immediate practice and to provide reinforcement of the concepts just learned.

STUDY TIP

When applying the properties of logarithms to a logarithmic function, you should be careful to check the domain of the function. For example, the domain of $f(x) = \ln x^2$ is all real $x \neq 0$, whereas the domain of $g(x) = 2 \ln x$ is all real $x > 0$.

STUDY TIPS

Scattered throughout the text, study tips address special cases, expand on concepts, and help you to avoid common errors.

Skills Review 2.7 The following warm-up exercises involve skills that were covered in earlier sections. You will use these skills in the exercise set for this section. For additional help, review Section 0.7.

In Exercises 1–10, perform the indicated operations and simplify the result.

1. $\dfrac{1}{x} + \dfrac{1}{1-x}$

2. $\dfrac{2}{x+3} - \dfrac{2}{x-3}$

3. $\dfrac{3}{x-2} - \dfrac{2}{x(x-2)}$

4. $\dfrac{x}{x-5} + \dfrac{1}{3}$

5. $(x-1)\left(\dfrac{1}{\sqrt{x^2-1}}\right)$

6. $\left(\dfrac{x}{x^2-4}\right)\left(\dfrac{x^2-x-2}{x^2}\right)$

7. $(x^2-4) \div \left(\dfrac{x+2}{5}\right)$

8. $\left(\dfrac{x}{x^2+3x-10}\right) \div \left(\dfrac{x^2+3x}{x^2+6x+5}\right)$

9. $\dfrac{(1/x)+5}{3-(1/x)}$

10. $\dfrac{(x/4)-(4/x)}{x-4}$

SKILLS REVIEW

These exercises at the beginning of each exercise set help you review skills covered in previous sections. The answers are provided at the back of the text to reinforce understanding of the skill sets learned.

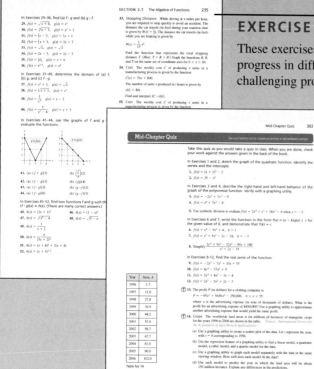

EXERCISE SETS

These exercises offer opportunities for practice and review. They progress in difficulty from skill-development problems to more challenging problems, to build confidence and understanding.

MID-CHAPTER QUIZ

Appearing in the middle of each chapter, this one-page test allows you to practice skills and concepts learned in the chapter. This opportunity for self-assessment will uncover any potentially weak areas that might require further review of the material.

CHAPTER TEST

Appearing at the end of each chapter, this test is designed to simulate an in-class exam. Taking these tests will help you to determine what concepts require further study and review.

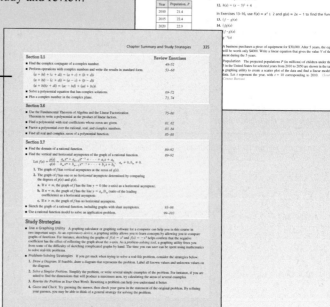

CHAPTER SUMMARY AND STUDY STRATEGIES

The *Summary* reviews the skills covered in the chapter and correlates each skill to the Review Exercises that test the skill. Following each *Chapter Summary* is a short list of *Study Strategies* for addressing topics or situations specific to the chapter.

Enhance Your Understanding Using Technology

TECHNOLOGY

Ⓣ There are several ways to use your graphing utility to locate the zeros of a polynomial function after listing the possible rational zeros. You can use the *table* feature by setting the increments of *x* to the smallest difference between possible rational zeros, or use the *table* feature in ASK mode. In either case the value in the function column will be 0 when *x* is a zero of the function. Another way to locate zeros is to graph the function. Be sure that your viewing window contains all the possible rational zeros.

TECHNOLOGY BOXES

These boxes appear throughout the text and provide guidance on using technology to facilitate lengthy calculations, present a graphical solution, or discuss where using technology can lead to misleading or wrong solutions.

TECHNOLOGY EXERCISES

Technology can help you visualize the math and develop a deeper understanding of mathematical concepts. Many of the exercises in the text can be solved using technology—giving you the opportunity to practice using these tools. The Ⓣ symbol identifies exercises for which you are specifically instructed to use a graphing calculator or a computer algebra system to solve the problem. Additionally, the Ⓢ symbol denotes exercises best solved by using a spreadsheet.

Ⓣ **92. Revenue** A company determines that the total revenue R (in hundreds of thousands of dollars) for the years 1997 to 2010 can be approximated by the function

$$R = -0.025t^3 + 0.8t^2 - 2.5t + 8.75, \quad 7 \le t \le 20$$

Ⓢ **60. Solar Energy** Photovoltaic cells convert light energy into electricity. The photovoltaic cell and module domestic shipments S (in peak kilowatts) for the years 1996 to 2005 are shown in the table. *(Source: Energy Information Administration)*

Year	Shipments, S	Year	Shipments, S
1996	13,016	2001	36,310
1997	12,561	2002	45,313
1998	15,069	2003	48,664
1999	21,225	2004	78,346
2000	19,838	2005	134,465

(a) Use a spreadsheet software program to create a scatter plot of the data. Let t represent the year, with $t = 6$ corresponding to 1996.

(b) Use the *regression* feature of a spreadsheet software program to find a cubic model and a quartic model for the data.

(c) Use each model to predict the year in which the shipments will be about 1,000,000 peak kilowatts. Then discuss the appropriateness of each model for predicting future values.

Prepare for Success in Applied Calculus and Beyond

Business Capsule

AP/Wide World Photos

SunPower Corporation develops and manufactures solar-electric power products. SunPower's new higher efficiency solar cells generate up to 50% more power than other solar technologies. SunPower's technology was developed by Dr. Richard Swanson and his students while he was Professor of Engineering at Stanford University. SunPower's 2006 revenues are projected to increase 300% from its 2005 revenues.

69. Research Project Use your campus library, the Internet, or some other reference source to find information about an alternative energy business experiencing strong growth similar to the example above. Write a brief report about the company or small business.

BUSINESS CAPSULES

Business Capsules appear at the ends of numerous sections. These capsules and their accompanying exercises deal with business situations that are related to the mathematical concepts covered in the chapter.

117. MAKE A DECISION You are a sales representative for an automobile manufacturer. You are paid an annual salary plus a bonus of 3% of your sales over $500,000. Consider the two functions given by

$$f(x) = x - 500,000$$

and

$$g(x) = 0.03x.$$

If x is greater than $500,000, does $f(g(x))$ or $g(f(x))$ represent your bonus? Explain.

MAKE A DECISION

These multi-step exercises reinforce your problem-solving skills and mastery of concepts, and take a real-life application further by testing what you know about a given problem to make a decision within the context of the problem.

Fundamental Concepts of Algebra

0

Jeff Schultz/AlaskaStock.com

The Iditarod Sled Dog Race includes a stop in McGrath, Alaska. Part of the challenge of this event is facing temperatures that reach well below zero. To find the range of a set of temperatures, you must find the distance between two numbers. (See Section 0.1, Exercise 81.)

Applications

The fundamental concepts of algebra have many real-life applications. The applications listed below represent a sample of the applications in this chapter.

- College Costs, Exercise 75, page 28
- Escape Velocity, Example 11, page 35
- Oxygen Level, Exercise 72, page 61

Section 0.1

Real Numbers: Order and Absolute Value

- Classify real numbers as natural numbers, integers, rational numbers, or irrational numbers.
- Order real numbers.
- Give a verbal description of numbers represented by an inequality.
- Use inequality notation to describe a set of real numbers.
- Interpret absolute value notation.
- Find the distance between two numbers on the real number line.
- Use absolute value to solve an application problem.

Real Numbers

The formal term that is used in mathematics to refer to a collection of objects is the word **set.** For instance, the set

$$\{1, 2, 3\}$$

contains the three numbers 1, 2, and 3. Note that a pair of braces { } is used to enclose the members of the set. In this text, a pair of braces will always indicate the members of a set. Parentheses ( ) and brackets [ ] are used to represent other ideas.

The set of numbers that is used in arithmetic is the set of **real numbers.** The term *real* distinguishes real numbers from *imaginary* or *complex* numbers.

A set A is called a **subset** of a set B if every member of A is also a member of B. Here are two examples.

- $\{1, 2, 3\}$ is a subset of $\{1, 2, 3, 4\}$.

- $\{0, 4\}$ is a subset of $\{0, 1, 2, 3, 4\}$.

One of the most commonly used subsets of real numbers is the set of **natural numbers** or **positive integers**

$$\{1, 2, 3, 4, \ldots\}. \qquad \text{Set of positive integers}$$

Note that the three dots indicate that the pattern continues. For instance, the set also contains the numbers 5, 6, 7, and so on.

Positive integers can be used to describe many quantities that you encounter in everyday life—for instance, you might be taking four classes this term, or you might be paying \$700 a month for rent. But even in everyday life, positive integers cannot describe some concepts accurately. For instance, you could have a zero balance in your checking account, or the temperature could be $-10°$ (10 degrees below zero). To describe such quantities, you need to expand the set of positive integers to include **zero** and the **negative integers.** The expanded set is called the set of **integers,** which can be written as follows.

$$\underbrace{\{\ldots, -3, -2, -1,}_{\text{Negative integers}} \quad \overset{\text{Zero}}{0}, \quad \underbrace{1, 2, 3, \ldots\}}_{\text{Positive integers}}$$

The set of integers is a subset of the set of real numbers. This means that every integer is a real number.

Even with the set of integers, there are still many quantities in everyday life that you cannot describe accurately. The costs of many items are not in whole dollar amounts, but in parts of dollars, such as $1.19 or $39.98. You might work $8\frac{1}{2}$ hours, or you might miss the first *half* of a movie. To describe such quantities, the set of integers is expanded to include **fractions.** The expanded set is called the set of **rational numbers.** Formally, a real number is called **rational** if it can be written as the ratio p/q of two integers, where $q \neq 0$. (The symbol $\neq$ means **not equal to.**) For instance,

$$2 = \frac{2}{1}, \quad 0.333\ldots = \frac{1}{3}, \quad 0.125 = \frac{1}{8}, \quad \text{and} \quad 1.126126\ldots = \frac{125}{111}$$

are rational numbers. Real numbers that cannot be written as the ratio of two integers are called **irrational.** For instance, the numbers

$$\sqrt{2} = 1.4142135\ldots \quad \text{and} \quad \pi = 3.1415926\ldots$$

are irrational numbers. The decimal representation of a rational number is either *terminating* or *repeating.* For instance, the decimal representation of

$$\tfrac{1}{4} = 0.25 \qquad\qquad\qquad \text{Terminating decimal}$$

is terminating, and the decimal representation of

$$\tfrac{4}{11} = 0.363636\ldots = 0.\overline{36} \qquad\qquad \text{Repeating decimal}$$

is repeating. (The line over "36" indicates which digits repeat.)

The decimal representation of an irrational number neither terminates nor repeats. When you perform calculations using decimal representations of nonterminating decimals, you usually use a decimal approximation that has been **rounded** to a certain number of decimal places. For instance, rounded to four decimal places, the decimal approximations of $\frac{2}{3}$ and π are

$$\frac{2}{3} \approx 0.6667 \quad \text{and} \quad \pi \approx 3.1416.$$

The symbol $\approx$ means **approximately equal to.**

The Venn diagram in Figure 0.1 shows the relationships between the real numbers and several commonly used subsets of the real numbers.

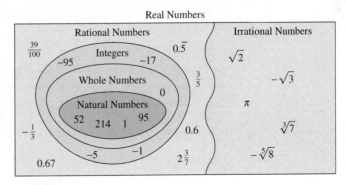

FIGURE 0.1

The Real Number Line and Ordering

The picture that is used to represent the real numbers is the **real number line.** It consists of a horizontal line with a point (the **origin**) labeled as 0 (zero). Points to the left of zero are associated with **negative numbers,** and points to the right of zero are associated with **positive numbers,** as shown in Figure 0.2. The real number zero is neither positive nor negative. So, when you want to talk about real numbers that might be positive *or* zero, you can use the term **nonnegative real numbers.**

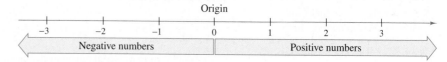

FIGURE 0.2   The Real Number Line

Each point on the real number line corresponds to exactly one real number, and each real number corresponds to exactly one point on the real number line, as shown in Figure 0.3. The number associated with a point on the real number line is the **coordinate** of the point.

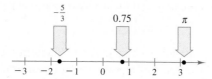

FIGURE 0.3   Every real number corresponds to a point on the real number line.

The real number line provides you with a way of comparing any two real numbers. For instance, if you choose any two (different) numbers on the real number line, one of the numbers must be to the left of the other number. The number to the left is **less than** the number to the right, and the number to the right is **greater than** the number to the left.

Definition of Order on the Real Number Line

If the real number a lies to the left of the real number b on the real number line, a is **less than** b, which is denoted by

$$a < b$$

as shown in Figure 0.4. This relationship can also be described by saying that b is **greater than** a and writing $b > a$. The inequality $a \leq b$ means that a is **less than or equal to** b, and the inequality $b \geq a$ means that b is **greater than or equal to** a.

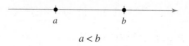

FIGURE 0.4 a is to the left of b.

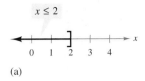

(a)

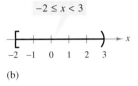

(b)

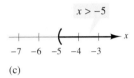

(c)

FIGURE 0.5

The symbols $<$, $>$, $\leq$, and $\geq$ are called **inequality symbols.** Inequalities are useful in denoting subsets of real numbers, as shown in Examples 1 and 2.

Example 1   Interpreting Inequalities

a. The inequality $x \leq 2$ denotes all real numbers that are less than or equal to 2, as shown in Figure 0.5(a).

b. The inequality $-2 \leq x < 3$ means that $x \geq -2$ *and* $x < 3$. This **double inequality** denotes all real numbers between -2 and 3, including -2 but *not* including 3, as shown in Figure 0.5(b).

c. The inequality $x > -5$ denotes all real numbers that are greater than -5, as shown in Figure 0.5(c).

✓**CHECKPOINT 1**

Give a verbal description of the subset of real numbers represented by $x \geq 7$. ◼

In Figure 0.5, notice that a bracket is used to *include* the endpoint of an interval and a parenthesis is used to *exclude* the endpoint.

Example 2   Inequalities and Sets of Real Numbers

a. "c is nonnegative" means that c is greater than or equal to zero, which you can write as $c \geq 0$.

b. "b is at most 5" can be written as $b \leq 5$.

c. "d is negative" can be written as $d < 0$, and "d is greater than -3" can be written as $-3 < d$. Combining these two inequalities produces $-3 < d < 0$.

d. "x is positive" can be written as $0 < x$, and "x is not more than 6" can be written as $x \leq 6$. Combining these two inequalities produces $0 < x \leq 6$.

✓**CHECKPOINT 2**

Use inequality notation to describe each subset of real numbers.

a. x is at least 5.

b. y is greater than 4, but no more than 11. ◼

The following property of real numbers is called the **Law of Trichotomy.** As the "tri" in its name suggests, this law tells you that for any two real numbers a and b, precisely one of *three* relationships is possible.

$$a < b, \quad a = b, \quad \text{or} \quad a > b \qquad \text{Law of Trichotomy}$$

Absolute Value and Distance

The **absolute value** of a real number is its *magnitude,* or its value disregarding its sign. For instance, the absolute value of -3, written $|-3|$, has the value of 3.

STUDY TIP

Be sure you see from the definition that the absolute value of a real number is never negative. For instance, if $a = -5$, then $|-5| = -(-5) = 5$.

Definition of Absolute Value

Let a be a real number. The **absolute value** of a, denoted by $|a|$, is

$$|a| = \begin{cases} a, & \text{if } a \geq 0 \\ -a, & \text{if } a < 0 \end{cases}.$$

The absolute value of any real number is either positive or zero. Moreover, 0 is the only real number whose absolute value is zero. That is, $|0| = 0$.

Example 3 Finding Absolute Value

a. $|-7| = 7$ **b.** $\left|\frac{1}{2}\right| = \frac{1}{2}$

c. $|-4.8| = 4.8$ **d.** $-|-9| = -(9) = -9$

✓ CHECKPOINT 3

Evaluate $|-12|$. ■

Example 4 Comparing Real Numbers

Place the correct symbol ($<$, $>$, or $=$) between the two real numbers.

a. $|-4|$ ░░ $|4|$ **b.** $|-5|$ ░░ 3 **c.** $-|-1|$ ░░ $|-1|$

SOLUTION

a. $|-4| = |4|$, because both are equal to 4.

b. $|-5| > 3$, because $|-5| = 5$ and 5 is greater than 3.

c. $-|-1| < |-1|$, because $-|-1| = -1$ and $|-1| = 1$.

✓ CHECKPOINT 4

Place the correct symbol ($<$, $>$, or $=$) between the two real numbers.

a. $-|-6|$ ░░ $-|6|$

b. $-|5|$ ░░ $|-5|$ ■

Properties of Absolute Value

Let a and b be real numbers. Then the following properties are true.

1. $|a| \geq 0$ **2.** $|-a| = |a|$

3. $|ab| = |a|\,|b|$ **4.** $\left|\dfrac{a}{b}\right| = \dfrac{|a|}{|b|}, \; b \neq 0$

Absolute value can be used to define the distance between two numbers on the real number line. To see how this is done, consider the numbers -3 and 4, as shown in Figure 0.6. To find the distance between these two numbers, subtract either number from the other and then take the absolute value of the difference.

$$(\text{Distance between } -3 \text{ and } 4) = |-3 - 4| = |-7| = 7$$

FIGURE 0.6 The distance between -3 and 4 is 7.

Distance Between Two Numbers

Let a and b be real numbers. The **distance between a and b** is given by

$$\text{Distance} = |b - a| = |a - b|.$$

| Example 5 | Finding the Distance Between Two Numbers |

a. The distance between 2 and 7 is $|2 - 7| = |-5| = 5$.

b. The distance between 0 and -4 is $|0 - (-4)| = |4| = 4$.

c. The statement "the distance between x and 2 is at least 3" can be written as $|x - 2| \geq 3$.

✓**CHECKPOINT 5**

Find the distance between -5 and 3. ■

Application

| Example 6 |
| MAKE A DECISION   Budget Variance |

You monitor monthly expenses for a home health care company. For each type of expense, the company wants the absolute value of the difference between the actual and budgeted amounts to be less than or equal to $500 *and* less than or equal to 5% of the budgeted amount. By letting a represent the actual expenses and b the budgeted expenses, these restrictions can be written as

$$|a - b| \leq 500 \quad \text{and} \quad |a - b| \leq 0.05b.$$

For travel, office supplies, and wages, the company budgeted $12,500, $750, and $84,600. The actual amounts paid for these expenses were $12,872.56, $704.15, and $85,143.95. Are these amounts within budget restrictions?

SOLUTION   One way to determine whether these three expenses are within budget restrictions is to create the table shown.

	Budgeted Expense, b	Actual Expense, a	$|a - b|$	$0.05b$
Travel	$12,500	$12,872.56	$372.56	$625.00
Office supplies	$750	$704.15	$45.85	$37.50
Wages	$84,600	$85,143.95	$543.95	$4230.00

From this table, you can see that travel expenses pass both tests, so they are within budget restrictions. Office supply expenses pass the first test but fail the second test, so they are *not* within budget restrictions. Wage expenses fail the first test and pass the second test, so they are *not* within budget restrictions.

SuperStock/Jupiter Images

Math plays an important part in keeping your personal finances in order as well as a company's expenses and budget.

✓**CHECKPOINT 6**

In Example 6, the company budgeted $28,000 for medical supplies, but actually paid $30,100. Is this within budget restrictions? ■

┌─ **CONCEPT CHECK** ─┐

Is the statement true? If not, explain why.

1. There are no integers in the set of rational numbers.

2. The set of integers is a subset of the set of natural numbers.

3. The expression $x < 5$ describes a subset of the set of rational numbers.

4. When a is negative, $|a| = -a$.

The symbol indicates an example that uses or is derived from real-life data.

Exercises 0.1

See www.CalcChat.com for worked-out solutions to odd-numbered exercises.

In Exercises 1–6, determine which numbers in the set are (a) natural numbers, (b) integers, (c) rational numbers, and (d) irrational numbers.

1. $\left\{-9, -\frac{7}{2}, 5, \frac{2}{3}, \sqrt{2}, 0.1\right\}$

2. $\left\{\sqrt{5}, -7, -\frac{7}{3}, 0, 3.12, \frac{5}{4}\right\}$

3. $\left\{12, -13, 1, \sqrt{4}, \sqrt{6}, \frac{3}{2}\right\}$

4. $\left\{3, -1, \frac{1}{3}, \frac{6}{3}, -\frac{1}{2}\sqrt{2}, -7.5\right\}$

5. $\left\{\frac{8}{2}, -\frac{8}{3}, \sqrt{10}, -4, 9, 14.2\right\}$

6. $\left\{25, -17, \frac{12}{5}, \sqrt{9}, \sqrt{8}, -\sqrt{8}\right\}$

In Exercises 7–10, use a calculator to find the decimal form of the rational number. If the number is a nonterminating decimal, write the repeating pattern.

7. $\dfrac{2}{3}$

8. $\dfrac{9}{40}$

9. $\dfrac{14}{111}$

10. $\dfrac{49}{160}$

In Exercises 11 and 12, approximate the two plotted numbers and place the correct symbol (< or >) between them.

11.

12.

In Exercises 13–18, plot the two real numbers on the real number line and place the appropriate inequality symbol (< or >) between them.

13. $\frac{3}{2}, 7$

14. $-4, -8$

15. $1, -3.5$

16. $\frac{16}{3}, 1$

17. $\frac{5}{6}, \frac{2}{3}$

18. $-\frac{8}{7}, -\frac{3}{7}$

In Exercises 19–22, use a calculator to order the numbers from least to greatest.

19. $\dfrac{7}{2}, 2\sqrt{3}, 3.45, \dfrac{204}{60}, \dfrac{31}{9}$

20. $\dfrac{559}{500}, 1.12, \dfrac{\sqrt{5}}{2}, \dfrac{115}{99}, \dfrac{23}{20}$

21. $\dfrac{7071}{5000}, \dfrac{584}{413}, \sqrt{2}, \dfrac{47}{33}, \dfrac{127}{90}$

22. $\dfrac{26}{15}, \sqrt{3}, 1.73\overline{20}, \dfrac{381}{220}, \sqrt{10} - \sqrt{2}$

In Exercises 23–26, write an inequality that describes the graph.

23.

24.

25.

26.

In Exercises 27–36, give a verbal description of the subset of real numbers that is represented by the inequality, and sketch the subset on the real number line.

27. $x < 0$

28. $x < 2$

29. $x \le 5$

30. $x \ge -2$

31. $x > 3$

32. $x \ge 4$

33. $-2 < x < 2$

34. $0 \le x \le 5$

35. $-1 \le x < 0$

36. $0 < x \le 6$

In Exercises 37–44, use inequality notation to describe the subset of real numbers.

37. x is positive.

38. t is no more than 20.

39. y is greater than 5 and less than or equal to 12.

40. m is at least -5 and at most 9.

41. The person's age A is at least 35.

42. The yield Y is no more than 42 bushels per acre.

43. The annual rate of inflation r is expected to be at least 3.5%, but no more than 6%.

44. The price p of unleaded gasoline is not expected to go below $2.13 per gallon during the coming year.

In Exercises 45–54, evaluate the expression.

45. $|-10|$

46. $|0|$

47. $-3 - |-3|$

48. $|-1| - |-2|$

49. $-3|-3|$

50. $-5|-5|$

51. $\dfrac{-5}{|-5|}$

52. $\dfrac{|-4|}{-4}$

53. $|3 - \pi|$

54. $|4 - \pi|$

In Exercises 55–60, place the correct symbol (<, >, or =) between the two real numbers.

55. $|-7|$ _____ $|7|$

56. -5 _____ $-|5|$

57. $|-3|$ _____ $-|-3|$

58. $-|-6|$ _____ $|-6|$

59. $-|-2|$ _____ $-|2|$

60. $-(-2)$ _____ -2

In Exercises 61–70, find the distance between a and b.

61. $a = -1$ $b = 3$

62. $a = -\frac{5}{2}$ $b = 0$

63. $a = -4, b = -\frac{3}{2}$

64. $a = \frac{1}{4}, b = \frac{11}{4}$

65. $a = -\frac{7}{2}, b = 0$

66. $a = \frac{3}{4}, b = \frac{9}{4}$

67. $a = 126, b = 75$

68. $a = -126, b = -75$

69. $a = \frac{16}{5}, b = \frac{112}{75}$

70. $a = 9.34, b = -5.65$

In Exercises 71–78, use absolute value notation to describe the sentence.

71. The distance between z and $\frac{3}{2}$ is greater than 1.

72. The distance between x and 5 is no more than 3.

73. The distance between x and -10 is at least 6.

74. The distance between z and 0 is less than 8.

75. y is at least six units from 0.

76. x is less than eight units from 0.

77. x is more than five units from m.

78. y is at most two units from a.

79. **Travel** While traveling on the Pennsylvania Turnpike, you pass milepost 57 near Pittsburgh, then milepost 236 near Gettysburg. How far do you travel during that time period?

80. **Travel** While traveling on the Pennsylvania Turnpike, you pass milepost 326 near Valley Forge, then milepost 351 near Philadelphia. How far do you travel during that time period?

Temperature In Exercises 81 and 82, the record January temperatures (in degrees Fahrenheit) for a city are given. Find the distance between the numbers to determine the range of temperatures for January.

81. McGrath, Alaska: lowest: $-75°F$

highest: $54°F$

82. Flagstaff, Arizona: lowest: $-22°F$

highest: $66°F$

MAKE A DECISION: BUDGET VARIANCE In Exercises 83–88, the accounting department of an Internet start-up company is checking to see whether various actual expenses differ from the budgeted expenses by more than $500 or by more than 5%. Complete the missing parts of the table. Then determine whether the actual expense passes the "budget variance test."

	Budgeted Expense, b	Actual Expense, a	$\|a - b\|$	$0.05b$
83.	$30,000	$29,123.45		
84.	$125,500	$126,347.85		
85.	$12,000	$11,735.68		
86.	$8300	$8632.59		
87.	$40,800	$39,862.17		
88.	$2625	$2196.89		

MAKE A DECISION: QUALITY CONTROL In Exercises 89–94, the quality control inspector for a tire factory is testing the rim diameters of various tires. A tire is rejected if its rim diameter varies too much from its expected measure. The diameter should not differ by more than 0.02 inch or by more than 0.12% of the expected diameter measure. Complete the missing parts of the table. Then determine whether the tire is passed or rejected according to the inspector's guide lines.

	Expected Diameter, b	Actual Diameter, a	$\|a - b\|$	$0.0012b$
89.	14 in.	13.998 in.		
90.	15 in.	15.012 in.		
91.	16 in.	15.973 in.		
92.	17 in.	16.992 in.		
93.	18 in.	18.027 in.		
94.	19 in.	18.993 in.		

95. **Think About It** Consider $|u + v|$ and $|u| + |v|$.

(a) Are the values of the expressions always equal? If not, under what conditions are they unequal?

(b) If the two expressions are not equal for certain values of u and v, is one of the expressions always greater than the other? Explain.

96. **Think About It** Is there a difference between saying that a real number is positive and saying that a real number is nonnegative? Explain.

97. Describe the differences among the sets of natural numbers, integers, rational numbers, and irrational numbers.

98. **Think About it** Can it ever be true that $|a| = -a$ for a real number a? Explain.

Section 0.2

The Basic Rules of Algebra

- Identify the terms of an algebraic expression.
- Evaluate an algebraic expression.
- Identify basic rules of algebra.
- Perform operations on real numbers.
- Use a calculator to evaluate an algebraic expression.

Algebraic Expressions

One of the basic characteristics of algebra is the use of letters (or combinations of letters) to represent numbers. The letters used to represent numbers are called **variables,** and combinations of letters and numbers are called **algebraic expressions.** Some examples of algebraic expressions are

$$5x, \quad 2x - 3, \quad \frac{4}{x^2 + 2}, \quad \text{and} \quad 7x + y.$$

Algebraic Expression

A collection of letters (called **variables**) and real numbers (called **constants**) that are combined using the operations of addition, subtraction, multiplication, and division is an **algebraic expression.** (Other operations can also be used to form an algebraic expression.)

The **terms** of an algebraic expression are those parts that are separated by addition. For example, the algebraic expression $x^2 - 5x + 8$ has three terms: x^2, $-5x$, and 8. Note that $-5x$, rather than $5x$, is a term, because

$$x^2 - 5x + 8 = x^2 + (-5x) + 8.$$

The terms x^2 and $-5x$ are the **variable terms** of the expression, and 8 is the **constant term** of the expression. The numerical factor of a variable term is the **coefficient** of the variable term. For instance, the coefficient of the variable term $-5x$ is -5, and the coefficient of the variable term x^2 is 1.

Example 1   Identifying the Terms of an Algebraic Expression

Algebraic Expression	Terms
a. $4x - 3$	$4x, -3$
b. $2x + 4y - 5$	$2x, 4y, -5$

✓ CHECKPOINT 1

Identify the terms of each algebraic expression.

a. $8 - 15x$

b. $4x^2 - 3y - 7$ ■

STUDY TIP

When you evaluate an expression with grouping symbols (such as parentheses), be careful to use the correct order of operations.

TECHNOLOGY

(T) To evaluate the expression $3 + 4x$ for the values 2 and 5, use the *last entry* feature of a graphing utility.

1. Evaluate $3 + 4 \cdot 2$.

2. Press (2nd) [ENTRY] (recalls previous expression to the home screen).

3. Cursor to 2, replace 2 with 5, and press (ENTER).

For specific keystrokes for the *last entry* feature, go to the text website at *college.hmco.com/ info/larsonapplied.*

✓ **CHECKPOINT 3**

Evaluate $3y + x^2$ when $x = 4$ and $y = -2$. ∎

Example 2 **Symbols of Grouping**

a. $7 - 3(4 - 2) = 7 - 3(2) = 7 - 6 = 1$

b. $(4 - 5) - (3 - 6) = (-1) - (-3) = -1 + 3 = 2$

✓ **CHECKPOINT 2**

Simplify the expression $5(7 - 3) + 9$. ∎

The **Substitution Principle** states, "If $a = b$, then a can be replaced by b in any expression involving a." You use this principle to **evaluate** an algebraic expression by substituting numerical values for each of the variables in the expression. In the first evaluation shown below, 3 is substituted for x in the expression $-3x + 5$.

Expression	Value of Variable	Substitution	Value of Expression
$-3x + 5$	$x = 3$	$-3(3) + 5$	$-9 + 5 = -4$
$3x^2 + 2x - 1$	$x = -1$	$3(-1)^2 + 2(-1) - 1$	$3 - 2 - 1 = 0$
$-2x(x + 4)$	$x = -2$	$-2(-2)(-2 + 4)$	$-2(-2)(2) = 8$
$\dfrac{1}{x - 2}$	$x = 2$	$\dfrac{1}{2 - 2}$	Undefined

Example 3 **Evaluating Algebraic Expressions**

Evaluate each algebraic expression when $x = -2$ and $y = 3$.

a. $4y - 2x$ **b.** $5 + x^2$ **c.** $5 - y^2$

SOLUTION

a. When $x = -2$ and $y = 3$, the expression $4y - 2x$ has a value of

$$4(3) - 2(-2) = 12 + 4 = 16.$$

b. When $x = -2$, the expression $5 + x^2$ has a value of

$$5 + (-2)^2 = 5 + 4 = 9.$$

c. When $y = 3$, the expression $5 - y^2$ has a value of

$$5 - (3)^2 = 5 - 9 = -4.$$

Basic Rules of Algebra

The four basic arithmetic operations are **addition, multiplication, subtraction,** and **division,** denoted by the symbols $+$, $\times$ or $\cdot$, $-$, and $\div$, respectively. Of these, addition and multiplication are considered to be the two primary arithmetic operations. Subtraction and division are defined as the inverse operations of addition and multiplication, as follows.

The symbol (T) indicates when to use graphing technology or a symbolic computer algebra system to solve a problem or an exercise. The solutions of other exercises may also be facilitated by use of appropriate technology.

Subtraction: Add the opposite   *Division:* Multiply by the reciprocal

$$a - b = a + (-b)$$ If $b \neq 0$, then $a \div b = a\left(\dfrac{1}{b}\right) = \dfrac{a}{b}.$

In these definitions, $-b$ is called the **additive inverse** (or opposite) of b, and $1/b$ is called the **multiplicative inverse** (or reciprocal) of b. In place of $a \div b$, you can use the fraction symbol a/b. In this fractional form, a is called the **numerator** of the fraction and b is called the **denominator.**

The **basic rules of algebra,** listed below, are true for variables and algebraic expressions as well as for real numbers.

Basic Rules of Algebra

Let a, b, and c be real numbers, variables, or algebraic expressions.

Property	*Example*
Commutative Property of Addition	
$a + b = b + a$	$4x + x^2 = x^2 + 4x$
Commutative Property of Multiplication	
$ab = ba$	$(4 - x)x^2 = x^2(4 - x)$
Associative Property of Addition	
$(a + b) + c = a + (b + c)$	$(-x + 5) + 2x^2 =$ $-x + (5 + 2x^2)$
Associative Property of Multiplication	
$(ab)c = a(bc)$	$(2x \cdot 3y)(8) = (2x)(3y \cdot 8)$
Distributive Property	
$a(b + c) = ab + ac$	$3x(5 + 2x) = 3x \cdot 5 + 3x \cdot 2x$
$(a + b)c = ac + bc$	$(y + 8)y = y \cdot y + 8 \cdot y$
Additive Identity Property	
$a + 0 = a$	$5y^2 + 0 = 5y^2$
Multiplicative Identity Property	
$a \cdot 1 = a$	$(4x^2)(1) = 4x^2$
Additive Inverse Property	
$a + (-a) = 0$	$5x^3 + (-5x^3) = 0$
Multiplicative Inverse Property	
$a \cdot \dfrac{1}{a} = 1, \quad a \neq 0$	$(x^2 + 4)\left(\dfrac{1}{x^2 + 4}\right) = 1$

Because subtraction is defined as "adding the opposite," the Distributive Property is also true for subtraction. For instance, the "subtraction form" of $a(b + c) = ab + ac$ is $a(b - c) = a[b + (-c)] = ab + a(-c) = ab - ac.$

| Example 4 | **Identifying the Basic Rules of Algebra** |

Identify the rule of algebra illustrated by each statement.

a. $(4x^2)5 = 5(4x^2)$

b. $(2y^3 + y) - (2y^3 + y) = 0$

c. $(4 + x^2) + 3x^2 = 4 + (x^2 + 3x^2)$

d. $(x - 5)7 + (x - 5)x = (x - 5)(7 + x)$

e. $2x \cdot \dfrac{1}{2x} = 1, \quad x \neq 0$

SOLUTION

a. This equation illustrates the Commutative Property of Multiplication.

b. This equation illustrates the Additive Inverse Property.

c. This equation illustrates the Associative Property of Addition.

d. This equation illustrates the Distributive Property in reverse order.

$$ab + ac = a(b + c) \qquad \text{Distributive Property}$$
$$(x - 5)7 + (x - 5)x = (x - 5)(7 + x)$$

e. This equation illustrates the Multiplicative Inverse Property. Note that it is important that x be a nonzero number. If x were allowed to be zero, you would be in trouble because the reciprocal of zero is undefined. ⎯⎯⎯

✓ **CHECKPOINT 4**

Identify the rule of algebra illustrated by each statement.

a. $3x^2 \cdot 1 = 3x^2$

b. $x^2 + 5 = 5 + x^2$ ■

The following three lists summarize the basic properties of negation, zero, and fractions. When you encounter such lists, you should not only *memorize* a verbal description of each property, but you should also try to gain an *intuitive feeling* for the validity of each.

Properties of Negation

Let a and b be real numbers, variables, or algebraic expressions.

Property	*Example*
1. $(-1)a = -a$	$(-1)7 = -7$
2. $-(-a) = a$	$-(-6) = 6$
3. $(-a)b = -(ab) = a(-b)$	$(-5)3 = -(5 \cdot 3) = 5(-3)$
4. $(-a)(-b) = ab$	$(-2)(-6) = 2 \cdot 6$
5. $-(a + b) = (-a) + (-b)$	$-(3 + 8) = (-3) + (-8)$

Be sure you see the difference between the opposite of a number and a negative number. If a is negative, then its opposite, $-a$, is positive. For instance, if $a = -5$, then $-a = -(-5) = 5$.

Properties of Zero

Let a and b be real numbers, variables, or algebraic expressions. Then the following properties are true.

1. $a + 0 = a$ and $a - 0 = a$

2. $a \cdot 0 = 0$

3. $\dfrac{0}{a} = 0$, $a \neq 0$

4. $\dfrac{a}{0}$ is undefined.

5. Zero-Factor Property: If $ab = 0$, then $a = 0$ or $b = 0$.

The "or" in the Zero-Factor Property includes the possibility that both factors are zero. This is called an **inclusive or,** and it is the way the word "or" is always used in mathematics.

Properties of Fractions

Let a, b, c, and d be real numbers, variables, or algebraic expressions such that $b \neq 0$ and $d \neq 0$. Then the following properties are true.

1. *Equivalent fractions:* $\dfrac{a}{b} = \dfrac{c}{d}$ if and only if $ad = bc$.

2. *Rules of signs:* $-\dfrac{a}{b} = \dfrac{-a}{b} = \dfrac{a}{-b}$ and $\dfrac{-a}{-b} = \dfrac{a}{b}$

3. *Generate equivalent fractions:* $\dfrac{a}{b} = \dfrac{ac}{bc}$, $c \neq 0$

4. *Add or subtract with like denominators:* $\dfrac{a}{b} \pm \dfrac{c}{b} = \dfrac{a \pm c}{b}$

5. *Add or subtract with unlike denominators:* $\dfrac{a}{b} \pm \dfrac{c}{d} = \dfrac{ad \pm bc}{bd}$

6. *Multiply fractions:* $\dfrac{a}{b} \cdot \dfrac{c}{d} = \dfrac{ac}{bd}$

7. *Divide fractions:* $\dfrac{a}{b} \div \dfrac{c}{d} = \dfrac{a}{b} \cdot \dfrac{d}{c} = \dfrac{ad}{bc}$, $c \neq 0$

In Property 1 (equivalent fractions) the phrase "if and only if" implies two statements. One statement is: If $a/b = c/d$, then $ad = bc$. The other statement is: If $ad = bc$, where $b \neq 0$ and $d \neq 0$, then $a/b = c/d$.

Example 5   **Properties of Zero and Properties of Fractions**

a. $x - \dfrac{0}{5} = x - 0 = x$ Properties 3 and 1 of zero

b. $\dfrac{x}{5} = \dfrac{3 \cdot x}{3 \cdot 5} = \dfrac{3x}{15}$ Generate equivalent fractions.

c. $\dfrac{x}{3} + \dfrac{2x}{5} = \dfrac{x \cdot 5 + 3 \cdot 2x}{15} = \dfrac{11x}{15}$ Add fractions with unlike denominators.

d. $\dfrac{7}{x} \div \dfrac{3}{2} = \dfrac{7}{x} \cdot \dfrac{2}{3} = \dfrac{14}{3x}$ Divide fractions.

✓ **CHECKPOINT 5**

Simplify the expression $\dfrac{x}{4} + \dfrac{2x}{3}$. ▪

If a, b, and c are integers such that $ab = c$, then a and b are **factors** or **divisors** of c. For example, 2 and 3 are factors of 6 because $2 \cdot 3 = 6$. A **prime number** is a positive integer that has exactly two factors: itself and 1. For example, 2, 3, 5, 7, and 11 are prime numbers, whereas 1, 4, 6, 8, 9, and 10 are not. The numbers 4, 6, 8, 9, and 10 are **composite** because they can be written as the products of two or more prime numbers. The number 1 is neither prime nor composite. The **Fundamental Theorem of Arithmetic** states that every positive integer greater than 1 is a prime number or can be written as the product of prime numbers in precisely one way (disregarding order). For instance, the *prime factorization* of 24 is

$$24 = 2 \cdot 2 \cdot 2 \cdot 3.$$

When you are adding or subtracting fractions that have unlike denominators, you can use Property 4 of fractions by rewriting both of the fractions so that they have the same denominator. This is called the **least common denominator** method.

STUDY TIP

To find the LCD, first factor the denominators.

$$15 = 3 \cdot 5$$
$$9 = 3^2$$
$$5 = 5$$

The LCD is the product of the prime factors, with each factor given the highest power of its occurrence in any denominator. So, the LCD is

$$3^2 \cdot 5 = 45.$$

Example 6   **Adding and Subtracting Fractions**

Evaluate $\dfrac{2}{15} - \dfrac{5}{9} + \dfrac{4}{5}$.

SOLUTION   Begin by factoring the denominators to find the least common denominator (LCD). Use the LCD, 45, to rewrite the fractions and simplify.

$$\frac{2}{15} - \frac{5}{9} + \frac{4}{5} = \frac{2 \cdot 3}{15 \cdot 3} - \frac{5 \cdot 5}{9 \cdot 5} + \frac{4 \cdot 9}{5 \cdot 9}$$

$$= \frac{6 - 25 + 36}{45}$$

$$= \frac{17}{45}$$

✓ **CHECKPOINT 6**

Evaluate $\dfrac{3}{4} + \dfrac{2}{3} - \dfrac{1}{2}$. ▪

Equations

An **equation** is a statement of equality between two expressions. So, the statement

$$a + b = c + d$$

means that the expressions $a + b$ and $c + d$ represent the same number. For instance, because $1 + 4$ and $3 + 2$ both represent the number 5, you can write $1 + 4 = 3 + 2$. Three important properties of equality follow.

Properties of Equality

Let a, b, and c be real numbers, variables, or algebraic expressions.

1. Reflexive: $a = a$

2. Symmetric: If $a = b$, then $b = a$.

3. Transitive: If $a = b$ and $b = c$, then $a = c$.

In algebra, you often rewrite expressions by making substitutions that are permitted under the Substitution Principle. Two important consequences of the Substitution Principle are the following rules.

1. If $a = b$, then $a + c = b + c$.   **2.** If $a = b$, then $ac = bc$.

The first rule allows you to add the same number to each side of an equation. The second allows you to multiply each side of an equation by the same number. The converses of these two rules are also true and are listed below.

1. If $a + c = b + c$, then $a = b$.   **2.** If $ac = bc$ and $c \neq 0$, then $a = b$.

So, you can also subtract the same number from each side of an equation as well as divide each side of an equation by the same nonzero number.

Calculators and Rounding

The table below shows keystrokes for several similar functions on a standard scientific calculator and a graphing calculator. These keystrokes may not be the same as those for your calculator. Consult your user's guide for specific keystrokes.

Graphing Calculator	Scientific Calculator
ENTER	=
(−)	+/−
^	y^x
x^{-1}	1/x

For example, you can evaluate 13^3 on a graphing calculator or a scientific calculator as follows.

Graphing Calculator	Scientific Calculator
13 ^ 3 ENTER	13 y^x 3 =

Example 7 **Using a Calculator**

Scientific Calculator

	Expression	Keystrokes	Display
a.	$7 - (5 \cdot 3)$	7 ⊖ 5 ⊗ 3 ⊜	-8
b.	$-12^2 - 100$	12 (x²) (+/−) ⊖ 100 ⊜	-244
c.	$24 \div 2^3$	24 ⊘ 2 (yˣ) 3 ⊜	3
d.	$3(10 - 4^2) \div 2$	3 ⊗ (◦ 10 ⊖ 4 (x²) ◦) ⊘ 2 ⊜	-9
e.	37% of 40	.37 ⊗ 40 ⊜	14.8

Graphing Calculator

	Expression	Keystrokes	Display
a.	$7 - (5 \cdot 3)$	7 ⊖ 5 ⊗ 3 (ENTER)	-8
b.	$-12^2 - 100$	(−) 12 (x²) ⊖ 100 (ENTER)	-244
c.	$24 \div 2^3$	24 ⊘ 2 (∧) 3 (ENTER)	3
d.	$3(10 - 4^2) \div 2$	3 (◦ 10 ⊖ 4 (x²) ◦) ⊘ 2 (ENTER)	-9
e.	37% of 40	.37 ⊗ 40 (ENTER)	14.8

TECHNOLOGY

(T) Be sure you see the difference between the change sign key (+/−) or (−) and the subtraction key ⊖, as used in Example 7(b).

✓CHECKPOINT 7

Write the keystrokes you can use to evaluate

$$6(8^3 - 481)$$

on a graphing calculator or a scientific calculator. ▪

When rounding decimals, look at the *decision digit* (the digit at the right of the last digit you want to keep). Round up when the decision digit is 5 or greater, and round down when it is 4 or less.

Example 8 **Rounding Decimal Numbers**

Number	Rounded to Three Decimal Places	
a. $\sqrt{2} = 1.4142135\ldots$	1.414	Round down because 2 < 4.
b. $\pi = 3.1415926\ldots$	3.142	Round up because 5 = 5.
c. $\dfrac{7}{9} = 0.7777777\ldots$	0.778	Round up because 7 > 5.

✓CHECKPOINT 8

Use a calculator to evaluate

$$4\left(\frac{2}{3} + \frac{4}{5}\right).$$

Then round the result to two decimal places. ▪

CONCEPT CHECK

1. Write an algebraic expression that contains a variable term, a constant term, and a coefficient. Identify the parts of your expression.

2. Is $(a - b) + c = a - (b + c)$ when a, b, and c are nonzero real numbers? Explain.

3. Is the expression $-x$ always negative? Explain.

4. Explain how to divide a/b by c/d when b, c, and d are nonzero real numbers.

Skills Review 0.2   The following warm-up exercises involve skills that were covered in earlier sections. You will use these skills in the exercise set for this section. For additional help, review Section 0.1.

In Exercises 1–4, place the correct inequality symbol ($<$ or $>$) between the two real numbers.

1. -4 ___ -2 **2.** 0 ___ -3 **3.** $\sqrt{3}$ ___ 1.73 **4.** $-\pi$ ___ -3

In Exercises 5–8, find the distance between the two real numbers.

5. $4, 6$ **6.** $-2, 2$ **7.** $0, -5$ **8.** $-1, 3$

In Exercises 9 and 10, evaluate the expression.

9. $|-7| + |7|$ **10.** $-|8 - 10|$

Exercises 0.2   See www.CalcChat.com for worked-out solutions to odd-numbered exercises.

In Exercises 1–6, identify the terms of the algebraic expression.

1. $7x + 4$ **2.** $-5 + 3x$

3. $x^2 - 4x + 8$ **4.** $4x^3 + x - 5$

5. $2x^2 - 9x + 13$ **6.** $3x^4 + 2x^3 - 1$

In Exercises 7–10, simplify the expression.

7. $(8 - 17) + 3$ **8.** $-3(5 - 2)$

9. $(4 - 7)(-2)$ **10.** $-5(-2 - 6)$

In Exercises 11–16, evaluate the expression for each value of x. (If not possible, state the reason.)

	Expression	Values
11.	$4x - 6$	(a) $x = -1$ (b) $x = 0$
12.	$5 - 3x$	(a) $x = -3$ (b) $x = 2$
13.	$x^2 - 3x + 4$	(a) $x = -2$ (b) $x = 2$
14.	$-x^3 + 2x - 1$	(a) $x = 0$ (b) $x = 2$
15.	$\dfrac{x}{x - 2}$	(a) $x = -2$ (b) $x = 2$
16.	$\dfrac{x + 3}{x - 3}$	(a) $x = 3$ (b) $x = -3$

In Exercises 17–22, evaluate the expression when $x = 3$, $y = -2$, and $z = 4$.

17. $x + 3y + z$ **18.** $6z + 5x - 3y$

19. $x^2 - 5y + 4z$ **20.** $z^2 + 6y - x$

21. $\dfrac{x - y}{5z}$ **22.** $\dfrac{4z - 2y}{20x}$

In Exercises 23–38, identify the rule(s) of algebra illustrated by the statement.

23. $3 + 4 = 4 + 3$ **24.** $x + 9 = 9 + x$

25. $-15 + 15 = 0$ **26.** $(x + 2) - (x + 2) = 0$

27. $2(x + 3) = 2x + 6$

28. $(5 + 11) \cdot 6 = 5 \cdot 6 + 11 \cdot 6$

29. $2\left(\frac{1}{2}\right) = 1$

30. $\dfrac{1}{h + 6}(h + 6) = 1$, $h \neq -6$

31. $h + 0 = h$ **32.** $(z - 2) + 0 = z - 2$

33. $57 \cdot 1 = 57$ **34.** $1 \cdot (1 + x) = 1 + x$

35. $6 + (7 + 8) = (6 + 7) + 8$

36. $x + (y + 10) = (x + y) + 10$

37. $x(3y) = (x \cdot 3)y = (3x)y$

38. $\frac{1}{7}(7 \cdot 12) = \left(\frac{1}{7} \cdot 7\right)12 = 1 \cdot 12 = 12$

In Exercises 39–42, write the prime factorization of the integer.

39. 48 **40.** 24

41. 240 **42.** 150

In Exercises 43–50, perform the indicated operation(s). (Write fractional answers in simplest form.)

43. $2\left(\dfrac{77}{-11}\right)$ **44.** $\dfrac{27 - 35}{4}$

45. $\frac{5}{8} + \frac{1}{4} - \frac{5}{6}$ **46.** $\frac{10}{11} + \frac{6}{33} - \frac{13}{66}$

47. $\frac{2}{5} \cdot \frac{7}{8}$ **48.** $\left(-\frac{2}{3}\right) \cdot \frac{5}{8} \cdot \frac{3}{4}$

49. $\frac{2}{3} \div 8$ **50.** $\left(\frac{3}{5} \div 3\right) - \left(6 \cdot \frac{4}{8}\right)$

In Exercises 51–54, use a calculator to evaluate the expression. (Round to two decimal places.)

51. $3\left(-\frac{5}{12} + \frac{3}{8}\right)$

52. $2\left(-7 + \frac{1}{6}\right)$

53. $\dfrac{11.46 - 5.37}{3.91}$

54. $\dfrac{-8.31 + 4.83}{7.65}$

In Exercises 55–58, use a calculator to solve.

55. 35% of 68

56. 35% of 820

57. 125% of 37

58. 147% of 22

59. Federal Government Expenses The circle graph shows the types of expenses for the federal government in 2005. *(Source: Office of Management and Budget)*

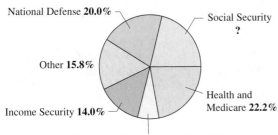

National Defense **20.0%**
Social Security **?**
Other **15.8%**
Income Security **14.0%**
Health and Medicare **22.2%**
Education and Veterans Benefits **6.8%**

(a) What percent of the total expenses was the amount spent on Social Security?

(b) The total of the 2005 expenses was $2,472,200,000,000. Find the amount spent for each category in the circle graph. (Round to the nearest billion dollars.)

60. Research Study The percent of people in a research study that have a particular health risk is 39.5%. The total number of people in the study is 12,857. How many people have the health risk?

61. Clinical Trial The percent of patients in a clinical trial of a cancer treatment showing a decrease in tumor size is 49.2%. There are 3445 patients in the trial. How many patients show a decrease in tumor size?

62. Calculator Keystrokes Write the algebraic expression that corresponds to each set of keystrokes.

(a) 5 ⨯ (2.7 − 9.4) =   Scientific

   5 (2.7 − 9.4) ENTER   Graphing

(b) 2 ⨯ (4 +/− + 2) =   Scientific

   2 ((−) 4 + 2) ENTER   Graphing

63. Calculator Keystrokes Write the keystrokes used to evaluate each algebraic expression on either a scientific or a graphing calculator.

(a) $5(18 - 2^3) \div 10$ (b) $-6^2 - [7 + (-2)^3]$

The symbol Ⓢ indicates an exercise in which you are instructed to use a spreadsheet.

Ⓢ In Exercises 64 and 65, a breakdown of pet spending for one year in the United States is given. Find the percent of total pet spending for each subcategory. Then use a spreadsheet software program to make a labeled circle graph for the percent data. *(Source: American Pet Products Manufacturers Association)*

64. Total pet spending (2005):   **$36.3 billion**

Food:	$14.7 billion
Vet care:	$8.7 billion
Supplies/OTC medicine:	$8.7 billion
Live animal purchases:	$1.7 billion
Grooming and boarding:	$2.5 billion

65. Total pet spending (2006):   **$38.4 billion**

Food:	$15.2 billion
Vet care:	$9.4 billion
Supplies/OTC medicine:	$9.3 billion
Live animal purchases:	$1.8 billion
Grooming and boarding:	$2.7 billion

Business Capsule

Tim Sloan/AFP/Getty Images

PetSmart, the largest U.S. pet store chain with 909 stores, has grown by offering pet lodging services in some stores. PetsHotels provide amenities such as supervised play areas with toys and slides, hypoallergenic lambskin blankets, TV, healthy pet snacks, and special fee services such as grooming, training, and phoning pet parents. These services are twice as profitable as retail sales, and they tend to attract greater sales as well. PetSmart's sales were 29% higher in stores with established PetsHotels than in those without them. In 2006, PetSmart had a goal to expand from 62 to 435 PetsHotels.

66. Research Project Use your campus library, the Internet, or some other reference source to find information about "special services" companies experiencing strong growth as in the example above. Write a brief report about one of these companies.

Section 0.3

Integer Exponents

- Use properties of exponents.
- Use scientific notation to represent real numbers.
- Use a calculator to raise a number to a power.
- Use interest formulas to solve an application problem.

Properties of Exponents

Repeated multiplication of a real number by itself can be written in **exponential form.** Here are some examples.

Repeated Multiplication	*Exponential Form*
$7 \cdot 7$	7^2
$a \cdot a \cdot a \cdot a \cdot a$	a^5
$(-4)(-4)(-4)$	$(-4)^3$
$(2x)(2x)(2x)(2x)$	$(2x)^4$

STUDY TIP

It is important to recognize the difference between exponential forms such as $(-2)^4$ and -2^4. In $(-2)^4$, the parentheses indicate that the exponent applies to the negative sign as well as to the 2, but in $-2^4 = -(2^4)$, the exponent applies only to the 2. Similarly, in $(5x)^3$, the parentheses indicate that the exponent applies to the 5 as well as to the x, whereas in $5x^3 = 5(x^3)$, the exponent applies only to the x.

Exponential Notation

Let a be a real number, a variable, or an algebraic expression, and let n be a positive integer. Then

$$a^n = \underbrace{a \cdot a \cdot a \cdots a}_{n \text{ factors}}$$

where n is the **exponent** and a is the **base.** The expression a^n is read as "a to the nth **power**" or simply "a to the nth."

When multiplying exponential expressions with the same base, *add* exponents.

$$a^m \cdot a^n = a^{m+n} \qquad \text{Add exponents when multiplying.}$$

For instance, to multiply 2^2 and 2^3, you can write

$$2^2 \cdot 2^3 = \overbrace{(2 \cdot 2)}^{\substack{\text{Two} \\ \text{factors}}} \cdot \overbrace{(2 \cdot 2 \cdot 2)}^{\substack{\text{Three} \\ \text{factors}}} = \overbrace{2 \cdot 2 \cdot 2 \cdot 2 \cdot 2}^{\substack{\text{Five} \\ \text{factors}}}$$

$$= 2^{2+3} = 2^5.$$

On the other hand, when dividing exponential expressions, *subtract* exponents. That is,

$$\frac{a^m}{a^n} = a^{m-n}, \quad a \neq 0. \qquad \text{Subtract exponents when dividing.}$$

These and other properties of exponents are summarized in the list on the following page.

Properties of Exponents

Let a and b be real numbers, variables, or algebraic expressions, and let m and n be integers. (Assume all denominators and bases are nonzero.)

Property		*Example*	
1. $a^m a^n = a^{m+n}$		$3^2 \cdot 3^4 = 3^{2+4} = 3^6$	Product of Powers
2. $\dfrac{a^m}{a^n} = a^{m-n}$		$\dfrac{x^7}{x^4} = x^{7-4} = x^3$	Quotient of Powers
3. $(ab)^m = a^m b^m$		$(5x)^3 = 5^3 x^3 = 125x^3$	Power of a Product
4. $\left(\dfrac{a}{b}\right)^m = \dfrac{a^m}{b^m}$		$\left(\dfrac{2}{x}\right)^3 = \dfrac{2^3}{x^3} = \dfrac{8}{x^3}$	Power of a Quotient
5. $(a^m)^n = a^{mn}$		$(y^3)^{-4} = y^{3(-4)} = y^{-12}$	Power of a Power
6. $a^{-n} = \dfrac{1}{a^n}$		$y^{-4} = \dfrac{1}{y^4}$	Definition of negative exponent
7. $a^0 = 1, \quad a \neq 0$		$(x^2 + 1)^0 = 1$	Definition of zero exponent
8. $\left(\dfrac{a}{b}\right)^{-n} = \left(\dfrac{b}{a}\right)^n, \quad a \neq 0, b \neq 0$		$\left(\dfrac{3}{2}\right)^{-3} = \left(\dfrac{2}{3}\right)^3$	
9. $|a^2| = |a|^2 = a^2$		$|2^2| = |2|^2 = 2^2$	

Notice that these properties of exponents apply for *all* integers m and n, not just positive integers. For instance, by the Quotient of Powers Property,

$$\frac{3^4}{3^{-5}} = 3^{4-(-5)} = 3^{4+5} = 3^9.$$

DISCOVERY

Using your calculator, find the values of 10^3, 10^2, 10^1, 10^0, 10^{-1}, and 10^{-2}. What do you observe?

Example 1   Using Properties of Exponents

a. $3^4 \cdot 3^{-1} = 3^{4-1} = 3^3 = 27$

b. $\dfrac{5^6}{5^4} = 5^{6-4} = 5^2 = 25$

c. $5\left(\dfrac{2}{5}\right)^3 = 5 \cdot \dfrac{2^3}{5^3} = 5 \cdot 5^{-3} \cdot 2^3 = 5^{-2} \cdot 2^3 = \dfrac{2^3}{5^2} = \dfrac{8}{25}$

d. $(-5 \cdot 2^3)^2 = (-5)^2 \cdot (2^3)^2 = 25 \cdot 2^6 = 25 \cdot 64 = 1600$

e. $(-3ab^4)(4ab^{-3}) = -3(4)(a)(a)(b^4)(b^{-3}) = -12a^2 b$

f. $3a(-4a^2)^0 = 3a(1) = 3a, \quad a \neq 0$

g. $\left(\dfrac{5x^3}{y}\right)^2 = \dfrac{5^2(x^3)^2}{y^2} = \dfrac{25x^6}{y^2}$

✓ CHECKPOINT 1

Evaluate the expression $4^2 \cdot 4^3$. ∎

The next example shows how expressions involving negative exponents can be rewritten using positive exponents.

Example 2   **Rewriting with Positive Exponents**

a. $x^{-1} = \dfrac{1}{x}$ Definition of negative exponent

b. $\dfrac{1}{3x^{-2}} = \dfrac{1(x^2)}{3} = \dfrac{x^2}{3}$ The exponent -2 applies only to x.

c. $\dfrac{12a^3b^{-4}}{4a^{-2}b} = \dfrac{12a^3 \cdot a^2}{4b \cdot b^4}$ Definition of negative exponent

$\phantom{\dfrac{12a^3b^{-4}}{4a^{-2}b}} = \dfrac{3a^5}{b^5}$ Product of Powers Property

d. $\left(\dfrac{3x^2}{y}\right)^{-2} = \dfrac{3^{-2}(x^2)^{-2}}{y^{-2}}$ Power of a Quotient and Power of a Product Properties

$\phantom{\left(\dfrac{3x^2}{y}\right)^{-2}} = \dfrac{3^{-2}x^{-4}}{y^{-2}}$ Power of a Power Property

$\phantom{\left(\dfrac{3x^2}{y}\right)^{-2}} = \dfrac{y^2}{3^2x^4}$ Definition of negative exponent

$\phantom{\left(\dfrac{3x^2}{y}\right)^{-2}} = \dfrac{y^2}{9x^4}$ Simplify.

✓ **CHECKPOINT 2**

Rewrite $\left(\dfrac{3}{x^{-2}z^4}\right)^{-3}$ with positive exponents and simplify. ■

Example 3   **Ratio of Volume to Surface Area**

The volume V and surface area S of a sphere are given by

$$V = \frac{4}{3}\pi r^3 \quad \text{and} \quad S = 4\pi r^2$$

where r is the radius of the sphere. A spherical weather balloon has a radius of 2 feet, as shown in Figure 0.7. Find the ratio of the volume to the surface area.

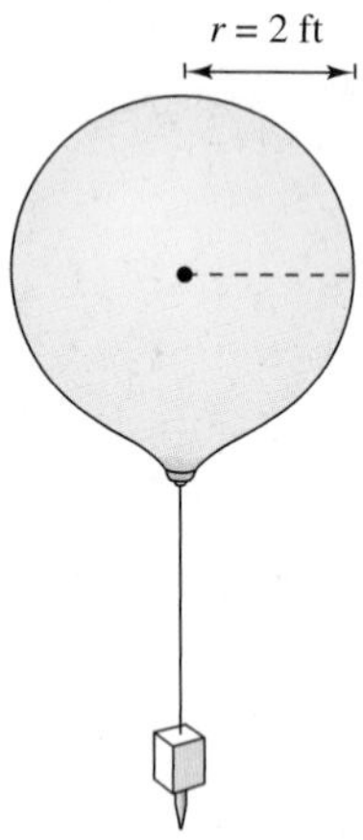

SOLUTION   To find the ratio, write the quotient V/S and simplify.

$$\frac{V}{S} = \frac{\frac{4}{3}\pi r^3}{4\pi r^2} = \frac{\frac{4}{3}\pi 2^3}{4\pi 2^2} = \frac{1}{3}(2) = \frac{2}{3}$$

✓ **CHECKPOINT 3**

Evaluate $\dfrac{\frac{5}{7}x^7}{25x^5}$ when $x = 7$. ■

$r = 2$ ft

FIGURE 0.7

Scientific Notation

Exponents provide an efficient way of writing and computing with very large (or very small) numbers. For instance, a drop of water contains more than 33 billion billion molecules—that is, 33 followed by 18 zeros.

33,000,000,000,000,000,000

It is convenient to write such numbers in **scientific notation.** This notation has the form $c \times 10^n$, where $1 \leq c < 10$ and n is an integer. So, the number of molecules in a drop of water can be written in scientific notation as

$3.3 \times 10,000,000,000,000,000,000 = 3.3 \times 10^{19}.$

The *positive* exponent 19 indicates that the number is *large* (10 or more) and that the decimal point has been moved 19 places. A *negative* exponent in scientific notation indicates that the number is *small* (less than 1). For instance, the mass (in grams) of one electron is approximately

$9.0 \times 10^{-28} = 0.0000000000000000000000000009.$

28 decimal places

Example 4   Converting to Scientific Notation

a. $0.0000572 = 5.72 \times 10^{-5}$ Number is less than 1.

b. $149,400,000 = 1.494 \times 10^8$ Number is greater than 10.

c. $32.675 = 3.2675 \times 10^1$ Number is greater than 10.

✓ CHECKPOINT 4

Write 0.00345 in scientific notation. ■

Example 5   Converting to Decimal Notation

a. $3.125 \times 10^2 = 312.5$ Number is greater than 10.

b. $3.73 \times 10^{-6} = 0.00000373$ Number is less than 1.

c. $7.91 \times 10^5 = 791,000$ Number is greater than 10.

✓ CHECKPOINT 5

Write 4.28×10^5 in decimal notation. ■

Most calculators automatically use scientific notation when showing large (or small) numbers that exceed the display range. Try multiplying $86,500,000 \times 6000$. If your calculator follows standard conventions, its display should be

5.19 11   or   5.19E11 .

This means that $c = 5.19$ and the exponent of 10 is $n = 11$, which implies that the number is 5.19×10^{11}. To *enter* numbers in scientific notation, your calculator should have an exponential entry key labeled (EXP) or (EE).

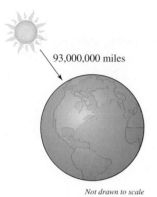

93,000,000 miles

Not drawn to scale

FIGURE 0.8

✓ **CHECKPOINT 6**

Evaluate the expression

$$\frac{4.6 \times 10^5}{2.3 \times 10^2}.$$ ∎

Example 6 **The Speed of Light**

The distance between Earth and the sun is approximately 93 million miles, as shown in Figure 0.8. How long does it take for light to travel from the sun to Earth? Use the fact that light travels at a rate of approximately 186,000 miles per second.

SOLUTION   Using the formula distance = (rate)(time), you find the time as follows.

$$\text{Time} = \frac{\text{distance}}{\text{rate}} = \frac{93 \text{ million miles}}{186,000 \text{ miles per second}}$$

$$= \frac{9.3 \times 10^7 \text{ miles}}{1.86 \times 10^5 \text{ miles/second}}$$

$$= 5 \times 10^2 \text{ seconds}$$

$$\approx 8.33 \text{ minutes}$$

Note that to convert 500 seconds to 8.33 minutes, you divide by 60, because there are 60 seconds in one minute.

Powers and Calculators

One of the most useful features of a calculator is its ability to evaluate exponential expressions. Consult your user's guide for specific keystrokes.

Example 7 **Using a Calculator to Raise a Number to a Power**

Scientific Calculator

	Expression	Keystrokes	Display
a.	$13^4 + 5$	13 [yˣ] 4 [+] 5 [=]	28566
b.	$3^{-2} + 4^{-1}$	3 [yˣ] 2 [+/−] [+] 4 [yˣ] 1 [+/−] [=]	0.361111111
c.	$\dfrac{3^5 + 1}{3^5 - 1}$	[(] 3 [yˣ] 5 [+] 1 [)] [÷] [(] 3 [yˣ] 5 [−] 1 [)] [=]	1.008264463

Graphing Calculator

	Expression	Keystrokes	Display
a.	$13^4 + 5$	13 [∧] 4 [+] 5 [ENTER]	28566
b.	$3^{-2} + 4^{-1}$	3 [∧] [(−)] 2 [+] 4 [∧] [(−)] 1 [ENTER]	.3611111111
c.	$\dfrac{3^5 + 1}{3^5 - 1}$	[(] 3 [∧] 5 [+] 1 [)] [÷] [(] 3 [∧] 5 [−] 1 [)] [ENTER]	1.008264463

✓ **CHECKPOINT 7**

Use a calculator to evaluate

$$\frac{4^4 - 6}{2^5 + 18}.$$ ∎

Applications

The formulas shown below can be used to find the balance in a savings account.

Balance in an Account

The balance A in an account that earns an annual interest rate r (in decimal form) for t years is given by one of the following.

$$A = P(1 + rt) \qquad \text{Simple interest}$$

$$A = P\left(1 + \frac{r}{n}\right)^{nt} \qquad \text{Compound interest}$$

In both formulas, P is the principal (or the initial deposit). In the formula for compound interest, n is the number of compoundings *per year*. Make sure you convert all units of time t to years. For instance, 6 months $= \frac{1}{2}$ year. So, $t = \frac{1}{2}$.

Example 8
MAKE A DECISION   **Finding the Balance in an Account**

You are trying to decide how to invest $5000 for 10 years. Which savings plan will earn more money?

a. 4% simple annual interest

b. 3.5% interest compounded quarterly

SOLUTION

a. The balance after 10 years is

$$A = P(1 + rt)$$
$$= 5000[1 + 0.04(10)]$$
$$= \$7000.$$

b. The balance after 10 years is

$$A = P\left(1 + \frac{r}{n}\right)^{nt}$$
$$= 5000\left(1 + \frac{0.035}{4}\right)^{(4)(10)}$$
$$\approx \$7084.54.$$

Savings plan (a) will earn $7000 - 5000 = \$2000$ and savings plan (b) will earn $7084.54 - 5000 = \$2084.54$. So, plan (b) will earn more money.

✓ **CHECKPOINT 8**

In Example 8, how much money would you earn in a savings plan with 3.4% annual interest compounded monthly? ■

In addition to finding the balance in an account, the compound interest formula can also be used to determine the rate of inflation. To apply the formula, you must know the cost of an item for two different years, as demonstrated in Example 9.

Example 9 Finding the Rate of Inflation

In 1984, the cost of a first-class postage stamp was $0.20. By 2007, the cost increased to $0.41, as shown in Figure 0.9. Find the average annual rate of inflation for first-class postage over this 23-year period.   *(Source: U.S. Postal Service)*

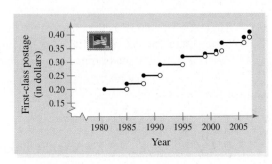

FIGURE 0.9

SOLUTION   To find the average annual rate of inflation, use the formula for compound interest with *annual* compounding. So, you need to find the value of r that will make the following equation true.

$$A = P\left(1 + \frac{r}{n}\right)^{nt}$$

$$0.41 = 0.20(1 + r)^{23}$$

You can begin by guessing that the average annual rate of inflation was 5%. Entering $r = 0.05$ in the formula, you find that $0.20(1 + 0.05)^{23} \approx 0.6143$. Because this result is more than 0.41, try some smaller values of r. Finally, you can discover that

$$0.20(1 + 0.032)^{23} \approx 0.41.$$

So, the average annual rate of inflation for first-class postage from 1984 to 2007 was about 3.2%.

✓ CHECKPOINT 9

The fee for a medical school application was $85. Three years later, the application fee is $95. What is the average annual rate of inflation over this three-year period? ▪

CONCEPT CHECK

1. Explain how to simplify the expression $a^{0.5}(a^{1.5})$.

2. Because $-2^3 = -8$ and $(-2)^3 = -8$, a student concludes that $-a^n = (-a)^n$, where n is an integer. Do you agree? Can you give an example where $-a^n \neq (-a)^n$?

3. A student claims "Any number with a zero exponent is equal to 1." Is the student correct? Explain.

4. Is 0.12×10^5 written in scientific notation? Explain.

Skills Review 0.3

The following warm-up exercises involve skills that were covered in earlier sections. You will use these skills in the exercise set for this section. For additional help, review Section 0.2.

In Exercises 1–10, perform the indicated operation(s) and simplify.

1. $\left(\frac{2}{3}\right)\left(\frac{3}{2}\right)$

2. $\left(\frac{1}{4}\right)(5)(4)$

3. $3\left(\frac{2}{7}\right) + 11\left(\frac{2}{7}\right)$

4. $11\left(\frac{1}{4}\right) + \frac{5}{4}$

5. $\frac{1}{2} \div 2$

6. $\frac{1}{3} \div \frac{1}{3}$

7. $\frac{1}{7} + \frac{1}{3} - \frac{1}{21}$

8. $\frac{1}{3} + \frac{1}{2} - \frac{5}{6}$

9. $\frac{1}{12} - \frac{1}{3} + \frac{1}{8}$

10. $\left(\frac{1}{2} - \frac{1}{3}\right) \div \frac{1}{6}$

Exercises 0.3

See www.CalcChat.com for worked-out solutions to odd-numbered exercises.

In Exercises 1–20, evaluate the expression. Write fractional answers in simplest form.

1. $2^2 \cdot 2^4$

2. $3 \cdot 3^5$

3. $\dfrac{2^6}{2^3}$

4. $\dfrac{5^7}{5^5}$

5. $(3^3)^2$

6. $(2^5)^3$

7. -3^4

8. $(-3)^4$

9. $8 \cdot 2^{-2} \cdot 4^{-1}$

10. $6 \cdot 2^{-3} \cdot 3^{-1}$

11. $\left(\frac{1}{2}\right)^{-3}$

12. $\left(\frac{2}{3}\right)^{-3}$

13. $5^{-1} - 2^{-1}$

14. $4^{-1} - 2^{-2}$

15. $(2^3 \cdot 3^2)^2$

16. $(-3 \cdot 4^2)^3$

17. $\left(-\frac{3}{5}\right)^3\left(\frac{5}{3}\right)^2$

18. $\left(\frac{-5}{4}\right)^3\left(\frac{4}{5}\right)^2$

19. 3^0

20. $(-2)^0$

In Exercises 21–24, evaluate the expression for the indicated value of x.

	Expression	Value
21.	$\dfrac{x^4}{2x^2}$	$x = -6$
22.	$4x^{-3}$	$x = 2$
23.	$7x^{-2}$	$x = 4$
24.	$8x^0 - (8x)^0$	$x = -7$

In Exercises 25–44, simplify the expression.

25. $(-5z)^3$

26. $(-2w)^5$

27. $(8x^4)(2x^3)$

28. $5x^4(x^2)$

29. $10(x^2)^2$

30. $(4x^4)^3$

31. $(-z)^3(3z^4)$

32. $(6y^2)(2y^3)^3$

33. $\dfrac{25y^8}{10y^4}$

34. $\dfrac{10x^9}{4x^6}$

35. $\left(\dfrac{4}{y}\right)^3\left(\dfrac{3}{y}\right)^4$

36. $\left(\dfrac{5}{z}\right)^2\left(\dfrac{2}{z}\right)^3$

37. $\dfrac{15(x+3)^3}{9(x+3)^2}$

38. $\dfrac{24(x-2)^2}{8(x-2)^4}$

39. $\dfrac{7x^2}{x^3}$

40. $\dfrac{5z^5}{z^7}$

41. $\dfrac{x^2 \cdot x^n}{x^3 \cdot x^n}$

42. $\dfrac{x^n \cdot x^{2n}}{x^{3n}}$

43. $3^n \cdot 3^{2n}$

44. $2^m \cdot 2^{3m}$

In Exercises 45–54, rewrite the expression with positive exponents and simplify.

45. $(2x^5)^0, \quad x \neq 0$

46. $(x+5)^0, \quad x \neq -5$

47. $(y+2)^{-2}(y+2)^{-1}$

48. $(x+y)^{-5}(x+y)^9$

49. $(4y^{-2})(8y^4)$

50. $(-2x^2)^3(4x^3)^{-1}$

51. $\left(\dfrac{x}{10}\right)^{-1}$

52. $\left(\dfrac{y}{5}\right)^{-2}$

53. $\left(\dfrac{x^{-3}y^4}{5}\right)^{-3}$

54. $\left(\dfrac{2z^2}{y}\right)^{-2}$

In Exercises 55–60, write the number in scientific notation.

55. Land Area of Earth: 57,300,000 square miles

56. Water Area of Earth: 139,500,000 square miles

57. Light Year: 9,460,000,000,000 kilometers

58. Mass of a Bacterium: 0.0000000000000003 gram

59. Thickness of a Soap Bubble: 0.0000001 meter

60. One Micron (millionth of a meter): 0.00003937 inch

In Exercises 61–64, write the number in decimal notation.

61. Number of Air Sacs in the Lungs: 3.5×10^8

62. Temperature of the Core of the Sun:
1.5×10^7 degrees Celsius

63. Charge of an Electron: 1.602×10^{-19} coulomb

64. Width of a Human Hair: 9.0×10^{-5} meter

In Exercises 65 and 66, evaluate each expression without using a calculator.

65. (a) $(1.2 \times 10^7)(5 \times 10^{-3})$ (b) $\dfrac{6.0 \times 10^8}{3.0 \times 10^{-3}}$

66. (a) $(9.8 \times 10^{-2})(3 \times 10^7)$ (b) $\dfrac{9.0 \times 10^5}{4.5 \times 10^{-2}}$

In Exercises 67–70, use a calculator to evaluate each expression. Write your answer in scientific notation. (Round to three decimal places.)

67. (a) $0.000345(8,900,000,000)$

(b) $\dfrac{67,000,000 + 93,000,000}{0.0052}$

68. (a) $0.000045(9,200,000)$

(b) $\dfrac{0.0000928 - 0.0000021}{0.0061}$

69. (a) $(9.3 \times 10^6)^3(6.1 \times 10^{-4})$ (b) $\dfrac{(2.414 \times 10^4)^6}{(1.68 \times 10^5)^5}$

70. (a) $(1.2 \times 10^2)^2(5.3 \times 10^{-5})$ (b) $\dfrac{(3.28 \times 10^{-6})^{10}}{(5.34 \times 10^{-3})^{22}}$

In Exercises 71 and 72, write each number in scientific notation. Perform the operation and write your answer in scientific notation.

71. (a) $48,000,000,000(250,000,000)$ (b) $\dfrac{0.000000012}{0.0000064}$

72. (a) $0.00000034(0.00000006)$ (b) $\dfrac{18,000,000,000}{2,400,000}$

73. Balance in an Account   You deposit \$10,000 in an account with an annual interest rate of 6.75% for 12 years. Determine the balance in the account when the interest is compounded (a) daily ($n = 365$), (b) weekly, (c) monthly, and (d) quarterly. How is the balance affected by the type of compounding?

74. Balance in an Account   You deposit \$2000 in an account with an annual interest rate of 7.5% for 15 years. Determine the balance in the account when the interest is compounded (a) daily ($n = 365$), (b) weekly, (c) monthly, and (d) quarterly. How is the balance affected by the type of compounding?

75. College Costs   The bar graph shows the average yearly costs of attending a public four-year college in the United States for the academic years 1998/1999 to 2004/2005. Find the average rate of inflation over this seven-year period. *(Source: U.S. National Center for Education Statistics)*

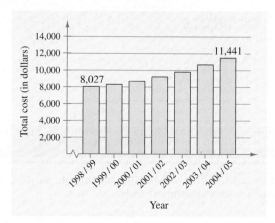

76. College Costs   The average yearly cost of tuition, fees, and room and board at private four-year colleges in the United States was \$19,929 for the academic year 1998/1999 and \$26,489 for the academic year 2004/2005. Find the average yearly rate of inflation over this seven-year period. *(Source: U.S. National Center for Education Statistics)*

77. Becoming a Millionaire   The compound interest formula can be rewritten as

$$P = \frac{A}{(1 + r/n)^{nt}}.$$

Find the principal amount P that would have had to have been invested on the day you were born at 7.5% annual interest compounded quarterly to make you a millionaire on your 21st birthday.

78. Electron Microscopes   Electron microscopes provide greater magnification than traditional light microscopes by using focused beams of electrons instead of visible light. It is the extremely short wavelengths of the electron beams that make electron microscopes so powerful. The wavelength λ (in meters) of any object in motion is given by $\lambda = \dfrac{6.626 \times 10^{-34}}{mv}$, where m is the mass (in kilograms) of the object and v is its velocity (in meters per second). Find the wavelength of an electron with a mass of 9.11×10^{-31} kilogram and a velocity of 5.9×10^6 meters per second.

(Submitted by Brian McIntyre, Senior Laboratory Engineer for the Optics Electron Microscopy Facility at the University of Rochester.)

Section 0.4

Radicals and Rational Exponents

- Simplify a radical.
- Rationalize a denominator.
- Use properties of rational exponents.
- Combine radicals.
- Use a calculator to evaluate a radical.
- Use a radical expression to solve an application problem.

Radicals and Properties of Radicals

A **square root** of a number is defined as one of its two equal factors. For example, 5 is a square root of 25 because 5 is one of the two equal factors of 25. In a similar way, a **cube root** of a number is one of its three equal factors. Here are some examples.

Number	*Equal Factors*	*Root*
$25 = (-5)^2$	$(-5)(-5)$	-5 (square root)
$-64 = (-4)^3$	$(-4)(-4)(-4)$	-4 (cube root)
$81 = 3^4$	$3 \cdot 3 \cdot 3 \cdot 3$	3 (fourth root)

Definition of *n*th Root of a Number

Let a and b be real numbers and let n be a positive integer. If

$$a = b^n$$

then b is an ***n*th root of a.** If $n = 2$, the root is a **square root,** and if $n = 3$, the root is a **cube root.**

From this definition, you can see that some numbers have more than one *n*th root. For example, both 5 and -5 are square roots of 25. The following definition distinguishes between these two roots.

Principal *n*th Root of a Number

Let a be a real number that has at least one real *n*th root. **The principal *n*th root of a** is the *n*th root that has the same sign as a, and it is denoted by the **radical symbol**

$$\sqrt[n]{a}.$$ Principal *n*th root

The positive integer n is the **index** (the plural of index is *indexes* or *indices*) of the radical, and the number a is the **radicand.** If $n = 2$, omit the index and write $\sqrt{a}$ rather than $\sqrt[2]{a}$.

Example 1   Evaluating Expressions Involving Radicals

a. The principal square root of 121 is $\sqrt{121} = 11$ because $11^2 = 121$.

b. The principal cube root of $\frac{125}{64}$ is $\sqrt[3]{\frac{125}{64}} = \frac{5}{4}$ because $\left(\frac{5}{4}\right)^3 = \frac{5^3}{4^3} = \frac{125}{64}$.

c. The principal fifth root of -32 is $\sqrt[5]{-32} = -2$ because $(-2)^5 = -32$.

d. $-\sqrt{49} = -7$ because $7^2 = 49$.

e. $\sqrt[4]{-81}$ is not a real number because there is no real number that can be raised to the fourth power to produce -81.

✓ CHECKPOINT 1

Evaluate $-\sqrt[3]{-8}$. ▪

From Example 1, you can make the following generalizations about nth roots of a real number.

1. If a is a positive real number and n is a positive *even* integer, then a has exactly two real nth roots, which are denoted by $\sqrt[n]{a}$ and $-\sqrt[n]{a}$.

2. If a is any real number and n is an *odd* integer, then a has only one (real) nth root. It is the principal nth root and is denoted by $\sqrt[n]{a}$.

3. If a is negative and n is an *even* integer, then a has no (real) nth root.

Integers such as 1, 4, 9, 16, 49, and 81 are called **perfect squares** because they have integer square roots. Similarly, integers such as 1, 8, 27, 64, and 125 are called **perfect cubes** because they have integer cube roots.

Properties of Radicals

Let a and b be real numbers, variables, or algebraic expressions such that the indicated roots are real numbers, and let m and n be positive integers. Then the following properties are true.

Property	Example
1. $\sqrt[n]{a^m} = \left(\sqrt[n]{a}\right)^m$	$\sqrt[3]{8^2} = \left(\sqrt[3]{8}\right)^2 = (2)^2 = 4$
2. $\sqrt[n]{a} \cdot \sqrt[n]{b} = \sqrt[n]{ab}$	$\sqrt{5} \cdot \sqrt{7} = \sqrt{5 \cdot 7} = \sqrt{35}$
3. $\dfrac{\sqrt[n]{a}}{\sqrt[n]{b}} = \sqrt[n]{\dfrac{a}{b}}, \quad b \neq 0$	$\dfrac{\sqrt[4]{27}}{\sqrt[4]{9}} = \sqrt[4]{\dfrac{27}{9}} = \sqrt[4]{3}$
4. $\sqrt[m]{\sqrt[n]{a}} = \sqrt[mn]{a}$	$\sqrt[3]{\sqrt{10}} = \sqrt[6]{10}$
5. $\left(\sqrt[n]{a}\right)^n = a$	$\left(\sqrt{3}\right)^2 = 3$
6. For n even, $\sqrt[n]{a^n} = \|a\|$.	$\sqrt{(-12)^2} = \|-12\| = 12$
For n odd, $\sqrt[n]{a^n} = a$.	$\sqrt[3]{(-12)^3} = -12$

A common special case of Property 6 is

$$\sqrt{a^2} = |a|.$$

Simplifying Radicals

An expression involving radicals is in **simplest form** when the following conditions are satisfied.

1. All possible factors have been removed from the radical.

2. All fractions have radical-free denominators (accomplished by a process called *rationalizing the denominator*).

3. The index of the radical has been reduced as far as possible.

To simplify a radical, factor the radicand into factors whose exponents are multiples of the index. The roots of these factors are written outside the radical, and the "leftover" factors make up the new radicand.

Example 2   **Simplifying Even Roots**

a. $\sqrt[4]{48} = \sqrt[4]{16 \cdot 3}$ Find largest fourth-power factor.

$\qquad = \sqrt[4]{2^4 \cdot 3}$ Rewrite.

$\qquad = 2\sqrt[4]{3}$ Find fourth root.

b. $\sqrt{75x^3} = \sqrt{25x^2 \cdot 3x}$ Find largest square factor.

$\qquad = \sqrt{(5x)^2 \cdot 3x}$ Rewrite.

$\qquad = 5x\sqrt{3x}, \quad x \geq 0$ Find root of perfect square.

c. $\sqrt[4]{(5x)^4} = |5x| = 5|x|$

✓ CHECKPOINT 2

Simplify $\sqrt{18x^5}$. ∎

In Example 2(c), note that the absolute value symbol is included in the answer because $\sqrt[4]{x^4} = |x|$.

Example 3   **Simplifying Odd Roots**

a. $\sqrt[3]{24} = \sqrt[3]{8 \cdot 3}$ Find largest cube factor.

$\qquad = \sqrt[3]{2^3 \cdot 3}$ Rewrite.

$\qquad = 2\sqrt[3]{3}$ Find root of perfect cube.

b. $\sqrt[5]{32a^{11}} = \sqrt[5]{32a^{10} \cdot a}$ Find largest fifth-power factor.

$\qquad = \sqrt[5]{(2a^2)^5 \cdot a}$ Rewrite.

$\qquad = 2a^2\sqrt[5]{a}$ Find fifth root.

c. $\sqrt[3]{-40x^6} = \sqrt[3]{(-8x^6) \cdot 5}$ Find largest cube factor.

$\qquad = \sqrt[3]{(-2x^2)^3 \cdot 5}$ Rewrite.

$\qquad = -2x^2\sqrt[3]{5}$ Find root of perfect cube.

✓ CHECKPOINT 3

Simplify $\sqrt[3]{54x^4}$. ∎

Some fractions have radicals in the denominator. To **rationalize a denominator** of the form $a + b\sqrt{m}$, multiply the numerator and denominator by the **conjugate** $a - b\sqrt{m}$.

$$a + b\sqrt{m} \text{ and } a - b\sqrt{m} \qquad \text{Conjugates}$$

When $a = 0$, the rationalizing factor of $\sqrt{m}$ is itself, $\sqrt{m}$.

Example 4 | Rationalizing Single-Term Denominators

a. To rationalize the denominator of the following fraction, multiply *both* the numerator and the denominator by $\sqrt{3}$ to obtain

$$\frac{5}{2\sqrt{3}} = \frac{5}{2\sqrt{3}} \cdot \frac{\sqrt{3}}{\sqrt{3}} = \frac{5\sqrt{3}}{2\sqrt{3^2}} = \frac{5\sqrt{3}}{2(3)} = \frac{5\sqrt{3}}{6}.$$

b. To rationalize the denominator of the following fraction, multiply *both* the numerator and the denominator by $\sqrt[3]{5^2}$. Note how this eliminates the radical from the denominator by producing a perfect *cube* in the radicand.

$$\frac{2}{\sqrt[3]{5}} = \frac{2}{\sqrt[3]{5}} \cdot \frac{\sqrt[3]{5^2}}{\sqrt[3]{5^2}} = \frac{2\sqrt[3]{5^2}}{\sqrt[3]{5^3}} = \frac{2\sqrt[3]{25}}{5}$$

✓ CHECKPOINT 4

Simplify $\dfrac{1}{\sqrt[3]{4}}$ by rationalizing the denominator. ■

Example 5 | Rationalizing a Denominator with Two Terms

$$\frac{2}{3 + \sqrt{7}} = \frac{2}{3 + \sqrt{7}} \cdot \frac{3 - \sqrt{7}}{3 - \sqrt{7}}$$ Multiply numerator and denominator by conjugate.

$$= \frac{2(3 - \sqrt{7})}{3^2 - (\sqrt{7})^2}$$ Multiply fractions.

$$= \frac{2(3 - \sqrt{7})}{9 - 7}$$ Simplify.

$$= \frac{2(3 - \sqrt{7})}{2}$$ Divide out like factors.

$$= 3 - \sqrt{7}$$ Simplify.

✓ CHECKPOINT 5

Simplify $\dfrac{6}{3 - \sqrt{3}}$ by rationalizing the denominator. ■

Don't confuse an expression such as $\sqrt{2} + \sqrt{7}$ with $\sqrt{2 + 7}$. In general, $\sqrt{x + y} \neq \sqrt{x} + \sqrt{y}$.

Rational Exponents

The definition on the following page shows how radicals are used to define **rational exponents**. Until now, work with exponents has been restricted to integer exponents.

The numerator of a rational exponent denotes the *power* to which the base is raised, and the denominator denotes the *index* or the *root* to be taken. It doesn't matter which operation is performed first, provided the nth root exists. Here is an example.

$$8^{2/3} = \left(\sqrt[3]{8}\right)^2 = 2^2 = 4$$

$$8^{2/3} = \sqrt[3]{8^2} = \sqrt[3]{64} = 4$$

Definition of Rational Exponents

If a is a real number and n is a positive integer such that the principal nth root of a exists, then $a^{1/n}$ is defined to be $a^{1/n} = \sqrt[n]{a}$.

If m is a positive integer that has no common factor with n, then
$$a^{m/n} = (a^{1/n})^m = \left(\sqrt[n]{a}\right)^m \text{ and } a^{m/n} = (a^m)^{1/n} = \sqrt[n]{a^m}.$$

The properties of exponents that were listed in Section 0.3 also apply to rational exponents (provided the roots indicated by the denominators exist). Some of those properties are relisted here, with different examples.

Properties of Exponents

Let r and s be rational numbers, and let a and b be real numbers, variables, or algebraic expressions. If the roots indicated by the rational exponents exist, then the following properties are true.

Property	*Example*
1. $a^r a^s = a^{r+s}$	$4^{1/2}(4^{1/3}) = 4^{5/6}$
2. $\dfrac{a^r}{a^s} = a^{r-s}, \quad a \neq 0$	$\dfrac{x^2}{x^{1/2}} = x^{2-(1/2)} = x^{3/2}, x \neq 0$
3. $(ab)^r = a^r b^r$	$(2x)^{1/2} = 2^{1/2}(x^{1/2})$
4. $\left(\dfrac{a}{b}\right)^r = \dfrac{a^r}{b^r}, \quad b \neq 0$	$\left(\dfrac{x}{3}\right)^{1/3} = \dfrac{x^{1/3}}{3^{1/3}}$
5. $(a^r)^s = a^{rs}$	$(x^3)^{1/3} = x$
6. $a^{-r} = \dfrac{1}{a^r}, \quad a \neq 0$	$4^{-1/2} = \dfrac{1}{4^{1/2}} = \dfrac{1}{2}$
7. $\left(\dfrac{a}{b}\right)^{-r} = \left(\dfrac{b}{a}\right)^r, \quad a \neq 0, \quad b \neq 0$	$\left(\dfrac{x}{4}\right)^{-1/2} = \left(\dfrac{4}{x}\right)^{1/2} = \dfrac{2}{x^{1/2}}$

Rational exponents can be tricky. Remember, the expression $b^{m/n}$ is not defined unless $\sqrt[n]{b}$ is a real number. For instance, $(-8)^{5/6}$ is not defined because $\sqrt[6]{-8}$ is not a real number. And yet, $(-8)^{2/3}$ is defined because $\sqrt[3]{-8} = -2$.

Rational exponents are particularly useful for evaluating roots of numbers on a calculator, for reducing the index of a radical, and for simplifying (and factoring) algebraic expressions. Examples 6 and 7 demonstrate some of these uses.

Example 6   Simplifying with Rational Exponents

a. $(27)^{1/3} = \sqrt[3]{27} = 3$

b. $(-32)^{-4/5} = \left(\sqrt[5]{-32}\right)^{-4} = (-2)^{-4} = \dfrac{1}{(-2)^4} = \dfrac{1}{16}$

c. $(-5x^{2/3})(3x^{-1/3}) = -15x^{(2/3)-(1/3)} = -15x^{1/3}, \quad x \neq 0$

✓CHECKPOINT 6

Simplify $(3^{1/2})(3^{3/2})$. ■

Example 7   **Reducing the Index of a Radical**

a. $\sqrt[6]{a^4} = a^{4/6} = a^{2/3} = \sqrt[3]{a^2}$

b. $\sqrt[3]{\sqrt{125}} = (125^{1/2})^{1/3}$ Rewrite with rational exponents.

$\qquad\qquad = (125)^{1/6}$ Multiply exponents.

$\qquad\qquad = (5^3)^{1/6}$ Rewrite base as perfect cube.

$\qquad\qquad = 5^{3/6}$ Multiply exponents.

$\qquad\qquad = 5^{1/2}$ Reduce exponent.

$\qquad\qquad = \sqrt{5}$ Rewrite as radical.

✓ CHECKPOINT 7

Use rational exponents to reduce the index of the radical $\sqrt[3]{2^6}$. ■

Radical expressions can be combined (added or subtracted) if they are **like radicals**—that is, if they have the same index and radicand. For instance, $2\sqrt{3x}$ and $\frac{1}{2}\sqrt{3x}$ are like radicals, but $\sqrt[3]{3x}$ and $2\sqrt{3x}$ are not like radicals.

Example 8   **Simplifying and Combining Like Radicals**

a. $2\sqrt{48} + 3\sqrt{27} = 2\sqrt{16 \cdot 3} + 3\sqrt{9 \cdot 3}$ Find square factors.

$\qquad\qquad\qquad = 8\sqrt{3} + 9\sqrt{3}$ Find square roots.

$\qquad\qquad\qquad = 17\sqrt{3}$ Combine like terms.

b. $\sqrt[3]{16x} - \sqrt[3]{54x} = \sqrt[3]{8 \cdot 2x} - \sqrt[3]{27 \cdot 2x}$ Find cube factors.

$\qquad\qquad\qquad = 2\sqrt[3]{2x} - 3\sqrt[3]{2x}$ Find cube roots.

$\qquad\qquad\qquad = -\sqrt[3]{2x}$ Combine like terms.

✓ CHECKPOINT 8

Simplify the expression $\sqrt{25x} + \sqrt{x}$. ■

Radicals and Calculators

You can use a calculator to evaluate radicals by using the square root key ⌐√⌐ , the cube root key ⌐∛⌐, or the xth root key ⌐ˣ√⌐. You can also use the exponential key ⌐^⌐ or ⌐yˣ⌐. To use these keys, first convert the radical to exponential form.

Example 9   **Evaluating a Cube Root with a Calculator**

Two ways to evaluate $\sqrt[3]{25}$ using a calculator are shown below.

25 ⌐yˣ⌐ ⌐(⌐ 1 ⌐÷⌐ 3 ⌐)⌐ ⌐=⌐   Exponential key

⌐∛⌐ 25 ⌐)⌐ ⌐ENTER⌐   Cube root key

Most calculators will display 2.924017738. So,

$$\sqrt[3]{25} \approx 2.924.$$

✓ CHECKPOINT 9

Use a calculator to approximate the value of $\sqrt[3]{18}$. ■

Example 10 **Evaluating Radicals with a Calculator**

a. Use the following keystroke sequence to evaluate $\sqrt[3]{-4}$.

4 (+/−) (yˣ) (() 1 (÷) 3 ()) (=)   Scientific

((−)) 4 (^) (() 1 (÷) 3 ()) (ENTER)   Graphing

The calculator display is -1.587401052, which implies that

$$\sqrt[3]{-4} \approx -1.587.$$

b. Use the following keystroke sequence to evaluate $(1.4)^{-2/5}$.

1.4 (yˣ) (() 2 (÷) 5 (+/−) ()) (=)   Scientific

1.4 (^) (() ((−)) 2 (÷) 5 ()) (ENTER)   Graphing

The calculator display is 0.874075175, which implies that

$$(1.4)^{-2/5} \approx 0.874.$$

✓ CHECKPOINT 10

Use a calculator to approximate the value of $2.2^{-1.2}$. Round to three decimal places. ■

Application

Example 11 **Escape Velocity** (R)

A rocket, launched vertically from Earth, has an initial velocity of 10,000 meters per second. All of the fuel is used for launching. The *escape velocity*, or the minimum initial velocity necessary for the rocket to escape the gravitational field of Earth, is

$$\sqrt{\frac{2(6.67 \times 10^{-11})(5.98 \times 10^{24})}{6.37 \times 10^{6}}}$$ meters per second.

Will the rocket escape Earth's gravitational field?

SOLUTION The escape velocity is

$$\sqrt{\frac{2(6.67 \times 10^{-11})(5.98 \times 10^{24})}{6.37 \times 10^{6}}} \approx 11,190.7 \text{ meters per second.}$$

The initial velocity of 10,000 meters per second is less than the escape velocity of 11,190.7 meters per second. So, the rocket will not escape Earth's gravitational field.

NASA

✓ CHECKPOINT 11

Will an object traveling at 10,000 meters per second exceed the escape velocity of Venus, which is

$$\sqrt{\frac{2(6.67 \times 10^{-11})(4.87 \times 10^{24})}{6.05 \times 10^{6}}}$$

meters per second? ■

CONCEPT CHECK

Let m and n be positive real numbers greater than 1.

1. Are the expressions $\left(\dfrac{m^2}{n}\right)^3$ and $\dfrac{m^5}{n^4}$ equivalent? Explain.

2. How many real cube roots does $-n$ have? Explain.

3. Is $3m\sqrt[4]{16mn^5}$ in simplest form? If not, simplify the expression.

4. Explain how to rationalize the denominator of $\dfrac{3}{\sqrt[5]{2}}$.

Skills Review 0.4

The following warm-up exercises involve skills that were covered in earlier sections. You will use these skills in the exercise set for this section. For additional help, review Section 0.3.

In Exercises 1–10, simplify the expression.

1. $\left(\frac{1}{3}\right)\left(\frac{2}{3}\right)^2$

2. $3(-4)^2$

3. $(-2x)^3$

4. $(-2x^3)(-3x^4)$

5. $(7x^5)(4x)$

6. $(5x^4)(25x^2)^{-1}$

7. $\dfrac{12z^6}{4z^2}$

8. $\left(\dfrac{2x}{5}\right)^2\left(\dfrac{2x}{5}\right)^{-4}$

9. $\left(\dfrac{3y^2}{x}\right)^0$, $\quad x \neq 0$, $y \neq 0$

10. $[(x+2)^2(x+2)^3]^2$

Exercises 0.4

See www.CalcChat.com for worked-out solutions to odd-numbered exercises.

In Exercises 1–12, fill in the missing form.

Radical Form	Rational Exponent Form
1. $\sqrt{9} = 3$	
2. $\sqrt[3]{125} = 5$	
3.	$32^{1/5} = 2$
4.	$-(144^{1/2}) = -12$
5.	$196^{1/2} = 14$
6.	$614.125^{1/3} = 8.5$
7. $\sqrt[3]{-216} = -6$	
8. $\sqrt[5]{-243} = -3$	
9. $\left(\sqrt[4]{81}\right)^3 = 27$	
10. $\sqrt[4]{81^3} = 27$	
11.	$125^{2/3} = 25$
12.	$16^{5/4} = 32$

In Exercises 13–30, evaluate the expression.

13. $\sqrt{9}$

14. $\sqrt[3]{64}$

15. $-\sqrt[3]{-27}$

16. $\sqrt[3]{0}$

17. $\dfrac{14}{\sqrt{49}}$

18. $\dfrac{\sqrt[4]{81}}{3}$

19. $\left(\sqrt[3]{-125}\right)^3$

20. $\sqrt[4]{562^4}$

21. $16^{1/2}$

22. $27^{1/3}$

23. $36^{3/2}$

24. $25^{3/2}$

25. $\sqrt{2} \cdot \sqrt{3}$

26. $\sqrt{2} \cdot \sqrt{5}$

27. $\left(\frac{16}{81}\right)^{-3/4}$

28. $\left(\frac{9}{4}\right)^{-1/2}$

29. $\left(-\frac{1}{64}\right)^{-1/3}$

30. $\left(-\frac{125}{27}\right)^{-1/3}$

In Exercises 31–36, simplify the expression.

31. $\sqrt[3]{16x^5}$

32. $\sqrt[4]{(3x^2)^4}$

33. $\sqrt{75x^{-2}y^4}$

34. $\sqrt[5]{96x^5}$

35. $\sqrt[5]{64y^{-5}}$

36. $\sqrt{8x^4y^3z^{-2}}$

In Exercises 37–40, evaluate the expression when $x = 2$, $y = 3$, and $z = 5$.

37. $\sqrt{2xy^4z^2}$

38. $\sqrt{3x^2yz^6}$

39. $\sqrt[4]{16x^{-4}y^8z^4}$

40. $\sqrt[5]{243x^5y^{-5}z^{15}}$

In Exercises 41–48, rewrite the expression by rationalizing the denominator. Simplify your answer.

41. $\dfrac{1}{\sqrt{5}}$

42. $\dfrac{5}{\sqrt{10}}$

43. $\dfrac{8}{\sqrt[3]{2}}$

44. $\dfrac{5}{\sqrt[3]{(5x)^2}}$

45. $\dfrac{2x}{5 - \sqrt{3}}$

46. $\dfrac{5x}{\sqrt{14} - 2}$

47. $\dfrac{3}{\sqrt{5} + \sqrt{6}}$

48. $\dfrac{5}{2\sqrt{10} - 5}$

In Exercises 49–60, simplify the expression.

49. $5^{1/2} \cdot 5^{3/2}$

50. $4^{1/3} \cdot 4^{5/3}$

51. $\dfrac{2^{3/2}}{2}$

52. $\dfrac{5^{1/2}}{5}$

53. $\dfrac{x^2}{x^{1/2}}$

54. $\dfrac{x \cdot x^{1/2}}{x^{3/2}}$

55. $\sqrt[3]{5} \cdot \sqrt[3]{5^2}$

56. $\sqrt[5]{3^7} \cdot \sqrt[5]{3^3}$

57. $(x^6x^3)^{1/3}$

58. $(x^3x^{12})^{1/5}$

59. $(16x^8y^4)^{3/4}$

60. $(27x^6y^9)^{2/3}$

In Exercises 61–66, use rational exponents to reduce the index of the radical.

61. $\sqrt{\sqrt{32}}$ **62.** $\sqrt{\sqrt{x^4}}$

63. $\sqrt[4]{3^2}$ **64.** $\sqrt[4]{(3x^2)^4}$

65. $\sqrt[9]{x^3}$ **66.** $\sqrt[6]{(x + 2)^4}$

In Exercises 67–72, simplify the expression.

67. $5\sqrt{x} - 3\sqrt{x}$ **68.** $3\sqrt{x + 1} + 10\sqrt{x + 1}$

69. $5\sqrt{50} + 3\sqrt{8}$ **70.** $2\sqrt{27} - \sqrt{75}$

71. $2\sqrt{4y} - 2\sqrt{9y}$ **72.** $2\sqrt{108} + \sqrt{147}$

In Exercises 73–80, use a calculator to approximate the number. (Round to three decimal places.)

73. $\sqrt[3]{45}$ **74.** $\sqrt{57}$

75. $5.7^{2/5}$ **76.** $24.7^{1.1}$

77. $0.26^{-0.8}$ **78.** $3.75^{-1/2}$

79. $\dfrac{3 - \sqrt{5}}{2}$ **80.** $\dfrac{-4 + \sqrt{12}}{4}$

81. Calculator Write the keystrokes you can use to evaluate $\dfrac{4 - \sqrt{7}}{3}$ in one step on your calculator.

82. Calculator Write the keystrokes you can use to evaluate $\sqrt[3]{(-5)^5}$ in one step on your calculator.

In Exercises 83–88, complete the statement with <, =, or >.

83. $\sqrt{5} + \sqrt{3}$ $\sqrt{5 + 3}$

84. $\sqrt{3} - \sqrt{2}$ $\sqrt{3 - 2}$

85. 5 $\sqrt{3^2 + 2^2}$

86. 5 $\sqrt{3^2 + 4^2}$

87. $\sqrt{3} \cdot \sqrt[4]{3}$ $\sqrt[8]{3}$

88. $\sqrt{\dfrac{3}{11}}$ $\dfrac{\sqrt{3}}{\sqrt{11}}$

89. Geometry Find the dimensions of a cube that has a volume of 15,625 cubic inches (see figure).

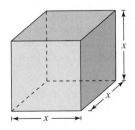

90. Geometry Find the dimensions of a square classroom that has 1100 square feet of floor space (see figure).

Declining Balances Depreciation In Exercises 91 and 92, find the annual depreciation rate r by using the declining balances formula

$$r = 1 - \left(\frac{S}{C}\right)^{1/n}$$

where n is the useful life of the item (in years), S is the salvage value (in dollars), and C is the original cost (in dollars).

91. A truck whose original cost is $75,000 is depreciated over an eight-year period, as shown in the bar graph.

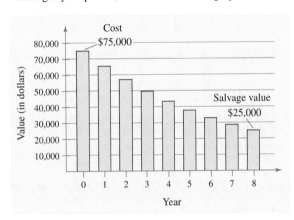

92. A printing press whose original cost is $125,000 is depreciated over a 10-year period, as shown in the bar graph.

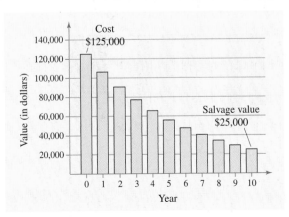

93. Escape Velocity The escape velocity (in meters per second) on the moon is

$$\sqrt{\frac{2(6.67 \times 10^{-11})(7.36 \times 10^{22})}{1.74 \times 10^{6}}}.$$

If all the fuel is consumed during launching, will a rocket with an initial velocity of 2000 meters per second escape the gravitational field of the moon?

94. Escape Velocity The escape velocity (in meters per second) on Mars is

$$\sqrt{\frac{2(6.67 \times 10^{-11})(6.42 \times 10^{23})}{3.37 \times 10^{6}}}.$$

Will an object traveling at 6000 meters per second escape the gravitational field of Mars?

95. Period of a Pendulum The period T (in seconds) of a pendulum is given by

$$T = 2\pi \sqrt{\frac{L}{32}}$$

where L is the length (in feet) of the pendulum. Find the period of a pendulum whose length is 4 feet.

96. Period of a Pendulum Use the formula given in Exercise 95 to find the period of a pendulum whose length is 2.5 feet.

97. Erosion A stream of water moving at the rate of v feet per second can carry particles of size $0.03\sqrt{v}$ inches. Find the size of the largest particle that can be carried by a stream flowing at the rate of $\frac{1}{2}$ foot per second.

98. Erosion A stream of water moving at the rate of v feet per second can carry particles of size $0.03\sqrt{v}$ inches. Find the size of the largest particle that can be carried by a stream flowing at the rate of $\frac{7}{9}$ foot per second.

Notes on a Musical Scale In Exercises 99–102, find the frequency of the indicated note on a piano (see figure). The musical note A above middle C has a frequency of 440 vibrations per second. If we denote this frequency by F_1, then the frequency of the next higher note is given by $F_2 = F_1 \cdot 2^{1/12}$. Similarly, the frequency of the next note is given by $F_3 = F_2 \cdot 2^{1/12}$.

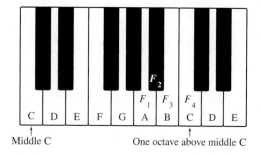

Middle C   One octave above middle C

99. Find the frequency of the musical note B above middle C.

100. Find the frequency of the musical note C that is one octave above middle C.

101. *MAKE A DECISION* Which note would you expect to have a higher frequency? Explain your reasoning.

(a) Musical note E one octave above middle C

(b) Musical note D one octave above middle C

102. *MAKE A DECISION* Assume the pattern shown on the piano continues. Which note would you expect to have a higher frequency?

(a) Musical note D one octave above middle C

(b) Musical note G one octave above middle C

Estimating Speed A formula used to help determine the speed of a car from its skid marks is $S = \sqrt{30Df}$, where S is the least possible speed (in miles per hour) of the car before its brakes are applied, D is the length of the car's skid marks (in feet) and f is the drag factor of the road surface. In Exercises 103 and 104, find the least possible speed of the car for the given conditions.

103. Skid marks: 60 feet, drag factor: 0.90

104. Skid marks: 100 feet, drag factor: 0.75

Wind Chill A wind chill temperature is a measure of how cold it *feels* outside. The wind chill temperature W (in degrees Fahrenheit) is given by

$$W = 35.75 + 0.6215T - 35.75v^{0.16} + 0.4275Tv^{0.16}$$

where T is the actual temperature (in degrees Fahrenheit) and v is the wind speed (in miles per hour). In Exercises 105 and 106, find the wind chill temperature for the given conditions. *(Source: NOAA's National Weather Service)*

105. Actual temperature: 30°F, wind speed: 20 mph

106. Actual temperature: 10°F, wind speed: 10 mph

107. Calculator Experiment Enter any positive real number in your calculator and repeatedly take the square root. What real number does the display appear to be approaching?

108. Calculator Experiment Square the real number $2/\sqrt{5}$ and note that the radical is eliminated from the denominator. Is this equivalent to rationalizing the denominator? Why or why not?

109. Think About It How can you show that $a^0 = 1, a \neq 0$? (*Hint:* Use the property of exponents $a^m/a^n = a^{m-n}$.)

110. Explain why $\sqrt{4x^2} \neq 2x$ for every real number x.

111. Explain why $\sqrt{2} + \sqrt{3} \neq \sqrt{5}$.

Mid-Chapter Quiz

See www.CalcChat.com for worked-out solutions to odd-numbered exercises.

© GI PhotoStock/RF/Alamy

When occupancy rates are not maximized, renters can sometimes negotiate for a lower rent. But when the market is overwhelmed by renters, rates are driven up.

Take this quiz as you would take a quiz in class. After you are done, check your work against the answers given in the back of the book.

In Exercises 1 and 2, place the correct symbol (<, >, or =) between the two real numbers.

1. $-|-7|$ ▢ $|-7|$

2. $-(-3)$ ▢ $|-3|$

In Exercises 3 and 4, use inequality notation to describe the subset of real numbers.

3. x is positive or x is equal to zero.

4. The apartment occupancy rate r will be at least 95% during the coming year.

5. Describe the subset of real numbers that is represented by the inequality $-2 \leq x < 3$, and sketch the subset on the real number line.

6. Identify the terms of the algebraic expression $3x^2 - 7x + 2$.

In Exercises 7–10, perform the indicated operation(s). Write fractional answers in simplest form.

7. $-4 - (-7)$

8. $\dfrac{31 - 5}{-2}$

9. $\dfrac{2}{3} \cdot \dfrac{5}{4} \cdot \dfrac{3}{7}$

10. $\dfrac{11}{15} \div \dfrac{3}{5}$

In Exercises 11–13, rewrite the expression with positive exponents and simplify.

11. $(-x)^3(2x^4)$

12. $\dfrac{5y^7}{15y^3}$

13. $\left(\dfrac{x^{-2}y^2}{3}\right)^{-3}$

14. You deposit $5000 in an account with an annual interest rate of 6.5%, compounded quarterly. Find the balance in the account after 10 years.

In Exercises 15 and 16, evaluate the expression.

15. $\dfrac{-\sqrt[4]{81}}{3}$

16. $\left(\sqrt[3]{-64}\right)^3$

In Exercises 17–19, simplify the expression.

17. $3^{1/2} \cdot 3^{3/2}$

18. $\sqrt[3]{81} - 4\sqrt[3]{3}$

19. $\sqrt[10]{12^5}$

20. Find the dimensions of a cube that has a volume of 10,648 cubic centimeters.

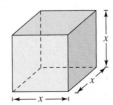

Polynomials and Special Products

- Write a polynomial in standard form.
- Add, subtract, and multiply polynomials.
- Use special products to multiply polynomials.
- Use polynomials to solve an application problem.

Polynomials

One of the simplest and most common types of algebraic expressions is a **polynomial.** Here are some examples.

$$2x + 5, \quad 3x^4 - 7x^2 + 2x + 4, \quad 5x^2y^2 - xy + 3$$

The first two are *polynomials in x* and the third is a *polynomial in x and y.* The terms of a polynomial in x have the form ax^k, where a is the **coefficient** and k is the **degree** of the term. Because a polynomial is defined as an algebraic sum, the coefficients take on the signs between the terms. For instance, the polynomial

$$2x^3 - 5x^2 + 1 = 2x^3 + (-5)x^2 + (0)x + 1$$

has coefficients 2, -5, 0, and 1.

Definition of a Polynomial in x

Let $a_n, \ldots, a_2, a_1, a_0$ be real numbers and let n be a *nonnegative integer.* A **polynomial in x** is an expression of the form

$$a_n x^n + \cdots + a_2 x^2 + a_1 x + a_0$$

where $a_n \neq 0$. The polynomial is of **degree** n, and the number a_n is the **leading coefficient.** The number a_0 is the **constant term.** The constant term is considered to have a degree of zero.

Note in the definition of a polynomial in x that the polynomial is written in descending powers of x. This is called the **standard form** of a polynomial.

Example 1   **Rewriting a Polynomial in Standard Form**

Polynomial	*Standard Form*	*Degree*
a. $4x^2 - 5x^3 - 2 + 3x$	$-5x^3 + 4x^2 + 3x - 2$	3
b. $4 - 9x^2$	$-9x^2 + 4$	2
c. 8	$8 (8 = 8x^0)$	0

✓**CHECKPOINT 1**

Rewrite the polynomial $7 - 9x^2 + 3x$ in standard form and state its degree. ■

Polynomials with one, two, and three terms are called **monomials, binomials,** and **trinomials,** respectively.

A polynomial that has all zero coefficients is called the **zero polynomial,** denoted by 0. This particular polynomial is not considered to have a degree.

Example 2   Identifying a Polynomial and Its Degree

a. $-2x^3 + x^2 + 3x - 2$ is a polynomial of degree 3.

b. $\sqrt{x^2 - 3x}$ is not a polynomial because the radical sign indicates a noninteger power of x.

c. $x^2 + 5x^{-1}$ is not a polynomial because of the negative exponent.

✓ **CHECKPOINT 2**

Determine whether the expression $\dfrac{2x + 5}{x}$ is a polynomial. If it is, state the degree. ▪

For a polynomial in more than one variable, the *degree of a term* is the sum of the powers of the variables in the term. The *degree of the polynomial* is the highest degree of all its terms. For instance, the polynomial

$$5x^3y - x^2y^2 + 2xy - 5$$

has two terms of degree 4, one term of degree 2, and one term of degree 0. The degree of the polynomial is 4.

Operations with Polynomials

You can **add** and **subtract** polynomials in much the same way that you add and subtract real numbers—you simply add or subtract the *like terms* (terms having the same variables to the same powers) by adding their coefficients. For instance, $-3x^2$ and $5x^2$ are like terms and their sum is given by

$$-3x^2 + 5x^2 = (-3 + 5)x^2 = 2x^2.$$

Example 3   Sums and Differences of Polynomials

a. $(5x^3 - 7x^2 - 3) + (x^3 + 2x^2 - x + 8)$

$\qquad = (5x^3 + x^3) + (2x^2 - 7x^2) - x + (8 - 3)$ Group like terms.

$\qquad = 6x^3 - 5x^2 - x + 5$ Combine like terms.

b. $(7x^4 - x^2 - 4x + 2) - (3x^4 - 4x^2 + 3x)$

$\qquad = 7x^4 - x^2 - 4x + 2 - 3x^4 + 4x^2 - 3x$ Distribute sign.

$\qquad = (7x^4 - 3x^4) + (4x^2 - x^2) + (-4x - 3x) + 2$ Group like terms.

$\qquad = 4x^4 + 3x^2 - 7x + 2$ Combine like terms.

✓ **CHECKPOINT 3**

Find the sum $(2x^2 + x + 3) + (4x + 1)$ and write the resulting polynomial in standard form. ▪

A common mistake is to fail to change the sign of *each* term inside parentheses preceded by a minus sign. For instance, note the following.

$$-(3x^4 - 4x^2 + 3x) = -3x^4 + 4x^2 - 3x$$ Correct

$$\cancel{-(3x^4 - 4x^2 + 3x) = -3x^4 - 4x^2 + 3x}$$ Common mistake

To find the **product** of two polynomials, you can use the left and right Distributive Properties. For example, if you treat $(5x + 7)$ as a single quantity, you can multiply $(3x - 2)$ by $(5x + 7)$ as follows.

$$(3x - 2)(5x + 7) = 3x(5x + 7) - 2(5x + 7)$$

$$= (3x)(5x) + (3x)(7) - (2)(5x) - (2)(7)$$

$$= 15x^2 + 21x - 10x - 14$$

Product of First terms	Product of Outer terms	Product of Inner terms	Product of Last terms

$$= 15x^2 + 11x - 14$$

You can use the four special products shown in the boxes above to write the product of two binomials in the FOIL form in just one step. This is called the **FOIL Method.**

Example 4 Using the FOIL Method

Use the FOIL Method to find the product of $2x - 4$ and $x + 5$.

SOLUTION

$$\overset{\text{F O I L}}{(2x - 4)(x + 5)} = 2x^2 + 10x - 4x - 20$$

$$= 2x^2 + 6x - 20$$

✓**CHECKPOINT 4**

Find the product of $3x + 1$ and $x - 1$. ■

When multiplying two polynomials, be sure to multiply *each* term of one polynomial by *each* term of the other. The following vertical pattern is a convenient way to multiply two polynomials.

Example 5 Using a Vertical Format to Multiply Polynomials

Multiply $(x^2 - 2x + 2)$ by $(x^2 + 2x + 2)$.

SOLUTION

$x^2 - 2x + 2$	Standard form
$x^2 + 2x + 2$	Standard form
$x^4 - 2x^3 + 2x^2$	$x^2(x^2 - 2x + 2)$
$2x^3 - 4x^2 + 4x$	$2x(x^2 - 2x + 2)$
$2x^2 - 4x + 4$	$2(x^2 - 2x + 2)$
$x^4 + 0x^3 + 0x^2 + 0x + 4 = x^4 + 4$	Combine like terms.

So, $(x^2 - 2x + 2)(x^2 + 2x + 2) = x^4 + 4$.

✓**CHECKPOINT 5**

Multiply $(x^2 + x + 4)$ by $(x^2 - 3x + 1)$. ■

Special Products

> **Special Products**
>
> Let u and v be real numbers, variables, or algebraic expressions.
>
> | *Special Product* | *Example* |
> |---|---|
> | **Sum and Difference of Two Terms** | |
> | $(u + v)(u - v) = u^2 - v^2$ | $(x + 4)(x - 4) = x^2 - 16$ |
> | **Square of a Binomial** | |
> | $(u + v)^2 = u^2 + 2uv + v^2$ | $(x + 3)^2 = x^2 + 6x + 9$ |
> | $(u - v)^2 = u^2 - 2uv + v^2$ | $(3x - 2)^2 = 9x^2 - 12x + 4$ |
> | **Cube of a Binomial** | |
> | $(u + v)^3 = u^3 + 3u^2v + 3uv^2 + v^3$ | $(x + 2)^3 = x^3 + 6x^2 + 12x + 8$ |
> | $(u - v)^3 = u^3 - 3u^2v + 3uv^2 - v^3$ | $(x - 1)^3 = x^3 - 3x^2 + 3x - 1$ |

Example 6   Sum and Difference of Two Terms

$$(5x + 9)(5x - 9) = (5x)^2 - 9^2 = 25x^2 - 81$$

✓**CHECKPOINT 6**

Find the product $(3 - x)(3 + x)$. ▪

Example 7   Square of a Binomial

$$(6x - 5)^2 = (6x)^2 - 2(6x)(5) + 5^2$$
$$= 36x^2 - 60x + 25$$

✓**CHECKPOINT 7**

Find the product $(x - 4)^2$. ▪

Example 8   Cube of a Binomial

✓**CHECKPOINT 8**

Find the product $(x - 3)^3$. ▪

$$(3x + 2)^3 = (3x)^3 + 3(3x)^2(2) + 3(3x)(2)^2 + 2^3$$
$$= 27x^3 + 54x^2 + 36x + 8$$

Example 9   The Product of Two Trinomials

✓**CHECKPOINT 9**

Find the product
$(x + 5 - y)(x + 5 + y)$. ▪

$$(x + y - 2)(x + y + 2) = [(x + y) - 2][(x + y) + 2]$$
$$= (x + y)^2 - 2^2$$
$$= x^2 + 2xy + y^2 - 4$$

Applications

Example 10 **A Savings Plan**

At the same time each year for five consecutive years, you deposit money in an account that earns 7% interest, compounded annually. The deposit amounts are $1500, $1800, $2400, $2600, and $3000. After the last deposit, is there enough money to pay a $12,000 tuition bill?

SOLUTION   Using the formula for compound interest, for *each* deposit you have

$$\text{Balance} = P\left(1 + \frac{r}{n}\right)^{nt} = P(1 + 0.07)^t = P(1.07)^t.$$

For the first deposit, $P = 1500$ and $t = 4$. For the second deposit, $P = 1800$ and $t = 3$, and so on. The balances for the five deposits are as follows.

Date	Deposit	Time in Account	Balance in Account
First Year	$1500	4 years	$1500(1.07)^4$
Second Year	$1800	3 years	$1800(1.07)^3$
Third Year	$2400	2 years	$2400(1.07)^2$
Fourth Year	$2600	1 year	$2600(1.07)$
Fifth Year	$3000	0 years	3000

By adding these five balances, you can find the total balance in the account to be

$$1500(1.07)^4 + 1800(1.07)^3 + 2400(1.07)^2 + 2600(1.07) + 3000.$$

Note that this expression is in polynomial form. By evaluating the expression, you can find the balance to be $12,701.03, as shown in Figure 0.10.

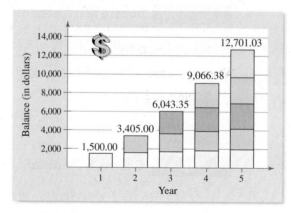

FIGURE 0.10

After the fifth deposit, there is enough money in the account to pay the college tuition bill.

✓ CHECKPOINT 10

In Example 10, suppose the account earns 5% interest. What is the balance of the account after the last deposit? ■

AP/Wide World Photos

Many families set up savings accounts to help pay their children's college expenses.

Example 11   **Geometry: Volume of a Box**

An open box is made by cutting squares from the corners of a piece of metal that measures 16 inches by 20 inches and turning up the sides, as shown in Figure 0.11. The sides of the cut-out squares are all x inches long, so the box is x inches tall. Write an expression for the volume of the box. Then find the volume when $x = 1$, $x = 2$, and $x = 3$ inches.

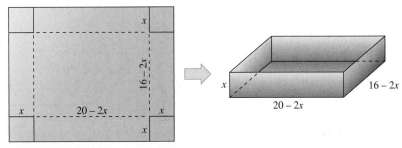

FIGURE 0.11

SOLUTION

Verbal Model: Volume = Length · Width · Height

Labels: Height = x (inches)
          Width = $16 - 2x$ (inches)
          Length = $20 - 2x$ (inches)

Equation: Volume = $(20 - 2x)(16 - 2x)(x)$

                = $(320 - 72x + 4x^2)(x)$

                = $320x - 72x^2 + 4x^3$

When $x = 1$ inch, the volume of the box is

    Volume = $320(1) - 72(1)^2 + 4(1)^3 = 252$ cubic inches.

When $x = 2$ inches, the volume of the box is

    Volume = $320(2) - 72(2)^2 + 4(2)^3 = 384$ cubic inches.

When $x = 3$ inches, the volume of the box is

    Volume = $320(3) - 72(3)^2 + 4(3)^3 = 420$ cubic inches.

✓**CHECKPOINT 11**

In Example 11, suppose the original piece of metal is 10 inches by 12 inches. Write an expression for the volume of the box. Then find the volume when $x = 2$ and $x = 3$. ▪

┌─ **CONCEPT CHECK** ─────────────────────────────

1. Is $2 - 3x + x^3 - x^5$ written in standard form? Explain.

2. How many terms are in the sum of $x^3 - 4x^2 + 3$ and $2x^2 - x$?

3. A student claims that $(x - 3)(x + 4) = x^2 - 12$. Is the student correct? Explain.

4. Describe how you would show that $\sqrt{a^2 + b^2} \neq a + b$, where $a, b \neq 0$, using an algebraic argument. Then give a numerical example.

Skills Review 0.5

The following warm-up exercises involve skills that were covered in earlier sections. You will use these skills in the exercise set for this section. For additional help, review Sections 0.3 and 0.4.

In Exercises 1–10, perform the indicated operation(s).

1. $(7x^2)(6x)$

2. $(10z^3)(-2z^{-1})$

3. $(-3x^2)^3$

4. $-3(x^2)^3$

5. $\dfrac{27z^5}{12z^2}$

6. $\sqrt{24} \cdot \sqrt{2}$

7. $\left(\dfrac{2x}{3}\right)^{-2}$

8. $16^{3/4}$

9. $\dfrac{4}{\sqrt{8}}$

10. $\sqrt[3]{-27x^3}$

Exercises 0.5

See www.CalcChat.com for worked-out solutions to odd-numbered exercises.

In Exercises 1–6, find the degree and leading coefficient of the polynomial.

1. $2x^2 - x + 1$

2. $-3x^4 + 2x^2 - 5$

3. $x^5 - 1$

4. 3

5. $3x^5 - 6x^4 + x - 2$

6. $-3x$

In Exercises 7–12, determine whether the algebraic expression is a polynomial. If it is, write the polynomial in standard form and state its degree.

7. $2x - 3x^3 + 8$

8. $2x^3 + x - 3x^{-1}$

9. $\dfrac{3x + 4}{x}$

10. $\dfrac{2x^2 + 5x - 3}{3}$

11. $w^2 - w^4 + 2w^3$

12. $\sqrt{y^2 - y^4}$

In Exercises 13–16, evaluate the polynomial for each value of x.

13. $4x + 5$ (a) $x = -2$ (b) $x = -1$
          (c) $x = 0$ (d) $x = 3$

14. $-x^2 + 3$ (a) $x = -3$ (b) $x = -2$
           (c) $x = 0$ (d) $x = 1$

15. $-2x^2 + 3x + 4$ (a) $x = -2$ (b) $x = -1$
                 (c) $x = 0$ (d) $x = 1$

16. $x^3 - 4x^2 + x$ (a) $x = -1$ (b) $x = 0$
                (c) $x = 1$ (d) $x = 2$

In Exercises 17–28, perform the indicated operation(s) and write the resulting polynomial in standard form.

17. $(6x + 5) - (8x + 15)$

18. $(3x^2 + 1) - (2x^2 - 2x + 3)$

19. $-(x^3 + 5) + (3x^3 - 4x)$

20. $-(5x^2 - 1) + (-3x^2 + 5)$

21. $(15x^2 - 6) - (-8x^3 - 14x^2 - 17)$

22. $(15x^4 - 18x - 19) - (13x^4 - 5x + 15)$

23. $3x(x^2 - 2x + 1)$ **24.** $z^2(2z^2 + 3z + 1)$

25. $-4x(3 - x^3)$ **26.** $-5y(2y - y^2)$

27. $-3x(-x)(3x - 7)$ **28.** $(2 - x^2)(-2x)(4x)$

In Exercises 29–54, find the product.

29. $(x + 3)(x + 4)$ **30.** $(x - 5)(x + 10)$

31. $(3x - 5)(2x + 1)$ **32.** $(7x - 2)(4x - 3)$

33. $(x + 5)(x - 5)$ **34.** $(3x + 2)(3x - 2)$

35. $(x + 6)^2$ **36.** $(3x - 2)^2$

37. $(2x - 5y)^2$ **38.** $(5 - 8x)^2$

39. $[(x - 3) + y]^2$ **40.** $[(x + 1) - y]^2$

41. $(x + 1)^3$ **42.** $(x - 2)^3$

43. $(2x - y)^3$ **44.** $(3x + 2y)^3$

45. $(3y^2 - 1)(3y^2 + 1)$ **46.** $(3x^2 - 4y^2)(3x^2 + 4y^2)$

47. $(m - 3 + n)(m - 3 - n)$

48. $(x + y + 1)(x + y - 1)$

49. $\left(\sqrt{x} + \sqrt{y}\right)\left(\sqrt{x} - \sqrt{y}\right)$

50. $\left(5 + \sqrt{x}\right)\left(5 - \sqrt{x}\right)$

51. $(x^2 - x + 1)(x^2 + x + 1)$

52. $(x^2 + 3x - 2)(x^2 - 3x - 2)$

53. $5x(x + 1) - 3x(x + 1)$

54. $(2x - 1)(x + 3) + 3(x + 3)$

55. Error Analysis A student claims that

$(x - 3)^2 = x^2 + 9.$

Describe and correct the student's error.

56. Error Analysis A student claims that

$(x - 3)(x + 3) = (x - 3)^2.$

Describe and correct the student's error.

57. Compound Interest After 3 years, an investment of $1000 earning an interest rate r compounded annually will be worth $1000(1 + r)^3$ dollars. Write this expression as a polynomial in standard form.

58. Compound Interest After 2 years, an investment of $800 earning an interest rate r compounded annually will be worth $800(1 + r)^2$ dollars. Write this expression as a polynomial in standard form.

59. Savings Plan At the same time each year for five consecutive years, you deposit money in an account that earns annually compounded interest. The deposits are $1500, $1700, $900, $2200, and $3000. Is there enough money in the account after the last deposit to pay a $10,000 college tuition bill at an interest rate of 6%? 5%? 4%?

60. Savings Plan You have an investment that pays an annual dividend. Each January for six consecutive years, you reinvest this dividend in an account that earns 6.25% interest, compounded annually. The dividends are shown in the table. Is there enough money in the account after the sixth deposit for a $7500 down payment on a car?

Year	Dividend
1	$920
2	$1000
3	$780
4	$1310
5	$1020
6	$1200

61. Federal Student Aid The total amount (in millions of dollars) of federal student aid disbursed in the years 1998 through 2005 can be approximated by

$453.11x^2 - 5546.7x + 55,833$

where x represents the year, with $x = 8$ corresponding to 1998. Evaluate the polynomial when $x = 14$ and $x = 15$. Then describe your results in everyday terms. *(Source: U.S. Department of Education)*

62. Federal Pell Grants The amount (in dollars) of the average Pell Grant awarded in the years 1998 through 2005 can be approximated by

$-4.874x^3 + 155.85x^2 - 1507.9x + 6443$

where x represents the year, with $x = 8$ corresponding to 1998. Evaluate the polynomial when $x = 14$ and $x = 15$. Then describe your results in everyday terms. *(Source: U.S. Department of Education)*

63. Geometry A box has a length of $(52 - 2x)$ inches, a width of $(42 - 2x)$ inches, and a height of x inches. Find the volume when $x = 3$, $x = 7$, and $x = 9$ inches. Which x-value gives the greatest volume?

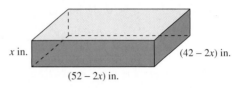

x in. $(42 - 2x)$ in.

$(52 - 2x)$ in.

64. Geometry A box has a length of $(57 - 2x)$ inches, a width of $(39 - 2x)$ inches, and a height of x inches. Find the volume when $x = 4$, $x = 6$, and $x = 10$ inches. Which x-value gives the greatest volume?

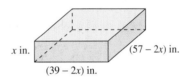

x in. $(57 - 2x)$ in.

$(39 - 2x)$ in.

65. Geometry Find the area of the shaded region in the figure. Write your answer as a polynomial in standard form.

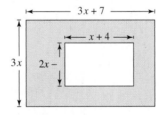

$3x + 7$

$x + 4$

$3x$ $2x -$

66. Geometry Find a polynomial that represents the total number of square feet in the floor plan.

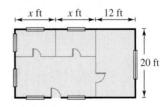

x ft x ft 12 ft

20 ft

67. Extended Application To work an extended application involving the population of the United States from 1990 to 2005, visit this text's website at *college.hmco.com*. *(Data Source: U.S. Census Bureau)*

Section 0.6

Factoring

- Factor a polynomial by removing common factors.
- Factor a polynomial in a special form.
- Factor a trinomial as the product of two binomials.
- Factor a polynomial by grouping.

Common Factors

The process of writing a polynomial as a product is called **factoring.** It is an important tool for solving equations and reducing fractional expressions.

A polynomial that cannot be factored using integer coefficients is called **prime** or **irreducible over the integers.** For instance, the polynomial $x^2 - 3$ is irreducible over the integers. [Over the *real numbers*, this polynomial can be factored as

$$x^2 - 3 = (x + \sqrt{3})(x - \sqrt{3}).]$$

A polynomial is **completely factored** when each of its factors is prime. For instance,

$$x^3 - x^2 + 4x - 4 = (x - 1)(x^2 + 4) \qquad \text{Completely factored}$$

is completely factored, but

$$x^3 - x^2 - 4x + 4 = (x - 1)(x^2 - 4) \qquad \text{Not completely factored}$$

is not completely factored. Its complete factorization is

$$x^3 - x^2 - 4x + 4 = (x - 1)(x + 2)(x - 2).$$

The simplest type of factoring involves a polynomial that can be written as the product of a monomial and another polynomial. To factor such a polynomial, you can use the Distributive Property in the *reverse* direction.

$$ab + ac = a(b + c) \qquad \text{\textit{a} is a common factor.}$$

Example 1   Removing Common Factors

Factor each expression.

a. $6x^3 - 4x$ **b.** $(x - 2)(2x) + (x - 2)(3)$

SOLUTION

a. Each term of this polynomial has $2x$ as a common factor.

$$6x^3 - 4x = 2x(3x^2) - 2x(2) = 2x(3x^2 - 2)$$

b. The binomial factor $(x - 2)$ is common to both terms.

$$(x - 2)(2x) + (x - 2)(3) = (x - 2)(2x + 3)$$

✓ CHECKPOINT 1

Factor the expression $(x + 1)^2 + 2x(x + 1)$. ■

Factoring Special Polynomial Forms

Factoring Special Polynomial Forms	
Factored Form	*Example*
Difference of Two Squares	
$u^2 - v^2 = (u + v)(u - v)$	$9x^2 - 4 = (3x + 2)(3x - 2)$
Perfect Square Trinomial	
$u^2 + 2uv + v^2 = (u + v)^2$	$x^2 + 6x + 9 = (x + 3)^2$
$u^2 - 2uv + v^2 = (u - v)^2$	$x^2 - 6x + 9 = (x - 3)^2$
Sum or Difference of Two Cubes	
$u^3 + v^3 = (u + v)(u^2 - uv + v^2)$	$x^3 + 8 = (x + 2)(x^2 - 2x + 4)$
$u^3 - v^3 = (u - v)(u^2 + uv + v^2)$	$27x^3 - 1 = (3x - 1)(9x^2 + 3x + 1)$

STUDY TIP

In Example 2, note that the first step in factoring a polynomial is to check for common factors. Once the common factor is removed, it is often possible to recognize patterns that were not obvious at first glance.

Example 2   **Removing a Common Factor First**

Factor the expression $3 - 12x^2$.

SOLUTION

$$3 - 12x^2 = 3(1 - 4x^2) \qquad \text{3 is a common factor.}$$
$$= 3[1^2 - (2x)^2] \qquad \text{Difference of two squares}$$
$$= 3(1 + 2x)(1 - 2x) \qquad \text{Completely factored}$$

✓**CHECKPOINT 2**

Factor the expression $x^3 - x$. ■

Example 3   **Factoring the Difference of Two Squares**

a. $(x + 2)^2 - y^2 = [(x + 2) + y][(x + 2) - y]$
$$= (x + 2 + y)(x + 2 - y)$$
$$= (x + y + 2)(x - y + 2)$$

b. You can factor $16x^4 - 81$ by applying the difference of two squares formula twice.

$$16x^4 - 81 = (4x^2)^2 - 9^2$$
$$= (4x^2 + 9)(4x^2 - 9) \qquad \text{First application}$$
$$= (4x^2 + 9)[(2x)^2 - 3^2]$$
$$= (4x^2 + 9)(2x + 3)(2x - 3) \qquad \text{Second application}$$

✓**CHECKPOINT 3**

Factor the expression $100 - 4y^2$. ■

A perfect square trinomial is the square of a binomial, and it has the following form. Note that the first and last terms of a perfect square trinomial are squares and the middle term is twice the product of u and v.

$$u^2 + 2uv + v^2 = (u + v)^2 \qquad \text{or} \qquad u^2 - 2uv + v^2 = (u - v)^2$$

Same sign   Same sign

Example 4   Factoring Perfect Square Trinomials

a. $16x^2 + 8x + 1 = (4x)^2 + 2(4x)(1) + 1^2 = (4x + 1)^2$

b. $x^2 - 10x + 25 = x^2 - 2(x)(5) + 5^2 = (x - 5)^2$

✓ CHECKPOINT 4

Factor the expression $x^2 - 12x + 36$. ■

The next two formulas show that sums and differences of cubes factor easily. Pay special attention to the signs of the terms.

Like signs   Like signs

$$u^3 + v^3 = (u + v)(u^2 - uv + v^2) \qquad u^3 - v^3 = (u - v)(u^2 + uv + v^2)$$

Unlike signs   Unlike signs

Example 5   Factoring the Sum and Difference of Cubes

Factor each expression.

a. $x^3 - 27$ **b.** $3x^3 + 192$

SOLUTION

a. $x^3 - 27 = x^3 - 3^3$ Rewrite 27 as 3^3.

            $= (x - 3)(x^2 + 3x + 9)$ Factor.

b. $3x^3 + 192 = 3(x^3 + 64)$ 3 is a common factor.

           $= 3(x^3 + 4^3)$ Rewrite 64 as 4^3.

           $= 3(x + 4)(x^2 - 4x + 16)$ Factor.

✓ CHECKPOINT 5

Factor the expression $y^3 - 1$. ■

Trinomials with Binomial Factors

To factor a trinomial of the form $ax^2 + bx + c$, use the following pattern.

Factors of a

$$ax^2 + bx + c = (\quad x + \quad)(\quad x + \quad)$$

Factors of c

The goal is to find a combination of factors of a and c such that the outer and inner products add up to the middle term bx. For instance, for the trinomial

$$6x^2 + 17x + 5$$

you can write

$$
\begin{array}{cccc}
\text{F} & \text{O} & \text{I} & \text{L} \\
\downarrow & \downarrow & \downarrow & \downarrow
\end{array}
$$

$$(2x + 5)(3x + 1) = 6x^2 + 2x + 15x + 5$$

$$
\begin{array}{c}
\text{O} + \text{I} \\
\downarrow
\end{array}
$$

$$= 6x^2 + 17x + 5.$$

Note that the outer (O) and inner (I) products add up to $17x$.

Example 6   Factoring a Trinomial: Leading Coefficient Is 1

Factor the trinomial $x^2 - 7x + 12$.

SOLUTION   For this trinomial, you have $a = 1$, $b = -7$, and $c = 12$. Because b is negative and c is positive, both factors of 12 must be negative. That is, $12 = (-2)(-6)$, $12 = (-1)(-12)$, or $12 = (-3)(-4)$. So, the possible factorizations of $x^2 - 7x + 12$ are

$$(x - 2)(x - 6), \quad (x - 1)(x - 12), \quad \text{and} \quad (x - 3)(x - 4).$$

Testing the middle term, you can find the correct factorization to be

$$x^2 - 7x + 12 = (x - 3)(x - 4).$$

✓ CHECKPOINT 6

Factor the trinomial $x^2 + x - 6$. ▪

Example 7   Factoring a Trinomial: Leading Coefficient Is Not 1

Factor the trinomial $2x^2 + x - 15$.

SOLUTION   For this trinomial, you have $a = 2$ and $c = -15$, which means that the factors of -15 must have unlike signs. The eight possible factorizations are as follows.

$$
\begin{array}{ll}
(2x - 1)(x + 15) & (2x + 1)(x - 15) \\
(2x - 3)(x + 5) & (2x + 3)(x - 5) \\
(2x - 5)(x + 3) & (2x + 5)(x - 3) \\
(2x - 15)(x + 1) & (2x + 15)(x - 1)
\end{array}
$$

Testing the middle term, you can find the correct factorization to be

$$2x^2 + x - 15 = (2x - 5)(x + 3).$$

✓ CHECKPOINT 7

Factor the trinomial $2x^2 - 5x + 3$. ▪

Factoring by Grouping

Sometimes polynomials with more than three terms can be **factored by grouping.**

Example 8   Factoring by Grouping

$$x^3 - 2x^2 - 3x + 6 = (x^3 - 2x^2) - (3x - 6) \qquad \text{Group terms.}$$
$$= x^2(x - 2) - 3(x - 2) \qquad \text{Factor groups.}$$
$$= (x - 2)(x^2 - 3) \qquad \text{Distributive Property}$$

✓ CHECKPOINT 8

Factor the polynomial $x^3 + x^2 + 5x + 5$. ▪

When factoring by grouping, sometimes several different groupings will work. For instance, a different grouping could have been used in Example 8.

$$x^3 - 2x^2 - 3x + 6 = (x^3 - 3x) - (2x^2 - 6)$$
$$= x(x^2 - 3) - 2(x^2 - 3)$$
$$= (x^2 - 3)(x - 2)$$

As you can see, you obtain the same result as in Example 8.

Factoring by grouping can save you some of the trial and error involved in factoring a trinomial. To factor a trinomial of the form $ax^2 + bx + c$ by grouping, rewrite the middle term as the sum of two factors of the product ac that add up to b. This technique is illustrated in Example 9.

Example 9   Factoring a Trinomial by Grouping

Use factoring by grouping to factor $2x^2 + 5x - 3$.

SOLUTION   In the trinomial $2x^2 + 5x - 3$, $a = 2$ and $c = -3$, so the product ac is -6. Notice that -6 factors as $(6)(-1)$, and $6 - 1 = 5 = b$. So, you can rewrite the middle term as $5x = 6x - x$. This produces the following.

$$2x^2 + 5x - 3 = 2x^2 + 6x - x - 3 \qquad \text{Rewrite middle term.}$$
$$= (2x^2 + 6x) - (x + 3) \qquad \text{Group terms.}$$
$$= 2x(x + 3) - (x + 3) \qquad \text{Factor groups.}$$
$$= (x + 3)(2x - 1) \qquad \text{Distributive Property}$$

✓ CHECKPOINT 9

Use factoring by grouping to factor $2x^2 + 5x - 12$. ▪

The trinomial factors as $2x^2 + 5x - 3 = (x + 3)(2x - 1)$.

CONCEPT CHECK

1. What is the common factor in the polynomial $3x^3 - 27x$?
2. Is $x^4 + 3x^3 - 8x - 24 = (x + 3)(x^3 - 8)$ factored completely? Explain.
3. Describe how you would show that $a^2 + b^2 \neq (a + b)^2$, where $a, b \neq 0$.
4. Can you factor $x^3 - 2x^2 - 8x + 24$ by grouping? Explain.

| **Skills Review 0.6** | The following warm-up exercises involve skills that were covered in earlier sections. You will use these skills in the exercise set for this section. For additional help, review Section 0.5. |

In Exercises 1–10, find the product.

1. $3x(5x - 2)$

2. $-2y(y + 1)$

3. $(2x + 3)^2$

4. $(3x - 8)^2$

5. $(2x - 3)(x + 8)$

6. $(4 - 5z)(1 + z)$

7. $(2y + 1)(2y - 1)$

8. $(x + a)(x - a)$

9. $(x + 4)^3$

10. $(2x - 3)^3$

Exercises 0.6

See www.CalcChat.com for worked-out solutions to odd-numbered exercises.

In Exercises 1–6, factor out the common factor.

1. $3x + 6$

2. $6y - 30$

3. $3x^3 - 6x$

4. $4x^3 - 6x^2 + 12x$

5. $(x - 1)^2 + 6(x - 1)$

6. $3x(x + 2) - 4(x + 2)$

In Exercises 7–12, factor the difference of two squares.

7. $x^2 - 36$

8. $x^2 - \frac{1}{9}$

9. $16x^2 - 9y^2$

10. $x^2 - 49y^2$

11. $(x - 1)^2 - 4$

12. $25 - (z + 5)^2$

In Exercises 13–18, factor the perfect square trinomial.

13. $x^2 - 4x + 4$

14. $x^2 + 10x + 25$

15. $4y^2 + 12y + 9$

16. $9x^2 - 12x + 4$

17. $y^2 - \frac{2}{3}y + \frac{1}{9}$

18. $z^2 + z + \frac{1}{4}$

In Exercises 19–24, factor the sum or difference of cubes.

19. $x^3 - 8$

20. $x^3 - 27$

21. $y^3 + 125$

22. $y^3 + 1000$

23. $8t^3 - 1$

24. $27x^3 + 8$

In Exercises 25–38, factor the trinomial.

25. $x^2 + x - 2$

26. $x^2 + 6x + 8$

27. $w^2 - 5w + 6$

28. $z^2 - z - 6$

29. $y^2 + y - 20$

30. $z^2 - 4z - 21$

31. $x^2 - 30x + 200$

32. $x^2 - 5x - 150$

33. $3x^2 - 5x + 2$

34. $2x^2 - x - 1$

35. $9x^2 - 3x - 2$

36. $12y^2 + 7y + 1$

37. $6x^2 + 37x + 6$

38. $5u^2 + 13u - 6$

In Exercises 39–44, factor by grouping.

39. $x^3 - x^2 + 2x - 2$

40. $x^3 + 5x^2 - 5x - 25$

41. $2x^3 - x^2 - 6x + 3$

42. $5x^3 - 10x^2 + 3x - 6$

43. $6 + 2y - 3y^3 - y^4$

44. $z^5 + 2z^3 + z^2 + 2$

In Exercises 45–68, completely factor the expression.

45. $4x^2 - 8x$

46. $12x^3 - 48x$

47. $y^3 - 9y$

48. $x^3 - 4x^2$

49. $3x^2 - 48$

50. $7y^2 - 63$

51. $x^2 - 2x + 1$

52. $9x^2 - 6x + 1$

53. $1 - 4x + 4x^2$

54. $16 + 6x - x^2$

55. $2y^3 - 7y^2 - 15y$

56. $3x^4 + x^3 - 10x^2$

57. $-2x^2 - 4x + 2x^3$

58. $13x + 6 + 5x^2$

59. $3x^3 + x^2 + 15x + 5$

60. $5 - x + 5x^2 - x^3$

61. $x^4 - 4x^3 + x^2 - 4x$

62. $3u - 2u^2 + 6 - u^3$

63. $25 - (x + 5)^2$

64. $(t - 1)^2 - 49$

65. $(x^2 + 1)^2 - 4x^2$

66. $(x^2 + 8)^2 - 36x^2$

67. $2t^3 - 16$

68. $3x^3 + 81$

Geometric Modeling In Exercises 69–72, make a "geometric factoring model" to represent the given factorization. For instance, a factoring model for $2x^2 + 5x + 2 = (2x + 1)(x + 2)$ is shown below.

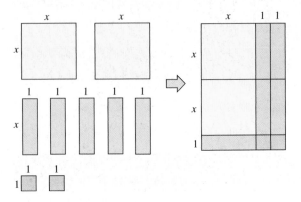

69. $x^2 + 3x + 2 = (x + 2)(x + 1)$

70. $x^2 + 4x + 3 = (x + 3)(x + 1)$

71. $2x^2 + 7x + 3 = (2x + 1)(x + 3)$

72. $3x^2 + 7x + 2 = (3x + 1)(x + 2)$

73. Geometry The room shown in the figure has a floor space of $(2x^2 - x - 3)$ square feet. If the width of the room is $(x + 1)$ feet, what is the length?

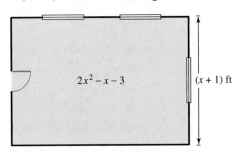

74. Geometry The room shown in the figure has a floor space of $(3x^2 + 8x + 4)$ square feet. If the width of the room is $(x + 2)$ feet, what is the length?

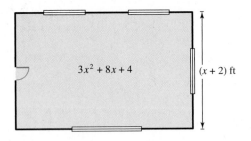

75. *MAKE A DECISION* Factor each trinomial. State whether you used factoring by grouping or factoring by trial and error.

(a) $x^2 + 11x + 24$ (b) $3x^2 + 7x - 20$

76. Find all integers b such that $x^2 + bx + 24$ can be factored. Describe how you found these values of b.

77. Find all integers $c > 0$ such that $x^2 + 8x + c$ can be factored. Describe how you found these values of c.

78. Think About It A student claims that

$$x^3 - 8 = (x - 2)^3.$$

Describe and correct the student's error.

79. Think About It Describe two different ways to factor $2x^2 - 7x - 15$.

80. Geometric Modeling The figure shows a large square with an area of a^2 that contains a smaller square with an area of b^2. If the smaller square is removed, the remaining figure has an area of $a^2 - b^2$. Rearrange the parts of the remaining figure to illustrate the factoring formula

$$a^2 - b^2 = (a - b)(a + b).$$

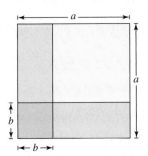

81. Geometric Modeling The figure shows a large cube with a volume of a^3 that contains a smaller cube with a volume of b^3. If the smaller cube is removed, the remaining solid has a volume of $a^3 - b^3$ and consists of the three rectangular boxes labeled Box 1, Box 2, and Box 3. Explain how you can use the figure to obtain the factoring formula

$$a^3 - b^3 = (a - b)(a^2 + ab + b^2).$$

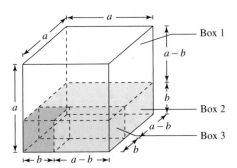

Section 0.7

Fractional Expressions

- Find the domain of an algebraic expression.
- Simplify a rational expression.
- Perform operations with rational expressions.
- Simplify a complex fraction.

Domain of an Expression

The set of all real numbers for which an algebraic expression is defined is called the **domain** of the expression. For instance, the domain of

$$\frac{1}{x}$$

is all real numbers other than $x = 0$. Two algebraic expressions are **equivalent** if they have the same domain and yield the same values for all numbers in their domain. For instance, the expressions

$$[(x + 1) + (x + 2)] \quad \text{and} \quad 2x + 3$$

are equivalent.

Example 1 Finding the Domain of an Algebraic Expression

a. The domain of the polynomial

$$2x^3 + 3x + 4$$

is the set of all real numbers. In fact, the domain of any polynomial is the set of all real numbers (unless the domain is specifically restricted).

b. The domain of the polynomial

$$x^2 + 5x + 2, \quad x > 0$$

is the set of positive real numbers, because the polynomial is specifically restricted to that set.

c. The domain of the radical expression

$$\sqrt{x}$$

is the set of nonnegative real numbers, because the square root of a negative number is not a real number.

d. The domain of the expression

$$\frac{x + 2}{x - 3}$$

is the set of all real numbers except $x = 3$, because the value $x = 3$ results in division by zero, which is undefined.

> **STUDY TIP**
>
> The domain of an algebraic expression does not include any value that creates *division by zero* or *the square root of a negative number*.

✓ CHECKPOINT 1

Find the domain of $\dfrac{1}{x - 5}$. ■

Simplifying Rational Expressions

The quotient of two algebraic expressions is a **fractional expression.** Moreover, the quotient of two *polynomials* such as

$$\frac{1}{x}, \quad \frac{2x-1}{x+1}, \quad \text{or} \quad \frac{x^2-1}{x^2+1}$$

is a **rational expression.** Recall that a fraction is in simplest form if its numerator and denominator have no factors in common aside from ± 1. To write a fraction in simplest form, divide out common factors.

$$\frac{a \cdot \overset{1}{\cancel{c}}}{b \cdot \underset{1}{\cancel{c}}} = \frac{a}{b}, \quad b \neq 0, \quad c \neq 0$$

The key to success in simplifying rational expressions lies in your ability to *factor* polynomials. For example,

$$\frac{18x^2 - 18}{6x - 6} = \frac{3(\overset{1}{\cancel{6}})(x+1)(\overset{1}{\cancel{x-1}})}{\underset{1}{\cancel{6}}(\underset{1}{\cancel{x-1}})} = 3(x+1), \quad x \neq 1.$$

Note that the original expression is undefined when $x = 1$ (because division by zero is undefined). Because this is not obvious in the simplified expression, you must add the domain restriction $x \neq 1$ to the simplified expression to make it *equivalent* to the original.

Example 2   Simplifying a Rational Expression

$$\frac{x^2 + 4x - 12}{3x - 6} = \frac{(x+6)(\overset{1}{\cancel{x-2}})}{3(\underset{1}{\cancel{x-2}})} \qquad \text{Factor completely.}$$

$$= \frac{x+6}{3}, \quad x \neq 2 \qquad \text{Divide out common factors.}$$

✓ CHECKPOINT 2

Write the expression $\dfrac{2x^2 - 2}{3x - 3}$ in simplest form. ▪

In Example 2, do not make the mistake of trying to simplify further by dividing out *terms.*

$$\frac{x+6}{3} = \frac{\cancel{x}+\overset{2}{\cancel{6}}}{\underset{1}{\cancel{3}}} = x+2$$

Remember that to simplify fractions, you divide out *factors*, not terms.

When simplifying rational expressions, be sure to factor each polynomial completely before concluding that the numerator and denominator have no factors in common. Moreover, changing the sign of a factor may allow further simplification, as demonstrated in part (b) of the next example.

Example 3 Simplifying Rational Expressions

a. $\dfrac{x^3 - 4x}{x^2 + x - 2} = \dfrac{x(x + 2)(x - 2)}{(x + 2)(x - 1)}$ Factor completely.

$\qquad\qquad\quad = \dfrac{x(x - 2)}{x - 1}, \quad x \neq -2$ Divide out common factors.

b. $\dfrac{12 + x - x^2}{2x^2 - 9x + 4} = \dfrac{(4 - x)(3 + x)}{(2x - 1)(x - 4)}$ Factor completely.

$\qquad\qquad\quad = \dfrac{-(x - 4)(3 + x)}{(2x - 1)(x - 4)}$ $4 - x = -(x - 4)$

$\qquad\qquad\quad = -\dfrac{3 + x}{2x - 1}, \quad x \neq 4$ Divide out common factors.

✓ CHECKPOINT 3

Write the expression $\dfrac{3 - 2x - x^2}{2x^2 - 2}$ in simplest form. ■

To multiply or divide rational expressions, use the properties of fractions (see Section 0.2). Recall that to divide fractions you invert the divisor and multiply.

Example 4 Multiplying Rational Expressions

$\dfrac{6x^2 - 6x}{x^2 + 2x - 3} \cdot \dfrac{x^2 + x - 6}{2x}$ Original product

$= \dfrac{6x(x - 1)(x + 3)(x - 2)}{(x - 1)(x + 3)(2x)}$ Factor and multiply.

$= \dfrac{3(2x)(x - 1)(x + 3)(x - 2)}{(x - 1)(x + 3)(2x)}$ Divide out common factors.

$= 3(x - 2), \quad x \neq -3, x \neq 0, x \neq 1$ Simplify.

✓ CHECKPOINT 4

Multiply and simplify:

$\dfrac{3}{x - 2} \cdot \dfrac{x - 2}{3x + 3}.$ ■

Example 5 Dividing Rational Expressions

$\dfrac{2x}{3x - 12} \div \dfrac{x^2 - 2x}{x^2 - 6x + 8} = \dfrac{2x}{3x - 12} \cdot \dfrac{x^2 - 6x + 8}{x^2 - 2x}$ Invert and multiply.

$= \dfrac{(2x)(x - 2)(x - 4)}{(3)(x - 4)(x)(x - 2)}$ Factor and multiply.

$= \dfrac{(2x)(x - 2)(x - 4)}{(3)(x - 4)(x)(x - 2)}$ Divide out common factors.

$= \dfrac{2}{3}, \quad x \neq 0, x \neq 2, x \neq 4$ Simplify.

✓ CHECKPOINT 5

Divide and simplify:

$\dfrac{4x + 4y}{5} \div \dfrac{x + y}{2}.$ ■

To add or subtract rational expressions, use the least common denominator (LCD) method or the following basic property of fractions that was covered on page 14.

$$\frac{a}{b} \pm \frac{c}{d} = \frac{ad \pm bc}{bd}, \quad b \neq 0, d \neq 0$$

This property is efficient for adding or subtracting *two* fractions that have no common factors in their denominators.

Example 6 Adding Rational Expressions

$$\frac{x}{x-3} + \frac{2}{3x+4} = \frac{x(3x+4) + 2(x-3)}{(x-3)(3x+4)} \qquad \frac{a}{b} + \frac{c}{d} = \frac{ad+bc}{bd}$$

$$= \frac{3x^2 + 4x + 2x - 6}{(x-3)(3x+4)} \qquad \text{Distributive Property}$$

$$= \frac{3x^2 + 6x - 6}{(x-3)(3x+4)} \qquad \text{Combine like terms.}$$

$$= \frac{3(x^2 + 2x - 2)}{(x-3)(3x+4)} \qquad \text{Factor.}$$

✓ CHECKPOINT 6

Subtract: $\dfrac{4}{x} - \dfrac{2x}{3}$. ∎

For fractions with a repeated factor in their denominators, the LCD method works well. Recall that the least common denominator of two or more fractions consists of the product of all prime factors in the denominators, with each factor given the highest power of its occurrence in any denominator.

Example 7 Combining Rational Expressions: The LCD Method

Perform the indicated operations and simplify.

$$\frac{3}{x-1} - \frac{2}{x} + \frac{x+3}{x^2-1}$$

SOLUTION Using the factored denominators $(x-1)$, x, and $(x+1)(x-1)$, you can see that the least common denominator is $x(x+1)(x-1)$.

$$\frac{3}{x-1} - \frac{2}{x} + \frac{x+3}{x^2-1}$$

$$= \frac{3(x)(x+1)}{x(x+1)(x-1)} - \frac{2(x+1)(x-1)}{x(x+1)(x-1)} + \frac{(x+3)(x)}{x(x+1)(x-1)}$$

$$= \frac{3(x)(x+1) - 2(x+1)(x-1) + (x+3)(x)}{x(x+1)(x-1)}$$

$$= \frac{3x^2 + 3x - 2x^2 + 2 + x^2 + 3x}{x(x+1)(x-1)}$$

$$= \frac{2x^2 + 6x + 2}{x(x+1)(x-1)} = \frac{2(x^2 + 3x + 1)}{x(x+1)(x-1)}$$

✓ CHECKPOINT 7

Perform the indicated operations and simplify:

$$\frac{5}{x} + \frac{4}{x-1} - \frac{4}{x(x-1)}. \quad ∎$$

Complex Fractions

Fractional expressions with separate fractions in the numerator or denominator are called **complex fractions.** Here are two examples.

$$\frac{\left(\dfrac{1}{x}\right)}{x^2 + 1} \quad \text{and} \quad \frac{\left(\dfrac{1}{x}\right)}{\left(\dfrac{1}{x^2 + 1}\right)}$$

A complex fraction can be simplified by combining the fractions in its numerator into a single fraction and then combining the fractions in its denominator into a single fraction. Then invert the denominator and multiply.

Example 8   **Simplifying a Complex Fraction**

$$\frac{\left(\dfrac{2}{x} - 3\right)}{\left(1 - \dfrac{1}{x-1}\right)} = \frac{\left[\dfrac{2 - 3(x)}{x}\right]}{\left[\dfrac{1(x-1) - 1}{x-1}\right]} \qquad \text{Combine fractions.}$$

$$= \frac{\left(\dfrac{2 - 3x}{x}\right)}{\left(\dfrac{x-2}{x-1}\right)} \qquad \text{Simplify.}$$

$$= \frac{2 - 3x}{x} \cdot \frac{x-1}{x-2} \qquad \text{Invert and multiply.}$$

$$= \frac{(2 - 3x)(x-1)}{x(x-2)}, \quad x \neq 1$$

✓ CHECKPOINT 8

Simplify the complex fraction

$$\frac{\left(\dfrac{x}{3} - 1\right)}{x - 3}.$$ ▪

Another way to simplify in Example 8 is to multiply its numerator and denominator by the LCD of all fractions in its numerator and denominator.

$$\frac{\left(\dfrac{2}{x} - 3\right)}{\left(1 - \dfrac{1}{x-1}\right)} = \frac{\left(\dfrac{2}{x} - 3\right)}{\left(1 - \dfrac{1}{x-1}\right)} \cdot \frac{x(x-1)}{x(x-1)} \qquad \text{LCD is } x(x-1).$$

$$= \frac{\left(\dfrac{2 - 3x}{x}\right) \cdot x(x-1)}{\left(\dfrac{x-2}{x-1}\right) \cdot x(x-1)} = \frac{(2 - 3x)(x-1)}{x(x-2)}, \quad x \neq 1$$

CONCEPT CHECK

1. Is $x \geq 0$ the domain of $\sqrt{x-2}$? Explain.

2. Explain why $\dfrac{x}{x^2 - 4} + \dfrac{5}{x + 2} \neq \dfrac{x + 5}{x^2 + x - 2}$.

3. In the expression $(3x - 2) \div (x + 1)$, explain why $x \neq -1$.

4. What is a complex fraction? Give an example.

Skills Review 0.7 The following warm-up exercises involve skills that were covered in earlier sections. You will use these skills in the exercise set for this section. For additional help, review Section 0.6.

In Exercises 1–10, completely factor the polynomial.

1. $5x^2 - 15x^3$

2. $16x^2 - 9$

3. $9x^2 - 6x + 1$

4. $9 + 12y + 4y^2$

5. $z^2 + 4z + 3$

6. $x^2 - 15x + 50$

7. $3 + 8x - 3x^2$

8. $3x^2 - 46x + 15$

9. $s^3 + s^2 - 4s - 4$

10. $y^3 + 64$

Exercises 0.7

See www.CalcChat.com for worked-out solutions to odd-numbered exercises.

In Exercises 1–4, determine if each value of x is in the domain of the expression.

1. $\dfrac{x + 2}{5x + 2}$ (a) $x = -\dfrac{2}{5}$ (b) $x = 2$

2. $\dfrac{2x + 3}{x - 4}$ (a) $x = -\dfrac{3}{2}$ (b) $x = 4$

3. $\sqrt{2x + 4}$ (a) $x = -2$ (b) $x = 2$

4. $\sqrt{3x - 9}$ (a) $x = -3$ (b) $x = 3$

In Exercises 5–12, find the domain of the expression.

5. $3x^2 - 4x + 7$

6. $6x^2 + 7x - 9,\ x > 0$

7. $\dfrac{1}{x - 2}$

8. $\dfrac{x + 1}{2x + 1}$

9. $\dfrac{x - 1}{x^2 - 4x}$

10. $\dfrac{4x + 3}{x^2 - 36}$

11. $\sqrt{x + 1}$

12. $\dfrac{1}{\sqrt{x + 1}}$

In Exercises 13–18, find the missing factor and state any domain restrictions necessary to make the two fractions equivalent.

13. $\dfrac{5}{2x} = \dfrac{5(\quad)}{6x^2}$

14. $\dfrac{3}{4} = \dfrac{3(\quad)}{4(x + 1)}$

15. $\dfrac{x + 1}{x} = \dfrac{(x + 1)(\quad)}{x(x - 2)}$

16. $\dfrac{3y - 4}{y + 1} = \dfrac{(3y - 4)(\quad)}{y^2 - 1}$

17. $\dfrac{3x}{x - 3} = \dfrac{3x(\quad)}{x^2 - x - 6}$

18. $\dfrac{1 - z}{z^2} = \dfrac{(1 - z)(\quad)}{z^3 + z^2}$

In Exercises 19–34, write the rational expression in simplest form.

19. $\dfrac{15x^2}{10x}$

20. $\dfrac{24y^3}{56y^7}$

21. $\dfrac{2x}{4x + 4}$

22. $\dfrac{9x^2 + 9x}{2x + 2}$

23. $\dfrac{x - 5}{10 - 2x}$

24. $\dfrac{3 - x}{8x - 24}$

25. $\dfrac{x^2 - 25}{5 - x}$

26. $\dfrac{x^2 - 16}{4 - x}$

27. $\dfrac{x^3 + 5x^2 + 6x}{x^2 - 4}$

28. $\dfrac{x^2 + 8x - 20}{x^2 + 11x + 10}$

29. $\dfrac{y^2 - 7y + 12}{y^2 + 3y - 18}$

30. $\dfrac{x + 1}{x^2 - 3x - 4}$

31. $\dfrac{2 - x + 2x^2 - x^3}{x - 2}$

32. $\dfrac{x^2 - 9}{x^3 + x^2 - 9x - 9}$

33. $\dfrac{z^3 - 27}{z^2 + 3z + 9}$

34. $\dfrac{y^3 - 2y^2 - 8y}{y^3 + 8}$

In Exercises 35–48, perform the indicated operations and simplify.

35. $\dfrac{5}{x - 1} \cdot \dfrac{x - 1}{25(x - 2)}$

36. $\dfrac{x + 13}{x^3(3 - x)} \cdot \dfrac{x(x - 3)}{5}$

37. $\dfrac{(x - 9)(x + 7)}{x + 1} \cdot \dfrac{x}{9 - x}$

38. $\dfrac{(x + 5)(x - 3)}{x + 2} \cdot \dfrac{1}{(x + 5)(x + 2)}$

39. $\dfrac{r}{r - 1} \cdot \dfrac{r^2 - 1}{r^2}$

40. $\dfrac{4y - 16}{5y + 15} \cdot \dfrac{2y + 6}{4 - y}$

41. $\dfrac{t^2 - t - 6}{t^2 + 6t + 9} \cdot \dfrac{t + 3}{t^2 - 4}$ **42.** $\dfrac{y^3 - 8}{2y^3} \cdot \dfrac{4y}{y^2 - 5y + 6}$

43. $\dfrac{x^2 + x - 2}{x^3 + x^2} \cdot \dfrac{x}{x^2 + 3x + 2}$

44. $\dfrac{x^3 - 8}{x + 1} \cdot \dfrac{x^2 - 1}{x^3 - 3x^2 + 2x}$

45. $\dfrac{3(x + y)}{4} \div \dfrac{x + y}{2}$

46. $\dfrac{x + 2}{5(x - 3)} \div \dfrac{x - 2}{5(x - 3)}$

47. $\dfrac{\left[\dfrac{x^2}{(x + 1)^2}\right]}{\left[\dfrac{x}{(x + 1)^3}\right]}$ **48.** $\dfrac{\left(\dfrac{x^2 - 1}{x}\right)}{\left[\dfrac{(x - 1)^2}{x}\right]}$

In Exercises 49–52, find the least common denominator of the expressions.

49. $\dfrac{1}{x^2}, \dfrac{1}{x - 1}, \dfrac{1}{x^2 - x}$ **50.** $\dfrac{1}{x}, \dfrac{1}{x^2 + 3x}, \dfrac{1}{x + 3}$

51. $\dfrac{10}{x + 5}, \dfrac{x + 4}{x - 7}, \dfrac{x + 5}{x^2 - 2x - 35}$

52. $\dfrac{x - 1}{x + 2}, \dfrac{8}{x^2 - x - 6}, \dfrac{x}{x - 3}$

In Exercises 53–62, perform the indicated operations and simplify.

53. $\dfrac{4x}{x - 2} + \dfrac{x}{x - 2}$ **54.** $\dfrac{3x - 2}{x + 1} + \dfrac{2 - x}{x + 1}$

55. $\dfrac{3x}{x - 4} + \dfrac{x}{4 - x}$ **56.** $\dfrac{4}{3 - x} + \dfrac{5}{x - 3}$

57. $4 - \dfrac{3}{x - 5}$ **58.** $\dfrac{4}{x + 2} - 6$

59. $\dfrac{2}{x^2 - 4} - \dfrac{1}{x^2 - 3x + 2}$

60. $\dfrac{x}{x^2 + x - 2} - \dfrac{1}{x + 2}$

61. $-\dfrac{1}{x} + \dfrac{2}{x^2 + 1} + \dfrac{1}{x^3 + x}$

62. $\dfrac{2}{x + 1} + \dfrac{2}{x - 1} + \dfrac{1}{x^2 - 1}$

In Exercises 63–68, simplify the complex fraction.

63. $\dfrac{\left(\dfrac{x}{2} - 1\right)}{(x - 2)}$ **64.** $\dfrac{(x - 3)}{\left(\dfrac{x}{4} - \dfrac{4}{x}\right)}$

65. $\dfrac{\left(\dfrac{1}{x} - \dfrac{1}{x + 1}\right)}{\left(\dfrac{1}{x + 1}\right)}$ **66.** $\dfrac{\left(\dfrac{5}{y} - \dfrac{6}{2y + 1}\right)}{\left(\dfrac{5}{y} + 4\right)}$

67. $\dfrac{\left(\sqrt{x} - \dfrac{1}{2\sqrt{x}}\right)}{\sqrt{x}}$ **68.** $\dfrac{\left(\dfrac{1}{\sqrt{2y}} + \sqrt{2y}\right)}{\sqrt{2y}}$

Monthly Payment In Exercises 69 and 70, use the formula for the approximate annual interest rate r of a monthly installment loan

$$r = \dfrac{\left[\dfrac{24(NM - P)}{N}\right]}{\left(P + \dfrac{NM}{12}\right)}$$

where N is the total number of payments, M is the monthly payment, and P is the amount financed.

69. (a) Approximate the annual interest rate r for a four-year car loan of $18,000 that has monthly payments of $475.

   (b) Simplify the expression for the annual interest rate r, and then rework part (a).

70. (a) Approximate the annual interest rate r for a five-year car loan of $20,000 that has monthly payments of $475.

   (b) Simplify the expression for the annual interest rate r, and then rework part (a).

71. Refrigeration When food is placed in a refrigerator, the time required for the food to cool depends on the amount of food, the air circulation in the refrigerator, the original temperature of the food, and the temperature of the refrigerator. One model for the temperature of food that starts at 75°F and is placed in a 40°F refrigerator is

$$T = 10\left(\dfrac{4t^2 + 16t + 75}{t^2 + 4t + 10}\right), \quad t \geq 0$$

where T is the temperature (in degrees Fahrenheit) and t is the time (in hours). Sketch a bar graph showing the temperature of the food when $t = 0, 1, 2, 3, 4,$ and 5 hours. According to the model, will the food reach a temperature of 40°F after 6 hours?

72. Oxygen Level The mathematical model

$$O = \dfrac{t^2 - t + 1}{t^2 + 1}, \quad t \geq 0$$

gives the percent of the normal level of oxygen in a pond, where t is the time in weeks after organic waste is dumped into the pond. Sketch a bar graph showing the oxygen level of the pond when $t = 0, 1, 2, 3, 4,$ and 5 weeks. What conclusions can you make from your bar graph?

Chapter Summary and Study Strategies

After studying this chapter, you should have acquired the following skills.
The exercise numbers are keyed to the Review Exercises that begin on page 64.
Answers to odd-numbered Review Exercises are given in the back of the text.*

Section 0.1 — Review Exercises

■ Classify real numbers as natural numbers, integers, rational numbers, or irrational numbers.	*1, 2*
■ Order real numbers.	*3, 4*
■ Use and interpret inequality notation.	*5–10*
■ Interpret absolute value notation.	*11–14, 19, 20*
■ Find the distance between two numbers on the real number line.	*15–18*

Section 0.2

■ Identify the terms of an algebraic expression.	*21–24*
■ Evaluate an algebraic expression.	*25, 26*
■ Identify basic rules of algebra.	*27–30*
■ Perform operations on real numbers.	*31–36*
■ Use the least common denominator method to add and subtract fractions.	*33, 34*
■ Use a calculator to evaluate an algebraic expression.	*37, 38*
■ Round decimal numbers.	*37, 38*

Section 0.3

■ Use properties of exponents to evaluate and simplify expressions with exponents.

$$a^m a^n = a^{m+n} \qquad \frac{a^m}{a^n} = a^{m-n} \qquad (ab)^m = a^m b^m$$

$$\left(\frac{a}{b}\right)^m = \frac{a^m}{b^m} \qquad (a^m)^n = a^{mn} \qquad a^{-n} = \frac{1}{a^n}$$

$$a^0 = 1 \qquad \left(\frac{a}{b}\right)^{-n} = \left(\frac{b}{a}\right)^n \qquad |a^2| = |a|^2 = a^2 \qquad \textit{39–46}$$

■ Use scientific notation.	*47–50*
■ Use a calculator to evaluate expressions involving powers.	*51, 52*
■ Use interest formulas to solve an application problem.	

Simple interest: $A = P(1 + rt)$

Compound interest: $A = P\left(1 + \dfrac{r}{n}\right)^{nt}$ *53, 54*

---

* Use a wide range of valuable study aids to help you master the material in this chapter. The *Student Solutions Guide* includes step-by-step solutions to all odd-numbered exercises to help you review and prepare. The student website at *college.hmco.com/info/larsonapplied* offers algebra help and a *Graphing Technology Guide*. The *Graphing Technology Guide* contains step-by-step commands and instructions for a wide variety of graphing calculators, including the most recent models.

Section 0.4   Review Exercises

■ Simplify and evaluate expressions involving radicals.

$$\sqrt[n]{a^m} = \left(\sqrt[n]{a}\right)^m \qquad \sqrt[n]{a} \cdot \sqrt[n]{b} = \sqrt[n]{ab} \qquad \frac{\sqrt[n]{a}}{\sqrt[n]{b}} = \sqrt[n]{\frac{a}{b}}$$

$$\sqrt[m]{\sqrt[n]{a}} = \sqrt[mn]{a} \qquad \left(\sqrt[n]{a}\right)^n = a \qquad \text{For } n \text{ even, } \sqrt[n]{a^n} = |a|$$

$$\text{For } n \text{ odd, } \sqrt[n]{a^n} = a \qquad \qquad 55\text{–}60$$

■ Rationalize a denominator by using its conjugate.   *61, 62*

■ Combine radicals.   *63–68*

■ Use properties of rational exponents.   *69, 70*

■ Use a calculator to evaluate a radical.   *71, 72*

Section 0.5

■ Write a polynomial in standard form.   *73–82*

■ Add and subtract polynomials by combining like terms.   *73–76*

■ Multiply polynomials using FOIL or a vertical format.   *77–82*

■ Use special products to multiply polynomials.

$$(u + v)(u - v) = u^2 - v^2 \qquad (u \pm v)^2 = u^2 \pm 2uv + v^2$$

$$(u \pm v)^3 = u^3 \pm 3u^2v + 3uv^2 \pm v^3 \qquad\qquad\qquad 78, 81, 82$$

■ Use polynomials to solve an application problem.   *83–86*

Section 0.6

■ Factor a polynomial by removing common factors.   *87, 89, 94*

■ Factor a polynomial in a special form.

$$u^2 - v^2 = (u + v)(u - v) \qquad u^2 \pm 2uv + v^2 = (u \pm v)^2$$

$$u^3 \pm v^3 = (u \pm v)(u^2 \mp uv + v^2) \qquad\qquad\qquad 87, 90\text{–}94$$

■ Factor a trinomial as the product of two binomials.   *88, 89*

■ Factor a polynomial by grouping.   *90, 93*

Section 0.7

■ Find the domain of an algebraic expression by finding values of the variable that
  make a denominator zero or a radicand negative.   *95–100*

■ Simplify a rational expression by dividing out common factors from the
  numerator and denominator.   *101–106*

■ Perform operations with rational expressions by using properties of fractions. *107–112*

■ Simplify a complex fraction.   *113–116*

Study Strategies

■ **Use the Skills Review Exercises**   Each section exercise set in this text (except the set for Section 0.1) begins with a set of skills review exercises. You should begin each homework session by quickly working through all of these exercises (all are answered in the back of the text). The "old" skills covered in these exercises are needed to master the "new" skills in the section exercise set. The skills review exercises remind you that mathematics is cumulative—to be successful in this course, you must retain "old" skills.

■ **Use the Additional Study Aids**   The additional study aids were prepared specifically to help you master the concepts discussed in the text. They are the *Student Solutions Manual*, the *Graphing Calculator Keystroke Guide*, and the *Instructional DVD*.

Review Exercises

See www.CalcChat.com for worked-out solutions to odd-numbered exercises.

In Exercises 1 and 2, determine which numbers in the set are (a) natural numbers, (b) integers, (c) rational numbers, and (d) irrational numbers.

1. $\left\{11, -14, -\frac{8}{9}, \frac{5}{2}, \sqrt{6}, 0.4\right\}$

2. $\left\{\sqrt{15}, -22, -\frac{10}{3}, 0, 5.2, \frac{3}{7}\right\}$

In Exercises 3 and 4, plot the two real numbers on the real number line and place the appropriate inequality sign (< or >) between them.

3. $-4, -3$

4. $\frac{1}{5}, \frac{1}{6}$

In Exercises 5 and 6, give a verbal description of the subset of real numbers that is represented by the inequality, and sketch the subset on the real number line.

5. $x \le -6$

6. $x > 5$

In Exercises 7–10, use inequality notation to describe the subset of real numbers.

7. x is nonnegative.

8. x is at most 7.

9. x is greater than 2 and less than or equal to 5.

10. x is less than or equal to -2 or x is greater than 2.

In Exercises 11 and 12, evaluate the expression.

11. $-|-14|$

12. $|-4 - 2|$

In Exercises 13 and 14, place the correct symbol (<, >, or =) between the two real numbers.

13. $|-12|$ ⬚ $-|12|$

14. $|9|$ ⬚ $|-9|$

In Exercises 15–18, find the distance between a and b.

15. $a = -14, \quad b = -18$

16. $a = -1, \quad b = -5$

17. $a = 2, \quad b = -8$

18. $a = 10, \quad b = -3$

In Exercises 19 and 20, use absolute value notation to describe the sentence.

19. The distance between x and 7 is at least 4.

20. The distance between x and -22 is no more than 10.

In Exercises 21–24, identify the terms of the algebraic expression.

21. $4 + x - 2x^2$

22. $16x^2 - 4$

23. $3x^3 + 7x - 4$

24. $3x^3 - 9x$

In Exercises 25 and 26, evaluate the expression for each value of x.

25. $-4x^2 - 6x$ (a) $x = -1$ (b) $x = 0$

26. $12 - 5x^2$ (a) $x = -2$ (b) $x = 3$

In Exercises 27–30, identify the rule of algebra illustrated by the statement.

27. $5(x^2 + x) = 5x^2 + 5x$

28. $x + (2x + 3) = (x + 2x) + 3$

29. $3x + 7 = 7 + 3x$

30. $(x^2 - 1)\left(\dfrac{1}{x^2 - 1}\right) = 1$

In Exercises 31–36, perform the indicated operation(s). Write fractional answers in simplest form.

31. $-3 - 2(4 - 5)$

32. $12(-3 + 5) - 20$

33. $\frac{1}{2} + \frac{1}{3} - \frac{1}{6}$

34. $\frac{5}{12} + \frac{3}{5}$

35. $5^2 \cdot 5^{-1}$

36. $(4^2)^2$

In Exercises 37 and 38, use a calculator to evaluate the expression. (Round to two decimal places.)

37. $4\left(\frac{1}{6} - \frac{1}{7}\right)$

38. $-2 + 3\left(\frac{1}{2} - \frac{1}{3}\right)$

In Exercises 39–42, evaluate the expression for the value of x.

39. $-2x^2, \quad x = -1$

40. $-\dfrac{(-x)^2}{6}, \quad x = -3$

41. $\dfrac{x^2 - 4}{x - 4}, \quad x = 3$

42. $\dfrac{2x^3 - x + 2}{x - 7}, \quad x = -2$

In Exercises 43–46, simplify the expression.

43. $\dfrac{(4x)^2}{2x}$

44. $(-x)^2(-3x)^3$

45. $\dfrac{10x^2}{2x^6}$

46. $2x(5x^2)^3$

In Exercises 47 and 48, write the number in scientific notation.

47. Population of the United States: 300,400,000
 (Source: U.S. Census Bureau)

48. Number of Meters in One Foot: 0.3048

In Exercises 49 and 50, write the number in decimal notation.

49. Diameter of the Sun: 8.644×10^5 miles

50. Length of an E. Coli Bacterium: 2×10^{-6} meter

In Exercises 51 and 52, use a calculator to evaluate the expression. (Round to three decimal places.)

51. (a) $1800(1 + 0.08)^{24}$

(b) $0.0024(7,658,400)$

52. (a) $50,000\left(1 + \dfrac{0.075}{12}\right)^{48}$

(b) $\dfrac{28,000,000 + 34,000,000}{87,000,000}$

In Exercises 53 and 54, complete the table by finding the balance.

53. Balance in an Account You deposit $1500 in an account with an annual interest rate of 6.5%, compounded monthly.

Year	5	10	15	20	25
Balance					

54. Balance in an Account You deposit $12,000 in an account with an annual interest rate of 6%, compounded quarterly.

Year	5	10	15	20	25
Balance					

In Exercises 55 and 56, fill in the missing form.

Radical Form   *Rational Exponent Form*

55. $\sqrt{16} = 4$

56. $16^{1/4} = 2$

In Exercises 57 and 58, evaluate the expression.

57. $\sqrt{169}$ **58.** $\sqrt[3]{125}$

In Exercises 59 and 60, simplify by removing all possible factors from the radical.

59. $\sqrt{4x^4}$ **60.** $\sqrt[3]{\dfrac{2x^3}{27}}$

In Exercises 61 and 62, rewrite the expression by rationalizing the denominator. Simplify your answer.

61. $\dfrac{1}{2 - \sqrt{3}}$

62. $\dfrac{2}{3 + \sqrt{5}}$

In Exercises 63–68, simplify the expression.

63. $2\sqrt{x} - 5\sqrt{x}$

64. $\sqrt{72} + \sqrt{128}$

65. $\sqrt{5}\sqrt{2}$

66. $\sqrt{3}\sqrt{4}$

67. $(64)^{-2/3}$

68. $4^{1/3} \cdot 4^{5/3}$

In Exercises 69 and 70, use rational exponents to reduce the index of the radical.

69. $\sqrt[4]{5^2}$ **70.** $\sqrt[8]{x^4}$

In Exercises 71 and 72, use a calculator to approximate the number. (Round your answer to three decimal places.)

71. $\sqrt{127}$ **72.** $\sqrt[3]{52}$

In Exercises 73–82, perform the indicated operation(s) and write the resulting polynomial in standard form.

73. $2(x - 3) - 4(2x - 8)$

74. $3(x^2 - 5x + 2) + 3x(2 - 4x)$

75. $x(x - 2) - 2(3x + 7)$

76. $2x(x + 1) + 3(x^2 - x)$

77. $(x + 1)(x - 2)$

78. $(2x - 5)(2x + 5)$

79. $(x + 4)(x^2 - 4x + 16)$

80. $(x - 2)(x^2 + 6x + 9)$

81. $(x + 4)^2$

82. $(2x + 1)^3$

83. Home Prices The average sale price (in thousands of dollars) of a newly manufactured residential mobile home in the United States from 2000 to 2005 can be represented by the polynomial

$3.17x + 45.7$

where x represents the year, with $x = 0$ corresponding to 2000. Evaluate the polynomial when $x = 5$. Then describe your result in everyday terms. *(Source: U.S. Census Bureau)*

84. Home Prices The median sale price (in thousands of dollars) of a new one-family home in the southern United States from 2000 to 2005 can be represented by the polynomial

$$9.38x + 145.4$$

where x represents the year, with $x = 0$ corresponding to 2000. Evaluate the polynomial when $x = 5$. Then describe your result in everyday terms. *(Source: U.S. Census Bureau and U.S. Department of Housing and Urban Development)*

85. Cell Phone Subscribers The numbers of cell phone subscribers (in millions) in the United States from 2000 to 2005 can be represented by the polynomial

$$19.18x + 106.6$$

where x represents the year, with $x = 0$ corresponding to 2000. Evaluate the polynomial when $x = 0$ and $x = 5$. Then describe your results in everyday terms. *(Source: Cellular Telecommunications & Internet Association)*

86. Cell Sites The numbers of cellular telecommunications sites in the United States from 2000 to 2005 can be represented by the polynomial

$$-1297.79x^2 + 22{,}637.7x + 104{,}230$$

where x represents the year, with $x = 0$ corresponding to 2000. Evaluate the polynomial when $x = 0$ and $x = 5$. Then describe your results in everyday terms. *(Source: Cellular Telecommunications & Internet Association)*

In Exercises 87–94, completely factor the expression.

87. $4x^2 - 36$

88. $x^2 - 4x - 5$

89. $-3x^2 - 6x + 3x^3$

90. $x^3 - 4x^2 - 2x + 8$

91. $x^3 - 16x$

92. $8x^3 - 125$

93. $x^3 - 2x^2 - 9x + 18$

94. $2x^5 - 16x^3$

In Exercises 95–98, find the domain of the expression.

95. $\dfrac{2x + 1}{x - 3}$

96. $\dfrac{x - 3}{x + 1}$

97. $2x^2 - 11x + 5$

98. $4\sqrt{2x}$

In Exercises 99 and 100, find the missing factor and state any domain restrictions necessary to make the two fractions equivalent.

99. $\dfrac{4}{3x} = \dfrac{4()}{9x^2}$

100. $\dfrac{5}{7} = \dfrac{5()}{7(x + 2)}$

In Exercises 101–106, write the rational expression in simplest form.

101. $\dfrac{x^2 - 4}{2x + 4}$

102. $\dfrac{2x^2 + 4x}{2x}$

103. $\dfrac{x^2 - 2x - 15}{x + 3}$

104. $\dfrac{x^3 + 2x^2 - 3x}{x - 1}$

105. $\dfrac{x^3 - 9x}{x^3 - 4x^2 + 3x}$

106. $\dfrac{x^3 + 64}{x^2 - x - 20}$

In Exercises 107–112, perform the operation and simplify.

107. $\dfrac{2x - 1}{x + 1} \cdot \dfrac{x^2 - 1}{2x^2 - 7x + 3}$

108. $\dfrac{x + 2}{x - 4} \div \dfrac{2x + 4}{8x}$

109. $\dfrac{x}{x - 1} + \dfrac{2x}{x - 2}$

110. $\dfrac{2}{x + 2} - \dfrac{3}{x - 2}$

111. $\dfrac{2}{x - 1} + \dfrac{4}{x + 1} + \dfrac{8}{x^2 - 1}$

112. $\dfrac{1}{x - 1} + \dfrac{2}{x} - \dfrac{1}{x^2 - x}$

In Exercises 113–116, simplify the complex fraction.

113. $\dfrac{\left(\dfrac{x^2 - 1}{x}\right)}{\left[\dfrac{(x - 1)^2}{x}\right]}$

114. $\dfrac{(x - 4)}{\left(\dfrac{x}{4} - \dfrac{4}{x}\right)}$

115. $\dfrac{\left(\dfrac{1}{x} - \dfrac{1}{y}\right)}{\left(\dfrac{1}{x} + \dfrac{1}{y}\right)}$

116. $\dfrac{\left(\dfrac{1}{2x - 3} - \dfrac{1}{2x + 3}\right)}{\left(\dfrac{1}{2x} - \dfrac{1}{2x + 3}\right)}$

Chapter Test

See www.CalcChat.com for worked-out solutions to odd-numbered exercises.

Take this test as you would take a test in class. When you are done, check your work against the answers given in the back of the book.

1. Evaluate the expression $-3x^2 - 5x$ when $x = -3$.

2. Complete the table at the left given that $4000 is deposited in an account with an annual interest rate of 7.5%, compounded monthly. What can you conclude from the table?

Year	Balance
5	
10	
15	
20	
25	

Table for 2

In Exercises 3–8, simplify the expression.

3. $8(-2x^2)^3$

4. $3\sqrt{x} - 7\sqrt{x}$

5. $5^{1/4} \cdot 5^{7/4}$

6. $\sqrt{48} - \sqrt{80}$

7. $\sqrt{12x^3}$

8. $\dfrac{2}{5 - \sqrt{7}}$

In Exercises 9 and 10, write the polynomial in standard form.

9. $(3x + 7)^2$

10. $3x(x + 5) - 2x(4x - 7)$

In Exercises 11–14, completely factor the expression.

11. $5x^2 - 80$

12. $4x^2 + 12x + 9$

13. $x^3 - 6x^2 - 3x + 18$

14. $x^3 + 2x^2 - 4x - 8$

15. Simplify: $\dfrac{x^2 - 16}{3x + 12}$.

16. Multiply and simplify: $\dfrac{3x - 5}{x + 3} \cdot \dfrac{x^2 + 7x + 12}{9x^2 - 25}$.

17. Add and simplify: $\dfrac{x}{x - 3} + \dfrac{3x}{x - 4}$.

18. Subtract and simplify: $\dfrac{3}{x + 5} - \dfrac{4}{x - 2}$.

In Exercises 19 and 20, find the domain of the expression.

19. $\sqrt{x - 2}$

20. $\dfrac{3}{x + 1}$

21. Simplify the complex fraction $\dfrac{\left(\dfrac{2x - 9}{x - 1}\right)}{\left(\dfrac{3}{x - 1} + \dfrac{1 - x}{x + 2}\right)}$.

22. Movie Price The average price of a movie ticket in the United States from 1995 to 2005 can be approximated by the polynomial $0.224x + 3.09$, where x is the year, with $x = 5$ corresponding to 1995. Evaluate the polynomial when $x = 5$ and $x = 15$. Then describe your results in everyday terms. *(Source: Exhibitor Relations Co., Inc.)*

1 Equations and Inequalities

© imagebroker/Alamy

The force of gravity on the moon is about one-sixth the force of gravity on Earth. So, objects fall at a different rate on the moon than on Earth. You can use a quadratic equation to model the height with respect to time of a falling object on the moon. (See Section 1.4, Example 6.)

Applications

Equations and inequalities are used to model and solve many real-life applications. The applications listed below represent a sample of the applications in this chapter.

- Blood Oxygen Level, Exercise 69, page 113
- Life Expectancy, Exercise 70, page 124
- Make a Decision: Company Profits, Exercise 65, page 146

Section 1.1

Linear Equations

- Classify an equation as an identity or a conditional equation.
- Solve a linear equation in one variable.
- Use a linear model to solve an application problem.

Equations and Solutions

An **equation** is a statement that two algebraic expressions are equal. Some examples of equations in x are

$$3x - 5 = 7, \quad x^2 - x - 6 = 0, \quad \text{and} \quad \sqrt{2x} = 4.$$

To **solve** an equation in x means to find all values of x for which the equation is **true.** Such values are called **solutions.** For instance, $x = 4$ is a solution of the equation $3x - 5 = 7$, because $3(4) - 5 = 7$ is a true statement.

An equation that is true for *every* real number in the domain of the variable is called an **identity.** Two examples of identities are

$$x^2 - 9 = (x + 3)(x - 3) \quad \text{and} \quad \frac{x}{3x^2} = \frac{1}{3x}, \quad x \neq 0.$$

The first equation is an identity because it is a true statement for all real values of x. The second is an identity because it is true for all nonzero real values of x.

An equation that is true for just *some* (or even none) of the real numbers in the domain of the variable is called a **conditional equation.** For example, the equation $x^2 - 9 = 0$ is conditional because $x = 3$ and $x = -3$ are the only values in the domain that satisfy the equation.

Example 1 Classifying Equations

Determine whether each equation is an identity or a conditional equation.

a. $2(x + 3) = 2x + 6$ **b.** $2(x + 3) = x + 6$ **c.** $2(x + 3) = 2x + 3$

SOLUTION

a. This equation is an identity because it is true for every real value of x.

b. This equation is a conditional equation because $x = 0$ is the only value in the domain for which the equation is true.

c. This equation is a conditional equation because there are no real number values of x for which the equation is true. ▬▬▬

✓CHECKPOINT 1

Determine whether the equation $4(x + 1) = 4x + 4$ is an identity or a conditional equation. ▪

Equations are used in algebra for two distinct purposes: (1) *identities* are usually used to state mathematical properties and (2) *conditional equations* are usually used to model and solve problems that occur in real life.

Linear Equations in One Variable

The most common type of conditional equation is a **linear equation.**

Definition of a Linear Equation

A **linear equation** in one variable x is an equation that can be written in the standard form

$$ax + b = 0$$

where a and b are real numbers with $a \neq 0$.

A linear equation in x has exactly one solution. To see this, consider the following steps. (Remember that $a \neq 0$.)

$ax + b = 0$	Original equation
$ax = -b$	Subtract b from each side.
$x = -\dfrac{b}{a}$	Divide each side by a.

So, the equation $ax + b = 0$ has exactly one solution, $x = -b/a$.

To solve a linear equation in x, you should isolate x by forming a sequence of **equivalent** (and usually simpler) equations, each having the same solution as the original equation. The operations that yield equivalent equations come from the basic rules of algebra reviewed in Section 0.2.

Forming Equivalent Equations

A given equation can be transformed into an equivalent equation by one or more of the following steps.

	Given Equation	*Equivalent Equation*
1. Remove symbols of grouping, combine like terms, or simplify one or both sides of the equation.	$2x - x = 4$ $3(x - 2) = 5$	$x = 4$ $3x - 6 = 5$
2. Add (or subtract) the same quantity to (from) *each* side of the equation.	$x + 1 = 6$	$x = 5$
3. Multiply (or divide) *each* side of the equation by the same *nonzero* quantity.	$2x = 6$	$x = 3$
4. Interchange sides of the equation.	$2 = x$	$x = 2$

The steps for solving a linear equation in x written in standard form are shown in Example 2.

Example 2   **Solving a Linear Equation**

Solve $3x - 6 = 0$.

SOLUTION

$3x - 6 = 0$	Write original equation.
$3x = 6$	Add 6 to each side.
$x = 2$	Divide each side by 3.

✓ **CHECKPOINT 2**

Solve $5 + 5x = 15$. ■

After solving an equation, you should **check each solution** in the *original* equation. For instance, in Example 2, you can check that 2 is a solution by substituting 2 for x in the original equation $3x - 6 = 0$, as follows.

CHECK

$3x - 6 = 0$	Write original equation.
$3(2) - 6 \overset{?}{=} 0$	Substitute 2 for x.
$6 - 6 = 0$	Solution checks. ✓

Example 3   **Solving a Linear Equation**

Solve $6(x - 1) + 4 = 3(7x + 1)$.

SOLUTION

$6(x - 1) + 4 = 3(7x + 1)$	Write original equation.
$6x - 6 + 4 = 21x + 3$	Distributive Property
$6x - 2 = 21x + 3$	Simplify.
$-15x = 5$	Add 2 to and subtract $21x$ from each side.
$x = -\frac{1}{3}$	Divide each side by -15.

The solution is $x = -\frac{1}{3}$. You can check this as follows.

CHECK

$6(x - 1) + 4 = 3(7x + 1)$	Write original equation.
$6\left(-\frac{1}{3} - 1\right) + 4 \overset{?}{=} 3\left[7\left(-\frac{1}{3}\right) + 1\right]$	Substitute $-\frac{1}{3}$ for x.
$6\left(-\frac{4}{3}\right) + 4 \overset{?}{=} 3\left(-\frac{7}{3} + 1\right)$	Add fractions.
$-8 + 4 \overset{?}{=} -7 + 3$	Simplify.
$-4 = -4$	Solution checks. ✓

✓ **CHECKPOINT 3**

Solve $2(x + 2) + 6 = 4(2x - 3)$. ■

STUDY TIP

You may think a solution to a problem looks easy when it is worked out in class, but you may not know where to begin when solving the problem on your own. Keep in mind that many problems involve some trial and error before a solution is found.

Some equations in one variable have *infinitely many solutions*. To recognize an equation of this type, perform the regular steps for solving the equation. If, when writing equivalent equations, you reach a statement that is true for all values in the domain of the variable, then the equation is an identity and has infinitely many solutions.

Example 4 An Equation with Infinitely Many Solutions

Solve $x + 4(x - 2) = 3x + 2(x - 4)$.

SOLUTION

$$x + 4(x - 2) = 3x + 2(x - 4) \qquad \text{Write original equation.}$$
$$x + 4x - 8 = 3x + 2x - 8 \qquad \text{Distributive Property}$$
$$5x - 8 = 5x - 8 \qquad \text{Simplify.}$$
$$-8 = -8 \qquad \text{Subtract } 5x \text{ from each side.}$$

Because the last equation is true for every real value of x, the original equation is an identity and you can conclude that it has infinitely many solutions.

✓ CHECKPOINT 4

Solve $x + 5 + 3(2x + 1) = 7x + 8$. ▪

It is also possible for an equation in one variable to have *no solution*. When solving an equation of this type, you will reach a statement that is not true for any value of the variable.

Example 5 An Equation with No Solution

Solve $4x - 9 + 2(x + 8) = 1 + 6(x + 4)$.

SOLUTION

$$4x - 9 + 2(x + 8) = 1 + 6(x + 4) \qquad \text{Write original equation.}$$
$$4x - 9 + 2x + 16 = 1 + 6x + 24 \qquad \text{Distributive Property}$$
$$6x + 7 = 6x + 25 \qquad \text{Simplify.}$$
$$7 \neq 25 \qquad \text{Subtract } 6x \text{ from each side.}$$

Because the statement $7 = 25$ is not true, you can conclude that the original equation has no solution.

✓ CHECKPOINT 5

Solve $1 + 4(x + 1) = 4(2 + x)$. ▪

Equations in one variable with infinitely many solutions or no solution are not linear because they cannot be written in the standard form $ax + b = 0$. Note that a linear equation in x has exactly one solution.

Equations Involving Fractional Expressions

To solve an equation involving fractional expressions, you can multiply every term in the equation by the least common denominator (LCD) of the terms.

Example 6 An Equation Involving Fractional Expressions

$$\frac{x}{3} + \frac{3x}{4} = 2 \qquad \text{Original equation}$$

$$(12)\frac{x}{3} + (12)\frac{3x}{4} = (12)2 \qquad \text{Multiply each term by least common denominator.}$$

$$4x + 9x = 24 \qquad \text{Simplify.}$$

$$13x = 24 \qquad \text{Combine like terms.}$$

$$x = \frac{24}{13} \qquad \text{Divide each side by 13.}$$

The solution is $x = \frac{24}{13}$. Check this in the original equation.

✓ **CHECKPOINT 6**

Solve $\dfrac{4x}{3} - \dfrac{x}{12} = 5$. ∎

When multiplying or dividing an equation by a *variable expression*, it is possible to introduce an **extraneous** solution—one that does not satisfy the original equation. In such cases a check is especially important.

Example 7 An Equation with an Extraneous Solution

Solve $\dfrac{1}{x-2} = \dfrac{3}{x+2} - \dfrac{6x}{x^2-4}$.

SOLUTION The least common denominator is $x^2 - 4 = (x+2)(x-2)$. Multiply each term by this LCD and simplify.

$$\frac{1}{x-2} = \frac{3}{x+2} - \frac{6x}{x^2-4} \qquad \text{Write original equation.}$$

$$\frac{1}{x-2}(x+2)(x-2) = \frac{3}{x+2}(x+2)(x-2) - \frac{6x}{x^2-4}(x+2)(x-2)$$

$$x + 2 = 3(x-2) - 6x, \quad x \neq \pm 2 \qquad \text{Simplify.}$$

$$x + 2 = 3x - 6 - 6x \qquad \text{Distributive Property}$$

$$4x = -8 \qquad \begin{array}{l}\text{Combine like terms}\\\text{and simplify.}\end{array}$$

$$x = -2 \qquad \text{Extraneous solution}$$

By checking $x = -2$, you can see that it yields a denominator of zero for the fraction $3/(x+2)$. So, $x = -2$ is extraneous, and the equation has *no solution*.

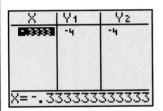
✓ **CHECKPOINT 7**

Solve $\dfrac{1}{x-4} + \dfrac{1}{x} = \dfrac{4}{x(x-4)}$. ∎

An equation with a *single fraction* on each side can be cleared of denominators by **cross-multiplying,** which is equivalent to multiplying each side of the equation by the least common denominator and then simplifying.

Example 8 **Cross–Multiplying to Solve an Equation**

Solve $\dfrac{3y - 2}{2y + 1} = \dfrac{6y - 9}{4y + 3}$.

SOLUTION

$$\dfrac{3y - 2}{2y + 1} = \dfrac{6y - 9}{4y + 3} \qquad \text{Write original equation.}$$

$$(3y - 2)(4y + 3) = (6y - 9)(2y + 1) \qquad \text{Cross-multiply.}$$

$$12y^2 + y - 6 = 12y^2 - 12y - 9 \qquad \text{Multiply.}$$

$$13y = -3 \qquad \text{Isolate } y\text{-term on left.}$$

$$y = -\dfrac{3}{13} \qquad \text{Divide each side by 13.}$$

The solution is $y = -\frac{3}{13}$. Check this in the original equation.

✓**CHECKPOINT 8**

Solve $\dfrac{3x - 6}{x + 10} = \dfrac{3}{4}$. ▪

Example 9 **Using a Calculator to Solve an Equation**

Solve $\dfrac{1}{9.38} - \dfrac{3}{x} = \dfrac{5}{0.3714}$.

SOLUTION Roundoff error will be minimized if you solve for x before performing any calculations. The least common denominator is $(9.38)(0.3714)(x)$.

$$\dfrac{1}{9.38} - \dfrac{3}{x} = \dfrac{5}{0.3714}$$

$$(9.38)(0.3714)(x)\left(\dfrac{1}{9.38} - \dfrac{3}{x}\right) = (9.38)(0.3714)(x)\left(\dfrac{5}{0.3714}\right)$$

$$0.3714x - 3(9.38)(0.3714) = 5(9.38)(x), \quad x \neq 0$$

$$[0.3714 - 5(9.38)]x = 3(9.38)(0.3714)$$

$$x = \dfrac{3(9.38)(0.3714)}{0.3714 - 5(9.38)}$$

$$x \approx -0.225 \qquad \text{Round to three decimal places.}$$

The solution is $x \approx -0.225$. Check this in the original equation.

✓**CHECKPOINT 9**

Solve $\dfrac{5}{x} + \dfrac{1}{2.7} = \dfrac{4}{0.6}$. ▪

Application

Example 10
MAKE A DECISION Hourly Earnings

The mean hourly earnings y (in dollars) of employees at outpatient care centers in the United States from 2000 to 2005 can be modeled by the linear equation

$$y = 0.782t + 15.20, \quad 0 \le t \le 5$$

where t represents the year, with $t = 0$ corresponding to 2000. Use the model to estimate the year in which the mean hourly earnings were $16.75. *(Source: U.S. Bureau of Labor Statistics)*

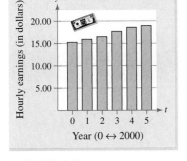

FIGURE 1.1

SOLUTION To determine when the mean hourly earnings were $16.75, solve the model for t when $y = 16.75$

$y = 0.782t + 15.20$	Write original model.
$16.75 = 0.782t + 15.20$	Substitute 16.75 for y so you can solve for t.
$1.55 = 0.782t$	Subtract 15.20 from each side.
$t = \dfrac{1.55}{0.782} \approx 2$	Divide each side by 0.782.

Because $t = 0$ corresponds to 2000, it follows that $t = 2$ corresponds to 2002. See Figure 1.1. So, mean hourly earnings were $16.75 in 2002. ———

✓**CHECKPOINT 10**

The mean hourly earnings y (in dollars) of the employees at a factory from 2000 to 2008 can be modeled by the linear equation

$$y = 0.825t + 18.60, \quad 0 \le t \le 8$$

where t represents the year, with $t = 0$ corresponding to 2000. Use the model to estimate the year in which the mean hourly earnings of the employees at the factory were $21.90. ■

CONCEPT CHECK

1. Is the equation $x(8 - x) = 15$ a linear equation? Explain.

2. Explain the difference between an identity and a conditional equation.

3. Can the equation $ax + b = 0$ have two solutions? Explain.

4. Does the equation $\dfrac{4x}{x - 3} = 8 + \dfrac{12}{x - 3}$ have an extraneous solution? Explain.

The symbol ⓡ indicates an example that uses or is derived from real-life data.

Skills Review 1.1   The following warm-up exercises involve skills that were covered in earlier sections. You will use these skills in the exercise set for this section. For additional help, review Sections 0.2 and 0.7.

In Exercises 1–10, perform the indicated operations and simplify your answer.

1. $(2x - 4) - (5x + 6)$

2. $(3x - 5) + (2x - 7)$

3. $2(x + 1) - (x + 2)$

4. $-3(2x - 4) + 7(x + 2)$

5. $\dfrac{x}{3} + \dfrac{x}{5}$

6. $x - \dfrac{x}{4}$

7. $\dfrac{1}{x + 1} - \dfrac{1}{x}$

8. $\dfrac{2}{x} + \dfrac{3}{x}$

9. $\dfrac{4}{x} + \dfrac{3}{x - 2}$

10. $\dfrac{1}{x + 1} - \dfrac{1}{x - 1}$

Exercises 1.1

See www.CalcChat.com for worked-out solutions to odd-numbered exercises.

In Exercises 1–6, determine whether the equation is an identity or a conditional equation.

1. $2(x - 1) = 2x - 2$

2. $3(x + 2) = 3x + 6$

3. $2(x - 1) = 3x + 4$

4. $3(x + 2) = 2x + 4$

5. $2(x + 1) = 2x + 1$

6. $3(x + 4) = 3x + 4$

In Exercises 7–16, determine whether each value of x is a solution of the equation.

	Equation		*Values*	
7.	$5x - 3 = 3x + 5$		(a) $x = 0$	(b) $x = -5$
			(c) $x = 4$	(d) $x = 10$
8.	$7 - 3x = 5x - 17$		(a) $x = -3$	(b) $x = 0$
			(c) $x = 8$	(d) $x = 3$
9.	$3x^2 + 2x - 5 = 2x^2 - 2$		(a) $x = -3$	(b) $x = 1$
			(c) $x = 4$	(d) $x = -5$
10.	$5x^3 + 2x - 3 = 4x^3 + 2x - 11$			
			(a) $x = 2$	(b) $x = -2$
			(c) $x = 0$	(d) $x = 10$
11.	$\dfrac{5}{2x} - \dfrac{4}{x} = 3$		(a) $x = -\dfrac{1}{2}$	(b) $x = 4$
			(c) $x = 0$	(d) $x = \dfrac{1}{4}$
12.	$3 + \dfrac{1}{x + 2} = 4$		(a) $x = -1$	(b) $x = -2$
			(c) $x = 0$	(d) $x = 5$
13.	$(x + 5)(x - 3) = 20$		(a) $x = 3$	(b) $x = -2$
			(c) $x = 0$	(d) $x = -7$

	Equation	*Values*	
14.	$(3x + 5)(2x - 7) = 0$	(a) $x = -\dfrac{5}{3}$	(b) $x = -\dfrac{2}{7}$
		(c) $x = \dfrac{2}{3}$	(d) $x = \dfrac{3}{2}$
15.	$\sqrt{2x - 3} = 3$	(a) $x = 6$	(b) $x = -3$
		(c) $x = -\dfrac{1}{3}$	(d) $x = -2$
16.	$\sqrt[3]{x - 8} = 3$	(a) $x = 2$	(b) $x = -5$
		(c) $x = 35$	(d) $x = 8$

In Exercises 17–54, solve the equation and check your solution. (Some equations have no solution.)

17. $x + 10 = 15$

18. $9 - x = 13$

19. $7 - 2x = 15$

20. $7x + 2 = 16$

21. $8x - 5 = 3x + 10$

22. $7x + 3 = 3x - 13$

23. $2(x + 5) - 7 = 3(x - 2)$

24. $2(13t - 15) + 3(t - 19) = 0$

25. $6[x - (2x + 3)] = 8 - 5x$

26. $3[2x - (x + 7)] = 5(x - 3)$

27. $\dfrac{5x}{4} + \dfrac{1}{2} = x - \dfrac{1}{2}$

28. $\dfrac{x}{5} - \dfrac{x}{2} = 3$

29. $\dfrac{3}{2}(z + 5) - \dfrac{1}{4}(z + 24) = 0$

30. $\dfrac{3x}{2} + \dfrac{1}{4}(x - 2) = 10$

31. $0.25x + 0.75(10 - x) = 3$

32. $0.60x + 0.40(100 - x) = 50$

33. $x + 8 = 2(x - 2) - x$

34. $3(x + 3) = 5(1 - x) - 1$

35. $\dfrac{100 - 4u}{3} = \dfrac{5u + 6}{4} + 6$

36. $\dfrac{17 + y}{y} + \dfrac{32 + y}{y} = 100$

37. $\dfrac{5x - 4}{5x + 4} = \dfrac{2}{3}$

38. $\dfrac{10x + 3}{5x + 6} = \dfrac{1}{2}$

39. $10 - \dfrac{13}{x} = 4 + \dfrac{5}{x}$

40. $\dfrac{15}{x} - 4 = \dfrac{6}{x} + 3$

41. $\dfrac{1}{x - 3} + \dfrac{1}{x + 3} = \dfrac{10}{x^2 - 9}$

42. $\dfrac{1}{x - 2} + \dfrac{3}{x + 3} = \dfrac{4}{x^2 + x - 6}$

43. $\dfrac{6}{(x - 3)(x - 1)} = \dfrac{3}{x - 3} + \dfrac{4}{x - 1}$

44. $\dfrac{2}{(x - 4)(x - 2)} = \dfrac{1}{x - 4} + \dfrac{2}{x - 2}$

45. $\dfrac{7}{2x + 1} - \dfrac{8x}{2x - 1} = -4$

46. $\dfrac{4}{u - 1} + \dfrac{6}{3u + 1} = \dfrac{15}{3u + 1}$

47. $\dfrac{3}{x(x - 3)} + \dfrac{4}{x} = \dfrac{1}{x - 3}$

48. $3 = 2 + \dfrac{2}{z + 2}$

49. $(x + 2)^2 + 5 = (x + 3)^2$

50. $(x + 1)^2 + 2(x - 2) = (x + 1)(x - 2)$

51. $(x + 2)^2 - x^2 = 4(x + 1)$

52. $4(x + 1) - 3x = x + 5$

53. $(2x + 1)^2 = 4(x^2 + x + 1)$

54. $(2x - 1)^2 = 4(x^2 - x + 6)$

55. A student states that the solution to the equation

$$\dfrac{2}{x(x - 2)} + \dfrac{5}{x} = \dfrac{1}{x - 2}$$

is $x = 2$. Describe and correct the student's error.

56. A student states that the equation

$$-3(x + 2) = -3x + 6$$

is an identity. Describe and correct the student's error.

57. Explain why a solution of an equation involving fractional expressions may be extraneous.

58. Describe two methods you can use to check a solution of an equation involving fractional expressions.

59. What is meant by "equivalent equations"? Give an example of two equivalent equations.

60. For what value(s) of b does the equation

$$7x + 3 = 7x + b$$

have infinitely many solutions? no solution?

In Exercises 61–66, use a calculator to solve the equation. (Round your solution to three decimal places.)

61. $0.275x + 0.725(500 - x) = 300$

62. $2.763 - 4.5(2.1x - 5.1432) = 6.32x + 5$

63. $\dfrac{x}{0.6321} + \dfrac{x}{0.0692} = 1000$

64. $(x + 5.62)^2 + 10.83 = (x + 7)^2$

65. $\dfrac{2}{7.398} - \dfrac{4.405}{x} = \dfrac{1}{x}$

66. $\dfrac{x}{2.625} + \dfrac{x}{4.875} = 1$

Ⓣ **67.** What method or methods would you recommend for checking the solutions to Exercises 61–66 using your graphing utility?

68. In Exercises 61–66, your answers are rounded to three decimal places. What effect does rounding have as you check a solution?

In Exercises 69–72, evaluate the expression in two ways. (a) Calculate entirely on your calculator using appropriate parentheses, and then round the answer to two decimal places. (b) Round both the numerator and the denominator to two decimal places before dividing, and then round the final answer to two decimal places. Does the second method introduce an additional roundoff error?

69. $\dfrac{1 + 0.73205}{1 - 0.73205}$

70. $\dfrac{1 + 0.86603}{1 - 0.86603}$

71. $\dfrac{333 + \dfrac{1.98}{0.74}}{4 + \dfrac{6.25}{3.15}}$

72. $\dfrac{1.73205 - 1.19195}{3 - (1.73205)(1.19195)}$

The symbol Ⓣ indicates when to use graphing technology or a symbolic computer algebra system to solve a problem or an exercise. The solutions of other exercises may also be facilitated by use of appropriate technology.

73. Personal Income   The per capita personal income in the United States from 1998 to 2005 can be approximated by the linear equation

$$y = 944.7t + 19,898, \quad 8 \le t \le 15$$

where t represents the year, with $t = 8$ corresponding to 1998. Use the model to estimate the year in which the per capita personal income was $32,000. *(Source: U.S. Department of Commerce, Bureau of Economic Analysis)*

74. Annual Sales   The annual sales S (in billions of dollars) of Microsoft Corporation from 1996 to 2006 can be approximated by the linear equation

$$S = 3.54t - 13.1, \quad 6 \le t \le 16$$

where t represents the year, with $t = 6$ corresponding to 1996. Use the model to estimate the year in which Microsoft's annual sales were about $20,000,000,000. *(Source: Microsoft Corporation)*

Human Height   In Exercises 75 and 76, use the following information. The relationship between the length of an adult's femur (thigh bone) and the height of the adult can be approximated by the linear equations

$$y = 0.432x - 10.44 \qquad \text{Female}$$

$$y = 0.449x - 12.15 \qquad \text{Male}$$

where y is the length of the femur in inches and x is the height of the adult in inches (see figure).

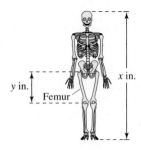

75. An anthropologist discovers a femur belonging to an adult human female. The bone is 15 inches long. Estimate the height of the female.

76. *MAKE A DECISION*   From the foot bones of an adult human male, an anthropologist estimates that the male was 65 inches tall. A few feet away from the site where the foot bones were discovered, the anthropologist discovers an adult male femur that is 17 inches long. Is it possible that the leg and foot bones came from the same person? Explain.

Consumer Credit   In Exercises 77 and 78, use the following information. From 1998 to 2005, the annual credit y (in billions of dollars) extended to consumers in the United States (other than real estate loans) can be approximated by the equation

$$y = 129.51t + 320.5, \quad 8 \le t \le 15$$

where t is the year, with $t = 8$ corresponding to 1998. *(Source: Federal Reserve Board)*

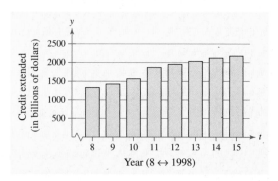

77. In which year was the credit extended to consumers about $2 trillion?

78. Use the model to predict the year in which the credit extended to consumers will be about $2.9 trillion.

Minimum Wage   In Exercises 79 and 80, use the following information. From 1997 to 2006, the federal minimum wage was $5.15 per hour. Adjusting for inflation, the federal minimum wage's *value in 1996 dollars* during these years can be approximated by the linear equation

$$y = -0.112t + 5.83, \quad 7 \le t \le 16$$

where t is the year, with $t = 7$ corresponding to 1997. *(Source: U.S. Department of Labor)*

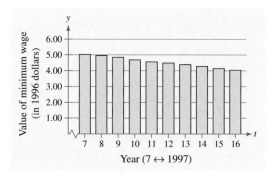

79. In which year was the value of the federal minimum wage about $4.60 in 1996 dollars?

80. According to the model, did the value of the federal minimum wage in 1996 dollars fall below $4.00 by 2007? Explain.

Section 1.2

Mathematical Modeling

- Construct a mathematical model from a verbal model.
- Model and solve percent and mixture problems.
- Use common formulas to solve geometry and simple interest problems.
- Develop a general problem-solving strategy.

Introduction to Problem Solving

In this section, you will use algebra to solve real-life problems. To do this, you will construct one or more equations that represent each real-life problem. This procedure is called **mathematical modeling.**

A good approach to mathematical modeling is to use two stages. First, use the verbal description of the problem to form a *verbal model.* Then, assign labels to each of the quantities in the verbal model and use the labels to form a *mathematical model* or an *algebraic equation.*

Verbal description		Verbal model		Algebraic equation

When you are trying to construct a verbal model, it is sometimes helpful to look for a *hidden equality*. For instance, in the following example the hidden equality equates your annual income to 24 pay periods and one bonus check.

Example 1 Using a Verbal Model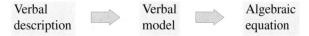

You accept a job with an annual income of $36,500. This includes your salary and a $500 year-end bonus. You are paid twice a month. What is your salary per pay period?

SOLUTION   Because there are 12 months in a year and you are paid twice a month, it follows that there are 24 pay periods during the year.

Verbal Model:

Income for year = 24 pay periods + Bonus
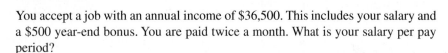

Labels: Income for year $= 36{,}500$ (dollars)
Salary for each pay period $= x$ (dollars)
Bonus $= 500$ (dollars)

Equation: $36{,}500 = 24x + 500$

Using the techniques discussed in Section 1.1, you can find that the solution is $x = \$1500$. Check whether a salary of $1500 per pay period is reasonable for the situation.

✓ CHECKPOINT 1

In Example 1, suppose you are paid weekly. What is your salary per pay period? ■

Translating Key Words and Phrases

Key Words and Phrases	Verbal Description	Algebraic Statement
Consecutive		
Next, subsequent	Consecutive integers	$n, n + 1$
Addition		
Sum, plus, greater,	The sum of 5 and x	$5 + x$
increased by, more than,	Seven more than y	$y + 7$
exceeds, total of		
Subtraction		
Difference, minus,	Four decreased by b	$4 - b$
less than, decreased by,	Three less than z	$z - 3$
subtracted from, reduced by,	Five subtracted	
the remainder	from w	$w - 5$
Multiplication		
Product, multiplied by,	Two times x	$2x$
twice, times, percent of		
Division		
Quotient, divided by, per	The quotient of x and 8	$\dfrac{x}{8}$

Example 2　**Constructing Mathematical Models**

a. A salary of $28,000 is increased by 9%. Write an equation that represents the new salary.

Verbal Model:　New salary = 9%(original salary) + Original salary

Labels:　Original salary = 28,000　　　　　　　　(dollars)
　　　　　New salary = S　　　　　　　　　　　(dollars)
　　　　　Percent = 0.09　　　　　　　(percent in decimal form)

Equation: $S = 0.09(28,000) + 28,000$

b. A laptop computer is marked down 20% to $1760. Write an equation you can use to find the original price.

Verbal Model:　Original price − 20%(original price) = Sale price

Labels:　Original price = p　　　　　　　　　(dollars)
　　　　　Sale price = 1760　　　　　　　　　　(dollars)
　　　　　Percent = 0.2　　　　　　　(percent in decimal form)

Equation: $p - 0.2p = 1760$

✓CHECKPOINT 2

A salary of $40,000 is increased by 5%. Write an equation that you can use to find the new salary. ■

Using Mathematical Models

Study the next several examples carefully. Your goal should be to develop a *general problem-solving strategy*.

Example 3 Finding the Percent of a Raise

You accept a job that pays $8 an hour. You are told that after a two-month probationary period, your hourly wage will be increased to $9 an hour. What percent raise will you receive after the two-month period?

SOLUTION

Verbal Model: Raise = Percent · Old wage

Labels:
Old wage = 8   (dollars)
Raise = 1   (dollar)
Percent = r (percent in decimal form)

Equation: $1 = r \cdot 8$

By solving this equation, you can find that you will receive a raise of $\frac{1}{8} = 0.125$, or 12.5%.

✓ **CHECKPOINT 3**

You buy stock at $25 per share. You sell the stock at $30 per share. What is the percent increase of the stock's value? ■

Example 4 Finding the Percent of a Salary

Your annual salary is $35,000. In addition to your salary, your employer also provides the following benefits. The total of this benefits package is equal to what percent of your annual salary?

Social Security (employer's portion):	6.2% of salary	$2170
Worker's compensation:	0.5% of salary	$175
Unemployment compensation:	0.75% of salary	$262.50
Medical insurance:	$2600 per year	$2600
Retirement contribution:	5% of salary	$1750

SOLUTION

Verbal Model: Benefits package = Percent · Salary

Labels:
Salary = 35,000   (dollars)
Benefits package = 6957.50   (dollars)
Percent = r (percent in decimal form)

Equation: $6957.50 = r \cdot 35{,}000$

By solving this equation, you can find that your benefits package is equal to $r = 6957.50/35{,}000$, or about 19.9% of your salary.

Charles Gupton/Getty Images
In 2005, 15.3% of the population of the United States had no health insurance. *(Source: Centers for Disease Control and Prevention, National Health Interview Survey)*

✓ **CHECKPOINT 4**

Your income last year was $42,000. Throughout that year you paid a total of $648 for parking fees. The total of the parking fees was equal to what percent of your income? ■

Example 5   Finding the Dimensions of a Room

A rectangular family room is twice as long as it is wide, and its perimeter is 84 feet. Find the dimensions of the family room.

SOLUTION   For this problem, it helps to sketch a diagram, as shown in Figure 1.2.

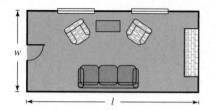

FIGURE 1.2

Verbal Model: $2 \cdot$ Length $+ 2 \cdot$ Width $=$ Perimeter

Labels:
Perimeter $= 84$ (feet)
Width $= w$ (feet)
Length $= l = 2w$ (feet)

Equation: $2(2w) + 2w = 84$

$$4w + 2w = 84$$

$$6w = 84$$

$$w = 14 \text{ feet}$$

$$l = 2w = 28 \text{ feet}$$

The dimensions of the room are 14 feet by 28 feet.

✓**CHECKPOINT 5**

A rectangular driveway is three times as long as it is wide, and its perimeter is 120 feet. Find the dimensions of the driveway. ■

Example 6
MAKE A DECISION   A Distance Problem

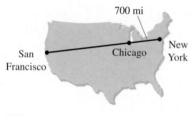

FIGURE 1.3

A plane travels nonstop from New York to San Francisco, a distance of 2600 miles. It takes 1.5 hours to fly from New York to Chicago, a distance of about 700 miles (see Figure 1.3). Assuming the plane flies at a constant speed, how long does the entire trip take? What time (EST) should the plane leave New York to arrive in San Francisco by 5 P.M. PST (8 P.M. EST)?

SOLUTION   To solve this problem, use the formula that relates distance, rate, and time. That is, (distance) = (rate)(time). Because it took the plane 1.5 hours to travel a distance of about 700 miles, you can conclude that its rate (or speed) is

$$\text{Rate} = \frac{\text{distance}}{\text{time}} = \frac{700 \text{ miles}}{1.5 \text{ hours}} \approx 466.67 \text{ miles per hour.}$$

Because the entire trip is about 2600 miles, the time for the entire trip is

$$\text{Time} = \frac{\text{distance}}{\text{rate}} = \frac{2600 \text{ miles}}{466.67 \text{ miles per hour}} \approx 5.57 \text{ hours.}$$

Because 0.57 hour represents about 34 minutes, you can conclude that the trip takes about 5 hours and 34 minutes. The plane must leave New York by 2:26 P.M. in order to arrive in San Francisco by 8 P.M. EST.

✓**CHECKPOINT 6**

A small boat travels at full speed to an island 11 miles away. It takes 0.3 hour to travel the first 3 miles. How long does the entire trip take? ■

Another way to solve the distance problem in Example 6 is to use the concept of **ratio and proportion.** To do this, let x represent the time required to fly from New York to San Francisco, set up the following proportion, and solve for x.

$$\frac{\text{Time to San Francisco}}{\text{Time to Chicago}} = \frac{\text{Distance to San Francisco}}{\text{Distance to Chicago}}$$

$$\frac{x}{1.5} = \frac{2600}{700}$$

$$x = 1.5 \cdot \frac{2600}{700}$$

$$x \approx 5.57$$

Notice how ratio and proportion are used with a property from geometry to solve the problem in the following example.

Example 7 **An Application Involving Similar Triangles**

To determine the height of Petronas Tower 1 (in Kuala Lumpur, Malaysia), you measure the shadow cast by the building to be 113 meters long, as shown in Figure 1.4. Then you measure the shadow cast by a 100-centimeter post and find that its shadow is 25 centimeters long. Use this information to determine the height of Petronas Tower 1.

SOLUTION   To find the height of the tower, you can use a property from geometry that states that the ratios of corresponding sides of similar triangles are equal.

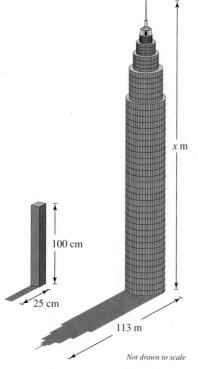

100 cm

25 cm

113 m

Not drawn to scale

FIGURE 1.4

Verbal Model:

$$\frac{\text{Height of tower}}{\text{Length of tower's shadow}} = \frac{\text{Height of post}}{\text{Length of post's shadow}}$$

Labels: Height of tower $= x$ (meters)
Length of tower's shadow $= 113$ (meters)
Height of post $= 100$ (centimeters)
Length of post's shadow $= 25$ (centimeters)

Equation: $\dfrac{x}{113} = \dfrac{100}{25}$

$$x = 113 \cdot \frac{100}{25}$$

$$x = 113 \cdot 4$$

$$x = 452 \text{ meters}$$

The Petronas Tower 1 is 452 meters high.

✓ CHECKPOINT 7

A tree casts a shadow that is 24 feet long. At the same time, a four-foot tall mailbox casts a shadow that is 3 feet long. How tall is the tree? ■

Mixture Problems

The next example is called a **mixture problem** because it involves two different unknown quantities that are *mixed* in a specific way. Watch for a *hidden product* in the verbal model.

Example 8   **A Simple Interest Problem**

You invested a total of $10,000 in accounts that earned $4\frac{1}{2}\%$ and $5\frac{1}{2}\%$ simple interest. In 1 year, the two accounts earned $508.75 in interest. How much did you invest in each account?

SOLUTION   The formula for simple interest is $I = Prt$, where I is the interest, P is the principal, r is the annual interest rate (in decimal form), and t is the time in years.

Verbal Model:	Interest from $4\frac{1}{2}\%$	+	Interest from $5\frac{1}{2}\%$	=	Total interest

You can let x represent the amount invested at $4\frac{1}{2}\%$. Because the total amount invested at $4\frac{1}{2}\%$ and $5\frac{1}{2}\%$ is $10,000, you can let $10,000 - x$ represent the amount invested at $5\frac{1}{2}\%$.

Labels: Amount invested at $4\frac{1}{2}\% = x$ (dollars)
Amount invested at $5\frac{1}{2}\% = 10{,}000 - x$ (dollars)
Interest from $4\frac{1}{2}\% = Prt = (x)(0.045)(1)$ (dollars)
Interest from $5\frac{1}{2}\% = Prt = (10{,}000 - x)(0.055)(1)$ (dollars)
Total interest $= 508.75$ (dollars)

Equation: $0.045x + 0.055(10{,}000 - x) = 508.75$

$$0.045x + 550 - 0.055x = 508.75$$
$$-0.01x = -41.25$$
$$x = \$4125$$

So, the amount invested at $4\frac{1}{2}\%$ is $4125 and the amount invested at $5\frac{1}{2}\%$ is

$$10{,}000 - x = 10{,}000 - 4125 = \$5875.$$

Check these results in the original statement of the problem, as follows.

CHECK

Interest from $4\frac{1}{2}\%$	Interest from $5\frac{1}{2}\%$	Total interest

$$0.045(4125) + 0.055(10{,}000 - 4125) \overset{?}{=} 508.75$$
$$185.625 + 323.125 \overset{?}{=} 508.75$$
$$508.75 = 508.75 \qquad \text{Solution checks. } \checkmark$$

In Example 8, did you recognize the hidden products in the two terms on the left side of the equation? Both hidden products come from the common formula

$$\text{Interest} = \text{Principal} \cdot \text{Rate} \cdot \text{Time}$$
$$I = Prt.$$

Common Formulas

Many common types of geometric, scientific, and investment problems use ready-made equations, called **formulas.** Knowing formulas such as those in the following lists will help you translate and solve a wide variety of real-life problems involving perimeter, area, volume, temperature, interest, and distance.

Common Formulas for Area, Perimeter, and Volume

Square

$A = s^2$

$P = 4s$

Rectangle

$A = lw$

$P = 2l + 2w$

Circle

$A = \pi r^2$

$C = 2\pi r$

Triangle

$A = \frac{1}{2}bh$

$P = a + b + c$

Cube

$V = s^3$

Rectangular Solid

$V = lwh$

Circular Cylinder

$V = \pi r^2 h$

Sphere

$V = \frac{4}{3}\pi r^3$

Miscellaneous Common Formulas

Temperature: F = degrees Fahrenheit, C = degrees Celsius

$$F = \frac{9}{5}C + 32$$

Simple interest: I = interest, P = principal, r = interest rate, t = time

$$I = Prt$$

Distance: d = distance traveled, r = rate, t = time

$$d = rt$$

When working with applied problems, you often need to rewrite common formulas. For instance, the formula

$$P = 2l + 2w$$

for the perimeter of a rectangle can be rewritten or solved for w to produce

$$w = \tfrac{1}{2}(P - 2l).$$

|← 4 cm →|

Cat Food

h

FIGURE 1.5

Example 9 Using a Formula

A cylindrical can has a volume of 200 cubic centimeters and a radius of 4 centimeters, as shown in Figure 1.5. Find the height of the can.

SOLUTION   The formula for the *volume of a cylinder* is $V = \pi r^2 h$. To find the height of the can, solve for h.

$$h = \frac{V}{\pi r^2}$$

Then, using $V = 200$ and $r = 4$, find the height.

$$h = \frac{200}{\pi (4)^2} \qquad \text{Substitute 200 for } V \text{ and 4 for } r.$$

$$= \frac{200}{16\pi} \qquad \text{Simplify denominator.}$$

$$\approx 3.98 \qquad \text{Use a calculator.}$$

So, the height of the can is about 3.98 centimeters. You can use unit analysis to check that your answer is reasonable.

$$\frac{200 \text{ cm}^3}{16\pi \text{ cm}^2} \approx 3.98 \text{ cm}$$

✓ **CHECKPOINT 9**

One cubic foot of water fills a cylindrical pipe with a radius of 0.5 foot. What is the height of the pipe? ■

Strategy for Solving Word Problems

1. *Search* for the hidden equality—two expressions said to be equal or known to be equal. A sketch may be helpful.

2. *Write* a verbal model that equates these two expressions. Identify any *hidden* products.

3. *Assign* numbers to the known quantities and letters (or algebraic expressions) to the unknown quantities.

4. *Rewrite* the verbal model as an algebraic equation using the assigned labels.

5. *Solve* the resulting algebraic equation.

6. *Check* to see that the answer satisfies the word problem as stated. (Remember that "solving for x" or some other variable may not completely answer the question.)

CONCEPT CHECK

1. Write a verbal model for the volume of a rectangular solid.

2. Describe and correct the error in the statement.
   The product of 10 and 5 less than x is $10(5 - x)$.

3. Two spherical balloons, each with radius r, are filled with air. Write an algebraic equation that represents the total volume of air in the balloons.

4. Using the formula for the volume of a rectangular solid, what information do you need to find the length of a block of ice?

The following warm-up exercises involve skills that were covered in earlier sections. You will use these skills in the exercise set for this section. For additional help, review Section 1.1.

In Exercises 1–10, solve the equation (if possible) and check your answer.

1. $3x - 42 = 0$

2. $64 - 16x = 0$

3. $2 - 3x = 14 + x$

4. $7 + 5x = 7x - 1$

5. $5[1 + 2(x + 3)] = 6 - 3(x - 1)$

6. $2 - 5(x - 1) = 2[x + 10(x - 1)]$

7. $\dfrac{x}{3} + \dfrac{x}{2} = \dfrac{1}{3}$

8. $\dfrac{2}{x} + \dfrac{2}{5} = 1$

9. $1 - \dfrac{2}{z} = \dfrac{z}{z + 3}$

10. $\dfrac{x}{x + 1} - \dfrac{1}{2} = \dfrac{4}{3}$

Exercises 1.2

See www.CalcChat.com for worked-out solutions to odd-numbered exercises.

Creating a Mathematical Model   In Exercises 1–10, write an algebraic expression for the verbal expression.

1. The sum of two consecutive natural numbers

2. The product of two natural numbers whose sum is 25

3. Distance Traveled   The distance traveled in t hours by a car traveling at 50 miles per hour

4. Travel Time   The travel time for a plane that is traveling at a rate of r miles per hour for 200 miles

5. Acid Solution   The amount of acid in x gallons of a 20% acid solution

6. Discount   The sale price of an item that is discounted by 20% of its list price L

7. Geometry   The perimeter of a rectangle whose width is x and whose length is twice the width

8. Geometry   The area of a triangle whose base is 20 inches and whose height is h inches

9. Total Cost   The total cost to buy x units at $25 per unit with a total shipping fee of $1200

10. Total Revenue   The total revenue obtained by selling x units at $3.59 per unit

In Exercises 11–16, write an equation that represents the statement.

11. The sum of 5 and x equals 8.

12. The difference of n and 7 is 4.

13. The quotient of r and 2 is 9.

14. The product of x and 6 equals -9.

15. The sum of a number n and twice the number is 15.

16. The product of 3 less than x and 8 is 40.

Using a Mathematical Model   In Exercises 17–22, write a mathematical model for the number problem, and solve the problem.

17. Find two consecutive numbers whose sum is 525.

18. Find three consecutive natural numbers whose sum is 804.

19. One positive number is five times another positive number. The difference between the two numbers is 148. Find the numbers.

20. One positive number is one-fifth of another number. The difference between the two numbers is 76. Find the numbers.

21. Find two consecutive integers whose product is five less than the square of the smaller number.

22. Find two consecutive natural numbers such that the difference of their reciprocals is one-fourth the reciprocal of the smaller number.

23. Weekly Paycheck   Your weekly paycheck is 12% *more* than your coworker's. Your two paychecks total $848. Find the amount of each paycheck.

24. Weekly Paycheck   Your weekly paycheck is 12% *less* than your coworker's. Your two paychecks total $848. Find the amount of each paycheck.

25. Monthly Profit   The profit for a company in February was 5% *higher* than it was in January. The total profit for the two months was $129,000. Find the profit for each month.

26. Monthly Profit   The profit for a company in February was 5% *lower* than it was in January. The total profit for the two months was $129,000. Find the profit for each month.

Movie Sequels   In Exercises 27–32, use the following information. The movie industry frequently releases sequels and/or prequels to successful movies. The revenue of each *Star Wars* movie is shown. Compare the revenue of the two given *Star Wars* movies by finding the percent increase or decrease in the domestic gross. *(Source: Infoplease.com)*

Movie	Domestic gross (in dollars)
Star Wars (1977)	$460,998,007
The Empire Strikes Back (1980)	$290,271,960
Return of the Jedi (1983)	$309,209,079
Episode I: The Phantom Menace (1999)	$431,088,295
Episode II: Attack of the Clones (2002)	$310,675,583
Episode III: Revenge of the Sith (2005)	$380,262,555

27. *Star Wars* (1977) to *The Empire Strikes Back* (1980)

28. *The Empire Strikes Back* (1980) to *Return of the Jedi* (1983)

29. *Return of the Jedi* (1983) to *Episode I: The Phantom Menace* (1999)

30. *Episode I: The Phantom Menace* (1999) to *Episode II: Attack of the Clones* (2002)

31. *Episode II: Attack of the Clones* (2002) to *Episode III: Revenge of the Sith* (2005)

32. *Star Wars* (1977) to *Episode III: Revenge of the Sith* (2005)

Size Inflation   In Exercises 33–36, use the following information. Restaurants tend to serve food in larger portions now than they have in the past. Several examples are shown in the table. Find the percent increase in size from the past to 2006 for the indicated food item. *(Source: The Portion Teller, McDonald's, Little Caesars, and Pizza Hut)*

Food or drink item	Past size	2006 size
Small soft drink (McDonald's)	7 fl oz	16 fl oz
Small French fries (McDonald's)	2.4 oz	2.6 oz
Large French fries (McDonald's)	3.5 oz	6 oz
Pizza (Little Caesars, Pizza Hut)	10 in.	12 in.

33. McDonald's small soft drink

34. McDonald's small French fries

35. McDonald's large French fries

36. Little Caesars or Pizza Hut standard pizza

37. Comparing Calories   A lunch consisting of a Big Mac, large fries, and large soft drink at McDonald's contains 1440 calories. A lunch consisting of a small hamburger, small fries, and a small soft drink at McDonald's contains 660 calories. Find the percent change in calories from the larger to the smaller lunch. *(Source: McDonald's Corporation)*

38. Comparing Calories   One slice (or one-tenth) of a 14-inch Little Caesars pizza with bacon, pepperoni, Italian sausage, and extra cheese has 315 calories. The same slice without the extra toppings has 200 calories. Find the percent change in calories from a slice with the extra toppings to a slice without them.   *(Source: Little Caesars)*

39. Salary   You accept a new job with a starting salary of $35,000. You receive an 8% raise at the start of your second year, a 7.8% raise at the start of your third year, and a 9.4% raise at the start of your fourth year.

(a) Find your salary for the second year.

(b) Find your salary for the third year.

(c) Find your salary for the fourth year.

40. Salary   You accept a new job with a starting salary of $48,000. You receive a 4% raise at the start of your second year, a 5.5% raise at the start of your third year, and an 11.4% raise at the start of your fourth year.

(a) Find your salary for the second year.

(b) Find your salary for the third year.

(c) Find your salary for the fourth year.

41. World Internet Users   The number of Internet users in the world reached 500 million in 2001. By the end of 2003, the number increased 43.8%. By the end of 2004, the number increased 13.6% from 2003. By the end of 2006 the number increased 33.8% from 2004. *(Source: Internet World Stats)*

(a) Find the number of users at the end of 2003.

(b) Find the number of users at the end of 2004.

(c) Find the number of users at the end of 2006.

(d) Find the percent increase in the number of users from 2001 to 2006.

42. Sporting Goods Sales   In 2002, the total sales of sporting goods in the United States was $77,726,000,000. In 2003, the total sales increased 2.6% from 2002. In 2004, the total sales increased 6.1% from 2003. In 2005, the total sales increased 2.5% from 2004. *(Source: National Sporting Goods Association)*

(a) Find the total sporting goods sales in 2003.

(b) Find the total sporting goods sales in 2004.

(c) Find the total sporting goods sales in 2005.

(d) Find the percent increase in total sales from 2002 to 2005.

43. Media Usage   It was projected that by 2009, the average person would spend 3555 hours per year using some type of media. Use the bar graph to determine the number of hours the average person will spend watching television, listening to the radio or recorded music, using the Internet, playing non-Internet video games, reading print media, and using other types of media in 2009. *(Source: Veronis Schuler Stevenson)*

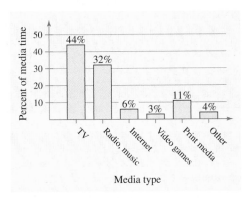

44. New Vehicle Sales   In 2005, the number of motor vehicles sold in the U.S. was about 17,445,000. Use the bar graph to determine how many cars, trucks, and light trucks were sold in 2005. *(Source: U.S. Bureau of Economic Analysis)*

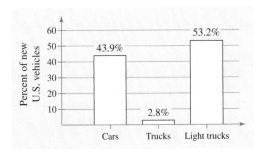

45. Geometry   A room is 1.5 times as long as it is wide, and its perimeter is 75 feet (see figure). Find the dimensions of the room.

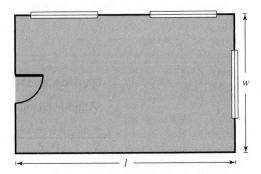

46. Geometry   A picture frame has a total perimeter of 3 feet (see figure). The width of the frame is 0.62 times its length. Find the dimensions of the frame.

47. Simple Interest   You invest $2500 at 7% simple interest. How many years will it take for the investment to earn $1000 in interest?

48. Simple Interest   An investment earns $3200 interest over a seven-year period. What is the rate of simple interest on a $4800 principal investment?

49. Course Grade   To get an A in a course, you need an average of 90% or better on four tests that are worth 100 points each. Your scores on the first three tests were 87, 92, and 84. What must you score on the fourth test to get an A for the course?

50. Course Grade   To get an A in a course, you need an average of 90% or better on four tests. The first three tests are worth 100 points each and the fourth is worth 200 points. Your scores on the first three tests are 87, 92, and 84. What must you score on the fourth test to get an A for the course?

51. List Price   The price of a swimming pool has been discounted 15%. The sale price is $1200. Find the original list price of the swimming pool.

52. List Price   The price of a home theater system has been discounted 10%. The sale price is $499. Find the original price of the system.

53. Discount Rate   A satellite radio system for your car has been discounted by $30. The sale price is $119. What percent of the original list price is the discount?

54. Discount Rate   The price of a shirt has been discounted by $20. The sale price is $29.95. What percent of the original list price is the discount?

55. Wholesale Price   A store marks up a power drill 60% from its wholesale price. In a clearance sale, the price is discounted by 25%. The sale price is $21.60. What was the wholesale price of the power drill?

56. Wholesale Price   A store marks up a picture frame 80% from its wholesale price. In a clearance sale, the price is discounted by 40%. The sale price is $28.08. What was the wholesale price of the picture frame?

Weekly Salary   In Exercises 57 and 58, use the following information to write a mathematical model and solve. Due to economic factors, your employer has reduced your weekly wage by 15%. Before the reduction, your weekly salary was $425.

57. What is your reduced salary?

58. What percent raise must you receive to bring your weekly salary back up to $425? Explain why the percent raise is different from the percent reduction.

59. Travel Time   You are driving to a college 150 miles from home. It takes 28 minutes to travel the first 30 miles. At this rate, how long is your entire trip?

60. Travel Time   Two friends fly from Denver to Orlando (a distance of 1526 miles). It takes 1 hour and 15 minutes to fly the first 500 miles. At this rate, how long is the entire flight?

61. Travel Time   Two cars start at the same time at a given point and travel in the same direction at constant speeds of 40 miles per hour and 55 miles per hour. After how long are the cars 5 miles apart?

62. Catch-Up Time   Students are traveling in two cars to a football game 135 miles away. One car travels at an average speed of 45 miles per hour. The second car starts $\frac{1}{2}$ hour later and travels at an average speed of 55 miles per hour. How long will it take the second car to catch up to the first car?

63. Radio Waves   Radio waves travel at the same speed as light, 3.0×10^8 meters per second. Find the time required for a radio wave to travel from mission control in Houston to NASA astronauts on the surface of the moon 3.84×10^8 meters away.

64. Distance to a Star   Find the distance (in miles) to a star that is 50 light years (distance traveled by light in 1 year) away. (Light travels at 186,000 miles per second.)

65. Height of a Tree   To determine the height of a tree, you measure its shadow and the shadow of a five-foot lamppost, as shown in the figure. How tall is the tree?

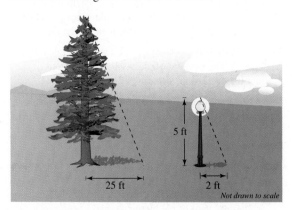

25 ft   5 ft   2 ft

Not drawn to scale

66. Height of a Building   To determine the height of a building, you measure the building's shadow and the shadow of a four-foot stake, as shown in the figure. How tall is the building?

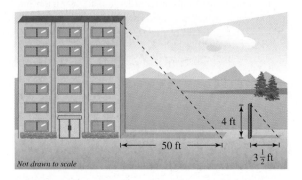

50 ft   4 ft $3\frac{1}{2}$ ft

Not drawn to scale

67. Projected Expenses   From January through May, a company's expenses totaled $325,450. If the monthly expenses continue at this rate, what will be the total expenses for the year?

68. Projected Revenue   From January through August, a company's revenues totaled $549,680. If the monthly revenue continues at this rate, what will be the total revenue for the year?

69. Investment Mix   You invest $15,000 in two funds paying 6.5% and 7.5% simple interest. The total annual interest is $1020. How much do you invest in each fund?

70. Investment Mix   You invest $30,000 in two funds paying 3% and $4\frac{1}{2}$% simple interest. The total annual interest is $1230. How much do you invest in each fund?

71. Stock Mix   You invest $5000 in two stocks. In one year, the value of stock A increases by 9.8% and the value of stock B increases by 6.2%. The total value of the stocks is now $5389.20. How much did you originally invest in each stock?

72. Stock Mix   You invest $4000 in two stocks. In one year, the value of stock A increases by 5.4% and the value of stock B increases by 12.8%. The total value of the stocks is now $4401. How much did you originally invest in each stock?

73. Comparing Investment Returns   You invest $12,000 in a fund paying $9\frac{1}{2}$% simple interest and $8000 in a fund for which the interest rate varies. At the end of the year the total interest for both funds is $2054.40. What simple interest rate yields the same interest amount as the variable rate fund?

74. Comparing Investment Returns   You have $10,000 in an account earning simple interest that is linked to the prime rate. The prime rate drops for the last quarter of the year, so your rate drops by $1\frac{1}{2}$% for the same period. Your total annual interest is $1112.50. What is your interest rate for the first three quarters and for the last quarter?

Production Limit   In Exercises 75 and 76, use the following information. *Variable costs* depend on the number of units produced. *Fixed costs* are the same regardless of how many units are produced. Find the greatest number of units the company can produce each month.

75. The company has fixed monthly costs of $15,000 and variable monthly costs of $8.75 per unit. The company has $90,000 available each month to cover costs.

76. The company has fixed monthly costs of $10,000 and variable monthly costs of $9.30 per unit. The company has $85,000 available each month to cover costs.

77. Length of a Tank   The diameter of a cylindrical propane gas tank is 4 feet (see figure). The total volume of the tank is 603.2 cubic feet. Find the length of the tank.

78. Water Depth   A trough is 12 feet long, 3 feet deep, and 3 feet wide (see figure). Find the depth of the water when the trough contains 70 gallons of water. (1 gallon ≈ 0.13368 cubic foot.)

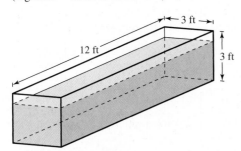

79. Mixture   A 55-gallon barrel contains a mixture with a concentration of 40%. How much of this mixture must be withdrawn and replaced by 100% concentrate to bring the mixture up to 75% concentration? (See figure.)

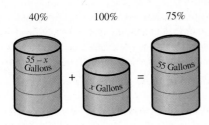

80. Mixture   A farmer mixes gasoline and oil to make 2 gallons of mixture for his two-cycle chain saw engine. This mixture is 32 parts gasoline and 1 part two-cycle oil. How much gasoline must be added to bring the mixture to 40 parts gasoline and 1 part oil?

New York City Marathon   In Exercises 81 and 82, the length of the New York City Marathon course is 26 miles, 385 yards. Find the average speed of the record holding runner. (Note that 1 mile = 5280 feet = 1760 yards.)

81. Men's record time: 2 hours, $7\frac{3}{4}$ minutes

82. Women's record time: 2 hours, $22\frac{1}{2}$ minutes

In Exercises 83–100, solve for the indicated variable.

83. Area of a Triangle
Solve for h in $A = \frac{1}{2}bh$.

84. Perimeter of a Rectangle
Solve for l in $P = 2l + 2w$.

85. Volume of a Rectangular Prism
Solve for l in $V = lwh$

86. Ideal Gas Law
Solve for T in $PV = nRT$.

87. Volume of a Right Circular Cylinder
Solve for h in $V = \pi r^2 h$.

88. Kinetic Energy
Solve for m in $E = \frac{1}{2}mv^2$.

89. Markup
Solve for C in $S = C + RC$.

90. Discount
Solve for L in $S = L - RL$.

91. Investment at Simple Interest
Solve for r in $A = P + Prt$.

92. Investment at Compound Interest
Solve for P in $A = P\left(1 + \frac{r}{n}\right)^{nt}$.

93. Area of a Trapezoid
Solve for b in $A = \frac{1}{2}(a + b)h$.

94. Area of a Sector of a Circle
Solve for θ in $A = \frac{\pi r^2 \theta}{360}$.

95. Arithmetic Progression
Solve for n in $L = a + (n - 1)d$.

96. Geometric Progression

Solve for r in $S = \dfrac{rL - a}{r - 1}$.

97. Lateral Surface Area of a Cylinder

Solve for h in $A = 2\pi rh$.

98. Surface Area of a Cone

Solve for l in $S = \pi r^2 + \pi rl$.

99. Lensmaker's Equation

Solve for R_1 in $\dfrac{1}{f} = (n - 1)\left(\dfrac{1}{R_1} - \dfrac{1}{R_2}\right)$

100. Capacitance in Series Circuits

Solve for C_1 in $C = \dfrac{1}{\dfrac{1}{C_1} + \dfrac{1}{C_2}}$

101. Monthly Sales   The table below shows the monthly sales of a sales team for the first quarter of the year. Find the average monthly sales for each salesperson. Then find the team's average sales for each month.

Name	January	February	March
Williams	$15,000	$18,800	$22,300
Gonzalez	$20,900	$17,500	$25,600
Walters	$18,600	$25,000	$16,400
Gilbert	$18,100	$18,700	$23,000
Hart	$13,000	$20,500	$20,000

102. Monthly Sales   The table below shows the monthly sales of a sales team for the second quarter of the year. Find the average monthly sales for each salesperson. Then find the team's average sales for each month.

Name	April	May	June
Williams	$25,000	$28,800	$21,000
Gonzalez	$26,200	$27,800	$29,500
Walters	$26,600	$23,400	$26,900
Gilbert	$27,100	$22,200	$29,000
Hart	$23,100	$27,400	$22,800

Ⓢ 103. Monthly Sales   The table below shows the monthly sales of a sales team for the third quarter of the year. Use a spreadsheet software program to find the average monthly sales for each salesperson. Then find the team's average sales for each month.

Name	July	August	September
Williams	$24,400	$29,500	$21,200
Gonzalez	$26,100	$22,900	$19,600
Walters	$29,200	$28,600	$18,400
Gilbert	$25,000	$27,600	$29,800
Hart	$31,400	$28,700	$24,200
Reyes	$27,300	$26,400	$21,200
Sanders	$8,200	$12,400	$20,300

Ⓢ 104. Monthly Sales   The table below shows the monthly sales of a sales team for the fourth quarter of the year. Use a spreadsheet software program to find the average monthly sales for each salesperson. Then find the team's average sales for each month.

Name	October	November	December
Williams	$20,000	$25,100	$23,900
Gonzalez	$24,200	$23,600	$18,500
Walters	$31,900	$23,800	$18,400
Gilbert	$24,600	$23,100	$30,700
Hart	$32,400	$19,100	$28,600
Reyes	$24,700	$24,500	$23,400
Sanders	$18,700	$22,100	$23,200

105. Applied problems in textbooks usually give just the amount of information that is necessary to solve the problem. In real life, however, you must often sort through the given information and discard facts that are irrelevant to the problem. Such an irrelevant fact is called a **red herring.** Find any red herrings in the following problem.

Beneath the surface of the ocean, pressure changes at a rate of approximately 4.4 pounds per square inch for every 10-foot change in depth. A diver takes 30 minutes to ascend 25 feet from a depth of 150 feet. What change in pressure does the diver experience?

The symbol Ⓢ indicates an exercise in which you are instructed to use a spreadsheet.

Section 1.3

Quadratic Equations

- Solve a quadratic equation by factoring.
- Solve a quadratic equation by extracting square roots.
- Construct and use a quadratic model to solve an application problem.

Solving Quadratic Equations by Factoring

In the first two sections of this chapter, you studied linear equations in one variable. In this and the next section, you will study quadratic equations.

Definition of a Quadratic Equation

A **quadratic equation** in x is an equation that can be written in the general form

$$ax^2 + bx + c = 0$$

where a, b, and c are real numbers with $a \neq 0$. Another name for a quadratic equation in x is a **second-degree polynomial equation in x.**

There are three basic techniques for solving quadratic equations: factoring, extracting square roots, and the *Quadratic Formula*. (The Quadratic Formula is discussed in the next section.) The first technique is based on the following property.

Zero-Factor Property

If $ab = 0$, then $a = 0$ or $b = 0$.

To use this property, rewrite the left side of the general form of a quadratic equation as the product of two linear factors. Then find the solutions of the quadratic equation by setting each linear factor equal to zero.

STUDY TIP

The Zero-Factor Property applies *only* to equations written in general form (in which one side of the equation is zero). So, be sure that all terms are collected on one side *before* factoring. For instance, in the equation

$$(x - 5)(x + 2) = 8$$

it is *incorrect* to set each factor equal to 8. Can you solve this equation correctly?

Example 1 Solving a Quadratic Equation by Factoring

Solve $x^2 - 3x - 10 = 0$.

SOLUTION

$x^2 - 3x - 10 = 0$		Write original equation.
$(x - 5)(x + 2) = 0$		Factor.
$x - 5 = 0 \implies x = 5$		Set 1st factor equal to 0.
$x + 2 = 0 \implies x = -2$		Set 2nd factor equal to 0.

The solutions are $x = 5$ and $x = -2$. Check these in the original equation.

✓ **CHECKPOINT 1**

Solve $x^2 - x - 12 = 0$. ■

Example 2 Solving a Quadratic Equation by Factoring

$6x^2 - 3x = 0$	Original equation
$3x(2x - 1) = 0$	Factor out common factor.
$3x = 0 \implies x = 0$	Set 1st factor equal to 0.
$2x - 1 = 0 \implies x = \frac{1}{2}$	Set 2nd factor equal to 0.

The solutions are $x = 0$ and $x = \frac{1}{2}$. Check these by substituting in the original equation, as follows.

CHECK

$6x^2 - 3x = 0$	Write original equation.
$6(0)^2 - 3(0) \overset{?}{=} 0$	Substitute 0 for x.
$0 - 0 = 0$	First solution checks. ✓
$6\left(\frac{1}{2}\right)^2 - 3\left(\frac{1}{2}\right) \overset{?}{=} 0$	Substitute $\frac{1}{2}$ for x.
$\frac{6}{4} - \frac{3}{2} = 0$	Second solution checks. ✓

✓ **CHECKPOINT 2**

Solve $4x^2 - 8x = 0$. ■

If the two factors of a quadratic expression are the same, the corresponding solution is a **double** or **repeated** solution.

TECHNOLOGY

Ⓣ To check the solution in Example 3 with your graphing utility, you should first write the equation in general form.

$$9x^2 - 6x + 1 = 0$$

Then enter the expression $9x^2 - 6x + 1$ into y_1 of the equation editor. Now you can use the ASK mode of the *table* feature of your graphing utility to check the solution. For instructions on how to use the *table* feature, see Appendix A; for specific keystrokes, go to the text website at *college.hmco.com/info/ larsonapplied.*

Example 3 A Quadratic Equation with a Repeated Solution

Solve $9x^2 - 6x = -1$.

SOLUTION

$9x^2 - 6x = -1$	Write original equation.
$9x^2 - 6x + 1 = 0$	Write in general form.
$(3x - 1)^2 = 0$	Factor.
$3x - 1 = 0$	Set repeated factor equal to 0.
$x = \frac{1}{3}$	Solution

The only solution is $x = \frac{1}{3}$. Check this by substituting in the original equation, as follows.

$9x^2 - 6x = -1$	Write original equation.
$9\left(\frac{1}{3}\right)^2 - 6\left(\frac{1}{3}\right) \overset{?}{=} -1$	Substitute $\frac{1}{3}$ for x.
$1 - 2 \overset{?}{=} -1$	Simplify.
$-1 = -1$	Solution checks. ✓

✓ **CHECKPOINT 3**

Solve $x^2 + 4x = -4$. ■

Extracting Square Roots

There is a shortcut for solving equations of the form $u^2 = d$, where $d > 0$. By factoring, you can see that this equation has two solutions.

$u^2 = d$	Write original equation.
$u^2 - d = 0$	Write in general form.
$(u + \sqrt{d})(u - \sqrt{d}) = 0$	Factor.
$u + \sqrt{d} = 0$ $\implies$ $u = -\sqrt{d}$	Set 1st factor equal to 0.
$u - \sqrt{d} = 0$ $\implies$ $u = \sqrt{d}$	Set 2nd factor equal to 0.

Solving an equation of the form $u^2 = d$ without going through the steps of factoring is called **extracting square roots.**

Extracting Square Roots

The equation $u^2 = d$, where $d > 0$, has exactly two solutions:

$$u = \sqrt{d} \quad \text{and} \quad u = -\sqrt{d}.$$

These solutions can also be written as $u = \pm\sqrt{d}$.

Example 4 **Extracting Square Roots**

Solve $4x^2 = 12$.

SOLUTION

$4x^2 = 12$	Write original equation.
$x^2 = 3$	Divide each side by 4.
$x = \pm\sqrt{3}$	Extract square roots.

The solutions are $x = \sqrt{3}$ and $x = -\sqrt{3}$. Check these in the original equation.

✓ CHECKPOINT 4

Solve $2x^2 = 8$. ▪

Example 5 **Extracting Square Roots**

Solve $(x - 3)^2 = 7$.

SOLUTION

$(x - 3)^2 = 7$	Write original equation.
$x - 3 = \pm\sqrt{7}$	Extract square roots.
$x = 3 \pm \sqrt{7}$	Add 3 to each side.

✓ CHECKPOINT 5

Solve $(x - 1)^2 = 16$. ▪

The solutions are $x = 3 \pm \sqrt{7}$. Check these in the original equation.

Applications

Quadratic equations often occur in problems dealing with area. Here is a simple example.

> A square room has an area of 144 square feet. Find the dimensions of the room.

To solve this problem, you can let x represent the length of each side of the room. Then, by solving the equation

$$x^2 = 144$$

you can conclude that each side of the room is 12 feet long. Note that although the equation $x^2 = 144$ has two solutions, $x = -12$ and $x = 12$, the negative solution makes no sense (for this problem), so you should choose the positive solution.

Example 6 Finding the Dimensions of a Room

A sunroom is 3 feet longer than it is wide (see Figure 1.6) and has an area of 154 square feet. Find the dimensions of the room.

SOLUTION You can begin by using the same type of problem-solving strategy that was presented in Section 1.2.

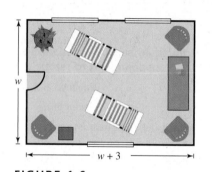

FIGURE 1.6

Verbal Model:	Width of room	·	Length of room	=	Area of room

Labels: Area of room = 154   (square feet)
        Width of room = w (feet)
        Length of room = $w + 3$ (feet)

Equation:
$$w(w + 3) = 154$$
$$w^2 + 3w - 154 = 0$$
$$(w - 11)(w + 14) = 0$$
$$w - 11 = 0 \implies w = 11$$
$$w + 14 = 0 \implies w = -14$$

Choosing the positive value, you can conclude that the width is 11 feet and the length is $w + 3 = 14$ feet. You can check this in the original statement of the problem as follows.

CHECK

The length of 14 feet is 3 feet more than the width of 11 feet. ✓

The area of the sunroom is $(11)(14) = 154$ square feet. ✓ _____

✓ CHECKPOINT 6

A rectangular kitchen is 8 feet longer than it is wide and has an area of 84 square feet. Find the dimensions of the kitchen. ■

Another application of quadratic equations involves an object that is falling (or is vertically projected into the air). The equation that gives the height of such an object is called a **position equation,** and on Earth's surface it has the form

$$s = -16t^2 + v_0t + s_0.$$

In this equation, s represents the height of the object (in feet), v_0 represents the initial velocity of the object (in feet per second), s_0 represents the initial height of the object (in feet), and t represents the time (in seconds).

The position equation shown above ignores air resistance. This implies that it is appropriate to use the position equation only to model falling objects that have little air resistance and that fall over short distances.

Example 7
MAKE A DECISION   Falling Object

A construction worker accidentally drops a wrench from a height of 235 feet and yells "Look out below!" (see Figure 1.7). Could a person at ground level hear this warning in time to get out of the way of the falling wrench?

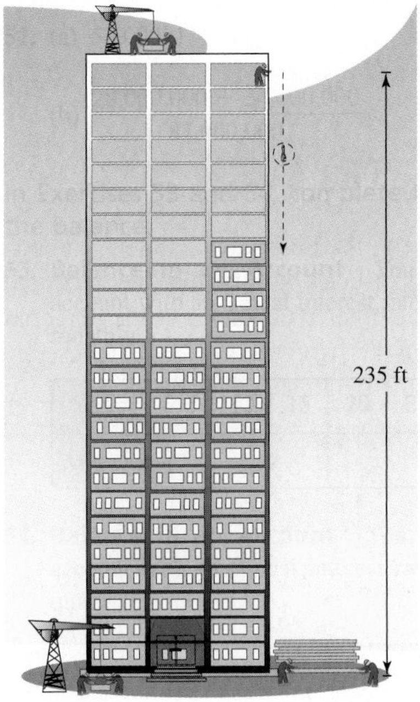

SOLUTION   Because sound travels at about 1100 feet per second, it follows that a person at ground level hears the warning within 1 second of the time the wrench is dropped. To set up a mathematical model for the height of the wrench, use the position equation

$$s = -16t^2 + v_0t + s_0.$$

Because the object is dropped rather than thrown, the initial velocity is $v_0 = 0$ feet per second. Moreover, because the initial height is $s_0 = 235$ feet, you have the following model.

$$s = -16t^2 + (0)t + 235 = -16t^2 + 235$$

After falling for 1 second, the height of the wrench is $-16(1)^2 + 235 = 219$ feet. After falling for 2 seconds, the height of the wrench is $-16(2)^2 + 235 = 171$ feet. To find the number of seconds it takes the wrench to hit the ground, let the height s be zero and solve the equation for t.

$s = -16t^2 + 235$	Write position equation.
$0 = -16t^2 + 235$	Substitute 0 for height.
$16t^2 = 235$	Add $16t^2$ to each side.
$t^2 = \dfrac{235}{16}$	Divide each side by 16.
$t = \dfrac{\sqrt{235}}{4}$	Extract positive square root.
$t \approx 3.83$	Use a calculator.

The wrench will take about 3.83 seconds to hit the ground. If the person hears the warning within 1 second after the wrench is dropped, the person still has almost 3 seconds to get out of the way.

235 ft

FIGURE 1.7

✓ **CHECKPOINT 7**

You drop a rock from a height of 144 feet. How long does it take the rock to hit the ground? ∎

A third type of application using a quadratic equation involves the hypotenuse of a right triangle. Recall from geometry that the sides of a right triangle are related by a formula called the **Pythagorean Theorem.** This theorem states that if a and b are the lengths of the legs of the triangle and c is the length of the hypotenuse (see Figure 1.8),

$$a^2 + b^2 = c^2. \qquad \text{Pythagorean Theorem}$$

Notice how this formula is used in the next example.

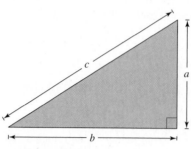

FIGURE 1.8

Example 8 **Cutting Across the Lawn**

An L-shaped sidewalk from the athletic center to the library on a college campus is shown in Figure 1.9. The sidewalk was constructed so that the length of one sidewalk forming the L is twice as long as the other. The length of the diagonal sidewalk that cuts across the grounds between the two buildings is 32 feet. How many feet does a person save by walking on the diagonal sidewalk?

SOLUTION   Using the Pythagorean Theorem, you have

$$a^2 + b^2 = c^2 \qquad \text{Pythagorean Theorem}$$
$$x^2 + (2x)^2 = 32^2 \qquad \text{Substitute for } a, b, \text{ and } c.$$
$$5x^2 = 1024 \qquad \text{Combine like terms.}$$
$$x^2 = 204.8 \qquad \text{Divide each side by 5.}$$
$$x = \pm\sqrt{204.8} \qquad \text{Take the square root of each side.}$$
$$x = \sqrt{204.8}. \qquad \text{Extract positive square root.}$$

The total distance covered by walking on the L-shaped sidewalk is

$$x + 2x = 3x$$
$$= 3\sqrt{204.8}$$
$$\approx 42.9 \text{ feet.}$$

Walking on the diagonal sidewalk saves a person about $42.9 - 32 = 10.9$ feet.

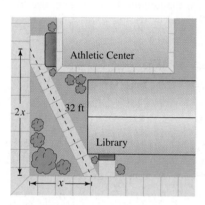

FIGURE 1.9

✓CHECKPOINT 8

In Example 8, suppose the length of one sidewalk forming the L is three times as long as the other. How many feet does a person save by walking on the 32-foot diagonal sidewalk? ▪

A fourth type of application of a quadratic equation is one in which a quantity y is changing over time t according to a quadratic model. In the next example, we exchange y for E, because E is a better descriptor in the model.

Example 9   Carbon Dioxide Emissions Ⓡ

From 2001 to 2005, yearly emissions E (in billions of metric tons) of carbon dioxide (CO_2) from energy consumption at power plants in the United States can be modeled by

$$E = 0.0053t^2 + 2.38, \quad 1 \le t \le 5$$

where t represents the year, with $t = 1$ corresponding to 2001 (see Figure 1.10). Use the model to approximate the year that CO_2 emissions were about 2,420,000,000 metric tons.   *(Source: Energy Information Administration)*

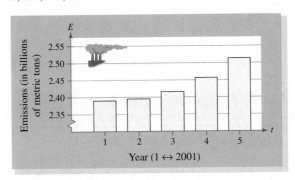

FIGURE 1.10

SOLUTION   To solve this problem, let the CO_2 emissions E be 2.42 billion and solve the equation for t.

$0.0053t^2 + 2.38 = 2.42$	Substitute 2.42 for E.
$0.0053t^2 = 0.04$	Subtract 2.38 from each side.
$t^2 \approx 7.547$	Divide each side by 0.0053.
$t \approx \sqrt{7.547}$	Extract positive square root.
$t \approx 2.747$	Simplify.

The solution is $t \approx 3$. Because $t = 1$ represents 2001, you can conclude that, according to the model, CO_2 emissions were about 2.42 billion metric tons in the year 2003.

✓**CHECKPOINT 9**

In Example 9, use the model to predict the year that CO_2 emissions will be about 3.0 billion metric tons. ■

CONCEPT CHECK

1. When using a quadratic model to solve an application problem, when can you reject one of the solutions?

2. Does the quadratic equation $x^2 = d$, where $d > 0$, have a repeated solution? Explain.

3. Which method would you use to solve the quadratic equation $(x - 5)^2 = 16$? Explain your reasoning.

4. Describe and correct the error in the solution:

$$x^2 - 2x = 3$$
$$x(x - 2) = 3$$
$$x = 3$$
$$(x - 2) = 3 \implies x = 5$$

Skills Review 1.3

The following warm-up exercises involve skills that were covered in earlier sections. You will use these skills in the exercise set for this section. For additional help, review Sections 0.4 and 0.6.

In Exercises 1–4, simplify the expression.

1. $\sqrt{\frac{7}{50}}$

2. $\sqrt{32}$

3. $\sqrt{7^2 + 3 \cdot 7^2}$

4. $\sqrt{\frac{1}{4} + \frac{3}{8}}$

In Exercises 5–10, factor the expression.

5. $3x^2 + 7x$

6. $4x^2 - 25$

7. $16 - (x - 11)^2$

8. $x^2 + 7x - 18$

9. $10x^2 + 13x - 3$

10. $6x^2 - 73x + 12$

Exercises 1.3

See www.CalcChat.com for worked-out solutions to odd-numbered exercises.

In Exercises 1–10, write the quadratic equation in general form.

1. $2x^2 = 3 - 5x$

2. $4x^2 - 2x = 9$

3. $x^2 = 25x$

4. $10x^2 = 90$

5. $(x - 3)^2 = 2$

6. $12 - 3(x + 7)^2 = 0$

7. $x(x + 2) = 3x^2 + 1$

8. $x(x + 5) = 2(x + 5)$

9. $\dfrac{3x^2 - 10}{5} = 12x$

10. $\dfrac{x^2 - 7}{3} = 2x$

In Exercises 11–22, solve the quadratic equation by factoring.

11. $x^2 - 2x - 8 = 0$

12. $x^2 - 10x + 9 = 0$

13. $6x^2 + 3x = 0$

14. $9x^2 - 1 = 0$

15. $x^2 + 10x + 25 = 0$

16. $16x^2 + 56x + 49 = 0$

17. $3 + 5x - 2x^2 = 0$

18. $2x^2 = 19x + 33$

19. $x^2 + 4x = 12$

20. $x^2 + 4x = 21$

21. $-x^2 - 7x = 10$

22. $-x^2 + 8x = 12$

In Exercises 23–40, solve the quadratic equation by extracting square roots. List both the exact answer *and* a decimal answer that has been rounded to two decimal places.

23. $x^2 = 16$

24. $x^2 = 144$

25. $x^2 = 7$

26. $x^2 = 27$

27. $3x^2 = 36$

28. $9x^2 = 25$

29. $(x - 12)^2 = 18$

30. $(x + 13)^2 = 21$

31. $(x + 2)^2 = 12$

32. $(x + 5)^2 = 20$

33. $12x^2 = 300$

34. $6x^2 = 250$

35. $5x^2 = 190$

36. $15x^2 = 620$

37. $3x^2 + 2(x^2 - 4) = 15$

38. $x^2 + 3(x^2 - 5) = 10$

39. $6x^2 - 3(x^2 + 1) = 23$

40. $2x^2 + 5(x^2 - 2) = 29$

In Exercises 41–62, solve the quadratic equation using any convenient method.

41. $x^2 = 64$

42. $7x^2 = 32$

43. $x^2 - 2x + 1 = 0$

44. $x^2 - 6x + 5 = 0$

45. $16x^2 - 9 = 0$

46. $11x^2 + 33x = 0$

47. $4x^2 - 12x + 9 = 0$

48. $x^2 - 14x + 49 = 0$

49. $(x + 4)^2 = 49$

50. $(x - 3)^2 = 36$

51. $4x = 4x^2 - 3$

52. $80 + 6x = 9x^2$

53. $50 + 5x = 3x^2$

54. $144 - 73x + 4x^2 = 0$

55. $12x = x^2 + 27$

56. $26x = 8x^2 + 15$

57. $50x^2 - 60x + 10 = 0$

58. $9x^2 + 12x + 3 = 0$

59. $(x + 3)^2 - 4 = 0$

60. $(x - 2)^2 - 9 = 0$

61. $(x + 1)^2 = x^2$

62. $(x + 1)^2 = 4x^2$

(T) **63.** Consider the expression $(x + 2)^2$. How would you convince someone in your class that $(x + 2)^2 \neq x^2 + 4$? Give an argument based on the rules of algebra. Give an argument using your graphing utility.

(T) **64.** Consider the expression $\sqrt{a^2 + b^2}$. How would you convince someone in your class that $\sqrt{a^2 + b^2} \neq a + b$? Give an argument based on the rules of algebra or geometry. Give an argument using your graphing utility.

65. Geometry A one-story building is 14 feet longer than it is wide (see figure). The building has 1632 square feet of floor space. What are the dimensions of the building?

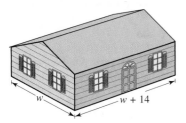

66. Geometry A billboard is 10 feet longer than it is high (see figure). The billboard has 336 square feet of advertising space. What are the dimensions of the billboard?

67. Geometry A triangular sign has a height that is equal to its base. The area of the sign is 4 square feet. Find the base and height of the sign.

68. Geometry The building lot shown in the figure has an area of 8000 square feet. What are the dimensions of the lot?

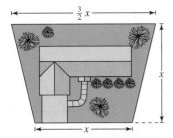

69. Geometry A rectangular garden that is 30 feet long and 20 feet wide is surrounded on all four sides by a rock path that is x feet wide. The total area of the garden and the rock path is 1200 square feet. What is the width of the path?

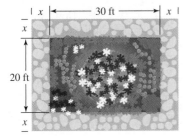

70. Geometry A rectangular pool is 30 feet wide and 40 feet long. It is surrounded on all four sides by a wooden deck that is x feet wide. The total area enclosed within the perimeter of the deck is 3000 square feet. What is the width of the deck?

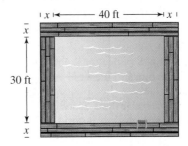

In Exercises 71–76, assume that air resistance is negligible, which implies that the position equation $s = -16t^2 + v_0t + s_0$ is a reasonable model.

71. Falling Object A rock is dropped from the top of a 200-foot cliff that overlooks the ocean. How long will it take for the rock to hit the water?

72. Royal Gorge Bridge The Royal Gorge Bridge near Canon City, Colorado is the highest suspension bridge in the world. The bridge is 1053 feet above the Arkansas river. A rock is dropped from the bridge. How long does it take the rock to hit the water?

73. Olympic Diver The high-dive platform in the Olympics is 10 meters above the water. A diver wants to perform an armstand dive, which means she will drop to the water from a handstand position. How long will the diver be in the air? (*Hint:* 1 meter $\approx$ 3.2808 feet)

74. The Owl and the Mouse An owl is circling a field and sees a mouse. The owl folds its wings and begins to dive. If the owl starts its dive from a height of 100 feet, how long does the mouse have to escape?

75. Wind Resistance At the same time a skydiver jumps from an airplane 13,000 feet above the ground, a steel ball is dropped from the plane. Because of air resistance, it takes the skydiver 67 seconds to freefall to a height of 3000 feet where the parachute opens. The steel ball has relatively no air resistance, so its height can be modeled by the position equation. How much faster does the ball reach a height of 3000 feet than the skydiver?

76. Wind Resistance At the same time a skydiver jumps from an airplane 8900 feet above the ground, a steel ball is dropped from the plane. Because of air resistance, it takes the skydiver 44 seconds to freefall to a height of 2500 feet where the parachute opens. The steel ball has relatively no air resistance and its height can be modeled by the position equation. How much faster does the ball reach a height of 2500 feet than the skydiver?

77. Geometry   The hypotenuse of an isosceles right triangle is 6 centimeters long. How long are the legs? (An isosceles right triangle is one whose two legs are of equal length.)

78. Geometry   An equilateral triangle has a height of 3 feet. How long are each of its legs? (*Hint:* Use the height of the triangle to partition the triangle into two right triangles of the same size.)

79. Flying Distance   A commercial jet flies to three cities whose locations form the vertices of a right triangle (see figure). The air distance from Atlanta to Buffalo is about 703 miles and the air distance from Atlanta to Chicago is about 583 miles. Approximate the air distance from Atlanta to Buffalo *by way of* Chicago.

In Exercises 80 and 81, use the following information. The sum of the angles of a triangle is 180°. Also, if two angles of a triangle are equal, the lengths of the sides opposite the angles are equal.

80. Depth of a Whale   The sonar of a research ship detects a whale that is 3000 feet from the ship. The angle formed by the ocean surface and a line from the ship to the whale is 45° (see figure). How deep is the whale?

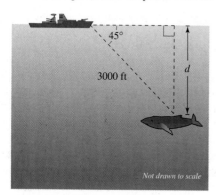

81. Depth of a Whale Shark   A research ship is tracking the movements of a whale shark that is 700 meters from the ship. The angle formed by the ocean surface and a line from the ship to the whale shark is 45°. How deep is the whale shark?

82. College Costs   The average yearly cost C of attending a private college full time for the academic years 1999/2000 to 2004/2005 in the United States can be approximated by the model

$$C = 45.6t^2 + 15,737, \quad 10 \le t \le 15$$

where $t = 10$ corresponds to the 1999/2000 academic year (see figure). Use the model to predict the year in which the average cost of attending a private college full time is about $30,000. (*Source: U.S. National Center for Education Statistics*)

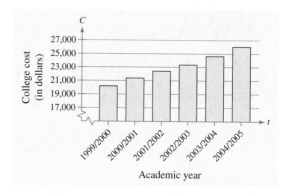

83. Total Revenue   The demand equation for a product is $p = 36 - 0.0003x$, where p is the price per unit and x is the number of units sold. The total revenue R for selling x units is given by

$$R = xp = x(36 - 0.0003x).$$

How many units must be sold to produce a revenue of $1,080,000?

84. Total Revenue   The demand equation for a product is $p = 40 - 0.0005x$, where p is the price per unit and x is the number of units sold. The total revenue R for selling x units is given by

$$R = xp = x(40 - 0.0005x).$$

How many units must be sold to produce a revenue of $800,000?

85. Production Cost   A company determines that the average monthly cost C (in dollars) of raw materials for manufacturing a product line can be modeled by

$$C = 35.65t^2 + 7205, \quad t \ge 0$$

where t is the year, with $t = 0$ corresponding to 2000. Use the model to estimate the year in which the average monthly cost reaches $12,000.

86. Monthly Cost A company determines that the average monthly cost C (in dollars) for staffing temporary positions can be modeled by

$$C = 135.47t^2 + 13,702, \quad t \geq 0$$

where t represents the year, with $t = 0$ corresponding to 2000. Use the model to predict the year in which the average monthly cost is about \$25,000.

87. *MAKE A DECISION: U.S. POPULATION* The resident population P (in thousands) of the United States from 1800 to 1890 can be approximated by the model

$$P = 694.59t^2 + 6179, \quad 0 \leq t \leq 9$$

where t represents the year, with $t = 0$ corresponding to 1800, $t = 1$ corresponding to 1810, and so on (see figure).

(a) Assume this model had continued to be valid up through the present time. In what year would the resident population of the United States have reached 250,000,000?

(b) Judging from the figure, would you say that this model is a good representation of the resident population through 1890?

(c) How about through 2006, when the United States resident population was approximately 300,000,000 people? *(Source: U.S. Census Bureau)*

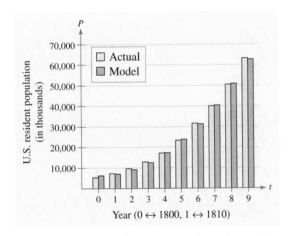

88. U.S. Population The resident population P (in thousands) of the United States from 1900 to 2000 can be approximated by the model

$$P = 1951.00t^2 + 97,551, \quad 0 \leq t \leq 10$$

where t represents the year, with $t = 0$ corresponding to 1900, $t = 1$ corresponding to 1910, and so on (see figure). Assume this model continues to be valid. In what year will the resident population of the United States reach 330,000,000? *(Source: U.S. Census Bureau)*

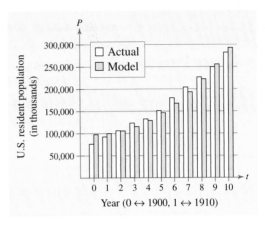

Figure for 88

(T) 89. *MAKE A DECISION* The U.S. Census Bureau predicts that the population in 2050 will be 419,854,000. Does the model in Exercise 88 appear to be a valid model for the year 2050?

90. *MAKE A DECISION* The enrollment E in an early childhood development program for a school district from 1995 to 2008 can be approximated by the model $E = 1.678t^2 + 1025$, $5 \leq t \leq 18$, where t represents the year, with $t = 5$ corresponding to 1995. Use the model to approximate the year in which the early childhood enrollment reached 1450 children. Can you use the model to estimate early childhood enrollment for the year 1980? Explain.

91. *MAKE A DECISION* The temperature T (in degrees Fahrenheit) during a certain day can be approximated by $T = 0.31t^2 + 32.9$, $7 \leq t \leq 15$, where t represents the hour of the day, with $t = 7$ corresponding to 7 A.M. Use the model to predict the time when the temperature was 85° F. Can you use this model to predict the temperature at 7 P.M.? Explain.

92. Hydrofluorocarbon Emissions From 2000 to 2005, yearly emissions E (in millions of metric tons) of hydrofluorocarbons (HFCs) in the United States can be modeled by $E = 1.26t^2 + 99.98$, $0 \leq t \leq 5$, where t represents the year, with $t = 0$ corresponding to 2000. Use the model to estimate the year in which HFC emissions were about 124,000,000 metric tons. *(Source: Energy Information Administration)*

93. Blue Oak The blue oak tree, native to California, is known for its slow rate of growth. Fencing enclosures protect seedlings from herbivore damage and promote faster growth. The height H (in inches) of an enclosed blue oak tree can be approximated by the model

$$H = 0.74t^2 + 25, \quad 0 \leq t \leq 5$$

where t represents the year, with $t = 0$ corresponding to 2000. Use the model to approximate the year in which the height of the tree was about 32 inches.

Section 1.4

The Quadratic Formula

- Develop the Quadratic Formula by completing the square.
- Use the discriminant to determine the number of real solutions of a quadratic equation.
- Solve a quadratic equation using the Quadratic Formula.
- Use the Quadratic Formula to solve an application problem.

Development of the Quadratic Formula

In Section 1.3 you studied two methods for solving quadratic equations. These two methods are efficient for special quadratic equations that are factorable or that can be solved by extracting square roots. There are, however, many quadratic equations that cannot be solved efficiently by either of these two techniques. Fortunately, there is a general formula that can be used to solve *any* quadratic equation. It is called the **Quadratic Formula.** This formula is derived using a process called **completing the square.**

$$ax^2 + bx + c = 0 \qquad \text{General form, } a \neq 0$$

$$ax^2 + bx = -c \qquad \text{Subtract } c \text{ from each side.}$$

$$x^2 + \frac{b}{a}x = -\frac{c}{a} \qquad \text{Divide each side by } a.$$

$$x^2 + \frac{b}{a}x + \left(\frac{b}{2a}\right)^2 = -\frac{c}{a} + \left(\frac{b}{2a}\right)^2 \qquad \text{Complete the square.}$$

$$\left(\text{half of } \frac{b}{a}\right)^2$$

$$\left(x + \frac{b}{2a}\right)^2 = \frac{b^2 - 4ac}{4a^2} \qquad \text{Simplify.}$$

$$x + \frac{b}{2a} = \pm \sqrt{\frac{b^2 - 4ac}{4a^2}} \qquad \text{Extract square roots.}$$

$$x = \frac{-b \pm \sqrt{b^2 - 4ac}}{2a} \qquad \text{Solutions}$$

STUDY TIP

The Quadratic Formula is one of the most important formulas in algebra, and you should memorize it. It might help to try to memorize a verbal statement of the rule. For instance, you might try to remember the following verbal statement of the Quadratic Formula: "The opposite of b, plus or minus the square root of b squared minus $4ac$, all divided by $2a$."

The Quadratic Formula

The solutions of

$$ax^2 + bx + c = 0, \quad a \neq 0$$

are given by the **Quadratic Formula,**

$$x = \frac{-b \pm \sqrt{b^2 - 4ac}}{2a}.$$

The Discriminant

In the Quadratic Formula, the quantity under the radical sign, $b^2 - 4ac$, is called the **discriminant** of the quadratic expression $ax^2 + bx + c$.

$$b^2 - 4ac \qquad \text{Discriminant}$$

It can be used to determine the number of the solutions of a quadratic equation.

Solutions of a Quadratic Equation

The solutions of a quadratic equation

$$ax^2 + bx + c = 0, \quad a \neq 0$$

can be classified by the discriminant, $b^2 - 4ac$, as follows.

1. If $b^2 - 4ac > 0$, the equation has *two* distinct real solutions.

2. If $b^2 - 4ac = 0$, the equation has *one* repeated real solution.

3. If $b^2 - 4ac < 0$, the equation has *no* real solutions.

If the discriminant of a quadratic equation is negative, as in case 3 above, then its square root is imaginary (not a real number) and the Quadratic Formula yields two complex solutions. You will study complex solutions in Section 3.5.

Example 1 Using the Discriminant

Use the discriminant to determine the number of real solutions of each of the following quadratic equations.

a. $4x^2 - 20x + 25 = 0$

b. $13x^2 + 7x + 1 = 0$

c. $5x^2 = 8x$

SOLUTION

a. Using $a = 4$, $b = -20$, and $c = 25$, the discriminant is

$$b^2 - 4ac = (-20)^2 - 4(4)(25) = 400 - 400 = 0.$$

Because $b^2 - 4ac = 0$, there is *one* repeated real solution.

b. Using $a = 13$, $b = 7$, and $c = 1$, the discriminant is

$$b^2 - 4ac = (7)^2 - 4(13)(1) = 49 - 52 = -3.$$

Because $b^2 - 4ac < 0$, there are *no* real solutions.

c. In general form, this equation is $5x^2 - 8x = 0$, with $a = 5$, $b = -8$, and $c = 0$, which implies that the discriminant is

$$b^2 - 4ac = (-8)^2 - 4(5)(0) = 64.$$

Because $b^2 - 4ac > 0$, there are *two* distinct real solutions.

✓CHECKPOINT 1

Use the discriminant to determine the number of real solutions of $x^2 + 6x + 9 = 0$. ∎

Using the Quadratic Formula

When using the Quadratic Formula, remember that *before* the formula can be applied, you must first write the quadratic equation in general form.

Example 2 Two Distinct Solutions

Solve $x^2 + 3x = 9$.

SOLUTION

$x^2 + 3x = 9$	Write original equation.
$x^2 + 3x - 9 = 0$	Write in general form.
$x = \dfrac{-3 \pm \sqrt{(3)^2 - 4(1)(-9)}}{2(1)}$	Quadratic Formula
$x = \dfrac{-3 \pm \sqrt{45}}{2}$	Simplify.
$x = \dfrac{-3 \pm 3\sqrt{5}}{2}$	Simplify.

The solutions are

$$x = \frac{-3 + 3\sqrt{5}}{2} \quad \text{and} \quad x = \frac{-3 - 3\sqrt{5}}{2}.$$

Check these in the original equation.

✓ CHECKPOINT 2

Solve $x^2 + 2x - 2 = 0$. ▪

Example 3 One Repeated Solution

Solve $8x^2 - 24x + 18 = 0$.

SOLUTION   Begin by dividing each side by the common factor 2.

$8x^2 - 24x + 18 = 0$	Write original equation.
$4x^2 - 12x + 9 = 0$	Divide each side by 2.
$x = \dfrac{-(-12) \pm \sqrt{(-12)^2 - 4(4)(9)}}{2(4)}$	Quadratic Formula
$x = \dfrac{12 \pm \sqrt{0}}{8}$	Simplify.
$x = \dfrac{3}{2}$	Repeated solution

The only solution is $x = \frac{3}{2}$. Check this in the original equation.

✓ CHECKPOINT 3

Solve $9x^2 - 6x = -1$. ▪

TECHNOLOGY

(T) You can check the solutions to Example 2 using a calculator.

```
((-3+3√(5))/2)²+
3((-3+3√(5))/2)
                 9
((-3-3√(5))/2)²+
3((-3-3√(5))/2)
                 9
```

The discriminant in Example 3 is a perfect square (zero in this case), and you could have factored the quadratic as

$$4x^2 - 12x + 9 = 0$$

$$(2x - 3)^2 = 0$$

and concluded that the solution is $x = \frac{3}{2}$. Because factoring is easier than applying the Quadratic Formula, try factoring first when solving a quadratic equation. If, however, factors cannot easily be found, then use the Quadratic Formula. For instance, try solving the quadratic equation

$$x^2 - x - 12 = 0$$

in two ways—by factoring and by the Quadratic Formula—to see that you get the same solutions either way.

When using a calculator with the Quadratic Formula, you should get in the habit of using the memory key to store intermediate steps. This will save steps and minimize roundoff error.

Example 4   Using a Calculator with the Quadratic Formula

Solve $16.3x^2 - 197.6x + 7.042 = 0$.

SOLUTION   In this case, $a = 16.3$, $b = -197.6$, $c = 7.042$, and you have

$$x = \frac{-(-197.6) \pm \sqrt{(-197.6)^2 - 4(16.3)(7.042)}}{2(16.3)}.$$

To evaluate these solutions, begin by calculating the square root of the discriminant, as follows.

Scientific Calculator Keystrokes

197.6 $\boxed{+/-}$ $\boxed{x^2}$ $\boxed{-}$ 4 $\boxed{\times}$ 16.3 $\boxed{\times}$ 7.042 $\boxed{=}$ $\boxed{\sqrt{}}$

Graphing Calculator Keystrokes

$\boxed{\sqrt{}}$ $\boxed{(}$ $\boxed{(}$ $\boxed{(-)}$ 197.6 $\boxed{)}$ $\boxed{x^2}$ $\boxed{-}$ 4 $\boxed{\times}$ 16.3 $\boxed{\times}$ 7.042 $\boxed{)}$ $\boxed{ENTER}$

In either case, the result is 196.434777. Storing this result and using the recall key, you can find the following two solutions.

$$x \approx \frac{197.6 + 196.434777}{2(16.3)} \approx 12.087 \qquad \text{Add stored value.}$$

$$x \approx \frac{197.6 - 196.434777}{2(16.3)} \approx 0.036 \qquad \text{Subtract stored value.}$$

✓CHECKPOINT 4

Solve $4.7x^2 - 3.2x - 5.9 = 0$. ■

TECHNOLOGY

Ⓣ Try to calculate the value of x in Example 4 by using additional parentheses instead of storing the intermediate result, 196.434777, in your calculator.

Applications

In Section 1.3, you studied four basic types of applications involving quadratic equations: area, falling bodies, the Pythagorean Theorem, and quadratic models. The solution to each of these types of problems can involve the Quadratic Formula. For instance, Example 5 shows how the Quadratic Formula can be used to analyze a quadratic model for a patient's blood oxygen level.

Example 5 | Blood Oxygen Level

Doctors treated a patient at an emergency room from 1:00 P.M. to 5:00 P.M. The patient's blood oxygen level L (in percent) during this time period can be modeled by

$$L = -0.25t^2 + 3.0t + 87, \quad 1 \le t \le 5$$

where t represents the time of day, with $t = 1$ corresponding to 1:00 P.M. Use the model to estimate the time when the patient's blood oxygen level was 95%.

SOLUTION   To find the hour when the patient's blood oxygen level was 95%, solve the equation

$$95 = -0.25t^2 + 3.0t + 87.$$

To begin, write the equation in general form.

$$-0.25t^2 + 3.0t - 8 = 0$$

Then apply the Quadratic Formula with $a = -0.25$, $b = 3.0$, and $c = -8$.

$$t = \frac{-3 \pm \sqrt{3^2 - 4(-0.25)(-8)}}{2(-0.25)}$$

$$= \frac{-3 \pm \sqrt{1}}{-0.5} = 4 \text{ or } 8$$

Of the two possible solutions, only $t = 4$ makes sense in the context of the problem, because $t = 8$ is not in the domain of L. Because $t = 1$ corresponds to 1:00 P.M., it follows that $t = 4$ corresponds to 4:00 P.M. So, from the model you can conclude that the patient's blood oxygen level was 95% at about 4:00 P.M. Figure 1.11 shows the patients oxygen level recorded every 30 minutes.

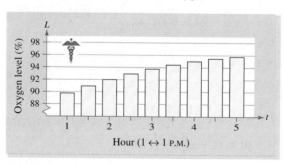

FIGURE 1.11

✓ **CHECKPOINT 5**

In Example 5, use the model to estimate the time when the patient's blood oxygen level was 92%. ∎

TECHNOLOGY

(T) A program can be written to solve equations using the Quadratic Formula. A program for several models of graphing utilities can be found on the website for this text at *college.hmco.com/info/larsonapplied.* Use a program to solve Example 5.

In Section 1.3, you learned that the position equation for a falling object is of the form

$$s = -16t^2 + v_0 t + s_0$$

where s is the height (in feet) of the object, v_0 is the initial velocity (in feet per second), t is the time (in seconds), and s_0 is the initial height (in feet). This equation is valid only for free-falling objects near Earth's surface. Because of differences in gravitational force, position equations are different on other planets or moons. The next example looks at a position equation for a falling object on our moon.

Example 6   Throwing an Object on the Moon

An astronaut standing on the surface of the moon throws a rock straight up at 27 feet per second from a height of 6 feet (see Figure 1.12). The height s (in feet) of the rock is given by

$$s = -2.7t^2 + 27t + 6$$

where t is the time (in seconds). How much time elapses before the rock strikes the lunar surface?

SOLUTION   Because s gives the height of the rock at any time t, you can find the time that the rock hits the surface of the moon by setting s equal to zero and solving for t.

$$-2.7t^2 + 27t + 6 = 0 \qquad \text{Substitute 0 for } s.$$

$$t = \frac{-27 \pm \sqrt{(27)^2 - 4(-2.7)(6)}}{2(-2.7)} \qquad \text{Quadratic Formula}$$

$$\approx 10.2 \text{ seconds} \qquad \text{Choose positive solution.}$$

So, about 10.2 seconds elapse before the rock hits the lunar surface.

FIGURE 1.12

✓**CHECKPOINT 6**

In Example 6, suppose the rock is thrown straight up at 13 feet per second from a height of 4 feet. The height s (in feet) of the rock is given by $s = -2.7t^2 + 13t + 4$. How much time (in seconds) elapses before the rock strikes the lunar surface? ▨

CONCEPT CHECK

1. When using the quadratic formula to solve $4x^2 = 2 - 3x$, what are the values of a, b, and c?

2. The quadratic equation $ax^2 + bx + c = 0$ has two distinct solutions. Does $b^2 - 4ac = 0$? Explain.

3. The area A (in square feet) of a parking lot is represented by $A = x^2 + 9x + 300$. Is it possible for the parking lot to have an area of 275 square feet? Explain.

4. The discriminants of two quadratic equations are 5 and -10. Can the equations have the same solution? Explain.

Skills Review 1.4

The following warm-up exercises involve skills that were covered in earlier sections. You will use these skills in the exercise set for this section. For additional help, review Sections 0.4 and 1.3.

In Exercises 1–4, simplify the expression.

1. $\sqrt{9 - 4(3)(-12)}$

2. $\sqrt{36 - 4(2)(3)}$

3. $\sqrt{12^2 - 4(3)(4)}$

4. $\sqrt{15^2 + 4(9)(12)}$

In Exercises 5–10, solve the quadratic equation by factoring.

5. $x^2 - x - 2 = 0$

6. $2x^2 + 3x - 9 = 0$

7. $x^2 - 4x = 5$

8. $2x^2 + 13x = 7$

9. $x^2 = 5x - 6$

10. $x(x - 3) = 4$

Exercises 1.4

See www.CalcChat.com for worked-out solutions to odd-numbered exercises.

In Exercises 1–8, use the discriminant to determine the number of real solutions of the quadratic equation.

1. $4x^2 - 4x + 1 = 0$

2. $2x^2 - x - 1 = 0$

3. $3x^2 + 4x + 1 = 0$

4. $x^2 + 2x + 4 = 0$

5. $2x^2 - 5x = -5$

6. $3 - 6x = -3x^2$

7. $\frac{1}{5}x^2 + \frac{6}{5}x - 8 = 0$

8. $\frac{1}{3}x^2 - 5x + 25 = 0$

In Exercises 9–30, use the Quadratic Formula to solve the quadratic equation.

9. $2x^2 + x - 1 = 0$

10. $2x^2 - x - 1 = 0$

11. $16x^2 + 8x - 3 = 0$

12. $25x^2 - 20x + 3 = 0$

13. $2 + 2x - x^2 = 0$

14. $x^2 - 10x + 22 = 0$

15. $x^2 + 14x + 44 = 0$

16. $6x = 4 - x^2$

17. $x^2 + 8x - 4 = 0$

18. $4x^2 - 4x - 4 = 0$

19. $12x - 9x^2 = -3$

20. $16x^2 + 22 = 40x$

21. $36x^2 + 24x = 7$

22. $3x + x^2 - 1 = 0$

23. $4x^2 + 4x = 7$

24. $16x^2 - 40x + 5 = 0$

25. $28x - 49x^2 = 4$

26. $9x^2 + 24x + 16 = 0$

27. $8t = 5 + 2t^2$

28. $25h^2 + 80h + 61 = 0$

29. $(y - 5)^2 = 2y$

30. $(x + 6)^2 = -2x$

In Exercises 31–36, use a calculator to solve the quadratic equation. (Round your answer to three decimal places.)

31. $5.1x^2 - 1.7x - 3.2 = 0$

32. $10.4x^2 + 8.6x + 1.2 = 0$

33. $7.06x^2 - 4.85x + 0.50 = 0$

34. $2x^2 - 2.50x - 0.42 = 0$

35. $-0.003x^2 + 0.025x - 0.98 = 0$

36. $-0.005x^2 + 0.101x - 0.193 = 0$

In Exercises 37–46, solve the quadratic equation using any convenient method.

37. $2x^2 + 7 = 2x^2 - x - 4$

38. $x^2 - 2x + 5 = x^2 - 5$

39. $4x^2 - 15 = 25$

40. $3x^2 - 16 = 38$

41. $x^2 + 3x + 1 = 0$

42. $x^2 + 3x - 4 = 0$

43. $(x - 1)^2 = 9$

44. $2x^2 - 4x - 6 = 0$

45. $3x^2 + 5x - 11 = 4(x - 2)$

46. $2x^2 + 4x - 9 = 2(x - 1)^2$

Writing Real-Life Problems In Exercises 47–50, solve the number problem *and* write a real-life problem that could be represented by this verbal model. For instance, an applied problem that could be represented by Exercise 47 is as follows.

The sum of the length and width of a one-story house is 100 feet. The house has 2500 square feet of floor space. What are the length and width of the house?

47. Find two numbers whose sum is 100 and whose product is 2500.

48. One number is 1 more than another number. The product of the two numbers is 72. Find the numbers.

49. One number is 1 more than another number. The sum of their squares is 113. Find the numbers.

50. One number is 2 more than another number. The product of the two numbers is 440. Find the numbers.

Cost Equation In Exercises 51–54, use the cost equation to find the number of units x that a manufacturer can produce for the cost C. (Round your answer to the nearest positive integer.)

51. $C = 0.125x^2 + 20x + 5000$ $C = \$14,000$

52. $C = 0.5x^2 + 15x + 5000$ $C = \$11,500$

53. $C = 800 + 0.04x + 0.002x^2$ $C = \$1680$

54. $C = 312.5 - 10x + 0.4x^2$ $C = \$900$

55. Seating Capacity A rectangular classroom seats 72 students. If the seats were rearranged with three more seats in each row, the classroom would have two fewer rows. Find the original number of seats in each row.

56. Geometry A rancher has 200 feet of fencing to enclose two adjacent rectangular corrals (see figure). Find the dimensions such that the total enclosed area will be 1400 square feet.

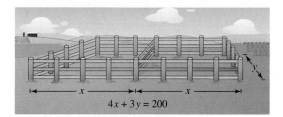

$4x + 3y = 200$

57. Geometry An open box is to be made from a square piece of material by cutting two-inch squares from the corners and turning up the sides (see figure). The volume of the finished box is to be 200 cubic inches. Find the size of the original piece of material.

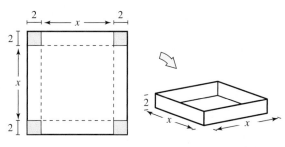

58. Geometry An open box (see figure) is to be constructed from 108 square inches of material. Find the dimensions of the square base. (*Hint:* The surface area is $S = x^2 + 4xh$.)

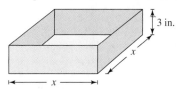

59. Eiffel Tower You throw a coin straight up from the top of the Eiffel Tower in Paris with a velocity of 20 miles per hour. The building has a height of 984 feet.

(a) Use the position equation to write a mathematical model for the height of the coin.

(b) Find the height of the coin after 4 seconds.

(c) How long will it take before the coin strikes the ground?

60. Sports Some Major League Baseball pitchers can throw a fastball at speeds of up to and over 100 miles per hour. Assume a Major League Baseball pitcher throws a baseball straight up into the air at 100 miles per hour from a height of 6 feet 3 inches.

(a) Use the position equation to write a mathematical model for the height of the baseball.

(b) Find the height of the baseball after 4 seconds, 5 seconds, and 6 seconds. What must have occurred sometime in the interval $4 \le t \le 6$? Explain.

(c) How many seconds is the baseball in the air?

61. On the Moon An astronaut on the moon throws a rock straight upward into space. The height s (in feet) of the rock is given by $s = -2.7t^2 + 40t + 5$, where the initial velocity is 40 feet per second, the initial height is 5 feet, and t is the time (in seconds). How long will it take the rock to hit the surface of the moon? If the rock had been thrown with the same initial velocity and height on Earth, how long would it have remained in the air?

62. Hot Air Balloon Two people are floating in a hot air balloon 200 feet above a lake. One person tosses out a coin with an initial velocity of 20 feet per second. One second later, the balloon is 20 feet higher and the other person drops another coin (see figure). The position equation for the first coin is $s = -16t^2 + 20t + 200$, and the position equation for the second coin is $s = -16t^2 + 220$. Which coin will hit the water first? (*Hint:* Remember that the first coin was tossed one second before the second coin was dropped.)

Falling Objects   In Exercises 63 and 64, use the following information. The position equation for falling objects on Earth is of the form

$$s = -16t^2 + v_0t + s_0$$

where s is the height of the object (in feet), v_0 is the initial velocity (in feet per second), t is the time (in seconds), and s_0 is the initial height (in feet).

63. *MAKE A DECISION* Would a rock thrown upward from an initial height of 6 feet with an initial velocity of 27 feet per second take longer to reach the ground on Earth or on the moon? (See Example 6.)

64. *MAKE A DECISION* Would a rock thrown downward from an initial height of 6 feet with an initial velocity of -14 feet per second take longer to reach the ground on Earth or on the moon? (See Example 6.)

65. Flying Distance   A small commuter airline flies to three cities whose locations form the vertices of a right triangle (see figure). The total flight distance (from City A to City B to City C and back to City A) is 1400 miles. It is 600 miles between the two cities that are farthest apart. Find the other two distances between cities.

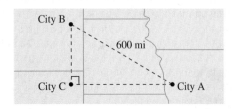

66. Distance from a Dock   A windlass is used to tow a boat to a dock. The figure shows a situation in which there is 75 feet of rope extended to the boat. How far is the boat from the dock?

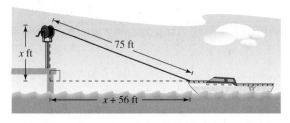

67. Starbucks   The total sales S (in millions of dollars) for Starbucks from 1996 to 2005 can be approximated by the model $S = 58.155t^2 - 612.3t + 2387.1$, $6 \le t \le 15$, where t represents the year, with $t = 6$ corresponding to 1996. The figure shows the actual sales and the sales represented by the model. (*Source: Starbucks Corporation*)

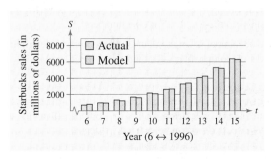

(a) Use the model to estimate the year when total sales were about $4 billion.

(b) Use the model to predict the year when the total sales were about $6.2 billion.

(c) Starbucks sales were expected to reach $9.45 billion in 2007. Does the model agree? Explain your reasoning.

68. Per Capita Income   The per capita income P (in dollars) in the United States from 1995 to 2005 can be approximated by the model $P = 7.14t^2 + 887.5t + 15,544$, $5 \le t \le 15$, where t represents the year, with $t = 5$ corresponding to 1995. The figure shows the actual per capita income and the per capita income represented by the model. (*Source: U.S. Bureau of Economic Analysis*)

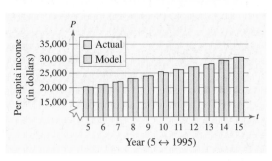

(a) Use the model to estimate the year in which the per capita income was about $26,500.

(b) Use the model to predict the year in which the per capita income is about $34,000.

69. Blood Oxygen Level   Doctors treated a patient at an emergency room from 2:00 P.M. to 7:00 P.M. The patient's blood oxygen level L (in percent) during this time period can be modeled by

$$L = -0.270t^2 + 3.59t + 83.1, \quad 2 \le t \le 7$$

where t represents the time of day, with $t = 2$ corresponding to 2:00 P.M. Use the model to estimate the time (rounded to the nearest hour) when the patient's blood oxygen level was 93%.

70. Prescription Drugs   The total amounts A (in billions of dollars) projected by the industry to be spent on prescription drugs in the United States from 2002 to 2012 can be approximated by the model.

$$A = 0.89t^2 + 15.9t + 126, \quad 2 \le t \le 12$$

where t represents the year, with $t = 2$ corresponding to 2002. Use the model to predict the year in which the total amount spent on prescription drugs will be about $374,000,000,000. *(Source: U.S. Center for Medicine and Medicaid Services)*

71. Flying Speed   Two planes leave simultaneously from the same airport, one flying due east and the other due south (see figure). The eastbound plane is flying 50 miles per hour faster than the southbound plane. After 3 hours the planes are 2440 miles apart. Find the speed of each plane.

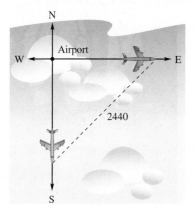

72. Flying Speed   Two planes leave simultaneously from the same airport, one flying due east and the other due south. The eastbound plane is flying 100 miles per hour faster than the southbound plane. After 2 hours the planes are 1500 miles apart. Find the speed of each plane.

73. Biology   The metabolic rate of an ectothermic organism increases with increasing temperature within a certain range. The graph shows experimental data for the oxygen consumption C (in microliters per gram per hour) of a beetle at certain temperatures. The data can be approximated by the model

$$C = 0.45x^2 - 1.65x + 50.75, \quad 10 \le x \le 25$$

where x is the temperature in degrees Celsius.

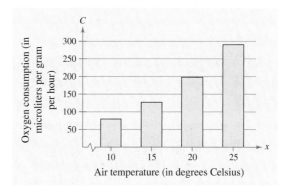

(a) The oxygen consumption is 150 microliters per gram per hour. What is the air temperature?

(b) The temperature is increased from 10°C to 20°C. The oxygen consumption is increased by approximately what factor?

74. Total Revenue   The demand equation for a product is

$$p = 60 - 0.0004x$$

where p is the price per unit and x is the number of units sold. The total revenue R for selling x units is given by

$$R = xp.$$

How many units must be sold to produce a revenue of $220,000?

75. Total Revenue   The demand equation for a product is

$$p = 50 - 0.0005x$$

where p is the price per unit and x is the number of units sold. The total revenue R for selling x units is given by

$$R = xp.$$

How many units must be sold to produce a revenue of $250,000?

76. When the Quadratic Formula is used to solve certain problems, such as the problem in Example 5 on page 108, why is only one solution used?

77. Extended Application   To work an extended application analyzing the sales per share of Starbucks Corporation from 1992 to 2005, visit this text's website at *college.hmco.com*. *(Data Source: Starbucks Corporation)*

Mid-Chapter Quiz

See www.CalcChat.com for worked-out solutions to odd-numbered exercises.

Take this quiz as you would take a quiz in class. When you are done, check your work against the answers given in the back of the book.

In Exercises 1–4, solve the equation and check your solution.

1. $3(x - 2) - 4(2x + 5) = 4$

2. $\dfrac{3x + 3}{5x - 2} = \dfrac{3}{4}$

3. $\dfrac{2}{x(x - 1)} + \dfrac{1}{x} = \dfrac{1}{x - 4}$

4. $(x + 3)^2 - x^2 = 6(x + 2)$

5. Describe how you can check your answers to Exercises 1–4 using your graphing utility.

In Exercises 6 and 7, solve the equation. (Round your solution to three decimal places.)

6. $\dfrac{x}{2.004} - \dfrac{x}{5.128} = 100$

7. $0.378x + 0.757(500 - x) = 215$

In Exercises 8 and 9, write an algebraic equation for the verbal description. Find the solution if possible and check.

8. A company has fixed costs of $30,000 per month and variable costs of $8.50 per unit manufactured. The company has $200,000 available each month to cover monthly costs. How many units can the company manufacture?

9. The demand equation for a product is $p = 75 - 0.0002x$, where p is the price per unit and x is the number of units sold. The total revenue R for selling x units is given by $R = xp$. How many units must be sold to produce a revenue of $300,000?

In Exercises 10–15, solve the quadratic equation by the indicated method.

10. *Factoring:* $3x^2 + 13x = 10$

11. *Extracting roots:* $3x^2 = 15$

12. *Extracting roots:* $(x + 3)^2 = 17$

13. *Quadratic Formula:* $2x + x^2 = 5$

14. *Quadratic Formula:* $3x^2 + 7x - 2 = 0$

15. *Quadratic Formula:* $3x^2 - 4.50x - 0.32 = 0$

In Exercises 16 and 17, use the discriminant to determine the number of real solutions of the quadratic equation.

16. $2x^2 - 4x + 9 = 0$

17. $4x^2 - 12x + 9 = 0$

18. Describe how you would convince a fellow student that $(x + 3)^2 = x^2 + 6x + 9$.

19. A rock is dropped from a height of 300 feet. How long will it take the rock to hit the ground?

20. An open box has a square base and a height of 6 inches. The volume of the box is 384 cubic inches. Find the dimensions of the box.

Section 1.5

Other Types of Equations

- Solve a polynomial equation by factoring.
- Rewrite and solve an equation involving radicals or rational exponents.
- Rewrite and solve an equation involving fractions or absolute values.
- Construct and use a nonquadratic model to solve an application problem.

Polynomial Equations

In this section you will extend the techniques for solving equations to nonlinear and nonquadratic equations. At this point in the text, you have only three basic methods for solving nonlinear equations—*factoring, extracting roots,* and the *Quadratic Formula.* So the main goal of this section is to learn to *rewrite* nonlinear equations in a form to which you can apply one of these methods.

STUDY TIP

When solving an equation, avoid dividing each side by a common variable factor to simplify. You may lose solutions. For instance, if you divide each side by x^2 in Example 1, you lose the solution $x = 0$. Also, when solving an equation by factoring, be sure to set each variable factor equal to zero to find all of the possible solutions.

Example 1   **Solving a Polynomial Equation by Factoring**

Solve $3x^4 = 48x^2$.

SOLUTION   The basic approach is first to write the polynomial equation in general form with zero on one side, then to factor the other side, and finally to set each factor equal to zero and solve.

$3x^4 = 48x^2$	Write original equation.
$3x^4 - 48x^2 = 0$	Write in general form.
$3x^2(x^2 - 16) = 0$	Factor out common factor.
$3x^2(x + 4)(x - 4) = 0$	Difference of two squares
$3x^2 = 0 \implies x = 0$	Set 1st factor equal to 0.
$x + 4 = 0 \implies x = -4$	Set 2nd factor equal to 0.
$x - 4 = 0 \implies x = 4$	Set 3rd factor equal to 0.

You can check these solutions by substituting in the original equation, as follows.

CHECK

$3x^4 = 48x^2$	Write original equation.
$3(0)^4 = 48(0)^2$	0 checks. ✓
$3(-4)^4 = 48(-4)^2$	-4 checks. ✓
$3(4)^4 = 48(4)^2$	4 checks. ✓

After checking, you can conclude that the solutions are $x = 0$, $x = -4$, and $x = 4$.

✓ CHECKPOINT 1

Solve $3x^3 = 3x$. ■

Example 2 Solving a Polynomial Equation by Factoring

Solve $x^3 - 3x^2 - 3x + 9 = 0$.

SOLUTION

$x^3 - 3x^2 - 3x + 9 = 0$	Write original equation.
$x^2(x - 3) - 3(x - 3) = 0$	Group terms.
$(x - 3)(x^2 - 3) = 0$	Factor by grouping.
$x - 3 = 0 \implies x = 3$	Set 1st factor equal to 0.
$x^2 - 3 = 0 \implies x = \pm\sqrt{3}$	Set 2nd factor equal to 0.

The solutions are $x = 3$, $x = \sqrt{3}$, and $x = -\sqrt{3}$. Check these in the original equation. Notice that this polynomial has a degree of 3 and has three solutions.

✓ CHECKPOINT 2

Solve $x^3 - x^2 - 2x + 2 = 0$. ■

DISCOVERY

What do you observe about the degrees of the polynomials in Examples 1, 2, and 3 and the possible numbers of solutions of the equations? Does your observation apply to the quadratic equations in Sections 1.3 and 1.4?

Occasionally, mathematical models involve equations that are of **quadratic type.** In general, an equation is of quadratic type if it can be written in the form

$$au^2 + bu + c = 0$$

where $a \neq 0$ and u is an algebraic expression.

Example 3 Solving an Equation of Quadratic Type

Solve $x^4 - 3x^2 + 2 = 0$.

SOLUTION This equation is of quadratic type with $u = x^2$.

$$(x^2)^2 - 3(x^2) + 2 = 0$$

To solve this equation, you can factor the left side of the equation as the product of two second-degree polynomials.

$x^4 - 3x^2 + 2 = 0$	Write original equation.
$(x^2 - 1)(x^2 - 2) = 0$	Partially factor.
$(x + 1)(x - 1)(x^2 - 2) = 0$	Completely factor.
$x + 1 = 0 \implies x = -1$	Set 1st factor equal to 0.
$x - 1 = 0 \implies x = 1$	Set 2nd factor equal to 0.
$x^2 - 2 = 0 \implies x = \pm\sqrt{2}$	Set 3rd factor equal to 0.

The solutions are $x = -1$, $x = 1$, $x = \sqrt{2}$, and $x = -\sqrt{2}$. Check these in the original equation. Notice that this polynomial has a degree of 4 and has four solutions.

✓ CHECKPOINT 3

Solve $x^4 - 5x^2 + 4 = 0$. ■

Solving Equations Involving Radicals

The steps involved in solving the remaining equations in this section will often introduce *extraneous solutions,* as discussed in Section 1.1. Operations such as squaring each side of an equation, raising each side of an equation to a rational power, or multiplying each side of an equation by a variable quantity all create this potential danger. So, when you use any of these operations, checking of solutions is crucial.

Example 4   An Equation Involving a Radical

Solve $\sqrt{2x + 7} - x = 2$.

SOLUTION

$\sqrt{2x + 7} - x = 2$	Write original equation.
$\sqrt{2x + 7} = x + 2$	Isolate the square root.
$2x + 7 = x^2 + 4x + 4$	Square each side.
$0 = x^2 + 2x - 3$	Write in general form.
$0 = (x + 3)(x - 1)$	Factor.
$x + 3 = 0 \implies x = -3$	Set 1st factor equal to 0.
$x - 1 = 0 \implies x = 1$	Set 2nd factor equal to 0.

By checking these values, you can determine that the only solution is $x = 1$.

✓ **CHECKPOINT 4**

Solve $\sqrt{3x} - 6 = 0$. ▪

STUDY TIP

The basic technique used in Example 5 is to isolate the factor with the rational exponent and raise each side to the reciprocal power. In Example 4, this is equivalent to isolating the square root and squaring each side.

Example 5   An Equation Involving a Rational Exponent

Solve $4x^{3/2} - 8 = 0$.

SOLUTION

$4x^{3/2} - 8 = 0$	Write original equation.
$4x^{3/2} = 8$	Add 8 to each side.
$x^{3/2} = 2$	Isolate $x^{3/2}$.
$x = 2^{2/3}$	Raise each side to the $\frac{2}{3}$ power.
$x \approx 1.587$	Round to three decimal places.

CHECK

$4x^{3/2} - 8 = 0$	Write original equation.
$4(2^{2/3})^{3/2} \overset{?}{=} 8$	Substitute $2^{2/3}$ for x.
$4(2) \overset{?}{=} 8$	Power of a Power Property
$8 = 8$	Solution checks. ✓

✓ **CHECKPOINT 5**

Solve $2x^{3/4} - 54 = 0$. ▪

Equations Involving Fractions or Absolute Values

In Section 1.1, you learned how to solve equations involving fractions. Recall that the first step is to multiply each term of the equation by the least common denominator (LCD).

Example 6   An Equation Involving Fractions

Solve $\dfrac{2}{x} = \dfrac{3}{x-2} - 1$.

SOLUTION   For this equation, the LCD of the three terms is $x(x-2)$, so begin by multiplying each term of the equation by this expression.

$$\frac{2}{x} = \frac{3}{x-2} - 1 \qquad\qquad \text{Write original equation.}$$

$$x(x-2)\frac{2}{x} = x(x-2)\frac{3}{x-2} - x(x-2)(1) \qquad \text{Multiply each term by LCD.}$$

$$2(x-2) = 3x - x(x-2), \quad x \neq 0, 2 \qquad \text{Simplify.}$$

$$2x - 4 = -x^2 + 5x \qquad\qquad \text{Distributive Property}$$

$$x^2 - 3x - 4 = 0 \qquad\qquad \text{Write in general form.}$$

$$(x-4)(x+1) = 0 \qquad\qquad \text{Factor.}$$

$$x - 4 = 0 \quad\Longrightarrow\quad x = 4 \qquad \text{Set 1st factor equal to 0.}$$

$$x + 1 = 0 \quad\Longrightarrow\quad x = -1 \qquad \text{Set 2nd factor equal to 0.}$$

Notice that the values $x = 0$ and $x = 2$ are excluded from the domains of the fractions because they result in division by zero. So, both $x = 4$ and $x = -1$ are possible solutions.

CHECK

$$\frac{2}{x} = \frac{3}{x-2} - 1 \qquad\qquad \text{Write original equation.}$$

$$\frac{2}{4} \stackrel{?}{=} \frac{3}{4-2} - 1 \qquad\qquad \text{Substitute 4 for } x.$$

$$\frac{1}{2} = \frac{3}{2} - 1 \qquad\qquad \text{4 checks. } \checkmark$$

$$\frac{2}{-1} \stackrel{?}{=} \frac{3}{-1-2} - 1 \qquad\qquad \text{Substitute } -1 \text{ for } x.$$

$$-2 = -1 - 1 \qquad\qquad -1 \text{ checks. } \checkmark$$

The solutions are $x = 4$ and $x = -1$.

✓ **CHECKPOINT 6**

Solve $\dfrac{3}{x} + \dfrac{1}{x-2} = 2$.

To solve an equation involving an absolute value, remember that the expression inside the absolute value signs can be positive or negative. This results in *two* separate equations, each of which must be solved. For instance, the equation

$$|x - 2| = 3$$

results in the two equations

$$x - 2 = 3 \quad \text{and} \quad -(x - 2) = 3$$

which implies that the original equation has two solutions: $x = 5$ and $x = -1$. When setting up the negative expression, it is important to remember to place parentheses around the entire expression that is inside the absolute value bars. After you set up the two equations, solve each one independently.

Example 7   An Equation Involving Absolute Value

Solve $|x^2 - 3x| = -4x + 6$.

SOLUTION   Because the variable expression inside the absolute value signs can be positive or negative, you must solve the following two equations.

First Equation

$x^2 - 3x = -4x + 6$	Use positive expression.
$x^2 + x - 6 = 0$	Write in general form.
$(x + 3)(x - 2) = 0$	Factor.
$x + 3 = 0 \implies x = -3$	Set 1st factor equal to 0.
$x - 2 = 0 \implies x = 2$	Set 2nd factor equal to 0.

Second Equation

$-(x^2 - 3x) = -4x + 6$	Use negative expression.
$x^2 - 7x + 6 = 0$	Write in general form.
$(x - 1)(x - 6) = 0$	Factor.
$x - 1 = 0 \implies x = 1$	Set 1st factor equal to 0.
$x - 6 = 0 \implies x = 6$	Set 2nd factor equal to 0.

CHECK

$|(-3)^2 - 3(-3)| = -4(-3) + 6$	-3 checks. ✓
$|(2)^2 - 3(2)| \neq -4(2) + 6$	2 does not check.
$|(1)^2 - 3(1)| = -4(1) + 6$	1 checks. ✓
$|(6)^2 - 3(6)| \neq -4(6) + 6$	6 does not check.

The solutions are $x = -3$ and $x = 1$.

✓ CHECKPOINT 7

Solve $|x^2 - 3| = 5x - 3$. ■

Applications

It would be virtually impossible to categorize all of the many different types of applications that involve nonlinear and nonquadratic models. However, from the few examples and exercises that follow, we hope you will gain some appreciation for the variety of applications that involve such models.

Example 8   Reduced Rates

A ski club charters a bus for a ski trip at a cost of $700. In an attempt to lower the bus fare per skier, the club invites five nonmembers to go along. As a result, the fare per skier decreases by $7. How many club members are going on the trip?

SOLUTION   Begin the solution by creating a verbal model and assigning labels, as follows.

Stockbyte/Getty Images

Verbal Model: $\boxed{\text{Cost per skier}} \cdot \boxed{\text{Number of skiers}} = \boxed{\text{Cost of trip}}$

Labels:
Cost of trip = 700   (dollars)
Number of ski club members = x (people)
Number of skiers = $x + 5$ (people)
Original cost per member = $\dfrac{700}{x}$ (dollars per person)
Cost per skier = $\dfrac{700}{x} - 7$ (dollars per person)

Equation:

$$\left(\frac{700}{x} - 7\right)(x + 5) = 700$$ Original equation

$$\left(\frac{700 - 7x}{x}\right)(x + 5) = 700$$ Rewrite first factor.

$$(700 - 7x)(x + 5) = 700x, \; x \neq 0$$ Multiply each side by x.

$$700x + 3500 - 7x^2 - 35x = 700x$$ Multiply factors.

$$-7x^2 - 35x + 3500 = 0$$ Subtract $700x$ from each side.

$$x^2 + 5x - 500 = 0$$ Divide each side by -7.

$$(x + 25)(x - 20) = 0$$ Factor left side of equation.

$$x + 25 = 0 \implies x = -25$$ Set 1st factor equal to 0.

$$x - 20 = 0 \implies x = 20$$ Set 2nd factor equal to 0.

Only the positive x-value makes sense in the context of the problem. So, you can conclude that 20 ski club members are going on the trip. Check this in the original statement of the problem. ───────

✓ CHECKPOINT 8

In Example 8, suppose the ski club invites eight nonmembers to join the trip. As a result, the fare per skier decreases by $10. How many club members are going on the trip? ∎

Interest earned on a savings account is calculated by one of three basic methods: simple interest, interest compounded n times per year, and interest compounded continuously. The next example uses the formula for interest that is compounded n times per year,

$$A = P\left(1 + \frac{r}{n}\right)^{nt}.$$

In this formula, A is the balance in the account, P is the principal (or original deposit), r is the annual interest rate (in decimal form), n is the number of compoundings per year, and t is the time in years. In Chapter 4, you will study the derivation of this formula for compound interest.

Example 9 Compound Interest

When you were born, your grandparents deposited $5000 in a savings account earning interest compounded quarterly. On your 25th birthday the balance of the account is $25,062.59. What is the average annual interest rate of the account?

SOLUTION

Formula: $A = P\left(1 + \dfrac{r}{n}\right)^{nt}$

Labels: Balance $= A = 25{,}062.59$ (dollars)
Principal $= P = 5000$ (dollars)
Time $= t = 25$ (years)
Compoundings per year $= n = 4$ (compoundings)
Annual interest rate $= r$ (percent in decimal form)

Equation: $25{,}062.59 = 5000\left(1 + \dfrac{r}{4}\right)^{4(25)}$ Substitute.

$\dfrac{25{,}062.59}{5000} = \left(1 + \dfrac{r}{4}\right)^{100}$ Divide each side by 5000.

$5.0125 \approx \left(1 + \dfrac{r}{4}\right)^{100}$ Use a calculator.

$(5.0125)^{1/100} \approx 1 + \dfrac{r}{4}$ Raise each side to reciprocal power.

$1.01625 \approx 1 + \dfrac{r}{4}$ Use a calculator.

$0.01625 \approx \dfrac{r}{4}$ Subtract 1 from each side.

$0.065 \approx r$ Multiply each side by 4.

The average annual interest rate is about $0.065 = 6.5\%$. Check this in the original statement of the problem.

✓ CHECKPOINT 9

You placed $1000 in an account earning interest compounded monthly. After 3 years, the account balance is $1144.25. What is the annual interest rate? ■

Example 10 **Market Research**

The marketing department of a publishing company is asked to determine the price of a book. The department determines that the demand for the book depends on the price of the book according to the model

$$p = 40 - \sqrt{0.0001x + 1}, \quad 0 \le x \le 15{,}990{,}000$$

where p is the price per book in dollars and x is the number of books sold at the given price. For instance, in Figure 1.13, note that if the price were \$39, then (according to the model) no one would be willing to buy the book. On the other hand, if the price were \$17.60, 5 million copies could be sold. The publisher set the price at \$12.95. How many copies can the publisher expect to sell?

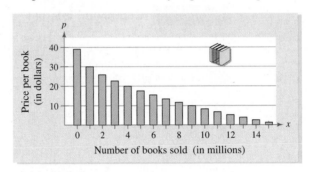

FIGURE 1.13

SOLUTION

$p = 40 - \sqrt{0.0001x + 1}$	Write given model.
$12.95 = 40 - \sqrt{0.0001x + 1}$	Set price at \$12.95.
$\sqrt{0.0001x + 1} = 27.05$	Isolate the radical.
$0.0001x + 1 = 731.7025$	Square each side.
$0.0001x = 730.7025$	Subtract 1 from each side.
$x = 7{,}307{,}025$	Divide each side by 0.0001.

So, by setting the book's price at \$12.95, the publisher can expect to sell about 7.3 million copies.

✓ **CHECKPOINT 10**

In Example 10, suppose the publisher set the price at \$19.95. How many copies can the publisher expect to sell? ■

┌───┐

CONCEPT CHECK

1. What method would you use to solve $x^3 + 3x^2 - 9x - 27 = 0$?

2. Explain why $x^6 + 2x^3 + 1 = 0$ is of the quadratic type, but $x^4 + 3x + 2 = 0$ is not.

3. How do you introduce an extraneous solution when solving $\sqrt{2x} + 4 = x$?

4. What two equations do you need to write in order to solve $|3x^2 - 5x| = 5$?

└───┘

Skills Review 1.5   The following warm-up exercises involve skills that were covered in earlier sections. You will use these skills in the exercise set for this section. For additional help, review Sections 1.3 and 1.4.

In Exercises 1–10, find the real solution(s) of the equation.

1. $x^2 - 22x + 121 = 0$

2. $x(x - 20) + 3(x - 20) = 0$

3. $(x + 20)^2 = 625$

4. $5x^2 + x = 0$

5. $3x^2 + 4x - 4 = 0$

6. $12x^2 + 8x - 55 = 0$

7. $x^2 + 4x - 5 = 0$

8. $4x^2 + 4x - 15 = 0$

9. $x^2 - 3x + 1 = 0$

10. $x^2 - 4x + 2 = 0$

Exercises 1.5

See www.CalcChat.com for worked-out solutions to odd-numbered exercises.

In Exercises 1–20, find the real solution(s) of the polynomial equation. Check your solutions.

1. $x^3 - 2x^2 - 3x = 0$

2. $20x^3 - 125x = 0$

3. $4x^4 - 18x^2 = 0$

4. $2x^4 - 15x^3 + 18x^2 = 0$

5. $x^4 - 81 = 0$

6. $x^6 - 64 = 0$

7. $5x^3 + 30x^2 + 45x = 0$

8. $9x^4 - 24x^3 + 16x^2 = 0$

9. $x^3 - 7x^2 - 4x + 28 = 0$

10. $x^3 + 2x^2 + 3x + 6 = 0$

11. $x^4 - x^3 + x - 1 = 0$

12. $x^4 + 2x^3 - 8x - 16 = 0$

13. $x^4 - 12x^2 + 11 = 0$

14. $x^4 - 29x^2 + 100 = 0$

15. $x^4 + 5x^2 - 36 = 0$

16. $x^4 - 4x^2 + 3 = 0$

17. $4x^4 - 65x^2 + 16 = 0$

18. $36t^4 + 29t^2 - 7 = 0$

19. $x^6 + 7x^3 - 8 = 0$

20. $x^6 + 3x^3 + 2 = 0$

In Exercises 21–34, find the real solution(s) of the radical equation. Check your solutions.

21. $\sqrt{2x} - 10 = 0$

22. $4\sqrt{x} - 3 = 0$

23. $\sqrt{x - 10} - 4 = 0$

24. $\sqrt{5 - x} - 3 = 0$

25. $\sqrt[3]{2x + 5} + 3 = 0$

26. $\sqrt[3]{3x + 1} - 5 = 0$

27. $2x + 9\sqrt{x} - 5 = 0$

28. $6x - 7\sqrt{x} - 3 = 0$

29. $x = \sqrt{11x - 30}$

30. $2x - \sqrt{15 - 4x} = 0$

31. $-\sqrt{26 - 11x} + 4 = x$

32. $x + \sqrt{31 - 9x} = 5$

33. $\sqrt{x + 1} - 3x = 1$

34. $\sqrt{2x + 1} + x = 7$

In Exercises 35–40, find the real solution(s) of the equation involving rational exponents. Check your solutions.

35. $(x - 5)^{2/3} = 16$

36. $(x + 3)^{4/3} = 16$

37. $(x + 3)^{3/2} = 8$

38. $(x^2 + 2)^{2/3} = 9$

39. $(x^2 - 5)^{2/3} = 16$

40. $(x^2 - x - 22)^{4/3} = 16$

In Exercises 41–48, find the real solution(s) of the equation involving fractions. Check your solutions.

41. $\dfrac{1}{x} - \dfrac{1}{x + 1} = 3$

42. $\dfrac{x}{x^2 - 4} + \dfrac{1}{x + 2} = 3$

43. $\dfrac{20 - x}{x} = x$

44. $\dfrac{4}{x} - \dfrac{5}{3} = \dfrac{x}{6}$

45. $\dfrac{1}{x} = \dfrac{4}{x - 1} + 1$

46. $x + \dfrac{9}{x + 1} = 5$

47. $\dfrac{4}{x + 1} - \dfrac{3}{x + 2} = 1$

48. $\dfrac{x + 1}{3} - \dfrac{x + 1}{x + 2} = 0$

In Exercises 49–56, find the real solution(s) of the equation involving absolute value. Check your solutions.

49. $|x + 1| = 2$

50. $|x - 2| = 3$

51. $|2x - 1| = 5$

52. $|3x + 2| = 7$

53. $|x| = x^2 + x - 3$

54. $|x^2 + 6x| = 3x + 18$

55. $|x - 10| = x^2 - 10x$

56. $|x + 1| = x^2 - 5$

57. Error Analysis   Find the error(s) in the solution.

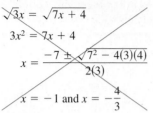

$$\sqrt{3}x = \sqrt{7x + 4}$$
$$3x^2 = 7x + 4$$
$$x = \frac{-7 \pm \sqrt{7^2 - 4(3)(4)}}{2(3)}$$
$$x = -1 \text{ and } x = -\frac{4}{3}$$

58. Error Analysis   Find the error(s) in the solution.

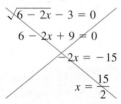

$$\sqrt{6 - 2x} - 3 = 0$$
$$6 - 2x + 9 = 0$$
$$-2x = -15$$
$$x = \frac{15}{2}$$

In Exercises 59–62, use a calculator to find the real solutions of the equation. (Round your answers to three decimal places.)

59. $3.2x^4 - 1.5x^2 - 2.1 = 0$

60. $7.08x^6 + 4.15x^3 - 9.6 = 0$

61. $1.8x - 6\sqrt{x} - 5.6 = 0$

62. $4x + 8\sqrt{x} + 3.6 = 0$

63. Sharing the Cost   A college charters a bus for $1700 to take a group of students to the Fiesta Bowl. When six more students join the trip, the cost per student decreases by $7.50. How many students were in the original group?

64. Sharing the Cost   Three students plan to share equally in the rent of an apartment. By adding a fourth person, each person could save $125 a month. How much is the monthly rent of the apartment?

65. Compound Interest   A deposit of $3000 reaches a balance of $4296.16 after 6 years. The interest on the account is compounded monthly. What is the annual interest rate for this investment?

66. Compound Interest   A sales representative describes a "guaranteed investment fund" that is offered to new investors. You are told that if you deposit $15,000 in the fund you will be guaranteed to receive a total of at least $40,000 after 20 years. (a) If after 20 years you received the minimum guarantee, what annual interest rate did you receive? (b) If after 20 years you received $48,000, what annual interest rate did you receive? (Assume that the interest in the fund is compounded quarterly.)

67. Borrowing Money   You borrow $300 from a friend and agree to pay the money back, plus $20 in interest, after 6 months. Assuming that the interest is compounded monthly, what annual interest rate are you paying?

68. Cash Advance   You take out a cash advance of $1000 on a credit card. After 2 months, you owe $1041.93. The interest is compounded monthly. What is the annual interest rate for this cash advance?

69. Airline Passengers   An airline offers daily flights between Chicago and Denver. The total monthly cost C (in millions of dollars) of these flights is modeled by

$$C = \sqrt{0.25x + 1}$$

where x is the number of passengers flying that month in thousands (see figure). The total cost of the flights for a month is 3.5 million dollars. Use the model to determine how many passengers flew that month.

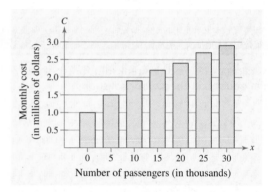

Number of passengers (in thousands)

70. Life Expectancy   The life expectancy of a person who is 16 to 25 years old can be modeled by

$$y = \sqrt{1.244x^2 - 161.16x + 6138.6}, \quad 16 \le x \le 25$$

where y represents the number of additional years the person is expected to live and x represents the person's current age. *(Source: U.S. National Center for Health Statistics)*

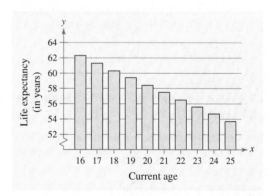

Current age

(a) Determine the life expectancies of persons who are 18, 20, and 22 years old.

(b) A person's life expectancy is 62 years. Use the model to determine the age of the person.

71. Life Expectancy   The life expectancy of a person who is 48 to 65 years old can be modeled by

$$y = \sqrt{0.874x^2 - 140.07x + 5752.5}, \quad 48 \le x \le 65$$

where y represents the number of additional years the person is expected to live and x represents the person's current age. A person's life expectancy is 20 years. How old is the person? *(Source: U.S. National Center for Health Statistics)*

72. New Homes   The number of new privately owned housing projects H (in thousands) started from 2000 to 2005 can be modeled by

$$H = -1993 + 204.9t + \frac{15{,}005}{t}, \quad 10 \le t \le 15$$

where t represents the year, with $t = 10$ corresponding to 2000 (see figure). Use the model to predict the year in which about 2,500,000 new housing projects were started. *(Source: U.S. Census Bureau)*

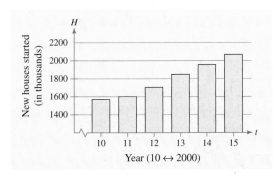

73. Market Research   The demand equation for a product is modeled by $p = 40 - \sqrt{0.01x + 1}$, where x is the number of units demanded per day and p is the price per unit. Find the demand when the price is set at $13.95. Explain why this model is only valid for $0 \le x \le 159{,}900$.

74. Power Line   A power station is on one side of a river that is $\frac{1}{2}$ mile wide. A factory is 6 miles downstream on the other side of the river. It costs $18 per foot to run power lines over land and $24 per foot to run them under water. The project's cost is $616,877.27. Find the length x as labeled in the figure.

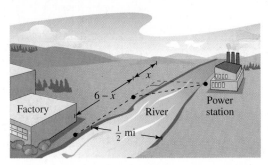

75. Sailboat Stays   Two stays for the mast on a sailboat are attached to the boat at two points, as shown in the figure. One point is 10 feet from the mast and the other point is 15 feet from the mast. The total length of the two stays is 35 feet. How high on the mast are the stays attached?

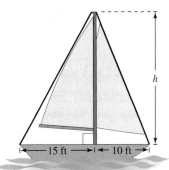

76. Flour Production   A company weighs each 16-ounce bag of flour it produces. After production, any bag that does not weigh within 0.4 ounce of 16 ounces cannot be sold. Solve the equation $|x - 16| = 0.4$ to find the least and greatest acceptable weights of a 16-ounce bag of flour.

77. Sugar Production   A company weighs each 80-ounce bag of sugar it produces. After production, any bag that does not weigh within 1.2 ounces of 80 ounces cannot be sold. Solve the equation $|x - 80| = 1.2$ to find the least and greatest acceptable weights of an 80-ounce bag of sugar.

78. Work Rate   With only the cold water valve open, it takes 8 minutes to fill the tub of a washing machine. With both the hot and cold water valves open, it takes 5 minutes. The time it takes for the tub to fill with only the hot water valve open can be modeled by the equation

$$\frac{1}{8} + \frac{1}{t} = \frac{1}{5}$$

where t is the time (in minutes) for the tub to fill. How long does it take for the tub of the washing machine to fill with only the hot water valve open?

79. Community Service   You and a friend volunteer to paint a small house as a community service project. Working alone, you can paint the house in 15 hours. Your friend can paint the house in 18 hours working alone. How long will it take both of you, working together, to paint the house?

80. Community Service   You and a friend volunteer to paint a large house as a community service project. Working alone, you can paint the house in 28 hours. Your friend can paint the house in 25 hours working alone. How long will it take both of you, working together, to paint the house?

Section 1.6

Linear Inequalities

■ Write bounded and unbounded intervals using inequalities or interval notation.
■ Solve and graph a linear inequality.
■ Construct and use a linear inequality to solve an application problem.

Introduction

Simple inequalities are used to *order* real numbers. The inequality symbols $<, \leq, >,$ and $\geq$ are used to compare two numbers and to denote subsets of real numbers. For instance, the simple inequality

$$x \geq 3$$

denotes all real numbers x that are greater than or equal to 3.

In this section you will expand your work with inequalities to include more involved statements such as

$$5x - 7 > 3x + 9 \quad \text{and} \quad -3 \leq 6x - 1 < 3.$$

As with an equation, you **solve an inequality** in the variable x by finding all values of x for which the inequality is true. Such values are **solutions** and are said to **satisfy** the inequality. The set of all real numbers that are solutions of an inequality is the **solution set** of the inequality. For instance, the solution set of

$$x + 3 > 4$$

is all real numbers that are greater than 1.

The set of all points on the real number line that represent the solution set of an inequality is the **graph** of the inequality. Graphs of many types of inequalities consist of intervals on the real number line. The four different types of **bounded** intervals are summarized below.

Bounded Intervals on the Real Number Line

Let a and b be real numbers such that $a < b$. The following intervals on the real number line are **bounded**. The numbers a and b are the **endpoints** of each interval.

Notation	Interval Type	Inequality	Graph
$[a, b]$	Closed	$a \leq x \leq b$	
(a, b)	Open	$a < x < b$	
$[a, b)$		$a \leq x < b$	
$(a, b]$		$a < x \leq b$	

Note that a closed interval contains both of its endpoints and an open interval does not contain either of its endpoints. Often, the solution of an inequality is an interval on the real line that is **unbounded.** For instance, the interval consisting of all positive numbers is unbounded. The symbols ∞, **positive infinity,** and $-\infty$, **negative infinity,** do not represent real numbers. They are simply convenient symbols used to describe the unboundedness of an interval such as $(1, \infty)$.

Unbounded Intervals on the Real Number Line

Let a and b be real numbers. The following intervals on the real number line are **unbounded.**

Notation	Interval Type	Inequality	Graph
$[a, \infty)$		$x \geq a$	
(a, ∞)	Open	$x > a$	
$(-\infty, b]$		$x \leq b$	
$(-\infty, b)$	Open	$x < b$	
$(-\infty, \infty)$	Entire real line	$-\infty < x < \infty$	

Example 1   **Intervals and Inequalities**

Write an inequality to represent each of the following intervals. Then state whether the interval is bounded or unbounded.

a. $(-3, 5]$

b. $(-3, \infty)$

c. $[0, 2]$

d. $(-\infty, 0)$

SOLUTION

a. $(-3, 5]$ corresponds to $-3 < x \leq 5$.   Bounded

b. $(-3, \infty)$ corresponds to $x > -3$.   Unbounded

c. $[0, 2]$ corresponds to $0 \leq x \leq 2$.   Bounded

d. $(-\infty, 0)$ corresponds to $x < 0$.   Unbounded

✓ **CHECKPOINT 1**

Write an inequality to represent each of the following intervals. Then state whether the interval is bounded or unbounded.

a. $[2, 7)$

b. $(-\infty, 3)$ ■

Properties of Inequalities

The procedures for solving linear inequalities in one variable are much like those for solving linear equations. To isolate the variable, you can make use of the **properties of inequalities.** These properties are similar to the properties of equality, but there are two important exceptions. When each side of an inequality is multiplied or divided by a negative number, the direction of the inequality symbol must be reversed. Here is an example.

$$-2 < 5 \qquad \text{Original inequality}$$

$$(-3)(-2) > (-3)(5) \qquad \text{Multiply each side by } -3 \text{ and reverse the inequality symbol.}$$

$$6 > -15 \qquad \text{Simplify.}$$

Two inequalities that have the same solution set are **equivalent.** For instance, the inequalities

$$x + 2 < 5 \quad \text{and} \quad x < 3$$

are equivalent. To obtain the second inequality from the first, you can subtract 2 from each side of the inequality. The following list describes operations that can be used to create equivalent inequalities.

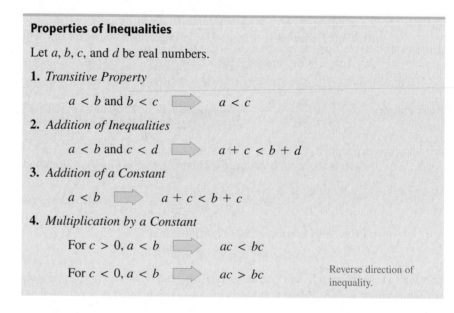

Properties of Inequalities

Let a, b, c, and d be real numbers.

1. *Transitive Property*

$$a < b \text{ and } b < c \quad \Longrightarrow \quad a < c$$

2. *Addition of Inequalities*

$$a < b \text{ and } c < d \quad \Longrightarrow \quad a + c < b + d$$

3. *Addition of a Constant*

$$a < b \quad \Longrightarrow \quad a + c < b + c$$

4. *Multiplication by a Constant*

$$\text{For } c > 0, a < b \quad \Longrightarrow \quad ac < bc$$

$$\text{For } c < 0, a < b \quad \Longrightarrow \quad ac > bc \qquad \text{Reverse direction of inequality.}$$

Each of the properties above is true if the symbol $<$ is replaced by $\leq$ and the symbol $>$ is replaced by $\geq$. For instance, another form of the multiplication property would be as follows.

$$\text{For } c > 0, a \leq b \quad \Longrightarrow \quad ac \leq bc.$$

$$\text{For } c < 0, a \leq b \quad \Longrightarrow \quad ac \geq bc.$$

On your own, try to verify each of the properties of inequalities by using several examples with real numbers.

Solving a Linear Inequality

The simplest type of inequality to solve is a **linear inequality** in a single variable. For instance, $2x + 3 > 4$ is a linear inequality in x.

As you read through the following examples, pay special attention to the steps in which the inequality symbol is reversed. Remember that when you multiply or divide by a negative number, you must reverse the inequality symbol.

Example 2 | Solving a Linear Inequality

Solve $5x - 7 > 3x + 9$.

SOLUTION

$5x - 7 > 3x + 9$	Write original inequality.
$2x - 7 > 9$	Subtract $3x$ from each side.
$x > 8$	Add 7 to each side and then divide each side by 2.

The solution set is all real numbers that are greater than 8, which is denoted by $(8, \infty)$. The graph is shown in Figure 1.14.

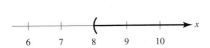

FIGURE 1.14 Solution Interval: $(8, \infty)$

✓ CHECKPOINT 2

Solve $3x < 2x + 1$. ■

Checking the solution set of an inequality is not as simple as checking the solutions of an equation. You can, however, get an indication of the validity of a solution set by substituting a few convenient values of x to see whether the original inequality is satisfied.

Example 3 | Solving a Linear Inequality

Solve $1 - \dfrac{3x}{2} \geq x - 4$.

SOLUTION

$1 - \dfrac{3x}{2} \geq x - 4$	Write original inequality.
$2 - 3x \geq 2x - 8$	Multiply each side by 2.
$2 - 5x \geq -8$	Subtract $2x$ from each side.
$-5x \geq -10$	Subtract 2 from each side.
$x \leq 2$	Divide each side by -5 and reverse inequality.

The solution set is all real numbers that are less than or equal to 2, which is denoted by $(-\infty, 2]$. The graph is shown in Figure 1.15.

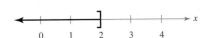

FIGURE 1.15 Solution Interval: $(-\infty, 2]$

✓ CHECKPOINT 3

Solve $-\dfrac{4x}{3} \leq 2 - x$. Then graph the solution set on the real number line. ■

Sometimes it is convenient to write two inequalities as a **double inequality.** For instance, you can write the two inequalities $-4 \le 5x - 2$ and $5x - 2 < 7$ more simply as

$$-4 \le 5x - 2 < 7.$$

This enables you to solve the two inequalities together, as demonstrated in Example 4.

Example 4   Solving a Double Inequality

Solve $-3 \le 6x - 1 < 3$.

SOLUTION   To solve a double inequality, you can isolate x as the middle term.

$-3 \le 6x - 1 < 3$	Write original inequality.
$-3 + 1 \le 6x - 1 + 1 < 3 + 1$	Add 1 to each part.
$-2 \le 6x < 4$	Simplify.
$\dfrac{-2}{6} \le \dfrac{6x}{6} < \dfrac{4}{6}$	Divide each part by 6.
$-\dfrac{1}{3} \le x < \dfrac{2}{3}$	Simplify.

The solution set is all real numbers that are greater than or equal to $-\frac{1}{3}$ and less than $\frac{2}{3}$. The interval notation for this solution set is

$$\left[-\tfrac{1}{3}, \tfrac{2}{3}\right).$$ Solution set

The graph of this solution set is shown in Figure 1.16.

FIGURE 1.16   Solution Interval: $\left[-\tfrac{1}{3}, \tfrac{2}{3}\right)$.

✓ CHECKPOINT 4

Solve $-1 < 3 - 2x \le 5$. Then graph the solution set on the real number line. ■

The double inequality in Example 4 could have been solved in two parts as follows.

$$-3 \le 6x - 1 \quad \text{and} \quad 6x - 1 < 3$$
$$-2 \le 6x \qquad\qquad\quad 6x < 4$$
$$-\frac{1}{3} \le x \qquad\qquad\quad x < \frac{2}{3}$$

The solution set consists of all real numbers that satisfy *both* inequalities. In other words, the solution set is the set of all values of x for which $-\frac{1}{3} \le x < \frac{2}{3}$.

When combining two inequalities to form a double inequality, be sure that the inequalities satisfy the Transitive Property. For instance, it is *incorrect* to combine the inequalities $3 < x$ and $x \le -1$ as $3 < x \le -1$. This "inequality" is obviously wrong because 3 is not less than -1.

Inequalities Involving Absolute Value

STUDY TIP

When an absolute value inequality is solved, the solution may consist of a bounded interval on the real number line (Example 5) or two unbounded intervals on the real number line (Example 6).

Solving an Absolute Value Inequality

Let x be a variable or an algebraic expression and let a be a real number such that $a \geq 0$.

1. The solutions of $|x| < a$ are all values of x that lie between $-a$ and a.

   $|x| < a$ if and only if $-a < x < a$.

2. The solutions of $|x| > a$ are all values of x that are less than $-a$ or greater than a.

   $|x| > a$ if and only if $x < -a$ or $x > a$.

These rules are also valid if $<$ is replaced by $\leq$ and $>$ is replaced by $\geq$.

Example 5 Solving an Absolute Value Inequality

Solve $|x - 5| < 2$.

SOLUTION

$\lvert x - 5\rvert < 2$	Write original inequality.
$-2 < x - 5 < 2$	Equivalent inequality
$-2 + 5 < x - 5 + 5 < 2 + 5$	Add 5 to each part.
$3 < x < 7$	Simplify.

The solution set consists of all real numbers that are greater than 3 and less than 7, which is denoted by $(3, 7)$. The graph is shown in Figure 1.17.

FIGURE 1.17

✓CHECKPOINT 5

Solve $|x + 2| \leq 7$. Then graph the solution set on the real number line.

Example 6 Solving an Absolute Value Inequality

Solve $|x + 3| \geq 7$.

SOLUTION

$\lvert x + 3\rvert \geq 7$			Write original inequality.
$x + 3 \leq -7$	or	$x + 3 \geq 7$	Equivalent inequalities
$x + 3 - 3 \leq -7 - 3$		$x + 3 - 3 \geq 7 - 3$	Subtract 3 from each side.
$x \leq -10$		$x \geq 4$	Simplify.

The solution set is all real numbers that are less than or equal to -10 *or* greater than or equal to 4, which is denoted by $(-\infty, -10] \cup [4, \infty)$ (see Figure 1.18). The symbol $\cup$ (union) means *or*.

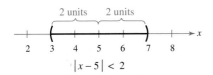

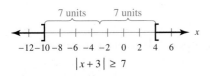

FIGURE 1.18

✓CHECKPOINT 6

Solve $|x + 1| > 3$. Then graph the solution set on the real number line.

Applications

Example 7 **Comparative Shopping**

The cost of renting a compact car from Company A is $200 per week with no extra charge for mileage. The cost of renting a similar car from Company B is $110 per week, plus $0.25 for each mile driven. How many miles must you drive in a week to make the rental fee for Company B more than that for Company A?

SOLUTION

Verbal Model:	Weekly cost for Company B	>	Weekly cost for Company A

Labels: Miles driven in one week $= m$ (miles)
            Weekly cost for Company A $= 200$ (dollars)
            Weekly cost for Company B $= 110 + 0.25m$ (dollars)

Inequality: $110 + 0.25m > 200$

$$0.25m > 90$$

$$m > 360$$

When you drive more than 360 miles in a week, the rental fee for Company B is more than the rental fee for Company A.

✓ **CHECKPOINT 7**

In Example 7, suppose the cost of renting a compact car from Company A is $250 per week with no extra charge for mileage. How many miles must you drive in a week to make the rental fee for Company B more than that for Company A? ■

Example 8 **Exercise Program**

A 225-pound man begins an exercise and diet program that is designed to reduce his weight by at least 2 pounds per week. Find the maximum number of weeks before the man's weight will reach his goal of 192 pounds.

SOLUTION

Verbal Model:	Desired weight	≤	Current weight	−	2 pounds per week	·	Number of weeks

Labels: Desired weight $= 192$ (pounds)
            Current weight $= 225$ (pounds)
            Number of weeks $= x$ (weeks)

Inequality: $192 \leq 225 - 2x$

$$-33 \leq -2x$$

$$16.5 \geq x$$

Losing at least 2 pounds per week, it will take at most $16\frac{1}{2}$ weeks for the man to reach his goal.

✓ **CHECKPOINT 8**

In Example 8, find the maximum number of weeks before the man's weight will reach 200 pounds. ■

Americans pay to be lean and fit. In 2005, Americans spent over $5 billion on exercise equipment. (*Source: National Sporting Goods Association*)

Example 9 **Accuracy of a Measurement**

You go to a candy store to buy chocolates that cost $9.89 per pound. The scale used in the store has a state seal of approval that indicates the scale is accurate to within half an ounce. According to the scale, your purchase weighs one-half pound and costs $4.95. How much might you have been undercharged or over charged due to an error in the scale?

SOLUTION To solve this problem, let x represent the *true* weight of the candy. Because the scale is accurate to within one-half an ounce (or $\frac{1}{32}$ of a pound), you can conclude that the absolute value of the difference between the exact weight (x) and the scale weight $\left(\frac{1}{2}\text{ of a pound}\right)$ is less than or equal to $\frac{1}{32}$ of a pound. That is,

$$\left| x - \frac{1}{2} \right| \le \frac{1}{32}.$$

You can solve this inequality as follows.

$$-\frac{1}{32} \le x - \frac{1}{2} \le \frac{1}{32}$$

$$\frac{15}{32} \le x \le \frac{17}{32}$$

$$0.46875 \le x \le 0.53125$$

In other words, your "one-half" pound of candy could have weighed as little as 0.46875 pound (which would have cost $0.46875 \cdot \$9.89 \approx \$4.64$) or as much as 0.53125 pound (which would have cost $0.53125 \cdot \$9.89 \approx \$5.25$). So, you could have been undercharged by as much as $0.30 or overcharged by as much as $0.31.

✓**CHECKPOINT 9**

You go to a grocery store to buy ground beef that costs $3.96 per pound. The scale used in the store is accurate to within $\frac{1}{3}$ ounce $\left(\text{or }\frac{1}{48}\text{ pound}\right)$. According to the scale, your purchase weighs 7.5 pounds and costs $29.70. How much might you have been undercharged or overcharged due to an error in the scale? ■

CONCEPT CHECK

1. Write an inequality for all values of x that lie between -6 and 8. Is the solution set bounded or unbounded? Explain.

2. Suppose $2x + 1 > 5$ and $y - 8 < 5$. Is it always true that $2x + 1 > y - 8$? Explain.

3. If $x < 12$, then $-x$ must be in what interval?

4. The solution set of an absolute value inequality is $(-\infty, -a] \cup [a, \infty)$. Is the inequality $|x| \le a$ or $|x| \ge a$?

Skills Review 1.6   The following warm-up exercises involve skills that were covered in earlier sections. You will use these skills in the exercise set for this section. For additional help, review Section 0.1.

In Exercises 1–4, determine which of the two numbers is larger.

1. $-\frac{1}{2}, -7$

2. $-\frac{1}{3}, -\frac{1}{6}$

3. $-\pi, -3$

4. $-6, -\frac{13}{2}$

In Exercises 5–8, use inequality notation to denote the statement.

5. x is nonnegative.

6. z is strictly between -3 and 10.

7. P is no more than 2.

8. W is at least 200.

In Exercises 9 and 10, evaluate the expression for the values of x.

9. $|x - 10|, x = 12, x = 3$

10. $|2x - 3|, x = \frac{3}{2}, x = 1$

Exercises 1.6

See www.CalcChat.com for worked-out solutions to odd-numbered exercises.

In Exercises 1–6, write an inequality that represents the interval. Then state whether the interval is bounded or unbounded.

1. $[-1, 5]$

2. $(2, 10]$

3. $(11, \infty)$

4. $[-5, \infty)$

5. $(-\infty, -2)$

6. $(-\infty, 7]$

In Exercises 7–14, match the inequality with its graph. [The graphs are labeled (a), (b), (c), (d), (e), (f), (g), and (h).]

7. $x < 4$

8. $x \geq 6$

9. $-2 < x \leq 5$

10. $0 \leq x \leq \frac{7}{2}$

11. $|x| < 4$

12. $|x| > 3$

13. $|x - 5| > 2$

14. $|x + 6| < 3$

(a)
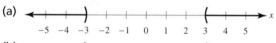

(b)

(c)

(d)

(e)

(f)

(g)

(h)

In Exercises 15–22, determine whether each value of x is a solution of the inequality.

15. $5x - 12 > 0$

  (a) $x = 3$ (b) $x = -3$ (c) $x = \frac{5}{2}$ (d) $x = \frac{3}{2}$

16. $x + 1 < \dfrac{2x}{3}$

  (a) $x = 0$ (b) $x = 4$ (c) $x = -4$ (d) $x = -3$

17. $0 < \dfrac{x - 2}{4} < 2$

  (a) $x = 4$ (b) $x = 10$ (c) $x = 0$ (d) $x = \frac{7}{2}$

18. $-1 < \dfrac{3 - x}{2} \leq 1$

  (a) $x = 0$ (b) $x = -5$ (c) $x = 1$ (d) $x = 5$

19. $|x - 10| \geq 3$

  (a) $x = 13$ (b) $x = -1$ (c) $x = 14$ (d) $x = 9$

20. $|3x + 5| > 7$

  (a) $x = -5$ (b) $x = -2$ (c) $x = \frac{1}{3}$ (d) $x = 10$

21. $|x + 2| \leq 10$

  (a) $x = -15$ (b) $x = -4$ (c) $x = 1$ (d) $x = 8$

22. $|2x - 3| < 15$

  (a) $x = -6$ (b) $x = 0$ (c) $x = 12$ (d) $x = 7$

In Exercises 23–30, copy and complete the statement using the correct inequality symbol.

23. If $2x > 6$, then x __?__ 3.

24. If $3x > 9$, then x __?__ 3.

25. If $2x \leq -8$, then x __?__ -4.

26. If $3x \leq -15$, then x __?__ -5.

27. If $2 - 4x > -10$, then x __?__ 3.

28. If $5 - 3x > -7$, then x __?__ 4.

29. If $-\dfrac{2}{3}x \geq -6$, then x __?__ 9.

30. If $-\dfrac{3}{4}x \geq -12$, then x __?__ 16.

In Exercises 31–70, solve the inequality. Then graph the solution set on the real number line.

31. $\dfrac{3}{2}x \geq 9$

32. $\dfrac{2}{5}x > 7$

33. $-10x < 40$

34. $-6x > 15$

35. $\dfrac{3}{5}x - 7 < 8$

36. $\dfrac{5}{4}x + 1 \leq 11$

37. $2x + 7 < 3 + 4x$

38. $6x - 4 \leq 2 + 8x$

39. $2x - 1 \geq 5x$

40. $3x + 1 \geq 2 + x$

41. $3(x + 2) + 7 < 2x - 5$

42. $2(x + 7) - 4 \geq 5(x - 3)$

43. $-3(x - 1) + 7 < 2x + 8$

44. $5 - 3x > -5(x + 4) + 6$

45. $3 \leq 2x - 1 < 7$

46. $3 > 1 - \dfrac{x}{2} > -3$

47. $1 < 2x + 3 < 9$

48. $-8 \leq 1 - 3(x - 2) < 13$

49. $-4 < \dfrac{2x - 3}{3} < 4$

50. $0 \leq \dfrac{x + 3}{2} < 5$

51. $\dfrac{3}{4} > x + 1 > \dfrac{1}{4}$

52. $-1 < -\dfrac{x}{3} < 1$

53. $|x| < 6$

54. $|x| > 8$

55. $\left|\dfrac{x}{2}\right| > 3$

56. $|5x| > 10$

57. $|x + 3| < 5$

58. $\left|\dfrac{2x + 1}{2}\right| < 6$

59. $|x - 20| \leq 4$

60. $|x - 7| < 6$

61. $|2x - 5| > 6$

62. $2|5 - 3x| + 7 < 21$

63. $\left|\dfrac{x - 3}{2}\right| \geq 5$

64. $\left|1 - \dfrac{2x}{3}\right| < 1$

65. $|9 - 2x| - 2 < -1$

66. $|x + 14| + 3 > 17$

67. $2|x + 10| \geq 9$

68. $3|4 - 5x| \leq 9$

69. $|x - 5| < 0$

70. $|x - 5| \geq 0$

In Exercises 71–78, use absolute value notation to define the solution set.

71.

72.

73.

74.

75. All real numbers at most 10 units from 12

76. All real numbers at least 5 units from 8

77. All real numbers whose distances from -3 are more than 5

78. All real numbers whose distances from -6 are no more than 7

79. Comparative Shopping   The cost of renting a midsize car from Company A is $279 per week with no extra charge for mileage. The cost of renting a similar car from Company B is $199 per week, plus 32 cents for each mile driven. How many miles must you drive in a week to make the rental fee for Company B greater than that for Company A?

80. Comparative Shopping   Your department sends its copying to a photocopy center. The photocopy center bills your department $0.08 per page. You are considering buying a departmental copier for $2500. With your own copier the cost per page would be $0.025. The expected life of the copier is 4 years. How many copies must you make in the four-year period to justify purchasing the copier?

81. Simple Interest   For $1500 to grow to more than $1890 in 3 years, what must the simple interest rate be?

82. Simple Interest   For $2000 to grow to more than $2500 in 2 years, what must the simple interest rate be?

83. Weight Loss Program   A person enrolls in a diet program that guarantees a loss of at least $1\frac{1}{2}$ pounds per week. The person's weight at the beginning of the program is 180 pounds. Find the maximum number of weeks before the person attains a weight of 130 pounds.

84. Salary Increase   You accept a new job with a starting salary of $28,800. You are told that you will receive an annual raise of at least $1500. What is the maximum number of years you must work before your annual salary will be $40,000?

85. Maximum Width An overnight delivery service will not accept any package whose combined length and girth (perimeter of a cross section) exceeds 132 inches. Suppose that you are sending a rectangular package that has square cross sections. If the length of the package is 68 inches, what is the maximum width of the sides of its square cross sections?

86. Maximum Width An overnight delivery service will not accept any package whose combined length and girth (perimeter of a cross section) exceeds 126 inches. Suppose that you are sending a rectangular package that has square cross sections. If the length of the package is 66 inches, what is the maximum width of the sides of its square cross sections?

87. Break-Even Analysis The revenue R for selling x units of a product is

$$R = 139.95x.$$

The cost C of producing x units is

$$C = 97x + 850.$$

In order to obtain a profit, the revenue must be greater than the cost.

(a) Complete the table.

x	10	20	30	40	50	60
R						
C						

(b) For what values of x will this product return a profit?

88. Break-Even Analysis The revenue R for selling x units of a product is $R = 25.95x$. The cost C of producing x units is

$$C = 13.95x + 125,000.$$

In order to obtain a profit, the revenue must be greater than the cost. For what values of x will this product return a profit?

89. Annual Operating Cost A utility company has a fleet of vans. The annual operating cost C per van is

$$C = 0.32m + 2500$$

where m is the number of miles traveled by a van in a year. What number of miles will yield an annual operating cost that is less than $12,000?

90. Daily Sales A doughnut shop sells a dozen doughnuts for $3.95. Beyond the fixed costs (rent, utilities, and insurance) of $165 per day, it costs $1.45 for enough materials (flour, sugar, and so on) and labor to produce a dozen doughnuts. The daily profit from doughnut sales varies between $100 and $400. Between what numbers of doughnuts (in dozens) do the daily sales vary?

91. IQ Scores The admissions office of a college wants to determine whether there is a relationship between IQ scores x and grade-point averages y after the first year of school. An equation that models the data obtained by the admissions office is

$$y = 0.068x - 4.753.$$

Estimate the values of x that predict a grade-point average of at least 3.0.

92. *MAKE A DECISION: WEIGHTLIFTING* You want to determine whether there is a relationship between an athlete's weight x (in pounds) and the athlete's maximum bench-press weight y (in pounds). An equation that models the data you obtained is

$$y = 1.4x - 39.$$

(a) Estimate the values of x that predict a maximum bench-press weight of at least 200 pounds.

(b) Do you think an athlete's weight is a good indicator of the athlete's maximum bench-press weight? What other factors might influence an individual's bench-press weight?

93. Baseball Salaries The average professional baseball player's salary S (in millions of dollars) from 1995 to 2006 can be modeled by

$$S = 0.1527t + 0.294, \quad 5 \le t \le 16$$

where t represents the year, with $t = 5$ corresponding to 1995 (see figure). Use the model to predict the year in which the average professional baseball player's salary exceeds $3,000,000. *(Source: Major League Baseball)*

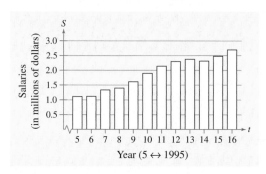

Year (5 ↔ 1995)

94. Public College Enrollment The projected public college enrollment E (in thousands) in the United States from 2005 to 2015 can be modeled by

$$E = 180.3t + 12,312, \quad 5 \le t \le 15$$

where t represents the year, with $t = 5$ corresponding to 2005 (see figure on next page). Use the model to predict the year in which public college enrollment will exceed 17,000,000. *(Source: U.S. National Center for Education Statistics)*

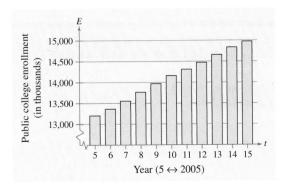

Figure for 94

95. Geometry You measure the side of a square as 10.4 inches with a possible error of $\frac{1}{16}$ inch. Using these measurements, determine the interval containing the possible areas of the square.

96. Geometry You measure the side of a square as 24.2 centimeters with a possible error of 0.25 centimeter. Using these measurements, determine the interval containing the possible areas of the square.

97. Accuracy of Measurement You buy six T-bone steaks that cost $7.99 per pound. The weight listed on the package is 5.72 pounds. The scale that weighed the package is accurate to within $\frac{1}{2}$ ounce. How much money might you have been undercharged or overcharged?

98. Accuracy of Measurement You stop at a self-service gas station to buy 15 gallons of 87-octane gasoline at $2.42 a gallon. The pump scale is accurate to within one-tenth of a gallon. How much money might you have been undercharged or overcharged?

99. Human Height The heights h of two-thirds of a population satisfy the inequality

$$|h - 68.5| \le 2.7$$

where h is measured in inches. Determine the interval on the real number line in which these heights lie.

100. Time Study A time study was conducted to determine the length of time required to perform a particular task in a manufacturing process. The times required by approximately two-thirds of the workers in the study satisfied the inequality

$$\left|\frac{t - 15.6}{1.9}\right| < 1$$

where t is time in minutes. Determine the interval on the real number line in which these times lie.

101. Humidity Control The specifications for an electronic device state that it is to be operated in a room with relative humidity h defined by $|h - 50| \le 30$. What are the minimum and maximum relative humidities for the operation of this device?

102. Body Temperature Physicians consider an adult's body temperature x (in degrees Fahrenheit) to be normal if it satisfies the inequality

$$|x - 98.6| \le 1.$$

Determine the range of temperatures that are considered to be normal.

103. Brand Name Drugs The average price B (in dollars) of brand name prescription drugs from 1998 to 2005 can be modeled by

$$B = 6.928t - 3.45, \quad 8 \le t \le 15$$

where t represents the year, with $t = 8$ corresponding to 1998. Use the model to find the year in which the price of the average brand name drug prescription exceeded $75. (Source: National Association of Chain Drug Stores)

104. Generic Drugs The average price G (in dollars) of generic prescription drugs from 1998 to 2005 can be modeled by

$$G = 2.005t + 0.40, \quad 8 \le t \le 15$$

where t represents the year, with $t = 8$ corresponding to 1998. Use the model to find the year in which the price of the average generic drug prescription exceeded $19. (Source: National Association of Chain Drug Stores)

105. Domestic Oil Demand The daily demand D (in thousands of barrels) for refined oil in the United States from 1995 to 2005 can be modeled by

$$D = 276.4t + 16,656, \quad 5 \le t \le 15$$

where t represents the year, with $t = 5$ corresponding to 1995. (Source: U.S. Energy Administration)

(a) Use the model to find the year in which the demand for U.S. oil exceeded 18 million barrels a day.

(b) Use the model to predict the year in which the demand for U.S. oil will exceed 22 million barrels a day.

106. Imported Oil The daily amount I (in thousands of barrels) of crude oil imported to the United States from 1995 to 2005 can be modeled by

$$I = 428.2t + 6976, \quad 5 \le t \le 15$$

where t represents the year, with $t = 5$ corresponding to 1995. (Source: U.S. Energy Administration)

(a) Use the model to find the year in which the amount of crude oil imported to the United States exceeded 12 million barrels a day.

(b) Use the model to predict the year in which the amount of oil imported to the United States will exceed 14 million barrels a day.

Section 1.7

Other Types of Inequalities

- ■ Use critical numbers to determine test intervals for a polynomial inequality.
- ■ Solve and graph a polynomial inequality.
- ■ Solve and graph a rational inequality.
- ■ Determine the domain of an expression involving a square root.
- ■ Construct and use a polynomial inequality to solve an application problem.

Polynomial Inequalities

To solve a polynomial inequality such as $x^2 - 2x - 3 < 0$, you can use the fact that a polynomial can change signs only at its **zeros** (the x-values that make the polynomial equal to zero). Between two consecutive zeros, a polynomial must be entirely positive or entirely negative. This means that when the real zeros of a polynomial are put in order, they divide the real number line into intervals in which the polynomial has no sign changes. These zeros are the **critical numbers** of the inequality, and the resulting intervals are the **test intervals** for the inequality. For example, the polynomial above factors as

$$x^2 - 2x - 3 = (x + 1)(x - 3)$$

and has two zeros, $x = -1$ and $x = 3$. These zeros divide the real number line into three test intervals:

$$(-\infty, -1), \quad (-1, 3), \quad \text{and} \quad (3, \infty). \qquad \text{(See Figure 1.19.)}$$

So, to solve the inequality $x^2 - 2x - 3 < 0$, you need only test one value from each of these test intervals.

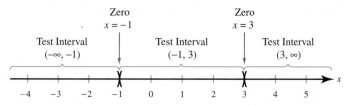

FIGURE 1.19   Three Test Intervals for $x^2 - 2x - 3 < 0$

STUDY TIP

If the value of the polynomial is negative at the representative x-value, the polynomial will have negative values for *every* x-value in the interval. If the value of the polynomial is positive, the polynomial will have positive values for *every* x-value in the interval.

Finding Test Intervals for a Polynomial

To determine the intervals on which the values of a polynomial are entirely negative or entirely positive, use the following steps.

1. Find all real zeros of the polynomial, and arrange the zeros in increasing order. These zeros are the **critical numbers** of the polynomial.

2. Use the critical numbers to determine the **test intervals.**

3. Choose one representative x-value in each test interval and evaluate the polynomial at that value.

Example 1   **Solving a Polynomial Inequality**

Solve $x^2 - x - 6 < 0$.

SOLUTION   By factoring the quadratic as

$$x^2 - x - 6 = (x + 2)(x - 3)$$

you can see that the critical numbers are $x = -2$ and $x = 3$. The boundaries between the numbers that satisfy the inequality and the numbers that do not satisfy the inequality always occur at critical numbers. So, the polynomial's test intervals are

$$(-\infty, -2), \quad (-2, 3), \quad \text{and} \quad (3, \infty). \qquad \text{Test intervals}$$

In each test interval, choose a representative x-value and evaluate the polynomial.

Test Interval	x-Value	Polynomial Value	Conclusion
$(-\infty, -2)$	$x = -3$	$(-3)^2 - (-3) - 6 = 6$	Positive
$(-2, 3)$	$x = 0$	$(0)^2 - (0) - 6 = -6$	Negative
$(3, \infty)$	$x = 4$	$(4)^2 - (4) - 6 = 6$	Positive

From this, you can conclude that the polynomial is positive for all x-values in $(-\infty, -2)$ and $(3, \infty)$, and is negative for all x-values in $(-2, 3)$. This implies that the solution of the inequality $x^2 - x - 6 < 0$ is the interval $(-2, 3)$, as shown in Figure 1.20.

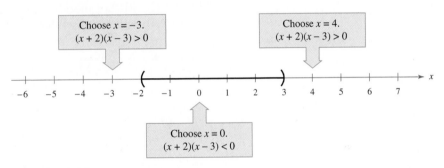

FIGURE 1.20

✓CHECKPOINT 1

Solve $x^2 + x - 2 < 0$. ▨

As with linear inequalities, you can check a solution interval of a polynomial inequality by substituting x-values into the original inequality. For instance, to check the solution found in Example 1, try substituting several x-values from the interval $(-2, 3)$ into the inequality

$$x^2 - x - 6 < 0.$$

Regardless of which x-values you choose, the inequality will be satisfied.

In Example 1, the polynomial inequality was given in general form. Whenever this is not the case, begin the solution process by writing the inequality in general form—with the polynomial on one side and zero on the other.

Example 2 Solving a Polynomial Inequality

Solve $x^3 - 3x^2 > 10x$.

SOLUTION

$$x^3 - 3x^2 > 10x \qquad \text{Write original inequality.}$$

$$x^3 - 3x^2 - 10x > 0 \qquad \text{Write in general form.}$$

$$x(x - 5)(x + 2) > 0 \qquad \text{Factor.}$$

You can see that the critical numbers are $x = -2$, $x = 0$, and $x = 5$, and the test intervals are $(-\infty, -2)$, $(-2, 0)$, $(0, 5)$, and $(5, \infty)$. In each test interval, choose a representative x-value and evaluate the polynomial.

Test Interval	x-Value	Polynomial Value	Conclusion
$(-\infty, -2)$	$x = -3$	$(-3)^3 - 3(-3)^2 - 10(-3) = -24$	Negative
$(-2, 0)$	$x = -1$	$(-1)^3 - 3(-1)^2 - 10(-1) = 6$	Positive
$(0, 5)$	$x = 2$	$2^3 - 3(2)^2 - 10(2) = -24$	Negative
$(5, \infty)$	$x = 6$	$6^3 - 3(6)^2 - 10(6) = 48$	Positive

From this, you can conclude that the inequality is satisfied on the open intervals $(-2, 0)$ and $(5, \infty)$. So, the solution set consists of all real numbers in the intervals $(-2, 0)$ and $(5, \infty)$, as shown in Figure 1.21.

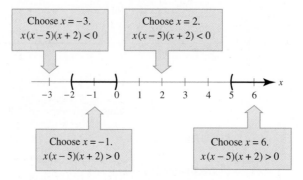

FIGURE 1.21

✓ CHECKPOINT 2

Solve $x^2 - 3x > -2$. ■

When solving a polynomial inequality, be sure to account for the type of inequality symbol given in the inequality. For instance, in Example 2, note that the solution consisted of two *open* intervals because the original inequality contained a "greater than" symbol. If the original inequality had been $x^3 - 3x^2 \geq 10x$, the solution would have consisted of the *closed* interval $[-2, 0]$ and the interval $[5, \infty)$.

Each of the polynomial inequalities in Examples 1 and 2 has a solution set that consists of a single interval or the union of two intervals. When solving the exercises for this section, you should watch for some unusual solution sets, as illustrated in Example 3.

Example 3 **Unusual Solution Sets**

What is unusual about the solution set for each inequality?

a. $x^2 + 2x + 4 > 0$

The solution set for this inequality consists of the entire set of real numbers, $(-\infty, \infty)$. In other words, the value of the quadratic $x^2 + 2x + 4$ is positive for every real value of x.

b. $x^2 + 2x + 1 \leq 0$

The solution set for this inequality consists of the single real number $\{-1\}$, because the quadratic $x^2 + 2x + 1$ has one critical number, $x = -1$, and it is the only value that satisfies the inequality.

c. $x^2 + 3x + 5 < 0$

The solution set for this inequality is empty. In other words, the quadratic $x^2 + 3x + 5$ is *not* less than zero for any value of x.

d. $x^2 - 4x + 4 > 0$

The solution set for this inequality consists of all real numbers *except* the number 2. In interval notation, this solution can be written as $(-\infty, 2) \cup (2, \infty)$.

✓ CHECKPOINT 3

What is unusual about the solution set for each inequality?

a. $x^2 + x + 3 \leq 0$

b. $x^2 - 2x + 1 > 0$ ∎

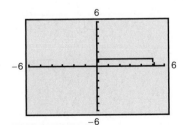

FIGURE 1.22

TECHNOLOGY

(T) **Graphs of Inequalities and Graphing Utilities**   Most graphing utilities can graph an inequality. Consult your user's guide for specific instructions. Once you know how to graph an inequality, you can check solutions by graphing. (Make sure you use an appropriate viewing window.) For example, the solution to

$$x^2 - 5x < 0$$

is the interval $(0, 5)$. When graphed, the solution occurs as an interval above the horizontal axis on the graphing utility, as shown in Figure 1.22. The graph does not indicate whether 0 and/or 5 are part of the solution. You must determine whether the endpoints are part of the solution based on the type of inequality.

Rational Inequalities

The concepts of critical numbers and test intervals can be extended to inequalities involving rational expressions. Use the fact that the value of a rational expression can change sign only at its *zeros* (the *x*-values for which its numerator is zero) and its *undefined values* (the *x*-values for which its denominator is zero). These two types of numbers make up the **critical numbers** of a rational inequality.

Example 4 Solving a Rational Inequality

Solve $\dfrac{2x - 7}{x - 5} \leq 3$.

SOLUTION

$$\frac{2x - 7}{x - 5} \leq 3 \qquad \text{Write original inequality.}$$

$$\frac{2x - 7}{x - 5} - 3 \leq 0 \qquad \text{Write in general form.}$$

$$\frac{2x - 7 - 3x + 15}{x - 5} \leq 0 \qquad \text{Add fractions.}$$

$$\frac{-x + 8}{x - 5} \leq 0 \qquad \text{Simplify.}$$

Critical numbers: $x = 5, x = 8$

Test intervals: $(-\infty, 5), (5, 8), (8, \infty)$

Test: Is $\dfrac{-x + 8}{x - 5} \leq 0$?

After testing these intervals, as shown in Figure 1.23, you can see that the inequality is satisfied on the open intervals $(-\infty, 5)$ and $(8, \infty)$. Moreover, because $(-x + 8)/(x - 5) = 0$ when $x = 8$, you can conclude that the solution set consists of all real numbers in the intervals $(-\infty, 5) \cup [8, \infty)$.

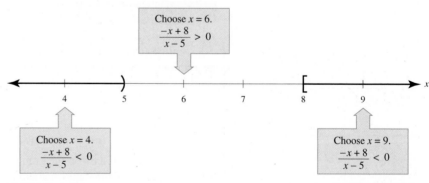

FIGURE 1.23

✓ **CHECKPOINT 4**

Solve $\dfrac{x - 1}{x - 3} \geq -1$. ∎

TECHNOLOGY

Ⓣ When using a graphing utility to check an inequality, always set your viewing window so that it includes all of the critical numbers.

Applications

One common application of inequalities comes from business and involves profit, revenue, and cost. The formula that relates these three quantities is

$$\text{Profit} = \text{Revenue} - \text{Cost}$$
$$P = R - C.$$

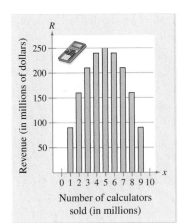

FIGURE 1.24

Example 5 Increasing the Profit for a Product

The marketing department of a calculator manufacturer has determined that the demand for a new model of calculator is given by

$$p = 100 - 10x, \quad 0 \le x \le 10 \qquad \text{Demand equation}$$

where p is the price per calculator in dollars and x represents the number of calculators sold, in millions. (If this model is accurate, no one would be willing to pay $100 for the calculator. At the other extreme, the company couldn't *give* away more than 10 million calculators.) The revenue, in millions of dollars, for selling x million calculators is given by

$$R = xp = x(100 - 10x). \qquad \text{Revenue equation}$$

See Figure 1.24. The total cost of producing x million calculators is $10 per calculator plus a one-time development cost of $2,500,000. So, the total cost, in millions of dollars, is

$$C = 10x + 2.5. \qquad \text{Cost equation}$$

What prices can the company charge per calculator to obtain a profit of at least $190,000,000?

SOLUTION

Verbal Model: $\quad \text{Profit} = \text{Revenue} - \text{Cost}$

Equation: $P = R - C$

$$P = 100x - 10x^2 - (10x + 2.5)$$
$$P = -10x^2 + 90x - 2.5$$

To answer the question, you must solve the inequality

$$-10x^2 + 90x - 2.5 \ge 190.$$

Using the techniques described in this section, you can find the solution set to be $3.5 \le x \le 5.5$, as shown in Figure 1.25. The prices that correspond to these x-values are given by

$$100 - 10(3.5) \ge p \ge 100 - 10(5.5)$$
$$45 \le p \le 65$$

The company can obtain a profit of $190,000,000 or better by charging at least $45 per calculator and at most $65 per calculator.

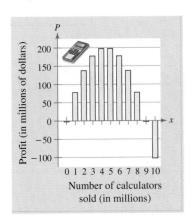

FIGURE 1.25

✓ CHECKPOINT 5

In Example 5, what prices can the company charge per calculator to obtain a profit of at least $160,000,000? ■

Another common application of inequalities is finding the domain of an expression that involves a square root, as shown in Example 6.

Example 6 Finding the Domain of an Expression

Find the domain of the expression $\sqrt{64 - 4x^2}$.

SOLUTION Remember that the domain of an expression is the set of all x-values for which the expression is defined. Because $\sqrt{64 - 4x^2}$ is defined (has real values) only if $64 - 4x^2$ is nonnegative, the domain is given by $64 - 4x^2 \geq 0$.

$$64 - 4x^2 \geq 0 \qquad \text{Write in general form.}$$

$$16 - x^2 \geq 0 \qquad \text{Divide each side by 4.}$$

$$(4 - x)(4 + x) \geq 0 \qquad \text{Factor.}$$

So, the inequality has two critical numbers: $x = -4$ and $x = 4$. You can use these two numbers to test the inequality as follows.

Critical numbers: $x = -4, x = 4$

Test intervals: $(-\infty, -4), (-4, 4), (4, \infty)$

Test: Is $(4 - x)(4 + x) \geq 0$?

A test shows that $64 - 4x^2$ is greater than or equal to 0 in the *closed interval* $[-4, 4]$. So, the domain of the expression $\sqrt{64 - 4x^2}$ is the interval $[-4, 4]$, as shown in Figure 1.26.

FIGURE 1.26

✓ CHECKPOINT 6

Find the domain of each expression.

a. $\sqrt{12 - 3x^2}$

b. $\sqrt[3]{x^2 - 2x - 8}$ ▪

CONCEPT CHECK

1. The test intervals for a polynomial inequality are $(-\infty, -2), (-2, 0), (0, 5)$, and $(5, \infty)$. What are the critical numbers of the polynomial?

2. Is -7 the only critical number of $\dfrac{x - 2}{x + 7} \geq 0$? Explain.

3. Describe and correct the error in the statement: The domain of the expression $\sqrt{(x - 3)(x + 3)}$ is all real numbers except -3 and 3.

4. Explain why the critical numbers of a polynomial inequality are not included in the test intervals.

Skills Review 1.7

The following warm-up exercises involve skills that were covered in earlier sections. You will use these skills in the exercise set for this section. For additional help, review Section 1.6.

In Exercises 1–10, solve the inequality.

1. $-\dfrac{y}{3} > 2$

2. $-6z < 27$

3. $-3 \le 2x + 3 < 5$

4. $-3x + 5 \ge 20$

5. $10 > 4 - 3(x + 1)$

6. $3 < 1 + 2(x - 4) < 7$

7. $2|x| \le 7$

8. $|x - 3| > 1$

9. $|x + 4| > 2$

10. $|2 - x| \le 4$

Exercises 1.7

See www.CalcChat.com for worked-out solutions to odd-numbered exercises.

In Exercises 1–6, find the test intervals of the inequality.

1. $x^2 - 25 < 0$

2. $x^2 - 6x + 8 > 0$

3. $2x^2 + 7x + 16 \ge 20$

4. $3x^2 - 26x + 25 \le 9$

5. $\dfrac{x - 3}{x - 1} < 2$

6. $\dfrac{x - 4}{2x + 3} \ge 1$

In Exercises 7–36, solve the inequality. Then graph the solution set on the real number line.

7. $x^2 \le 9$

8. $x^2 < 5$

9. $x^2 > 4$

10. $(x - 3)^2 \ge 1$

11. $(x + 2)^2 < 25$

12. $(x + 6)^2 \le 8$

13. $x^2 + 4x + 4 \ge 9$

14. $x^2 - 6x + 9 < 16$

15. $x^2 + x < 6$

16. $x^2 + 2x > 3$

17. $3(x - 1)(x + 1) > 0$

18. $6(x + 2)(x - 1) < 0$

19. $x^2 + 2x - 3 < 0$

20. $x^2 - 4x - 1 > 0$

21. $4x^3 - 6x^2 < 0$

22. $4x^3 - 12x^2 > 0$

23. $x^3 - 4x \ge 0$

24. $2x^3 - x^4 \le 0$

25. $x^3 - 2x^2 - x + 2 \ge 0$

26. $x^3 + 5x^2 - 4x - 20 \le 0$

27. $\dfrac{1}{x} > x$

28. $\dfrac{1}{x} < 4$

29. $\dfrac{x + 6}{x + 1} < 2$

30. $\dfrac{x + 12}{x + 2} \ge 3$

31. $\dfrac{3x - 5}{x - 5} > 4$

32. $\dfrac{5 + 7x}{1 + 2x} < 4$

33. $\dfrac{4}{x + 5} > \dfrac{1}{2x + 3}$

34. $\dfrac{5}{x - 6} > \dfrac{3}{x + 2}$

35. $\dfrac{1}{x - 3} \le \dfrac{9}{4x + 3}$

36. $\dfrac{1}{x} \ge \dfrac{1}{x + 3}$

In Exercises 37–46, find the domain of the expression.

37. $\sqrt{x^2 - 9}$

38. $\sqrt{x^2 - 4}$

39. $\sqrt[4]{6 + x^2}$

40. $\sqrt{x^2 + 4}$

41. $\sqrt{81 - 4x^2}$

42. $\sqrt{147 - 3x^2}$

43. $\sqrt{x^2 - 7x + 10}$

44. $\sqrt{12 - x - x^2}$

45. $\sqrt{x^2 - 3x + 3}$

46. $\sqrt[4]{-x^2 + 2x - 2}$

In Exercises 47 and 48, consider the domains of the expressions $\sqrt[3]{x^2 - 7x + 12}$ and $\sqrt{x^2 - 7x + 12}$.

47. Explain why the domain of $\sqrt[3]{x^2 - 7x + 12}$ consists of all real numbers.

48. Explain why the domain of $\sqrt{x^2 - 7x + 12}$ is different from the domain of $\sqrt[3]{x^2 - 7x + 12}$.

In Exercises 49–54, solve the inequality and write the solution set in interval notation.

49. $6x^3 - 10x^2 > 0$

50. $25x^3 - 10x^2 < 0$

51. $x^3 - 9x \le 0$

52. $4x^3 - x^4 \ge 0$

53. $(x - 1)^2(x + 2)^3 \ge 0$

54. $x^4(x - 3) \le 0$

In Exercises 55–60, use a calculator to solve the inequality. (Round each number in your answer to two decimal places.)

55. $0.4x^2 + 5.26 < 10.2$

56. $-1.3x^2 + 3.78 > 2.12$

57. $-0.5x^2 + 12.5x + 1.6 > 0$

58. $1.2x^2 + 4.8x + 3.1 < 5.3$

59. $\dfrac{1}{2.3x - 5.2} > 3.4$

60. $\dfrac{2}{3.1x - 3.7} > 5.8$

61. Height of a Projectile　A projectile is fired straight upward from ground level with an initial velocity of 200 feet per second. During what time period will its height exceed 400 feet?

62. Height of a Projectile　A projectile is fired straight upward from ground level with an initial velocity of 160 feet per second. During what time period will its height be less than 384 feet?

63. Geometry　A rectangular playing field with a perimeter of 100 meters is to have an area of at least 500 square meters (see figure). Within what bounds must the length be?

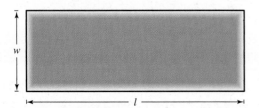

64. Geometry　A rectangular room with a perimeter of 50 feet is to have an area of at least 120 square feet. Within what bounds must the length be?

65. *MAKE A DECISION: COMPANY PROFITS*　The revenue R and cost C for a product are given by $R = x(50 - 0.0002x)$ and $C = 12x + 150,000$, where R and C are measured in dollars and x represents the number of units sold (see figure).

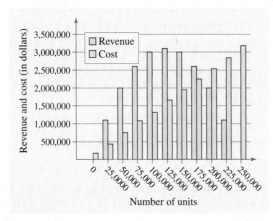

(a) How many units must be sold to obtain a profit of at least $1,650,000?

(b) The demand equation for the product is

$$p = 50 - 0.0002x$$

where p is the price per unit. What prices will produce a profit of at least $1,650,000?

(c) As the number of units increases, the revenue eventually decreases. After this point, at what number of units is the revenue approximately equal to the cost? How should this affect the company's decision about the level of production?

66. *MAKE A DECISION: COMPANY PROFITS*　The revenue R and cost C for a product are given by $R = x(75 - 0.0005x)$ and $C = 30x + 250,000$, where R and C are measured in dollars and x represents the number of units sold (see figure).

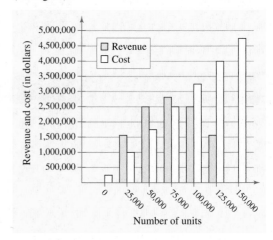

(a) How many units must be sold to obtain a profit of at least $750,000?

(b) The demand equation for the product is

$$p = 75 - 0.0005x$$

where p is the price per unit. What prices will produce a profit of at least $750,000?

(c) As the number of units increases, the revenue eventually decreases. After this point, at what number of units is the revenue approximately equal to the cost? How should this affect the company's decision about the level of production?

67. Compound Interest　P dollars, invested at interest rate r compounded annually, increases to an amount

$$A = P(1 + r)^3$$

in 3 years. For an investment of $1000 to increase to an amount greater than $1500 in 3 years, the interest rate must be greater than what percent?

68. Compound Interest　P dollars, invested at interest rate r compounded annually, increases to an amount

$$A = P(1 + r)^2$$

in 2 years. For an investment of $2000 to increase to an amount greater than $2350 in 2 years, the interest rate must be greater than what percent?

69. World Population The world population P (in millions) from 1995 to 2006 can be modeled by

$$P = -0.18t^2 + 80.30t + 5288, \quad 5 \leq t \leq 16$$

where t represents the year, with $t = 5$ corresponding to 1995 (see figure). Use the model to predict the year in which the world population will exceed 7,000,000,000. *(Source: U.S. Census Bureau)*

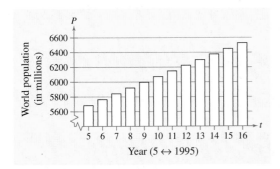

Year (5 ↔ 1995)

70. Higher Education The average yearly cost C of higher education at public institutions in the United States for the academic years 1995/1996 to 2004/2005 can be modeled by

$$C = 30.57t^2 - 259.6t + 6828, \quad 6 \leq t \leq 15$$

where t represents the year, with $t = 6$ corresponding to the 1995/1996 school year (see figure). Use the model to predict the academic year in which the average yearly cost of higher education at public institutions exceeds $12,000. *(Source: U.S. Department of Education)*

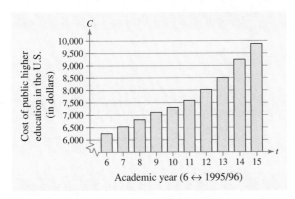

Academic year (6 ↔ 1995/96)

71. Higher Education The average yearly cost C of higher education at private institutions in the United States for the academic years 1995/1996 to 2004/2005 can be modeled by

$$C = 42.93t^2 + 68.0t + 15,309, \quad 6 \leq t \leq 15$$

where t represents the year, with $t = 6$ corresponding to the academic year 1995/1996 (see figure). Use the model to predict the academic year in which the average yearly cost of higher education at private institutions exceeds $32,000. *(Source: U.S. Department of Education.)*

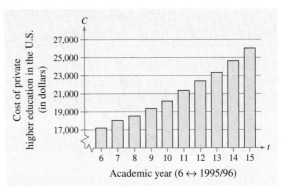

Academic year (6 ↔ 1995/96)

Figure for 71

72. Sales The total sales S (in millions of dollars) for Univision Communications from 1997 to 2005 can be modeled by

$$S = 18.471t^2 - 221.96t + 1152.6, \quad 7 \leq t \leq 15$$

where t represents the year, with $t = 7$ corresponding to 1997 (see figure). Univision Communications predicts sales will exceed $2.7 billion between 2009 and 2011. Does the model support this prediction? Explain your reasoning. *(Source: Univision Communications)*

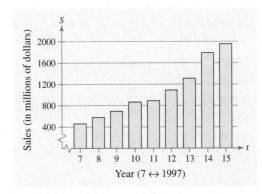

Year (7 ↔ 1997)

73. Resistors When two resistors of resistances R_1 and R_2 are connected in parallel (see figure), the total resistance R satisfies the equation

$$\frac{1}{R} = \frac{1}{R_1} + \frac{1}{R_2}.$$

Find R_1 for a parallel circuit in which $R_2 = 2$ ohms and R must be at least 1 ohm.

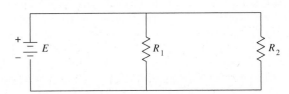

Chapter Summary and Study Strategies

After studying this chapter, you should have acquired the following skills.
The exercise numbers are keyed to the Review Exercises that begin on page 150.
Answers to odd-numbered Review Exercises are given in the back of the text.*

Section 1.1   Review Exercises

- Classify an equation as an identity or a conditional equation.   *1, 2*
- Determine whether a given value is a solution.   *3, 4*
- Solve a linear equation in one variable.   *5–8*
    Can be written in the general form: $ax + b = 0$.
- Solve an equation involving fractions.   *9–12*
- Use a calculator to solve an equation.   *13–16*

Section 1.2

- Use mathematical models to solve word problems.   *17, 19, 27–30*
- Model and solve percent and mixture problems.   *18, 20, 25, 26, 31, 32*
- Use common formulas to solve geometry and simple interest problems.   *21–24*

  Square: $A = s^2$,   Rectangle: $A = lw$,
  $\quad\quad P = 4s$ $\quad\quad P = 2l + 2w$

  Circle: $A = \pi r^2$,   Triangle: $A = \frac{1}{2}bh$,
  $\quad\quad C = 2\pi r$ $\quad\quad P = a + b + c$

  Cube: $V = s^3$

  Rectangular Solid: $V = lwh$

  Circular Cylinder: $V = \pi r^2 h$

  Sphere: $V = \frac{4}{3}\pi r^3$

  Temperature: $F = \frac{9}{5}C + 32$

  Simple Interest: $I = Prt$

  Distance: $d = rt$

Section 1.3

- Solve a quadratic equation by factoring.   *33–36*
    Can be written in the general form: $ax^2 + bx + c = 0$.
    Zero-Factor Property: If $ab = 0$, then $a = 0$ or $b = 0$.
- Solve a quadratic equation by extracting square roots.   *37–40*
- Analyze a quadratic equation.   *41, 42*
- Construct and use a quadratic model to solve area problems, falling   *43–46*
  object problems, right triangle problems, and other applications.

---

* Use a wide range of valuable study aids to help you master the material in this chapter. The *Student
Solutions Guide* includes step-by-step solutions to all odd-numbered exercises to help you review
and prepare. The student website at *college.hmco.com/info/larsonapplied* offers algebra help and a
Graphing Technology Guide. The *Graphing Technology Guide* contains step-by-step commands
and instructions for a wide variety of graphing calculators, including the most recent models.

Section 1.4

	Review Exercises
■ Use the discriminant to determine the number of real solutions of a quadratic equation.	47, 48

If $b^2 - 4ac > 0$, the equation has two distinct real solutions.

If $b^2 - 4ac = 0$, the equation has one repeated real solution.

If $b^2 - 4ac < 0$, the equation has no real solutions.

■ Solve a quadratic equation using the Quadratic Formula.   49–58

Quadratic Formula: $x = \dfrac{-b \pm \sqrt{b^2 - 4ac}}{2a}$

■ Use the Quadratic Formula to solve an application problem.   59, 60

Section 1.5

■ Solve a polynomial equation by factoring.   61, 62

■ Solve an equation of quadratic type.   63, 64

■ Rewrite and solve an equation involving radicals or rational exponents.   65–70

■ Rewrite and solve an equation involving fractions or absolute values.   71–74

■ Construct and use a nonquadratic model to solve an application problem.   75, 76, 78

■ Solve a compound interest problem.   77

Section 1.6

■ Solve and graph a linear inequality.   79–82

Transitive Property: $a < b$ and $b < c \implies a < c$

Addition of Inequalities: $a < b$ and $c < d \implies a + c < b + d$

Addition of a Constant: $a < b \implies a + c < b + c$

Multiplication by a Constant: For $c > 0, a < b \implies ac < bc$

For $c < 0, a < b \implies ac > bc$

■ Solve and graph inequalities involving absolute value.   83, 84

$|x| < a$ if and only if $-a < x < a$

$|x| > a$ if and only if $x < -a$ or $x > a$

■ Construct and use a linear inequality to solve an application problem.   85, 86

Section 1.7

■ Solve and graph a polynomial inequality.   87–89, 93, 94

■ Solve and graph a rational inequality.   90–92, 95, 96

■ Determine the domain of an expression involving a radical.   97–102

■ Construct and use a polynomial inequality to solve an application problem.   103–113

Study Strategies

■ **Check Your Answers** Because of the number of steps involved in solving an equation or inequality, there are many ways to make mistakes. So, always check your answers. In some cases, you may even want to check your answers in more than one way, just to be sure.

■ **Using Test Intervals** Make sure that you understand how to use critical numbers to determine test intervals for inequalities. The logic and mathematical reasoning involved in this concept can be applied in many real-life situations.

Review Exercises

See www.CalcChat.com for worked-out solutions to odd-numbered exercises.

In Exercises 1 and 2, determine whether the equation is an identity or a conditional equation.

1. $5(x - 3) = 2x + 9$ **2.** $3(x + 2) = 3x + 6$

In Exercises 3 and 4, determine whether each value of x is a solution of the equation.

3. $3x^2 + 7x + 5 = x^2 + 9$

   (a) $x = 0$ (b) $x = \frac{1}{2}$ (c) $x = -4$ (d) $x = -1$

4. $6 + \dfrac{3}{x - 4} = 5$

   (a) $x = 5$ (b) $x = 0$ (c) $x = -2$ (d) $x = 7$

In Exercises 5–12, solve the equation (if possible) and check your solution.

5. $x + 7 = 20$

6. $2x + 15 = 43$

7. $4(x + 3) - 3 = 2(4 - 3x) - 4$

8. $(x + 3) + 2(x - 4) = 5(x + 3)$

9. $\dfrac{3x - 2}{5x - 1} = \dfrac{3}{4}$

10. $\dfrac{3}{x - 4} + \dfrac{8}{2x + 5} = \dfrac{11}{2x^2 - 3x - 20}$

11. $\dfrac{x}{x + 3} - \dfrac{4}{x + 3} + 2 = 0$

12. $7 - \dfrac{3}{x} = 8 + \dfrac{5}{x}$

In Exercises 13–16, use a calculator to solve the equation. (Round your solution to three decimal places.)

13. $0.375x - 0.75(300 - x) = 200$

14. $0.235x + 2.6(-x - 4) = 30$

15. $\dfrac{x}{0.055} + \dfrac{x}{0.085} = 1$

16. $\dfrac{x}{0.0645} + \dfrac{x}{0.098} = 2$

17. Three consecutive even integers have a sum of 42. Find the smallest of these integers.

18. Annual Salary Your annual salary is $28,900. You receive a 7% raise. What is your new annual salary?

19. Fitness When using a pull-up weight machine, the amount you set is subtracted from your weight and you pull the remaining amount. Write a model that describes the weight x that must be set if a person weighing 150 pounds wishes to pull 120 pounds. Solve for x.

20. Oil Imports The United States imported 1738 million barrels of crude oil from members of OPEC (Organization of the Petroleum Exporting Countries) in 2005. Use the bar graph to determine the amount imported from each of the four top contributing countries. *(Source: U.S. Energy Information Administration)*

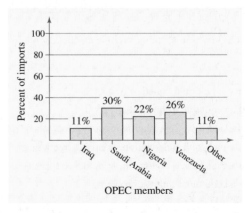

21. Geometry A volleyball court is twice as long as it is wide, and its perimeter is 177 feet. Find the dimensions of the volleyball court.

22. Geometry A room is 1.25 times as long as it is wide, and its perimeter is 90 feet. Find the dimensions of the room.

23. Simple Interest You deposit $500 in a savings account earning 4% simple interest. How much interest will you have earned after 1 year?

24. Simple Interest You deposit $800 in a money market account. One year later the account balance is $862.40. What was the simple interest rate?

25. List Price The price of an outdoor barbeque grill has been discounted 15%. The sale price is $139. Find the original price of the grill.

26. Discount Rate The price of a three-station home gym is discounted by $300. The sale price is $599.99. What percent of the original price is the discount?

27. Travel Time Two cars start at the same time at a given point and travel in the same direction at average speeds of 45 miles per hour and 50 miles per hour. After how long are the cars 10 miles apart?

28. Exercise Two bicyclists start at the same time at a given point and travel in the same direction at average speeds of 8 miles per hour and 10 miles per hour. After how long are the bicyclists 5 miles apart?

29. Projected Revenue From January through June, a company's revenues have totaled $375,832. If the monthly revenues continue at this rate, what will be the total revenue for the year?

30. Projected Profit From January through March, a company's profits have totaled $425,345. If the monthly profits continue at this rate, what will be the total profit for the year?

31. Mixture A car radiator contains 10 quarts of a 10% antifreeze solution. The car's owner wishes to create a 10-quart solution that is 30% antifreeze. How many quarts will have to be replaced with pure antifreeze?

32. Mixture A three-gallon acid solution contains 3% boric acid. How many gallons of 20% boric acid solution should be added to make a final solution that is 8% boric acid?

In Exercises 33–36, solve the quadratic equation by factoring. Check your solutions.

33. $6x^2 = 5x + 4$ **34.** $-x^2 = 15x + 36$

35. $x^2 - 11x + 24 = 0$ **36.** $4 - 4x + x^2 = 0$

In Exercises 37–40, solve the quadratic equation by extracting square roots. List both the exact answer and a decimal answer that has been rounded to two decimal places.

37. $x^2 = 11$ **38.** $16x^2 = 25$

39. $(x + 4)^2 = 18$ **40.** $(x - 1)^2 = 5$

Ⓣ **41.** Describe at least two ways you can use a graphing utility to check a solution of a quadratic equation.

42. Error Analysis A student solves Exercise 37 by extracting square roots and states that the exact and rounded solutions are $x = \sqrt{11}$ and $x \approx 3.32$. What error has the student made? Give an analytical argument to persuade the student that there are two *different* solutions to Exercise 37.

43. Geometry A billboard is 12 feet longer than it is high. The billboard has 405 square feet of advertising space. What are the dimensions of the billboard? Use a diagram to help answer the question.

44. Grand Canyon The Grand Canyon is 6000 feet deep at its deepest part. A rock is dropped over the deepest part of the canyon. How long does the rock take to hit the water in the Colorado River below?

45. Total Revenue The demand equation for a product is

$$p = 60 - 0.0001x$$

where p is the price per unit and x is the number of units sold. The total revenue R for selling x units is given by

$$R = xp = x(60 - 0.0001x).$$

How many units must be sold to produce a revenue of $8,000,000?

46. Depth of an Underwater Cable A ship's sonar locates a cable 2000 feet from the ship (see figure). The angle between the surface of the water and a line from the ship to the cable is 45°. How deep is the cable?

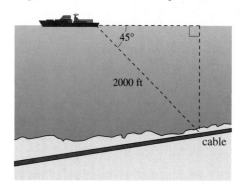

In Exercises 47 and 48, use the discriminant to determine the number of real solutions of the quadratic equation.

47. $x^2 + 11x + 24 = 0$

48. $x^2 + 5x + 12 = 0$

In Exercises 49–54, use the Quadratic Formula to solve the quadratic equation. Check your solutions.

49. $x^2 - 12x + 30 = 0$

50. $5x^2 + 16x - 12 = 0$

51. $(y + 7)^2 = -5y$

52. $6x = 7 - 2x^2$

53. $x^2 + 6x - 3 = 0$

54. $10x^2 - 11x = 2$

In Exercises 55–58, use a calculator to solve the quadratic equation. (Round your answers to three decimal places.)

55. $3.6x^2 - 5.7x - 1.9 = 0$

56. $2.3x^2 + 6.6x - 3.9 = 0$

57. $34x^2 - 296x + 47 = 0$

58. $39x^2 + 75x - 21 = 0$

59. On the Moon An astronaut standing on the edge of a cliff on the moon drops a rock over the cliff. The height s of the rock after t seconds is given by

$$s = -2.7t^2 + 200.$$

The rock's initial velocity is 0 feet per second and the initial height is 200 feet. Determine how long it will take the rock to hit the lunar surface. If the rock were dropped from a similar cliff on Earth, how long would it remain in the air?

60. Geometry An open box is to be made from a square piece of material by cutting three-inch squares from the corners and turning up the sides (see figure). The volume of the finished box is to be 363 cubic inches. Find the size of the original piece of material.

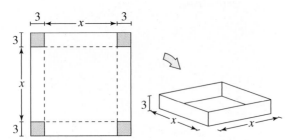

In Exercises 61–74, find the real solutions of the equation. Check your solutions.

61. $3x^3 - 9x^2 - 12x = 0$

62. $x^4 + 3x^3 - 5x - 15 = 0$

63. $x^4 - 5x^2 + 4 = 0$

64. $x^6 + 26x^3 - 27 = 0$

65. $2\sqrt{x} - 5 = 0$ **66.** $\sqrt{3x - 2} + x = 4$

67. $2\sqrt{x - 3} - 4 = 3x$ **68.** $\sqrt[3]{3x + 5} = 5$

69. $(x^2 - 5)^{2/3} = 9$ **70.** $(x^2 - 5x - 6)^{4/3} = 16$

71. $|5x + 4| = 11$ **72.** $|x^2 + 4x| - 2x = 8$

73. $\dfrac{5}{x + 1} + \dfrac{3}{x + 3} = 1$ **74.** $x + \dfrac{3}{x + 2} = 2$

75. Sharing the Cost Three students are planning to share the expense of renting a condominium at a resort for 1 week. By adding a fourth person to the group, each person could save $75 in rental fees. How much is the rent for the week?

76. Sharing the Cost A college charters a bus for $1800 to take a group to a museum. When four more students join the trip, the cost per student decreases by $5. How many students were in the original group?

77. Cash Advance You take out a cash advance of $500 on a credit card. After 3 months, the amount you owe is $535.76. What is the annual percentage rate for this cash advance? (Assume that the interest is compounded monthly and that you made no payments yet.)

78. Market Research The demand equation for a product is given by

$$p = 45 - \sqrt{0.002x + 1}$$

where x is the number of units demanded per day and p is the price per unit. Find the demand when the price is set at $19.95.

In Exercises 79–84, solve the inequality and graph the solution set on the real number line.

79. $3(x - 1) < 2x + 8$

80. $-5 \leq 2 - 4(x + 2) \leq 6$

81. $-3 < \dfrac{2x + 1}{4} < 3$

82. $-1 \leq -5 - 3x < 4$

83. $|x + 10| + 3 < 5$

84. $|2x - 3| - 4 > 2$

85. *MAKE A DECISION: BREAK-EVEN ANALYSIS* The revenue R for selling x units of a product is

$$R = 89.95x.$$

The cost C of producing x units is

$$C = 35x + 2500.$$

In order to obtain a profit, the revenue must be greater than the cost. What are the numbers of units the company can produce in order to return a profit?

86. Accuracy of Measurement You buy a 16-inch gold chain that costs $9.95 per inch. If the chain is measured accurately to within $\frac{1}{16}$ of an inch, how much money might you have been undercharged or overcharged?

In Exercises 87–92, solve the inequality and graph the solution set on the real number line.

87. $5(x + 1)(x - 3) < 0$ **88.** $(x + 4)^2 \leq 4$

89. $x^3 - 9x < 0$ **90.** $\dfrac{x + 5}{x + 8} \geq 2$

91. $\dfrac{2 + 3x}{4 - x} < 2$ **92.** $\dfrac{1}{x + 1} \geq \dfrac{1}{x + 5}$

In Exercises 93–96, use a calculator to solve the inequality. (Round each number in your answer to two decimal places.)

93. $-1.2x^2 + 4.76 > 1.32$

94. $3.5x^2 + 4.9x - 6.1 < 2.4$

95. $\dfrac{1}{3.7x - 6.1} > 2.9$

96. $\dfrac{3}{5.4x - 2.7} < 8.9$

In Exercises 97–102, find the domain of the expression.

97. $\sqrt{x - 10}$ **98.** $\sqrt[4]{2x + 5}$

99. $\sqrt[3]{2x - 1}$ **100.** $\sqrt[5]{x^2 - 4}$

101. $\sqrt{x^2 - 15x + 54}$ **102.** $\sqrt{81 - 4x^2}$

103. Height of a Projectile A projectile is fired straight upward from ground level with an initial velocity of 134 feet per second. During what time period will its height exceed 276 feet?

104. Height of a Flare A flare is fired straight upward from ground level with an initial velocity of 100 feet per second. During what time period will its height exceed 150 feet?

105. Path of a Soccer Ball The path of a soccer ball kicked from the ground can be modeled by

$$y = -0.054x^2 + 1.43x$$

where x is the horizontal distance (in feet) from where the ball was kicked and y is the corresponding height (in feet).

(a) A soccer goal is 8 feet high. Write an inequality to determine for what values of x the ball is low enough to go into the goal.

(b) Solve the inequality from part (a).

(c) A soccer player kicks the ball toward the goal from a distance of 15 feet. No one is blocking the goal. Will the player score a goal? Explain your reasoning.

106. Geometry A rectangular field with a perimeter of 80 meters is to have an area of at least 380 square meters (see figure). Within what bounds must the length be?

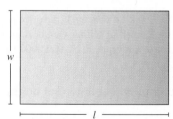

107. Geometry A rectangular room with a perimeter of 60 feet is to have an area of at least 150 square feet. Within what bounds must the length be?

108. Compound Interest P dollars, invested at interest rate r compounded annually, increases to an amount

$$A = P(1 + r)^5$$

in 5 years. An investment of $1000 increases to an amount greater than $1400 in 5 years. The interest rate must be greater than what percent?

109. Compound Interest P dollars, invested at an interest rate r compounded semiannually, increases to an amount

$$A = P(1 + r/2)^{2 \cdot 8}$$

in 8 years. An investment of $2000 increases to an amount greater than $4200 in 8 years. The interest rate must be greater than what percent?

110. Company Profits The revenue R and cost C for a product are given by

$$R = x(75 - 0.0005x) \quad \text{and} \quad C = 25x + 100{,}000$$

where R and C are measured in dollars and x represents the number of units sold. How many units must be sold to obtain a profit of at least $500,000?

111. Price of a Product In Exercise 110, the revenue equation is

$$R = x(75 - 0.0005x)$$

which implies that the demand equation is

$$p = 75 - 0.0005x$$

where p is the price per unit. What prices per unit can the company set to obtain a profit of at least $1,000,000?

112. Mail Order Sales The total sales S (in billions of dollars) of prescription drugs by mail order in the United States from 1998 to 2005 can be approximated by the model

$$S = 4.37t - 21.4, \quad 8 \le t \le 15$$

where t represents the year, with $t = 8$ corresponding to 1998. *(Source: National Center for Health Statistics)*

(a) Complete the table.

t	8	11	13	15
S				

(b) Use the model to predict the year in which mail order drug sales will be at least $60 billion

113. Revenue The revenue per share R (in dollars) for the Sonic Corporation from 1996 to 2005 can be approximated by the model

$$R = 0.0399t^2 - 0.244t + 1.61, \quad 6 \le t \le 15$$

where t represents the year, with $t = 6$ corresponding to 1996. *(Source: Sonic Corporation)*

(a) Complete the table. Round each value of R to the nearest cent.

t	6	10	13	15
R				

(b) In 2006, Sonic predicted that their revenue per share would be at least $8.80 in 2007. Does the model support this prediction? Explain.

(c) Sonic also predicted their revenue per share will be at least $11.10 sometime between 2009 and 2011. Does the model support this prediction? Explain.

Chapter Test

See www.CalcChat.com for worked-out solutions to odd-numbered exercises.

Take this test as you would take a test in class. When you are done, check your work against the answers given in the back of the book.

1. Solve the equation $3(x + 2) - 8 = 4(2 - 5x) + 7$.

2. Find the domain of (a) $\sqrt[3]{2x + 3}$ and (b) $\sqrt{9 - x^2}$.

3. In May, the total profit for a company was 8% less than it was in April. The total profit for the 2 months was $625,509.12. Find the profit for each month.

In Exercises 4–13, solve the equation. Check your solution(s).

4. *Factoring:* $6x^2 + 7x = 5$

5. *Factoring:* $12 + 5x - 2x^2 = 0$

6. *Extracting roots:* $x^2 - 5 = 10$

7. *Quadratic Formula:* $(x + 5)^2 = -3x$

8. *Quadratic Formula:* $3x^2 - 11x = 2$

9. *Quadratic Formula:* $5.4x^2 - 3.2x - 2.5 = 0$

10. $|2x - 3| = 10$

11. $\sqrt{x - 3} + x = 5$

12. $x^4 - 10x^2 + 9 = 0$

13. $(x^2 - 9)^{2/3} = 9$

14. The demand equation for a product is $p = 40 - 0.0001x$, where p is the price per unit and x is the number of units sold. The total revenue R for selling x units is given by $R = xp$. How many units must be sold to produce a revenue of $2,000,000? Explain your reasoning.

In Exercises 15–18, solve the inequality and graph the solution set on the real number line.

15. $\dfrac{3x + 1}{5} < 2$

16. $|4 - 5x| \geq 24$

17. $\dfrac{x + 3}{x + 7} > 2$

18. $3x^3 - 12x \leq 0$

19. The revenue R and cost C for a product are given by

$$R = x(90 - 0.0004x) \quad \text{and} \quad C = 25x + 300,000$$

where R and C are measured in dollars and x represents the number of units sold. How many units must be sold to obtain a profit of at least $800,000?

20. The average annual cost C (in dollars) to stay in a college dormitory from 2000 to 2005 can be approximated by the model

$$C = 7.71t^2 + 136.9t + 2433, \quad 0 \leq t \leq 5$$

where t represents the year, with $t = 0$ corresponding to 2000. Use the model to predict the year in which the average dormitory cost exceeds $4000. *(Source: U.S. National Center for Education Statistics)*

Cumulative Test: Chapters 0–1

See www.CalcChat.com for worked-out solutions to odd-numbered exercises.

Take this test as you would take a test in class. When you are done, check your work against the answers given in the back of the book.

In Exercises 1–3, simplify the expression.

1. $4(-2x^2)^3$

2. $\sqrt{18x^5}$

3. $\dfrac{2}{3 - \sqrt{5}}$

4. Factor completely: $x^3 - 6x^2 - 3x + 18$.

5. Simplify: $\dfrac{x^2 - 16}{5x - 20}$.

6. Simplify: $\dfrac{\frac{1}{x} - \frac{1}{y}}{\frac{1}{y} + \frac{1}{x}}$.

7. The average monthly retail sales C (in billions of dollars) in the United States from 2000 to 2005 can be approximated by the model

$$C = 11.9t + 243, \quad 0 \le t \le 5$$

where t represents the year, with $t = 0$ corresponding to 2000. *(Source: U.S. Council of Economic Advisors)*

(a) Estimate the average monthly retail sales in 2005.

(b) Use the model to predict the first year in which the average monthly retail sales will exceed $360,000,000,000.

In Exercises 8–13, solve the equation.

8. *Factoring:* $2x^2 - 11x = -5$

9. *Quadratic Formula:* $5.2x^2 + 1.5x - 3.9 = 0$

10. $|3x + 1| = 9$

11. $\sqrt{2x - 1} + x = 4$

12. $x^4 - 17x^2 = -16$

13. $(x^2 - 14)^{3/2} = 8$

In Exercises 14–16, solve the inequality and graph the solution set on the real number line.

14. $-2 < \dfrac{1 - 3x}{5} < 2$

15. $2x^3 - 16x \ge 0$

16. $|5 - 3x| \le 21$

17. The revenue R and cost C for a product are given by

$$R = x(120 - 0.0002x) \quad \text{and} \quad C = 40x + 200,000$$

where R and C are measured in dollars and x represents the number of units sold. How many units must be sold to obtain a profit of at least $600,000?

18. The per capita gross domestic product D (in dollars) in the United States from 2000 to 2005 can be approximated by the model

$$D = 228.57t^2 + 323.3t + 34,808, \quad 0 \le t \le 5$$

where t represents the year, with $t = 0$ corresponding to 2000. According to this model, in what year will per capita gross domestic product first exceed $50,000? *(Source: U.S. Bureau of Economic Analysis)*

2

Functions and Graphs

alixeudo images/Getty Images

The first Ferris wheel stood about 264 feet tall. It was designed by George Washington Gale Ferris Jr. for the World's Columbian Exposition in Chicago, Illinois, in 1893. You can use the standard form of the equation of a circle to model the shape of a Ferris wheel. (See Section 2.1, Exercises 111 and 112.)

Applications

Functions and graphs are used to model and solve many real-life applications. The applications listed below represent a sample of the applications in this chapter.

- Make a Decision: Yahoo! Inc. Revenue, Exercise 101, page 181
- Path of a Salmon, Exercise 74, page 205
- Earnings-Dividend Ratio, Exercise 73, page 247

Graphs of Equations

- Plot points in the Cartesian plane.
- Use the Distance Formula to find the distance between two points in the coordinate plane.
- Use the Midpoint Formula to find the midpoint of a line segment joining two points.
- Determine whether a point is a solution of an equation.
- Sketch the graph of an equation using a table of values.
- Find the *x*- and *y*-intercepts of the graph of an equation.
- Determine the symmetry of a graph.
- Write the equation of a circle in standard form.

The Cartesian Plane

Just as you can represent real numbers by points on a real number line, you can represent ordered pairs of real numbers by points in a plane. This plane is called the **rectangular coordinate system,** or the **Cartesian plane,** named after the French mathematician René Descartes (1596–1650).

The Cartesian plane is formed by using two real number lines intersecting at right angles, as shown in Figure 2.1. The horizontal real number line is usually called the *x*-**axis,** and the vertical real number line is usually called the *y*-**axis.** The point of intersection of these two axes is the **origin,** and the two axes divide the plane into four parts called **quadrants.**

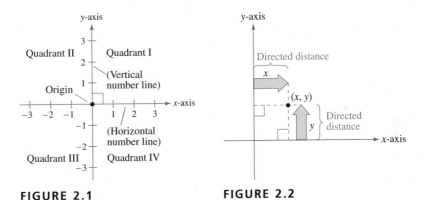

FIGURE 2.1   **FIGURE 2.2**

Each point in the plane corresponds to an **ordered pair** (x, y) of real numbers x and y, called the **coordinates** of the point. The *x*-**coordinate** represents the directed distance from the *y*-axis to the point, and the *y*-**coordinate** represents the directed distance from the *x*-axis to the point, as shown in Figure 2.2.

$$\underset{\substack{\text{Directed distance} \\ \text{from }y\text{-axis}}}{\qquad} (x, y) \underset{\substack{\text{Directed distance} \\ \text{from }x\text{-axis}}}{\qquad}$$

The notation (x, y) denotes both a point in the plane and an open interval on the real number line. The context will tell you which meaning is intended.

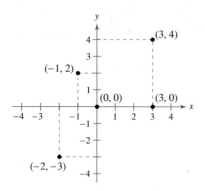

FIGURE 2.3

Example 1 **Plotting Points in the Cartesian Plane**

Plot the points $(-1, 2)$, $(3, 4)$, $(0, 0)$, $(3, 0)$, and $(-2, -3)$.

SOLUTION To plot the point $(-1, 2)$, imagine a vertical line through -1 on the x-axis and a horizontal line through 2 on the y-axis. The intersection of these two lines is the point $(-1, 2)$. The other four points can be plotted in a similar way, as shown in Figure 2.3.

✓**CHECKPOINT 1**

Plot the points $(2, 3)$ and $(-4, 1)$. ∎

The Distance and Midpoint Formulas

Recall from the Pythagorean Theorem that, for a right triangle with hypotenuse of length c and legs of lengths a and b, you have $a^2 + b^2 = c^2$, as shown in Figure 2.4. (The converse is also true. That is, if $a^2 + b^2 = c^2$, then the triangle is a right triangle.)

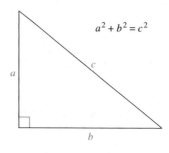

FIGURE 2.4

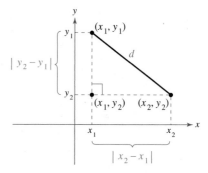

FIGURE 2.5

Suppose you want to determine the distance d between two points (x_1, y_1) and (x_2, y_2) that do not lie on the same horizontal or vertical line. With these two points, a right triangle can be formed, as shown in Figure 2.5. The length of the vertical side of the triangle is $|y_2 - y_1|$, and the length of the horizontal side is $|x_2 - x_1|$. By the Pythagorean Theorem, you can write

$$d^2 = |x_2 - x_1|^2 + |y_2 - y_1|^2$$
$$d = \sqrt{|x_2 - x_1|^2 + |y_2 - y_1|^2} \qquad \text{Choose positive square root.}$$
$$d = \sqrt{(x_2 - x_1)^2 + (y_2 - y_1)^2}.$$

The result is the **Distance Formula.**

The Distance Formula

The distance d between the points (x_1, y_1) and (x_2, y_2) in the coordinate plane is

$$d = \sqrt{(x_2 - x_1)^2 + (y_2 - y_1)^2}.$$

The following formula shows how to find the *midpoint* of the line segment that joins two points.

> **The Midpoint Formula**
>
> The midpoint of the line segment joining the points (x_1, y_1) and (x_2, y_2) in the coordinate plane is
>
> $$\left(\frac{x_1 + x_2}{2}, \frac{y_1 + y_2}{2} \right).$$

Example 2 Using the Distance and Midpoint Formulas

Find (a) the distance between, and (b) the midpoint of the line segment joining, the points $(-2, 1)$ and $(3, 4)$.

SOLUTION

a. Let $(x_1, y_1) = (-2, 1)$ and $(x_2, y_2) = (3, 4)$, and apply the Distance Formula.

$$
\begin{aligned}
d &= \sqrt{(x_2 - x_1)^2 + (y_2 - y_1)^2} && \text{Distance Formula} \\
&= \sqrt{[3 - (-2)]^2 + (4 - 1)^2} && \text{Substitute for } x_1, x_2, y_1, \text{ and } y_2. \\
&= \sqrt{5^2 + 3^2} && \text{Simplify.} \\
&= \sqrt{34} \approx 5.83 && \text{Simplify.}
\end{aligned}
$$

See Figure 2.6.

b. By the Midpoint Formula, you have

$$
\begin{aligned}
\text{Midpoint} &= \left(\frac{x_1 + x_2}{2}, \frac{y_1 + y_2}{2} \right) && \text{Midpoint Formula} \\
&= \left(\frac{-2 + 3}{2}, \frac{1 + 4}{2} \right) && \text{Substitute for } x_1, x_2, y_1, \text{ and } y_2. \\
&= \left(\frac{1}{2}, \frac{5}{2} \right). && \text{Simplify.}
\end{aligned}
$$

See Figure 2.7.

✓ **CHECKPOINT 2**

Find (a) the distance between, and (b) the midpoint of the line segment joining, the points $(-2, 2)$ and $(6, -4)$. ■

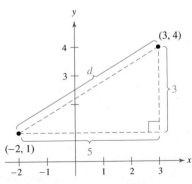

FIGURE 2.6 $d = \sqrt{34} \approx 5.83$

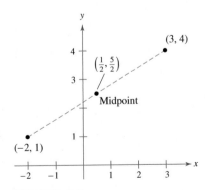

FIGURE 2.7

The Graph of an Equation

Frequently, a relationship between two quantities is written in the form of an equation. In the remainder of this section, you will study a procedure for sketching the graph of an equation. For an equation in the variables x and y, a point (a, b) is a **solution** if the substitution $x = a$ and $y = b$ satisfies the equation.

Example 3 Solution of an Equation

Determine whether $(-1, 0)$ is a solution of the equation $y = 2x^2 - 4x - 6$.

SOLUTION

$$y = 2x^2 - 4x - 6 \qquad \text{Write original equation.}$$

$$0 \stackrel{?}{=} 2(-1)^2 - 4(-1) - 6 \qquad \text{Substitute } -1 \text{ for } x \text{ and } 0 \text{ for } y.$$

$$0 = 0 \qquad \text{Simplify.}$$

Both sides of the equation are equivalent, so the point $(-1, 0)$ is a solution.

✓**CHECKPOINT 3**

Determine whether $(-1, 3)$ is a solution of the equation $y = x + 4$. ■

Most equations have *infinitely* many solutions. The **graph of an equation** is the set of all points that are solutions of the equation.

Example 4 Sketching the Graph of an Equation

Sketch the graph of $3x + y = 5$.

SOLUTION   First rewrite the equation as $y = 5 - 3x$ with y isolated on the left. Next, construct a table of values by choosing several values of x and calculating the corresponding values of y.

x	-1	0	1	2	3
$y = 5 - 3x$	8	5	2	-1	-4

From the table, it follows that $(-1, 8)$, $(0, 5)$, $(1, 2)$, $(2, -1)$, and $(3, -4)$ are solution points of the equation. After plotting these points and connecting them, you can see that they appear to lie on a line, as shown in Figure 2.8.

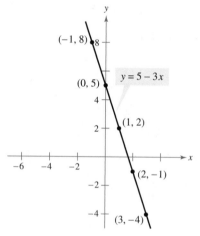

FIGURE 2.8

✓**CHECKPOINT 4**

Sketch the graph of $y - x = 3$. ■

The Point-Plotting Method of Graphing

1. If possible, isolate one of the variables.

2. Construct a table of values showing several solution points.

3. Plot these points on a rectangular coordinate system.

4. Connect the points with a smooth curve or line.

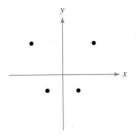

FIGURE 2.9

Step 4 of the point-plotting method can be difficult. For instance, how would you connect the four points in Figure 2.9? Without further information about the equation, any one of the three graphs in Figure 2.10 would be reasonable. These graphs show that with too few solution points, you can misrepresent the graph of an equation. Throughout this course, you will study many ways to improve your graphing techniques. For now, you should plot enough points to reveal the essential behavior of the graph. It is important to use negative values, zero, and positive values for x when constructing a table.

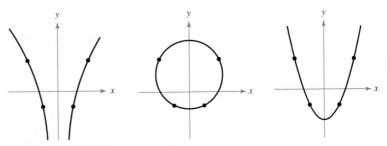

FIGURE 2.10

> **Example 5** **Sketching the Graph of an Equation**

Sketch the graph of $y = x^2 - 2$.

SOLUTION   First, construct a table of values by choosing several convenient values of x and calculating the corresponding values of y.

x	-3	-2	-1	0	1	2	3
$y = x^2 - 2$	7	2	-1	-2	-1	2	7

Next, plot the corresponding solution points. Finally, connect the points with a smooth curve, as shown in Figure 2.11.

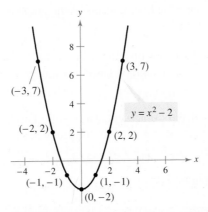

FIGURE 2.11

✓ **CHECKPOINT 5**

Sketch the graph of $y = x^2 + 3$. ▨

Intercepts of a Graph

When you are sketching a graph, points for which either the y-coordinate or the x-coordinate is zero are especially useful.

Definition of Intercepts

1. The **x-intercepts** of a graph are the points at which the graph intersects the x-axis. To find the x-intercepts, let y equal zero and solve for x.

2. The **y-intercepts** of a graph are the points at which the graph intersects the y-axis. To find the y-intercepts, let x equal zero and solve for y.

Some texts denote the x-intercept as the x-coordinate of the point $(a, 0)$ rather than the point itself. Unless it is necessary to make a distinction, we will use the term *intercept* to mean either the point or the coordinate.

A graph may have no intercepts, one intercept, or several intercepts. For instance, consider the three graphs in Figure 2.12.

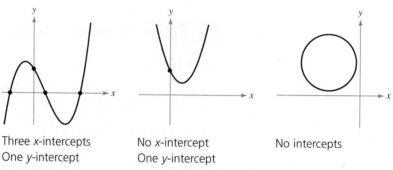

Three x-intercepts   No x-intercept   No intercepts
One y-intercept   One y-intercept

FIGURE 2.12

Example 6   **Finding x- and y-Intercepts**

Find the x- and y-intercepts of the graph of

$$y^2 - 3 = x.$$

SOLUTION   To find the x-intercept, let $y = 0$. This produces $-3 = x$, which implies that the graph has one x-intercept, which occurs at

$$(-3, 0). \qquad \textit{x-intercept}$$

To find the y-intercept, let $x = 0$. This produces $y^2 - 3 = 0$, which has two solutions: $y = \pm\sqrt{3}$. So, the graph has two y-intercepts, which occur at

$$\left(0, \sqrt{3}\right) \quad \text{and} \quad \left(0, -\sqrt{3}\right). \qquad \textit{y-intercepts}$$

See Figure 2.13.

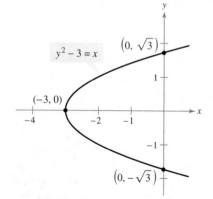

FIGURE 2.13

✓ CHECKPOINT 6

Find the x- and y-intercepts of the graph of $y = 3x + 2$. ∎

Symmetry

Symmetry with respect to the x-axis means that if the Cartesian plane were folded along the x-axis, the portion of the graph above the x-axis would coincide with the portion below the x-axis. Symmetry with respect to the y-axis can be described in a similar manner. Symmetry with respect to the origin means that when the graph rotates $180°$ about the origin, it looks the same. (See Figure 2.14.)

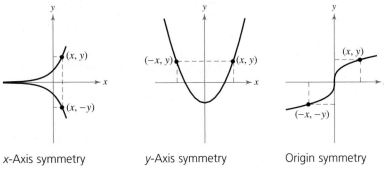

x-Axis symmetry   y-Axis symmetry   Origin symmetry

FIGURE 2.14

Knowing the symmetry of a graph *before* attempting to sketch it is helpful, because then you need only half as many solution points to sketch the graph. The three basic types of symmetry are described as follows.

Definition of Symmetry

1. A graph is **symmetric with respect to the x-axis** if, whenever (x, y) is on the graph, $(x, -y)$ is also on the graph.

2. A graph is **symmetric with respect to the y-axis** if, whenever (x, y) is on the graph, $(-x, y)$ is also on the graph.

3. A graph is **symmetric with respect to the origin** if, whenever (x, y) is on the graph, $(-x, -y)$ is also on the graph.

You can apply this definition of symmetry to the graph of the equation $y = x^2 - 1$. Replacing x with $-x$ produces the following.

$y = x^2 - 1$ Write original equation.

$y = (-x)^2 - 1$ Replace x with $-x$.

$y = x^2 - 1$ Replacement yields equivalent equation.

Because the substitution did not change the equation, it follows that if (x, y) is a solution of the equation, then $(-x, y)$ must also be a solution. So, the graph of $y = x^2 - 1$ is symmetric with respect to the y-axis. By plotting the points in the table below, you can confirm that the graph is symmetric with respect to the y-axis, as shown in Figure 2.15.

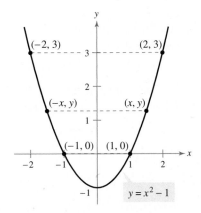

FIGURE 2.15   *y*-Axis Symmetry

x	-2	-1	1	2
y	3	0	0	3

Tests for Symmetry

1. The graph of an equation is symmetric with respect to the x-axis if replacing y with $-y$ yields an equivalent equation.

2. The graph of an equation is symmetric with respect to the y-axis if replacing x with $-x$ yields an equivalent equation.

3. The graph of an equation is symmetric with respect to the *origin* if replacing x with $-x$ *and* y with $-y$ yields an equivalent equation.

Example 7 **Using Symmetry as a Sketching Aid**

Describe the symmetry of the graph of $x - y^2 = 1$.

SOLUTION Of the three tests for symmetry, the only one that is satisfied by this equation is the test for x-axis symmetry.

$$x - y^2 = 1 \qquad \text{Write original equation.}$$
$$x - (-y)^2 = 1 \qquad \text{Replace } y \text{ with } -y.$$
$$x - y^2 = 1 \qquad \text{Replacement yields equivalent equation.}$$

So, the graph is symmetric with respect to the x-axis. To sketch the graph, plot the points above the x-axis and use symmetry to complete the graph, as shown in Figure 2.16.

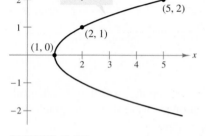

FIGURE 2.16

✓**CHECKPOINT 7**

Describe the symmetry of the graph of $y = |x|$. ▪

The Equation of a Circle

In this section, you have studied the point-plotting method and two additional concepts (intercepts and symmetry) that can be used to streamline the graphing procedure. Another graphing aid is *equation recognition,* which is the ability to recognize the general shape of a graph simply by looking at its equation.

Figure 2.17 shows a circle of *radius* r with *center* at the point (h, k). The point (x, y) is on this circle if and only if its distance from the center (h, k) is r. This means that a **circle** in the plane consists of all points (x, y) that are a given positive distance r from a fixed point (h, k). Using the Distance Formula, you can conclude that the point (x, y) lies on the circle if and only if

$$\sqrt{(x - h)^2 + (y - k)^2} = r.$$

By squaring each side of this equation, you obtain the **standard form of the equation of a circle.** For example, a circle with its center at the origin, $(h, k) = (0, 0)$, and radius $r = 4$ is given by

$$\sqrt{(x - 0)^2 + (y - 0)^2} = 4 \qquad \text{Substitute for } h, k, \text{ and } r.$$
$$\sqrt{x^2 + y^2} = 4 \qquad \text{Simplify.}$$
$$x^2 + y^2 = 16. \qquad \text{Square each side.}$$

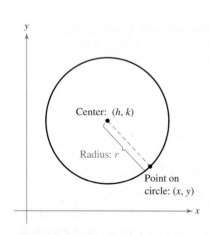

FIGURE 2.17

> **Standard Form of the Equation of a Circle**
>
> The **standard form of the equation of a circle** is
>
> $$(x - h)^2 + (y - k)^2 = r^2.$$
>
> The point (h, k) is called the **center** of the circle, and the positive number r is called the **radius** of the circle. The standard form of the equation of a circle whose center is the *origin* is $x^2 + y^2 = r^2$.

Example 8   Finding the Equation of a Circle

The point $(3, 4)$ lies on a circle whose center is at $(-1, 2)$, as shown in Figure 2.18. Write the standard form of the equation of this circle.

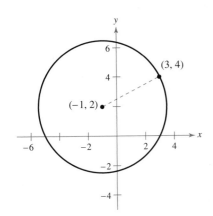

FIGURE 2.18

SOLUTION   The radius of the circle is the distance between the center $(-1, 2)$ and the point $(3, 4)$.

$$r = \sqrt{(x - h)^2 + (y - k)^2} \qquad \text{Distance Formula}$$
$$r = \sqrt{[3 - (-1)]^2 + (4 - 2)^2} \qquad \text{Substitute for } x, y, h, \text{ and } k.$$
$$= \sqrt{4^2 + 2^2} \qquad \text{Simplify.}$$
$$= \sqrt{16 + 4} \qquad \text{Simplify.}$$
$$= \sqrt{20} \qquad \text{Radius}$$

Using $(h, k) = (-1, 2)$ and $r = \sqrt{20}$, the equation of the circle is

$$(x - h)^2 + (y - k)^2 = r^2 \qquad \text{Equation of circle}$$
$$[x - (-1)]^2 + (y - 2)^2 = \left(\sqrt{20}\right)^2 \qquad \text{Substitute for } h, k, \text{ and } r.$$
$$(x + 1)^2 + (y - 2)^2 = 20. \qquad \text{Standard form}$$

✓CHECKPOINT 8

The point $(4, 4)$ lies on a circle whose center is at $(0, 1)$. Write the standard form of the equation of this circle. ■

If you expand the standard equation in Example 8, you obtain the following.

$$(x + 1)^2 + (y - 2)^2 = 20 \qquad \text{Standard form}$$
$$x^2 + 2x + 1 + y^2 - 4y + 4 = 20 \qquad \text{Expand terms.}$$
$$x^2 + y^2 + 2x - 4y - 15 = 0 \qquad \text{General form}$$

The last equation is in the **general form of the equation of a circle,**

$$Ax^2 + Ay^2 + Dx + Ey + F = 0, \quad A \neq 0.$$

The general form of the equation of a circle is less useful than the standard form. For instance, it is not immediately apparent from the general equation shown above that the center is $(-1, 2)$ and the radius is $\sqrt{20}$. To graph the equation of a circle, it is best to write the equation in standard form. You can do this by **completing the square,** as demonstrated in Example 9.

| Example 9 | Completing the Square to Sketch a Circle |

Sketch the circle given by $4x^2 + 4y^2 + 20x - 16y + 37 = 0$.

SOLUTION Begin by writing the original equation in standard form by completing the square for both the *x*-terms *and* the *y*-terms.

$$4x^2 + 4y^2 + 20x - 16y + 37 = 0 \qquad \text{Write original equation.}$$

$$x^2 + y^2 + 5x - 4y + \frac{37}{4} = 0 \qquad \text{Divide by 4.}$$

$$\left(x^2 + 5x + \phantom{\rule{1em}{0ex}}\right) + \left(y^2 - 4y + \phantom{\rule{1em}{0ex}}\right) = -\frac{37}{4} \qquad \text{Group terms.}$$

$$\left[x^2 + 5x + \left(\frac{5}{2}\right)^2\right] + (y^2 - 4y + 2^2) = -\frac{37}{4} + \frac{25}{4} + 4 \quad \text{Complete the square.}$$

$$\left(x + \frac{5}{2}\right)^2 + (y - 2)^2 = 1 \qquad \text{Standard form}$$

So, the center of the circle is $\left(-\frac{5}{2}, 2\right)$ and the radius of the circle is 1. Using this information, you can sketch the circle, as shown in Figure 2.19.

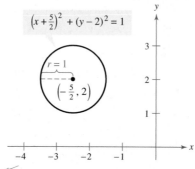

FIGURE 2.19

STUDY TIP

Recall that to complete the square, you add the square of half the coefficient of the linear term to each side.

✓ **CHECKPOINT 9**

Sketch the circle given by $x^2 + y^2 - 2x - 4y + 1 = 0$. ∎

CONCEPT CHECK

1. Determine whether the following statement is true or false. Explain your reasoning.

    The points (3, 4) and (−4, 3) both lie on the same circle whose center is the origin.

2. Explain how to find the *x*- and *y*-intercepts of the graph of an equation.

3. For every point (*x*, *y*) on a graph, the point (−*x*, *y*) is also on the graph. What type of symmetry must the graph have? Explain.

4. Is the point (0, 0) on the circle whose equation in standard form is $(x - 0)^2 + (y - 0)^2 = 4$? Explain.

Skills Review 2.1	The following warm-up exercises involve skills that were covered in earlier sections. You will use these skills in the exercise set for this section. For additional help, review Sections 0.2, 0.4, and 1.5.

In Exercises 1–6, simplify the expression.

1. $\sqrt{(2-6)^2 + [1-(-2)]^2}$

2. $\sqrt{(1-4)^2 + (-2-1)^2}$

3. $\dfrac{4+(-2)}{2}$

4. $\dfrac{-1+(-3)}{2}$

5. $\sqrt{18} + \sqrt{45}$

6. $\sqrt{12} + \sqrt{44}$

In Exercises 7–10, solve the equation.

7. $\sqrt{(4-x)^2 + (5-2)^2} = \sqrt{58}$

8. $\sqrt{(8-6)^2 + (y-5)^2} = 2\sqrt{5}$

9. $x^3 - 9x = 0$

10. $x^4 - 8x^2 + 16 = 0$

Exercises 2.1

See www.CalcChat.com for worked-out solutions to odd-numbered exercises.

In Exercises 1–12, (a) plot the points, (b) find the distance between the points, and (c) find the midpoint of the line segment joining the points.

1. $(2, -5), (-6, 1)$

2. $(1, 12), (6, 0)$

3. $(3, -11), (-12, -3)$

4. $(-7, 3), (2, -9)$

5. $(-1, 2), (5, 4)$

6. $(2, 10), (10, 2)$

7. $\left(\frac{1}{2}, 1\right), \left(-\frac{5}{2}, \frac{4}{3}\right)$

8. $\left(-\frac{1}{3}, -\frac{1}{3}\right), \left(-\frac{1}{6}, -\frac{1}{2}\right)$

9. $(1.8, 7.5), (-2.5, 2.1)$

10. $(37.5, -12.3), (-6.2, 5.9)$

11. $(-36, -18), (48, -72)$

12. $(1.451, 3.051), (5.906, 11.360)$

In Exercises 13–16, find the length of the hypotenuse in two ways: (a) using the Pythagorean Theorem and (b) using the Distance Formula.

13.
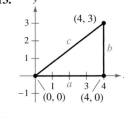

14.
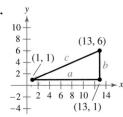

15.
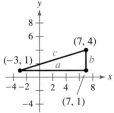

16.
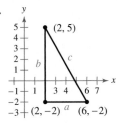

In Exercises 17 and 18, find x such that the distance between the points is 15.

17. $(3, -4), (x, 5)$

18. $(x, 8), (-9, -4)$

In Exercises 19 and 20, find y such that the distance between the points is 20.

19. $(-15, y), (-3, -7)$

20. $(6, -1), (-10, y)$

In Exercises 21–24, determine whether each point is a solution of the equation.

	Equation	Points	
21.	$2x - 3y + 11 = 0$	(a) $(2, 5)$	(b) $(3, 2)$
22.	$y = 2x^2 - 7x + 3$	(a) $(1, -1)$	(b) $(3, 0)$
23.	$y = \sqrt{x - 5}$	(a) $(9, 2)$	(b) $(21, 4)$
24.	$y = \dfrac{x + 1}{5 - x}$	(a) $\left(1, \frac{1}{2}\right)$	(b) $(0, 1)$

In Exercises 25 and 26, complete the table. Use the resulting solution points to sketch the graph of the equation.

25. $y = \frac{3}{4}x - 1$

x	-2	0	1	$\frac{4}{3}$	2
y					

26. $y = 5 - x^2$

x	-2	-1	0	1	2
y					

In Exercises 27–34, find the *x*- and *y*-intercepts of the graph of the equation.

27. $y = 2x - 1$

28. $y = (x - 4)(x + 2)$

29. $y = x^2 + x - 2$

30. $y = 4 - x^2$

31. $y = x\sqrt{x + 2}$

32. $y = x\sqrt{x + 5}$

33. $2y - xy + 3x = 4$

34. $x^2y - x^2 + 4y = 0$

35. Use your knowledge of the Cartesian plane and intercepts to explain why you let *y* equal zero when you are finding the *x*-intercepts of the graph of an equation, and why you let *x* equal zero when you are finding the *y*-intercepts of the graph of an equation.

36. Is it possible for a graph to have no *x*-intercepts? no *y*-intercepts? no *x*-intercepts and no *y*-intercepts? Give examples to support your answers.

In Exercises 37–48, check for symmetry with respect to both axes and the origin.

37. $x^4 - 2y = 0$

38. $y = x^4 - x^2 + 3$

39. $x - y^2 = 0$

40. $y^2 = x + 2$

41. $y = \sqrt{16 - x^2}$

42. $y = \sqrt{4 - x^2}$

43. $xy = 2$

44. $x^3y = 1$

45. $y = \dfrac{x}{x^2 - 4}$

46. $y = \dfrac{x}{x^2 + 1}$

47. $x^2 + y^2 = 25$

48. $x^2 + y^2 = 9$

In Exercises 49–52, use symmetry to complete the graph of the equation.

49. *y*-axis symmetry

$y = -x^2 + 4$

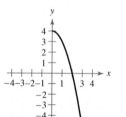

50. *x*-axis symmetry

$y^2 = -x + 4$

51. Origin symmetry

$y = -x^3 + x$

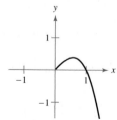

52. *y*-axis symmetry

$y = |x| - 2$

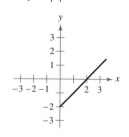

In Exercises 53–58, match the equation with its graph. [The graphs are labeled (a), (b), (c), (d), (e), and (f).]

(a)
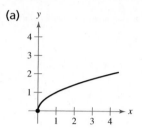

(b)
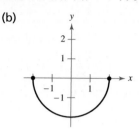

(c)
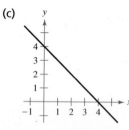

(d)
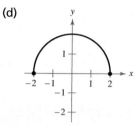

(e)
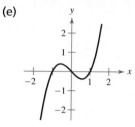

(f)
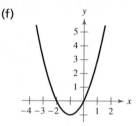

53. $y = 4 - x$

54. $y = \sqrt{4 - x^2}$

55. $y = x^2 + 2x$

56. $y = \sqrt{x}$

57. $y = x^3 - x$

58. $y = -\sqrt{4 - x^2}$

In Exercises 59–78, sketch the graph of the equation. Identify any intercepts and test for symmetry.

59. $y = 5 - 3x$

60. $y = 2x - 3$

61. $y = 1 - x^2$

62. $y = x^2 - 1$

63. $y = x^2 - 4x + 3$

64. $y = -x^2 - 4x$

65. $y = x^3 + 2$

66. $y = x^3 - 1$

67. $y = \dfrac{8}{x^2 + 4}$

68. $y = \dfrac{4}{x^2 + 1}$

69. $y = \sqrt{x + 1}$

70. $y = \sqrt{1 - x}$

71. $y = \sqrt[3]{x}$

72. $y = \sqrt[3]{x + 1}$

73. $y = |x - 4|$

74. $y = |x| - 3$

75. $x = y^2 - 1$

76. $x = y^2 - 4$

77. $x^2 + y^2 = 4$

78. $x^2 + y^2 = 16$

In Exercises 79–82, find the radius of the circle given the center and a point on the circle.

79.

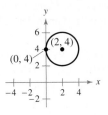

80.

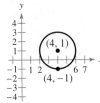

81.

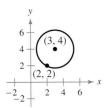

82.

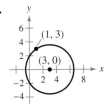

In Exercises 83 and 84, find the center and radius of the circle given the endpoints of the diameter of the circle.

83.

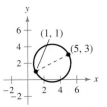

84.

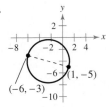

In Exercises 85–92, find the standard form of the equation of the specified circle.

85. Center: $(0, 0)$; radius: 3

86. Center: $(0, 0)$; radius: 5

87. Center: $(-4, 1)$; radius: $\sqrt{2}$

88. Center: $\left(0, \frac{1}{2}\right)$; radius: $\frac{2}{3}$

89. Center: $(-1, 2)$; point on circle: $(0, 0)$

90. Center: $(3, -2)$; point on circle: $(-1, 1)$

91. Endpoints of a diameter: $(-3, 4)$, $(5, -2)$

92. Endpoints of a diameter: $(-4, -1)$, $(4, 1)$

In Exercises 93–100, write the equation of the circle in standard form. Then sketch the circle.

93. $x^2 + y^2 - 6x + 4y - 3 = 0$

94. $x^2 + y^2 - 2x + 6y - 15 = 0$

95. $x^2 + y^2 - 4x + 6y + 9 = 0$

96. $5x^2 + 5y^2 + 10x + 1 = 0$

97. $2x^2 + 2y^2 - 2x - 2y - 3 = 0$

98. $4x^2 + 4y^2 - 4x + 2y - 1 = 0$

99. $16x^2 + 16y^2 + 16x + 40y - 7 = 0$

100. $x^2 + y^2 - 4x + 2y + 3 = 0$

In Exercises 101 and 102, an equation of a circle is written in standard form. Indicate the coordinates of the center of the circle and determine the radius of the circle. Rewrite the equation of the circle in general form.

101. $(x - 3)^2 + (y + 1)^2 = 25$

102. $\left(x - \frac{1}{2}\right)^2 + (y - 2)^2 = 7$

Gold Prices In Exercises 103 and 104, use the figure below, which shows the average prices of gold for the years 1975 to 2005. *(Sources: U.S. Bureau of Mines; U.S. Geological Survey)*

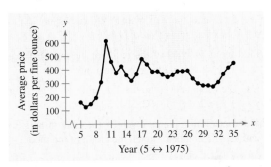

103. What is the highest price of gold shown in the graph? When did this occur?

104. What is the lowest price of gold shown in the graph? When did this occur?

Vehicle Distance In Exercises 105 and 106, use the figure below, which shows the total number of miles traveled by vehicles in the United States each year from 1994 to 2005. *(Source: National Highway Traffic Safety Administration)*

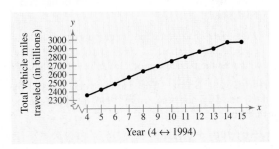

105. Estimate the percent increase in miles traveled by vehicles from 1994 to 2000.

106. Estimate the percent increase in miles traveled by vehicles from 2000 to 2005.

107. Population   The population y (in millions of people) of North America from 1980 to 2050 can be modeled by

$$y = 5.3x + 377, \quad -20 \le x \le 50$$

where x represents the year, with $x = 50$ corresponding to 2050.   (*Source: U.S. Census Bureau*)

(a) Find the y-intercept of the graph of the model. What does it represent in the given situation?

(b) Construct a table of values for $x = -20, -10, 0, 10, 20, 30, 40,$ and 50.

(c) Plot the solution points given by the table in part (b) and use the points to sketch the graph of the model.

108. Profit   The annual profits y (in millions of dollars) of UnitedHealth Group from 1997 to 2006 can be modeled by

$$y = 58.86x^2 + 228.0x + 677, \quad -3 \le x \le 6$$

where x represents the year, with $x = 6$ corresponding to 2006.   (*Source: UnitedHealth Group*)

(a) Find the y-intercept of the graph of the model. What does it represent in the given situation?

(b) Use the model to complete the table of values for $x = -3, -2, -1, 0, 1, 2, 3, 4, 5,$ and 6.

(c) Plot the solution points given by the table in part (b) and use the points to sketch the graph of the model.

109. Earnings per Share   The earnings per share y (in dollars) for Dollar Tree Stores from 1996 to 2005 can be modeled by

$$y = -0.0082t^2 + 0.318t - 1.28, \quad 6 \le t \le 15$$

where t represents the year, with $t = 6$ corresponding to 1996.   (*Source: Dollar Tree Stores*)

(a) Sketch a graph of the equation.

(b) In 2005, Dollar Tree predicted that its earnings per share would be $1.75 in 2006 and $1.95 in 2007. Use the model to predict the earnings per share for these years. How well does the model support Dollar Tree's predictions?

(c) Dollar Tree also predicted its earnings per share to reach $3.00 sometime in 2009, 2010, or 2011. How well does the model support Dollar Tree's prediction?

110. Earnings per Share   The earnings per share y (in dollars) for Paychex, Inc. from 1996 to 2005 can be modeled by

$$y = -0.0014t^2 + 0.123t - 0.57, \quad 6 \le t \le 15$$

where t represents the year, with $t = 6$ corresponding to 1996.   (*Source: Paychex, Inc.*)

(a) Sketch a graph of the equation.

(b) In 2005, Paychex predicted that its earnings per share would be $1.22 in 2006 and $1.40 in 2007. Use the model to predict the earnings per share for these years. How well does the model support Paychex's predictions?

(c) Paychex also predicted its earnings per share to reach $1.90 sometime in 2009, 2010, or 2011. How well does the model support Paychex's prediction?

111. The London Eye   The London Eye is a Ferris wheel that opened in 1999 in London, England as the tallest in the world. It remained the tallest Ferris wheel in the world until the Star of Nanchang was opened in 2006. The London Eye stands 135 meters tall. Use the diagram below to write an equation that models the circular shape of the London Eye wheel.

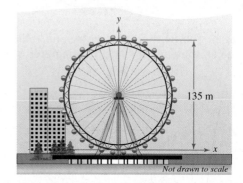

Not drawn to scale

112. Star of Nanchang   In 2006, a Ferris wheel called the Star of Nanchang was opened in the Jiangxi province of China to replace the London Eye as the world's largest Ferris wheel. The Star of Nanchang stands 525 feet tall. Use the diagram below to write an equation that models the circular shape of the Star of Nanchang wheel.

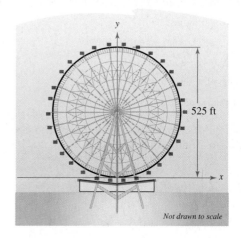

Not drawn to scale

Section 2.2

Lines in the Plane

- Find the slope of a line passing through two points.
- Use the point-slope form to find the equation of a line.
- Use the slope-intercept form to sketch a line.
- Use slope to determine if lines are parallel or perpendicular, and write the equation of a line parallel or perpendicular to a given line.

The Slope of a Line

The **slope** of a nonvertical line is a measure of the steepness of the line. The slope represents the number of units the line rises or falls vertically for each unit of horizontal change from left to right. For instance, consider the two points (x_1, y_1) and (x_2, y_2) on the line shown in Figure 2.20. As you move from left to right along this line, a change of $y_2 - y_1$ units in the vertical direction corresponds to a change of $x_2 - x_1$ units in the horizontal direction. That is,

$$y_2 - y_1 = \text{the change in } y \quad \text{and} \quad x_2 - x_1 = \text{the change in } x.$$

The slope of the line is defined as the quotient of these two changes.

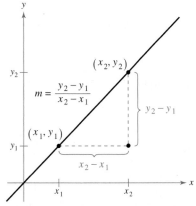

FIGURE 2.20

Definition of the Slope of a Line

The **slope** m of the nonvertical line passing through the points (x_1, y_1) and (x_2, y_2) is

$$m = \frac{y_2 - y_1}{x_2 - x_1} = \frac{\text{change in } y}{\text{change in } x}$$

where $x_1 \neq x_2$.

The change in x is sometimes called the *run* and the change in y is sometimes called the *rise*.

When this formula is used for slope, the *order of subtraction* is important. Given two points on a line, you are free to label either one of them as (x_1, y_1) and the other as (x_2, y_2). However, once this is done, you must form the numerator and denominator using the same order of subtraction.

$$m = \frac{y_2 - y_1}{x_2 - x_1} \qquad m = \frac{y_1 - y_2}{x_1 - x_2} \qquad m = \frac{y_2 - y_1}{x_1 - x_2}$$

Correct · Correct · Incorrect

In real-life problems, such as finding the steepness of a ramp or the increase in the value of a product, the slope of a line can be interpreted as either a *ratio* or a *rate*. If the x-axis and the y-axis have the same units of measure, then the slope has no units and is a *ratio*. If the x-axis and the y-axis have different units of measure, then the slope is a *rate* or *rate of change*. You will learn more about rates of change in Section 2.3.

| Example 1 | Finding the Slope of a Line Through Two Points |

Find the slope of the line passing through each pair of points.

a. $(-2, 0)$ and $(3, 1)$ **b.** $(-1, 2)$ and $(2, 2)$

c. $(0, 4)$ and $(1, -1)$ **d.** $(3, 4)$ and $(3, 1)$

SOLUTION

a. Letting $(x_1, y_1) = (-2, 0)$ and $(x_2, y_2) = (3, 1)$, you obtain a slope of

$$m = \frac{y_2 - y_1}{x_2 - x_1}$$

⬅ Difference in y-values

⬅ Difference in x-values

$$= \frac{1 - 0}{3 - (-2)}$$

$$= \frac{1}{5}.$$

b. The slope of the line passing through $(-1, 2)$ and $(2, 2)$ is

$$m = \frac{2 - 2}{2 - (-1)} = \frac{0}{3} = 0.$$

c. The slope of the line passing through $(0, 4)$ and $(1, -1)$ is

$$m = \frac{-1 - 4}{1 - 0} = \frac{-5}{1} = -5.$$

d. The slope of the line passing through $(3, 4)$ and $(3, 1)$ is undefined. Applying the formula for slope, you have

$$m = \frac{1 - 4}{3 - 3} = \frac{-3}{0}.$$ Division by zero is undefined.

Because division by zero is not defined, the slope of a vertical line is not defined.

The graphs of the four lines are shown in Figure 2.21.

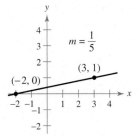

(a) If m is positive, the line rises from left to right.

(b) If m is zero, the line is horizontal.

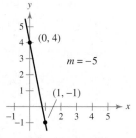

(c) If m is negative, the line falls from left to right.

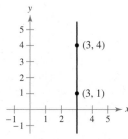

(d) If the line is vertical, the slope is undefined.

FIGURE 2.21

✓ **CHECKPOINT 1**

Find the slope of the line passing through the points $(-1, 2)$ and $(3, 4)$. ■

From the slopes of the lines shown in Example 1, you can make the following generalizations about the slope of a line.

Slope of a Line

1. A line with positive slope $(m > 0)$ *rises* from left to right.

2. A line with negative slope $(m < 0)$ *falls* from left to right.

3. A line with zero slope $(m = 0)$ is *horizontal*.

4. A line with undefined slope is *vertical*.

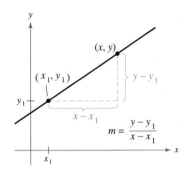

FIGURE 2.22   Any two points on a line can be used to determine the slope of the line.

The Point-Slope Form

If you know the slope of a line and the coordinates of one point on the line, you can find an equation for the line. For instance, in Figure 2.22, let (x_1, y_1) be a given point on the line whose slope is m. If (x, y) is *any other* point on the line, it follows that

$$\frac{y - y_1}{x - x_1} = m.$$

This equation in the variables x and y can be rewritten to produce the following **point-slope form** of the equation of a line.

Point-Slope Form of the Equation of a Line

The **point-slope form** of the equation of the line that passes through the point (x_1, y_1) and has a slope of m is

$$y - y_1 = m(x - x_1).$$

Example 2   **The Point-Slope Form of the Equation of a Line**

Find an equation of the line that passes through $(1, -2)$ and has a slope of 3.

SOLUTION   Use the point-slope form with $(x_1, y_1) = (1, -2)$ and $m = 3$.

$$y - y_1 = m(x - x_1) \qquad \text{Point-slope form}$$

$$y - (-2) = 3(x - 1) \qquad \text{Substitute } y_1 = -2, x_1 = 1, \text{ and } m = 3.$$

$$y + 2 = 3x - 3 \qquad \text{Simplify.}$$

$$y = 3x - 5 \qquad \text{Equation of line}$$

The graph of this line is shown in Figure 2.23.

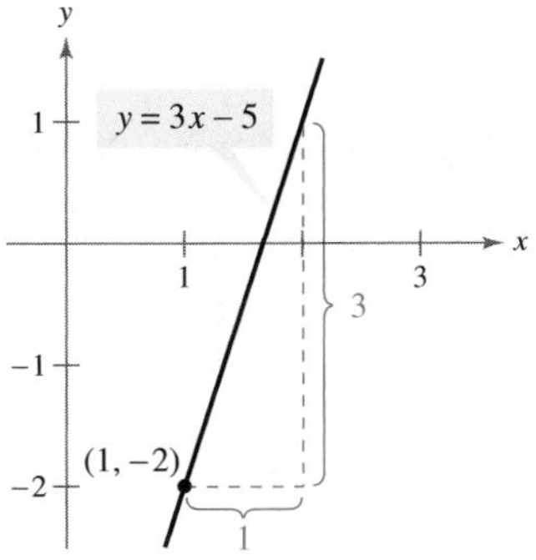

FIGURE 2.23

✓**CHECKPOINT 2**

Find an equation of the line that passes through the given point and has the given slope.

a. $(2, 4), m = -2$

b. $(-8, -3), m = \frac{3}{2}$ ■

The point-slope form can be used to find the equation of a line passing through two points (x_1, y_1) and (x_2, y_2). First, use the formula for the slope of a line passing through two points. Then, use the point-slope form to obtain

$$y - y_1 = \frac{y_2 - y_1}{x_2 - x_1}(x - x_1).$$

This is sometimes called the **two-point form** of the equation of a line.

Example 3 A Linear Model for Sales Prediction

During the first two quarters of the year, a jewelry company had sales of \$3.4 million and \$3.7 million, respectively. (a) Write a linear equation giving the sales y in terms of the quarter x. (b) Use the equation to predict the sales during the fourth quarter. Can you assume that sales will follow this linear pattern?

SOLUTION

a. Let $(1, 3.4)$ and $(2, 3.7)$ be two points on the line representing the total sales. Use the two-point form to find an equation of the line.

$$y - 3.4 = \frac{3.7 - 3.4}{2 - 1}(x - 1) \qquad \text{Substitute for } x_1, y_1, x_2, \text{ and } y_2 \text{ in two-point form.}$$

$$y - 3.4 = 0.3(x - 1) \qquad \text{Simplify quotient.}$$

$$y = 0.3x + 3.1 \qquad \text{Equation of line}$$

b. Using the equation from part (a), the fourth-quarter sales $(x = 4)$ should be $y = 0.3(4) + 3.1 = \$4.3$ million. See Figure 2.24. Without more data, you cannot assume that the sales pattern will be linear. Many factors, such as seasonal demand and past sales history, help to determine the sales pattern.

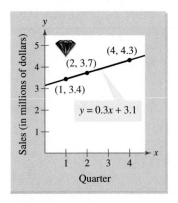

FIGURE 2.24

✓ CHECKPOINT 3

A company has sales of \$1.2 million and \$1.4 million in its first two years. Write a linear equation giving the sales y in terms of the year x. ■

The estimation method illustrated in Example 3 is called **linear extrapolation.** Note in Figure 2.25(a) that for linear extrapolation, the estimated point lies to the *right* of the given points. When the estimated point lies *between* two given points, the procedure is called **linear interpolation,** as shown in Figure 2.25(b).

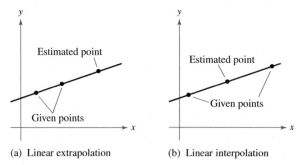

(a) Linear extrapolation   (b) Linear interpolation

FIGURE 2.25

Sketching Graphs of Lines

You have seen that to *find the equation of a line* it is convenient to use the point-slope form. This formula, however, is not particularly useful for *sketching the graph of a line*. The form that is better suited to graphing linear equations is the **slope-intercept form** of the equation of a line. To derive the slope-intercept form, write the following.

$$y - y_1 = m(x - x_1) \qquad \text{Point-slope form}$$
$$y = mx - mx_1 + y_1 \qquad \text{Solve for } y.$$
$$y = mx + (y_1 - mx_1) \qquad \text{Commutative Property of Addition}$$
$$y = mx + b \qquad \text{Slope-intercept form}$$

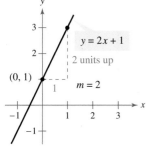

(a) When m is positive, the line rises from left to right.

Slope-Intercept Form of the Equation of a Line

The graph of the equation

$$y = mx + b$$

is a line whose slope is m and whose y-intercept is $(0, b)$.

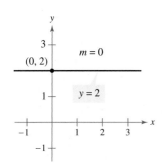

(b) When m is zero, the line is horizontal.

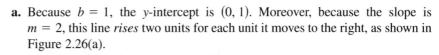 **Example 4** **Sketching the Graphs of Linear Equations**

Sketch the graph of each linear equation.

a. $y = 2x + 1$

b. $y = 2$

c. $x + y = 2$

SOLUTION

a. Because $b = 1$, the y-intercept is $(0, 1)$. Moreover, because the slope is $m = 2$, this line *rises* two units for each unit it moves to the right, as shown in Figure 2.26(a).

b. By writing the equation $y = 2$ in the form

$$y = (0)x + 2$$

you can see that the y-intercept is $(0, 2)$ and the slope is zero. A zero slope implies that the line is horizontal, as shown in Figure 2.26(b).

c. By writing the equation $x + y = 2$ in slope-intercept form

$$y = -x + 2$$

you can see that the y-intercept is $(0, 2)$. Moreover, because the slope is $m = -1$, this line *falls* one unit for each unit it moves to the right, as shown in Figure 2.26(c).

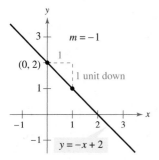

(c) When m is negative, the line falls from left to right.

FIGURE 2.26

✓ CHECKPOINT 4

Sketch the graph of the linear equation $y - 2x = -3$. ■

From the slope-intercept form of the equation of a line, you can see that a horizontal line ($m = 0$) has an equation of the form

$$y = (0)x + b \quad \text{or} \quad y = b. \qquad \text{Horizontal line}$$

This is consistent with the fact that each point on a horizontal line through $(0, b)$ has a y-coordinate of b, as shown in Figure 2.27.

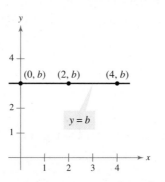

FIGURE 2.27   Horizontal Line   **FIGURE 2.28**   Vertical Line

Similarly, each point on a vertical line through $(a, 0)$ has an x-coordinate of a, as shown in Figure 2.28. So, a vertical line has an equation of the form

$$x = a. \qquad\qquad\qquad\qquad \text{Vertical line}$$

This equation cannot be written in slope-intercept form because the slope of a vertical line is undefined. However, *every* line has an equation that can be written in the **general form**

$$Ax + By + C = 0 \qquad\qquad \text{General form}$$

where A and B are not *both* zero. If $A = 0$ (and $B \neq 0$), the general equation can be reduced to the form $y = b$, which represents a horizontal line. If $B = 0$ (and $A \neq 0$), the general equation can be reduced to the form $x = a$, which represents a vertical line.

Summary of Equations of Lines

1. General form: $Ax + By + C = 0$

2. Vertical line: $x = a$

3. Horizontal line: $y = b$

4. Slope-intercept form: $y = mx + b$

5. Point-slope form: $y - y_1 = m(x - x_1)$

DISCOVERY

Use a graphing utility to graph each equation in the same viewing window.

$$y_1 = \tfrac{3}{2}x - 1 \qquad\qquad y_2 = \tfrac{3}{2}x \qquad\qquad y_3 = \tfrac{3}{2}x + 2$$

What is true about the graphs? What do you notice about the slopes of the equations?

Parallel and Perpendicular Lines

The slope of a line is a convenient tool for determining whether two lines are parallel, perpendicular, or neither.

Parallel Lines

Two distinct nonvertical lines are **parallel** if and only if their slopes are equal.

Example 5   Equations of Parallel Lines

Find an equation of the line that passes through the point $(2, -1)$ and is parallel to the line $2x - 3y = 5$, as shown in Figure 2.29.

SOLUTION   Start by rewriting the equation in slope-intercept form.

$$2x - 3y = 5 \qquad \text{Write original equation.}$$

$$-3y = -2x + 5 \qquad \text{Subtract } 2x \text{ from each side.}$$

$$y = \frac{2}{3}x - \frac{5}{3} \qquad \text{Write in slope-intercept form.}$$

So, the given line has a slope of $m = \frac{2}{3}$. Because any line parallel to the given line must also have a slope of $\frac{2}{3}$, the required line through $(2, -1)$ has the following equation.

$$y - y_1 = m(x - x_1) \qquad \text{Point-slope form}$$

$$y - (-1) = \frac{2}{3}(x - 2) \qquad \text{Substitute for } y_1, x_1, \text{ and } m.$$

$$y + 1 = \frac{2}{3}x - \frac{4}{3} \qquad \text{Simplify.}$$

$$y = \frac{2}{3}x - \frac{4}{3} - 1 \qquad \text{Solve for } y.$$

$$y = \frac{2}{3}x - \frac{7}{3} \qquad \text{Write in slope-intercept form.}$$

Notice the similarity between the slope-intercept form of the original equation and the slope-intercept form of the parallel equation.

✓CHECKPOINT 5

Find an equation that passes through the point $(2, 4)$ and is parallel to the line $2y - 6x = 2$. ■

You have seen that two nonvertical lines are parallel if and only if they have the same slope. Two nonvertical lines are *perpendicular* if and only if their slopes are negative reciprocals of each other. For instance, the lines $y = 2x$ and $y = -\frac{1}{2}x$ are perpendicular because one has a slope of $2 = \frac{2}{1}$ and the other has a slope of $-\frac{1}{2}$.

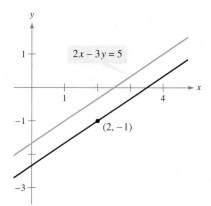

FIGURE 2.29

DISCOVERY

Use a graphing utility to graph each equation in the same viewing window.

$$y_1 = \tfrac{2}{3}x + \tfrac{5}{2}$$

$$y_2 = -\tfrac{3}{2}x + 2$$

When you examine the graphs with a square setting, what do you observe? What do you notice about the slopes of the two lines?

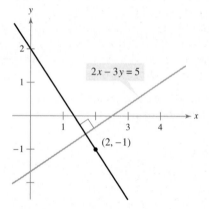

FIGURE 2.30

Perpendicular Lines

Two nonvertical lines are **perpendicular** if and only if their slopes are negative reciprocals of each other. That is,

$$m_1 = -\frac{1}{m_2}.$$

Example 6   **Equations of Perpendicular Lines**

Find an equation of the line that passes through the point $(2, -1)$ and is perpendicular to the line $2x - 3y = 5$, as shown in Figure 2.30.

SOLUTION   By writing the equation of the original line in slope-intercept form

$$y = \frac{2}{3}x - \frac{5}{3}$$

you can see that the line has a slope of $\tfrac{2}{3}$. So, any line that is perpendicular to this line must have a slope of $-\tfrac{3}{2}$ (because $-\tfrac{3}{2}$ is the negative reciprocal of $\tfrac{2}{3}$). The required line through the point $(2, -1)$ has the following equation.

$y - y_1 = m(x - x_1)$	Point-slope form
$y - (-1) = -\dfrac{3}{2}(x - 2)$	Substitute for y_1, x_1, and m.
$y + 1 = -\dfrac{3}{2}x + 3$	Simplify.
$y = -\dfrac{3}{2}x + 3 - 1$	Solve for y.
$y = -\dfrac{3}{2}x + 2$	Write in slope-intercept form.

✓**CHECKPOINT 6**

Find an equation of the line that passes through the point $(-2, 12)$ and is perpendicular to the line $y = \tfrac{1}{4}x - 2$. ■

CONCEPT CHECK

1. What is the slope of a line that falls five units for each two units it moves to the right?

2. What is an equation of a horizontal line that passes through the point (a, b)?

3. Why is it convenient to use the slope-intercept form when sketching the graph of a linear equation?

4. Line A and line B are perpendicular to each other and the slope of line A is 1/2. What is the slope of line B?

Skills Review 2.2   The following warm-up exercises involve skills that were covered in earlier sections. You will use these skills in the exercise set for this section. For additional help, review Sections 0.2 and 1.1.

In Exercises 1–4, simplify the expression.

1. $\dfrac{4 - (-5)}{-3 - (-1)}$

2. $\dfrac{-5 - 8}{0 - (-3)}$

3. Find $-1/m$ for $m = 4/5$.

4. Find $-1/m$ for $m = -2$.

In Exercises 5–10, solve for y in terms of x.

5. $2x - 3y = 5$

6. $4x + 2y = 0$

7. $y - (-4) = 3[x - (-1)]$

8. $y - 7 = \dfrac{2}{3}(x - 3)$

9. $y - (-1) = \dfrac{3 - (-1)}{2 - 4}(x - 4)$

10. $y - 5 = \dfrac{3 - 5}{0 - 2}(x - 2)$

Exercises 2.2

See www.CalcChat.com for worked-out solutions to odd-numbered exercises.

In Exercises 1–6, estimate the slope of the line.

1.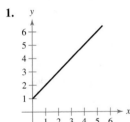

2.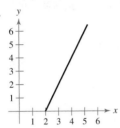

3.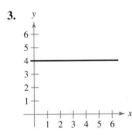

4.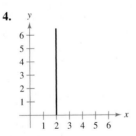

5.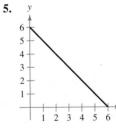

6.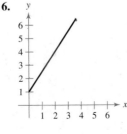

In Exercises 7 and 8, determine if a line with the following description has a positive slope, a negative slope, or an undefined slope.

7. Line rises from left to right

8. Vertical line

In Exercises 9 and 10, sketch the lines through the point with the indicated slopes on the same set of coordinate axes.

	Point	Slopes	
9.	$(-3, 4)$	(a) -2	(b) $\frac{2}{3}$
		(c) 0	(d) Undefined
10.	$(-2, -5)$	(a) -1	(b) $\frac{3}{4}$
		(c) 0	(d) Undefined

In Exercises 11–16, plot the points and find the slope of the line passing through the points.

11. $(6, 9), (-4, -1)$

12. $(2, 4), (4, -4)$

13. $(-6, -1), (-6, 4)$

14. $(0, -10), (-4, 0)$

15. $\left(-\frac{1}{3}, 1\right), \left(-\frac{2}{3}, \frac{5}{6}\right)$

16. $\left(\frac{7}{8}, \frac{3}{4}\right), \left(\frac{5}{4}, -\frac{1}{4}\right)$

In Exercises 17–24, use the point on the line and the slope of the line to find three additional points through which the line passes. (There are many correct answers.)

	Point	Slope
17.	$(5, -2)$	$m = 0$
18.	$(-3, 4)$	$m = 0$
19.	$(2, -5)$	m is undefined.
20.	$(-1, 3)$	m is undefined.
21.	$(5, -6)$	$m = 1$
22.	$(10, -6)$	$m = -1$
23.	$(-6, -1)$	$m = \frac{1}{2}$
24.	$(7, -5)$	$m = -\frac{2}{3}$

In Exercises 25–38, find an equation of the line that passes through the point and has the indicated slope. Then sketch the line.

Point	Slope
25. $(7, 0)$	$m = 1$
26. $(0, -4)$	$m = -1$
27. $(-2, 0)$	$m = -4$
28. $(1, 3)$	$m = 3$
29. $(-3, 6)$	$m = -2$
30. $(-8, 3)$	$m = -\frac{1}{2}$
31. $(4, 0)$	$m = -\frac{1}{3}$
32. $(-2, -5)$	$m = \frac{3}{4}$
33. $(6, -1)$	m is undefined.
34. $(3, -2)$	m is undefined.
35. $(-2, -7)$	$m = 0$
36. $(-10, 4)$	$m = 0$
37. $\left(4, \frac{5}{2}\right)$	$m = \frac{4}{3}$
38. $\left(-\frac{1}{2}, \frac{3}{2}\right)$	$m = -3$

In Exercises 39–48, find the slope and y-intercept (if possible) of the line specified by the equation. Then sketch the line.

39. $y = 2x - 1$ **40.** $y = 3 - x$

41. $4x - y - 6 = 0$ **42.** $2x + 3y - 9 = 0$

43. $8 - 3x = 0$ **44.** $2x + 5 = 0$

45. $7x + 6y - 30 = 0$ **46.** $x - y - 10 = 0$

47. $2y - 7 = 0$ **48.** $8 - 5y = 0$

In Exercises 49–60, find an equation of the line passing through the points.

49. $(2, 5), (-1, -4)$ **50.** $(6, -1), (-2, 1)$

51. $(7, -4), (-7, 3)$ **52.** $(4, 3), (-4, -4)$

53. $(-9, 11), (-9, 14)$ **54.** $(3, 5), (3, -2)$

55. $(-1, 7), (3, 7)$ **56.** $(3, -2), (-8, -2)$

57. $\left(2, \frac{1}{2}\right), \left(\frac{1}{2}, \frac{5}{4}\right)$ **58.** $\left(1, 1\right), \left(6, -\frac{2}{3}\right)$

59. $(1, 0.6), (-2, -0.6)$ **60.** $(-8, 0.6), (2, -2.4)$

61. A fellow student does not understand why the slope of a vertical line is undefined. Describe how you would help this student understand the concept of undefined slope.

62. Another student overhears your conversation in Exercise 61 and states, "I do not understand why a horizontal line has zero slope and how that is different from undefined or no slope." Describe how you would explain the concepts of zero slope and undefined slope and how they are different from each other.

In Exercises 63–68, use the *intercept form* to find the equation of the line with the given intercepts. The intercept form of the equation of a line with intercepts $(a, 0)$ and $(0, b)$ is

$$\frac{x}{a} + \frac{y}{b} = 1, \quad a \neq 0, \quad b \neq 0.$$

63. x-intercept: $(1, 0)$ **64.** x-intercept: $(-3, 0)$
  y-intercept: $(0, -4)$ y-intercept: $(0, 4)$

65. x-intercept: $(-2, 0)$ **66.** x-intercept: $(5, 0)$
  y-intercept: $(0, -2)$ y-intercept: $(0, 1)$

67. x-intercept: $\left(-\frac{1}{6}, 0\right)$ **68.** x-intercept: $\left(-\frac{2}{3}, 0\right)$
  y-intercept: $\left(0, -\frac{2}{3}\right)$ y-intercept: $\left(0, \frac{1}{2}\right)$

In Exercises 69–76, the equations of two lines are given. Determine if lines L_1 and L_2 are parallel, perpendicular, or neither.

69. $L_1: y = 3x + 4$; $L_2: y = x - \frac{1}{4}$

70. $L_1: y = \frac{3}{4}x + 1$; $L_2: y = -\frac{4}{3}x + 3$

71. $L_1: 2x - y = 1$; $L_2: x + 2y = -1$

72. $L_1: x - 5y = -2$; $L_2: -3x + 15y = 6$

73. $L_1: x - 3y = -3$; $L_2: 2x - 6y = 6$

74. $L_1: 4x - y = -2$; $L_2: 8x - 2y = 6$

75. $L_1: 2x - 3y - 15 = 0$; $L_2: 3x + 2y + 8 = 0$

76. $L_1: x - 4y - 12 = 0$; $L_2: 3x - 4y - 8 = 0$

In Exercises 77–84, determine if the lines L_1 and L_2 passing through the indicated pairs of points are parallel, perpendicular, or neither.

77. $L_1: (-5, 0), (-2, 1)$; $L_2: (0, 1), (3, 2)$

78. $L_1: (-1, 6), (1, 4)$; $L_2: (3, -3), (6, -9)$

79. $L_1: (0, -1), (5, 9)$; $L_2: (0, 3), (4, 1)$

80. $L_1: (3, 6), (-6, 0)$; $L_2: (0, -1), \left(5, \frac{7}{3}\right)$

81. $L_1: (-2, -1), (1, 5)$; $L_2: (1, 3), (5, -5)$

82. $L_1: (4, 8), (-4, 2)$; $L_2: (3, -5), \left(-1, \frac{1}{3}\right)$

83. $L_1: (-1, 7), (-6, 4)$; $L_2: (0, 1), (5, 4)$

84. $L_1: (-1, 3), (2, -5)$; $L_2: (3, 0), (2, -7)$

In Exercises 85–90, write equations of the lines through the point (a) parallel to the given line and (b) perpendicular to the given line.

Point	Line
85. $(6, 2)$	$y - 2x = -1$
86. $(-5, 4)$	$x + y = 8$
87. $\left(\frac{1}{4}, -\frac{2}{3}\right)$	$2x - 3y = 5$

88. $\left(\frac{7}{8}, \frac{3}{4}\right)$ $\qquad$ $5x + 3y = 0$

89. $(-1, 0)$ $\qquad$ $y = -3$

90. $(2, 5)$ $\qquad$ $x = 4$

91. Temperature Find an equation of the line that gives the relationship between the temperature in degrees Celsius C and the temperature in degrees Fahrenheit F. Remember that water freezes at $0°$ Celsius ($32°$ Fahrenheit) and boils at $100°$ Celsius ($212°$ Fahrenheit).

92. Temperature Use the result of Exercise 91 to complete the table. Is there a temperature for which the Fahrenheit reading is the same as the Celsius reading? If so, what is it?

C		$-10°$	$10°$			$177°$
F	$0°$			$68°$	$90°$	

93. Simple Interest A person deposits P dollars in an account that pays simple interest. After 2 months, the balance in the account is $813 and after 3 months, the balance in the account is $819.50. Find an equation that gives the relationship between the balance A and the time t in months.

94. Simple Interest Use the result of Exercise 93 to complete the table.

A			$813.00	$819.50			
t	0	1			4	5	6

95. Wheelchair Ramp The maximum recommended slope of a wheelchair ramp is $\frac{1}{12}$. A business is installing a wheelchair ramp that rises 34 inches over a horizontal length of 30 feet. Is the ramp steeper than recommended? *(Source: Americans with Disabilities Act Handbook)*

96. Revenue A line representing daily revenues y in terms of time x in days has a slope of $m = 100$. Interpret the change in daily revenues for a one-day increase in time.

97. College Enrollment A small college had 3125 students in 2005 and 3582 students in 2008. The enrollment follows a linear growth pattern. How many students will the college have in 2012?

98. Annual Salary Your salary was $30,200 in 2007 and $33,500 in 2009. Your salary follows a linear growth pattern. What salary will you be making in 2012?

99. *MAKE A DECISION: FOURTH-QUARTER SALES* During the first and second quarters of the year, a business had sales of $158,000 and $165,000. From these data, can you assume that the sales follow a linear growth pattern? If the pattern is linear, what will the sales be during the fourth quarter?

100. Fatal Crashes In 1998, there were 37,107 motor vehicle traffic crashes involving fatalities in the United States. In 2005, there were 39,189 such crashes. Assume that the trend is linear. Predict the number of crashes with fatalities in 2007. *(Source: National Highway Traffic Safety Administration)*

101. *MAKE A DECISION: YAHOO! INC. REVENUE* In 2000, Yahoo! Inc. had revenues of $1110.2 million. In 2003, their revenues were $1625.1 million. Assume the revenue followed a linear trend. What would the approximate revenue have been in 2005? The actual revenue in 2005 was $5257.7 million. Do you think the yearly revenue followed a linear trend? Explain your reasoning. *(Source: Yahoo! Inc.)*

102. Applebee's Revenue Applebee's is one of the largest casual dining chains in the United States. In 2000, Applebee's had revenues of $690.2 million. In 2004, their revenues were $1111.6 million. Assume the yearly revenue followed a linear trend. What would the approximate revenue have been in 2005? The actual revenue in 2005 was $1216.6 million. From these data, is it possible that Applebee's yearly revenue followed a linear trend? Explain your reasoning. *(Source: Applebee's International, Inc.)*

103. Scuba Diving The pressure (in atmospheres) exerted on a scuba diver's body has a linear relationship with the diver's depth. At sea level (or a depth of 0 feet), the pressure exerted on a diver is 1 atmosphere. At a depth of 99 feet, the pressure exerted on a diver is 4 atmospheres. Write a linear equation to describe the pressure p (in atmospheres) in terms of the depth d (in feet) below the surface of the sea. What is the rate of change of pressure with respect to depth? *(Source: PADI Open Water Diver Manual)*

104. Stone Cutting A stone cutter is making a 6-foot tall memorial stone. The diagram shows coordinates labeled in feet. The stone cutter plans to make the cut indicated by the dashed line. This cut follows a line perpendicular to one side of the stone that passes through the point labeled $(-1, 6)$. Find an equation of the line of the cut.

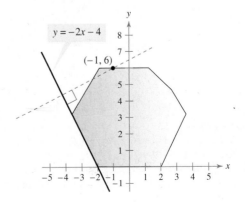

Section 2.3

Linear Modeling and Direct Variation

- Use a mathematical model to approximate a set of data points.
- Construct a linear model to relate quantities that vary directly.
- Construct and use a linear model with slope as the rate of change.
- Use a scatter plot to find a linear model that fits a set of data.

Introduction

The primary objective of applied mathematics is to find equations or **mathematical models** that describe real-world situations. In developing a mathematical model to represent actual data, you should strive for two (often conflicting) goals—accuracy and simplicity. That is, you want the model to be simple enough to be workable, yet accurate enough to produce meaningful results.

You have already studied some techniques for fitting models to data. For instance, in Section 2.2, you learned how to find the equation of a line that passes through two points. In this section, you will study other techniques for fitting models to data: *direct variation, rates of change,* and *linear regression.*

Image Source Pink/Getty Images

For most breeds, the body weight of a dog increases at an approximately constant rate through the first several months of life.

Example 1   A Mathematical Model

The weight of a puppy recorded every two months is shown in the table.

Age (in months)	2	4	6	8	10	12
Weight (in pounds)	24	45	67	93	117	130

A linear model that approximates the puppy's weight w (in pounds) in month t is

$$w = 11.03t + 2.1, \quad 2 \le t \le 12.$$

How closely does the model represent the data?

SOLUTION   By graphing the data points with the linear model (see Figure 2.31), you can see that the model is a "good fit" for the actual data. The table shows how each actual weight w compares with the weight $w*$ given by the model.

✓ **CHECKPOINT 1**

In Example 1, what are the best and worst approximations given by the model? ■

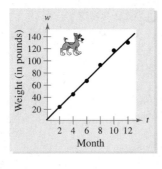

FIGURE 2.31

t	w	$w*$
2	24	24.16
4	45	46.22
6	67	68.28
8	93	90.34
10	117	112.4
12	130	134.46

Direct Variation

There are two basic types of linear models in x and y. The more general model has a y-intercept that is nonzero: $y = mx + b$, $b \neq 0$. The simpler model, $y = mx$, has a y-intercept that is zero. In the simpler model, y is said to **vary directly** as x, or to be **directly proportional** to x.

Direct Variation

The following statements are equivalent.

1. y **varies directly** as x.

2. y is **directly proportional** to x.

3. $y = mx$ for some nonzero constant m, where m is the **constant of variation** or the **constant of proportionality.**

Example 2 **State Income Tax**

In Colorado, state income tax is directly proportional to *taxable income*. For a taxable income of $30,000, the Colorado state income tax is $1389. Find a mathematical model that gives the Colorado state income tax in terms of taxable income.

SOLUTION

Verbal Model: $\dfrac{\text{State}}{\text{income tax}} = m \cdot \dfrac{\text{Taxable}}{\text{income}}$

Labels: State income tax $= y$ (dollars)
Taxable income $= x$ (dollars)
Income tax rate $= m$ (percent in decimal form)

Equation: $y = mx$

Find m by substituting the given information into the equation $y = mx$.

$y = mx$ Direct variation model

$1389 = m(30,000)$ Substitute $y = 1389$ and $x = 30,000$.

$0.0463 = m$ Income tax rate

An equation (or model) for state income tax in Colorado is

$y = 0.0463x.$

So, Colorado has a state income tax rate of 4.63% of taxable income. The graph of this equation is shown in Figure 2.32. ▬▬▬

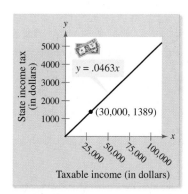

FIGURE 2.32

✓ CHECKPOINT 2

You buy a flash drive for $14.50 and pay sales tax of $0.87. The sales tax is directly proportional to the price. Find a mathematical model that gives the sales tax in terms of the price. ■

Most measurements in the English system and metric system are directly proportional. The next example shows how to use a direct proportion to convert between miles per hour and kilometers per hour.

Example 3   The English and Metric Systems

While driving, your speedometer indicates that your speed is 64 miles per hour or 103 kilometers per hour. Use this information to find a mathematical model that relates miles per hour to kilometers per hour.

SOLUTION   Let y represent the speed in miles per hour and let x represent the speed in kilometers per hour. Then y and x are related by the equation

$$y = mx.$$

Use the fact that $y = 64$ when $x = 103$ to find the value of m.

$$y = mx \qquad \text{Direct variation model}$$

$$64 = m(103) \qquad \text{Substitute } y = 64 \text{ and } x = 103.$$

$$\frac{64}{103} = m \qquad \text{Divide each side by 103.}$$

$$0.62136 \approx m \qquad \text{Use a calculator.}$$

So, the conversion factor from kilometers per hour to miles per hour is approximately 0.62136, and the model is

$$y = 0.62136x.$$

The graph of this equation is shown in Figure 2.33.

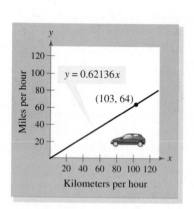

FIGURE 2.33

✓ CHECKPOINT 3

You buy an ice bucket with a capacity of 44 ounces, or 1.3 liters. Write a mathematical model that relates ounces to liters. ■

You can use the model from Example 3 to convert any speed in kilometers per hour to miles per hour, as shown in the table.

Kilometers per hour	Miles per hour
20	12.4
40	24.9
60	37.3
80	49.7
100	62.1
120	74.6

The conversion equation $y = 0.62136x$ can be approximated by the simpler equation $y = \frac{5}{8}x$ because $\frac{5}{8} = 0.625$.

Rates of Change

A second common type of linear model is one that involves a known rate of change. In the linear equation

$$y = mx + b$$

you know that m represents the slope of the line. In real-life problems, the slope can often be interpreted as the **rate of change** of y with respect to x. Rates of change should always be listed in appropriate units of measure.

Example 4 **Mountain Climbing**

A mountain climber is climbing up a 500-foot cliff. At 1 P.M., the climber is 115 feet up the cliff. By 4 P.M., the climber has reached a height of 280 feet, as shown in Figure 2.34.

a. Find the average rate of change of the climber. Use this rate of change to find an equation that relates the height of the climber to the time.

b. Use the equation to estimate the time when the climber reaches the top of the cliff.

SOLUTION

a. Let y represent the climber's height on the cliff and let t represent the time. Then the two points that represent the climber's two positions are

$$(t_1, y_1) = (1, 115) \quad \text{and} \quad (t_2, y_2) = (4, 280).$$

So, the average rate of change of the climber is

$$\text{Average rate of change} = \frac{y_2 - y_1}{t_2 - t_1}$$

$$= \frac{280 - 115}{4 - 1}$$

$$= 55 \text{ feet per hour.}$$

An equation that relates the height of the climber to the time is

$$y - y_1 = m(t - t_1) \qquad \text{Point-slope form}$$

$$y - 115 = 55(t - 1) \qquad \text{Substitute } y_1 = 115, t_1 = 1, \text{ and } m = 55.$$

$$y = 55t + 60. \qquad \text{Linear model}$$

If you had chosen to use the point (t_2, y_2) to determine the equation, you would have obtained a different equation initially: $y - 280 = 55(t - 4)$. However, simplifying this equation yields the same linear model $y = 55t + 60$.

b. To estimate the time when the climber reaches the top of the cliff, let $y = 500$ and solve for t to obtain $t = 8$. Because $t = 8$ corresponds to 8 P.M., at the average rate of change, the climber will reach the top at 8 P.M.

✓CHECKPOINT 4

How long does it take the climber in Example 4 to climb 275 feet? ■

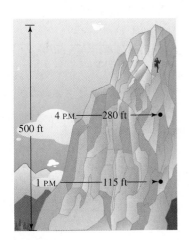

FIGURE 2.34

Example 5   Population of Orlando, Florida

Between 1990 and 2005, the population of Orlando, Florida increased at an average rate of approximately 3233 people per year. In 1990, the population was about 164,700. Find a mathematical model that gives the population of Orlando in terms of the year, and use the model to estimate the population in 2010. *(Source: U.S. Census Bureau)*

SOLUTION   Let y represent the population of Orlando, and let t represent the calendar year, with $t = 0$ corresponding to 1990. It is convenient to let $t = 0$ correspond to 1990 because you were given the population in 1990. Now, using the rate of change of 3233 people per year, you have

Rate of change	1990 population
↓	↓

$$y = \quad mt + b$$

$$y = 3233t + 164,700.$$

Using this model, you can predict the 2010 population to be

$$2010 \text{ population} = 3233(20) + 164,700$$

$$= 229,360.$$

The graph is shown in Figure 2.35.

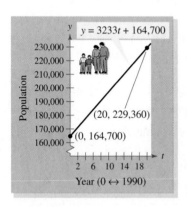

FIGURE 2.35

✓ **CHECKPOINT 5**

Use the model in Example 5 to predict the population of Orlando in 2012. ▨

Example 6   Straight-Line Depreciation

A racing team buys a $4750 welder that has a useful life of 10 years. The salvage value of the welder at the end of the 10 years is $400. Write a linear equation that describes the value of the welder throughout its useable life.

SOLUTION   Let V represent the value of the welder (in dollars) at the end of the year t. You can represent the initial value of the welder by the ordered pair $(0, 4750)$ and the salvage value by the ordered pair $(10, 400)$. The slope of the line is

$$m = \frac{400 - 4750}{10 - 0}$$

$$= -435$$

which represents the annual depreciation in *dollars per year.* Using the slope-intercept form, you can write the equation of the line as follows.

$$V = -435t + 4750 \qquad \text{Slope-intercept form}$$

The graph of the equation is shown in Figure 2.36.

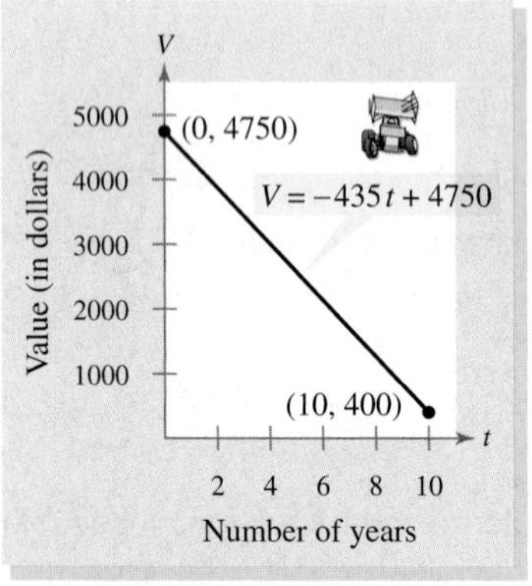

FIGURE 2.36

✓ **CHECKPOINT 6**

Write a linear equation to model the value of a new machine that costs $2300 and is worth $350 after 10 years. ▨

(T) When you use the *regression* feature of your graphing utility, you may obtain an "*r*-value," which gives a measure of how well the model fits the data (see figure).

```
LinReg
 y=ax+b
 a=18.48351648
 b=500.4175824
 r²=.9987189141
 r=.9993592518
■
```

The closer the value of |*r*| is to 1, the better the fit. For the data in Example 7, *r* ≈ 0.999, which implies that the model is a good fit. For instructions on how to use the *regression* feature, see Appendix A; for specific keystrokes, go to the text website at *college.hmco.com/info/ larsonapplied.*

Scatter Plots and Regression Analysis

Another type of linear modeling is a graphical approach that is commonly used in statistics. To find a mathematical model that approximates a set of actual data points, plot the points on a rectangular coordinate system. This collection of points is called a **scatter plot.** You can use the statistical features of a graphing utility to calculate the equation of the best-fitting line for the data in your scatter plot. The statistical method of fitting a line to a collection of points is called **linear regression.** A discussion of linear regression is beyond the scope of this text, but the program in most graphing utilities is easy to use and allows you to analyze linear data that may not be convenient to graph by hand.

Example 7 Dentistry

The table shows the numbers of employees *y* (in thousands) in dentist offices and clinics in the United States in the years 1993 to 2005. *(Source: U.S. Bureau of Labor Statistics)*

Year	x	Employees, y
1993	3	556
1994	4	574
1995	5	592
1996	6	611
1997	7	629
1998	8	646
1999	9	667

Year	x	Employees, y
2000	10	688
2001	11	705
2002	12	725
2003	13	744
2004	14	760
2005	15	771

a. Use the *regression* feature of a graphing utility to find a linear model for the data. Let *x* = 3 represent 1993.

b. Use a graphing utility to graph the linear model along with a scatter plot of the data.

c. Use the linear model to estimate the number of employees in 2007.

SOLUTION

a. Enter the data into a graphing utility. Then, using the *regression* feature of the graphing utility, you should obtain a linear model for the data that can be rounded to the following.

$$y = 18.48x + 500.4, \quad 3 \leq x \leq 15$$

b. The graph of the equation and the scatter plot are shown in Figure 2.37.

c. Substituting *x* = 17 into the equation *y* = 18.48*x* + 500.4, you get *y* = 814.56. So, according to the model, there will be about 815,000 employees in dentist offices and clinics in the United States in 2007.

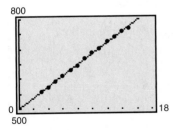

FIGURE 2.37

✓ CHECKPOINT 7

Redo Example 7 using only the data for the years 2000–2005. ■

Example 8 Prize Money at the Indianapolis 500

The total prize money p (in millions of dollars) awarded at the Indianapolis 500 in each year from 1995 to 2006 is shown in the table. Construct a scatter plot that represents the data and find a linear model that approximates the data. *(Source: Indianapolis 500)*

Year	1995	1996	1997	1998	1999	2000
p	8.06	8.11	8.61	8.72	9.05	9.48

Year	2001	2002	2003	2004	2005	2006
p	9.61	10.03	10.15	10.25	10.30	10.52

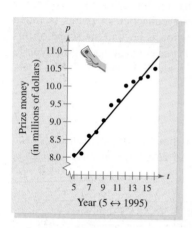

FIGURE 2.38

SOLUTION   Let $t = 5$ represent 1995. The scatter plot of the data is shown in Figure 2.38. Draw a line on the scatter plot that approximates the data. To find an equation of the line, approximate two points on the line: $(5, 8)$ and $(9, 9)$. So, the slope of the line is

$$m \approx \frac{p_2 - p_1}{t_2 - t_1} = \frac{9 - 8}{9 - 5} = 0.25.$$

Using the point-slope form, you can determine that an equation of the line is

$$p - 8 = 0.25(t - 5) \qquad \text{Point-slope form}$$

$$p = 0.25t + 6.75. \qquad \text{Slope-intercept form}$$

To check this model, compare the actual p-values with the p-values given by the model (these values are labeled as p^* in the table at the left). ───

✓ CHECKPOINT 8

Redo Example 8 using only the data for 2001 to 2006. ▪

t	p	p^*
5	8.06	8.00
6	8.11	8.25
7	8.61	8.50
8	8.72	8.75
9	9.05	9.00
10	9.48	9.25
11	9.61	9.50
12	10.03	9.75
13	10.15	10.00
14	10.25	10.25
15	10.30	10.5
16	10.52	10.75

CONCEPT CHECK

1. Name a point that is on the graph of any direct variation equation.

2. What does the constant of variation tell you about the graph of a direct variation equation?

3. The cost y (in dollars) of producing x units of a product is modeled by
   $$y = 30x + 240.$$
   Explain what the rate of change represents in this situation.

4. A girl grows at a rate of 2 inches per year from the time she is 2 years old until she is 10 years old. What other information do you need to write an equation that models the girl's height during this time period? Explain.

Skills Review 2.3

The following warm-up exercises involve skills that were covered in earlier sections. You will use these skills in the exercise set for this section. For additional help, review Sections 2.1 and 2.2.

In Exercises 1–4, sketch the line.

1. $y = 2x$

2. $y = \frac{1}{2}x$

3. $y = 2x + 1$

4. $y = \frac{1}{2}x + 1$

In Exercises 5 and 6, find an equation of the line that has the given slope and y-intercept.

5. Slope: 1; y-intercept: $(0, 2)$

6. Slope: $\frac{3}{2}$; y-intercept: $(0, 3)$

In Exercises 7–10, find an equation of the line that passes through the two points.

7. $(1, 3)$ and $(6, 8)$

8. $(0, 4)$ and $(7, 10)$

9. $(1, 5.2)$ and $(5, 4.7)$

10. $(2, 6.5)$ and $(8, 3.6)$

Exercises 2.3

See www.CalcChat.com for worked-out solutions to odd-numbered exercises.

1. Dog Growth The weight of a puppy recorded every two months is shown in the table.

Age (in months)	2	4	6
Weight (in pounds)	21	44	63

Age (in months)	8	10	12
Weight (in pounds)	82	92	101

A linear model that approximates the puppy's weight w (in pounds) in month t is $w = 8.0t + 11$, $2 \le t \le 12$. Plot the actual data with the model. How closely does the model represent the data?

2. Non-Wage Earners The numbers of working-age civilians (in millions) in the United States that were not involved in the labor force from 1995 to 2005 are given by the following ordered pairs.

(1995, 66.3)   (1996, 66.6)   (1997, 66.8)
(1998, 67.5)   (1999, 68.4)   (2000, 70.0)
(2001, 71.4)   (2002, 72.7)   (2003, 74.7)
(2004, 76.0)   (2005, 76.8)

A linear model that approximates the data is $y = 1.16t + 59.1$, $5 \le t \le 15$, where y is the number of civilians (in millions) and $t = 5$ represents 1995. Plot the actual data with the model. How closely does the model represent the data? *(Source: U.S. Bureau of Labor Statistics)*

T 3. UPS Revenue The yearly revenues (in billions of dollars) of UPS from 1997 to 2005 are given by the following ordered pairs.

(1997, 22.5)   (1998, 24.8)   (1999, 27.1)
(2000, 29.8)   (2001, 30.6)   (2002, 31.3)
(2003, 33.5)   (2004, 36.6)   (2005, 42.6)

Use a graphing utility to create a scatter plot of the data. Let $x = 7$ represent 1997. Then use the *regression* feature of the graphing utility to find a best-fitting line for the data. Graph the model and the data together. How closely does the model represent the data? *(Source: United Parcel Service)*

T 4. Consumer Price Index For urban consumers of educational and communication materials, the Consumer Price Index giving the dollar amount equal to the buying power of $100 in December 1997 is given for each year from 1994 to 2005 by the following ordered pairs.

(1994, 88.8)   (1995, 92.2)   (1996, 95.3)
(1997, 98.4)   (1998, 100.3)   (1999, 101.2)
(2000, 102.5)   (2001, 105.2)   (2002, 107.9)
(2003, 109.8)   (2004, 111.6)   (2005, 113.7)

Use a graphing utility to create a scatter plot of the data. Let $x = 4$ represent 1994. Then use the *regression* feature of the graphing utility to find a best-fitting line for the data. Graph the model and the data together. How closely does the model represent the data? *(Source: U.S. Bureau of Labor Statistics)*

Direct Variation In Exercises 5–10, *y* is proportional to *x*. Use the *x*- and *y*-values to find a linear model that relates *y* and *x*.

5. $x = 8$, $y = 3$ **6.** $x = 5$, $y = 9$

7. $x = 15$, $y = 300$ **8.** $x = 12$, $y = 204$

9. $x = 7$, $y = 3.2$ **10.** $x = 11$, $y = 1.5$

Direct Variation In Exercises 11–14, write a linear model that relates the variables.

11. *H* varies directly as *p*; $H = 27$ when $p = 9$

12. *s* is proportional to *t*; $s = 32$ when $t = 4$

13. *c* is proportional to *d*; $c = 12$ when $d = 20$

14. *r* varies directly as *s*; $r = 25$ when $s = 40$

15. Simple Interest The simple interest received from an investment is directly proportional to the amount of the investment. By investing $2500 in a bond issue, you obtain an interest payment of $187.50 at the end of 1 year. Find a mathematical model that gives the interest *I* at the end of 1 year in terms of the amount invested *P*.

16. Simple Interest The simple interest received from an investment is directly proportional to the amount of the investment. By investing $5000 in a municipal bond, you obtain interest of $337.50 at the end of 1 year. Find a mathematical model that gives the interest *I* at the end of 1 year in terms of the amount invested *P*.

17. Property Tax Your property tax is based on the assessed value of your property. (The assessed value is often lower than the actual value of the property.) A house that has an assessed value of $150,000 has a property tax of $5520.

  (a) Find a mathematical model that gives the amount of property tax *y* in terms of the assessed value *x* of the property.

  (b) Use the model to find the property tax on a house that has an assessed value of $185,000.

18. State Sales Tax An item that sells for $145.99 has a sales tax of $10.22.

  (a) Find a mathematical model that gives the amount of sales tax *y* in terms of the retail price *x*.

  (b) Use the model to find the sales tax on a purchase that has a retail price of $540.50.

19. Centimeters and Inches On a yardstick, you notice that 13 inches is the same length as 33 centimeters.

  (a) Use this information to find a mathematical model that relates centimeters to inches.

  (b) Use the model to complete the table.

Inches	5	10	20	25	30
Centimeters					

20. Liters and Gallons You are buying gasoline and notice that 14 gallons of gasoline is the same as 53 liters.

  (a) Use this information to find a mathematical model that relates gallons to liters.

  (b) Use the model to complete the table.

Gallons	5	10	20	25	30
Liters					

In Exercises 21–26, you are given the 2005 value of a product *and* the rate at which the value is expected to change during the next 5 years. Use this information to write a linear equation that gives the dollar value of the product in terms of the year. (Let $t = 5$ represent 2005.)

	2005 Value	*Rate*
21.	$2540	$125 increase per year
22.	$156	$4.50 increase per year
23.	$20,400	$2000 decrease per year
24.	$45,000	$2800 decrease per year
25.	$154,000	$12,500 increase per year
26.	$245,000	$5600 increase per year

27. Parachuting After opening the parachute, the descent of a parachutist follows a linear model. At 2:08 P.M., the height of the parachutist is 7000 feet. At 2:10 P.M., the height is 4600 feet.

  (a) Write a linear equation that gives the height of the parachutist in terms of the time *t*. (Let $t = 0$ represent 2:08 P.M. and let *t* be measured in seconds.)

  (b) Use the equation in part (a) to find the time when the parachutist will reach the ground.

28. Distance Traveled by a Car You are driving at a constant speed. At 4:30 P.M., you drive by a sign that gives the distance to Montgomery, Alabama as 84 miles. At 4:59 P.M., you drive by another sign that gives the distance to Montgomery as 56 miles.

  (a) Write a linear equation that gives your distance from Montgomery in terms of time *t*. (Let $t = 0$ represent 4:30 P.M. and let *t* be measured in minutes.)

  (b) Use the equation in part (a) to find the time when you will reach Montgomery.

29. Straight-Line Depreciation A business purchases a piece of equipment for $875. After 5 years the equipment will have no value. Write a linear equation giving the value *V* of the equipment during the 5 years.

30. Straight-Line Depreciation   A business purchases a piece of equipment for $25,000. The equipment will be replaced in 10 years, at which time its salvage value is expected to be $2000. Write a linear equation giving the value *V* of the equipment during the 10 years.

31. Sale Price and List Price   A store is offering a 15% discount on all items. Write a linear equation giving the sale price *S* for an item with a list price *L*.

32. Sale Price and List Price   A store is offering a 25% discount on all shirts. Write a linear equation giving the sale price *S* for a shirt with a list price *L*.

33. Hourly Wages   A manufacturer pays its assembly line workers $11.50 per hour. In addition, workers receive a piecework rate of $0.75 per unit produced. Write a linear equation for the hourly wages *W* in terms of the number of units *x* produced per hour.

34. Sales Commission   A salesperson receives a monthly salary of $2500 plus a commission of 7% of sales. Write a linear equation for the salesperson's monthly wage *W* in terms of the person's monthly sales *S*.

35. Deer Population   A forest region had a population of 1300 deer in the year 2000. During the next 8 years, the deer population increased by about 60 deer per year.

(a) Write a linear equation giving the deer population *P* in terms of the year *t*. Let *t* = 0 represent 2000.

(b) The deer population keeps growing at this constant rate. Predict the number of deer in 2012.

36. Pest Management   The cost of implementing an invasive species management system in a forest is related to the area of the forest. It costs $630 to implement the system in a forest area of 10 acres. It costs $1070 in a forest area of 18 acres.

(a) Write a linear equation giving the cost of the invasive species management system in terms of the number of acres *x* of forest.

(b) Use the equation in part (a) to find the cost of implementing the system in a forest area of 30 acres.

In Exercises 37–40, match the description with one of the graphs. Also find the slope of the graph and describe how it is interpreted in the real-life situation. [The graphs are labeled (a), (b), (c), and (d).]

(a)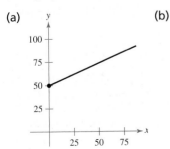

(b)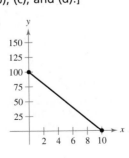

(c)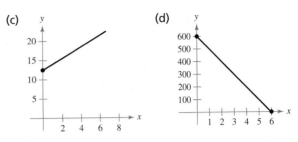

(d)

37. A person is paying $10 per week to a friend to repay a $100 loan.

38. An employee is paid $12.50 per hour plus $1.50 for each unit produced per hour.

39. A sales representative receives $50 per day for food, plus $0.48 for each mile traveled.

40. You purchased a digital camera for $600 that depreciates $100 per year.

41. Think About It   You begin a video game with 100 points and earn 10 points for each coin you collect. Does this description match graph (b) in Exercises 37–40? Explain.

42. Think About It   You start with $1.50 and save $12.50 per week. Does this description match graph (c) in Exercises 37–40? Explain.

In Exercises 43–48, can the data be approximated by a linear model? If so, sketch the line that best approximates the data. Then find an equation of the line.

43.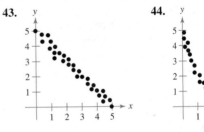

44.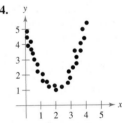

45.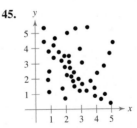

46.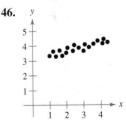

47.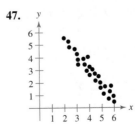

48.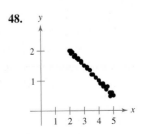

(T) **49. Advertising** The estimated annual amounts *A* (in millions of dollars) spent on cable TV advertising for the years 1996 to 2005 are shown in the table. *(Source: Universal McCann)*

Year	1996	1997	1998	1999
Advertising, A	7778	8750	10,340	12,570

Year	2000	2001	2002
Advertising, A	15,455	15,536	16,297

Year	2003	2004	2005
Advertising, A	18,814	21,527	24,501

(a) Use a graphing utility to create a scatter plot of the data. Let *t* = 6 represent 1996. Do the data appear linear?

(b) Use the *regression* feature of a graphing utility to find a linear model for the data.

(c) State the slope of the graph of the linear model from part (b) and interpret its meaning in the context of the problem.

(d) Use the linear model to estimate the amounts spent on cable TV advertising in 2006 and 2007. Are your estimates reasonable?

(T) **50. Japan** The population of Japan is expected to drop by 30% over the next 50 years as the percent of its citizens that are elderly increases. Projections for Japan's population through 2050 are shown in the table.

Year, t	2005	2010	2020
Population, P (in millions)	127.8	127.5	124.1

Year, t	2030	2040	2050
Population, P (in millions)	117.6	109.3	100.6

(a) Use a graphing utility to create a scatter plot of the data. Let *t* = 5 represent 2005. Do the data appear linear?

(b) Use the *regression* feature of a graphing utility to find a linear model for the data.

(c) Identify the slope of the model from part (b) and interpret its meaning in the context of the problem.

(d) Use the linear model to predict the populations in 2015, 2035, and 2060. Are these predictions reasonable?

(T) **51. Yearly Revenue** The yearly revenues (in millions of dollars) for Sonic Corporation for the years 1996 to 2005 are given by the following ordered pairs. *(Source: Sonic Corporation)*

(1996, 151.1)	(1997, 184.0)	(1998, 219.1)
(1999, 257.6)	(2000, 280.1)	(2001, 330.6)
(2002, 400.2)	(2003, 446.6)	(2004, 536.4)
(2005, 623.1)		

(a) Use a graphing utility to create a scatter plot of the data. Let *t* = 6 represent 1996.

(b) Use two points on the scatter plot to find an equation of a line that approximates the data.

(c) Use the *regression* feature of a graphing utility to find a linear model for the data. Use this model and the model from part (b) to predict the revenues in 2006 and 2007.

(d) Sonic Corporation projected its revenues in 2006 and 2007 to be $695 million and $765 million. How close are these projections to the predictions from the models?

(e) Sonic Corporation also expected their yearly revenue to reach $965 million in 2009, 2010, or 2011. Do the models from parts (b) and (c) support this? Explain your reasoning.

(T) **52. Revenue per Share** The revenues per share of stock (in dollars) for Sonic Corporation for the years 1996 to 2005 are given by the following ordered pairs. *(Source: Sonic Corporation)*

(1996, 1.48)	(1997, 1.90)	(1998, 2.29)
(1999, 2.74)	(2000, 3.15)	(2001, 3.64)
(2002, 4.48)	(2003, 5.06)	(2004, 6.01)
(2005, 7.00)		

(a) Use a graphing utility to create a scatter plot of the data. Let *t* = 6 represent 1996.

(b) Use two points on the scatter plot to find an equation of a line that approximates the data.

(c) Use the *regression* feature of a graphing utility to find a linear model for the data. Use this model and the model from part (b) to predict the revenues per share in 2006 and 2007.

(d) Sonic projected the revenues per share in 2006 and 2007 to be $8.00 and $8.80. How close are these projections to the predictions from the models?

(e) Sonic also expected the revenue per share to reach $11.10 in 2009, 2010, or 2011. Do the models from parts (b) and (c) support this? Explain your reasoning.

(S) 53. **Purchasing Power**   The value (in 1982 dollars) of each dollar received by producers in each of the years from 1991 to 2005 in the United States is represented by the following ordered pairs. *(Source: U.S. Bureau of Labor Statistics)*

(1991, 0.822)	(1992, 0.812)	(1993, 0.802)
(1994, 0.797)	(1995, 0.782)	(1996, 0.762)
(1997, 0.759)	(1998, 0.765)	(1999, 0.752)
(2000, 0.725)	(2001, 0.711)	(2002, 0.720)
(2003, 0.698)	(2004, 0.673)	(2005, 0.642)

(a) Use a spreadsheet software program to generate a scatter plot of the data. Let $t = 1$ represent 1991. Do the data appear to be linear?

(b) Use the *regression* feature of a spreadsheet software program to find a linear model for the data.

(c) Use the model to estimate the value (in 1982 dollars) of 1 dollar received by producers in 2007 and in 2008. Discuss the reliability of your estimates based on your scatter plot and the graph of your linear model for the data.

(S) 54. **Purchasing Power**   The value (in 1982–1984 dollars) of each dollar paid by consumers in each of the years from 1991 to 2005 in the United States is represented by the following ordered pairs. *(Source: U.S. Bureau of Labor Statistics)*

(1991, 0.734)	(1992, 0.713)	(1993, 0.692)
(1994, 0.675)	(1995, 0.656)	(1996, 0.638)
(1997, 0.623)	(1998, 0.614)	(1999, 0.600)
(2000, 0.581)	(2001, 0.565)	(2002, 0.556)
(2003, 0.544)	(2004, 0.530)	(2005, 0.512)

(a) Use a spreadsheet software program to generate a scatter plot of the data. Let $t = 1$ represent 1991. Do the data appear to be linear?

(b) Use the *regression* feature of a spreadsheet software program to find a linear model for the data.

(c) Use the model to estimate the value (in 1982–1984 dollars) of 1 dollar paid by consumers in 2007 and in 2008. Discuss the reliability of your estimates based on your scatter plot and the graph of your linear model for the data.

(T) 55. **Health Services**   The numbers of employees E (in thousands) in the health services industry for the years 2000 to 2005 are shown in the table. *(Source: U.S. Department of Health and Human Services)*

Year	2000	2001	2002
Employees, E	12,718	13,134	13,556

Year	2003	2004	2005
Employees, E	13,893	14,190	14,523

(a) Use a graphing utility to create a scatter plot of the data. Let $t = 0$ represent 2000. Do the data appear to be linear?

(b) Use the *regression* feature of a graphing utility to find a linear model for the data.

(c) Use the model to estimate the numbers of employees in 2007 and 2009.

(d) Graph the linear model along with the scatter plot of the data. Comparing the data with the model, are the predictions in part (c) most likely to be high, low, or just about right? Explain your reasoning.

(T) 56. **Health Care**   The total yearly health care expenditures E (in billions of dollars) in the United States for the years 1996 to 2005 are shown in the table. *(Source: U.S. Centers for Medicare and Medicaid Services)*

Year	1996	1997	1998	1999
Expenditures, E	1073	1125	1191	1265

Year	2000	2001	2002
Expenditures, E	1353	1470	1603

Year	2003	2004	2005
Expenditures, E	1733	1859	1988

(a) Use a graphing utility to create a scatter plot of the data. Let $t = 6$ represent 1996. Do the data appear to be linear?

(b) Use the *regression* feature of a graphing utility to find a linear model for the data.

(c) Use the model to estimate the health care expenditures in 2006, 2007, and 2008.

(d) Graph the linear model along with the scatter plot. Use the trend in the scatter plot to explain why the predictions from the model differ from the following 2007 government projections for the same expenditures: $2164 billion in 2006, $2320 billion in 2007, and $2498 billion in 2008.

57. **Think About It**   Annual data from three years are used to create linear models for the population and the yearly snowfall of Reno, Nevada. Which model is more likely to give better predictions for future years? Discuss the appropriateness of using only three data points in each situation.

Section 2.4

Functions

- Determine if an equation or a set of ordered pairs represents a function.
- Use function notation and evaluate a function.
- Find the domain of a function.
- Write a function that relates quantities in an application problem.

Introduction to Functions

Many everyday phenomena involve two quantities that are related to each other by some rule of correspondence. Here are some examples.

1. The simple interest I earned on $1000 for 1 year is related to the annual interest rate r by the formula $I = 1000r$.

2. The distance d traveled on a bicycle in 2 hours is related to the speed s of the bicycle by the formula $d = 2s$.

3. The area A of a circle is related to its radius r by the formula $A = \pi r^2$.

Not all correspondences between two quantities have simple mathematical formulas. For instance, people commonly match up athletes with jersey numbers and hours of the day with temperatures. In each of these cases, however, there is some rule of correspondence that matches each item from one set with exactly one item from a different set. Such a rule of correspondence is called a **function.**

Definition of a Function

A **function** f from a set A to a set B is a rule of correspondence that assigns to each element x in the set A exactly one element y in the set B. The set A is the **domain** (or set of inputs) of the function f, and the set B contains the **range** (or set of outputs).

To get a better idea of this definition, look at the function that relates the time of day to the temperature in Figure 2.39. This function can be represented by the following set of ordered pairs.

$$\{(1, 9°), (2, 13°), (3, 15°), (4, 15°), (5, 12°), (6, 4°)\}$$

In each ordered pair, the first coordinate (x-value) is the input and the second coordinate (y-value) is the output. In this example, note the following characteristics of a function.

1. Each element of A (the domain) must be matched with an element of B (the range).

2. Some elements of B may not be matched with any element of A.

3. Two or more elements of A may be matched with the same element of B.

4. An element of A cannot be matched with two different elements of B.

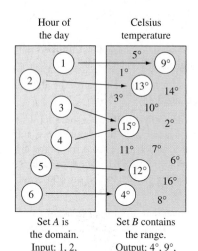

Hour of the day

Celsius temperature

Set A is the domain.
Input: 1, 2, 3, 4, 5, 6

Set B contains the range.
Output: 4°, 9°, 12°, 13°, 15°

FIGURE 2.39 Function from Set A to Set B

In the following two examples, you are asked to decide whether different correspondences are functions. To do this, you must decide whether each element of the domain A is matched with exactly one element of the range B. If any element of A is matched with two or more elements of B, the correspondence is not a function. For example, people are not a function of their birthday month because many people are born in any given month.

Example 1 Testing for Functions

Let $A = \{a, b, c\}$ and $B = \{1, 2, 3, 4, 5\}$. Which of the following sets of ordered pairs or figures represent functions from set A to set B?

a. $\{(a, 2), (b, 3), (c, 4)\}$

b. $\{(a, 4), (b, 5)\}$

c.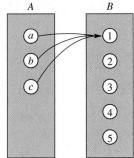

d.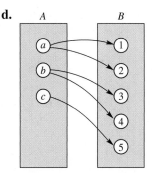

SOLUTION

a. This collection of ordered pairs *does* represent a function from A to B. Each element of A is matched with exactly one element of B.

b. This collection of ordered pairs *does not* represent a function from A to B. Not every element of A is matched with an element of B.

c. This figure *does* represent a function from A to B. It does not matter that each element of A is matched with the same element of B.

d. This figure *does not* represent a function from A to B. The element a of A is matched with *two* elements of B. This is also true of the element b.

✓ CHECKPOINT 1

Let $A = \{a, b, c, d\}$ and $B = \{1, 3, 5, 7\}$. Does the set of ordered pairs $\{(a, 3), (b, 7), (c, 1), (d, 3)\}$ represent a function from set A to set B? ▨

Representing functions by sets of ordered pairs is a common practice in *discrete mathematics*. In algebra, however, it is more common to represent functions by equations or formulas involving two variables. For instance, the equation

$$y = x^2 \qquad \text{\textit{y} is a function of \textit{x}.}$$

represents the variable y as a function of the variable x. In this equation, x is the **independent variable** and y is the **dependent variable.** The domain of the function is the set of all values taken on by the independent variable x, and the range of the function is the set of all values taken on by the dependent variable y.

Example 2   **Testing for Functions Represented by Equations**

Which of the equations represent(s) y as a function of x?

a. $x^2 + y = 1$

b. $-x + y^2 = 1$

SOLUTION   To determine whether y is a function of x, try to solve for y in terms of x.

a. Solving for y yields

$$x^2 + y = 1 \qquad \text{Write original equation.}$$
$$y = 1 - x^2. \qquad \text{Solve for } y.$$

To each value of x there corresponds exactly one value of y. So, y *is* a function of x.

b. Solving for y yields

$$-x + y^2 = 1 \qquad \text{Write original equation.}$$
$$y^2 = 1 + x \qquad \text{Add } x \text{ to each side.}$$
$$y = \pm\sqrt{1 + x}. \qquad \text{Solve for } y.$$

The $\pm$ indicates that to a given value of x there correspond two values of y. So, y *is not* a function of x.

✓ **CHECKPOINT 2**

Does the equation $y - 2 = x^2$ represent y as a function of x? ▪

Function Notation

When an equation is used to represent a function, it is convenient to name the function so that it can be referenced easily. For example, you know that the equation $y = 1 - x^2$ describes y as a function of x. Suppose you give this function the name "f." Then you can use the following **function notation.**

Input	*Output*	*Equation*
x	$f(x)$	$f(x) = 1 - x^2$

The symbol $f(x)$ is read as the **value of f at x** or simply **f of x.** The symbol $f(x)$ corresponds to the y-value for a given x. So, you can write $y = f(x)$. Keep in mind that f is the *name* of the function, whereas $f(x)$ is the *value* of the function at x. For instance, the function given by

$$f(x) = 3 - 2x$$

has *function values* denoted by $f(-1)$, $f(0)$, $f(2)$, and so on. To find these values, substitute the specified input values into the given equation.

For $x = -1$, $f(-1) = 3 - 2(-1) = 3 + 2 = 5.$

For $x = 0$, $f(0) = 3 - 2(0) = 3 - 0 = 3.$

For $x = 2$, $f(2) = 3 - 2(2) = 3 - 4 = -1.$

Although f is often used as a convenient function name and x is often used as the independent variable, you can use other letters. For instance,

$$f(x) = x^2 - 4x + 7, \quad f(t) = t^2 - 4t + 7, \quad \text{and} \quad g(s) = s^2 - 4s + 7$$

all define the same function. In fact, the role of the independent variable in a function is simply that of a "placeholder." Consequently, the function above could be described by the form

$$f(\quad) = (\quad)^2 - 4(\quad) + 7.$$

Example 3   Evaluating a Function

Let $g(x) = -x^2 + 4x + 1$. Find the following.

a. $g(2)$ **b.** $g(t)$ **c.** $g(x + 2)$

SOLUTION

a. Replacing x with 2 in $g(x) = -x^2 + 4x + 1$ yields the following.

$$g(2) = -(2)^2 + 4(2) + 1 = -4 + 8 + 1 = 5$$

b. Replacing x with t yields the following.

$$g(t) = -(t)^2 + 4(t) + 1 = -t^2 + 4t + 1$$

c. Replacing x with $x + 2$ yields the following.

$$\begin{aligned} g(x + 2) &= -(x + 2)^2 + 4(x + 2) + 1 \\ &= -(x^2 + 4x + 4) + 4x + 8 + 1 \\ &= -x^2 - 4x - 4 + 4x + 8 + 1 \\ &= -x^2 + 5 \end{aligned}$$

> **STUDY TIP**
>
> In Example 3(c), note that $g(x + 2)$ is not equal to $g(x) + g(2)$. In general, $g(u + v) \neq g(u) + g(v)$.

✓ **CHECKPOINT 3**

Let $h(x) = 2x^2 + x - 4$. Find $h(-1)$. ■

A function defined by two or more equations over a specified domain is called a **piecewise-defined function.**

Example 4   A Piecewise-Defined Function

Evaluate the function when $x = -1, 0,$ and 1.

$$f(x) = \begin{cases} x^2 + 1, & x < 0 \\ x - 1, & x \geq 0 \end{cases}$$

SOLUTION   Because $x = -1$ is less than 0, use $f(x) = x^2 + 1$ to obtain

$$f(-1) = (-1)^2 + 1 = 2.$$

For $x = 0$, use $f(x) = x - 1$ to obtain $f(0) = (0) - 1 = -1$. For $x = 1$, use $f(x) = x - 1$ to obtain $f(1) = (1) - 1 = 0$. The graph of the function is shown in Figure 2.40.

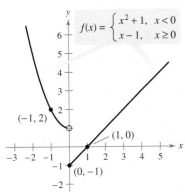

FIGURE 2.40

✓ **CHECKPOINT 4**

Evaluate the function in Example 4 when $x = -3$ and 3. ■

DISCOVERY

Use a graphing utility to graph $y = \sqrt{4 - x^2}$. What is the domain of this function? Then graph $y = \sqrt{x^2 - 4}$. What is the domain of this function? Do the domains of these two functions overlap? If so, for what values?

Finding the Domain of a Function

The domain of a function can be described explicitly or it can be *implied* by the expression used to define the function. The **implied domain** is the set of all real numbers for which the expression is defined. For instance, the function given by

$$f(x) = \frac{1}{x^2 - 4}$$
Domain excludes x-values that result in division by zero.

has an implied domain that consists of all real x other than $x = \pm 2$. These two values are excluded from the domain because division by zero is undefined. Another common type of implied domain results from the restrictions needed to avoid even roots of negative numbers. For example, the function given by

$$f(x) = \sqrt{x}$$
Domain excludes x-values that result in even roots of negative numbers.

is defined only for $x \geq 0$. So, its implied domain is the interval $[0, \infty)$. In general, the domain of a function *excludes* values that would cause division by zero *or* result in the even root of a negative number.

Example 5 Finding the Domain of a Function

Find the domain of each function.

a. f: $\{(-3, 0), (-1, 4), (0, 2), (2, 2), (4, -1)\}$ **b.** $g(x) = \dfrac{1}{x + 5}$

c. Volume of a sphere: $V = \frac{4}{3}\pi r^3$ **d.** $h(x) = \sqrt{4 - x^2}$

e. $r(x) = \sqrt[3]{x + 3}$

SOLUTION

a. The domain of f consists of all first coordinates in the set of ordered pairs.

Domain $= \{-3, -1, 0, 2, 4\}$

b. Excluding x-values that yield zero in the denominator, the domain of g is the set of all real numbers x such that $x \neq -5$.

c. Because this function represents the volume of a sphere, the values of the radius r must be positive. So, the domain is the set of all real numbers r such that $r > 0$.

d. This function is defined only for x-values for which $4 - x^2 \geq 0$. Using the methods described in Section 1.7, you can conclude that $-2 \leq x \leq 2$. So, the domain of h is the interval $[-2, 2]$.

e. Because the cube root of any real number is defined, the domain of r is the set of all real numbers, or $(-\infty, \infty)$.

✓ CHECKPOINT 5

Find the domain of the function $f(x) = 6 - x^3$. ∎

In Example 5(c), note that the domain of a function may be implied by the physical context. For instance, from the equation $V = \frac{4}{3}\pi r^3$, you would have no reason to restrict r to positive values, but the physical context implies that a sphere cannot have a negative or zero radius.

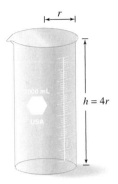

FIGURE 2.41

Applications

Example 6   **The Dimensions of a Container**

You are working with a cylindrical beaker in a chemistry lab experiment. The height of the beaker is 4 times the radius, as shown in Figure 2.41.

a. Write the volume of the beaker as a function of the radius r.

b. Write the volume of the beaker as a function of the height h.

SOLUTION

a. $V = \pi r^2 h = \pi r^2 (4r) = 4\pi r^3$ V is a function of r.

b. $V = \pi \left(\dfrac{h}{4}\right)^2 h = \dfrac{\pi h^3}{16}$ V is a function of h.

✓CHECKPOINT 6

In Example 6, suppose the radius is twice the height. Write the volume of the beaker as a function of the height h. ▪

Example 7   **The Path of a Baseball**

A baseball is hit 3 feet above home plate at a velocity of 100 feet per second and an angle of 45°. The path of the baseball is given by the function

$$y = -0.0032x^2 + x + 3$$

where y and x are measured in feet. Will the baseball clear a 10-foot fence located 300 feet from home plate?

SOLUTION   When $x = 300$, the height of the baseball is given by

$$y = -0.0032(300)^2 + 300 + 3 = 15 \text{ feet.}$$

The ball will clear the fence, as shown in Figure 2.42.

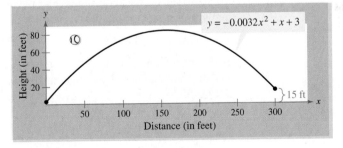

FIGURE 2.42

Notice that in Figure 2.42, the baseball is not at the point $(0, 0)$ before it is hit. This is because the original problem states that the baseball was hit 3 feet above the ground.

✓CHECKPOINT 7

In Example 7, will the baseball clear a 35-foot fence located 280 feet from home plate? ▪

Example 8 Patents

The number P (in thousands) of patents issued increased in a linear pattern from 1998 to 2001. Then, in 2002, the pattern changed from a linear to a quadratic pattern (see Figure 2.43). These two patterns can be approximated by the function

$$P = \begin{cases} 6.96t + 106.9, & 8 \leq t \leq 11 \\ -6.550t^2 + 168.27t - 892.1, & 12 \leq t \leq 15 \end{cases}$$

with $t = 8$ corresponding to 1998. Use this function to approximate the total number of patents issued between 1998 and 2005. *(Source: U.S. Patent and Trademark Office)*

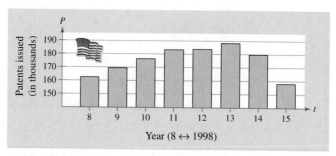

FIGURE 2.43

SOLUTION For 1998 to 2001, use the equation $P = 6.96t + 106.9$ to approximate the number of patents issued, as shown in the table. For 2002 to 2005, use the equation $P = -6.550t^2 + 168.27t - 892.1$ to approximate the number of patents issued, as shown in the table.

t	8	9	10	11	12	13	14	15
P	162.6	169.5	176.5	183.5	183.9	188.5	179.9	158.2

$$P = 6.96t + 106.9 \qquad\qquad P = -6.550t^2 + 168.27t - 892.1$$

To approximate the total number of patents issued from 1998 to 2005, add the amounts for each of the years, as follows.

$$162.6 + 169.5 + 176.5 + 183.5 + 183.9 + 188.5 + 179.9 + 158.2 = 1402.6$$

Because the number of patents issued is measured in thousands, you can conclude that the total number of patents issued between 1998 and 2005 was approximately 1,402,600.

✓**CHECKPOINT 8**

The number of cat cadavers purchased for dissection in a biology class from 2000 to 2008 can be modeled by the function

$$C = \begin{cases} 2t + 48, & 0 \leq t \leq 3 \\ 4t + 42, & 4 \leq t \leq 8 \end{cases}$$

with $t = 0$ corresponding to 2000. Use the function to approximate the total number of cat cadavers purchased from 2000 to 2008. ■

Summary of Function Terminology

Function: A **function** is a relationship between two variables such that to each value of the independent variable there corresponds exactly one value of the dependent variable.

For instance, let $A = \{a, b, c\}$ and $B = \{1, 2, 3, 4\}$.

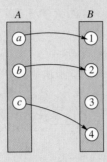   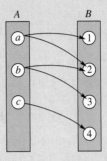

The set of ordered pairs $\{(a, 1), (b, 2), (c, 4)\}$ *is a function.*   The set of ordered pairs $\{(a, 1), (a, 2), (b, 2), (b, 3), (c, 4)\}$ *is not a function.*

Function Notation: $y = f(x)$

   f is the **name** of the function.

   y is the **dependent variable.**

   x is the **independent variable.**

   $f(x)$ is the **value of the function at x.**

Domain: The **domain** of a function is the set of all values (inputs) of the independent variable for which the function is defined. If x is in the domain of f, then f is said to be **defined** at x. If x is not in the domain of f, then f is said to be **undefined** at x.

Range: The **range** of a function is the set of all values (outputs) assumed by the dependent variable (that is, the set of all function values).

Implied Domain: If f is defined by an algebraic expression and the domain is not specified, the **implied domain** consists of all real numbers for which the expression is defined.

CONCEPT CHECK

1. Let $A = \{0, 2, 4, 6\}$ and $B = \{1, 3, 5, 7, 9\}$. Give an example of a set of ordered pairs that represent a function from set A to set B.

2. Is $f(2)$ equivalent to $2 \cdot f(x)$ for every function f? Explain.

3. Give an example of a function whose domain is the set of all real numbers x such that $x \neq 6$.

4. You want to write the area of a rectangle as a function of the width w. What information is needed? Explain.

Skills Review 2.4

The following warm-up exercises involve skills that were covered in earlier sections. You will use these skills in the exercise set for this section. For additional help, review Sections 0.2, 1.1, 1.5, and 1.7.

In Exercises 1–4, simplify the expression.

1. $2(-3)^3 + 4(-3) - 7$

2. $4(-1)^2 - 5(-1) + 4$

3. $(x + 1)^2 + 3(x + 1) - 4 - (x^2 + 3x - 4)$

4. $(x - 2)^2 - 4(x - 2) - (x^2 - 4)$

In Exercises 5 and 6, solve for y in terms of x.

5. $2x + 5y - 7 = 0$

6. $y^2 = x^2$

In Exercises 7–10, solve the inequality.

7. $x^2 - 4 \geq 0$

8. $9 - x^2 \geq 0$

9. $x^2 + 2x + 1 \geq 0$

10. $x^2 - 3x + 2 \geq 0$

Exercises 2.4

See www.CalcChat.com for worked-out solutions to odd-numbered exercises.

In Exercises 1–4, decide whether the set of figures represents a function from A to B.

$A = \{a, b, c\}$ and $B = \{1, 2, 3, 4\}$

Give reasons for your answers.

1.   **2.**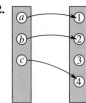

3. 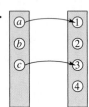   **4.**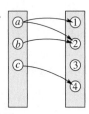

In Exercises 5–8, decide whether the set of ordered pairs represents a function from A to B.

$A = \{0, 1, 2, 3\}$ and $B = \{-2, -1, 0, 1, 2\}$

Give reasons for your answers.

5. $\{(0, 1), (1, -2), (2, 0), (3, 2)\}$

6. $\{(0, -1), (2, 2), (1, -2), (3, 0), (1, 1)\}$

7. $\{(0, 0), (1, 0), (2, 0), (3, 0)\}$

8. $\{(0, 2), (3, 0), (1, 1)\}$

In Exercises 9–12, decide whether the set of ordered pairs represents a function from A to B.

$A = \{a, b, c\}$ and $B = \{0, 1, 2, 3\}$

Give reasons for your answers.

9. $\{(a, 1), (c, 2), (c, 3), (b, 3)\}$

10. $\{(a, 1), (b, 2), (c, 3)\}$

11. $\{(1, a), (0, a), (2, c), (3, b)\}$

12. $\{(c, 0), (b, 0), (a, 3)\}$

In Exercises 13–16, the domain of f is the set

$A = \{-2, -1, 0, 1, 2\}$.

Write the function as a set of ordered pairs.

13. $f(x) = x^2$

14. $f(x) = \dfrac{2x}{x^2 + 1}$

15. $f(x) = \sqrt{x + 2}$

16. $f(x) = |x + 1|$

In Exercises 17–26, determine whether the equation represents y as a function of x.

17. $x^2 + y^2 = 4$

18. $x = y^2$

19. $x^2 + y = 4$

20. $x + y^2 = 4$

21. $2x + 3y = 4$

22. $x^2 + y^2 - 2x - 4y + 1 = 0$

23. $y^2 = x^2 - 1$

24. $y = \sqrt{x + 5}$

25. $x^2y - x^2 + 4y = 0$

26. $xy - y - x - 2 = 0$

In Exercises 27–30, fill in the blank and simplify.

27. $f(x) = 6 - 4x$

(a) $f(3) = 6 - 4(\quad)$

(b) $f(-7) = 6 - 4(\quad)$

(c) $f(t) = 6 - 4(\quad)$

(d) $f(c + 1) = 6 - 4(\quad)$

28. $f(s) = \dfrac{1}{s + 1}$

(a) $f(4) = \dfrac{1}{(\quad) + 1}$

(b) $f(0) = \dfrac{1}{(\quad) + 1}$

(c) $f(4x) = \dfrac{1}{(\quad) + 1}$

(d) $f(x + 1) = \dfrac{1}{(\quad) + 1}$

29. $g(x) = \dfrac{1}{x^2 - 2x}$

(a) $g(1) = \dfrac{1}{(\quad)^2 - 2(\quad)}$

(b) $g(-3) = \dfrac{1}{(\quad)^2 - 2(\quad)}$

(c) $g(t) = \dfrac{1}{(\quad)^2 - 2(\quad)}$

(d) $g(t + 1) = \dfrac{1}{(\quad)^2 - 2(\quad)}$

30. $f(t) = \sqrt{25 - t^2}$

(a) $f(3) = \sqrt{25 - (\quad)^2}$

(b) $f(5) = \sqrt{25 - (\quad)^2}$

(c) $f(x + 5) = \sqrt{25 - (\quad)^2}$

(d) $f(2x) = \sqrt{25 - (\quad)^2}$

In Exercises 31–44, evaluate the function at each specified value of the independent variable and simplify.

31. $f(x) = 2x - 3$

(a) $f(1)$ (b) $f(-3)$

(c) $f(x - 1)$ (d) $f\left(\tfrac{1}{4}\right)$

32. $g(y) = 7 - 3y$

(a) $g(0)$ (b) $g\left(\tfrac{7}{3}\right)$

(c) $g(s)$ (d) $g(s + 2)$

33. $h(t) = t^2 - 2t$

(a) $h(2)$ (b) $h(-1)$

(c) $h(x + 2)$ (d) $h(1.5)$

34. $k(b) = 2b^2 + 7b + 3$

(a) $k(0)$ (b) $k\left(-\tfrac{1}{2}\right)$

(c) $k(a)$ (d) $k(x + 2)$

35. $V(r) = \tfrac{4}{3}\pi r^3$

(a) $V(3)$ (b) $V(0)$

(c) $V\left(\tfrac{3}{2}\right)$ (d) $V(2r)$

36. $A(s) = \dfrac{\sqrt{3}s^2}{4}$

(a) $A(1)$ (b) $A(0)$

(c) $A(2x)$ (d) $A(3)$

37. $f(y) = 3 - \sqrt{y}$

(a) $f(4)$ (b) $f(100)$

(c) $f(4x^2)$ (d) $f(0.25)$

38. $f(x) = \sqrt{x + 3} - 2$

(a) $f(-3)$ (b) $f(1)$

(c) $f(x - 3)$ (d) $f(x + 4)$

39. $c(x) = \dfrac{1}{x^2 - 16}$

(a) $c(4)$ (b) $c(0)$

(c) $c(y + 2)$ (d) $c(y - 2)$

40. $q(t) = \dfrac{2t^2 + 3}{t^2}$

(a) $q(2)$ (b) $q(0)$

(c) $q(x)$ (d) $q(-x)$

41. $f(x) = \dfrac{|x|}{x}$

(a) $f(2)$ (b) $f(-2)$

(c) $f(x^2)$ (d) $f(x - 1)$

42. $f(x) = |x| + 4$

(a) $f(2)$ (b) $f(-2)$

(c) $f(x^2)$ (d) $f(x + 2)$

43. $f(x) = \begin{cases} 3x - 1, & x < 0 \\ 2x + 3, & x \geq 0 \end{cases}$

(a) $f(-1)$ (b) $f(0)$

(c) $f(-2)$ (d) $f(2)$

44. $f(x) = \begin{cases} x^2 + 1, & x \leq 1 \\ 2x - 3, & x > 1 \end{cases}$

(a) $f(-2)$ (b) $f(1)$

(c) $f\left(\tfrac{3}{2}\right)$ (d) $f(0)$

In Exercises 45–52, find all real values of x such that $f(x) = 0$.

45. $f(x) = 15 - 3x$

46. $f(x) = \dfrac{2x - 5}{3}$

47. $f(x) = x^2 - 9$

48. $f(x) = 2x^2 - 11x + 5$

49. $f(x) = x^3 - x$

50. $f(x) = x^3 - 3x^2 - 4x + 12$

51. $f(x) = \dfrac{3}{x - 1} + \dfrac{4}{x - 2}$

52. $f(x) = 3 + \dfrac{2}{x - 1}$

In Exercises 53–66, find the domain of the function.

53. $g(x) = 1 - 2x^2$

54. $f(x) = 5x^2 + 2x - 1$

55. $h(t) = \dfrac{4}{t}$

56. $s(y) = \dfrac{3y}{y + 5}$

57. $g(y) = \sqrt[3]{y - 10}$

58. $f(t) = \sqrt[3]{t + 4}$

59. $f(x) = \sqrt[4]{1 - x^2}$

60. $g(x) = \sqrt{x + 1}$

61. $g(x) = \dfrac{1}{x} - \dfrac{3}{x + 2}$

62. $h(x) = \dfrac{10}{x^2 - 2x}$

63. $f(x) = \dfrac{\sqrt{x + 1}}{x - 2}$

64. $f(s) = \dfrac{\sqrt{s - 1}}{s - 4}$

65. $f(x) = \dfrac{x - 4}{\sqrt{x}}$

66. $f(x) = \dfrac{x - 5}{\sqrt{x^2 - 9}}$

67. Consider $f(x) = \sqrt{x - 2}$ and $g(x) = \sqrt[3]{x - 2}$. Why are the domains of f and g different?

68. A student says that the domain of

$$f(x) = \frac{\sqrt{x + 1}}{x - 3}$$

is all real numbers except $x = 3$. Is the student correct? Explain.

69. Volume of a Box   An open box is to be made from a square piece of material 18 inches on a side by cutting equal squares from the corners and turning up the sides (see figure).

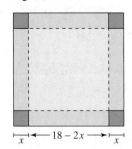
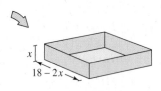

(a) Write the volume V of the box as a function of its height x.

(b) What is the domain of the function?

(c) Determine the volume of a box with a height of 4 inches.

70. Height of a Balloon   A balloon carrying a transmitter ascends vertically from a point 2000 feet from the receiving station (see figure). Let d be the distance between the balloon and the receiving station. Write the height h of the balloon as a function of d. What is the domain of this function?

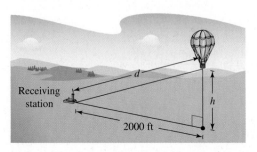

71. Cost, Revenue, and Profit   A company produces a product for which the variable cost is \$11.75 per unit and the fixed costs are \$112,000. The product sells for \$21.95 per unit. Let x be the number of units produced and sold.

(a) Add the variable cost and the fixed costs to write the total cost C as a function of the number of units produced.

(b) Write the revenue R as a function of the number of units sold.

(c) Use the formula

$$P = R - C$$

to write the profit P as a function of the number of units sold.

72. Cost, Revenue, and Profit   A company produces a product for which the variable cost is \$9.85 per unit and the fixed costs are \$85,000. The product sells for \$19.95 per unit. Let x be the number of units produced and sold.

(a) Add the variable cost and the fixed costs to write the total cost C as a function of the number of units produced.

(b) Write the revenue R as a function of the number of units sold.

(c) Use the formula

$$P = R - C$$

to write the profit P as a function of the number of units sold.

73. Path of a Ball The height y (in feet) of a baseball thrown by a child is given by

$$y = -\frac{1}{10}x^2 + 3x + 6$$

where x is the horizontal distance (in feet) from where the ball was thrown. Will the ball fly over the head of another child 30 feet away trying to catch the ball? (Assume that the child who is trying to catch the ball holds a baseball glove at a height of 5 feet.)

74. Path of a Salmon Part of the life cycle of a salmon is migration for reproduction. Salmon are anadromous fish. This means that they swim from the ocean to fresh water streams to lay their eggs. During migration, salmon must jump waterfalls to reach their destination. The path of a jumping salmon is given by

$$h = -0.42x^2 + 2.52x$$

where h is the height (in feet) and x is the horizontal distance (in feet) from where the salmon left the water. Will the salmon clear a waterfall that is 3 feet high if it leaves the water 4 feet from the waterfall?

75. National Defense The national defense budget expenses for veterans V (in billions of dollars) in the United States from 1990 to 2005 can be approximated by the model

$$V = \begin{cases} -0.326t^2 + 3.40t + 28.7, & 0 \le t \le 6 \\ 0.441t^2 - 6.23t + 62.6, & 7 \le t \le 15 \end{cases}$$

where t represents the year, with $t = 0$ corresponding to 1990. Use the model to find total veteran expenses in 1995 and 2005. *(Source: U.S. Office of Management and Budget)*

76. Mobile Homes The number N (in thousands) of mobile homes manufactured for residential use in the United States from 1991 to 2005 can be approximated by the model

$$N = \begin{cases} 29.08t + 157.0, & 1 \le t \le 7 \\ 4.902t^2 - 151.70t + 1289.2, & 8 \le t \le 15 \end{cases}$$

where t represents the year, with $t = 1$ corresponding to 1991. Use the model to find the total number of mobile homes manufactured between 1991 and 2005. *(Source: U.S. Census Bureau)*

Ⓣ **77. Total Sales** The total sales S (in millions of dollars) for the Cheesecake Factory for the years 1999 to 2005 are shown in the table. *(Source: Cheesecake Factory)*

Year	1999	2000	2001	2002
Sales, S	347.5	438.3	539.1	652.0

Year	2003	2004	2005
Sales, S	773.8	969.2	1177.6

(a) Use a graphing utility to create a scatter plot of the data. Let t represent the year, with $t = 9$ corresponding to 1999.

(b) Use the *regression* feature of a graphing utility to find a linear model and a quadratic model for the data.

(c) Use each model to approximate the total sales for each year from 1999 to 2005. Compare the values generated by each model with the actual values shown in the table. Which model is a better fit? Justify your answer.

Ⓣ **78. Book Value per Share** The book values per share B (in dollars) for Analog Devices for the years 1996 to 2005 are shown in the table. *(Source: Analog Devices)*

Year	BV/share, B	Year	BV/share, B
1996	2.72	2001	7.83
1997	3.36	2002	7.99
1998	3.52	2003	8.88
1999	4.62	2004	10.11
2000	6.44	2005	10.06

(a) Use a graphing utility to create a scatter plot of the data. Let t represent the year, with $t = 6$ corresponding to 1996.

(b) Use the *regression* feature of a graphing utility to find a linear model and a quadratic model for the data.

(c) Use each model to approximate the book value per share for each year from 1996 to 2005. Compare the values generated by each model with the actual values shown in the table. Which model is a better fit? Justify your answer.

79. Average Cost The inventor of a new game determines that the variable cost of producing the game is $2.95 per unit and the fixed costs are $8000. The inventor sells each game for $8.79. Let x be the number of games sold.

(a) Write the total cost C as a function of the number of games sold.

(b) Write the average cost per unit $\overline{C} = C/x$ as a function of x.

(c) Complete the table.

x	100	1000	10,000	100,000
$\overline{C}$				

(d) Write a paragraph analyzing the data in the table. What do you observe about the average cost per unit as x gets larger?

80. Average Cost A manufacturer determines that the variable cost for a new product is $2.05 per unit and the fixed costs are $57,000. The product is to be sold for $5.89 per unit. Let x be the number of units sold.

(a) Write the total cost C as a function of the number of units sold.

(b) Write the average cost per unit $\overline{C} = C/x$ as a function of x.

(c) Complete the table.

x	100	1000	10,000	100,000
$\overline{C}$				

(d) Write a paragraph analyzing the data in the table. What do you observe about the average cost per unit as x gets larger?

81. Charter Bus Fares For groups of 80 or more people, a charter bus company determines the rate per person (in dollars) according to the formula

$$\text{Rate} = 8 - 0.05(n - 80) \quad n \geq 80$$

where n is the number of people in the group.

(a) Write the total revenue R for the bus company as a function of n.

(b) Complete the table.

n	90	100	110	120	130	140	150
R							

(c) Write a paragraph analyzing the data in the table.

82. Ripples in a Pond A stone is thrown into the middle of a calm pond, causing ripples to form in concentric circles. The radius r of the outermost ripple increases at the rate of 0.75 foot per second.

(a) Write a function for the radius r of the circle formed by the outermost ripple in terms of time t.

(b) Write a function for the area A enclosed by the outermost ripple. Complete the table.

Time, t	1	2	3	4	5
Radius, r (in feet)					
Area, A (in square feet)					

(c) Compare the ratios $A(2)/A(1)$ and $A(4)/A(2)$. What do you observe? Based on your observation, predict the area when $t = 8$. Verify by checking $t = 8$ in the area function.

83. MAKE A DECISION: DIVIDENDS The dividends D (in dollars) per share declared by Coca-Cola for the years 1990 to 2005 are shown in the table. *(Source: Coca-Cola Company)*

Year	Dividend, D	Year	Dividend, D
1990	0.20	1998	0.60
1991	0.24	1999	0.64
1992	0.28	2000	0.68
1993	0.34	2001	0.72
1994	0.39	2002	0.80
1995	0.44	2003	0.88
1996	0.50	2004	1.00
1997	0.56	2005	1.12

(a) Use a graphing utility to create a scatter plot of the data. Let t represent the year, with $t = 0$ corresponding to 1990.

(b) Use the *regression* feature of a graphing utility to find a linear model and a quadratic model for the data.

(c) Use the graphing utility to graph each model from part (b) with the data.

(d) Which model do you think better fits the data? Explain your reasoning.

(e) Use the model you selected in part (d) to estimate the dividends per share in 2006 and 2007. Coca-Cola predicts the dividends per share in 2006 and 2007 will be $1.24 and $1.32, respectively. How well do your estimates match the ones given by Coca-Cola?

MAKE A DECISION In Exercises 84 and 85, determine whether the statements use the word *function* in ways that are *mathematically* correct. Explain your reasoning.

84. (a) The sales tax on a purchased item is a function of the selling price.

(b) Your score on the next algebra exam is a function of the number of hours you study for the exam.

85. (a) The amount in your savings account is a function of your salary.

(b) The speed at which a free-falling baseball strikes the ground is a function of its initial height.

86. Extended Application To work an extended application analyzing the sales per share of St. Jude Medical, Inc. for the years 1991 to 2005, visit this text's website at *college.hmco.com/info/larsonapplied*. *(Source: St. Jude Medical, Inc.)*

Mid-Chapter Quiz

See www.CalcChat.com for worked-out solutions to odd-numbered exercises.

Take this quiz as you would take a quiz in class. When you are done, check your work against the answers given in the back of the book.

In Exercises 1–3, (a) plot the points, (b) find the distance between the points, and (c) find the midpoint of the line segment joining the points.

1. $(-3, 2),\ (4, -5)$ **2.** $(1.3, -4.5),\ (-3.7, 0.7)$ **3.** $(4, -2), \left(-1, -\frac{5}{2}\right)$

4. A city had a population of 233,134 in 2004 and 244,288 in 2007. Predict the population in 2009. Explain your reasoning.

In Exercises 5–8, find an equation of the line that passes through the given point and has the indicated slope. Then sketch the line.

Point	Slope		Point	Slope
5. $(3, 5)$	$m = \frac{2}{3}$		**6.** $(-2, 4)$	$m = 0$
7. $(2, -3)$	m is undefined.		**8.** $(-2, -5)$	$m = -2$

In Exercises 9–11, sketch the graph of the equation. Identify any intercepts and symmetry.

9. $y = 9 - x^2$ **10.** $y = x\sqrt{x + 4}$ **11.** $y = |x - 3|$

In Exercises 12 and 13, find the standard form of the equation of the circle.

12. Center: $(2, -3)$; radius: 4

13. Center: $\left(0, -\frac{1}{2}\right)$; point on circle: $\left(-1, \frac{3}{2}\right)$

14. Write the equation $x^2 + y^2 - 2x + 4y - 4 = 0$ in standard form. Then sketch the circle.

In Exercises 15 and 16, evaluate the function as indicated and simplify.

15. $f(x) = 3(x + 2) - 4$
  (a) $f(0)$ (b) $f(-3)$

16. $g(t) = 2t^3 - t^2$
  (a) $g(1)$ (b) $g(-2)$

In Exercises 17 and 18, find the domain of the function.

17. $h(x) = \sqrt{x - 4}$

18. $f(x) = \dfrac{x}{x + 2}$

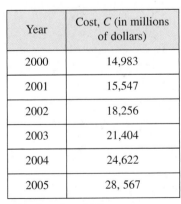

Year	Cost, C (in millions of dollars)
2000	14,983
2001	15,547
2002	18,256
2003	21,404
2004	24,622
2005	28, 567

Table for 19 and 20

(T) In Exercises 19 and 20, use the U.S. Department of Agriculture's estimates for the federal costs C of food stamps (in millions of dollars) shown in the table. *(Source: U.S. Department of Agriculture)*

19. Let $t = 0$ represent 2000. Use a graphing utility to create a scatter plot of the data and use the *regression* feature to find a linear model and a quadratic model for the data.

20. Use each model you found in Exercise 19 to predict the federal costs of food stamps in 2006 and 2007.

21. Write the area A of a circle as a function of its circumference C.

Section 2.5

Graphs of Functions

■ Find the domain and range using the graph of a function.
■ Identify the graph of a function using the Vertical Line Test.
■ Describe the increasing and decreasing behavior of a function.
■ Find the relative minima and relative maxima of the graph of a function.
■ Classify a function as even or odd.
■ Identify six common graphs and use them to sketch the graph of a function.

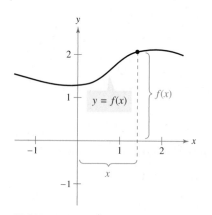

FIGURE 2.44

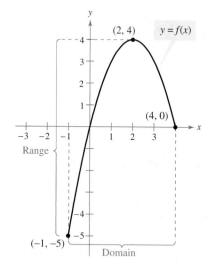

FIGURE 2.45

The Graph of a Function

In Section 2.4, you studied functions from an algebraic point of view. In this section, you will study functions from a graphical perspective.

The **graph of a function** f is the collection of ordered pairs $(x, f(x))$ such that x is in the domain of f. As you study this section, remember that

$$x = \text{the directed distance from the } y\text{-axis}$$

$$f(x) = \text{the directed distance from the } x\text{-axis}$$

as shown in Figure 2.44. If the graph of a function has an x-intercept at $(a, 0)$, then a is a **zero** of the function. In other words, the zeros of a function are the values of x for which $f(x) = 0$. For instance, the function given by $f(x) = x^2 - 4$ has two zeros: -2 and 2.

The **range** of a function (the set of values assumed by the dependent variable) is often easier to determine graphically than algebraically. This technique is illustrated in Example 1.

Example 1 Finding the Domain and Range of a Function

Use the graph of the function f, shown in Figure 2.45, to find (a) the domain of f, (b) the function values $f(-1)$ and $f(2)$, and (c) the range of f.

SOLUTION

a. Because the graph does not extend beyond $x = -1$ (on the left) and $x = 4$ (on the right), the domain of f is all x in the interval $[-1, 4]$.

b. Because $(-1, -5)$ is a point on the graph of f, it follows that

$$f(-1) = -5.$$

Similarly, because $(2, 4)$ is a point on the graph of f, it follows that

$$f(2) = 4.$$

c. Because the graph does not extend below $f(-1) = -5$ or above $f(2) = 4$, the range of f is the interval $[-5, 4]$.

✓ CHECKPOINT 1

Use the graph of $f(x) = x^2 - 3$ to find the domain and range of f. ■

By the definition of a function, at most one y-value corresponds to a given x-value. This means that the graph of a function cannot have two or more different points with the same x-coordinate, and no two points on the graph of a function can be vertically above or below each other. It follows, then, that a vertical line can intersect the graph of a function at most once. This observation provides a convenient visual test called the **Vertical Line Test** for functions.

Vertical Line Test for Functions

A set of points in a coordinate plane is the graph of y as a function of x if and only if no vertical line intersects the graph at more than one point.

Example 2 **Vertical Line Test for Functions**

Use the Vertical Line Test to decide whether the graphs in Figure 2.46 represent y as a function of x.

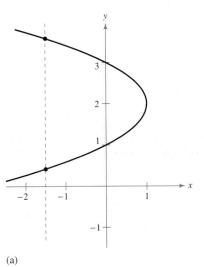

(a)

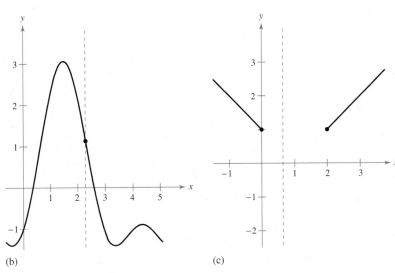

(b)   (c)

FIGURE 2.46

SOLUTION

a. This *is not* a graph of y as a function of x because you can find a vertical line that intersects the graph twice. That is, for a particular input x, there is more than one output y.

b. This *is* a graph of y as a function of x because every vertical line intersects the graph at most once. That is, for a particular input x, there is at most one output y.

c. This *is* a graph of y as a function of x. That is, for a particular input x, there is at most one output y. Note that if a vertical line does not intersect the graph, it simply means that the function is undefined for that particular value of x.

✓ CHECKPOINT 2

Use the Vertical Line Test to decide whether the graph of $x^2 + y = 2$ represents y as a function of x. ▪

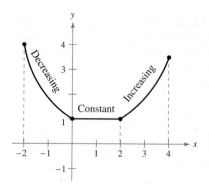

FIGURE 2.47

Increasing and Decreasing Functions

The more you know about the graph of a function, the more you know about the function itself. Consider the graph that is shown in Figure 2.47, for example. As you move from *left to right,* this graph decreases, then is constant, and then increases.

Increasing, Decreasing, and Constant Functions

A function f is **increasing** on an interval if, for any x_1 and x_2 in the interval, $x_1 < x_2$ implies $f(x_1) < f(x_2)$.

A function f is **decreasing** on an interval if, for any x_1 and x_2 in the interval, $x_1 < x_2$ implies $f(x_1) > f(x_2)$.

A function f is **constant** on an interval if, for any x_1 and x_2 in the interval, $f(x_1) = f(x_2)$.

Example 3　**Increasing and Decreasing Functions**

Describe the increasing or decreasing behavior of each function shown in Figure 2.48.

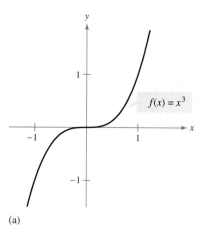

(a)

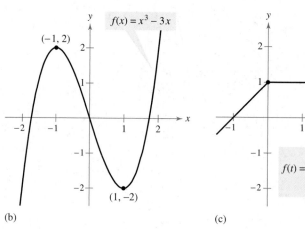

(b)　　　　　　　　　　　　(c)

FIGURE 2.48

SOLUTION

a. This function is increasing over the entire real line.

b. This function is increasing on the interval $(-\infty, -1)$, decreasing on the interval $(-1, 1)$, and increasing on the interval $(1, \infty)$.

c. This function is increasing on the interval $(-\infty, 0)$, constant on the interval $(0, 2)$, and decreasing on the interval $(2, \infty)$.

✓ **CHECKPOINT 3**

Describe the increasing or decreasing behavior of the function $f(x) = x^2 + 3x$. ■

The points at which a function changes its increasing, decreasing, or constant behavior are helpful in determining the **relative minimum** or **relative maximum** values of the function.

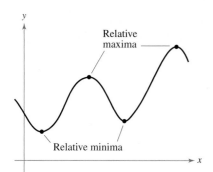

FIGURE 2.49

Definition of Relative Minimum and Relative Maximum

A function value $f(a)$ is called a **relative minimum** of f if there exists an interval (x_1, x_2) that contains a such that

$$x_1 < x < x_2 \quad \text{implies} \quad f(a) \leq f(x).$$

A function value $f(a)$ is called a **relative maximum** of f if there exists an interval (x_1, x_2) that contains a such that

$$x_1 < x < x_2 \quad \text{implies} \quad f(a) \geq f(x).$$

Figure 2.49 shows several examples of relative minima and relative maxima. In Section 3.1, you will study a technique for finding the *exact point* at which a second-degree polynomial function has a relative minimum or relative maximum. For the time being, however, you can use a graphing utility to find reasonable approximations of these points.

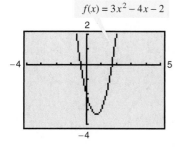

FIGURE 2.50

Example 4   **Approximating a Relative Minimum**

Use a graphing utility to approximate the relative minimum of the function given by $f(x) = 3x^2 - 4x - 2$.

SOLUTION   The graph of f is shown in Figure 2.50. By using the *zoom* and *trace* features of a graphing utility, you can estimate that the function has a relative minimum at the point

$(0.67, -3.33)$.   Relative minimum

Later, in Section 3.1, you will be able to determine that the exact point at which the relative minimum occurs is $\left(\frac{2}{3}, -\frac{10}{3}\right)$.

✓ CHECKPOINT 4

Use a graphing utility to approximate the relative maximum of the function given by $f(x) = -x^2 + 4x - 2$. ∎

You can also use the *table* feature of a graphing utility to approximate numerically the relative minimum of the function in Example 4. Using a table that begins at 0.6 and increments the value of x by 0.01, you can approximate the minimum of $f(x) = 3x^2 - 4x - 2$ to be -3.33, which occurs at $(0.67, -3.33)$. A third way to find the relative minimum is to use the *minimum* feature of a graphing utility.

TECHNOLOGY

Ⓣ   For instructions on how to use the *table* feature and the *minimum* feature, see Appendix A; for specific keystrokes, go to the text website at *college.hmco.com/info/larsonapplied.*

TECHNOLOGY

Ⓣ   If you use a graphing utility to estimate the x- and y-values of a relative minimum or relative maximum, the *zoom* feature will often produce graphs that are nearly flat. To overcome this problem, you can manually change the vertical setting of the viewing window. The graph will stretch vertically if the values of Ymin and Ymax are closer together.

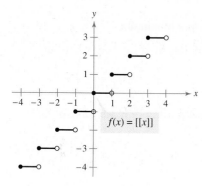

FIGURE 2.51   Greatest Integer Function

Step Functions

The **greatest integer function** is denoted by $[\![x]\!]$ and is defined as

$$f(x) = [\![x]\!] = \text{the greatest integer less than or equal to } x.$$

The graph of this function is shown in Figure 2.51. Note that the graph of the greatest integer function jumps vertically one unit at each integer and is constant (a horizontal line segment) between each pair of consecutive integers. Because of the jumps in its graph, the greatest integer function is an example of a type of function called a **step function.** Some values of the greatest integer function are as follows.

$$[\![-1]\!] = -1 \qquad [\![-0.5]\!] = -1$$
$$[\![0]\!] = 0 \qquad [\![0.5]\!] = 0$$
$$[\![1]\!] = 1 \qquad [\![1.5]\!] = 1$$

The range of the greatest integer function is the set of all integers.

If you use a graphing utility to graph a step function, you should set the utility to *dot* mode rather than *connected* mode.

Example 5   The Price of a Telephone Call

The cost of a long-distance telephone call is $0.10 for up to, but not including, the first minute and $0.05 for each additional minute (or portion of a minute). The greatest integer function

$$C = 0.10 + 0.05[\![t]\!], \quad t > 0$$

can be used to model the cost of this call, where C is the total cost of the call (in dollars) and t is the length of the call (in minutes).

a. Sketch the graph of this function.

b. How long can you talk without spending more than $1?

SOLUTION

a. For calls up to, but not including, 1 minute, the cost is $0.10. For calls between 1 and 2 minutes, the cost is $0.15, and so on.

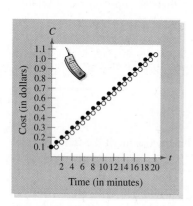

FIGURE 2.52

Length of call, t	Cost of call, C
$0 < t < 1$	$0.10
$1 \le t < 2$	$0.15
$2 \le t < 3$	$0.20
$\vdots$	$\vdots$
$19 \le t < 20$	$1.05

Using these and other values, you can sketch the graph shown in Figure 2.52.

b. From the graph, you can see that your phone call must be less than 19 minutes to avoid spending more than $1.

✓**CHECKPOINT 5**

In Example 5, suppose the cost of each additional minute (or portion of a minute) is $0.07. Sketch the graph of this function. How long can you talk without spending more than $1? ▓

Even and Odd Functions

In Section 2.1, you studied different types of symmetry of a graph. A function is said to be **even** if its graph is symmetric with respect to the y-axis and **odd** if its graph is symmetric with respect to the origin. The symmetry tests in Section 2.1 yield the following tests for even and odd functions. Even though symmetry with respect to the x-axis is introduced in Section 2.1, it will not be discussed here because a graph that is symmetric about the x-axis is not a function.

Tests for Even and Odd Functions

A function given by $y = f(x)$ is even if, for each x in the domain of f,

$$f(-x) = f(x).$$

A function given by $y = f(x)$ is odd if, for each x in the domain of f,

$$f(-x) = -f(x).$$

Example 6   Even and Odd Functions

Decide whether each function is even, odd, or neither.

a. $g(x) = x^3 - x$ **b.** $h(x) = x^2 + 1$

SOLUTION

a. The function given by $g(x) = x^3 - x$ is odd because

$$g(-x) = (-x)^3 - (-x) = -x^3 + x = -(x^3 - x) = -g(x).$$

b. The function given by $h(x) = x^2 + 1$ is even because

$$h(-x) = (-x)^2 + 1 = x^2 + 1 = h(x).$$

The graphs of the two functions are shown in Figure 2.53.

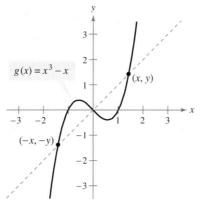

(a) Odd function (symmetric about origin)

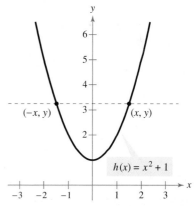

(b) Even function (symmetric about y-axis)

FIGURE 2.53

✓ **CHECKPOINT 6**

Decide whether the function $f(x) = -2x^2 + x - 1$ is even, odd, or neither. ■

Common Graphs

Figure 2.54 shows the graphs of six common functions. You need to be familiar with these graphs. They can be used as an aid when sketching other graphs. For instance, the graph of the absolute value function given by

$$f(x) = |x - 2|$$

is $\bigvee$-shaped.

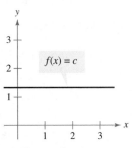

(a) Constant function

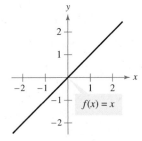

(b) Identity function

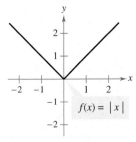

(c) Absolute value function

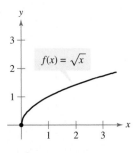

(d) Square root function

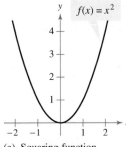

(e) Squaring function

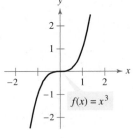

(f) Cubing function

FIGURE 2.54

CONCEPT CHECK

In Exercises 1 and 2, determine whether the statement is true or false. Justify your answer.

1. If $a < 0$, then $f(0)$ is the relative maximum of the function $f(x) = ax^2$.

2. The graph of the greatest integer function is increasing over its entire domain.

3. Is the function represented by the following set of ordered pairs *even*, *odd*, or *neither*?

    $\{(1, 4), (-1, 4)\}$

4. The line $x = 1$ does not intersect the graph of f. Can you conclude that f is a function? Explain.

Skills Review 2.5

The following warm-up exercises involve skills that were covered in earlier sections. You will use these skills in the exercise set for this section. For additional help, review Sections 1.4, 1.5, and 2.4.

1. Find $f(2)$ for $f(x) = -x^3 + 5x$.

2. Find $f(6)$ for $f(x) = x^2 - 6x$.

3. Find $f(-x)$ for $f(x) = 3/x$.

4. Find $f(-x)$ for $f(x) = x^2 + 3$.

In Exercises 5 and 6, solve the equation.

5. $x^3 - 16x = 0$

6. $2x^2 - 3x + 1 = 0$

In Exercises 7–10, find the domain of the function.

7. $g(x) = 4(x - 4)^{-1}$

8. $f(x) = 2x/(x^2 - 9x + 20)$

9. $h(t) = \sqrt[4]{5 - 3t}$

10. $f(t) = t^3 + 3t - 5$

Exercises 2.5

See www.CalcChat.com for worked-out solutions to odd-numbered exercises.

In Exercises 1–8, find the domain and range of the function. Then evaluate f at the given x-value.

1. $f(x) = \sqrt{x - 1}$,

$x = 1$

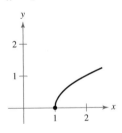

2. $f(x) = \sqrt{x^2 - 4}$,

$x = -2$

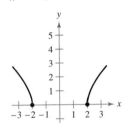
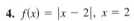

3. $f(x) = 4 - x^2$, $x = 0$

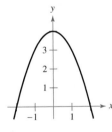

4. $f(x) = |x - 2|$, $x = 2$

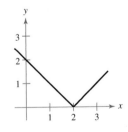

5. $f(x) = x^3 - 1$, $x = 0$

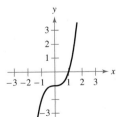

6. $f(x) = \dfrac{|x|}{x}$, $x = 5$

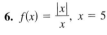
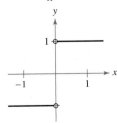

7. $f(x) = \sqrt{25 - x^2}$,

$x = 0$

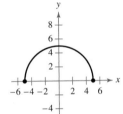

8. $f(x) = \sqrt{x^2 - 9}$,

$x = 3$

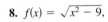
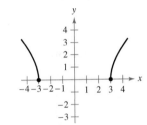

In Exercises 9–12, use the Vertical Line Test to decide whether y is a function of x.

9. $y = x^2$

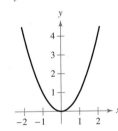

10. $x - y^2 = 0$

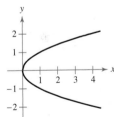

11. $x^2 + y^2 = 9$

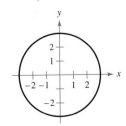

12. $x^2 = xy - 1$

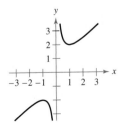

In Exercises 13–20, describe the increasing and decreasing behavior of the function. Find the point or points where the behavior of the function changes.

13. $f(x) = 2x$

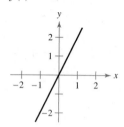

14. $f(x) = x^2 - 2x$

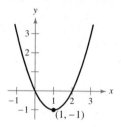

15. $f(x) = x^3 - 3x^2$

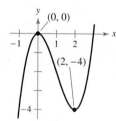

16. $f(x) = \sqrt{x^2 - 4}$

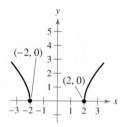

17. $f(x) = 3x^4 - 6x^2$

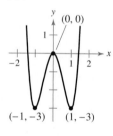

18. $f(x) = x^{2/3}$

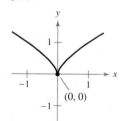

19. $y = x\sqrt{x + 3}$

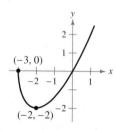

20. $y = |x + 1| + |x - 1|$

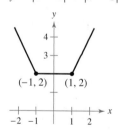

(T) In Exercises 21–26, use a graphing utility to graph the function, approximate the relative minimum or maximum of the function, and estimate the open intervals on which the function is increasing or decreasing.

21. $f(x) = x^2 - 4x + 1$

22. $f(x) = -x^2 + 6x + 3$

23. $f(x) = x^3 - 3x^2$

24. $f(x) = -x^3 + 3x + 1$

25. $f(x) = \frac{1}{4}(-4x^4 - 5x^3 + 10x^2 + 8x + 6)$

26. $f(x) = \frac{1}{4}(x^4 + x^3 - 10x^2 + 2x - 15)$

In Exercises 27–32, decide whether the function is even, odd, or neither.

27. $f(x) = x^6 - 2x^2 + 3$ **28.** $f(t) = t^2 + 3t - 10$

29. $g(x) = x^3 - 5x$ **30.** $h(x) = x^3 + 3$

31. $f(x) = x\sqrt{4 - x^2}$ **32.** $g(s) = 4s^{2/3}$

In Exercises 33–36, evaluate the function at each specified value of the independent variable.

33. $f(x) = [\![x]\!]$

   (a) $f(2)$ (b) $f(2.5)$

   (c) $f(-2.5)$ (d) $f(-4)$

34. $f(x) = [\![-x]\!]$

   (a) $f(3)$ (b) $f(6.1)$

   (c) $f(-5.9)$ (d) $f(-9)$

35. $f(x) = [\![x - 1.8]\!]$

   (a) $f(4)$ (b) $f(3.7)$

   (c) $f(-5.8)$ (d) $f(-6.3)$

36. $f(x) = [\![x + 0.3]\!]$

   (a) $f(2.9)$ (b) $f(4.6)$

   (c) $f(-2.3)$ (d) $f(-4.2)$

In Exercises 37–50, sketch the graph of the function and determine whether the function is even, odd, or neither.

37. $f(x) = 3$

38. $g(x) = x$

39. $f(x) = 5 - 3x$

40. $h(x) = x^2 - 4$

41. $g(s) = \dfrac{s^3}{4}$

42. $f(t) = -t^4$

43. $f(x) = \sqrt{1 - x}$

44. $g(t) = \sqrt[3]{t - 1}$

45. $f(x) = x^{3/2}$

46. $f(x) = |x + 2|$

47. $f(x) = \begin{cases} x^2 + 1, & x \le 1 \\ 3x - 1, & x > 1 \end{cases}$

48. $f(x) = \begin{cases} 2x - 1, & x \le -1 \\ x^2 - 1, & x > -1 \end{cases}$

49. $f(x) = \begin{cases} x + 1, & x \le 0 \\ 4, & 0 < x \le 2 \\ 3x - 1, & x > 2 \end{cases}$

50. $f(x) = \begin{cases} 2x - 1, & x \le 1 \\ 3, & 1 < x \le 3 \\ 2x + 1, & x > 3 \end{cases}$

In Exercises 51–64, sketch the graph of the function.

51. $f(x) = 4 - x$

52. $f(x) = 4x + 2$

53. $f(x) = x^2 - 9$

54. $f(x) = x^2 - 4x$

55. $f(x) = 1 - x^4$

56. $f(x) = x^4 - 4x^2$

57. $f(x) = \frac{1}{3}(3 + |x|)$

58. $f(x) = -1(1 + |x|)$

59. $f(x) = \sqrt{x + 3}$

60. $f(x) = \sqrt{x - 1}$

61. $f(x) = -[\![x]\!]$

62. $f(x) = 2[\![x]\!]$

63. $f(x) = [\![x - 1]\!]$

64. $f(x) = [\![x + 1]\!]$

65. *MAKE A DECISION: PRICE OF GOLD*   The price P (in dollars) of an ounce of gold from 1995 to 2005 can be approximated by the model

$$P = -0.203513t^4 + 8.27786t^3 - 115.1479t^2 + 635.832t - 819.60, \quad 5 \le t \le 15$$

where t represents the year, with $t = 5$ corresponding to 1995. Use the graph of P to find the maximum price of gold between 1995 and 2005. During which years was the price decreasing? During which years was the price increasing? Is it realistic to assume that the price of gold will continue to follow this model? *(Source: World Gold Council)*

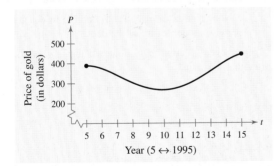

Year (5 ↔ 1995)

66. *MAKE A DECISION: SALES*   The sales S (in billions of dollars) of petroleum and coal products from 1997 to 2005 can be approximated by the model

$$S = 1.34668t^4 - 57.7219t^3 + 918.390t^2 - 6355.84t + 16,367.4, \quad 7 \le t \le 15$$

where t represents the year, with $t = 7$ corresponding to 1997. Use the graph of S to find the maximum sales of these products between 1997 and 2005. During which years were sales decreasing? During which years were sales increasing? Is it realistic to assume that sales will continue to follow this model? *(Source: U.S. Census Bureau)*

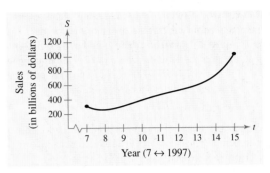

Year (7 ↔ 1997)

(T) 67. **Lung Volume**   The change in volume V (in milliliters) of the lungs as they expand and contract during a breath can be approximated by the model

$$V = (-6.549s^2 + 26.20s - 3.8)^2, \quad 0 \le s \le 4$$

where s represents the number of seconds. Graph the volume function with a graphing utility and use the *trace* feature to estimate the number of seconds in which the volume is increasing and in which the volume is decreasing. Find the maximum change in volume between 0 and 4 seconds.

68. **Book Value**   For the years 1990 to 2005, the book value B (in dollars) of a share of Wells Fargo stock can be approximated by the model

$$B = 0.0272t^2 + 0.268t + 1.71, \quad 0 \le t \le 15$$

where t represents the year, with $t = 0$ corresponding to 1990 (see figure). *(Source: Wells Fargo)*

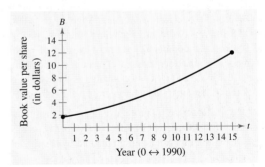

Year (0 ↔ 1990)

(a) Estimate the maximum book value per share from 1990 to 2005.

(b) Estimate the minimum book value per share from 1990 to 2005.

(T) (c) Verify your estimates from parts (a) and (b) with a graphing utility.

69. **Reasoning**   When finding a maximum or minimum value in Exercises 65–68, why should you also check the endpoints of the function?

70. **Reasoning**   Assume that the book value B in Exercise 68 continues to follow the model through 2007. In which year is B at a maximum?

71. Maximum Profit The marketing department of a company estimates that the demand for a product is given by $p = 100 - 0.0001x$, where p is the price per unit and x is the number of units. The cost C of producing x units is given by $C = 350,000 + 30x$, and the profit P for producing and selling x units is given by

$$P = R - C = xp - C.$$

Sketch the graph of the profit function and estimate the number of units that would produce a maximum profit. Verify your estimate using a graphing utility.

72. Maximum Profit The marketing department of a company estimates that the demand for a product is given by $p = 125 - 0.0002x$, where p is the price per unit and x is the number of units. The cost C of producing x units is given by $C = 225,000 + 80x$, and the profit P for producing and selling x units is given by

$$P = R - C = xp - C.$$

Sketch the graph of the profit function and estimate the number of units that would produce a maximum profit. Verify your estimate using a graphing utility.

73. Cost of Overnight Delivery The cost of sending an overnight package from New York to Atlanta is $9.80 for up to, but not including, the first pound and $2.50 for each additional pound (or portion of a pound). A model for the total cost C of sending the package is $C = 9.8 + 2.5 \llbracket x \rrbracket$, $x > 0$, where x is the weight of the package (in pounds). Sketch the graph of this function.

74. Cost of Overnight Delivery The cost of sending an overnight package from Los Angeles to Miami is $10.75 for up to, but not including, the first pound and $3.95 for each additional pound (or portion of a pound). A model for the total cost C of sending the package is $C = 10.75 + 3.95 \llbracket x \rrbracket$, $x > 0$, where x is the weight of the package (in pounds). Sketch the graph of this function.

75. Strategic Reserve The total volume V (in millions of barrels) of the Strategic Oil Reserve R in the United States from 1995 to 2005 can be approximated by the model

$$V = \begin{cases} -2.722t^3 + 61.18t^2 - 451.5t + 1660, & 5 \le t \le 10 \\ 34.7t + 179, & 11 \le t \le 15 \end{cases}$$

where t represents the year, with $t = 5$ corresponding to 1995. Sketch the graph of this function. *(Source: U.S. Energy Information Administration)*

76. Grade Level Salaries The 2007 salary S (in dollars) for federal employees at the Step 1 level can be approximated by the model

$$S = \begin{cases} 2904.3x + 12,155, & x = 1, 2, \ldots, 10 \\ 11,499.2x - 81,008, & x = 11, \ldots, 15 \end{cases}$$

where x represents the "GS" grade. Sketch a *bar graph* that represents this function. *(Source: U.S. Office of Personnel Management)*

77. Air Travel The total numbers (in thousands) of U.S. airline delays, cancellations, and diversions for the years 1995 to 2005 are given by the following ordered pairs. *(Source: U.S. Bureau of Transportation Statistics)*

(1995, 5327.4) (1996, 5352.0) (1997, 5411.8)
(1998, 5384.7) (1999, 5527.9) (2000, 5683.0)
(2001, 5967.8) (2002, 5271.4) (2003, 6488.5)
(2004, 7129.3) (2005, 7140.6)

(a) Use the *regression* feature of a graphing utility to find a quadratic model for the data from 1995 to 2001. Let t represent the year, with $t = 5$ corresponding to 1995.

(b) Use the *regression* feature of a graphing utility to find a quadratic model for the data from 2002 to 2005. Let t represent the year, with $t = 12$ corresponding to 2002.

(c) Use your results from parts (a) and (b) to construct a piecewise model for all of the data.

78. Revenues The revenues of Symantec Corporation (in millions of dollars) from 1996 to 2005 are given by the following ordered pairs. *(Source: Symantec Corporation)*

(1996, 472.2) (1997, 578.4) (1998, 633.8)
(1999, 745.7) (2000, 853.6) (2001, 1071.4)
(2002, 1406.9) (2003, 1870.1) (2004, 2582.8)
(2005, 4143.4)

(a) Use the *regression* feature of a graphing utility to find a linear model for the data from 1996 to 2000. Let t represent the year, with $t = 6$ corresponding to 1996.

(b) Use the *regression* feature of a graphing utility to find a quadratic model for the data from 2001 to 2005. Let t represent the year, with $t = 11$ corresponding to 2001.

(c) Use your results from parts (a) and (b) to construct a piecewise model for all of the data.

79. If f is an even function, determine whether g is even, odd, or neither. Explain.

(a) $g(x) = -f(x)$

(b) $g(x) = f(-x)$

(c) $g(x) = f(x) - 2$

(d) $g(x) = f(x - 2)$

Think About It In Exercises 80–83, find the coordinates of a second point on the graph of a function f if the given point is on the graph and the function is (a) even and (b) odd.

80. $\left(-\frac{3}{2}, 4\right)$

81. $\left(-\frac{5}{3}, -7\right)$

82. $(4, 9)$

83. $(5, -1)$

Section 2.6

Transformations of Functions

- Use vertical and horizontal shifts to sketch graphs of functions.
- Use reflections to sketch graphs of functions.
- Use nonrigid transformations to sketch graphs of functions.

Vertical and Horizontal Shifts

Many functions have graphs that are simple transformations of the common graphs that are summarized on page 214. For example, you can obtain the graph of $h(x) = x^2 + 2$ by shifting the graph of $f(x) = x^2$ *upward* two units, as shown in Figure 2.55. In function notation, h and f are related as follows.

$$h(x) = x^2 + 2 = f(x) + 2 \qquad \text{Upward shift of two units}$$

Similarly, you can obtain the graph of $g(x) = (x - 2)^2$ by shifting the graph of $f(x) = x^2$ to the *right* two units, as shown in Figure 2.56. In this case, the functions g and f have the following relationship.

$$g(x) = (x - 2)^2 = f(x - 2) \qquad \text{Right shift of two units}$$

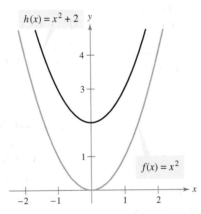

FIGURE 2.55 Vertical Shift Upward

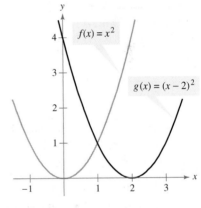

FIGURE 2.56 Horizontal Shift to the Right

The following list summarizes this discussion about horizontal and vertical shifts.

STUDY TIP

In items 3 and 4, be sure you see that $h(x) = f(x - c)$ corresponds to a *right* shift and $h(x) = f(x + c)$ corresponds to a *left* shift.

Vertical and Horizontal Shifts

Let c be a positive real number. **Vertical** and **horizontal shifts** of the graph of $y = f(x)$ are represented as follows.

1. Vertical shift c units **upward:** $h(x) = f(x) + c$
2. Vertical shift c units **downward:** $h(x) = f(x) - c$
3. Horizontal shift c units to the **right:** $h(x) = f(x - c)$
4. Horizontal shift c units to the **left:** $h(x) = f(x + c)$

Some graphs can be obtained from a combination of vertical and horizontal shifts, as demonstrated in Example 1(b). Vertical and horizontal shifts generate a *family of functions,* each with the same shape but at different locations in the plane.

Example 1   Shifts in the Graph of a Function

Use the graph of $f(x) = x^3$ to sketch the graph of each function.

a. $g(x) = x^3 + 1$ **b.** $h(x) = (x + 2)^3 + 1$

SOLUTION

a. Relative to the graph of $f(x) = x^3$, the graph of $g(x) = x^3 + 1$ is an upward shift of one unit, as shown in Figure 2.57(a).

b. Relative to the graph of $f(x) = x^3$, the graph of $h(x) = (x + 2)^3 + 1$ involves a left shift of two units *and* an upward shift of one unit, as shown in Figure 2.57(b).

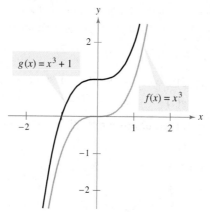

(a) Vertical shift: one unit upward

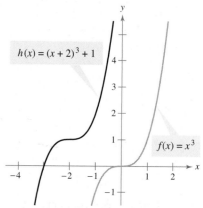

(b) Horizontal shift: two units left;
    Vertical shift: one unit upward

FIGURE 2.57

Note that the functions f, g, and h belong to the family of cubic functions.

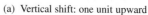

✓**CHECKPOINT 1**

Use the graph of $f(x) = \sqrt{x}$ to sketch the graph of $g(x) = \sqrt{x - 1} - 1$. ▪

TECHNOLOGY

Ⓣ Graphing utilities are ideal tools for exploring transformations of functions. Try to predict how the graphs of g and h relate to the graph of f. Graph f, g, and h in the same viewing window to confirm your prediction.

a. $f(x) = x^2$, $g(x) = (x - 4)^2$, $h(x) = (x - 4)^2 + 3$

b. $f(x) = x^2$, $g(x) = (x + 1)^2$, $h(x) = (x + 1)^2 - 2$

c. $f(x) = x^2$, $g(x) = (x + 4)^2$, $h(x) = (x + 4)^2 + 2$

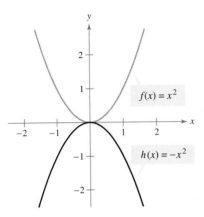

FIGURE 2.58 Reflection

Reflections

The second common type of transformation is a **reflection.** For instance, if you consider the x-axis to be a mirror, the graph of $h(x) = -x^2$ is the mirror image (or reflection) of the graph of $f(x) = x^2$, as shown in Figure 2.58.

Reflections in the Coordinate Axes

Reflections in the coordinate axes of the graph of $y = f(x)$ are represented as follows.

1. **Reflection in the x-axis:** $g(x) = -f(x)$
2. **Reflection in the y-axis:** $h(x) = f(-x)$

Example 2 **Reflections of the Graph of a Function**

Compare the graph of each function with the graph of $f(x) = \sqrt{x}$.

a. $g(x) = -\sqrt{x}$ **b.** $h(x) = \sqrt{-x}$

SOLUTION

a. The graph of g is a reflection of the graph of f in the x-axis because

$$g(x) = -\sqrt{x} = -f(x). \qquad \text{See Figure 2.59(a).}$$

b. The graph of h is a reflection of the graph of f in the y-axis because

$$h(x) = \sqrt{-x} = f(-x). \qquad \text{See Figure 2.59(b).}$$

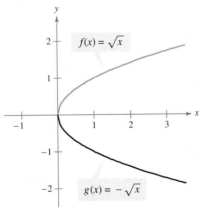

(a) Reflection in x-axis

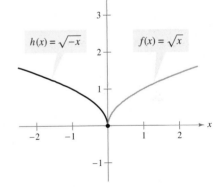

(b) Reflection in y-axis

FIGURE 2.59

When sketching the graph of a function involving square roots, remember that the domain must be restricted to exclude numbers that make the radicand negative. For instance, here are the domains of the functions in Example 2.

$$\text{Domain of } g(x) = -\sqrt{x}: \quad x \geq 0$$
$$\text{Domain of } h(x) = \sqrt{-x}: \quad x \leq 0$$

✓ **CHECKPOINT 2**

Compare the graph of each function with the graph of $f(x) = |x|$.

a. $g(x) = -|x|$

b. $h(x) = |-x|$ ∎

Example 3 **Reflections and Shifts**

Use the graph of $f(x) = x^2$ to sketch the graph of each function.

a. $g(x) = -(x - 3)^2$ **b.** $h(x) = -x^2 + 2$

SOLUTION

a. To sketch the graph of $g(x) = -(x - 3)^2$, first shift the graph of $f(x) = x^2$ to the right three units. Then reflect the result in the x-axis.

b. To sketch the graph of $h(x) = -x^2 + 2$, first reflect the graph of $f(x) = x^2$ in the x-axis. Then shift the result upward two units.

The graphs of both functions are shown in Figure 2.60.

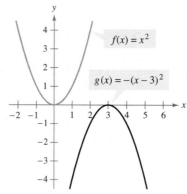
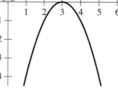

(a) Shift and then reflect in x-axis

(b) Reflect in x-axis and then shift

FIGURE 2.60

✓ **CHECKPOINT 3**

Use the graph of $f(x) = x^3$ to sketch the graph of each function.

a. $g(x) = -(x + 2)^3$ **b.** $h(x) = -x^3 - 3$ ∎

Example 4 **Finding Equations from Graphs**

The graphs labeled g and h in Figure 2.61(a) are transformations of the graph of $f(x) = x^4$. Find an equation for each function.

SOLUTION The graph of g is a reflection in the x-axis *followed* by a downward shift of two units of the graph of $f(x) = x^4$. So, the equation for g is $g(x) = -x^4 - 2$. The graph of h is a horizontal shift of one unit to the left *followed* by a reflection in the x-axis of the graph of $f(x) = x^4$. So, the equation for h is $h(x) = -(x + 1)^4$.

Can you think of another way to find an equation for g in Example 4? If you were to shift the graph of f upward two units and then reflect the graph in the x-axis, you would obtain the equation $g(x) = -(x^4 + 2)$. The Distributive Property yields $g(x) = -x^4 - 2$, which is the same equation obtained in Example 4.

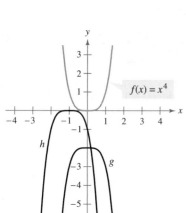
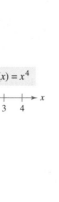

(a)

(b)

FIGURE 2.61

✓ **CHECKPOINT 4**

The graph labeled h in Figure 2.61(b) is a transformation of the graph of $f(x) = (x + 2)^2$. Find an equation for the function h. ∎

Nonrigid Transformations

Horizontal shifts, vertical shifts, and reflections are **rigid** transformations because the basic shape of the graph is unchanged. These transformations change only the *position* of the graph in the *xy*-plane. A **nonrigid** transformation is one that causes a *distortion*—a change in the shape of the original graph. For instance, a nonrigid transformation of the graph of $y = f(x)$ is represented by $g(x) = cf(x)$, where the transformation is a **vertical stretch** if $|c| > 1$ and a **vertical shrink** if $0 < |c| < 1$.

Example 5 Nonrigid Transformations

Compare the graph of each function with the graph of $f(x) = |x|$.

a. $h(x) = 3|x|$ **b.** $g(x) = \frac{1}{3}|x|$

SOLUTION

a. Relative to the graph of $f(x) = |x|$, the graph of

$$h(x) = 3|x| = 3f(x)$$

is a vertical stretch (each *y*-value is multiplied by 3) of the graph of *f*.

b. Similarly, the graph of

$$g(x) = \frac{1}{3}|x| = \frac{1}{3}f(x)$$

is a vertical shrink $\left(\text{each } y\text{-value is multiplied by } \frac{1}{3}\right)$ of the graph of *f*.

The graphs of both functions are shown in Figure 2.62.

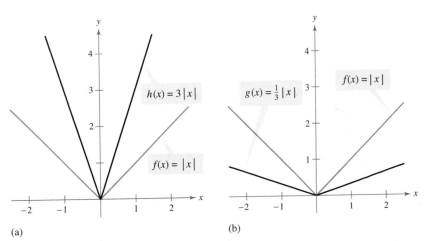

(a)   (b)

FIGURE 2.62

✓ CHECKPOINT 5

Compare the graph of each function with the graph of $f(x) = \sqrt{x}$.

a. $g(x) = 4\sqrt{x}$ **b.** $h(x) = \frac{1}{4}\sqrt{x}$

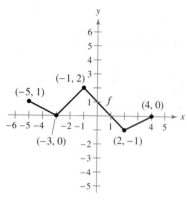

FIGURE 2.63

| Example 6 | **Rigid and Nonrigid Transformations** |

Use the graph of f shown in Figure 2.63 to sketch each graph.

a. $g(x) = f(x - 2) + 1$

b. $h(x) = \frac{1}{2}f(x)$

SOLUTION

a. The graph of g is a horizontal shift to the right two units and a vertical shift upward one unit of the graph of f. The graph of g is shown in Figure 2.64(a).

b. The graph of h is a vertical shrink of the graph of f. The graph of h is shown in Figure 2.64(b).

For $x = -5$, $h(-5) = \frac{1}{2}f(-5) = \frac{1}{2}(1) = \frac{1}{2}$.

For $x = -3$, $h(-3) = \frac{1}{2}f(-3) = \frac{1}{2}(0) = 0$.

For $x = -1$, $h(-1) = \frac{1}{2}f(-1) = \frac{1}{2}(2) = 1$.

For $x = 2$, $h(2) = \frac{1}{2}f(2) = \frac{1}{2}(-1) = -\frac{1}{2}$.

For $x = 4$, $h(4) = \frac{1}{2}f(4) = \frac{1}{2}(0) = 0$.

✓ CHECKPOINT 6

Use the graph of g shown in Figure 2.64(a) to sketch the graph of $p(x) = 2g(x) - 1$. ∎

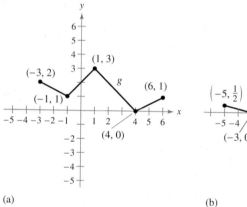

(a)

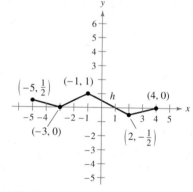

(b)

FIGURE 2.64

┌─ **CONCEPT CHECK** ─────────────────────

In Exercises 1–4, determine whether the statement is true or false. Explain your reasoning.

1. A rigid transformation preserves the basic shape of a graph.

2. The graph of $g(x) = x^2 + 5$ is a vertical shift downward five units of the graph of $f(x) = x^2$.

3. The graph of $g(x) = (x - 1)^2$ is a horizontal shift to the left one unit of the graph of $f(x) = x^2$.

4. The graph of $g(x) = 2x^2$ is an example of a nonrigid transformation of the graph of $f(x) = x^2$.

Skills Review 2.6   The following warm-up exercises involve skills that were covered in earlier sections. You will use these skills in the exercise set for this section. For additional help, review Sections 1.4, 1.5, 2.4, and 2.5.

In Exercises 1 and 2, evaluate the function at the indicated value.

1. Find $f(3)$ for $f(x) = x^2 - 4x + 15$.

2. Find $f(-x)$ for $f(x) = 2x/(x - 3)$.

In Exercises 3 and 4, solve the equation.

3. $-x^3 + 10x = 0$

4. $3x^2 + 2x - 8 = 0$

In Exercises 5–10, sketch the graph of the function.

5. $f(x) = -2$

6. $f(x) = -x$

7. $f(x) = x + 5$

8. $f(x) = 2 - x$

9. $f(x) = 3x - 4$

10. $f(x) = 9x + 10$

Exercises 2.6

See www.CalcChat.com for worked-out solutions to odd-numbered exercises.

In Exercises 1–8, describe the sequence of transformations from $f(x) = x^2$ to g. Then sketch the graph of g by hand. Verify with a graphing utility.

1. $g(x) = x^2 - 4$ **2.** $g(x) = x^2 + 1$

3. $g(x) = (x + 2)^2$ **4.** $g(x) = (x - 3)^2$

5. $g(x) = (x - 2)^2 + 2$

6. $g(x) = (x + 1)^2 - 3$

7. $g(x) = -x^2 + 1$

8. $g(x) = -(x - 2)^2$

In Exercises 9–16, describe the sequence of transformations from $f(x) = |x|$ to g. Then sketch the graph of g by hand. Verify with a graphing utility.

9. $g(x) = |x| + 2$ **10.** $g(x) = |x| - 3$

11. $g(x) = |x - 1|$ **12.** $g(x) = |x + 4|$

13. $g(x) = -|x| + 3$

14. $g(x) = 5 - |x - 1|$

15. $g(x) = |x + 1| - 3$

16. $g(x) = |x - 2| + 2$

In Exercises 17–24, describe the sequence of transformations from $f(x) = \sqrt{x}$ to g. Then sketch the graph of g by hand. Verify with a graphing utility.

17. $g(x) = \sqrt{x - 3}$ **18.** $g(x) = \sqrt{x + 4}$

19. $g(x) = \sqrt{x - 3} + 1$ **20.** $g(x) = \sqrt{x + 5} - 2$

21. $g(x) = \sqrt{2x}$ **22.** $g(x) = \sqrt{2x} - 5$

23. $g(x) = 2 - \sqrt{x - 4}$ **24.** $g(x) = \sqrt{-x} + 1$

In Exercises 25–34, describe the sequence of transformations from $f(x) = \sqrt[3]{x}$ to y. Then sketch the graph of y by hand. Verify with a graphing utility.

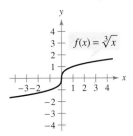

25. $y = \sqrt[3]{-x}$ **26.** $y = -\sqrt[3]{x}$

27. $y = \sqrt[3]{x} - 1$ **28.** $y = \sqrt[3]{x + 1}$

29. $y = 2 - \sqrt[3]{x + 1}$ **30.** $y = -\sqrt[3]{x - 1} - 4$

31. $y = \sqrt[3]{x + 1} - 1$ **32.** $y = 2\sqrt[3]{x - 2} + 1$

33. $y = \frac{1}{2}\sqrt[3]{x}$ **34.** $y = \frac{1}{2}\sqrt[3]{x} - 3$

In Exercises 35–40, identify the transformation shown in the graph and the associated common function. Write the equation of the graphed function.

35.

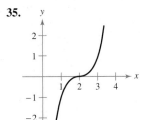

36.

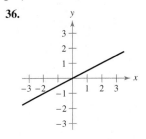

37.

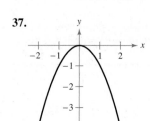

38.

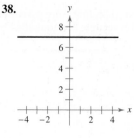

39.

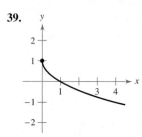

40.

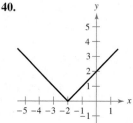

41. Use a graphing utility to graph f for $c = -2, 0,$ and 2 in the same viewing window.

(a) $f(x) = \frac{1}{2}x + c$

(b) $f(x) = \frac{1}{2}(x - c)$

(c) $f(x) = \frac{1}{2}(cx)$

In each case, compare the graph with the graph of $y = \frac{1}{2}x$.

42. Use a graphing utility to graph f for $c = -2, 0,$ and 2 in the same viewing window.

(a) $f(x) = x^3 + c$

(b) $f(x) = (x - c)^3$

(c) $f(x) = (x - 2)^3 + c$

In each case, compare the graph with the graph of $y = x^3$.

43. Use the graph of $f(x) = x^2$ to write equations for the functions whose graphs are shown.

(a)

(b)

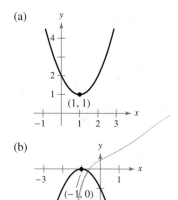

44. Use the graph of $f(x) = x^3$ to write equations for the functions whose graphs are shown.

(a)

(b)

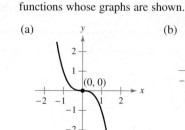

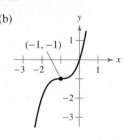

45. Use the graph of f (see figure) to sketch each graph.

(a) $y = f(x) + 2$ (b) $y = -f(x)$

(c) $y = f(x - 2)$ (d) $y = f(x + 3)$

(e) $y = 2f(x)$ (f) $y = f(-x)$

46. Use the graph of f (see figure) to sketch each graph.

(a) $y = f(x) - 1$ (b) $y = f(x + 1)$

(c) $y = f(x - 1)$ (d) $y = -f(x - 2)$

(e) $y = f(-x)$ (f) $y = \frac{1}{2}f(x)$

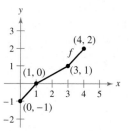

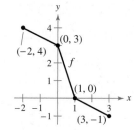

Figure for 45   Figure for 46

47. Use the graph of f (see figure) to sketch each graph.

(a) $y = f(-x)$ (b) $y = f(x) + 4$

(c) $y = 2f(x)$ (d) $y = -f(x - 4)$

(e) $y = f(x) - 3$ (f) $y = -f(x) - 1$

48. Use the graph of f (see figure) to sketch each graph.

(a) $y = f(x - 5)$ (b) $y = -f(x) + 3$

(c) $y = \frac{1}{3}f(x)$ (d) $y = -f(x + 1)$

(e) $y = f(-x)$ (f) $y = f(x) - 5$

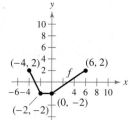

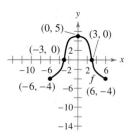

Figure for 47   Figure for 48

In Exercises 49–52, consider the graph of $f(x) = x^3$. Use your knowledge of rigid and nonrigid transformations to write an equation for each of the following descriptions. Verify with a graphing utility.

49. The graph of f is shifted two units downward.

50. The graph of f is shifted three units to the left.

51. The graph of f is vertically stretched by a factor of 4.

52. The graph of f is vertically shrunk by a factor of $\frac{1}{3}$.

In Exercises 53–56, consider the graph of $f(x) = |x|$. Use your knowledge of rigid and nonrigid transformations to write an equation for each of the following descriptions. Verify with a graphing utility.

53. The graph of f is shifted three units to the right and two units upward.

54. The graph of f is reflected in the x-axis, shifted two units to the left, and shifted three units upward.

55. The graph of f is vertically stretched by a factor of 4 and reflected in the x-axis.

56. The graph of f is vertically shrunk by a factor of $\frac{1}{3}$ and shifted two units to the left.

In Exercises 57–60, consider the graph of $g(x) = \sqrt{x}$. Use your knowledge of rigid and nonrigid transformations to write an equation for each of the following descriptions. Verify with a graphing utility.

57. The graph of g is shifted four units to the right and three units downward.

58. The graph of g is reflected in the x-axis, shifted two units to the left, and shifted one unit upward.

59. The graph of g is vertically shrunk by a factor of $\frac{1}{2}$ and shifted three units to the right.

60. The graph of g is vertically stretched by a factor of 2, reflected in the x-axis, and shifted three units upward.

In Exercises 61 and 62, use the graph of $f(x) = x^3 - 3x^2$ to write an equation for the function g shown in the graph.

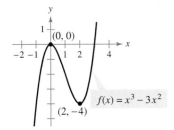

61.

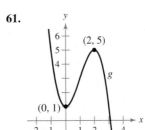

62.

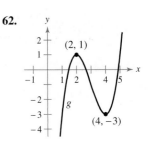

63. The point $(3, 9)$ on the graph of $f(x) = x^2$ has been shifted to the point $(4, 7)$ after a rigid transformation. Identify the shift and write the new function g in terms of f.

64. The point $(8, 2)$ on the graph of $f(x) = \sqrt[3]{x}$ has been shifted to the point $(5, 0)$ after a rigid transformation. Identify the shift and write the new function h in terms of f.

65. Profit A company's weekly profit P (in hundreds of dollars) from a product is given by the model

$$P(x) = 80 + 20x - 0.5x^2, \quad 0 \le x \le 20$$

where x is the amount (in hundreds of dollars) spent on advertising.

(T) (a) Use a graphing utility to graph the profit function.

(b) The company estimates that taxes and operating costs will increase by an average of \$2500 per week during the next year. Rewrite the profit equation to reflect this expected decrease in profits. Identify the type of transformation applied to the graph of the equation.

(c) Rewrite the profit equation so that x measures advertising expenditures in dollars. [Find $P(x/100)$.] Identify the type of transformation applied to the graph of the profit function.

66. Automobile Aerodynamics The number of horsepower H required to overcome wind drag on an automobile is approximated by

$$H(x) = 0.002x^2 + 0.005x - 0.029, \quad 10 \le x \le 100$$

where x is the speed of the car (in miles per hour).

(T) (a) Use a graphing utility to graph the function.

(b) Rewrite the horsepower function so that x represents the speed in kilometers per hour. [Find $H(x/1.6)$.] Identify the type of transformation applied to the graph of the horsepower function.

(T) **67. Exploration** Use a graphing utility to graph the six functions below in the same viewing window. Describe any similarities and differences you observe among the graphs.

(a) $y = x$ (b) $y = x^2$ (c) $y = x^3$

(d) $y = x^4$ (e) $y = x^5$ (f) $y = x^6$

68. Reasoning Use the results of Exercise 67 to make a conjecture about the shapes of the graphs of $y = x^7$ and $y = x^8$. Use a graphing utility to verify your conjecture.

Section 2.7

The Algebra of Functions

- Find the sum, difference, product, and quotient of two functions.
- Form the composition of two functions and determine its domain.
- Identify a function as the composition of two functions.
- Use combinations and compositions of functions to solve application problems.

Arithmetic Combinations of Functions

Just as two real numbers can be combined by the operations of addition, subtraction, multiplication, and division to form other real numbers, two functions can be combined to create new functions. For example, the functions given by $f(x) = 2x - 3$ and $g(x) = x^2 - 1$ can be combined as follows.

$$f(x) + g(x) = (2x - 3) + (x^2 - 1) = x^2 + 2x - 4 \qquad \text{Sum}$$

$$f(x) - g(x) = (2x - 3) - (x^2 - 1) = -x^2 + 2x - 2 \qquad \text{Difference}$$

$$f(x)g(x) = (2x - 3)(x^2 - 1) = 2x^3 - 3x^2 - 2x + 3 \qquad \text{Product}$$

$$\frac{f(x)}{g(x)} = \frac{2x - 3}{x^2 - 1}, \quad x \neq \pm 1, \quad g(x) \neq 0 \qquad \text{Quotient}$$

The domain of an arithmetic combination of the functions f and g consists of all real numbers that are common to the domains of f and g.

Sum, Difference, Product, and Quotient of Functions

Let f and g be two functions with overlapping domains. Then, for all x common to both domains, the **sum, difference, product,** and **quotient** of f and g are defined as follows.

1. *Sum:* $\qquad (f + g)(x) = f(x) + g(x)$

2. *Difference:* $\qquad (f - g)(x) = f(x) - g(x)$

3. *Product:* $\qquad (fg)(x) = f(x) \cdot g(x)$

4. *Quotient:* $\qquad \left(\dfrac{f}{g}\right)(x) = \dfrac{f(x)}{g(x)}, \quad g(x) \neq 0$

Example 1   **Finding the Sum of Two Functions**

Given $f(x) = 2x + 1$ and $g(x) = x^2 + 2x - 1$, find $(f + g)(x)$.

SOLUTION

$$(f + g)(x) = f(x) + g(x) = (2x + 1) + (x^2 + 2x - 1) = x^2 + 4x$$

✓ **CHECKPOINT 1**

Given $f(x) = x^2 - 4$ and $h(x) = x^2 + x + 3$, find $(f + h)(x)$. ∎

Example 2   Finding the Difference of Two Functions

Given the functions

$$f(x) = 2x + 1 \quad \text{and} \quad g(x) = x^2 + 2x - 1$$

find $(f - g)(x)$. Then evaluate the difference when $x = 2$.

SOLUTION   The difference of the functions f and g is given by

$$(f - g)(x) = f(x) - g(x) \qquad \text{Definition of difference of two functions}$$

$$= (2x + 1) - (x^2 + 2x - 1) \quad \text{Substitute for } f(x) \text{ and } g(x).$$

$$= -x^2 + 2. \qquad \text{Simplify.}$$

When $x = 2$, the value of this difference is

$$(f - g)(2) = -(2)^2 + 2 = -2.$$

STUDY TIP

Note that in Example 2, $(f - g)(2)$ can also be evaluated as follows.

$$(f - g)(2) = f(2) - g(2)$$

$$= [2(2) + 1]$$

$$\quad - [2^2 + 2(2) - 1]$$

$$= 5 - 7$$

$$= -2$$

✓ CHECKPOINT 2

Given $f(x) = x^2 - 4$ and $h(x) = x^2 + x + 3$, find $(f - h)(x)$. Then evaluate the difference when $x = 3$. ▪

In Examples 1 and 2, both f and g have domains that consist of all real numbers. So, the domains of $(f + g)$ and $(f - g)$ are also the set of all real numbers. Remember that any restrictions on the domains of f and g must be considered when forming the sum, difference, product, or quotient of f and g.

Example 3   The Quotient of Two Functions

Find the domains of $\left(\dfrac{f}{g}\right)(x)$ and $\left(\dfrac{g}{f}\right)(x)$ for the functions

$$f(x) = \sqrt{x} \quad \text{and} \quad g(x) = \sqrt{4 - x^2}.$$

SOLUTION   The quotient of f and g is given by

$$\left(\frac{f}{g}\right)(x) = \frac{f(x)}{g(x)} = \frac{\sqrt{x}}{\sqrt{4 - x^2}}$$

and the quotient of g and f is given by

$$\left(\frac{g}{f}\right)(x) = \frac{g(x)}{f(x)} = \frac{\sqrt{4 - x^2}}{\sqrt{x}}.$$

The domain of f is $[0, \infty)$ and the domain of g is $[-2, 2]$. The intersection of these two domains is $[0, 2]$, which implies that the domains of f/g and g/f are as follows. Notice that the domains differ slightly.

$$\text{Domain of } \frac{f}{g}: \ [0, 2) \qquad \text{Domain of } \frac{g}{f}: \ (0, 2]$$

✓ CHECKPOINT 3

Find the domains of $\left(\dfrac{f}{h}\right)(x)$ and $\left(\dfrac{h}{f}\right)(x)$ for the functions $f(x) = x - 1$ and $h(x) = x - 3$. ▪

Composition of Functions

Another way to combine two functions is to form the **composition** of one with the other. For instance, if $f(x) = x^2$ and $g(x) = x + 1$, the composition of f with g is given by

$$f(g(x)) = f(x + 1) = (x + 1)^2.$$

This composition is denoted as $f \circ g$ and is read as "f composed with g."

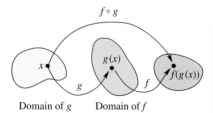

$f \circ g$

Domain of g Domain of f

FIGURE 2.65

Definition of the Composition of Two Functions

The **composition** of the functions f and g is given by

$$(f \circ g)(x) = f(g(x)).$$

The domain of $f \circ g$ is the set of all x in the domain of g such that $g(x)$ is in the domain of f. (See Figure 2.65.)

From the definition above, it follows that the domain of $f \circ g$ is always a subset of the domain of g, and the range of $f \circ g$ is always a subset of the range of f.

Example 4 **Composition of Functions**

Given $f(x) = x + 2$ and $g(x) = 4 - x^2$, find the following.

a. $(f \circ g)(x)$

b. $(g \circ f)(x)$

SOLUTION

a. The composition of f with g is as follows.

$$\begin{aligned}
(f \circ g)(x) &= f(g(x)) &&\text{Definition of } f \circ g \\
&= f(4 - x^2) &&\text{Definition of } g(x) \\
&= (4 - x^2) + 2 &&\text{Definition of } f(x) \\
&= -x^2 + 6 &&\text{Simplify.}
\end{aligned}$$

b. The composition of g with f is as follows.

$$\begin{aligned}
(g \circ f)(x) &= g(f(x)) &&\text{Definition of } g \circ f \\
&= g(x + 2) &&\text{Definition of } f(x) \\
&= 4 - (x + 2)^2 &&\text{Definition of } g(x) \\
&= 4 - (x^2 + 4x + 4) &&\text{Expand.} \\
&= -x^2 - 4x &&\text{Simplify.}
\end{aligned}$$

Note that, in this case, $(f \circ g)(x) \neq (g \circ f)(x)$.

✓**CHECKPOINT 4**

Given $f(x) = x^2 - 2$ and $g(x) = x + 1$, find $(f \circ g)(x)$. ■

TECHNOLOGY

(T) In Example 5, the domain of the composite function is $[-3, 3]$. To convince yourself of this, use a graphing utility to graph

$$y = \left(\sqrt{9 - x^2}\right)^2 - 9$$

as shown in the figure below. Notice that the graphing utility does not extend the graph to the left of $x = -3$ or to the right of $x = 3$.

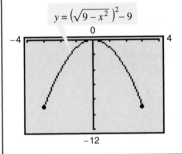

$y = \left(\sqrt{9-x^2}\ \right)^2 - 9$

Example 5   Finding the Domain of a Composite Function

Find the composition $(f \circ g)(x)$ for the functions given by

$$f(x) = x^2 - 9 \quad \text{and} \quad g(x) = \sqrt{9 - x^2}.$$

Then find the domain of $f \circ g$.

SOLUTION   The composition of the functions is as follows.

$$(f \circ g)(x) = f(g(x))$$
$$= f\left(\sqrt{9 - x^2}\right)$$
$$= \left(\sqrt{9 - x^2}\right)^2 - 9$$
$$= 9 - x^2 - 9$$
$$= -x^2$$

From this result, it might appear that the domain of the composition is the set of all real numbers. However, because the domain of f is the set of all real numbers and the domain of g is $[-3, 3]$, the domain of $f \circ g$ is $[-3, 3]$. _____

✓ CHECKPOINT 5

Find the composition $(f \circ g)(x)$ for the functions given by $f(x) = \sqrt{x}$ and $g(x) = 3 - x^2$. Then find the domain of $f \circ g$. ■

In Examples 4 and 5, you formed the composition of two functions. To "decompose" a composite function, look for an "inner" function and an "outer" function. For instance, the function h given by

$$h(x) = (3x - 5)^3$$

is the composition of f with g, where $f(x) = x^3$ and $g(x) = 3x - 5$. That is,

$$h(x) = (3x - 5)^3 = [g(x)]^3 = f(g(x)).$$

In the function h, $g(x) = 3x - 5$ is the *inner* function and $f(x) = x^3$ is the *outer* function.

Example 6   Identifying a Composite Function

Write the function given by $h(x) = \dfrac{1}{(x - 2)^2}$ as a composition of two functions.

SOLUTION   One way to write h as a composition of two functions is to take the inner function to be $g(x) = x - 2$ and the outer function to be

$$f(x) = \frac{1}{x^2} = x^{-2}.$$

Then you can write

$$h(x) = \frac{1}{(x - 2)^2} = (x - 2)^{-2} = f(x - 2) = f(g(x)).$$

✓ CHECKPOINT 6

Write the function given by $h(x) = (x - 1)^2 + 2$ as a composition of two functions. ■

Applications

Andy Williams/Getty Images

The Capitol building in Washington, D.C. is where each state's Congressional representatives convene. In recent years, no party has had a strong majority, which can make it difficult to pass legislation.

Example 7   **Political Makeup of the U.S. Senate**

Consider three functions R, D, and I that represent the numbers of Republicans, Democrats, and Independents, respectively, in the U.S. Senate from 1967 to 2005. Sketch the graphs of R, D, and I and the sum of R, D, and I in the same coordinate plane. The numbers of senators from each political party are shown below.

Year	R	D	I	Year	R	D	I
1967	36	64	0	1987	45	55	0
1969	42	58	0	1989	45	55	0
1971	44	54	2	1991	44	56	0
1973	42	56	2	1993	43	57	0
1975	37	61	2	1995	52	48	0
1977	38	61	1	1997	55	45	0
1979	41	58	1	1999	55	45	0
1981	53	46	1	2001	50	50	0
1983	54	46	0	2003	51	48	1
1985	53	47	0	2005	55	44	1

SOLUTION   The graphs of R, D, and I are shown in Figure 2.66. Note that the sum of R, D, and I is the constant function $R + D + I = 100$. This follows from the fact that the number of senators in the United States is 100 (two from each state).

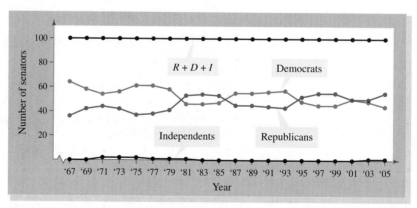

FIGURE 2.66   Numbers of U.S. Senators by Political Party

✓CHECKPOINT 7

In Example 7, consider the function f given by $f = 100 - (R + D)$. What does f represent in the context of the real-life situation? ■

Example 8 Bacteria Count

The number of bacteria in a certain food is given by

$$N(T) = 20T^2 - 80T + 500, \quad 2 \le T \le 14$$

where T is the temperature of the food in degrees Celsius. When the food is removed from refrigeration, the temperature of the food is given by

$$T(t) = 4t + 2, \quad 0 \le t \le 3$$

where t is the time in hours. Find (a) the composition $N(T(t))$, (b) the number of bacteria in the food when $t = 2$ hours, and (c) how long the food can remain unrefrigerated before the bacteria count reaches 2000.

SOLUTION

a. $N(T(t)) = 20(4t + 2)^2 - 80(4t + 2) + 500$

$$= 20(16t^2 + 16t + 4) - 320t - 160 + 500$$

$$= 320t^2 + 320t + 80 - 320t - 160 + 500$$

$$= 320t^2 + 420$$

b. When $t = 2$, the number of bacteria is

$$N(T(2)) = 320(2)^2 + 420 = 1280 + 420 = 1700.$$

c. The bacteria count will reach $N = 2000$ when $320t^2 + 420 = 2000$. By solving this equation, you can determine that the bacteria count will reach 2000 when $t \approx 2.2$ hours. So, the food can remain unrefrigerated for about 2 hours and 12 minutes.

✓ CHECKPOINT 8

In Example 8, how long can the food remain unrefrigerated before the bacteria count reaches 1000? ▨

CONCEPT CHECK

1. Given $g(x) = x^2 + 3x$ and $f(x) = 2x + 3$, describe and correct the error in finding $(g - f)(x)$.

$$\overline{(g - f)(x)} = x^2 + 3x - 2x + 3$$
$$= x^2 + x + 3$$

2. Given $f(x) = x^2$ and $g(x) = 2x - 1$, describe and correct the error in finding $(f \circ g)(x)$.

$$\overline{(f \circ g)(x)} = f(g(x)) = 2(x^2) - 1 = 2x^2 - 1$$

3. Explain why the domain of the composition $f \circ g$ is a subset of the domain of g.

4. Are the domains of the functions given by $h(x) = \sqrt{x - 3}$ and $g(x) = \dfrac{1}{\sqrt{x - 3}}$ the same? Explain.

In Exercises 1–10, perform the indicated operations and simplify the result.

1. $\dfrac{1}{x} + \dfrac{1}{1 - x}$

2. $\dfrac{2}{x + 3} - \dfrac{2}{x - 3}$

3. $\dfrac{3}{x - 2} - \dfrac{2}{x(x - 2)}$

4. $\dfrac{x}{x - 5} + \dfrac{1}{3}$

5. $(x - 1)\left(\dfrac{1}{\sqrt{x^2 - 1}}\right)$

6. $\left(\dfrac{x}{x^2 - 4}\right)\left(\dfrac{x^2 - x - 2}{x^2}\right)$

7. $(x^2 - 4) \div \left(\dfrac{x + 2}{5}\right)$

8. $\left(\dfrac{x}{x^2 + 3x - 10}\right) \div \left(\dfrac{x^2 + 3x}{x^2 + 6x + 5}\right)$

9. $\dfrac{(1/x) + 5}{3 - (1/x)}$

10. $\dfrac{(x/4) - (4/x)}{x - 4}$

Exercises 2.7

In Exercises 1–4, use the graphs of f and g to graph $h(x) = (f + g)(x)$.

1.

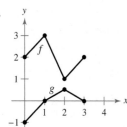

2.

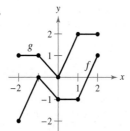

3.

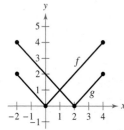

4.

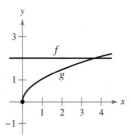

In Exercises 5–12, find (a) $(f + g)(x)$, (b) $(f - g)(x)$, (c) $(fg)(x)$, and (d) $(f/g)(x)$. What is the domain of f/g?

5. $f(x) = x + 1, \quad g(x) = x - 1$

6. $f(x) = 2x - 3, \quad g(x) = 1 - x$

7. $f(x) = x^2, \quad g(x) = 1 - x$

8. $f(x) = 2x + 3, \quad g(x) = x^2 - 1$

9. $f(x) = x^2 + 5, \quad g(x) = \sqrt{1 - x}$

10. $f(x) = \sqrt{x^2 - 4}, \quad g(x) = \dfrac{x^2}{x^2 + 1}$

11. $f(x) = \dfrac{1}{x}, \quad g(x) = \dfrac{1}{x^2}$

12. $f(x) = \dfrac{x}{x + 1}, \quad g(x) = x^3$

In Exercises 13–24, evaluate the function for $f(x) = 2x + 1$ and $g(x) = x^2 - 2$.

13. $(f + g)(3)$

14. $(f - g)(-2)$

15. $(f - g)(2t)$

16. $(f + g)(t - 1)$

17. $(fg)(-2)$

18. $(fg)(-6)$

19. $\left(\dfrac{f}{g}\right)(5)$

20. $\left(\dfrac{f}{g}\right)(0)$

21. $(f - g)(0)$

22. $(f + g)(1)$

23. $\left(\dfrac{f}{g}\right)(-1) - g(3)$

24. $(2f)(5) + (3g)(-4)$

In Exercises 25–28, find (a) $f \circ g$, (b) $g \circ f$, and (c) $f \circ f$.

25. $f(x) = 3x, \quad g(x) = 2x + 5$

26. $f(x) = 2x - 1, \quad g(x) = 7 - x$

27. $f(x) = x^2, \quad g(x) = 3x + 1$

28. $f(x) = x^3, \quad g(x) = \dfrac{1}{x}$

In Exercises 29–36, find (a) $f \circ g$ and (b) $g \circ f$.

29. $f(x) = \sqrt{x + 4}, \quad g(x) = x^2$

30. $f(x) = \sqrt[3]{x - 1}, \quad g(x) = x^3 + 1$

31. $f(x) = \frac{1}{3}x - 3, \quad g(x) = 3x + 1$

32. $f(x) = \frac{1}{2}x + 1, \quad g(x) = 2x + 3$

33. $f(x) = \sqrt{x}, \quad g(x) = \sqrt{x}$

34. $f(x) = 2x - 3, \quad g(x) = 2x - 3$

35. $f(x) = |x|, \quad g(x) = x + 6$

36. $f(x) = x^{2/3}, \quad g(x) = x^6$

In Exercises 37–40, determine the domain of (a) f, (b) g, and (c) $f \circ g$.

37. $f(x) = x^2 + 3, \quad g(x) = \sqrt{x}$

38. $f(x) = \sqrt[3]{x + 1}, \quad g(x) = x^3$

39. $f(x) = \dfrac{1}{x^2}, \quad g(x) = x - 2$

40. $f(x) = \dfrac{5}{x^2 - 4}, \quad g(x) = x + 3$

In Exercises 41–44, use the graphs of f and g to evaluate the functions.

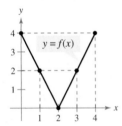
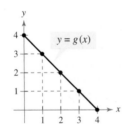

41. (a) $(f + g)(3)$ \qquad (b) $\left(\dfrac{f}{g}\right)(2)$

42. (a) $(f - g)(1)$ \qquad (b) $(fg)(4)$

43. (a) $(f \circ g)(2)$ \qquad (b) $(g \circ f)(2)$

44. (a) $(f \circ g)(0)$ \qquad (b) $(g \circ f)(3)$

In Exercises 45–52, find two functions f and g such that $(f \circ g)(x) = h(x)$. (There are many correct answers.)

45. $h(x) = (2x + 1)^2$ \qquad **46.** $h(x) = (1 - x)^3$

47. $h(x) = \sqrt[3]{x^2 - 4}$ \qquad **48.** $h(x) = \sqrt{9 - x}$

49. $h(x) = \dfrac{1}{x + 2}$

50. $h(x) = \dfrac{4}{(5x + 2)^2}$

51. $h(x) = (x + 4)^2 + 2(x + 4)$

52. $h(x) = (x + 3)^{3/2}$

53. Stopping Distance While driving at x miles per hour, you are required to stop quickly to avoid an accident. The distance the car travels (in feet) during your reaction time is given by $R(x) = \frac{3}{4}x$. The distance the car travels (in feet) while you are braking is given by

$$B(x) = \frac{1}{15}x^2.$$

Find the function that represents the total stopping distance T. (*Hint: $T = R + B$.*) Graph the functions R, B, and T on the same set of coordinate axes for $0 \le x \le 60$.

54. Cost The weekly cost C of producing x units in a manufacturing process is given by the function

$$C(x) = 70x + 800.$$

The number of units x produced in t hours is given by

$$x(t) = 40t.$$

Find and interpret $(C \circ x)(t)$.

55. Cost The weekly cost C of producing x units in a manufacturing process is given by the function

$$C(x) = 50x + 495.$$

The number of units x produced in t hours is given by

$$x(t) = 30t.$$

Find and interpret $(C \circ x)(t)$.

56. Comparing Profits A company has two manufacturing plants, one in New Jersey and the other in California. From 2000 to 2008, the profits for the manufacturing plant in New Jersey were decreasing according to the function

$$P_1 = 18.97 - 0.55t, \quad t = 0, 1, 2, 3, 4, 5, 6, 7, 8$$

where P_1 represents the profits (in millions of dollars) and t represents the year, with $t = 0$ corresponding to 2000. On the other hand, the profits for the manufacturing plant in California were increasing according to the function

$$P_2 = 15.85 + 0.67t, \quad t = 0, 1, 2, 3, 4, 5, 6, 7, 8.$$

Write a function that represents the overall company profits during the nine-year period. Use the *stacked bar graph* in the figure, which represents the total profits for the company during this nine-year period, to determine whether the overall company profits were increasing or decreasing.

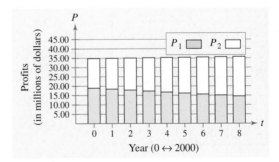

57. Comparing Sales You own two fast-food restaurants. During the years 2000 to 2008, the sales for the first restaurant were decreasing according to the function

$$R_1 = 525 - 15.2t, \quad t = 0, 1, 2, 3, 4, 5, 6, 7, 8$$

where R_1 represents the sales (in thousands of dollars) and t represents the year, with $t = 0$ corresponding to 2000. During the same nine-year period, the sales for the second restaurant were increasing according to the function

$$R_2 = 392 + 8.5t, \quad t = 0, 1, 2, 3, 4, 5, 6, 7, 8.$$

Write a function that represents the total sales for the two restaurants. Use the *stacked bar graph* in the figure, which represents the total sales during this nine-year period, to determine whether the total sales were increasing or decreasing.

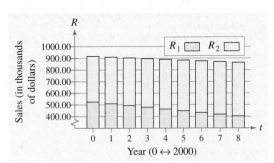

Year (0 ↔ 2000)

(T) 58. Female Labor Force The table shows the marital status of women in the civilian labor force for the years 1995 to 2005. The numbers (in millions) of working women whose status is single, married, or other (widowed, divorced, or separated) are represented by the variables y_1, y_2, and y_3, respectively. *(Source: U.S. Bureau of Labor Statistics)*

Year	y_1	y_2	y_3
1995	15.5	33.4	12.1
1996	15.8	33.6	12.4
1997	16.5	33.8	12.7
1998	17.1	33.9	12.8
1999	17.6	34.4	12.9
2000	17.8	35.1	13.3
2001	18.0	35.2	13.6
2002	18.2	35.5	13.7
2003	18.4	36.0	13.8
2004	18.6	35.8	14.0
2005	19.2	35.9	14.2

(T) (a) Create a stacked bar graph for the data.

(b) Use the *regression* feature of a graphing utility to find linear models for y_1, y_2, and y_3. Let t represent the year, with $t = 5$ corresponding to 1995.

(c) Use a graphing utility to graph the models for y_1, y_2, y_3, and

$$y_4 = y_1 + y_2 + y_3$$

in the same viewing window. Use y_4 to predict the total number of women in the work force in 2007 and 2009.

59. Cost, Revenue, and Profit The table shows the revenues y_1 (in thousands of dollars) and total costs y_2 (in thousands of dollars) for a sports memorabilia store for the years 1998 to 2008.

Year	y_1	y_2	Year	y_1	y_2
1998	40.9	29.8	2004	71.0	51.1
1999	46.3	32.9	2005	75.7	53.7
2000	51.3	36.5	2006	80.8	57.6
2001	55.9	39.9	2007	85.6	62.1
2002	60.8	43.8	2008	90.7	68.7
2003	65.9	46.9			

(a) Use the *regression* feature of a graphing utility to find linear models for y_1 and y_2. Let t represent the year, with $t = 8$ corresponding to 1998.

(b) Use a graphing utility to graph the models for y_1, y_2, and

$$y_3 = y_1 - y_2$$

in the same viewing window. What does y_3 represent in the context of the problem? Determine the value of y_3 in 2010.

(c) Create a stacked bar graph for y_2 and y_3. What do the heights of the bars represent?

60. Bacteria Count The number of bacteria in a certain food product is given by

$$N(T) = 10T^2 - 20T + 600, \quad 1 \le T \le 20$$

where T is the temperature of the food. When the food is removed from the refrigerator, the temperature of the food is given by

$$T(t) = 3t + 1$$

where t is the time in hours. Find (a) the composite function $N(T(t))$ and (b) the time when the bacteria count reaches 1500.

61. Bacteria Count   The number of bacteria in a certain food product is given by

$$N(T) = 25T^2 - 50T + 300, \quad 2 \le T \le 20$$

where T is the temperature of the food. When the food is removed from the refrigerator, the temperature of the food is given by

$$T(t) = 2t + 1$$

where t is the time in hours. Find (a) the composite function $N(T(t))$ and (b) the time when the bacteria count reaches 750.

62. Troubled Waters   A pebble is dropped into a calm pond, causing ripples in the form of concentric circles. The radius (in feet) of the outermost ripple is given by

$$r(t) = 0.6t$$

where t is time in seconds after the pebble strikes the water. The area of the outermost circle is given by the function

$$A(r) = \pi r^2.$$

Find and interpret $(A \circ r)(t)$.

63. Consumer Awareness   The suggested retail price of a new hybrid car is p dollars. The dealership advertises a factory rebate of $2000 and a 10% discount.

(a) Write a function R in terms of p giving the cost of the hybrid car after receiving the rebate from the factory.

(b) Write a function S in terms of p giving the cost of the hybrid car after receiving the dealership discount.

(c) Form the composite functions $(R \circ S)(p)$ and $(S \circ R)(p)$ and interpret each.

(d) Find $(R \circ S)(20,500)$ and $(S \circ R)(20,500)$. Which yields the lower cost for the hybrid car? Explain.

Price-Earnings Ratio   In Exercises 64 and 65, the average annual price-earnings ratio for a corporation's stock is defined as the average price of the stock divided by the earnings per share. The average price of a corporation's stock is given as the function P and the earnings per share is given as the function E. Find the price-earnings ratios, P/E, for the years 2001 to 2005.

64. *Cheesecake Factory*

Year	2001	2002	2003	2004	2005
P	$18.34	$23.17	$23.63	$29.04	$33.90
E	$0.53	$0.64	$0.75	$0.88	$1.09

(Source: Cheesecake Factory)

65. *Jack in the Box*

Year	2001	2002	2003	2004	2005
P	$27.22	$28.19	$19.38	$25.20	$36.21
E	$2.11	$2.33	$2.04	$2.27	$2.48

(Source: Jack in the Box)

66. Find the domains of $(f/g)(x)$ and $(g/f)(x)$ for the functions

$$f(x) = \sqrt{x} \quad \text{and} \quad g(x) = \sqrt{9 - x^2}.$$

Why do the two domains differ?

True or False?   In Exercises 67 and 68, determine whether the statement is true or false. Justify your answer.

67. If $f(x) = x + 1$ and $g(x) = 6x$, then

$$(f \circ g)(x) = (g \circ f)(x).$$

68. If you are given two functions $f(x)$ and $g(x)$, you can calculate $(f \circ g)(x)$ if and only if the range of g is a subset of the domain of f.

Business Capsule

AP/Wide World Photos

SunPower Corporation develops and manufactures solar-electric power products. SunPower's new higher efficiency solar cells generate up to 50% more power than other solar technologies. SunPower's technology was developed by Dr. Richard Swanson and his students while he was Professor of Engineering at Stanford University. SunPower's 2006 revenues are projected to increase 300% from its 2005 revenues.

69. Research Project   Use your campus library, the Internet, or some other reference source to find information about an alternative energy business experiencing strong growth similar to the example above. Write a brief report about the company or small business.

Section 2.8

Inverse Functions

- Determine if a function has an inverse function.
- Find the inverse function of a function.
- Graph a function and its inverse function.

Inverse Functions

Recall from Section 2.4 that a function can be represented by a set of ordered pairs. For instance, the function $f(x) = x + 4$ from the set $A = \{1, 2, 3, 4\}$ to the set $B = \{5, 6, 7, 8\}$ can be written as follows.

$$f(x) = x + 4: \quad \{(1, 5), (2, 6), (3, 7), (4, 8)\}$$

By interchanging the first and second coordinates of each of these ordered pairs, you can form the **inverse function** of f, which is denoted by f^{-1}. It is a function from the set B to the set A and can be written as follows.

$$f^{-1}(x) = x - 4: \quad \{(5, 1), (6, 2), (7, 3), (8, 4)\}$$

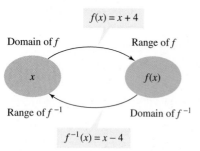

FIGURE 2.67

Note that the domain of f is equal to the range of f^{-1} and vice versa, as shown in Figure 2.67. Also note that the functions f and f^{-1} have the effect of "undoing" each other. In other words, when you form the composition of f with f^{-1} or the composition of f^{-1} with f, you obtain the identity function, as follows.

$$f(f^{-1}(x)) = f(x - 4) = (x - 4) + 4 = x$$
$$f^{-1}(f(x)) = f^{-1}(x + 4) = (x + 4) - 4 = x$$

Example 1   Finding Inverse Functions Informally

Find the inverse function of $f(x) = 4x$. Then verify that both $f(f^{-1}(x))$ and $f^{-1}(f(x))$ are equal to the identity function.

SOLUTION   The given function *multiplies* each input by 4. To "undo" this function, you need to *divide* each input by 4. So, the inverse function of $f(x) = 4x$ is

$$f^{-1}(x) = \frac{x}{4}.$$

You can verify that both $f(f^{-1}(x))$ and $f^{-1}(f(x))$ are equal to the identity function as follows.

$$f(f^{-1}(x)) = f\left(\frac{x}{4}\right) = 4\left(\frac{x}{4}\right) = x$$

$$f^{-1}(f(x)) = f^{-1}(4x) = \frac{4x}{4} = x$$

✓ CHECKPOINT 1

Find the inverse function of $f(x) = \dfrac{x}{6}$. Then verify that both $f(f^{-1}(x))$ and $f^{-1}(f(x))$ are equal to the identity function. ∎

| Example 2 | Finding Inverse Functions Informally

Find the inverse function of

$$f(x) = x - 6.$$

Then verify that both $f(f^{-1}(x))$ and $f^{-1}(f(x))$ are equal to the identity function.

SOLUTION   The given function *subtracts* 6 from each input. To "undo" this function, you need to *add* 6 to each input. So, the inverse function of $f(x) = x - 6$ is

$$f^{-1}(x) = x + 6.$$

You can verify that both $f(f^{-1}(x))$ and $f^{-1}(f(x))$ are equal to the identity function as follows.

$$f(f^{-1}(x)) = f(x + 6)$$ Substitute $x + 6$ for $f^{-1}(x)$.

$$= (x + 6) - 6$$ Substitute $x + 6$ into $f(x)$.

$$= x$$ Identity function

$$f^{-1}(f(x)) = f^{-1}(x - 6)$$ Substitute $x - 6$ for $f(x)$.

$$= (x - 6) + 6$$ Substitute $x - 6$ into $f^{-1}(x)$.

$$= x$$ Identity function

✓ **CHECKPOINT 2**

Find the inverse function of $f(x) = x + 10$. Then verify that both $f(f^{-1}(x))$ and $f^{-1}(f(x))$ are equal to the identity function. ∎

The formal definition of inverse function is as follows.

Definition of Inverse Function

Let f and g be two functions such that

$$f(g(x)) = x$$ for every x in the domain of g

and

$$g(f(x)) = x$$ for every x in the domain of f.

Under these conditions, the function g is the **inverse function** of the function f. The function g is denoted by f^{-1} (read "f-inverse"). So,

$$f(f^{-1}(x)) = x \quad \text{and} \quad f^{-1}(f(x)) = x.$$

The domain of f must be equal to the range of f^{-1}, and the range of f must be equal to the domain of f^{-1}.

Don't be confused by the use of -1 to denote the inverse function f^{-1}. In this text, f^{-1} *always* refers to the inverse function of the function f and *not* to the reciprocal of $f(x)$. That is,

$$f^{-1}(x) \neq \frac{1}{f(x)}.$$

If the function g is the inverse function of the function f, it must also be true that the function f is the inverse function of the function g. For this reason, you can say that the functions f and g are *inverse functions of each other*.

Example 3 **Verifying Inverse Functions**

Show that the following functions are inverse functions.

$$f(x) = 2x^3 - 1 \quad \text{and} \quad g(x) = \sqrt[3]{\frac{x+1}{2}}$$

SOLUTION

$$f(g(x)) = f\left(\sqrt[3]{\frac{x+1}{2}}\right) = 2\left(\sqrt[3]{\frac{x+1}{2}}\right)^3 - 1$$

$$= 2\left(\frac{x+1}{2}\right) - 1$$

$$= x + 1 - 1$$

$$= x$$

$$g(f(x)) = g(2x^3 - 1) = \sqrt[3]{\frac{(2x^3 - 1) + 1}{2}}$$

$$= \sqrt[3]{\frac{2x^3}{2}}$$

$$= \sqrt[3]{x^3}$$

$$= x$$

✓ CHECKPOINT 3

Show that the following functions are inverse functions.

$$f(x) = x^3 + 6 \quad \text{and} \quad g(x) = \sqrt[3]{x - 6} \quad \blacksquare$$

Example 4 **Verifying Inverse Functions**

Which of the functions given by

$$g(x) = \frac{x - 2}{5} \quad \text{and} \quad h(x) = \frac{5}{x} + 2$$

is the inverse function of $f(x) = \frac{5}{x - 2}$?

SOLUTION By forming the composition of f with g, you can see that

$$f(g(x)) = f\left(\frac{x - 2}{5}\right) = \frac{5}{[(x - 2)/5] - 2} = \frac{25}{x - 12} \neq x.$$

Because this composition is not equal to the identity function x, it follows that g is *not* the inverse function of f. By forming the composition of f with h, you have

$$f(h(x)) = f\left(\frac{5}{x} + 2\right) = \frac{5}{[(5/x) + 2] - 2} = \frac{5}{5/x} = x.$$

So, it appears that h is the inverse function of f. You can confirm this result by showing that the composition of h with f is also equal to the identity function. (Try doing this.)

DISCOVERY

Graph the equations from Example 3 and the equation $y = x$ on a graphing utility using a square viewing window.

$$y_1 = 2x^3 - 1$$

$$y_2 = \sqrt[3]{\frac{x+1}{2}}$$

$$y_3 = x$$

What do you observe about the graphs of y_1 and y_2?

✓ CHECKPOINT 4

Which of the functions given by $g(x) = \frac{x + 4}{3}$ and $h(x) = \frac{x}{3} + 4$ is the inverse function of $f(x) = 3x - 4$? ∎

Finding Inverse Functions

For simple functions (such as the ones in Examples 1 and 2), you can find inverse functions by inspection. For more complicated functions it is best to use the following guidelines. The key step in these guidelines is switching the roles of x and y. This step corresponds to the fact that inverse functions have ordered pairs with the coordinates reversed.

STUDY TIP

Note in Step 3 of the guidelines for finding inverse functions that it is possible for a function to have no inverse function. For instance, the function given by $f(x) = x^2$ has no inverse function.

Finding Inverse Functions

1. In the equation for $f(x)$, replace $f(x)$ by y.

2. Interchange the roles of x and y.

3. Solve the new equation for y. If the new equation does not represent y as a function of x, the function f does not have an inverse function. If the new equation does represent y as a function of x, continue to Step 4.

4. Replace y by $f^{-1}(x)$ in the new equation.

5. Verify that f and f^{-1} are inverse functions of each other by showing that the domain of f is equal to the range of f^{-1}, the range of f is equal to the domain of f^{-1}, and $f(f^{-1}(x)) = x = f^{-1}(f(x))$.

Example 5 **Finding Inverse Functions**

Find the inverse function of $f(x) = \dfrac{5 - 3x}{2}$.

SOLUTION

$$f(x) = \frac{5 - 3x}{2} \qquad \text{Write original function.}$$

$$y = \frac{5 - 3x}{2} \qquad \text{Replace } f(x) \text{ by } y.$$

$$x = \frac{5 - 3y}{2} \qquad \text{Interchange } x \text{ and } y.$$

$$2x = 5 - 3y \qquad \text{Multiply each side by 2.}$$

$$3y = 5 - 2x \qquad \text{Isolate the } y\text{-term.}$$

$$y = \frac{5 - 2x}{3} \qquad \text{Solve for } y.$$

$$f^{-1}(x) = \frac{5 - 2x}{3} \qquad \text{Replace } y \text{ by } f^{-1}(x).$$

Note that both f and f^{-1} have domains and ranges that consist of the entire set of real numbers. Check that $f(f^{-1}(x)) = x$ and $f^{-1}(f(x)) = x$.

✓ **CHECKPOINT 5**

Find the inverse function of $f(x) = 4x + 5$. ■

The Graph of an Inverse Function

The graphs of a function *f* and its inverse function f^{-1} are related to each other in the following way. If the point (a, b) lies on the graph of *f*, then the point (b, a) must lie on the graph of f^{-1}, and vice versa. This means that the graph of f^{-1} is a *reflection* of the graph of *f* in the line $y = x$, as shown in Figure 2.68.

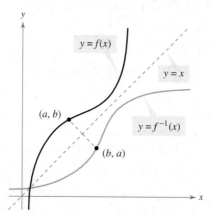

The graph of f^{-1} is a reflection of the graph of *f* in the line $y = x$.

FIGURE 2.68

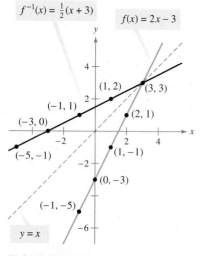

FIGURE 2.69

Example 6 **The Graphs of f and f⁻¹**

Sketch the graphs of the inverse functions given by

$$f(x) = 2x - 3 \quad \text{and} \quad f^{-1}(x) = \tfrac{1}{2}(x + 3)$$

in the same coordinate plane and show that the graphs are reflections of each other in the line $y = x$.

SOLUTION The graphs of *f* and f^{-1} are shown in Figure 2.69. Visually, it appears that the graphs are reflections of each other in the line $y = x$. You can further verify this reflective property by testing a few points on each graph. Note in the following list that if the point (a, b) is on the graph of *f*, then the point (b, a) is on the graph of f^{-1}.

Graph of $f(x) = 2x - 3$	Graph of $f^{-1}(x) = \tfrac{1}{2}(x + 3)$
$(0, -3)$	$(-3, 0)$
$(1, -1)$	$(-1, 1)$
$(2, 1)$	$(1, 2)$
$(3, 3)$	$(3, 3)$

✓**CHECKPOINT 6**

Sketch the graphs of the inverse functions given by $f(x) = \tfrac{2}{5}x + 2$ and $f^{-1}(x) = \tfrac{5}{2}x - 5$ in the same coordinate plane and show that the graphs are reflections of each other in the line $y = x$. ∎

The Study Tip on page 241 mentioned that the function given by

$$f(x) = x^2$$

has no inverse function. What this means is that, *assuming the domain of f is the entire real line,* the function given by $f(x) = x^2$ has no inverse function. If the domain of f is restricted to the nonnegative real numbers, however, then f does have an inverse function, as demonstrated in Example 7.

Example 7   The Graphs of *f* and *f*⁻¹

Sketch the graphs of the inverse functions given by

$$f(x) = x^2, \quad x \geq 0, \quad \text{and} \quad f^{-1}(x) = \sqrt{x}$$

in the same coordinate plane and show that the graphs are reflections of each other in the line $y = x$.

SOLUTION   The graphs of f and f^{-1} are shown in Figure 2.70. Visually, it appears that the graphs are reflections of each other in the line $y = x$. You can further verify this reflective property by testing a few points on each graph. Note in the following list that if the point (a, b) is on the graph of f, then the point (b, a) is on the graph of f^{-1}.

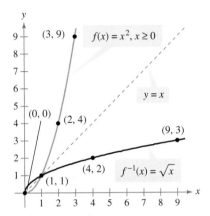

FIGURE 2.70

Graph of $f(x) = x^2, \quad x \geq 0$	*Graph of* $f^{-1}(x) = \sqrt{x}$
$(0, 0)$	$(0, 0)$
$(1, 1)$	$(1, 1)$
$(2, 4)$	$(4, 2)$
$(3, 9)$	$(9, 3)$

You can verify algebraically that the functions are inverse functions of each other by showing that $f(f^{-1}(x)) = x$ and $f^{-1}(f(x)) = x$ as follows.

$$f(f^{-1}(x)) = f(\sqrt{x}) = (\sqrt{x})^2 = x, \quad \text{if } x \geq 0$$
$$f^{-1}(f(x)) = f^{-1}(x^2) = \sqrt{x^2} = x, \quad \text{if } x \geq 0$$

✓ CHECKPOINT 7

Sketch the graphs of the inverse functions given by $f(x) = x^2 + 3, \ x \geq 0$, and $f^{-1}(x) = \sqrt{x - 3}$ in the same coordinate plane and show that the graphs are reflections of each other in the line $y = x$. ∎

The guidelines for finding the inverse function of a function include an *algebraic* test for determining whether a function has an inverse function. The reflective property of the graphs of inverse functions gives you a *geometric* test for determining whether a function has an inverse function. This test is called the **Horizontal Line Test** for inverse functions.

Horizontal Line Test for Inverse Functions

A function f has an inverse function if and only if no *horizontal* line intersects the graph of f at more than one point.

Example 8 **Applying the Horizontal Line Test**

Use the graph of f to determine whether the function has an inverse function.

a. $f(x) = x^3 - 1$ **b.** $f(x) = x^2 - 1$

SOLUTION

a. The graph of the function given by

$$f(x) = x^3 - 1$$

is shown in Figure 2.71(a). Because no horizontal line intersects the graph of f at more than one point, you can conclude that f does have an inverse function.

b. The graph of the function given by

$$f(x) = x^2 - 1$$

is shown in Figure 2.71(b). Because it is possible to find a horizontal line that intersects the graph of f at more than one point, you can conclude that f does not have an inverse function.

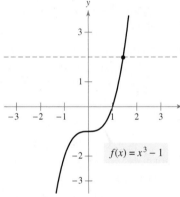

(a)

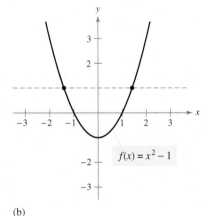

(b)

FIGURE 2.71

✓**CHECKPOINT 8**

Use the graph of f to determine whether the function has an inverse function.

a. $f(x) = |x|$

b. $f(x) = \sqrt{x}$ ▨

┌─────(**CONCEPT CHECK**)─────────────────────────┐

1. What can you say about the functions m and n given that $m(n(x)) = x$ for every x in the domain of n and $n(m(x)) = x$ for every x in the domain of m?

2. Given that the functions g and h are inverses of each other and (a, b) is a point on the graph of g, name a point on the graph of h.

3. Explain how to find an inverse function algebraically.

4. The line $y = 2$ intersects the graph of $f(x)$ at two points. Does f have an inverse? Explain.

└──┘

| **Skills Review 2.8** | The following warm-up exercises involve skills that were covered in earlier sections. You will use these skills in the exercise set for this section. For additional help, review Sections 0.2, 0.4, 1.1, 1.5, and 2.4. |

In Exercises 1–4, find the domain of the function.

1. $f(x) = \sqrt[3]{x + 1}$

2. $f(x) = \sqrt{x + 1}$

3. $g(x) = \dfrac{2}{x^2 - 2x}$

4. $h(x) = \dfrac{x}{3x + 5}$

In Exercises 5–8, simplify the expression.

5. $2\left(\dfrac{x + 5}{2}\right) - 5$

6. $7 - 10\left(\dfrac{7 - x}{10}\right)$

7. $\sqrt[3]{2\left(\dfrac{x^3}{2} - 2\right) + 4}$

8. $\sqrt[5]{(x + 2)^5} - 2$

In Exercises 9 and 10, solve for x in terms of y.

9. $y = \dfrac{2x - 6}{3}$

10. $y = \sqrt[3]{2x - 4}$

Exercises 2.8

See www.CalcChat.com for worked-out solutions to odd-numbered exercises.

In Exercises 1–4, find the inverse function of the function f given by the set of ordered pairs.

1. $\{(1, 4), (2, 5), (3, 6), (4, 7)\}$

2. $\{(6, 2), (5, 3), (4, 4), (3, 5)\}$

3. $\{(-1, 1), (-2, 2), (-3, 3), (-4, 4)\}$

4. $\{(6, -2), (5, -3), (4, -4), (3, -5)\}$

Ⓣ In Exercises 5–8, find the inverse function informally. Verify that $f(f^{-1}(x)) = x$ and $f^{-1}(f(x)) = x$.

5. $f(x) = 2x$

6. $f(x) = -\dfrac{x}{4}$

7. $f(x) = x - 5$

8. $f(x) = x + 7$

In Exercises 9–16, show that f and g are inverse functions by (a) using the definition of inverse functions and (b) graphing the functions. Make sure you test a few points, as shown in Examples 6 and 7.

9. $f(x) = 5x + 1$, $g(x) = \dfrac{x - 1}{5}$

10. $f(x) = 3 - 4x$, $g(x) = \dfrac{3 - x}{4}$

11. $f(x) = x^3$, $g(x) = \sqrt[3]{x}$

12. $f(x) = \dfrac{1}{x}$, $g(x) = \dfrac{1}{x}$

13. $f(x) = \sqrt{x - 4}$, $g(x) = x^2 + 4$, $x \geq 0$

14. $f(x) = 9 - x^2$, $x \geq 0$
  $g(x) = \sqrt{9 - x}$, $x \leq 9$

15. $f(x) = 1 - x^3$, $g(x) = \sqrt[3]{1 - x}$

16. $f(x) = \dfrac{1}{1 + x}$, $x \geq 0$
  $g(x) = \dfrac{1 - x}{x}$, $0 < x \leq 1$

In Exercises 17–20, use the graph of f to complete the table and to sketch the graph of f^{-1}.

17.

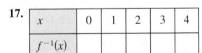

x	0	1	2	3	4
$f^{-1}(x)$					

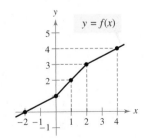

18.

x	0	2	4	6
$f^{-1}(x)$				

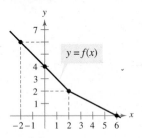

19.

x	-2	0	2	3
$f^{-1}(x)$				

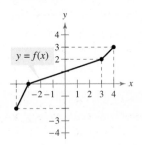

20.

x	1	2	3	4
$f^{-1}(x)$				

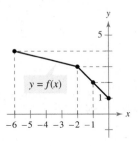

In Exercises 21–36, determine whether the function has an inverse function. If it does, find its inverse function.

21. $f(x) = x^4$

22. $f(x) = \dfrac{1}{x^2}$

23. $g(x) = \dfrac{x}{8}$

24. $f(x) = 3x + 5$

25. $p(x) = -4$

26. $f(x) = \dfrac{3x + 4}{5}$

27. $f(x) = (x + 3)^2, \ x \geq -3$

28. $q(x) = (x - 5)^2$

29. $h(x) = \dfrac{1}{x}$

30. $f(x) = |x - 2|, \ x \leq 2$

31. $f(x) = \sqrt{2x + 3}$

32. $f(x) = \sqrt{x - 2}$

33. $g(x) = x^2 - x^4$

34. $f(x) = \dfrac{x^2}{x^2 + 1}$

35. $f(x) = 25 - x^2, \ x \leq 0$ **36.** $f(x) = 36 + x^2, \ x \leq 0$

Error Analysis   In Exercises 37 and 38, a student has handed in the answer to a problem on a quiz. Find the error(s) in each solution and discuss how to explain each error to the student.

37. Find the inverse function f^{-1} of $f(x) = \sqrt{2x - 5}$.

$$f(x) = \sqrt{2x - 5}, \text{ so}$$
$$f^{-1}(x) = \frac{1}{\sqrt{2x - 5}}$$

38. Find the inverse function f^{-1} of $f(x) = \frac{3}{5}x + \frac{1}{3}$.

$$f(x) = \tfrac{3}{5}x + \tfrac{1}{3}, \text{ so}$$
$$f^{-1}(x) = \tfrac{5}{3}x - 3$$

Ⓣ In Exercises 39–48, find the inverse function f^{-1} of the function f. Then, using a graphing utility, graph both f and f^{-1} in the same viewing window.

39. $f(x) = 2x - 3$ **40.** $f(x) = 5x + 2$

41. $f(x) = x^5$ **42.** $f(x) = x^3 + 1$

43. $f(x) = \sqrt{x}$ **44.** $f(x) = x^2, \ x \geq 0$

45. $f(x) = \sqrt{16 - x^2}, \ 0 \leq x \leq 4$

46. $f(x) = \dfrac{3}{x + 1}$

47. $f(x) = \sqrt[3]{x + 2}$ **48.** $f(x) = x^{3/5} - 2$

In Exercises 49–52, does the function have an inverse function? Explain your reasoning.

49.

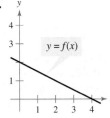

50.

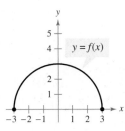

51.

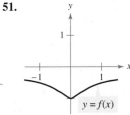

52.

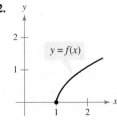

In Exercises 53–58, graph the function and use the Horizontal Line Test to determine whether the function has an inverse function.

53. $g(x) = \dfrac{5 - 2x}{3}$ **54.** $f(x) = 10$

55. $h(x) = |x - 5|$ **56.** $g(x) = (x - 3)^2$

57. $f(x) = -\sqrt{9 - x^2}$ **58.** $f(x) = (x - 1)^3$

In Exercises 59–62, use the functions given by

$$f(x) = \frac{1}{8}x - 3 \quad \text{and} \quad g(x) = x^3$$

to find the value.

59. $(f^{-1} \circ g^{-1})(1)$ **60.** $(g^{-1} \circ f^{-1})(-3)$

61. $(f^{-1} \circ f^{-1})(6)$ **62.** $(g^{-1} \circ g^{-1})(-4)$

In Exercises 63–66, use the functions given by

$$f(x) = x + 4 \quad \text{and} \quad g(x) = 2x - 5$$

to find the composition of functions.

63. $g^{-1} \circ f^{-1}$ **64.** $f^{-1} \circ g^{-1}$

65. $(f \circ g)^{-1}$ **66.** $(g \circ f)^{-1}$

67. Cost With fixed daily costs of $1500, the cost C for a T-shirt business to make x personalized T-shirts is given by $C(x) = 7.50x + 1500$. Find the inverse function $C^{-1}(x)$ and explain what it represents. Describe the domains of $C(x)$ and $C^{-1}(x)$.

68. Profit A company's profit P for producing x units is given by $P(x) = 47x - 5736$. Find the inverse function $P^{-1}(x)$ and explain what it represents. Describe the domains of $P(x)$ and $P^{-1}(x)$.

(T) 69. Movie Theaters The average prices of admission y (in dollars) to a movie theater for the years 1998 to 2005 are shown in the table. *(Source: Motion Picture Association of America, Inc.)*

Year	1998	1999	2000	2001
Admission price, y	4.69	5.08	5.39	5.66

Year	2002	2003	2004	2005
Admission price, y	5.81	6.03	6.21	6.41

(a) Use a graphing utility to create a scatter plot of the data. Let t represent the year, with $t = 8$ corresponding to 1998.

(b) Use the *regression* feature of a graphing utility to find a linear model for the data.

(c) Algebraically find the inverse function of the model in part (b). Explain what this inverse function represents in a real-life context.

(d) Use the inverse function you found in part (c) to estimate the year in which the average admission price to a movie theater will reach $8.00.

(T) 70. Lead Exposure A project is conducted to study the amount of lead accumulated in the bones of humans. The concentration L (in micrograms per gram of bone mineral) of lead found in the tibia of a man is measured every five years. The results are shown in the table.

Age	15	20	25	30	35	40
Lead, L	3.2	5.4	9.2	12.2	13.8	16.0

(a) Use a graphing utility to create a scatter plot of the data. Let x represent the age (in years) of the man.

(b) Use the *regression* feature of a graphing utility to find a linear model for the data.

(c) Algebraically find the inverse function of the model in part (b). Explain what this inverse function represents in a real-life context.

(d) Use the inverse function you found in part (c) to estimate the age of the man when the concentration of lead in his tibia reaches 25 micrograms per gram of bone mineral.

71. Reasoning You are helping a friend to find the inverse function of a one-to-one function. He states that interchanging the roles of x and y is "cheating." Explain how you would use the graphs of $f(x) = x^2 + 1$, $x \geq 0$, and $f^{-1}(x) = \sqrt{x - 1}$ to justify that particular step in the process of finding an inverse function.

72. Diesel Mechanics The function given by

$$y = 0.03x^2 + 245.5, \quad 0 < x < 100$$

approximates the exhaust temperature y for a diesel engine in degrees Fahrenheit, where x is the percent load for the diesel engine. Solve the equation for x in terms of y and use the result to find the percent load for a diesel engine when the exhaust temperature is 410°F.

73. Earnings-Dividend Ratio From 1995 to 2005, the earnings per share for Wal-Mart Stores were approximately related to the dividends per share by the function given by

$$f(x) = \sqrt{0.0161x^3 + 0.008}, \quad 0.6 \leq x \leq 2.63$$

where f represents the dividends per share (in dollars) and x represents the earnings per share (in dollars). In 2004, Wal-Mart paid dividends of $0.48 per share. Find the inverse function of f and use the inverse function to approximate the earnings per share in 2004. *(Source: Wal-Mart Stores, Inc.)*

Chapter Summary and Study Strategies

After studying this chapter, you should have acquired the following skills.
The exercise numbers are keyed to the Review Exercises that begin on page 250.
Answers to odd-numbered Review Exercises are given in the back of the text.*

Section 2.1
                                                                            Review Exercises

■ Plot points in the Cartesian plane, find the distance between two points,   *1–6*
and find the midpoint of a line segment joining two points.

$$d = \sqrt{(x_2 - x_1)^2 + (y_2 - y_1)^2} \quad \text{Midpoint} = \left(\frac{x_1 + x_2}{2}, \frac{y_1 + y_2}{2}\right)$$

■ Determine whether a point is a solution of an equation.   *7, 8*

■ Sketch the graph of an equation using a table of values.   *9, 10*

■ Find the x- and y-intercepts, and determine the symmetry, of   *11–16*
the graph of an equation.

■ Write the equation of a circle in standard form.   *17–20*

$$(x - h)^2 + (y - k)^2 = r^2$$

Section 2.2

■ Find the slope of a line passing through two points.   *21–24*

$$m = \frac{y_2 - y_1}{x_2 - x_1}$$

■ Use the point-slope form to find the equation of a line.   *25–28*

$$y - y_1 = m(x - x_1)$$

■ Use the slope-intercept form to sketch a line.   *29–32*

$$y = mx + b$$

■ Use slope to determine if lines are parallel or perpendicular, and write the   *33–40*
equation of a line parallel or perpendicular to a given line.

Parallel lines: $m_1 = m_2$

Perpendicular lines: $m_1 = -\dfrac{1}{m_2}$

Section 2.3

■ Construct and use a linear model to relate quantities that vary directly.   *41–50*

Direct variation: $y = mx$

■ Construct and use a linear model with slope as the rate of change.   *51–53*

■ Use a scatter plot to find a linear model that fits a set of data.   *54*

---

* Use a wide range of valuable study aids to help you master the material in this chapter. The *Student
Solutions Guide* includes step-by-step solutions to all odd-numbered exercises to help you review
and prepare. The student website at *college.hmco.com/info/larsonapplied* offers algebra help and a
Graphing Technology Guide. The *Graphing Technology Guide* contains step-by-step commands
and instructions for a wide variety of graphing calculators, including the most recent models.

Section 2.4

	Review Exercises
■ Determine if an equation or a set of ordered pairs represents a function.	*55–60*
■ Use function notation, evaluate a function, and find the domain of a function.	*61–69*
■ Write a function that relates quantities in an application problem.	*70–72*

Section 2.5

■ Find the domain and range using the graph of a function.	*73–76*
■ Identify the graph of a function using the Vertical Line Test.	*77–82*
■ Describe the increasing and decreasing behavior of a function.	*73–76, 92*
■ Find the relative maxima and relative minima of the graph of a function.	*73–76*
■ Classify a function as even or odd.	*73–76*

In an even function, $f(-x) = f(x)$

In an odd function, $f(-x) = -f(x)$

■ Identify six common graphs and use them to sketch the graph of a function.	*83–91*

Section 2.6

■ Use vertical and horizontal shifts, reflections, and nonrigid transformations to sketch graphs of functions.	*93–100*

Section 2.7

■ Find the sum, difference, product, and quotient of two functions.	*101–106*
■ Form the composition of two functions and determine its domain.	*107–110*
■ Identify a function as the composition of two functions.	*111–114*
■ Use combinations and compositions of functions to solve application problems.	*115–118*

Section 2.8

■ Verify that two functions are inverse functions of each other.	*119, 120, 125–128*

$f(f^{-1}(x)) = x$

$f^{-1}(f(x)) = x$

■ Determine if a function has an inverse function.	*119–129*
■ Find the inverse function of a function.	*121–128*
■ Graph a function and its inverse function.	*121–128*
■ Find and use an inverse function in an application problem.	*129*

Study Strategies

■ **To Memorize or Not to Memorize?** When studying mathematics, you often need to memorize formulas, rules, and properties. The formulas that you use most often can become committed to memory through practice. Some formulas, however, are used infrequently or may be easily forgotten. When you are unsure of a formula, you may be able to *derive* it using other information that you know. For instance, if you forget the standard form of the equation of a circle, you can use the Distance Formula and properties of a circle to derive it, as shown on pages 164 and 165. If you also forget the Distance Formula, you can depict the distance between two generic points graphically and use the Pythagorean Theorem to derive the formula, as shown on page 158.

■ **Choose Convenient Values for Yearly Data** When you work with data involving years, you may want to reassign simpler values to represent the years. For instance, you might represent the years 1992 to 2009 by the *x*-values 2 to 19. If you sketch a graph of these data, be sure to account for this in the *x*-axis title: Year (2 ↔ 1992).

Review Exercises

See www.CalcChat.com for worked-out solutions to odd-numbered exercises.

In Exercises 1–4, (a) plot the points, (b) find the distance between the points, and (c) find the midpoint of the line segment joining the points.

1. $(3, 2), (-3, -5)$

2. $(-9, 3), (5, 7)$

3. $(3.45, 6.55), (-1.06, -3.87)$

4. $(-6.7, -3.9), (5.1, 8.2)$

In Exercises 5 and 6, find x such that the distance between the points is 25.

5. $(10, 10), (x, -5)$

6. $(x, -5), (-15, 10)$

In Exercises 7 and 8, determine whether the point is a solution of the equation.

7. $y = 2x^2 - 7x - 15$

   (a) $(5, 0)$ (b) $(-2, 7)$

8. $y = \sqrt{16 - x^2}$

   (a) $(1, 5)$ (b) $(4, 0)$

In Exercises 9 and 10, complete the table. Use the resulting solution points to sketch the graph of the equation.

9. $y = -\frac{1}{2}x + 2$

x	-2	0	2	3	4
y					

10. $y = x^2 - 3x$

x	-1	0	1	2	3
y					

In Exercises 11–16, sketch the graph of the equation. Identify any intercepts and test for symmetry.

11. $y = x^2 + 3$ **12.** $y^2 = x$

13. $y = 3x - 4$ **14.** $y = \sqrt{9 - x}$

15. $y = x^3 + 1$ **16.** $y = |x - 3|$

In Exercises 17 and 18, find the standard form of the equation of the specified circle.

17. Center: $(-1, 2)$; radius: 6

18. Endpoints of the diameter: $(-2, -3), (4, 5)$

In Exercises 19 and 20, write the equation of the circle in standard form and sketch its graph.

19. $x^2 + y^2 - 4x + 6y - 12 = 0$

20. $4x^2 + 4y^2 - 4x - 8y - 11 = 0$

In Exercises 21–24, plot the points and find the slope of the line passing through the points.

21. $(3, 7), (2, -1)$ **22.** $(3, -2), (-1, -2)$

23. $(3, 4), (3, -2)$ **24.** $(-1, 5), (2, -3)$

In Exercises 25–28, find an equation of the line that passes through the point and has the indicated slope. Sketch the line.

	Point	Slope
25.	$(0, -5)$	$m = \frac{3}{2}$
26.	$(3, 0)$	$m = -\frac{2}{3}$
27.	$(-2, 6)$	$m = 0$
28.	$(5, 4)$	m is undefined.

In Exercises 29–32, find the slope and y-intercept (if possible) of the line specified by the equation. Then sketch the line.

29. $5x - 4y + 11 = 0$

30. $3y - 2 = 0$

31. $17 - 5x = 10$

32. $16x + 12y - 24 = 0$

In Exercises 33–36, determine whether the lines L_1 and L_2 passing through the pairs of points are parallel, perpendicular, or neither.

33. L_1: $(0, 3), (-2, 1)$; L_2: $(-8, -3), (4, 9)$

34. L_1: $(-3, -1), (2, 5)$; L_2: $(2, 1), (8, 6)$

35. L_1: $(3, 6), (-1, -5)$; L_2: $(-2, 3), (4, 7)$

36. L_1: $(-1, 2), (-1, 4)$; L_2: $(7, 3), (4, 7)$

In Exercises 37–40, write an equation of the line through the point (a) parallel to the given line and (b) perpendicular to the given line.

	Point	Line
37.	$(3, -2)$	$5x - 4y = 8$
38.	$(-8, 3)$	$2x + 3y = 5$
39.	$(-1, -2)$	$y = 2$
40.	$(0, 5)$	$x = -3$

Direct Variation   In Exercises 41–44, *y* is proportional to *x*. Use the *x*- and *y*-values to find a linear model that relates *x* and *y*.

41. $x = 3, y = 7$

42. $x = 5, y = 3.8$

43. $x = 10, y = 3480$

44. $x = 14, y = 1.95$

Direct Variation   In Exercises 45–48, write a linear model that relates the variables.

45. *A* varies directly as *r*; $A = 30$ when $r = 6$.

46. *y* varies directly as *z*; $y = 7$ when $z = 14$.

47. *a* is proportional to *b*; $a = 15$ when $b = 20$.

48. *m* varies directly as *n*; $m = 12$ when $n = 21$.

49. Property Tax   The property tax in a city is based on the assessed value of the property. A house that has an assessed value of $80,000 has a property tax of $2920. Find a mathematical model that gives the amount of property tax *y* in terms of the assessed value of the property *x*. Use the model to find the property tax on a house that has an assessed value of $102,000.

50. Feet and Meters   You are driving and you notice a billboard that indicates it is 1000 feet or 305 meters to the next restaurant of a national fast-food chain. Use this information to find a linear model that relates feet to meters. Use the model to complete the table.

Feet	20	50	100	120
Meters				

51. Fourth-Quarter Sales   During the second and third quarters of the year, a business had sales of $275,000 and $305,500, respectively. Assume the growth of the sales follows a linear pattern. What will sales be during the fourth quarter?

52. Dollar Value   The dollar value of a product in 2008 is $75 and the item is expected to increase in value at a rate of $5.95 per year. Write a linear equation that gives the dollar value of the product in terms of the year. Use this model to predict the dollar value of the product in 2010. (Let $t = 8$ represent 2008.)

53. Straight-Line Depreciation   A small business purchases a piece of equipment for $135,000. After 10 years, the equipment will have to be replaced. Its salvage value at that time is expected to be $5500. Write a linear equation giving the value *V* of the equipment during the 10 years it will be used.

(T) 54. Sales   The sales *S* (in millions of dollars) for Intuit Corporation for the years 2000 to 2005 are shown in the table. *(Source: Intuit Corporation)*

Year	Sales S (in millions of dollars)
2000	1093.8
2001	1261.5
2002	1358.3
2003	1650.7
2004	1867.7
2005	2079.9

(a) Use a graphing utility to create a scatter plot of the data. Let *t* represent the year, with $t = 0$ corresponding to 2000. Do the data appear to be linear?

(b) Use the *regression* feature of a graphing utility to find a linear model for the data.

(c) Use the linear model from part (b) to predict sales in 2006 and 2007.

(d) Intuit Corporation predicts sales of $2325 million for 2006 and $2500 million for 2007. Do your estimates from part (c) agree with those of Intuit Corporation? Which set of estimates do you think is more reasonable? Explain.

In Exercises 55–58, decide whether the equation represents *y* as a function of *x*.

55. $3x - 4y = 12$

56. $y^2 = x^2 - 9$

57. $y = \sqrt{x + 3}$

58. $x^2 + y^2 - 6x + 8y = 0$

In Exercises 59 and 60, decide whether the set of ordered pairs represents a function from *A* to *B*.

$A = \{1, 2, 3\}$ $B = \{-3, -4, -7\}$

Give reasons for your answer.

59. $\{(1, -3), (2, -7), (3, -3)\}$

60. $\{(1, -4), (2, -3), (3, -9)\}$

In Exercises 61 and 62, evaluate the function at each specified value of the independent variable and simplify.

61. $f(x) = \sqrt{x + 4} - 5$

(a) $f(5)$ (b) $f(0)$ (c) $f(-4)$ (d) $f(x + 3)$

62. $f(x) = \begin{cases} 2x - 1, & x \le 1 \\ x^2 + 2, & x > 1 \end{cases}$

(a) $f(0)$ (b) $f(1)$ (c) $f(3)$ (d) $f(-4)$

In Exercises 63–68, find the domain of the function.

63. $f(x) = 2x^2 + 7x + 3$

64. $g(t) = \dfrac{3}{t^2 - 4}$

65. $h(x) = \sqrt{x + 5}$

66. $f(t) = \sqrt[3]{t - 3}$

67. $g(t) = \dfrac{\sqrt{t - 1}}{t - 4}$

68. $h(x) = \sqrt[4]{16 - x^2}$

(T) 69. Reasoning   A student has difficulty understanding why the domains of

$$h(x) = \dfrac{x^2 - 4}{x} \quad \text{and} \quad k(x) = \dfrac{x}{x^2 - 4}$$

are different. How would you explain their respective domains algebraically? How could you use a graphing utility to explain their domains?

70. Volume of a Box   An open box is to be made from a square piece of material 20 inches on a side by cutting equal squares from the corners and turning the sides up (see figure).

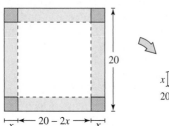
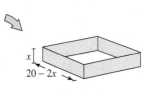

(a) Write the volume V of the box as a function of its height x.

(b) What is the domain of this function?

(T) (c) Use a graphing utility to graph the function.

71. Balance in an Account   A person deposits $6500 in an account that pays 6.85% interest compounded quarterly.

(a) Write the balance of the account in terms of the time t that the principal is left in the account.

(b) What is the domain of this function?

72. Vertical Motion   The velocity v (in feet per second) of a ball thrown vertically upward from ground level is given by

$$v(t) = -32t + 80$$

where t is the time (in seconds).

(a) Find the velocity when $t = 1$.

(b) Find the time when the ball reaches its maximum height. [*Hint:* Find the time when $v(t) = 0$.]

(c) Find the velocity when $t = 3$.

In Exercises 73–76, (a) determine the domain and range of the function, (b) determine the intervals over which the function is increasing, decreasing, or constant, (c) determine if the function is even, odd, or neither, and (d) approximate any relative minimum or relative maximum values of the function.

73. $f(x) = x^2 + 1$

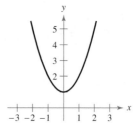

74. $f(x) = \sqrt{x^2 - 9}$

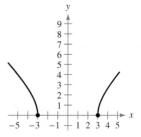

75. $f(x) = x^3 - 4x^2$

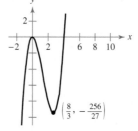

76. $f(x) = |x - 2|$

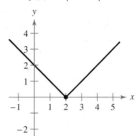

In Exercises 77–82, use the Vertical Line Test to decide whether y is a function of x.

77. $y = \frac{1}{2}x^2$

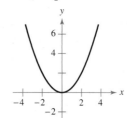

78. $y = \frac{1}{4}x^3$

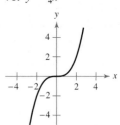

79. $x - y^2 = 1$

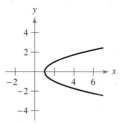

80. $x^2 + y^2 = 25$

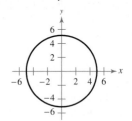

81. $x^2 = 2xy - 1$

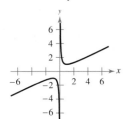

82. $x = |y + 2|$

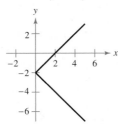

In Exercises 83–90, sketch the graph of the function.

83. $f(x) = |x + 3|$

84. $g(x) = \sqrt{x^2 - 16}$

85. $h(x) = 2[\![x]\!] + 1$

86. $f(x) = 3$

87. $g(x) = \begin{cases} x + 2, & x < 0 \\ 2, & x = 0 \\ x^2 + 2, & x > 0 \end{cases}$

88. $g(x) = \begin{cases} 3x + 1, & x < -1 \\ x^2 - 3, & x \geq -1 \end{cases}$

89. $h(x) = x^2 - 3x$

90. $f(x) = \sqrt{9 - x^2}$

91. Cost of Overnight Delivery   The cost of sending an overnight package from Los Angeles to Dallas is $10.25 for up to, but not including, the first pound and $2.75 for each additional pound (or portion of a pound). A model for the total cost C of sending the package is

$$C = 10.25 + 2.75[\![x]\!], \quad x > 0$$

where x is the weight of the package (in pounds). Sketch the graph of this function.

Ⓣ 92. Revenue   A company determines that the total revenue R (in hundreds of thousands of dollars) for the years 1997 to 2010 can be approximated by the function

$$R = -0.025t^3 + 0.8t^2 - 2.5t + 8.75, \quad 7 \leq t \leq 20$$

where t represents the year, with $t = 7$ corresponding to 1997. Graph the revenue function using a graphing utility and use the *trace* feature to estimate the years during which the revenue was increasing and the years during which the revenue was decreasing.

In Exercises 93 and 94, describe the sequence of transformations from $f(x) = x^2$ to g. Then sketch the graph of g.

93. $g(x) = -(x - 1)^2 - 2$

94. $g(x) = -x^2 + 3$

In Exercises 95 and 96, describe the sequence of transformations from $f(x) = \sqrt{x}$ to g. Then sketch the graph of g.

95. $g(x) = -\sqrt{x - 2}$

96. $g(x) = \sqrt{x} + 2$

In Exercises 97 and 98, describe the sequence of transformations from $f(x) = \sqrt[3]{x}$ to g. Then sketch the graph of g.

97. $g(x) = \sqrt[3]{x + 2}$

98. $g(x) = 2\sqrt[3]{x}$

In Exercises 99 and 100, identify the transformation shown in the graph and the associated common function. Write the equation of the graphed function.

99.

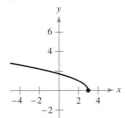

100.

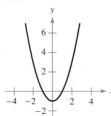

In Exercises 101 and 102, find $(f + g)(x)$, $(f - g)(x)$, $(fg)(x)$, and $(f/g)(x)$. What is the domain of f/g?

101. $f(x) = 3x - 1$, $g(x) = x^2 + 2x$

102. $f(x) = 3x$, $g(x) = \sqrt{x^2 + 1}$

In Exercises 103–106, evaluate the function for $f(x) = x^2 + 3x$ and $g(x) = 2x - 5$.

103. $(f + g)(2)$

104. $(f - g)(-1)$

105. $(fg)(3)$

106. $\left(\dfrac{f}{g}\right)(0)$

In Exercises 107–110, find and determine the domains of (a) $f \circ g$ and (b) $g \circ f$.

107. $f(x) = x^2$, $g(x) = x + 3$

108. $f(x) = 2x - 5$, $g(x) = x^2 + 2$

109. $f(x) = \dfrac{1}{x}$, $g(x) = 3x + x^2$

110. $f(x) = \dfrac{1}{x^2}$, $g(x) = x^3$

In Exercises 111–114, find two functions f and g such that $(f \circ g)(x) = h(x)$. (There are many correct answers.)

111. $h(x) = (6x - 5)^2$

112. $h(x) = \sqrt[3]{x + 2}$

113. $h(x) = \dfrac{1}{(x - 1)^2}$

114. $h(x) = (x - 3)^3 + 2(x - 3)$

115. *MAKE A DECISION: COMPARING SALES*   You own two dry cleaning establishments. From 2000 to 2008, the sales for one of the establishments were increasing according to the function

$$R_1 = 499.7 - 0.3t + 0.2t^2, \quad t = 0, 1, 2, 3, 4, 5, 6, 7, 8$$

where R_1 represents the sales (in thousands of dollars) and t represents the year, with $t = 0$ corresponding to 2000. During the same nine-year period, the sales for the second establishment were decreasing according to the function

$$R_2 = 300.8 - 0.62t, \quad t = 0, 1, 2, 3, 4, 5, 6, 7, 8.$$

Write a function that represents the total sales for the two establishments. Make a stacked bar graph to represent the total sales during this nine-year period. Were total sales increasing or decreasing?

116. Area   A square concrete foundation is prepared as a base for a large cylindrical aquatic tank that is to be used in ecology experiments (see figure).

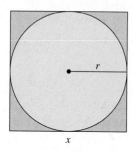

(a) Write the radius r of the tank as a function of the length x of the sides of the square.

(b) Write the area A of the circular base of the tank as a function of the radius r.

(c) Find and interpret $(A \circ r)(x)$.

117. *MAKE A DECISION*   You are a sales representative for an automobile manufacturer. You are paid an annual salary plus a bonus of 3% of your sales over $500,000. Consider the two functions given by

$$f(x) = x - 500,000$$

and

$$g(x) = 0.03x.$$

If x is greater than $500,000, does $f(g(x))$ or $g(f(x))$ represent your bonus? Explain.

118. Bacteria   The number N of bacteria is given by $N(T) = 8T^2 - 14T + 200$, where T is the temperature (in degrees Fahrenheit). The temperature is $T(t) = 2t + 2$, where t is the time in hours. Find and interpret $(N \circ T)(t)$.

In Exercises 119 and 120, show that f and g are inverse functions of each other.

119. $f(x) = 3x + 5, \quad g(x) = \dfrac{x - 5}{3}$

120. $f(x) = \sqrt[3]{x - 3}, \quad g(x) = x^3 + 3$

In Exercises 121–124, determine whether the function has an inverse function. If it does, find the inverse function and graph f and f^{-1} in the same coordinate plane.

121. $f(x) = 3x^2$

122. $f(x) = \sqrt[3]{x + 1}$

123. $f(x) = \dfrac{1}{x}$

124. $f(x) = \dfrac{x^2}{x^2 - 9}$

In Exercises 125–128, (a) find f^{-1}, (b) sketch the graphs of f and f^{-1} on the same coordinate plane, and (c) verify that $f^{-1}(f(x)) = x$ and $f(f^{-1}(x)) = x$.

125. $f(x) = \frac{1}{2}x - 3$

126. $f(x) = \sqrt{x + 1}$

127. $f(x) = x^2, \quad x \geq 0$

128. $f(x) = \sqrt[3]{x - 1}$

(T) 129. Federal Student Aid   The average awards A (in dollars) of federal financial aid (including grants and loans) for the years 2000 to 2005 are shown in the table. *(Source: U.S. Department of Education)*

Year	Average award, A (in dollars)
2000	2925
2001	2982
2002	3089
2003	3208
2004	3316
2005	3425

(a) Use a graphing utility to create a scatter plot of the data. Let t represent the year, with $t = 0$ corresponding to 2000.

(b) Use the *regression* feature of a graphing utility to find a linear model for the data.

(c) If the data can be modeled by a one-to-one function, find the inverse function of the model and use it to predict in what year the average award will be $3600.

Chapter Test

See www.CalcChat.com for worked-out solutions to odd-numbered exercises.

Take this test as you would take a test in class. When you are done, check your work against the answers given in the back of the book.

In Exercises 1 and 2, find the distance between the points and the midpoint of the line segment connecting the points.

1. $(-3, 2), (5, -2)$ **2.** $(3.25, 7.05), (-2.37, 1.62)$

3. Find the intercepts of the graph of $y = (x + 5)(x - 3)$.

4. Describe the symmetry of the graph of $y = \dfrac{x}{x^2 - 4}$.

5. Find an equation of the line through $(-3, 5)$ with a slope of $\frac{2}{3}$.

6. Write the equation of the circle in standard form and sketch its graph.

$$x^2 + y^2 - 6x + 4y - 3 = 0$$

In Exercises 7 and 8, decide whether the statement is true or false. Explain.

7. The equation $2x - 3y = 5$ identifies y as a function of x.

8. If $A = \{3, 4, 5\}$ and $B = \{-1, -2, -3\}$, the set $\{(3, -9), (4, -2), (5, -3)\}$ represents a function from A to B.

In Exercises 9 and 10, (a) find the domain and range of the function, (b) determine the intervals over which the function is increasing, decreasing, or constant, (c) determine whether the function is even or odd, and (d) approximate any relative minimum or relative maximum values of the function.

9. $f(x) = 2 - x^2$ (See figure.)   **10.** $g(x) = \sqrt{x^2 - 4}$ (See figure.)

In Exercises 11 and 12, sketch the graph of the function.

11. $g(x) = \begin{cases} x + 1, & x < 0 \\ 1, & x = 0 \\ x^2 + 1, & x > 0 \end{cases}$

12. $h(x) = (x - 3)^2 + 4$

In Exercises 13–16, use $f(x) = x^2 + 2$ and $g(x) = 2x - 1$ to find the function.

13. $(f - g)(x)$

14. $(fg)(x)$

15. $(f \circ g)(x)$

16. $g^{-1}(x)$

17. A business purchases a piece of equipment for $30,000. After 5 years, the equipment will be worth only $4000. Write a linear equation that gives the value V of the equipment during the 5 years.

(T) **18. Population** The projected populations P (in millions) of children under the age of 5 in the United States for selected years from 2010 to 2050 are shown in the table. Use a graphing utility to create a scatter plot of the data and find a linear model for the data. Let t represent the year, with $t = 10$ corresponding to 2010. *(Source: U.S. Census Bureau)*

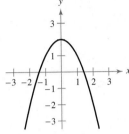

Figure for 9

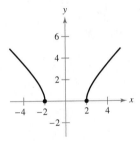

Figure for 10

Year	Population, P
2010	21.4
2015	22.4
2020	22.9
2025	23.5
2030	24.3
2035	25.3
2040	26.3
2045	27.2
2050	28.1

Table for 18

3

Polynomial and Rational Functions

©Bettmann/CORBIS

Many professional athletes sign contracts with sportswear companies to promote clothing lines and footwear. Quadratic functions are often used to model real-life phenomena, such as the profit from selling a line of sportswear. You can use a quadratic model to determine how much money a company can spend on advertising to obtain a certain profit. (See Section 3.4, Example 9.)

Applications

Polynomial and rational functions are used to model and solve many real-life applications. The applications listed below represent a sample of the applications in this chapter.

Section 3.1

Quadratic Functions and Models

- Sketch the graph of a quadratic function and identify its vertex and intercepts.
- Find a quadratic function given its vertex and a point on its graph.
- Construct and use a quadratic model to solve an application problem.

The Graph of a Quadratic Function

In this and the next section, you will study the graphs of polynomial functions.

Definition of a Polynomial Function

Let n be a nonnegative integer and let $a_n, a_{n-1}, \ldots, a_2, a_1, a_0$ be real numbers with $a_n \neq 0$. The function given by

$$f(x) = a_n x^n + a_{n-1} x^{n-1} + \cdots + a_2 x^2 + a_1 x + a_0$$

is called a **polynomial function of x with degree n.**

Polynomial functions are classified by degree. Recall that the degree of a polynomial is the highest degree of its terms. For instance, the polynomial function given by

$$f(x) = a, \quad a \neq 0 \qquad \text{Constant function}$$

has degree 0 and is called a **constant function.** In Chapter 2, you learned that the graph of this type of function is a horizontal line. The polynomial function given by

$$f(x) = ax + b, \quad a \neq 0 \qquad \text{Linear function}$$

has degree 1 and is called a **linear function.** In Chapter 2, you learned that the graph of the linear function given by $f(x) = ax + b$ is a line whose slope is a and whose y-intercept is $(0, b)$. In this section, you will study second-degree polynomial functions, which are called **quadratic functions.**

For instance, each of the following functions is a quadratic function.

$$f(x) = x^2 + 6x + 2 \quad g(x) = 2(x + 1)^2 - 3 \quad h(x) = (x - 2)(x + 1)$$

Definition of a Quadratic Function

Let a, b, and c be real numbers with $a \neq 0$. The function of x given by

$$f(x) = ax^2 + bx + c \qquad \text{Quadratic function}$$

is called a quadratic function.

The graph of a quadratic function is called a **parabola.** It is " $\cup$ "-shaped and can open upward or downward.

All parabolas are symmetric with respect to a line called the **axis of symmetry,** or simply the **axis** of the parabola. The point at which the axis intersects the parabola is the **vertex** of the parabola, as shown in Figure 3.1. If the leading coefficient is positive, the graph of $f(x) = ax^2 + bx + c$ is a parabola that opens upward, and if the leading coefficient is negative, the graph of $f(x) = ax^2 + bx + c$ is a parabola that opens downward.

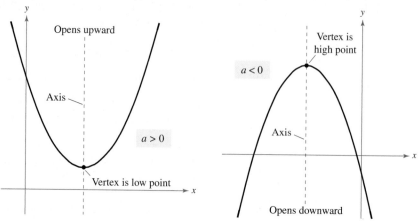

FIGURE 3.1

The simplest type of quadratic function is

$$f(x) = ax^2.$$

Its graph is a parabola whose vertex is $(0, 0)$. When $a > 0$, the vertex is the point with the *minimum* y-value on the graph, and when $a < 0$, the vertex is the point with the *maximum* y-value on the graph, as shown in Figure 3.2.

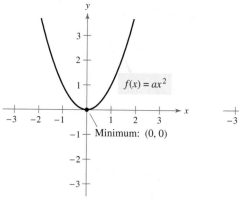
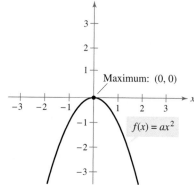

$a > 0$: Parabola opens upward $a < 0$: Parabola opens downward
FIGURE 3.2

When sketching the graph of $f(x) = ax^2$, it is helpful to use the graph of $y = x^2$ as a reference, as discussed in Section 2.6. There you saw that when $a > 1$, the graph of $y = af(x)$ is a vertical stretch of the graph of $y = f(x)$. When $0 < a < 1$, the graph of $y = af(x)$ is a vertical shrink of the graph of $y = f(x)$. This is demonstrated again in Example 1.

Example 1   **Sketching the Graph of a Quadratic Function**

a. Compared with the graph of $y = x^2$, each output of $f(x) = \frac{1}{3}x^2$ vertically "shrinks" the graph by a factor of $\frac{1}{3}$, creating the wider parabola shown in Figure 3.3(a).

b. Compared with the graph of $y = x^2$, each output of $g(x) = 2x^2$ vertically "stretches" the graph by a factor of 2, creating the narrower parabola shown in Figure 3.3(b).

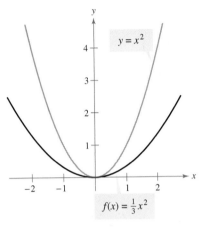
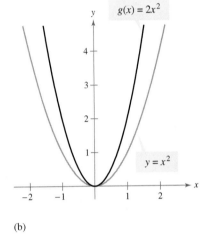

(a)

(b)

FIGURE 3.3

✓ **CHECKPOINT 1**

Sketch the graph of $f(x) = 4x^2$. Then compare the graph with the graph of $y = x^2$. ■

In Example 1, note that the coefficient a determines how widely the parabola given by $f(x) = ax^2$ opens. If $|a|$ is small, the parabola opens more widely than if $|a|$ is large.

Recall from Section 2.6 that the graphs of $y = f(x \pm c)$, $y = f(x) \pm c$, $y = -f(x)$, and $y = f(-x)$ are rigid transformations of the graph of $y = f(x)$. For instance, in Figure 3.4, notice how the graph of $y = x^2$ can be transformed to produce the graphs of $f(x) = -x^2 + 1$ and $g(x) = (x + 2)^2 - 3$.

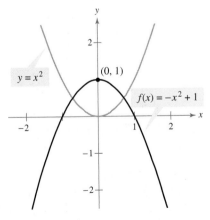
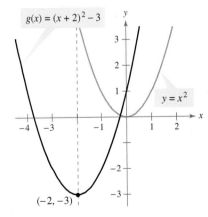

FIGURE 3.4

The Standard Form of a Quadratic Function

The **standard form** of a quadratic function is

$$f(x) = a(x - h)^2 + k.$$

This form is especially convenient for sketching a parabola because it identifies the vertex of the parabola.

Standard Form of a Quadratic Function

The quadratic function given by

$$f(x) = a(x - h)^2 + k, \quad a \neq 0$$

is said to be in **standard form.** The graph of f is a parabola whose axis is the vertical line $x = h$ and whose vertex is the point (h, k). If $a > 0$, the parabola opens upward, and if $a < 0$, the parabola opens downward.

To write a quadratic function in standard form, you can use the process of *completing the square,* as illustrated in Example 2.

Example 2   Graphing a Parabola in Standard Form

Sketch the graph of $f(x) = 2x^2 + 8x + 7$ and identify the vertex.

SOLUTION   Begin by writing the quadratic function in standard form. The first step in completing the square is to factor out any coefficient of x^2 that is not 1.

$$f(x) = 2x^2 + 8x + 7 \qquad \text{Write original function.}$$

$$= 2(x^2 + 4x) + 7 \qquad \text{Factor 2 out of } x \text{ terms.}$$

$$= 2(x^2 + 4x + 4 - 4) + 7 \qquad \begin{array}{l}\text{Add and subtract 4 within parentheses} \\ \text{to complete the square.}\end{array}$$

$$(4/2)^2$$

After adding and subtracting 4 within the parentheses, you must now regroup the terms to form a perfect square trinomial. The -4 can be removed from inside the parentheses. But, because of the 2 outside the parentheses, you must multiply -4 by 2 as shown below.

$$f(x) = 2(x^2 + 4x + 4) - 2(4) + 7 \qquad \text{Regroup terms.}$$

$$= 2(x^2 + 4x + 4) - 8 + 7 \qquad \text{Simplify.}$$

$$= 2(x + 2)^2 - 1 \qquad \text{Standard form}$$

From this form, you can see that the graph of f is a parabola that opens upward with vertex $(-2, -1)$. This corresponds to a left shift of two units and a downward shift of one unit relative to the graph of $y = 2x^2$, as shown in Figure 3.5.

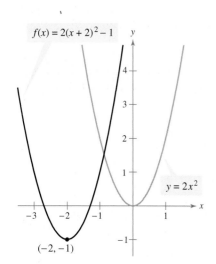

$f(x) = 2(x + 2)^2 - 1$

$y = 2x^2$

$(-2, -1)$

FIGURE 3.5

✓ CHECKPOINT 2

Sketch the graph of $f(x) = 2x^2 - 12x + 20$ and identify the vertex. ■

Example 3   Graphing a Parabola in Standard Form

Sketch the graph of $f(x) = -x^2 + 6x - 8$ and identify the vertex.

SOLUTION   As in Example 2, begin by writing the quadratic function in standard form.

$$f(x) = -x^2 + 6x - 8 \qquad \text{Write original function.}$$

$$= -(x^2 - 6x) - 8 \qquad \text{Factor } -1 \text{ out of } x \text{ terms.}$$

$$= -(x^2 - 6x + 9 - 9) - 8 \qquad \begin{array}{l}\text{Add and subtract 9 within parentheses}\\ \text{to complete the square.}\end{array}$$

$$(-6/2)^2$$

$$= -(x^2 - 6x + 9) - (-9) - 8 \qquad \text{Regroup terms.}$$

$$= -(x - 3)^2 + 1 \qquad \text{Standard form}$$

So, the graph of f is a parabola that opens downward with vertex at $(3, 1)$, as shown in Figure 3.6.

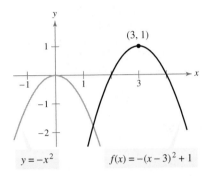

$y = -x^2$ $f(x) = -(x-3)^2 + 1$

FIGURE 3.6

✓ **CHECKPOINT 3**

Sketch the graph of $f(x) = -3x^2 + 12x + 1$ and identify the vertex. ▨

Example 4   Finding an Equation of a Parabola

Find an equation of the parabola whose vertex is $(1, 2)$ and that passes through the point $(0, 0)$, as shown in Figure 3.7.

SOLUTION   Because the parabola has a vertex at $(h, k) = (1, 2)$, the equation must have the form

$$f(x) = a(x - 1)^2 + 2. \qquad \text{Standard form}$$

Because the parabola passes through the point $(0, 0)$, it follows that when $x = 0$, $f(x)$ must equal 0. Substitute 0 for x and 0 for $f(x)$ to obtain the equation

$$0 = a(0 - 1)^2 + 2.$$

This equation can be solved easily for a, and you can see that

$$a = -2.$$

You can now write an equation of the parabola.

$$f(x) = -2(x - 1)^2 + 2 \qquad \text{Substitute for } a, h, \text{ and } k \text{ in standard form.}$$

$$= -2x^2 + 4x \qquad \text{Simplify.}$$

To find the x-intercepts of the graph of $f(x) = ax^2 + bx + c$, you must solve the equation

$$ax^2 + bx + c = 0.$$

If the equation $ax^2 + bx + c$ does not factor, you can use the Quadratic Formula to determine the x-intercepts. Remember, however, that a parabola may have no x-intercepts.

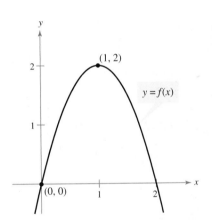

$y = f(x)$

FIGURE 3.7

✓ **CHECKPOINT 4**

Find an equation of the parabola whose vertex is $(3, 4)$ and that passes through the point $(2, 5)$. ▨

Applications

Many applications involve finding the maximum or minimum value of a quadratic function. By writing $f(x) = ax^2 + bx + c$ in standard form, you can determine that the vertex occurs at $x = -b/2a$.

Example 5   The Maximum Height of a Baseball

A baseball is hit 3 feet above the ground at a velocity of 100 feet per second and at an angle of 45° with respect to the ground. The path of the baseball is given by

$$f(x) = -0.0032x^2 + x + 3$$

where $f(x)$ is the height of the baseball (in feet) and x is the distance from home plate (in feet). What is the maximum height reached by the baseball?

SOLUTION   For this quadratic function, you have

$$f(x) = ax^2 + bx + c$$
$$= -0.0032x^2 + x + 3.$$

So, $a = -0.0032$ and $b = 1$. Because the function has a maximum when $x = -b/2a$, the baseball reaches its maximum height when it is

$$x = -\frac{b}{2a} = -\frac{1}{2(-0.0032)} = 156.25 \text{ feet}$$

from home plate. At this distance, the maximum height is

$$f(156.25) = -0.0032(156.25)^2 + 156.25 + 3 = 81.125 \text{ feet}.$$

The path of the baseball is shown in Figure 3.8.

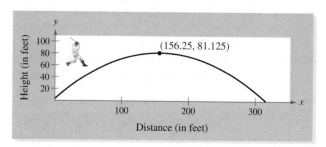

FIGURE 3.8

✓ CHECKPOINT 5

In Example 5, suppose the baseball is hit at a velocity of 70 feet per second. The path of the baseball is given by $f(x) = -0.007x^2 + x + 4$, where $f(x)$ is the height of the baseball (in feet) and x is the distance from home plate (in feet). What is the maximum height reached by the baseball? ■

In Section 2.3 you plotted data points in the coordinate plane and estimated the best-fitting line. Fitting a quadratic model by this same process would be complicated. Most graphing utilities have a built-in statistical program that easily calculates the best-fitting quadratic model for a set of data points. Refer to the user's guide of your graphing utility for the required steps.

Example 6 Fitting a Quadratic Function to Data   (R)

Sparrow Population   The table shows the numbers N of sparrows in a nature preserve for the years 1993 to 2008. Use a graphing utility to plot the data and find the quadratic model that best fits the data. Find the vertex of the graph of the quadratic model and interpret its meaning in the context of the problem. Let $x = 3$ represent the year 1993.

Year	1993	1994	1995	1996	1997	1998	1999	2000
x	3	4	5	6	7	8	9	10
Number, N	211	187	148	120	95	76	67	62

Year	2001	2002	2003	2004	2005	2006	2007	2008
x	11	12	13	14	15	16	17	18
Number, N	66	71	92	107	128	145	167	197

SOLUTION   Begin by entering the data into your graphing utility and displaying the scatter plot. From the scatter plot that is shown in Figure 3.9(a) you can see that the points have a parabolic trend. Use the *quadratic regression* feature to find the quadratic function that best fits the data. The quadratic equation that best fits the data is

$$N = 2.53x^2 - 53.5x + 350, \quad 3 \le x \le 18.$$

Graph the data and the equation in the same viewing window, as shown in Figure 3.9(b). By using the *minimum* feature of your graphing utility, you can see that the vertex of the graph is approximately $(10.6, 67.2)$, as shown in Figure 3.9(c). The vertex corresponds to the year in which the number of sparrows in the nature preserve was the least. So, in 2001, the number of sparrows in the nature preserve reached a minimum.

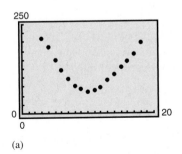

(a)

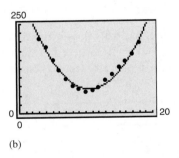

(b)

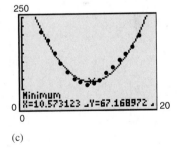

(c)

FIGURE 3.9

✓**CHECKPOINT 6**

In Example 6, use the model to predict the number of sparrows in the nature preserve in 2011. ■

Example 7 **Charitable Contributions**

The percent of their income that a family gives to charities is related to their income level. For families with annual incomes between $5000 and $100,000, the percent P can be modeled by

$$P(x) = 0.0014x^2 - 0.1529x + 5.855, \quad 5 \le x \le 100$$

where x is the annual income (in thousands of dollars). Use the model to estimate the income that corresponds to the minimum percent of income given to charities.

SOLUTION   One way to answer the question is to sketch the graph of the quadratic function, as shown in Figure 3.10. From this graph, it appears that the minimum percent corresponds to an income level of about $55,000.

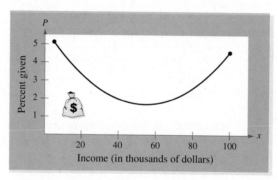

FIGURE 3.10

Another way to answer the question is to use the fact that the minimum point of the parabola occurs when $x = -b/2a$.

$$x = -\frac{b}{2a} = -\frac{-0.1529}{2(0.0014)} \approx 54.6$$

From this x-value, you can conclude that the minimum percent corresponds to an income level of about $54,600.

✓ **CHECKPOINT 7**

A manufacturer has daily production costs C (in dollars per unit) of $C = 0.15x^2 - 9x + 700$ where x is the number of units produced. How many units should be produced each day to yield a minimum cost per unit? ■

CONCEPT CHECK

1. Does the vertex of the graph of $f(x) = -3(x + 1)^2 - 1$ contain a minimum y-value or a maximum y-value? Explain.

2. Is the quadratic function given by $f(x) = 2(x - 1)^2 + 3$ written in standard form? Explain.

3. Write an equation of a parabola that is the graph of $y = x^2$ shifted right three units, downward one unit, and vertically stretched by a factor of 2.

4. The graph of the quadratic function given by $f(x) = a(x - 1)^2 + 3$ has two x-intercepts. What can you conclude about the value of a?

Skills Review 3.1

The following warm-up exercises involve skills that were covered in earlier sections. You will use these skills in the exercise set for this section. For additional help, review Sections 1.3 and 1.4.

In Exercises 1–4, solve the quadratic equation by factoring.

1. $2x^2 + 11x - 6 = 0$

2. $5x^2 - 12x - 9 = 0$

3. $3 + x - 2x^2 = 0$

4. $x^2 + 20x + 100 = 0$

In Exercises 5–10, use the Quadratic Formula to solve the quadratic equation.

5. $x^2 - 6x + 4 = 0$

6. $x^2 + 4x + 1 = 0$

7. $2x^2 - 16x + 25 = 0$

8. $3x^2 + 30x + 74 = 0$

9. $x^2 + 3x + 1 = 0$

10. $x^2 + 3x - 3 = 0$

Exercises 3.1

See www.CalcChat.com for worked-out solutions to odd-numbered exercises.

In Exercises 1–8, match the quadratic function with its graph. [The graphs are labeled (a), (b), (c), (d), (e), (f), (g), and (h).]

(a)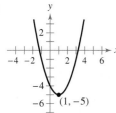

(b)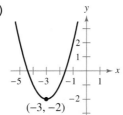

(c)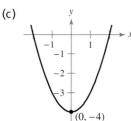

(d)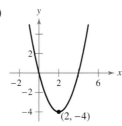

(e)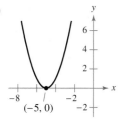

(f)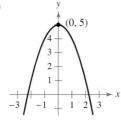

(g)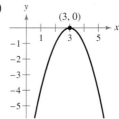

(h)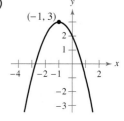

1. $f(x) = -(x - 3)^2$

2. $f(x) = (x + 5)^2$

3. $f(x) = x^2 - 4$

4. $f(x) = 5 - x^2$

5. $f(x) = (x + 3)^2 - 2$

6. $f(x) = (x - 1)^2 - 5$

7. $f(x) = -(x + 1)^2 + 3$

8. $f(x) = (x - 2)^2 - 4$

In Exercises 9–14, find an equation of the parabola.

9.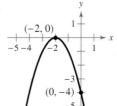

10.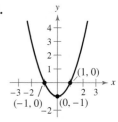

11.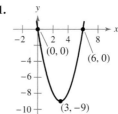

12.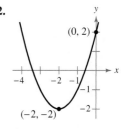

13.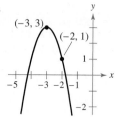

14.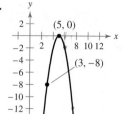

In Exercises 15–18, compare the graph of the quadratic function with the graph of $y = x^2$.

15. $f(x) = 5x^2$

16. $f(x) = -\frac{1}{4}x^2$

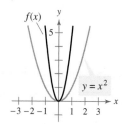

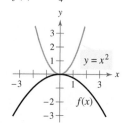

17. $f(x) = -(x + 1)^2 + 1$

18. $f(x) = 3(x - 2)^2 - 1$

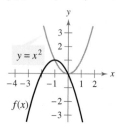

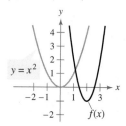

In Exercises 19–36, sketch the graph of the quadratic function. Identify the vertex and intercepts.

19. $f(x) = 3x^2$

20. $f(x) = -2x^2$

21. $f(x) = 16 - x^2$

22. $h(x) = x^2 - 9$

23. $f(x) = (x + 5)^2 - 6$

24. $f(x) = (x - 6)^2 + 3$

25. $g(x) = x^2 + 2x + 1$

26. $h(x) = x^2 - 4x + 2$

27. $f(x) = -(x^2 + 2x - 3)$

28. $f(x) = -(x^2 + 6x - 3)$

29. $f(x) = x^2 - x + \frac{5}{4}$

30. $f(x) = x^2 + 3x + \frac{1}{4}$

31. $f(x) = -x^2 + 2x + 5$

32. $f(x) = -x^2 - 4x + 1$

33. $h(x) = 4x^2 - 4x + 21$

34. $f(x) = 2x^2 - x + 1$

35. $f(x) = \frac{1}{4}(x^2 - 16x + 32)$

36. $g(x) = \frac{1}{2}(x^2 + 4x - 2)$

In Exercises 37–40, find an equation of the parabola that has the indicated vertex and whose graph passes through the given point.

37. Vertex: $(2, -1)$; point: $(4, -3)$

38. Vertex: $(-3, 5)$; point: $(-6, -1)$

39. Vertex: $(5, 12)$; point: $(7, 15)$

40. Vertex: $(-2, -2)$; point: $(-1, 0)$

In Exercises 41–46, find two quadratic functions whose graphs have the given x-intercepts. Find one function whose graph opens upward and another whose graph opens downward. (There are many correct answers.)

41. $(2, 0), (-1, 0)$

42. $(-4, 0), (0, 0)$

43. $(0, 0), (10, 0)$

44. $(4, 0), (8, 0)$

45. $(-3, 0), \left(-\frac{1}{2}, 0\right)$

46. $\left(-\frac{5}{2}, 0\right), (2, 0)$

47. Optimal Area The perimeter of a rectangle is 200 feet. Let x represent the width of the rectangle. Write a quadratic function for the area of the rectangle in terms of its width. Find the vertex of the graph of the quadratic function and interpret its meaning in the context of the problem.

48. Optimal Area The perimeter of a rectangle is 540 feet. Let x represent the width of the rectangle. Write a quadratic function for the area of the rectangle in terms of its width. Find the vertex of the graph of the quadratic function and interpret its meaning in the context of the problem.

49. Optimal Area A rancher has 1200 feet of fencing with which to enclose two adjacent rectangular corrals (see figure). What measurements will produce a maximum enclosed area?

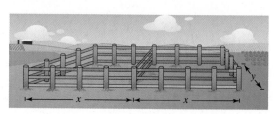

50. Optimal Area An indoor physical-fitness room consists of a rectangular region with a semicircle on each end (see figure). The perimeter of the room is to be a 200-meter running track. What measurements will produce a maximum area of the rectangle?

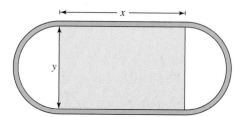

Optimal Revenue In Exercises 51 and 52, find the number of units that produces a maximum revenue. The revenue R is measured in dollars and x is the number of units produced.

51. $R = 1000x - 0.02x^2$

52. $R = 80x - 0.0001x^2$

53. Optimal Cost A manufacturer of lighting fixtures has daily production costs C (in dollars per unit) of

$$C(x) = 800 - 10x + 0.25x^2$$

where x is the number of units produced. How many fixtures should be produced each day to yield a minimum cost per unit?

54. Optimal Profit The profit P (in dollars) for a manufacturer of sound systems is given by

$$P(x) = -0.0003x^2 + 150x - 375,000$$

where x is the number of units produced. What production level will yield a maximum profit?

(T) 55. Maximum Height of a Diver The path of a diver is given by

$$y = -\frac{4}{9}x^2 + \frac{24}{9}x + 10$$

where y is the height (in feet) and x is the horizontal distance from the end of the diving board (in feet) (see figure). Use a graphing utility and the *trace* or *maximum* feature to find the maximum height of the diver.

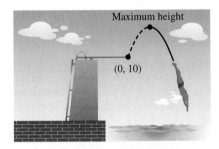

Maximum height

$(0, 10)$

(T) 56. Maximum Height The winning men's shot put in the 2004 Summer Olympics was recorded by Yuriy Belonog of Ukraine. The path of his winning toss is approximately given by

$$y = -0.011x^2 + 0.65x + 8.3$$

where y is the height of the shot (in feet) and x is the horizontal distance (in feet). Use a graphing utility and the *trace* or *maximum* feature to find the length of the winning toss and the maximum height of the shot.

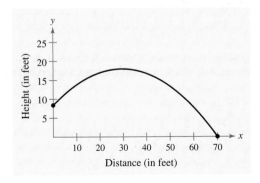

(T) 57. Cable TV Subscribers The table shows the average numbers S (in millions) of basic cable subscribers for the years 1995 to 2005. *(Source: Kagan Research, LLC)*

Year	1995	1996	1997	1998
Subscribers, S	60.6	62.3	63.6	64.7

Year	1999	2000	2001	2002
Subscribers, S	65.5	66.3	66.7	66.5

Year	2003	2004	2005
Subscribers, S	66.1	65.7	65.3

(a) Use a graphing utility to create a scatter plot of the data. Let t represent the year, with $t = 5$ corresponding to 1995.

(b) Use the *regression* feature of a graphing utility to find a quadratic model for the data.

(c) Use a graphing utility to graph the model from part (b) in the same viewing window as the scatter plot.

(d) Use the graph of the model from part (c) to estimate when the number of basic cable subscribers was the greatest. Does this result agree with the actual data?

(T) 58. Price of Gold The table shows the average annual prices P (in dollars) of gold for the years 1996 to 2005. *(Source: World Gold Council)*

Year	1996	1997	1998	1999
Price of gold, P	387.82	330.98	294.12	278.55

Year	2000	2001	2002	2003
Price of gold, P	279.10	272.67	309.66	362.91

Year	2004	2005
Price of gold, P	409.17	444.47

(a) Use a graphing utility to create a scatter plot of the data. Let t represent the year, with $t = 6$ corresponding to 1996.

(b) Use the *regression* feature of a graphing utility to find a quadratic model for the data.

(c) Use a graphing utility to graph the model from part (b) in the same viewing window as the scatter plot.

(d) Use the graph of the model from part (c) to estimate when the price of gold was the lowest. Does this result agree with the actual data?

(T) 59. Tuition and Fees   The table shows the average values of tuition and fees F (in dollars) for in-state students at public institutions of higher education in the years 1996 to 2005. *(Source: U.S. National Center for Educational Statistics)*

Year	1996	1997	1998	1999	2000
Tuition and fees, F	2179	2271	2360	2430	2506

Year	2001	2002	2003	2004	2005
Tuition and fees, F	2562	2700	2903	3319	3638

(a) Use a graphing utility to create a scatter plot of the data. Let t represent the year, with $t = 6$ corresponding to 1996.

(b) Use the *regression* feature of a graphing utility to find a quadratic model for the data.

(c) Use a graphing utility to graph the model from part (b) in the same viewing window as the scatter plot of the data.

(d) Use the graph of the model from part (c) to predict the average value of tuition and fees in 2008.

(T) 60. Liver Transplants   The table shows the numbers L of liver transplant procedures performed in the United States in the years 1995 to 2005.   *(Source: U.S. Department of Health and Human Services)*

Year	1995	1996	1997	1998
Transplants, L	3818	3918	4005	4356

Year	1999	2000	2001	2002
Transplants, L	4586	4816	5177	5326

Year	2003	2004	2005
Transplants, L	5671	6168	6444

(a) Use a graphing utility to create a scatter plot of the data. Let t represent the year, with $t = 5$ corresponding to 1995.

(b) Use the *regression* feature of a graphing utility to find a quadratic model for the data.

(c) Use a graphing utility to graph the model from part (b) in the same viewing window as the scatter plot of the data.

(d) Use the graph of the model from part (c) to predict the number of liver transplant procedures performed in 2008.

(T) 61. Regression Problem   Let x be the number of units (in tens of thousands) that a computer company produces and let $p(x)$ be the profit (in hundreds of thousands of dollars). The table shows the profits for different levels of production.

Units, x	2	4	6	8	10
Profit, $p(x)$	270.5	307.8	320.1	329.2	325.0

Units, x	12	14	16	18	20
Profit, $p(x)$	311.2	287.8	254.8	212.2	160.0

(a) Use a graphing utility to create a scatter plot of the data.

(b) Use the *regression* feature of a graphing utility to find a quadratic model for $p(x)$.

(c) Use a graphing utility to graph your model for $p(x)$ with the scatter plot of the data.

(d) Find the vertex of the graph of the model from part (c). Interpret its meaning in the context of the problem.

(e) With these data and this model, the profit begins to decrease. Discuss how it is possible for production to increase and profit to decrease.

(T) 62. Regression Problem   Let x be the angle (in degrees) at which a baseball is hit with no spin at an initial speed of 40 meters per second and let $d(x)$ be the distance (in meters) the ball travels. The table shows the distances for the different angles at which the ball is hit.   *(Source: The Physics of Sports)*

Angle, x	10	15	30	36	42
Distance, $d(x)$	58.3	79.7	126.9	136.6	140.6

Angle, x	44	45	48	54	60
Distance, $d(x)$	140.9	140.9	139.3	132.5	120.5

(a) Use a graphing utility to create a scatter plot of the data.

(b) Use the *regression* feature of a graphing utility to find a quadratic model for $d(x)$.

(c) Use a graphing utility to graph your model for $d(x)$ with the scatter plot of the data.

(d) Find the vertex of the graph of the model from part (c). Interpret its meaning in the context of the problem.

63. Write the quadratic function

$$f(x) = ax^2 + bx + c$$

in standard form to verify that the vertex occurs at

$$\left(-\frac{b}{2a},\ f\left(-\frac{b}{2a}\right)\right).$$

Polynomial Functions of Higher Degree

- Sketch a transformation of a monomial function.
- Determine right-hand and left-hand behavior of graphs of polynomial functions.
- Find the real zeros of a polynomial function.
- Sketch the graph of a polynomial function.
- Use a polynomial model to solve an application problem.

Graphs of Polynomial Functions

In this section, you will study basic characteristics of the graphs of polynomial functions. The first characteristic is that the graph of a polynomial function is **continuous.** Essentially, this means that the graph of a polynomial function has no breaks, as shown in Figure 3.11(a). Functions with graphs that are not continuous are not polynomial functions, as shown in Figure 3.11(b).

> **STUDY TIP**
>
> The graphs of polynomial functions of degree greater than 2 are more complicated than those of degree 0, 1, or 2. However, using the characteristics presented in this section, together with point plotting, intercepts, and symmetry, you should be able to make reasonably accurate sketches *by hand.* Of course, if you have a graphing utility, the task is easier.

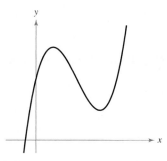

(a) Continuous

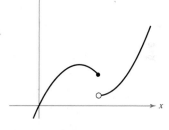

(b) Not continuous

FIGURE 3.11

The second characteristic is that the graph of a polynomial function has only smooth, rounded turns, as shown in Figure 3.12(a). A polynomial function cannot have a sharp turn, as shown in Figure 3.12(b).

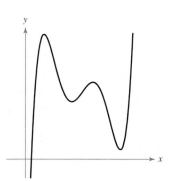

(a) Polynomial functions have smooth, rounded turns.

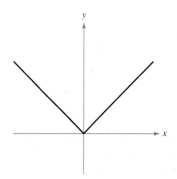

(b) Polynomial functions cannot have sharp turns.

FIGURE 3.12

The polynomial functions that have the simplest graphs are monomial functions of the form $f(x) = x^n$, where n is an integer greater than zero. From Figure 3.13, you can see that when n is *even*, the graph is similar to the graph of $f(x) = x^2$, and when n is *odd*, the graph is similar to the graph of $f(x) = x^3$. Moreover, the greater the value of n, the flatter the graph near the origin.

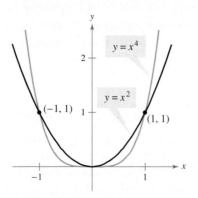

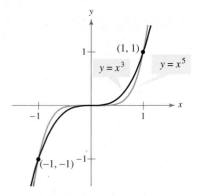

(a) If n is even, the graph of $y = x^n$ touches the axis at the x-intercept.

(b) If n is odd, the graph of $y = x^n$ crosses the axis at the x-intercept.

FIGURE 3.13

Example 1　**Sketching Transformations of Monomial Functions**

Sketch the graph of each function.

a. $f(x) = -x^5$　　**b.** $h(x) = (x + 1)^4$

SOLUTION

a. Because the degree of $f(x) = -x^5$ is odd, its graph is similar to the graph of $y = x^3$. In Figure 3.14(a), note that the negative coefficient has the effect of reflecting the graph about the x-axis.

b. The graph of $h(x) = (x + 1)^4$ is a left shift, by one unit, of the graph of $y = x^4$, as shown in Figure 3.14(b).

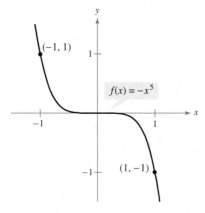

(a)

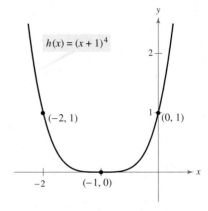

(b)

FIGURE 3.14

✓ **CHECKPOINT 1**

Sketch the graph of $f(x) = (x - 3)^3$. ∎

The Leading Coefficient Test

In Example 1, note that both graphs eventually rise or fall without bound as x moves to the right. Whether the graph of a polynomial function eventually rises or falls can be determined by the function's degree (even or odd) and by its leading coefficient (positive or negative), as indicated in the **Leading Coefficient Test.**

DISCOVERY

For each function below, identify the degree of the function and whether it is even or odd. Identify the leading coefficient, and whether the leading coefficient is positive or negative. Use a graphing utility to graph each function. Describe the relationship between the function's degree and the sign of its leading coefficient and the right-hand and left-hand behavior of the graph of the function.

a. $y = x^3 - 2x^2 - x + 1$

b. $y = 2x^5 + 2x^2 - 5x + 1$

c. $y = -2x^5 - x^2 + 5x + 3$

d. $y = -x^3 + 5x - 2$

e. $y = 2x^2 + 3x - 4$

f. $y = x^4 - 3x^2 + 2x - 1$

g. $y = x^2 + 3x + 2$

h. $y = -x^6 - x^2 - 5x + 4$

Leading Coefficient Test

As x moves without bound to the left or to the right, the graph of the polynomial function given by

$$f(x) = a_n x^n + \cdots + a_1 x + a_0$$

eventually rises or falls in the following manner.

1. When n is *odd:*

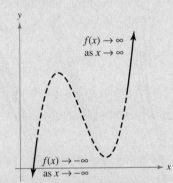

If the leading coefficient is positive ($a_n > 0$), the graph falls to the left and rises to the right.

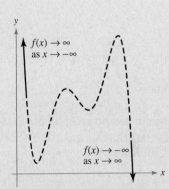

If the leading coefficient is negative ($a_n < 0$), the graph rises to the left and falls to the right.

2. When n is *even:*

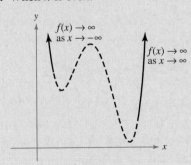

If the leading coefficient is positive ($a_n > 0$), the graph rises to the left and right.

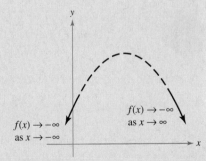

If the leading coefficient is negative ($a_n < 0$), the graph falls to the left and right.

The dashed portions of the graphs indicate that the test determines *only* the right-hand and left-hand behavior of the graph.

The notation "$f(x) \to -\infty$ as $x \to -\infty$" indicates that the graph falls to the left. The notation "$f(x) \to \infty$ as $x \to \infty$" indicates that the graph rises to the right.

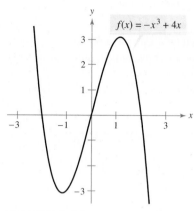

$f(x) = -x^3 + 4x$

FIGURE 3.15

| Example 2 | **Applying the Leading Coefficient Test** |

Describe the right-hand and left-hand behavior of the graph of $f(x) = -x^3 + 4x$.

SOLUTION   Because the degree is odd and the leading coefficient is negative, the graph rises to the left and falls to the right, as shown in Figure 3.15.

✓ **CHECKPOINT 2**

Describe the right-hand and left-hand behavior of the graph of
$f(x) = -2x^4 + x$. ■

In Example 2, note that the Leading Coefficient Test tells you only whether the graph *eventually* rises or falls to the right or left. Other characteristics of the graph, such as intercepts, relative minima, and relative maxima, must be determined by other tests. For example, later you will use the number of real zeros of a polynomial function to determine how many times the graph of the function crosses the x-axis.

STUDY TIP

The function in Example 3 part (a) is a fourth-degree polynomial function. This can also be referred to as a *quartic* function.

| Example 3 | **Applying the Leading Coefficient Test** |

Describe the right-hand and left-hand behavior of the graph of each function.

a. $f(x) = x^4 - 5x^2 + 4$ **b.** $f(x) = x^5 - x$

SOLUTION

a. Because the degree is even and the leading coefficient is positive, the graph rises to the left and right, as shown in Figure 3.16(a).

b. Because the degree is odd and the leading coefficient is positive, the graph falls to the left and rises to the right, as shown in Figure 3.16(b).

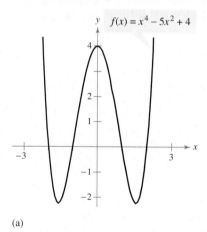

(a)   (b)

$f(x) = x^4 - 5x^2 + 4$

$f(x) = x^5 - x$

FIGURE 3.16

✓ **CHECKPOINT 3**

Describe the right-hand and left-hand behavior of the graph of
$f(x) = x^4 - 4x^2$. ■

Real Zeros of Polynomial Functions

It can be shown that for a polynomial function f of degree n, the following statements are true. Remember that the **zeros** of a function are the x-values for which the function is zero.

1. The graph of f has, at most, $n - 1$ *turning points*. Turning points are points at which the graph changes from increasing to decreasing or vice versa. For instance, the graph of $f(x) = x^4 - 1$ has at most $4 - 1 = 3$ turning points.

2. The function f has, at most, n real zeros. For instance, the function given by $f(x) = x^4 - 1$ has at most $n = 4$ real zeros. (You will study this result in detail in Section 3.6 on the Fundamental Theorem of Algebra.)

Finding the zeros of polynomial functions is one of the most important problems in algebra. There is a strong interplay between graphical and algebraic approaches to this problem. Sometimes you can use information about the graph of a function to help find its zeros, and in other cases you can use information about the zeros of a function to help sketch its graph.

Real Zeros of Polynomial Functions

If f is a polynomial function and a is a real number, then the following statements are equivalent.

1. $x = a$ is a zero of the function f.

2. $x = a$ is a solution of the polynomial equation $f(x) = 0$.

3. $(x - a)$ is a factor of the polynomial $f(x)$.

4. $(a, 0)$ is an x-intercept of the graph of f.

In the equivalent statements above, notice that finding zeros of polynomial functions is closely related to factoring and finding x-intercepts.

Example 4   Finding Zeros of a Polynomial Function

Find all real zeros of $f(x) = x^3 - x^2 - 2x$.

SOLUTION   By factoring, you obtain the following.

$$f(x) = x^3 - x^2 - 2x \qquad \text{Write original function.}$$
$$= x(x^2 - x - 2) \qquad \text{Remove common monomial factor.}$$
$$= x(x - 2)(x + 1) \qquad \text{Factor completely.}$$

So, the real zeros are $x = 0$, $x = 2$, and $x = -1$, and the corresponding x-intercepts are $(0, 0)$, $(2, 0)$, and $(-1, 0)$, as shown in Figure 3.17. Note that the graph in the figure has two turning points. This is consistent with the fact that the graph of a third-degree polynomial function can have *at most* $3 - 1 = 2$ turning points.

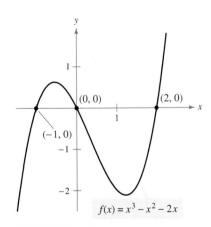

FIGURE 3.17

✓ **CHECKPOINT 4**

Find all real zeros of $f(x) = x^2 - 4$. ∎

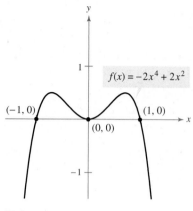

$f(x) = -2x^4 + 2x^2$

$(-1, 0)$ $(1, 0)$ $(0, 0)$

FIGURE 3.18

Example 5   **Finding Zeros of a Polynomial Function**

Find all real zeros of $f(x) = -2x^4 + 2x^2$.

SOLUTION   In this case, the polynomial factors as follows.

$$f(x) = -2x^2(x^2 - 1) = -2x^2(x - 1)(x + 1)$$

So, the real zeros are $x = 0$, $x = 1$, and $x = -1$, and the corresponding x-intercepts are $(0, 0)$, $(1, 0)$, and $(-1, 0)$, as shown in Figure 3.18. Note that the graph in the figure has three turning points, which is consistent with the fact that the graph of a fourth-degree polynomial function can have *at most* three turning points.

✓ **CHECKPOINT 5**

Find all real zeros of $f(x) = x^3 - x$. ∎

In Example 5, the real zero arising from $-2x^2 = 0$ is called a **repeated zero.** In general, a factor $(x - a)^k$ yields a repeated zero $x = a$ of **multiplicity** k. If k is odd, the graph *crosses* the x-axis at $x = a$. If k is even, the graph *touches* (but does not cross) the x-axis at $x = a$. This is illustrated in Figure 3.18.

Example 6   **Sketching the Graph of a Polynomial Function**

Sketch the graph of $f(x) = 3x^4 - 4x^3$.

SOLUTION   Because the leading coefficient is positive and the degree is even, you know that the graph eventually rises to the left and right, as shown in Figure 3.19(a). By factoring $f(x) = 3x^4 - 4x^3$ as $f(x) = x^3(3x - 4)$, you can see that the zeros of f are $x = 0$ and $x = \frac{4}{3}$ (both of odd multiplicity). So, the x-intercepts occur at $(0, 0)$ and $\left(\frac{4}{3}, 0\right)$. To sketch the graph by hand, find a few additional points, as shown in the table. Then plot the points and draw a continuous curve through the points to complete the graph, as shown in Figure 3.19(b). If you are unsure of the shape of a portion of a graph, plot some additional points.

x	$f(x)$
-1	7
0.5	-0.3125
1	-1
1.5	1.6875

✓ **CHECKPOINT 6**

Sketch the graph of
$f(x) = 2x^3 - 3x^2$. ∎

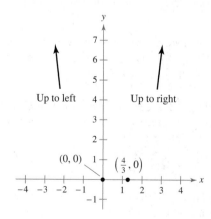

(a)

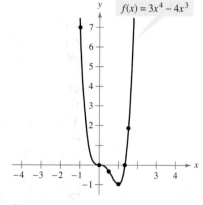

(b)

FIGURE 3.19

Application

Example 7 Charitable Contributions Revisited

Example 7 in Section 3.1 discussed the model

$$P(x) = 0.0014x^2 - 0.1529x + 5.855, \quad 5 \le x \le 100$$

where P is the percent of annual income given to charities and x is the annual income (in thousands of dollars). Note that this model gives the charitable contributions as a *percent* of annual income. To find the *amount* that a family gives to charity, you can multiply the given model by the income $1000x$ (and divide by 100 to change from percent to decimal form) to obtain

$$A(x) = 0.014x^3 - 1.529x^2 + 58.55x, \quad 5 \le x \le 100$$

where A represents the amount of charitable contributions (in dollars). Sketch the graph of this function and use the graph to estimate the annual salary of a family that gives $1000 a year to charities.

SOLUTION Because the leading coefficient is positive and the degree is odd, you know that the graph eventually falls to the left and rises to the right. To sketch the graph by hand, find a few points, as shown in the table. Then plot the points and complete the graph, as shown in Figure 3.20.

x	5	25	45	65	86	100
$A(x)$	256.28	726.88	814.28	1190.48	2527.48	4565.00

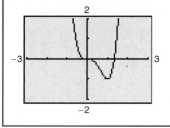

From the graph you can see that an annual contribution of $1000 corresponds to an annual income of about $59,000.

✓ CHECKPOINT 7

The median prices P (in thousands of dollars) of new privately owned homes in housing developments from 1998 to 2008 can be approximated by the model

$$P(t) = 0.139t^3 - 4.42t^2 + 51.1t - 39$$

where t represents the year, with $t = 8$ corresponding to 1998. Sketch the graph of this function and use the graph to estimate the year in which the median price of a new privately owned home was about $195,000. ▪

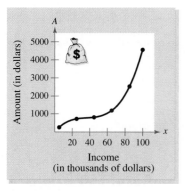

FIGURE 3.20

CONCEPT CHECK

1. Write a function whose graph is a downward shift, by one unit, and a reflection in the x-axis, of the graph of $y = x^4$.

2. The graph of a fifth-degree polynomial function rises to the left. Describe the right-hand behavior of the graph.

3. Name a zero of the function f given that $(x - 5)$ is a factor of the polynomial $f(x)$.

4. Does the graph of every function with real zeros cross the x-axis? Explain.

Skills Review 3.2

The following warm-up exercises involve skills that were covered in earlier sections. You will use these skills in the exercise set for this section. For additional help, review Sections 0.6, 1.3, 1.4, and 1.5.

In Exercises 1–6, factor the expression completely.

1. $12x^2 + 7x - 10$

2. $25x^3 - 60x^2 + 36x$

3. $12z^4 + 17z^3 + 5z^2$

4. $y^3 + 125$

5. $x^3 + 3x^2 - 4x - 12$

6. $x^3 + 2x^2 + 3x + 6$

In Exercises 7–10, find all real solutions of the equation.

7. $5x^2 + 8 = 0$

8. $x^2 - 6x + 4 = 0$

9. $4x^2 + 4x - 11 = 0$

10. $x^4 - 18x^2 + 81 = 0$

Exercises 3.2

See www.CalcChat.com for worked-out solutions to odd-numbered exercises.

In Exercises 1–8, match the polynomial function with its graph. [The graphs are labeled (a), (b), (c), (d), (e), (f), (g), and (h).]

1. $f(x) = \frac{1}{2}(x^3 + 2x^2 - 3x)$

2. $f(x) = x^2 - 2x$

3. $f(x) = \frac{1}{3}x^4 - x^2$

4. $f(x) = -3x^4 - 4x^3$

5. $f(x) = 3x^3 - 9x + 1$

6. $f(x) = x^5 - 5x^3 + 4x$

7. $f(x) = -\frac{1}{3}x^3 + x - \frac{2}{3}$

8. $f(x) = -x^5 + 5x^3 - 4x$

(a)
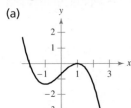

(b)
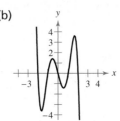

(c)
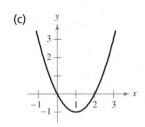

(d)
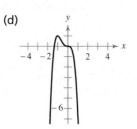

(e)
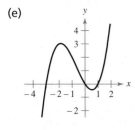

(f)
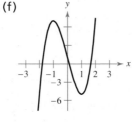

(g)
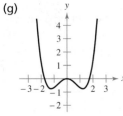

(h)
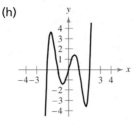

In Exercises 9–12, use the graph of $y = x^3$ to sketch the graph of the function.

9. $f(x) = x^3 - 2$

10. $f(x) = (x + 3)^3$

11. $f(x) = (x + 1)^3 - 4$

12. $f(x) = -(x - 2)^3 + 2$

In Exercises 13–16, use the graph of $y = x^4$ to sketch the graph of the function.

13. $f(x) = (x + 3)^4$

14. $f(x) = x^4 - 4$

15. $f(x) = 3 - x^4$

16. $f(x) = \frac{1}{2}(x - 1)^4$

In Exercises 17–26, describe the right-hand and left-hand behavior of the graph of the polynomial function.

17. $f(x) = -x^3 + 1$

18. $f(x) = \frac{1}{3}x^3 + 5x$

19. $g(x) = 6 - 4x^2 + x - 3x^5$

20. $f(x) = 2x^5 - 5x + 7.5$

21. $f(x) = 4x^8 - 2$

22. $h(x) = 1 - x^6$

23. $f(x) = 2 + 5x - x^2 - x^3 + 2x^4$

24. $f(x) = \dfrac{3x^4 - 2x + 5}{4}$

25. $h(t) = -\frac{2}{3}(t^2 - 5t + 3)$

26. $f(s) = -\frac{7}{8}(s^3 + 5s^2 - 7s + 1)$

In Exercises 27–30, determine (a) the maximum number of turning points of the graph of the function and (b) the maximum number of real zeros of the function.

27. $f(x) = x^2 - 4x + 1$ **28.** $f(x) = -3x^4 + 1$

29. $f(x) = -x^5 + x^4 - x$ **30.** $f(x) = 2x^3 + x^2 + 1$

Algebraic and Graphical Approaches   In Exercises 31–46, find all real zeros of the function algebraically. Then use a graphing utility to confirm your results.

31. $f(x) = 9 - x^2$ **32.** $f(x) = x^2 - 25$

33. $h(t) = t^2 + 8t + 16$ **34.** $f(x) = x^2 - 12x + 36$

35. $f(x) = \frac{1}{3}x^2 + \frac{1}{3}x - \frac{2}{3}$ **36.** $f(x) = \frac{1}{2}x^2 + \frac{5}{2}x - \frac{3}{2}$

37. $f(x) = 2x^2 + 4x + 6$

38. $g(x) = -5(x^2 + 2x - 4)$

39. $f(t) = t^3 - 4t^2 + 4t$

40. $f(x) = x^4 - x^3 - 20x^2$

41. $g(t) = \frac{1}{2}t^4 - \frac{1}{2}$

42. $f(x) = \frac{1}{3} - \frac{1}{3}x^2$

43. $f(x) = 2x^4 - 2x^2 - 40$

44. $g(t) = t^5 - 6t^3 + 9t$

45. $f(x) = x^3 - 3x^2 + 2x - 6$

46. $f(x) = x^3 - 4x^2 - 25x + 100$

Analyzing a Graph   In Exercises 47–58, analyze the graph of the function algebraically and use the results to sketch the graph *by hand*. Then use a graphing utility to confirm your sketch.

47. $f(x) = \frac{2}{3}x + 5$ **48.** $h(x) = -\frac{3}{4}x + 2$

49. $f(t) = \frac{1}{2}(t^2 - 4t - 1)$

50. $g(x) = -x^2 + 10x - 16$

51. $f(x) = 4x^2 - x^3$ **52.** $f(x) = 1 - x^3$

53. $f(x) = x^3 - 9x$ **54.** $f(x) = \frac{1}{4}x^4 - 2x^2$

55. $g(t) = -\frac{1}{4}(t - 2)^2(t + 2)^2$

56. $f(x) = x(x - 2)^2(x + 1)$

57. $f(x) = 1 - x^6$ **58.** $g(x) = 1 - (x + 1)^6$

59. Modeling Polynomials   Sketch the graph of a polynomial function that is of fourth degree, has a zero of multiplicity 2, and has a negative leading coefficient. Sketch another graph under the same conditions but with a positive leading coefficient.

60. Modeling Polynomials   Sketch the graph of a polynomial function that is of fifth degree, has a zero of multiplicity 2, and has a negative leading coefficient. Sketch another graph under the same conditions but with a positive leading coefficient.

61. Modeling Polynomials   Determine the equation of the fourth-degree polynomial function f whose graph is shown.

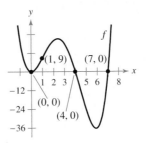

62. Modeling Polynomials   Determine the equation of the third-degree polynomial function g whose graph is shown.

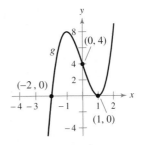

(T) **63. Credit Cards**   The numbers of active American Express cards C (in millions) in the years 1997 to 2006 are shown in the table.   *(Source: American Express)*

Year	1997	1998	1999	2000	2001
Cards, C	42.7	42.7	46.0	51.7	55.2

Year	2002	2003	2004	2005	2006
Cards, C	57.3	60.5	65.4	71.0	78.0

(a) Use a graphing utility to create a scatter plot of the data. Let t represent the year, with $t = 7$ corresponding to 1997.

(b) Use what you know about end behavior and the scatter plot from part (a) to predict the sign of the leading coefficient of a quartic model for C.

(c) Use the *regression* feature of a graphing utility to find a quartic model for C. Does your model agree with your answer from part (b)?

(d) Use a graphing utility to graph the model from part (c). Use the graph to predict the year in which the number of active American Express cards would be about 92 million. Is your prediction reasonable?

(T) **64. Population** The immigrant population P (in millions) living in the United States at the beginning of each decade from 1900 to 2000 is shown in the table. *(Source: Center of Immigration Studies)*

Year	1900	1910	1920	1930
Population, P	10.3	13.5	13.9	14.2

Year	1940	1950	1960	1970
Population, P	11.6	10.3	9.7	9.6

Year	1980	1990	2000
Population, P	14.1	19.8	30.0

(a) Use a graphing utility to create a scatter plot of the data. Let $t = 0$ correspond to 1900.

(b) Use what you know about end behavior and the scatter plot from part (a) to predict the sign of the leading coefficient of a cubic model for P.

(c) Use the *regression* feature of a graphing utility to find a cubic model for P. Does your model agree with your answer from part (b)?

(d) Use a graphing utility to graph the model from part (c). Use the graph to predict the year in which the immigrant population will be about 45 million. Is your prediction reasonable?

65. Advertising Expenses The total revenue R (in millions of dollars) for a soft-drink company is related to its advertising expenses by the function

$$R = \frac{1}{50,000}(-x^3 + 600x^2), \quad 0 \le x \le 400$$

where x is the amount spent on advertising (in tens of thousands of dollars). Use the graph of R to estimate the point on the graph at which the function is increasing most rapidly. This point is called the *point of diminishing returns* because any expenditure above this amount will yield less return per dollar invested in advertising.

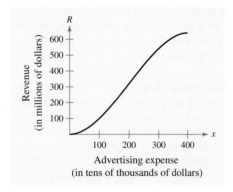

66. Advertising Expenses The total revenue R (in millions of dollars) for a hotel corporation is related to its advertising expenses by the function

$$R = -0.148x^3 + 4.889x^2 - 17.778x + 125.185,$$
$$0 \le x \le 20$$

where x is the amount spent on advertising (in millions of dollars). Use the graph of R to estimate the point on the graph at which the function is increasing most rapidly. This point is called the *point of diminishing returns* because any expenditure above this amount will yield less return per dollar invested in advertising.

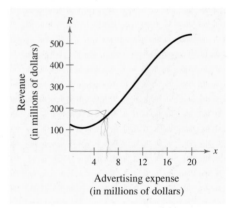

67. Maximum Value An open box with locking tabs is to be made from a square piece of material 24 inches on a side. This is to be done by cutting equal squares from the corners and folding along the dashed lines shown in the figure. Verify that the volume of the box is given by the function $V(x) = 8x(6 - x)(12 - x)$. Determine the domain of the function V. Then sketch a graph of the function and estimate the value of x for which $V(x)$ is maximum.

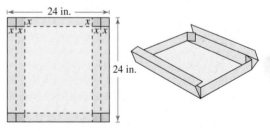

(T) **68. Comparing Graphs** Use a graphing utility to graph the functions given by $f(x) = x^2$, $g(x) = x^4$, and $h(x) = x^6$. Do the three functions have a common shape? Are their graphs identical? Why or why not?

(T) **69. Comparing Graphs** Use a graphing utility to graph the functions given by $f(x) = x^3$, $g(x) = x^5$, and $h(x) = x^7$. Do the three functions have a common shape? Are their graphs identical? Why or why not?

Section 3.3

Polynomial Division

- Divide one polynomial by a second polynomial using long division.
- Simplify a rational expression using long division.
- Use synthetic division to divide two polynomials.
- Use the Remainder Theorem and synthetic division to evaluate a polynomial.
- Use the Factor Theorem to factor a polynomial.
- Use polynomial division to solve an application problem.

Long Division of Polynomials

In this section, you will study two procedures for *dividing* polynomials. These procedures are especially valuable in factoring polynomials and finding the zeros of polynomial functions. To begin, suppose you are given the graph of

$$f(x) = 6x^3 - 19x^2 + 16x - 4.$$

Notice that a zero of f occurs at $x = 2$, as shown in Figure 3.21. Because $x = 2$ is a zero of the polynomial function f, you know that $(x - 2)$ is a factor of $f(x)$. This means that there exists a second-degree polynomial $q(x)$ such that

$$f(x) = (x - 2) \cdot q(x).$$

To find $q(x)$, you can use **long division,** as illustrated in Example 1.

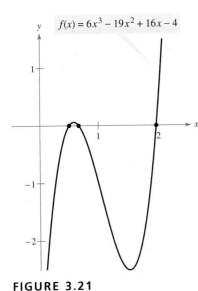

$f(x) = 6x^3 - 19x^2 + 16x - 4$

FIGURE 3.21

Example 1   Long Division of Polynomials

Divide the polynomial $6x^3 - 19x^2 + 16x - 4$ by $x - 2$, and use the result to factor the polynomial completely.

SOLUTION

$$
\require{enclose}
\begin{array}{r}
6x^2 - 7x + 2 \\[-3pt]
x - 2 \enclose{longdiv}{6x^3 - 19x^2 + 16x - 4} \\
\end{array}
$$

$6x^3 - 12x^2$	Multiply: $6x^2$ by $x - 2$.
$-7x^2 + 16x$	Subtract and bring down $16x$.
$-7x^2 + 14x$	Multiply: $-7x$ by $x - 2$
$2x - 4$	Subtract and bring down -4.
$2x - 4$	Multiply: 2 by $x - 2$.
0	Subtract.

From this division, you can conclude that

$$6x^3 - 19x^2 + 16x - 4 = (x - 2)(6x^2 - 7x + 2)$$

and by factoring the quadratic $6x^2 - 7x + 2$, you have

$$6x^3 - 19x^2 + 16x - 4 = (x - 2)(2x - 1)(3x - 2).$$

✓ **CHECKPOINT 1**

Divide $x^3 - 6x^2 + 5x + 12$ by $x - 4$, and use the result to factor the polynomial completely. ■

Note that the factorization shown in Example 1 agrees with the graph shown in Figure 3.21 in that the three x-intercepts occur at $x = 2$, $x = \frac{1}{2}$, and $x = \frac{2}{3}$.

In Example 1, $x - 2$ is a factor of the polynomial $6x^3 - 19x^2 + 16x - 4$, and the long division process produces a remainder of zero. Often, long division will produce a nonzero remainder. For instance, when you divide $x^2 + 3x + 5$ by $x + 1$, you obtain the following.

$$
\begin{array}{r}
x + 2 \quad \longleftarrow \text{Quotient}\\
x + 1 \overline{)\ x^2 + 3x + 5} \quad \longleftarrow \text{Dividend}\\
\underline{x^2 + \ x}\\
2x + 5\\
\underline{2x + 2}\\
3 \quad \longleftarrow \text{Remainder}
\end{array}
$$

Divisor $\Longrightarrow$

In fractional form, you can write this result as follows.

$$
\underbrace{\frac{\overbrace{x^2 + 3x + 5}^{\text{Dividend}}}{\underbrace{x + 1}_{\text{Divisor}}}} = \overbrace{x + 2}^{\text{Quotient}} + \overset{\text{Remainder}}{\frac{3}{\underbrace{x + 1}_{\text{Divisor}}}}
$$

This implies that

$$
x^2 + 3x + 5 = (x + 1)(x + 2) + 3 \qquad \text{Multiply each side by } (x + 1).
$$

which illustrates the following well-known theorem called the **Division Algorithm.**

The Division Algorithm

If $f(x)$ and $d(x)$ are polynomials such that $d(x) \neq 0$, and the degree of $d(x)$ is less than or equal to the degree of $f(x)$, there exist unique polynomials $q(x)$ and $r(x)$ such that

$$
f(x) = d(x)q(x) + r(x)
$$

$$
\underset{\text{Dividend}}{\uparrow} \quad \underset{\text{Divisor}}{\uparrow} \quad \underset{\text{Quotient}}{\uparrow} \quad \underset{\text{Remainder}}{\uparrow}
$$

where $r(x) = 0$ or the degree of $r(x)$ is less than the degree of $d(x)$. If the remainder $r(x)$ is zero, $d(x)$ **divides evenly** into $f(x)$.

The Division Algorithm can also be written as

$$
\frac{f(x)}{d(x)} = q(x) + \frac{r(x)}{d(x)}.
$$

In the Division Algorithm, the rational expression $f(x)/d(x)$ is **improper** because the degree of $f(x)$ is greater than or equal to the degree of $d(x)$. On the other hand, the rational expression $r(x)/d(x)$ is **proper** because the degree of $r(x)$ is less than the degree of $d(x)$.

Before you apply the Division Algorithm, follow these steps.

1. Write the dividend and divisor in descending powers of the variable.

2. Insert placeholders with zero coefficients for missing powers of the variable.

Example 2   Long Division of Polynomials

Divide $x^3 - 1$ by $x - 1$.

SOLUTION   Because there is no x^2-term or x-term in the dividend, you need to line up the subtraction by using zero coefficients (or leaving spaces) for the missing terms.

$$
\begin{array}{r}
x^2 + x + 1 \\
x - 1 \overline{\smash{)}\ x^3 + 0x^2 + 0x - 1} \\
\underline{x^3 - x^2} \\
x^2 + 0x \\
\underline{x^2 - x} \\
x - 1 \\
\underline{x - 1} \\
0
\end{array}
$$

Insert $0x^2$ and $0x$.
Multiply x^2 by $x - 1$.
Subtract and bring down $0x$.
Multiply x by $x - 1$.
Subtract and bring down -1.
Multiply 1 by $x - 1$.
Subtract.

So, $x - 1$ divides evenly into $x^3 - 1$ and you can write

$$\frac{x^3 - 1}{x - 1} = x^2 + x + 1.$$

✓ CHECKPOINT 2

Divide $x^3 + 8$ by $x + 2$. ■

You can check the result of a division problem by multiplying. For instance, in Example 2, try checking that $(x - 1)(x^2 + x + 1) = x^3 - 1$.

Example 3   Long Division of Polynomials

Divide $2x^4 + 4x^3 - 5x^2 + 3x - 2$ by $x^2 + 2x - 3$.

SOLUTION

$$
\begin{array}{r}
2x^2 + 1 \\
x^2 + 2x - 3 \overline{\smash{)}\ 2x^4 + 4x^3 - 5x^2 + 3x - 2} \\
\underline{2x^4 + 4x^3 - 6x^2} \\
x^2 + 3x - 2 \\
\underline{x^2 + 2x - 3} \\
x + 1
\end{array}
$$

Multiply $2x^2$ by $x^2 + 2x - 3$.
Subtract and bring down $3x - 2$.
Multiply 1 by $x^2 + 2x - 3$.
Subtract.

✓ CHECKPOINT 3

Divide

$$5x^4 + 10x^3 - 7x^2 + 28x - 39$$

by

$$x^2 + 2x - 4. \quad ■$$

Note that the first subtraction eliminated two terms from the dividend. When this happens, the quotient skips a term. So, you can write

$$\frac{2x^4 + 4x^3 - 5x^2 + 3x - 2}{x^2 + 2x - 3} = 2x^2 + 1 + \frac{x + 1}{x^2 + 2x - 3}.$$

Synthetic Division

There is a nice shortcut for long division of polynomials when dividing by divisors of the form $x - k$. This shortcut is called **synthetic division.** We summarize the pattern for synthetic division of a cubic polynomial as follows. (The pattern for higher-degree polynomials is similar.)

Synthetic Division (for a Cubic Polynomial)

To divide $ax^3 + bx^2 + cx + d$ by $x - k$, use the following pattern.

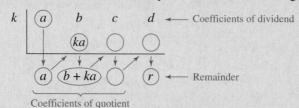

Vertical pattern: Add terms in columns.
Diagonal pattern: Multiply results by k.

Example 4 Using Synthetic Division

Use synthetic division to divide $x^4 - 10x^2 - 2x + 4$ by $x + 3$.

SOLUTION   You should set up the array as follows. Note that a zero is included for the missing x^3-term in the dividend.

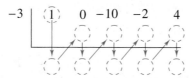

Then, use the synthetic division pattern by adding terms in columns and multiplying the results by -3.

Divisor: $x + 3$ Dividend: $x^4 - 10x^2 - 2x + 4$

$$
\begin{array}{r|rrrrr}
-3 & 1 & 0 & -10 & -2 & 4 \\
   &   & -3 & 9 & 3 & -3 \\
\hline
   & 1 & -3 & -1 & 1 & 1
\end{array}
$$

$-3(1)$ $-3(-3)$ $-3(-1)$ $-3(1)$ Remainder: 1

Quotient: $x^3 - 3x^2 - x + 1$

So, you have $\dfrac{x^4 - 10x^2 - 2x + 4}{x + 3} = x^3 - 3x^2 - x + 1 + \dfrac{1}{x + 3}$.

✓ CHECKPOINT 4

Use synthetic division to divide $2x^3 - 7x^2 - 80$ by $x - 5$. ∎

Remainder and Factor Theorems

The remainder obtained in the synthetic division process has an important interpretation, as described in the Remainder Theorem.

The Remainder Theorem

If a polynomial $f(x)$ is divided by $x - k$, the remainder is

$r = f(k)$.

The Remainder Theorem tells you that synthetic division can be used to evaluate a polynomial function. That is, to evaluate a polynomial function f at $x = k$, divide $f(x)$ by $x - k$. The remainder will be $f(k)$, as illustrated in Example 5.

TECHNOLOGY

(T) Remember, you can also evaluate a function with your graphing utility by entering the function in the equation editor and using the *table* feature in ASK mode. For instructions on how to use the *table* feature, see Appendix A; for specific keystrokes, go to the text website at *college.hmco.com/info/ larsonapplied*.

Example 5   Using the Remainder Theorem

Use the Remainder Theorem to evaluate the following function when $x = -2$.

$$f(x) = 3x^3 + 8x^2 + 5x - 7$$

SOLUTION   Using synthetic division, you obtain the following.

$$
\begin{array}{r|rrrr}
-2 & 3 & 8 & 5 & -7 \\
   &   & -6 & -4 & -2 \\
\hline
   & 3 & 2 & 1 & -9
\end{array}
$$

Because the remainder is $r = -9$, you can conclude that

$$f(-2) = -9.$$

This means that $(-2, -9)$ is a point on the graph of f. You can check this by substituting $x = -2$ in the original function.

CHECK

$$f(-2) = 3(-2)^3 + 8(-2)^2 + 5(-2) - 7$$
$$= 3(-8) + 8(4) - 10 - 7 = -9$$

✓ **CHECKPOINT 5**

Use the Remainder Theorem to evaluate $f(x) = 4x^3 + 6x^2 + 4x + 5$ when $x = -1$. ■

Another important theorem is the Factor Theorem, which is stated below.

Factor Theorem

A polynomial $f(x)$ has a factor $(x - k)$ if and only if $f(k) = 0$.

You can think of the Factor Theorem as stating that if $(x - k)$ is a factor of $f(x)$, then $f(k) = 0$. Conversely, if $f(k) = 0$, then $(x - k)$ is a factor of $f(x)$.

Example 6   **Factoring a Polynomial: Repeated Division**

Show that $(x - 2)$ and $(x + 3)$ are factors of the polynomial

$$f(x) = 2x^4 + 7x^3 - 4x^2 - 27x - 18.$$

Then find the remaining factors of $f(x)$.

SOLUTION   Using synthetic division with the factor $(x - 2)$, you obtain the following.

$$
\begin{array}{r|rrrrr}
2 & 2 & 7 & -4 & -27 & -18 \\
  &   & 4 & 22 & 36 & 18 \\
\hline
  & 2 & 11 & 18 & 9 & 0
\end{array}
$$

⟹ 0 remainder, so $f(2) = 0$ and $(x - 2)$ is a factor.

Take the result of this division and perform synthetic division again using the factor $(x + 3)$.

$$
\begin{array}{r|rrrr}
-3 & 2 & 11 & 18 & 9 \\
   &   & -6 & -15 & -9 \\
\hline
   & 2 & 5 & 3 & 0
\end{array}
$$

⟹ 0 remainder, so $f(-3) = 0$ and $(x + 3)$ is a factor.

Quadratic: $2x^2 + 5x + 3$

Because the resulting quadratic expression factors as

$$2x^2 + 5x + 3 = (2x + 3)(x + 1)$$

the complete factorization of $f(x)$ is

$$f(x) = (x - 2)(x + 3)(2x + 3)(x + 1).$$

Note that this factorization implies that f has four real zeros:

$$2, -3, -\tfrac{3}{2}, \text{ and } -1.$$

This is confirmed by the graph of f, which is shown in Figure 3.22.

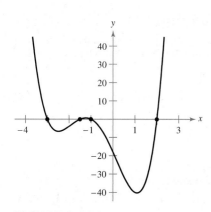

FIGURE 3.22

✓ **CHECKPOINT 6**

Show that $(x + 2)$ and $(x + 4)$ are factors of the polynomial $f(x) = x^4 + 6x^3 + 7x^2 - 6x - 8$. Then find the remaining factors of $f(x)$. ▪

Uses of The Remainder in Synthetic Division

The remainder r obtained in the synthetic division of $f(x)$ by $x - k$ provides the following information.

1. The remainder r gives the value of f at $x = k$. That is, $r = f(k)$.

2. If $r = 0$, $(x - k)$ is a factor of $f(x)$.

3. If $r = 0$, $(k, 0)$ is an x-intercept of the graph of f.

Throughout this text, the importance of developing several problem-solving strategies is emphasized. In the exercises for this section, try using more than one strategy to solve several of the exercises. For instance, if you find that $x - k$ divides evenly into $f(x)$ (with no remainder), try sketching the graph of f. You should find that $(k, 0)$ is an x-intercept of the graph.

Application

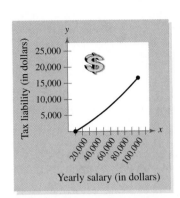

FIGURE 3.23

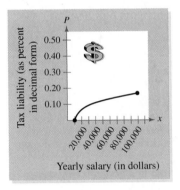

FIGURE 3.24

Example 7   **Tax Liability**

The 2005 federal income tax liability for an employee who was single and claimed no dependents is given by the function

$$y = 0.00000066x^2 + 0.113x - 1183, \quad 10{,}000 \le x \le 100{,}000$$

where y represents the tax liability (in dollars) and x represents the employee's yearly salary (in dollars) (see Figure 3.23). *(Source: U.S. Department of the Treasury)*

a. Find a function that gives the tax liability as a *percent* of the yearly salary.

b. Graph the function from part (a). What conclusions can you make from the graph?

SOLUTION

a. Because the yearly salary is given by x and the tax liability is given by y, the percent (in decimal form) of yearly salary that the person owes in federal income tax is

$$P = \frac{y}{x}$$

$$= \frac{0.00000066x^2 + 0.113x - 1183}{x}$$

$$= 0.00000066x + 0.113 - \frac{1183}{x}.$$

b. The graph of the function P is shown in Figure 3.24. From the graph you can see that as a person's yearly salary increases, the percent that he or she must pay in federal income tax also increases.

✓ CHECKPOINT 7

Using the function P from part (a) of Example 7, what percent of a $39,000 yearly salary does a person owe in federal income tax? ■

┌─ **CONCEPT CHECK** ─────────────────────────────────┐

1. How should you write the dividend $x^5 - 3x + 10$ to apply the Division Algorithm?

2. Describe and correct the error in using synthetic division to divide $x^3 + 4x^2 - x - 4$ by $x + 4$.

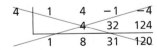

3. A factor of the polynomial $f(x)$ is $(x - 3)$. What is the value of $f(3)$?

4. A fourth-degree polynomial is divided by a first-degree polynomial. What is the degree of the quotient?

└──┘

In Exercises 1–4, write the expression in standard polynomial form.

1. $(x - 1)(x^2 + 2) + 5$

2. $(x^2 - 3)(2x + 4) + 8$

3. $(x^2 + 1)(x^2 - 2x + 3) - 10$

4. $(x + 6)(2x^3 - 3x) - 5$

In Exercises 5–10, factor the polynomial.

5. $x^2 - 4x + 3$

6. $8x^2 - 24x - 80$

7. $3x^2 + 2x - 5$

8. $9x^2 - 24x + 16$

9. $4x^3 - 10x^2 + 6x$

10. $6x^3 + 7x^2 + 2x$

Exercises 3.3

See www.CalcChat.com for worked-out solutions to odd-numbered exercises.

In Exercises 1–18, use long division to divide.

Dividend	Divisor
1. $3x^2 - 7x + 4$	$x - 1$
2. $5x^2 - 17x - 12$	$x - 4$
3. $2x^2 + 10x + 12$	$x + 3$
4. $2x^2 + x - 11$	$x + 5$
5. $2x^3 + 6x^2 - x - 3$	$2x^2 - 1$
6. $3x^3 - 12x^2 - 2x + 8$	$3x^2 - 2$
7. $x^4 + 5x^3 + 6x^2 - x - 2$	$x + 2$
8. $x^4 + 2x^3 - 3x^2 - 8x - 4$	$x^2 - 4$
9. $7x + 3$	$x + 4$
10. $8x - 5$	$2x + 3$
11. $6x^3 + 10x^2 + x + 8$	$2x^2 + 1$
12. $2x^3 - 8x^2 + 3x - 9$	$x - 4$
13. $x^3 - 27$	$x^2 - 1$
14. $x^3 - 9$	$x^2 + 1$
15. $x^3 - 4x^2 + 5x - 2$	$x + 2$
16. $x^3 - x^2 + 2x - 8$	$x - 2$
17. $2x^5 - 8x^3 + 4x - 1$	$x^2 - 2x + 1$
18. $x^5 + 7$	$x^3 - 1$

In Exercises 19–36, use synthetic division to divide.

Dividend	Divisor
19. $2x^3 + 5x^2 - 7x + 20$	$x + 4$
20. $3x^3 - 23x^2 - 12x + 32$	$x - 8$

Dividend	Divisor
21. $4x^3 - 9x + 8x^2 - 18$	$x + 2$
22. $9x^3 - 16x - 18x^2 + 32$	$x - 2$
23. $-x^3 + 75x - 250$	$x + 10$
24. $3x^3 - 16x^2 - 72$	$x - 6$
25. $x^4 - 4x^3 - 7x^2 + 22x + 24$	$x + 3$
26. $6x^4 - 15x^3 - 11x$	$x + 2$
27. $10x^4 - 50x^3 - 800$	$x - 6$
28. $x^5 - 13x^4 - 120x + 80$	$x + 3$
29. $2x^5 - 30x^3 - 37x + 13$	$x - 4$
30. $5x^3$	$x + 3$
31. $-3x^4$	$x - 2$
32. $2x^5$	$x + 3$
33. $5 - 3x + 2x^2 - x^3$	$x + 1$
34. $180x - x^4$	$x - 6$
35. $4x^3 + 16x^2 - 23x - 15$	$x + \frac{1}{2}$
36. $3x^3 - 4x^2 + 5$	$x - \frac{3}{2}$

In Exercises 37–44, write the function in the form

$$f(x) = (x - k)q(x) + r$$

for the given value of k, and demonstrate that $f(k) = r$.

37. $f(x) = x^3 + x^2 - 12x + 20, \quad k = 3$

38. $f(x) = x^3 - 2x^2 - 15x + 7, \quad k = -4$

39. $f(x) = 3x^3 + 2x^2 + 5x - 2, \quad k = \frac{1}{3}$

40. $f(x) = 4x^4 + 6x^3 + 4x^2 - 5x + 13, \quad k = -\frac{1}{2}$

41. $f(x) = x^3 + 2x^2 - 3x - 12$, $k = \sqrt{3}$

42. $f(x) = x^3 + 3x^2 - 7x - 6$, $k = -\sqrt{2}$

43. $f(x) = 2x^3 + x^2 - 14x - 10$, $k = 1 + \sqrt{3}$

44. $f(x) = 3x^3 - 19x^2 + 27x - 7$, $k = 3 - \sqrt{2}$

In Exercises 45–50, use synthetic division to find each function value.

45. $f(x) = 2x^5 - 3x^2 - 4x - 1$

   (a) $f(-2)$ (b) $f(-4)$

   (c) $f(1)$ (d) $f(3)$

46. $g(x) = x^6 - 4x^4 + 3x^2 + 2$

   (a) $g(2)$ (b) $g(-4)$

   (c) $g(7)$ (d) $g(-1)$

47. $f(x) = 2x^3 - 3x^2 + 8x - 14$

   (a) $f(2)$ (b) $f(-1)$

   (c) $f(1.1)$ (d) $f(3)$

48. $f(x) = 3x^4 - 7x^3 + 5x - 12$

   (a) $f(1)$ (b) $f(4)$

   (c) $f(-3)$ (d) $f(-1.2)$

49. $f(x) = 1.2x^3 - 0.5x^2 - 2.1x - 2.4$

   (a) $f(2)$ (b) $f(-6)$

   (c) $f\left(\frac{2}{3}\right)$ (d) $f(1)$

50. $f(x) = 0.4x^4 - 1.6x^3 + 0.7x^2 - 2$

   (a) $f(1)$ (b) $f(-2)$

   (c) $f(5)$ (d) $f(-10)$

In Exercises 51–56, (a) verify the given factors of $f(x)$, (b) find the remaining factor of $f(x)$, (c) use your results to write the complete factorization of $f(x)$, (d) list all real zeros of f, and (e) confirm your results by using a graphing utility to graph the function.

   Function   *Factors*

51. $f(x) = x^3 - 12x - 16$ $(x + 2), (x - 4)$

52. $f(x) = x^3 - 28x - 48$ $(x + 4), (x - 6)$

53. $f(x) = 3x^3 + 10x^2 - 27x - 10$ $(3x + 1), (x - 2)$

54. $f(x) = 5x^3 - 11x^2 - 38x + 8$ $(5x - 1), (x - 4)$

55. $f(x) = x^3 + 2x^2 - 3x - 6$ $\left(x - \sqrt{3}\right), (x + 2)$

56. $f(x) = x^3 + 2x^2 - 2x - 4$ $\left(x - \sqrt{2}\right), (x + 2)$

(T) 57. You divide a polynomial by another polynomial. The remainder is zero. What conclusion(s) can you make?

(T) 58. Suppose that the remainder obtained in a polynomial division by $x - k$ is zero. How is the divisor related to the graph of the dividend?

In Exercises 59–64, match the function with its graph and use the result to find all real solutions of $f(x) = 0$. [The graphs are labeled (a), (b), (c), (d), (e), and (f).]

(a) 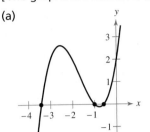   (b)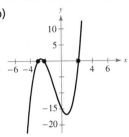

(c) 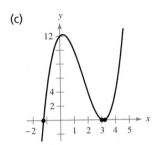   (d)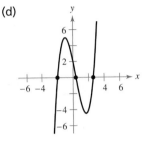

(e) 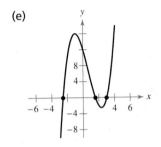   (f)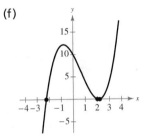

59. $f(x) = x^3 - 2x^2 - 7x + 12$

60. $f(x) = x^3 - x^2 - 5x + 2$

61. $f(x) = x^3 + 5x^2 + 6x + 2$

62. $f(x) = x^3 - 5x^2 + 2x + 12$

63. $f(x) = x^3 + 3x^2 - 5x - 15$

64. $f(x) = x^3 - 2x^2 - 5x + 10$

65. **Modeling Polynomials** A third-degree polynomial function f has real zeros -1, 2, and $\frac{10}{3}$. Find two different polynomial functions, one with a positive leading coefficient and one with a negative leading coefficient, that could be f. How many different polynomial functions are possible for f?

66. **Modeling Polynomials** A fourth-degree polynomial function g has real zeros -2, 0, 1, and 5. Find two different polynomial functions, one with a positive leading coefficient and one with a negative leading coefficient, that could be g. How many different polynomial functions are possible for g?

In Exercises 67–74, simplify the rational expression.

67. $\dfrac{x^3 - 10x^2 + 31x - 30}{x - 3}$

68. $\dfrac{x^3 + 15x^2 + 68x + 96}{x + 4}$

69. $\dfrac{6x^3 + x^2 - 21x - 10}{2x + 1}$

70. $\dfrac{3x^3 - 5x^2 - 34x + 24}{3x - 2}$

71. $\dfrac{x^4 - 5x^3 + 14x^2 - 120x}{x^2 + x + 20}$

72. $\dfrac{x^4 + x^3 + 3x^2 + 10x}{x^2 - x + 5}$

73. $\dfrac{x^4 + 4x^3 - 6x^2 - 36x - 27}{x^2 - 9}$

74. $\dfrac{x^4 + x^3 - 13x^2 - x + 12}{x^2 + x - 12}$

75. Examination Room A rectangular examination room in a veterinary clinic has a volume of

$$x^3 + 11x^2 + 34x + 24$$

cubic feet. The height of the room is $x + 1$ feet (see figure). Find the number of square feet of floor space in the examination room.

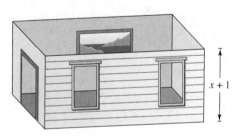

76. Veterinary Clinic A rectangular veterinary clinic has a volume of

$$x^3 + 55x^2 + 650x + 2000$$

cubic feet (the space in the attic is not counted). The height of the clinic is $x + 5$ feet (see figure). Find the number of square feet of floor space *on the first floor* of the clinic.

77. Profit A company making fishing poles estimated that the profit P (in dollars) from selling a particular fishing pole was

$$P = -140.75x^3 + 5348.3x^2 - 76{,}560, \quad 0 \le x \le 35$$

where x was the advertising expense (in tens of thousands of dollars). For this fishing pole, the advertising expense was \$300,000 ($x = 30$) and the profit was \$936,660.

(a) From the graph shown in the figure, it appears that the company could have obtained the same profit by spending less on advertising. Use the graph to estimate another amount the company could have spent on advertising that would have produced the same profit.

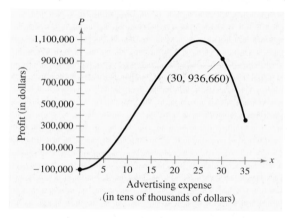

(b) Use synthetic division to confirm the result of part (a) algebraically.

78. Profit A company that produces calculators estimated that the profit P (in dollars) from selling a particular model of calculator was

$$P = -152x^3 + 7545x^2 - 169{,}625, \quad 0 \le x \le 45$$

where x was the advertising expense (in tens of thousands of dollars). For this model of calculator, the advertising expense was \$400,000 ($x = 40$) and the profit was \$2,174,375.

Ⓣ (a) Use a graphing utility to graph the profit function.

(b) Could the company have obtained the same profit by spending less on advertising? Explain your reasoning.

79. Writing Briefly explain what it means for a divisor to divide evenly into a dividend.

80. Writing Briefly explain how to check polynomial division, and justify your answer. Give an example.

Exploration In Exercises 81 and 82, find the constant c such that the denominator will divide evenly into the numerator.

81. $\dfrac{x^3 + 4x^2 - 3x + c}{x - 5}$

82. $\dfrac{x^5 - 2x^2 + x + c}{x + 2}$

Section 3.4

Real Zeros of Polynomial Functions

- Find all possible rational zeros of a function using the Rational Zero Test.
- Find all real zeros of a function.
- Approximate the real zeros of a polynomial function using the Intermediate Value Theorem.
- Approximate the real zeros of a polynomial function using a graphing utility.
- Apply techniques for approximating real zeros to solve an application problem.

The Rational Zero Test

The **Rational Zero Test** relates the possible rational zeros of a polynomial function (having integer coefficients) to the leading coefficient and to the constant term of the polynomial.

STUDY TIP

When the leading coefficient is 1, the possible rational zeros are simply the factors of the constant term.

The Rational Zero Test

If the polynomial function given by

$$f(x) = a_n x^n + a_{n-1} x^{n-1} + \cdots + a_2 x^2 + a_1 x + a_0$$

has *integer* coefficients, then every rational zero of f has the form

$$\text{Rational zeros} = \frac{\text{a factor of the constant term } a_0}{\text{a factor of the leading coefficient } a_n} = \frac{p}{q}$$

where p and q have no common factors other than 1.

Make a list of *possible rational zeros*. Then use a trial-and-error method to determine which, if any, are actual zeros of the polynomial function.

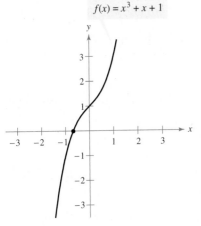

$f(x) = x^3 + x + 1$

FIGURE 3.25

| Example 1 | **Rational Zero Test with Leading Coefficient of 1**

Find the rational zeros of $f(x) = x^3 + x + 1$.

SOLUTION   Because the leading coefficient is 1, the possible rational zeros are the factors of the constant term, 1 and -1. By testing these possible zeros, you can see that neither checks.

$$f(1) = (1)^3 + 1 + 1 = 3 \qquad f(-1) = (-1)^3 + (-1) + 1 = -1$$

So, you can conclude that the given function has *no* rational zeros. Note from the graph of f in Figure 3.25 that f does have one real zero (between -1 and 0). By the Rational Zero Test, you know that this real zero is *not* a rational number.

✓**CHECKPOINT 1**

Find the rational zeros of $f(x) = x^3 + 2x^2 + 1$. ∎

Example 2   **Rational Zero Test with Leading Coefficient of 1**

Find the rational zeros of

$$f(x) = x^4 - x^3 + x^2 - 3x - 6.$$

SOLUTION   Because the leading coefficient is 1, the possible rational zeros are the factors of the constant term.

Possible rational zeros: $\pm 1, \pm 2, \pm 3, \pm 6$

Test each possible rational zero. The test shows $x = -1$ and $x = 2$ are the only two rational zeros of the function.

✓ CHECKPOINT 2

Find the rational zeros of $f(x) = x^4 + 2x^3 + x^2 - 4.$ ■

If the leading coefficient of a polynomial is not 1, the list of possible rational zeros can increase dramatically. In such cases, the search can be shortened in several ways: (1) a programmable calculator can be used to speed up the calculations; (2) a graph, created either by hand or with a graphing utility, can give a good estimate of the locations of the zeros; and (3) synthetic division can be used to test the possible rational zeros.

TECHNOLOGY

Ⓣ   There are several ways to use your graphing utility to locate the zeros of a polynomial function after listing the possible rational zeros. You can use the *table* feature by setting the increments of x to the smallest difference between possible rational zeros, or use the *table* feature in ASK mode. In either case the value in the function column will be 0 when x is a zero of the function. Another way to locate zeros is to graph the function. Be sure that your viewing window contains all the possible rational zeros.

To see how to use synthetic division to test the possible rational zeros, let's take another look at the function given by

$$f(x) = x^4 - x^3 + x^2 - 3x - 6$$

from Example 2. To test that $x = -1$ and $x = 2$ are zeros of f, you can apply synthetic division, as follows.

$$
\begin{array}{r|rrrrr}
-1 & 1 & -1 & 1 & -3 & -6 \\
   &   & -1 & 2 & -3 & 6 \\
\hline
   & 1 & -2 & 3 & -6 & 0
\end{array}
\qquad
\begin{array}{r|rrrr}
2 & 1 & -2 & 3 & -6 \\
  &   & 2 & 0 & 6 \\
\hline
  & 1 & 0 & 3 & 0
\end{array}
$$

So, you have

$$f(x) = (x + 1)(x - 2)(x^2 + 3).$$

Because the factor $(x^2 + 3)$ produces no real zeros, you can conclude that $x = -1$ and $x = 2$ are the only *real* zeros of f. This is verified in the graph of f shown in Figure 3.26.

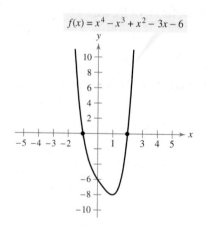

$f(x) = x^4 - x^3 + x^2 - 3x - 6$

FIGURE 3.26

Finding the first zero is often the hardest part. After that, the search is simplified by using the lower-degree polynomial obtained in synthetic division. Once the lower-degree polynomial is quadratic, either factoring or the Quadratic Formula can be used to find the remaining zeros.

Example 3 Using the Rational Zero Test

Find the rational zeros of $f(x) = 2x^3 + 3x^2 - 8x + 3$.

SOLUTION The leading coefficient is 2 and the constant term is 3.

$$\textit{Possible rational zeros:} \quad \frac{\text{Factors of 3}}{\text{Factors of 2}} = \frac{\pm 1, \pm 3}{\pm 1, \pm 2} = \pm 1, \pm 3, \pm \frac{1}{2}, \pm \frac{3}{2}$$

By synthetic division, you can determine that $x = 1$ is a rational zero.

$$
\begin{array}{r|rrrr}
1 & 2 & 3 & -8 & 3 \\
  &   & 2 & 5 & -3 \\
\hline
  & 2 & 5 & -3 & 0
\end{array}
$$

So, $f(x)$ factors as

$$f(x) = (x - 1)(2x^2 + 5x - 3)$$
$$= (x - 1)(2x - 1)(x + 3)$$

and you can conclude that the rational zeros of f are $x = 1$, $x = \frac{1}{2}$, and $x = -3$.

✓ CHECKPOINT 3

Find the rational zeros of $f(x) = 2x^3 + 5x^2 + x - 2$. ▪

Example 4 Using the Rational Zero Test

Find all the real zeros of $f(x) = 10x^3 - 15x^2 - 16x + 12$.

SOLUTION The leading coefficient is 10 and the constant term is 12.

$$\textit{Possible rational zeros:} \quad \frac{\text{Factors of 12}}{\text{Factors of 10}} = \frac{\pm 1, \pm 2, \pm 3, \pm 4, \pm 6, \pm 12}{\pm 1, \pm 2, \pm 5, \pm 10}$$

With so many possibilities (32, in fact), it is worth your time to stop and sketch a graph. From Figure 3.27, it looks like three reasonable choices would be $x = -\frac{6}{5}$, $x = \frac{1}{2}$, and $x = 2$. Testing these by synthetic division shows that only $x = 2$ checks. So, you have

$$f(x) = (x - 2)(10x^2 + 5x - 6).$$

Using the Quadratic Formula, you find that the two additional zeros are irrational numbers.

$$x = \frac{-5 + \sqrt{265}}{20} \approx 0.5639 \quad \text{and} \quad x = \frac{-5 - \sqrt{265}}{20} \approx -1.0639$$

You can conclude that the real zeros of f are $x = 2$, $x \approx 0.5639$, and $x \approx -1.0639$.

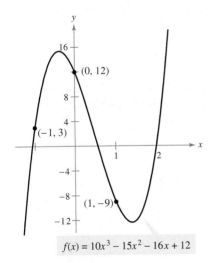

$$f(x) = 10x^3 - 15x^2 - 16x + 12$$

FIGURE 3.27

✓ CHECKPOINT 4

Find the rational zero of $f(x) = 3x^3 + 2x^2 - 5x + 6$. ▪

The Intermediate Value Theorem

The next theorem, called the **Intermediate Value Theorem,** tells you of the existence of real zeros of polynomial functions. The theorem implies that if $(a, f(a))$ and $(b, f(b))$ are two points on the graph of a polynomial function such that $f(a) \neq f(b)$, then for any number d between $f(a)$ and $f(b)$ there must be a number c between a and b such that $f(c) = d$. (See Figure 3.28.)

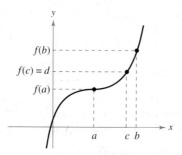

FIGURE 3.28

Intermediate Value Theorem

Let a and b be real numbers such that $a < b$. If f is a polynomial function such that $f(a) \neq f(b)$, then, in the interval $[a, b]$, f takes on every value between $f(a)$ and $f(b)$.

The Intermediate Value Theorem helps you locate the real zeros of a polynomial function in the following way. If you can find a value $x = a$ where a polynomial function is positive, and another value $x = b$ where it is negative, you can conclude that the function has at least one real zero between these two values. For example, the function given by

$$f(x) = x^3 + x^2 + 1$$

is negative when $x = -2$ and positive when $x = -1$. So, it follows from the Intermediate Value Theorem that f must have a real zero somewhere between -2 and -1, as shown in Figure 3.29.

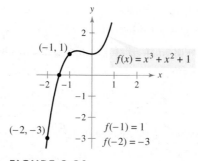

FIGURE 3.29

By continuing this line of reasoning, you can approximate any real zeros of a polynomial function to any desired level of accuracy. This concept is further demonstrated in Example 5.

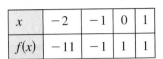 **Example 5** Approximating a Zero of a Polynomial Function

Use the Intermediate Value Theorem to approximate a real zero of

$$f(x) = x^3 - x^2 + 1.$$

SOLUTION Begin by computing a few function values, as follows.

x	-2	-1	0	1
$f(x)$	-11	-1	1	1

Because $f(-1)$ is negative and $f(0)$ is positive, you can apply the Intermediate Value Theorem to conclude that the function has a zero between -1 and 0. To pinpoint this zero more closely, divide the interval $[-1, 0]$ into tenths and evaluate the function at each point. When you do this, you will find that

$$f(-0.8) = -0.152$$

and

$$f(-0.7) = 0.167.$$

So, f must have a zero between -0.8 and -0.7, as shown in Figure 3.30. By continuing this process, you can approximate this zero to any desired level of accuracy.

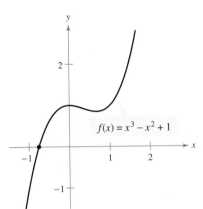

FIGURE 3.30 f has a zero between -8 and -0.7.

✓**CHECKPOINT 5**

Use the Intermediate Value Theorem to approximate a real zero of $f(x) = x^3 + x - 4.$ ∎

Approximating Zeros of Polynomial Functions

There are several different techniques for approximating the zeros of a polynomial function. All such techniques are better suited to computers or graphing utilities than they are to "hand calculations." In this section, you will study two techniques that can be used with a graphing utility. The first is called the **zoom-and-trace** technique.

Zoom-and-Trace Technique

To approximate a real zero of a function with a graphing utility, use the following steps.

1. Graph the function so that the real zero you want to approximate appears as an x-intercept on the screen.

2. Move the cursor near the x-intercept and use the *zoom* feature to zoom in to get a better look at the intercept.

3. Use the *trace* feature to find the x-values that occur just before and just after the x-intercept. If the difference between these values is sufficiently small, use their average as the approximation. If not, continue zooming in until the approximation reaches the desired level of accuracy.

The amount that a graphing utility zooms in is determined by the *zoom factor*. The zoom factor is a positive number greater than or equal to 1 that gives the ratio of the larger screen to the smaller screen. For instance, if you zoom in with a zoom factor of 2, you will obtain a screen in which the *x*- and *y*-values are half their original values. This text uses a zoom factor of 4.

Example 6   Approximating a Zero of a Polynomial Function

Approximate a real zero of

$$f(x) = x^3 + 4x + 2$$

to the nearest thousandth.

SOLUTION   To begin, use a graphing utility to graph the function, as shown in Figure 3.31(a). Set the X-scale to 0.001 and zoom in several times until the tick marks on the *x*-axis become visible. The final screen should be similar to the one shown in Figure 3.31(b).

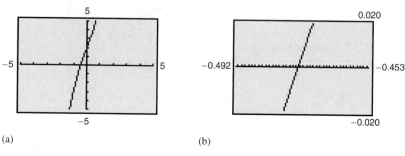

(a)   (b)

FIGURE 3.31

At this point, you can use the *trace* feature to determine that the *x*-values just to the left and right of the *x*-intercept are

$$x \approx -0.4735 \quad \text{and} \quad x \approx -0.4733.$$

So, to the nearest thousandth, you can approximate the zero of the function to be

$$x \approx -0.473.$$

To check this, try substituting -0.473 into the function. You should obtain a result that is approximately zero.

✓ CHECKPOINT 6

Approximate a real zero of $f(x) = 2x^3 - x + 3$ to the nearest thousandth. ▪

In Example 6, the cubic polynomial function has only one real zero. Remember that functions can have two or more real zeros. In such cases, you can use the zoom-and-trace technique for each zero separately. For instance, the function given by

$$f(x) = x^3 - 4x^2 + x + 2$$

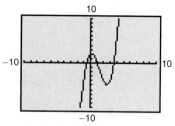

FIGURE 3.32

has three real zeros, as shown in Figure 3.32. Using a zoom-and-trace approach for each real zero, you can approximate the real zeros to be

$$x \approx -0.562, \quad x = 1.000, \quad \text{and} \quad x \approx 3.562.$$

The second technique that can be used with some graphing utilities is to employ the graphing utility's *zero* or *root* feature. The name of this feature differs with different calculators. Consult your user's guide to determine if this feature is available.

Example 7   Approximating the Zeros of a Polynomial Function

Approximate the real zeros of $f(x) = x^3 - 2x^2 - x + 1$.

SOLUTION   To begin, use a graphing utility to graph the function, as shown in the first screen in Figure 3.33. Notice that the graph has three x-intercepts. To approximate the leftmost intercept, find an appropriate viewing window and use the zero feature, as shown below. The calculator should display an approximation of $x \approx -0.8019377$, which is accurate to seven decimal places.

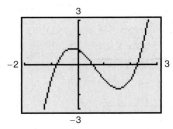

Find an appropriate viewing window, then use the *zero* feature.

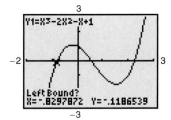

Move the cursor to the left of the intercept and press "Enter."

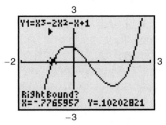

Move the cursor to the right of the intercept and press "Enter."

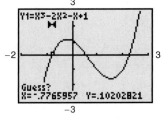

Move the cursor near the intercept and press "Enter."

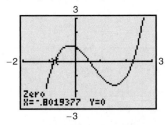

FIGURE 3.33

TECHNOLOGY

For instructions on how to use the *zoom*, *trace*, *zero*, and *root* features, see Appendix A; for specific keystrokes, go to the text website at *college.hmco.com/info/larsonapplied.*

By repeating this process, you can determine that the other two zeros are $x \approx 0.555$ and $x \approx 2.247$.

✓ **CHECKPOINT 7**

Approximate the real zeros of $f(x) = x^3 - 4x^2 + 3x + 1$. ■

You may be wondering why so much time is spent in algebra trying to find the zeros of a function. The reason is that if you have a technique that will enable you to solve the equation $f(x) = 0$, you can use the same technique to solve the more general equation

$$f(x) = c$$

where c is any real number. This procedure is demonstrated in Example 8.

Example 8 Solving the Equation $f(x) = c$

Find a value of x such that $f(x) = 30$ for the function given by

$$f(x) = x^3 - 4x + 4.$$

SOLUTION The graph of

$$f(x) = x^3 - 4x + 4$$

is shown in Figure 3.34. Note from the graph that $f(x) = 30$ when x is about 3.5. To use the zoom-and-trace technique to approximate this x-value more closely, consider the equation

$$x^3 - 4x + 4 = 30$$
$$x^3 - 4x - 26 = 0.$$

So, the *solutions* of the equation $f(x) = 30$ are precisely the same x-values as the *zeros* of

$$g(x) = x^3 - 4x - 26.$$

Using the graph of g, as shown in Figure 3.35, you can approximate the zero of g to be

$$x \approx 3.41.$$

You can check this value by substituting $x = 3.41$ into the original function.

$$f(3.41) = (3.41)^3 - 4(3.41) + 4$$
$$\approx 30.01 \checkmark$$

Remember that with decimal approximations, a check usually will not produce an exact value.

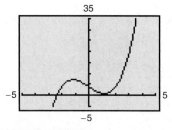

FIGURE 3.34

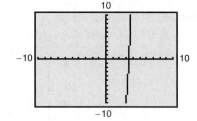

FIGURE 3.35

✓CHECKPOINT 8

Find a value of x such that $f(x) = 20$ for the function given by $f(x) = x^3 + 4x^2 - 1$. ■

Application

Example 9
MAKE A DECISION **Profit and Advertising Expenses**

A company that produces sports clothes estimates that the profit from selling a particular line of sportswear is given by

$$P = -0.014x^3 + 0.752x^2 - 40, \quad 0 \le x \le 50$$

where P is the profit (in tens of thousands of dollars) and x is the advertising expense (in tens of thousands of dollars). According to this model, how much money should the company spend on advertising to obtain a profit of $2,750,000?

SOLUTION   From Figure 3.36, it appears that there are two different values of x between 0 and 50 that will produce a profit of $2,750,000. However, because of the context of the problem, it is clear that the better answer is the smaller of the two numbers. So, to solve the equation

$$-0.014x^3 + 0.752x^2 - 40 = 275$$

$$-0.014x^3 + 0.752x^2 - 315 = 0$$

find the zeros of the function

$$g(x) = -0.014x^3 + 0.752x^2 - 315.$$

Using the zoom-and-trace technique, you can find that the leftmost zero is

$$x \approx 32.8.$$

You can check this solution by substituting

$$x = 32.8$$

into the original function.

$$P = -0.014(32.8)^3 + 0.752(32.8)^2 - 40$$

$$\approx 275$$

The company should spend about $328,000 on advertising for the line of sportswear.

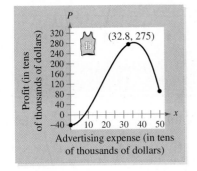

FIGURE 3.36

✓ CHECKPOINT 9

In Example 9, how much should the company spend on advertising to obtain a profit of $2,500,000? ■

CONCEPT CHECK

1. Use the Rational Zero Test to explain why $\frac{3}{2}$ is not a possible rational zero of $f(x) = 3x^2 - x + 2$.

2. Can you use the *zero* feature of a graphing utility to find rational zeros of a function? Irrational zeros? Imaginary zeros? Explain your reasoning.

3. Is it possible for a polynomial function to have no real zeros? Explain your reasoning.

4. Explain how to use the Intermediate Value Theorem to approximate the real zeros of a function.

Skills Review 3.4

The following warm-up exercises involve skills that were covered in earlier sections. You will use these skills in the exercise set for this section. For additional help, review Sections 1.5 and 3.3.

In Exercises 1 and 2, find a polynomial function with integer coefficients having the given zeros.

1. $-1, \frac{2}{3}, 3$

2. $-2, 0, \frac{3}{4}, 2$

In Exercises 3 and 4, use synthetic division to divide.

3. $\dfrac{x^5 - 9x^3 + 5x + 18}{x + 3}$

4. $\dfrac{3x^4 + 17x^3 + 10x^2 - 9x - 8}{x + \frac{2}{3}}$

In Exercises 5–8, use the given zero to find all the real zeros of f.

5. $f(x) = 2x^3 + 11x^2 + 2x - 4, \ x = \frac{1}{2}$

6. $f(x) = 6x^3 - 47x^2 - 124x - 60, \ x = 10$

7. $f(x) = 4x^3 - 13x^2 - 4x + 6, \ x = -\frac{3}{4}$

8. $f(x) = 10x^3 + 51x^2 + 48x - 28, \ x = \frac{2}{5}$

In Exercises 9 and 10, find all real solutions of the equation.

9. $x^4 - 3x^2 + 2 = 0$

10. $x^4 - 7x^2 + 12 = 0$

Exercises 3.4

See www.CalcChat.com for worked-out solutions to odd-numbered exercises.

(T) In Exercises 1 and 2, use the Rational Zero Test to list all possible rational zeros of f. Then use a graphing utility to graph the function. Use the graph to help determine which of the possible rational zeros are actual zeros of the function.

1. $f(x) = x^3 + x^2 - 4x - 4$

2. $f(x) = 2x^4 - x^2 - 6$

In Exercises 3–6, find the rational zeros of the polynomial function.

3. $f(x) = x^3 - \frac{3}{2}x^2 - \frac{23}{2}x + 6$

4. $f(x) = x^3 + 3x^2 - x - 3$

5. $f(x) = 4x^4 - 17x^2 + 4$

6. $f(x) = -2x^4 + 13x^3 - 21x^2 + 2x + 8$

In Exercises 7–14, find all real zeros of the function.

7. $f(x) = x^3 - 6x^2 + 11x - 6$

8. $g(x) = x^3 - 4x^2 - x + 4$

9. $h(t) = t^3 + 12t^2 + 21t + 10$

10. $f(x) = x^3 - 4x^2 + 5x - 2$

11. $C(x) = 2x^3 + 3x^2 - 1$

12. $f(x) = 3x^3 - 19x^2 + 33x - 9$

13. $f(x) = x^4 - 11x^2 + 18$

14. $P(t) = t^4 - 19t^2 + 48$

In Exercises 15–20, find all real solutions of the polynomial equation.

15. $z^4 - z^3 - 2z - 4 = 0$

16. $x^4 - 13x^2 - 12x = 0$

17. $2y^4 + 7y^3 - 26y^2 + 23y - 6 = 0$

18. $2x^4 - 11x^3 - 6x^2 + 64x + 32 = 0$

19. $x^5 - x^4 - 3x^3 + 5x^2 - 2x = 0$

20. $x^5 - 7x^4 + 10x^3 + 14x^2 - 24x = 0$

In Exercises 21 and 22, (a) list the possible rational zeros of f, (b) sketch the graph of f so that some of the possible zeros in part (a) can be discarded, and (c) determine all real zeros of f.

21. $f(x) = 32x^3 - 52x^2 + 17x + 3$

22. $f(x) = 4x^3 + 7x^2 - 11x - 18$

In Exercises 23–26, use the Intermediate Value Theorem to show that the function has at least one zero in the interval $[a, b]$. (You do not have to approximate the zero.)

23. $f(x) = x^3 + 2x - 5$, $\ [1, 2]$

24. $f(x) = x^5 - 3x + 3$, $\ [-2, -1]$

25. $f(x) = x^4 - 3x^2 - 10$, $\ [2, 3]$

26. $f(x) = -x^3 + 2x^2 + 7x - 3$, $\ [3, 4]$

In Exercises 27–30, use the Intermediate Value Theorem to approximate the zero of f in the interval $[a, b]$. Give your approximation to the nearest tenth. (If you have a graphing utility, use it to help you approximate the zero.)

27. $f(x) = x^3 + x - 1$, $\ [0, 1]$

28. $f(x) = x^5 + x + 1$, $\ [-1, 0]$

29. $f(x) = x^4 - 10x^2 - 11$, $\ [3, 4]$

30. $f(x) = -x^3 + 3x^2 + 9x - 2$, $\ [4, 5]$

In Exercises 31–36, match the function with its graph. Then approximate the real zeros of the function to three decimal places. [The graphs are labeled (a), (b), (c), (d), (e), and (f).]

(a)
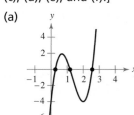

(b)
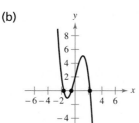

(c)
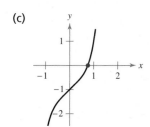

(d)
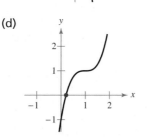

(e)
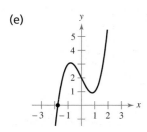

(f)
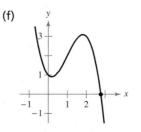

31. $f(x) = x^3 - 2x + 2$ **32.** $f(x) = x^5 + x - 1$

33. $f(x) = 2x^3 - 6x^2 + 6x - 1$

34. $f(x) = 5x^3 - 20x^2 + 20x - 4$

35. $f(x) = -x^3 + 3x^2 - x + 1$

36. $f(x) = -x^3 + 4x + 2$

(T) In Exercises 37–40, use the *zoom* and *trace* features of a graphing utility to approximate the real zeros of f. Give your approximations to the nearest thousandth.

37. $f(x) = x^4 - x - 3$ **38.** $f(x) = 4x^3 + 14x - 8$

39. $f(x) = x^3 - 3.9x^2 + 4.79x - 1.881$

40. $f(x) = -x^3 + 2x^2 + 4x + 5$

(T) In Exercises 41–44, use the *zero* or *root* feature of a graphing utility to approximate the real zeros of f. Give your approximations to the nearest thousandth.

41. $f(x) = x^4 + x - 3$

42. $f(x) = -x^4 + 2x^3 + 4$

43. $f(x) = 7x^4 - 42x^3 + 43x^2 + 216x - 324$

44. $f(x) = 3x^4 - 12x^3 + 27x^2 + 4x - 4$

In Exercises 45–48, match the cubic function with the numbers of rational and irrational zeros.

(a) Rational zeros: 0;   Irrational zeros: 1

(b) Rational zeros: 3;   Irrational zeros: 0

(c) Rational zeros: 1;   Irrational zeros: 2

(d) Rational zeros: 1;   Irrational zeros: 0

45. $f(x) = x^3 - 1$ **46.** $f(x) = x^3 - 2$

47. $f(x) = x^3 - x$ **48.** $f(x) = x^3 - 2x$

49. Dimensions of a Box   An open box is to be made from a rectangular piece of material, 18 inches by 15 inches, by cutting equal squares from the corners and turning up the sides (see figure).

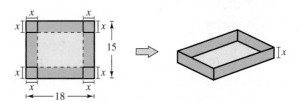

(a) Write the volume V of the box as a function of x. Determine the domain of the function.

(b) Sketch the graph of the function and approximate the dimensions of the box that yield a maximum volume.

(c) Find values of x such that $V = 108$. Which of these values is a physical impossibility in the construction of the box? Explain.

(d) What value of x should you use to make the tallest possible box with a volume of 108 cubic inches?

50. Dimensions of a Box   An open box is to be made from a rectangular piece of material, 16 inches by 12 inches, by cutting equal squares from the corners and turning up the sides (see figure).

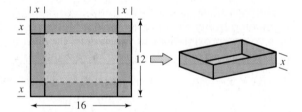

(a) Write the volume V of the box as a function of x. Determine the domain of the function.

(b) Sketch the graph of the function and approximate the dimensions of the box that yield a maximum volume.

(c) Find values of x such that $V = 120$. Which of these values is a physical impossibility in the construction of the box? Explain.

(d) What value of x should you use to make the tallest possible box with a volume of 120 cubic inches?

51. Dimensions of a Terrarium   A rectangular terrarium with a square cross section has a combined length and girth (perimeter of a cross section) of 108 inches (see figure). Find the dimensions of the terrarium, given that the volume is 11,664 cubic inches.

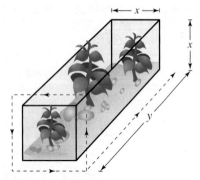

Figure for 51 and 52

52. Dimensions of a Terrarium   A rectangular terrarium has a combined length and girth (perimeter of a cross section) of 120 inches (see figure). Find the dimensions of the terrarium, given that the volume is 16,000 cubic inches.

53. Geometry   A bulk food storage bin with dimensions 2 feet by 3 feet by 4 feet needs to be increased in size to hold five times as much food as the current bin. (Assume each dimension is increased by the same amount.)

(a) Write a function that represents the volume V of the new bin.

(b) Find the dimensions of the new bin.

54. Geometry   A rancher wants to enlarge an existing rectangular corral such that the total area of the new corral is 1.5 times that of the original corral. The current corral's dimensions are 250 feet by 160 feet. The rancher wants to increase each dimension by the same amount.

(a) Write a function that represents the area A of the new corral.

(b) Find the dimensions of the new corral.

(c) A rancher wants to add a length to the sides of the corral that are 160 feet, and twice the length to the sides that are 250 feet, such that the total area of the new corral is 1.5 times that of the original corral. Repeat parts (a) and (b). Explain your results.

55. Medicine   The concentration C of a chemical in the bloodstream t hours after injection into muscle tissue is given by

$$C = \frac{3t^2 + t}{t^3 + 50}, \quad t \geq 0.$$

The concentration is greatest when

$$3t^4 + 2t^3 - 300t - 50 = 0.$$

Approximate this time to the nearest hundredth of an hour.

56. Transportation Cost   The transportation cost C (in thousands of dollars) of the components used in manufacturing prefabricated homes is given by

$$C = 100\left(\frac{200}{x^2} + \frac{x}{x + 30}\right), \quad x \geq 1$$

where x is the order size (in hundreds). The cost is a minimum when

$$3x^3 - 40x^2 - 2400x - 36,000 = 0.$$

Approximate the optimal order size to the nearest unit.

(T) **57. Online Sales**   The revenues per share R (in dollars) for Amazon.com for the years 1996 to 2005 are shown in the table.   *(Source: Amazon.com)*

Year	Revenue per share, R	Year	Revenue per share, R
1996	0.07	2001	8.37
1997	0.51	2002	10.14
1998	1.92	2003	13.05
1999	4.75	2004	17.16
2000	7.73	2005	20.41

(a) Use a graphing utility to create a scatter plot of the data. Let t represent the year, with $t = 6$ corresponding to 1996.

(b) Use the *regression* feature of a graphing utility to find a linear model, a quadratic model, a cubic model, and a quartic model for the data.

(c) Use a graphing utility to graph each model separately with the data in the same viewing window. How well does each model fit the data?

(d) Use each model to predict the year in which the revenue per share is about $37. Explain any differences in the predictions.

(T) **58. Population** The numbers P (in millions) of people age 18 and over in the United States for the years 1996 to 2005 are shown in the table. *(Source: U.S. Census Bureau)*

Year	Population, P	Year	Population, P
1996	199.2	2001	212.5
1997	201.7	2002	215.1
1998	204.4	2003	217.8
1999	207.1	2004	220.4
2000	209.1	2005	222.9

(a) Use a graphing utility to create a scatter plot of the data. Let $t = 6$ correspond to 1996.

(b) Use the *regression* feature of a graphing utility to find a linear model, a quadratic model, and a cubic model for the data.

(c) Use a graphing utility to graph each model separately with the data in the same viewing window. How well does each model fit the data?

(d) Use each model to predict the year in which the population is about 231,000,000. Explain any differences in the predictions.

(S) **59. Cost of Dental Care** The amount that $100 worth of dental care at 1982–1984 prices would cost in a different year is given by a CPI (Consumer Price Index). The CPIs for dental care in the United States for the years 1996 to 2005 are shown in the table. *(Source: U.S. Bureau of Labor Statistics)*

Year	CPI	Year	CPI
1996	216.5	2001	269.0
1997	226.6	2002	281.0
1998	236.2	2003	292.5
1999	247.2	2004	306.9
2000	258.5	2005	324.0

(a) Use a spreadsheet software program to create a scatter plot of the data. Let t represent the year, with $t = 6$ corresponding to 1996.

(b) Use the *regression* feature of a spreadsheet software program to find a linear model, a quadratic model, a cubic model, and a quartic model for the data.

(c) Use each model to predict the year in which the CPI for dental care will be about $400. Then discuss the appropriateness of each model for predicting future values.

(S) **60. Solar Energy** Photovoltaic cells convert light energy into electricity. The photovoltaic cell and module domestic shipments S (in peak kilowatts) for the years 1996 to 2005 are shown in the table. *(Source: Energy Information Administration)*

Year	Shipments, S	Year	Shipments, S
1996	13,016	2001	36,310
1997	12,561	2002	45,313
1998	15,069	2003	48,664
1999	21,225	2004	78,346
2000	19,838	2005	134,465

(a) Use a spreadsheet software program to create a scatter plot of the data. Let t represent the year, with $t = 6$ corresponding to 1996.

(b) Use the *regression* feature of a spreadsheet software program to find a cubic model and a quartic model for the data.

(c) Use each model to predict the year in which the shipments will be about 1,000,000 peak kilowatts. Then discuss the appropriateness of each model for predicting future values.

61. Advertising Cost A company that produces video games estimates that the profit P (in dollars) from selling a new game is given by

$$P = -82x^3 + 7250x^2 - 450,000, \quad 0 \le x \le 80$$

where x is the advertising expense (in tens of thousands of dollars). Using this model, how much should the company spend on advertising to obtain a profit of $5,900,000?

62. Advertising Cost A company that manufactures hydroponic gardening systems estimates that the profit P (in dollars) from selling a new system is given by

$$P = -35x^3 + 2700x^2 - 300,000, \quad 0 \le x \le 70$$

where x is the advertising expense (in tens of thousands of dollars). Using this model, how much should the company spend on advertising to obtain a profit of $1,800,000?

63. *MAKE A DECISION: DEMAND FUNCTION* A company that produces cell phones estimates that the demand D for a new model of phone is given by

$$D = -x^3 + 54x^2 - 140x - 3000, \quad 10 \le x \le 50$$

where x is the price of the phone (in dollars).

(T) (a) Use a graphing utility to graph D. Use the *trace* feature to determine the values of x for which the demand is 14,400 phones.

(b) You may also determine the values of x for which the demand is 14,400 phones by setting D equal to 14,400 and solving for x with a graphing utility. Discuss this alternative solution method. Of the solutions that lie within the given interval, what price would you recommend the company charge for the phones?

64. *MAKE A DECISION: DEMAND FUNCTION* A company that produces hand-held organizers estimates that the demand D for a new model of organizer is given by

$$D = -0.005x^3 + 2.65x^2 - 70x - 2500, \quad 50 \le x \le 500$$

where x is the price of the organizer (in dollars).

(T) (a) Use a graphing utility to graph D. Use the *trace* feature to determine the values of x for which the demand will be 80,000 organizers.

(b) You may also determine the values of x for which the demand will be 80,000 organizers by setting D equal to 80,000 and solving for x with a graphing utility. Discuss this alternative solution method. Of the solutions that lie within the given interval, what price would you recommend the company charge for the new organizers?

65. Height of a Baseball A baseball is launched upward from ground level with an initial velocity of 48 feet per second, and its height h (in feet) is

$$h(t) = -16t^2 + 48t, \quad 0 \le t \le 3$$

where t is the time (in seconds). You are told the ball reaches a height of 64 feet. Is this possible?

(T) **66. Exploration** Use a graphing utility to graph the function

$$f(x) = x^4 - 4x^2 + k$$

for different values of k. Find the values of k such that the zeros of f satisfy the specified characteristics. (Some parts do not have unique answers.)

(a) Four real zeros

(b) Two real zeros and two complex roots

67. Reasoning Is it possible that a second-degree polynomial function with integer coefficients has one rational zero and one irrational zero? If so, give an example.

68. Reasoning Is it possible that a third-degree polynomial function with integer coefficients has one rational zero and two irrational zeros? If so, give an example.

69. Use the information in the table.

Interval	Value of $f(x)$
$(-\infty, -2)$	Positive
$(-2, 1)$	Negative
$(1, 4)$	Negative
$(4, \infty)$	Positive

(a) What are the three real zeros of the polynomial function f?

(b) What can be said about the behavior of the graph of f at $x = 1$?

(c) What is the least possible degree of f? Explain. Can the degree of f ever be odd? Explain.

(d) Is the leading coefficient of f positive or negative? Explain.

(e) Write an equation for f. (There are many correct answers.)

(f) Sketch a graph of the equation you wrote in part (e).

70. Graphical Reasoning The graph of one of the following functions is shown below. Identify the function shown in the graph. Explain why each of the others is not the correct function. Use a graphing utility to verify your result.

(a) $f(x) = x^2(x + 2)(x - 3.5)$

(b) $g(x) = (x + 2)(x - 3.5)$

(c) $h(x) = (x + 2)(x - 3.5)(x^2 + 1)$

(d) $k(x) = (x + 1)(x + 2)(x - 3.5)$

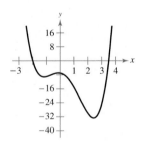

71. Extended Application To work an extended application analyzing the sales per share of Best Buy, visit this text's website at *college.hmco.com*. (*Source: Best Buy*)

Mid-Chapter Quiz

See www.CalcChat.com for worked-out solutions to odd-numbered exercises.

Take this quiz as you would take a quiz in class. When you are done, check your work against the answers given in the back of the book.

In Exercises 1 and 2, sketch the graph of the quadratic function. Identify the vertex and the intercepts.

1. $f(x) = (x + 1)^2 - 2$

2. $f(x) = 25 - x^2$

In Exercises 3 and 4, describe the right-hand and left-hand behavior of the graph of the polynomial function. Verify with a graphing utility.

3. $f(x) = -2x^3 + 7x^2 - 9$

4. $f(x) = x^4 + 7x^2 - 8$

5. Use synthetic division to evaluate $f(x) = 2x^4 + x^3 + 18x^2 - 4$ when $x = -3$.

In Exercises 6 and 7, write the function in the form $f(x) = (x - k)q(x) + r$ for the given value of k, and demonstrate that $f(k) = r$.

6. $f(x) = x^4 - 5x^2 + 4$, $k = 1$

7. $f(x) = x^3 + 5x^2 - 2x - 24$, $k = -3$

8. Simplify $\dfrac{2x^4 + 9x^3 - 32x^2 - 99x + 180}{x^2 + 2x - 15}$.

In Exercises 9–12, find the real zeros of the function.

9. $f(x) = -2x^3 - 7x^2 + 10x + 35$

10. $f(x) = 4x^4 - 37x^2 + 9$

11. $f(x) = 3x^4 + 4x^3 - 3x - 4$

12. $f(x) = 2x^3 - 3x^2 + 2x - 3$

(T) **13.** The profit P (in dollars) for a clothing company is

$$P = -95x^3 + 5650x^2 - 250{,}000, \quad 0 \le x \le 55$$

where x is the advertising expense (in tens of thousands of dollars). What is the profit for an advertising expense of $450,000? Use a graphing utility to approximate another advertising expense that would yield the same profit.

(T) **14. Crops** The worldwide land areas A (in millions of hectares) of transgenic crops for the years 1996 to 2006 are shown in the table. *(Source: International Service for the Acquisition of Agri-Biotech Applications)*

(a) Use a graphing utility to create a scatter plot of the data. Let t represent the year, with $t = 6$ corresponding to 1996.

(b) Use the *regression* feature of a graphing utility to find a linear model, a quadratic model, a cubic model, and a quartic model for the data.

(c) Use a graphing utility to graph each model separately with the data in the same viewing window. How well does each model fit the data?

(d) Use each model to predict the year in which the land area will be about 150 million hectares. Explain any differences in the predictions.

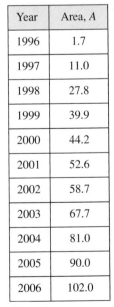

Year	Area, A
1996	1.7
1997	11.0
1998	27.8
1999	39.9
2000	44.2
2001	52.6
2002	58.7
2003	67.7
2004	81.0
2005	90.0
2006	102.0

Table for 14

Section 3.5

Complex Numbers

- ■ Perform operations with complex numbers and write the results in standard form.
- ■ Find the complex conjugate of a complex number.
- ■ Solve a polynomial equation that has complex solutions.
- ■ Plot a complex number in the complex plane.
- ■ Determine whether a complex number is in the Mandelbrot Set.

The Imaginary Unit *i*

Some quadratic equations have no real solutions. For instance, the quadratic equations

$$x^2 + 1 = 0 \quad \text{and} \quad x^2 = -5 \qquad \text{Equations with no real solutions}$$

have no real solutions because there is no real number x that can be squared to produce a negative number. To overcome this deficiency, mathematicians utilized an expanded system of numbers that used the **imaginary unit *i*,** which is defined as

$$i = \sqrt{-1} \qquad \text{Imaginary unit}$$

where $i^2 = -1$. By adding real numbers to real multiples of this imaginary unit, we obtain the set of **complex numbers.** Each complex number can be written in the **standard form** $a + bi$.

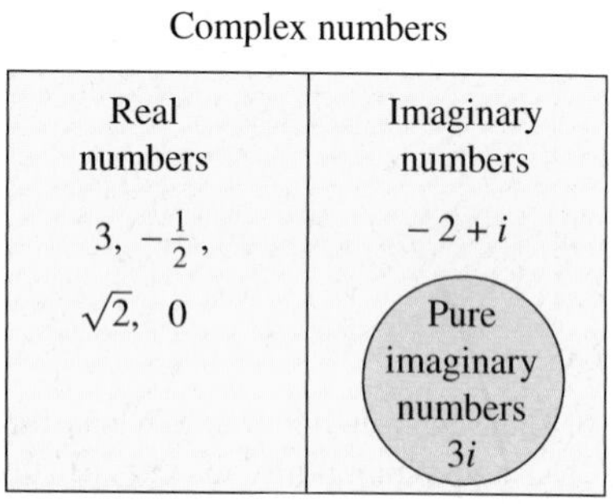

Complex numbers

Real numbers	Imaginary numbers
$3, -\frac{1}{2},$ $\sqrt{2}, 0$	$-2 + i$ Pure imaginary numbers $3i$

FIGURE 3.37

Definition of a Complex Number

If a and b are real numbers, the number $a + bi$ is called a **complex number,** and it is said to be written in **standard form.** If $b = 0$, the number $a + bi = a$ is a real number. If $b \neq 0$, the number $a + bi$ is called an **imaginary number.** A number of the form bi, where $b \neq 0$, is called a **pure imaginary number.**

The set of real numbers is a subset of the set of complex numbers, as shown in Figure 3.37. This is true because every real number a can be written as a complex number using $b = 0$. That is, for every real number a, we can write $a = a + 0i$.

Equality of Complex Numbers

Two complex numbers $a + bi$ and $c + di$ written in standard form are **equal** to each other,

$$a + bi = c + di \qquad \text{Equality of two complex numbers}$$

if and only if $a = c$ and $b = d$.

Operations with Complex Numbers

To add (or subtract) two complex numbers, you add (or subtract) the real and imaginary parts of the numbers separately.

Addition and Subtraction of Complex Numbers

If $a + bi$ and $c + di$ are two complex numbers written in standard form, their sum and difference are defined as follows.

Sum: $(a + bi) + (c + di) = (a + c) + (b + d)i$

Difference: $(a + bi) - (c + di) = (a - c) + (b - d)i$

The **additive identity** in the complex number system is zero (the same as in the real number system). Furthermore, the **additive inverse** of the complex number $a + bi$ is

$-(a + bi) = -a - bi.$ Additive inverse

So, you have

$(a + bi) + (-a - bi) = 0 + 0i = 0.$

Example 1   **Adding and Subtracting Complex Numbers**

Perform the operation(s) and write each result in standard form.

a. $(3 - i) + (2 + 3i)$ **b.** $2i + (-4 - 2i)$ **c.** $3 - (-2 + 3i) + (-5 + i)$

SOLUTION

a. $(3 - i) + (2 + 3i) = 3 - i + 2 + 3i$ Remove parentheses.

$= 3 + 2 - i + 3i$ Group like terms.

$= (3 + 2) + (-1 + 3)i$

$= 5 + 2i$ Write in standard form.

b. $2i + (-4 - 2i) = 2i - 4 - 2i$ Remove parentheses.

$= -4 + 2i - 2i$ Group like terms.

$= -4$ Write in standard form.

c. $3 - (-2 + 3i) + (-5 + i) = 3 + 2 - 3i - 5 + i$

$= 3 + 2 - 5 - 3i + i$

$= 0 - 2i$

$= -2i$

✓**CHECKPOINT 1**

Perform the operation(s) and write each result in standard form.

a. $(4 + 7i) + (1 - 6i)$ **b.** $3i - (-2 + 3i) - (2 + 5i)$ ∎

Note in Example 1(b) that the sum of two imaginary numbers can be a real number.

Many of the properties of real numbers are valid for complex numbers as well. Here are some examples.

Associative Properties of Addition and Multiplication

Commutative Properties of Addition and Multiplication

Distributive Property of Multiplication Over Addition

Notice how these properties are used when two complex numbers are multiplied.

$$(a + bi)(c + di) = a(c + di) + bi(c + di) \qquad \text{Distributive Property}$$
$$= ac + (ad)i + (bc)i + (bd)i^2 \qquad \text{Distributive Property}$$
$$= ac + (ad)i + (bc)i + (bd)(-1) \qquad i^2 = -1$$
$$= ac - bd + (ad)i + (bc)i \qquad \text{Commutative Property}$$
$$= (ac - bd) + (ad + bc)i \qquad \text{Associative Property}$$

Rather than trying to memorize this multiplication rule, you should simply remember how the Distributive Property is used to multiply two complex numbers. The procedure is similar to multiplying two binomials and combining like terms (as in the FOIL Method).

Example 2 Multiplying Complex Numbers

Find each product.

a. $(i)(-3i)$ **b.** $(2 - i)(4 + 3i)$ **c.** $(3 + 2i)(3 - 2i)$ **d.** $(3 + 2i)^2$

SOLUTION

a. $(i)(-3i) = -3i^2$ Multiply.

$ = -3(-1)$ $i^2 = -1$

$ = 3$ Simplify.

b. $(2 - i)(4 + 3i) = 8 + 6i - 4i - 3i^2$ Distributive Property

$ = 8 + 6i - 4i - 3(-1)$ $i^2 = -1$

$ = 8 + 3 + 6i - 4i$ Group like terms.

$ = 11 + 2i$ Write in standard form.

c. $(3 + 2i)(3 - 2i) = 9 - 6i + 6i - 4i^2$ Distributive Property

$ = 9 - 4(-1)$ $i^2 = -1$

$ = 9 + 4$ Simplify.

$ = 13$ Write in standard form.

d. $(3 + 2i)^2 = 9 + 6i + 6i + 4i^2$ Distributive Property

$ = 9 + 4(-1) + 12i$ $i^2 = -1$

$ = 9 - 4 + 12i$ Simplify.

$ = 5 + 12i$ Write in standard form.

DISCOVERY

Fill in the blanks:

$i^1 = i$ $i^5 = \rule{1cm}{0.4pt}$ $i^9 = \rule{1cm}{0.4pt}$

$i^2 = -1$ $i^6 = \rule{1cm}{0.4pt}$ $i^{10} = \rule{1cm}{0.4pt}$

$i^3 = -i$ $i^7 = \rule{1cm}{0.4pt}$ $i^{11} = \rule{1cm}{0.4pt}$

$i^4 = 1$ $i^8 = \rule{1cm}{0.4pt}$ $i^{12} = \rule{1cm}{0.4pt}$

What pattern do you see? Write a brief description of how you would find i raised to any positive integer power.

✓CHECKPOINT 2

Find each product.

a. $4(-2 + 3i)$

b. $(5 - 3i)^2$ ▪

Complex Conjugates

Notice in Example 2(c) that the product of two complex numbers can be a real number. This occurs with pairs of complex numbers of the form $a + bi$ and $a - bi$, called **complex conjugates.** In general, the product of two complex conjugates can be written as follows.

$$(a + bi)(a - bi) = a^2 - abi + abi - b^2i^2$$
$$= a^2 - b^2(-1) = a^2 + b^2$$

Complex conjugates can be used to write the quotient of $a + bi$ and $c + di$ in standard form, where c and d are not both zero. To do this, multiply the numerator and denominator by the complex conjugate of the denominator to obtain

$$\frac{a + bi}{c + di} = \frac{a + bi}{c + di}\left(\frac{c - di}{c - di}\right) = \frac{(ac + bd) + (bc - ad)i}{c^2 + d^2}.$$

TECHNOLOGY

Ⓣ Some graphing utilities can perform operations with complex numbers. For specific keystrokes, go to the text website at *college.hmco.com/ info/larsonapplied.*

Example 3 **Writing Quotients of Complex Numbers in Standard Form**

Write each quotient in standard form.

a. $\dfrac{1}{1 + i}$ b. $\dfrac{2 + 3i}{4 - 2i}$

SOLUTION

a.
$$\frac{1}{1 + i} = \frac{1}{1 + i}\left(\frac{1 - i}{1 - i}\right) \qquad \text{Multiply numerator and denominator by complex conjugate of denominator.}$$

$$= \frac{1 - i}{1^2 - i^2} \qquad \text{Expand.}$$

$$= \frac{1 - i}{1 - (-1)} \qquad i^2 = -1$$

$$= \frac{1 - i}{2} \qquad \text{Simplify.}$$

$$= \frac{1}{2} - \frac{1}{2}i \qquad \text{Write in standard form.}$$

b.
$$\frac{2 + 3i}{4 - 2i} = \frac{2 + 3i}{4 - 2i}\left(\frac{4 + 2i}{4 + 2i}\right) \qquad \text{Multiply numerator and denominator by complex conjugate of denominator.}$$

$$= \frac{8 + 4i + 12i + 6i^2}{16 - 4i^2} \qquad \text{Expand.}$$

$$= \frac{8 - 6 + 16i}{16 + 4} \qquad i^2 = -1$$

$$= \frac{2 + 16i}{20} \qquad \text{Simplify.}$$

$$= \frac{1}{10} + \frac{4}{5}i \qquad \text{Write in standard form.}$$

✓ CHECKPOINT 3

Write $\dfrac{6 - 7i}{1 - 2i}$ in standard form. ▪

Complex Solutions

When using the Quadratic Formula to solve a quadratic equation, you often obtain a result such as $\sqrt{-3}$, which you know is not a real number. By factoring out $i = \sqrt{-1}$, you can write this number in standard form.

$$\sqrt{-3} = \sqrt{3(-1)} = \sqrt{3}\sqrt{-1} = \sqrt{3}\,i$$

The number $\sqrt{3}\,i$ is called the *principal square root* of -3.

STUDY TIP

The definition of principal square root uses the rule

$$\sqrt{ab} = \sqrt{a}\sqrt{b}$$

for $a > 0$ and $b < 0$. This rule is not valid if *both* a and b are negative. For example,

$$\sqrt{(-5)(-5)} = \sqrt{25} = 5.$$

whereas

$$\sqrt{-5}\sqrt{-5} = 5i^2 = -5$$

To avoid problems with multiplying square roots of negative numbers, be sure to convert complex numbers to standard form *before* multiplying.

Principal Square Root of a Negative Number

If a is a positive number, the **principal square root** of the negative number $-a$ is defined as

$$\sqrt{-a} = \sqrt{a}\,i.$$

Example 4 **Writing Complex Numbers in Standard Form**

a. $\sqrt{-3}\sqrt{-12} = \sqrt{3}\,i\sqrt{12}\,i = \sqrt{36}\,i^2 = 6(-1) = -6$

b. $\sqrt{-48} - \sqrt{-27} = \sqrt{48}\,i - \sqrt{27}\,i = 4\sqrt{3}\,i - 3\sqrt{3}\,i = \sqrt{3}\,i$

c. $\left(-1 + \sqrt{-3}\right)^2 = \left(-1 + \sqrt{3}\,i\right)^2$

$$= (-1)^2 - 2\sqrt{3}\,i + \left(\sqrt{3}\right)^2(i^2)$$

$$= 1 - 2\sqrt{3}\,i + 3(-1)$$

$$= -2 - 2\sqrt{3}\,i$$

✓ CHECKPOINT 4

Write $4 + \sqrt{-9}$ in standard form. ■

Example 5 **Complex Solutions of a Quadratic Equation**

Solve $3x^2 - 2x + 5 = 0$.

SOLUTION

$$x = \frac{-(-2) \pm \sqrt{(-2)^2 - 4(3)(5)}}{2(3)} \qquad \text{Quadratic Formula}$$

$$= \frac{2 \pm \sqrt{-56}}{6} \qquad \text{Simplify.}$$

$$= \frac{2 \pm 2\sqrt{14}\,i}{6} \qquad \text{Write in } i\text{-form.}$$

$$= \frac{1}{3} \pm \frac{\sqrt{14}}{3}\,i \qquad \text{Write in standard form.}$$

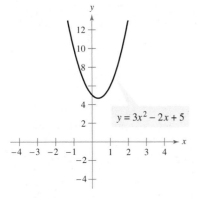

FIGURE 3.38

✓ CHECKPOINT 5

Solve $x^2 + 3x + 4 = 0$. ■

The graph of $f(x) = 3x^2 - 2x + 5$, shown in Figure 3.38, does not touch or cross the x-axis. This confirms that the equation in Example 5 has no real solution.

Applications

Most applications involving complex numbers are either theoretical (see the next section) or very technical, and so are not appropriate for inclusion in this text. However, to give you some idea of how complex numbers can be used in applications, a general description of their use in **fractal geometry** is presented.

To begin, consider a coordinate system called the **complex plane.** Just as every real number corresponds to a point on the real number line, every complex number corresponds to a point in the complex plane, as shown in Figure 3.39. In this figure, note that the vertical axis is the **imaginary axis** and the horizontal axis is the real axis. The point that corresponds to the complex number $a + bi$ is (a, b).

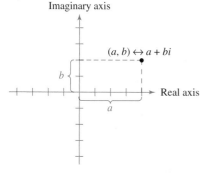

Imaginary axis

$(a, b) \leftrightarrow a + bi$

b

Real axis

a

FIGURE 3.39

Complex number $a + bi$ ⟹   Ordered pair (a, b)

From Figure 3.39, you can see that i is called the imaginary unit because it is located one unit from the origin on the imaginary axis of the complex plane.

Example 6 **Plotting Complex Numbers in the Complex Plane**

Plot each complex number in the complex plane.

a. $2 + 3i$ **b.** $-1 + 2i$ **c.** 4

SOLUTION

a. To plot the complex number $2 + 3i$, move (from the origin) two units to the right on the real axis and then three units upward. See Figure 3.40. In other words, plotting the complex number $2 + 3i$ in the complex plane is comparable to plotting the point $(2, 3)$ in the Cartesian plane.

b. The complex number $-1 + 2i$ corresponds to the point $(-1, 2)$. See Figure 3.40.

c. The complex number 4 corresponds to the point $(4, 0)$. See Figure 3.40.

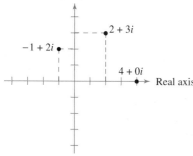

Imaginary axis

$2 + 3i$

$-1 + 2i$

$4 + 0i$

Real axis

FIGURE 3.40

✓**CHECKPOINT 6**

Plot $-3i$ in the complex plane. ■

In the hands of a person who understands "fractal geometry," the complex plane can become an easel on which stunning pictures, called **fractals,** can be drawn. The most famous such picture is called the **Mandelbrot Set,** named after the Polish-born mathematician Benoit Mandelbrot. To draw the Mandelbrot Set, consider the following sequence of numbers.

$$c, \quad c^2 + c, \quad (c^2 + c)^2 + c, \quad [(c^2 + c)^2 + c]^2 + c, \quad \ldots$$

The behavior of this sequence depends on the value of the complex number c. For some values of c, this sequence is **bounded,** which means that the absolute value of each number $\left(|a + bi| = \sqrt{a^2 + b^2}\right)$ in the sequence is less than some fixed number N. For other values of c, this sequence is **unbounded,** which means that the absolute values of the terms of the sequence become infinitely large. If the sequence is bounded, the complex number c is in the Mandelbrot Set, and if the sequence is unbounded, the complex number c is not in the Mandelbrot Set.

Example 7

MAKE A DECISION **Members of the Mandelbrot Set**

Decide whether each complex number is a member of the Mandelbrot Set.

a. -2 **b.** i **c.** $1 + i$

SOLUTION

a. For $c = -2$, the corresponding Mandelbrot sequence is

$$-2, \quad 2, \quad 2, \quad 2, \quad 2, \quad 2, \ldots$$

Because the sequence is bounded, the complex number -2 is in the Mandelbrot Set.

b. For $c = i$, the corresponding Mandelbrot sequence is

$$i, \quad -1 + i, \quad -i, \quad -1 + i, \quad -i, \quad -1 + i, \ldots$$

Because the sequence is bounded, the complex number i is in the Mandelbrot Set.

c. For $c = 1 + i$, the corresponding Mandelbrot sequence is

$$1 + i, \quad 1 + 3i, \quad -7 + 7i, \quad 1 - 97i, \quad -9407 - 193i,$$
$$88454401 + 3631103i, \ldots$$

Because the sequence is unbounded, the complex number $1 + i$ is *not* in the Mandelbrot Set.

✓ **CHECKPOINT 7**

Decide whether -3 is in the Mandelbrot Set. Explain your reasoning. ▪

With this definition, a picture of the Mandelbrot Set would have only two colors: one color for points that are in the set (the sequence is bounded) and one color for points that are outside the set (the sequence is unbounded). Figure 3.41 shows a black and yellow picture of the Mandelbrot Set. The points that are black are in the Mandelbrot Set and the points that are yellow are not.

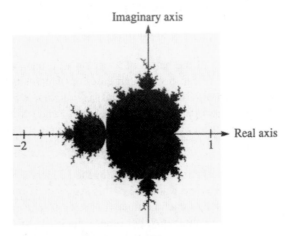

FIGURE 3.41 Mandelbrot Set

To add more interest to the picture, computer scientists discovered that the points that are not in the Mandelbrot Set can be assigned a variety of colors, depending on "how quickly" their sequences diverge. Figure 3.42 shows three different appendages of the Mandelbrot Set using a spectrum of colors. (The colored portions of the picture represent points that are *not* in the Mandelbrot Set.)

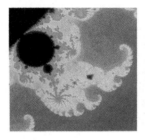 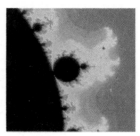 

American Mathematical Society

FIGURE 3.42

Figures 3.43, 3.44, and 3.45 show other types of fractal sets. From these pictures, you can see why fractals have fascinated people since their discovery (around 1980).

Fred Espenak/Photo Researchers, Inc.

FIGURE 3.43

Gregory Sams/Photo Researchers, Inc.

FIGURE 3.44

Francoise Sauze/Photo Researchers, Inc.

FIGURE 3.45

CONCEPT CHECK

1. Is $3 + \sqrt{-4}$ written in standard form? Explain.

2. Is $-(m + ni)$ the complex conjugate of $(m + ni)$? Use multiplication to justify your answer.

3. Is $-2\sqrt{2}$ the principal square root of -8? Explain.

4. Can the difference of two imaginary numbers be a real number? Justify your answer with an example.

Skills Review 3.5   The following warm-up exercises involve skills that were covered in earlier sections. You will use these skills in the exercise set for this section. For additional help, review Sections 0.4 and 1.4.

In Exercises 1–8, simplify the expression.

1. $\sqrt{12}$

2. $\sqrt{500}$

3. $\sqrt{20} - \sqrt{5}$

4. $\sqrt{27} - \sqrt{243}$

5. $\sqrt{24}\sqrt{6}$

6. $2\sqrt{18}\sqrt{32}$

7. $\dfrac{1}{\sqrt{3}}$

8. $\dfrac{2}{\sqrt{2}}$

In Exercises 9 and 10, solve the quadratic equation.

9. $x^2 + x - 1 = 0$

10. $x^2 + 2x - 1 = 0$

Exercises 3.5

See www.CalcChat.com for worked-out solutions to odd-numbered exercises.

1. Write out the first 16 positive integer powers of i $(i, i^2, i^3, \ldots, i^{16})$, and write each as i, $-i$, 1, or -1. What pattern do you observe?

2. Use the pattern you found in Exercise 1 to help you write each power of i as i, $-i$, 1, or -1.

   (a) i^{28} (b) i^{37} (c) i^{127} (d) i^{82}

In Exercises 3–6, find the real numbers a and b such that the equation is true.

3. $a + bi = 7 + 12i$

4. $a + bi = -2 - 5i$

5. $(a + 3) + (b - 1)i = 7 - 4i$

6. $(a + 6) + 2bi = 6 - 5i$

In Exercises 7–18, write the complex number in standard form and find its complex conjugate.

7. $9 + \sqrt{-16}$

8. $2 + \sqrt{-25}$

9. $-3 - \sqrt{-12}$

10. $1 + \sqrt{-8}$

11. -21

12. 45

13. $-6i + i^2$

14. $4i^2 - 2i^3$

15. $-5i^5$

16. $(-i)^3$

17. $\left(\sqrt{-6}\right)^2 + 3$

18. $\left(\sqrt{-4}\right)^2 - 5$

In Exercises 19–44, perform the indicated operation and write the result in standard form.

19. $(-4 + 3i) + (6 - 2i)$

20. $(13 - 2i) + (-5 + 6i)$

21. $(12 + 5i) - (7 - i)$

22. $(3 + 2i) - (6 + 13i)$

23. $\left(-2 + \sqrt{-8}\right) + \left(5 - \sqrt{-50}\right)$

24. $\left(5 + \sqrt{-18}\right) - \left(3 + \sqrt{-32}\right)$

25. $-\left(\frac{3}{2} + \frac{5}{2}i\right) + \left(\frac{5}{3} + \frac{11}{3}i\right)$

26. $(1.6 + 3.2i) + (-5.8 + 4.3i)$

27. $(3 + 4i)^2 + (3 - 4i)^2$

28. $(2 - 5i)^2 - (2 + 5i)^2$

29. $\sqrt{-3} \cdot \sqrt{-8}$

30. $\sqrt{-5} \cdot \sqrt{-10}$

31. $\left(\sqrt{-10}\right)^2$

32. $\left(\sqrt{-75}\right)^3$

33. $(2 + 3i)(1 - i)$

34. $(6 - 5i)(1 + i)$

35. $(3 + 4i)(3 - 4i)$

36. $(8 + 3i)(8 - 3i)$

37. $5i(4 - 6i)$

38. $-2i(7 + 9i)$

39. $(5 + 6i)^2$

40. $(3 - 7i)^2$

41. $\left(\sqrt{5} - \sqrt{3}i\right)\left(\sqrt{5} + \sqrt{3}i\right)$

42. $\left(\sqrt{14} + \sqrt{10}i\right)\left(\sqrt{14} - \sqrt{10}i\right)$

43. $\left(2 - \sqrt{-8}\right)\left(8 + \sqrt{-6}\right)$

44. $\left(3 + \sqrt{-5}\right)\left(7 - \sqrt{-10}\right)$

In Exercises 45–56, write the quotient in standard form.

45. $\dfrac{3 - i}{3 + i}$

46. $\dfrac{8 - 5i}{1 - 3i}$

47. $\dfrac{5}{4 - 2i}$

48. $\dfrac{3}{1 + 2i}$

49. $\dfrac{7 + 10i}{i}$

50. $\dfrac{8 + 15i}{3i}$

51. $\dfrac{1}{(2i)^3}$

52. $\dfrac{1}{(3i)^3}$

53. $\dfrac{4}{(1 - 2i)^3}$

54. $\dfrac{3}{(5 - 2i)^2}$

55. $\dfrac{(21 - 7i)(4 + 3i)}{2 - 5i}$

56. $\dfrac{(3 - i)(2 + 5i)}{4 + 3i}$

Error Analysis   In Exercises 57 and 58, a student has handed in the specified problem. Find the error(s) and discuss how to explain the error(s) to the student.

57. Write $\dfrac{5}{3 - 2i}$ in standard form.

$$\dfrac{5}{3 - 2i} \cdot \dfrac{3 + 2i}{3 + 2i} = \dfrac{15 + 10i}{9 - 4} = 3 + 2i$$

58. Multiply $\left(\sqrt{-4} + 3\right)\left(i - \sqrt{-3}\right)$.

$$\left(\sqrt{-4} + 3\right)\left(i - \sqrt{-3}\right)$$
$$= i\sqrt{-4} - \sqrt{-4}\sqrt{-3} + 3i - 3\sqrt{-3}$$
$$= -2i - \sqrt{12} + 3i - 3i\sqrt{3}$$
$$= \left(1 - 3\sqrt{3}\right)i - 2\sqrt{3}$$

In Exercises 59–66, solve the quadratic equation.

59. $x^2 - 2x + 2 = 0$

60. $x^2 + 6x + 10 = 0$

61. $4x^2 + 16x + 17 = 0$

62. $9x^2 - 6x + 37 = 0$

63. $4x^2 + 16x + 15 = 0$

64. $9x^2 - 6x + 35 = 0$

65. $16t^2 - 4t + 3 = 0$

66. $5s^2 + 6s + 3 = 0$

(T) In Exercises 67–70, solve the quadratic equation and then use a graphing utility to graph the related quadratic function in the standard viewing window. Discuss how the graph of the quadratic function relates to the solutions of the quadratic equation.

Equation	Function
67. $x^2 + x + 2 = 0$	$y = x^2 + x + 2$
68. $-x^2 + 3x - 5 = 0$	$y = -x^2 + 3x - 5$
69. $x^2 + 3x - 5 = 0$	$y = x^2 + 3x - 5$
70. $-x^2 - 3x + 4 = 0$	$y = -x^2 - 3x + 4$

In Exercises 71–76, plot the complex number.

71. 3

72. i

73. $-2 + i$

74. $-2 - 3i$

75. $1 - 2i$

76. $-2i$

In Exercises 77–82, decide whether the number is in the Mandelbrot Set. Explain your reasoning.

77. $c = 0$

78. $c = 2$

79. $c = 1$

80. $c = -1$

81. $c = \frac{1}{2}i$

82. $c = -i$

In Exercises 83 and 84, determine whether the statement is *true* or *false*. Explain.

83. There is no complex number that is equal to its conjugate.

84. The conjugate of the sum of two complex numbers is equal to the sum of the conjugates of the two complex numbers.

Business Capsule

© Greg Smith/CORBIS

Fractal Graphics, established in 1992, built a world-class reputation as a leader in application of 3-D visualization technology as applied to the interpretation of complex geoscientific models. In 2002, Fractal Graphics split to form the software development group Fractal Technologies Pty Ltd and the geological consulting group Fractal Geoscience. Fractal Technologies develops dimensional data management and visualization software for the geosciences. One of Fractal Technologies' product suites is FracSIS, which stores geological, geochemical, and geophysical data with an interactive visualization environment.

85. Research Project   Use your campus library, the Internet, or some other reference source to find information about a company that uses algorithms to generate 3-D images or gaming software. Write a brief paper about such a company or small business.

Section 3.6

The Fundamental Theorem of Algebra

- Use the Fundamental Theorem of Algebra and the Linear Factorization Theorem to write a polynomial as the product of linear factors.
- Find a polynomial with real coefficients whose zeros are given.
- Factor a polynomial over the rational, real, and complex numbers.
- Find all real and complex zeros of a polynomial function.

The Fundamental Theorem of Algebra

You have been using the fact that an nth-degree polynomial function can have at most n real zeros. In the complex number system, this statement can be improved. That is, in the complex number system, every nth-degree polynomial function has *precisely* n zeros. This important result is derived from the **Fundamental Theorem of Algebra,** first proved by the famous German mathematician Carl Friedrich Gauss (1777–1855).

The Fundamental Theorem of Algebra

If $f(x)$ is a polynomial of degree n, where $n > 0$, then f has at least one zero in the complex number system.

Using the Fundamental Theorem of Algebra and the equivalence of zeros and factors, you obtain the following theorem.

Linear Factorization Theorem

If $f(x)$ is a polynomial of degree n

$$f(x) = a_n x^n + a_{n-1} x^{n-1} + \cdots + a_1 x + a_0$$

where $n > 0$, then $f(x)$ has precisely n linear factors

$$f(x) = a_n(x - c_1)(x - c_2) \cdots (x - c_n)$$

where $c_1, c_2, \ldots, c_n$ are complex numbers and a_n is the leading coefficient of $f(x)$.

Note that neither the Fundamental Theorem of Algebra nor the Linear Factorization Theorem tells you *how* to find the zeros or factors of a polynomial. Such theorems are called **existence theorems.** To find the zeros of a polynomial function, you still must rely on the techniques developed in the earlier parts of the text.

Remember that the n zeros of a polynomial function can be real or complex, and they may be repeated. Example 1 illustrates several cases.

Example 1 **Zeros of Polynomial Functions**

Determine the number of zeros of each polynomial function. Then list the zeros.

a. $f(x) = x - 2$ **b.** $f(x) = x^2 - 6x + 9$

c. $f(x) = x^3 + 4x$ **d.** $f(x) = x^4 - 1$

SOLUTION

a. The first-degree polynomial function given by $f(x) = x - 2$ has exactly *one* zero: $x = 2$.

b. Counting multiplicity, the second-degree polynomial function given by

$$f(x) = x^2 - 6x + 9 = (x - 3)(x - 3)$$

has exactly *two* zeros: $x = 3$ and $x = 3$.

c. The third-degree polynomial function given by

$$f(x) = x^3 + 4x = x(x - 2i)(x + 2i)$$

has exactly *three* zeros: $x = 0, x = 2i$, and $x = -2i$.

d. The fourth-degree polynomial function given by

$$f(x) = x^4 - 1 = (x - 1)(x + 1)(x - i)(x + i)$$

has exactly *four* zeros: $x = 1, x = -1, x = i$, and $x = -i$. _____

✓**CHECKPOINT 1**

Determine the number of zeros of $f(x) = x^4 - 36$. Then list the zeros. ■

TECHNOLOGY

Remember that when you use a graphing utility to locate the zeros of a function, the only zeros that appear as x-intercepts are the *real zeros*. Compare the graphs below with the four polynomial functions in Example 1. Which zeros appear on the graphs?

(a)
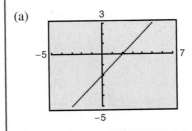

(b)
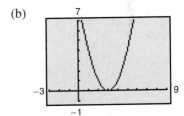

(c)
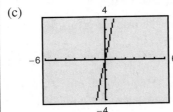

(d)
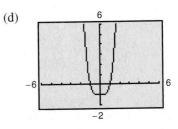

Example 2 shows how you can use the methods described in Sections 3.3 and 3.4 (the Rational Zero Test, synthetic division, and factoring) to find all the zeros of a polynomial function, including the complex zeros.

Example 2 Finding the Zeros of a Polynomial Function

Find all of the zeros of

$$f(x) = x^5 + x^3 + 2x^2 - 12x + 8$$

and write the polynomial as a product of linear factors.

SOLUTION From the Rational Zero Test, the possible rational zeros are ± 1, $\pm 2, \pm 4$, and ± 8. Synthetic division produces the following.

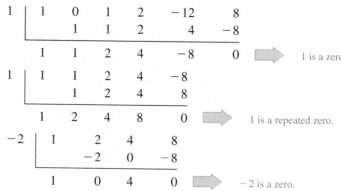

1	1	0	1	2	-12	8		
		1	1	2	4	-8		
	1	1	2	4	-8	0		1 is a zero.

1	1	1	2	4	-8		
		1	2	4	8		
	1	2	4	8	0		1 is a repeated zero.

-2	1	2	4	8		
		-2	0	-8		
	1	0	4	0		-2 is a zero.

So, you have

$$f(x) = x^5 + x^3 + 2x^2 - 12x + 8$$
$$= (x - 1)(x - 1)(x + 2)(x^2 + 4).$$

By factoring $x^2 + 4$ as the difference of two squares over the imaginary numbers

$$x^2 - (-4) = \left(x - \sqrt{-4}\right)\left(x + \sqrt{-4}\right)$$
$$= (x - 2i)(x + 2i)$$

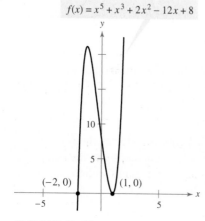

$$f(x) = x^5 + x^3 + 2x^2 - 12x + 8$$

FIGURE 3.46

you obtain

$$f(x) = (x - 1)(x - 1)(x + 2)(x - 2i)(x + 2i)$$

which gives the following five zeros of f.

$$1, \quad 1, \quad -2, \quad 2i, \quad \text{and} \quad -2i$$

Note from the graph of f shown in Figure 3.46 that the *real* zeros are the only ones that appear as x-intercepts.

✓ CHECKPOINT 2

Find all the zeros of each function and write the polynomial as the product of linear factors.

a. $f(x) = x^4 + 8x^2 - 9$

b. $g(x) = x^5 - 5x^4 + 4x^3 + 4x^2 + 3x + 9$ ∎

DISCOVERY

Use a graphing utility to graph

$$f(x) = x^3 + x^2 - 2x + 1$$

and

$$g(x) = x^3 + x^2 - 2x - 1.$$

How many zeros does f have? How many zeros does g have? Is it possible for an odd-degree polynomial function with real coefficients to have no real zeros (only complex zeros)? Can an even-degree polynomial function with real coefficients have only imaginary zeros? If so, how does the graph of such a polynomial function behave?

Conjugate Pairs

In Example 2, note that the two imaginary zeros are **conjugates.** That is, they are of the form $a + bi$ and $a - bi$.

Complex Zeros Occur in Conjugate Pairs

Let f be a polynomial function that has *real coefficients*. If $a + bi$, where $b \neq 0$, is a zero of the function, then the conjugate $a - bi$ is also a zero of the function.

Be sure you see that this result is true only if the polynomial function has *real coefficients.* For instance, the result applies to the function given by $f(x) = x^2 + 1$, but not to the function given by $g(x) = x - i$.

You have been using the Rational Zero Test, synthetic division, and factoring to find the zeros of polynomial functions. The Linear Factorization Theorem enables you to reverse this process and find a polynomial function when its zeros are given.

Example 3   Finding a Polynomial Function with Given Zeros

Find a *fourth-degree* polynomial function with real coefficients that has -1, -1, and $3i$ as zeros.

SOLUTION   Because $3i$ is a zero *and* the function is stated to have real coefficients, you know that the conjugate $-3i$ must also be a zero. So, -1, -1, $3i$, and $-3i$ are the four zeros and from the Linear Factorization Theorem, $f(x)$ can be written as a product of linear factors, as shown.

$$f(x) = a(x + 1)(x + 1)(x - 3i)(x + 3i)$$

For simplicity, let $a = 1$. Then multiply the factors with real coefficients to get $(x^2 + 2x + 1)$ and multiply the complex conjugates to get $(x^2 + 9)$. So, you obtain the following fourth-degree polynomial function.

$$f(x) = (x^2 + 2x + 1)(x^2 + 9)$$
$$= x^4 + 2x^3 + 10x^2 + 18x + 9$$

✓ CHECKPOINT 3

Find a *fourth-degree* polynomial function with real coefficients that has -3, 3, and $2i$ as zeros. ▪

Factoring a Polynomial

The Linear Factorization Theorem shows that you can write any nth-degree polynomial as the product of n linear factors.

$$f(x) = a_n(x - c_1)(x - c_2)(x - c_3) \cdots (x - c_n)$$

However, this result includes the possibility that some of the values of c_i are complex. The following result implies that even if you do not want to get involved with "imaginary factors," you can still write $f(x)$ as the product of linear and/or quadratic factors.

> **Factors of a Polynomial**
>
> Every polynomial of degree $n > 0$ with real coefficients can be written as the product of linear and quadratic factors with real coefficients, where the quadratic factors have no real zeros.

A quadratic factor with no real zeros is said to be **irreducible over the reals.** Be sure you see that this is not the same as being *irreducible over the rationals.* For example, the quadratic

$$x^2 + 1 = (x - i)(x + i)$$

is irreducible over the reals (and therefore over the rationals). On the other hand, the quadratic

$$x^2 - 2 = \left(x - \sqrt{2}\right)\left(x + \sqrt{2}\right)$$

is irreducible over the rationals, but it is *reducible* over the reals.

Example 4   Factoring a Polynomial

Use the polynomial $x^4 - x^2 - 20$ to complete the following.

a. Write the polynomial as the product of factors that are irreducible over the *rationals.*

b. Write the polynomial as the product of linear factors and quadratic factors that are irreducible over the *reals.*

c. Write the polynomial in completely factored form.

d. How many of the zeros are rational, irrational, or imaginary?

SOLUTION

a. Begin by factoring the polynomial into the product of two quadratic polynomials.

$$x^4 - x^2 - 20 = (x^2 - 5)(x^2 + 4)$$

Both of these factors are irreducible over the rationals.

b. By factoring over the reals, you have

$$x^4 - x^2 - 20 = \left(x + \sqrt{5}\right)\left(x - \sqrt{5}\right)(x^2 + 4)$$

where the quadratic factor is irreducible over the reals.

c. In completely factored form, you have

$$x^4 - x^2 - 20 = \left(x + \sqrt{5}\right)\left(x - \sqrt{5}\right)(x - 2i)(x + 2i).$$

d. Using the completely factored form, you can conclude that there are no rational zeros, two irrational zeros $\left(\pm\sqrt{5}\right)$, and two imaginary zeros $(\pm 2i)$.

✓ CHECKPOINT 4

In Example 4, complete parts (a)–(d) using the polynomial $x^4 + x^2 - 12.$ ∎

Example 5   **Finding the Zeros of a Polynomial Function**

Find all the zeros of $f(x) = x^4 - 3x^3 + 6x^2 + 2x - 60$, given that $1 + 3i$ is a zero of f.

SOLUTION   Because imaginary zeros occur in conjugate pairs, you know that $1 - 3i$ is also a zero of f. This means that both

$$[x - (1 + 3i)] \quad \text{and} \quad [x - (1 - 3i)]$$

are factors of $f(x)$. Multiplying these two factors produces

$$[x - (1 + 3i)][x - (1 - 3i)] = [(x - 1) - 3i][(x - 1) + 3i]$$
$$= (x - 1)^2 - 9i^2$$
$$= x^2 - 2x + 10.$$

Using long division, you can divide $x^2 - 2x + 10$ into $f(x)$ to obtain the following.

$$
\require{enclose}
\begin{array}{r}
x^2 - x - 6 \\[-3pt]
x^2 - 2x + 10 \enclose{longdiv}{x^4 - 3x^3 + 6x^2 + 2x - 60} \\
\end{array}
$$

$$
\begin{aligned}
&\underline{x^4 - 2x^3 + 10x^2}\\
&-x^3 - 4x^2 + 2x\\
&\underline{-x^3 + 2x^2 - 10x}\\
&-6x^2 + 12x - 60\\
&\underline{-6x^2 + 12x - 60}\\
&0
\end{aligned}
$$

So, you have

$$f(x) = (x^2 - 2x + 10)(x^2 - x - 6)$$
$$= (x^2 - 2x + 10)(x - 3)(x + 2)$$

and you can conclude that the zeros of f are $1 + 3i$, $1 - 3i$, 3, and -2.

✓CHECKPOINT 5

Find all the zeros of $f(x) = 3x^3 - 5x^2 + 48x - 80$, given that $4i$ is a zero of f. ■

CONCEPT CHECK

1. Given that $2 + 3i$ is a zero of a polynomial function f with real coefficients, name another zero of f.

2. Explain how to find a second-degree polynomial function with real coefficients that has $-i$ as a zero.

3. Explain the difference between a polynomial that is irreducible over the rationals and a polynomial that is irreducible over the reals. Justify your answer with examples.

4. Does the Fundamental Theorem of Algebra indicate that a cubic function must have at least one imaginary zero? Explain.

Skills Review 3.6

The following warm-up exercises involve skills that were covered in earlier sections. You will use these skills in the exercise set for this section. For additional help, review Section 3.5.

In Exercises 1–4, write the complex number in standard form and find its complex conjugate.

1. $4 - \sqrt{-29}$ **2.** $-5 - \sqrt{-144}$ **3.** $-1 + \sqrt{-32}$ **4.** $6 + \sqrt{-1/4}$

In Exercises 5–10, perform the indicated operation and write the result in standard form.

5. $(-3 + 6i) - (10 - 3i)$

6. $(12 - 4i) + 20i$

7. $(4 - 2i)(3 + 7i)$

8. $(2 - 5i)(2 + 5i)$

9. $\dfrac{1 + i}{1 - i}$

10. $(3 + 2i)^3$

Exercises 3.6

See www.CalcChat.com for worked-out solutions to odd-numbered exercises.

In Exercises 1–6, determine the number of zeros of the polynomial function.

1. $f(x) = x - 7$ **2.** $g(x) = x^4 - 256$

3. $h(x) = -x^3 + 2x^2 - 5$ **4.** $f(t) = -2t^5 - 3t^3 + 1$

5. $f(x) = 6x - x^4$

6. $f(x) = 3 - 7x^2 - 5x^4 + 9x^6$

In Exercises 7–34, find all the zeros of the function and write the polynomial as a product of linear factors.

7. $f(x) = x^2 + 16$ **8.** $f(x) = x^2 + 36$

9. $h(x) = x^2 - 5x + 5$ **10.** $g(x) = x^2 + 10x + 23$

11. $f(x) = x^4 - 81$ **12.** $f(t) = t^4 - 625$

13. $g(x) = x^3 + 5x$ **14.** $g(x) = x^3 + 7x$

15. $h(x) = x^3 - 11x^2 - 15x + 325$

16. $h(x) = x^3 - 3x^2 + 4x - 2$

17. $g(x) = x^3 - 6x^2 + 13x - 10$

18. $f(x) = x^3 - 2x^2 - 11x + 52$

19. $f(t) = t^3 - 3t^2 - 15t + 125$

20. $f(x) = x^3 + 8x^2 + 20x + 13$

21. $f(x) = x^3 + 24x^2 + 214x + 740$

22. $h(x) = x^3 - x + 6$

23. $h(x) = x^3 + 9x^2 + 27x + 35$

24. $f(s) = 2s^3 - 5s^2 + 12s - 5$

25. $f(x) = 16x^3 - 20x^2 - 4x + 15$

26. $f(x) = 9x^3 - 15x^2 + 11x - 5$

27. $f(x) = 5x^3 - 9x^2 + 28x + 6$

28. $g(x) = 3x^3 - 4x^2 + 8x + 8$

29. $g(x) = x^4 - 4x^3 + 8x^2 - 16x + 16$

30. $h(x) = x^4 + 6x^3 + 10x^2 + 6x + 9$

31. $f(x) = x^4 + 10x^2 + 9$

32. $f(x) = x^4 + 29x^2 + 100$

33. $f(t) = t^5 + 5t^4 - 7t^3 - 43t^2 - 8t - 48$

34. $g(x) = x^5 - 8x^4 + 28x^3 - 56x^2 + 64x - 32$

In Exercises 35–44, find a polynomial with real coefficients that has the given zeros. (There are many correct answers.)

35. $-2, 3i, -3i$ **36.** $5, 2i, -2i$

37. $1, 2 + i, 2 - i$ **38.** $6, -5 + 2i, -5 - 2i$

39. $-4, 3i, -3i, 2i, -2i$ **40.** $2, 2, 2, 4i, -4i$

41. $-5, -5, 1 + \sqrt{3}\,i$ **42.** $0, 0, 4, 1 + i$

43. $\frac{2}{3}, -1, 3 + \sqrt{2}\,i$ **44.** $\frac{3}{4}, -2, -\frac{1}{2} + i$

In Exercises 45–48, write the polynomial (a) as the product of factors that are irreducible over the *rationals,* (b) as the product of linear and quadratic factors that are irreducible over the *reals,* and (c) in completely factored form.

45. $x^4 - 7x^2 - 8$

46. $x^4 - 6x^2 - 72$

47. $x^4 - 5x^3 + 4x^2 + x - 15$
(*Hint:* One factor is $x^2 - 2x + 3$.)

48. $x^4 + x^3 + 8x^2 + 9x - 9$
(*Hint:* One factor is $x^2 + 9$.)

In Exercises 49–58, use the given zero of f to find all the zeros of f.

49. $f(x) = 3x^3 - 7x^2 + 12x - 28,\ 2i$

50. $f(x) = 3x^3 - x^2 + 27x - 9,\ 3i$

51. $f(x) = x^4 - 2x^3 + 37x^2 - 72x + 36,\quad 6i$

52. $f(x) = x^3 - 7x^2 - x + 87,\quad 5 + 2i$

53. $f(x) = 4x^3 + 23x^2 + 34x - 10,\quad -3 + i$

54. $f(x) = 3x^3 - 10x^2 + 31x + 26,\ 2 + 3i$

55. $f(x) = x^4 + 3x^3 - 5x^2 - 21x + 22,\ -3 + \sqrt{2}\,i$

56. $f(x) = 2x^3 - 13x^2 + 34x - 35,\ 2 - \sqrt{3}\,i$

57. $f(x) = 8x^3 - 14x^2 + 18x - 9,\ \frac{1}{2}(1 - \sqrt{5}\,i)$

58. $f(x) = 25x^3 - 55x^2 - 54x - 18,\ \frac{1}{5}(-2 + \sqrt{2}\,i)$

(T) 59. Graphic Reasoning   Solve $x^4 - 5x^2 + 4 = 0$. Then use a graphing utility to graph

$$y = x^4 - 5x^2 + 4.$$

What is the connection between the solutions you found and the intercepts of the graph?

(T) 60. Graphical Reasoning   Solve $x^4 + 5x^2 + 4 = 0$. Then use a graphing utility to graph

$$y = x^4 + 5x^2 + 4.$$

What is the connection between the solutions you found and the intercepts of the graph?

(T) 61. Graphical Analysis   Find a fourth-degree polynomial function that has (a) four real zeros, (b) two real zeros, and (c) no real zeros. Use a graphing utility to graph the functions and describe the similarities and differences among them.

(T) 62. Graphical Analysis   Find a sixth-degree polynomial function that has (a) six real zeros, (b) four real zeros, (c) two real zeros, and (d) no real zeros. Use a graphing utility to graph the functions and describe the similarities and differences among them.

(T) 63. Profit   The demand and cost equations for a stethoscope are given by

$$p = 140 - 0.0001x$$

and

$$C = 80x + 150,000$$

where p is the unit price (in dollars), C is the total cost (in dollars), and x is the number of units. The total profit P (in dollars) obtained by producing and selling x units is given by

$$P = R - C = xp - C.$$

Try to determine a price p that would yield a profit of $9 million, and then use a graphing utility to explain why this is not possible.

(T) 64. Revenue   The demand equation for a stethoscope is given by

$$p = 140 - 0.0001x$$

where p is the unit price (in dollars) and x is the number of units sold. The total revenue R obtained by producing and selling x units is given by

$$R = xp.$$

Try to determine a price p that would yield a revenue of $50 million, and then use a graphing utility to explain why this is not possible.

65. Reasoning   The imaginary number $2i$ is a zero of

$$f(x) = x^3 - 2ix^2 - 4x + 8i$$

but the complex conjugate of $2i$ is not a zero of $f(x)$. Is this a contradiction of the conjugate pairs statement on page 317? Explain.

66. Reasoning   The imaginary number $1 - 2i$ is a zero of

$$f(x) = x^3 - (1 - 2i)x^2 - 9x + 9(1 - 2i)$$

but $1 + 2i$ is not a zero of $f(x)$. Is this a contradiction of the conjugate pairs statement on page 317? Explain.

67. Reasoning   Let f be a fourth-degree polynomial function with real coefficients. Three of the zeros of f are

$$3, 1 + i, \text{ and } 1 - i.$$

Explain how you know that the fourth zero must be a real number.

68. Reasoning   Let f be a fourth-degree polynomial function with real coefficients. Three of the zeros of f are

$$-1, 2, \text{ and } 3 + 2i.$$

What is the fourth zero? Explain.

69. Reasoning   Let f be a third-degree polynomial function with real coefficients. Explain how you know that f must have at least one zero that is a real number.

70. Reasoning   Let f be a fifth-degree polynomial function with real coefficients. Explain how you know that f must have at least one zero that is a real number.

(T) 71. Think About It   A student claims that a third-degree polynomial function with real coefficients can have three complex zeros. Describe how you could use a graphing utility and the Leading Coefficient Test (Section 3.2) to convince the student otherwise.

72. Think About It   A student claims that the polynomial

$$x^4 - 7x^2 + 12$$

may be factored over the rational numbers as

$$\left(x - \sqrt{3}\right)\left(x + \sqrt{3}\right)(x - 2)(x + 2).$$

Do you agree with this claim? Explain your answer.

Rational Functions

- Find the domain of a rational function.
- Find the vertical and horizontal asymptotes of the graph of a rational function.
- Sketch the graph of a rational function.
- Sketch the graph of a rational function that has a slant asymptote.
- Use a rational function model to solve an application problem.

Introduction

A **rational function** is one that can be written in the form

$$f(x) = \frac{p(x)}{q(x)}$$

where $p(x)$ and $q(x)$ are polynomials and $q(x)$ is not the zero polynomial. In this section, assume that $p(x)$ and $q(x)$ have no common factors. Unlike polynomial functions, whose domains consist of all real numbers, rational functions often have restricted domains. In general, the *domain* of a rational function of x includes all real numbers except x-values that make the denominator zero.

Example 1 Finding the Domain of a Rational Function

Find the domain of $f(x) = \dfrac{1}{x}$ and discuss its behavior near any excluded x-values.

SOLUTION The domain of f is all real numbers except $x = 0$. To determine the behavior of f near this x-value, evaluate $f(x)$ to the left and right of $x = 0$.

				x approaches 0 from the left		
x	-1	-0.5	-0.1	-0.01	-0.001	$\longrightarrow 0$
$f(x)$	-1	-2	-10	-100	-1000	$\longrightarrow -\infty$

			x approaches 0 from the right			
x	$0 \longleftarrow$	0.001	0.01	0.1	0.5	1
$f(x)$	$\infty \longleftarrow$	1000	100	10	2	1

Note that as x approaches 0 *from the left,* $f(x)$ decreases without bound. In contrast, as x approaches 0 *from the right,* $f(x)$ increases without bound. The graph of f is shown in Figure 3.47.

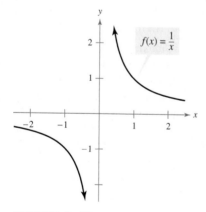

FIGURE 3.47

✓ CHECKPOINT 1

Find the domain of $f(x) = \dfrac{1}{x-1}$ and discuss the behavior of f near any excluded x-values. ■

Horizontal and Vertical Asymptotes

In Example 1, the behavior of f near $x = 0$ is denoted as follows.

$$f(x) \to -\infty \text{ as } x \to 0^-$$

$f(x)$ decreases without bound
as x approaches 0 from the left.

$$f(x) \to \infty \text{ as } x \to 0^+$$

$f(x)$ increases without bound
as x approaches 0 from the right.

The line $x = 0$ is a **vertical asymptote** of the graph of f, as shown in Figure 3.48. In this figure, note that the graph of f also has a **horizontal asymptote**—the line $y = 0$. The behavior of f near $y = 0$ is denoted as follows.

$$f(x) \to 0 \text{ as } x \to -\infty$$

$f(x)$ approaches 0 as x
decreases without bound.

$$f(x) \to 0 \text{ as } x \to \infty$$

$f(x)$ approaches 0 as x
increases without bound.

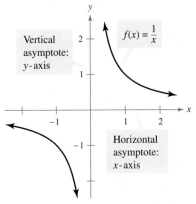

Vertical
asymptote:
y-axis

$f(x) = \dfrac{1}{x}$

Horizontal
asymptote:
x-axis

FIGURE 3.48

Definition of Vertical and Horizontal Asymptotes

1. The line $x = a$ is a **vertical asymptote** of the graph of f if

$$f(x) \to \infty \quad \text{or} \quad f(x) \to -\infty$$

   as $x \to a$, either from the right or from the left.

2. The line $y = b$ is a **horizontal asymptote** of the graph of f if

$$f(x) \to b$$

   as $x \to \infty$ or $x \to -\infty$.

The graph of a rational function can never intersect its vertical asymptote. It may or may not intersect its horizontal asymptote. In either case, the distance between the horizontal asymptote and the points on the graph must approach zero (as $x \to \infty$ or $x \to -\infty$). Figure 3.49 shows the horizontal and vertical asymptotes of the graphs of three rational functions.

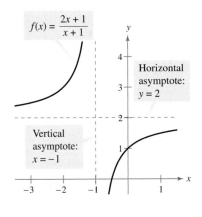

$f(x) = \dfrac{2x + 1}{x + 1}$

Horizontal
asymptote:
$y = 2$

Vertical
asymptote:
$x = -1$

(a)

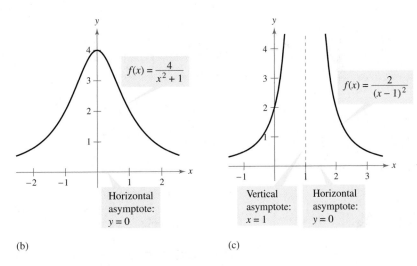

$f(x) = \dfrac{4}{x^2 + 1}$

Horizontal
asymptote:
$y = 0$

(b)

$f(x) = \dfrac{2}{(x - 1)^2}$

Vertical
asymptote:
$x = 1$

Horizontal
asymptote:
$y = 0$

(c)

FIGURE 3.49

The graphs of $f(x) = 1/x$ in Figure 3.48 and $f(x) = (2x + 1)/(x + 1)$ in Figure 3.49(a) are **hyperbolas.**

Asymptotes of a Rational Function

Let f be the rational function given by

$$f(x) = \frac{p(x)}{q(x)} = \frac{a_n x^n + a_{n-1} x^{n-1} + \cdots + a_1 x + a_0}{b_m x^m + b_{m-1} x^{m-1} + \cdots + b_1 x + b_0}, \quad a_n \neq 0, b_m \neq 0.$$

1. The graph of f has *vertical* asymptotes at the zeros of $q(x)$.

2. The graph of f has one or no *horizontal* asymptote determined by comparing the degrees of $p(x)$ and $q(x)$.

   a. If $n < m$, the graph of f has the line $y = 0$ (the x-axis) as a horizontal asymptote.

   b. If $n = m$, the graph of f has the line $y = a_n/b_m$ (ratio of the leading coefficients) as a horizontal asymptote.

   c. If $n > m$, the graph of f has no horizontal asymptote.

Example 2   Finding Horizontal and Vertical Asymptotes

Find all horizontal and vertical asymptotes of the graph of each rational function.

a. $f(x) = \dfrac{2x}{3x^2 + 1}$ **b.** $f(x) = \dfrac{2x^2}{x^2 - 1}$

SOLUTION

a. For this rational function, the degree of the numerator is *less than* the degree of the denominator, so the graph has the line $y = 0$ as a horizontal asymptote. To find any vertical asymptotes, set the denominator equal to zero and solve the resulting equation for x. Because the equation $3x^2 + 1 = 0$ has no real solutions, you can conclude that the graph has no vertical asymptote. The graph of the function is shown in Figure 3.50(a).

b. For this rational function, the degree of the numerator is *equal to* the degree of the denominator. The leading coefficient of the numerator is 2 and the leading coefficient of the denominator is 1, so the graph has the line $y = 2$ as a horizontal asymptote. To find any vertical asymptotes, set the denominator equal to zero and solve the resulting equation for x.

$$x^2 - 1 = 0 \qquad \text{Set denominator equal to zero.}$$

$$(x + 1)(x - 1) = 0 \qquad \text{Factor.}$$

$$x + 1 = 0 \implies x = -1 \qquad \text{Set 1st factor equal to 0.}$$

$$x - 1 = 0 \implies x = 1 \qquad \text{Set 2nd factor equal to 0.}$$

This equation has two real solutions $x = -1$ and $x = 1$, so the graph has the lines $x = -1$ and $x = 1$ as vertical asymptotes. The graph of the function is shown in Figure 3.50(b).

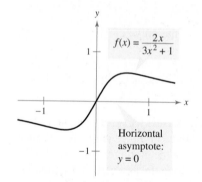

(a)

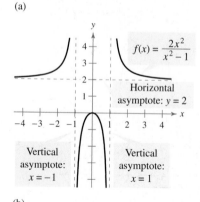

(b)

FIGURE 3.50

✓ CHECKPOINT 2

Find all horizontal and vertical asymptotes of the graph of $f(x) = \dfrac{x^2}{x^2 + 1}$. ▪

Sketching the Graph of a Rational Function

DISCOVERY

Consider the rational function

$$f(x) = \frac{x^2 - 4}{x - 2}.$$

Is $x = 2$ in the domain of f? Graph f on a graphing utility. Is there a vertical asymptote at $x = 2$? Describe the graph of f. Factor the numerator and reduce the rational function. Describe the resulting function. Under what conditions will a rational function have no vertical asymptote?

Guidelines for Graphing Rational Functions

Let $f(x) = p(x)/q(x)$, where $p(x)$ and $q(x)$ are polynomials with no common factors.

1. Find and plot the y-intercept (if any) by evaluating $f(0)$.

2. Find the zeros of the numerator (if any) by solving the equation $p(x) = 0$. Then plot the corresponding x-intercepts.

3. Find the zeros of the denominator (if any) by solving the equation $q(x) = 0$. Then sketch the corresponding vertical asymptotes.

4. Find and sketch the horizontal asymptote (if any) by using the rule for finding the horizontal asymptote of a rational function.

5. Test for symmetry.

6. Plot at least one point both *between and beyond* each x-intercept and vertical asymptote.

7. Use smooth curves to complete the graph between and beyond the vertical asymptotes.

Testing for symmetry can be useful, especially for simple rational functions. For example, the graph of $f(x) = 1/x$ is symmetric with respect to the origin, and the graph of $g(x) = 1/x^2$ is symmetric with respect to the y-axis.

Example 3   Sketching the Graph of a Rational Function

Sketch the graph of $g(x) = \dfrac{3}{x - 2}$.

SOLUTION Begin by noting that the numerator and denominator have no common factors.

y-intercept:	$\left(0, -\frac{3}{2}\right)$, because $g(0) = -\frac{3}{2}$
x-intercept:	None, numerator has no zeros.
Vertical asymptote:	$x = 2$, zero of denominator
Horizontal asymptote:	$y = 0$, degree of $p(x) <$ degree of $q(x)$

Additional points:

x	-4	1	2	3	5
$g(x)$	-0.5	-3	Undefined	3	1

By plotting the intercepts, asymptotes, and a few additional points, you can obtain the graph shown in Figure 3.51. In the figure, note that the graph of g is a vertical stretch and a right shift of the graph of $y = 1/x$.

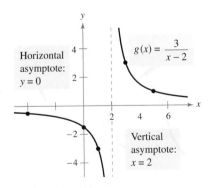

Horizontal asymptote: $y = 0$

$g(x) = \dfrac{3}{x - 2}$

Vertical asymptote: $x = 2$

FIGURE 3.51

✓ CHECKPOINT 3

Sketch the graph of $f(x) = \dfrac{1}{x + 2}$. ∎

Note that in the examples in this section, the vertical asymptotes are included in the table of additional points. This is done to emphasize numerically the behavior of the graph of the function.

Example 4 | Sketching the Graph of a Rational Function

Sketch the graph of $f(x) = \dfrac{x}{x^2 - x - 2}$.

SOLUTION Factor the denominator to determine more easily the zeros of the denominator.

$$f(x) = \frac{x}{x^2 - x - 2} = \frac{x}{(x + 1)(x - 2)}$$

y-intercept: $(0, 0)$, because $f(0) = 0$

x-intercept: $(0, 0)$

Vertical asymptotes: $x = -1, x = 2$, zeros of denominator

Horizontal asymptote: $y = 0$, degree of $p(x) <$ degree of $q(x)$

Additional points:

x	-3	-1	-0.5	1	2	3
$f(x)$	-0.3	Undefined	0.4	-0.5	Undefined	0.75

The graph is shown in Figure 3.52. Confirm the graph with your graphing utility.

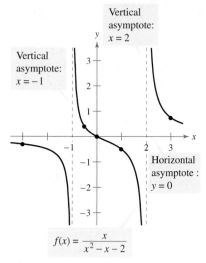

Vertical asymptote: $x = -1$

Vertical asymptote: $x = 2$

Horizontal asymptote: $y = 0$

$f(x) = \dfrac{x}{x^2 - x - 2}$

FIGURE 3.52

✓ CHECKPOINT 4

Sketch the graph of $f(x) = \dfrac{3x}{x^2 + x - 6}$. ■

Example 5 | Sketching the Graph of a Rational Function

Sketch the graph of $f(x) = \dfrac{2(x^2 - 9)}{x^2 - 4}$.

SOLUTION By factoring the numerator and denominator, you have

$$f(x) = \frac{2(x^2 - 9)}{x^2 - 4} = \frac{2(x - 3)(x + 3)}{(x - 2)(x + 2)}.$$

y-intercept: $\left(0, \frac{9}{2}\right)$, because $f(0) = \frac{9}{2}$

x-intercepts: $(-3, 0)$ and $(3, 0)$

Vertical asymptotes: $x = -2, x = 2$, zeros of denominator

Horizontal asymptote: $y = 2$, degree of $p(x) =$ degree of $q(x)$

Symmetry: With respect to y-axis, because $f(-x) = f(x)$

Additional points:

x	-2	0.5	2	2.5	6
$f(x)$	Undefined	4.67	Undefined	-2.44	1.6875

The graph is shown in Figure 3.53.

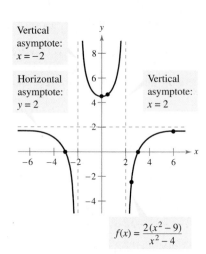

Vertical asymptote: $x = -2$

Horizontal asymptote: $y = 2$

Vertical asymptote: $x = 2$

$f(x) = \dfrac{2(x^2 - 9)}{x^2 - 4}$

FIGURE 3.53

✓ CHECKPOINT 5

Sketch the graph of

$$f(x) = \frac{5(x^2 - 1)}{x^2 - 9}. \quad ■$$

Slant Asymptotes

Consider a rational function whose denominator is of degree 1 or greater. If the degree of the numerator is exactly *one more* than the degree of the denominator, the graph of the function has a **slant** (or **oblique**) **asymptote.** For example, the graph of

$$f(x) = \frac{x^2 - x}{x + 1}$$

has a slant asymptote, as shown in Figure 3.54. To find the equation of a slant asymptote, use long division. For instance, by dividing $x + 1$ into $x^2 - x$, you have

$$f(x) = \frac{x^2 - x}{x + 1} = \underbrace{x - 2}_{\substack{\text{Slant asymptote} \\ (y = x - 2)}} + \frac{2}{x + 1}.$$

As x increases or decreases without bound, the remainder term $2/(x + 1)$ approaches 0, so the graph of f approaches the line $y = x - 2$, as shown in Figure 3.54.

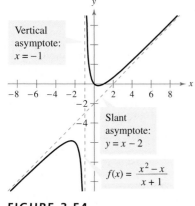

Vertical asymptote: $x = -1$

Slant asymptote: $y = x - 2$

$f(x) = \dfrac{x^2 - x}{x + 1}$

FIGURE 3.54

Example 6 A Rational Function with a Slant Asymptote

Sketch the graph of $f(x) = \dfrac{x^2 - x - 2}{x - 1}$.

SOLUTION First write f in two different ways. Factoring the numerator enables you to recognize the x-intercepts.

$$f(x) = \frac{x^2 - x - 2}{x - 1} = \frac{(x - 2)(x + 1)}{x - 1}$$

Then long division enables you to recognize that the line $y = x$ is a slant asymptote of the graph.

$$f(x) = \frac{x^2 - x - 2}{x - 1} = x - \frac{2}{x - 1}$$

y-intercept: $(0, 2)$, because $f(0) = 2$

x-intercepts: $(-1, 0)$ and $(2, 0)$

Vertical asymptote: $x = 1$, zero of denominator

Horizontal asymptote: None; degree of $p(x) >$ degree of $q(x)$

Slant asymptote: $y = x$

Additional points:

x	-2	0.5	1	1.5	3
$f(x)$	$-1.\overline{3}$	4.5	Undefined	-2.5	2

The graph is shown in Figure 3.55.

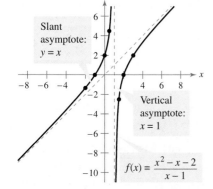

Slant asymptote: $y = x$

Vertical asymptote: $x = 1$

$f(x) = \dfrac{x^2 - x - 2}{x - 1}$

FIGURE 3.55

✓ CHECKPOINT 6

Sketch the graph of $f(x) = \dfrac{x^2 + 3x + 2}{x - 1}$. ∎

Applications

There are many examples of asymptotic behavior in business and biology. For instance, the following example describes the asymptotic behavior related to the cost of removing smokestack emissions.

Example 7   Cost-Benefit Model ®

A utility company burns coal to generate electricity. The cost of removing a certain *percent* of the pollutants from the stack emissions is typically not a linear function. That is, if it costs C dollars to remove 25% of the pollutants, it would cost more than 2C dollars to remove 50% of the pollutants. As the percent of pollutants removed approaches 100%, the cost tends to become prohibitive. The cost C (in dollars) of removing p percent of the smokestack pollutants is given by

$$C = \frac{80,000p}{100 - p}.$$

Suppose that you are a member of a state legislature that is considering a law that will require utility companies to remove 90% of the pollutants from their smokestack emissions. The current law requires 85% removal.

a. How much additional expense is the new law asking the utility company to incur?

b. According to the model, would it be possible to remove 100% of the pollutants?

SOLUTION

a. The graph of this function is shown in Figure 3.56. Note that the graph has a vertical asymptote at $p = 100$. Because the current law requires 85% removal, the current cost to the utility company is

$$C = \frac{80,000(85)}{100 - 85} \qquad \text{Substitute 85 for } p.$$

$$\approx \$453,333. \qquad \text{Use a calculator.}$$

If the new law increases the percent removal to 90%, the cost to the utility company will be

$$C = \frac{80,000(90)}{100 - 90} \qquad \text{Substitute 90 for } p.$$

$$= \$720,000. \qquad \text{Use a calculator.}$$

The new law would require the utility company to spend an additional

$$\$720,000 - \$453,333 = \$266,667.$$

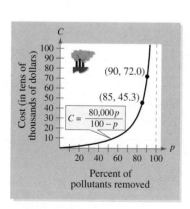

FIGURE 3.56

b. From Figure 3.56, you can see that the graph has a vertical asymptote at $p = 100$. Because the graph of a rational function can never intersect its vertical asymptote, you can conclude that it is not possible for the company to remove 100% of the pollutants from the stack emissions. ▬▬▬

✓ CHECKPOINT 7

In Example 7, suppose the new law will require utility companies to remove 95% of the pollutants. Find the additional cost to the utility company. ▪

Example 8 **Per Capita Land Area**

A model for the population P (in millions) of the United States from 1960 to 2005 is $P = 2.5049t + 179.214$, where t represents the year, with $t = 0$ corresponding to 1960. A model for the land area A (in millions of acres) of the United States from 1960 to 2005 is $A = 2263.960$. Construct a rational function for per capita land area L (in acres per person). Sketch a graph of the rational function. Use the model to predict the per capita land area in 2013.   *(Source: U.S. Census Bureau)*

SOLUTION   The rational function for the per capita land area L is

$$L = \frac{A}{P} = \frac{2263.960}{2.5049t + 179.214}.$$

The graph of the function is shown in Figure 3.57. To find the per capita land area in 2013, substitute $t = 53$ into L.

$$L = \frac{2263.960}{2.5049t + 179.214} = \frac{2263.960}{2.5049(53) + 179.214} \approx \frac{2263.960}{311.974} \approx 7.26$$

The per capita land area will be approximately 7.3 acres per person in 2013.

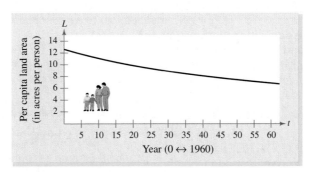

FIGURE 3.57

✓ CHECKPOINT 8

In Example 8, use the model to predict the per capita land area in 2020. ▪

CONCEPT CHECK

In Exercises 1–4, determine whether the statement is true or false. Justify your answer.

1. The domain of $f(x) = \dfrac{x^2 + 2x - 8}{x^2 - 9}$ is all real numbers except $x = -3$ and $x = 3$.

2. The graph of $g(x) = \dfrac{x^2 - 1}{x^2 + 4x + 4}$ has vertical asymptotes $x = -1$ and $x = 1$.

3. The graph of every rational function has a horizontal asymptote.

4. A rational function f has a numerator of degree n. The graph of f has a slant asymptote. So, the denominator has degree n.

The following warm-up exercises involve skills that were covered in earlier sections. You will use these skills in the exercise set for this section. For additional help, review Sections 0.6 and 2.2.

In Exercises 1–6, factor the polynomial.

1. $x^2 - 4x$

2. $2x^3 - 6x$

3. $x^2 - 3x - 10$

4. $x^2 - 7x + 10$

5. $x^3 + 4x^2 + 3x$

6. $x^3 - 4x^2 - 2x + 8$

In Exercises 7–10, sketch the graph of the equation.

7. $y = 2$

8. $x = -1$

9. $y = x - 2$

10. $y = -x + 1$

Exercises 3.7

In Exercises 1–8, find the domain of the function and identify any horizontal and vertical asymptotes.

1. $f(x) = \dfrac{3x}{x + 1}$

2. $f(x) = \dfrac{x}{x - 2}$

3. $f(x) = \dfrac{x - 7}{5 - x}$

4. $f(x) = \dfrac{1 - 5x}{1 + 2x}$

5. $f(x) = \dfrac{3x^2 + 1}{x^2 + 9}$

6. $f(x) = \dfrac{3x^2 + x - 5}{x^2 + 1}$

7. $f(x) = \dfrac{5}{(x + 4)^2}$

8. $f(x) = \dfrac{1}{(x - 1)^2}$

In Exercises 9–12, find any (a) vertical, (b) horizontal, and (c) slant asymptotes of the graph of the function. Then sketch the graph of f.

9. $f(x) = \dfrac{x^2 - 7x + 12}{x - 3}$

10. $f(x) = \dfrac{x + 3}{x^2 - 9}$

11. $f(x) = \dfrac{x^2}{x + 1}$

12. $f(x) = \dfrac{x^3 + x}{x^2 - 1}$

In Exercises 13–18, match the function with its graph. [The graphs are labeled (a), (b), (c), (d), (e), and (f).]

(a)
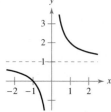

(b)
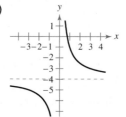

(c)
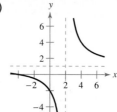

(d)
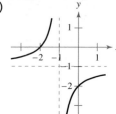

(e)
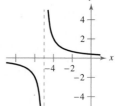

(f)
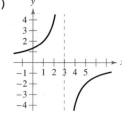

13. $f(x) = -\dfrac{4}{x - 3}$

14. $f(x) = \dfrac{2}{x + 5}$

15. $f(x) = \dfrac{x + 1}{x}$

16. $f(x) = \dfrac{3 - 4x}{x}$

17. $f(x) = \dfrac{x + 4}{x - 2}$

18. $f(x) = -\dfrac{x + 2}{x + 1}$

In Exercises 19–22, compare the graph of $f(x) = 1/x$ with the graph of g.

19. $g(x) = f(x) - 2 = \dfrac{1}{x} - 2$

20. $g(x) = f(x + 1) = \dfrac{1}{x + 1}$

21. $g(x) = -f(x) = -\dfrac{1}{x}$

22. $g(x) = -f(x + 1) = -\dfrac{1}{x + 1}$

In Exercises 23–26, compare the graph of $f(x) = 4/x^2$ with the graph of g.

23. $g(x) = f(x) + 3 = \dfrac{4}{x^2} + 3$

24. $g(x) = f(x - 1) = \dfrac{4}{(x - 1)^2}$

25. $g(x) = -f(x) = -\dfrac{4}{x^2}$

26. $g(x) = \dfrac{1}{8}f(x) = \dfrac{1}{2x^2}$

In Exercises 27–30, compare the graph of $f(x) = 8/x^3$ with the graph of g.

27. $g(x) = f(x) + 5 = \dfrac{8}{x^3} + 5$

28. $g(x) = f(x - 3) = \dfrac{8}{(x - 3)^3}$

29. $g(x) = -f(x) = -\dfrac{8}{x^3}$

30. $g(x) = \dfrac{1}{4}f(x) = \dfrac{2}{x^3}$

In Exercises 31–56, sketch the graph of the rational function. To aid in sketching the graphs, check for intercepts, symmetry, vertical asymptotes, and horizontal asymptotes.

31. $f(x) = \dfrac{1}{x + 3}$

32. $f(x) = \dfrac{1}{x - 3}$

33. $f(x) = \dfrac{1}{x - 4}$

34. $f(x) = \dfrac{1}{x + 6}$

35. $f(x) = \dfrac{-1}{x + 1}$

36. $f(x) = \dfrac{-2}{x - 3}$

37. $f(x) = \dfrac{x + 4}{x - 5}$

38. $f(x) = \dfrac{x - 2}{x - 3}$

39. $f(x) = \dfrac{2 + x}{1 - x}$

40. $f(x) = \dfrac{3 - x}{2 - x}$

41. $f(t) = \dfrac{3t + 1}{t}$

42. $f(t) = \dfrac{1 - 2t}{t}$

43. $C(x) = \dfrac{5 + 2x}{1 + x}$

44. $P(x) = \dfrac{1 - 3x}{1 - x}$

45. $g(x) = \dfrac{1}{x + 2} + 2$

46. $h(x) = \dfrac{1}{x - 3} + 1$

47. $f(x) = \dfrac{1}{x^2} + 2$

48. $f(x) = 2 - \dfrac{3}{x^2}$

49. $h(x) = \dfrac{x^2}{x^2 - 9}$

50. $h(t) = \dfrac{3t^2}{t^2 - 4}$

51. $g(s) = \dfrac{s}{s^2 + 1}$

52. $g(x) = \dfrac{x}{x^2 + 3}$

53. $f(x) = \dfrac{x}{x^2 - 3x - 4}$

54. $f(x) = \dfrac{-x}{x^2 + x - 6}$

55. $f(x) = \dfrac{3x}{x^2 - x - 2}$

56. $f(x) = \dfrac{2x}{x^2 + x - 2}$

In Exercises 57–60, write a rational function f that has the specified characteristics. (There are many correct answers.)

57. Vertical asymptote: None
    Horizontal asymptote: $y = 2$

58. Vertical asymptotes: $x = 0, x = \dfrac{5}{2}$
    Horizontal asymptote: $y = -3$

59. Vertical asymptotes: $x = -2, x = 1$
    Horizontal asymptote: None

60. Vertical asymptote: $x = 3$
    Horizontal asymptote: x-axis

(T) In Exercises 61–64, find a counterexample to show that the statement is incorrect.

61. Every rational function has a vertical asymptote.

62. Every rational function has at least one asymptote.

63. A rational function can have only one vertical asymptote.

64. The graph of a rational function with a slant asymptote cannot cross its slant asymptote.

(T) **65.** Is it possible for a rational function to have all three types of asymptotes (vertical, horizontal, and slant)? Why or why not?

(T) **66.** Is it possible for a rational function to have more than one horizontal asymptote? Why or why not?

67. *MAKE A DECISION: SEIZURE OF ILLEGAL DRUGS* The cost C (in millions of dollars) for the federal government to seize p percent of an illegal drug as it enters the country is

$$C = \frac{528p}{100 - p}, \quad 0 \le p < 100.$$

(a) Find the cost of seizing 25% of the drug.

(b) Find the cost of seizing 50% of the drug.

(c) Find the cost of seizing 75% of the drug.

(d) According to this model, would it be possible to seize 100% of the drug? Explain.

68. *MAKE A DECISION: WATER POLLUTION* The cost C (in millions of dollars) of removing p percent of the industrial and municipal pollutants discharged into a river is

$$C = \frac{255p}{100 - p}, \quad 0 \le p < 100.$$

(a) Find the cost of removing 15% of the pollutants.

(b) Find the cost of removing 50% of the pollutants.

(c) Find the cost of removing 80% of the pollutants.

(d) According to the model, would it be possible to remove 100% of the pollutants? Explain.

69. Population of Deer The Game Commission introduces 100 deer into newly acquired state game lands. The population N of the herd is given by

$$N = \frac{25(4 + 2t)}{1 + 0.02t}, \quad t \ge 0$$

where t is time (in years).

(a) Find the populations when t is 5, 10, and 25.

(b) What is the limiting size of the herd as time progresses?

70. Population of Elk The Game Commission introduces 40 elk into newly acquired state game lands. The population N of the herd is given by

$$N = \frac{10(4 + 2t)}{1 + 0.03t}, \quad t \ge 0$$

where t is time (in years).

(a) Find the populations when t is 5, 10, and 25.

(b) What is the limiting size of the herd as time progresses?

71. Defense The table shows the national defense outlays D (in billions of dollars) from 1997 to 2005. The data can be modeled by

$$D = \frac{1.493t^2 - 39.06t + 273.5}{0.0051t^2 - 0.1398t + 1}, \quad 7 \le t \le 15$$

where t is the year, with $t = 7$ corresponding to 1997. *(Source: U.S. Office of Management and Budget)*

Year	Defense outlays	Year	Defense outlays
1997	270.5	2002	348.6
1998	268.5	2003	404.9
1999	274.9	2004	455.9
2000	294.5	2005	465.9
2001	305.5		

(a) Use a graphing utility to plot the data and graph the model in the same viewing window. How well does the model represent the data?

(b) Use the model to predict the national defense outlays for the years 2010, 2015, and 2020. Are the predictions reasonable?

(c) Determine the horizontal asymptote of the graph of the model. What does it represent in the context of the situation?

72. Average Cost The cost C (in dollars) of producing x basketballs is $C = 375,000 + 4x$. The average cost $\overline{C}$ per basketball is

$$\overline{C} = \frac{C}{x} = \frac{375,000 + 4x}{x}, \quad x > 0.$$

(a) Sketch the graph of the average cost function.

(b) Find the average costs of producing 1000, 10,000, and 100,000 basketballs.

(c) Find the horizontal asymptote and explain its meaning in the context of the problem.

73. Human Memory Model Psychologists have developed mathematical models to predict memory performance as a function of the number of trials n of a certain task. Consider the learning curve modeled by

$$P = \frac{0.6 + 0.95(n - 1)}{1 + 0.95(n - 1)}, \quad n > 0$$

where P is the percent of correct responses (in decimal form) after n trials.

(a) Complete the table.

n	1	2	3	4	5	6	7	8	9	10
P										

(b) According to this model, what is the limiting percent of correct responses as n increases?

74. Human Memory Model Consider the learning curve modeled by

$$P = \frac{0.55 + 0.87(n - 1)}{1 + 0.87(n - 1)}, \quad n > 0$$

where P is the percent of correct responses (in decimal form) after n trials.

(a) Complete the table.

n	1	2	3	4	5	6	7	8	9	10
P										

(b) According to this model, what is the limiting percent of correct responses as n increases?

75. Average Recycling Cost The cost C (in dollars) of recycling a waste product is

$$C = 450,000 + 6x, \quad x > 0$$

where x is the number of pounds of waste. The average recycling cost $\overline{C}$ per pound is

$$\overline{C} = \frac{C}{x} = \frac{450,000 + 6x}{x}, \quad x > 0.$$

(T) (a) Use a graphing utility to graph $\overline{C}$.

(b) Find the average costs of recycling 10,000, 100,000, 1,000,000, and 10,000,000 pounds of waste. What can you conclude?

76. Drug Concentration The concentration C of a medication in the bloodstream t minutes after sublingual (under the tongue) application is given by

$$C(t) = \frac{3t - 1}{2t^2 + 5}, \quad t > 0.$$

(T) (a) Use a graphing utility to graph the function. Estimate when the concentration is greatest.

(b) Does this function have a horizontal asymptote? If so, discuss the meaning of the asymptote in terms of the concentration of the medication.

77. Domestic Demand The U.S. domestic demand D (in millions of barrels) for refined oil products from 1995 to 2005 can be modeled by

$$D = 100.9708t + 6083.999, \quad 5 \le t \le 15$$

where t represents the year, with $t = 5$ corresponding to 1995. The population P (in millions) of the United States from 1995 to 2005 can be modeled by

$$P = 3.0195t + 251.817, \quad 5 \le t \le 15$$

where t represents the year, with $t = 5$ corresponding to 1995. *(Sources: U.S. Energy Information Administration and the U.S. Census Bureau)*

(a) Construct a rational function B to describe the per capita demand for refined oil products.

(T) (b) Use a graphing utility to graph the rational function B.

(c) Use the model to predict the per capita demand for refined oil products in 2010.

78. Health Care Spending The total health care spending H (in millions of dollars) in the United States from 1995 to 2005 can be modeled by

$$H = 6136.36t^2 - 22,172.7t + 979,909, \quad 5 \le t \le 15$$

where t represents the year, with $t = 5$ corresponding to 1995. The population P (in millions) of the United States from 1995 to 2005 can be modeled by

$$P = 3.0195t + 251.817, \quad 5 \le t \le 15$$

where t represents the year, with $t = 5$ corresponding to 1995. *(Sources: U.S. Centers for Medicare and Medicaid Services and the U.S. Census Bureau)*

(a) Construct a rational function S to describe the per capita health spending.

(T) (b) Use a graphing utility to graph the rational function S.

(c) Use the model to predict the per capita health care spending in 2010.

79. 100-Meter Freestyle The winning times for the men's 100-meter freestyle swim at the Olympics from 1952 to 2004 can be approximated by the quadratic model

$$y = 86.24 - 0.752t + 0.0037t^2, \quad 52 \le t \le 104$$

where y is the winning time (in seconds) and t represents the year, with $t = 52$ corresponding to 1952. *(Sources: The World Almanac and Book of Facts 2005)*

(T) (a) Use a graphing utility to graph the model.

(b) Use the model to predict the winning times in 2008 and 2012.

(c) Does this model have a horizontal asymptote? Do you think that a model for this type of data should have a horizontal asymptote?

80. 3000-Meter Speed Skating The winning times for the women's 3000-meter speed skating race at the Olympics from 1960 to 2006 can be approximated by the quadratic model

$$y = 0.0202t^2 - 5.066t + 550.24, \quad 60 \le t \le 106$$

where y is the winning time (in seconds) and t represents the year, with $t = 60$ corresponding to 1960. *(Sources: World Almanac and Book of Facts 2005 and NBC)*

(a) Use a graphing utility to graph the model.

(T) (b) Use the model to predict the winning times in 2010 and 2014.

(c) Does this model have a horizontal asymptote? Do you think that a model for this type of data should have a horizontal asymptote?

Chapter Summary and Study Strategies

After studying this chapter, you should have acquired the following skills.
The exercise numbers are keyed to the Review Exercises that begin on page 336.
Answers to odd-numbered Review Exercises are given in the back of the text.*

Section 3.1	Review Exercises
■ Sketch the graph of a quadratic function and identify its vertex and intercepts.	*1–4*
■ Find a quadratic function given its vertex and a point on its graph.	*5, 6*
■ Construct and use a quadratic model to solve an application problem.	*7–12*

Section 3.2

■ Determine right-hand and left-hand behavior of graphs of polynomial functions. *13–16*

When n is odd and the leading coefficient is positive,

$$f(x) \to -\infty \text{ as } x \to -\infty \text{ and } f(x) \to \infty \text{ as } x \to \infty.$$

When n is odd and the leading coefficient is negative,

$$f(x) \to \infty \text{ as } x \to -\infty \text{ and } f(x) \to -\infty \text{ as } x \to \infty.$$

When n is even and the leading coefficient is positive,

$$f(x) \to \infty \text{ as } x \to -\infty \text{ and } f(x) \to \infty \text{ as } x \to \infty.$$

When n is even and the leading coefficient is negative,

$$f(x) \to -\infty \text{ as } x \to -\infty \text{ and } f(x) \to -\infty \text{ as } x \to \infty.$$

■ Find the real zeros of a polynomial function. *17–20*

Section 3.3

■ Divide one polynomial by a second polynomial using long division.	*21, 22*
■ Simplify a rational expression using long division.	*23, 24*
■ Use synthetic division to divide two polynomials.	*25, 26, 31, 32*
■ Use the Remainder Theorem and synthetic division to evaluate a polynomial.	*27, 28*
■ Use the Factor Theorem to factor a polynomial.	*29, 30*

Section 3.4

■ Find all possible rational zeros of a function using the Rational Zero Test.	*33, 34*
■ Find all real zeros of a function.	*35–42*
■ Approximate the real zeros of a polynomial function using the Intermediate Value Theorem.	*43, 44*
■ Approximate the real zeros of a polynomial function using a graphing utility.	*45, 46*
■ Apply techniques for approximating real zeros to solve an application problem.	*47, 48*

---

* Use a wide range of valuable study aids to help you master the material in this chapter. The *Student Solutions Guide* includes step-by-step solutions to all odd-numbered exercises to help you review and prepare. The student website at *college.hmco.com/info/larsonapplied* offers algebra help and a *Graphing Technology Guide*. The *Graphing Technology Guide* contains step-by-step commands and instructions for a wide variety of graphing calculators, including the most recent models.

Section 3.5

Review Exercises

- Find the complex conjugate of a complex number.   *49–52*
- Perform operations with complex numbers and write the results in standard form.   *53–68*

$$(a + bi) + (c + di) = (a + c) + (b + d)i$$
$$(a + bi) - (c + di) = (a - c) + (b - d)i$$
$$(a + bi)(c + di) = (ac - bd) + (ad + bc)i$$

- Solve a polynomial equation that has complex solutions.   *69–72*
- Plot a complex number in the complex plane.   *73, 74*

Section 3.6

- Use the Fundamental Theorem of Algebra and the Linear Factorization   *75–80*
  Theorem to write a polynomial as the product of linear factors.
- Find a polynomial with real coefficients whose zeros are given.   *81, 82*
- Factor a polynomial over the rational, real, and complex numbers.   *83, 84*
- Find all real and complex zeros of a polynomial function.   *85–88*

Section 3.7

- Find the domain of a rational function.   *89–92*
- Find the vertical and horizontal asymptotes of the graph of a rational function.   *89–92*

$$\text{Let } f(x) = \frac{p(x)}{q(x)} = \frac{a_n x^n + a_{n-1}x^{n-1} + \cdots + a_1 x + a_0}{b_m x^m + b_{m-1}x^{m-1} + \cdots + b_1 x + b_0}, \quad a_n \neq 0, b_m \neq 0.$$

  1. The graph of f has *vertical* asymptotes at the zeros of $q(x)$.

  2. The graph of f has one or no *horizontal* asymptote determined by comparing
  the degrees of $p(x)$ and $q(x)$.

  a. If $n < m$, the graph of f has the line $y = 0$ (the x-axis) as a horizontal asymptote.

  b. If $n = m$, the graph of f has the line $y = a_n / b_m$ (ratio of the leading
  coefficients) as a horizontal asymptote.

  c. If $n > m$, the graph of f has no horizontal asymptote.

- Sketch the graph of a rational function, including graphs with slant asymptotes.   *93–98*
- Use a rational function model to solve an application problem.   *99–103*

Study Strategies

- **Use a Graphing Utility**   A graphing calculator or graphing software for a computer can help you in this course in
  two important ways. As an *exploratory device*, a graphing utility allows you to learn concepts by allowing you to compare
  graphs of functions. For instance, sketching the graphs of $f(x) = x^3$ and $f(x) = -x^3$ helps confirm that the negative
  coefficient has the effect of reflecting the graph about the x-axis. As a *problem-solving tool*, a graphing utility frees you
  from some of the difficulty of sketching complicated graphs by hand. The time you can save can be spent using mathematics
  to solve real-life problems.

- **Problem-Solving Strategies**   If you get stuck when trying to solve a real-life problem, consider the strategies below.

  1. *Draw a Diagram.* If feasible, draw a diagram that represents the problem. Label all known values and unknown values on
     the diagram.

  2. *Solve a Simpler Problem.* Simplify the problem, or write several simple examples of the problem. For instance, if you are
     asked to find the dimensions that will produce a maximum area, try calculating the areas of several examples.

  3. *Rewrite the Problem in Your Own Words.* Rewriting a problem can help you understand it better.

  4. *Guess and Check.* Try guessing the answer, then check your guess in the statement of the original problem. By refining
     your guesses, you may be able to think of a general strategy for solving the problem.

Review Exercises

See www.CalcChat.com for worked-out solutions to odd-numbered exercises.

In Exercises 1–4, sketch the graph of the quadratic function. Identify the vertex and intercepts.

1. $f(x) = (x + 3)^2 - 5$

2. $g(x) = -(x - 1)^2 + 3$

3. $h(x) = 3x^2 - 12x + 11$

4. $f(x) = \frac{1}{3}(x^2 + 5x - 4)$

In Exercises 5 and 6, find an equation of the parabola that has the indicated vertex and whose graph passes through the given point.

5. Vertex: $(-5, -1)$; point: $(-2, 6)$

6. Vertex: $(2, 5)$; point: $(4, 7)$

7. Optimal Area The perimeter of a rectangular archaeological dig site is 500 feet. Let x represent the width of the dig site. Write a quadratic function for the area of the rectangle in terms of its width. Of all possible dig sites with perimeters of 500 feet, what are the measurements of the one with the greatest area?

8. Optimal Revenue Find the number of units that produces a maximum revenue R (in dollars) for

$$R = 900x - 0.015x^2$$

where x is the number of units produced.

(T) 9. Optimal Cost A manufacturer of retinal imaging systems has daily production costs C (in dollars per unit) of

$$C = 25{,}000 - 50x + 0.065x^2$$

where x is the number of units produced.

(a) Use a graphing utility to graph the cost function.

(b) Graphically estimate the number of units that should be produced to yield a minimum cost per unit.

(c) Explain how to confirm the result of part (b) algebraically.

(T) 10. Optimal Profit The profit P (in dollars) for an electronics company is given by

$$P = -0.00015x^2 + 155x - 450{,}000$$

where x is the number of units produced.

(a) Use a graphing utility to graph the profit function.

(b) Graphically estimate the number of units that should be produced to yield a maximum profit.

(c) Explain how to confirm the result of part (b) algebraically.

(T) 11. Regression Problem Let x be the angle (in degrees) at which a baseball is hit with a 30-hertz backspin at an initial speed of 40 meters per second and let $d(x)$ be the distance (in meters) the ball travels. The table shows the distances traveled for the different angles at which the ball is hit. *(Source: The Physics of Sports)*

x	10	15	30	36	42	43
$d(x)$	61.2	83.0	130.4	139.4	143.2	143.3

x	44	45	48	54	60
$d(x)$	142.8	142.7	140.7	132.8	119.7

(a) Use a graphing utility to create a scatter plot of the data.

(b) Use the *regression* feature of a graphing utility to find a quadratic model for the data.

(c) Use a graphing utility to graph the model from part (b) in the same viewing window as the scatter plot of the data.

(d) Find the vertex of the graph of the model from part (c). Interpret its meaning in the context of the problem.

12. Doctorates in Science The numbers of non-U.S. citizens from Thailand with temporary visas that were awarded doctorates in science for the years 2000 to 2005 are shown in the table. *(Source: National Science Foundation)*

Year	2000	2001	2002	2003
Number, D	90	118	130	138

Year	2004	2005
Number, D	128	115

(a) Use a graphing utility to create a scatter plot of the data. Let t represent the year, with $t = 0$ corresponding to 2000.

(b) Use the *regression* feature of a graphing utility to find a quadratic model for the data.

(c) Use a graphing utility to graph the model from part (b) in the same viewing window as the scatter plot of the data.

(d) Find the vertex of the graph of the model from part (c). Interpret its meaning in the context of the problem.

In Exercises 13–16, describe the right-hand and left-hand behavior of the graph of the polynomial function.

13. $f(x) = \frac{1}{2}x^3 + 2x$

14. $f(x) = 5 + 4x^3 - x^5$

15. $f(x) = -x^6 + 3x^4 - x^2 + 6$

16. $f(x) = \frac{3}{4}(x^4 + 3x^2 + 2)$

In Exercises 17–20, find all real zeros of the function.

17. $f(x) = 16 - x^2$

18. $f(x) = x^4 - 6x^2 + 8$

19. $f(x) = x^3 - 7x^2 + 10x$

20. $f(x) = x^3 - 6x^2 - 3x + 18$

In Exercises 21 and 22, use long division to divide.

Dividend	Divisor
21. $2x^3 - 5x^2 - x$	$2x + 1$
22. $x^4 - 5x^3 + 10x^2 - 12$	$x^2 - 2x + 4$

In Exercises 23 and 24, simplify the rational expression.

23. $\dfrac{x^3 + 9x^2 + 2x - 48}{x - 2}$

24. $\dfrac{x^4 + 5x^3 - 20x - 16}{x^2 - 4}$

In Exercises 25 and 26, use synthetic division to divide.

Dividend	Divisor
25. $x^3 - 6x + 9$	$x + 3$
26. $x^5 - x^4 + x^3 - 13x^2 + x + 6$	$x - 2$

In Exercises 27 and 28, use synthetic division to find each function value.

27. $f(x) = 6 + 2x^2 - 3x^3$ (a) $f(2)$ (b) $f(-1)$

28. $f(x) = 2x^4 + 3x^3 + 6$ (a) $f\left(\frac{1}{2}\right)$ (b) $f(-1)$

In Exercises 29 and 30, (a) verify the given factors of $f(x)$, (b) find the remaining factors of $f(x)$, (c) use your results to write the complete factorization of $f(x)$, (d) list all real zeros of f, and (e) confirm your results by using a graphing utility to graph the function.

Function	Factors
29. $f(x) = x^3 - 4x^2 - 11x + 30$	$(x - 5), (x + 3)$
30. $f(x) = 3x^3 + 23x^2 + 37x - 15$	$(3x - 1), (x + 5)$

(T) **31. Profit** The profit P (in dollars) from selling a motorcycle is given by

$$P = -42x^3 + 3000x^2 - 6000, \quad 0 \le x \le 65$$

where x is the advertising expense (in tens of thousands of dollars). For this motorcycle, the advertising expense was $600,000 ($x = 60$) and the profit was $1,722,000.

(a) Use a graphing utility to graph the function and use the result to find another advertising expense that would have produced the same profit.

(b) Use synthetic division to confirm the result of part (a) algebraically.

(T) **32. Profit** The profit P (in dollars) from selling a novel is given by

$$P = -150x^3 + 7500x^2 - 450,000, \quad 0 \le x \le 45$$

where x is the advertising expense (in tens of thousands of dollars). For this novel, the advertising expense was $400,000 ($x = 40$), and the profit was $1,950,000.

(a) Use a graphing utility to graph the function and use the result to find another advertising expense that would have produced the same profit.

(b) Use synthetic division to confirm the result of part (a) algebraically.

(T) In Exercises 33 and 34, use the Rational Zero Test to list all possible rational zeros of f. Then use a graphing utility to graph the function. Use the graph to help determine which of the possible rational zeros are actual zeros of the function.

33. $f(x) = -4x^3 + 8x^2 - 3x + 15$

34. $f(x) = 3x^4 + 4x^3 - 5x^2 + 10x - 8$

In Exercises 35–42, find all real zeros of the function.

35. $f(x) = x^3 + 2x^2 - 5x - 6$

36. $g(x) = 2x^3 - 15x^2 + 24x + 16$

37. $h(x) = 3x^4 - 27x^2 + 60$

38. $f(x) = x^5 - 4x^3 + 3x$

39. $B(x) = 6x^3 - 19x^2 + 11x + 6$

40. $C(x) = 3x^4 + 3x^3 - 7x^2 - x + 2$

41. $p(x) = x^4 - x^3 - 2x - 4$

42. $q(x) = x^5 - 2x^4 + 2x^3 - 4x^2 - 3x + 6$

In Exercises 43 and 44, use the Intermediate Value Theorem to approximate the zero of f in the interval $[a, b]$. Give your approximation to the nearest tenth.

43. $f(x) = x^3 - 4x + 3, \quad [-3, -2]$

44. $f(x) = x^5 + 5x^2 + x - 1, \quad [0, 1]$

(T) In Exercises 45 and 46, use a graphing utility to approximate the real zeros of f. Give your approximations to the nearest thousandth.

45. $f(x) = 5x^3 - 11x - 3$

46. $f(x) = 2x^4 - 9x^3 - 5x^2 + 10x + 12$

(T) **47. Wholesale Revenue** The revenues R (in millions of dollars) for Costco Wholesale for the years 1996 to 2005 are shown in the table. *(Source: Costco Wholesale)*

Year	Revenue, R	Year	Revenue, R
1996	19,566	2001	34,797
1997	21,874	2002	38,762
1998	24,270	2003	42,546
1999	27,456	2004	48,107
2000	32,164	2005	52,935

(a) Use a graphing utility to create a scatter plot of the data. Let t represent the year, with $t = 6$ corresponding to 1996.

(b) Use the *regression* feature of a graphing utility to find a linear model, a quadratic model, a cubic model, and a quartic model for the data.

(c) Use a graphing utility to graph each model separately with the data in the same viewing window. How well does each model fit the data?

(d) Use each model to predict the year in which the revenue will be about \$65 billion. Explain any differences in the predictions.

(T) **48. Shoe Sales** The sales S (in millions of dollars) for Steve Madden for the years 1996 to 2005 are shown in the table. *(Source: Steve Madden, LTD)*

Year	Sales, S	Year	Sales, S
1996	45.8	2001	243.4
1997	59.3	2002	326.1
1998	85.8	2003	324.2
1999	163.0	2004	338.1
2000	205.1	2005	375.8

(a) Use a graphing utility to create a scatter plot of the data. Let t represent the year, with $t = 6$ corresponding to 1996.

(b) Use the *regression* feature of a graphing utility to find a linear model, a quadratic model, and a quartic model for the data.

(c) Use a graphing utility to graph each model separately with the data in the same viewing window. How well does each model fit the data?

(d) Use each model to predict the year in which the sales will be about \$500 million. Explain any differences in the predictions.

In Exercises 49–52, write the complex number in standard form and find its complex conjugate.

49. $\sqrt{-32}$　　　　　　**50.** 12

51. $-3 + \sqrt{-16}$　　　　**52.** $2 - \sqrt{-18}$

In Exercises 53–64, perform the indicated operation and write the result in standard form.

53. $(7 - 4i) + (-2 + 5i)$

54. $(14 + 6i) - (-1 - 2i)$

55. $\left(1 + \sqrt{-12}\right)\left(5 - \sqrt{-3}\right)$

56. $\left(3 - \sqrt{-4}\right)\left(4 - \sqrt{-49}\right)$

57. $(5 + 8i)(5 - 8i)$　　　**58.** $\left(\frac{1}{2} + \frac{3}{4}i\right)\left(\frac{1}{2} - \frac{3}{4}i\right)$

59. $-2i(4 - 5i)$　　　　　**60.** $-3(-2 + 4i)$

61. $(3 + 4i)^2$　　　　　　**62.** $(2 - 5i)^2$

63. $(3 + 2i)^2 + (3 - 2i)^2$　　**64.** $(1 + i)^2 - (1 - i)^2$

In Exercises 65–68, write the quotient in standard form.

65. $\dfrac{8 - i}{2 + i}$　　　　　　**66.** $\dfrac{3 - 4i}{1 - 5i}$

67. $\dfrac{4 - 3i}{i}$　　　　　　**68.** $\dfrac{2}{(1 + i)^2}$

In Exercises 69–72, solve the equation.

69. $2x^2 - x + 3 = 0$

70. $3x^2 + 6x + 11 = 0$

71. $4x^2 + 11x + 3 = 0$

72. $9x^2 - 2x + 5 = 0$

In Exercises 73 and 74, plot the complex number.

73. $-3 + 2i$　　　　　　**74.** $-1 - 4i$

In Exercises 75–80, find all the zeros of the function and write the polynomial as a product of linear factors.

75. $f(x) = x^4 - 81$

76. $h(x) = 2x^3 - 5x^2 + 4x - 10$

77. $f(t) = t^3 + 5t^2 + 3t + 15$

78. $h(x) = x^4 + 17x^2 + 16$

79. $g(x) = 4x^3 - 8x^2 + 9x - 18$

80. $f(x) = x^5 - 2x^4 + x^3 - x^2 + 2x - 1$

In Exercises 81 and 82, find a polynomial with real coefficients that has the given zeros. (There are many correct answers.)

81. $1, 3i, -3i$

82. $1, -2, 1 - 3i, 1 + 3i$

In Exercises 83 and 84, write the polynomial (a) as the product of factors that are irreducible over the *rationals*, (b) as the product of linear and quadratic factors that are irreducible over the *reals*, and (c) in completely factored form.

83. $x^4 + 5x^2 - 24$

84. $x^4 - 2x^3 - 2x^2 - 14x - 63$

(*Hint:* One factor is $x^2 + 7$.)

In Exercises 85–88, use the given zero of f to find all the zeros of f.

85. $f(x) = 4x^3 - x^2 + 64x - 16$, $-4i$

86. $f(x) = 50 - 75x + 2x^2 - 3x^3$, $5i$

87. $f(x) = x^4 + 7x^3 + 24x^2 + 58x + 40$, $-1 + 3i$

88. $f(x) = x^4 + 4x^3 + 8x^2 + 4x + 7$, $-2 - \sqrt{3}i$

In Exercises 89–92, find the domain of the function and identify any horizontal or vertical asymptotes.

89. $f(x) = \dfrac{-3}{x + 2}$

90. $f(x) = \dfrac{3x^2 + 7x - 5}{x^2 + 1}$

91. $f(x) = \dfrac{2x^2}{x^2 - 9}$

92. $f(x) = \dfrac{3x}{x^2 + x - 6}$

In Exercises 93–96, sketch the graph of the rational function. As sketching aids, check for intercepts, symmetry, vertical asymptotes, and horizontal asymptotes.

93. $P(x) = \dfrac{3 - x}{x + 2}$

94. $f(x) = \dfrac{4}{(x - 1)^2}$

95. $g(x) = \dfrac{1}{x^2 - 4} + 2$

96. $h(x) = \dfrac{-3x}{2x^2 + 3x - 5}$

In Exercises 97 and 98, find all possible asymptotes (vertical, horizontal, and/or slant) of the given function. Sketch the graph of f.

97. $f(x) = \dfrac{x^2 - 16}{x - 4}$

98. $f(x) = \dfrac{x^3}{x^2 - 5}$

99. Average Cost   The cost C (in dollars) of producing x charcoal grills is $C = 125,000 + 9.65x$. The average cost $\overline{C}$ per charcoal grill is

$$\overline{C} = \frac{C}{x} = \frac{125,000 + 9.65x}{x}, \quad x > 0.$$

(a) Sketch the graph of the average cost function.

(b) Find the average cost of producing 1000, 10,000, 100,000, and 1,000,000 charcoal grills. What can you conclude?

100. Average Recycling Cost   The cost C (in dollars) of recycling a waste product is

$$C = 325,000 + 8.5x, \quad x > 0$$

where x is the number of pounds of waste. The average recycling cost $\overline{C}$ per pound is

$$\overline{C} = \frac{C}{x} = \frac{325,000 + 8.5x}{x}, \quad x > 0.$$

(a) Sketch the graph of $\overline{C}$.

(b) Find the average cost of recycling 1000, 10,000, 100,000, and 1,000,000 pounds of waste. What can you conclude?

101. Population of Fish   The Wildlife Commission introduces 60,000 game fish into a large lake. The population N (in thousands) of the fish is

$$N = \frac{20(3 + 5t)}{1 + 0.06t}, \quad t \geq 0$$

where t is time (in years).

(a) Find the populations when $t = 5$, 10, and 25.

(b) What is the limiting number of fish in the lake as time progresses?

102. Human Memory Model   Consider the learning curve modeled by

$$P = \frac{0.7 + 0.65(n - 1)}{1 + 0.65(n - 1)}, \quad n \geq 0$$

where P is the percent of correct responses (in decimal form) after n trials.

(a) Complete the table.

n	1	2	3	4	5	6	7	8	9	10
P										

(b) According to this model, what is the limiting percent of correct responses as n increases?

103. Smokestack Emissions   The cost C (in dollars) of removing p percent of the air pollutants in the stack emissions of a utility company that burns coal to generate electricity is

$$C = \frac{105,000p}{100 - p}, \quad 0 \leq p < 100.$$

(a) Find the cost of removing 25% of the pollutants.

(b) Find the cost of removing 60% of the pollutants.

(c) Find the cost of removing 99% of the pollutants.

(d) According to the model, would it be possible to remove 100% of the pollutants? Explain.

Chapter Test

See www.CalcChat.com for worked-out solutions to odd-numbered exercises.

Take this test as you would take a test in class. When you are done, check your work against the answers given in the back of the book.

1. Sketch the graph of the quadratic function given by

   $f(x) = -\frac{1}{2}(x - 1)^2 - 5.$

   Identify the vertex and intercepts.

2. Describe the right-hand and left-hand behavior of the graph of f.

   (a) $f(x) = 12x^3 - 5x^2 - 49x + 15$

   (b) $f(x) = 5x^4 - 3x^3 + 2x^2 + 11x + 12$

3. Simplify $\dfrac{x^4 + 4x^3 - 19x^2 - 106x - 120}{x^2 - 3x - 10}$.

4. List all the possible rational zeros of

   $f(x) = 4x^4 - 16x^3 + 3x^2 + 36x - 27.$

   Use synthetic division to show that $x = -\frac{3}{2}$ and $x = \frac{3}{2}$ are zeros of f. Using these results, completely factor the polynomial.

5. The sales per share S (in dollars) for Cost Plus, Inc. for the years 1996 to 2005 are shown in the table at the left. *(Source: Cost Plus Inc.)*

   (a) Use a graphing utility to create a scatter plot of the data. Let t represent the year, with $t = 6$ corresponding to 1996.

   (b) Use the *regression* feature of a graphing utility to find a linear model, a quadratic model, and a cubic model for the data.

   (c) Use a graphing utility to graph each model separately with the data in the same viewing window. How well does each model fit the data?

   (d) Use each model to predict the year in which the sales per share will be about $50. Then discuss the appropriateness of each model for predicting future values.

Year	Sales per share, S
1996	11.79
1997	13.33
1998	15.81
1999	19.60
2000	23.50
2001	26.38
2002	32.12
2003	36.73
2004	41.62
2005	43.99

Table for 5

In Exercises 6–9, perform the indicated operation and write the result in standard form.

6. $(12 + 3i) + (4 - 6i)$

7. $(10 - 2i) - (3 + 7i)$

8. $\left(5 + \sqrt{-12}\right)\left(3 - \sqrt{-12}\right)$

9. $(4 + 3i)(2 - 5i)$

10. Write the quotient in standard form: $\dfrac{1 + i}{1 - i}$.

In Exercises 11 and 12, solve the quadratic equation.

11. $x^2 + 5x + 7 = 0$

12. $2x^2 - 5x + 11 = 0$

13. Find a polynomial with real coefficients that has 2, 5, $3i$, and $-3i$ as zeros.

14. Find all the zeros of $f(x) = x^3 + 2x^2 + 5x + 10$, given that $\sqrt{5}\,i$ is a zero.

15. Sketch the graph of $f(x) = \dfrac{3x}{x - 2}$. Label any intercepts and asymptotes. What is the domain of f?

Exponential and Logarithmic Functions

© Tetra Images/Alamy

Some scientists believe the duration of short-term memory is less than a minute. In contrast, the duration of long-term memory is theoretically unlimited. You can use logarithmic functions to model long-term memory to see how well humans retain information over time. (See Section 4.2, Example 10.)

Applications

Exponential and logarithmic functions are used to model and solve many real-life applications. The applications listed below represent a sample of the applications in this chapter.

- Population Growth, Exercises 65 and 66, page 352
- Bone Graft Procedures, Example 11, page 379
- Super Bowl Ad Revenue, Exercise 43, page 393

4

Section 4.1

Exponential Functions

- Evaluate an exponential expression.
- Sketch the graph of an exponential function.
- Evaluate and sketch the graph of the natural exponential function.
- Use the compound interest formulas.
- Use an exponential model to solve an application problem.

Exponential Functions

So far, this text has dealt only with **algebraic functions,** which include polynomial functions and rational functions. In this chapter, you will study two types of nonalgebraic functions—*exponential* functions and *logarithmic* functions. These functions are examples of **transcendental functions.**

> **Definition of Exponential Function**
>
> The **exponential function f with base a** is denoted by
>
> $$f(x) = a^x$$
>
> where $a > 0$, $a \neq 1$, and x is any real number.

The base $a = 1$ is excluded because it yields

$$f(x) = 1^x = 1.$$

This is a constant function, not an exponential function.

You already know how to evaluate a^x for integer and rational values of x. For example, you know that $4^3 = 64$ and $4^{1/2} = 2$. However, to evaluate 4^x for any real number x, you need to interpret forms with *irrational* exponents. For the purposes of this text, it is sufficient to think of

$$a^{\sqrt{2}} \quad \left(\text{where } \sqrt{2} \approx 1.414214\right)$$

as that value having the successively closer approximations

$$a^{1.4}, a^{1.41}, a^{1.414}, a^{1.4142}, a^{1.41421}, a^{1.414214}, \ldots.$$

Example 1   Evaluating an Exponential Expression

Scientific Calculator

Number	*Keystrokes*	*Display*
$2^{-\pi}$	2 $\boxed{y^2}$ π $\boxed{+/-}$ $\boxed{=}$	0.113314732

Graphing Calculator

Number	*Keystrokes*	*Display*
$2^{-\pi}$	2 $\boxed{\wedge}$ $\boxed{(-)}$ π $\boxed{\text{ENTER}}$	.1133147323

✓ **CHECKPOINT 1**

Use a calculator to evaluate $(2.2)^{1.8}$. Round your result to three decimal places. ■

Graphs of Exponential Functions

The graphs of all exponential functions have similar characteristics, as shown in Examples 2, 3, and 4.

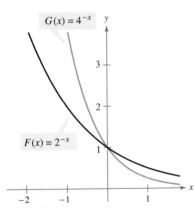

FIGURE 4.1

Example 2 Graphs of $y = a^x$

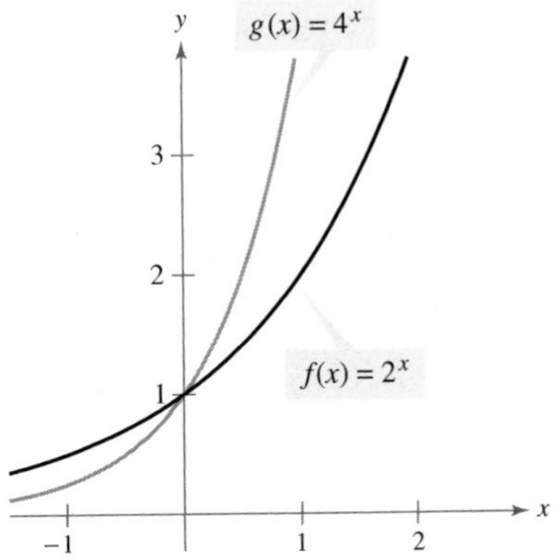

In the same coordinate plane, sketch the graph of each function.

a. $f(x) = 2^x$ **b.** $g(x) = 4^x$

SOLUTION The table below lists some values for each function, and Figure 4.1 shows their graphs. Note that both graphs are increasing. Moreover, the graph of $g(x) = 4^x$ is increasing more rapidly than the graph of $f(x) = 2^x$.

x	-2	-1	0	1	2	3
$f(x) = 2^x$	$\frac{1}{4}$	$\frac{1}{2}$	1	2	4	8
$g(x) = 4^x$	$\frac{1}{16}$	$\frac{1}{4}$	1	4	16	64

✓**CHECKPOINT 2**

Sketch the graph of $f(x) = 5^x$. ■

Example 3 Graphs of $y = a^{-x}$

In the same coordinate plane, sketch the graph of each function.

a. $F(x) = 2^{-x}$ **b.** $G(x) = 4^{-x}$

SOLUTION The table below lists some values for each function, and Figure 4.2 shows their graphs. Note that both graphs are decreasing. Moreover, the graph of $G(x) = 4^{-x}$ is decreasing more rapidly than the graph of $F(x) = 2^{-x}$.

FIGURE 4.2

x	-3	-2	-1	0	1	2
$F(x) = 2^{-x}$	8	4	2	1	$\frac{1}{2}$	$\frac{1}{4}$
$G(x) = 4^{-x}$	64	16	4	1	$\frac{1}{4}$	$\frac{1}{16}$

✓**CHECKPOINT 3**

Sketch the graph of $F(x) = 5^{-x}$. ■

The tables in Examples 2 and 3 were evaluated by hand. You could, of course, use the *table* feature of a graphing utility to construct tables with even more values.

In Example 3, note that the functions given by $F(x) = 2^{-x}$ and $G(x) = 4^{-x}$ can be rewritten with positive exponents.

$$F(x) = 2^{-x} = \left(\tfrac{1}{2}\right)^x \quad \text{and} \quad G(x) = 4^{-x} = \left(\tfrac{1}{4}\right)^x$$

Comparing the functions in Examples 2 and 3, observe that

$$F(x) = 2^{-x} = f(-x) \quad \text{and} \quad G(x) = 4^{-x} = g(-x).$$

Consequently, the graph of F is a reflection (in the y-axis) of the graph of f. The graphs of G and g have the same relationship.

The graphs in Figures 4.1 and 4.2 are typical of the exponential functions $y = a^x$ and $y = a^{-x}$. They have one y-intercept and one horizontal asymptote (the x-axis), and they are continuous. The basic characteristics of these exponential functions are summarized in Figures 4.3 and 4.4.

Characteristics of Exponential Functions

Graph of $y = a^x$, $a > 1$

- Domain: $(-\infty, \infty)$
- Range: $(0, \infty)$
- Intercept: $(0, 1)$
- Increasing
- x-axis is a horizontal asymptote
    $(a^x \to 0$ as $x \to -\infty)$
- Continuous

Graph of $y = a^{-x}$, $a > 1$

- Domain: $(-\infty, \infty)$
- Range: $(0, \infty)$
- Intercept: $(0, 1)$
- Decreasing
- x-axis is a horizontal asymptote
    $(a^{-x} \to 0$ as $x \to \infty)$
- Continuous
- Reflection of graph of $y = a^x$ about y-axis

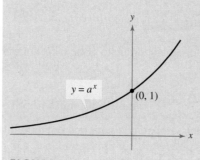

FIGURE 4.3

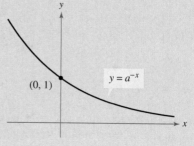

FIGURE 4.4

DISCOVERY

Use a graphing utility to graph

$$y = a^x$$

for $a = 3$, 5, and 7 in the same viewing window. (Use a viewing window in which $-2 \le x \le 1$ and $0 \le y \le 2$.) How do the graphs compare with each other? Which graph is on the top in the interval $(-\infty, 0)$? Which is on the bottom? Which graph is on the top in the interval $(0, \infty)$? Which is on the bottom?

Repeat this experiment with the graphs of $y = b^x$ for $b = \frac{1}{3}, \frac{1}{5},$ and $\frac{1}{7}$. (Use a viewing window in which $-1 \le x \le 2$ and $0 \le y \le 2$.) What can you conclude about the shape of the graph of $y = b^x$ and the value of b?

In the following example, notice how the graph of $y = a^x$ is used to sketch the graphs of functions of the form $f(x) = b \pm a^{x+c}$.

Example 4 **Transformations of Graphs of Exponential Functions**

Each of the following graphs is a transformation of the graph of $f(x) = 3^x$, as shown in Figure 4.5.

a. Because $g(x) = 3^{x+1} = f(x + 1)$, the graph of g can be obtained by shifting the graph of f one unit to the *left*.

b. Because $h(x) = 3^x - 2 = f(x) - 2$, the graph of h can be obtained by shifting the graph of f *downward* two units.

c. Because $k(x) = -3^x = -f(x)$, the graph of k can be obtained by *reflecting* the graph of f in the x-axis.

d. Because $j(x) = 3^{-x} = f(-x)$, the graph of j can be obtained by *reflecting* the graph of f in the y-axis.

STUDY TIP

Notice in Example 4(b) that shifting the graph downward two units also shifts the horizontal asymptote of $f(x) = 3^x$ from the x-axis $(y = 0)$ to the line $y = -2$.

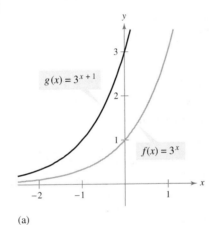

(a)

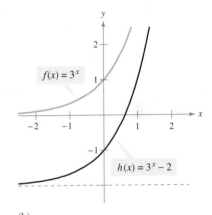

(b)

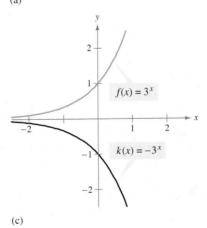

(c)

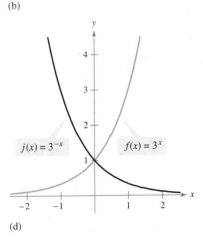

(d)

FIGURE 4.5

✓ **CHECKPOINT 4**

Sketch the graph of $f(x) = 2^{x-1}$. ∎

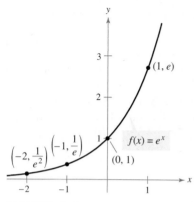

FIGURE 4.6

The Natural Base e

In many applications, the most convenient choice for a base is the irrational number

$$e = 2.718281828 . . .$$

called the **natural base**. The function given by $f(x) = e^x$ is called the **natural exponential function**. Its graph is shown in Figure 4.6. The graph of the natural exponential function has the same basic characteristics as the graph of the exponential function given by $f(x) = a^x$ (see page 344). Be sure you see that for the exponential function given by $f(x) = e^x$, e is the constant $2.718281828 . . .$, whereas x is the variable.

Example 5   Evaluating the Natural Exponential Function

Use a calculator to evaluate the function given by $f(x) = e^x$ when $x = 2$ and $x = -1$.

SOLUTION

Scientific Calculator

Number	Keystrokes	Display
e^2	2 [2nd] [eˣ]	7.389056099
e^{-1}	1 [+/−] [2nd] [eˣ]	0.367879441

Graphing Calculator

Number	Keystrokes	Display
e^2	[2nd] [eˣ] 2 [)] [ENTER]	7.389056099
e^{-1}	[2nd] [eˣ] [(−)] 1 [)] [ENTER]	.3678794412

✓**CHECKPOINT 5**

Use a calculator to evaluate $f(x) = e^x$ when $x = 6$. ■

Example 6   Graphing Natural Exponential Functions

Sketch the graph of each natural exponential function.

a. $f(x) = 2e^{0.24x}$ **b.** $g(x) = \frac{1}{2}e^{-0.58x}$

SOLUTION   To sketch these two graphs, you can use a calculator to plot several points on each graph, as shown in the table. Then, connect the points with smooth curves, as shown in Figure 4.7. Note that the graph in part (a) is increasing, whereas the graph in part (b) is decreasing.

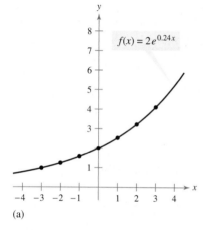

(a)

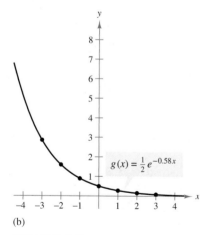

(b)

FIGURE 4.7

x	-3	-2	-1	0	1	2	3
$f(x) = 2e^{0.24x}$	0.974	1.238	1.573	2	2.542	3.232	4.109
$g(x) = \frac{1}{2}e^{-0.58x}$	2.849	1.595	0.893	0.5	0.280	0.157	0.088

✓**CHECKPOINT 6**

Sketch the graph of $f(x) = e^{0.5x}$. ■

Compound Interest

One of the most familiar examples of exponential growth is that of an investment earning **continuously compounded interest.** The formula for the balance in an account that is compounded n times per year is $A = P(1 + r/n)^{nt}$, where A is the balance in the account, P is the initial deposit, r is the annual interest rate (in decimal form), and t is the number of years. Using exponential functions, you will *develop* this formula and show how it leads to continuous compounding.

Suppose a principal P is invested at an annual interest rate r, compounded once a year. The principal at the end of the first year, P_1, is equal to the initial deposit P plus the interest earned, Pr. So,

$$P_1 = P + Pr.$$

This can be rewritten by factoring out P from each term as follows.

$$P_1 = P + Pr$$
$$= P(1 + r)$$

This pattern of multiplying the previous principal by $1 + r$ is then repeated each successive year, as shown below.

Year	*Balance After Each Compounding*
0	$P = P$
1	$P_1 = P(1 + r)$
2	$P_2 = P_1(1 + r) = P(1 + r)(1 + r) = P(1 + r)^2$
3	$P_3 = P_2(1 + r) = P(1 + r)^2(1 + r) = P(1 + r)^3$
$\vdots$	
t	$P_t = P(1 + r)^t$

To accommodate more frequent (quarterly, monthly, or daily) compounding of interest, let n be the number of compoundings per year and let t be the number of years. Then the rate per compounding is r/n and the account balance after t years is

$$A = P\left(1 + \frac{r}{n}\right)^{nt}. \qquad \text{Amount (balance) with } n \text{ compoundings per year}$$

If you let the number of compoundings n increase without bound, the process approaches what is called **continuous compounding.** In the formula for n compoundings per year, let $m = n/r$. This produces

$$A = P\left(1 + \frac{r}{n}\right)^{nt} \qquad \text{Amount with } n \text{ compoundings per year}$$

$$= P\left(1 + \frac{1}{m}\right)^{mrt} \qquad \text{Substitute } mr \text{ for } n \text{ and simplify.}$$

$$= P\left[\left(1 + \frac{1}{m}\right)^{m}\right]^{rt}. \qquad \text{Property of exponents}$$

As m increases without bound, it can be shown that $[1 + (1/m)]^m$ approaches e. From this, you can conclude that the formula for continuous compounding is $A = Pe^{rt}$.

DISCOVERY

Use a calculator and the formula $A = P(1 + r/n)^{nt}$ to calculate the amount in an account when $P = \$3000$, $r = 6\%$, t is 10 years, and the number of compoundings is (1) by the day, (2) by the hour, (3) by the minute, and (4) by the second. Use these results to present an argument that increasing the number of compoundings does not mean unlimited growth of the amount in the account.

Formulas for Compound Interest

After t years, the balance A in an account with principal P and annual interest rate r (in decimal form) is given by the following formulas.

1. For n compoundings per year: $A = P\left(1 + \dfrac{r}{n}\right)^{nt}$

2. For continuous compounding: $A = Pe^{rt}$

Be sure that the annual interest rate is written in decimal form. For instance, 6% should be written as 0.06 when using compound interest formulas.

Example 7
MAKE A DECISION   **Compound Interest**

You invest $12,000 at an annual rate of 3%. Find the balance after 5 years when the interest is compounded (a) quarterly, (b) monthly, and (c) continuously. Which type of compounding earns the most money?

SOLUTION

a. For quarterly compounding, you have $n = 4$. So, in 5 years at 3%, the balance is

$$A = P\left(1 + \frac{r}{n}\right)^{nt} \qquad \text{Formula for compound interest}$$

$$= 12{,}000\left(1 + \frac{0.03}{4}\right)^{4(5)} \qquad \text{Substitute for } P, r, n, \text{ and } t.$$

$$\approx \$13{,}934.21. \qquad \text{Use a calculator.}$$

b. For monthly compounding, you have $n = 12$. So, in 5 years at 3%, the balance is

$$A = P\left(1 + \frac{r}{n}\right)^{nt} \qquad \text{Formula for compound interest}$$

$$= 12{,}000\left(1 + \frac{0.03}{12}\right)^{12(5)} \qquad \text{Substitute for } P, r, n \text{ and } t.$$

$$\approx \$13{,}939.40 \qquad \text{Use a calculator.}$$

c. For continuous compounding, the balance is

$$A = Pe^{rt} \qquad \text{Formula for continuous compounding}$$

$$= 12{,}000e^{0.03(5)} \qquad \text{Substitute for } P, r, \text{ and } t.$$

$$\approx \$13{,}942.01 \qquad \text{Use a calculator.}$$

Note that continuous compounding yields more than quarterly and monthly compounding. This is typical of the two types of compounding. That is, for a given principal, interest rate, and time, continuous compounding will always yield a larger balance than compounding n times a year.

✓ **CHECKPOINT 7**

You invest $6000 at an annual rate of 4%. Find the balance after 7 years when the interest is compounded continuously. ▪

Another Application

> **Example 8**
> **MAKE A DECISION   Radioactive Decay**

In 1986, a nuclear reactor accident occurred in Chernobyl in what was then the Soviet Union. The explosion spread highly toxic radioactive chemicals, such as plutonium, over hundreds of square miles, and the government evacuated the city and the surrounding area. Consider the model

$$P = 10\left(\tfrac{1}{2}\right)^{t/24,100}$$

which represents the amount of plutonium P that remains (from an initial amount of 10 pounds) after t years. Sketch the graph of this function over the interval from $t = 0$ to $t = 100,000$, where $t = 0$ represents 1986. How much of the 10 pounds of plutonium will remain in the year 2010? How much of the 10 pounds will remain after 100,000 years? Why is this city uninhabited?

SOLUTION   The graph of this function is shown in Figure 4.8. Note from this graph that plutonium has a *half-life* of about 24,100 years. That is, after 24,100 years, *half* of the original amount of plutonium will remain. After another 24,100 years, one-quarter of the original amount will remain, and so on. In the year 2010 ($t = 24$), there will still be

$$P = 10\left(\tfrac{1}{2}\right)^{24/24,100} \approx 10\left(\tfrac{1}{2}\right)^{0.0009959} \approx 9.993 \text{ pounds}$$

of the original amount of plutonium remaining. After 100,000 years, there will still be

$$P = 10\left(\tfrac{1}{2}\right)^{100,000/24,100} \approx 10\left(\tfrac{1}{2}\right)^{4.149} \approx 0.564 \text{ pound}$$

of the original amount of plutonium remaining. This city is uninhabited because much of the original amount of radioactive plutonium still remains in the city.

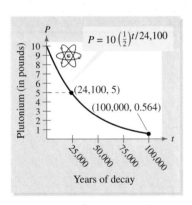

FIGURE 4.8

✓**CHECKPOINT 8**

In Example 8, how much of the initial 10 pounds of plutonium will remain in the year 2086? ■

┌─ **CONCEPT CHECK** ─────────────────────────────────

1. Is the value of 8^x when $x = 3$ equivalent to the value of 8^{-x} when $x = -3$? Explain.

2. What formula would you use to find the balance A of an account after t years with a principal of \$1000 earning an annual interest rate of 5% compounded continuously?

3. What is the range of the graph of $f(x) = 5^x - 1$?

4. Write a natural exponential function whose graph is the graph of $y = e^x$ shifted two units to the left and three units upward.

Skills Review 4.1

The following warm-up exercises involve skills that were covered in earlier sections. You will use these skills in the exercise set for this section. For additional help, review Sections 0.3 and 0.4.

In Exercises 1–12, use the properties of exponents to simplify the expression.

1. $5^{2x}(5^{-x})$

2. $3^{-x}(3^{3x})$

3. $\dfrac{4^{5x}}{4^{2x}}$

4. $\dfrac{10^{2x}}{10^x}$

5. $(4^x)^2$

6. $(4^{2x})^5$

7. $\left(\dfrac{2^x}{3^x}\right)^{-1}$

8. $(4^{6x})^{1/2}$

9. $(2^{3x})^{-1/3}$

10. $\left(\dfrac{3^{4x}}{5^{4x}}\right)^{1/4}$

11. $(16^x)^{1/4}$

12. $(27^x)^{1/3}$

Exercises 4.1

See www.CalcChat.com for worked-out solutions to odd-numbered exercises.

In Exercises 1–10, use a calculator to evaluate the expression. Round your result to three decimal places.

1. $(2.6)^{1.3}$

2. $(1.07)^{50}$

3. $100(1.03)^{-1.4}$

4. $1500(2^{-5/2})$

5. $6^{-\sqrt{2}}$

6. $1.3^{\sqrt{5}}$

7. e^4

8. e^{-5}

9. $e^{2/3}$

10. $e^{-2.7}$

In Exercises 11–18, match the function with its graph. [The graphs are labeled (a), (b), (c), (d), (e), (f), (g), and (h).]

(a)
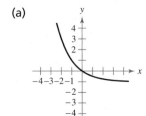

(b)
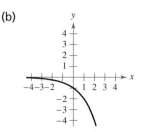

(c)
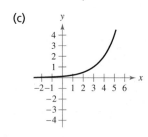

(d)
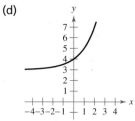

(e)
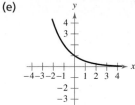

(f)
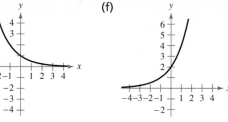

(g)
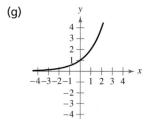

(h)
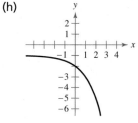

11. $f(x) = 2^x$

12. $f(x) = 2^{-x}$

13. $f(x) = -2^x$

14. $f(x) = -2^x - 1$

15. $f(x) = 2^x + 3$

16. $f(x) = 2^{-x} - 1$

17. $f(x) = 2^{x+1}$

18. $f(x) = 2^{x-3}$

In Exercises 19–36, sketch the graph of the function.

19. $g(x) = 4^x$

20. $f(x) = \left(\tfrac{3}{2}\right)^x$

21. $f(x) = 4^{-x}$

22. $h(x) = \left(\tfrac{3}{2}\right)^{-x}$

23. $h(x) = 4^{x-3}$

24. $g(x) = \left(\tfrac{3}{2}\right)^{x+2}$

25. $g(x) = 4^{-x} - 2$

26. $f(x) = \left(\tfrac{3}{2}\right)^{-x} + 2$

27. $y = 2^{-x^2}$

28. $y = 3^{-x^2}$

29. $y = e^{-0.1x}$

30. $y = e^{0.2x}$

31. $f(x) = 2e^{0.12x}$

32. $f(x) = 3e^{-0.2x}$

33. $f(x) = e^{2x}$

34. $h(x) = e^{x-2}$

35. $g(x) = 1 + e^{-x}$

36. $N(t) = 1000e^{-0.2t}$

Compound Interest   In Exercises 37–40, complete the table to find the balance A for P dollars invested at rate r for t years, compounded n times per year.

n	1	2	4	12	365	Continuous
A						

37. $P = \$5000$, $r = 8\%$, $t = 5$ years

38. $P = \$1000$, $r = 10\%$, $t = 10$ years

39. $P = \$2500$, $r = 12\%$, $t = 20$ years

40. $P = \$1000$, $r = 10\%$, $t = 40$ years

Compound Interest   In Exercises 41–44, complete the table to find the amount P that must be invested at rate r to obtain a balance of $A = \$100,000$ in t years.

t	1	10	20	30	40	50
P						

41. $r = 9\%$, compounded continuously

42. $r = 12\%$, compounded continuously

43. $r = 10\%$, compounded monthly

44. $r = 7\%$, compounded daily

45. Compound Interest   A bank offers two types of interest-bearing accounts. The first account pays 5% interest compounded quarterly. The second account pays 3% interest compounded continuously. Which account earns more money? Why?

46. Compound Interest   A bank offers two types of interest-bearing accounts. The first account pays 6% interest compounded monthly. The second account pays 5% interest compounded continuously. Which account earns more money? Why?

47. *MAKE A DECISION: CASH SETTLEMENT*   You invest a cash settlement of \$10,000 for 5 years. You have a choice between an account that pays 6.25% interest compounded monthly with a monthly online access fee of \$5 and an account that pays 5.25% interest compounded continuously with free online access. Which account should you choose? Explain your reasoning.

48. *MAKE A DECISION: SALES COMMISSION*   You invest a sales commission of \$12,000 for 6 years. You have a choice between an account that pays 4.85% interest compounded monthly with a monthly online access fee of \$3 and an account that pays 4.25% interest compounded continuously with free online access. Which account should you choose? Explain your reasoning.

Compound Interest   On the day a child was born, a lump sum P was deposited in a trust fund paying 6.5% interest compounded continuously. In Exercises 49–52, use the balance A of the fund on the child's 25th birthday to find P.

49. $A = \$100,000$

50. $A = \$500,000$

51. $A = \$750,000$

52. $A = \$1,000,000$

Compound Interest   On the day you were born, a lump sum P was deposited in a trust fund paying 7.5% interest compounded continuously. In Exercises 53–56, use the balance A of the fund, which is the balance on your 21st birthday, to find P.

53. $A = \$100,000$

54. $A = \$500,000$

55. $A = \$750,000$

56. $A = \$1,000,000$

(T) 57. Demand Function   The demand function for a limited edition comic book is given by

$$p = 3000\left(1 - \frac{5}{5 + e^{-0.015x}}\right).$$

(a) Find the price p for a demand of $x = 75$ units.

(b) Find the price p for a demand of $x = 200$ units.

(c) Use a graphing utility to graph the demand function.

(d) Use the graph from part (c) to approximate the demand when the price is \$100.

(T) 58. Demand Function   The demand function for a home theater sound system is given by

$$p = 7500\left(1 - \frac{7}{7 + e^{-0.003x}}\right).$$

(a) Find the price p for a demand of $x = 200$ units.

(b) Find the price p for a demand of $x = 900$ units.

(c) Use a graphing utility to graph the demand function.

(d) Use the graph from part (c) to approximate the demand when the price is \$400.

59. Bacteria Growth   The number of a certain type of bacteria increases according to the model

$$P(t) = 100e^{0.01896t}$$

where t is time (in hours).

(a) Find $P(0)$.

(b) Find $P(5)$.

(c) Find $P(10)$.

(d) Find $P(24)$.

60. Bacteria Growth   As a result of a medical treatment, the number of a certain type of bacteria decreases according to the model

$$P(t) = 100e^{-0.685t}$$

where t is time (in hours).

(a) Find $P(0)$.

(b) Find $P(5)$.

(c) Find $P(10)$.

(d) Find $P(24)$.

Present Value   The present value of money is the principal P you need to invest today so that it will grow to an amount A at the end of a specified time. The present value formula

$$P = A\left(1 + \frac{r}{n}\right)^{-nt}$$

is obtained by solving the compound interest formula

$$A = P\left(1 + \frac{r}{n}\right)^{nt}$$

for P. Recall that t is the number of years, r is the interest rate per year, and n is the number of compoundings per year. In Exercises 61–64, find the present value of amount A invested at rate r for t years, compounded n times per year.

61. $A = \$10,000$, $r = 6\%$, $t = 5$ years, $n = 4$

62. $A = \$50,000$, $r = 7\%$, $t = 10$ years, $n = 12$

63. $A = \$20,000$, $r = 8\%$, $t = 6$ years, $n = 4$

64. $A = \$1,000,000$, $r = 8\%$, $t = 20$ years, $n = 2$

65. Population Growth   The population P of a town increases according to the model

$$P(t) = 4500e^{0.0272t}$$

where t represents the year, with $t = 0$ corresponding to 2000. Use the model to predict the population in each year.

(a) 2010   (b) 2012

(c) 2015   (d) 2020

66. Population Growth   The population P of a small city increases according to the model

$$P(t) = 36,000e^{0.0156t}$$

where t represents the year, with $t = 0$ corresponding to 2000. Use the model to predict the population in each year.

(a) 2009   (b) 2011

(c) 2015   (d) 2018

67. Radioactive Decay   Strontium-90 has a half-life of 29.1 years. The amount S of 100 kilograms of strontium-90 present after t years is given by

$$S = 100e^{-0.0238t}.$$

How much of the 100 kilograms will remain after 50 years?

68. Radioactive Decay   Neptunium-237 has a half-life of 2.1 million years. The amount N of 200 kilograms of neptunium-237 present after t years is given by

$$N = 200e^{-0.00000033007t}.$$

How much of the 200 kilograms will remain after 20,000 years?

Ⓣ **69. Radioactive Decay**   Five pounds of the element plutonium (^{230}Pu) is released in a nuclear accident. The amount of plutonium P that is present after t months is given by $P = 5e^{-0.1507t}$.

(a) Use a graphing utility to graph this function over the interval from $t = 0$ to $t = 10$.

(b) How much of the 5 pounds of plutonium will remain after 10 months?

(c) Use the graph to estimate the half-life of ^{230}Pu. Explain your reasoning.

Ⓣ **70. Radioactive Decay**   One hundred grams of radium (^{226}Ra) is stored in a container. The amount of radium R present after t years is given by $R = 100e^{-0.0004335t}$.

(a) Use a graphing utility to graph this function over the interval from $t = 0$ to $t = 10,000$.

(b) How much of the 100 grams of radium will remain after 10,000 years?

(c) Use the graph to estimate the half-life of ^{226}Ra. Explain your reasoning.

71. Guitar Sales   The sales S (in millions of dollars) for Guitar Center, Inc. from 1996 to 2005 can be modeled by

$$S = 63.7e^{0.2322t}, \quad 6 \le t \le 15$$

where t represents the year, with $t = 6$ corresponding to 1996.   *(Source: Guitar Center, Inc.)*

(a) Use the graph to estimate *graphically* the sales for Guitar Center, Inc. in 1998, 2000, and 2005.

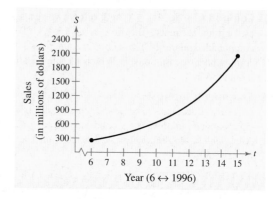

(b) Use the model to confirm *algebraically* the estimates obtained in part (a).

72. Digital Cinema Screens   The numbers y of digital cinema screens in the world from 2000 to 2005 can be modeled by

$$y = 28.7e^{0.6577t}, \quad 0 \le t \le 5$$

where t represents the year, with $t = 0$ corresponding to 2000.   *(Source: Screen Digest)*

(a) Use the graph to estimate *graphically* the numbers of digital cinema screens in 2001 and 2004.

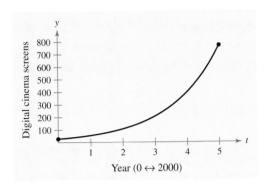

Year (0 ↔ 2000)

(b) Use the model to confirm *algebraically* the estimates obtained in part (a).

73. Age at First Marriage From 1980 to 2005, the median age A of an American woman at her first marriage can be approximated by the model

$$A = 17.91 + \frac{7.88}{1 + e^{-0.1117t - 0.1138}}, \quad 0 \le t \le 25$$

where t represents the year, with $t = 0$ corresponding to 1980. *(Source: U.S. Census Bureau)*

(a) Use the graph to estimate *graphically* the median age of an American woman at her first marriage in 1980, 1990, 2000, and 2005.

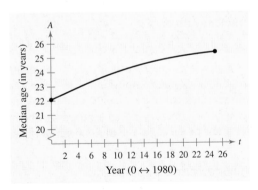

Year (0 ↔ 1980)

(b) Use the model to confirm *algebraically* the estimates obtained in part (a).

74. Age at First Marriage From 1980 to 2005, the median age A of an American man at his first marriage can be approximated by the model

$$A = 22.55 + \frac{4.74}{1 + e^{-0.1412t - 0.1513}}, \quad 0 \le t \le 25$$

where t represents the year, with $t = 0$ corresponding to 1980. *(Source: U.S. Census Bureau)*

(a) Use the graph to estimate *graphically* the median age of an American man at his first marriage in 1980, 1990, 2000, and 2005.

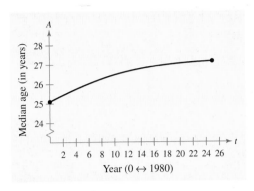

Year (0 ↔ 1980)

(b) Use the model to confirm *algebraically* the estimates obtained in part (a).

75. Compare Compare the results of Exercises 73 and 74. What can you conclude about the differences in men's and women's ages at first marriage?

Ⓣ **76. Hospital Employment** The numbers of people E (in thousands) employed in hospitals from 1999 to 2005 can be modeled by

$$E = 3331(1.0182)^t, \quad 9 \le t \le 15$$

where t represents the year, with $t = 9$ corresponding to 1999. *(Source: U.S. Bureau of Labor Statistics)*

(a) Use a graphing utility to graph E for the years 1999 to 2005.

(b) Use the graph from part (a) to estimate the numbers of hospital employees in 2000, 2002, and 2005.

Ⓣ **77. Prescriptions** The numbers of prescriptions P (in millions) filled in the United States from 1998 to 2005 can be modeled by

$$P = -11{,}415 + \frac{15{,}044}{1 + e^{-0.2166t - 0.7667}}, \quad 8 \le t \le 15$$

where t represents the year, with $t = 8$ corresponding to 1998. *(Source: National Association of Chain Drug Stores)*

(a) Use a graphing utility to graph P for the years 1998 to 2005.

(b) Use the graph from part (a) to estimate the numbers of prescriptions filled in 1999, 2002, and 2005.

78. Writing Determine whether $e = \dfrac{271{,}801}{99{,}990}$. Justify your answer.

79. Extended Application To work an extended application involving the healing rate of a wound, visit this text's website at *college.hmco.com/info/larsonapplied.*

Logarithmic Functions

- Recognize and evaluate a logarithmic function with base a.
- Sketch the graph of a logarithmic function.
- Recognize and evaluate the natural logarithmic function.
- Use a logarithmic model to solve an application problem.

Logarithmic Functions

In Section 2.8, you studied inverse functions. There, you learned that if a function has the property that no horizontal line intersects the graph of the function more than once, the function must have an inverse function. By looking back at the graphs of the exponential functions introduced in Section 4.1, you will see that every function of the form $f(x) = a^x$ (where $a > 0$ and $a \neq 1$) passes the Horizontal Line Test and therefore must have an inverse function. This inverse function is called the **logarithmic function with base a.**

Definition of a Logarithmic Function

For $x > 0$, $a > 0$, and $a \neq 1$,

$$y = \log_a x \text{ if and only if } x = a^y.$$

The function given by

$$f(x) = \log_a x$$

is called the **logarithmic function with base a.**

The equations $y = \log_a x$ and $x = a^y$ are equivalent. The first equation is in logarithmic form and the second is in exponential form.

When evaluating logarithms, remember that *a logarithm is an exponent.* This means that $\log_a x$ is the exponent to which a must be raised to obtain x. For instance, $\log_2 8 = 3$ because 2 must be raised to the third power to obtain 8.

Example 1　Evaluating Logarithmic Expressions

a. $\log_2 32 = 5$　　　because　　$2^5 = 32.$

b. $\log_4 2 = \dfrac{1}{2}$　　　because　　$4^{1/2} = \sqrt{4} = 2.$

c. $\log_{10} \dfrac{1}{100} = -2$　　because　　$10^{-2} = \dfrac{1}{10^2} = \dfrac{1}{100}.$

d. $\log_3 1 = 0$　　　because　　$3^0 = 1.$

✓**CHECKPOINT 1**

Evaluate the expression $\log_7 \dfrac{1}{49}.$ ▪

The logarithmic function with base 10 is called the **common logarithmic function.** On most calculators, this function is denoted by $\boxed{\text{LOG}}$.

STUDY TIP

Because $\log_a x$ is the inverse function of $y = a^x$, it follows that the domain of $y = \log_a x$ is the range of $y = a^x$, $(0, \infty)$. In other words, $y = \log_a x$ is defined only if x is positive.

Example 2 Evaluating Logarithmic Expressions on a Calculator

Scientific Calculator

Number	Keystrokes	Display
a. $\log_{10} 10$	10 $\boxed{\text{LOG}}$	1
b. $2 \log_{10} 2.5$	2.5 $\boxed{\text{LOG}}$ $\boxed{\times}$ 2 $\boxed{=}$	0.795880017
c. $\log_{10}(-2)$	2 $\boxed{+/-}$ $\boxed{\text{LOG}}$	ERROR

Graphing Calculator

Number	Keystrokes	Display
a. $\log_{10} 10$	$\boxed{\text{LOG}}$ 10 $\boxed{)}$ $\boxed{\text{ENTER}}$	1
b. $2 \log_{10} 2.5$	2 $\boxed{\text{LOG}}$ 2.5 $\boxed{)}$ $\boxed{\text{ENTER}}$	.7958800173
c. $\log_{10}(-2)$	$\boxed{\text{LOG}}$ $\boxed{(-)}$ 2 $\boxed{)}$ $\boxed{\text{ENTER}}$	ERROR

Many calculators display an error message (or a complex number) when you try to evaluate $\log_{10}(-2)$. This is because the domain of every logarithmic function is the set of *positive real numbers*. In other words, there is no real number power to which 10 can be raised to obtain -2.

✓ CHECKPOINT 2

Use a calculator to evaluate the expression $\log_{10} 200$. Round your result to three decimal places. ▪

The following properties follow directly from the definition of the logarithmic function with base a.

Properties of Logarithms

1. $\log_a 1 = 0$ because $a^0 = 1$.

2. $\log_a a = 1$ because $a^1 = a$.

3. $\log_a a^x = x$ and $a^{\log_a x} = x$ Inverse Properties

4. If $\log_a x = \log_a y$, then $x = y$.   One-to-One Property

---

Example 3 Using Properties of Logarithms

a. Solve the equation $\log_2 x = \log_2 3$ for x.

b. Solve the equation $\log_5 x = 1$ for x.

SOLUTION

a. Using the One-to-One Property (Property 4), you can conclude that $x = 3$.

b. Using Property 2, you can conclude that $x = 5$.

✓ CHECKPOINT 3

Solve the equation $\log_4 1 = x$ for x. ▪

Graphs of Logarithmic Functions

To sketch the graph of $y = \log_a x$, you can use the fact that the graphs of inverse functions are reflections of each other in the line $y = x$.

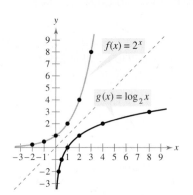

FIGURE 4.9 Inverse Functions

Example 4   Graphs of Exponential and Logarithmic Functions

In the same coordinate plane, sketch the graph of each function.

a. $f(x) = 2^x$

b. $g(x) = \log_2 x$

SOLUTION

a. For $f(x) = 2^x$, construct a table of values, as follows.

x	-2	-1	0	1	2	3
$f(x) = 2^x$	$\frac{1}{4}$	$\frac{1}{2}$	1	2	4	8

By plotting these points and connecting them with a smooth curve, you obtain the graph shown in Figure 4.9.

b. Because $g(x) = \log_2 x$ is the inverse function of $f(x) = 2^x$, the graph of g is obtained by plotting the points $(f(x), x)$ and connecting them with a smooth curve. The graph of g is a reflection of the graph of f in the line $y = x$, as shown in Figure 4.9.

Before you can confirm the result of Example 4 with a graphing utility, you need to know how to enter $\log_2 x$. You will learn how to do this using the *change-of-base formula* discussed in Section 4.3.

✓ **CHECKPOINT 4**

In the same coordinate plane, sketch the graph of each function.

a. $f(x) = 4^x$

b. $g(x) = \log_4 x$ ∎

Example 5   Sketching the Graph of a Logarithmic Function

Sketch the graph of the common logarithmic function given by $f(x) = \log_{10} x$.

SOLUTION Begin by constructing a table of values. Note that some of the values can be obtained without a calculator by using the Inverse Property of logarithms. Others require a calculator. Next, plot the points and connect them with a smooth curve, as shown in Figure 4.10.

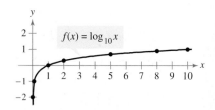

FIGURE 4.10

	Without Calculator				With Calculator		
x	$\frac{1}{100}$	$\frac{1}{10}$	1	10	2	5	8
$f(x) = \log_{10} x$	-2	-1	0	1	0.301	0.699	0.903

✓ **CHECKPOINT 5**

Sketch the graph of the function given by $f(x) = 2 \log_{10} x$. ∎

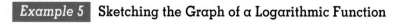

The nature of the graph in Figure 4.10 is typical of functions of the form $f(x) = \log_a x$, $a > 1$. They have one x-intercept and one vertical asymptote. Notice how slowly the graph rises for $x > 1$. The basic characteristics of logarithmic graphs are summarized in Figure 4.11. Note that the vertical asymptote occurs at $x = 0$, where $\log_a x$ is *undefined*.

Characteristics of Logarithmic Functions

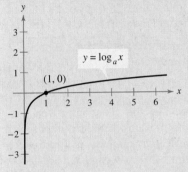

Graph of $y = \log_a x$, $a > 1$

- Domain: $(0, \infty)$
- Range: $(-\infty, \infty)$
- Intercept: $(1, 0)$
- Increasing
- One-to-one; therefore has an inverse function
- y-axis is a vertical asymptote $(\log_a x \to -\infty$ as $x \to 0^+)$
- Continuous
- Reflection of graph of $y = a^x$ about the line $y = x$

FIGURE 4.11

Example 6   Sketching the Graphs of Logarithmic Functions

The graph of each function below is similar to the graph of $f(x) = \log_{10} x$.

a. Because $g(x) = \log_{10}(x - 1) = f(x - 1)$, the graph of g can be obtained by shifting the graph of f one unit to the *right*. See Figure 4.12(a).

b. Because $h(x) = 2 + \log_{10} x = 2 + f(x)$, the graph of h can be obtained by shifting the graph of f two units *upward*. See Figure 4.12(b).

STUDY TIP

Notice in Example 6(a) that shifting the graph of $f(x)$ one unit to the right also shifts the vertical asymptote from the y-axis ($x = 0$) to the line $x = 1$.

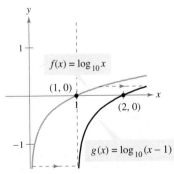

(a) Right shift of one unit

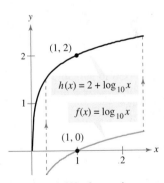

(b) Upward shift of two units

FIGURE 4.12

✓**CHECKPOINT 6**

Sketch the graph of $f(x) = \log_{10}(x + 3)$. ◼

The Natural Logarithmic Function

By looking back at the graph of the natural exponential function introduced in Section 4.1, you will see that $f(x) = e^x$ is one-to-one and so has an inverse function. This inverse function is called the **natural logarithmic function** and is denoted by the special symbol $\ln x$, read as "el en of x."

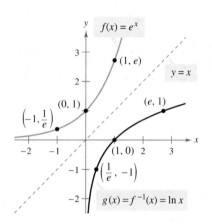

FIGURE 4.13

The Natural Logarithmic Function

The function defined by

$$f(x) = \log_e x = \ln x, \quad x > 0$$

is called the **natural logarithmic function.**

Because the functions given by $f(x) = e^x$ and $g(x) = \ln x$ are inverse functions of each other, their graphs are reflections of each other in the line $y = x$. This reflective property is illustrated in Figure 4.13. The four properties of logarithms listed on page 355 are also valid for natural logarithms.

Properties of Natural Logarithms

1. $\ln 1 = 0$ because $e^0 = 1.$

2. $\ln e = 1$ because $e^1 = e.$

3. $\ln e^x = x$ and $e^{\ln x} = x$ Inverse Properties

4. If $\ln x = \ln y$, then $x = y.$ One-to-One Property

Example 7   **Using Properties of Natural Logarithms**

Evaluate each logarithmic expression.

a. $\ln \dfrac{1}{e}$ **b.** $e^{\ln 5}$ **c.** $\dfrac{\ln 1}{3}$ **d.** $2 \ln e$

SOLUTION

a. $\ln \dfrac{1}{e} = \ln e^{-1} = -1$ Inverse Property

b. $e^{\ln 5} = 5$ Inverse Property

c. $\dfrac{\ln 1}{3} = \dfrac{0}{3} = 0$ Property 1

d. $2 \ln e = 2(1) = 2$ Property 2

✓**CHECKPOINT 7**

Evaluate each logarithmic expression.

a. $\ln e^7$ **b.** $5 \ln 1$ ■

On most calculators, the natural logarithm is denoted by $\boxed{\text{LN}}$, as illustrated in Example 8.

Example 8   Evaluating the Natural Logarithmic Function

Use a calculator to evaluate each expression.

a. $\ln 2$ **b.** $\ln 0.3$ **c.** $\ln e^2$ **d.** $\ln(-1)$

SOLUTION

Scientific Calculator

	Number	*Keystrokes*	*Display*
a.	$\ln 2$	2 $\boxed{\text{LN}}$	0.69314718
b.	$\ln 0.3$	.3 $\boxed{\text{LN}}$	-1.203972804
c.	$\ln e^2$	2 $\boxed{\text{2nd}}$ [e^x] $\boxed{\text{LN}}$	2
d.	$\ln(-1)$	1 $\boxed{+/-}$ $\boxed{\text{LN}}$	ERROR

Graphing Calculator

	Number	*Keystrokes*	*Display*
a.	$\ln 2$	$\boxed{\text{LN}}$ 2 $\boxed{)}$ $\boxed{\text{ENTER}}$	.6931471806
b.	$\ln 0.3$	$\boxed{\text{LN}}$ .3 $\boxed{)}$ $\boxed{\text{ENTER}}$	-1.203972804
c.	$\ln e^2$	$\boxed{\text{LN}}$ $\boxed{\text{2nd}}$ [e^x] 2 $\boxed{)}$ $\boxed{)}$ $\boxed{\text{ENTER}}$	2
d.	$\ln(-1)$	$\boxed{\text{LN}}$ $\boxed{(-)}$ 1 $\boxed{)}$ $\boxed{\text{ENTER}}$	ERROR

✓ **CHECKPOINT 8**

Use a calculator to evaluate the expression $\ln 0.1$. Round your result to three decimal places. ◼

In Example 8, note that $\ln(-1)$ gives an error message on most calculators. This occurs because the domain of $\ln x$ is the set of positive real numbers (see Figure 4.13). So, $\ln(-1)$ is undefined.

Example 9   Finding the Domains of Logarithmic Functions

Find the domain of each function.

a. $f(x) = \ln(x - 2)$ **b.** $g(x) = \ln(2 - x)$ **c.** $h(x) = \ln x^2$

SOLUTION

a. Because $\ln(x - 2)$ is defined only if $x - 2 > 0$, it follows that the domain of f is $(2, \infty)$.

b. Because $\ln(2 - x)$ is defined only if $2 - x > 0$, it follows that the domain of g is $(-\infty, 2)$. The graph of g is shown in Figure 4.14.

c. Because $\ln x^2$ is defined only if $x^2 > 0$, it follows that the domain of h is all real numbers except $x = 0$.

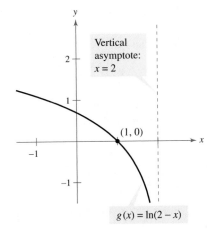

FIGURE 4.14

✓ **CHECKPOINT 9**

Find the domain of the function given by $f(x) = \ln(x + 5)$. ◼

Application

Example 10   Human Memory Model

A group of students participating in a psychological experiment attended several lectures on a subject. Every month for a year after that, the group of students were tested to see how much of the material they remembered. The average scores for the group are given by the *human memory model*

$$f(t) = 75 - 6 \ln(t + 1), \quad 0 \le t \le 12$$

where t is the time (in months). Based on the results of the experiment, how many months can a student wait before retaking the exam and still expect to score 60 or better? (Do not count portions of months.)

SOLUTION   To determine how many months a student can wait before retaking the exam and still expect to score 60 or better, use the model to create a table of values showing the scores for several months.

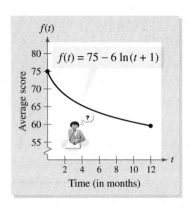

FIGURE 4.15

Month, t	0	1	2	3	4	5	6
Score, $f(t)$	75	70.84	68.41	66.68	65.34	64.25	63.32

Month, t	7	8	9	10	11	12
Score, $f(t)$	62.52	61.82	61.18	60.61	60.09	59.61

From the table, you can see that a student would need to retake the exam by the 11th month in order to score 60 or better. The graph of f is shown in Figure 4.15.

✓ CHECKPOINT 10

Biologists have found that an alligator's length l (in inches) can be approximated by the model

$$l = 27.1 \ln w - 32.8$$

where w is the weight (in pounds) of the alligator. Find the lengths of alligators for which $w = 150, 225, 380, 450,$ and 625 pounds. Round your results to the nearest tenth of an inch. ■

CONCEPT CHECK

1. Is $\log_c b = a$ equivalent to $a^b = c$ when a, b, and c are greater than 0, $a \ne 1$, and $c \ne 1$? Explain.

2. What property would you use to solve $\log_x 7 = 1$ for x?

3. Explain how you can use the graph of an exponential function to sketch the graph of $f(x) = \log_5 x$.

4. How is the graph of $g(x) = -\ln x$ related to the graph of $f(x) = \ln x$?

Skills Review 4.2   The following warm-up exercises involve skills that were covered in earlier sections. You will use these skills in the exercise set for this section. For additional help, review Sections 2.6 and 4.1.

In Exercises 1–4, solve for *x*.

1. $2^x = 8$ **2.** $4^x = 1$ **3.** $10^x = 0.1$ **4.** $e^x = e$

In Exercises 5 and 6, evaluate the expression. (Round the result to three decimal places.)

5. e^2 **6.** e^{-1}

In Exercises 7–10, describe how the graph of *g* is related to the graph of *f*.

7. $g(x) = f(x + 2)$ **8.** $g(x) = -f(x)$ **9.** $g(x) = -1 + f(x)$ **10.** $g(x) = f(-x)$

Exercises 4.2

In Exercises 1–6, match the logarithmic equation with its exponential form. [The exponential forms are labeled (a), (b), (c), (d), (e), and (f).]

1. $\log_4 16 = 2$ (a) $4^{1/2} = 2$

2. $\log_2 16 = 4$ (b) $2^{-4} = \frac{1}{16}$

3. $\log_2 \frac{1}{16} = -4$ (c) $4^2 = 16$

4. $\log_4 \frac{1}{16} = -2$ (d) $4^{-2} = \frac{1}{16}$

5. $\log_4 2 = \frac{1}{2}$ (e) $16^{1/2} = 4$

6. $\log_{16} 4 = \frac{1}{2}$ (f) $2^4 = 16$

In Exercises 7–16, use the definition of a logarithm to write the equation in logarithmic form. For example, the logarithmic form of $2^3 = 8$ is $\log_2 8 = 3$.

7. $4^4 = 256$ **8.** $7^3 = 343$

9. $81^{1/4} = 3$ **10.** $9^{3/2} = 27$

11. $6^{-2} = \frac{1}{36}$ **12.** $10^{-3} = 0.001$

13. $e^1 = e$ **14.** $e^4 = 54.5981\ldots$

15. $e^x = 4$ **16.** $e^{-x} = 2$

In Exercises 17–26, use the definition of a logarithm to write the equation in exponential form. For example, the exponential form of $\log_5 125 = 3$ is $5^3 = 125$.

17. $\log_4 16 = 2$ **18.** $\log_{10} 1000 = 3$

19. $\log_2 \frac{1}{2} = -1$ **20.** $\log_3 \frac{1}{9} = -2$

21. $\ln e = 1$ **22.** $\ln \frac{1}{e} = -1$

23. $\log_5 0.2 = -1$ **24.** $\log_{10} 0.1 = -1$

25. $\log_{27} 3 = \frac{1}{3}$ **26.** $\log_8 2 = \frac{1}{3}$

In Exercises 27–42, evaluate the expression without using a calculator.

27. $\log_3 9$ **28.** $\log_5 125$

29. $\log_2 \frac{1}{16}$ **30.** $\log_6 \frac{1}{36}$

31. $\log_8 2$ **32.** $\log_{64} 4$

33. $\log_7 7$ **34.** $\log_{12} 1$

35. $\log_{10} 0.0001$ **36.** $\log_{10} 100$

37. $\ln e$ **38.** $\ln e^{10}$

39. $\ln e^{-4}$ **40.** $\ln \frac{1}{e^3}$

41. $\log_a a^5$ **42.** $\log_a 1$

In Exercises 43–54, use a calculator to evaluate the logarithm. Round your result to three decimal places.

43. $\log_{10} 345$ **44.** $\log_{10} 163$

45. $\log_{10} \frac{4}{5}$ **46.** $\log_{10} \frac{3}{4}$

47. $\log_{10} \sqrt{8}$ **48.** $\log_{10} \sqrt{3}$

49. $\ln 7$ **50.** $2 \ln 9$

51. $\ln 18.42$ **52.** $\ln 36.7$

53. $\ln \sqrt{6}$ **54.** $\ln \sqrt{10}$

In Exercises 55–58, sketch the graphs of *f* and *g* in the same coordinate plane.

55. $f(x) = 3^x$, $g(x) = \log_3 x$

56. $f(x) = 5^x$, $g(x) = \log_5 x$

57. $f(x) = e^x$, $g(x) = \ln x$

58. $f(x) = 10^x$, $g(x) = \log_{10} x$

In Exercises 59–64, match the function with its graph. [The graphs are labeled (a), (b), (c), (d), (e), and (f).]

(a)
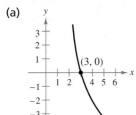

(b)
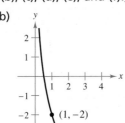

(c)
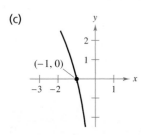

(d)
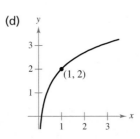

(e)
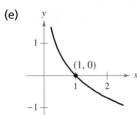

(f)
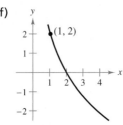

59. $f(x) = \ln x + 2$

60. $f(x) = -\ln x$

61. $f(x) = -3 \ln(x - 2)$

62. $f(x) = 4 \ln(-x)$

63. $f(x) = -3 \ln x + 2$

64. $f(x) = -3 \ln x - 2$

In Exercises 65–74, find the domain, vertical asymptote, and x-intercept of the logarithmic function. Then sketch its graph.

65. $f(x) = \log_2 x$

66. $g(x) = \log_4 x$

67. $h(x) = \log_2(x + 4)$

68. $f(x) = \log_4(x - 3)$

69. $f(x) = -\log_2 x$

70. $h(x) = -\log_4(x - 1)$

71. $g(x) = \ln(-x)$

72. $f(x) = \ln(3 - x)$

73. $h(x) = \ln(x + 1)$

74. $f(x) = 3 + \ln x$

(T) In Exercises 75–80, use a graphing utility to graph the function. Be sure to use an appropriate viewing window.

75. $f(x) = \log(x + 1)$

76. $f(x) = \log(x - 1)$

77. $f(x) = \ln(x - 1)$

78. $f(x) = \ln(x + 2)$

79. $f(x) = \ln x + 1$

80. $f(x) = 3 \ln x - 1$

81. Population Growth The population of a town will double in

$$t = \frac{8 \ln 3}{\ln 63 - \ln 45}$$

years. Find t.

82. Work The work W (in foot-pounds) done in compressing a volume of 9 cubic feet at a pressure of 15 pounds per square inch to a volume of 3 cubic feet is $W = 19{,}440(\ln 9 - \ln 3)$. Find W.

83. Human Memory Model Students in a mathematics class were given an exam and then retested monthly with an equivalent exam. The average score g for the class can be approximated by the human memory model

$$g(t) = 78 - 14 \log_{10}(t + 1), \quad 0 \le t \le 12$$

where t is the time (in months).

(a) What was the average score on the original exam?

(b) What was the average score after 4 months?

(c) When did the average score drop below 70?

84. Human Memory Model Students in a seventh-grade class were given an exam. During the next 2 years, the same students were retested several times. The average score g can be approximated by the model

$$g(t) = 87 - 16 \log_{10}(t + 1), \quad 0 \le t \le 24$$

where t is the time (in months).

(a) What was the average score on the original exam?

(b) What was the average score after 6 months?

(c) When did the average score drop below 70?

85. Investment Time A principal P, invested at 5.25% interest and compounded continuously, increases to an amount that is K times the principal after t years, where t is given by

$$t = \frac{\ln K}{0.0525}.$$

(a) Complete the table.

K	1	2	4	6	8	10	12
t							

(b) Use the table in part (a) to graph the function.

(T) **86. Investment Time** A principal P, invested at 4.85% interest and compounded continuously, increases to an amount that is K times the principal after t years, where t is given by

$$t = \frac{\ln K}{0.0485}.$$

Use a graphing utility to graph this function.

Skill Retention Model In Exercises 87 and 88, participants in an industrial psychology study were taught a simple mechanical task and tested monthly on this mechanical task for a period of 1 year. The average scores for the participants are given by the model

$$f(t) = 98 - 14 \log_{10}(t + 1), \quad 0 \le t \le 12$$

where t is the time (in months).

87. Use a graphing utility to graph the function. Use the graph to discuss the domain and range of the function.

88. Think About It Based on the graph of f, do you think the study's participants practiced the simple mechanical task very often? Cite the behavior of the graph to justify your answer.

Productivity In Exercises 89 and 90, the productivity of a new employee (in units produced per day) is given by the model

$$g(t) = 2 + 12 \ln t, \quad 1 \le t \le 15$$

where t is the time (in work days).

89. Use a graphing utility to graph the function. Use the graph to discuss the domain and range of the function.

90. Think About It Based on the graph of g, do you think the new employee will reach a benchmark of 40 units produced per day by the end of three work weeks? Explain.

91. Median Age of U.S. Population The model

$$A = 15.68 - 0.037t + 6.131 \ln t, \quad 10 \le t \le 80$$

approximates the median age A of the United States population from 1980 to 2050. In the model, t represents the year, with $t = 10$ corresponding to 1980 (see figure). *(Source: U.S. Census Bureau)*

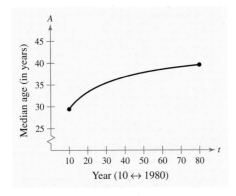

Year (10 ↔ 1980)

(a) Use the model to approximate the median age in the United States in 1980.

(b) Use the model to approximate the median age in the United States in 1990.

(c) Use the model to approximate the change in median age in the United States from 1980 to 2000.

(d) Use the model to project the change in median age in the United States from 1980 to 2050.

92. Monthly Payment The model

$$t = 12.542 \ln\left(\frac{x}{x - 1000}\right), \quad x > 1000$$

approximates the length of a home mortgage of $150,000 at 8% interest in terms of the monthly payment. In the model, t is the length of the mortgage (in years) and x is the monthly payment (in dollars) (see figure).

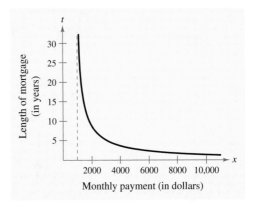

Monthly payment (in dollars)

(a) Use the model to approximate the length of a $150,000 mortgage at 8% interest when the monthly payment is $1100.65 and when the monthly payment is $1254.68.

(b) Approximate the total amount paid over the term of the mortgage with a monthly payment of $1100.65 and with a monthly payment of $1254.68.

(c) Approximate the total interest charge for a monthly payment of $1100.65 and for a monthly payment of $1254.68.

(d) What is the vertical asymptote of the model? Interpret its meaning in the context of the problem.

93. Think About It The table of values was obtained by evaluating a function. Determine which of the statements may be true and which must be false.

x	1	2	8
y	0	1	3

(a) y is an exponential function of x.

(b) y is a logarithmic function of x.

(c) x is an exponential function of y.

(d) y is a linear function of x.

Properties of Logarithms

- Evaluate a logarithm using the change-of-base formula.
- Use properties of logarithms to evaluate or rewrite a logarithmic expression.
- Use properties of logarithms to expand or condense a logarithmic expression.
- Use logarithmic functions to model and solve real-life applications.

Change of Base

Most calculators have only two types of log keys, one for common logarithms (base 10) and one for natural logarithms (base e). Although common logs and natural logs are the most frequently used, you may occasionally need to evaluate logarithms with other bases. To do this, you can use the following **change-of-base formula.**

Change-of-Base Formula

Let a, b, and x be positive real numbers such that $a \neq 1$ and $b \neq 1$. Then $\log_a x$ can be converted to a different base as follows.

Base b	*Base 10*	*Base e*
$\log_a x = \dfrac{\log_b x}{\log_b a}$	$\log_a x = \dfrac{\log_{10} x}{\log_{10} a}$	$\log_a x = \dfrac{\ln x}{\ln a}$

One way to look at the change-of-base formula is that logarithms to base a are simply *constant multiples* of logarithms to base b. The constant multiplier is $1/(\log_b a)$.

Example 1 **Changing Bases Using Common Logarithms**

a. $\log_4 30 = \dfrac{\log_{10} 30}{\log_{10} 4} \approx \dfrac{1.47712}{0.60206} \approx 2.4534$

b. $\log_2 14 = \dfrac{\log_{10} 14}{\log_{10} 2} \approx \dfrac{1.14613}{0.30103} \approx 3.8074$

Example 2 **Changing Bases Using Natural Logarithms**

a. $\log_4 30 = \dfrac{\ln 30}{\ln 4} \approx \dfrac{3.40120}{1.386294} \approx 2.4534$

b. $\log_2 14 = \dfrac{\ln 14}{\ln 2} \approx \dfrac{2.63906}{0.693147} \approx 3.8074$

Notice in Examples 1 and 2 that the result is the same whether common logarithms or natural logarithms are used in the change-of-base formula.

✓**CHECKPOINT 1**

Evaluate $\log_8 56$ using common logarithms. Round your result to three decimal places. ■

✓**CHECKPOINT 2**

Evaluate $\log_8 56$ using natural logarithms. Round your result to three decimal places. ■

Properties of Logarithms

You know from the preceding section that the logarithmic function with base a is the *inverse function* of the exponential function with base a. So, it makes sense that the properties of exponents should have corresponding properties involving logarithms. For instance, the exponential property $a^0 = 1$ has the corresponding logarithmic property $\log_a 1 = 0$.

> ### Properties of Logarithms
>
> Let a be a positive number such that $a \neq 1$, and let n be a real number. If u and v are positive real numbers, then the following properties are true.
>
> | *Logarithm with Base a* | *Natural Logarithm* | |
> |---|---|---|
> | $\log_a(uv) = \log_a u + \log_a v$ | $\ln(uv) = \ln u + \ln v$ | Product Rule |
> | $\log_a \dfrac{u}{v} = \log_a u - \log_a v$ | $\ln \dfrac{u}{v} = \ln u - \ln v$ | Quotient Rule |
> | $\log_a u^n = n \log_a u$ | $\ln u^n = n \ln u$ | Power Rule |

STUDY TIP

There is no general property that can be used to rewrite $\log_a(u \pm v)$. Specifically, $\log_a(x + y)$ is *not* equal to $\log_a x + \log_a y$.

Example 3 Using Properties of Logarithms

Write each logarithm in terms of $\ln 2$ and $\ln 3$.

a. $\ln 6$ **b.** $\ln \dfrac{2}{27}$

SOLUTION

a. $\ln 6 = \ln(2 \cdot 3)$ Rewrite 6 as $2 \cdot 3$.

   $ = \ln 2 + \ln 3$ Product Rule

b. $\ln \dfrac{2}{27} = \ln 2 - \ln 27$ Quotient Rule

   $\phantom{\ln \dfrac{2}{27}} = \ln 2 - \ln 3^3$ Rewrite 27 as 3^3.

   $\phantom{\ln \dfrac{2}{27}} = \ln 2 - 3 \ln 3$ Power Rule

✓ CHECKPOINT 3

Write $\log_{10} \frac{25}{3}$ in terms of $\log_{10} 3$ and $\log_{10} 5$. ▪

Example 4 Using Properties of Logarithms

Use the properties of logarithms to verify that $-\log_{10} \frac{1}{100} = \log_{10} 100$.

SOLUTION

$$-\log_{10} \frac{1}{100} = -\log_{10}(100^{-1}) = -(-1)\log_{10} 100 = \log_{10} 100$$

Try checking this result on your calculator.

✓ CHECKPOINT 4

Use the properties of logarithms to verify that $-\ln \dfrac{2}{e} = 1 - \ln 2$. ▪

Rewriting Logarithmic Expressions

The properties of logarithms are useful for rewriting logarithmic expressions in forms that simplify the operations of algebra. This is true because these properties convert complicated products, quotients, and exponential forms into simpler sums, differences, and products, respectively.

DISCOVERY

Use a calculator to approximate $\ln \sqrt[3]{e^2}$. Now find the exact value by rewriting $\ln \sqrt[3]{e^2}$ with a rational exponent using the properties of logarithms. How do the two values compare?

Example 5 **Expanding Logarithmic Expressions**

Expand each logarithmic expression.

a. $\log_4 5x^3y$ **b.** $\ln \dfrac{\sqrt{3x-5}}{7}$

SOLUTION

a. $\log_4 5x^3y = \log_4 5 + \log_4 x^3 + \log_4 y$ Product Rule

$\qquad\qquad = \log_4 5 + 3 \log_4 x + \log_4 y$ Power Rule

b. $\ln \dfrac{\sqrt{3x-5}}{7} = \ln \dfrac{(3x-5)^{1/2}}{7}$ Rewrite using rational exponent.

$\qquad\qquad = \ln(3x-5)^{1/2} - \ln 7$ Quotient Rule

$\qquad\qquad = \dfrac{1}{2}\ln(3x-5) - \ln 7$ Power Rule

✓ **CHECKPOINT 5**

Expand the expression $\ln 2mn^2$. ■

In Example 5, the properties of logarithms were used to *expand* logarithmic expressions. In Example 6, this procedure is reversed and the properties of logarithms are used to *condense* logarithmic expressions.

STUDY TIP

When applying the properties of logarithms to a logarithmic function, you should be careful to check the domain of the function. For example, the domain of $f(x) = \ln x^2$ is all real $x \neq 0$, whereas the domain of $g(x) = 2 \ln x$ is all real $x > 0$.

Example 6 **Condensing Logarithmic Expressions**

Condense each logarithmic expression.

a. $\frac{1}{2} \log_{10} x + 3 \log_{10}(x+1)$ **b.** $2 \ln(x+2) - \ln x$

c. $\frac{1}{3}[\log_2 x + \log_2(x+1)]$

SOLUTION

a. $\frac{1}{2} \log_{10} x + 3 \log_{10}(x+1) = \log_{10} x^{1/2} + \log_{10}(x+1)^3$ Power Rule

$\qquad\qquad\qquad = \log_{10}\left[\sqrt{x}(x+1)^3\right]$ Product Rule

b. $2 \ln(x+2) - \ln x = \ln(x+2)^2 - \ln x$ Power Rule

$\qquad\qquad\qquad = \ln \dfrac{(x+2)^2}{x}$ Quotient Rule

c. $\frac{1}{3}[\log_2 x + \log_2(x+1)] = \frac{1}{3}\{\log_2[x(x+1)]\}$ Product Rule

$\qquad\qquad\qquad = \log_2[x(x+1)]^{1/3}$ Power Rule

$\qquad\qquad\qquad = \log_2 \sqrt[3]{x(x+1)}$ Rewrite with a radical.

✓ **CHECKPOINT 6**

Condense the expression
$2 \log_{10}(x+1) - 3 \log_{10}(x-1)$. ■

Applications

One method of determining how the x- and y-values of a set of nonlinear data are related begins by taking the natural logarithm of each of the x- and y-values. If you graph the points $(\ln x, \ln y)$ and they fall in a straight line, then you can determine that the x- and y-values are related by the equation

$$\ln y = m \ln x$$

where m is the slope of the straight line.

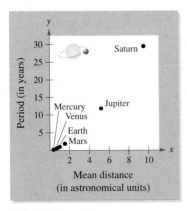

FIGURE 4.16

Example 7 Finding a Mathematical Model

The table shows the mean distance from the sun x and the period (the time it takes a planet to orbit the sun) y for each of the six planets that are closest to the sun. In the table, the mean distance is given in astronomical units (where Earth's mean distance is defined as 1.0), and the period is given in years. Find an equation that relates y and x.

Planet	Mean distance, x	Period, y
Mercury	0.387	0.241
Venus	0.723	0.615
Earth	1.000	1.000
Mars	1.524	1.881
Jupiter	5.203	11.863
Saturn	9.537	29.447

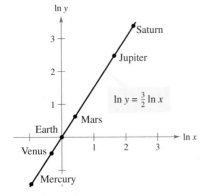

FIGURE 4.17

SOLUTION   The points in the table are plotted in Figure 4.16. From this figure it is not clear how to find an equation that relates y and x. To solve this problem, take the natural logarithm of each of the x- and y-values in the table. This produces the following results.

Planet	Mercury	Venus	Earth	Mars	Jupiter	Saturn
$\ln x$	-0.949	-0.324	0.000	0.421	1.649	2.255
$\ln y$	-1.423	-0.486	0.000	0.632	2.473	3.383

Now, by plotting the points in the second table, you can see that all six of the points appear to lie in a line (see Figure 4.17). Choose any two points to determine the slope of the line. Using the two points $(0.421, 0.632)$ and $(0, 0)$, you can determine that the slope of the line is

$$m = \frac{0.632 - 0}{0.421 - 0} \approx 1.5 = \frac{3}{2}.$$

By the point-slope form, the equation of the line is $Y = \frac{3}{2}X$, where $Y = \ln y$ and $X = \ln x$. You can therefore conclude that $\ln y = \frac{3}{2} \ln x$.

✓ **CHECKPOINT 7**

Find a logarithmic equation that relates y and x.

x	1	2	3	4
y	1	1.414	1.732	2

© Grain Belt Pictures/Alamy

Example 8   **Sound Intensity**

The level of sound L (in decibels) with an intensity of I (in watts per square meter) is given by

$$L = 10 \log_{10} \frac{I}{I_0}$$

where I_0 represents the faintest sound that can be heard by the human ear, and is approximately equal to 10^{-12} watt per square meter. You and your roommate are playing your stereos at the same time and at the same intensity. How much louder is the music when both stereos are playing compared with when just one stereo is playing?

SOLUTION   Let L_1 represent the level of sound when one stereo is playing and let L_2 represent the level of sound when both stereos are playing. Using the formula for level of sound, you can express L_1 as

$$L_1 = 10 \log_{10} \frac{I}{10^{-12}}.$$

For L_2, multiply I by 2 as shown below

$$L_2 = 10 \log_{10} \frac{2I}{10^{-12}}$$

because L_2 represents the level of sound when *two* stereos are playing at the same intensity I. To determine the increase in loudness, subtract L_1 from L_2 as follows.

$$L_2 - L_1 = 10 \log_{10} \frac{2I}{10^{-12}} - 10 \log_{10} \frac{I}{10^{-12}}$$

$$= 10 \left(\log_{10} \frac{2I}{10^{-12}} - \log_{10} \frac{I}{10^{-12}} \right)$$

$$= 10 \left(\log_{10} 2 + \log_{10} \frac{I}{10^{-12}} - \log_{10} \frac{I}{10^{-12}} \right)$$

$$= 10 \log_{10} 2 \approx 3$$

So, the music is about 3 decibels louder. Notice that the variable I drops out of the equation when it is simplified. This means that the loudness increases by 3 decibels when both stereos are played at the same intensity, regardless of the individual intensities of the stereos.

✓ **CHECKPOINT 8**

Two sounds have intensities of $I_1 = 10^{-6}$ watt per square meter and $I_2 = 10^{-9}$ watt per square meter. Use the formula for the level of sound in Example 8 to find the difference in loudness between the two sounds. ▓

(CONCEPT CHECK)

In Exercises 1–4, let x and y be positive real numbers. Determine whether the statement is true or false. Explain your reasoning.

1. The expression $\log_{10} 25$ can be rewritten as $2 \log_{10} 5$.

2. The expression $\ln(x + y)$ can be rewritten as $\ln x + \ln y$.

3. The expression $2 \ln(x + 1) - \ln y$ can be rewritten as $\ln \frac{(x + 1)^2}{y}$.

4. The expressions $\frac{\log_{10} x}{\log_{10} y}$ and $\frac{\ln x}{\ln y}$ are equivalent.

Skills Review 4.3

The following warm-up exercises involve skills that were covered in earlier sections. You will use these skills in the exercise set for this section. For additional help, review Sections 0.3, 0.4, and 4.2.

In Exercises 1–4, evaluate the expression without using a calculator.

1. $\log_7 49$ **2.** $\log_2 \dfrac{1}{32}$ **3.** $\ln \dfrac{1}{e^2}$ **4.** $\log_{10} 0.001$

In Exercises 5–8, simplify the expression.

5. $e^2 e^3$ **6.** $\dfrac{e^2}{e^3}$ **7.** $(e^2)^3$ **8.** $(e^2)^0$

In Exercises 9–12, rewrite the equation in exponential form.

9. $y = \dfrac{1}{x^2}$ **10.** $y = \sqrt{x}$ **11.** $\log_4 64 = 3$ **12.** $\log_{16} 4 = \dfrac{1}{2}$

Exercises 4.3

See www.CalcChat.com for worked-out solutions to odd-numbered exercises.

In Exercises 1–12, write the logarithm in terms of common logarithms.

1. $\log_5 8$ **2.** $\log_7 12$

3. $\ln 30$ **4.** $\ln 20$

5. $\log_3 n$ **6.** $\log_4 m$

7. $\log_{1/5} x$ **8.** $\log_{1/3} x$

9. $\log_x \frac{3}{10}$ **10.** $\log_x \frac{3}{4}$

11. $\log_{2.6} x$ **12.** $\log_{7.1} x$

In Exercises 13–24, write the logarithm in terms of natural logarithms.

13. $\log_5 8$ **14.** $\log_7 12$

15. $\log_{10} 5$ **16.** $\log_{10} 20$

17. $\log_3 n$ **18.** $\log_2 m$

19. $\log_{1/5} x$ **20.** $\log_{1/3} x$

21. $\log_x \frac{3}{10}$ **22.** $\log_x \frac{3}{4}$

23. $\log_{2.6} x$ **24.** $\log_{7.1} x$

In Exercises 25–36, evaluate the logarithm. Round your result to three decimal places.

25. $\log_2 6$ **26.** $\log_8 3$

27. $\log_{27} 35$ **28.** $\log_{19} 42$

29. $\log_{15} 1250$ **30.** $\log_{20} 1575$

31. $\log_5 \frac{1}{3}$ **32.** $\log_9 \frac{3}{5}$

33. $\log_{1/4} 10$ **34.** $\log_{1/3} 5$

35. $\log_{1/2} 0.2$ **36.** $\log_{1/6} 0.025$

In Exercises 37–50, approximate the logarithm using the properties of logarithms, given $\log_b 2 \approx 0.3562$, $\log_b 3 \approx 0.5646$, and $\log_b 5 \approx 0.8271$.

37. $\log_b 10$ **38.** $\log_b 15$

39. $\log_b \frac{2}{3}$ **40.** $\log_b \frac{3}{5}$

41. $\log_b 8$ **42.** $\log_b 81$

43. $\log_b \sqrt{2}$ **44.** $\log_b \sqrt{5}$

45. $\log_b 40$ **46.** $\log_b 45$

47. $\log_b (2b)^{-2}$ **48.** $\log_b (3b^2)$

49. $\log_b \sqrt[3]{4b}$ **50.** $\log_b \sqrt[3]{3b}$

In Exercises 51–56, find the *exact* value of the logarithmic expression without using a calculator.

51. $\log_4 \sqrt[3]{4}$ **52.** $\log_8 \sqrt[4]{8}$

53. $\ln \dfrac{1}{\sqrt{e}}$ **54.** $\ln \sqrt[4]{e^3}$

55. $\log_5 \frac{1}{125}$ **56.** $\log_7 \frac{49}{343}$

In Exercises 57–64, use the properties of logarithms to simplify the given logarithmic expression.

57. $\log_9 \frac{1}{18}$ **58.** $\log_5 \frac{1}{15}$

59. $\log_7 \sqrt{70}$ **60.** $\log_5 \sqrt{75}$

61. $\log_5 \frac{1}{250}$ **62.** $\log_{10} \frac{9}{300}$

63. $\ln(5e^6)$ **64.** $\ln \dfrac{6}{e^2}$

In Exercises 65–86, use the properties of logarithms to expand the expression as a sum, difference, and/or multiple of logarithms. (Assume all variables are positive.)

65. $\log_2(4^3 \cdot 3^5)$

66. $\log_3(3^2 \cdot 4^2)$

67. $\log_3 4n$

68. $\log_6 6x$

69. $\log_5 \dfrac{x}{25}$

70. $\log_{10} \dfrac{y}{2}$

71. $\log_2 x^4$

72. $\log_2 z^{-3}$

73. $\ln \sqrt{z}$

74. $\ln \sqrt[3]{t}$

75. $\ln xyz$

76. $\ln \dfrac{xy}{z}$

77. $\ln \sqrt{a - 1}, \quad a > 1$

78. $\ln \sqrt[3]{y - 2}, \quad y > 2$

79. $\ln \left[\dfrac{(z - 1)^2}{z} \right]$

80. $\ln \left(\dfrac{x}{\sqrt{x^2 + 1}} \right)$

81. $\ln \dfrac{z}{\sqrt[3]{z + 3}}$

82. $\log_9 \dfrac{\sqrt{y}}{z^2}$

83. $\ln \sqrt[3]{\dfrac{x}{y}}$

84. $\ln \sqrt{\dfrac{x^2}{y^3}}$

85. $\ln \sqrt[4]{x^3(x^2 + 3)}$

86. $\ln \sqrt{x^2(x + 2)}$

In Exercises 87–102, condense the expression to the logarithm of a single quantity.

87. $\log_3 x + \log_3 5$

88. $\log_5 y + \log_5 x$

89. $\log_4 8 - \log_4 x$

90. $\log_{10} 4 - \log_{10} z$

91. $2 \log_{10}(x + 4)$

92. $-4 \log_{10} 2x$

93. $-\ln x - 3 \ln 6$

94. $2 \ln 8 + 5 \ln z$

95. $\frac{1}{3} \ln 5x - \ln(x + 1)$

96. $\frac{3}{2} \ln(z - 2) + \ln z$

97. $\log_8(x - 2) - \log_8(x + 2)$

98. $3 \log_7 x + 2 \log_7 y - 4 \log_7 z$

99. $2 \ln 3 - \frac{1}{2} \ln(x^2 + 1)$

100. $\frac{3}{2} \ln t^6 - \frac{3}{4} \ln t^4$

101. $\ln x - \ln(x + 2) - \ln(x - 2)$

102. $\ln(x + 1) + 2 \ln(x - 1) + 3 \ln x$

Curve Fitting In Exercises 103–106, find a logarithmic equation that relates y and x. Explain the steps used to find the equation.

103.

x	1	2	3	4	5	6
y	1	1.189	1.316	1.414	1.495	1.565

104.

x	1	2	3	4	5	6
y	1	1.587	2.080	2.520	2.924	3.302

105.

x	1	2	3	4	5	6
y	2.5	2.102	1.9	1.768	1.672	1.597

106.

x	1	2	3	4	5	6
y	0.5	2.828	7.794	16	27.951	44.091

107. Nail Length The approximate lengths and diameters (in inches) of common nails are shown in the table. Find a logarithmic equation that relates the diameter y of a common nail to its length x.

Length, x	Diameter, y	Length, x	Diameter, y
1	0.070	4	0.176
2	0.111	5	0.204
3	0.146	6	0.231

108. Galloping Speeds of Animals Four-legged animals run with two different types of motion: trotting and galloping. An animal that is trotting has at least one foot on the ground at all times, whereas an animal that is galloping has all four feet off the ground at some point in its stride. The number of strides per minute at which an animal breaks from a trot to a gallop depends on the weight of the animal. Use the table to find a logarithmic equation that relates an animal's lowest galloping speed y (in strides per minute) to its weight x (in pounds).

Weight, x	Galloping speed, y	Weight, x	Galloping speed, y
25	191.5	75	164.2
35	182.7	500	125.9
50	173.8	1000	114.2

109. Sound Intensity Use the equation for the level of sound in Example 8 to find the difference in loudness between an average office and a broadcast studio with the intensities given below.

Office: 1.26×10^{-7} watt per square meter

Broadcast studio: 3.16×10^{-10} watt per square meter

Algebra

Factors and Zeros of Polynomials

Given the polynomial $p(x) = a_n x^n + a_{n-1}x^{n-1} + \cdots + a_1 x + a_0$. If $p(b) = 0$, then b is a *zero* of the polynomial and a solution of the equation $p(x) = 0$. Furthermore, $(x - b)$ is a *factor* of the polynomial.

Fundamental Theorem of Algebra

An nth degree polynomial has n (not necessarily distinct) zeros.

Quadratic Formula

If $p(x) = ax^2 + bx + c$, $a \neq 0$ and $b^2 - 4ac \geq 0$, then the real zeros of p are $x = \left(-b \pm \sqrt{b^2 - 4ac}\right)/2a$.

Example

If $p(x) = x^2 + 3x - 1$, then $p(x) = 0$ if

$$x = \frac{-3 \pm \sqrt{13}}{2}.$$

Special Factors

$$x^2 - a^2 = (x - a)(x + a)$$
$$x^3 - a^3 = (x - a)(x^2 + ax + a^2)$$
$$x^3 + a^3 = (x + a)(x^2 - ax + a^2)$$
$$x^4 - a^4 = (x - a)(x + a)(x^2 + a^2)$$
$$x^4 + a^4 = \left(x^2 + \sqrt{2}ax + a^2\right)\left(x^2 - \sqrt{2}ax + a^2\right)$$
$$x^n - a^n = (x - a)(x^{n-1} + ax^{n-2} + \cdots + a^{n-1}), \text{ for } n \text{ odd}$$
$$x^n + a^n = (x + a)(x^{n-1} - ax^{n-2} + \cdots + a^{n-1}), \text{ for } n \text{ odd}$$
$$x^{2n} - a^{2n} = (x^n - a^n)(x^n + a^n)$$

Examples

$$x^2 - 9 = (x - 3)(x + 3)$$
$$x^3 - 8 = (x - 2)(x^2 + 2x + 4)$$
$$x^3 + 4 = \left(x + \sqrt[3]{4}\right)\left(x^2 - \sqrt[3]{4}x + \sqrt[3]{16}\right)$$
$$x^4 - 4 = \left(x - \sqrt{2}\right)\left(x + \sqrt{2}\right)(x^2 + 2)$$
$$x^4 + 4 = (x^2 + 2x + 2)(x^2 - 2x + 2)$$
$$x^5 - 1 = (x - 1)(x^4 + x^3 + x^2 + x + 1)$$
$$x^7 + 1 = (x + 1)(x^6 - x^5 + x^4 - x^3 + x^2 - x + 1)$$
$$x^6 - 1 = (x^3 - 1)(x^3 + 1)$$

Binomial Theorem

$$(x + a)^2 = x^2 + 2ax + a^2$$
$$(x - a)^2 = x^2 - 2ax + a^2$$
$$(x + a)^3 = x^3 + 3ax^2 + 3a^2x + a^3$$
$$(x - a)^3 = x^3 - 3ax^2 + 3a^2x - a^3$$
$$(x + a)^4 = x^4 + 4ax^3 + 6a^2x^2 + 4a^3x + a^4$$
$$(x - a)^4 = x^4 - 4ax^3 + 6a^2x^2 - 4a^3x + a^4$$
$$(x + a)^n = x^n + nax^{n-1} + \frac{n(n-1)}{2!}a^2x^{n-2} + \cdots + na^{n-1}x + a^n$$
$$(x - a)^n = x^n - nax^{n-1} + \frac{n(n-1)}{2!}a^2x^{n-2} - \cdots \pm na^{n-1}x \mp a^n$$

Examples

$$(x + 3)^2 = x^2 + 6x + 9$$
$$(x^2 - 5)^2 = x^4 - 10x^2 + 25$$
$$(x + 2)^3 = x^3 + 6x^2 + 12x + 8$$
$$(x - 1)^3 = x^3 - 3x^2 + 3x - 1$$
$$\left(x + \sqrt{2}\right)^4 = x^4 + 4\sqrt{2}x^3 + 12x^2 + 8\sqrt{2}x + 4$$
$$(x - 4)^4 = x^4 - 16x^3 + 96x^2 - 256x + 256$$
$$(x + 1)^5 = x^5 + 5x^4 + 10x^3 + 10x^2 + 5x + 1$$
$$(x - 1)^6 = x^6 - 6x^5 + 15x^4 - 20x^3 + 15x^2 - 6x + 1$$

Rational Zero Test

If $p(x) = a_n x^n + a_{n-1}x^{n-1} + \cdots + a_1 x + a_0$ has integer coefficients, then every *rational* zero of $p(x) = 0$ is of the form $x = r/s$, where r is a factor of a_0 and s is a factor of a_n.

Example

If $p(x) = 2x^4 - 7x^3 + 5x^2 - 7x + 3$, then the only possible *rational* zeros are $x = \pm 1, \pm\frac{1}{2}, \pm 3$, and $\pm\frac{3}{2}$. By testing, you find the two rational zeros to be $\frac{1}{2}$ and 3.

Factoring by Grouping

$$acx^3 + adx^2 + bcx + bd = ax^2(cx + d) + b(cx + d)$$
$$= (ax^2 + b)(cx + d)$$

Example

$$3x^3 - 2x^2 - 6x + 4 = x^2(3x - 2) - 2(3x - 2)$$
$$= (x^2 - 2)(3x - 2)$$

Algebra

Arithmetic Operations

$$ab + ac = a(b + c)$$

$$\frac{a}{b} + \frac{c}{d} = \frac{ad + bc}{bd}$$

$$\frac{a + b}{c} = \frac{a}{c} + \frac{b}{c}$$

$$\frac{\left(\dfrac{a}{b}\right)}{\left(\dfrac{c}{d}\right)} = \frac{ad}{bc}$$

$$a\left(\frac{b}{c}\right) = \frac{ab}{c}$$

$$\frac{a - b}{c - d} = \frac{b - a}{d - c}$$

$$\frac{ab + ac}{a} = b + c, \ a \neq 0$$

$$\frac{\left(\dfrac{a}{b}\right)}{c} = \frac{a}{bc}$$

$$\frac{a}{\left(\dfrac{b}{c}\right)} = \frac{ac}{b}$$

Exponents and Radicals

$$a^0 = 1, \ a \neq 0$$

$$\frac{a^x}{a^y} = a^{x-y}$$

$$\left(\frac{a}{b}\right)^x = \frac{a^x}{b^x}$$

$$\sqrt[n]{a^m} = a^{m/n} = \left(\sqrt[n]{a}\right)^m$$

$$a^{-x} = \frac{1}{a^x}$$

$$(a^x)^y = a^{xy}$$

$$\sqrt{a} = a^{1/2}$$

$$\sqrt[n]{ab} = \sqrt[n]{a}\,\sqrt[n]{b}$$

$$a^x a^y = a^{x+y}$$

$$(ab)^x = a^x b^x$$

$$\sqrt[n]{a} = a^{1/n}$$

$$\sqrt[n]{\left(\frac{a}{b}\right)} = \frac{\sqrt[n]{a}}{\sqrt[n]{b}}$$

Algebraic Errors to Avoid

$$\frac{a}{x + b} \neq \frac{a}{x} + \frac{a}{b}$$

(To see this error, let $a = b = x = 1$.)

$$\sqrt{x^2 + a^2} \neq x + a$$

(To see this error, let $x = 3$ and $a = 4$.)

$$a - b(x - 1) \neq a - bx - b$$

[Remember to distribute negative signs. The equation should be $a - b(x - 1) = a - bx + b$.]

$$\frac{\left(\dfrac{x}{a}\right)}{b} \neq \frac{bx}{a}$$

[To divide fractions, invert and multiply. The equation should be
$$\frac{\left(\dfrac{x}{a}\right)}{b} = \frac{\left(\dfrac{x}{a}\right)}{\left(\dfrac{b}{1}\right)} = \left(\frac{x}{a}\right)\left(\frac{1}{b}\right) = \frac{x}{ab}.]$$

$$\sqrt{-x^2 + a^2} \neq -\sqrt{x^2 - a^2}$$

(The negative sign cannot be factored out of the square root.)

$$\frac{a + bx}{a} \neq 1 + bx$$

(This is one of many examples of incorrect cancellation. The equation should be
$$\frac{a + bx}{a} = \frac{a}{a} + \frac{bx}{a} = 1 + \frac{bx}{a}.)$$

$$\frac{1}{x^{1/2} - x^{1/3}} \neq x^{-1/2} - x^{-1/3}$$

(This error is a more complex version of the first error.)

$$(x^2)^3 \neq x^5$$

[This equation should be $(x^2)^3 = x^2 x^2 x^2 = x^6$.]

Conversion Table

1 centimeter $\approx$ 0.394 inch	1 joule $\approx$ 0.738 foot-pound	1 mile $\approx$ 1.609 kilometers
1 meter $\approx$ 39.370 inches	1 gram $\approx$ 0.035 ounce	1 gallon $\approx$ 3.785 liters
$\approx$ 3.281 feet	1 kilogram $\approx$ 2.205 pounds	1 pound $\approx$ 4.448 newtons
1 kilometer $\approx$ 0.621 mile	1 inch $\approx$ 2.540 centimeters	1 foot-lb $\approx$ 1.356 joules
1 liter $\approx$ 0.264 gallon	1 foot $\approx$ 30.480 centimeters	1 ounce $\approx$ 28.350 grams
1 newton $\approx$ 0.225 pound	$\approx$ 0.305 meter	1 pound $\approx$ 0.454 kilogram

Formulas from Geometry

Triangle

$h = a \sin \theta$

$\text{Area} = \dfrac{1}{2}bh$

(Law of Cosines)
$c^2 = a^2 + b^2 - 2ab \cos \theta$

Right Triangle

(Pythagorean Theorem)
$c^2 = a^2 + b^2$

Equilateral Triangle

$h = \dfrac{\sqrt{3}s}{2}$

$\text{Area} = \dfrac{\sqrt{3}s^2}{4}$

Parallelogram

$\text{Area} = bh$

Trapezoid

$\text{Area} = \dfrac{h}{2}(a + b)$

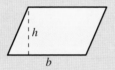

Circle

$\text{Area} = \pi r^2$

$\text{Circumference} = 2\pi r$

Sector of Circle

(θ in radians)

$\text{Area} = \dfrac{\theta r^2}{2}$

$s = r\theta$

Circular Ring

(p = average radius,
w = width of ring)

$\text{Area} = \pi(R^2 - r^2)$

$\quad = 2\pi pw$

Sector of Circular Ring

(p = average radius,
w = width of ring,
θ in radians)

$\text{Area} = \theta pw$

Ellipse

$\text{Area} = \pi ab$

$\text{Circumference} \approx 2\pi \sqrt{\dfrac{a^2 + b^2}{2}}$

Cone

(A = area of base)

$\text{Volume} = \dfrac{Ah}{3}$

Right Circular Cone

$\text{Volume} = \dfrac{\pi r^2 h}{3}$

$\text{Lateral Surface Area} = \pi r \sqrt{r^2 + h^2}$

Frustum of Right Circular Cone

$\text{Volume} = \dfrac{\pi(r^2 + rR + R^2)h}{3}$

$\text{Lateral Surface Area} = \pi s(R + r)$

Right Circular Cylinder

$\text{Volume} = \pi r^2 h$

$\text{Lateral Surface Area} = 2\pi rh$

Sphere

$\text{Volume} = \dfrac{4}{3}\pi r^3$

$\text{Surface Area} = 4\pi r^2$

Wedge

(A = area of upper face,
B = area of base)

$A = B \sec \theta$

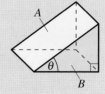

Graphs of Common Functions

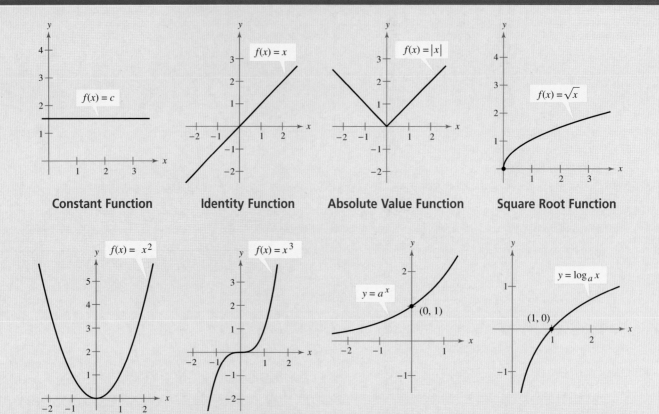

Constant Function   **Identity Function**   **Absolute Value Function**   **Square Root Function**

Squaring Function   **Cubing Function**   **Exponential Function**   **Logarithmic Function**

Symmetry

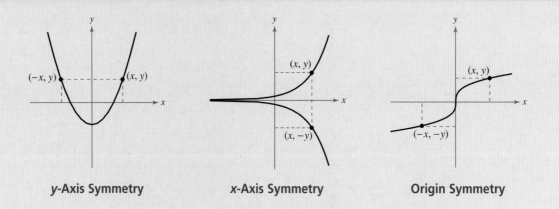

y-Axis Symmetry   **x-Axis Symmetry**   **Origin Symmetry**

110. Sound Intensity Use the equation for the level of sound in Example 8 to find the difference in loudness between a bird singing and rustling leaves with the intensities given below.

Bird singing: 10^{-8} watt per square meter

Rustling leaves: 10^{-10} watt per square meter

(T) 111. Graphical Analysis Use a graphing utility to graph

$$f(x) = \ln 5x \quad \text{and} \quad g(x) = \ln 5 + \ln x$$

in the same viewing window. What do you observe about the two graphs? What property of logarithms is being demonstrated graphically?

(T) 112. Graphical Analysis You are helping another student learn the properties of logarithms. How would you use a graphing utility to demonstrate to this student the logarithmic property

$$\log_a u^v = v \log_a u$$

(u is a positive number, v is a real number, and a is a positive number such that $a \neq 1$)? What two functions could you use? Briefly describe your explanation of this property using these functions and their graphs.

(T) 113. Reasoning An algebra student claims that the following is true:

$$\log_a \frac{x}{y} = \frac{\log_a x}{\log_a y} = \log_a x - \log_a y.$$

Discuss how you would use a graphing utility to demonstrate that this claim is not true. Describe how to demonstrate the actual property of logarithms that is hidden in this faulty claim.

(T) 114. Reasoning A classmate claims that the following is true:

$$\ln(x + y) = \ln x + \ln y = \ln xy.$$

Discuss how you would use a graphing utility to demonstrate that this claim is not true. Describe how to demonstrate the actual property of logarithms that is hidden in this faulty claim.

115. Complete the proof of the logarithmic property

$$\log_a uv = \log_a u + \log_a v.$$

Let $\log_a u = x$ and $\log_a v = y$.

$a^x = \boxed{}$ and $a^y = \boxed{}$ Rewrite in exponential form.

$u \cdot v = \boxed{} \cdot \boxed{} = a^{\boxed{}}$ Multiply and substitute for u and v.

$\boxed{} = x + y$ Rewrite in logarithmic form.

$\log_a uv = \boxed{} + \boxed{}$ Substitute for x and y.

116. Complete the proof of the logarithmic property

$$\log_a \frac{u}{v} = \log_a u - \log_a v.$$

Let $\log_a u = x$ and $\log_a v = y$.

$a^x = \boxed{}$ and $a^y = \boxed{}$ Rewrite in exponential form.

$\dfrac{u}{v} = \dfrac{\boxed{}}{\boxed{}} = a^{\boxed{}}$ Divide and substitute for u and v.

$\boxed{} = x - y$ Rewrite in logarithmic form.

$\log_a \dfrac{u}{v} = \boxed{} - \boxed{}$ Substitute for x and y.

Business Capsule

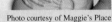

Homes of hospitality for expectant women.

www.maggiesplace.org

Photo courtesy of Maggie's Place

Co-founded by five recent college graduates, Maggie's Place is a community of homes that provides hospitality for pregnant women who are alone or living on the streets. Maggie's Place provides for immediate needs such as shelter, clothing, food, and community support. Expectant mothers are connected to community resources such as prenatal care, education programs, and low-cost housing. Maggie's Place opened its first home, the Magdalene House, on May 13, 2000 in Phoenix, Arizona and has since expanded to four homes. Maggie's Place is working with other local and national groups to develop homes in other communities.

117. Research Project Use your campus library, the Internet, or some other reference source to find information about a nonprofit group or company whose growth can be modeled by a logarithmic function. Write a brief report about the growth of the group or company.

Mid-Chapter Quiz

See www.CalcChat.com for worked-out solutions to odd-numbered exercises.

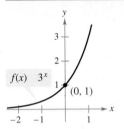

Figure for 1–4

Take this quiz as you would take a quiz in class. When you are done, check your work against the answers given in the back of the book.

In Exercises 1–4, use the graph of $f(x) = 3^x$ to sketch the graph of the function.

1. $g(x) = 3^x - 2$

2. $h(x) = 3^{-x}$

3. $k(x) = \log_3 x$

4. $j(x) = \log_3 (x - 1)$

5. For $P = \$10,000$, $r = 5.15\%$, and $t = 8$ years, find the balance in an account when interest is compounded (a) monthly and (b) continuously.

6. The numbers of children C (in millions) participating in the Federal School Breakfast Program from 1997 to 2005 can be approximated by the model

$$C = 5.26(1.04)^t, \quad 7 \le t \le 15$$

where t represents the year, with $t = 7$ corresponding to 1997. *(Source: U.S. Department of Agriculture)*

(a) Use the model to estimate the numbers of children participating in the Federal School Breakfast Program in 2000 and in 2005.

(b) Use the model to predict the numbers of children that will participate in the Federal School Breakfast Program in 2009 and in 2010.

7. The size of a bacteria population is modeled by

$$P(t) = 100e^{0.2154t}$$

where t is the time in hours. Find (a) $P(0)$, (b) $P(6)$, and (c) $P(12)$.

8. Use the demand function

$$p = 4000\left(1 - \frac{8}{8 + e^{-0.003x}}\right)$$

to find the price for a demand of $x = 500$ MP3 players.

In Exercises 9–12, evaluate the expression without using a calculator.

9. $\log_{10} 100$

10. $\ln e^4$

11. $\log_4 \frac{1}{16}$

12. $\ln 1$

13. Sketch the graphs of $f(x) = 4^x$ and $g(x) = \log_4 x$ in the same coordinate plane. Identify the domains of f and g. Discuss the special relationship between f and g that is demonstrated by their graphs.

In Exercises 14 and 15, find the *exact* value of the logarithm.

14. $\log_7 \sqrt{343}$

15. $\ln \sqrt[5]{e^6}$

In Exercises 16 and 17, expand the logarithmic expression.

16. $\log_{10} \sqrt[3]{\dfrac{xy}{z}}$

17. $\ln\left(\dfrac{x^2 + 3}{x^3}\right)$

In Exercises 18 and 19, condense the logarithmic expression.

18. $\ln x + \ln y - \ln 3$

19. $-3 \log_{10} 4 - 3 \log_{10} x$

20. Use the values in the table at the left to find a logarithmic equation that relates y and x.

x	y
1	1
2	1.260
3	1.442
4	1.587
6	1.817
8	2.000

Table for 20

Section 4.4

Solving Exponential and Logarithmic Equations

- Solve an exponential equation.
- Solve a logarithmic equation.
- Use an exponential or a logarithmic model to solve an application problem.

Introduction

So far in this chapter, you have studied the definitions, graphs, and properties of exponential and logarithmic functions. In this section, you will study procedures for *solving equations* involving these exponential and logarithmic functions.

There are two basic strategies for solving exponential or logarithmic equations. The first is based on the One-to-One Properties and the second is based on the Inverse Properties. For $a > 0$ and $a \neq 1$, the following properties are true for all x and y for which $\log_a x$ and $\log_a y$ are defined.

One-to-One Properties	*Inverse Properties*
$a^x = a^y$ if and only if $x = y$.	$a^{\log_a x} = x$
$\log_a x = \log_a y$ if and only if $x = y$.	$\log_a a^x = x$

Example 1 Solving Simple Equations

Original Equation	*Rewritten Equation*	*Solution*	*Property*
a. $2^x = 32$	$2^x = 2^5$	$x = 5$	One-to-One
b. $\ln x - \ln 3 = 0$	$\ln x = \ln 3$	$x = 3$	One-to-One
c. $e^x = 7$	$\ln e^x = \ln 7$	$x = \ln 7$	Inverse
d. $\ln x = -3$	$e^{\ln x} = e^{-3}$	$x = e^{-3}$	Inverse
e. $\log_{10} x = -1$	$10^{\log_{10} x} = 10^{-1}$	$x = 10^{-1} = \frac{1}{10}$	Inverse

✓ CHECKPOINT 1

Solve each equation for x.

a. $3^x = 81$ **b.** $\log_6 x = 3$ ■

Strategies for Solving Exponential and Logarithmic Equations

1. Rewrite the original equation in a form that allows the use of the One-to-One Property of exponential or logarithmic functions.

2. Rewrite an *exponential* equation in logarithmic form and apply the Inverse Property of logarithmic functions.

3. Rewrite a *logarithmic* equation in exponential form and apply the Inverse Property of exponential functions.

Solving Exponential Equations

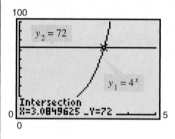
✓ **CHECKPOINT 3**

Solve $62 - 10^x = 24$ and approximate the result to three decimal places. ■

Example 2   **Solving Exponential Equations**

Solve each equation and approximate the result to three decimal places.

a. $4^x = 72$ **b.** $3(2^x) = 42$

SOLUTION

a.

$4^x = 72$	Write original equation.
$\log_4 4^x = \log_4 72$	Take log (base 4) of each side.
$x = \log_4 72$	Inverse Property
$x = \dfrac{\ln 72}{\ln 4} \approx 3.085$	Change-of-base formula

The solution is $x = \log_4 72 \approx 3.085$. Check this in the original equation.

b.

$3(2^x) = 42$	Write original equation.
$2^x = 14$	Divide each side by 3 to isolate the exponential expression.
$\log_2 2^x = \log_2 14$	Take log (base 2) of each side.
$x = \log_2 14$	Inverse Property
$x = \dfrac{\ln 14}{\ln 2} \approx 3.807$	Change-of-base formula

The solution is $x = \log_2 14 \approx 3.807$. Check this in the original equation.

✓ **CHECKPOINT 2**

Solve $6^x = 84$ and approximate the result to three decimal places. ■

In Example 2(a), the exact solution is $x = \log_4 72$ and the approximate solution is $x \approx 3.085$. An exact answer is preferred when the solution is an intermediate step in a larger problem. For a final answer, an approximate solution in decimal form is easier to comprehend.

Example 3   **Solving an Exponential Equation**

Solve $e^x + 5 = 60$ and approximate the result to three decimal places.

SOLUTION

$e^x + 5 = 60$	Write original equation.
$e^x = 55$	Subtract 5 from each side to isolate the exponential expression.
$\ln e^x = \ln 55$	Take natural log of each side.
$x = \ln 55$	Inverse Property
$x \approx 4.007$	Use a calculator.

The solution is $x = \ln 55 \approx 4.007$. Check this in the original equation.

Example 4   **Solving an Exponential Equation**

Solve $2(3^{2t-5}) - 4 = 11$ and approximate the result to three decimal places.

SOLUTION

$2(3^{2t-5}) - 4 = 11$	Write original equation.
$2(3^{2t-5}) = 15$	Add 4 to each side.
$3^{2t-5} = \dfrac{15}{2}$	Divide each side by 2.
$\log_3 3^{2t-5} = \log_3 \dfrac{15}{2}$	Take log (base 3) of each side.
$2t - 5 = \log_3 \dfrac{15}{2}$	Inverse Property
$2t = 5 + \log_3 7.5$	Add 5 to each side.
$t = \dfrac{5}{2} + \dfrac{1}{2}\log_3 7.5$	Divide each side by 2.
$t \approx 3.417$	Use a calculator.

The solution is $t = \frac{5}{2} + \frac{1}{2}\log_3 7.5 \approx 3.417$. Check this in the original equation.

✓ **CHECKPOINT 4**

Solve $4(4^{2t-7}) + 14 = 110$ and approximate the result to three decimal places. ▦

When an equation involves two or more exponential expressions, you can still use a procedure similar to that demonstrated in Examples 2, 3, and 4. However, the algebra is a bit more complicated.

Example 5   **Solving an Exponential Equation of Quadratic Type**

Solve $e^{2x} - 3e^x + 2 = 0$.

SOLUTION

$e^{2x} - 3e^x + 2 = 0$	Write original equation.
$(e^x)^2 - 3e^x + 2 = 0$	Write in quadratic form.
$(e^x - 2)(e^x - 1) = 0$	Factor.
$e^x - 2 = 0 \quad \Longrightarrow \quad x = \ln 2$	Set 1st factor equal to 0.
$e^x - 1 = 0 \quad \Longrightarrow \quad x = 0$	Set 2nd factor equal to 0.

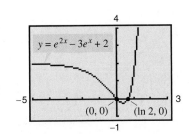

FIGURE 4.18

The solutions are $x = \ln 2$ and $x = 0$. Check these in the original equation. Or, check by graphing $y = e^{2x} - 3e^x + 2$ using a graphing utility. The graph should have two x-intercepts: $x = \ln 2$ and $x = 0$, as shown in Figure 4.18.

✓ **CHECKPOINT 5**

Solve $e^{2x} - 7e^x + 12 = 0$. ▦

Solving Logarithmic Equations

To solve a logarithmic equation such as

$$\ln x = 3 \qquad \text{Logarithmic form}$$

write the equation in exponential form as follows.

$$e^{\ln x} = e^3 \qquad \text{Exponentiate each side.}$$

$$x = e^3 \qquad \text{Exponential form}$$

This procedure is called **exponentiating** each side of an equation.

Example 6   Solving Logarithmic Equations

a. Solve $\ln x = 2$.

b. Solve $2 \log_5 3x = 4$.

SOLUTION

a. $\ln x = 2$ Write original equation.

$\quad e^{\ln x} = e^2$ Exponentiate each side.

$\quad\quad x = e^2$ Inverse Property

The solution is $x = e^2$. Check this in the original equation.

b. $2 \log_5 3x = 4$ Write original equation.

$\quad \log_5 3x = 2$ Divide each side by 2.

$\quad 5^{\log_5 3x} = 5^2$ Exponentiate each side (base 5).

$\quad\quad 3x = 25$ Inverse Property

$\quad\quad x = \dfrac{25}{3}$ Divide each side by 3.

The solution is $x = \frac{25}{3}$. Check this in the original equation.

✓ **CHECKPOINT 6**

Solve $\log_3 2x = 4$. ■

Example 7   Solving a Logarithmic Equation

Solve $\log_3(5x - 1) = \log_3(x + 7)$.

SOLUTION

$$\log_3(5x - 1) = \log_3(x + 7) \qquad \text{Write original equation.}$$

$$5x - 1 = x + 7 \qquad \text{One-to-One Property}$$

$$4x = 8 \qquad \text{Add } -x \text{ and 1 to each side.}$$

$$x = 2 \qquad \text{Divide each side by 4.}$$

The solution is $x = 2$. Check this in the original equation.

✓ **CHECKPOINT 7**

Solve $\ln(3x + 2) = \ln(x + 8)$. ■

(T) You can use a graphing
utility to verify that the
equation in Example 9 has
$x = 5$ as its only solution.
Graph

$$y_1 = \log_{10} 5x$$
$$+ \log_{10}(x - 1)$$

and

$$y_2 = 2$$

in the same viewing window.
From the graph shown below,
it appears that the graphs of the
two equations intersect at one
point. Use the *intersect* feature
or the *zoom* and *trace* features
to determine that $x = 5$ is the
solution. You can verify this
algebraically by substituting
$x = 5$ into the original
equation.

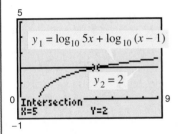

For instructions on how to use
the *zoom* and *trace* features,
see Appendix A; for specific
keystrokes, go to the text
website at *college.hmco.com/
info/larsonapplied.*

✓ **CHECKPOINT 9**

Solve $\log_6 x + \log_6(x + 5) = 2$. ■

Example 8   **Solving a Logarithmic Equation**

Solve $5 + 2 \ln x = 4$ and approximate the result to three decimal places.

SOLUTION

$5 + 2 \ln x = 4$	Write original equation.
$2 \ln x = -1$	Subtract 5 from each side.
$\ln x = -\dfrac{1}{2}$	Divide each side by 2.
$e^{\ln x} = e^{-1/2}$	Exponentiate each side.
$x = e^{-1/2}$	Inverse Property
$x \approx 0.607$	Use a calculator.

The solution is $x = e^{-1/2} \approx 0.607$. Check this in the original equation.

✓ **CHECKPOINT 8**

Solve $4 + 3 \ln x = 16$ and approximate the result to three decimal places. ■

Because the domain of a logarithmic function generally does not include all real numbers, be sure to check for extraneous solutions of logarithmic equations.

Example 9   **Checking for Extraneous Solutions**

Solve $\log_{10} 5x + \log_{10}(x - 1) = 2$.

SOLUTION

$\log_{10} 5x + \log_{10}(x - 1) = 2$	Write original equation.
$\log_{10}[5x(x - 1)] = 2$	Product Rule of logarithms
$10^{\log_{10}(5x^2 - 5x)} = 10^2$	Exponentiate each side (base 10).
$5x^2 - 5x = 100$	Inverse Property
$x^2 - x - 20 = 0$	Write in general form.
$(x - 5)(x + 4) = 0$	Factor.
$x - 5 = 0$	Set 1st factor equal to 0.
$x = 5$	Solution
$x + 4 = 0$	Set 2nd factor equal to 0.
$x = -4$	Solution

The solutions appear to be $x = 5$ and $x = -4$. However, when you check these in the original equation, you can see that $x = 5$ is the only solution.

In Example 9, the domain of $\log_{10} 5x$ is $x > 0$ and the domain of $\log_{10}(x - 1)$ is $x > 1$, so the domain of the original equation is $x > 1$. Because the domain is all real numbers greater than 1, the solution $x = -4$ is extraneous.

Applications

Example 10   Doubling and Tripling an Investment

You deposit $500 in an account that pays 6.75% interest, compounded continuously.

a. How long will it take your money to double?

b. How long will it take your money to triple?

SOLUTION Using the formula for compound interest with continuous compounding, you can find that the balance in the account is given by

$$A = Pe^{rt}$$

$$= 500e^{0.0675t}.$$

a. To find the time required for the balance to double, let $A = 1000$ and solve the resulting equation for t.

$500e^{0.0675t} = 1000$	Write original equation.
$e^{0.0675t} = 2$	Divide each side by 500.
$\ln e^{0.0675t} = \ln 2$	Take natural log of each side.
$0.0675t = \ln 2$	Inverse Property
$t = \dfrac{1}{0.0675} \ln 2$	Divide each side by 0.0675.
$t \approx 10.27$	Use a calculator.

The balance in the account will double after approximately 10.27 years.

b. To find the time required for the balance to triple, let $A = 1500$ and solve the resulting equation for t.

$500e^{0.0675t} = 1500$	Write original equation.
$e^{0.0675t} = 3$	Divide each side by 500.
$\ln e^{0.0675t} = \ln 3$	Take natural log of each side.
$0.0675t = \ln 3$	Inverse Property
$t = \dfrac{1}{0.0675} \ln 3$	Divide each side by 0.0675.
$t \approx 16.28$	Use a calculator.

The balance in the account will triple after approximately 16.28 years.

Notice that it took 10.27 years to earn the first $500 and only 6.01 years to earn the second $500. This result is graphically demonstrated in Figure 4.19.

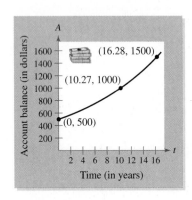

FIGURE 4.19

✓ CHECKPOINT 10

In Example 10, how long will it take for the account balance to reach $600? ▪

Example 11 Bone Graft Procedures

From 1998 to 2005, the numbers of bone graft procedures y (in thousands) performed in the United States can be approximated by

$$y = 144.32 e^{0.164t}$$

where t represents the year, with $t = 8$ corresponding to 1998 (see Figure 4.20). Use the model to estimate the year in which the number of bone graft procedures reached about 880,000. *(Source: U.S. Department of Health and Human Services)*

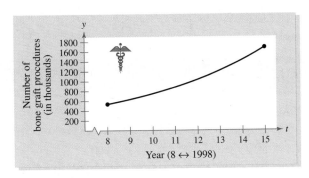

FIGURE 4.20

SOLUTION

$144.32 e^{0.164t} = y$	Write original model.
$144.32 e^{0.164t} = 880$	Substitute 880 for y.
$e^{0.164t} \approx 6.098$	Divide each side by 144.32.
$\ln e^{0.164t} \approx \ln 6.098$	Take natural log of each side.
$0.164t \approx 1.808$	Inverse Property.
$t \approx 11$	Divide each side by 0.164.

The solution is $t \approx 11$. Because $t = 8$ represents 1998, it follows that there were about 880,000 bone graft procedures performed in 2001.

✓ CHECKPOINT 11

Use the model in Example 11 to estimate the year in which the number of bone graft procedures performed in the United States reached about 1,440,000. ■

CONCEPT CHECK

1. What property would you use to solve the equation $3^x = 81$? Explain.

2. What strategy would you use for solving an equation of the form $\log_a x = b$?

3. Describe and correct the error in solving the equation.

$$\log_8 x = 10$$
$$8^{\log_8 x} = 10^8$$
$$x = 100,000,000$$

4. Discuss the steps you would take to solve the equation $\ln(x + 1) - \ln x = 100$.

Skills Review 4.4

The following warm-up exercises involve skills that were covered in earlier sections. You will use these skills in the exercise set for this section. For additional help, review Sections 1.4, 3.5, and 4.2.

In Exercises 1–6, solve for x.

1. $x \ln 2 = \ln 3$

2. $(x - 1) \ln 4 = 2$

3. $2xe^2 = e^3$

4. $4xe^{-1} = 8$

5. $x^2 - 4x + 5 = 0$

6. $2x^2 - 3x + 1 = 0$

In Exercises 7–10, simplify the expression.

7. $\log_{10} 10^x$

8. $\log_{10} 10^{2x}$

9. $\ln e^{2x}$

10. $\ln e^{-x^2}$

Exercises 4.4

See www.CalcChat.com for worked-out solutions to odd-numbered exercises.

In Exercises 1–10, solve for x.

1. $5^x = 125$

2. $2^x = 64$

3. $7^x = \frac{1}{49}$

4. $4^x = \frac{1}{256}$

5. $4^{2x-1} = 64$

6. $3^{x-1} = 27$

7. $\log_4 x = 3$

8. $\log_5 5x = 2$

9. $\log_{10} x = -1$

10. $\ln(2x - 1) = 0$

In Exercises 11–22, apply the Inverse Property of logarithmic or exponential functions to simplify the expression.

11. $\ln e^{x^2}$

12. $\ln e^{2x-1}$

13. $\log_{10} 10^{x^2} + 1$

14. $\log_{10} 10^{2x+3}$

15. $\log_5 5^{x^3} - 7$

16. $\log_8 8^{x^5} + 1$

17. $-8 + e^{\ln x^3}$

18. $-1 + \ln e^{2x}$

19. $10^{\log_{10}(x+5)}$

20. $10^{\log_{10}(x^2+7x+10)}$

21. $2^{\log_2 x^2}$

22. $9^{\log_9(3x+7)}$

In Exercises 23–60, solve the exponential equation algebraically. Approximate the result to three decimal places.

23. $3e^x = 9$

24. $5e^x = 20$

25. $2(3^x) = 16$

26. $3(4^x) = 81$

27. $e^x - 9 = 19$

28. $6^x + 10 = 47$

29. $3^{2x} = 80$

30. $6^{5x} = 3000$

31. $5^{-t/2} = 0.20$

32. $4^{-t/3} = 0.15$

33. $3^{x-1} = 28$

34. $2^{x-3} = 31$

35. $2^{3-x} = 565$

36. $8^{-2-x} = 431$

37. $8(10^{3x}) = 12$

38. $5(10^{x-6}) = 7$

39. $3(5^{x-1}) = 21$

40. $8(3^{6-x}) = 40$

41. $e^{3x} = 12$

42. $e^{2x} = 50$

43. $500e^{-x} = 300$

44. $1000e^{-4x} = 75$

45. $7 - 2e^x = 6$

46. $-14 + 3e^x = 11$

47. $6(2^{3x-1}) - 7 = 9$

48. $8(4^{6-2x}) + 13 = 41$

49. $e^{2x} - 8e^x + 12 = 0$

50. $e^{2x} - 5e^x + 6 = 0$

51. $e^{2x} - 3e^x - 4 = 0$

52. $e^{2x} - 9e^x - 36 = 0$

53. $\dfrac{500}{100 - e^{x/2}} = 20$

54. $\dfrac{400}{1 + e^{-x}} = 350$

55. $\dfrac{3000}{2 + e^{2x}} = 2$

56. $\dfrac{119}{e^{6x} - 14} = 7$

57. $\left(1 + \dfrac{0.065}{365}\right)^{365t} = 4$

58. $\left(1 + \dfrac{0.075}{4}\right)^{4t} = 5$

59. $\left(1 + \dfrac{0.10}{12}\right)^{12t} = 2$

60. $\left(1 + \dfrac{0.0825}{26}\right)^{26t} = 9$

In Exercises 61–90, solve the logarithmic equation algebraically. Approximate the result to three decimal places.

61. $\log_{10} x = 4$

62. $\ln x = 5$

63. $\ln x = -3$

64. $\log_{10} x = -5$

65. $\ln 2x = 2.4$

66. $\ln 4x = 1$

67. $\log_{10} 2x = 7$

68. $\log_{10} 3z = 2$

69. $5 \log_3(x + 1) = 12$

70. $5 \log_{10}(x - 2) = 11$

71. $3 \ln 5x = 10$

72. $2 \ln x = 7$

73. $\ln \sqrt{x + 2} = 1$

74. $\ln \sqrt{x - 8} = 5$

75. $7 + 3 \ln x = 5$

76. $2 - 6 \ln x = 10$

77. $\ln x - \ln(x + 1) = 2$

78. $\ln x - \ln(x + 2) = 3$

79. $\ln x + \ln(x - 2) = 1$

80. $\ln x + \ln(x + 3) = 1$

81. $\ln(x + 5) = \ln(x - 1) - \ln(x + 1)$

82. $\ln(x + 1) - \ln(x - 2) = \ln x$

83. $\log_2(2x - 3) = \log_2(x + 4)$

84. $\log_3(x + 8) = \log_3(3x + 2)$

85. $\log_{10}(x + 4) - \log_{10} x = \log_{10}(x + 2)$

86. $\log_{10} x + \log_{10}(x + 1) = \log_{10}(x + 3)$

87. $\log_4 x - \log_4(x - 1) = \frac{1}{2}$

88. $\log_3 x + \log_3(x - 8) = 2$

89. $\log_{10} 8x - \log_{10}\left(1 + \sqrt{x}\right) = 2$

90. $\log_{10} 4x - \log_{10}\left(12 + \sqrt{x}\right) = 2$

In Exercises 91–94, solve for *y* in terms of *x*.

91. $\ln y = \ln(2x + 1) + \ln 1$

92. $\ln y = 2 \ln x + \ln(x - 3)$

93. $\log_{10} y = 2 \log_{10}(x - 1) - \log_{10}(x + 2)$

94. $\log_{10}(y - 4) + \log_{10} x = 3 \log_{10} x$

(T) In Exercises 95–98, use a graphing utility to solve the equation. Approximate the result to three decimal places. Verify your result algebraically.

95. $2^x - 7 = 0$

96. $500 - 1500e^{-x/2} = 0$

97. $3 - \ln x = 0$

98. $10 - 4 \ln(x - 2) = 0$

Compound Interest In Exercises 99 and 100, find the time required for a $1000 investment to double at interest rate *r*, compounded continuously.

99. $r = 0.0625$

100. $r = 0.085$

Compound Interest In Exercises 101 and 102, find the time required for a $1000 investment to triple at interest rate *r*, compounded continuously.

101. $r = 0.0725$

102. $r = 0.0875$

103. Suburban Wildlife The number *V* of varieties of suburban nondomesticated wildlife in a community is approximated by the model

$$V = 15 \cdot 10^{0.02x}, \quad 0 \le x \le 36$$

where *x* is the number of months since the development of the community was completed. Use this model to approximate the number of months since the development was completed when $V = 50$.

104. Native Prairie Grasses The number *A* of varieties of native prairie grasses per acre within a farming region is approximated by the model

$$A = 10.5 \cdot 10^{0.04x}, \quad 0 \le x \le 24$$

where *x* is the number of months since the farming region was plowed. Use this model to approximate the number of months since the region was plowed using a test acre for which $A = 70$.

105. Demand Function The demand function for a special limited edition coin set is given by

$$p = 1000\left(1 - \frac{5}{5 + e^{-0.001x}}\right).$$

(a) Find the demand *x* for a price of $p = \$139.50$.

(b) Find the demand *x* for a price of $p = \$99.99$.

(T) (c) Use a graphing utility to confirm graphically the results found in parts (a) and (b).

106. Demand Function The demand function for a hot tub spa is given by

$$p = 105,000\left(1 - \frac{3}{3 + e^{-0.002x}}\right).$$

(a) Find the demand *x* for a price of $p = \$25,000$.

(b) Find the demand *x* for a price of $p = \$21,000$.

(T) (c) Use a graphing utility to confirm graphically the results found in parts (a) and (b).

(T) **107. Forest Yield** The yield *V* (in millions of cubic feet per acre) for a forest at age *t* years is given by

$$V = 6.7e^{-48.1/t}, \quad t > 0.$$

(a) Use a graphing utility to find the time necessary to obtain a yield of 1.3 million cubic feet per acre.

(b) Use a graphing utility to find the time necessary to obtain a yield of 2 million cubic feet per acre.

108. Human Memory Model In a group project on learning theory, a mathematical model for the percent *P* (in decimal form) of correct responses after *n* trials was found to be

$$P = \frac{0.98}{1 + e^{-0.3n}}, \quad n \ge 0.$$

(a) After how many trials will 80% of the responses be correct? (That is, for what value of *n* will $P = 0.8$?)

(T) (b) Use a graphing utility to graph the memory model and confirm the result found in part (a).

(c) Write a paragraph describing the memory model.

(T) **109. U.S. Currency** The value *y* (in billions of dollars) of U.S. currency in circulation (outside the U.S. Treasury and not held by banks) from 1996 to 2005 can be approximated by the model

$$y = -302 + 374 \ln t, \quad 6 \le t \le 15$$

where *t* represents the year, with $t = 6$ corresponding to 1996. *(Source: Board of Governors of the Federal Reserve System)*

(a) Use a graphing utility to graph the model.

(b) Use a graphing utility to estimate the year when the value of U.S. currency in circulation exceeded $600 billion.

(c) Verify your answer to part (b) algebraically.

(T) **110. Retail Trade** The average monthly sales y (in billions of dollars) in retail trade in the United States from 1996 to 2005 can be approximated by the model

$$y = -22 + 117 \ln t, \quad 6 \le t \le 15$$

where t represents the year, with $t = 6$ corresponding to 1996. *(Source: U.S. Council of Economic Advisors)*

(a) Use a graphing utility to graph the model.

(b) Use a graphing utility to estimate the year in which the average monthly sales first exceeded $270 billion.

(c) Verify your answer to part (b) algebraically.

111. *MAKE A DECISION: AUTOMOBILES* Automobiles are designed with crumple zones that help protect their occupants in crashes. The crumple zones allow the occupants to move short distances when the automobiles come to abrupt stops. The greater the distance moved, the fewer g's the crash victims experience. (One g is equal to the acceleration due to gravity. For very short periods of time, humans have withstood as much as 40 g's.) In crash tests with vehicles moving at 90 kilometers per hour, analysts measured the numbers of g's experienced during deceleration by crash dummies that were permitted to move x meters during impact. The data are shown in the table.

x	0.2	0.4	0.6	0.8	1.0
g's	158	80	53	40	32

A model for these data is given by

$$y = -3.00 + 11.88 \ln x + \frac{36.94}{x}$$

where y is the number of g's.

(a) Complete the table using the model.

x	0.2	0.4	0.6	0.8	1.0
y					

(T) (b) Use a graphing utility to graph the data points and the model in the same viewing window. How do they compare?

(c) Use the model to estimate the least distance traveled during impact for which the passenger does not experience more than 30 g's.

(d) Do you think it is practical to lower the number of g's experienced during impact to fewer than 23? Explain your reasoning.

112. Average Heights The percent m of American males (between 20 and 29 years old) who are less than x inches tall is approximated by

$$m = 0.027 + \frac{0.986}{1 + e^{-0.5857(x - 70.38)}}, \quad 65 \le x \le 78$$

and the percent f of American females (between 20 and 29 years old) who are less than x inches tall is approximated by

$$f = -0.023 + \frac{1.031}{1 + e^{-0.6500(x - 64.34)}}, \quad 60 \le x \le 74$$

where m and f are the percents (in decimal form) and x is the height (in inches) (see figure). *(Source: U.S. National Center for Health Statistics)*

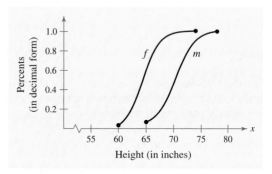

(a) What is the median height for each sex between 20 and 29 years old? (In other words, for what values of x are m and f equal to 0.5?)

(b) Write a paragraph describing each height model.

In Exercises 113–116, rewrite each verbal statement as an equation. Then decide whether the statement is true or false. Justify your answer.

113. The logarithm of the product of two numbers is equal to the sum of the logarithms of the numbers.

114. The logarithm of the sum of two numbers is equal to the product of the logarithms of the numbers.

115. The logarithm of the difference of two numbers is equal to the difference of the logarithms of the numbers.

116. The logarithm of the quotient of two numbers is equal to the difference of the logarithms of the numbers.

117. Think About It Is it possible for a logarithmic equation to have more than one extraneous solution? Explain.

118. Think About It Are the times required for the investments in Exercises 99 and 100 to quadruple twice as long as the times for them to double? Give a reason for your answer and verify your answer algebraically.

Section 4.5

Exponential and Logarithmic Models

- Construct and use a model for exponential growth or exponential decay.
- Use a Gaussian model to solve an application problem.
- Use a logistic growth model to solve an application problem.
- Use a logarithmic model to solve an application problem.
- Choose an appropriate model involving exponential or logarithmic functions for a real-life situation.

Introduction

The five most common types of mathematical models involving exponential functions and logarithmic functions are as follows.

1. Exponential growth model: $y = ae^{bx}, \quad b > 0$

2. Exponential decay model: $y = ae^{-bx}, \quad b > 0$

3. Gaussian model: $y = ae^{-(x-b)^2/c}$

4. Logistic growth model: $y = \dfrac{a}{1 + be^{-rx}}$

5. Logarithmic models: $y = a + b \ln x, \quad y = a + b \log_{10} x$

The basic shape of each of these graphs is shown in Figure 4.21.

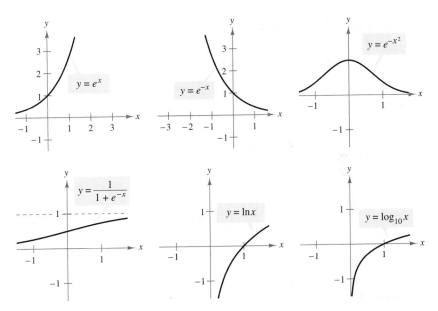

FIGURE 4.21

You can often gain quite a bit of insight into a situation modeled by an exponential or logarithmic function by identifying and interpreting the function's asymptotes. Use the graphs in Figure 4.21 to identify the asymptote(s) of the graph of each function.

Exponential Growth and Decay

Example 1   Population Increase

The world populations (in millions) for each year from 1996 through 2005 are shown in the table.   *(Source: U.S. Census Bureau)*

Year	1996	1997	1998	1999	2000
Population	5763	5842	5920	5997	6073

Year	2001	2002	2003	2004	2005
Population	6149	6224	6299	6375	6451

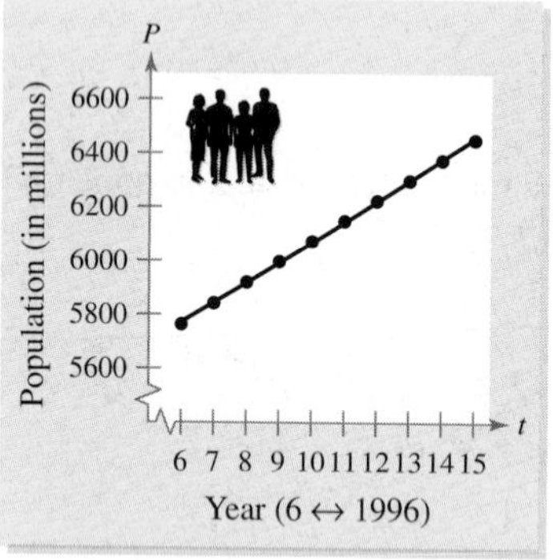

An exponential growth model that approximates these data is

$$P = 5356e^{0.012469t}, \quad 6 \le t \le 15$$

where P is the population (in millions) and $t = 6$ represents 1996. Compare the estimates given by the model with the values given by the U.S. Census Bureau. Use the model to predict the year in which the world population reaches 6.7 billion.

SOLUTION   The following table compares the two sets of population figures. The graph of the model and the original data values are shown in Figure 4.22.

Year	1996	1997	1998	1999	2000	2001
Population	5763	5842	5920	5997	6073	6149
Model	5772	5845	5918	5992	6067	6143

Year	2002	2003	2004	2005
Population	6224	6299	6375	6451
Model	6221	6299	6378	6458

To find the year in which the world population reaches 6.7 billion, let $P = 6700$ in the model and solve for t.

$5356e^{0.012469t} = P$	Write original model.
$5356e^{0.012469t} = 6700$	Let $P = 6700$.
$e^{0.012469t} \approx 1.25093$	Divide each side by 5356.
$\ln e^{0.012469t} \approx \ln 1.25093$	Take natural log of each side.
$0.012469t \approx 0.223890$	Inverse Property
$t \approx 18.0$	Divide each side by 0.012469.

According to the model, the world population reaches 6.7 billion in 2008.

P

Population (in millions)

6600
6400
6200
6000
5800
5600

6 7 8 9 10 11 12 13 14 15
Year (6 ↔ 1996)

FIGURE 4.22

TECHNOLOGY

(T) Some graphing utilities have an *exponential regression* feature that can be used to find an exponential model that represents data. If you have such a graphing utility, try using it to find a model for the data given in Example 1. How does your model compare with the model given in Example 1? For instructions on how to use the *regression* feature, see Appendix A; for specific keystrokes, go to the text website at *college.hmco.com/info/larsonapplied.*

✓ **CHECKPOINT 1**

Use the model in Example 1 to predict the year in which the world population will reach 7.3 billion. ∎

The exponential model in Example 1 increases (or decreases) by the same percent each year. What is the annual percent increase for this exponential model?

In Example 1, you were given the exponential growth model. But suppose this model were not given; how could you find such a model? If you are given a set of data, as in Example 1, but you are not given the exponential growth model that fits the data, you can choose any two of the points and substitute them in the general exponential growth model $y = ae^{bx}$. This technique is demonstrated in Example 2.

Example 2   Finding an Exponential Growth Model

Find an exponential growth model whose graph passes through the points $(0, 4453)$ and $(7, 5024)$, as shown in Figure 4.23(a).

SOLUTION   The general form of the model is

$$y = ae^{bx}.$$

From the fact that the graph passes through the point $(0, 4453)$, you know that $y = 4453$ when $x = 0$. By substituting these values into the general model, you have

$$4453 = ae^0 \quad \Longrightarrow \quad a = 4453.$$

In a similar way, from the fact that the graph passes through the point $(7, 5024)$, you know that $y = 5024$ when $x = 7$. By substituting these values into the model, you obtain

$$5024 = 4453e^{7b} \quad \Longrightarrow \quad b = \frac{1}{7} \ln \frac{5024}{4453} \approx 0.01724.$$

So, the exponential growth model is

$$y = 4453e^{0.01724x}.$$

The graph of the model is shown in Figure 4.23(b).

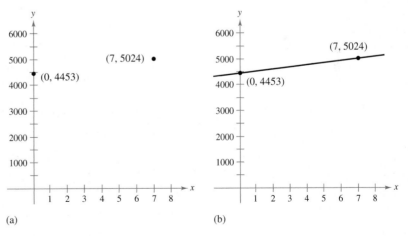

(a)

(b)

FIGURE 4.23

✓ CHECKPOINT 2

Find an exponential growth model whose graph passes through the points $(0, 3)$ and $(5, 8)$. ▨

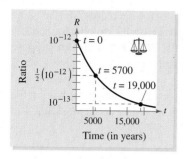

FIGURE 4.24

© Louie Psihoyos/CORBIS

In 1960, Willard Libby of the University of Chicago won the Nobel Prize for Chemistry for the carbon 14 method, a valuable tool for estimating the ages of ancient materials.

✓**CHECKPOINT 3**

The ratio of carbon 14 to carbon 12 in a newly discovered fossil is

$$R = \frac{1}{9^{13}}.$$

Estimate the age of the fossil. ■

In living organic material, the ratio of the number of radioactive carbon isotopes (carbon 14) to the number of nonradioactive carbon isotopes (carbon 12) is about 1 to 10^{12}. When organic material dies, its carbon 12 content remains fixed, whereas its radioactive carbon 14 begins to decay with a half-life of about 5700 years. To estimate the age of dead organic material, scientists use the following formula, which denotes the ratio of carbon 14 to carbon 12 present at any time t (in years).

$$R = \frac{1}{10^{12}} e^{-t/8223} \qquad \text{Carbon dating model}$$

In Figure 4.24, note that R decreases as the time t increases. Any material that is composed of carbon, such as wood, bone, hair, pottery, paper, and water, can be dated.

Example 3 **Carbon Dating**

The ratio of carbon 14 to carbon 12 in a newly discovered fossil is

$$R = \frac{1}{10^{13}}.$$

Estimate the age of the fossil.

SOLUTION In the carbon dating model, substitute the given value of R to obtain the following.

$$\frac{1}{10^{12}} e^{-t/8223} = R \qquad \text{Write original model.}$$

$$\frac{e^{-t/8223}}{10^{12}} = \frac{1}{10^{13}} \qquad \text{Substitute } \frac{1}{10^{13}} \text{ for } R.$$

$$e^{-t/8223} = \frac{1}{10} \qquad \text{Multiply each side by } 10^{12}.$$

$$\ln e^{-t/8223} = \ln \frac{1}{10} \qquad \text{Take natural log of each side.}$$

$$-\frac{t}{8223} \approx -2.3026 \qquad \text{Inverse Property}$$

$$t \approx 18{,}934 \qquad \text{Multiply each side by } -8223.$$

So, you can estimate the age of the fossil to be about 19,000 years.

An exponential model can be used to determine the *decay* of radioactive isotopes. For instance, to find how much of an initial 10 grams of radioactive radium (^{226}Ra) with a half-life of 1599 years is left after 500 years, you would use the exponential decay model, as follows.

$$y = ae^{-bt} \implies \frac{1}{2}(10) = 10e^{-b(1599)} \implies \ln\frac{1}{2} = -1599b \implies -\frac{\ln\frac{1}{2}}{1599} = b$$

Using the value of b found above, $a = 10$, and $t = 500$, the amount left is

$$y = 10e^{-[-\ln(1/2)/1599](500)} \approx 8.05 \text{ grams}$$

Gaussian Models

As mentioned at the beginning of this section, Gaussian models are of the form

$$y = ae^{-(x-b)^2/c}.$$

This type of model is commonly used in probability and statistics to represent populations that are **normally distributed.** For *standard* normal distributions, the model takes the form

$$y = \frac{1}{\sqrt{2\pi}} e^{-x^2/2}.$$

The graph of a Gaussian model is called a **bell-shaped curve.** Try to sketch the standard normal distribution curve with a graphing utility. Can you see why it is called a bell-shaped curve?

The **average value** of a population can be found from the bell-shaped curve by observing where the maximum y-value of the function occurs. The x-value corresponding to the maximum y-value of the function represents the average value of the independent variable, x.

Example 4   SAT Scores

In 2006, the SAT (Scholastic Aptitude Test) mathematics scores for college-bound seniors in the United States roughly followed a normal distribution given by

$$y = 0.0035e^{-(x-518)^2/26,450}, \quad 200 \leq x \leq 800$$

where x is the SAT score for mathematics. Sketch the graph of this function. From the graph, estimate the average SAT score.   *(Source: College Board)*

SOLUTION   The graph of the function is shown in Figure 4.25. On this bell-shaped curve, the x-value corresponding to the maximum value of the curve represents the average score. From the graph, you can estimate that the average mathematics score for college-bound seniors in 2006 was 518.

✓ CHECKPOINT 4

In 2006, the SAT reading scores for college-bound seniors in the United States roughly followed a normal distribution given by

$$y = 0.0035e^{-(x-503)^2/25,538},$$
$$200 \leq x \leq 800$$

where x is the SAT score for reading. Sketch the graph of this function. From the graph, estimate the average SAT score.   *(Source: College Board)* ▪

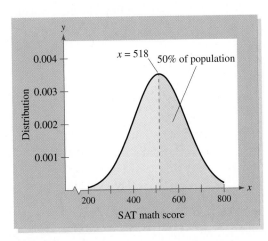

FIGURE 4.25

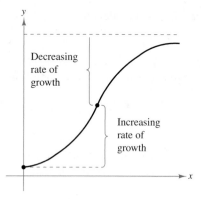

FIGURE 4.26   Logistic Curve

Logistic Growth Models

Some populations initially have rapid growth, followed by a declining rate of growth, as shown by the graph in Figure 4.26. One model for describing this type of growth pattern is the **logistic curve** given by the function

$$y = \frac{a}{1 + be^{-rx}}$$

where y is the population size and x is the time. An example is a bacteria culture that is initially allowed to grow under ideal conditions, followed by less favorable conditions that inhibit growth. A logistic growth curve is also called a **sigmoidal curve.**

| *Example 5*   **Spread of a Virus** | |

On a college campus of 5000 students, one student returned from vacation with a contagious flu virus. The spread of the virus through the student population is given by

$$y = \frac{5000}{1 + 4999e^{-0.8t}}, \quad t \geq 0$$

where y is the total number of students infected after t days. The college will cancel classes when 40% or more of the students become infected.

a. How many students are infected after 5 days?

b. After how many days will the college cancel classes?

SOLUTION

a. After 5 days, the number of students infected is

$$y = \frac{5000}{1 + 4999e^{-0.8(5)}} = \frac{5000}{1 + 4999e^{-4}} \approx 54.$$

b. Classes are cancelled when the number infected is $(0.40)(5000) = 2000$. So, solve for t in the following equation.

$$2000 = \frac{5000}{1 + 4999e^{-0.8t}}$$

$$1 + 4999e^{-0.8t} = 2.5$$

$$e^{-0.8t} = \frac{1.5}{4999}$$

$$\ln e^{-0.8t} = \ln \frac{1.5}{4999}$$

$$-0.8t = \ln \frac{1.5}{4999}$$

$$t \approx 10.1$$

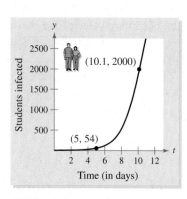

FIGURE 4.27

So, after 10 days, at least 40% of the students will become infected, and the college will cancel classes. The graph of the function is shown in Figure 4.27.

✓ **CHECKPOINT 5**

In Example 5, how many days does it take for 25% of the students to become infected? ■

Logarithmic Models

© Warrick Page/CORBIS

The severity of the destruction caused by an earthquake depends on its magnitude and duration. Earthquakes can destroy buildings, and can cause landslides and tsunamis.

Example 6 Magnitudes of Earthquakes

On the Richter scale, the magnitude R of an earthquake of intensity I per unit of area is given by

$$R = \log_{10}\left(\frac{I}{I_0}\right)$$

where $I_0 = 1$ is the minimum intensity used for comparison. Find the intensity per unit of area for each earthquake. (Intensity is a measure of the wave energy of an earthquake.)

a. Prince William Sound, Alaska, in 1964; $R = 9.2$

b. Off the coast of Northern California in 2005; $R = 7.2$

SOLUTION

a. Because $I_0 = 1$ and $R = 9.2$,

$$9.2 = \log_{10} I$$
$$I = 10^{9.2} \approx 1,584,893,000.$$

b. For $R = 7.2$,

$$7.2 = \log_{10} I$$
$$I = 10^{7.2} \approx 15,849,000$$

Note that an increase of 2.0 units on the Richter scale (from 7.2 to 9.2) represents an intensity change by a factor of

$$\frac{1,584,893,000}{15,849,000} \approx 100.$$

In other words, the Prince William Sound earthquake in 1964 had a magnitude about 100 times greater than that of the earthquake off the coast of Northern California in 2005.

✓ **CHECKPOINT 6**

Find the intensity I per unit of area of an earthquake measuring $R = 6.4$ on the Richter scale. (Let $I_0 = 1$.) ∎

Example 7 pH Levels

Acidity, or pH level, is a measure of the hydrogen ion concentration $[H^+]$ (measured in moles of hydrogen per liter) of a solution. Use the model given by

$$pH = -\log_{10}[H^+]$$

to determine the hydrogen ion concentration of milk of magnesia, which has a pH of 10.5.

SOLUTION

$pH = -\log_{10}[H^+]$	Write original model.
$10.5 = -\log_{10}[H^+]$	Substitute 10.5 for pH.
$-10.5 = \log_{10}[H^+]$	Multiply each side by -1.
$10^{-10.5} = 10^{\log_{10}[H^+]}$	Exponentiate each side (base 10).
$3.16 \times 10^{-11} = [H^+]$	Simplify.

So, the hydrogen ion concentration of milk of magnesia is 3.16×10^{-11} mole of hydrogen per liter.

✓ **CHECKPOINT 7**

Use the model in Example 7 to determine the hydrogen ion concentration of coffee, which has a pH of 5.0 ∎

Comparing Models

So far you have been given the type of model to use for a data set. Now you will use the general trends of the graphs of the five models presented in this section to choose appropriate models for real-life situations.

Example 8
MAKE A DECISION **Choosing an Appropriate Model**

Decide whether to use an exponential growth model or a logistic growth model to represent each data set.

a.

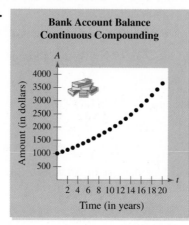

b.

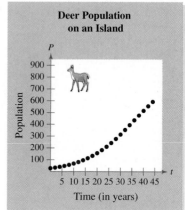

✓ CHECKPOINT 8

Decide whether to use an exponential growth model or a logistic growth model to represent the data set for the fish population of a lake in the figure below.

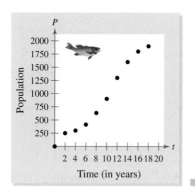

SOLUTION

a. As long as withdrawals and deposits are not made and the interest rate remains constant, the bank account balance will grow exponentially. So, an exponential growth model is an appropriate model.

b. The growth of the deer population will slow as the population approaches the carrying capacity of the island. So, a logistic growth model is an appropriate model.

┌───┐
│ **CONCEPT CHECK**

1. What type of model is the function $y = 8e^{-0.5x^2}$?

2. Does a Gaussian model generally represent population growth well? Explain your reasoning.

3. Explain why the growth of a population of bacteria in a petri dish can be modeled by a logistic growth model.

4. Is it possible for the graph of an exponential decay model to pass through the points $(0, 220)$ and $(-4, 736)$? Explain.
└───┘

Skills Review 4.5

The following warm-up exercises involve skills that were covered in earlier sections. You will use these skills in the exercise set for this section. For additional help, review Sections 4.1, 4.2, and 4.4.

In Exercises 1–6, sketch the graph of the equation.

1. $y = e^{0.1x}$

2. $y = e^{-0.25x}$

3. $y = e^{-x^2/5}$

4. $y = \dfrac{2}{1 + e^{-x}}$

5. $y = \log_{10} 2x$

6. $y = \ln 4x$

In Exercises 7 and 8, solve the equation algebraically.

7. $3e^{2x} = 7$

8. $4 \ln 5x = 14$

In Exercises 9 and 10, solve the equation graphically.

9. $2e^{-0.2x} = 0.002$

10. $6 \ln 2x = 12$

Exercises 4.5

See www.CalcChat.com for worked-out solutions to odd-numbered exercises.

Compound Interest In Exercises 1–10, complete the table for a savings account in which interest is compounded continuously.

Initial Investment	Annual % Rate	Time to Double	Amount After 10 Years
1. $5000	7%		
2. $1000	$9\frac{1}{4}\%$		
3. $500		10 yr	
4. $10,000		5 yr	
5. $1000			$2281.88
6. $2000			$3000
7.	11%		$19,205
8.	8%		$20,000
9. $5000			$11,127.70
10. $250			$600

Radioactive Decay In Exercises 11–16, complete the table for the radioactive isotope.

Isotope	Half-Life (Years)	Initial Quantity	Amount After 1000 Years
11. ^{226}Ra	1599	4 g	
12. ^{226}Ra	1599		0.15 g
13. ^{14}C	5715		3.5 g
14. ^{14}C	5715	8 g	
15. ^{239}Pu	24,100		1.6 g
16. ^{239}Pu	24,100		0.38 g

In Exercises 17–20, classify the model as an exponential growth model or an exponential decay model.

17. $y = 3e^{0.5t}$

18. $y = 2e^{-0.6t}$

19. $y = 20e^{-1.5t}$

20. $y = 4e^{0.07t}$

In Exercises 21–24, find the constants C and k such that the exponential function $y = Ce^{kt}$ passes through the points on the graph.

21.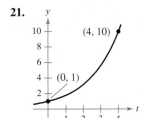

22.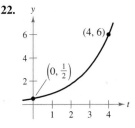

23.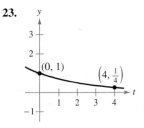

24.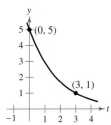

25. Population The population P of a city is given by

$$P = 120,000e^{0.016t}$$

where t represents the year, with $t = 0$ corresponding to 2000. Sketch the graph of this equation. Use the model to predict the year in which the population of the city will reach about 180,000.

26. Population The population P of a city is given by

$$P = 25{,}000e^{kt}$$

where t represents the year, with $t = 0$ corresponding to 2000. In 1980, the population was 15,000. Find the value of k and use this result to predict the population in the year 2010.

27. Bacteria Growth The number N of bacteria in a culture is given by the model $N = 100e^{kt}$, where t is the time (in hours), with $t = 0$ corresponding to the time when $N = 100$. When $t = 6$, there are 140 bacteria. How long does it take the bacteria population to double in size? To triple in size?

28. Bacteria Growth The number N of bacteria in a culture is given by the model $N = 250e^{kt}$, where t is the time (in hours), with $t = 0$ corresponding to the time when $N = 250$. When $t = 10$, there are 320 bacteria. How long does it take the bacteria population to double in size? To triple in size?

29. Carbon Dating The ratio of carbon 14 to carbon 12 in a piece of wood discovered in a cave is $R = 1/8^{14}$. Estimate the age of the piece of wood.

30. Carbon Dating The ratio of carbon 14 to carbon 12 in a piece of paper buried in a tomb is $R = 1/13^{11}$. Estimate the age of the piece of paper.

31. Radioactive Decay What percent of a present amount of radioactive cesium (^{137}Cs) will remain after 100 years? Use the fact that radioactive cesium has a half-life of 30 years.

32. Radioactive Decay Find the half-life of radioactive iodine (^{131}I) if, after 20 days, 0.53 kilogram of an initial 3 kilograms remains.

33. Learning Curve The management at a factory has found that the maximum number of units a worker can produce in a day is 40. The learning curve for the number of units N produced per day after a new employee has worked t days is given by

$$N = 40(1 - e^{kt}).$$

After 20 days on the job, a particular worker produced 25 units in 1 day.

(a) Find the learning curve for this worker (first find the value of k).

(b) How many days should pass before this worker is producing 35 units per day?

34. Learning Curve The management at a customer service center has found that the maximum number of customer calls an employee can process effectively in a day is 90. The learning curve for the number N of calls processed per day after a new employee has worked t days is given by

$$N = 90(1 - e^{kt}).$$

After 15 days on the job, a particular employee processed 60 calls in 1 day.

(a) Find the learning curve for this worker (first find the value of k).

(b) How many days should pass before this employee will process 80 calls per day?

35. Motorola The sales per share S (in dollars) for Motorola from 1992 to 2005 can be approximated by the function

$$S = \begin{cases} 2.33 - 0.909t + 10.394 \ln t, & 2 \le t \le 10 \\ 0.6157t^2 - 15.597t + 110.25, & 11 \le t \le 15 \end{cases}$$

where t represents the year, with $t = 2$ corresponding to 1992. *(Source: Motorola)*

(T) (a) Use a graphing utility to graph the function.

(b) Describe the change in sales per share that occurred in 2001.

36. Intel The sales per share S (in dollars) for Intel from 1992 to 2005 can be approximated by the function

$$S = \begin{cases} -1.48 + 2.65 \ln t, & 2 \le t \le 10 \\ 0.1586t^2 - 3.465t + 22.87, & 11 \le t \le 15 \end{cases}$$

where t represents the year, with $t = 2$ corresponding to 1992. *(Source: Intel)*

(T) (a) Use a graphing utility to graph the function.

(b) Describe the change in sales per share that occurred in 2001.

(T) **37. Women's Heights** The distribution of heights of American women (between 30 and 39 years of age) can be approximated by the function

$$p = 0.163e^{-(x - 64.9)^2/12.03}, \quad 60 \le x \le 74$$

where x is the height (in inches) and p is the percent (in decimal form). Use a graphing utility to graph the function. Then determine the average height of women in this age bracket. *(Source: U.S. National Center for Health Statistics)*

(T) **38. Men's Heights** The distribution of heights of American men (between 30 and 39 years of age) can be approximated by the function

$$p = 0.131e^{-(x - 69.9)^2/18.66}, \quad 63 \le x \le 77$$

where x is the height (in inches) and p is the percent (in decimal form). Use a graphing utility to graph the function. Then determine the average height of men in this age bracket. *(Source: U.S. National Center for Health Statistics)*

39. Stocking a Lake with Fish A lake is stocked with 500 fish, and the fish population P increases according to the logistic curve

$$P = \frac{10{,}000}{1 + 19e^{-t/5}}, \quad t \ge 0$$

where t is the time (in months).

(a) Use a graphing utility to graph the logistic curve.

(b) Find the fish population after 5 months.

(c) After how many months will the fish population reach 2000?

40. Endangered Species   A conservation organization releases 100 animals of an endangered species into a game preserve. The organization believes that the preserve has a carrying capacity of 1000 animals and that the growth of the herd will be modeled by the logistic curve

$$p = \frac{1000}{1 + 9e^{-kt}}, \quad t \geq 0$$

where p is the number of animals and t is the time (in years). The herd size is 134 after 2 years. Find k. Then find the population after 5 years.

(T) 41. Aged Population   The table shows the projected U.S. populations P (in thousands) of people who are 85 years old or older for several years from 2010 to 2050.   *(Source: U.S. Census Bureau)*

Year	85 years and older
2010	6123
2015	6822
2020	7269
2025	8011
2030	9603
2035	12,430
2040	15,409
2045	18,498
2050	20,861

(a) Use a graphing utility to create a scatter plot of the data. Let t represent the year, with $t = 10$ corresponding to 2010.

(b) Use the *regression* feature of a graphing utility to find an exponential model for the data. Use the Inverse Property

$$b = e^{\ln b}$$

to rewrite the model as an exponential model in base e.

(c) Use a graphing utility to graph the exponential model in base e.

(d) Use the exponential model in base e to estimate the populations of people who are 85 years old or older in 2022 and in 2042.

(T) 42. Super Bowl Ad Cost   The table shows the costs C (in millions of dollars) of a 30-second TV ad during the Super Bowl for several years from 1987 to 2006.   *(Source: TNS Media Intelligence)*

Year	Cost
1987	0.6
1992	0.9
1997	1.2
2002	2.2
2006	2.5

(a) Use a graphing utility to create a scatter plot of the data. Let t represent the year, with $t = 7$ corresponding to 1987.

(b) Use the *regression* feature of a graphing utility to find an exponential model for the data. Use the Inverse Property $b = e^{\ln b}$ to rewrite the model as an exponential model in base e.

(c) Use a graphing utility to graph the exponential model in base e.

(d) Use the exponential model in base e to predict the costs of a 30-second ad during the Super Bowl in 2009 and in 2010.

(S) 43. Super Bowl Ad Revenue   The table shows Super Bowl TV ad revenues R (in millions of dollars) for several years from 1987 to 2006.   *(Source: TNS Media Intelligence)*

Year	Revenue
1987	31.5
1992	48.2
1997	72.2
2002	134.2
2006	162.5

(a) Use a spreadsheet software program to create a scatter plot of the data. Let t represent the year, with $t = 7$ corresponding to 1987.

(b) Use the *regression* feature of a spreadsheet software program to find an exponential model for the data. Use the Inverse Property $b = e^{\ln b}$ to rewrite the model as an exponential model in base e.

(c) Use a spreadsheet software program to graph the exponential model in base e.

(d) Use the exponential model in base e to predict the Super Bowl ad revenues in 2009 and in 2010.

44. Domestic Demand The domestic demands D (in thousands of barrels) for refined oil products in the United States from 1995 to 2005 are shown in the table. *(Source: U.S. Energy Information Administration)*

Year	Demand
1995	6,469,625
1996	6,701,094
1997	6,796,300
1998	6,904,705
1999	7,124,435
2000	7,210,566

Year	Demand
2001	7,171,885
2002	7,212,765
2003	7,312,410
2004	7,587,546
2005	7,539,440

(a) Use a spreadsheet software program to create a scatter plot of the data. Let t represent the year, with $t = 5$ corresponding to 1995.

(b) Use the *regression* feature of a spreadsheet software program to find an exponential model for the data. Use the Inverse Property $b = e^{\ln b}$ to rewrite the model as an exponential model in base e.

(c) Use the *regression* feature of a spreadsheet software program to find a logarithmic model $(y = a + b \ln x)$ for the data.

(d) Use a spreadsheet software program to graph the exponential model in base e and the logarithmic model with the scatter plot.

(e) Use both models to predict domestic demands in 2008, 2009, and 2010. Do both models give reasonable predictions? Explain.

45. Population The populations P of the United States (in thousands) from 1990 to 2005 are shown in the table. *(Source: U.S. Census Bureau)*

Year	Population
1990	250,132
1991	253,493
1992	256,894
1993	260,255
1994	263,436
1995	266,557
1996	269,667
1997	272,912

Year	Population
1998	276,115
1999	279,295
2000	282,403
2001	285,335
2002	288,216
2003	291,089
2004	293,908
2005	296,639

(a) Use a graphing utility to create a scatter plot of the data. Let t represent the year, with $t = 0$ corresponding to 1990.

(b) Use the *regression* feature of a graphing utility to find an exponential model for the data. Use the Inverse Property $b = e^{\ln b}$ to rewrite the model as an exponential model in base e.

(c) Use the *regression* feature of a graphing utility to find a linear model and a quadratic model for the data.

(d) Use a graphing utility to graph the exponential model in base e and the models in part (c) with the scatter plot.

(e) Use each model to predict the populations in 2008, 2009, and 2010. Do all models give reasonable predictions? Explain.

46. Population The population P of the United States officially reached 300 million at about 7:46 A.M. E.S.T. on Tuesday, October 17, 2006. The table shows the U.S. populations (in millions) since 1900. *(Source: U.S. Census Bureau)*

Year	Population
1900	76
1910	92
1920	106
1930	123
1940	132
1950	151

Year	Population
1960	179
1970	203
1980	227
1990	250
2000	282
2006	300

(a) Use a graphing utility to create a scatter plot of the data. Let t represent the year, with $t = 0$ corresponding to 1900.

(b) Use the *regression* feature of a graphing utility to find an exponential model for the data. Use the Inverse Property $b = e^{\ln b}$ to rewrite the model as an exponential model in base e.

(c) Graph the exponential model in base e with the scatter plot of the data. What appears to be happening to the relationship between the data points and the regression curve at $t = 100$ and $t = 106$?

(d) Use the *regression* feature of a graphing utility to find a logistic growth model for the data. Graph each model using the window settings shown below. Which model do you think will give more accurate predictions of the population well beyond 2006?

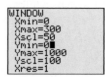

```
WINDOW
 Xmin=0
 Xmax=300
 Xscl=50
 Ymin=0
 Ymax=1000
 Yscl=100
 Xres=1
```

(e) The U.S. Census Bureau predicts that the U.S. population will be about 420 million in 2050. Use each model to predict the population in 2050. Which model gives an estimate closer to the prediction of 420 million?

Earthquake Magnitudes In Exercises 47 and 48, use the Richter scale (see Example 6) for measuring the magnitudes of earthquakes.

47. Find the magnitude R (on the Richter scale) of an earthquake of intensity I. (Let $I_0 = 1$.)

(a) $I = 80,500,000$ (b) $I = 48,275,000$

48. Find the intensity I of an earthquake measuring R on the Richter scale. (Let $I_0 = 1$.)

(a) Vanuatu Islands in 2002, $R = 7.3$

(b) Near coast of Peru in 2001, $R = 8.4$

Intensity of Sound In Exercises 49 and 50, find the level of sound using the following information for determining sound intensity. The level of sound L (in decibels) of a sound with an intensity of I is given by

$$L = 10 \log_{10} \frac{I}{I_0}$$

where I_0 is an intensity of 10^{-12} watt per square meter, corresponding roughly to the faintest sound that can be heard by the human ear.

49. (a) $I = 10^{-10}$ watt per square meter (quiet room)

(b) $I = 10^{-5}$ watt per square meter (busy street corner)

50. (a) $I = 10^{-3}$ watt per square meter (loud car horn)

(b) $I \approx 10^{0}$ watt per square meter (threshold of pain)

pH Levels In Exercises 51–54, use the acidity model given in Example 7.

51. Compute $[\text{H}^+]$ for a solution for which pH $= 5.8$.

52. Compute $[\text{H}^+]$ for a solution for which pH $= 7.3$.

53. A grape has a pH of 3.5, and baking soda has a pH of 8.0. The hydrogen ion concentration of the grape is how many times that of the baking soda?

54. The pH of a solution is decreased by one unit. The hydrogen ion concentration is increased by what factor?

55. Estimating the Time of Death At 8:30 A.M., a coroner was called to the home of a person who had died during the night. In order to estimate the time of death, the coroner took the person's temperature twice. At 9:00 A.M. the temperature was 85.7°F, and at 11:00 A.M. the temperature was 82.8°F.

From these two temperature readings, the coroner was able to determine that the time elapsed since death and the body temperature are related by the formula

$$t = -10 \ln \frac{T - 70}{98.6 - 70}$$

where t is the time (in hours) elapsed since the person died and T is the temperature (in degrees Fahrenheit) of the person's body. The coroner assumed that the person had a normal body temperature of 98.6°F at death, and that the room temperature was a constant 70°F. Use this formula to estimate the time of death of the person.

56. Thawing a Package of Steaks You take a three-pound package of steaks out of the freezer at 11 A.M. and place it in the refrigerator. Will the steaks be thawed in time to be grilled at 6 P.M.? Assume that the refrigerator temperature is 40°F and that the freezer temperature is 0°F. Use the formula for Newton's Law of Cooling

$$t = -5.05 \ln \frac{T - 40}{0 - 40}$$

where t is the time in hours (with $t = 0$ corresponding to 11 A.M.) and T is the temperature of the package of steaks (in degrees Fahrenheit).

Ⓣ **57.** *MAKE A DECISION: WORKER'S PRODUCTIVITY* The numbers n of units per day that a new worker can produce after t days on the job are listed in the table. Use a graphing utility to create a scatter plot of the data. Do the data fit an exponential model or a logarithmic model? Use the *regression* feature of the graphing utility to find the model. Graph the model with the original data. Is the model a good fit? Can you think of a better model to use for these data? Explain.

Days, t	5	10	15	20	25
Units, n	6	13	22	34	56

Ⓣ **58. Chemical Reaction** The table shows the yield y (in milligrams) of a chemical reaction after x minutes. Use a graphing utility to create a scatter plot of the data. Do the data fit an exponential model or a logarithmic model? Use the *regression* feature of the graphing utility to find the model. Graph the model with the original data. Is this model a good fit for the data?

Minutes, x	1	2	3	4	5
Yield, y	1.5	7.4	10.2	13.4	15.8

Minutes, x	6	7	8
Yield, y	16.3	18.2	18.3

Chapter Summary and Study Strategies

After studying this chapter, you should have acquired the following skills.
The exercise numbers are keyed to the Review Exercises that begin on page 398.
Answers to odd-numbered Review Exercises are given in the back of the text.*

Section 4.1   Review Exercises

■ Sketch the graph of an exponential function.   *1–4, 9–16*

    Characteristics of Exponential Functions

    Graph of $y = a^x$, $a > 1$ Graph of $y = a^{-x}$, $a > 1$

    • Domain: $(-\infty, \infty)$ • Domain: $(-\infty, \infty)$
    • Range: $(0, \infty)$ • Range: $(0, \infty)$
    • Intercept: $(0, 1)$ • Intercept: $(0, 1)$
    • Increasing   • Decreasing
    • x-axis is a horizontal asymptote: • x-axis is a horizontal asymptote:
      $(a^x \to 0$ as $x \to -\infty)$ $(a^{-x} \to 0$ as $x \to \infty)$
    • Continuous   • Continuous
                           • Reflection of graph of $y = a^x$
                            about y-axis

■ Use the compound interest formulas.   *17–20*

    For n compoundings per year: $A = P(1 + r/n)^{nt}$

    For continuous compounding: $A = Pe^{rt}$

■ Use an exponential model to solve an application problem.   *21, 22*

Section 4.2

■ Recognize and evaluate a logarithmic function.   *23–36*

    $y = \log_a x$ if and only if $x = a^y$

    $y = \log_e x = \ln x$

■ Sketch the graph of a logarithmic function.   *5–8, 37–42*

    Characteristics of Logarithmic Functions

    Graph of $y = \log_a x$, $a > 1$

    • Domain: $(0, \infty)$
    • Range: $(-\infty, \infty)$
    • Intercept: $(1, 0)$
    • Increasing
    • One-to-one; therefore has an inverse function
    • y-axis is a vertical asymptote ($\log_a x \to -\infty$ as $x \to 0^+$)
    • Continuous
    • Reflection of graph of $y = a^x$ about the line $y = x$

■ Use a logarithmic model to solve an application problem.   *43–46*

---

* Use a wide range of valuable study aids to help you master the material in this chapter. The *Student
Solutions Guide* includes step-by-step solutions to all odd-numbered exercises to help you review
and prepare. The student website at *college.hmco.com/info/larsonapplied* offers algebra help and a
Graphing Technology Guide. The *Graphing Technology Guide* contains step-by-step commands
and instructions for a wide variety of graphing calculators, including the most recent models.

Section 4.3

Review Exercises

■ Evaluate a logarithm using the change-of-base formula.   *47–50*

$$\log_a x = \frac{\log_b x}{\log_b a}, \quad \log_a x = \frac{\ln x}{\ln a}$$

■ Use properties of logarithms to evaluate or rewrite a logarithmic expression.   *51–58*

$$\log_a(uv) = \log_a u + \log_a v \qquad \ln(uv) = \ln u + \ln v$$

$$\log_a \frac{u}{v} = \log_a u - \log_a v \qquad \ln \frac{u}{v} = \ln u - \ln v$$

$$\log_a u^n = n \log_a u \qquad \ln u^n = n \ln u$$

■ Use properties of logarithms to expand or condense a logarithmic expression.   *59–70*

■ Use logarithmic functions to model and solve real-life applications.   *71, 72*

Section 4.4

■ Solve an exponential equation.   *73–78*

■ Solve a logarithmic equation.   *79–86*

■ Use an exponential or a logarithmic model to solve an application problem.   *87, 88*

Section 4.5

■ Construct and use a model for exponential growth or exponential decay.   *89–95*

$$y = ae^{bx}, \quad b > 0 \qquad\qquad y = ae^{-bx}, \quad b > 0$$

■ Use a Gaussian model to solve an application problem.   *96*

$$y = ae^{-(x-b)^2/c}$$

■ Use a logistic growth model to solve an application problem.   *97*

$$y = \frac{a}{1 + be^{-rx}}$$

■ Use a logarithmic model to solve an application problem.   *98, 99*

$$y = a + b \ln x, \quad y = a + b \log_{10} x$$

■ Choose an appropriate model involving exponential or logarithmic functions for a real-life situation.   *100*

Study Strategies

■ **Solve Problems Analytically or Graphically** When solving an exponential or logarithmic equation, you could use a variety of problem-solving strategies. For instance, if you were asked to solve the logarithmic equation

$$\ln(x + 4) - \ln x = 1$$

you could solve the equation *analytically*. That is, you could use the properties of logarithms to rewrite the equation, exponentiate each side, use the Inverse Property, and solve the resulting equation to determine that $x \approx 2.328$.

You could also solve the equation *graphically*. That is, you could use a graphing utility to graph $y_1 = \ln(x + 4) - \ln x$ and $y_2 = 1$ in the same viewing window. Then you could use the *intersect* feature or the *zoom* and *trace* features to determine that the solution of the original equation is $x \approx 2.328$.

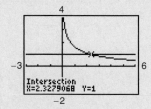

Review Exercises

See www.CalcChat.com for worked-out solutions to odd-numbered exercises.

In Exercises 1–8, match the function with its graph. [The graphs are labeled (a), (b), (c), (d), (e), (f), (g), and (h).]

(a)

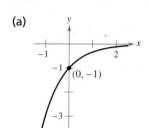

(b)

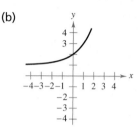

(c)

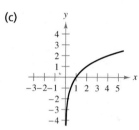

(d)

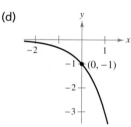

(e)

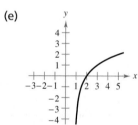

(f)

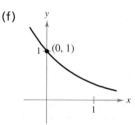

(g)

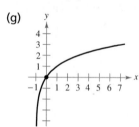

(h)

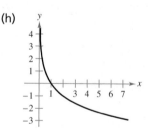

1. $f(x) = -3^x$

2. $f(x) = 3^{-x}$

3. $f(x) = -3^{-x}$

4. $f(x) = 2^x + 1$

5. $f(x) = \log_2 x$

6. $f(x) = \log_2(x - 1)$

7. $f(x) = -\log_2 x$

8. $f(x) = \log_2(x + 1)$

In Exercises 9–16, sketch the graph of the function.

9. $f(x) = 4^x$

10. $f(x) = 4^{x-1}$

11. $f(x) = \left(\frac{1}{2}\right)^x$

12. $f(x) = \left(\frac{1}{2}\right)^{x+1}$

13. $f(x) = 3e^{0.2x}$

14. $f(x) = 10e^{-0.1x}$

15. $f(x) = 3^{-x^2}$

16. $f(x) = 2^{1-x^2}$

Compound Interest In Exercises 17 and 18, complete the table to find the balance A for P dollars invested at a rate r for t years, compounded n times per year.

n	1	2	4	12	365	Continuous
A						

17. $P = \$5000$, $r = 8.75\%$, $t = 12$ years

18. $P = \$8000$, $r = 6.5\%$, $t = 25$ years

Compound Interest In Exercises 19 and 20, complete the table to determine the amount P that should be invested at a rate r to produce a final balance of $A = \$200,000$ in t years.

t	1	10	20	30	40	50
P						

19. $r = 7.5\%$, compounded continuously

20. $r = 9.5\%$, compounded quarterly

21. Investment Plan   You deposit \$6000 in a fund that yields 5.75% interest, compounded continuously. How much money will be in the fund after 6 years?

22. Population   The population P of a town increases according to the model

$$P(t) = 15,000e^{0.025t}$$

where t is the time in years, with $t = 8$ corresponding to 2008. Use the model to approximate the population in 2009 and 2011.

In Exercises 23–26, use the definition of a logarithm to write the equation in logarithmic form.

23. $4^3 = 64$

24. $25^{3/2} = 125$

25. $e^2 = 7.3890\ldots$

26. $e^x = 8$

In Exercises 27–30, use the definition of a logarithm to write the equation in exponential form.

27. $\log_3 81 = 4$

28. $\log_5 0.2 = -1$

29. $\ln 1 = 0$

30. $\ln 4 = 1.3862\ldots$

In Exercises 31–36, evaluate the expression without using a calculator.

31. $\log_2 32$

32. $\log_9 3$

33. $\ln e^7$

34. $\log_4 \frac{1}{4}$

35. $\ln e^{-1/2}$

36. $\ln 1$

In Exercises 37 and 38, use the fact that f and g are inverse functions of each other to sketch their graphs in the same coordinate plane.

37. $f(x) = 10^x$, $g(x) = \log_{10} x$

38. $f(x) = e^x$, $g(x) = \ln x$

In Exercises 39–42, find the domain, vertical asymptote, and x-intercept of the logarithmic function. Then sketch its graph.

39. $f(x) = \log_2(x - 3)$

40. $f(x) = 5 - 2 \log_{10} x$

41. $g(x) = 2 \ln x$

42. $g(x) = \ln(4 - x)$

43. Human Memory Model   Students in a sociology class were given an exam and then retested monthly for 6 months with an equivalent exam. The average score for the class is given by the human memory model

$$f(t) = 82 - 16 \log_{10}(t + 1), \quad 0 \le t \le 6$$

where t is the time (in months). How did the average score change over the six-month period?

44. Investment Time   A principal P, invested at 5.85% interest and compounded continuously, increases to an amount that is K times the principal after t years, where t is given by

$$t = \frac{\ln K}{0.0585}.$$

Complete the table and describe the result.

K	1	2	3	4	6	8	10
t							

45. Antler Spread   The antler spread a (in inches) and shoulder height h (in inches) of an adult American elk are related by the model

$$h = 116 \log_{10}(a + 40) - 176.$$

(a) Approximate the shoulder height of an elk with an antler spread of 55 inches.

(T) (b) Use a graphing utility to graph the model.

46. Snow Removal   The number of miles s of roads cleared of snow is approximated by the model

$$s = 25 - \frac{13 \ln(h/12)}{\ln 3}, \quad 2 \le h \le 15$$

where h is the depth of the snow in inches.

(a) Use the model to find s when $h = 10$ inches.

(T) (b) Use a graphing utility to graph the model.

In Exercises 47–50, evaluate the logarithm using the change-of-base formula. Do each problem twice, once with common logarithms and once with natural logarithms. (Round your answer to three decimal places.)

47. $\log_3 10$

48. $\log_{1/4} 7$

49. $\log_{12} 200$

50. $\log_3 0.28$

In Exercises 51–54, approximate the logarithm using the properties of logarithms, given $\log_b 2 \approx 0.3562$, $\log_b 3 \approx 0.5646$, and $\log_b 5 \approx 0.8271$.

51. $\log_b 6$

52. $\log_b \frac{4}{25}$

53. $\log_b \sqrt{3}$

54. $\log_b 30$

In Exercises 55–58, find the *exact* value of the logarithm.

55. $\log_7 49$

56. $\log_6 \frac{1}{36}$

57. $\ln e^{3.2}$

58. $\ln \sqrt[5]{e^3}$

In Exercises 59–64, use the properties of logarithms to expand the expression as a sum, difference, and/or multiple of logarithms. (Assume that all variables are positive.)

59. $\log_{10} \dfrac{x}{y}$

60. $\log_{10} \dfrac{xy^3}{z^2}$

61. $\ln\left(x\sqrt{x - 3}\right)$

62. $\ln \sqrt[3]{\dfrac{x^3}{y^2}}$

63. $\log_5(y - 3)^4$

64. $\log_2 2xy^2 z$

In Exercises 65–70, condense the expression to the logarithm of a single quantity.

65. $\log_4 2 + \log_4 3$

66. $\ln y + 2 \ln z$

67. $\frac{1}{2} \ln x$

68. $4 \log_3 x + \log_3 y - 2 \log_3 z$

69. $\ln x - \ln(x - 3) - \ln(x + 1)$

70. $\log_{10}(x + 2) + 2 \log_{10} x - 3 \log_{10}(x + 4)$

71. Curve Fitting   Find a logarithmic equation that relates y and x (see figure).

x	1	2	3	4	5	6
y	1	2.520	4.327	6.350	8.550	10.903

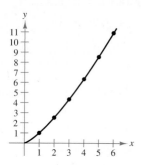

72. Human Memory Model   Students in a learning theory study were given an exam and then retested monthly for 6 months with an equivalent exam. The average scores for the class are shown in the table, with $t = 1$ representing 1 month after the initial exam. Use the table to find a logarithmic equation that relates s and t.

Month, t	Score, s
1	87.9
2	79.7
3	74.8
4	71.3
5	68.6
6	66.5

In Exercises 73–86, solve the equation. Approximate the result to three decimal places.

73. $e^x = 8$

74. $2e^{x+1} = 7$

75. $3^{4x-1} - 4 = 23$

76. $2^{3x+1} + 5 = 133$

77. $e^{2x} - 3e^x - 4 = 0$

78. $e^{2x} - 8e^x + 12 = 0$

79. $\ln 3x = 8.2$

80. $2 \log_3 4x = 15$

81. $-2 + \ln 5x = 0$

82. $\ln 4x^2 = 21$

83. $\ln x - \ln 3 = 2$

84. $\log_3 x - \log_3 4 = 5$

85. $\log_2 \sqrt[3]{x + 1} = 1$

86. $\ln \sqrt{x + 1} = 2$

87. Demand Function   The demand function for a desk is given by

$$p = 6000\left(1 - \frac{5}{5 + e^{-0.004x}}\right).$$

Find the demand x for each price p.

(a) $p = \$500$

(b) $p = \$400$.

88. Demand Function   The demand function for a bicycle is given by

$$p = 4000\left(1 - \frac{3}{3 + e^{-0.004x}}\right).$$

Find the demand x for each price p.

(a) $p = \$700$

(b) $p = \$400$.

Radioactive Decay   In Exercises 89 and 90, complete the table for the radioactive isotope.

Isotope	Half-Life (Years)	Initial Quantity	Amount After 1000 Years
89. ^{14}C	5715	12 g	
90. ^{239}Pu	24,100		3.1 g

91. Population   The population P of a city is given by

$$P = 185,000e^{0.018t}$$

where t represents the year, with $t = 8$ corresponding to 2008.

(T) (a) Use a graphing utility to graph this equation.

(b) Use the model to predict the year in which the population of the city will reach 250,000.

92. Population   The population P of a city is given by

$$P = 50,000e^{kt}$$

where t represents the year, with $t = 0$ corresponding to 2000. In 1990, the population was 34,500.

(a) Find the value of k and use this result to predict the population in the year 2030.

(T) (b) Use a graphing utility to confirm the result of part (a).

93. Bacteria Growth   The number of bacteria N in a culture is given by the model

$$N = 250e^{kt}$$

where t is the time (in hours), with $t = 0$ corresponding to the time when $N = 250$. When $t = 6$, there are 380 bacteria. How long does it take the bacteria population to double in size? To triple in size?

94. Bacteria Growth   The number of bacteria N in a culture is given by the model

$$N = 200e^{kt}$$

where t is the time (in hours), with $t = 0$ corresponding to the time when $N = 200$. When $t = 5$, there are 325 bacteria. How long does it take for the bacteria population to double in size? To triple in size?

95. Learning Curve   The management at a factory has found that the maximum number of units a worker can produce in a day is 50. The learning curve for the number of units N produced per day after a new employee has worked t days is given by

$$N = 50(1 - e^{kt}).$$

After 20 days on the job, a particular worker produced 31 units in 1 day.

(a) Find the learning curve for this worker.

(b) How many days should pass before this worker is producing 45 units per day?

96. Test Scores   The scores on a general aptitude test roughly follow a normal distribution given by

$$y = 0.0040e^{-[(x-300)^2]/20,000}, \quad 100 \le x \le 500.$$

Sketch the graph of this function. Estimate the average score on this test.

97. Wildlife Management   A state parks and wildlife department releases 100 deer into a wilderness area. The department believes that the carrying capacity of the area is 400 deer and that the growth of the herd will be modeled by the logistic curve

$$P = \frac{400}{1 + 3e^{-kt}}, \quad t \ge 0$$

where P is the number of deer and t is the time (in years).

(a) The herd size is 135 after 2 years. Find k.

(b) Find the populations after 5 years, after 10 years, and after 20 years.

98. Earthquake Magnitudes   On the Richter scale, the magnitude R of an earthquake of intensity I is given by

$$R = \log_{10} \frac{I}{I_0}$$

where $I_0 = 1$ is the minimum intensity used for comparison. Find the intensity per unit of area for each value of R.

(a) $R = 8.4$ (b) $R = 6.85$ (c) $R = 9.1$

99. Thawing a Package of Steaks   You take a package of steaks out of a freezer at 10 A.M. and place it in the refrigerator. Will the steaks be thawed in time to be grilled at 6 P.M.? Assume that the refrigerator temperature is 40°F and the freezer temperature is 0°F. Use the formula

$$t = -3.95 \ln \frac{T - 40}{0 - 40}$$

where t is the time in hours (with $t = 0$ corresponding to 10 A.M.) and T is the temperature of the package of steaks (in degrees Fahrenheit).

(T) **100.** *MAKE A DECISION: COSTCO REVENUES*   The annual revenues R (in millions of dollars) for the Costco Wholesale Corporation from 1996 to 2005 are shown in the table.   *(Source: Costco Wholesale Corporation)*

Year	Revenue, R
1996	19,566
1997	21,874
1998	24,270
1999	27,456
2000	32,164
2001	34,797
2002	38,762
2003	42,546
2004	48,107
2005	52,935

(a) Use a graphing utility to create a scatter plot of the data. Let t represent the year, with $t = 6$ corresponding to 1996.

(b) Use the *regression* feature of a graphing utility to find an exponential model for the data. Use the Inverse Property $b = e^{\ln b}$ to rewrite the model as an exponential model in base e.

(c) Use the *regression* feature of a graphing utility to find a logarithmic model for the data.

(d) Use the exponential model in base e and the logarithmic model to predict revenues in 2006 and in 2007. It is projected that revenues in 2006 and in 2007 will be $59,050 million and $64,500 million. Do the predictions from the two models agree with these projections? Explain.

(e) Use the exponential model in base e and the logarithmic model to predict revenues in 2009, 2010, and 2011. It is projected that revenue will reach $81,000 million during the period from 2009 to 2011. Does the prediction from each model agree with this projection? Explain.

Chapter Test

See www.CalcChat.com for worked-out solutions to odd-numbered exercises.

Take this test as you would take a test in class. When you are done, check your work against the answers given in the back of the book.

In Exercises 1–4, sketch the graph of the function.

1. $y = 2^x$

2. $y = e^{-2x}$

3. $y = \ln x$

4. $y = \log_3(x - 1)$

In Exercises 5 and 6, students in a psychology class were given an exam and then retested monthly with an equivalent exam. The average score for the class is given by the human memory model

$$f(t) = 87 - 15 \log_{10}(t + 1), \quad 0 \le t \le 4$$

where t is the time (in months).

5. What was the average score on the original exam? After 2 months? After 4 months?

6. The students in this psychology class participated in a study that required that they continue taking an equivalent exam every 6 months for 2 years. Use the model to predict the average score after 12 months and after 18 months. What could this indicate about human memory?

In Exercises 7–10, expand the logarithmic expression.

7. $\ln \dfrac{x^2 y^3}{z}$

8. $\log_{10} 3xyz^2$

9. $\log_2\left(x \sqrt[3]{x - 2}\right)$

10. $\log_8 \sqrt[5]{x^2 + 1}$

In Exercises 11 and 12, condense the logarithmic expression.

11. $2 \ln x + 3 \ln y - \ln z$

12. $\frac{2}{3}(\log_{10} x + \log_{10} y)$

In Exercises 13–16, solve the equation. Approximate the result to three decimal places.

13. $2^{4x} = 21$

14. $e^{2x} - 8e^x + 12 = 0$

15. $\log_2(x + 1) - 7 = 0$

16. $\ln \sqrt{x + 2} = 3$

17. You deposit $40,000 in a fund that pays 6.75% interest, compounded continuously. When will the balance be greater than $120,000?

18. The population P of a city is given by

$$P = 85,000e^{0.025t}$$

where t represents the year, with $t = 8$ corresponding to 2008. When will the city have a population of 125,000? Explain.

19. The number of bacteria N in a culture is given by

$$N = 100e^{kt}$$

where t is the time (in hours), with $t = 0$ corresponding to the time when $N = 100$. When $t = 8$, $N = 175$. How long does it take the bacteria population to double?

20. Carbon 14 has a half-life of 5715 years. You have an initial quantity of 10 grams. How many grams will remain after 10,000 years? After 20,000 years?

21. If you are given the annual bear population on a small Alaskan island for the past decade, would you expect the bear population to grow exponentially or logistically? Explain your reasoning.

Take this test as you would take a test in class. When you are done, check your work against the answers given in the back of the book.

In Exercises 1–6, use the functions given by $f(x) = x^2 + 1$ and $g(x) = 3x - 5$ to find the indicated function.

1. $(f + g)(x)$ **2.** $(f - g)(x)$ **3.** $(fg)(x)$

4. $\left(\dfrac{f}{g}\right)(x)$ **5.** $(f \circ g)(x)$ **6.** $(g \circ f)(x)$

In Exercises 7–11, sketch the graph of the function. Describe the domain and range of the function.

7. $f(x) = (x - 2)^2 + 3$ **8.** $g(x) = \dfrac{2}{x - 3}$ **9.** $h(x) = 2^{-x}$

10. $f(x) = \log_3(x + 1)$ **11.** $g(x) = \begin{cases} x + 5, & x < 0 \\ 5, & x = 0 \\ x^2 + 5, & x > 0 \end{cases}$

12. The profit P (in dollars) for a software company is given by

$$P = -0.001x^2 + 150x - 175{,}000$$

where x is the number of units produced. What production level will yield a maximum profit?

In Exercises 13–15, perform the indicated operation and write the result in standard form.

13. $(5 + 3i)(6 - 5i)$ **14.** $(4 + 5i)^2$

15. Write the quotient in standard form: $\dfrac{1 - 3i}{5 + 2i}$.

16. Use the Quadratic Formula to solve $3x^2 - 5x + 7 = 0$.

17. Find all the zeros of

$$f(x) = x^4 + 17x^2 + 16$$

given that $4i$ is a zero. Explain your reasoning.

18. Use long division or synthetic division to divide.

   (a) $(6x^3 - 4x^2) \div (2x^2 + 1)$
   (b) $(2x^4 + 3x^3 - 6x + 5) \div (x + 2)$

In Exercises 19 and 20, solve the equation. Approximate the result to three decimal places.

19. $e^{2x} - 3e^x - 18 = 0$

20. $\frac{1}{3} \ln(x - 3) = 4$

(T) **21.** The IQ scores for adults roughly follow the normal distribution given by

$$y = 0.0266e^{-(x - 100)^2/450}, \quad 70 \le x \le 130$$

where x is the IQ score. Use a graphing utility to graph the function. From the graph, estimate the average IQ score.

5 Systems of Equations and Inequalities

Paul Grebliunas/Getty Images

Ancient Greeks and Romans used naturally occurring substances to control insects and protect crops. Today, farmers use chemicals to protect crops from insects. You can use a system of equations to find the amounts of chemicals needed to obtain a desired mixture. (See Section 5.3, Exercise 57.)

Applications

Systems of equations and inequalities are used to model and solve many real-life applications. The applications listed below represent a sample of the applications in this chapter.

- Atmosphere, Exercise 62, page 426
- Peregrine Falcons, Exercise 64, page 450
- Investments, Exercise 47, page 460

Section 5.1

Solving Systems Using Substitution

- Solve a system of equations by the method of substitution.
- Solve a system of equations graphically.
- Construct and use a system of equations to solve an application problem.

The Method of Substitution

Up to this point in the text, most problems have involved either a function of one variable or a single equation in two variables. However, many problems in science, business, and engineering involve two or more equations in two or more variables. To solve such a problem, you need to find the solutions of a **system of equations.** Here is an example of a system of two equations in x and y.

$$\begin{cases} 2x + y = 5 & \text{Equation 1} \\ 3x - 2y = 4 & \text{Equation 2} \end{cases}$$

A **solution** of this system is an ordered pair that satisfies each equation in the system. For instance, the ordered pair $(2, 1)$ is a solution of this system. To check this, you can substitute 2 for x and 1 for y in *each* equation.

$$2x + y = 5 \qquad \text{Write Equation 1.}$$
$$2(2) + 1 \overset{?}{=} 5 \qquad \text{Substitute 2 for } x \text{ and 1 for } y.$$
$$4 + 1 = 5 \qquad \text{Solution checks in Equation 1.} \checkmark$$

$$3x - 2y = 4 \qquad \text{Write Equation 2.}$$
$$3(2) - 2(1) \overset{?}{=} 4 \qquad \text{Substitute 2 for } x \text{ and 1 for } y.$$
$$6 - 2 = 4 \qquad \text{Solution checks in Equation 2.} \checkmark$$

Finding the set of all solutions is called **solving the system of equations.** There are several different ways to solve systems of equations. In this chapter, you will study four of the most common techniques: *the method of substitution, the graphical approach, the method of elimination,* and *Gaussian elimination.* This section begins with the **method of substitution.**

Method of Substitution

1. *Solve* one of the equations for one variable in terms of the other.

2. *Substitute* the expression found in Step 1 into the other equation to obtain an equation in one variable.

3. *Solve* the equation obtained in Step 2.

4. *Back-substitute* the value found in Step 3 into the expression obtained in Step 1 to find the value of the other variable.

5. *Check* that the solution satisfies *each* of the original equations.

✓ **CHECKPOINT 1**

Solve the system of equations.

$$\begin{cases} x + y = 6 \\ x - y = 4 \end{cases}$$ ■

When using the method of substitution to solve a system of equations, it does not matter which variable you solve for first. You will obtain the same solution regardless. When making your choice, you should choose the variable that is easier to work with. For instance, solve for a variable that has a coefficient of 1 or -1 to avoid working with fractions.

Example 1 **Solving a System of Two Equations by Substitution**

Solve the system of equations.

$$\begin{cases} x + y = 4 & \text{Equation 1} \\ x - y = 2 & \text{Equation 2} \end{cases}$$

SOLUTION Begin by solving for y in Equation 1.

$$y = 4 - x \qquad \text{Revised Equation 1}$$

Next, substitute this expression for y into Equation 2 and solve the resulting single-variable equation for x.

$x - y = 2$	Write Equation 2.
$x - (4 - x) = 2$	Substitute $4 - x$ for y.
$x - 4 + x = 2$	Distributive Property
$2x = 6$	Combine like terms.
$x = 3$	Divide each side by 2.

Finally, you can solve for y by *back-substituting* $x = 3$ into the equation $y = 4 - x$, to obtain

$y = 4 - x$	Write revised Equation 1.
$y = 4 - 3$	Substitute 3 for x.
$y = 1$	Solve for y.

The solution is the ordered pair $(3, 1)$. You can check this as follows.

CHECK

Substitute $(3, 1)$ into Equation 1:

$x + y = 4$	Write Equation 1.
$3 + 1 \stackrel{?}{=} 4$	Substitute for x and y.
$4 = 4$	Solution checks in Equation 1. ✓

Substitute $(3, 1)$ into Equation 2:

$x - y = 2$	Write Equation 2.
$3 - 1 \stackrel{?}{=} 2$	Substitute for x and y.
$2 = 2$	Solution checks in Equation 2.

The term *back-substitution* implies that you work *backwards*. First you solve for one of the variables, and then you substitute that value *back* into one of the equations in the system to find the value of the other variable.

Example 2 **Solving a System by Substitution**

A total of $12,000 is invested in two funds paying 9% and 11% simple interest. (Recall that the formula for simple interest is $I = Prt$, where P is the principal, r is the annual interest rate, and t is time.) The total annual interest is $1180. How much is invested at each rate?

SOLUTION

Verbal Model:

$$\boxed{\begin{array}{c}9\% \\ \text{fund}\end{array}} + \boxed{\begin{array}{c}11\% \\ \text{fund}\end{array}} = \boxed{\begin{array}{c}\text{Total} \\ \text{investment}\end{array}}$$

$$\boxed{\begin{array}{c}9\% \\ \text{interest}\end{array}} + \boxed{\begin{array}{c}11\% \\ \text{interest}\end{array}} = \boxed{\begin{array}{c}\text{Total} \\ \text{interest}\end{array}}$$

Labels:

Amount in 9% fund $= x$	(dollars)
Interest for 9% fund $= 0.09x$	(dollars)
Amount in 11% fund $= y$	(dollars)
Interest for 11% fund $= 0.11y$	(dollars)
Total investment $= 12,000$	(dollars)
Total interest $= 1180$	(dollars)

System:
$$\begin{cases} x + \quad y = 12{,}000 & \text{Equation 1} \\ 0.09x + 0.11y = \quad 1180 & \text{Equation 2} \end{cases}$$

To begin, it is convenient to multiply each side of Equation 2 by 100. This eliminates the need to work with decimals.

$$100(0.09x + 0.11y) = 100(1180) \qquad \text{Multiply each side by 100.}$$

$$9x + 11y = 118{,}000 \qquad \text{Revised Equation 2}$$

To solve this system, you can begin by solving for x in Equation 1.

$$x = 12{,}000 - y \qquad \text{Revised Equation 1}$$

Then, substitute this expression for x into revised Equation 2 and solve the resulting equation for y.

$$9x + 11y = 118{,}000 \qquad \text{Write revised Equation 2.}$$

$$9(12{,}000 - y) + 11y = 118{,}000 \qquad \text{Substitute } 12{,}000 - y \text{ for } x.$$

$$108{,}000 - 9y + 11y = 118{,}000 \qquad \text{Distributive Property}$$

$$2y = 10{,}000 \qquad \text{Combine like terms.}$$

$$y = 5000 \qquad \text{Divide each side by 2.}$$

Next, back-substitute the value $y = 5000$ to solve for x.

$$x = 12{,}000 - y \qquad \text{Write revised Equation 1.}$$

$$x = 12{,}000 - 5000 \qquad \text{Substitute 5000 for } y.$$

$$x = 7000 \qquad \text{Solve for } x.$$

The solution is (7000, 5000). So, $7000 is invested at 9% and $5000 is invested at 11%. Check this in the original system.

✓ CHECKPOINT 2

In Example 2, suppose a total of $15,000 is invested in the same two funds. The total annual interest is $1420. How much is invested at each rate? ■

The equations in Examples 1 and 2 are linear. The method of substitution can also be used to solve systems in which one or both of the equations are nonlinear.

Example 3 Substitution: Two-Solution Case

Solve the system of equations.

$$\begin{cases} x^2 - x - y = 1 & \text{Equation 1} \\ -x + y = -1 & \text{Equation 2} \end{cases}$$

SOLUTION Begin by solving for y in Equation 2 to obtain $y = x - 1$. Next, substitute this expression for y into Equation 1 and solve for x.

$x^2 - x - y = 1$	Write Equation 1.
$x^2 - x - (x - 1) = 1$	Substitute for y.
$x^2 - 2x + 1 = 1$	Simplify.
$x^2 - 2x = 0$	General form
$x(x - 2) = 0$	Factor.
$x = 0$	Set 1st factor equal to 0.
$x - 2 = 0 \implies x = 2$	Set 2nd factor equal to 0.

Back-substituting these values of x to solve for the corresponding values of y produces the two solutions $(0, -1)$ and $(2, 1)$. Check these solutions in the original system.

✓ **CHECKPOINT 3**

Solve the system of equations.

$$\begin{cases} x^2 + 4x - y = 7 \\ 2x - y = -1 \end{cases}$$ ▩

When using the method of substitution, you may encounter an equation that has no solution, as shown in Example 4.

Example 4 Substitution: No-Real-Solution Case

Solve the system of equations.

$$\begin{cases} -x + y = 4 & \text{Equation 1} \\ x^2 + y = 3 & \text{Equation 2} \end{cases}$$

SOLUTION Begin by solving for y in Equation 1 to obtain $y = x + 4$. Next, substitute this expression for y into Equation 2 and solve for x.

$x^2 + y = 3$	Write Equation 2.
$x^2 + (x + 4) = 3$	Substitute $x + 4$ for y.
$x^2 + x + 1 = 0$	Simplify.
$x = \dfrac{-1 \pm \sqrt{1^2 - 4(1)(1)}}{2(1)}$	Use the Quadratic Formula.
$x = \dfrac{-1 \pm \sqrt{-3}}{2}$	Simplify.

✓ **CHECKPOINT 4**

Solve the system of equations.

$$\begin{cases} 2x^2 - y = 1 \\ x + y = -2 \end{cases}$$ ▩

Because the discriminant is negative, the equation $x^2 + x + 1 = 0$ has no (real) solution. So, this system has no (real) solution.

Graphical Approach to Finding Solutions

From Examples 1, 3, and 4, you can see that a system of two equations in two unknowns can have exactly one solution, more than one solution, or no solution. In practice, you can gain insight about the location and number of solutions of a system of equations by graphing each of the equations in the same coordinate plane. The solution(s) of the system correspond to the **point(s) of intersection** of the graphs. For instance, the graph of the system in Example 1 is two lines with a *single point* of intersection, as shown in Figure 5.1(a). The graph of the system in Example 3 is a parabola and a line with *two points* of intersection, as shown in Figure 5.1(b). The graph of the system in Example 4 is a line and a parabola that have *no points* of intersection, as shown in Figure 5.1(c).

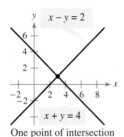

One point of intersection

(a) One solution

Two points of intersection

(b) Two solutions

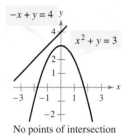

No points of intersection

(c) No solution

FIGURE 5.1

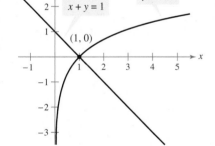

FIGURE 5.2

✓ CHECKPOINT 5

Solve the system of equations.

$$\begin{cases} 2x + 2y = 8 \\ \quad\quad y = \ln \frac{1}{4}x \end{cases}$$ ■

Example 5   **Solving a System of Equations Graphically**

Solve the system of equations.

$$\begin{cases} \quad\quad y = \ln x & \text{Equation 1} \\ x + y = 1 & \text{Equation 2} \end{cases}$$

SOLUTION   The graph of each equation is shown in Figure 5.2. From the graph, you can see that there is only one point of intersection. So, it appears that $(1, 0)$ is the solution point. You can confirm this by substituting 1 for x and 0 for y in *both* equations.

CHECK

Equation 1: $0 = \ln 1$ ✓

Equation 2: $1 + 0 = 1$ ✓

TECHNOLOGY

Ⓣ   Your graphing utility may have an *intersect* feature that approximates the point(s) of intersection of two graphs. Use the *intersect* feature to verify the solution of Example 5. For instructions on how to use the *intersect* feature, see Appendix A; for specific keystrokes, go to the text website at *college.hmco.com/info/larsonapplied.*

© Jeff Greenberg/Alamy

In 2005, the average price for cross-training shoes in the United States was $46.34. *(Source: National Sporting Goods Association)*

Applications

The total cost C of producing x units of a product typically has two components—the initial cost and the cost per unit. When enough units have been sold so that the total revenue R equals the total cost C, the sales are said to have reached the **break-even point.** You will find that the break-even point corresponds to the point of intersection of the cost and revenue graphs.

Example 6   Break–Even Analysis

A shoe company invests $300,000 in equipment to produce cross-training shoes. Each pair of shoes costs $3 to produce and is sold for $60. How many pairs of shoes must be sold before the business breaks even?

SOLUTION   The total cost of producing x units is

$$\frac{\text{Total}}{\text{cost}} = \frac{\text{Cost per}}{\text{unit}} \cdot \frac{\text{Number}}{\text{of units}} + \frac{\text{Initial}}{\text{cost}}$$

$$C = 3x + 300{,}000.$$

The total revenue obtained by selling x units is

$$\frac{\text{Total}}{\text{revenue}} = \frac{\text{Price per}}{\text{unit}} \cdot \frac{\text{Number}}{\text{of units}}$$

$$R = 60x.$$

Because the break-even point occurs when $R = C$, you have $C = 60x$, and the system of equations to solve is

$$\begin{cases} C = 3x + 300{,}000 & \text{Equation 1} \\ C = 60x & \text{Equation 2} \end{cases}.$$

Now you can solve by substitution.

$$60x = 3x + 300{,}000 \qquad \text{Substitute } 60x \text{ for } C \text{ in Equation 1.}$$

$$57x = 300{,}000 \qquad \text{Subtract } 3x \text{ from each side.}$$

$$x = \frac{300{,}000}{57} \qquad \text{Divide each side by 57.}$$

$$x \approx 5263 \qquad \text{Use a calculator.}$$

The company must sell about 5263 pairs of shoes to break even. Note in Figure 5.3 that sales less than the break-even point correspond to an overall loss, whereas sales greater than the break-even point correspond to a profit.

FIGURE 5.3

✓ CHECKPOINT 6

In Example 6, suppose each pair of shoes costs $5 to produce. How many pairs of shoes must be sold before the business breaks even? ▪

Another way to view the solution in Example 6 is to consider the profit function $P = R - C$. The break-even point occurs when the profit is 0, which is the same as saying that $R = C$.

Example 7
MAKE A DECISION Long–Distance Phone Plans

You are choosing between two long-distance telephone plans. Plan A charges $0.05 per minute plus a basic monthly fee of $7.50. Plan B charges $0.075 per minute plus a basic monthly fee of $4.25. After how many long-distance minutes are the costs of the two plans equal? Which plan should you choose if you use 100 long-distance minutes each month?

SOLUTION Models for each long-distance phone plan are

$$C = 0.05m + 7.5 \qquad \text{Plan A}$$
$$C = 0.075m + 4.25 \qquad \text{Plan B}$$

where C is the monthly phone cost and m is the number of monthly long-distance minutes used. (See Figure 5.4.) Because the first equation has already been solved for C in terms of m, substitute this value into the second equation and solve for m, as follows.

$$0.05m + 7.5 = 0.075m + 4.25$$
$$0.05m - 0.075m = 4.25 - 7.5$$
$$-0.025m = -3.25$$
$$m = 130$$

So, the costs of the two plans are equal after 130 long-distance minutes. Because Plan B costs less than Plan A when you use less than 130 long-distance minutes, you should choose Plan B.

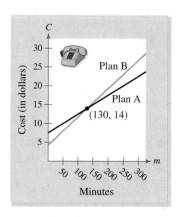

FIGURE 5.4

✓ CHECKPOINT 7

In Example 7, suppose Plan A charges $0.045 per minute plus a basic monthly fee of $7.49. Which plan should you choose if you use 150 long-distance minutes each month? ▪

CONCEPT CHECK

1. The ordered pair $(2, -3)$ is a solution to $x - 2y = 8$. Give values of a, b, and c so that $(2, -3)$ is a solution of the system
$$\begin{cases} x - 2y = 8 \\ ax + by = c \end{cases}$$
where a, b, and c are real numbers.

2. When solving a system of quadratic equations using substitution, the resulting equation is not factorable. Explain your next step.

3. A system of equations consists of a linear equation and a cubic equation. what is the greatest number of possible solutions? Explain.

4. Explain why you can set a cost equation equal to a revenue equation when finding the break-even point.

Skills Review 5.1

The following warm-up exercises involve skills that were covered in earlier sections. You will use these skills in the exercise set for this section. For additional help, review Sections 0.5, 1.1, 1.3, and 2.1.

In Exercises 1–4, sketch the graph of the equation.

1. $y = -\frac{1}{3}x + 6$ **2.** $y = 2(x - 3)$

3. $x^2 + y^2 = 4$ **4.** $y = 5 - (x - 3)^2$

In Exercises 5–8, perform the indicated operations and simplify.

5. $(3x + 2y) - 2(x + y)$ **6.** $(-10u + 3v) + 5(2u - 8v)$

7. $x^2 + (x - 3)^2 + 6x$ **8.** $y^2 - (y + 1)^2 + 2y$

In Exercises 9 and 10, solve the equation.

9. $3x + (x - 5) = 15 + 4$ **10.** $y^2 + (y - 2)^2 = 2$

Exercises 5.1   See www.CalcChat.com for worked-out solutions to odd-numbered exercises.

In Exercises 1–6, determine whether each ordered pair is a solution of the system of equations.

1. $\begin{cases} x + 4y = -3 \\ 5x - y = 6 \end{cases}$

  (a) $(-1, -1)$

  (b) $(1, -1)$

2. $\begin{cases} 2x - y = 2 \\ x + 3y = 8 \end{cases}$

  (a) $(2, 1)$

  (b) $(2, 2)$

3. $\begin{cases} 2x + 5y = -5 \\ 2x - y^2 = 1 \end{cases}$

  (a) $(5, -3)$

  (b) $(0, -1)$

4. $\begin{cases} 4x^2 + y = 3 \\ -x - y = 11 \end{cases}$

  (a) $(-2, -9)$

  (b) $(2, -13)$

5. $\begin{cases} y = -2e^x \\ 3x - y = 2 \end{cases}$

  (a) $(-2, 0)$

  (b) $(-1, 2)$

6. $\begin{cases} -\log_{10} x + 3 = y \\ \frac{1}{9}x + y = \frac{28}{9} \end{cases}$

  (a) $(1, 3)$

  (b) $\left(9, \frac{37}{9}\right)$

In Exercises 7–16, solve the system by the method of substitution. Then use the graph to confirm your solution.

7. $\begin{cases} x + y = -1 \\ -2x + y = -7 \end{cases}$

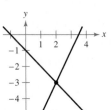

8. $\begin{cases} x - y = -5 \\ x + 2y = 4 \end{cases}$

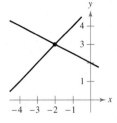

9. $\begin{cases} x - y = -3 \\ x^2 - y = -1 \end{cases}$

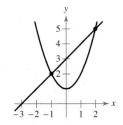

10. $\begin{cases} x^2 - y = 0 \\ x^2 - 4x + y = 0 \end{cases}$

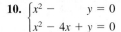

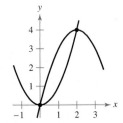

11. $\begin{cases} x - y = 0 \\ x^3 - 5x + y = 0 \end{cases}$

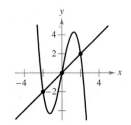

12. $\begin{cases} y = x^3 - 3x^2 + 3 \\ 2x + y = 3 \end{cases}$

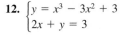

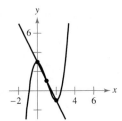

13. $\begin{cases} 3x + y = 4 \\ x^2 + y^2 = 16 \end{cases}$ **14.** $\begin{cases} 3x - 4y = 18 \\ x^2 + y^2 = 36 \end{cases}$

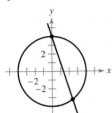   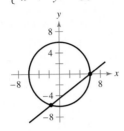

In Exercises 39–50, solve the system graphically.

39. $\begin{cases} -x + 2y = 2 \\ 3x + y = 15 \end{cases}$ **40.** $\begin{cases} x + y = 0 \\ 3x - 2y = 10 \end{cases}$

41. $\begin{cases} x - 3y = -2 \\ 5x + 3y = 17 \end{cases}$ **42.** $\begin{cases} -x + 2y = 1 \\ x - y = 2 \end{cases}$

43. $\begin{cases} x + y = 4 \\ x^2 + y^2 - 4x = 0 \end{cases}$

44. $\begin{cases} -x + y = 3 \\ x^2 - 6x - 27 + y^2 = 0 \end{cases}$

45. $\begin{cases} x - y + 3 = 0 \\ x^2 - 4x + 7 = y \end{cases}$

46. $\begin{cases} y^2 - 4x + 11 = 0 \\ -\frac{1}{2}x + y = -\frac{1}{2} \end{cases}$

47. $\begin{cases} 7x + 8y = 24 \\ x - 8y = 8 \end{cases}$ **48.** $\begin{cases} x - y = 0 \\ 5x - 2y = 6 \end{cases}$

49. $\begin{cases} 3x - 2y = 0 \\ x^2 - y^2 = 4 \end{cases}$ **50.** $\begin{cases} 2x - y + 3 = 0 \\ x^2 + y^2 - 4x = 0 \end{cases}$

15. $\begin{cases} y = -x^2 + 1 \\ y = x^2 - 1 \end{cases}$ **16.** $\begin{cases} y = x^2 - 3x - 4 \\ y = -x^2 + 3x + 4 \end{cases}$

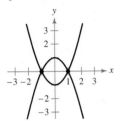   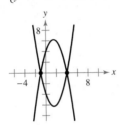

In Exercises 17–38, solve the system by the method of substitution.

17. $\begin{cases} 2x - y = -3 \\ -3x - 4y = -1 \end{cases}$ **18.** $\begin{cases} x + 2y = 1 \\ 5x - 4y = -23 \end{cases}$

19. $\begin{cases} 2x - y + 2 = 0 \\ 4x + y - 5 = 0 \end{cases}$ **20.** $\begin{cases} 6x - 3y - 4 = 0 \\ x + 2y - 4 = 0 \end{cases}$

21. $\begin{cases} x - y = 7 \\ 2x + y = 23 \end{cases}$ **22.** $\begin{cases} x - 2y = -2 \\ 3x - y = 6 \end{cases}$

23. $\begin{cases} 0.3x - 0.4y - 0.33 = 0 \\ 0.1x + 0.2y - 0.21 = 0 \end{cases}$

24. $\begin{cases} 1.5x + 0.8y = 2.3 \\ 0.3x - 0.2y = 0.1 \end{cases}$

25. $\begin{cases} \frac{1}{5}x + \frac{1}{2}y = 8 \\ x + y = 20 \end{cases}$ **26.** $\begin{cases} \frac{1}{2}x + \frac{3}{4}y = 10 \\ \frac{3}{2}x - y = 4 \end{cases}$

27. $\begin{cases} 6x + 5y = -3 \\ -x - \frac{5}{6}y = -7 \end{cases}$ **28.** $\begin{cases} -\frac{2}{3}x + y = 2 \\ 2x - 3y = 6 \end{cases}$

29. $\begin{cases} y = 2x \\ y = x^2 - 1 \end{cases}$ **30.** $\begin{cases} x + y = 4 \\ x^2 - y = 2 \end{cases}$

31. $\begin{cases} 3x - 7y + 6 = 0 \\ x^2 - y^2 = 4 \end{cases}$ **32.** $\begin{cases} x^2 + y^2 = 25 \\ 2x + y = 10 \end{cases}$

33. $\begin{cases} x - 2y = 4 \\ x^2 - y = 0 \end{cases}$ **34.** $\begin{cases} x^2 + y^2 = 9 \\ x - y = -5 \end{cases}$

35. $\begin{cases} y = x^4 - 2x^2 + 1 \\ y = 1 - x^2 \end{cases}$ **36.** $\begin{cases} y = x^3 - 2x^2 + x - 1 \\ y = -x^2 + 3x - 1 \end{cases}$

37. $\begin{cases} xy - 2 = 0 \\ y = \sqrt{x - 1} \end{cases}$ **38.** $\begin{cases} xy = 3 \\ y = \sqrt{x - 2} \end{cases}$

In Exercises 51–56, use a graphing utility to determine whether the system of equations has *one solution*, *two solutions*, or *no solution*.

51. $\begin{cases} y = -5x + 1 \\ y = x + 3 \end{cases}$ **52.** $\begin{cases} -\frac{1}{2}x + y = -1 \\ 7x + y = 2 \end{cases}$

53. $\begin{cases} y = x^2 + 2x - 1 \\ y = 2x + 5 \end{cases}$

54. $\begin{cases} x^2 + 3x + y = 4 \\ 3x + y = -5 \end{cases}$

55. $\begin{cases} y = x^2 + 3x + 7 \\ y = -x^2 - 3x + 1 \end{cases}$

56. $\begin{cases} -10x + y = 2 \\ -10x + y = -3 \end{cases}$

In Exercises 57–64, use a graphing utility to find the point(s) of intersection of the graphs. Then confirm your solution algebraically.

57. $\begin{cases} y = x^2 + 3x - 1 \\ y = -x^2 - 2x + 2 \end{cases}$ **58.** $\begin{cases} y = -2x^2 + x - 1 \\ y = x^2 - 2x - 1 \end{cases}$

59. $\begin{cases} x - y + 3 = 0 \\ x^2 - 4x + 7 = y \end{cases}$ **60.** $\begin{cases} x - y = 3 \\ x - y^2 = 1 \end{cases}$

61. $\begin{cases} y = e^x \\ x - y + 1 = 0 \end{cases}$ **62.** $\begin{cases} y = \sqrt{x} \\ y = x \end{cases}$

63. $\begin{cases} 4x^2 - y^2 - 32x - 2y = -59 \\ 2x + y - 7 = 0 \end{cases}$

64. $\begin{cases} x^2 + y^2 = 8 \\ y = x^2 + 4 \end{cases}$

Break-Even Analysis In Exercises 65–68, find the sales necessary to break even ($R = C$) for the cost C of producing x units and the revenue R obtained by selling x units. (Round your answer to the nearest whole unit.)

65. $C = 8650x + 250,000$; $R = 9950x$

66. $C = 5.5\sqrt{x} + 10,000$; $R = 3.29x$

67. $C = 2.65x + 350,000$; $R = 4.15x$

68. $C = 0.08x + 50,000$; $R = 0.25x$

69. Break-Even Analysis You invest $18,000 in equipment to make CDs. The CDs can be produced for $1.95 each and will be sold for $13.95 each. How many CDs must you sell to break even?

70. Break-Even Analysis You invest $3000 in a fishing lure business. A lure costs $1.06 to produce and will be sold for $5.86. How many lures must you sell to break even?

71. Comparing Populations From 1995 to 2005, the population of Kentucky grew more slowly than that of Colorado. Models that represent the populations of the two states are given by

$$\begin{cases} P = 27.9t + 3757 & \text{Kentucky} \\ P = 86.1t + 3425 & \text{Colorado} \end{cases}$$

where P is the population (in thousands) and t represents the year, with $t = 5$ corresponding to 1995. Use the models to estimate when the population of Colorado first exceeded the population of Kentucky. *(Source: U.S. Census Bureau)*

72. Comparing Populations From 1995 to 2005, the population of Maryland grew more slowly than that of Arizona. Models that represent the populations of the two states are given by

$$\begin{cases} P = 55.6t + 4771 & \text{Maryland} \\ P = 145.9t + 3703 & \text{Arizona} \end{cases}$$

where P is the population (in thousands) and t represents the year, with $t = 5$ corresponding to 1995. Use the models to estimate when the population of Arizona first exceeded the population of Maryland. *(Source: U.S. Census Bureau)*

(T) 73. Body Mass Index Body mass index (BMI) is a measure of body fat based on height and weight. The 75th percentile BMI for females, ages 9 to 20, grew more slowly than that of males of the same age range. Models that represent the 75th percentile BMI for males and females, ages 9 to 20, are given by

$$\begin{cases} B = 0.73a + 11 & \text{Males} \\ B = 0.61a + 12.8 & \text{Females} \end{cases}$$

where B is the BMI (kg/m^2) and a represents the age, with $a = 9$ corresponding to 9 years old. Use a graphing utility to determine whether the BMI for males will exceed the BMI for females. *(Source: National Center for Health Statistics)*

(T) 74. Clothing Sales From 1996 to 2005, the sales of Abercrombie & Fitch Company grew faster than those of Timberland Company. Models that represent the sales of the two companies are given by

$$\begin{cases} S = 235.1t - 1126 & \text{Abercrombie \& Fitch Company} \\ S = 97.7t + 88 & \text{Timberland Company} \end{cases}$$

where S is the sales (in millions) and t represents the year, with $t = 6$ corresponding to 1996. Use a graphing utility to determine whether the sales of Abercrombie & Fitch Company will exceed the sales of Timberland Company. *(Source: Abercrombie & Fitch Company and Timberland Company)*

75. A total of $35,000 is invested in two funds paying 8.5% and 12% simple interest. The total annual interest is $3675. How much is invested at each rate?

76. A total of $35,000 is invested in two funds paying 8% and 10.5% simple interest. The total annual interest is $3275. How much is invested at each rate?

77. Job Choices You are offered two different jobs. Company A offers an annual salary of $30,000 plus a year-end bonus of 2.5% of your total sales. Company B offers a salary of $24,000 plus a year-end bonus of 6.5% of your total sales. What is the amount you must sell in one year to earn the same salary working for either company?

78. Camping You are choosing between camping outfitters. Outfitter A charges a reservation fee of $150 plus a daily guide fee of $70. Outfitter B charges a reservation fee of $75 plus a daily guide fee of $90. Estimate when the cost of Outfitter A equals the cost of Outfitter B.

(T) 79. Financial Aid The average award for Federal Pell Grants and Federal Perkins Loans from 1995 to 2005 can be approximated by

$$\begin{cases} A = -2.051t^3 + 56.87t^2 - 376.7t + 2238 & \text{Federal Pell Grant} \\ A = -1.810t^3 + 56.64t^2 - 476.4t + 2711 & \text{Federal Perkins Loan} \end{cases}$$

where A is the award (in dollars) and t represents the year, with $t = 5$ corresponding to 1995. Use a graphing utility to determine whether Federal Perkins Loan awards will exceed Federal Pell Grant awards. Do you think these models will continue to be accurate? Explain your reasoning. *(Source: U.S. Department of Education)*

(T) 80. SAT or ACT? The number of participants in SAT and ACT testing from 1995 to 2005 can be approximated by

$$\begin{cases} y = 0.68t^2 + 28.1t + 903 & \text{SAT} \\ y = -0.485t^3 + 14.88t^2 - 115.1t + 1201 & \text{ACT} \end{cases}$$

where y is the number of participants (in thousands) and t represents the year, with $t = 5$ corresponding to 1995. Use a graphing utility to determine whether the number of participants in ACT testing will exceed the number of participants in SAT testing. Do you think these models will continue to be accurate? Explain your reasoning. *(Source: College Board; ACT, Inc.)*

Section 5.2

Solving Systems Using Elimination

- Solve a linear system by the method of elimination.
- Interpret the solution of a linear system graphically.
- Construct and use a linear system to solve an application problem.

The Method of Elimination

In Section 5.1, you studied two methods for solving a system of equations: substitution and graphing. In this section, you will study a third method called the **method of elimination.** The key step in the method of elimination is to obtain, for one of the variables, coefficients that differ only in sign, so that *adding* the two equations eliminates this variable. The following system provides an example.

$$3x + 5y = 7 \qquad \text{Equation 1}$$

$$\underline{-3x - 2y = -1} \qquad \text{Equation 2}$$

$$3y = 6 \qquad \text{Add equations.}$$

Note that by adding the two equations, you eliminate the variable x and obtain a single equation in y. Solving this equation for y produces $y = 2$, which you can then back-substitute into one of the original equations to solve for x.

Example 1   The Method of Elimination

Solve the system of linear equations.

$$\begin{cases} 3x + 2y = 4 & \text{Equation 1} \\ 5x - 2y = 8 & \text{Equation 2} \end{cases}$$

SOLUTION   Because the coefficients of the y-terms differ only in sign, you can eliminate the y-terms by adding the two equations. This leaves you with a single equation in x.

$$3x + 2y = 4 \qquad \text{Write Equation 1.}$$

$$\underline{5x - 2y = 8} \qquad \text{Write Equation 2.}$$

$$8x = 12 \qquad \text{Add equations.}$$

So, $x = \frac{3}{2}$. By back-substituting this value into Equation 1, you can solve for y.

$$3x + 2y = 4 \qquad \text{Write Equation 1.}$$

$$3\left(\tfrac{3}{2}\right) + 2y = 4 \qquad \text{Substitute } \tfrac{3}{2} \text{ for } x.$$

$$y = -\tfrac{1}{4} \qquad \text{Solve for } y.$$

The solution is $\left(\frac{3}{2}, -\frac{1}{4}\right)$. Check this in the original system.

✓ CHECKPOINT 1

Solve the system of linear equations.

$$\begin{cases} 2x - 3y = 5 \\ 5x + 3y = 9 \end{cases} \blacksquare$$

STUDY TIP

The method of substitution can also be used to solve the system in Example 1. Use substitution to solve the system. Which method do you think is easier?

To obtain coefficients (for one of the variables) that differ only in sign, you may need to multiply one or both of the equations by a suitable constant, as demonstrated in Example 2.

Example 2 The Method of Elimination

Solve the system of linear equations.

$$\begin{cases} 2x - 3y = -7 & \text{Equation 1} \\ 3x + y = -5 & \text{Equation 2} \end{cases}$$

SOLUTION For this system, you can obtain coefficients that differ only in sign by multiplying Equation 2 by 3. Then, by adding the two equations, you can eliminate the y-terms. This leaves you with a single equation in x.

$$
\begin{array}{llll}
2x - 3y = -7 & \Longrightarrow & 2x - 3y = -7 & \text{Write Equation 1.} \\
\underline{3x + y = -5} & \Longrightarrow & \underline{9x + 3y = -15} & \text{Multiply Equation 2 by 3.} \\
& & 11x = -22 & \text{Add equations.}
\end{array}
$$

By dividing each side by 11, you can see that $x = -2$. By back-substituting this value of x into Equation 1, you can solve for y.

$$
\begin{array}{ll}
2x - 3y = -7 & \text{Write Equation 1.} \\
2(-2) - 3y = -7 & \text{Substitute } -2 \text{ for } x. \\
-3y = -3 & \text{Add 4 to each side.} \\
y = 1 & \text{Solve for } y.
\end{array}
$$

The solution is $(-2, 1)$. Check this in the original system, as follows.

CHECK

$$
\begin{array}{ll}
2(-2) - 3(1) \overset{?}{=} -7 & \text{Substitute into Equation 1.} \\
-4 - 3 = -7 & \text{Equation 1 checks.} \checkmark \\
3(-2) + 1 \overset{?}{=} -5 & \text{Substitute into Equation 2.} \\
-6 + 1 = -5 & \text{Equation 2 checks.} \checkmark
\end{array}
$$

✓ CHECKPOINT 2

Solve the system of linear equations.

$$\begin{cases} 3x - 5y = -1 \\ x - 2y = -1 \end{cases} \blacksquare$$

In Example 2, the two systems of linear equations

$$\begin{cases} 2x - 3y = -7 \\ 3x + y = -5 \end{cases} \quad \text{and} \quad \begin{cases} 2x - 3y = -7 \\ 9x + 3y = -15 \end{cases}$$

are called **equivalent systems** because they have precisely the same solution set. The operations that can be performed on a system of linear equations to produce an equivalent system are (1) interchanging any two equations, (2) multiplying an equation by a nonzero constant, and (3) adding a multiple of one equation to any other equation in the system.

The Method of Elimination

To use the **method of elimination** to solve a system of two linear equations in x and y, use the following steps.

1. Examine the system to determine which variable is easiest to eliminate.

2. Obtain coefficients of x (or y) that differ only in sign by multiplying all terms of one or both equations by suitably chosen constants.

3. Add the equations to eliminate one variable and solve the resulting equation.

4. Back-substitute the value obtained in Step 3 into either of the original equations and solve for the other variable.

5. Check your solution in both of the original equations.

Example 3 **The Method of Elimination**

Solve the system of linear equations.

$$\begin{cases} 5x + 3y = 9 & \text{Equation 1} \\ 2x - 4y = 14 & \text{Equation 2} \end{cases}$$

SOLUTION You can obtain coefficients of y that differ only in sign by multiplying Equation 1 by 4 and multiplying Equation 2 by 3.

$$5x + 3y = 9 \quad \Longrightarrow \quad 20x + 12y = 36 \qquad \text{Multiply Equation 1 by 4.}$$

$$\underline{2x - 4y = 14} \quad \Longrightarrow \quad \underline{6x - 12y = 42} \qquad \text{Multiply Equation 2 by 3.}$$

$$26x = 78 \qquad \text{Add equations.}$$

From this equation, you can see that $x = 3$. By back-substituting this value of x into Equation 2, you can solve for y, as follows.

$$2x - 4y = 14 \qquad \text{Write Equation 2.}$$

$$2(3) - 4y = 14 \qquad \text{Substitute 3 for } x.$$

$$-4y = 8 \qquad \text{Subtract 6 from each side.}$$

$$y = -2 \qquad \text{Solve for } y.$$

The solution is $(3, -2)$. Check this in the original system. —————

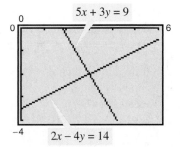

FIGURE 5.5

✓ CHECKPOINT 3

Solve the system of linear equations.

$$\begin{cases} -2x + 7y = -15 \\ 3x - 5y = 6 \end{cases} \blacksquare$$

Remember that you can check the solution of a system of equations graphically. For instance, to check the solution found in Example 3, graph both equations in the same viewing window, as shown in Figure 5.5. Notice that the two lines intersect at $(3, -2)$.

Example 4 illustrates a strategy for solving a system of linear equations that has decimal coefficients.

Example 4 A Linear System Having Decimal Coefficients

Solve the system of linear equations.

$$\begin{cases} 0.02x - 0.05y = -0.38 & \text{Equation 1} \\ 0.03x + 0.04y = 1.04 & \text{Equation 2} \end{cases}$$

SOLUTION Because the coefficients in this system have two decimal places, you can begin by multiplying each equation by 100. (This produces a system in which the coefficients are all integers.)

$$\begin{cases} 2x - 5y = -38 & \text{Revised Equation 1} \\ 3x + 4y = 104 & \text{Revised Equation 2} \end{cases}$$

Now, to obtain coefficients of x that differ only in sign, multiply revised Equation 1 by 3 and revised Equation 2 by -2.

$$2x - 5y = -38 \implies 6x - 15y = -114 \quad \text{Multiply by 3.}$$
$$3x + 4y = 104 \implies -6x - 8y = -208 \quad \text{Multiply by } -2.$$
$$\overline{-23y = -322} \quad \text{Add equations.}$$

So, you can conclude that $y = \dfrac{-322}{-23} = 14$. Now, back-substitute $y = 14$ into any of the original or revised equations of the system that contain the variable y. Back-substituting this value into revised Equation 2 produces the following.

$3x + 4y = 104$	Write revised Equation 2.
$3x + 4(14) = 104$	Substitute 14 for y.
$3x = 48$	Subtract 56 from each side.
$x = 16$	Solve for x.

The solution is $(16, 14)$. Check this in the original system. ▔▔▔▔▔

✓ CHECKPOINT 4

Solve the system of linear equations.

$$\begin{cases} 0.03x + 0.04y = -0.13 \\ -0.04x + 0.05y = -0.24 \end{cases} \blacksquare$$

DISCOVERY

Rewrite each system of equations in slope-intercept form and graph the system using a graphing utility. What is the relationship between the slopes of the two lines and the number of points of intersection?

a. $\begin{cases} 2x + 4y = 8 \\ 4x - 3y = -6 \end{cases}$ **b.** $\begin{cases} -x + 5y = 15 \\ 2x - 10y = -7 \end{cases}$ **c.** $\begin{cases} x - y = 9 \\ 2x - 2y = 18 \end{cases}$

Graphical Interpretation of Solutions

STUDY TIP

Keep in mind that the terminology and methods discussed in this section and the following section apply only to systems of linear equations.

It is possible for a *general* system of equations to have exactly one solution, two or more solutions, or no solution. If a system of *linear* equations has two different solutions, it must have an *infinite* number of solutions. To see why this is true, consider the following graphical interpretations of systems of two linear equations in two variables. (Remember that the graph of a linear equation in two variables is a line.)

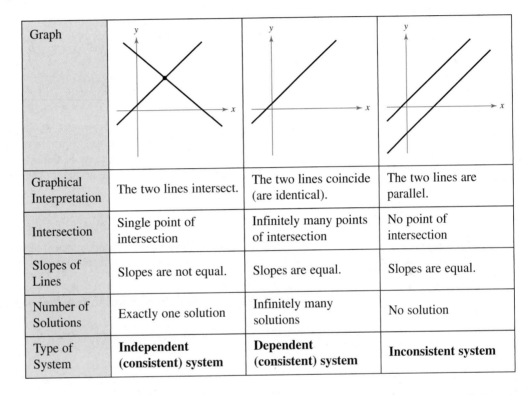

Graph			
Graphical Interpretation	The two lines intersect.	The two lines coincide (are identical).	The two lines are parallel.
Intersection	Single point of intersection	Infinitely many points of intersection	No point of intersection
Slopes of Lines	Slopes are not equal.	Slopes are equal.	Slopes are equal.
Number of Solutions	Exactly one solution	Infinitely many solutions	No solution
Type of System	**Independent (consistent) system**	**Dependent (consistent) system**	**Inconsistent system**

A system of linear equations is *consistent* if it has at least one solution. A consistent system with exactly one solution is *independent*, whereas a consistent system with infinitely many solutions is *dependent*. A system is *inconsistent* if it has no solution.

From the graphs above, you can see that a comparison of the slopes and *y*-intercepts of two lines is helpful in determining the number of solutions of the corresponding system of equations. For instance:

Independent (consistent) systems have lines with slopes that are not equal.
Dependent (consistent) systems have lines with equal slopes and the same *y*-intercept.
Inconsistent systems have lines with equal slopes, but different *y*-intercepts.

So, when solving a system of linear equations graphically, it is helpful to know the slope of each line. Writing each linear equation in the slope-intercept form

$$y = mx + b \qquad \text{Slope-intercept form}$$

enables you to identify the slopes quickly.

In Examples 5 and 6, note how you can use the method of elimination to determine that a linear system has no solution or infinitely many solutions.

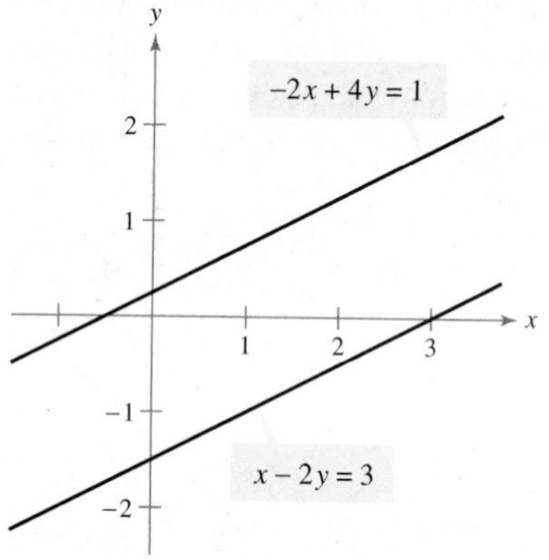

Example 5   The Method of Elimination: No–Solution Case

Solve the system of linear equations.

$$\begin{cases} x - 2y = 3 & \text{Equation 1} \\ -2x + 4y = 1 & \text{Equation 2} \end{cases}$$

SOLUTION   Obtain coefficients that differ only in sign, as follows.

$$x - 2y = 3 \implies 2x - 4y = 6 \qquad \text{Multiply Equation 1 by 2.}$$
$$\underline{-2x + 4y = 1} \implies \underline{-2x + 4y = 1} \qquad \text{Write Equation 2.}$$
$$0 = 7 \qquad \text{False statement}$$

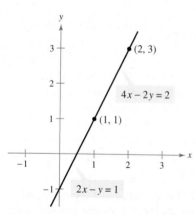

FIGURE 5.6   No Solution

Because there are no values of x and y for which $0 = 7$, you can conclude that the system is inconsistent and has no solution. The graphs of the equations are shown in Figure 5.6. Note that the two lines have equal slopes, but different y-intercepts. Therefore, the lines are parallel and have no point of intersection.

✓ **CHECKPOINT 5**

Solve the system of linear equations.

$$\begin{cases} -3x + 6y = 5 \\ x - 2y = 2 \end{cases}$$ ▨

In Example 5, note that the occurrence of a false statement, such as $0 = 7$, indicates that the system has no solution. In the next example, note that the occurrence of a statement that is true for all values of the variables, such as $0 = 0$, indicates that the system has infinitely many solutions.

Example 6   The Method of Elimination: Many–Solutions Case

Solve the system of linear equations.

$$\begin{cases} 2x - y = 1 & \text{Equation 1} \\ 4x - 2y = 2 & \text{Equation 2} \end{cases}$$

SOLUTION   Obtain coefficients that differ only in sign, as follows.

$$2x - y = 1 \implies 2x - y = 1 \qquad \text{Write Equation 1.}$$
$$\underline{4x - 2y = 2} \implies \underline{-2x + y = -1} \qquad \text{Multiply Equation 2 by } -\tfrac{1}{2}.$$
$$0 = 0 \qquad \text{Add equations.}$$

Because the two equations are equivalent (have the same solution set), you can conclude that the system is consistent and has infinitely many solutions. The solution set consists of all points (x, y) lying on the line $2x - y = 1$, as shown in Figure 5.7. To represent the solution set as an ordered pair, let $x = a$, where a is any real number. Then $y = 2a - 1$ and the solution set can be written as $(a, 2a - 1)$.

FIGURE 5.7   Infinite Number of Solutions

✓ **CHECKPOINT 6**

Solve the system of linear equations.

$$\begin{cases} -x - y = 5 \\ 4x + 4y = -20 \end{cases}$$ ▨

Applications

At this point, you may be asking the question, "How can I tell which application problems can be solved using a system of linear equations?" The answer comes from the following considerations.

1. Does the problem involve more than one unknown quantity?

2. Are there two (or more) equations or conditions to be satisfied?

If one or both of these conditions occur, the appropriate mathematical model for the problem may be a system of linear equations. Example 7 shows how to construct such a model.

Example 7   An Application of a Linear System

An airplane flying into a headwind travels the 2000-mile flying distance between Wilmington, Delaware and Tucson, Arizona in 4 hours and 24 minutes. On the return flight, the same distance is traveled in 4 hours. Find the air speed of the plane and the speed of the wind, assuming that both remain constant.

SOLUTION   The two unknown quantities are the speeds of the wind and the plane. If r_1 is the air speed of the plane and r_2 is the speed of the wind, then

$$r_1 - r_2 = \text{speed of the plane } against \text{ the wind}$$

$$r_1 + r_2 = \text{speed of the plane } with \text{ the wind}$$

as shown in Figure 5.8. Using the formula distance = (rate)(time) for these two speeds, you obtain the following equations.

$$2000 = (r_1 - r_2)\left(4 + \frac{24}{60}\right)$$

$$2000 = (r_1 + r_2)(4)$$

These two equations simplify as follows.

$$\begin{cases} 5000 = 11r_1 - 11r_2 & \text{Equation 1} \\ 500 = r_1 + r_2 & \text{Equation 2} \end{cases}$$

To solve this system by elimination, multiply Equation 2 by 11.

$5000 = 11r_1 - 11r_2$	$5000 = 11r_1 - 11r_2$	Write Equation 1.
$500 = r_1 + r_2$	$5500 = 11r_1 + 11r_2$	Multiply Equation 2 by 11.
	$10{,}500 = 22r_1$	Add equations.

The solution is

$$r_1 = \frac{10{,}500}{22} = \frac{5250}{11} \approx 477.27$$

$$r_2 = 500 - \frac{5250}{11} = \frac{250}{11} \approx 22.73$$

So, the air speed of the plane is about 477.27 miles per hour and the speed of the wind is about 22.73 miles per hour. Check this solution in the original statement of the problem.

Original flight

$r_1 - r_2$

Return flight

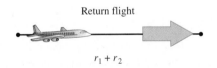

$r_1 + r_2$

FIGURE 5.8

✓ CHECKPOINT 7

In Example 7, suppose the return flight takes 4 hours and 6 minutes. Find the air speed of the plane and the speed of the wind, assuming that both remain constant. ■

In a free market, the demands for many products are related to the prices of the products. As the prices decrease, the demands by consumers increase and the amounts that producers are able or willing to supply decrease.

Example 8   Finding the Point of Equilibrium

The demand and supply equations for a DVD are given by

$$\begin{cases} p = 35 - 0.0001x & \text{Demand equation} \\ p = 8 + 0.0001x & \text{Supply equation} \end{cases}$$

where p is the price (in dollars) and x represents the number of DVDs. For how many units will the quantity demanded equal the quantity supplied? What price corresponds to this value?

SOLUTION   To obtain coefficients of p that differ only in sign, multiply the demand equation by -1.

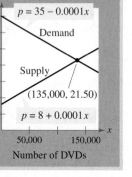

FIGURE 5.9

$$\begin{array}{llll} p = 35 - 0.0001x & \Longrightarrow & -p = -35 + 0.0001x & \text{Multiply demand equation by } -1. \\ \underline{p = 8 + 0.0001x} & \Longrightarrow & \underline{p = 8 + 0.0001x} & \text{Write supply equation.} \\ & & 0 = -27 + 0.0002x & \text{Add equations.} \end{array}$$

By solving the equation $0 = -27 + 0.0002x$, you get $x = 135,000$. So, the quantity demanded equals the quantity supplied for 135,000 units (see Figure 5.9). The price that corresponds to this x-value is obtained by back-substituting $x = 135,000$ into either of the original equations. For instance, back-substituting into the demand equation produces

$$p = 35 - 0.0001(135,000) = 35 - 13.5 = \$21.50.$$

Back-substitute $x = 135,000$ into the supply equation to see that you obtain the same price. The solution $(135,000, 21.50)$ is called the *point of equilibrium*. The **point of equilibrium** is the price p and the number of units x that satisfy both the demand and supply equations.

✓CHECKPOINT 8

In Example 8, suppose the supply equation is $p = 9 + 0.0001x$. Find the point of equilibrium. ■

CONCEPT CHECK

1. Two systems have infinitely many solutions. Are the systems equivalent? Explain.

2. Using the method of elimination, you reduce a system to $0 = 5$. What can you conclude about the system?

3. Can the graphs of the equations in an inconsistent system intersect? Explain.

4. Can the graphs of the equations in an independent (consistent) system have the same y-intercept? Can they have different y-intercepts? Explain.

Skills Review 5.2   The following warm-up exercises involve skills that were covered in earlier sections. You will use these skills in the exercise set for this section. For additional help, review Section 2.2.

In Exercises 1 and 2, sketch the graph of the equation.

1. $2x + y = 4$

2. $5x - 2y = 3$

In Exercises 3 and 4, find an equation of the line passing through the two points.

3. $(-1, 3), (4, 8)$

4. $(2, 6), (5, 1)$

In Exercises 5 and 6, determine the slope of the line.

5. $3x + 6y = 4$

6. $7x - 4y = 10$

In Exercises 7–10, determine whether the lines represented by the pair of equations are parallel, perpendicular, or neither.

7. $2x - 3y = -10$
$3x + 2y = 11$

8. $4x - 12y = 5$
$-2x + 6y = 3$

9. $5x + y = 2$
$3x + 2y = 1$

10. $x - 3y = 2$
$6x + 2y = 4$

Exercises 5.2

See www.CalcChat.com for worked-out solutions to odd-numbered exercises.

In Exercises 1–10, solve the system by elimination. Then use the graph to confirm your solution. Copy the graph and label each line with the appropriate equation.

1. $\begin{cases} 3x - 2y = 2 \\ x + 2y = 6 \end{cases}$

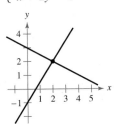

2. $\begin{cases} -x + 3y = 2 \\ x - 4y = -4 \end{cases}$

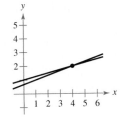

3. $\begin{cases} x - 4y = 2 \\ 2x + y = 4 \end{cases}$

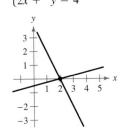

4. $\begin{cases} 2x - y = 2 \\ 4x + 3y = 24 \end{cases}$

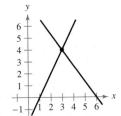

5. $\begin{cases} x - y = 1 \\ -2x + 2y = 5 \end{cases}$

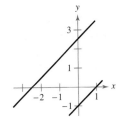

6. $\begin{cases} 3x + 2y = 2 \\ 6x + 4y = 14 \end{cases}$

7. $\begin{cases} 3x - 2y = 6 \\ -6x + 4y = -12 \end{cases}$

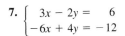

8. $\begin{cases} 2x + 4y = 8 \\ 6x + 12y = 24 \end{cases}$

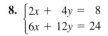
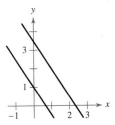
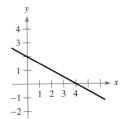

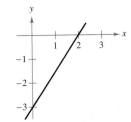

9. $\begin{cases} 9x - 3y = -1 \\ 3x + 6y = -5 \end{cases}$ **10.** $\begin{cases} 5x + 3y = 18 \\ 2x - 7y = -1 \end{cases}$

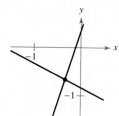

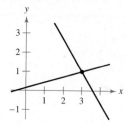

In Exercises 11–32, solve the system by elimination. Then state whether the system is consistent or inconsistent.

11. $\begin{cases} x + 2y = 3 \\ x - 2y = 1 \end{cases}$ **12.** $\begin{cases} 2x - 3y = 4 \\ -2x - y = 4 \end{cases}$

13. $\begin{cases} 4x - 3y = 11 \\ -6x + 3y = 3 \end{cases}$ **14.** $\begin{cases} 3x - 5y = 2 \\ 2x + 5y = 13 \end{cases}$

15. $\begin{cases} 3x - y = 17 \\ 5x + 5y = -5 \end{cases}$ **16.** $\begin{cases} x + 7y = 12 \\ 3x - 5y = 10 \end{cases}$

17. $\begin{cases} 3x + 2y = 10 \\ 2x + 5y = 3 \end{cases}$ **18.** $\begin{cases} 8r + 16s = 20 \\ 16r + 50s = 55 \end{cases}$

19. $\begin{cases} 2u + v = 120 \\ u + 2v = 120 \end{cases}$ **20.** $\begin{cases} 5u + 6v = 24 \\ 3u + 5v = 18 \end{cases}$

21. $\begin{cases} 4b + 3m = 3 \\ 3b + 11m = 13 \end{cases}$ **22.** $\begin{cases} 3b + 3m = 7 \\ 3b + 5m = 3 \end{cases}$

23. $\begin{cases} 6r - 5s = 3 \\ -1.2r + s = 0.5 \end{cases}$ **24.** $\begin{cases} 1.8x + 1.2y = 4 \\ 9x + 6y = 3 \end{cases}$

25. $\begin{cases} \dfrac{x}{4} + \dfrac{y}{6} = 1 \\ x - y = 3 \end{cases}$ **26.** $\begin{cases} \dfrac{1}{6}x - \dfrac{2}{3}y = 3 \\ 3x + y = 15 \end{cases}$

27. $\begin{cases} \dfrac{x+3}{4} + \dfrac{y-1}{3} = 1 \\ x - y = 3 \end{cases}$

28. $\begin{cases} \dfrac{x-1}{2} + \dfrac{y+2}{3} = 4 \\ x - 2y = 5 \end{cases}$

29. $\begin{cases} 2.5x - 3y = 1.5 \\ 10x - 12y = 6 \end{cases}$

30. $\begin{cases} 1.5x + 2y = 3.75 \\ 7.5x + 10y = 18.75 \end{cases}$

31. $\begin{cases} 0.05x - 0.03y = 0.21 \\ 0.07x + 0.02y = 0.16 \end{cases}$

32. $\begin{cases} 0.02x - 0.05y = -0.19 \\ 0.03x + 0.04y = 0.52 \end{cases}$

In Exercises 33 and 34, the graphs of the two equations appear to be parallel. Are they? Justify your answer by using elimination to solve the system.

33. $\begin{cases} 200y - x = 200 \\ 199y - x = -198 \end{cases}$ **34.** $\begin{cases} 25x - 24y = 0 \\ 13x - 12y = 120 \end{cases}$

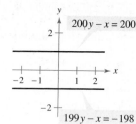

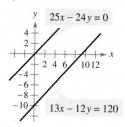

In Exercises 35–38, use the given statements to write a system of equations. Solve the system by elimination.

35. The sum of a number x and a number y is 13. The difference of x and y is 3.

36. The sum of a number a and a number b is 43. The difference of a and b is -27.

37. The sum of twice a number r and a number s is 8. The difference of r and s is 7.

38. The difference of a number m and twice a number n is 1. The sum of two times m and n is 22.

39. Airplane Speed An airplane flying into a headwind travels the 1800-mile flying distance between Los Angeles, California and South Bend, Indiana in 3 hours and 36 minutes. On the return flight, the distance is traveled in 3 hours. Find the air speed of the plane and the speed of the wind, assuming that both remain constant.

40. Airplane Speed Two planes start from the same airport and fly in opposite directions. The second plane starts $\frac{1}{2}$ hour after the first plane, but its speed is 50 miles per hour faster. Find the air speed of each plane if, 2 hours after the first plane departs, the planes are 2000 miles apart.

41. Acid Mixture Ten gallons of a 30% acid solution is obtained by mixing a 20% solution with a 50% solution. How much of each solution is required to obtain the specified concentration of the final mixture?

42. Fuel Mixture Five hundred gallons of 89-octane gasoline is obtained by mixing 87-octane gasoline with 92-octane gasoline. How much of each type of gasoline is required to obtain the specified mixture? (Octane ratings can be interpreted as percents.)

43. Investment Portfolio A total of $25,000 is invested in two corporate bonds that pay 9.5% and 14% simple interest. The total annual interest is $3050. How much is invested in each bond?

44. Investment Portfolio A total of $50,000 is invested in two municipal bonds that pay 6.75% and 8.25% simple interest. The total annual interest is $3900. How much is invested in each bond?

45. Ticket Sales You are the manager of a theater. On Saturday morning you are going over the ticket sales for Friday evening. A total of 740 tickets were sold. The tickets for adults and children sold for $8.50 and $4.00, respectively, and the total receipts for the performance were $4688. However, your assistant manager did not record how many of each type of ticket were sold. From the information you have, can you determine how many of each type were sold? Explain your reasoning.

46. Shoe Sales You are the manager of a shoe store. On Sunday morning you are going over the receipts for the previous week's sales. A total of 320 pairs of cross-training shoes were sold. One style sold for $56.95 and the other sold for $72.95. The total receipts were $21,024. The cash register that was supposed to keep track of the number of each type of shoe sold malfunctioned. Can you recover the information? If so, how many of each type were sold?

Supply and Demand In Exercises 47–50, find the point of equilibrium for the pair of demand and supply equations.

Demand	*Supply*
47. $p = 56 - 0.0001x$	$p = 22 + 0.00001x$
48. $p = 60 - 0.00001x$	$p = 15 + 0.00004x$
49. $p = 140 - 0.00002x$	$p = 80 + 0.00001x$
50. $p = 400 - 0.0002x$	$p = 225 + 0.0005x$

51. Restaurants The total sales y (in billions of dollars) for fast-food and full-service restaurants for the years 1999 to 2005 are shown in the table. *(Source: National Restaurant Association)*

Year	Fast-food	Full-service
1999	103.0	125.4
2000	107.1	133.8
2001	111.6	139.9
2002	115.1	141.9
2003	120.5	148.3
2004	129.4	157.0
2005	135.6	164.9

(S) (a) Use a spreadsheet software program to create a scatter plot of the data for fast-food sales and use the *regression* feature to find a linear model. Let x represent the year, with $x = 9$ corresponding to 1999. Repeat the procedure for the data for full-service sales.

(b) Assuming that the amounts for the given 7 years are representative of future years, will fast-food sales ever equal full-service sales?

52. Prescriptions The numbers of prescriptions y (in thousands) filled at two pharmacies in the years 2002 to 2008 are shown in the table.

Year	Pharmacy A	Pharmacy B
2002	18.1	19.5
2003	18.6	19.9
2004	19.2	20.4
2005	19.6	20.8
2006	20.0	21.1
2007	20.4	21.4
2008	21.3	22.0

(S) (a) Use a spreadsheet software program to create a scatter plot of the data for pharmacy A and use the *regression* feature to find a linear model. Let x represent the year, with $x = 2$ corresponding to 2002. Repeat the procedure for the data for pharmacy B.

(b) Assuming the amounts for the given 7 years are representative of future years, will the number of prescriptions filled at pharmacy A ever exceed the number of prescriptions filled at pharmacy B?

53. Supply and Demand The supply and demand equations for a small LCD television are given by

$$\begin{cases} p + 0.53x = 1542 & \text{Demand} \\ p - 0.37x = 300 & \text{Supply} \end{cases}$$

where p is the price (in dollars) and x represents the number of televisions. For how many units will the quantity demanded equal the quantity supplied? What price corresponds to this value?

54. Supply and Demand The supply and demand equations for a microscope are given by

$$\begin{cases} p + 0.85x = 650 & \text{Demand} \\ p - 0.4x = 75 & \text{Supply} \end{cases}$$

where p is the price (in dollars) and x represents the number of microscopes. For how many units will the quantity demanded equal the quantity supplied? What price corresponds to this value?

Fitting a Line to Data In Exercises 55–60, find the least squares regression line $y = ax + b$ for the points $(x_1, y_1), (x_2, y_2), \ldots, (x_n, y_n)$ by solving the system for a and b. (If you are unfamiliar with summation notation, look at the discussion in Section 7.1.)

$$\begin{cases} nb + \left(\sum_{i=1}^{n} x_i\right)a = \sum_{i=1}^{n} y_i \\ \left(\sum_{i=1}^{n} x_i\right)b + \left(\sum_{i=1}^{n} x_i^2\right)a = \sum_{i=1}^{n} x_i y_i \end{cases}$$

55. $\begin{cases} 5b + 10a = 20.2 \\ 10b + 30a = 50.1 \end{cases}$ **56.** $\begin{cases} 5b + 10a = 11.7 \\ 10b + 30a = 25.6 \end{cases}$

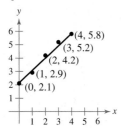
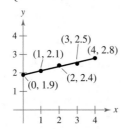
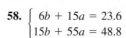

57. $\begin{cases} 7b + 21a = 35.1 \\ 21b + 91a = 114.2 \end{cases}$ **58.** $\begin{cases} 6b + 15a = 23.6 \\ 15b + 55a = 48.8 \end{cases}$

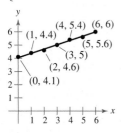
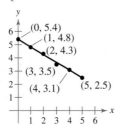

59. $(0, 4), (1, 3), (1, 1), (2, 0)$

60. $(1, 0), (2, 0), (3, 0), (3, 1), (4, 1), (4, 2), (5, 2), (6, 2)$

Ⓣ **61. Classic Cars** The numbers of cars y sold at Barrett-Jackson Collector Car Auction in Scottsdale in the years 2003 to 2007 are shown in the table. *(Source: Barrett-Jackson Auction Company)*

Year	2003	2004	2005	2006	2007
t	0	1	2	3	4
Cars, y	655	727	877	1105	1271

(a) Solve the following system for a and b to find the least squares regression line $y = at + b$ for the data. Let t represent the year, with $t = 0$ corresponding to 2003.

$$\begin{cases} 5b + 10a = 4635 \\ 10b + 30a = 10{,}880 \end{cases}$$

(b) Use a graphing utility to graph the regression line and estimate the number of cars that will be sold in 2009.

(c) Use the *regression* feature of a graphing utility to find a linear model for the data. Compare this model with the one you found in part (a).

Ⓣ **62. Atmosphere** The concentration y (in parts per million) of carbon dioxide in the atmosphere is measured at the Mauna Loa Observatory in Hawaii. The greatest monthly carbon dioxide concentrations for the years 2002 to 2006 are shown in the table. *(Source: Scripps CO2 Program)*

Year	t	Concentration, y
2002	0	375.55
2003	1	378.35
2004	2	380.63
2005	3	382.26
2006	4	384.92

(a) Solve the following system for a and b to find the least squares regression line $y = at + b$ for the data. Let t represent the year, with $t = 0$ corresponding to 2002.

$$\begin{cases} 5b + 10a = 1901.71 \\ 10b + 30a = 3826.07 \end{cases}$$

(b) Use a graphing utility to graph the regression line and predict the largest monthly carbon dioxide concentration in 2012.

(c) Use the *regression* feature of a graphing utility to find a linear model for the data. Compare this model with the one you found in part (a).

63. Reasoning Design a system of two linear equations with infinitely many solutions. Solve the system algebraically and explain how the solution indicates that there are infinitely many solutions.

64. Reasoning Design a system of two linear equations with no solution. Solve the system algebraically and explain how the solution indicates that there is no solution.

65. Think About It For the system below, find the value(s) of k for which the system is (a) inconsistent and (b) consistent (dependent). Explain how you found your answers.

$$\begin{cases} 3x - 12y = 9 \\ x - 4y = k \end{cases}$$

66. Think About It For the system in Exercise 65, can you find a value of k for which the system is consistent (independent)? Explain.

Section 5.3

Linear Systems in Three or More Variables

■ Solve a linear system in row-echelon form using back-substitution.
■ Use Gaussian elimination to solve a linear system.
■ Solve a nonsquare linear system.
■ Construct and use a linear system in three or more variables to solve an application problem.
■ Find the equation of a circle or a parabola using a linear system in three or more variables.

Row-Echelon Form and Back-Substitution

The method of elimination can be applied to a system of linear equations in more than two variables. In fact, this method easily adapts to computer use for solving linear systems with dozens of variables.

When elimination is used to solve a system of linear equations, the goal is to rewrite the system in a form to which back-substitution can be applied. To see how this works, consider the following two systems of linear equations.

$$\begin{cases} x - 2y + 3z = 9 \\ -x + 3y = -4 \\ 2x - 5y + 5z = 17 \end{cases}$$ System of Three Linear Equations in Three Variables

$$\begin{cases} x - 2y + 3z = 9 \\ y + 3z = 5 \\ z = 2 \end{cases}$$ Equivalent System in Row-Echelon Form

The second system is said to be in **row-echelon form,** which means that it has a "stair-step" pattern with leading coefficients of 1. After comparing the two systems, it should be clear that it is easier to solve the second system.

Example 1   Using Back-Substitution

Solve the system of linear equations.

$$\begin{cases} x - 2y + 3z = 9 & \text{Equation 1} \\ y + 3z = 5 & \text{Equation 2} \\ z = 2 & \text{Equation 3} \end{cases}$$

SOLUTION   From Equation 3, you know the value of z. To solve for y, substitute $z = 2$ into Equation 2 to obtain

$$y + 3(2) = 5 \quad \Longrightarrow \quad y = -1.$$

Finally, substitute $y = -1$ and $z = 2$ into Equation 1 to obtain

$$x - 2(-1) + 3(2) = 9 \quad \Longrightarrow \quad x = 1.$$

The solution is $x = 1$, $y = -1$, and $z = 2$, which can be written as the **ordered triple** $(1, -1, 2)$. Check this in the original system of equations.

✓**CHECKPOINT 1**

Solve the system of linear equations.

$$\begin{cases} 2x + y - 3z = 10 \\ y + z = 4 \\ z = 2 \end{cases}$$ ■

Gaussian Elimination

Two systems of equations are **equivalent** if they have the same solution set. To solve a system that is not in row-echelon form, first convert it to an *equivalent* system that is in row-echelon form. To see how this is done, let's take another look at the method of elimination, as applied to a system of two linear equations.

Example 2 The Method of Elimination

Solve the system of linear equations.

$$\begin{cases} 3x - 2y = -1 & \text{Equation 1} \\ x - y = 0 & \text{Equation 2} \end{cases}$$

SOLUTION An easy way of obtaining a leading coefficient of 1 is to interchange the two equations.

$$\begin{cases} x - y = 0 \\ 3x - 2y = -1 \end{cases}$$ Interchange two equations in the system.

$$\begin{cases} -3x + 3y = 0 \\ 3x - 2y = -1 \end{cases}$$ Multiply the first equation by -3.

$$\begin{array}{r} -3x + 3y = 0 \\ \underline{3x - 2y = -1} \\ y = -1 \end{array}$$ Add the multiple of the first equation to the second equation to obtain a new second equation.

$$\begin{cases} x - y = 0 \\ y = -1 \end{cases}$$ New system in row-echelon form

Now, using back-substitution, you can determine that the solution is $y = -1$ and $x = -1$, which can be written as the ordered pair $(-1, -1)$. Check this in the original system of equations.

✓ CHECKPOINT 2

Solve the system of linear equations.

$$\begin{cases} 2x + 4y = 1 \\ x + y = 0 \end{cases}$$ ■

The process of rewriting a system of equations in row-echelon form by using the three basic row operations is called **Gaussian elimination,** after the German mathematician Carl Friedrich Gauss. Example 2 shows the chain of equivalent systems used to solve a linear system in two variables.

Operations That Produce Equivalent Systems

Each of the following **row operations** on a system of linear equations produces an *equivalent* system of linear equations.

1. Interchange two equations.

2. Multiply one of the equations by a nonzero constant.

3. Add a multiple of one of the equations to another equation to replace the latter equation.

Example 3   **Using Elimination to Solve a System**

Solve the system of linear equations.

$$\begin{cases} x - 2y + 3z = 9 & \text{Equation 1} \\ -x + 3y = -4 & \text{Equation 2} \\ 2x - 5y + 5z = 17 & \text{Equation 3} \end{cases}$$

SOLUTION   Because the leading coefficient of Equation 1 is 1, you can begin by saving the x in the upper left position and eliminating the other x-terms from the first column.

$$\begin{cases} x - 2y + 3z = 9 \\ y + 3z = 5 \\ 2x - 5y + 5z = 17 \end{cases}$$

Adding the first equation to the second equation produces a new second equation.

$$\begin{cases} x - 2y + 3z = 9 \\ y + 3z = 5 \\ -y - z = -1 \end{cases}$$

Adding -2 times the first equation to the third equation produces a new third equation.

Now that all but the first x have been eliminated from the first column, work on the second column. (You need to eliminate y from the third equation.)

$$\begin{cases} x - 2y + 3z = 9 \\ y + 3z = 5 \\ 2z = 4 \end{cases}$$

Adding the second equation to the third equation produces a new third equation.

Finally, you need a coefficient of 1 for z in the third equation.

$$\begin{cases} x - 2y + 3z = 9 \\ y + 3z = 5 \\ z = 2 \end{cases}$$

Multiplying the third equation by $\frac{1}{2}$ produces a new third equation.

This is the same system that was solved in Example 1, and, as in that example, you can conclude that the solution is

$$x = 1, \quad y = -1, \quad \text{and} \quad z = 2.$$

✓ CHECKPOINT 3

Solve the system of linear equations.

$$\begin{cases} x + y + z = 6 \\ 2x - y + z = 3 \\ 3x - z = 0 \end{cases}$$ ∎

In Example 3, you can check the solution by substituting $x = 1$, $y = -1$, and $z = 2$ into each original equation, as follows.

Equation 1: $(1) - 2(-1) + 3(2) = 9$ ✓

Equation 2: $-(1) + 3(-1) = -4$ ✓

Equation 3: $2(1) - 5(-1) + 5(2) = 17$ ✓

The next example involves an inconsistent system—one that has no solution. The key to recognizing an inconsistent system is that at some stage in the elimination process, you obtain a false statement such as $0 = -2$.

Example 4 An Inconsistent System

Solve the system of linear equations.

$$\begin{cases} x - 3y + z = 1 & \text{Equation 1} \\ 2x - y - 2z = 2 & \text{Equation 2} \\ x + 2y - 3z = -1 & \text{Equation 3} \end{cases}$$

SOLUTION

$$\begin{cases} x - 3y + z = 1 \\ 5y - 4z = 0 \\ x + 2y - 3z = -1 \end{cases}$$

Adding -2 times the first equation to the second equation produces a new second equation.

$$\begin{cases} x - 3y + z = 1 \\ 5y - 4z = 0 \\ 5y - 4z = -2 \end{cases}$$

Adding -1 times the first equation to the third equation produces a new third equation.

$$\begin{cases} x - 3y + z = 1 \\ 5y - 4z = 0 \\ 0 = -2 \end{cases}$$

Adding -1 times the second equation to the third equation produces a new third equation.

Because $0 = -2$ is a false statement, you can conclude that this system is inconsistent and therefore has no solution. Moreover, because this system is equivalent to the original system, you can conclude that the original system also has no solution.

✓ CHECKPOINT 4

Solve the system of linear equations.

$$\begin{cases} 2x + y - z = 7 \\ x - 2y + 2z = -9 \\ 3x - y + z = 5 \end{cases}$$ ■

As with a system of linear equations in two variables, the solution(s) of a system of linear equations in more than two variables must fall into one of three categories. Because an equation in three variables represents a plane in space, the possible solutions can be shown graphically. See Figure 5.10.

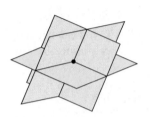

(a) Solution: one point

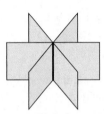

(b) Solution: one line

(c) Solution: one plane

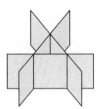

(d) Solution: none

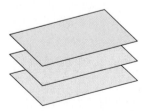

(e) Solution: none

FIGURE 5.10

The Number of Solutions of a Linear System

For a system of linear equations, exactly one of the following is true.

1. There is exactly one solution. [See Figure 5.10(a).]

2. There are infinitely many solutions. [See Figures 5.10(b) and (c).]

3. There is no solution. [See Figures 5.10(d) and (e).]

Example 5   **A System with Infinitely Many Solutions**

Solve the system of linear equations.

$$\begin{cases} x + y - 3z = -1 & \text{Equation 1} \\ y - z = 0 & \text{Equation 2} \\ -x + 2y = 1 & \text{Equation 3} \end{cases}$$

SOLUTION

$$\begin{cases} x + y - 3z = -1 \\ y - z = 0 \\ 3y - 3z = 0 \end{cases}$$

> Adding the first equation to the third equation produces a new third equation.

$$\begin{cases} x + y - 3z = -1 \\ y - z = 0 \\ 0 = 0 \end{cases}$$

> Adding -3 times the second equation to the third equation produces a new third equation.

This means that Equation 3 depends on Equations 1 and 2 in the sense that it gives us no additional information about the variables. Because $0 = 0$ is a true statement, you can conclude that this system has infinitely many solutions. So, the original system is equivalent to the system

$$\begin{cases} x + y - 3z = -1 \\ y - z = 0 \end{cases}.$$

In this last equation, solve for y in terms of z to obtain $y = z$. Back-substituting for y into the previous equation produces $x = 2z - 1$. Finally, letting $z = a$, the solutions to the original system are all of the form

$$x = 2a - 1, \quad y = a, \quad \text{and} \quad z = a$$

where a is a real number. So, every ordered triple of the form

$$(2a - 1, a, a), \quad a \text{ is a real number}$$

is a solution of the system.

✓ CHECKPOINT 5

Solve the system of linear equations.

$$\begin{cases} 2x + y + 3z = 1 \\ 2x + 6y + 12z = 3 \\ 6x + 8y + 18z = 5 \end{cases}$$ ▪

In Example 5, there are other ways to write the same infinite set of solutions. For instance, the solutions could have been written as

$$\left(b, \tfrac{1}{2}(b + 1), \tfrac{1}{2}(b + 1)\right), \quad b \text{ is a real number.}$$

To convince yourself that this description produces the same set of solutions, consider the following.

Substitution	*Solution*
$a = 0$	$(2(0) - 1, 0, 0) = (-1, 0, 0)$
$b = -1$	$\left(-1, \tfrac{1}{2}(-1 + 1), \tfrac{1}{2}(-1 + 1)\right) = (-1, 0, 0)$
$a = 1$	$(2(1) - 1, 1, 1) = (1, 1, 1)$
$b = 1$	$\left(1, \tfrac{1}{2}(1 + 1), \tfrac{1}{2}(1 + 1)\right) = (1, 1, 1)$

In both cases, you obtain the same ordered triples. So, when comparing descriptions of an infinite solution set, keep in mind that there is more than one way to describe the set.

Nonsquare Systems

So far, each system of linear equations you have looked at has been **square,** which means that the number of equations is equal to the number of variables. In a **nonsquare** system, the number of equations differs from the number of variables. A system of linear equations cannot have a unique solution unless there are at least as many equations as there are variables in the system.

Example 6 **A System with Fewer Equations than Variables**

Solve the system of linear equations.

$$\begin{cases} x - 2y + z = 2 & \text{Equation 1} \\ 2x - y - z = 1 & \text{Equation 2} \end{cases}$$

SOLUTION Begin by rewriting the system in row-echelon form, as follows.

$$\begin{cases} x - 2y + z = 2 \\ \quad\quad 3y - 3z = -3 \end{cases}$$

> Adding -2 times the first equation to the second equation produces a new second equation.

$$\begin{cases} x - 2y + z = 2 \\ \quad\quad y - z = -1 \end{cases}$$

> Multiplying the second equation by $\frac{1}{3}$ produces a new second equation.

Solving for y in terms of z, you obtain $y = z - 1$. Back-substitution into Equation 1 yields

$$x - 2(z - 1) + z = 2$$
$$x - 2z + 2 + z = 2$$
$$x = z.$$

Finally, by letting $z = a$, you have the solution

$$x = a, \quad y = a - 1, \quad \text{and} \quad z = a$$

where a is a real number. So, every ordered triple of the form

$$(a, a - 1, a), \quad a \text{ is a real number}$$

is a solution of the system. Because there were originally three variables and only two equations, the system cannot have a unique solution. ──────

✓CHECKPOINT 6

Solve the system of linear equations.

$$\begin{cases} 2x - 2y + 5z = 2 \\ 4x \quad\quad\quad - z = 0 \end{cases} \blacksquare$$

In Example 6, try choosing some values of a to obtain different solutions of the system, such as $(1, 0, 1)$, $(2, 1, 2)$, and $(3, 2, 3)$. Then check each of the solutions in the original system. For example, you can check the solution $(1, 0, 1)$ as follows.

Equation 1: $1 - 2(0) + 1 = 2$ ✓

Equation 2: $2(1) - 0 - 1 = 1$ ✓

Applications

Example 7
MAKE A DECISION   **An Investment Portfolio**

You have a portfolio totaling $450,000 and want to invest in (1) certificates of deposit, (2) municipal bonds, (3) blue-chip stocks, and (4) growth or speculative stocks. The certificates pay 9% simple annual interest, and the municipal bonds pay 6% simple annual interest. You expect the blue-chip stocks to return 10% simple annual interest and the growth stocks to return 15% simple annual interest. You want a combined annual return of 8%, and you also want to have only one-third of the portfolio invested in stocks. How much should be allocated to each type of investment?

SOLUTION   Let C, M, B, and G represent the amounts in the four types of investments. Because the total investment is $450,000, you can write the equation

$$C + M + B + G = 450{,}000.$$

A second equation can be derived from the fact that the combined annual return should be 8%.

$$0.09C + 0.06M + 0.10B + 0.15G = 0.08(450{,}000)$$

Finally, because only one-third of the total investment should be allocated to stocks, you can write

$$B + G = \tfrac{1}{3}(450{,}000).$$

These three equations make up the following system.

$$\begin{cases} C + M + B + G = 450{,}000 & \text{Equation 1} \\ 0.09C + 0.06M + 0.10B + 0.15G = 36{,}000 & \text{Equation 2} \\ B + G = 150{,}000 & \text{Equation 3} \end{cases}$$

Using elimination, you find that the system has infinitely many solutions, which can be written as follows.

$$C = -\tfrac{5}{3}a + 100{,}000$$

$$M = \tfrac{5}{3}a + 200{,}000$$

$$B = -a + 150{,}000$$

$$G = a$$

So, you have many different options. One possible solution is to choose $a = 30{,}000$, which yields the following portfolio.

1. Certificates of deposit: $50,000

2. Municipal bonds: $250,000

3. Blue-chip stocks: $120,000

4. Growth or speculative stocks: $30,000

_____

✓ CHECKPOINT 7

In Example 7, suppose the total investment is $360,000. How much should be allocated to each type of investment? ■

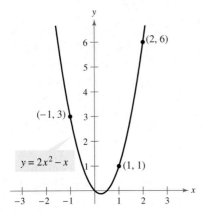

FIGURE 5.11

DISCOVERY

The total numbers of sides and diagonals of regular polygons with three, four, and five sides are three, six, and ten, respectively, as shown in the figure.

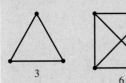

3

6

10

15

Find a quadratic function, $y = ax^2 + bx + c$, where y represents the total number of sides and diagonals and x represents the number of sides, that fits these data. Check to see if the quadratic function gives the correct answers for a polygon with six sides.

Example 8 **Data Analysis: Curve-Fitting**

Find a quadratic equation, $y = ax^2 + bx + c$, whose graph passes through the points $(-1, 3)$, $(1, 1)$, and $(2, 6)$.

SOLUTION Because the graph of $y = ax^2 + bx + c$ passes through the points $(-1, 3)$, $(1, 1)$, and $(2, 6)$, you can write the following.

When $x = -1$, $y = 3$: $\qquad a(-1)^2 + b(-1) + c = 3$

When $x = 1$, $y = 1$: $\qquad a(1)^2 + b(1) + c = 1$

When $x = 2$, $y = 6$: $\qquad a(2)^2 + b(2) + c = 6$

This produces the following system of linear equations.

$$\begin{cases} a - b + c = 3 & \text{Equation 1} \\ a + b + c = 1 & \text{Equation 2} \\ 4a + 2b + c = 6 & \text{Equation 3} \end{cases}$$

The solution of this system is $a = 2$, $b = -1$, and $c = 0$. So, the equation of the parabola is $y = 2x^2 - x$, as shown in Figure 5.11.

✓ **CHECKPOINT 8**

Find a quadratic equation, $y = ax^2 + bx + c$, whose graph passes through the points $(-1, 7)$, $(1, 3)$, and $(2, 7)$. ▦

(CONCEPT CHECK)

1. The ordered triple (a, b, c) is the solution of system A and system B. Are the systems equivalent? Explain.

2. Using Gaussian elimination to solve a system of three linear equations produces $0 = -2$. What does this tell you about the graphs of the equations in the system?

3. Describe the solution set of a system of equations with two equations in three variables.

4. The graph of the quadratic equation $y = 2x^2 + 3x - 1$ passes through the points (x_1, y_1), (x_2, y_2), and (x_3, y_3). Describe the solution set of the given system.

$$\begin{cases} ax_1^2 + bx_1 + c = y_1 \\ ax_2^2 + bx_2 + c = y_2 \\ ax_3^2 + bx_3 + c = y_3 \end{cases}$$

Skills Review 5.3

The following warm-up exercises involve skills that were covered in earlier sections. You will use these skills in the exercise set for this section. For additional help, review Sections 1.2, 2.1, 5.1, and 5.2.

In Exercises 1–4, solve the system of linear equations.

1. $\begin{cases} x + y = 25 \\ y = 10 \end{cases}$

2. $\begin{cases} 2x - 3y = 4 \\ 6x = -12 \end{cases}$

3. $\begin{cases} x + y = 32 \\ x - y = 24 \end{cases}$

4. $\begin{cases} 2r - s = 5 \\ r + 2s = 10 \end{cases}$

In Exercises 5–8, determine whether the ordered triple is a solution of the equation.

5. $5x - 3y + 4z = 2$

   $(-1, -2, 1)$

6. $x - 2y + 12z = 9$

   $(6, 3, 2)$

7. $2x - 5y + 3z = -9$

   $(a - 2, a + 1, a)$

8. $-5x + y + z = 21$

   $(a - 4, 4a + 1, a)$

In Exercises 9 and 10, solve for x in terms of a.

9. $x + 2y - 3z = 4$

   $y = 1 - a, z = a$

10. $x - 3y + 5z = 4$

    $y = 2a + 3, z = a$

Exercises 5.3

See www.CalcChat.com for worked-out solutions to odd-numbered exercises.

In Exercises 1–4, match each system of equations with its solution. [The solutions are labeled (a), (b), (c), and (d).]

1. $\begin{cases} -2x + 3y - 2z = 5 \\ 3x - 4y + z = -1 \\ x + 2y + 5z = -11 \end{cases}$

2. $\begin{cases} 5x + 2y - 4z = -17 \\ -8x + y - 5z = -7 \\ 4x + 3y - z = -7 \end{cases}$

3. $\begin{cases} 2x + 5y - 7z = 36 \\ x + 6y - 10z = 38 \\ x - 4y + 8z = -18 \end{cases}$

4. $\begin{cases} -x - 2y + 5z = 23 \\ -3x + y + 6z = 17 \\ 9x + 2y - 7z = -1 \end{cases}$

(a) $(-1, 0, 3)$

(b) $(6, 2, -2)$

(c) $(2, 1, -3)$

(d) $(4, -1, 5)$

In Exercises 5–8, determine whether the system of equations is in row–echelon form. Justify your answer.

5. $\begin{cases} x + 3y - 7z = -11 \\ y - 2z = -3 \\ z = 2 \end{cases}$

6. $\begin{cases} x - y + 3z = -11 \\ y + 8z = -12 \\ z = -2 \end{cases}$

7. $\begin{cases} x - 9y + z = 22 \\ 2y + z = -3 \\ z = 1 \end{cases}$

8. $\begin{cases} x - y - 8z = 12 \\ 2y - 2z = 2 \\ 7z = -7 \end{cases}$

In Exercises 9 and 10, use back-substitution to solve the system of linear equations.

9. $\begin{cases} x - y + z = 4 \\ 2y + z = -6 \\ z = -2 \end{cases}$

10. $\begin{cases} 4x - 2y + z = 8 \\ -y + z = 4 \\ z = 2 \end{cases}$

In Exercises 11–36, solve the system of equations.

11. $\begin{cases} 4x + y - 3z = 11 \\ 2x - 3y + 2z = 9 \\ x + y + z = -3 \end{cases}$

12. $\begin{cases} 6y + 4z = -12 \\ 3x + 3y = 9 \\ 2x - 3z = 10 \end{cases}$

13. $\begin{cases} 3x + 2z = 13 \\ x + 2y + z = -5 \\ - 3y - z = 10 \end{cases}$

14. $\begin{cases} 2x + 3y + z = -4 \\ 2x - 4y + 3z = 18 \\ 3x - 2y + 2z = 9 \end{cases}$

15. $\begin{cases} 3x - 2y + 4z = 1 \\ x + y - 2z = 3 \\ 2x - 3y + 6z = 8 \end{cases}$

16. $\begin{cases} 5x - 3y + 2z = 3 \\ 2x + 4y - z = 7 \\ x - 11y + 4z = 3 \end{cases}$

17. $\begin{cases} 3x + 3y + 5z = 1 \\ 3x + 5y + 9z = 0 \\ 5x + 9y + 17z = 0 \end{cases}$

18. $\begin{cases} 2x + y - z = 13 \\ x + 2y + z = 2 \\ 8x - 3y + 4z = -2 \end{cases}$

19. $\begin{cases} x + 2y - 7z = -4 \\ 2x + y + z = 13 \\ 3x + 9y - 36z = -33 \end{cases}$

20. $\begin{cases} 2x + y - 3z = 4 \\ 4x + 2z = 10 \\ -2x + 3y - 13z = -8 \end{cases}$

21. $\begin{cases} x + 4z = 13 \\ 4x - 2y + z = 7 \\ 2x - 2y - 7z = -19 \end{cases}$

22. $\begin{cases} 4x - y + 5z = 11 \\ x + 2y - z = 5 \\ 5x - 8y + 13z = 7 \end{cases}$

23. $\begin{cases} x + 4z = 1 \\ x + y + 10z = 10 \\ 2x - y + 2z = -5 \end{cases}$

24. $\begin{cases} 3x - 2y - 6z = 4 \\ -3x + 2y + 6z = 1 \\ x - y - 5z = 3 \end{cases}$

25. $\begin{cases} 4x + 3y + 5z = 10 \\ 5x + 2y + 10z = 13 \\ 3x + y - 2z = -9 \end{cases}$

26. $\begin{cases} 2x + 5y = 25 \\ 3x - 2y + 4z = 1 \\ 4x - 3y + z = 9 \end{cases}$

27. $\begin{cases} 2x + 3y - z = 1 \\ x - 2y + z = 7 \\ 3x + y + 2z = 12 \end{cases}$

28. $\begin{cases} 2x + 3y = 0 \\ 4x + 3y - z = 0 \\ 8x + 3y + 3z = 0 \end{cases}$

29. $\begin{cases} 12x + 5y + z = 0 \\ 12x + 4y - z = 0 \end{cases}$

30. $\begin{cases} x - 2y + 5z = 2 \\ 3x + 2y - z = -2 \end{cases}$

31. $\begin{cases} x - 3y + 2z = 18 \\ 5x - 13y + 12z = 80 \end{cases}$

32. $\begin{cases} 2x - 3y + z = -2 \\ -4x + 9y = 7 \end{cases}$

33. $\begin{cases} 3x - 3y + 6z = 7 \\ -x + y - 2z = 3 \\ 2x + 3y - 4z = 8 \end{cases}$

34. $\begin{cases} 4x + 3y = 7 \\ x - 2y + z = 0 \\ -2x + 4y - 2z = 13 \end{cases}$

35. $\begin{cases} x + 3w = 4 \\ 2y - z - w = 0 \\ 3y - 2w = 1 \\ 2x - y + 4z = 5 \end{cases}$

36. $\begin{cases} x + y + z + w = 6 \\ 2x + 3y - w = 0 \\ -3x + 4y + z + 2w = 4 \\ x + 2y - z + w = 0 \end{cases}$

In Exercises 37–40, find two systems of equations that have the ordered triple as a solution. (There are many correct answers.)

37. $(3, -1, 2)$

38. $\left(-\dfrac{1}{2}, -2, 4\right)$

39. $(1, -5, -3)$

40. $\left(0, 2, \dfrac{1}{2}\right)$

In Exercises 41–44, write three ordered triples of the given form.

41. $\left(a, a - 5, \dfrac{2}{3}a + 1\right)$

42. $(3a, 5 - a, a)$

43. $\left(\dfrac{1}{2}a, 3a, 5\right)$

44. $\left(-\dfrac{1}{2}a + 5, -1, a\right)$

In Exercises 45–48, find the equation of the parabola $y = ax^2 + bx + c$ that passes through the points.

45.
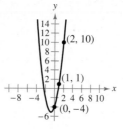

46.
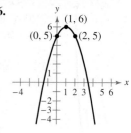

47.
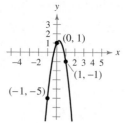

48.
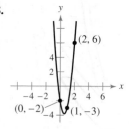

In Exercises 49–52, find the equation of the circle $x^2 + y^2 + Dx + Ey + F = 0$ that passes through the points.

49.
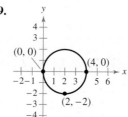

50.
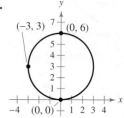

51.
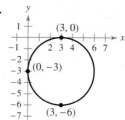

52.
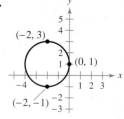

53. Investment   A real estate company borrows $1,500,000. Some of the money is borrowed at 7%, some at 8%, and some at 10% simple annual interest. How much is borrowed at each rate when the total annual interest is $117,000 and the amount borrowed at 8% is the same as the amount borrowed at 10%?

54. Investment   A clothing company borrows $700,000. Some of the money is borrowed at 8%, some at 9%, and some at 10% simple annual interest. How much is borrowed at each rate when the total annual interest is $60,500 and the amount borrowed at 8% is three times the amount borrowed at 10%?

55. Candles A candle company sells three types of candles for $15, $10, and $5 per unit. In one year, the total revenue for the three products was $550,000, which corresponded to the sale of 50,000 units. The company sold half as many units of the $15 candles as units of the $10 candles. How many units of each type of candle were sold?

56. Hair Products A hair product company sells three types of hair products for $30, $20, and $10 per unit. In one year, the total revenue for the three products was $800,000, which corresponded to the sale of 40,000 units. The company sold half as many units of the $30 product as units of the $20 product. How many units of each product were sold?

57. Crop Spraying A mixture of 5 gallons of chemical A, 8 gallons of chemical B, and 12 gallons of chemical C is required to kill a crop destroying insect. Commercial spray X contains 1, 2, and 3 parts of these chemicals, respectively. Commercial spray Y contains only chemical C. Commercial spray Z contains chemicals A, B, and C in equal amounts. How much of each type of commercial spray is needed to obtain the desired mixture?

58. Acid Mixture A chemist needs 10 liters of a 25% acid solution. The solution is to be mixed from three solutions whose acid concentrations are 10%, 20%, and 50%. How many liters of each solution should the chemist use to satisfy the following?

(a) Use as little as possible of the 50% solution.

(b) Use as much as possible of the 50% solution.

(c) Use 2 liters of the 50% solution.

MAKE A DECISION: INVESTMENT PORTFOLIO In Exercises 59 and 60, you have a total of $500,000 that is to be invested in (1) certificates of deposit, (2) municipal bonds, (3) blue-chip stocks, and (4) growth or speculative stocks. How much should be put in each type of investment?

59. The certificates of deposit pay 2.5% simple annual interest, and the municipal bonds pay 10% simple annual interest. Over a five-year period, you expect the blue-chip stocks to return 12% simple annual interest and the growth stocks to return 18% simple annual interest. You want a combined annual return of 10% and you also want to have only one-fourth of the portfolio invested in stocks.

60. The certificates of deposit pay 3% simple annual interest, and the municipal bonds pay 10% simple annual interest. Over a five-year period, you expect the blue-chip stocks to return 12% simple annual interest and the growth stocks to return 15% simple annual interest. You want a combined annual return of 10% and you also want to have only one-fourth of the portfolio invested in stocks.

Fitting a Parabola to Data In Exercises 61–64, find the least squares regression parabola

$$y = ax^2 + bx + c$$

for the points $(x_1, y_1), (x_2, y_2), \ldots, (x_n, y_n)$ by solving the system of linear equations for a, b, and c.

61.
$$\begin{cases} 5c & + 10a = 15.5 \\ 10b & = 6.3 \\ 10c & + 34a = 32.1 \end{cases}$$

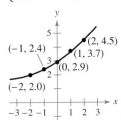

62.
$$\begin{cases} 5c & + 10a = 15.0 \\ 10b & = 17.3 \\ 10c & + 34a = 34.5 \end{cases}$$

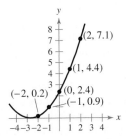

63.
$$\begin{cases} 6c + 3b + 19a = 23.9 \\ 3c + 19b + 27a = -7.2 \\ 19c + 27b + 115a = 48.8 \end{cases}$$

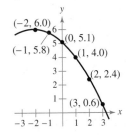

64.
$$\begin{cases} 6c + 3b + 19a = 13.1 \\ 3c + 19b + 27a = -2.6 \\ 19c + 27b + 115a = 29.0 \end{cases}$$

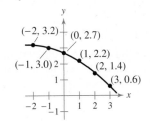

65. Sailboats   The total numbers y (in thousands) of sailboats purchased in the United States in the years 2001 to 2005 are shown in the table. In the table, x represents the year, with $x = 0$ corresponding to 2003.   *(Source: National Marine Manufacturers Association)*

Year, x	Number, y
-2	18.6
-1	15.8
0	15.0
1	14.3
2	14.4

(a) Find the least squares regression parabola $y = ax^2 + bx + c$ for the data by solving the following system.

$$\begin{cases} 5c & + 10a = & 78.1 \\ 10b & = & -9.9 \\ 10c & + 34a = & 162.1 \end{cases}$$

(T) (b) Use the *regression* feature of a graphing utility to find a quadratic model for the data. Compare the quadratic model with the model found in part (a).

66. Genetically Modified Soybeans   The global areas y (in millions of hectares) of genetically modified soybean crops planted in the years 2002 to 2006 are shown in the table. In the table, x represents the year, with $x = 0$ corresponding to 2004.   *(Source: ISAAA, Clive James, 2006)*

Year, x	Area, y
-2	36.5
-1	41.4
0	48.4
1	54.4
2	58.6

(a) Find the least squares regression parabola $y = ax^2 + bx + c$ for the data by solving the following system.

$$\begin{cases} 5c & + 10a = 239.3 \\ 10b & = 57.2 \\ 10c & + 34a = 476.2 \end{cases}$$

(T) (b) Use the *regression* feature of a graphing utility to find a quadratic model for the data. Compare the quadratic model with the model found in part (a).

67. Federal Debt   The values of the federal debt of the United States as percents of the Gross Domestic Product (GDP) for the years 2001 to 2005 are shown in the table. In the table, x represents the year, with $x = 0$ corresponding to 2002.   *(Source: U.S. Office of Management and Budget)*

Year, x	% of GDP
-1	57.4
0	59.7
1	62.6
2	63.7
3	64.3

(a) Find the least squares regression parabola $y = ax^2 + bx + c$ for the data by solving the following system.

$$\begin{cases} 5c + 5b + 15a = 307.7 \\ 5c + 15b + 35a = 325.5 \\ 15c + 35b + 99a = 953.5 \end{cases}$$

(T) (b) Use the *regression* feature of a graphing utility to find a quadratic model for the data. Compare the quadratic model with the model found in part (a).

(c) Use either model to predict the federal debt as a percent of the GDP in 2007.

68. Revenues Per Share   The revenues per share (in dollars) for Panera Bread Company for the years 2002 to 2006 are shown in the table. In the table, x represents the year, with $x = 0$ corresponding to 2003.   *(Source: Panera Bread Company)*

Year, x	Revenues per share
-1	9.47
0	11.85
1	15.72
2	20.49
3	26.11

(a) Find the least squares regression parabola $y = ax^2 + bx + c$ for the data by solving the following system.

$$\begin{cases} 5c + 5b + 15a = 83.64 \\ 5c + 15b + 35a = 125.56 \\ 15c + 35b + 99a = 342.14 \end{cases}$$

(b) Use the *regression* feature of a graphing utility to find a quadratic model for the data. Compare the quadratic model with the model found in part (a).

(c) Use either model to predict the revenues per share in 2008 and 2009.

69. *MAKE A DECISION: STOPPING DISTANCE*   In testing of the new braking system of an automobile, the speed (in miles per hour) and the stopping distance (in feet) were recorded in the table below.

Speed, x	Stopping distance, y
30	54
40	116
50	203
60	315
70	452

(a) Find the least squares regression parabola $y = ax^2 + bx + c$ for the data by solving the following system.

$$\begin{cases} 5c + 250b + 13{,}500a = 1140 \\ 250c + 13{,}500b + 775{,}000a = 66{,}950 \\ 13{,}500c + 775{,}000b + 46{,}590{,}000a = 4{,}090{,}500 \end{cases}$$

(b) Use the *regression* feature of a graphing utility to check your answer to part (a).

(c) A car design specification requires the car to stop within 520 feet when traveling 75 miles per hour. Does the new braking system meet this specification?

70. Sound Recordings   The percents of sound recordings purchased over the Internet (not including digital downloads) in the years 1999 to 2005 are shown in the table. In the table, x represents the year, with $x = 0$ corresponding to 2000. *(Source: The Recording Industry Association of America)*

Year, x	Percent of sound recordings, y
-1	2.4
0	3.2
1	2.9
2	3.4
3	5.0
4	5.9
5	8.2

(a) Find the least squares regression parabola $y = ax^2 + bx + c$ for the data by solving the following system.

$$\begin{cases} 7c + 14b + 56a = 31.0 \\ 14c + 56b + 224a = 86.9 \\ 56c + 224b + 980a = 363.3 \end{cases}$$

(b) Use the *regression* feature of a graphing utility to find a quadratic model for the data. Compare the quadratic model with the model found in part (a).

(c) Use either model to predict the percent of Internet sales in 2008. Does your result seem reasonable? Explain.

71. Reasoning   Is it possible for a square linear system to have no solution? Explain.

72. Reasoning   Is it possible for a square linear system to have infinitely many solutions? Explain.

73. One solution for Exercise 30 is $(-a, 2a - 1, a)$. A student gives $(b, -2b - 1, -b)$ as a solution to the same exercise. Explain why both solutions are correct.

74. Extended Application   To work an extended application analyzing the sales per share of Wal-Mart, visit this text's website at *college.hmco.com*.   *(Source: Wal-Mart Stores, Inc.)*

Business Capsule

© C.Devan/zefa/CORBIS

SAS is a leader in business software and services. Using SAS forecasting technologies, customers can accurately analyze and forecast processes that take place over time. SAS/ETS software contains popular forecasting methods such as regression analysis and trend extrapolation.

75. Research Project   Use your campus library, the Internet, or some other reference source to find information about a company or small business that generates software which uses regression analysis to predict trends. Write a brief paper about the company or small business.

Mid-Chapter Quiz

See www.CalcChat.com for worked-out solutions to odd-numbered exercises.

Take this quiz as you would take a quiz in class. When you are done, check your work against the answers given in the back of the book.

In Exercises 1–4, solve the system algebraically. Use a graphing utility to verify your solution.

1. $\begin{cases} 3x + y = 11 \\ x - 2y = -8 \end{cases}$

2. $\begin{cases} 4x + 8y = 8 \\ x + 2y = 6 \end{cases}$

3. $\begin{cases} x + y = 4 \\ y = 2\sqrt{x} + 1 \end{cases}$

4. $\begin{cases} x^2 + y^2 = 9 \\ y = 2x + 1 \end{cases}$

In Exercises 5 and 6, find the number of units x that need to be sold to break even.

5. $C = 10.50x + 9000, R = 16.50x$

6. $C = 3.79x + 400{,}000, R = 4.59x$

In Exercises 7 and 8, solve the system by substitution or elimination.

7. $\begin{cases} 2.5x - y = 6 \\ 3x + 4y = 2 \end{cases}$

8. $\begin{cases} \frac{1}{2}x + \frac{1}{3}y = 1 \\ x - 2y = -2 \end{cases}$

9. Find the point of equilibrium for the pair of supply and demand equations. Verify your solution graphically.

   Demand: $p = 50 - 0.002x$

   Supply: $p = 20 + 0.004x$

10. The total numbers y (in millions) of Medicare enrollees in the years 2001 to 2005 are shown in the table at the left. In the table, x represents the year, with $x = 0$ corresponding to 2001. Solve the following system for a and b to find the least squares regression line $y = ax + b$ for the data. *(Source: U.S. Centers for Medicare and Medicaid Services)*

    $\begin{cases} 5b + 10a = 206.2 \\ 10b + 30a = 418.6 \end{cases}$

Year, x	Number, y
0	40.1
1	40.5
2	41.2
3	41.9
4	42.5

Table for 10

In Exercises 11–13, solve the system of equations.

11. $\begin{cases} 2x + 3y - z = -7 \\ x + 3z = 10 \\ 2y + z = -1 \end{cases}$

12. $\begin{cases} x + y - 2z = 12 \\ 2x - y - z = 6 \\ y - z = 6 \end{cases}$

13. $\begin{cases} 3x + 2y + z = 17 \\ -x + y + z = 4 \\ x - y - z = 3 \end{cases}$

14. The average prices y (in dollars) of retail prescription drugs for the years 2001 to 2005 are shown in the table at the left. In the table, x represents the year, with $x = 0$ corresponding to 2003. Solve the following system for a, b, and c to find the least squares regression parabola $y = ax^2 + bx + c$ for the data. *(Source: National Association of Chain Drug Stores)*

    $\begin{cases} 5c + 10a = 293.40 \\ 10b = 37.82 \\ 10c + 34a = 578.64 \end{cases}$

Year, x	Average price, y
-2	50.06
-1	55.37
0	59.52
1	63.59
2	64.86

Table for 14

Section 5.4

Systems of Inequalities

- Sketch the graph of an inequality in two variables.
- Solve a system of inequalities.
- Construct and use a system of inequalities to solve an application problem.

The Graph of an Inequality

The following statements are inequalities in two variables:

$$3x - 2y < 6 \quad \text{and} \quad 2x^2 + 3y^2 \geq 6.$$

An ordered pair (a, b) is a **solution of an inequality** in x and y if the inequality is true when a and b are substituted for x and y, respectively. The **graph of an inequality** is the collection of all solutions of the inequality. To sketch the graph of an inequality, begin by sketching the graph of the *corresponding equation*. The graph of the equation will normally separate the plane into two or more regions. In each such region, one of the following must be true.

1. *All* points in the region are solutions of the inequality.

2. *No* point in the region is a solution of the inequality.

So, you can determine whether the points in an entire region satisfy the inequality simply by testing *one* point in the region.

Sketching the Graph of an Inequality in Two Variables

1. Replace the inequality sign by an equal sign, and sketch the graph of the resulting equation. (Use a dashed line for $<$ or $>$ and a solid line for $\leq$ or $\geq$.)

2. Test one point in each of the regions formed by the graph in Step 1. If the point satisfies the inequality, shade the entire region to denote that every point in the region satisfies the inequality.

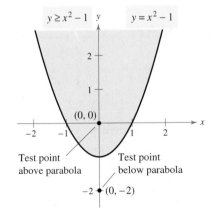

$y \geq x^2 - 1$ $y = x^2 - 1$

$(0, 0)$

Test point above parabola

Test point below parabola

$(0, -2)$

FIGURE 5.12

Example 1   Sketching the Graph of an Inequality

Sketch the graph of the inequality $y \geq x^2 - 1$.

SOLUTION   The graph of the corresponding *equation* $y = x^2 - 1$ is a parabola, as shown in Figure 5.12. By testing a point *above* the parabola $(0, 0)$ and a point *below* the parabola $(0, -2)$, you can see that the points that satisfy the inequality are those lying above (or on) the parabola.

✓ **CHECKPOINT 1**

Sketch the graph of $y < x^2 + 2$. ▪

The inequality in Example 1 is a nonlinear inequality in two variables. Most of the following examples involve **linear inequalities** such as $ax + by < c$ (a and b are not both zero). The graph of a linear inequality is a half-plane lying on one side of the line $ax + by = c$. The simplest linear inequalities are those corresponding to horizontal or vertical lines, as shown in Example 2.

Example 2 Sketching the Graph of a Linear Inequality

Sketch the graph of each linear inequality.

a. $x > -2$ **b.** $y \le 3$

SOLUTION

a. The graph of the corresponding equation $x = -2$ is a vertical line. The points that satisfy the inequality $x > -2$ are those lying to the right of this line, as shown in Figure 5.13.

b. The graph of the corresponding equation $y = 3$ is a horizontal line. The points that satisfy the inequality $y \le 3$ are those lying below (or on) this line, as shown in Figure 5.14.

STUDY TIP

To graph a linear inequality, it can help to write the inequality in slope-intercept form. For instance, by writing $x - y < 2$ in the form

$$y > x - 2$$

you can see that the solution points lie above the line $x - y = 2$ (or $y = x - 2$), as shown in Figure 5.15.

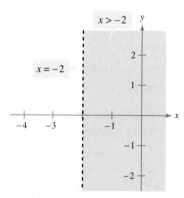

FIGURE 5.13

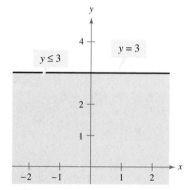

FIGURE 5.14

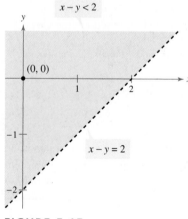

FIGURE 5.15

✓ **CHECKPOINT 2**

Sketch the graph of $x \ge -1$. ∎

Example 3 Sketching the Graph of a Linear Inequality

Sketch the graph of $x - y < 2$.

SOLUTION The graph of the corresponding equation $x - y = 2$ is a line, as shown in Figure 5.15. Because the origin $(0, 0)$ satisfies the inequality, the graph consists of the half-plane lying above the line. (Try checking a point below the line. Regardless of which point you choose, you will see that it does not satisfy the inequality.)

✓ **CHECKPOINT 3**

Sketch the graph of $x + y \le 1$. ∎

Systems of Inequalities

A **solution** of a system of inequalities in x and y is a point (x, y) that satisfies each inequality in the system. To solve a system of inequalities in two variables, first sketch the graph of each individual inequality (on the same coordinate system) and then find the region that is *common* to every graph in the system. This region represents the **solution set** of the system. For systems of *linear* inequalities, it is helpful to find the vertices of the solution region.

Example 4   Solving a System of Inequalities

Solve the system of linear inequalities.

$$\begin{cases} x - y < 2 & \text{Inequality 1} \\ x > -2 & \text{Inequality 2} \\ y \le 3 & \text{Inequality 3} \end{cases}$$

SOLUTION   To solve the system, sketch the graph of the solution set. The graphs of these inequalities are shown in Figures 5.15, 5.13, and 5.14 on page 442. The triangular region common to all three graphs can be found by superimposing the graphs on the same coordinate plane, as shown in Figure 5.16. To find the vertices of the region, solve the three systems of equations obtained by taking the *pairs* of equations representing the boundaries of the individual regions.

Vertex A: $(-2, -4)$
Obtained by solving the system

$$\begin{cases} x - y = 2 \\ x = -2 \end{cases}$$

Vertex B: $(5, 3)$
Obtained by solving the system

$$\begin{cases} x - y = 2 \\ y = 3 \end{cases}$$

Vertex C: $(-2, 3)$
Obtained by solving the system

$$\begin{cases} x = -2 \\ y = 3 \end{cases}$$

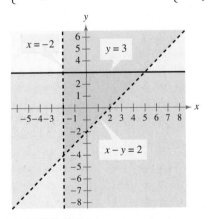
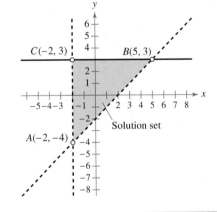

FIGURE 5.16

✓ **CHECKPOINT 4**

Solve the system of linear inequalities.

$$\begin{cases} 2x - y > -3 \\ x \le 2 \\ y > -2 \end{cases}$$

For the triangular region shown in Figure 5.16, each point of intersection of a pair of boundary lines corresponds to a vertex. With more complicated regions, two border lines can sometimes intersect at a point that is *not* a vertex of the region, as shown in Figure 5.17. In order to determine which points of intersection are actually vertices of the region, you should sketch the region and refer to your sketch as you find each point of intersection.

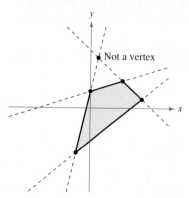

FIGURE 5.17   Boundary lines can intersect at a point that is not a vertex.

Example 5 **Solving a System of Inequalities**

Solve the system of inequalities.

$$\begin{cases} x^2 - y \le 1 & \text{Inequality 1} \\ -x + y \le 1 & \text{Inequality 2} \end{cases}$$

SOLUTION   To solve the system, sketch the graph of the solution set. As shown in Figure 5.18, the points that satisfy the inequality $x^2 - y \le 1$ are the points lying above (or on) the parabola given by

$$y = x^2 - 1. \qquad \text{Parabola}$$

The points satisfying the inequality $-x + y \le 1$ are the points lying below (or on) the line given by

$$y = x + 1. \qquad \text{Line}$$

To find the points of intersection of the parabola and the line, solve the system of corresponding equations.

$$\begin{cases} x^2 - y = 1 \\ -x + y = 1 \end{cases}$$

Using the method of substitution, you can find the points of intersection to be $(-1, 0)$ and $(2, 3)$. The graph of the solution set of the system is shown in Figure 5.18.

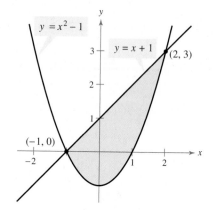

FIGURE 5.18

✓ **CHECKPOINT 5**

Solve the system of linear inequalities.

$$\begin{cases} x^2 + y < 3 \\ x + y > -3 \end{cases} \ ■$$

When solving a system of inequalities, you should be aware that the system might have no solution. For instance, the system

$$\begin{cases} x + y > 3 \\ x + y < -1 \end{cases}$$

has no solution points, because the quantity $(x + y)$ cannot be less than -1 *and* greater than 3, as shown in Figure 5.19.

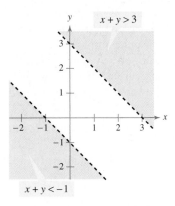

FIGURE 5.19   No Solution   **FIGURE 5.20**   Unbounded Region

Another possibility is that the solution set of a system of inequalities can be unbounded. For instance, the solution set of

$$\begin{cases} x + y < 3 \\ x + 2y > 3 \end{cases}$$

forms an *infinite wedge,* as shown in Figure 5.20.

TECHNOLOGY

Inequalities and Graphing Utilities

Ⓣ A graphing utility can be used to graph an inequality. The graph of $y \geq x^2 - 2$ is shown below.

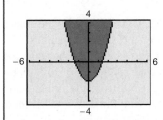

Use your graphing utility to graph each inequality given below. For specific keystrokes on how to graph inequalities and systems of inequalities, go to the text website at *college.hmco.com/info/larsonapplied.*

a. $y \leq 2x + 2$ **b.** $y \geq \frac{1}{2}x^2 - 4$ **c.** $y \leq x^3 - 4x^2 + 4$

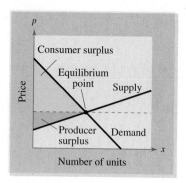

FIGURE 5.21

Applications

Example 8 in Section 5.2 discussed the *point of equilibrium* for a demand function and a supply function. The next example discusses two related concepts that economists call **consumer surplus** and **producer surplus.** As shown in Figure 5.21, the consumer surplus is defined as the area of the region that lies *below* the demand graph, *above* the horizontal line passing through the equilibrium point, and to the right of the p-axis. Similarly, the producer surplus is defined as the area of the region that lies *above* the supply graph, *below* the horizontal line passing through the equilibrium point, and to the right of the p-axis. The consumer surplus is a measure of the amount that consumers would have been willing to pay *above what they actually paid,* whereas the producer surplus is a measure of the amount that producers would have been willing to receive *below what they actually received.*

Example 6 Consumer and Producer Surpluses

The demand and supply equations for a DVD are given by

$$\begin{cases} p = 35 - 0.0001x & \text{Demand equation} \\ p = 8 + 0.0001x & \text{Supply equation} \end{cases}$$

where p is the price (in dollars) and x represents the number of DVDs. Find the consumer surplus and producer surplus for these two equations.

SOLUTION   In Example 8 in Section 5.2, you saw that the point of equilibrium for these equations is

$$(135,000, 21.50).$$

So, the horizontal line passing through this point is $p = 21.50$. Now you can determine that the consumer surplus and producer surplus are the areas of the triangular regions given by the following systems of inequalities, respectively.

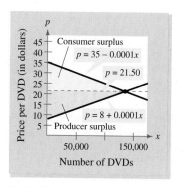

FIGURE 5.22

Consumer Surplus	*Producer Surplus*
$\begin{cases} p \le 35 - 0.0001x \\ p \ge 21.50 \\ x \ge 0 \end{cases}$	$\begin{cases} p \ge 8 + 0.0001x \\ p \le 21.50 \\ x \ge 0 \end{cases}$

In Figure 5.22, you can see that the consumer and producer surpluses are defined as the areas of the shaded triangles. The base of the triangle representing the consumer surplus is 135,000 because the x-value of the point of equilibrium is 135,000. To find the height of this triangle, subtract the p-value of the point of equilibrium, 21.50, from the p-intercept of the demand equation, 35, to obtain 13.50. You can find the base and height of the triangle representing the producer surplus in a similar manner.

$$\text{Consumer surplus} = \tfrac{1}{2}(\text{base})(\text{height}) = \tfrac{1}{2}(135,000)(13.50) = \$911,250$$

$$\text{Producer surplus} = \tfrac{1}{2}(\text{base})(\text{height}) = \tfrac{1}{2}(135,000)(13.50) = \$911,250.$$

✓ CHECKPOINT 6

In Example 6, suppose the supply equation is given by $p = 9 + 0.0001x$ and the new point of equilibrium is $(130,000, 22)$. Find the consumer surplus and producer surplus. ■

© Michael Keller/CORBIS

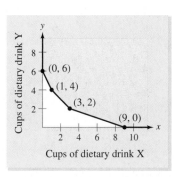

FIGURE 5.23

Example 7 **Nutrition**

The liquid portion of a diet requires at least 300 calories, 36 units of vitamin A, and 90 units of vitamin C daily. A cup of dietary drink X provides 60 calories, 12 units of vitamin A, and 10 units of vitamin C. A cup of dietary drink Y provides 60 calories, 6 units of vitamin A, and 30 units of vitamin C.

a. Set up a system of linear inequalities that describes how many cups of each drink should be consumed each day to meet or exceed the minimum daily requirements for calories and vitamins.

b. A nutritionist normally gives a patient 6 cups of dietary drink X and 1 cup of dietary drink Y per day. Supplies on dietary drink X are running low. Use the graph of the system of linear inequalities to determine other combinations of drinks X and Y that can be given that will meet the minimum daily requirements.

SOLUTION

a. Begin by letting x represent the number of cups of drink X and letting y be the number of cups of drink Y. To meet or exceed the minimum daily requirements, the following inequalities must be satisfied.

$$\begin{cases} 60x + 60y \geq 300 & \text{Calories} \\ 12x + \ 6y \geq \ 36 & \text{Vitamin A} \\ 10x + 30y \geq \ 90 & \text{Vitamin C} \\ \qquad\quad x \geq \ \ 0 \\ \qquad\quad y \geq \ \ 0 \end{cases}$$

The last two inequalities are included because x and y cannot be negative. The graph of this system of inequalities is shown in Figure 5.23. (More is said about this application in Example 5 in Section 5.5.)

b. From Figure 5.23, there are many different possible substitutions that the nutritionist can make. Because supplies are running low on dietary drink X, the nutritionist should choose a combination that contains a small amount of drink X. For instance, 1 cup of dietary drink X and 4 cups of dietary drink Y will also meet the minimum requirements.

✓CHECKPOINT 7

In Example 7, should the nutritionist give a patient 4 cups of dietary drink X and 1 cup of dietary drink Y? Explain. ■

CONCEPT CHECK

1. How do solid lines and dashed lines differ in representing the solution set of the graph of an inequality?

2. When sketching the graph of $y > x$, does testing the point (1, 1) help you determine which region to shade? Explain.

3. Under what circumstances are the vertices of the graph of a solution set represented by open dots?

4. Can a system of inequalities have only one solution? Justify your answer.

Skills Review 5.4

The following warm-up exercises involve skills that were covered in earlier sections. You will use these skills in the exercise set for this section. For additional help, review Sections 2.1, 2.2, 3.1, 5.1, and 5.2.

In Exercises 1–6, classify the graph of the equation as a line, a parabola, or a circle.

1. $x + y = 3$

2. $y = x^2 - 4$

3. $x^2 + y^2 = 9$

4. $y = -x^2 + 1$

5. $4x - y = 8$

6. $y^2 = 16 - x^2$

In Exercises 7–10, solve the system of equations.

7. $\begin{cases} x + 2y = 3 \\ 4x - 7y = -3 \end{cases}$

8. $\begin{cases} 2x - 3y = 4 \\ x + 5y = 2 \end{cases}$

9. $\begin{cases} x^2 + y = 5 \\ 2x - 4y = 0 \end{cases}$

10. $\begin{cases} x^2 + y^2 = 13 \\ x + y = 5 \end{cases}$

Exercises 5.4

See www.CalcChat.com for worked-out solutions to odd-numbered exercises.

In Exercises 1–6, match the inequality with its graph. [The graphs are labeled (a), (b), (c), (d), (e), and (f).]

(a)

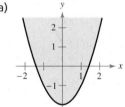

(b)

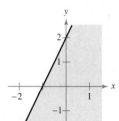

(c)

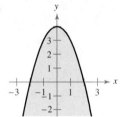

(d)

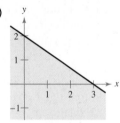

(e)

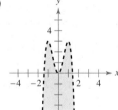

(f)

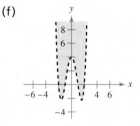

1. $2x + 3y \le 6$

2. $2x - y \ge -2$

3. $x^2 - y \le 2$

4. $y \le 4 - x^2$

5. $y > x^4 - 5x^2 + 4$

6. $3x^4 + y < 6x^2$

In Exercises 7–22, sketch the graph of the inequality.

7. $x \ge 2$

8. $x < 4$

9. $y + 2x^2 > 0$

10. $y^2 - x < 0$

11. $y > -1$

12. $y \le 3$

13. $y < 2 - x$

14. $y > 2x - 4$

15. $2y - x \ge 4$

16. $5x + 3y \ge -15$

17. $y < \ln x$

18. $y \ge -\ln x + 1$

19. $x^2 + y^2 \le 4$

20. $x^2 + y^2 > 4$

21. $x^2 + (y - 2)^2 < 16$

22. $y - (x - 3)^3 \ge 0$

In Exercises 23–44, graph the solution set of the system of inequalities.

23. $\begin{cases} x + y \le 2 \\ -x + y \le 2 \\ y \ge 0 \end{cases}$

24. $\begin{cases} 3x + 2y < 6 \\ x > 1 \\ y > 0 \end{cases}$

25. $\begin{cases} x + y \le 5 \\ x \ge 2 \\ y \ge 0 \end{cases}$

26. $\begin{cases} 2x + y \ge 2 \\ x \le 2 \\ y \le 1 \end{cases}$

27. $\begin{cases} -3x + 2y < 6 \\ x + 4y < -2 \\ 2x + y < 3 \end{cases}$

28. $\begin{cases} x - 7y > -36 \\ 5x + 2y > 5 \\ 6x - 5y > 6 \end{cases}$

29. $\begin{cases} x^2 + y \le 6 \\ x \ge -1 \\ y \ge 0 \end{cases}$

30. $\begin{cases} 2x^2 + y > 4 \\ x < 0 \\ y < 2 \end{cases}$

31. $\begin{cases} 2x + y > 2 \\ 6x + 3y < 2 \end{cases}$ **32.** $\begin{cases} 5x - 3y > -6 \\ 5x - 3y < -9 \end{cases}$

33. $\begin{cases} y \geq -3 \\ y \leq 1 - x^2 \end{cases}$ **34.** $\begin{cases} x - y^2 > 0 \\ y > (x - 3)^2 - 4 \end{cases}$

35. $\begin{cases} x^2 + y^2 \leq 16 \\ x^2 + y^2 < 1 \end{cases}$ **36.** $\begin{cases} x^2 + y^2 \leq 25 \\ x^2 + y^2 \geq 9 \end{cases}$

37. $\begin{cases} x > y^2 \\ x < y + 2 \end{cases}$ **38.** $\begin{cases} x < 2y - y^2 \\ 0 < x + y \end{cases}$

39. $\begin{cases} y \leq \sqrt{3x} + 1 \\ y \geq x + 1 \end{cases}$ **40.** $\begin{cases} y < \sqrt{2x} + 3 \\ y > x + 3 \end{cases}$

41. $\begin{cases} y < x^3 - 2x + 1 \\ y > -2x \\ x \leq 1 \end{cases}$ **42.** $\begin{cases} x^2 + y \leq 4 \\ y \geq 2x \\ x \geq -1 \end{cases}$

43. $\begin{cases} y \leq e^x \\ y \geq \ln x \\ x \geq \frac{1}{2} \\ x \leq 2 \end{cases}$ **44.** $\begin{cases} y \leq e^{-x^2/2} \\ y \geq 0 \\ x \geq -1 \\ x \leq 0 \end{cases}$

In Exercises 45–50, write a system of inequalities that corresponds to the solution set shown in the graph.

45. Rectangle

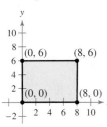

46. Parallelogram

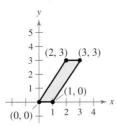

47. Triangle

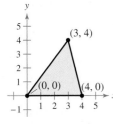

48. Triangle

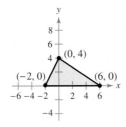

49. Sector of a circle

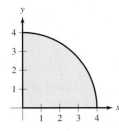

50. Sector of a circle

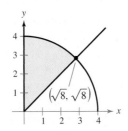

51. Furniture Production   A furniture company produces tables and chairs. Each table requires 2 hours in the assembly center and $1\frac{1}{2}$ hours in the finishing center. Each chair requires $1\frac{1}{2}$ hours in the assembly center and $1\frac{1}{2}$ hours in the finishing center. The company's assembly center is available 18 hours per day, and its finishing center is available 15 hours per day. Let x and y be the numbers of tables and chairs produced per day, respectively. (a) Find a system of inequalities describing all possible production levels, and (b) sketch the graph of the system.

52. Kayak Inventory   A store sells two models of kayaks. Because of the demand, it is necessary to stock at least twice as many units of model A as units of model B. The costs to the store for the two models are $500 and $700, respectively. The management does not want more than $30,000 in kayak inventory at any one time, and it wants at least six model A kayaks and three model B kayaks in inventory at all times. (a) Find a system of inequalities describing all possible inventory levels, and (b) sketch the graph of the system.

Consumer and Producer Surpluses   In Exercises 53–56, find the consumer surplus and producer surplus for the pair of demand and supply equations.

	Demand	Supply
53.	$p = 56 - 0.0001x$	$p = 22 + 0.00001x$
54.	$p = 60 - 0.00001x$	$p = 15 + 0.00004x$
55.	$p = 140 - 0.00002x$	$p = 80 + 0.00001x$
56.	$p = 600 - 0.0002x$	$p = 125 + 0.0006x$

57. Think About It   Under what circumstances are the consumer surplus and producer surplus equal for a pair of linear supply and demand equations? Explain.

58. Think About It   Under what circumstances is the consumer surplus greater than the producer surplus for a pair of linear supply and demand equations? Explain.

59. Investment   You plan to invest up to $30,000 in two different interest-bearing accounts. Each account is to contain at least $6000. Moreover, one account should have at least twice the amount that is in the other account. (a) Find a system of inequalities that describes the amounts that you can invest in each account, and (b) sketch the graph of the system.

60. Concert Ticket Sales   Two types of tickets are to be sold for a concert. One type costs $20 per ticket and the other type costs $30 per ticket. The promoter of the concert must sell at least 20,000 tickets, including at least 8000 of the $20 tickets and at least 5000 of the $30 tickets. Moreover, the gross receipts must total at least $480,000 in order for the concert to be held. (a) Find a system of inequalities describing the different numbers of tickets that must be sold, and (b) sketch the graph of the system.

61. MAKE A DECISION: DIET SUPPLEMENT A dietitian designs a special diet supplement using two different foods. Each ounce of food X contains 20 units of calcium, 10 units of iron, and 15 units of vitamin B. Each ounce of food Y contains 15 units of calcium, 20 units of iron, and 20 units of vitamin B. The minimum daily requirements for the diet are 400 units of calcium, 250 units of iron, and 220 units of vitamin B.

(a) Find a system of inequalities describing the different amounts of food X and food Y that the dietitian can use in the diet.

(b) Sketch the graph of the system.

(c) A nutritionist normally gives a patient 18 ounces of food X and 3.5 ounces of food Y per day. Supplies of food X are running low. What other combinations of foods X and Y can be given to the patient to meet the minimum daily requirements?

62. MAKE A DECISION: DIET SUPPLEMENT A dietitian designs a special diet supplement using two different foods. Each ounce of food X contains 12 units of calcium, 10 units of iron, and 20 units of vitamin B. Each ounce of food Y contains 15 units of calcium, 20 units of iron, and 12 units of vitamin B. The minimum daily requirements for the diet are 300 units of calcium, 280 units of iron, and 300 units of vitamin B.

(a) Find a system of inequalities describing the different amounts of food X and food Y that the dietitian can use in the diet.

(b) Sketch the graph of the system.

(c) A nutritionist normally gives a patient 10 ounces of food X and 12 ounces of food Y per day. Supplies of food Y are running low. What other combinations of foods X and Y can be given to the patient to meet the minimum daily requirements?

63. Health A person's maximum heart rate is $220 - x$, where x is the person's age in years for $20 \leq x \leq 70$. When a person exercises, it is recommended that the person strive for a heart rate that is at least 50% of the maximum and at most 75% of the maximum. *(Source: American Heart Association)*

(a) Write a system of inequalities that describes the exercise target heart rate region. Let y represent a person's heart rate.

(b) Sketch a graph of the region in part (a).

(c) Find two solutions to the system and interpret their meanings in the context of the problem.

64. Peregrine Falcons The numbers of nesting pairs y of peregrine falcons in Yellowstone National Park from 2001 to 2005 can be approximated by the linear model

$$y = 3.4t + 13, \quad 1 \leq t \leq 5$$

where t represents the year, with $t = 1$ corresponding to 2001. *(Source: Yellowstone Bird Report 2005)*

(T) (a) The *total* number of nesting pairs during this five-year period can be approximated by finding the area of the trapezoid represented by the following system.

$$\begin{cases} y \leq 3.4t + 13 \\ y \geq 0 \\ t \geq 0.5 \\ t \leq 5.5 \end{cases}$$

Graph this region using a graphing utility.

(b) Use the formula for the area of a trapezoid to approximate the total number of nesting pairs.

65. Computers The sales y (in billions of dollars) for Dell Inc. from 1996 to 2005 can be approximated by the linear model

$$y = 5.07t - 22.4, \quad 6 \leq t \leq 15$$

where t represents the year, with $t = 6$ corresponding to 1996. *(Source: Dell Inc.)*

(T) (a) The *total* sales during this ten-year period can be approximated by finding the area of the trapezoid represented by the following system.

$$\begin{cases} y \leq 5.07t - 22.4 \\ y \geq 0 \\ t \geq 5.5 \\ t \leq 15.5 \end{cases}$$

Graph this region using a graphing utility.

(b) Use the formula for the area of a trapezoid to approximate the total sales.

66. Write a system of inequalities whose graphed solution set is a right triangle.

67. Write a system of inequalities whose graphed solution set is an isosceles triangle.

68. Writing Explain the difference between the graphs of the inequality $x \leq 4$ on the real number line and on the rectangular coordinate system.

69. Graphical Reasoning Two concentric circles have radii x and y, where $y > x$. The area between the circles must be at least 10 square units.

(a) Find a system of inequalities describing the constraints on the circles.

(T) (b) Use a graphing utility to graph the system of inequalities in part (a). Graph the line $y = x$ in the same viewing window.

(c) Identify the graph of the line in relation to the boundary of the inequality. Explain its meaning in the context of the problem.

Linear Programming

- Use linear programming to minimize or maximize an objective function.
- Use linear programming to optimize an application.

Linear Programming: A Graphical Approach

Many applications in business and economics involve a process called **optimization,** in which you are asked to find the minimum cost, the maximum profit, or the minimum use of resources. In this section you will study an optimization strategy called **linear programming.**

A two-dimensional linear programming problem consists of a linear **objective function** and a system of linear inequalities called **constraints.** The objective function gives the quantity that is to be maximized (or minimized), and the constraints determine the set of **feasible solutions.** For example, consider a linear programming problem in which you are asked to maximize the value of

$$z = ax + by \qquad \text{Objective function}$$

subject to a set of constraints that determines the region in Figure 5.24. Because every point in the region satisfies each constraint, it is not clear how you should go about finding the point that yields a maximum value of z. Fortunately, it can be shown that if there is an optimal solution, it must occur at one of the vertices of the region. This means that *you can find the maximum value by testing z at each of the vertices.*

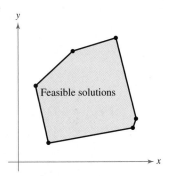

Feasible solutions

FIGURE 5.24

Optimal Solution of a Linear Programming Problem

If a linear programming problem has a solution, it must occur at a vertex of the set of feasible solutions. If the problem has more than one solution, then at least one solution must occur at a vertex of the set of feasible solutions. In either case, the value of the objective function is unique.

The process for solving a linear programming problem in two variables is shown in Example 1 on the next page.

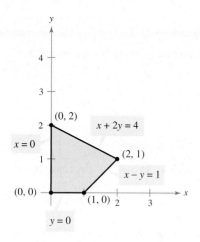

FIGURE 5.25

Example 1   Solving a Linear Programming Problem

Find the maximum value of

$$z = 3x + 2y \qquad \text{Objective function}$$

subject to the following constraints.

$$\left. \begin{aligned} x &\geq 0 \\ y &\geq 0 \\ x + 2y &\leq 4 \\ x - y &\leq 1 \end{aligned} \right\} \qquad \text{Constraints}$$

SOLUTION   The constraints form the region shown in Figure 5.25. At the four vertices of this region, the objective function has the following values.

At $(0, 0)$: $z = 3(0) + 2(0) = 0$

At $(1, 0)$: $z = 3(1) + 2(0) = 3$

At $(2, 1)$: $z = 3(2) + 2(1) = 8$ Maximum value of z

At $(0, 2)$: $z = 3(0) + 2(2) = 4$

So, the maximum value of z is 8, which occurs when $x = 2$ and $y = 1$.

✓ CHECKPOINT 1

Find the maximum value of

$$z = 2x + 3y$$

subject to the following constraints.

$$\left. \begin{aligned} x &\geq 0 \\ y &\geq 0 \\ x + y &\leq 3 \\ x - y &\leq 2 \end{aligned} \right\}$$

In Example 1, try testing some of the *interior* points of the region. You will see that the corresponding values of z are less than 8. Here are some examples.

At $(1, 1)$: $z = 3(1) + 2(1) = 5$

At $\left(1, \tfrac{1}{2}\right)$: $z = 3(1) + 2\left(\tfrac{1}{2}\right) = 4$

To see why the maximum value of the objective function in Example 1 must occur at a vertex, consider writing the objective function in slope-intercept form

$$y = -\frac{3}{2}x + \frac{z}{2} \qquad \text{Family of lines}$$

where $z/2$ is the y-intercept of the objective function. This equation represents a family of lines, each of slope $-\tfrac{3}{2}$. Of these infinitely many lines, you want the one that has the largest z-value while still intersecting the region determined by the constraints. In other words, of all the lines whose slope is $-\tfrac{3}{2}$, you want the one that has the largest y-intercept *and* intersects the given region, as shown in Figure 5.26. It should be clear that such a line will pass through one (or more) of the vertices of the region.

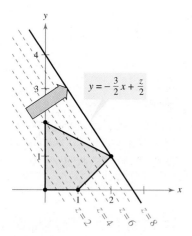

FIGURE 5.26

Solving a Linear Programming Problem

To solve a linear programming problem involving two variables by the graphical method, use the following steps.

1. Sketch the region corresponding to the system of constraints.

2. Find the vertices of the region.

3. Test the objective function at each of the vertices and select the values of the variables that optimize the objective function. For a bounded region, both a minimum and a maximum value will exist. (For an unbounded region, *if* an optimal solution exists, it will occur at a vertex.)

The guidelines above will work whether the objective function is to be maximized or minimized. For instance, the same test used in Example 1 to find the maximum value of z can be used to conclude that the minimum value of z is 0 and that this value occurs at the vertex $(0, 0)$.

Example 2 Solving a Linear Programming Problem

Find (a) the maximum value and (b) the minimum value of

$$z = 4x + 6y \qquad \text{Objective function}$$

subject to the following constraints.

$$\left. \begin{array}{r} x \geq 0 \\ y \geq 0 \\ -x + y \leq 11 \\ x + y \leq 27 \\ 2x + 5y \leq 90 \end{array} \right\} \qquad \text{Constraints}$$

SOLUTION

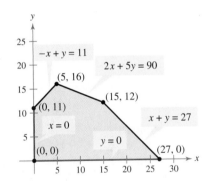

FIGURE 5.27

a. The region bounded by the constraints is shown in Figure 5.27. By testing the objective function at each vertex, you obtain the following.

$$\text{At } (0, 0): \quad z = 4(0) + 6(0) = 0$$
$$\text{At } (0, 11): \quad z = 4(0) + 6(11) = 66$$
$$\text{At } (5, 16): \quad z = 4(5) + 6(16) = 116$$
$$\text{At } (15, 12): \quad z = 4(15) + 6(12) = 132 \qquad \text{Maximum value of } z$$
$$\text{At } (27, 0): \quad z = 4(27) + 6(0) = 108$$

So, the maximum value of z is 132, which occurs when $x = 15$ and $y = 12$.

b. Using the values of z at the vertices in part (a), you can conclude that the minimum value of z is 0, and that this value occurs when $x = 0$ and $y = 0$.

✓ **CHECKPOINT 2**

Find (a) the maximum value and (b) the minimum value of $z = 5x + 2y$ subject to the same constraints as in Example 2. ■

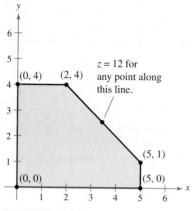

FIGURE 5.28

It is possible for the maximum (or minimum) value in a linear programming problem to occur at *two* different vertices. For instance, at the vertices of the region shown in Figure 5.28, the objective function

$$z = 2x + 2y \qquad \text{Objective function}$$

has the following values.

At $(0, 0)$: $z = 2(0) + 2(0) = 0$

At $(0, 4)$: $z = 2(0) + 2(4) = 8$

At $(2, 4)$: $z = 2(2) + 2(4) = 12$ Maximum value of z

At $(5, 1)$: $z = 2(5) + 2(1) = 12$ Maximum value of z

At $(5, 0)$: $z = 2(5) + 2(0) = 10$

In this case, you can conclude that the objective function has a maximum value not only at the vertices $(2, 4)$ and $(5, 1)$; it also has a maximum value (of 12) at *any point on the line segment connecting these two vertices*. Note that the objective function $y = -x + \frac{1}{2}z$ has the same slope as the line through the vertices $(2, 4)$ and $(5, 1)$.

Some linear programming problems have no optimal solution. This can occur if the region determined by the constraints is *unbounded*. Example 3 illustrates such a problem.

Example 3   **An Unbounded Region**

Find the maximum value of

$$z = 4x + 2y \qquad \text{Objective function}$$

where $x \geq 0$ and $y \geq 0$, subject to the following constraints.

$$\left. \begin{array}{r} x + 2y \geq 4 \\ 3x + y \geq 7 \\ -x + 2y \leq 7 \end{array} \right\} \qquad \text{Constraints}$$

SOLUTION   The region determined by the constraints is shown in Figure 5.29. For this unbounded region, there is no maximum value of z. To see this, note that the point $(x, 0)$ lies in the region for all values of $x \geq 4$. By choosing x to be large, you can obtain values of

$$z = 4(x) + 2(0) = 4x$$

that are as large as you want. So, there is no maximum value of z. For this problem, there *is* a minimum value of z, $z = 10$, which occurs at the vertex $(2, 1)$, as shown below.

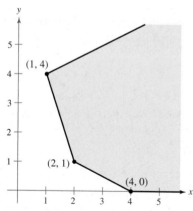

FIGURE 5.29

At $(1, 4)$: $z = 4(1) + 2(4) = 12$

At $(2, 1)$: $z = 4(2) + 2(1) = 10$ Minimum value of z

At $(4, 0)$: $z = 4(4) + 2(0) = 16$

CHECKPOINT 3

Find the maximum value of the objective function $z = x + 8y$ where $x \geq 0$ and $y \geq 0$, subject to the same constraints as in Example 3. ∎

Applications

Example 4 shows how linear programming can be used to find the maximum profit in a business application.

Example 4 Optimal Profit ®

A manufacturer wants to maximize the profit for two laboratory products. Product I yields a profit of $1.50 per unit, and product II yields a profit of $2.00 per unit. Market tests and available resources have indicated the following constraints.

1. The combined production level should not exceed 1200 units per month.

2. The demand for product II is no more than half the demand for product I.

3. The production level of product I is less than or equal to 600 units plus three times the production level of product II.

What is the optimal production level for each product?

SOLUTION Let x be the number of units of product I and let y be the number of units of product II. The objective function (for the combined profit) is given by

$$P = 1.5x + 2y. \qquad \text{\small Objective function}$$

The three constraints translate into the following linear inequalities.

1. $x + y \le 1200$ ⟹ $x + y \le 1200$

2. $y \le \frac{1}{2}x$ ⟹ $-x + 2y \le 0$

3. $x \le 3y + 600$ ⟹ $x - 3y \le 600$

Because neither x nor y can be negative, you also have the two additional constraints of $x \ge 0$ and $y \ge 0$. Figure 5.30 shows the region determined by the constraints. To find the maximum profit, test the value of P at each vertex of the region.

At $(0, 0)$: $\quad P = 1.5(0) \quad + 2(0) \quad = \quad 0$

At $(800, 400)$: $\quad P = 1.5(800) \quad + 2(400) = 2000 \qquad \text{\small Maximum profit}$

At $(1050, 150)$: $P = 1.5(1050) + 2(150) = 1875$

At $(600, 0)$: $\quad P = 1.5(600) \quad + 2(0) \quad = \quad 900$

So, the maximum profit is $2000, and it occurs when the monthly production levels are 800 units of product I and 400 units of product II.

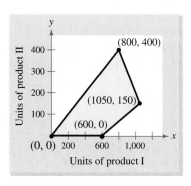

FIGURE 5.30

✓ CHECKPOINT 4

In Example 4, suppose the manufacturer improved the production of product I so that it yielded a profit of $2.50 per unit. How would this improvement affect the optimal number of units the manufacturer should sell in order to obtain a maximum profit? ▪

Example 5 shows how linear programming can be used to find the optimal cost in a real-life application.

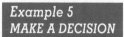
Optimal Cost

The liquid portion of a diet requires at least 300 calories, 36 units of vitamin A, and 90 units of vitamin C daily. A cup of dietary drink X costs $0.12 and provides 60 calories, 12 units of vitamin A, and 10 units of vitamin C. A cup of dietary drink Y costs $0.15 and provides 60 calories, 6 units of vitamin A, and 30 units of vitamin C. How many cups of each drink should be consumed each day to obtain an optimal cost and still meet the daily requirements?

SOLUTION As in Example 7 on page 447, let x be the number of cups of dietary drink X and let y be the number of cups of dietary drink Y.

$$
\left.
\begin{array}{lrl}
\text{For calories:} & 60x + 60y & \geq 300 \\
\text{For vitamin A:} & 12x + 6y & \geq 36 \\
\text{For vitamin C:} & 10x + 30y & \geq 90 \\
& x & \geq 0 \\
& y & \geq 0
\end{array}
\right\} \quad \text{Constraints}
$$

The cost C is given by

$$C = 0.12x + 0.15y. \qquad \text{Objective function}$$

The graph of the region corresponding to the constraints is shown in Figure 5.31. Because you want to incur as little cost as possible, you want to determine the *minimum* cost. To determine the minimum cost, test C at each vertex of the region, as follows.

At $(0, 6)$: $C = 0.12(0) + 0.15(6) = 0.90$

At $(1, 4)$: $C = 0.12(1) + 0.15(4) = 0.72$

At $(3, 2)$: $C = 0.12(3) + 0.15(2) = 0.66$ Minimum value of C

At $(9, 0)$: $C = 0.12(9) + 0.15(0) = 1.08$

So, the minimum cost is $0.66 per day, and this cost occurs when 3 cups of drink X and 2 cups of drink Y are consumed each day.

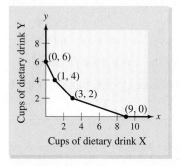

FIGURE 5.31

✓ **CHECKPOINT 5**

In Example 5, suppose a cup of dietary drink Y costs $0.11. How would this affect the number of cups of each drink that should be consumed each day to obtain an optimal cost and still meet the daily requirements? ■

CONCEPT CHECK

1. Does every linear programming problem have an optimal solution? Explain.

2. Can a linear programming problem have a maximum value but no minimum value? Explain.

3. Can a linear programming problem have a minimum value at two vertices and a maximum value at two vertices? Justify your answer.

4. Suppose that a linear programming problem has a minimum value of a and a maximum value of b. Write an interval for the value of the objective function at any point in the interior of the region determined by the constraints. Explain your reasoning.

Skills Review 5.5

The following warm-up exercises involve skills that were covered in earlier sections. You will use these skills in the exercise set for this section. For additional help, review Section 2.2, 5.1, and 5.4.

In Exercises 1–4, sketch the graph of the linear equation.

1. $y + x = 3$

2. $y - x = 12$

3. $x = 0$

4. $y = 4$

In Exercises 5–8, find the point of intersection of the two lines.

5. $x + y = 4$
   $x = 0$

6. $x + 2y = 12$
   $y = 0$

7. $x + y = 4$
   $2x + 3y = 9$

8. $x + 2y = 12$
   $2x + y = 9$

In Exercises 9 and 10, sketch the graph of the inequality.

9. $2x + 3y \geq 18$

10. $4x + 3y \geq 12$

Exercises 5.5

See www.CalcChat.com for worked-out solutions to odd-numbered exercises.

In Exercises 1–8, find the minimum and maximum values of the objective function and where they occur, subject to the indicated constraints. (For each exercise, the graph of the region determined by the constraints is provided.)

1. *Objective function:*

   $z = 6x + 5y$

   Constraints:

   $x \geq 0$
   $y \geq 0$
   $x + y \leq 6$

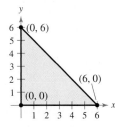

2. *Objective function:*

   $z = 2x + 8y$

   Constraints:

   $x \geq 0$
   $y \geq 0$
   $2x + y \leq 4$

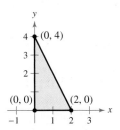

3. *Objective function:*

   $z = 8x + 7y$

   Constraints:

   See Exercise 1.

4. *Objective function:*

   $z = 7x + 3y$

   Constraints:

   See Exercise 2.

5. *Objective function:*

   $z = 3x + 2y$

   Constraints:

   $x \geq 0$
   $y \geq 0$
   $x + 3y \leq 15$
   $4x + y \leq 16$

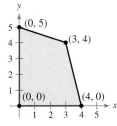

6. *Objective function:*

   $z = 5x + 4y$

   Constraints:

   $x \geq 0$
   $y \geq 0$
   $2x + 3y \geq 6$
   $3x - 2y \leq 9$
   $x + 5y \leq 20$

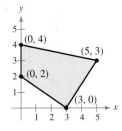

7. *Objective function:*

   $z = 5x + 0.5y$

   Constraints:

   See Exercise 5.

8. *Objective function:*

   $z = x + 6y$

   Constraints:

   See Exercise 6.

In Exercises 9–20, sketch the region determined by the constraints. Then find the minimum and maximum values of the objective function and where they occur, subject to the indicated constraints.

9. *Objective function:*

$z = 6x + 10y$

Constraints:

$x \geq 0$

$y \geq 0$

$3x + 5y \leq 15$

10. *Objective function:*

$z = 7x + 8y$

Constraints:

$x \geq 0$

$y \geq 0$

$x + 2y \leq 8$

11. *Objective function:*

$z = 9x + 4y$

Constraints:

See Exercise 9.

12. *Objective function:*

$z = 7x + 2y$

Constraints:

See Exercise 10.

13. *Objective function:*

$z = 4x + 5y$

Constraints:

$x \geq 0$

$y \geq 0$

$x + y \geq 8$

$3x + 5y \geq 30$

14. *Objective function:*

$z = 4x + 5y$

Constraints:

$x \geq 0$

$y \geq 0$

$x + y \leq 5$

$x + 2y \leq 6$

15. *Objective function:*

$z = 2x + 7y$

Constraints:

See Exercise 13.

16. *Objective function:*

$z = 2x - y$

Constraints:

See Exercise 14.

17. *Objective function:*

$z = x + 2y$

Constraints:

$x \geq 0$

$y \geq 0$

$x + 2y \leq 40$

$x + y \leq 30$

$2x + 3y \leq 65$

18. *Objective function:*

$z = x$

Constraints:

$x \geq 0$

$y \geq 0$

$2x + 3y \leq 60$

$2x + y \leq 28$

$4x + y \leq 48$

19. *Objective function:*

$z = x + y$

Constraints:

See Exercise 17.

20. *Objective function:*

$z = y$

Constraints:

See Exercise 18.

In Exercises 21–24, maximize the objective function subject to the constraints $3x + y \leq 15$, $4x + 3y \leq 30$, $x \geq 0$, and $y \geq 0$.

21. $z = 2x + y$

22. $z = 5x + y$

23. $z = 4x + 3y$

24. $z = 3x + y$

In Exercises 25–28, maximize the objective function subject to the constraints $x + 4y \leq 20$, $x + y \leq 8$, $3x + 2y \leq 21$, $x \geq 0$, and $y \geq 0$.

25. $z = 2x + 5y$

26. $z = 3x + 5y$

27. $z = 12x + 5y$

28. $z = 15x + 8y$

Think About It In Exercises 29–36, find an objective function that has a maximum or minimum value at the indicated vertex of the constraint region shown. (There are many correct answers.)

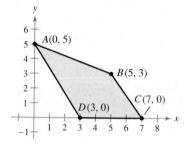

29. The maximum occurs at vertex *A*.

30. The maximum occurs at vertex *B*.

31. The minimum occurs at vertex *C*.

32. The minimum occurs at vertex *D*.

33. The maximum occurs at vertices *A* and *B*.

34. The maximum occurs at vertices *B* and *C*.

35. The minimum occurs at vertices *A* and *D*.

36. The minimum occurs at vertices *C* and *D*.

37. Optimal Profit A fruit grower raises crops A and B. The profit is $185 per acre for crop A and $245 per acre for crop B. Research and available resources indicate the following constraints.

- The fruit grower has 150 acres of land for raising the crops.
- It takes 1 day to trim an acre of crop A and 2 days to trim an acre of crop B, and there are 240 days per year available for trimming.
- It takes 0.3 day to pick an acre of crop A and 0.1 day to pick an acre of crop B, and there are 30 days per year available for picking.

What is the optimal acreage for each fruit? What is the optimal profit?

38. Optimal Profit   The costs to a store for two models of Global Positioning System (GPS) receivers are $80 and $100. The $80 model yields a profit of $25 and the $100 model yields a profit of $30. Market tests and available resources indicate the following constraints.

- The merchant estimates that the total monthly demand will not exceed 200 units.
- The merchant does not want to invest more than $18,000 in GPS receiver inventory.

What is the optimal inventory level for each model? What is the optimal profit?

39. Optimal Cost   A farming cooperative mixes two brands of cattle feed. Brand X costs $30 per bag, and brand Y costs $25 per bag. Research and available resources have indicated the following constraints.

- Brand X contains two units of nutritional element A, two units of element B, and two units of element C.
- Brand Y contains one unit of nutritional element A, nine units of element B, and three units of element C.
- The minimum requirements for nutrients A, B, and C are 12 units, 36 units, and 24 units, respectively.

What is the optimal number of bags of each brand that should be mixed? What is the optimal cost?

40. Optimal Cost   A humanitarian agency can use two models of vehicles for a refugee rescue mission. Each model A vehicle costs $1000 and each model B vehicle costs $1500. Mission strategies and objectives indicate the following constraints.

- A total of at least 20 vehicles must be used.
- A model A vehicle can hold 45 boxes of supplies. A model B vehicle can hold 30 boxes of supplies. The agency must deliver at least 690 boxes of supplies to the refugee camp.
- A model A vehicle can hold 20 refugees. A model B vehicle can hold 32 refugees. The agency must rescue at least 520 refugees.

What is the optimal number of vehicles of each model that should be used? What is the optimal cost?

41. Optimal Profit   A manufacturer produces two models of bicycles. The times (in hours) required for assembling, painting, and packaging each model are shown in the table.

Process	Model A	Model B
Assembling	2	2.5
Painting	4	1
Packaging	1	0.75

The total times available for assembling, painting, and packaging are 4000 hours, 4800 hours, and 1500 hours, respectively. The profits per unit are $50 for model A and $75 for model B. What is the optimal production level for each model? What is the optimal profit?

42. Optimal Profit   A company makes two models of doghouses. The times (in hours) required for assembling, painting, and packaging are shown in the table.

Process	Model A	Model B
Assembling	2.5	3
Painting	2	1
Packaging	0.75	1.25

The total times available for assembling, painting, and packaging are 4000 hours, 2500 hours, and 1500 hours, respectively. The profits per unit are $60 for model A and $75 for model B. What is the optimal production level for each model? What is the optimal profit?

43. Optimal Revenue   An accounting firm charges $2500 for an audit and $350 for a tax return. Research and available resources have indicated the following constraints.

- The firm has 900 hours of staff time available each week.
- The firm has 155 hours of review time available each week.
- Each audit requires 75 hours of staff time and 10 hours of review time.
- Each tax return requires 12.5 hours of staff time and 2.5 hours of review time.

What numbers of audits and tax returns will bring in an optimal revenue?

44. Optimal Revenue   The accounting firm in Exercise 43 lowers its charge for an audit to $2000. What numbers of audits and tax returns will bring in an optimal revenue?

45. Media Selection   A company has budgeted a maximum of $1,000,000 for national advertising of an allergy medication. Each minute of television time costs $100,000 and each one-page newspaper ad costs $20,000. Each television ad is expected to be viewed by 20 million viewers, and each newspaper ad is expected to be seen by 5 million readers. The company's market research department recommends that at most 80% of the advertising budget be spent on television ads. What is the optimal amount that should be spent on each type of ad? What is the optimal total audience?

46. Optimal Profit   A fruit juice company makes two drinks by blending apple and pineapple juices. The percents of apple juice and pineapple juice in each drink are shown in the table.

Mixture	Drink A	Drink B
Apple juice	30%	60%
Pineapple juice	70%	40%

There are 1000 liters of apple juice and 1500 liters of pineapple juice available. The profit for drink A is $0.70 per liter and the profit for drink B is $0.60 per liter. What is the optimal production level for each type of drink? What is the optimal profit?

47. Investments   An investor has up to $250,000 to invest in two types of investments. Type A investments pay 7% annually and type B pay 12% annually. To have a well-balanced portfolio, the investor imposes the following conditions. At least one-fourth of the total portfolio is to be allocated to type A investments and at least one-fourth is to be allocated to type B investments. What is the optimal amount that should be invested in each type of investment? What is the optimal return?

48. Investments   An investor has up to $450,000 to invest in two types of investments. Type A investments pay 8% annually and type B pay 14% annually. To have a well-balanced portfolio, the investor imposes the following conditions. At least one-half of the total portfolio is to be allocated to type A investments and at least one-fourth is to be allocated to type B investments. What is the optimal amount that should be invested in each type of investment? What is the optimal return?

49. Optimal Profit   A company makes two models of a patio furniture set. The times for assembling, finishing, and packaging model A are 3 hours, 2.5 hours, and 0.6 hour, respectively. The times for model B are 2.75 hours, 1 hour, and 1.25 hours. The total times available for assembling, finishing, and packaging are 3000 hours, 2400 hours, and 1200 hours, respectively. The profit per unit for model A is $100 and the profit per unit for model B is $85. What is the optimal production level for each model? What is the optimal profit?

50. Optimal Profit   A manufacturer produces two models of elliptical cross-training exercise machines. The times for assembling, finishing, and packaging model A are 3 hours, 3 hours, and 0.8 hour, respectively. The times for model B are 4 hours, 2.5 hours, and 0.4 hour. The total times available for assembling, finishing, and packaging are 6000 hours, 4200 hours, and 950 hours, respectively. The profits per unit are $300 for model A and $375 for model B. What is the optimal production level for each model? What is the optimal profit?

In Exercises 51–56, the given linear programming problem has an unusual characteristic. Sketch a graph of the solution region for the problem and describe the unusual characteristic. Find the maximum value of the objective function and where it occurs.

51. *Objective function:*

$$z = 2.5x + y$$

Constraints:

$$x \geq 0$$
$$y \geq 0$$
$$3x + 5y \leq 15$$
$$5x + 2y \leq 10$$

52. *Objective function:*

$$z = x + y$$

Constraints:

$$x \geq 0$$
$$y \geq 0$$
$$-x + y \leq 1$$
$$-x + 2y \leq 4$$

53. *Objective function:*

$$z = -x + 2y$$

Constraints:

$$x \geq 0$$
$$y \geq 0$$
$$x \leq 10$$
$$x + y \leq 7$$

54. *Objective function:*

$$z = x + y$$

Constraints:

$$x \geq 0$$
$$y \geq 0$$
$$-x + y \leq 1$$
$$-3x + y \geq 3$$

55. *Objective function:*

$$z = 3x + 4y$$

Constraints:

$$x \geq 0$$
$$y \geq 0$$
$$x + y \leq 1$$
$$2x + y \leq 4$$

56. *Objective function:*

$$z = x + 2y$$

Constraints:

$$x \geq 0$$
$$y \geq 0$$
$$x + 2y \leq 4$$
$$2x + y \leq 4$$

57. Reasoning   An objective function has a maximum value at the vertices $(0, 14)$ and $(3, 8)$.

(a) Can you conclude that it also has a maximum value at the point $(1, 12)$? Explain.

(b) Can you conclude that it also has a maximum value at the point $(4, 6)$? Explain.

(c) Find another point that maximizes the objective function.

58. Reasoning   An objective function has a minimum value at the vertex $(20, 0)$. Can you conclude that it also has a minimum value at the point $(0, 0)$? Explain.

59. Reasoning   When solving a linear programming problem, you find that the objective function has a maximum value at more than one vertex. Can you assume that there are an infinite number of points that will produce the maximum value? Explain your reasoning.

Chapter Summary and Study Strategies

After studying this chapter, you should have acquired the following skills.
The exercise numbers are keyed to the Review Exercises that begin on page 462.
Answers to odd-numbered Review Exercises are given in the back of the text.*

Section 5.1	Review Exercises
■ Solve a system of equations by the method of substitution.	*1–6*
■ Solve a system of equations graphically.	*7, 8, 12*
■ Construct and use a system of equations to solve an application problem.	*9–12*

Section 5.2	
■ Solve a linear system by the method of elimination.	*13–20, 41*
■ Interpret the solution of a linear system graphically.	*21, 22*
■ Construct and use a linear system to solve an application problem.	*23–26*

Section 5.3	
■ Solve a linear system in row-echelon form using back-substitution.	*27, 28*
■ Use Gaussian elimination to solve a linear system.	*29, 30, 33, 34*
■ Solve a nonsquare linear system.	*31, 32*
■ Construct and use a linear system in three or more variables to solve an application problem.	*39, 40, 43*
■ Find the equation of a circle or a parabola using a linear system in three or more variables.	*35–38, 42*

Section 5.4	
■ Sketch the graph of an inequality in two variables.	*44–49*
■ Solve a system of inequalities.	*50–57*
■ Construct and use a system of inequalities to solve an application problem.	*58–63*

Section 5.5	
■ Use linear programming to minimize or maximize an objective function.	*64–71*
■ Use linear programming to optimize an application.	*72–76*

Study Strategies

■ **Units of Variables in Applied Problems**   When using systems of equations to solve real-life applications, be sure to keep track of the unit(s) assigned to each variable. This will allow you to write correctly each equation of the system based on the constraints given in the application.

---

* Use a wide range of valuable study aids to help you master the material in this chapter. The *Student Solutions Guide* includes step-by-step solutions to all odd-numbered exercises to help you review and prepare. The student website at *college.hmco.com/info/larsonapplied* offers algebra help and a *Graphing Technology Guide*. The *Graphing Technology Guide* contains step-by-step commands and instructions for a wide variety of graphing calculators, including the most recent models.

Review Exercises

See www.CalcChat.com for worked-out solutions to odd-numbered exercises.

In Exercises 1–6, solve the system by the method of substitution.

1. $\begin{cases} x + 3y = 10 \\ 4x - 5y = -28 \end{cases}$

2. $\begin{cases} 3x - y - 13 = 0 \\ 4x + 3y - 26 = 0 \end{cases}$

3. $\begin{cases} \frac{1}{2}x + \frac{3}{5}y = -2 \\ 2x + y = 6 \end{cases}$

4. $\begin{cases} 1.3x + 0.9y = 7.5 \\ 0.4x - 0.5y = -0.8 \end{cases}$

5. $\begin{cases} x^2 + y^2 = 100 \\ x + 2y = 20 \end{cases}$

6. $\begin{cases} y = x^3 - 2x^2 - 2x - 3 \\ y = -x^2 + 4x - 3 \end{cases}$

(T) In Exercises 7 and 8, use a graphing utility to find the point(s) of intersection of the graphs.

7. $\begin{cases} y = x^2 - 3x + 11 \\ y = -x^2 + 2x + 8 \end{cases}$

8. $\begin{cases} y = \sqrt{9 - x^2} \\ y = e^x + 1 \end{cases}$

9. Break-Even Analysis You invest $5000 in a greenhouse. The planter, potting soil, and seed for each plant costs $6.43 and the selling price is $12.68. How many plants must you sell to break even?

10. Break-Even Analysis You are setting up a basket-weaving business and have made an initial investment of $20,000. The cost of each basket is $3.25 and the selling price is $6.95. How many baskets must you sell to break even? (Round to the nearest whole unit.)

11. Choice of Newscasts Television Stations A and B are competing for the 6 P.M. newscast audience. Station A is implementing a new newscast format for the 6 P.M. audience. Models that represent the numbers of 6 P.M. viewers each month for the two stations are given by

$\begin{cases} y = 950x + 10,000 \quad \text{Station A (new format)} \\ y = -875x + 18,000 \quad \text{Station B} \end{cases}$

where y is the number of viewers and x represents the month, with $x = 1$ corresponding to the first month of the new format. Use the models to estimate when the number of viewers for Station A's 6 P.M. newscast will exceed the number of viewers for Station B's 6 P.M. newscast.

(T) **12. Comparing Populations** From 2000 to 2005, the population of Vermont grew more slowly than that of Alaska. Models that represent the populations of the two states are given by

$\begin{cases} P = 7.7t + 626 \quad \text{Alaska} \\ P = 2.8t + 610 \quad \text{Vermont} \end{cases}$

where P is the population (in thousands) and t represents the year, with $t = 0$ corresponding to 2000. Use a graphing utility to determine whether the population of Vermont will exceed that of Alaska. *(Source: U.S. Census Bureau)*

In Exercises 13–20, solve the system by elimination.

13. $\begin{cases} 2x - 3y = 21 \\ 3x + y = 4 \end{cases}$

14. $\begin{cases} 3u + 5v = 9 \\ 12u + 10v = 22 \end{cases}$

15. $\begin{cases} 4x - 3y = 10 \\ 8x - 6y = 20 \end{cases}$

16. $\begin{cases} 3x + 4y = 18 \\ 6x + 8y = 18 \end{cases}$

17. $\begin{cases} 1.25x - 2y = 3.5 \\ 5x - 8y = 14 \end{cases}$

18. $\begin{cases} 1.5x + 2.5y = 8.5 \\ 6x + 10y = 24 \end{cases}$

19. $\begin{cases} \dfrac{x-2}{3} + \dfrac{y+3}{4} = 5 \\ 2x - y = 7 \end{cases}$

20. $\begin{cases} \dfrac{3}{5}x + \dfrac{2}{7}y = 10 \\ x + 2y = 38 \end{cases}$

In Exercises 21 and 22, describe the graph of the solution of the linear system.

21. $\begin{cases} 2x + y = -1 \\ 3x - 2y = -5 \end{cases}$

22. $\begin{cases} x - 2y = -1 \\ -2x + 4y = 2 \end{cases}$

23. Acid Mixture Twelve gallons of a 25% acid solution is obtained by mixing a 10% solution with a 50% solution.

(T) (a) Write a system of equations that represents the problem and use a graphing utility to graph the equations in the same viewing window.

(b) How much of each solution is required to obtain the specified concentration of the final mixture?

24. Acid Mixture Twenty gallons of a 30% acid solution is obtained by mixing a 12% solution with a 60% solution.

(T) (a) Write a system of equations that represents the problem and use a graphing utility to graph the equations in the same viewing window.

(b) How much of each solution is required to obtain the specified concentration of the final mixture?

Supply and Demand In Exercises 25 and 26, find the point of equilibrium for the pair of demand and supply equations.

Demand	*Supply*
25. $p = 37 - 0.0002x$	$p = 22 + 0.00001x$
26. $p = 120 - 0.0001x$	$p = 45 + 0.0002x$

In Exercises 27–34, solve the system of equations.

27. $\begin{cases} 4x - 3y + 2z = 1 \\ 2y - 4z = 2 \\ z = 2 \end{cases}$

28. $\begin{cases} 2x + y - 4z = 6 \\ 3y + z = 2 \\ z = -4 \end{cases}$

29. $\begin{cases} 2x + y + z = 6 \\ x - 4y - z = 3 \\ x + y + z = 4 \end{cases}$

30. $\begin{cases} x + 3y - z = 13 \\ 2x \quad\quad - 5z = 23 \\ 4x - y - 2z = 4 \end{cases}$

31. $\begin{cases} x + y + z = 10 \\ -2x + 3y + 4z = 22 \end{cases}$

32. $\begin{cases} 5x - 12y + 7z = 16 \\ 3x - 7y + 4z = 9 \end{cases}$

33. $\begin{cases} 2x + 6y - z = 1 \\ x - 3y + z = 2 \\ \frac{3}{2}x + \frac{3}{2}y \quad = 6 \end{cases}$

34. $\begin{cases} x + y + z + w = 8 \\ 4y + 5z - 2w = 3 \\ 2x + 3y - z \quad = -2 \\ 3x + 2y \quad - 4w = -20 \end{cases}$

In Exercises 35 and 36, find the equation of the parabola $y = ax^2 + bx + c$ that passes through the points.

35.

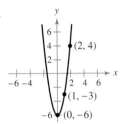

36.

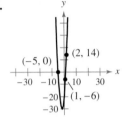

In Exercises 37 and 38, find the equation of the circle $x^2 + y^2 + Dx + Ey + F = 0$ that passes through the points.

37.

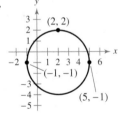

38.

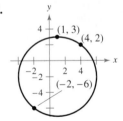

39. Investment Portfolio   An investor allocates a portfolio totaling $500,000 among the following types of investments: (1) certificates of deposit, (2) municipal bonds, (3) blue-chip stocks, and (4) growth or speculative stocks. The certificates of deposit pay 5% simple annual interest, and the municipal bonds pay 8% simple annual interest. Over a five-year period, the investor expects the blue-chip stocks to return 10% simple annual interest and the growth stocks to return 15% simple annual interest. The investor wishes a combined return of 9.45% and also wants to have only two-fifths of the portfolio invested in stocks. How much should the investor allocate to each type of investment if the amount invested in certificates of deposit is twice that invested in municipal bonds?

40. Investment   You receive $8580 a year in simple annual interest from three investments. The interest rates for the three investments are 6%, 8%, and 10%. The value of the 10% investment is two times that of the 6% investment, and the 8% investment is $1000 more than the 6% investment. What is the amount of each investment?

41. Fitting a Line to Data   Find the least squares regression line $y = ax + b$ for the points $(0, 1.6)$, $(1, 2.4)$, $(2, 3.6)$, $(3, 4.7)$, and $(4, 5.5)$ by solving the following system of linear equations for a and b.

$$\begin{cases} 5b + 10a = 17.8 \\ 10b + 30a = 45.7 \end{cases}$$

42. Fitting a Parabola to Data   Find the least squares regression parabola $y = ax^2 + bx + c$ for the points

$(-2, 0.4)$, $(-1, 0.9)$, $(0, 1.9)$, $(1, 2.1)$, and $(2, 3.8)$

by solving the following system of linear equations for a, b, and c.

$$\begin{cases} 5c \quad\quad + 10a = 9.1 \\ \quad 10b \quad\quad = 8.0 \\ 10c \quad + 34a = 19.8 \end{cases}$$

Ⓣ **43. Revenue**   The revenues y (in billions of dollars) for McDonald's Corporation for the years 2001 to 2005 are shown in the table, where t represents the year, with $t = 0$ corresponding to 2002. (*Source: McDonald's Corporation*)

Year, t	Revenue, y
-1	14.9
0	15.4
1	17.1
2	19.1
3	20.5

(a) Use a graphing utility to create a scatter plot of the data.

(b) Solve the following system for a and b to find the least squares regression line $y = at + b$ for the data.

$$\begin{cases} 5b + 5a = 87.0 \\ 5b + 15a = 101.9 \end{cases}$$

(c) Solve the following system for a, b, and c to find the least squares regression parabola $y = at^2 + bt + c$ for the data.

$$\begin{cases} 5c + 5b + 15a = 87.0 \\ 5c + 15b + 35a = 101.9 \\ 15c + 35b + 99a = 292.9 \end{cases}$$

(d) Use the *regression* feature of a graphing utility to find linear and quadratic models for the data. Compare them with the least squares regression models found in parts (b) and (c).

(e) Use a graphing utility to graph the linear and quadratic models. Use the models to predict the revenues in 2006 and 2007. Compare the predictions for each year.

In Exercises 44–49, sketch the graph of the inequality.

44. $x \geq -4$

45. $y < 5$

46. $y \leq 5 - \frac{1}{2}x$

47. $3y - x \geq 7$

48. $y - 4x^2 > -1$

49. $y \leq \dfrac{3}{x^2 + 2}$

In Exercises 50–57, graph the solution set of the system of inequalities.

50. $\begin{cases} 2x + 3y < 9 \\ \quad\quad x > 0 \\ \quad\quad y > 0 \end{cases}$

51. $\begin{cases} 2x - y > \;\; 6 \\ \quad\;\; x < \;\; 5 \\ \quad\;\; y \geq -2 \end{cases}$

52. $\begin{cases} 3x - y > -4 \\ 2x + y > -1 \\ 7x + y < \;\; 4 \end{cases}$

53. $\begin{cases} x + y > 4 \\ 3x + y < 10 \\ x - y \leq 0 \end{cases}$

54. $\begin{cases} \quad x^2 + y^2 \leq 9 \\ x^2 - x - 2 \leq y \end{cases}$

55. $\begin{cases} \quad x^2 + y^2 \leq 4 \\ -2x^2 + 2 < y \end{cases}$

56. $\begin{cases} \ln x < \;\; y \\ \quad\; y > -1 \\ \quad\; x < \;\; 4 \end{cases}$

57. $\begin{cases} \quad \ln x \geq \;\; y \\ -x + y < -2 \\ \quad\;\; x > \;\; 2 \end{cases}$

In Exercises 58 and 59, write a system of inequalities that corresponds to the solution set that is shown in the graph.

58. Parallelogram

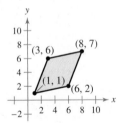

59. Triangle

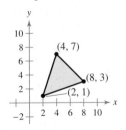

Consumer and Producer Surpluses In Exercises 60 and 61, find the consumer surplus and producer surplus for the pair of demand and supply equations.

Demand	Supply
60. $p = 160 - 0.0001x$	$p = 70 + 0.0002x$
61. $p = 130 - 0.0002x$	$p = 30 + 0.0003x$

62. Movie Player Inventory A store sells two models of Blu-ray Disc™ players (BDPs). Because of the demand, it is necessary to stock at least twice as many units of model Y as units of model Z. The costs to the store for the two models are $200 and $300, respectively. The management does not want more than $4000 in BDP inventory at any one time, and it wants at least four model Y BDPs and two model Z BDPs in inventory at all times. Find a system of inequalities that describes all possible inventory levels. Sketch the graph of the system.

63. Concert Ticket Sales Two types of tickets are to be sold for a concert. One type costs $30 per ticket and the other type costs $50 per ticket. The promoter of the concert must sell at least 15,000 tickets, including at least 8000 of the $30 tickets and at least 4000 of the $50 tickets. Moreover, the gross receipts must total at least $550,000 in order for the concert to be held. Find a system of inequalities describing the different numbers of tickets that must be sold. Sketch the graph of the system.

In Exercises 64–67, find the minimum and maximum values of the objective function and where they occur, subject to the indicated constraints. (For each exercise, the graph of the region determined by the constraints is provided.)

64. *Objective function:*

$z = 5x + 6y$

Constraints:

$x \geq 0$

$y \geq 0$

$x + y \leq 8$

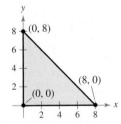

65. *Objective function:*

$z = 15x + 12y$

Constraints:

$x \geq \;\; 0$

$y \geq \;\; 0$

$x + 3y \leq 12$

$3x + 2y \leq 15$

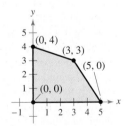

66. *Objective function:*

$z = 8x + 10y$

Constraints:

$0 \le x \le 50$

$0 \le y \le 35$

$4x + 5y \le 275$

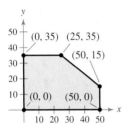

67. *Objective function:*

$z = 50x + 60y$

Constraints:

$x \ge 0$

$y \ge 0$

$3x + 4y \ge 1200$

$5x + 6y \le 3000$

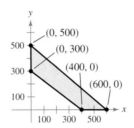

In Exercises 68–71, sketch the region determined by the constraints. Then find the minimum and maximum values of the objective function and where they occur, subject to the indicated constraints.

68. *Objective function:*

$z = 6x + 8y$

Constraints:

$x \ge 0$

$y \ge 0$

$x + 4y \le 16$

$3x + 2y \le 18$

69. *Objective function:*

$z = 5x + 8y$

Constraints:

$0 \le x \le 5$

$y \ge 0$

$x + 2y \le 12$

$2x + 3y \le 19$

70. *Objective function:*

$z = 8x + 3y$

Constraints:

$0 \le x \le 5$

$0 \le y \le 7$

$x + y \le 9$

$3x + y \le 17$

71. *Objective function:*

$z = 10x + 11y$

Constraints:

$x \ge 0$

$y \ge 0$

$2x + 5y \le 30$

$x + y \ge 3$

$2x + y \le 14$

72. Optimal Profit A company makes two models of desks. The times (in hours) required for assembling, finishing, and packaging each model are shown in the table.

Process	Model A	Model B
Assembling	3.5	8
Finishing	2.5	2
Packaging	1.3	0.7

The total times available for assembling, finishing, and packaging are 5600 hours, 2000 hours, and 910 hours, respectively. The profits per unit are $100 for model A and $150 for model B. What is the optimal production level for each model? What is the optimal profit?

73. Optimal Profit A factory manufactures two television set models: a basic model that yields $100 profit and a deluxe model that yields a profit of $180. The times (in hours) required for assembling, finishing, and packaging each model are shown in the table.

Process	Basic model	Deluxe model
Assembling	2	5
Finishing	1	2
Packaging	1	1

The total times available for assembling, finishing, and packaging are 3000 hours, 1300 hours, and 1000 hours, respectively. What is the optimal production level for each model? What is the optimal profit?

74. Optimal Profit The costs to a merchant for two models of digital camcorders are $525 and $675. The $525 model yields a profit of $75 and the $675 model yields a profit of $125. The merchant estimates that the total monthly demand will not exceed 350 units. There should be no more than $206,250 in digital camcorder inventory. Find the number of units of each model that should be stocked in order to optimize profit. What is the optimal profit?

75. Optimal Profit The costs to a merchant for two models of home theater systems are $270 and $455. The $270 model yields a profit of $30 and the $455 model yields a profit of $45. The merchant estimates that the total monthly demand will not exceed 100 units. There should be no more than $36,250 in home theater system inventory. Find the number of units of each model that should be stocked in order to optimize profit. What is the optimal profit?

76. Optimal Revenue An accounting firm has 800 hours of staff time and 90 hours of review time available each week. The firm charges $2500 for an audit and $200 for a tax return. Each audit requires 100 hours of staff time and 10 hours of review time. Each tax return requires 10 hours of staff time and 2 hours of review time. What numbers of audits and tax returns will bring in an optimal revenue? What is the optimal revenue?

Chapter Test

See www.CalcChat.com for worked-out solutions to odd-numbered exercises.

Take this test as you would take a test in class. When you are done, check your work against the answers given in the back of the book.

In Exercises 1–6, solve the system of equations using the indicated method.

1. *Substitution*

$$\begin{cases} 5x - 7y = -18 \\ 4x + 3y = 20 \end{cases}$$

2. *Substitution*

$$\begin{cases} x + y = 3 \\ x^2 + y = 9 \end{cases}$$

3. *Graphing*

$$\begin{cases} 5x - y = 6 \\ 2x^2 + y = 8 \end{cases}$$

4. *Graphing*

$$\begin{cases} 1.5x - 2.25y = 8 \\ 2.5x + 2y = 5.75 \end{cases}$$

5. *Elimination*

$$\begin{cases} 2x - 4y + z = 11 \\ x + 2y + 3z = 9 \\ 3y + 5z = 12 \end{cases}$$

6. *Elimination*

$$\begin{cases} 3x - 2y + z = 16 \\ 5x - z = 6 \\ 2x - y - z = 3 \end{cases}$$

7. A total of $80,000 is invested in two funds paying 9% and 9.5% simple interest. The total annual interest is $7300. How much is invested in each fund?

8. Find the point of equilibrium for a system that has a demand equation of $p = 49 - 0.0003x$ and a supply equation of $p = 33 + 0.00002x$.

9. The numbers y of adults (in millions) who participated in baking as a leisure activity in the years 2001 to 2005 are shown in the table at the left. Find the least squares regression parabola $y = at^2 + bt + c$ for the data by solving the following system. *(Source: Mediamark Research, Inc.)*

$$\begin{cases} 5c + 10a = 188.4 \\ 10b = 5.1 \\ 10c + 34a = 383.1 \end{cases}$$

Use the model to predict the number of adults who participated in baking as a leisure activity in 2006.

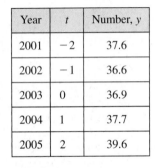

Year	t	Number, y
2001	-2	37.6
2002	-1	36.6
2003	0	36.9
2004	1	37.7
2005	2	39.6

Table for 9

In Exercises 10–13, sketch the graph of the inequality.

10. $x \geq 0$

11. $y \geq 0$

12. $x + 3y \leq 12$

13. $3x + 2y \leq 15$

14. Sketch the solution set of the system of inequalities composed of the inequalities in Exercises 10–13.

15. Find the minimum and maximum values of the objective function $z = 6x + 7y$, subject to the constraints given in Exercises 10–13.

16. A manufacturer produces two models of stair climbers. The times required for assembling, painting, and packaging each model are as follows.

- Assembling: 3.5 hours for model A; 8 hours for model B
- Painting: 2.5 hours for model A; 2 hours for model B
- Packaging: 1.3 hours for model A; 0.9 hour for model B

The total times available for assembling, painting, and packaging are 5600 hours, 2000 hours, and 900 hours, respectively. The profits per unit are $200 for model A and $275 for model B. What is the optimal production level for each model? What is the optimal profit? Explain your reasoning.

Matrices and Determinants

6

© Jupiter Images/Comstock Images/Alamy

The Bank of New York, which opened for business on June 9, 1784, is the oldest bank in the United States. In 1789, Alexander Hamilton negotiated the first loan given to the government for $200,000. You can use matrices to find amounts of money borrowed at various interest rates. (See Section 6.1, Exercises 83 and 84.)

Applications

Matrices are used to solve many real-life applications. The applications listed below represent a sample of the applications in this chapter.

- Contract Bonuses, Exercise 67, page 495
- Raw Materials, Exercises 75–78, page 505
- Gypsy Moths, Exercise 15, page 525

Matrices and Linear Systems

- Determine the order of a matrix.
- Perform elementary row operations on a matrix in order to write the matrix in row-echelon form or reduced row-echelon form.
- Solve a system of linear equations using Gaussian elimination.
- Solve a system of linear equations using Gauss-Jordan elimination.

Matrices

In this section, you will study a streamlined technique for solving systems of linear equations. This technique involves the use of a rectangular array of real numbers called a **matrix.** The plural of matrix is **matrices.**

Definition of a Matrix

If m and n are positive integers, an $m \times n$ matrix (read "m by n") is a rectangular array

$$\begin{bmatrix} a_{11} & a_{12} & a_{13} & \cdots & a_{1n} \\ a_{21} & a_{22} & a_{23} & \cdots & a_{2n} \\ a_{31} & a_{32} & a_{33} & \cdots & a_{3n} \\ \vdots & \vdots & \vdots & & \vdots \\ a_{m1} & a_{m2} & a_{m3} & \cdots & a_{mn} \end{bmatrix} \Bigg\} m \text{ rows}$$

$$\underbrace{\hspace{4cm}}_{n \text{ columns}}$$

in which each **entry,** a_{ij}, of the matrix is a number. An $m \times n$ matrix has m **rows** (horizontal lines) and n **columns** (vertical lines).

The entry in the ith row and jth column is denoted by the *double subscript* notation a_{ij}. That is, a_{21} refers to the entry in row 2, column 1. A matrix having m rows and n columns is said to be of **order** $m \times n$. If $m = n$, the matrix is **square** of order n. For a square matrix, the entries $a_{11}, a_{22}, a_{33}, \ldots$ are the **main diagonal** entries.

Example 1 Orders of Matrices

The following matrices have the indicated orders.

a. *Order:* 1×4

$$\begin{bmatrix} 1 & -3 & 0 & \frac{1}{2} \end{bmatrix}$$

b. *Order:* 2×2

$$\begin{bmatrix} 0 & 0 \\ 0 & 0 \end{bmatrix}$$

c. *Order:* 3×2

$$\begin{bmatrix} 5 & 0 \\ 2 & -2 \\ -7 & 4 \end{bmatrix}$$

✓ **CHECKPOINT 1**

Determine the order of the matrix.

$$\begin{bmatrix} 0 & 8 & 5 \\ 3 & 1 & -2 \end{bmatrix}$$ ■

A matrix that has only one row [such as the matrix in Example 1(a)] is called a **row matrix,** and a matrix that has only one column is called a **column matrix.**

A matrix derived from a system of linear equations (each written in standard form with the constant term on the right) is the **augmented matrix** of the system. Moreover, the matrix derived from the coefficients of the system (but that does not include the constant terms) is the **coefficient matrix** of the system. Note in the matrices below the use of 0 for the coefficient of the missing y-variable in the third equation. Also note that the fourth column (the column of constant terms) in the augmented matrix is separated from the coefficients of the linear system by vertical dots.

$$
\begin{array}{ccc}
\textit{System} & \textit{Augmented Matrix} & \textit{Coefficient Matrix} \\
\begin{cases} x - 4y + 3z = 5 \\ -x + 3y - z = -3 \\ 2x \qquad - 4z = 6 \end{cases} &
\left[\begin{array}{ccc:c} 1 & -4 & 3 & 5 \\ -1 & 3 & -1 & -3 \\ 2 & 0 & -4 & 6 \end{array}\right] &
\left[\begin{array}{ccc} 1 & -4 & 3 \\ -1 & 3 & -1 \\ 2 & 0 & -4 \end{array}\right]
\end{array}
$$

When forming either the coefficient matrix or the augmented matrix of a system, you should begin by vertically aligning the variables in the equations and using zeros for the coefficients of any missing variables.

$$
\begin{array}{ccc}
\textit{Original System} & \textit{Line Up Variables} & \textit{Form Augmented Matrix} \\
\begin{cases} x + 3y = 9 \\ -y + 4z = -2 \\ x - 5z = 0 \end{cases} &
\begin{cases} x + 3y \qquad = 9 \\ \quad -y + 4z = -2 \\ x \qquad - 5z = 0 \end{cases} &
\left[\begin{array}{ccc:c} 1 & 3 & 0 & 9 \\ 0 & -1 & 4 & -2 \\ 1 & 0 & -5 & 0 \end{array}\right]
\end{array}
$$

Elementary Row Operations

In Section 5.3, you studied three operations that can be used on a system of linear equations to produce an equivalent system.

1. Interchange two equations.

2. Multiply an equation by a nonzero constant.

3. Add a multiple of an equation to another equation.

In matrix terminology, these three operations correspond to **elementary row operations.** An elementary row operation on an augmented matrix of a given system of linear equations produces a new augmented matrix corresponding to a new (but equivalent) system of linear equations. Two matrices are **row-equivalent** if one can be obtained from the other by a sequence of elementary row operations.

Elementary Row Operations

1. Interchange two rows.

2. Multiply a row by a nonzero constant.

3. Add a multiple of a row to another row.

Although elementary row operations are simple to perform, they involve a lot of arithmetic. Because it is easy to make a mistake, you should get in the habit of noting in each step, next to the row you are changing, the elementary row operation performed, so that you can go back and check your work.

The next example demonstrates each of the elementary row operations that can be performed on a matrix to produce a row-equivalent matrix.

Example 2 Elementary Row Operations

a. Interchange the first and second rows.

Original Matrix

$$\begin{bmatrix} 0 & 1 & 3 & 4 \\ -1 & 2 & 0 & 3 \\ 2 & -3 & 4 & 1 \end{bmatrix}$$

New Row-Equivalent Matrix

$$\begin{matrix} R_2 \\ R_1 \end{matrix} \begin{bmatrix} -1 & 2 & 0 & 3 \\ 0 & 1 & 3 & 4 \\ 2 & -3 & 4 & 1 \end{bmatrix}$$

b. Multiply the first row by $\frac{1}{2}$.

Original Matrix

$$\begin{bmatrix} 2 & -4 & 6 & -2 \\ 1 & 3 & -3 & 0 \\ 5 & -2 & 1 & 2 \end{bmatrix}$$

New Row-Equivalent Matrix

$$\frac{1}{2}R_1 \rightarrow \begin{bmatrix} 1 & -2 & 3 & -1 \\ 1 & 3 & -3 & 0 \\ 5 & -2 & 1 & 2 \end{bmatrix}$$

c. Add -2 times the first row to the third row.

Original Matrix

$$\begin{bmatrix} 1 & 2 & -4 & 3 \\ 0 & 3 & -2 & -1 \\ 2 & 1 & 5 & -2 \end{bmatrix}$$

New Row-Equivalent Matrix

$$-2R_1 + R_3 \rightarrow \begin{bmatrix} 1 & 2 & -4 & 3 \\ 0 & 3 & -2 & -1 \\ 0 & -3 & 13 & -8 \end{bmatrix}$$

Note that the elementary row operation is written beside *the row that is changing*.

✓CHECKPOINT 2

Identify the elementary row operation being performed.

Original Matrix

$$\begin{bmatrix} 1 & 0 & 2 \\ 0 & 3 & 6 \end{bmatrix}$$

New Row-Equivalent Matrix

$$\begin{bmatrix} 1 & 0 & 2 \\ 0 & 1 & 2 \end{bmatrix}$$ ■

TECHNOLOGY

(T) Most graphing utilities can perform elementary row operations on matrices. The screens below show how one graphing utility displays each new row-equivalent matrix from Example 2. For specific instructions on how to use the elementary row operations features of a graphing utility, go to the text website at *college.hmco.com/info/larsonapplied*.

a. Interchange the first and second rows.

b. Multiply the first row by $\frac{1}{2}$.

c. Add -2 times the first row to the third row.

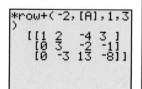

```
rowSwap([A],1,2)
   [[-1 2  0 3]
    [0  1  3 4]
    [2 -3 4 1]]
```

```
*row(.5,[A],1)
   [[1 -2 3  -1]
    [1  3 -3  0 ]
    [5 -2 1   2 ]]
```

```
*row+(-2,[A],1,3)
)
   [[1 2  -4 3 ]
    [0 3  -2 -1]
    [0 -3 13 -8]]
```

In Example 3 in Section 5.3, you used Gaussian elimination with back-substitution to solve a system of linear equations. The next example demonstrates the matrix version of Gaussian elimination. The two methods are essentially the same. The basic difference is that with matrices you do not need to keep writing the variables.

Example 3   Comparing Linear Systems and Matrix Operations

Linear System

$$\begin{cases} x - 2y + 3z = 9 \\ -x + 3y \quad\quad = -4 \\ 2x - 5y + 5z = 17 \end{cases}$$

Add the first equation to the second equation.

$$\begin{cases} x - 2y + 3z = 9 \\ \quad\quad y + 3z = 5 \\ 2x - 5y + 5z = 17 \end{cases}$$

Add -2 times the first equation to the third equation.

$$\begin{cases} x - 2y + 3z = 9 \\ \quad\quad y + 3z = 5 \\ \quad\quad -y - z = -1 \end{cases}$$

Add the second equation to the third equation.

$$\begin{cases} x - 2y + 3z = 9 \\ \quad\quad y + 3z = 5 \\ \quad\quad\quad 2z = 4 \end{cases}$$

Multiply the third equation by $\frac{1}{2}$.

$$\begin{cases} x - 2y + 3z = 9 \\ \quad\quad y + 3z = 5 \\ \quad\quad\quad z = 2 \end{cases}$$

Associated Augmented Matrix

$$\begin{bmatrix} 1 & -2 & 3 & \vdots & 9 \\ -1 & 3 & 0 & \vdots & -4 \\ 2 & -5 & 5 & \vdots & 17 \end{bmatrix}$$

Add the first row to the second row $(R_1 + R_2)$.

$$R_1 + R_2 \rightarrow \begin{bmatrix} 1 & -2 & 3 & \vdots & 9 \\ 0 & 1 & 3 & \vdots & 5 \\ 2 & -5 & 5 & \vdots & 17 \end{bmatrix}$$

Add -2 times the first row to the third row $(-2R_1 + R_3)$.

$$-2R_1 + R_3 \rightarrow \begin{bmatrix} 1 & -2 & 3 & \vdots & 9 \\ 0 & 1 & 3 & \vdots & 5 \\ 0 & -1 & -1 & \vdots & -1 \end{bmatrix}$$

Add the second row to the third row $(R_2 + R_3)$.

$$R_2 + R_3 \rightarrow \begin{bmatrix} 1 & -2 & 3 & \vdots & 9 \\ 0 & 1 & 3 & \vdots & 5 \\ 0 & 0 & 2 & \vdots & 4 \end{bmatrix}$$

Multiply the third row by $\frac{1}{2}$.

$$\tfrac{1}{2}R_3 \rightarrow \begin{bmatrix} 1 & -2 & 3 & \vdots & 9 \\ 0 & 1 & 3 & \vdots & 5 \\ 0 & 0 & 1 & \vdots & 2 \end{bmatrix}$$

At this point, you can use back-substitution to find that the solution is

$$x = 1, \, y = -1, \text{ and } z = 2$$

as shown in Example 1 in Section 5.3.

✓ CHECKPOINT 3

Write the system of equations represented by the augmented matrix. Use back-substitution to find the solution. (Use the variables x, y, and z.)

$$\begin{bmatrix} 1 & -2 & 5 & \vdots & 3 \\ 0 & 1 & 4 & \vdots & -3 \\ 0 & 0 & 1 & \vdots & 2 \end{bmatrix}$$ ∎

Remember that you can check a solution by substituting the values of x, y, and z into each equation in the original system. For example, you can check the solution to Example 3 as follows.

Equation 1: $1 - 2(-1) + 3(2) = 9$ ✓

Equation 2: $-1 + 3(-1) = -4$ ✓

Equation 3: $2(1) - 5(-1) + 5(2) = 17$ ✓

The last matrix in Example 3 is said to be in *row-echelon form*. The term *echelon* refers to the stair-step pattern formed by the nonzero entries of the matrix. To be in this form, a matrix must have the following properties.

Row-Echelon Form and Reduced Row-Echelon Form

A matrix in **row-echelon form** has the following properties.

1. Any rows consisting entirely of zeros occur at the bottom of the matrix.

2. For each row that does not consist entirely of zeros, the first nonzero entry is 1 (called a **leading 1**).

3. For two successive (nonzero) rows, the leading 1 in the higher row is farther to the left than the leading 1 in the lower row.

A matrix in *row-echelon form* is in **reduced row-echelon form** if every column that has a leading 1 has zeros in every position above and below the leading 1.

TECHNOLOGY

Some graphing utilities can automatically transform a matrix to row-echelon form and reduced row-echelon form. For specific keystrokes, go to the text website at *college.hmco.com/ info/larsonapplied.*

✓ **CHECKPOINT 4**

Determine whether the matrix is in row-echelon form. If it is, determine whether the matrix is in reduced row-echelon form.

$$\begin{bmatrix} 1 & 0 & 3 & -2 \\ 0 & 1 & 0 & -3 \\ 0 & 0 & 1 & 4 \end{bmatrix}$$

Example 4 Row-Echelon Form

Determine whether each matrix is in row-echelon form. If it is, determine whether the matrix is in reduced row-echelon form.

a. $\begin{bmatrix} 1 & 2 & -1 & 4 \\ 0 & 1 & 0 & 3 \\ 0 & 0 & 1 & -2 \end{bmatrix}$ **b.** $\begin{bmatrix} 1 & 2 & -1 & 2 \\ 0 & 0 & 0 & 0 \\ 0 & 1 & 2 & -4 \end{bmatrix}$

c. $\begin{bmatrix} 1 & -5 & 2 & -1 & 3 \\ 0 & 0 & 1 & 3 & -2 \\ 0 & 0 & 0 & 1 & 4 \\ 0 & 0 & 0 & 0 & 1 \end{bmatrix}$ **d.** $\begin{bmatrix} 1 & 0 & 0 & -1 \\ 0 & 1 & 0 & 2 \\ 0 & 0 & 1 & 3 \\ 0 & 0 & 0 & 0 \end{bmatrix}$

e. $\begin{bmatrix} 1 & 2 & -3 & 4 \\ 0 & 2 & 1 & -1 \\ 0 & 0 & 1 & -3 \end{bmatrix}$ **f.** $\begin{bmatrix} 0 & 1 & 0 & 5 \\ 0 & 0 & 1 & 3 \\ 0 & 0 & 0 & 0 \end{bmatrix}$

SOLUTION The matrices in (a), (c), (d), and (f) are in row-echelon form. The matrices in (d) and (f) are in reduced row-echelon form because every column that has a leading 1 has zeros in every position above and below the leading 1. The matrix in (b) is not in row-echelon form because the row of all zeros does not occur at the bottom of the matrix. The matrix in (e) is not in row-echelon form because the first nonzero entry in row 2 is not 1.

Every matrix can be converted to a row-equivalent matrix that is in row-echelon form. For instance, in Example 4, you can change the matrix in part (e) to row-echelon form by multiplying its second row by $\frac{1}{2}$, as shown below.

Original Matrix

$$\begin{bmatrix} 1 & 2 & -3 & 4 \\ 0 & 2 & 1 & -1 \\ 0 & 0 & 1 & -3 \end{bmatrix}$$

Row-Echelon Form

$$\frac{1}{2}R_2 \rightarrow \begin{bmatrix} 1 & 2 & -3 & 4 \\ 0 & 1 & \frac{1}{2} & -\frac{1}{2} \\ 0 & 0 & 1 & -3 \end{bmatrix}$$

Gaussian Elimination with Back-Substitution

Gaussian elimination with back-substitution works well for solving systems of linear equations by hand or with a computer. For this algorithm, the order in which the elementary row operations are performed is important. You should operate from *left to right* by columns, using elementary row operations to obtain zeros in all entries directly below the leading 1's.

Example 5 Gaussian Elimination with Back-Substitution

Solve the system.

$$\begin{cases} y + z - 2w = -3 \\ x + 2y - z = 2 \\ 2x + 4y + z - 3w = -2 \\ x - 4y - 7z - w = -19 \end{cases}$$

SOLUTION

$$\begin{matrix} R_2 \\ R_1 \end{matrix} \begin{bmatrix} 1 & 2 & -1 & 0 & \vdots & 2 \\ 0 & 1 & 1 & -2 & \vdots & -3 \\ 2 & 4 & 1 & -3 & \vdots & -2 \\ 1 & -4 & -7 & -1 & \vdots & -19 \end{bmatrix}$$ Interchange R_1 and R_2 so that there is a leading 1 in the upper left corner.

$$\begin{matrix} \\ \\ -2R_1 + R_3 \rightarrow \\ -R_1 + R_4 \rightarrow \end{matrix} \begin{bmatrix} 1 & 2 & -1 & 0 & \vdots & 2 \\ 0 & 1 & 1 & -2 & \vdots & -3 \\ 0 & 0 & 3 & -3 & \vdots & -6 \\ 0 & -6 & -6 & -1 & \vdots & -21 \end{bmatrix}$$ Perform operations on R_3 and R_4 so that the first column has zeros below the leading 1.

$$\begin{matrix} \\ \\ \\ 6R_2 + R_4 \rightarrow \end{matrix} \begin{bmatrix} 1 & 2 & -1 & 0 & \vdots & 2 \\ 0 & 1 & 1 & -2 & \vdots & -3 \\ 0 & 0 & 3 & -3 & \vdots & -6 \\ 0 & 0 & 0 & -13 & \vdots & -39 \end{bmatrix}$$ Perform operations on R_4 so that the second column has zeros below the leading 1.

$$\begin{matrix} \\ \\ \frac{1}{3}R_3 \rightarrow \\ \\ \end{matrix} \begin{bmatrix} 1 & 2 & -1 & 0 & \vdots & 2 \\ 0 & 1 & 1 & -2 & \vdots & -3 \\ 0 & 0 & 1 & -1 & \vdots & -2 \\ 0 & 0 & 0 & -13 & \vdots & -39 \end{bmatrix}$$ Multiply R_3 by $\frac{1}{3}$ so that the third row has a leading 1.

$$\begin{matrix} \\ \\ \\ -\frac{1}{13}R_4 \rightarrow \end{matrix} \begin{bmatrix} 1 & 2 & -1 & 0 & \vdots & 2 \\ 0 & 1 & 1 & -2 & \vdots & -3 \\ 0 & 0 & 1 & -1 & \vdots & -2 \\ 0 & 0 & 0 & 1 & \vdots & 3 \end{bmatrix}$$ Multiply R_4 by $-\frac{1}{13}$ so that the fourth row has a leading 1.

The matrix is now in row-echelon form, and the corresponding system is

$$\begin{cases} x + 2y - z = 2 \\ y + z - 2w = -3 \\ z - w = -2 \\ w = 3 \end{cases}.$$

Using back-substitution, you can determine that the solution is $x = -1$, $y = 2$, $z = 1$, and $w = 3$. Check this in the original system of equations.

✓ **CHECKPOINT 5**

Solve the system.

$$\begin{cases} y + 2z - w = -5 \\ x - 3y - z = 0 \\ 2x - 6y + z + 3w = 6 \\ 4x + 4y - 2z + w = 1 \end{cases}$$

Gaussian Elimination with Back-Substitution

1. Write the augmented matrix of the system of linear equations.

2. Use elementary row operations to rewrite the augmented matrix in row-echelon form.

3. Write the system of linear equations corresponding to the matrix in row-echelon form, and use back-substitution to find the solution.

When solving a system of linear equations, remember that it is possible for the system to have no solution. If, in the elimination process, you obtain a row with zeros except for the last entry, it is unnecessary to continue the elimination process. You can conclude that the system has no solution, or is inconsistent.

Example 6 **A System with No Solution**

Solve the system.

$$\begin{cases} 3x + 2y - z = 1 \\ x \quad\quad + z = 6 \\ 2x - 3y + 5z = 4 \\ x - y + 2z = 4 \end{cases}$$

SOLUTION

$$\begin{bmatrix} 3 & 2 & -1 & \vdots & 1 \\ 1 & 0 & 1 & \vdots & 6 \\ 2 & -3 & 5 & \vdots & 4 \\ 1 & -1 & 2 & \vdots & 4 \end{bmatrix}$$

$$\begin{matrix} R_4 \\ \\ \\ R_1 \end{matrix} \begin{bmatrix} 1 & -1 & 2 & \vdots & 4 \\ 1 & 0 & 1 & \vdots & 6 \\ 2 & -3 & 5 & \vdots & 4 \\ 3 & 2 & -1 & \vdots & 1 \end{bmatrix}$$

$$\begin{matrix} \\ -R_1 + R_2 \rightarrow \\ -2R_1 + R_3 \rightarrow \\ -3R_1 + R_4 \rightarrow \end{matrix} \begin{bmatrix} 1 & -1 & 2 & \vdots & 4 \\ 0 & 1 & -1 & \vdots & 2 \\ 0 & -1 & 1 & \vdots & -4 \\ 0 & 5 & -7 & \vdots & -11 \end{bmatrix}$$

$$\begin{matrix} \\ \\ R_2 + R_3 \rightarrow \\ \end{matrix} \begin{bmatrix} 1 & -1 & 2 & \vdots & 4 \\ 0 & 1 & -1 & \vdots & 2 \\ 0 & 0 & 0 & \vdots & -2 \\ 0 & 5 & -7 & \vdots & -11 \end{bmatrix}$$

Note that the third row of this matrix consists of zeros except for the last entry. This means that the original system of linear equations is *inconsistent*. You can see why this is true by converting back to a system of linear equations.

$$\begin{cases} x - y + 2z = \quad 4 \\ y - z = \quad 2 \\ 0 = -2 \\ 5y - 7z = -11 \end{cases}$$

Because $0 = -2$ is a false statement, the system has no solution.

✓ CHECKPOINT 6

Solve the system.

$$\begin{cases} x + 2y - z = 3 \\ -x - y + 3z = 2 \\ 2x + 3y - 4z = 0 \\ 3x + 2y + z = -4 \end{cases}$$ ∎

Gauss-Jordan Elimination

With Gaussian elimination, elementary row operations are applied to a matrix to obtain a (row-equivalent) row-echelon form of the matrix. A second method of elimination, called **Gauss-Jordan elimination** after Carl Friedrich Gauss and Wilhelm Jordan (1842–1899), continues the reduction process until a *reduced* row-echelon form is obtained. This procedure is demonstrated in Example 7.

STUDY TIP

Either Gaussian elimination or Gauss-Jordan elimination can be used to solve a system of equations. The method you use depends on your preference.

Example 7   Gauss-Jordan Elimination

Use Gauss-Jordan elimination to solve the system.

$$\begin{cases} x - 2y + 3z = 9 \\ -x + 3y = -4 \\ 2x - 5y + 5z = 17 \end{cases}$$

SOLUTION   In Example 3, Gaussian elimination was used to obtain the row-echelon form

$$\begin{bmatrix} 1 & -2 & 3 & \vdots & 9 \\ 0 & 1 & 3 & \vdots & 5 \\ 0 & 0 & 1 & \vdots & 2 \end{bmatrix}.$$

Now, apply elementary row operations until you obtain a matrix in reduced row-echelon form. To do this, you must produce zeros above each of the leading 1's, as follows.

$$2R_2 + R_1 \longrightarrow \begin{bmatrix} 1 & 0 & 9 & \vdots & 19 \\ 0 & 1 & 3 & \vdots & 5 \\ 0 & 0 & 1 & \vdots & 2 \end{bmatrix}$$

Perform operations on R_1 so that the second column has a zero above the leading 1.

$$\begin{matrix} -9R_3 + R_1 \longrightarrow \\ -3R_3 + R_2 \longrightarrow \end{matrix} \begin{bmatrix} 1 & 0 & 0 & \vdots & 1 \\ 0 & 1 & 0 & \vdots & -1 \\ 0 & 0 & 1 & \vdots & 2 \end{bmatrix}$$

Perform operations on R_1 and R_2 so that the third column has zeros above the leading 1.

✓CHECKPOINT 7

Use Gauss-Jordan elimination to solve the system.

$$\begin{cases} x - 3y + 2z = 1 \\ -x - y + 3z = 4 \\ y - 2z = -5 \end{cases}$$

Now, converting back to a system of linear equations, you have

$$\begin{cases} x = 1 \\ y = -1. \\ z = 2 \end{cases}$$

An advantage of Gauss-Jordan elimination is that you can read the solution from the matrix in reduced row-echelon form.

The elimination procedures described in this section sometimes result in fractional coefficients. For instance, in the elimination procedure for the system

$$\begin{cases} 2x - 5y + 5z = 17 \\ 3x - 2y + 3z = 11 \\ -3x + 3y = 6 \end{cases}$$

you may be inclined to multiply the first row by $\frac{1}{2}$ to produce a leading 1, which will result in working with fractional coefficients. You can sometimes avoid fractions by judiciously choosing the order in which you apply elementary row operations.

Example 8 Comparing Row-Echelon Forms

Compare the row-echelon form obtained below with the one found in Example 3. Is it the same? Does it yield the same solution?

$$\begin{cases} x - 2y + 3z = 9 \\ -x + 3y \quad\;\; = -4 \\ 2x - 5y + 5z = 17 \end{cases}$$

$$\begin{bmatrix} 1 & -2 & 3 & \vdots & 9 \\ -1 & 3 & 0 & \vdots & -4 \\ 2 & -5 & 5 & \vdots & 17 \end{bmatrix}$$

$$\begin{matrix} R_2 \\ R_1 \end{matrix} \begin{bmatrix} -1 & 3 & 0 & \vdots & -4 \\ 1 & -2 & 3 & \vdots & 9 \\ 2 & -5 & 5 & \vdots & 17 \end{bmatrix}$$

$$-R_1 \rightarrow \begin{bmatrix} 1 & -3 & 0 & \vdots & 4 \\ 1 & -2 & 3 & \vdots & 9 \\ 2 & -5 & 5 & \vdots & 17 \end{bmatrix}$$

$$\begin{matrix} -R_1 + R_2 \rightarrow \\ -2R_1 + R_3 \rightarrow \end{matrix} \begin{bmatrix} 1 & -3 & 0 & \vdots & 4 \\ 0 & 1 & 3 & \vdots & 5 \\ 0 & 1 & 5 & \vdots & 9 \end{bmatrix}$$

$$-R_2 + R_3 \rightarrow \begin{bmatrix} 1 & -3 & 0 & \vdots & 4 \\ 0 & 1 & 3 & \vdots & 5 \\ 0 & 0 & 2 & \vdots & 4 \end{bmatrix}$$

$$\tfrac{1}{2}R_3 \rightarrow \begin{bmatrix} 1 & -3 & 0 & \vdots & 4 \\ 0 & 1 & 3 & \vdots & 5 \\ 0 & 0 & 1 & \vdots & 2 \end{bmatrix}$$

SOLUTION This row-echelon form is different from the one that was obtained in Example 3. The corresponding system of linear equations for this matrix is

$$\begin{cases} x - 3y \quad\;\; = 4 \\ \quad\;\; y + 3z = 5. \\ \quad\quad\quad z = 2 \end{cases}$$

Using back-substitution on this system, you obtain the solution

$$x = 1, y = -1, \text{ and } z = 2$$

which is the same solution that was obtained in Example 3. This row-echelon form is not the same as the one found in Example 3, but both forms yield the same solution.

✓**CHECKPOINT 8**

Compare the row-echelon form below with the one found in Example 8. Is it the same? Does it yield the same solution?

$$\begin{bmatrix} 1 & -2 & 0 & \vdots & 3 \\ 0 & 1 & 0 & \vdots & -1 \\ 0 & 0 & 1 & \vdots & 2 \end{bmatrix}$$

In Example 8, you discovered that the row-echelon form of a matrix is *not* unique. Two different sequences of elementary row operations may yield different row-echelon forms. However, the *reduced* row-echelon form of a given matrix *is* unique. Try applying Gauss-Jordan elimination to the row-echelon matrix in Example 8 to see that you obtain the same reduced row-echelon form as in Example 7.

Example 9   A System with an Infinite Number of Solutions

Solve the system.

$$\begin{cases} 2x + 4y - 2z = 0 \\ 3x + 5y \quad\quad = 1 \end{cases}$$

SOLUTION

$$\begin{bmatrix} 2 & 4 & -2 & \vdots & 0 \\ 3 & 5 & 0 & \vdots & 1 \end{bmatrix} \qquad \frac{1}{2}R_1 \rightarrow \begin{bmatrix} 1 & 2 & -1 & \vdots & 0 \\ 3 & 5 & 0 & \vdots & 1 \end{bmatrix}$$

$$-3R_1 + R_2 \rightarrow \begin{bmatrix} 1 & 2 & -1 & \vdots & 0 \\ 0 & -1 & 3 & \vdots & 1 \end{bmatrix}$$

$$-R_2 \rightarrow \begin{bmatrix} 1 & 2 & -1 & \vdots & 0 \\ 0 & 1 & -3 & \vdots & -1 \end{bmatrix}$$

$$-2R_2 + R_1 \rightarrow \begin{bmatrix} 1 & 0 & 5 & \vdots & 2 \\ 0 & 1 & -3 & \vdots & -1 \end{bmatrix}$$

The corresponding system of equations is

$$\begin{cases} x \quad\quad + 5z = \quad 2 \\ \quad y - 3z = -1 \end{cases}.$$

Solving for x and y in terms of z, you have $x = -5z + 2$ and $y = 3z - 1$. To write a solution of the system that does not use any of the three variables of the system, let a represent any real number and let $z = a$. Now, substitute a for z in the equations for x and y.

$$x = -5z + 2 = -5a + 2$$

$$y = 3z - 1 = 3a - 1$$

So, the solution set has the form

$$(-5a + 2, 3a - 1, a)$$

where a is a real number. Try substituting values for a to obtain a few solutions. Then check each solution in the original system of equations.

✓ CHECKPOINT 9

Solve the system. $\begin{cases} \quad\quad y + 5z = -2 \\ -x + y - 4z = \quad 8 \end{cases}$ ▨

CONCEPT CHECK

1. A matrix has four columns and three rows. Is the order of the matrix 4×3? Explain.

2. Can every matrix be written in row-echelon form? Explain.

3. When solving a system of equations using Gaussian elimination, you obtain the statement $0 = 4$. What can you conclude? Explain.

4. Explain the difference between using Gaussian elimination and using Gauss-Jordan elimination when solving a system of linear equations.

Skills Review 6.1

The following warm-up exercises involve skills that were covered in earlier sections. You will use these skills in the exercise set for this section. For additional help, review Sections 0.2, 5.1, and 5.3.

In Exercises 1–4, evaluate the expression.

1. $2(-1) - 3(5) + 7(2)$

2. $-4(-3) + 6(7) + 8(-3)$

3. $11\left(\frac{1}{2}\right) - 7\left(-\frac{3}{2}\right) - 5(2)$

4. $\frac{2}{3}\left(\frac{1}{2}\right) + \frac{4}{3}\left(-\frac{1}{3}\right)$

In Exercises 5 and 6, determine whether $x = 1$, $y = 3$, and $z = -1$ is a solution of the system.

5. $\begin{cases} 4x - 2y + 3z = -5 \\ x + 3y - z = 11 \\ -x + 2y = 5 \end{cases}$

6. $\begin{cases} -x + 2y + z = 4 \\ 2x - 3z = 5 \\ 3x + 5y - 2z = 21 \end{cases}$

In Exercises 7–10, use back-substitution to solve the system of linear equations.

7. $\begin{cases} 2x - 3y = 4 \\ y = 2 \end{cases}$

8. $\begin{cases} 5x + 4y = 0 \\ y = -3 \end{cases}$

9. $\begin{cases} x - 3y + z = 0 \\ y - 3z = 8 \\ z = 2 \end{cases}$

10. $\begin{cases} 2x - 5y + 3z = -2 \\ y - 4z = 0 \\ z = 1 \end{cases}$

Exercises 6.1

See www.CalcChat.com for worked-out solutions to odd-numbered exercises.

In Exercises 1–8, determine the order of the matrix.

1. $\begin{bmatrix} 0 & -3 & 0 \\ 9 & 2 & -7 \end{bmatrix}$

2. $[-7 \quad 21]$

3. $\begin{bmatrix} 6 & 4 & 1 \\ 8 & 3 & 0 \\ -1 & 2 & 1 \\ 1 & 5 & 4 \end{bmatrix}$

4. $\begin{bmatrix} 1 \\ 0 \\ 3 \\ 5 \\ 6 \end{bmatrix}$

5. $\begin{bmatrix} 33 & 45 \\ -9 & 20 \\ 12 & 15 \\ 16 & -2 \end{bmatrix}$

6. $\begin{bmatrix} 12 & -2 & 4 \\ -3 & 4 & 0 \\ -8 & 12 & 2 \end{bmatrix}$

7. $\begin{bmatrix} 2 & 7 & 11 & -3 \\ -1 & 10 & -5 & 0 \end{bmatrix}$

8. $[-11]$

In Exercises 9–12, fill in the blank(s) to form a new row-equivalent matrix.

Original Matrix   *New Row-Reduced Matrix*

9. $\begin{bmatrix} 1 & 1 & 1 \\ 5 & -2 & 4 \end{bmatrix}$ $\begin{bmatrix} 1 & 1 & 1 \\ 0 & & -1 \end{bmatrix}$

10. $\begin{bmatrix} -3 & 3 & 12 \\ 18 & -8 & 4 \end{bmatrix}$ $\begin{bmatrix} 1 & -1 & \\ 18 & -8 & 4 \end{bmatrix}$

Original Matrix   *New Row-Reduced Matrix*

11. $\begin{bmatrix} 1 & 5 & 4 & -1 \\ 0 & 1 & -2 & 2 \\ 0 & 0 & 1 & -7 \end{bmatrix}$ $\begin{bmatrix} 1 & 0 & & \\ 0 & 1 & -2 & 2 \\ 0 & 0 & 1 & -7 \end{bmatrix}$

12. $\begin{bmatrix} 1 & 0 & 6 & 1 \\ 0 & -1 & 0 & 7 \\ 0 & 0 & -1 & 3 \end{bmatrix}$ $\begin{bmatrix} 1 & 0 & 6 & 1 \\ 0 & 1 & 0 & \\ 0 & 0 & 1 & \end{bmatrix}$

In Exercises 13–16, identify the elementary row operation(s) being performed to obtain the new row-equivalent matrix.

Original Matrix   *New Row-Equivalent Matrix*

13. $\begin{bmatrix} -2 & 5 & 1 \\ 3 & -1 & -8 \end{bmatrix}$ $\begin{bmatrix} 13 & 0 & -39 \\ 3 & -1 & -8 \end{bmatrix}$

14. $\begin{bmatrix} 3 & -1 & -4 \\ -4 & 3 & 7 \end{bmatrix}$ $\begin{bmatrix} 3 & -1 & -4 \\ 5 & 0 & -5 \end{bmatrix}$

15. $\begin{bmatrix} 0 & -1 & -5 & 5 \\ -1 & 3 & -7 & 6 \\ 4 & -5 & 1 & 3 \end{bmatrix}$ $\begin{bmatrix} -1 & 3 & -7 & 6 \\ 0 & -1 & -5 & 5 \\ 0 & 7 & -27 & 27 \end{bmatrix}$

Original Matrix *New Row-Equivalent Matrix*

16. $\begin{bmatrix} -1 & -2 & 3 & -2 \\ 2 & -5 & 1 & -7 \\ 5 & 4 & -7 & 6 \end{bmatrix}$ $\begin{bmatrix} -1 & -2 & 3 & -2 \\ 0 & -9 & 7 & -11 \\ 0 & -6 & 8 & -4 \end{bmatrix}$

In Exercises 17–22, determine whether the matrix is in row-echelon form. If it is, determine if it is also in *reduced* row-echelon form.

17. $\begin{bmatrix} 1 & 0 & 0 & 0 \\ 0 & 1 & 1 & 5 \\ 0 & 0 & 0 & 0 \end{bmatrix}$ **18.** $\begin{bmatrix} 1 & 0 & 2 & 1 \\ 0 & 1 & -3 & 10 \\ 0 & 0 & 1 & 0 \end{bmatrix}$

19. $\begin{bmatrix} 2 & 0 & 4 & 0 \\ 0 & -1 & 3 & 6 \\ 0 & 0 & 1 & 5 \end{bmatrix}$ **20.** $\begin{bmatrix} 0 & 0 & 0 & 0 \\ 0 & 1 & 0 & 5 \\ 0 & 0 & 1 & 3 \end{bmatrix}$

21. $\begin{bmatrix} 1 & 3 & 0 & 0 & 0 & 0 \\ 0 & 0 & 1 & 8 & 1 & 0 \\ 0 & 0 & 0 & 0 & 1 & 1 \\ 0 & 0 & 0 & 0 & 1 & 1 \end{bmatrix}$

22. $\begin{bmatrix} 1 & 0 & 0 & 10 \\ 0 & 1 & 3 & 9 \\ 0 & 0 & 0 & 1 \\ 0 & 0 & 0 & 0 \end{bmatrix}$

Ⓣ **23.** Use a graphing utility to perform the sequence of row operations in parts (a) through (d) to reduce the matrix to row-echelon form.

$\begin{bmatrix} 1 & 1 & 2 \\ 3 & 4 & -3 \\ 2 & -1 & 1 \end{bmatrix}$

(a) Add -3 times R_1 to R_2.

(b) Add -2 times R_1 to R_3.

(c) Add 3 times R_2 to R_3.

(d) Multiply R_3 by $-\frac{1}{30}$.

Ⓣ **24.** Use a graphing utility to perform the sequence of row operations in parts (a) through (f) to reduce the matrix to *reduced* row-echelon form.

$\begin{bmatrix} 7 & 1 \\ 0 & 2 \\ -3 & 4 \\ 4 & 1 \end{bmatrix}$

(a) Add R_3 to R_4.

(b) Interchange R_1 and R_4.

(c) Add 3 times R_1 to R_3.

(d) Add -7 times R_1 to R_4.

(e) Multiply R_2 by $\frac{1}{2}$.

(f) Add the appropriate multiple of R_2 to R_1, R_3, and R_4.

In Exercises 25–28, write the matrix in row-echelon form. (Note: Row-echelon forms are not unique.)

25. $\begin{bmatrix} 1 & 2 & -1 & 5 \\ 3 & 2 & 1 & 11 \\ 4 & 8 & 1 & 10 \end{bmatrix}$

26. $\begin{bmatrix} 1 & 2 & -1 & 3 \\ 3 & 7 & -5 & 14 \\ -2 & -1 & -3 & 8 \end{bmatrix}$

27. $\begin{bmatrix} 1 & -1 & -1 & 1 \\ 5 & -4 & 1 & 8 \\ -6 & 8 & 18 & 0 \end{bmatrix}$

28. $\begin{bmatrix} 1 & -3 & 0 & -7 \\ -3 & 10 & 1 & 23 \\ 1 & 0 & 1 & 12 \\ 4 & -10 & 2 & -24 \end{bmatrix}$

In Exercises 29–34, write the matrix in reduced row-echelon form.

29. $\begin{bmatrix} 4 & 4 & 8 \\ 1 & 2 & 2 \\ -3 & 6 & -9 \end{bmatrix}$ **30.** $\begin{bmatrix} 1 & 3 & 2 \\ 5 & 15 & 9 \\ 2 & 6 & 10 \end{bmatrix}$

31. $\begin{bmatrix} 1 & 0 & 2 & 1 \\ 0 & 3 & -6 & 6 \\ 2 & 0 & 5 & -4 \\ 0 & 1 & 0 & 1 \end{bmatrix}$ **32.** $\begin{bmatrix} 1 & 2 & 3 & -5 \\ 1 & 2 & 4 & -9 \\ -2 & -4 & -4 & 3 \\ 4 & 8 & 11 & -14 \end{bmatrix}$

33. $\begin{bmatrix} 2 & 1 \\ 1 & 4 \\ -2 & -1 \end{bmatrix}$ **34.** $\begin{bmatrix} 1 & -3 \\ -1 & 8 \\ 0 & 4 \\ -2 & 10 \end{bmatrix}$

In Exercises 35–38, write the system of linear equations represented by the augmented matrix. (Use the variables x, y, z, and w.)

35. $\begin{bmatrix} 2 & 4 & \vdots & 6 \\ -1 & 3 & \vdots & -8 \end{bmatrix}$

36. $\begin{bmatrix} 7 & -2 & \vdots & 7 \\ -8 & 3 & \vdots & -3 \end{bmatrix}$

37. $\begin{bmatrix} 1 & 0 & 2 & \vdots & -10 \\ 0 & 3 & -1 & \vdots & 5 \\ 4 & 2 & 0 & \vdots & 3 \end{bmatrix}$

38. $\begin{bmatrix} 5 & 8 & 2 & 0 & \vdots & -1 \\ -2 & 15 & 5 & 1 & \vdots & 9 \\ 1 & 6 & -7 & 0 & \vdots & -3 \end{bmatrix}$

In Exercises 39–44, write the augmented matrix for the system of linear equations.

39. $\begin{cases} 2x - y = 3 \\ 5x + 7y = 12 \end{cases}$ **40.** $\begin{cases} 8x + 3y = 25 \\ 3x - 9y = 12 \end{cases}$

41. $\begin{cases} x + 10y - 3z = 2 \\ 5x - 3y + 4z = 0 \\ 2x + 4y = 6 \end{cases}$ **42.** $\begin{cases} 2x + 3y - z = 8 \\ y + 2z = -10 \\ x - 2y - 3z = 21 \end{cases}$

43. $\begin{cases} 9w - 3x + 20y + z = 13 \\ 12w - 8y = 5 \\ w + 2x + 3y - 4z = -2 \\ -w - x + y + z = 1 \end{cases}$

44. $\begin{cases} w + 2x - 3y + z = 18 \\ 3w - 5y = 8 \\ w + x + y + 2z = 15 \\ -w - x + 2y + z = -3 \end{cases}$

In Exercises 45–48, write the system of equations represented by the augmented matrix. Use back-substitution to find the solution. (Use x, y, z, and w.)

45. $\begin{bmatrix} 1 & -5 & \vdots & 6 \\ 0 & 1 & \vdots & -2 \end{bmatrix}$

46. $\begin{bmatrix} 1 & 2 & -1 & \vdots & 3 \\ 0 & 1 & -2 & \vdots & -3 \\ 0 & 0 & 1 & \vdots & 4 \end{bmatrix}$

47. $\begin{bmatrix} 1 & 3 & -1 & \vdots & 15 \\ 0 & 1 & 4 & \vdots & -12 \\ 0 & 0 & 1 & \vdots & -5 \end{bmatrix}$

48. $\begin{bmatrix} 1 & 2 & -2 & 0 & \vdots & -1 \\ 0 & 1 & 1 & 2 & \vdots & 9 \\ 0 & 0 & 1 & 0 & \vdots & 2 \\ 0 & 0 & 0 & 1 & \vdots & -3 \end{bmatrix}$

In Exercises 49–54, an augmented matrix that represents a system of linear equations (in variables x, y, and z) has been reduced using Gauss-Jordan elimination. Write the solution represented by the augmented matrix.

49. $\begin{bmatrix} 1 & 0 & \vdots & -4 \\ 0 & 1 & \vdots & 6 \end{bmatrix}$ **50.** $\begin{bmatrix} 1 & 0 & \vdots & 9 \\ 0 & 1 & \vdots & -3 \end{bmatrix}$

51. $\begin{bmatrix} 1 & 0 & 0 & \vdots & -4 \\ 0 & 1 & 0 & \vdots & -8 \\ 0 & 0 & 1 & \vdots & 2 \end{bmatrix}$

52. $\begin{bmatrix} 1 & 0 & 0 & \vdots & 3 \\ 0 & 1 & 0 & \vdots & -1 \\ 0 & 0 & 1 & \vdots & 0 \end{bmatrix}$

53. $\begin{bmatrix} 1 & 0 & 2 & \vdots & -4 \\ 0 & 1 & 1 & \vdots & 6 \\ 0 & 0 & 0 & \vdots & 0 \end{bmatrix}$

54. $\begin{bmatrix} 1 & 0 & 2 & \vdots & 9 \\ 0 & 1 & 5 & \vdots & -3 \\ 0 & 0 & 0 & \vdots & 0 \end{bmatrix}$

In Exercises 55–78, use matrices to solve the system of equations (if possible). Use Gaussian elimination with back-substitution or Gauss-Jordan elimination.

55. $\begin{cases} x + 2y = 7 \\ 2x + y = 8 \end{cases}$ **56.** $\begin{cases} 2x + 6y = 16 \\ 2x + 3y = 7 \end{cases}$

57. $\begin{cases} -3x + 5y = -22 \\ 3x + 4y = 4 \\ 4x - 8y = 32 \end{cases}$ **58.** $\begin{cases} x + 2y = 0 \\ x + y = 6 \\ 3x - 2y = 8 \end{cases}$

59. $\begin{cases} 8x - 4y = 7 \\ 5x + 2y = 1 \end{cases}$ **60.** $\begin{cases} x - 3y = 5 \\ -2x + 6y = -10 \end{cases}$

61. $\begin{cases} -x + 2y = 1.5 \\ 2x - 4y = 3 \end{cases}$ **62.** $\begin{cases} 2x - y = -0.1 \\ 3x + 2y = 1.6 \end{cases}$

63. $\begin{cases} 2x + 2y - z = 2 \\ x - 3y + z = -28 \\ -x + y = 14 \end{cases}$ **64.** $\begin{cases} -x + y - z = -14 \\ 2x - y + z = 21 \\ 3x + 2y + z = 19 \end{cases}$

65. $\begin{cases} 2x + 3z = 3 \\ 4x - 3y + 7z = 5 \\ 8x - 9y + 15z = 9 \end{cases}$ **66.** $\begin{cases} 2x - y + 3z = 24 \\ 2y - z = 14 \\ 7x - 5y = 6 \end{cases}$

67. $\begin{cases} x + y - 5z = 3 \\ x - 2z = 1 \\ 2x - y - z = 1 \end{cases}$ **68.** $\begin{cases} x - 3z = -2 \\ 3x + y - 2z = 5 \\ 2x + 2y + z = 4 \end{cases}$

69. $\begin{cases} x + 2y + z = 8 \\ 3x + 7y + 6z = 26 \end{cases}$ **70.** $\begin{cases} x + y + 4z = 5 \\ 2x + y - z = 9 \end{cases}$

71. $\begin{cases} 3x + 3y + 12z = 6 \\ x + y + 4z = 2 \\ 2x + 5y + 20z = 10 \\ -x + 2y + 8z = 4 \end{cases}$ **72.** $\begin{cases} 2x + 10y + 2z = 6 \\ x + 5y + 2z = 6 \\ x + 5y + z = 3 \\ -3x + 15y - 3z = -9 \end{cases}$

73. $\begin{cases} 4x + 12y - 7z - 20w = 22 \\ 3x + 9y - 5z - 28w = 30 \end{cases}$

74. $\begin{cases} x + 2y + 2z + 4w = 11 \\ 3x + 6y + 5z + 12w = 30 \end{cases}$

75. $\begin{cases} x + 2y = 0 \\ -x - y = 0 \end{cases}$ **76.** $\begin{cases} x + 2y = 0 \\ 2x + 4y = 0 \end{cases}$

77. $\begin{cases} x + y + z = 0 \\ 2x + 3y + z = 0 \\ 3x + 5y + z = 0 \end{cases}$ **78.** $\begin{cases} x - 2y + z + 3w = 0 \\ x - y + w = 0 \\ y - z + 2w = 0 \end{cases}$

In Exercises 79–82, determine whether the two systems of linear equations yield the same solution. If so, find the solution using matrices.

79. (a) $\begin{cases} x - 2y + z = -6 \\ y - 5z = 16 \\ z = -3 \end{cases}$ (b) $\begin{cases} x + y - 2z = 6 \\ y + 3z = -8 \\ z = -3 \end{cases}$

80. (a) $\begin{cases} x - 3y + 4z = -11 \\ \quad\ y - z = -4 \\ \qquad\qquad z = 2 \end{cases}$ (b) $\begin{cases} x + 4y \qquad = -11 \\ \quad\ y + 3z = 4 \\ \qquad\qquad z = 2 \end{cases}$

81. (a) $\begin{cases} x - 4y + 5z = 27 \\ \quad\ y - 7z = -54 \\ \qquad\qquad z = 8 \end{cases}$ (b) $\begin{cases} x - 6y + z = 15 \\ \quad\ y + 5z = 42 \\ \qquad\qquad z = 8 \end{cases}$

82. (a) $\begin{cases} x + 3y - z = 19 \\ \quad\ y + 6z = -18 \\ \qquad\qquad z = -4 \end{cases}$ (b) $\begin{cases} x - y + 3z = -15 \\ \quad\ y - 2z = 14 \\ \qquad\qquad z = -4 \end{cases}$

83. Breeding Facility   A city zoo borrowed $2,000,000 at simple annual interest to construct a breeding facility. Some of the money was borrowed at 8%, some at 9%, and some at 12%. Use a system of equations to determine how much was borrowed at each rate if the total annual interest was $186,000 and the amount borrowed at 8% was twice the amount borrowed at 12%. Solve the system using matrices.

84. Museum   A natural history museum borrowed $2,000,000 at simple annual interest to purchase new exhibits. Some of the money was borrowed at 7%, some at 8.5%, and some at 9.5%. Use a system of equations to determine how much was borrowed at each rate if the total annual interest was $169,750 and the amount borrowed at 8.5% was four times the amount borrowed at 9.5%. Solve the system using matrices.

85. You and a friend solve the following system of equations independently.

$$\begin{cases} 2x - 4y - 3z = 3 \\ \quad\ x + 3y + z = -1 \\ 5x + \ y - 2z = 2 \end{cases}$$

You write your solution set as

$(a, -a, 2a - 1)$

where a is any real number. Your friend's solution set is

$\left(\tfrac{1}{2}b + \tfrac{1}{2}, -\tfrac{1}{2}b - \tfrac{1}{2}, b\right)$

where b is any real number. Are you both correct? Explain. If you let $a = 3$, what value of b must be selected so that you both have the same ordered triple?

86. Describe how you would explain to another student that the augmented matrix below represents a dependent system of equations. Describe a way to write the infinitely many solutions of this system.

$$\begin{bmatrix} 1 & -2 & 3 & \vdots & -6 \\ 0 & 1 & 2 & \vdots & 5 \\ 0 & 0 & 0 & \vdots & 0 \end{bmatrix}$$

87. Health and Wellness   From 1997 to 2008, the number of new cases of a waterborne disease in a small city increased in a pattern that was approximately linear (see figure). Find the least squares regression line

$y = at + b$

for the data shown in the figure by solving the following system using matrices. Let t represent the year, with $t = 0$ corresponding to 1997.

$$\begin{cases} 12b + 66a = 831 \\ 66b + 506a = 5643 \end{cases}$$

Use the result to predict the number of new cases of the waterborne disease in 2011. Is the estimate reasonable? Explain.

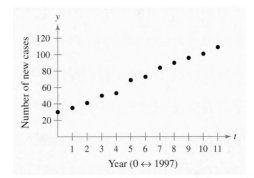

Year (0 ↔ 1997)

88. Energy Imports   From 1994 to 2005, the total energy imports y (in quadrillions of Btu's) to the United States increased in a pattern that was approximately linear (see figure). Find the least squares regression line

$y = at + b$

for the data shown in the figure by solving the following system using matrices. Let t represent the year, with $t = 0$ corresponding to 1994.

$$\begin{cases} 12b + 66a = 334.80 \\ 66b + 506a = 1999.91 \end{cases}$$

Use the result to predict the total energy imports in 2010. Is the estimate reasonable? Explain. *(Source: Energy Information Administration)*

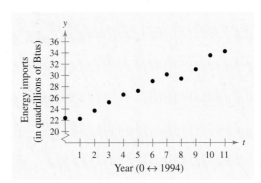

Year (0 ↔ 1994)

Section 6.2

Operations with Matrices

■ Determine whether two matrices are equal.
■ Add or subtract two matrices and multiply a matrix by a scalar.
■ Find the product of two matrices.
■ Solve a matrix equation.
■ Use matrix multiplication to solve an application problem.

Equality of Matrices

In Section 6.1, you used matrices to solve systems of linear equations. Matrices, however, can do much more than this. There is a rich mathematical theory of matrices, with numerous applications. This section and the next introduce some fundamentals of matrix theory. It is standard mathematical convention to represent matrices in any of the following three ways.

1. A matrix can be denoted by an uppercase letter such as A, B, or C.

2. A matrix can be denoted by a representative element enclosed in brackets, such as $[a_{ij}]$, $[b_{ij}]$, or $[c_{ij}]$.

3. A matrix can be denoted by a rectangular array of numbers such as

$$A = [a_{ij}] = \begin{bmatrix} a_{11} & a_{12} & a_{13} & \cdots & a_{1n} \\ a_{21} & a_{22} & a_{23} & \cdots & a_{2n} \\ a_{31} & a_{32} & a_{33} & \cdots & a_{3n} \\ \vdots & \vdots & \vdots & & \vdots \\ a_{m1} & a_{m2} & a_{m3} & \cdots & a_{mn} \end{bmatrix}.$$

Two matrices $A = [a_{ij}]$ and $B = [b_{ij}]$ are **equal** if they have the same order $(m \times n)$ and if $a_{ij} = b_{ij}$ for $1 \leq i \leq m$ and $1 \leq j \leq n$. In other words, two matrices are equal if their corresponding entries are equal.

Example 1 Equality of Matrices

Solve for a_{11}, a_{12}, a_{21}, and a_{22} in the matrix equation.

$$\begin{bmatrix} a_{11} & a_{12} \\ a_{21} & a_{22} \end{bmatrix} = \begin{bmatrix} 2 & -1 \\ -3 & 0 \end{bmatrix}$$

SOLUTION Because two matrices are equal only if their corresponding entries are equal, you can conclude that

$$a_{11} = 2, \quad a_{12} = -1, \quad a_{21} = -3, \quad \text{and} \quad a_{22} = 0.$$

✓ CHECKPOINT 1

Solve for a_{11}, a_{12}, a_{21}, and a_{22} in the matrix equation.

$$\begin{bmatrix} a_{11} & a_{12} \\ a_{21} & a_{22} \end{bmatrix} = \begin{bmatrix} 5 & 2 \\ -1 & 3 \end{bmatrix} \quad ■$$

Matrix Addition and Scalar Multiplication

You can **add** two matrices (of the same order) by adding their corresponding entries.

Definition of Matrix Addition

If $A = [a_{ij}]$ and $B = [b_{ij}]$ are matrices of order $m \times n$, their **sum** is the $m \times n$ matrix given by

$$A + B = [a_{ij} + b_{ij}].$$

The sum of two matrices of different orders is undefined.

Example 2   **Addition of Matrices**

a. $\begin{bmatrix} -1 & 2 \\ 0 & 1 \end{bmatrix} + \begin{bmatrix} 1 & 3 \\ -1 & 2 \end{bmatrix} = \begin{bmatrix} -1+1 & 2+3 \\ 0+(-1) & 1+2 \end{bmatrix} = \begin{bmatrix} 0 & 5 \\ -1 & 3 \end{bmatrix}$

b. $\begin{bmatrix} 0 & 1 & -2 \\ 1 & 2 & 3 \end{bmatrix} + \begin{bmatrix} 0 & 0 & 0 \\ 0 & 0 & 0 \end{bmatrix} = \begin{bmatrix} 0 & 1 & -2 \\ 1 & 2 & 3 \end{bmatrix}$

c. $\begin{bmatrix} 1 \\ -3 \\ -2 \end{bmatrix} + \begin{bmatrix} -1 \\ 3 \\ 2 \end{bmatrix} = \begin{bmatrix} 0 \\ 0 \\ 0 \end{bmatrix}$

d. The sum of

$$A = \begin{bmatrix} 2 & 1 & 0 \\ 4 & 0 & -1 \\ 3 & -2 & 2 \end{bmatrix} \quad \text{and} \quad B = \begin{bmatrix} 0 & 1 \\ -1 & 3 \\ 2 & 4 \end{bmatrix}$$

is undefined because A is of order 3×3 and B is of order 3×2.

✓**CHECKPOINT 2**

Find $\begin{bmatrix} 2 & -7 \\ -1 & 3 \end{bmatrix} + \begin{bmatrix} 4 & 5 \\ -1 & -6 \end{bmatrix}$. ■

In operations with matrices, numbers are usually referred to as **scalars**. In this text, scalars will always be real numbers. You can multiply a matrix A by a scalar c by multiplying each entry in A by c, as shown below.

Scalar   Matrix

$$3 \begin{bmatrix} -1 & 2 \\ 6 & 5 \end{bmatrix} = \begin{bmatrix} 3(-1) & 3(2) \\ 3(6) & 3(5) \end{bmatrix} = \begin{bmatrix} -3 & 6 \\ 18 & 15 \end{bmatrix}$$

Definition of Scalar Multiplication

If $A = [a_{ij}]$ is an $m \times n$ matrix and c is a scalar, the **scalar multiple** of A by c is the $m \times n$ matrix given by

$$cA = [ca_{ij}].$$

The symbol $-A$ represents the negation of A, or the scalar product $(-1)A$. Moreover, if A and B are of the same order, then $A - B$ represents the sum of A and $(-1)B$. That is,

$$A - B = A + (-1)B. \qquad \text{Subtraction of matrices}$$

Example 3 Scalar Multiplication and Matrix Subtraction

For the following matrices, find (a) $3A$, (b) $-A$, and (c) $3A - B$.

$$A = \begin{bmatrix} 2 & 2 & 4 \\ -3 & 0 & -1 \\ 2 & 1 & 2 \end{bmatrix} \quad \text{and} \quad B = \begin{bmatrix} 2 & 0 & 0 \\ 1 & -4 & 3 \\ -1 & 3 & 2 \end{bmatrix}$$

SOLUTION

a. $3A = 3\begin{bmatrix} 2 & 2 & 4 \\ -3 & 0 & -1 \\ 2 & 1 & 2 \end{bmatrix}$ Scalar multiplication

$$= \begin{bmatrix} 3(2) & 3(2) & 3(4) \\ 3(-3) & 3(0) & 3(-1) \\ 3(2) & 3(1) & 3(2) \end{bmatrix} \qquad \text{Multiply each entry by 3.}$$

$$= \begin{bmatrix} 6 & 6 & 12 \\ -9 & 0 & -3 \\ 6 & 3 & 6 \end{bmatrix} \qquad \text{Simplify.}$$

b. $-A = (-1)\begin{bmatrix} 2 & 2 & 4 \\ -3 & 0 & -1 \\ 2 & 1 & 2 \end{bmatrix}$ Definition of negation

$$= \begin{bmatrix} -2 & -2 & -4 \\ 3 & 0 & 1 \\ -2 & -1 & -2 \end{bmatrix} \qquad \text{Multiply each entry by } -1.$$

c. $3A - B = \begin{bmatrix} 6 & 6 & 12 \\ -9 & 0 & -3 \\ 6 & 3 & 6 \end{bmatrix} - \begin{bmatrix} 2 & 0 & 0 \\ 1 & -4 & 3 \\ -1 & 3 & 2 \end{bmatrix}$ Matrix subtraction

$$= \begin{bmatrix} 4 & 6 & 12 \\ -10 & 4 & -6 \\ 7 & 0 & 4 \end{bmatrix} \qquad \text{Subtract corresponding entries.}$$

STUDY TIP

The order of operations for matrix expressions is similar to that for real numbers. In particular, you perform scalar multiplication before matrix addition and subtraction, as shown in Example 3(c).

✓ CHECKPOINT 3

For the following matrices, find (a) $2A$ and (b) $2A - B$.

$$A = \begin{bmatrix} 2 & 4 & -1 \\ 0 & 1 & 3 \\ -3 & 2 & 5 \end{bmatrix} \quad \text{and}$$

$$B = \begin{bmatrix} 0 & 6 & 3 \\ 7 & -4 & 1 \\ 2 & 0 & -2 \end{bmatrix} \ ■$$

It is often convenient to rewrite the scalar multiple cA by factoring c out of every entry in the matrix. For instance, in the first matrix below, the scalar $\frac{1}{2}$ has been factored out of the matrix, and in the second matrix the scalar -2 has been factored out of the matrix.

$$\begin{bmatrix} \frac{1}{2} & -\frac{3}{2} \\ \frac{5}{2} & \frac{1}{2} \end{bmatrix} = \frac{1}{2}\begin{bmatrix} 1 & -3 \\ 5 & 1 \end{bmatrix}$$

$$\begin{bmatrix} -4 & -20 \\ -10 & -2 \end{bmatrix} = -2\begin{bmatrix} 2 & 10 \\ 5 & 1 \end{bmatrix}$$

The properties of matrix addition and scalar multiplication are similar to those of addition and multiplication of real numbers.

Properties of Matrix Addition and Scalar Multiplication

If A, B, and C are $m \times n$ matrices and c and d are scalars, then the following properties are true.

1. $A + B = B + A$ Commutative Property of Matrix Addition

2. $A + (B + C) = (A + B) + C$ Associative Property of Matrix Addition

3. $(cd)A = c(dA)$ Associative Property of Scalar Multiplication

4. $1A = A$ Scalar Identity Property

5. $c(A + B) = cA + cB$ Distributive Property

6. $(c + d)A = cA + dA$ Distributive Property

Note that the Associative Property of Matrix Addition allows you to write expressions such as $A + B + C$ without ambiguity, because you obtain the same sum regardless of how the matrices are grouped. In other words, you obtain the same sum whether you group $A + B + C$ as $(A + B) + C$ or as $A + (B + C)$. This same reasoning applies to sums of four or more matrices.

Example 4   Addition of More than Two Matrices

Add the following four matrices.

$$\begin{bmatrix} 1 \\ 2 \\ -3 \end{bmatrix}, \begin{bmatrix} -1 \\ -1 \\ 2 \end{bmatrix}, \begin{bmatrix} 0 \\ 1 \\ 4 \end{bmatrix}, \begin{bmatrix} 2 \\ -3 \\ -2 \end{bmatrix}$$

SOLUTION   By adding corresponding entries, you obtain the following sum of four matrices.

$$\begin{bmatrix} 1 \\ 2 \\ -3 \end{bmatrix} + \begin{bmatrix} -1 \\ -1 \\ 2 \end{bmatrix} + \begin{bmatrix} 0 \\ 1 \\ 4 \end{bmatrix} + \begin{bmatrix} 2 \\ -3 \\ -2 \end{bmatrix} = \begin{bmatrix} 2 \\ -1 \\ 1 \end{bmatrix}$$

✓ **CHECKPOINT 4**

Add the following three matrices.

$$\begin{bmatrix} 2 & -9 \\ 0 & 1 \end{bmatrix}, \begin{bmatrix} -1 & 4 \\ 7 & -3 \end{bmatrix},$$

$$\begin{bmatrix} 0 & 6 \\ 5 & -2 \end{bmatrix}$$ ■

TECHNOLOGY

(T) Most graphing utilities can add and subtract matrices and multiply matrices by scalars. Use your graphing utility to find (a) $A + B$, (b) $A - B$, (c) $4A$, and (d) $4A + B$. For specific keystrokes on how to perform matrix operations using a graphing utility, go to the text website at *college.hmco.com/info/larsonapplied*.

$$A = \begin{bmatrix} 2 & -3 \\ -1 & 0 \end{bmatrix} \quad \text{and} \quad B = \begin{bmatrix} -1 & 4 \\ 2 & -5 \end{bmatrix}$$

One important property of addition of real numbers is that the number 0 is the additive identity. That is, $c + 0 = c$ for any real number c. For matrices, a similar property holds. That is, if A is an $m \times n$ matrix and O is the $m \times n$ **zero matrix** consisting entirely of zeros, then $A + O = A$.

In other words, O is the **additive identity** for the set of all $m \times n$ matrices. For example, the following matrices are the additive identities for the sets of all 2×3 and 2×2 matrices, respectively.

$$O = \begin{bmatrix} 0 & 0 & 0 \\ 0 & 0 & 0 \end{bmatrix} \quad \text{and} \quad O = \begin{bmatrix} 0 & 0 \\ 0 & 0 \end{bmatrix}.$$

2×3 zero matrix $\qquad$ 2×2 zero matrix

The algebra of real numbers and the algebra of matrices have many similarities. For example, compare the following solutions.

Real Numbers *(Solve for x.)*	$m \times n$ *Matrices* *(Solve for X.)*
$x + a = b$	$X + A = B$
$x + a + (-a) = b + (-a)$	$X + A + (-A) = B + (-A)$
$x + 0 = b - a$	$X + O = B - A$
$x = b - a$	$X = B - A$

This means that you can apply some of your knowledge of solving real number equations to solving matrix equations. It is often easier to complete the algebraic steps first, and then substitute the matrices into the equation, as illustrated in Example 5.

Example 5 Solving a Matrix Equation

Solve for X in the equation $3X + A = B$, where

$$A = \begin{bmatrix} 1 & -2 \\ 0 & 3 \end{bmatrix} \quad \text{and} \quad B = \begin{bmatrix} -3 & 4 \\ 2 & 1 \end{bmatrix}.$$

SOLUTION Begin by solving the equation for X to obtain

$$3X = B - A$$

$$X = \frac{1}{3}(B - A).$$

Now, using the matrices A and B, you have

$$X = \frac{1}{3}\left(\begin{bmatrix} -3 & 4 \\ 2 & 1 \end{bmatrix} - \begin{bmatrix} 1 & -2 \\ 0 & 3 \end{bmatrix} \right) \qquad \text{Substitute the matrices.}$$

$$= \frac{1}{3}\begin{bmatrix} -4 & 6 \\ 2 & -2 \end{bmatrix} \qquad \text{Subtract matrix } A \text{ from matrix } B.$$

$$= \begin{bmatrix} -\frac{4}{3} & 2 \\ \frac{2}{3} & -\frac{2}{3} \end{bmatrix}. \qquad \text{Multiply the resulting matrix by } \frac{1}{3}.$$

✓ **CHECKPOINT 5**

Solve for X in the equation $2X - A = B$, where

$$A = \begin{bmatrix} 7 & 0 \\ -1 & 2 \end{bmatrix} \text{ and}$$

$$B = \begin{bmatrix} 3 & 1 \\ 2 & 4 \end{bmatrix}. \quad \blacksquare$$

TECHNOLOGY

Ⓣ Some graphing utilities can multiply two matrices. Use your graphing utility to find the product AB.

$$A = \begin{bmatrix} 1 & 2 & 3 \\ 2 & -5 & 1 \end{bmatrix}$$

$$B = \begin{bmatrix} -3 & 2 & 1 \\ 4 & -2 & 0 \\ 1 & 2 & 3 \end{bmatrix}$$

Now use your graphing utility to find the product BA. What is the result of this operation? For specific keystrokes, go to the text website at *college.hmco.com/info/larsonapplied.*

Matrix Multiplication

The third basic matrix operation is **matrix multiplication.** At first glance the definition may seem unusual. You will see later, however, that this definition of the product of two matrices has many practical applications.

Definition of Matrix Multiplication

If $A = [a_{ij}]$ is an $m \times n$ matrix and $B = [b_{ij}]$ is an $n \times p$ matrix, the **product** AB is an $m \times p$ matrix

$$AB = [c_{ij}]$$

where $c_{ij} = a_{i1}b_{1j} + a_{i2}b_{2j} + a_{i3}b_{3j} + \cdots + a_{in}b_{nj}.$

The definition of matrix multiplication indicates a *row-by-column* multiplication, where the entry in the ith row and jth column of the product AB is obtained by multiplying the entries in the ith row of A by the corresponding entries in the jth column of B and then adding the results. The general pattern for matrix multiplication is as follows.

$$\begin{bmatrix} a_{11} & a_{12} & a_{13} & \cdots & a_{1n} \\ a_{21} & a_{22} & a_{23} & \cdots & a_{2n} \\ a_{31} & a_{32} & a_{33} & \cdots & a_{3n} \\ \vdots & \vdots & \vdots & & \vdots \\ a_{i1} & a_{i2} & a_{i3} & \cdots & a_{in} \\ \vdots & \vdots & \vdots & & \vdots \\ a_{m1} & a_{m2} & a_{m3} & \cdots & a_{mn} \end{bmatrix} \begin{bmatrix} b_{11} & b_{12} & \cdots & b_{1j} & \cdots & b_{1p} \\ b_{21} & b_{22} & \cdots & b_{2j} & \cdots & b_{2p} \\ b_{31} & b_{32} & \cdots & b_{3j} & \cdots & b_{3p} \\ \vdots & \vdots & & \vdots & & \vdots \\ b_{n1} & b_{n2} & \cdots & b_{nj} & \cdots & b_{np} \end{bmatrix} = \begin{bmatrix} c_{11} & c_{12} & \cdots & c_{1j} & \cdots & c_{1p} \\ c_{21} & c_{22} & \cdots & c_{2j} & \cdots & c_{2p} \\ \vdots & \vdots & & & & \\ c_{i1} & c_{i2} & \cdots & c_{ij} & \cdots & c_{ip} \\ \vdots & \vdots & & & & \\ c_{m1} & c_{m2} & \cdots & c_{mj} & \cdots & c_{mp} \end{bmatrix}$$

$$a_{i1}b_{1j} + a_{i2}b_{2j} + a_{i3}b_{3j} + \cdots + a_{in}b_{nj} = c_{ij}$$

Example 6   Finding the Product of Two Matrices

Find the product AB using $A = \begin{bmatrix} -1 & 3 \\ 4 & -2 \\ 5 & 0 \end{bmatrix}$ and $B = \begin{bmatrix} -3 & 2 \\ -4 & 1 \end{bmatrix}.$

SOLUTION   First, note that the product AB is defined because the number of *columns* of A is equal to the number of *rows* of B. Moreover, the product AB has order 3×2. To find the entries of the product, multiply each row of A by each column of B.

$$AB = \begin{bmatrix} -1 & 3 \\ 4 & -2 \\ 5 & 0 \end{bmatrix} \begin{bmatrix} -3 & 2 \\ -4 & 1 \end{bmatrix}$$

$$= \begin{bmatrix} (-1)(-3) + (3)(-4) & (-1)(2) + (3)(1) \\ (4)(-3) + (-2)(-4) & (4)(2) + (-2)(1) \\ (5)(-3) + (0)(-4) & (5)(2) + (0)(1) \end{bmatrix}$$

$$= \begin{bmatrix} -9 & 1 \\ -4 & 6 \\ -15 & 10 \end{bmatrix}$$

✓**CHECKPOINT 6**

Find the product AB using

$$A = \begin{bmatrix} -2 & 2 \\ 0 & 4 \\ 3 & -1 \end{bmatrix} \text{ and }$$

$$B = \begin{bmatrix} -3 & 0 \\ 1 & -4 \end{bmatrix}.$$

DISCOVERY

Use a graphing utility to multiply the matrices

$$A = \begin{bmatrix} 1 & 2 \\ 3 & 4 \end{bmatrix} \text{ and}$$

$$B = \begin{bmatrix} 0 & 1 \\ 2 & 3 \end{bmatrix}.$$

Do you obtain the same result for the product AB as for the product BA? What does this tell you about matrix multiplication and commutativity?

Be sure you understand that for the product of two matrices to be defined, the number of *columns* of the first matrix must equal the number of *rows* of the second matrix. That is, the middle two indices must be the same and the outside two indices give the order of the product, as shown below.

$$\underset{m \times n}{A} \qquad \underset{n \times p}{B} \qquad = \qquad \underset{m \times p}{AB}$$

Equal

Order of AB

Example 7   Finding the Product of Two Matrices

Find the product AB using $A = \begin{bmatrix} 6 & 2 & 0 \\ 3 & -1 & 2 \\ 1 & 4 & 6 \end{bmatrix}$ and $B = \begin{bmatrix} 1 & 0 \\ 2 & 7 \\ -3 & 5 \end{bmatrix}$.

SOLUTION   Note that the order of A is 3×3 and the order of B is 3×2. So, the product AB is defined and is of order 3×2.

$$AB = \begin{bmatrix} 6 & 2 & 0 \\ 3 & -1 & 2 \\ 1 & 4 & 6 \end{bmatrix} \begin{bmatrix} 1 & 0 \\ 2 & 7 \\ -3 & 5 \end{bmatrix}$$

$$= \begin{bmatrix} 6(1) + 2(2) + 0(-3) & 6(0) + 2(7) + 0(5) \\ 3(1) + (-1)(2) + 2(-3) & 3(0) + (-1)(7) + 2(5) \\ 1(1) + 4(2) + 6(-3) & 1(0) + 4(7) + 6(5) \end{bmatrix}$$

$$= \begin{bmatrix} 10 & 14 \\ -5 & 3 \\ -9 & 58 \end{bmatrix}$$

✓**CHECKPOINT 7**

Find the product AB using

$$A = \begin{bmatrix} 0 & 4 & -3 \\ 2 & 1 & 7 \\ 3 & -2 & 1 \end{bmatrix} \text{ and}$$

$$B = \begin{bmatrix} -2 & 0 \\ 0 & -4 \\ 1 & 2 \end{bmatrix}. \ \blacksquare$$

Example 8   Patterns in Matrix Multiplication

a. $\begin{bmatrix} 1 & 0 & 3 \\ 2 & -1 & -2 \end{bmatrix} \begin{bmatrix} -2 & 4 & 2 \\ 1 & 0 & 0 \\ -1 & 1 & -1 \end{bmatrix} = \begin{bmatrix} -5 & 7 & -1 \\ -3 & 6 & 6 \end{bmatrix}$

$\quad \quad 2 \times 3 \qquad\qquad 3 \times 3 \qquad\qquad\quad 2 \times 3$

b. $\begin{bmatrix} 3 & 4 \\ -2 & 5 \end{bmatrix} \begin{bmatrix} 1 & 0 \\ 0 & 1 \end{bmatrix} = \begin{bmatrix} 3 & 4 \\ -2 & 5 \end{bmatrix}$

$\quad\quad 2 \times 2 \qquad 2 \times 2 \qquad 2 \times 2$

c. The product AB for the following matrices is not defined.

$$A = \begin{bmatrix} -2 & 1 \\ 1 & -3 \\ 1 & 4 \end{bmatrix} \text{ and } B = \begin{bmatrix} -2 & 3 & 1 & 4 \\ 0 & 1 & -1 & 2 \\ 2 & -1 & 0 & 1 \end{bmatrix}$$

$\quad\quad\quad\quad 3 \times 2 \qquad\qquad\qquad\quad 3 \times 4$

✓**CHECKPOINT 8**

Find AB, if possible, using $A = \begin{bmatrix} 2 & 0 & 1 \\ 6 & 1 & -3 \end{bmatrix}$ and $B = \begin{bmatrix} -5 & 3 \\ 3 & 0 \end{bmatrix}. \ \blacksquare$

Example 9   **Patterns in Matrix Multiplication**

a. $\begin{bmatrix} 1 & -2 \end{bmatrix} \begin{bmatrix} 2 \\ -1 \end{bmatrix} = \begin{bmatrix} 4 \end{bmatrix}$ **b.** $\begin{bmatrix} 2 \\ -1 \end{bmatrix} \begin{bmatrix} 1 & -2 \end{bmatrix} = \begin{bmatrix} 2 & -4 \\ -1 & 2 \end{bmatrix}$

$\quad 1 \times 2 \quad 2 \times 1 \quad 1 \times 1 \qquad\qquad 2 \times 1 \quad 1 \times 2 \qquad\qquad 2 \times 2$

✓ **CHECKPOINT 9**

Find AB and BA using $A = \begin{bmatrix} 3 & -1 \end{bmatrix}$ and $B = \begin{bmatrix} 1 \\ -3 \end{bmatrix}$. ▪

In Example 9, note that the two products are different. Even if AB and BA are defined, matrix multiplication is not, in general, commutative. That is, for most matrices, $AB \neq BA$.

Properties of Matrix Multiplication

Let A, B, and C be matrices and let c be a scalar.

1. $A(BC) = (AB)C$ \qquad\qquad Associative Property of Matrix Multiplication

2. $A(B + C) = AB + AC$ \qquad\quad Left Distributive Property

3. $(A + B)C = AC + BC$ \qquad\quad Right Distributive Property

4. $c(AB) = (cA)B = A(cB)$ \qquad Associative Property of Scalar Multiplication

Definition of the Identity Matrix

The $n \times n$ matrix that consists of 1's on its main diagonal and 0's elsewhere is called the **identity matrix of order n** and is denoted by

$$I_n = \begin{bmatrix} 1 & 0 & 0 & \cdots & 0 \\ 0 & 1 & 0 & \cdots & 0 \\ 0 & 0 & 1 & \cdots & 0 \\ \vdots & \vdots & \vdots & & \vdots \\ 0 & 0 & 0 & \cdots & 1 \end{bmatrix}.$$ Identity matrix

Note that an identity matrix must be *square*. When the order is understood to be n, you can denote I_n simply by I.

If A is an $n \times n$ matrix, the identity matrix has the property that $AI_n = A$ and $I_n A = A$. For example,

$$\begin{bmatrix} 3 & -2 & 5 \\ 1 & 0 & 4 \\ -1 & 2 & -3 \end{bmatrix} \begin{bmatrix} 1 & 0 & 0 \\ 0 & 1 & 0 \\ 0 & 0 & 1 \end{bmatrix} = \begin{bmatrix} 3 & -2 & 5 \\ 1 & 0 & 4 \\ -1 & 2 & -3 \end{bmatrix} \quad AI = A$$

and

$$\begin{bmatrix} 1 & 0 & 0 \\ 0 & 1 & 0 \\ 0 & 0 & 1 \end{bmatrix} \begin{bmatrix} 3 & -2 & 5 \\ 1 & 0 & 4 \\ -1 & 2 & -3 \end{bmatrix} = \begin{bmatrix} 3 & -2 & 5 \\ 1 & 0 & 4 \\ -1 & 2 & -3 \end{bmatrix}. \quad IA = A$$

Applications

One application of matrix multiplication is the representation of a system of linear equations. Note how the system

$$\begin{cases} a_{11}x_1 + a_{12}x_2 + a_{13}x_3 = b_1 \\ a_{21}x_1 + a_{22}x_2 + a_{23}x_3 = b_2 \\ a_{31}x_1 + a_{32}x_2 + a_{33}x_3 = b_3 \end{cases}$$

can be written as the matrix equation $AX = B$, where A is the *coefficient matrix* of the system, and X and B are column matrices.

$$\begin{bmatrix} a_{11} & a_{12} & a_{13} \\ a_{21} & a_{22} & a_{23} \\ a_{31} & a_{32} & a_{33} \end{bmatrix} \begin{bmatrix} x_1 \\ x_2 \\ x_3 \end{bmatrix} = \begin{bmatrix} b_1 \\ b_2 \\ b_3 \end{bmatrix}$$

$$\quad A \qquad\qquad \times\; X \;=\; B$$

STUDY TIP

The column matrix B is also called a *constant matrix*. Its entries are the constant terms in the system of equations.

Example 10   Solving a System of Linear Equations

Consider the system of linear equations.

$$\begin{cases} x_1 - 2x_2 + x_3 = -4 \\ x_2 + 2x_3 = 4 \\ 2x_1 + 3x_2 - 2x_3 = 2 \end{cases}$$

a. Write this system as a matrix equation $AX = B$.

b. Use Gauss-Jordan elimination on $[A \,\vdots\, B]$ to solve for the matrix X.

SOLUTION

a. In matrix form $AX = B$, the system is written as

$$\begin{bmatrix} 1 & -2 & 1 \\ 0 & 1 & 2 \\ 2 & 3 & -2 \end{bmatrix} \begin{bmatrix} x_1 \\ x_2 \\ x_3 \end{bmatrix} = \begin{bmatrix} -4 \\ 4 \\ 2 \end{bmatrix}.$$

Coefficient matrix   Constant matrix

b. The augmented matrix is formed by adjoining matrix B to matrix A.

$$[A \,\vdots\, B] = \begin{bmatrix} 1 & -2 & 1 & \vdots & -4 \\ 0 & 1 & 2 & \vdots & 4 \\ 2 & 3 & -2 & \vdots & 2 \end{bmatrix}$$

Using Gauss-Jordan elimination, you can rewrite this matrix as

$$[I \,\vdots\, X] = \begin{bmatrix} 1 & 0 & 0 & \vdots & -1 \\ 0 & 1 & 0 & \vdots & 2 \\ 0 & 0 & 1 & \vdots & 1 \end{bmatrix}.$$

So, the solution of the matrix equation is

$$X = \begin{bmatrix} x_1 \\ x_2 \\ x_3 \end{bmatrix} = \begin{bmatrix} -1 \\ 2 \\ 1 \end{bmatrix}.$$

STUDY TIP

The notation $[A \,\vdots\, B]$ represents the augmented matrix formed when matrix B is adjoined to matrix A. The notation $[I \,\vdots\, X]$ represents the reduced row-echelon form of the augmented matrix that yields the solution of the system.

✓CHECKPOINT 10

Write the system of linear equations as a matrix equation $AX = B$. Then use Gauss-Jordan elimination on the augmented matrix $[A \,\vdots\, B]$ to solve for the matrix X.

$$\begin{cases} -2x_1 - 3x_2 = -4 \\ 6x_1 + x_2 = -36 \end{cases}$$ ■

Example 11   Long-Distance Phone Plans

The charges (in dollars per minute) of two long-distance telephone companies are shown in the table.

	Company A	Company B
In-state	0.07	0.095
State-to-state	0.10	0.08
International	0.28	0.25

You plan to use 120 minutes on in-state long-distance calls, 80 minutes on state-to-state calls, and 20 minutes on international calls. Use matrices to determine which company you should choose to be your long-distance carrier.

SOLUTION   The charges C and amounts of time T spent on the phone can be written in matrix form as

$$C = \begin{bmatrix} 0.07 & 0.095 \\ 0.10 & 0.08 \\ 0.28 & 0.25 \end{bmatrix} \quad \text{and} \quad T = \begin{bmatrix} 120 & 80 & 20 \end{bmatrix}.$$

The total amount that each company charges is given by the product

$$TC = \begin{bmatrix} 120 & 80 & 20 \end{bmatrix} \begin{bmatrix} 0.07 & 0.095 \\ 0.10 & 0.08 \\ 0.28 & 0.25 \end{bmatrix} = \begin{bmatrix} 22 & 22.8 \end{bmatrix}.$$

Company A charges \$22 for the calls and Company B charges \$22.80. Company A charges less for the calling pattern, so you should choose Company A. Notice that you cannot find the total amount that each company charges using the product CT because CT is not defined. That is, the number of columns of C does not equal the number of rows of T.

✓CHECKPOINT 11

In Example 11, suppose you plan to use 100 minutes on in-state long-distance calls, 70 minutes on state-to-state calls, and 40 minutes on international calls. Use matrices to determine which company you should choose to be your long-distance carrier. ▨

CONCEPT CHECK

1. Under what conditions are matrices $A = [a_{ij}]$ and $B = [b_{ij}]$ equal?
2. What is the sum of a matrix A and the negation of A?
3. Discuss the similarities and differences between solving real number equations and solving matrix equations.
4. Explain why AB is not defined and BA is defined when matrix A is of order 1×3 and matrix B is of order 2×1.

Skills Review 6.2

The following warm-up exercises involve skills that were covered in earlier sections. You will use these skills in the exercise set for this section. For additional help, review Sections 0.2 and 6.1.

In Exercises 1 and 2, evaluate the expression.

1. $-3\left(-\frac{5}{6}\right) + 10\left(-\frac{3}{4}\right)$

2. $-22\left(\frac{5}{2}\right) + 6(8)$

In Exercises 3 and 4, determine whether the matrix is in *reduced* row-echelon form.

3. $\begin{bmatrix} 0 & 1 & 0 & -5 \\ 1 & 0 & 3 & 2 \\ 0 & 0 & 1 & 0 \end{bmatrix}$

4. $\begin{bmatrix} 1 & 0 & 0 & 2 & 3 \\ 0 & 0 & 0 & 0 & 0 \\ 0 & 1 & 1 & 3 & 10 \end{bmatrix}$

In Exercises 5 and 6, write the augmented matrix for the system of linear equations.

5. $\begin{cases} -5x + 10y = 12 \\ 7x - 3y = 0 \end{cases}$

6. $\begin{cases} 10x + 15y - 9z = 42 \\ 6x - 5y = 0 \end{cases}$

In Exercises 7–10, solve the system of linear equations represented by the augmented matrix.

7. $\begin{bmatrix} 1 & 0 & \vdots & 0 \\ 0 & 1 & \vdots & 2 \end{bmatrix}$

8. $\begin{bmatrix} 1 & 0 & -1 & \vdots & 2 \\ 0 & 1 & 1 & \vdots & 3 \end{bmatrix}$

9. $\begin{bmatrix} 1 & 2 & 1 & \vdots & 0 \\ 0 & 0 & 1 & \vdots & -1 \\ 0 & 0 & 0 & \vdots & 0 \end{bmatrix}$

10. $\begin{bmatrix} 1 & -1 & 0 & \vdots & 3 \\ 0 & 1 & -2 & \vdots & 1 \\ 0 & 0 & 1 & \vdots & -1 \end{bmatrix}$

Exercises 6.2

See www.CalcChat.com for worked-out solutions to odd-numbered exercises.

In Exercises 1–4, find x and y.

1. $\begin{bmatrix} 4 & x \\ -1 & y \end{bmatrix} = \begin{bmatrix} 4 & -3 \\ -1 & 2 \end{bmatrix}$

2. $\begin{bmatrix} x & -7 \\ 9 & y \end{bmatrix} = \begin{bmatrix} 5 & -7 \\ 9 & -8 \end{bmatrix}$

3. $\begin{bmatrix} -4 & 3 \\ 6 & -1 \\ 8 & 2 \\ 5 & 9 \end{bmatrix} = \begin{bmatrix} x-2 & 3 \\ 6 & -1 \\ 8 & -x \\ 5 & 2y-1 \end{bmatrix}$

4. $\begin{bmatrix} x+2 & 8 & -3 \\ 1 & 2y & 2x \\ 7 & -2 & y+2 \end{bmatrix} = \begin{bmatrix} 2x+6 & 8 & -3 \\ 1 & 18 & -8 \\ 7 & -2 & 11 \end{bmatrix}$

In Exercises 5–10, find (a) $A + B$, (b) $A - B$, (c) $3A$, and (d) $3A - 2B$.

5. $A = \begin{bmatrix} 5 & -2 \\ 3 & 1 \end{bmatrix}$, $B = \begin{bmatrix} 3 & 1 \\ -2 & 6 \end{bmatrix}$

6. $A = \begin{bmatrix} 7 & 4 \\ -4 & 5 \end{bmatrix}$, $B = \begin{bmatrix} -3 & 1 \\ 8 & -4 \end{bmatrix}$

7. $A = \begin{bmatrix} 6 & -1 \\ 2 & 4 \\ -3 & 5 \end{bmatrix}$, $B = \begin{bmatrix} 1 & 4 \\ -1 & 5 \\ 1 & 10 \end{bmatrix}$

8. $A = \begin{bmatrix} 6 & 8 & -3 & 2 & 1 \\ -4 & 2 & 1 & 5 & -2 \end{bmatrix}$,

$B = \begin{bmatrix} 6 & 0 & 4 & -1 & 3 \\ 4 & 5 & -2 & 1 & 2 \end{bmatrix}$

9. $A = \begin{bmatrix} 2 & 2 & -1 \\ 1 & 1 & -2 \\ 1 & -1 & 3 \end{bmatrix}$, $B = \begin{bmatrix} 1 & 1 & -1 \\ -3 & 4 & 9 \\ 0 & -7 & 8 \end{bmatrix}$

10. $A = \begin{bmatrix} 3 \\ 2 \\ -1 \end{bmatrix}$, $B = \begin{bmatrix} -4 \\ 6 \\ 2 \end{bmatrix}$

In Exercises 11–16, evaluate the expression.

11. $\begin{bmatrix} -5 & 0 \\ 3 & -6 \end{bmatrix} + \begin{bmatrix} 7 & 1 \\ -2 & -1 \end{bmatrix} + \begin{bmatrix} -10 & -8 \\ 14 & 6 \end{bmatrix}$

12. $\begin{bmatrix} 6 & 8 \\ -1 & 0 \end{bmatrix} + \begin{bmatrix} 0 & 5 \\ -3 & -1 \end{bmatrix} + \begin{bmatrix} -11 & -7 \\ 2 & -1 \end{bmatrix}$

13. $4\left(\begin{bmatrix} -4 & 0 & 1 \\ 0 & 2 & 3 \end{bmatrix} - \begin{bmatrix} 2 & 1 & -2 \\ 3 & -6 & 0 \end{bmatrix} \right)$

14. $\frac{1}{2}([5 \quad -2 \quad 4 \quad 0] + [14 \quad 6 \quad -18 \quad 9])$

15. $-3\left(\begin{bmatrix} 0 & -3 \\ 7 & 2 \end{bmatrix} + \begin{bmatrix} -6 & 3 \\ 8 & 1 \end{bmatrix} \right) - 2\begin{bmatrix} 4 & -4 \\ 7 & -9 \end{bmatrix}$

16. $-1\begin{bmatrix} 4 & 11 \\ -2 & -1 \\ 9 & 3 \end{bmatrix} + \frac{1}{6}\left(\begin{bmatrix} -5 & -1 \\ 3 & 4 \\ 0 & 13 \end{bmatrix} + \begin{bmatrix} 7 & 5 \\ -9 & -1 \\ 6 & -1 \end{bmatrix} \right)$

Ⓣ In Exercises 17–20, use the matrix capabilities of a graphing utility to evaluate the expression. Round your results to three decimal places, if necessary.

17. $\frac{3}{7}\begin{bmatrix} 2 & 5 \\ -1 & -4 \end{bmatrix} + 6\begin{bmatrix} -3 & 0 \\ 2 & 2 \end{bmatrix}$

18. $55\left(\begin{bmatrix} 14 & -11 \\ -22 & 19 \end{bmatrix} + \begin{bmatrix} -22 & 20 \\ 13 & 6 \end{bmatrix} \right)$

19. $-\begin{bmatrix} 3.211 & 6.829 \\ -1.004 & 4.914 \\ 0.055 & -3.889 \end{bmatrix} - \begin{bmatrix} -1.630 & -3.090 \\ 5.256 & 8.335 \\ -9.768 & 4.251 \end{bmatrix}$

20. $-12\left(\begin{bmatrix} 6 & 20 \\ 1 & -9 \\ -2 & 5 \end{bmatrix} + \begin{bmatrix} 14 & -15 \\ -8 & -6 \\ 7 & 0 \end{bmatrix} + \begin{bmatrix} -31 & -19 \\ 16 & 10 \\ 24 & -10 \end{bmatrix} \right)$

In Exercises 21–24, solve for X when

$$A = \begin{bmatrix} -2 & -1 \\ 1 & 0 \\ 3 & -4 \end{bmatrix} \quad \text{and} \quad B = \begin{bmatrix} 0 & 3 \\ 2 & 0 \\ -4 & -1 \end{bmatrix}.$$

21. $X = 3A - 2B$

22. $2X = 2A - B$

23. $2X + 3A = B$

24. $2A + 4B = -2X$

In Exercises 25–32, find AB, if possible.

25. $A = \begin{bmatrix} 3 & -2 \\ 4 & 5 \\ 1 & -1 \end{bmatrix}, \ B = \begin{bmatrix} -1 & 4 & -2 & 5 \\ 2 & 1 & 3 & -1 \end{bmatrix}$

26. $A = \begin{bmatrix} 0 & -1 & 0 \\ 4 & 0 & 2 \\ 8 & -1 & 7 \end{bmatrix}, \ B = \begin{bmatrix} 2 & 1 \\ -3 & 4 \\ 1 & 6 \end{bmatrix}$

27. $A = \begin{bmatrix} -1 & 3 \\ 4 & -5 \\ 0 & 2 \end{bmatrix}, \ B = \begin{bmatrix} 1 & 2 \\ 0 & 7 \end{bmatrix}$

28. $A = \begin{bmatrix} 1 & 0 & 0 \\ 0 & 4 & 0 \\ 0 & 0 & -2 \end{bmatrix}, \ B = \begin{bmatrix} 3 & 0 & 0 \\ 0 & -1 & 0 \\ 0 & 0 & 5 \end{bmatrix}$

29. $A = \begin{bmatrix} 5 & 0 & 0 \\ 0 & -8 & 0 \\ 0 & 0 & 7 \end{bmatrix}, \ B = \begin{bmatrix} \frac{1}{5} & 0 & 0 \\ 0 & -\frac{1}{8} & 0 \\ 0 & 0 & \frac{1}{2} \end{bmatrix}$

30. $A = \begin{bmatrix} 6 \\ -2 \\ 1 \\ 6 \end{bmatrix}, \ B = [10 \quad 12]$

31. $A = \begin{bmatrix} 0 & 1 & 0 \\ 3 & 0 & 2 \\ 5 & 0 & 0 \end{bmatrix}, \ B = \begin{bmatrix} 4 \\ -2 \\ 0 \\ 1 \end{bmatrix}$

32. $A = \begin{bmatrix} 1 & 0 & 3 & -2 & 4 \\ 6 & 13 & 8 & -17 & 10 \end{bmatrix}, \ B = \begin{bmatrix} 1 & 6 \\ 4 & 2 \end{bmatrix}$

Ⓣ In Exercises 33–38, use the matrix capabilities of a graphing utility to find AB, if possible.

33. $A = \begin{bmatrix} 5 & 6 & -3 \\ -2 & 5 & 1 \\ 10 & -5 & 5 \end{bmatrix}, \ B = \begin{bmatrix} 1 & -1 & 2 \\ 8 & 1 & 4 \\ 4 & -2 & 9 \end{bmatrix}$

34. $A = \begin{bmatrix} 11 & -12 & 4 \\ 14 & 10 & 12 \\ 6 & -2 & 9 \end{bmatrix}, \ B = \begin{bmatrix} 12 & 10 \\ -5 & 12 \\ 15 & 16 \end{bmatrix}$

35. $A = \begin{bmatrix} -3 & 8 & -6 & 8 \\ -12 & 15 & 9 & 6 \\ 5 & -1 & 1 & 5 \end{bmatrix}, \ B = \begin{bmatrix} 3 & 1 & 6 \\ 24 & 15 & 14 \\ 16 & 10 & 21 \\ 8 & -4 & 10 \end{bmatrix}$

36. $A = \begin{bmatrix} 15 & -18 \\ -4 & 12 \\ -8 & 22 \end{bmatrix}, \ B = \begin{bmatrix} -7 & 22 & 1 \\ 8 & 16 & 24 \end{bmatrix}$

37. $A = \begin{bmatrix} -2 & 4 & 8 \\ 21 & 5 & 6 \\ 13 & 2 & 6 \end{bmatrix}, \ B = \begin{bmatrix} 2 & 0 \\ -7 & 15 \\ 32 & 14 \\ 0.5 & 1.6 \end{bmatrix}$

38. $A = \begin{bmatrix} 9 & 10 & -38 & 18 \\ 100 & -50 & 250 & 75 \end{bmatrix}, \ B = \begin{bmatrix} 52 & -85 & 27 & 45 \\ 40 & -35 & 60 & 82 \end{bmatrix}$

In Exercises 39–44, find (a) AB, (b) BA, and, if possible, (c) A^2. (Note: $A^2 = AA$.)

39. $A = \begin{bmatrix} 1 & 2 \\ 4 & 2 \end{bmatrix}, \ B = \begin{bmatrix} 2 & -1 \\ -1 & 8 \end{bmatrix}$

40. $A = \begin{bmatrix} 2 & -1 \\ 1 & 4 \end{bmatrix}$, $B = \begin{bmatrix} 0 & 0 \\ 3 & -3 \end{bmatrix}$

41. $A = \begin{bmatrix} -1 & 2 & 3 \\ 4 & 1 & -1 \end{bmatrix}$, $B = \begin{bmatrix} 1 & 3 \\ -1 & -2 \\ 2 & 4 \end{bmatrix}$

42. $A = \begin{bmatrix} 1 & -1 & 7 \\ 2 & -1 & 8 \\ 3 & 1 & -1 \end{bmatrix}$, $B = \begin{bmatrix} 1 & 1 & 2 \\ 2 & 1 & 1 \\ 1 & -3 & 2 \end{bmatrix}$

43. $A = \begin{bmatrix} -4 & 2 & 3 \end{bmatrix}$, $B = \begin{bmatrix} 1 \\ 0 \\ 5 \end{bmatrix}$

44. $A = \begin{bmatrix} 3 & 2 & 1 & 0 \end{bmatrix}$, $B = \begin{bmatrix} 2 \\ 3 \\ 1 \\ 0 \end{bmatrix}$

In Exercises 45–50, (a) write the system of linear equations as a matrix equation $AX = B$, and (b) use Gauss-Jordan elimination on the augmented matrix $[A \vdots B]$ to solve for the matrix X.

45. $\begin{cases} -x + y = 4 \\ -2x + y = 0 \end{cases}$

46. $\begin{cases} 2x + 3y = 5 \\ x + 4y = 10 \end{cases}$

47. $\begin{cases} x + 2y = 3 \\ 3x - y = 2 \end{cases}$

48. $\begin{cases} 2x - 4y + z = 0 \\ -x + 3y + z = 1 \\ x + y = 3 \end{cases}$

49. $\begin{cases} x - 2y + 3z = 9 \\ -x + 3y - z = -6 \\ 2x - 5y + 5z = 17 \end{cases}$

50. $\begin{cases} x + y - 3z = -1 \\ -x + 2y = 1 \\ x - y + z = 2 \end{cases}$

51. Factory Production A corporation that makes sunglasses has four factories, each of which manufactures two products. The number of units of product i produced at factory j in one day is represented by a_{ij} in the matrix

$$A = \begin{bmatrix} 100 & 120 & 60 & 40 \\ 140 & 160 & 200 & 80 \end{bmatrix}.$$

Find the production levels if production is increased by 10%. (*Hint:* Because an increase of 10% corresponds to $100\% + 10\%$, multiply the matrix by 1.10.)

52. Factory Production A tire corporation has three factories, each of which manufactures two products. The number of units of product i produced at factory j in one day is represented by a_{ij} in the matrix

$$A = \begin{bmatrix} 80 & 120 & 140 \\ 40 & 100 & 80 \end{bmatrix}.$$

Find the production levels if production is decreased by 5%. (*Hint:* Because a decrease of 5% corresponds to $100\% - 5\%$, multiply the matrix by 0.95.)

53. Hotel Pricing A convention planning service has identified three suitable hotels for a convention. The quoted room rates are for single, double, triple, and quadruple occupancy. The current rates for the four types of rooms at the three hotels are represented by the matrix A.

$$A = \begin{array}{c} \\ \\ \begin{bmatrix} 85 & 92 & 110 \\ 100 & 120 & 130 \\ 110 & 130 & 140 \\ 110 & 140 & 155 \end{bmatrix} \end{array} \begin{array}{l} \text{Single} \\ \text{Double} \\ \text{Triple} \\ \text{Quadruple} \end{array}$$

with column headers: Hotel x, Hotel y, Hotel z, and braced "Occupancy".

If room rates are guaranteed not to increase by more than 15% by the time of the convention, what is the maximum rate per room per hotel?

54. Vacation Packages A vacation service has identified four resort hotels with a special all-inclusive package (room and meals included) at a popular travel destination. The quoted room rates are for double and family (maximum of four people) occupancy for 5 days and 4 nights. The current rates for the two types of rooms at the four hotels are represented by the matrix A.

$$A = \begin{bmatrix} 615 & 670 & 740 & 990 \\ 995 & 1030 & 1180 & 1105 \end{bmatrix} \begin{array}{l} \text{Double} \\ \text{Family} \end{array}$$

with column headers: Hotel w, Hotel x, Hotel y, Hotel z, and braced "Occupancy".

If room rates are guaranteed not to increase by more than 12% by next season, what is the maximum rate per package per hotel?

55. Inventory Levels A company sells five different models of computers through three retail outlets. The inventories of the five models at the three outlets are given by the matrix S.

$$S = \begin{array}{c} \\ \begin{bmatrix} 3 & 2 & 2 & 3 & 0 \\ 0 & 2 & 3 & 4 & 3 \\ 4 & 2 & 1 & 3 & 2 \end{bmatrix} \end{array} \begin{array}{l} 1 \\ 2 \\ 3 \end{array} \text{Outlet}$$

with column headers A, B, C, D, E under "Model".

The wholesale and retail prices for each model are given by the matrix T.

Price

$$T = \begin{bmatrix} \$900 & \$1200 \\ \$1200 & \$1450 \\ \$1400 & \$1650 \\ \$2650 & \$3250 \\ \$3050 & \$3375 \end{bmatrix} \begin{matrix} A \\ B \\ C \\ D \\ E \end{matrix} \Big\} \text{Model}$$

Wholesale   Retail

(a) What is the total retail price of the inventory at Outlet 1?

(b) What is the total wholesale price of the inventory at Outlet 3?

(c) Compute the product ST and interpret the result in the context of the problem.

56. Labor/Wage Requirements A company that manufactures boats has the following labor-hour and wage requirements.

Labor-Hour Requirements (per boat)

Department

$$S = \begin{bmatrix} 1.0 \text{ hour} & 0.5 \text{ hour} & 0.2 \text{ hour} \\ 1.6 \text{ hours} & 1.0 \text{ hour} & 0.2 \text{ hour} \\ 2.5 \text{ hours} & 2.0 \text{ hours} & 0.4 \text{ hour} \end{bmatrix} \begin{matrix} \text{Small} \\ \text{Medium} \\ \text{Large} \end{matrix} \Big\} \begin{matrix} \text{Boat} \\ \text{size} \end{matrix}$$

Cutting   Assembly   Packaging

Wage Requirements (per hour)

Plant

$3 \times 3 \quad 3 \times 2$

$$T = \begin{bmatrix} \$8 & \$15 \\ \$13 & \$12 \\ \$10 & \$11 \end{bmatrix} \begin{matrix} \text{Cutting} \\ \text{Assembly} \\ \text{Packaging} \end{matrix} \Big\} \text{Department}$$

A   B

(a) What is the labor cost for a medium boat at Plant B?

(b) What is the labor cost for a large boat at Plant A?

(c) Compute ST and interpret the result.

57. Exercise The numbers of calories burned by individuals of different body weights while performing different types of aerobic exercises for a 20-minute time period are shown in the matrix A.

Calories burned

$$A = \begin{bmatrix} 109 & 136 \\ 127 & 159 \\ 64 & 79 \end{bmatrix} \begin{matrix} \text{Bicycling} \\ \text{Jogging} \\ \text{Walking} \end{matrix}$$

120-lb   150-lb
person   person

(a) A 120-pound person and a 150-pound person bicycled for 40 minutes, jogged for 10 minutes, and walked for 60 minutes. Organize the times spent exercising in a matrix B.

(b) Compute BA and interpret the result.

58. Agriculture A fruit grower raises apples and peaches, which are shipped to three different outlets. The numbers of units of apples and peaches that are shipped to the three outlets are shown in the matrix A.

Outlet

$$A = \begin{bmatrix} 125 & 100 & 75 \\ 100 & 175 & 125 \end{bmatrix} \begin{matrix} \text{Apples} \\ \text{Peaches} \end{matrix} \Big\} \text{Units shipped}$$

X   Y   Z

(a) The profit per unit of apples is \$3.50 and the profit per unit of peaches is \$6. Organize the profits per unit in a matrix B.

(b) Compute BA and interpret the result.

Think About It   In Exercises 59–66, let matrices A, B, C, and D be of orders 2×3, 2×3, 3×2, and 2×2, respectively. Determine whether the matrices are of proper order to perform the operation(s). If so, give the order of the answer.

59. $A + 2C$

60. $B - 3C$

61. AB

62. BC

63. $BC - D$

64. $CB - D$

65. $D(A - 3B)$

66. $(BC - D)A$

67. Contract Bonuses Professional athletes frequently have bonus or incentive clauses in their contracts. For example, a defensive football player might receive bonuses for defensive plays such as sacks, interceptions, and/or key tackles. In one contract, a sack is worth \$2000, an interception is worth \$1000, and a key tackle is worth \$800. The table shows the numbers of sacks, interceptions, and key tackles for three players.

Player	Sacks	Interceptions	Key tackles
Player X	3	0	4
Player Y	1	2	5
Player Z	2	3	3

(a) Write a matrix D that represents the number of each type of defensive play i made by each player j using the data from the table. State what each entry d_{ij} of the matrix represents.

(b) Write a matrix B that represents the bonus amount received for each type of defensive play. State what each entry b_{ij} of the matrix represents.

(c) Find the product BD of the two matrices and state what each entry of matrix BD represents.

(d) Which player receives the largest bonus?

68. Long-Distance Plans You are choosing between two monthly long-distance phone plans offered by two different companies. Company A charges $0.05 per minute for in-state calls, $0.12 per minute for state-to-state calls, and $0.30 per minute for international calls. Company B charges $0.085 per minute for in-state calls, $0.10 per minute for state-to-state calls, and $0.25 per minute for international calls. In a month, you normally use 20 minutes on in-state calls, 60 minutes on state-to-state calls, and 30 minutes on international calls.

(a) Write a matrix C that represents the charges for each type of call i by each company j. State what each entry c_{ij} of the matrix represents.

(b) Write a matrix T that represents the times spent on the phone for each type of call. State what each entry of the matrix represents.

(c) Find the product TC and state what each entry of the matrix represents.

(d) Which company should you choose? Explain.

(T) **69. Voting Preference** The matrix

$$
\begin{array}{c}
\text{From} \\
\overbrace{\begin{array}{ccc} R & D & I \end{array}} \\
P = \begin{bmatrix} 0.6 & 0.1 & 0.1 \\ 0.2 & 0.7 & 0.1 \\ 0.2 & 0.2 & 0.8 \end{bmatrix} \begin{array}{c} R \\ D \\ I \end{array} \Big\} \text{To}
\end{array}
$$

is called a *stochastic matrix*. Each entry p_{ij} $(i \neq j)$ represents the proportion of the voting population that changes from Party i to Party j, and p_{ii} represents the proportion that remains loyal to the party from one election to the next. Use a graphing utility to find P^2. (This matrix gives the transition probabilities from the first election to the third.)

(T) **70. Voting Preference** Use a graphing utility to find P^3, P^4, P^5, P^6, P^7, and P^8 for the matrix given in Exercise 69. Can you detect a pattern as P is raised to higher and higher powers?

In Exercises 71 and 72, find a matrix B such that AB is the identity matrix. Is there more than one correct result?

71. $A = \begin{bmatrix} 1 & 3 \\ 1 & 2 \end{bmatrix}$

72. $A = \begin{bmatrix} 2 & 1 \\ 5 & 2 \end{bmatrix}$

73. If a, b, and c are real numbers such that $c \neq 0$ and $ac = bc$, then $a = b$. However, if A, B, and C are matrices such that $AC = BC$, then A is *not* necessarily equal to B. Illustrate this using the following matrices.

$$
A = \begin{bmatrix} 1 & 2 & 3 \\ 0 & 5 & 4 \\ 3 & -2 & 1 \end{bmatrix}, B = \begin{bmatrix} 4 & -6 & 3 \\ 5 & 4 & 4 \\ -1 & 0 & 1 \end{bmatrix},
$$

and $C = \begin{bmatrix} 0 & 0 & 0 \\ 0 & 0 & 0 \\ 4 & -2 & 3 \end{bmatrix}$

74. If a and b are real numbers such that $ab = 0$, then $a = 0$ or $b = 0$. However, if A and B are matrices such that $AB = O$, then it is *not* necessarily true that $A = O$ or $B = O$. Illustrate this using the following matrices.

$$
A = \begin{bmatrix} 3 & 3 \\ 4 & 4 \end{bmatrix} \quad \text{and} \quad B = \begin{bmatrix} 1 & -1 \\ -1 & 1 \end{bmatrix}
$$

Find another example of two nonzero matrices whose product is the zero matrix.

In Exercises 75 and 76, determine whether the statement is true or false. Justify your answer.

75. $\begin{bmatrix} 3 & 2 \\ 1 & -4 \end{bmatrix}\begin{bmatrix} 1 & 0 \\ 0 & 1 \end{bmatrix} = \begin{bmatrix} 1 & 0 \\ 0 & 1 \end{bmatrix}\begin{bmatrix} 3 & 2 \\ 1 & -4 \end{bmatrix}$

76. $\begin{bmatrix} -6 & -2 \\ 2 & -6 \end{bmatrix}\begin{bmatrix} 4 & 0 \\ 0 & -1 \end{bmatrix} = \begin{bmatrix} 4 & 0 \\ 0 & -1 \end{bmatrix}\begin{bmatrix} -6 & -2 \\ 2 & -6 \end{bmatrix}$

77. Cable Television Two competing companies offer cable television to a city with 100,000 households. Gold Cable Company has 25,000 subscribers and Galaxy Cable Company has 30,000 subscribers. (The other 45,000 households do not subscribe.) The percent changes in cable subscriptions each year are shown in the matrix below.

	Percent Changes		
	From Gold	From Galaxy	From Non-subscriber
To Gold	0.70	0.15	0.15
To Galaxy	0.20	0.80	0.15
To Nonsubscriber	0.10	0.05	0.70

(a) Find the number of subscribers each company will have in one year using matrix multiplication. Explain how you obtained your answer.

(b) Find the number of subscribers each company will have in two years using matrix multiplication. Explain how you obtained your answer.

(c) Find the number of subscribers each company will have in three years using matrix multiplication. Explain how you obtained your answer.

(d) What is happening to the number of subscribers to each company? What is happening to the number of nonsubscribers?

78. Extended Application To work an extended application analyzing airline routes with matrices, visit this text's website at *college.hmco.com*.

Section 6.3

The Inverse of a Square Matrix

- Verify that a matrix is the inverse of a given matrix.
- Find the inverse of a matrix.
- Find the inverse of a 2 × 2 matrix using a formula.
- Use an inverse matrix to solve a system of linear equations.

The Inverse of a Matrix

This section further develops the algebra of matrices. To begin, consider the real number equation $ax = b$. To solve this equation for x, multiply each side of the equation by a^{-1} (provided $a \neq 0$).

$$ax = b$$
$$(a^{-1}a)x = a^{-1}b$$
$$(1)x = a^{-1}b$$
$$x = a^{-1}b$$

The number a^{-1} is called the *multiplicative inverse of a* because $a^{-1}a = 1$. The definition of the multiplicative inverse of a matrix is similar.

Definition of the Inverse of a Square Matrix

Let A be an $n \times n$ matrix and let I_n be the $n \times n$ identity matrix. If there exists a matrix A^{-1} such that

$$AA^{-1} = I_n = A^{-1}A$$

then A^{-1} is called the **inverse** of A. (The symbol A^{-1} is read "A inverse.")

Example 1   The Inverse of a Matrix

Show that B is the inverse of A, where

$$A = \begin{bmatrix} -1 & 2 \\ -1 & 1 \end{bmatrix} \quad \text{and} \quad B = \begin{bmatrix} 1 & -2 \\ 1 & -1 \end{bmatrix}.$$

SOLUTION   To show that B is the inverse of A, show that $AB = I = BA$, as follows.

$$AB = \begin{bmatrix} -1 & 2 \\ -1 & 1 \end{bmatrix}\begin{bmatrix} 1 & -2 \\ 1 & -1 \end{bmatrix} = \begin{bmatrix} -1+2 & 2-2 \\ -1+1 & 2-1 \end{bmatrix} = \begin{bmatrix} 1 & 0 \\ 0 & 1 \end{bmatrix}$$

$$BA = \begin{bmatrix} 1 & -2 \\ 1 & -1 \end{bmatrix}\begin{bmatrix} -1 & 2 \\ -1 & 1 \end{bmatrix} = \begin{bmatrix} -1+2 & 2-2 \\ -1+1 & 2-1 \end{bmatrix} = \begin{bmatrix} 1 & 0 \\ 0 & 1 \end{bmatrix}$$

STUDY TIP

Recall that it is not always true that $AB = BA$, even if both products are defined. However, if A and B are both square matrices and $AB = I_n$, it can be shown that $BA = I_n$. So, in Example 1, you need only check that $AB = I_2$.

✓ **CHECKPOINT 1**

Show that B is the inverse of A, where $A = \begin{bmatrix} 3 & 5 \\ -1 & -2 \end{bmatrix}$ and $B = \begin{bmatrix} 2 & 5 \\ -1 & -3 \end{bmatrix}$. ∎

If a matrix A has an inverse, A is called **invertible** (or **nonsingular**); otherwise, A is called **singular.** A nonsquare matrix cannot have an inverse. To see this, note that if A is of order $m \times n$ and B is of order $n \times m$ (where $m \neq n$), the products AB and BA are of different orders and therefore cannot be equal to each other. Not all square matrices have inverses (see the matrix at the bottom of page 500). If, however, a matrix does have an inverse, that inverse is unique. The following example shows how to use a system of equations to find an inverse.

Example 2   Finding the Inverse of a Matrix

Find the inverse of the matrix

$$A = \begin{bmatrix} 1 & 4 \\ -1 & -3 \end{bmatrix}.$$

SOLUTION   To find the inverse of A, try to solve the matrix equation $AX = I$ for X.

$$\begin{array}{ccc} AX & = & I \end{array}$$

$$\begin{bmatrix} 1 & 4 \\ -1 & -3 \end{bmatrix} \begin{bmatrix} x_{11} & x_{12} \\ x_{21} & x_{22} \end{bmatrix} = \begin{bmatrix} 1 & 0 \\ 0 & 1 \end{bmatrix} \qquad \text{Write matrix equation.}$$

$$\begin{bmatrix} x_{11} + 4x_{21} & x_{12} + 4x_{22} \\ -x_{11} - 3x_{21} & -x_{12} - 3x_{22} \end{bmatrix} = \begin{bmatrix} 1 & 0 \\ 0 & 1 \end{bmatrix} \qquad \text{Multiply } A \text{ and } X.$$

Equating corresponding entries, you obtain the following two systems of linear equations.

$$\begin{cases} x_{11} + 4x_{21} = 1 \\ -x_{11} - 3x_{21} = 0 \end{cases} \qquad \begin{cases} x_{12} + 4x_{22} = 0 \\ -x_{12} - 3x_{22} = 1 \end{cases}$$

You can solve these systems using the methods learned in Chapter 5. From the first system you can determine that $x_{11} = -3$ and $x_{21} = 1$, and from the second system you can determine that $x_{12} = -4$ and $x_{22} = 1$. So, the inverse of A is

$$X = A^{-1}$$

$$= \begin{bmatrix} -3 & -4 \\ 1 & 1 \end{bmatrix}.$$

You can use matrix multiplication to check this result.

CHECK

$$AA^{-1} = \begin{bmatrix} 1 & 4 \\ -1 & -3 \end{bmatrix} \begin{bmatrix} -3 & -4 \\ 1 & 1 \end{bmatrix}$$

$$= \begin{bmatrix} 1 & 0 \\ 0 & 1 \end{bmatrix} \checkmark$$

$$A^{-1}A = \begin{bmatrix} -3 & -4 \\ 1 & 1 \end{bmatrix} \begin{bmatrix} 1 & 4 \\ -1 & -3 \end{bmatrix}$$

$$= \begin{bmatrix} 1 & 0 \\ 0 & 1 \end{bmatrix} \checkmark$$

✓ **CHECKPOINT 2**

Find the inverse of the matrix

$$A = \begin{bmatrix} -4 & -1 \\ 5 & 1 \end{bmatrix}.$$

Finding Inverse Matrices

In Example 2, note that the two systems of linear equations have the *same coefficient matrix A*. Rather than solve the two systems represented by

$$\begin{bmatrix} 1 & 4 & \vdots & 1 \\ -1 & -3 & \vdots & 0 \end{bmatrix} \quad \text{and} \quad \begin{bmatrix} 1 & 4 & \vdots & 0 \\ -1 & -3 & \vdots & 1 \end{bmatrix}$$

separately, you can solve them simultaneously by adjoining the identity matrix to the coefficient matrix to obtain

$$\begin{array}{cc} A & I \\ \begin{bmatrix} 1 & 4 & \vdots & 1 & 0 \\ -1 & -3 & \vdots & 0 & 1 \end{bmatrix}. \end{array}$$

Then, by applying Gauss-Jordan elimination to this matrix, you can solve *both* systems with a single elimination process, as follows.

$$\begin{bmatrix} 1 & 4 & \vdots & 1 & 0 \\ -1 & -3 & \vdots & 0 & 1 \end{bmatrix}$$

$$R_1 + R_2 \rightarrow \begin{bmatrix} 1 & 4 & \vdots & 1 & 0 \\ 0 & 1 & \vdots & 1 & 1 \end{bmatrix}$$

$$-4R_2 + R_1 \rightarrow \begin{bmatrix} 1 & 0 & \vdots & -3 & -4 \\ 0 & 1 & \vdots & 1 & 1 \end{bmatrix}$$

So, from the "doubly augmented" matrix $[A \; \vdots \; I]$, you obtain $[I \; \vdots \; A^{-1}]$.

$$\begin{array}{cccc} A & I & I & A^{-1} \\ \begin{bmatrix} 1 & 4 & \vdots & 1 & 0 \\ -1 & -3 & \vdots & 0 & 1 \end{bmatrix} & \Longrightarrow & \begin{bmatrix} 1 & 0 & \vdots & -3 & -4 \\ 0 & 1 & \vdots & 1 & 1 \end{bmatrix} \end{array}$$

This procedure (or algorithm) works for an arbitrary square matrix that has an inverse.

Finding an Inverse Matrix

Let A be a square matrix of order n.

1. Write the $n \times 2n$ matrix that consists of the given matrix A on the left and the $n \times n$ identity matrix I on the right to obtain $[A \; \vdots \; I]$. Note that the matrices A and I are separated by a dotted line. This process is called **adjoining** the matrices A and I.

2. If possible, row reduce A to I using elementary row operations on the *entire* matrix $[A \; \vdots \; I]$. The result will be the matrix $[I \; \vdots \; A^{-1}]$. If this is not possible, A is not invertible.

3. Check your work by multiplying to see that $AA^{-1} = I = A^{-1}A$.

DISCOVERY

Select two 2×2 matrices A and B that have inverses. Calculate $(AB)^{-1}$ and then calculate $B^{-1}A^{-1}$ and $A^{-1}B^{-1}$. Make a conjecture about the inverse of the product of two invertible matrices.

Example 3   Finding the Inverse of a Matrix

Find the inverse of the matrix $A = \begin{bmatrix} 1 & -1 & 0 \\ 1 & 0 & -1 \\ 6 & -2 & -3 \end{bmatrix}$.

SOLUTION   Begin by adjoining the identity matrix to A to form the matrix

$$[A \vdots I] = \begin{bmatrix} 1 & -1 & 0 & \vdots & 1 & 0 & 0 \\ 1 & 0 & -1 & \vdots & 0 & 1 & 0 \\ 6 & -2 & -3 & \vdots & 0 & 0 & 1 \end{bmatrix}.$$

Use elementary row operations to obtain the matrix $[I \vdots A^{-1}]$, as follows.

$$\begin{matrix} \\ -R_1 + R_2 \rightarrow \\ -6R_1 + R_3 \rightarrow \end{matrix} \begin{bmatrix} 1 & -1 & 0 & \vdots & 1 & 0 & 0 \\ 0 & 1 & -1 & \vdots & -1 & 1 & 0 \\ 0 & 4 & -3 & \vdots & -6 & 0 & 1 \end{bmatrix}$$

$$\begin{matrix} R_2 + R_1 \rightarrow \\ \\ -4R_2 + R_3 \rightarrow \end{matrix} \begin{bmatrix} 1 & 0 & -1 & \vdots & 0 & 1 & 0 \\ 0 & 1 & -1 & \vdots & -1 & 1 & 0 \\ 0 & 0 & 1 & \vdots & -2 & -4 & 1 \end{bmatrix}$$

$$\begin{matrix} R_3 + R_1 \rightarrow \\ R_3 + R_2 \rightarrow \\ \end{matrix} \begin{bmatrix} 1 & 0 & 0 & \vdots & -2 & -3 & 1 \\ 0 & 1 & 0 & \vdots & -3 & -3 & 1 \\ 0 & 0 & 1 & \vdots & -2 & -4 & 1 \end{bmatrix}$$

So, the matrix A is invertible and its inverse is

$$A^{-1} = \begin{bmatrix} -2 & -3 & 1 \\ -3 & -3 & 1 \\ -2 & -4 & 1 \end{bmatrix}.$$

✓ **CHECKPOINT 3**

Find the inverse of the matrix

$$A = \begin{bmatrix} 1 & -2 & -1 \\ 0 & -1 & 2 \\ 1 & -2 & 0 \end{bmatrix}.$$

Confirm this result by multiplying A and A^{-1} to obtain I, as follows.

$$AA^{-1} = \begin{bmatrix} 1 & -1 & 0 \\ 1 & 0 & -1 \\ 6 & -2 & -3 \end{bmatrix} \begin{bmatrix} -2 & -3 & 1 \\ -3 & -3 & 1 \\ -2 & -4 & 1 \end{bmatrix} = \begin{bmatrix} 1 & 0 & 0 \\ 0 & 1 & 0 \\ 0 & 0 & 1 \end{bmatrix} = I$$

The process shown in Example 3 applies to any $n \times n$ matrix A. If A has an inverse, this process will find it. When using this process, if the matrix A does not reduce to the identity matrix, then A does not have an inverse.

To confirm that matrix A shown below has no inverse, begin by adjoining the identity matrix to A to form the following.

$$A = \begin{bmatrix} 1 & 2 & 0 \\ 3 & -1 & 2 \\ -2 & 3 & -2 \end{bmatrix} \implies [A \vdots I] = \begin{bmatrix} 1 & 2 & 0 & \vdots & 1 & 0 & 0 \\ 3 & -1 & 2 & \vdots & 0 & 1 & 0 \\ -2 & 3 & -2 & \vdots & 0 & 0 & 1 \end{bmatrix}$$

Then use elementary row operations to obtain

$$\begin{bmatrix} 1 & 2 & 0 & \vdots & 1 & 0 & 0 \\ 0 & -7 & 2 & \vdots & -3 & 1 & 0 \\ 0 & 0 & 0 & \vdots & -2 & 1 & 1 \end{bmatrix}.$$

At this point in the elimination process, you can see that it is impossible to obtain the identity matrix I on the left. So, A is not invertible.

The Inverse of a 2 × 2 Matrix (Quick Method)

Using Gauss-Jordan elimination to find the inverse of a matrix works well (even as a computer technique) for matrices of order 3 × 3 or greater. For 2 × 2 matrices, however, many people prefer to use a formula for the inverse rather than Gauss-Jordan elimination. This simple formula, which works *only* for 2 × 2 matrices, is explained as follows. If A is a 2 × 2 matrix given by

$$A = \begin{bmatrix} a & b \\ c & d \end{bmatrix}$$

then A is invertible if and only if $ad - bc \neq 0$. Moreover, if $ad - bc \neq 0$, the inverse is given by

$$A^{-1} = \frac{1}{ad - bc} \begin{bmatrix} d & -b \\ -c & a \end{bmatrix}.$$

Try verifying this inverse by multiplication.

The denominator $ad - bc$ is called the **determinant** of the 2 × 2 matrix A. You will study determinants in the next section.

Example 4 Finding the Inverse of a 2 × 2 Matrix

If possible, find the inverse of each matrix.

a. $A = \begin{bmatrix} 3 & -1 \\ -2 & 2 \end{bmatrix}$ **b.** $B = \begin{bmatrix} 3 & -1 \\ -6 & 2 \end{bmatrix}$

SOLUTION

a. For the matrix A, begin by applying the formula for the determinant of a 2 × 2 matrix to obtain

$$ad - bc = 3(2) - (-1)(-2) = 4.$$

Because this quantity is not zero, the matrix is invertible. The inverse is formed by interchanging the entries on the main diagonal, changing the signs of the other two entries, and multiplying by the scalar $\frac{1}{4}$, as follows.

$$A^{-1} = \frac{1}{ad - bc} \begin{bmatrix} d & -b \\ -c & a \end{bmatrix}$$ Formula for inverse of a 2 × 2 matrix

$$= \frac{1}{4} \begin{bmatrix} 2 & 1 \\ 2 & 3 \end{bmatrix}$$ Substitute for a, b, c, d, and the determinant.

$$= \begin{bmatrix} \frac{1}{4}(2) & \frac{1}{4}(1) \\ \frac{1}{4}(2) & \frac{1}{4}(3) \end{bmatrix}$$ Multiply by the scalar $\frac{1}{4}$.

$$= \begin{bmatrix} \frac{1}{2} & \frac{1}{4} \\ \frac{1}{2} & \frac{3}{4} \end{bmatrix}$$ Simplify.

b. For the matrix B, you have

$$ad - bc = 3(2) - (-1)(-6)$$
$$= 0.$$

Because $ad - bc = 0$, B is not invertible.

✓**CHECKPOINT 4**

Find the inverse of the matrix

$$A = \begin{bmatrix} -2 & 3 \\ 4 & -1 \end{bmatrix}.$$

Systems of Linear Equations

You know that a system of linear equations can have exactly one solution, infinitely many solutions, or no solution. If the coefficient matrix A of a *square* system (a system that has the same number of equations as variables) is invertible, then the system has a unique solution, which is defined as follows.

A System of Equations with a Unique Solution

If A is an invertible matrix, then the system of linear equations represented by $AX = B$ has a unique solution given by $X = A^{-1}B$.

TECHNOLOGY

(T) To solve a system of equations with a graphing utility, enter the matrices A and B in the *matrix editor*. Then, using the inverse key, solve for X.

$$A \; \boxed{x^{-1}} \; B \; \boxed{\text{ENTER}}$$

The screen will display the solution, matrix X.

Example 5   Solving a System of Equations Using an Inverse Matrix

Use an inverse matrix to solve the system.

$$\begin{cases} 2x + 3y + z = -1 \\ 3x + 3y + z = 1 \\ 2x + 4y + z = -2 \end{cases}$$

SOLUTION   Begin by writing the system in the matrix form $AX = B$.

$$\begin{bmatrix} 2 & 3 & 1 \\ 3 & 3 & 1 \\ 2 & 4 & 1 \end{bmatrix} \begin{bmatrix} x \\ y \\ z \end{bmatrix} = \begin{bmatrix} -1 \\ 1 \\ -2 \end{bmatrix}$$

Next, use Gauss-Jordan elimination to find A^{-1}.

$$A^{-1} = \begin{bmatrix} -1 & 1 & 0 \\ -1 & 0 & 1 \\ 6 & -2 & -3 \end{bmatrix}$$

Finally, multiply B by A^{-1} on the left to obtain the solution.

$$X = A^{-1}B = \begin{bmatrix} -1 & 1 & 0 \\ -1 & 0 & 1 \\ 6 & -2 & -3 \end{bmatrix} \begin{bmatrix} -1 \\ 1 \\ -2 \end{bmatrix} = \begin{bmatrix} 2 \\ -1 \\ -2 \end{bmatrix}$$

So, the solution is $x = 2$, $y = -1$, and $z = -2$.

✓ **CHECKPOINT 5**

Use an inverse matrix to solve the system.

$$\begin{cases} -x + y + z = 4 \\ 2x - y - 3z = -7 \\ -2x + 3y + 2z = 10 \end{cases}$$ ■

CONCEPT CHECK

1. What is the product of a square matrix of order n and its inverse?

2. Matrix A is a singular matrix of order n. Does a matrix B exist such that $AB = I$? Explain.

3. Consider the matrix $A = \begin{bmatrix} x_{11} & x_{12} \\ x_{21} & x_{22} \end{bmatrix}$, where $x_{11} \cdot x_{22} = x_{12} \cdot x_{21}$. Is A invertible? Explain.

4. Matrix A is nonsingular. Can a system of linear equations represented by $AX = B$ have infinitely many solutions? Explain.

Skills Review 6.3

The following warm-up exercises involve skills that were covered in earlier sections. You will use these skills in the exercise set for this section. For additional help, review Sections 6.1 and 6.2.

In Exercises 1–8, perform the indicated matrix operations.

1. $4\begin{bmatrix} 1 & 6 \\ 0 & -4 \\ 12 & 2 \end{bmatrix}$

2. $\dfrac{1}{2}\begin{bmatrix} 11 & 10 & 48 \\ 1 & 0 & 16 \\ 0 & 2 & 8 \end{bmatrix}$

3. $\begin{bmatrix} 1 & -10 & 3 \\ 4 & 1 & 0 \end{bmatrix} - 2\begin{bmatrix} 3 & -4 & 8 \\ 0 & 7 & 1 \end{bmatrix}$

4. $\begin{bmatrix} 5 & 20 \\ -7 & 15 \end{bmatrix} - 3\begin{bmatrix} 6 & 3 \\ 4 & -2 \end{bmatrix}$

5. $\begin{bmatrix} 1 & -2 \\ -1 & 3 \end{bmatrix}\begin{bmatrix} 3 & 2 \\ 1 & 1 \end{bmatrix}$

6. $\begin{bmatrix} 1 & 0 \\ 0 & 1 \end{bmatrix}\begin{bmatrix} 6 & 5 \\ 3 & -2 \end{bmatrix}$

7. $\begin{bmatrix} 2 & 0 & 0 \\ 0 & -1 & 0 \\ 0 & 0 & 3 \end{bmatrix}\begin{bmatrix} \frac{1}{2} & 0 & 0 \\ 0 & -1 & 0 \\ 0 & 0 & \frac{1}{3} \end{bmatrix}$

8. $\begin{bmatrix} 1 & 2 & 3 \\ 3 & -1 & -2 \\ 3 & 1 & 1 \end{bmatrix}\begin{bmatrix} 1 & 1 & -1 \\ -9 & -8 & 11 \\ 6 & 5 & -7 \end{bmatrix}$

In Exercises 9 and 10, rewrite the matrix in reduced row-echelon form.

9. $\begin{bmatrix} 3 & -2 & 1 & 0 \\ 4 & -3 & 0 & 1 \end{bmatrix}$

10. $\begin{bmatrix} 1 & 1 & 2 & 1 & 0 & 0 \\ -1 & 0 & 3 & 0 & 1 & 0 \\ 1 & 2 & 8 & 0 & 0 & 1 \end{bmatrix}$

Exercises 6.3

See www.CalcChat.com for worked-out solutions to odd-numbered exercises.

In Exercises 1–10, show that B is the inverse of A.

1. $A = \begin{bmatrix} 7 & 4 \\ 5 & 3 \end{bmatrix}$, $B = \begin{bmatrix} 3 & -4 \\ -5 & 7 \end{bmatrix}$

2. $A = \begin{bmatrix} -4 & 1 \\ -9 & 2 \end{bmatrix}$, $B = \begin{bmatrix} 2 & -1 \\ 9 & -4 \end{bmatrix}$

3. $A = \begin{bmatrix} 2 & -1 \\ 5 & -4 \end{bmatrix}$, $B = \begin{bmatrix} \frac{4}{3} & -\frac{1}{3} \\ \frac{5}{3} & -\frac{2}{3} \end{bmatrix}$

4. $A = \begin{bmatrix} 1 & -2 \\ 3 & -10 \end{bmatrix}$, $B = \begin{bmatrix} \frac{5}{2} & -\frac{1}{2} \\ \frac{3}{4} & -\frac{1}{4} \end{bmatrix}$

5. $A = \begin{bmatrix} -2 & 2 & 3 \\ 1 & -1 & 0 \\ 0 & 1 & 4 \end{bmatrix}$, $B = \dfrac{1}{3}\begin{bmatrix} -4 & -5 & 3 \\ -4 & -8 & 3 \\ 1 & 2 & 0 \end{bmatrix}$

6. $A = \begin{bmatrix} -1 & 0 & 2 \\ 1 & -2 & 0 \\ 1 & 0 & 3 \end{bmatrix}$, $B = \dfrac{1}{10}\begin{bmatrix} -6 & 0 & 4 \\ -3 & -5 & 2 \\ 2 & 0 & 2 \end{bmatrix}$

7. $A = \begin{bmatrix} 2 & -17 & 11 \\ -1 & 11 & -7 \\ 0 & 3 & -2 \end{bmatrix}$, $B = \begin{bmatrix} 1 & 1 & 2 \\ 2 & 4 & -3 \\ 3 & 6 & -5 \end{bmatrix}$

8. $A = \begin{bmatrix} -1 & 1 & -3 \\ 2 & -1 & 4 \\ -1 & 1 & -2 \end{bmatrix}$, $B = \begin{bmatrix} 2 & 1 & -1 \\ 0 & 1 & 2 \\ -1 & 0 & 1 \end{bmatrix}$

9. $A = \begin{bmatrix} 2 & 0 & 2 & 1 \\ 3 & 0 & 0 & 1 \\ -1 & 1 & -2 & 1 \\ 3 & -1 & 1 & 0 \end{bmatrix}$,

$B = \dfrac{1}{3}\begin{bmatrix} -1 & 3 & -2 & -2 \\ -2 & 9 & -7 & -10 \\ 1 & 0 & -1 & -1 \\ 3 & -6 & 6 & 6 \end{bmatrix}$

10. $A = \begin{bmatrix} -1 & 1 & 0 & -1 \\ 1 & -1 & 2 & 0 \\ -1 & 1 & 2 & 0 \\ 0 & -1 & 1 & 1 \end{bmatrix}$,

$B = \dfrac{1}{4}\begin{bmatrix} -4 & 1 & 1 & -4 \\ -4 & -1 & 3 & -4 \\ 0 & 1 & 1 & 0 \\ -4 & -2 & 2 & 0 \end{bmatrix}$

In Exercises 11–30, find the inverse of the matrix (if it exists).

11. $\begin{bmatrix} 1 & 2 \\ 3 & 7 \end{bmatrix}$ **12.** $\begin{bmatrix} -1 & 1 \\ -2 & 1 \end{bmatrix}$

13. $\begin{bmatrix} 11 & 1 \\ -1 & 0 \end{bmatrix}$ **14.** $\begin{bmatrix} -7 & 33 \\ 4 & -19 \end{bmatrix}$

15. $\begin{bmatrix} 8 & 4 \\ -2 & -2 \end{bmatrix}$ **16.** $\begin{bmatrix} 2 & 3 \\ 1 & 4 \end{bmatrix}$

17. $\begin{bmatrix} 0 & 4 \\ -3 & 6 \end{bmatrix}$ **18.** $\begin{bmatrix} 2 & 3 \\ 6 & 9 \end{bmatrix}$

19. $\begin{bmatrix} 2 & 7 & 1 \\ -3 & -9 & 2 \end{bmatrix}$ **20.** $\begin{bmatrix} -2 & 5 \\ 6 & -15 \\ 0 & 1 \end{bmatrix}$

21. $\begin{bmatrix} 1 & 1 & 1 \\ 3 & 5 & 4 \\ 3 & 6 & 5 \end{bmatrix}$ **22.** $\begin{bmatrix} 1 & 2 & 2 \\ 3 & 7 & 9 \\ -1 & -4 & -7 \end{bmatrix}$

23. $\begin{bmatrix} 1 & 1 & 2 \\ 3 & 1 & 0 \\ -2 & 0 & 3 \end{bmatrix}$ **24.** $\begin{bmatrix} 3 & 2 & 2 \\ 2 & 2 & 2 \\ -4 & 4 & 3 \end{bmatrix}$

25. $\begin{bmatrix} 3 & 0 & 0 \\ 0 & -2 & 0 \\ 0 & 0 & 4 \end{bmatrix}$ **26.** $\begin{bmatrix} 2 & 0 & 0 \\ 0 & 3 & 0 \\ 0 & 0 & 5 \end{bmatrix}$

27. $\begin{bmatrix} 1 & 0 & 0 \\ 3 & 4 & 0 \\ 2 & 5 & 5 \end{bmatrix}$ **28.** $\begin{bmatrix} 1 & 0 & 0 \\ 3 & 0 & 0 \\ 2 & 5 & 5 \end{bmatrix}$

29. $\begin{bmatrix} 1 & 0 & 3 & 0 \\ 0 & 2 & 0 & 4 \\ 1 & 0 & 3 & 0 \\ 0 & 2 & 0 & 4 \end{bmatrix}$ **30.** $\begin{bmatrix} -1 & 0 & 1 & 0 \\ 0 & 2 & 0 & -1 \\ 2 & 0 & -1 & 0 \\ 0 & -1 & 0 & 1 \end{bmatrix}$

ⓉIn Exercises 31–38, use the matrix capabilities of a graphing utility to find the inverse of the matrix (if it exists).

31. $\begin{bmatrix} 1 & 2 & -1 \\ 3 & 7 & -10 \\ -5 & -7 & -15 \end{bmatrix}$ **32.** $\begin{bmatrix} 10 & 5 & -7 \\ -5 & 1 & 4 \\ 3 & 2 & -2 \end{bmatrix}$

33. $\begin{bmatrix} 0.1 & 0.2 & 0.3 \\ -0.3 & 0.2 & 0.2 \\ 0.5 & 0.4 & 0.4 \end{bmatrix}$ **34.** $\begin{bmatrix} 0.6 & 0 & -0.3 \\ 0.7 & -1 & 0.2 \\ 1 & 0 & -0.9 \end{bmatrix}$

35. $\begin{bmatrix} 1 & -3 & 2 & -1 \\ 0 & 4 & -12 & 8 \\ 3 & 0 & 5 & -2 \\ 0 & -3 & 9 & -6 \end{bmatrix}$ **36.** $\begin{bmatrix} 1 & 3 & -2 & 0 \\ 0 & 2 & 4 & 6 \\ 0 & 0 & -2 & 1 \\ 0 & 0 & 0 & 5 \end{bmatrix}$

37. $\begin{bmatrix} 1 & -2 & -1 & -2 \\ 3 & -5 & -2 & -3 \\ 2 & -5 & -2 & -5 \\ -1 & 4 & 4 & 11 \end{bmatrix}$

38. $\begin{bmatrix} 4 & 8 & -7 & 14 \\ 2 & 5 & -4 & 6 \\ 0 & 2 & 1 & -7 \\ 3 & 6 & -5 & 10 \end{bmatrix}$

In Exercises 39–44, use the formula on page 501 to find the inverse of the matrix (if it exists).

39. $\begin{bmatrix} 5 & -2 \\ 2 & 3 \end{bmatrix}$ **40.** $\begin{bmatrix} 7 & 12 \\ -8 & -5 \end{bmatrix}$

41. $\begin{bmatrix} -4 & -6 \\ 2 & 3 \end{bmatrix}$ **42.** $\begin{bmatrix} -12 & 3 \\ 5 & -2 \end{bmatrix}$

43. $\begin{bmatrix} \frac{7}{2} & -\frac{3}{4} \\ \frac{1}{5} & \frac{4}{5} \end{bmatrix}$ **44.** $\begin{bmatrix} -\frac{1}{4} & \frac{9}{4} \\ \frac{5}{3} & \frac{8}{9} \end{bmatrix}$

In Exercises 45–48, use the inverse matrix found in Exercise 11 to solve the system of linear equations.

45. $\begin{cases} x + 2y = 0 \\ 3x + 7y = 1 \end{cases}$ **46.** $\begin{cases} x + 2y = -5 \\ 3x + 7y = -16 \end{cases}$

47. $\begin{cases} x + 2y = 8 \\ 3x + 7y = 26 \end{cases}$ **48.** $\begin{cases} x + 2y = -6 \\ 3x + 7y = -21 \end{cases}$

In Exercises 49–52, use the inverse matrix found in Exercise 16 to solve the system of linear equations.

49. $\begin{cases} 2x + 3y = 5 \\ x + 4y = 10 \end{cases}$ **50.** $\begin{cases} 2x + 3y = 0 \\ x + 4y = 3 \end{cases}$

51. $\begin{cases} 2x + 3y = 4 \\ x + 4y = 2 \end{cases}$ **52.** $\begin{cases} 2x + 3y = 1 \\ x + 4y = -2 \end{cases}$

In Exercises 53 and 54, use the inverse matrix found in Exercise 21 to solve the system of linear equations.

53. $\begin{cases} x + y + z = 0 \\ 3x + 5y + 4z = 5 \\ 3x + 6y + 5z = 2 \end{cases}$ **54.** $\begin{cases} x + y + z = -1 \\ 3x + 5y + 4z = 2 \\ 3x + 6y + 5z = 0 \end{cases}$

In Exercises 55 and 56, use the inverse matrix found in Exercise 37 to solve the system of linear equations.

55. $\begin{cases} x_1 - 2x_2 - x_3 - 2x_4 = 0 \\ 3x_1 - 5x_2 - 2x_3 - 3x_4 = 1 \\ 2x_1 - 5x_2 - 2x_3 - 5x_4 = -1 \\ -x_1 + 4x_2 + 4x_3 + 11x_4 = 2 \end{cases}$

56. $\begin{cases} x_1 - 2x_2 - x_3 - 2x_4 = 1 \\ 3x_1 - 5x_2 - 2x_3 - 3x_4 = -2 \\ 2x_1 - 5x_2 - 2x_3 - 5x_4 = 0 \\ -x_1 + 4x_2 + 4x_3 + 11x_4 = -3 \end{cases}$

In Exercises 57–64, use an inverse matrix to solve (if possible) the system of linear equations.

57. $\begin{cases} 3x + 4y = -2 \\ 5x + 3y = 4 \end{cases}$

58. $\begin{cases} 18x + 12y = 13 \\ 30x + 24y = 23 \end{cases}$

59. $\begin{cases} -0.4x + 0.8y = 1.6 \\ 2x - 4y = 5 \end{cases}$

60. $\begin{cases} 0.2x - 0.6y = 2.4 \\ -x + 1.4y = -8.8 \end{cases}$

61. $\begin{cases} -\frac{1}{4}x + \frac{3}{8}y = -2 \\ \frac{3}{2}x + \frac{3}{4}y = -12 \end{cases}$

62. $\begin{cases} \frac{5}{6}x - y = -20 \\ \frac{4}{3}x - \frac{7}{2}y = -51 \end{cases}$

63. $\begin{cases} 4x - y + z = -5 \\ 2x + 2y + 3z = 10 \\ 5x - 2y + 6z = 1 \end{cases}$

64. $\begin{cases} 4x - 2y + 3z = -2 \\ 2x + 2y + 5z = 16 \\ 8x - 5y - 2z = 4 \end{cases}$

Ⓣ In Exercises 65 and 66, use the matrix capabilities of a graphing utility to solve (if possible) the system of linear equations.

65. $\begin{cases} 7x - 3y + 2w = 41 \\ -2x + y - w = -13 \\ 4x + z - 2w = 12 \\ -x + y - w = -8 \end{cases}$

66. $\begin{cases} 2x + 5y + w = 11 \\ x + 4y + 2z - 2w = -7 \\ 2x - 2y + 5z + w = 3 \\ x - 3w = -1 \end{cases}$

In Exercises 67 and 68, develop for the given matrix a system of equations that has the given solution. Use an inverse matrix to verify that the system of equations has the desired solution.

67. $\begin{bmatrix} 2 & 1 & 3 \\ 4 & 0 & -2 \\ 0 & 3 & 2 \end{bmatrix}$ $\begin{aligned} x &= 2 \\ y &= -3 \\ z &= 5 \end{aligned}$

68. $\begin{bmatrix} 1 & 0 & 2 \\ 1 & 1 & 1 \\ 2 & -1 & 0 \end{bmatrix}$ $\begin{aligned} x &= 5 \\ y &= -2 \\ z &= 1 \end{aligned}$

Bond Investment   In Exercises 69–72, you invest in AAA-rated bonds, A-rated bonds, and B-rated bonds. Your average yield is 9% on AAA bonds, 7% on A bonds, and 8% on B bonds. You invest twice as much in B bonds as in A bonds. The desired system of linear equations (where x, y, and z represent the amounts invested in AAA, A, and B bonds, respectively) is as follows.

$\begin{cases} x + y + z = \text{(total investment)} \\ 0.09x + 0.07y + 0.08z = \text{(annual return)} \\ 2y - z = 0 \end{cases}$

Use the inverse of the coefficient matrix of this system to find the amount invested in each type of bond for the given total investment and annual return.

69. Total investment = $35,000; annual return = $2950

70. Total investment = $50,000; annual return = $4180

71. Total investment = $36,000; annual return = $3040

72. Total investment = $45,000; annual return = $3770

Circuit Analysis   In Exercises 73 and 74, consider the circuit shown in the figure. The currents I_1, I_2, and I_3, in amperes, are the solution of the system of linear equations

$\begin{cases} 2I_1 + 4I_3 = E_1 \\ I_2 + 4I_3 = E_2 \\ I_1 + I_2 - I_3 = 0 \end{cases}$

where E_1 and E_2 are voltages. Use the inverse of the coefficient matrix of this system to find the unknown currents for the given voltages.

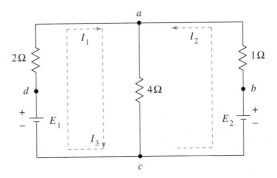

73. $E_1 = 28$ volts, $E_2 = 21$ volts

74. $E_1 = 24$ volts, $E_2 = 23$ volts

Raw Materials   In Exercises 75–78, consider a company that specializes in potting soil. Each bag of potting soil for seedlings requires 2 units of sand, 1 unit of loam, and 1 unit of peat moss. Each bag of potting soil for general potting requires 1 unit of sand, 2 units of loam, and 1 unit of peat moss. Each bag of potting soil for hardwood plants requires 2 units of sand, 2 units of loam, and 2 units of peat moss. Find the numbers of bags of the three types of potting soil that the company can produce with the given amounts of raw materials.

75. 500 units of sand
    500 units of loam
    400 units of peat moss

76. 500 units of sand
    750 units of loam
    450 units of peat moss

77. 350 units of sand
    445 units of loam
    345 units of peat moss

78. 975 units of sand
    1050 units of loam
    725 units of peat moss

79. Child Support The total values y (in billions of dollars) of child support collections from 1998 to 2005 are shown in the figure. The least squares regression parabola $y = at^2 + bt + c$ for these data is found by solving the system

$$\begin{cases} 8c + 92b + 1100a = 153.3 \\ 92c + 1100b + 13{,}616a = 1813.9. \\ 1100c + 13{,}616b + 173{,}636a = 22{,}236.7 \end{cases}$$

Let t represent the year, with $t = 8$ corresponding to 1998. *(Source: U.S. Department of Health and Human Services)*

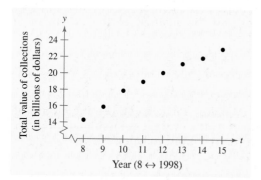

Year (8 ↔ 1998)

(T) (a) Use a graphing utility to find an inverse matrix to solve this system, and find the equation of the least squares regression parabola.

(b) Use the result from part (a) to estimate the value of child support collections in 2007.

(c) An analyst predicted that the value of child support collections in 2007 would be $24.0 billion. How does this value compare with your estimate in part (b)? Do both estimates seem reasonable?

80. Alaskan Fishing The total annual profits y (in thousands of dollars) for an Alaskan fishing captain from 2000 to 2008 are shown in the figure. The least squares regression parabola $y = at^2 + bt + c$ for these data is found by solving the system

$$\begin{cases} 9c + 36b + 204a = 1152 \\ 36c + 204b + 1296a = 4399. \\ 204c + 1296b + 8772a = 24{,}597 \end{cases}$$

Let t represent the year, with $t = 0$ corresponding to 2000.

(T) (a) Use a graphing utility to find an inverse matrix to solve this system, and find the equation of the least squares regression parabola.

(b) Use the result from part (a) to predict the captain's profit in 2010.

(c) Due to increased competition, the captain projects profits of $115,000 in 2010. How does this value compare with your prediction in part (b)?

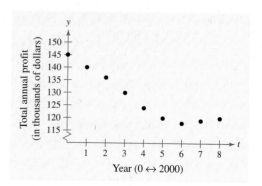

Year (0 ↔ 2000)

Figure for 80

In Exercises 81 and 82, use the following matrices.

$$A = \begin{bmatrix} 4 & 3 \\ -2 & 1 \end{bmatrix}, B = \begin{bmatrix} 1 & -2 \\ 3 & 4 \end{bmatrix}, C = \begin{bmatrix} 13 & 4 \\ 1 & 8 \end{bmatrix}$$

81. Find AB and BA. What do you observe about the two products?

82. Find $C^{-1}, A^{-1} \cdot B^{-1}$, and $B^{-1} \cdot A^{-1}$. What do you observe about the three resulting matrices?

In Exercises 83 and 84, find a value of k that makes the matrix invertible and then find a value of k that makes the matrix singular. (There are many correct answers.)

83. $\begin{bmatrix} 4 & 3 \\ -2 & k \end{bmatrix}$ **84.** $\begin{bmatrix} 2k+1 & 3 \\ -7 & 1 \end{bmatrix}$

In Exercises 85 and 86, determine whether the statement is true or false. Justify your answer.

85. There exists a matrix A such that $A = A^{-1}$.

86. Multiplication of a nonsingular matrix and its inverse is commutative.

87. If A is a 2×2 matrix $A = \begin{bmatrix} a & b \\ c & d \end{bmatrix}$, then A is invertible if and only if $ad - bc \neq 0$. If $ad - bc \neq 0$, verify that the inverse is

$$A^{-1} = \frac{1}{ad - bc} \begin{bmatrix} d & -b \\ -c & a \end{bmatrix}.$$

88. Exploration Consider matrices of the form

$$A = \begin{bmatrix} a_{11} & 0 & 0 & 0 & \cdots & 0 \\ 0 & a_{22} & 0 & 0 & \cdots & 0 \\ 0 & 0 & a_{33} & 0 & \cdots & 0 \\ \vdots & \vdots & \vdots & \vdots & \cdots & \vdots \\ 0 & 0 & 0 & 0 & \cdots & a_{nn} \end{bmatrix}$$

(a) Write a 2×2 matrix and a 3×3 matrix of the form of A. Find the inverse of each.

(b) Use the result from part (a) to make a conjecture about the inverses of matrices of the form of A.

Mid-Chapter Quiz

See www.CalcChat.com for worked-out solutions to odd-numbered exercises.

Take this quiz as you would take a quiz in class. When you are done, check your work against the answers given in the back of the book.

In Exercises 1 and 2, write a matrix of the given order.

1. 4×3

2. 3×1

In Exercises 3 and 4, write the augmented matrix for the system of equations.

3. $\begin{cases} 3x + 2y = -2 \\ 5x - y = 19 \end{cases}$

4. $\begin{cases} x + 3z = -5 \\ x + 2y - z = 3 \\ 3x + 4z = 0 \end{cases}$

5. Use Gaussian elimination with back-substitution to solve the augmented matrix found in Exercise 3.

6. Use Gauss-Jordan elimination to solve the augmented matrix found in Exercise 4.

In Exercises 7–12, use the following matrices to find the indicated matrix (if possible).

$$A = \begin{bmatrix} -1 & 4 \\ -2 & 6 \end{bmatrix}, \quad B = \begin{bmatrix} -1 & 2 & -3 \\ 2 & 0 & 5 \end{bmatrix}, \quad C = \begin{bmatrix} 0 & -2 \\ 3 & 1 \end{bmatrix}$$

7. $2A + 3C$

8. AB

9. $A - 3C$

10. C^2

11. A^{-1}

12. B^{-1}

In Exercises 13 and 14, solve for X using matrices A and C from Exercises 7–12.

13. $X = 3A - 2C$

14. $2X + 4A = 2C$

In Exercises 15–18, a hang glider manufacturer has the labor-hour and wage requirements indicated at the left.

15. What is the labor cost for model A at Plant 1?

16. What is the labor cost for model B at Plant 2?

17. What is the labor cost for model C at Plant 2?

18. Compute LW and interpret the result.

In Exercises 19 and 20, use an inverse matrix to solve the system of linear equations.

19. $\begin{cases} x - 3y = 10 \\ -2x + y = -10 \end{cases}$

20. $\begin{cases} 2x - y + z = 3 \\ 3x - z = 15 \\ 4y + 3z = -1 \end{cases}$

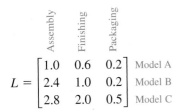

$$L = \begin{bmatrix} 1.0 & 0.6 & 0.2 \\ 2.4 & 1.0 & 0.2 \\ 2.8 & 2.0 & 0.5 \end{bmatrix} \begin{matrix} \text{Model A} \\ \text{Model B} \\ \text{Model C} \end{matrix}$$

with columns labeled Assembly, Finishing, Packaging

Labor-Hour Requirements
(in hours per hang glider)

$$W = \begin{bmatrix} 15 & 12 \\ 10 & 11 \\ 9 & 8 \end{bmatrix} \begin{matrix} \text{Assembly} \\ \text{Finishing} \\ \text{Packaging} \end{matrix}$$

with columns labeled Plant 1, Plant 2

Wage Requirements
(in dollars per hour)

Matrices for 15–18

Section 6.4

The Determinant of a Square Matrix

- Evaluate the determinant of a 2 × 2 matrix.
- Find the minors and cofactors of a matrix.
- Find the determinant of a square matrix.
- Find the determinant of a triangular matrix.

The Determinant of a 2 × 2 Matrix

Every *square* matrix can be associated with a real number called its **determinant.** Determinants have many uses, and several will be discussed in this and the next section. The use of determinants is derived from special number patterns that occur when systems of linear equations are solved. For instance, the system

$$\begin{cases} a_1x + b_1y = c_1 \\ a_2x + b_2y = c_2 \end{cases}$$

has a solution given by

$$x = \frac{c_1b_2 - c_2b_1}{a_1b_2 - a_2b_1} \quad \text{and} \quad y = \frac{a_1c_2 - a_2c_1}{a_1b_2 - a_2b_1}$$

provided that $a_1b_2 - a_2b_1 \neq 0$. Note that the denominator of each fraction is the same. This denominator is called the *determinant* of the coefficient matrix of the system.

Coefficient Matrix   *Determinant*

$$A = \begin{bmatrix} a_1 & b_1 \\ a_2 & b_2 \end{bmatrix} \qquad \det(A) = a_1b_2 - a_2b_1$$

The determinant of the matrix A can also be denoted by vertical bars on both sides of the matrix, as indicated in the following definition.

Definition of the Determinant of a 2 × 2 Matrix

The **determinant** of the matrix

$$A = \begin{bmatrix} a_1 & b_1 \\ a_2 & b_2 \end{bmatrix}$$

is given by

$$\det(A) = |A| = \begin{vmatrix} a_1 & b_1 \\ a_2 & b_2 \end{vmatrix} = a_1b_2 - a_2b_1.$$

In this text, $\det(A)$ and $|A|$ are used interchangeably to represent the determinant of A. Although vertical bars are also used to denote the absolute value of a real number, the context will show which use is intended.

A convenient method for remembering the formula for the determinant of a 2 × 2 matrix is shown in the following diagram.

$$\det(A) = \begin{vmatrix} a_1 & b_1 \\ a_2 & b_2 \end{vmatrix} = a_1b_2 - a_2b_1$$

Note that the determinant is the difference of the products of the two diagonals of the matrix. In Example 1 you will see that the determinant of a matrix can be positive, zero, or negative.

Example 1   The Determinant of a 2 × 2 Matrix

Find the determinant of each matrix.

a. $A = \begin{bmatrix} 2 & -3 \\ 1 & 2 \end{bmatrix}$ **b.** $B = \begin{bmatrix} 2 & 1 \\ 4 & 2 \end{bmatrix}$ **c.** $C = \begin{bmatrix} 0 & 3 \\ 2 & 4 \end{bmatrix}$

SOLUTION   Use the formula $\det(A) = \begin{vmatrix} a_1 & b_1 \\ a_2 & b_2 \end{vmatrix} = a_1 b_2 - a_2 b_1$.

a. $\det(A) = \begin{vmatrix} 2 & -3 \\ 1 & 2 \end{vmatrix} = 2(2) - 1(-3) = 4 + 3 = 7$

b. $\det(B) = \begin{vmatrix} 2 & 1 \\ 4 & 2 \end{vmatrix} = 2(2) - 4(1) = 4 - 4 = 0$

c. $\det(C) = \begin{vmatrix} 0 & 3 \\ 2 & 4 \end{vmatrix} = 0(4) - 2(3) = 0 - 6 = -6$

STUDY TIP

The determinant of a matrix of order 1×1 is defined simply as the entry of the matrix. For instance, if $A = [-2]$, then $\det(A) = -2$.

✓ **CHECKPOINT 1**

Find the determinant of $A = \begin{bmatrix} 1 & 2 \\ 3 & -1 \end{bmatrix}$. ▪

TECHNOLOGY

Ⓣ   Most graphing utilities can evaluate the determinant of a matrix. Use a graphing utility to find the determinant of matrix A from Example 1. The result should be 7, as shown below. For specific keystrokes on how to use a graphing utility to evaluate the determinant of a matrix, go to the text website at *college.hmco.com/info/larsonapplied*.

```
det([A])
              7
```

Try evaluating the determinant of B with your graphing utility.

$$B = \begin{bmatrix} 3 & -1 & 1 \\ 0 & 2 & 1 \end{bmatrix}$$

What happens when you try to evaluate the determinant of a nonsquare matrix?

Minors and Cofactors

To define the determinant of a square matrix of order 3×3 or higher, it is convenient to introduce the concepts of **minors** and **cofactors.**

Sign Pattern for Cofactors

$$\begin{bmatrix} + & - & + \\ - & + & - \\ + & - & + \end{bmatrix}$$

3×3 matrix

$$\begin{bmatrix} + & - & + & - \\ - & + & - & + \\ + & - & + & - \\ - & + & - & + \end{bmatrix}$$

4×4 matrix

$$\begin{bmatrix} + & - & + & - & + \cdots \\ - & + & - & + & - \cdots \\ + & - & + & - & + \cdots \\ - & + & - & + & - \cdots \\ + & - & + & - & + \cdots \\ \vdots & \vdots & \vdots & \vdots & \vdots \end{bmatrix}$$

$n \times n$ matrix

Minors and Cofactors of a Square Matrix

If A is a square matrix, the **minor** M_{ij} of the entry a_{ij} is the determinant of the matrix obtained by deleting the ith row and jth column of A. The **cofactor** C_{ij} of the entry a_{ij} is given by $C_{ij} = (-1)^{i+j} M_{ij}$.

In the sign pattern for cofactors at the left, notice that *odd* positions (where $i + j$ is odd) have negative signs and *even* positions (where $i + j$ is even) have positive signs.

Example 2 Finding the Minors and Cofactors of a Matrix

Find all the minors and cofactors of $A = \begin{bmatrix} 0 & 2 & 1 \\ 3 & -1 & 2 \\ 4 & 0 & 1 \end{bmatrix}$.

SOLUTION To find the minor M_{11}, delete the first row and first column of A and evaluate the determinant of the resulting matrix.

$$\begin{bmatrix} 0 & 2 & 1 \\ 3 & -1 & 2 \\ 4 & 0 & 1 \end{bmatrix}, \quad M_{11} = \begin{vmatrix} -1 & 2 \\ 0 & 1 \end{vmatrix} = -1(1) - 0(2) = -1$$

Similarly, to find M_{12}, delete the first row and second column.

$$\begin{bmatrix} 0 & 2 & 1 \\ 3 & -1 & 2 \\ 4 & 0 & 1 \end{bmatrix}, \quad M_{12} = \begin{vmatrix} 3 & 2 \\ 4 & 1 \end{vmatrix} = 3(1) - 4(2) = -5$$

Continuing this pattern, you obtain the following minors.

$$\begin{array}{lll} M_{11} = -1 & M_{12} = -5 & M_{13} = 4 \\ M_{21} = 2 & M_{22} = -4 & M_{23} = -8 \\ M_{31} = 5 & M_{32} = -3 & M_{33} = -6 \end{array}$$

Now, to find the cofactors, combine the minors above with the checkerboard pattern of signs for a 3×3 matrix shown at the upper left.

$$\begin{array}{lll} C_{11} = -1 & C_{12} = 5 & C_{13} = 4 \\ C_{21} = -2 & C_{22} = -4 & C_{23} = 8 \\ C_{31} = 5 & C_{32} = 3 & C_{33} = -6 \end{array}$$

✓CHECKPOINT 2

Find all the minors and cofactors of $A = \begin{bmatrix} 1 & 2 & 3 \\ 0 & -1 & 5 \\ 2 & 1 & 4 \end{bmatrix}$.

The Determinant of a Square Matrix

The definition below is called **inductive** because it uses determinants of matrices of order $n - 1$ to define determinants of matrices of order n.

Determinant of a Square Matrix

If A is a square matrix (of order 2×2 or greater), then the determinant of A is the sum of the entries in any row (or column) of A multiplied by their respective cofactors. For instance, expanding along the first row yields

$$|A| = a_{11}C_{11} + a_{12}C_{12} + \cdots + a_{1n}C_{1n}.$$

Applying this definition to find a determinant is called **expanding by cofactors.**

Try checking that for a 2×2 matrix

$$A = \begin{bmatrix} a_1 & b_1 \\ a_2 & b_2 \end{bmatrix}$$

this definition of the determinant yields

$$|A| = a_1 b_2 - a_2 b_1$$

as previously defined.

Example 3   **The Determinant of a Matrix of Order 3 × 3**

Find the determinant of

$$A = \begin{bmatrix} 0 & 2 & 1 \\ 3 & -1 & 2 \\ 4 & 0 & 1 \end{bmatrix}.$$

SOLUTION   Note that this is the same matrix that was given in Example 2. There you found the cofactors of the entries in the first row to be

$$C_{11} = -1, \quad C_{12} = 5, \quad \text{and} \quad C_{13} = 4.$$

So, by the definition of a determinant, you have

$$\begin{aligned} |A| &= a_{11}C_{11} + a_{12}C_{12} + a_{13}C_{13} \qquad \text{First-row expansion} \\ &= 0(-1) + 2(5) + 1(4) \\ &= 14. \end{aligned}$$

✓**CHECKPOINT 3**

Find the determinant of

$$A = \begin{bmatrix} 3 & 0 & 2 \\ 1 & 2 & -1 \\ 2 & 0 & -4 \end{bmatrix}.$$

In Example 3, the determinant was found by expanding by the cofactors in the first row. You could have used any row or column. For instance, you could have expanded along the second row to obtain

$$\begin{aligned} |A| &= a_{21}C_{21} + a_{22}C_{22} + a_{23}C_{23} \qquad \text{Second-row expansion} \\ &= 3(-2) + (-1)(-4) + 2(8) \\ &= 14. \end{aligned}$$

When expanding by cofactors, you do not need to find cofactors of zero entries, because zero times its cofactor is zero.

$$a_{ij}C_{ij} = (0)C_{ij} = 0$$

So, the row (or column) containing the most zeros is usually the best choice for expansion by cofactors. This is demonstrated in the next example.

Example 4   The Determinant of a Matrix of Order 4 × 4

Find the determinant of

$$A = \begin{bmatrix} 1 & -2 & 3 & 0 \\ -1 & 1 & 0 & 2 \\ 0 & 2 & 0 & 3 \\ 3 & 4 & 0 & 2 \end{bmatrix}.$$

SOLUTION   After inspecting this matrix, you can see that three of the entries in the third column are zeros. So, you can eliminate some of the work in the expansion by using the third column.

$$|A| = 3(C_{13}) + 0(C_{23}) + 0(C_{33}) + 0(C_{43})$$

Because C_{23}, C_{33}, and C_{43} have zero coefficients, you only need to find the cofactor C_{13}. To do this, delete the first row and third column of A and evaluate the determinant of the resulting matrix.

$$C_{13} = (-1)^{1+3} \begin{vmatrix} -1 & 1 & 2 \\ 0 & 2 & 3 \\ 3 & 4 & 2 \end{vmatrix} \qquad \text{Delete 1st row and 3rd column.}$$

$$= \begin{vmatrix} -1 & 1 & 2 \\ 0 & 2 & 3 \\ 3 & 4 & 2 \end{vmatrix} \qquad \text{Simplify.}$$

Expanding by minors in the second row yields

$$C_{13} = 0(-1)^3 \begin{vmatrix} 1 & 2 \\ 4 & 2 \end{vmatrix} + 2(-1)^4 \begin{vmatrix} -1 & 2 \\ 3 & 2 \end{vmatrix} + 3(-1)^5 \begin{vmatrix} -1 & 1 \\ 3 & 4 \end{vmatrix}$$

$$= 0 + 2(1)(-8) + 3(-1)(-7)$$

$$= 5.$$

So, you obtain

$$|A| = 3C_{13} = 3(5) = 15.$$

✓ CHECKPOINT 4

Find the determinant of $A = \begin{bmatrix} 3 & 0 & 7 & 0 \\ 2 & 6 & 0 & 11 \\ 4 & 1 & 0 & 2 \\ 1 & 5 & 0 & 10 \end{bmatrix}.$ ∎

Try using a graphing utility to confirm the result of Example 4.

There is an alternative method that is commonly used to evaluate the determinant of a 3×3 matrix A. This method works *only* for 3×3 matrices. To apply this method, copy the first and second columns of A to form fourth and fifth columns. The determinant of A is then obtained by adding the products of the three "downward diagonals" and subtracting the products of the three "upward diagonals," as shown in the following diagram.

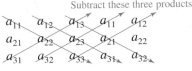

Subtract these three products

Add these three products

So, the determinant of the 3×3 matrix A is given by

$$|A| = a_{11}a_{22}a_{33} + a_{12}a_{23}a_{31} + a_{13}a_{21}a_{32}$$
$$- a_{31}a_{22}a_{13} - a_{32}a_{23}a_{11} - a_{33}a_{21}a_{12}.$$

Example 5   The Determinant of a 3×3 Matrix

Find the determinant of

$$A = \begin{bmatrix} 0 & 2 & 1 \\ 3 & -1 & 2 \\ 4 & -4 & 1 \end{bmatrix}.$$

SOLUTION   Because A is a 3×3 matrix, you can use the alternative procedure for finding $|A|$. Begin by copying the first and second columns to form fourth and fifth columns. Then compute the six diagonal products, as follows.

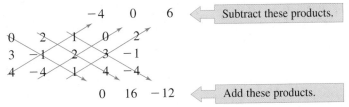

Subtract these products.

Add these products.

Now, by adding the lower three products and subtracting the upper three products, you find the determinant of A to be

$$|A| = 0 + 16 + (-12) - (-4) - 0 - 6$$
$$= 2.$$

✓ CHECKPOINT 5

Find the determinant of $A = \begin{bmatrix} -3 & -4 & 0 \\ 4 & 1 & -2 \\ 0 & 2 & 3 \end{bmatrix}.$ ∎

Be sure you understand that the diagonal process illustrated in Example 5 is valid *only* for matrices of order 3×3. For matrices of higher orders, another method must be used, such as expansion by cofactors or a graphing utility.

Triangular Matrices

Evaluating determinants of matrices of order 4 or higher can be tedious. There is, however, an important exception: the determinant of a **triangular** matrix. A triangular matrix is a square matrix with all *zero* entries either below or above its main diagonal. A square matrix is **upper triangular** if it has all zero entries below its main diagonal and **lower triangular** if it has all zero entries above its main diagonal. A matrix that is both upper and lower triangular is called **diagonal.** That is, a diagonal matrix is one in which all entries above and below the main diagonal are zero.

Upper Triangular Matrix

$$\begin{bmatrix} a_{11} & a_{12} & a_{13} & \cdots & a_{1n} \\ 0 & a_{22} & a_{23} & \cdots & a_{2n} \\ 0 & 0 & a_{33} & \cdots & a_{3n} \\ \vdots & \vdots & \vdots & & \vdots \\ 0 & 0 & 0 & \cdots & a_{nn} \end{bmatrix}$$

Lower Triangular Matrix

$$\begin{bmatrix} a_{11} & 0 & 0 & \cdots & 0 \\ a_{21} & a_{22} & 0 & \cdots & 0 \\ a_{31} & a_{32} & a_{33} & \cdots & 0 \\ \vdots & \vdots & \vdots & & \vdots \\ a_{n1} & a_{n2} & a_{n3} & \cdots & a_{nn} \end{bmatrix}$$

To find the determinant of a triangular matrix of any order, simply find the product of the entries on the main diagonal.

Example 6 The Determinant of a Triangular Matrix

a.
$$\begin{vmatrix} 2 & 0 & 0 & 0 \\ 4 & -2 & 0 & 0 \\ -5 & 6 & 1 & 0 \\ 1 & 5 & 3 & 3 \end{vmatrix} = 2(-2)(1)(3) = -12$$

b.
$$\begin{vmatrix} -1 & 0 & 0 & 0 & 0 \\ 0 & 3 & 0 & 0 & 0 \\ 0 & 0 & 2 & 0 & 0 \\ 0 & 0 & 0 & 4 & 0 \\ 0 & 0 & 0 & 0 & -2 \end{vmatrix} = -1(3)(2)(4)(-2) = 48$$

✓ **CHECKPOINT 6**

Find the determinant of

$$A = \begin{bmatrix} -1 & -2 & -1 & 6 \\ 0 & 5 & 4 & 2 \\ 0 & 0 & 2 & 5 \\ 0 & 0 & 0 & 3 \end{bmatrix}.$$

CONCEPT CHECK

1. Explain the difference between $\begin{bmatrix} 0 & 3 \\ 1 & 4 \end{bmatrix}$ and $\begin{vmatrix} 0 & 3 \\ 1 & 4 \end{vmatrix}$.

2. Explain the difference between the minors and cofactors of a square matrix.

3. Consider the matrix $A = \begin{bmatrix} 2 & 4 & 1 \\ 1 & 0 & 2 \\ 3 & 0 & 5 \end{bmatrix}$. When expanding by cofactors, which column reduces the amount of work in finding the determinant? Explain your reasoning.

4. What is the determinant of any identity matrix? Explain.

Skills Review 6.4

The following warm-up exercises involve skills that were covered in earlier sections. You will use these skills in the exercise set for this section. For additional help, review Sections 0.2, 0.3, and 6.2.

In Exercises 1–4, perform the indicated matrix operations.

1. $\begin{bmatrix} 1 & -2 \\ 0 & 3 \end{bmatrix} + \begin{bmatrix} 2 & 7 \\ 4 & -3 \end{bmatrix}$

2. $\begin{bmatrix} -2 & 5 \\ 3 & -2 \end{bmatrix} - \begin{bmatrix} 0 & -3 \\ 1 & 2 \end{bmatrix}$

3. $3\begin{bmatrix} 3 & -4 & 2 \\ 1 & 0 & -1 \\ 0 & 1 & -2 \end{bmatrix}$

4. $4\begin{bmatrix} 0 & 2 & 3 \\ -1 & 2 & 3 \\ -2 & 1 & -2 \end{bmatrix}$

In Exercises 5–10, perform the indicated arithmetic operations.

5. $[(1)(3) + (-3)(2)] - [(1)(4) + (3)(5)]$

6. $[(4)(4) + (-1)(-3)] - [(-1)(2) + (-2)(7)]$

7. $\dfrac{4(7) - 1(-2)}{(-5)(-2) - 3(4)}$

8. $\dfrac{3(6) - 2(7)}{6(-5) - 2(1)}$

9. $-5(-1)^2[6(-2) - 7(-3)]$

10. $4(-1)^3[3(6) - 2(7)]$

Exercises 6.4

See www.CalcChat.com for worked-out solutions to odd-numbered exercises.

In Exercises 1–14, find the determinant of the matrix.

1. $[-5]$

2. $[6]$

3. $\begin{bmatrix} 1 & 3 \\ 2 & 7 \end{bmatrix}$

4. $\begin{bmatrix} -3 & 4 \\ -2 & 1 \end{bmatrix}$

5. $\begin{bmatrix} 5 & 6 \\ 2 & 3 \end{bmatrix}$

6. $\begin{bmatrix} -7 & -4 \\ 8 & 7 \end{bmatrix}$

7. $\begin{bmatrix} 9 & 3 \\ 12 & 4 \end{bmatrix}$

8. $\begin{bmatrix} -5 & -2 \\ 10 & 4 \end{bmatrix}$

9. $\begin{bmatrix} 2 & 1 \\ 3 & 4 \end{bmatrix}$

10. $\begin{bmatrix} -3 & 1 \\ 5 & 2 \end{bmatrix}$

11. $\begin{bmatrix} \frac{2}{3} & 0 \\ -1 & 6 \end{bmatrix}$

12. $\begin{bmatrix} 9 & -\frac{1}{4} \\ 8 & 0 \end{bmatrix}$

13. $\begin{bmatrix} -\frac{1}{2} & \frac{1}{3} \\ -6 & \frac{1}{3} \end{bmatrix}$

14. $\begin{bmatrix} \frac{2}{3} & \frac{4}{3} \\ -1 & -\frac{1}{3} \end{bmatrix}$

(T) In Exercises 15–20, use the matrix capabilities of a graphing utility to find the determinant of the matrix.

15. $\begin{bmatrix} 0.1 & 0.3 & 0.2 \\ -0.3 & -0.2 & 0.1 \\ 1 & 2 & 3 \end{bmatrix}$

16. $\begin{bmatrix} 0.2 & -0.1 & -0.3 \\ 0.1 & -0.1 & 0.4 \\ -0.5 & -0.2 & -0.1 \end{bmatrix}$

17. $\begin{bmatrix} 0.9 & 0.7 & 0 \\ -0.1 & 0.3 & 1.3 \\ 2.2 & 4.2 & 6.1 \end{bmatrix}$

18. $\begin{bmatrix} 0.1 & 0.1 & -4.3 \\ 7.5 & 6.2 & 0.7 \\ 0.3 & 0.6 & -1.2 \end{bmatrix}$

19. $\begin{bmatrix} 5 & -3 & 2 \\ 7 & 5 & -7 \\ 0 & 6 & -1 \end{bmatrix}$

20. $\begin{bmatrix} 2 & 3 & 1 \\ 0 & 5 & -2 \\ 0 & 0 & -2 \end{bmatrix}$

In Exercises 21–28, find all (a) minors and (b) cofactors of the matrix.

21. $\begin{bmatrix} 3 & 4 \\ 2 & -5 \end{bmatrix}$

22. $\begin{bmatrix} 11 & 0 \\ -3 & 2 \end{bmatrix}$

23. $\begin{bmatrix} 3 & 1 \\ -2 & -4 \end{bmatrix}$

24. $\begin{bmatrix} -6 & 5 \\ 7 & -2 \end{bmatrix}$

25. $\begin{bmatrix} 4 & 0 & 2 \\ -3 & 2 & 1 \\ 1 & -1 & 1 \end{bmatrix}$

26. $\begin{bmatrix} 1 & -1 & 0 \\ 3 & 2 & 5 \\ 4 & -6 & 4 \end{bmatrix}$

27. $\begin{bmatrix} 3 & -2 & 8 \\ 3 & 2 & -6 \\ -1 & 3 & 6 \end{bmatrix}$

28. $\begin{bmatrix} -2 & 9 & 4 \\ 7 & -6 & 0 \\ 6 & 7 & -6 \end{bmatrix}$

In Exercises 29–34, find the determinant of the matrix by the method of expansion by cofactors. Expand using the indicated row or column.

29. $\begin{bmatrix} 4 & 1 & -3 \\ 6 & 5 & -2 \\ -1 & 3 & -4 \end{bmatrix}$

(a) Row 3

(b) Column 2

30. $\begin{bmatrix} -3 & 4 & 2 \\ 6 & 3 & 1 \\ 4 & -7 & -8 \end{bmatrix}$

(a) Row 2

(b) Column 3

31. $\begin{bmatrix} 7 & 0 & -4 \\ 2 & -3 & 0 \\ 5 & 8 & 1 \end{bmatrix}$

(a) Row 1

(b) Column 3

32. $\begin{bmatrix} 10 & -5 & 5 \\ 30 & 0 & 10 \\ 0 & 10 & 1 \end{bmatrix}$

(a) Row 3

(b) Column 1

33. $\begin{bmatrix} 6 & 0 & -3 & 5 \\ 4 & 13 & 6 & -8 \\ -1 & 0 & 7 & 4 \\ 8 & 6 & 0 & 2 \end{bmatrix}$

(a) Row 2   (b) Column 2

34. $\begin{bmatrix} 10 & 8 & 3 & -7 \\ 4 & 0 & 5 & -6 \\ 0 & 3 & 2 & 7 \\ 1 & 0 & -3 & 2 \end{bmatrix}$

(a) Row 3   (b) Column 1

In Exercises 35–52, find the determinant of the matrix. Expand by cofactors on the row or column that appears to make the computations easiest. Use a graphing utility to confirm your result.

35. $\begin{bmatrix} 1 & 4 & -2 \\ 3 & 2 & 0 \\ -1 & 4 & 3 \end{bmatrix}$

36. $\begin{bmatrix} 2 & -1 & 3 \\ 1 & 4 & 4 \\ 1 & 0 & 2 \end{bmatrix}$

37. $\begin{bmatrix} 2 & 4 & 6 \\ 0 & 3 & 1 \\ 0 & 0 & -5 \end{bmatrix}$

38. $\begin{bmatrix} -3 & 0 & 0 \\ 7 & 11 & 0 \\ 1 & 2 & 2 \end{bmatrix}$

39. $\begin{bmatrix} 2 & -1 & 0 \\ 4 & 2 & 1 \\ 4 & 2 & 1 \end{bmatrix}$

40. $\begin{bmatrix} -2 & 2 & 3 \\ 1 & -1 & 0 \\ 0 & 1 & 4 \end{bmatrix}$

41. $\begin{bmatrix} 1 & 4 & -2 \\ 3 & 6 & -6 \\ -2 & 1 & 4 \end{bmatrix}$

42. $\begin{bmatrix} -1 & 3 & 1 \\ 4 & 2 & 5 \\ -2 & 1 & 6 \end{bmatrix}$

43. $\begin{bmatrix} 0.3 & 0.2 & 0.2 \\ 0.2 & 0.2 & 0.2 \\ -0.4 & 0.4 & 0.3 \end{bmatrix}$

44. $\begin{bmatrix} 0.1 & 0.2 & 0.3 \\ -0.3 & 0.2 & 0.2 \\ 0.5 & 0.4 & 0.4 \end{bmatrix}$

45. $\begin{bmatrix} 6 & 3 & -7 \\ 0 & 0 & 0 \\ 4 & -6 & 3 \end{bmatrix}$

46. $\begin{bmatrix} 5 & 0 & 3 \\ -4 & 0 & 8 \\ 3 & 0 & -6 \end{bmatrix}$

47. $\begin{bmatrix} 3 & 6 & -5 & 4 \\ -2 & 0 & 6 & 0 \\ 1 & 1 & 2 & 2 \\ 0 & 3 & -1 & -1 \end{bmatrix}$

48. $\begin{bmatrix} 2 & 6 & 6 & 2 \\ 2 & 7 & 3 & 6 \\ 1 & 5 & 0 & 1 \\ 3 & 7 & 0 & 7 \end{bmatrix}$

49. $\begin{bmatrix} 5 & 3 & 0 & 6 \\ 4 & 6 & 4 & 12 \\ 0 & 2 & -3 & 4 \\ 0 & 1 & -2 & 2 \end{bmatrix}$

50. $\begin{bmatrix} 1 & 4 & 3 & 2 \\ -5 & 6 & 2 & 1 \\ 0 & 0 & 0 & 0 \\ 3 & -2 & 1 & 5 \end{bmatrix}$

51. $\begin{bmatrix} 3 & 2 & 4 & -1 & 5 \\ -2 & 0 & 1 & 3 & 2 \\ 1 & 0 & 0 & 4 & 0 \\ 6 & 0 & 2 & -1 & 0 \\ 3 & 0 & 5 & 1 & 0 \end{bmatrix}$

52. $\begin{bmatrix} 5 & 2 & 0 & 0 & -2 \\ 0 & 1 & 4 & 3 & 2 \\ 0 & 0 & 2 & 6 & 3 \\ 0 & 0 & 3 & 4 & 1 \\ 0 & 0 & 0 & 0 & 2 \end{bmatrix}$

Ⓣ In Exercises 53–60, use the matrix capabilities of a graphing utility to evaluate the determinant.

53. $\begin{vmatrix} 3 & 8 & -7 \\ 0 & -5 & 4 \\ 8 & 1 & 6 \end{vmatrix}$

54. $\begin{vmatrix} 5 & -8 & 0 \\ 9 & 7 & 4 \\ -8 & 7 & 1 \end{vmatrix}$

55. $\begin{vmatrix} 7 & 0 & -14 \\ -2 & 5 & 4 \\ -6 & 2 & 12 \end{vmatrix}$

56. $\begin{vmatrix} 3 & 0 & 0 \\ -2 & 5 & 0 \\ 12 & 5 & 7 \end{vmatrix}$

57. $\begin{vmatrix} 1 & -1 & 8 & 4 \\ 2 & 6 & 0 & -4 \\ 2 & 0 & 2 & 6 \\ 0 & 2 & 8 & 0 \end{vmatrix}$

58. $\begin{vmatrix} 0 & -3 & 8 & 3 \\ 8 & 1 & -1 & 6 \\ -4 & 6 & 0 & 9 \\ -7 & 0 & 0 & 14 \end{vmatrix}$

59. $\begin{vmatrix} 3 & -2 & 4 & 3 & 1 \\ -1 & 0 & 2 & 1 & 0 \\ 5 & -1 & 0 & 3 & 2 \\ 4 & 7 & -8 & 0 & 0 \\ 1 & 2 & 3 & 0 & 2 \end{vmatrix}$

60. $\begin{vmatrix} -2 & 0 & 0 & 0 & 0 \\ 0 & 3 & 0 & 0 & 0 \\ 0 & 0 & -1 & 0 & 0 \\ 0 & 0 & 0 & 2 & 0 \\ 0 & 0 & 0 & 0 & -4 \end{vmatrix}$

In Exercises 61–70, evaluate the determinant of the matrix. Do not use a graphing utility.

61. $\begin{bmatrix} 2 & 0 & 0 \\ 4 & -3 & 0 \\ 6 & 5 & 1 \end{bmatrix}$

62. $\begin{bmatrix} 1 & 0 & 0 \\ -4 & -1 & 0 \\ 5 & 1 & 5 \end{bmatrix}$

63. $\begin{bmatrix} 2 & 3 & -1 & -1 \\ 0 & -1 & -3 & 5 \\ 0 & 0 & -2 & 7 \\ 0 & 0 & 0 & -4 \end{bmatrix}$

64. $\begin{bmatrix} 4 & 0 & 0 & 0 \\ 1 & -4 & 0 & 0 \\ 2 & 1 & -1 & 0 \\ 6 & -2 & 3 & -1 \end{bmatrix}$

65.
$$\begin{bmatrix} 1 & 0 & 0 & 0 & 0 \\ 0 & 2 & 0 & 0 & 0 \\ 0 & 0 & 3 & 0 & 0 \\ 0 & 0 & 0 & 4 & 0 \\ 0 & 0 & 0 & 0 & 5 \end{bmatrix}$$

66.
$$\begin{bmatrix} -2 & 0 & 0 & 0 & 0 \\ 0 & 3 & 0 & 0 & 0 \\ 0 & 0 & -1 & 0 & 0 \\ 0 & 0 & 0 & 2 & 0 \\ 0 & 0 & 0 & 0 & -4 \end{bmatrix}$$

67.
$$\begin{bmatrix} 4 & 0 & 0 & 0 \\ 6 & -5 & 0 & 0 \\ 1 & 3 & 2 & 0 \\ 1 & 2 & 7 & -1 \end{bmatrix}$$

68.
$$\begin{bmatrix} 5 & 3 & 6 & 1 \\ 0 & -10 & 4 & 3 \\ 0 & 0 & 5 & 2 \\ 0 & 0 & 0 & 8 \end{bmatrix}$$

69.
$$\begin{bmatrix} -6 & 7 & 2 & 0 & 5 \\ 0 & -1 & 3 & 4 & -3 \\ 0 & 0 & -7 & 0 & 4 \\ 0 & 0 & 0 & -2 & 1 \\ 0 & 0 & 0 & 0 & -2 \end{bmatrix}$$

70.
$$\begin{bmatrix} -3 & 0 & 0 & 0 & 0 \\ 4 & 1 & 0 & 0 & 0 \\ 7 & -8 & 7 & 0 & 0 \\ 6 & 4 & 0 & -2 & 0 \\ 1 & 5 & 1 & -10 & 6 \end{bmatrix}$$

In Exercises 71–74, find the determinant of the matrix. Tell which method you used.

71.
$$\begin{bmatrix} 2 & 1 & 3 \\ 7 & 3 & -2 \\ 4 & 1 & 1 \end{bmatrix}$$

72.
$$\begin{bmatrix} 6 & -5 & 2 \\ 0 & 5 & -3 \\ 0 & 0 & 2 \end{bmatrix}$$

73.
$$\begin{bmatrix} 3 & 0 & 0 \\ 4 & -2 & 0 \\ 5 & 4 & 3 \end{bmatrix}$$

74.
$$\begin{bmatrix} 3 & 2 & -4 \\ -1 & 5 & -3 \\ 0 & 1 & 0 \end{bmatrix}$$

In Exercises 75–82, find (a) $|A|$, (b) $|B|$, (c) AB, and (d) $|AB|$.

75. $A = \begin{bmatrix} -1 & 0 \\ 0 & 3 \end{bmatrix}$, $B = \begin{bmatrix} 2 & 0 \\ 0 & -1 \end{bmatrix}$

76. $A = \begin{bmatrix} -2 & 1 \\ 4 & -2 \end{bmatrix}$, $B = \begin{bmatrix} 1 & 2 \\ 0 & -1 \end{bmatrix}$

77. $A = \begin{bmatrix} 4 & 0 \\ 3 & -2 \end{bmatrix}$, $B = \begin{bmatrix} -1 & 1 \\ -2 & 2 \end{bmatrix}$

78. $A = \begin{bmatrix} 5 & 4 \\ 3 & -1 \end{bmatrix}$, $B = \begin{bmatrix} 0 & 6 \\ 1 & -2 \end{bmatrix}$

79. $A = \begin{bmatrix} 0 & 1 & 2 \\ -3 & -2 & 1 \\ 0 & 4 & 1 \end{bmatrix}$, $B = \begin{bmatrix} 3 & -2 & 0 \\ 1 & -1 & 2 \\ 3 & 1 & 1 \end{bmatrix}$

80. $A = \begin{bmatrix} 3 & 2 & 0 \\ -1 & -3 & 4 \\ -2 & 0 & 1 \end{bmatrix}$, $B = \begin{bmatrix} -3 & 0 & 1 \\ 0 & 2 & -1 \\ -2 & -1 & 1 \end{bmatrix}$

81. $A = \begin{bmatrix} -1 & 2 & 1 \\ 1 & 0 & 1 \\ 0 & 1 & 0 \end{bmatrix}$, $B = \begin{bmatrix} -1 & 0 & 0 \\ 0 & 2 & 0 \\ 0 & 0 & 3 \end{bmatrix}$

82. $A = \begin{bmatrix} 2 & 0 & 1 \\ 1 & -1 & 2 \\ 3 & 1 & 0 \end{bmatrix}$, $B = \begin{bmatrix} 2 & -1 & 4 \\ 0 & 1 & 3 \\ 3 & -2 & 1 \end{bmatrix}$

In Exercises 83–86, find a 4 × 4 *upper* triangular matrix whose determinant is equal to the given value and a 4 × 4 *lower* triangular matrix whose determinant is equal to the given value. Use a graphing utility to confirm your results.

83. -18

84. -40

85. 28

86. 36

In Exercises 87–90, explain why the determinant of the matrix is equal to zero.

87.
$$\begin{bmatrix} 3 & 4 & -2 & 7 \\ 1 & 3 & -1 & 2 \\ 0 & 5 & 7 & 1 \\ 1 & 3 & -1 & 2 \end{bmatrix}$$

88.
$$\begin{bmatrix} 3 & 2 & -1 \\ -6 & -4 & 2 \\ 5 & -7 & 9 \end{bmatrix}$$

89.
$$\begin{bmatrix} 3 & 0 & 1 & 7 \\ 2 & -1 & 4 & 3 \\ 11 & 5 & -7 & 8 \\ -6 & 3 & -12 & -9 \end{bmatrix}$$

90.
$$\begin{bmatrix} -1 & 3 & 2 \\ 5 & 7 & 0 \\ -1 & 3 & 2 \end{bmatrix}$$

In Exercises 91 and 92, determine whether the statement is true or false. Justify your answer.

91. If a square matrix has an entire row of zeros, the determinant will always be zero.

92. If two columns of a square matrix are the same, the determinant of the matrix will be zero.

In Exercises 93–96, evaluate the determinant(s) to verify the equation.

93. $\begin{vmatrix} w & x \\ y & z \end{vmatrix} = -\begin{vmatrix} y & z \\ w & x \end{vmatrix}$

94. $\begin{vmatrix} w & cx \\ y & cz \end{vmatrix} = c\begin{vmatrix} w & x \\ y & z \end{vmatrix}$

95. $\begin{vmatrix} w & x \\ y & z \end{vmatrix} = \begin{vmatrix} w & x + cw \\ y & z + cy \end{vmatrix}$

96. $\begin{vmatrix} w & x \\ cw & cx \end{vmatrix} = 0$

Section 6.5

Applications of Matrices and Determinants

- Find the area of a triangle using a determinant.
- Determine whether three points are collinear using a determinant.
- Use a determinant to find an equation of a line.
- Encode and decode a cryptogram using a matrix.

Area of a Triangle

In this section, you will study some additional applications of matrices and determinants. The first involves a formula for finding the area of a triangle whose vertices are given by three points on a rectangular coordinate system.

> **Area of a Triangle**
>
> The area of a triangle with vertices (x_1, y_1), (x_2, y_2), and (x_3, y_3) is given by
>
> $$\text{Area} = \pm \frac{1}{2} \begin{vmatrix} x_1 & y_1 & 1 \\ x_2 & y_2 & 1 \\ x_3 & y_3 & 1 \end{vmatrix}$$
>
> where the symbol $(\pm)$ indicates that the appropriate sign should be chosen to yield a positive area.

Example 1 **Finding the Area of a Triangle**

Find the area of the triangle whose vertices are $(1, 0)$, $(2, 2)$, and $(4, 3)$, as shown in Figure 6.1.

SOLUTION Let $(x_1, y_1) = (1, 0)$, $(x_2, y_2) = (2, 2)$, and $(x_3, y_3) = (4, 3)$. Then, to find the area of the triangle, evaluate the determinant

$$\begin{vmatrix} x_1 & y_1 & 1 \\ x_2 & y_2 & 1 \\ x_3 & y_3 & 1 \end{vmatrix} = \begin{vmatrix} 1 & 0 & 1 \\ 2 & 2 & 1 \\ 4 & 3 & 1 \end{vmatrix}$$

$$= 1(-1)^2 \begin{vmatrix} 2 & 1 \\ 3 & 1 \end{vmatrix} + 0(-1)^3 \begin{vmatrix} 2 & 1 \\ 4 & 1 \end{vmatrix} + 1(-1)^4 \begin{vmatrix} 2 & 2 \\ 4 & 3 \end{vmatrix}$$

$$= 1(-1) + 0 + 1(-2) = -3.$$

Using this value, you can conclude that the area of the triangle is

$$\text{Area} = -\frac{1}{2} \begin{vmatrix} 1 & 0 & 1 \\ 2 & 2 & 1 \\ 4 & 3 & 1 \end{vmatrix} = -\frac{1}{2}(-3) = \frac{3}{2}.$$

Choose $(-)$ so that the area is positive.

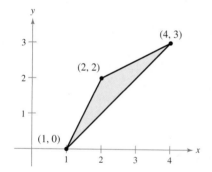

FIGURE 6.1

✓**CHECKPOINT 1**

Find the area of the triangle whose vertices are $(2, 1)$, $(3, 5)$, and $(10, 5)$. ■

Lines in the Plane

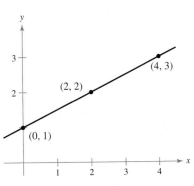

FIGURE 6.2

Suppose the three points in Example 1 had been on the same line. What would have happened had the area formula been applied to three such points? The answer is that the determinant would have been zero. Consider, for instance, the three collinear points $(0, 1)$, $(2, 2)$, and $(4, 3)$, as shown in Figure 6.2. The area of the "triangle" that has these three points as vertices is

$$\frac{1}{2}\begin{vmatrix} 0 & 1 & 1 \\ 2 & 2 & 1 \\ 4 & 3 & 1 \end{vmatrix} = \frac{1}{2}\left[0(-1)^2\begin{vmatrix} 2 & 1 \\ 3 & 1 \end{vmatrix} + 1(-1)^3\begin{vmatrix} 2 & 1 \\ 4 & 1 \end{vmatrix} + 1(-1)^4\begin{vmatrix} 2 & 2 \\ 4 & 3 \end{vmatrix} \right]$$

$$= \frac{1}{2}[0(-1) - 1(-2) + 1(-2)] = 0.$$

This result is generalized as follows.

Test for Collinear Points

Three points (x_1, y_1), (x_2, y_2), and (x_3, y_3) are collinear (lie on the same line) if and only if

$$\begin{vmatrix} x_1 & y_1 & 1 \\ x_2 & y_2 & 1 \\ x_3 & y_3 & 1 \end{vmatrix} = 0.$$

Example 2   **Testing for Collinear Points**

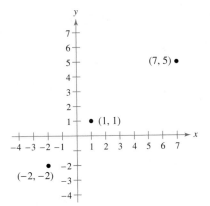

FIGURE 6.3

Determine whether the points $(-2, -2)$, $(1, 1)$, and $(7, 5)$ are collinear. (See Figure 6.3.)

SOLUTION   Letting $(x_1, y_1) = (-2, -2)$, $(x_2, y_2) = (1, 1)$, and $(x_3, y_3) = (7, 5)$, you have

$$\begin{vmatrix} x_1 & y_1 & 1 \\ x_2 & y_2 & 1 \\ x_3 & y_3 & 1 \end{vmatrix} = \begin{vmatrix} -2 & -2 & 1 \\ 1 & 1 & 1 \\ 7 & 5 & 1 \end{vmatrix}$$

$$= -2(-1)^2\begin{vmatrix} 1 & 1 \\ 5 & 1 \end{vmatrix} + (-2)(-1)^3\begin{vmatrix} 1 & 1 \\ 7 & 1 \end{vmatrix} + 1(-1)^4\begin{vmatrix} 1 & 1 \\ 7 & 5 \end{vmatrix}$$

$$= -2(-4) + 2(-6) + 1(-2)$$

$$= -6.$$

Because the value of this determinant is *not* zero, you can conclude that the three points *do not* lie on the same line and are not collinear.

✓ CHECKPOINT 2

Determine whether the points $(-2, 4)$, $(3, 0)$, and $(6, -4)$ are collinear. ■

Another way to test for collinear points in Example 2 is to find the slope of the line between $(-2, -2)$ and $(1, 1)$ and the slope of the line between $(-2, -2)$ and $(7, 5)$. Try doing this. If the slopes are equal, then the points are collinear. If the slopes are not equal, then the points are not collinear.

The test for collinear points can be adapted to another use. That is, if you are given two points on a rectangular coordinate system, you can find an equation of the line passing through the two points, as follows.

Two-Point Form of the Equation of a Line

An equation of the line passing through the distinct points (x_1, y_1) and (x_2, y_2) is given by

$$\begin{vmatrix} x & y & 1 \\ x_1 & y_1 & 1 \\ x_2 & y_2 & 1 \end{vmatrix} = 0.$$

Example 3 Finding an Equation of a Line

Find an equation of the line passing through the two points $(2, 4)$ and $(-1, 3)$, as shown in Figure 6.4.

SOLUTION   Let $(x_1, y_1) = (2, 4)$ and $(x_2, y_2) = (-1, 3)$. Applying the determinant formula for the equation of a line produces

$$\begin{vmatrix} x & y & 1 \\ 2 & 4 & 1 \\ -1 & 3 & 1 \end{vmatrix} = 0.$$

To evaluate this determinant, you can expand by cofactors along the first row to obtain the following.

$$x(-1)^2 \begin{vmatrix} 4 & 1 \\ 3 & 1 \end{vmatrix} + y(-1)^3 \begin{vmatrix} 2 & 1 \\ -1 & 1 \end{vmatrix} + 1(-1)^4 \begin{vmatrix} 2 & 4 \\ -1 & 3 \end{vmatrix} = 0$$

$$x(1) - y(3) + (1)(10) = 0$$

$$x - 3y + 10 = 0$$

So, an equation of the line is

$$x - 3y + 10 = 0.$$

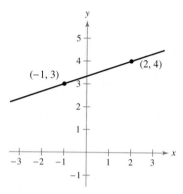

FIGURE 6.4

✓ **CHECKPOINT 3**

Find an equation of the line passing through the two points $(-3, -1)$ and $(3, 5)$. ■

Note that this method of finding the equation of a line works for all lines, including horizontal and vertical lines. For instance, the equation of the vertical line through $(2, 0)$ and $(2, 2)$ is

$$\begin{vmatrix} x & y & 1 \\ 2 & 0 & 1 \\ 2 & 2 & 1 \end{vmatrix} = 0$$

$$4 - 2x = 0$$

$$x = 2.$$

Cryptography

A **cryptogram** is a message written according to a secret code. (The Greek word *kryptos* means "hidden.") Matrix multiplication can be used to **encode** and **decode** messages. To begin, you need to assign a number to each letter in the alphabet (with 0 assigned to a blank space), as follows.

0 = __	9 = I	18 = R
1 = A	10 = J	19 = S
2 = B	11 = K	20 = T
3 = C	12 = L	21 = U
4 = D	13 = M	22 = V
5 = E	14 = N	23 = W
6 = F	15 = O	24 = X
7 = G	16 = P	25 = Y
8 = H	17 = Q	26 = Z

© CORBIS

During World War II, Navajo soldiers created a code using their native language to send messages between batallions. Native words were assigned to represent characters in the English alphabet, and they created a number of expressions for important military terms, like iron-fish to mean submarine. Without the Navajo Code Talkers, the Second World War might have had a very different outcome.

The message is then converted to numbers and partitioned into **uncoded row matrices,** each having n entries, as demonstrated in Example 4.

Example 4 Forming Uncoded Row Matrices

Write the uncoded row matrices of order 1×3 for the message

MEET ME MONDAY.

SOLUTION Partitioning the message (including blank spaces, but ignoring punctuation) into groups of three produces the following uncoded row matrices.

$$\begin{bmatrix} 13 & 5 & 5 \end{bmatrix} \begin{bmatrix} 20 & 0 & 13 \end{bmatrix} \begin{bmatrix} 5 & 0 & 13 \end{bmatrix} \begin{bmatrix} 15 & 14 & 4 \end{bmatrix} \begin{bmatrix} 1 & 25 & 0 \end{bmatrix}$$
$$\quad\; M \;\; E \;\; E \qquad T \qquad M \;\; E \qquad M \;\; O \qquad N \;\; D \qquad A \;\; Y$$

Note that a blank space is used to fill out the last uncoded row matrix.

✓CHECKPOINT 4

Write the uncoded row matrices of order 1×3 for the message

OWLS ARE NOCTURNAL. ▪

To encode a message, choose an $n \times n$ invertible matrix A and multiply the uncoded row matrices by A to obtain **coded row matrices.** The uncoded matrix should be on the left, whereas the encoding matrix A should be on the right. Here is an example.

$$\underset{\text{Uncoded Matrix}}{\begin{bmatrix} 13 & 5 & 5 \end{bmatrix}} \quad \underset{\text{Encoding Matrix } A}{\begin{bmatrix} 1 & -2 & 2 \\ -1 & 1 & 3 \\ 1 & -1 & -4 \end{bmatrix}} = \underset{\text{Coded Matrix}}{\begin{bmatrix} 13 & -26 & 21 \end{bmatrix}}$$

This technique is further illustrated in Example 5.

Example 5 **Encoding a Message**

Use the following matrix to encode the message MEET ME MONDAY.

$$A = \begin{bmatrix} 1 & -2 & 2 \\ -1 & 1 & 3 \\ 1 & -1 & -4 \end{bmatrix}$$

SOLUTION The coded row matrices are obtained by multiplying each of the uncoded row matrices found in Example 4 by the matrix A, as follows.

Uncoded Matrix	Encoding Matrix A	Coded Matrix

$$\begin{bmatrix} 13 & 5 & 5 \end{bmatrix} \begin{bmatrix} 1 & -2 & 2 \\ -1 & 1 & 3 \\ 1 & -1 & -4 \end{bmatrix} = \begin{bmatrix} 13 & -26 & 21 \end{bmatrix}$$

$$\begin{bmatrix} 20 & 0 & 13 \end{bmatrix} \begin{bmatrix} 1 & -2 & 2 \\ -1 & 1 & 3 \\ 1 & -1 & -4 \end{bmatrix} = \begin{bmatrix} 33 & -53 & -12 \end{bmatrix}$$

$$\begin{bmatrix} 5 & 0 & 13 \end{bmatrix} \begin{bmatrix} 1 & -2 & 2 \\ -1 & 1 & 3 \\ 1 & -1 & -4 \end{bmatrix} = \begin{bmatrix} 18 & -23 & -42 \end{bmatrix}$$

$$\begin{bmatrix} 15 & 14 & 4 \end{bmatrix} \begin{bmatrix} 1 & -2 & 2 \\ -1 & 1 & 3 \\ 1 & -1 & -4 \end{bmatrix} = \begin{bmatrix} 5 & -20 & 56 \end{bmatrix}$$

$$\begin{bmatrix} 1 & 25 & 0 \end{bmatrix} \begin{bmatrix} 1 & -2 & 2 \\ -1 & 1 & 3 \\ 1 & -1 & -4 \end{bmatrix} = \begin{bmatrix} -24 & 23 & 77 \end{bmatrix}$$

✓**CHECKPOINT 5**

Use the following matrix to encode the message OWLS ARE NOCTURNAL.

$$A = \begin{bmatrix} 1 & -1 & 0 \\ 1 & 0 & -1 \\ 6 & -2 & -3 \end{bmatrix} ■$$

So, the sequence of coded row matrices is

$$\begin{bmatrix} 13 & -26 & 21 \end{bmatrix} \begin{bmatrix} 33 & -53 & -12 \end{bmatrix} \begin{bmatrix} 18 & -23 & -42 \end{bmatrix} \begin{bmatrix} 5 & -20 & 56 \end{bmatrix} \begin{bmatrix} -24 & 23 & 77 \end{bmatrix}.$$

Finally, removing the matrix notation produces the following cryptogram.

$$13 \ -26 \ 21 \ 33 \ -53 \ -12 \ 18 \ -23 \ -42 \ 5 \ -20 \ 56 \ -24 \ 23 \ 77$$

For those who do not know the encoding matrix A, decoding the cryptogram found in Example 5 is difficult. But for an authorized receiver who knows the encoding matrix A, decoding is simple. The receiver only needs to multiply the coded row matrices by A^{-1} (on the right) to retrieve the uncoded row matrices. Here is an example.

$$\underbrace{\begin{bmatrix} 13 & -26 & 21 \end{bmatrix}}_{\text{Coded}} \underbrace{\begin{bmatrix} -1 & -10 & -8 \\ -1 & -6 & -5 \\ 0 & -1 & -1 \end{bmatrix}}_{A^{-1}} = \underbrace{\begin{bmatrix} 13 & 5 & 5 \end{bmatrix}}_{\text{Uncoded}}$$

The receiver could then easily refer to the number code chart on page 521 and translate $\begin{bmatrix} 13 & 5 & 5 \end{bmatrix}$ into the letters M E E.

Example 6   Decoding a Message

Use the inverse of the matrix $A = \begin{bmatrix} 1 & -2 & 2 \\ -1 & 1 & 3 \\ 1 & -1 & -4 \end{bmatrix}$ to decode the cryptogram.

$$13 \ -26 \ 21 \ 33 \ -53 \ -12 \ 18 \ -23 \ -42 \ 5 \ -20 \ 56 \ -24 \ 23 \ 77$$

SOLUTION   First find A^{-1} by using the techniques demonstrated in Section 6.3. A^{-1} is the decoding matrix. Then partition the message into groups of three to form the coded row matrices. Multiply each coded row matrix on the right by A^{-1} to obtain the decoded row matrices.

Coded Matrix	Decoding Matrix A^{-1}	Decoded Matrix
$\begin{bmatrix} 13 & -26 & 21 \end{bmatrix}$	$\begin{bmatrix} -1 & -10 & -8 \\ -1 & -6 & -5 \\ 0 & -1 & -1 \end{bmatrix} =$	$\begin{bmatrix} 13 & 5 & 5 \end{bmatrix}$
$\begin{bmatrix} 33 & -53 & -12 \end{bmatrix}$	$\begin{bmatrix} -1 & -10 & -8 \\ -1 & -6 & -5 \\ 0 & -1 & -1 \end{bmatrix} =$	$\begin{bmatrix} 20 & 0 & 13 \end{bmatrix}$
$\begin{bmatrix} 18 & -23 & -42 \end{bmatrix}$	$\begin{bmatrix} -1 & -10 & -8 \\ -1 & -6 & -5 \\ 0 & -1 & -1 \end{bmatrix} =$	$\begin{bmatrix} 5 & 0 & 13 \end{bmatrix}$
$\begin{bmatrix} 5 & -20 & 56 \end{bmatrix}$	$\begin{bmatrix} -1 & -10 & -8 \\ -1 & -6 & -5 \\ 0 & -1 & -1 \end{bmatrix} =$	$\begin{bmatrix} 15 & 14 & 4 \end{bmatrix}$
$\begin{bmatrix} -24 & 23 & 77 \end{bmatrix}$	$\begin{bmatrix} -1 & -10 & -8 \\ -1 & -6 & -5 \\ 0 & -1 & -1 \end{bmatrix} =$	$\begin{bmatrix} 1 & 25 & 0 \end{bmatrix}$

So, the message is as follows.

$$\begin{bmatrix} 13 & 5 & 5 \end{bmatrix} \ \begin{bmatrix} 20 & 0 & 13 \end{bmatrix} \ \begin{bmatrix} 5 & 0 & 13 \end{bmatrix} \ \begin{bmatrix} 15 & 14 & 4 \end{bmatrix} \ \begin{bmatrix} 1 & 25 & 0 \end{bmatrix}$$
$$\text{M E E} \quad \text{T} \quad \text{M} \quad \text{E} \quad \text{M} \quad \text{O N D} \quad \text{A Y}$$

✓ CHECKPOINT 6

Use the inverse of the matrix

$$A = \begin{bmatrix} 1 & -1 & 0 \\ 1 & 0 & -1 \\ 6 & -2 & -3 \end{bmatrix}$$

to decode the cryptogram.

$110, -39, -59, 25, -21, -3, 23,$
$-18, -5, 47, -20, -24, 149,$
$-56, -75, 87, -38, -37$ ■

CONCEPT CHECK

1. The area of a triangle with vertices (x_1, y_1), (x_2, y_2), and (x_3, y_3) is 4. What are the possible values of the determinant in the area formula? Explain.

2. Suppose the matrix formed by three points is not invertible. What does this tell you about the points?

3. You are finding the equation of a line given two points using the determinant formula. By expanding by cofactors along the first row and second column, you find that the determinant is 0. What does this tell you about the line?

4. Can you use the matrix $A = \begin{bmatrix} 1 & 0 & 0 \\ 3 & 0 & 0 \\ 2 & 5 & 5 \end{bmatrix}$ to encode a message? Explain.

Skills Review 6.5

The following warm-up exercises involve skills that were covered in earlier sections. You will use these skills in the exercise set for this section. For additional help, review Sections 6.2, 6.3, and 6.4.

In Exercises 1–6, evaluate the determinant.

1. $\begin{vmatrix} 4 & 3 \\ -3 & -2 \end{vmatrix}$

2. $\begin{vmatrix} 10 & -20 \\ -1 & 2 \end{vmatrix}$

3. $\begin{vmatrix} 4 & 0 \\ -3 & -2 \end{vmatrix}$

4. $\begin{vmatrix} x & x^2 \\ 1 & 2x \end{vmatrix}$

5. $\begin{vmatrix} 4 & 0 & -2 \\ 3 & 1 & 2 \\ -8 & 0 & 6 \end{vmatrix}$

6. $\begin{vmatrix} 3 & 2 & 5 \\ 0 & 0 & -4 \\ -6 & 1 & 1 \end{vmatrix}$

In Exercises 7 and 8, find the inverse of the matrix.

7. $A = \begin{bmatrix} 1 & 3 \\ 2 & 7 \end{bmatrix}$

8. $A = \begin{bmatrix} 10 & 5 & -2 \\ -4 & -2 & 1 \\ 1 & 1 & 0 \end{bmatrix}$

In Exercises 9 and 10, perform the indicated matrix multiplication.

9. $\begin{bmatrix} 0.1 & 0.2 \\ 0.4 & 0.3 \end{bmatrix}\begin{bmatrix} 0.4 \\ 0.5 \end{bmatrix}$

10. $\begin{bmatrix} 2 & 5 \end{bmatrix}\begin{bmatrix} 1 & 2 \\ 1 & 2 \end{bmatrix}$

Exercises 6.5

See www.CalcChat.com for worked-out solutions to odd-numbered exercises.

In Exercises 1–10, use a determinant to find the area of the triangle with the given vertices.

1.

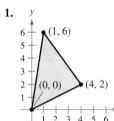

2.

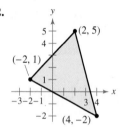

3.

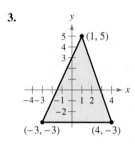

4.

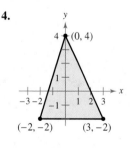

5.

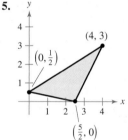

6.

7. $(-2, 4), (2, 3), (-1, 5)$

8. $(0, -2), (-1, 4), (3, 5)$

9. $(-3, 5), (2, 6), (3, -5)$

10. $(-2, 4), (1, 5), (3, -2)$

In Exercises 11 and 12, find a value of y such that the triangle with the given vertices has an area of 4 square units.

11. $(-5, 1), (0, 2), (-2, y)$

12. $(-4, 2), (-3, 5), (-1, y)$

In Exercises 13 and 14, find a value of y such that the triangle with the given vertices has an area of 6 square units.

13. $(-2, -3), (1, -1), (-8, y)$

14. $(1, 0), (5, -3), (-3, y)$

(T) 15. Gypsy Moths A large region of forest has been infested with gypsy moths. The region is roughly triangular, as shown in the figure. From the northernmost vertex *A* of the region, the distances to the other vertices are 30 miles south and 15 miles east (for vertex *B*), and 25 miles south and 33 miles east (for vertex *C*). Use a graphing utility to approximate the number of square miles in this region.

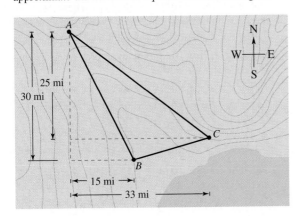

(T) 16. Botany A botanist is studying the plants growing in a triangular tract of land, as shown in the figure. To estimate the number of square feet in the tract, the botanist starts at one vertex, walks 70 feet east and 55 feet north to the second vertex, and then walks 90 feet west and 35 feet north to the third vertex. Use a graphing utility to determine how many square feet there are in the tract of land.

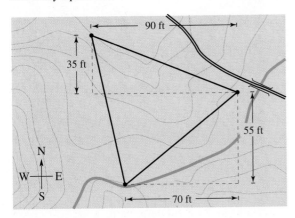

In Exercises 17–24, use a determinant to determine whether the points are collinear.

17. $(-4, -7), (0, -4), (4, -1)$

18. $(2, 4), (4, 5), (-2, 2)$

19. $(-1, -7), (0, -3), (1, 2)$

20. $(1, 7), (0, 4), (-1, 2)$

21. $(-2, -11), (4, 13), (2, 5)$

22. $(4, 3), (3, 1), (2, -1)$

23. $(-2, 3), (2, -1), (7, -4)$

24. $(-3, -4), (-1, -1), (5, 5)$

In Exercises 25 and 26, find *y* such that the points are collinear.

25. $(2, -5), (3, y), (5, -2)$

26. $(-6, 2), (-4, y), (-3, 5)$

In Exercises 27 and 28, find *x* such that the points are collinear.

27. $(-4, -1), (-1, 2), (x, 6)$

28. $(1, 5), (5, -1), (x, 3)$

In Exercises 29–36, use a determinant to find an equation of the line passing through the points.

29. $(-1, 2), (5, 3)$ **30.** $(3, 1), (-2, -5)$

31. $(-4, 3), (2, 1)$ **32.** $(10, 7), (-2, -7)$

33. $(-4, 5), (-4, -2)$ **34.** $(3, 3), (6, 3)$

35. $\left(-\frac{1}{2}, 3\right), \left(\frac{5}{2}, 1\right)$ **36.** $\left(\frac{2}{3}, 4\right), (6, 12)$

In the remaining exercises for this section, use the number code chart on page 521.

In Exercises 37 and 38, find the uncoded 1 × 2 row matrices for the message. Then encode the message using the encoding matrix.

	Message	Encoding Matrix

37. COME HOME SOON $\begin{bmatrix} 1 & 2 \\ 3 & 5 \end{bmatrix}$

38. HELP IS ON THE WAY $\begin{bmatrix} -2 & 3 \\ -1 & 1 \end{bmatrix}$

In Exercises 39 and 40, find the uncoded 1 × 3 row matrices for the message. Then encode the message using the encoding matrix.

	Message	Encoding Matrix

39. CALL ME TOMORROW $\begin{bmatrix} 1 & -1 & 0 \\ 1 & 0 & -1 \\ -6 & 2 & 3 \end{bmatrix}$

40. PLEASE SEND MONEY $\begin{bmatrix} 4 & 2 & 1 \\ -3 & -3 & -1 \\ 3 & 2 & 1 \end{bmatrix}$

In Exercises 41–46, write a cryptogram for the message using the matrix

$$A = \begin{bmatrix} 1 & 2 & 2 \\ 3 & 7 & 9 \\ -1 & -4 & -7 \end{bmatrix}.$$

41. LANDING SUCCESSFUL

42. BEAM ME UP SCOTTY

43. HAPPY BIRTHDAY

44. OPERATION OVERLORD

45. CONTACT AT DAWN

46. HEAD DUE WEST

In Exercises 47–50, use A^{-1} to decode the cryptogram.

47. $A = \begin{bmatrix} 1 & 2 \\ 3 & 5 \end{bmatrix}$

11, 21, 64, 112, 25, 50, 29, 53, 23, 46, 40, 75, 55, 92

48. $A = \begin{bmatrix} 2 & 3 \\ 3 & 4 \end{bmatrix}$

19, 26, 41, 57, 28, 42, 78, 109, 64, 87, 62, 83, 63, 87, 28, 42, 73, 102, 46, 69

49. $A = \begin{bmatrix} 4 & 2 & 1 \\ -3 & -3 & -1 \\ 3 & 2 & 1 \end{bmatrix}$

94, 35, 25, 44, 16, 10, 4, −10, 1, 27, 15, 9, 71, 43, 22

50. $A = \begin{bmatrix} 1 & -1 & 0 \\ 1 & 0 & -1 \\ -6 & 2 & 3 \end{bmatrix}$

9, −1, −9, 38, −19, −19, 28, −9, −19, −80, 25, 41, −64, 21, 31, −7, −4, 7

In Exercises 51 and 52, decode the cryptogram by using the inverse of the matrix A.

$A = \begin{bmatrix} 1 & 2 & 2 \\ 3 & 7 & 9 \\ -1 & -4 & -7 \end{bmatrix}$

51. 20, 17, −15, −9, −44, −83, 64, 136, 157, 24, 31, 12, 4, −37, −102

52. −10, −57, −111, 74, 168, 209, 35, 75, 85, 16, 35, 42, 34, 55, 43

53. The following cryptogram was encoded with a 2×2 matrix.

8, 21, −15, −10, −13, −13, 5, 10, 5, 25, 5, 19, −1, 6, 20, 40, −18, −18, 1, 16

The last word of the message is __RON. What is the message?

54. The following cryptogram was encoded with a 2×2 matrix.

5, 2, 25, 11, −2, −7, −15, −15, 32, 14, −8, −13, 38, 19, −19, −19, 37, 16

The last word of the message is __SUE. What is the message?

55. Cryptography   A code breaker intercepted the encoded message below.

45, −35, 38, −30, 18, −18, 35, −30, 81, −60, 42, −28, 75, −55, 2, −2, 22, −21, 15, −10

Let $A^{-1} = \begin{bmatrix} w & x \\ y & z \end{bmatrix}$. You know that $[45 \ -35]A^{-1} = [10 \ 15]$ and that $[38 \ -30]A^{-1} = [8 \ 14]$, where A^{-1} is the inverse of the encoding matrix A. Explain how you can find the values of w, x, y, and z. Decode the message.

56. Cryptography   Your biology professor gives you the encoded message below.

−204, 47, −231, 53, −265, 61, −223, 51, −9, 2, −117, 28, −117, 26, −166, 37, −265, 61, −145, 34, −112, 25, −76, 19

Let $A^{-1} = \begin{bmatrix} w & x \\ y & z \end{bmatrix}$. You know that $[-204 \ 47]A^{-1} = [15 \ 16]$ and that $[-231 \ 53]A^{-1} = [15 \ 19]$, where A^{-1} is the inverse of the encoding matrix A. Explain how you can find the values of w, x, y, and z. Decode the message.

Business Capsule

Kei Uesugi/Getty Images

Voltage Security, Inc. is a leader in secure business communications and data protection. The company provides the most scalable enterprise key management and encryption capabilities for securing data. Invented by Dr. Dan Boneh and Dr. Matt Franklin in 2001, Identity-Based Encryption or IBE is a breakthrough in cryptography. IBE enables users to simply use an identity, such as an e-mail address, to secure business communications.

57. Research Project   Use your campus library, the Internet, or some other reference source to find information about a company that generates software which uses cryptography to secure data. Write a brief paper about such a company or small business.

Chapter Summary and Study Strategies

After studying this chapter, you should have acquired the following skills.
The exercise numbers are keyed to the Review Exercises that begin on page 529.
Answers to odd-numbered Review Exercises are given in the back of the book.*

Section 6.1	Review Exercises
■ Determine the order of a matrix.	1, 2

$$\begin{bmatrix} a_{11} & a_{12} & a_{13} & \cdots & a_{1n} \\ a_{21} & a_{22} & a_{23} & \cdots & a_{2n} \\ a_{31} & a_{32} & a_{33} & \cdots & a_{3n} \\ \vdots & \vdots & \vdots & & \vdots \\ a_{m1} & a_{m2} & a_{m3} & \cdots & a_{mn} \end{bmatrix} \Bigg\} \; m \text{ rows}$$

$\underbrace{\hspace{3cm}}_{n \text{ columns}}$

A matrix having m rows and n columns is of order $m \times n$.

■ Perform elementary row operations on a matrix in order to write the matrix in row-echelon form or reduced row-echelon form.	3–6
■ Solve a system of linear equations using Gaussian elimination or Gauss-Jordan elimination.	7–16

Section 6.2

■ Add or subtract two matrices and multiply a matrix by a scalar.	17–20, 33, 34

If $A = [a_{ij}]$ and $B = [b_{ij}]$ are $m \times n$ matrices and c is a scalar, then

$$A + B = [a_{ij} + b_{ij}] \text{ and } cA = [ca_{ij}].$$

■ Find the product of two matrices.	25–32

If $A = [a_{ij}]$ is an $m \times n$ matrix and $B = [b_{ij}]$ is an $n \times p$ matrix, then AB is an $m \times p$ matrix

$$AB = [c_{ij}]$$

where $c_{ij} = a_{i1}b_{1j} + a_{i2}b_{2j} + a_{i3}b_{3j} + \cdots + a_{in}b_{nj}$.

■ Solve a matrix equation.	21–24
■ Use matrix multiplication to solve an application problem.	35, 36

Section 6.3

■ Verify that a matrix is the inverse of a given matrix.	37, 38
■ Find the inverse of a matrix.	39, 40
■ Find the inverse of a 2×2 matrix using a formula.	41, 42

$$A^{-1} = \frac{1}{ad - bc} \begin{bmatrix} d & -b \\ -c & a \end{bmatrix}$$

■ Use an inverse matrix to solve a system of linear equations.	43–54

---

* Use a wide range of valuable study aids to help you master the material in this chapter. The *Student Solutions Guide* includes step-by-step solutions to all odd-numbered exercises to help you review and prepare. The student website at *college.hmco.com/info/larsonapplied* offers algebra help and a *Graphing Technology Guide*. The *Graphing Technology Guide* contains step-by-step commands and instructions for a wide variety of graphing calculators, including the most recent models.

Section 6.4

- Evaluate the determinant of a 2×2 matrix.

$$\det(A) = |A| = \begin{vmatrix} a_1 & b_1 \\ a_2 & b_2 \end{vmatrix} = a_1 b_2 - a_2 b_1$$

55–58

- Find the minors of a matrix.

59–62

- Find the cofactors of a matrix.

59–62

- Find the determinant of a square matrix.

63–68

$$|A| = a_{11}C_{11} + a_{12}C_{12} + \cdots + a_{1n}C_{1n}$$

- Find the determinant of a triangular matrix.

65, 66

Section 6.5

- Find the area of a triangle using a determinant.

69–72

$$\text{Area} = \pm\frac{1}{2}\begin{vmatrix} x_1 & y_1 & 1 \\ x_2 & y_2 & 1 \\ x_3 & y_3 & 1 \end{vmatrix}$$

- Determine whether three points are collinear using a determinant.

73–76

$$\begin{vmatrix} x_1 & y_1 & 1 \\ x_2 & y_2 & 1 \\ x_3 & y_3 & 1 \end{vmatrix} = 0$$

- Use a determinant to find an equation of a line.

77–80

$$\begin{vmatrix} x & y & 1 \\ x_1 & y_1 & 1 \\ x_2 & y_2 & 1 \end{vmatrix} = 0$$

- Encode and decode a cryptogram using a matrix.

81–85

Study Strategies

- **Variety of Approaches**   You can use a variety of approaches when finding the determinant of a square matrix.

    1. For a 2×2 matrix, you can use the definition $|A| = \begin{vmatrix} a_1 & b_1 \\ a_2 & b_2 \end{vmatrix} = a_1 b_2 - a_2 b_1$.

    2. For a 3×3 matrix, you can use the diagonal process shown in Example 5 on page 513.

    3. For any square matrix (of order 2×2 or greater), you can use expansion by cofactors. Be sure you choose the row or column that makes the computations the easiest.

    4. You can always use the matrix capabilities of a graphing utility.

- **Using Technology**   Performing operations with matrices can be tedious. You can use a graphing utility to accomplish the following.

    • Perform elementary row operations on matrices.

    • Reduce matrices to row-echelon form and reduced row-echelon form.

    • Add and subtract matrices.

    • Multiply matrices.

    • Multiply matrices by scalars.

    • Find inverses of matrices.

    • Solve systems of equations using matrices.

    • Evaluate determinants of matrices.

Review Exercises

See www.CalcChat.com for worked-out solutions to odd-numbered exercises.

In Exercises 1 and 2, determine the order of the matrix.

1. $\begin{bmatrix} 3 & 7 & 4 & -2 \\ 1 & 8 & 6 & 1 \end{bmatrix}$ **2.** $\begin{bmatrix} 5 \\ -1 \\ 2 \\ 4 \end{bmatrix}$

In Exercises 3 and 4, write the matrix in row-echelon form.

3. $\begin{bmatrix} 1 & 3 & 0 & 2 \\ 3 & 10 & 1 & 8 \\ 2 & 3 & 3 & 10 \end{bmatrix}$

4. $\begin{bmatrix} 1 & 2 & -1 & 0 \\ -2 & -3 & 3 & 4 \\ 4 & 0 & 1 & 3 \end{bmatrix}$

In Exercises 5 and 6, write the matrix in *reduced* row-echelon form.

5. $\begin{bmatrix} 1 & 2 & 3 \\ -2 & 0 & 2 \\ 2 & 1 & 2 \end{bmatrix}$

6. $\begin{bmatrix} 2 & 3 & 1 & -5 \\ 1 & 0 & 5 & 2 \\ -1 & 4 & 3 & 6 \\ 0 & -2 & 6 & -8 \end{bmatrix}$

In Exercises 7–14, use matrices to solve the system of equations (if possible). Use Gaussian elimination with back-substitution or Gauss-Jordan elimination.

7. $\begin{cases} 4x - 3y = 18 \\ x + y = 1 \end{cases}$ **8.** $\begin{cases} 2x + 4y = 16 \\ -x + 3y = 17 \end{cases}$

9. $\begin{cases} 2x + 3y - z = 13 \\ 3x + z = 8 \\ x - 2y + 3z = -4 \end{cases}$ **10.** $\begin{cases} 3x + 4y + 2z = 5 \\ 2x + 3y = 7 \\ 2y - 3z = 12 \end{cases}$

11. $\begin{cases} x + 2y + 2z = 10 \\ 2x + 3y + 5z = 20 \end{cases}$ **12.** $\begin{cases} 3x + 10y + 4z = 20 \\ x + 3y - 2z = 8 \end{cases}$

13. $\begin{cases} 2x + y - 3z = 4 \\ x + 2y + 2z = 10 \\ x - 2z = 12 \\ x + y + z = 6 \end{cases}$ **14.** $\begin{cases} 2x + 4y + 2z = 10 \\ x + 3z = 9 \\ 3x - 2y = 4 \\ x + y + z = 8 \end{cases}$

15. Biology A school district borrowed $200,000 at simple annual interest to upgrade microbiology equipment. Some of the money was borrowed at 8%, some at 10%, and some at 12%. Use a system of equations to determine how much was borrowed at each rate if the total annual interest was $20,000 and the amount borrowed at 10% was three times the amount borrowed at 8%. Solve the system using matrices.

16. Amusement Park An amusement park borrowed $650,000 at simple annual interest to renovate a roller coaster. Some of the money was borrowed at 8.5%, some at 9.5%, and some at 10%. Use a system of equations to determine how much was borrowed at each rate if the total annual interest was $58,250 and the amount borrowed at 8.5% was four times the amount borrowed at 10%. Solve the system using matrices.

In Exercises 17–20, find (a) $A + B$, (b) $A - B$, (c) $4A$, and (d) $4A - 3B$.

17. $A = \begin{bmatrix} -1 & 5 \\ 2 & 1 \end{bmatrix}$, $B = \begin{bmatrix} 4 & 2 \\ -6 & 3 \end{bmatrix}$

18. $A = \begin{bmatrix} 1 & 0 & 2 \\ -1 & 3 & 5 \\ 2 & -2 & 3 \end{bmatrix}$, $B = \begin{bmatrix} 2 & 0 & 1 \\ 3 & -4 & 6 \\ 1 & 2 & -3 \end{bmatrix}$

19. $A = \begin{bmatrix} 1 & 3 & -2 & 6 \\ 0 & 1 & 3 & 2 \end{bmatrix}$,

$B = \begin{bmatrix} 2 & 1 & 4 & -5 \\ 3 & -6 & 3 & -2 \end{bmatrix}$

20. $A = \begin{bmatrix} 3 \\ -2 \\ 3 \end{bmatrix}$, $B = \begin{bmatrix} -1 \\ 4 \\ 5 \end{bmatrix}$

In Exercises 21–24, solve for X when

$A = \begin{bmatrix} 1 & -2 \\ 0 & 1 \\ 2 & 3 \end{bmatrix}$ and $B = \begin{bmatrix} 0 & 1 \\ 1 & 1 \\ 3 & 5 \end{bmatrix}$.

21. $X = 4A - 3B$ **22.** $X = 5B + 2A$

23. $2X - 3A = B$ **24.** $4X - 8B = 4A$

In Exercises 25–30, find AB, if possible.

25. $A = \begin{bmatrix} 1 & 4 \\ -2 & -1 \\ 3 & 2 \end{bmatrix}$, $B = \begin{bmatrix} -4 \\ 3 \end{bmatrix}$

26. $A = \begin{bmatrix} 3 \\ 2 \\ 4 \\ 6 \end{bmatrix}$, $B = \begin{bmatrix} 2 & 0 & -1 \end{bmatrix}$

27. $A = \begin{bmatrix} 4 & 0 & 0 \\ 0 & 3 & 0 \\ 0 & 0 & -2 \end{bmatrix}$, $B = \begin{bmatrix} \frac{1}{4} & 0 & 0 \\ 0 & \frac{1}{3} & 0 \\ 0 & 0 & -\frac{1}{2} \end{bmatrix}$

28. $A = \begin{bmatrix} 0 & 0 & 2 \\ 1 & 0 & 6 \\ 0 & 2 & 2 \end{bmatrix}$, $B = \begin{bmatrix} 3 & 4 & 0 & 1 \\ 2 & 1 & 0 & 0 \\ 0 & 0 & 1 & 1 \end{bmatrix}$

29. $A = \begin{bmatrix} 3 & 1 \\ 4 & 7 \\ 1 & 1 \end{bmatrix}$, $B = \begin{bmatrix} 1 & 2 & -2 \\ 3 & 4 & 0 \\ 0 & 1 & 0 \end{bmatrix}$

30. $A = \begin{bmatrix} 1 & 2 & 3 & 6 & -1 \\ 2 & 8 & 0 & 0 & 2 \end{bmatrix}$, $B = \begin{bmatrix} 3 & 2 \\ 4 & -1 \end{bmatrix}$

In Exercises 31 and 32, find (a) AB, (b) BA, and, if possible, (c) A^2. (*Note:* $A^2 = AA$.)

31. $A = \begin{bmatrix} 1 & -3 & 4 \end{bmatrix}$, $B = \begin{bmatrix} 2 \\ -2 \\ -1 \end{bmatrix}$

32. $A = \begin{bmatrix} 1 & 0 & 2 \\ 3 & 1 & -2 \\ 1 & 1 & 1 \end{bmatrix}$, $B = \begin{bmatrix} 2 & 0 & 0 \\ 1 & -2 & 1 \\ 5 & 4 & -2 \end{bmatrix}$

33. Factory Production A window corporation has four factories, each of which manufactures three products. The number of units of product i produced at factory j in one day is represented by a_{ij} in the matrix

$$A = \begin{bmatrix} 80 & 120 & 20 & 40 \\ 40 & 60 & 80 & 20 \\ 140 & 60 & 100 & 80 \end{bmatrix}.$$

Find the production levels if production is increased by 20%.

34. Factory Production An electronics manufacturer has three factories, each of which manufactures four products. The number of units of product i produced at factory j in one day is represented by a_{ij} in the matrix

$$A = \begin{bmatrix} 120 & 140 & 60 \\ 80 & 100 & 40 \\ 40 & 160 & 80 \\ 20 & 120 & 100 \end{bmatrix}.$$

Find the production levels if production is decreased by 10%.

35. Inventory Levels A company sells four different models of car sound systems through three retail outlets. The inventories of the four models at the three outlets are given by matrix S.

	Model			
	A	B	C	D	
$S =$	3	2	1	4	1
	1	3	4	3	2 } Outlet
	5	3	2	2	3

The wholesale and retail prices of the four models are given by matrix T.

	Price		
	Wholesale	Retail	
	\$350	\$600	A
$T =$	\$425	\$705	B
	\$300	\$455	C } Model
	\$750	\$1150	D

(a) What is the total retail price of the inventory at Outlet 3?

(b) What is the total wholesale price of the inventory at Outlet 1?

(c) Compute ST and interpret the result in the context of the problem.

36. Labor/Wage Requirements A company that manufactures racing bicycles has the following labor-hour and wage requirements.

Labor-Hour Requirements (per bicycle)

	Department			
	Cutting	Assembly	Packaging	
	0.9 hour	0.8 hour	0.2 hour	Basic
$S =$	1.5 hours	1.0 hour	0.4 hour	Light } Models
	3.5 hours	3.0 hours	0.5 hour	Ultra-light

Wage Requirements (per hour)

	Plant		
	A	B	
	\$12.00	\$13.00	Cutting
$T =$	\$9.00	\$8.50	Assembly } Department
	\$7.50	\$8.00	Packaging

(a) What is the labor cost for a light racing bicycle at Plant A?

(b) What is the labor cost for an ultra-light racing bicycle at Plant B?

(c) Compute ST and interpret the result.

In Exercises 37 and 38, show that B is the inverse of A.

37. $A = \begin{bmatrix} 1 & 2 & 1 \\ 3 & 6 & 4 \\ 0 & 1 & 3 \end{bmatrix}$, $B = \begin{bmatrix} -14 & 5 & -2 \\ 9 & -3 & 1 \\ -3 & 1 & 0 \end{bmatrix}$

38. $A = \begin{bmatrix} 2 & 0 & 1 & 2 \\ 3 & 0 & 0 & 1 \\ -1 & 1 & 2 & 0 \\ 0 & -1 & 2 & 2 \end{bmatrix}$,

$B = \dfrac{1}{9}\begin{bmatrix} -4 & 6 & 1 & 1 \\ 10 & -6 & 2 & -7 \\ -7 & 6 & 4 & 4 \\ 12 & -9 & -3 & -3 \end{bmatrix}$

In Exercises 39 and 40, find the inverse of the matrix.

39. $\begin{bmatrix} -1 & 0 & 0 \\ 0 & 2 & 0 \\ 0 & 0 & 4 \end{bmatrix}$ **40.** $\begin{bmatrix} 3 & 2 & 2 \\ 0 & 2 & 1 \\ 1 & 0 & 1 \end{bmatrix}$

In Exercises 41 and 42, use the formula on page 501 to find the inverse of the matrix.

41. $\begin{bmatrix} 1 & 3 \\ 2 & 5 \end{bmatrix}$ **42.** $\begin{bmatrix} -2 & 1 \\ 4 & 3 \end{bmatrix}$

In Exercises 43 and 44, use the inverse matrix found in Exercise 41 to solve the system of linear equations.

43. $\begin{cases} x + 3y = 15 \\ 2x + 5y = 26 \end{cases}$ **44.** $\begin{cases} x + 3y = 7 \\ 2x + 5y = 11 \end{cases}$

In Exercises 45 and 46, use the inverse matrix found in Exercise 40 to solve the system of linear equations.

45. $\begin{cases} 3x + 2y + 2z = 13 \\ 2y + z = 4 \\ x + z = 5 \end{cases}$ **46.** $\begin{cases} 3x + 2y + 2z = 12 \\ 2y + z = 13 \\ x + z = 3 \end{cases}$

In Exercises 47–50, use an inverse matrix to solve the system of linear equations.

47. $\begin{cases} -3x + 10y = 8 \\ 5x - 17y = -13 \end{cases}$

48. $\begin{cases} 5x - y = 13 \\ -9x + 2y = -24 \end{cases}$

49. $\begin{cases} 3x + 2y - z = 6 \\ x - y + 2z = -1 \\ 5x + y + z = 7 \end{cases}$

50. $\begin{cases} -x + 4y - 2z = 12 \\ 2x - 9y + 5z = -25 \\ -x + 5y - 4z = 10 \end{cases}$

Raw Materials In Exercises 51 and 52, you are making three types of windshield washer fluid in chemistry class. Fluid X requires 9 cups of water, 1 cup of isopropyl alcohol, and 1 tablespoon of detergent. Fluid Y requires 10 cups of water, 3 cups of isopropyl alcohol, and 1 tablespoon of detergent. Fluid Z requires 14 cups of water, 2 cups of isopropyl alcohol, and 2 tablespoons of detergent. A system of linear equations (where *x*, *y*, and *z* represent fluids X, Y, and Z, respectively) is as follows.

$\begin{cases} 9x + 10y + 14z = (cups\ of\ water) \\ x + 3y + 2z = (cups\ of\ isopropyl\ alcohol) \\ x + y + 2z = (tablespoons\ of\ detergent) \end{cases}$

Use the inverse of the coefficient matrix of this system to find the numbers of units of fluids X, Y, and Z that you can produce with the given amounts of ingredients.

51. 240 cups of water
44 cups of isopropyl alcohol
28 tablespoons of detergent

52. 235 cups of water
41 cups of isopropyl alcohol
29 tablespoons of detergent

53. Field of Study The percent *y* of U.S. college freshmen who identified computer science as their probable field of study from 2001 to 2005 decreased in a pattern that was approximately parabolic. The least squares regression parabola $y = at^2 + bt + c$ for the data is found by solving the system

$\begin{cases} 5c + 15b + 55a = 9.7 \\ 15c + 55b + 225a = 23.9. \\ 55c + 225b + 979a = 77.3 \end{cases}$

Let *t* represent the year, with $t = 1$ corresponding to 2001. *(Source: The Higher Education Research Institute)*

(T) (a) Use a graphing utility to find an inverse matrix with which to solve the system, and find the equation of the least squares regression parabola.

(b) Use the result of part (a) to estimate the percent in 2000.

(c) The actual percent in 2000 was 3.7. How does this value compare with your estimate in part (b)?

54. Carnivorous Plants A Venus flytrap is grown in a greenhouse, and the size *y* (in millimeters) of its traps is measured at the end of each year for 5 years. The least squares regression parabola $y = at^2 + bt + c$ for the data is found by solving the system

$\begin{cases} 5c + 15b + 55a = 53.3 \\ 15c + 55b + 225a = 190.6. \\ 55c + 225b + 979a = 755.8 \end{cases}$

Let *t* represent the year, with $t = 1$ corresponding to the first year.

(T) (a) Use a graphing utility to find an inverse matrix with which to solve the system, and find the equation of the least squares regression parabola.

(b) Use the result of part (a) to estimate the sizes of the traps after the first and third years.

(c) The actual sizes of the traps were 2.5 millimeters after the first year and 12.8 millimeters after the third year. How do these values compare with your estimates in part (b)?

In Exercises 55–58, find the determinant of the matrix.

55. $\begin{bmatrix} 8 & 4 \\ 3 & 2 \end{bmatrix}$ **56.** $\begin{bmatrix} 7 & 2 \\ 9 & -3 \end{bmatrix}$

57. $\begin{bmatrix} 5 & 2 \\ 0 & 0 \end{bmatrix}$ **58.** $\begin{bmatrix} 3 & 0 \\ 0 & -7 \end{bmatrix}$

In Exercises 59–62, find all (a) minors and (b) cofactors of the matrix.

59. $\begin{bmatrix} 2 & -1 \\ 7 & 4 \end{bmatrix}$

60. $\begin{bmatrix} 3 & 6 \\ 5 & -4 \end{bmatrix}$

61. $\begin{bmatrix} 3 & 2 & -1 \\ -2 & 5 & 0 \\ 1 & 8 & 6 \end{bmatrix}$

62. $\begin{bmatrix} 8 & 3 & 4 \\ 6 & 5 & -9 \\ -4 & 1 & 2 \end{bmatrix}$

In Exercises 63–68, find the determinant of the matrix. Tell whether you used expansion by cofactors, the product of the entries on the main diagonal, or upward and downward diagonals.

63. $\begin{bmatrix} 1 & 2 & 3 \\ 8 & 6 & 7 \\ 0 & 2 & -1 \end{bmatrix}$

64. $\begin{bmatrix} -2 & 3 & 3 \\ -1 & 0 & 5 \\ 1 & 2 & -1 \end{bmatrix}$

65. $\begin{bmatrix} 1 & 3 & 2 & 4 \\ 0 & -1 & 2 & 2 \\ 0 & 0 & 3 & 0 \\ 0 & 0 & 0 & 4 \end{bmatrix}$

66. $\begin{bmatrix} 2 & 0 & 0 & 0 \\ 3 & 4 & 0 & 0 \\ 5 & 1 & -2 & 0 \\ 6 & 3 & 1 & 1 \end{bmatrix}$

67. $\begin{bmatrix} -2 & 4 & 1 \\ 6 & 1 & 2 \\ 5 & 3 & 4 \end{bmatrix}$

68. $\begin{bmatrix} 4 & 7 & -1 \\ 2 & -3 & 4 \\ -5 & 1 & -1 \end{bmatrix}$

In Exercises 69–72, use a determinant to find the area of the triangle with the given vertices.

69.

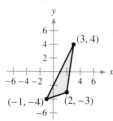

70.

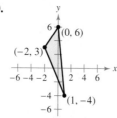

71.

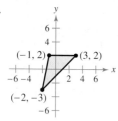

72.

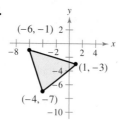

In Exercises 73–76, use a determinant to determine whether the points are collinear.

73. $(0, 3), (1, 5), (2, 8)$

74. $(2, 6), (-2, 3), (0, 5)$

75. $(-4, 1), (6, 6), (0, 3)$

76. $(-3, -1), (0, 5), (-4, -3)$

In Exercises 77–80, use a determinant to find an equation of the line passing through the points.

77. $(-7, 3), (8, 2)$ **78.** $(5, -4), (-3, 2)$

79. $(2, 4), (2, -7)$ **80.** $(-5, -1), (1, -1)$

In the remaining exercises for this review, use the number code chart on page 521.

In Exercises 81 and 82, find the uncoded row matrices for the message. Then encode the message using the encoding matrix.

Message	*Encoding Matrix*

81. TRANSMIT NOW $\begin{bmatrix} 2 & 3 \\ 3 & 4 \end{bmatrix}$

82. CALL AT MIDNIGHT $\begin{bmatrix} 1 & 2 & 2 \\ 3 & 7 & 9 \\ -1 & -4 & -7 \end{bmatrix}$

In Exercises 83 and 84, use A^{-1} to decode the cryptogram.

83. $A = \begin{bmatrix} 1 & 2 \\ -1 & 3 \end{bmatrix}$

$14, 53, -17, 96, 5, 10, 12, 64, 5, 10, 3, 11, 25, 50$

84. $A = \begin{bmatrix} 1 & -1 & 0 \\ 1 & 0 & -1 \\ -6 & 2 & 3 \end{bmatrix}$

$-14, -1, 10, -38, 2, 27, -94, 18, 57, 7, -11, -1, -96,$
$20, 57, -74, 23, 35, 17, -12, -5$

85. Cryptography   A family sends the encoded message below to a relative overseas.

$-57, -13, 91, 26, 97, 29, -76, -19, 5, 5, -84, -21, 55,$
$16, -28, -7, 97, 28, -8, -2$

Let $A^{-1} = \begin{bmatrix} w & x \\ y & z \end{bmatrix}$.

(a) You know that $\begin{bmatrix} -57 & -13 \end{bmatrix} A^{-1} = \begin{bmatrix} 23 & 5 \end{bmatrix}$ and that $\begin{bmatrix} 91 & 26 \end{bmatrix} A^{-1} = \begin{bmatrix} 0 & 13 \end{bmatrix}$, where A^{-1} is the inverse of the encoding matrix A. Explain how you can find the values of $w, x, y,$ and z.

(b) Decode the message.

Chapter Test

See www.CalcChat.com for worked-out solutions to odd-numbered exercises.

Take this test as you would take a test in class. When you are done, check your work against the answers given in the back of the book.

In Exercises 1 and 2, write the augmented matrix for the system of linear equations.

1. $\begin{cases} 2x + y + 4z = 2 \\ x + 4y - z = 0 \\ -x + 3y + 3z = -1 \end{cases}$

2. $\begin{cases} 3x + 4y + 2z = 4 \\ 2x + 3y = -2 \\ 2y - 3z = -13 \end{cases}$

In Exercises 3–5, use matrices to solve the system of equations.

3. $\begin{cases} x + 2y + 3z = 16 \\ 5x + 4y - z = 22 \end{cases}$

4. $\begin{cases} x - 2y + z = 14 \\ y - 3z = 2 \\ z = -6 \end{cases}$

5. $\begin{cases} 2x - 3y + z = 14 \\ x + 2y = -4 \\ y - z = -4 \end{cases}$

In Exercises 6–9, use the matrices to find the indicated matrix.

$$A = \begin{bmatrix} 1 & 3 \\ 2 & 4 \end{bmatrix}, \ B = \begin{bmatrix} 2 & -1 & 3 \\ 4 & 0 & 1 \end{bmatrix}, \ C = \begin{bmatrix} 0 & -2 \\ 3 & 5 \end{bmatrix}, \ D = \begin{bmatrix} 3 \\ 2 \\ -1 \end{bmatrix}$$

6. $2A + C$ **7.** CA **8.** BD **9.** A^2

In Exercises 10–12, find the inverse of the matrix.

10. $A = \begin{bmatrix} 2 & -1 \\ -3 & 4 \end{bmatrix}$

11. $A = \begin{bmatrix} 1 & 0 \\ 0 & 1 \end{bmatrix}$

12. $A = \begin{bmatrix} 3 & 4 & 2 \\ 2 & 3 & 0 \\ 0 & 2 & -3 \end{bmatrix}$

In Exercises 13–15, find the determinant of the matrix.

13. $\begin{bmatrix} -5 & 2 \\ 1 & 3 \end{bmatrix}$

14. $\begin{bmatrix} 3 & 2 & -1 \\ 1 & 0 & 2 \\ 4 & 5 & 2 \end{bmatrix}$

15. $\begin{bmatrix} 2 & 0 & 0 \\ 0 & 5 & 0 \\ 0 & 0 & -2 \end{bmatrix}$

16. Use the inverse matrix found in Exercise 12 to solve the system in Exercise 2.

17. Find two nonzero matrices whose product is a zero matrix.

18. Find the area of the triangle whose vertices are $(-3, 1)$, $(0, 4)$, and $(5, 2)$.

19. Use a determinant to decide whether $(2, 1)$, $(-3, -14)$, and $(4, 7)$ are collinear.

20. Use a determinant to find an equation of the line passing through the points $(1, -2)$ and $(5, 2)$.

21. A manufacturer produces three models of a product, which are shipped to two warehouses. The number of units i that are shipped to warehouse j is represented by a_{ij} in matrix A below. The prices per unit are represented by matrix B. Find the product BA and interpret the result.

$$A = \begin{bmatrix} 1500 & 4000 \\ 3000 & 4500 \\ 5500 & 7000 \end{bmatrix}$$

$$B = [\$55 \quad \$40 \quad \$33]$$

7

Limits and Derivatives

AP/Wide World Photos

A graph showing changes in a company's earnings and other financial indicators can depict the company's general financial trends over time. (See Section 7.4, Example 10.)

Applications

Limits and derivatives have many real-life applications. The applications listed below represent a sample of the applications in this chapter.

Section 7.1

Limits

- Find limits of functions graphically and numerically.
- Use the properties of limits to evaluate limits of functions.
- Use different analytic techniques to evaluate limits of functions.
- Evaluate one-sided limits.
- Recognize unbounded behavior of functions.

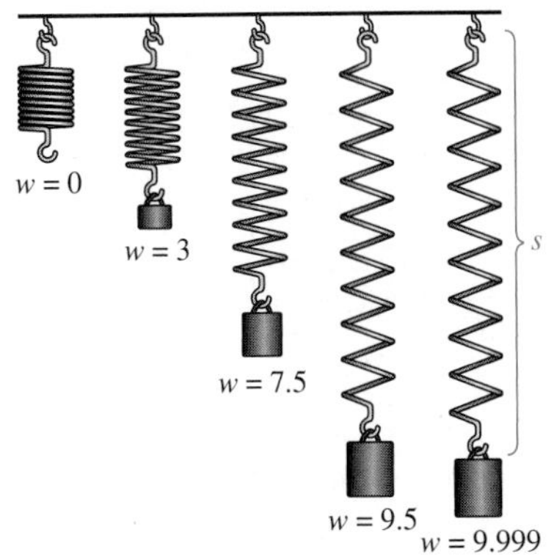

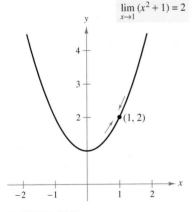

FIGURE 7.1   What is the limit of s as w approaches 10 pounds?

The Limit of a Function

In everyday language, people refer to a speed limit, a wrestler's weight limit, the limit of one's endurance, or stretching a spring to its limit. These phrases all suggest that a limit is a bound, which on some occasions may not be reached but on other occasions may be reached or exceeded.

Consider a spring that will break only if a weight of 10 pounds or more is attached. To determine how far the spring will stretch without breaking, you could attach increasingly heavier weights and measure the spring length s for each weight w, as shown in Figure 7.1. If the spring length approaches a value of L, then it is said that "the limit of s as w approaches 10 is L." A mathematical limit is much like the limit of a spring. The notation for a limit is

$$\lim_{x \to c} f(x) = L$$

which is read as "the limit of $f(x)$ as x approaches c is L."

Example 1   Finding a Limit

Find the limit: $\lim_{x \to 1} (x^2 + 1)$.

SOLUTION   Let $f(x) = x^2 + 1$. From the graph of f in Figure 7.2, it appears that $f(x)$ approaches 2 as x approaches 1 from either side, and you can write

$$\lim_{x \to 1} (x^2 + 1) = 2.$$

The table yields the same conclusion. Notice that as x gets closer and closer to 1, $f(x)$ gets closer and closer to 2.

$\lim_{x \to 1} (x^2 + 1) = 2$

FIGURE 7.2

	x approaches 1.				*x* approaches 1.		
x	0.900	0.990	0.999	1.000	1.001	1.010	1.100
$f(x)$	1.810	1.980	1.998	2.000	2.002	2.020	2.210

$f(x)$ approaches 2. $f(x)$ approaches 2.

✓ CHECKPOINT 1

Find the limit: $\lim_{x \to 1} (2x + 4)$. ■

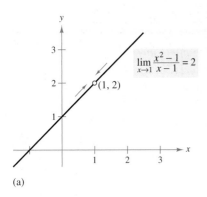

$$\lim_{x \to 1} \frac{x^2 - 1}{x - 1} = 2$$

(a)

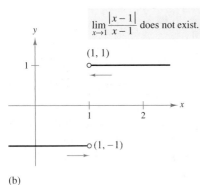

$$\lim_{x \to 1} \frac{|x - 1|}{x - 1} \text{ does not exist.}$$

(b)

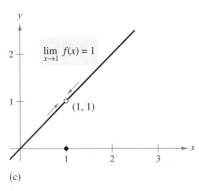

$$\lim_{x \to 1} f(x) = 1$$

(c)

FIGURE 7.3

Example 2 **Finding Limits Graphically and Numerically**

Find the limit: $\lim\limits_{x \to 1} f(x)$.

a. $f(x) = \dfrac{x^2 - 1}{x - 1}$ **b.** $f(x) = \dfrac{|x - 1|}{x - 1}$ **c.** $f(x) = \begin{cases} x, & x \neq 1 \\ 0, & x = 1 \end{cases}$

SOLUTION

a. From the graph of f, in Figure 7.3(a), it appears that $f(x)$ approaches 2 as x approaches 1 from either side. A missing point is denoted by the open dot on the graph. This conclusion is reinforced by the table. Be sure you see that *it does not matter that $f(x)$ is undefined when $x = 1$. The limit depends only on values of $f(x)$ near 1, not at 1.*

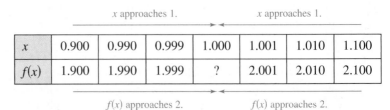

x	0.900	0.990	0.999	1.000	1.001	1.010	1.100
$f(x)$	1.900	1.990	1.999	?	2.001	2.010	2.100

$f(x)$ approaches 2. $f(x)$ approaches 2.

b. From the graph of f, in Figure 7.3(b), you can see that $f(x) = -1$ for all values to the left of $x = 1$ and $f(x) = 1$ for all values to the right of $x = 1$. So, $f(x)$ is approaching a different value from the left of $x = 1$ than it is from the right of $x = 1$. In such situations, we say that *the limit does not exist.* This conclusion is reinforced by the table.

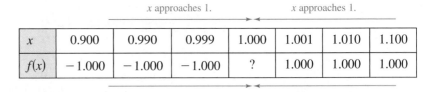

x	0.900	0.990	0.999	1.000	1.001	1.010	1.100
$f(x)$	-1.000	-1.000	-1.000	?	1.000	1.000	1.000

$f(x)$ approaches -1. $f(x)$ approaches 1.

c. From the graph of f, in Figure 7.3(c), it appears that $f(x)$ approaches 1 as x approaches 1 from either side. This conclusion is reinforced by the table. It does not matter that $f(1) = 0$. The limit depends only on values of $f(x)$ near 1, not at 1.

x	0.900	0.990	0.999	1.000	1.001	1.010	1.100
$f(x)$	0.900	0.990	0.999	?	1.001	1.010	1.100

$f(x)$ approaches 1. $f(x)$ approaches 1.

✓ CHECKPOINT 2

Find the limit: $\lim\limits_{x \to 2} f(x)$.

a. $f(x) = \dfrac{x^2 - 4}{x - 2}$ **b.** $f(x) = \dfrac{|x - 2|}{x - 2}$ **c.** $f(x) = \begin{cases} x^2, & x \neq 2 \\ 0, & x = 2 \end{cases}$ ∎

TECHNOLOGY

(T) Try using a graphing utility to determine the following limit.

$$\lim_{x \to 1} \frac{x^3 + 4x - 5}{x - 1}$$

You can do this by graphing

$$f(x) = \frac{x^3 + 4x - 5}{x - 1}$$

and zooming in near $x = 1$. From the graph, what does the limit appear to be?

There are three important ideas to learn from Examples 1 and 2.

1. Saying that the limit of $f(x)$ approaches L as x approaches c means that the value of $f(x)$ may be made *arbitrarily close* to the number L by choosing x closer and closer to c.

2. For a limit to exist, you must allow x to approach c from *either side* of c. If $f(x)$ approaches a different number as x approaches c from the left than it does as x approaches c from the right, then the limit *does not exist*. [See Example 2(b).]

3. The value of $f(x)$ when $x = c$ has no bearing on the existence or nonexistence of the limit of $f(x)$ as x approaches c. For instance, in Example 2(a), the limit of $f(x)$ exists as x approaches 1 even though the function f is not defined at $x = 1$.

Definition of the Limit of a Function

If $f(x)$ becomes arbitrarily close to a single number L as x approaches c from either side, then

$$\lim_{x \to c} f(x) = L$$

which is read as "the **limit** of $f(x)$ as x approaches c is L."

Properties of Limits

Many times the limit of $f(x)$ as x approaches c is simply $f(c)$, as shown in Example 1. Whenever the limit of $f(x)$ as x approaches c is

$$\lim_{x \to c} f(x) = f(c) \qquad \text{Substitute } c \text{ for } x.$$

the limit can be evaluated by **direct substitution.** (In the next section, you will learn that a function that has this property is *continuous at c*.) It is important that you learn to recognize the types of functions that have this property. Some basic ones are given in the following list.

Properties of Limits

Let b and c be real numbers, and let n be a positive integer.

1. $\lim\limits_{x \to c} b = b$

2. $\lim\limits_{x \to c} x = c$

3. $\lim\limits_{x \to c} x^n = c^n$

4. $\lim\limits_{x \to c} \sqrt[n]{x} = \sqrt[n]{c}$

In Property 4, if n is even, then c must be positive.

By combining the properties of limits with the rules for operating with limits shown below, you can find limits for a wide variety of algebraic functions.

Operations with Limits

Let b and c be real numbers, let n be a positive integer, and let f and g be functions with the following limits.

$$\lim_{x \to c} f(x) = L \quad \text{and} \quad \lim_{x \to c} g(x) = K$$

1. Scalar multiple: $\lim_{x \to c} [bf(x)] = bL$

2. Sum or difference: $\lim_{x \to c} [f(x) \pm g(x)] = L \pm K$

3. Product: $\lim_{x \to c} [f(x) \cdot g(x)] = LK$

4. Quotient: $\lim_{x \to c} \dfrac{f(x)}{g(x)} = \dfrac{L}{K}$, provided $K \neq 0$

5. Power: $\lim_{x \to c} [f(x)]^n = L^n$

6. Radical: $\lim_{x \to c} \sqrt[n]{f(x)} = \sqrt[n]{L}$

In Property 6, if n is even, then L must be positive.

DISCOVERY

Use a graphing utility to graph $y_1 = 1/x^2$. Does y_1 approach a limit as x approaches 0? Evaluate $y_1 = 1/x^2$ at several positive and negative values of x near 0 to confirm your answer. Does $\lim_{x \to 1} 1/x^2$ exist?

Example 3 **Finding the Limit of a Polynomial Function**

Find the limit: $\lim_{x \to 2} (x^2 + 2x - 3)$.

$$\lim_{x \to 2} (x^2 + 2x - 3) = \lim_{x \to 2} x^2 + \lim_{x \to 2} 2x - \lim_{x \to 2} 3 \qquad \text{Apply Property 2.}$$
$$= 2^2 + 2(2) - 3 \qquad \text{Use direct substitution.}$$
$$= 4 + 4 - 3 \qquad \text{Simplify.}$$
$$= 5$$

✓**CHECKPOINT 3**

Find the limit: $\lim_{x \to 1} (2x^2 - x + 4)$. ▪

Example 3 is an illustration of the following important result, which states that the limit of a polynomial function can be evaluated by direct substitution.

The Limit of a Polynomial Function

If p is a polynomial function and c is any real number, then

$$\lim_{x \to c} p(x) = p(c).$$

Techniques for Evaluating Limits

Many techniques for evaluating limits are based on the following important theorem. Basically, the theorem states that if two functions agree at all but a single point c, then they have identical limit behavior at $x = c$.

The Replacement Theorem

Let c be a real number and let $f(x) = g(x)$ for all $x \neq c$. If the limit of $g(x)$ exists as $x \to c$, then the limit of $f(x)$ also exists and

$$\lim_{x \to c} f(x) = \lim_{x \to c} g(x).$$

To apply the Replacement Theorem, you can use a result from algebra which states that for a polynomial function p, $p(c) = 0$ if and only if $(x - c)$ is a factor of $p(x)$. This concept is demonstrated in Example 4.

Example 4   Finding the Limit of a Function

Find the limit: $\lim_{x \to 1} \dfrac{x^3 - 1}{x - 1}$.

SOLUTION   Note that the numerator and denominator are zero when $x = 1$. This implies that $x - 1$ is a factor of both, and you can divide out this like factor.

$$\frac{x^3 - 1}{x - 1} = \frac{(x - 1)(x^2 + x + 1)}{x - 1} \qquad \text{Factor numerator.}$$

$$= \frac{(x - 1)(x^2 + x + 1)}{x - 1} \qquad \text{Divide out like factor.}$$

$$= x^2 + x + 1, \quad x \neq 1 \qquad \text{Simplify.}$$

So, the rational function $(x^3 - 1)/(x - 1)$ and the polynomial function $x^2 + x + 1$ agree for all values of x other than $x = 1$, and you can apply the Replacement Theorem.

$$\lim_{x \to 1} \frac{x^3 - 1}{x - 1} = \lim_{x \to 1} (x^2 + x + 1) = 1^2 + 1 + 1 = 3$$

Figure 7.4 illustrates this result graphically. Note that the two graphs are identical except that the graph of g contains the point $(1, 3)$, whereas this point is missing on the graph of f. (In the graph of f in Figure 7.4, the missing point is denoted by an open dot.)

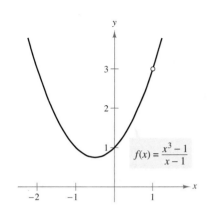

$f(x) = \dfrac{x^3 - 1}{x - 1}$

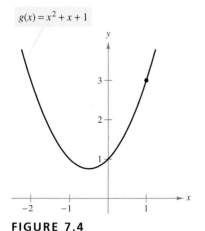

$g(x) = x^2 + x + 1$

FIGURE 7.4

✓ CHECKPOINT 4

Find the limit: $\lim_{x \to 2} \dfrac{x^3 - 8}{x - 2}$. ■

The technique used to evaluate the limit in Example 4 is called the **dividing out** technique. This technique is further demonstrated in the next example.

DISCOVERY

Using the graphs in Figure 7.4, what is the domain of $f(x)$? of $g(x)$?

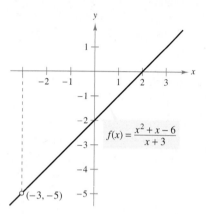

FIGURE 7.5 f is undefined when $x = -3$.

Example 5 Using the Dividing Out Technique

Find the limit: $\displaystyle\lim_{x \to -3} \frac{x^2 + x - 6}{x + 3}$.

SOLUTION Direct substitution fails because both the numerator and the denominator are zero when $x = -3$.

$$\lim_{x \to -3} \frac{x^2 + x - 6}{x + 3} \qquad \begin{array}{l} \longleftarrow \lim_{x \to -3} (x^2 + x - 6) = 0 \\ \longleftarrow \lim_{x \to -3} (x + 3) = 0 \end{array}$$

However, because the limits of both the numerator and denominator are zero, you know that they have a *common factor* of $x + 3$. So, for all $x \neq -3$, you can divide out this factor to obtain the following.

$$\lim_{x \to -3} \frac{x^2 + x - 6}{x + 3} = \lim_{x \to -3} \frac{(x - 2)(x + 3)}{x + 3} \qquad \text{Factor numerator.}$$

$$= \lim_{x \to -3} \frac{(x - 2)(x + 3)}{x + 3} \qquad \text{Divide out like factor.}$$

$$= \lim_{x \to -3} (x - 2) \qquad \text{Simplify.}$$

$$= -5 \qquad \text{Direct substitution}$$

This result is shown graphically in Figure 7.5. Note that the graph of f coincides with the graph of $g(x) = x - 2$, except that the graph of f has a hole at $(-3, -5)$.

✓ CHECKPOINT 5

Find the limit: $\displaystyle\lim_{x \to 3} \frac{x^2 + x - 12}{x - 3}$.

Example 6 Finding a Limit of a Function

Find the limit: $\displaystyle\lim_{x \to 0} \frac{\sqrt{x + 1} - 1}{x}$.

SOLUTION Direct substitution fails because both the numerator and the denominator are zero when $x = 0$. In this case, you can rewrite the fraction by rationalizing the numerator.

$$\frac{\sqrt{x + 1} - 1}{x} = \left(\frac{\sqrt{x + 1} - 1}{x} \right)\left(\frac{\sqrt{x + 1} + 1}{\sqrt{x + 1} + 1} \right)$$

$$= \frac{(x + 1) - 1}{x(\sqrt{x + 1} + 1)}$$

$$= \frac{x}{x(\sqrt{x + 1} + 1)} = \frac{1}{\sqrt{x + 1} + 1}, \quad x \neq 0$$

Now, using the Replacement Theorem, you can evaluate the limit as shown.

$$\lim_{x \to 0} \frac{\sqrt{x + 1} - 1}{x} = \lim_{x \to 0} \frac{1}{\sqrt{x + 1} + 1} = \frac{1}{1 + 1} = \frac{1}{2}$$

✓ CHECKPOINT 6

Find the limit: $\displaystyle\lim_{x \to 0} \frac{\sqrt{x + 4} - 2}{x}$.

One-Sided Limits

In Example 2(b), you saw that one way in which a limit can fail to exist is when a function approaches a different value from the left of c than it approaches from the right of c. This type of behavior can be described more concisely with the concept of a **one-sided limit.**

$$\lim_{x \to c^-} f(x) = L \qquad \text{Limit from the left}$$

$$\lim_{x \to c^+} f(x) = L \qquad \text{Limit from the right}$$

The first of these two limits is read as "the limit of $f(x)$ as x approaches c from the left is L." The second is read as "the limit of $f(x)$ as x approaches c from the right is L."

Example 7   Finding One-Sided Limits

Find the limit as $x \to 0$ from the left and the limit as $x \to 0$ from the right for the function

$$f(x) = \frac{|2x|}{x}.$$

SOLUTION   From the graph of f, shown in Figure 7.6, you can see that $f(x) = -2$ for all $x < 0$. So, the limit from the left is

$$\lim_{x \to 0^-} \frac{|2x|}{x} = -2. \qquad \text{Limit from the left}$$

Because $f(x) = 2$ for all $x > 0$, the limit from the right is

$$\lim_{x \to 0^+} \frac{|2x|}{x} = 2. \qquad \text{Limit from the right}$$

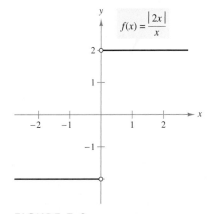

$f(x) = \dfrac{|2x|}{x}$

FIGURE 7.6

✓ **CHECKPOINT 7**

Find each limit.   (a) $\displaystyle\lim_{x \to 2^-} \frac{|x - 2|}{x - 2}$ (b) $\displaystyle\lim_{x \to 2^+} \frac{|x - 2|}{x - 2}$ ■

In Example 7, note that the function approaches different limits from the left and from the right. In such cases, the limit of $f(x)$ as $x \to c$ does not exist. For the limit of a function to exist as $x \to c$, *both* one-sided limits must exist and must be equal.

Existence of a Limit

If f is a function and c and L are real numbers, then

$$\lim_{x \to c} f(x) = L$$

if and only if both the left and right limits are equal to L.

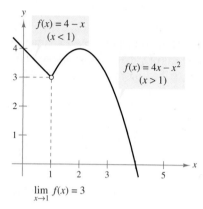

$f(x) = 4 - x$
$(x < 1)$

$f(x) = 4x - x^2$
$(x > 1)$

$\lim_{x \to 1} f(x) = 3$

FIGURE 7.7

✓ **CHECKPOINT 8**

Find the limit of $f(x)$ as x approaches 0.

$$f(x) = \begin{cases} x^2 + 1, & x < 0 \\ 2x + 1, & x > 0 \end{cases}$$ ∎

| Example 8 | **Finding One-Sided Limits** |

Find the limit of $f(x)$ as x approaches 1.

$$f(x) = \begin{cases} 4 - x, & x < 1 \\ 4x - x^2, & x > 1 \end{cases}$$

SOLUTION   Remember that you are concerned about the value of f near $x = 1$ rather than at $x = 1$. So, for $x < 1, f(x)$ is given by $4 - x$, and you can use direct substitution to obtain

$$\lim_{x \to 1^-} f(x) = \lim_{x \to 1^-} (4 - x)$$
$$= 4 - 1 = 3.$$

For $x > 1, f(x)$ is given by $4x - x^2$, and you can use direct substitution to obtain

$$\lim_{x \to 1^+} f(x) = \lim_{x \to 1^+} (4x - x^2)$$
$$= 4(1) - 1^2 = 4 - 1 = 3.$$

Because both one-sided limits exist and are equal to 3, it follows that

$$\lim_{x \to 1} f(x) = 3.$$

The graph in Figure 7.7 confirms this conclusion.

| Example 9 | **Comparing One-Sided Limits** | |

An overnight delivery service charges \$12 for the first pound and \$2 for each additional pound. Let x represent the weight of a parcel and let $f(x)$ represent the shipping cost.

$$f(x) = \begin{cases} 12, & 0 < x \le 1 \\ 14, & 1 < x \le 2 \\ 16, & 2 < x \le 3 \end{cases}$$

Show that the limit of $f(x)$ as $x \to 2$ does not exist.

SOLUTION   The graph of f is shown in Figure 7.8. The limit of $f(x)$ as x approaches 2 from the left is

$$\lim_{x \to 2^-} f(x) = 14$$

whereas the limit of $f(x)$ as x approaches 2 from the right is

$$\lim_{x \to 2^+} f(x) = 16.$$

Because these one-sided limits are not equal, the limit of $f(x)$ as $x \to 2$ does not exist.

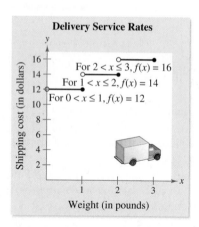

Delivery Service Rates

For $2 < x \le 3, f(x) = 16$
For $1 < x \le 2, f(x) = 14$
For $0 < x \le 1, f(x) = 12$

Shipping cost (in dollars)

Weight (in pounds)

FIGURE 7.8   Demand Curve

✓ **CHECKPOINT 9**

Show that the limit of $f(x)$ as $x \to 1$ does not exist in Example 9. ∎

Unbounded Behavior

Example 9 shows a limit that fails to exist because the limits from the left and right differ. Another important way in which a limit can fail to exist is when $f(x)$ increases or decreases without bound as x approaches c.

Example 10 An Unbounded Function

Find the limit (if possible).

$$\lim_{x \to 2} \frac{3}{x - 2}$$

SOLUTION From Figure 7.9, you can see that $f(x)$ decreases without bound as x approaches 2 from the left and $f(x)$ increases without bound as x approaches 2 from the right. Symbolically, you can write this as

$$\lim_{x \to 2^-} \frac{3}{x - 2} = -\infty$$

and

$$\lim_{x \to 2^+} \frac{3}{x - 2} = \infty.$$

Because f is unbounded as x approaches 2, the limit does not exist.

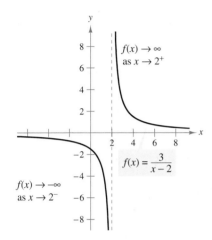

$f(x) \to \infty$
as $x \to 2^+$

$f(x) = \dfrac{3}{x-2}$

$f(x) \to -\infty$
as $x \to 2^-$

FIGURE 7.9

DISCOVERY

Using the graph in Figure 7.9, what is the domain of $f(x)$? the range?

✓ CHECKPOINT 10

Find the limit (if possible): $\lim\limits_{x \to -2} \dfrac{5}{x + 2}$. ▪

STUDY TIP

The equal sign in the statement $\lim\limits_{x \to c^+} f(x) = \infty$ does not mean that the limit exists. On the contrary, it tells you how the limit *fails to exist* by denoting the unbounded behavior of $f(x)$ as x approaches c.

CONCEPT CHECK

1. If $\lim\limits_{x \to c^-} f(x) \neq \lim\limits_{x \to c^+} f(x)$, what can you conclude about $\lim\limits_{x \to c} f(x)$?
2. Describe how to find the limit of a polynomial function $p(x)$ as x approaches c.
3. Is the limit of $f(x)$ as x approaches c always equal to $f(c)$? Why or why not?
4. If f is undefined at $x = c$, can you conclude that the limit of $f(x)$ as x approaches c does not exist? Explain.

Skills Review 7.1	The following warm-up exercises involve skills that were covered in earlier sections. You will use these skills in the exercise set for this section. For additional help, review Sections 2.4 and 2.5.

In Exercises 1–4, evaluate the expression and simplify.

1. $f(x) = x^2 - 3x + 3$

    (a) $f(-1)$ (b) $f(c)$ (c) $f(x + h)$

2. $f(x) = \begin{cases} 2x - 2, & x < 1 \\ 3x + 1, & x \geq 1 \end{cases}$

    (a) $f(-1)$ (b) $f(3)$ (c) $f(t^2 + 1)$

3. $f(x) = x^2 - 2x + 2$ $\dfrac{f(1 + h) - f(1)}{h}$

4. $f(x) = 4x$ $\dfrac{f(2 + h) - f(2)}{h}$

In Exercises 5–8, find the domain and range of the function and sketch its graph.

5. $h(x) = -\dfrac{5}{x}$

6. $g(x) = \sqrt{36 - x^2}$

7. $f(x) = |x - 3|$

8. $f(x) = \dfrac{|x|}{2x}$

In Exercises 9 and 10, determine whether y is a function of x.

9. $9x^2 + 4y^2 = 49$

10. $2x^2y + 8x = 7y$

Exercises 7.1

See www.CalcChat.com for worked-out solutions to odd-numbered exercises.

In Exercises 1–8, complete the table and use the result to estimate the limit. Use a graphing utility to graph the function to confirm your result.

1. $\lim\limits_{x \to 2} (2x + 5)$

x	1.9	1.99	1.999	2	2.001	2.01	2.1
$f(x)$				?			

2. $\lim\limits_{x \to 2} (x^2 - 3x + 1)$

x	1.9	1.99	1.999	2	2.001	2.01	2.1
$f(x)$				?			

3. $\lim\limits_{x \to 2} \dfrac{x - 2}{x^2 - 4}$

x	1.9	1.99	1.999	2	2.001	2.01	2.1
$f(x)$				?			

4. $\lim\limits_{x \to 2} \dfrac{x - 2}{x^2 - 3x + 2}$

x	1.9	1.99	1.999	2	2.001	2.01	2.1
$f(x)$				?			

5. $\lim\limits_{x \to 0} \dfrac{\sqrt{x + 1} - 1}{x}$

x	-0.1	-0.01	-0.001	0	0.001	0.01	0.1
$f(x)$				?			

6. $\lim\limits_{x \to 0} \dfrac{\sqrt{x + 2} - \sqrt{2}}{x}$

x	-0.1	-0.01	-0.001	0	0.001	0.01	0.1
$f(x)$				?			

7. $\lim\limits_{x \to 0^-} \dfrac{\dfrac{1}{x+4} - \dfrac{1}{4}}{x}$

x	-0.5	-0.1	-0.01	-0.001	0
$f(x)$					?

8. $\lim\limits_{x \to 0^+} \dfrac{\dfrac{1}{2+x} - \dfrac{1}{2}}{2x}$

x	0.5	0.1	0.01	0.001	0
$f(x)$					?

In Exercises 9–12, use the graph to find the limit (if it exists).

9.

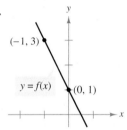

(a) $\lim\limits_{x \to 0} f(x)$

(b) $\lim\limits_{x \to -1} f(x)$

10.

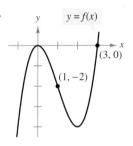

(a) $\lim\limits_{x \to 1} f(x)$

(b) $\lim\limits_{x \to 3} f(x)$

11.

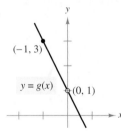

(a) $\lim\limits_{x \to 0} g(x)$

(b) $\lim\limits_{x \to -1} g(x)$

12.

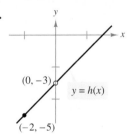

(a) $\lim\limits_{x \to -2} h(x)$

(b) $\lim\limits_{x \to 0} h(x)$

In Exercises 13 and 14, find the limit of (a) $f(x) + g(x)$, (b) $f(x)g(x)$, and (c) $f(x)/g(x)$, as x approaches c.

13. $\lim\limits_{x \to c} f(x) = 3$

$\lim\limits_{x \to c} g(x) = 9$

14. $\lim\limits_{x \to c} f(x) = \frac{3}{2}$

$\lim\limits_{x \to c} g(x) = \frac{1}{2}$

In Exercises 15 and 16, find the limit of (a) $\sqrt{f(x)}$, (b) $[3f(x)]$, and (c) $[f(x)]^2$, as x approaches c.

15. $\lim\limits_{x \to c} f(x) = 16$

16. $\lim\limits_{x \to c} f(x) = 9$

In Exercises 17–22, use the graph to find the limit (if it exists).

(a) $\lim\limits_{x \to c^+} f(x)$ (b) $\lim\limits_{x \to c^-} f(x)$ (c) $\lim\limits_{x \to c} f(x)$

17.

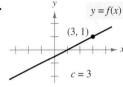

18.

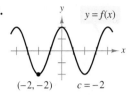

19.

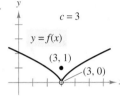

20.

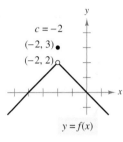

21.

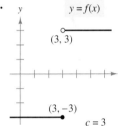

22.

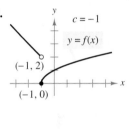

In Exercises 23–40, find the limit.

23. $\lim\limits_{x \to 2} x^2$

24. $\lim\limits_{x \to -2} x^3$

25. $\lim\limits_{x \to -3} (2x + 5)$

26. $\lim\limits_{x \to 0} (3x - 2)$

27. $\lim\limits_{x \to 1} (1 - x^2)$

28. $\lim\limits_{x \to 2} (-x^2 + x - 2)$

29. $\lim\limits_{x \to 3} \sqrt{x + 6}$

30. $\lim\limits_{x \to 4} \sqrt[3]{x + 4}$

31. $\lim\limits_{x \to -3} \dfrac{2}{x + 2}$

32. $\lim\limits_{x \to -2} \dfrac{3x + 1}{2 - x}$

33. $\lim\limits_{x \to -2} \dfrac{x^2 - 1}{2x}$

34. $\lim\limits_{x \to -1} \dfrac{4x - 5}{3 - x}$

35. $\lim\limits_{x \to 7} \dfrac{5x}{x + 2}$

36. $\lim\limits_{x \to 3} \dfrac{\sqrt{x + 1}}{x - 4}$

37. $\lim\limits_{x \to 3} \dfrac{\sqrt{x + 1} - 1}{x}$

38. $\lim\limits_{x \to 5} \dfrac{\sqrt{x + 4} - 2}{x}$

39. $\lim\limits_{x \to 1} \dfrac{\dfrac{1}{x + 4} - \dfrac{1}{4}}{x}$

40. $\lim\limits_{x \to 2} \dfrac{\dfrac{1}{x + 2} - \dfrac{1}{2}}{x}$

In Exercises 41–60, find the limit (if it exists).

41. $\lim\limits_{x \to 1} \dfrac{x^2 - 1}{x - 1}$

42. $\lim\limits_{x \to -1} \dfrac{2x^2 - x - 3}{x + 1}$

43. $\lim\limits_{x \to 2} \dfrac{x - 2}{x^2 - 4x + 4}$

44. $\lim\limits_{x \to 2} \dfrac{2 - x}{x^2 - 4}$

45. $\lim\limits_{t \to 4} \dfrac{t + 4}{t^2 - 16}$

46. $\lim\limits_{t \to 1} \dfrac{t^2 + t - 2}{t^2 - 1}$

47. $\lim\limits_{x \to -2} \dfrac{x^3 + 8}{x + 2}$

48. $\lim\limits_{x \to -1} \dfrac{x^3 - 1}{x + 1}$

49. $\lim\limits_{x \to -2} \dfrac{|x + 2|}{x + 2}$

50. $\lim\limits_{x \to 2} \dfrac{|x - 2|}{x - 2}$

51. $\lim\limits_{x \to 2} f(x)$, where $f(x) = \begin{cases} 4 - x, & x \neq 2 \\ 0 & x = 2 \end{cases}$

52. $\lim\limits_{x \to 1} f(x)$, where $f(x) = \begin{cases} x^2 + 2, & x \neq 1 \\ 1, & x = 1 \end{cases}$

53. $\lim\limits_{x \to 3} f(x)$, where $f(x) = \begin{cases} \frac{1}{3}x - 2, & x \leq 3 \\ -2x + 5, & x > 3 \end{cases}$

54. $\lim\limits_{s \to 1} f(s)$, where $f(s) = \begin{cases} s, & s \leq 1 \\ 1 - s, & s > 1 \end{cases}$

55. $\lim\limits_{\Delta x \to 0} \dfrac{2(x + \Delta x) - 2x}{\Delta x}$

56. $\lim\limits_{\Delta x \to 0} \dfrac{4(x + \Delta x) - 5 - (4x - 5)}{\Delta x}$

57. $\lim\limits_{\Delta x \to 0} \dfrac{\sqrt{x + 2 + \Delta x} - \sqrt{x + 2}}{\Delta x}$

58. $\lim\limits_{\Delta x \to 0} \dfrac{\sqrt{x + \Delta x} - \sqrt{x}}{\Delta x}$

59. $\lim\limits_{\Delta t \to 0} \dfrac{(t + \Delta t)^2 - 5(t + \Delta t) - (t^2 - 5t)}{\Delta t}$

60. $\lim\limits_{\Delta t \to 0} \dfrac{(t + \Delta t)^2 - 4(t + \Delta t) + 2 - (t^2 - 4t + 2)}{\Delta t}$

(T) **Graphical, Numerical, and Analytic Analysis** In Exercises 61–64, use a graphing utility to graph the function and estimate the limit. Use a table to reinforce your conclusion. Then find the limit by analytic methods.

61. $\lim\limits_{x \to 1^-} \dfrac{2}{x^2 - 1}$

62. $\lim\limits_{x \to 1^+} \dfrac{5}{1 - x}$

63. $\lim\limits_{x \to -2^-} \dfrac{1}{x + 2}$

64. $\lim\limits_{x \to 0^-} \dfrac{x + 1}{x}$

(T) In Exercises 65–68, use a graphing utility to estimate the limit (if it exists).

65. $\lim\limits_{x \to 2} \dfrac{x^2 - 5x + 6}{x^2 - 4x + 4}$

66. $\lim\limits_{x \to 1} \dfrac{x^2 + 6x - 7}{x^3 - x^2 + 2x - 2}$

67. $\lim\limits_{x \to -4} \dfrac{x^3 + 4x^2 + x + 4}{2x^2 + 7x - 4}$

68. $\lim\limits_{x \to -2} \dfrac{4x^3 + 7x^2 + x + 6}{3x^2 - x - 14}$

69. Environment The cost (in dollars) of removing $p\%$ of the pollutants from the water in a small lake is given by

$$C = \dfrac{25{,}000p}{100 - p}, \quad 0 \leq p < 100$$

where C is the cost and p is the percent of pollutants.

(a) Find the cost of removing 50% of the pollutants.

(b) What percent of the pollutants can be removed for $100,000?

(c) Evaluate $\lim\limits_{p \to 100^-} C$. Explain your results.

70. Compound Interest You deposit $2000 in an account that is compounded quarterly at an annual rate of r (in decimal form). The balance A after 10 years is

$$A = 2000\left(1 + \dfrac{r}{4}\right)^{40}.$$

(a) Complete the table.

r	0.059	0.0599	0.06	0.0601	0.061
A					

(b) Does the limit of A exist as the interest rate approaches 6%? If so, what is the limit?

(T) **71. Compound Interest** Consider a certificate of deposit that pays 10% (annual percentage rate) on an initial deposit of $1000. The balance A after 10 years is

$$A = 1000(1 + 0.1x)^{10/x}$$

where x is the length of the compounding period (in years).

(a) Use a graphing utility to graph A, where $0 \leq x \leq 1$.

(b) Use the *zoom* and *trace* features to estimate the balance for quarterly compounding and daily compounding.

(c) Use the *zoom* and *trace* features to estimate

$$\lim\limits_{x \to 0^+} A.$$

What do you think this limit represents? Explain your reasoning.

72. The limit of $f(x) = (1 + x)^{1/x}$ is a natural base for many business applications, as you will see in Section 10.2.

$$\lim\limits_{x \to 0} (1 + x)^{1/x} = e \approx 2.718$$

(a) Show the reasonableness of this limit by completing the table.

x	-0.01	-0.001	-0.0001	0	0.0001	0.001	0.01
$f(x)$							

(T) (b) Use a graphing utility to graph f and to confirm the answer in part (a).

(c) Find the domain and range of the function.

Section 7.2

Continuity

- Determine the continuity of functions.
- Determine the continuity of functions on a closed interval.
- Use the greatest integer function to model and solve real-life problems.
- Use compound interest models to solve real-life problems.

Continuity

In mathematics, the term "continuous" has much the same meaning as it does in everyday use. To say that a function is continuous at $x = c$ means that there is no interruption in the graph of f at c. The graph of f is unbroken at c, and there are no holes, jumps, or gaps. As simple as this concept may seem, its precise definition eluded mathematicians for many years. In fact, it was not until the early 1800's that a precise definition was finally developed.

Before looking at this definition, consider the function whose graph is shown in Figure 7.10. This figure identifies three values of x at which the function f is not continuous.

1. At $x = c_1$, $f(c_1)$ is not defined.

2. At $x = c_2$, $\lim\limits_{x \to c_2} f(x)$ does not exist.

3. At $x = c_3$, $f(c_3) \neq \lim\limits_{x \to c_3} f(x)$.

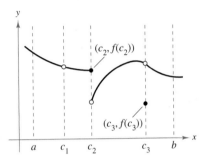

FIGURE 7.10 f is not continuous when $x = c_1, c_2, c_3$.

At all other points in the interval (a, b), the graph of f is uninterrupted, which implies that the function f is continuous at all other points in the interval (a, b).

Definition of Continuity

Let c be a number in the interval (a, b), and let f be a function whose domain contains the interval (a, b). The function f is **continuous at the point c** if the following conditions are true.

1. $f(c)$ is defined.

2. $\lim\limits_{x \to c} f(x)$ exists.

3. $\lim\limits_{x \to c} f(x) = f(c)$.

If f is continuous at every point in the interval (a, b), then it is **continuous on an open interval (a, b).**

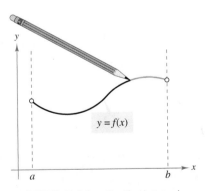

FIGURE 7.11 On the interval (a, b), the graph of f can be traced with a pencil.

Roughly, you can say that a function is continuous on an interval if its graph on the interval can be traced using a pencil and paper without lifting the pencil from the paper, as shown in Figure 7.11.

In Section 7.1, you studied several types of functions that meet the three conditions for continuity. Specifically, if *direct substitution* can be used to evaluate the limit of a function at c, then the function is continuous at c. Two types of functions that have this property are polynomial functions and rational functions.

Continuity of Polynomial and Rational Functions

1. A polynomial function is continuous at every real number.

2. A rational function is continuous at every number in its domain.

Example 1 **Determining Continuity of a Polynomial Function**

Discuss the continuity of each function.

a. $f(x) = x^2 - 2x + 3$

b. $f(x) = x^3 - x$

SOLUTION Each of these functions is a *polynomial function*. So, each is continuous on the entire real line, as indicated in Figure 7.12.

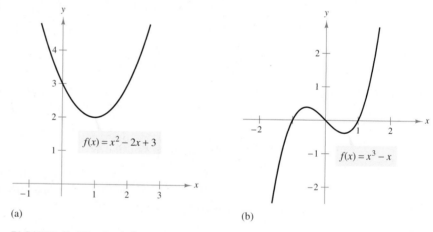

(a)   (b)

FIGURE 7.12 Both functions are continuous on $(-\infty, \infty)$.

✓**CHECKPOINT 1**

Discuss the continuity of each function.

a. $f(x) = x^2 + x + 1$ **b.** $f(x) = x^3 + x$ ▪

Polynomial functions are one of the most important types of functions used in calculus. Be sure you see from Example 1 that the graph of a polynomial function is continuous on the entire real line, and therefore has no holes, jumps, or gaps. Rational functions, on the other hand, need not be continuous on the entire real line, as shown in Example 2.

Example 2 Determining Continuity of a Rational Function

Discuss the continuity of each function.

a. $f(x) = 1/x$ **b.** $f(x) = (x^2 - 1)/(x - 1)$ **c.** $f(x) = 1/(x^2 + 1)$

SOLUTION Each of these functions is a rational function and is therefore continuous at every number in its domain.

a. The domain of $f(x) = 1/x$ consists of all real numbers except $x = 0$. So, this function is continuous on the intervals $(-\infty, 0)$ and $(0, \infty)$. [See Figure 7.13(a).]

b. The domain of $f(x) = (x^2 - 1)/(x - 1)$ consists of all real numbers except $x = 1$. So, this function is continuous on the intervals $(-\infty, 1)$ and $(1, \infty)$. [See Figure 7.13(b).]

c. The domain of $f(x) = 1/(x^2 + 1)$ consists of all real numbers. So, this function is continuous on the entire real line. [See Figure 7.13(c).]

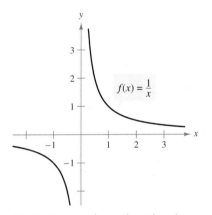

(a) Continuous on $(-\infty, 0)$ and $(0, \infty)$.

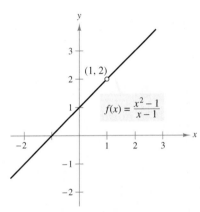

(b) Continuous on $(-\infty, 1)$ and $(1, \infty)$.

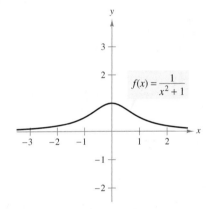

(c) Continuous on $(-\infty, \infty)$.

FIGURE 7.13

✓ CHECKPOINT 2

Discuss the continuity of each function.

a. $f(x) = \dfrac{1}{x - 1}$ **b.** $f(x) = \dfrac{x^2 - 4}{x - 2}$ **c.** $f(x) = \dfrac{1}{x^2 + 2}$ ∎

Consider an open interval I that contains a real number c. If a function f is defined on I (except possibly at c), and f is not continuous at c, then f is said to have a **discontinuity** at c. Discontinuities fall into two categories: **removable** and **nonremovable.** A discontinuity at c is called removable if f can be made continuous by appropriately defining (or redefining) $f(c)$. For instance, the function in Example 2(b) has a removable discontinuity at $(1, 2)$. To remove the discontinuity, all you need to do is redefine the function so that $f(1) = 2$.

A discontinuity at $x = c$ is nonremovable if the function cannot be made continuous at $x = c$ by defining or redefining the function at $x = c$. For instance, the function in Example 2(a) has a nonremovable discontinuity at $x = 0$.

Continuity on a Closed Interval

The intervals discussed in Examples 1 and 2 are open. To discuss continuity on a closed interval, you can use the concept of one-sided limits, as defined in Section 7.1.

Definition of Continuity on a Closed Interval

Let f be defined on a closed interval $[a, b]$. If f is continuous on the open interval (a, b) and

$$\lim_{x \to a^+} f(x) = f(a) \qquad \text{and} \qquad \lim_{x \to b^-} f(x) = f(b)$$

then f is **continuous on the closed interval** $[a, b]$. Moreover, f is **continuous from the right** at a and **continuous from the left** at b.

Similar definitions can be made to cover continuity on intervals of the form $(a, b]$ and $[a, b)$, or on infinite intervals. For example, the function

$$f(x) = \sqrt{x}$$

is continuous on the infinite interval $[0, \infty)$.

Example 3 **Examining Continuity at an Endpoint**

Discuss the continuity of

$$f(x) = \sqrt{3 - x}.$$

SOLUTION Notice that the domain of f is the set $(-\infty, 3]$. Moreover, f is continuous from the left at $x = 3$ because

$$\lim_{x \to 3^-} f(x) = \lim_{x \to 3^-} \sqrt{3 - x}$$
$$= 0$$
$$= f(3).$$

For all $x < 3$, the function f satisfies the three conditions for continuity. So, you can conclude that f is continuous on the interval $(-\infty, 3]$, as shown in Figure 7.14.

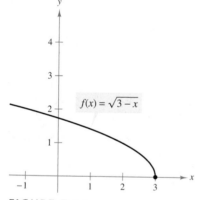

$f(x) = \sqrt{3 - x}$

FIGURE 7.14

✓ **CHECKPOINT 3**

Discuss the continuity of $f(x) = \sqrt{x - 2}$. ■

STUDY TIP

When working with radical functions of the form

$$f(x) = \sqrt{g(x)}$$

remember that the domain of f coincides with the solution of $g(x) \geq 0$.

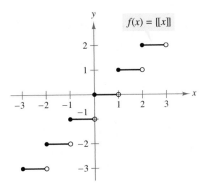

FIGURE 7.15

Example 4 Examining Continuity on a Closed Interval

Discuss the continuity of $g(x) = \begin{cases} 5 - x, & -1 \le x \le 2 \\ x^2 - 1, & 2 < x \le 3 \end{cases}$.

SOLUTION The polynomial functions $5 - x$ and $x^2 - 1$ are continuous on the intervals $[-1, 2]$ and $(2, 3]$, respectively. So, to conclude that g is continuous on the entire interval $[-1, 3]$, you only need to check the behavior of g when $x = 2$. You can do this by taking the one-sided limits when $x = 2$.

$$\lim_{x \to 2^-} g(x) = \lim_{x \to 2^-} (5 - x) = 3 \qquad \text{Limit from the left}$$

and

$$\lim_{x \to 2^+} g(x) = \lim_{x \to 2^+} (x^2 - 1) = 3 \qquad \text{Limit from the right}$$

Because these two limits are equal,

$$\lim_{x \to 2} g(x) = g(2) = 3.$$

So, g is continuous at $x = 2$ and, consequently, it is continuous on the entire interval $[-1, 3]$. The graph of g is shown in Figure 7.15.

✓ **CHECKPOINT 4**

Discuss the continuity of $f(x) = \begin{cases} x + 2, & -1 \le x < 3 \\ 14 - x^2, & 3 \le x \le 5 \end{cases}$. ■

The Greatest Integer Function

Many functions that are used in business applications are **step functions.** For instance, the function in Example 9 in Section 7.1 is a step function. The **greatest integer function** is another example of a step function. This function is denoted by

$$[\![x]\!] = \text{greatest integer less than or equal to } x.$$

For example,

$$[\![-2.1]\!] = \text{greatest integer less than or equal to } -2.1 = -3$$
$$[\![-2]\!] = \text{greatest integer less than or equal to } -2 = -2$$
$$[\![1.5]\!] = \text{greatest integer less than or equal to } 1.5 = 1.$$

Note that the graph of the greatest integer function (Figure 7.16) jumps up one unit at each integer. This implies that the function is not continuous at each integer.

In real-life applications, the domain of the greatest integer function is often restricted to nonnegative values of x. In such cases this function serves the purpose of **truncating** the decimal portion of x. For example, 1.345 is truncated to 1 and 3.57 is truncated to 3. That is,

$$[\![1.345]\!] = 1 \qquad \text{and} \qquad [\![3.57]\!] = 3.$$

FIGURE 7.16 Greatest Integer Function

TECHNOLOGY

Ⓣ Use a graphing utility to calculate the following.

a. $[\![3.5]\!]$ **b.** $[\![-3.5]\!]$ **c.** $[\![0]\!]$

AP/Wide World Photos

R. R. Donnelley & Sons Company is one of the world's largest commercial printers. It prints and binds a major share of the national publications in the United States, including *Time*, *Newsweek*, and *TV Guide*.

Example 5   Modeling a Cost Function

A bookbinding company produces 10,000 books in an eight-hour shift. The fixed cost *per shift* amounts to $5000, and the unit cost per book is $3. Using the greatest integer function, you can write the cost of producing x books as

$$C = 5000\left(1 + \left[\!\left[\frac{x-1}{10,000}\right]\!\right]\right) + 3x.$$

Sketch the graph of this cost function.

SOLUTION   Note that during the first eight-hour shift

$$\left[\!\left[\frac{x-1}{10,000}\right]\!\right] = 0, \quad 1 \le x \le 10,000$$

which implies

$$C = 5000\left(1 + \left[\!\left[\frac{x-1}{10,000}\right]\!\right]\right) + 3x = 5000 + 3x.$$

During the second eight-hour shift

$$\left[\!\left[\frac{x-1}{10,000}\right]\!\right] = 1, \quad 10,001 \le x \le 20,000$$

which implies

$$C = 5000\left(1 + \left[\!\left[\frac{x-1}{10,000}\right]\!\right]\right) + 3x$$

$$= 10,000 + 3x.$$

The graph of C is shown in Figure 7.17. Note the graph's discontinuities.

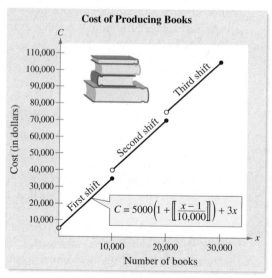

FIGURE 7.17

✓ **CHECKPOINT 5**

Use a graphing utility to graph the cost function in Example 5. ■

TECHNOLOGY

Step Functions and Compound Functions

Ⓣ To graph a step function or compound function with a graphing utility, you must be familiar with the utility's programming language. For instance, different graphing utilities have different "integer truncation" functions. One is IPart(x), and it yields the truncated integer part of x. For example, IPart(-1.2) $= -1$ and IPart(3.4) $= 3$. The other function is Int(x), which is the greatest integer function. The graphs of these two functions are shown below. When graphing a step function, you should set your graphing utility to *dot mode*.

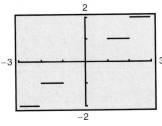

Graph of $f(x) = $ IPart(x)

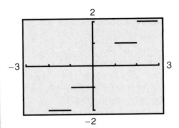

Graph of $f(x) = $ Int(x)

On some graphing utilities, you can graph a piecewise-defined function such as

$$f(x) = \begin{cases} x^2 - 4, & x \leq 2 \\ -x + 2, & 2 < x \end{cases}.$$

The graph of this function is shown below.

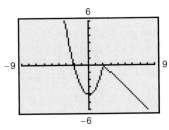

Consult the user's guide for your graphing utility for specific keystrokes you can use to graph these functions.

Quarterly Compounding

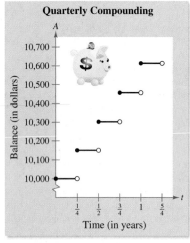

FIGURE 7.18

Extended Application: Compound Interest

Banks and other financial institutions differ on how interest is paid to an account. If the interest is added to the account so that future interest is paid on previously earned interest, then the interest is said to be compounded. Suppose, for example, that you deposited $10,000 in an account that pays 6% interest, compounded quarterly. Because the 6% is the annual interest rate, the quarterly rate is $\frac{1}{4}(0.06) = 0.015$ or 1.5%. The balances during the first five quarters are shown below.

Quarter	Balance
1st	$10,000.00
2nd	$10,000.00 + (0.015)(10,000.00) = \$10,150.00$
3rd	$10,150.00 + (0.015)(10,150.00) = \$10,302.25$
4th	$10,302.25 + (0.015)(10,302.25) = \$10,456.78$
5th	$10,456.78 + (0.015)(10,456.78) = \$10,613.63$

Example 6 Graphing Compound Interest ®

Sketch the graph of the balance in the account described above.

SOLUTION Let A represent the balance in the account and let t represent the time, in years. You can use the greatest integer function to represent the balance, as shown.

$$A = 10,000(1 + 0.015)^{[\![4t]\!]}$$

From the graph shown in Figure 7.18, notice that the function has a discontinuity at each quarter.

✓**CHECKPOINT 6**

Write an equation that gives the balance of the account in Example 6 if the annual interest rate is 8%. ∎

CONCEPT CHECK

1. Describe the continuity of a polynomial function.
2. Describe the continuity of a rational function.
3. If a function f is continuous at every point in the interval (a, b), then what can you say about f on an open interval (a, b)?
4. Describe in your own words what it means to say that a function f is continuous at $x = c$.

Skills Review 7.2

The following warm-up exercises involve skills that were covered in earlier sections. You will use these skills in the exercise set for this section. For additional help, review Sections 0.7, 1.3, 1.5, and 7.1.

In Exercises 1–4, simplify the expression.

1. $\dfrac{x^2 + 6x + 8}{x^2 - 6x - 16}$

2. $\dfrac{x^2 - 5x - 6}{x^2 - 9x + 18}$

3. $\dfrac{2x^2 - 2x - 12}{4x^2 - 24x + 36}$

4. $\dfrac{x^3 - 16x}{x^3 + 2x^2 - 8x}$

In Exercises 5–8, solve for x.

5. $x^2 + 7x = 0$

6. $x^2 + 4x - 5 = 0$

7. $3x^2 + 8x + 4 = 0$

8. $x^3 + 5x^2 - 24x = 0$

In Exercises 9 and 10, find the limit.

9. $\lim\limits_{x \to 3} (2x^2 - 3x + 4)$

10. $\lim\limits_{x \to -2} (3x^3 - 8x + 7)$

Exercises 7.2

See www.CalcChat.com for worked-out solutions to odd-numbered exercises.

In Exercises 1–10, determine whether the function is continuous on the entire real line. Explain your reasoning.

1. $f(x) = 5x^3 - x^2 + 2$

2. $f(x) = (x^2 - 1)^3$

3. $f(x) = \dfrac{1}{x^2 - 4}$

4. $f(x) = \dfrac{1}{9 - x^2}$

5. $f(x) = \dfrac{1}{4 + x^2}$

6. $f(x) = \dfrac{3x}{x^2 + 1}$

7. $f(x) = \dfrac{2x - 1}{x^2 - 8x + 15}$

8. $f(x) = \dfrac{x + 4}{x^2 - 6x + 5}$

9. $g(x) = \dfrac{x^2 - 4x + 4}{x^2 - 4}$

10. $g(x) = \dfrac{x^2 - 9x + 20}{x^2 - 16}$

In Exercises 11–34, describe the interval(s) on which the function is continuous. Explain why the function is continuous on the interval(s). If the function has a discontinuity, identify the conditions of continuity that are not satisfied.

11. $f(x) = \dfrac{x^2 - 1}{x}$

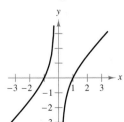

12. $f(x) = \dfrac{1}{x^2 - 4}$

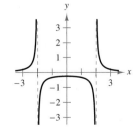

13. $f(x) = \dfrac{x^2 - 1}{x + 1}$

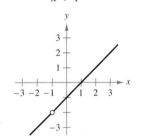

14. $f(x) = \dfrac{x^3 - 8}{x - 2}$

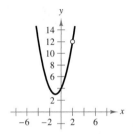

15. $f(x) = x^2 - 2x + 1$

16. $f(x) = 3 - 2x - x^2$

17. $f(x) = \dfrac{x}{x^2 - 1}$

18. $f(x) = \dfrac{x - 3}{x^2 - 9}$

19. $f(x) = \dfrac{x}{x^2 + 1}$

20. $f(x) = \dfrac{1}{x^2 + 1}$

21. $f(x) = \dfrac{x - 5}{x^2 - 9x + 20}$

22. $f(x) = \dfrac{x - 1}{x^2 + x - 2}$

23. $f(x) = [\![2x]\!] + 1$

24. $f(x) = \dfrac{[\![x]\!]}{2} + x$

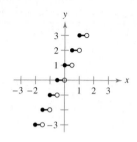

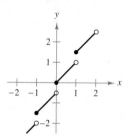

25. $f(x) = \begin{cases} -2x + 3, & x < 1 \\ x^2, & x \geq 1 \end{cases}$

26. $f(x) = \begin{cases} 3 + x, & x \leq 2 \\ x^2 + 1, & x > 2 \end{cases}$

27. $f(x) = \begin{cases} \frac{1}{2}x + 1, & x \leq 2 \\ 3 - x, & x > 2 \end{cases}$

28. $f(x) = \begin{cases} x^2 - 4, & x \leq 0 \\ 3x + 1, & x > 0 \end{cases}$

29. $f(x) = \dfrac{|x + 1|}{x + 1}$

30. $f(x) = \dfrac{|4 - x|}{4 - x}$

31. $f(x) = [\![x - 1]\!]$

32. $f(x) = x - [\![x]\!]$

33. $h(x) = f(g(x)), \quad f(x) = \dfrac{1}{\sqrt{x}}, \quad g(x) = x - 1, x > 1$

34. $h(x) = f(g(x)), \quad f(x) = \dfrac{1}{x - 1}, \quad g(x) = x^2 + 5$

In Exercises 35–38, discuss the continuity of the function on the closed interval. If there are any discontinuities, determine whether they are removable.

Function	Interval
35. $f(x) = x^2 - 4x - 5$	$[-1, 5]$
36. $f(x) = \dfrac{5}{x^2 + 1}$	$[-2, 2]$
37. $f(x) = \dfrac{1}{x - 2}$	$[1, 4]$
38. $f(x) = \dfrac{x}{x^2 - 4x + 3}$	$[0, 4]$

In Exercises 39–44, sketch the graph of the function and describe the interval(s) on which the function is continuous.

39. $f(x) = \dfrac{x^2 - 16}{x - 4}$

40. $f(x) = \dfrac{2x^2 + x}{x}$

41. $f(x) = \dfrac{x^3 + x}{x}$

42. $f(x) = \dfrac{x - 3}{4x^2 - 12x}$

43. $f(x) = \begin{cases} x^2 + 1, & x < 0 \\ x - 1, & x \geq 0 \end{cases}$

44. $f(x) = \begin{cases} x^2 - 4, & x \leq 0 \\ 2x + 4, & x > 0 \end{cases}$

In Exercises 45 and 46, find the constant a (Exercise 45) and the constants a and b (Exercise 46) such that the function is continuous on the entire real line.

45. $f(x) = \begin{cases} x^3, & x \leq 2 \\ ax^2, & x > 2 \end{cases}$

46. $f(x) = \begin{cases} 2, & x \leq -1 \\ ax + b, & -1 < x < 3 \\ -2, & x \geq 3 \end{cases}$

Ⓣ In Exercises 47–52, use a graphing utility to graph the function. Use the graph to determine any x-value(s) at which the function is not continuous. Explain why the function is not continuous at the x-value(s).

47. $h(x) = \dfrac{1}{x^2 - x - 2}$

48. $k(x) = \dfrac{x - 4}{x^2 - 5x + 4}$

49. $f(x) = \begin{cases} 2x - 4, & x \leq 3 \\ x^2 - 2x, & x > 3 \end{cases}$

50. $f(x) = \begin{cases} 3x - 1, & x \leq 1 \\ x + 1, & x > 1 \end{cases}$

51. $f(x) = x - 2[\![x]\!]$

52. $f(x) = [\![2x - 1]\!]$

In Exercises 53–56, describe the interval(s) on which the function is continuous.

53. $f(x) = \dfrac{x}{x^2 + 1}$

54. $f(x) = x\sqrt{x + 3}$

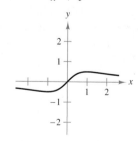

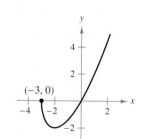

55. $f(x) = \dfrac{1}{2}[\![2x]\!]$

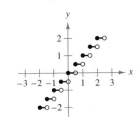

56. $f(x) = \dfrac{x+1}{\sqrt{x}}$

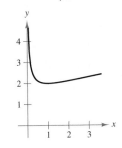

 Writing In Exercises 57 and 58, use a graphing utility to graph the function on the interval $[-4, 4]$. Does the graph of the function appear to be continuous on this interval? Is the function in fact continuous on $[-4, 4]$? Write a short paragraph about the importance of examining a function analytically as well as graphically.

57. $f(x) = \dfrac{x^2 + x}{x}$

58. $f(x) = \dfrac{x^3 - 8}{x - 2}$

59. Compound Interest A deposit of \$7500 is made in an account that pays 6% compounded quarterly. The amount A in the account after t years is

$$A = 7500(1.015)^{[\![4t]\!]}, \quad t \geq 0.$$

(a) Sketch the graph of A. Is the graph continuous? Explain your reasoning.

(b) What is the balance after 7 years?

60. Environmental Cost The cost C (in millions of dollars) of removing x percent of the pollutants emitted from the smokestack of a factory can be modeled by

$$C = \dfrac{2x}{100 - x}.$$

(a) What is the implied domain of C? Explain your reasoning.

 (b) Use a graphing utility to graph the cost function. Is the function continuous on its domain? Explain your reasoning.

(c) Find the cost of removing 75% of the pollutants from the smokestack.

 61. Consumer Awareness A shipping company's charge for sending an overnight package from New York to Atlanta is \$12.80 for the first pound and \$2.50 for each additional pound or fraction thereof. Use the greatest integer function to create a model for the charge C for overnight delivery of a package weighing x pounds. Use a graphing utility to graph the function, and discuss its continuity.

62. Consumer Awareness The United States Postal Service first class mail rates are \$0.41 for the first ounce and \$0.17 for each additional ounce or fraction thereof up to 3.5 ounces. A model for the cost C (in dollars) of a first class mailing that weighs 3.5 ounces or less is given below. *(Source: United States Postal Service)*

$$C(x) = \begin{cases} 0.41, & 0 \leq x \leq 1 \\ 0.58, & 1 < x \leq 2 \\ 0.75, & 2 < x \leq 3 \\ 0.92, & 3 < x \leq 3.5 \end{cases}$$

 (a) Use a graphing utility to graph the function and discuss its continuity. At what values is the function not continuous? Explain your reasoning.

(b) Find the cost of mailing a 2.5-ounce letter.

63. Salary Contract A union contract guarantees a 9% yearly increase for 5 years. For a current salary of \$28,500, the salaries for the next 5 years are given by

$$S = 28,500(1.09)^{[\![t]\!]}$$

where $t = 0$ represents the present year.

 (a) Use the greatest integer function of a graphing utility to graph the salary function, and discuss its continuity.

(b) Find the salary during the fifth year (when $t = 5$).

64. Inventory Management The number of units in inventory in a small company is

$$N = 25\left(2\left[\!\left[\dfrac{t+2}{2}\right]\!\right] - t\right), \quad 0 \leq t \leq 12$$

where the real number t is the time in months.

 (a) Use the greatest integer function of a graphing utility to graph this function, and discuss its continuity.

(b) How often must the company replenish its inventory?

65. Owning a Franchise You have purchased a franchise. You have determined a linear model for your revenue as a function of time. Is the model a continuous function? Would your actual revenue be a continuous function of time? Explain your reasoning.

66. Biology The gestation period of rabbits is about 29 to 35 days. Therefore, the population of a form (rabbits' home) can increase dramatically in a short period of time. The table gives the population of a form, where t is the time in months and N is the rabbit population.

t	0	1	2	3	4	5	6
N	2	8	10	14	10	15	12

 (a) Use a graphing utility to graph the population as a function of time.

(b) Find any points of discontinuity in the function. Explain your reasoning.

Section 7.3

The Derivative and the Slope of a Graph

- Identify tangent lines to a graph at a point.
- Approximate the slopes of tangent lines to graphs at points.
- Use the limit definition to find the slopes of graphs at points.
- Use the limit definition to find the derivatives of functions.
- Describe the relationship between differentiability and continuity.

Tangent Line to a Graph

Calculus is a branch of mathematics that studies rates of change of functions. In this course, you will learn that rates of change have many applications in real life. In Section 2.2, you learned how the slope of a line indicates the rate at which the line rises or falls. For a line, this rate (or slope) is the same at every point on the line. For graphs other than lines, the rate at which the graph rises or falls changes from point to point. For instance, in Figure 7.19, the parabola is rising more quickly at the point (x_1, y_1) than it is at the point (x_2, y_2). At the vertex (x_3, y_3), the graph levels off, and at the point (x_4, y_4), the graph is falling.

To determine the rate at which a graph rises or falls at a *single point*, you can find the slope of the **tangent line** at the point. In simple terms, the tangent line to the graph of a function f at a point $P(x_1, y_1)$ is the line that best approximates the graph at that point, as shown in Figure 7.19. Figure 7.20 shows other examples of tangent lines.

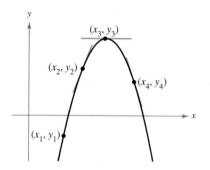

FIGURE 7.19 The slope of a nonlinear graph changes from one point to another.

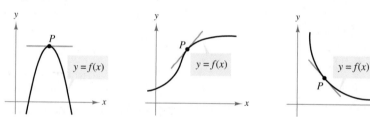

FIGURE 7.20 Tangent Line to a Graph at a Point

When Isaac Newton (1642–1727) was working on the "tangent line problem," he realized that it is difficult to define precisely what is meant by a tangent to a general curve. From geometry, you know that a line is tangent to a circle if the line intersects the circle at only one point, as shown in Figure 7.21. Tangent lines to a noncircular graph, however, can intersect the graph at more than one point. For instance, in the second graph in Figure 7.20, if the tangent line were extended, it would intersect the graph at a point other than the point of tangency. In this section, you will see how the notion of a limit can be used to define a general tangent line.

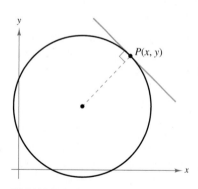

FIGURE 7.21 Tangent Line to a Circle

DISCOVERY

Use a graphing utility to graph $f(x) = 2x^3 - 4x^2 + 3x - 5$. On the same screen, sketch the graphs of $y = x - 5$, $y = 2x - 5$, and $y = 3x - 5$. Which of these lines, if any, appears to be tangent to the graph of f at the point $(0, -5)$? Explain your reasoning.

Slope of a Graph

Because a tangent line approximates the graph at a point, the problem of finding the slope of a graph at a point becomes one of finding the slope of the tangent line at the point.

Example 1   Approximating the Slope of a Graph

Use the graph in Figure 7.22 to approximate the slope of the graph of $f(x) = x^2$ at the point $(1, 1)$.

SOLUTION   From the graph of $f(x) = x^2$, you can see that the tangent line at $(1, 1)$ rises approximately two units for each unit change in x. So, the slope of the tangent line at $(1, 1)$ is given by

$$\text{Slope} = \frac{\text{change in } y}{\text{change in } x} \approx \frac{2}{1} = 2.$$

Because the tangent line at the point $(1, 1)$ has a slope of about 2, you can conclude that the graph has a slope of about 2 at the point $(1, 1)$.

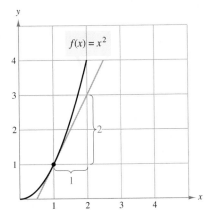

FIGURE 7.22

✓ **CHECKPOINT 1**

Use the graph to approximate the slope of the graph of $f(x) = x^3$ at the point $(1, 1)$.

STUDY TIP

When visually approximating the slope of a graph, note that the scales on the horizontal and vertical axes may differ. When this happens (as it frequently does in applications), the slope of the tangent line is distorted, and you must be careful to account for the difference in scales.

Example 2   Interpreting Slope

Figure 7.23 graphically depicts the average monthly temperature (in degrees Fahrenheit) in Duluth, Minnesota. Estimate the slope of this graph at the indicated point and give a physical interpretation of the result.   *(Source: National Oceanic and Atmospheric Administration)*

SOLUTION   From the graph, you can see that the tangent line at the given point falls approximately 28 units for each two-unit change in x. So, you can estimate the slope at the given point to be

$$\text{Slope} = \frac{\text{change in } y}{\text{change in } x} \approx \frac{-28}{2}$$

$$= -14 \text{ degrees per month.}$$

This means that you can expect the average daily temperatures in November to be about 14 degrees *lower* than the corresponding temperatures in October.

✓ **CHECKPOINT 2**

For which months do the slopes of the tangent lines appear to be positive? Negative? Interpret these slopes in the context of the problem.   ■

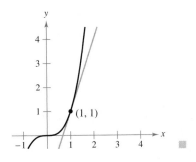

FIGURE 7.23

Slope and the Limit Process

In Examples 1 and 2, you approximated the slope of a graph at a point by making a careful graph and then "eyeballing" the tangent line at the point of tangency. A more precise method of approximating the slope of a tangent line makes use of a **secant line** through the point of tangency and a second point on the graph, as shown in Figure 7.24. If $(x, f(x))$ is the point of tangency and $(x + \Delta x, f(x + \Delta x))$ is a second point on the graph of f, then the slope of the secant line through the two points is

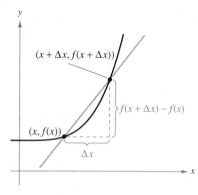

FIGURE 7.24 The Secant Line Through the Two Points $(x, f(x))$ and $(x + \Delta x, f(x + \Delta x))$

$$m_{\text{sec}} = \frac{f(x + \Delta x) - f(x)}{\Delta x}. \qquad \text{Slope of secant line}$$

The right side of this equation is called the **difference quotient.** The denominator Δx is the **change in x,** and the numerator is the **change in y.** The beauty of this procedure is that you obtain more and more accurate approximations of the slope of the tangent line by choosing points closer and closer to the point of tangency, as shown in Figure 7.25. Using the limit process, you can find the *exact* slope of the tangent line at $(x, f(x))$, which is also the slope of the graph of f at $(x, f(x))$.

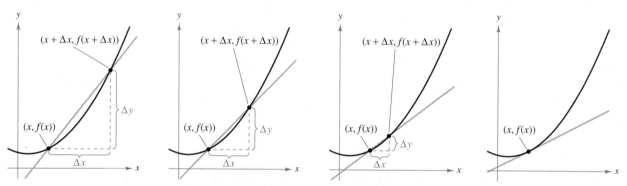

FIGURE 7.25   As Δx approaches 0, the secant lines approach the tangent line.

Definition of the Slope of a Graph

The **slope** m of the graph of f at the point $(x, f(x))$ is equal to the slope of its tangent line at $(x, f(x))$, and is given by

$$m = \lim_{\Delta x \to 0} m_{\text{sec}} = \lim_{\Delta x \to 0} \frac{f(x + \Delta x) - f(x)}{\Delta x}$$

provided this limit exists.

STUDY TIP

Δx is used as a variable to represent the change in x in the definition of the slope of a graph. Other variables may also be used. For instance, this definition is sometimes written as

$$m = \lim_{h \to 0} \frac{f(x + h) - f(x)}{h}.$$

Algebra Review

For help in evaluating the expressions in Examples 3–6, see the review of simplifying fractional expressions on page 617.

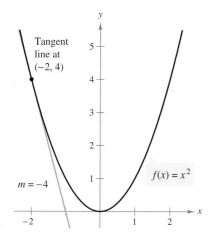

Tangent line at $(-2, 4)$

$m = -4$

$f(x) = x^2$

FIGURE 7.26

| Example 3 | **Finding Slope by the Limit Process** |

Find the slope of the graph of $f(x) = x^2$ at the point $(-2, 4)$.

SOLUTION   Begin by finding an expression that represents the slope of a secant line at the point $(-2, 4)$.

$$m_{sec} = \frac{f(-2 + \Delta x) - f(-2)}{\Delta x} \qquad \text{Set up difference quotient.}$$

$$= \frac{(-2 + \Delta x)^2 - (-2)^2}{\Delta x} \qquad \text{Use } f(x) = x^2.$$

$$= \frac{4 - 4\,\Delta x + (\Delta x)^2 - 4}{\Delta x} \qquad \text{Expand terms.}$$

$$= \frac{-4\,\Delta x + (\Delta x)^2}{\Delta x} \qquad \text{Simplify.}$$

$$= \frac{\Delta x(-4 + \Delta x)}{\Delta x} \qquad \text{Factor and divide out.}$$

$$= -4 + \Delta x, \quad \Delta x \neq 0 \qquad \text{Simplify.}$$

Next, take the limit of m_{sec} as $\Delta x \to 0$.

$$m = \lim_{\Delta x \to 0} m_{sec} = \lim_{\Delta x \to 0} (-4 + \Delta x) = -4$$

So, the graph of f has a slope of -4 at the point $(-2, 4)$, as shown in Figure 7.26.

✓ CHECKPOINT 3

Find the slope of the graph of $f(x) = x^2$ at the point $(2, 4)$. ■

| Example 4 | **Finding the Slope of a Graph** |

Find the slope of the graph of $f(x) = -2x + 4$.

SOLUTION   You know from your study of linear functions that the line given by $f(x) = -2x + 4$ has a slope of -2, as shown in Figure 7.27. This conclusion is consistent with the limit definition of slope.

$$m = \lim_{\Delta x \to 0} \frac{f(x + \Delta x) - f(x)}{\Delta x}$$

$$= \lim_{\Delta x \to 0} \frac{[-2(x + \Delta x) + 4] - [-2x + 4]}{\Delta x}$$

$$= \lim_{\Delta x \to 0} \frac{-2x - 2\,\Delta x + 4 + 2x - 4}{\Delta x}$$

$$= \lim_{\Delta x \to 0} \frac{-2\Delta x}{\Delta x} = -2$$

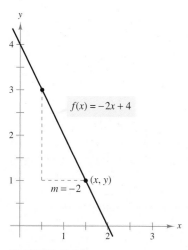

$f(x) = -2x + 4$

(x, y)

$m = -2$

FIGURE 7.27

✓ CHECKPOINT 4

Find the slope of the graph of $f(x) = 2x + 5$. ■

DISCOVERY

Use a graphing utility to graph the function $y_1 = x^2 + 1$ and the three lines $y_2 = 3x - 1$, $y_3 = 4x - 3$, and $y_4 = 5x - 5$. Which of these lines appears to be tangent to y_1 at the point $(2, 5)$? Confirm your answer by showing that the graphs of y_1 and its tangent line have only one point of intersection, whereas the graphs of y_1 and the other lines each have two points of intersection.

It is important that you see the distinction between the ways the difference quotients were set up in Examples 3 and 4. In Example 3, you were finding the slope of a graph at a specific point $(c, f(c))$. To find the slope, you can use the following form of a difference quotient.

$$m = \lim_{\Delta x \to 0} \frac{f(c + \Delta x) - f(c)}{\Delta x} \qquad \text{Slope at specific point}$$

In Example 4, however, you were finding a formula for the slope at *any* point on the graph. In such cases, you should use x, rather than c, in the difference quotient.

$$m = \lim_{\Delta x \to 0} \frac{f(x + \Delta x) - f(x)}{\Delta x} \qquad \text{Formula for slope}$$

Except for linear functions, this form will always produce a function of x, which can then be evaluated to find the slope at any desired point.

Example 5 Finding a Formula for the Slope of a Graph

Find a formula for the slope of the graph of $f(x) = x^2 + 1$. What are the slopes at the points $(-1, 2)$ and $(2, 5)$?

SOLUTION

$$
\begin{aligned}
m_{\text{sec}} &= \frac{f(x + \Delta x) - f(x)}{\Delta x} & \text{Set up difference quotient.}\\[2mm]
&= \frac{[(x + \Delta x)^2 + 1] - (x^2 + 1)}{\Delta x} & \text{Use } f(x) = x^2 + 1.\\[2mm]
&= \frac{x^2 + 2x\,\Delta x + (\Delta x)^2 + 1 - x^2 - 1}{\Delta x} & \text{Expand terms.}\\[2mm]
&= \frac{2x\,\Delta x + (\Delta x)^2}{\Delta x} & \text{Simplify.}\\[2mm]
&= \frac{\cancel{\Delta x}(2x + \Delta x)}{\cancel{\Delta x}} & \text{Factor and divide out.}\\[2mm]
&= 2x + \Delta x, \quad \Delta x \neq 0 & \text{Simplify.}
\end{aligned}
$$

Next, take the limit of m_{sec} as $\Delta x \to 0$.

$$
\begin{aligned}
m &= \lim_{\Delta x \to 0} m_{\text{sec}}\\
&= \lim_{\Delta x \to 0} (2x + \Delta x)\\
&= 2x
\end{aligned}
$$

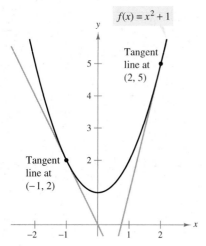

$f(x) = x^2 + 1$

Tangent line at $(2, 5)$

Tangent line at $(-1, 2)$

FIGURE 7.28

Using the formula $m = 2x$, you can find the slopes at the specified points. At $(-1, 2)$ the slope is $m = 2(-1) = -2$, and at $(2, 5)$ the slope is $m = 2(2) = 4$. The graph of f is shown in Figure 7.28.

✓ CHECKPOINT 5

Find a formula for the slope of the graph of $f(x) = 4x^2 + 1$. What are the slopes at the points $(0, 1)$ and $(1, 5)$?

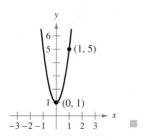

STUDY TIP

The slope of the graph of $f(x) = x^2 + 1$ varies for different values of x. For what value of x is the slope equal to 0?

The Derivative of a Function

In Example 5, you started with the function $f(x) = x^2 + 1$ and used the limit process to derive another function, $m = 2x$, that represents the slope of the graph of f at the point $(x, f(x))$. This derived function is called the **derivative** of f at x. It is denoted by $f'(x)$, which is read as "f prime of x."

STUDY TIP

The notation dy/dx is read as "the derivative of y with respect to x," and using limit notation, you can write

$$\frac{dy}{dx} = \lim_{\Delta x \to 0} \frac{\Delta y}{\Delta x}$$

$$= \lim_{\Delta x \to 0} \frac{f(x + \Delta x) - f(x)}{\Delta x}$$

$$= f'(x).$$

Definition of the Derivative

The **derivative of f at x** is given by

$$f'(x) = \lim_{\Delta x \to 0} \frac{f(x + \Delta x) - f(x)}{\Delta x}$$

provided this limit exists. A function is **differentiable** at x if its derivative exists at x. The process of finding derivatives is called **differentiation**.

In addition to $f'(x)$, other notations can be used to denote the derivative of $y = f(x)$. The most common are

$$\frac{dy}{dx}, \quad y', \quad \frac{d}{dx}[f(x)], \quad \text{and} \quad D_x[y].$$

Example 6 Finding a Derivative

Find the derivative of $f(x) = 3x^2 - 2x$.

SOLUTION

$$f'(x) = \lim_{\Delta x \to 0} \frac{f(x + \Delta x) - f(x)}{\Delta x}$$

$$= \lim_{\Delta x \to 0} \frac{[3(x + \Delta x)^2 - 2(x + \Delta x)] - (3x^2 - 2x)}{\Delta x}$$

$$= \lim_{\Delta x \to 0} \frac{3x^2 + 6x \Delta x + 3(\Delta x)^2 - 2x - 2 \Delta x - 3x^2 + 2x}{\Delta x}$$

$$= \lim_{\Delta x \to 0} \frac{6x \Delta x + 3(\Delta x)^2 - 2 \Delta x}{\Delta x}$$

$$= \lim_{\Delta x \to 0} \frac{\Delta x(6x + 3 \Delta x - 2)}{\Delta x}$$

$$= \lim_{\Delta x \to 0} (6x + 3 \Delta x - 2)$$

$$= 6x - 2$$

So, the derivative of $f(x) = 3x^2 - 2x$ is $f'(x) = 6x - 2$.

✓ CHECKPOINT 6

Find the derivative of $f(x) = x^2 - 5x$. ■

In many applications, it is convenient to use a variable other than x as the independent variable. Example 7 shows a function that uses t as the independent variable.

TECHNOLOGY

(T) You can use a graphing utility to confirm the result given in Example 7. One way to do this is to choose a point on the graph of $y = 2/t$, such as $(1, 2)$, and find the equation of the tangent line at that point. Using the derivative found in the example, you know that the slope of the tangent line when $t = 1$ is $m = -2$. This means that the tangent line at the point $(1, 2)$ is

$$y - y_1 = m(t - t_1)$$

$$y - 2 = -2(t - 1) \text{ or}$$

$$y = -2t + 4.$$

By graphing $y = 2/t$ and $y = -2t + 4$ in the same viewing window, as shown below, you can confirm that the line is tangent to the graph at the point $(1, 2)$.*

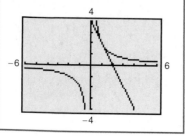

Example 7   Finding a Derivative

Find the derivative of y with respect to t for the function

$$y = \frac{2}{t}.$$

SOLUTION   Consider $y = f(t)$, and use the limit process as shown.

$$\frac{dy}{dt} = \lim_{\Delta t \to 0} \frac{f(t + \Delta t) - f(t)}{\Delta t} \qquad \text{Set up difference quotient.}$$

$$= \lim_{\Delta t \to 0} \frac{\dfrac{2}{t + \Delta t} - \dfrac{2}{t}}{\Delta t} \qquad \text{Use } f(t) = 2/t.$$

$$= \lim_{\Delta t \to 0} \frac{\dfrac{2t - 2t - 2\,\Delta t}{t(t + \Delta t)}}{\Delta t} \qquad \text{Expand terms.}$$

$$= \lim_{\Delta t \to 0} \frac{-2\,\Delta t}{t(\Delta t)(t + \Delta t)} \qquad \text{Factor and divide out.}$$

$$= \lim_{\Delta t \to 0} \frac{-2}{t(t + \Delta t)} \qquad \text{Simplify.}$$

$$= -\frac{2}{t^2} \qquad \text{Evaluate the limit.}$$

So, the derivative of y with respect to t is

$$\frac{dy}{dt} = -\frac{2}{t^2}.$$

Remember that the derivative of a function gives you a formula for finding the slope of the tangent line at any point on the graph of the function. For example, the slope of the tangent line to the graph of f at the point $(1, 2)$ is given by

$$f'(1) = -\frac{2}{1^2} = -2.$$

To find the slopes of the graph at other points, substitute the t-coordinate of the point into the derivative, as shown below.

Point	t-Coordinate	Slope
$(2, 1)$	$t = 2$	$m = f'(2) = -\dfrac{2}{2^2} = -\dfrac{1}{2}$
$(-2, -1)$	$t = -2$	$m = f'(-2) = -\dfrac{2}{(-2)^2} = -\dfrac{1}{2}$

✓**CHECKPOINT 7**

Find the derivative of y with respect to t for the function $y = 4/t$. ∎

---

*Specific calculator keystroke instructions for operations in this and other technology boxes can be found at *college.hmco.com/info/larsonapplied*.

Differentiability and Continuity

Not every function is differentiable. Figure 7.29 shows some common situations in which a function will not be differentiable at a point—vertical tangent lines, discontinuities, and sharp turns in the graph. Each of the functions shown in Figure 7.29 is differentiable at every value of x *except* $x = 0$.

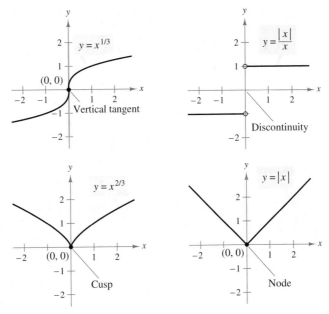

FIGURE 7.29   Functions That Are Not Differentiable at $x = 0$

In Figure 7.29, you can see that all but one of the functions are continuous at $x = 0$ but none are differentiable there. This shows that continuity is not a strong enough condition to guarantee differentiability. On the other hand, if a function is differentiable at a point, then it must be continuous at that point. This important result is stated in the following theorem.

Differentiability Implies Continuity

If a function f is differentiable at $x = c$, then f is continuous at $x = c$.

> ### CONCEPT CHECK
>
> 1. What is the name of the line that best approximates the slope of a graph at a point?
> 2. What is the name of a line through the point of tangency and a second point on the graph?
> 3. Sketch a graph of a function whose derivative is always negative.
> 4. Sketch a graph of a function whose derivative is always positive.

| **Skills Review 7.3** | The following warm-up exercises involve skills that were covered in earlier sections. You will use these skills in the exercise set for this section. For additional help, review Sections 2.2, 2.4, and 7.1. |

In Exercises 1–3, find an equation of the line containing P and Q.

1. $P(2, 1)$, $Q(2, 4)$ **2.** $P(2, 2)$, $Q(-5, 2)$ **3.** $P(2, 0)$, $Q(3, -1)$

In Exercises 4–7, find the limit.

4. $\lim\limits_{\Delta x \to 0} \dfrac{2x\Delta x + (\Delta x)^2}{\Delta x}$

5. $\lim\limits_{\Delta x \to 0} \dfrac{3x^2\Delta x + 3x(\Delta x)^2 + (\Delta x)^3}{\Delta x}$

6. $\lim\limits_{\Delta x \to 0} \dfrac{1}{x(x + \Delta x)}$

7. $\lim\limits_{\Delta x \to 0} \dfrac{(x + \Delta x)^2 - x^2}{\Delta x}$

In Exercises 8–10, find the domain of the function.

8. $f(x) = \dfrac{1}{x - 1}$ **9.** $f(x) = \dfrac{1}{5}x^3 - 2x^2 + \dfrac{1}{3}x - 1$ **10.** $f(x) = \dfrac{6x}{x^3 + x}$

Exercises 7.3

See www.CalcChat.com for worked-out solutions to odd-numbered exercises.

In Exercises 1–4, trace the graph and sketch the tangent lines at (x_1, y_1) and (x_2, y_2).

1.

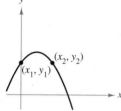

2.

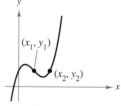

3.

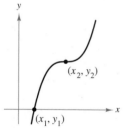

4.

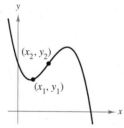

In Exercises 5–10, estimate the slope of the graph at the point (x, y). (Each square on the grid is 1 unit by 1 unit.)

5.

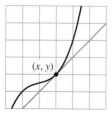

6.

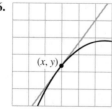

7.

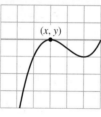

8.

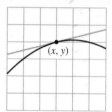

9.

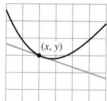

10.

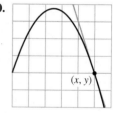

11. Revenue The graph represents the revenue R (in millions of dollars per year) for Polo Ralph Lauren from 1999 through 2005, where t represents the year, with $t = 9$ corresponding to 1999. Estimate the slopes of the graph for the years 2002 and 2004. *(Source: Polo Ralph Lauren Corp.)*

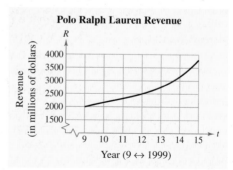

12. Sales The graph represents the sales S (in millions of dollars per year) for Scotts Miracle-Gro Company from 1999 through 2005, where t represents the year, with $t = 9$ corresponding to 1999. Estimate the slopes of the graph for the years 2001 and 2004. *(Source: Scotts Miracle-Gro Company)*

Scotts Miracle-Gro Company

13. Consumer Trends The graph shows the number of visitors V to a national park in hundreds of thousands during a one-year period, where $t = 1$ corresponds to January. Estimate the slopes of the graph at $t = 1$, 8, and 12.

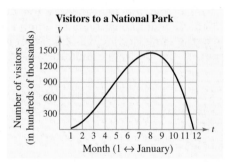

Visitors to a National Park

14. Athletics Two long distance runners starting out side by side begin a 10,000-meter run. Their distances are given by $s = f(t)$ and $s = g(t)$, where s is measured in thousands of meters and t is measured in minutes.

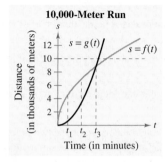

10,000-Meter Run

(a) Which runner is running faster at t_1?

(b) What conclusion can you make regarding their rates at t_2?

(c) What conclusion can you make regarding their rates at t_3?

(d) Which runner finishes the race first? Explain.

In Exercises 15–24, use the limit definition to find the slope of the tangent line to the graph of f at the given point.

15. $f(x) = 6 - 2x$; $(2, 2)$ **16.** $f(x) = 2x + 4$; $(1, 6)$

17. $f(x) = -1$; $(0, -1)$ **18.** $f(x) = 6$; $(-2, 6)$

19. $f(x) = x^2 - 1$; $(2, 3)$ **20.** $f(x) = 4 - x^2$; $(2, 0)$

21. $f(x) = x^3 - x$; $(2, 6)$

22. $f(x) = x^3 + 2x$; $(1, 3)$

23. $f(x) = 2\sqrt{x}$; $(4, 4)$

24. $f(x) = \sqrt{x + 1}$; $(8, 3)$

In Exercises 25–38, use the limit definition to find the derivative of the function.

25. $f(x) = 3$ **26.** $f(x) = -2$

27. $f(x) = -5x$ **28.** $f(x) = 4x + 1$

29. $g(s) = \frac{1}{3}s + 2$ **30.** $h(t) = 6 - \frac{1}{2}t$

31. $f(x) = x^2 - 4$ **32.** $f(x) = 1 - x^2$

33. $h(t) = \sqrt{t - 1}$ **34.** $f(x) = \sqrt{x + 2}$

35. $f(t) = t^3 - 12t$ **36.** $f(t) = t^3 + t^2$

37. $f(x) = \dfrac{1}{x + 2}$ **38.** $g(s) = \dfrac{1}{s - 1}$

In Exercises 39–46, use the limit definition to find an equation of the tangent line to the graph of f at the given point. Then verify your results by using a graphing utility to graph the function and its tangent line at the point.

39. $f(x) = \frac{1}{2}x^2$; $(2, 2)$ **40.** $f(x) = -x^2$; $(-1, -1)$

41. $f(x) = (x - 1)^2$; $(-2, 9)$ **42.** $f(x) = 2x^2 - 1$; $(0, -1)$

43. $f(x) = \sqrt{x} + 1$; $(4, 3)$ **44.** $f(x) = \sqrt{x + 2}$; $(7, 3)$

45. $f(x) = \dfrac{1}{x}$; $(1, 1)$ **46.** $f(x) = \dfrac{1}{x - 1}$; $(2, 1)$

In Exercises 47–50, find an equation of the line that is tangent to the graph of f and parallel to the given line.

Function	*Line*
47. $f(x) = -\frac{1}{4}x^2$	$x + y = 0$
48. $f(x) = x^2 + 1$	$2x + y = 0$
49. $f(x) = -\frac{1}{2}x^3$	$6x + y + 4 = 0$
50. $f(x) = x^2 - x$	$x + 2y - 6 = 0$

In Exercises 51–58, describe the x-values at which the function is differentiable. Explain your reasoning.

51. $y = |x + 3|$

52. $y = |x^2 - 9|$

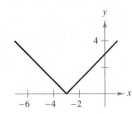

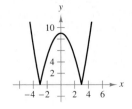

53. $y = (x - 3)^{2/3}$

54. $y = x^{2/5}$

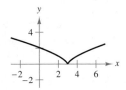

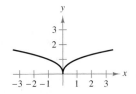

55. $y = \sqrt{x - 1}$

56. $y = \dfrac{x^2}{x^2 - 4}$

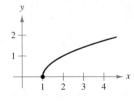

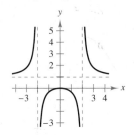

57. $y = \begin{cases} x^3 + 3, & x < 0 \\ x^3 - 3, & x \geq 0 \end{cases}$

58. $y = \begin{cases} x^2, & x \leq 1 \\ -x^2, & x > 1 \end{cases}$

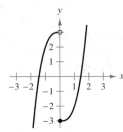

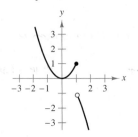

In Exercises 59 and 60, describe the x-values at which f is differentiable.

59. $f(x) = \dfrac{1}{x - 1}$

60. $f(x) = \begin{cases} x^2 - 3, & x \leq 0 \\ 3 - x^2, & x > 0 \end{cases}$

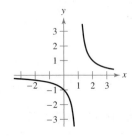

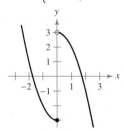

In Exercises 61 and 62, identify a function f that has the given characteristics. Then sketch the function.

61. $f(0) = 2; f'(x) = -3, -\infty < x < \infty$

62. $f(-2) = f(4) = 0; f'(1) = 0, f'(x) < 0$ for $x < 1; f'(x) > 0$ for $x > 1$

(T) **Graphical, Numerical, and Analytic Analysis**   In Exercises 63–66, use a graphing utility to graph f on the interval $[-2, 2]$. Complete the table by graphically estimating the slopes of the graph at the given points. Then evaluate the slopes analytically and compare your results with those obtained graphically.

x	-2	$-\frac{3}{2}$	-1	$-\frac{1}{2}$	0	$\frac{1}{2}$	1	$\frac{3}{2}$	2
$f(x)$									
$f'(x)$									

63. $f(x) = \frac{1}{4}x^3$

64. $f(x) = \frac{1}{2}x^2$

65. $f(x) = -\frac{1}{2}x^3$

66. $f(x) = -\frac{3}{2}x^2$

(T) In Exercises 67–70, find the derivative of the given function f. Then use a graphing utility to graph f and its derivative in the same viewing window. What does the x-intercept of the derivative indicate about the graph of f?

67. $f(x) = x^2 - 4x$

68. $f(x) = 2 + 6x - x^2$

69. $f(x) = x^3 - 3x$

70. $f(x) = x^3 - 6x^2$

True or False?   In Exercises 71–74, determine whether the statement is true or false. If it is false, explain why or give an example that shows it is false.

71. The slope of the graph of $y = x^2$ is different at every point on the graph of f.

72. If a function is continuous at a point, then it is differentiable at that point.

73. If a function is differentiable at a point, then it is continuous at that point.

74. A tangent line to a graph can intersect the graph at more than one point.

(T) **75. Writing**   Use a graphing utility to graph the two functions $f(x) = x^2 + 1$ and $g(x) = |x| + 1$ in the same viewing window. Use the *zoom* and *trace* features to analyze the graphs near the point $(0, 1)$. What do you observe? Which function is differentiable at this point? Write a short paragraph describing the geometric significance of differentiability at a point.

Section 7.4

Some Rules for Differentiation

- Find the derivatives of functions using the Constant Rule.
- Find the derivatives of functions using the Power Rule.
- Find the derivatives of functions using the Constant Multiple Rule.
- Find the derivatives of functions using the Sum and Difference Rules.
- Use derivatives to answer questions about real-life situations.

The Constant Rule

In Section 7.3, you found derivatives by the limit process. This process is tedious, even for simple functions, but fortunately there are rules that greatly simplify differentiation. These rules allow you to calculate derivatives without the *direct* use of limits.

The Constant Rule

The derivative of a constant function is zero. That is,

$$\frac{d}{dx}[c] = 0, \quad c \text{ is a constant.}$$

PROOF   Let $f(x) = c$. Then, by the limit definition of the derivative, you can write

$$f'(x) = \lim_{\Delta x \to 0} \frac{f(x + \Delta x) - f(x)}{\Delta x} = \lim_{\Delta x \to 0} \frac{c - c}{\Delta x} = \lim_{\Delta x \to 0} 0 = 0.$$

So, $\frac{d}{dx}[c] = 0$.

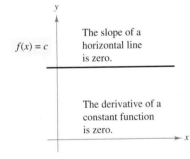

$f(x) = c$

The slope of a horizontal line is zero.

The derivative of a constant function is zero.

FIGURE 7.30

STUDY TIP

Note in Figure 7.30 that the Constant Rule is equivalent to saying that the slope of a horizontal line is zero.

STUDY TIP

An interpretation of the Constant Rule says that the tangent line to a constant function is the function itself. Find an equation of the tangent line to $f(x) = -4$ at $x = 3$.

Example 1   Finding Derivatives of Constant Functions

a. $\frac{d}{dx}[7] = 0$

b. If $f(x) = 0$, then $f'(x) = 0$.

c. If $y = 2$, then $\frac{dy}{dx} = 0$.

d. If $g(t) = -\frac{3}{2}$, then $g'(t) = 0$.

✓ **CHECKPOINT 1**

Find the derivative of each function.

a. $f(x) = -2$ **b.** $y = \pi$ **c.** $g(w) = \sqrt{5}$ **d.** $s(t) = 320.5$ ■

The Power Rule

STUDY TIP

For more information on binomial expansions, see Section 16.6.

The binomial expansion process is used to prove the Power Rule.

$$(x + \Delta x)^2 = x^2 + 2x \, \Delta x + (\Delta x)^2$$

$$(x + \Delta x)^3 = x^3 + 3x^2 \, \Delta x + 3x(\Delta x)^2 + (\Delta x)^3$$

$$(x + \Delta x)^n = x^n + nx^{n-1} \, \Delta x + \underbrace{\frac{n(n-1)x^{n-2}}{2}(\Delta x)^2 + \cdots + (\Delta x)^n}_{(\Delta x)^2 \text{ is a factor of these terms.}}$$

The (Simple) Power Rule

$$\frac{d}{dx}[x^n] = nx^{n-1}, \qquad n \text{ is any real number.}$$

PROOF   We prove only the case in which n is a positive integer. Let $f(x) = x^n$. Using the binomial expansion, you can write

$$f'(x) = \lim_{\Delta x \to 0} \frac{f(x + \Delta x) - f(x)}{\Delta x} \qquad \text{Definition of derivative}$$

$$= \lim_{\Delta x \to 0} \frac{(x + \Delta x)^n - x^n}{\Delta x}$$

$$= \lim_{\Delta x \to 0} \frac{x^n + nx^{n-1} \, \Delta x + \dfrac{n(n-1)x^{n-2}}{2}(\Delta x)^2 + \cdots + (\Delta x)^n - x^n}{\Delta x}$$

$$= \lim_{\Delta x \to 0} \left[nx^{n-1} + \frac{n(n-1)x^{n-2}}{2}(\Delta x) + \cdots + (\Delta x)^{n-1} \right]$$

$$= nx^{n-1} + 0 + \cdots + 0 = nx^{n-1}.$$

For the Power Rule, the case in which $n = 1$ is worth remembering as a separate differentiation rule. That is,

$$\frac{d}{dx}[x] = 1. \qquad \text{The derivative of } x \text{ is 1.}$$

This rule is consistent with the fact that the slope of the line given by $y = x$ is 1. (See Figure 7.31.)

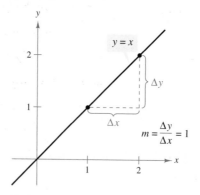

FIGURE 7.31   The slope of the line $y = x$ is 1.

Example 2   **Applying the Power Rule**

Find the derivative of each function.

	Function	*Derivative*
a.	$f(x) = x^3$	$f'(x) = 3x^2$
b.	$y = \dfrac{1}{x^2} = x^{-2}$	$\dfrac{dy}{dx} = (-2)x^{-3} = -\dfrac{2}{x^3}$
c.	$g(t) = t$	$g'(t) = 1$
d.	$R = x^4$	$\dfrac{dR}{dx} = 4x^3$

✓**CHECKPOINT 2**

Find the derivative of each function.

a. $f(x) = x^4$ **b.** $y = \dfrac{1}{x^3}$

c. $g(w) = w^2$ **d.** $s(t) = \dfrac{1}{t}$ ∎

In Example 2(b), note that *before* differentiating, you should rewrite $1/x^2$ as x^{-2}. Rewriting is the first step in *many* differentiation problems.

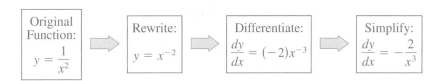

Remember that the derivative of a function f is another function that gives the slope of the graph of f at any point at which f is differentiable. So, you can use the derivative to find slopes, as shown in Example 3.

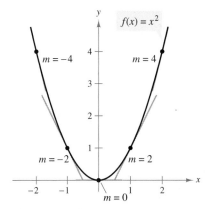

FIGURE 7.32

Example 3   **Finding the Slope of a Graph**

Find the slopes of the graph of

$$f(x) = x^2 \qquad \text{Original function}$$

when $x = -2, -1, 0, 1,$ and 2.

SOLUTION   Begin by using the Power Rule to find the derivative of f.

$$f'(x) = 2x \qquad \text{Derivative}$$

You can use the derivative to find the slopes of the graph of f, as shown.

x-Value	*Slope of Graph of f*
$x = -2$	$m = f'(-2) = 2(-2) = -4$
$x = -1$	$m = f'(-1) = 2(-1) = -2$
$x = 0$	$m = f'(0) = 2(0) = 0$
$x = 1$	$m = f'(1) = 2(1) = 2$
$x = 2$	$m = f'(2) = 2(2) = 4$

The graph of f is shown in Figure 7.32.

✓**CHECKPOINT 3**

Find the slopes of the graph of $f(x) = x^3$ when $x = -1, 0,$ and 1.

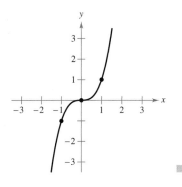

The Constant Multiple Rule

To prove the Constant Multiple Rule, the following property of limits is used.

$$\lim_{x \to a} cg(x) = c\left[\lim_{x \to a} g(x)\right]$$

The Constant Multiple Rule

If f is a differentiable function of x, and c is a real number, then

$$\frac{d}{dx}[cf(x)] = cf'(x), \qquad c \text{ is a constant.}$$

PROOF   Apply the definition of the derivative to produce

$$\frac{d}{dx}[cf(x)] = \lim_{\Delta x \to 0} \frac{cf(x + \Delta x) - cf(x)}{\Delta x} \qquad \text{Definition of derivative}$$

$$= \lim_{\Delta x \to 0} c\left[\frac{f(x + \Delta x) - f(x)}{\Delta x}\right]$$

$$= c\left[\lim_{\Delta x \to 0} \frac{f(x + \Delta x) - f(x)}{\Delta x}\right] = cf'(x).$$

Informally, the Constant Multiple Rule states that constants can be factored out of the differentiation process.

$$\frac{d}{dx}[cf(x)] = c\frac{d}{dx}[(\ \ )f(x)] = cf'(x)$$

The usefulness of this rule is often overlooked, especially when the constant appears in the denominator, as shown below.

$$\frac{d}{dx}\left[\frac{f(x)}{c}\right] = \frac{d}{dx}\left[\frac{1}{c}f(x)\right] = \frac{1}{c}\left(\frac{d}{dx}[(\ \ )f(x)]\right) = \frac{1}{c}f'(x)$$

To use the Constant Multiple Rule efficiently, look for constants that can be factored out *before* differentiating. For example,

$$\frac{d}{dx}[5x^2] = 5\frac{d}{dx}[x^2] \qquad \text{Factor out 5.}$$

$$= 5(2x) \qquad \text{Differentiate.}$$

$$= 10x \qquad \text{Simplify.}$$

and

$$\frac{d}{dx}\left[\frac{x^2}{5}\right] = \frac{1}{5}\left(\frac{d}{dx}[x^2]\right) \qquad \text{Factor out } \frac{1}{5}.$$

$$= \frac{1}{5}(2x) \qquad \text{Differentiate.}$$

$$= \frac{2}{5}x. \qquad \text{Simplify.}$$

Example 4   Using the Power and Constant Multiple Rules

Differentiate each function.

a. $y = 2x^{1/2}$ **b.** $f(t) = \dfrac{4t^2}{5}$

SOLUTION

a. Using the Constant Multiple Rule and the Power Rule, you can write

$$\frac{dy}{dx} = \frac{d}{dx}[2x^{1/2}] = 2\underbrace{\frac{d}{dx}[x^{1/2}]}_{\text{Constant Multiple Rule}} = 2\underbrace{\left(\frac{1}{2}x^{-1/2}\right)}_{\text{Power Rule}} = x^{-1/2} = \frac{1}{\sqrt{x}}.$$

b. Begin by rewriting $f(t)$ as

$$f(t) = \frac{4t^2}{5} = \frac{4}{5}t^2.$$

Then, use the Constant Multiple Rule and the Power Rule to obtain

$$f'(t) = \frac{d}{dt}\left[\frac{4}{5}t^2\right] = \frac{4}{5}\left[\frac{d}{dt}(t^2)\right] = \frac{4}{5}(2t) = \frac{8}{5}t.$$

✓ **CHECKPOINT 4**

Differentiate each function.

a. $y = 4x^2$

b. $f(x) = 16x^{1/2}$ ▪

You may find it helpful to combine the Constant Multiple Rule and the Power Rule into one combined rule.

$$\frac{d}{dx}[cx^n] = cnx^{n-1}, \quad n \text{ is a real number, } c \text{ is a constant.}$$

For instance, in Example 4(b), you can apply this combined rule to obtain

$$\frac{d}{dt}\left[\frac{4}{5}t^2\right] = \left(\frac{4}{5}\right)(2)(t) = \frac{8}{5}t.$$

The three functions in the next example are simple, yet errors are frequently made in differentiating functions involving constant multiples of the first power of x. Keep in mind that

$$\frac{d}{dx}[cx] = c, \quad c \text{ is a constant.}$$

Example 5   Applying the Constant Multiple Rule

Find the derivative of each function.

✓ **CHECKPOINT 5**

Find the derivative of each function.

a. $y = \dfrac{t}{4}$

b. $y = -\dfrac{2x}{5}$ ▪

Original Function	*Derivative*
a. $y = -\dfrac{3x}{2}$	$y' = -\dfrac{3}{2}$
b. $y = 3\pi x$	$y' = 3\pi$
c. $y = -\dfrac{x}{2}$	$y' = -\dfrac{1}{2}$

Parentheses can play an important role in the use of the Constant Multiple Rule and the Power Rule. In Example 6, be sure you understand the mathematical conventions involving the use of parentheses.

Example 6   Using Parentheses When Differentiating

Find the derivative of each function.

a. $y = \dfrac{5}{2x^3}$ **b.** $y = \dfrac{5}{(2x)^3}$ **c.** $y = \dfrac{7}{3x^{-2}}$ **d.** $y = \dfrac{7}{(3x)^{-2}}$

SOLUTION

Function	Rewrite	Differentiate	Simplify
a. $y = \dfrac{5}{2x^3}$	$y = \dfrac{5}{2}(x^{-3})$	$y' = \dfrac{5}{2}(-3x^{-4})$	$y' = -\dfrac{15}{2x^4}$
b. $y = \dfrac{5}{(2x)^3}$	$y = \dfrac{5}{8}(x^{-3})$	$y' = \dfrac{5}{8}(-3x^{-4})$	$y' = -\dfrac{15}{8x^4}$
c. $y = \dfrac{7}{3x^{-2}}$	$y = \dfrac{7}{3}(x^2)$	$y' = \dfrac{7}{3}(2x)$	$y' = \dfrac{14x}{3}$
d. $y = \dfrac{7}{(3x)^{-2}}$	$y = 63(x^2)$	$y' = 63(2x)$	$y' = 126x$

✓CHECKPOINT 6

Find the derivative of each function.

a. $y = \dfrac{9}{4x^2}$ **b.** $y = \dfrac{9}{(4x)^2}$ ■

Example 7   Differentiating Radical Functions

Find the derivative of each function.

a. $y = \sqrt{x}$ **b.** $y = \dfrac{1}{2\sqrt[3]{x^2}}$ **c.** $y = \sqrt{2x}$

SOLUTION

Function	Rewrite	Differentiate	Simplify
a. $y = \sqrt{x}$	$y = x^{1/2}$	$y' = \left(\dfrac{1}{2}\right)x^{-1/2}$	$y' = \dfrac{1}{2\sqrt{x}}$
b. $y = \dfrac{1}{2\sqrt[3]{x^2}}$	$y = \dfrac{1}{2}x^{-2/3}$	$y' = \dfrac{1}{2}\left(-\dfrac{2}{3}\right)x^{-5/3}$	$y' = -\dfrac{1}{3x^{5/3}}$
c. $y = \sqrt{2x}$	$y = \sqrt{2}(x^{1/2})$	$y' = \sqrt{2}\left(\dfrac{1}{2}\right)x^{-1/2}$	$y' = \dfrac{1}{\sqrt{2x}}$

STUDY TIP

When differentiating functions involving radicals, you should rewrite the function with rational exponents. For instance, you should rewrite $y = \sqrt[3]{x}$ as $y = x^{1/3}$, and you should rewrite

$$y = \dfrac{1}{\sqrt[3]{x^4}} \quad \text{as } y = x^{-4/3}.$$

✓CHECKPOINT 7

Find the derivative of each function.

a. $y = \sqrt{5x}$

b. $y = \sqrt[3]{x}$ ■

The Sum and Difference Rules

The next two rules are ones that you might expect to be true, and you may have used them without thinking about it. For instance, if you were asked to differentiate $y = 3x + 2x^3$, you would probably write

$$y' = 3 + 6x^2$$

without questioning your answer. The validity of differentiating a sum or difference of functions term by term is given by the Sum and Difference Rules.

The Sum and Difference Rules

The derivative of the sum or difference of two differentiable functions is the sum or difference of their derivatives.

$$\frac{d}{dx}[f(x) + g(x)] = f'(x) + g'(x) \qquad \text{Sum Rule}$$

$$\frac{d}{dx}[f(x) - g(x)] = f'(x) - g'(x) \qquad \text{Difference Rule}$$

PROOF   Let $h(x) = f(x) + g(x)$. Then, you can prove the Sum Rule as shown.

$$h'(x) = \lim_{\Delta x \to 0} \frac{h(x + \Delta x) - h(x)}{\Delta x} \qquad \text{Definition of derivative}$$

$$= \lim_{\Delta x \to 0} \frac{f(x + \Delta x) + g(x + \Delta x) - f(x) - g(x)}{\Delta x}$$

$$= \lim_{\Delta x \to 0} \frac{f(x + \Delta x) - f(x) + g(x + \Delta x) - g(x)}{\Delta x}$$

$$= \lim_{\Delta x \to 0} \left[\frac{f(x + \Delta x) - f(x)}{\Delta x} + \frac{g(x + \Delta x) - g(x)}{\Delta x} \right]$$

$$= \lim_{\Delta x \to 0} \frac{f(x + \Delta x) - f(x)}{\Delta x} + \lim_{\Delta x \to 0} \frac{g(x + \Delta x) - g(x)}{\Delta x}$$

$$= f'(x) + g'(x)$$

So,

$$\frac{d}{dx}[f(x) + g(x)] = f'(x) + g'(x).$$

The Difference Rule can be proved in a similar manner.

The Sum and Difference Rules can be extended to the sum or difference of any finite number of functions. For instance, if $y = f(x) + g(x) + h(x)$, then $y' = f'(x) + g'(x) + h'(x)$.

STUDY TIP

Look back at Example 6 on page 563. Notice that the example asks for the derivative of the difference of two functions. Verify this result by using the Difference Rule.

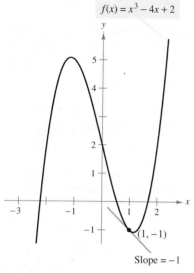

$f(x) = x^3 - 4x + 2$

(1, −1)

Slope = −1

FIGURE 7.33

With the four differentiation rules listed in this section, you can differentiate *any* polynomial function.

Example 8   Using the Sum and Difference Rules

Find the slope of the graph of $f(x) = x^3 - 4x + 2$ at the point $(1, -1)$.

SOLUTION   The derivative of $f(x)$ is

$$f'(x) = 3x^2 - 4.$$

So, the slope of the graph of f at $(1, -1)$ is

$$\text{Slope} = f'(1) = 3(1)^2 - 4 = -1$$

as shown in Figure 7.33.

✓ **CHECKPOINT 8**

Find the slope of the graph of $f(x) = x^2 - 5x + 1$ at the point $(2, -5)$. ▪

Example 8 illustrates the use of the derivative for determining the shape of a graph. A rough sketch of the graph of $f(x) = x^3 - 4x + 2$ might lead you to think that the point $(1, -1)$ is a minimum point of the graph. After finding the slope at this point to be -1, however, you can conclude that the minimum point (where the slope is 0) is farther to the right. (You will study techniques for finding minimum and maximum points in Section 8.5.)

Example 9   Using the Sum and Difference Rules

Find an equation of the tangent line to the graph of

$$g(x) = -\frac{1}{2}x^4 + 3x^3 - 2x$$

at the point $\left(-1, -\frac{3}{2}\right)$.

SOLUTION   The derivative of $g(x)$ is $g'(x) = -2x^3 + 9x^2 - 2$, which implies that the slope of the graph at the point $\left(-1, -\frac{3}{2}\right)$ is

$$\begin{aligned} \text{Slope} = g'(-1) &= -2(-1)^3 + 9(-1)^2 - 2 \\ &= 2 + 9 - 2 \\ &= 9 \end{aligned}$$

as shown in Figure 7.34. Using the point-slope form, you can write the equation of the tangent line at $\left(-1, -\frac{3}{2}\right)$ as shown.

$$y - \left(-\frac{3}{2}\right) = 9[x - (-1)]$$ Point-slope form

$$y = 9x + \frac{15}{2}$$ Equation of tangent line

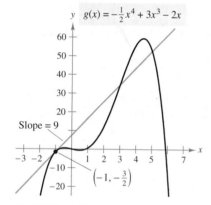

$g(x) = -\frac{1}{2}x^4 + 3x^3 - 2x$

Slope = 9

$\left(-1, -\frac{3}{2}\right)$

FIGURE 7.34

✓ **CHECKPOINT 9**

Find an equation of the tangent line to the graph of $f(x) = -x^2 + 3x - 2$ at the point $(2, 0)$. ▪

Application

Example 10 Modeling Revenue

From 2000 through 2005, the revenue R (in millions of dollars per year) for Microsoft Corporation can be modeled by

$$R = -110.194t^3 + 993.98t^2 + 1155.6t + 23,036, \quad 0 \le t \le 5$$

where t represents the year, with $t = 0$ corresponding to 2000. At what rate was Microsoft's revenue changing in 2001?   *(Source: Microsoft Corporation)*

SOLUTION   One way to answer this question is to find the derivative of the revenue model with respect to time.

$$\frac{dR}{dt} = -330.582t^2 + 1987.96t + 1155.6, \quad 0 \le t \le 5$$

In 2001 (when $t = 1$), the rate of change of the revenue with respect to time is given by

$$-330.582(1)^2 + 1987.96(1) + 1155.6 \approx 2813.$$

Because R is measured in millions of dollars and t is measured in years, it follows that the derivative dR/dt is measured in millions of dollars per year. So, at the end of 2001, Microsoft's revenue was increasing at a rate of about \$2813 million per year, as shown in Figure 7.35.

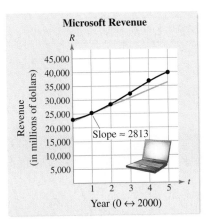

Microsoft Revenue

Slope ≈ 2813

FIGURE 7.35

✓ CHECKPOINT 10

From 1998 through 2005, the revenue per share R (in dollars) for McDonald's Corporation can be modeled by

$$R = 0.0598t^2 - 0.379t + 8.44, \quad 8 \le t \le 15$$

where t represents the year, with $t = 8$ corresponding to 1998. At what rate was McDonald's revenue per share changing in 2003?   *(Source: McDonald's Corporation)* ▪

CONCEPT CHECK

1. What is the derivative of any constant function?
2. Write a verbal description of the Power Rule.
3. According to the Sum Rule, the derivative of the sum of two differentiable functions is equal to what?
4. According to the Difference Rule, the derivative of the difference of two differentiable functions is equal to what?

Skills Review 7.4

The following warm-up exercises involve skills that were covered in earlier sections. You will use these skills in the exercise set for this section. For additional help, review Sections 0.3, 0.6, 1.3, and 1.5.

In Exercises 1 and 2, evaluate each expression when $x = 2$.

1. (a) $2x^2$ (b) $(2x)^2$ (c) $2x^{-2}$

2. (a) $\dfrac{1}{(3x)^2}$ (b) $\dfrac{1}{4x^3}$ (c) $\dfrac{(2x)^{-3}}{4x^{-2}}$

In Exercises 3–6, simplify the expression.

3. $4(3)x^3 + 2(2)x$

4. $\frac{1}{2}(3)x^2 - \frac{3}{2}x^{1/2}$

5. $\left(\frac{1}{4}\right)x^{-3/4}$

6. $\frac{1}{3}(3)x^2 - 2\left(\frac{1}{2}\right)x^{-1/2} + \frac{1}{3}x^{-2/3}$

In Exercises 7–10, solve the equation.

7. $3x^2 + 2x = 0$

8. $x^3 - x = 0$

9. $x^2 + 8x - 20 = 0$

10. $x^2 - 10x - 24 = 0$

Exercises 7.4

See www.CalcChat.com for worked-out solutions to odd-numbered exercises.

In Exercises 1–4, find the slope of the tangent line to $y = x^n$ at the point $(1, 1)$.

1. (a) $y = x^2$ (b) $y = x^{1/2}$

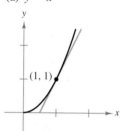 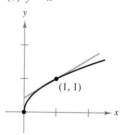

2. (a) $y = x^{3/2}$ (b) $y = x^3$

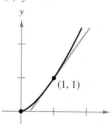 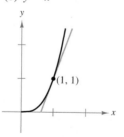

3. (a) $y = x^{-1}$ (b) $y = x^{-1/3}$

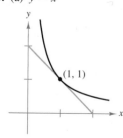 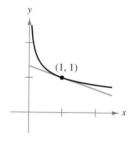

4. (a) $y = x^{-1/2}$ (b) $y = x^{-2}$

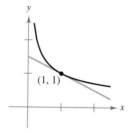 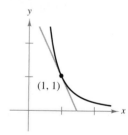

In Exercises 5–22, find the derivative of the function.

5. $y = 3$ **6.** $f(x) = -2$

7. $y = x^4$ **8.** $h(x) = 2x^5$

9. $f(x) = 4x + 1$ **10.** $g(x) = 3x - 1$

11. $g(x) = x^2 + 5x$ **12.** $y = t^2 - 6$

13. $f(t) = -3t^2 + 2t - 4$

14. $y = x^3 - 9x^2 + 2$

15. $s(t) = t^3 - 2t + 4$

16. $y = 2x^3 - x^2 + 3x - 1$

17. $y = 4t^{4/3}$

18. $h(x) = x^{5/2}$

19. $f(x) = 4\sqrt{x}$

20. $g(x) = 4\sqrt[3]{x} + 2$

21. $y = 4x^{-2} + 2x^2$

22. $s(t) = 4t^{-1} + 1$

In Exercises 23–28, use Example 6 as a model to find the derivative.

Function	Rewrite	Differentiate	Simplify
23. $y = \dfrac{1}{x^3}$			
24. $y = \dfrac{2}{3x^2}$			
25. $y = \dfrac{1}{(4x)^3}$			
26. $y = \dfrac{\pi}{(3x)^2}$			
27. $y = \dfrac{\sqrt{x}}{x}$			
28. $y = \dfrac{4x}{x^{-3}}$			

In Exercises 29–34, find the value of the derivative of the function at the given point.

Function	Point
29. $f(x) = \dfrac{1}{x}$	$(1, 1)$
30. $f(t) = 4 - \dfrac{4}{3t}$	$\left(\dfrac{1}{2}, \dfrac{4}{3}\right)$
31. $f(x) = -\dfrac{1}{2}x(1 + x^2)$	$(1, -1)$
32. $y = 3x\left(x^2 - \dfrac{2}{x}\right)$	$(2, 18)$
33. $y = (2x + 1)^2$	$(0, 1)$
34. $f(x) = 3(5 - x)^2$	$(5, 0)$

In Exercises 35–48, find $f'(x)$.

35. $f(x) = x^2 - \dfrac{4}{x} - 3x^{-2}$

36. $f(x) = x^2 - 3x - 3x^{-2} + 5x^{-3}$

37. $f(x) = x^2 - 2x - \dfrac{2}{x^4}$
38. $f(x) = x^2 + 4x + \dfrac{1}{x}$

39. $f(x) = x(x^2 + 1)$
40. $f(x) = (x^2 + 2x)(x + 1)$

41. $f(x) = (x + 4)(2x^2 - 1)$

42. $f(x) = (3x^2 - 5x)(x^2 + 2)$

43. $f(x) = \dfrac{2x^3 - 4x^2 + 3}{x^2}$
44. $f(x) = \dfrac{2x^2 - 3x + 1}{x}$

45. $f(x) = \dfrac{4x^3 - 3x^2 + 2x + 5}{x^2}$

46. $f(x) = \dfrac{-6x^3 + 3x^2 - 2x + 1}{x}$

47. $f(x) = x^{4/5} + x$
48. $f(x) = x^{1/3} - 1$

(T) In Exercises 49–52, (a) find an equation of the tangent line to the graph of the function at the given point, (b) use a graphing utility to graph the function and its tangent line at the point, and (c) use the *derivative* feature of a graphing utility to confirm your results.

Function	Point
49. $y = -2x^4 + 5x^2 - 3$	$(1, 0)$
50. $y = x^3 + x$	$(-1, -2)$
51. $f(x) = \sqrt[3]{x} + \sqrt[5]{x}$	$(1, 2)$
52. $f(x) = \dfrac{1}{\sqrt[3]{x^2}} - x$	$(-1, 2)$

In Exercises 53–56, determine the point(s), if any, at which the graph of the function has a horizontal tangent line.

53. $y = -x^4 + 3x^2 - 1$
54. $y = x^3 + 3x^2$

55. $y = \dfrac{1}{2}x^2 + 5x$
56. $y = x^2 + 2x$

In Exercises 57 and 58, (a) sketch the graphs of f and g, (b) find $f'(1)$ and $g'(1)$, (c) sketch the tangent line to each graph when $x = 1$, and (d) explain the relationship between f' and g'.

57. $f(x) = x^3$
   $g(x) = x^3 + 3$

58. $f(x) = x^2$
   $g(x) = 3x^2$

59. Use the Constant Rule, the Constant Multiple Rule, and the Sum Rule to find $h'(1)$ given that $f'(1) = 3$.

(a) $h(x) = f(x) - 2$ (b) $h(x) = 2f(x)$

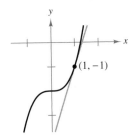
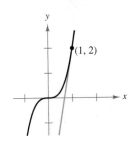

(c) $h(x) = -f(x)$ (d) $h(x) = -1 + 2f(x)$

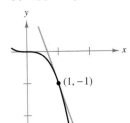
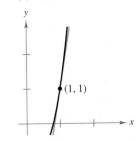

60. Revenue The revenue R (in millions of dollars per year) for Polo Ralph Lauren from 1999 through 2005 can be modeled by

$$R = 0.59221t^4 - 18.0042t^3 + 175.293t^2 - 316.42t - 116.5$$

where t is the year, with $t = 9$ corresponding to 1999. *(Source: Polo Ralph Lauren Corp.)*

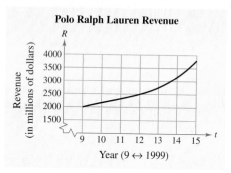

Polo Ralph Lauren Revenue

(a) Find the slopes of the graph for the years 2002 and 2004.

(b) Compare your results with those obtained in Exercise 11 in Section 7.3.

(c) What are the units for the slope of the graph? Interpret the slope of the graph in the context of the problem.

61. Sales The sales S (in millions of dollars per year) for Scotts Miracle-Gro Company from 1999 through 2005 can be modeled by

$$S = -1.29242t^4 + 69.9530t^3 - 1364.615t^2 + 11,511.47t - 33,932.9$$

where t is the year, with $t = 9$ corresponding to 1999. *(Source: Scotts Miracle-Gro Company)*

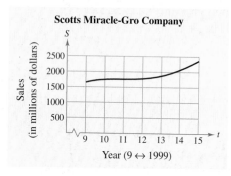

Scotts Miracle-Gro Company

(a) Find the slopes of the graph for the years 2001 and 2004.

(b) Compare your results with those obtained in Exercise 12 in Section 7.3.

(c) What are the units for the slope of the graph? Interpret the slope of the graph in the context of the problem.

62. Cost The variable cost for manufacturing an electrical component is $7.75 per unit, and the fixed cost is $500. Write the cost C as a function of x, the number of units produced. Show that the derivative of this cost function is a constant and is equal to the variable cost.

63. Political Fundraiser A politician raises funds by selling tickets to a dinner for $500. The politician pays $150 for each dinner and has fixed costs of $7000 to rent a dining hall and wait staff. Write the profit P as a function of x, the number of dinners sold. Show that the derivative of the profit function is a constant and is equal to the increase in profit from each dinner sold.

Ⓑ **64. Psychology: Migraine Prevalence** The graph illustrates the prevalence of migraine headaches in males and females in selected income groups. *(Source: Adapted from Sue/Sue/Sue, Understanding Abnormal Behavior, Seventh Edition)*

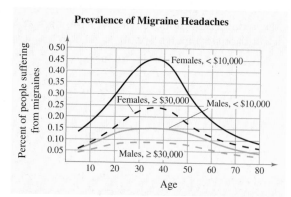

Prevalence of Migraine Headaches

(a) Write a short paragraph describing your general observations about the prevalence of migraines in females and males with respect to age group and income bracket.

(b) Describe the graph of the derivative of each curve, and explain the significance of each derivative. Include an explanation of the units of the derivatives, and indicate the time intervals in which the derivatives would be positive and negative.

Ⓣ In Exercises 65 and 66, use a graphing utility to graph f and f' over the given interval. Determine any points at which the graph of f has horizontal tangents.

Function	Interval
65. $f(x) = 4.1x^3 - 12x^2 + 2.5x$	$[0, 3]$
66. $f(x) = x^3 - 1.4x^2 - 0.96x + 1.44$	$[-2, 2]$

True or False? In Exercises 67 and 68, determine whether the statement is true or false. If it is false, explain why or give an example that shows it is false.

67. If $f'(x) = g'(x)$, then $f(x) = g(x)$.

68. If $f(x) = g(x) + c$, then $f'(x) = g'(x)$.

Mid-Chapter Quiz

See www.CalcChat.com for worked-out solutions to odd-numbered exercises.

Take this test as you would take a test in class. When you are done, check your work against the answers given in the back of the book.

In Exercises 1–6, find the limit (if it exists).

1. $\lim\limits_{x \to 2} (5x + 4)$

2. $\lim\limits_{x \to 3^-} \sqrt{x + 1}$

3. $\lim\limits_{x \to -3} \dfrac{x + 1}{x + 3}$

4. $\lim\limits_{x \to 2} \dfrac{x^2 + 3x - 10}{x - 2}$

5. $\lim\limits_{x \to 0} \dfrac{4 - \sqrt{x + 16}}{x}$

6. $\lim\limits_{x \to 0^+} [\![x]\!]$

In Exercises 7–10, describe the interval(s) on which the function is continuous. Explain why the function is continuous on the interval(s). If the function has a discontinuity at a point, identify all conditions of continuity that are not satisfied.

7. $f(x) = \dfrac{x}{x^2 + 2}$

8. $f(x) = \dfrac{x}{(x + 2)^2}$

9. $f(x) = \dfrac{x + 3}{x^2 + 2x - 3}$

10. $f(x) = \begin{cases} x^2, & x < 0 \\ x^3, & x \geq 0 \end{cases}$

In Exercises 11 and 12, use the limit definition to find the derivative of the function. Then find the slope of the tangent line to the graph of f at the given point.

11. $f(x) = -x + 2; (2, 0)$ **12.** $f(x) = \dfrac{4}{x}; (1, 4)$

In Exercises 13–18, find the derivative of the function.

13. $f(x) = 12$

14. $f(x) = 19x + 9$

15. $f(x) = 5 - 3x^2$

16. $f(x) = 12x^{1/4}$

17. $f(x) = 4x^{-2}$

18. $f(x) = 2\sqrt{x}$

(T) In Exercises 19 and 20, find an equation of the tangent line to the graph of f at the given point. Then use a graphing utility to graph the function and the equation of the tangent line in the same viewing window.

19. $f(x) = 5x^2 + 6x - 1; (-1, -2)$

20. $f(x) = x^{4/3} + x; (0, 0)$

21. From 2000 through 2005, the sales per share S (in dollars) for CVS Corporation can be modeled by

$$S = 0.18390t^3 - 0.8242t^2 + 3.492t + 25.60, \; 0 \leq t \leq 5$$

where t represents the year, with $t = 0$ corresponding to 2000. *(Source: CVS Corporation)*

(a) Find the rate of change of the sales per share with respect to the year.

(b) At what rate were the sales per share changing in 2001? in 2004? in 2005?

Section 7.5

Rates of Change: Velocity and Marginals

- Find the average rates of change of functions over intervals.
- Find the instantaneous rates of change of functions at points.
- Find the marginal revenues, marginal costs, and marginal profits for products.

Average Rate of Change

In Sections 7.3 and 7.4, you studied the two primary applications of derivatives.

1. **Slope** The derivative of f is a function that gives the slope of the graph of f at a point $(x, f(x))$.

2. **Rate of Change** The derivative of f is a function that gives the rate of change of $f(x)$ with respect to x at the point $(x, f(x))$.

In this section, you will see that there are many real-life applications of rates of change. A few are velocity, acceleration, population growth rates, unemployment rates, production rates, and water flow rates. Although rates of change often involve change with respect to time, you can investigate the rate of change of one variable with respect to any other related variable.

When determining the rate of change of one variable with respect to another, you must be careful to distinguish between *average* and *instantaneous* rates of change. The distinction between these two rates of change is comparable to the distinction between the slope of the secant line through two points on a graph and the slope of the tangent line at one point on the graph.

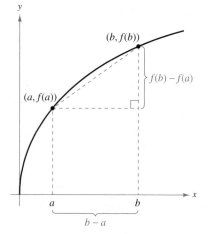

FIGURE 7.36

Definition of Average Rate of Change

If $y = f(x)$, then the **average rate of change** of y with respect to x on the interval $[a, b]$ is

$$\text{Average rate of change} = \frac{f(b) - f(a)}{b - a}$$

$$= \frac{\Delta y}{\Delta x}.$$

Note that $f(a)$ is the value of the function at the *left* endpoint of the interval, $f(b)$ is the value of the function at the *right* endpoint of the interval, and $b - a$ is the width of the interval, as shown in Figure 7.36.

STUDY TIP

In real-life problems, it is important to list the units of measure for a rate of change. The units for $\Delta y / \Delta x$ are "y-units" per "x-units." For example, if y is measured in miles and x is measured in hours, then $\Delta y / \Delta x$ is measured in *miles per hour.*

Example 1 **Medicine**

The concentration C (in milligrams per milliliter) of a drug in a patient's bloodstream is monitored over 10-minute intervals for 2 hours, where t is measured in minutes, as shown in the table. Find the average rate of change over each interval.

a. $[0, 10]$ **b.** $[0, 20]$ **c.** $[100, 110]$

t	0	10	20	30	40	50	60	70	80	90	100	110	120
C	0	2	17	37	55	73	89	103	111	113	113	103	68

STUDY TIP

In Example 1, the average rate of change is positive when the concentration increases and negative when the concentration decreases, as shown in Figure 7.37.

SOLUTION

a. For the interval $[0, 10]$, the average rate of change is

Value of C at right endpoint
Value of C at left endpoint

$$\frac{\Delta C}{\Delta t} = \frac{2 - 0}{10 - 0} = \frac{2}{10} = 0.2 \text{ milligram per milliliter per minute.}$$

Width of interval

b. For the interval $[0, 20]$, the average rate of change is

$$\frac{\Delta C}{\Delta t} = \frac{17 - 0}{20 - 0} = \frac{17}{20} = 0.85 \text{ milligram per milliliter per minute.}$$

c. For the interval $[100, 110]$, the average rate of change is

$$\frac{\Delta C}{\Delta t} = \frac{103 - 113}{110 - 100} = \frac{-10}{10} = -1 \text{ milligram per milliliter per minute.}$$

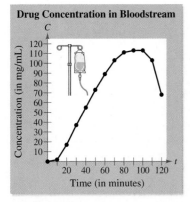

FIGURE 7.37

✓CHECKPOINT 1

Use the table in Example 1 to find the average rate of change over each interval.

a. $[0, 120]$ **b.** $[90, 100]$ **c.** $[90, 120]$ ■

The rates of change in Example 1 are in milligrams per milliliter per minute because the concentration is measured in milligrams per milliliter and the time is measured in minutes.

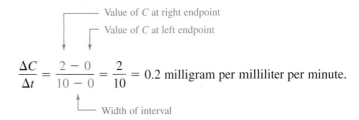

Concentration is measured in milligrams per milliliter.

Rate of change is measured in milligrams per milliliter per minute.

$$\frac{\Delta C}{\Delta t} = \frac{2 - 0}{10 - 0} = \frac{2}{10} = 0.2 \text{ milligram per milliliter per minute}$$

Time is measured in minutes.

A common application of an average rate of change is to find the **average velocity** of an object that is moving in a straight line. That is,

$$\text{Average velocity} = \frac{\text{change in distance}}{\text{change in time}}.$$

This formula is demonstrated in Example 2.

Example 2 Finding an Average Velocity

If a free-falling object is dropped from a height of 100 feet, and *air resistance is neglected*, the height h (in feet) of the object at time t (in seconds) is given by

$$h = -16t^2 + 100. \qquad \text{(See Figure 7.38.)}$$

Find the average velocity of the object over each interval.

a. $[1, 2]$ **b.** $[1, 1.5]$ **c.** $[1, 1.1]$

SOLUTION You can use the position equation $h = -16t^2 + 100$ to determine the heights at $t = 1$, $t = 1.1$, $t = 1.5$, and $t = 2$, as shown in the table.

t (in seconds)	0	1	1.1	1.5	2
h (in feet)	100	84	80.64	64	36

a. For the interval $[1, 2]$, the object falls from a height of 84 feet to a height of 36 feet. So, the average velocity is

$$\frac{\Delta h}{\Delta t} = \frac{36 - 84}{2 - 1} = \frac{-48}{1} = -48 \text{ feet per second.}$$

b. For the interval $[1, 1.5]$, the average velocity is

$$\frac{\Delta h}{\Delta t} = \frac{64 - 84}{1.5 - 1} = \frac{-20}{0.5} = -40 \text{ feet per second.}$$

c. For the interval $[1, 1.1]$, the average velocity is

$$\frac{\Delta h}{\Delta t} = \frac{80.64 - 84}{1.1 - 1} = \frac{-3.36}{0.1} = -33.6 \text{ feet per second.}$$

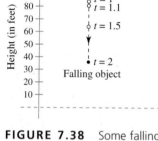

h

100
90
80
70
60
50
40
30
20
10

Height (in feet)

$t = 0$
$t = 1$
$t = 1.1$
$t = 1.5$

$t = 2$
Falling object

FIGURE 7.38 Some falling objects have considerable air resistance. Other falling objects have negligible air resistance. When modeling a falling-body problem, you must decide whether to account for air resistance or neglect it.

✓ **CHECKPOINT 2**

The height h (in feet) of a free-falling object at time t (in seconds) is given by $h = -16t^2 + 180$. Find the average velocity of the object over each interval.

a. $[0, 1]$ **b.** $[1, 2]$ **c.** $[2, 3]$ ▣

STUDY TIP

In Example 2, the average velocities are negative because the object is moving downward.

Instantaneous Rate of Change and Velocity

Suppose in Example 2 you wanted to find the rate of change of h at the instant $t = 1$ second. Such a rate is called an **instantaneous rate of change.** You can approximate the instantaneous rate of change at $t = 1$ by calculating the average rate of change over smaller and smaller intervals of the form $[1, 1 + \Delta t]$, as shown in the table. From the table, it seems reasonable to conclude that the instantaneous rate of change of the height when $t = 1$ is -32 feet per second.

Δt approaches 0.

Δt	1	0.5	0.1	0.01	0.001	0.0001	0
$\dfrac{\Delta h}{\Delta t}$	-48	-40	-33.6	-32.16	-32.016	-32.0016	-32

$\dfrac{\Delta h}{\Delta t}$ approaches -32.

STUDY TIP

The limit in this definition is the same as the limit in the definition of the derivative of f at x. This is the second major interpretation of the derivative—as an *instantaneous rate of change in one variable with respect to another.* Recall that the first interpretation of the derivative is as the slope of the graph of f at x.

Definition of Instantaneous Rate of Change

The **instantaneous rate of change** (or simply **rate of change**) of $y = f(x)$ at x is the limit of the average rate of change on the interval $[x, x + \Delta x]$, as Δx approaches 0.

$$\lim_{\Delta x \to 0} \frac{\Delta y}{\Delta x} = \lim_{\Delta x \to 0} \frac{f(x + \Delta x) - f(x)}{\Delta x}$$

If y is a distance and x is time, then the rate of change is a **velocity.**

Example 3   Finding an Instantaneous Rate of Change

Find the velocity of the object in Example 2 when $t = 1$.

SOLUTION   From Example 2, you know that the height of the falling object is given by

$h = -16t^2 + 100.$ Position function

By taking the derivative of this position function, you obtain the velocity function.

$h'(t) = -32t$ Velocity function

The velocity function gives the velocity at *any* time. So, when $t = 1$, the velocity is

$h'(1) = -32(1)$

$= -32$ feet per second.

✓ CHECKPOINT 3

Find the velocities of the object in Checkpoint 2 when $t = 1.75$ and $t = 2$. ■

The general **position function** for a free-falling object, neglecting air resistance, is

$$h = -16t^2 + v_0 t + h_0 \qquad \text{Position function}$$

where h is the height (in feet), t is the time (in seconds), v_0 is the initial velocity (in feet per second), and h_0 is the initial height (in feet). Remember that the model assumes that positive velocities indicate upward motion and negative velocities indicate downward motion. The derivative $h' = -32t + v_0$ is the **velocity function.** The absolute value of the velocity is the **speed** of the object.

Example 4   Finding the Velocity of a Diver

At time $t = 0$, a diver jumps from a diving board that is 32 feet high, as shown in Figure 7.39. Because the diver's initial velocity is 16 feet per second, his position function is

$$h = -16t^2 + 16t + 32. \qquad \text{Position function}$$

a. When does the diver hit the water?

b. What is the diver's velocity at impact?

SOLUTION

a. To find the time at which the diver hits the water, let $h = 0$ and solve for t.

$$
\begin{aligned}
-16t^2 + 16t + 32 &= 0 && \text{Set } h \text{ equal to } 0. \\
-16(t^2 - t - 2) &= 0 && \text{Factor out common factor.} \\
-16(t + 1)(t - 2) &= 0 && \text{Factor.} \\
t = -1 \text{ or } t &= 2 && \text{Solve for } t.
\end{aligned}
$$

The solution $t = -1$ does not make sense in the problem because it would mean the diver hits the water 1 second before he jumps. So, you can conclude that the diver hits the water when $t = 2$ seconds.

b. The velocity at time t is given by the derivative

$$h' = -32t + 16. \qquad \text{Velocity function}$$

The velocity at time $t = 2$ is $-32(2) + 16 = -48$ feet per second.

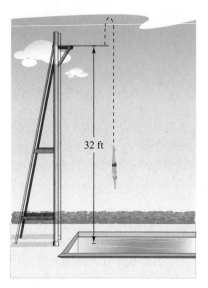

FIGURE 7.39

✓ CHECKPOINT 4

Give the position function of a diver who jumps from a board 12 feet high with initial velocity 16 feet per second. Then find the diver's velocity function.

In Example 4, note that the diver's initial velocity is $v_0 = 16$ feet per second (upward) and his initial height is $h_0 = 32$ feet.

Initial velocity is 16 feet per second.

Initial height is 32 feet.

$$h = -16t^2 + 16t + 32$$

Rates of Change in Economics: Marginals

Another important use of rates of change is in the field of economics. Economists refer to *marginal profit*, *marginal revenue*, and *marginal cost* as the rates of change of the profit, revenue, and cost with respect to the number x of units produced or sold. An equation that relates these three quantities is

$$P = R - C$$

where P, R, and C represent the following quantities.

P = total profit

R = total revenue

and

C = total cost

The derivatives of these quantities are called the **marginal profit, marginal revenue,** and **marginal cost,** respectively.

$$\frac{dP}{dx} = \text{marginal profit}$$

$$\frac{dR}{dx} = \text{marginal revenue}$$

$$\frac{dC}{dx} = \text{marginal cost}$$

In many business and economics problems, the number of units produced or sold is restricted to positive integer values, as indicated in Figure 7.40(a). (Of course, it could happen that a sale involves half or quarter units, but it is hard to conceive of a sale involving $\sqrt{2}$ units.) The variable that denotes such units is called a **discrete variable.** To analyze a function of a discrete variable x, you can temporarily assume that x is a **continuous variable** and is able to take on any real value in a given interval, as indicated in Figure 7.40(b). Then, you can use the methods of calculus to find the x-value that corresponds to the marginal revenue, maximum profit, minimum cost, or whatever is called for. Finally, you should round the solution to the nearest sensible x-value—cents, dollars, units, or days, depending on the context of the problem.

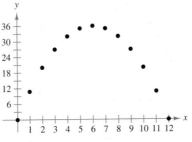

(a) Function of a Discrete Variable

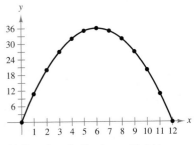

(b) Function of a Continuous Variable

FIGURE 7.40

Example 5   **Finding the Marginal Profit**

The profit derived from selling x units of an alarm clock is given by

$$P = 0.0002x^3 + 10x.$$

a. Find the marginal profit for a production level of 50 units.

b. Compare this with the actual gain in profit obtained by increasing the production level from 50 to 51 units.

SOLUTION

a. Because the profit is $P = 0.0002x^3 + 10x$, the marginal profit is given by the derivative

$$dP/dx = 0.0006x^2 + 10.$$

When $x = 50$, the marginal profit is

$$0.0006(50)^2 + 10 = 1.5 + 10$$
$$= \$11.50 \text{ per unit.} \qquad \text{Marginal profit for } x = 50$$

b. For $x = 50$, the actual profit is

$$P = (0.0002)(50)^3 + 10(50) \qquad \text{Substitute 50 for } x.$$
$$= 25 + 500$$
$$= \$525.00 \qquad \text{Actual profit for } x = 50$$

and for $x = 51$, the actual profit is

$$P = (0.0002)(51)^3 + 10(51) \qquad \text{Substitute 51 for } x.$$
$$\approx 26.53 + 510$$
$$= \$536.53. \qquad \text{Actual profit for } x = 51$$

So, the additional profit obtained by increasing the production level from 50 to 51 units is

$$536.53 - 525.00 = \$11.53. \qquad \text{Extra profit for one unit}$$

Note that the actual profit increase of \$11.53 (when x increases from 50 to 51 units) can be approximated by the marginal profit of \$11.50 per unit (when $x = 50$), as shown in Figure 7.41.

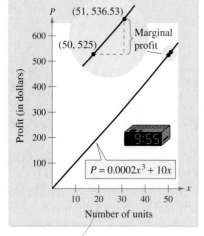

Marginal Profit

FIGURE 7.41

✓ CHECKPOINT 5

Use the profit function in Example 5 to find the marginal profit for a production level of 100 units. Compare this with the actual gain in profit by increasing production from 100 to 101 units. ∎

STUDY TIP

The reason the marginal profit gives a good approximation of the actual change in profit is that the graph of P is nearly straight over the interval $50 \le x \le 51$. You will study more about the use of marginals to approximate actual changes in Section 9.5.

The profit function in Example 5 is unusual in that the profit continues to increase as long as the number of units sold increases. In practice, it is more common to encounter situations in which sales can be increased only by lowering the price per item. Such reductions in price will ultimately cause the profit to decline.

The number of units x that consumers are willing to purchase at a given price per unit p is given by the **demand function**

$$p = f(x).$$ Demand function

The total revenue R is then related to the price per unit and the quantity demanded (or sold) by the equation

$$R = xp.$$ Revenue function

Example 6   Finding a Demand Function

A business sells 2000 items per month at a price of $10 each. It is estimated that monthly sales will increase 250 units for each $0.25 reduction in price. Use this information to find the demand function and total revenue function.

SOLUTION   From the given estimate, x increases 250 units each time p drops $0.25 from the original cost of $10. This is described by the equation

$$x = 2000 + 250\left(\frac{10 - p}{0.25}\right)$$

$$= 2000 + 10{,}000 - 1000p$$

$$= 12{,}000 - 1000p.$$

Solving for p in terms of x produces

$$p = 12 - \frac{x}{1000}.$$ Demand function

This, in turn, implies that the revenue function is

$$R = xp$$ Formula for revenue

$$= x\left(12 - \frac{x}{1000}\right)$$

$$= 12x - \frac{x^2}{1000}.$$ Revenue function

The graph of the demand function is shown in Figure 7.42. Notice that as the price decreases, the quantity demanded increases.

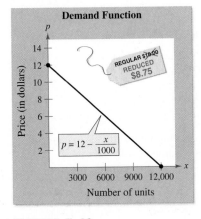

FIGURE 7.42

✓ CHECKPOINT 6

Find the demand function in Example 6 if monthly sales increase 200 units for each $0.10 reduction in price. ■

TECHNOLOGY

Modeling a Demand Function

(T) To model a demand function, you need data that indicate how many units of a product will sell at a given price. As you might imagine, such data are not easy to obtain for a new product. After a product has been on the market awhile, however, its sales history can provide the necessary data.

As an example, consider the two bar graphs shown below. From these graphs, you can see that from 2001 through 2005, the number of prerecorded DVDs sold increased from about 300 million to about 1100 million. During that time, the price per unit dropped from an average price of about $18 to an average price of about $15. *(Source: Kagan Research, LLC)*

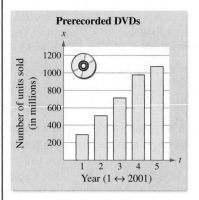
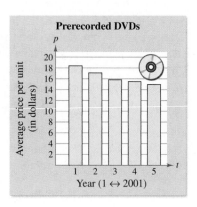

The information in the two bar graphs is combined in the table, where x represents the units sold (in millions) and p represents the price (in dollars).

t	1	2	3	4	5
x	291.5	507.5	713.0	976.6	1072.4
p	18.40	17.11	15.83	15.51	14.94

By entering the ordered pairs (x, p) into a graphing utility, you can find that the power model for the demand for prerecorded DVDs is: $p = 44.55x^{-0.155}$, $291.5 \leq x \leq 1072.4$. A graph of this demand function and its data points is shown below.

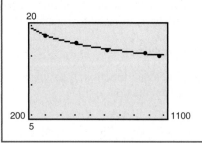

Example 7 **Finding the Marginal Revenue**

A fast-food restaurant has determined that the monthly demand for its hamburgers is given by

$$p = \frac{60,000 - x}{20,000}.$$

Figure 7.43 shows that as the price decreases, the quantity demanded increases. The table shows the demands for hamburgers at various prices.

x	60,000	50,000	40,000	30,000	20,000	10,000	0
p	$0.00	$0.50	$1.00	$1.50	$2.00	$2.50	$3.00

Find the increase in revenue per hamburger for monthly sales of 20,000 hamburgers. In other words, find the marginal revenue when $x = 20,000$.

SOLUTION Because the demand is given by

$$p = \frac{60,000 - x}{20,000}$$

and the revenue is given by $R = xp$, you have

$$R = xp = x\left(\frac{60,000 - x}{20,000}\right)$$

$$= \frac{1}{20,000}(60,000x - x^2).$$

By differentiating, you can find the marginal revenue to be

$$\frac{dR}{dx} = \frac{1}{20,000}(60,000 - 2x).$$

So, when $x = 20,000$, the marginal revenue is

$$\frac{1}{20,000}[60,000 - 2(20,000)] = \frac{20,000}{20,000} = \$1 \text{ per unit.}$$

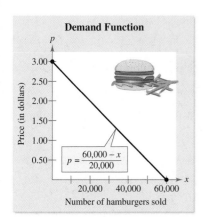

Demand Function

$$p = \frac{60,000 - x}{20,000}$$

FIGURE 7.43 As the price decreases, more hamburgers are sold.

✓CHECKPOINT 7

Find the revenue function and marginal revenue for a demand function of $p = 2000 - 4x$. ▪

STUDY TIP

Writing a demand function in the form $p = f(x)$ is a convention used in economics. From a consumer's point of view, it might seem more reasonable to think that the quantity demanded is a function of the price. Mathematically, however, the two points of view are equivalent because a typical demand function is one-to-one and so has an inverse function. For instance, in Example 7, you could write the demand function as $x = 60,000 - 20,000p$.

> **Example 8** **Finding the Marginal Profit**

Suppose that in Example 7, the cost of producing x hamburgers is

$$C = 5000 + 0.56x, \quad 0 \le x \le 50,000.$$

Find the profit and the marginal profit for each production level.

a. $x = 20,000$ **b.** $x = 24,400$ **c.** $x = 30,000$

SOLUTION From Example 7, you know that the total revenue from selling x hamburgers is

$$R = \frac{1}{20,000}(60,000x - x^2).$$

Because the total profit is given by $P = R - C$, you have

$$P = \frac{1}{20,000}(60,000x - x^2) - (5000 + 0.56x)$$

$$= 3x - \frac{x^2}{20,000} - 5000 - 0.56x$$

$$= 2.44x - \frac{x^2}{20,000} - 5000. \qquad \text{See Figure 7.44.}$$

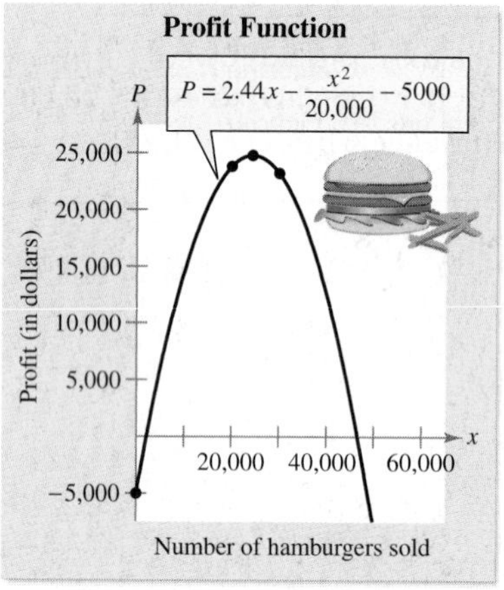

So, the marginal profit is

$$\frac{dP}{dx} = 2.44 - \frac{x}{10,000}.$$

Using these formulas, you can compute the profit and marginal profit.

Profit Function

$P = 2.44x - \dfrac{x^2}{20,000} - 5000$

Profit (in dollars): 25,000 · 20,000 · 15,000 · 10,000 · 5,000 · −5,000

Number of hamburgers sold: 20,000 40,000 60,000

FIGURE 7.44 Demand Curve

Production	Profit	Marginal Profit
a. $x = 20,000$	$P = \$23,800.00$	$2.44 - \dfrac{20,000}{10,000} = \$0.44$ per unit
b. $x = 24,400$	$P = \$24,768.00$	$2.44 - \dfrac{24,400}{10,000} = \$0.00$ per unit
c. $x = 30,000$	$P = \$23,200.00$	$2.44 - \dfrac{30,000}{10,000} = -\$0.56$ per unit

✓ CHECKPOINT 8

From Example 8, compare the marginal profit when 10,000 units are produced with the actual increase in profit from 10,000 units to 10,001 units. ∎

CONCEPT CHECK

1. You are asked to find the rate of change of a function over a certain interval. Should you find the average rate of change or the instantaneous rate of change?

2. You are asked to find the rate of change of a function at a certain instant. Should you find the average rate of change or the instantaneous rate of change?

3. If a variable can take on any real value in a given interval, is the variable discrete or continuous?

4. What does a demand function represent?

Skills Review 7.5

The following warm-up exercises involve skills that were covered in earlier sections. You will use these skills in the exercise set for this section. For additional help, review Sections 7.3 and 7.4.

In Exercises 1 and 2, evaluate the expression.

1. $\dfrac{-63 - (-105)}{21 - 7}$

2. $\dfrac{-37 - 54}{16 - 3}$

In Exercises 3–10, find the derivative of the function.

3. $y = 4x^2 - 2x + 7$

4. $y = -3t^3 + 2t^2 - 8$

5. $s = -16t^2 + 24t + 30$

6. $y = -16x^2 + 54x + 70$

7. $A = \frac{1}{10}(-2r^3 + 3r^2 + 5r)$

8. $y = \frac{1}{9}(6x^3 - 18x^2 + 63x - 15)$

9. $y = 12x - \dfrac{x^2}{5000}$

10. $y = 138 + 74x - \dfrac{x^3}{10,000}$

Exercises 7.5

See www.CalcChat.com for worked-out solutions to odd-numbered exercises.

1. Research and Development The table shows the amounts A (in billions of dollars per year) spent on R&D in the United States from 1980 through 2004, where t is the year, with $t = 0$ corresponding to 1980. Approximate the average rate of change of A during each period. *(Source: U.S. National Science Foundation)*

(a) 1980–1985   (b) 1985–1990   (c) 1990–1995

(d) 1995–2000   (e) 1980–2004   (f) 1990–2004

t	0	1	2	3	4	5	6
A	63	72	81	90	102	115	120

t	7	8	9	10	11	12
A	126	134	142	152	161	165

t	13	14	15	16	17	18
A	166	169	184	197	212	228

t	19	20	21	22	23	24
A	245	267	277	276	292	312

2. Trade Deficit The graph shows the values I (in billions of dollars per year) of goods imported to the United States and the values E (in billions of dollars per year) of goods exported from the United States from 1980 through 2005. Approximate each indicated average rate of change. *(Source: U.S. International Trade Administration)*

(a) Imports: 1980–1990   (b) Exports: 1980–1990

(c) Imports: 1990–2000   (d) Exports: 1990–2000

(e) Imports: 1980–2005   (f) Exports: 1980–2005

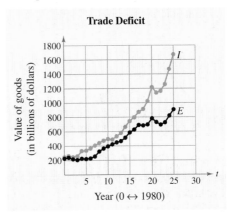

Figure for 2

(T) In Exercises 3–12, use a graphing utility to graph the function and find its average rate of change on the interval. Compare this rate with the instantaneous rates of change at the endpoints of the interval.

3. $f(t) = 3t + 5;\ [1, 2]$

4. $h(x) = 2 - x;\ [0, 2]$

5. $h(x) = x^2 - 4x + 2;\ [-2, 2]$

6. $f(x) = x^2 - 6x - 1;\ [-1, 3]$

7. $f(x) = 3x^{4/3};\ [1, 8]$

8. $f(x) = x^{3/2};\ [1, 4]$

9. $f(x) = \dfrac{1}{x};\ [1, 4]$

10. $f(x) = \dfrac{1}{\sqrt{x}};\ [1, 4]$

11. $g(x) = x^4 - x^2 + 2;\ [1, 3]$

12. $g(x) = x^3 - 1;\ [-1, 1]$

13. Consumer Trends The graph shows the number of visitors V to a national park in hundreds of thousands during a one-year period, where $t = 1$ represents January.

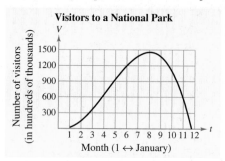

Visitors to a National Park

(a) Estimate the rate of change of V over the interval $[9, 12]$ and explain your results.

(b) Over what interval is the average rate of change approximately equal to the rate of change at $t = 8$? Explain your reasoning.

14. Medicine The graph shows the estimated number of milligrams of a pain medication M in the bloodstream t hours after a 1000-milligram dose of the drug has been given.

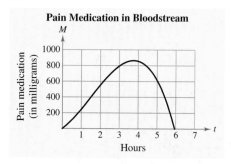

Pain Medication in Bloodstream

(a) Estimate the one-hour interval over which the average rate of change is the greatest.

(b) Over what interval is the average rate of change approximately equal to the rate of change at $t = 4$? Explain your reasoning.

15. Medicine The effectiveness E (on a scale from 0 to 1) of a pain-killing drug t hours after entering the bloodstream is given by

$$E = \frac{1}{27}(9t + 3t^2 - t^3), \quad 0 \le t \le 4.5.$$

Find the average rate of change of E on each indicated interval and compare this rate with the instantaneous rates of change at the endpoints of the interval.

(a) $[0, 1]$ (b) $[1, 2]$ (c) $[2, 3]$ (d) $[3, 4]$

16. Chemistry: Wind Chill At 0° Celsius, the heat loss H (in kilocalories per square meter per hour) from a person's body can be modeled by

$$H = 33\left(10\sqrt{v} - v + 10.45\right)$$

where v is the wind speed (in meters per second).

(a) Find $\dfrac{dH}{dv}$ and interpret its meaning in this situation.

(b) Find the rates of change of H when $v = 2$ and when $v = 5$.

17. Velocity The height s (in feet) at time t (in seconds) of a silver dollar dropped from the top of the Washington Monument is given by

$$s = -16t^2 + 555.$$

(a) Find the average velocity on the interval $[2, 3]$.

(b) Find the instantaneous velocities when $t = 2$ and when $t = 3$.

(c) How long will it take the dollar to hit the ground?

(d) Find the velocity of the dollar when it hits the ground.

Ⓑ **18. Physics: Velocity** A racecar travels northward on a straight, level track at a constant speed, traveling 0.750 kilometer in 20.0 seconds. The return trip over the same track is made in 25.0 seconds.

(a) What is the average velocity of the car in meters per second for the first leg of the run?

(b) What is the average velocity for the total trip?

(Source: Shipman/Wilson/Todd, An Introduction to Physical Science, Eleventh Edition)

Marginal Cost In Exercises 19–22, find the marginal cost for producing x units. (The cost is measured in dollars.)

19. $C = 4500 + 1.47x$ **20.** $C = 205,000 + 9800x$

21. $C = 55,000 + 470x - 0.25x^2, \quad 0 \le x \le 940$

22. $C = 100\left(9 + 3\sqrt{x}\right)$

Marginal Revenue In Exercises 23–26, find the marginal revenue for producing x units. (The revenue is measured in dollars.)

23. $R = 50x - 0.5x^2$ **24.** $R = 30x - x^2$

25. $R = -6x^3 + 8x^2 + 200x$ **26.** $R = 50(20x - x^{3/2})$

Marginal Profit In Exercises 27–30, find the marginal profit for producing x units. (The profit is measured in dollars.)

27. $P = -2x^2 + 72x - 145$

28. $P = -0.25x^2 + 2000x - 1,250,000$

29. $P = -0.00025x^2 + 12.2x - 25,000$

30. $P = -0.5x^3 + 30x^2 - 164.25x - 1000$

31. Marginal Cost The cost C (in dollars) of producing x units of a product is given by

$$C = 3.6\sqrt{x} + 500.$$

(a) Find the additional cost when the production increases from 9 to 10 units.

(b) Find the marginal cost when $x = 9$.

(c) Compare the results of parts (a) and (b).

32. Marginal Revenue The revenue R (in dollars) from renting x apartments can be modeled by

$$R = 2x(900 + 32x - x^2).$$

(a) Find the additional revenue when the number of rentals is increased from 14 to 15.

(b) Find the marginal revenue when $x = 14$.

(c) Compare the results of parts (a) and (b).

33. Marginal Profit The profit P (in dollars) from selling x units of calculus textbooks is given by

$$P = -0.05x^2 + 20x - 1000.$$

(a) Find the additional profit when the sales increase from 150 to 151 units.

(b) Find the marginal profit when $x = 150$.

(c) Compare the results of parts (a) and (b).

34. Population Growth The population P (in thousands) of Japan can be modeled by

$$P = -14.71t^2 + 785.5t + 117,216$$

where t is time in years, with $t = 0$ corresponding to 1980. *(Source: U.S. Census Bureau)*

(a) Evaluate P for $t = 0$, 10, 15, 20, and 25. Explain these values.

(b) Determine the population growth rate, dP/dt.

(c) Evaluate dP/dt for the same values as in part (a). Explain your results.

35. Health The temperature T (in degrees Fahrenheit) of a person during an illness can be modeled by the equation $T = -0.0375t^2 + 0.3t + 100.4$, where t is time in hours since the person started to show signs of a fever.

(T) (a) Use a graphing utility to graph the function. Be sure to choose an appropriate window.

(b) Do the slopes of the tangent lines appear to be positive or negative? What does this tell you?

(c) Evaluate the function for $t = 0$, 4, 8, and 12.

(d) Find dT/dt and explain its meaning in this situation.

(e) Evaluate dT/dt for $t = 0$, 4, 8, and 12.

36. Marginal Profit The profit P (in dollars) from selling x units of a product is given by

$$P = 36,000 + 2048\sqrt{x} - \frac{1}{8x^2}, \quad 150 \le x \le 275.$$

Find the marginal profit for each of the following sales.

(a) $x = 150$ (b) $x = 175$ (c) $x = 200$

(d) $x = 225$ (e) $x = 250$ (f) $x = 275$

37. Profit The monthly demand function and cost function for x newspapers at a newsstand are given by $p = 5 - 0.001x$ and $C = 35 + 1.5x$.

(a) Find the monthly revenue R as a function of x.

(b) Find the monthly profit P as a function of x.

(c) Complete the table.

x	600	1200	1800	2400	3000
dR/dx					
dP/dx					
P					

(B) **38. Economics** Use the table to answer the questions below.

Quantity produced and sold (Q)	Price (p)	Total revenue (TR)	Marginal revenue (MR)
0	160	0	—
2	140	280	130
4	120	480	90
6	100	600	50
8	80	640	10
10	60	600	−30

(T) (a) Use the *regression* feature of a graphing utility to find a quadratic model that relates the total revenue (TR) to the quantity produced and sold (Q).

(b) Using derivatives, find a model for marginal revenue from the model you found in part (a).

(c) Calculate the marginal revenue for all values of Q using your model in part (b), and compare these values with the actual values given. How good is your model?

(Source: Adapted from Taylor, Economics, Fifth Edition)

39. Marginal Profit When the price of a glass of lemonade at a lemonade stand was \$1.75, 400 glasses were sold. When the price was lowered to \$1.50, 500 glasses were sold. Assume that the demand function is linear and that the variable and fixed costs are \$0.10 and \$25, respectively.

(a) Find the profit P as a function of x, the number of glasses of lemonade sold.

(T) (b) Use a graphing utility to graph P, and comment about the slopes of P when $x = 300$ and when $x = 700$.

(c) Find the marginal profits when 300 glasses of lemonade are sold and when 700 glasses of lemonade are sold.

40. Marginal Cost The cost C of producing x units is modeled by $C = v(x) + k$, where v represents the variable cost and k represents the fixed cost. Show that the marginal cost is independent of the fixed cost.

41. Marginal Profit When the admission price for a baseball game was $6 per ticket, 36,000 tickets were sold. When the price was raised to $7, only 33,000 tickets were sold. Assume that the demand function is linear and that the variable and fixed costs for the ballpark owners are $0.20 and $85,000, respectively.

(a) Find the profit P as a function of x, the number of tickets sold.

(T) (b) Use a graphing utility to graph P, and comment about the slopes of P when $x = 18,000$ and when $x = 36,000$.

(c) Find the marginal profits when 18,000 tickets are sold and when 36,000 tickets are sold.

42. Marginal Profit In Exercise 41, suppose ticket sales decreased to 30,000 when the price increased to $7. How would this change the answers?

43. Profit The demand function for a product is given by $p = 50/\sqrt{x}$ for $1 \le x \le 8000$, and the cost function is given by $C = 0.5x + 500$ for $0 \le x \le 8000$.

Find the marginal profits for (a) $x = 900$, (b) $x = 1600$, (c) $x = 2500$, and (d) $x = 3600$.

If you were in charge of setting the price for this product, what price would you set? Explain your reasoning.

44. Inventory Management The annual inventory cost for a manufacturer is given by

$$C = 1{,}008{,}000/Q + 6.3Q$$

where Q is the order size when the inventory is replenished. Find the change in annual cost when Q is increased from 350 to 351, and compare this with the instantaneous rate of change when $Q = 350$.

45. *MAKE A DECISION: FUEL COST* A car is driven 15,000 miles a year and gets x miles per gallon. Assume that the average fuel cost is $2.95 per gallon. Find the annual cost of fuel C as a function of x and use this function to complete the table.

x	10	15	20	25	30	35	40
C							
dC/dx							

Who would benefit more from a 1 mile per gallon increase in fuel efficiency—the driver who gets 15 miles per gallon or the driver who gets 35 miles per gallon? Explain.

46. Gasoline Sales The number N of gallons of regular unleaded gasoline sold by a gasoline station at a price of p dollars per gallon is given by $N = f(p)$.

(a) Describe the meaning of $f'(2.959)$

(b) Is $f'(2.959)$ usually positive or negative? Explain.

47. Dow Jones Industrial Average The table shows the year-end closing prices p of the Dow Jones Industrial Average (DJIA) from 1992 through 2006, where t is the year, and $t = 2$ corresponds to 1992. *(Source: Dow Jones Industrial Average)*

t	2	3	4	5	6
p	3301.11	3754.09	3834.44	5117.12	6448.26

t	7	8	9	10	11
p	7908.24	9181.43	11,497.12	10,786.85	10,021.50

t	12	13	14	15	16
p	8341.63	10,453.92	10,783.01	10,717.50	12,463.15

(a) Determine the average rate of change in the value of the DJIA from 1992 to 2006.

(b) Estimate the instantaneous rate of change in 1998 by finding the average rate of change from 1996 to 2000.

(c) Estimate the instantaneous rate of change in 1998 by finding the average rate of change from 1997 to 1999.

(d) Compare your answers for parts (b) and (c). Which interval do you think produced the best estimate for the instantaneous rate of change in 1998?

(B) **48. Biology** Many populations in nature exhibit logistic growth, which consists of four phases, as shown in the figure. Describe the rate of growth of the population in each phase, and give possible reasons as to why the rates might be changing from phase to phase. *(Source: Adapted from Levine/Miller, Biology: Discovering Life, Second Edition)*

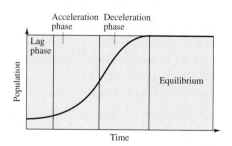

Section 7.6

The Product and Quotient Rules

■ Find the derivatives of functions using the Product Rule.
■ Find the derivatives of functions using the Quotient Rule.
■ Simplify derivatives.
■ Use derivatives to answer questions about real-life situations.

The Product Rule

In Section 7.4, you saw that the derivative of a sum or difference of two functions is simply the sum or difference of their derivatives. The rules for the derivative of a product or quotient of two functions are not as simple.

STUDY TIP

Rather than trying to remember the formula for the Product Rule, it can be more helpful to remember its verbal statement: *the first function times the derivative of the second plus the second function times the derivative of the first.*

The Product Rule

The derivative of the product of two differentiable functions is equal to the first function times the derivative of the second plus the second function times the derivative of the first.

$$\frac{d}{dx}[f(x)g(x)] = f(x)g'(x) + g(x)f'(x)$$

PROOF Some mathematical proofs, such as the proof of the Sum Rule, are straightforward. Others involve clever steps that may not appear to follow clearly from a prior step. The proof below involves such a step—adding and subtracting the same quantity. (This step is shown in color.) Let $F(x) = f(x)g(x)$.

$$F'(x) = \lim_{\Delta x \to 0} \frac{F(x + \Delta x) - F(x)}{\Delta x}$$

$$= \lim_{\Delta x \to 0} \frac{f(x + \Delta x)g(x + \Delta x) - f(x)g(x)}{\Delta x}$$

$$= \lim_{\Delta x \to 0} \frac{f(x + \Delta x)g(x + \Delta x) - f(x + \Delta x)g(x) + f(x + \Delta x)g(x) - f(x)g(x)}{\Delta x}$$

$$= \lim_{\Delta x \to 0} \left[f(x + \Delta x) \frac{g(x + \Delta x) - g(x)}{\Delta x} + g(x) \cdot \frac{f(x + \Delta x) - f(x)}{\Delta x} \right]$$

$$= \lim_{\Delta x \to 0} f(x + \Delta x) \frac{g(x + \Delta x) - g(x)}{\Delta x} + \lim_{\Delta x \to 0} g(x) \cdot \frac{f(x + \Delta x) - f(x)}{\Delta x}$$

$$= \left[\lim_{\Delta x \to 0} f(x + \Delta x) \right] \left[\lim_{\Delta x \to 0} \frac{g(x + \Delta x) - g(x)}{\Delta x} \right]$$

$$+ \left[\lim_{\Delta x \to 0} g(x) \right] \left[\lim_{\Delta x \to 0} \frac{f(x + \Delta x) - f(x)}{\Delta x} \right]$$

$$= f(x)g'(x) + g(x)f'(x)$$

| Example 1 | **Finding the Derivative of a Product** |

Find the derivative of $y = (3x - 2x^2)(5 + 4x)$.

SOLUTION Using the Product Rule, you can write

$$\frac{dy}{dx} = \overbrace{(3x - 2x^2)}^{\text{First}} \overbrace{\frac{d}{dx}[5 + 4x]}^{\substack{\text{Derivative} \\ \text{of second}}} + \overbrace{(5 + 4x)}^{\text{Second}} \overbrace{\frac{d}{dx}[3x - 2x^2]}^{\substack{\text{Derivative} \\ \text{of first}}}$$

$$= (3x - 2x^2)(4) + (5 + 4x)(3 - 4x)$$

$$= (12x - 8x^2) + (15 - 8x - 16x^2)$$

$$= 15 + 4x - 24x^2.$$

✓ **CHECKPOINT 1**

Find the derivative of $y = (4x + 3x^2)(6 - 3x)$. ∎

STUDY TIP

In general, the derivative of the product of two functions is not equal to the product of the derivatives of the two functions. To see this, compare the product of the derivatives of $f(x) = 3x - 2x^2$ and $g(x) = 5 + 4x$ with the derivative found in Example 1.

In the next example, notice that the first step in differentiating is *rewriting the original function*.

TECHNOLOGY

Ⓣ If you have access to a symbolic differentiation utility, try using it to confirm several of the derivatives in this section. The form of the derivative can depend on how you use software.

| Example 2 | **Finding the Derivative of a Product** |

Find the derivative of

$$f(x) = \left(\frac{1}{x} + 1\right)(x - 1). \qquad \text{Original function}$$

SOLUTION Rewrite the function. Then use the Product Rule to find the derivative.

$$f(x) = (x^{-1} + 1)(x - 1) \qquad \text{Rewrite function.}$$

$$f'(x) = (x^{-1} + 1)\frac{d}{dx}[x - 1] + (x - 1)\frac{d}{dx}[x^{-1} + 1] \qquad \text{Product Rule}$$

$$= (x^{-1} + 1)(1) + (x - 1)(-x^{-2})$$

$$= \frac{1}{x} + 1 - \frac{x - 1}{x^2}$$

$$= \frac{x + x^2 - x + 1}{x^2} \qquad \text{Write with common denominator.}$$

$$= \frac{x^2 + 1}{x^2} \qquad \text{Simplify.}$$

✓ **CHECKPOINT 2**

Find the derivative of

$$f(x) = \left(\frac{1}{x} + 1\right)(2x + 1). \quad ∎$$

You now have two differentiation rules that deal with products—the Constant Multiple Rule and the Product Rule. The difference between these two rules is that the Constant Multiple Rule deals with the product of a constant and a variable quantity:

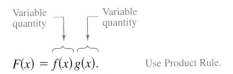

Constant ⌐ Variable quantity

$$F(x) = c\,f(x)$$ Use Constant Multiple Rule.

whereas the Product Rule deals with the product of two variable quantities:

Variable quantity   Variable quantity

$$F(x) = f(x)\,g(x).$$ Use Product Rule.

The next example compares these two rules.

Example 3   Comparing Differentiation Rules

Find the derivative of each function.

a. $y = 2x(x^2 + 3x)$

b. $y = 2(x^2 + 3x)$

SOLUTION

a. By the Product Rule,

$$\frac{dy}{dx} = (2x)\frac{d}{dx}[x^2 + 3x] + (x^2 + 3x)\frac{d}{dx}[2x]$$ Product Rule

$$= (2x)(2x + 3) + (x^2 + 3x)(2)$$

$$= 4x^2 + 6x + 2x^2 + 6x$$

$$= 6x^2 + 12x.$$

b. By the Constant Multiple Rule,

$$\frac{dy}{dx} = 2\frac{d}{dx}[x^2 + 3x]$$ Constant Multiple Rule

$$= 2(2x + 3)$$

$$= 4x + 6.$$

✓ **CHECKPOINT 3**

Find the derivative of each function.

a. $y = 3x(2x^2 + 5x)$

b. $y = 3(2x^2 + 5x)$ ■

The Product Rule can be extended to products that have more than two factors. For example, if f, g, and h are differentiable functions of x, then

$$\frac{d}{dx}[f(x)g(x)h(x)] = f'(x)g(x)h(x) + f(x)g'(x)h(x) + f(x)g(x)h'(x).$$

The Quotient Rule

In Section 7.4, you saw that by using the Constant Rule, the Power Rule, the Constant Multiple Rule, and the Sum and Difference Rules, you were able to differentiate any polynomial function. By combining these rules with the Quotient Rule, you can now differentiate any *rational* function.

The Quotient Rule

The derivative of the quotient of two differentiable functions is equal to the denominator times the derivative of the numerator minus the numerator times the derivative of the denominator, all divided by the square of the denominator.

$$\frac{d}{dx}\left[\frac{f(x)}{g(x)}\right] = \frac{g(x)f'(x) - f(x)g'(x)}{[g(x)]^2}, \quad g(x) \neq 0$$

STUDY TIP

From this differentiation rule, you can see that the derivative of a quotient is not, in general, the quotient of the derivatives. That is,

$$\frac{d}{dx}\left[\frac{f(x)}{g(x)}\right] \neq \frac{f'(x)}{g'(x)}.$$

PROOF Let $F(x) = f(x)/g(x)$. As in the proof of the Product Rule, a key step in this proof is adding and subtracting the same quantity.

$$F'(x) = \lim_{\Delta x \to 0} \frac{F(x + \Delta x) - F(x)}{\Delta x}$$

$$= \lim_{\Delta x \to 0} \frac{\dfrac{f(x + \Delta x)}{g(x + \Delta x)} - \dfrac{f(x)}{g(x)}}{\Delta x}$$

$$= \lim_{\Delta x \to 0} \frac{g(x)f(x + \Delta x) - f(x)g(x + \Delta x)}{\Delta x g(x)g(x + \Delta x)}$$

$$= \lim_{\Delta x \to 0} \frac{g(x)f(x + \Delta x) - f(x)g(x) + f(x)g(x) - f(x)g(x + \Delta x)}{\Delta x g(x)g(x + \Delta x)}$$

$$= \frac{\displaystyle\lim_{\Delta x \to 0} \frac{g(x)[f(x + \Delta x) - f(x)]}{\Delta x} - \lim_{\Delta x \to 0} \frac{f(x)[g(x + \Delta x) - g(x)]}{\Delta x}}{\displaystyle\lim_{\Delta x \to 0} [g(x)g(x + \Delta x)]}$$

$$= \frac{g(x)\left[\displaystyle\lim_{\Delta x \to 0} \frac{f(x + \Delta x) - f(x)}{\Delta x}\right] - f(x)\left[\displaystyle\lim_{\Delta x \to 0} \frac{g(x + \Delta x) - g(x)}{\Delta x}\right]}{\displaystyle\lim_{\Delta x \to 0} [g(x)g(x + \Delta x)]}$$

$$= \frac{g(x)f'(x) - f(x)g'(x)}{[g(x)]^2}$$

STUDY TIP

As suggested for the Product Rule, it can be more helpful to remember the verbal statement of the Quotient Rule rather than trying to remember the formula for the rule.

Algebra Review

When applying the Quotient Rule, it is suggested that you enclose all factors and derivatives in symbols of grouping, such as parentheses. Also, pay special attention to the subtraction required in the numerator. For help in evaluating expressions like the one in Example 4, see the *Chapter 7 Algebra Review* on page 618, Example 2(d).

Example 4 **Finding the Derivative of a Quotient**

Find the derivative of $y = \dfrac{x - 1}{2x + 3}$.

SOLUTION Apply the Quotient Rule, as shown.

$$\frac{dy}{dx} = \frac{(2x + 3)\dfrac{d}{dx}[x - 1] - (x - 1)\dfrac{d}{dx}[2x + 3]}{(2x + 3)^2}$$

$$= \frac{(2x + 3)(1) - (x - 1)(2)}{(2x + 3)^2}$$

$$= \frac{2x + 3 - 2x + 2}{(2x + 3)^2}$$

$$= \frac{5}{(2x + 3)^2}$$

✓**CHECKPOINT 4**

Find the derivative of $y = \dfrac{x + 4}{5x - 2}$. ∎

Example 5 **Finding an Equation of a Tangent Line**

Find an equation of the tangent line to the graph of

$$y = \frac{2x^2 - 4x + 3}{2 - 3x}$$

when $x = 1$.

SOLUTION Apply the Quotient Rule, as shown.

$$\frac{dy}{dx} = \frac{(2 - 3x)\dfrac{d}{dx}[2x^2 - 4x + 3] - (2x^2 - 4x + 3)\dfrac{d}{dx}[2 - 3x]}{(2 - 3x)^2}$$

$$= \frac{(2 - 3x)(4x - 4) - (2x^2 - 4x + 3)(-3)}{(2 - 3x)^2}$$

$$= \frac{-12x^2 + 20x - 8 - (-6x^2 + 12x - 9)}{(2 - 3x)^2}$$

$$= \frac{-12x^2 + 20x - 8 + 6x^2 - 12x + 9}{(2 - 3x)^2}$$

$$= \frac{-6x^2 + 8x + 1}{(2 - 3x)^2}$$

When $x = 1$, the value of the function is $y = -1$ and the slope is $m = 3$. Using the point-slope form of a line, you can find the equation of the tangent line to be $y = 3x - 4$. The graph of the function and the tangent line is shown in Figure 7.45.

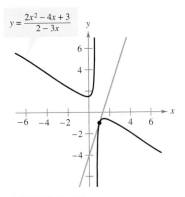

$y = \dfrac{2x^2 - 4x + 3}{2 - 3x}$

FIGURE 7.45

✓**CHECKPOINT 5**

Find an equation of the tangent line to the graph of

$$y = \frac{x^2 - 4}{2x + 5} \quad \text{when } x = 0.$$

Sketch the line tangent to the graph at $x = 0$. ∎

Example 6   **Finding the Derivative of a Quotient**

Find the derivative of

$$y = \frac{3 - (1/x)}{x + 5}.$$

SOLUTION   Begin by rewriting the original function. Then apply the Quotient Rule and simplify the result.

$$y = \frac{3 - (1/x)}{x + 5} \qquad \text{Write original function.}$$

$$= \frac{3x - 1}{x(x + 5)} \qquad \text{Multiply numerator and denominator by } x.$$

$$= \frac{3x - 1}{x^2 + 5x} \qquad \text{Rewrite.}$$

$$\frac{dy}{dx} = \frac{(x^2 + 5x)(3) - (3x - 1)(2x + 5)}{(x^2 + 5x)^2} \qquad \text{Apply Quotient Rule.}$$

$$= \frac{(3x^2 + 15x) - (6x^2 + 13x - 5)}{(x^2 + 5x)^2}$$

$$= \frac{-3x^2 + 2x + 5}{(x^2 + 5x)^2} \qquad \text{Simplify.}$$

✓**CHECKPOINT 6**

Find the derivative of $y = \dfrac{3 - (2/x)}{x + 4}$. ▪

Not every quotient needs to be differentiated by the Quotient Rule. For instance, each of the quotients in the next example can be considered as the product of a constant and a function of x. In such cases, the Constant Multiple Rule is more efficient than the Quotient Rule.

STUDY TIP

To see the efficiency of using the Constant Multiple Rule in Example 7, try using the Quotient Rule to find the derivatives of the four functions.

Example 7   **Rewriting Before Differentiating**

Find the derivative of each function.

	Original Function	Rewrite	Differentiate	Simplify
a.	$y = \dfrac{x^2 + 3x}{6}$	$y = \dfrac{1}{6}(x^2 + 3x)$	$y' = \dfrac{1}{6}(2x + 3)$	$y' = \dfrac{1}{3}x + \dfrac{1}{2}$
b.	$y = \dfrac{5x^4}{8}$	$y = \dfrac{5}{8}x^4$	$y' = \dfrac{5}{8}(4x^3)$	$y' = \dfrac{5}{2}x^3$
c.	$y = \dfrac{-3(3x - 2x^2)}{7x}$	$y = -\dfrac{3}{7}(3 - 2x)$	$y' = -\dfrac{3}{7}(-2)$	$y' = \dfrac{6}{7}$
d.	$y = \dfrac{9}{5x^2}$	$y = \dfrac{9}{5}(x^{-2})$	$y' = \dfrac{9}{5}(-2x^{-3})$	$y' = -\dfrac{18}{5x^3}$

✓**CHECKPOINT 7**

Find the derivative of each function.

a. $y = \dfrac{x^2 + 4x}{5}$ **b.** $y = \dfrac{3x^4}{4}$ ▪

Simplifying Derivatives

Example 8 **Combining the Product and Quotient Rules**

Find the derivative of

$$y = \frac{(1 - 2x)(3x + 2)}{5x - 4}.$$

SOLUTION This function contains a product within a quotient. You could first multiply the factors in the numerator and then apply the Quotient Rule. However, to gain practice in using the Product Rule within the Quotient Rule, try differentiating as shown.

$$y' = \frac{(5x - 4)\dfrac{d}{dx}[(1 - 2x)(3x + 2)] - (1 - 2x)(3x + 2)\dfrac{d}{dx}[5x - 4]}{(5x - 4)^2}$$

$$= \frac{(5x - 4)[(1 - 2x)(3) + (3x + 2)(-2)] - (1 - 2x)(3x + 2)(5)}{(5x - 4)^2}$$

$$= \frac{(5x - 4)(-12x - 1) - (1 - 2x)(15x + 10)}{(5x - 4)^2}$$

$$= \frac{(-60x^2 + 43x + 4) - (-30x^2 - 5x + 10)}{(5x - 4)^2}$$

$$= \frac{-30x^2 + 48x - 6}{(5x - 4)^2}$$

✓**CHECKPOINT 8**

Find the derivative of $y = \dfrac{(1 + x)(2x - 1)}{x - 1}$. ▪

In the examples in this section, much of the work in obtaining the final form of the derivative occurs *after* the differentiation. As summarized in the list below, direct application of differentiation rules often yields results that are not in simplified form. Note that two characteristics of simplified form are the absence of negative exponents and the combining of like terms.

	f'(x) After Differentiating	*f'(x) After Simplifying*
Example 1	$(3x - 2x^2)(4) + (5 + 4x)(3 - 4x)$	$15 + 4x - 24x^2$
Example 2	$(x^{-1} + 1)(1) + (x - 1)(-x^{-2})$	$\dfrac{x^2 + 1}{x^2}$
Example 5	$\dfrac{(2 - 3x)(4x - 4) - (2x^2 - 4x + 3)(-3)}{(2 - 3x)^2}$	$\dfrac{-6x^2 + 8x + 1}{(2 - 3x)^2}$
Example 8	$\dfrac{(5x - 4)[(1 - 2x)(3) + (3x + 2)(-2)] - (1 - 2x)(3x + 2)(5)}{(5x - 4)^2}$	$\dfrac{-30x^2 + 48x - 6}{(5x - 4)^2}$

Application

| Example 9 | **Rate of Change of Systolic Blood Pressure** | |

As blood moves from the heart through the major arteries out to the capillaries and back through the veins, the systolic blood pressure continuously drops. Consider a person whose systolic blood pressure P (in millimeters of mercury) is given by

$$P = \frac{25t^2 + 125}{t^2 + 1}, \quad 0 \le t \le 10$$

where t is measured in seconds. At what rate is the blood pressure changing 5 seconds after blood leaves the heart?

SOLUTION Begin by applying the Quotient Rule.

$$\frac{dP}{dt} = \frac{(t^2 + 1)(50t) - (25t^2 + 125)(2t)}{(t^2 + 1)^2} \qquad \text{Quotient Rule}$$

$$= \frac{50t^3 + 50t - 50t^3 - 250t}{(t^2 + 1)^2}$$

$$= -\frac{200t}{(t^2 + 1)^2} \qquad \text{Simplify.}$$

When $t = 5$, the rate of change is

$$-\frac{200(5)}{26^2} \approx -1.48 \text{ millimeters per second.}$$

So, the pressure is *dropping* at a rate of 1.48 millimeters per second when $t = 5$ seconds.

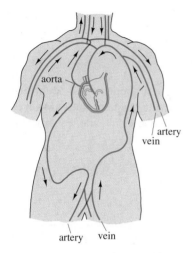

aorta

artery
vein

artery   vein

✓ CHECKPOINT 9

In Example 9, find the rate at which systolic blood pressure is changing at each time shown in the table below. Describe the changes in blood pressure as the blood moves away from the heart.

t	0	1	2	3	4	5	6	7
$\dfrac{dP}{dt}$								

CONCEPT CHECK

1. Write a verbal statement that represents the Product Rule.

2. Write a verbal statement that represents the Quotient Rule.

3. Is it possible to find the derivative of $f(x) = \dfrac{x^3 + 5x}{2}$ without using the Quotient Rule? If so, what differentiation rule can you use to find f'? (You do not need to find the derivative.)

4. Complete the following: In general, you can use the Product Rule to differentiate the _____ of two variable quantities and the Quotient Rule to differentiate any _____ function.

Skills Review 7.6

The following warm-up exercises involve skills that were covered in earlier sections. You will use these skills in the exercise set for this section. For additional help, review Sections 0.3, 0.6, 0.7, and 7.4.

In Exercises 1–10, simplify the expression.

1. $(x^2 + 1)(2) + (2x + 7)(2x)$

2. $(2x - x^3)(8x) + (4x^2)(2 - 3x^2)$

3. $x(4)(x^2 + 2)^3(2x) + (x^2 + 4)(1)$

4. $x^2(2)(2x + 1)(2) + (2x + 1)^4(2x)$

5. $\dfrac{(2x + 7)(5) - (5x + 6)(2)}{(2x + 7)^2}$

6. $\dfrac{(x^2 - 4)(2x + 1) - (x^2 + x)(2x)}{(x^2 - 4)^2}$

7. $\dfrac{(x^2 + 1)(2) - (2x + 1)(2x)}{(x^2 + 1)^2}$

8. $\dfrac{(1 - x^4)(4) - (4x - 1)(-4x^3)}{(1 - x^4)^2}$

9. $(x^{-1} + x)(2) + (2x - 3)(-x^{-2} + 1)$

10. $\dfrac{(1 - x^{-1})(1) - (x - 4)(x^{-2})}{(1 - x^{-1})^2}$

In Exercises 11–14, find $f'(2)$.

11. $f(x) = 3x^2 - x + 4$

12. $f(x) = -x^3 + x^2 + 8x$

13. $f(x) = \dfrac{1}{x}$

14. $f(x) = x^2 - \dfrac{1}{x^2}$

Exercises 7.6

See www.CalcChat.com for worked-out solutions to odd-numbered exercises.

In Exercises 1–16, find the value of the derivative of the function at the given point. State which differentiation rule you used to find the derivative.

Function	Point
1. $f(x) = x(x^2 + 3)$	$(2, 14)$
2. $g(x) = (x - 4)(x + 2)$	$(4, 0)$
3. $f(x) = x^2(3x^3 - 1)$	$(1, 2)$
4. $f(x) = (x^2 + 1)(2x + 5)$	$(-1, 6)$
5. $f(x) = \frac{1}{3}(2x^3 - 4)$	$\left(0, -\frac{4}{3}\right)$
6. $f(x) = \frac{1}{7}(5 - 6x^2)$	$\left(1, -\frac{1}{7}\right)$
7. $g(x) = (x^2 - 4x + 3)(x - 2)$	$(4, 6)$
8. $g(x) = (x^2 - 2x + 1)(x^3 - 1)$	$(1, 0)$
9. $h(x) = \dfrac{x}{x - 5}$	$(6, 6)$
10. $h(x) = \dfrac{x^2}{x + 3}$	$\left(-1, \frac{1}{2}\right)$
11. $f(t) = \dfrac{2t^2 - 3}{3t + 1}$	$\left(3, \frac{3}{2}\right)$
12. $f(x) = \dfrac{3x}{x^2 + 4}$	$\left(-1, -\frac{3}{5}\right)$
13. $g(x) = \dfrac{2x + 1}{x - 5}$	$(6, 13)$
14. $f(x) = \dfrac{x + 1}{x - 1}$	$(2, 3)$
15. $f(t) = \dfrac{t^2 - 1}{t + 4}$	$(1, 0)$
16. $g(x) = \dfrac{4x - 5}{x^2 - 1}$	$(0, 5)$

In Exercises 17–24, find the derivative of the function. Use Example 7 as a model.

Function	Rewrite	Differentiate	Simplify
17. $y = \dfrac{x^2 + 2x}{x}$			
18. $y = \dfrac{4x^{3/2}}{x}$			
19. $y = \dfrac{7}{3x^3}$			
20. $y = \dfrac{4}{5x^2}$			
21. $y = \dfrac{4x^2 - 3x}{8\sqrt{x}}$			
22. $y = \dfrac{3x^2 - 4x}{6x}$			
23. $y = \dfrac{x^2 - 4x + 3}{x - 1}$			
24. $y = \dfrac{x^2 - 4}{x + 2}$			

In Exercises 25–40, find the derivative of the function. State which differentiation rule(s) you used to find the derivative.

25. $f(x) = (x^3 - 3x)(2x^2 + 3x + 5)$

26. $h(t) = (t^5 - 1)(4t^2 - 7t - 3)$

27. $g(t) = (2t^3 - 1)^2$

28. $h(p) = (p^3 - 2)^2$

29. $f(x) = \sqrt[3]{x}(\sqrt{x} + 3)$

30. $f(x) = \sqrt[3]{x}(x + 1)$

31. $f(x) = \dfrac{3x - 2}{2x - 3}$

32. $f(x) = \dfrac{x^3 + 3x + 2}{x^2 - 1}$

33. $f(x) = \dfrac{3 - 2x - x^2}{x^2 - 1}$

34. $f(x) = (x^5 - 3x)\left(\dfrac{1}{x^2}\right)$

35. $f(x) = x\left(1 - \dfrac{2}{x + 1}\right)$

36. $h(t) = \dfrac{t + 2}{t^2 + 5t + 6}$

37. $g(s) = \dfrac{s^2 - 2s + 5}{\sqrt{s}}$

38. $f(x) = \dfrac{x + 1}{\sqrt{x}}$

39. $g(x) = \left(\dfrac{x - 3}{x + 4}\right)(x^2 + 2x + 1)$

40. $f(x) = (3x^3 + 4x)(x - 5)(x + 1)$

(T) In Exercises 41–46, find an equation of the tangent line to the graph of the function at the given point. Then use a graphing utility to graph the function and the tangent line in the same viewing window.

Function	*Point*
41. $f(x) = (x - 1)^2(x - 2)$	$(0, -2)$
42. $h(x) = (x^2 - 1)^2$	$(-2, 9)$
43. $f(x) = \dfrac{x - 2}{x + 1}$	$\left(1, -\tfrac{1}{2}\right)$
44. $f(x) = \dfrac{2x + 1}{x - 1}$	$(2, 5)$
45. $f(x) = \left(\dfrac{x + 5}{x - 1}\right)(2x + 1)$	$(0, -5)$
46. $g(x) = (x + 2)\left(\dfrac{x - 5}{x + 1}\right)$	$(0, -10)$

In Exercises 47–50, find the point(s), if any, at which the graph of f has a horizontal tangent.

47. $f(x) = \dfrac{x^2}{x - 1}$

48. $f(x) = \dfrac{x^2}{x^2 + 1}$

49. $f(x) = \dfrac{x^4}{x^3 + 1}$

50. $f(x) = \dfrac{x^4 + 3}{x^2 + 1}$

(T) In Exercises 51–54, use a graphing utility to graph f and f' on the interval $[-2, 2]$.

51. $f(x) = x(x + 1)$

52. $f(x) = x^2(x + 1)$

53. $f(x) = x(x + 1)(x - 1)$

54. $f(x) = x^2(x + 1)(x - 1)$

Demand   In Exercises 55 and 56, use the demand function to find the rate of change in the demand x for the given price p.

55. $x = 275\left(1 - \dfrac{3p}{5p + 1}\right)$, $p = \$4$

56. $x = 300 - p - \dfrac{2p}{p + 1}$, $p = \$3$

57. Environment   The model

$$f(t) = \frac{t^2 - t + 1}{t^2 + 1}$$

measures the level of oxygen in a pond, where t is the time (in weeks) after organic waste is dumped into the pond. Find the rates of change of f with respect to t when (a) $t = 0.5$, (b) $t = 2$, and (c) $t = 8$.

58. Physical Science   The temperature T (in degrees Fahrenheit) of food placed in a refrigerator is modeled by

$$T = 10\left(\frac{4t^2 + 16t + 75}{t^2 + 4t + 10}\right)$$

where t is the time (in hours). What is the initial temperature of the food? Find the rates of change of T with respect to t when (a) $t = 1$, (b) $t = 3$, (c) $t = 5$, and (d) $t = 10$.

59. Population Growth   A population of bacteria is introduced into a culture. The number of bacteria P can be modeled by

$$P = 500\left(1 + \frac{4t}{50 + t^2}\right)$$

where t is the time (in hours). Find the rate of change of the population when $t = 2$.

60. Quality Control   The percent P of defective parts produced by a new employee t days after the employee starts work can be modeled by

$$P = \frac{t + 1750}{50(t + 2)}.$$

Find the rates of change of P when (a) $t = 1$ and (b) $t = 10$.

61. *MAKE A DECISION: NEGOTIATING A PRICE*   You decide to form a partnership with another business. Your business determines that the demand x for your product is inversely proportional to the square of the price for $x \geq 5$.

(a) The price is \$1000 and the demand is 16 units. Find the demand function.

(b) Your partner determines that the product costs \$250 per unit and the fixed cost is \$10,000. Find the cost function.

(c) Find the profit function and use a graphing utility to graph it. From the graph, what price would you negotiate with your partner for this product? Explain your reasoning.

(T) **62. Managing a Store** You are managing a store and have been adjusting the price of an item. You have found that you make a profit of $50 when 10 units are sold, $60 when 12 units are sold, and $65 when 14 units are sold.

(a) Fit these data to the model $P = ax^2 + bx + c$.

(b) Use a graphing utility to graph P.

(c) Find the point on the graph at which the marginal profit is zero. Interpret this point in the context of the problem.

(T) **63. Demand Function** Given $f(x) = x + 1$, which function would most likely represent a demand function? Explain your reasoning. Use a graphing utility to graph each function, and use each graph as part of your explanation.

(a) $p = f(x)$ (b) $p = xf(x)$ (c) $p = -f(x) + 5$

(T) **64. Cost** The cost of producing x units of a product is given by

$$C = x^3 - 15x^2 + 87x - 73, \quad 4 \le x \le 9.$$

(a) Use a graphing utility to graph the marginal cost function and the average cost function, C/x, in the same viewing window.

(b) Find the point of intersection of the graphs of dC/dx and C/x. Does this point have any significance?

65. *MAKE A DECISION: INVENTORY REPLENISHMENT* The ordering and transportation cost C per unit (in thousands of dollars) of the components used in manufacturing a product is given by

$$C = 100\left(\frac{200}{x^2} + \frac{x}{x + 30}\right), \quad 1 \le x$$

where x is the order size (in hundreds). Find the rate of change of C with respect to x for each order size. What do these rates of change imply about increasing the size of an order? Of the given order sizes, which would you choose? Explain.

(a) $x = 10$ (b) $x = 15$ (c) $x = 20$

66. Inventory Replenishment The ordering and transportation cost C per unit for the components used in manufacturing a product is

$$C = (375,000 + 6x^2)/x, \quad x \ge 1$$

where C is measured in dollars and x is the order size. Find the rate of change of C with respect to x when (a) $x = 200$, (b) $x = 250$, and (c) $x = 300$. Interpret the meaning of these values.

67. Consumer Awareness The prices per pound of lean and extra lean ground beef in the United States from 1998 to 2005 can be modeled by

$$P = \frac{1.755 - 0.2079t + 0.00673t^2}{1 - 0.1282t + 0.00434t^2}, \quad 8 \le t \le 15$$

where t is the year, with $t = 8$ corresponding to 1998. Find dP/dt and evaluate it for $t = 8, 10, 12,$ and 14. Interpret the meaning of these values. *(Source: U.S. Bureau of Labor Statistics)*

68. Sales Analysis The monthly sales of memberships M at a newly built fitness center are modeled by

$$M(t) = \frac{300t}{t^2 + 1} + 8$$

where t is the number of months since the center opened.

(a) Find $M'(t)$.

(b) Find $M(3)$ and $M'(3)$ and interpret the results.

(c) Find $M(24)$ and $M'(24)$ and interpret the results.

In Exercises 69–72, use the given information to find $f'(2)$.

$$g(2) = 3 \quad \text{and} \quad g'(2) = -2$$
$$h(2) = -1 \quad \text{and} \quad h'(2) = 4$$

69. $f(x) = 2g(x) + h(x)$ **70.** $f(x) = 3 - g(x)$

71. $f(x) = g(x) + h(x)$ **72.** $f(x) = \dfrac{g(x)}{h(x)}$

Business Capsule

AP/Wide World Photos

In 1978 Ben Cohen and Jerry Greenfield used their combined life savings of $8000 to convert an abandoned gas station in Burlington, Vermont into their first ice cream shop. Today, Ben & Jerry's Homemade Holdings, Inc. has over 600 scoop shops in 16 countries. The company's three-part mission statement emphasizes product quality, economic reward, and a commitment to the community. Ben & Jerry's contributes a minimum of $1.1 million annually through corporate philanthropy that is primarily employee led.

73. Research Project Use your school's library, the Internet, or some other reference source to find information on a company that is noted for its philanthropy and community commitment. (One such business is described above.) Write a short paper about the company.

Section 7.7

The Chain Rule

- Find derivatives using the Chain Rule.
- Find derivatives using the General Power Rule.
- Write derivatives in simplified form.
- Use derivatives to answer questions about real-life situations.
- Use the differentiation rules to differentiate algebraic functions.

The Chain Rule

In this section, you will study one of the most powerful rules of differential calculus—the **Chain Rule.** This differentiation rule deals with composite functions and adds versatility to the rules presented in Sections 7.4 and 7.6. For example, compare the functions below. Those on the left can be differentiated without the Chain Rule, whereas those on the right are best done with the Chain Rule.

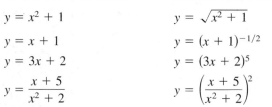

Without the Chain Rule	*With the Chain Rule*
$y = x^2 + 1$	$y = \sqrt{x^2 + 1}$
$y = x + 1$	$y = (x + 1)^{-1/2}$
$y = 3x + 2$	$y = (3x + 2)^5$
$y = \dfrac{x + 5}{x^2 + 2}$	$y = \left(\dfrac{x + 5}{x^2 + 2}\right)^2$

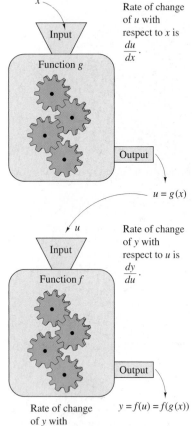

x

Rate of change of u with respect to x is $\dfrac{du}{dx}$.

Input

Function g

Output

$u = g(x)$

u

Rate of change of y with respect to u is $\dfrac{dy}{du}$.

Input

Function f

Output

$y = f(u) = f(g(x))$

Rate of change of y with respect to x is $\dfrac{dy}{dx} = \dfrac{dy}{du}\dfrac{du}{dx}$.

FIGURE 7.46

> ### The Chain Rule
>
> If $y = f(u)$ is a differentiable function of u, and $u = g(x)$ is a differentiable function of x, then $y = f(g(x))$ is a differentiable function of x, and
>
> $$\frac{dy}{dx} = \frac{dy}{du} \cdot \frac{du}{dx}$$
>
> or, equivalently,
>
> $$\frac{d}{dx}[f(g(x))] = f'(g(x))g'(x).$$

Basically, the Chain Rule states that if y changes dy/du times as fast as u, and u changes du/dx times as fast as x, then y changes

$$\frac{dy}{du} \cdot \frac{du}{dx}$$

times as fast as x, as illustrated in Figure 7.46. One advantage of the dy/dx notation for derivatives is that it helps you remember differentiation rules, such as the Chain Rule. For instance, in the formula

$$dy/dx = (dy/du)(du/dx)$$

you can imagine that the du's divide out.

When applying the Chain Rule, it helps to think of the composite function $y = f(g(x))$ or $y = f(u)$ as having two parts—an *inside* and an *outside*—as illustrated below.

$$y = \underset{\text{Outside}}{\overset{\overset{\text{Inside}}{\downarrow\quad\downarrow}}{f(g(x))}} = f(u)$$

The Chain Rule tells you that the derivative of $y = f(u)$ is the derivative of the outer function (at the inner function u) *times* the derivative of the inner function. That is,

$$y' = f'(u) \cdot u'.$$

✓ **CHECKPOINT 1**

Write each function as the composition of two functions, where $y = f(g(x))$.

a. $y = \dfrac{1}{\sqrt{x + 1}}$

b. $y = (x^2 + 2x + 5)^3$ ◼

Example 1   Decomposing Composite Functions

Write each function as the composition of two functions.

a. $y = \dfrac{1}{x + 1}$ **b.** $y = \sqrt{3x^2 - x + 1}$

SOLUTION   There is more than one correct way to decompose each function. One way for each is shown below.

	$y = f(g(x))$	$u = g(x)$ *(inside)*	$y = f(u)$ *(outside)*
a.	$y = \dfrac{1}{x + 1}$	$u = x + 1$	$y = \dfrac{1}{u}$
b.	$y = \sqrt{3x^2 - x + 1}$	$u = 3x^2 - x + 1$	$y = \sqrt{u}$

Example 2   Using the Chain Rule

Find the derivative of $y = (x^2 + 1)^3$.

SOLUTION   To apply the Chain Rule, you need to identify the inside function u.

$$y = \overset{u}{\overbrace{(x^2 + 1)}}^3 = u^3$$

By the Chain Rule, you can write the derivative as shown.

$$\frac{dy}{dx} = \overset{\frac{dy}{du}}{\overbrace{3(x^2 + 1)^2}}\,\overset{\frac{du}{dx}}{\overbrace{(2x)}} = 6x(x^2 + 1)^2$$

STUDY TIP

Try checking the result of Example 2 by expanding the function to obtain

$$y = x^6 + 3x^4 + 3x^2 + 1$$

and finding the derivative. Do you obtain the same answer?

✓ **CHECKPOINT 2**

Find the derivative of $y = (x^3 + 1)^2$. ◼

The General Power Rule

The function in Example 2 illustrates one of the most common types of composite functions—a power function of the form

$$y = [u(x)]^n.$$

The rule for differentiating such functions is called the **General Power Rule,** and it is a special case of the Chain Rule.

The General Power Rule

If $y = [u(x)]^n$, where u is a differentiable function of x and n is a real number, then

$$\frac{dy}{dx} = n[u(x)]^{n-1}\frac{du}{dx}$$

or, equivalently,

$$\frac{d}{dx}[u^n] = nu^{n-1}u'.$$

PROOF   Apply the Chain Rule and the Simple Power Rule as shown.

$$\frac{dy}{dx} = \frac{dy}{du} \cdot \frac{du}{dx}$$

$$= \frac{d}{du}[u^n]\frac{du}{dx}$$

$$= nu^{n-1}\frac{du}{dx}$$

TECHNOLOGY

(T) If you have access to a symbolic differentiation utility, try using it to confirm the result of Example 3.

Example 3   **Using the General Power Rule**

Find the derivative of

$$f(x) = (3x - 2x^2)^3.$$

SOLUTION   The inside function is $u = 3x - 2x^2$. So, by the General Power Rule,

$$f'(x) = \overset{n}{3}\overbrace{(3x - 2x^2)^2}^{u^{n-1}}\overbrace{\frac{d}{dx}[3x - 2x^2]}^{u'}$$

$$= 3(3x - 2x^2)^2(3 - 4x)$$

$$= (9 - 12x)(3x - 2x^2)^2.$$

✓ **CHECKPOINT 3**

Find the derivative of $y = (x^2 + 3x)^4$. ▪

Example 4 Rewriting Before Differentiating

Find the tangent line to the graph of

$$y = \sqrt[3]{(x^2 + 4)^2}$$ Original function

when $x = 2$.

SOLUTION Begin by rewriting the function in rational exponent form.

$$y = (x^2 + 4)^{2/3}$$ Rewrite original function.

Then, using the inside function, $u = x^2 + 4$, apply the General Power Rule.

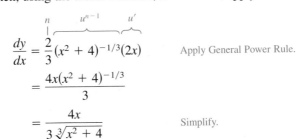

$$\frac{dy}{dx} = \frac{2}{3}(x^2 + 4)^{-1/3}(2x)$$ Apply General Power Rule.

$$= \frac{4x(x^2 + 4)^{-1/3}}{3}$$

$$= \frac{4x}{3\sqrt[3]{x^2 + 4}}$$ Simplify.

When $x = 2$, $y = 4$ and the slope of the line tangent to the graph at $(2, 4)$ is $\frac{4}{3}$. Using the point-slope form, you can find the equation of the tangent line to be $y = \frac{4}{3}x + \frac{4}{3}$. The graph of the function and the tangent line is shown in Figure 7.47.

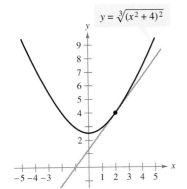

$y = \sqrt[3]{(x^2 + 4)^2}$

FIGURE 7.47

✓ CHECKPOINT 4

Find the tangent line to the graph of $y = \sqrt[3]{(x + 4)^2}$ when $x = 4$. Sketch the line tangent to the graph at $x = 4$. ■

STUDY TIP

The derivative of a quotient can sometimes be found more easily with the General Power Rule than with the Quotient Rule. This is especially true when the numerator is a constant, as shown in Example 5.

Example 5 Finding the Derivative of a Quotient

Find the derivative of each function.

a. $y = \dfrac{3}{x^2 + 1}$ **b.** $y = \dfrac{3}{(x + 1)^2}$

SOLUTION

a. Begin by rewriting the function as

$$y = 3(x^2 + 1)^{-1}.$$ Rewrite original function.

Then apply the General Power Rule to obtain

$$\frac{dy}{dx} = -3(x^2 + 1)^{-2}(2x) = -\frac{6x}{(x^2 + 1)^2}.$$ Apply General Power Rule.

b. Begin by rewriting the function as

$$y = 3(x + 1)^{-2}.$$ Rewrite original function.

Then apply the General Power Rule to obtain

$$\frac{dy}{dx} = -6(x + 1)^{-3}(1) = -\frac{6}{(x + 1)^3}.$$ Apply General Power Rule.

✓ CHECKPOINT 5

Find the derivative of each function.

a. $y = \dfrac{4}{2x + 1}$

b. $y = \dfrac{2}{(x - 1)^3}$ ■

Simplification Techniques

Throughout this chapter, writing derivatives in simplified form has been emphasized. The reason for this is that most applications of derivatives require a simplified form. The next two examples illustrate some useful simplification techniques.

Algebra Review

In Example 6, note that you subtract exponents when factoring. That is, when $(1 - x^2)^{-1/2}$ is factored out of $(1 - x^2)^{1/2}$, the *remaining* factor has an exponent of $\frac{1}{2} - \left(-\frac{1}{2}\right) = 1$. So,

$$(1 - x^2)^{1/2} = (1 - x^2)^{-1/2}$$
$$(1 - x^2)^{1}.$$

For help in evaluating expressions like the one in Example 6, see the *Chapter 7 Algebra Review* on pages 617 and 618.

Example 6 Simplifying by Factoring Out Least Powers

Find the derivative of $y = x^2 \sqrt{1 - x^2}$.

$$y = x^2 \sqrt{1 - x^2} \qquad \text{Write original function.}$$
$$= x^2 (1 - x^2)^{1/2} \qquad \text{Rewrite function.}$$
$$y' = x^2 \frac{d}{dx}\left[(1 - x^2)^{1/2}\right] + (1 - x^2)^{1/2} \frac{d}{dx}[x^2] \qquad \text{Product Rule}$$
$$= x^2\left[\frac{1}{2}(1 - x^2)^{-1/2}(-2x)\right] + (1 - x^2)^{1/2}(2x) \qquad \text{Power Rule}$$
$$= -x^3(1 - x^2)^{-1/2} + 2x(1 - x^2)^{1/2}$$
$$= x(1 - x^2)^{-1/2}\left[-x^2(1) + 2(1 - x^2)\right] \qquad \text{Factor.}$$
$$= x(1 - x^2)^{-1/2}(2 - 3x^2)$$
$$= \frac{x(2 - 3x^2)}{\sqrt{1 - x^2}} \qquad \text{Simplify.}$$

✓ **CHECKPOINT 6**

Find and simplify the derivative of $y = x^2 \sqrt{x^2 + 1}$. ∎

STUDY TIP

In Example 7, try to find $f'(x)$ by applying the Quotient Rule to

$$f(x) = \frac{(3x - 1)^2}{(x^2 + 3)^2}.$$

Which method do you prefer?

Example 7 Differentiating a Quotient Raised to a Power

Find the derivative of

$$f(x) = \left(\frac{3x - 1}{x^2 + 3}\right)^2.$$

SOLUTION

$$f'(x) = \overset{n}{2}\overset{u^{n-1}}{\left(\frac{3x - 1}{x^2 + 3}\right)}\overset{u'}{\frac{d}{dx}\left[\frac{3x - 1}{x^2 + 3}\right]}$$
$$= \left[\frac{2(3x - 1)}{x^2 + 3}\right]\left[\frac{(x^2 + 3)(3) - (3x - 1)(2x)}{(x^2 + 3)^2}\right]$$
$$= \frac{2(3x - 1)(3x^2 + 9 - 6x^2 + 2x)}{(x^2 + 3)^3}$$
$$= \frac{2(3x - 1)(-3x^2 + 2x + 9)}{(x^2 + 3)^3}$$

✓ **CHECKPOINT 7**

Find the derivative of

$$f(x) = \left(\frac{x + 1}{x - 5}\right)^2.$$ ∎

Example 8   **Finding Rates of Change**

From 1996 through 2005, the revenue per share R (in dollars) for U.S. Cellular can be modeled by $R = (-0.009t^2 + 0.54t - 0.1)^2$ for $6 \le t \le 15$, where t is the year, with $t = 6$ corresponding to 1996. Use the model to approximate the rates of change in the revenue per share in 1997, 1999, and 2003. If you had been a U.S. Cellular stockholder from 1996 through 2005, would you have been satisfied with the performance of this stock?   *(Source: U.S. Cellular)*

SOLUTION   The rate of change in R is given by the derivative dR/dt. You can use the General Power Rule to find the derivative.

$$\frac{dR}{dt} = 2(-0.009t^2 + 0.54t - 0.1)^1(-0.018t + 0.54)$$

$$= (-0.036t + 1.08)(-0.009t^2 + 0.54t - 0.1)$$

In 1997, the revenue per share was changing at a rate of

$$[-0.036(7) + 1.08][-0.009(7)^2 + 0.54(7) - 0.1] \approx \$2.68 \text{ per year.}$$

In 1999, the revenue per share was changing at a rate of

$$[-0.036(9) + 1.08][-0.009(9)^2 + 0.54(9) - 0.1] \approx \$3.05 \text{ per year.}$$

In 2003, the revenue per share was changing at a rate of

$$[-0.036(13) + 1.08][-0.009(13)^2 + 0.54(13) - 0.1] \approx \$3.30 \text{ per year.}$$

The graph of the revenue per share function is shown in Figure 7.48. For most investors, the performance of U.S. Cellular stock would be considered to be good.

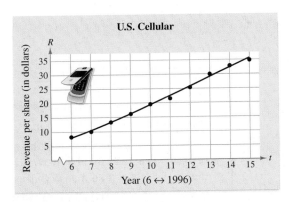

FIGURE 7.48

✓ **CHECKPOINT 8**

From 1996 through 2005, the sales per share (in dollars) for Dollar Tree can be modeled by $S = (-0.002t^2 + 0.39t + 0.1)^2$ for $6 \le t \le 15$, where t is the year, with $t = 6$ corresponding to 1996. Use the model to approximate the rate of change in sales per share in 2003.   *(Source: Dollar Tree Stores, Inc.)* ■

Summary of Differentiation Rules

You now have all the rules you need to differentiate *any* algebraic function. For your convenience, they are summarized below.

Summary of Differentiation Rules

Let u and v be differentiable functions of x.

1. Constant Rule
$$\frac{d}{dx}[c] = 0, \quad c \text{ is a constant.}$$

2. Constant Multiple Rule
$$\frac{d}{dx}[cu] = c\frac{du}{dx}, \quad c \text{ is a constant.}$$

3. Sum and Difference Rules
$$\frac{d}{dx}[u \pm v] = \frac{du}{dx} \pm \frac{dv}{dx}$$

4. Product Rule
$$\frac{d}{dx}[uv] = u\frac{dv}{dx} + v\frac{du}{dx}$$

5. Quotient Rule
$$\frac{d}{dx}\left[\frac{u}{v}\right] = \frac{v\dfrac{du}{dx} - u\dfrac{dv}{dx}}{v^2}$$

6. Power Rules
$$\frac{d}{dx}[x^n] = nx^{n-1}$$
$$\frac{d}{dx}[u^n] = nu^{n-1}\frac{du}{dx}$$

7. Chain Rule
$$\frac{dy}{dx} = \frac{dy}{du} \cdot \frac{du}{dx}$$

CONCEPT CHECK

1. Write a verbal statement that represents the Chain Rule.

2. Write a verbal statement that represents the General Power Rule.

3. Complete the following: When the numerator of a quotient is a constant, you may be able to find the derivative of the quotient more easily with the _____ _____ Rule than with the Quotient Rule.

4. In the expression $f(g(x))$, f is the outer function and g is the inner function. Write a verbal statement of the Chain Rule using the words "inner" and "outer."

Skills Review 7.7

The following warm-up exercises involve skills that were covered in earlier sections. You will use these skills in the exercise set for this section. For additional help, review Sections 0.4 and 0.6.

In Exercises 1–6, rewrite the expression with rational exponents.

1. $\sqrt[5]{(1 - 5x)^2}$

2. $\sqrt[4]{(2x - 1)^3}$

3. $\dfrac{1}{\sqrt{4x^2 + 1}}$

4. $\dfrac{1}{\sqrt[3]{x - 6}}$

5. $\dfrac{\sqrt{x}}{\sqrt[3]{1 - 2x}}$

6. $\dfrac{\sqrt{(3 - 7x)^3}}{2x}$

In Exercises 7–10, factor the expression.

7. $3x^3 - 6x^2 + 5x - 10$

8. $5x\sqrt{x} - x - 5\sqrt{x} + 1$

9. $4(x^2 + 1)^2 - x(x^2 + 1)^3$

10. $-x^5 + 3x^3 + x^2 - 3$

Exercises 7.7

See www.CalcChat.com for worked-out solutions to odd-numbered exercises.

In Exercises 1–8, identify the inside function, $u = g(x)$, and the outside function, $y = f(u)$.

$y = f(g(x))$	$u = g(x)$	$y = f(u)$
1. $y = (6x - 5)^4$		
2. $y = (x^2 - 2x + 3)^3$		
3. $y = (4 - x^2)^{-1}$		
4. $y = (x^2 + 1)^{4/3}$		
5. $y = \sqrt{5x - 2}$		
6. $y = \sqrt{1 - x^2}$		
7. $y = (3x + 1)^{-1}$		
8. $y = (x + 2)^{-1/2}$		

In Exercises 9–14, find dy/du, du/dx, and dy/dx.

9. $y = u^2, u = 4x + 7$

10. $y = u^3, u = 3x^2 - 2$

11. $y = \sqrt{u}, u = 3 - x^2$

12. $y = 2\sqrt{u}, u = 5x + 9$

13. $y = u^{2/3}, u = 5x^4 - 2x$

14. $y = u^{-1}, u = x^3 + 2x^2$

In Exercises 15–22, match the function with the rule that you would use to find the derivative *most efficiently*.

(a) Simple Power Rule

(b) Constant Rule

(c) General Power Rule

(d) Quotient Rule

15. $f(x) = \dfrac{2}{1 - x^3}$

16. $f(x) = \dfrac{2x}{1 - x^3}$

17. $f(x) = \sqrt[3]{8^2}$

18. $f(x) = \sqrt[3]{x^2}$

19. $f(x) = \dfrac{x^2 + 2}{x}$

20. $f(x) = \dfrac{x^4 - 2x + 1}{\sqrt{x}}$

21. $f(x) = \dfrac{2}{x - 2}$

22. $f(x) = \dfrac{5}{x^2 + 1}$

In Exercises 23–40, use the General Power Rule to find the derivative of the function.

23. $y = (2x - 7)^3$

24. $y = (2x^3 + 1)^2$

25. $g(x) = (4 - 2x)^3$

26. $h(t) = (1 - t^2)^4$

27. $h(x) = (6x - x^3)^2$

28. $f(x) = (4x - x^2)^3$

29. $f(x) = (x^2 - 9)^{2/3}$

30. $f(t) = (9t + 2)^{2/3}$

31. $f(t) = \sqrt{t + 1}$

32. $g(x) = \sqrt{5 - 3x}$

33. $s(t) = \sqrt{2t^2 + 5t + 2}$

34. $y = \sqrt[3]{3x^3 + 4x}$

35. $y = \sqrt[3]{9x^2 + 4}$

36. $y = 2\sqrt{4 - x^2}$

37. $f(x) = -3\sqrt[4]{2 - 9x}$

38. $f(x) = (25 + x^2)^{-1/2}$

39. $h(x) = (4 - x^3)^{-4/3}$

40. $f(x) = (4 - 3x)^{-5/2}$

In Exercises 41–46, find an equation of the tangent line to the graph of f at the point $(2, f(2))$. Use a graphing utility to check your result by graphing the original function and the tangent line in the same viewing window.

41. $f(x) = 2(x^2 - 1)^3$

42. $f(x) = 3(9x - 4)^4$

43. $f(x) = \sqrt{4x^2 - 7}$

44. $f(x) = x\sqrt{x^2 + 5}$

45. $f(x) = \sqrt{x^2 - 2x + 1}$

46. $f(x) = (4 - 3x^2)^{-2/3}$

Ⓣ In Exercises 47–50, use a symbolic differentiation utility to find the derivative of the function. Graph the function and its derivative in the same viewing window. Describe the behavior of the function when the derivative is zero.

47. $f(x) = \dfrac{\sqrt{x} + 1}{x^2 + 1}$

48. $f(x) = \sqrt{\dfrac{2x}{x + 1}}$

49. $f(x) = \sqrt{\dfrac{x + 1}{x}}$

50. $f(x) = \sqrt{x}(2 - x^2)$

In Exercises 51–66, find the derivative of the function. State which differentiation rule(s) you used to find the derivative.

51. $y = \dfrac{1}{x - 2}$

52. $s(t) = \dfrac{1}{t^2 + 3t - 1}$

53. $y = -\dfrac{4}{(t + 2)^2}$

54. $f(x) = \dfrac{3}{(x^3 - 4)^2}$

55. $f(x) = \dfrac{1}{(x^2 - 3x)^2}$

56. $y = \dfrac{1}{\sqrt{x + 2}}$

57. $g(t) = \dfrac{1}{t^2 - 2}$

58. $g(x) = \dfrac{3}{\sqrt[3]{x^3 - 1}}$

59. $f(x) = x(3x - 9)^3$

60. $f(x) = x^3(x - 4)^2$

61. $y = x\sqrt{2x + 3}$

62. $y = t\sqrt{t + 1}$

63. $y = t^2\sqrt{t - 2}$

64. $y = \sqrt{x}(x - 2)^2$

65. $y = \left(\dfrac{6 - 5x}{x^2 - 1}\right)^2$

66. $y = \left(\dfrac{4x^2}{3 - x}\right)^3$

(T) In Exercises 67–72, find an equation of the tangent line to the graph of the function at the given point. Then use a graphing utility to graph the function and the tangent line in the same viewing window.

Function	*Point*
67. $f(t) = \dfrac{36}{(3 - t)^2}$	$(0, 4)$
68. $s(x) = \dfrac{1}{\sqrt{x^2 - 3x + 4}}$	$\left(3, \tfrac{1}{2}\right)$
69. $f(t) = (t^2 - 9)\sqrt{t + 2}$	$(-1, -8)$
70. $y = \dfrac{2x}{\sqrt{x + 1}}$	$(3, 3)$
71. $f(x) = \dfrac{x + 1}{\sqrt{2x - 3}}$	$(2, 3)$
72. $y = \dfrac{x}{\sqrt{25 + x^2}}$	$(0, 0)$

73. Compound Interest You deposit $1000 in an account with an annual interest rate of r (in decimal form) compounded monthly. At the end of 5 years, the balance is

$$A = 1000\left(1 + \frac{r}{12}\right)^{60}.$$

Find the rates of change of A with respect to r when (a) $r = 0.08$, (b) $r = 0.10$, and (c) $r = 0.12$.

74. Environment An environmental study indicates that the average daily level P of a certain pollutant in the air, in parts per million, can be modeled by the equation

$$P = 0.25\sqrt{0.5n^2 + 5n + 25}$$

where n is the number of residents of the community, in thousands. Find the rate at which the level of pollutant is increasing when the population of the community is 12,000.

75. Biology The number N of bacteria in a culture after t days is modeled by

$$N = 400\left[1 - \frac{3}{(t^2 + 2)^2}\right].$$

Complete the table. What can you conclude?

t	0	1	2	3	4
dN/dt					

76. Depreciation The value V of a machine t years after it is purchased is inversely proportional to the square root of $t + 1$. The initial value of the machine is $10,000.

(a) Write V as a function of t.

(b) Find the rate of depreciation when $t = 1$.

(c) Find the rate of depreciation when $t = 3$.

77. Depreciation Repeat Exercise 76 given that the value of the machine t years after it is purchased is inversely proportional to the cube root of $t + 1$.

(T) **78. Credit Card Rate** The average annual rate r (in percent form) for commercial bank credit cards from 2000 through 2005 can be modeled by

$$r = \sqrt{-1.7409t^4 + 18.070t^3 - 52.68t^2 + 10.9t + 249}$$

where t represents the year, with $t = 0$ corresponding to 2000. *(Source: Federal Reserve Bulletin)*

(a) Find the derivative of this model. Which differentiation rule(s) did you use?

(b) Use a graphing utility to graph the derivative on the interval $0 \le t \le 5$.

(c) Use the *trace* feature to find the years during which the finance rate was changing the most.

(d) Use the *trace* feature to find the years during which the finance rate was changing the least.

True or False? In Exercises 79 and 80, determine whether the statement is true or false. If it is false, explain why or give an example that shows it is false.

79. If $y = (1 - x)^{1/2}$, then $y' = \tfrac{1}{2}(1 - x)^{-1/2}$.

80. If y is a differentiable function of u, u is a differentiable function of v, and v is a differentiable function of x, then

$$\frac{dy}{dx} = \frac{dy}{du} \cdot \frac{du}{dv} \cdot \frac{dv}{dx}.$$

81. Given that $f(x) = h(g(x))$, find $f'(2)$ for each of the following.

(a) $g(2) = -6$ and $g'(2) = 5$, $h(5) = 4$ and $h'(-6) = 3$

(b) $g(2) = -1$ and $g'(2) = -2$, $h(2) = 4$ and $h'(-1) = 5$

Algebra Review

Simplifying Algebraic Expressions

To be successful in using derivatives, you must be good at simplifying algebraic expressions. Here are some helpful simplification techniques.

1. Combine *like terms*. This may involve expanding an expression by multiplying factors.

2. Divide out *like factors* in the numerator and denominator of an expression.

3. Factor an expression.

4. Rationalize a denominator.

5. Add, subtract, multiply, or divide fractions.

Example 1 Simplifying a Fractional Expression

a.
$$\frac{(x + \Delta x)^2 - x^2}{\Delta x} = \frac{x^2 + 2x(\Delta x) + (\Delta x)^2 - x^2}{\Delta x}$$ Expand expression.

$$= \frac{2x(\Delta x) + (\Delta x)^2}{\Delta x}$$ Combine like terms.

$$= \frac{\Delta x(2x + \Delta x)}{\Delta x}$$ Factor.

$$= 2x + \Delta x, \qquad \Delta x \neq 0$$ Divide out like factors.

b.
$$\frac{(x^2 - 1)(-2 - 2x) - (3 - 2x - x^2)(2)}{(x^2 - 1)^2}$$

$$= \frac{(-2x^2 - 2x^3 + 2 + 2x) - (6 - 4x - 2x^2)}{(x^2 - 1)^2}$$ Expand expression.

$$= \frac{-2x^2 - 2x^3 + 2 + 2x - 6 + 4x + 2x^2}{(x^2 - 1)^2}$$ Remove parentheses.

$$= \frac{-2x^3 + 6x - 4}{(x^2 - 1)^2}$$ Combine like terms.

c.
$$2\left(\frac{2x + 1}{3x}\right)\left[\frac{3x(2) - (2x + 1)(3)}{(3x)^2}\right]$$

$$= 2\left(\frac{2x + 1}{3x}\right)\left[\frac{6x - (6x + 3)}{(3x)^2}\right]$$ Multiply factors.

$$= \frac{2(2x + 1)(6x - 6x - 3)}{(3x)^3}$$ Multiply fractions and remove parentheses.

$$= \frac{2(2x + 1)(-3)}{3(9)x^3}$$ Combine like terms and factor.

$$= \frac{-2(2x + 1)}{9x^3}$$ Divide out like factors.

| Example 2 | **Simplifying an Expression with Powers or Radicals** |

a. $(2x + 1)^2(6x + 1) + (3x^2 + x)(2)(2x + 1)(2)$

$= (2x + 1)[(2x + 1)(6x + 1) + (3x^2 + x)(2)(2)]$ Factor.

$= (2x + 1)[12x^2 + 8x + 1 + (12x^2 + 4x)]$ Multiply factors.

$= (2x + 1)(12x^2 + 8x + 1 + 12x^2 + 4x)$ Remove parentheses.

$= (2x + 1)(24x^2 + 12x + 1)$ Combine like terms.

b. $(-1)(6x^2 - 4x)^{-2}(12x - 4)$

$= \dfrac{(-1)(12x - 4)}{(6x^2 - 4x)^2}$ Rewrite as a fraction.

$= \dfrac{(-1)(4)(3x - 1)}{(6x^2 - 4x)^2}$ Factor.

$= \dfrac{-4(3x - 1)}{(6x^2 - 4x)^2}$ Multiply factors.

c. $(x)\left(\dfrac{1}{2}\right)(2x + 3)^{-1/2} + (2x + 3)^{1/2}(1)$

$= (2x + 3)^{-1/2}\left(\dfrac{1}{2}\right)[x + (2x + 3)(2)]$ Factor.

$= \dfrac{x + 4x + 6}{(2x + 3)^{1/2}(2)}$ Rewrite as a fraction.

$= \dfrac{5x + 6}{2(2x + 3)^{1/2}}$ Combine like terms.

d. $\dfrac{x^2\left(\frac{1}{2}\right)(2x)(x^2 + 1)^{-1/2} - (x^2 + 1)^{1/2}(2x)}{x^4}$

$= \dfrac{(x^3)(x^2 + 1)^{-1/2} - (x^2 + 1)^{1/2}(2x)}{x^4}$ Multiply factors.

$= \dfrac{(x^2 + 1)^{-1/2}(x)[x^2 - (x^2 + 1)(2)]}{x^4}$ Factor.

$= \dfrac{x[x^2 - (2x^2 + 2)]}{(x^2 + 1)^{1/2}x^4}$ Write with positive exponents.

$= \dfrac{x^2 - 2x^2 - 2}{(x^2 + 1)^{1/2}x^3}$ Divide out like factors and remove parentheses.

$= \dfrac{-x^2 - 2}{(x^2 + 1)^{1/2}x^3}$ Combine like terms.

All but one of the expressions in this Algebra Review are derivatives. Can you see what the original function is? Explain your reasoning.

Chapter Summary and Study Strategies

After studying this chapter, you should have acquired the following skills.
The exercise numbers are keyed to the Review Exercises that begin on page 621.
Answers to odd-numbered Review Exercises are given in the back of the text.

Section 7.1	Review Exercises
■ Determine whether limits exist. If they do, find the limits.	*1–18*
■ Use a table to estimate one-sided limits.	*19, 20*
■ Determine whether statements about limits are true or false.	*21–26*

Section 7.2

■ Determine whether functions are continuous at a point, on an open interval, and on a closed interval.	*27–34*
■ Determine the constant such that f is continuous.	*35, 36*
■ Use analytic and graphical models of real-life data to solve real-life problems.	*37–40*

Section 7.3

■ Approximate the slope of the tangent line to a graph at a point.	*41–44*
■ Interpret the slope of a graph in a real-life setting.	*45–48*
■ Use the limit definition to find the derivative of a function and the slope of a graph at a point.	*49–56*

$$f'(x) = \lim_{\Delta x \to 0} \frac{f(x + \Delta x) - f(x)}{\Delta x}$$

■ Use the derivative to find the slope of a graph at a point.	*57–64*
■ Use the graph of a function to recognize points at which the function is not differentiable.	*65–68*

Section 7.4

■ Use the Constant Multiple Rule for differentiation.	*69, 70*

$$\frac{d}{dx}[cf(x)] = cf'(x)$$

■ Use the Sum and Difference Rules for differentiation.	*71–78*

$$\frac{d}{dx}[f(x) \pm g(x)] = f'(x) \pm g'(x)$$

Section 7.5 | Review Exercises

■ Find the average rate of change of a function over an interval and the instantaneous   79, 80
rate of change at a point.

$$\text{Average rate of change} = \frac{f(b) - f(a)}{b - a}$$

$$\text{Instantaneous rate of change} = \lim_{\Delta x \to 0} \frac{f(x + \Delta x) - f(x)}{\Delta x}$$

■ Find the velocity of an object that is moving in a straight line.   81, 82

■ Find the average and instantaneous rates of change of a quantity in a real-life problem.   83, 84

■ Create mathematical models for the revenue, cost, and profit for a product.   85, 86

$$P = R - C, \quad R = xp$$

■ Find the marginal revenue, marginal cost, and marginal profit for a product.   87–96

Section 7.6

■ Use the Product Rule for differentiation.   97–100

$$\frac{d}{dx}[f(x)g(x)] = f(x)g'(x) + g(x)f'(x)$$

■ Use the Quotient Rule for differentiation.   101, 102

$$\frac{d}{dx}\left[\frac{f(x)}{g(x)}\right] = \frac{g(x)f'(x) - f(x)g'(x)}{[g(x)]^2}$$

Section 7.7

■ Use the General Power Rule for differentiation.   103–106

$$\frac{d}{dx}[u^n] = nu^{n-1}u'$$

■ Use differentiation rules efficiently to find the derivative of any algebraic function,   107–116
then simplify the result.

■ Use derivatives to answer questions about real-life situations.   (Sections 7.3–7.7)   117, 118

Study Strategies

■ **Simplify Your Derivatives** Often our students ask if they have to simplify their derivatives. Our answer is "Yes, if you expect to use them." In the next two chapters, you will see that almost all applications of derivatives require that the derivatives be written in simplified form. It is not difficult to see the advantage of a derivative in simplified form. Consider, for instance, the derivative of

$$f(x) = \frac{x}{\sqrt{x^2 + 1}}.$$

The "raw form" produced by the Quotient and Chain Rules

$$f'(x) = \frac{(x^2 + 1)^{1/2}(1) - (x)(\frac{1}{2})(x^2 + 1)^{-1/2}(2x)}{\left(\sqrt{x^2 + 1}\right)^2}$$

is obviously much more difficult to use than the simplified form

$$f'(x) = \frac{1}{(x^2 + 1)^{3/2}}.$$

■ **List Units of Measure in Applied Problems** When using derivatives in real-life applications, be sure to list the units of measure for each variable. For instance, if R is measured in dollars and t is measured in years, then the derivative dR/dt is measured in dollars per year.

Review Exercises

See www.CalcChat.com for worked-out solutions to odd-numbered exercises.

In Exercises 1–18, find the limit (if it exists).

1. $\lim\limits_{x \to 2} (5x - 3)$

2. $\lim\limits_{x \to 2} (2x + 9)$

3. $\lim\limits_{x \to 2} (5x - 3)(2x + 3)$

4. $\lim\limits_{x \to 2} \dfrac{5x - 3}{2x + 9}$

5. $\lim\limits_{t \to 3} \dfrac{t^2 + 1}{t}$

6. $\lim\limits_{t \to 0} \dfrac{t^2 + 1}{t}$

7. $\lim\limits_{t \to 1} \dfrac{t + 1}{t - 2}$

8. $\lim\limits_{t \to 2} \dfrac{t + 1}{t - 2}$

9. $\lim\limits_{x \to -2} \dfrac{x + 2}{x^2 - 4}$

10. $\lim\limits_{x \to 3^-} \dfrac{x^2 - 9}{x - 3}$

11. $\lim\limits_{x \to 0^+} \left(x - \dfrac{1}{x} \right)$

12. $\lim\limits_{x \to 1/2} \dfrac{2x - 1}{6x - 3}$

13. $\lim\limits_{x \to 0} \dfrac{[1/(x - 2)] - 1}{x}$

14. $\lim\limits_{x \to 0} \dfrac{[1/(x - 4)] - (1/4)}{x}$

15. $\lim\limits_{t \to 0} \dfrac{(1/\sqrt{t + 4}) - (1/2)}{t}$

16. $\lim\limits_{s \to 0} \dfrac{(1/\sqrt{1 + s}) - 1}{s}$

17. $\lim\limits_{\Delta x \to 0} \dfrac{(x + \Delta x)^3 - (x + \Delta x) - (x^3 - x)}{\Delta x}$

18. $\lim\limits_{\Delta x \to 0} \dfrac{1 - (x + \Delta x)^2 - (1 - x^2)}{\Delta x}$

In Exercises 19 and 20, use a table to estimate the limit.

19. $\lim\limits_{x \to 1^+} \dfrac{\sqrt{2x + 1} - \sqrt{3}}{x - 1}$

20. $\lim\limits_{x \to 1^+} \dfrac{1 - \sqrt[3]{x}}{x - 1}$

True or False? In Exercises 21–26, determine whether the statement is true or false. If it is false, explain why or give an example that shows it is false.

21. $\lim\limits_{x \to 0} \dfrac{|x|}{x} = 1$

22. $\lim\limits_{x \to 0} x^3 = 0$

23. $\lim\limits_{x \to 0} \sqrt{x} = 0$

24. $\lim\limits_{x \to 0} \sqrt[3]{x} = 0$

25. $\lim\limits_{x \to 2} f(x) = 3, \quad f(x) = \begin{cases} 3, & x \le 2 \\ 0, & x > 2 \end{cases}$

26. $\lim\limits_{x \to 3} f(x) = 1, \quad f(x) = \begin{cases} x - 2, & x \le 3 \\ -x^2 + 8x - 14, & x > 3 \end{cases}$

In Exercises 27–34, describe the interval(s) on which the function is continuous. Explain why the function is continuous on the interval(s). If the function has a discontinuity, identify the conditions of continuity that are not satisfied.

27. $f(x) = \dfrac{1}{(x + 4)^2}$

28. $f(x) = \dfrac{x + 2}{x}$

29. $f(x) = \dfrac{3}{x + 1}$

30. $f(x) = \dfrac{x + 1}{2x + 2}$

31. $f(x) = [\![x + 3]\!]$

32. $f(x) = [\![x]\!] - 2$

33. $f(x) = \begin{cases} x, & x \le 0 \\ x + 1, & x > 0 \end{cases}$

34. $f(x) = \begin{cases} x, & x \le 0 \\ x^2, & x > 0 \end{cases}$

In Exercises 35 and 36, find the constant a such that f is continuous on the entire real line.

35. $f(x) = \begin{cases} -x + 1, & x \le 3 \\ ax - 8, & x > 3 \end{cases}$

36. $f(x) = \begin{cases} x + 1, & x < 1 \\ 2x + a, & x \ge 1 \end{cases}$

37. Consumer Awareness The cost C (in dollars) of making x photocopies at a copy shop is given below.

$$C(x) = \begin{cases} 0.15x, & 0 < x \le 25 \\ 0.10x, & 25 < x \le 100 \\ 0.07x, & 100 < x \le 500 \\ 0.05x, & x > 500 \end{cases}$$

(T) (a) Use a graphing utility to graph the function and discuss its continuity. At what values is the function not continuous? Explain your reasoning.

(b) Find the cost of making 100 copies.

38. Salary Contract A union contract guarantees a 10% salary increase yearly for 3 years. For a current salary of $28,000, the salary S (in thousands of dollars) for the next 3 years is given by

$$S(t) = \begin{cases} 28.00, & 0 < t \le 1 \\ 30.80, & 1 < t \le 2 \\ 33.88, & 2 < t \le 3 \end{cases}$$

where $t = 0$ represents the present year. Does the limit of S exist as t approaches 2? Explain your reasoning.

39. Recycling A recycling center pays $0.50 for each pound of aluminum cans. Twenty-four aluminum cans weigh one pound. A mathematical model for the amount A paid by the recycling center is

$$A = \dfrac{1}{2} \left[\!\left[\dfrac{x}{24} \right]\!\right]$$

where x is the number of cans.

(T) (a) Use a graphing utility to graph the function and then discuss its continuity.

(b) How much does the recycling center pay out for 1500 cans?

(T) **40. Consumer Awareness** A pay-as-you-go cellular phone charges $1 for the first time you access the phone and $0.10 for each additional minute or fraction thereof. Use the greatest integer function to create a model for the cost C of a phone call lasting t minutes. Use a graphing utility to graph the function, and discuss its continuity.

In Exercises 41–44, approximate the slope of the tangent line to the graph at (x, y).

41.

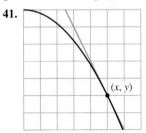

42.

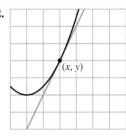

43.

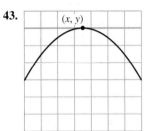

44.

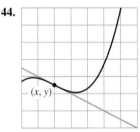

45. Sales The graph approximates the annual sales S (in millions of dollars per year) of Home Depot for the years 1999 through 2005, where t is the year, with $t = 9$ corresponding to 1999. Estimate the slopes of the graph when $t = 10$, $t = 13$, and $t = 15$. Interpret each slope in the context of the problem. *(Source: The Home Depot, Inc.)*

46. Consumer Trends The graph approximates the number of subscribers S (in millions per year) of cellular telephones for the years 1996 through 2005, where t is the year, with $t = 6$ corresponding to 1996. Estimate the slopes of the graph when $t = 7$, $t = 11$, and $t = 15$. Interpret each slope in the context of the problem. *(Source: Cellular Telecommunications & Internet Association)*

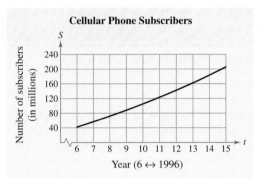

Figure for 46

47. Medicine The graph shows the estimated number of milligrams of a pain medication M in the bloodstream t hours after a 1000-milligram dose of the drug has been given. Estimate the slopes of the graph at $t = 0, 4$, and 6.

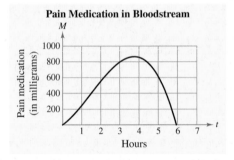

48. White-Water Rafting Two white-water rafters leave a campsite simultaneously and start downstream on a 9-mile trip. Their distances from the campsite are given by $s = f(t)$ and $s = g(t)$, where s is measured in miles and t is measured in hours.

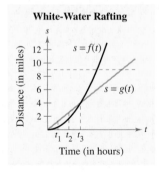

(a) Which rafter is traveling at a greater rate at t_1?

(b) What can you conclude about their rates at t_2? at t_3?

(c) Which rafter finishes the trip first? Explain your reasoning.

In Exercises 49–56, use the limit definition to find the derivative of the function. Then use the limit definition to find the slope of the tangent line to the graph of f at the given point.

49. $f(x) = -3x - 5$; $(-2, 1)$ **50.** $f(x) = 7x + 3$; $(-1, 4)$

51. $f(x) = x^2 - 4x$; $(1, -3)$ **52.** $f(x) = x^2 + 10$; $(2, 14)$

53. $f(x) = \sqrt{x + 9}$; $(-5, 2)$ **54.** $f(x) = \sqrt{x - 1}$; $(10, 3)$

55. $f(x) = \dfrac{1}{x - 5}$; $(6, 1)$ **56.** $f(x) = \dfrac{1}{x + 4}$; $(-3, 1)$

In Exercises 57–64, find the slope of the graph of f at the given point.

57. $f(x) = 5 - 3x$; $(1, -2)$ **58.** $f(x) = 1 - 4x$; $(2, -7)$

59. $f(x) = -\frac{1}{2}x^2 + 2x$; $(2, 2)$ **60.** $f(x) = 4 - x^2$; $(-1, 3)$

61. $f(x) = \sqrt{x} + 2$; $(9, 5)$ **62.** $f(x) = 2\sqrt{x} + 1$; $(4, 5)$

63. $f(x) = \dfrac{5}{x}$; $(1, 5)$ **64.** $f(x) = \dfrac{2}{x} - 1$; $\left(\dfrac{1}{2}, 3\right)$

In Exercises 65–68, determine the x-value at which the function is not differentiable.

65. $y = \dfrac{x + 1}{x - 1}$

66. $y = -|x| + 3$

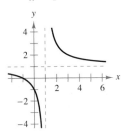

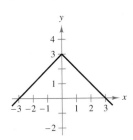

67. $y = \begin{cases} -x - 2, & x \le 0 \\ x^3 + 2, & x > 0 \end{cases}$

68. $y = (x + 1)^{2/3}$

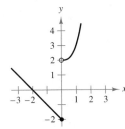

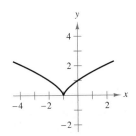

(T) In Exercises 69–78, find the equation of the tangent line at the given point. Then use a graphing utility to graph the function and the equation of the tangent line in the same viewing window.

69. $g(t) = \dfrac{2}{3t^2}$, $\left(1, \dfrac{2}{3}\right)$

70. $h(x) = \dfrac{2}{(3x)^2}$, $\left(2, \dfrac{1}{18}\right)$

71. $f(x) = x^2 + 3$, $(1, 4)$

72. $f(x) = 2x^2 - 3x + 1$, $(2, 3)$

73. $y = 11x^4 - 5x^2 + 1$, $(-1, 7)$

74. $y = x^3 - 5 + \dfrac{3}{x^3}$, $(-1, -9)$

75. $f(x) = \sqrt{x} - \dfrac{1}{\sqrt{x}}$, $(1, 0)$

76. $f(x) = 2x^{-3} + 4 - \sqrt{x}$, $(1, 5)$

77. $f(x) = \dfrac{x^2 + 3}{x}$, $(1, 4)$

78. $f(x) = -x^2 - 4x - 4$, $(-4, -4)$

In Exercises 79 and 80, find the average rate of change of the function over the indicated interval. Then compare the average rate of change with the instantaneous rates of change at the endpoints of the interval.

79. $f(x) = x^2 + 3x - 4$; $[0, 1]$

80. $f(x) = x^3 + x$; $[-2, 2]$

81. Velocity A rock is dropped from a tower on the Brooklyn Bridge, 276 feet above the East River. Let t represent the time in seconds.

(a) Write a model for the position function (assume that air resistance is negligible).

(b) Find the average velocity during the first 2 seconds.

(c) Find the instantaneous velocities when $t = 2$ and $t = 3$.

(d) How long will it take for the rock to hit the water?

(e) When it hits the water, what is the rock's speed?

82. Velocity The straight-line distance s (in feet) traveled by an accelerating bicyclist can be modeled by

$$s = 2t^{3/2}, \quad 0 \le t \le 8$$

where t is the time (in seconds). Complete the table, showing the velocity of the bicyclist at two-second intervals.

Time, t	0	2	4	6	8
Velocity					

83. Sales The annual sales S (in millions of dollars per year) of Home Depot for the years 1999 through 2005 can be modeled by

$$S = 123.833t^3 - 4319.55t^2 + 56{,}278.0t - 208{,}517$$

where t is the time in years, with $t = 9$ corresponding to 1999. A graph of this model appears in Exercise 5. *(Source: The Home Depot, Inc.)*

(a) Find the average rate of change for the interval from 1999 through 2005.

(b) Find the instantaneous rates of change of the model for 1999 and 2005.

(c) Interpret the results of parts (a) and (b) in the context of the problem.

84. Consumer Trends The numbers of subscribers S (in millions per year) of cellular telephones for the years 1996 through 2005 can be modeled by

$$S = \frac{-33.2166 + 11.6732t}{1 - 0.0207t}$$

where t is the time in years, with $t = 6$ corresponding to 1996. A graph of this model appears in Exercise 6. *(Source: Cellular Telecommunications & Internet Association)*

(a) Find the average rate of change for the interval from 2000 through 2005.

(b) Find the instantaneous rates of change of the model for 2000 and 2005.

(c) Interpret the results of parts (a) and (b) in the context of the problem.

85. Cost, Revenue, and Profit The fixed cost of operating a small flower shop is $2500 per month. The average cost of a floral arrangement is $15 and the average price is $27.50. Write the monthly revenue, cost, and profit functions for the floral shop in terms of x, the number of arrangements sold.

86. Profit The weekly demand and cost functions for a product are given by

$$p = 1.89 - 0.0083x \quad \text{and} \quad C = 21 + 0.65x.$$

Write the profit function for this product.

Marginal Cost In Exercises 87–90, find the marginal cost function.

87. $C = 2500 + 320x$ **88.** $C = 225x + 4500$

89. $C = 370 + 2.55\sqrt{x}$ **90.** $C = 475 + 5.25x^{2/3}$

Marginal Revenue In Exercises 91–94, find the marginal revenue function.

91. $R = 200x - \frac{1}{5}x^2$ **92.** $R = 150x - \frac{3}{4}x^2$

93. $R = \frac{35x}{\sqrt{x - 2}}, \quad x \geq 6$ **94.** $R = x\left(5 + \frac{10}{\sqrt{x}}\right)$

Marginal Profit In Exercises 95 and 96, find the marginal profit function.

95. $P = -0.0002x^3 + 6x^2 - x - 2000$

96. $P = -\frac{1}{15}x^3 + 4000x^2 - 120x - 144{,}000$

In Exercises 97–116, find the derivative of the function. Simplify your result. State which differentiation rule(s) you used to find the derivative.

97. $f(x) = x^3(5 - 3x^2)$ **98.** $y = (3x^2 + 7)(x^2 - 2x)$

99. $y = (4x - 3)(x^3 - 2x^2)$ **100.** $s = \left(4 - \frac{1}{t^2}\right)(t^2 - 3t)$

101. $f(x) = \frac{6x - 5}{x^2 + 1}$ **102.** $f(x) = \frac{x^2 + x - 1}{x^2 - 1}$

103. $f(x) = (5x^2 + 2)^3$ **104.** $f(x) = \sqrt[3]{x^2 - 1}$

105. $h(x) = \frac{2}{\sqrt{x + 1}}$

106. $g(x) = \sqrt{x^6 - 12x^3 + 9}$

107. $g(x) = x\sqrt{x^2 + 1}$

108. $g(t) = \frac{t}{(1 - t)^3}$

109. $f(x) = x(1 - 4x^2)^2$

110. $f(x) = \left(x^2 + \frac{1}{x}\right)^5$

111. $h(x) = [x^2(2x + 3)]^3$

112. $f(x) = [(x - 2)(x + 4)]^2$

113. $f(x) = x^2(x - 1)^5$

114. $f(s) = s^3(s^2 - 1)^{5/2}$

115. $h(t) = \frac{\sqrt{3t + 1}}{(1 - 3t)^2}$

116. $g(x) = \frac{(3x + 1)^2}{(x^2 + 1)^2}$

117. Physical Science The temperature T (in degrees Fahrenheit) of food placed in a freezer can be modeled by

$$T = \frac{1300}{t^2 + 2t + 25}$$

where t is the time (in hours).

(a) Find the rates of change of T when $t = 1$, $t = 3$, $t = 5$, and $t = 10$.

Ⓣ (b) Graph the model on a graphing utility and describe the rate at which the temperature is changing.

118. Forestry According to the *Doyle Log Rule*, the volume V (in board-feet) of a log of length L (feet) and diameter D (inches) at the small end is

$$V = \left(\frac{D - 4}{4}\right)^2 L.$$

Find the rates at which the volume is changing with respect to D for a 12-foot-long log whose smallest diameter is (a) 8 inches, (b) 16 inches, (c) 24 inches, and (d) 36 inches.

Take this test as you would take a test in class. When you are done, check your work against the answers given in the back of the book.

In Exercises 1–4, find the limit (if it exists).

1. $\lim\limits_{x \to 0} \dfrac{x + 5}{x - 5}$

2. $\lim\limits_{x \to 5} \dfrac{x + 5}{x - 5}$

3. $\lim\limits_{x \to -3} \dfrac{x^2 + 2x - 3}{x^2 + 4x + 3}$

4. $\lim\limits_{x \to 0} \dfrac{\sqrt{x + 9} - 3}{x}$

In Exercises 5–8, describe the interval(s) on which the function is continuous. Explain why the function is continuous on the interval(s). If the function has a discontinuity at a point, identify all conditions of continuity that are not satisfied.

5. $f(x) = x^2 - 2x + 4$

6. $f(x) = \dfrac{x^2 - 16}{x - 4}$

7. $f(x) = \sqrt{5 - x}$

8. $f(x) = \begin{cases} 1 - x, & x < 1 \\ x - x^2, & x \geq 1 \end{cases}$

In Exercises 9 and 10, use the limit definition to find the derivative of the function. Then find the slope of the tangent line to the graph of f at the given point.

9. $f(x) = x^2 + 1; \ (2, 5)$

10. $f(x) = \sqrt{x} - 2; \ (4, 0)$

In Exercises 11–19, find the derivative of the function. Simplify your result.

11. $f(t) = t^3 + 2t$

12. $f(x) = 4x^2 - 8x + 1$

13. $f(x) = x^{3/2}$

14. $f(x) = (x + 3)(x - 3)$

15. $f(x) = -3x^{-3}$

16. $f(x) = \sqrt{x}(5 + x)$

17. $f(x) = (3x^2 + 4)^2$

18. $f(x) = \sqrt{1 - 2x}$

19. $f(x) = \dfrac{(5x - 1)^3}{x}$

(T) **20.** Find an equation of the tangent line to the graph of $f(x) = x - \dfrac{1}{x}$ at the point $(1, 0)$. Then use a graphing utility to graph the function and the tangent line in the same viewing window.

21. The annual sales S (in millions of dollars per year) of Bausch & Lomb for the years 1999 through 2005 can be modeled by

$$S = -2.9667t^3 + 135.008t^2 - 1824.42t + 9426.3, \ 9 \leq t \leq 15$$

where t represents the year, with $t = 9$ corresponding to 1999. *(Source: Bausch & Lomb, Inc.)*

(a) Find the average rate of change for the interval from 2001 through 2005.

(b) Find the instantaneous rates of change of the model for 2001 and 2005.

(c) Interpret the results of parts (a) and (b) in the context of the problem.

22. The monthly demand and cost functions for a product are given by

$$p = 1700 - 0.016x \quad \text{and} \quad C = 715{,}000 + 240x.$$

Write the profit function for this product.

8 Applications of the Derivative

© Schlegelmilch/Corbis

Higher-order derivatives are used to determine the acceleration function of a sports car. The acceleration function shows the changes in the car's velocity. As the car reaches its "cruising" speed, is the acceleration increasing or decreasing? (See Section 8.1, Exercise 45.)

Applications

Derivatives have many real-life applications. The applications listed below represent a sample of the applications in this chapter.

Higher-Order Derivatives

- Find higher-order derivatives.
- Find and use the position functions to determine the velocity and acceleration of moving objects.

Second, Third, and Higher-Order Derivatives

The derivative of f' is the **second derivative** of f and is denoted by f''.

$$\frac{d}{dx}[f'(x)] = f''(x) \qquad \text{Second derivative}$$

The derivative of f'' is the **third derivative** of f and is denoted by f'''.

$$\frac{d}{dx}[f''(x)] = f'''(x) \qquad \text{Third derivative}$$

By continuing this process, you obtain **higher-order derivatives** of f. Higher-order derivatives are denoted as follows.

STUDY TIP

In the context of higher-order derivatives, the "standard" derivative f' is often called the **first derivative** of f.

DISCOVERY

For each function, find the indicated higher-order derivative.

a. $y = x^2$ **b.** $y = x^3$

y'' y'''

c. $y = x^4$ **d.** $y = x^n$

$y^{(4)}$ $y^{(n)}$

Notation for Higher-Order Derivatives

1. 1st derivative:	y',	$f'(x)$,	$\dfrac{dy}{dx}$,	$\dfrac{d}{dx}[f(x)]$,	$D_x[y]$
2. 2nd derivative:	y'',	$f''(x)$,	$\dfrac{d^2y}{dx^2}$,	$\dfrac{d^2}{dx^2}[f(x)]$,	$D_x^2[y]$
3. 3rd derivative:	y''',	$f'''(x)$,	$\dfrac{d^3y}{dx^3}$,	$\dfrac{d^3}{dx^3}[f(x)]$,	$D_x^3[y]$
4. 4th derivative:	$y^{(4)}$,	$f^{(4)}(x)$,	$\dfrac{d^4y}{dx^4}$,	$\dfrac{d^4}{dx^4}[f(x)]$,	$D_x^4[y]$
5. nth derivative:	$y^{(n)}$,	$f^{(n)}(x)$,	$\dfrac{d^ny}{dx^n}$,	$\dfrac{d^n}{dx^n}[f(x)]$,	$D_x^n[y]$

Example 1 Finding Higher-Order Derivatives

Find the first five derivatives of $f(x) = 2x^4 - 3x^2$.

$f(x) = 2x^4 - 3x^2$	Write original function.
$f'(x) = 8x^3 - 6x$	First derivative
$f''(x) = 24x^2 - 6$	Second derivative
$f'''(x) = 48x$	Third derivative
$f^{(4)}(x) = 48$	Fourth derivative
$f^{(5)}(x) = 0$	Fifth derivative

✓ **CHECKPOINT 1**

Find the first four derivatives of $f(x) = 6x^3 - 2x^2 + 1$. ■

Example 2   **Finding Higher-Order Derivatives**

Find the value of $g'''(2)$ for the function

$$g(t) = -t^4 + 2t^3 + t + 4.$$ Original function

SOLUTION   Begin by differentiating three times.

$$g'(t) = -4t^3 + 6t^2 + 1$$ First derivative

$$g''(t) = -12t^2 + 12t$$ Second derivative

$$g'''(t) = -24t + 12$$ Third derivative

Then, evaluate the third derivative of g at $t = 2$.

$$g'''(2) = -24(2) + 12$$

$$= -36$$ Value of third derivative

TECHNOLOGY

(T) Higher-order derivatives of nonpolynomial functions can be difficult to find by hand. If you have access to a symbolic differentiation utility, try using it to find higher-order derivatives.

✓ **CHECKPOINT 2**

Find the value of $g'''(1)$ for $g(x) = x^4 - x^3 + 2x$. ■

Examples 1 and 2 show how to find higher-order derivatives of *polynomial* functions. Note that with each successive differentiation, the degree of the polynomial drops by one. Eventually, higher-order derivatives of polynomial functions degenerate to a constant function. Specifically, the nth-order derivative of an nth-degree polynomial function

$$f(x) = a_n x^n + a_{n-1} x^{n-1} + \cdots + a_1 x + a_0$$

is the constant function

$$f^{(n)}(x) = n! a_n$$

where $n! = 1 \cdot 2 \cdot 3 \cdots n$. Each derivative of order higher than n is the zero function. Polynomial functions are the *only* functions with this characteristic. For other functions, successive differentiation never produces a constant function.

Example 3   **Finding Higher-Order Derivatives**

Find the first four derivatives of $y = x^{-1}$.

$$y = x^{-1} = \frac{1}{x}$$ Write original function.

$$y' = (-1)x^{-2} = -\frac{1}{x^2}$$ First derivative

$$y'' = (-1)(-2)x^{-3} = \frac{2}{x^3}$$ Second derivative

✓ **CHECKPOINT 3**

Find the fourth derivative of

$$y = \frac{1}{x^2}. ■$$

$$y''' = (-1)(-2)(-3)x^{-4} = -\frac{6}{x^4}$$ Third derivative

$$y^{(4)} = (-1)(-2)(-3)(-4)x^{-5} = \frac{24}{x^5}$$ Fourth derivative

Acceleration

In Section 7.5, you saw that the velocity of an object moving in a straight path (neglecting air resistance) is given by the derivative of its position function. In other words, the rate of change of the position with respect to time is defined to be the velocity. In a similar way, the rate of change of the velocity with respect to time is defined to be the **acceleration** of the object.

$$s = f(t)$$ 　　　　　　Position function

$$\frac{ds}{dt} = f'(t)$$ 　　　　　Velocity function

$$\frac{d^2s}{dt^2} = f''(t)$$ 　　　　Acceleration function

To find the position, velocity, or acceleration at a particular time t, substitute the given value of t into the appropriate function, as illustrated in Example 4.

Example 4   Finding Acceleration

A ball is thrown upward from the top of a 160-foot cliff, as shown in Figure 8.1. The initial velocity of the ball is 48 feet per second, which implies that the position function is

$$s = -16t^2 + 48t + 160$$

where the time t is measured in seconds. Find the height, the velocity, and the acceleration of the ball when $t = 3$.

SOLUTION   Begin by differentiating to find the velocity and acceleration functions.

$$s = -16t^2 + 48t + 160$$ 　　Position function

$$\frac{ds}{dt} = -32t + 48$$ 　　　Velocity function

$$\frac{d^2s}{dt^2} = -32$$ 　　　　Acceleration function

To find the height, velocity, and acceleration when $t = 3$, substitute $t = 3$ into each of the functions above.

Height $= -16(3)^2 + 48(3) + 160 = 160$ feet

Velocity $= -32(3) + 48 = -48$ feet per second

Acceleration $= -32$ feet per second squared

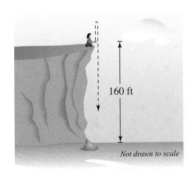

160 ft

Not drawn to scale

FIGURE 8.1

✓CHECKPOINT 4

A ball is thrown upward from the top of an 80-foot cliff with an initial velocity of 64 feet per second. Give the position function. Then find the velocity and acceleration functions. ■

The acceleration due to gravity on the surface of the moon is only about one-sixth that exerted by Earth. So, if you were on the moon and threw an object into the air, it would rise to a greater height than it would on Earth's surface.

In Example 4, notice that the acceleration of the ball is -32 feet per second squared at any time t. This constant acceleration is due to the gravitational force of Earth and is called the **acceleration due to gravity.** Note that the negative value indicates that the ball is being pulled *down*—toward Earth.

Although the acceleration exerted on a falling object is relatively constant near Earth's surface, it varies greatly throughout our solar system. Large planets exert a much greater gravitational pull than do small planets or moons. The next example describes the motion of a free-falling object on the moon.

Example 5 Finding Acceleration on the Moon

An astronaut standing on the surface of the moon throws a rock into the air. The height s (in feet) of the rock is given by

$$s = -\frac{27}{10}t^2 + 27t + 6$$

where t is measured in seconds. How does the acceleration due to gravity on the moon compare with that on Earth?

SOLUTION

$$s = -\frac{27}{10}t^2 + 27t + 6 \qquad \text{Position function}$$

$$\frac{ds}{dt} = -\frac{27}{5}t + 27 \qquad \text{Velocity function}$$

$$\frac{d^2s}{dt^2} = -\frac{27}{5} \qquad \text{Acceleration function}$$

So, the acceleration at any time is

$$-\frac{27}{5} = -5.4 \text{ feet per second squared}$$

—about one-sixth of the acceleration due to gravity on Earth.

✓ **CHECKPOINT 5**

The position function on Earth, where s is measured in meters, t is measured in seconds, v_0 is the initial velocity in meters per second, and h_0 is the initial height in meters, is

$$s = -4.9t^2 + v_0 t + h_0.$$

If the initial velocity is 2.2 and the initial height is 3.6, what is the acceleration due to gravity on Earth in meters per second per second? ■

The position function described in Example 5 neglects air resistance, which is appropriate because the moon has no atmosphere—and *no air resistance.* This means that the position function for any free-falling object on the moon is given by

$$s = -\frac{27}{10}t^2 + v_0 t + h_0$$

where s is the height (in feet), t is the time (in seconds), v_0 is the initial velocity, and h_0 is the initial height. For instance, the rock in Example 5 was thrown upward with an initial velocity of 27 feet per second and had an initial height of 6 feet. This position function is valid for all objects, whether heavy ones such as hammers or light ones such as feathers.

In 1971, astronaut David R. Scott demonstrated the lack of atmosphere on the moon by dropping a hammer and a feather from the same height. Both took exactly the same time to fall to the ground. If they were dropped from a height of 6 feet, how long did each take to hit the ground?

Example 6 **Finding Velocity and Acceleration**

The velocity v (in feet per second) of a certain automobile starting from rest is

$$v = \frac{80t}{t + 5}$$ Velocity function

where t is the time (in seconds). The positions of the automobile at 10-second intervals are shown in Figure 8.2. Find the velocity and acceleration of the automobile at 10-second intervals from $t = 0$ to $t = 60$.

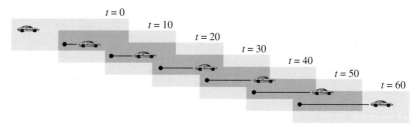

FIGURE 8.2

SOLUTION   To find the acceleration function, differentiate the velocity function.

$$\frac{dv}{dt} = \frac{(t + 5)(80) - (80t)(1)}{(t + 5)^2}$$

$$= \frac{400}{(t + 5)^2}$$ Acceleration function

t (seconds)	0	10	20	30	40	50	60
v (ft/sec)	0	53.5	64.0	68.6	71.1	72.7	73.8
$\dfrac{dv}{dt}$ (ft/sec²)	16	1.78	0.64	0.33	0.20	0.13	0.09

✓ **CHECKPOINT 6**

Use a graphing utility to graph the velocity function and acceleration function in Example 6 in the same viewing window. Compare the graphs with the table at the right. As the velocity levels off, what does the acceleration approach? ■

In the table, note that the acceleration approaches zero as the velocity levels off. This observation should agree with your experience—when riding in an accelerating automobile, you do not feel the velocity, but you do feel the acceleration. In other words, you feel changes in velocity.

CONCEPT CHECK

1. Use mathematical notation to write the third derivative of $f(x)$.

2. Give a verbal description of what is meant by $\dfrac{d^2y}{dx^2}$.

3. Complete the following: If $f(x)$ is an nth-degree polynomial, then $f^{(n+1)}(x)$ is equal to _____.

4. If the velocity of an object is constant, what is its acceleration?

Skills Review 8.1

The following warm-up exercises involve skills that were covered in earlier sections. You will use these skills in the exercise set for this section. For additional help, review Sections 1.3, 1.4, 2.5, and 7.6.

In Exercises 1–4, solve the equation.

1. $-16t^2 + 24t = 0$

2. $-16t^2 + 80t + 224 = 0$

3. $-16t^2 + 128t + 320 = 0$

4. $-16t^2 + 9t + 1440 = 0$

In Exercises 5–8, find dy/dx.

5. $y = x^2(2x + 7)$

6. $y = (x^2 + 3x)(2x^2 - 5)$

7. $y = \dfrac{x^2}{2x + 7}$

8. $y = \dfrac{x^2 + 3x}{2x^2 - 5}$

In Exercises 9 and 10, find the domain and range of f.

9. $f(x) = x^2 - 4$

10. $f(x) = \sqrt{x - 7}$

Exercises 8.1

See www.CalcChat.com for worked-out solutions to odd-numbered exercises.

In Exercises 1–16, find the second derivative of the function.

1. $f(x) = 9 - 2x$

2. $f(x) = 4x + 15$

3. $f(x) = x^2 + 7x - 4$

4. $f(x) = 3x^2 + 4x$

5. $g(t) = \frac{1}{3}t^3 - 4t^2 + 2t$

6. $f(x) = 4(x^2 - 1)^2$

7. $f(t) = \dfrac{3}{4t^2}$

8. $g(t) = 32t^{-2}$

9. $f(x) = 3(2 - x^2)^3$

10. $f(x) = x\sqrt[3]{x}$

11. $y = (x^3 - 2x)^4$

12. $y = 4(x^2 + 5x)^3$

13. $f(x) = \dfrac{x + 1}{x - 1}$

14. $g(t) = -\dfrac{4}{(t + 2)^2}$

15. $y = x^2(x^2 + 4x + 8)$

16. $h(s) = s^3(s^2 - 2s + 1)$

In Exercises 17–22, find the third derivative of the function.

17. $f(x) = x^5 - 3x^4$

18. $f(x) = x^4 - 2x^3$

19. $f(x) = 5x(x + 4)^3$

20. $f(x) = (x^3 - 6)^4$

21. $f(x) = \dfrac{3}{16x^2}$

22. $f(x) = \dfrac{1}{x}$

In Exercises 23–28, find the given value.

Function	Value
23. $g(t) = 5t^4 + 10t^2 + 3$	$g''(2)$
24. $f(x) = 9 - x^2$	$f''\left(-\sqrt{5}\right)$
25. $f(x) = \sqrt{4 - x}$	$f'''(-5)$
26. $f(t) = \sqrt{2t + 3}$	$f'''\left(\frac{1}{2}\right)$
27. $f(x) = x^2(3x^2 + 3x - 4)$	$f'''(-2)$
28. $g(x) = 2x^3(x^2 - 5x + 4)$	$g'''(0)$

In Exercises 29–34, find the higher-order derivative.

Given	Derivative
29. $f'(x) = 2x^2$	$f''(x)$
30. $f''(x) = 20x^3 - 36x^2$	$f'''(x)$
31. $f'''(x) = (3x - 1)/x$	$f^{(4)}(x)$
32. $f'''(x) = 2\sqrt{x - 1}$	$f^{(4)}(x)$
33. $f^{(4)}(x) = (x^2 + 1)^2$	$f^{(6)}(x)$
34. $f''(x) = 2x^2 + 7x - 12$	$f^{(5)}(x)$

In Exercises 35–42, find the second derivative and solve the equation $f''(x) = 0$.

35. $f(x) = x^3 - 9x^2 + 27x - 27$

36. $f(x) = 3x^3 - 9x + 1$

37. $f(x) = (x + 3)(x - 4)(x + 5)$

38. $f(x) = (x + 2)(x - 2)(x + 3)(x - 3)$

39. $f(x) = x\sqrt{x^2 - 1}$

40. $f(x) = x\sqrt{4 - x^2}$

41. $f(x) = \dfrac{x}{x^2 + 3}$

42. $f(x) = \dfrac{x}{x - 1}$

43. Velocity and Acceleration   A ball is propelled straight upward from ground level with an initial velocity of 144 feet per second.

(a) Write the position, velocity, and acceleration functions of the ball.

(b) When is the ball at its highest point? How high is this point?

(c) How fast is the ball traveling when it hits the ground? How is this speed related to the initial velocity?

44. Velocity and Acceleration   A brick becomes dislodged from the top of the Empire State Building (at a height of 1250 feet) and falls to the sidewalk below.

(a) Write the position, velocity, and acceleration functions of the brick.

(b) How long does it take the brick to hit the sidewalk?

(c) How fast is the brick traveling when it hits the sidewalk?

45. Velocity and Acceleration   The velocity (in feet per second) of an automobile starting from rest is modeled by

$$\frac{ds}{dt} = \frac{90t}{t+10}.$$

Create a table showing the velocity and acceleration at 10-second intervals during the first minute of travel. What can you conclude?

46. Stopping Distance   A car is traveling at a rate of 66 feet per second (45 miles per hour) when the brakes are applied. The position function for the car is given by $s = -8.25t^2 + 66t$, where s is measured in feet and t is measured in seconds. Create a table showing the position, velocity, and acceleration for each given value of t. What can you conclude?

(T) In Exercises 47 and 48, use a graphing utility to graph f, f', and f'' in the same viewing window. What is the relationship among the degree of f and the degrees of its successive derivatives? In general, what is the relationship among the degree of a polynomial function and the degrees of its successive derivatives?

47. $f(x) = x^2 - 6x + 6$ **48.** $f(x) = 3x^3 - 9x$

In Exercises 49 and 50, the graphs of f, f', and f'' are shown on the same set of coordinate axes. Which is which? Explain your reasoning.

49.

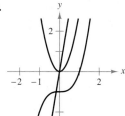

50.

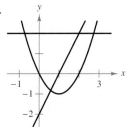

(T) **51. Modeling Data**   The table shows the retail values y (in billions of dollars) of motor homes sold in the United States for 2000 to 2005, where t is the year, with $t = 0$ corresponding to 2000. *(Source: Recreation Vehicle Industry Association)*

t	0	1	2	3	4	5
y	9.5	8.6	11.0	12.1	14.7	14.4

(a) Use a graphing utility to find a cubic model for the total retail value $y(t)$ of the motor homes.

(b) Use a graphing utility to graph the model and plot the data in the same viewing window. How well does the model fit the data?

(c) Find the first and second derivatives of the function.

(d) Show that the retail value of motor homes was increasing from 2001 to 2004.

(e) Find the year when the retail value was increasing at the greatest rate by solving $y''(t) = 0$.

(f) Explain the relationship among your answers for parts (c), (d), and (e).

52. Projectile Motion   An object is thrown upward from the top of a 64-foot building with an initial velocity of 48 feet per second.

(a) Write the position, velocity, and acceleration functions of the object.

(b) When will the object hit the ground?

(c) When is the velocity of the object zero?

(d) How high does the object go?

(T) (e) Use a graphing utility to graph the position, velocity, and acceleration functions in the same viewing window. Write a short paragraph that describes the relationship among these functions.

True or False?   In Exercises 53–56, determine whether the statement is true or false. If it is false, explain why or give an example that shows it is false.

53. If $y = f(x)g(x)$, then $y' = f'(x)g'(x)$.

54. If $y = (x+1)(x+2)(x+3)(x+4)$, then $\dfrac{d^5y}{dx^5} = 0$.

55. If $f'(c)$ and $g'(c)$ are zero and $h(x) = f(x)g(x)$, then $h'(c) = 0$.

56. The second derivative represents the rate of change of the first derivative.

57. Finding a Pattern   Develop a general rule for $[x\,f(x)]^{(n)}$ where f is a differentiable function of x.

58. Extended Application   To work an extended application analyzing the median prices of new privately owned U.S. homes in the South for 1980 through 2005, visit this text's website at *college.hmco.com*. *(Data Source: U.S. Census Bureau)*

Section 8.2

Implicit Differentiation

■ Find derivatives explicitly.
■ Find derivatives implicitly.
■ Use derivatives to answer questions about real-life situations.

Explicit and Implicit Functions

So far in this text, most functions involving two variables have been expressed in the **explicit form** $y = f(x)$. That is, one of the two variables has been explicitly given in terms of the other. For example, in the equation

$$y = 3x - 5 \qquad \text{Explicit form}$$

the variable y is explicitly written as a function of x. Some functions, however, are not given explicitly and are only implied by a given equation, as shown in Example 1.

Example 1   Finding a Derivative Explicitly

Find dy/dx for the equation

$$xy = 1.$$

SOLUTION   In this equation, y is **implicitly** defined as a function of x. One way to find dy/dx is first to solve the equation for y, then differentiate as usual.

$$xy = 1 \qquad \text{Write original equation.}$$

$$y = \frac{1}{x} \qquad \text{Solve for } y.$$

$$= x^{-1} \qquad \text{Rewrite.}$$

$$\frac{dy}{dx} = -x^{-2} \qquad \text{Differentiate with respect to } x.$$

$$= -\frac{1}{x^2} \qquad \text{Simplify.}$$

✓CHECKPOINT 1

Find dy/dx for the equation $x^2 y = 1$. ■

The procedure shown in Example 1 works well whenever you can easily write the given function explicitly. You cannot, however, use this procedure when you are unable to solve for y as a function of x. For instance, how would you find dy/dx in the equation

$$x^2 - 2y^3 + 4y = 2$$

where it is very difficult to express y as a function of x explicitly? To do this, you can use a procedure called **implicit differentiation.**

Implicit Differentiation

To understand how to find dy/dx implicitly, you must realize that the differentiation is taking place *with respect to x*. This means that when you differentiate terms involving x alone, you can differentiate as usual. *But* when you differentiate terms involving y, you must apply the Chain Rule because you are assuming that y is defined implicitly as a differentiable function of x. Study the next example carefully. Note in particular how the Chain Rule is used to introduce the dy/dx factors in Examples 2(b) and 2(d).

Example 2 Applying the Chain Rule

Differentiate each expression with respect to x.

a. $3x^2$ **b.** $2y^3$ **c.** $x + 3y$ **d.** xy^2

SOLUTION

a. The only variable in this expression is x. So, to differentiate with respect to x, you can use the Simple Power Rule and the Constant Multiple Rule to obtain

$$\frac{d}{dx}[3x^2] = 6x.$$

b. This case is different. The variable in the expression is y, and yet you are asked to differentiate with respect to x. To do this, assume that y is a differentiable function of x and use the Chain Rule.

$$\frac{d}{dx}[2y^3] = \overset{c}{2}\,\overset{n}{(3)}\,\overset{u^{n-1}}{y^2}\,\overset{u'}{\frac{dy}{dx}} \qquad \text{Chain Rule}$$

$$= 6y^2\frac{dy}{dx}$$

c. This expression involves both x and y. By the Sum Rule and the Constant Multiple Rule, you can write

$$\frac{d}{dx}[x + 3y] = 1 + 3\frac{dy}{dx}.$$

d. By the Product Rule and the Chain Rule, you can write

$$\frac{d}{dx}[xy^2] = x\frac{d}{dx}[y^2] + y^2\frac{d}{dx}[x] \qquad \text{Product Rule}$$

$$= x\left(2y\frac{dy}{dx}\right) + y^2(1) \qquad \text{Chain Rule}$$

$$= 2xy\frac{dy}{dx} + y^2.$$

✓CHECKPOINT 2

Differentiate each expression with respect to x.

a. $4x^3$ **b.** $3y^2$ **c.** $x + 5y$ **d.** xy^3 ∎

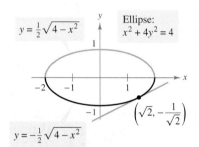

$y = \frac{1}{2}\sqrt{4 - x^2}$

Ellipse: $x^2 + 4y^2 = 4$

$\left(\sqrt{2}, -\frac{1}{\sqrt{2}}\right)$

$y = -\frac{1}{2}\sqrt{4 - x^2}$

FIGURE 8.3 Slope of tangent line is $\frac{1}{2}$.

STUDY TIP

An ellipse is an example of a conic section. For more information on conic sections, see Appendix B.

✓ CHECKPOINT 3

Find the slope of the tangent line to the circle $x^2 + y^2 = 25$ at the point $(3, -4)$.

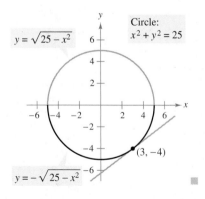

$y = \sqrt{25 - x^2}$

Circle: $x^2 + y^2 = 25$

$(3, -4)$

$y = -\sqrt{25 - x^2}$

Implicit Differentiation

Consider an equation involving x and y in which y is a differentiable function of x. You can use the steps below to find dy/dx.

1. Differentiate both sides of the equation *with respect to x*.

2. Write the result so that all terms involving dy/dx are on the left side of the equation and all other terms are on the right side of the equation.

3. Factor dy/dx out of the terms on the left side of the equation.

4. Solve for dy/dx by dividing both sides of the equation by the left-hand factor that does not contain dy/dx.

In Example 3, note that implicit differentiation can produce an expression for dy/dx that contains both x and y.

Example 3 Finding the Slope of a Graph Implicitly

Find the slope of the tangent line to the ellipse given by $x^2 + 4y^2 = 4$ at the point $\left(\sqrt{2}, -1/\sqrt{2}\right)$, as shown in Figure 8.3.

SOLUTION

$$x^2 + 4y^2 = 4 \qquad \text{Write original equation.}$$

$$\frac{d}{dx}[x^2 + 4y^2] = \frac{d}{dx}[4] \qquad \text{Differentiate with respect to } x.$$

$$2x + 8y\left(\frac{dy}{dx}\right) = 0 \qquad \text{Implicit differentiation}$$

$$8y\left(\frac{dy}{dx}\right) = -2x \qquad \text{Subtract } 2x \text{ from each side.}$$

$$\frac{dy}{dx} = \frac{-2x}{8y} \qquad \text{Divide each side by } 8y.$$

$$\frac{dy}{dx} = -\frac{x}{4y} \qquad \text{Simplify.}$$

To find the slope at the given point, substitute $x = \sqrt{2}$ and $y = -1/\sqrt{2}$ into the derivative, as shown below.

$$-\frac{\sqrt{2}}{4\left(-1/\sqrt{2}\right)} = \frac{1}{2}$$

STUDY TIP

To see the benefit of implicit differentiation, try reworking Example 3 using the explicit function

$$y = -\frac{1}{2}\sqrt{4 - x^2}.$$

The graph of this function is the lower half of the ellipse.

Example 4   **Using Implicit Differentiation**

Find dy/dx for the equation $y^3 + y^2 - 5y - x^2 = -4$.

SOLUTION

$$y^3 + y^2 - 5y - x^2 = -4 \qquad \text{Write original equation.}$$

$$\frac{d}{dx}[y^3 + y^2 - 5y - x^2] = \frac{d}{dx}[-4] \qquad \text{Differentiate with respect to } x.$$

$$3y^2\frac{dy}{dx} + 2y\frac{dy}{dx} - 5\frac{dy}{dx} - 2x = 0 \qquad \text{Implicit differentiation}$$

$$3y^2\frac{dy}{dx} + 2y\frac{dy}{dx} - 5\frac{dy}{dx} = 2x \qquad \text{Collect } dy/dx \text{ terms.}$$

$$\frac{dy}{dx}(3y^2 + 2y - 5) = 2x \qquad \text{Factor.}$$

$$\frac{dy}{dx} = \frac{2x}{3y^2 + 2y - 5}$$

The graph of the original equation is shown in Figure 8.4. What are the slopes of the graph at the points $(1, -3)$, $(2, 0)$, and $(1, 1)$?

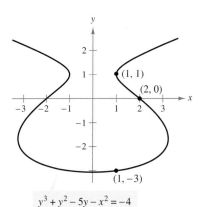

$y^3 + y^2 - 5y - x^2 = -4$

FIGURE 8.4

✓**CHECKPOINT 4**

Find dy/dx for the equation $y^2 + x^2 - 2y - 4x = 4$. ■

Example 5   **Finding the Slope of a Graph Implicitly**

Find the slope of the graph of $2x^2 - y^2 = 1$ at the point $(1, 1)$.

SOLUTION   Begin by finding dy/dx implicitly.

$$2x^2 - y^2 = 1 \qquad \text{Write original equation.}$$

$$4x - 2y\left(\frac{dy}{dx}\right) = 0 \qquad \text{Differentiate with respect to } x.$$

$$-2y\left(\frac{dy}{dx}\right) = -4x \qquad \text{Subtract } 4x \text{ from each side.}$$

$$\frac{dy}{dx} = \frac{2x}{y} \qquad \text{Divide each side by } -2y.$$

At the point $(1, 1)$, the slope of the graph is

$$\frac{2(1)}{1} = 2$$

as shown in Figure 8.5. The graph is called a **hyperbola.**

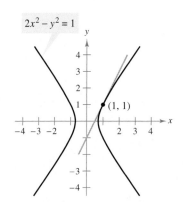

$2x^2 - y^2 = 1$

FIGURE 8.5   Hyperbola

✓**CHECKPOINT 5**

Find the slope of the graph of $x^2 - 9y^2 = 16$ at the point $(5, 1)$. ■

Application

Demand Function

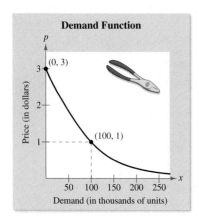

Price (in dollars) — Demand (in thousands of units)

(0, 3)

(100, 1)

50 100 150 200 250

FIGURE 8.6

Example 6 **Using a Demand Function**

The demand function for a product is modeled by

$$p = \frac{3}{0.000001x^3 + 0.01x + 1}$$

where p is measured in dollars and x is measured in thousands of units, as shown in Figure 8.6. Find the rate of change of the demand x with respect to the price p when $x = 100$.

SOLUTION To simplify the differentiation, begin by rewriting the function. Then, differentiate *with respect to p.*

$$p = \frac{3}{0.000001x^3 + 0.01x + 1}$$

$$0.000001x^3 + 0.01x + 1 = \frac{3}{p}$$

$$0.000003x^2\frac{dx}{dp} + 0.01\frac{dx}{dp} = -\frac{3}{p^2}$$

$$(0.000003x^2 + 0.01)\frac{dx}{dp} = -\frac{3}{p^2}$$

$$\frac{dx}{dp} = -\frac{3}{p^2(0.000003x^2 + 0.01)}$$

When $x = 100$, the price is

$$p = \frac{3}{0.000001(100)^3 + 0.01(100) + 1} = \$1.$$

So, when $x = 100$ and $p = 1$, the rate of change of the demand with respect to the price is

$$-\frac{3}{(1)^2[0.000003(100)^2 + 0.01]} = -75.$$

This means that when $x = 100$, the demand is dropping at the rate of 75 thousand units for each dollar increase in price.

✓CHECKPOINT 6

The demand function for a product is given by

$$p = \frac{2}{0.001x^2 + x + 1}.$$

Find dx/dp implicitly. ▪

CONCEPT CHECK

1. Complete the following: The equation $x + y = 1$ is written in _____ form and the equation $y = 1 - x$ is written in _____ form.

2. Complete the following: When you are asked to find dy/dt, you are being asked to find the derivative of _____ with respect to _____.

3. Describe the difference between the explicit form of a function and an implicit equation. Give an example of each.

4. In your own words, state the guidelines for implicit differentiation.

Skills Review 8.2	The following warm-up exercises involve skills that were covered in earlier sections. You will use these skills in the exercise set for this section. For additional help, review Sections 0.2, 1.1, 1.5, and 2.1.

In Exercises 1–6, solve the equation for y.

1. $x - \dfrac{y}{x} = 2$

2. $\dfrac{4}{x - 3} = \dfrac{1}{y}$

3. $xy - x + 6y = 6$

4. $12 + 3y = 4x^2 + x^2 y$

5. $x^2 + y^2 = 5$

6. $x = \pm\sqrt{6 - y^2}$

In Exercises 7–10, evaluate the expression at the given point.

7. $\dfrac{3x^2 - 4}{3y^2}$, $(2, 1)$

8. $\dfrac{x^2 - 2}{1 - y}$, $(0, -3)$

9. $\dfrac{5x}{3y^2 - 12y + 5}$, $(-1, 2)$

10. $\dfrac{1}{y^2 - 2xy + x^2}$, $(4, 3)$

Exercises 8.2

See www.CalcChat.com for worked-out solutions to odd-numbered exercises.

In Exercises 1–12, find dy/dx.

1. $xy = 4$

2. $3x^2 - y = 8x$

3. $y^2 = 1 - x^2$, $0 \le x \le 1$

4. $4x^2 y - \dfrac{3}{y} = 0$

5. $x^2 y^2 - 2x = 3$

6. $xy^2 + 4xy = 10$

7. $4y^2 - xy = 2$

8. $2xy^3 - x^2 y = 2$

9. $\dfrac{2y - x}{y^2 - 3} = 5$

10. $\dfrac{xy - y^2}{y - x} = 1$

11. $\dfrac{x + y}{2x - y} = 1$

12. $\dfrac{2x + y}{x - 5y} = 1$

In Exercises 13–24, find dy/dx by implicit differentiation and evaluate the derivative at the given point.

Equation	Point
13. $x^2 + y^2 = 16$	$(0, 4)$
14. $x^2 - y^2 = 25$	$(5, 0)$
15. $y + xy = 4$	$(-5, -1)$
16. $x^3 - y^2 = 0$	$(1, 1)$
17. $x^3 - xy + y^2 = 4$	$(0, -2)$
18. $x^2 y + y^2 x = -2$	$(2, -1)$
19. $x^3 y^3 - y = x$	$(0, 0)$
20. $x^3 + y^3 = 2xy$	$(1, 1)$
21. $x^{1/2} + y^{1/2} = 9$	$(16, 25)$
22. $\sqrt{xy} = x - 2y$	$(4, 1)$
23. $x^{2/3} + y^{2/3} = 5$	$(8, 1)$
24. $(x + y)^3 = x^3 + y^3$	$(-1, 1)$

In Exercises 25–30, find the slope of the graph at the given point.

25. $3x^2 - 2y + 5 = 0$

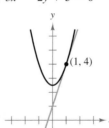

26. $4x^2 + 2y - 1 = 0$

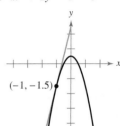

27. $x^2 + y^2 = 4$

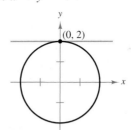

28. $4x^2 + y^2 = 4$

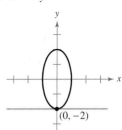

29. $4x^2 + 9y^2 = 36$

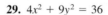
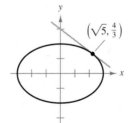

30. $x^2 - y^3 = 0$

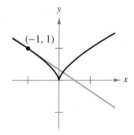

In Exercises 31–34, find dy/dx implicitly and explicitly (the explicit functions are shown on the graph) and show that the results are equivalent. Use the graph to estimate the slope of the tangent line at the labeled point. Then verify your result analytically by evaluating dy/dx at the point.

31. $x^2 + y^2 = 25$

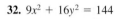
$y = \sqrt{25 - x^2}$

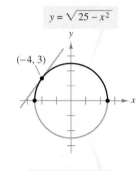
$y = -\sqrt{25 - x^2}$

32. $9x^2 + 16y^2 = 144$

$y = \dfrac{\sqrt{144 - 9x^2}}{4}$

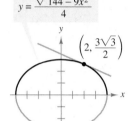

$y = -\dfrac{\sqrt{144 - 9x^2}}{4}$

33. $x - y^2 - 1 = 0$

$y = \sqrt{x - 1}$

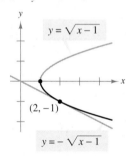
$y = -\sqrt{x - 1}$

34. $4y^2 - x^2 = 7$

$y = \dfrac{\sqrt{x^2 + 7}}{2}$

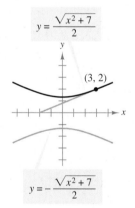

$y = -\dfrac{\sqrt{x^2 + 7}}{2}$

(T) In Exercises 35–40, find equations of the tangent lines to the graph at the given points. Use a graphing utility to graph the equation and the tangent lines in the same viewing window.

Equation	Points
35. $x^2 + y^2 = 100$	$(8, 6)$ and $(-6, 8)$
36. $x^2 + y^2 = 9$	$(0, 3)$ and $\left(2, \sqrt{5}\right)$
37. $y^2 = 5x^3$	$\left(1, \sqrt{5}\right)$ and $\left(1, -\sqrt{5}\right)$
38. $4xy + x^2 = 5$	$(1, 1)$ and $(5, -1)$
39. $x^3 + y^3 = 8$	$(0, 2)$ and $(2, 0)$
40. $y^2 = \dfrac{x^3}{4 - x}$	$(2, 2)$ and $(2, -2)$

Demand In Exercises 41–44, find the rate of change of x with respect to p.

41. $p = \dfrac{2}{0.00001x^3 + 0.1x}$ $x \geq 0$

42. $p = \dfrac{4}{0.000001x^2 + 0.05x + 1}$ $x \geq 0$

43. $p = \sqrt{\dfrac{200 - x}{2x}}$, $0 < x \leq 200$

44. $p = \sqrt{\dfrac{500 - x}{2x}}$, $0 < x \leq 500$

45. Production Let x represent the units of labor and y the capital invested in a manufacturing process. When 135,540 units are produced, the relationship between labor and capital can be modeled by $100x^{0.75}y^{0.25} = 135{,}540$.

(a) Find the rate of change of y with respect to x when $x = 1500$ and $y = 1000$.

(T) (b) The model used in the problem is called the *Cobb-Douglas production function*. Graph the model on a graphing utility and describe the relationship between labor and capital.

46. Production Repeat Exercise 45(a) by finding the rate of change of y with respect to x when $x = 3000$ and $y = 125$.

47. Health: U.S. HIV/AIDS Epidemic The numbers (in thousands) of cases y of HIV/AIDS reported in the years 2001 through 2005 can be modeled by

$$y^2 - 1141.6 = 24.9099t^3 - 183.045t^2 + 452.79t$$

where t represents the year, with $t = 1$ corresponding to 2001. *(Source: U.S. Centers for Disease Control and Prevention)*

(T) (a) Use a graphing utility to graph the model and describe the results.

(b) Use the graph to estimate the year during which the number of reported cases was increasing at the greatest rate.

(c) Complete the table to estimate the year during which the number of reported cases was increasing at the greatest rate. Compare this estimate with your answer in part (b).

t	1	2	3	4	5
y					
y'					

Section 8.3
Related Rates

- Examine related variables.
- Solve related-rate problems.

Related Variables

In this section, you will study problems involving variables that are changing with respect to time. If two or more such variables are related to each other, then their rates of change with respect to time are also related.

For instance, suppose that x and y are related by the equation $y = 2x$. If both variables are changing with respect to time, then their rates of change will also be related.

x and y are related.   The rates of change of x and y are related.

$$y = 2x \implies \frac{dy}{dt} = 2\frac{dx}{dt}$$

In this simple example, you can see that because y always has twice the value of x, it follows that the rate of change of y with respect to time is always twice the rate of change of x with respect to time.

Example 1   Examining Two Rates That Are Related

The variables x and y are differentiable functions of t and are related by the equation

$$y = x^2 + 3.$$

When $x = 1$, $dx/dt = 2$. Find dy/dt when $x = 1$.

SOLUTION   Use the Chain Rule to differentiate both sides of the equation with respect to t.

$$y = x^2 + 3 \qquad \text{Write original equation.}$$

$$\frac{d}{dt}[y] = \frac{d}{dt}[x^2 + 3] \qquad \text{Differentiate with respect to } t.$$

$$\frac{dy}{dt} = 2x\frac{dx}{dt} \qquad \text{Apply Chain Rule.}$$

When $x = 1$ and $dx/dt = 2$, you have

$$\frac{dy}{dt} = 2(1)(2)$$

$$= 4.$$

✓ CHECKPOINT 1

When $x = 1$, $dx/dt = 3$. Find dy/dt when $x = 1$ if $y = x^3 + 2$. ■

Solving Related-Rate Problems

In Example 1, you were *given* the mathematical model.

$$\text{Given equation: } y = x^2 + 3$$

$$\text{Given rate: } \frac{dx}{dt} = 2 \text{ when } x = 1$$

$$\text{Find: } \frac{dy}{dt} \text{ when } x = 1$$

In the next example, you are asked to *create* a similar mathematical model.

Example 2 Changing Area

A pebble is dropped into a calm pool of water, causing ripples in the form of concentric circles, as shown in the photo. The radius r of the outer ripple is increasing at a constant rate of 1 foot per second. When the radius is 4 feet, at what rate is the total area A of the disturbed water changing?

SOLUTION   The variables r and A are related by the equation for the area of a circle, $A = \pi r^2$. To solve this problem, use the fact that the rate of change of the radius is given by dr/dt.

$$\text{Equation: } A = \pi r^2$$

$$\text{Given rate: } \frac{dr}{dt} = 1 \text{ when } r = 4$$

$$\text{Find: } \frac{dA}{dt} \text{ when } r = 4$$

Using this model, you can proceed as in Example 1.

$A = \pi r^2$	Write original equation.
$\dfrac{d}{dt}[A] = \dfrac{d}{dt}\left[\pi r^2\right]$	Differentiate with respect to t.
$\dfrac{dA}{dt} = 2\pi r \dfrac{dr}{dt}$	Apply Chain Rule.

When $r = 4$ and $dr/dt = 1$, you have

$$\frac{dA}{dt} = 2\pi(4)(1) = 8\pi \qquad \text{Substitute 4 for } r \text{ and 1 for } dr/dt.$$

When the radius is 4 feet, the area is changing at a rate of 8π square feet per second.

© Randy Faris/Corbis

Total area increases as the outer radius increases.

✓ **CHECKPOINT 2**

If the radius r of the outer ripple in Example 2 is increasing at a rate of 2 feet per second, at what rate is the total area changing when the radius is 3 feet? ■

STUDY TIP

In Example 2, note that the radius changes at a *constant* rate ($dr/dt = 1$ for all t), but the area changes at a *nonconstant* rate.

When $r = 1$ ft	When $r = 2$ ft	When $r = 3$ ft	When $r = 4$ ft
$\dfrac{dA}{dt} = 2\pi$ ft^2/sec	$\dfrac{dA}{dt} = 4\pi$ ft^2/sec	$\dfrac{dA}{dt} = 6\pi$ ft^2/sec	$\dfrac{dA}{dt} = 8\pi$ ft^2/sec

The solution shown in Example 2 illustrates the steps for solving a related-rate problem.

Guidelines for Solving a Related-Rate Problem

1. Identify all *given* quantities and all quantities *to be determined.* If possible, make a sketch and label the quantities.

2. Write an equation that relates all variables whose rates of change are either given or to be determined.

3. Use the Chain Rule to differentiate both sides of the equation *with respect to time.*

4. Substitute into the resulting equation all known values of the variables and their rates of change. Then solve for the required rate of change.

STUDY TIP

Be sure you notice the order of Steps 3 and 4 in the guidelines. Do not substitute the known values for the variables until after you have differentiated.

In Step 2 of the guidelines, note that you must write an equation that relates the given variables. To help you with this step, reference tables that summarize many common formulas are included in the appendices. For instance, the volume of a sphere of radius r is given by the formula

$$V = \frac{4}{3}\pi r^3$$

as listed in Appendix D.

The table below shows the mathematical models for some common rates of change that can be used in the first step of the solution of a related-rate problem.

Verbal statement	Mathematical model
The velocity of a car after traveling for 1 hour is 50 miles per hour.	$x =$ distance traveled $\frac{dx}{dt} = 50$ when $t = 1$
Water is being pumped into a swimming pool at the rate of 10 cubic feet per minute.	$V =$ volume of water in pool $\frac{dV}{dt} = 10 \text{ ft}^3/\text{min}$
A population of bacteria is increasing at the rate of 2000 per hour.	$x =$ number in population $\frac{dx}{dt} = 2000$ bacteria per hour
Revenue is increasing at the rate of $4000 per month.	$R =$ revenue $\frac{dR}{dt} = 4000$ dollars per month

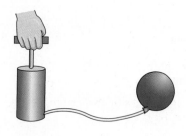

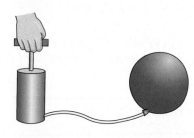

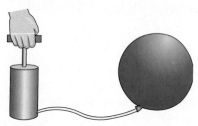

FIGURE 8.7   Expanding Balloon

| Example 3 | Changing Volume | |

Air is being pumped into a spherical balloon at the rate of 4.5 cubic inches per minute. See Figure 8.7. Find the rate of change of the radius when the radius is 2 inches.

SOLUTION   Let V represent the volume of the balloon and let r represent the radius. Because the volume is increasing at the rate of 4.5 cubic inches per minute, you know that $dV/dt = 4.5$. An equation that relates V and r is $V = \frac{4}{3}\pi r^3$. So, the problem can be represented by the model shown below.

$$\text{Equation: } V = \frac{4}{3}\pi r^3$$

$$\text{Given rate: } \frac{dV}{dt} = 4.5$$

$$\text{Find: } \frac{dr}{dt} \text{ when } r = 2$$

By differentiating the equation, you obtain

$$V = \frac{4}{3}\pi r^3 \qquad \text{Write original equation.}$$

$$\frac{d}{dt}[V] = \frac{d}{dt}\left[\frac{4}{3}\pi r^3\right] \qquad \text{Differentiate with respect to } t.$$

$$\frac{dV}{dt} = \frac{4}{3}\pi(3r^2)\frac{dr}{dt} \qquad \text{Apply Chain Rule.}$$

$$\frac{1}{4\pi r^2}\frac{dV}{dt} = \frac{dr}{dt}. \qquad \text{Solve for } dr/dt.$$

When $r = 2$ and $dV/dt = 4.5$, the rate of change of the radius is

$$\frac{dr}{dt} = \frac{1}{4\pi(2^2)}(4.5)$$

$$\approx 0.09 \text{ inch per minute.}$$

✓ CHECKPOINT 3

If the radius of a spherical balloon increases at a rate of 1.5 inches per minute, find the rate at which the surface area changes when the radius is 6 inches. (Formula for surface area of a sphere: $S = 4\pi r^2$) ∎

In Example 3, note that the volume is increasing at a *constant rate* but the radius is increasing at a *variable* rate. In this particular example, the radius is increasing more and more slowly as t increases. This is illustrated in the table below.

t	1	3	5	7	9	11
$V = 4.5t$	4.5	13.5	22.5	31.5	40.5	49.5
$t = \sqrt[3]{\dfrac{3V}{4\pi}}$	1.02	1.48	1.75	1.96	2.13	2.28
$\dfrac{dr}{dt}$	0.34	0.16	0.12	0.09	0.08	0.07

Example 4   **Analyzing a Profit Function**

A company's profit P (in dollars) from selling x units of a product can be modeled by

$$P = 500x - \left(\frac{1}{4}\right)x^2. \qquad \text{Model for profit}$$

The sales are increasing at a rate of 10 units per day. Find the rate of change in the profit (in dollars per day) when 500 units have been sold.

SOLUTION   Because you are asked to find the rate of change in dollars per day, you should differentiate the given equation with respect to the time t.

$$P = 500x - \left(\frac{1}{4}\right)x^2 \qquad \text{Write model for profit.}$$

$$\frac{dP}{dt} = 500\left(\frac{dx}{dt}\right) - 2\left(\frac{1}{4}\right)(x)\left(\frac{dx}{dt}\right) \qquad \text{Differentiate with respect to } t.$$

The sales are increasing at a constant rate of 10 units per day, so

$$\frac{dx}{dt} = 10.$$

When $x = 500$ units and $dx/dt = 10$, the rate of change in the profit is

$$\frac{dP}{dt} = 500(10) - 2\left(\frac{1}{4}\right)(500)(10)$$

$$= 5000 - 2500$$

$$= \$2500 \text{ per day.} \qquad \text{Simplify.}$$

The graph of the profit function (in terms of x) is shown in Figure 8.8.

STUDY TIP

In Example 4, note that one of the keys to successful use of calculus in applied problems is the interpretation of a rate of change as a derivative.

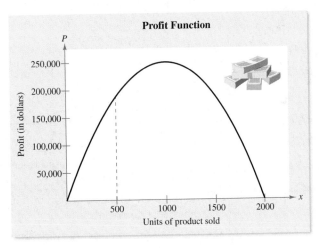

FIGURE 8.8

✓CHECKPOINT 4

Find the rate of change in profit (in dollars per day) when 50 units have been sold, sales have increased at a rate of 10 units per day, and $P = 200x - \frac{1}{2}x^2$. ■

Example 5
MAKE A DECISION   **Increasing Production**

A company is increasing the production of a product at the rate of 200 units per week. The weekly demand function is modeled by

$$p = 100 - 0.001x$$

where p is the price per unit and x is the number of units produced in a week. Find the rate of change of the revenue with respect to time when the weekly production is 2000 units. Will the rate of change of the revenue be greater than $20,000 per week?

SOLUTION

$$\textit{Equation: } R = xp = x(100 - 0.001x) = 100x - 0.001x^2$$

$$\textit{Given rate: } \frac{dx}{dt} = 200$$

$$\textit{Find: } \frac{dR}{dt} \text{ when } x = 2000$$

By differentiating the equation, you obtain

$$R = 100x - 0.001x^2 \qquad \text{Write original equation.}$$

$$\frac{d}{dt}[R] = \frac{d}{dt}[100x - 0.001x^2] \qquad \text{Differentiate with respect to } t.$$

$$\frac{dR}{dt} = (100 - 0.002x)\frac{dx}{dt}. \qquad \text{Apply Chain Rule.}$$

Using $x = 2000$ and $dx/dt = 200$, you have

$$\frac{dR}{dt} = [100 - 0.002(2000)](200)$$

$$= \$19,200 \text{ per week.}$$

No, the rate of change of the revenue will not be greater than $20,000 per week.

✓ **CHECKPOINT 5**

Find the rate of change of revenue with respect to time for the company in Example 5 if the weekly demand function is

$$p = 150 - 0.002x. \ \blacksquare$$

⌐ **CONCEPT CHECK** ⌐

1. Complete the following. Two variables x and y are changing with respect to _____. If x and y are related to each other, then their rates of change with respect to time are also _____.

2. The volume V of an object is a differentiable function of time t. Describe what dV/dt represents.

3. The area A of an object is a differentiable function of time t. Describe what dA/dt represents.

4. In your own words, state the guidelines for solving related-rate problems.

The following warm-up exercises involve skills that were covered in earlier sections. You will use these skills in the exercise set for this section. For additional help, review Section 8.2.

In Exercises 1–6, write a formula for the given quantity.

1. Area of a circle

2. Volume of a sphere

3. Surface area of a cube

4. Volume of a cube

5. Volume of a cone

6. Area of a triangle

In Exercises 7–10, find *dy/dx* by implicit differentiation.

7. $x^2 + y^2 = 9$

8. $3xy - x^2 = 6$

9. $x^2 + 2y + xy = 12$

10. $x + xy^2 - y^2 = xy$

Exercises 8.3

In Exercises 1–4, use the given values to find *dy/dt* and *dx/dt*.

Equation	Find	Given
1. $y = \sqrt{x}$	(a) $\dfrac{dy}{dt}$	$x = 4, \dfrac{dx}{dt} = 3$
	(b) $\dfrac{dx}{dt}$	$x = 25, \dfrac{dy}{dt} = 2$
2. $y = 2(x^2 - 3x)$	(a) $\dfrac{dy}{dt}$	$x = 3, \dfrac{dx}{dt} = 2$
	(b) $\dfrac{dx}{dt}$	$x = 1, \dfrac{dy}{dt} = 5$
3. $xy = 4$	(a) $\dfrac{dy}{dt}$	$x = 8, \dfrac{dx}{dt} = 10$
	(b) $\dfrac{dx}{dt}$	$x = 1, \dfrac{dy}{dt} = -6$
4. $x^2 + y^2 = 25$	(a) $\dfrac{dy}{dt}$	$x = 3, y = 4, \dfrac{dx}{dt} = 8$
	(b) $\dfrac{dx}{dt}$	$x = 4, y = 3, \dfrac{dy}{dt} = -2$

5. Area The radius *r* of a circle is increasing at a rate of 3 inches per minute. Find the rates of change of the area when (a) *r* = 6 inches and (b) *r* = 24 inches.

6. Volume The radius *r* of a sphere is increasing at a rate of 3 inches per minute. Find the rates of change of the volume when (a) *r* = 6 inches and (b) *r* = 24 inches.

7. Area Let *A* be the area of a circle of radius *r* that is changing with respect to time. If *dr/dt* is constant, is *dA/dt* constant? Explain your reasoning.

8. Volume Let *V* be the volume of a sphere of radius *r* that is changing with respect to time. If *dr/dt* is constant, is *dV/dt* constant? Explain your reasoning.

9. Volume A spherical balloon is inflated with gas at a rate of 10 cubic feet per minute. How fast is the radius of the balloon changing at the instant the radius is (a) 1 foot and (b) 2 feet?

10. Volume The radius *r* of a right circular cone is increasing at a rate of 2 inches per minute. The height *h* of the cone is related to the radius by *h* = 3*r*. Find the rates of change of the volume when (a) *r* = 6 inches and (b) *r* = 24 inches.

11. Cost, Revenue, and Profit A company that manufactures sport supplements calculates that its costs and revenue can be modeled by the equations

$$C = 125{,}000 + 0.75x \quad \text{and} \quad R = 250x - \frac{1}{10}x^2$$

where *x* is the number of units of sport supplements produced in 1 week. If production in one particular week is 1000 units and is increasing at a rate of 150 units per week, find:

(a) the rate at which the cost is changing.

(b) the rate at which the revenue is changing.

(c) the rate at which the profit is changing.

12. Cost, Revenue, and Profit A company that manufactures pet toys calculates that its costs and revenue can be modeled by the equations

$$C = 75{,}000 + 1.05x \quad \text{and} \quad R = 500x - \frac{x^2}{25}$$

where *x* is the number of toys produced in 1 week. If production in one particular week is 5000 toys and is increasing at a rate of 250 toys per week, find:

(a) the rate at which the cost is changing.

(b) the rate at which the revenue is changing.

(c) the rate at which the profit is changing.

13. **Volume** All edges of a cube are expanding at a rate of 3 centimeters per second. How fast is the volume changing when each edge is (a) 1 centimeter and (b) 10 centimeters?

14. **Surface Area** All edges of a cube are expanding at a rate of 3 centimeters per second. How fast is the surface area changing when each edge is (a) 1 centimeter and (b) 10 centimeters?

15. **Moving Point** A point is moving along the graph of $y = x^2$ such that dx/dt is 2 centimeters per minute. Find dy/dt for each value of x.

   (a) $x = -3$ (b) $x = 0$ (c) $x = 1$ (d) $x = 3$

16. **Moving Point** A point is moving along the graph of $y = 1/(1 + x^2)$ such that dx/dt is 2 centimeters per minute. Find dy/dt for each value of x.

   (a) $x = -2$ (b) $x = 2$ (c) $x = 0$ (d) $x = 10$

17. **Moving Ladder** A 25-foot ladder is leaning against a house (see figure). The base of the ladder is pulled away from the house at a rate of 2 feet per second. How fast is the top of the ladder moving down the wall when the base is (a) 7 feet, (b) 15 feet, and (c) 24 feet from the house?

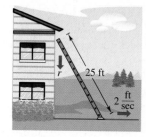
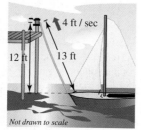

Figure for 17   Figure for 18

18. **Boating** A boat is pulled by a winch on a dock, and the winch is 12 feet above the deck of the boat (see figure). The winch pulls the rope at a rate of 4 feet per second. Find the speed of the boat when 13 feet of rope is out. What happens to the speed of the boat as it gets closer and closer to the dock?

19. **Air Traffic Control** An air traffic controller spots two airplanes at the same altitude converging to a point as they fly at right angles to each other. One airplane is 150 miles from the point and has a speed of 450 miles per hour. The other is 200 miles from the point and has a speed of 600 miles per hour.

   (a) At what rate is the distance between the planes changing?

   (b) How much time does the controller have to get one of the airplanes on a different flight path?

20. **Air Traffic Control** An airplane flying at an altitude of 6 miles passes directly over a radar antenna (see figure). When the airplane is 10 miles away ($s = 10$), the radar detects that the distance s is changing at a rate of 240 miles per hour. What is the speed of the airplane?

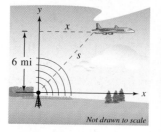
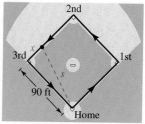

Figure for 20   Figure for 21

21. **Baseball** A (square) baseball diamond has sides that are 90 feet long (see figure). A player 26 feet from third base is running at a speed of 30 feet per second. At what rate is the player's distance from home plate changing?

22. **Advertising Costs** A retail sporting goods store estimates that weekly sales S and weekly advertising costs x are related by the equation $S = 2250 + 50x + 0.35x^2$. The current weekly advertising costs are $1500, and these costs are increasing at a rate of $125 per week. Find the current rate of change of weekly sales.

23. **Environment** An accident at an oil drilling platform is causing a circular oil slick. The slick is 0.08 foot thick, and when the radius of the slick is 150 feet, the radius is increasing at the rate of 0.5 foot per minute. At what rate (in cubic feet per minute) is oil flowing from the site of the accident?

Ⓣ 24. **Profit** A company is increasing the production of a product at the rate of 25 units per week. The demand and cost functions for the product are given by $p = 50 - 0.01x$ and $C = 4000 + 40x - 0.02x^2$. Find the rate of change of the profit with respect to time when the weekly sales are $x = 800$ units. Use a graphing utility to graph the profit function, and use the *zoom* and *trace* features of the graphing utility to verify your result.

25. **Sales** The profit for a product is increasing at a rate of $5600 per week. The demand and cost functions for the product are given by $p = 6000 - 25x$ and $C = 2400x + 5200$. Find the rate of change of sales with respect to time when the weekly sales are $x = 44$ units.

Ⓣ 26. **Cost** The annual cost (in millions of dollars) for a government agency to seize $p\%$ of an illegal drug is given by

$$C = \frac{528p}{100 - p}, \quad 0 \le p < 100.$$

The agency's goal is to increase p by 5% per year. Find the rates of change of the cost when (a) $p = 30\%$ and (b) $p = 60\%$. Use a graphing utility to graph C. What happens to the graph of C as p approaches 100?

Mid-Chapter Quiz

Take this test as you would take a test in class. When you are done, check your work against the answers given in the back of the book.

In Exercises 1–4, find the second derivative of the function. Simplify your result.

1. $f(x) = x^3 - x^2 + 2x - 1$

2. $h(x) = \dfrac{1}{\sqrt[3]{x - 2}}$

3. $g(x) = (x^2 + 1)^3$

4. $f(x) = \dfrac{x - 5}{2x + 5}$

In Exercises 5–7, find the given value.

Function	*Value*
5. $f(x) = \sqrt{x}$	$f''(4)$
6. $f(x) = x^5 - 4x^3 + \frac{3}{2}x^2 + 19$	$f^{(4)}(-1)$
7. $f(x) = \dfrac{1}{x}$	$f'''\left(\dfrac{1}{2}\right)$

8. An object is thrown upward from the top of an 800-foot building with an initial velocity of 80 feet per second. Find the height, the velocity, and the acceleration of the object when $t = 1$.

In Exercises 9–12, use implicit differentiation to find dy/dx.

9. $x^2 + 3y = x$

10. $\sqrt{y} = x^3$

11. $xy = x + y$

12. $y^3 + y - 2x^2y = 12$

Ⓣ **13.** Use implicit differentiation to find an equation of the tangent line to the graph of $-2xy + 3x^2 = 1$ at the point $(-1, -1)$. Use a graphing utility to graph the equation and the tangent line in the same viewing window.

In Exercises 14 and 15, use the given values to find dy/dt.

Equation	*Given*
14. $y = 2x^2 + 5$	$x = 1, \dfrac{dx}{dt} = \dfrac{1}{2}$
15. $x^2 - y^2 = \dfrac{16}{y}$	$x = \sqrt{12}, y = -4, \dfrac{dx}{dt} = 1$

16. A company that manufactures a type of automobile part calculates that its costs and revenue can be modeled by the equations

$$C = 200{,}000 + 0.95x \quad \text{and} \quad R = 300x - \frac{1}{75}x^2$$

where x is the number of parts produced in 1 week. If production in one particular week is 7500 parts and is increasing at a rate of 200 parts per week, find the rate of change of (a) the cost, (b) the revenue, and (c) the profit.

Increasing and Decreasing Functions

- Test for increasing and decreasing functions.
- Find the critical numbers of functions and find the open intervals on which functions are increasing or decreasing.
- Use increasing and decreasing functions to model and solve real-life problems.

Increasing and Decreasing Functions

A function is **increasing** if its graph moves up as x moves to the right and **decreasing** if its graph moves down as x moves to the right. The following definition states this more formally.

Definition of Increasing and Decreasing Functions

A function f is **increasing** on an interval if for any two numbers x_1 and x_2 in the interval

$$x_2 > x_1 \quad \text{implies} \quad f(x_2) > f(x_1).$$

A function f is **decreasing** on an interval if for any two numbers x_1 and x_2 in the interval

$$x_2 > x_1 \quad \text{implies} \quad f(x_2) < f(x_1).$$

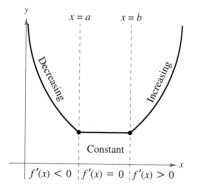

FIGURE 8.9

The function in Figure 8.9 is decreasing on the interval $(-\infty, a)$, constant on the interval (a, b), and increasing on the interval (b, ∞). Actually, from the definition of increasing and decreasing functions, the function shown in Figure 8.9 is decreasing on the interval $(-\infty, a]$ and increasing on the interval $[b, \infty)$. This text restricts the discussion to finding *open* intervals on which a function is increasing or decreasing.

The derivative of a function can be used to determine whether the function is increasing or decreasing on an interval.

Test for Increasing and Decreasing Functions

Let f be differentiable on the interval (a, b).

1. If $f'(x) > 0$ for all x in (a, b), then f is increasing on (a, b).

2. If $f'(x) < 0$ for all x in (a, b), then f is decreasing on (a, b).

3. If $f'(x) = 0$ for all x in (a, b), then f is constant on (a, b).

STUDY TIP

The conclusions in the first two cases of testing for increasing and decreasing functions are valid even if $f'(x) = 0$ at a finite number of x-values in (a, b).

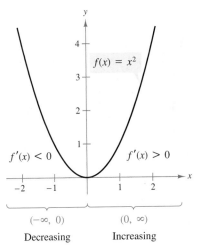

$f(x) = x^2$

$f'(x) < 0$ $f'(x) > 0$

$(-\infty, 0)$ $(0, \infty)$

Decreasing   Increasing

FIGURE 8.10

✓ **CHECKPOINT 2**

From 1995 through 2004, the consumption W of bottled water in the United States (in gallons per person per year) can be modeled by

$$W = 0.058t^2 + 0.19t + 9.2,$$

$$5 \le t \le 14$$

where $t = 5$ corresponds to 1995. Show that the consumption of bottled water was increasing from 1995 to 2004.   *(Source: U.S. Department of Agriculture)* ∎

Example 1 **Testing for Increasing and Decreasing Functions**

Show that the function

$$f(x) = x^2$$

is decreasing on the open interval $(-\infty, 0)$ and increasing on the open interval $(0, \infty)$.

SOLUTION   The derivative of f is

$$f'(x) = 2x.$$

On the open interval $(-\infty, 0)$, the fact that x is negative implies that $f'(x) = 2x$ is also negative. So, by the test for a decreasing function, you can conclude that f is *decreasing* on this interval. Similarly, on the open interval $(0, \infty)$, the fact that x is positive implies that $f'(x) = 2x$ is also positive. So, it follows that f is *increasing* on this interval, as shown in Figure 8.10.

✓ **CHECKPOINT 1**

Show that the function $f(x) = x^4$ is decreasing on the open interval $(-\infty, 0)$ and increasing on the open interval $(0, \infty)$. ∎

Example 2 **Modeling Consumption**

From 1997 through 2004, the consumption C of Italian cheeses in the United States (in pounds per person per year) can be modeled by

$$C = -0.0333t^2 + 0.996t + 5.40, \quad 7 \le t \le 14$$

where $t = 7$ corresponds to 1997 (see Figure 8.11). Show that the consumption of Italian cheeses was increasing from 1997 to 2004.   *(Source: U.S. Department of Agriculture)*

SOLUTION   The derivative of this model is $dC/dt = -0.0666t + 0.996$. For the open interval $(7, 14)$, the derivative is positive. So, the function is increasing, which implies that the consumption of Italian cheeses was increasing during the given time period.

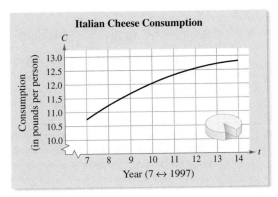

FIGURE 8.11

Critical Numbers and Their Use

In Example 1, you were given two intervals: one on which the function was decreasing and one on which it was increasing. Suppose you had been asked to determine these intervals. To do this, you could have used the fact that for a continuous function, $f'(x)$ can change signs only at x-values where $f'(x) = 0$ or at x-values where $f'(x)$ is undefined, as shown in Figure 8.12. These two types of numbers are called the **critical numbers** of f.

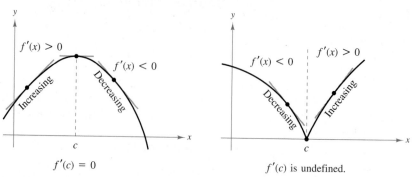

FIGURE 8.12

Definition of Critical Number

If f is defined at c, then c is a critical number of f if $f'(c) = 0$ or if $f'(c)$ is undefined.

STUDY TIP

This definition requires that a critical number be in the domain of the function. For example, $x = 0$ is not a critical number of the function $f(x) = 1/x$.

To determine the intervals on which a continuous function is increasing or decreasing, you can use the guidelines below.

Guidelines for Applying Increasing/Decreasing Test

1. Find the derivative of f.

2. Locate the critical numbers of f and use these numbers to determine test intervals. That is, find all x for which $f'(x) = 0$ or $f'(x)$ is undefined.

3. Test the sign of $f'(x)$ at an arbitrary number in each of the test intervals.

4. Use the test for increasing and decreasing functions to decide whether f is increasing or decreasing on each interval.

Example 3 **Finding Increasing and Decreasing Intervals**

Find the open intervals on which the function is increasing or decreasing.

$$f(x) = x^3 - \frac{3}{2}x^2$$

SOLUTION Begin by finding the derivative of f. Then set the derivative equal to zero and solve for the critical numbers.

$$f'(x) = 3x^2 - 3x \qquad \text{Differentiate original function.}$$
$$3x^2 - 3x = 0 \qquad \text{Set derivative equal to 0.}$$
$$3(x)(x - 1) = 0 \qquad \text{Factor.}$$
$$x = 0, x = 1 \qquad \text{Critical numbers}$$

Because there are no x-values for which f' is undefined, it follows that $x = 0$ and $x = 1$ are the *only* critical numbers. So, the intervals that need to be tested are $(-\infty, 0)$, $(0, 1)$, and $(1, \infty)$. The table summarizes the testing of these three intervals.

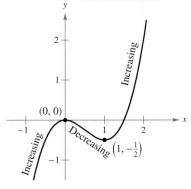

$f(x) = x^3 - \frac{3}{2}x^2$

FIGURE 8.13

Interval	$-\infty < x < 0$	$0 < x < 1$	$1 < x < \infty$
Test value	$x = -1$	$x = \frac{1}{2}$	$x = 2$
Sign of $f'(x)$	$f'(-1) = 6 > 0$	$f'\left(\frac{1}{2}\right) = -\frac{3}{4} < 0$	$f'(2) = 6 > 0$
Conclusion	Increasing	Decreasing	Increasing

✓ **CHECKPOINT 3**

Find the open intervals on which the function $f(x) = x^3 - 12x$ is increasing or decreasing. ■

The graph of f is shown in Figure 8.13. Note that the test values in the intervals were chosen for convenience—other x-values could have been used.

TECHNOLOGY

Ⓣ You can use the *trace* feature of a graphing utility to confirm the result of Example 3. Begin by graphing the function, as shown at the right. Then activate the *trace* feature and move the cursor from left to right. In intervals on which the function is increasing, note that the y-values increase as the x-values increase, whereas in intervals on which the function is decreasing, the y-values decrease as the x-values increase.*

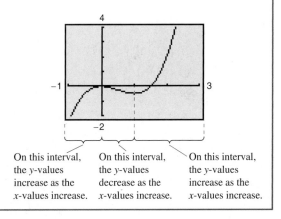

On this interval, the y-values increase as the x-values increase.

On this interval, the y-values decrease as the x-values increase.

On this interval, the y-values increase as the x-values increase.

---

*Specific calculator keystroke instructions for operations in this and other technology boxes can be found at *college.hmco.com/info/larsonapplied*.

Not only is the function in Example 3 continuous on the entire real line, it is also differentiable there. For such functions, the only critical numbers are those for which $f'(x) = 0$. The next example considers a continuous function that has *both* types of critical numbers—those for which $f'(x) = 0$ and those for which $f'(x)$ is undefined.

Algebra Review

For help on the algebra in Example 4, see Example 2(c) in the *Chapter 8 Algebra Review*, on page 680.

Example 4 **Finding Increasing and Decreasing Intervals**

Find the open intervals on which the function

$$f(x) = (x^2 - 4)^{2/3}$$

is increasing or decreasing.

SOLUTION Begin by finding the derivative of the function.

$$f'(x) = \frac{2}{3}(x^2 - 4)^{-1/3}(2x) \qquad \text{Differentiate.}$$

$$= \frac{4x}{3(x^2 - 4)^{1/3}} \qquad \text{Simplify.}$$

From this, you can see that the derivative is zero when $x = 0$ and the derivative is undefined when $x = \pm 2$. So, the critical numbers are

$$x = -2, \quad x = 0, \quad \text{and} \quad x = 2. \qquad \text{Critical numbers}$$

This implies that the test intervals are

$$(-\infty, -2), \quad (-2, 0), \quad (0, 2), \quad \text{and} \quad (2, \infty). \qquad \text{Test intervals}$$

The table summarizes the testing of these four intervals, and the graph of the function is shown in Figure 8.14.

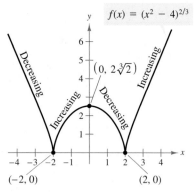

FIGURE 8.14

Interval	$-\infty < x < -2$	$-2 < x < 0$	$0 < x < 2$	$2 < x < \infty$
Test value	$x = -3$	$x = -1$	$x = 1$	$x = 3$
Sign of $f'(x)$	$f'(-3) < 0$	$f'(-1) > 0$	$f'(1) < 0$	$f'(3) > 0$
Conclusion	Decreasing	Increasing	Decreasing	Increasing

✓ CHECKPOINT 4

Find the open intervals on which the function $f(x) = x^{2/3}$ is increasing or decreasing. ■

STUDY TIP

To test the intervals in the table, it is not necessary to *evaluate* $f'(x)$ at each test value—you only need to determine its sign. For example, you can determine the sign of $f'(-3)$ as shown.

$$f'(-3) = \frac{4(-3)}{3(9 - 4)^{1/3}} = \frac{\text{negative}}{\text{positive}} = \text{negative}$$

The functions in Examples 1 through 4 are continuous on the entire real line. If there are isolated x-values at which a function is not continuous, then these x-values should be used along with the critical numbers to determine the test intervals. For example, the function

$$f(x) = \frac{x^4 + 1}{x^2}$$

is not continuous when $x = 0$. Because the derivative of f

$$f'(x) = \frac{2(x^4 - 1)}{x^3}$$

is zero when $x = \pm 1$, you should use the following numbers to determine the test intervals.

$x = -1, x = 1$ Critical numbers

$x = 0$ Discontinuity

After testing $f'(x)$, you can determine that the function is decreasing on the intervals $(-\infty, -1)$ and $(0, 1)$, and increasing on the intervals $(-1, 0)$ and $(1, \infty)$, as shown in Figure 8.15.

The converse of the test for increasing and decreasing functions is *not* true. For instance, it is possible for a function to be increasing on an interval even though its derivative is not positive at every point in the interval.

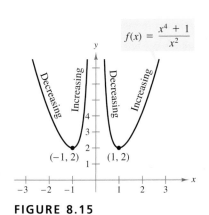

$f(x) = \dfrac{x^4 + 1}{x^2}$

FIGURE 8.15

Example 5   Testing an Increasing Function

Show that

$$f(x) = x^3 - 3x^2 + 3x$$

is increasing on the entire real line.

SOLUTION From the derivative of f

$$f'(x) = 3x^2 - 6x + 3 = 3(x - 1)^2$$

you can see that the only critical number is $x = 1$. So, the test intervals are $(-\infty, 1)$ and $(1, \infty)$. The table summarizes the testing of these two intervals. From Figure 8.16, you can see that f is increasing on the entire real line, even though $f'(1) = 0$. To convince yourself of this, look back at the definition of an increasing function.

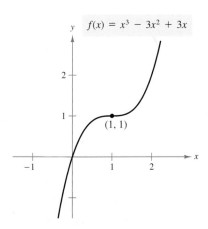

$f(x) = x^3 - 3x^2 + 3x$

FIGURE 8.16

Interval	$-\infty < x < 1$	$1 < x < \infty$
Test value	$x = 0$	$x = 2$
Sign of $f'(x)$	$f'(0) = 3(-1)^2 > 0$	$f'(2) = 3(1)^2 > 0$
Conclusion	Increasing	Increasing

✓ **CHECKPOINT 5**

Show that $f(x) = -x^3 + 2$ is decreasing on the entire real line. ▪

Application

| Example 6 | **Profit Analysis**

A national toy distributor determines the cost and revenue models for one of its games.

$$C = 2.4x - 0.0002x^2, \quad 0 \le x \le 6000$$
$$R = 7.2x - 0.001x^2, \quad 0 \le x \le 6000$$

Determine the interval on which the profit function is increasing.

SOLUTION The profit for producing x games is

$$
\begin{aligned}
P &= R - C \\
&= (7.2x - 0.001x^2) - (2.4x - 0.0002x^2) \\
&= 4.8x - 0.0008x^2.
\end{aligned}
$$

To find the interval on which the profit is increasing, set the marginal profit P' equal to zero and solve for x.

$$P' = 4.8 - 0.0016x \qquad \text{Differentiate profit function.}$$
$$4.8 - 0.0016x = 0 \qquad \text{Set } P' \text{ equal to 0.}$$
$$-0.0016x = -4.8 \qquad \text{Subtract 4.8 from each side.}$$
$$x = \frac{-4.8}{-0.0016} \qquad \text{Divide each side by } -0.0016.$$
$$x = 3000 \text{ games} \qquad \text{Simplify.}$$

On the interval $(0, 3000)$, P' is positive and the profit is *increasing*. On the interval $(3000, 6000)$, P' is negative and the profit is *decreasing*. The graphs of the cost, revenue, and profit functions are shown in Figure 8.17.

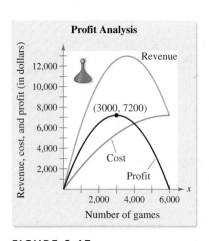

Profit Analysis

(3000, 7200)

Revenue

Cost

Profit

Number of games

Revenue, cost, and profit (in dollars)

FIGURE 8.17

✓ CHECKPOINT 6

A national distributor of pet toys determines the cost and revenue functions for one of its toys.

$$C = 1.2x - 0.0001x^2, \quad 0 \le x \le 6000$$

$$R = 3.6x - 0.0005x^2, \quad 0 \le x \le 6000$$

Determine the interval on which the profit function is increasing. ▪

(CONCEPT CHECK)

1. Write a verbal description of (a) the graph of an increasing function and (b) the graph of a decreasing function.

2. Complete the following: If $f'(x) > 0$ for all x in (a, b), then f is _____ on (a, b). [Assume f is differentiable on (a, b).]

3. If f is defined at c, under what condition(s) is c a critical number of f?

4. In your own words, state the guidelines for determining the intervals on which a continuous function is increasing or decreasing.

Skills Review 8.4

The following warm-up exercises involve skills that were covered in earlier sections. You will use these skills in the exercise set for this section. For additional help, review Sections 0.2, 0.7, 1.3, and 1.5.

In Exercises 1–4, solve the equation.

1. $x^2 = 8x$

2. $15x = \dfrac{5}{8}x^2$

3. $\dfrac{x^2 - 25}{x^3} = 0$

4. $\dfrac{2x}{\sqrt{1 - x^2}} = 0$

In Exercises 5–8, find the domain of the expression.

5. $\dfrac{x + 3}{x - 3}$

6. $\dfrac{2}{\sqrt{1 - x}}$

7. $\dfrac{2x + 1}{x^2 - 3x - 10}$

8. $\dfrac{3x}{\sqrt{9 - 3x^2}}$

In Exercises 9–12, evaluate the expression when $x = -2$, 0, and 2.

9. $-2(x + 1)(x - 1)$

10. $4(2x + 1)(2x - 1)$

11. $\dfrac{2x + 1}{(x - 1)^2}$

12. $\dfrac{-2(x + 1)}{(x - 4)^2}$

Exercises 8.4

See www.CalcChat.com for worked-out solutions to odd-numbered exercises.

In Exercises 1–4, evaluate the derivative of the function at the indicated points on the graph.

1. $f(x) = \dfrac{x^2}{x^2 + 4}$

2. $f(x) = x + \dfrac{32}{x^2}$

In Exercises 5–8, use the derivative to identify the open intervals on which the function is increasing or decreasing. Verify your result with the graph of the function.

5. $f(x) = -(x + 1)^2$

6. $f(x) = \dfrac{x^3}{4} - 3x$

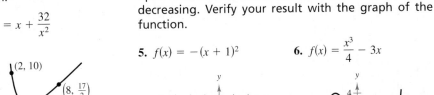

3. $f(x) = (x + 2)^{2/3}$

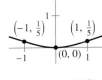

4. $f(x) = -3x\sqrt{x + 1}$

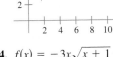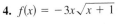

7. $f(x) = x^4 - 2x^2$

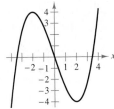

8. $f(x) = \dfrac{x^2}{x + 1}$

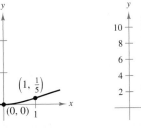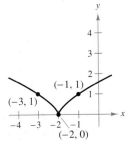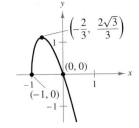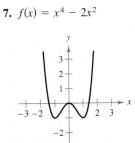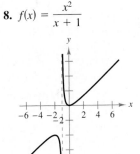

(T) In Exercises 9–32, find the critical numbers and the open intervals on which the function is increasing or decreasing. Then use a graphing utility to graph the function.

9. $f(x) = 2x - 3$

10. $f(x) = 5 - 3x$

11. $g(x) = -(x - 1)^2$

12. $g(x) = (x + 2)^2$

13. $y = x^2 - 6x$

14. $y = -x^2 + 2x$

15. $y = x^3 - 6x^2$

16. $y = (x - 2)^3$

17. $f(x) = \sqrt{x^2 - 1}$

18. $f(x) = \sqrt{9 - x^2}$

19. $y = x^{1/3} + 1$

20. $y = x^{2/3} - 4$

21. $g(x) = (x - 1)^{1/3}$

22. $g(x) = (x - 1)^{2/3}$

23. $f(x) = -2x^2 + 4x + 3$

24. $f(x) = x^2 + 8x + 10$

25. $y = 3x^3 + 12x^2 + 15x$

26. $y = x^3 - 3x + 2$

27. $f(x) = x\sqrt{x + 1}$

28. $h(x) = x\sqrt[3]{x - 1}$

29. $f(x) = x^4 - 2x^3$

30. $f(x) = \frac{1}{4}x^4 - 2x^2$

31. $f(x) = \dfrac{x}{x^2 + 4}$

32. $f(x) = \dfrac{x^2}{x^2 + 4}$

In Exercises 33–38, find the critical numbers and the open intervals on which the function is increasing or decreasing. (*Hint:* Check for discontinuities.) Sketch the graph of the function.

33. $f(x) = \dfrac{2x}{16 - x^2}$

34. $f(x) = \dfrac{x}{x + 1}$

35. $y = \begin{cases} 4 - x^2, & x \le 0 \\ -2x, & x > 0 \end{cases}$

36. $y = \begin{cases} 2x + 1, & x \le -1 \\ x^2 - 2, & x > -1 \end{cases}$

37. $y = \begin{cases} 3x + 1, & x \le 1 \\ 5 - x^2, & x > 1 \end{cases}$

38. $y = \begin{cases} -x^3 + 1, & x \le 0 \\ -x^2 + 2x, & x > 0 \end{cases}$

(T) **39. Cost** The ordering and transportation cost C (in hundreds of dollars) for an automobile dealership is modeled by

$$C = 10\left(\frac{1}{x} + \frac{x}{x + 3}\right), \quad x \ge 1$$

where x is the number of automobiles ordered.

(a) Find the intervals on which C is increasing or decreasing.

(b) Use a graphing utility to graph the cost function.

(c) Use the *trace* feature to determine the order sizes for which the cost is \$900. Assuming that the revenue function is increasing for $x \ge 0$, which order size would you use? Explain your reasoning.

(B) **40. Chemistry: Molecular Velocity** Plots of the relative numbers of N_2 (nitrogen) molecules that have a given velocity at each of three temperatures (in kelvins) are shown in the figure. Identify the differences in the average

velocities (indicated by the peaks of the curves) for the three temperatures, and describe the intervals on which the velocity is increasing and decreasing for each of the three temperatures. (*Source: Adapted from Zumdahl, Chemistry, Seventh Edition*)

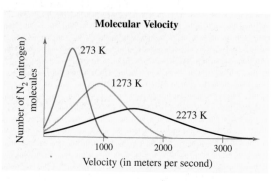

Molecular Velocity

41. Medical Degrees The number y of medical degrees conferred in the United States from 1970 through 2004 can be modeled by

$$y = 0.813t^3 - 55.70t^2 + 1185.2t + 7752, \quad 0 \le t \le 34$$

where t is the time in years, with $t = 0$ corresponding to 1970. (*Source: U.S. National Center for Education Statistics*)

(T) (a) Use a graphing utility to graph the model. Then graphically estimate the years during which the model is increasing and the years during which it is decreasing.

(b) Use the test for increasing and decreasing functions to verify the result of part (a).

42. MAKE A DECISION: PROFIT The profit P made by a cinema from selling x bags of popcorn can be modeled by

$$P = 2.36x - \frac{x^2}{25,000} - 3500, \quad 0 \le x \le 50,000.$$

(a) Find the intervals on which P is increasing and decreasing.

(b) If you owned the cinema, what price would you charge to obtain a maximum profit for popcorn? Explain your reasoning.

43. Profit Analysis A fast-food restaurant determines the cost and revenue models for its hamburgers.

$$C = 0.6x + 7500, \quad 0 \le x \le 50,000$$

$$R = \frac{1}{20,000}(65,000x - x^2), \quad 0 \le x \le 50,000$$

(a) Write the profit function for this situation.

(b) Determine the intervals on which the profit function is increasing and decreasing.

(c) Determine how many hamburgers the restaurant needs to sell to obtain a maximum profit. Explain your reasoning.

Extrema and the First-Derivative Test

- Recognize the occurrence of relative extrema of functions.
- Use the First-Derivative Test to find the relative extrema of functions.
- Find absolute extrema of continuous functions on a closed interval.
- Find minimum and maximum values of real-life models and interpret the results in context.

Relative Extrema

You have used the derivative to determine the intervals on which a function is increasing or decreasing. In this section, you will examine the points at which a function changes from increasing to decreasing, or vice versa. At such a point, the function has a **relative extremum.** (The plural of extremum is *extrema.*) The **relative extrema** of a function include the **relative minima** and **relative maxima** of the function. For instance, the function shown in Figure 8.18 has a relative maximum at the left point and a relative minimum at the right point.

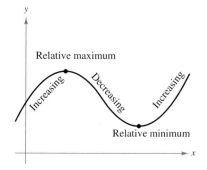

FIGURE 8.18

Definition of Relative Extrema

Let f be a function defined at c.

1. $f(c)$ is a **relative maximum** of f if there exists an interval (a, b) containing c such that $f(x) \leq f(c)$ for all x in (a, b).
2. $f(c)$ is a **relative minimum** of f if there exists an interval (a, b) containing c such that $f(x) \geq f(c)$ for all x in (a, b).

If $f(c)$ is a relative extremum of f, then the relative extremum is said to occur at $x = c$.

For a continuous function, the relative extrema must occur at critical numbers of the function, as shown in Figure 8.19.

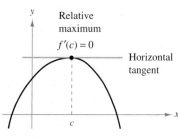

FIGURE 8.19

Occurrences of Relative Extrema

If f has a relative minimum or relative maximum when $x = c$, then c is a critical number of f. That is, either $f'(c) = 0$ or $f'(c)$ is undefined.

The First-Derivative Test

The discussion on the preceding page implies that in your search for relative extrema of a continuous function, you only need to test the critical numbers of the function. Once you have determined that c is a critical number of a function f, the **First-Derivative Test** for relative extrema enables you to classify $f(c)$ as a relative minimum, a relative maximum, or neither.

First-Derivative Test for Relative Extrema

Let f be continuous on the interval (a, b) in which c is the only critical number. If f is differentiable on the interval (except possibly at c), then $f(c)$ can be classified as a relative minimum, a relative maximum, or neither, as shown.

1. On the interval (a, b), if $f'(x)$ is negative to the left of $x = c$ and positive to the right of $x = c$, then $f(c)$ is a relative minimum.

2. On the interval (a, b), if $f'(x)$ is positive to the left of $x = c$ and negative to the right of $x = c$, then $f(c)$ is a relative maximum.

3. On the interval (a, b), if $f'(x)$ is positive on both sides of $x = c$ or negative on both sides of $x = c$, then $f(c)$ is not a relative extremum of f.

A graphical interpretation of the First-Derivative Test is shown in Figure 8.20.

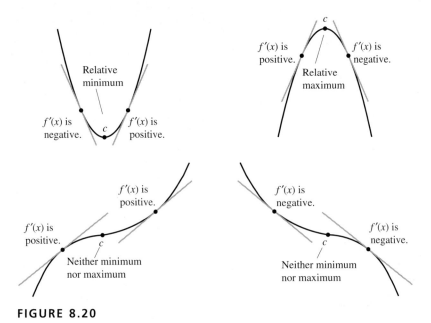

FIGURE 8.20

Example 1   **Finding Relative Extrema**

Find all relative extrema of the function

$$f(x) = 2x^3 - 3x^2 - 36x + 14.$$

SOLUTION   Begin by finding the critical numbers of f.

$f'(x) = 6x^2 - 6x - 36$	Find derivative of f.
$6x^2 - 6x - 36 = 0$	Set derivative equal to 0.
$6(x^2 - x - 6) = 0$	Factor out common factor.
$6(x - 3)(x + 2) = 0$	Factor.
$x = -2, x = 3$	Critical numbers

Because $f'(x)$ is defined for all x, the only critical numbers of f are $x = -2$ and $x = 3$. Using these numbers, you can form the three test intervals $(-\infty, -2)$, $(-2, 3)$, and $(3, \infty)$. The testing of the three intervals is shown in the table.

Interval	$-\infty < x < -2$	$-2 < x < 3$	$3 < x < \infty$
Test value	$x = -3$	$x = 0$	$x = 4$
Sign of $f'(x)$	$f'(-3) = 36 > 0$	$f'(0) = -36 < 0$	$f'(4) = 36 > 0$
Conclusion	Increasing	Decreasing	Increasing

Using the First-Derivative Test, you can conclude that the critical number -2 yields a relative maximum [$f'(x)$ changes sign from positive to negative], and the critical number 3 yields a relative minimum [$f'(x)$ changes sign from negative to positive].

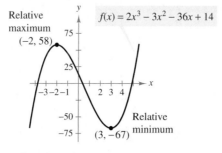

Relative maximum $(-2, 58)$

Relative minimum $(3, -67)$

$f(x) = 2x^3 - 3x^2 - 36x + 14$

FIGURE 8.21

The graph of f is shown in Figure 8.21. The relative maximum is $f(-2) = 58$ and the relative minimum is $f(3) = -67$.

✓**CHECKPOINT 1**

Find all relative extrema of $f(x) = 2x^3 - 6x + 1$. ■

In Example 1, both critical numbers yielded relative extrema. In the next example, only one of the two critical numbers yields a relative extremum.

> ### Algebra Review
>
> For help on the algebra in Example 2, see Example 2(b) in the *Chapter 8 Algebra Review*, on page 680.

Example 2 Finding Relative Extrema

Find all relative extrema of the function $f(x) = x^4 - x^3$.

SOLUTION From the derivative of the function

$$f'(x) = 4x^3 - 3x^2 = x^2(4x - 3)$$

you can see that the function has only two critical numbers: $x = 0$ and $x = \frac{3}{4}$. These numbers produce the test intervals $(-\infty, 0)$, $(0, \frac{3}{4})$, and $(\frac{3}{4}, \infty)$, which are tested in the table.

Interval	$-\infty < x < 0$	$0 < x < \frac{3}{4}$	$\frac{3}{4} < x < \infty$
Test value	$x = -1$	$x = \frac{1}{2}$	$x = 1$
Sign of $f'(x)$	$f'(-1) = -7 < 0$	$f'(\frac{1}{2}) = -\frac{1}{4} < 0$	$f'(1) = 1 > 0$
Conclusion	Decreasing	Decreasing	Increasing

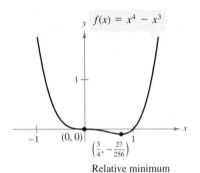

$f(x) = x^4 - x^3$

$\left(\frac{3}{4}, -\frac{27}{256}\right)$ Relative minimum

FIGURE 8.22

By the First-Derivative Test, it follows that f has a relative minimum when $x = \frac{3}{4}$, as shown in Figure 8.22. The relative minimum is $f(\frac{3}{4}) = -\frac{27}{256}$. Note that the critical number $x = 0$ does not yield a relative extremum.

✓ **CHECKPOINT 2**

Find all relative extrema of $f(x) = x^4 - 4x^3$. ■

Example 3 Finding Relative Extrema

Find all relative extrema of the function

$$f(x) = 2x - 3x^{2/3}.$$

SOLUTION From the derivative of the function

$$f'(x) = 2 - \frac{2}{x^{1/3}} = \frac{2(x^{1/3} - 1)}{x^{1/3}}$$

you can see that $f'(1) = 0$ and f' is undefined at $x = 0$. So, the function has two critical numbers: $x = 1$ and $x = 0$. These numbers produce the test intervals $(-\infty, 0)$, $(0, 1)$, and $(1, \infty)$. By testing these intervals, you can conclude that f has a relative maximum at $(0, 0)$ and a relative minimum at $(1, -1)$, as shown in Figure 8.23.

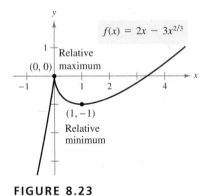

$f(x) = 2x - 3x^{2/3}$

$(0, 0)$ Relative maximum

$(1, -1)$ Relative minimum

FIGURE 8.23

✓ **CHECKPOINT 3**

Find all relative extrema of $f(x) = 3x^{2/3} - 2x$. ■

TECHNOLOGY

(T) There are several ways to use technology to find relative extrema of a function. One way is to use a graphing utility to graph the function, and then use the *zoom* and *trace* features to find the relative minimum and relative maximum points. For instance, consider the graph of

$$f(x) = 3.1x^3 - 7.3x^2 + 1.2x + 2.5$$

as shown below.

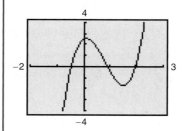

From the graph, you can see that the function has one relative maximum and one relative minimum. You can approximate these values by zooming in and using the *trace* feature, as shown below.

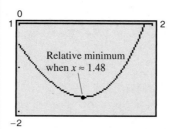
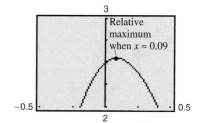

STUDY TIP

Some graphing calculators have a special feature that allows you to find the minimum or maximum of a function on an interval. Consult the user's manual for information on the *minimum value* and *maximum value* features of your graphing utility.

A second way to use technology to find relative extrema is to perform the First-Derivative Test with a symbolic differentiation utility. You can use the utility to differentiate the function, set the derivative equal to zero, and then solve the resulting equation. After obtaining the critical numbers, 1.48288 and 0.0870148, you can graph the function and observe that the first yields a relative minimum and the second yields a relative maximum. Compare the two ways shown above with doing the calculations by hand, as shown below.

$$f(x) = 3.1x^3 - 7.3x^2 + 1.2x + 2.5 \qquad \text{Write original function.}$$

$$f'(x) = \frac{d}{dx}[3.1x^3 - 7.3x^2 + 1.2x + 2.5] \qquad \text{Differentiate with respect to } x.$$

$$f'(x) = 9.3x^2 - 14.6x + 1.2 \qquad \text{First derivative}$$

$$9.3x^2 - 14.6x + 1.2 = 0 \qquad \text{Set derivative equal to 0.}$$

$$x = \frac{73 \pm \sqrt{4213}}{93} \qquad \text{Solve for } x.$$

$$x \approx 1.48288,\ x \approx 0.0870148 \qquad \text{Approximate.}$$

Absolute Extrema

The terms *relative minimum* and *relative maximum* describe the *local* behavior of a function. To describe the *global* behavior of the function on an entire interval, you can use the terms **absolute maximum** and **absolute minimum.**

Definition of Absolute Extrema

Let f be defined on an interval I containing c.

1. $f(c)$ is an **absolute minimum of f** on I if $f(c) \leq f(x)$ for every x in I.

2. $f(c)$ is an **absolute maximum of f** on I if $f(c) \geq f(x)$ for every x in I.

The absolute minimum and absolute maximum values of a function on an interval are sometimes simply called the **minimum** and **maximum** of f on I.

Be sure that you understand the distinction between relative extrema and absolute extrema. For instance, in Figure 8.24, the function has a relative minimum that also happens to be an absolute minimum on the interval $[a, b]$. The relative maximum of f, however, is not the absolute maximum on the interval $[a, b]$. The next theorem points out that if a continuous function has a closed interval as its domain, then it *must* have both an absolute minimum and an absolute maximum on the interval. From Figure 8.24, note that these extrema can occur at endpoints of the interval.

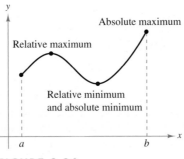

FIGURE 8.24

Extreme Value Theorem

If f is continuous on $[a, b]$, then f has both a minimum value and a maximum value on $[a, b]$.

Although a continuous function has just one minimum and one maximum value on a closed interval, either of these values can occur for more than one x-value. For instance, on the interval $[-3, 3]$, the function $f(x) = 9 - x^2$ has a minimum value of zero when $x = -3$ *and* when $x = 3$, as shown in Figure 8.25.

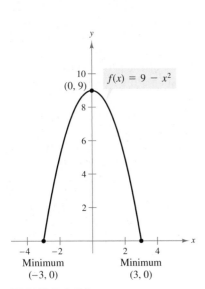

FIGURE 8.25

When looking for extrema of a function on a *closed* interval, remember that you must consider the values of the function at the endpoints as well as at the critical numbers of the function. You can use the guidelines below to find extrema on a closed interval.

Guidelines for Finding Extrema on a Closed Interval

To find the extrema of a continuous function f on a closed interval $[a, b]$, use the steps below.

1. Evaluate f at each of its critical numbers in (a, b).

2. Evaluate f at each endpoint, a and b.

3. The least of these values is the minimum, and the greatest is the maximum.

Example 4 Finding Extrema on a Closed Interval

Find the minimum and maximum values of

$$f(x) = x^2 - 6x + 2$$

on the interval $[0, 5]$.

SOLUTION   Begin by finding the critical numbers of the function.

$$f'(x) = 2x - 6 \qquad \text{Find derivative of } f.$$
$$2x - 6 = 0 \qquad \text{Set derivative equal to 0.}$$
$$2x = 6 \qquad \text{Add 6 to each side.}$$
$$x = 3 \qquad \text{Solve for } x.$$

From this, you can see that the only critical number of f is $x = 3$. Because this number lies in the interval under question, you should test the values of $f(x)$ at this number *and* at the endpoints of the interval, as shown in the table.

x-value	Endpoint: $x = 0$	Critical number: $x = 3$	Endpoint: $x = 5$
$f(x)$	$f(0) = 2$	$f(3) = -7$	$f(5) = -3$
Conclusion	Maximum is 2	Minimum is -7	Neither maximum nor minimum

From the table, you can see that the minimum of f on the interval $[0, 5]$ is $f(3) = -7$. Moreover, the maximum of f on the interval $[0, 5]$ is $f(0) = 2$. This is confirmed by the graph of f, as shown in Figure 8.26.

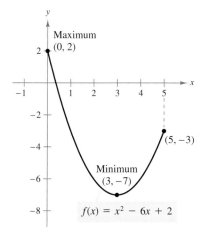

FIGURE 8.26

✓ **CHECKPOINT 4**

Find the minimum and maximum values of $f(x) = x^2 - 8x + 10$ on the interval $[0, 7]$. Sketch the graph of $f(x)$ and label the minimum and maximum values. ■

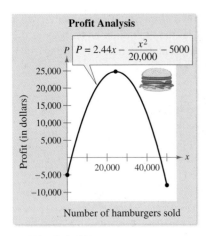

FIGURE 8.27

✓**CHECKPOINT 5**

Verify the results of Example 5 by completing the table.

x (units)	24,000	24,200
P (profit)		

x (units)	24,300	24,400
P (profit)		

x (units)	24,500	24,600
P (profit)		

x (units)	24,800	25,000
P (profit)		

Applications of Extrema

Finding the minimum and maximum values of a function is one of the most common applications of calculus.

Example 5 **Finding the Maximum Profit**

Recall the fast-food restaurant in Examples 7 and 8 in Section 7.5. The restaurant's profit function for hamburgers is given by

$$P = 2.44x - \frac{x^2}{20,000} - 5000, \quad 0 \le x \le 50,000.$$

Find the sales level that yields a maximum profit.

SOLUTION To begin, find an equation for marginal profit. Then set the marginal profit equal to zero and solve for x.

$$P' = 2.44 - \frac{x}{10,000} \qquad \text{Find marginal profit.}$$

$$2.44 - \frac{x}{10,000} = 0 \qquad \text{Set marginal profit equal to 0.}$$

$$-\frac{x}{10,000} = -2.44 \qquad \text{Subtract 2.44 from each side.}$$

$$x = 24,400 \text{ hamburgers} \qquad \text{Critical number}$$

From Figure 8.27, you can see that the critical number $x = 24,400$ corresponds to the sales level that yields a maximum profit. To find the maximum profit, substitute $x = 24,400$ into the profit function.

$$P = 2.44x - \frac{x^2}{20,000} - 5000$$

$$= 2.44(24,400) - \frac{(24,400)^2}{20,000} - 5000$$

$$= \$24,768$$

CONCEPT CHECK

1. Complete the following: The relative extrema of a function include the relative _____ and the relative _____.

2. Let f be continuous on the open interval (a, b) in which c is the only critical number and assume f is differentiable on the interval (except possibly at c). According to the First-Derivative Test, what are the three possible classifications for $f(c)$?

3. Let f be defined on an interval I containing c. The value $f(c)$ is an absolute minimum of f on I if what is true?

4. In your own words, state the guidelines for finding the extrema of a continuous function f on a closed interval $[a, b]$.

Skills Review 8.5

The following warm-up exercises involve skills that were covered in earlier sections. You will use these skills in the exercise set for this section. For additional help, review Sections 7.4, 7.6, and 8.4.

In Exercises 1–6, solve the equation $f'(x) = 0$.

1. $f(x) = 4x^4 - 2x^2 + 1$

2. $f(x) = \frac{1}{3}x^3 - \frac{3}{2}x^2 - 10x$

3. $f(x) = 5x^{4/5} - 4x$

4. $f(x) = \frac{1}{2}x^2 - 3x^{5/3}$

5. $f(x) = \dfrac{x+4}{x^2+1}$

6. $f(x) = \dfrac{x-1}{x^2+4}$

In Exercises 7–10, use $g(x) = -x^5 - 2x^4 + 4x^3 + 2x - 1$ to determine the sign of the derivative.

7. $g'(-4)$

8. $g'(0)$

9. $g'(1)$

10. $g'(3)$

In Exercises 11 and 12, decide whether the function is increasing or decreasing on the given interval.

11. $f(x) = 2x^2 - 11x - 6$, $(3, 6)$

12. $f(x) = x^3 + 2x^2 - 4x - 8$, $(-2, 0)$

Exercises 8.5

See www.CalcChat.com for worked-out solutions to odd-numbered exercises.

In Exercises 1–4, use a table similar to that in Example 1 to find all relative extrema of the function.

1. $f(x) = -2x^2 + 4x + 3$

2. $f(x) = x^2 + 8x + 10$

3. $f(x) = x^2 - 6x$

4. $f(x) = -4x^2 + 4x + 1$

In Exercises 5–12, find all relative extrema of the function.

5. $g(x) = 6x^3 - 15x^2 + 12x$

6. $g(x) = \frac{1}{5}x^5 - x$

7. $h(x) = -(x + 4)^3$

8. $h(x) = 2(x - 3)^3$

9. $f(x) = x^3 - 6x^2 + 15$

10. $f(x) = x^4 - 32x + 4$

11. $f(x) = x^4 - 2x^3 + x + 1$

12. $f(x) = x^4 - 12x^3$

(T) In Exercises 13–18, use a graphing utility to graph the function. Then find all relative extrema of the function.

13. $f(x) = (x - 1)^{2/3}$

14. $f(t) = (t - 1)^{1/3}$

15. $g(t) = t - \dfrac{1}{2t^2}$

16. $f(x) = x + \dfrac{1}{x}$

17. $f(x) = \dfrac{x}{x + 1}$

18. $h(x) = \dfrac{4}{x^2 + 1}$

In Exercises 19–30, find the absolute extrema of the function on the closed interval. Use a graphing utility to verify your results.

19. $f(x) = 2(3 - x)$, $[-1, 2]$

20. $f(x) = \frac{1}{3}(2x + 5)$, $[0, 5]$

21. $f(x) = 5 - 2x^2$, $[0, 3]$

22. $f(x) = x^2 + 2x - 4$, $[-1, 1]$

23. $f(x) = x^3 - 3x^2$, $[-1, 3]$

24. $f(x) = x^3 - 12x$, $[0, 4]$

25. $h(s) = \dfrac{1}{3 - s}$, $[0, 2]$

26. $h(t) = \dfrac{t}{t - 2}$, $[3, 5]$

27. $f(x) = 3x^{2/3} - 2x$, $[-1, 2]$

28. $g(t) = \dfrac{t^2}{t^2 + 3}$, $[-1, 1]$

29. $h(t) = (t - 1)^{2/3}$, $[-7, 2]$

30. $g(x) = 4\left(1 + \dfrac{1}{x} + \dfrac{1}{x^2}\right)$, $[-4, 5]$

In Exercises 31 and 32, approximate the critical numbers of the function shown in the graph. Determine whether the function has a relative maximum, a relative minimum, an absolute maximum, an absolute minimum, or none of these at each critical number on the interval shown.

31.

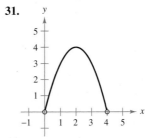

32.

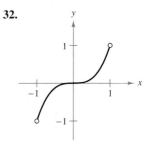

(T) In Exercises 33–36, use a graphing utility to find graphically the absolute extrema of the function on the closed interval.

33. $f(x) = 0.4x^3 - 1.8x^2 + x - 3$, $[0, 5]$

34. $f(x) = 3.2x^5 + 5x^3 - 3.5x$, $[0, 1]$

35. $f(x) = \frac{4}{3}x\sqrt{3 - x}$, $[0, 3]$

36. $f(x) = 4\sqrt{x} - 2x + 1$, $[0, 6]$

In Exercises 37–40, find the absolute extrema of the function on the interval $[0, \infty)$.

37. $f(x) = \dfrac{4x}{x^2 + 1}$

38. $f(x) = \dfrac{8}{x + 1}$

39. $f(x) = \dfrac{2x}{x^2 + 4}$

40. $f(x) = 8 - \dfrac{4x}{x^2 + 1}$

In Exercises 41 and 42, find the maximum value of $|f''(x)|$ on the closed interval. (You will use this skill in Section 12.4 to estimate the error in the Trapezoidal Rule.)

41. $f(x) = \sqrt{1 + x^3}$, $[0, 2]$

42. $f(x) = \dfrac{1}{x^2 + 1}$, $[0, 3]$

In Exercises 43 and 44, find the maximum value of $|f^{(4)}(x)|$ on the closed interval. (You will use this skill in Section 12.4 to estimate the error in Simpson's Rule.)

43. $f(x) = (x + 1)^{2/3}$, $[0, 2]$

44. $f(x) = \dfrac{1}{x^2 + 1}$, $[-1, 1]$

In Exercises 45 and 46, graph a function on the interval $[-2, 5]$ having the given characteristics.

45. Absolute maximum at $x = -2$

Absolute minimum at $x = 1$

Relative maximum at $x = 3$

46. Relative minimum at $x = -1$

Critical number at $x = 0$, but no extrema

Absolute maximum at $x = 2$

Absolute minimum at $x = 5$

47. **Cost** A retailer has determined the cost C for ordering and storing x units of a product to be modeled by

$$C = 3x + \frac{20{,}000}{x}, \quad 0 < x \le 200.$$

The delivery truck can bring at most 200 units per order. Find the order size that will minimize the cost. Use a graphing utility to verify your result.

48. **Profit** The quantity demanded x for a product is inversely proportional to the cube of the price p for $p > 1$. When the price is $10 per unit, the quantity demanded is eight units. The initial cost is $100 and the cost per unit is $4. What price will yield a maximum profit?

(T) 49. **Profit** When soft drinks were sold for $1.00 per can at football games, approximately 6000 cans were sold. When the price was raised to $1.20 per can, the quantity demanded dropped to 5600. The initial cost is $5000 and the cost per unit is $0.50. Assuming that the demand function is linear, use the *table* feature of a graphing utility to determine the price that will yield a maximum profit.

50. **Medical Science** Coughing forces the trachea (windpipe) to contract, which in turn affects the velocity of the air through the trachea. The velocity of the air during coughing can be modeled by $v = k(R - r)r^2$, $0 \le r < R$, where k is a constant, R is the normal radius of the trachea, and r is the radius during coughing. What radius r will produce the maximum air velocity?

51. **Population** The resident population P (in millions) of the United States from 1790 through 2000 can be modeled by $P = 0.00000583t^3 + 0.005003t^2 + 0.13776t + 4.658$, $-10 \le t \le 200$, where $t = 0$ corresponds to 1800. *(Source: U.S. Census Bureau)*

(a) Make a conjecture about the maximum and minimum populations in the U.S. from 1790 to 2000.

(b) Analytically find the maximum and minimum populations over the interval.

(c) Write a brief paragraph comparing your conjecture with your results in part (b).

52. **Biology: Fertility Rates** The graph of the United States fertility rate shows the number of births per 1000 women in their lifetime according to the birth rate in that particular year. *(Source: U.S. National Center for Health Statistics)*

(a) Around what year was the fertility rate the highest, and to how many births per 1000 women did this rate correspond?

(b) During which time periods was the fertility rate increasing most rapidly? Most slowly?

(c) During which time periods was the fertility rate decreasing most rapidly? Most slowly?

(d) Give some possible real-life reasons for fluctuations in the fertility rate.

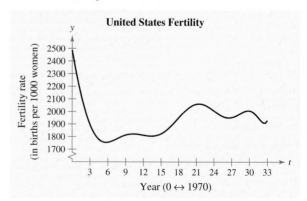

United States Fertility

Fertility rate (in births per 1000 women)

Year (0 ↔ 1970)

Section 8.6

Concavity and the Second-Derivative Test

- Determine the intervals on which the graphs of functions are concave upward or concave downward.
- Find the points of inflection of the graphs of functions.
- Use the Second-Derivative Test to find the relative extrema of functions.
- Find the points of diminishing returns of input-output models.

Concavity

You already know that locating the intervals over which a function f increases or decreases is helpful in determining its graph. In this section, you will see that locating the intervals on which f' increases or decreases can determine where the graph of f is curving upward or curving downward. This property of curving upward or downward is defined formally as the **concavity** of the graph of the function.

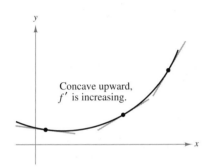

Concave upward,
f' is increasing.

Definition of Concavity

Let f be differentiable on an open interval I. The graph of f is

1. **concave upward** on I if f' is increasing on the interval.

2. **concave downward** on I if f' is decreasing on the interval.

From Figure 8.28, you can observe the following graphical interpretation of concavity.

1. A curve that is concave upward lies *above* its tangent line.

2. A curve that is concave downward lies *below* its tangent line.

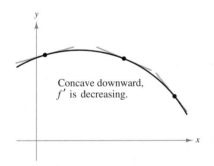

Concave downward,
f' is decreasing.

This visual test for concavity is useful when the graph of a function is given. To determine concavity without seeing a graph, you need an analytic test. It turns out that you can use the second derivative to determine these intervals in much the same way that you use the first derivative to determine the intervals on which f is increasing or decreasing.

FIGURE 8.28

Test for Concavity

Let f be a function whose second derivative exists on an open interval I.

1. If $f''(x) > 0$ for all x in I, then f is concave upward on I.

2. If $f''(x) < 0$ for all x in I, then f is concave downward on I.

For a *continuous* function f, you can find the open intervals on which the graph of f is concave upward and concave downward as follows. [For a function that is not continuous, the test intervals should be formed using points of discontinuity, along with the points at which $f''(x)$ is zero or undefined.]

Guidelines for Applying Concavity Test

1. Locate the x-values at which $f''(x) = 0$ or $f''(x)$ is undefined.

2. Use these x-values to determine the test intervals.

3. Test the sign of $f''(x)$ in each test interval.

Example 1   Applying the Test for Concavity

a. The graph of the function

$$f(x) = x^2 \qquad \text{Original function}$$

is concave upward on the entire real line because its second derivative

$$f''(x) = 2 \qquad \text{Second derivative}$$

is positive for all x. (See Figure 8.29.)

b. The graph of the function

$$f(x) = \sqrt{x} \qquad \text{Original function}$$

is concave downward for $x > 0$ because its second derivative

$$f''(x) = -\frac{1}{4}x^{-3/2} \qquad \text{Second derivative}$$

is negative for all $x > 0$. (See Figure 8.30.)

✓ CHEC⌐

of

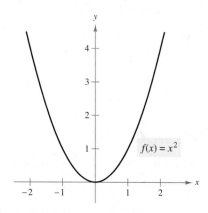

FIGURE 8.29   Concave Upward

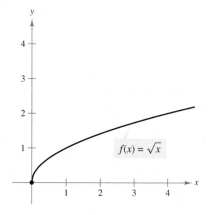

FIGURE 8.30   Concave Downward

Example 2 **Determining Concavity**

Determine the open intervals on which the graph of the function is concave upward or concave downward.

$$f(x) = \frac{6}{x^2 + 3}$$

SOLUTION   Begin by finding the second derivative of f.

$$f(x) = 6(x^2 + 3)^{-1}$$ Rewrite original function.

$$f'(x) = (-6)(2x)(x^2 + 3)^{-2}$$ Chain Rule

$$= \frac{-12x}{(x^2 + 3)^2}$$ Simplify.

$$f''(x) = \frac{(x^2 + 3)^2(-12) - (-12x)(2)(2x)(x^2 + 3)}{(x^2 + 3)^4}$$ Quotient Rule

$$= \frac{-12(x^2 + 3) + (48x^2)}{(x^2 + 3)^3}$$ Simplify.

$$= \frac{36(x^2 - 1)}{(x^2 + 3)^3}$$ Simplify.

From this, you can see that $f''(x)$ is defined for all real numbers and $f''(x) = 0$ when $x = \pm 1$. So, you can test the concavity of f by testing the intervals $(-\infty, -1)$, $(-1, 1)$, and $(1, \infty)$, as shown in the table. The graph of f is shown in Figure 8.31.

Interval	$-\infty < x < -1$	$-1 < x < 1$	$1 < x < \infty$
Test value	$x = -2$	$x = 0$	$x = 2$
Sign of $f''(x)$	$f''(-2) > 0$	$f''(0) < 0$	$f''(2) > 0$
Conclusion	Concave upward	Concave downward	Concave upward

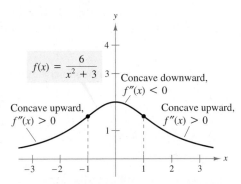

$$f(x) = \frac{6}{x^2 + 3}$$

Concave upward, $f''(x) > 0$

Concave downward, $f''(x) < 0$

Concave upward, $f''(x) > 0$

FIGURE 8.31

STUDY TIP

In Example 2, f' is increasing on the interval $(1, \infty)$ even though f is decreasing there. Be sure you see that the increasing or decreasing of f' does not necessarily correspond to the increasing or decreasing of f.

✓ **CHECKPOINT 2**

Determine the intervals on which the graph of the function is concave upward or concave downward.

$$f(x) = \frac{12}{x^2 + 4}$$ ■

Points of Inflection

If the tangent line to a graph exists at a point at which the concavity changes, then the point is a **point of inflection.** Three examples of inflection points are shown in Figure 8.32. (Note that the third graph has a vertical tangent line at its point of inflection.)

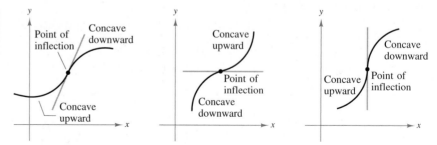

FIGURE 8.32   The graph *crosses* its tangent line at a point of inflection.

> **STUDY TIP**
>
> As shown in Figure 8.32, a graph crosses its tangent line at a point of inflection.

Definition of Point of Inflection

If the graph of a continuous function has a tangent line at a point where its concavity changes from upward to downward (or downward to upward), then the point is a **point of inflection.**

DISCOVERY

Use a graphing utility to graph

$$f(x) = x^3 - 6x^2 + 12x - 6 \quad \text{and} \quad f''(x) = 6x - 12$$

in the same viewing window. At what x-value does $f''(x) = 0$? At what x-value does the point of inflection occur? Repeat this analysis for

$$g(x) = x^4 - 5x^2 + 7 \quad \text{and} \quad g''(x) = 12x^2 - 10.$$

Make a general statement about the relationship of the point of inflection of a function and the second derivative of the function.

Because a point of inflection occurs where the concavity of a graph changes, it must be true that at such points the sign of f'' changes. So, to locate possible points of inflection, you only need to determine the values of x for which $f''(x) = 0$ or for which $f''(x)$ does not exist. This parallels the procedure for locating the relative extrema of f by determining the critical numbers of f.

Property of Points of Inflection

If $(c, f(c))$ is a point of inflection of the graph of f, then either $f''(c) = 0$ or $f''(c)$ is undefined.

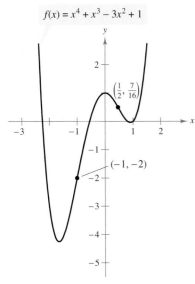

$f(x) = x^4 + x^3 - 3x^2 + 1$

FIGURE 8.33   Two Points of Inflection

Example 3   **Finding Points of Inflection**

Discuss the concavity of the graph of f and find its points of inflection.

$$f(x) = x^4 + x^3 - 3x^2 + 1$$

SOLUTION   Begin by finding the second derivative of f.

$$f(x) = x^4 + x^3 - 3x^2 + 1 \qquad \text{Write original function.}$$
$$f'(x) = 4x^3 + 3x^2 - 6x \qquad \text{Find first derivative.}$$
$$f''(x) = 12x^2 + 6x - 6 \qquad \text{Find second derivative.}$$
$$= 6(2x - 1)(x + 1) \qquad \text{Factor.}$$

From this, you can see that the possible points of inflection occur at $x = \frac{1}{2}$ and $x = -1$. After testing the intervals $(-\infty, -1)$, $\left(-1, \frac{1}{2}\right)$, and $\left(\frac{1}{2}, \infty\right)$, you can determine that the graph is concave upward on $(-\infty, -1)$, concave downward on $\left(-1, \frac{1}{2}\right)$, and concave upward on $\left(\frac{1}{2}, \infty\right)$. Because the concavity changes at $x = -1$ and $x = \frac{1}{2}$, you can conclude that the graph has points of inflection at these x-values, as shown in Figure 8.33.

✓ CHECKPOINT 3

Discuss the concavity of the graph of f and find its points of inflection.

$$f(x) = x^4 - 2x^3 + 1 \quad ■$$

It is possible for the second derivative to be zero at a point that is *not* a point of inflection. For example, compare the graphs of $f(x) = x^3$ and $g(x) = x^4$, as shown in Figure 8.34. Both second derivatives are zero when $x = 0$, but only the graph of f has a point of inflection at $x = 0$. This shows that before concluding that a point of inflection exists at a value of x for which $f''(x) = 0$, you must test to be certain that the concavity actually changes at that point.

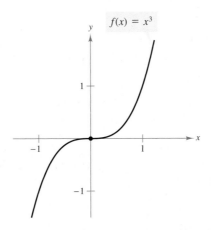

$f''(0) = 0$, and $(0, 0)$ is a point of inflection.

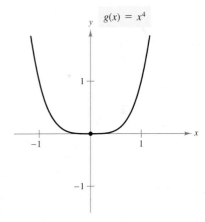

$g''(0) = 0$, but $(0, 0)$ is not a point of inflection.

FIGURE 8.34

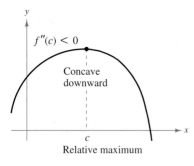

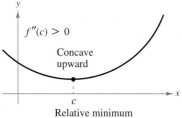

FIGURE 8.35

The Second-Derivative Test

The second derivative can be used to perform a simple test for relative minima and relative maxima. If f is a function such that $f'(c) = 0$ and the graph of f is concave upward at $x = c$, then $f(c)$ is a relative minimum of f. Similarly, if f is a function such that $f'(c) = 0$ and the graph of f is concave downward at $x = c$, then $f(c)$ is a relative maximum of f, as shown in Figure 8.35.

Second-Derivative Test

Let $f'(c) = 0$, and let f'' exist on an open interval containing c.

1. If $f''(c) > 0$, then $f(c)$ is a relative minimum.

2. If $f''(c) < 0$, then $f(c)$ is a relative maximum.

3. If $f''(c) = 0$, then the test fails. In such cases, you can use the First-Derivative Test to determine whether $f(c)$ is a relative minimum, a relative maximum, or neither.

Example 4   **Using the Second-Derivative Test**

Find the relative extrema of

$$f(x) = -3x^5 + 5x^3.$$

SOLUTION   Begin by finding the first derivative of f.

$$f'(x) = -15x^4 + 15x^2$$
$$= 15x^2(1 - x^2)$$

From this derivative, you can see that $x = 0$, $x = -1$, and $x = 1$ are the only critical numbers of f. Using the second derivative

$$f''(x) = -60x^3 + 30x$$

you can apply the Second-Derivative Test, as shown.

Point	Sign of $f''(x)$	Conclusion
$(-1, -2)$	$f''(-1) = 30 > 0$	Relative minimum
$(0, 0)$	$f''(0) = 0$	Test fails.
$(1, 2)$	$f''(1) = -30 < 0$	Relative maximum

Because the test fails at $(0, 0)$, you can apply the First-Derivative Test to conclude that the point $(0, 0)$ is neither a relative minimum nor a relative maximum—a test for concavity would show that this point is a point of inflection. The graph of f is shown in Figure 8.36.

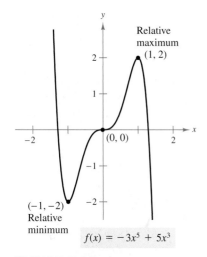

FIGURE 8.36

✓ **CHECKPOINT 4**

Find all relative extrema of $f(x) = x^4 - 4x^3 + 1$. ■

Extended Application: Diminishing Returns

In economics, the notion of concavity is related to the concept of **diminishing returns.** Consider a function

Output ⌐   ⌐Input

$$y = f(x)$$

where x measures input (in dollars) and y measures output (in dollars). In Figure 8.37, notice that the graph of this function is concave upward on the interval (a, c) and is concave downward on the interval (c, b). On the interval (a, c), each additional dollar of input returns more than the previous input dollar. By contrast, on the interval (c, b), each additional dollar of input returns less than the previous input dollar. The point $(c, f(c))$ is called the **point of diminishing returns.** An increased investment beyond this point is usually considered a poor use of capital.

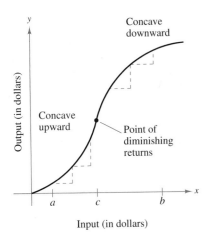

FIGURE 8.37

Example 5 **Exploring Diminishing Returns**

By increasing its advertising cost x (in thousands of dollars) for a product, a company discovers that it can increase the sales y (in thousands of dollars) according to the model

$$y = -\frac{1}{10}x^3 + 6x^2 + 400, \quad 0 \le x \le 40.$$

Find the point of diminishing returns for this product.

SOLUTION Begin by finding the first and second derivatives.

$$y' = 12x - \frac{3x^2}{10} \qquad \text{First derivative}$$

$$y'' = 12 - \frac{3x}{5} \qquad \text{Second derivative}$$

The second derivative is zero only when $x = 20$. By testing the intervals $(0, 20)$ and $(20, 40)$, you can conclude that the graph has a point of diminishing returns when $x = 20$, as shown in Figure 8.38. So, the point of diminishing returns for this product occurs when \$20,000 is spent on advertising. ▬

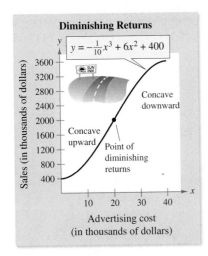

FIGURE 8.38

✓ CHECKPOINT 5

Find the point of diminishing returns for the model below, where R is the revenue (in thousands of dollars) and x is the advertising cost (in thousands of dollars).

$$R = \frac{1}{20,000}(450x^2 - x^3),$$

$$0 \le x \le 300 \quad ■$$

┌─ **CONCEPT CHECK** ─┐

1. Let f be differentiable on an open interval I. If the graph of f is concave upward on I, what can you conclude about the behavior of f' on the interval?

2. Let f be a function whose second derivative exists on an open interval I and $f''(x) > 0$ for all x in I. Is f concave upward or concave downward on I?

3. Let $f'(c) = 0$, and let f'' exist on an open interval containing c. According to the Second-Derivative Test, what are the possible classifications for $f(c)$?

4. A newspaper headline states that "The rate of growth of the national deficit is decreasing." What does this mean? What does it imply about the graph of the deficit as a function of time?

Skills Review 8.6

The following warm-up exercises involve skills that were covered in earlier sections. You will use these skills in the exercise set for this section. For additional help, review Sections 7.6, 7.7, 8.1, and 8.4.

In Exercises 1–6, find the second derivative of the function.

1. $f(x) = 4x^4 - 9x^3 + 5x - 1$

2. $g(s) = (s^2 - 1)(s^2 - 3s + 2)$

3. $g(x) = (x^2 + 1)^4$

4. $f(x) = (x - 3)^{4/3}$

5. $h(x) = \dfrac{4x + 3}{5x - 1}$

6. $f(x) = \dfrac{2x - 1}{3x + 2}$

In Exercises 7–10, find the critical numbers of the function.

7. $f(x) = 5x^3 - 5x + 11$

8. $f(x) = x^4 - 4x^3 - 10$

9. $g(t) = \dfrac{16 + t^2}{t}$

10. $h(x) = \dfrac{x^4 - 50x^2}{8}$

Exercises 8.6

See www.CalcChat.com for worked-out solutions to odd-numbered exercises.

In Exercises 1–8, analytically find the open intervals on which the graph is concave upward and those on which it is concave downward.

1. $y = x^2 - x - 2$

2. $y = -x^3 + 3x^2 - 2$

3. $f(x) = \dfrac{x^2 - 1}{2x + 1}$

4. $f(x) = \dfrac{x^2 + 4}{4 - x^2}$

5. $f(x) = \dfrac{24}{x^2 + 12}$

6. $f(x) = \dfrac{x^2}{x^2 + 1}$

7. $y = -x^3 + 6x^2 - 9x - 1$

8. $y = x^5 + 5x^4 - 40x^2$

In Exercises 9–22, find all relative extrema of the function. Use the Second-Derivative Test when applicable.

9. $f(x) = 6x - x^2$

10. $f(x) = (x - 5)^2$

11. $f(x) = x^3 - 5x^2 + 7x$

12. $f(x) = x^4 - 4x^3 + 2$

13. $f(x) = x^{2/3} - 3$

14. $f(x) = x + \dfrac{4}{x}$

15. $f(x) = \sqrt{x^2 + 1}$

16. $f(x) = \sqrt{2x^2 + 6}$

17. $f(x) = \sqrt{9 - x^2}$

18. $f(x) = \sqrt{4 - x^2}$

19. $f(x) = \dfrac{8}{x^2 + 2}$

20. $f(x) = \dfrac{18}{x^2 + 3}$

21. $f(x) = \dfrac{x}{x - 1}$

22. $f(x) = \dfrac{x}{x^2 - 1}$

(T) In Exercises 23–26, use a graphing utility to estimate graphically all relative extrema of the function.

23. $f(x) = \frac{1}{2}x^4 - \frac{1}{3}x^3 - \frac{1}{2}x^2$

24. $f(x) = -\frac{1}{3}x^5 - \frac{1}{2}x^4 + x$

25. $f(x) = 5 + 3x^2 - x^3$

26. $f(x) = 3x^3 + 5x^2 - 2$

In Exercises 27–30, state the signs of $f'(x)$ and $f''(x)$ on the interval (0, 2).

27.
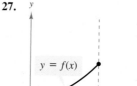

28.
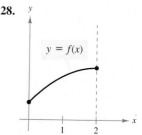

29.
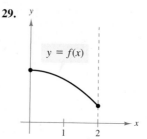

30.
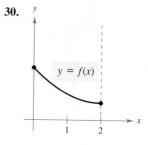

In Exercises 31–38, find the point(s) of inflection of the graph of the function.

31. $f(x) = x^3 - 9x^2 + 24x - 18$

32. $f(x) = x(6 - x)^2$

33. $f(x) = (x - 1)^3(x - 5)$

34. $f(x) = x^4 - 18x^2 + 5$

35. $g(x) = 2x^4 - 8x^3 + 12x^2 + 12x$

36. $f(x) = -4x^3 - 8x^2 + 32$

37. $h(x) = (x - 2)^3(x - 1)$

38. $f(t) = (1 - t)(t - 4)(t^2 - 4)$

Ⓣ In Exercises 39–50, use a graphing utility to graph the function and identify all relative extrema and points of inflection.

39. $f(x) = x^3 - 12x$

40. $f(x) = x^3 - 3x$

41. $f(x) = x^3 - 6x^2 + 12x$

42. $f(x) = x^3 - \frac{3}{2}x^2 - 6x$

43. $f(x) = \frac{1}{4}x^4 - 2x^2$

44. $f(x) = 2x^4 - 8x + 3$

45. $g(x) = (x - 2)(x + 1)^2$

46. $g(x) = (x - 6)(x + 2)^3$

47. $g(x) = x\sqrt{x + 3}$

48. $g(x) = x\sqrt{9 - x}$

49. $f(x) = \dfrac{4}{1 + x^2}$

50. $f(x) = \dfrac{2}{x^2 - 1}$

In Exercises 51–54, sketch a graph of a function f having the given characteristics.

51. $f(2) = f(4) = 0$
$f'(x) < 0$ if $x < 3$
$f'(3) = 0$
$f'(x) > 0$ if $x > 3$
$f''(x) > 0$

52. $f(2) = f(4) = 0$
$f'(x) > 0$ if $x < 3$
$f'(3)$ is undefined.
$f'(x) < 0$ if $x > 3$
$f''(x) > 0, x \neq 3$

53. $f(0) = f(2) = 0$
$f'(x) > 0$ if $x < 1$
$f'(1) = 0$
$f'(x) < 0$ if $x > 1$
$f''(x) < 0$

54. $f(0) = f(2) = 0$
$f'(x) < 0$ if $x < 1$
$f'(1) = 0$
$f'(x) > 0$ if $x > 1$
$f''(x) > 0$

In Exercises 55 and 56, use the graph to sketch the graph of f'. Find the intervals on which (a) $f'(x)$ is positive, (b) $f'(x)$ is negative, (c) f' is increasing, and (d) f' is decreasing. For each of these intervals, describe the corresponding behavior of f.

55.

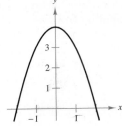

56.

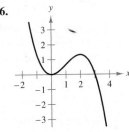

In Exercises 57–60, you are given f'. Find the intervals on which (a) $f'(x)$ is increasing or decreasing and (b) the graph of f is concave upward or concave downward. (c) Find the relative extrema and inflection points of f. (d) Then sketch a graph of f.

57. $f'(x) = 2x + 5$

58. $f'(x) = 3x^2 - 2$

59. $f'(x) = -x^2 + 2x - 1$

60. $f'(x) = x^2 + x - 6$

Point of Diminishing Returns In Exercises 61 and 62, identify the point of diminishing returns for the input-output function. For each function, R is the revenue and x is the amount spent on advertising. Use a graphing utility to verify your results.

61. $R = \dfrac{1}{50,000}(600x^2 - x^3), \quad 0 \le x \le 400$

62. $R = -\frac{4}{9}(x^3 - 9x^2 - 27), \quad 0 \le x \le 5$

Average Cost In Exercises 63 and 64, you are given the total cost of producing x units. Find the production level that minimizes the average cost per unit. Use a graphing utility to verify your results.

63. $C = 0.5x^2 + 15x + 5000$

64. $C = 0.002x^3 + 20x + 500$

Productivity In Exercises 65 and 66, consider a college student who works from 7 P.M. to 11 P.M. assembling mechanical components. The number N of components assembled after t hours is given by the function. At what time is the student assembling components at the greatest rate?

65. $N = -0.12t^3 + 0.54t^2 + 8.22t, \quad 0 \le t \le 4$

66. $N = \dfrac{20t^2}{4 + t^2}, \quad 0 \le t \le 4$

Sales Growth In Exercises 67 and 68, find the time t in years when the annual sales x of a new product are increasing at the greatest rate. Use a graphing utility to verify your results.

67. $x = \dfrac{10,000t^2}{9 + t^2}$

68. $x = \dfrac{500,000t^2}{36 + t^2}$

Ⓣ In Exercises 69–72, use a graphing utility to graph f, f', and f'' in the same viewing window. Graphically locate the relative extrema and points of inflection of the graph of f. State the relationship between the behavior of f and the signs of f' and f''.

69. $f(x) = \frac{1}{2}x^3 - x^2 + 3x - 5, \quad [0, 3]$

70. $f(x) = -\frac{1}{20}x^5 - \frac{1}{12}x^2 - \frac{1}{3}x + 1, \quad [-2, 2]$

71. $f(x) = \dfrac{2}{x^2 + 1}, \quad [-3, 3]$ **72.** $f(x) = \dfrac{x^2}{x^2 + 1}, \quad [-3, 3]$

73. Average Cost A manufacturer has determined that the total cost C of operating a factory is $C = 0.5x^2 + 10x + 7200$, where x is the number of units produced. At what level of production will the average cost per unit be minimized? (The average cost per unit is C/x.)

74. Inventory Cost The cost C for ordering and storing x units is $C = 2x + 300,000/x$. What order size will produce a minimum cost?

75. Phishing Phishing is a criminal activity used by an individual or group to fraudulently acquire information by masquerading as a trustworthy person or business in an electronic communication. Criminals create spoof sites on the Internet to trick victims into giving them information. The sites are designed to copy the exact look and feel of a "real" site. A model for the number of reported spoof sites from November 2005 through October 2006 is

$$f(t) = 88.253t^3 - 1116.16t^2 + 4541.4t + 4161, \quad 0 \le t \le 11$$

where t represents the number of months since November 2005. *(Source: Anti-Phishing Working Group)*

(T) (a) Use a graphing utility to graph the model on the interval $[0, 11]$.

(b) Use the graph in part (a) to estimate the month corresponding to the absolute minimum number of spoof sites.

(c) Use the graph in part (a) to estimate the month corresponding to the absolute maximum number of spoof sites.

(d) During approximately which month was the rate of increase of the number of spoof sites the greatest? the least?

76. Dow Jones Industrial Average The graph shows the Dow Jones Industrial Average y on Black Monday, October 19, 1987, where $t = 0$ corresponds to 9:30 A.M., when the market opens, and $t = 6.5$ corresponds to 4 P.M., the closing time. *(Source: Wall Street Journal)*

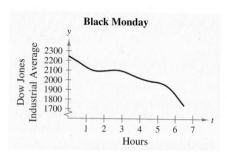

Black Monday

(a) Estimate the relative extrema and absolute extrema of the graph. Interpret your results in the context of the problem.

(b) Estimate the point of inflection of the graph on the interval $[1, 3]$. Interpret your result in the context of the problem.

77. Think About It Let S represent monthly sales of a new digital audio player. Write a statement describing S' and S'' for each of the following.

(a) The rate of change of sales is increasing.

(b) Sales are increasing, but at a greater rate.

(c) The rate of change of sales is steady.

(d) Sales are steady.

(e) Sales are declining, but at a lower rate.

(f) Sales have bottomed out and have begun to rise.

78. Medicine The spread of a virus can be modeled by

$$N = -t^3 + 12t^2, \quad 0 \le t \le 12$$

where N is the number of people infected (in hundreds), and t is the time (in weeks).

(a) What is the maximum number of people projected to be infected?

(b) When will the virus be spreading most rapidly?

(T) (c) Use a graphing utility to graph the model and to verify your results.

Business Capsule

Photo courtesy of Pat Alexander Sanford

In 1985, Pat Alexander Sanford started Alexander Perry, Inc., in Philadelphia, Pennsylvania. The company specializes in providing interior architecture and space planning to corporations, educational institutions, and private residences. Sanford started the company using about $5000 from her personal savings and a grant from the Women's Enterprise Center in Philadelphia. The company was incorporated in 1992. Revenues for the company topped $714,000 in 2004 and contracts for 2006 totaled about $6 million. Projected sales are currently expected to approach $10 million.

79. Research Project Use your school's library, the Internet, or some other reference source to research the financial history of a small company like the one above. Gather the data on the company's costs and revenues over a period of time, and use a graphing utility to graph a scatter plot of the data. Fit models to the data. Do the models appear to be concave upward or downward? Do they appear to be increasing or decreasing? Discuss the implications of your answers.

Algebra Review

(T) The equations in Example 1 are solved algebraically. Most graphing utilities have a "solve" key that allows you to solve equations graphically. If you have a graphing utility, try using it to solve graphically the equations in Example 1.

Solving Equations

Much of the algebra in Chapter 8 involves simplifying algebraic expressions (see pages 617 and 618) and solving algebraic equations, as illustrated in the following examples. In Example 1, you can review some of the basic techniques for solving equations. In Example 2 on the next page, you can review some of the more complicated techniques for solving equations.

When solving an equation, remember that your basic goal is to isolate the variable on one side of the equation.

1. To solve a *linear equation*, you can add or subtract the same quantity from each side of the equation. You can also multiply or divide each side of the equation by the same *nonzero* quantity.

2. To solve a *quadratic equation*, you can take the square root of each side, use factoring, or use the Quadratic Formula.

3. To solve a *radical equation*, isolate the radical on one side of the equation and square each side of the equation.

STUDY TIP

Remember, solving radical equations can sometimes lead to *extraneous solutions* (those that do not satisfy the original equation). For example, squaring both sides of the following equation yields two possible solutions, one of which is extraneous.

$$\sqrt{x} = x - 2$$
$$x = x^2 - 4x + 4$$
$$0 = x^2 - 5x + 4$$
$$= (x - 4)(x - 1)$$
$$x - 4 = 0 \implies x = 4$$
$$\text{(solution)}$$
$$x - 1 = 0 \implies x = 1$$
$$\text{(extraneous)}$$

Example 1   Solving Equations

Solve each equation.

a. $3x - 3 = 5x - 7$ **b.** $2x^2 = 10$

c. $2x^2 + 5x - 6 = 6$ **d.** $\sqrt{2x - 7} = 5$

SOLUTION

a. $3x - 3 = 5x - 7$		Write original (linear) equation.
$-3 = 2x - 7$		Subtract $3x$ from each side.
$4 = 2x$		Add 7 to each side.
$2 = x$		Divide each side by 2.
b. $2x^2 = 10$		Write original (quadratic) equation.
$x^2 = 5$		Divide each side by 2.
$x = \pm\sqrt{5}$		Take the square root of each side.
c. $2x^2 + 5x - 6 = 6$		Write original (quadratic) equation.
$2x^2 + 5x - 12 = 0$		Write in general form.
$(2x - 3)(x + 4) = 0$		Factor.
$2x - 3 = 0 \implies x = \frac{3}{2}$		Set first factor equal to zero.
$x + 4 = 0 \implies x = -4$		Set second factor equal to zero.
d. $\sqrt{2x - 7} = 5$		Write original (radical) equation.
$2x - 7 = 25$		Square each side.
$2x = 32$		Add 7 to each side.
$x = 16$		Divide each side by 2.

| Example 2 | Solving an Equation |

Solve each equation.

a. $\dfrac{36(x^2 - 1)}{(x^2 + 3)^3} = 0$ **b.** $x^2(4x - 3) = 0$

c. $\dfrac{4x}{3(x^2 - 4)^{1/3}} = 0$ **d.** $g'(x) = 0$, where $g(x) = (x - 2)(x + 1)^2$

SOLUTION

a. $\dfrac{36(x^2 - 1)}{(x^2 + 3)^3} = 0$ Example 2, page 671

$\quad 36(x^2 - 1) = 0$ A fraction is zero only if its numerator is zero.

$\quad\quad\quad x^2 - 1 = 0$ Divide each side by 36.

$\quad\quad\quad\quad\quad x^2 = 1$ Add 1 to each side.

$\quad\quad\quad\quad\quad x = \pm 1$ Take the square root of each side.

b. $x^2(4x - 3) = 0$ Example 2, page 662

$\quad\quad x^2 = 0 \implies x = 0$ Set first factor equal to zero.

$\quad 4x - 3 = 0 \implies x = \tfrac{3}{4}$ Set second factor equal to zero.

c. $\dfrac{4x}{3(x^2 - 4)^{1/3}} = 0$ Example 4, page 654

$\quad\quad\quad 4x = 0$ A fraction is zero only if its numerator is zero.

$\quad\quad\quad\quad x = 0$ Divide each side by 4.

d. $g(x) = (x - 2)(x + 1)^2$ Exercise 45, page 677

$\quad (x - 2)(2)(x + 1) + (x + 1)^2(1) = 0$ Find derivative and set equal to zero.

$\quad\quad (x + 1)[2(x - 2) + (x + 1)] = 0$ Factor.

$\quad\quad\quad (x + 1)(2x - 4 + x + 1) = 0$ Multiply factors.

$\quad\quad\quad\quad (x + 1)(3x - 3) = 0$ Combine like terms.

$\quad\quad\quad\quad\quad x + 1 = 0 \implies x = -1$ Set first factor equal to zero.

$\quad\quad\quad\quad 3x - 3 = 0 \implies x = 1$ Set second factor equal to zero.

Chapter Summary and Study Strategies

After studying this chapter, you should have acquired the following skills.
The exercise numbers are keyed to the Review Exercises that begin on page 683.
Answers to odd-numbered Review Exercises are given in the back of the text.

Section 8.1

	Review Exercises
■ Find higher-order derivatives.	*1–12, 16*
■ Find and use the position function to determine the velocity and acceleration of a moving object.	*13–15*

Section 8.2

■ Find derivatives implicitly.	*17–28*

Section 8.3

■ Solve related-rate problems.	*29–32*

Section 8.4

■ Find the critical numbers of a function.	*33–36*
$\quad c$ is a critical number of f if $f'(c) = 0$ or $f'(c)$ is undefined.	
■ Find the open intervals on which a function is increasing or decreasing.	*37–40*
$\quad$ Increasing if $f'(x) > 0$	
$\quad$ Decreasing if $f'(x) < 0$	
■ Find intervals on which a real-life model is increasing or decreasing, and interpret the results in context.	*41–44*

Section 8.5

■ Use the First-Derivative Test to find the relative extrema of a function.	*45–54*
■ Find the absolute extrema of a continuous function on a closed interval.	*55–64*
■ Find minimum and maximum values of a real-life model and interpret the results in context.	*65–70*

Section 8.6

■ Find the open intervals on which the graph of a function is concave upward or concave downward.	*71–74*
$\quad$ Concave upward if $f''(x) > 0$	
$\quad$ Concave downward if $f''(x) < 0$	
■ Find the points of inflection of the graph of a function.	*75–78, 83*
■ Use the Second-Derivative Test to find the relative extrema of a function.	*79–82, 84*
■ Find the point of diminishing returns of an input-output model.	*85, 86*

Study Strategies

- **Solve Problems Graphically, Analytically, and Numerically**   When analyzing the graph of a function, use a variety of problem-solving strategies. For instance, if you were asked to analyze the graph of

$$f(x) = x^3 - 4x^2 + 5x - 4$$

you could begin *graphically*. That is, you could use a graphing utility to find a viewing window that appears to show the important characteristics of the graph. From the graph shown below, the function appears to have one relative minimum, one relative maximum, and one point of inflection.

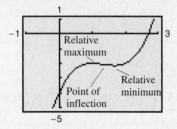

Next, you could use calculus to *analyze* the graph. Because the derivative of f is

$$f'(x) = 3x^2 - 8x + 5 = (3x - 5)(x - 1)$$

the critical numbers of f are $x = \frac{5}{3}$ and $x = 1$. By the First-Derivative Test, you can conclude that $x = \frac{5}{3}$ yields a relative minimum and $x = 1$ yields a relative maximum. Because

$$f''(x) = 6x - 8$$

you can conclude that $x = \frac{4}{3}$ yields a point of inflection. Finally, you could analyze the graph *numerically*. For instance, you could construct a table of values and observe that f is increasing on the interval $(-\infty, 1)$, decreasing on the interval $\left(1, \frac{5}{3}\right)$, and increasing on the interval $\left(\frac{5}{3}, \infty\right)$.

Review Exercises

See www.CalcChat.com for worked-out solutions to odd-numbered exercises.

In Exercises 1–8, find the higher-order derivative.

1. Given $f(x) = 3x^2 + 7x + 1$, find $f''(x)$.

2. Given $f'(x) = 5x^4 - 6x^2 + 2x$, find $f'''(x)$.

3. Given $f'''(x) = -\dfrac{6}{x^4}$, find $f^{(5)}(x)$.

4. Given $f(x) = \sqrt{x}$, find $f^{(4)}(x)$.

5. Given $f'(x) = 7x^{5/2}$, find $f''(x)$.

6. Given $f(x) = x^2 + \dfrac{3}{x}$, find $f''(x)$.

7. Given $f''(x) = 6\sqrt[3]{x}$, find $f'''(x)$.

8. Given $f'''(x) = 20x^4 - \dfrac{2}{x^3}$, find $f^{(5)}(x)$.

In Exercises 9–12, find the given value.

Function	Value
9. $f(x) = x^2 + 3x + 4$	$f''(1)$
10. $f(x) = \dfrac{1}{x}$	$f'''(3)$
11. $f(x) = \sqrt{16x + 9}$	$f'''(0)$
12. $f(x) = x^2(x - 2)^2$	$f''(-1)$

13. **Athletics** A person dives from a 30-foot platform with an initial velocity of 5 feet per second (upward).

  (a) Find the position function of the diver.

  (b) How long will it take for the diver to hit the water?

  (c) What is the diver's velocity at impact?

  (d) What is the diver's acceleration at impact?

14. **Projectile Motion** An object is thrown upward from the top of a 96-foot building with an initial velocity of 80 feet per second.

  (a) Write the position, velocity, and acceleration functions of the object.

  (b) When will the object hit the ground?

  (c) When is the velocity of the object zero?

  (d) How high does the object go?

  (T) (e) Use a graphing utility to graph the position, velocity, and acceleration functions in the same viewing window.

15. **Velocity and Acceleration** The position function of a particle is given by

$$s = \frac{1}{t^2 + 2t + 1}$$

where s is the height (in feet) and t is the time (in seconds). Find the velocity and acceleration functions.

(T) 16. **Modeling Data** The table shows the utilized productions y of citrus fruits (in millions of pounds) in the United States for the years 2000 through 2005, where t is the year, with $t = 0$ corresponding to 2000. *(Source: U.S. Department of Agriculture)*

t	0	1	2	3	4	5
y	8355	8331	8256	8442	8156	7366

  (a) Use a graphing utility to find a cubic model for the data.

  (b) Use a graphing utility to graph the model and plot the data in the same viewing window. How well does the model fit the data?

  (c) Find the first and second derivatives of the function.

  (d) Show that the utilized production was decreasing from 2003 to 2005.

  (e) Find the year in which the utilized production was increasing at the greatest rate by solving $y''(t) = 0$.

  (f) Explain the relationship among your answers for parts (c), (d), and (e).

In Exercises 17–20, use implicit differentiation to find dy/dx.

17. $x^2 + 3xy + y^3 = 10$

18. $x^2 + 9xy + y^2 = 0$

19. $y^2 - x^2 + 8x - 9y - 1 = 0$

20. $y^2 + x^2 - 6y - 2x - 5 = 0$

In Exercises 21–24, use implicit differentiation to find the slope of the graph at the given point.

21. $5x^2 - 2y - 3 = 0$

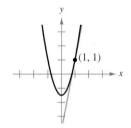

22. $3x^2 + 2y - 1 = 0$

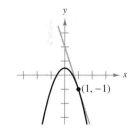

23. $x^2 + y^2 = 1$

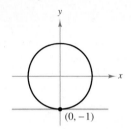

24. $x^2 + 4y^2 = 16$

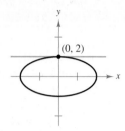

In Exercises 25–28, use implicit differentiation to find an equation of the tangent line at the given point.

Equation	Point
25. $y^2 = x - y$	$(2, 1)$
26. $2\sqrt[3]{x} + 3\sqrt{y} = 10$	$(8, 4)$
27. $y^2 - 2x = xy$	$(1, 2)$
28. $y^3 - 2x^2y + 3xy^2 = -1$	$(0, -1)$

29. Water Level A swimming pool is 40 feet long, 20 feet wide, 4 feet deep at the shallow end, and 9 feet deep at the deep end (see figure). Water is being pumped into the pool at the rate of 10 cubic feet per minute. How fast is the water level rising when there is 4 feet of water in the deep end?

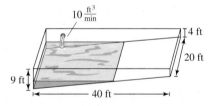

30. Water Level A trough is 12 feet long and 3 feet across the top (see figure). Its ends are isosceles triangles with heights of 3 feet.

(a) If water is being pumped into the trough at 2 cubic feet per minute, how fast is the water level rising when h is 1 foot deep?

(b) If the water is rising at a rate of $\frac{3}{8}$ inch per minute when $h = 2$, determine the rate at which water is being pumped into the trough.

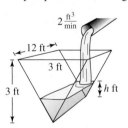

31. Sales The profit for a product is increasing at a rate of \$6384 per week. The demand and cost functions for the product are given by $p = 6000 - 0.4x^2$ and $C = 2400x + 5200$.

(a) Write the profit function for this product.

(b) Find the profit when the weekly sales are $x = 44$ units.

(c) Find the rate of change of sales with respect to time when the weekly sales are $x = 44$ units.

32. Electricity The combined electrical resistance R of R_1 and R_2, connected in parallel, is given by

$$\frac{1}{R} = \frac{1}{R_1} + \frac{1}{R_2}$$

where R, R_1, and R_2 are measured in ohms. R_1 and R_2 are increasing at rates of 1 and 1.5 ohms per second, respectively. At what rate is R changing when $R_1 = 50$ ohms and $R_2 = 75$ ohms?

In Exercises 33–36, find the critical numbers of the function.

33. $f(x) = -x^2 + 2x + 4$

34. $g(x) = (x - 1)^2(x - 3)$

35. $h(x) = \sqrt{x}(x - 3)$

36. $f(x) = (x + 3)^2$

In Exercises 37–40, determine the open intervals on which the function is increasing or decreasing. Verify your result with the graph of the function.

37. $f(x) = x^2 + x - 2$

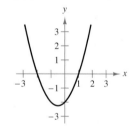

38. $g(x) = (x + 2)^3$

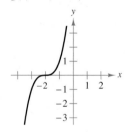

39. $h(x) = \dfrac{x^2 - 3x - 4}{x - 3}$

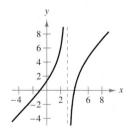

40. $f(x) = -x^3 + 3x^2 - 2$

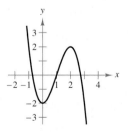

41. Meteorology The monthly normal temperature T (in degrees Fahrenheit) for New York City can be modeled by

$$T = 0.0380t^4 - 1.092t^3 + 9.23t^2 - 19.6t + 44$$

where $1 \le t \le 12$ and $t = 1$ corresponds to January. *(Source: National Climatic Data Center)*

(a) Find the interval(s) on which the model is increasing.

(b) Find the interval(s) on which the model is decreasing.

(c) Interpret the results of parts (a) and (b).

(T) (d) Use a graphing utility to graph the model.

42. CD Shipments The number S of manufacturer unit shipments (in millions) of CDs in the United States from 2000 through 2005 can be modeled by

$$S = -4.17083t^4 + 40.3009t^3 - 110.524t^2$$
$$+ 19.40t + 941.6$$

where $0 \le t \le 5$ and $t = 0$ corresponds to 2000. *(Source: Recording Industry Association of America)*

(a) Find the interval(s) on which the model is increasing.

(b) Find the interval(s) on which the model is decreasing.

(c) Interpret the results of parts (a) and (b).

(T) (d) Use a graphing utility to graph the model.

43. Consumer Trends The average number of hours N (per person per year) of TV usage in the United States from 2000 through 2005 can be modeled by

$$N = -0.382t^3 - 0.97t^2 + 30.5t + 1466, \quad 0 \le t \le 5$$

where $t = 0$ corresponds to 2000. *(Source: Veronis Suhler Stevenson)*

(a) Find the intervals on which dN/dt is increasing and decreasing.

(b) Find the limit of N as $t \to \infty$.

(c) Briefly explain your results for parts (a) and (b).

(T) **44. Revenue Per Share** The revenues per share R (in dollars) for the Walt Disney Company for the years 1994 through 2005 are shown in the table. *(Source: The Walt Disney Company)*

Year, t	4	5	6	7	8	9
Revenue per share, R	6.40	7.70	10.50	11.10	11.21	11.34

Year, t	10	11	12	13	14	15
Revenue per share, R	12.09	12.52	12.40	13.23	15.05	15.91

(a) Use a graphing utility to create a scatter plot of the data, where t is the time in years, with $t = 4$ corresponding to 1994.

(b) Describe any trends and/or patterns in the data.

(c) A model for the data is

$$R = \frac{5.75 - 2.043t + 0.1959t^2}{1 - 0.378t + 0.0438t^2 - 0.00117t^3}, \quad 4 \le t \le 15.$$

Graph the model and the data in the same viewing window.

(d) Find the years in which the revenue per share was increasing and decreasing.

(e) Find the years in which the rate of change of the revenue per share was increasing and decreasing.

(f) Briefly explain your results for parts (d) and (e).

In Exercises 45–54, use the First-Derivative Test to find the relative extrema of the function. Then use a graphing utility to verify your result.

45. $f(x) = 4x^3 - 6x^2 - 2$ **46.** $f(x) = \frac{1}{4}x^4 - 8x$

47. $g(x) = x^2 - 16x + 12$ **48.** $h(x) = 4 + 10x - x^2$

49. $h(x) = 2x^2 - x^4$ **50.** $s(x) = x^4 - 8x^2 + 3$

51. $f(x) = \dfrac{6}{x^2 + 1}$ **52.** $f(x) = \dfrac{2}{x^2 - 1}$

53. $h(x) = \dfrac{x^2}{x - 2}$ **54.** $g(x) = x - 6\sqrt{x}, \quad x > 0$

In Exercises 55–64, find the absolute extrema of the function on the closed interval. Then use a graphing utility to confirm your result.

55. $f(x) = x^2 + 5x + 6; \quad [-3, 0]$

56. $f(x) = x^4 - 2x^3; \quad [0, 2]$

57. $f(x) = x^3 - 12x + 1; \quad [-4, 4]$

58. $f(x) = x^3 + 2x^2 - 3x + 4; \quad [-3, 2]$

59. $f(x) = 4\sqrt{x} - x^2; \quad [0, 3]$

60. $f(x) = 2\sqrt{x} - x; \quad [0, 9]$

61. $f(x) = \dfrac{x}{\sqrt{x^2 + 1}}; \quad [0, 2]$

62. $f(x) = -x^4 + x^2 + 2; \quad [0, 2]$

63. $f(x) = \dfrac{2x}{x^2 + 1}; \quad [-1, 2]$

64. $f(x) = \dfrac{8}{x} + x; \quad [1, 4]$

65. Twins The number y pairs of twins born (per 1,000 live births) in the United States from 1971 through 2004 can be modeled by

$$y = 0.0143t^2 - 0.074t + 18, \quad 1 \le t \le 34$$

where $t = 1$ corresponds to 1971. When were the fewest pairs of twins born? *(Source: U.S. Department of Health and Human Services)*

66. Newspaper Circulation   The total number N of daily newspapers in circulation (in millions) in the United States from 1970 through 2005 can be modeled by

$$N = 0.022t^3 - 1.27t^2 + 9.7t + 1746$$

where $0 \le t \le 35$ and $t = 0$ corresponds to 1970. *(Source: Editor and Publisher Company)*

(a) Find the absolute maximum and minimum over the time period.

(b) Find the year in which the circulation was changing at the greatest rate.

(c) Briefly explain your results for parts (a) and (b).

Ⓣ **67. Biology**   The growth of a red oak tree is approximated by the model

$$y = -0.003x^3 + 0.137x^2 + 0.458x - 0.839,$$

$$2 \le x \le 34$$

where y is the height of the tree in feet and x is its age in years. Find the age of the tree when it is growing most rapidly. Then use a graphing utility to graph the function and to verify your result. (*Hint:* Use the viewing window $2 \le x \le 34$ and $-10 \le y \le 60$.)

68. Environment   When organic waste is dumped into a pond, the decomposition of the waste consumes oxygen. A model for the oxygen level O (where 1 is the normal level) of a pond as waste material oxidizes is

$$O = \frac{t^2 - t + 1}{t^2 + 1}, \quad 0 \le t$$

where t is the time in weeks.

(a) When is the oxygen level lowest? What is this level?

(b) When is the oxygen level highest? What is this level?

(c) Describe the oxygen level as t increases.

69. Bloodstream   The concentration C (in milligrams per milliliter) of a chemical in the bloodstream t hours after injection into muscle tissue can be modeled by

$$C = \frac{3t}{27 + t^3}, \quad t \ge 0.$$

(a) Complete the table and use it to approximate the time when the concentration reached a maximum.

t	0	0.5	1	1.5	2	2.5	3
$C(t)$							

Ⓣ (b) Use a graphing utility to graph the concentration function. Use the *trace* feature to approximate the time when the concentration reached a maximum.

(c) Determine analytically the time when the concentration reached a maximum.

Ⓣ **70. Surface Area**   A right circular cylinder of radius r and height h has a volume of 25 cubic inches. The total surface area of the cylinder in terms of r is given by

$$S = 2\pi r \left(r + \frac{25}{\pi r^2} \right).$$

Use a graphing utility to graph S and S' and find the value of r that yields the minimum surface area.

In Exercises 71–74, determine the open intervals on which the graph of the function is concave upward or concave downward. Then use a graphing utility to confirm your result.

71. $f(x) = (x - 2)^3$ **72.** $h(x) = x^5 - 10x^2$

73. $g(x) = \frac{1}{4}(-x^4 + 8x^2 - 12)$ **74.** $h(x) = x^3 - 6x$

In Exercises 75–78, find the points of inflection of the graph of the function.

75. $f(x) = \frac{1}{2}x^4 - 4x^3$ **76.** $f(x) = \frac{1}{4}x^4 - 2x^2 - x$

77. $f(x) = x^3(x - 3)^2$ **78.** $f(x) = (x - 1)^2(x - 3)$

In Exercises 79–82, use the Second-Derivative Test to find the relative extrema of the function.

79. $f(x) = x^5 - 5x^3$ **80.** $f(x) = x(x^2 - 3x - 9)$

81. $f(x) = 2x^2(1 - x^2)$ **82.** $f(x) = x - 4\sqrt{x + 1}$

83. High School Dropouts   From 2000 through 2005, the number d of high school dropouts not in the labor force (in thousands) can be modeled by

$$d = -20.444t^3 + 152.33t^2 - 266.6t + 1162$$

where t is the year, with $t = 0$ corresponding to 2000. *(Source: U.S. Bureau of Labor Statistics)*

Ⓣ (a) Use a graphing utility to graph the model.

(b) Use the second derivative to determine the concavity of d.

(c) Find the point(s) of inflection of the graph of d.

(d) Interpret the meaning of the inflection point(s) of the graph of d.

84. Medicine: Poiseuille's Law   The speed of blood that is r centimeters from the center of an artery is modeled by

$$s(r) = c(R^2 - r^2), \quad c > 0$$

where c is a constant, R is the radius of the artery, and s is measured in centimeters per second. Show that the speed is a maximum at the center of an artery.

Take this test as you would take a test in class. When you are done, check your work against the answers given in the back of the book.

In Exercises 1–3, find the third derivative of the function. Simplify your result.

1. $f(x) = 2x^2 + 3x + 1$ **2.** $f(x) = \sqrt{3 - x}$ **3.** $f(x) = \dfrac{2x + 1}{2x - 1}$

In Exercises 4–6, use implicit differentiation to find dy/dx.

4. $x + xy = 6$ **5.** $y^2 + 2x - 2y + 1 = 0$ **6.** $x^2 - 2y^2 = 4$

In Exercises 7–9, find the critical numbers of the function and the open intervals on which the function is increasing or decreasing.

7. $f(x) = 3x^2 - 4$ **8.** $f(x) = x^3 - 12x$ **9.** $f(x) = (x - 5)^4$

Ⓣ In Exercises 10–12, use a graphing utility to graph the function. Then use the First-Derivative Test to find all relative extrema of the function.

10. $f(x) = \dfrac{1}{3}x^3 - 9x + 4$ **11.** $f(x) = 2x^4 - 4x^2 - 5$ **12.** $f(x) = \dfrac{5}{x^2 + 2}$

In Exercises 13 and 14, find the absolute extrema of the function on the closed interval.

13. $f(x) = x^2 + 6x + 8$, $[-4, 0]$ **14.** $f(x) = 12\sqrt{x} - 4x$, $[0, 5]$

In Exercises 15 and 16, determine the open intervals on which the graph of the function is concave upward or concave downward.

15. $f(x) = x^5 - 4x^2$ **16.** $f(x) = \dfrac{20}{3x^2 + 8}$

In Exercises 17 and 18, find the point(s) of inflection of the graph of the function.

17. $f(x) = x^3 - 6x^2 + 7x$ **18.** $f(x) = \dfrac{1}{5}x^5 - 4x^2$

In Exercises 19 and 20, use the Second-Derivative Test to find all relative extrema of the function.

19. $f(x) = x^3 - 6x^2 - 24x + 12$ **20.** $f(x) = 0.6x^5 - 9x^3$

21. The radius r of a right circular cylinder is increasing at a rate of 0.25 centimeter per minute. The height h of the cylinder is related to the radius by $h = 20r$. Find the rate of change of the volume when (a) $r = 0.5$ centimeter and (b) $r = 1$ centimeter.

22. The resident population P (in thousands) of the District of Columbia from 1999 through 2005 can be modeled by

$$P = 0.2694t^3 - 2.048t^2 - 0.73t + 571.9$$

where $-1 \le t \le 5$ and $t = 0$ corresponds to 2000. *(Source: U.S. Census Bureau)*

(a) During which year, from 1999 through 2005, was the population the greatest? the least?

(b) During which year(s) was the population increasing? decreasing?

9

Further Applications of the Derivative

Still Images/Getty Images

Designers use the derivative to find the dimensions of a container that will minimize cost. (See Section 9.1, Exercise 28.)

Applications

Derivatives have many real-life applications in addition to those discussed in Chapter 8. The applications listed below represent a sample of the applications in this chapter.

Section 9.1

Optimization Problems

■ Solve real-life optimization problems.

Solving Optimization Problems

One of the most common applications of calculus is the determination of optimum (minimum or maximum) values. Before learning a general method for solving optimization problems, consider the next example.

Example 1 Finding the Maximum Volume

A manufacturer wants to design an open box that has a square base and a surface area of 108 square inches, as shown in Figure 9.1. What dimensions will produce a box with a maximum volume?

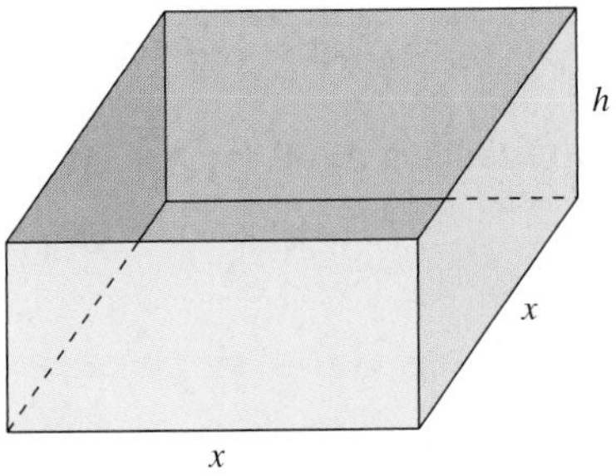

SOLUTION   Because the base of the box is square, the volume is

$$V = x^2 h. \qquad \text{Primary equation}$$

This equation is called the **primary equation** because it gives a formula for the quantity to be optimized. The surface area of the box is

$$S = (\text{area of base}) + (\text{area of four sides})$$
$$108 = x^2 + 4xh. \qquad \text{Secondary equation}$$

Because V is to be optimized, it helps to express V as a function of just one variable. To do this, solve the secondary equation for h in terms of x to obtain

$$h = \frac{108 - x^2}{4x}$$

and substitute into the primary equation.

$$V = x^2 h = x^2 \left(\frac{108 - x^2}{4x} \right) = 27x - \frac{1}{4}x^3 \qquad \text{Function of one variable}$$

Before finding which x-value yields a maximum value of V, you need to determine the *feasible domain* of the function. That is, what values of x make sense in the problem? Because x must be nonnegative and the area of the base $(A = x^2)$ is at most 108, you can conclude that the feasible domain is

$$0 \le x \le \sqrt{108}. \qquad \text{Feasible domain}$$

Using the techniques described in Sections 8.4 through 8.6, you can determine that $\left(\text{on the interval } 0 \le x \le \sqrt{108} \right)$ this function has an absolute maximum when $x = 6$ inches and $h = 3$ inches.

FIGURE 9.1   Open Box with Square Base: $S = x^2 + 4xh = 108$

Algebra Review

For help on the algebra in Example 1, see Example 1(c) in the *Chapter 9 Algebra Review*, on page 737.

✓**CHECKPOINT 1**

Use a graphing utility to graph the volume function $V = 27x - \frac{1}{4}x^3$ on $0 \le x \le \sqrt{108}$ from Example 1. Verify that the function has an absolute maximum when $x = 6$. What is the maximum volume? ■

In studying Example 1, be sure that you understand the basic question that it asks. Some students have trouble with optimization problems because they are too eager to start solving the problem by using a standard formula. For instance, in Example 1, you should realize that there are infinitely many open boxes having 108 square inches of surface area. You might begin to solve this problem by asking yourself which basic shape would seem to yield a maximum volume. Should the box be tall, squat, or nearly cubical? You might even try calculating a few volumes, as shown in Figure 9.2, to see if you can get a good feeling for what the optimum dimensions should be.

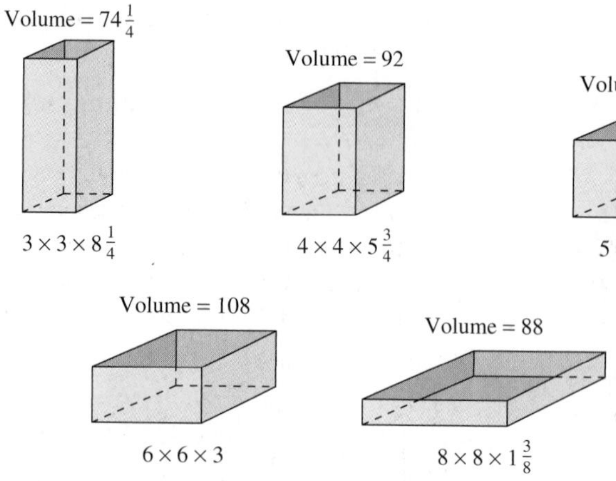

Volume $= 74\frac{1}{4}$

$3 \times 3 \times 8\frac{1}{4}$

Volume $= 92$

$4 \times 4 \times 5\frac{3}{4}$

Volume $= 103\frac{3}{4}$

$5 \times 5 \times 4\frac{3}{20}$

Volume $= 108$

$6 \times 6 \times 3$

Volume $= 88$

$8 \times 8 \times 1\frac{3}{8}$

FIGURE 9.2   Which box has the greatest volume?

STUDY TIP

Remember that you are not ready to begin solving an optimization problem until you have clearly identified what the problem is. Once you are sure you understand what is being asked, you are ready to begin considering a method for solving the problem.

There are several steps in the solution of Example 1. The first step is to sketch a diagram and identify all *known* quantities and all quantities *to be determined*. The second step is to write a primary equation for the quantity to be optimized. Then, a secondary equation is used to rewrite the primary equation as a function of one variable. Finally, calculus is used to determine the optimum value. These steps are summarized below.

STUDY TIP

When performing Step 5, remember that to determine the maximum or minimum value of a continuous function f on a closed interval, you need to compare the values of f at its critical numbers with the values of f at the endpoints of the interval. The greatest of these values is the desired maximum and the least is the desired minimum.

Guidelines for Solving Optimization Problems

1. Identify all given quantities and all quantities to be determined. If possible, make a sketch.

2. Write a **primary equation** for the quantity that is to be maximized or minimized. (A summary of several common formulas is given in Appendix D.)

3. Reduce the primary equation to one having a single independent variable. This may involve the use of a **secondary equation** that relates the independent variables of the primary equation.

4. Determine the feasible domain of the primary equation. That is, determine the values for which the stated problem makes sense.

5. Determine the desired maximum or minimum value by the calculus techniques discussed in Sections 8.4 through 8.6.

Algebra Review

For help on the algebra in Example 2, see Example 1(a) in the *Chapter 9 Algebra Review*, on page 737.

Example 2 **Finding a Minimum Sum**

The product of two positive numbers is 288. Minimize the sum of the second number and twice the first number.

SOLUTION

1. Let x be the first number, y the second, and S the sum to be minimized.

2. Because you want to minimize S, the primary equation is

   $$S = 2x + y. \qquad \text{Primary equation}$$

3. Because the product of the two numbers is 288, you can write the secondary equation as

   $$xy = 288 \qquad \text{Secondary equation}$$

   $$y = \frac{288}{x}.$$

   Using this result, you can rewrite the primary equation as a function of one variable.

   $$S = 2x + \frac{288}{x} \qquad \text{Function of one variable}$$

4. Because the numbers are positive, the feasible domain is

   $$x > 0. \qquad \text{Feasible domain}$$

5. To find the minimum value of S, begin by finding its critical numbers.

   $$\frac{dS}{dx} = 2 - \frac{288}{x^2} \qquad \text{Find derivative of } S.$$

   $$0 = 2 - \frac{288}{x^2} \qquad \text{Set derivative equal to 0.}$$

   $$x^2 = 144 \qquad \text{Simplify.}$$

   $$x = \pm 12 \qquad \text{Critical numbers}$$

   Choosing the positive x-value, you can use the First-Derivative Test to conclude that S is decreasing on the interval $(0, 12)$ and increasing on the interval $(12, \infty)$, as shown in the table. So, $x = 12$ yields a minimum, and the two numbers are

   $$x = 12 \quad \text{and} \quad y = \frac{288}{12} = 24.$$

TECHNOLOGY

After you have written the primary equation as a function of a single variable, you can estimate the optimum value by graphing the function. For instance, the graph of

$$S = 2x + \frac{288}{x}$$

shown below indicates that the minimum value of S occurs when x is about 12.

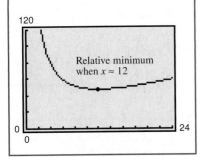

Relative minimum when $x \approx 12$

Interval	$0 < x < 12$	$12 < x < \infty$
Test value	$x = 1$	$x = 13$
Sign of $\dfrac{dS}{dx}$	$\dfrac{dS}{dx} < 0$	$\dfrac{dS}{dx} > 0$
Conclusion	S is decreasing.	S is increasing.

✓CHECKPOINT 2

The product of two numbers is 72. Minimize the sum of the second number and twice the first number. ■

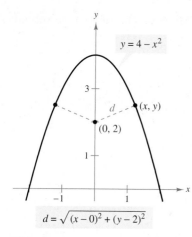

FIGURE 9.3

$$d = \sqrt{(x-0)^2 + (y-2)^2}$$

Example 3 Finding a Minimum Distance

Find the points on the graph of

$$y = 4 - x^2$$

that are closest to $(0, 2)$.

SOLUTION

1. Figure 9.3 indicates that there are two points at a minimum distance from the point $(0, 2)$.

2. You are asked to minimize the distance d. So, you can use the Distance Formula to obtain a primary equation.

$$d = \sqrt{(x-0)^2 + (y-2)^2} \qquad \text{Primary equation}$$

3. Using the secondary equation $y = 4 - x^2$, you can rewrite the primary equation as a function of a single variable.

$$d = \sqrt{x^2 + (4 - x^2 - 2)^2} \qquad \text{Substitute } 4 - x^2 \text{ for } y.$$
$$= \sqrt{x^4 - 3x^2 + 4} \qquad \text{Simplify.}$$

Because d is smallest when the expression under the radical is smallest, you simplify the problem by finding the minimum value of $f(x) = x^4 - 3x^2 + 4$.

4. The domain of f is the entire real line.

5. To find the minimum value of $f(x)$, first find the critical numbers of f.

$$f'(x) = 4x^3 - 6x \qquad \text{Find derivative of } f.$$
$$0 = 4x^3 - 6x \qquad \text{Set derivative equal to 0.}$$
$$0 = 2x(2x^2 - 3) \qquad \text{Factor.}$$
$$x = 0, x = \sqrt{\tfrac{3}{2}}, x = -\sqrt{\tfrac{3}{2}} \qquad \text{Critical numbers}$$

By the First-Derivative Test, you can conclude that $x = 0$ yields a relative maximum, whereas both $\sqrt{3/2}$ and $-\sqrt{3/2}$ yield a minimum. So, on the graph of $y = 4 - x^2$, the points that are closest to the point $(0, 2)$ are

$$\left(\sqrt{\tfrac{3}{2}}, \tfrac{5}{2}\right) \quad \text{and} \quad \left(-\sqrt{\tfrac{3}{2}}, \tfrac{5}{2}\right).$$

✓ **CHECKPOINT 3**

Find the points on the graph of $y = 4 - x^2$ that are closest to $(0, 3)$. ∎

Algebra Review

For help on the algebra in Example 3, see Example 1(b) in the *Chapter 9 Algebra Review*, on page 737.

STUDY TIP

To confirm the result in Example 3, try computing the distances between several points on the graph of $y = 4 - x^2$ and the point $(0, 2)$. For instance, the distance between $(1, 3)$ and $(0, 2)$ is

$$d = \sqrt{(0-1)^2 + (2-3)^2} = \sqrt{2} \approx 1.414.$$

Note that this is greater than the distance between $\left(\sqrt{3/2}, 5/2\right)$ and $(0, 2)$, which is

$$d = \sqrt{\left(0 - \sqrt{\tfrac{3}{2}}\right)^2 + \left(2 - \tfrac{5}{2}\right)^2} = \sqrt{\tfrac{7}{4}} \approx 1.323.$$

Example 4 **Finding a Minimum Area**

A rectangular page will contain 24 square inches of print. The margins at the top and bottom of the page are $1\frac{1}{2}$ inches wide. The margins on each side are 1 inch wide. What should the dimensions of the page be to minimize the amount of paper used?

SOLUTION

1. A diagram of the page is shown in Figure 9.4.

2. Letting A be the area to be minimized, the primary equation is

$$A = (x + 3)(y + 2).$$ Primary equation

3. The printed area inside the margins is given by

$$24 = xy.$$ Secondary equation

Solving this equation for y produces

$$y = \frac{24}{x}.$$

By substituting this into the primary equation, you obtain

$$A = (x + 3)\left(\frac{24}{x} + 2\right)$$ Write as a function of one variable.

$$= (x + 3)\left(\frac{24 + 2x}{x}\right)$$ Rewrite second factor as a single fraction.

$$= \frac{2x^2}{x} + \frac{30x}{x} + \frac{72}{x}$$ Multiply and separate into terms.

$$= 2x + 30 + \frac{72}{x}.$$ Simplify.

4. Because x must be positive, the feasible domain is $x > 0$.

5. To find the minimum area, begin by finding the critical numbers of A.

$$\frac{dA}{dx} = 2 - \frac{72}{x^2}$$ Find derivative of A.

$$0 = 2 - \frac{72}{x^2}$$ Set derivative equal to 0.

$$-2 = -\frac{72}{x^2}$$ Subtract 2 from each side.

$$x^2 = 36$$ Simplify.

$$x = \pm 6$$ Critical numbers

Because $x = -6$ is not in the feasible domain, you only need to consider the critical number $x = 6$. Using the First-Derivative Test, it follows that A is a minimum when $x = 6$. So, the dimensions of the page should be

$$x + 3 = 6 + 3 = 9 \text{ inches by } y + 2 = \frac{24}{6} + 2 = 6 \text{ inches.}$$

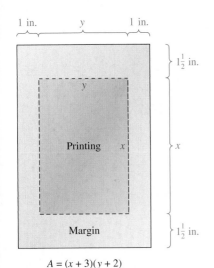

1 in.   y   1 in.

$1\frac{1}{2}$ in.

y

Printing   x   x

Margin $1\frac{1}{2}$ in.

$A = (x + 3)(y + 2)$

FIGURE 9.4

✓ **CHECKPOINT 4**

A rectangular page will contain 54 square inches of print. The margins at the top and bottom of the page are $1\frac{1}{2}$ inches wide. The margins on each side are 1 inch wide. What should the dimensions of the page be to minimize the amount of paper used? ■

As applications go, the four examples described in this section are fairly simple, and yet the resulting primary equations are quite complicated. Real-life applications often involve equations that are at least as complex as these four. Remember that one of the main goals of this course is to enable you to use the power of calculus to analyze equations that at first glance seem formidable.

Also remember that once you have found the primary equation, you can use the graph of the equation to help solve the problem. For instance, the graphs of the primary equations in Examples 1 through 4 are shown in Figure 9.5.

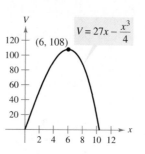

Example 1

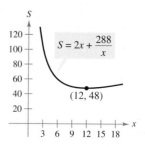

Example 2

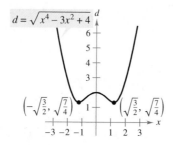

Example 3

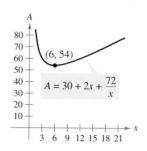

Example 4

FIGURE 9.5

CONCEPT CHECK

1. Complete the following: In an optimization problem, the formula that represents the quantity to be optimized is called the _____ _____.

2. Explain what is meant by the term *feasible domain*.

3. Explain the difference between a primary equation and a secondary equation.

4. In your own words, state the guidelines for solving an optimization problem.

Skills Review 9.1

The following warm-up exercises involve skills that were covered in earlier sections. You will use these skills in the exercise set for this section. For additional help, review Section 8.4.

In Exercises 1–4, write a formula for the written statement.

1. The sum of one number and half a second number is 12

2. The product of one number and twice another is 24.

3. The area of a rectangle is 24 square units.

4. The distance between two points is 10 units.

In Exercises 5–10, find the critical numbers of the function.

5. $y = x^2 + 6x - 9$

6. $y = 2x^3 - x^2 - 4x$

7. $y = 5x + \dfrac{125}{x}$

8. $y = 3x + \dfrac{96}{x^2}$

9. $y = \dfrac{x^2 + 1}{x}$

10. $y = \dfrac{x}{x^2 + 9}$

Exercises 9.1

See www.CalcChat.com for worked-out solutions to odd-numbered exercises.

In Exercises 1–6, find two positive numbers satisfying the given requirements.

1. The sum is 120 and the product is a maximum.

2. The sum is S and the product is a maximum.

3. The sum of the first and twice the second is 36 and the product is a maximum.

4. The sum of the first and twice the second is 100 and the product is a maximum.

5. The product is 192 and the sum is a minimum.

6. The product is 192 and the sum of the first plus three times the second is a minimum.

In Exercises 7 and 8, find the length and width of a rectangle that has the given perimeter and a maximum area.

7. Perimeter: 100 meters

8. Perimeter: P units

In Exercises 9 and 10, find the length and width of a rectangle that has the given area and a minimum perimeter.

9. Area: 64 square feet

10. Area: A square centimeters

11. Maximum Area A rancher has 200 feet of fencing to enclose two adjacent rectangular corrals (see figure). What dimensions should be used so that the enclosed area will be a maximum?

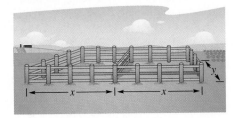

12. Area A dairy farmer plans to enclose a rectangular pasture adjacent to a river. To provide enough grass for the herd, the pasture must contain 180,000 square meters. No fencing is required along the river. What dimensions will use the least amount of fencing?

13. Maximum Volume

(a) Verify that each of the rectangular solids shown in the figure has a surface area of 150 square inches.

(b) Find the volume of each solid.

(c) Determine the dimensions of a rectangular solid (with a square base) of maximum volume if its surface area is 150 square inches.

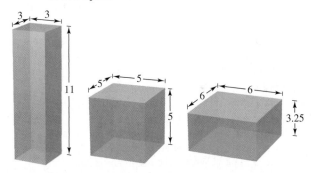

14. Maximum Volume Determine the dimensions of a rectangular solid (with a square base) with maximum volume if its surface area is 337.5 square centimeters.

15. Minimum Cost A storage box with a square base must have a volume of 80 cubic centimeters. The top and bottom cost $0.20 per square centimeter and the sides cost $0.10 per square centimeter. Find the dimensions that will minimize cost.

16. Maximum Area A Norman window is constructed by adjoining a semicircle to the top of an ordinary rectangular window (see figure). Find the dimensions of a Norman window of maximum area if the total perimeter is 16 feet.

17. Minimum Surface Area A net enclosure for golf practice is open at one end (see figure). The volume of the enclosure is $83\frac{1}{3}$ cubic meters. Find the dimensions that require the least amount of netting.

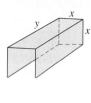
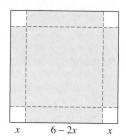

Figure for 17   Figure for 18

18. Volume An open box is to be made from a six-inch by six-inch square piece of material by cutting equal squares from the corners and turning up the sides (see figure). Find the volume of the largest box that can be made.

19. Volume An open box is to be made from a two-foot by three-foot rectangular piece of material by cutting equal squares from the corners and turning up the sides. Find the volume of the largest box that can be made in this manner.

20. Maximum Yield A home gardener estimates that 16 apple trees will have an average yield of 80 apples per tree. But because of the size of the garden, for each additional tree planted the yield will decrease by four apples per tree. How many trees should be planted to maximize the total yield of apples? What is the maximum yield?

21. Area A rectangular page is to contain 36 square inches of print. The margins at the top and bottom and on each side are to be $1\frac{1}{2}$ inches. Find the dimensions of the page that will minimize the amount of paper used.

22. Area A rectangular page is to contain 30 square inches of print. The margins at the top and bottom of the page are to be 2 inches wide. The margins on each side are to be 1 inch wide. Find the dimensions of the page such that the least amount of paper is used.

23. Maximum Area A rectangle is bounded by the x- and y-axes and the graph of $y = (6 - x)/2$ (see figure). What length and width should the rectangle have so that its area is a maximum?

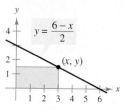
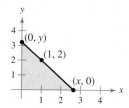

Figure for 23   Figure for 24

24. Minimum Length A right triangle is formed in the first quadrant by the x- and y-axes and a line through the point $(1, 2)$ (see figure).

(a) Write the length L of the hypotenuse as a function of x.

(T) (b) Use a graphing utility to approximate x graphically such that the length of the hypotenuse is a minimum.

(c) Find the vertices of the triangle such that its area is a minimum.

25. Maximum Area A rectangle is bounded by the x-axis and the semicircle

$$y = \sqrt{25 - x^2}$$

(see figure). What length and width should the rectangle have so that its area is a maximum?

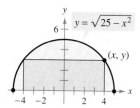

26. Area Find the dimensions of the largest rectangle that can be inscribed in a semicircle of radius r. (See Exercise 25.)

27. Volume You are designing a soft drink container that has the shape of a right circular cylinder. The container is supposed to hold 12 fluid ounces (1 fluid ounce is approximately 1.80469 cubic inches). Find the dimensions that will use a minimum amount of construction material.

28. Minimum Cost An energy drink container of the shape described in Exercise 27 must have a volume of 16 fluid ounces. The cost per square inch of constructing the top and bottom is twice the cost of constructing the sides. Find the dimensions that will minimize cost.

In Exercises 29–32, find the points on the graph of the function that are closest to the given point.

29. $f(x) = x^2$, $\left(2, \dfrac{1}{2}\right)$

30. $f(x) = (x + 1)^2$, $(5, 3)$

31. $f(x) = \sqrt{x}$, $(4, 0)$

32. $f(x) = \sqrt{x - 8}$, $(2, 0)$

33. Maximum Volume   A rectangular package to be sent by a postal service can have a maximum combined length and girth (perimeter of a cross section) of 108 inches. Find the dimensions of the package with maximum volume. Assume that the package's dimensions are x by x by y (see figure).

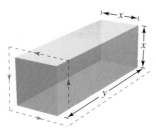

34. Minimum Surface Area   A solid is formed by adjoining two hemispheres to the ends of a right circular cylinder. The total volume of the solid is 12 cubic inches. Find the radius of the cylinder that produces the minimum surface area.

35. Minimum Cost   An industrial tank of the shape described in Exercise 34 must have a volume of 3000 cubic feet. The hemispherical ends cost twice as much per square foot of surface area as the sides. Find the dimensions that will minimize cost.

36. Minimum Area   The sum of the perimeters of a circle and a square is 16. Find the dimensions of the circle and square that produce a minimum total area.

37. Minimum Area   The sum of the perimeters of an equilateral triangle and a square is 10. Find the dimensions of the triangle and square that produce a minimum total area.

38. Minimum Time   You are in a boat 2 miles from the nearest point on the coast. You are to go to point Q, located 3 miles down the coast and 1 mile inland (see figure). You can row at a rate of 2 miles per hour and you can walk at a rate of 4 miles per hour. Toward what point on the coast should you row in order to reach point Q in the least time?

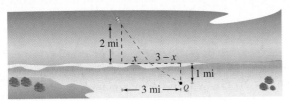

39. Maximum Area   An indoor physical fitness room consists of a rectangular region with a semicircle on each end. The perimeter of the room is to be a 200-meter running track. Find the dimensions that will make the area of the rectangular region as large as possible.

40. Farming   A strawberry farmer will receive $30 per bushel of strawberries during the first week of harvesting. Each week after that, the value will drop $0.80 per bushel. The farmer estimates that there are approximately 120 bushels of strawberries in the fields, and that the crop is increasing at a rate of four bushels per week. When should the farmer harvest the strawberries to maximize their value? How many bushels of strawberries will yield the maximum value? What is the maximum value of the strawberries?

41. Beam Strength   A wooden beam has a rectangular cross section of height h and width w (see figure). The strength S of the beam is directly proportional to its width and the square of its height. What are the dimensions of the strongest beam that can be cut from a round log of diameter 24 inches?   (*Hint:* $S = kh^2w$, where $k > 0$ is the proportionality constant.)

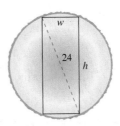

42. Area   Four feet of wire is to be used to form a square and a circle.

(a) Express the sum of the areas of the square and the circle as a function A of the side of the square x.

(b) What is the domain of A?

(T) (c) Use a graphing utility to graph A on its domain.

(d) How much wire should be used for the square and how much for the circle in order to enclose the least total area? the greatest total area?

43. Profit   The profit P (in thousands of dollars) for a company spending an amount s (in thousands of dollars) on advertising is

$$P = -\frac{1}{10}s^3 + 6s^2 + 400.$$

(a) Find the amount of money the company should spend on advertising in order to yield a maximum profit.

(b) Find the point of diminishing returns.

Section 9.2

Business and Economics Applications

■ Solve business and economics optimization problems.
■ Find the price elasticity of demand for demand functions.
■ Recognize basic business terms and formulas.

Optimization in Business and Economics

The problems in this section are primarily optimization problems, so the five-step procedure used in Section 9.1 is an appropriate strategy to follow.

Example 1   Finding the Maximum Revenue

A company has determined that its total revenue (in dollars) for a product can be modeled by

$$R = -x^3 + 450x^2 + 52,500x$$

where x is the number of units produced (and sold). What production level will yield a maximum revenue?

SOLUTION

1. A sketch of the revenue function is shown in Figure 9.6.

2. The primary equation is the given revenue function.

$$R = -x^3 + 450x^2 + 52,500x \qquad \text{Primary equation}$$

3. Because R is already given as a function of one variable, you do not need a secondary equation.

4. The feasible domain of the primary equation is

$$0 \le x \le 546. \qquad \text{Feasible domain}$$

This is determined by finding the x-intercepts of the revenue function, as shown in Figure 9.6.

5. To maximize the revenue, find the critical numbers.

$$\frac{dR}{dx} = -3x^2 + 900x + 52,500 = 0 \qquad \text{Set derivative equal to 0.}$$

$$-3(x - 350)(x + 50) = 0 \qquad \text{Factor.}$$

$$x = 350, \ x = -50 \qquad \text{Critical numbers}$$

The only critical number in the feasible domain is $x = 350$. From the graph of the function, you can see that the production level of 350 units corresponds to a maximum revenue.

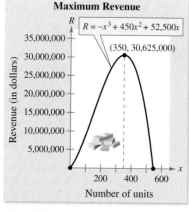

FIGURE 9.6 Maximum revenue occurs when $dR/dx = 0$.

✓ CHECKPOINT 1

Find the number of units that must be produced to maximize the revenue function $R = -x^3 + 150x^2 + 9375x$. What is the maximum revenue? ■

To study the effects of production levels on cost, economists use the **average cost function** $\overline{C}$, which is defined as

$$\overline{C} = \frac{C}{x}$$ Average cost function

where $C = f(x)$ is the total cost function and x is the number of units produced.

Example 2 **Finding the Minimum Average Cost**

A company estimates that the cost (in dollars) of producing x units of a product can be modeled by $C = 800 + 0.04x + 0.0002x^2$. Find the production level that minimizes the average cost per unit.

SOLUTION

1. C represents the total cost, x represents the number of units produced, and $\overline{C}$ represents the average cost per unit.

2. The primary equation is

$$\overline{C} = \frac{C}{x}.$$ Primary equation

3. Substituting the given equation for C produces

$$\overline{C} = \frac{800 + 0.04x + 0.0002x^2}{x}$$ Substitute for C.

$$= \frac{800}{x} + 0.04 + 0.0002x.$$ Function of one variable

4. The feasible domain for this function is

$$x > 0.$$ Feasible domain

5. You can find the critical numbers as shown.

$$\frac{d\overline{C}}{dx} = -\frac{800}{x^2} + 0.0002 = 0$$ Set derivative equal to 0.

$$0.0002 = \frac{800}{x^2}$$

$$x^2 = \frac{800}{0.0002}$$ Multiply each side by x^2 and divide each side by 0.0002.

$$x^2 = 4{,}000{,}000$$

$$x = \pm 2000$$ Critical numbers

By choosing the positive value of x and sketching the graph of $\overline{C}$, as shown in Figure 9.7, you can see that a production level of $x = 2000$ minimizes the average cost per unit.

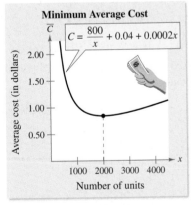

Minimum Average Cost

$$C = \frac{800}{x} + 0.04 + 0.0002x$$

FIGURE 9.7   Minimum average cost occurs when $d\overline{C}/dx = 0$.

✓**CHECKPOINT 2**

Find the production level that minimizes the average cost per unit for the cost function $C = 400 + 0.05x + 0.0025x^2$. ■

| Example 3 | Finding the Maximum Revenue | |

A business sells 2000 units of a product per month at a price of $10 each. It can sell 250 more items per month for each $0.25 reduction in price. What price per unit will maximize the monthly revenue?

SOLUTION

1. Let x represent the number of units sold in a month, let p represent the price per unit, and let R represent the monthly revenue.

2. Because the revenue is to be maximized, the primary equation is

$$R = xp. \qquad \text{Primary equation}$$

3. A price of $p = \$10$ corresponds to $x = 2000$, and a price of $p = \$9.75$ corresponds to $x = 2250$. Using this information, you can use the point-slope form to create the demand equation.

$$p - 10 = \frac{10 - 9.75}{2000 - 2250}(x - 2000) \qquad \text{Point-slope form}$$

$$p - 10 = -0.001(x - 2000) \qquad \text{Simplify.}$$

$$p = -0.001x + 12 \qquad \text{Secondary equation}$$

Substituting this value into the revenue equation produces

$$R = x(-0.001x + 12) \qquad \text{Substitute for } p.$$

$$= -0.001x^2 + 12x. \qquad \text{Function of one variable}$$

4. The feasible domain of the revenue function is

$$0 \le x \le 12{,}000. \qquad \text{Feasible domain}$$

5. To maximize the revenue, find the critical numbers.

$$\frac{dR}{dx} = 12 - 0.002x = 0 \qquad \text{Set derivative equal to 0.}$$

$$-0.002x = -12$$

$$x = 6000 \qquad \text{Critical number}$$

From the graph of R in Figure 9.8, you can see that this production level yields a maximum revenue. The price that corresponds to this production level is

$$p = 12 - 0.001x \qquad \text{Demand function}$$

$$= 12 - 0.001(6000) \qquad \text{Substitute 6000 for } x.$$

$$= \$6. \qquad \text{Price per unit}$$

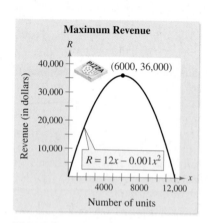

Maximum Revenue

$R = 12x - 0.001x^2$

(6000, 36,000)

Revenue (in dollars) / Number of units

FIGURE 9.8

STUDY TIP

In Example 3, the revenue function was written as a function of x. It could also have been written as a function of p. That is, $R = 1000(12p - p^2)$. By finding the critical numbers of this function, you can determine that the maximum revenue occurs when $p = 6$.

✓ **CHECKPOINT 3**

Find the price per unit that will maximize the monthly revenue for the business in Example 3 if it can sell only 200 more items per month for each $0.25 reduction in price. ■

Algebra Review

For help on the algebra in Example 4, see Example 2(b) in the *Chapter 9 Algebra Review*, on page 738.

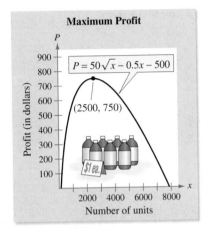

Maximum Profit

$P = 50\sqrt{x} - 0.5x - 500$

(2500, 750)

$1 ea.

FIGURE 9.9

✓ CHECKPOINT 4

Find the price that will maximize profit for the demand and cost functions.

$$p = \frac{40}{\sqrt{x}} \text{ and } C = 2x + 50 \ ■$$

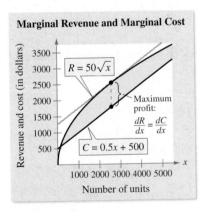

Marginal Revenue and Marginal Cost

$R = 50\sqrt{x}$

Maximum profit:

$$\frac{dR}{dx} = \frac{dC}{dx}$$

$C = 0.5x + 500$

FIGURE 9.10

Example 4 **Finding the Maximum Profit**

The marketing department of a business has determined that the demand for a product can be modeled by

$$p = \frac{50}{\sqrt{x}}.$$

The cost of producing x units is given by $C = 0.5x + 500$. What price will yield a maximum profit?

SOLUTION

1. Let R represent the revenue, P the profit, p the price per unit, x the number of units, and C the total cost of producing x units.

2. Because you are maximizing the profit, the primary equation is

$$P = R - C. \qquad \text{Primary equation}$$

3. Because the revenue is $R = xp$, you can write the profit function as

$$P = R - C$$
$$= xp - (0.5x + 500) \qquad \text{Substitute for } R \text{ and } C.$$
$$= x\left(\frac{50}{\sqrt{x}}\right) - 0.5x - 500 \qquad \text{Substitute for } p.$$
$$= 50\sqrt{x} - 0.5x - 500. \qquad \text{Function of one variable}$$

4. The feasible domain of the function is $127 < x \le 7872$. (When x is less than 127 or greater than 7872, the profit is negative.)

5. To maximize the profit, find the critical numbers.

$$\frac{dP}{dx} = \frac{25}{\sqrt{x}} - 0.5 = 0 \qquad \text{Set derivative equal to 0.}$$
$$\sqrt{x} = 50 \qquad \text{Isolate } x\text{-term on one side.}$$
$$x = 2500 \qquad \text{Critical number}$$

From the graph of the profit function shown in Figure 9.9, you can see that a maximum profit occurs when $x = 2500$. The price that corresponds to $x = 2500$ is

$$p = \frac{50}{\sqrt{x}} = \frac{50}{\sqrt{2500}} = \frac{50}{50} = \$1.00. \qquad \text{Price per unit}$$

STUDY TIP

To find the maximum profit in Example 4, the equation $P = R - C$ was differentiated and set equal to zero. From the equation

$$\frac{dP}{dx} = \frac{dR}{dx} - \frac{dC}{dx} = 0$$

it follows that the maximum profit occurs when the marginal revenue is equal to the marginal cost, as shown in Figure 9.10.

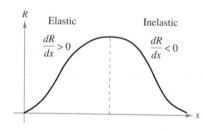

FIGURE 9.11   Revenue Curve

Price Elasticity of Demand

One way economists measure the responsiveness of consumers to a change in the price of a product is with **price elasticity of demand.** For example, a drop in the price of vegetables might result in a much greater demand for vegetables; such a demand is called **elastic.** On the other hand, the demand for items such as milk and water is relatively unresponsive to changes in price; the demand for such items is called **inelastic.**

More formally, the elasticity of demand is the percent change of a quantity demanded x, divided by the percent change in its price p. You can develop a formula for price elasticity of demand using the approximation

$$\frac{\Delta p}{\Delta x} \approx \frac{dp}{dx}$$

which is based on the definition of the derivative. Using this approximation, you can write

$$\text{Price elasticity of demand} = \frac{\text{rate of change in demand}}{\text{rate of change in price}}$$

$$= \frac{\Delta x / x}{\Delta p / p}$$

$$= \frac{p / x}{\Delta p / \Delta x}$$

$$\approx \frac{p / x}{dp / dx}.$$

Definition of Price Elasticity of Demand

If $p = f(x)$ is a differentiable function, then the **price elasticity of demand** is given by

$$\eta = \frac{p / x}{dp / dx}$$

where η is the lowercase Greek letter eta. For a given price, the demand is **elastic** if $|\eta| > 1$, the demand is **inelastic** if $|\eta| < 1$, and the demand has **unit elasticity** if $|\eta| = 1$.

Price elasticity of demand is related to the total revenue function, as indicated in Figure 9.11 and the list below.

1. If the demand is *elastic*, then a decrease in price is accompanied by an increase in unit sales sufficient to increase the total revenue.

2. If the demand is *inelastic*, then a decrease in price is not accompanied by an increase in unit sales sufficient to increase the total revenue.

Comparing Elasticity and Revenue Ⓡ

The demand function for a product is modeled by $p = 18 - 1.5\sqrt{x}$, $0 \le x \le 144$, as shown in Figure 9.12(a).

a. Find the intervals on which the demand is elastic, inelastic, and of unit elasticity.

b. Use the result of part (a) to describe the behavior of the revenue function.

SOLUTION

a. The price elasticity of demand is given by

$$\eta = \frac{p/x}{dp/dx}$$ Formula for price elasticity of demand

$$= \frac{\dfrac{18 - 1.5\sqrt{x}}{x}}{\dfrac{-3}{4\sqrt{x}}}$$ Substitute for p/x and dp/dx.

$$= \frac{-24\sqrt{x} + 2x}{x}$$ Multiply numerator and denominator by $-\dfrac{4\sqrt{x}}{3}$.

$$= -\frac{24\sqrt{x}}{x} + 2.$$ Rewrite as two fractions and simplify.

The demand is of unit elasticity when $|\eta| = 1$. In the interval $[0, 144]$, the only solution of the equation

$$|\eta| = \left| -\frac{24\sqrt{x}}{x} + 2 \right| = 1$$ Unit elasticity

is $x = 64$. So, the demand is of unit elasticity when $x = 64$. For x-values in the interval $(0, 64)$,

$$|\eta| = \left| -\frac{24\sqrt{x}}{x} + 2 \right| > 1, \quad 0 < x < 64$$ Elastic

which implies that the demand is elastic when $0 < x < 64$. For x-values in the interval $(64, 144)$,

$$|\eta| = \left| -\frac{24\sqrt{x}}{x} + 2 \right| < 1, \quad 64 < x < 144$$ Inelastic

which implies that the demand is inelastic when $64 < x < 144$.

b. From part (a), you can conclude that the revenue function R is increasing on the open interval $(0, 64)$, is decreasing on the open interval $(64, 144)$, and is a maximum when $x = 64$, as indicated in Figure 9.12(b).

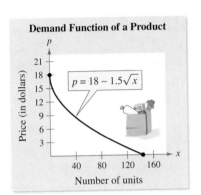

Demand Function of a Product

$p = 18 - 1.5\sqrt{x}$

Price (in dollars) vs. Number of units

(a)

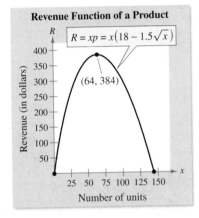

Revenue Function of a Product

$R = xp = x(18 - 1.5\sqrt{x})$

(64, 384)

Revenue (in dollars) vs. Number of units

(b)

FIGURE 9.12

Algebra Review

For help on the algebra in Example 5, see Example 2(c) in the *Chapter 9 Algebra Review*, on page 738.

✓**CHECKPOINT 5**

Find the intervals on which the demand function $p = 36 - 2\sqrt{x}$, $0 \le x \le 324$, is elastic, inelastic, and of unit elasticity. ■

STUDY TIP

In the discussion of price elasticity of demand, the price is assumed to decrease as the quantity demanded increases. So, the demand function $p = f(x)$ is decreasing and dp/dx is negative.

Business Terms and Formulas

This section concludes with a summary of the basic business terms and formulas used in this section. A summary of the graphs of the demand, revenue, cost, and profit functions is shown in Figure 9.13.

Summary of Business Terms and Formulas

x = number of units produced (or sold)	η = price elasticity of demand
p = price per unit	$\quad = (p/x)/(dp/dx)$
R = total revenue from selling x units = xp	dR/dx = marginal revenue
C = total cost of producing x units	dC/dx = marginal cost
P = total profit from selling x units = $R - C$	dP/dx = marginal profit
$\overline{C}$ = average cost per unit = $\dfrac{C}{x}$	

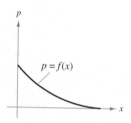

Demand function

Quantity demanded increases as price decreases.

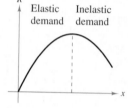

Revenue function

The low prices required to sell more units eventually result in a decreasing revenue.

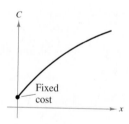

Cost function

The total cost to produce x units includes the fixed cost.

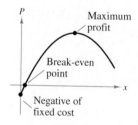

Profit function

The break-even point occurs when $R = C$.

FIGURE 9.13

CONCEPT CHECK

1. In the average cost function $\overline{C} = \dfrac{C}{x}$, what does C represent? What does x represent?

2. After a drop in the price of tomatoes, the demand for tomatoes increased. This is an example of what type of demand?

3. Even though the price of gasoline rose, the demand for gasoline was the same. This is an example of what type of demand?

4. Explain how price elasticity of demand is related to the total revenue function.

Skills Review 9.2   The following warm-up exercises involve skills that were covered in earlier sections. You will use these skills in the exercise set for this section. For additional help, review Sections 0.1, 0.3, 0.4, 0.7, and 7.5.

In Exercises 1–4, evaluate the expression for $x = 150$.

1. $\left| -\dfrac{300}{x} + 3 \right|$

2. $\left| -\dfrac{600}{5x} + 2 \right|$

3. $\left| \dfrac{(20x^{-1/2})/x}{-10x^{-3/2}} \right|$

4. $\left| \dfrac{(4000/x^2)/x}{-8000x^{-3}} \right|$

In Exercises 5–10, find the marginal revenue, marginal cost, or marginal profit.

5. $C = 650 + 1.2x + 0.003x^2$

6. $P = 0.01x^2 + 11x$

7. $R = 14x - \dfrac{x^2}{2000}$

8. $R = 3.4x - \dfrac{x^2}{1500}$

9. $P = -0.7x^2 + 7x - 50$

10. $C = 1700 + 4.2x + 0.001x^3$

Exercises 9.2

See www.CalcChat.com for worked-out solutions to odd-numbered exercises.

In Exercises 1–4, find the number of units x that produces a maximum revenue R.

1. $R = 800x - 0.2x^2$

2. $R = 48x^2 - 0.02x^3$

3. $R = 400x - x^2$

4. $R = 30x^{2/3} - 2x$

In Exercises 5–8, find the number of units x that produces the minimum average cost per unit $\overline{C}$.

5. $C = 0.125x^2 + 20x + 5000$

6. $C = 0.001x^3 + 5x + 250$

7. $C = 2x^2 + 255x + 5000$

8. $C = 0.02x^3 + 55x^2 + 1380$

In Exercises 9–12, find the price per unit p that produces the maximum profit P.

Cost Function	Demand Function
9. $C = 100 + 30x$	$p = 90 - x$
10. $C = 0.5x + 500$	$p = \dfrac{50}{\sqrt{x}}$
11. $C = 8000 + 50x + 0.03x^2$	$p = 70 - 0.01x$
12. $C = 35x + 500$	$p = 50 - 0.1\sqrt{x}$

(T) **Average Cost** In Exercises 13 and 14, use the cost function to find the production level for which the average cost is a minimum. For this production level, show that the marginal cost and average cost are equal. Use a graphing utility to graph the average cost function and verify your results.

13. $C = 2x^2 + 5x + 18$

14. $C = x^3 - 6x^2 + 13x$

15. Maximum Profit A commodity has a demand function modeled by $p = 100 - 0.5x$, and a total cost function modeled by $C = 40x + 37.5$.

(a) What price yields a maximum profit?

(b) When the profit is maximized, what is the average cost per unit?

16. Maximum Profit How would the answer to Exercise 15 change if the marginal cost rose from $40 per unit to $50 per unit? In other words, rework Exercise 15 using the cost function $C = 50x + 37.5$.

Maximum Profit In Exercises 17 and 18, find the amount s of advertising that maximizes the profit P. (s and P are measured in thousands of dollars.) Find the point of diminishing returns.

17. $P = -2s^3 + 35s^2 - 100s + 200$

18. $P = -0.1s^3 + 6s^2 + 400$

19. Maximum Profit The cost per unit of producing a type of digital audio player is $60. The manufacturer charges $90 per unit for orders of 100 or less. To encourage large orders, however, the manufacturer reduces the charge by $0.10 per player for each order in excess of 100 units. For instance, an order of 101 players would be $89.90 per player, an order of 102 players would be $89.80 per player, and so on. Find the largest order the manufacturer should allow to obtain a maximum profit.

20. Maximum Profit   A real estate office handles a 50-unit apartment complex. When the rent is $580 per month, all units are occupied. For each $40 increase in rent, however, an average of one unit becomes vacant. Each occupied unit requires an average of $45 per month for service and repairs. What rent should be charged to obtain a maximum profit?

21. Maximum Revenue   When a wholesaler sold a product at $40 per unit, sales were 300 units per week. After a price increase of $5, however, the average number of units sold dropped to 275 per week. Assuming that the demand function is linear, what price per unit will yield a maximum total revenue?

22. Maximum Profit   Assume that the amount of money deposited in a bank is proportional to the square of the interest rate the bank pays on the money. Furthermore, the bank can reinvest the money at 12% simple interest. Find the interest rate the bank should pay to maximize its profit.

(T) **23. Minimum Cost**   A power station is on one side of a river that is 0.5 mile wide, and a factory is 6 miles downstream on the other side of the river (see figure). It costs $18 per foot to run overland power lines and $25 per foot to run underwater power lines. Write a cost function for running the power lines from the power station to the factory. Use a graphing utility to graph your function. Estimate the value of x that minimizes the cost. Explain your results.

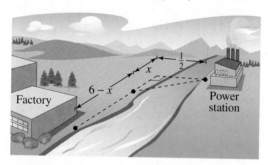

24. Minimum Cost   An offshore oil well is 1 mile off the coast. The oil refinery is 2 miles down the coast. Laying pipe in the ocean is twice as expensive as laying it on land. Find the most economical path for the pipe from the well to the oil refinery.

Minimum Cost   In Exercises 25 and 26, find the speed v, in miles per hour, that will minimize costs on a 110-mile delivery trip. The cost per hour for fuel is C dollars, and the driver is paid W dollars per hour. (Assume there are no costs other than wages and fuel.)

25. Fuel cost: $C = \dfrac{v^2}{300}$

Driver: $W = \$12$

26. Fuel cost: $C = \dfrac{v^2}{500}$

Driver: $W = \$9.50$

(T) **Elasticity**   In Exercises 27–32, find the price elasticity of demand for the demand function at the indicated x-value. Is the demand elastic, inelastic, or of unit elasticity at the indicated x-value? Use a graphing utility to graph the revenue function, and identify the intervals of elasticity and inelasticity.

Demand Function	Quantity Demanded
27. $p = 600 - 5x$	$x = 30$
28. $p = 400 - 3x$	$x = 20$
29. $p = 5 - 0.03x$	$x = 100$
30. $p = 20 - 0.0002x$	$x = 30$
31. $p = \dfrac{500}{x + 2}$	$x = 23$
32. $p = \dfrac{100}{x^2} + 2$	$x = 10$

33. Elasticity   The demand function for a product is given by
$$p = 20 - 0.02x, \quad 0 < x < 1000.$$

(a) Find the price elasticity of demand when $x = 560$.

(b) Find the values of x and p that maximize the total revenue.

(c) For the value of x found in part (b), show that the price elasticity of demand has unit elasticity.

34. Elasticity   The demand function for a product is given by
$$p = 800 - 4x, \quad 0 < x < 200.$$

(a) Find the price elasticity of demand when $x = 150$.

(b) Find the values of x and p that maximize the total revenue.

(c) For the value of x found in part (b), show that the price elasticity of demand has unit elasticity.

(T) **35. Minimum Cost**   The shipping and handling cost C of a manufactured product is modeled by
$$C = 4\left(\frac{25}{x^2} - \frac{x}{x - 10}\right), \quad 0 < x < 10$$
where C is measured in thousands of dollars and x is the number of units shipped (in hundreds). Find the shipment size that minimizes the cost. (*Hint:* Use the *root* feature of a graphing utility.)

(T) **36. Minimum Cost**   The ordering and transportation cost C of the components used in manufacturing a product is modeled by
$$C = 8\left(\frac{2500}{x^2} - \frac{x}{x - 100}\right), \quad 0 < x < 100$$
where C is measured in thousands of dollars and x is the order size in hundreds. Find the order size that minimizes the cost. (*Hint:* Use the *root* feature of a graphing utility.)

37. *MAKE A DECISION: REVENUE* The demand for a car wash is $x = 600 - 50p$, where the current price is \$5. Can revenue be increased by lowering the price and thus attracting more customers? Use price elasticity of demand to determine your answer.

38. Revenue Repeat Exercise 37 for a demand function of $x = 800 - 40p$.

39. Sales The sales S (in billions of dollars per year) for Procter & Gamble for the years 2001 through 2006 can be modeled by

$$S = 1.09312t^2 - 1.8682t + 39.831, \quad 1 \le t \le 6$$

where t represents the year, with $t = 1$ corresponding to 2001. *(Source: Procter & Gamble Company)*

(a) During which year, from 2001 through 2006, were Procter & Gamble's sales increasing most rapidly?

(b) During which year were the sales increasing at the lowest rate?

(c) Find the rate of increase or decrease for each year in parts (a) and (b).

(T) (d) Use a graphing utility to graph the sales function. Then use the *zoom* and *trace* features to confirm the results in parts (a), (b), and (c).

40. Revenue The revenue R (in millions of dollars per year) for Papa John's from 1996 to 2005 can be modeled by

$$R = \frac{-485.0 + 116.68t}{1 - 0.12t + 0.0097t^2}, \quad 6 \le t \le 15$$

where t represents the year, with $t = 6$ corresponding to 1996. *(Source: Papa John's Int'l.)*

(a) During which year, from 1996 through 2005, was Papa John's revenue the greatest? the least?

(b) During which year was the revenue increasing at the greatest rate? decreasing at the greatest rate?

(T) (c) Use a graphing utility to graph the revenue function, and confirm your results in parts (a) and (b).

41. Match each graph with the function it best represents— a demand function, a revenue function, a cost function, or a profit function. Explain your reasoning. (The graphs are labeled *a*–*d*.)

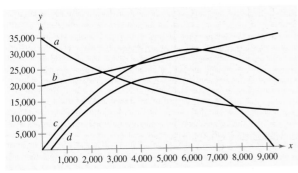

42. Demand A demand function is modeled by $x = a/p^m$, where a is a constant and $m > 1$. Show that $\eta = -m$. In other words, show that a 1% increase in price results in an $m\%$ decrease in the quantity demanded.

43. Think About It Throughout this text, it is assumed that demand functions are decreasing. Can you think of a product that has an increasing demand function? That is, can you think of a product that becomes more in demand as its price increases? Explain your reasoning, and sketch a graph of the function.

44. Extended Application To work an extended application analyzing the sales per share for Lowe's from 1990 through 2005, visit this text's website at *college.hmco.com*. *(Data Source: Lowe's Companies)*

Business Capsule

Photo courtesy of Jim Bell

Illinois native Jim Bell moved to California in 1996 to pursue his dream of working in the skateboarding industry. After a string of sales jobs with several skate companies, Bell started San Diego-based Jim Bell Skateboard Ramps in 2004 with an initial cash outlay of \$50. His custom-built skateboard ramp business brought in sales of \$250,000 the following year. His latest product, the U-Built-It Skateboard Ramp, is expected to nearly double his annual sales. Bell marketed his new product by featuring it at trade shows. He backed it up by showing pictures of the hundreds of ramps he has built. So, Bell was able to prove the demand existed, as well as the quality and customer satisfaction his work boasted.

45. Research Project Choose an innovative product like the one described above. Use your school's library, the Internet, or some other reference source to research the history of the product or service. Collect data about the revenue that the product or service has generated, and find a mathematical model of the data. Summarize your findings.

Section 9.3

Asymptotes

■ Find the vertical asymptotes of functions and find infinite limits.
■ Find the horizontal asymptotes of functions and find limits at infinity.
■ Use asymptotes to answer questions about real-life situations.

Vertical Asymptotes and Infinite Limits

In Sections 8.4 through 8.6, you studied ways in which you can use calculus to help analyze the graph of a function. In this section, you will study another valuable aid to curve sketching: the determination of vertical and horizontal asymptotes.

Recall from Section 7.1, Example 10, that the function

$$f(x) = \frac{3}{x - 2}$$

is unbounded as x approaches 2 (see Figure 9.14). This type of behavior is described by saying that the line $x = 2$ is a **vertical asymptote** of the graph of f. The type of limit in which $f(x)$ approaches infinity (or negative infinity) as x approaches c from the left or from the right is an **infinite limit.** The infinite limits for the function $f(x) = 3/(x - 2)$ can be written as

$$\lim_{x \to 2^-} \frac{3}{x - 2} = -\infty$$

and

$$\lim_{x \to 2^+} \frac{3}{x - 2} = \infty.$$

$f(x) = \dfrac{3}{x - 2}$

$\dfrac{3}{x - 2} \to \infty$ as $x \to 2^+$

$\dfrac{3}{x - 2} \to -\infty$ as $x \to 2^-$

$x = 2$ is a vertical asymptote.

FIGURE 9.14

Definition of Vertical Asymptote

If $f(x)$ approaches infinity (or negative infinity) as x approaches c from the right or from the left, then the line $x = c$ is a **vertical asymptote** of the graph of f.

TECHNOLOGY

When you use a graphing utility to graph a function that has a vertical asymptote, the utility may try to connect separate branches of the graph. For instance, the figure at the right shows the graph of

$$f(x) = \frac{3}{x - 2}$$

on a graphing calculator.

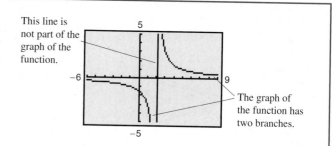

This line is not part of the graph of the function.

The graph of the function has two branches.

x Approaches 1 from the Left

x	$f(x) = 1/(x - 1)$
0	-1
0.9	-10
0.99	-100
0.999	-1000
0.9999	$-10,000$

x Approaches 1 from the Right

x	$f(x) = 1/(x - 1)$
2	1
1.1	10
1.01	100
1.001	1000
1.0001	10,000

✓ **CHECKPOINT 1**

Find each limit.

a. *Limit from the left*

$$\lim_{x \to 2^-} \frac{1}{x - 2}$$

Limit from the right

$$\lim_{x \to 2^+} \frac{1}{x - 2}$$

b. *Limit from the left*

$$\lim_{x \to -3^-} \frac{-1}{x + 3}$$

Limit from the right

$$\lim_{x \to -3^+} \frac{-1}{x + 3}$$ ■

One of the most common instances of a vertical asymptote is the graph of a *rational function*—that is, a function of the form $f(x) = p(x)/q(x)$, where $p(x)$ and $q(x)$ are polynomials. If c is a real number such that $q(c) = 0$ and $p(c) \neq 0$, the graph of f has a vertical asymptote at $x = c$. Example 1 shows four cases.

Example 1   Finding Infinite Limits

Find each limit.

Limit from the left	*Limit from the right*	
a. $\lim\limits_{x \to 1^-} \dfrac{1}{x - 1} = -\infty$	$\lim\limits_{x \to 1^+} \dfrac{1}{x - 1} = \infty$	See Figure 9.15(a).
b. $\lim\limits_{x \to 1^-} \dfrac{-1}{x - 1} = \infty$	$\lim\limits_{x \to 1^+} \dfrac{-1}{x - 1} = -\infty$	See Figure 9.15(b).
c. $\lim\limits_{x \to 1^-} \dfrac{-1}{(x - 1)^2} = -\infty$	$\lim\limits_{x \to 1^+} \dfrac{-1}{(x - 1)^2} = -\infty$	See Figure 9.15(c).
d. $\lim\limits_{x \to 1^-} \dfrac{1}{(x - 1)^2} = \infty$	$\lim\limits_{x \to 1^+} \dfrac{1}{(x - 1)^2} = \infty$	See Figure 9.15(d).

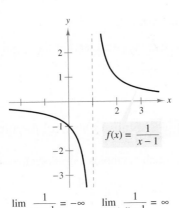

$$\lim_{x \to 1^-} \frac{1}{x - 1} = -\infty \qquad \lim_{x \to 1^+} \frac{1}{x - 1} = \infty$$

(a)

$$\lim_{x \to 1^-} \frac{-1}{x - 1} = \infty \qquad \lim_{x \to 1^+} \frac{-1}{x - 1} = -\infty$$

(b)

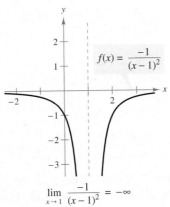

$$\lim_{x \to 1} \frac{-1}{(x - 1)^2} = -\infty$$

(c)

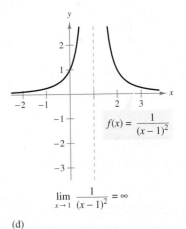

$$\lim_{x \to 1} \frac{1}{(x - 1)^2} = \infty$$

(d)

FIGURE 9.15

Each of the graphs in Example 1 has only one vertical asymptote. As shown in the next example, the graph of a rational function can have more than one vertical asymptote.

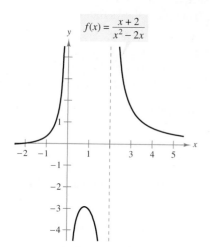

$$f(x) = \frac{x + 2}{x^2 - 2x}$$

FIGURE 9.16 Vertical Asymptotes at $x = 0$ and $x = 2$

Example 2 Finding Vertical Asymptotes

Find the vertical asymptotes of the graph of

$$f(x) = \frac{x + 2}{x^2 - 2x}.$$

SOLUTION The possible vertical asymptotes correspond to the x-values for which the denominator is zero.

$$x^2 - 2x = 0 \qquad \text{Set denominator equal to 0.}$$
$$x(x - 2) = 0 \qquad \text{Factor.}$$
$$x = 0, x = 2 \qquad \text{Zeros of denominator}$$

Because the numerator of f is not zero at either of these x-values, you can conclude that the graph of f has two vertical asymptotes—one at $x = 0$ and one at $x = 2$, as shown in Figure 9.16.

✓CHECKPOINT 2

Find the vertical asymptote(s) of the graph of

$$f(x) = \frac{x + 4}{x^2 - 4x}. \blacksquare$$

Example 3 Finding Vertical Asymptotes

Find the vertical asymptotes of the graph of

$$f(x) = \frac{x^2 + 2x - 8}{x^2 - 4}.$$

SOLUTION First factor the numerator and denominator. Then divide out like factors.

$$f(x) = \frac{x^2 + 2x - 8}{x^2 - 4} \qquad \text{Write original function.}$$

$$= \frac{(x + 4)(x - 2)}{(x + 2)(x - 2)} \qquad \text{Factor numerator and denominator.}$$

$$= \frac{(x + 4)(x - 2)}{(x + 2)(x - 2)} \qquad \text{Divide out like factors.}$$

$$= \frac{x + 4}{x + 2}, \quad x \neq 2 \qquad \text{Simplify.}$$

For all values of x other than $x = 2$, the graph of this simplified function is the same as the graph of f. So, you can conclude that the graph of f has only one vertical asymptote. This occurs at $x = -2$, as shown in Figure 9.17.

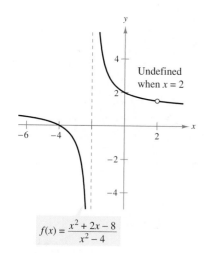

$$f(x) = \frac{x^2 + 2x - 8}{x^2 - 4}$$

FIGURE 9.17 Vertical Asymptote at $x = -2$

✓CHECKPOINT 3

Find the vertical asymptotes of the graph of

$$f(x) = \frac{x^2 + 4x + 3}{x^2 - 9}. \blacksquare$$

From Example 3, you know that the graph of

$$f(x) = \frac{x^2 + 2x - 8}{x^2 - 4}$$

has a vertical asymptote at $x = -2$. This implies that the limit of $f(x)$ as $x \to -2$ from the right (or from the left) is either ∞ or $-\infty$. But without looking at the graph, how can you determine that the limit from the left is *negative* infinity and the limit from the right is *positive* infinity? That is, why is the limit from the left

$$\lim_{x \to -2^-} \frac{x^2 + 2x - 8}{x^2 - 4} = -\infty \qquad \text{Limit from the left}$$

and why is the limit from the right

$$\lim_{x \to -2^+} \frac{x^2 + 2x - 8}{x^2 - 4} = \infty? \qquad \text{Limit from the right}$$

It is cumbersome to determine these limits analytically, and you may find the graphical method shown in Example 4 to be more efficient.

Example 4   Determining Infinite Limit

Find the limits.

$$\lim_{x \to 1^-} \frac{x^2 - 3x}{x - 1} \quad \text{and} \quad \lim_{x \to 1^+} \frac{x^2 - 3x}{x - 1}$$

SOLUTION   Begin by considering the function

$$f(x) = \frac{x^2 - 3x}{x - 1}.$$

Because the denominator is zero when $x = 1$ and the numerator is not zero when $x = 1$, it follows that the graph of the function has a vertical asymptote at $x = 1$. This implies that each of the given limits is either ∞ or $-\infty$. To determine which, use a graphing utility to graph the function, as shown in Figure 9.18. From the graph, you can see that the limit from the left is positive infinity and the limit from the right is negative infinity. That is,

$$\lim_{x \to 1^-} \frac{x^2 - 3x}{x - 1} = \infty \qquad \text{Limit from the left}$$

and

$$\lim_{x \to 1^+} \frac{x^2 - 3x}{x - 1} = -\infty. \qquad \text{Limit from the right}$$

From the left, $f(x)$ approaches positive infinity.

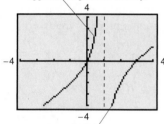

From the right, $f(x)$ approaches negative infinity.

FIGURE 9.18

STUDY TIP

In Example 4, try evaluating $f(x)$ at x-values that are just barely to the left of 1. You will find that you can make the values of $f(x)$ arbitrarily large by choosing x sufficiently close to 1. For instance, $f(0.99999) = 199,999$.

✓ CHECKPOINT 4

Find the limits.

$$\lim_{x \to 2^-} \frac{x^2 - 4x}{x - 2} \quad \text{and} \quad \lim_{x \to 2^+} \frac{x^2 - 4x}{x - 2}$$

Then verify your solution by graphing the function. ■

Horizontal Asymptotes and Limits at Infinity

Another type of limit, called a **limit at infinity,** specifies a finite value approached by a function as x increases (or decreases) without bound.

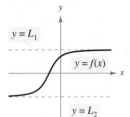

> **Definition of Horizontal Asymptote**
>
> If f is a function and L_1 and L_2 are real numbers, the statements
>
> $$\lim_{x \to \infty} f(x) = L_1 \quad \text{and} \quad \lim_{x \to -\infty} f(x) = L_2$$
>
> denote **limits at infinity.** The lines $y = L_1$ and $y = L_2$ are **horizontal asymptotes** of the graph of f.

Figure 9.19 shows two ways in which the graph of a function can approach one or more horizontal asymptotes. Note that it is possible for the graph of a function to cross its horizontal asymptote.

Limits at infinity share many of the properties of limits discussed in Section 7.1. When finding horizontal asymptotes, you can use the property that

$$\lim_{x \to \infty} \frac{1}{x^r} = 0, \quad r > 0 \quad \text{and} \quad \lim_{x \to -\infty} \frac{1}{x^r} = 0, \quad r > 0.$$

FIGURE 9.19

(The second limit assumes that x^r is defined when $x < 0$.)

Example 5 **Finding Limits at Infinity**

Find the limit: $\displaystyle \lim_{x \to \infty} \left(5 - \frac{2}{x^2}\right).$

SOLUTION

$$\lim_{x \to \infty} \left(5 - \frac{2}{x^2}\right) = \lim_{x \to \infty} 5 - \lim_{x \to \infty} \frac{2}{x^2} \qquad \lim_{x \to \infty} [f(x) - g(x)] = \lim_{x \to \infty} f(x) - \lim_{x \to \infty} g(x)$$

$$= \lim_{x \to \infty} 5 - 2\left(\lim_{x \to \infty} \frac{1}{x^2}\right) \qquad \lim_{x \to \infty} cf(x) = c \lim_{x \to \infty} f(x)$$

$$= 5 - 2(0)$$

$$= 5$$

You can verify this limit by sketching the graph of

$$f(x) = 5 - \frac{2}{x^2}$$

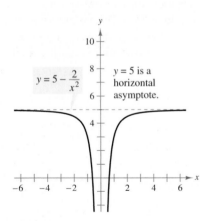

$y = 5 - \dfrac{2}{x^2}$

$y = 5$ is a horizontal asymptote.

FIGURE 9.20

as shown in Figure 9.20. Note that the graph has $y = 5$ as a horizontal asymptote to the right. By evaluating the limit of $f(x)$ as $x \to -\infty$, you can show that this line is also a horizontal asymptote to the left.

✓**CHECKPOINT 5**

Find the limit: $\displaystyle \lim_{x \to \infty} \left(2 + \frac{5}{x^2}\right).$ ∎

There is an easy way to determine whether the graph of a *rational* function has a horizontal asymptote. This shortcut is based on a comparison of the degrees of the numerator and denominator of the rational function.

Horizontal Asymptotes of Rational Functions

Let $f(x) = p(x)/q(x)$ be a rational function.

1. If the degree of the numerator is less than the degree of the denominator, then $y = 0$ is a horizontal asymptote of the graph of f (to the left and to the right).

2. If the degree of the numerator is equal to the degree of the denominator, then $y = a/b$ is a horizontal asymptote of the graph of f (to the left and to the right), where a and b are the leading coefficients of $p(x)$ and $q(x)$, respectively.

3. If the degree of the numerator is greater than the degree of the denominator, then the graph of f has no horizontal asymptote.

✓ CHECKPOINT 6

Find the horizontal asymptote of the graph of each function.

a. $y = \dfrac{2x + 1}{4x^2 + 5}$

b. $y = \dfrac{2x^2 + 1}{4x^2 + 5}$

c. $y = \dfrac{2x^3 + 1}{4x^2 + 5}$ ∎

Example 6   Finding Horizontal Asymptotes

Find the horizontal asymptote of the graph of each function.

a. $y = \dfrac{-2x + 3}{3x^2 + 1}$ **b.** $y = \dfrac{-2x^2 + 3}{3x^2 + 1}$ **c.** $y = \dfrac{-2x^3 + 3}{3x^2 + 1}$

SOLUTION

a. Because the degree of the numerator is less than the degree of the denominator, $y = 0$ is a horizontal asymptote. [See Figure 9.21(a).]

b. Because the degree of the numerator is equal to the degree of the denominator, the line $y = -\frac{2}{3}$ is a horizontal asymptote. [See Figure 9.21(b).]

c. Because the degree of the numerator is greater than the degree of the denominator, the graph has no horizontal asymptote. [See Figure 9.21(c).]

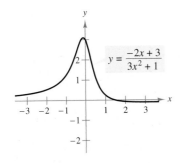

(a) $y = 0$ is a horizontal asymptote.

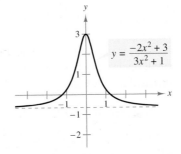

(b) $y = -\frac{2}{3}$ is a horizontal asymptote.

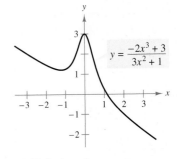

(c) No horizontal asymptote

FIGURE 9.21

Applications of Asymptotes

There are many examples of asymptotic behavior in real life. For instance, Example 7 describes the asymptotic behavior of an average cost function.

Example 7   Modeling Average Cost

A small business invests $5000 in a new product. In addition to this initial investment, the product will cost $0.50 per unit to produce. Find the average cost per unit if 1000 units are produced, if 10,000 units are produced, and if 100,000 units are produced. What is the limit of the average cost as the number of units produced increases?

SOLUTION   From the given information, you can model the total cost C (in dollars) by

$$C = 0.5x + 5000 \qquad \text{Total cost function}$$

where x is the number of units produced. This implies that the average cost function is

$$\overline{C} = \frac{C}{x} = 0.5 + \frac{5000}{x}. \qquad \text{Average cost function}$$

If only 1000 units are produced, then the average cost per unit is

$$\overline{C} = 0.5 + \frac{5000}{1000} = \$5.50. \qquad \text{Average cost for 1000 units}$$

If 10,000 units are produced, then the average cost per unit is

$$\overline{C} = 0.5 + \frac{5000}{10,000} = \$1.00. \qquad \text{Average cost for 10,000 units}$$

If 100,000 units are produced, then the average cost per unit is

$$\overline{C} = 0.5 + \frac{5000}{100,000} = \$0.55. \qquad \text{Average cost for 100,000 units}$$

As x approaches infinity, the limiting average cost per unit is

$$\lim_{x \to \infty} \left(0.5 + \frac{5000}{x} \right) = \$0.50.$$

As shown in Figure 9.22, this example points out one of the major problems of small businesses. That is, it is difficult to have competitively low prices when the production level is low.

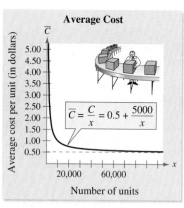

FIGURE 9.22   As $x \to \infty$, the average cost per unit approaches $0.50.

✓ CHECKPOINT 7

A small business invests $25,000 in a new product. In addition, the product will cost $0.75 per unit to produce. Find the cost function and the average cost function. What is the limit of the average cost function as production increases?

Since the 1980s, industries in the United States have spent billions of dollars to reduce air pollution.

Example 8   Modeling Smokestack Emission

A manufacturing plant has determined that the cost C (in dollars) of removing $p\%$ of the smokestack pollutants of its main smokestack is modeled by

$$C = \frac{80,000p}{100 - p}, \quad 0 \le p < 100.$$

What is the vertical asymptote of this function? What does the vertical asymptote mean to the plant owners?

SOLUTION   The graph of the cost function is shown in Figure 9.23. From the graph, you can see that $p = 100$ is the vertical asymptote. This means that as the plant attempts to remove higher and higher percents of the pollutants, the cost increases dramatically. For instance, the cost of removing 85% of the pollutants is

$$C = \frac{80,000(85)}{100 - 85} \approx \$453,333 \qquad \text{Cost for 85\% removal}$$

but the cost of removing 90% is

$$C = \frac{80,000(90)}{100 - 90} = \$720,000. \qquad \text{Cost for 90\% removal}$$

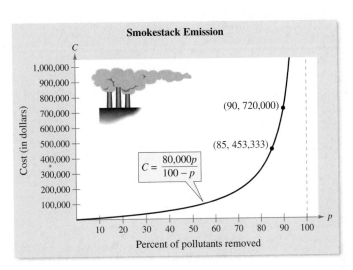

FIGURE 9.23

✓ CHECKPOINT 8

According to the cost function in Example 8, is it possible to remove 100% of the smokestack pollutants? Why or why not? ▪

CONCEPT CHECK

1. Complete the following: If $f(x) \to \pm\infty$ as $x \to c$ from the right or the left, then the line $x = c$ is a _____ _____ of the graph of f.
2. Describe in your own words what is meant by $\lim\limits_{x \to \infty} f(x) = 4$.
3. Describe in your own words what is meant by $\lim\limits_{x \to -\infty} f(x) = 2$.
4. Complete the following: Given a rational function f, if the degree of the numerator is less than the degree of the denominator, then _____ is a horizontal asymptote of the graph of f (to the left and to the right).

| **Skills Review 9.3** | The following warm-up exercises involve skills that were covered in earlier sections. You will use these skills in the exercise set for this section. For additional help, review Sections 7.1, 7.5, and 9.2. |

In Exercises 1–8, find the limit.

1. $\lim_{x \to 2} (x + 1)$

2. $\lim_{x \to -1} (3x + 4)$

3. $\lim_{x \to -3} \dfrac{2x^2 + x - 15}{x + 3}$

4. $\lim_{x \to 2} \dfrac{3x^2 - 8x + 4}{x - 2}$

5. $\lim_{x \to 2^+} \dfrac{x^2 - 5x + 6}{x^2 - 4}$

6. $\lim_{x \to 1^-} \dfrac{x^2 - 6x + 5}{x^2 - 1}$

7. $\lim_{x \to 0^+} \sqrt{x}$

8. $\lim_{x \to 1^+} \left(x + \sqrt{x - 1}\right)$

In Exercises 9–12, find the average cost and the marginal cost.

9. $C = 150 + 3x$

10. $C = 1900 + 1.7x + 0.002x^2$

11. $C = 0.005x^2 + 0.5x + 1375$

12. $C = 760 + 0.05x$

Exercises 9.3

See www.CalcChat.com for worked-out solutions to odd-numbered exercises.

In Exercises 1–8, find the vertical and horizontal asymptotes. Write the asymptotes as equations of lines.

1. $f(x) = \dfrac{x^2 + 1}{x^2}$

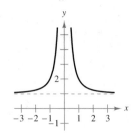

2. $f(x) = \dfrac{4}{(x - 2)^3}$

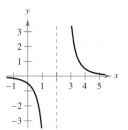

3. $f(x) = \dfrac{x^2 - 2}{x^2 - x - 2}$

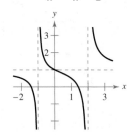

4. $f(x) = \dfrac{2 + x}{1 - x}$

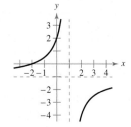

5. $f(x) = \dfrac{3x^2}{2(x^2 + 1)}$

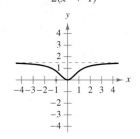

6. $f(x) = \dfrac{-4x}{x^2 + 4}$

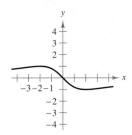

7. $f(x) = \dfrac{x^2 - 1}{2x^2 - 8}$

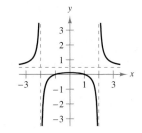

8. $f(x) = \dfrac{x^2 + 1}{x^3 - 8}$

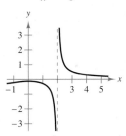

In Exercises 9–12, match the function with its graph. Use horizontal asymptotes as an aid. [The graphs are labeled (a)–(d).]

(a)

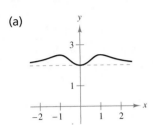

(b)

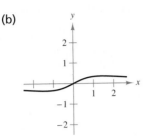

(c)

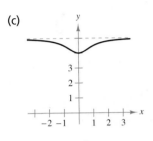

(d)

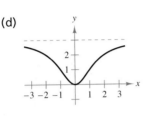

9. $f(x) = \dfrac{3x^2}{x^2 + 2}$

10. $f(x) = \dfrac{x}{x^2 + 2}$

11. $f(x) = 2 + \dfrac{x^2}{x^4 + 1}$

12. $f(x) = 5 - \dfrac{1}{x^2 + 1}$

In Exercises 13–20, find the limit.

13. $\lim\limits_{x \to -2^-} \dfrac{1}{(x + 2)^2}$

14. $\lim\limits_{x \to -2^-} \dfrac{1}{x + 2}$

15. $\lim\limits_{x \to 3^+} \dfrac{x - 4}{x - 3}$

16. $\lim\limits_{x \to 1^+} \dfrac{2 + x}{1 - x}$

17. $\lim\limits_{x \to 4^-} \dfrac{x^2}{x^2 - 16}$

18. $\lim\limits_{x \to 4} \dfrac{x^2}{x^2 + 16}$

19. $\lim\limits_{x \to 0^-} \left(1 + \dfrac{1}{x}\right)$

20. $\lim\limits_{x \to 0^-} \left(x^2 - \dfrac{1}{x}\right)$

(T)(S) In Exercises 21–24, use a graphing utility or spreadsheet software program to complete the table. Then use the result to estimate the limit of $f(x)$ as x approaches infinity.

x	10^0	10^1	10^2	10^3	10^4	10^5	10^6
$f(x)$							

21. $f(x) = \dfrac{x + 1}{x\sqrt{x}}$

22. $f(x) = \dfrac{2x^2}{x + 1}$

23. $f(x) = \dfrac{x^2 - 1}{0.02x^2}$

24. $f(x) = \dfrac{3x^2}{0.1x^2 + 1}$

(T)(S) In Exercises 25 and 26, use a graphing utility or a spreadsheet software program to complete the table and use the result to estimate the limit of $f(x)$ as x approaches infinity and as x approaches negative infinity.

x	-10^6	-10^4	-10^2	10^0	10^2	10^4	10^6
$f(x)$							

25. $f(x) = \dfrac{2x}{\sqrt{x^2 + 4}}$

26. $f(x) = x - \sqrt{x(x - 1)}$

In Exercises 27 and 28, find $\lim\limits_{x \to \infty} h(x)$, if possible.

27. $f(x) = 5x^3 - 3$

(a) $h(x) = \dfrac{f(x)}{x^2}$ (b) $h(x) = \dfrac{f(x)}{x^3}$ (c) $h(x) = \dfrac{f(x)}{x^4}$

28. $f(x) = 3x^2 + 7$

(a) $h(x) = \dfrac{f(x)}{x}$ (b) $h(x) = \dfrac{f(x)}{x^2}$ (c) $h(x) = \dfrac{f(x)}{x^3}$

In Exercises 29 and 30, find each limit, if possible.

29. (a) $\lim\limits_{x \to \infty} \dfrac{x^2 + 2}{x^3 - 1}$

(b) $\lim\limits_{x \to \infty} \dfrac{x^2 + 2}{x^2 - 1}$

(c) $\lim\limits_{x \to \infty} \dfrac{x^2 + 2}{x - 1}$

30. (a) $\lim\limits_{x \to \infty} \dfrac{3 - 2x}{3x^3 - 1}$

(b) $\lim\limits_{x \to \infty} \dfrac{3 - 2x}{3x - 1}$

(c) $\lim\limits_{x \to \infty} \dfrac{3 - 2x^2}{3x - 1}$

In Exercises 31–40, find the limit.

31. $\lim\limits_{x \to \infty} \dfrac{4x - 3}{2x + 1}$

32. $\lim\limits_{x \to \infty} \dfrac{5x^3 + 1}{10x^3 - 3x^2 + 7}$

33. $\lim\limits_{x \to \infty} \dfrac{3x}{4x^2 - 1}$

34. $\lim\limits_{x \to -\infty} \dfrac{2x^2 - 5x - 12}{1 - 6x - 8x^2}$

35. $\lim\limits_{x \to -\infty} \dfrac{5x^2}{x + 3}$

36. $\lim\limits_{x \to \infty} \dfrac{x^3 - 2x^2 + 3x + 1}{x^2 - 3x + 2}$

37. $\lim\limits_{x \to \infty} (2x - x^{-2})$

38. $\lim\limits_{x \to \infty} (2 - x^{-3})$

39. $\lim\limits_{x \to -\infty} \left(\dfrac{2x}{x - 1} + \dfrac{3x}{x + 1}\right)$

40. $\lim\limits_{x \to \infty} \left(\dfrac{2x^2}{x - 1} + \dfrac{3x}{x + 1}\right)$

In Exercises 41–58, sketch the graph of the equation. Use intercepts, extrema, and asymptotes as sketching aids.

41. $y = \dfrac{3x}{1 - x}$

42. $y = \dfrac{x - 3}{x - 2}$

43. $f(x) = \dfrac{x^2}{x^2 + 9}$

44. $f(x) = \dfrac{x}{x^2 + 4}$

45. $g(x) = \dfrac{x^2}{x^2 - 16}$

46. $g(x) = \dfrac{x}{x^2 - 4}$

47. $xy^2 = 4$

48. $x^2y = 4$

49. $y = \dfrac{2x}{1-x}$

50. $y = \dfrac{2x}{1-x^2}$

51. $y = 1 - 3x^{-2}$

52. $y = 1 + x^{-1}$

53. $f(x) = \dfrac{1}{x^2 - x - 2}$

54. $f(x) = \dfrac{x-2}{x^2 - 4x + 3}$

55. $g(x) = \dfrac{x^2 - x - 2}{x-2}$

56. $g(x) = \dfrac{x^2 - 9}{x+3}$

57. $y = \dfrac{2x^2 - 6}{(x-1)^2}$

58. $y = \dfrac{x}{(x+1)^2}$

59. Cost The cost C (in dollars) of producing x units of a product is $C = 1.35x + 4570$.

(a) Find the average cost function $\overline{C}$.

(b) Find $\overline{C}$ when $x = 100$ and when $x = 1000$.

(c) What is the limit of $\overline{C}$ as x approaches infinity?

60. Average Cost A business has a cost (in dollars) of $C = 0.5x + 500$ for producing x units.

(a) Find the average cost function $\overline{C}$.

(b) Find $\overline{C}$ when $x = 250$ and when $x = 1250$.

(c) What is the limit of $\overline{C}$ as x approaches infinity?

61. Average Cost The cost function for a certain model of personal digital assistant (PDA) is given by $C = 13.50x + 45{,}750$, where C is measured in dollars and x is the number of PDAs produced.

(a) Find the average cost function $\overline{C}$.

(b) Find $\overline{C}$ when $x = 100$ and $x = 1000$.

(c) Determine the limit of the average cost function as x approaches infinity. Interpret the limit in the context of the problem.

62. Average Cost The cost function for a company to recycle x tons of material is given by $C = 1.25x + 10{,}500$, where C is measured in dollars.

(a) Find the average cost function $\overline{C}$.

(b) Find the average costs of recycling 100 tons of material and 1000 tons of material.

(c) Determine the limit of the average cost function as x approaches infinity. Interpret the limit in the context of the problem.

63. Seizing Drugs The cost C (in millions of dollars) for the federal government to seize $p\%$ of a type of illegal drug as it enters the country is modeled by

$C = 528p/(100 - p), \quad 0 \le p < 100.$

(a) Find the costs of seizing 25%, 50%, and 75%.

(b) Find the limit of C as $p \to 100^-$. Interpret the limit in the context of the problem. Use a graphing utility to verify your result.

64. Removing Pollutants The cost C (in dollars) of removing $p\%$ of the air pollutants in the stack emission of a utility company that burns coal is modeled by

$C = 80{,}000p/(100 - p), \quad 0 \le p < 100.$

(a) Find the costs of removing 15%, 50%, and 90%.

(b) Find the limit of C as $p \to 100^-$. Interpret the limit in the context of the problem. Use a graphing utility to verify your result.

65. Learning Curve Psychologists have developed mathematical models to predict performance P (the percent of correct responses) as a function of n, the number of times a task is performed. One such model is

$$P = \dfrac{0.5 + 0.9(n-1)}{1 + 0.9(n-1)}, \quad 0 < n.$$

Ⓢ (a) Use a spreadsheet software program to complete the table for the model.

n	1	2	3	4	5	6	7	8	9	10
P										

(b) Find the limit as n approaches infinity.

Ⓣ (c) Use a graphing utility to graph this learning curve, and interpret the graph in the context of the problem.

66. Biology: Wildlife Management The state game commission introduces 30 elk into a new state park. The population N of the herd is modeled by

$N = [10(3 + 4t)]/(1 + 0.1t)$

where t is the time in years.

(a) Find the size of the herd after 5, 10, and 25 years.

(b) According to this model, what is the limiting size of the herd as time progresses?

67. Average Profit The cost and revenue functions for a product are $C = 34.5x + 15{,}000$ and $R = 69.9x$.

(a) Find the average profit function $\overline{P} = (R - C)/x$.

(b) Find the average profits when x is 1000, 10,000, and 100,000.

(c) What is the limit of the average profit function as x approaches infinity? Explain your reasoning.

68. Average Profit The cost and revenue functions for a product are $C = 25.5x + 1000$ and $R = 75.5x$.

(a) Find the average profit function $\overline{P} = \dfrac{R - C}{x}$.

(b) Find the average profits when x is 100, 500, and 1000.

(c) What is the limit of the average profit function as x approaches infinity? Explain your reasoning.

Mid-Chapter Quiz

See www.CalcChat.com for worked-out solutions to odd-numbered exercises.

Take this quiz as you would take a quiz in class. When you are done, check your work against the answers given in the back of the book.

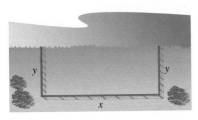

Figure for 1

1. A gardener has 200 feet of fencing to enclose a rectangular garden adjacent to a river (see figure). No fencing is needed along the river.

   (a) What dimensions should be used so that the area of the garden wil be a maximum?

   (b) Find the maximum area.

2. A rectangular page is to contain 48 square inches of print. The margins at the top and bottom of the page are to be 1 inch wide. The margins on each side are to be $\frac{3}{4}$ inch wide. Find the dimensions of the page that will minimize the amount of paper used.

In Exercises 3 and 4, find the number of units x that produces the minimum average cost per unit $\overline{C}$.

3. $C = 0.06x^2 + 12x + 9600$ 4. $C = 0.003x^3 + 8x + 2058$

In Exercises 5 and 6, find the price per unit p that yields the maximum profit P.

Cost Function	Demand Function
5. $C = 200 + 26x$	$p = 100 - x$
6. $C = 0.4x + 300$	$p = \dfrac{48}{\sqrt{x}}$

Ⓣ In Exercises 7 and 8, (a) find the price elasticity of demand for the demand function at the indicated x-value, (b) determine whether the demand is elastic, inelastic, or of unit elasticity at the indicated x-value, (c) use a graphing utility to graph the revenue function, and (d) identify the intervals of elasticity and inelasticity.

Demand Function	Quantity Demanded
7. $p = 500 - 4x$	$x = 250$
8. $p = 15 - \sqrt{x}$	$x = 900$

In Exercises 9–14, find the limit, if possible.

9. $\displaystyle\lim_{x \to 5^+} \frac{5 - x}{x - 5}$ 10. $\displaystyle\lim_{x \to 2^-} \frac{x^2}{x^2 + 2x - 8}$ 11. $\displaystyle\lim_{x \to 0^+} \frac{x}{x^2 + 0.1x}$

12. $\displaystyle\lim_{x \to \infty} \frac{x}{3x + 2}$ 13. $\displaystyle\lim_{x \to -\infty} \left(\frac{3}{x^2} - \frac{2}{x} - 1 \right)$ 14. $\displaystyle\lim_{x \to \infty} \frac{x^2 - 9}{x + 3}$

Ⓣ In Exercises 15–17, find any vertical and horizontal asymptotes of the graph. Then use a graphing utility to graph the function.

15. $f(x) = \dfrac{2x + 1}{x - 1}$ 16. $f(x) = \dfrac{3}{x^2 - 2x}$ 17. $f(x) = \dfrac{x^2 - 4}{x - 3}$

Curve Sketching: A Summary

■ Analyze the graphs of functions.

■ Recognize the graphs of simple polynomial functions.

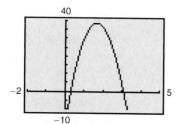

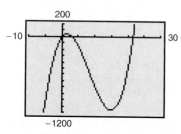

FIGURE 9.24

Summary of Curve-Sketching Techniques

It would be difficult to overstate the importance of using graphs in mathematics. Descartes's introduction of analytic geometry contributed significantly to the rapid advances in calculus that began during the mid-seventeenth century.

So far, you have studied several concepts that are useful in analyzing the graph of a function.

- x-intercepts and y-intercepts   (Section 2.1)
- Domain and range   (Section 2.4)
- Continuity   (Section 7.2)
- Differentiability   (Section 7.3)
- Relative extrema   (Section 8.5)
- Concavity   (Section 8.6)
- Points of inflection   (Section 8.6)
- Vertical asymptotes   (Section 9.3)
- Horizontal asymptotes   (Section 9.3)

When you are sketching the graph of a function, either by hand or with a graphing utility, remember that you cannot normally show the *entire* graph. The decision as to which part of the graph to show is crucial. For instance, which of the viewing windows in Figure 9.24 better represents the graph of

$$f(x) = x^3 - 25x^2 + 74x - 20?$$

The lower viewing window gives a more complete view of the graph, but the context of the problem might indicate that the upper view is better. Here are some guidelines for analyzing the graph of a function.

Guidelines for Analyzing the Graph of a Function

1. Determine the domain and range of the function. If the function models a real-life situation, consider the context.

2. Determine the intercepts and asymptotes of the graph.

3. Locate the x-values where $f'(x)$ and $f''(x)$ are zero or undefined. Use the results to determine where the relative extrema and the points of inflection occur.

In these guidelines, note the importance of *algebra* (as well as calculus) for solving the equations $f(x) = 0$, $f'(x) = 0$, and $f''(x) = 0$.

| Example 1 | **Analyzing a Graph**

Analyze the graph of

$$f(x) = x^3 + 3x^2 - 9x + 5.$$ Original function

SOLUTION   Begin by finding the intercepts of the graph. This function factors as

$$f(x) = (x - 1)^2(x + 5).$$ Factored form

So, the x-intercepts occur when $x = 1$ and $x = -5$. The derivative is

$$f'(x) = 3x^2 + 6x - 9$$ First derivative
$$= 3(x - 1)(x + 3).$$ Factored form

So, the critical numbers of f are $x = 1$ and $x = -3$. The second derivative of f is

$$f''(x) = 6x + 6$$ Second derivative
$$= 6(x + 1)$$ Factored form

which implies that the second derivative is zero when $x = -1$. By testing the values of $f'(x)$ and $f''(x)$, as shown in the table, you can see that f has one relative minimum, one relative maximum, and one point of inflection. The graph of f is shown in Figure 9.25.

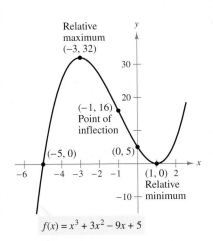

Relative maximum $(-3, 32)$

$(-1, 16)$ Point of inflection

$(-5, 0)$ $(0, 5)$

$(1, 0)$ Relative minimum

$f(x) = x^3 + 3x^2 - 9x + 5$

FIGURE 9.25

	$f(x)$	$f'(x)$	$f''(x)$	Characteristics of graph
x in $(-\infty, -3)$		$+$	$-$	Increasing, concave downward
$x = -3$	32	0	$-$	Relative maximum
x in $(-3, -1)$		$-$	$-$	Decreasing, concave downward
$x = -1$	16	$-$	0	Point of inflection
x in $(-1, 1)$		$-$	$+$	Decreasing, concave upward
$x = 1$	0	0	$+$	Relative minimum
x in $(1, \infty)$		$+$	$+$	Increasing, concave upward

✓ **CHECKPOINT 1**

Analyze the graph of $f(x) = -x^3 + 3x^2 + 9x - 27$. ■

TECHNOLOGY

Ⓣ In Example 1, you are able to find the zeros of f, f', and f'' algebraically (by factoring). When this is not feasible, you can use a graphing utility to find the zeros. For instance, the function

$$g(x) = x^3 + 3x^2 - 9x + 6$$

is similar to the function in the example, but it does not factor with integer coefficients. Using a graphing utility, you can determine that the function has only one x-intercept, $x \approx -5.0275$.

Example 2   Analyzing a Graph

Analyze the graph of

$$f(x) = x^4 - 12x^3 + 48x^2 - 64x.$$ Original function

SOLUTION   Begin by finding the intercepts of the graph. This function factors as

$$f(x) = x(x^3 - 12x^2 + 48x - 64)$$
$$= x(x - 4)^3.$$ Factored form

So, the x-intercepts occur when $x = 0$ and $x = 4$. The derivative is

$$f'(x) = 4x^3 - 36x^2 + 96x - 64$$ First derivative
$$= 4(x - 1)(x - 4)^2.$$ Factored form

So, the critical numbers of f are $x = 1$ and $x = 4$. The second derivative of f is

$$f''(x) = 12x^2 - 72x + 96$$ Second derivative
$$= 12(x - 4)(x - 2)$$ Factored form

which implies that the second derivative is zero when $x = 2$ and $x = 4$. By testing the values of $f'(x)$ and $f''(x)$, as shown in the table, you can see that f has one relative minimum and two points of inflection. The graph is shown in Figure 9.26.

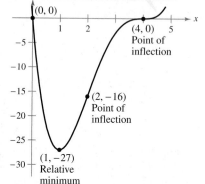

FIGURE 9.26

	$f(x)$	$f'(x)$	$f''(x)$	Characteristics of graph
x in $(-\infty, 1)$		$-$	$+$	Decreasing, concave upward
$x = 1$	-27	0	$+$	Relative minimum
x in $(1, 2)$		$+$	$+$	Increasing, concave upward
$x = 2$	-16	$+$	0	Point of inflection
x in $(2, 4)$		$+$	$-$	Increasing, concave downward
$x = 4$	0	0	0	Point of inflection
x in $(4, \infty)$		$+$	$+$	Increasing, concave upward

✓**CHECKPOINT 2**

Analyze the graph of $f(x) = x^4 - 4x^3 + 5$. ∎

DISCOVERY

A polynomial function of degree n can have at most $n - 1$ relative extrema and at most $n - 2$ points of inflection. For instance, the third-degree polynomial in Example 1 has two relative extrema and one point of inflection. Similarly, the fourth-degree polynomial function in Example 2 has one relative extremum and two points of inflection. Is it possible for a third-degree function to have no relative extrema? Is it possible for a fourth-degree function to have no relative extrema?

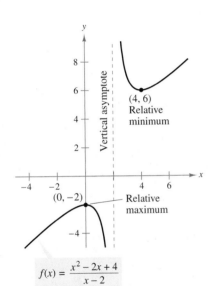

$$f(x) = \frac{x^2 - 2x + 4}{x - 2}$$

FIGURE 9.27

Example 3 Analyzing a Graph

Analyze the graph of

$$f(x) = \frac{x^2 - 2x + 4}{x - 2}.$$ Original function

SOLUTION The y-intercept occurs at $(0, -2)$. Using the Quadratic Formula on the numerator, you can see that there are no x-intercepts. Because the denominator is zero when $x = 2$ (and the numerator is not zero when $x = 2$), it follows that $x = 2$ is a vertical asymptote of the graph. There are no horizontal asymptotes because the degree of the numerator is greater than the degree of the denominator. The derivative is

$$f'(x) = \frac{(x - 2)(2x - 2) - (x^2 - 2x + 4)}{(x - 2)^2}$$ First derivative

$$= \frac{x(x - 4)}{(x - 2)^2}.$$ Factored form

So, the critical numbers of f are $x = 0$ and $x = 4$. The second derivative is

$$f''(x) = \frac{(x - 2)^2(2x - 4) - (x^2 - 4x)(2)(x - 2)}{(x - 2)^4}$$ Second derivative

$$= \frac{(x - 2)(2x^2 - 8x + 8 - 2x^2 + 8x)}{(x - 2)^4}$$

$$= \frac{8}{(x - 2)^3}.$$ Factored form

Because the second derivative has no zeros and because $x = 2$ is not in the domain of the function, you can conclude that the graph has no points of inflection. By testing the values of $f'(x)$ and $f''(x)$, as shown in the table, you can see that f has one relative minimum and one relative maximum. The graph of f is shown in Figure 9.27.

	$f(x)$	$f'(x)$	$f''(x)$	Characteristics of graph
x in $(-\infty, 0)$		+	−	Increasing, concave downward
$x = 0$	−2	0	−	Relative maximum
x in $(0, 2)$		−	−	Decreasing, concave downward
$x = 2$	Undef.	Undef.	Undef.	Vertical asymptote
x in $(2, 4)$		−	+	Decreasing, concave upward
$x = 4$	6	0	+	Relative minimum
x in $(4, \infty)$		+	+	Increasing, concave upward

✓ CHECKPOINT 3

Analyze the graph of $f(x) = \dfrac{x^2}{x - 1}$. ∎

Example 4 **Analyzing a Graph**

Analyze the graph of

$$f(x) = \frac{2(x^2 - 9)}{x^2 - 4}.$$ Original function

SOLUTION Begin by writing the function in factored form.

$$f(x) = \frac{2(x - 3)(x + 3)}{(x - 2)(x + 2)}$$ Factored form

The y-intercept is $\left(0, \frac{9}{2}\right)$, and the x-intercepts are $(-3, 0)$ and $(3, 0)$. The graph of f has vertical asymptotes at $x = \pm 2$ and a horizontal asymptote at $y = 2$. The first derivative is

$$f'(x) = \frac{2[(x^2 - 4)(2x) - (x^2 - 9)(2x)]}{(x^2 - 4)^2}$$ First derivative

$$= \frac{20x}{(x^2 - 4)^2}.$$ Factored form

So, the critical number of f is $x = 0$. The second derivative of f is

$$f''(x) = \frac{(x^2 - 4)^2(20) - (20x)(2)(2x)(x^2 - 4)}{(x^2 - 4)^4}$$ Second derivative

$$= \frac{20(x^2 - 4)(x^2 - 4 - 4x^2)}{(x^2 - 4)^4}$$

$$= -\frac{20(3x^2 + 4)}{(x^2 - 4)^3}.$$ Factored form

Because the second derivative has no zeros and $x = \pm 2$ are not in the domain of the function, you can conclude that the graph has no points of inflection. By testing the values of $f'(x)$ and $f''(x)$, as shown in the table, you can see that f has one relative minimum. The graph of f is shown in Figure 9.28.

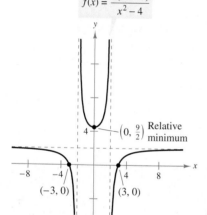

$f(x) = \dfrac{2(x^2 - 9)}{x^2 - 4}$

$\left(0, \frac{9}{2}\right)$ Relative minimum

$(-3, 0)$ $(3, 0)$

FIGURE 9.28

	$f(x)$	$f'(x)$	$f''(x)$	Characteristics of graph
x in $(-\infty, -2)$		$-$	$-$	Decreasing, concave downward
$x = -2$	Undef.	Undef.	Undef.	Vertical asymptote
x in $(-2, 0)$		$-$	$+$	Decreasing, concave upward
$x = 0$	$\frac{9}{2}$	0	$+$	Relative minimum
x in $(0, 2)$		$+$	$+$	Increasing, concave upward
$x = 2$	Undef.	Undef.	Undef.	Vertical asymptote
x in $(2, \infty)$		$+$	$-$	Increasing, concave downward

✓**CHECKPOINT 4**

Analyze the graph of $f(x) = \dfrac{x^2 + 1}{x^2 - 1}$. ∎

Example 5   **Analyzing a Graph**

Analyze the graph of

$$f(x) = 2x^{5/3} - 5x^{4/3}.$$ Original function

SOLUTION   Begin by writing the function in factored form.

$$f(x) = x^{4/3}(2x^{1/3} - 5)$$ Factored form

One of the intercepts is $(0, 0)$. A second x-intercept occurs when $2x^{1/3} = 5$.

$$2x^{1/3} = 5$$

$$x^{1/3} = \tfrac{5}{2}$$

$$x = \left(\tfrac{5}{2}\right)^3$$

$$x = \tfrac{125}{8}$$

The first derivative is

$$f'(x) = \tfrac{10}{3}x^{2/3} - \tfrac{20}{3}x^{1/3}$$ First derivative

$$= \tfrac{10}{3}x^{1/3}(x^{1/3} - 2).$$ Factored form

So, the critical numbers of f are $x = 0$ and $x = 8$. The second derivative is

$$f''(x) = \tfrac{20}{9}x^{-1/3} - \tfrac{20}{9}x^{-2/3}$$ Second derivative

$$= \tfrac{20}{9}x^{-2/3}(x^{1/3} - 1)$$

$$= \frac{20(x^{1/3} - 1)}{9x^{2/3}}.$$ Factored form

So, possible points of inflection occur when $x = 1$ and when $x = 0$. By testing the values of $f'(x)$ and $f''(x)$, as shown in the table, you can see that f has one relative maximum, one relative minimum, and one point of inflection. The graph of f is shown in Figure 9.29.

Algebra Review

For help on the algebra in Example 5, see Example 2(a) in the *Chapter 9 Algebra Review*, on page 738.

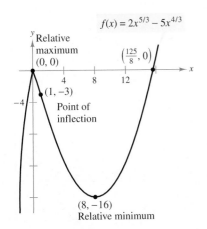

$$f(x) = 2x^{5/3} - 5x^{4/3}$$

Relative maximum $(0, 0)$

$\left(\tfrac{125}{8}, 0\right)$

$(1, -3)$

Point of inflection

$(8, -16)$ Relative minimum

FIGURE 9.29

	$f(x)$	$f'(x)$	$f''(x)$	Characteristics of graph
x in $(-\infty, 0)$		$+$	$-$	Increasing, concave downward
$x = 0$	0	0	Undef.	Relative maximum
x in $(0, 1)$		$-$	$-$	Decreasing, concave downward
$x = 1$	-3	$-$	0	Point of inflection
x in $(1, 8)$		$-$	$+$	Decreasing, concave upward
$x = 8$	-16	0	$+$	Relative minimum
x in $(8, \infty)$		$+$	$+$	Increasing, concave upward

✓ **CHECKPOINT 5**

Analyze the graph of

$$f(x) = 2x^{3/2} - 6x^{1/2}. \ \blacksquare$$

Summary of Simple Polynomial Graphs

A summary of the graphs of polynomial functions of degrees 0, 1, 2, and 3 is shown in Figure 9.30. Because of their simplicity, lower-degree polynomial functions are commonly used as mathematical models.

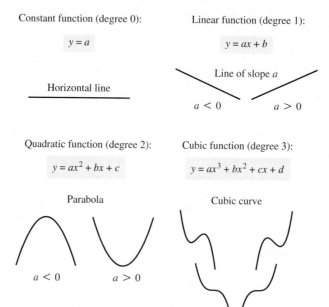

Constant function (degree 0):

$$y = a$$

Horizontal line

Linear function (degree 1):

$$y = ax + b$$

Line of slope a

$a < 0$ $a > 0$

Quadratic function (degree 2):

$$y = ax^2 + bx + c$$

Parabola

$a < 0$ $a > 0$

Cubic function (degree 3):

$$y = ax^3 + bx^2 + cx + d$$

Cubic curve

$a < 0$ $a > 0$

FIGURE 9.30

STUDY TIP

The graph of any cubic polynomial has one point of inflection. The slope of the graph at the point of inflection may be zero or nonzero.

CONCEPT CHECK

1. A fourth-degree polynomial can have at most how many relative extrema?

2. A fourth-degree polynomial can have at most how many points of inflection?

3. Complete the following: A polynomial function of degree n can have at most _____ relative extrema.

4. Complete the following: A polynomial function of degree n can have at most _____ points of inflection.

Skills Review 9.4

The following warm-up exercises involve skills that were covered in earlier sections. You will use these skills in the exercise set for this section. For additional help, review Sections 8.4 and 9.3.

In Exercises 1–4, find the vertical and horizontal asymptotes of the graph.

1. $f(x) = \dfrac{1}{x^2}$

2. $f(x) = \dfrac{8}{(x-2)^2}$

3. $f(x) = \dfrac{40x}{x+3}$

4. $f(x) = \dfrac{x^2 - 3}{x^2 - 4x + 3}$

In Exercises 5–10, determine the open intervals on which the function is increasing or decreasing.

5. $f(x) = x^2 + 4x + 2$

6. $f(x) = -x^2 - 8x + 1$

7. $f(x) = x^3 - 3x + 1$

8. $f(x) = \dfrac{-x^3 + x^2 - 1}{x^2}$

9. $f(x) = \dfrac{x-2}{x-1}$

10. $f(x) = -x^3 - 4x^2 + 3x + 2$

Exercises 9.4

See www.CalcChat.com for worked-out solutions to odd-numbered exercises.

In Exercises 1–22, sketch the graph of the function. Choose a scale that allows all relative extrema and points of inflection to be identified on the graph.

1. $y = -x^2 - 2x + 3$

2. $y = 2x^2 - 4x + 1$

3. $y = x^3 - 4x^2 + 6$

4. $y = -x^3 + x - 2$

5. $y = 2 - x - x^3$

6. $y = x^3 + 3x^2 + 3x + 2$

7. $y = 3x^3 - 9x + 1$

8. $y = -4x^3 + 6x^2$

9. $y = 3x^4 + 4x^3$

10. $y = x^4 - 2x^2$

11. $y = x^3 - 6x^2 + 3x + 10$

12. $y = -x^3 + 3x^2 + 9x - 2$

13. $y = x^4 - 8x^3 + 18x^2 - 16x + 5$

14. $y = x^4 - 4x^3 + 16x - 16$

15. $y = x^4 - 4x^3 + 16x$

16. $y = x^5 + 1$

17. $y = x^5 - 5x$

18. $y = (x-1)^5$

19. $y = \dfrac{x^2 + 1}{x}$

20. $y = \dfrac{x+2}{x}$

21. $y = \begin{cases} x^2 + 1, & x \le 0 \\ 1 - 2x, & x > 0 \end{cases}$

22. $y = \begin{cases} x^2 + 4, & x < 0 \\ 4 - x, & x \ge 0 \end{cases}$

(T) In Exercises 23–34, use a graphing utility to graph the function. Choose a window that allows all relative extrema and points of inflection to be identified on the graph.

23. $y = \dfrac{x^2}{x^2 + 3}$

24. $y = \dfrac{x}{x^2 + 1}$

25. $y = 3x^{2/3} - 2x$

26. $y = 3x^{2/3} - x^2$

27. $y = 1 - x^{2/3}$

28. $y = (1 - x)^{2/3}$

29. $y = x^{1/3} + 1$

30. $y = x^{-1/3}$

31. $y = x^{5/3} - 5x^{2/3}$

32. $y = x^{4/3}$

33. $y = x\sqrt{x^2 - 9}$

34. $y = \dfrac{x}{\sqrt{x^2 - 4}}$

In Exercises 35–44, sketch the graph of the function. Label the intercepts, relative extrema, points of inflection, and asymptotes. Then state the domain of the function.

35. $y = \dfrac{5 - 3x}{x - 2}$

36. $y = \dfrac{x^2 + 1}{x^2 - 9}$

37. $y = \dfrac{2x}{x^2 - 1}$

38. $y = \dfrac{x^2 - 6x + 12}{x - 4}$

39. $y = x\sqrt{4 - x}$

40. $y = x\sqrt{4 - x^2}$

41. $y = \dfrac{x - 3}{x}$

42. $y = x + \dfrac{32}{x^2}$

43. $y = \dfrac{x^3}{x^3 - 1}$

44. $y = \dfrac{x^4}{x^4 - 1}$

In Exercises 45 and 46, find values of a, b, c, and d such that the graph of $f(x) = ax^3 + bx^2 + cx + d$ will resemble the given graph. Then use a graphing utility to verify your result. (There are many correct answers.)

45.

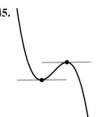

46.

In Exercises 47–50, use the graph of f' or f'' to sketch the graph of f. (There are many correct answers.)

47.

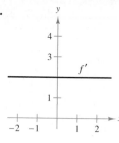

48.

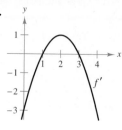

49.

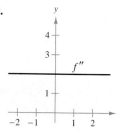

50.

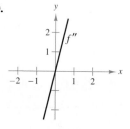

In Exercises 51 and 52, sketch a graph of a function f having the given characteristics. (There are many correct answers.)

51. $f(-2) = f(0) = 0$

$f'(x) > 0$ if $x < -1$

$f'(x) < 0$ if $-1 < x < 0$

$f'(x) > 0$ if $x > 0$

$f'(-1) = f'(0) = 0$

52. $f(-1) = f(3) = 0$

$f'(1)$ is undefined.

$f'(x) < 0$ if $x < 1$

$f'(x) > 0$ if $x > 1$

$f''(x) < 0, x \neq 1$

$\lim_{x \to \infty} f(x) = 4$

In Exercises 53 and 54, create a function whose graph has the given characteristics. (There are many correct answers.)

53. Vertical asymptote: $x = 5$

Horizontal asymptote: $y = 0$

54. Vertical asymptote: $x = -3$

Horizontal asymptote: None

55. *MAKE A DECISION: SOCIAL SECURITY* The table lists the average monthly Social Security benefits B (in dollars) for retired workers aged 62 and over from 1998 through 2005. A model for the data is

$$B = \frac{582.6 + 38.38t}{1 + 0.025t - 0.0009t^2}, \quad 8 \leq t \leq 15$$

where $t = 8$ corresponds to 1998. *(Source: U.S. Social Security Administration)*

t	8	9	10	11	12	13	14	15
B	780	804	844	874	895	922	955	1002

(T) (a) Use a graphing utility to create a scatter plot of the data and graph the model in the same viewing window. How well does the model fit the data?

(b) Use the model to predict the average monthly benefit in 2008.

(c) Should this model be used to predict the average monthly Social Security benefits in future years? Why or why not?

56. Cost An employee of a delivery company earns $10 per hour driving a delivery van in an area where gasoline costs $2.80 per gallon. When the van is driven at a constant speed s (in miles per hour, with $40 \leq s \leq 65$), the van gets $700/s$ miles per gallon.

(a) Find the cost C as a function of s for a 100-mile trip on an interstate highway.

(T) (b) Use a graphing utility to graph the function found in part (a) and determine the most economical speed.

57. *MAKE A DECISION: PROFIT* The management of a company is considering three possible models for predicting the company's profits from 2003 through 2008. Model I gives the expected annual profits if the current trends continue. Models II and III give the expected annual profits for various combinations of increased labor and energy costs. In each model, p is the profit (in billions of dollars) and $t = 0$ corresponds to 2003.

Model I: $p = 0.03t^2 - 0.01t + 3.39$

Model II: $p = 0.08t + 3.36$

Model III: $p = -0.07t^2 + 0.05t + 3.38$

(T) (a) Use a graphing utility to graph all three models in the same viewing window.

(b) For which models are profits increasing during the interval from 2003 through 2008?

(c) Which model is the most optimistic? Which is the most pessimistic? Which model would you choose? Explain.

(T) **58. Meteorology** The monthly normal temperature T (in degrees Fahrenheit) for Pittsburgh, Pennsylvania can be modeled by

$$T = \frac{22.329 - 0.7t + 0.029t^2}{1 - 0.203t + 0.014t^2}, \quad 1 \leq t \leq 12$$

where t is the month, with $t = 1$ corresponding to January. Use a graphing utility to graph the model and find all absolute extrema. Interpret the meaning of these values in the context of the problem. *(Source: National Climatic Data Center)*

(T) **Writing** In Exercises 59 and 60, use a graphing utility to graph the function. Explain why there is no vertical asymptote when a superficial examination of the function may indicate that there should be one.

59. $h(x) = \dfrac{6 - 2x}{3 - x}$

60. $g(x) = \dfrac{x^2 + x - 2}{x - 1}$

Differentials and Marginal Analysis

- Find the differentials of functions.
- Use differentials to approximate changes in functions.
- Use differentials to approximate changes in real-life models.

Differentials

When the derivative was defined in Section 7.3 as the limit of the ratio $\Delta y/\Delta x$, it seemed natural to retain the quotient symbolism for the limit itself. So, the derivative of y with respect to x was denoted by

$$\frac{dy}{dx} = \lim_{\Delta x \to 0} \frac{\Delta y}{\Delta x}$$

even though we did not interpret dy/dx as the quotient of two separate quantities. In this section, you will see that the quantities dy and dx can be assigned meanings in such a way that their quotient, when $dx \neq 0$, is equal to the derivative of y with respect to x.

STUDY TIP

In this definition, dx can have any nonzero value. In most applications, however, dx is chosen to be small and this choice is denoted by $dx = \Delta x$.

Definition of Differentials

Let $y = f(x)$ represent a differentiable function. The **differential of x** (denoted by dx) is any nonzero real number. The **differential of y** (denoted by dy) is $dy = f'(x)\, dx$.

One use of differentials is in approximating the change in $f(x)$ that corresponds to a change in x, as shown in Figure 9.31. This change is denoted by

$$\Delta y = f(x + \Delta x) - f(x). \qquad \text{Change in } y$$

In Figure 9.31, notice that as Δx gets smaller and smaller, the values of dy and Δy get closer and closer. That is, when Δx is small, $dy \approx \Delta y$.

STUDY TIP

Note in Figure 9.31 that near the point of tangency, the graph of f is very close to the tangent line. This is the essence of the approximations used in this section. In other words, near the point of tangency, $dy \approx \Delta y$.

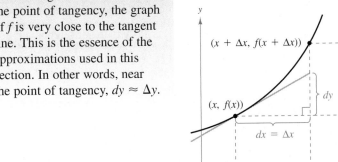

FIGURE 9.31

This **tangent line approximation** is the basis for most applications of differentials.

Example 1 **Interpreting Differentials Graphically**

Consider the function given by

$$f(x) = x^2. \qquad \text{Original function}$$

Find the value of dy when $x = 1$ and $dx = 0.01$. Compare this with the value of Δy when $x = 1$ and $\Delta x = 0.01$. Interpret the results graphically.

SOLUTION   Begin by finding the derivative of f.

$$f'(x) = 2x \qquad \text{Derivative of } f$$

When $x = 1$ and $dx = 0.01$, the value of the differential dy is

$$
\begin{aligned}
dy &= f'(x)\,dx & &\text{Differential of } y \\
&= f'(1)(0.01) & &\text{Substitute 1 for } x \text{ and 0.01 for } dx. \\
&= 2(1)(0.01) & &\text{Use } f'(x) = 2x. \\
&= 0.02. & &\text{Simplify.}
\end{aligned}
$$

When $x = 1$ and $\Delta x = 0.01$, the value of Δy is

$$
\begin{aligned}
\Delta y &= f(x + \Delta x) - f(x) & &\text{Change in } y \\
&= f(1.01) - f(1) & &\text{Substitute 1 for } x \text{ and 0.01 for } \Delta x. \\
&= (1.01)^2 - (1)^2 \\
&= 0.0201. & &\text{Simplify.}
\end{aligned}
$$

Note that $dy \approx \Delta y$, as shown in Figure 9.32.

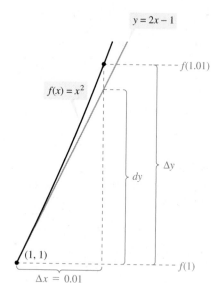

$y = 2x - 1$

$f(1.01)$

$f(x) = x^2$

Δy

dy

$(1, 1)$

$f(1)$

$\Delta x = 0.01$

FIGURE 9.32

✓ **CHECKPOINT 1**

Find the value of dy when $x = 2$ and $dx = 0.01$ for $f(x) = x^4$. Compare this with the value of Δy when $x = 2$ and $\Delta x = 0.01$. ▨

STUDY TIP

Find an equation of the tangent line $y = g(x)$ to the graph of $f(x) = x^2$ at the point $x = 1$. Evaluate $g(1.01)$ and $f(1.01)$.

The validity of the approximation

$$dy \approx \Delta y, \quad dx \neq 0$$

stems from the definition of the derivative. That is, the existence of the limit

$$f'(x) = \lim_{\Delta x \to 0} \frac{f(x + \Delta x) - f(x)}{\Delta x}$$

implies that when Δx is close to zero, then $f'(x)$ is close to the difference quotient. So, you can write

$$\frac{f(x + \Delta x) - f(x)}{\Delta x} \approx f'(x)$$

$$f(x + \Delta x) - f(x) \approx f'(x)\,\Delta x$$

$$\Delta y \approx f'(x)\,\Delta x.$$

Substituting dx for Δx and dy for $f'(x)\,dx$ produces

$$\Delta y \approx dy.$$

Marginal Analysis

Differentials are used in economics to approximate changes in revenue, cost, and profit. Suppose that $R = f(x)$ is the total revenue for selling x units of a product. When the number of units increases by 1, the change in x is $\Delta x = 1$, and the change in R is

$$\Delta R = f(x + \Delta x) - f(x) \approx dR = \frac{dR}{dx} \, dx.$$

In other words, you can use the differential dR to approximate the change in the revenue that accompanies the sale of one additional unit. Similarly, the differentials dC and dP can be used to approximate the changes in cost and profit that accompany the sale (or production) of one additional unit.

Example 2   **Using Marginal Analysis**

The demand function for a product is modeled by

$$p = 400 - x, \quad 0 \le x \le 400.$$

Use differentials to approximate the change in revenue as sales increase from 149 units to 150 units. Compare this with the actual change in revenue.

SOLUTION   Begin by finding the marginal revenue, dR/dx.

$R = xp$	Formula for revenue
$\quad = x(400 - x)$	Use $p = 400 - x$
$\quad = 400x - x^2$	Multiply.
$\dfrac{dR}{dx} = 400 - 2x$	Power Rule

When $x = 149$ and $dx = \Delta x = 1$, you can approximate the change in the revenue to be

$$[400 - 2(149)](1) = \$102.$$

When x increases from 149 to 150, the actual change in revenue is

$$\Delta R = \left[400(150) - 150^2\right] - \left[400(149) - 149^2\right]$$
$$= 37{,}500 - 37{,}399$$
$$= \$101$$

✓ CHECKPOINT 2

The demand function for a product is modeled by

$$p = 200 - x, \quad 0 \le x \le 200.$$

Use differentials to approximate the change in revenue as sales increase from 89 to 90 units. Compare this with the actual change in revenue. ▪

Example 3
MAKE A DECISION   Using Marginal Analysis

The profit derived from selling x units of an item is modeled by

$$P = 0.0002x^3 + 10x.$$

Use the differential dP to approximate the change in profit when the production level changes from 50 to 51 units. Compare this with the actual gain in profit obtained by increasing the production level from 50 to 51 units. Will the gain in profit exceed $11?

SOLUTION   The marginal profit is

$$\frac{dP}{dx} = 0.0006x^2 + 10.$$

When $x = 50$ and $dx = 1$, the differential is

$$[0.0006(50)^2 + 10](1) = \$11.50.$$

When x changes from 50 to 51 units, the actual change in profit is

$$\Delta P = [(0.0002)(51)^3 + 10(51)] - [(0.0002)(50)^3 + 10(50)]$$
$$\approx 536.53 - 525.00$$
$$= \$11.53.$$

These values are shown graphically in Figure 9.33. Note that the gain in profit will exceed $11.

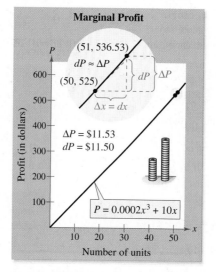

FIGURE 9.33

✓ **CHECKPOINT 3**

Use the differential dP to approximate the change in profit for the profit function in Example 3 when the production level changes from 40 to 41 units. Compare this with the actual gain in profit obtained by increasing the production level from 40 to 41 units. ■

Formulas for Differentials

You can use the definition of differentials to rewrite each differentiation rule in **differential form.** For example, if u and v are differentiable functions of x, then $du = (du/dx)\, dx$ and $dv = (dv/dx)\, dx$, which implies that you can write the Product Rule in the following differential form.

$$d[uv] = \frac{d}{dx}[uv]\, dx \qquad\qquad \text{Differential of } uv$$

$$= \left[u\frac{dv}{dx} + v\frac{du}{dx} \right] dx \qquad \text{Product Rule}$$

$$= u\frac{dv}{dx}\, dx + v\frac{du}{dx}\, dx$$

$$= u\, dv + v\, du \qquad\qquad \text{Differential form of Product Rule}$$

The following summary gives the differential forms of the differentiation rules presented so far in the text.

Differential Forms of Differentiation Rules

Constant Multiple Rule:	$d[cu] = c\, du$
Sum or Difference Rule:	$d[u \pm v] = du \pm dv$
Product Rule:	$d[uv] = u\, dv + v\, du$
Quotient Rule:	$d\left[\dfrac{u}{v}\right] = \dfrac{v\, du - u\, dv}{v^2}$
Constant Rule:	$d[c] = 0$
Power Rule:	$d[x^n] = nx^{n-1}\, dx$

The next example compares the derivatives and differentials of several simple functions.

Example 4 Finding Differentials

Find the differential dy of each function.

Function	*Derivative*	*Differential*
a. $y = x^2$	$\dfrac{dy}{dx} = 2x$	$dy = 2x\, dx$
b. $y = \dfrac{3x + 2}{5}$	$\dfrac{dy}{dx} = \dfrac{3}{5}$	$dy = \dfrac{3}{5}\, dx$
c. $y = 2x^2 - 3x$	$\dfrac{dy}{dx} = 4x - 3$	$dy = (4x - 3)\, dx$
d. $y = \dfrac{1}{x}$	$\dfrac{dy}{dx} = -\dfrac{1}{x^2}$	$dy = -\dfrac{1}{x^2}\, dx$

✓ **CHECKPOINT 4**

Find the differential dy of each function.

a. $y = 4x^3$

b. $y = \dfrac{2x + 1}{3}$

c. $y = 3x^2 - 2x$

d. $y = \dfrac{1}{x^2}$

Error Propagation

A common use of differentials is the estimation of errors that result from inaccuracies of physical measuring devices. This is shown in Example 5.

Example 5 Estimating Measurement Errors

The radius of a ball bearing is measured to be 0.7 inch, as shown in Figure 9.34. This implies that the volume of the ball bearing is $\frac{4}{3}\pi(0.7)^3 \approx 1.4368$ cubic inches. You are told that the measurement of the radius is correct to within 0.01 inch. How far off could the calculation of the volume be?

SOLUTION Because the value of r can be off by 0.01 inch, it follows that

$$-0.01 \le \Delta r \le 0.01. \qquad \text{Possible error in measuring}$$

Using $\Delta r = dr$, you can estimate the possible error in the volume.

$$V = \tfrac{4}{3}\pi r^3 \qquad \text{Formula for volume}$$

$$dV = \frac{dV}{dr}\,dr = 4\pi r^2\,dr \qquad \text{Formula for differential of } V$$

The possible error in the volume is

$$4\pi r^2\,dr = 4\pi(0.7)^2(\pm 0.01) \qquad \text{Substitute for } r \text{ and } dr.$$

$$\approx \pm 0.0616 \text{ cubic inch.} \qquad \text{Possible error}$$

So, the volume of the ball bearing could range between

$$(1.4368 - 0.0616) = 1.3752 \text{ cubic inches}$$

and

$$(1.4368 + 0.0616) = 1.4984 \text{ cubic inches.}$$

In Example 5, the **relative error** in the volume is defined to be the ratio of dV to V. This ratio is

$$\frac{dV}{V} \approx \frac{\pm 0.0616}{1.4368} \approx \pm 0.0429.$$

This corresponds to a **percentage error** of 4.29%.

FIGURE 9.34

✓ **CHECKPOINT 5**

Find the surface area of the ball bearing in Example 5. How far off could your calculation of the surface area be? The surface area of a sphere is given by $S = 4\pi r^2$. ■

┌─ **CONCEPT CHECK** ─────────────────────────────

1. Given a differentiable function $y = f(x)$, what is the differential of x?
2. Given a differentiable function $y = f(x)$, write an expression for the differential of y.
3. Write the differential form of the Quotient Rule.
4. When using differentials, what is meant by the terms *relative error* and *percentage error*?

Skills Review 9.5   The following warm-up exercises involve skills that were covered in earlier sections. You will use these skills in the exercise set for this section. For additional help, review Sections 7.4 and 7.6.

In Exercises 1–12, find the derivative.

1. $C = 44 + 0.09x^2$

2. $C = 250 + 0.15x$

3. $R = x(1.25 + 0.02\sqrt{x})$

4. $R = x(15.5 - 1.55x)$

5. $P = -0.03x^{1/3} + 1.4x - 2250$

6. $P = -0.02x^2 + 25x - 1000$

7. $A = \frac{1}{4}\sqrt{3}x^2$

8. $A = 6x^2$

9. $C = 2\pi r$

10. $P = 4w$

11. $S = 4\pi r^2$

12. $P = 2x + \sqrt{2}x$

In Exercises 13–16, write a formula for the quantity.

13. Area A of a circle of radius r

14. Area A of a square of side x

15. Volume V of a cube of edge x

16. Volume V of a sphere of radius r

Exercises 9.5

See www.CalcChat.com for worked-out solutions to odd-numbered exercises.

In Exercises 1–6, find the differential dy.

1. $y = 3x^2 - 4$

2. $y = 3x^{2/3}$

3. $y = (4x - 1)^3$

4. $y = \dfrac{x + 1}{2x - 1}$

5. $y = \sqrt{9 - x^2}$

6. $y = \sqrt[3]{6x^2}$

In Exercises 7–10, let $x = 1$ and $\Delta x = 0.01$. Find Δy.

7. $f(x) = 5x^2 - 1$

8. $f(x) = \sqrt{3x}$

9. $f(x) = \dfrac{4}{\sqrt[3]{x}}$

10. $f(x) = \dfrac{x}{x^2 + 1}$

In Exercises 11–14, compare the values of dy and Δy.

11. $y = 0.5x^3$ $x = 2$ $\Delta x = dx = 0.1$

12. $y = 1 - 2x^2$ $x = 0$ $\Delta x = dx = -0.1$

13. $y = x^4 + 1$ $x = -1$ $\Delta x = dx = 0.01$

14. $y = 2x + 1$ $x = 2$ $\Delta x = dx = 0.01$

In Exercises 15–20, let $x = 2$ and complete the table for the function.

$dx = \Delta x$	dy	Δy	$\Delta y - dy$	$dy/\Delta y$
1.000				
0.500				
0.100				
0.010				
0.001				

15. $y = x^2$

16. $y = x^5$

17. $y = \dfrac{1}{x^2}$

18. $y = \dfrac{1}{x}$

19. $y = \sqrt[4]{x}$

20. $y = \sqrt{x}$

In Exercises 21–24, find an equation of the tangent line to the function at the given point. Then find the function values and the tangent line values at $f(x + \Delta x)$ and $y(x + \Delta x)$ for $\Delta x = -0.01$ and 0.01.

Function	*Point*
21. $f(x) = 2x^3 - x^2 + 1$	$(-2, -19)$
22. $f(x) = 3x^2 - 1$	$(2, 11)$
23. $f(x) = \dfrac{x}{x^2 + 1}$	$(0, 0)$
24. $f(x) = \sqrt{25 - x^2}$	$(3, 4)$

25. Profit   The profit P for a company producing x units is

$$P = (500x - x^2) - \left(\frac{1}{2}x^2 - 77x + 3000\right).$$

Approximate the change and percent change in profit as production changes from $x = 115$ to $x = 120$ units.

26. Revenue   The revenue R for a company selling x units is

$$R = 900x - 0.1x^2.$$

Use differentials to approximate the change in revenue if sales increase from $x = 3000$ to $x = 3100$ units.

27. Demand The demand function for a product is modeled by

$p = 75 - 0.25x$.

(a) If x changes from 7 to 8, what is the corresponding change in p? Compare the values of Δp and dp.

(b) Repeat part (a) when x changes from 70 to 71 units.

28. Biology: Wildlife Management A state game commission introduces 50 deer into newly acquired state game lands. The population N of the herd can be modeled by

$$N = \frac{10(5 + 3t)}{1 + 0.04t}$$

where t is the time in years. Use differentials to approximate the change in the herd size from $t = 5$ to $t = 6$.

Ⓣ **Marginal Analysis** In Exercises 29–34, use differentials to approximate the change in cost, revenue, or profit corresponding to an increase in sales of one unit. For instance, in Exercise 29, approximate the change in cost as x increases from 12 units to 13 units. Then use a graphing utility to graph the function, and use the *trace* feature to verify your result.

Function	x-Value
29. $C = 0.05x^2 + 4x + 10$	$x = 12$
30. $C = 0.025x^2 + 8x + 5$	$x = 10$
31. $R = 30x - 0.15x^2$	$x = 75$
32. $R = 50x - 1.5x^2$	$x = 15$
33. $P = -0.5x^3 + 2500x - 6000$	$x = 50$
34. $P = -x^2 + 60x - 100$	$x = 25$

35. Marginal Analysis A retailer has determined that the monthly sales x of a watch are 150 units when the price is $50, but decrease to 120 units when the price is $60. Assume that the demand is a linear function of the price. Find the revenue R as a function of x and approximate the change in revenue for a one-unit increase in sales when $x = 141$. Make a sketch showing dR and ΔR.

36. Marginal Analysis A manufacturer determines that the demand x for a product is inversely proportional to the square of the price p. When the price is $10, the demand is 2500. Find the revenue R as a function of x and approximate the change in revenue for a one-unit increase in sales when $x = 3000$. Make a sketch showing dR and ΔR.

37. Marginal Analysis The demand x for a web camera is 30,000 units per month when the price is $25 and 40,000 units when the price is $20. The initial investment is $275,000 and the cost per unit is $17. Assume that the demand is a linear function of the price. Find the profit P as a function of x and approximate the change in profit for a one-unit increase in sales when $x = 28,000$. Make a sketch showing dP and ΔP.

38. Marginal Analysis The variable cost for the production of a calculator is $14.25 and the initial investment is $110,000. Find the total cost C as a function of x, the number of units produced. Then use differentials to approximate the change in the cost for a one-unit increase in production when $x = 50,000$. Make a sketch showing dC and ΔC. Explain why $dC = \Delta C$ in this problem.

39. Area The side of a square is measured to be 12 inches, with a possible error of $\frac{1}{64}$ inch. Use differentials to approximate the possible error and the relative error in computing the area of the square.

40. Volume The radius of a sphere is measured to be 6 inches, with a possible error of 0.02 inch. Use differentials to approximate the possible error and the relative error in computing the volume of the sphere.

41. Economics: Gross Domestic Product The gross domestic product (GDP) of the United States for 2001 through 2005 is modeled by

$G = 0.0026x^2 - 7.246x + 14,597.85$

where G is the GDP (in billions of dollars) and x is the capital outlay (in billions of dollars). Use differentials to approximate the change in the GDP when the capital outlays change from $2100 billion to $2300 billion. (*Source: U.S. Bureau of Economic Analysis, U.S. Office of Management and Budget*)

42. Medical Science The concentration C (in milligrams per milliliter) of a drug in a patient's bloodstream t hours after injection into muscle tissue is modeled by

$$C = \frac{3t}{27 + t^3}.$$

Use differentials to approximate the change in the concentration when t changes from $t = 1$ to $t = 1.5$.

43. Physiology: Body Surface Area The body surface area (BSA) of a 180-centimeter-tall (about six-feet-tall) person is modeled by

$B = 0.1\sqrt{5w}$

where B is the BSA (in square meters) and w is the weight (in kilograms). Use differentials to approximate the change in the person's BSA when the person's weight changes from 90 kilograms to 95 kilograms.

True or False? In Exercises 44 and 45, determine whether the statement is true or false. If it is false, explain why or give an example that shows it is false.

44. If $y = x + c$, then $dy = dx$.

45. If $y = ax + b$, then $\Delta y/\Delta x = dy/dx$.

Algebra Review

Solving Equations

Example 1 on page 679 illustrates some of the basic techniques for solving equations. Example 2 on page 680 illustrates some of the more complicated techniques. In the examples that follow, you can further review some of the more complicated techniques for solving equations. Note in Example 2(c) that with an *absolute value* equation, the definition of absolute value is used to rewrite the equation as two equations.

Remember that when solving an equation, your basic goal is to isolate the variable on one side of the equation. To do this, you use inverse operations. For instance, to get rid of the *subtract* 2 in

$$x - 2 = 0$$

you *add* 2 to each side of the equation. Similarly, to get rid of the *square root* in

$$\sqrt{x + 3} = 2$$

you *square* both sides of the equation.

Example 1 Solving an Equation

Solve each equation.

a. $0 = 2 - \dfrac{288}{x^2}$ **b.** $0 = 2x(2x^2 - 3)$ **c.** $V' = 0$, where $V = 27x - \dfrac{1}{4}x^2$

SOLUTION

a. $0 = 2 - \dfrac{288}{x^2}$ Example 2, page 691

$-2 = -\dfrac{288}{x^2}$ Subtract 2 from each side.

$1 = \dfrac{144}{x^2}$ Divide each side by -2.

$x^2 = 144$ Multiply each side by x^2.

$x = \pm 12$ Take the square root of each side.

b. $0 = 2x(2x^2 - 3)$ Example 3, page 692

$2x = 0 \implies x = 0$ Set first factor equal to zero.

$2x^2 - 3 = 0 \implies x = \pm\sqrt{\tfrac{3}{2}}$ Set second factor equal to zero.

c. $V = 27x - \dfrac{1}{4}x^3$ Example 1, page 689

$27 - \dfrac{3}{4}x^2 = 0$ Find derivative and set equal to zero.

$27 = \dfrac{3}{4}x^2$ Add $\tfrac{3}{4}x^2$ to each side.

$36 = x^2$ Divide each side by $\tfrac{3}{4}$.

$\pm 6 = x$ Take the square root of each side.

Example 2 Solving an Equation

Solve each equation.

a. $\dfrac{20(x^{1/3} - 1)}{9x^{2/3}} = 0$

Example 5, page 725

$20(x^{1/3} - 1) = 0$

A fraction is zero only if its numerator is zero.

$x^{1/3} - 1 = 0$

Divide each side by 20.

$x^{1/3} = 1$

Add 1 to each side.

$x = 1$

Cube each side.

b. $\dfrac{25}{\sqrt{x}} - 0.5 = 0$

Example 4, page 701

$\dfrac{25}{\sqrt{x}} = 0.5$

Add 0.5 to each side.

$25 = 0.5\sqrt{x}$

Multiply each side by

$50 = \sqrt{x}$

Divide each side by 0.5.

$2500 = x$

Square both sides.

c. $\left| -\dfrac{24\sqrt{x}}{x} + 2 \right| = 1$

Example 5, page 703

First Equation

$-\dfrac{24\sqrt{x}}{x} + 2 = 1$

Use positive expression.

$-\dfrac{24\sqrt{x}}{x} = -1$

Subtract 2 from each side.

$24\sqrt{x} = x$

Multiply each side by $-x$.

$576x = x^2$

Square both sides.

$0 = x(x - 576)$

Subtract 576x from both sides and factor.

$x = 0$

Set first factor equal to zero (extraneous solution).

$x = 576$

Set second factor equal to zero.

Second Equation

$-\left(-\dfrac{24\sqrt{x}}{x} + 2 \right) = 1$

Use negative expression.

$\dfrac{24\sqrt{x}}{x} - 2 = 1$

Rewrite without parentheses.

$\dfrac{24\sqrt{x}}{x} = 3$

Add 2 to each side.

$24\sqrt{x} = 3x$

Multiply each side by x.

$576x = 9x^2$

Square both sides.

$0 = 9x(x - 64)$

Subtract 576x from both sides and factor.

$x = 0$

Set first factor equal to zero (extraneous solution).

$x = 64$

Set second factor equal to zero.

Chapter Summary and Study Strategies

After studying this chapter, you should have acquired the following skills.
The exercise numbers are keyed to the Review Exercises that begin on page 740.
Answers to odd-numbered Review Exercises are given in the back of the text.

Section 9.1	Review Exercises
■ Solve real-life optimization problems.	*1–8*

Section 9.2	
■ Solve business and economics optimization problems.	*9–14*
■ Find the price elasticity of demand for a demand function.	*15–18*

Section 9.3	
■ Find infinite limits and limits at infinity.	*19–26*
■ Find the vertical and horizontal asymptotes of a function and sketch its graph.	*27–36*
■ Use asymptotes to answer questions about real life.	*37–40*

Section 9.4	
■ Analyze the graph of a function.	*41–48*

Section 9.5	
■ Find the differential of a function.	*49–52*
■ Use differentials to approximate changes in a function.	*53–56*
■ Use differentials to approximate changes in real-life models.	*57–60*

Study Strategies

■ **Problem-Solving Strategies**   If you get stuck when trying to solve an optimization problem, consider the strategies below.

1. *Draw a Diagram.*   If feasible, draw a diagram that represents the problem. Label all known values and unknown values on the diagram.

2. *Solve a Simpler Problem.*   Simplify the problem, or write several simple examples of the problem. For instance, if you are asked to find the dimensions that will produce a maximum area, try calculating the areas of several examples.

3. *Rewrite the Problem in Your Own Words.*   Rewriting a problem can help you understand it better.

4. *Guess and Check.*   Try guessing the answer, then check your guess in the statement of the original problem. By refining your guesses, you may be able to think of a general strategy for solving the problem.

Review Exercises

(T) 1. Minimum Sum Find two positive numbers whose product is 169 and whose sum is a minimum. Solve the problem analytically, and use a graphing utility to solve the problem graphically.

(T) 2. Maximum Product The sum of a positive number and three times another positive number is 768. Find the two numbers if their product is a maximum. Solve the problem analytically, and use a graphing utility to solve the problem graphically.

3. Maximum Volume A rectangular solid with a square base has a surface area of 9600 square inches.

(a) Determine the dimensions that yield the maximum volume.

(b) Find the maximum volume.

4. Minimum Cost A fence is to be built to enclose a rectangular region of 4800 square feet. The fencing material along three sides costs $3 per foot. The fencing material along the fourth side costs $4 per foot.

(a) Find the most economical dimensions of the region.

(b) How would the result of part (a) change if the fencing material costs for all sides increased by $1 per foot?

5. Maximum Yield A citrus grower estimates that 90 orange trees per acre will have an average yield of 700 oranges per tree. For each additional tree per acre, the yield will decrease by 25 oranges per tree.

(a) How many trees should be planted per acre to maximize the yield of oranges?

(b) What is the maximum yield per acre?

6. Maximum Volume A solid is formed by adjoining a hemisphere to one end of a right circular cylinder. The total surface area of the solid is 1000 square centimeters. Find the radius of the cylinder that produces the maximum volume.

7. Minimum Length Two posts, one 12 feet high and the other 28 feet high, stand 30 feet apart. They are to be secured by two wires, attached to a single stake, running from ground level to the top of each post (see figure). Where should the stake be placed to use the least amount of wire?

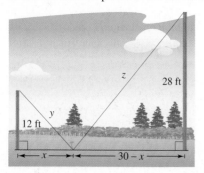

8. Minimum Length The wall of a building is to be braced by a beam that must pass over a five-foot fence that is parallel to the building and 4 feet from the building. Find the length of the shortest beam that can be used.

9. Profit The demand and cost functions for a product are $p = 36 - 4x$ and $C = 2x^2 + 6$.

(a) What level of production will produce a maximum profit?

(b) What level of production will produce a minimum average cost per unit?

10. Revenue For groups of 20 or more, a theater determines the ticket price p according to the formula

$$p = 15 - 0.1(n - 20), \quad 20 \le n \le N$$

where n is the number in the group. What should the value of N be? Explain your reasoning.

11. Minimum Cost The cost of fuel to run a locomotive is proportional to the $\frac{3}{2}$ power of the speed. At a speed of 25 miles per hour, the cost of fuel is $50 per hour. Other costs amount to $100 per hour. Find the speed that will minimize the cost per mile.

(B) 12. Economics: Revenue Consider the following cost and demand information for a monopoly (in dollars). Complete the table, and then use the information to answer the questions. *(Source: Adapted from Taylor, Economics, Fifth Edition)*

Quantity of output	Price	Total revenue	Marginal revenue
1	14.00		
2	12.00		
3	10.00		
4	8.50		
5	7.00		
6	5.50		

(T) (a) Use the *regression* feature of a graphing utility to find a quadratic model for the total revenue data.

(b) From the total revenue model you found in part (a), use derivatives to find an equation for the marginal revenue. Now use the values for output in the table and compare the results with the values in the marginal revenue column of the table. How close was your model?

(c) What quantity maximizes total revenue for the monopoly?

13. Inventory Cost The cost C of inventory modeled by

$$C = \left(\frac{Q}{x}\right)s + \left(\frac{x}{2}\right)r$$

depends on ordering and storage costs, where Q is the number of units sold per year, r is the cost of storing one unit for 1 year, s is the cost of placing an order, and x is the number of units in the order. Determine the order size that will minimize the cost when $Q = 10{,}000$, $s = 4.5$, and $r = 5.76$.

14. Profit The demand and cost functions for a product are given by

$$p = 600 - 3x$$

and

$$C = 0.3x^2 + 6x + 600$$

where p is the price per unit, x is the number of units, and C is the total cost. The profit for producing x units is given by

$$P = xp - C - xt$$

where t is the excise tax per unit. Find the maximum profits for excise taxes of $t = \$5$, $t = \$10$, and $t = \$20$.

In Exercises 15–18, find the intervals on which the demand is elastic, inelastic, and of unit elasticity.

15. $p = 30 - 0.2x, \quad 0 \le x \le 150$

16. $p = 60 - 0.04x, \quad 0 \le x \le 1500$

17. $p = 300 - x, \quad 0 \le x \le 300$

18. $p = 960 - x, \quad 0 \le x \le 960$

In Exercises 19–26, find the limit, if it exists.

19. $\displaystyle \lim_{x \to 0^+} \left(x - \frac{1}{x^3}\right)$

20. $\displaystyle \lim_{x \to 0^-} \left(3 + \frac{1}{x}\right)$

21. $\displaystyle \lim_{x \to -1^+} \frac{x^2 - 2x + 1}{x + 1}$

22. $\displaystyle \lim_{x \to 3^-} \frac{3x^2 + 1}{x^2 - 9}$

23. $\displaystyle \lim_{x \to \infty} \frac{2x^2}{3x^2 + 5}$

24. $\displaystyle \lim_{x \to \infty} \frac{3x^2 - 2x + 3}{x + 1}$

25. $\displaystyle \lim_{x \to -\infty} \frac{3x}{x^2 + 1}$

26. $\displaystyle \lim_{x \to -\infty} \left(\frac{x}{x - 2} + \frac{2x}{x + 2}\right)$

In Exercises 27–30, find the vertical and horizontal asymptotes. Write the asymptotes as equations of lines.

27. $f(x) = \dfrac{1}{2x} - 2$

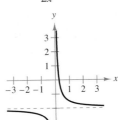

28. $g(x) = \dfrac{x + 1}{x - 2}$

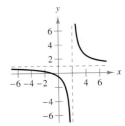

29. $h(x) = \dfrac{\sqrt{4x^2 + 1}}{x}$

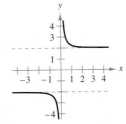

30. $f(x) = \dfrac{x^2}{x^2 - x - 2}$

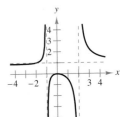

(T) In Exercises 31–36, find any vertical and horizontal asymptotes of the graph. Then use a graphing utility to graph the function.

31. $h(x) = \dfrac{2x + 3}{x - 4}$

32. $g(x) = \dfrac{3}{x} - 2$

33. $f(x) = \dfrac{x + 10}{x^2 + 3x - 10}$

34. $h(x) = \dfrac{3x}{\sqrt{x^2 + 2}}$

35. $f(x) = \dfrac{4}{x^2 + 1}$

36. $h(x) = \dfrac{2x^2 + 3x - 5}{x - 1}$

37. Temperature The graph shows the temperature T (in degrees Fahrenheit) of an apple pie t seconds after it is removed from an oven and placed on a cooling rack.

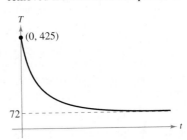

(a) Find $\displaystyle \lim_{t \to 0^+} T$. What does this limit represent?

(b) Find $\displaystyle \lim_{t \to \infty} T$. What does this limit represent?

38. Health For a person with sensitive skin, the maximum amount T (in hours) of exposure to the sun that can be tolerated before skin damage occurs can be modeled by

$$T = \frac{-0.03s + 33.6}{s}, \quad 0 < s \le 120$$

where s is the Sunsor Scale reading. *(Source: Sunsor, Inc.)*

Ⓣ (a) Use a graphing utility to graph the model. Compare your result with the graph below.

(b) Describe the value of T as s increases.

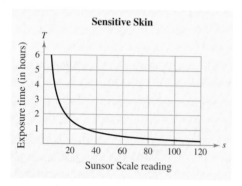

Sensitive Skin

39. Average Cost and Profit The cost and revenue functions for a product are given by

$$C = 10{,}000 + 48.9x$$

and

$$R = 68.5x.$$

(a) Find the average cost function.

(b) What is the limit of the average cost as x approaches infinity?

(c) Find the average profits when x is 1 million, 2 million, and 10 million.

(d) What is the limit of the average profit as x increases without bound?

40. Average Cost and Profit Repeat Exercise 39 if the cost and revenue functions are given by $C = 16{,}500 + 0.63x$ and $R = 1.16x$. Interpret your results to parts (b) and (d) in the context of the problem.

Ⓣ In Exercises 41–48, use a graphing utility to graph the function. Use the graph to approximate any intercepts, relative extrema, points of inflection, and asymptotes. State the domain of the function.

41. $f(x) = 4x - x^2$

42. $f(x) = 4x^3 - x^4$

43. $f(x) = x\sqrt{16 - x^2}$

44. $f(x) = x^2\sqrt{9 - x^2}$

45. $f(x) = \dfrac{x + 1}{x - 1}$

46. $f(x) = \dfrac{x - 1}{3x^2 + 1}$

47. $f(x) = x^2 + \dfrac{2}{x}$

48. $f(x) = x^{4/5}$

In Exercises 49–52, find the differential dy.

49. $y = x(1 - x)$

50. $y = (3x^2 - 2)^3$

51. $y = \sqrt{36 - x^2}$

52. $y = \dfrac{2 - x}{x + 5}$

In Exercises 53–56, use differentials to approximate the change in cost, revenue, or profit corresponding to an increase in sales of one unit.

53. $C = 40x^2 + 1225, \quad x = 10$

54. $C = 1.5\sqrt[3]{x} + 500, \quad x = 125$

55. $R = 6.25x + 0.4x^{3/2}, \quad x = 225$

56. $P = 0.003x^2 + 0.019x - 1200, \quad x = 750$

57. Area The area A of a square of side x is $A = x^2$.

(a) Compare dA and ΔA in terms of x and Δx.

(b) In the figure, identify the region whose area is dA.

(c) Identify the region whose area is $\Delta A - dA$.

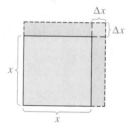

58. Surface Area and Volume The diameter of a sphere is measured to be 18 inches with a possible error of 0.05 inch. Use differentials to approximate the possible error in the surface area and the volume of the sphere.

59. Demand A company finds that the demand for its product is modeled by $p = 85 - 0.125x$. If x changes from 7 to 8, what is the corresponding change in p? Compare the values of Δp and dp.

60. Aquaculture The recommended daily percent p of biomass (plant matter) to be included in a fish's diet can be modeled by

$$p = 0.000235w^2 - 0.054w + 7.1$$

where w is the weight of the fish in grams. *(Source: Food and Agriculture Organization of the United Nations)*

(a) Use differentials to approximate the change in the recommended percent of biomass when the fish's weight changes from 10 grams to 20 grams.

(b) Use differentials to approximate the change in the recommended percent of biomass when the fish's weight changes from 40 grams to 60 grams.

Chapter Test

See www.CalcChat.com for worked-out solutions to odd-numbered exercises.

Take this test as you would take a test in class. When you are done, check your work against the answers given in the back of the book.

In Exercises 1 and 2, find any vertical and horizontal asymptotes of the graph. Write the asymptotes as equations of lines.

1. $f(x) = \dfrac{2x - 5}{x - 1}$

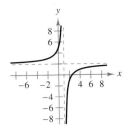

2. $f(x) = \dfrac{2x}{x^2 + 3}$

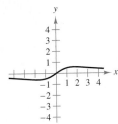

(T) In Exercises 3 and 4, find the vertical and horizontal asymptotes of the graph. Then use a graphing utility to graph the function.

3. $f(x) = \dfrac{3x + 2}{x - 5}$

4. $f(x) = \dfrac{x^2 + 2x + 3}{x^2 - 1}$

In Exercises 5–10, find the limit, if possible.

5. $\displaystyle\lim_{x \to 1^-} \frac{x + 1}{x - 1}$

6. $\displaystyle\lim_{x \to 2^+} \frac{x}{x^2 - 4}$

7. $\displaystyle\lim_{x \to -1^-} \frac{x^2 + 1}{x^2 - 1}$

8. $\displaystyle\lim_{x \to \infty} \left(\frac{3}{x} + 1 \right)$

9. $\displaystyle\lim_{x \to \infty} \frac{3x^2 - 4x + 1}{x - 7}$

10. $\displaystyle\lim_{x \to -\infty} \frac{6x^2 + x - 5}{2x^2 - 5x}$

(T) In Exercises 11 and 12, use a graphing utility to graph the function. Find any intercepts, relative extrema, and points of inflection. State the domain of the function.

11. $f(x) = x^3 + x^2 - 4x - 4$

12. $f(x) = 4x\sqrt{1 - x}$

In Exercises 13–15, find the differential *dy*.

13. $y = 5x^2 - 3$

14. $y = \dfrac{1 - x}{x + 3}$

15. $y = (x + 4)^3$

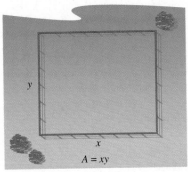

$A = xy$

Figure for 16

16. An ecologist has 500 meters of fencing to enclose a rectangular study plot (see figure). What should the dimensions of the plot be to maximize the enclosed area?

17. A rectangular solid with a square base has a volume of 8000 cubic inches.

   (a) Determine the dimensions that yield the minimum surface area.

   (b) Find the minimum surface area.

18. The demand function for a product is modeled by $p = 250 - 0.4x$, $0 \le x \le 625$, where p is the price at which x units of the product are demanded by the market. Find the interval of inelasticity for the function.

10

Exponential and Logarithmic Functions

AP/Wide World Photos

On May 26, 2006, Java, Indonesia experienced an earthquake measuring 6.3 on the Richter scale, a logarithmic function that serves as one way to calculate an earthquake's magnitude. (See Section 10.5, Exercise 87.)

Applications

Exponential and logarithmic functions have many real-life applications. The applications listed below represent a sample of the applications in this chapter.

- Make a Decision: Median Sales Prices, Exercise 37, page 750
- Bacteria Growth, Exercise 47, page 759
- Learning Theory, Exercise 88, page 777
- Consumer Trends, Exercise 85, page 786
- Make a Decision: Revenue, Exercise 41, page 795

Section 10.1

Exponential Functions

- Use the properties of exponents to evaluate and simplify exponential expressions.
- Sketch the graphs of exponential functions.

Exponential Functions

You are already familiar with the behavior of algebraic functions such as

$$f(x) = x^2, \quad g(x) = \sqrt{x} = x^{1/2}, \quad \text{and} \quad h(x) = \frac{1}{x} = x^{-1}$$

each of which involves a variable raised to a constant power. By interchanging roles and raising a constant to a variable power, you obtain another important class of functions called **exponential functions.** Some simple examples are

$$f(x) = 2^x, \quad g(x) = \left(\frac{1}{10}\right)^x = \frac{1}{10^x}, \quad \text{and} \quad h(x) = 3^{2x} = 9^x.$$

In general, you can use any positive base $a \neq 1$ as the base of an exponential function.

Definition of Exponential Function

If $a > 0$ and $a \neq 1$, then the **exponential function** with base a is given by

$$f(x) = a^x.$$

STUDY TIP

In the definition above, the base $a = 1$ is excluded because it yields $f(x) = 1^x = 1$. This is a constant function, not an exponential function.

When working with exponential functions, the properties of exponents, shown below, are useful.

Properties of Exponents

Let a and b be positive numbers.

1. $a^0 = 1$ **2.** $a^x a^y = a^{x+y}$ **3.** $\dfrac{a^x}{a^y} = a^{x-y}$

4. $(a^x)^y = a^{xy}$ **5.** $(ab)^x = a^x b^x$ **6.** $\left(\dfrac{a}{b}\right)^x = \dfrac{a^x}{b^x}$

7. $a^{-x} = \dfrac{1}{a^x}$

| Example 1 | Applying Properties of Exponents |

Simplify each expression using the properties of exponents.

a. $(2^2)(2^3)$ **b.** $(2^2)(2^{-3})$ **c.** $(3^2)^3$

d. $\left(\dfrac{1}{3}\right)^{-2}$ **e.** $\dfrac{3^2}{3^3}$ **f.** $(2^{1/2})(3^{1/2})$

SOLUTION

a. $(2^2)(2^3) = 2^{2+3} = 2^5 = 32$ Apply Property 2.

b. $(2^2)(2^{-3}) = 2^{2-3} = 2^{-1} = \frac{1}{2}$ Apply Properties 2 and 7.

c. $(3^2)^3 = 3^{2(3)} = 3^6 = 729$ Apply Property 4.

d. $\left(\dfrac{1}{3}\right)^{-2} = \dfrac{1}{(1/3)^2} = \left(\dfrac{1}{1/3}\right)^2 = 3^2 = 9$ Apply Properties 6 and 7.

e. $\dfrac{3^2}{3^3} = 3^{2-3} = 3^{-1} = \dfrac{1}{3}$ Apply Properties 3 and 7.

f. $(2^{1/2})(3^{1/2}) = [(2)(3)]^{1/2} = 6^{1/2} = \sqrt{6}$ Apply Property 5.

✓ **CHECKPOINT 1**

Simplify each expression using the properties of exponents.

a. $(3^2)(3^3)$ **b.** $(3^2)(3^{-1})$

c. $(2^3)^2$ **d.** $\left(\dfrac{1}{2}\right)^{-3}$

e. $\dfrac{2^2}{2^3}$ **f.** $(2^{1/2})(5^{1/2})$ ■

Although Example 1 demonstrates the properties of exponents with integer and rational exponents, it is important to realize that the properties hold for *all* real exponents. With a calculator, you can obtain approximations of a^x for any base a and any real exponent x. Here are some examples.

$$2^{-0.6} \approx 0.660, \qquad \pi^{0.75} \approx 2.360, \qquad (1.56)^{\sqrt{2}} \approx 1.876$$

| Example 2 | Dating Organic Material | ®

In living organic material, the ratio of radioactive carbon isotopes to the total number of carbon atoms is about 1 to 10^{12}. When organic material dies, its radioactive carbon isotopes begin to decay, with a half-life of about 5715 years. This means that after 5715 years, the ratio of isotopes to atoms will have decreased to one-half the original ratio, after a second 5715 years the ratio will have decreased to one-fourth of the original, and so on. Figure 10.1 shows this decreasing ratio. The formula for the ratio R of carbon isotopes to carbon atoms is

$$R = \left(\dfrac{1}{10^{12}}\right)\left(\dfrac{1}{2}\right)^{t/5715}$$

where t is the time in years. Find the value of R for each period of time.

a. 10,000 years   **b.** 20,000 years   **c.** 25,000 years

SOLUTION

a. $R = \left(\dfrac{1}{10^{12}}\right)\left(\dfrac{1}{2}\right)^{10,000/5715} \approx 2.973 \times 10^{-13}$ Ratio for 10,000 years

b. $R = \left(\dfrac{1}{10^{12}}\right)\left(\dfrac{1}{2}\right)^{20,000/5715} \approx 8.842 \times 10^{-14}$ Ratio for 20,000 years

c. $R = \left(\dfrac{1}{10^{12}}\right)\left(\dfrac{1}{2}\right)^{25,000/5715} \approx 4.821 \times 10^{-14}$ Ratio for 25,000 years

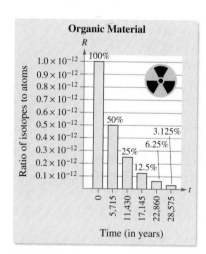

FIGURE 10.1

✓ **CHECKPOINT 2**

Use the formula for the ratio of carbon isotopes to carbon atoms in Example 2 to find the value of R for each period of time.

a. 5,000 years

b. 15,000 years

c. 30,000 years ■

Graphs of Exponential Functions

The basic nature of the graph of an exponential function can be determined by the point-plotting method or by using a graphing utility.

Example 3 Graphing Exponential Functions

Sketch the graph of each exponential function.

a. $f(x) = 2^x$ **b.** $g(x) = \left(\frac{1}{2}\right)^x = 2^{-x}$ **c.** $h(x) = 3^x$

SOLUTION To sketch these functions by hand, you can begin by constructing a table of values, as shown below.

x	-3	-2	-1	0	1	2	3	4
$f(x) = 2^x$	$\frac{1}{8}$	$\frac{1}{4}$	$\frac{1}{2}$	1	2	4	8	16
$g(x) = 2^{-x}$	8	4	2	1	$\frac{1}{2}$	$\frac{1}{4}$	$\frac{1}{8}$	$\frac{1}{16}$
$h(x) = 3^x$	$\frac{1}{27}$	$\frac{1}{9}$	$\frac{1}{3}$	1	3	9	27	81

The graphs of the three functions are shown in Figure 10.2. Note that the graphs of $f(x) = 2^x$ and $h(x) = 3^x$ are increasing, whereas the graph of $g(x) = 2^{-x}$ is decreasing.

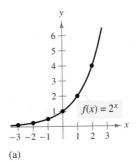
(a)

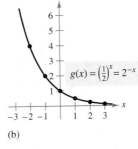
(b)

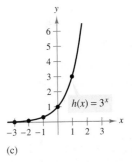
(c)

FIGURE 10.2

✓ CHECKPOINT 3

Complete the table of values for $f(x) = 5^x$. Sketch the graph of the exponential function.

x	-3	-2	-1	0
$f(x)$				

x	1	2	3
$f(x)$			

TECHNOLOGY

(T) Try graphing the functions $f(x) = 2^x$ and $h(x) = 3^x$ in the same viewing window, as shown at the right. From the display, you can see that the graph of h is increasing more rapidly than the graph of f.*

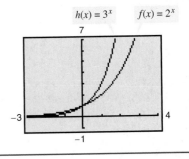

*Specific calculator keystroke instructions for operations in this and other technology boxes can be found at *college.hmco.com/info/larsonapplied.*

The forms of the graphs in Figure 10.2 are typical of the graphs of the exponential functions $y = a^{-x}$ and $y = a^x$, where $a > 1$. The basic characteristics of such graphs are summarized in Figure 10.3.

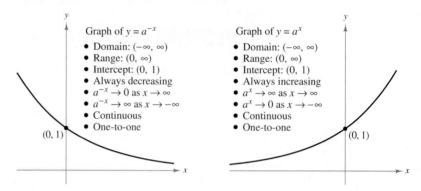

FIGURE 10.3 Characteristics of the Exponential Functions $y = a^{-x}$ and $y = a^x (a > 1)$

Example 4 **Graphing an Exponential Function**

Sketch the graph of $f(x) = 3^{-x} - 1$.

SOLUTION Begin by creating a table of values, as shown below.

x	-2	-1	0	1	2
$f(x)$	$3^2 - 1 = 8$	$3^1 - 1 = 2$	$3^0 - 1 = 0$	$3^{-1} - 1 = -\frac{2}{3}$	$3^{-2} - 1 = -\frac{8}{9}$

From the limit

$$\lim_{x \to \infty} (3^{-x} - 1) = \lim_{x \to \infty} 3^{-x} - \lim_{x \to \infty} 1$$

$$= \lim_{x \to \infty} \frac{1}{3^x} - \lim_{x \to \infty} 1$$

$$= 0 - 1$$

$$= -1$$

you can see that $y = -1$ is a horizontal asymptote of the graph. The graph is shown in Figure 10.4.

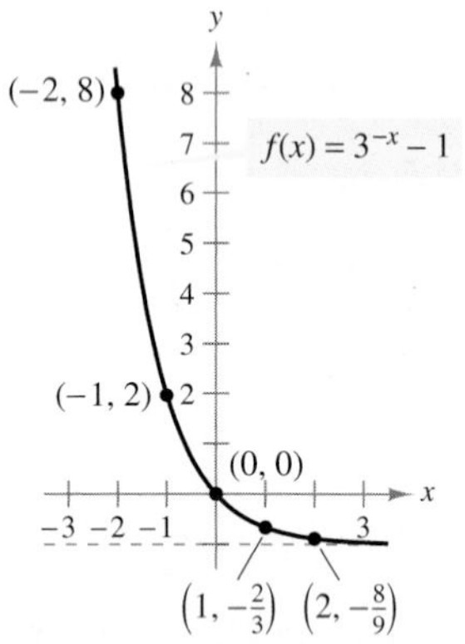

FIGURE 10.4

(Graph showing $f(x) = 3^{-x} - 1$ with points $(-2, 8)$, $(-1, 2)$, $(0, 0)$, $\left(1, -\frac{2}{3}\right)$, $\left(2, -\frac{8}{9}\right)$.)

✓ **CHECKPOINT 4**

Complete the table of values for $f(x) = 2^{-x} + 1$. Sketch the graph of the function. Determine the horizontal asymptote of the graph.

x	-3	-2	-1	0
$f(x)$				

x	1	2	3
$f(x)$			

CONCEPT CHECK

1. Complete the following: If $a > 0$ and $a \neq 1$, then $f(x) = a^x$ is a(n) _____ function.

2. Identify the domain and range of the exponential functions (a) $y = a^{-x}$ and (b) $y = a^x$. (Assume $a > 1$.)

3. As x approaches ∞, what does a^{-x} approach? (Assume $a > 1$.)

4. Explain why 1^x is *not* an exponential function.

Skills Review 10.1

The following warm-up exercises involve skills that were covered in earlier sections. You will use these skills in the exercise set for this section. For additional help, review Sections 1.1, 1.3, 2.6, and 7.2.

In Exercises 1–6, describe how the graph of g is related to the graph of f.

1. $g(x) = f(x + 2)$

2. $g(x) = -f(x)$

3. $g(x) = -1 + f(x)$

4. $g(x) = f(-x)$

5. $g(x) = f(x - 1)$

6. $g(x) = f(x) + 2$

In Exercises 7–10, discuss the continuity of the function.

7. $f(x) = \dfrac{x^2 + 2x - 1}{x + 4}$

8. $f(x) = \dfrac{x^2 - 3x + 1}{x^2 + 2}$

9. $f(x) = \dfrac{x^2 - 3x - 4}{x^2 - 1}$

10. $f(x) = \dfrac{x^2 - 5x + 4}{x^2 + 1}$

In Exercises 11–16, solve for x.

11. $2x - 6 = 4$

12. $3x + 1 = 5$

13. $(x + 4)^2 = 25$

14. $(x - 2)^2 = 8$

15. $x^2 + 4x - 5 = 0$

16. $2x^2 - 3x + 1 = 0$

Exercises 10.1

See www.CalcChat.com for worked-out solutions to odd-numbered exercises.

In Exercises 1 and 2, evaluate each expression.

1. (a) $5(5^3)$

  (b) $27^{2/3}$

  (c) $64^{3/4}$

  (d) $81^{1/2}$

  (e) $25^{3/2}$

  (f) $32^{2/5}$

2. (a) $\left(\frac{1}{5}\right)^3$

  (b) $\left(\frac{1}{8}\right)^{1/3}$

  (c) $64^{2/3}$

  (d) $\left(\frac{5}{8}\right)^2$

  (e) $100^{3/2}$

  (f) $4^{5/2}$

In Exercises 3–6, use the properties of exponents to simplify the expression.

3. (a) $(5^2)(5^3)$

  (b) $(5^2)(5^{-3})$

  (c) $(5^2)^2$

  (d) 5^{-3}

4. (a) $\dfrac{5^3}{5^6}$

  (b) $\left(\dfrac{1}{5}\right)^{-2}$

  (c) $(8^{1/2})(2^{1/2})$

  (d) $(32^{3/2})\left(\frac{1}{2}\right)^{3/2}$

5. (a) $\dfrac{5^3}{25^2}$

  (b) $(9^{2/3})(3)(3^{2/3})$

  (c) $[(25^{1/2})(5^2)]^{1/3}$

  (d) $(8^2)(4^3)$

6. (a) $(4^3)(4^2)$

  (b) $\left(\frac{1}{4}\right)^2(4^2)$

  (c) $(4^6)^{1/2}$

  (d) $[(8^{-1})(8^{2/3})]^3$

Ⓣ In Exercises 7–10, evaluate the function. If necessary, use a graphing utility, rounding your answers to three decimal places.

7. $f(x) = 2^{x-1}$

  (a) $f(3)$ (b) $f\left(\frac{1}{2}\right)$ (c) $f(-2)$ (d) $f\left(-\frac{3}{2}\right)$

8. $f(x) = 3^{x+2}$

  (a) $f(-4)$ (b) $f\left(-\frac{1}{2}\right)$ (c) $f(2)$ (d) $f\left(-\frac{5}{2}\right)$

9. $g(x) = 1.05^x$

  (a) $g(-2)$ (b) $g(120)$ (c) $g(12)$ (d) $g(5.5)$

10. $g(x) = 1.075^x$

  (a) $g(1.2)$ (b) $g(180)$ (c) $g(60)$ (d) $g(12.5)$

11. Radioactive Decay After t years, the remaining mass y (in grams) of 16 grams of a radioactive element whose half-life is 30 years is given by

$$y = 16\left(\frac{1}{2}\right)^{t/30}, \quad t \ge 0.$$

How much of the initial mass remains after 90 years?

12. Radioactive Decay After t years, the remaining mass y (in grams) of 23 grams of a radioactive element whose half-life is 45 years is given by

$$y = 23\left(\frac{1}{2}\right)^{t/45}, \quad t \ge 0.$$

How much of the initial mass remains after 150 years?

In Exercises 13–18, match the function with its graph.
[The graphs are labeled (a)–(f).]

(a)

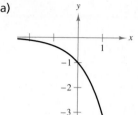

(b)

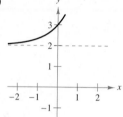

(c)

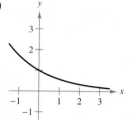

(d)

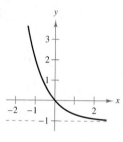

(e)

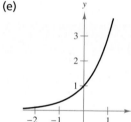

(f)

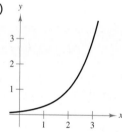

13. $f(x) = 3^x$

14. $f(x) = 3^{-x/2}$

15. $f(x) = -3^x$

16. $f(x) = 3^{x-2}$

17. $f(x) = 3^{-x} - 1$

18. $f(x) = 3^x + 2$

(T) In Exercises 19–30, use a graphing utility to graph the function.

19. $f(x) = 6^x$

20. $f(x) = 4^x$

21. $f(x) = \left(\frac{1}{5}\right)^x = 5^{-x}$

22. $f(x) = \left(\frac{1}{4}\right)^x = 4^{-x}$

23. $y = 2^{x-1}$

24. $y = 4^x + 3$

25. $y = -2^x$

26. $y = -5^x$

27. $y = 3^{-x^2}$

28. $y = 2^{-x^2}$

29. $s(t) = \frac{1}{4}(3^{-t})$

30. $s(t) = 2^{-t} + 3$

31. Population Growth The population P (in millions) of the United States from 1992 through 2005 can be modeled by the exponential function $P(t) = 252.12(1.011)^t$, where t is the time in years, with $t = 2$ corresponding to 1992. Use the model to estimate the population in the years (a) 2008 and (b) 2012. *(Source: U.S. Census Bureau)*

32. Sales The sales S (in millions of dollars) for Starbucks from 1996 through 2005 can be modeled by the exponential function $S(t) = 182.34(1.272)^t$, where t is the time in

years, with $t = 6$ corresponding to 1996. Use the model to estimate the sales in the years (a) 2008 and (b) 2014. *(Source: Starbucks Corp.)*

33. Property Value Suppose that the value of a piece of property doubles every 15 years. If you buy the property for $64,000, its value t years after the date of purchase should be $V(t) = 64,000(2)^{t/15}$. Use the model to approximate the value of the property (a) 5 years and (b) 20 years after it is purchased.

34. Depreciation After t years, the value of a car that originally cost $16,000 depreciates so that each year it is worth $\frac{3}{4}$ of its value for the previous year. Find a model for $V(t)$, the value of the car after t years. Sketch a graph of the model and determine the value of the car 4 years after it was purchased.

35. Inflation Rate Suppose that the annual rate of inflation averages 4% over the next 10 years. With this rate of inflation, the approximate cost C of goods or services during any year in that decade will be given by

$$C(t) = P(1.04)^t, \quad 0 \le t \le 10$$

where t is time in years and P is the present cost. If the price of an oil change for your car is presently $24.95, estimate the price 10 years from now.

36. Inflation Rate Repeat Exercise 35 assuming that the annual rate of inflation is 10% over the next 10 years and the approximate cost C of goods or services will be given by

$$C(t) = P(1.10)^t, \quad 0 \le t \le 10.$$

(T) **37.** *MAKE A DECISION: MEDIAN SALES PRICES* For the years 1998 through 2005, the median sales prices y (in dollars) of one-family homes in the United States are shown in the table. *(Source: U.S. Census Bureau and U.S. Department of Housing and Urban Development)*

Year	1998	1999	2000	2001
Price	152,500	161,000	169,000	175,200

Year	2002	2003	2004	2005
Price	187,600	195,000	221,000	240,900

A model for this data is given by $y = 90,120(1.0649)^t$, where t represents the year, with $t = 8$ corresponding to 1998.

(a) Compare the actual prices with those given by the model. Does the model fit the data? Explain your reasoning.

(b) Use a graphing utility to graph the model.

(c) Use the *zoom* and *trace* features of a graphing utility to predict during which year the median sales price of one-family homes will reach $300,000.

Section 10.2

Natural Exponential Functions

■ Evaluate and graph functions involving the natural exponential function.
■ Solve compound interest problems.
■ Solve present value problems.

Natural Exponential Functions

In Section 10.1, exponential functions were discussed using an unspecified base a. In calculus, the most convenient (or natural) choice for a base is the irrational number e, whose decimal approximation is

$$e \approx 2.71828182846.$$

Although this choice of base may seem unusual, its convenience will become apparent as the rules for differentiating exponential functions are developed in Section 10.3. In that development, you will encounter the limit used in the definition of e.

> **TECHNOLOGY**
>
> (T) Try graphing
> $y = (1 + x)^{1/x}$ with a
> graphing utility. Then use the
> *zoom* and *trace* features to find
> values of y near $x = 0$. You will
> find that the y-values get closer
> and closer to the number
> $e \approx 2.71828$.

Limit Definition of e

The irrational number e is defined to be the limit of $(1 + x)^{1/x}$ as $x \to 0$.
That is,

$$\lim_{x \to 0} (1 + x)^{1/x} = e.$$

Example 1   Graphing the Natural Exponential Function

Sketch the graph of $f(x) = e^x$.

SOLUTION   Begin by evaluating the function for several values of x, as shown in the table.

x	-2	-1	0	1	2
$f(x)$	$e^{-2} \approx 0.135$	$e^{-1} \approx 0.368$	$e^0 \approx 1$	$e^1 \approx 2.718$	$e^2 \approx 7.389$

The graph of $f(x) = e^x$ is shown in Figure 10.5. Note that e^x is positive for all values of x. Moreover, the graph has the x-axis as a horizontal asymptote to the left. That is,

$$\lim_{x \to -\infty} e^x = 0.$$

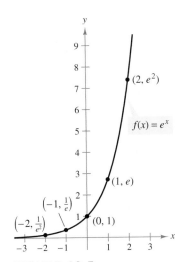

FIGURE 10.5

✓ **CHECKPOINT 1**

Complete the table of values for $f(x) = e^{-x}$. Sketch the graph of the function.

x	-2	-1	0	1	2
$f(x)$					

Exponential functions are often used to model the growth of a quantity or a population. When the quantity's growth *is not* restricted, an exponential model is often used. When the quantity's growth *is* restricted, the best model is often a **logistic growth function** of the form

$$f(t) = \frac{a}{1 + be^{-kt}}.$$

Graphs of both types of population growth models are shown in Figure 10.6.

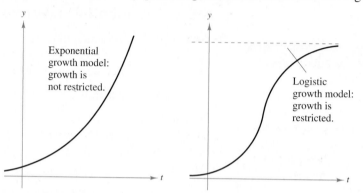

When a culture is grown in a dish, the size of the dish and the available food limit the culture's growth.

FIGURE 10.6

Example 2
MAKE A DECISION   Modeling a Population

A bacterial culture is growing according to the *logistic growth model*

$$y = \frac{1.25}{1 + 0.25e^{-0.4t}}, \quad t \ge 0$$

where y is the culture weight (in grams) and t is the time (in hours). Find the weight of the culture after 0 hours, 1 hour, and 10 hours. What is the limit of the model as t increases without bound? According to the model, will the weight of the culture reach 1.5 grams?

SOLUTION

$$y = \frac{1.25}{1 + 0.25e^{-0.4(0)}} = 1 \text{ gram} \qquad \text{Weight when } t = 0$$

$$y = \frac{1.25}{1 + 0.25e^{-0.4(1)}} \approx 1.071 \text{ grams} \qquad \text{Weight when } t = 1$$

$$y = \frac{1.25}{1 + 0.25e^{-0.4(10)}} \approx 1.244 \text{ grams} \qquad \text{Weight when } t = 10$$

As t approaches infinity, the limit of y is

$$\lim_{t \to \infty} \frac{1.25}{1 + 0.25e^{-0.4t}} = \lim_{t \to \infty} \frac{1.25}{1 + (0.25/e^{0.4t})} = \frac{1.25}{1 + 0} = 1.25.$$

So, as t increases without bound, the weight of the culture approaches 1.25 grams. According to the model, the weight of the culture will not reach 1.5 grams. The graph of the model is shown in Figure 10.7.

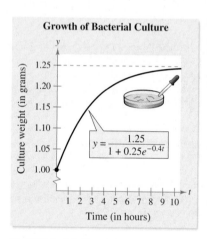

FIGURE 10.7

✓ CHECKPOINT 2

A bacterial culture is growing according to the model

$$y = \frac{1.50}{1 + 0.2e^{-0.5t}}, \quad t \ge 0$$

where y is the culture weight (in grams) and t is the time (in hours). Find the weight of the culture after 0 hours, 1 hour, and 10 hours. What is the limit of the model as t increases without bound? ∎

Extended Application: Compound Interest

If P dollars is deposited in an account at an annual interest rate of r (in decimal form), what is the balance after 1 year? The answer depends on the number of times the interest is compounded, according to the formula

$$A = P\left(1 + \frac{r}{n}\right)^n$$

where n is the number of compoundings per year. The balances for a deposit of $1000 at 8%, at various compounding periods, are shown in the table.

Number of times compounded per year, n	Balance (in dollars), A
Annually, $n = 1$	$A = 1000\left(1 + \frac{0.08}{1}\right)^1 = \$1080.00$
Semiannually, $n = 2$	$A = 1000\left(1 + \frac{0.08}{2}\right)^2 = \$1081.60$
Quarterly, $n = 4$	$A = 1000\left(1 + \frac{0.08}{4}\right)^4 \approx \$1082.43$
Monthly, $n = 12$	$A = 1000\left(1 + \frac{0.08}{12}\right)^{12} \approx \$1083.00$
Daily, $n = 365$	$A = 1000\left(1 + \frac{0.08}{365}\right)^{365} \approx \$1083.28$

You may be surprised to discover that as n increases, the balance A approaches a limit, as indicated in the following development. In this development, let $x = r/n$. Then $x \to 0$ as $n \to \infty$, and you have

$$A = \lim_{n \to \infty} P\left(1 + \frac{r}{n}\right)^n$$

$$= P \lim_{n \to \infty} \left[\left(1 + \frac{r}{n}\right)^{n/r}\right]^r$$

$$= P\left[\lim_{x \to 0} (1 + x)^{1/x}\right]^r \qquad \text{Substitute } x \text{ for } r/n.$$

$$= Pe^r.$$

This limit is the balance after 1 year of **continuous compounding**. So, for a deposit of $1000 at 8%, compounded continuously, the balance at the end of the year would be

$$A = 1000e^{0.08}$$

$$\approx \$1083.29.$$

Summary of Compound Interest Formulas

Let P be the amount deposited, t the number of years, A the balance, and r the annual interest rate (in decimal form).

1. Compounded n times per year: $A = P\left(1 + \frac{r}{n}\right)^{nt}$

2. Compounded continuously: $A = Pe^{rt}$

The average interest rates paid by banks on savings accounts have varied greatly during the past 30 years. At times, savings accounts have earned as much as 12% annual interest and at times they have earned as little as 3%. The next example shows how the annual interest rate can affect the balance of an account.

Example 3
MAKE A DECISION Finding Account Balances

You are creating a trust fund for your newborn nephew. You deposit $12,000 in an account, with instructions that the account be turned over to your nephew on his 25th birthday. Compare the balances in the account for each situation. Which account should you choose?

a. 7%, compounded continuously

b. 7%, compounded quarterly

c. 11%, compounded continuously

d. 11%, compounded quarterly

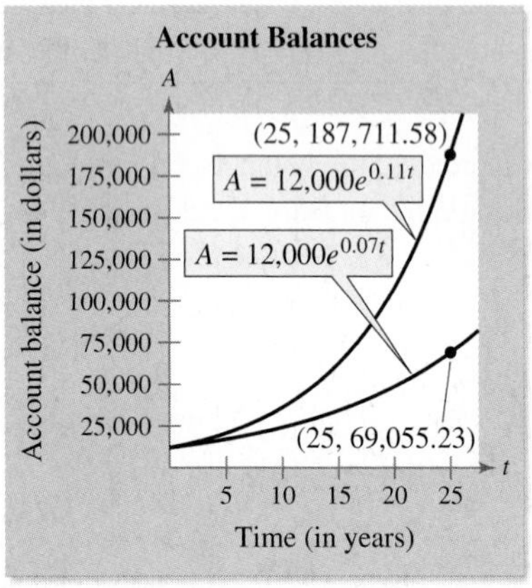

SOLUTION

a. $12,000e^{0.07(25)} \approx 69,055.23$ 7%, compounded continuously

b. $12,000\left(1 + \dfrac{0.07}{4}\right)^{4(25)} \approx 68,017.87$ 7%, compounded quarterly

c. $12,000e^{0.11(25)} \approx 187,711.58$ 11%, compounded continuously

d. $12,000\left(1 + \dfrac{0.11}{4}\right)^{4(25)} \approx 180,869.07$ 11%, compounded quarterly

The growth of the account for parts (a) and (c) is shown in Figure 10.8. Notice the dramatic difference between the balances at 7% and 11%. You should choose the account described in part (c) because it earns more money than the other accounts.

Account Balances

A = 12,000e^{0.11t}

A = 12,000e^{0.07t}

(25, 187,711.58)

(25, 69,055.23)

FIGURE 10.8

✓ CHECKPOINT 3

Find the balance in an account if $2000 is deposited for 10 years at an interest rate of 9%, compounded as follows. Compare the results and make a general statement about compounding.

a. quarterly   **b.** monthly

c. daily   **d.** continuously ■

In Example 3, note that the interest earned depends on the frequency with which the interest is compounded. The annual percentage rate is called the **stated rate** or **nominal rate**. However, the nominal rate does not reflect the actual rate at which interest is earned, which means that the compounding produced an **effective rate** that is larger than the nominal rate. In general, the effective rate corresponding to a nominal rate of r that is compounded n times per year is

$$\text{Effective rate} = r_{eff} = \left(1 + \frac{r}{n}\right)^n - 1.$$

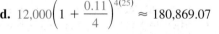

Example 4   **Finding the Effective Rate of Interest**

Find the effective rate of interest corresponding to a nominal rate of 6% per year compounded (a) annually, (b) quarterly, and (c) monthly.

SOLUTION

a. $r_{eff} = \left(1 + \dfrac{r}{n}\right)^n - 1$ Formula for effective rate of interest

$\qquad = \left(1 + \dfrac{0.06}{1}\right)^1 - 1$ Substitute for r and n.

$\qquad = 1.06 - 1$ Simplify.

$\qquad = 0.06$

So, the effective rate is 6% per year.

b. $r_{eff} = \left(1 + \dfrac{r}{n}\right)^n - 1$ Formula for effective rate of interest

$\qquad = \left(1 + \dfrac{0.06}{4}\right)^4 - 1$ Substitute for r and n.

$\qquad = (1.015)^4 - 1$ Simplify.

$\qquad \approx 0.0614$

So, the effective rate is about 6.14% per year.

c. $r_{eff} = \left(1 + \dfrac{r}{n}\right)^n - 1$ Formula for effective rate of interest

$\qquad = \left(1 + \dfrac{0.06}{12}\right)^{12} - 1$ Substitute for r and n.

$\qquad = (1.005)^{12} - 1$ Simplify.

$\qquad \approx 0.0617$

So, the effective rate is about 6.17% per year.

✓ **CHECKPOINT 4**

Find the effective rate of interest corresponding to a nominal rate of 7% per year compounded (a) semiannually and (b) daily. ▧

Present Value

In planning for the future, this problem often arises: "How much money P should be deposited now, at a fixed rate of interest r, in order to have a balance of A, t years from now?" The answer to this question is given by the **present value** of A.

To find the present value of a future investment, use the formula for compound interest as shown.

$$A = P\left(1 + \frac{r}{n}\right)^{nt}$$ Formula for compound interest

Solving for P gives a present value of

$$P = \frac{A}{\left(1 + \dfrac{r}{n}\right)^{nt}} \quad \text{or} \quad P = \frac{A}{(1 + i)^N}$$

where $i = r/n$ is the interest rate per compounding period and $N = nt$ is the total number of compounding periods. You will learn another way to find the present value of a future investment in Section 12.1.

Example 5   Finding Present Value

An investor is purchasing a 12-year certificate of deposit that pays an annual percentage rate of 8%, compounded monthly. How much should the person invest in order to obtain a balance of $15,000 at maturity?

SOLUTION   Here, $A = 15,000$, $r = 0.08$, $n = 12$, and $t = 12$. Using the formula for present value, you obtain

$$P = \frac{15,000}{\left(1 + \dfrac{0.08}{12}\right)^{12(12)}} \qquad \text{Substitute for } A, r, n, \text{ and } t.$$

$$\approx 5761.72. \qquad \text{Simplify.}$$

So, the person should invest $5761.72 in the certificate of deposit.

✓ CHECKPOINT 5

How much money should be deposited in an account paying 6% interest compounded monthly in order to have a balance of $20,000 after 3 years? ▣

CONCEPT CHECK

1. Can the number e be written as the ratio of two integers? Explain.

2. When a quantity's growth is not restricted, which model is more often used: an exponential model or a logistic growth model?

3. When a quantity's growth is restricted, which model is more often used: an exponential model or a logistic growth model?

4. Write the formula for the balance A in an account after t years with principal P and an annual interest rate r compounded continuously.

The following warm-up exercises involve skills that were covered in earlier sections. You will use these skills in the exercise set for this section. For additional help, review Sections 7.2 and 9.3.

In Exercises 1–4, discuss the continuity of the function.

1. $f(x) = \dfrac{3x^2 + 2x + 1}{x^2 + 1}$

2. $f(x) = \dfrac{x + 1}{x^2 - 4}$

3. $f(x) = \dfrac{x^2 - 6x + 5}{x^2 - 3}$

4. $g(x) = \dfrac{x^2 - 9x + 20}{x - 4}$

In Exercises 5–12, find the limit.

5. $\displaystyle\lim_{x \to \infty} \frac{25}{1 + 4x}$

6. $\displaystyle\lim_{x \to \infty} \frac{16x}{3 + x^2}$

7. $\displaystyle\lim_{x \to \infty} \frac{8x^3 + 2}{2x^3 + x}$

8. $\displaystyle\lim_{x \to \infty} \frac{x}{2x}$

9. $\displaystyle\lim_{x \to \infty} \frac{3}{2 + (1/x)}$

10. $\displaystyle\lim_{x \to \infty} \frac{6}{1 + x^{-2}}$

11. $\displaystyle\lim_{x \to \infty} 2^{-x}$

12. $\displaystyle\lim_{x \to \infty} \frac{7}{1 + 5x}$

Exercises 10.2

In Exercises 1–4, use the properties of exponents to simplify the expression.

1. (a) $(e^3)(e^4)$ (b) $(e^3)^4$

   (c) $(e^3)^{-2}$ (d) e^0

2. (a) $\left(\dfrac{1}{e}\right)^{-2}$ (b) $\left(\dfrac{e^5}{e^2}\right)^{-1}$

   (c) $\dfrac{e^5}{e^3}$ (d) $\dfrac{1}{e^{-3}}$

3. (a) $(e^2)^{5/2}$ (b) $(e^2)(e^{1/2})$

   (c) $(e^{-2})^{-3}$ (d) $\dfrac{e^5}{e^{-2}}$

4. (a) $(e^{-3})^{2/3}$ (b) $\dfrac{e^4}{e^{-1/2}}$

   (c) $(e^{-2})^{-4}$ (d) $(e^{-4})(e^{-3/2})$

In Exercises 5–10, match the function with its graph. [The graphs are labeled (a)–(f).]

(a)

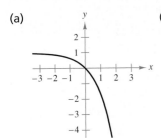

(b)

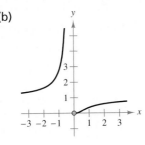

(c)

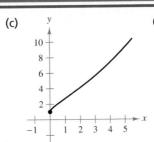

(d)

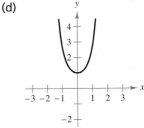

(e)

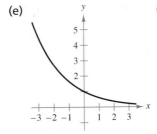

(f)

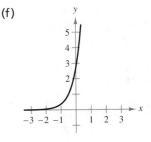

5. $f(x) = e^{2x+1}$ **6.** $f(x) = e^{-x/2}$

7. $f(x) = e^{x^2}$ **8.** $f(x) = e^{-1/x}$

9. $f(x) = e^{\sqrt{x}}$ **10.** $f(x) = -e^x + 1$

In Exercises 11–14, sketch the graph of the function.

11. $h(x) = e^{x-3}$ **12.** $f(x) = e^{2x}$

13. $g(x) = e^{1-x}$ **14.** $j(x) = e^{-x+2}$

Ⓣ In Exercises 15–18, use a graphing utility to graph the function. Be sure to choose an appropriate viewing window.

15. $N(t) = 500e^{-0.2t}$

16. $A(t) = 500e^{0.15t}$

17. $g(x) = \dfrac{2}{1 + e^{x^2}}$

18. $g(x) = \dfrac{10}{1 + e^{-x}}$

Ⓣ In Exercises 19–22, use a graphing utility to graph the function. Determine whether the function has any horizontal asymptotes and discuss the continuity of the function.

19. $f(x) = \dfrac{e^x + e^{-x}}{2}$

20. $f(x) = \dfrac{e^x - e^{-x}}{2}$

21. $f(x) = \dfrac{2}{1 + e^{1/x}}$

22. $f(x) = \dfrac{2}{1 + 2e^{-0.2x}}$

Ⓣ **23.** Use a graphing utility to graph $f(x) = e^x$ and the given function in the same viewing window. How are the two graphs related?

(a) $g(x) = e^{x-2}$

(b) $h(x) = -\dfrac{1}{2}e^x$

(c) $q(x) = e^x + 3$

Ⓣ **24.** Use a graphing utility to graph the function. Describe the shape of the graph for very large and very small values of x.

(a) $f(x) = \dfrac{8}{1 + e^{-0.5x}}$

(b) $g(x) = \dfrac{8}{1 + e^{-0.5/x}}$

Ⓢ **Compound Interest**   In Exercises 25–28, use a spreadsheet to complete the table to determine the balance A for P dollars invested at rate r for t years, compounded n times per year.

n	1	2	4	12	365	Continuous compounding
A						

25. $P = \$1000$, $r = 3\%$, $t = 10$ years

26. $P = \$2500$, $r = 2.5\%$, $t = 20$ years

27. $P = \$1000$, $r = 4\%$, $t = 20$ years

28. $P = \$2500$, $r = 5\%$, $t = 40$ years

Ⓢ **Compound Interest**   In Exercises 29–32, use a spreadsheet to complete the table to determine the amount of money P that should be invested at rate r to produce a final balance of $\$100,000$ in t years.

t	1	10	20	30	40	50
P						

29. $r = 4\%$, compounded continuously

30. $r = 3\%$, compounded continuously

31. $r = 5\%$, compounded monthly

32. $r = 6\%$, compounded daily

33. Trust Fund   On the day of a child's birth, a deposit of $\$20,000$ is made in a trust fund that pays 8% interest, compounded continuously. Determine the balance in this account on the child's 21st birthday.

34. Trust Fund   A deposit of $\$10,000$ is made in a trust fund that pays 7% interest, compounded continuously. It is specified that the balance will be given to the college from which the donor graduated after the money has earned interest for 50 years. How much will the college receive?

35. Effective Rate   Find the effective rate of interest corresponding to a nominal rate of 9% per year compounded (a) annually, (b) semiannually, (c) quarterly, and (d) monthly.

36. Effective Rate   Find the effective rate of interest corresponding to a nominal rate of 7.5% per year compounded (a) annually, (b) semiannually, (c) quarterly, and (d) monthly.

37. Present Value   How much should be deposited in an account paying 7.2% interest compounded monthly in order to have a balance of $\$15,503.77$ three years from now?

38. Present Value   How much should be deposited in an account paying 7.8% interest compounded monthly in order to have a balance of $\$21,154.03$ four years from now?

39. Future Value   Find the future value of an $\$8000$ investment if the interest rate is 4.5% compounded monthly for 2 years.

40. Future Value   Find the future value of a $\$6500$ investment if the interest rate is 6.25% compounded monthly for 3 years.

41. Demand   The demand function for a product is modeled by

$$p = 5000\left(1 - \frac{4}{4 + e^{-0.002x}}\right).$$

Find the price of the product if the quantity demanded is (a) $x = 100$ units and (b) $x = 500$ units. What is the limit of the price as x increases without bound?

42. Demand   The demand function for a product is modeled by

$$p = 10{,}000\left(1 - \frac{3}{3 + e^{-0.001x}}\right).$$

Find the price of the product if the quantity demanded is (a) $x = 1000$ units and (b) $x = 1500$ units. What is the limit of the price as x increases without bound?

43. Probability The average time between incoming calls at a switchboard is 3 minutes. If a call has just come in, the probability that the next call will come within the next t minutes is $P(t) = 1 - e^{-t/3}$. Find the probability of each situation.

(a) A call comes in within $\frac{1}{2}$ minute.

(b) A call comes in within 2 minutes.

(c) A call comes in within 5 minutes.

44. Consumer Awareness An automobile gets 28 miles per gallon at speeds of up to and including 50 miles per hour. At speeds greater than 50 miles per hour, the number of miles per gallon drops at the rate of 12% for each 10 miles per hour. If s is the speed (in miles per hour) and y is the number of miles per gallon, then $y = 28e^{0.6 - 0.012s}$, $s > 50$. Use this information and a spreadsheet to complete the table. What can you conclude?

Speed (s)	50	55	60	65	70
Miles per gallon (y)					

45. MAKE A DECISION: SALES The sales S (in millions of dollars) for Avon Products from 1998 through 2005 are shown in the table. *(Source: Avon Products Inc.)*

t	8	9	10	11
S	5212.7	5289.1	5673.7	5952.0

t	12	13	14	15
S	6170.6	6804.6	7656.2	8065.2

A model for this data is given by $S = 2962.6e^{0.0653t}$, where t represents the year, with $t = 8$ corresponding to 1998.

(a) How well does the model fit the data?

(b) Find a linear model for the data. How well does the linear model fit the data? Which model, exponential or linear, is a better fit?

(c) Use the exponential growth model and the linear model from part (b) to predict when the sales will exceed 10 billion dollars.

46. Population The population P (in thousands) of Las Vegas, Nevada from 1960 through 2005 can be modeled by $P = 68.4e^{0.0467t}$, where t is the time in years, with $t = 0$ corresponding to 1960. *(Source: U.S. Census Bureau)*

(a) Find the populations in 1960, 1970, 1980, 1990, 2000, and 2005.

(b) Explain why the data do not fit a linear model.

(c) Use the model to estimate when the population will exceed 900,000.

47. Biology The population y of a bacterial culture is modeled by the logistic growth function $y = 925/(1 + e^{-0.3t})$, where t is the time in days.

(a) Use a graphing utility to graph the model.

(b) Does the population have a limit as t increases without bound? Explain your answer.

(c) How would the limit change if the model were $y = 1000/(1 + e^{-0.3t})$? Explain your answer. Draw some conclusions about this type of model.

48. Biology: Cell Division Suppose that you have a single imaginary bacterium able to divide to form two new cells every 30 seconds. Make a table of values for the number of individuals in the population over 30-second intervals up to 5 minutes. Graph the points and use a graphing utility to fit an exponential model to the data. *(Source: Adapted from Levine/Miller, Biology: Discovering Life, Second Edition)*

49. Learning Theory In a learning theory project, the proportion P of correct responses after n trials can be modeled by

$$P = \frac{0.83}{1 + e^{-0.2n}}.$$

(a) Use a graphing utility to estimate the proportion of correct responses after 10 trials. Verify your result analytically.

(b) Use a graphing utility to estimate the number of trials required to have a proportion of correct responses of 0.75.

(c) Does the proportion of correct responses have a limit as n increases without bound? Explain your answer.

50. Learning Theory In a typing class, the average number N of words per minute typed after t weeks of lessons can be modeled by

$$N = \frac{95}{1 + 8.5e^{-0.12t}}.$$

(a) Use a graphing utility to estimate the average number of words per minute typed after 10 weeks. Verify your result analytically.

(b) Use a graphing utility to estimate the number of weeks required to achieve an average of 70 words per minute.

(c) Does the number of words per minute have a limit as t increases without bound? Explain your answer.

51. MAKE A DECISION: CERTIFICATE OF DEPOSIT You want to invest $5000 in a certificate of deposit for 12 months. You are given the options below. Which would you choose? Explain.

(a) $r = 5.25\%$, quarterly compounding

(b) $r = 5\%$, monthly compounding

(c) $r = 4.75\%$, continuous compounding

Section 10.3

Derivatives of Exponential Functions

■ Find the derivatives of natural exponential functions.
■ Use calculus to analyze the graphs of functions that involve the natural exponential function.
■ Explore the normal probability density function.

Derivatives of Exponential Functions

In Section 10.2, it was stated that the most convenient base for exponential functions is the irrational number e. The convenience of this base stems primarily from the fact that the function $f(x) = e^x$ *is its own derivative.* You will see that this is not true of other exponential functions of the form $y = a^x$ where $a \neq e$. To verify that $f(x) = e^x$ is its own derivative, notice that the limit

$$\lim_{\Delta x \to 0} (1 + \Delta x)^{1/\Delta x} = e$$

implies that for small values of Δx, $e \approx (1 + \Delta x)^{1/\Delta x}$, or $e^{\Delta x} \approx 1 + \Delta x$. This approximation is used in the following derivation.

$$f'(x) = \lim_{\Delta x \to 0} \frac{f(x + \Delta x) - f(x)}{\Delta x} \qquad \text{Definition of derivative}$$

$$= \lim_{\Delta x \to 0} \frac{e^{x + \Delta x} - e^x}{\Delta x} \qquad \text{Use } f(x) = e^x.$$

$$= \lim_{\Delta x \to 0} \frac{e^x(e^{\Delta x} - 1)}{\Delta x} \qquad \text{Factor numerator.}$$

$$= \lim_{\Delta x \to 0} \frac{e^x[(1 + \Delta x) - 1]}{\Delta x} \qquad \text{Substitute } 1 + \Delta x \text{ for } e^{\Delta x}.$$

$$= \lim_{\Delta x \to 0} \frac{e^x(\Delta x)}{\Delta x} \qquad \text{Divide out like factor.}$$

$$= \lim_{\Delta x \to 0} e^x \qquad \text{Simplify.}$$

$$= e^x \qquad \text{Evaluate limit.}$$

If u is a function of x, you can apply the Chain Rule to obtain the derivative of e^u with respect to x. Both formulas are summarized below.

Derivative of the Natural Exponential Function

Let u be a differentiable function of x.

1. $\dfrac{d}{dx}[e^x] = e^x$ **2.** $\dfrac{d}{dx}[e^u] = e^u \dfrac{du}{dx}$

DISCOVERY

Use a spreadsheet software program to compare the expressions $e^{\Delta x}$ and $1 + \Delta x$ for values of Δx near 0.

Δx	$e^{\Delta x}$	$1 + \Delta x$
0.1		
0.01		
0.001		

What can you conclude? Explain how this result is used in the development of the derivative of $f(x) = e^x$.

TECHNOLOGY

(T) Let $f(x) = e^x$. Use a graphing utility to evaluate $f(x)$ and the numerical derivative of $f(x)$ at each x-value. Explain the results.

a. $x = -2$ **b.** $x = 0$ **c.** $x = 2$

Example 1   Interpreting a Derivative

Find the slopes of the tangent lines to

$$f(x) = e^x \qquad \text{Original function}$$

at the points $(0, 1)$ and $(1, e)$. What conclusion can you make?

SOLUTION   Because the derivative of f is

$$f'(x) = e^x \qquad \text{Derivative}$$

it follows that the slope of the tangent line to the graph of f is

$$f'(0) = e^0 = 1 \qquad \text{Slope at point } (0, 1)$$

at the point $(0, 1)$ and

$$f'(1) = e^1 = e \qquad \text{Slope at point } (1, e)$$

at the point $(1, e)$, as shown in Figure 10.9. From this pattern, you can see that the slope of the tangent line to the graph of $f(x) = e^x$ at any point (x, e^x) is equal to the y-coordinate of the point.

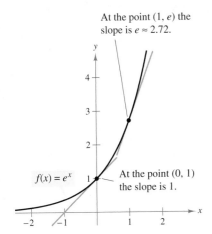

At the point $(1, e)$ the slope is $e \approx 2.72$.

$f(x) = e^x$

At the point $(0, 1)$ the slope is 1.

FIGURE 10.9

✓ CHECKPOINT 1

Find the equations of the tangent lines to $f(x) = e^x$ at the points $(0, 1)$ and $(1, e)$. ■

STUDY TIP
In Example 2, notice that when you differentiate an exponential function, the exponent does not change. For instance, the derivative of $y = e^{3x}$ is $y' = 3e^{3x}$. In both the function and its derivative, the exponent is $3x$.

✓ CHECKPOINT 2

Differentiate each function.

a. $f(x) = e^{3x}$

b. $f(x) = e^{-2x^3}$

c. $f(x) = 4e^{x^2}$

d. $f(x) = e^{-2x}$ ■

Example 2   Differentiating Exponential Functions

Differentiate each function.

a. $f(x) = e^{2x}$ **b.** $f(x) = e^{-3x^2}$

c. $f(x) = 6e^{x^3}$ **d.** $f(x) = e^{-x}$

SOLUTION

a. Let $u = 2x$. Then $du/dx = 2$, and you can apply the Chain Rule.

$$f'(x) = e^u \frac{du}{dx} = e^{2x}(2) = 2e^{2x}$$

b. Let $u = -3x^2$. Then $du/dx = -6x$, and you can apply the Chain Rule.

$$f'(x) = e^u \frac{du}{dx} = e^{-3x^2}(-6x) = -6xe^{-3x^2}$$

c. Let $u = x^3$. Then $du/dx = 3x^2$, and you can apply the Chain Rule.

$$f'(x) = 6e^u \frac{du}{dx} = 6e^{x^3}(3x^2) = 18x^2e^{x^3}$$

d. Let $u = -x$. Then $du/dx = -1$, and you can apply the Chain Rule.

$$f'(x) = e^u \frac{du}{dx} = e^{-x}(-1) = -e^{-x}$$

The differentiation rules that you studied in Chapter 7 can be used with exponential functions, as shown in Example 3.

Example 3 Differentiating Exponential Functions

Differentiate each function.

a. $f(x) = xe^x$ **b.** $f(x) = \dfrac{e^x - e^{-x}}{2}$

c. $f(x) = \dfrac{e^x}{x}$ **d.** $f(x) = xe^x - e^x$

SOLUTION

a. $f(x) = xe^x$ Write original function.

     $f'(x) = xe^x + e^x(1)$ Product Rule

     $= xe^x + e^x$ Simplify.

b. $f(x) = \dfrac{e^x - e^{-x}}{2}$ Write original function.

     $= \frac{1}{2}(e^x - e^{-x})$ Rewrite.

     $f'(x) = \frac{1}{2}(e^x + e^{-x})$ Constant Multiple Rule

c. $f(x) = \dfrac{e^x}{x}$ Write original function.

     $f'(x) = \dfrac{xe^x - e^x(1)}{x^2}$ Quotient Rule

     $= \dfrac{e^x(x - 1)}{x^2}$ Simplify.

d. $f(x) = xe^x - e^x$ Write original function.

     $f'(x) = [xe^x + e^x(1)] - e^x$ Product and Difference Rules

     $= xe^x + e^x - e^x$

     $= xe^x$ Simplify.

✓ CHECKPOINT 3

Differentiate each function.

a. $f(x) = x^2 e^x$ **b.** $f(x) = \dfrac{e^x + e^{-x}}{2}$

c. $f(x) = \dfrac{e^x}{x^2}$ **d.** $f(x) = x^2 e^x - e^x$

TECHNOLOGY

(T) If you have access to a symbolic differentiation utility, try using it to find the derivatives of the functions in Example 3.

Applications

In Chapter 8 and Chapter 9, you learned how to use derivatives to analyze the graphs of functions. The next example applies those techniques to a function composed of exponential functions. In the example, notice that $e^a = e^b$ implies that $a = b$.

Example 4 Analyzing a Catenary

When a telephone wire is hung between two poles, the wire forms a U-shaped curve called a **catenary.** For instance, the function

$$y = 30(e^{x/60} + e^{-x/60}), \quad -30 \le x \le 30$$

models the shape of a telephone wire strung between two poles that are 60 feet apart (x and y are measured in feet). Show that the lowest point on the wire is midway between the two poles. How much does the wire sag between the two poles?

SOLUTION   The derivative of the function is

$$y' = 30\left[e^{x/60}\left(\tfrac{1}{60}\right) + e^{-x/60}\left(-\tfrac{1}{60}\right)\right]$$
$$= \tfrac{1}{2}(e^{x/60} - e^{-x/60}).$$

To find the critical numbers, set the derivative equal to zero.

$\tfrac{1}{2}(e^{x/60} - e^{-x/60}) = 0$	Set derivative equal to 0.
$e^{x/60} - e^{-x/60} = 0$	Multiply each side by 2.
$e^{x/60} = e^{-x/60}$	Add $e^{-x/60}$ to each side.
$\dfrac{x}{60} = -\dfrac{x}{60}$	If $e^a = e^b$, then $a = b$.
$x = -x$	Multiply each side by 60.
$2x = 0$	Add x to each side.
$x = 0$	Divide each side by 2.

Using the First-Derivative Test, you can determine that the critical number $x = 0$ yields a relative minimum of the function. From the graph in Figure 10.10, you can see that this relative minimum is actually a minimum on the interval $[-30, 30]$. To find how much the wire sags between the two poles, you can compare its height at each pole with its height at the midpoint.

$y = 30(e^{-30/60} + e^{-(-30)/60}) \approx 67.7$ feet	Height at left pole
$y = 30(e^{0/60} + e^{-(0)/60}) = 60$ feet	Height at midpoint
$y = 30(e^{30/60} + e^{-(30)/60}) \approx 67.7$ feet	Height at right pole

From this, you can see that the wire sags about 7.7 feet.

Utility wires strung between poles have the shape of a catenary.

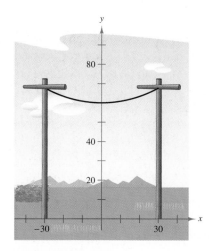

FIGURE 10.10

✓ CHECKPOINT 4

Use a graphing utility to graph the function in Example 4. Verify the minimum value. Use the information in the example to choose an appropriate viewing window. ■

Example 5   **Finding a Maximum Revenue**

The demand function for a product is modeled by

$$p = 56e^{-0.000012x}$$ Demand function ·

where p is the price per unit (in dollars) and x is the number of units. What price will yield a maximum revenue?

SOLUTION   The revenue function is

$$R = xp = 56xe^{-0.000012x}.$$ Revenue function

To find the maximum revenue *analytically*, you would set the marginal revenue, dR/dx, equal to zero and solve for x. In this problem, it is easier to use a *graphical* approach. After experimenting to find a reasonable viewing window, you can obtain a graph of R that is similar to that shown in Figure 10.11. Using the *zoom* and *trace* features, you can conclude that the maximum revenue occurs when x is about 83,300 units. To find the price that corresponds to this production level, substitute $x \approx 83,300$ into the demand function.

$$p \approx 56e^{-0.000012(83,300)} \approx \$20.61.$$

So, a price of about \$20.61 will yield a maximum revenue.

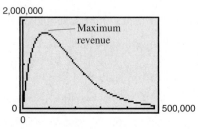

FIGURE 10.11   Use the *zoom* and *trace* features to approximate the *x*-value that corresponds to the maximum revenue.

✓**CHECKPOINT 4**

The demand function for a product is modeled by

$$p = 50e^{-0.0000125x}$$

where p is the price per unit in dollars and x is the number of units. What price will yield a maximum revenue? ■

STUDY TIP

Try solving the problem in Example 5 analytically. When you do this, you obtain

$$\frac{dR}{dx} = 56xe^{-0.000012x}(-0.000012) + e^{-0.000012x}(56) = 0.$$

Explain how you would solve this equation. What is the solution?

The Normal Probability Density Function

If you take a course in statistics or quantitative business analysis, you will spend quite a bit of time studying the characteristics and use of the **normal probability density function** given by

$$f(x) = \frac{1}{\sigma\sqrt{2\pi}}e^{-(x-\mu)^2/2\sigma^2}$$

where σ is the lowercase Greek letter sigma, and μ is the lowercase Greek letter mu. In this formula, σ represents the *standard deviation* of the probability distribution, and μ represents the *mean* of the probability distribution.

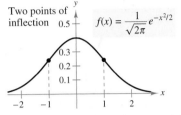

Two points of inflection

$$f(x) = \frac{1}{\sqrt{2\pi}}e^{-x^2/2}$$

FIGURE 10.12   The graph of the normal probability density function is bell-shaped.

Example 6   Exploring a Probability Density Function

Show that the graph of the normal probability density function

$$f(x) = \frac{1}{\sqrt{2\pi}}e^{-x^2/2} \qquad \text{Original function}$$

has points of inflection at $x = \pm 1$.

SOLUTION   Begin by finding the second derivative of the function.

$$f'(x) = \frac{1}{\sqrt{2\pi}}(-x)e^{-x^2/2} \qquad \text{First derivative}$$

$$f''(x) = \frac{1}{\sqrt{2\pi}}[(-x)(-x)e^{-x^2/2} + (-1)e^{-x^2/2}] \qquad \text{Second derivative}$$

$$= \frac{1}{\sqrt{2\pi}}(e^{-x^2/2})(x^2 - 1) \qquad \text{Simplify.}$$

By setting the second derivative equal to 0, you can determine that $x = \pm 1$. By testing the concavity of the graph, you can then conclude that these x-values yield points of inflection, as shown in Figure 10.12.

✓ CHECKPOINT 6

Graph the normal probability density function

$$f(x) = \frac{1}{4\sqrt{2\pi}}e^{-x^2/32}$$

and approximate the points of inflection. ∎

CONCEPT CHECK

1. What is the derivative of $f(x) = e^x$?
2. What is the derivative of $f(x) = e^u$? (Assume that u is a differentiable function of x.)
3. If $e^a = e^b$, then a is equal to what?
4. In the normal probability density function given by

$$f(x) = \frac{1}{\sigma\sqrt{2\pi}}e^{-(x-\mu)^2/2\sigma^2}$$

   identify what is represented by (a) σ and (b) μ.

| **Skills Review 10.3** | The following warm-up exercises involve skills that were covered in earlier sections. You will use these skills in the exercise set for this section. For additional help, review Sections 0.6, 7.4, 7.6, and 8.5. |

In Exercises 1–4, factor the expression.

1. $x^2 e^x - \frac{1}{2}e^x$

2. $(xe^{-x})^{-1} + e^x$

3. $xe^x - e^{2x}$

4. $e^x - xe^{-x}$

In Exercises 5–8, find the derivative of the function.

5. $f(x) = \dfrac{3}{7x^2}$

6. $g(x) = 3x^2 - \dfrac{x}{6}$

7. $f(x) = (4x - 3)(x^2 + 9)$

8. $f(t) = \dfrac{t - 2}{\sqrt{t}}$

In Exercises 9 and 10, find the relative extrema of the function.

9. $f(x) = \frac{1}{8}x^3 - 2x$

10. $f(x) = x^4 - 2x^2 + 5$

Exercises 10.3

See www.CalcChat.com for worked-out solutions to odd-numbered exercises.

In Exercises 1–4, find the slope of the tangent line to the exponential function at the point (0, 1).

1. $y = e^{3x}$

2. $y = e^{2x}$

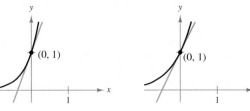

3. $y = e^{-x}$

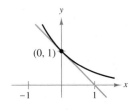

4. $y = e^{-2x}$

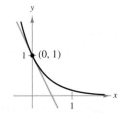

In Exercises 5–16, find the derivative of the function.

5. $y = e^{5x}$

6. $y = e^{1-x}$

7. $y = e^{-x^2}$

8. $f(x) = e^{1/x}$

9. $f(x) = e^{-1/x^2}$

10. $g(x) = e^{\sqrt{x}}$

11. $f(x) = (x^2 + 1)e^{4x}$

12. $y = 4x^3 e^{-x}$

13. $f(x) = \dfrac{2}{(e^x + e^{-x})^3}$

14. $f(x) = \dfrac{(e^x + e^{-x})^4}{2}$

15. $y = xe^x - 4e^{-x}$

16. $y = x^2 e^x - 2xe^x + 2e^x$

In Exercises 17–22, determine an equation of the tangent line to the function at the given point.

17. $y = e^{-2x+x^2}$, $(2, 1)$

18. $g(x) = e^{x^3}$, $\left(-1, \dfrac{1}{e}\right)$

19. $y = x^2 e^{-x}$, $\left(2, \dfrac{4}{e^2}\right)$

20. $y = \dfrac{x}{e^{2x}}$, $\left(1, \dfrac{1}{e^2}\right)$

21. $y = (e^{2x} + 1)^3$, $(0, 8)$

22. $y = (e^{4x} - 2)^2$, $(0, 1)$

In Exercises 23–26, find dy/dx implicitly.

23. $xe^y - 10x + 3y = 0$

24. $x^2 y - e^y - 4 = 0$

25. $x^2 e^{-x} + 2y^2 - xy = 0$

26. $e^{xy} + x^2 - y^2 = 10$

In Exercises 27–30, find the second derivative.

27. $f(x) = 2e^{3x} + 3e^{-2x}$

28. $f(x) = (1 + 2x)e^{4x}$

29. $f(x) = 5e^{-x} - 2e^{-5x}$

30. $f(x) = (3 + 2x)e^{-3x}$

In Exercises 31–34, graph and analyze the function. Include extrema, points of inflection, and asymptotes in your analysis.

31. $f(x) = \dfrac{1}{2 - e^{-x}}$

32. $f(x) = \dfrac{e^x - e^{-x}}{2}$

33. $f(x) = x^2 e^{-x}$

34. $f(x) = xe^{-x}$

(T) In Exercises 35 and 36, use a graphing utility to graph the function. Determine any asymptotes of the graph.

35. $f(x) = \dfrac{8}{1 + e^{-0.5x}}$

36. $g(x) = \dfrac{8}{1 + e^{-0.5/x}}$

In Exercises 37–40, solve the equation for x.

37. $e^{-3x} = e$ **38.** $e^x = 1$

39. $e^{\sqrt{x}} = e^3$ **40.** $e^{-1/x} = e^{1/2}$

Depreciation   In Exercises 41 and 42, the value V (in dollars) of an item is a function of the time t (in years).

(a) Sketch the function over the interval $[0, 10]$. Use a graphing utility to verify your graph.

(b) Find the rate of change of V when $t = 1$.

(c) Find the rate of change of V when $t = 5$.

(d) Use the values $(0, V(0))$ and $(10, V(10))$ to find the linear depreciation model for the item.

(e) Compare the exponential function and the model from part (d). What are the advantages of each?

41. $V = 15{,}000e^{-0.6286t}$ **42.** $V = 500{,}000e^{-0.2231t}$

43. Learning Theory   The average typing speed N (in words per minute) after t weeks of lessons is modeled by

$$N = \frac{95}{1 + 8.5e^{-0.12t}}.$$

Find the rates at which the typing speed is changing when (a) $t = 5$ weeks, (b) $t = 10$ weeks, and (c) $t = 30$ weeks.

44. Compound Interest   The balance A (in dollars) in a savings account is given by $A = 5000e^{0.08t}$, where t is measured in years. Find the rates at which the balance is changing when (a) $t = 1$ year, (b) $t = 10$ years, and (c) $t = 50$ years.

45. Ebbinghaus Model   The *Ebbinghaus Model* for human memory is $p = (100 - a)e^{-bt} + a$, where p is the percent retained after t weeks. (The constants a and b vary from one person to another.) If $a = 20$ and $b = 0.5$, at what rate is information being retained after 1 week? After 3 weeks?

46. Agriculture   The yield V (in pounds per acre) for an orchard at age t (in years) is modeled by

$$V = 7955.6e^{-0.0458/t}.$$

At what rate is the yield changing when (a) $t = 5$ years, (b) $t = 10$ years, and (c) $t = 25$ years?

Ⓣ **47. Employment**   From 1996 through 2005, the numbers y (in millions) of employed people in the United States can be modeled by

$$y = 98.020 + 6.2472t - 0.24964t^2 + 0.000002e^t$$

where t represents the year, with $t = 6$ corresponding to 1996.   *(Source: U.S. Bureau of Labor Statistics)*

(a) Use a graphing utility to graph the model.

(b) Use the graph to estimate the rates of change in the number of employed people in 1996, 2000, and 2005.

(c) Confirm the results of part (b) analytically.

Ⓣ **48. Cell Sites**   A cell site is a site where electronic communications equipment is placed in a cellular network for the use of mobile phones. From 1985 through 2006, the numbers y of cell sites can be modeled by

$$y = \frac{222{,}827}{1 + 2677e^{-0.377t}}$$

where t represents the year, with $t = 5$ corresponding to 1985.   *(Source: Cellular Telecommunications & Internet Association)*

(a) Use a graphing utility to graph the model.

(b) Use the graph to estimate when the rate of change in the number of cell cites began to decrease.

(c) Confirm the result of part (b) analytically.

49. Probability   A survey of high school seniors from a certain school district who took the SAT has determined that the mean score on the mathematics portion was 650 with a standard deviation of 12.5.

(a) Assuming the data can be modeled by a normal probability density function, find a model for these data.

Ⓣ (b) Use a graphing utility to graph the model. Be sure to choose an appropriate viewing window.

(c) Find the derivative of the model.

(d) Show that $f' > 0$ for $x < \mu$ and $f' < 0$ for $x > \mu$.

50. Probability   A survey of a college freshman class has determined that the mean height of females in the class is 64 inches with a standard deviation of 3.2 inches.

(a) Assuming the data can be modeled by a normal probability density function, find a model for these data.

Ⓣ (b) Use a graphing utility to graph the model. Be sure to choose an appropriate viewing window.

(c) Find the derivative of the model.

(d) Show that $f' > 0$ for $x < \mu$ and $f' < 0$ for $x > \mu$.

Ⓣ **51.** Use a graphing utility to graph the normal probability density function with $\mu = 0$ and $\sigma = 2, 3$, and 4 in the same viewing window. What effect does the standard deviation σ have on the function? Explain your reasoning.

Ⓣ **52.** Use a graphing utility to graph the normal probability density function with $\sigma = 1$ and $\mu = -2, 1$, and 3 in the same viewing window. What effect does the mean μ have on the function? Explain your reasoning.

Ⓣ **53.** Use Example 6 as a model to show that the graph of the normal probability density function with $\mu = 0$

$$f(x) = \frac{1}{\sigma\sqrt{2\pi}}e^{-x^2/2\sigma^2}$$

has points of inflection at $x = \pm\sigma$. What is the maximum value of the function? Use a graphing utility to verify your answer by graphing the function for several values of σ.

Mid-Chapter Quiz

See www.CalcChat.com for worked-out solutions to odd-numbered exercises.

Take this quiz as you would take a quiz in class. When you are done, check your work against the answers given in the back of the book.

In Exercises 1–4, evaluate each expression.

1. $4(4^2)$

2. $\left(\dfrac{2}{3}\right)^3$

3. $81^{1/3}$

4. $\left(\dfrac{4}{9}\right)^2$

In Exercises 5–12, use properties of exponents to simplify the expression.

5. $4^3(4^2)$

6. $\left(\dfrac{1}{6}\right)^{-3}$

7. $\dfrac{3^8}{3^5}$

8. $(5^{1/2})(3^{1/2})$

9. $(e^2)(e^5)$

10. $(e^{2/3})(e^3)$

11. $\dfrac{e^2}{e^{-4}}$

12. $(e^{-1})^{-3}$

Ⓣ In Exercises 13–18, use a graphing utility to graph the function.

13. $f(x) = 3^x - 2$

14. $f(x) = 5^{-x} + 2$

15. $f(x) = 6^{x-3}$

16. $f(x) = e^{x+2}$

17. $f(x) = 250e^{0.15x}$

18. $f(x) = \dfrac{5}{1 + e^x}$

19. Suppose that the annual rate of inflation averages 4.5% over the next 10 years. With this rate of inflation, the approximate cost C of goods or services during any year in that decade will be given by

$$C(t) = P(1.045)^t, \quad 0 \le t \le 10$$

where t is time in years and P is the present cost. If the price of a baseball game ticket is presently $14.95, estimate the price 10 years from now.

20. For $P = \$3000$, $r = 3.5\%$, and $t = 5$ years, find the balance in an account if interest is compounded (a) monthly and (b) continuously.

In Exercises 21–24, find the derivative of the function.

21. $y = e^{5x}$

22. $y = e^{x-4}$

23. $y = 5e^{x+2}$

24. $y = 3e^x - xe^x$

25. Determine an equation of the tangent line to $y = e^{-2x}$ at the point $(0, 1)$.

26. Graph and analyze the function $f(x) = 0.5x^2e^{-0.5x}$. Include extrema, points of inflection, and asymptotes in your analysis.

Section 10.4

Logarithmic Functions

- Sketch the graphs of natural logarithmic functions.
- Use properties of logarithms to simplify, expand, and condense logarithmic expressions.
- Use inverse properties of exponential and logarithmic functions to solve exponential and logarithmic equations.
- Use properties of natural logarithms to answer questions about real-life situations.

The Natural Logarithmic Function

From your previous algebra courses, you should be somewhat familiar with logarithms. For instance, the **common logarithm** $\log_{10} x$ is defined as

$$\log_{10} x = b \quad \text{if and only if} \quad 10^b = x.$$

The base of common logarithms is 10. In calculus, the most useful base for logarithms is the number e.

Definition of the Natural Logarithmic Function

The **natural logarithmic function,** denoted by $\ln x$, is defined as

$$\ln x = b \quad \text{if and only if} \quad e^b = x.$$

$\ln x$ is read as "el en of x" or as "the natural log of x."

This definition implies that the natural logarithmic function and the natural exponential function are inverse functions. So, every logarithmic equation can be written in an equivalent exponential form and every exponential equation can be written in logarithmic form. Here are some examples.

Logarithmic form:	*Exponential form:*
$\ln 1 = 0$	$e^0 = 1$
$\ln e = 1$	$e^1 = e$
$\ln \dfrac{1}{e} = -1$	$e^{-1} = \dfrac{1}{e}$
$\ln 2 \approx 0.693$	$e^{0.693} \approx 2$

Because the functions $f(x) = e^x$ and $g(x) = \ln x$ are inverse functions, their graphs are reflections of each other in the line $y = x$. This reflective property is illustrated in Figure 10.13. The figure also contains a summary of several properties of the graph of the natural logarithmic function.

Notice that the domain of the natural logarithmic function is the set of *positive real numbers*—be sure you see that $\ln x$ is not defined for zero or for negative numbers. You can test this on your calculator. If you try evaluating $\ln(-1)$ or $\ln 0$, your calculator should indicate that the value is not a real number.

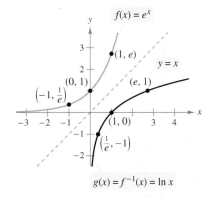

$g(x) = \ln x$

- Domain: $(0, \infty)$
- Range: $(-\infty, \infty)$
- Intercept: $(1, 0)$
- Always increasing
- $\ln x \to \infty$ as $x \to \infty$
- $\ln x \to -\infty$ as $x \to 0^+$
- Continuous
- One-to-one

FIGURE 10.13

Example 1 Graphing Logarithmic Functions

Sketch the graph of each function.

a. $f(x) = \ln(x + 1)$ **b.** $f(x) = 2\ln(x - 2)$

SOLUTION

a. Because the natural logarithmic function is defined only for positive values, the domain of the function is $x + 1 > 0$, or

$$x > -1. \qquad \text{Domain}$$

To sketch the graph, begin by constructing a table of values, as shown below. Then plot the points in the table and connect them with a smooth curve, as shown in Figure 10.14(a).

x	-0.5	0	0.5	1	1.5	2
$\ln(x + 1)$	-0.693	0	0.405	0.693	0.916	1.099

b. The domain of this function is $x - 2 > 0$, or

$$x > 2. \qquad \text{Domain}$$

A table of values for the function is shown below, and its graph is shown in Figure 10.14(b).

x	2.5	3	3.5	4	4.5	5
$2\ln(x - 2)$	-1.386	0	0.811	1.386	1.833	2.197

✓ CHECKPOINT 1

Use a graphing utility to complete the table and graph the function.

$$f(x) = \ln(x + 2)$$

x	-1.5	-1	-0.5
$f(x)$			

x	0	0.5	1
$f(x)$			

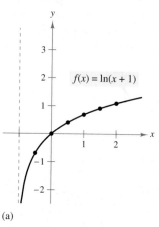

(a)

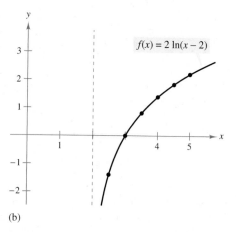

(b)

FIGURE 10.14

STUDY TIP

How does the graph of $f(x) = \ln(x + 1)$ relate to the graph of $y = \ln x$? The graph of f is a translation of the graph of $y = \ln x$ one unit to the left.

Properties of Logarithmic Functions

Recall from Section 2.8 that inverse functions have the property that

$$f(f^{-1}(x)) = x \quad \text{and} \quad f^{-1}(f(x)) = x.$$

The properties listed below follow from the fact that the natural logarithmic function and the natural exponential function are inverse functions.

Inverse Properties of Logarithms and Exponents

1. $\ln e^x = x$ **2.** $e^{\ln x} = x$

Example 2 **Applying Inverse Properties**

Simplify each expression.

a. $\ln e^{\sqrt{2}}$ **b.** $e^{\ln 3x}$

SOLUTION

a. Because $\ln e^x = x$, it follows that

$$\ln e^{\sqrt{2}} = \sqrt{2}.$$

b. Because $e^{\ln x} = x$, it follows that

$$e^{\ln 3x} = 3x.$$

✓ CHECKPOINT 2

Simplify each expression.

a. $\ln e^3$ **b.** $e^{\ln(x+1)}$ ■

Most of the properties of exponential functions can be rewritten in terms of logarithmic functions. For instance, the property

$$e^x e^y = e^{x+y}$$

states that you can multiply two exponential expressions by adding their exponents. In terms of logarithms, this property becomes

$$\ln xy = \ln x + \ln y.$$

This property and two other properties of logarithms are summarized below.

STUDY TIP

There is no general property that can be used to rewrite $\ln(x + y)$. Specifically, $\ln(x + y)$ is not equal to $\ln x + \ln y$.

Properties of Logarithms

1. $\ln xy = \ln x + \ln y$ **2.** $\ln \dfrac{x}{y} = \ln x - \ln y$

3. $\ln x^n = n \ln x$

Rewriting a logarithm of a single quantity as the sum, difference, or multiple of logarithms is called *expanding* the logarithmic expression. The reverse procedure is called *condensing* a logarithmic expression.

TECHNOLOGY

(T) Try using a graphing utility to verify the results of Example 3(b). That is, try graphing the functions

$$y = \ln \sqrt{x^2 + 1}$$

and

$$y = \frac{1}{2} \ln(x^2 + 1).$$

Because these two functions are equivalent, their graphs should coincide.

Example 3 **Expanding Logarithmic Expressions**

Use the properties of logarithms to rewrite each expression as a sum, difference, or multiple of logarithms. (Assume $x > 0$ and $y > 0$.)

a. $\ln \dfrac{10}{9}$ **b.** $\ln \sqrt{x^2 + 1}$ **c.** $\ln \dfrac{xy}{5}$ **d.** $\ln[x^2(x + 1)]$

SOLUTION

a. $\ln \frac{10}{9} = \ln 10 - \ln 9$ Property 2

b. $\ln \sqrt{x^2 + 1} = \ln(x^2 + 1)^{1/2}$ Rewrite with rational exponent.

$\qquad\qquad\quad = \frac{1}{2} \ln(x^2 + 1)$ Property 3

c. $\ln \frac{xy}{5} = \ln(xy) - \ln 5$ Property 2

$\qquad\quad = \ln x + \ln y - \ln 5$ Property 1

d. $\ln[x^2(x + 1)] = \ln x^2 + \ln(x + 1)$ Property 1

$\qquad\qquad\quad = 2 \ln x + \ln(x + 1)$ Property 3

✓ **CHECKPOINT 3**

Use the properties of logarithms to rewrite each expression as a sum, difference, or multiple of logarithms. (Assume $x > 0$ and $y > 0$.)

a. $\ln \dfrac{2}{5}$ **b.** $\ln \sqrt[3]{x + 2}$ **c.** $\ln \dfrac{x}{5y}$ **d.** $\ln x(x + 1)^2$ ■

Example 4 **Condensing Logarithmic Expressions**

Use the properties of logarithms to rewrite each expression as the logarithm of a single quantity. (Assume $x > 0$ and $y > 0$.)

a. $\ln x + 2 \ln y$

b. $2 \ln(x + 2) - 3 \ln x$

SOLUTION

a. $\ln x + 2 \ln y = \ln x + \ln y^2$ Property 3

$\qquad\qquad\qquad = \ln xy^2$ Property 1

b. $2 \ln(x + 2) - 3 \ln x = \ln(x + 2)^2 - \ln x^3$ Property 3

$\qquad\qquad\qquad\qquad = \ln \dfrac{(x + 2)^2}{x^3}$ Property 2

✓ **CHECKPOINT 4**

Use the properties of logarithms to rewrite each expression as the logarithm of a single quantity. (Assume $x > 0$ and $y > 0$.)

a. $4 \ln x + 3 \ln y$

b. $\ln(x + 1) - 2 \ln(x + 3)$ ■

Solving Exponential and Logarithmic Equations

The inverse properties of logarithms and exponents can be used to solve exponential and logarithmic equations, as shown in the next two examples.

STUDY TIP

In the examples on this page, note that the key step in solving an exponential equation is to take the log of each side, and the key step in solving a logarithmic equation is to exponentiate each side.

Example 5 **Solving Exponential Equations**

Solve each equation.

a. $e^x = 5$ **b.** $10 + e^{0.1t} = 14$

SOLUTION

a.

$e^x = 5$	Write original equation.
$\ln e^x = \ln 5$	Take natural log of each side.
$x = \ln 5$	Inverse property: $\ln e^x = x$

b.

$10 + e^{0.1t} = 14$	Write original equation.
$e^{0.1t} = 4$	Subtract 10 from each side.
$\ln e^{0.1t} = \ln 4$	Take natural log of each side.
$0.1t = \ln 4$	Inverse property: $\ln e^{0.1t} = 0.1t$
$t = 10 \ln 4$	Multiply each side by 10.

✓**CHECKPOINT 5**

Solve each equation.

a. $e^x = 6$ **b.** $5 + e^{0.2t} = 10$ ▪

Example 6 **Solving Logarithmic Equations**

Solve each equation.

a. $\ln x = 5$ **b.** $3 + 2 \ln x = 7$

SOLUTION

a.

$\ln x = 5$	Write original equation.
$e^{\ln x} = e^5$	Exponentiate each side.
$x = e^5$	Inverse property: $e^{\ln x} = x$

b.

$3 + 2 \ln x = 7$	Write original equation.
$2 \ln x = 4$	Subtract 3 from each side.
$\ln x = 2$	Divide each side by 2.
$e^{\ln x} = e^2$	Exponentiate each side.
$x = e^2$	Inverse property: $e^{\ln x} = x$

✓**CHECKPOINT 6**

Solve each equation.

a. $\ln x = 4$ **b.** $4 + 5 \ln x = 19$ ▪

Example 7   Finding Doubling Time

You deposit P dollars in an account whose annual interest rate is r, compounded continuously. How long will it take for your balance to double?

SOLUTION   The balance in the account after t years is

$$A = Pe^{rt}.$$

So, the balance will have doubled when $Pe^{rt} = 2P$. To find the "doubling time," solve this equation for t.

$Pe^{rt} = 2P$	Balance in account has doubled.
$e^{rt} = 2$	Divide each side by P.
$\ln e^{rt} = \ln 2$	Take natural log of each side.
$rt = \ln 2$	Inverse property:
$t = \dfrac{1}{r} \ln 2$	Divide each side by r.

From this result, you can see that the time it takes for the balance to double is inversely proportional to the interest rate r. The table shows the doubling times for several interest rates. Notice that the doubling time decreases as the rate increases. The relationship between doubling time and the interest rate is shown graphically in Figure 10.15.

Doubling Account Balances

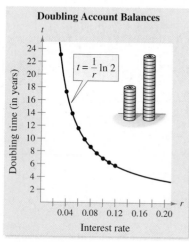

$t = \dfrac{1}{r} \ln 2$

FIGURE 10.15

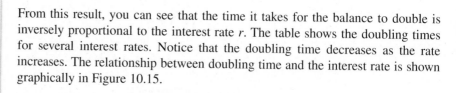

r	3%	4%	5%	6%	7%	8%	9%	10%	11%	12%
t	23.1	17.3	13.9	11.6	9.9	8.7	7.7	6.9	6.3	5.8

✓ **CHECKPOINT 7**

Use the equation found in Example 7 to determine the amount of time it would take for your balance to double at an interest rate of 8.75%. ■

```
CONCEPT CHECK

1. What are common logarithms and natural logarithms?
2. Write "logarithm of x with base 3" symbolically.
3. What are the domain and range of f(x) = ln x?
4. Explain the relationship between the functions f(x) = ln x and g(x) = eˣ.
```

In Exercises 1–8, use the properties of exponents to simplify the expression.

1. $(4^2)(4^{-3})$

2. $(2^3)^2$

3. $\dfrac{3^4}{3^{-2}}$

4. $\left(\dfrac{3}{2}\right)^{-3}$

5. e^0

6. $(3e)^4$

7. $\left(\dfrac{2}{e^3}\right)^{-1}$

8. $\left(\dfrac{4e^2}{25}\right)^{-3/2}$

In Exercises 9–12, solve for x.

9. $0 < x + 4$

10. $0 < x^2 + 1$

11. $0 < \sqrt{x^2 - 1}$

12. $0 < x - 5$

In Exercises 13 and 14, find the balance in the account after 10 years.

13. $P = \$1900$, $r = 6\%$, compounded continuously

14. $P = \$2500$, $r = 3\%$, compounded continuously

Exercises 10.4

See www.CalcChat.com for worked-out solutions to odd-numbered exercises.

In Exercises 1–8, write the logarithmic equation as an exponential equation, or vice versa.

1. $\ln 2 = 0.6931\ldots$

2. $\ln 9 = 2.1972\ldots$

3. $\ln 0.2 = -1.6094\ldots$

4. $\ln 0.05 = -2.9957\ldots$

5. $e^0 = 1$

6. $e^2 = 7.3891\ldots$

7. $e^{-3} = 0.0498\ldots$

8. $e^{0.25} = 1.2840\ldots$

In Exercises 9–12, match the function with its graph. [The graphs are labeled (a)–(d).]

(a)
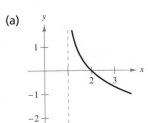

(b)
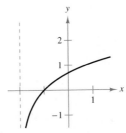

(c)
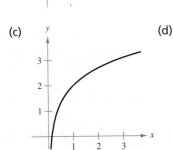

(d)
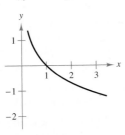

9. $f(x) = 2 + \ln x$

10. $f(x) = -\ln x$

11. $f(x) = \ln(x + 2)$

12. $f(x) = -\ln(x - 1)$

In Exercises 13–18, sketch the graph of the function.

13. $y = \ln(x - 1)$

14. $y = \ln|x|$

15. $y = \ln 2x$

16. $y = 5 + \ln x$

17. $y = 3 \ln x$

18. $y = \frac{1}{4} \ln x$

(T) In Exercises 19–22, analytically show that the functions are inverse functions. Then use a graphing utility to show this graphically.

19. $f(x) = e^{2x}$
$g(x) = \ln \sqrt{x}$

20. $f(x) = e^x - 1$
$g(x) = \ln(x + 1)$

21. $f(x) = e^{2x - 1}$
$g(x) = \frac{1}{2} + \ln \sqrt{x}$

22. $f(x) = e^{x/3}$
$g(x) = \ln x^3$

In Exercises 23–28, apply the inverse properties of logarithmic and exponential functions to simplify the expression.

23. $\ln e^{x^2}$

24. $\ln e^{2x - 1}$

25. $e^{\ln(5x + 2)}$

26. $e^{\ln \sqrt{x}}$

27. $-1 + \ln e^{2x}$

28. $-8 + e^{\ln x^3}$

In Exercises 29 and 30, use the properties of logarithms and the fact that $\ln 2 \approx 0.6931$ and $\ln 3 \approx 1.0986$ to approximate the logarithm. Then use a calculator to confirm your approximation.

29. (a) $\ln 6$ (b) $\ln \frac{3}{2}$ (c) $\ln 81$ (d) $\ln \sqrt{3}$

30. (a) $\ln 0.25$ (b) $\ln 24$ (c) $\ln \sqrt[3]{12}$ (d) $\ln \frac{1}{72}$

In Exercises 31–40, use the properties of logarithms to write the expression as a sum, difference, or multiple of logarithms.

31. $\ln \frac{2}{3}$

32. $\ln \frac{1}{5}$

33. $\ln 2xy$

34. $\ln \frac{xy}{2}$

35. $\ln \sqrt{x^2 + 1}$

36. $\ln \sqrt{\dfrac{x^3}{x + 1}}$

37. $\ln [z(z - 1)^2]$

38. $\ln\left(x \sqrt[3]{x^2 + 1}\right)$

39. $\ln \dfrac{3x(x + 1)}{(2x + 1)^2}$

40. $\ln \dfrac{2x}{\sqrt{x^2 - 1}}$

In Exercises 41–50, write the expression as the logarithm of a single quantity.

41. $\ln(x - 2) - \ln(x + 2)$

42. $\ln(2x + 1) + \ln(2x - 1)$

43. $3 \ln x + 2 \ln y - 4 \ln z$

44. $2 \ln 3 - \frac{1}{2} \ln(x^2 + 1)$

45. $3[\ln x + \ln(x + 3) - \ln(x + 4)]$

46. $\frac{1}{3}[2 \ln(x + 3) + \ln x - \ln(x^2 - 1)]$

47. $\frac{3}{2}[\ln x(x^2 + 1) - \ln(x + 1)]$

48. $2\left[\ln x + \frac{1}{4} \ln(x + 1)\right]$

49. $\frac{1}{3} \ln(x + 1) - \frac{2}{3} \ln(x - 1)$

50. $\frac{1}{2} \ln(x - 2) + \frac{3}{2} \ln(x + 2)$

In Exercises 51–74, solve for x or t.

51. $e^{\ln x} = 4$

52. $e^{\ln x^2} - 9 = 0$

53. $\ln x = 0$

54. $2 \ln x = 4$

55. $\ln 2x = 1.2$

56. $\ln 5x = 1$

57. $3 \ln 5x = 8$

58. $2 \ln 4x = 7$

59. $e^{x+1} = 4$

60. $e^{-0.5x} = 0.075$

61. $300e^{-0.2t} = 700$

62. $400e^{-0.0174t} = 1000$

63. $4e^{2x-1} - 1 = 5$

64. $2e^{-x+1} - 5 = 9$

65. $\dfrac{10}{1 + 4e^{-0.01x}} = 2.5$

66. $\dfrac{50}{1 + 12e^{-0.02x}} = 10.5$

67. $5^{2x} = 15$

68. $2^{1-x} = 6$

69. $500(1.07)^t = 1000$

70. $400(1.06)^t = 1300$

71. $\left(1 + \dfrac{0.07}{12}\right)^{12t} = 3$

72. $\left(1 + \dfrac{0.06}{12}\right)^{12t} = 5$

73. $\left(16 - \dfrac{0.878}{26}\right)^{3t} = 30$

74. $\left(4 - \dfrac{2.471}{40}\right)^{9t} = 21$

In Exercises 75 and 76, $3000 is invested in an account at interest rate r, compounded continuously. Find the time required for the amount to (a) double and (b) triple.

75. $r = 0.085$

76. $r = 0.12$

77. Compound Interest A deposit of $1000 is made in an account that earns interest at an annual rate of 5%. How long will it take for the balance to double if the interest is compounded (a) annually, (b) monthly, (c) daily, and (d) continuously?

Ⓢ **78. Compound Interest** Use a spreadsheet to complete the table, which shows the time t necessary for P dollars to triple if the interest is compounded continuously at the rate of r.

r	2%	4%	6%	8%	10%	12%	14%
t							

79. Demand The demand function for a product is given by

$$p = 5000\left(1 - \dfrac{4}{4 + e^{-0.002x}}\right)$$

where p is the price per unit and x is the number of units sold. Find the numbers of units sold for prices of (a) $p = \$200$ and (b) $p = \$800$.

80. Demand The demand function for a product is given by

$$p = 10{,}000\left(1 - \dfrac{3}{3 + e^{-0.001x}}\right)$$

where p is the price per unit and x is the number of units sold. Find the numbers of units sold for prices of (a) $p = \$500$ and (b) $p = \$1500$.

81. Population Growth The population P (in thousands) of Orlando, Florida from 1980 through 2005 can be modeled by

$$P = 131e^{0.019t}$$

where $t = 0$ corresponds to 1980. *(Source: U.S. Census Bureau)*

(a) According to this model, what was the population of Orlando in 2005?

(b) According to this model, in what year will Orlando have a population of 300,000?

82. Population Growth The population P (in thousands) of Houston, Texas from 1980 through 2005 can be modeled by $P = 1576e^{0.01t}$, where $t = 0$ corresponds to 1980. *(Source: U.S. Census Bureau)*

(a) According to this model, what was the population of Houston in 2005?

(b) According to this model, in what year will Houston have a population of 2,500,000?

Carbon Dating In Exercises 83–86, you are given the ratio of carbon atoms in a fossil. Use the information to estimate the age of the fossil. In living organic material, the ratio of radioactive carbon isotopes to the total number of carbon atoms is about 1 to 10^{12}. (See Example 2 in Section 10.1.) When organic material dies, its radioactive carbon isotopes begin to decay, with a half-life of about 5715 years. So, the ratio R of carbon isotopes to carbon-14 atoms is modeled by $R = 10^{-12}\left(\frac{1}{2}\right)^{t/5715}$, where t is the time (in years) and $t = 0$ represents the time when the organic material died.

83. $R = 0.32 \times 10^{-12}$ **84.** $R = 0.27 \times 10^{-12}$

85. $R = 0.22 \times 10^{-12}$ **86.** $R = 0.13 \times 10^{-12}$

87. Learning Theory Students in a mathematics class were given an exam and then retested monthly with equivalent exams. The average scores S (on a 100-point scale) for the class can be modeled by $S = 80 - 14 \ln(t + 1)$, $0 \le t \le 12$, where t is the time in months.

(a) What was the average score on the original exam?

(b) What was the average score after 4 months?

(c) After how many months was the average score 46?

(T) 88. Learning Theory In a group project in learning theory, a mathematical model for the proportion P of correct responses after n trials was found to be

$$P = \frac{0.83}{1 + e^{-0.2n}}.$$

(a) Use a graphing utility to graph the function.

(b) Use the graph to determine any horizontal asymptotes of the graph of the function. Interpret the meaning of the upper asymptote in the context of the problem.

(c) After how many trials will 60% of the responses be correct?

(T) 89. Agriculture The yield V (in pounds per acre) for an orchard at age t (in years) is modeled by

$$V = 7955.6 e^{-0.0458/t}.$$

(a) Use a graphing utility to graph the function.

(b) Determine the horizontal asymptote of the graph of the function. Interpret its meaning in the context of the problem.

(c) Find the time necessary to obtain a yield of 7900 pounds per acre.

90. MAKE A DECISION: FINANCE You are investing P dollars at an annual interest rate of r, compounded continuously, for t years, Which of the following options would you choose to get the highest value of the investment? Explain your reasoning.

(a) Double the amount you invest.

(b) Double your interest rate.

(c) Double the number of years.

(S) 91. Demonstrate that

$$\frac{\ln x}{\ln y} \ne \ln \frac{x}{y} = \ln x - \ln y$$

by using a spreadsheet to complete the table.

x	y	$\dfrac{\ln x}{\ln y}$	$\ln \dfrac{x}{y}$	$\ln x - \ln y$
1	2			
3	4			
10	5			
4	0.5			

(S) 92. Use a spreadsheet to complete the table using $f(x) = \dfrac{\ln x}{x}$.

x	1	5	10	10^2	10^4	10^6
$f(x)$						

(a) Use the table to estimate the limit: $\lim\limits_{x \to \infty} f(x)$.

(T) (b) Use a graphing utility to estimate the relative extrema of f.

(T) In Exercises 93 and 94, use a graphing utility to verify that the functions are equivalent for $x > 0$.

93. $f(x) = \ln \dfrac{x^2}{4}$ **94.** $f(x) = \ln \sqrt{x(x^2 + 1)}$

     $g(x) = 2 \ln x - \ln 4$ $g(x) = \frac{1}{2}[\ln x + \ln(x^2 + 1)]$

True or False? In Exercises 95–100, determine whether the statement is true or false given that $f(x) = \ln x$. If it is false, explain why or give an example that shows it is false.

95. $f(0) = 0$

96. $f(ax) = f(a) + f(x), \quad a > 0, x > 0$

97. $f(x - 2) = f(x) - f(2), \quad x > 2$

98. $\sqrt{f(x)} = \frac{1}{2} f(x)$

99. If $f(u) = 2f(v)$, then $v = u^2$.

100. If $f(x) < 0$, then $0 < x < 1$.

(T) 101. Research Project Use a graphing utility to graph

$$y = 10 \ln\left(\frac{10 + \sqrt{100 - x^2}}{10}\right) - \sqrt{100 - x^2}$$

over the interval $(0, 10]$. This graph is called a *tractrix* or *pursuit curve*. Use your school's library, the Internet, or some other reference source to find information about a tractrix. Explain how such a curve can arise in a real-life setting.

Section 10.5

Derivatives of Logarithmic Functions

■ Find derivatives of natural logarithmic functions.
■ Use calculus to analyze the graphs of functions that involve the natural logarithmic function.
■ Use the definition of logarithms and the change-of-base formula to evaluate logarithmic expressions involving other bases.
■ Find derivatives of exponential and logarithmic functions involving other bases.

Derivatives of Logarithmic Functions

Implicit differentiation can be used to develop the derivative of the natural logarithmic function.

$$y = \ln x \qquad \text{Natural logarithmic function}$$

$$e^y = x \qquad \text{Write in exponential form.}$$

$$\frac{d}{dx}[e^y] = \frac{d}{dx}[x] \qquad \text{Differentiate with respect to } x.$$

$$e^y \frac{dy}{dx} = 1 \qquad \text{Chain Rule}$$

$$\frac{dy}{dx} = \frac{1}{e^y} \qquad \text{Divide each side by } e^y.$$

$$\frac{dy}{dx} = \frac{1}{x} \qquad \text{Substitute } x \text{ for } e^y.$$

This result and its Chain Rule version are summarized below.

Derivative of the Natural Logarithmic Function

Let u be a differentiable function of x.

1. $\dfrac{d}{dx}[\ln x] = \dfrac{1}{x}$ **2.** $\dfrac{d}{dx}[\ln u] = \dfrac{1}{u}\dfrac{du}{dx}$

Example 1 **Differentiating a Logarithmic Function**

Find the derivative of

$$f(x) = \ln 2x.$$

SOLUTION Let $u = 2x$. Then $du/dx = 2$, and you can apply the Chain Rule as shown.

$$f'(x) = \frac{1}{u}\frac{du}{dx} = \frac{1}{2x}(2) = \frac{1}{x}$$

✓**CHECKPOINT 1**

Find the derivative of $f(x) = \ln 5x$. ■

Example 2 Differentiating Logarithmic Functions

Find the derivative of each function.

a. $f(x) = \ln(2x^2 + 4)$ **b.** $f(x) = x \ln x$ **c.** $f(x) = \dfrac{\ln x}{x}$

SOLUTION

a. Let $u = 2x^2 + 4$. Then $du/dx = 4x$, and you can apply the Chain Rule.

$$f'(x) = \frac{1}{u}\frac{du}{dx}$$ Chain Rule

$$= \frac{1}{2x^2 + 4}(4x)$$

$$= \frac{2x}{x^2 + 2}$$ Simplify.

b. Using the Product Rule, you can find the derivative.

$$f'(x) = x\frac{d}{dx}[\ln x] + (\ln x)\frac{d}{dx}[x]$$ Product Rule

$$= x\left(\frac{1}{x}\right) + (\ln x)(1)$$

$$= 1 + \ln x$$ Simplify.

c. Using the Quotient Rule, you can find the derivative.

$$f'(x) = \frac{x\dfrac{d}{dx}[\ln x] - (\ln x)\dfrac{d}{dx}[x]}{x^2}$$ Quotient Rule

$$= \frac{x\left(\dfrac{1}{x}\right) - \ln x}{x^2}$$

$$= \frac{1 - \ln x}{x^2}$$ Simplify.

STUDY TIP

When you are differentiating logarithmic functions, it is often helpful to use the properties of logarithms to rewrite the function *before* differentiating. To see the advantage of rewriting before differentiating, try using the Chain Rule to differentiate $f(x) = \ln\sqrt{x + 1}$ and compare your work with that shown in Example 3.

✓ CHECKPOINT 2

Find the derivative of each function.

a. $f(x) = \ln(x^2 - 4)$

b. $f(x) = x^2 \ln x$

c. $f(x) = -\dfrac{\ln x}{x^2}$ ∎

Example 3 Rewriting Before Differentiating

Find the derivative of $f(x) = \ln\sqrt{x + 1}$.

SOLUTION

$$f(x) = \ln\sqrt{x + 1}$$ Write original function.

$$= \ln(x + 1)^{1/2}$$ Rewrite with rational exponent.

$$= \frac{1}{2}\ln(x + 1)$$ Property of logarithms

$$f'(x) = \frac{1}{2}\left(\frac{1}{x + 1}\right)$$ Differentiate.

$$= \frac{1}{2(x + 1)}$$ Simplify.

✓ CHECKPOINT 3

Find the derivative of $f(x) = \ln\sqrt[3]{x + 1}$.   ∎

The next example is an even more dramatic illustration of the benefit of rewriting a function before differentiating.

Example 4 Rewriting Before Differentiating

Find the derivative of $f(x) = \ln[x(x^2 + 1)^2]$.

SOLUTION

$$f(x) = \ln[x(x^2 + 1)^2] \qquad \text{Write original function.}$$
$$= \ln x + \ln(x^2 + 1)^2 \qquad \text{Logarithmic properties}$$
$$= \ln x + 2\ln(x^2 + 1) \qquad \text{Logarithmic properties}$$
$$f'(x) = \frac{1}{x} + 2\left(\frac{2x}{x^2 + 1}\right) \qquad \text{Differentiate.}$$
$$= \frac{1}{x} + \frac{4x}{x^2 + 1} \qquad \text{Simplify.}$$

✓CHECKPOINT 4

Find the derivative of $f(x) = \ln[x^2\sqrt{x^2 + 1}]$. ▪

Applications

Example 5   Analyzing a Graph

Analyze the graph of the function $f(x) = \dfrac{x^2}{2} - \ln x$.

SOLUTION   From Figure 10.16, it appears that the function has a minimum at $x = 1$. To find the minimum analytically, find the critical numbers by setting the derivative of f equal to zero and solving for x.

$$f(x) = \frac{x^2}{2} - \ln x \qquad \text{Write original function.}$$

$$f'(x) = x - \frac{1}{x} \qquad \text{Differentiate.}$$

$$x - \frac{1}{x} = 0 \qquad \text{Set derivative equal to 0.}$$

$$x = \frac{1}{x} \qquad \text{Add } 1/x \text{ to each side.}$$

$$x^2 = 1 \qquad \text{Multiply each side by } x.$$

$$x = \pm 1 \qquad \text{Take square root of each side.}$$

Of these two possible critical numbers, only the positive one lies in the domain of f. By applying the First-Derivative Test, you can confirm that the function has a relative minimum when $x = 1$.

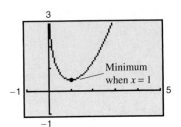

FIGURE 10.16

✓ CHECKPOINT 5

Determine the relative extrema of the function

$$f(x) = x - 2 \ln x. \ \blacksquare$$

Example 6   Finding a Rate of Change

A group of 200 college students was tested every 6 months over a four-year period. The group was composed of students who took Spanish during the fall semester of their freshman year and did not take subsequent Spanish courses. The average test score p (in percent) is modeled by

$$p = 91.6 - 15.6 \ln(t + 1), \quad 0 \le t \le 48$$

where t is the time in months, as shown in Figure 10.17. At what rate was the average score changing after 1 year?

SOLUTION   The rate of change is

$$\frac{dp}{dt} = -\frac{15.6}{t + 1}.$$

When $t = 12$, $dp/dt = -1.2$, which means that the average score was decreasing at the rate of 1.2% per month.

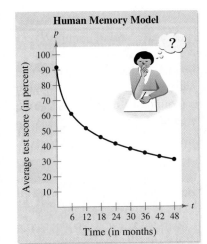

Human Memory Model

FIGURE 10.17

✓ CHECKPOINT 6

Suppose the average test score p in Example 6 was modeled by $p = 92.3 - 16.9 \ln(t + 1)$, where t is the time in months. How would the rate at which the average test score changed after 1 year compare with that of the model in Example 6? ■

Other Bases

This chapter began with a definition of a general exponential function

$$f(x) = a^x$$

where a is a positive number such that $a \neq 1$. The corresponding **logarithm to the base a** is defined by

$$\log_a x = b \quad \text{if and only if} \quad a^b = x.$$

As with the natural logarithmic function, the domain of the logarithmic function to the base a is the set of positive numbers.

TECHNOLOGY

(T) Use a graphing utility to graph the three functions $y_1 = \log_2 x = \ln x / \ln 2$, $y_2 = 2^x$, and $y_3 = x$ in the same viewing window. Explain why the graphs of y_1 and y_2 are reflections of each other in the line $y_3 = x$.

Example 7 Evaluating Logarithms

Evaluate each logarithm without using a calculator.

a. $\log_2 8$ **b.** $\log_{10} 100$ **c.** $\log_{10} \frac{1}{10}$ **d.** $\log_3 81$

SOLUTION

a. $\log_2 8 = 3$ $2^3 = 8$

b. $\log_{10} 100 = 2$ $10^2 = 100$

c. $\log_{10} \frac{1}{10} = -1$ $10^{-1} = \frac{1}{10}$

d. $\log_3 81 = 4$ $3^4 = 81$

✓**CHECKPOINT 7**

Evaluate each logarithm without using a calculator.

a. $\log_2 16$

b. $\log_{10} \frac{1}{100}$

c. $\log_2 \frac{1}{32}$

d. $\log_5 125$ ▨

Logarithms to the base 10 are called **common logarithms.** Most calculators have only two logarithm keys—a natural logarithm key denoted by (LN) and a common logarithm key denoted by (LOG). Logarithms to other bases can be evaluated with the following change-of-base formula.

$$\log_a x = \frac{\ln x}{\ln a} \qquad \text{Change-of-base formula}$$

Example 8 Evaluating Logarithms

Use the change-of-base formula and a calculator to evaluate each logarithm.

a. $\log_2 3$ **b.** $\log_3 6$ **c.** $\log_2(-1)$

SOLUTION   In each case, use the change-of-base formula and a calculator.

a. $\log_2 3 = \dfrac{\ln 3}{\ln 2} \approx 1.585$ $\log_a x = \dfrac{\ln x}{\ln a}$

b. $\log_3 6 = \dfrac{\ln 6}{\ln 3} \approx 1.631$ $\log_a x = \dfrac{\ln x}{\ln a}$

c. $\log_2(-1)$ is not defined.

✓**CHECKPOINT 8**

Use the change-of-base formula and a calculator to evaluate each logarithm.

a. $\log_2 5$

b. $\log_3 18$

c. $\log_4 80$

d. $\log_{16} 0.25$ ▨

To find derivatives of exponential or logarithmic functions to bases other than e, you can either convert to base e or use the differentiation rules shown on the next page.

STUDY TIP

Remember that you can convert to base e using the formulas

$$a^x = e^{(\ln a)x}$$

and

$$\log_a x = \left(\frac{1}{\ln a}\right) \ln x.$$

Other Bases and Differentiation

Let u be a differentiable function of x.

1. $\dfrac{d}{dx}[a^x] = (\ln a)a^x$ **2.** $\dfrac{d}{dx}[a^u] = (\ln a)a^u \dfrac{du}{dx}$

3. $\dfrac{d}{dx}[\log_a x] = \left(\dfrac{1}{\ln a}\right)\dfrac{1}{x}$ **4.** $\dfrac{d}{dx}[\log_a u] = \left(\dfrac{1}{\ln a}\right)\left(\dfrac{1}{u}\right)\dfrac{du}{dx}$

PROOF   By definition, $a^x = e^{(\ln a)x}$. So, you can prove the first rule by letting $u = (\ln a)x$ and differentiating with base e to obtain

$$\frac{d}{dx}[a^x] = \frac{d}{dx}[e^{(\ln a)x}] = e^u \frac{du}{dx} = e^{(\ln a)x}(\ln a) = (\ln a)a^x.$$

Example 9   Finding a Rate of Change

Radioactive carbon isotopes have a half-life of 5715 years. If 1 gram of the isotopes is present in an object now, the amount A (in grams) that will be present after t years is

$$A = \left(\frac{1}{2}\right)^{t/5715}.$$

At what rate is the amount changing when $t = 10{,}000$ years?

SOLUTION   The derivative of A with respect to t is

$$\frac{dA}{dt} = \left(\ln \frac{1}{2}\right)\left(\frac{1}{2}\right)^{t/5715}\left(\frac{1}{5715}\right).$$

When $t = 10{,}000$, the rate at which the amount is changing is

$$\left(\ln \frac{1}{2}\right)\left(\frac{1}{2}\right)^{10{,}000/5715}\left(\frac{1}{5715}\right) \approx -0.000036$$

which implies that the amount of isotopes in the object is decreasing at the rate of 0.000036 gram per year.

✓ CHECKPOINT 9

Use a graphing utility to graph the model in Example 9. Describe the rate at which the amount is changing as time t increases. ∎

┌─ **CONCEPT CHECK** ─┐

1. What is the derivative of $f(x) = \ln x$?

2. What is the derivative of $f(x) = \ln u$? (Assume u is a differentiable function of x.)

3. Complete the following: The change-of-base formula for base e is given by $\log_a x = $ _____.

4. Logarithms to the base e are called natural logarithms. What are logarithms to the base 10 called?

Skills Review 10.5   The following warm-up exercises involve skills that were covered in earlier sections. You will use these skills in the exercise set for this section. For additional help, review Sections 8.1, 8.2, and 10.4.

In Exercises 1–6, expand the logarithmic expression.

1. $\ln(x + 1)^2$

2. $\ln x(x + 1)$

3. $\ln \dfrac{x}{x + 1}$

4. $\ln\left(\dfrac{x}{x - 3}\right)^3$

5. $\ln \dfrac{4x(x - 7)}{x^2}$

6. $\ln x^3(x + 1)$

In Exercises 7 and 8, find dy/dx implicitly.

7. $y^2 + xy = 7$

8. $x^2 y - xy^2 = 3x$

In Exercises 9 and 10, find the second derivative of f.

9. $f(x) = x^2(x + 1) - 3x^3$

10. $f(x) = -\dfrac{1}{x^2}$

Exercises 10.5

See www.CalcChat.com for worked-out solutions to odd-numbered exercises.

In Exercises 1–4, find the slope of the tangent line to the graph of the function at the point (1, 0).

1. $y = \ln x^3$

2. $y = \ln x^{5/2}$

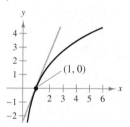

3. $y = \ln x^2$

4. $y = \ln x^{1/2}$

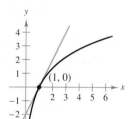

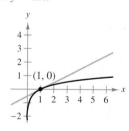

In Exercises 5–26, find the derivative of the function.

5. $y = \ln x^2$

6. $f(x) = \ln 2x$

7. $y = \ln(x^2 + 3)$

8. $f(x) = \ln(1 - x^2)$

9. $y = \ln \sqrt{x - 4}$

10. $y = \ln(1 - x)^{3/2}$

11. $y = (\ln x)^4$

12. $y = (\ln x^2)^2$

13. $f(x) = 2x \ln x$

14. $y = \dfrac{\ln x}{x^2}$

15. $y = \ln\left(x\sqrt{x^2 - 1}\right)$

16. $y = \ln \dfrac{x}{x^2 + 1}$

17. $y = \ln \dfrac{x}{x + 1}$

18. $y = \ln \dfrac{x^2}{x^2 + 1}$

19. $y = \ln \sqrt[3]{\dfrac{x - 1}{x + 1}}$

20. $y = \ln \sqrt{\dfrac{x + 1}{x - 1}}$

21. $y = \ln \dfrac{\sqrt{4 + x^2}}{x}$

22. $y = \ln\left(x\sqrt{4 + x^2}\right)$

23. $g(x) = e^{-x} \ln x$

24. $f(x) = x \ln e^{x^2}$

25. $g(x) = \ln \dfrac{e^x + e^{-x}}{2}$

26. $f(x) = \ln \dfrac{1 + e^x}{1 - e^x}$

In Exercises 27–30, write the expression with base e.

27. 2^x

28. 3^x

29. $\log_4 x$

30. $\log_3 x$

In Exercises 31–38, use a calculator to evaluate the logarithm. Round to three decimal places.

31. $\log_4 7$

32. $\log_6 10$

33. $\log_2 48$

34. $\log_5 12$

35. $\log_3 \frac{1}{2}$

36. $\log_7 \frac{2}{9}$

37. $\log_{1/5} 31$

38. $\log_{2/3} 32$

In Exercises 39–48, find the derivative of the function.

39. $y = 3^x$

40. $y = \left(\frac{1}{4}\right)^x$

41. $f(x) = \log_2 x$

42. $g(x) = \log_5 x$

43. $h(x) = 4^{2x-3}$

44. $y = 6^{5x}$

45. $y = \log_{10}(x^2 + 6x)$

46. $f(x) = 10^{x^2}$

47. $y = x2^x$

48. $y = x3^{x+1}$

In Exercises 49–52, determine an equation of the tangent line to the function at the given point.

Function	Point
49. $y = x \ln x$	$(1, 0)$
50. $y = \dfrac{\ln x}{x}$	$\left(e, \dfrac{1}{e}\right)$
51. $y = \log_3 x$	$(27, 3)$
52. $g(x) = \log_{10} 2x$	$(5, 1)$

In Exercises 53–56, find dy/dx implicitly.

53. $x^2 - 3 \ln y + y^2 = 10$

54. $\ln xy + 5x = 30$

55. $4x^3 + \ln y^2 + 2y = 2x$

56. $4xy + \ln(x^2 y) = 7$

In Exercises 57 and 58, use implicit differentiation to find an equation of the tangent line to the graph at the given point.

57. $x + y - 1 = \ln(x^2 + y^2)$, $(1, 0)$

58. $y^2 + \ln(xy) = 2$, $(e, 1)$

In Exercises 59–64, find the second derivative of the function.

59. $f(x) = x \ln \sqrt{x} + 2x$

60. $f(x) = 3 + 2 \ln x$

61. $f(x) = 2 + x \ln x$

62. $f(x) = \dfrac{\ln x}{x} + x$

63. $f(x) = 5^x$

64. $f(x) = \log_{10} x$

65. Sound Intensity The relationship between the number of decibels β and the intensity of a sound I in watts per square centimeter is given by

$$\beta = 10 \log_{10}\left(\frac{I}{10^{-16}}\right).$$

Find the rate of change in the number of decibels when the intensity is 10^{-4} watt per square centimeter.

66. Chemistry The temperatures T (°F) at which water boils at selected pressures p (pounds per square inch) can be modeled by

$$T = 87.97 + 34.96 \ln p + 7.91 \sqrt{p}.$$

Find the rate of change of the temperature when the pressure is 60 pounds per square inch.

In Exercises 67–72, find the slope of the graph at the indicated point. Then write an equation of the tangent line to the graph of the function at the given point.

67. $f(x) = 1 + 2x \ln x$, $(1, 1)$

68. $f(x) = 2 \ln x^3$, $(e, 6)$

69. $f(x) = \ln \dfrac{5(x + 2)}{x}$, $(-2.5, 0)$

70. $f(x) = \ln\left(x\sqrt{x + 3}\right)$, $(1.2, 0.9)$

71. $f(x) = x \log_2 x$, $(1, 0)$ **72.** $f(x) = x^2 \log_3 x$, $(1, 0)$

In Exercises 73–78, graph and analyze the function. Include any relative extrema and points of inflection in your analysis. Use a graphing utility to verify your results.

73. $y = x - \ln x$

74. $y = \dfrac{x}{\ln x}$

75. $y = \dfrac{\ln x}{x}$

76. $y = x \ln x$

77. $y = x^2 \ln \dfrac{x}{4}$

78. $y = (\ln x)^2$

Demand In Exercises 79 and 80, find dx/dp for the demand function. Interpret this rate of change when the price is $10.

79. $x = \ln \dfrac{1000}{p}$

80. $x = \dfrac{500}{\ln(p^2 + 1)}$

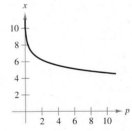

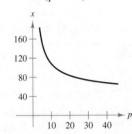

81. Demand Solve the demand function in Exercise 79 for p. Use the result to find dp/dx. Then find the rate of change when $p = \$10$. What is the relationship between this derivative and dx/dp?

82. Demand Solve the demand function in Exercise 80 for p. Use the result to find dp/dx. Then find the rate of change when $p = \$10$. What is the relationship between this derivative and dx/dp?

83. Minimum Average Cost The cost of producing x units of a product is modeled by

$$C = 500 + 300x - 300 \ln x, \quad x \geq 1.$$

(a) Find the average cost function $\overline{C}$.

(b) Analytically find the minimum average cost. Use a graphing utility to confirm your result.

84. Minimum Average Cost The cost of producing x units of a product is modeled by

$$C = 100 + 25x - 120 \ln x, \quad x \geq 1.$$

(a) Find the average cost function $\overline{C}$.

(b) Analytically find the minimum average cost. Use a graphing utility to confirm your result.

85. Consumer Trends The retail sales S (in billions of dollars per year) of e-commerce companies in the United States from 1999 through 2004 are shown in the table.

t	9	10	11	12	13	14
S	14.5	27.8	34.5	45.0	56.6	70.9

The data can be modeled by $S = -254.9 + 121.95 \ln t$, where $t = 9$ corresponds to 1999. *(Source: U.S. Census Bureau)*

(T) (a) Use a graphing utility to plot the data and graph S over the interval $[9, 14]$.

(b) At what rate were the sales changing in 2002?

86. Home Mortgage The term t (in years) of a $200,000 home mortgage at 7.5% interest can be approximated by

$$t = -13.375 \ln \frac{x - 1250}{x}, \quad x > 1250$$

where x is the monthly payment in dollars.

(T) (a) Use a graphing utility to graph the model.

(b) Use the model to approximate the term of a home mortgage for which the monthly payment is $1398.43. What is the total amount paid?

(c) Use the model to approximate the term of a home mortgage for which the monthly payment is $1611.19. What is the total amount paid?

(d) Find the instantaneous rate of change of t with respect to x when $x = \$1398.43$ and $x = \$1611.19$.

(e) Write a short paragraph describing the benefit of the higher monthly payment.

87. Earthquake Intensity On the Richter scale, the magnitude R of an earthquake of intensity I is given by

$$R = \frac{\ln I - \ln I_0}{\ln 10}$$

where I_0 is the minimum intensity used for comparison. Assume $I_0 = 1$.

(a) Find the intensity of the 1906 San Francisco earthquake for which $R = 8.3$.

(b) Find the intensity of the May 26, 2006 earthquake in Java, Indonesia for which $R = 6.3$.

(c) Find the factor by which the intensity is increased when the value of R is doubled.

(d) Find dR/dI.

88. Learning Theory Students in a learning theory study were given an exam and then retested monthly for 6 months with an equivalent exam. The data obtained in the study are shown in the table, where t is the time in months after the initial exam and s is the average score for the class.

t	1	2	3	4	5	6
s	84.2	78.4	72.1	68.5	67.1	65.3

(a) Use these data to find a logarithmic equation that relates t and s.

(T) (b) Use a graphing utility to plot the data and graph the model. How well does the model fit the data?

(c) Find the rate of change of s with respect to t when $t = 2$. Interpret the meaning in the context of the problem.

Business Capsule

AP/Wide World Photos

Lillian Vernon Corporation is a leading national catalog and online retailer that markets gift, household, children's, and fashion accessory products. Lilly Menasche founded the company in Mount Vernon, New York in 1951 using $2000 of wedding gift money. Today, headquartered in Virginia Beach, Virginia, Lillian Vernon's annual sales exceed $287 million. More than 3.3 million packages were shipped in 2006.

89. Research Project Use your school's library, the Internet, or some other reference source to research information about a mail-order or e-commerce company, such as that mentioned above. Collect data about the company (sales or membership over a 20-year period, for example) and find a mathematical model to represent the data.

Section 10.6

Exponential Growth and Decay

■ Use exponential growth and decay to model real-life situations.

Exponential Growth and Decay

In this section, you will learn to create models of *exponential growth and decay.* Real-life situations that involve exponential growth and decay deal with a substance or population whose *rate of change at any time t is proportional to the amount of the substance present at that time.* For example, the rate of decomposition of a radioactive substance is proportional to the amount of radioactive substance at a given instant. In its simplest form, this relationship is described by the equation below.

Rate of change of y is   proportional to y.

$$\frac{dy}{dt} = ky$$

In this equation, k is a constant and y is a function of t. The solution of this equation is shown below.

Law of Exponential Growth and Decay

If y is a positive quantity whose rate of change with respect to time is proportional to the quantity present at any time t, then y is of the form

$$y = Ce^{kt}$$

where C is the **initial value** and k is the **constant of proportionality.** **Exponential growth** is indicated by $k > 0$ and **exponential decay** by $k < 0$.

DISCOVERY

Use a graphing utility to graph $y = Ce^{2t}$ for $C = 1, 2,$ and 5. How does the value of C affect the shape of the graph? Now graph $y = 2e^{kt}$ for $k = -2, -1, 0, 1,$ and 2. How does the value of k affect the shape of the graph? Which function grows faster, $y = e^x$ or $y = x^{10}$?

PROOF   Because the rate of change of y is proportional to y, you can write

$$\frac{dy}{dt} = ky.$$

You can see that $y = Ce^{kt}$ is a solution of this equation by differentiating to obtain $dy/dt = kCe^{kt}$ and substituting

$$\frac{dy}{dt} = kCe^{kt} = k(Ce^{kt}) = ky.$$

STUDY TIP

In the model $y = Ce^{kt}$, C is called the "initial value" because when $t = 0$

$$y = Ce^{k(0)} = C(1) = C.$$

Applications

Much of the cost of nuclear energy is the cost of disposing of radioactive waste. Because of the long half-life of the waste, it must be stored in containers that will remain undisturbed for thousands of years.

Radioactive decay is measured in terms of **half-life,** the number of years required for half of the atoms in a sample of radioactive material to decay. The half-lives of some common radioactive isotopes are as shown.

Uranium (^{238}U)	4,470,000,000 years
Plutonium (^{239}Pu)	24,100 years
Carbon (^{14}C)	5,715 years
Radium (^{226}Ra)	1,599 years
Einsteinium (^{254}Es)	276 days
Nobelium (^{257}No)	25 seconds

Example 1
MAKE A DECISION   Modeling Radioactive Decay

A sample contains 1 gram of radium. Will more than 0.5 gram of radium remain after 1000 years?

SOLUTION   Let y represent the mass (in grams) of the radium in the sample. Because the rate of decay is proportional to y, you can apply the Law of Exponential Decay to conclude that y is of the form $y = Ce^{kt}$, where t is the time in years. From the given information, you know that $y = 1$ when $t = 0$. Substituting these values into the model produces

$$1 = Ce^{k(0)} \qquad \text{Substitute 1 for } y \text{ and 0 for } t.$$

which implies that $C = 1$. Because radium has a half-life of 1599 years, you know that $y = \frac{1}{2}$ when $t = 1599$. Substituting these values into the model allows you to solve for k.

$$y = e^{kt} \qquad \text{Exponential decay model}$$
$$\frac{1}{2} = e^{k(1599)} \qquad \text{Substitute } \tfrac{1}{2} \text{ for } y \text{ and 1599 for } t.$$
$$\ln \tfrac{1}{2} = 1599k \qquad \text{Take natural log of each side.}$$
$$\tfrac{1}{1599} \ln \tfrac{1}{2} = k \qquad \text{Divide each side by 1599.}$$

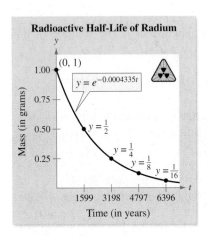

Radioactive Half-Life of Radium

$(0, 1)$

$y = e^{-0.0004335t}$

$y = \frac{1}{2}$

$y = \frac{1}{4}$

$y = \frac{1}{8}$

$y = \frac{1}{16}$

Time (in years)

Mass (in grams)

FIGURE 10.18

So, $k \approx -0.0004335$, and the exponential decay model is $y = e^{-0.0004335t}$. To find the amount of radium remaining in the sample after 1000 years, substitute $t = 1000$ into the model. This produces

$$y = e^{-0.0004335(1000)} \approx 0.648 \text{ gram.}$$

Yes, more than 0.5 gram of radium will remain after 1000 years. The graph of the model is shown in Figure 10.18.

✔ **CHECKPOINT 1**

Use the model in Example 1 to determine the number of years required for a one-gram sample of radium to decay to 0.4 gram. ∎

Note: Instead of approximating the value of k in Example 1, you could leave the value exact and obtain

$$y = e^{\ln[(1/2)^{(t/1599)}]} = \frac{1}{2}^{(t/1599)}.$$

This version of the model clearly shows the "half-life." When $t = 1599$, the value of y is $\frac{1}{2}$. When $t = 2(1599)$, the value of y is $\frac{1}{4}$, and so on.

Guidelines for Modeling Exponential Growth and Decay

1. Use the given information to write *two* sets of conditions involving y and t.

2. Substitute the given conditions into the model $y = Ce^{kt}$ and use the results to solve for the constants C and k. (If one of the conditions involves $t = 0$, substitute that value first to solve for C.)

3. Use the model $y = Ce^{kt}$ to answer the question.

Example 2   **Modeling Population Growth**

In a research experiment, a population of fruit flies is increasing in accordance with the exponential growth model. After 2 days, there are 100 flies, and after 4 days, there are 300 flies. How many flies will there be after 5 days?

Algebra Review

For help with the algebra in Example 2, see Example 1(c) in the *Chapter 10 Algebra Review* on page 796.

SOLUTION   Let y be the number of flies at time t. From the given information, you know that $y = 100$ when $t = 2$ and $y = 300$ when $t = 4$. Substituting this information into the model $y = Ce^{kt}$ produces

$$100 = Ce^{2k} \quad \text{and} \quad 300 = Ce^{4k}.$$

To solve for k, solve for C in the first equation and substitute the result into the second equation.

$$300 = Ce^{4k} \qquad \text{Second equation}$$

$$300 = \left(\frac{100}{e^{2k}}\right)e^{4k} \qquad \text{Substitute } 100/e^{2k} \text{ for } C.$$

$$\frac{300}{100} = e^{2k} \qquad \text{Divide each side by 100.}$$

$$\ln 3 = 2k \qquad \text{Take natural log of each side.}$$

$$\frac{1}{2}\ln 3 = k \qquad \text{Solve for } k.$$

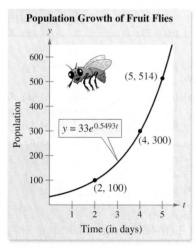

Population Growth of Fruit Flies

$y = 33e^{0.5493t}$

(5, 514)

(4, 300)

(2, 100)

Time (in days)

FIGURE 10.19

Using $k = \frac{1}{2}\ln 3 \approx 0.5493$, you can determine that $C \approx 100/e^{2(0.5493)} \approx 33$. So, the exponential growth model is

$$y = 33e^{0.5493t}$$

as shown in Figure 10.19. This implies that, after 5 days, the population is

$$y = 33e^{0.5493(5)} \approx 514 \text{ flies.}$$

✓CHECKPOINT 2

Find the exponential growth model if a population of fruit flies is 100 after 2 days and 400 after 4 days. ■

| Example 3 | **Modeling Compound Interest** | |

Money is deposited in an account for which the interest is compounded continuously. The balance in the account doubles in 6 years. What is the annual interest rate?

SOLUTION   The balance A in an account with continuously compounded interest is given by the exponential growth model

$$A = Pe^{rt} \qquad \text{Exponential growth model}$$

where P is the original deposit, r is the annual interest rate (in decimal form), and t is the time (in years). From the given information, you know that $A = 2P$ when $t = 6$, as shown in Figure 10.20. Use this information to solve for r.

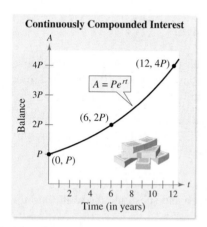

Continuously Compounded Interest

FIGURE 10.20

$$A = Pe^{rt} \qquad \text{Exponential growth model}$$
$$2P = Pe^{r(6)} \qquad \text{Substitute } 2P \text{ for } A \text{ and } 6 \text{ for } t.$$
$$2 = e^{6r} \qquad \text{Divide each side by } P.$$
$$\ln 2 = 6r \qquad \text{Take natural log of each side.}$$
$$\tfrac{1}{6} \ln 2 = r \qquad \text{Divide each side by } 6.$$

So, the annual interest rate is

$$r = \tfrac{1}{6} \ln 2$$
$$\approx 0.1155$$

or about 11.55%.

✓ CHECKPOINT 3

Find the annual interest rate if the balance in an account doubles in 8 years where the interest is compounded continuously. ▪

Each of the examples in this section uses the exponential growth model in which the base is e. Exponential growth, however, can be modeled with *any* base. That is, the model

$$y = Ca^{bt}$$

also represents exponential growth. (To see this, note that the model can be written in the form $y = Ce^{(\ln a)bt}$.) In some real-life settings, bases other than e are more convenient. For instance, in Example 1, knowing that the half-life of radium is 1599 years, you can immediately write the exponential decay model as

$$y = \left(\frac{1}{2}\right)^{t/1599}.$$

Using this model, the amount of radium left in the sample after 1000 years is

$$y = \left(\frac{1}{2}\right)^{1000/1599} \approx 0.648 \text{ gram}$$

which is the same answer obtained in Example 1.

STUDY TIP

Can you see why you can immediately write the model $y = \left(\frac{1}{2}\right)^{t/1599}$ for the radioactive decay described in Example 1? Notice that when $t = 1599$, the value of y is $\frac{1}{2}$, when $t = 3198$, the value of y is $\frac{1}{4}$, and so on.

TECHNOLOGY

Fitting an Exponential Model to Data

(T) Most graphing utilities have programs that allow you to find the *least squares regression exponential model* for data. Depending on the type of graphing utility, you can fit the data to a model of the form

$y = ab^x$ Exponential model with base b

or

$y = ae^{bx}$.   Exponential model with base e

To see how to use such a program, consider the example below.

The cash flow per share y for Harley-Davidson, Inc. from 1998 through 2005 is shown in the table.   *(Source: Harley-Davidson, Inc.)*

x	8	9	10	11	12	13	14	15
y	$0.98	$1.26	$1.59	$1.95	$2.50	$3.18	$3.75	$4.25

In the table, $x = 8$ corresponds to 1998. To fit an exponential model to these data, enter the coordinates listed below into the statistical data bank of a graphing utility.

(8, 0.98), (9, 1.26), (10, 1.59), (11, 1.95),

(12, 2.50), (13, 3.18), (14, 3.75), (15, 4.25)

After running the exponential regression program with a graphing utility that uses the model $y = ab^x$, the display should read $a \approx 0.183$ and $b \approx 1.2397$. (The coefficient of determination of $r^2 \approx 0.993$ tells you that the fit is very good.) So, a model for the data is

$y = 0.183(1.2397)^x$.   Exponential model with base b

If you use a graphing utility that uses the model $y = ae^{bx}$, the display should read $a \approx 0.183$ and $b \approx 0.2149$. The corresponding model is

$y = 0.183e^{0.2149x}$.   Exponential model with base e

The graph of the second model is shown at the right. Notice that one way to interpret the model is that the cash flow per share increased by about 21.5% each year from 1998 through 2005.

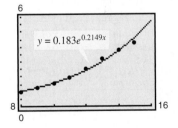

You can use either model to predict the cash flow per share in future years. For instance, in 2006 ($x = 16$), the cash flow per share is predicted to be

$y = 0.183e^{(0.2149)(16)}$

$\approx \$5.70$.

Graph the model $y = 0.183(1.2397)^x$ and use the model to predict the cash flow for 2006. Compare your results with those obtained using the model $y = 0.183e^{0.2149x}$. What do you notice?

Example 4   **Modeling Sales**

Algebra Review

For help with the algebra in Example 4, see Example 1(b) in the *Chapter 10 Algebra Review* on page 796.

Four months after discontinuing advertising on national television, a manufacturer notices that sales have dropped from 100,000 MP3 players per month to 80,000 MP3 players. If the sales follow an exponential pattern of decline, what will they be after another 4 months?

SOLUTION   Let y represent the number of MP3 players, let t represent the time (in months), and consider the exponential decay model

$$y = Ce^{kt}.$$ Exponential decay model

From the given information, you know that $y = 100{,}000$ when $t = 0$. Using this information, you have

$$100{,}000 = Ce^0$$

which implies that $C = 100{,}000$. To solve for k, use the fact that $y = 80{,}000$ when $t = 4$.

$y = 100{,}000e^{kt}$	Exponential decay model
$80{,}000 = 100{,}000e^{k(4)}$	Substitute 80,000 for y and 4 for t.
$0.8 = e^{4k}$	Divide each side by 100,000.
$\ln 0.8 = 4k$	Take natural log of each side.
$\frac{1}{4}\ln 0.8 = k$	Divide each side by 4.

So, $k = \frac{1}{4}\ln 0.8 \approx -0.0558$, which means that the model is

$$y = 100{,}000e^{-0.0558t}.$$

After four more months ($t = 8$), you can expect sales to drop to

$$y = 100{,}000e^{-0.0558(8)}$$
$$\approx 64{,}000 \text{ MP3 players}$$

as shown in Figure 10.21.

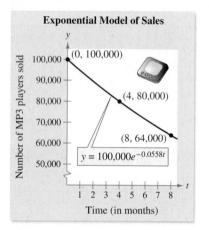

Exponential Model of Sales

(0, 100,000)

(4, 80,000)

(8, 64,000)

$y = 100{,}000e^{-0.0558t}$

Number of MP3 players sold

Time (in months)

FIGURE 10.21

✓**CHECKPOINT 4**

Use the model in Example 4 to determine when sales drop to 50,000 MP3 players. ■

CONCEPT CHECK

1. Describe what the values of C and k represent in the exponential growth and decay model, $y = Ce^{kt}$.

2. For what values of k is $y = Ce^{kt}$ an exponential growth model? an exponential decay model?

3. Can the base used in an exponential growth model be a number other than e?

4. In exponential growth, is the rate of growth constant? Explain why or why not.

Skills Review 10.6

The following warm-up exercises involve skills that were covered in earlier sections. You will use these skills in the exercise set for this section. For additional help, review Sections 10.3 and 10.4.

In Exercises 1–4, solve the equation for k.

1. $12 = 24e^{4k}$

2. $10 = 3e^{5k}$

3. $25 = 16e^{-0.01k}$

4. $22 = 32e^{-0.02k}$

In Exercises 5–8, find the derivative of the function.

5. $y = 32e^{0.23t}$

6. $y = 18e^{0.072t}$

7. $y = 24e^{-1.4t}$

8. $y = 25e^{-0.001t}$

In Exercises 9–12, simplify the expression.

9. $e^{\ln 4}$

10. $4e^{\ln 3}$

11. $e^{\ln(2x+1)}$

12. $e^{\ln(x^2+1)}$

Exercises 10.6

See www.CalcChat.com for worked-out solutions to odd-numbered exercises.

In Exercises 1–6, find the exponential function $y = Ce^{kt}$ that passes through the two given points.

1. $y = Ce^{kt}$

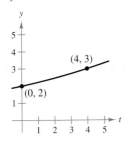

2. $y = Ce^{kt}$

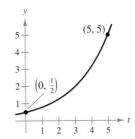

3. $y = Ce^{kt}$

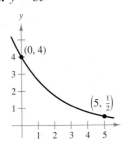

4. $y = Ce^{kt}$

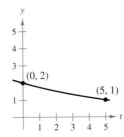

5. $y = Ce^{kt}$

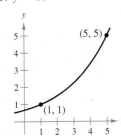

6. $y = Ce^{kt}$

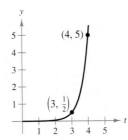

In Exercises 7–10, use the given information to write an equation for y. Confirm your result analytically by showing that the function satisfies the equation $dy/dt = Cy$. Does the function represent exponential growth or exponential decay?

7. $\dfrac{dy}{dt} = 2y$, $y = 10$ when $t = 0$

8. $\dfrac{dy}{dt} = -\dfrac{2}{3}y$, $y = 20$ when $t = 0$

9. $\dfrac{dy}{dt} = -4y$, $y = 30$ when $t = 0$

10. $\dfrac{dy}{dt} = 5.2y$, $y = 18$ when $t = 0$

Radioactive Decay In Exercises 11–16, complete the table for each radioactive isotope.

Isotope	Half-life (in years)	Initial quantity	Amount after 1000 years	Amount after 10,000 years
11. ^{226}Ra	1599	10 grams		
12. ^{226}Ra	1599		1.5 grams	
13. ^{14}C	5715			2 grams
14. ^{14}C	5715	3 grams		
15. ^{239}Pu	24,100		2.1 grams	
16. ^{239}Pu	24,100			0.4 gram

17. Radioactive Decay What percent of a present amount of radioactive radium (^{226}Ra) will remain after 900 years?

18. Radioactive Decay Find the half-life of a radioactive material if after 1 year 99.57% of the initial amount remains.

19. Carbon Dating ^{14}C dating assumes that the carbon dioxide on the Earth today has the same radioactive content as it did centuries ago. If this is true, then the amount of ^{14}C absorbed by a tree that grew several centuries ago should be the same as the amount of ^{14}C absorbed by a similar tree today. A piece of ancient charcoal contains only 15% as much of the radioactive carbon as a piece of modern charcoal. How long ago was the tree burned to make the ancient charcoal? (The half-life of ^{14}C is 5715 years.)

20. Carbon Dating   Repeat Exercise 19 for a piece of charcoal that contains 30% as much radioactive carbon as a modern piece.

In Exercises 21 and 22, find exponential models

$$y_1 = Ce^{k_1 t} \quad \text{and} \quad y_2 = C(2)^{k_2 t}$$

that pass through the points. Compare the values of k_1 and k_2. Briefly explain your results.

21. $(0, 5), (12, 20)$ **22.** $(0, 8), \left(20, \frac{1}{2}\right)$

23. Population Growth   The number of a certain type of bacteria increases continuously at a rate proportional to the number present. There are 150 present at a given time and 450 present 5 hours later.

   (a) How many will there be 10 hours after the initial time?

   (b) How long will it take for the population to double?

   (c) Does the answer to part (b) depend on the starting time? Explain your reasoning.

24. School Enrollment   In 1970, the total enrollment in public universities and colleges in the United States was 5.7 million students. By 2004, enrollment had risen to 13.7 million students. Assume enrollment can be modeled by exponential growth. *(Source: U.S. Census Bureau)*

   (a) Estimate the total enrollments in 1980, 1990, and 2000.

   (b) How many years until the enrollment doubles from the 2004 figure?

   (c) By what percent is the enrollment increasing each year?

Compound Interest   In Exercises 25–32, complete the table for an account in which interest is compounded continuously.

Initial investment	Annual rate	Time to double	Amount after 10 years	Amount after 25 years
25. $1,000	12%			
26. $20,000	$10\frac{1}{2}$%			
27. $750		8 years		
28. $10,000		10 years		

Initial investment	Annual rate	Time to double	Amount after 10 years	Amount after 25 years
29. $500			$1292.85	
30. $2,000				$6008.33
31.	4.5%		$10,000.00	
32.	2%		$2000.00	

In Exercises 33 and 34, determine the principal P that must be invested at interest rate r, compounded continuously, so that $1,000,000 will be available for retirement in t years.

33. $r = 7.5\%, t = 40$ **34.** $r = 10\%, t = 25$

35. Effective Yield   The effective yield is the annual rate i that will produce the same interest per year as the nominal rate r compounded n times per year.

   (a) For a rate r that is compounded n times per year, show that the effective yield is

$$i = \left(1 + \frac{r}{n}\right)^n - 1.$$

   (b) Find the effective yield for a nominal rate of 6%, compounded monthly.

36. Effective Yield   The effective yield is the annual rate i that will produce the same interest per year as the nominal rate r.

   (a) For a rate r that is compounded continuously, show that the effective yield is $i = e^r - 1$.

   (b) Find the effective yield for a nominal rate of 6%, compounded continuously.

Effective Yield   In Exercises 37 and 38, use the results of Exercises 35 and 36 to complete the table showing the effective yield for a nominal rate of r.

Number of compoundings per year	4	12	365	Continuous
Effective yield				

37. $r = 5\%$ **38.** $r = 7\frac{1}{2}\%$

39. Investment: Rule of 70   Verify that the time necessary for an investment to double its value is approximately $70/r$, where r is the annual interest rate entered as a percent.

40. Investment: Rule of 70   Use the Rule of 70 from Exercise 39 to approximate the times necessary for an investment to double in value if (a) $r = 10\%$ and (b) $r = 7\%$.

41. *MAKE A DECISION: REVENUE* The revenues for Sonic Corporation were $151.1 million in 1996 and $693.3 million in 2006. *(Source: Sonic Corporation)*

(a) Use an exponential growth model to estimate the revenue in 2011.

(b) Use a linear model to estimate the 2011 revenue.

(T) (c) Use a graphing utility to graph the models from parts (a) and (b). Which model is more accurate?

(T) **42.** *MAKE A DECISION: SALES* The sales for exercise equipment in the United States were $1824 million in 1990 and $5112 million in 2005. *(Source: National Sporting Goods Association)*

(a) Use the *regression* feature of a graphing utility to find an exponential growth model and a linear model for the data.

(b) Use the exponential growth model to estimate the sales in 2011.

(c) Use the linear model to estimate the sales in 2011.

(d) Use a graphing utility to graph the models from part (a). Which model is more accurate?

43. **Sales** The cumulative sales S (in thousands of units) of a new product after it has been on the market for t years are modeled by

$$S = Ce^{k/t}.$$

During the first year, 5000 units were sold. The saturation point for the market is 30,000 units. That is, the limit of S as $t \to \infty$ is 30,000.

(a) Solve for C and k in the model.

(b) How many units will be sold after 5 years?

(T) (c) Use a graphing utility to graph the sales function.

44. **Sales** The cumulative sales S (in thousands of units) of a new product after it has been on the market for t years are modeled by

$$S = 30(1 - 3^{kt}).$$

During the first year, 5000 units were sold.

(a) Solve for k in the model.

(b) What is the saturation point for this product?

(c) How many units will be sold after 5 years?

(T) (d) Use a graphing utility to graph the sales function.

45. **Learning Curve** The management of a factory finds that the maximum number of units a worker can produce in a day is 30. The learning curve for the number of units N produced per day after a new employee has worked t days is modeled by $N = 30(1 - e^{kt})$. After 20 days on the job, a worker is producing 19 units in a day. How many days should pass before this worker is producing 25 units per day?

46. **Learning Curve** The management in Exercise 45 requires that a new employee be producing at least 20 units per day after 30 days on the job.

(a) Find a learning curve model that describes this minimum requirement.

(b) Find the number of days before a minimal achiever is producing 25 units per day.

47. **Profit** Because of a slump in the economy, a company finds that its annual profits have dropped from $742,000 in 1998 to $632,000 in 2000. If the profit follows an exponential pattern of decline, what is the expected profit for 2003? (Let $t = 0$ correspond to 1998.)

48. **Revenue** A small business assumes that the demand function for one of its new products can be modeled by $p = Ce^{kx}$. When $p = \$45$, $x = 1000$ units, and when $p = \$40$, $x = 1200$ units.

(a) Solve for C and k.

(b) Find the values of x and p that will maximize the revenue for this product.

49. **Revenue** Repeat Exercise 48 given that when $p = \$5$, $x = 300$ units, and when $p = \$4$, $x = 400$ units.

50. **Forestry** The value V (in dollars) of a tract of timber can be modeled by $V = 100,000e^{0.75\sqrt{t}}$, where $t = 0$ corresponds to 1990. If money earns interest at a rate of 4%, compounded continuously, then the present value A of the timber at any time t is $A = Ve^{-0.04t}$. Find the year in which the timber should be harvested to maximize the present value.

51. **Forestry** Repeat Exercise 50 using the model

$$V = 100,000e^{0.6\sqrt{t}}.$$

(T) **52.** *MAKE A DECISION: MODELING DATA* The table shows the population P (in millions) of the United States from 1960 through 2005. *(Source: U.S. Census Bureau)*

Year	1960	1970	1980	1990	2000	2005
Population, P	181	205	228	250	282	297

(a) Use the 1960 and 1970 data to find an exponential model P_1 for the data. Let $t = 0$ represent 1960.

(b) Use a graphing utility to find an exponential model P_2 for the data. Let $t = 0$ represent 1960.

(c) Use a graphing utility to plot the data and graph both models in the same viewing window. Compare the actual data with the predictions. Which model is more accurate?

53. **Extended Application** To work an extended application analyzing the revenue per share for Target Corporation from 1990 through 2005, visit this text's website at *college.hmco.com*. *(Data Source: Target Corporation)*

Algebra Review

Solving Exponential and Logarithmic Equations

To find the extrema or points of inflection of an exponential or logarithmic function, you must know how to solve exponential and logarithmic equations. A few examples are given on page 773. Some additional examples are presented in this Algebra Review.

As with all equations, remember that your basic goal is to isolate the variable on one side of the equation. To do this, you use inverse operations. For instance, to get rid of an exponential expression such as e^{2x}, take the natural log of each side and use the property $\ln e^{2x} = 2x$. Similarly, to get rid of a logarithmic expression such as $\log_2 3x$, exponentiate each side and use the property $2^{\log_2 3x} = 3x$.

Example 1   Solving Exponential Equations

Solve each exponential equation.

a. $25 = 5e^{7t}$ **b.** $80{,}000 = 100{,}000e^{k(4)}$ **c.** $300 = \left(\dfrac{100}{e^{2k}}\right)e^{4k}$

SOLUTION

a.

$25 = 5e^{7t}$	Write original equation.
$5 = e^{7t}$	Divide each side by 5.
$\ln 5 = \ln e^{7t}$	Take natural log of each side.
$\ln 5 = 7t$	Apply the property $\ln e^a = a$.
$\frac{1}{7}\ln 5 = t$	Divide each side by 7.

b.

$80{,}000 = 100{,}000e^{k(4)}$	Example 4, page 792
$0.8 = e^{4k}$	Divide each side by 100,000.
$\ln 0.8 = \ln e^{4k}$	Take natural log of each side.
$\ln 0.8 = 4k$	Apply the property $\ln e^a = a$.
$\frac{1}{4}\ln 0.8 = k$	Divide each side by 4.

c.

$300 = \left(\dfrac{100}{e^{2k}}\right)e^{4k}$	Example 2, page 789
$300 = (100)\dfrac{e^{4k}}{e^{2k}}$	Rewrite product.
$300 = 100e^{4k-2k}$	To divide powers, subtract exponents.
$300 = 100e^{2k}$	Simplify.
$3 = e^{2k}$	Divide each side by 100.
$\ln 3 = \ln e^{2k}$	Take natural log of each side.
$\ln 3 = 2k$	Apply the property $\ln e^a = a$.
$\frac{1}{2}\ln 3 = k$	Divide each side by 2.

Example 2   Solving Logarithmic Equations

Solve each logarithmic equation.

a. $\ln x = 8$ **b.** $3 + 2 \ln x = 2$

c. $2 \ln 3x = 4$ **d.** $\ln x - \ln(x - 1) = 1$

SOLUTION

a.
$\ln x = 8$	Write original equation.
$e^{\ln x} = e^8$	Exponentiate each side.
$x = e^8$	Apply the property $e^{\ln a} = a$.

b.
$3 + 2 \ln x = 2$	Write original equation.
$2 \ln x = -1$	Subtract 3 from each side.
$\ln x = -\dfrac{1}{2}$	Divide each side by 2.
$e^{\ln x} = e^{-1/2}$	Exponentiate each side.
$x = e^{-1/2}$	Apply the property $e^{\ln a} = a$.

c.
$2 \ln 3x = 4$	Write original equation.
$\ln 3x = 2$	Divide each side by 2.
$e^{\ln 3x} = e^2$	Exponentiate each side.
$3x = e^2$	Apply the property $e^{\ln a} = a$.
$x = \frac{1}{3}e^2$	Divide each side by 3.

d.
$\ln x - \ln(x - 1) = 1$	Write original equation.
$\ln \dfrac{x}{x - 1} = 1$	$\ln m - \ln n = \ln(m/n)$
$e^{\ln(x/x-1)} = e^1$	Exponentiate each side.
$\dfrac{x}{x - 1} = e^1$	Apply the property $e^{\ln a} = a$.
$x = ex - e$	Multiply each side by $x - 1$.
$x - ex = -e$	Subtract ex from each side.
$x(1 - e) = -e$	Factor.
$x = \dfrac{-e}{1 - e}$	Divide each side by $1 - e$.
$x = \dfrac{e}{e - 1}$	Simplify.

STUDY TIP

Because the domain of a logarithmic function generally does not include all real numbers, be sure to check for extraneous solutions.

Chapter Summary and Study Strategies

After studying this chapter, you should have acquired the following skills.
The exercise numbers are keyed to the Review Exercises that begin on page 800.
Answers to odd-numbered Review Exercises are given in the back of the text.

Section 10.1	Review Exercises
■ Use the properties of exponents to evaluate and simplify exponential expressions and functions.	*1–16*

$$a^0 = 1, \qquad a^x a^y = a^{x+y}, \qquad \frac{a^x}{a^y} = a^{x-y}, \qquad (a^x)^y = a^{xy}$$

$$(ab)^x = a^x b^x, \qquad \left(\frac{a}{b}\right)^x = \frac{a^x}{b^x}, \qquad a^{-x} = \frac{1}{a^x}$$

■ Use properties of exponents to answer questions about real life.	*17, 18*

Section 10.2

■ Sketch the graphs of exponential functions.	*19–28*
■ Evaluate limits of exponential functions in real life.	*29, 30*
■ Evaluate and graph functions involving the natural exponential function.	*31–34*
■ Graph logistic growth functions.	*35, 36*
■ Solve compound interest problems.	*37–40*

$$A = P(1 + r/n)^{nt}, \quad A = Pe^{rt}$$

■ Solve effective rate of interest problems.	*41, 42*

$$r_{eff} = (1 + r/n)^n - 1$$

■ Solve present value problems.	*43, 44*

$$P = \frac{A}{(1 + r/n)^{nt}}$$

■ Answer questions involving the natural exponential function as a real-life model.	*45, 46*

Section 10.3

■ Find the derivatives of natural exponential functions.	*47–54*

$$\frac{d}{dx}[e^x] = e^x, \quad \frac{d}{dx}[e^u] = e^u \frac{du}{dx}$$

■ Use calculus to analyze the graphs of functions that involve the natural exponential function.	*55–62*

Section 10.4

■ Use the definition of the natural logarithmic function to write exponential equations in logarithmic form, and vice versa.	*63–66*

$$\ln x = b \quad \text{if and only if} \quad e^b = x.$$

Section 10.4 (continued)

$$\ln xy = \ln x + \ln y, \quad \ln \frac{x}{y} = \ln x - \ln y, \quad \ln x^n = n \ln x$$

$$\ln e^x = x, \quad e^{\ln x} = x$$

Section 10.5

$$\frac{d}{dx}[\ln x] = \frac{1}{x}, \quad \frac{d}{dx}[\ln u] = \frac{1}{u}\frac{du}{dx}$$

$$\log_a x = b \quad \text{if and only if} \quad a^b = x$$

$$\log_a x = \frac{\ln x}{\ln a}$$

$$\frac{d}{dx}[a^x] = (\ln a)a^x, \quad \frac{d}{dx}[a^u] = (\ln a)a^u\frac{du}{dx}$$

$$\frac{d}{dx}[\log_a x] = \left(\frac{1}{\ln a}\right)\frac{1}{x}, \quad \frac{d}{dx}[\log_a u] = \left(\frac{1}{\ln a}\right)\left(\frac{1}{u}\right)\frac{du}{dx}$$

Section 10.6

Study Strategies

■ **Classifying Differentiation Rules**   Differentiation rules fall into two basic classes: (1) general rules that apply to all differentiable functions; and (2) specific rules that apply to special types of functions. At this point in the course, you have studied six general rules: the Constant Rule, the Constant Multiple Rule, the Sum Rule, the Difference Rule, the Product Rule, and the Quotient Rule. Although these rules were introduced in the context of algebraic functions, remember that they can also be used with exponential and logarithmic functions. You have also studied three specific rules: the Power Rule, the derivative of the natural exponential function, and the derivative of the natural logarithmic function. Each of these rules comes in two forms: the "simple" version, such as $D_x[e^x] = e^x$, and the Chain Rule version, such as $D_x[e^u] = e^u(du/dx)$.

■ **To Memorize or Not to Memorize?**   When studying mathematics, you need to memorize some formulas and rules. Much of this will come from practice—the formulas that you use most often will be committed to memory. Some formulas, however, are used only infrequently. With these, it is helpful to be able to *derive* the formula from a *known* formula. For instance, knowing the Log Rule for differentiation and the change-of-base formula, $\log_a x = (\ln x)/(\ln a)$, allows you to derive the formula for the derivative of a logarithmic function to base a.

Review Exercises

See www.CalcChat.com for worked-out solutions to odd-numbered exercises.

In Exercises 1–4, evaluate the expression.

1. $32^{3/5}$

2. $25^{3/2}$

3. $\left(\frac{1}{16}\right)^{-3/2}$

4. $\left(\frac{27}{8}\right)^{-1/3}$

In Exercises 5–12, use the properties of exponents to simplify the expression.

5. $\left(\frac{9}{16}\right)^{0}$

6. $(9^{1/3})(3^{1/3})$

7. $\frac{6^3}{36^2}$

8. $\frac{1}{4}\left(\frac{1}{2}\right)^{-3}$

9. $(e^2)^5$

10. $\frac{e^6}{e^4}$

11. $(e^{-1})(e^4)$

12. $(e^{1/2})(e^3)$

(T) In Exercises 13–16, evaluate the function for the indicated value of x. If necessary, use a graphing utility, rounding your answers to three decimal places.

13. $f(x) = 2^{x+3}$, $x = 4$

14. $f(x) = 4^{x-1}$, $x = -2$

15. $f(x) = 1.02^x$, $x = 10$

16. $f(x) = 1.12^x$, $x = 1.3$

17. Revenue   The revenues R (in millions of dollars) for California Pizza Kitchen from 1999 through 2005 can be modeled by

$$R = 39.615(1.183)^t$$

where $t = 9$ corresponds to 1999.   *(Source: California Pizza Kitchen, Inc.)*

(a) Use this model to estimate the net profits in 1999, 2003, and 2005.

(b) Do you think the model will be valid for years beyond 2005? Explain your reasoning.

18. Property Value   Suppose that the value of a piece of property doubles every 12 years. If you buy the property for $55,000, its value t years after the date of purchase should be

$$V(t) = 55,000(2)^{t/12}.$$

Use the model to approximate the value of the property (a) 4 years and (b) 25 years after it is purchased.

In Exercises 19–28, sketch the graph of the function.

19. $f(x) = 9^{x/2}$

20. $g(x) = 16^{3x/2}$

21. $f(t) = \left(\frac{1}{6}\right)^t$

22. $g(t) = \left(\frac{1}{3}\right)^{-t}$

23. $f(x) = \left(\frac{1}{2}\right)^{2x} + 4$

24. $g(x) = \left(\frac{2}{3}\right)^{2x} + 1$

25. $f(x) = e^{-x} + 1$

26. $g(x) = e^{2x} - 1$

27. $f(x) = 1 - e^x$

28. $g(x) = 2 + e^{x-1}$

29. Demand   The demand function for a product is given by

$$p = 12,500 - \frac{10,000}{2 + e^{-0.001x}}$$

where p is the price per unit and x is the number of units produced (see figure). What is the limit of the price as x increases without bound? Explain what this means in the context of the problem.

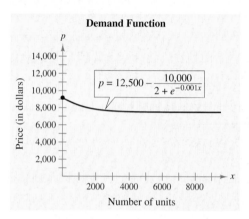

Demand Function

30. Biology: Endangered Species   Biologists consider a species of a plant or animal to be endangered if it is expected to become extinct in less than 20 years. The population y of a certain species is modeled by

$$y = 1096e^{-0.39t}$$

(see figure). Is this species endangered? Explain your reasoning.

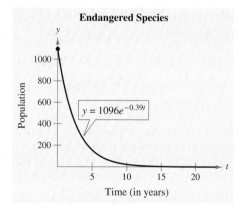

Endangered Species

In Exercises 31–34, evaluate the function at each indicated value.

31. $f(x) = 5e^{x-1}$

    (a) $x = 2$ (b) $x = \frac{1}{2}$ (c) $x = 10$

32. $f(t) = e^{4t} - 2$

    (a) $t = 0$ (b) $t = 2$ (c) $t = -\frac{3}{4}$

33. $g(t) = 6e^{-0.2t}$

    (a) $t = 17$ (b) $t = 50$ (c) $t = 100$

34. $g(x) = \dfrac{24}{1 + e^{-0.3x}}$

    (a) $x = 0$ (b) $x = 300$ (c) $x = 1000$

35. Biology A lake is stocked with 500 fish and the fish population P begins to increase according to the logistic growth model

$$P = \frac{10,000}{1 + 19e^{-t/5}}, \quad t \geq 0$$

where t is measured in months.

(T) (a) Use a graphing utility to graph the function.

    (b) Estimate the number of fish in the lake after 4 months.

    (c) Does the population have a limit as t increases without bound? Explain your reasoning.

    (d) After how many months is the population increasing most rapidly? Explain your reasoning.

36. Medicine On a college campus of 4000 students, the spread of a flu virus through the student body is modeled by

$$P = \frac{4000}{1 + 3999e^{-0.8t}}, \quad t \geq 0$$

where P is the total number of infected people and t is the time, measured in days.

(T) (a) Use a graphing utility to graph the function.

    (b) How many students will be infected after 5 days?

    (c) According to this model, will all the students on campus become infected with the flu? Explain your reasoning.

In Exercises 37 and 38, complete the table to determine the balance A when P dollars is invested at an annual rate of r for t years, compounded n times per year.

n	1	2	4	12	365	Continuous compounding
A						

37. $P = \$1000$, $r = 4\%$, $t = 5$ years

38. $P = \$7000$, $r = 6\%$, $t = 20$ years

In Exercises 39 and 40, \$2000 is deposited in an account. Decide which account, (a) or (b), will have the greater balance after 10 years.

39. (a) 5%, compounded continuously

    (b) 6%, compounded quarterly

40. (a) $6\frac{1}{2}\%$, compounded monthly

    (b) $6\frac{1}{4}\%$, compounded continuously

Effective Rate In Exercises 41 and 42, find the effective rate of interest corresponding to a nominal rate r, compounded (a) quarterly and (b) monthly.

41. $r = 6\%$ **42.** $r = 8.25\%$

43. Present Value How much should be deposited in an account paying 5% interest compounded quarterly in order to have a balance of \$12,000 three years from now?

44. Present Value How much should be deposited in an account paying 8% interest compounded monthly in order to have a balance of \$20,000 five years from now?

45. Vital Statistics The population P (in millions) of people 65 years old and over in the United States from 1990 through 2005 can be modeled by

$$P = 29.7e^{0.01t}, \quad 0 \leq t \leq 15$$

where $t = 0$ corresponds to 1990. Use this model to estimate the populations of people 65 years old and over in 1990, 2000, and 2005. *(Source: U.S. Census Bureau)*

46. Revenue The revenues R (in millions of dollars per year) for Papa John's International from 1998 through 2005 can be modeled by

$$R = -6310 + 1752.5t - 139.23t^2 + 3.634t^3$$
$$+ 0.000017e^t, \quad 8 \leq t \leq 15$$

where $t = 8$ corresponds to 1998. Use this model to estimate the revenues for Papa John's in 1998, 2002, and 2005. *(Source: Papa John's International)*

In Exercises 47–54, find the derivative of the function.

47. $y = 4e^{x^2}$ **48.** $y = 4e^{\sqrt{x}}$

49. $y = \dfrac{x}{e^{2x}}$ **50.** $y = x^2e^x$

51. $y = \sqrt{4e^{4x}}$ **52.** $y = \sqrt[3]{2e^{3x}}$

53. $y = \dfrac{5}{1 + e^{2x}}$ **54.** $y = \dfrac{10}{1 - 2e^x}$

In Exercises 55–62, graph and analyze the function. Include any relative extrema, points of inflection, and asymptotes in your analysis.

55. $f(x) = 4e^{-x}$ **56.** $f(x) = 2e^{x^2}$

57. $f(x) = x^3e^x$ **58.** $f(x) = \dfrac{e^x}{x^2}$

59. $f(x) = \dfrac{1}{xe^x}$

60. $f(x) = \dfrac{x^2}{e^x}$

61. $f(x) = xe^{2x}$

62. $f(x) = xe^{-2x}$

In Exercises 63 and 64, write the logarithmic equation as an exponential equation.

63. $\ln 12 = 2.4849 \ldots$

64. $\ln 0.6 = -0.5108 \ldots$

In Exercises 65 and 66, write the exponential equation as a logarithmic equation.

65. $e^{1.5} = 4.4816 \ldots$

66. $e^{-4} = 0.0183 \ldots$

In Exercises 67–70, sketch the graph of the function.

67. $y = \ln(4 - x)$

68. $y = 5 + \ln x$

69. $y = \ln \dfrac{x}{3}$

70. $y = -2 \ln x$

In Exercises 71–76, use the properties of logarithms to write the expression as a sum, difference, or multiple of logarithms.

71. $\ln \sqrt{x^2(x - 1)}$

72. $\ln \sqrt[3]{x^2 - 1}$

73. $\ln \dfrac{x^2}{(x + 1)^3}$

74. $\ln \dfrac{x^2}{x^2 + 1}$

75. $\ln\left(\dfrac{1 - x}{3x}\right)^3$

76. $\ln\left(\dfrac{x - 1}{x + 1}\right)^2$

In Exercises 77–92, solve the equation for x.

77. $e^{\ln x} = 3$

78. $e^{\ln(x+2)} = 5$

79. $\ln x = 3e^{-1}$

80. $\ln x = 2e^5$

81. $\ln 2x - \ln(3x - 1) = 0$

82. $\ln x - \ln(x + 1) = 2$

83. $e^{2x-1} - 6 = 0$

84. $4e^{2x-3} - 5 = 0$

85. $\ln x + \ln(x - 3) = 0$

86. $2 \ln x + \ln(x - 2) = 0$

87. $e^{-1.386x} = 0.25$

88. $e^{-0.01x} - 5.25 = 0$

89. $100(1.21)^x = 110$

90. $500(1.075)^{120x} = 100{,}000$

91. $\dfrac{40}{1 - 5e^{-0.01x}} = 200$

92. $\dfrac{50}{1 - 2e^{-0.001x}} = 1000$

93. *MAKE A DECISION: HOME MORTGAGE* The monthly payment M for a home mortgage of P dollars for t years at an annual interest rate r is given by

$$M = P\left\{\dfrac{\dfrac{r}{12}}{1 - \left[\dfrac{1}{(r/12) + 1}\right]^{12t}}\right\}.$$

(T) (a) Use a graphing utility to graph the model when $P = \$150{,}000$ and $r = 0.075$.

(b) You are given a choice of a 20-year term or a 30-year term. Which would you choose? Explain your reasoning.

94. Hourly Wages The average hourly wages w in the United States from 1990 through 2005 can be modeled by

$$w = 8.25 + 0.681t - 0.0105t^2 + 1.94366e^{-t}$$

where $t = 0$ corresponds to 1990. *(Source: U.S. Bureau of Labor Statistics)*

(T) (a) Use a graphing utility to graph the model.

(b) Use the model to determine the year in which the average hourly wage was $12.

(c) For how many years past 2005 do you think this equation might be a good model for the average hourly wage? Explain your reasoning.

In Exercises 95–108, find the derivative of the function.

95. $f(x) = \ln 3x^2$

96. $y = \ln \sqrt{x}$

97. $y = \ln \dfrac{x(x - 1)}{x - 2}$

98. $y = \ln \dfrac{x^2}{x + 1}$

99. $f(x) = \ln e^{2x+1}$

100. $f(x) = \ln e^{x^2}$

101. $y = \dfrac{\ln x}{x^3}$

102. $y = \dfrac{x^2}{\ln x}$

103. $y = \ln(x^2 - 2)^{2/3}$

104. $y = \ln \sqrt[3]{x^3 + 1}$

105. $f(x) = \ln\left(x^2 \sqrt{x + 1}\right)$

106. $f(x) = \ln \dfrac{x}{\sqrt{x + 1}}$

107. $y = \ln \dfrac{e^x}{1 + e^x}$

108. $y = \ln\left(e^{2x}\sqrt{e^{2x} - 1}\right)$

In Exercises 109–112, graph and analyze the function. Include any relative extrema and points of inflection in your analysis.

109. $y = \ln(x + 3)$

110. $y = \dfrac{8 \ln x}{x^2}$

111. $y = \ln \dfrac{10}{x + 2}$

112. $y = \ln \dfrac{x^2}{9 - x^2}$

In Exercises 113–116, evaluate the logarithm.

113. $\log_8 64$

114. $\log_2 64$

115. $\log_{10} 1$

116. $\log_4 \frac{1}{64}$

In Exercises 117–120, use the change-of-base formula to evaluate the logarithm. Round the result to three decimal places.

117. $\log_5 13$

118. $\log_4 18$

119. $\log_{16} 64$

120. $\log_4 125$

In Exercises 121–124, find the derivative of the function.

121. $y = \log_3(2x - 1)$

122. $y = \log_{10} \frac{3}{x}$

123. $y = \log_2 \frac{1}{x^2}$

124. $y = \log_{16}(x^2 - 3x)$

125. Depreciation After t years, the value V of a car purchased for $25,000 is given by

$$V = 25,000(0.75)^t.$$

(a) Sketch a graph of the function and determine the value of the car 2 years after it was purchased.

(b) Find the rates of change of V with respect to t when $t = 1$ and when $t = 4$.

(c) After how many years will the car be worth $5000?

126. Inflation Rate If the annual rate of inflation averages 4% over the next 10 years, then the approximate cost of goods or services C during any year in that decade will be given by

$$C = P(1.04)^t$$

where t is the time in years and P is the present cost.

(a) The price of an oil change is presently $24.95. Estimate the price of an oil change 10 years from now.

(b) Find the rate of change of C with respect to t when $t = 1$.

127. Medical Science A medical solution contains 500 milligrams of a drug per milliliter when the solution is prepared. After 40 days, it contains only 300 milligrams per milliliter. Assuming that the rate of decomposition is proportional to the concentration present, find an equation giving the concentration A after t days.

128. Population Growth A population is growing continuously at the rate of $2\frac{1}{2}\%$ per year. Find the time necessary for the population to (a) double in size and (b) triple in size.

129. Radioactive Decay A sample of radioactive waste is taken from a nuclear plant. The sample contains 50 grams of strontium-90 at time $t = 0$ years and 42.031 grams after 7 years. What is the half-life of strontium-90?

130. Radioactive Decay The half-life of cobalt-60 is 5.2 years. Find the time it would take for a sample of 0.5 gram of cobalt-60 to decay to 0.1 gram.

131. Profit The profit P (in millions of dollars) for Affiliated Computer Services, Inc. was $23.8 million in 1996 and $406.9 million in 2005 (see figure). Use an exponential growth model to predict the profit in 2008. *(Source: Affiliated Computer Services, Inc.)*

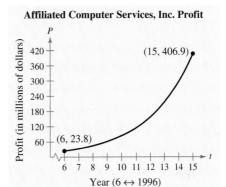

Affiliated Computer Services, Inc. Profit

132. Profit The profit P (in millions of dollars) for Bank of America was $2375 million in 1996 and $16,465 million in 2005 (see figure). Use an exponential growth model to predict the profit in 2008. *(Source: Bank of America)*

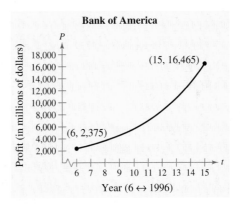

Bank of America

Chapter Test

See www.CalcChat.com for worked-out solutions to odd-numbered exercises.

Take this test as you would take a test in class. When you are done, check your work against the answers given in the back of the book.

In Exercises 1–4, use properties of exponents to simplify the expression.

1. $3^2(3^{-2})$ **2.** $\left(\dfrac{2^3}{2^{-5}}\right)^{-1}$ **3.** $(e^{1/2})(e^4)$ **4.** $(e^3)(e^{-1})$

(T) In Exercises 5–10, use a graphing utility to graph the function.

5. $f(x) = 5^{x-2}$ **6.** $f(x) = 4^{-x}$ **7.** $f(x) = 3^{x-3}$

8. $f(x) = 8 + \ln x^2$ **9.** $f(x) = \ln(x - 5)$ **10.** $f(x) = 0.5 \ln x$

In Exercises 11–13, use the properties of logarithms to write the expression as a sum, difference , or multiple of logarithms.

11. $\ln \dfrac{3}{2}$ **12.** $\ln \sqrt{x + y}$ **13.** $\ln \dfrac{x + 1}{y}$

In Exercises 14–16, condense the logarithmic expression.

14. $\ln y + \ln(x + 1)$ **15.** $3 \ln 2 - 2 \ln(x - 1)$

16. $2 \ln x + \ln y - \ln(z + 4)$

In Exercises 17–19, solve the equation.

17. $e^{x-1} = 9$ **18.** $10e^{2x+1} = 900$ **19.** $50(1.06)^x = 1500$

20. A deposit of $500 is made to an account that earns interest at an annual rate of 4%. How long will it take for the balance to double if the interest is compounded (a) annually, (b) monthly, (c) daily, and (d) continuously?

In Exercises 21–24, find the derivative of the function.

21. $y = e^{-3x} + 5$ **22.** $y = 7e^{x+2} + 2x$

23. $y = \ln(3 + x^2)$ **24.** $y = \ln \dfrac{5x}{x + 2}$

25. The gross revenues R (in millions of dollars) of symphony orchestras in the United States from 1997 through 2004 can be modeled by

$R = -93.4 + 349.36 \ln t$

where $t = 7$ corresponds to 1997. *(Source: American Symphony Orchestra League, Inc.)*

(a) Use this model to estimate the gross revenues in 2004.

(b) At what rate were the gross revenues changing in 2004?

26. What percent of a present amount of radioactive radium (^{226}Ra) will remain after 1200 years? (The half-life of ^{266}Ra is 1599 years.)

27. A population is growing continuously at the rate of 1.75% per year. Find the time necessary for the population to double in size.

Integration and Its Applications

Integration can be used to solve business problems, such as estimating the surface area of an oil spill. (See Chapter 11 Review Exercises, Exercise 101.)

Natalie Fobes/Getty Images

Applications

Integration has many real-life applications. The applications listed below represent a sample of the applications in this chapter.

- Make a Decision: Internet Users, Exercise 79, page 816
- Average Nurse's Salary, Exercise 61, page 832
- Biology: Fishing Population/Population Extinction, Exercise 97, page 845
- Make a Decision: Job Offers, Exercise 45, page 853
- Make a Decision: Budget Deficits, Exercise 46, page 853
- Spread of Disease, Exercise 50, page 854
- Consumer Trends, Exercise 51, page 854

Section 11.1

Antiderivatives and Indefinite Integrals

■ Understand the definition of antiderivative.
■ Use indefinite integral notation for antiderivatives.
■ Use basic integration rules to find antiderivatives.
■ Use initial conditions to find particular solutions of indefinite integrals.
■ Use antiderivatives to solve real-life problems.

Antiderivatives

Up to this point in the text, you have been concerned primarily with this problem: given a function, find its derivative. Many important applications of calculus involve the inverse problem: given the derivative of a function, find the function. For example, suppose you are given

$$f'(x) = 2, \quad g'(x) = 3x^2, \quad \text{and} \quad s'(t) = 4t.$$

Your goal is to determine the functions f, g, and s. By making educated guesses, you might come up with the following functions.

$$f(x) = 2x \quad \text{because} \quad \frac{d}{dx}[2x] = 2.$$

$$g(x) = x^3 \quad \text{because} \quad \frac{d}{dx}[x^3] = 3x^2.$$

$$s(t) = 2t^2 \quad \text{because} \quad \frac{d}{dt}[2t^2] = 4t.$$

This operation of determining the original function from its derivative is the inverse operation of differentiation. It is called **antidifferentiation.**

Definition of Antiderivative

A function F is an **antiderivative** of a function f if for every x in the domain of f, it follows that $F'(x) = f(x)$.

If $F(x)$ is an antiderivative of $f(x)$, then $F(x) + C$, where C is any constant, is also an antiderivative of $f(x)$. For example,

$$F(x) = x^3, \quad G(x) = x^3 - 5, \quad \text{and} \quad H(x) = x^3 + 0.3$$

are all antiderivatives of $3x^2$ because the derivative of each is $3x^2$. As it turns out, *all* antiderivatives of $3x^2$ are of the form $x^3 + C$. So, the process of antidifferentiation does not determine a single function, but rather a *family* of functions, each differing from the others by a constant.

STUDY TIP

In this text, the phrase "$F(x)$ is an antiderivative of $f(x)$" is used synonymously with "F is an antiderivative of f."

Notation for Antiderivatives and Indefinite Integrals

The antidifferentiation process is also called **integration** and is denoted by the symbol

$$\int$$ Integral sign

which is called an **integral sign.** The symbol

$$\int f(x)\,dx$$ Indefinite integral

is the **indefinite integral** of $f(x)$, and it denotes the family of antiderivatives of $f(x)$. That is, if $F'(x) = f(x)$ for all x, then you can write

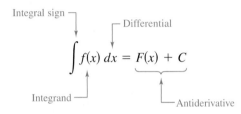

where $f(x)$ is the **integrand** and C is the **constant of integration.** The differential dx in the indefinite integral identifies the variable of integration. That is, the symbol $\int f(x)\,dx$ denotes the "antiderivative of f *with respect to* x" just as the symbol dy/dx denotes the "derivative of y *with respect to x.*"

DISCOVERY

Verify that $F_1(x) = x^2 - 2x$, $F_2(x) = x^2 - 2x - 1$, and $F_3(x) = (x - 1)^2$ are all antiderivatives of $f(x) = 2x - 2$. Use a graphing utility to graph F_1, F_2, and F_3 in the same coordinate plane. How are their graphs related? What can you say about the graph of any other antiderivative of f?

Integral Notation of Antiderivatives

The notation

$$\int f(x)\,dx = F(x) + C$$

where C is an arbitrary constant, means that F is an antiderivative of f. That is, $F'(x) = f(x)$ for all x in the domain of f.

Example 1 Notation for Antiderivatives

Using integral notation, you can write the three antiderivatives from the beginning of this section as shown.

a. $\displaystyle\int 2\,dx = 2x + C$ **b.** $\displaystyle\int 3x^2\,dx = x^3 + C$ **c.** $\displaystyle\int 4t\,dt = 2t^2 + C$

✓CHECKPOINT 1

Rewrite each antiderivative using integral notation.

a. $\dfrac{d}{dx}[3x] = 3$ **b.** $\dfrac{d}{dx}[x^2] = 2x$ **c.** $\dfrac{d}{dt}[3t^3] = 9t^2$

Finding Antiderivatives

The inverse relationship between the operations of integration and differentiation can be shown symbolically, as follows.

$$\frac{d}{dx}\left[\int f(x)\,dx\right] = f(x)$$ Differentiation is the inverse of integration.

$$\int f'(x)\,dx = f(x) + C$$ Integration is the inverse of differentiation.

This inverse relationship between integration and differentiation allows you to obtain integration formulas directly from differentiation formulas. The following summary lists the integration formulas that correspond to some of the differentiation formulas you have studied.

Basic Integration Rules

1. $\int k\,dx = kx + C,\quad k$ is a constant.   Constant Rule

2. $\int kf(x)\,dx = k\int f(x)\,dx$ Constant Multiple Rule

3. $\int [f(x) + g(x)]\,dx = \int f(x)\,dx + \int g(x)\,dx$ Sum Rule

4. $\int [f(x) - g(x)]\,dx = \int f(x)\,dx - \int g(x)\,dx$ Difference Rule

5. $\int x^n\,dx = \dfrac{x^{n+1}}{n+1} + C,\quad n \neq -1$ Simple Power Rule

STUDY TIP

You will study the General Power Rule for integration in Section 11.2 and the Exponential and Log Rules in Section 11.3.

Be sure you see that the Simple Power Rule has the restriction that n cannot be -1. So, you *cannot* use the Simple Power Rule to evaluate the integral

$$\int \frac{1}{x}\,dx.$$

To evaluate this integral, you need the Log Rule, which is described in Section 11.3.

STUDY TIP

In Example 2(b), the integral $\int 1\,dx$ is usually shortened to the form $\int dx$.

✓ CHECKPOINT 2

Find each indefinite integral.

a. $\displaystyle\int 5\,dx$

b. $\displaystyle\int -1\,dr$

c. $\displaystyle\int 2\,dt$

Example 2 Finding Indefinite Integrals

Find each indefinite integral.

a. $\displaystyle\int \frac{1}{2}\,dx$ b. $\displaystyle\int 1\,dx$ c. $\displaystyle\int -5\,dt$

SOLUTION

a. $\displaystyle\int \frac{1}{2}\,dx = \frac{1}{2}x + C$ b. $\displaystyle\int 1\,dx = x + C$ c. $\displaystyle\int -5\,dt = -5t + C$

Example 3 **Finding an Indefinite Integral**

Find $\displaystyle\int 3x\, dx$.

SOLUTION

$$\int 3x\, dx = 3\int x\, dx \qquad \text{Constant Multiple Rule}$$

$$= 3\int x^1\, dx \qquad \text{Rewrite } x \text{ as } x^1.$$

$$= 3\left(\frac{x^2}{2}\right) + C \qquad \text{Simple Power Rule with } n = 1$$

$$= \frac{3}{2}x^2 + C \qquad \text{Simplify.}$$

✓ **CHECKPOINT 3**

Find $\displaystyle\int 5x\, dx$.

In finding indefinite integrals, a strict application of the basic integration rules tends to produce cumbersome constants of integration. For instance, in Example 3, you could have written

$$\int 3x\, dx = 3\int x\, dx = 3\left(\frac{x^2}{2} + C\right) = \frac{3}{2}x^2 + 3C.$$

However, because C represents *any* constant, it is unnecessary to write $3C$ as the constant of integration. You can simply write $\frac{3}{2}x^2 + C$.

In Example 3, note that the general pattern of integration is similar to that of differentiation.

STUDY TIP

Remember that you can check your answer to an antidifferentiation problem by differentiating. For instance, in Example 4(b), you can check that $\frac{2}{3}x^{3/2}$ is the correct antiderivative by differentiating to obtain

$$\frac{d}{dx}\left[\frac{2}{3}x^{3/2}\right] = \left(\frac{2}{3}\right)\left(\frac{3}{2}\right)x^{1/2}$$

$$= \sqrt{x}.$$

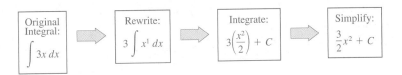

Original Integral: $\int 3x\, dx$ → Rewrite: $3\int x^1\, dx$ → Integrate: $3\left(\frac{x^2}{2}\right) + C$ → Simplify: $\frac{3}{2}x^2 + C$

Example 4 **Rewriting Before Integrating**

Find each indefinite integral.

a. $\displaystyle\int \frac{1}{x^3}\, dx$

b. $\displaystyle\int \sqrt{x}\, dx$

SOLUTION

	Original Integral	Rewrite	Integrate	Simplify
a.	$\displaystyle\int \frac{1}{x^3}\, dx$	$\displaystyle\int x^{-3}\, dx$	$\dfrac{x^{-2}}{-2} + C$	$-\dfrac{1}{2x^2} + C$
b.	$\displaystyle\int \sqrt{x}\, dx$	$\displaystyle\int x^{1/2}\, dx$	$\dfrac{x^{3/2}}{3/2} + C$	$\dfrac{2}{3}x^{3/2} + C$

✓ **CHECKPOINT 4**

Find each indefinite integral.

a. $\displaystyle\int \frac{1}{x^2}\, dx$ **b.** $\displaystyle\int \sqrt[3]{x}\, dx$

With the five basic integration rules, you can integrate *any* polynomial function, as demonstrated in the next example.

✓ **CHECKPOINT 5**

Find each indefinite integral.

a. $\displaystyle\int (x + 4)\, dx$

b. $\displaystyle\int (4x^3 - 5x + 2)\, dx$

Example 5 **Integrating Polynomial Functions**

Find each indefinite integral.

a. $\displaystyle\int (x + 2)\, dx$ **b.** $\displaystyle\int (3x^4 - 5x^2 + x)\, dx$

SOLUTION

a. $\displaystyle\int (x + 2)\, dx = \int x\, dx + \int 2\, dx$ Apply Sum Rule.

$\displaystyle\qquad\qquad\qquad = \frac{x^2}{2} + C_1 + 2x + C_2$ Integrate.

$\displaystyle\qquad\qquad\qquad = \frac{x^2}{2} + 2x + C$ $C = C_1 + C_2$

The second line in the solution is usually omitted.

b. Try to identify each basic integration rule used to evaluate this integral.

$\displaystyle\int (3x^4 - 5x^2 + x)\, dx = 3\left(\frac{x^5}{5}\right) - 5\left(\frac{x^3}{3}\right) + \frac{x^2}{2} + C$

$\displaystyle\qquad\qquad\qquad\qquad = \frac{3}{5}x^5 - \frac{5}{3}x^3 + \frac{1}{2}x^2 + C$

STUDY TIP

When integrating quotients, remember *not* to integrate the numerator and denominator separately. For instance, in Example 6, be sure you understand that

$\displaystyle\int \frac{x + 1}{\sqrt{x}}\, dx = \frac{2}{3}\sqrt{x}(x + 3) + C$

is not the same as

$\displaystyle\frac{\int (x + 1)\, dx}{\int \sqrt{x}\, dx} = \frac{\frac{1}{2}x^2 + x + C_1}{\frac{2}{3}x\sqrt{x} + C_2}.$

Example 6 **Rewriting Before Integrating**

Find $\displaystyle\int \frac{x + 1}{\sqrt{x}}\, dx.$

SOLUTION Begin by rewriting the quotient in the integrand as a sum. Then rewrite each term using rational exponents.

$\displaystyle\int \frac{x + 1}{\sqrt{x}}\, dx = \int \left(\frac{x}{\sqrt{x}} + \frac{1}{\sqrt{x}}\right) dx$ Rewrite as a sum.

$\displaystyle\qquad\qquad\quad = \int (x^{1/2} + x^{-1/2})\, dx$ Rewrite using rational exponents.

$\displaystyle\qquad\qquad\quad = \frac{x^{3/2}}{3/2} + \frac{x^{1/2}}{1/2} + C$ Apply Power Rule.

$\displaystyle\qquad\qquad\quad = \frac{2}{3}x^{3/2} + 2x^{1/2} + C$ Simplify.

$\displaystyle\qquad\qquad\quad = \frac{2}{3}\sqrt{x}(x + 3) + C$ Factor.

Algebra Review

For help on the algebra in Example 6, see Example 1(a) in the *Chapter 11 Algebra Review*, on page 861.

✓ **CHECKPOINT 6**

Find $\displaystyle\int \frac{x + 2}{\sqrt{x}}\, dx.$

Particular Solutions

You have already seen that the equation $y = \int f(x)\, dx$ has many solutions, each differing from the others by a constant. This means that the graphs of any two antiderivatives of f are vertical translations of each other. For example, Figure 11.1 shows the graphs of several antiderivatives of the form

$$y = F(x) = \int (3x^2 - 1)\, dx = x^3 - x + C$$

for various integer values of C. Each of these antiderivatives is a solution of the *differential equation* $dy/dx = 3x^2 - 1$. A **differential equation** in x and y is an equation that involves x, y, and derivatives of y. The **general solution** of $dy/dx = 3x^2 - 1$ is $F(x) = x^3 - x + C$.

In many applications of integration, you are given enough information to determine a **particular solution.** To do this, you only need to know the value of $F(x)$ for one value of x. (This information is called an **initial condition.**) For example, in Figure 11.1, there is only one curve that passes through the point $(2, 4)$. To find this curve, use the information below.

$$F(x) = x^3 - x + C \qquad \text{General solution}$$
$$F(2) = 4 \qquad \text{Initial condition}$$

By using the initial condition in the general solution, you can determine that $F(2) = 2^3 - 2 + C = 4$, which implies that $C = -2$. So, the particular solution is

$$F(x) = x^3 - x - 2. \qquad \text{Particular solution}$$

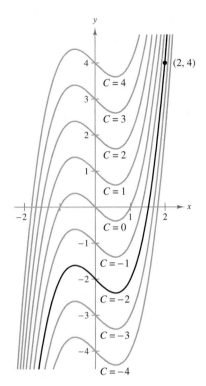

$$F(x) = x^3 - x + C$$

FIGURE 11.1

Example 7 Finding a Particular Solution

Find the general solution of

$$F'(x) = 2x - 2$$

and find the particular solution that satisfies the initial condition $F(1) = 2$.

SOLUTION Begin by integrating to find the general solution.

$$F(x) = \int (2x - 2)\, dx \qquad \text{Integrate } F'(x) \text{ to obtain } F(x).$$
$$= x^2 - 2x + C \qquad \text{General solution}$$

Using the initial condition $F(1) = 2$, you can write

$$F(1) = 1^2 - 2(1) + C = 2$$

which implies that $C = 3$. So, the particular solution is

$$F(x) = x^2 - 2x + 3. \qquad \text{Particular solution}$$

This solution is shown graphically in Figure 11.2. Note that each of the gray curves represents a solution of the equation $F'(x) = 2x - 2$. The black curve, however, is the only solution that passes through the point $(1, 2)$, which means that $F(x) = x^2 - 2x + 3$ is the only solution that satisfies the initial condition.

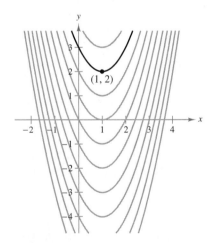

FIGURE 11.2

✓ CHECKPOINT 7

Find the general solution of $F'(x) = 4x + 2$, and find the particular solution that satisfies the initial condition $F(1) = 8$. ■

Applications

In Chapter 7 and Chapter 8, you used the general position function (neglecting air resistance) for a falling object

$$s(t) = -16t^2 + v_0 t + s_0$$

where $s(t)$ is the height (in feet) and t is the time (in seconds). In the next example, integration is used to *derive* this function.

Example 8

MAKE A DECISION   **Deriving a Position Function**

A ball is thrown upward with an initial velocity of 64 feet per second from an initial height of 80 feet, as shown in Figure 11.3. Derive the position function giving the height s (in feet) as a function of the time t (in seconds). Will the ball be in the air for more than 5 seconds?

SOLUTION   Let $t = 0$ represent the initial time. Then the two given conditions can be written as

$$s(0) = 80 \qquad \text{Initial height is 80 feet.}$$
$$s'(0) = 64. \qquad \text{Initial velocity is 64 feet per second.}$$

Because the acceleration due to gravity is -32 feet per second per second, you can integrate the acceleration function to find the velocity function as shown.

$$s''(t) = -32 \qquad \text{Acceleration due to gravity}$$
$$s'(t) = \int -32 \, dt \qquad \text{Integrate } s''(t) \text{ to obtain } s'(t).$$
$$= -32t + C_1 \qquad \text{Velocity function}$$

Using the initial velocity, you can conclude that $C_1 = 64$.

$$s'(t) = -32t + 64 \qquad \text{Velocity function}$$
$$s(t) = \int (-32t + 64) \, dt \qquad \text{Integrate } s'(t) \text{ to obtain } s(t).$$
$$= -16t^2 + 64t + C_2 \qquad \text{Position function}$$

Using the initial height, it follows that $C_2 = 80$. So, the position function is given by

$$s(t) = -16t^2 + 64t + 80. \qquad \text{Position function}$$

To find the time when the ball hits the ground, set the position function equal to 0 and solve for t.

$$-16t^2 + 64t + 80 = 0 \qquad \text{Set } s(t) \text{ equal to zero.}$$
$$-16(t + 1)(t - 5) = 0 \qquad \text{Factor.}$$
$$t = -1, \quad t = 5 \qquad \text{Solve for } t.$$

Because the time must be positive, you can conclude that the ball hits the ground 5 seconds after it is thrown. No, the ball was not in the air for more than 5 seconds.

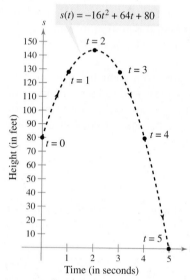

$s(t) = -16t^2 + 64t + 80$

FIGURE 11.3

✓CHECKPOINT 8

Derive the position function if a ball is thrown upward with an initial velocity of 32 feet per second from an initial height of 48 feet. When does the ball hit the ground? With what velocity does the ball hit the ground? ∎

Example 9 Finding a Cost Function

The marginal cost for producing x units of a product is modeled by

$$\frac{dC}{dx} = 32 - 0.04x. \qquad\qquad \text{Marginal cost}$$

It costs $50 to produce one unit. Find the total cost of producing 200 units.

SOLUTION   To find the cost function, integrate the marginal cost function.

$$C = \int (32 - 0.04x)\, dx \qquad\qquad \text{Integrate } \frac{dC}{dx} \text{ to obtain } C.$$

$$= 32x - 0.04\left(\frac{x^2}{2}\right) + K$$

$$= 32x - 0.02x^2 + K \qquad\qquad \text{Cost function}$$

To solve for K, use the initial condition that $C = 50$ when $x = 1$.

$$50 = 32(1) - 0.02(1)^2 + K \qquad \text{Substitute 50 for } C \text{ and 1 for } x.$$
$$18.02 = K \qquad\qquad\qquad\qquad \text{Solve for } K.$$

So, the total cost function is given by

$$C = 32x - 0.02x^2 + 18.02 \qquad\qquad \text{Cost function}$$

which implies that the cost of producing 200 units is

$$C = 32(200) - 0.02(200)^2 + 18.02$$
$$= \$5618.02.$$

✓ CHECKPOINT 9

The marginal cost function for producing x units of a product is modeled by

$$\frac{dC}{dx} = 28 - 0.02x.$$

It costs $40 to produce one unit. Find the cost of producing 200 units. ∎

CONCEPT CHECK

1. How can you check your answer to an antidifferentiation problem?

2. Write what is meant by the symbol $\int f(x)\, dx$ in words.

3. Given $\int (2x + 1)\, dx = x^2 + x + C$, identify (a) the integrand and (b) the antiderivative.

4. True or false: The antiderivative of a second-degree polynomial function is a third-degree polynomial function.

The following warm-up exercises involve skills that were covered in earlier sections. You will use these skills in the exercise set for this section. For additional help, review Sections 0.4 and 2.1.

In Exercises 1–6, rewrite the expression using rational exponents.

1. $\dfrac{\sqrt{x}}{x}$

2. $\sqrt[3]{2x}(2x)$

3. $\sqrt{5x^3} + \sqrt{x^5}$

4. $\dfrac{1}{\sqrt{x}} + \dfrac{1}{\sqrt[3]{x^2}}$

5. $\dfrac{(x + 1)^3}{\sqrt{x + 1}}$

6. $\dfrac{\sqrt{x}}{\sqrt[3]{x}}$

In Exercises 7–10, let $(x, y) = (2, 2)$, and solve the equation for C.

7. $y = x^2 + 5x + C$

8. $y = 3x^3 - 6x + C$

9. $y = -16x^2 + 26x + C$

10. $y = -\frac{1}{4}x^4 - 2x^2 + C$

Exercises 11.1

In Exercises 1–8, verify the statement by showing that the derivative of the right side is equal to the integrand of the left side.

1. $\displaystyle\int \left(-\dfrac{9}{x^4}\right) dx = \dfrac{3}{x^3} + C$

2. $\displaystyle\int \dfrac{4}{\sqrt{x}}\, dx = 8\sqrt{x} + C$

3. $\displaystyle\int \left(4x^3 - \dfrac{1}{x^2}\right) dx = x^4 + \dfrac{1}{x} + C$

4. $\displaystyle\int \left(1 - \dfrac{1}{\sqrt[3]{x^2}}\right) dx = x - 3\sqrt[3]{x} + C$

5. $\displaystyle\int 2\sqrt{x}(x - 3)\, dx = \dfrac{4x^{3/2}(x - 5)}{5} + C$

6. $\displaystyle\int 4\sqrt{x}(x^2 - 2)\, dx = \dfrac{8x^{3/2}(3x^2 - 14)}{21} + C$

7. $\displaystyle\int (x - 2)(x + 2)\, dx = \dfrac{1}{3}x^3 - 4x + C$

8. $\displaystyle\int \dfrac{x^2 - 1}{x^{3/2}}\, dx = \dfrac{2(x^2 + 3)}{3\sqrt{x}} + C$

In Exercises 9–20, find the indefinite integral and check your result by differentiation.

9. $\displaystyle\int 6\, dx$

10. $\displaystyle\int -4\, dx$

11. $\displaystyle\int 5t^2\, dt$

12. $\displaystyle\int 3t^4\, dt$

13. $\displaystyle\int 5x^{-3}\, dx$

14. $\displaystyle\int 4y^{-3}\, dy$

15. $\displaystyle\int du$

16. $\displaystyle\int dr$

17. $\displaystyle\int e\, dt$

18. $\displaystyle\int e^3\, dy$

19. $\displaystyle\int y^{3/2}\, dy$

20. $\displaystyle\int v^{-1/2}\, dv$

In Exercises 21–26, complete the table.

	Original Integral	Rewrite	Integrate	Simplify
21.	$\displaystyle\int \sqrt[3]{x}\, dx$			
22.	$\displaystyle\int \dfrac{1}{x^2}\, dx$			
23.	$\displaystyle\int \dfrac{1}{x\sqrt{x}}\, dx$			
24.	$\displaystyle\int x(x^2 + 3)\, dx$			
25.	$\displaystyle\int \dfrac{1}{2x^3}\, dx$			
26.	$\displaystyle\int \dfrac{1}{(3x)^2}\, dx$			

In Exercises 27–38, find the indefinite integral and check your result by differentiation.

27. $\displaystyle\int (x + 3)\, dx$

28. $\displaystyle\int (5 - x)\, dx$

29. $\displaystyle\int (x^3 + 2)\, dx$

30. $\displaystyle\int (x^3 - 4x + 2)\, dx$

31. $\displaystyle\int \left(\sqrt[3]{x} - \dfrac{1}{2\sqrt[3]{x}}\right) dx$

32. $\displaystyle\int \left(\sqrt{x} + \dfrac{1}{2\sqrt{x}}\right) dx$

33. $\int \sqrt[3]{x^2}\, dx$

34. $\int \left(\sqrt[4]{x^3} + 1 \right) dx$

35. $\int \frac{1}{x^4}\, dx$

36. $\int \frac{1}{4x^2}\, dx$

37. $\int \frac{2x^3 + 1}{x^3}\, dx$

38. $\int \frac{t^2 + 2}{t^2}\, dt$

T In Exercises 39–44, use a symbolic integration utility to find the indefinite integral.

39. $\int u(3u^2 + 1)\, du$

40. $\int \sqrt{x}(x + 1)\, dx$

41. $\int (x + 1)(3x - 2)\, dx$

42. $\int (2t^2 - 1)^2\, dt$

43. $\int y^2 \sqrt{y}\, dy$

44. $\int (1 + 3t)t^2\, dt$

In Exercises 45–48, the graph of the derivative of a function is given. Sketch the graphs of two functions that have the given derivative. (There is more than one correct answer.)

45.
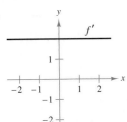

46.
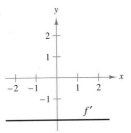

47.
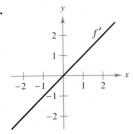

48.
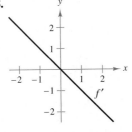

In Exercises 49–54, find the particular solution $y = f(x)$ that satisfies the differential equation and initial condition.

49. $f'(x) = 4x$; $f(0) = 6$

50. $f'(x) = \frac{1}{5}x - 2$; $f(10) = -10$

51. $f'(x) = 2(x - 1)$; $f(3) = 2$

52. $f'(x) = (2x - 3)(2x + 3)$; $f(3) = 0$

53. $f'(x) = \frac{2 - x}{x^3}$, $x > 0$; $f(2) = \frac{3}{4}$

54. $f'(x) = \frac{x^2 - 5}{x^2}$, $x > 0$; $f(1) = 2$

In Exercises 55 and 56, find the equation for y, given the derivative and the indicated point on the curve.

55. $\frac{dy}{dx} = -5x - 2$

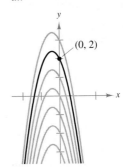

56. $\frac{dy}{dx} = 2(x - 1)$

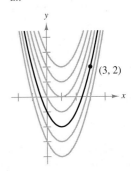

In Exercises 57 and 58, find the equation of the function f whose graph passes through the point.

Derivative	*Point*
57. $f'(x) = 2x$	$(-2, -2)$
58. $f'(x) = 2\sqrt{x}$	$(4, 12)$

In Exercises 59–62, find a function f that satisfies the conditions.

59. $f''(x) = 2$, $f'(2) = 5$, $f(2) = 10$

60. $f''(x) = x^2$, $f'(0) = 6$, $f(0) = 3$

61. $f''(x) = x^{-2/3}$, $f'(8) = 6$, $f(0) = 0$

62. $f''(x) = x^{-3/2}$, $f'(1) = 2$, $f(9) = -4$

Cost In Exercises 63–66, find the cost function for the marginal cost and fixed cost.

Marginal Cost	*Fixed Cost* $(x = 0)$
63. $\frac{dC}{dx} = 85$	$\$5500$
64. $\frac{dC}{dx} = \frac{1}{50}x + 10$	$\$1000$
65. $\frac{dC}{dx} = \frac{1}{20\sqrt{x}} + 4$	$\$750$
66. $\frac{dC}{dx} = \frac{\sqrt[4]{x}}{10} + 10$	$\$2300$

Demand Function In Exercises 67 and 68, find the revenue and demand functions for the given marginal revenue. (Use the fact that $R = 0$ when $x = 0$.)

67. $\frac{dR}{dx} = 225 - 3x$

68. $\frac{dR}{dx} = 310 - 4x$

Profit In Exercises 69–72, find the profit function for the given marginal profit and initial condition.

Marginal Profit	Initial Condition
69. $\dfrac{dP}{dx} = -18x + 1650$	$P(15) = \$22{,}725$
70. $\dfrac{dP}{dx} = -40x + 250$	$P(5) = \$650$
71. $\dfrac{dP}{dx} = -24x + 805$	$P(12) = \$8000$
72. $\dfrac{dP}{dx} = -30x + 920$	$P(8) = \$6500$

Vertical Motion In Exercises 73 and 74, use $a(t) = -32$ feet per second per second as the acceleration due to gravity.

73. The Grand Canyon is 6000 feet deep at the deepest part. A rock is dropped from this height. Express the height s of the rock as a function of the time t (in seconds). How long will it take the rock to hit the canyon floor?

74. With what initial velocity must an object be thrown upward from the ground to reach the height of the Washington Monument (550 feet)?

75. **Cost** A company produces a product for which the marginal cost of producing x units is modeled by $dC/dx = 2x - 12$, and the fixed costs are $125.

(a) Find the total cost function and the average cost function.

(b) Find the total cost of producing 50 units.

(c) In part (b), how much of the total cost is fixed? How much is variable? Give examples of fixed costs associated with the manufacturing of a product. Give examples of variable costs.

76. **Tree Growth** An evergreen nursery usually sells a certain shrub after 6 years of growth and shaping. The growth rate during those 6 years is approximated by $dh/dt = 1.5t + 5$, where t is the time in years and h is the height in centimeters. The seedlings are 12 centimeters tall when planted $(t = 0)$.

(a) Find the height after t years.

(b) How tall are the shrubs when they are sold?

77. **MAKE A DECISION: POPULATION GROWTH** The growth rate of Horry County in South Carolina can be modeled by $dP/dt = 105.46t + 2642.7$, where t is the time in years, with $t = 0$ corresponding to 1970. The county's population was 226,992 in 2005. *(Source: U.S. Census Bureau)*

(a) Find the model for Horry County's population.

(b) Use the model to predict the population in 2012. Does your answer seem reasonable? Explain your reasoning.

78. **MAKE A DECISION: VITAL STATISTICS** The rate of increase of the number of married couples M (in thousands) in the United States from 1970 to 2005 can be modeled by

$$\frac{dM}{dt} = 1.218t^2 - 44.72t + 709.1$$

where t is the time in years, with $t = 0$ corresponding to 1970. The number of married couples in 2005 was 59,513 thousand. *(Source: U.S. Census Bureau)*

(a) Find the model for the number of married couples in the United States.

(b) Use the model to predict the number of married couples in the United States in 2012. Does your answer seem reasonable? Explain your reasoning.

79. **MAKE A DECISION: INTERNET USERS** The rate of growth of the number of Internet users I (in millions) in the world from 1991 to 2004 can be modeled by

$$\frac{dI}{dt} = -0.25t^3 + 5.319t^2 - 19.34t + 21.03$$

where t is the time in years, with $t = 1$ corresponding to 1991. The number of Internet users in 2004 was 863 million. *(Source: International Telecommunication Union)*

(a) Find the model for the number of Internet users in the world.

(b) Use the model to predict the number of Internet users in the world in 2012. Does your answer seem reasonable? Explain your reasoning.

(T)(B) 80. **Economics: Marginal Benefits and Costs** The table gives the marginal benefit and marginal cost of producing x units of a product for a given company. Plot the points in each column and use the *regression* feature of a graphing utility to find a linear model for marginal benefit and a quadratic model for marginal cost. Then use integration to find the benefit B and cost C equations. Assume $B(0) = 0$ and $C(0) = 425$. Finally, find the intervals in which the benefit exceeds the cost of producing x units, and make a recommendation for how many units the company should produce based on your findings. *(Source: Adapted from Taylor, Economics, Fifth Edition)*

Number of units	1	2	3	4	5
Marginal benefit	330	320	290	270	250
Marginal cost	150	120	100	110	120

Number of units	6	7	8	9	10
Marginal benefit	230	210	190	170	160
Marginal cost	140	160	190	250	320

Section 11.2

Integration by Substitution and the General Power Rule

- Use the General Power Rule to find indefinite integrals.
- Use substitution to find indefinite integrals.
- Use the General Power Rule to solve real-life problems.

The General Power Rule

In Section 11.1, you used the Simple Power Rule

$$\int x^n \, dx = \frac{x^{n+1}}{n+1} + C, \quad n \neq -1$$

to find antiderivatives of functions expressed as powers of x alone. In this section, you will study a technique for finding antiderivatives of more complicated functions.

To begin, consider how you might find the antiderivative of $2x(x^2 + 1)^3$. Because you are hunting for a function whose derivative is $2x(x^2 + 1)^3$, you might discover the antiderivative as shown.

$$\frac{d}{dx}[(x^2 + 1)^4] = 4(x^2 + 1)^3(2x) \qquad \text{Use Chain Rule.}$$

$$\frac{d}{dx}\left[\frac{(x^2 + 1)^4}{4}\right] = (x^2 + 1)^3(2x) \qquad \text{Divide both sides by 4.}$$

$$\frac{(x^2 + 1)^4}{4} + C = \int 2x(x^2 + 1)^3 \, dx \qquad \text{Write in integral form.}$$

The key to this solution is the presence of the factor $2x$ in the integrand. In other words, this solution works because $2x$ is precisely the derivative of $(x^2 + 1)$. Letting $u = x^2 + 1$, you can write

$$\int \overbrace{(x^2 + 1)^3}^{u^3} \underbrace{2x \, dx}_{du} = \int u^3 \, du$$

$$= \frac{u^4}{4} + C.$$

This is an example of the **General Power Rule** for integration.

General Power Rule for Integration

If u is a differentiable function of x, then

$$\int u^n \frac{du}{dx} \, dx = \int u^n \, du = \frac{u^{n+1}}{n+1} + C, \quad n \neq -1.$$

When using the General Power Rule, you must first identify a factor u of the integrand that is raised to a power. Then, you must show that its derivative du/dx is also a factor of the integrand. This is demonstrated in Example 1.

Example 1 **Applying the General Power Rule**

Find each indefinite integral.

a. $\int 3(3x - 1)^4 \, dx$ **b.** $\int (2x + 1)(x^2 + x) \, dx$

c. $\int 3x^2 \sqrt{x^3 - 2} \, dx$ **d.** $\int \frac{-4x}{(1 - 2x^2)^2} \, dx$

SOLUTION

STUDY TIP

Example 1(b) illustrates a case of the General Power Rule that is sometimes overlooked—when the power is $n = 1$. In this case, the rule takes the form

$$\int u \frac{du}{dx} \, dx = \frac{u^2}{2} + C.$$

a. $\int 3(3x - 1)^4 \, dx = \int \overbrace{(3x - 1)^4}^{u^n} \overbrace{(3)}^{\frac{du}{dx}} \, dx$ Let $u = 3x - 1$.

$= \frac{(3x - 1)^5}{5} + C$ General Power Rule

b. $\int (2x + 1)(x^2 + x) \, dx = \int \overbrace{(x^2 + x)}^{u^n} \overbrace{(2x + 1)}^{\frac{du}{dx}} \, dx$ Let $u = x^2 + x$.

$= \frac{(x^2 + x)^2}{2} + C$ General Power Rule

c. $\int 3x^2 \sqrt{x^3 - 2} \, dx = \int \overbrace{(x^3 - 2)^{1/2}}^{u^n} \overbrace{(3x^2)}^{\frac{du}{dx}} \, dx$ Let $u = x^3 - 2$.

$= \frac{(x^3 - 2)^{3/2}}{3/2} + C$ General Power Rule

$= \frac{2}{3}(x^3 - 2)^{3/2} + C$ Simplify.

STUDY TIP

Remember that you can verify the result of an indefinite integral by differentiating the function. Check the answer to Example 1(d) by differentiating the function

$$F(x) = -\frac{1}{1 - 2x^2} + C.$$

$$\frac{d}{dx}\left[-\frac{1}{1 - 2x^2} + C\right]$$

$$= \frac{-4x}{(1 - 2x^2)^2}$$

d. $\int \frac{-4x}{(1 - 2x^2)^2} \, dx = \int \overbrace{(1 - 2x^2)^{-2}}^{u^n} \overbrace{(-4x)}^{\frac{du}{dx}} \, dx$ Let $u = 1 - 2x^2$.

$= \frac{(1 - 2x^2)^{-1}}{-1} + C$ General Power Rule

$= -\frac{1}{1 - 2x^2} + C$ Simplify.

✓CHECKPOINT 1

Find each indefinite integral.

a. $\int (3x^2 + 6)(x^3 + 6x)^2 \, dx$ **b.** $\int 2x\sqrt{x^2 - 2} \, dx$

Many times, part of the derivative du/dx is missing from the integrand, and in *some* cases you can make the necessary adjustments to apply the General Power Rule.

Algebra Review

For help on the algebra in Example 2, see Example 1(b) in the *Chapter 11 Algebra Review*, on page 861.

Example 2 **Multiplying and Dividing by a Constant**

Find $\displaystyle\int x(3 - 4x^2)^2 \, dx$.

SOLUTION Let $u = 3 - 4x^2$. To apply the General Power Rule, you need to create $du/dx = -8x$ as a factor of the integrand. You can accomplish this by multiplying and dividing by the constant -8.

$$\int x(3 - 4x^2)^2 \, dx = \int \left(-\frac{1}{8}\right) \overbrace{(3 - 4x^2)^2}^{u^n} \overbrace{(-8x)}^{\frac{du}{dx}} \, dx \qquad \text{Multiply and divide by } -8.$$

$$= -\frac{1}{8}\int (3 - 4x^2)^2(-8x) \, dx \qquad \text{Factor } -\tfrac{1}{8} \text{ out of integrand.}$$

$$= \left(-\frac{1}{8}\right)\frac{(3 - 4x^2)^3}{3} + C \qquad \text{General Power Rule}$$

$$= -\frac{(3 - 4x^2)^3}{24} + C \qquad \text{Simplify.}$$

STUDY TIP

Try using the Chain Rule to check the result of Example 2. After differentiating $-\frac{1}{24}(3 - 4x^2)^3$ and simplifying, you should obtain the original integrand.

✓**CHECKPOINT 2**

Find $\displaystyle\int x^3(3x^4 + 1)^2 \, dx$.

STUDY TIP

In Example 3, be sure you see that you cannot factor variable quantities outside the integral sign. After all, if this were permissible, then you could move the entire integrand outside the integral sign and eliminate the need for all integration rules except the rule $\int dx = x + C$.

Example 3 **A Failure of the General Power Rule**

Find $\displaystyle\int -8(3 - 4x^2)^2 \, dx$.

SOLUTION Let $u = 3 - 4x^2$. As in Example 2, to apply the General Power Rule you must create $du/dx = -8x$ as a factor of the integrand. In Example 2, you could do this by multiplying and dividing by a constant, and then factoring that constant out of the integrand. This strategy doesn't work with variables. That is,

$$\int -8(3 - 4x^2)^2 \, dx \neq \frac{1}{x}\int (3 - 4x^2)^2(-8x) \, dx.$$

To find this indefinite integral, you can expand the integrand and use the Simple Power Rule.

$$\int -8(3 - 4x^2)^2 \, dx = \int (-72 + 192x^2 - 128x^4) \, dx$$

$$= -72x + 64x^3 - \frac{128}{5}x^5 + C$$

✓**CHECKPOINT 3**

Find $\displaystyle\int 2(3x^4 + 1)^2 \, dx$.

When an integrand contains an extra constant factor that is not needed as part of du/dx, you can simply move the factor outside the integral sign, as shown in the next example.

Example 4   Applying the General Power Rule

Find $\displaystyle\int 7x^2\sqrt{x^3 + 1}\ dx$.

SOLUTION   Let $u = x^3 + 1$. Then you need to create $du/dx = 3x^2$ by multiplying and dividing by 3. The constant factor $\frac{7}{3}$ is not needed as part of du/dx, and can be moved outside the integral sign.

$$\int 7x^2\sqrt{x^3 + 1}\ dx = \int 7x^2(x^3 + 1)^{1/2}\ dx \qquad \text{Rewrite with rational exponent.}$$

$$= \int \frac{7}{3}(x^3 + 1)^{1/2}(3x^2)\ dx \qquad \text{Multiply and divide by 3.}$$

$$= \frac{7}{3}\int (x^3 + 1)^{1/2}(3x^2)\ dx \qquad \text{Factor } \tfrac{7}{3} \text{ outside integral.}$$

$$= \frac{7}{3}\frac{(x^3 + 1)^{3/2}}{3/2} + C \qquad \text{General Power Rule}$$

$$= \frac{14}{9}(x^3 + 1)^{3/2} + C \qquad \text{Simplify.}$$

✓CHECKPOINT 4

Find $\displaystyle\int 5x\sqrt{x^2 - 1}\ dx$.

Algebra Review

For help on the algebra in Example 4, see Example 1(c) in the *Chapter 11 Algebra Review*, on page 861.

TECHNOLOGY

(T) If you use a symbolic integration utility to find indefinite integrals, you should be in for some surprises. This is true because integration is not nearly as straightforward as differentiation. By trying different integrands, you should be able to find several that the program cannot solve: in such situations, it may list a new indefinite integral. You should also be able to find several that have horrendous antiderivatives, some with functions that you may not recognize.

Substitution

The integration technique used in Examples 1, 2, and 4 depends on your ability to recognize or create an integrand of the form $u^n \, du/dx$. With more complicated integrands, it is difficult to recognize the steps needed to fit the integrand to a basic integration formula. When this occurs, an alternative procedure called **substitution** or **change of variables** can be helpful. With this procedure, you completely rewrite the integral in terms of u and du. That is, if $u = f(x)$, then $du = f'(x) \, dx$, and the General Power Rule takes the form

$$\int u^n \frac{du}{dx} \, dx = \int u^n \, du. \qquad \text{General Power Rule}$$

DISCOVERY

Calculate the derivative of each function. Which one is the anti-derivative of $f(x) = \sqrt{1 - 3x}$?

$$F(x) = (1 - 3x)^{3/2} + C$$

$$F(x) = \tfrac{2}{3}(1 - 3x)^{3/2} + C$$

$$F(x) = -\tfrac{2}{9}(1 - 3x)^{3/2} + C$$

Example 5   Integrating by Substitution

Find $\displaystyle\int \sqrt{1 - 3x} \, dx$.

SOLUTION   Begin by letting $u = 1 - 3x$. Then, $du/dx = -3$ and $du = -3 \, dx$. This implies that $dx = -\tfrac{1}{3} \, du$, and you can find the indefinite integral as shown.

$$\int \sqrt{1 - 3x} \, dx = \int (1 - 3x)^{1/2} \, dx \qquad \text{Rewrite with rational exponent.}$$

$$= \int u^{1/2} \left(-\frac{1}{3} \, du \right) \qquad \text{Substitute for } x \text{ and } dx.$$

$$= -\frac{1}{3} \int u^{1/2} \, du \qquad \text{Factor } -\tfrac{1}{3} \text{ out of integrand.}$$

$$= -\frac{1}{3} \frac{u^{3/2}}{3/2} + C \qquad \text{Apply Power Rule.}$$

$$= -\frac{2}{9} u^{3/2} + C \qquad \text{Simplify.}$$

$$= -\frac{2}{9}(1 - 3x)^{3/2} + C \qquad \text{Substitute } 1 - 3x \text{ for } u.$$

✓ **CHECKPOINT 5**

Find $\int \sqrt{1 - 2x} \, dx$ by the method of substitution. ■

The basic steps for integration by substitution are outlined in the guidelines below.

Guidelines for Integration by Substitution

1. Let u be a function of x (usually part of the integrand).

2. Solve for x and dx in terms of u and du.

3. Convert the entire integral to u-variable form.

4. After integrating, rewrite the antiderivative as a function of x.

5. Check your answer by differentiating.

Example 6   **Integration by Substitution**

Find $\int x\sqrt{x^2 - 1}\, dx$.

SOLUTION   Consider the substitution $u = x^2 - 1$, which produces $du = 2x\, dx$. To create $2x\, dx$ as part of the integral, multiply and divide by 2.

$$\int x\sqrt{x^2 - 1}\, dx = \frac{1}{2} \int \overbrace{(x^2 - 1)^{1/2}}^{u^{1/n}} \overbrace{2x\, dx}^{du} \qquad \text{Multiply and divide by 2.}$$

$$= \frac{1}{2} \int u^{1/2}\, du \qquad \text{Substitute for } x \text{ and } dx.$$

$$= \frac{1}{2} \frac{u^{3/2}}{3/2} + C \qquad \text{Apply Power Rule.}$$

$$= \frac{1}{3} u^{3/2} + C \qquad \text{Simplify.}$$

$$= \frac{1}{3}(x^2 - 1)^{3/2} + C \qquad \text{Substitute for } u.$$

You can check this result by differentiating.

$$\frac{d}{dx}\left[\frac{1}{3}(x^2 - 1)^{3/2} + C\right] = \frac{1}{3}\left(\frac{3}{2}\right)(x^2 - 1)^{1/2}(2x)$$

$$= \frac{1}{2}(2x)(x^2 - 1)^{1/2}$$

$$= x\sqrt{x^2 - 1}$$

✓**CHECKPOINT 6**

Find $\int x\sqrt{x^2 + 4}\, dx$ by the method of substitution. ▪

To become efficient at integration, you should learn to use *both* techniques discussed in this section. For simpler integrals, you should use pattern recognition and create du/dx by multiplying and dividing by an appropriate constant. For more complicated integrals, you should use a formal change of variables, as shown in Examples 5 and 6. For the integrals in this section's exercise set, try working several of the problems twice—once with pattern recognition and once using formal substitution.

DISCOVERY

Suppose you were asked to evaluate the integrals below. Which one would you choose? Explain your reasoning.

$$\int \sqrt{x^2 + 1}\, dx \quad \text{or} \quad \int x\sqrt{x^2 + 1}\, dx$$

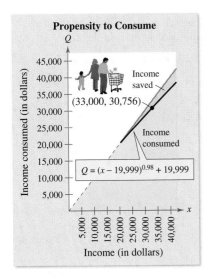

Propensity to Consume

$Q = (x - 19,999)^{0.98} + 19,999$

(33,000, 30,756)

Income saved

Income consumed

Income consumed (in dollars) — vertical axis: 5,000 to 45,000

Income (in dollars) — horizontal axis: 5,000 to 40,000

FIGURE 11.4

STUDY TIP

When you use the initial condition to find the value of C in Example 7, you substitute 20,000 for Q and 20,000 for x.

$$Q = (x - 19,999)^{0.98} + C$$

$$20,000 = (20,000 - 19,999)^{0.98} + C$$

$$20,000 = 1 + C$$

$$19,999 = C$$

✓ CHECKPOINT 7

According to the model in Example 7, at what income level would a family of four consume $30,000? ■

Extended Application: Propensity to Consume

In 2005, the U.S. poverty level for a family of four was about $20,000. Families at or below the poverty level tend to consume 100% of their income—that is, they use all their income to purchase necessities such as food, clothing, and shelter. As income level increases, the average consumption tends to drop below 100%. For instance, a family earning $22,000 may be able to save $440 and so consume only $21,560 (98%) of their income. As the income increases, the ratio of consumption to savings tends to decrease. The rate of change of consumption with respect to income is called the **marginal propensity to consume.** *(Source: U.S. Census Bureau)*

Example 7
MAKE A DECISION   Analyzing Consumption

For a family of four in 2005, the marginal propensity to consume income x can be modeled by

$$\frac{dQ}{dx} = \frac{0.98}{(x - 19,999)^{0.02}}, \quad x \geq 20,000$$

where Q represents the income consumed. Use the model to estimate the amount consumed by a family of four whose 2005 income was $33,000. Would the family have consumed more than $30,000?

SOLUTION   Begin by integrating dQ/dx to find a model for the consumption Q. Use the initial condition that $Q = 20,000$ and $x = 20,000$.

$$\frac{dQ}{dx} = \frac{0.98}{(x - 19,999)^{0.02}}$$ Marginal propensity to consume

$$Q = \int \frac{0.98}{(x - 19,999)^{0.02}} \, dx$$ Integrate to obtain Q.

$$= \int 0.98(x - 19,999)^{-0.02} \, dx$$ Rewrite.

$$= (x - 19,999)^{0.98} + C$$ General Power Rule

$$= (x - 19,999)^{0.98} + 19,999$$ Use initial condition to find C.

Using this model, you can estimate that a family of four with an income of $x = 33,000$ consumed about $30,756. So, a family of four would have consumed more than $30,000. The graph of Q is shown in Figure 11.4. ▬▬▬

CONCEPT CHECK

1. When using the General Power Rule for an integrand that contains an extra constant factor that is not needed as part of du/dx, what can you do with the factor?

2. Write the General Power Rule for integration.

3. Write the guidelines for integration by substitution.

4. Explain why the General Power Rule works for finding $\int 2x\sqrt{x^2 + 1} \, dx$, but not for finding $\int 2\sqrt{x^2 + 1} \, dx$.

Skills Review 11.2

The following warm-up exercises involve skills that were covered in earlier sections. You will use these skills in the exercise set for this section. For additional help, review Sections 0.3, 0.4, 0.7, and 11.1.

In Exercises 1–10, find the indefinite integral.

1. $\int (2x^3 + 1)\, dx$

2. $\int (x^{1/2} + 3x - 4)\, dx$

3. $\int \frac{1}{x^2}\, dx$

4. $\int \frac{1}{3t^3}\, dt$

5. $\int (1 + 2t)t^{3/2}\, dt$

6. $\int \sqrt{x}(2x - 1)\, dx$

7. $\int \frac{5x^3 + 2}{x^2}\, dx$

8. $\int \frac{2x^2 - 5}{x^4}\, dx$

9. $\int (x^2 + 1)^2\, dx$

10. $\int (x^3 - 2x + 1)^2\, dx$

In Exercises 11–14, simplify the expression.

11. $\left(-\frac{5}{4}\right)\frac{(x - 2)^4}{4}$

12. $\left(\frac{1}{6}\right)\frac{(x - 1)^{-2}}{-2}$

13. $(6)\frac{(x^2 + 3)^{2/3}}{2/3}$

14. $\left(\frac{5}{2}\right)\frac{(1 - x^3)^{-1/2}}{-1/2}$

Exercises 11.2

See www.CalcChat.com for worked-out solutions to odd-numbered exercises.

In Exercises 1–8, identify u and du/dx for the integral $\int u^n(du/dx)\, dx$.

1. $\int (5x^2 + 1)^2(10x)\, dx$

2. $\int (3 - 4x^2)^3(-8x)\, dx$

3. $\int \sqrt{1 - x^2}(-2x)\, dx$

4. $\int 3x^2\sqrt{x^3 + 1}\, dx$

5. $\int \left(4 + \frac{1}{x^2}\right)^5\left(\frac{-2}{x^3}\right) dx$

6. $\int \frac{1}{(1 + 2x)^2}(2)\, dx$

7. $\int (1 + \sqrt{x})^3\left(\frac{1}{2\sqrt{x}}\right) dx$

8. $\int (4 - \sqrt{x})^2\left(\frac{-1}{2\sqrt{x}}\right) dx$

In Exercises 9–28, find the indefinite integral and check the result by differentiation.

9. $\int (1 + 2x)^4(2)\, dx$

10. $\int (x^2 - 1)^3(2x)\, dx$

11. $\int \sqrt{4x^2 - 5}(8x)\, dx$

12. $\int \sqrt[3]{1 - 2x^2}(-4x)\, dx$

13. $\int (x - 1)^4\, dx$

14. $\int (x - 3)^{5/2}\, dx$

15. $\int 2x(x^2 - 1)^7\, dx$

16. $\int x(1 - 2x^2)^3\, dx$

17. $\int \frac{x^2}{(1 + x^3)^2}\, dx$

18. $\int \frac{x^2}{(x^3 - 1)^2}\, dx$

19. $\int \frac{x + 1}{(x^2 + 2x - 3)^2}\, dx$

20. $\int \frac{6x}{(1 + x^2)^3}\, dx$

21. $\int \frac{x - 2}{\sqrt{x^2 - 4x + 3}}\, dx$

22. $\int \frac{4x + 6}{(x^2 + 3x + 7)^3}\, dx$

23. $\int 5u\sqrt[3]{1 - u^2}\, du$

24. $\int u^3\sqrt{u^4 + 2}\, du$

25. $\int \frac{4y}{\sqrt{1 + y^2}}\, dy$

26. $\int \frac{3x^2}{\sqrt{1 - x^3}}\, dx$

27. $\int \frac{-3}{\sqrt{2t + 3}}\, dt$

28. $\int \frac{t + 2t^2}{\sqrt{t}}\, dt$

Ⓣ In Exercises 29–34, use a symbolic integration utility to find the indefinite integral.

29. $\int \frac{x^3}{\sqrt{1 - x^4}}\, dx$

30. $\int \frac{3x}{\sqrt{1 - 4x^2}}\, dx$

31. $\int \left(1 + \frac{4}{t^2}\right)^2\left(\frac{1}{t^3}\right) dt$

32. $\int \left(1 + \frac{1}{t}\right)^3\left(\frac{1}{t^2}\right) dt$

33. $\int (x^3 + 3x + 9)(x^2 + 1)\, dx$

34. $\int (7 - 3x - 3x^2)(2x + 1)\, dx$

In Exercises 35–42, use formal substitution (as illustrated in Examples 5 and 6) to find the indefinite integral.

35. $\displaystyle\int 12x(6x^2 - 1)^3 \, dx$

36. $\displaystyle\int 3x^2(1 - x^3)^2 \, dx$

37. $\displaystyle\int x^2(2 - 3x^3)^{3/2} \, dx$

38. $\displaystyle\int t\sqrt{t^2 + 1} \, dt$

39. $\displaystyle\int \frac{x}{\sqrt{x^2 + 25}} \, dx$

40. $\displaystyle\int \frac{3}{\sqrt{2x + 1}} \, dx$

41. $\displaystyle\int \frac{x^2 + 1}{\sqrt{x^3 + 3x + 4}} \, dx$

42. $\displaystyle\int \sqrt{x}\,(4 - x^{3/2})^2 \, dx$

In Exercises 43–46, (a) perform the integration in two ways: once using the Simple Power Rule and once using the General Power Rule. (b) Explain the difference in the results. (c) Which method do you prefer? Explain your reasoning.

43. $\displaystyle\int (x - 1)^2 \, dx$

44. $\displaystyle\int (3 - x)^2 \, dx$

45. $\displaystyle\int x(x^2 - 1)^2 \, dx$

46. $\displaystyle\int x(2x^2 + 1)^2 \, dx$

47. Find the equation of the function f whose graph passes through the point $\left(0, \frac{4}{3}\right)$ and whose derivative is
$$f'(x) = x\sqrt{1 - x^2}.$$

48. Find the equation of the function f whose graph passes through the point $\left(0, \frac{7}{3}\right)$ and whose derivative is
$$f'(x) = x\sqrt{1 - x^2}.$$

49. Cost The marginal cost of a product is modeled by
$$\frac{dC}{dx} = \frac{4}{\sqrt{x + 1}}. \text{ When } x = 15, C = 50.$$

(a) Find the cost function.

(T) (b) Use a graphing utility to graph dC/dx and C in the same viewing window.

50. Cost The marginal cost of a product is modeled by
$$\frac{dC}{dx} = \frac{12}{\sqrt[3]{12x + 1}}.$$
When $x = 13$, $C = 100$.

(a) Find the cost function.

(T) (b) Use a graphing utility to graph dC/dx and C in the same viewing window.

Supply Function In Exercises 51 and 52, find the supply function $x = f(p)$ that satisfies the initial conditions.

51. $\dfrac{dx}{dp} = p\sqrt{p^2 - 25}, \quad x = 600 \text{ when } p = \$13$

52. $\dfrac{dx}{dp} = \dfrac{10}{\sqrt{p - 3}}, \quad x = 100 \text{ when } p = \$3$

Demand Function In Exercises 53 and 54, find the demand function $x = f(p)$ that satisfies the initial conditions.

53. $\dfrac{dx}{dp} = -\dfrac{6000p}{(p^2 - 16)^{3/2}}, \quad x = 5000 \text{ when } p = \$5$

54. $\dfrac{dx}{dp} = -\dfrac{400}{(0.02p - 1)^3}, \quad x = 10{,}000 \text{ when } p = \$100$

55. Gardening An evergreen nursery usually sells a type of shrub after 5 years of growth and shaping. The growth rate during those 5 years is approximated by
$$\frac{dh}{dt} = \frac{17.6t}{\sqrt{17.6t^2 + 1}}$$
where t is time in years and h is height in inches. The seedlings are 6 inches tall when planted $(t = 0)$.

(a) Find the height function.

(b) How tall are the shrubs when they are sold?

56. Cash Flow The rate of disbursement dQ/dt of a $4 million federal grant is proportional to the square of $100 - t$, where t is the time (in days, $0 \le t \le 100$) and Q is the amount that remains to be disbursed. Find the amount that remains to be disbursed after 50 days. Assume that the entire grant will be disbursed after 100 days.

(T)(S) **Marginal Propensity to Consume** In Exercises 57 and 58, (a) use the marginal propensity to consume, dQ/dx, to write Q as a function of x, where x is the income (in dollars) and Q is the income consumed (in dollars). Assume that 100% of the income is consumed for families that have annual incomes of $25,000 or less. (b) Use the result of part (a) and a spreadsheet to complete the table showing the income consumed and the income saved, $x - Q$, for various incomes. (c) Use a graphing utility to represent graphically the income consumed and saved.

x	25,000	50,000	100,000	150,000
Q				
$x - Q$				

57. $\dfrac{dQ}{dx} = \dfrac{0.95}{(x - 24{,}999)^{0.05}}, \quad x \ge 25{,}000$

58. $\dfrac{dQ}{dx} = \dfrac{0.93}{(x - 24{,}999)^{0.07}}, \quad x \ge 25{,}000$

(T) In Exercises 59 and 60, use a symbolic integration utility to find the indefinite integral. Verify the result by differentiating.

59. $\displaystyle\int \frac{1}{\sqrt{x} + \sqrt{x + 1}} \, dx$

60. $\displaystyle\int \frac{x}{\sqrt{3x + 2}} \, dx$

Section 11.3

Exponential and Logarithmic Integrals

■ Use the Exponential Rule to find indefinite integrals.
■ Use the Log Rule to find indefinite integrals.

Using the Exponential Rule

Each of the differentiation rules for exponential functions has its corresponding integration rule.

Integrals of Exponential Functions

Let u be a differentiable function of x.

$$\int e^x \, dx = e^x + C \qquad\qquad \text{Simple Exponential Rule}$$

$$\int e^u \frac{du}{dx} \, dx = \int e^u \, du = e^u + C \qquad\qquad \text{General Exponential Rule}$$

> **Example 1** **Integrating Exponential Functions**

Find each indefinite integral.

a. $\displaystyle\int 2e^x \, dx$ **b.** $\displaystyle\int 2e^{2x} \, dx$ **c.** $\displaystyle\int (e^x + x) \, dx$

SOLUTION

a. $\displaystyle\int 2e^x \, dx = 2 \int e^x \, dx$ Constant Multiple Rule

$\qquad\qquad = 2e^x + C$ Simple Exponential Rule

b. $\displaystyle\int 2e^{2x} \, dx = \int e^{2x}(2) \, dx$ Let $u = 2x$, then $\dfrac{du}{dx} = 2$.

$\qquad\qquad = \displaystyle\int e^u \frac{du}{dx} \, dx$

$\qquad\qquad = e^{2x} + C$ General Exponential Rule

c. $\displaystyle\int (e^x + x) \, dx = \int e^x \, dx + \int x \, dx$ Sum Rule

$\qquad\qquad = e^x + \dfrac{x^2}{2} + C$ Simple Exponential and Power Rules

You can check each of these results by differentiating.

✓ **CHECKPOINT 1**

Find each indefinite integral.

a. $\displaystyle\int 3e^x \, dx$

b. $\displaystyle\int 5e^{5x} \, dx$

c. $\displaystyle\int (e^x - x) \, dx$

If you use a symbolic integration utility to find antiderivatives of exponential or logarithmic functions, you can easily obtain results that are beyond the scope of this course. For instance, the antiderivative of e^{x^2} involves the imaginary unit i and the probability function called ERF. In this course, you are not expected to interpret or use such results. You can simply state that the function cannot be integrated using elementary functions.

Example 2 Integrating an Exponential Function

Find $\displaystyle\int e^{3x+1}\,dx$.

SOLUTION Let $u = 3x + 1$, then $du/dx = 3$. You can introduce the missing factor of 3 in the integrand by multiplying and dividing by 3.

$$\int e^{3x+1}\,dx = \frac{1}{3}\int e^{3x+1}(3)\,dx \qquad \text{Multiply and divide by 3.}$$

$$= \frac{1}{3}\int e^{u}\frac{du}{dx}\,dx \qquad \text{Substitute } u \text{ and } du/dx.$$

$$= \frac{1}{3}e^{u} + C \qquad \text{General Exponential Rule}$$

$$= \frac{1}{3}e^{3x+1} + C \qquad \text{Substitute for } u.$$

✓**CHECKPOINT 2**

Find $\displaystyle\int e^{2x+3}\,dx$.

Algebra Review

For help on the algebra in Example 3, see Example 1(d) in the *Chapter 11 Algebra Review*, on page 861.

Example 3 Integrating an Exponential Function

Find $\displaystyle\int 5xe^{-x^2}\,dx$.

SOLUTION Let $u = -x^2$, then $du/dx = -2x$. You can create the factor $-2x$ in the integrand by multiplying and dividing by -2.

$$\int 5xe^{-x^2}\,dx = \int\left(-\frac{5}{2}\right)e^{-x^2}(-2x)\,dx \qquad \text{Multiply and divide by } -2.$$

$$= -\frac{5}{2}\int e^{-x^2}(-2x)\,dx \qquad \text{Factor } -\tfrac{5}{2} \text{ out of the integrand.}$$

$$= -\frac{5}{2}\int e^{u}\frac{du}{dx}\,dx \qquad \text{Substitute } u \text{ and } \frac{du}{dx}.$$

$$= -\frac{5}{2}e^{u} + C \qquad \text{General Exponential Rule}$$

$$= -\frac{5}{2}e^{-x^2} + C \qquad \text{Substitute for } u.$$

STUDY TIP

Remember that you cannot introduce a missing *variable* in the integrand. For instance, you cannot find $\int e^{x^2}\,dx$ by multiplying and dividing by $2x$ and then factoring $1/(2x)$ out of the integral. That is,

$$\int e^{x^2}\,dx \neq \frac{1}{2x}\int e^{x^2}(2x)\,dx.$$

✓**CHECKPOINT 3**

Find $\displaystyle\int 4xe^{x^2}\,dx$.

Using the Log Rule

When the Power Rules for integration were introduced in Sections 11.1 and 11.2, you saw that they work for powers other than $n = -1$.

$$\int x^n \, dx = \frac{x^{n+1}}{n+1} + C, \qquad n \neq -1 \qquad \text{Simple Power Rule}$$

$$\int u^n \frac{du}{dx} \, dx = \int u^n \, du = \frac{u^{n+1}}{n+1} + C, \qquad n \neq -1 \qquad \text{General Power Rule}$$

The Log Rules for integration allow you to integrate functions of the form $\int x^{-1} \, dx$ and $\int u^{-1} \, du$.

Integrals of Logarithmic Functions

Let u be a differentiable function of x.

$$\int \frac{1}{x} \, dx = \ln|x| + C \qquad \text{Simple Logarithmic Rule}$$

$$\int \frac{du/dx}{u} \, dx = \int \frac{1}{u} \, du = \ln|u| + C \qquad \text{General Logarithmic Rule}$$

You can verify each of these rules by differentiating. For instance, to verify that $d/dx[\ln|x|] = 1/x$, notice that

$$\frac{d}{dx}[\ln x] = \frac{1}{x} \quad \text{and} \quad \frac{d}{dx}[\ln(-x)] = \frac{-1}{-x} = \frac{1}{x}.$$

Example 4   Integrating Logarithmic Functions

Find each indefinite integral.

a. $\displaystyle\int \frac{4}{x} \, dx$ **b.** $\displaystyle\int \frac{2x}{x^2} \, dx$ **c.** $\displaystyle\int \frac{3}{3x+1} \, dx$

SOLUTION

a. $\displaystyle\int \frac{4}{x} \, dx = 4 \int \frac{1}{x} \, dx$ Constant Multiple Rule

$\qquad\qquad = 4\ln|x| + C$ Simple Logarithmic Rule

b. $\displaystyle\int \frac{2x}{x^2} \, dx = \int \frac{du/dx}{u} \, dx$ Let $u = x^2$, then $\dfrac{du}{dx} = 2x$.

$\qquad\qquad = \ln|u| + C$ General Logarithmic Rule

$\qquad\qquad = \ln x^2 + C$ Substitute for u.

c. $\displaystyle\int \frac{3}{3x+1} \, dx = \int \frac{du/dx}{u} \, dx$ Let $u = 3x+1$, then $\dfrac{du}{dx} = 3$.

$\qquad\qquad = \ln|u| + C$ General Logarithmic Rule

$\qquad\qquad = \ln|3x+1| + C$ Substitute for u.

Example 5   **Using the Log Rule**

Find $\displaystyle\int \frac{1}{2x-1}\,dx$.

SOLUTION   Let $u = 2x - 1$, then $du/dx = 2$. You can create the necessary factor of 2 in the integrand by multiplying and dividing by 2.

$$\int \frac{1}{2x-1}\,dx = \frac{1}{2}\int \frac{2}{2x-1}\,dx \qquad \text{Multiply and divide by 2.}$$

$$= \frac{1}{2}\int \frac{du/dx}{u}\,dx \qquad \text{Substitute } u \text{ and } \frac{du}{dx}.$$

$$= \frac{1}{2}\ln|u| + C \qquad \text{General Log Rule}$$

$$= \frac{1}{2}\ln|2x-1| + C \qquad \text{Substitute for } u.$$

✓**CHECKPOINT 5**

Find $\displaystyle\int \frac{1}{4x+1}\,dx$. ▪

Example 6   **Using the Log Rule**

Find $\displaystyle\int \frac{6x}{x^2+1}\,dx$.

SOLUTION   Let $u = x^2 + 1$, then $du/dx = 2x$. You can create the necessary factor of $2x$ in the integrand by factoring a 3 out of the integrand.

$$\int \frac{6x}{x^2+1}\,dx = 3\int \frac{2x}{x^2+1}\,dx \qquad \text{Factor 3 out of integrand.}$$

$$= 3\int \frac{du/dx}{u}\,dx \qquad \text{Substitute } u \text{ and } \frac{du}{dx}.$$

$$= 3\ln|u| + C \qquad \text{General Log Rule}$$

$$= 3\ln(x^2+1) + C \qquad \text{Substitute for } u.$$

✓**CHECKPOINT 6**

Find $\displaystyle\int \frac{3x}{x^2+4}\,dx$. ▪

Integrals to which the Log Rule can be applied are often given in disguised form. For instance, if a rational function has a numerator of degree greater than or equal to that of the denominator, you should use long division to rewrite the integrand. Here is an example.

$$\int \frac{x^2+6x+1}{x^2+1}\,dx = \int \left(1 + \frac{6x}{x^2+1}\right)\,dx$$

$$= x + 3\ln(x^2+1) + C$$

Algebra Review

For help on the algebra in the integral at the right, see Example 2(d) in the *Chapter 11 Algebra Review*, on page 862.

The next example summarizes some additional situations in which it is helpful to rewrite the integrand in order to recognize the antiderivative.

Algebra Review

For help on the algebra in Example 7, see Example 2(a)–(c) in the *Chapter 11 Algebra Review*, on page 862.

Example 7 Rewriting Before Integrating

Find each indefinite integral.

a. $\int \dfrac{3x^2 + 2x - 1}{x^2}\, dx$ **b.** $\int \dfrac{1}{1 + e^{-x}}\, dx$ **c.** $\int \dfrac{x^2 + x + 1}{x - 1}\, dx$

SOLUTION

a. Begin by rewriting the integrand as the sum of three fractions.

$$\int \frac{3x^2 + 2x - 1}{x^2}\, dx = \int \left(\frac{3x^2}{x^2} + \frac{2x}{x^2} - \frac{1}{x^2}\right) dx$$
$$= \int \left(3 + \frac{2}{x} - \frac{1}{x^2}\right) dx$$
$$= 3x + 2\ln|x| + \frac{1}{x} + C$$

b. Begin by rewriting the integrand by multiplying and dividing by e^x.

$$\int \frac{1}{1 + e^{-x}}\, dx = \int \left(\frac{e^x}{e^x}\right)\frac{1}{1 + e^{-x}}\, dx$$
$$= \int \frac{e^x}{e^x + 1}\, dx$$
$$= \ln(e^x + 1) + C$$

c. Begin by dividing the numerator by the denominator.

$$\int \frac{x^2 + x + 1}{x - 1}\, dx = \int \left(x + 2 + \frac{3}{x - 1}\right) dx$$
$$= \frac{x^2}{2} + 2x + 3\ln|x - 1| + C$$

✓CHECKPOINT 7

Find each indefinite integral.

a. $\int \dfrac{4x^2 - 3x + 2}{x^2}\, dx$

b. $\int \dfrac{2}{e^{-x} + 1}\, dx$

c. $\int \dfrac{x^2 + 2x + 4}{x + 1}\, dx$ ∎

STUDY TIP

The Exponential and Log Rules are necessary to solve certain real-life problems, such as population growth. You will see such problems in the exercise set for this section.

CONCEPT CHECK

1. Write the General Exponential Rule for integration.
2. Write the General Logarithmic Rule for integration.
3. Which integration rule allows you to integrate functions of the form $\int e^u \dfrac{du}{dx}\, dx$?
4. Which integration rule allows you to integrate $\int x^{-1}\, dx$?

Skills Review 11.3	The following warm-up exercises involve skills that were covered in earlier sections. You will use these skills in the exercise set for this section. For additional help, review Sections 3.3, 10.4, 11.1, and 11.2.

In Exercises 1 and 2, find the domain of the function.

1. $y = \ln(2x - 5)$

2. $y = \ln(x^2 - 5x + 6)$

In Exercises 3–6, use long division to rewrite the quotient.

3. $\dfrac{x^2 + 4x + 2}{x + 2}$

4. $\dfrac{x^2 - 6x + 9}{x - 4}$

5. $\dfrac{x^3 + 4x^2 - 30x - 4}{x^2 - 4x}$

6. $\dfrac{x^4 - x^3 + x^2 + 15x + 2}{x^2 + 5}$

In Exercises 7–10, evaluate the integral.

7. $\displaystyle \int \left(x^3 + \frac{1}{x^2} \right) dx$

8. $\displaystyle \int \frac{x^2 + 2x}{x}\, dx$

9. $\displaystyle \int \frac{x^3 + 4}{x^2}\, dx$

10. $\displaystyle \int \frac{x + 3}{x^3}\, dx$

Exercises 11.3

See www.CalcChat.com for worked-out solutions to odd-numbered exercises.

In Exercises 1–12, use the Exponential Rule to find the indefinite integral.

1. $\displaystyle \int 2e^{2x}\, dx$

2. $\displaystyle \int -3e^{-3x}\, dx$

3. $\displaystyle \int e^{4x}\, dx$

4. $\displaystyle \int e^{-0.25x}\, dx$

5. $\displaystyle \int 9xe^{-x^2}\, dx$

6. $\displaystyle \int 3xe^{0.5x^2}\, dx$

7. $\displaystyle \int 5x^2 e^{x^3}\, dx$

8. $\displaystyle \int (2x + 1)e^{x^2 + x}\, dx$

9. $\displaystyle \int (x^2 + 2x)e^{x^3 + 3x^2 - 1}\, dx$

10. $\displaystyle \int 3(x - 4)e^{x^2 - 8x}\, dx$

11. $\displaystyle \int 5e^{2-x}\, dx$

12. $\displaystyle \int 3e^{-(x+1)}\, dx$

In Exercises 13–28, use the Log Rule to find the indefinite integral.

13. $\displaystyle \int \frac{1}{x + 1}\, dx$

14. $\displaystyle \int \frac{1}{x - 5}\, dx$

15. $\displaystyle \int \frac{1}{3 - 2x}\, dx$

16. $\displaystyle \int \frac{1}{6x - 5}\, dx$

17. $\displaystyle \int \frac{2}{3x + 5}\, dx$

18. $\displaystyle \int \frac{5}{2x - 1}\, dx$

19. $\displaystyle \int \frac{x}{x^2 + 1}\, dx$

20. $\displaystyle \int \frac{x^2}{3 - x^3}\, dx$

21. $\displaystyle \int \frac{x^2}{x^3 + 1}\, dx$

22. $\displaystyle \int \frac{x}{x^2 + 4}\, dx$

23. $\displaystyle \int \frac{x + 3}{x^2 + 6x + 7}\, dx$

24. $\displaystyle \int \frac{x^2 + 2x + 3}{x^3 + 3x^2 + 9x + 1}\, dx$

25. $\displaystyle \int \frac{1}{x \ln x}\, dx$

26. $\displaystyle \int \frac{1}{x(\ln x)^2}\, dx$

27. $\displaystyle \int \frac{e^{-x}}{1 - e^{-x}}\, dx$

28. $\displaystyle \int \frac{e^x}{1 + e^x}\, dx$

Ⓣ In Exercises 29–38, use a symbolic integration utility to find the indefinite integral.

29. $\displaystyle \int \frac{1}{x^2} e^{2/x}\, dx$

30. $\displaystyle \int \frac{1}{x^3} e^{1/4x^2}\, dx$

31. $\displaystyle \int \frac{1}{\sqrt{x}} e^{\sqrt{x}}\, dx$

32. $\displaystyle \int \frac{e^{1/\sqrt{x}}}{x^{3/2}}\, dx$

33. $\displaystyle \int (e^x - 2)^2\, dx$

34. $\displaystyle \int (e^x - e^{-x})^2\, dx$

35. $\displaystyle \int \frac{e^{-x}}{1 + e^{-x}}\, dx$

36. $\displaystyle \int \frac{3e^x}{2 + e^x}\, dx$

37. $\displaystyle \int \frac{4e^{2x}}{5 - e^{2x}}\, dx$

38. $\displaystyle \int \frac{-e^{3x}}{2 - e^{3x}}\, dx$

In Exercises 39–54, use any basic integration formula or formulas to find the indefinite integral. State which integration formula(s) you used to find the integral.

39. $\displaystyle \int \frac{e^{2x} + 2e^x + 1}{e^x} \, dx$

40. $\displaystyle \int (6x + e^x)\sqrt{3x^2 + e^x} \, dx$

41. $\displaystyle \int e^x \sqrt{1 - e^x} \, dx$

42. $\displaystyle \int \frac{2(e^x - e^{-x})}{(e^x + e^{-x})^2} \, dx$

43. $\displaystyle \int \frac{1}{(x - 1)^2} \, dx$

44. $\displaystyle \int \frac{1}{\sqrt{x + 1}} \, dx$

45. $\displaystyle \int 4e^{2x-1} \, dx$

46. $\displaystyle \int (5e^{-2x} + 1) \, dx$

47. $\displaystyle \int \frac{x^3 - 8x}{2x^2} \, dx$

48. $\displaystyle \int \frac{x - 1}{4x} \, dx$

49. $\displaystyle \int \frac{2}{1 + e^{-x}} \, dx$

50. $\displaystyle \int \frac{3}{1 + e^{-3x}} \, dx$

51. $\displaystyle \int \frac{x^2 + 2x + 5}{x - 1} \, dx$

52. $\displaystyle \int \frac{x - 3}{x + 3} \, dx$

53. $\displaystyle \int \frac{1 + e^{-x}}{1 + xe^{-x}} \, dx$

54. $\displaystyle \int \frac{5}{e^{-5x} + 7} \, dx$

In Exercises 55 and 56, find the equation of the function *f* whose graph passes through the point.

55. $f'(x) = \dfrac{x^2 + 4x + 3}{x - 1}; \quad (2, 4)$

56. $f'(x) = \dfrac{x^3 - 4x^2 + 3}{x - 3}; \quad (4, -1)$

57. Biology A population of bacteria is growing at the rate of

$$\frac{dP}{dt} = \frac{3000}{1 + 0.25t}$$

where *t* is the time in days. When $t = 0$, the population is 1000.

(a) Write an equation that models the population *P* in terms of the time *t*.

(b) What is the population after 3 days?

(c) After how many days will the population be 12,000?

58. Biology Because of an insufficient oxygen supply, the trout population in a lake is dying. The population's rate of change can be modeled by

$$\frac{dP}{dt} = -125e^{-t/20}$$

where *t* is the time in days. When $t = 0$, the population is 2500.

(a) Write an equation that models the population *P* in terms of the time *t*.

(b) What is the population after 15 days?

(c) According to this model, how long will it take for the entire trout population to die?

(T) **59. Demand** The marginal price for the demand of a product can be modeled by $dp/dx = 0.1e^{-x/500}$, where *x* is the quantity demanded. When the demand is 600 units, the price is $30.

(a) Find the demand function, $p = f(x)$.

(b) Use a graphing utility to graph the demand function. Does price increase or decrease as demand increases?

(c) Use the *zoom* and *trace* features of the graphing utility to find the quantity demanded when the price is $22.

(T) **60. Revenue** The marginal revenue for the sale of a product can be modeled by

$$\frac{dR}{dx} = 50 - 0.02x + \frac{100}{x + 1}$$

where *x* is the quantity demanded.

(a) Find the revenue function.

(b) Use a graphing utility to graph the revenue function.

(c) Find the revenue when 1500 units are sold.

(d) Use the *zoom* and *trace* features of the graphing utility to find the number of units sold when the revenue is $60,230.

61. Average Salary From 2000 through 2005, the average salary for public school nurses *S* (in dollars) in the United States changed at the rate of

$$\frac{dS}{dt} = 1724.1e^{-t/4.2}$$

where $t = 0$ corresponds to 2000. In 2005, the average salary for public school nurses was $40,520. *(Source: Educational Research Service)*

(a) Write a model that gives the average salary for public school nurses per year.

(b) Use the model to find the average salary for public school nurses in 2002.

62. Sales The rate of change in sales for The Yankee Candle Company from 1998 through 2005 can be modeled by

$$\frac{dS}{dt} = 0.528t + \frac{597.2099}{t}$$

where *S* is the sales (in millions) and $t = 8$ corresponds to 1998. In 1999, the sales for The Yankee Candle Company were $256.6 million. *(Source: The Yankee Candle Company)*

(a) Find a model for sales from 1998 through 2005.

(b) Find The Yankee Candle Company's sales in 2004.

True or False? In Exercises 63 and 64, determine whether the statement is true or false. If it is false, explain why or give an example that shows it is false.

63. $(\ln x)^{1/2} = \frac{1}{2}(\ln x)$

64. $\displaystyle \int \ln x = \left(\frac{1}{x}\right) + C$

Mid-Chapter Quiz

See www.CalcChat.com for worked-out solutions to odd-numbered exercises.

Take this quiz as you would take a quiz in class. When you are done, check your work against the answers given in the back of the book.

In Exercises 1–9, find the indefinite integral and check your result by differentiation.

1. $\displaystyle\int 3\,dx$

2. $\displaystyle\int 10x\,dx$

3. $\displaystyle\int \frac{1}{x^5}\,dx$

4. $\displaystyle\int (x^2 - 2x + 15)\,dx$

5. $\displaystyle\int x(x + 4)\,dx$

6. $\displaystyle\int (6x + 1)^3(6)\,dx$

7. $\displaystyle\int (x^2 - 5x)(2x - 5)\,dx$

8. $\displaystyle\int \frac{3x^2}{(x^3 + 3)^3}\,dx$

9. $\displaystyle\int \sqrt{5x + 2}\,dx$

In Exercises 10 and 11, find the particular solution $y = f(x)$ that satisfies the differential equation and initial condition.

10. $f'(x) = 16x; f(0) = 1$

11. $f'(x) = 9x^2 + 4; f(1) = 5$

12. The marginal cost function for producing x units of a product is modeled by

$$\frac{dC}{dx} = 16 - 0.06x.$$

It costs \$25 to produce one unit. Find (a) the fixed cost (when $x = 0$) and (b) the total cost of producing 500 units.

13. Find the equation of the function f whose graph passes through the point $(0, 1)$ and whose derivative is

$$f'(x) = 2x^2 + 1.$$

In Exercises 14–16, use the Exponential Rule to find the indefinite integral. Check your result by differentiation.

14. $\displaystyle\int 5e^{5x+4}\,dx$

15. $\displaystyle\int (x + 2e^{2x})\,dx$

16. $\displaystyle\int 3x^2 e^{x^3}\,dx$

In Exercises 17–19, use the Log Rule to find the indefinite integral.

17. $\displaystyle\int \frac{2}{2x - 1}\,dx$

18. $\displaystyle\int \frac{-2x}{x^2 + 3}\,dx$

19. $\displaystyle\int \frac{3(3x^2 + 4x)}{x^3 + 2x^2}\,dx$

20. The number of bolts B produced by a foundry changes according to the model

$$\frac{dB}{dt} = \frac{250t}{\sqrt{t^2 + 36}}, \quad 0 \le t \le 40$$

where t is measured in hours. Find the number of bolts produced in (a) 8 hours and (b) 40 hours.

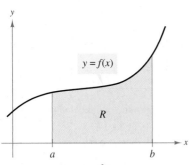

Section 11.4

Area and the Fundamental Theorem of Calculus

■ Evaluate definite integrals.
■ Evaluate definite integrals using the Fundamental Theorem of Calculus.
■ Use definite integrals to solve marginal analysis problems.
■ Find the average values of functions over closed intervals.
■ Use properties of even and odd functions to help evaluate definite integrals.
■ Find the amounts of annuities.

Area and Definite Integrals

From your study of geometry, you know that area is a number that defines the size of a bounded region. For simple regions, such as rectangles, triangles, and circles, area can be found using geometric formulas.

In this section, you will learn how to use calculus to find the areas of nonstandard regions, such as the region R shown in Figure 11.5.

FIGURE 11.5 $\displaystyle\int_a^b f(x)\,dx =$ Area

Definition of a Definite Integral

Let f be nonnegative and continuous on the closed interval $[a, b]$. The area of the region bounded by the graph of f, the x-axis, and the lines $x = a$ and $x = b$ is denoted by

$$\text{Area} = \int_a^b f(x)\,dx.$$

The expression $\int_a^b f(x)\,dx$ is called the **definite integral** from a to b, where a is the **lower limit of integration** and b is the **upper limit of integration.**

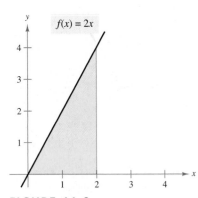

FIGURE 11.6

Example 1 Evaluating a Definite Integral

Evaluate $\displaystyle\int_0^2 2x\,dx.$

SOLUTION This definite integral represents the area of the region bounded by the graph of $f(x) = 2x$, the x-axis, and the line $x = 2$, as shown in Figure 11.6. The region is triangular, with a height of four units and a base of two units.

$$\int_0^2 2x\,dx = \frac{1}{2}(\text{base})(\text{height}) \qquad \text{Formula for area of triangle}$$

$$= \frac{1}{2}(2)(4) = 4 \qquad \text{Simplify.}$$

✓**CHECKPOINT 1**

Evaluate the definite integral using a geometric formula. Illustrate your answer with an appropriate sketch.

$$\int_0^3 4x\,dx$$

The Fundamental Theorem of Calculus

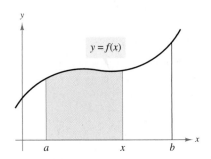

Consider the function A, which denotes the area of the region shown in Figure 11.7. To discover the relationship between A and f, let x increase by an amount Δx. This increases the area by ΔA. Let $f(m)$ and $f(M)$ denote the minimum and maximum values of f on the interval $[x, x + \Delta x]$.

FIGURE 11.7 $A(x)$ = Area from a to x

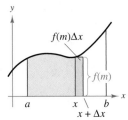 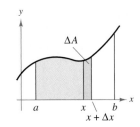 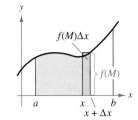

FIGURE 11.8

As indicated in Figure 11.8, you can write the inequality below.

$$f(m) \, \Delta x \leq \quad \Delta A \quad \leq f(M) \, \Delta x \qquad \text{See Figure 11.8.}$$

$$f(m) \leq \quad \frac{\Delta A}{\Delta x} \quad \leq f(M) \qquad \text{Divide each term by } \Delta x.$$

$$\lim_{\Delta x \to 0} f(m) \leq \lim_{\Delta x \to 0} \frac{\Delta A}{\Delta x} \leq \lim_{\Delta x \to 0} f(M) \qquad \text{Take limit of each term.}$$

$$f(x) \leq \quad A'(x) \quad \leq f(x) \qquad \text{Definition of derivative of } A(x)$$

So, $f(x) = A'(x)$, and $A(x) = F(x) + C$, where $F'(x) = f(x)$. Because $A(a) = 0$, it follows that $C = -F(a)$. So, $A(x) = F(x) - F(a)$, which implies that

$$A(b) = \int_a^b f(x) \, dx = F(b) - F(a).$$

This equation tells you that *if you can find an antiderivative for f,* then you can use the antiderivative to evaluate the definite integral $\int_a^b f(x) \, dx$. This result is called the **Fundamental Theorem of Calculus.**

The Fundamental Theorem of Calculus

If f is nonnegative and continuous on the closed interval $[a, b]$, then

$$\int_a^b f(x) \, dx = F(b) - F(a)$$

where F is any function such that $F'(x) = f(x)$ for all x in $[a, b]$.

STUDY TIP

There are two basic ways to introduce the Fundamental Theorem of Calculus. One way uses an area function, as shown here. The other uses a summation process, as shown in Appendix D.

Guidelines for Using the Fundamental Theorem of Calculus

1. The Fundamental Theorem of Calculus describes a way of *evaluating* a definite integral, not a procedure for finding antiderivatives.

2. In applying the Fundamental Theorem, it is helpful to use the notation

$$\int_a^b f(x)\, dx = F(x)\Big]_a^b = F(b) - F(a).$$

3. The constant of integration C can be dropped because

$$\int_a^b f(x)\, dx = \left[F(x) + C\right]_a^b$$
$$= [F(b) + C] - [F(a) + C]$$
$$= F(b) - F(a) + C - C$$
$$= F(b) - F(a).$$

In the development of the Fundamental Theorem of Calculus, f was assumed to be nonnegative on the closed interval $[a, b]$. As such, the definite integral was defined as an area. Now, with the Fundamental Theorem, the definition can be extended to include functions that are negative on all or part of the closed interval $[a, b]$. Specifically, if f is *any* function that is continuous on a closed interval $[a, b]$, then the **definite integral** of $f(x)$ from a to b is defined to be

$$\int_a^b f(x)\, dx = F(b) - F(a)$$

where F is an antiderivative of f. Remember that definite integrals do not necessarily represent areas and can be negative, zero, or positive.

STUDY TIP

Be sure you see the distinction between indefinite and definite integrals. The *indefinite integral*

$$\int f(x)\, dx$$

denotes a family of *functions*, each of which is an antiderivative of f, whereas the *definite integral*

$$\int_a^b f(x)\, dx$$

is a *number*.

Properties of Definite Integrals

Let f and g be continuous on the closed interval $[a, b]$.

1. $\displaystyle\int_a^b k f(x)\, dx = k \int_a^b f(x)\, dx, \quad k$ is a constant.

2. $\displaystyle\int_a^b \left[f(x) \pm g(x)\right] dx = \int_a^b f(x)\, dx \pm \int_a^b g(x)\, dx$

3. $\displaystyle\int_a^b f(x)\, dx = \int_a^c f(x)\, dx + \int_c^b f(x)\, dx, \quad a < c < b$

4. $\displaystyle\int_a^a f(x)\, dx = 0$

5. $\displaystyle\int_a^b f(x)\, dx = -\int_b^a f(x)\, dx$

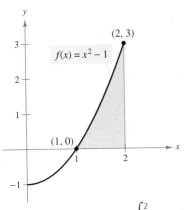

FIGURE 11.9   Area $= \int_1^2 (x^2 - 1)\, dx$

✓ **CHECKPOINT 2**

Find the area of the region bounded by the x-axis and the graph of $f(x) = x^2 + 1$, $2 \le x \le 3$. ▪

Example 2   **Finding Area by the Fundamental Theorem**

Find the area of the region bounded by the x-axis and the graph of

$$f(x) = x^2 - 1, \quad 1 \le x \le 2.$$

SOLUTION   Note that $f(x) \ge 0$ on the interval $1 \le x \le 2$, as shown in Figure 11.9. So, you can represent the area of the region by a definite integral. To find the area, use the Fundamental Theorem of Calculus.

$$\text{Area} = \int_1^2 (x^2 - 1)\, dx \qquad \text{Definition of definite integral}$$

$$= \left(\frac{x^3}{3} - x \right)_1^2 \qquad \text{Find antiderivative.}$$

$$= \left(\frac{2^3}{3} - 2 \right) - \left(\frac{1^3}{3} - 1 \right) \qquad \text{Apply Fundamental Theorem.}$$

$$= \frac{4}{3} \qquad \text{Simplify.}$$

So, the area of the region is $\frac{4}{3}$ square units.

STUDY TIP

It is easy to make errors in signs when evaluating definite integrals. To avoid such errors, enclose the values of the antiderivative at the upper and lower limits of integration in separate sets of parentheses, as shown above.

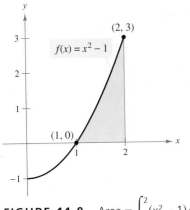

FIGURE 11.10

✓ **CHECKPOINT 3**

Evaluate $\int_0^1 (2t + 3)^3\, dt$. ▪

Example 3   **Evaluating a Definite Integral**

Evaluate the definite integral

$$\int_0^1 (4t + 1)^2\, dt$$

and sketch the region whose area is represented by the integral.

SOLUTION

$$\int_0^1 (4t + 1)^2\, dt = \frac{1}{4} \int_0^1 (4t + 1)^2 (4)\, dt \qquad \text{Multiply and divide by 4.}$$

$$= \frac{1}{4} \left[\frac{(4t + 1)^3}{3} \right]_0^1 \qquad \text{Find antiderivative.}$$

$$= \frac{1}{4} \left[\left(\frac{5^3}{3} \right) - \left(\frac{1}{3} \right) \right] \qquad \text{Apply Fundamental Theorem.}$$

$$= \frac{31}{3} \qquad \text{Simplify.}$$

The region is shown in Figure 11.10.

Example 4 **Evaluating Definite Integrals**

Evaluate each definite integral.

a. $\displaystyle\int_0^3 e^{2x}\,dx$ **b.** $\displaystyle\int_1^2 \frac{1}{x}\,dx$ **c.** $\displaystyle\int_1^4 -3\sqrt{x}\,dx$

SOLUTION

a. $\displaystyle\int_0^3 e^{2x}\,dx = \frac{1}{2}e^{2x}\Big]_0^3 = \frac{1}{2}(e^6 - e^0) \approx 201.21$

b. $\displaystyle\int_1^2 \frac{1}{x}\,dx = \ln x\Big]_1^2 = \ln 2 - \ln 1 = \ln 2 \approx 0.69$

c. $\displaystyle\int_1^4 -3\sqrt{x}\,dx = -3\int_1^4 x^{1/2}\,dx$ Rewrite with rational exponent.

$\displaystyle\qquad\qquad = -3\left[\frac{x^{3/2}}{3/2}\right]_1^4$ Find antiderivative.

$\displaystyle\qquad\qquad = -2x^{3/2}\Big]_1^4$

$\displaystyle\qquad\qquad = -2(4^{3/2} - 1^{3/2})$ Apply Fundamental Theorem.

$\displaystyle\qquad\qquad = -2(8 - 1)$

$\displaystyle\qquad\qquad = -14$ Simplify.

✓**CHECKPOINT 4**

Evaluate each definite integral.

a. $\displaystyle\int_0^1 e^{4x}\,dx$

b. $\displaystyle\int_2^5 -\frac{1}{x}\,dx$ ∎

STUDY TIP

In Example 4(c), note that the value of a definite integral can be negative.

Example 5 **Interpreting Absolute Value**

Evaluate $\displaystyle\int_0^2 |2x - 1|\,dx$.

SOLUTION The region represented by the definite integral is shown in Figure 11.11. From the definition of absolute value, you can write

$$|2x - 1| = \begin{cases} -(2x - 1), & x < \frac{1}{2} \\ 2x - 1, & x \geq \frac{1}{2} \end{cases}.$$

Using Property 3 of definite integrals, you can rewrite the integral as two definite integrals.

$$\int_0^2 |2x - 1|\,dx = \int_0^{1/2} -(2x - 1)\,dx + \int_{1/2}^2 (2x - 1)\,dx$$

$$= \left[-x^2 + x\right]_0^{1/2} + \left[x^2 - x\right]_{1/2}^2$$

$$= \left(-\frac{1}{4} + \frac{1}{2}\right) - (0 + 0) + (4 - 2) - \left(\frac{1}{4} - \frac{1}{2}\right) = \frac{5}{2}$$

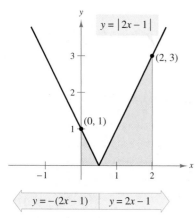

$y = |2x - 1|$

(2, 3)

(0, 1)

$y = -(2x - 1)$ $y = 2x - 1$

FIGURE 11.11

✓**CHECKPOINT 5**

Evaluate $\displaystyle\int_0^5 |x - 2|\,dx$. ∎

Marginal Analysis

You have already studied *marginal analysis* in the context of derivatives and differentials (Sections 7.5 and 9.5). There, you were given a cost, revenue, or profit function, and you used the derivative to approximate the additional cost, revenue, or profit obtained by selling one additional unit. In this section, you will examine the reverse process. That is, you will be given the marginal cost, marginal revenue, or marginal profit and will be asked to use a definite integral to find the exact increase or decrease in cost, revenue, or profit obtained by selling one or several additional units.

For instance, suppose you wanted to find the additional revenue obtained by increasing sales from x_1 to x_2 units. If you knew the revenue function R you could simply subtract $R(x_1)$ from $R(x_2)$. If you didn't know the revenue function, but did know the marginal revenue function, you could still find the additional revenue by using a definite integral, as shown.

$$\int_{x_1}^{x_2} \frac{dR}{dx} \, dx = R(x_2) - R(x_1)$$

Example 6   Analyzing a Profit Function

The marginal profit for a product is modeled by $\dfrac{dP}{dx} = -0.0005x + 12.2$.

a. Find the change in profit when sales increase from 100 to 101 units.

b. Find the change in profit when sales increase from 100 to 110 units.

SOLUTION

a. The change in profit obtained by increasing sales from 100 to 101 units is

$$\int_{100}^{101} \frac{dP}{dx} \, dx = \int_{100}^{101} (-0.0005x + 12.2) \, dx$$
$$= \left[-0.00025x^2 + 12.2x \right]_{100}^{101}$$
$$\approx \$12.15.$$

b. The change in profit obtained by increasing sales from 100 to 110 units is

$$\int_{100}^{110} \frac{dP}{dx} \, dx = \int_{100}^{110} (-0.0005x + 12.2) \, dx$$
$$= \left[-0.00025x^2 + 12.2x \right]_{100}^{110}$$
$$\approx \$121.48$$

✓ **CHECKPOINT 6**

The marginal profit for a product is modeled by

$$\frac{dP}{dx} = -0.0002x + 14.2.$$

a. Find the change in profit when sales increase from 100 to 101 units.

b. Find the change in profit when sales increase from 100 to 110 units. ∎

TECHNOLOGY

Ⓣ Symbolic integration utilities can be used to evaluate definite integrals as well as indefinite integrals. If you have access to such a program, try using it to evaluate several of the definite integrals in this section.

Average Value

The *average value* of a function on a closed interval is defined below.

Definition of the Average Value of a Function

If f is continuous on $[a, b]$, then the **average value** of f on $[a, b]$ is

$$\text{Average value of } f \text{ on } [a, b] = \frac{1}{b-a} \int_a^b f(x) \, dx.$$

In Section 9.2, you studied the effects of production levels on cost using an average cost function. In the next example, you will study the effects of time on cost by using integration to find the average cost.

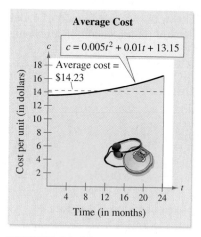

Average Cost

$c = 0.005t^2 + 0.01t + 13.15$

Average cost = $14.23

Cost per unit (in dollars)

Time (in months)

FIGURE 11.12

Example 7

MAKE A DECISION **Finding the Average Cost**

The cost per unit c of producing CD players over a two-year period is modeled by

$$c = 0.005t^2 + 0.01t + 13.15, \quad 0 \le t \le 24$$

where t is the time in months. Approximate the average cost per unit over the two-year period. Will the average cost per unit be less than $15?

SOLUTION   The average cost can be found by integrating c over the interval $[0, 24]$.

$$\text{Average cost per unit} = \frac{1}{24} \int_0^{24} (0.005t^2 + 0.01t + 13.15) \, dt$$

$$= \frac{1}{24} \left[\frac{0.005t^3}{3} + \frac{0.01t^2}{2} + 13.15t \right]_0^{24}$$

$$= \frac{1}{24}(341.52)$$

$$= \$14.23 \quad \text{(See Figure 11.12.)}$$

Yes, the average cost per unit will be less than $15.

✓ **CHECKPOINT 7**

Find the average cost per unit over a two-year period if the cost per unit c of roller blades is given by $c = 0.005t^2 + 0.02t + 12.5$, for $0 \le t \le 24$, where t is the time in months. ∎

To check the reasonableness of the average value found in Example 7, assume that one unit is produced each month, beginning with $t = 0$ and ending with $t = 24$. When $t = 0$, the cost is

$$c = 0.005(0)^2 + 0.01(0) + 13.15$$

$$= \$13.15.$$

Similarly, when $t = 1$, the cost is

$$c = 0.005(1)^2 + 0.01(1) + 13.15$$

$$\approx \$13.17.$$

Each month, the cost increases, and the average of the 25 costs is

$$\frac{13.15 + 13.17 + 13.19 + 13.23 + \cdots + 16.27}{25} \approx \$14.25.$$

Even and Odd Functions

Several common functions have graphs that are symmetric with respect to the y-axis or the origin, as shown in Figure 11.13. If the graph of f is symmetric with respect to the y-axis, as in Figure 11.13(a), then

$$f(-x) = f(x) \qquad \text{Even function}$$

and f is called an **even** function. If the graph of f is symmetric with respect to the origin, as in Figure 11.13(b), then

$$f(-x) = -f(x) \qquad \text{Odd function}$$

and f is called an **odd** function.

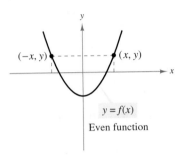

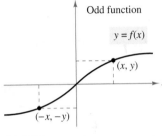

(a) y-axis symmetry

(b) Origin symmetry

FIGURE 11.13

Integration of Even and Odd Functions

1. If f is an *even* function, then $\displaystyle\int_{-a}^{a} f(x)\,dx = 2\int_{0}^{a} f(x)\,dx.$

2. If f is an *odd* function, then $\displaystyle\int_{-a}^{a} f(x)\,dx = 0.$

Example 8   Integrating Even and Odd Functions

Evaluate each definite integral.

a. $\displaystyle\int_{-2}^{2} x^2\,dx$ **b.** $\displaystyle\int_{-2}^{2} x^3\,dx$

SOLUTION

a. Because $f(x) = x^2$ is even,

$$\int_{-2}^{2} x^2\,dx = 2\int_{0}^{2} x^2\,dx = 2\left[\frac{x^3}{3}\right]_{0}^{2} = 2\left(\frac{8}{3} - 0\right) = \frac{16}{3}.$$

b. Because $f(x) = x^3$ is odd,

$$\int_{-2}^{2} x^3\,dx = 0.$$

✓ **CHECKPOINT 8**

Evaluate each definite integral.

a. $\displaystyle\int_{-1}^{1} x^4\,dx$

b. $\displaystyle\int_{-1}^{1} x^5\,dx$

Annuity

A sequence of equal payments made at regular time intervals over a period of time is called an **annuity.** Some examples of annuities are payroll savings plans, monthly home mortgage payments, and individual retirement accounts. The **amount of an annuity** is the sum of the payments plus the interest earned and can be found as shown below.

Amount of an Annuity

If c represents a continuous income function in dollars per year (where t is the time in years), r represents the interest rate compounded continuously, and T represents the term of the annuity in years, then the **amount of an annuity** is

$$\text{Amount of an annuity} = e^{rT} \int_0^T c(t)e^{-rt}\, dt.$$

Example 9   Finding the Amount of an Annuity

You deposit $2000 each year for 15 years in an individual retirement account (IRA) paying 5% interest. How much will you have in your IRA after 15 years?

SOLUTION   The income function for your deposit is $c(t) = 2000$. So, the amount of the annuity after 15 years will be

$$
\begin{aligned}
\text{Amount of an annuity} &= e^{rT} \int_0^T c(t)e^{-rt}\, dt \\
&= e^{(0.05)(15)} \int_0^{15} 2000 e^{-0.05t}\, dt \\
&= 2000 e^{0.75} \left[-\frac{e^{-0.05t}}{0.05} \right]_0^{15} \\
&\approx \$44{,}680.00.
\end{aligned}
$$

✓ **CHECKPOINT 9**

If you deposit $1000 in a savings account every year, paying 4% interest, how much will be in the account after 10 years? ■

CONCEPT CHECK

1. Complete the following: The indefinite integral $\int f(x)\, dx$ denotes a family of _____ , each of which is a(n) _____ of f, whereas the definite integral $\int_a^b f(x)\, dx$ is a _____ .

2. If f is an odd function, then $\int_{-a}^{a} f(x)\, dx$ equals what?

3. State the Fundamental Theorem of Calculus.

4. What is an annuity?

In Exercises 1–4, find the indefinite integral.

1. $\int (3x + 7)\, dx$

2. $\int \left(x^{3/2} + 2\sqrt{x}\right) dx$

3. $\int \frac{1}{5x}\, dx$

4. $\int e^{-6x}\, dx$

In Exercises 5 and 6, evaluate the expression when $a = 5$ and $b = 3$.

5. $\left(\dfrac{a}{5} - a\right) - \left(\dfrac{b}{5} - b\right)$

6. $\left(6a - \dfrac{a^3}{3}\right) - \left(6b - \dfrac{b^3}{3}\right)$

In Exercises 7–10, integrate the marginal function.

7. $\dfrac{dC}{dx} = 0.02x^{3/2} + 29{,}500$

8. $\dfrac{dR}{dx} = 9000 + 2x$

9. $\dfrac{dP}{dx} = 25{,}000 - 0.01x$

10. $\dfrac{dC}{dx} = 0.03x^2 + 4600$

Exercises 11.4

 In Exercises 1 and 2, use a graphing utility to graph the integrand. Use the graph to determine whether the definite integral is positive, negative, or zero.

1. $\displaystyle\int_0^3 \frac{5x}{x^2 + 1}\, dx$

2. $\displaystyle\int_{-2}^2 x\sqrt{x^2 + 1}\, dx$

In Exercises 3–12, sketch the region whose area is represented by the definite integral. Then use a geometric formula to evaluate the integral.

3. $\displaystyle\int_0^2 3\, dx$

4. $\displaystyle\int_0^3 4\, dx$

5. $\displaystyle\int_0^4 x\, dx$

6. $\displaystyle\int_0^4 \frac{x}{2}\, dx$

7. $\displaystyle\int_0^5 (x + 1)\, dx$

8. $\displaystyle\int_0^3 (2x + 1)\, dx$

9. $\displaystyle\int_{-2}^3 |x - 1|\, dx$

10. $\displaystyle\int_{-1}^4 |x - 2|\, dx$

11. $\displaystyle\int_{-3}^3 \sqrt{9 - x^2}\, dx$

12. $\displaystyle\int_0^2 \sqrt{4 - x^2}\, dx$

In Exercises 13 and 14, use the values $\int_0^5 f(x)\, dx = 6$ and $\int_0^5 g(x)\, dx = 2$ to evaluate the definite integral.

13. (a) $\displaystyle\int_0^5 [f(x) + g(x)]\, dx$

(b) $\displaystyle\int_0^5 [f(x) - g(x)]\, dx$

(c) $\displaystyle\int_0^5 -4f(x)\, dx$

(d) $\displaystyle\int_0^5 [f(x) - 3g(x)]\, dx$

14. (a) $\displaystyle\int_0^5 2g(x)\, dx$

(b) $\displaystyle\int_5^0 f(x)\, dx$

(c) $\displaystyle\int_5^5 f(x)\, dx$

(d) $\displaystyle\int_0^5 [f(x) - f(x)]\, dx$

In Exercises 15–22, find the area of the region.

15. $y = x - x^2$

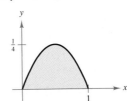

16. $y = 1 - x^4$

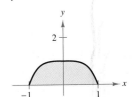

17. $y = \dfrac{1}{x^2}$

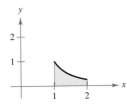

18. $y = \dfrac{2}{\sqrt{x}}$

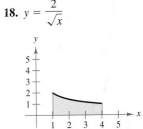

19. $y = 3e^{-x/2}$

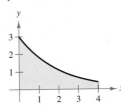

20. $y = 2e^{x/2}$

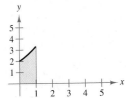

21. $y = \dfrac{x^2 + 4}{x}$

22. $y = \dfrac{x - 2}{x}$

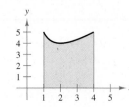

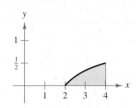

In Exercises 23–46, evaluate the definite integral.

23. $\displaystyle\int_0^1 2x\, dx$

24. $\displaystyle\int_2^7 3v\, dv$

25. $\displaystyle\int_{-1}^0 (x - 2)\, dx$

26. $\displaystyle\int_2^5 (-3x + 4)\, dx$

27. $\displaystyle\int_{-1}^1 (2t - 1)^2\, dt$

28. $\displaystyle\int_0^1 (1 - 2x)^2\, dx$

29. $\displaystyle\int_0^3 (x - 2)^3\, dx$

30. $\displaystyle\int_2^2 (x - 3)^4\, dx$

31. $\displaystyle\int_{-1}^1 \left(\sqrt[3]{t} - 2\right) dt$

32. $\displaystyle\int_1^4 \sqrt{\dfrac{2}{x}}\, dx$

33. $\displaystyle\int_1^4 \dfrac{u - 2}{\sqrt{u}}\, du$

34. $\displaystyle\int_0^1 \dfrac{x - \sqrt{x}}{3}\, dx$

35. $\displaystyle\int_{-1}^0 (t^{1/3} - t^{2/3})\, dt$

36. $\displaystyle\int_0^4 (x^{1/2} + x^{1/4})\, dx$

37. $\displaystyle\int_0^4 \dfrac{1}{\sqrt{2x + 1}}\, dx$

38. $\displaystyle\int_0^2 \dfrac{x}{\sqrt{1 + 2x^2}}\, dx$

39. $\displaystyle\int_0^1 e^{-2x}\, dx$

40. $\displaystyle\int_1^2 e^{1-x}\, dx$

41. $\displaystyle\int_1^3 \dfrac{e^{3/x}}{x^2}\, dx$

42. $\displaystyle\int_{-1}^1 (e^x - e^{-x})\, dx$

43. $\displaystyle\int_0^1 e^{2x}\sqrt{e^{2x} + 1}\, dx$

44. $\displaystyle\int_0^1 \dfrac{e^{-x}}{\sqrt{e^{-x} + 1}}\, dx$

45. $\displaystyle\int_0^2 \dfrac{x}{1 + 4x^2}\, dx$

46. $\displaystyle\int_0^1 \dfrac{e^{2x}}{e^{2x} + 1}\, dx$

In Exercises 47–50, evaluate the definite integral by the most convenient method. Explain your approach.

47. $\displaystyle\int_{-1}^1 |4x|\, dx$

48. $\displaystyle\int_0^3 |2x - 3|\, dx$

49. $\displaystyle\int_0^4 (2 - |x - 2|)\, dx$

50. $\displaystyle\int_{-4}^4 (4 - |x|)\, dx$

(T) In Exercises 51–54, evaluate the definite integral by hand. Then use a symbolic integration utility to evaluate the definite integral. Briefly explain any differences in your results.

51. $\displaystyle\int_{-1}^2 \dfrac{x}{x^2 - 9}\, dx$

52. $\displaystyle\int_2^3 \dfrac{x + 1}{x^2 + 2x - 3}\, dx$

53. $\displaystyle\int_0^3 \dfrac{2e^x}{2 + e^x}\, dx$

54. $\displaystyle\int_1^2 \dfrac{(2 + \ln x)^3}{x}\, dx$

(T) In Exercises 55–60, evaluate the definite integral by hand. Then use a graphing utility to graph the region whose area is represented by the integral.

55. $\displaystyle\int_1^3 (4x - 3)\, dx$

56. $\displaystyle\int_0^2 (x + 4)\, dx$

57. $\displaystyle\int_0^1 (x - x^3)\, dx$

58. $\displaystyle\int_0^2 (2 - x)\sqrt{x}\, dx$

59. $\displaystyle\int_2^4 \dfrac{3x^2}{x^3 - 1}\, dx$

60. $\displaystyle\int_0^{\ln 6} \dfrac{e^x}{2}\, dx$

In Exercises 61–64, find the area of the region bounded by the graphs of the equations. Use a graphing utility to verify your results.

61. $y = 3x^2 + 1$, $\quad y = 0$, $\quad x = 0$, and $\quad x = 2$

62. $y = 1 + \sqrt{x}$, $\quad y = 0$, $\quad x = 0$, and $\quad x = 4$

63. $y = 4/x$ $\quad y = 0$, $\quad x = 1$, and $x = 3$

64. $y = e^x$, $\quad y = 0$, $\quad x = 0$, and $\quad x = 2$

(T) In Exercises 65–72, use a graphing utility to graph the function over the interval. Find the average value of the function over the interval. Then find all x-values in the interval for which the function is equal to its average value.

Function	Interval
65. $f(x) = 4 - x^2$	$[-2, 2]$
66. $f(x) = x - 2\sqrt{x}$	$[0, 4]$
67. $f(x) = 2e^x$	$[-1, 1]$
68. $f(x) = e^{x/4}$	$[0, 4]$
69. $f(x) = x\sqrt{4 - x^2}$	$[0, 2]$
70. $f(x) = \dfrac{1}{(x - 3)^2}$	$[0, 2]$
71. $f(x) = \dfrac{6x}{x^2 + 1}$	$[0, 7]$
72. $f(x) = \dfrac{4x}{x^2 + 1}$	$[0, 1]$

In Exercises 73–76, state whether the function is even, odd, or neither.

73. $f(x) = 3x^4$

74. $g(x) = x^3 - 2x$

75. $g(t) = 2t^5 - 3t^2$

76. $f(t) = 5t^4 + 1$

77. Use the value $\displaystyle\int_0^1 x^2\, dx = \dfrac{1}{3}$ to evaluate each definite integral. Explain your reasoning.

(a) $\displaystyle\int_{-1}^0 x^2\, dx$ (b) $\displaystyle\int_{-1}^1 x^2\, dx$ (c) $\displaystyle\int_0^1 -x^2\, dx$

78. Use the value $\int_0^2 x^3 \, dx = 4$ to evaluate each definite integral. Explain your reasoning.

(a) $\displaystyle\int_{-2}^0 x^3 \, dx$ (b) $\displaystyle\int_{-2}^2 x^3 \, dx$ (c) $\displaystyle\int_0^2 3x^3 \, dx$

Marginal Analysis   In Exercises 79–84, find the change in cost C, revenue R, or profit P, for the given marginal. In each case, assume that the number of units x increases by 3 from the specified value of x.

Marginal	Number of Units, x
79. $\dfrac{dC}{dx} = 2.25$	$x = 100$
80. $\dfrac{dC}{dx} = \dfrac{20{,}000}{x^2}$	$x = 10$
81. $\dfrac{dR}{dx} = 48 - 3x$	$x = 12$
82. $\dfrac{dR}{dx} = 75\left(20 + \dfrac{900}{x}\right)$	$x = 500$
83. $\dfrac{dP}{dx} = \dfrac{400 - x}{150}$	$x = 200$
84. $\dfrac{dP}{dx} = 12.5\left(40 - 3\sqrt{x}\right)$	$x = 125$

Annuity   In Exercises 85–88, find the amount of an annuity with income function $c(t)$, interest rate r, and term T.

85. $c(t) = \$250$, $r = 8\%$, $T = 6$ years

86. $c(t) = \$500$, $r = 7\%$, $T = 4$ years

87. $c(t) = \$1500$, $r = 2\%$, $T = 10$ years

88. $c(t) = \$2000$, $r = 3\%$, $T = 15$ years

Capital Accumulation   In Exercises 89–92, you are given the rate of investment dI/dt. Find the capital accumulation over a five-year period by evaluating the definite integral

$$\text{Capital accumulation} = \int_0^5 \frac{dI}{dt} \, dt$$

where t is the time in years.

89. $\dfrac{dI}{dt} = 500$

90. $\dfrac{dI}{dt} = 100t$

91. $\dfrac{dI}{dt} = 500\sqrt{t + 1}$

92. $\dfrac{dI}{dt} = \dfrac{12{,}000t}{(t^2 + 2)^2}$

93. Cost   The total cost of purchasing and maintaining a piece of equipment for x years can be modeled by

$$C = 5000\left(25 + 3\int_0^x t^{1/4} \, dt\right).$$

Find the total cost after (a) 1 year, (b) 5 years, and (c) 10 years.

94. Depreciation   A company purchases a new machine for which the rate of depreciation can be modeled by

$$\frac{dV}{dt} = 10{,}000(t - 6), \quad 0 \le t \le 5$$

where V is the value of the machine after t years. Set up and evaluate the definite integral that yields the total loss of value of the machine over the first 3 years.

95. Compound Interest   A deposit of \$2250 is made in a savings account at an annual interest rate of 6%, compounded continuously. Find the average balance in the account during the first 5 years.

96. Mortgage Debt   The rate of change of mortgage debt outstanding for one- to four-family homes in the United States from 1998 through 2005 can be modeled by

$$\frac{dM}{dt} = 5.142t^2 - 283{,}426.2e^{-t}$$

where M is the mortgage debt outstanding (in billions of dollars) and t is the year, with $t = 8$ corresponding to 1998. In 1998, the mortgage debt outstanding in the United States was \$4259 billion. *(Source: Board of Governors of the Federal Reserve System)*

(a) Write a model for the debt as a function of t.

(b) What was the average mortgage debt outstanding for 1998 through 2005?

97. Biology   In the North Sea, cod fish are in danger of becoming extinct because a large proportion of the catch is being taken before the cod can reach breeding age. The fishing quotas set in the United Kingdom from the years 1999 through 2006 can be approximated by the equation

$$y = -0.7020t^3 + 29.802t^2 - 422.77t + 2032.9$$

where y is the total catch weight (in thousands of kilograms) and t is the year, with $t = 9$ corresponding to 1999. Determine the average recommended quota during the years 1995 through 2006. *(Source: International Council for Exploration of the Sea)*

98. Blood Flow   The velocity v of the flow of blood at a distance r from the center of an artery of radius R can be modeled by

$$v = k(R^2 - r^2), \quad k > 0$$

where k is a constant. Find the average velocity along a radius of the artery. (Use 0 and R as the limits of integration.)

(T) In Exercises 99–102, use a symbolic integration utility to evaluate the definite integral.

99. $\displaystyle\int_3^6 \frac{x}{3\sqrt{x^2 - 8}} \, dx$

100. $\displaystyle\int_{1/2}^1 (x + 1)\sqrt{1 - x} \, dx$

101. $\displaystyle\int_2^5 \left(\frac{1}{x^2} - \frac{1}{x^3}\right) dx$

102. $\displaystyle\int_0^1 x^3(x^3 + 1)^3 \, dx$

Section 11.5

The Area of a Region Bounded by Two Graphs

- Find the areas of regions bounded by two graphs.
- Find consumer and producer surpluses.
- Use the areas of regions bounded by two graphs to solve real-life problems.

Area of a Region Bounded by Two Graphs

With a few modifications, you can extend the use of definite integrals from finding the area of a region *under a graph* to finding the area of a region *bounded by two graphs*. To see how this is done, consider the region bounded by the graphs of f, g, $x = a$, and $x = b$, as shown in Figure 11.14. If the graphs of both f and g lie above the x-axis, then you can interpret the area of the region between the graphs as the area of the region under the graph of g subtracted from the area of the region under the graph of f, as shown in Figure 11.14.

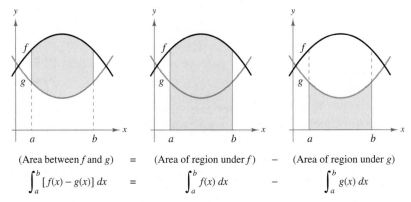

$$\text{(Area between } f \text{ and } g) \quad = \quad (\text{Area of region under } f) \quad - \quad (\text{Area of region under } g)$$

$$\int_a^b [f(x) - g(x)]\, dx \quad = \quad \int_a^b f(x)\, dx \quad - \quad \int_a^b g(x)\, dx$$

FIGURE 11.14

Although Figure 11.14 depicts the graphs of f and g lying above the x-axis, this is not necessary, and the same integrand $[f(x) - g(x)]$ can be used as long as both functions are continuous and $g(x) \leq f(x)$ on the interval $[a, b]$.

Area of a Region Bounded by Two Graphs

If f and g are continuous on $[a, b]$ and $g(x) \leq f(x)$ for all x in the interval, then the area of the region bounded by the graphs of f, g, $x = a$, and $x = b$ is given by

$$A = \int_a^b [f(x) - g(x)]\, dx.$$

DISCOVERY

Sketch the graph of $f(x) = x^3 - 4x$ and shade in the regions bounded by the graph of f and the x-axis. Write the appropriate integral(s) for this area.

Example 1   Finding the Area Bounded by Two Graphs

Find the area of the region bounded by the graphs of

$$y = x^2 + 2 \quad \text{and} \quad y = x$$

for $0 \le x \le 1$.

SOLUTION   Begin by sketching the graphs of both functions, as shown in Figure 11.15. From the figure, you can see that $x \le x^2 + 2$ for all x in $[0, 1]$. So, you can let $f(x) = x^2 + 2$ and $g(x) = x$. Then compute the area as shown.

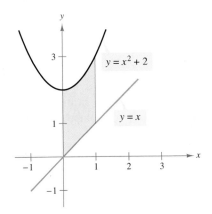

FIGURE 11.15

$$\text{Area} = \int_a^b [f(x) - g(x)]\, dx \qquad \text{Area between } f \text{ and } g$$

$$= \int_0^1 [(x^2 + 2) - (x)]\, dx \qquad \text{Substitute for } f \text{ and } g.$$

$$= \int_0^1 (x^2 - x + 2)\, dx$$

$$= \left[\frac{x^3}{3} - \frac{x^2}{2} + 2x \right]_0^1 \qquad \text{Find antiderivative.}$$

$$= \frac{11}{6} \text{ square units} \qquad \text{Apply Fundamental Theorem.}$$

✓ **CHECKPOINT 1**

Find the area of the region bounded by the graphs of $y = x^2 + 1$ and $y = x$ for $0 \le x \le 2$. Sketch the region bounded by the graphs. ▪

Example 2   Finding the Area Between Intersecting Graphs

Find the area of the region bounded by the graphs of

$$y = 2 - x^2 \quad \text{and} \quad y = x.$$

SOLUTION   In this problem, the values of a and b are not given and you must compute them by finding the points of intersection of the two graphs. To do this, equate the two functions and solve for x. When you do this, you will obtain $x = -2$ and $x = 1$. In Figure 11.16, you can see that the graph of $f(x) = 2 - x^2$ lies above the graph of $g(x) = x$ for all x in the interval $[-2, 1]$.

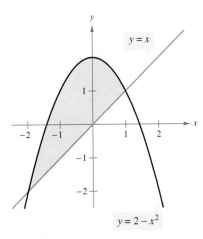

FIGURE 11.16

✓ **CHECKPOINT 2**

Find the area of the region bounded by the graphs of $y = 3 - x^2$ and $y = 2x$. ▪

$$\text{Area} = \int_a^b [f(x) - g(x)]\, dx \qquad \text{Area between } f \text{ and } g$$

$$= \int_{-2}^1 [(2 - x^2) - (x)]\, dx \qquad \text{Substitute for } f \text{ and } g.$$

$$= \int_{-2}^1 (-x^2 - x + 2)\, dx$$

$$= \left[-\frac{x^3}{3} - \frac{x^2}{2} + 2x \right]_{-2}^1 \qquad \text{Find antiderivative.}$$

$$= \frac{9}{2} \text{ square units} \qquad \text{Apply Fundamental Theorem.}$$

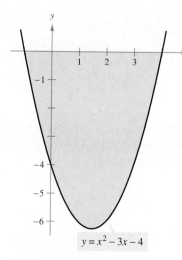

$$y = x^2 - 3x - 4$$

FIGURE 11.17

STUDY TIP

When finding the area of a region bounded by two graphs, be sure to use the integrand $[f(x) - g(x)]$. Be sure you realize that you cannot interchange $f(x)$ and $g(x)$. For instance, when solving Example 3, if you subtract $f(x)$ from $g(x)$, you will obtain an answer of $-\frac{125}{6}$, which is not correct.

Example 3   **Finding an Area Below the x-Axis**

Find the area of the region bounded by the graph of

$$y = x^2 - 3x - 4$$

and the x-axis.

SOLUTION   Begin by finding the x-intercepts of the graph. To do this, set the function equal to zero and solve for x.

$$x^2 - 3x - 4 = 0 \qquad \text{Set function equal to 0.}$$
$$(x - 4)(x + 1) = 0 \qquad \text{Factor.}$$
$$x = 4, x = -1 \qquad \text{Solve for } x.$$

From Figure 11.17, you can see that $x^2 - 3x - 4 \leq 0$ for all x in the interval $[-1, 4]$. So, you can let $f(x) = 0$ and $g(x) = x^2 - 3x - 4$, and compute the area as shown.

$$\text{Area} = \int_a^b [f(x) - g(x)]\, dx \qquad \text{Area between } f \text{ and } g$$
$$= \int_{-1}^4 [(0) - (x^2 - 3x - 4)]\, dx \qquad \text{Substitute for } f \text{ and } g.$$
$$= \int_{-1}^4 (-x^2 + 3x + 4)\, dx$$
$$= \left[-\frac{x^3}{3} + \frac{3x^2}{2} + 4x \right]_{-1}^4 \qquad \text{Find antiderivative.}$$
$$= \frac{125}{6} \text{ square units} \qquad \text{Apply Fundamental Theorem.}$$

✓ CHECKPOINT 3

Find the area of the region bounded by the graph of $y = x^2 - x - 2$ and the x-axis. ∎

TECHNOLOGY

(T)   Most graphing utilities can display regions that are bounded by two graphs. For instance, to graph the region in Example 3, set the viewing window to $-1 \leq x \leq 4$ and $-7 \leq y \leq 1$. Consult your user's manual for specific keystrokes on how to shade the graph. You should obtain the graph shown at the right.*

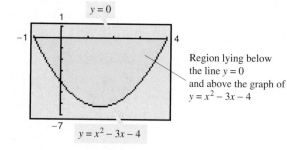

Region lying below the line $y = 0$ and above the graph of $y = x^2 - 3x - 4$

$y = x^2 - 3x - 4$

*Specific calculator keystroke instructions for operations in this and other technology boxes can be found at *college.hmco.com/info/larsonapplied*.

Sometimes two graphs intersect at more than two points. To determine the area of the region bounded by two such graphs, you must find *all* points of intersection and check to see which graph is above the other in each interval determined by the points.

Example 4   Using Multiple Points of Intersection

Find the area of the region bounded by the graphs of

$$f(x) = 3x^3 - x^2 - 10x \quad \text{and} \quad g(x) = -x^2 + 2x.$$

SOLUTION   To find the points of intersection of the two graphs, set the functions equal to each other and solve for x.

$f(x) = g(x)$	Set $f(x)$ equal to $g(x)$.
$3x^3 - x^2 - 10x = -x^2 + 2x$	Substitute for $f(x)$ and $g(x)$.
$3x^3 - 12x = 0$	Write in general form.
$3x(x^2 - 4) = 0$	
$3x(x - 2)(x + 2) = 0$	Factor.
$x = 0, x = 2, x = -2$	Solve for x.

These three points of intersection determine two intervals of integration: $[-2, 0]$ and $[0, 2]$. In Figure 11.18, you can see that $g(x) \leq f(x)$ in the interval $[-2, 0]$, and that $f(x) \leq g(x)$ in the interval $[0, 2]$. So, you must use two integrals to determine the area of the region bounded by the graphs of f and g: one for the interval $[-2, 0]$ and one for the interval $[0, 2]$.

$$\begin{aligned}
\text{Area} &= \int_{-2}^{0} [f(x) - g(x)] \, dx + \int_{0}^{2} [g(x) - f(x)] \, dx \\
&= \int_{-2}^{0} (3x^3 - 12x) \, dx + \int_{0}^{2} (-3x^3 + 12x) \, dx \\
&= \left[\frac{3x^4}{4} - 6x^2 \right]_{-2}^{0} + \left[-\frac{3x^4}{4} + 6x^2 \right]_{0}^{2} \\
&= (0 - 0) - (12 - 24) + (-12 + 24) - (0 + 0) \\
&= 24
\end{aligned}$$

So, the region has an area of 24 square units.

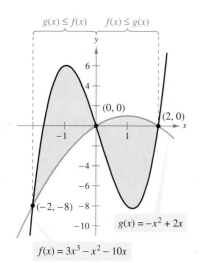

$g(x) \leq f(x)$ $f(x) \leq g(x)$

$(0, 0)$

$(2, 0)$

$(-2, -8)$

$g(x) = -x^2 + 2x$

$f(x) = 3x^3 - x^2 - 10x$

FIGURE 11.18

✓ CHECKPOINT 4

Find the area of the region bounded by the graphs of $f(x) = x^3 + 2x^2 - 3x$ and $g(x) = x^2 + 3x$. Sketch a graph of the region. ∎

STUDY TIP

It is easy to make an error when calculating areas such as that in Example 4. To give yourself some idea about the reasonableness of your solution, you could make a careful sketch of the region on graph paper and then use the grid on the graph paper to approximate the area. Try doing this with the graph shown in Figure 11.18. Is your approximation close to 24 square units?

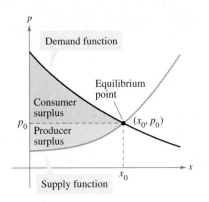

FIGURE 11.19

Consumer Surplus and Producer Surplus

In Section 7.5, you learned that a demand function relates the price of a product to the consumer demand. A supply function relates the price of a product to producers' willingness to supply the product. The point (x_0, p_0) at which a demand function $p = D(x)$ and a supply function $p = S(x)$ intersect is the equilibrium point.

Economists call the area of the region bounded by the graph of the demand function, the horizontal line $p = p_0$, and the vertical line $x = 0$ the **consumer surplus.** Similarly, the area of the region bounded by the graph of the supply function, the horizontal line $p = p_0$, and the vertical line $x = 0$ is called the **producer surplus,** as shown in Figure 11.19.

Example 5 Finding Surpluses

The demand and supply functions for a product are modeled by

Demand: $p = -0.36x + 9$ and *Supply:* $p = 0.14x + 2$

where x is the number of units (in millions). Find the consumer and producer surpluses for this product.

SOLUTION By equating the demand and supply functions, you can determine that the point of equilibrium occurs when $x = 14$ (million) and the price is $3.96 per unit.

$$\text{Consumer surplus} = \int_0^{14} (\text{demand function} - \text{price})\, dx$$

$$= \int_0^{14} [(-0.36x + 9) - 3.96]\, dx$$

$$= \left[-0.18x^2 + 5.04x \right]_0^{14}$$

$$= 35.28$$

$$\text{Producer surplus} = \int_0^{14} (\text{price} - \text{supply function})\, dx$$

$$= \int_0^{14} [3.96 - (0.14x + 2)]\, dx$$

$$= \left[-0.07x^2 + 1.96x \right]_0^{14}$$

$$= 13.72$$

The consumer surplus and producer surplus are shown in Figure 11.20.

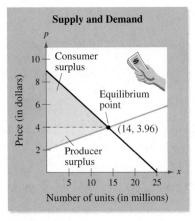

FIGURE 11.20

✓ CHECKPOINT 5

The demand and supply functions for a product are modeled by

Demand: $p = -0.2x + 8$ and *Supply:* $p = 0.1x + 2$

where x is the number of units (in millions). Find the consumer and producer surpluses for this product. ■

AP/Wide World Photos

In 2005, the United States consumed about 40.4 quadrillion Btu of petroleum.

Application

In addition to consumer and producer surpluses, there are many other types of applications involving the area of a region bounded by two graphs. Example 6 shows one of these applications.

Example 6   Modeling Petroleum Consumption

In the *Annual Energy Outlook*, the U.S. Energy Information Administration projected the consumption C (in quadrillions of Btu per year) of petroleum to follow the model

$$C_1 = 0.004t^2 + 0.330t + 38.3, \quad 0 \le t \le 30$$

where $t = 0$ corresponds to 2000. If the actual consumption more closely followed the model

$$C_2 = 0.005t^2 + 0.301t + 38.2, \quad 0 \le t \le 30$$

how much petroleum would be saved?

SOLUTION   The petroleum saved can be represented as the area of the region between the graphs of C_1 and C_2, as shown in Figure 11.21.

$$\text{Petroleum saved} = \int_0^{30} (C_1 - C_2) \, dt$$
$$= \int_0^{30} (-0.001t^2 + 0.029t + 0.1) \, dt$$
$$= \left[-\frac{0.001}{3} t^3 + \frac{0.029}{2} t^2 + 0.1t \right]_0^{30}$$
$$\approx 7.1$$

So, about 7.1 quadrillion Btu of petroleum would be saved.

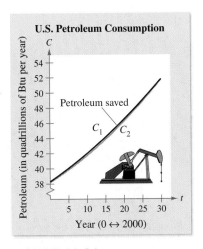

FIGURE 11.21

✓ CHECKPOINT 6

The projected fuel cost C (in millions of dollars per year) for a trucking company from 2008 through 2020 is $C_1 = 5.6 + 2.21t$, $8 \le t \le 20$, where $t = 8$ corresponds to 2008. If the company purchases more efficient truck engines, fuel cost is expected to decrease and to follow the model $C_2 = 4.7 + 2.04t$, $8 \le t \le 20$. How much can the company save with the more efficient engines? ■

CONCEPT CHECK

1. When finding the area of a region bounded by two graphs, you use the integrand $[f(x) - g(x)]$. Identify what f and g represent.

2. Consider the functions f and g, where f and g are continuous on $[a, b]$ and $g(x) \le f(x)$ for all x in the interval. How can you find the area of the region bounded by the graphs of f, g, $x = a$, and $x = b$?

3. Describe the characteristics of typical demand and supply functions.

4. Suppose that the demand and supply functions for a product do not intersect. What can you conclude?

Skills Review 11.5   The following warm-up exercises involve skills that were covered in earlier sections. You will use these skills in the exercise set for this section. For additional help, review Sections 0.5, 2.5, 4.1 and 5.1.

In Exercises 1–4, simplify the expression.

1. $(-x^2 + 4x + 3) - (x + 1)$

2. $(-2x^2 + 3x + 9) - (-x + 5)$

3. $(-x^3 + 3x^2 - 1) - (x^2 - 4x + 4)$

4. $(3x + 1) - (-x^3 + 9x + 2)$

In Exercises 5–10, find the points of intersection of the graphs.

5. $f(x) = x^2 - 4x + 4$, $g(x) = 4$

6. $f(x) = -3x^2$, $g(x) = 6 - 9x$

7. $f(x) = x^2$, $g(x) = -x + 6$

8. $f(x) = \frac{1}{2}x^3$, $g(x) = 2x$

9. $f(x) = x^2 - 3x$, $g(x) = 3x - 5$

10. $f(x) = e^x$, $g(x) = e$

Exercises 11.5

See www.CalcChat.com for worked-out solutions to odd-numbered exercises.

In Exercises 1–6, find the area of the region.

1. $f(x) = x^2 - 6x$
   $g(x) = 0$

2. $f(x) = x^2 + 2x + 1$
   $g(x) = 2x + 5$

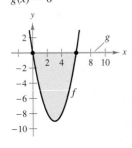

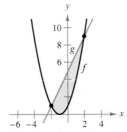

3. $f(x) = x^2 - 4x + 3$
   $g(x) = -x^2 + 2x + 3$

4. $f(x) = x^2$
   $g(x) = x^3$

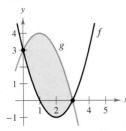

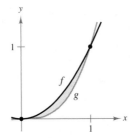

5. $f(x) = 3(x^3 - x)$
   $g(x) = 0$

6. $f(x) = (x - 1)^3$
   $g(x) = x - 1$

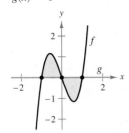

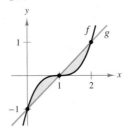

In Exercises 7–12, the integrand of the definite integral is a difference of two functions. Sketch the graph of each function and shade the region whose area is represented by the integral.

7. $\displaystyle\int_0^4 \left[(x + 1) - \tfrac{1}{2}x\right] dx$

8. $\displaystyle\int_{-1}^1 \left[(1 - x^2) - (x^2 - 1)\right] dx$

9. $\displaystyle\int_{-2}^2 \left[2x^2 - (x^4 - 2x^2)\right] dx$

10. $\displaystyle\int_{-4}^0 \left[(x - 6) - (x^2 + 5x - 6)\right] dx$

11. $\displaystyle\int_{-1}^2 \left[(y^2 + 2) - 1\right] dy$ **12.** $\displaystyle\int_{-2}^3 \left[(y + 6) - y^2\right] dy$

Think About It In Exercises 13 and 14, determine which value best approximates the area of the region bounded by the graphs of *f* and *g*. (Make your selection on the basis of a sketch of the region and not by performing any calculations.)

13. $f(x) = x + 1$, $g(x) = (x - 1)^2$

   (a) -2 (b) 2   (c) 10   (d) 4   (e) 8

14. $f(x) = 2 - \dfrac{1}{2}x$, $g(x) = 2 - \sqrt{x}$

   (a) 1   (b) 6   (c) -3 (d) 3   (e) 4

In Exercises 15–30, sketch the region bounded by the graphs of the functions and find the area of the region.

15. $y = \dfrac{1}{x^2}, y = 0, x = 1, x = 5$

16. $y = x^3 - 2x + 1, y = -2x, x = 1$

17. $f(x) = \sqrt[3]{x}$, $g(x) = x$

18. $f(x) = \sqrt{3x} + 1$, $g(x) = x + 1$

19. $y = x^2 - 4x + 3$, $y = 3 + 4x - x^2$

20. $y = 4 - x^2$, $y = x^2$

21. $y = xe^{-x^2}$, $y = 0$, $x = 0$, $x = 1$

22. $y = \dfrac{e^{1/x}}{x^2}$, $y = 0$, $x = 1$, $x = 3$

23. $y = \dfrac{8}{x}$, $y = x^2$, $y = 0$, $x = 1$, $x = 4$

24. $y = \dfrac{1}{x}$, $y = x^3$, $x = \dfrac{1}{2}$, $x = 1$

25. $f(x) = e^{0.5x}$, $g(x) = -\dfrac{1}{x}$, $x = 1$, $x = 2$

26. $f(x) = \dfrac{1}{x}$, $g(x) = -e^x$, $x = \dfrac{1}{2}$, $x = 1$

27. $f(y) = y^2$, $g(y) = y + 2$

28. $f(y) = y(2 - y)$, $g(y) = -y$

29. $f(y) = \sqrt{y}$, $y = 9$, $x = 0$

30. $f(y) = y^2 + 1$, $g(y) = 4 - 2y$

Ⓣ In Exercises 31–34, use a graphing utility to graph the region bounded by the graphs of the functions. Write the definite integrals that represent the area of the region. (*Hint:* Multiple integrals may be necessary.)

31. $f(x) = 2x$, $g(x) = 4 - 2x$, $h(x) = 0$

32. $f(x) = x(x^2 - 3x + 3)$, $g(x) = x^2$

33. $y = \dfrac{4}{x}$, $y = x$, $x = 1$, $x = 4$

34. $y = x^3 - 4x^2 + 1$, $y = x - 3$

Ⓣ In Exercises 35–38, use a graphing utility to graph the region bounded by the graphs of the functions, and find the area of the region.

35. $f(x) = x^2 - 4x$, $g(x) = 0$

36. $f(x) = 3 - 2x - x^2$, $g(x) = 0$

37. $f(x) = x^2 + 2x + 1$, $g(x) = x + 1$

38. $f(x) = -x^2 + 4x + 2$, $g(x) = x + 2$

In Exercises 39 and 40, use integration to find the area of the triangular region having the given vertices.

39. $(0, 0), (4, 0), (4, 4)$

40. $(0, 0), (4, 0), (6, 4)$

Consumer and Producer Surpluses In Exercises 41–44, find the consumer and producer surpluses.

Demand Function	Supply Function
41. $p_1(x) = 50 - 0.5x$	$p_2(x) = 0.125x$
42. $p_1(x) = 300 - x$	$p_2(x) = 100 + x$
43. $p_1(x) = 200 - 0.4x$	$p_2(x) = 100 + 1.6x$
44. $p_1(x) = 975 - 23x$	$p_2(x) = 42x$

45. *MAKE A DECISION: JOB OFFERS*   A college graduate has two job offers. The starting salary for each is \$32,000, and after 8 years of service each will pay \$54,000. The salary increase for each offer is shown in the figure. From a strictly monetary viewpoint, which is the better offer? Explain.

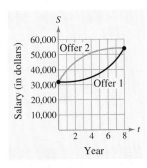 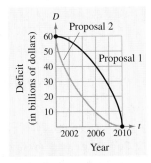

Figure for 45   Figure for 46

46. *MAKE A DECISION: BUDGET DEFICITS*   A state legislature is debating two proposals for eliminating the annual budget deficits by the year 2010. The rate of decrease of the deficits for each proposal is shown in the figure. From the viewpoint of minimizing the cumulative state deficit, which is the better proposal? Explain.

Revenue   In Exercises 47 and 48, two models, R_1 and R_2, are given for revenue (in billions of dollars per year) for a large corporation. Both models are estimates of revenues for 2007 through 2011, with $t = 7$ corresponding to 2007. Which model is projecting the greater revenue? How much more total revenue does that model project over the five-year period?

47. $R_1 = 7.21 + 0.58t$, $R_2 = 7.21 + 0.45t$

48. $R_1 = 7.21 + 0.26t + 0.02t^2$, $R_2 = 7.21 + 0.1t + 0.01t^2$

49. **Fuel Cost**   The projected fuel cost C (in millions of dollars per year) for an airline company from 2007 through 2013 is $C_1 = 568.5 + 7.15t$, where $t = 7$ corresponds to 2007. If the company purchases more efficient airplane engines, fuel cost is expected to decrease and to follow the model $C_2 = 525.6 + 6.43t$. How much can the company save with the more efficient engines? Explain your reasoning.

50. Health An epidemic was spreading such that t weeks after its outbreak it had infected

$$N_1(t) = 0.1t^2 + 0.5t + 150, \quad 0 \le t \le 50$$

people. Twenty-five weeks after the outbreak, a vaccine was developed and administered to the public. At that point, the number of people infected was governed by the model

$$N_2(t) = -0.2t^2 + 6t + 200.$$

Approximate the number of people that the vaccine prevented from becoming ill during the epidemic.

51. Consumer Trends For the years 1996 through 2004, the per capita consumption of fresh pineapples (in pounds per year) in the United States can be modeled by

$$C(t) = \begin{cases} -0.046t^2 + 1.07t - 2.9, & 6 \le t \le 10 \\ -0.164t^2 + 4.53t - 26.8, & 10 < t \le 14 \end{cases}$$

where t is the year, with $t = 6$ corresponding to 1996. *(Source: U.S. Department of Agriculture)*

(T) (a) Use a graphing utility to graph this model.

(b) Suppose the fresh pineapple consumption from 2001 through 2004 had continued to follow the model for 1996 through 2000. How many more or fewer pounds of fresh pineapples would have been consumed from 2001 through 2004?

52. Consumer and Producer Surpluses Factory orders for an air conditioner are about 6000 units per week when the price is $331 and about 8000 units per week when the price is $303. The supply function is given by $p = 0.0275x$. Find the consumer and producer surpluses. (Assume the demand function is linear.)

53. Consumer and Producer Surpluses Repeat Exercise 52 with a demand of about 6000 units per week when the price is $325 and about 8000 units per week when the price is $300. Find the consumer and producer surpluses. (Assume the demand function is linear.)

54. Cost, Revenue, and Profit The revenue from a manufacturing process (in millions of dollars per year) is projected to follow the model $R = 100$ for 10 years. Over the same period of time, the cost (in millions of dollars per year) is projected to follow the model $C = 60 + 0.2t^2$, where t is the time (in years). Approximate the profit over the 10-year period.

55. Cost, Revenue, and Profit Repeat Exercise 54 for revenue and cost models given by $R = 100 + 0.08t$ and $C = 60 + 0.2t^2$.

56. Lorenz Curve Economists use *Lorenz curves* to illustrate the distribution of income in a country. Letting x represent the percent of families in a country and y the percent of total income, the model $y = x$ would represent a country in which each family had the same income. The Lorenz curve, $y = f(x)$, represents the actual income distribution. The area between these two models, for

$0 \le x \le 100$, indicates the "income inequality" of a country. In 2005, the Lorenz curve for the United States could be modeled by

$$y = (0.00061x^2 + 0.0218x + 1.723)^2, \quad 0 \le x \le 100$$

where x is measured from the poorest to the wealthiest families. Find the income inequality for the United States in 2005. *(Source: U.S. Census Bureau)*

(S) **57. Income Distribution** Using the Lorenz curve in Exercise 56 and a spreadsheet, complete the table, which lists the percent of total income earned by each quintile in the United States in 2005.

Quintile	Lowest	2nd	3rd	4th	Highest
Percent					

58. Extended Application To work an extended application analyzing the receipts and expenditures for the Old-Age and Survivors Insurance Trust Fund (Social Security Trust Fund) from 1990 through 2005, visit this text's website at *college.hmco.com*. *(Data Source: Social Security Administration)*

Business Capsule

Photo courtesy of Avis Yates Rivers

After losing her job as an account executive in 1985, Avis Yates Rivers used $2500 to start a word processing business from the basement of her home. In 1996, as a spin-off from her word processing business, Rivers established Technology Concepts Group. Today, this Somerset, New Jersey-based firm provides information technology management consulting, e-business solutions, and network and desktop support for corporate and government customers. Annual revenue is currently $1.1 million.

59. Research Project Use your school's library, the Internet, or some other reference source to research a small company similar to that described above. Describe the impact of different factors, such as start-up capital and market conditions, on a company's revenue.

Section 11.6

The Definite Integral as the Limit of a Sum

■ Use the Midpoint Rule to approximate definite integrals.
■ Use a symbolic integration utility to approximate definite integrals.

The Midpoint Rule

In Section 11.4, you learned that you cannot use the Fundamental Theorem of Calculus to evaluate a definite integral unless you can find an antiderivative of the integrand. In cases where this cannot be done, you can approximate the value of the integral using an approximation technique. One such technique is called the **Midpoint Rule.** (Two other techniques are discussed in Section 12.4.)

Example 1   Approximating the Area of a Plane Region

Use the five rectangles in Figure 11.22 to approximate the area of the region bounded by the graph of $f(x) = -x^2 + 5$, the x-axis, and the lines $x = 0$ and $x = 2$.

SOLUTION   You can find the heights of the five rectangles by evaluating f at the midpoint of each of the following intervals.

$$\left[0, \frac{2}{5}\right], \quad \left[\frac{2}{5}, \frac{4}{5}\right], \quad \left[\frac{4}{5}, \frac{6}{5}\right], \quad \left[\frac{6}{5}, \frac{8}{5}\right], \quad \left[\frac{8}{5}, \frac{10}{5}\right]$$

Evaluate f at the midpoints of these intervals.

The width of each rectangle is $\frac{2}{5}$. So, the sum of the five areas is

$$\begin{aligned}
\text{Area} &\approx \frac{2}{5}f\left(\frac{1}{5}\right) + \frac{2}{5}f\left(\frac{3}{5}\right) + \frac{2}{5}f\left(\frac{5}{5}\right) + \frac{2}{5}f\left(\frac{7}{5}\right) + \frac{2}{5}f\left(\frac{9}{5}\right) \\
&= \frac{2}{5}\left[f\left(\frac{1}{5}\right) + f\left(\frac{3}{5}\right) + f\left(\frac{5}{5}\right) + f\left(\frac{7}{5}\right) + f\left(\frac{9}{5}\right)\right] \\
&= \frac{2}{5}\left(\frac{124}{25} + \frac{116}{25} + \frac{100}{25} + \frac{76}{25} + \frac{44}{25}\right) \\
&= \frac{920}{125} \\
&= 7.36.
\end{aligned}$$

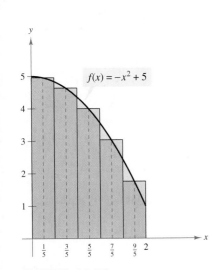

$f(x) = -x^2 + 5$

FIGURE 11.22

✓CHECKPOINT 1

Use four rectangles to approximate the area of the region bounded by the graph of $f(x) = x^2 + 1$, the x-axis, $x = 0$ and $x = 2$.   ■

For the region in Example 1, you can find the exact area with a definite integral. That is,

$$\text{Area} = \int_0^2 (-x^2 + 5)\, dx = \frac{22}{3} \approx 7.33.$$

Ⓣ The easiest way to use the Midpoint Rule to approximate the definite integral $\int_a^b f(x)\, dx$ is to program it into a computer or programmable calculator. For instance, the pseudocode below will help you write a program to evaluate the Midpoint Rule. (Appendix H lists this program for several models of graphing utilities.)

Program

- *Prompt for value of a.*
- *Input value of a.*
- *Prompt for value of b.*
- *Input value of b.*
- *Prompt for value of n.*
- *Input value of n.*
- *Initialize sum of areas.*
- *Calculate width of subinterval.*
- *Initialize counter.*
- *Begin loop.*
- *Calculate left endpoint.*
- *Calculate right endpoint.*
- *Calculate midpoint of subinterval.*
- *Add area to sum.*
- *Test counter.*
- *End loop.*
- *Display approximation.*

Before executing the program, enter the function. When the program is executed, you will be prompted to enter the lower and upper limits of integration and the number of subintervals you want to use.

The approximation procedure used in Example 1 is the **Midpoint Rule.** You can use the Midpoint Rule to approximate *any* definite integral—not just those representing area. The basic steps are summarized below.

Guidelines for Using the Midpoint Rule

To approximate the definite integral $\int_a^b f(x)\, dx$ with the Midpoint Rule, use the steps below.

1. Divide the interval $[a, b]$ into n subintervals, each of width

$$\Delta x = \frac{b - a}{n}.$$

2. Find the midpoint of each subinterval.

$$\text{Midpoints} = \{x_1, x_2, x_3, \ldots, x_n\}$$

3. Evaluate f at each midpoint and form the sum as shown.

$$\int_a^b f(x)\, dx \approx \frac{b - a}{n}[f(x_1) + f(x_2) + f(x_3) + \cdots + f(x_n)]$$

An important characteristic of the Midpoint Rule is that the approximation tends to improve as n increases. The table below shows the approximations for the area of the region described in Example 1 for various values of n. For example, for $n = 10$, the Midpoint Rule yields

$$\int_0^2 (-x^2 + 5)\, dx \approx \frac{2}{10}\left[f\left(\frac{1}{10}\right) + f\left(\frac{3}{10}\right) + \cdots + f\left(\frac{19}{10}\right)\right]$$
$$= 7.34.$$

n	5	10	15	20	25	30
Approximation	7.3600	7.3400	7.3363	7.3350	7.3344	7.3341

Note that as n increases, the approximation gets closer and closer to the exact value of the integral, which was found to be

$$\frac{22}{3} \approx 7.3333.$$

STUDY TIP

In Example 1, the Midpoint Rule is used to approximate an integral whose exact value can be found with the Fundamental Theorem of Calculus. This was done to illustrate the accuracy of the rule. In practice, of course, you would use the Midpoint Rule to approximate the values of definite integrals for which you cannot find an antiderivative. Examples 2 and 3 illustrate such integrals.

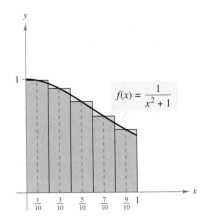

FIGURE 11.23

Example 2 Using the Midpoint Rule

Use the Midpoint Rule with $n = 5$ to approximate $\int_0^1 \dfrac{1}{x^2 + 1}\, dx$.

SOLUTION   With $n = 5$, the interval $[0, 1]$ is divided into five subintervals.

$$\left[0, \frac{1}{5}\right], \quad \left[\frac{1}{5}, \frac{2}{5}\right], \quad \left[\frac{2}{5}, \frac{3}{5}\right], \quad \left[\frac{3}{5}, \frac{4}{5}\right], \quad \left[\frac{4}{5}, 1\right]$$

The midpoints of these intervals are $\frac{1}{10}, \frac{3}{10}, \frac{5}{10}, \frac{7}{10}$, and $\frac{9}{10}$. Because each subinterval has a width of $\Delta x = (1 - 0)/5 = \frac{1}{5}$, you can approximate the value of the definite integral as shown.

$$\int_0^1 \frac{1}{x^2 + 1}\, dx \approx \frac{1}{5}\left(\frac{1}{1.01} + \frac{1}{1.09} + \frac{1}{1.25} + \frac{1}{1.49} + \frac{1}{1.81}\right)$$
$$\approx 0.786$$

The region whose area is represented by the definite integral is shown in Figure 11.23. The actual area of this region is $\pi/4 \approx 0.785$. So, the approximation is off by only 0.001.

✓ **CHECKPOINT 2**

Use the Midpoint Rule with $n = 4$ to approximate the area of the region bounded by the graph of $f(x) = 1/(x^2 + 2)$, the x-axis, and the lines $x = 0$ and $x = 1$. ▪

Example 3 Using the Midpoint Rule

Use the Midpoint Rule with $n = 10$ to approximate $\int_1^3 \sqrt{x^2 + 1}\, dx$.

SOLUTION   Begin by dividing the interval $[1, 3]$ into 10 subintervals. The midpoints of these intervals are

$$\frac{11}{10}, \quad \frac{13}{10}, \quad \frac{3}{2}, \quad \frac{17}{10}, \quad \frac{19}{10}, \quad \frac{21}{10}, \quad \frac{23}{10}, \quad \frac{5}{2}, \quad \frac{27}{10}, \quad \text{and} \quad \frac{29}{10}.$$

Because each subinterval has a width of $\Delta x = (3 - 1)/10 = \frac{1}{5}$, you can approximate the value of the definite integral as shown.

$$\int_1^3 \sqrt{x^2 + 1}\, dx \approx \frac{1}{5}\left[\sqrt{(1.1)^2 + 1} + \sqrt{(1.3)^2 + 1} + \cdots + \sqrt{(2.9)^2 + 1}\right]$$
$$\approx 4.504$$

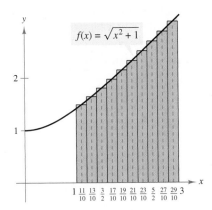

FIGURE 11.24

The region whose area is represented by the definite integral is shown in Figure 11.24. Using techniques that are not within the scope of this course, it can be shown that the actual area is

$$\tfrac{1}{2}\left[3\sqrt{10} + \ln\left(3 + \sqrt{10}\right) - \sqrt{2} - \ln\left(1 + \sqrt{2}\right)\right] \approx 4.505.$$

So, the approximation is off by only 0.001.

✓ **CHECKPOINT 3**

Use the Midpoint Rule with $n = 4$ to approximate the area of the region bounded by the graph of $f(x) = \sqrt{x^2 - 1}$, the x-axis, and the lines $x = 2$ and $x = 4$. ▪

STUDY TIP

The Midpoint Rule is necessary for solving certain real-life problems, such as measuring irregular areas like bodies of water (see Exercise 38).

The Definite Integral as the Limit of a Sum

Consider the closed interval $[a, b]$, divided into n subintervals whose midpoints are x_i and whose widths are $\Delta x = (b - a)/n$. In this section, you have seen that the midpoint approximation

$$\int_a^b f(x)\, dx \approx f(x_1)\, \Delta x + f(x_2)\, \Delta x + f(x_3)\, \Delta x + \cdots + f(x_n)\, \Delta x$$

$$= \left[f(x_1) + f(x_2) + f(x_3) + \cdots + f(x_n) \right] \Delta x$$

becomes better and better as n increases. In fact, the limit of this sum as n approaches infinity is exactly equal to the definite integral. That is,

$$\int_a^b f(x)\, dx = \lim_{n \to \infty} \left[f(x_1) + f(x_2) + f(x_3) + \cdots + f(x_n) \right] \Delta x.$$

It can be shown that this limit is valid as long as x_i is *any* point in the ith interval.

Example 4 **Approximating a Definite Integral**

Use a computer, programmable calculator, or symbolic integration utility to approximate the definite integral

$$\int_0^1 e^{-x^2}\, dx.$$

SOLUTION   Using the program on page 856, with $n = 10, 20, 30, 40,$ and 50, it appears that the value of the integral is approximately 0.7468. If you have access to a computer or calculator with a built-in program for approximating definite integrals, try using it to approximate this integral. When a computer with such a built-in program approximated the integral, it returned a value of 0.746824.

✓CHECKPOINT 4

Use a computer, programmable calculator, or symbolic integration utility to approximate the definite integral

$$\int_0^1 e^{x^2}\, dx. \quad \blacksquare$$

┌─ **CONCEPT CHECK** ─────────────────────────────┐

1. Complete the following: In cases where the Fundamental Theorem of Calculus cannot be used to evaluate a definite integral, you can approximate the value of the integral using the _____ _____.

2. True or false: The Midpoint Rule can be used to approximate any definite integral.

3. In the Midpoint Rule, as the number of subintervals n increases, does the approximation of a definite integral become better or worse?

4. State the guidelines for using the Midpoint Rule.

└──┘

Skills Review 11.6

The following warm-up exercises involve skills that were covered in earlier sections. You will use these skills in the exercise set for this section. For additional help, review Sections 2.1, 7.1, and 9.3.

In Exercises 1–6, find the midpoint of the interval.

1. $\left[0, \frac{1}{3}\right]$

2. $\left[\frac{1}{10}, \frac{2}{10}\right]$

3. $\left[\frac{3}{20}, \frac{4}{20}\right]$

4. $\left[1, \frac{7}{6}\right]$

5. $\left[2, \frac{31}{15}\right]$

6. $\left[\frac{26}{9}, 3\right]$

In Exercises 7–10, find the limit.

7. $\lim\limits_{x \to \infty} \dfrac{2x^2 + 4x - 1}{3x^2 - 2x}$

8. $\lim\limits_{x \to \infty} \dfrac{4x + 5}{7x - 5}$

9. $\lim\limits_{x \to \infty} \dfrac{x - 7}{x^2 + 1}$

10. $\lim\limits_{x \to \infty} \dfrac{5x^3 + 1}{x^3 + x^2 + 4}$

Exercises 11.6

See www.CalcChat.com for worked-out solutions to odd-numbered exercises.

In Exercises 1–4, use the Midpoint Rule with $n = 4$ to approximate the area of the region. Compare your result with the exact area obtained with a definite integral.

In Exercises 5–16, use the Midpoint Rule with $n = 4$ to approximate the area of the region bounded by the graph of f and the x-axis over the interval. Compare your result with the exact area. Sketch the region.

1. $f(x) = -2x + 3$, $[0, 1]$ 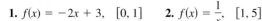 **2.** $f(x) = \dfrac{1}{x}$, $[1, 5]$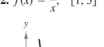

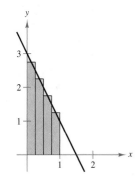

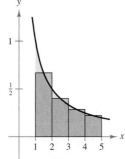
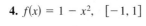

3. $f(x) = \sqrt{x}$, $[0, 1]$

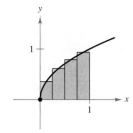

4. $f(x) = 1 - x^2$, $[-1, 1]$

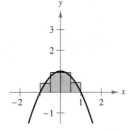

	Function	Interval
5.	$f(x) = 4 - x^2$	$[0, 2]$
6.	$f(x) = 4x^2$	$[0, 2]$
7.	$f(x) = x^2 + 3$	$[-1, 1]$
8.	$f(x) = 4 - x^2$	$[-2, 2]$
9.	$f(x) = 2x^2$	$[1, 3]$
10.	$f(x) = 3x^2 + 1$	$[-1, 3]$
11.	$f(x) = 2x - x^3$	$[0, 1]$
12.	$f(x) = x^2 - x^3$	$[0, 1]$
13.	$f(x) = x^2 - x^3$	$[-1, 0]$
14.	$f(x) = x(1 - x)^2$	$[0, 1]$
15.	$f(x) = x^2(3 - x)$	$[0, 3]$
16.	$f(x) = x^2 + 4x$	$[0, 4]$

Ⓣ In Exercises 17–22, use a program similar to that on page 856 to approximate the area of the region. How large must n be to obtain an approximation that is correct to within 0.01?

17. $\displaystyle\int_0^4 (2x^2 + 3)\, dx$

18. $\displaystyle\int_0^4 (2x^3 + 3)\, dx$

19. $\displaystyle\int_1^2 (2x^2 - x + 1)\, dx$

20. $\displaystyle\int_1^2 (x^3 - 1)\, dx$

21. $\displaystyle\int_1^4 \dfrac{1}{x + 1}\, dx$

22. $\displaystyle\int_1^2 \sqrt{x + 2}\, dx$

In Exercises 23–26, use the Midpoint Rule with $n = 4$ to approximate the area of the region. Compare your result with the exact area obtained with a definite integral.

23. $f(y) = \frac{1}{4}y$, $[2, 4]$

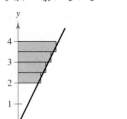

24. $f(y) = 2y$, $[0, 2]$

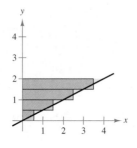

25. $f(y) = y^2 + 1$, $[0, 4]$

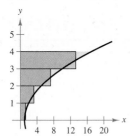

26. $f(y) = 4y - y^2$, $[0, 4]$

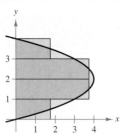

Trapezoidal Rule In Exercises 27 and 28, use the Trapezoidal Rule with $n = 8$ to approximate the definite integral. Compare the result with the exact value and the approximation obtained with $n = 8$ and the Midpoint Rule. Which approximation technique appears to be better? Let f be continuous on $[a, b]$ and let n be the number of equal subintervals (see figure). Then the Trapezoidal Rule for approximating $\int_a^b f(x)\, dx$ is

$$\frac{b - a}{2n}[f(x_0) + 2f(x_1) + \cdots + 2f(x_{n-1}) + f(x_n)]$$

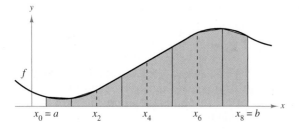

27. $\int_0^2 x^3 \, dx$

28. $\int_1^3 \frac{1}{x^2} \, dx$

In Exercises 29–32, use the Trapezoidal Rule with $n = 4$ to approximate the definite integral.

29. $\int_0^2 \frac{1}{x + 1} \, dx$

30. $\int_0^4 \sqrt{1 + x^2} \, dx$

31. $\int_{-1}^1 \frac{1}{x^2 + 1} \, dx$

32. $\int_1^5 \frac{\sqrt{x - 1}}{x} \, dx$

(T) In Exercises 33 and 34, use a computer or programmable calculator to approximate the definite integral using the Midpoint Rule and the Trapezoidal Rule for $n = 4$, 8, 12, 16, and 20.

33. $\int_0^4 \sqrt{2 + 3x^2} \, dx$

34. $\int_0^2 \frac{5}{x^3 + 1} \, dx$

In Exercises 35 and 36, use the Trapezoidal Rule with $n = 10$ to approximate the area of the region bounded by the graphs of the equations.

35. $y = \sqrt{\dfrac{x^3}{4 - x}}$, $y = 0$, $x = 3$

36. $y = x\sqrt{\dfrac{4 - x}{4 + x}}$, $y = 0$, $x = 4$

37. Surface Area Estimate the surface area of the golf green shown in the figure using (a) the Midpoint Rule and (b) the Trapezoidal Rule.

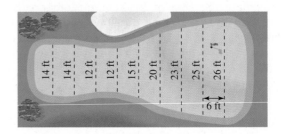

38. Surface Area To estimate the surface area of a pond, a surveyor takes several measurements, as shown in the figure. Estimate the surface area of the pond using (a) the Midpoint Rule and (b) the Trapezoidal Rule.

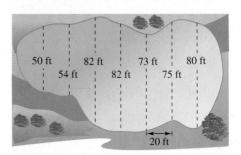

(T) **39. Numerical Approximation** Use the Midpoint Rule and the Trapezoidal Rule with $n = 4$ to approximate π where

$$\pi = \int_0^1 \frac{4}{1 + x^2} \, dx.$$

Then use a graphing utility to evaluate the definite integral. Compare all of your results.

Algebra Review

"Unsimplifying" an Algebraic Expression

In algebra it is often helpful to write an expression in simplest form. In this chapter, you have seen that the reverse is often true in integration. That is, to fit an integrand to an integration formula, it often helps to "unsimplify" the expression. To do this, you use the same algebraic rules, but your goal is different. Here are some examples.

Example 1 **Rewriting an Algebraic Expression**

Rewrite each algebraic expression as indicated in the example.

a. $\dfrac{x+1}{\sqrt{x}}$ Example 6, page 810

b. $x(3-4x^2)^2$ Example 2, page 819

c. $7x^2\sqrt{x^3+1}$ Example 4, page 820

d. $5xe^{-x^2}$ Example 3, page 827

SOLUTION

a. $\dfrac{x+1}{\sqrt{x}} = \dfrac{x}{\sqrt{x}} + \dfrac{1}{\sqrt{x}}$

Example 6, page 810
Rewrite as two fractions.

$= \dfrac{x^1}{x^{1/2}} + \dfrac{1}{x^{1/2}}$

Rewrite with rational exponents.

$= x^{1-1/2} + x^{-1/2}$

Properties of exponents

$= x^{1/2} + x^{-1/2}$

Simplify exponent.

b. $x(3-4x^2)^2 = \dfrac{-8}{-8}x(3-4x^2)^2$

Example 2, page 819
Multiply and divide by -8.

$= \left(-\dfrac{1}{8}\right)(-8)x(3-4x^2)^2$

Regroup.

$= \left(-\dfrac{1}{8}\right)(3-4x^2)^2(-8x)$

Regroup.

c. $7x^2\sqrt{x^3+1} = 7x^2(x^3+1)^{1/2}$

Example 4, page 820
Rewrite with rational exponent.

$= \dfrac{3}{3}(7x^2)(x^3+1)^{1/2}$

Multiply and divide by 3.

$= \dfrac{7}{3}(3x^2)(x^3+1)^{1/2}$

Regroup.

$= \dfrac{7}{3}(x^3+1)^{1/2}(3x^2)$

Regroup.

d. $5xe^{-x^2} = \dfrac{-2}{-2}(5x)e^{-x^2}$

Example 3, page 827
Multiply and divide by -2.

$= \left(-\dfrac{5}{2}\right)(-2x)e^{-x^2}$

Regroup.

$= \left(-\dfrac{5}{2}\right)e^{-x^2}(-2x)$

Regroup.

| Example 2 | **Rewriting an Algebraic Expression** |

Rewrite each algebraic expression.

a. $\dfrac{3x^2 + 2x - 1}{x^2}$ **b.** $\dfrac{1}{1 + e^{-x}}$

c. $\dfrac{x^2 + x + 1}{x - 1}$ **d.** $\dfrac{x^2 + 6x + 1}{x^2 + 1}$

SOLUTION

a. $\dfrac{3x^2 + 2x - 1}{x^2} = \dfrac{3x^2}{x^2} + \dfrac{2x}{x^2} - \dfrac{1}{x^2}$ Example 7(a), page 830
Rewrite as separate fractions.

$= 3 + \dfrac{2}{x} - x^{-2}$ Properties of exponents.

$= 3 + 2\left(\dfrac{1}{x}\right) - x^{-2}$ Regroup.

b. $\dfrac{1}{1 + e^{-x}} = \left(\dfrac{e^x}{e^x}\right)\dfrac{1}{1 + e^{-x}}$ Example 7(b), page 830
Multiply and divide by e^x.

$= \dfrac{e^x}{e^x + e^x(e^{-x})}$ Multiply.

$= \dfrac{e^x}{e^x + e^{x-x}}$ Property of exponents

$= \dfrac{e^x}{e^x + e^0}$ Simplify exponent.

$= \dfrac{e^x}{e^x + 1}$ $e^0 = 1$

c. $\dfrac{x^2 + x + 1}{x - 1} = x + 2 + \dfrac{3}{x - 1}$ Example 7(c), page 830
Use long division as shown below.

$$
\begin{array}{r}
x + 2 \\
x - 1 \overline{)\, x^2 + x + 1} \\
\underline{x^2 - x } \\
2x + 1 \\
\underline{2x - 2} \\
3
\end{array}
$$

d. $\dfrac{x^2 + 6x + 1}{x^2 + 1} = 1 + \dfrac{6x}{x^2 + 1}$ Bottom of page 829.
Use long division as shown below.

$$
\begin{array}{r}
1 \\
x^2 + 1 \overline{)\, x^2 + 6x + 1} \\
\underline{x^2 + 1} \\
6x
\end{array}
$$

Chapter Summary and Study Strategies

After studying this chapter, you should have acquired the following skills.
The exercise numbers are keyed to the Review Exercises that begin on page 865.
Answers to odd-numbered Review Exercises are given in the back of the text.

Section 11.1

Review Exercises

■ Use basic integration rules to find indefinite integrals. *1–10*

$$\int k \, dx = kx + C \qquad\qquad \int [f(x) - g(x)] \, dx = \int f(x) \, dx - \int g(x) \, dx$$

$$\int kf(x) \, dx = k \int f(x) \, dx \qquad\qquad \int x^n \, dx = \frac{x^{n+1}}{n+1} + C, \quad n \neq -1$$

$$\int [f(x) + g(x)] \, dx = \int f(x) \, dx + \int g(x) \, dx$$

■ Use initial conditions to find particular solutions of indefinite integrals. *11–14*

■ Use antiderivatives to solve real-life problems. *15, 16*

Section 11.2

■ Use the General Power Rule or integration by substitution to find indefinite integrals. *17–24*

$$\int u^n \frac{du}{dx} \, dx = \int u^n \, du = \frac{u^{n+1}}{n+1} + C, \quad n \neq -1$$

■ Use the General Power Rule or integration by substitution to solve real-life problems. *25, 26*

Section 11.3

■ Use the Exponential and Log Rules to find indefinite integrals. *27–32*

$$\int e^x \, dx = e^x + C \qquad\qquad \int \frac{1}{x} \, dx = \ln|x| + C$$

$$\int e^u \frac{du}{dx} \, dx = \int e^u \, du = e^u + C \qquad \int \frac{du/dx}{u} \, dx = \int \frac{1}{u} \, du = \ln|u| + C$$

■ Use a symbolic integration utility to find indefinite integrals. *33, 34*

Section 11.4

■ Find the areas of regions using a geometric formula. *35, 36*

■ Find the areas of regions bounded by the graph of a function and the *x*-axis. *37–44*

■ Use properties of definite integrals. *45, 46*

Section 11.4 (continued)

Review Exercises

■ Use the Fundamental Theorem of Calculus to evaluate definite integrals.   *47–64*

$$\int_a^b f(x)\, dx = F(x)\Big]_a^b = F(b) - F(a), \quad \text{where} \quad F'(x) = f(x)$$

■ Use definite integrals to solve marginal analysis problems.   *65, 66*

■ Find average values of functions over closed intervals.   *67–70*

$$\text{Average value} = \frac{1}{b-a}\int_a^b f(x)\, dx$$

■ Use average values to solve real-life problems.   *71–74*

■ Find amounts of annuities.   *75, 76*

■ Use properties of even and odd functions to help evaluate definite integrals.   *77–80*

Even function: $f(-x) = f(x)$ *Odd function:* $f(-x) = -f(x)$

If f is an *even* function, then $\displaystyle\int_{-a}^a f(x)\, dx = 2\int_0^a f(x)\, dx.$

If f is an *odd* function, then $\displaystyle\int_{-a}^a f(x)\, dx = 0.$

Section 11.5

■ Find areas of regions bounded by two (or more) graphs.   *81–90*

$$A = \int_a^b [f(x) - g(x)]\, dx$$

■ Find consumer and producer surpluses.   *91, 92*

■ Use the areas of regions bounded by two graphs to solve real-life problems.   *93–96*

Section 11.6

■ Use the Midpoint Rule to approximate values of definite integrals.   *97–100*

$$\int_a^b f(x)\, dx \approx \frac{b-a}{n}[f(x_1) + f(x_2) + f(x_3) + \cdots + f(x_n)]$$

■ Use the Midpoint Rule to solve real-life problems.   *101, 102*

Study Strategies

■ **Indefinite and Definite Integrals**   When evaluating integrals, remember that an indefinite integral is a *family of antiderivatives*, each differing by a constant C, whereas a definite integral is a number.

■ **Checking Antiderivatives by Differentiating**   When finding an antiderivative, remember that you can check your result by differentiating. For example, you can check that the antiderivative

$$\int (3x^3 - 4x)\, dx = \frac{3}{4}x^4 - 2x^2 + C \quad \text{is correct by differentiating to obtain} \quad \frac{d}{dx}\left[\frac{3}{4}x^4 - 2x^2 + C\right] = 3x^3 - 4x.$$

Because the derivative is equal to the original integrand, you know that the antiderivative is correct.

■ **Grouping Symbols and the Fundamental Theorem**   When using the Fundamental Theorem of Calculus to evaluate a definite integral, you can avoid sign errors by using grouping symbols. Here is an example.

$$\int_1^3 (x^3 - 9x)\, dx = \left[\frac{x^4}{4} - \frac{9x^2}{2}\right]_1^3 = \left[\frac{3^4}{4} - \frac{9(3^2)}{2}\right] - \left[\frac{1^4}{4} - \frac{9(1^2)}{2}\right] = \frac{81}{4} - \frac{81}{2} - \frac{1}{4} + \frac{9}{2} = -16$$

Review Exercises

See www.CalcChat.com for worked-out solutions to odd-numbered exercises.

In Exercises 1–10, find the indefinite integral.

1. $\int 16\, dx$

2. $\int \frac{3}{5}x\, dx$

3. $\int (2x^2 + 5x)\, dx$

4. $\int (5 - 6x^2)\, dx$

5. $\int \frac{2}{3\sqrt[3]{x}}\, dx$

6. $\int 6x^2 \sqrt{x}\, dx$

7. $\int \left(\sqrt[3]{x^4} + 3x \right) dx$

8. $\int \left(\frac{4}{\sqrt{x}} + \sqrt{x} \right) dx$

9. $\int \frac{2x^4 - 1}{\sqrt{x}}\, dx$

10. $\int \frac{1 - 3x}{x^2}\, dx$

In Exercises 11–14, find the particular solution, $y = f(x)$, that satisfies the conditions.

11. $f'(x) = 3x + 1,\quad f(2) = 6$

12. $f'(x) = x^{-1/3} - 1,\quad f(8) = 4$

13. $f''(x) = 2x^2,\quad f'(3) = 10,\quad f(3) = 6$

14. $f''(x) = \dfrac{6}{\sqrt{x}} + 3,\quad f'(1) = 12,\quad f(4) = 56$

15. Vertical Motion An object is projected upward from the ground with an initial velocity of 80 feet per second.

(a) How long does it take the object to rise to its maximum height?

(b) What is the maximum height?

(c) When is the velocity of the object half of its initial velocity?

(d) What is the height of the object when its velocity is one-half the initial velocity?

16. Revenue The weekly revenue for a new product has been increasing. The rate of change of the revenue can be modeled by

$$\frac{dR}{dt} = 0.675t^{3/2},\quad 0 \le t \le 225$$

where t is the time (in weeks). When $t = 0$, $R = 0$.

(a) Find a model for the revenue function.

(b) When will the weekly revenue be $27,000?

In Exercises 17–24, find the indefinite integral.

17. $\int (1 + 5x)^2\, dx$

18. $\int (x - 6)^{4/3}\, dx$

19. $\int \frac{1}{\sqrt{5x - 1}}\, dx$

20. $\int \frac{4x}{\sqrt{1 - 3x^2}}\, dx$

21. $\int x(1 - 4x^2)\, dx$

22. $\int \frac{x^2}{(x^3 - 4)^2}\, dx$

23. $\int (x^4 - 2x)(2x^3 - 1)\, dx$

24. $\int \frac{\sqrt{x}}{(1 - x^{3/2})^3}\, dx$

25. Production The output P (in board-feet) of a small sawmill changes according to the model

$$\frac{dP}{dt} = 2t(0.001t^2 + 0.5)^{1/4},\quad 0 \le t \le 40$$

where t is measured in hours. Find the numbers of board-feet produced in (a) 6 hours and (b) 12 hours.

26. Cost The marginal cost for a catering service to cater to x people can be modeled by

$$\frac{dC}{dx} = \frac{5x}{\sqrt{x^2 + 1000}}.$$

When $x = 225$, the cost is $1136.06. Find the costs of catering to (a) 500 people and (b) 1000 people.

In Exercises 27–32, find the indefinite integral.

27. $\int 3e^{-3x}\, dx$

28. $\int (2t - 1)e^{t^2 - t}\, dt$

29. $\int (x - 1)e^{x^2 - 2x}\, dx$

30. $\int \frac{4}{6x - 1}\, dx$

31. $\int \frac{x^2}{1 - x^3}\, dx$

32. $\int \frac{x - 4}{x^2 - 8x}\, dx$

(T) In Exercises 33 and 34, use a symbolic integration utility to find the indefinite integral.

33. $\int \frac{(\sqrt{x} + 1)^2}{\sqrt{x}}\, dx$

34. $\int \frac{e^{5x}}{5 + e^{5x}}\, dx$

In Exercises 35 and 36, sketch the region whose area is given by the definite integral. Then use a geometric formula to evaluate the integral.

35. $\int_0^5 (5 - |x - 5|)\, dx$

36. $\int_{-4}^4 \sqrt{16 - x^2}\, dx$

In Exercises 37–44, find the area of the region.

37. $f(x) = 4 - 2x$

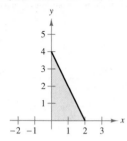

38. $f(x) = 3x + 6$

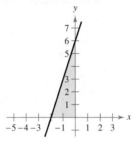

39. $f(x) = 4 - x^2$

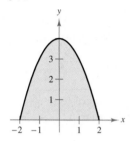

40. $f(x) = 9 - x^2$

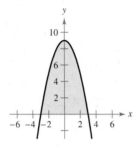

41. $f(y) = (y - 2)^2$

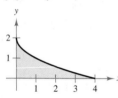

42. $f(x) = \sqrt{9 - x^2}$

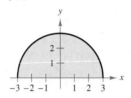

43. $f(x) = \dfrac{2}{x + 1}$

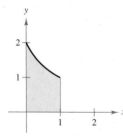

44. $f(x) = 2xe^{x^2 - 4}$

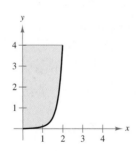

45. Given $\displaystyle\int_2^6 f(x)\,dx = 10$ and $\displaystyle\int_2^6 g(x)\,dx = 3$, evaluate the definite integral.

(a) $\displaystyle\int_2^6 [f(x) + g(x)]\,dx$

(b) $\displaystyle\int_2^6 [f(x) - g(x)]\,dx$

(c) $\displaystyle\int_2^6 [2f(x) - 3g(x)]\,dx$

(d) $\displaystyle\int_2^6 5f(x)\,dx$

46. Given $\displaystyle\int_0^3 f(x)\,dx = 4$ and $\displaystyle\int_3^6 f(x)\,dx = -1$, evaluate the definite integral.

(a) $\displaystyle\int_0^6 f(x)\,dx$

(b) $\displaystyle\int_6^3 f(x)\,dx$

(c) $\displaystyle\int_4^4 f(x)\,dx$

(d) $\displaystyle\int_3^6 -10f(x)\,dx$

In Exercises 47–60, use the Fundamental Theorem of Calculus to evaluate the definite integral.

47. $\displaystyle\int_0^4 (2 + x)\,dx$

48. $\displaystyle\int_{-1}^1 (t^2 + 2)\,dt$

49. $\displaystyle\int_4^9 x\sqrt{x}\,dx$

50. $\displaystyle\int_1^4 2x\sqrt{x}\,dx$

51. $\displaystyle\int_{-1}^1 (4t^3 - 2t)\,dt$

52. $\displaystyle\int_{-2}^2 (x^4 + 2x^2 - 5)\,dx$

53. $\displaystyle\int_0^3 \dfrac{1}{\sqrt{1 + x}}\,dx$

54. $\displaystyle\int_3^6 \dfrac{x}{3\sqrt{x^2 - 8}}\,dx$

55. $\displaystyle\int_1^2 \left(\dfrac{1}{x^2} - \dfrac{1}{x^3}\right)\,dx$

56. $\displaystyle\int_0^1 x^2(x^3 + 1)^3\,dx$

57. $\displaystyle\int_1^3 \dfrac{(3 + \ln x)}{x}\,dx$

58. $\displaystyle\int_0^{\ln 5} e^{x/5}\,dx$

59. $\displaystyle\int_{-1}^1 3xe^{x^2 - 1}\,dx$

60. $\displaystyle\int_1^3 \dfrac{1}{x(\ln x + 2)^2}\,dx$

In Exercises 61–64, sketch the graph of the region whose area is given by the integral, and find the area.

61. $\displaystyle\int_1^3 (2x - 1)\,dx$

62. $\displaystyle\int_0^2 (x + 4)\,dx$

63. $\displaystyle\int_3^4 (x^2 - 9)\,dx$

64. $\displaystyle\int_{-1}^2 (-x^2 + x + 2)\,dx$

65. Cost The marginal cost of serving an additional typical client at a law firm can be modeled by

$$\dfrac{dC}{dx} = 675 + 0.5x$$

where x is the number of clients. How does the cost C change when x increases from 50 to 51 clients?

66. Profit The marginal profit obtained by selling x dollars of automobile insurance can be modeled by

$$\frac{dP}{dx} = 0.4\left(1 - \frac{5000}{x}\right), \quad x \geq 5000.$$

Find the change in the profit when x increases from $75,000 to $100,000.

In Exercises 67–70, find the average value of the function on the closed interval. Then find all x-values in the interval for which the function is equal to its average value.

67. $f(x) = \dfrac{1}{\sqrt{x}}, \quad [4, 9]$

68. $f(x) = \dfrac{20 \ln x}{x}, \quad [2, 10]$

69. $f(x) = e^{5-x}, \quad [2, 5]$

70. $f(x) = x^3, \quad [0, 2]$

71. Compound Interest An interest-bearing checking account yields 4% interest compounded continuously. If you deposit $500 in such an account, and never write checks, what will the average value of the account be over a period of 2 years? Explain your reasoning.

72. Consumer Awareness Suppose the price p of gasoline can be modeled by

$$p = 0.0782t^2 - 0.352t + 1.75$$

where $t = 1$ corresponds to January 1, 2001. Find the cost of gasoline for an automobile that is driven 15,000 miles per year and gets 33 miles per gallon from 2001 through 2006. *(Source: U.S. Department of Energy)*

73. Consumer Trends The rates of change of lean and extra lean beef prices (in dollars per pound) in the United States from 1999 through 2006 can be modeled by

$$\frac{dB}{dt} = -0.0391t + 0.6108$$

where t is the year, with $t = 9$ corresponding to 1999. The price of 1 pound of lean and extra lean beef in 2006 was $2.95. *(Source: U.S. Bureau of Labor Statistics)*

(a) Find the price function in terms of the year.

(b) If the price of beef per pound continues to change at this rate, in what year does the model predict the price per pound of lean and extra lean beef will surpass $3.25? Explain your reasoning.

74. Medical Science The volume V (in liters) of air in the lungs during a five-second respiratory cycle is approximated by the model

$$V = 0.1729t + 0.1522t^2 - 0.0374t^3$$

where t is time in seconds.

(T) (a) Use a graphing utility to graph the equation on the interval $[0, 5]$.

(b) Determine the intervals on which the function is increasing and decreasing.

(c) Determine the maximum volume during the respiratory cycle.

(d) Determine the average volume of air in the lungs during one cycle.

(e) Briefly explain your results for parts (a) through (d).

Annuity In Exercises 75 and 76, find the amount of an annuity with income function $c(t)$, interest rate r, and term T.

75. $c(t) = \$3000, r = 6\%, T = 5$ years

76. $c(t) = \$1200, r = 7\%, T = 8$ years

In Exercises 77–80, explain how the given value can be used to evaluate the second integral.

77. $\displaystyle\int_0^2 6x^5 \, dx = 64, \quad \int_{-2}^2 6x^5 \, dx$

78. $\displaystyle\int_0^3 (x^4 + x^2) \, dx = 57.6, \quad \int_{-3}^3 (x^4 + x^2) \, dx$

79. $\displaystyle\int_1^2 \frac{4}{x^2} \, dx = 2, \quad \int_{-2}^{-1} \frac{4}{x^2} \, dx$

80. $\displaystyle\int_0^1 (x^3 - x) \, dx = -\frac{1}{4}, \quad \int_{-1}^0 (x^3 - x) \, dx$

In Exercises 81–88, sketch the region bounded by the graphs of the equations. Then find the area of the region.

81. $y = \dfrac{1}{x^2}, y = 0, x = 1, x = 5$

82. $y = \dfrac{1}{x^2}, y = 4, x = 5$

83. $y = x, y = x^3$

84. $y = 1 - \dfrac{1}{2}x, y = x - 2, y = 1$

85. $y = \dfrac{4}{\sqrt{x + 1}}, y = 0, x = 0, x = 8$

86. $y = \sqrt{x}(x - 1), y = 0$

87. $y = (x - 3)^2, y = 8 - (x - 3)^2$

88. $y = 4 - x, y = x^2 - 5x + 8, x = 0$

(T) In Exercises 89 and 90, use a graphing utility to graph the region bounded by the graphs of the equations. Then find the area of the region.

89. $y = x, y = 2 - x^2$

90. $y = x, y = x^5$

Consumer and Producer Surpluses In Exercises 91 and 92, find the consumer surplus and producer surplus for the demand and supply functions.

91. Demand function: $p_2(x) = 500 - x$

Supply function: $p_1(x) = 1.25x + 162.5$

92. Demand function: $p_2(x) = \sqrt{100,000 - 0.15x^2}$

Supply function: $p_1(x) = \sqrt{0.01x^2 + 36,000}$

93. Sales The sales S (in millions of dollars per year) for Avon from 1996 through 2001 can be modeled by

$S = 12.73t^2 + 4379.7, \quad 6 \le t \le 11$

where $t = 6$ corresponds to 1996. The sales for Avon from 2002 through 2005 can be modeled by

$S = 24.12t^2 + 2748.7, \quad 11 < t \le 15.$

If sales for Avon had followed the first model from 1996 through 2005, how much more or less sales would there have been for Avon? *(Source: Avon Products, Inc.)*

94. Revenue The revenues (in millions of dollars per year) for Telephone & Data Systems, U.S. Cellular, and IDT from 2001 through 2005 can be modeled by

$R = -35.643t^2 + 561.68t + 2047.0$ Telephone & Data Systems

$R = -23.307t^2 + 433.37t + 1463.4$ U.S. Cellular

$R = -1.321t^2 + 323.96t + 899.2$ IDT

where $1 \le t \le 5$ corresponds to the five-year period from 2001 through 2005. *(Source: Telephone & Data Systems Inc., U.S. Cellular Corp., and IDT Corp.)*

(a) From 2001 through 2005, how much more was Telephone & Data Systems' revenue than U.S. Cellular's revenue?

(b) From 2001 through 2005, how much more was U.S. Cellular's revenue than IDT's revenue?

95. Revenue The revenues (in millions of dollars per year) for The Men's Wearhouse from 1996 through 1999 can be modeled by

$R = 67.800t^2 - 792.36t + 2811.5, \quad 6 \le t \le 9$

where $t = 6$ corresponds to 1996. From 2000 through 2005, the revenues can be modeled by

$R = 30.738t^2 - 686.29t + 5113.9, \quad 9 < t \le 15.$

If sales for The Men's Wearhouse had followed the first model from 1996 through 2005, how much more or less revenues would there have been for The Men's Wearhouse? *(Source: The Men's Wearhouse, Inc.)*

Ⓑ **96. Psychology: Sleep Patterns** The graph shows three areas, representing awake time, REM (rapid eye movement) sleep time, and non-REM sleep time, over a typical individual's lifetime. Make generalizations about the amount of total sleep, non-REM sleep, and REM sleep an individual

gets as he or she gets older. If you wanted to estimate mathematically the amount of non-REM sleep an individual gets between birth and age 50, how would you do so? How would you mathematically estimate the amount of REM sleep an individual gets during this interval? *(Source: Adapted from Bernstein/Clarke-Stewart/Roy/Wickens, Psychology, Seventh Edition)*

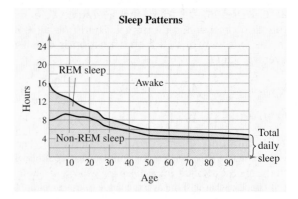

Sleep Patterns

Ⓣ In Exercises 97–100, use the Midpoint Rule with $n = 4$ to approximate the definite integral. Then use a programmable calculator or computer to approximate the definite integral with $n = 20$. Compare the two approximations.

97. $\int_0^2 (x^2 + 1)^2 \, dx$

98. $\int_{-1}^1 \sqrt{1 - x^2} \, dx$

99. $\int_0^1 \frac{1}{x^2 + 1} \, dx$

100. $\int_{-1}^1 e^{3 - x^2} \, dx$

101. Surface Area Use the Midpoint Rule to estimate the surface area of the oil spill shown in the figure.

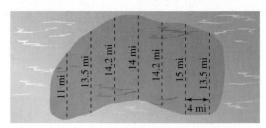

102. Velocity and Acceleration The table lists the velocity v (in feet per second) of an accelerating car over a 20-second interval. Approximate the distance in feet that the car travels during the 20 seconds using (a) the Midpoint Rule and (b) the Trapezoidal Rule. (The distance is given by $s = \int_0^{20} v \, dt.$)

Time, t	0	5	10	15	20
Velocity, v	0.0	29.3	51.3	66.0	73.3

Chapter Test

See www.CalcChat.com for worked-out solutions to odd-numbered exercises.

Take this test as you would take a test in class. When you are done, check your work against the answers given in the back of the book.

In Exercises 1–6, find the indefinite integral.

1. $\int (9x^2 - 4x + 13)\, dx$ **2.** $\int (x + 1)^2\, dx$ **3.** $\int 4x^3\sqrt{x^4 - 7}\, dx$

4. $\int \dfrac{5x - 6}{\sqrt{x}}\, dx$ **5.** $\int 15e^{3x}\, dx$ **6.** $\int \dfrac{3x^2 - 11}{x^3 - 11x}\, dx$

In Exercises 7 and 8, find the particular solution $y = f(x)$ that satisfies the differential equation and initial condition.

7. $f'(x) = e^x + 1;\, f(0) = 1$ **8.** $f'(x) = \dfrac{1}{x};\, f(-1) = 2$

In Exercises 9–14, evaluate the definite integral.

9. $\displaystyle\int_0^1 16x\, dx$ **10.** $\displaystyle\int_{-3}^3 (3 - 2x)\, dx$ **11.** $\displaystyle\int_{-1}^1 (x^3 + x^2)\, dx$

12. $\displaystyle\int_{-1}^2 \dfrac{2x}{\sqrt{x^2 + 1}}\, dx$ **13.** $\displaystyle\int_0^3 e^{4x}\, dx$ **14.** $\displaystyle\int_{-2}^3 \dfrac{1}{x + 3}\, dx$

15. The rate of change in sales for PetSmart, Inc. from 1998 through 2005 can be modeled by

$$\frac{dS}{dt} = 15.7e^{0.23t}$$

where S is the sales (in millions of dollars) and $t = 8$ corresponds to 1998. In 1998, the sales for PetSmart were $2109.3 million.   *(Source: PetSmart, Inc.)*

(a) Write a model for the sales as a function of t.

(b) What were the average sales for 1998 through 2005?

Ⓣ In Exercises 16 and 17, use a graphing utility to graph the region bounded by the graphs of the functions. Then find the area of the region.

16. $f(x) = 6,\ g(x) = x^2 - x - 6$ **17.** $f(x) = \sqrt[3]{x},\ g(x) = x^2$

18. The demand and supply functions for a product are modeled by

Demand: $p_1(x) = -0.625x + 10$ and   *Supply:* $p_2(x) = 0.25x + 3$

where x is the number of units (in millions). Find the consumer and producer surpluses for this product.

In Exercises 19 and 20, use the Midpoint Rule with $n = 4$ to approximate the area of the region bounded by the graph of f and the x-axis over the interval. Compare your result with the exact area. Sketch the region.

19. $f(x) = 3x^2,\, [0, 1]$

20. $f(x) = x^2 + 1,\, [-1, 1]$

12 Techniques of Integration

Integration can be used to find the amount of lumber used per year for residential upkeep and improvements. (See Section 12.4, Exercise 51.)

Applications

Integration has many real-life applications. The applications listed below represent a sample of the applications in this chapter.

© Pedar Björkegren/Etsa/Corbis

Section 12.1

Integration by Parts and Present Value

- Use integration by parts to find indefinite and definite integrals.
- Find the present value of future income.

Integration by Parts

In this section, you will study an integration technique called **integration by parts.** This technique is particularly useful for integrands involving the products of algebraic and exponential or logarithmic functions, such as

$$\int x^2 e^x \, dx \quad \text{and} \quad \int x \ln x \, dx.$$

Integration by parts is based on the Product Rule for differentiation.

$$\frac{d}{dx}[uv] = u\frac{dv}{dx} + v\frac{du}{dx} \qquad \text{Product Rule}$$

$$uv = \int u\frac{dv}{dx}\,dx + \int v\frac{du}{dx}\,dx \qquad \text{Integrate each side.}$$

$$uv = \int u\,dv + \int v\,du \qquad \text{Write in differential form.}$$

$$\int u\,dv = uv - \int v\,du \qquad \text{Rewrite.}$$

Integration by Parts

Let u and v be differentiable functions of x.

$$\int u\,dv = uv - \int v\,du$$

Note that the formula for integration by parts expresses the original integral in terms of another integral. Depending on the choices for u and dv, it may be easier to evaluate the second integral than the original one.

STUDY TIP

When using integration by parts, note that you can first choose dv or first choose u. After you choose, however, the choice of the other factor is determined—it must be the remaining portion of the integrand. Also note that dv *must* contain the differential dx of the original integral.

Guidelines for Integration by Parts

1. Let dv be the most complicated portion of the integrand that fits a basic integration formula. Let u be the remaining factor.

2. Let u be the portion of the integrand whose derivative is a function simpler than u. Let dv be the remaining factor.

| Example 1 | **Integration by Parts**

Find $\int xe^x \, dx$.

SOLUTION To apply integration by parts, you must rewrite the original integral in the form $\int u \, dv$. That is, you must break $xe^x \, dx$ into two factors—one "part" representing u and the other "part" representing dv. There are several ways to do this.

$$\int \underbrace{(x)}_{u}\underbrace{(e^x \, dx)}_{dv} \qquad \int \underbrace{(e^x)}_{u}\underbrace{(x \, dx)}_{dv} \qquad \int \underbrace{(1)}_{u}\underbrace{(xe^x \, dx)}_{dv} \qquad \int \underbrace{(xe^x)}_{u}\underbrace{(dx)}_{dv}$$

Following the guidelines, you should choose the first option because $dv = e^x \, dx$ is the most complicated portion of the integrand that fits a basic integration formula *and* because the derivative of $u = x$ is simpler than x.

$$dv = e^x \, dx \qquad \Longrightarrow \qquad v = \int dv = \int e^x \, dx = e^x$$

$$u = x \qquad \Longrightarrow \qquad du = dx$$

With these substitutions, you can apply the integration by parts formula as shown.

$$\int xe^x \, dx = xe^x - \int e^x \, dx \qquad \int u \, dv = uv - \int v \, du$$

$$= xe^x - e^x + C \qquad \text{Integrate } \int e^x \, dx.$$

✓ **CHECKPOINT 1**

Find $\int xe^{2x} \, dx$. ▪

STUDY TIP

In Example 1, notice that you do not need to include a constant of integration when solving $v = \int e^x \, dx = e^x$. To see why this is true, try replacing e^x by $e^x + C_1$ in the solution.

$$\int xe^x \, dx = x(e^x + C_1) - \int (e^x + C_1) \, dx$$

After integrating, you can see that the terms involving C_1 subtract out.

TECHNOLOGY

Ⓣ If you have access to a symbolic integration utility, try using it to solve several of the exercises in this section. Note that the form of the integral may be slightly different from what you obtain when solving the exercise by hand.

STUDY TIP

To remember the integration by parts formula, you might like to use the "Z" pattern below. The top row represents the original integral, the diagonal row represents uv, and the bottom row represents the new integral.

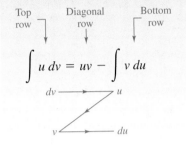

Top row   Diagonal row   Bottom row

$$\int u \, dv = uv - \int v \, du$$

$dv \longrightarrow u$

$v \longrightarrow du$

Example 2   **Integration by Parts**

Find $\displaystyle\int x^2 \ln x \, dx$.

SOLUTION   For this integral, x^2 is more easily integrated than $\ln x$. Furthermore, the derivative of $\ln x$ is simpler than $\ln x$. So, you should choose $dv = x^2 \, dx$.

$$dv = x^2 \, dx \qquad \Longrightarrow \qquad v = \int dv = \int x^2 \, dx = \frac{x^3}{3}$$

$$u = \ln x \qquad \Longrightarrow \qquad du = \frac{1}{x} \, dx$$

Using these substitutions, apply the integration by parts formula as shown.

$$\int x^2 \ln x \, dx = \frac{x^3}{3} \ln x - \int \left(\frac{x^3}{3}\right)\left(\frac{1}{x}\right) dx \qquad \text{\small $\int u \, dv = uv - \int v \, du$}$$

$$= \frac{x^3}{3} \ln x - \frac{1}{3} \int x^2 \, dx \qquad \text{\small Simplify.}$$

$$= \frac{x^3}{3} \ln x - \frac{x^3}{9} + C \qquad \text{\small Integrate.}$$

✓**CHECKPOINT 2**

Find $\displaystyle\int x \ln x \, dx$.

Example 3   **Integrating by Parts with a Single Factor**

Find $\displaystyle\int \ln x \, dx$.

SOLUTION   This integral is unusual because it has only one factor. In such cases, you should choose $dv = dx$ and choose u to be the single factor.

$$dv = dx \qquad \Longrightarrow \qquad v = \int dv = \int dx = x$$

$$u = \ln x \qquad \Longrightarrow \qquad du = \frac{1}{x} \, dx$$

Using these substitutions, apply the integration by parts formula as shown.

✓**CHECKPOINT 3**

Differentiate $y = x \ln x - x + C$ to show that it is the antiderivative of $\ln x$. ▇

$$\int \ln x \, dx = x \ln x - \int (x)\left(\frac{1}{x}\right) dx \qquad \text{\small $\int u \, dv = uv - \int v \, du$}$$

$$= x \ln x - \int dx \qquad \text{\small Simplify.}$$

$$= x \ln x - x + C \qquad \text{\small Integrate.}$$

Example 4   **Using Integration by Parts Repeatedly**

Find $\int x^2 e^x \, dx$.

SOLUTION   Using the guidelines, notice that the derivative of x^2 becomes simpler, whereas the derivative of e^x does not. So, you should let $u = x^2$ and let $dv = e^x \, dx$.

$$dv = e^x \, dx \implies v = \int dv = \int e^x \, dx = e^x$$

$$u = x^2 \implies du = 2x \, dx$$

Using these substitutions, apply the integration by parts formula as shown.

$$\int x^2 e^x \, dx = x^2 e^x - \int 2x e^x \, dx \qquad \text{First application of integration by parts}$$

To evaluate the new integral on the right, apply integration by parts a second time, using the substitutions below.

$$dv = e^x \, dx \implies v = \int dv = \int e^x \, dx = e^x$$

$$u = 2x \implies du = 2 \, dx$$

Using these substitutions, apply the integration by parts formula as shown.

$$\int x^2 e^x \, dx = x^2 e^x - \int 2x e^x \, dx \qquad \text{First application of integration by parts}$$

$$= x^2 e^x - \left(2x e^x - \int 2e^x \, dx \right) \qquad \text{Second application of integration by parts}$$

$$= x^2 e^x - 2x e^x + 2e^x + C \qquad \text{Integrate.}$$

$$= e^x(x^2 - 2x + 2) + C \qquad \text{Simplify.}$$

You can confirm this result by differentiating.

✓ **CHECKPOINT 4**

Find $\int x^3 e^x \, dx$. ■

STUDY TIP

Remember that you can check an indefinite integral by differentiating. For instance, in Example 4, try differentiating the antiderivative

$$e^x(x^2 - 2x + 2) + C$$

to check that you obtain the original integrand, $x^2 e^x$.

When making repeated applications of integration by parts, be careful not to interchange the substitutions in successive applications. For instance, in Example 4, the first substitutions were $dv = e^x \, dx$ and $u = x^2$. If in the second application you had switched to $dv = 2x \, dx$ and $u = e^x$, you would have reversed the previous integration and returned to the *original* integral.

$$\int x^2 e^x \, dx = x^2 e^x - \left(x^2 e^x - \int x^2 e^x \, dx \right)$$

$$= \int x^2 e^x \, dx$$

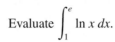

Evaluating a Definite Integral

Evaluate $\displaystyle\int_{1}^{e} \ln x \, dx$.

SOLUTION Integration by parts was used to find the antiderivative of $\ln x$ in Example 3. Using this result, you can evaluate the definite integral as shown.

$$\int_{1}^{e} \ln x \, dx = \left[x \ln x - x \right]_{1}^{e} \qquad \text{Use result of Example 3.}$$
$$= (e \ln e - e) - (1 \ln 1 - 1) \qquad \text{Apply Fundamental Theorem.}$$
$$= (e - e) - (0 - 1)$$
$$= 1 \qquad \text{Simplify.}$$

The area represented by this definite integral is shown in Figure 12.1.

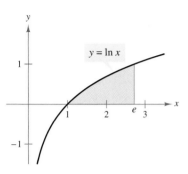

FIGURE 12.1

✓ CHECKPOINT 5

Evaluate $\displaystyle\int_{0}^{1} x^2 e^x \, dx$. ▪

Before starting the exercises in this section, remember that it is not enough to know *how* to use the various integration techniques. You also must know *when* to use them. Integration is first and foremost a problem of recognition—recognizing which formula or technique to apply to obtain an antiderivative. Often, a slight alteration of an integrand will necessitate the use of a different integration technique. Here are some examples.

Integral	*Technique*	*Antiderivative*
$\displaystyle\int x \ln x \, dx$	Integration by parts	$\dfrac{x^2}{2} \ln x - \dfrac{x^2}{4} + C$
$\displaystyle\int \dfrac{\ln x}{x} \, dx$	Power Rule: $\displaystyle\int u^n \dfrac{du}{dx} \, dx$	$\dfrac{(\ln x)^2}{2} + C$
$\displaystyle\int \dfrac{1}{x \ln x} \, dx$	Log Rule: $\displaystyle\int \dfrac{1}{u} \dfrac{du}{dx} \, dx$	$\ln|\ln x| + C$

As you gain experience with integration by parts, your skill in determining u and dv will improve. The summary below gives suggestions for choosing u and dv.

Summary of Common Uses of Integration by Parts

1. $\displaystyle\int x^n e^{ax} \, dx$ Let $u = x^n$ and $dv = e^{ax} \, dx$. (Examples 1 and 4)

2. $\displaystyle\int x^n \ln x \, dx$ Let $u = \ln x$ and $dv = x^n \, dx$. (Examples 2 and 3)

Present Value

Recall from Section 10.2 that the present value of a future payment is the amount that would have to be deposited today to produce the future payment. What is the present value of a future payment of $1000 one year from now? Because of inflation, $1000 today buys more than $1000 will buy a year from now. The definition below considers only the effect of inflation.

STUDY TIP

According to this definition, if the rate of inflation were 4%, then the present value of $1000 one year from now is just $980.26.

Present Value

If c represents a continuous income function in dollars per year and the annual rate of inflation is r, then the actual total income over t_1 years is

$$\text{Actual income over } t_1 \text{ years} = \int_0^{t_1} c(t)\, dt$$

and its **present value** is

$$\text{Present value} = \int_0^{t_1} c(t)e^{-rt}\, dt.$$

Ignoring inflation, the equation for present value also applies to an interest-bearing account where the annual interest rate r is compounded continuously and c is an income function in dollars per year.

Example 6 Finding Present Value

You have just won a state lottery for $1,000,000. You will be paid an annuity of $50,000 a year for 20 years. Assuming an annual inflation rate of 6%, what is the present value of this income?

SOLUTION The income function for your winnings is given by $c(t) = 50,000$. So,

$$\text{Actual income} = \int_0^{20} 50,000\, dt = \left[50,000t \right]_0^{20} = \$1,000,000.$$

Because you do not receive this entire amount now, its present value is

$$\text{Present value} = \int_0^{20} 50,000e^{-0.06t}\, dt = \left[\frac{50,000}{-0.06}e^{-0.06t} \right]_0^{20} \approx \$582,338.$$

This present value represents the amount that the state must deposit now to cover your payments over the next 20 years. This shows why state lotteries are so profitable—for the states!

AP/Wide World Photos

On February 18, 2006, a group of eight coworkers at a meat processing plant in Nebraska won the largest lottery jackpot in the world. They chose to receive a lump sum payment of $177.3 million instead of an annuity that would have paid $365 million over a 29-year period. The odds of winning the PowerBall jackpot are about 1 in 146.1 million.

✓**CHECKPOINT 6**

Find the present value of the income from the lottery ticket in Example 6 if the inflation rate is 7%. ◼

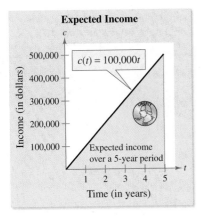

Expected Income

Income (in dollars)

$c(t) = 100,000t$

Expected income over a 5-year period

Time (in years)

(a)

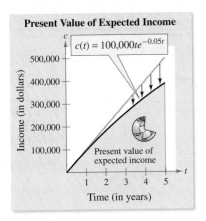

Present Value of Expected Income

Income (in dollars)

$c(t) = 100,000te^{-0.05t}$

Present value of expected income

Time (in years)

(b)

FIGURE 12.2

Example 7
MAKE A DECISION **Finding Present Value**

A company expects its income during the next 5 years to be given by

$$c(t) = 100,000t, \quad 0 \le t \le 5.$$ See Figure 12.2(a).

Assuming an annual inflation rate of 5%, can the company claim that the present value of this income is at least $1 million?

SOLUTION The present value is

$$\text{Present value} = \int_0^5 100,000te^{-0.05t}\, dt = 100,000 \int_0^5 te^{-0.05t}\, dt.$$

Using integration by parts, let $dv = e^{-0.05t}\, dt$.

$$dv = e^{-0.05t}\, dt \implies v = \int dv = \int e^{-0.05t}\, dt = -20e^{-0.05t}$$

$$u = t \implies du = dt$$

This implies that

$$\int te^{-0.05t}\, dt = -20te^{-0.05t} + 20 \int e^{-0.05t}\, dt$$

$$= -20te^{-0.05t} - 400e^{-0.05t}$$

$$= -20e^{-0.05t}(t + 20).$$

So, the present value is

$$\text{Present value} = 100,000 \int_0^5 te^{-0.05t}\, dt$$ See Figure 12.2(b).

$$= 100,000 \left[-20e^{-0.05t}(t + 20) \right]_0^5$$

$$\approx \$1,059,961.$$

Yes, the company can claim that the present value of its expected income during the next 5 years is at least $1 million.

✓ CHECKPOINT 7

A company expects its income during the next 10 years to be given by $c(t) = 20,000t$, for $0 \le t \le 10$. Assuming an annual inflation rate of 5%, what is the present value of this income? ■

CONCEPT CHECK

1. Integration by parts is based on what differentiation rule?
2. Write the formula for integration by parts.
3. State the guidelines for integration by parts.
4. Without integrating, which formula or technique of integration would you use to find $\int xe^{4x}\, dx$? Explain your reasoning.

The following warm-up exercises involve skills that were covered in earlier sections. You will use these skills in the exercise set for this section. For additional help, review Sections 10.3, 10.5, and 11.5.

In Exercises 1–6, find $f'(x)$.

1. $f(x) = \ln(x + 1)$ **2.** $f(x) = \ln(x^2 - 1)$ **3.** $f(x) = e^{x^3}$

4. $f(x) = e^{-x^2}$ **5.** $f(x) = x^2 e^x$ **6.** $f(x) = xe^{-2x}$

In Exercises 7–10, find the area between the graphs of f and g.

7. $f(x) = -x^2 + 4$, $g(x) = x^2 - 4$ **8.** $f(x) = -x^2 + 2$, $g(x) = 1$

9. $f(x) = 4x$, $g(x) = x^2 - 5$ **10.** $f(x) = x^3 - 3x^2 + 2$, $g(x) = x - 1$

Exercises 12.1

In Exercises 1–4, identify u and dv for finding the integral using integration by parts. (Do not evaluate the integral.)

1. $\displaystyle\int xe^{3x}\, dx$ **2.** $\displaystyle\int x^2 e^{3x}\, dx$

3. $\displaystyle\int x \ln 2x\, dx$ **4.** $\displaystyle\int \ln 4x\, dx$

In Exercises 5–10, use integration by parts to find the indefinite integral.

5. $\displaystyle\int xe^{3x}\, dx$ **6.** $\displaystyle\int xe^{-x}\, dx$

7. $\displaystyle\int x^2 e^{-x}\, dx$ **8.** $\displaystyle\int x^2 e^{2x}\, dx$

9. $\displaystyle\int \ln 2x\, dx$ **10.** $\displaystyle\int \ln x^2\, dx$

In Exercises 11–38, find the indefinite integral. (*Hint:* Integration by parts is not required for all the integrals.)

11. $\displaystyle\int e^{4x}\, dx$ **12.** $\displaystyle\int e^{-2x}\, dx$

13. $\displaystyle\int xe^{4x}\, dx$ **14.** $\displaystyle\int xe^{-2x}\, dx$

15. $\displaystyle\int xe^{x^2}\, dx$ **16.** $\displaystyle\int x^2 e^{x^3}\, dx$

17. $\displaystyle\int \frac{x}{e^x}\, dx$ **18.** $\displaystyle\int \frac{2x}{e^x}\, dx$

19. $\displaystyle\int 2x^2 e^x\, dx$ **20.** $\displaystyle\int \frac{1}{2}x^3 e^x\, dx$

21. $\displaystyle\int t \ln(t + 1)\, dt$ **22.** $\displaystyle\int x^3 \ln x\, dx$

23. $\displaystyle\int (x - 1)e^x\, dx$ **24.** $\displaystyle\int x^4 \ln x\, dx$

25. $\displaystyle\int \frac{e^{1/t}}{t^2}\, dt$ **26.** $\displaystyle\int \frac{1}{x(\ln x)^3}\, dx$

27. $\displaystyle\int x(\ln x)^2\, dx$ **28.** $\displaystyle\int \ln 3x\, dx$

29. $\displaystyle\int \frac{(\ln x)^2}{x}\, dx$

30. $\displaystyle\int \frac{1}{x \ln x}\, dx$

31. $\displaystyle\int \frac{\ln x}{x^2}\, dx$

32. $\displaystyle\int \frac{\ln 2x}{x^2}\, dx$

33. $\displaystyle\int x\sqrt{x - 1}\, dx$

34. $\displaystyle\int \frac{x}{\sqrt{x - 1}}\, dx$

35. $\displaystyle\int x(x + 1)^2\, dx$

36. $\displaystyle\int \frac{x}{\sqrt{2 + 3x}}\, dx$

37. $\displaystyle\int \frac{xe^{2x}}{(2x + 1)^2}\, dx$

38. $\displaystyle\int \frac{x^3 e^{x^2}}{(x^2 + 1)^2}\, dx$

In Exercises 39–46, evaluate the definite integral.

39. $\displaystyle\int_{1}^{2} x^2 e^x \, dx$

40. $\displaystyle\int_{0}^{2} \frac{x^2}{e^x} \, dx$

41. $\displaystyle\int_{0}^{4} \frac{x}{e^{x/2}} \, dx$

42. $\displaystyle\int_{1}^{2} x^2 \ln x \, dx$

43. $\displaystyle\int_{1}^{e} x^5 \ln x \, dx$

44. $\displaystyle\int_{1}^{e} 2x \ln x \, dx$

45. $\displaystyle\int_{-1}^{0} \ln(x + 2) \, dx$

46. $\displaystyle\int_{0}^{1} \ln(1 + 2x) \, dx$

(T) In Exercises 47–50, find the area of the region bounded by the graphs of the equations. Then use a graphing utility to graph the region and verify your answer.

47. $y = x^3 e^x, \; y = 0, \; x = 0, \; x = 2$

48. $y = (x^2 - 1)e^x, \; y = 0, \; x = -1, \; x = 1$

49. $y = x^2 \ln x, \; y = 0, \; x = 1, \; x = e$

50. $y = \dfrac{\ln x}{x^2}, \; y = 0, \; x = 1, \; x = e$

In Exercises 51 and 52, use integration by parts to verify the formula.

51. $\displaystyle\int x^n \ln x \, dx = \frac{x^{n+1}}{(n+1)^2}[-1 + (n+1)\ln x] + C,$

$n \neq -1$

52. $\displaystyle\int x^n e^{ax} \, dx = \frac{x^n e^{ax}}{a} - \frac{n}{a}\int x^{n-1} e^{ax} \, dx, \quad n > 0$

In Exercises 53–56, use the results of Exercises 51 and 52 to find the indefinite integral.

53. $\displaystyle\int x^2 e^{5x} \, dx$

54. $\displaystyle\int x e^{-3x} \, dx$

55. $\displaystyle\int x^{-2} \ln x \, dx$

56. $\displaystyle\int x^{1/2} \ln x \, dx$

In Exercises 57–60, find the area of the region bounded by the graphs of the given equations.

57. $y = xe^{-x},$
$y = 0, \; x = 4$

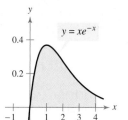

58. $y = \frac{1}{9}xe^{-x/3},$
$y = 0, \; x = 0, \; x = 3$

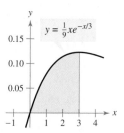

59. $y = x \ln x,$
$y = 0, \; x = e$

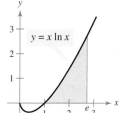

60. $y = x^{-3} \ln x,$
$y = 0, \; x = e$

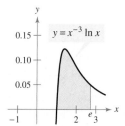

(T) In Exercises 61–64, use a symbolic integration utility to evaluate the integral.

61. $\displaystyle\int_{0}^{2} t^3 e^{-4t} \, dt$

62. $\displaystyle\int_{1}^{4} \ln x(x^2 + 4) \, dx$

63. $\displaystyle\int_{0}^{5} x^4(25 - x^2)^{3/2} \, dx$

64. $\displaystyle\int_{1}^{e} x^9 \ln x \, dx$

65. Demand A manufacturing company forecasts that the demand x (in units per year) for its product over the next 10 years can be modeled by $x = 500(20 + te^{-0.1t})$ for $0 \leq t \leq 10$, where t is the time in years.

(T) (a) Use a graphing utility to decide whether the company is forecasting an increase or a decrease in demand over the decade.

(b) According to the model, what is the total demand over the next 10 years?

(c) Find the average annual demand during the 10-year period.

66. Capital Campaign The board of trustees of a college is planning a five-year capital gifts campaign to raise money for the college. The goal is to have an annual gift income I that is modeled by $I = 2000(375 + 68te^{-0.2t})$ for $0 \le t \le 5$, where t is the time in years.

(T) (a) Use a graphing utility to decide whether the board of trustees expects the gift income to increase or decrease over the five-year period.

(b) Find the expected total gift income over the five-year period.

(c) Determine the average annual gift income over the five-year period. Compare the result with the income given when $t = 3$.

67. Memory Model A model for the ability M of a child to memorize, measured on a scale from 0 to 10, is

$$M = 1 + 1.6t \ln t, \quad 0 < t \le 4$$

where t is the child's age in years. Find the average value of this model between

(a) the child's first and second birthdays.

(b) the child's third and fourth birthdays.

68. Revenue A company sells a seasonal product. The revenue R (in dollars per year) generated by sales of the product can be modeled by

$$R = 410.5t^2 e^{-t/30} + 25,000, \quad 0 \le t \le 365$$

where t is the time in days.

(a) Find the average daily receipts during the first quarter, which is given by $0 \le t \le 90$.

(b) Find the average daily receipts during the fourth quarter, which is given by $274 \le t \le 365$.

(c) Find the total daily receipts during the year.

Present Value In Exercises 69–74, find the present value of the income c (measured in dollars) over t_1 years at the given annual inflation rate r.

69. $c = 5000, \ r = 4\%, \ t_1 = 4$ years

70. $c = 450, \ r = 4\%, \ t_1 = 10$ years

71. $c = 100,000 + 4000t, \ r = 5\%, \ t_1 = 10$ years

72. $c = 30,000 + 500t, \ r = 7\%, \ t_1 = 6$ years

73. $c = 1000 + 50e^{t/2}, \ r = 6\%, \ t_1 = 4$ years

74. $c = 5000 + 25te^{t/10}, \ r = 6\%, \ t_1 = 10$ years

75. Present Value A company expects its income c during the next 4 years to be modeled by

$$c = 150,000 + 75,000t.$$

(a) Find the actual income for the business over the 4 years.

(b) Assuming an annual inflation rate of 4%, what is the present value of this income?

76. Present Value A professional athlete signs a three-year contract in which the earnings can be modeled by

$$c = 300,000 + 125,000t.$$

(a) Find the actual value of the athlete's contract.

(b) Assuming an annual inflation rate of 3%, what is the present value of the contract?

Future Value In Exercises 77 and 78, find the future value of the income (in dollars) given by $f(t)$ over t_1 years at the annual interest rate of r. If the function f represents a continuous investment over a period of t_1 years at an annual interest rate of r (compounded continuously), then the future value of the investment is given by

$$\text{Future value} = e^{rt_1} \int_0^{t_1} f(t)e^{-rt} \, dt.$$

77. $f(t) = 3000, \ r = 8\%, \ t_1 = 10$ years

78. $f(t) = 3000e^{0.05t}, \ r = 10\%, \ t_1 = 5$ years

(B) **79. Finance: Future Value** Use the equation from Exercises 77 and 78 to calculate the following. *(Source: Adapted from Garman/Forgue, Personal Finance, Eighth Edition)*

(a) The future value of $1200 saved each year for 10 years earning 7% interest.

(b) A person who wishes to invest $1200 each year finds one investment choice that is expected to pay 9% interest per year and another, riskier choice that may pay 10% interest per year. What is the difference in return (future value) if the investment is made for 15 years?

80. MAKE A DECISION: COLLEGE TUITION FUND In 2006, the total cost of attending Pennsylvania State University for 1 year was estimated to be $20,924. Assume your grandparents had continuously invested in a college fund according to the model

$$f(t) = 400t$$

for 18 years, at an annual interest rate of 10%. Will the fund have grown enough to allow you to cover 4 years of expenses at Pennsylvania State University? *(Source: Pennsylvania State University)*

(T) **81.** Use a program similar to the Midpoint Rule program on page 856 with $n = 10$ to approximate

$$\int_1^4 \frac{4}{\sqrt{x} + \sqrt[3]{x}} \, dx.$$

(T) **82.** Use a program similar to the Midpoint Rule program on page 856 with $n = 12$ to approximate the area of the region bounded by the graphs of

$$y = \frac{10}{\sqrt{x}e^x}, \ y = 0, \ x = 1, \text{ and } x = 4.$$

Section 12.2

Partial Fractions and Logistic Growth

■ Use partial fractions to find indefinite integrals.
■ Use logistic growth functions to model real-life situations.

Partial Fractions

In Sections 11.2 and 12.1, you studied integration by substitution and by parts. In this section you will study a third technique called **partial fractions.** This technique involves the decomposition of a rational function into the sum of two or more simple rational functions. For instance, suppose you know that

$$\frac{x + 7}{x^2 - x - 6} = \frac{2}{x - 3} - \frac{1}{x + 2}.$$

Knowing the "partial fractions" on the right side would allow you to integrate the left side as shown.

$$\int \frac{x + 7}{x^2 - x - 6}\, dx = \int \left(\frac{2}{x - 3} - \frac{1}{x + 2}\right) dx$$

$$= 2 \int \frac{1}{x - 3}\, dx - \int \frac{1}{x + 2}\, dx$$

$$= 2 \ln|x - 3| - \ln|x + 2| + C$$

This method depends on the ability to factor the denominator of the original rational function *and* on finding the partial fraction decomposition of the function.

STUDY TIP

Recall that finding the partial fraction decomposition of a rational function is a *precalculus* topic. Explain how you could verify that

$$\frac{1}{x - 1} + \frac{2}{x + 2}$$

is the partial fraction decomposition of

$$\frac{3x}{x^2 + x - 2}.$$

Partial Fractions

To find the partial fraction decomposition of the *proper* rational function $p(x)/q(x)$, factor $q(x)$ and write an equation that has the form

$$\frac{p(x)}{q(x)} = (\text{sum of partial fractions}).$$

For each *distinct* linear factor $ax + b$, the right side should include a term of the form

$$\frac{A}{ax + b}.$$

For each *repeated* linear factor $(ax + b)^n$, the right side should include n terms of the form

$$\frac{A_1}{ax + b} + \frac{A_2}{(ax + b)^2} + \cdots + \frac{A_n}{(ax + b)^n}.$$

STUDY TIP

A rational function $p(x)/q(x)$ is *proper* if the degree of the numerator is less than the degree of the denominator.

Example 1 Finding a Partial Fraction Decomposition

Write the partial fraction decomposition for

$$\frac{x + 7}{x^2 - x - 6}.$$

SOLUTION   Begin by factoring the denominator as $x^2 - x - 6 = (x - 3)(x + 2)$. Then, write the partial fraction decomposition as

$$\frac{x + 7}{x^2 - x - 6} = \frac{A}{x - 3} + \frac{B}{x + 2}.$$

To solve this equation for A and B, multiply each side of the equation by the least common denominator $(x - 3)(x + 2)$. This produces the **basic equation** as shown.

$$x + 7 = A(x + 2) + B(x - 3) \qquad \text{Basic equation}$$

Because this equation is true for all x, you can substitute any convenient values of x into the equation. The x-values that are especially convenient are the ones that make particular factors equal to 0.

To solve for B, substitute $x = -2$:

$$x + 7 = A(x + 2) + B(x - 3) \qquad \text{Write basic equation.}$$
$$-2 + 7 = A(-2 + 2) + B(-2 - 3) \qquad \text{Substitute } -2 \text{ for } x.$$
$$5 = A(0) + B(-5) \qquad \text{Simplify.}$$
$$-1 = B \qquad \text{Solve for } B.$$

To solve for A, substitute $x = 3$:

$$x + 7 = A(x + 2) + B(x - 3) \qquad \text{Write basic equation.}$$
$$3 + 7 = A(3 + 2) + B(3 - 3) \qquad \text{Substitute 3 for } x.$$
$$10 = A(5) + B(0) \qquad \text{Simplify.}$$
$$2 = A \qquad \text{Solve for } A.$$

Now that you have solved the basic equation for A and B, you can write the partial fraction decomposition as

$$\frac{x + 7}{x^2 - x - 6} = \frac{2}{x - 3} - \frac{1}{x + 2}$$

as indicated at the beginning of this section.

> **Algebra Review**
>
> You can check the result in Example 1 by subtracting the partial fractions to obtain the original fraction, as shown in Example 1(a) in the *Chapter 12 Algebra Review*, on page 922.

✓**CHECKPOINT 1**

Write the partial fraction decomposition for $\dfrac{x + 8}{x^2 + 7x + 12}$. ▨

STUDY TIP

Be sure you see that the substitutions for x in Example 1 are chosen for their convenience in solving for A and B. The value $x = -2$ is selected because it eliminates the term $A(x + 2)$, and the value $x = 3$ is chosen because it eliminates the term $B(x - 3)$.

Example 2 **Integrating with Repeated Factors**

Find $\int \dfrac{5x^2 + 20x + 6}{x^3 + 2x^2 + x} \, dx.$

SOLUTION Begin by factoring the denominator as $x(x + 1)^2$. Then, write the partial fraction decomposition as

$$\frac{5x^2 + 20x + 6}{x(x + 1)^2} = \frac{A}{x} + \frac{B}{x + 1} + \frac{C}{(x + 1)^2}.$$

To solve this equation for A, B, and C, multiply each side of the equation by the least common denominator $x(x + 1)^2$.

$$5x^2 + 20x + 6 = A(x + 1)^2 + Bx(x + 1) + Cx \qquad \text{Basic equation}$$

Now, solve for A and C by substituting $x = -1$ and $x = 0$ into the basic equation.

Substitute $x = -1$:

$$5(-1)^2 + 20(-1) + 6 = A(-1 + 1)^2 + B(-1)(-1 + 1) + C(-1)$$
$$-9 = A(0) + B(0) - C$$
$$9 = C \qquad \text{Solve for } C.$$

Substitute $x = 0$:

$$5(0)^2 + 20(0) + 6 = A(0 + 1)^2 + B(0)(0 + 1) + C(0)$$
$$6 = A(1) + B(0) + C(0)$$
$$6 = A \qquad \text{Solve for } A.$$

At this point, you have exhausted the convenient choices for x and have yet to solve for B. When this happens, you can use *any* other x-value along with the known values of A and C.

Substitute $x = 1$, $A = 6$, and $C = 9$:

$$5(1)^2 + 20(1) + 6 = (6)(1 + 1)^2 + B(1)(1 + 1) + (9)(1)$$
$$31 = 6(4) + B(2) + 9(1)$$
$$-1 = B \qquad \text{Solve for } B.$$

Now that you have solved for A, B, and C, you can use the partial fraction decomposition to integrate.

$$\int \frac{5x^2 + 20x + 6}{x^3 + 2x^2 + x} \, dx = \int \left(\frac{6}{x} - \frac{1}{x + 1} + \frac{9}{(x + 1)^2} \right) dx$$

$$= 6 \ln|x| - \ln|x + 1| + 9 \frac{(x + 1)^{-1}}{-1} + C$$

$$= \ln\left| \frac{x^6}{x + 1} \right| - \frac{9}{x + 1} + C$$

✓ CHECKPOINT 2

Find $\int \dfrac{3x^2 + 7x + 4}{x^3 + 4x^2 + 4x} \, dx.$ ∎

You can use the partial fraction decomposition technique outlined in Examples 1 and 2 only with a *proper* rational function—that is, a rational function whose numerator is of lower degree than its denominator. If the numerator is of equal or greater degree, you must divide first. For instance, the rational function

$$\frac{x^3}{x^2 + 1}$$

is improper because the degree of the numerator is greater than the degree of the denominator. Before applying partial fractions to this function, you should divide the denominator into the numerator to obtain

$$\frac{x^3}{x^2 + 1} = x - \frac{x}{x^2 + 1}.$$

Example 3   Integrating an Improper Rational Function

Find $\displaystyle\int \frac{x^5 + x - 1}{x^4 - x^3}\, dx.$

SOLUTION   This rational function is improper—its numerator has a degree greater than that of its denominator. So, you should begin by dividing the denominator into the numerator to obtain

$$\frac{x^5 + x - 1}{x^4 - x^3} = x + 1 + \frac{x^3 + x - 1}{x^4 - x^3}.$$

Now, applying partial fraction decomposition produces

$$\frac{x^3 + x - 1}{x^3(x - 1)} = \frac{A}{x} + \frac{B}{x^2} + \frac{C}{x^3} + \frac{D}{x - 1}.$$

Multiplying both sides by the least common denominator $x^3(x - 1)$ produces the basic equation.

$$x^3 + x - 1 = Ax^2(x - 1) + Bx(x - 1) + C(x - 1) + Dx^3 \qquad \text{Basic equation}$$

Using techniques similar to those in the first two examples, you can solve for A, B, C, and D to obtain

$$A = 0, \quad B = 0, \quad C = 1, \quad \text{and} \quad D = 1.$$

So, you can integrate as shown.

$$\int \frac{x^5 + x - 1}{x^4 - x^3}\, dx = \int \left(x + 1 + \frac{x^3 + x - 1}{x^4 - x^3} \right) dx$$

$$= \int \left(x + 1 + \frac{1}{x^3} + \frac{1}{x - 1} \right) dx$$

$$= \frac{x^2}{2} + x - \frac{1}{2x^2} + \ln|x - 1| + C$$

> ### Algebra Review
>
> You can check the partial fraction decomposition in Example 3 by combining the partial fractions to obtain the original fraction, as shown in Example 2(a) in the *Chapter 12 Algebra Review*, on page 923.

✓ **CHECKPOINT 3**

Find $\displaystyle\int \frac{x^4 - x^3 + 2x^2 + x + 1}{x^3 + x^2}.$ ∎

Logistic Growth Function

In Section 10.6, you saw that exponential growth occurs in situations for which the rate of growth is proportional to the quantity present at any given time. That is, if y is the quantity at time t, then $dy/dt = ky$. The general solution of this differential equation is $y = Ce^{kt}$. Exponential growth is unlimited. As long as C and k are positive, the value of Ce^{kt} can be made arbitrarily large by choosing sufficiently large values of t.

In many real-life situations, however, the growth of a quantity is limited and cannot increase beyond a certain size L, as shown in Figure 12.3. This upper limit L is called the **carrying capacity,** which is the maximum population $y(t)$ that can be sustained or supported as time t increases. A model that is often used for this type of growth is the **logistic differential equation**

$$\frac{dy}{dt} = ky\left(1 - \frac{y}{L}\right) \qquad \text{Logistic differential equation}$$

where k and L are positive constants. A population that satisfies this equation does not grow without bound, but approaches L as t increases. The general solution of this differential equation is called the *logistic growth model* and is derived in Example 4.

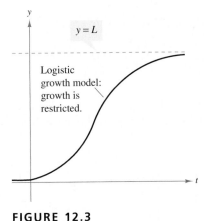

FIGURE 12.3

STUDY TIP

The graph of

$$y = \frac{L}{1 + be^{-kt}}$$

is called the *logistic curve,* as shown in Figure 12.3.

Algebra Review

For help with the algebra used to solve for y in Example 4, see Example 2(c) in the *Chapter 12 Algebra Review,* on page 923.

✓ **CHECKPOINT 4**

Show that if

$$y = \frac{1}{1 + be^{-kt}}, \text{ then}$$

$$\frac{dy}{dt} = ky(1 - y).$$

[*Hint:* First find $ky(1 - y)$ in terms of t, then find dy/dt and show that they are equivalent.] ∎

Example 4   **Deriving the Logistic Growth Model**

Solve the equation $\dfrac{dy}{dt} = ky\left(1 - \dfrac{y}{L}\right)$.

SOLUTION

$$\frac{dy}{dt} = ky\left(1 - \frac{y}{L}\right) \qquad \text{Write differential equation.}$$

$$\frac{1}{y(1 - y/L)}\, dy = k\, dt \qquad \text{Write in differential form.}$$

$$\int \frac{1}{y(1 - y/L)}\, dy = \int k\, dt \qquad \text{Integrate each side.}$$

$$\int \left(\frac{1}{y} + \frac{1}{L - y}\right) dy = \int k\, dt \qquad \text{Rewrite left side using partial fractions.}$$

$$\ln|y| - \ln|L - y| = kt + C \qquad \text{Find antiderivative of each side.}$$

$$\ln\left|\frac{L - y}{y}\right| = -kt - C \qquad \text{Multiply each side by } -1 \text{ and simplify.}$$

$$\left|\frac{L - y}{y}\right| = e^{-kt - C} = e^{-C}e^{-kt} \qquad \text{Exponentiate each side.}$$

$$\frac{L - y}{y} = be^{-kt} \qquad \text{Let } \pm e^{-C} = b.$$

Solving this equation for y produces the **logistic growth model** $y = \dfrac{L}{1 + be^{-kt}}$.

Example 5   Comparing Logistic Growth Functions

Use a graphing utility to investigate the effects of the values of L, b, and k on the graph of

$$y = \frac{L}{1 + be^{-kt}}.$$ Logistic growth function ($L > 0, b > 0, k > 0$)

SOLUTION   The value of L determines the horizontal asymptote of the graph to the right. In other words, as t increases without bound, the graph approaches a limit of L (see Figure 12.4).

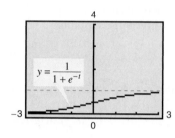 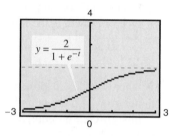 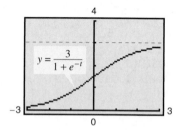

FIGURE 12.4

The value of b determines the point of inflection of the graph. When $b = 1$, the point of inflection occurs when $t = 0$. If $b > 1$, the point of inflection is to the right of the y-axis. If $0 < b < 1$, the point of inflection is to the left of the y-axis (see Figure 12.5).

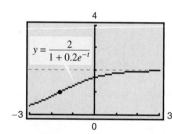 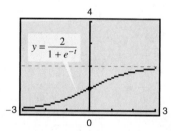 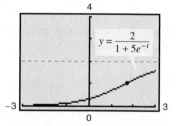

FIGURE 12.5

The value of k determines the rate of growth of the graph. For fixed values of b and L, larger values of k correspond to higher rates of growth (see Figure 12.6).

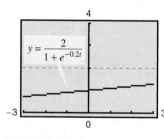 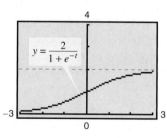 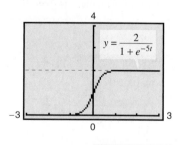

FIGURE 12.6

✓**CHECKPOINT 5**

Find the horizontal asymptote of the graph of $y = \dfrac{4}{1 + 5e^{-6t}}$. ∎

Daniel J. Cox/Getty Images

Example 6 | Modeling a Population

The state game commission releases 100 deer into a game preserve. During the first 5 years, the population increases to 432 deer. The commission believes that the population can be modeled by logistic growth with a limit of 2000 deer. Write a logistic growth model for this population. Then use the model to create a table showing the size of the deer population over the next 30 years.

SOLUTION   Let y represent the number of deer in year t. Assuming a logistic growth model means that the rate of change in the population is proportional to both y and $(1 - y/2000)$. That is

$$\frac{dy}{dt} = ky\left(1 - \frac{y}{2000}\right), \quad 100 \le y \le 2000.$$

The solution of this equation is

$$y = \frac{2000}{1 + be^{-kt}}.$$

Using the fact that $y = 100$ when $t = 0$, you can solve for b.

$$100 = \frac{2000}{1 + be^{-k(0)}} \quad \Longrightarrow \quad b = 19$$

Then, using the fact that $y = 432$ when $t = 5$, you can solve for k.

$$432 = \frac{2000}{1 + 19e^{-k(5)}} \quad \Longrightarrow \quad k \approx 0.33106$$

So, the logistic growth model for the population is

$$y = \frac{2000}{1 + 19e^{-0.33106t}}. \qquad \text{Logistic growth model}$$

The population, in five-year intervals, is shown in the table.

Time, t	0	5	10	15	20	25	30
Population, y	100	432	1181	1766	1951	1990	1998

✓ **CHECKPOINT 6**

Write the logistic growth model for the population of deer in Example 6 if the game preserve could contain a limit of 4000 deer.

╭─ **CONCEPT CHECK** ─────────────────────────────────────╮

1. Complete the following: The technique of partial fractions involves the decomposition of a _____ function into the _____ of two or more simple _____ functions.

2. What is a proper rational function?

3. Before applying partial fractions to an improper rational function, what should you do?

4. Describe what the value of L represents in the logistic growth function
   $$y = \frac{L}{1 + be^{-kt}}.$$

╰──╯

Skills Review 12.2

The following warm-up exercises involve skills that were covered in earlier sections. You will use these skills in the exercise set for this section. For additional help, review Sections 0.6 and 3.3.

In Exercises 1–8, factor the expression.

1. $x^2 - 16$ **2.** $x^2 - 25$ **3.** $x^2 - x - 12$ **4.** $x^2 + x - 30$

5. $x^3 - x^2 - 2x$ **6.** $x^3 - 4x^2 + 4x$ **7.** $x^3 - 4x^2 + 5x - 2$ **8.** $x^3 - 5x^2 + 7x - 3$

In Exercises 9–14, rewrite the improper rational expression as the sum of a proper rational expression and a polynomial.

9. $\dfrac{x^2 - 2x + 1}{x - 2}$ **10.** $\dfrac{2x^2 - 4x + 1}{x - 1}$ **11.** $\dfrac{x^3 - 3x^2 + 2}{x - 2}$

12. $\dfrac{x^3 + 2x - 1}{x + 1}$ **13.** $\dfrac{x^3 + 4x^2 + 5x + 2}{x^2 - 1}$ **14.** $\dfrac{x^3 + 3x^2 - 4}{x^2 - 1}$

Exercises 12.2

See www.CalcChat.com for worked-out solutions to odd-numbered exercises.

In Exercises 1–12, write the partial fraction decomposition for the expression.

1. $\dfrac{2(x + 20)}{x^2 - 25}$ **2.** $\dfrac{3x + 11}{x^2 - 2x - 3}$

3. $\dfrac{8x + 3}{x^2 - 3x}$ **4.** $\dfrac{10x + 3}{x^2 + x}$

5. $\dfrac{4x - 13}{x^2 - 3x - 10}$ **6.** $\dfrac{7x + 5}{6(2x^2 + 3x + 1)}$

7. $\dfrac{3x^2 - 2x - 5}{x^3 + x^2}$ **8.** $\dfrac{3x^2 - x + 1}{x(x + 1)^2}$

9. $\dfrac{x + 1}{3(x - 2)^2}$ **10.** $\dfrac{3x - 4}{(x - 5)^2}$

11. $\dfrac{8x^2 + 15x + 9}{(x + 1)^3}$ **12.** $\dfrac{6x^2 - 5x}{(x + 2)^3}$

In Exercises 13–32, use partial fractions to find the indefinite integral.

13. $\displaystyle\int \dfrac{1}{x^2 - 1}\,dx$ **14.** $\displaystyle\int \dfrac{4}{x^2 - 4}\,dx$

15. $\displaystyle\int \dfrac{-2}{x^2 - 16}\,dx$ **16.** $\displaystyle\int \dfrac{-4}{x^2 - 4}\,dx$

17. $\displaystyle\int \dfrac{1}{2x^2 - x}\,dx$ **18.** $\displaystyle\int \dfrac{2}{x^2 - 2x}\,dx$

19. $\displaystyle\int \dfrac{10}{x^2 - 10x}\,dx$ **20.** $\displaystyle\int \dfrac{5}{x^2 + x - 6}\,dx$

21. $\displaystyle\int \dfrac{3}{x^2 + x - 2}\,dx$ **22.** $\displaystyle\int \dfrac{1}{4x^2 - 9}\,dx$

23. $\displaystyle\int \dfrac{5 - x}{2x^2 + x - 1}\,dx$ **24.** $\displaystyle\int \dfrac{x + 1}{x^2 + 4x + 3}\,dx$

25. $\displaystyle\int \dfrac{x^2 - 4x - 4}{x^3 - 4x}\,dx$ **26.** $\displaystyle\int \dfrac{x^2 + 12x + 12}{x^3 - 4x}\,dx$

27. $\displaystyle\int \dfrac{x + 2}{x^2 - 4x}\,dx$ **28.** $\displaystyle\int \dfrac{4x^2 + 2x - 1}{x^3 + x^2}\,dx$

29. $\displaystyle\int \dfrac{2x - 3}{(x - 1)^2}\,dx$ **30.** $\displaystyle\int \dfrac{x^4}{(x - 1)^3}\,dx$

31. $\displaystyle\int \dfrac{3x^2 + 3x + 1}{x(x^2 + 2x + 1)}\,dx$ **32.** $\displaystyle\int \dfrac{3x}{x^2 - 6x + 9}\,dx$

In Exercises 33–40, evaluate the definite integral.

33. $\displaystyle\int_4^5 \dfrac{1}{9 - x^2}\,dx$ **34.** $\displaystyle\int_0^1 \dfrac{3}{2x^2 + 5x + 2}\,dx$

35. $\displaystyle\int_1^5 \dfrac{x - 1}{x^2(x + 1)}\,dx$ **36.** $\displaystyle\int_0^1 \dfrac{x^2 - x}{x^2 + x + 1}\,dx$

37. $\displaystyle\int_0^1 \dfrac{x^3}{x^2 - 2}\,dx$ **38.** $\displaystyle\int_0^1 \dfrac{x^3 - 1}{x^2 - 4}\,dx$

39. $\displaystyle\int_1^2 \dfrac{x^3 - 4x^2 - 3x + 3}{x^2 - 3x}\,dx$ **40.** $\displaystyle\int_2^4 \dfrac{x^4 - 4}{x^2 - 1}\,dx$

In Exercises 41–44, find the area of the shaded region.

41. $y = \dfrac{14}{16 - x^2}$ **42.** $y = \dfrac{-4}{x^2 - x - 6}$

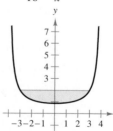

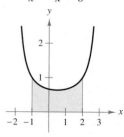

43. $y = \dfrac{x + 1}{x^2 - x}$

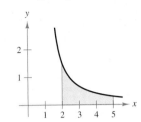

44. $y = \dfrac{x^2 + 2x - 1}{x^2 - 4}$

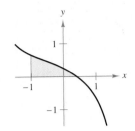

In Exercises 45 and 46, find the area of the region bounded by the graphs of the given equations.

45. $y = \dfrac{12}{x^2 + 5x + 6}, y = 0, x = 0, x = 1$

46. $y = \dfrac{-24}{x^2 - 16}, y = 0, x = 1, x = 3$

In Exercises 47–50, write the partial fraction decomposition for the rational expression. Check your result algebraically. Then assign a value to the constant *a* and use a graphing utility to check the result graphically.

47. $\dfrac{1}{a^2 - x^2}$

48. $\dfrac{1}{x(x + a)}$

49. $\dfrac{1}{x(a - x)}$

50. $\dfrac{1}{(x + 1)(a - x)}$

51. Writing What is the first step when integrating

$\displaystyle \int \dfrac{x^2}{x - 5} \, dx$? Explain. (Do not integrate.)

52. Writing State the method you would use to evaluate each integral. Explain why you chose that method. (Do not integrate.)

(a) $\displaystyle \int \dfrac{2x + 1}{x^2 + x - 8} \, dx$

(b) $\displaystyle \int \dfrac{7x + 4}{x^2 + 2x - 8} \, dx$

53. Biology A conservation organization releases 100 animals of an endangered species into a game preserve. During the first 2 years, the population increases to 134 animals. The organization believes that the preserve has a capacity of 1000 animals and that the herd will grow according to a logistic growth model. That is, the size *y* of the herd will follow the equation

$\displaystyle \int \dfrac{1}{y(1 - y/1000)} \, dy = \int k \, dt$

where *t* is measured in years. Find this logistic curve. (To solve for the constant of integration *C* and the proportionality constant *k*, assume $y = 100$ when $t = 0$ and $y = 134$ when $t = 2$.) Use a graphing utility to graph your solution.

54. Health: Epidemic A single infected individual enters a community of 500 individuals susceptible to the disease. The disease spreads at a rate proportional to the product of the total number infected and the number of susceptible individuals not yet infected. A model for the time it takes for the disease to spread to *x* individuals is

$t = 5010 \displaystyle \int \dfrac{1}{(x + 1)(500 - x)} \, dx$

where *t* is the time in hours.

(a) Find the time it takes for 75% of the population to become infected (when $t = 0, x = 1$).

(b) Find the number of people infected after 100 hours.

55. Marketing After test-marketing a new menu item, a fast-food restaurant predicts that sales of the new item will grow according to the model

$\dfrac{dS}{dt} = \dfrac{2t}{(t + 4)^2}$

where *t* is the time in weeks and *S* is the sales (in thousands of dollars). Find the sales of the menu item at 10 weeks.

56. Biology One gram of a bacterial culture is present at time $t = 0$, and 10 grams is the upper limit of the culture's weight. The time required for the culture to grow to *y* grams is modeled by

$kt = \displaystyle \int \dfrac{1}{y(1 - y/10)} \, dy$

where *y* is the weight of the culture (in grams) and *t* is the time in hours.

(a) Verify that the weight of the culture at time *t* is modeled by

$y = \dfrac{10}{1 + 9e^{-kt}}.$

Use the fact that $y = 1$ when $t = 0$.

(b) Use the graph to determine the constant *k*.

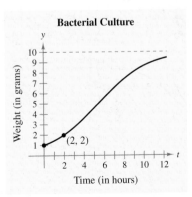

Bacterial Culture

Weight (in grams)

(2, 2)

Time (in hours)

57. Revenue The revenue R (in millions of dollars per year) for Symantec Corporation from 1997 through 2005 can be modeled by

$$R = \frac{1340t^2 + 24{,}044t + 22{,}704}{-6t^2 + 94t + 100}$$

where $t = 7$ corresponds to 1997. Find the total revenue from 1997 through 2005. Then find the average revenue during this time period. *(Source: Symantec Corporation)*

58. Environment The predicted cost C (in hundreds of thousands of dollars) for a company to remove $p\%$ of a chemical from its waste water is shown in the table.

p	0	10	20	30	40
C	0	0.7	1.0	1.3	1.7

p	50	60	70	80	90
C	2.0	2.7	3.6	5.5	11.2

A model for the data is given by

$$C = \frac{124p}{(10 + p)(100 - p)}, \quad 0 \le p < 100.$$

Use the model to find the average cost for removing between 75% and 80% of the chemical.

B 59. Biology: Population Growth The graph shows the logistic growth curves for two species of the single-celled Paramecium in a laboratory culture. During which time intervals is the rate of growth of each species increasing? During which time intervals is the rate of growth of each species decreasing? Which species has a higher limiting population under these conditions? *(Source: Adapted from Levine/Miller, Biology: Discovering Life, Second Edition)*

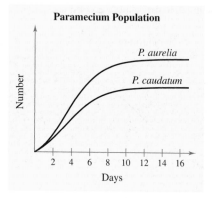

Paramecium Population

60. Population Growth The population of the United States was 76 million people in 1900 and reached 300 million people in 2006. From 1900 through 2006, assume the population of the United States can be modeled by logistic growth with a limit of 839.1 million people. *(Source: U.S. Census Bureau)*

(a) Write a differential equation of the form

$$\frac{dy}{dt} = ky\left(1 - \frac{y}{L}\right)$$

where y represents the population of the United States (in millions of people) and t represents the number of years since 1900.

(b) Find the logistic growth model $y = \dfrac{L}{1 + be^{-kt}}$ for this population.

T (c) Use a graphing utility to graph the model from part (b). Then estimate the year in which the population of the United States will reach 400 million people.

Business Capsule

Photo courtesy of Susie Wang and Ric Kostick

Susie Wang and Ric Kostick graduated from the University of California at Berkeley with degrees in mathematics. In 1999, Wang used $10,000 to start Aqua Dessa Spa Therapy, a high-end cosmetics company that uses natural ingredients in its products. Now, the company run by Wang and Kostick has annual sales of over $10 million, operates under several brand names, including 100% Pure, and has a global customer base. Wang and Kostick attribute the success of their business to applying what they learned from their studies.

61. Research Project Use your school's library, the Internet, or some other reference source to research the opportunity cost of attending graduate school for 2 years to receive a Masters of Business Administration (MBA) degree rather than working for 2 years with a bachelor's degree. Write a short paper describing these costs.

Integration Tables

- Use integration tables to find indefinite integrals.
- Use reduction formulas to find indefinite integrals.

Integration Tables

You have studied several integration techniques that can be used with the basic integration formulas. Certainly these techniques and formulas do not cover every possible method for finding an antiderivative, but they do cover most of the important ones.

In this section, you will expand the list of integration formulas to form a table of integrals. As you add new integration formulas to the basic list, two effects occur. On one hand, it becomes increasingly difficult to memorize, or even become familiar with, the entire list of formulas. On the other hand, with a longer list you need fewer techniques for fitting an integral to one of the formulas on the list. The procedure of integrating by means of a long list of formulas is called **integration by tables.** (The table in this section constitutes only a partial listing of integration formulas. Much longer lists exist, some of which contain several hundred formulas.)

Integration by tables should not be considered a trivial task. It requires considerable thought and insight, and it often requires substitution. Many people find a table of integrals to be a valuable supplement to the integration techniques discussed in this text. We encourage you to gain competence in the use of integration tables, as well as to continue to improve in the use of the various integration techniques. In doing so, you should find that a combination of techniques and tables is the most versatile approach to integration.

Each integration formula in the table on the next three pages can be developed using one or more of the techniques you have studied. You should try to verify several of the formulas. For instance, Formula 4

$$\int \frac{u}{(a + bu)^2}\, du = \frac{1}{b^2}\left(\frac{a}{a + bu} + \ln|a + bu|\right) + C \qquad \text{Formula 4}$$

can be verified using partial fractions, Formula 17

$$\int \frac{\sqrt{a + bu}}{u}\, du = 2\sqrt{a + bu} + a\int \frac{1}{u\sqrt{a + bu}}\, du \qquad \text{Formula 17}$$

can be verified using integration by parts, and Formula 37

$$\int \frac{1}{1 + e^u}\, du = u - \ln(1 + e^u) + C \qquad \text{Formula 37}$$

can be verified using substitution.

STUDY TIP

A symbolic integration utility consists, in part, of a database of integration tables. The primary difference between using a symbolic integration utility and using a table of integrals is that with a symbolic integration utility the computer searches through the database to find a fit. With a table of integrals, *you* must do the searching.

In the table of integrals below and on the next two pages, the formulas have been grouped into eight different types according to the form of the integrand.

Forms involving u^n

Forms involving $a + bu$

Forms involving $\sqrt{a + bu}$

Forms involving $\sqrt{u^2 \pm a^2}$

Forms involving $u^2 - a^2$

Forms involving $\sqrt{a^2 - u^2}$

Forms involving e^u

Forms involving $\ln u$

Table of Integrals

Forms involving u^n

1. $\displaystyle \int u^n \, du = \frac{u^{n+1}}{n+1} + C, \quad n \neq -1$

2. $\displaystyle \int \frac{1}{u} \, du = \ln|u| + C$

Forms involving $a + bu$

3. $\displaystyle \int \frac{u}{a + bu} \, du = \frac{1}{b^2}(bu - a \ln|a + bu|) + C$

4. $\displaystyle \int \frac{u}{(a + bu)^2} \, du = \frac{1}{b^2}\left(\frac{a}{a + bu} + \ln|a + bu|\right) + C$

5. $\displaystyle \int \frac{u}{(a + bu)^n} \, du = \frac{1}{b^2}\left[\frac{-1}{(n-2)(a + bu)^{n-2}} + \frac{a}{(n-1)(a + bu)^{n-1}}\right] + C, \quad n \neq 1, 2$

6. $\displaystyle \int \frac{u^2}{a + bu} \, du = \frac{1}{b^3}\left[-\frac{bu}{2}(2a - bu) + a^2 \ln|a + bu|\right] + C$

7. $\displaystyle \int \frac{u^2}{(a + bu)^2} \, du = \frac{1}{b^3}\left(bu - \frac{a^2}{a + bu} - 2a \ln|a + bu|\right) + C$

8. $\displaystyle \int \frac{u^2}{(a + bu)^3} \, du = \frac{1}{b^3}\left[\frac{2a}{a + bu} - \frac{a^2}{2(a + bu)^2} + \ln|a + bu|\right] + C$

9. $\displaystyle \int \frac{u^2}{(a + bu)^n} \, du = \frac{1}{b^3}\left[\frac{-1}{(n-3)(a + bu)^{n-3}} + \frac{2a}{(n-2)(a + bu)^{n-2}} - \frac{a^2}{(n-1)(a + bu)^{n-1}}\right] + C, \quad n \neq 1, 2, 3$

10. $\displaystyle \int \frac{1}{u(a + bu)} \, du = \frac{1}{a} \ln\left|\frac{u}{a + bu}\right| + C$

11. $\displaystyle \int \frac{1}{u(a + bu)^2} \, du = \frac{1}{a}\left(\frac{1}{a + bu} + \frac{1}{a} \ln\left|\frac{u}{a + bu}\right|\right) + C$

12. $\displaystyle \int \frac{1}{u^2(a + bu)} \, du = -\frac{1}{a}\left(\frac{1}{u} + \frac{b}{a} \ln\left|\frac{u}{a + bu}\right|\right) + C$

13. $\displaystyle \int \frac{1}{u^2(a + bu)^2} \, du = -\frac{1}{a^2}\left[\frac{a + 2bu}{u(a + bu)} + \frac{2b}{a} \ln\left|\frac{u}{a + bu}\right|\right] + C$

Table of Integrals (*continued*)

Forms involving $\sqrt{a + bu}$

14. $\displaystyle \int u^n \sqrt{a + bu}\, du = \frac{2}{b(2n + 3)}\left[u^n(a + bu)^{3/2} - na \int u^{n-1} \sqrt{a + bu}\, du \right]$

15. $\displaystyle \int \frac{1}{u\sqrt{a + bu}}\, du = \frac{1}{\sqrt{a}} \ln\left| \frac{\sqrt{a + bu} - \sqrt{a}}{\sqrt{a + bu} + \sqrt{a}} \right| + C, \quad a > 0$

16. $\displaystyle \int \frac{1}{u^n \sqrt{a + bu}}\, du = \frac{-1}{a(n-1)}\left[\frac{\sqrt{a + bu}}{u^{n-1}} + \frac{(2n - 3)b}{2} \int \frac{1}{u^{n-1}\sqrt{a + bu}}\, du \right], \quad n \neq 1$

17. $\displaystyle \int \frac{\sqrt{a + bu}}{u}\, du = 2\sqrt{a + bu} + a \int \frac{1}{u\sqrt{a + bu}}\, du$

18. $\displaystyle \int \frac{\sqrt{a + bu}}{u^n}\, du = \frac{-1}{a(n-1)}\left[\frac{(a + bu)^{3/2}}{u^{n-1}} + \frac{(2n - 5)b}{2} \int \frac{\sqrt{a + bu}}{u^{n-1}}\, du \right], \quad n \neq 1$

19. $\displaystyle \int \frac{u}{\sqrt{a + bu}}\, du = -\frac{2(2a - bu)}{3b^2}\sqrt{a + bu} + C$

20. $\displaystyle \int \frac{u^n}{\sqrt{a + bu}}\, du = \frac{2}{(2n + 1)b}\left(u^n \sqrt{a + bu} - na \int \frac{u^{n-1}}{\sqrt{a + bu}}\, du \right)$

Forms involving $\sqrt{u^2 \pm a^2}, \quad a > 0$

21. $\displaystyle \int \sqrt{u^2 \pm a^2}\, du = \frac{1}{2}\left(u\sqrt{u^2 \pm a^2} \pm a^2 \ln\left| u + \sqrt{u^2 \pm a^2}\right| \right) + C$

22. $\displaystyle \int u^2 \sqrt{u^2 \pm a^2}\, du = \frac{1}{8}\left[u(2u^2 \pm a^2)\sqrt{u^2 \pm a^2} - a^4 \ln\left| u + \sqrt{u^2 \pm a^2}\right| \right] + C$

23. $\displaystyle \int \frac{\sqrt{u^2 + a^2}}{u}\, du = \sqrt{u^2 + a^2} - a \ln\left| \frac{a + \sqrt{u^2 + a^2}}{u} \right| + C$

24. $\displaystyle \int \frac{\sqrt{u^2 \pm a^2}}{u^2}\, du = \frac{-\sqrt{u^2 \pm a^2}}{u} + \ln\left| u + \sqrt{u^2 \pm a^2}\right| + C$

25. $\displaystyle \int \frac{1}{\sqrt{u^2 \pm a^2}}\, du = \ln\left| u + \sqrt{u^2 \pm a^2}\right| + C$

26. $\displaystyle \int \frac{1}{u\sqrt{u^2 + a^2}}\, du = -\frac{1}{a}\ln\left| \frac{a + \sqrt{u^2 + a^2}}{u} \right| + C$

27. $\displaystyle \int \frac{u^2}{\sqrt{u^2 \pm a^2}}\, du = \frac{1}{2}\left(u\sqrt{u^2 \pm a^2} \mp a^2 \ln\left| u + \sqrt{u^2 \pm a^2}\right| \right) + C$

28. $\displaystyle \int \frac{1}{u^2 \sqrt{u^2 \pm a^2}}\, du = \mp \frac{\sqrt{u^2 \pm a^2}}{a^2 u} + C$

Table of Integrals (*continued*)

Forms involving $u^2 - a^2$, $a > 0$

29. $\int \dfrac{1}{u^2 - a^2}\, du = -\int \dfrac{1}{a^2 - u^2}\, du = \dfrac{1}{2a} \ln\left|\dfrac{u - a}{u + a}\right| + C$

30. $\int \dfrac{1}{(u^2 - a^2)^n}\, du = \dfrac{-1}{2a^2(n - 1)}\left[\dfrac{u}{(u^2 - a^2)^{n-1}} + (2n - 3)\int \dfrac{1}{(u^2 - a^2)^{n-1}}\, du\right],\quad n \neq 1$

Forms involving $\sqrt{a^2 - u^2}$, $a > 0$

31. $\int \dfrac{\sqrt{a^2 - u^2}}{u}\, du = \sqrt{a^2 - u^2} - a \ln\left|\dfrac{a + \sqrt{a^2 - u^2}}{u}\right| + C$

32. $\int \dfrac{1}{u\sqrt{a^2 - u^2}}\, du = -\dfrac{1}{a} \ln\left|\dfrac{a + \sqrt{a^2 - u^2}}{u}\right| + C$

33. $\int \dfrac{1}{u^2 \sqrt{a^2 - u^2}}\, du = \dfrac{-\sqrt{a^2 - u^2}}{a^2 u} + C$

Forms involving e^u

34. $\int e^u\, du = e^u + C$

35. $\int u e^u\, du = (u - 1)e^u + C$

36. $\int u^n e^u\, du = u^n e^u - n\int u^{n-1} e^u\, du$

37. $\int \dfrac{1}{1 + e^u}\, du = u - \ln(1 + e^u) + C$

38. $\int \dfrac{1}{1 + e^{nu}}\, du = u - \dfrac{1}{n} \ln(1 + e^{nu}) + C$

Forms involving $\ln u$

39. $\int \ln u\, du = u(-1 + \ln u) + C$

40. $\int u \ln u\, du = \dfrac{u^2}{4}(-1 + 2\ln u) + C$

41. $\int u^n \ln u\, du = \dfrac{u^{n+1}}{(n + 1)^2}[-1 + (n + 1)\ln u] + C,\quad n \neq -1$

42. $\int (\ln u)^2\, du = u[2 - 2\ln u + (\ln u)^2] + C$

43. $\int (\ln u)^n\, du = u(\ln u)^n - n\int (\ln u)^{n-1}\, du$

Example 1 **Using Integration Tables**

Find $\displaystyle\int \frac{x}{\sqrt{x-1}}\,dx$.

SOLUTION Because the expression inside the radical is linear, you should consider forms involving $\sqrt{a+bu}$, as in Formula 19.

$$\int \frac{u}{\sqrt{a+bu}}\,du = -\frac{2(2a-bu)}{3b^2}\sqrt{a+bu}+C \qquad \text{Formula 19}$$

Using this formula, let $a=-1$, $b=1$, and $u=x$. Then $du=dx$, and you obtain

$$\int \frac{x}{\sqrt{x-1}}\,dx = -\frac{2(-2-x)}{3}\sqrt{x-1}+C \qquad \begin{array}{l}\text{Substitute values of}\\ a,\ b,\ \text{and}\ u.\end{array}$$

$$= \frac{2}{3}(2+x)\sqrt{x-1}+C. \qquad \text{Simplify.}$$

✓ CHECKPOINT 1

Use the integration table to find $\displaystyle\int \frac{x}{\sqrt{2+x}}\,dx$. ■

Example 2 **Using Integration Tables**

Find $\displaystyle\int x\sqrt{x^4-9}\,dx$.

SOLUTION Because it is not clear which formula to use, you can begin by letting $u=x^2$ and $du=2x\,dx$. With these substitutions, you can write the integral as shown.

$$\int x\sqrt{x^4-9}\,dx = \frac{1}{2}\int \sqrt{(x^2)^2-9}\,(2x)\,dx \qquad \text{Multiply and divide by 2.}$$

$$= \frac{1}{2}\int \sqrt{u^2-9}\,du \qquad \text{Substitute } u \text{ and } du.$$

Now, it appears that you can use Formula 21.

$$\int \sqrt{u^2-a^2}\,du = \frac{1}{2}\left(u\sqrt{u^2-a^2}-a^2\ln\left|u+\sqrt{u^2-a^2}\right|\right)+C$$

Letting $a=3$, you obtain

$$\int x\sqrt{x^4-9}\,dx = \frac{1}{2}\int \sqrt{u^2-a^2}\,du$$

$$= \frac{1}{2}\left[\frac{1}{2}\left(u\sqrt{u^2-a^2}-a^2\ln\left|u+\sqrt{u^2-a^2}\right|\right)\right]+C$$

$$= \frac{1}{4}\left(x^2\sqrt{x^4-9}-9\ln\left|x^2+\sqrt{x^4-9}\right|\right)+C.$$

✓ CHECKPOINT 2

Use the integration table to find

$$\int \frac{\sqrt{x^2+16}}{x}\,dx.$$ ■

Using Integration Tables

Find $\displaystyle\int \frac{1}{x\sqrt{x+1}}\,dx$.

SOLUTION Considering forms involving $\sqrt{a+bu}$, where $a = 1$, $b = 1$, and $u = x$, you can use Formula 15.

$$\int \frac{1}{u\sqrt{a+bu}}\,du = \frac{1}{\sqrt{a}}\ln\left|\frac{\sqrt{a+bu}-\sqrt{a}}{\sqrt{a+bu}+\sqrt{a}}\right| + C, \quad a > 0$$

So,

$$\int \frac{1}{x\sqrt{x+1}}\,dx = \int \frac{1}{u\sqrt{a+bu}}\,du = \frac{1}{\sqrt{a}}\ln\left|\frac{\sqrt{a+bu}-\sqrt{a}}{\sqrt{a+bu}+\sqrt{a}}\right| + C$$

$$= \ln\left|\frac{\sqrt{x+1}-1}{\sqrt{x+1}+1}\right| + C.$$

✓ **CHECKPOINT 3**

Use the integration table to find

$$\int \frac{1}{x^2-4}\,dx.$$ ▪

Example 4 **Using Integration Tables**

Evaluate $\displaystyle\int_0^2 \frac{x}{1+e^{-x^2}}\,dx$.

SOLUTION Of the forms involving e^u, Formula 37

$$\int \frac{1}{1+e^u}\,du = u - \ln(1+e^u) + C$$

seems most appropriate. To use this formula, let $u = -x^2$ and $du = -2x\,dx$.

$$\int \frac{x}{1+e^{-x^2}}\,dx = -\frac{1}{2}\int \frac{1}{1+e^{-x^2}}(-2x)\,dx = -\frac{1}{2}\int \frac{1}{1+e^u}\,du$$

$$= -\frac{1}{2}[u - \ln(1+e^u)] + C$$

$$= -\frac{1}{2}[-x^2 - \ln(1+e^{-x^2})] + C$$

$$= \frac{1}{2}[x^2 + \ln(1+e^{-x^2})] + C$$

So, the value of the definite integral is

$$\int_0^2 \frac{x}{1+e^{-x^2}}\,dx = \frac{1}{2}\left[x^2 + \ln(1+e^{-x^2})\right]_0^2 \approx 1.66$$

as shown in Figure 12.7.

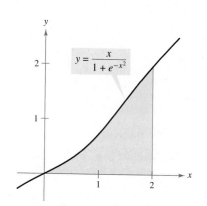

$$y = \frac{x}{1+e^{-x^2}}$$

FIGURE 12.7

✓ **CHECKPOINT 4**

Use the integration table to evaluate $\displaystyle\int_0^1 \frac{x^2}{1+e^{x^3}}\,dx$. ▪

Reduction Formulas

Several of the formulas in the integration table have the form

$$\int f(x)\, dx = g(x) + \int h(x)\, dx$$

where the right side contains an integral. Such integration formulas are called **reduction formulas** because they reduce the original integral to the sum of a function and a simpler integral.

Algebra Review

For help on the algebra in Example 5, see Example 2(b) in the *Chapter 12 Algebra Review*, on page 923.

Example 5   Using a Reduction Formula

Find $\int x^2 e^x\, dx$.

SOLUTION   Using Formula 36

$$\int u^n e^u\, du = u^n e^u - n \int u^{n-1} e^u\, du$$

you can let $u = x$ and $n = 2$. Then $du = dx$, and you can write

$$\int x^2 e^x\, dx = x^2 e^x - 2 \int x e^x\, dx.$$

Then, using Formula 35

$$\int u e^u\, du = (u - 1)e^u + C$$

you can write

$$\begin{aligned}
\int x^2 e^x\, dx &= x^2 e^x - 2 \int x e^x\, dx \\
&= x^2 e^x - 2(x - 1)e^x + C \\
&= x^2 e^x - 2x e^x + 2e^x + C \\
&= e^x(x^2 - 2x + 2) + C.
\end{aligned}$$

✓ **CHECKPOINT 5**

Use the integration table to find the indefinite integral $\int (\ln x)^2\, dx$.   ■

TECHNOLOGY

(T)   You have now studied two ways to find the indefinite integral in Example 5. Example 5 uses an integration table, and Example 4 in Section 12.1 uses integration by parts. A third way would be to use a symbolic integration utility.

Application

Researchers such as psychologists use definite integrals to represent the probability that an event will occur. For instance, a probability of 0.5 means that an event will occur about 50% of the time.

Integration can be used to find the probability that an event will occur. In such an application, the real-life situation is modeled by a *probability density function f*, and the probability that x will lie between a and b is represented by

$$P(a \leq x \leq b) = \int_a^b f(x)\, dx.$$

The probability $P(a \leq x \leq b)$ must be a number between 0 and 1.

Example 6 Finding a Probability

A psychologist finds that the probability that a participant in a memory experiment will recall between a and b percent (in decimal form) of the material is

$$P(a \leq x \leq b) = \int_a^b \frac{1}{e-2}\, x^2 e^x\, dx, \quad 0 \leq a \leq b \leq 1.$$

Find the probability that a randomly chosen participant will recall between 0% and 87.5% of the material.

SOLUTION   You can use the Constant Multiple Rule to rewrite the integral as

$$\frac{1}{e-2}\int_a^b x^2 e^x\, dx.$$

Note that the integrand is the same as the one in Example 5. Use the result of Example 5 to find the probability with $a = 0$ and $b = 0.875$.

$$\frac{1}{e-2}\int_0^{0.875} x^2 e^x\, dx = \frac{1}{e-2}\left[e^x \left(x^2 - 2x + 2 \right) \right]_0^{0.875} \approx 0.608$$

So, the probability is about 60.8%, as indicated in Figure 12.8. ———

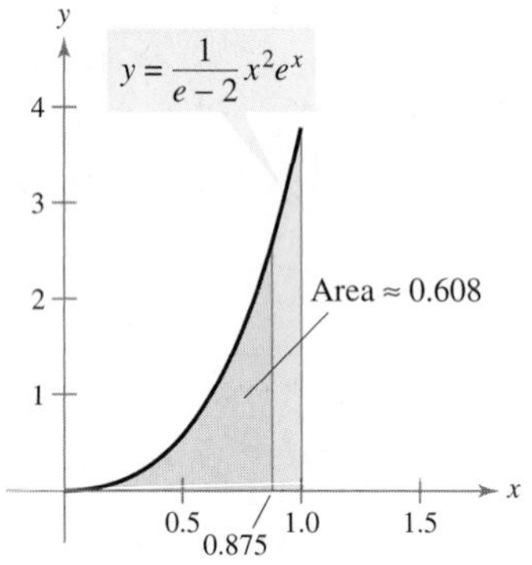

The graph shows $y = \dfrac{1}{e-2} x^2 e^x$ with shaded region labeled Area ≈ 0.608 between $x = 0$ and $x = 0.875$.

FIGURE 12.8

✓CHECKPOINT 6

Use Example 6 to find the probability that a participant will recall between 0% and 62.5% of the material. ▪

CONCEPT CHECK

1. Which integration formula would you use to find $\displaystyle\int \frac{1}{e^x + 1}\, dx$? (Do not integrate.)

2. Which integration formula would you use to find $\displaystyle\int \sqrt{x^2 + 4}\, dx$? (Do not integrate.)

3. True or false: When using a table of integrals, you may have to make substitutions to rewrite your integral in the form in which it appears in the table.

4. Describe what is meant by a reduction formula. Give an example.

Skills Review 12.3

The following warm-up exercises involve skills that were covered in earlier sections. You will use these skills in the exercise set for this section. For additional help, review Sections 0.5, 12.1, and 12.2.

In Exercises 1–4, expand the expression.

1. $(x + 5)^2$

2. $(x - 1)^2$

3. $\left(x + \frac{1}{2}\right)^2$

4. $\left(x - \frac{1}{3}\right)^2$

In Exercises 5–8, write the partial fraction decomposition for the expression.

5. $\dfrac{4}{x(x + 2)}$

6. $\dfrac{3}{x(x - 4)}$

7. $\dfrac{x + 4}{x^2(x - 2)}$

8. $\dfrac{3x^2 + 4x - 8}{x(x - 2)(x + 1)}$

In Exercises 9 and 10, use integration by parts to find the indefinite integral.

9. $\displaystyle \int 2xe^x \, dx$

10. $\displaystyle \int 3x^2 \ln x \, dx$

Exercises 12.3

See www.CalcChat.com for worked-out solutions to odd-numbered exercises.

In Exercises 1–8, use the indicated formula from the table of integrals in this section to find the indefinite integral.

1. $\displaystyle \int \frac{x}{(2 + 3x)^2} \, dx$, Formula 4

2. $\displaystyle \int \frac{1}{x(2 + 3x)^2} \, dx$, Formula 11

3. $\displaystyle \int \frac{x}{\sqrt{2 + 3x}} \, dx$, Formula 19

4. $\displaystyle \int \frac{4}{x^2 - 9} \, dx$, Formula 29

5. $\displaystyle \int \frac{2x}{\sqrt{x^4 - 9}} \, dx$, Formula 25

6. $\displaystyle \int x^2 \sqrt{x^2 + 9} \, dx$, Formula 22

7. $\displaystyle \int x^3 e^{x^2} \, dx$, Formula 35

8. $\displaystyle \int \frac{x}{1 + e^{x^2}} \, dx$, Formula 37

In Exercises 9–36, use the table of integrals in this section to find the indefinite integral.

9. $\displaystyle \int \frac{1}{x(1 + x)} \, dx$

10. $\displaystyle \int \frac{1}{x(1 + x)^2} \, dx$

11. $\displaystyle \int \frac{1}{x\sqrt{x^2 + 9}} \, dx$

12. $\displaystyle \int \frac{1}{\sqrt{x^2 - 1}} \, dx$

13. $\displaystyle \int \frac{1}{x\sqrt{4 - x^2}} \, dx$

14. $\displaystyle \int \frac{\sqrt{x^2 - 9}}{x^2} \, dx$

15. $\displaystyle \int x \ln x \, dx$

16. $\displaystyle \int (\ln 5x)^2 \, dx$

17. $\displaystyle \int \frac{6x}{1 + e^{3x^2}} \, dx$

18. $\displaystyle \int \frac{1}{1 + e^x} \, dx$

19. $\displaystyle \int x\sqrt{x^4 - 4} \, dx$

20. $\displaystyle \int \frac{x}{x^4 - 9} \, dx$

21. $\displaystyle \int \frac{t^2}{(2 + 3t)^3} \, dt$

22. $\displaystyle \int \frac{\sqrt{3 + 4t}}{t} \, dt$

23. $\displaystyle \int \frac{s}{s^2\sqrt{3 + s}} \, ds$

24. $\displaystyle \int \sqrt{3 + x^2} \, dx$

25. $\displaystyle \int \frac{x^2}{1 + x} \, dx$

26. $\displaystyle \int \frac{1}{1 + e^{2x}} \, dx$

27. $\displaystyle \int \frac{x^2}{(3 + 2x)^5} \, dx$

28. $\displaystyle \int \frac{1}{x^2\sqrt{x^2 - 4}} \, dx$

29. $\displaystyle \int \frac{1}{x^2\sqrt{1 - x^2}} \, dx$

30. $\displaystyle \int \frac{2x}{(1 - 3x)^2} \, dx$

31. $\displaystyle \int x^2 \ln x \, dx$

32. $\displaystyle \int xe^{x^2} \, dx$

33. $\displaystyle \int \frac{x^2}{(3x - 5)^2} \, dx$

34. $\displaystyle \int \frac{1}{2x^2(2x - 1)^2} \, dx$

35. $\displaystyle \int \frac{\ln x}{x(4 + 3 \ln x)} \, dx$

36. $\displaystyle \int (\ln x)^3 \, dx$

(T) In Exercises 37–42, use the table of integrals to find the exact area of the region bounded by the graphs of the equations. Then use a graphing utility to graph the region and approximate the area.

37. $y = \dfrac{x}{\sqrt{x+1}}$, $y = 0$, $x = 8$

38. $y = \dfrac{2}{1 + e^{4x}}$, $y = 0$, $x = 0$, $x = 1$

39. $y = \dfrac{x}{1 + e^{x^2}}$, $y = 0$, $x = 2$

40. $y = \dfrac{-e^x}{1 - e^{2x}}$, $y = 0$, $x = 1$, $x = 2$

41. $y = x^2\sqrt{x^2 + 4}$, $y = 0$, $x = \sqrt{5}$

42. $y = x \ln x^2$, $y = 0$, $x = 4$

In Exercises 43–50, evaluate the definite integral.

43. $\displaystyle\int_0^1 \dfrac{x}{\sqrt{1+x}}\,dx$

44. $\displaystyle\int_0^5 \dfrac{x}{\sqrt{5+2x}}\,dx$

45. $\displaystyle\int_0^5 \dfrac{x}{(4+x)^2}\,dx$

46. $\displaystyle\int_2^4 \dfrac{x^2}{(3x-5)}\,dx$

47. $\displaystyle\int_0^4 \dfrac{6}{1 + e^{0.5x}}\,dx$

48. $\displaystyle\int_2^4 \sqrt{3 + x^2}\,dx$

49. $\displaystyle\int_1^4 x \ln x\,dx$

50. $\displaystyle\int_1^3 x^2 \ln x\,dx$

In Exercises 51–54, find the indefinite integral (a) using the integration table and (b) using the specified method.

Integral	Method
51. $\displaystyle\int x^2 e^x\,dx$	Integration by parts
52. $\displaystyle\int x^4 \ln x\,dx$	Integration by parts
53. $\displaystyle\int \dfrac{1}{x^2(x+1)}\,dx$	Partial fractions
54. $\displaystyle\int \dfrac{1}{x^2 - 75}\,dx$	Partial fractions

55. Probability   The probability of recall in an experiment is modeled by

$$P(a \le x \le b) = \int_a^b \dfrac{75}{14}\left(\dfrac{x}{\sqrt{4+5x}}\right)dx, \quad 0 \le a \le b \le 1$$

where x is the percent of recall (see figure).

(a) What is the probability of recalling between 40% and 80%?

(b) What is the probability of recalling between 0% and 50%?

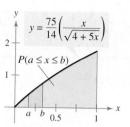

Figure for 55

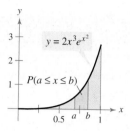

Figure for 56

56. Probability   The probability of finding between a and b percent iron in ore samples is modeled by

$$P(a \le x \le b) = \int_a^b 2x^3 e^{x^2}\,dx, \quad 0 \le a \le b \le 1$$

(see figure). Find the probabilities that a sample will contain between (a) 0% and 25% and (b) 50% and 100% iron.

(T) **Population Growth**   In Exercises 57 and 58, use a graphing utility to graph the growth function. Use the table of integrals to find the average value of the growth function over the interval, where N is the size of a population and t is the time in days.

57. $N = \dfrac{5000}{1 + e^{4.8 - 1.9t}}$, $[0, 2]$

58. $N = \dfrac{375}{1 + e^{4.20 - 0.25t}}$, $[21, 28]$

59. Revenue   The revenue (in dollars per year) for a new product is modeled by

$$R = 10{,}000\left[1 - \dfrac{1}{(1 + 0.1t^2)^{1/2}}\right]$$

where t is the time in years. Estimate the total revenue from sales of the product over its first 2 years on the market.

60. Consumer and Producer Surpluses   Find the consumer surplus and the producer surplus for a product with the given demand and supply functions.

$$\text{Demand: } p = \dfrac{60}{\sqrt{x^2 + 81}}, \quad \text{Supply: } p = \dfrac{x}{3}$$

61. Profit   The net profits P (in billions of dollars per year) for The Hershey Company from 2002 through 2005 can be modeled by

$$P = \sqrt{0.00645t^2 + 0.1673}, \quad 2 \le t \le 5$$

where t is time in years, with $t = 2$ corresponding to 2002. Find the average net profit over that time period.   (*Source: The Hershey Co.*)

62. Extended Application   To work an extended application analyzing the purchasing power of the dollar from 1983 through 2005, visit this text's website at *college.hmco.com*. (*Data Source: U.S. Bureau of Labor Statistics*)

See www.CalcChat.com for worked-out solutions to odd-numbered exercises.

Take this quiz as you would take a quiz in class. When you are done, check your work against the answers given in the back of the book.

In Exercises 1–6, use integration by parts to find the indefinite integral.

1. $\displaystyle\int xe^{5x}\,dx$

2. $\displaystyle\int \ln x^3\,dx$

3. $\displaystyle\int (x+1)\ln x\,dx$

4. $\displaystyle\int x\sqrt{x+3}\,dx$

5. $\displaystyle\int x\ln\sqrt{x}\,dx$

6. $\displaystyle\int x^2 e^{-2x}\,dx$

7. A small business expects its income during the next 7 years to be given by

$$c(t) = 32{,}000t, \quad 0 \le t \le 7.$$

Assuming an annual inflation rate of 3.3%, can the business claim that the present value of its income during the next 7 years is at least $650,000?

In Exercises 8–10, use partial fractions to find the indefinite integral.

8. $\displaystyle\int \frac{10}{x^2-25}\,dx$

9. $\displaystyle\int \frac{x-14}{x^2+2x-8}\,dx$

10. $\displaystyle\int \frac{5x-1}{(x+1)^2}\,dx$

11. The population of a colony of bees can be modeled by logistic growth. The capacity of the colony's hive is 100,000 bees. One day in the early spring, there are 25,000 bees in the hive. Thirteen days later, the population of the hive increases to 28,000 bees. Write a logistic growth model for the colony.

In Exercises 12–17, use the table of integrals in Section 12.3 to find the indefinite integral.

12. $\displaystyle\int \frac{x}{1+2x}\,dx$

13. $\displaystyle\int \frac{1}{x(0.1+0.2x)}\,dx$

14. $\displaystyle\int \frac{\sqrt{x^2-16}}{x^2}\,dx$

15. $\displaystyle\int \frac{1}{x\sqrt{4+9x}}\,dx$

16. $\displaystyle\int \frac{2x}{1+e^{4x^2}}\,dx$

17. $\displaystyle\int 2x(x^2+1)e^{x^2+1}\,dx$

18. The number of Kohl's Corporation stores in the United States from 1999 through 2006 can be modeled by

$$N(t) = 75.0 + 1.07t^2 \ln t, \quad 9 \le t \le 16$$

where t is the year, with $t = 9$ corresponding to 1999. Find the average number of Kohl's stores in the U.S. from 1999 through 2006.   *(Source: Kohl's Corporation)*

In Exercises 19–24, evaluate the definite integral.

19. $\displaystyle\int_{-2}^{0} xe^{x/2}\,dx$

20. $\displaystyle\int_{1}^{e} (\ln x)^2\,dx$

21. $\displaystyle\int_{1}^{4} \frac{3x+1}{x(x+1)}\,dx$

22. $\displaystyle\int_{4}^{5} \frac{120}{(x-3)(x+5)}\,dx$

23. $\displaystyle\int_{2}^{3} \frac{1}{x^2\sqrt{9-x^2}}\,dx$

24. $\displaystyle\int_{4}^{6} \frac{2x}{x^4-4}\,dx$

Numerical Integration

■ Use the Trapezoidal Rule to approximate definite integrals.
■ Use Simpson's Rule to approximate definite integrals.
■ Analyze the sizes of the errors when approximating definite integrals with the Trapezoidal Rule and Simpson's Rule.

Trapezoidal Rule

In Section 11.6, you studied one technique for approximating the value of a *definite* integral—the Midpoint Rule. In this section, you will study two other approximation techniques: the **Trapezoidal Rule** and **Simpson's Rule.**

To develop the Trapezoidal Rule, consider a function f that is nonnegative and continuous on the closed interval $[a, b]$. To approximate the area represented by $\int_a^b f(x)dx$, partition the interval into n subintervals, each of width

$$\Delta x = \frac{b - a}{n}. \qquad \text{Width of each subinterval}$$

Next, form n trapezoids, as shown in Figure 12.9. As you can see in Figure 12.10, the area of the first trapezoid is

$$\text{Area of first trapezoid} = \left(\frac{b - a}{n}\right)\left[\frac{f(x_0) + f(x_1)}{2}\right].$$

The areas of the other trapezoids follow a similar pattern, and the sum of the n areas is

$$\left(\frac{b - a}{n}\right)\left[\frac{f(x_0) + f(x_1)}{2} + \frac{f(x_1) + f(x_2)}{2} + \cdots + \frac{f(x_{n-1}) + f(x_n)}{2}\right]$$

$$= \left(\frac{b - a}{2n}\right)[f(x_0) + f(x_1) + f(x_1) + f(x_2) + \cdots + f(x_{n-1}) + f(x_n)]$$

$$= \left(\frac{b - a}{2n}\right)[f(x_0) + 2f(x_1) + 2f(x_2) + \cdots + 2f(x_{n-1}) + f(x_n)].$$

Although this development assumes f to be continuous *and* nonnegative on $[a, b]$, the resulting formula is valid as long as f is continuous on $[a, b]$.

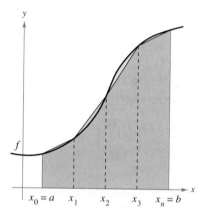

FIGURE 12.9 The area of the region can be approximated using four trapezoids.

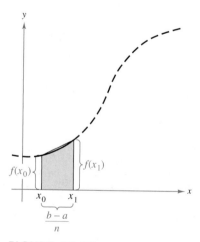

FIGURE 12.10

The Trapezoidal Rule

If f is continuous on $[a, b]$, then

$$\int_a^b f(x)dx \approx \left(\frac{b - a}{2n}\right)[f(x_0) + 2f(x_1) + \cdots + 2f(x_{n-1}) + f(x_n)].$$

STUDY TIP

The coefficients in the Trapezoidal Rule have the pattern

$$1 \quad 2 \quad 2 \quad 2 \ldots 2 \quad 2 \quad 1.$$

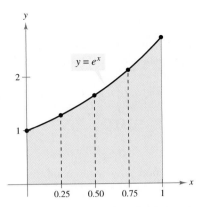

FIGURE 12.11 Four Subintervals

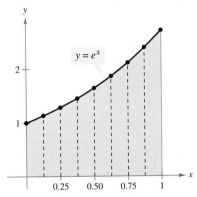

FIGURE 12.12 Eight Subintervals

Use the Trapezoidal Rule with $n = 4$ to approximate

$$\int_0^1 e^{2x} \, dx.$$

Example 1   Using the Trapezoidal Rule

Use the Trapezoidal Rule to approximate $\int_0^1 e^x \, dx$. Compare the results for $n = 4$ and $n = 8$.

SOLUTION   When $n = 4$, the width of each subinterval is

$$\frac{1 - 0}{4} = \frac{1}{4}$$

and the endpoints of the subintervals are

$$x_0 = 0, \quad x_1 = \frac{1}{4}, \quad x_2 = \frac{1}{2}, \quad x_3 = \frac{3}{4}, \quad \text{and} \quad x_4 = 1$$

as indicated in Figure 12.11. So, by the Trapezoidal Rule,

$$\int_0^1 e^x \, dx = \frac{1}{8}(e^0 + 2e^{0.25} + 2e^{0.5} + 2e^{0.75} + e^1)$$

$$\approx 1.7272. \qquad \text{Approximation using } n = 4$$

When $n = 8$, the width of each subinterval is

$$\frac{1 - 0}{8} = \frac{1}{8}$$

and the endpoints of the subintervals are

$$x_0 = 0, \quad x_1 = \frac{1}{8}, \quad x_2 = \frac{1}{4}, \quad x_3 = \frac{3}{8}, \quad x_4 = \frac{1}{2}$$

$$x_5 = \frac{5}{8}, \quad x_6 = \frac{3}{4}, \quad x_7 = \frac{7}{8}, \quad \text{and} \quad x_8 = 1$$

as indicated in Figure 12.12. So, by the Trapezoidal Rule,

$$\int_0^1 e^x \, dx = \frac{1}{16}(e^0 + 2e^{0.125} + 2e^{0.25} + \cdots + 2e^{0.875} + e^1)$$

$$\approx 1.7205. \qquad \text{Approximation using } n = 8$$

Of course, for *this particular* integral, you could have found an antiderivative and used the Fundamental Theorem of Calculus to find the exact value of the definite integral. The exact value is

$$\int_0^1 e^x \, dx = e - 1 \approx 1.718282. \qquad \text{Exact value}$$

There are two important points that should be made concerning the Trapezoidal Rule. First, the approximation tends to become more accurate as n increases. For instance, in Example 1, if $n = 16$, the Trapezoidal Rule yields an approximation of 1.7188. Second, although you could have used the Fundamental Theorem of Calculus to evaluate the integral in Example 1, this theorem cannot be used to evaluate an integral as simple as $\int_0^1 e^{x^2} \, dx$, because e^{x^2} has no elementary function as an antiderivative. Yet the Trapezoidal Rule can be easily applied to this integral.

TECHNOLOGY

Ⓣ A graphing utility can also evaluate a definite integral that does not have an elementary function as an antiderivative. Use the integration capabilities of a graphing utility to approximate the integral $\int_0^1 e^{x^2} \, dx$.*

*Specific calculator keystroke instructions for operations in this and other technology boxes can be found at *college.hmco.com/info/larsonapplied*.

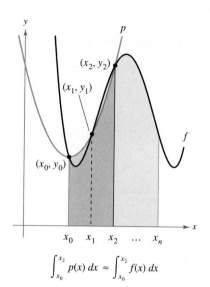

$$\int_{x_0}^{x_2} p(x)\, dx \approx \int_{x_0}^{x_2} f(x)\, dx$$

FIGURE 12.13

Simpson's Rule

One way to view the Trapezoidal Rule is to say that on each subinterval, f is approximated by a first-degree polynomial. In Simpson's Rule, f is approximated by a second-degree polynomial on each subinterval.

To develop Simpson's Rule, partition the interval $[a, b]$ into an *even number* n of subintervals, each of width

$$\Delta x = \frac{b - a}{n}.$$

On the subinterval $[x_0, x_2]$, approximate the function f by the second-degree polynomial $p(x)$ that passes through the points

$$(x_0, f(x_0)), \quad (x_1, f(x_1)), \quad \text{and} \quad (x_2, f(x_2))$$

as shown in Figure 12.13. The Fundamental Theorem of Calculus can be used to show that

$$\int_{x_0}^{x_2} f(x)\, dx \approx \int_{x_0}^{x_2} p(x)\, dx$$

$$= \left(\frac{x_2 - x_0}{6}\right)\left[p(x_0) + 4p\left(\frac{x_0 + x_2}{2}\right) + p(x_2)\right]$$

$$= \frac{2[(b - a)/n]}{6}[p(x_0) + 4p(x_1) + p(x_2)]$$

$$= \left(\frac{b - a}{3n}\right)[f(x_0) + 4f(x_1) + f(x_2)].$$

Repeating this process on the subintervals $[x_{i-2}, x_i]$ produces

$$\int_{a}^{b} f(x)\, dx \approx \left(\frac{b - a}{3n}\right)[f(x_0) + 4f(x_1) + f(x_2) + f(x_2) + 4f(x_3) +$$

$$f(x_4) + \cdots + f(x_{n-2}) + 4f(x_{n-1}) + f(x_n)].$$

By grouping like terms, you can obtain the approximation shown below, which is known as Simpson's Rule. This rule is named after the English mathematician Thomas Simpson (1710–1761).

Simpson's Rule (*n* Is Even)

If f is continuous on $[a, b]$, then

$$\int_{a}^{b} f(x)\, dx \approx \left(\frac{b - a}{3n}\right)[f(x_0) + 4f(x_1) + 2f(x_2) + 4f(x_3) +$$

$$\cdots + 4f(x_{n-1}) + f(x_n)].$$

In Example 1, the Trapezoidal Rule was used to estimate the value of

$$\int_0^1 e^x \, dx.$$

The next example uses Simpson's Rule to approximate the same integral.

Example 2 Using Simpson's Rule

Use Simpson's Rule to approximate

$$\int_0^1 e^x \, dx.$$

Compare the results for $n = 4$ and $n = 8$.

SOLUTION   When $n = 4$, the width of each subinterval is $(1 - 0)/4 = \frac{1}{4}$ and the endpoints of the subintervals are

$$x_0 = 0, \quad x_1 = \frac{1}{4}, \quad x_2 = \frac{1}{2}, \quad x_3 = \frac{3}{4}, \quad \text{and} \quad x_4 = 1$$

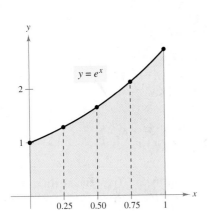

FIGURE 12.14   Four Subintervals

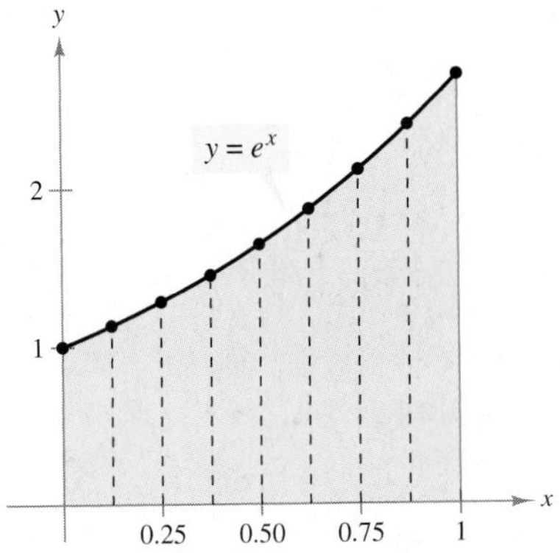

as indicated in Figure 12.14. So, by Simpson's Rule,

$$\int_0^1 e^x dx = \frac{1}{12}(e^0 + 4e^{0.25} + 2e^{0.5} + 4e^{0.75} + e^1)$$

$$\approx 1.718319. \qquad \text{Approximation using } n = 4$$

When $n = 8$, the width of each subinterval is $(1 - 0)/8 = \frac{1}{8}$ and the endpoints of the subintervals are

$$x_0 = 0, \quad x_1 = \frac{1}{8}, \quad x_2 = \frac{1}{4}, \quad x_3 = \frac{3}{8}, \quad x_4 = \frac{1}{2}$$

$$x_5 = \frac{5}{8}, \quad x_6 = \frac{3}{4}, \quad x_7 = \frac{7}{8}, \quad \text{and} \quad x_8 = 1$$

FIGURE 12.15   Eight Subintervals

as indicated in Figure 12.15. So, by Simpson's Rule,

$$\int_0^1 e^x dx = \frac{1}{24}(e^0 + 4e^{0.125} + 2e^{0.25} + \cdots + 4e^{0.875} + e^1)$$

$$\approx 1.718284. \qquad \text{Approximation using } n = 8$$

Recall that the exact value of this integral is

$$\int_0^1 e^x dx = e - 1 \approx 1.718282. \qquad \text{Exact value}$$

So, with only eight subintervals, you obtained an approximation that is correct to the nearest 0.000002—an impressive result.

STUDY TIP

Comparing the results of Examples 1 and 2, you can see that for a given value of n, Simpson's Rule tends to be more accurate than the Trapezoidal Rule.

✓ CHECKPOINT 2

Use Simpson's Rule with $n = 4$ to approximate $\int_0^1 e^{2x} \, dx$.

Programming Simpson's Rule

In Section 11.6, you saw how to program the Midpoint Rule into a computer or programmable calculator. The pseudocode below can be used to write a program that will evaluate Simpson's Rule. (Appendix E lists this program for several models of graphing utilities.)

Program

- *Prompt for value of a.*
- *Input value of a.*
- *Prompt for value of b.*
- *Input value of b.*
- *Prompt for value of n/2.*
- *Input value of n/2.*
- *Initialize sum of areas.*
- *Calculate width of subinterval.*
- *Initialize counter.*
- *Begin loop.*
- *Calculate left endpoint.*
- *Calculate right endpoint.*
- *Calculate midpoint of subinterval.*
- *Store left endpoint.*
- *Evaluate f(x) at left endpoint.*
- *Store midpoint of subinterval.*
- *Evaluate f(x) at midpoint.*
- *Store right endpoint.*
- *Evaluate f(x) at right endpoint.*
- *Store Simpson's Rule.*
- *Check value of index.*
- *End loop.*
- *Display approximation.*

Before executing the program, enter the function. When the program is executed, you will be prompted to enter the lower and upper limits of integration, and *half* the number of subintervals you want to use.

Error Analysis

In Examples 1 and 2, you were able to calculate the exact value of the integral and compare that value with the approximations to see how good they were. In practice, you need to have a different way of telling how good an approximation is: such a way is provided in the next result.

Errors in the Trapezoidal Rule and Simpson's Rule

The errors E in approximating $\int_a^b f(x)\,dx$ are as shown.

Trapezoidal Rule: $|E| \leq \dfrac{(b-a)^3}{12n^2}\left[\max|f''(x)|\right], \quad a \leq x \leq b$

Simpson's Rule: $|E| \leq \dfrac{(b-a)^5}{180n^4}\left[\max|f^{(4)}(x)|\right], \quad a \leq x \leq b$

This result indicates that the errors generated by the Trapezoidal Rule and Simpson's Rule have upper bounds dependent on the extreme values of $f''(x)$ and $f^{(4)}(x)$ in the interval $[a, b]$. Furthermore, the bounds for the errors can be made arbitrarily small by *increasing n*. To determine what value of n to choose, consider the steps below.

Trapezoidal Rule

1. Find $f''(x)$.

2. Find the maximum of $|f''(x)|$ on the interval $[a, b]$.

3. Set up the inequality

$$|E| \leq \frac{(b-a)^3}{12n^2}\left[\max|f''(x)|\right].$$

4. For an error less than ϵ, solve for n in the inequality

$$\frac{(b-a)^3}{12n^2}\left[\max|f''(x)|\right] < \epsilon.$$

5. Partition $[a, b]$ into n subintervals and apply the Trapezoidal Rule.

Simpson's Rule

1. Find $f^{(4)}(x)$.

2. Find the maximum of $|f^{(4)}(x)|$ on the interval $[a, b]$.

3. Set up the inequality

$$|E| \leq \frac{(b-a)^5}{180n^4}\left[\max|f^{(4)}(x)|\right].$$

4. For an error less than ϵ, solve for n in the inequality

$$\frac{(b-a)^5}{180n^4}\left[\max|f^{(4)}(x)|\right] < \epsilon.$$

5. Partition $[a, b]$ into n subintervals and apply Simpson's Rule.

Example 3   **Using the Trapezoidal Rule**

Use the Trapezoidal Rule to estimate the value of $\displaystyle\int_0^1 e^{-x^2}\,dx$ such that the approximation error is less than 0.01.

SOLUTION

1. Begin by finding the second derivative of $f(x) = e^{-x^2}$.

$$f(x) = e^{-x^2}$$
$$f'(x) = -2xe^{-x^2}$$
$$f''(x) = 4x^2e^{-x^2} - 2e^{-x^2}$$
$$= 2e^{-x^2}(2x^2 - 1)$$

2. f'' has only one critical number in the interval $[0, 1]$, and the maximum value of $|f''(x)|$ on this interval is $|f''(0)| = 2$.

3. The error E using the Trapezoidal Rule is bounded by

$$|E| \le \frac{(b - a)^3}{12n^2}(2) = \frac{1}{12n^2}(2) = \frac{1}{6n^2}.$$

4. To ensure that the approximation has an error of less than 0.01, you should choose n such that

$$\frac{1}{6n^2} < 0.01.$$

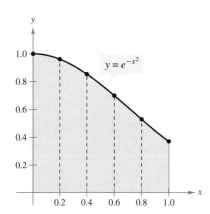

FIGURE 12.16

Solving for n, you can determine that n must be 5 or more.

5. Partition $[0, 1]$ into five subintervals, as shown in Figure 12.16. Then apply the Trapezoidal Rule to obtain

$$\int_0^1 e^{-x^2}\,dx = \frac{1}{10}\left(\frac{1}{e^0} + \frac{2}{e^{0.04}} + \frac{2}{e^{0.16}} + \frac{2}{e^{0.36}} + \frac{2}{e^{0.64}} + \frac{1}{e^1}\right)$$
$$\approx 0.744.$$

So, with an error no larger than 0.01, you know that

$$0.734 \le \int_0^1 e^{-x^2}\,dx \le 0.754.$$

✓ CHECKPOINT 3

Use the Trapezoidal Rule to estimate the value of

$$\int_0^1 \sqrt{1 + x^2}\,dx$$

such that the approximation error is less than 0.01. ■

⸨ CONCEPT CHECK ⸩

1. For the Trapezoidal Rule, the number of subintervals n can be odd or even. For Simpson's Rule, n must be what?

2. As the number of subintervals n increases, does an approximation given by the Trapezoidal Rule or Simpson's Rule tend to become less accurate or more accurate?

3. Write the formulas for (a) the Trapezoidal Rule and (b) Simpson's Rule.

4. The Trapezoidal Rule and Simpson's Rule yield approximations of a definite integral $\int_a^b f(x)\,dx$ based on polynomial approximations of f. What degree polynomial is used for each?

Skills Review 12.4

The following warm-up exercises involve skills that were covered in earlier sections. You will use these skills in the exercise set for this section. For additional help, review Sections 1.7, 7.4, 8.1, 8.5, 10.3, and 10.5.

In Exercises 1–6, find the indicated derivative.

1. $f(x) = \dfrac{1}{x}, \; f''(x)$

2. $f(x) = \ln(2x + 1), \; f^{(4)}(x)$

3. $f(x) = 2 \ln x, \; f^{(4)}(x)$

4. $f(x) = x^3 - 2x^2 + 7x - 12, \; f''(x)$

5. $f(x) = e^{2x}, \; f^{(4)}(x)$

6. $f(x) = e^{x^2}, \; f''(x)$

In Exercises 7 and 8, find the absolute maximum of f on the interval.

7. $f(x) = -x^2 + 6x + 9, \, [0, 4]$

8. $f(x) = \dfrac{8}{x^3}, \, [1, 2]$

In Exercises 9 and 10, solve for n.

9. $\dfrac{1}{4n^2} < 0.001$

10. $\dfrac{1}{16n^4} < 0.0001$

Exercises 12.4

See www.CalcChat.com for worked-out solutions to odd-numbered exercises.

In Exercises 1–14, use the Trapezoidal Rule and Simpson's Rule to approximate the value of the definite integral for the indicated value of n. Compare these results with the exact value of the definite integral. Round your answers to four decimal places.

1. $\displaystyle\int_0^2 x^2 \, dx, \; n = 4$

2. $\displaystyle\int_0^1 \left(\dfrac{x^2}{2} + 1\right) dx, \; n = 4$

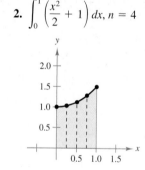

3. $\displaystyle\int_0^2 (x^4 + 1) \, dx, \; n = 4$

4. $\displaystyle\int_1^2 \dfrac{1}{x} \, dx, \; n = 4$

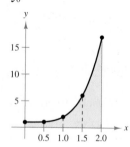
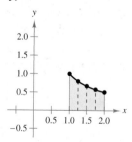

5. $\displaystyle\int_0^2 x^3 \, dx, \; n = 8$

6. $\displaystyle\int_1^3 (4 - x^2) \, dx, \; n = 4$

7. $\displaystyle\int_1^2 \dfrac{1}{x} \, dx, \; n = 8$

8. $\displaystyle\int_1^2 \dfrac{1}{x^2} \, dx, \; n = 4$

9. $\displaystyle\int_0^4 \sqrt{x} \, dx, \; n = 8$

10. $\displaystyle\int_0^2 \sqrt{1 + x} \, dx, \; n = 4$

11. $\displaystyle\int_4^9 \sqrt{x} \, dx, \; n = 8$

12. $\displaystyle\int_0^8 \sqrt[3]{x} \, dx, \; n = 8$

13. $\displaystyle\int_0^1 \dfrac{1}{1 + x} \, dx, \; n = 4$

14. $\displaystyle\int_0^2 x\sqrt{x^2 + 1} \, dx, \; n = 4$

In Exercises 15–24, approximate the integral using (a) the Trapezoidal Rule and (b) Simpson's Rule for the indicated value of n. (Round your answers to three significant digits.)

15. $\displaystyle\int_0^1 \dfrac{1}{1 + x^2} \, dx, \; n = 4$

16. $\displaystyle\int_0^2 \dfrac{1}{\sqrt{1 + x^3}} \, dx, \; n = 4$

17. $\displaystyle\int_0^2 \sqrt{1 + x^3} \, dx, \; n = 4$

18. $\displaystyle\int_0^1 \sqrt{1 - x} \, dx, \; n = 4$

19. $\displaystyle\int_0^1 \sqrt{1 - x^2} \, dx, \; n = 4$

20. $\displaystyle\int_0^1 \sqrt{1 - x^2} \, dx, \; n = 8$

21. $\displaystyle\int_0^2 e^{-x^2} \, dx, \; n = 2$

22. $\displaystyle\int_0^2 e^{-x^2} \, dx, \; n = 4$

23. $\displaystyle\int_0^3 \dfrac{1}{2 - 2x + x^2} \, dx, \; n = 6$

24. $\displaystyle\int_0^3 \dfrac{x}{2 + x + x^2} \, dx, \; n = 6$

(T) **Present Value** In Exercises 25 and 26, use a program similar to the Simpson's Rule program on page 906 with $n = 8$ to approximate the present value of the income $c(t)$ over t_1 years at the given annual interest rate r. Then use the integration capabilities of a graphing utility to approximate the present value. Compare the results. (Present value is defined in Section 12.1.)

25. $c(t) = 6000 + 200\sqrt{t}$, $r = 7\%$, $t_1 = 4$

26. $c(t) = 200{,}000 + 15{,}000\sqrt[3]{t}$, $r = 10\%$, $t_1 = 8$

(T) **Marginal Analysis** In Exercises 27 and 28, use a program similar to the Simpson's Rule program on page 906 with $n = 4$ to approximate the change in revenue from the marginal revenue function dR/dx. In each case, assume that the number of units sold x increases from 14 to 16.

27. $\dfrac{dR}{dx} = 5\sqrt{8000 - x^3}$ **28.** $\dfrac{dR}{dx} = 50\sqrt{x}\sqrt{20 - x}$

(T) **Probability** In Exercises 29–32, use a program similar to the Simpson's Rule program on page 906 with $n = 6$ to approximate the indicated normal probability. The standard normal probability density function is $f(x) = \left(1/\sqrt{2\pi}\right)e^{-x^2/2}$. If x is chosen at random from a population with this density, then the probability that x lies in the interval $[a, b]$ is $P(a \le x \le b) = \int_a^b f(x)\,dx$.

29. $P(0 \le x \le 1)$ **30.** $P(0 \le x \le 2)$

31. $P(0 \le x \le 4)$ **32.** $P(0 \le x \le 1.5)$

(T) **Surveying** In Exercises 33 and 34, use a program similar to the Simpson's Rule program on page 906 to estimate the number of square feet of land in the lot, where x and y are measured in feet, as shown in the figures. In each case, the land is bounded by a stream and two straight roads.

33.

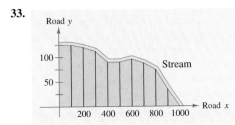

x	0	100	200	300	400	500
y	125	125	120	112	90	90

x	600	700	800	900	1000
y	95	88	75	35	0

34.

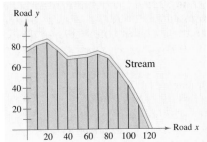

x	0	10	20	30	40	50	60
y	75	81	84	76	67	68	69

x	70	80	90	100	110	120
y	72	68	56	42	23	0

In Exercises 35–38, use the error formulas to find bounds for the error in approximating the integral using (a) the Trapezoidal Rule and (b) Simpson's Rule. (Let $n = 4$.)

35. $\displaystyle\int_0^2 x^3\,dx$ **36.** $\displaystyle\int_0^1 \frac{1}{x+1}\,dx$

37. $\displaystyle\int_0^1 e^{x^3}\,dx$ **38.** $\displaystyle\int_0^1 e^{-x^2}\,dx$

In Exercises 39–42, use the error formulas to find n such that the error in the approximation of the definite integral is less than 0.0001 using (a) the Trapezoidal Rule and (b) Simpson's Rule.

39. $\displaystyle\int_0^1 x^3\,dx$ **40.** $\displaystyle\int_1^3 \frac{1}{x}\,dx$

41. $\displaystyle\int_1^3 e^{2x}\,dx$ **42.** $\displaystyle\int_3^5 \ln x\,dx$

(T) In Exercises 43–46, use a program similar to the Simpson's Rule program on page 906 to approximate the integral. Use $n = 100$.

43. $\displaystyle\int_1^4 x\sqrt{x+4}\,dx$ **44.** $\displaystyle\int_1^4 x^2\sqrt{x+4}\,dx$

45. $\displaystyle\int_2^5 10xe^{-x}\,dx$ **46.** $\displaystyle\int_2^5 10x^2e^{-x}\,dx$

(T) In Exercises 47 and 48, use a program similar to the Simpson's Rule program on page 906 with $n = 4$ to find the area of the region bounded by the graphs of the equations.

47. $y = x\sqrt[3]{x+4}$, $y = 0$, $x = 1$, $x = 5$

48. $y = \sqrt{2 + 3x^2}$, $y = 0$, $x = 1$, $x = 3$

In Exercises 49 and 50, use the definite integral below to find the required arc length. If f has a continuous derivative, then the arc length of f between the points $(a, f(a))$ and $(b, f(b))$ is

$$\int_b^a \sqrt{1 + [f'(x)]^2} \, dx.$$

(T) **49. Arc Length**   The suspension cable on a bridge that is 400 feet long is in the shape of a parabola whose equation is $y = x^2/800$ (see figure). Use a program similar to the Simpson's Rule program on page 906 with $n = 12$ to approximate the length of the cable. Compare this result with the length obtained by using the table of integrals in Section 12.3 to perform the integration.

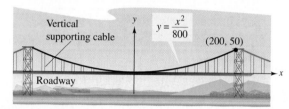

50. Arc Length   A fleeing hare leaves its burrow $(0, 0)$ and moves due north (up the y-axis). At the same time, a pursuing lynx leaves from 1 yard east of the burrow $(1, 0)$ and always moves toward the fleeing hare (see figure). If the lynx's speed is twice that of the hare's, the equation of the lynx's path is

$$y = \frac{1}{3}(x^{3/2} - 3x^{1/2} + 2).$$

Find the distance traveled by the lynx by integrating over the interval $[0, 1]$.

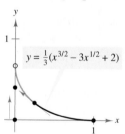

51. Lumber Use   The table shows the amounts of lumber used for residential upkeep and improvements (in billions of board-feet per year) for the years 1997 through 2005. (*Source: U.S. Forest Service*)

Year	1997	1998	1999	2000	2001
Amount	15.1	14.7	15.1	16.4	17.0

Year	2002	2003	2004	2005
Amount	17.8	18.3	20.0	20.6

(a) Use Simpson's Rule to estimate the average number of board-feet (in billions) used per year over the time period.

(b) A model for the data is

$$L = 6.613 + 0.93t + 2095.7e^{-t}, \quad 7 \le t \le 15$$

where L is the amount of lumber used and t is the year, with $t = 7$ corresponding to 1997. Use integration to find the average number of board-feet (in billions) used per year over the time period.

(c) Compare the results of parts (a) and (b).

52. Median Age   The table shows the median ages of the U.S. resident population for the years 1997 through 2005. (*Source: U.S. Census Bureau*)

Year	1997	1998	1999	2000	2001
Median age	34.7	34.9	35.2	35.3	35.6

Year	2002	2003	2004	2005
Median age	35.7	35.9	36.0	36.2

(a) Use Simpson's Rule to estimate the average age over the time period.

(b) A model for the data is $A = 31.5 + 1.21\sqrt{t}$, $7 \le t \le 15$, where A is the median age and t is the year, with $t = 7$ corresponding to 1997. Use integration to find the average age over the time period.

(c) Compare the results of parts (a) and (b).

53. Medicine   A body assimilates a 12-hour cold tablet at a rate modeled by $dC/dt = 8 - \ln(t^2 - 2t + 4)$, $0 \le t \le 12$, where dC/dt is measured in milligrams per hour and t is the time in hours. Use Simpson's Rule with $n = 8$ to estimate the total amount of the drug absorbed into the body during the 12 hours.

54. Medicine   The concentration M (in grams per liter) of a six-hour allergy medicine in a body is modeled by $M = 12 - 4\ln(t^2 - 4t + 6)$, $0 \le t \le 6$, where t is the time in hours since the allergy medication was taken. Use Simpson's Rule with $n = 6$ to estimate the average level of concentration in the body over the six-hour period.

55. Consumer Trends   The rate of change S in the number of subscribers to a newly introduced magazine is modeled by $dS/dt = 1000t^2e^{-t}$, $0 \le t \le 6$, where t is the time in years. Use Simpson's Rule with $n = 12$ to estimate the total increase in the number of subscribers during the first 6 years.

56. Prove that Simpson's Rule is exact when used to approximate the integral of a cubic polynomial function, and demonstrate the result for $\int_0^1 x^3 \, dx$, $n = 2$.

Section 12.5

Improper Integrals

■ Recognize improper integrals.
■ Evaluate improper integrals with infinite limits of integration.
■ Evaluate improper integrals with infinite integrands.
■ Use improper integrals to solve real-life problems.
■ Find the present value of a perpetuity.

Improper Integrals

The definition of the definite integral

$$\int_a^b f(x)\, dx$$

includes the requirements that the interval $[a, b]$ be finite and that f be continuous on $[a, b]$. In this section, you will study integrals that do not satisfy these requirements because of one of the conditions below.

1. One or both of the limits of integration are infinite.

2. f has an infinite discontinuity in the interval $[a, b]$.

Integrals having either of these characteristics are called **improper integrals.** For instance, the integrals

$$\int_0^\infty e^{-x}\, dx \quad \text{and} \quad \int_{-\infty}^\infty \frac{1}{x^2 + 1}\, dx$$

are improper because one or both limits of integration are infinite, as indicated in Figure 12.17. Similarly, the integrals

$$\int_1^5 \frac{1}{\sqrt{x - 1}}\, dx \quad \text{and} \quad \int_{-2}^2 \frac{1}{(x + 1)^2}\, dx$$

are improper because their integrands have an **infinite discontinuity**—that is, they approach infinity somewhere in the interval of integration, as indicated in Figure 12.18.

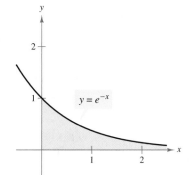

$y = e^{-x}$

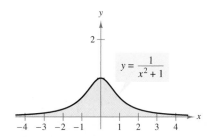

$y = \dfrac{1}{x^2 + 1}$

FIGURE 12.17

DISCOVERY

Use a graphing utility to calculate the definite integral $\int_0^b e^{-x}\, dx$ for $b = 10$ and for $b = 20$. What is the area of the region bounded by the graph of $y = e^{-x}$ and the two coordinate axes?

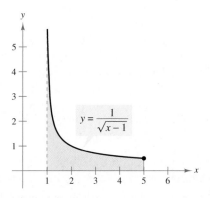

$y = \dfrac{1}{\sqrt{x - 1}}$

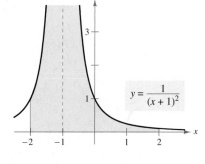

$y = \dfrac{1}{(x + 1)^2}$

FIGURE 12.18

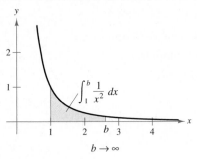

FIGURE 12.19

Integrals with Infinite Limits of Integration

To see how to evaluate an improper integral, consider the integral shown in Figure 12.19. As long as b is a real number that is greater than 1 (no matter how large), this is a definite integral whose value is

$$\int_1^b \frac{1}{x^2}\, dx = \left[-\frac{1}{x}\right]_1^b$$

$$= -\frac{1}{b} + 1$$

$$= 1 - \frac{1}{b}.$$

The table shows the values of this integral for several values of b.

b	2	5	10	100	1000	10,000
$\int_1^b \frac{1}{x^2}\, dx = 1 - \frac{1}{b}$	0.5000	0.8000	0.9000	0.9900	0.9990	0.9999

From this table, it appears that the value of the integral is approaching a limit as b increases without bound. This limit is denoted by the *improper integral* shown below.

$$\int_1^\infty \frac{1}{x^2}\, dx = \lim_{b\to\infty} \int_1^b \frac{1}{x^2}\, dx$$

$$= \lim_{b\to\infty}\left(1 - \frac{1}{b}\right)$$

$$= 1$$

Improper Integrals (Infinite Limits of Integration)

1. If f is continuous on the interval $[a, \infty)$, then

$$\int_a^\infty f(x)\, dx = \lim_{b\to\infty} \int_a^b f(x)\, dx.$$

2. If f is continuous on the interval $(-\infty, b]$, then

$$\int_{-\infty}^b f(x)\, dx = \lim_{a\to-\infty} \int_a^b f(x)\, dx.$$

3. If f is continuous on the interval $(-\infty, \infty)$, then

$$\int_{-\infty}^\infty f(x)\, dx = \int_{-\infty}^c f(x)\, dx + \int_c^\infty f(x)\, dx$$

where c is any real number.

In the first two cases, if the limit exists, then the improper integral **converges;** otherwise, the improper integral **diverges.** In the third case, the integral on the left will diverge if either one of the integrals on the right diverges.

Example 1   Evaluating an Improper Integral

Determine the convergence or divergence of $\int_{1}^{\infty} \frac{1}{x}\, dx$.

SOLUTION   Begin by applying the definition of an improper integral.

$$\int_{1}^{\infty} \frac{1}{x}\, dx = \lim_{b \to \infty} \int_{1}^{b} \frac{1}{x}\, dx \qquad \text{Definition of improper integral}$$

$$= \lim_{b \to \infty} \left[\ln x \right]_{1}^{b} \qquad \text{Find antiderivative.}$$

$$= \lim_{b \to \infty} (\ln b - 0) \qquad \text{Apply Fundamental Theorem.}$$

$$= \infty \qquad \text{Evaluate limit.}$$

Because the limit is infinite, the improper integral diverges.

✓**CHECKPOINT 1**

Determine the convergence or divergence of each improper integral.

a. $\int_{1}^{\infty} \frac{1}{x^3}\, dx$ **b.** $\int_{1}^{\infty} \frac{1}{\sqrt{x}}\, dx$ ▪

As you begin to work with improper integrals, you will find that integrals that appear to be similar can have very different values. For instance, consider the two improper integrals

$$\int_{1}^{\infty} \frac{1}{x}\, dx = \infty \qquad \text{Divergent integral}$$

and

$$\int_{1}^{\infty} \frac{1}{x^2}\, dx = 1. \qquad \text{Convergent integral}$$

The first integral diverges and the second converges to 1. Graphically, this means that the areas shown in Figure 12.20 are very different. The region lying between the graph of $y = 1/x$ and the x-axis (for $x \geq 1$) has an *infinite* area, and the region lying between the graph of $y = 1/x^2$ and the x-axis (for $x \geq 1$) has a *finite* area.

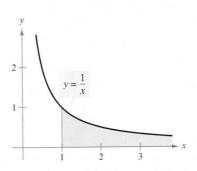

Diverges (infinite area)

FIGURE 12.20

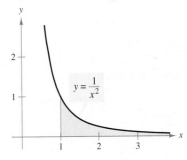

Converges (finite area)

Example 2 **Evaluating an Improper Integral**

Evaluate the improper integral.

$$\int_{-\infty}^{0} \frac{1}{(1 - 2x)^{3/2}} \, dx$$

SOLUTION Begin by applying the definition of an improper integral.

$$\int_{-\infty}^{0} \frac{1}{(1 - 2x)^{3/2}} \, dx = \lim_{a \to -\infty} \int_{a}^{0} \frac{1}{(1 - 2x)^{3/2}} \, dx \qquad \text{Definition of improper integral}$$

$$= \lim_{a \to -\infty} \left[\frac{1}{\sqrt{1 - 2x}} \right]_{a}^{0} \qquad \text{Find antiderivative.}$$

$$= \lim_{a \to -\infty} \left(1 - \frac{1}{\sqrt{1 - 2a}} \right) \qquad \text{Apply Fundamental Theorem.}$$

$$= 1 - 0 \qquad \text{Evaluate limit.}$$

$$= 1 \qquad \text{Simplify.}$$

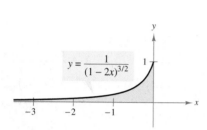

$$y = \frac{1}{(1 - 2x)^{3/2}}$$

FIGURE 12.21

So, the improper integral converges to 1. As shown in Figure 12.21, this implies that the region lying between the graph of $y = 1/(1 - 2x)^{3/2}$ and the x-axis (for $x \le 0$) has an area of 1 square unit.

✓**CHECKPOINT 2**

Evaluate the improper integral, if possible.

$$\int_{-\infty}^{0} \frac{1}{(x - 1)^2} \, dx$$

Example 3 **Evaluating an Improper Integral**

Evaluate the improper integral.

$$\int_{0}^{\infty} 2xe^{-x^2} \, dx$$

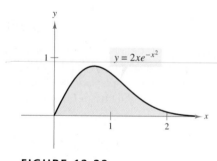

$$y = 2xe^{-x^2}$$

FIGURE 12.22

SOLUTION Begin by applying the definition of an improper integral.

$$\int_{0}^{\infty} 2xe^{-x^2} \, dx = \lim_{b \to \infty} \int_{0}^{b} 2xe^{-x^2} \, dx \qquad \text{Definition of improper integral}$$

$$= \lim_{b \to \infty} \left[-e^{-x^2} \right]_{0}^{b} \qquad \text{Find antiderivative.}$$

$$= \lim_{b \to \infty} \left(-e^{-b^2} + 1 \right) \qquad \text{Apply Fundamental Theorem.}$$

$$= 0 + 1 \qquad \text{Evaluate limit.}$$

$$= 1 \qquad \text{Simplify.}$$

✓**CHECKPOINT 3**

Evaluate the improper integral, if possible.

$$\int_{-\infty}^{0} e^{2x} \, dx$$

So, the improper integral converges to 1. As shown in Figure 12.22, this implies that the region lying between the graph of $y = 2xe^{-x^2}$ and the x-axis (for $x \ge 0$) has an area of 1 square unit.

Integrals with Infinite Integrands

Improper Integrals (Infinite Integrands)

1. If f is continuous on the interval $[a, b)$ and approaches infinity at b, then

$$\int_a^b f(x)\, dx = \lim_{c \to b^-} \int_a^c f(x)\, dx.$$

2. If f is continuous on the interval $(a, b]$ and approaches infinity at a, then

$$\int_a^b f(x)\, dx = \lim_{c \to a^+} \int_c^b f(x)\, dx.$$

3. If f is continuous on the interval $[a, b]$, except for some c in (a, b) at which f approaches infinity, then

$$\int_a^b f(x)\, dx = \int_a^c f(x)\, dx + \int_c^b f(x)\, dx.$$

In the first two cases, if the limit exists, then the improper integral **converges;** otherwise, the improper integral **diverges.** In the third case, the improper integral on the left diverges if either of the improper integrals on the right diverges.

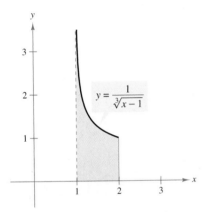

$y = \dfrac{1}{\sqrt[3]{x - 1}}$

FIGURE 12.23

TECHNOLOGY

(T) Use a graphing utility to verify the result of Example 4 by calculating each definite integral.

$$\int_{1.01}^2 \frac{1}{\sqrt[3]{x - 1}}\, dx$$

$$\int_{1.001}^2 \frac{1}{\sqrt[3]{x - 1}}\, dx$$

$$\int_{1.0001}^2 \frac{1}{\sqrt[3]{x - 1}}\, dx$$

Example 4   **Evaluating an Improper Integral**

Evaluate $\displaystyle\int_1^2 \frac{1}{\sqrt[3]{x - 1}}\, dx$.

SOLUTION

$$\int_1^2 \frac{1}{\sqrt[3]{x - 1}}\, dx = \lim_{c \to 1^+} \int_c^2 \frac{1}{\sqrt[3]{x - 1}}\, dx \qquad \text{Definition of improper integral}$$

$$= \lim_{c \to 1^+} \left[\frac{3}{2}(x - 1)^{2/3} \right]_c^2 \qquad \text{Find antiderivative.}$$

$$= \lim_{c \to 1^+} \left[\frac{3}{2} - \frac{3}{2}(c - 1)^{2/3} \right] \qquad \text{Apply Fundamental Theorem.}$$

$$= \frac{3}{2} - 0 \qquad \text{Evaluate limit.}$$

$$= \frac{3}{2} \qquad \text{Simplify.}$$

So, the integral converges to $\frac{3}{2}$. This implies that the region shown in Figure 12.23 has an area of $\frac{3}{2}$ square units.

✓**CHECKPOINT 4**

Evaluate $\displaystyle\int_1^2 \frac{1}{\sqrt{x - 1}}\, dx$. ∎

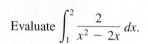

Example 5 Evaluating an Improper Integral

Evaluate $\displaystyle\int_1^2 \frac{2}{x^2 - 2x}\, dx$.

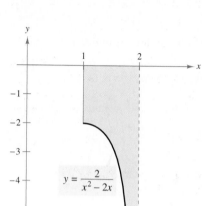

FIGURE 12.24

SOLUTION

$$\int_1^2 \frac{2}{x^2 - 2x}\, dx = \int_1^2 \left(\frac{1}{x - 2} - \frac{1}{x}\right) dx \qquad \text{Use partial fractions.}$$

$$= \lim_{c \to 2^-} \int_1^c \left(\frac{1}{x - 2} - \frac{1}{x}\right) dx \qquad \text{Definition of improper integral}$$

$$= \lim_{c \to 2^-} \left[\ln|x - 2| - \ln|x|\right]_1^c \qquad \text{Find antiderivative.}$$

$$= -\infty \qquad \text{Evaluate limit.}$$

So, you can conclude that the integral diverges. This implies that the region shown in Figure 12.24 has an infinite area.

✓ **CHECKPOINT 5**

Evaluate $\displaystyle\int_1^3 \frac{3}{x^2 - 3x}\, dx$. ■

Example 6 Evaluating an Improper Integral

Evaluate $\displaystyle\int_{-1}^2 \frac{1}{x^3}\, dx$.

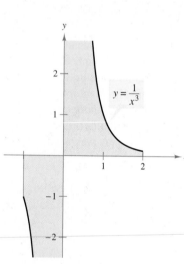

FIGURE 12.25

SOLUTION This integral is improper because the integrand has an infinite discontinuity at the interior value $x = 0$, as shown in Figure 12.25. So, you can write

$$\int_{-1}^2 \frac{1}{x^3}\, dx = \int_{-1}^0 \frac{1}{x^3}\, dx + \int_0^2 \frac{1}{x^3}\, dx.$$

By applying the definition of an improper integral, you can show that each of these integrals diverges. So, the original improper integral also diverges.

✓ **CHECKPOINT 6**

Evaluate $\displaystyle\int_{-1}^1 \frac{1}{x^2}\, dx$. ■

STUDY TIP

Had you not recognized that the integral in Example 6 was improper, you would have obtained the incorrect result

$$\int_{-1}^2 \frac{1}{x^3}\, dx = \left[-\frac{1}{2x^2}\right]_{-1}^2 = -\frac{1}{8} + \frac{1}{2} = \frac{3}{8}. \qquad \text{Incorrect}$$

Improper integrals in which the integrand has an infinite discontinuity *between* the limits of integration are often overlooked, so keep alert for such possibilities. Even symbolic integrators can have trouble with this type of integral, and can give the same incorrect result.

Application

In Section 10.3, you studied the graph of the *normal probability density function*

$$f(x) = \frac{1}{\sigma\sqrt{2\pi}}e^{-(x-\mu)^2/2\sigma^2}.$$

This function is used in statistics to represent a population that is normally distributed with a mean of μ and a standard deviation of σ. Specifically, if an outcome x is chosen at random from the population, the probability that x will have a value between a and b is

$$P(a \le x \le b) = \int_a^b \frac{1}{\sigma\sqrt{2\pi}}e^{-(x-\mu)^2/2\sigma^2}\,dx.$$

As shown in Figure 12.26, the probability $P(-\infty < x < \infty)$ is

$$P(-\infty < x < \infty) = \int_{-\infty}^{\infty} \frac{1}{\sigma\sqrt{2\pi}}e^{-(x-\mu)^2/2\sigma^2}\,dx = 1.$$

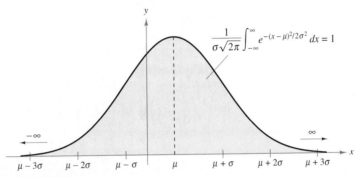

FIGURE 12.26

Example 7 Finding a Probability ®

The mean height of American men (from 20 to 29 years old) is 70 inches, and the standard deviation is 3 inches. A 20- to 29-year-old man is chosen at random from the population. What is the probability that he is 6 feet tall or taller? *(Source: U.S. National Center for Health Statistics)*

SOLUTION Using a mean of $\mu = 70$ and a standard deviation of $\sigma = 3$, the probability $P(72 \le x < \infty)$ is given by the improper integral

$$P(72 \le x < \infty) = \int_{72}^{\infty} \frac{1}{3\sqrt{2\pi}}e^{-(x-70)^2/18}\,dx.$$

Using a symbolic integration utility, you can approximate the value of this integral to be 0.252. So, the probability that the man is 6 feet tall or taller is about 25.2%.

✓ CHECKPOINT 7

Use Example 7 to find the probability that a 20- to 29-year-old man chosen at random from the population is 6 feet 6 inches tall or taller. ■

Victor Baldizon/NBAE via Getty Images

Many professional basketball players are over $6\frac{1}{2}$ feet tall. If a man is chosen at random from the population, the probability that he is $6\frac{1}{2}$ feet tall or taller is less than half of one percent.

Present Value of a Perpetuity

Recall from Section 12.1 that for an interest-bearing account, the present value over t_1 years is

$$\text{Present value} = \int_0^{t_1} c(t)e^{-rt}\,dt$$

where c represents a continuous income function (in dollars per year) and the annual interest rate r is compounded continuously. If the size of an annuity's payment is a constant number of dollars P, then $c(t)$ is equal to P and the present value is

$$\text{Present value} = \int_0^{t_1} Pe^{-rt}\,dt = P\int_0^{t_1} e^{-rt}\,dt. \qquad \text{Present value of an annuity with payment } P$$

Suppose you wanted to start an annuity, such as a scholarship fund, that pays the same amount each year *forever*? Because the annuity continues indefinitely, the number of years t_1 approaches infinity. Such an annuity is called a **perpetual annuity** or a **perpetuity.** This situation can be represented by the following improper integral.

$$\text{Present value} = P\int_0^{\infty} e^{-rt}\,dt \qquad \text{Present value of a perpetuity with payment } P$$

This integral is simplified as follows.

$$P\int_0^{\infty} e^{-rt}\,dt = P\lim_{b\to\infty}\int_0^{b} e^{-rt}\,dt \qquad \text{Definition of improper integral}$$

$$= P\lim_{b\to\infty}\left[-\frac{e^{-rt}}{r}\right]_0^{b} \qquad \text{Find antiderivative.}$$

$$= P\lim_{b\to\infty}\left(-\frac{e^{-rb}}{r}+\frac{1}{r}\right) \qquad \text{Apply Fundamental Theorem.}$$

$$= P\left(0+\frac{1}{r}\right) \qquad \text{Evaluate limit.}$$

$$= \frac{P}{r} \qquad \text{Simplify.}$$

So, the improper integral converges to P/r. As shown in Figure 12.27, this implies that the region lying between the graph of $y = Pe^{-rt}$ and the t-axis for $t \geq 0$ has an area equal to the annual payment P divided by the annual interest rate r.

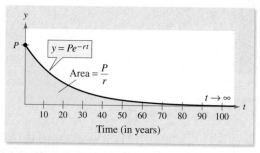

FIGURE 12.27

The present value of a perpetuity is defined as follows.

Present Value of a Perpetuity

If P represents the size of each annual payment in dollars and the annual interest rate is r (compounded continuously), then the present value of a perpetuity is

$$\text{Present value} = P \int_0^\infty e^{-rt}\, dt = \frac{P}{r}.$$

This definition is useful in determining the amount of money needed to start an endowment, such as a scholarship fund, as shown in Example 8.

Example 8
MAKE A DECISION   **Finding Present Value**

You want to start a scholarship fund at your alma mater. You plan to give one $9000 scholarship annually beginning one year from now, and you have at most $120,000 to start the fund. You also want the scholarship to be given out indefinitely. Assuming an annual interest rate of 8% (compounded continuously), do you have enough money for the scholarship fund?

SOLUTION   To answer this question, you must find the present value of the scholarship fund. Because the scholarship is to be given out each year indefinitely, the time period is infinite. The fund is a perpetuity with $P = 9000$ and $r = 0.08$. The present value is

$$\begin{aligned} \text{Present value} &= \frac{P}{r} \\[6pt] &= \frac{9000}{0.08} \\[6pt] &= 112{,}500. \end{aligned}$$

The amount you need to start the scholarship fund is $112,500. Yes, you have enough money to start the scholarship fund.   ———

✓ CHECKPOINT 8

In Example 8, do you have enough money to start a scholarship fund that pays $10,000 annually? Explain why or why not. ▪

CONCEPT CHECK

1. Integrals are improper integrals if they have either of what two characteristics?

2. Describe the different types of improper integrals.

3. Define the term *converges* when working with improper integrals.

4. Define the term *diverges* when working with improper integrals.

Skills Review 12.5

The following warm-up exercises involve skills that were covered in earlier sections. You will use these skills in the exercise set for this section. For additional help, review Sections 0.2, 7.1, 10.1, and 10.4.

In Exercises 1–6, find the limit.

1. $\lim\limits_{x \to 2} (2x + 5)$

2. $\lim\limits_{x \to 1} \left(\dfrac{1}{x} + 2x^2 \right)$

3. $\lim\limits_{x \to -4} \dfrac{x + 4}{x^2 - 16}$

4. $\lim\limits_{x \to 0} \dfrac{x^2 - 2x}{x^3 + 3x^2}$

5. $\lim\limits_{x \to 1} \dfrac{1}{\sqrt{x - 1}}$

6. $\lim\limits_{x \to -3} \dfrac{x^2 + 2x - 3}{x + 3}$

In Exercises 7–10, evaluate the expression (a) when $x = b$ and (b) when $x = 0$.

7. $\dfrac{4}{3}(2x - 1)^3$

8. $\dfrac{1}{x - 5} + \dfrac{3}{(x - 2)^2}$

9. $\ln(5 - 3x^2) - \ln(x + 1)$

10. $e^{3x^2} + e^{-3x^2}$

Exercises 12.5

See www.CalcChat.com for worked-out solutions to odd-numbered exercises.

In Exercises 1–4, decide whether the integral is improper. Explain your reasoning.

1. $\displaystyle\int_0^1 \dfrac{dx}{3x - 2}$

2. $\displaystyle\int_1^3 \dfrac{dx}{x^2}$

3. $\displaystyle\int_0^1 \dfrac{2x - 5}{x^2 - 5x + 6}\, dx$

4. $\displaystyle\int_1^\infty x^2\, dx$

In Exercises 5–10, explain why the integral is improper and determine whether it diverges or converges. Evaluate the integral if it converges.

5. $\displaystyle\int_0^4 \dfrac{1}{\sqrt{x}}\, dx$

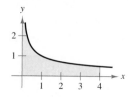

6. $\displaystyle\int_3^4 \dfrac{1}{\sqrt{x - 3}}\, dx$

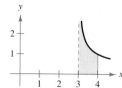

7. $\displaystyle\int_0^2 \dfrac{1}{(x - 1)^{2/3}}\, dx$

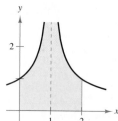

8. $\displaystyle\int_0^2 \dfrac{1}{(x - 1)^2}\, dx$

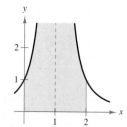

9. $\displaystyle\int_0^\infty e^{-x}\, dx$

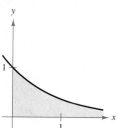

10. $\displaystyle\int_{-\infty}^0 e^{2x}\, dx$

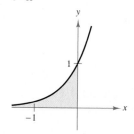

In Exercises 11–22, determine whether the improper integral diverges or converges. Evaluate the integral if it converges.

11. $\displaystyle\int_1^\infty \dfrac{1}{x^2}\, dx$

12. $\displaystyle\int_1^\infty \dfrac{1}{\sqrt[3]{x}}\, dx$

13. $\displaystyle\int_0^\infty e^{x/3}\, dx$

14. $\displaystyle\int_0^\infty \dfrac{5}{e^{2x}}\, dx$

15. $\displaystyle\int_5^\infty \dfrac{x}{\sqrt{x^2 - 16}}\, dx$

16. $\displaystyle\int_{1/2}^\infty \dfrac{1}{\sqrt{2x - 1}}\, dx$

17. $\displaystyle\int_{-\infty}^0 e^{-x}\, dx$

18. $\displaystyle\int_{-\infty}^{-1} \dfrac{1}{x^2}\, dx$

19. $\displaystyle\int_1^\infty \dfrac{e^{\sqrt{x}}}{\sqrt{x}}\, dx$

20. $\displaystyle\int_{-\infty}^0 \dfrac{x}{x^2 + 1}\, dx$

21. $\displaystyle\int_{-\infty}^\infty 2xe^{-3x^2}\, dx$

22. $\displaystyle\int_{-\infty}^\infty x^2 e^{-x^3}\, dx$

In Exercises 23–32, determine whether the improper integral diverges or converges. Evaluate the integral if it converges, and check your results with the results obtained by using the integration capabilities of a graphing utility.

23. $\int_0^1 \frac{1}{1 - x}\, dx$

24. $\int_0^{27} \frac{5}{\sqrt[3]{x}}\, dx$

25. $\int_0^9 \frac{1}{\sqrt{9 - x}}\, dx$

26. $\int_0^2 \frac{x}{\sqrt{4 - x^2}}\, dx$

27. $\int_0^1 \frac{1}{x^2}\, dx$

28. $\int_0^1 \frac{1}{x}\, dx$

29. $\int_0^2 \frac{1}{\sqrt[3]{x - 1}}\, dx$

30. $\int_0^2 \frac{1}{(x - 1)^{4/3}}\, dx$

31. $\int_3^4 \frac{1}{\sqrt{x^2 - 9}}\, dx$

32. $\int_3^5 \frac{1}{x^2\sqrt{x^2 - 9}}\, dx$

In Exercises 33 and 34, consider the region satisfying the inequalities. Find the area of the region.

33. $y \leq \frac{1}{x^2}, y \geq 0, x \geq 1$

34. $y \leq e^{-x}, y \geq 0, x \geq 0$

Ⓢ In Exercises 35–38, use a spreadsheet to complete the table for the specified values of a and n to demonstrate that

$$\lim_{x \to \infty} x^n e^{-ax} = 0, \quad a > 0, n > 0.$$

x	1	10	25	50
$x^n e^{-ax}$				

35. $a = 1, n = 1$

36. $a = 2, n = 4$

37. $a = \frac{1}{2}, n = 2$

38. $a = \frac{1}{2}, n = 5$

In Exercises 39–42, use the results of Exercises 35–38 to evaluate the improper integral.

39. $\int_0^\infty x^2 e^{-x}\, dx$

40. $\int_0^\infty (x - 1)e^{-x}\, dx$

41. $\int_0^\infty x e^{-2x}\, dx$

42. $\int_0^\infty x e^{-x}\, dx$

43. Women's Height The mean height of American women between the ages of 30 and 39 is 64.5 inches, and the standard deviation is 2.7 inches. Find the probability that a 30- to 39-year-old woman chosen at random is

(a) between 5 and 6 feet tall.

(b) 5 feet 8 inches or taller.

(c) 6 feet or taller.

(Source: U.S. National Center for Health Statistics)

44. Quality Control A company manufactures wooden yardsticks. The lengths of the yardsticks are normally distributed with a mean of 36 inches and a standard deviation of 0.2 inch. Find the probability that a yardstick is

(a) longer than 35.5 inches.   (b) longer than 35.9 inches.

Endowment In Exercises 45 and 46, determine the amount of money required to set up a charitable endowment that pays the amount P each year indefinitely for the annual interest rate r compounded continuously.

45. $P = \$5000, r = 7.5\%$

46. $P = \$12,000, r = 6\%$

47. *MAKE A DECISION: SCHOLARSHIP FUND*   You want to start a scholarship fund at your alma mater. You plan to give one $18,000 scholarship annually beginning one year from now and you have at most $400,000 to start the fund. You also want the scholarship to be given out indefinitely. Assuming an annual interest rate of 5% compounded continuously, do you have enough money for the scholarship fund?

48. *MAKE A DECISION: CHARITABLE FOUNDATION*   A charitable foundation wants to help schools buy computers. The foundation plans to donate $35,000 each year to one school beginning one year from now, and the foundation has at most $500,000 to start the fund. The foundation wants the donation to be given out indefinitely. Assuming an annual interest rate of 8% compounded continuously, does the foundation have enough money to fund the donation?

49. Present Value A business is expected to yield a continuous flow of profit at the rate of $500,000 per year. If money will earn interest at the nominal rate of 9% per year compounded continuously, what is the present value of the business (a) for 20 years and (b) forever?

50. Present Value Repeat Exercise 49 for a farm that is expected to produce a profit of $75,000 per year. Assume that money will earn interest at the nominal rate of 8% compounded continuously.

Capitalized Cost In Exercises 51 and 52, find the capitalized cost C of an asset (a) for $n = 5$ years, (b) for $n = 10$ years, and (c) forever. The capitalized cost is given by

$$C = C_0 + \int_0^n c(t)e^{-rt}\, dt$$

where C_0 is the original investment, t is the time in years, r is the annual interest rate compounded continuously, and $c(t)$ is the annual cost of maintenance (measured in dollars). [*Hint:* For part (c), see Exercises 35–38.]

51. $C_0 = \$650,000, c(t) = 25,000, r = 10\%$

52. $C_0 = \$650,000, c(t) = 25,000(1 + 0.08t), r = 12\%$

Algebra Review

Algebra and Integration Techniques

Integration techniques involve many different algebraic skills. Study the examples in this Algebra Review. Be sure that you understand the algebra used in each step.

Example 1   Algebra and Integration Techniques

Perform each operation and simplify.

a. $\dfrac{2}{x-3} - \dfrac{1}{x+2}$ Example 1, page 882

$= \dfrac{2(x+2)}{(x-3)(x+2)} - \dfrac{(x-3)}{(x-3)(x+2)}$ Rewrite with common denominator.

$= \dfrac{2(x+2) - (x-3)}{(x-3)(x+2)}$ Rewrite as single fraction.

$= \dfrac{2x+4-x+3}{x^2-x-6}$ Multiply factors.

$= \dfrac{x+7}{x^2-x-6}$ Combine like terms.

b. $\dfrac{6}{x} - \dfrac{1}{x+1} + \dfrac{9}{(x+1)^2}$ Example 2, page 883

$= \dfrac{6(x+1)^2}{x(x+1)^2} - \dfrac{x(x+1)}{x(x+1)^2} + \dfrac{9x}{x(x+1)^2}$ Rewrite with common denominator.

$= \dfrac{6(x+1)^2 - x(x+1) + 9x}{x(x+1)^2}$ Rewrite as single fraction.

$= \dfrac{6x^2 + 12x + 6 - x^2 - x + 9x}{x^3 + 2x^2 + x}$ Multiply factors.

$= \dfrac{5x^2 + 20x + 6}{x^3 + 2x^2 + x}$ Combine like terms.

c. $6\ln|x| - \ln|x+1| + 9\dfrac{(x+1)^{-1}}{-1}$ Example 2, page 883

$= \ln|x|^6 - \ln|x+1| + 9\dfrac{(x+1)^{-1}}{-1}$ $m\ln n = \ln n^m$

$= \ln|x^6| - \ln|x+1| + 9\dfrac{(x+1)^{-1}}{-1}$ Property of absolute value

$= \ln\dfrac{|x^6|}{|x+1|} + 9\dfrac{(x+1)^{-1}}{-1}$ $\ln m - \ln n = \ln\dfrac{m}{n}$

$= \ln\left|\dfrac{x^6}{x+1}\right| + 9\dfrac{(x+1)^{-1}}{-1}$ $\dfrac{|a|}{|b|} = \left|\dfrac{a}{b}\right|$

$= \ln\left|\dfrac{x^6}{x+1}\right| - 9(x+1)^{-1}$ Rewrite sum as difference.

$= \ln\left|\dfrac{x^6}{x+1}\right| - \dfrac{9}{x+1}$ Rewrite with positive exponent.

Example 2   Algebra and Integration Techniques

Perform each operation and simplify.

a. $x + 1 + \dfrac{1}{x^3} + \dfrac{1}{x - 1}$

b. $x^2 e^x - 2(x - 1)e^x$

c. Solve for y: $\ln |y| - \ln |L - y| = kt + C$

SOLUTION

a. $x + 1 + \dfrac{1}{x^3} + \dfrac{1}{x - 1}$ Example 3, page 884

$= \dfrac{(x + 1)(x^3)(x - 1)}{x^3(x - 1)} + \dfrac{x - 1}{x^3(x - 1)} + \dfrac{x^3}{x^3(x - 1)}$

$= \dfrac{(x + 1)(x^3)(x - 1) + (x - 1) + x^3}{x^3(x - 1)}$ Rewrite as single fraction.

$= \dfrac{(x^2 - 1)(x^3) + x - 1 + x^3}{x^3(x - 1)}$ $(x + 1)(x - 1) = x^2 - 1$

$= \dfrac{x^5 - x^3 + x - 1 + x^3}{x^4 - x^3}$ Multiply factors.

$= \dfrac{x^5 + x - 1}{x^4 - x^3}$ Combine like terms.

b. $x^2 e^x - 2(x - 1)e^x$ Example 5, page 897

$= x^2 e^x - 2(xe^x - e^x)$ Multiply factors.

$= x^2 e^x - 2xe^x + 2e^x$ Multiply factors.

$= e^x(x^2 - 2x + 2)$ Factor.

c. $\ln |y| - \ln |L - y| = kt + C$ Example 4, page 885

$-\ln |y| + \ln |L - y| = -kt - C$ Multiply each side by -1.

$\ln \left| \dfrac{L - y}{y} \right| = -kt - C$ $\ln x - \ln y = \ln \dfrac{x}{y}$

$\left| \dfrac{L - y}{y} \right| = e^{-kt - C}$ Exponentiate each side.

$\left| \dfrac{L - y}{y} \right| = e^{-C}e^{-kt}$ $x^{n+m} = x^n x^m$

$\dfrac{L - y}{y} = \pm e^{-C}e^{-kt}$ Property of absolute value

$L - y = be^{-kt}y$ Let $\pm e^{-C} = b$ and multiply each side by y.

$L = y + be^{-kt}y$ Add y to each side.

$L = y(1 + be^{-kt})$ Factor.

$\dfrac{L}{1 + be^{-kt}} = y$ Divide.

Chapter Summary and Study Strategies

After studying this chapter, you should have acquired the following skills.
The exercise numbers are keyed to the Review Exercises that begin on page 926.
Answers to odd-numbered Review Exercises are given in the back of the text.

Section 12.1   **Review Exercises**

- Use integration by parts to find indefinite integrals.   *1–4*

$$\int u \, dv = uv - \int v \, du$$

- Use integration by parts repeatedly to find indefinite integrals.   *5, 6*
- Find the present value of future income.   *7–14*

Section 12.2

- Use partial fractions to find indefinite integrals.   *15–20*
- Use logistic growth functions to model real-life situations.   *21, 22*

$$y = \frac{L}{1 + be^{-kt}}$$

Section 12.3

- Use integration tables to find indefinite and definite integrals.   *23–30*
- Use reduction formulas to find indefinite integrals.   *31–34*
- Use integration tables to solve real-life problems.   *35, 36*

Section 12.4

- Use the Trapezoidal Rule to approximate definite integrals.   *37–40*

$$\int_a^b f(x) \, dx \approx \left(\frac{b-a}{2n}\right)[f(x_0) + 2f(x_1) + \cdots + 2f(x_{n-1}) + f(x_n)]$$

- Use Simpson's Rule to approximate definite integrals.   *41–44*

$$\int_a^b f(x) \, dx \approx \left(\frac{b-a}{3n}\right)[f(x_0) + 4f(x_1) + 2f(x_2) + 4f(x_3) + \cdots + 4f(x_{n-1}) + f(x_n)]$$

- Analyze the sizes of the errors when approximating definite integrals with the   *45, 46*
  Trapezoidal Rule.

$$|E| \le \frac{(b-a)^3}{12n^2}[\max|f''(x)|], \quad a \le x \le b$$

- Analyze the sizes of the errors when approximating definite integrals with Simpson's Rule.   *47, 48*

$$|E| \le \frac{(b-a)^5}{180n^4}[\max|f^{(4)}(x)|], \quad a \le x \le b$$

Section 12.5

- Evaluate improper integrals with infinite limits of integration.

$$\int_a^\infty f(x)\,dx = \lim_{b\to\infty} \int_a^b f(x)\,dx, \qquad \int_{-\infty}^b f(x)\,dx = \lim_{a\to-\infty}\int_a^b f(x)\,dx,$$

$$\int_{-\infty}^\infty f(x)\,dx = \int_{-\infty}^c f(x)\,dx + \int_c^\infty f(x)\,dx$$

- Evaluate improper integrals with infinite integrands.

$$\int_a^b f(x)\,dx = \lim_{c\to b^-}\int_a^c f(x)\,dx, \qquad \int_a^b f(x)\,dx = \lim_{c\to a^+}\int_c^b f(x)\,dx,$$

$$\int_a^b f(x)\,dx = \int_a^c f(x)\,dx + \int_c^b f(x)\,dx$$

- Use improper integrals to solve real-life problems.

Study Strategies

- **Use a Variety of Approaches**   To be efficient at finding antiderivatives, you need to use a variety of approaches.

  1. Check to see whether the integral fits one of the basic integration formulas—you should have these formulas memorized.

  2. Try an integration technique such as substitution, integration by parts, or partial fractions to rewrite the integral in a form that fits one of the basic integration formulas.

  3. Use a table of integrals.

  4. Use a symbolic integration utility.

- **Use Numerical Integration**   When solving a definite integral, remember that you cannot apply the Fundamental Theorem of Calculus unless you can find an antiderivative of the integrand. This is not always possible—even with a symbolic integration utility. In such cases, you can use a numerical technique such as the Midpoint Rule, the Trapezoidal Rule, or Simpson's Rule to approximate the value of the integral.

- **Improper Integrals**   When solving integration problems, remember that the symbols used to denote definite integrals are the same as those used to denote improper integrals. Evaluating an improper integral as a definite integral can lead to an incorrect value. For instance, if you evaluated the integral

$$\int_{-2}^1 \frac{1}{x^2}\,dx$$

as though it were a definite integral, you would obtain a value of $-\frac{3}{2}$. This is not, however, correct. This integral is actually a divergent improper integral.

Review Exercises

See www.CalcChat.com for worked-out solutions to odd-numbered exercises.

In Exercises 1–4, use integration by parts to find the indefinite integral.

1. $\displaystyle\int \frac{\ln x}{\sqrt{x}}\, dx$

2. $\displaystyle\int \sqrt{x}\,\ln x\, dx$

3. $\displaystyle\int (x + 1)e^x\, dx$

4. $\displaystyle\int \ln\!\left(\frac{x}{x + 1}\right) dx$

In Exercises 5 and 6, use integration by parts repeatedly to find the indefinite integral. Use a symbolic integration utility to verify your answer.

5. $\displaystyle\int 2x^2 e^{2x}\, dx$

6. $\displaystyle\int (\ln x)^3\, dx$

Present Value   In Exercises 7–10, find the present value of the income given by $c(t)$ (measured in dollars) over t_1 years at the given annual inflation rate r.

7. $c(t) = 20,000,\ r = 4\%,\ t_1 = 5$ years

8. $c(t) = 10,000\ + 1500t,\ r = 6\%,\ t_1 = 10$ years

9. $c(t) = 24,000t,\ r = 5\%,\ t_1 = 10$ years

10. $c(t) = 20,000 + 100e^{t/2},\ r = 5\%,\ t_1 = 5$ years

Ⓑ **11. Economics: Present Value**   Calculate the present value of each scenario.

(a) $2000 per year for 5 years at interest rates of 5%, 10%, and 15%

(b) A lottery ticket that pays $200,000 per year after taxes over 20 years, assuming an inflation rate of 8%

(Source: Adapted from Boyes/Melvin, Economics, Third Edition)

Ⓑ **12. Finance: Present Value**   You receive $2000 at the end of each year for the next 3 years to help with college expenses. Assuming an annual interest rate of 6%, what is the present value of that stream of payments?   *(Source: Adapted from Garman/Forgue, Personal Finance, Eighth Edition)*

Ⓑ **13. Finance: Present Value**   Determine the amount a person planning for retirement would need to deposit today to be able to withdraw $12,000 each year for the next 10 years from an account earning 6% interest.   *(Source: Adapted from Garman/Forgue, Personal Finance, Eighth Edition)*

Ⓑ **14. Finance: Present Value**   A person invests $100,000 earning 6% interest. If $10,000 is withdrawn each year, use present value to determine how many years it will take for the fund to run out.   *(Source: Adapted from Garman/ Forgue, Personal Finance, Eighth Edition)*

In Exercises 15–20, use partial fractions to find the indefinite integral.

15. $\displaystyle\int \frac{1}{x(x + 5)}\, dx$

16. $\displaystyle\int \frac{4x - 2}{3(x - 1)^2}\, dx$

17. $\displaystyle\int \frac{x - 28}{x^2 - x - 6}\, dx$

18. $\displaystyle\int \frac{4x^2 - x - 5}{x^2(x + 5)}\, dx$

19. $\displaystyle\int \frac{x^2}{x^2 + 2x - 15}\, dx$

20. $\displaystyle\int \frac{x^2 + 2x - 12}{x(x + 3)}\, dx$

21. Sales   A new product initially sells 1250 units per week. After 24 weeks, the number of sales increases to 6500. The sales can be modeled by logistic growth with a limit of 10,000 units per week.

(a) Find a logistic growth model for the number of units.

(b) Use the model to complete the table.

Time, t	0	3	6	12	24
Sales, y					

(c) Use the graph shown below to approximate the time t when sales will be 7500.

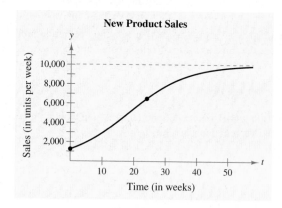

New Product Sales

22. Biology   A conservation society has introduced a population of 300 ring-necked pheasants into a new area. After 5 years, the population has increased to 966. The population can be modeled by logistic growth with a limit of 2700 pheasants.

(a) Find a logistic growth model for the population of ring-necked pheasants.

(b) How many pheasants were present after 4 years?

(c) How long will it take to establish a population of 1750 pheasants?

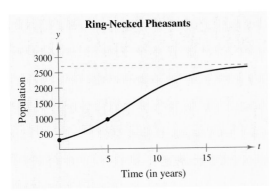

Figure for 22

In Exercises 23–30, use the table of integrals in Section 12.3 to find or evaluate the integral.

23. $\displaystyle\int \frac{x}{(2 + 3x)^2}\, dx$

24. $\displaystyle\int \frac{x}{\sqrt{2 + 3x}}\, dx$

25. $\displaystyle\int \frac{\sqrt{x^2 + 25}}{x}\, dx$

26. $\displaystyle\int \frac{1}{x(4 + 3x)}\, dx$

27. $\displaystyle\int \frac{1}{x^2 - 4}\, dx$

28. $\displaystyle\int (\ln 3x)^2\, dx$

29. $\displaystyle\int_0^3 \frac{x}{\sqrt{1 + x}}\, dx$

30. $\displaystyle\int_1^3 \frac{1}{x^2\sqrt{16 - x^2}}\, dx$

In Exercises 31–34, use a reduction formula from the table of integrals in Section 12.3 to find the indefinite integral.

31. $\displaystyle\int \frac{\sqrt{1 + x}}{x}\, dx$

32. $\displaystyle\int \frac{1}{(x^2 - 9)^2}\, dx$

33. $\displaystyle\int (x - 5)^3 e^{x-5}\, dx$

34. $\displaystyle\int (\ln x)^4\, dx$

35. Probability The probability of recall in an experiment is found to be

$$P(a \le x \le b) = \int_a^b \frac{96}{11}\left(\frac{x}{\sqrt{9 + 16x}}\right) dx, \quad 0 \le a \le b \le 1$$

where x represents the percent of recall (see figure).

(a) Find the probability that a randomly chosen individual will recall between 0% and 80% of the material.

(b) Find the probability that a randomly chosen individual will recall between 0% and 50% of the material.

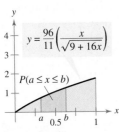

Figure for 35

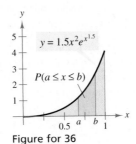

Figure for 36

36. Probability The probability of locating between a and b percent of oil and gas deposits in a region is

$$P(a \le x \le b) = \int_a^b 1.5x^2 e^{x^{1.5}}\, dx \text{ (see figure).}$$

(a) Find the probability that between 40% and 60% of the deposits will be found.

(b) Find the probability that between 0% and 50% of the deposits will be found.

In Exercises 37–40, use the Trapezoidal Rule to approximate the definite integral.

37. $\displaystyle\int_1^3 \frac{1}{x^2}\, dx, \ n = 4$

38. $\displaystyle\int_0^2 (x^2 + 1)\, dx, \ n = 4$

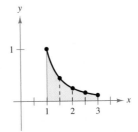

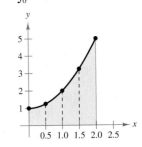

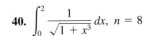

39. $\displaystyle\int_1^2 \frac{1}{1 + \ln x}\, dx, \ n = 4$

40. $\displaystyle\int_0^2 \frac{1}{\sqrt{1 + x^3}}\, dx, \ n = 8$

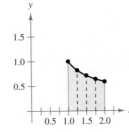

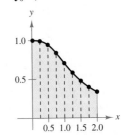

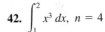

In Exercises 41–44, use Simpson's Rule to approximate the definite integral.

41. $\displaystyle\int_1^2 \frac{1}{x^3}\, dx, \ n = 4$

42. $\displaystyle\int_1^2 x^3\, dx, \ n = 4$

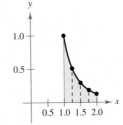

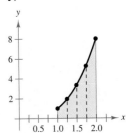

43. $\int_0^1 \dfrac{x^{3/2}}{2 - x^2}\, dx$, $n = 4$ **44.** $\int_0^1 e^{x^2}\, dx$, $n = 6$

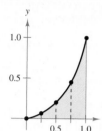

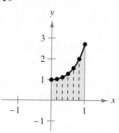

In Exercises 45 and 46, use the error formula to find bounds for the error in approximating the integral using the Trapezoidal Rule.

45. $\int_0^2 e^{2x}\, dx$, $n = 4$ **46.** $\int_0^2 e^{2x}\, dx$, $n = 8$

In Exercises 47 and 48, use the error formula to find bounds for the error in approximating the integral using Simpson's Rule.

47. $\int_2^4 \dfrac{1}{x - 1}\, dx$, $n = 4$ **48.** $\int_2^4 \dfrac{1}{x - 1}\, dx$, $n = 8$

In Exercises 49–56, determine whether the improper integral diverges or converges. Evaluate the integral if it converges.

49. $\int_0^\infty 4xe^{-2x^2}\, dx$ **50.** $\int_{-\infty}^0 \dfrac{3}{(1 - 3x)^{2/3}}\, dx$

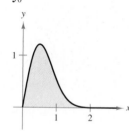

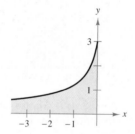

51. $\int_{-\infty}^0 \dfrac{1}{3x^2}\, dx$ **52.** $\int_0^\infty 2x^2 e^{-x^3}\, dx$

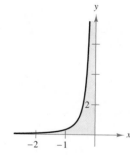

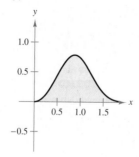

53. $\int_0^4 \dfrac{1}{\sqrt{4x}}\, dx$ **54.** $\int_1^2 \dfrac{x}{16(x - 1)^2}\, dx$

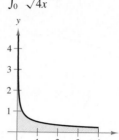

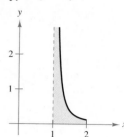

55. $\int_2^3 \dfrac{1}{\sqrt{x - 2}}\, dx$ **56.** $\int_0^2 \dfrac{x + 2}{(x - 1)^2}\, dx$

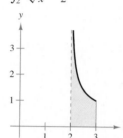

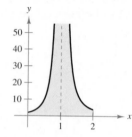

57. Present Value   You are considering buying a franchise that yields a continuous income stream of $100,000 per year. Find the present value of the franchise (a) for 15 years and (b) forever. Assume that money earns 6% interest per year, compounded continuously.

58. Capitalized Cost   A company invests $1.5 million in a new manufacturing plant that will cost $75,000 per year in maintenance. Find the capitalized cost for (a) 20 years and (b) forever. Assume that money earns 6% interest, compounded continuously.

59. SAT Scores   In 2006, the Scholastic Aptitude Test (SAT) math scores for college-bound seniors roughly followed a normal distribution

$$y = 0.0035e^{-(x - 518)^2/26,450}, \quad 200 \le x \le 800$$

where x is the SAT score for mathematics. Find the probability that a senior chosen at random had an SAT score (a) between 500 and 650, (b) 650 or better, and (c) 750 or better.   *(Source: College Board)*

60. ACT Scores   In 2006, the ACT composite scores for college-bound seniors followed a normal distribution

$$y = 0.0831e^{-(x - 21.1)^2/46.08}, \quad 1 \le x \le 36$$

where x is the composite ACT score. Find the probability that a senior chosen at random had an ACT score (a) between 16.3 and 25.9, (b) 25.9 or better, and (c) 30.7 or better.   *(Source: ACT, Inc.)*

Chapter Test

See www.CalcChat.com for worked-out solutions to odd-numbered exercises.

Take this test as you would take a test in class. When you are done, check your work against the answers given in the back of the book.

In Exercises 1–3, use integration by parts to find the indefinite integral.

1. $\displaystyle\int xe^{x+1}\,dx$

2. $\displaystyle\int 9x^2 \ln x\,dx$

3. $\displaystyle\int x^2 e^{-x/3}\,dx$

4. The earnings per share E (in dollars) for Home Depot from 2000 through 2006 can be modeled by

$$E = -2.62 + 0.495\sqrt{t}\,\ln t, \quad 10 \le t \le 16$$

where t is the year, with $t = 10$ corresponding to 2000. Find the average earnings per share for the years 2000 through 2006.   *(Source: The Home Depot, Inc.)*

In Exercises 5–7, use partial fractions to find the indefinite integral.

5. $\displaystyle\int \frac{18}{x^2 - 81}\,dx$

6. $\displaystyle\int \frac{3x}{(3x + 1)^2}\,dx$

7. $\displaystyle\int \frac{x + 4}{x^2 + 2x}\,dx$

In Exercises 8–10, use the table of integrals in Section 12.3 to find the indefinite integral.

8. $\displaystyle\int \frac{x}{(7 + 2x)^2}\,dx$

9. $\displaystyle\int \frac{3x^2}{1 + e^{x^3}}\,dx$

10. $\displaystyle\int \frac{2x^3}{\sqrt{1 + 5x^2}}\,dx$

In Exercises 11–13, evaluate the definite integral.

11. $\displaystyle\int_0^1 \ln(3 - 2x)\,dx$

12. $\displaystyle\int_5^{10} \frac{28}{x^2 - x - 12}\,dx$

13. $\displaystyle\int_{-3}^{-1} \frac{\sqrt{x^2 + 16}}{x}\,dx$

14. Use the Trapezoidal Rule with $n = 4$ to approximate $\displaystyle\int_1^2 \frac{1}{x^2\sqrt{x^2 + 4}}\,dx$. Compare your result with the exact value of the definite integral.

15. Use Simpson's Rule with $n = 4$ to approximate $\displaystyle\int_0^1 9xe^{3x}\,dx$. Compare your result with the exact value of the definite integral.

In Exercises 16–18, determine whether the improper integral converges or diverges. Evaluate the integral if it converges.

16. $\displaystyle\int_0^\infty e^{-3x}\,dx$

17. $\displaystyle\int_0^9 \frac{2}{\sqrt{x}}\,dx$

18. $\displaystyle\int_{-\infty}^0 \frac{1}{(4x - 1)^{2/3}}\,dx$

19. A magazine publisher offers two subscription plans. Plan A is a one-year subscription for $19.95. Plan B is a lifetime subscription (lasting indefinitely) for $149.

(a) A subscriber considers using plan A indefinitely. Assuming an annual inflation rate of 4%, find the present value of the money the subscriber will spend using plan A.

(b) Based on your answer to part (a), which plan should the subscriber use? Explain.

13

Functions of Several Variables

A spherical building can be represented by an equation involving three variables. (See Section 13.1, Exercise 61.)

Applications

Functions of several variables have many real-life applications. The applications listed below represent a sample of the applications in this chapter.

- Modeling Data: Milk Consumption, Exercise 59, page 947
- Make a Decision: Monthly Mortgage Payments, Exercise 51, page 956
- Shareholder's Equity, Exercise 66, page 967
- Medicine: Dosage and Duration of Infection, Exercise 50, page 976
- Make a Decision: Revenue, Exercise 33, page 996

© Chuck Savage/Corbis

The Three-Dimensional Coordinate System

- Plot points in space.
- Find distances between points in space and find midpoints of line segments in space.
- Write the standard forms of the equations of spheres and find the centers and radii of spheres.
- Sketch the coordinate plane traces of surfaces.

The Three-Dimensional Coordinate System

Recall from Section 2.1 that the Cartesian plane is determined by two perpendicular number lines called the x-axis and the y-axis. These axes together with their point of intersection (the origin) allow you to develop a two-dimensional coordinate system for identifying points in a plane. To identify a point in space, you must introduce a third dimension to the model. The geometry of this three-dimensional model is called **solid analytic geometry.**

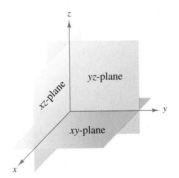

FIGURE 13.1

You can construct a **three-dimensional coordinate system** by passing a z-axis perpendicular to both the x- and y-axes at the origin. Figure 13.1 shows the positive portion of each coordinate axis. Taken as pairs, the axes determine three **coordinate planes:** the **xy-plane,** the **xz-plane,** and the **yz-plane.** These three coordinate planes separate the three-dimensional coordinate system into eight **octants.** The first octant is the one for which all three coordinates are positive. In this three-dimensional system, a point P in space is determined by an ordered triple (x, y, z), where x, y, and z are as follows.

$x = $ directed distance from yz-plane to P

$y = $ directed distance from xz-plane to P

$z = $ directed distance from xy-plane to P

A three-dimensional coordinate system can have either a **left-handed** or a **right-handed** orientation. To determine the orientation of a system, imagine that you are standing at the origin, with your arms pointing in the direction of the positive x- and y-axes, and with the z-axis pointing up, as shown in Figure 13.2. The system is right-handed or left-handed depending on which hand points along the x-axis. In this text, you will work exclusively with the right-handed system.

Right-handed system   Left-handed system

FIGURE 13.2

Example 1   **Plotting Points in Space**

Plot each point in space.

a. $(2, -3, 3)$

b. $(-2, 6, 2)$

c. $(1, 4, 0)$

d. $(2, 2, -3)$

SOLUTION   To plot the point $(2, -3, 3)$, notice that $x = 2$, $y = -3$, and $z = 3$. To help visualize the point (see Figure 13.3), locate the point $(2, -3)$ in the xy-plane (denoted by a cross). The point $(2, -3, 3)$ lies three units above the cross. The other three points are also shown in the figure.

✓ **CHECKPOINT 1**

Plot each point on the three-dimensional coordinate system.

a. $(2, 5, 1)$

b. $(-2, -4, 3)$

c. $(4, 0, -5)$ ■

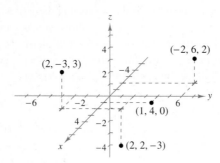

FIGURE 13.3

The Distance and Midpoint Formulas

Many of the formulas established for the two-dimensional coordinate system can be extended to three dimensions. For example, to find the distance between two points in space, you can use the Pythagorean Theorem twice, as shown in Figure 13.4. By doing this, you will obtain the formula for the distance between two points in space.

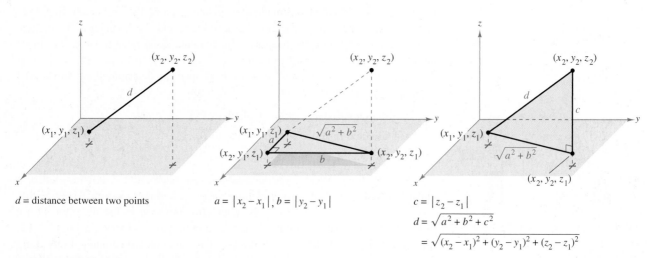

d = distance between two points

$a = |x_2 - x_1|,\ b = |y_2 - y_1|$

$c = |z_2 - z_1|$

$d = \sqrt{a^2 + b^2 + c^2}$

$\quad = \sqrt{(x_2 - x_1)^2 + (y_2 - y_1)^2 + (z_2 - z_1)^2}$

FIGURE 13.4

Distance Formula in Space

The distance between the points (x_1, y_1, z_1) and (x_2, y_2, z_2) is

$$d = \sqrt{(x_2 - x_1)^2 + (y_2 - y_1)^2 + (z_2 - z_1)^2}.$$

Example 2   Finding the Distance Between Two Points

Find the distance between $(1, 0, 2)$ and $(2, 4, -3)$.

SOLUTION

$$\begin{aligned} d &= \sqrt{(x_2 - x_1)^2 + (y_2 - y_1)^2 + (z_2 - z_1)^2} &&\text{Write Distance Formula.} \\ &= \sqrt{(2 - 1)^2 + (4 - 0)^2 + (-3 - 2)^2} &&\text{Substitute.} \\ &= \sqrt{1 + 16 + 25} &&\text{Simplify.} \\ &= \sqrt{42} &&\text{Simplify.} \end{aligned}$$

✓ **CHECKPOINT 2**

Find the distance between $(2, 3, -1)$ and $(0, 5, 3)$. ■

Notice the similarity between the Distance Formula in the plane and the Distance Formula in space. The Midpoint Formulas in the plane and in space are also similar.

Midpoint Formula in Space

The midpoint of the line segment joining the points (x_1, y_1, z_1) and (x_2, y_2, z_2) is

$$\text{Midpoint} = \left(\frac{x_1 + x_2}{2}, \frac{y_1 + y_2}{2}, \frac{z_1 + z_2}{2} \right).$$

Example 3   Using the Midpoint Formula

Find the midpoint of the line segment joining $(5, -2, 3)$ and $(0, 4, 4)$.

SOLUTION   Using the Midpoint Formula, the midpoint is

$$\left(\frac{5 + 0}{2}, \frac{-2 + 4}{2}, \frac{3 + 4}{2} \right) = \left(\frac{5}{2}, 1, \frac{7}{2} \right)$$

as shown in Figure 13.5.

✓ **CHECKPOINT 3**

Find the midpoint of the line segment joining $(3, -2, 0)$ and $(-8, 6, -4)$. ■

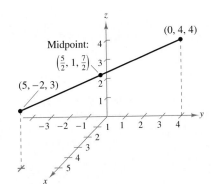

FIGURE 13.5

The Equation of a Sphere

A **sphere** with center at (h, k, l) and radius r is defined to be the set of all points (x, y, z) such that the distance between (x, y, z) and (h, k, l) is r, as shown in Figure 13.6. Using the Distance Formula, this condition can be written as

$$\sqrt{(x - h)^2 + (y - k)^2 + (z - l)^2} = r.$$

By squaring both sides of this equation, you obtain the standard equation of a sphere.

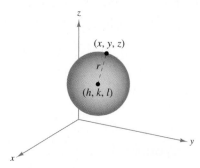

FIGURE 13.6   Sphere: Radius r, Center (h, k, l)

Standard Equation of a Sphere

The **standard equation of a sphere** whose center is (h, k, l) and whose radius is r is

$$(x - h)^2 + (y - k)^2 + (z - l)^2 = r^2.$$

Example 4   Finding the Equation of a Sphere

Find the standard equation of the sphere whose center is $(2, 4, 3)$ and whose radius is 3. Does this sphere intersect the xy-plane?

SOLUTION

$(x - h)^2 + (y - k)^2 + (z - l)^2 = r^2$	Write standard equation.
$(x - 2)^2 + (y - 4)^2 + (z - 3)^2 = 3^2$	Substitute.
$(x - 2)^2 + (y - 4)^2 + (z - 3)^2 = 9$	Simplify.

From the graph shown in Figure 13.7, you can see that the center of the sphere lies three units above the xy-plane. Because the sphere has a radius of 3, you can conclude that it does intersect the xy-plane—at the point $(2, 4, 0)$.

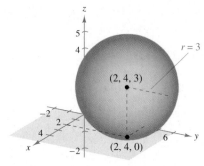

FIGURE 13.7

✓**CHECKPOINT 4**

Find the standard equation of the sphere whose center is $(4, 3, 2)$ and whose radius is 5. ■

Example 5 **Finding the Equation of a Sphere**

Find the equation of the sphere that has the points $(3, -2, 6)$ and $(-1, 4, 2)$ as endpoints of a diameter.

SOLUTION By the Midpoint Formula, the center of the sphere is

$$(h, k, l) = \left(\frac{3 + (-1)}{2}, \frac{-2 + 4}{2}, \frac{6 + 2}{2}\right)$$ Apply Midpoint Formula.

$$= (1, 1, 4).$$ Simplify.

By the Distance Formula, the radius is

$$r = \sqrt{(3 - 1)^2 + (-2 - 1)^2 + (6 - 4)^2}$$

$$= \sqrt{17}.$$ Simplify.

So, the standard equation of the sphere is

$$(x - h)^2 + (y - k)^2 + (z - l)^2 = r^2$$ Write formula for a sphere.

$$(x - 1)^2 + (y - 1)^2 + (z - 4)^2 = 17.$$ Substitute.

✓ **CHECKPOINT 5**

Find the equation of the sphere that has the points $(-2, 5, 7)$ and $(4, 1, -3)$ as endpoints of a diameter. ■

Example 6 **Finding the Center and Radius of a Sphere**

Find the center and radius of the sphere whose equation is

$$x^2 + y^2 + z^2 - 2x + 4y - 6z + 8 = 0.$$

SOLUTION You can obtain the standard equation of the sphere by completing the square. To do this, begin by grouping terms with the same variable. Then add "the square of half the coefficient of each linear term" to each side of the equation. For instance, to complete the square of $(x^2 - 2x)$, add $\left[\frac{1}{2}(-2)\right]^2 = 1$ to each side.

$$x^2 + y^2 + z^2 - 2x + 4y - 6z + 8 = 0$$

$$(x^2 - 2x + \quad) + (y^2 + 4y + \quad) + (z^2 - 6z + \quad) = -8$$

$$(x^2 - 2x + 1) + (y^2 + 4y + 4) + (z^2 - 6z + 9) = -8 + 1 + 4 + 9$$

$$(x - 1)^2 + (y + 2)^2 + (z - 3)^2 = 6$$

So, the center of the sphere is $(1, -2, 3)$, and its radius is $\sqrt{6}$, as shown in Figure 13.8.

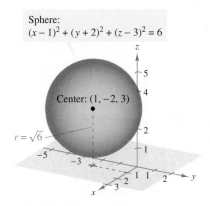

Sphere:
$(x - 1)^2 + (y + 2)^2 + (z - 3)^2 = 6$

Center: $(1, -2, 3)$

$r = \sqrt{6}$

FIGURE 13.8

✓ **CHECKPOINT 6**

Find the center and radius of the sphere whose equation is

$$x^2 + y^2 + z^2 + 6x - 8y + 2z - 10 = 0. ■$$

Note in Example 6 that the points satisfying the equation of the sphere are "surface points," not "interior points." In general, the collection of points satisfying an equation involving x, y, and z is called a **surface in space.**

Traces of Surfaces

Finding the intersection of a surface with one of the three coordinate planes (or with a plane parallel to one of the three coordinate planes) helps visualize the surface. Such an intersection is called a **trace** of the surface. For example, the xy-trace of a surface consists of all points that are common to both the surface *and* the xy-plane. Similarly, the xz-trace of a surface consists of all points that are common to both the surface and the xz-plane.

Example 7	**Finding a Trace of a Surface**

Sketch the xy-trace of the sphere whose equation is

$$(x - 3)^2 + (y - 2)^2 + (z + 4)^2 = 5^2.$$

SOLUTION   To find the xy-trace of this surface, use the fact that every point in the xy-plane has a z-coordinate of zero. This means that if you substitute $z = 0$ into the original equation, the resulting equation will represent the intersection of the surface with the xy-plane.

$$(x - 3)^2 + (y - 2)^2 + (z + 4)^2 = 5^2 \qquad \text{Write original equation.}$$
$$(x - 3)^2 + (y - 2)^2 + (0 + 4)^2 = 25 \qquad \text{Let } z = 0 \text{ to find } xy\text{-trace.}$$
$$(x - 3)^2 + (y - 2)^2 + 16 = 25$$
$$(x - 3)^2 + (y - 2)^2 = 9$$
$$(x - 3)^2 + (y - 2)^2 = 3^2 \qquad \text{Equation of circle}$$

From this equation, you can see that the xy-trace is a circle of radius 3, as shown in Figure 13.9.

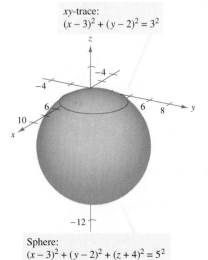

xy-trace:
$(x - 3)^2 + (y - 2)^2 = 3^2$

Sphere:
$(x - 3)^2 + (y - 2)^2 + (z + 4)^2 = 5^2$

FIGURE 13.9

✓ **CHECKPOINT 7**

Find the equation of the xy-trace of the sphere whose equation is

$$(x + 1)^2 + (y - 2)^2 + (z + 3)^2 = 5^2. \quad ■$$

CONCEPT CHECK

1. Name the three coordinate planes of a three-dimensional coordinate system formed by passing a z-axis perpendicular to both the x- and y-axes at the origin.

2. A point in the three-dimensional coordinate system has coordinates (x_1, y_1, z_1). Describe what each coordinate measures.

3. Give the formula for the distance between the points (x_1, y_1, z_1) and (x_2, y_2, z_2).

4. Give the standard equation of a sphere of radius r centered at (h, k, l).

In Exercises 1–4, find the distance between the points.

1. $(5, 1), (3, 5)$ **2.** $(2, 3), (-1, -1)$ **3.** $(-5, 4), (-5, -4)$ **4.** $(-3, 6), (-3, -2)$

In Exercises 5–8, find the midpoint of the line segment connecting the points.

5. $(2, 5), (6, 9)$ **6.** $(-1, -2), (3, 2)$ **7.** $(-6, 0), (6, 6)$ **8.** $(-4, 3), (2, -1)$

In Exercises 9 and 10, write the standard form of the equation of the circle.

9. Center: $(2, 3)$; radius: 2   **10.** Endpoints of a diameter: $(4, 0), (-2, 8)$

Exercises 13.1

In Exercises 1–4, plot the points on the same three-dimensional coordinate system.

1. (a) $(2, 1, 3)$
   (b) $(-1, 2, 1)$

2. (a) $(3, -2, 5)$
   (b) $\left(\frac{3}{2}, 4, -2\right)$

3. (a) $(5, -2, 2)$
   (b) $(5, -2, -2)$

4. (a) $(0, 4, -5)$
   (b) $(4, 0, 5)$

In Exercises 5 and 6, approximate the coordinates of the points.

5.

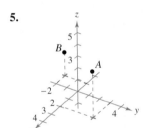

6.

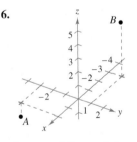

In Exercises 7–10, find the coordinates of the point.

7. The point is located three units behind the yz-plane, four units to the right of the xz-plane, and five units above the xy-plane.

8. The point is located seven units in front of the yz-plane, two units to the left of the xz-plane, and one unit below the xy-plane.

9. The point is located on the x-axis, 10 units in front of the yz-plane.

10. The point is located in the yz-plane, three units to the right of the xz-plane, and two units above the xy-plane.

11. Think About It What is the z-coordinate of any point in the xy-plane?

12. Think About It What is the x-coordinate of any point in the yz-plane?

In Exercises 13–16, find the distance between the two points.

13. $(4, 1, 5), (8, 2, 6)$ **14.** $(-4, -1, 1), (2, -1, 5)$

15. $(-1, -5, 7), (-3, 4, -4)$ **16.** $(8, -2, 2), (8, -2, 4)$

In Exercises 17–20, find the coordinates of the midpoint of the line segment joining the two points.

17. $(6, -9, 1), (-2, -1, 5)$ **18.** $(4, 0, -6), (8, 8, 20)$

19. $(-5, -2, 5), (6, 3, -7)$ **20.** $(0, -2, 5), (4, 2, 7)$

In Exercises 21–24, find (x, y, z).

21.

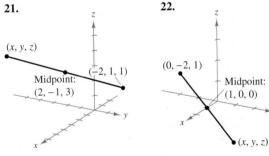

22.

23.

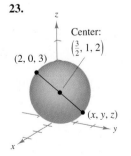

24.

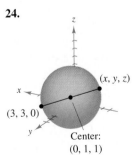

In Exercises 25–28, find the lengths of the sides of the triangle with the given vertices, and determine whether the triangle is a right triangle, an isosceles triangle, or neither of these.

25. $(0, 0, 0)$, $(2, 2, 1)$, $(2, -4, 4)$

26. $(5, 3, 4)$, $(7, 1, 3)$, $(3, 5, 3)$

27. $(-2, 2, 4)$, $(-2, 2, 6)$, $(-2, 4, 8)$

28. $(5, 0, 0)$, $(0, 2, 0)$, $(0, 0, -3)$

29. Think About It   The triangle in Exercise 25 is translated five units upward along the z-axis. Determine the coordinates of the translated triangle.

30. Think About It   The triangle in Exercise 26 is translated three units to the right along the y-axis. Determine the coordinates of the translated triangle.

In Exercises 31–40, find the standard equation of the sphere.

31.
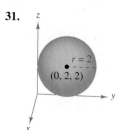

32.
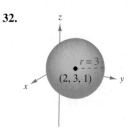

33.
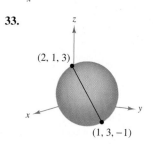

34.
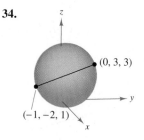

35. Center: $(1, 1, 5)$; radius: 3

36. Center: $(4, -1, 1)$; radius: 5

37. Endpoints of a diameter: $(2, 0, 0)$, $(0, 6, 0)$

38. Endpoints of a diameter: $(1, 0, 0)$, $(0, 5, 0)$

39. Center: $(-2, 1, 1)$; tangent to the xy-plane

40. Center: $(1, 2, 0)$; tangent to the yz-plane

In Exercises 41–46, find the sphere's center and radius.

41. $x^2 + y^2 + z^2 - 5x = 0$

42. $x^2 + y^2 + z^2 - 8y = 0$

43. $x^2 + y^2 + z^2 - 2x + 6y + 8z + 1 = 0$

44. $x^2 + y^2 + z^2 - 4y + 6z + 4 = 0$

45. $2x^2 + 2y^2 + 2z^2 - 4x - 12y - 8z + 3 = 0$

46. $4x^2 + 4y^2 + 4z^2 - 8x + 16y + 11 = 0$

In Exercises 47–50, sketch the xy-trace of the sphere.

47. $(x - 1)^2 + (y - 3)^2 + (z - 2)^2 = 25$

48. $(x + 1)^2 + (y + 2)^2 + (z - 2)^2 = 16$

49. $x^2 + y^2 + z^2 - 6x - 10y + 6z + 30 = 0$

50. $x^2 + y^2 + z^2 - 4y + 2z - 60 = 0$

In Exercises 51–54, sketch the yz-trace of the sphere.

51. $x^2 + (y + 3)^2 + z^2 = 25$

52. $(x + 2)^2 + (y - 3)^2 + z^2 = 9$

53. $x^2 + y^2 + z^2 - 4x - 4y - 6z - 12 = 0$

54. $x^2 + y^2 + z^2 - 6x - 10y + 6z + 30 = 0$

In Exercises 55–58, sketch the trace of the intersection of each plane with the given sphere.

55. $x^2 + y^2 + z^2 = 25$

   (a) $z = 3$ (b) $x = 4$

56. $x^2 + y^2 + z^2 = 169$

   (a) $x = 5$ (b) $y = 12$

57. $x^2 + y^2 + z^2 - 4x - 6y + 9 = 0$

   (a) $x = 2$ (b) $y = 3$

58. $x^2 + y^2 + z^2 - 8x - 6z + 16 = 0$

   (a) $x = 4$ (b) $z = 3$

59. Geology   Crystals are classified according to their symmetry. Crystals shaped like cubes are classified as isometric. The vertices of an isometric crystal mapped onto a three-dimensional coordinate system are shown in the figure. Determine (x, y, z).

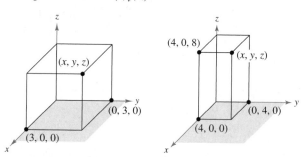

Figure for 59   Figure for 60

60. Crystals   Crystals shaped like rectangular prisms are classified as tetragonal. The vertices of a tetragonal crystal mapped onto a three-dimensional coordinate system are shown in the figure. Determine (x, y, z).

61. Architecture   A spherical building has a diameter of 165 feet. The center of the building is placed at the origin of a three-dimensional coordinate system. What is the equation of the sphere?

- Sketch planes in space.
- Draw planes in space with different numbers of intercepts.
- Classify quadric surfaces in space.

Equations of Planes in Space

In Section 13.1, you studied one type of surface in space—a sphere. In this section, you will study a second type—a plane in space. The **general equation of a plane** in space is

$$ax + by + cz = d.$$ General equation of a plane

Note the similarity of this equation to the general equation of a line in the plane. In fact, if you intersect the plane represented by this equation with each of the three coordinate planes, you will obtain traces that are lines, as shown in Figure 13.10.

In Figure 13.10, the points where the plane intersects the three coordinate axes are the x-, y-, and z-intercepts of the plane. By connecting these three points, you can form a triangular region, which helps you visualize the plane in space.

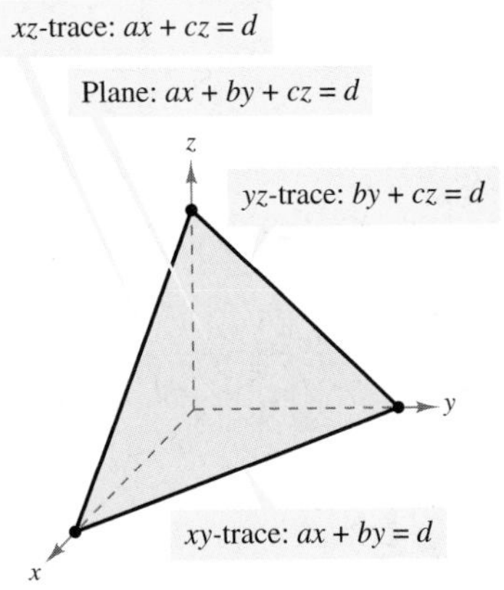

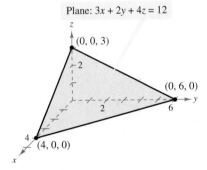

xz-trace: $ax + cz = d$

Plane: $ax + by + cz = d$

yz-trace: $by + cz = d$

xy-trace: $ax + by = d$

FIGURE 13.10

Example 1   Sketching a Plane in Space

Find the x-, y-, and z-intercepts of the plane given by

$$3x + 2y + 4z = 12.$$

Then sketch the plane.

SOLUTION   To find the x-intercept, let both y and z be zero.

$$3x + 2(0) + 4(0) = 12$$ Substitute 0 for y and z.

$$3x = 12$$ Simplify.

$$x = 4$$ Solve for x.

So, the x-intercept is $(4, 0, 0)$. To find the y-intercept, let x and z be zero and conclude that $y = 6$. So, the y-intercept is $(0, 6, 0)$. Similarly, by letting x and y be zero, you can determine that $z = 3$ and that the z-intercept is $(0, 0, 3)$. Figure 13.11 shows the triangular portion of the plane formed by connecting the three intercepts.

Plane: $3x + 2y + 4z = 12$

$(0, 0, 3)$

$(0, 6, 0)$

$(4, 0, 0)$

FIGURE 13.11   Sketch Made by Connecting Intercepts: $(4, 0, 0)$, $(0, 6, 0)$, $(0, 0, 3)$

✓ CHECKPOINT 1

Find the x-, y-, and z-intercepts of the plane given by

$$2x + 4y + z = 8.$$

Then sketch the plane. ■

Drawing Planes in Space

The planes shown in Figures 13.10 and 13.11 have three intercepts. When this occurs, we suggest that you draw the plane by sketching the triangular region formed by connecting the three intercepts.

It is possible for a plane in space to have fewer than three intercepts. This occurs when one or more of the coefficients in the equation $ax + by + cz = d$ is zero. Figure 13.12 shows some planes in space that have only one intercept, and Figure 13.13 shows some that have only two intercepts. In each figure, note the use of dashed lines and shading to give the illusion of three dimensions.

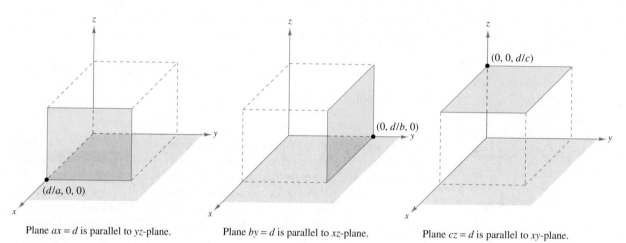

Plane $ax = d$ is parallel to yz-plane.   Plane $by = d$ is parallel to xz-plane.   Plane $cz = d$ is parallel to xy-plane.

FIGURE 13.12   Planes Parallel to Coordinate Planes

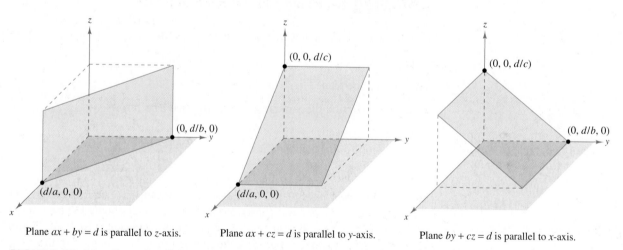

Plane $ax + by = d$ is parallel to z-axis.   Plane $ax + cz = d$ is parallel to y-axis.   Plane $by + cz = d$ is parallel to x-axis.

FIGURE 13.13   Planes Parallel to Coordinate Axes

DISCOVERY

What is the equation of each plane?

a. xy-plane   **b.** xz-plane   **c.** yz-plane

Quadric Surfaces

A third common type of surface in space is a **quadric surface.** Every quadric surface has an equation of the form

$$Ax^2 + By^2 + Cz^2 + Dx + Ey + Fz + G = 0. \qquad \text{Second-degree equation}$$

There are six basic types of quadric surfaces.

1. Elliptic cone

2. Elliptic paraboloid

3. Hyperbolic paraboloid

4. Ellipsoid

5. Hyperboloid of one sheet

6. Hyperboloid of two sheets

The six types are summarized on pages 942 and 943. Notice that each surface is pictured with two types of three-dimensional sketches. The computer-generated sketches use traces with hidden lines to give the illusion of three dimensions. The artist-rendered sketches use shading to create the same illusion.

All of the quadric surfaces on pages 942 and 943 are centered at the origin and have axes along the coordinate axes. Moreover, only one of several possible orientations of each surface is shown. If the surface has a different center or is oriented along a different axis, then its standard equation will change accordingly. For instance, the ellipsoid

$$\frac{x^2}{1^2} + \frac{y^2}{3^2} + \frac{z^2}{2^2} = 1$$

has $(0, 0, 0)$ as its center, but the ellipsoid

$$\frac{(x - 2)^2}{1^2} + \frac{(y + 1)^2}{3^2} + \frac{(z - 4)^2}{2^2} = 1$$

has $(2, -1, 4)$ as its center. A computer-generated graph of the first ellipsoid is shown in Figure 13.14.

DISCOVERY

One way to help visualize a quadric surface is to determine the intercepts of the surface with the coordinate axes. What are the intercepts of the ellipsoid in Figure 13.14?

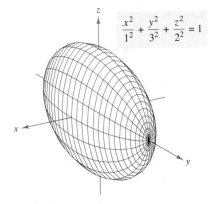

$$\frac{x^2}{1^2} + \frac{y^2}{3^2} + \frac{z^2}{2^2} = 1$$

FIGURE 13.14

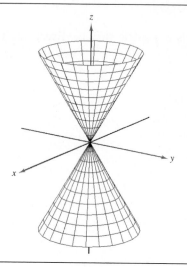

Elliptic Cone

$$\frac{x^2}{a^2} + \frac{y^2}{b^2} - \frac{z^2}{c^2} = 0$$

Trace	*Plane*
Ellipse	Parallel to xy-plane
Hyperbola	Parallel to xz-plane
Hyperbola	Parallel to yz-plane

The axis of the cone corresponds to the variable whose coefficient is negative. The traces in the coordinate planes parallel to this axis are intersecting lines.

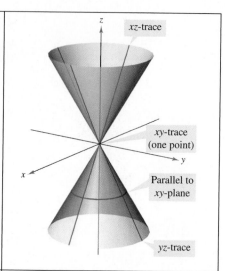

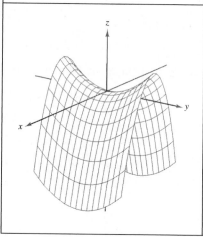

Elliptic Paraboloid

$$z = \frac{x^2}{a^2} + \frac{y^2}{b^2}$$

Trace	*Plane*
Ellipse	Parallel to xy-plane
Parabola	Parallel to xz-plane
Parabola	Parallel to yz-plane

The axis of the paraboloid corresponds to the variable raised to the first power.

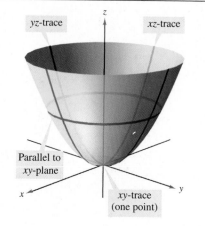

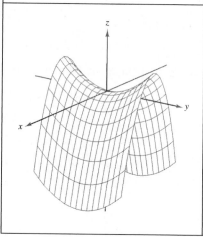

Hyperbolic Paraboloid

$$z = \frac{y^2}{b^2} - \frac{x^2}{a^2}$$

Trace	*Plane*
Hyperbola	Parallel to xy-plane
Parabola	Parallel to xz-plane
Parabola	Parallel to yz-plane

The axis of the paraboloid corresponds to the variable raised to the first power.

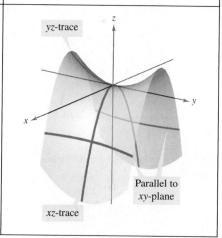

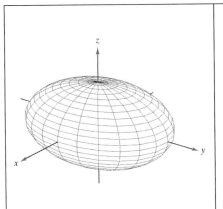

Ellipsoid

$$\frac{x^2}{a^2} + \frac{y^2}{b^2} + \frac{z^2}{c^2} = 1$$

Trace	*Plane*
Ellipse	Parallel to *xy*-plane
Ellipse	Parallel to *xz*-plane
Ellipse	Parallel to *yz*-plane

The surface is a sphere if the coefficients *a*, *b*, and *c* are equal and nonzero.

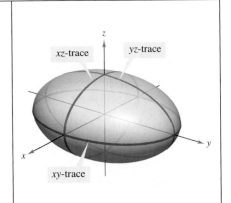

Hyperboloid of One Sheet

$$\frac{x^2}{a^2} + \frac{y^2}{b^2} - \frac{z^2}{c^2} = 1$$

Trace	*Plane*
Ellipse	Parallel to *xy*-plane
Hyperbola	Parallel to *xz*-plane
Hyperbola	Parallel to *yz*-plane

The axis of the hyperboloid corresponds to the variable whose coefficient is negative.

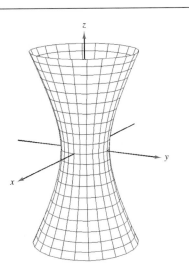

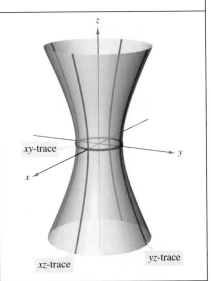

Hyperboloid of Two Sheets

$$\frac{z^2}{c^2} - \frac{x^2}{a^2} - \frac{y^2}{b^2} = 1$$

Trace	*Plane*
Ellipse	Parallel to *xy*-plane
Hyperbola	Parallel to *xz*-plane
Hyperbola	Parallel to *yz*-plane

The axis of the hyperboloid corresponds to the variable whose coefficient is positive. There is no trace in the coordinate plane perpendicular to this axis.

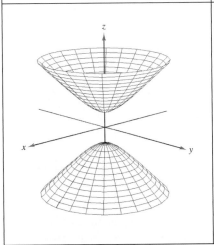

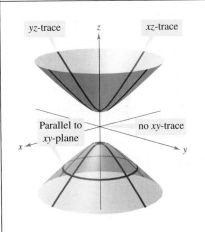

When classifying quadric surfaces, note that the two types of paraboloids have one variable raised to the first power. The other four types of quadric surfaces have equations that are of second degree in *all* three variables.

Example 2 Classifying a Quadric Surface

Classify the surface given by $x - y^2 - z^2 = 0$. Describe the traces of the surface in the xy-plane, the xz-plane, and the plane given by $x = 1$.

SOLUTION Because x is raised only to the first power, the surface is a paraboloid whose axis is the x-axis, as shown in Figure 13.15. In standard form, the equation is

$$x = y^2 + z^2.$$

The traces in the xy-plane, the xz-plane, and the plane given by $x = 1$ are as shown.

Trace in xy-plane $(z = 0)$:	$x = y^2$	Parabola
Trace in xz-plane $(y = 0)$:	$x = z^2$	Parabola
Trace in plane $x = 1$:	$y^2 + z^2 = 1$	Circle

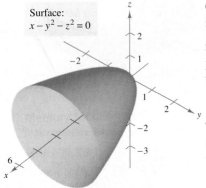

Surface:
$x - y^2 - z^2 = 0$

FIGURE 13.15 Elliptic Paraboloid

These three traces are shown in Figure 13.16. From the traces, you can see that the surface is an elliptic (or circular) paraboloid. If you have access to a three-dimensional graphing utility, try using it to graph this surface. If you do this, you will discover that sketching surfaces in space is not a simple task—even with a graphing utility.

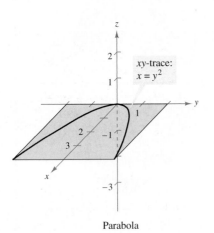

xy-trace:
$x = y^2$

Parabola

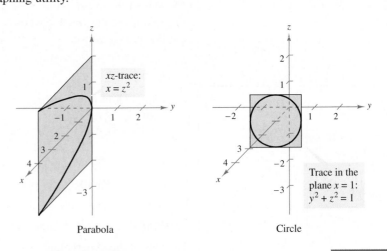

xz-trace:
$x = z^2$

Parabola

Trace in the plane $x = 1$:
$y^2 + z^2 = 1$

Circle

FIGURE 13.16

✓ **CHECKPOINT 2**

Classify the surface given by $x^2 + y^2 - z^2 = 1$. Describe the traces of the surface in the xy-plane, the yz-plane, the xz-plane, and the plane given by $z = 3$. ∎

Example 3   Classifying Quadric Surfaces

Classify the surface given by each equation.

a. $x^2 - 4y^2 - 4z^2 - 4 = 0$

b. $x^2 + 4y^2 + z^2 - 4 = 0$

SOLUTION

a. The equation $x^2 - 4y^2 - 4z^2 - 4 = 0$ can be written in standard form as

$$\frac{x^2}{4} - y^2 - z^2 = 1. \qquad \text{Standard form}$$

From the standard form, you can see that the graph is a hyperboloid of two sheets, with the x-axis as its axis, as shown in Figure 13.17(a).

b. The equation $x^2 + 4y^2 + z^2 - 4 = 0$ can be written in standard form as

$$\frac{x^2}{4} + y^2 + \frac{z^2}{4} = 1. \qquad \text{Standard form}$$

From the standard form, you can see that the graph is an ellipsoid, as shown in Figure 13.17(b).

✓ **CHECKPOINT 3**

Write each quadric surface in standard form and classify each equation.

a. $4x^2 + 9y^2 - 36z = 0$

b. $36x^2 + 16y^2 - 144z^2 = 0$ ▪

Surface:
$x^2 - 4y^2 - 4z^2 - 4 = 0$

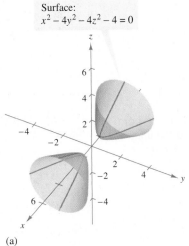

(a)

Surface:
$x^2 + 4y^2 + z^2 - 4 = 0$

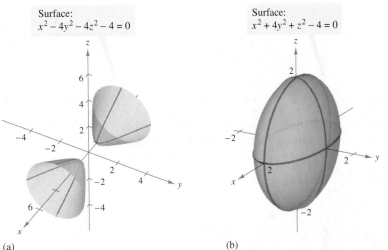

(b)

FIGURE 13.17

┌─ **CONCEPT CHECK** ─┐

1. Give the general equation of a plane in space.

2. List the six basic types of quadric surfaces.

3. Which types of quadric surfaces have equations that are of second degree in *all* three variables? Which types of quadric surfaces have equations that have one variable raised to the first power?

4. Is it possible for a plane in space to have fewer than three intercepts? If so, when does this occur?

Skills Review 13.2   The following warm-up exercises involve skills that were covered in earlier sections. You will use these skills in the exercise set for this section. For additional help, review Sections 2.1 and 13.1.

In Exercises 1–4, find the x- and y-intercepts of the function.

1. $3x + 4y = 12$ **2.** $6x + y = -8$ **3.** $-2x + y = -2$ **4.** $-x - y = 5$

In Exercises 5–8, rewrite the expression by completing the square.

5. $x^2 + y^2 + z^2 - 2x - 4y - 6z + 15 = 0$ **6.** $x^2 + y^2 - z^2 - 8x + 4y - 6z + 11 = 0$

7. $z - 2 = x^2 + y^2 + 2x - 2y$ **8.** $x^2 + y^2 + z^2 - 6x + 10y + 26z = -202$

In Exercises 9 and 10, write the equation of the sphere in standard form.

9. $16x^2 + 16y^2 + 16z^2 = 4$ **10.** $9x^2 + 9y^2 + 9z^2 = 36$

Exercises 13.2

See www.CalcChat.com for worked-out solutions to odd-numbered exercises.

In Exercises 1–12, find the intercepts and sketch the graph of the plane.

1. $4x + 2y + 6z = 12$ **2.** $3x + 6y + 2z = 6$

3. $3x + 3y + 5z = 15$ **4.** $x + y + z = 3$

5. $2x - y + 3z = 4$ **6.** $2x - y + z = 4$

7. $z = 8$ **8.** $x = 5$

9. $y + z = 5$ **10.** $x + 2y = 4$

11. $x + y - z = 0$ **12.** $x - 3z = 3$

In Exercises 13–20, find the distance between the point and the plane (see figure). The distance D between a point (x_0, y_0, z_0) and the plane $ax + by + cz + d = 0$ is

$$D = \frac{|ax_0 + by_0 + cz_0 + d|}{\sqrt{a^2 + b^2 + c^2}}$$

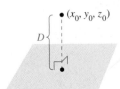

Plane:
$ax + by + cz + d = 0$

13. $(0, 0, 0), 2x + 3y + z = 12$

14. $(0, 0, 0), 8x - 4y + z = 8$

15. $(1, 5, -4), 3x - y + 2z = 6$

16. $(3, 2, 1), x - y + 2z = 4$

17. $(1, 0, -1), 2x - 4y + 3z = 12$

18. $(2, -1, 0), 3x + 3y + 2z = 6$

19. $(3, 2, -1), 2x - 3y + 4z = 24$

20. $(-2, 1, 0), 2x + 5y - z = 20$

In Exercises 21–30, determine whether the planes $a_1x + b_1y + c_1z = d_1$ and $a_2x + b_2y + c_2z = d_2$ are parallel, perpendicular, or neither. The planes are parallel if there exists a nonzero constant k such that $a_1 = ka_2$, $b_1 = kb_2$, and $c_1 = kc_2$, and are perpendicular if $a_1a_2 + b_1b_2 + c_1c_2 = 0$.

21. $5x - 3y + z = 4, x + 4y + 7z = 1$

22. $3x + y - 4z = 3, -9x - 3y + 12z = 4$

23. $x - 5y - z = 1, 5x - 25y - 5z = -3$

24. $x + 3y + 2z = 6, 4x - 12y + 8z = 24$

25. $x + 2y = 3, 4x + 8y = 5$

26. $x + 3y + z = 7, x - 5z = 0$

27. $2x + y = 3, 3x - 5z = 0$

28. $2x - z = 1, 4x + y + 8z = 10$

29. $x = 6, y = -1$

30. $x = -2, y = 4$

In Exercises 31–36, match the equation with its graph. [The graphs are labeled (a)–(f).]

(a)

(b)

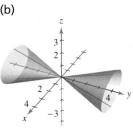

(c)

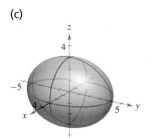

(d)

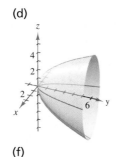

(e)

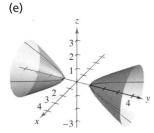

(f)

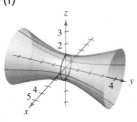

31. $\dfrac{x^2}{9} + \dfrac{y^2}{16} + \dfrac{z^2}{9} = 1$ **32.** $15x^2 - 4y^2 + 15z^2 = -4$

33. $4x^2 - y^2 + 4z^2 = 4$ **34.** $y^2 = 4x^2 + 9z^2$

35. $4x^2 - 4y + z^2 = 0$ **36.** $4x^2 - y^2 + 4z = 0$

In Exercises 37–40, describe the traces of the surface in the given planes.

Surface	*Planes*
37. $x^2 - y - z^2 = 0$	xy-plane, $y = 1$, yz-plane
38. $y = x^2 + z^2$	xy-plane, $y = 1$, yz-plane
39. $\dfrac{x^2}{4} + y^2 + z^2 = 1$	xy-plane, xz-plane, yz-plane
40. $y^2 + z^2 - x^2 = 1$	xy-plane, xz-plane, yz-plane

In Exercises 41–54, identify the quadric surface.

41. $x^2 + \dfrac{y^2}{4} + z^2 = 1$ **42.** $\dfrac{x^2}{9} + \dfrac{y^2}{16} + \dfrac{z^2}{16} = 1$

43. $25x^2 + 25y^2 - z^2 = 5$ **44.** $9x^2 + 4y^2 - 8z^2 = 72$

45. $x^2 - y + z^2 = 0$ **46.** $z = 4x^2 + y^2$

47. $x^2 - y^2 + z = 0$ **48.** $z^2 - x^2 - \dfrac{y^2}{4} = 1$

49. $2x^2 - y^2 + 2z^2 = -4$ **50.** $z^2 = x^2 + \dfrac{y^2}{4}$

51. $z^2 = 9x^2 + y^2$ **52.** $4y = x^2 + z^2$

53. $3z = -y^2 + x^2$ **54.** $z^2 = 2x^2 + 2y^2$

Think About It In Exercises 55–58, each figure is a graph of the quadric surface $z = x^2 + y^2$. Match each of the four graphs with the point in space from which the paraboloid is viewed. The four points are $(0, 0, 20)$, $(0, 20, 0)$, $(20, 0, 0)$, and $(10, 10, 20)$.

55.

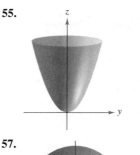

56.

57.

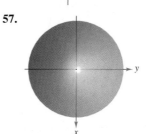

58.

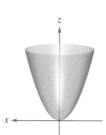

59. Modeling Data Per capita consumptions (in gallons) of different types of plain milk in the United States from 1999 through 2004 are shown in the table. Consumption of reduced-fat (1%) and skim milks, reduced-fat milk (2%), and whole milk are represented by the variables x, y, and z, respectively. *(Source: U.S. Department of Agriculture)*

Year	1999	2000	2001	2002	2003	2004
x	6.2	6.1	5.9	5.8	5.6	5.5
y	7.3	7.1	7.0	7.0	6.9	6.9
z	7.8	7.7	7.4	7.3	7.2	6.9

A model for the data in the table is given by $-1.25x + 0.125y + z = 0.95$.

(a) Complete a fourth row of the table using the model to approximate z for the given values of x and y. Compare the approximations with the actual values of z.

(b) According to this model, increases in consumption of milk types y and z would correspond to what kind of change in consumption of milk type x?

60. Physical Science Because of the forces caused by its rotation, Earth is actually an oblate ellipsoid rather than a sphere. The equatorial radius is 3963 miles and the polar radius is 3950 miles. Find an equation of the ellipsoid. Assume that the center of Earth is at the origin and the xy-trace $(z = 0)$ corresponds to the equator.

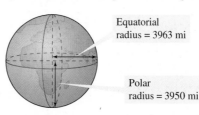

Equatorial radius = 3963 mi

Polar radius = 3950 mi

Functions of Several Variables

- Evaluate functions of several variables.
- Find the domains and ranges of functions of several variables.
- Read contour maps and sketch level curves of functions of two variables.
- Use functions of several variables to answer questions about real-life situations.

Functions of Several Variables

So far in this text, you have studied functions of a single independent variable. Many quantities in science, business, and technology, however, are functions not of one, but of two or more variables. For instance, the demand function for a product is often dependent on the price *and* the advertising, rather than on the price alone. The notation for a function of two or more variables is similar to that for a function of a single variable. Here are two examples.

$$z = f(x, y) = x^2 + xy \qquad \text{Function of two variables}$$

$$\underbrace{}_{\text{2 variables}}$$

and

$$w = f(x, y, z) = x + 2y - 3z \qquad \text{Function of three variables}$$

$$\underbrace{}_{\text{3 variables}}$$

Definition of a Function of Two Variables

Let D be a set of ordered pairs of real numbers. If to each ordered pair (x, y) in D there corresponds a unique real number $f(x, y)$, then f is called a **function of x and y.** The set D is the **domain** of f, and the corresponding set of z-values is the **range** of f. Functions of three, four, or more variables are defined similarly.

Example 1 Evaluating Functions of Several Variables

a. For $f(x, y) = 2x^2 - y^2$, you can evaluate $f(2, 3)$ as shown.

$$f(2, 3) = 2(2)^2 - (3)^2$$
$$= 8 - 9$$
$$= -1$$

b. For $f(x, y, z) = e^x(y + z)$, you can evaluate $f(0, -1, 4)$ as shown.

$$f(0, -1, 4) = e^0(-1 + 4)$$
$$= (1)(3)$$
$$= 3$$

✓**CHECKPOINT 1**

Find the function values of $f(x, y)$.

a. For $f(x, y) = x^2 + 2xy$, find $f(2, -1)$.

b. For $f(x, y, z) = \dfrac{2x^2z}{y^3}$, find $f(-3, 2, 1)$. ■

The Graph of a Function of Two Variables

A function of two variables can be represented graphically as a surface in space by letting $z = f(x, y)$. When sketching the graph of a function of x and y, remember that even though the graph is three-dimensional, the domain of the function is two-dimensional—it consists of the points in the xy-plane for which the function is defined. As with functions of a single variable, unless specifically restricted, the domain of a function of two variables is assumed to be the set of all points (x, y) for which the defining equation has meaning. In other words, to each point (x, y) in the domain of f there corresponds a point (x, y, z) on the surface, and conversely, to each point (x, y, z) on the surface there corresponds a point (x, y) in the domain of f.

Hemisphere:
$f(x, y) = \sqrt{64 - x^2 - y^2}$

Domain: $x^2 + y^2 \leq 64$
Range: $0 \leq z \leq 8$

FIGURE 13.18

Example 2 Finding the Domain and Range of a Function

Find the domain and range of the function

$$f(x, y) = \sqrt{64 - x^2 - y^2}.$$

SOLUTION Because no restrictions are given, the domain is assumed to be the set of all points for which the defining equation makes sense.

$$64 - x^2 - y^2 \geq 0 \qquad \text{Quantity inside radical must be nonnegative.}$$

$$x^2 + y^2 \leq 64 \qquad \text{Domain of the function}$$

So, the domain is the set of all points that lie on or inside the circle given by $x^2 + y^2 = 8^2$. The range of f is the set

$$0 \leq z \leq 8. \qquad \text{Range of the function}$$

As shown in Figure 13.18, the graph of the function is a hemisphere.

✓ **CHECKPOINT 2**

Find the domain and range of the function

$$f(x, y) = \sqrt{9 - x^2 - y^2}. \blacksquare$$

TECHNOLOGY

(T) Some three-dimensional graphing utilities can graph equations in x, y, and z. Others are programmed to graph only functions of x and y. A surface in space represents the graph of a function of x and y only if each vertical line intersects the surface at most once. For instance, the surface shown in Figure 13.18 passes this vertical line test, but the surface at the right (drawn by *Mathematica*) does not represent the graph of a function of x and y.

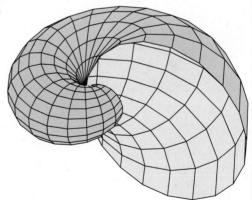

Some vertical lines intersect this surface more than once. So, the surface does not pass the Vertical Line Test and is not a function of x and y.

Contour Maps and Level Curves

A **contour map** of a surface is created by *projecting* traces, taken in evenly spaced planes that are parallel to the *xy*-plane, onto the *xy*-plane. Each projection is a **level curve** of the surface.

Contour maps are used to create weather maps, topographical maps, and population density maps. For instance, Figure 13.19(a) shows a graph of a "mountain and valley" surface given by $z = f(x, y)$. Each of the level curves in Figure 13.19(b) represents the intersection of the surface $z = f(x, y)$ with a plane $z = c$, where $c = 828, 830, . . . , 854$.

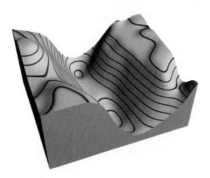

(a) Surface

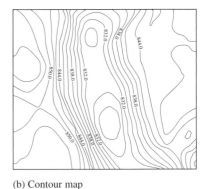

(b) Contour map

FIGURE 13.19

| *Example 3* | Reading a Contour Map | |

The "contour map" in Figure 13.20 was computer generated using data collected by satellite instrumentation. Color is used to show the "ozone hole" in Earth's atmosphere. The purple and blue areas represent the lowest levels of ozone and the green areas represent the highest level. Describe the areas that have the lowest levels of ozone.   *(Source: National Aeronautics and Space Administration)*

SOLUTION   The lowest levels of ozone are over Antarctica and the Antarctic Ocean. The ozone layer acts to protect life on Earth by blocking harmful ultraviolet rays from the sun. The "ozone hole" in the polar region of the Southern Hemisphere is an area in which there is a severe depletion of the ozone levels in the atmosphere. It is primarily caused by compounds that release chlorine and bromine gases into the atmosphere.   ━━━━━━━

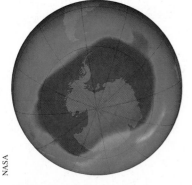

NASA

FIGURE 13.20

✓ CHECKPOINT 3

When the level curves of a contour map are close together, is the surface represented by the contour map steep or nearly level? When the level curves of a contour map are far apart, is the surface represented by the contour map steep or nearly level? ■

Example 4   **Reading a Contour Map**

The contour map shown in Figure 13.21 represents the economy of the United States. Discuss the use of color to represent the level curves.   *(Source: U.S. Census Bureau)*

SOLUTION   You can see from the key that the light yellow regions are mainly used in crop production. The gray areas represent regions that are unproductive. Manufacturing centers are denoted by large red dots and mineral deposits are denoted by small black dots.

One advantage of such a map is that it allows you to "see" the components of the country's economy at a glance. From the map it is clear that the Midwest is responsible for most of the crop production in the United States.

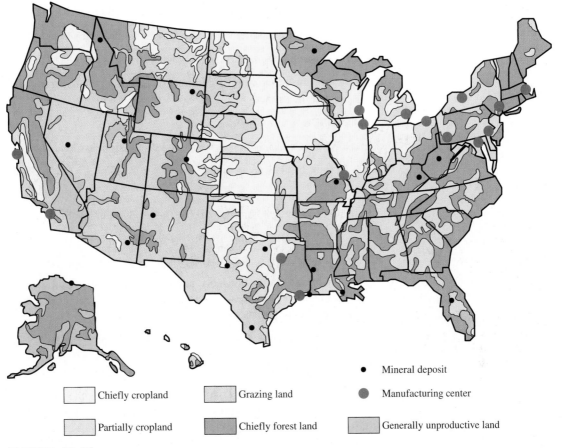

Chiefly cropland	Grazing land	• Mineral deposit
Partially cropland	Chiefly forest land	Manufacturing center
		Generally unproductive land

FIGURE 13.21

✓ **CHECKPOINT 4**

Use Figure 13.21 to describe how Alaska contributes to the U.S. economy. Does Alaska contain any manufacturing centers? Does Alaska contain any mineral deposits? ▪

Applications

The **Cobb-Douglas production function** is used in economics to represent the numbers of units produced by varying amounts of labor and capital. Let x represent the number of units of labor and let y represent the number of units of capital. Then, the number of units produced is modeled by

$$f(x, y) = Cx^a y^{1-a}$$

where C is a constant and $0 < a < 1$.

Example 5 | Using a Production Function

A manufacturer estimates that its production (measured in units of a product) can be modeled by $f(x, y) = 100x^{0.6}y^{0.4}$, where the labor x is measured in person-hours and the capital y is measured in thousands of dollars.

a. What is the production level when $x = 1000$ and $y = 500$?

b. What is the production level when $x = 2000$ and $y = 1000$?

c. How does doubling the amounts of labor and capital from part (a) to part (b) affect the production?

SOLUTION

a. When $x = 1000$ and $y = 500$, the production level is

$$f(1000, 500) = 100(1000)^{0.6}(500)^{0.4}$$
$$\approx 75{,}786 \text{ units.}$$

b. When $x = 2000$ and $y = 1000$, the production level is

$$f(2000, 1000) = 100(2000)^{0.6}(1000)^{0.4}$$
$$\approx 151{,}572 \text{ units.}$$

c. When the amounts of labor and capital are doubled, the production level also doubles. In Exercise 42, you are asked to show that this is characteristic of the Cobb-Douglas production function.

A contour graph of this function is shown in Figure 13.22. _____

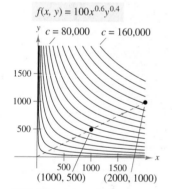

FIGURE 13.22 Level Curves (at Increments of 10,000)

✓ CHECKPOINT 5

Use the Cobb-Douglas production function in Example 5 to find the production levels when $x = 1500$ and $y = 1000$ and when $x = 1000$ and $y = 1500$. Use your results to determine which variable has a greater influence on production.

STUDY TIP

In Figure 13.22, note that the level curves of the function

$$f(x, y) = 100x^{0.6}y^{0.4}$$

occur at increments of 10,000.

Kayte M. Deioma/PhotoEdit

For many Americans, buying a house is the largest single purchase they will ever make. During the 1970s, 1980s, and 1990s, the annual interest rate on home mortgages varied drastically. It was as high as 18% and as low as 5%. Such variations can change monthly payments by hundreds of dollars.

Example 6   Finding Monthly Payments

The monthly payment M for an installment loan of P dollars taken out over t years at an annual interest rate of r is given by

$$M = f(P, r, t) = \frac{\dfrac{Pr}{12}}{1 - \left[\dfrac{1}{1 + (r/12)}\right]^{12t}}.$$

a. Find the monthly payment for a home mortgage of $100,000 taken out for 30 years at an annual interest rate of 7%.

b. Find the monthly payment for a car loan of $22,000 taken out for 5 years at an annual interest rate of 8%.

SOLUTION

a. If $P = \$100{,}000$, $r = 0.07$, and $t = 30$, then the monthly payment is

$$M = f(100{,}000, 0.07, 30)$$

$$= \frac{\dfrac{(100{,}000)(0.07)}{12}}{1 - \left[\dfrac{1}{1 + (0.07/12)}\right]^{12(30)}}$$

$$\approx \$665.30.$$

b. If $P = \$22{,}000$, $r = 0.08$, and $t = 5$, then the monthly payment is

$$M = f(22{,}000, 0.08, 5)$$

$$= \frac{\dfrac{(22{,}000)(0.08)}{12}}{1 - \left[\dfrac{1}{1 + (0.08/12)}\right]^{12(5)}}$$

$$\approx \$446.08.$$

✓ CHECKPOINT 6

a. Find the monthly payment M for a home mortgage of $100,000 taken out for 30 years at an annual interest rate of 8%.

b. Find the total amount of money you will pay for the mortgage. ■

CONCEPT CHECK

1. The function $f(x, y) = x + y$ is a function of how many variables?
2. What is a graph of a function of two variables?
3. Give a description of the domain of a function of two variables.
4. How is a contour map created? What is a level curve?

The following warm-up exercises involve skills that were covered in earlier sections. You will use these skills in the exercise set for this section. For additional help, review Sections 0.4, 0.5, and 2.4.

In Exercises 1–4, evaluate the function when $x = -3$.

1. $f(x) = 5 - 2x$ **2.** $f(x) = -x^2 + 4x + 5$ **3.** $y = \sqrt{4x^2 - 3x + 4}$ **4.** $y = \sqrt[3]{34 - 4x + 2x^2}$

In Exercises 5–8, find the domain of the function.

5. $f(x) = 5x^2 + 3x - 2$ **6.** $g(x) = \dfrac{1}{2x} - \dfrac{2}{x+3}$ **7.** $h(y) = \sqrt{y - 5}$ **8.** $f(y) = \sqrt{y^2 - 5}$

In Exercises 9 and 10, evaluate the expression.

9. $(476)^{0.65}$ **10.** $(251)^{0.35}$

Exercises 13.3

See www.CalcChat.com for worked-out solutions to odd-numbered exercises.

In Exercises 1–14, find the function values.

1. $f(x, y) = \dfrac{x}{y}$

  (a) $f(3, 2)$ (b) $f(-1, 4)$ (c) $f(30, 5)$

  (d) $f(5, y)$ (e) $f(x, 2)$ (f) $f(5, t)$

2. $f(x, y) = 4 - x^2 - 4y^2$

  (a) $f(0, 0)$ (b) $f(0, 1)$ (c) $f(2, 3)$

  (d) $f(1, y)$ (e) $f(x, 0)$ (f) $f(t, 1)$

3. $f(x, y) = xe^y$

  (a) $f(5, 0)$ (b) $f(3, 2)$ (c) $f(2, -1)$

  (d) $f(5, y)$ (e) $f(x, 2)$ (f) $f(t, t)$

4. $g(x, y) = \ln|x + y|$

  (a) $g(2, 3)$ (b) $g(5, 6)$ (c) $g(e, 0)$

  (d) $g(0, 1)$ (e) $g(2, -3)$ (f) $g(e, e)$

5. $h(x, y, z) = \dfrac{xy}{z}$

  (a) $h(2, 3, 9)$ (b) $h(1, 0, 1)$

6. $f(x, y, z) = \sqrt{x + y + z}$

  (a) $f(0, 5, 4)$ (b) $f(6, 8, -3)$

7. $V(r, h) = \pi r^2 h$

  (a) $V(3, 10)$ (b) $V(5, 2)$

8. $F(r, N) = 500\left(1 + \dfrac{r}{12}\right)^N$

  (a) $F(0.09, 60)$ (b) $F(0.14, 240)$

9. $A(P, r, t) = P\left[\left(1 + \dfrac{r}{12}\right)^{12t} - 1\right]\left(1 + \dfrac{12}{r}\right)$

  (a) $A(100, 0.10, 10)$ (b) $A(275, 0.0925, 40)$

10. $A(P, r, t) = Pe^{rt}$

  (a) $A(500, 0.10, 5)$ (b) $A(1500, 0.12, 20)$

11. $f(x, y) = \displaystyle\int_x^y (2t - 3)\, dt$

  (a) $f(1, 2)$ (b) $f(1, 4)$

12. $g(x, y) = \displaystyle\int_x^y \dfrac{1}{t}\, dt$

  (a) $g(4, 1)$ (b) $g(6, 3)$

13. $f(x, y) = x^2 - 2y$

  (a) $f(x + \Delta x, y)$ (b) $\dfrac{f(x, y + \Delta y) - f(x, y)}{\Delta y}$

14. $f(x, y) = 3xy + y^2$

  (a) $f(x + \Delta x, y)$ (b) $\dfrac{f(x, y + \Delta y) - f(x, y)}{\Delta y}$

In Exercises 15–18, describe the region R in the xy-plane that corresponds to the domain of the function, and find the range of the function.

15. $f(x, y) = \sqrt{16 - x^2 - y^2}$

16. $f(x, y) = x^2 + y^2 - 1$

17. $f(x, y) = e^{x/y}$

18. $f(x, y) = \ln(x + y)$

In Exercises 19–28, describe the region R in the xy-plane that corresponds to the domain of the function.

19. $z = \sqrt{4 - x^2 - y^2}$ **20.** $z = \sqrt{4 - x^2 - 4y^2}$

21. $f(x, y) = x^2 + y^2$ **22.** $f(x, y) = \dfrac{x}{y}$

23. $f(x, y) = \dfrac{1}{xy}$

24. $g(x, y) = \dfrac{1}{x - y}$

25. $h(x, y) = x\sqrt{y}$

26. $f(x, y) = \sqrt{xy}$

27. $g(x, y) = \ln(4 - x - y)$

28. $f(x, y) = ye^{1/x}$

In Exercises 29–32, match the graph of the surface with one of the contour maps. [The contour maps are labeled (a)–(d).]

(a)

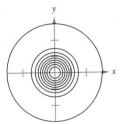

(b)

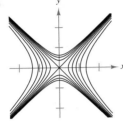

(c)

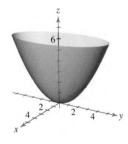

(d)

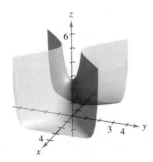

29. $f(x, y) = x^2 + \dfrac{y^2}{4}$

30. $f(x, y) = e^{1 - x^2 + y^2}$

31. $f(x, y) = e^{1 - x^2 - y^2}$

32. $f(x, y) = \ln|y - x^2|$

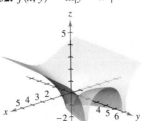

In Exercises 33–40, describe the level curves of the function. Sketch the level curves for the given c-values.

Function	c-Values
33. $z = x + y$	$c = -1, 0, 2, 4$
34. $z = 6 - 2x - 3y$	$c = 0, 2, 4, 6, 8, 10$
35. $z = \sqrt{25 - x^2 - y^2}$	$c = 0, 1, 2, 3, 4, 5$
36. $f(x, y) = x^2 + y^2$	$c = 0, 2, 4, 6, 8$
37. $f(x, y) = xy$	$c = \pm 1, \pm 2, \ldots, \pm 6$
38. $z = e^{xy}$	$c = 1, 2, 3, 4, \frac{1}{2}, \frac{1}{3}, \frac{1}{4}$
39. $f(x, y) = \dfrac{x}{x^2 + y^2}$	$c = \pm\frac{1}{2}, \pm 1, \pm\frac{3}{2}, \pm 2$
40. $f(x, y) = \ln(x - y)$	$c = 0, \pm\frac{1}{2}, \pm 1, \pm\frac{3}{2}, \pm 2$

41. Cobb-Douglas Production Function A manufacturer estimates the Cobb-Douglas production function to be given by
$$f(x, y) = 100x^{0.75}y^{0.25}.$$
Estimate the production levels when $x = 1500$ and $y = 1000$.

42. Cobb-Douglas Production Function Use the Cobb-Douglas production function (Example 5) to show that if both the number of units of labor and the number of units of capital are doubled, the production level is also doubled.

43. Profit A sporting goods manufacturer produces regulation soccer balls at two plants. The costs of producing x_1 units at location 1 and x_2 units at location 2 are given by
$$C_1(x_1) = 0.02x_1^2 + 4x_1 + 500$$
and
$$C_2(x_2) = 0.05x_2^2 + 4x_2 + 275$$
respectively. If the product sells for \$50 per unit, then the profit function for the product is given by
$$P(x_1, x_2) = 50(x_1 + x_2) - C_1(x_1) - C_2(x_2).$$
Find (a) $P(250, 150)$ and (b) $P(300, 200)$.

44. Queuing Model The average amount of time that a customer waits in line for service is given by
$$W(x, y) = \dfrac{1}{x - y}, \quad y < x$$
where y is the average arrival rate and x is the average service rate (x and y are measured in the number of customers per hour). Evaluate W at each point.

(a) $(15, 10)$ (b) $(12, 9)$ (c) $(12, 6)$ (d) $(4, 2)$

Ⓢ **45. Investment**   In 2008, an investment of $1000 was made in a bond earning 10% compounded annually. The investor pays tax at rate R, and the annual rate of inflation is I. In the year 2018, the value V of the bond in constant 2008 dollars is given by

$$V(I, R) = 1000\left[\frac{1 + 0.10(1 - R)}{1 + I}\right]^{10}.$$

Use this function of two variables and a spreadsheet to complete the table.

	Inflation Rate		
Tax Rate	0	0.03	0.05
0			
0.28			
0.35			

Ⓢ **46. Investment**   A principal of $1000 is deposited in a savings account that earns an interest rate of r (written as a decimal), compounded continuously. The amount $A(r, t)$ after t years is $A(r, t) = 1000\, e^{rt}$. Use this function of two variables and a spreadsheet to complete the table.

	Number of Years			
Rate	5	10	15	20
0.02				
0.04				
0.06				
0.08				

47. Meteorology   Meteorologists measure the atmospheric pressure in millibars. From these observations they create weather maps on which the curves of equal atmospheric pressure (isobars) are drawn (see figure). On the map, the closer the isobars the higher the wind speed. Match points A, B, and C with (a) highest pressure, (b) lowest pressure, and (c) highest wind velocity.

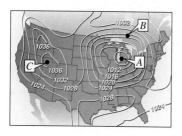

Ⓑ **48. Geology**   The contour map below represents color-coded seismic amplitudes of a fault horizon and a projected contour map, which is used in earthquake studies. *(Source: Adapted from Shipman/Wilson/Todd, An Introduction to Physical Science, Tenth Edition)*

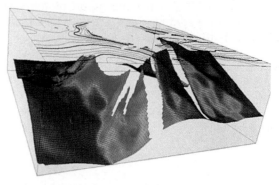

Shipman, *An Introduction to Physical Science* 10/e, 2003, Houghton Mifflin Company

(a) Discuss the use of color to represent the level curves.

(b) Do the level curves correspond to equally spaced amplitudes? Explain your reasoning.

49. Earnings per Share   The earnings per share z (in dollars) for Starbucks Corporation from 1998 through 2006 can be modeled by $z = 0.106x - 0.036y - 0.005$, where x is sales (in billions of dollars) and y is the shareholder's equity (in billions of dollars). *(Source: Starbucks Corporation)*

(a) Find the earnings per share when $x = 8$ and $y = 5$.

(b) Which of the two variables in this model has the greater influence on the earnings per share? Explain.

50. Shareholder's Equity   The shareholder's equity z (in billions of dollars) for Wal-Mart Corporation from 2000 to 2006 can be modeled by $z = 0.205x - 0.073y - 0.728$, where x is net sales (in billions of dollars) and y is the total assets (in billions of dollars). *(Source: Wal-Mart Corporation)*

(a) Find the shareholder's equity when $x = 300$ and $y = 130$.

(b) Which of the two variables in this model has the greater influence on shareholder's equity? Explain.

51. MAKE A DECISION: MONTHLY PAYMENTS   You are taking out a home mortgage for $120,000, and you are given the options below. Which option would you choose? Explain your reasoning.

(a) A fixed annual rate of 8%, over a term of 20 years.

(b) A fixed annual rate of 7%, over a term of 30 years.

(c) An adjustable annual rate of 7%, over a term of 20 years. The annual rate can fluctuate—each year it is set at 1% above the prime rate.

(d) A fixed annual rate of 7%, over a term of 15 years.

Section 13.4

Partial Derivatives

- Find the first partial derivatives of functions of two variables.
- Find the slopes of surfaces in the *x*- and *y*-directions and use partial derivatives to answer questions about real-life situations.
- Find the partial derivatives of functions of several variables.
- Find higher-order partial derivatives.

Functions of Two Variables

Real-life applications of functions of several variables are often concerned with how changes in one of the variables will affect the values of the functions. For instance, an economist who wants to determine the effect of a tax increase on the economy might make calculations using different tax rates while holding all other variables, such as unemployment, constant.

You can follow a similar procedure to find the rate of change of a function *f* with respect to one of its independent variables. That is, you find the derivative of *f* with respect to one independent variable, while holding the other variable(s) constant. This process is called **partial differentiation,** and each derivative is called a **partial derivative.** A function of several variables has as many partial derivatives as it has independent variables.

STUDY TIP

Note that this definition indicates that partial derivatives of a function of two variables are determined by temporarily considering one variable to be fixed. For instance, if $z = f(x, y)$, then to find $\partial z/\partial x$, you consider y to be constant and differentiate with respect to x. Similarly, to find $\partial z/\partial y$, you consider x to be constant and differentiate with respect to y.

Partial Derivatives of a Function of Two Variables

If $z = f(x, y)$, then the **first partial derivatives of f with respect to x and y** are the functions $\partial z/\partial x$ and $\partial z/\partial y$, defined as shown.

$$\frac{\partial z}{\partial x} = \lim_{\Delta x \to 0} \frac{f(x + \Delta x, y) - f(x, y)}{\Delta x} \qquad \text{\textit{y} is held constant.}$$

$$\frac{\partial z}{\partial y} = \lim_{\Delta y \to 0} \frac{f(x, y + \Delta y) - f(x, y)}{\Delta y} \qquad \text{\textit{x} is held constant.}$$

Example 1 Finding Partial Derivatives

Find $\partial z/\partial x$ and $\partial z/\partial y$ for the function $z = 3x - x^2 y^2 + 2x^3 y$.

SOLUTION

$$\frac{\partial z}{\partial x} = 3 - 2xy^2 + 6x^2 y \qquad \text{Hold \textit{y} constant and differentiate with respect to \textit{x}.}$$

$$\frac{\partial z}{\partial y} = -2x^2 y + 2x^3 \qquad \text{Hold \textit{x} constant and differentiate with respect to \textit{y}.}$$

✓**CHECKPOINT 1**

Find $\dfrac{\partial z}{\partial x}$ and $\dfrac{\partial z}{\partial y}$ for $z = 2x^2 - 4x^2 y^3 + y^4$. ■

> **Notation for First Partial Derivatives**
>
> The first partial derivatives of $z = f(x, y)$ are denoted by
>
> $$\frac{\partial z}{\partial x} = f_x(x, y) = z_x = \frac{\partial}{\partial x}[f(x, y)]$$
>
> and
>
> $$\frac{\partial z}{\partial y} = f_y(x, y) = z_y = \frac{\partial}{\partial y}[f(x, y)].$$
>
> The values of the first partial derivatives at the point (a, b) are denoted by
>
> $$\left.\frac{\partial z}{\partial x}\right|_{(a, b)} = f_x(a, b) \quad \text{and} \quad \left.\frac{\partial z}{\partial y}\right|_{(a, b)} = f_y(a, b).$$

TECHNOLOGY

(T) Symbolic differentiation utilities can be used to find partial derivatives of a function of two variables. Try using a symbolic differentiation utility to find the first partial derivatives of the function in Example 2.

Example 2 **Finding and Evaluating Partial Derivatives**

Find the first partial derivatives of $f(x, y) = xe^{x^2y}$ and evaluate each at the point $(1, \ln 2)$.

SOLUTION   To find the first partial derivative with respect to x, hold y constant and differentiate using the Product Rule.

$$f_x(x, y) = x\frac{\partial}{\partial x}[e^{x^2y}] + e^{x^2y}\frac{\partial}{\partial x}[x] \qquad \text{Apply Product Rule.}$$

$$= x(2xy)e^{x^2y} + e^{x^2y} \qquad \text{y is held constant.}$$

$$= e^{x^2y}(2x^2y + 1) \qquad \text{Simplify.}$$

At the point $(1, \ln 2)$, the value of this derivative is

$$f_x(1, \ln 2) = e^{(1)^2(\ln 2)}[2(1)^2(\ln 2) + 1] \qquad \text{Substitute for x and y.}$$

$$= 2(2 \ln 2 + 1) \qquad \text{Simplify.}$$

$$\approx 4.773. \qquad \text{Use a calculator.}$$

To find the first partial derivative with respect to y, hold x constant and differentiate to obtain

$$f_y(x, y) = x(x^2)e^{x^2y} \qquad \text{Apply Constant Multiple Rule.}$$

$$= x^3e^{x^2y}. \qquad \text{Simplify.}$$

At the point $(1, \ln 2)$, the value of this derivative is

$$f_y(1, \ln 2) = (1)^3e^{(1)^2(\ln 2)} \qquad \text{Substitute for x and y.}$$

$$= 2. \qquad \text{Simplify.}$$

✓**CHECKPOINT 2**

Find the first partial derivatives of $f(x, y) = x^2y^3$ and evaluate each at the point $(1, 2)$. ■

Graphical Interpretation of Partial Derivatives

At the beginning of this course, you studied graphical interpretations of the derivative of a function of a single variable. There, you found that $f'(x_0)$ represents the slope of the tangent line to the graph of $y = f(x)$ at the point (x_0, y_0). The partial derivatives of a function of two variables also have useful graphical interpretations. Consider the function

$z = f(x, y).$ Function of two variables

As shown in Figure 13.23(a), the graph of this function is a surface in space. If the variable y is fixed, say at $y = y_0$, then

$z = f(x, y_0)$ Function of one variable

is a function of one variable. The graph of this function is the curve that is the intersection of the plane $y = y_0$ and the surface $z = f(x, y)$. On this curve, the partial derivative

$f_x(x, y_0)$ Slope in x-direction

represents the slope in the plane $y = y_0$, as shown in Figure 13.23(a). In a similar way, if the variable x is fixed, say at $x = x_0$, then

$z = f(x_0, y)$ Function of one variable

is a function of one variable. Its graph is the intersection of the plane $x = x_0$ and the surface $z = f(x, y)$. On this curve, the partial derivative

$f_y(x_0, y)$ Slope in y-direction

represents the slope in the plane $x = x_0$, as shown in Figure 13.23(b).

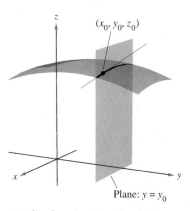

(a) $f_x(x, y_0)$ = slope in x-direction

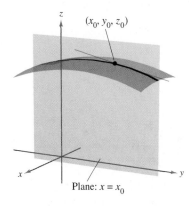

(b) $f_y(x_0, y)$ = slope in y-direction

FIGURE 13.23

DISCOVERY

How can partial derivatives be used to find *relative extrema* of graphs of functions of two variables?

Example 3 **Finding Slopes in the x- and y-Directions**

Find the slopes of the surface given by

$$f(x, y) = -\frac{x^2}{2} - y^2 + \frac{25}{8}$$

at the point $\left(\frac{1}{2}, 1, 2\right)$ in (a) the x-direction and (b) the y-direction.

SOLUTION

a. To find the slope in the x-direction, hold y constant and differentiate with respect to x to obtain

$$f_x(x, y) = -x. \qquad \text{Partial derivative with respect to } x$$

At the point $\left(\frac{1}{2}, 1, 2\right)$, the slope in the x-direction is

$$f_x\left(\frac{1}{2}, 1\right) = -\frac{1}{2} \qquad \text{Slope in } x\text{-direction}$$

as shown in Figure 13.24(a).

b. To find the slope in the y-direction, hold x constant and differentiate with respect to y to obtain

$$f_y(x, y) = -2y. \qquad \text{Partial derivative with respect to } y$$

At the point $\left(\frac{1}{2}, 1, 2\right)$, the slope in the y-direction is

$$f_y\left(\frac{1}{2}, 1\right) = -2 \qquad \text{Slope in } y\text{-direction}$$

as shown in Figure 13.24(b).

 **CHECKPOINT 3**

Find the slopes of the surface given by

$$f(x, y) = 4x^2 + 9y^2 + 36$$

at the point $(1, -1, 49)$ in the x-direction and the y-direction.

DISCOVERY

Find the partial derivatives f_x and f_y at $(0, 0)$ for the function in Example 3. What are the slopes of f in the x- and y-directions at $(0, 0)$? Describe the shape of the graph of f at this point.

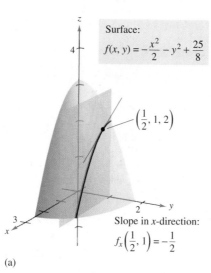

Surface:

$$f(x, y) = -\frac{x^2}{2} - y^2 + \frac{25}{8}$$

$\left(\frac{1}{2}, 1, 2\right)$

Slope in x-direction:

$$f_x\left(\frac{1}{2}, 1\right) = -\frac{1}{2}$$

(a)

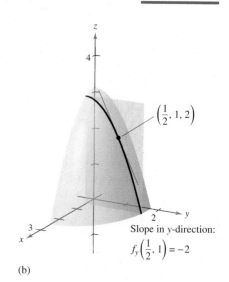

$\left(\frac{1}{2}, 1, 2\right)$

Slope in y-direction:

$$f_y\left(\frac{1}{2}, 1\right) = -2$$

(b)

FIGURE 13.24

Consumer products in the same market or in related markets can be classified as **complementary** or **substitute products.** If two products have a complementary relationship, an increase in the sale of one product will be accompanied by an increase in the sale of the other product. For instance, DVD players and DVDs have a complementary relationship.

If two products have a substitute relationship, an increase in the sale of one product will be accompanied by a decrease in the sale of the other product. For instance, videocassette recorders and DVD players both compete in the same home entertainment market and you would expect a drop in the price of one to be a deterrent to the sale of the other.

| Example 4 | Examining Demand Functions | |

The demand functions for two products are represented by

$$x_1 = f(p_1, p_2) \quad \text{and} \quad x_2 = g(p_1, p_2)$$

where p_1 and p_2 are the prices per unit for the two products, and x_1 and x_2 are the numbers of units sold. The graphs of two different demand functions for x_1 are shown below. Use them to classify the products as complementary or substitute products.

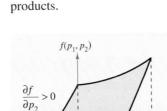

(a)

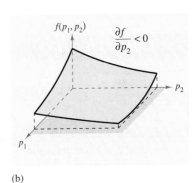

(b)

FIGURE 13.25

SOLUTION

a. Notice that Figure 13.25(a) represents the demand for the *first product.* From the graph of this function, you can see that for a fixed price p_1, an increase in p_2 results in an increase in the demand for the first product. Remember that an increase in p_2 will also result in a decrease in the demand for the second product. So, if $\partial f/\partial p_2 > 0$, the two products have a *substitute* relationship.

b. Notice that Figure 13.25(b) represents a different demand for the *first product.* From the graph of this function, you can see that for a fixed price p_1, an increase in p_2 results in a decrease in the demand for the first product. Remember that an increase in p_2 will also result in a decrease in the demand for the second product. So, if $\partial f/\partial p_2 < 0$, the two products have a *complementary* relationship.

✓ **CHECKPOINT 4**

Determine if the demand functions below describe a complementary or a substitute product relationship.

$$x_1 = 100 - 2p_1 + 1.5p_2$$
$$x_2 = 145 + \tfrac{1}{2}p_1 - \tfrac{3}{4}p_2 \quad \blacksquare$$

AP/World Wide Photos

In 2007, Subway was chosen as the number one franchise by Entrepreneur Magazine. By the end of the year 2006, Subway had a total of 26,197 franchises worldwide. What type of product would be complementary to a Subway sandwich? What type of product would be a substitute?

Functions of Three Variables

The concept of a partial derivative can be extended naturally to functions of three or more variables. For instance, the function $w = f(x, y, z)$ has three partial derivatives, each of which is formed by considering two of the variables to be constant. That is, to define the partial derivative of w with respect to x, consider y *and* z to be constant and write

$$\frac{\partial w}{\partial x} = f_x(x, y, z) = \lim_{\Delta x \to 0} \frac{f(x + \Delta x, y, z) - f(x, y, z)}{\Delta x}.$$

To define the partial derivative of w with respect to y, consider x *and* z to be constant and write

$$\frac{\partial w}{\partial y} = f_y(x, y, z) = \lim_{\Delta y \to 0} \frac{f(x, y + \Delta y, z) - f(x, y, z)}{\Delta y}.$$

To define the partial derivative of w with respect to z, consider x *and* y to be constant and write

$$\frac{\partial w}{\partial z} = f_z(x, y, z) = \lim_{\Delta z \to 0} \frac{f(x, y, z + \Delta z) - f(x, y, z)}{\Delta z}.$$

Example 5   Finding Partial Derivatives of a Function

Find the three partial derivatives of the function

$$w = xe^{xy + 2z}.$$

SOLUTION   Holding y and z constant, you obtain

$$
\begin{aligned}
\frac{\partial w}{\partial x} &= x\frac{\partial}{\partial x}[e^{xy + 2z}] + e^{xy + 2z}\frac{\partial}{\partial x}[x] && \text{Apply Product Rule.}\\
&= x(ye^{xy + 2z}) + e^{xy + 2z}(1) && \text{Hold } y \text{ and } z \text{ constant.}\\
&= (xy + 1)e^{xy + 2z}. && \text{Simplify.}
\end{aligned}
$$

Holding x and z constant, you obtain

$$
\begin{aligned}
\frac{\partial w}{\partial y} &= x(x)e^{xy + 2z} && \text{Hold } x \text{ and } z \text{ constant.}\\
&= x^2 e^{xy + 2z}. && \text{Simplify.}
\end{aligned}
$$

Holding x and y constant, you obtain

$$
\begin{aligned}
\frac{\partial w}{\partial z} &= x(2)e^{xy + 2z} && \text{Hold } x \text{ and } y \text{ constant.}\\
&= 2xe^{xy + 2z}. && \text{Simplify.}
\end{aligned}
$$

✓ CHECKPOINT 5

Find the three partial derivatives of the function

$$w = x^2 y \ln(xz). \quad ∎$$

TECHNOLOGY

(T) A symbolic differentiation utility can be used to find the partial derivatives of a function of three or more variables. Try using a symbolic differentiation utility to find the partial derivative $f_y(x, y, z)$ for the function in Example 5.

STUDY TIP

Note that in Example 5 the Product Rule was used only when finding the partial derivative with respect to x. Can you see why?

Higher-Order Partial Derivatives

As with ordinary derivatives, it is possible to take second, third, and higher partial derivatives of a function of several variables, provided such derivatives exist. Higher-order derivatives are denoted by the order in which the differentiation occurs. For instance, there are four different ways to find a second partial derivative of $z = f(x, y)$.

$$\frac{\partial}{\partial x}\left(\frac{\partial f}{\partial x}\right) = \frac{\partial^2 f}{\partial x^2} = f_{xx} \qquad \text{Differentiate twice with respect to } x.$$

$$\frac{\partial}{\partial y}\left(\frac{\partial f}{\partial y}\right) = \frac{\partial^2 f}{\partial y^2} = f_{yy} \qquad \text{Differentiate twice with respect to } y.$$

$$\frac{\partial}{\partial y}\left(\frac{\partial f}{\partial x}\right) = \frac{\partial^2 f}{\partial y \partial x} = f_{xy} \qquad \begin{array}{l}\text{Differentiate first with respect to } x \text{ and} \\ \text{then with respect to } y.\end{array}$$

$$\frac{\partial}{\partial x}\left(\frac{\partial f}{\partial y}\right) = \frac{\partial^2 f}{\partial x \partial y} = f_{yx} \qquad \begin{array}{l}\text{Differentiate first with respect to } y \text{ and} \\ \text{then with respect to } x.\end{array}$$

The third and fourth cases are **mixed partial derivatives.** Notice that with the two types of notation for mixed partials, different conventions are used for indicating the order of differentiation. For instance, the partial derivative

$$\frac{\partial}{\partial y}\left(\frac{\partial f}{\partial x}\right) = \frac{\partial^2 f}{\partial y \partial x} \qquad \text{Right-to-left order}$$

indicates differentiation with respect to x first, but the partial derivative

$$(f_y)_x = f_{yx} \qquad \text{Left-to-right order}$$

indicates differentiation with respect to y first. To remember this, note that in each case you differentiate first with respect to the variable "nearest" f.

Example 6   Finding Second Partial Derivatives

Find the second partial derivatives of

$$f(x, y) = 3xy^2 - 2y + 5x^2y^2$$

and determine the value of $f_{xy}(-1, 2)$.

SOLUTION   Begin by finding the first partial derivatives.

$$f_x(x, y) = 3y^2 + 10xy^2 \qquad f_y(x, y) = 6xy - 2 + 10x^2y$$

Then, differentiating with respect to x and y produces

$$f_{xx}(x, y) = 10y^2, \qquad f_{yy}(x, y) = 6x + 10x^2$$
$$f_{xy}(x, y) = 6y + 20xy, \qquad f_{yx}(x, y) = 6y + 20xy.$$

Finally, the value of $f_{xy}(x, y)$ at the point $(-1, 2)$ is

$$f_{xy}(-1, 2) = 6(2) + 20(-1)(2) = 12 - 40 = -28.$$

✓CHECKPOINT 6

Find the second partial derivatives of

$$f(x, y) = 4x^2y^2 + 2x + 4y^2. \quad \blacksquare$$

A function of two variables has two first partial derivatives and four second partial derivatives. For a function of three variables, there are three first partials

$$f_x, f_y, \quad \text{and} \quad f_z$$

and nine second partials

$$f_{xx}, f_{xy}, f_{xz}, f_{yx}, f_{yy}, f_{yz}, f_{zx}, f_{zy}, \quad \text{and} \quad f_{zz}$$

of which six are mixed partials. To find partial derivatives of order three and higher, follow the same pattern used to find second partial derivatives. For instance, if $z = f(x, y)$, then

$$z_{xxx} = \frac{\partial}{\partial x}\left(\frac{\partial^2 f}{\partial x^2}\right) = \frac{\partial^3 f}{\partial x^3} \quad \text{and} \quad z_{xxy} = \frac{\partial}{\partial y}\left(\frac{\partial^2 f}{\partial x^2}\right) = \frac{\partial^3 f}{\partial y \partial x^2}.$$

Example 7   **Finding Second Partial Derivatives**

Find the second partial derivatives of

$$f(x, y, z) = ye^x + x \ln z.$$

SOLUTION   Begin by finding the first partial derivatives.

$$f_x(x, y, z) = ye^x + \ln z, \qquad f_y(x, y, z) = e^x, \qquad f_z(x, y, z) = \frac{x}{z}$$

Then, differentiate with respect to x, y, and z to find the nine second partial derivatives.

$$f_{xx}(x, y, z) = ye^x, \qquad f_{xy}(x, y, z) = e^x, \qquad f_{xz}(x, y, z) = \frac{1}{z}$$

$$f_{yx}(x, y, z) = e^x, \qquad f_{yy}(x, y, z) = 0, \qquad f_{yz}(x, y, z) = 0$$

$$f_{zx}(x, y, z) = \frac{1}{z}, \qquad f_{zy}(x, y, z) = 0, \qquad f_{zz}(x, y, z) = -\frac{x}{z^2}$$

✓**CHECKPOINT 7**

Find the second partial derivatives of $f(x, y, z) = xe^y + 2xz + y^2$. ▪

╭─ **CONCEPT CHECK** ─────────────────────────────────╮

1. Write the notation that denotes the first partial derivative of $z = f(x, y)$ with respect to x.

2. Write the notation that denotes the first partial derivative of $z = f(x, y)$ with respect to y.

3. Let f be a function of two variables x and y. Describe the procedure for finding the first partial derivatives.

4. Define the first partial derivatives of a function f of two variables x and y.

╰───╯

Skills Review 13.4

The following warm-up exercises involve skills that were covered in earlier sections. You will use these skills in the exercise set for this section. For additional help, review Sections 7.4, 7.6, 7.7, 10.3, and 10.5.

In Exercises 1–8, find the derivative of the function.

1. $f(x) = \sqrt{x^2 + 3}$

2. $g(x) = (3 - x^2)^3$

3. $g(t) = te^{2t+1}$

4. $f(x) = e^{2x}\sqrt{1 - e^{2x}}$

5. $f(x) = \ln(3 - 2x)$

6. $u(t) = \ln\sqrt{t^3 - 6t}$

7. $g(x) = \dfrac{5x^2}{(4x - 1)^2}$

8. $f(x) = \dfrac{(x + 2)^3}{(x^2 - 9)^2}$

In Exercises 9 and 10, evaluate the derivative at the point $(2, 4)$.

9. $f(x) = x^2e^{x-2}$

10. $g(x) = x\sqrt{x^2 - x + 2}$

Exercises 13.4

See www.CalcChat.com for worked-out solutions to odd-numbered exercises.

In Exercises 1–14, find the first partial derivatives with respect to x and with respect to y.

1. $z = 3x + 5y - 1$

2. $z = x^2 - 2y$

3. $f(x, y) = 3x - 6y^2$

4. $f(x, y) = x + 4y^{3/2}$

5. $f(x, y) = \dfrac{x}{y}$

6. $z = x\sqrt{y}$

7. $f(x, y) = \sqrt{x^2 + y^2}$

8. $f(x, y) = \dfrac{xy}{x^2 + y^2}$

9. $z = x^2e^{2y}$

10. $z = xe^{x+y}$

11. $h(x, y) = e^{-(x^2+y^2)}$

12. $g(x, y) = e^{x/y}$

13. $z = \ln\dfrac{x + y}{x - y}$

14. $g(x, y) = \ln(x^2 + y^2)$

In Exercises 15–20, let $f(x, y) = 3x^2ye^{x-y}$ and $g(x, y) = 3xy^2e^{y-x}$. Find each of the following.

15. $f_x(x, y)$

16. $f_y(x, y)$

17. $g_x(x, y)$

18. $g_y(x, y)$

19. $f_x(1, 1)$

20. $g_x(-2, -2)$

In Exercises 21–28, evaluate f_x and f_y at the point.

Function	Point
21. $f(x, y) = 3x^2 + xy - y^2$	$(2, 1)$
22. $f(x, y) = x^2 - 3xy + y^2$	$(1, -1)$
23. $f(x, y) = e^{3xy}$	$(0, 4)$
24. $f(x, y) = e^xy^2$	$(0, 2)$
25. $f(x, y) = \dfrac{xy}{x - y}$	$(2, -2)$
26. $f(x, y) = \dfrac{4xy}{\sqrt{x^2 + y^2}}$	$(1, 0)$
27. $f(x, y) = \ln(x^2 + y^2)$	$(1, 0)$
28. $f(x, y) = \ln\sqrt{xy}$	$(-1, -1)$

In Exercises 29–32, find the first partial derivatives with respect to x, y, and z.

29. $w = xyz$

30. $w = x^2 - 3xy + 4yz + z^3$

31. $w = \dfrac{2z}{x + y}$

32. $w = \sqrt{x^2 + y^2 + z^2}$

In Exercises 33–38, evaluate w_x, w_y, and w_z at the point.

Function	Point
33. $w = \sqrt{x^2 + y^2 + z^2}$	$(2, -1, 2)$
34. $w = \dfrac{xy}{x + y + z}$	$(1, 2, 0)$
35. $w = \ln\sqrt{x^2 + y^2 + z^2}$	$(3, 0, 4)$
36. $w = \dfrac{1}{\sqrt{1 - x^2 - y^2 - z^2}}$	$(0, 0, 0)$
37. $w = 2xz^2 + 3xyz - 6y^2z$	$(1, -1, 2)$
38. $w = xye^{z^2}$	$(2, 1, 0)$

In Exercises 39–42, find values of x and y such that $f_x(x, y) = 0$ and $f_y(x, y) = 0$ simultaneously.

39. $f(x, y) = x^2 + 4xy + y^2 - 4x + 16y + 3$

40. $f(x, y) = 3x^3 - 12xy + y^3$

41. $f(x, y) = \dfrac{1}{x} + \dfrac{1}{y} + xy$

42. $f(x, y) = \ln(x^2 + y^2 + 1)$

In Exercises 43–46, find the slope of the surface at the given point in (a) the x-direction and (b) the y-direction.

43. $z = xy$

$(1, 2, 2)$

44. $z = \sqrt{25 - x^2 - y^2}$

$(3, 0, 4)$

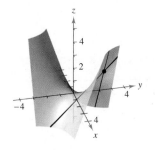
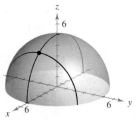

45. $z = 4 - x^2 - y^2$

$(1, 1, 2)$

46. $z = x^2 - y^2$

$(-2, 1, 3)$

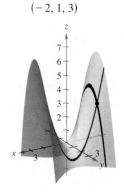

In Exercises 47–54, find the four second partial derivatives. Observe that the second mixed partials are equal.

47. $z = x^2 - 2xy + 3y^2$

48. $z = y^3 - 4xy^2 - 1$

49. $z = \dfrac{e^{2xy}}{4x}$

50. $z = \dfrac{x^2 - y^2}{2xy}$

51. $z = x^3 - 4y^2$

52. $z = \sqrt{9 - x^2 - y^2}$

53. $z = \dfrac{1}{x - y}$

54. $z = \dfrac{x}{x + y}$

In Exercises 55–58, evaluate the second partial derivatives f_{xx}, f_{xy}, f_{yy}, and f_{yx} at the point.

Function	Point
55. $f(x, y) = x^4 - 3x^2y^2 + y^2$	$(1, 0)$
56. $f(x, y) = \sqrt{x^2 + y^2}$	$(0, 2)$
57. $f(x, y) = \ln(x - y)$	$(2, 1)$
58. $f(x, y) = x^2e^y$	$(-1, 0)$

59. Marginal Cost A company manufactures two models of bicycles: a mountain bike and a racing bike. The cost function for producing x mountain bikes and y racing bikes is given by

$$C = 10\sqrt{xy} + 149x + 189y + 675.$$

(a) Find the marginal costs ($\partial C/\partial x$ and $\partial C/\partial y$) when $x = 120$ and $y = 160$.

(b) When additional production is required, which model of bicycle results in the cost increasing at a higher rate? How can this be determined from the cost model?

60. Marginal Revenue A pharmaceutical corporation has two plants that produce the same over-the-counter medicine. If x_1 and x_2 are the numbers of units produced at plant 1 and plant 2, respectively, then the total revenue for the product is given by

$$R = 200x_1 + 200x_2 - 4x_1^2 - 8x_1x_2 - 4x_2^2.$$

When $x_1 = 4$ and $x_2 = 12$, find

(a) the marginal revenue for plant 1, $\partial R/\partial x_1$.

(b) the marginal revenue for plant 2, $\partial R/\partial x_2$.

61. Marginal Productivity Consider the Cobb-Douglas production function $f(x, y) = 200x^{0.7}y^{0.3}$. When $x = 1000$ and $y = 500$, find

(a) the marginal productivity of labor, $\partial f/\partial x$.

(b) the marginal productivity of capital, $\partial f/\partial y$.

62. Marginal Productivity Repeat Exercise 61 for the production function given by $f(x, y) = 100x^{0.75}y^{0.25}$.

Complementary and Substitute Products In Exercises 63 and 64, determine whether the demand functions describe complementary or substitute product relationships. Using the notation of Example 4, let x_1 and x_2 be the demands for products p_1 and p_2, respectively.

63. $x_1 = 150 - 2p_1 - \frac{5}{2}p_2$, $x_2 = 350 - \frac{3}{2}p_1 - 3p_2$

64. $x_1 = 150 - 2p_1 + 1.8p_2$, $x_2 = 350 + \frac{3}{4}p_1 - 1.9p_2$

65. Milk Consumption A model for the per capita consumptions (in gallons) of different types of plain milk in the United States from 1999 through 2004 is

$$z = 1.25x - 0.125y + 0.95.$$

Consumption of reduced-fat (1%) and skim milks, reduced-fat milk (2%), and whole milk are represented by variables x, y, and z, respectively. *(Source: U.S. Department of Agriculture)*

(a) Find $\dfrac{\partial z}{\partial x}$ and $\dfrac{\partial z}{\partial y}$.

(b) Interpret the partial derivatives in the context of the problem.

66. Shareholder's Equity The shareholder's equity z (in billions of dollars) for Wal-Mart Corporation from 2000 through 2006 can be modeled by

$$z = 0.205x - 0.073y - 0.728$$

where x is net sales (in billions of dollars) and y is the total assets (in billions of dollars). *(Source: Wal-Mart Corporation)*

(a) Find $\dfrac{\partial z}{\partial x}$ and $\dfrac{\partial z}{\partial y}$.

(b) Interpret the partial derivatives in the context of the problem.

Ⓑ 67. Psychology Early in the twentieth century, an intelligence test called the *Stanford-Binet Test* (more commonly known as the *IQ test*) was developed. In this test, an individual's mental age M is divided by the individual's chronological age C and the quotient is multiplied by 100. The result is the individual's *IQ*.

$$IQ(M, C) = \frac{M}{C} \times 100$$

Find the partial derivatives of *IQ* with respect to M and with respect to C. Evaluate the partial derivatives at the point $(12, 10)$ and interpret the result. *(Source: Adapted from Bernstein/Clark-Stewart/Roy/Wickens, Psychology, Fourth Edition)*

68. Investment The value of an investment of $1000 earning 10% compounded annually is

$$V(I, R) = 1000\left[\frac{1 + 0.10(1 - R)}{1 + I}\right]^{10}$$

where I is the annual rate of inflation and R is the tax rate for the person making the investment. Calculate $V_I(0.03, 0.28)$ and $V_R(0.03, 0.28)$. Determine whether the tax rate or the rate of inflation is the greater "negative" factor on the growth of the investment.

69. Think About It Let N be the number of applicants to a university, p the charge for food and housing at the university, and t the tuition. Suppose that N is a function of p and t such that $\partial N/\partial p < 0$ and $\partial N/\partial t < 0$. How would you interpret the fact that both partials are negative?

70. Marginal Utility The utility function $U = f(x, y)$ is a measure of the utility (or satisfaction) derived by a person from the consumption of two products x and y. Suppose the utility function is given by $U = -5x^2 + xy - 3y^2$.

(a) Determine the marginal utility of product x.

(b) Determine the marginal utility of product y.

(c) When $x = 2$ and $y = 3$, should a person consume one more unit of product x or one more unit of product y? Explain your reasoning.

Ⓣ (d) Use a three-dimensional graphing utility to graph the function. Interpret the marginal utilities of products x and y graphically.

Business Capsule

Photo courtesy of Izzy and Coco Tihanyi

In 1996, twin sisters Izzy and Coco Tihanyi started Surf Diva, a surf school and apparel company for women and girls, in La Jolla, California. To advertise their business, they would donate surf lessons and give the surf report on local radio stations in exchange for air time. Today, they have schools in Japan and Costa Rica, and their clothing line can be found in surf and specialty shops, sporting goods stores, and airport gift shops. Sales from their surf schools have increased nearly 13% per year, and product sales are expected to double each year.

71. Research Project Use your school's library, the Internet, or some other reference source to research a company that increased the demand for its product by creative advertising. Write a paper about the company. Use graphs to show how a change in demand is related to a change in the marginal utility of a product or service.

Extrema of Functions of Two Variables

- Understand the relative extrema of functions of two variables.
- Use the First-Partials Test to find the relative extrema of functions of two variables.
- Use the Second-Partials Test to find the relative extrema of functions of two variables.
- Use relative extrema to answer questions about real-life situations.

Relative Extrema

Earlier in the text, you learned how to use derivatives to find the relative minimum and relative maximum values of a function of a single variable. In this section, you will learn how to use partial derivatives to find the relative minimum and relative maximum values of a function of two variables.

Relative Extrema of a Function of Two Variables

Let f be a function defined on a region containing (x_0, y_0). The function f has a **relative maximum** at (x_0, y_0) if there is a circular region R centered at (x_0, y_0) such that

$$f(x, y) \leq f(x_0, y_0) \qquad \text{\scriptsize f has a relative maximum at (x_0, y_0).}$$

for all (x, y) in R. The function f has a **relative minimum** at (x_0, y_0) if there is a circular region R centered at (x_0, y_0) such that

$$f(x, y) \geq f(x_0, y_0) \qquad \text{\scriptsize f has a relative minimum at (x_0, y_0).}$$

for all (x, y) in R.

To say that f has a relative maximum at (x_0, y_0) means that the point (x_0, y_0, z_0) is at least as high as all nearby points on the graph of $z = f(x, y)$. Similarly, f has a relative minimum at (x_0, y_0) if (x_0, y_0, z_0) is at least as low as all nearby points on the graph. (See Figure 13.26.)

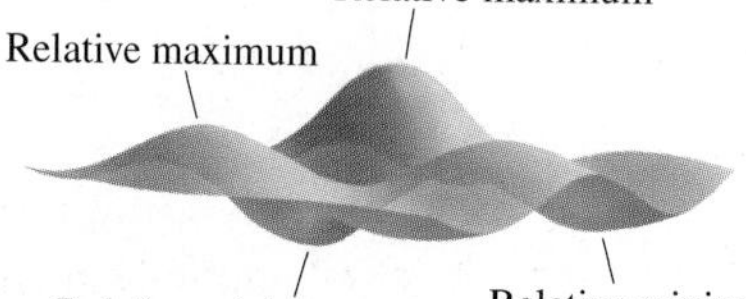

Relative maximum

Relative maximum

Relative minimum   Relative minimum

FIGURE 13.26   Relative Extrema

As in single-variable calculus, you need to distinguish between relative extrema and absolute extrema of a function of two variables. The number $f(x_0, y_0)$ is an absolute maximum of f in the region R if it is greater than or equal to all other function values in the region. For instance, the function $f(x, y) = -(x^2 + y^2)$ graphs as a paraboloid, opening downward, with vertex at $(0, 0, 0)$. (See Figure 13.27.) The number $f(0, 0) = 0$ is an absolute maximum of the function over the entire xy-plane. An absolute minimum of f in a region is defined similarly.

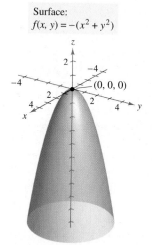

Surface:
$f(x, y) = -(x^2 + y^2)$

$(0, 0, 0)$

FIGURE 13.27 f has an absolute maximum at $(0, 0, 0)$.

The First-Partials Test for Relative Extrema

To locate the relative extrema of a function of two variables, you can use a procedure that is similar to the First-Derivative Test used for functions of a single variable.

> ### First-Partials Test for Relative Extrema
>
> If f has a relative extremum at (x_0, y_0) on an open region R in the xy-plane, and the first partial derivatives of f exist in R, then
>
> $$f_x(x_0, y_0) = 0$$
>
> and
>
> $$f_y(x_0, y_0) = 0$$
>
> as shown in Figure 13.28.

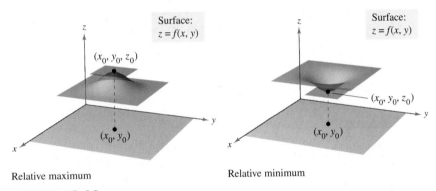

Relative maximum   Relative minimum

FIGURE 13.28

An *open* region in the xy-plane is similar to an open interval on the real number line. For instance, the region R consisting of the interior of the circle $x^2 + y^2 = 1$ is an open region. If the region R consists of the interior of the circle *and* the points on the circle, then it is a *closed* region.

A point (x_0, y_0) is a **critical point** of f if $f_x(x_0, y_0)$ or $f_y(x_0, y_0)$ is undefined or if

$$f_x(x_0, y_0) = 0 \quad \text{and} \quad f_y(x_0, y_0) = 0. \qquad \text{Critical point}$$

The First-Partials Test states that if the first partial derivatives exist, then you need only examine values of $f(x, y)$ at critical points to find the relative extrema. As is true for a function of a single variable, however, the critical points of a function of two variables do not always yield relative extrema. For instance, the point $(0, 0)$ is a critical point of the surface shown in Figure 13.29, but $f(0, 0)$ is not a relative extremum of the function. Such points are called **saddle points** of the function.

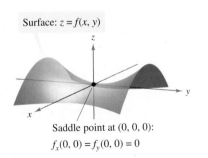

Surface: $z = f(x, y)$

Saddle point at $(0, 0, 0)$:
$f_x(0, 0) = f_y(0, 0) = 0$

FIGURE 13.29

Surface:

$f(x, y) = 2x^2 + y^2 + 8x - 6y + 20$

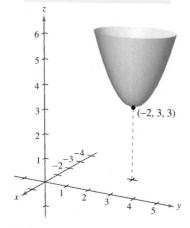

FIGURE 13.30

Surface:

$f(x, y) = 1 - (x^2 + y^2)^{1/3}$

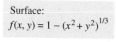
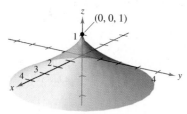

FIGURE 13.31 $f_x(x, y)$ and $f_y(x, y)$ are undefined at (0, 0).

✓**CHECKPOINT 2**

Find the relative extrema of

$$f(x, y) = \sqrt{1 - \frac{x^2}{16} - \frac{y^2}{4}}.$$ ■

Example 1 **Finding Relative Extrema**

Find the relative extrema of

$$f(x, y) = 2x^2 + y^2 + 8x - 6y + 20.$$

SOLUTION Begin by finding the first partial derivatives of f.

$$f_x(x, y) = 4x + 8 \quad \text{and} \quad f_y(x, y) = 2y - 6$$

Because these partial derivatives are defined for all points in the xy-plane, the only critical points are those for which both first partial derivatives are zero. To locate these points, set $f_x(x, y)$ and $f_y(x, y)$ equal to 0, and solve the resulting system of equations.

$$4x + 8 = 0 \qquad \text{Set } f_x(x, y) \text{ equal to 0.}$$
$$2y - 6 = 0 \qquad \text{Set } f_y(x, y) \text{ equal to 0.}$$

The solution of this system is $x = -2$ and $y = 3$. So, the point $(-2, 3)$ is the only critical number of f. From the graph of the function, shown in Figure 13.30, you can see that this critical point yields a relative minimum of the function. So, the function has only one relative extremum, which is

$$f(-2, 3) = 3. \qquad \text{Relative minimum}$$

✓**CHECKPOINT 1**

Find the relative extrema of $f(x, y) = x^2 + 2y^2 + 16x - 8y + 8.$ ■

Example 1 shows a relative minimum occurring at one type of critical point—the type for which both $f_x(x, y)$ and $f_y(x, y)$ are zero. The next example shows a relative maximum that occurs at the other type of critical point—the type for which either $f_x(x, y)$ or $f_y(x, y)$ is undefined.

Example 2 **Finding Relative Extrema**

Find the relative extrema of

$$f(x, y) = 1 - (x^2 + y^2)^{1/3}.$$

SOLUTION Begin by finding the first partial derivatives of f.

$$f_x(x, y) = -\frac{2x}{3(x^2 + y^2)^{2/3}} \quad \text{and} \quad f_y(x, y) = -\frac{2y}{3(x^2 + y^2)^{2/3}}$$

These partial derivatives are defined for all points in the xy-plane *except* the point $(0, 0)$. So, $(0, 0)$ is a critical point of f. Moreover, this is the only critical point, because there are no other values of x and y for which either partial is undefined or for which both partials are zero. From the graph of the function, shown in Figure 13.31, you can see that this critical point yields a relative maximum of the function. So, the function has only one relative extremum, which is

$$f(0, 0) = 1. \qquad \text{Relative maximum}$$

The Second-Partials Test for Relative Extrema

For functions such as those in Examples 1 and 2, you can determine the *types* of extrema at the critical points by sketching the graph of the function. For more complicated functions, a graphical approach is not so easy to use. The **Second-Partials Test** is an analytical test that can be used to determine whether a critical number yields a relative minimum, a relative maximum, or neither.

Second-Partials Test for Relative Extrema

Let f have continuous second partial derivatives on an open region containing (a, b) for which $f_x(a, b) = 0$ and $f_y(a, b) = 0$. To test for relative extrema of f, consider the quantity

$$d = f_{xx}(a, b)f_{yy}(a, b) - [f_{xy}(a, b)]^2.$$

1. If $d > 0$ and $f_{xx}(a, b) > 0$, then f has a **relative minimum** at (a, b).

2. If $d > 0$ and $f_{xx}(a, b) < 0$, then f has a **relative maximum** at (a, b).

3. If $d < 0$, then $(a, b, f(a, b))$ is a **saddle point.**

4. The test gives no information if $d = 0$.

Example 3   Applying the Second-Partials Test

Find the relative extrema and saddle points of $f(x, y) = xy - \frac{1}{4}x^4 - \frac{1}{4}y^4$.

SOLUTION   Begin by finding the critical points of f. Because $f_x(x, y) = y - x^3$ and $f_y(x, y) = x - y^3$ are defined for all points in the xy-plane, the only critical points are those for which both first partial derivatives are zero. By solving the equations $y - x^3 = 0$ and $x - y^3 = 0$ simultaneously, you can determine that the critical points are $(1, 1)$, $(-1, -1)$, and $(0, 0)$. Furthermore, because

$$f_{xx}(x, y) = -3x^2, \quad f_{yy}(x, y) = -3y^2, \quad \text{and} \quad f_{xy}(x, y) = 1$$

you can use the quantity $d = f_{xx}(a, b)f_{yy}(a, b) - [f_{xy}(a, b)]^2$ to classify the critical points as shown.

Critical Point	d	$f_{xx}(x, y)$	Conclusion
$(1, 1)$	$(-3)(-3) - 1 = 8$	-3	Relative maximum
$(-1, -1)$	$(-3)(-3) - 1 = 8$	-3	Relative maximum
$(0, 0)$	$(0)(0) - 1 = -1$	0	Saddle point

The graph of f is shown in Figure 13.32.

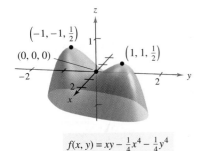

$$f(x, y) = xy - \frac{1}{4}x^4 - \frac{1}{4}y^4$$

FIGURE 13.32

✓ CHECKPOINT 3

Find the relative extrema and saddle points of $f(x, y) = \dfrac{y^2}{16} - \dfrac{x^2}{4}$. ∎

Application of Extrema

Example 4 Finding a Maximum Profit

A company makes two substitute products whose demand functions are given by

$$x_1 = 200(p_2 - p_1)$$ Demand for product 1

$$x_2 = 500 + 100p_1 - 180p_2$$ Demand for product 2

where p_1 and p_2 are the prices per unit (in dollars) and x_1 and x_2 are the numbers of units sold. The costs of producing the two products are $0.50 and $0.75 per unit, respectively. Find the prices that will yield a maximum profit.

SOLUTION The cost and revenue functions are as shown.

$$C = 0.5x_1 + 0.75x_2$$ Write cost function.

$$= 0.5(200)(p_2 - p_1) + 0.75(500 + 100p_1 - 180p_2)$$ Substitute.

$$= 375 - 25p_1 - 35p_2$$ Simplify.

$$R = p_1x_1 + p_2x_2$$ Write revenue function.

$$= p_1(200)(p_2 - p_1) + p_2(500 + 100p_1 - 180p_2)$$ Substitute.

$$= -200p_1{}^2 - 180p_2{}^2 + 300p_1p_2 + 500p_2$$ Simplify.

This implies that the profit function is

$$P = R - C$$ Write profit function.

$$= -200p_1{}^2 - 180p_2{}^2 + 300p_1p_2 + 500p_2 - (375 - 25p_1 - 35p_2)$$

$$= -200p_1{}^2 - 180p_2{}^2 + 300p_1p_2 + 25p_1 + 535p_2 - 375.$$

The maximum profit occurs when the two first partial derivatives are zero.

$$\frac{\partial P}{\partial p_1} = -400p_1 + 300p_2 + 25 = 0$$

$$\frac{\partial P}{\partial p_2} = 300p_1 - 360p_2 + 535 = 0$$

By solving this system simultaneously, you can conclude that the solution is $p_1 = \$3.14$ and $p_2 = \$4.10$. From the graph of the function shown in Figure 13.33, you can see that this critical number yields a maximum. So, the maximum profit is

$$P(3.14, 4.10) = \$761.48.$$

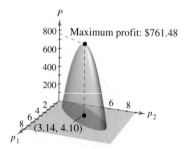

FIGURE 13.33

✓ **CHECKPOINT 4**

Find the prices that will yield a maximum profit for the products in Example 4 if the costs of producing the two products are $0.75 and $0.50 per unit, respectively. ▪

Algebra Review

For help in solving the system of equations

$$y(24 - 12x - 4y) = 0$$

$$x(24 - 6x - 8y) = 0$$

in Example 5, see Example 2(a) in the *Chapter 13 Algebra Review*, on page 1014.

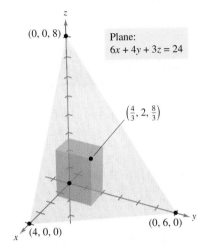

FIGURE 13.34

Example 5 **Finding a Maximum Volume**

Consider all possible rectangular boxes that are resting on the xy-plane with one vertex at the origin and the opposite vertex in the plane $6x + 4y + 3z = 24$, as shown in Figure 13.34. Of all such boxes, which has the greatest volume?

SOLUTION Because one vertex of the box lies in the plane given by $6x + 4y + 3z = 24$ or $z = \frac{1}{3}(24 - 6x - 4y)$, you can write the volume of the box as

$$V = xyz \qquad \text{Volume} = (\text{width})(\text{length})(\text{height})$$

$$= xy\left(\tfrac{1}{3}\right)(24 - 6x - 4y) \qquad \text{Substitute for } z.$$

$$= \tfrac{1}{3}(24xy - 6x^2y - 4xy^2). \qquad \text{Simplify.}$$

To find the critical numbers, set the first partial derivatives equal to zero.

$$V_x = \tfrac{1}{3}(24y - 12xy - 4y^2) \qquad \text{Partial with respect to } x$$

$$= \tfrac{1}{3}y(24 - 12x - 4y) = 0 \qquad \text{Factor and set equal to 0.}$$

$$V_y = \tfrac{1}{3}(24x - 6x^2 - 8xy) \qquad \text{Partial with respect to } y$$

$$= \tfrac{1}{3}x(24 - 6x - 8y) = 0 \qquad \text{Factor and set equal to 0.}$$

The four solutions of this system are $(0, 0)$, $(0, 6)$, $(4, 0)$, and $\left(\frac{4}{3}, 2\right)$. Using the Second-Partials Test, you can determine that the maximum volume occurs when the width is $x = \frac{4}{3}$ and the length is $y = 2$. For these values, the height of the box is

$$z = \tfrac{1}{3}\left[24 - 6\left(\tfrac{4}{3}\right) - 4(2)\right] = \tfrac{8}{3}.$$

So, the maximum volume is

$$V = xyz = \left(\tfrac{4}{3}\right)(2)\left(\tfrac{8}{3}\right) = \tfrac{64}{9} \text{ cubic units.}$$

✓**CHECKPOINT 5**

Find the maximum volume of a box that is resting on the xy-plane with one vertex at the origin and the opposite vertex in the plane $2x + 4y + z = 8$. ■

CONCEPT CHECK

1. Given a function of two variables f, state how you can determine whether (x_0, y_0) is a critical point of f.

2. The point $(a, b, f(a, b))$ is a saddle point if what is true?

3. If $d > 0$ and $f_{xx}(a, b) > 0$, then what does f have at (a, b): a relative minimum or a relative maximum?

4. If $d > 0$ and $f_{xx}(a, b) < 0$, then what does f have at (a, b): a relative minimum or a relative maximum?

Skills Review 13.5	The following warm-up exercises involve skills that were covered in earlier sections. You will use these skills in the exercise set for this section. For additional help, review Sections 5.1, 5.2, and 13.4.

In Exercises 1–8, solve the system of equations.

1. $\begin{cases} 5x = 15 \\ 3x - 2y = 5 \end{cases}$
2. $\begin{cases} \frac{1}{2}y = 3 \\ -x + 5y = 19 \end{cases}$
3. $\begin{cases} x + y = 5 \\ x - y = -3 \end{cases}$
4. $\begin{cases} x + y = 8 \\ 2x - y = 4 \end{cases}$

5. $\begin{cases} 2x - y = 8 \\ 3x - 4y = 7 \end{cases}$
6. $\begin{cases} 2x - 4y = 14 \\ 3x + y = 7 \end{cases}$
7. $\begin{cases} x^2 + x = 0 \\ 2yx + y = 0 \end{cases}$
8. $\begin{cases} 3y^2 + 6y = 0 \\ xy + x + 2 = 0 \end{cases}$

In Exercises 9–14, find all first and second partial derivatives of the function.

9. $z = 4x^3 - 3y^2$
10. $z = 2x^5 - y^3$
11. $z = x^4 - \sqrt{xy} + 2y$

12. $z = 2x^2 - 3xy + y^2$
13. $z = ye^{xy^2}$
14. $z = xe^{xy}$

Exercises 13.5

See www.CalcChat.com for worked-out solutions to odd-numbered exercises.

In Exercises 1–4, find any critical points and relative extrema of the function.

1. $f(x, y) = x^2 - y^2 + 4x - 8y - 11$

2. $f(x, y) = x^2 + y^2 + 2x - 6y + 6$

3. $f(x, y) = \sqrt{x^2 + y^2 + 1}$

4. $f(x, y) = \sqrt{25 - (x - 2)^2 - y^2}$

In Exercises 5–20, examine the function for relative extrema and saddle points.

5. $f(x, y) = (x - 1)^2 + (y - 3)^2$

6. $f(x, y) = 9 - (x - 3)^2 - (y + 2)^2$

7. $f(x, y) = 2x^2 + 2xy + y^2 + 2x - 3$

8. $f(x, y) = -x^2 - 5y^2 + 8x - 10y - 13$

9. $f(x, y) = -5x^2 + 4xy - y^2 + 16x + 10$

10. $f(x, y) = x^2 + 6xy + 10y^2 - 4y + 4$

11. $f(x, y) = 3x^2 + 2y^2 - 12x - 4y + 7$

12. $f(x, y) = -3x^2 - 2y^2 + 3x - 4y + 5$

13. $f(x, y) = x^2 - y^2 + 4x - 4y - 8$

14. $f(x, y) = x^2 - 3xy - y^2$

15. $f(x, y) = \frac{1}{2}xy$

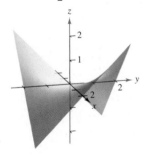

16. $f(x, y) = x + y + 2xy - x^2 - y^2$

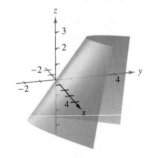

17. $f(x, y) = (x + y)e^{1 - x^2 - y^2}$

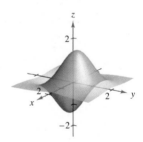

18. $f(x, y) = 3e^{-(x^2 + y^2)}$

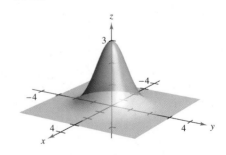

19. $f(x, y) = 4e^{xy}$

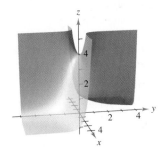

20. $f(x, y) = -\dfrac{3}{x^2 + y^2 + 1}$

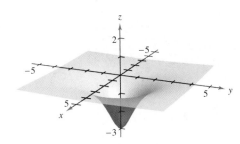

Think About It In Exercises 21–24, determine whether there is a relative maximum, a relative minimum, a saddle point, or insufficient information to determine the nature of the function $f(x, y)$ at the critical point (x_0, y_0).

21. $f_{xx}(x_0, y_0) = 9$, $f_{yy}(x_0, y_0) = 4$, $f_{xy}(x_0, y_0) = 6$

22. $f_{xx}(x_0, y_0) = -3$, $f_{yy}(x_0, y_0) = -8$, $f_{xy}(x_0, y_0) = 2$

23. $f_{xx}(x_0, y_0) = -9$, $f_{yy}(x_0, y_0) = 6$, $f_{xy}(x_0, y_0) = 10$

24. $f_{xx}(x_0, y_0) = 25$, $f_{vv}(x_0, y_0) = 8$, $f_{xv}(x_0, y_0) = 10$

In Exercises 25–30, find the critical points and test for relative extrema. List the critical points for which the Second-Partials Test fails.

25. $f(x, y) = (xy)^2$

26. $f(x, y) = \sqrt{x^2 + y^2}$

27. $f(x, y) = x^3 + y^3$

28. $f(x, y) = x^3 + y^3 - 3x^2 + 6y^2 + 3x + 12y + 7$

29. $f(x, y) = x^{2/3} + y^{2/3}$

30. $f(x, y) = (x^2 + y^2)^{2/3}$

In Exercises 31 and 32, find the critical points of the function and, from the form of the function, determine whether a relative maximum or a relative minimum occurs at each point.

31. $f(x, y, z) = (x - 1)^2 + (y + 3)^2 + z^2$

32. $f(x, y, z) = 6 - [x(y + 2)(z - 1)]^2$

In Exercises 33–36, find three positive numbers x, y, and z that satisfy the given conditions.

33. The sum is 30 and the product is a maximum.

34. The sum is 32 and $P = xy^2z$ is a maximum.

35. The sum is 30 and the sum of the squares is a minimum.

36. The sum is 1 and the sum of the squares is a minimum.

37. Revenue A company manufactures two types of sneakers: running shoes and basketball shoes. The total revenue from x_1 units of running shoes and x_2 units of basketball shoes is

$$R = -5x_1^2 - 8x_2^2 - 2x_1x_2 + 42x_1 + 102x_2$$

where x_1 and x_2 are in thousands of units. Find x_1 and x_2 so as to maximize the revenue.

38. Revenue A retail outlet sells two types of riding lawn mowers, the prices of which are p_1 and p_2. Find p_1 and p_2 so as to maximize total revenue, where $R = 515p_1 + 805p_2 + 1.5p_1p_2 - 1.5p_1^2 - p_2^2$.

Revenue In Exercises 39 and 40, find p_1 and p_2 so as to maximize the total revenue $R = x_1p_1 + x_2p_2$ for a retail outlet that sells two competitive products with the given demand functions.

39. $x_1 = 1000 - 2p_1 + p_2$, $x_2 = 1500 + 2p_1 - 1.5p_2$

40. $x_1 = 1000 - 4p_1 + 2p_2$, $x_2 = 900 + 4p_1 - 3p_2$

41. Profit A corporation manufactures a high-performance automobile engine product at two locations. The cost of producing x_1 units at location 1 is

$$C_1 = 0.05x_1^2 + 15x_1 + 5400$$

and the cost of producing x_2 units at location 2 is

$$C_2 = 0.03x_2^2 + 15x_2 + 6100.$$

The demand function for the product is

$$p = 225 - 0.4(x_1 + x_2)$$

and the total revenue function is

$$R = [225 - 0.4(x_1 + x_2)](x_1 + x_2).$$

Find the production levels at the two locations that will maximize the profit

$$P = R - C_1 - C_2.$$

42. Profit A corporation manufactures candles at two locations. The cost of producing x_1 units at location 1 is

$$C_1 = 0.02x_1^2 + 4x_1 + 500$$

and the cost of producing x_2 units at location 2 is

$$C_2 = 0.05x_2^2 + 4x_2 + 275.$$

The candles sell for \$15 per unit. Find the quantity that should be produced at each location to maximize the profit

$$P = 15(x_1 + x_2) - C_1 - C_2.$$

43. Volume Find the dimensions of a rectangular package of maximum volume that may be sent by a shipping company assuming that the sum of the length and the girth (perimeter of a cross section) cannot exceed 96 inches.

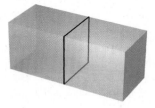

44. Volume Repeat Exercise 43 assuming that the sum of the length and the girth cannot exceed 144 inches.

45. Cost A manufacturer makes a wooden storage crate that has an open top. The volume of each crate is 6 cubic feet. Material costs are $0.15 per square foot for the base of the crate and $0.10 per square foot for the sides. Find the dimensions that minimize the cost of each crate. What is the minimum cost?

46. Cost A home improvement contractor is painting the walls and ceiling of a rectangular room. The volume of the room is 668.25 cubic feet. The cost of wall paint is $0.06 per square foot and the cost of ceiling paint is $0.11 per square foot. Find the room dimensions that result in a minimum cost for the paint. What is the minimum cost for the paint?

47. Hardy-Weinberg Law Common blood types are determined genetically by the three alleles A, B, and O. (An allele is any of a group of possible mutational forms of a gene.) A person whose blood type is AA, BB, or OO is homozygous. A person whose blood type is AB, AO, or BO is heterozygous. The Hardy-Weinberg Law states that the proportion P of heterozygous individuals in any given population is modeled by

$$P(p, q, r) = 2pq + 2pr + 2qr$$

where p represents the percent of allele A in the population, q represents the percent of allele B in the population, and r represents the percent of allele O in the population. Use the fact that $p + q + r = 1$ (the sum of the three must equal 100%) to show that the maximum proportion of heterozygous individuals in any population is $\frac{2}{3}$.

48. Biology A lake is to be stocked with smallmouth and largemouth bass. Let x represent the number of smallmouth bass and let y represent the number of largemouth bass in the lake. The weight of each fish is dependent on the population densities. After a six-month period, the weight of a single smallmouth bass is given by

$$W_1 = 3 - 0.002x - 0.001y$$

and the weight of a single largemouth bass is given by

$$W_2 = 4.5 - 0.004x - 0.005y.$$

Assuming that no fish die during the six-month period, how many smallmouth and largemouth bass should be stocked in the lake so that the *total* weight T of bass in the lake is a maximum?

Steve & Dave Maslowski/Photo Researchers, Inc.

Bass help to keep a pond healthy. A suitable quantity of bass keeps other fish populations in check and helps balance the food chain.

49. Cost An automobile manufacturer has determined that its annual labor and equipment cost (in millions of dollars) can be modeled by

$$C(x, y) = 2x^2 + 3y^2 - 15x - 20y + 4xy + 39$$

where x is the amount spent per year on labor and y is the amount spent per year on equipment (both in millions of dollars). Find the values of x and y that minimize the annual labor and equipment cost. What is this cost?

50. Medicine In order to treat a certain bacterial infection, a combination of two drugs is being tested. Studies have shown that the duration of the infection in laboratory tests can be modeled by

$$D(x, y) = x^2 + 2y^2 - 18x - 24y + 2xy + 120$$

where x is the dosage in hundreds of milligrams of the first drug and y is the dosage in hundreds of milligrams of the second drug. Determine the partial derivatives of D with respect to x and with respect to y. Find the amount of each drug necessary to minimize the duration of the infection.

True or False? In Exercises 51 and 52, determine whether the statement is true or false. If it is false, explain why or give an example that shows it is false.

51. A saddle point always occurs at a critical point.

52. If $f(x, y)$ has a relative maximum at (x_0, y_0, z_0), then $f_x(x_0, y_0) = f_y(x_0, y_0) = 0$.

Mid-Chapter Quiz

See www.CalcChat.com for worked-out solutions to odd-numbered exercises.

Take this quiz as you would take a quiz in class. When you are done, check your work against the answers given in the back of the book.

In Exercises 1–3, (a) plot the points on a three-dimensional coordinate system, (b) find the distance between the points, and (c) find the coordinates of the midpoint of the line segment joining the points.

1. $(1, 3, 2), (-1, 2, 0)$ **2.** $(-1, 4, 3), (5, 1, -6)$ **3.** $(0, -3, 3), (3, 0, -3)$

In Exercises 4 and 5, find the standard equation of the sphere.

4. Center: $(2, -1, 3)$; radius: 4

5. Endpoints of a diameter: $(0, 3, 1), (2, 5, -5)$

6. Find the center and radius of the sphere whose equation is

$$x^2 + y^2 + z^2 - 8x - 2y - 6z - 23 = 0.$$

In Exercises 7–9, find the intercepts and sketch the graph of the plane.

7. $2x + 3y + z = 6$ **8.** $x - 2z = 4$ **9.** $z = -5$

In Exercises 10–12, identify the quadric surface.

10. $\dfrac{x^2}{4} + \dfrac{y^2}{9} + \dfrac{z^2}{16} = 1$ **11.** $z^2 - x^2 - y^2 = 25$ **12.** $81z - 9x^2 - y^2 = 0$

In Exercises 13–15, find $f(1, 0)$ and $f(4, -1)$.

13. $f(x, y) = x - 9y^2$ **14.** $f(x, y) = \sqrt{4x^2 + y}$ **15.** $f(x, y) = \ln(x + 3y)$

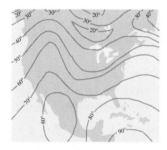

Figure for 16

16. The contour map shows level curves of equal temperature (isotherms), measured in degrees Fahrenheit, across North America on a spring day. Use the map to find the approximate range of temperatures in (a) the Great Lakes region, (b) the United States, and (c) Mexico.

In Exercises 17 and 18, find f_x and f_y and evaluate each at the point $(-2, 3)$.

17. $f(x, y) = x^2 + 2y^2 - 3x - y + 1$ **18.** $f(x, y) = \dfrac{3x - y^2}{x + y}$

In Exercises 19 and 20, find any critical points, relative extrema, and saddle points of the function.

19. $f(x, y) = 3x^2 + y^2 - 2xy - 6x + 2y$ **20.** $f(x, y) = -x^3 + 4xy - 2y^2 + 1$

21. A company manufactures two types of wood burning stoves: a freestanding model and a fireplace-insert model. The total cost (in thousands of dollars) for producing x freestanding stoves and y fireplace-insert stoves can be modeled by

$$C(x, y) = \tfrac{1}{16}x^2 + y^2 - 10x - 40y + 820.$$

Find the values of x and y that minimize the total cost. What is this cost?

22. Physical Science Assume that Earth is a sphere with a radius of 3963 miles. If the center of Earth is placed at the origin of a three-dimensional coordinate system, what is the equation of the sphere? Lines of longitude that run north-south could be represented by what trace(s)? What shape would each of these traces form? Why? Lines of latitude that run east-west could be represented by what trace(s)? Why? What shape would each of these traces form? Why?

Section 13.6

Lagrange Multipliers

- Use Lagrange multipliers with one constraint to find extrema of functions of several variables and to answer questions about real-life situations.
- Use Lagrange multipliers with two constraints to find extrema of functions of several variables.

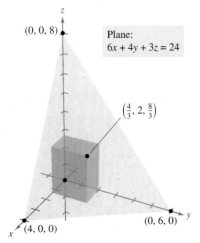

Plane:
$6x + 4y + 3z = 24$

$(0, 0, 8)$

$\left(\frac{4}{3}, 2, \frac{8}{3}\right)$

$(0, 6, 0)$

$(4, 0, 0)$

FIGURE 13.35

Lagrange Multipliers with One Constraint

In Example 5 in Section 13.5, you were asked to find the dimensions of the rectangular box of maximum volume that would fit in the first octant beneath the plane

$$6x + 4y + 3z = 24$$

as shown again in Figure 13.35. Another way of stating this problem is to say that you are asked to find the maximum of

$$V = xyz \qquad \text{Objective function}$$

subject to the constraint

$$6x + 4y + 3z - 24 = 0. \qquad \text{Constraint}$$

This type of problem is called a **constrained optimization** problem. In Section 13.5, you answered this question by solving for z in the constraint equation and then rewriting V as a function of two variables.

In this section, you will study a different (and often better) way to solve constrained optimization problems. This method involves the use of variables called **Lagrange multipliers,** named after the French mathematician Joseph Louis Lagrange (1736–1813).

STUDY TIP

When using the Method of Lagrange Multipliers for functions of three variables, F has the form

$F(x, y, z, \lambda) = f(x, y, z) - \lambda g(x, y, z).$

The system of equations used in Step 1 are as follows.

$F_x(x, y, z, \lambda) = 0$

$F_y(x, y, z, \lambda) = 0$

$F_z(x, y, z, \lambda) = 0$

$F_\lambda(x, y, z, \lambda) = 0$

Method of Lagrange Multipliers

If $f(x, y)$ has a maximum or minimum subject to the constraint $g(x, y) = 0$, then it will occur at one of the critical numbers of the function F defined by

$$F(x, y, \lambda) = f(x, y) - \lambda g(x, y).$$

The variable λ (the lowercase Greek letter lambda) is called a **Lagrange multiplier.** To find the minimum or maximum of f, use the following steps.

1. Solve the following system of equations.

$$F_x(x, y, \lambda) = 0 \qquad F_y(x, y, \lambda) = 0 \qquad F_\lambda(x, y, \lambda) = 0$$

2. Evaluate f at each solution point obtained in the first step. The greatest value yields the maximum of f subject to the constraint $g(x, y) = 0$, and the least value yields the minimum of f subject to the constraint $g(x, y) = 0$.

The Method of Lagrange Multipliers gives you a way of finding critical points but does not tell you whether these points yield minima, maxima, or neither. To make this distinction, you must rely on the context of the problem.

| Example 1 | Using Lagrange Multipliers: One Constraint |

Find the maximum of

$$V = xyz \qquad \text{Objective function}$$

subject to the constraint

$$6x + 4y + 3z - 24 = 0. \qquad \text{Constraint}$$

SOLUTION   First, let $f(x, y, z) = xyz$ and $g(x, y, z) = 6x + 4y + 3z - 24$. Then, define a new function F as

$$\begin{aligned} F(x, y, z, \lambda) &= f(x, y, z) - \lambda g(x, y, z) \\ &= xyz - \lambda(6x + 4y + 3z - 24). \end{aligned}$$

To find the critical numbers of F, set the partial derivatives of F with respect to x, y, z, and λ equal to zero and obtain

$$\begin{aligned} F_x(x, y, z, \lambda) &= yz - 6\lambda = 0 \\ F_y(x, y, z, \lambda) &= xz - 4\lambda = 0 \\ F_z(x, y, z, \lambda) &= xy - 3\lambda = 0 \\ F_\lambda(x, y, z, \lambda) &= -6x - 4y - 3z + 24 = 0. \end{aligned}$$

Solving for λ in the first equation and substituting into the second and third equations produces the following.

$$xz - 4\left(\frac{yz}{6}\right) = 0 \implies y = \frac{3}{2}x$$

$$xy - 3\left(\frac{yz}{6}\right) = 0 \implies z = 2x$$

Next, substitute for y and z in the equation $F_\lambda(x, y, z, \lambda) = 0$ and solve for x.

$$\begin{aligned} F_\lambda(x, y, z, \lambda) &= 0 \\ -6x - 4\left(\tfrac{3}{2}x\right) - 3(2x) + 24 &= 0 \\ -18x &= -24 \\ x &= \tfrac{4}{3} \end{aligned}$$

Using this x-value, you can conclude that the critical values are $x = \frac{4}{3}$, $y = 2$, and $z = \frac{8}{3}$, which implies that the maximum is

$$\begin{aligned} V &= xyz && \text{Write objective function.} \\ &= \left(\frac{4}{3}\right)(2)\left(\frac{8}{3}\right) && \text{Substitute values of } x, y, \text{ and } z. \\ &= \frac{64}{9} \text{ cubic units.} && \text{Maximum volume} \end{aligned}$$

✓ **CHECKPOINT 1**

Find the maximum volume of $V = xyz$ subject to the constraint $2x + 4y + z - 8 = 0$. ∎

AP/Wide World Photos

For many industrial applications, a simple robot can cost more than a year's wages and benefits for one employee. So, manufacturers must carefully balance the amount of money spent on labor and capital.

Example 2
MAKE A DECISION Finding a Maximum Production Level

A manufacturer's production is modeled by the Cobb-Douglas function

$$f(x, y) = 100x^{3/4}y^{1/4} \qquad \text{Objective function}$$

where x represents the units of labor and y represents the units of capital. Each labor unit costs \$150 and each capital unit costs \$250. The total expenses for labor and capital cannot exceed \$50,000. Will the maximum production level exceed 16,000 units?

SOLUTION Because total labor and capital expenses cannot exceed \$50,000, the constraint is

$$150x + 250y = 50,000 \qquad \text{Constraint}$$
$$150x + 250y - 50,000 = 0. \qquad \text{Write in standard form.}$$

To find the maximum production level, begin by writing the function

$$F(x, y, \lambda) = 100x^{3/4}y^{1/4} - \lambda(150x + 250y - 50,000).$$

Next, set the partial derivatives of this function equal to zero.

$$F_x(x, y, \lambda) = 75x^{-1/4}y^{1/4} - 150\lambda = 0 \qquad \text{Equation 1}$$
$$F_y(x, y, \lambda) = 25x^{3/4}y^{-3/4} - 250\lambda = 0 \qquad \text{Equation 2}$$
$$F_\lambda(x, y, \lambda) = -150x - 250y + 50,000 = 0 \qquad \text{Equation 3}$$

The strategy for solving such a system must be customized to the particular system. In this case, you can solve for λ in the first equation, substitute into the second equation, solve for x, substitute into the third equation, and solve for y.

$$75x^{-1/4}y^{1/4} - 150\lambda = 0 \qquad \text{Equation 1}$$
$$\lambda = \tfrac{1}{2}x^{-1/4}y^{1/4} \qquad \text{Solve for } \lambda.$$
$$25x^{3/4}y^{-3/4} - 250\left(\tfrac{1}{2}\right)x^{-1/4}y^{1/4} = 0 \qquad \text{Substitute in Equation 2.}$$
$$25x - 125y = 0 \qquad \text{Multiply by } x^{1/4}y^{3/4}.$$
$$x = 5y \qquad \text{Solve for } x.$$
$$-150(5y) - 250y + 50,000 = 0 \qquad \text{Substitute in Equation 3.}$$
$$-1000y = -50,000 \qquad \text{Simplify.}$$
$$y = 50 \qquad \text{Solve for } y.$$

Using this value for y, it follows that $x = 5(50) = 250$. So, the maximum production level of

$$f(250, 50) = 100(250)^{3/4}(50)^{1/4} \qquad \text{Substitute for } x \text{ and } y.$$
$$\approx 16,719 \text{ units} \qquad \text{Maximum production}$$

occurs when $x = 250$ units of labor and $y = 50$ units of capital. Yes, the maximum production level will exceed 16,000 units.

TECHNOLOGY

You can use a spreadsheet to solve constrained optimization problems. Spreadsheet software programs have a built-in algorithm that finds absolute extrema of functions. Be sure you enter each constraint and the objective function into the spreadsheet. You should also enter initial values of the variables you are working with. Try using a spreadsheet to solve the problem in Example 2. What is your result? (Consult the user's manual of a spreadsheet software program for specific instructions on how to solve a constrained optimization problem.)

✓ CHECKPOINT 2

In Example 2, suppose that each labor unit costs \$200 and each capital unit costs \$250. Find the maximum production level if labor and capital cannot exceed \$50,000. ■

Economists call the Lagrange multiplier obtained in a production function the **marginal productivity of money.** For instance, in Example 2, the marginal productivity of money when $x = 250$ and $y = 50$ is

$$\lambda = \tfrac{1}{2}x^{-1/4}y^{1/4} = \tfrac{1}{2}(250)^{-1/4}(50)^{1/4} \approx 0.334.$$

This means that if one additional dollar is spent on production, approximately 0.334 additional unit of the product can be produced.

Example 3   Finding a Maximum Production Level

In Example 2, suppose that $70,000 is available for labor and capital. What is the maximum number of units that can be produced?

SOLUTION   You could rework the entire problem, as demonstrated in Example 2. However, because the only change in the problem is the availability of additional money to spend on labor and capital, you can use the fact that the marginal productivity of money is

$$\lambda \approx 0.334.$$

Because an additional $20,000 is available and the maximum production in Example 2 was 16,719 units, you can conclude that the maximum production is now

$$16{,}719 + (0.334)(20{,}000) \approx 23{,}400 \text{ units}.$$

Try using the procedure demonstrated in Example 2 to confirm this result.

✓ CHECKPOINT 3

In Example 3, suppose that $80,000 is available for labor and capital. What is the maximum number of units that can be produced? ▪

TECHNOLOGY

(T) You can use a three-dimensional graphing utility to confirm graphically the results of Examples 2 and 3. Begin by graphing the surface $f(x, y) = 100x^{3/4}y^{1/4}$. Then graph the vertical plane given by $150x + 250y = 50{,}000$. As shown at the right, the maximum production level corresponds to the highest point on the intersection of the surface and the plane.

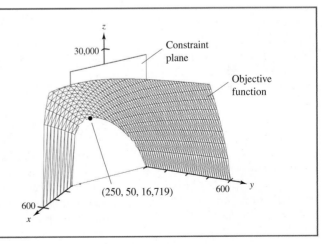

In Example 4 in Section 13.5, you found the maximum profit for two substitute products whose demand functions are given by

$$x_1 = 200(p_2 - p_1) \qquad \text{Demand for product 1}$$
$$x_2 = 500 + 100p_1 - 180p_2. \qquad \text{Demand for product 2}$$

With this model, the total demand, $x_1 + x_2$, is completely determined by the prices p_1 and p_2. In many real-life situations, this assumption is too simplistic; regardless of the prices of the substitute brands, the annual total demands for some products, such as toothpaste, are relatively constant. In such situations, the total demand is **limited,** and variations in price do not affect the total demand as much as they affect the market share of the substitute brands.

Example 4 Finding a Maximum Profit

A company makes two substitute products whose demand functions are given by

$$x_1 = 200(p_2 - p_1) \qquad \text{Demand for product 1}$$
$$x_2 = 500 + 100p_1 - 180p_2 \qquad \text{Demand for product 2}$$

where p_1 and p_2 are the prices per unit (in dollars) and x_1 and x_2 are the numbers of units sold. The costs of producing the two products are \$0.50 and \$0.75 per unit, respectively. The total demand is limited to 200 units per year. Find the prices that will yield a maximum profit.

SOLUTION   From Example 4 in Section 13.5, the profit function is modeled by

$$P = -200p_1^2 - 180p_2^2 + 300p_1p_2 + 25p_1 + 535p_2 - 375.$$

The total demand for the two products is

$$x_1 + x_2 = 200(p_2 - p_1) + 500 + 100p_1 - 180p_2$$
$$= -100p_1 + 20p_2 + 500.$$

Because the total demand is limited to 200 units,

$$-100p_1 + 20p_2 + 500 = 200. \qquad \text{Constraint}$$

Using Lagrange multipliers, you can determine that the maximum profit occurs when $p_1 = \$3.94$ and $p_2 = \$4.69$. This corresponds to an annual profit of \$712.21.

✓**CHECKPOINT 4**

In Example 4, suppose the total demand is limited to 250 units per year. Find the prices that will yield a maximum profit. ■

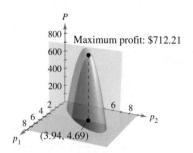

FIGURE 13.36

STUDY TIP

The constrained optimization problem in Example 4 is represented graphically in Figure 13.36. The graph of the objective function is a paraboloid and the graph of the constraint is a vertical plane. In the "unconstrained" optimization problem on page 972, the maximum profit occurred at the vertex of the paraboloid. In this "constrained" problem, however, the maximum profit corresponds to the highest point on the curve that is the intersection of the paraboloid and the vertical "constraint" plane.

Lagrange Multipliers with Two Constraints

In Examples 1 through 4, each of the optimization problems contained only one constraint. When an optimization problem has two constraints, you need to introduce a second Lagrange multiplier. The customary symbol for this second multiplier is μ, the Greek letter mu.

Example 5   Using Lagrange Multipliers: Two Constraints

Find the minimum value of

$$f(x, y, z) = x^2 + y^2 + z^2 \qquad \text{Objective function}$$

subject to the constraints

$$x + y - 3 = 0 \qquad \text{Constraint 1}$$
$$x + z - 5 = 0. \qquad \text{Constraint 2}$$

SOLUTION   Begin by forming the function

$$F(x, y, z, \lambda, \mu) = x^2 + y^2 + z^2 - \lambda(x + y - 3) - \mu(x + z - 5).$$

Next, set the five partial derivatives equal to zero, and solve the resulting system of equations for x, y, and z.

$$F_x(x, y, z, \lambda, \mu) = 2x - \lambda - \mu = 0 \qquad \text{Equation 1}$$
$$F_y(x, y, z, \lambda, \mu) = 2y - \lambda = 0 \qquad \text{Equation 2}$$
$$F_z(x, y, z, \lambda, \mu) = 2z - \mu = 0 \qquad \text{Equation 3}$$
$$F_\lambda(x, y, z, \lambda, \mu) = -x - y + 3 = 0 \qquad \text{Equation 4}$$
$$F_\mu(x, y, z, \lambda, \mu) = -x - z + 5 = 0 \qquad \text{Equation 5}$$

Solving this system of equations produces $x = \frac{8}{3}$, $y = \frac{1}{3}$, and $z = \frac{7}{3}$. So, the minimum value of $f(x, y, z)$ is

$$f\left(\frac{8}{3}, \frac{1}{3}, \frac{7}{3}\right) = \left(\frac{8}{3}\right)^2 + \left(\frac{1}{3}\right)^2 + \left(\frac{7}{3}\right)^2$$
$$= \frac{38}{3}.$$

✓ **CHECKPOINT 5**

Find the minimum value of $f(x, y, z) = x^2 + y^2 + z^2$ subject to the constraints

$$x + y - 2 = 0$$
$$x + z - 4 = 0. \ \blacksquare$$

┌─ **CONCEPT CHECK** ─┐

1. Lagrange multipliers are named after what French mathematician?
2. What do economists call the Lagrange multiplier obtained in a production function?
3. Explain what is meant by constrained optimization problems.
4. Explain the Method of Lagrange Multipliers for solving constrained optimization problems.

Skills Review 13.6

The following warm-up exercises involve skills that were covered in earlier sections. You will use these skills in the exercise set for this section. For additional help, review Sections 5.1, 5.2, 5.3, and 13.4.

In Exercises 1–6, solve the system of linear equations.

1. $\begin{cases} 4x - 6y = 3 \\ 2x + 3y = 2 \end{cases}$

2. $\begin{cases} 6x - 6y = 5 \\ -3x - y = 1 \end{cases}$

3. $\begin{cases} 5x - y = 25 \\ x - 5y = 15 \end{cases}$

4. $\begin{cases} 4x - 9y = 5 \\ -x + 8y = -2 \end{cases}$

5. $\begin{cases} 2x - y + z = 3 \\ 2x + 2y + z = 4 \\ -x + 2y + 3z = -1 \end{cases}$

6. $\begin{cases} -x - 4y + 6z = -2 \\ x - 3y - 3z = 4 \\ 3x + y + 3z = 0 \end{cases}$

In Exercises 7–10, find all first partial derivatives.

7. $f(x, y) = x^2y + xy^2$

8. $f(x, y) = 25(xy + y^2)^2$

9. $f(x, y, z) = x(x^2 - 2xy + yz)$

10. $f(x, y, z) = z(xy + xz + yz)$

Exercises 13.6

See www.CalcChat.com for worked-out solutions to odd-numbered exercises.

In Exercises 1–12, use Lagrange multipliers to find the given extremum. In each case, assume that x and y are positive.

Objective Function	*Constraint*
1. Maximize $f(x, y) = xy$	$x + y = 10$
2. Maximize $f(x, y) = xy$	$2x + y = 4$
3. Minimize $f(x, y) = x^2 + y^2$	$x + y - 4 = 0$
4. Minimize $f(x, y) = x^2 + y^2$	$-2x - 4y + 5 = 0$
5. Maximize $f(x, y) = x^2 - y^2$	$2y - x^2 = 0$
6. Minimize $f(x, y) = x^2 - y^2$	$x - 2y + 6 = 0$
7. Maximize $f(x, y) = 2x + 2xy + y$	$2x + y = 100$
8. Minimize $f(x, y) = 3x + y + 10$	$x^2y = 6$
9. Maximize $f(x, y) = \sqrt{6 - x^2 - y^2}$	$x + y - 2 = 0$
10. Minimize $f(x, y) = \sqrt{x^2 + y^2}$	$2x + 4y - 15 = 0$
11. Maximize $f(x, y) = e^{xy}$	$x^2 + y^2 - 8 = 0$
12. Minimize $f(x, y) = 2x + y$	$xy = 32$

In Exercises 13–18, use Lagrange multipliers to find the given extremum. In each case, assume that x, y, and z are positive.

13. Minimize $f(x, y, z) = 2x^2 + 3y^2 + 2z^2$

Constraint: $x + y + z - 24 = 0$

14. Maximize $f(x, y, z) = xyz$

Constraint: $x + y + z - 6 = 0$

15. Minimize $f(x, y, z) = x^2 + y^2 + z^2$

Constraint: $x + y + z = 1$

16. Minimize $f(x, y) = x^2 - 8x + y^2 - 12y + 48$

Constraint: $x + y = 8$

17. Maximize $f(x, y, z) = x + y + z$

Constraint: $x^2 + y^2 + z^2 = 1$

18. Maximize $f(x, y, z) = x^2y^2z^2$

Constraint: $x^2 + y^2 + z^2 = 1$

In Exercises 19–22, use Lagrange multipliers to find the given extremum of f subject to two constraints. In each case, assume that x, y, and z are nonnegative.

19. Maximize $f(x, y, z) = xyz$

Constraints: $x + y + z = 32$, $x - y + z = 0$

20. Minimize $f(x, y, z) = x^2 + y^2 + z^2$

Constraints: $x + 2z = 6$, $x + y = 12$

21. Maximize $f(x, y, z) = xyz$

Constraints: $x^2 + z^2 = 5$, $x - 2y = 0$

22. Maximize $f(x, y, z) = xy + yz$

Constraints: $x + 2y = 6$, $x - 3z = 0$

Ⓢ In Exercises 23 and 24, use a spreadsheet to find the given extremum. In each case, assume that x, y, and z are nonnegative.

23. Maximize $f(x, y, z) = xyz$

Constraints: $x + 3y = 6$, $x - 2z = 0$

24. Minimize $f(x, y, z) = x^2 + y^2 + z^2$

Constraints: $x + 2y = 8$, $x + z = 4$

In Exercises 25–28, find three positive numbers x, y, and z that satisfy the given conditions.

25. The sum is 120 and the product is maximum.

26. The sum is 120 and the sum of the squares is minimum.

27. The sum is S and the product is maximum.

28. The sum is S and the sum of the squares is minimum.

In Exercises 29–32, find the minimum distance from the curve or surface to the given point. (*Hint:* Start by minimizing the square of the distance.)

29. Line: $x + y = 6$, $(0, 0)$

Minimize $d^2 = x^2 + y^2$

30. Circle: $(x - 4)^2 + y^2 = 4$, $(0, 10)$

Minimize $d^2 = x^2 + (y - 10)^2$

31. Plane: $x + y + z = 1$, $(2, 1, 1)$

Minimize $d^2 = (x - 2)^2 + (y - 1)^2 + (z - 1)^2$

32. Cone: $z = \sqrt{x^2 + y^2}$, $(4, 0, 0)$

Minimize $d^2 = (x - 4)^2 + y^2 + z^2$

33. Volume Find the dimensions of the rectangular package of largest volume subject to the constraint that the sum of the length and the girth cannot exceed 108 inches (see figure). (*Hint:* Maximize $V = xyz$ subject to the constraint $x + 2y + 2z = 108$.)

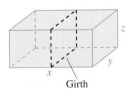

Figure for 33   Figure for 34

34. Cost In redecorating an office, the cost for new carpeting is $3 per square foot and the cost of wallpapering a wall is $1 per square foot. Find the dimensions of the largest office that can be redecorated for $1296 (see figure). (*Hint:* Maximize $V = xyz$ subject to $3xy + 2xz + 2yz = 1296$.)

35. Cost A cargo container (in the shape of a rectangular solid) must have a volume of 480 cubic feet. Use Lagrange multipliers to find the dimensions of the container of this size that has a minimum cost, if the bottom will cost $5 per square foot to construct and the sides and top will cost $3 per square foot to construct.

36. Cost A manufacturer has an order for 1000 units of fine paper that can be produced at two locations. Let x_1 and x_2 be the numbers of units produced at the two plants. Find the number of units that should be produced at each plant to minimize the cost if the cost function is given by

$$C = 0.25x_1^2 + 25x_1 + 0.05x_2^2 + 12x_2.$$

37. Cost A manufacturer has an order for 2000 units of all-terrain vehicle tires that can be produced at two locations. Let x_1 and x_2 be the numbers of units produced at the two plants. The cost function is modeled by

$$C = 0.25x_1^2 + 10x_1 + 0.15x_2^2 + 12x_2.$$

Find the number of units that should be produced at each plant to minimize the cost.

38. Hardy-Weinberg Law Repeat Exercise 47 in Section 13.5 using Lagrange multipliers—that is, maximize

$$P(p, q, r) = 2pq + 2pr + 2qr$$

subject to the constraint

$$p + q + r = 1.$$

39. Least-Cost Rule The production function for a company is given by

$$f(x, y) = 100x^{0.25}y^{0.75}$$

where x is the number of units of labor and y is the number of units of capital. Suppose that labor costs $48 per unit, capital costs $36 per unit, and management sets a production goal of 20,000 units.

(a) Find the numbers of units of labor and capital needed to meet the production goal while minimizing the cost.

(b) Show that the conditions of part (a) are met when

$$\frac{\text{Marginal productivity of labor}}{\text{Marginal productivity of capital}} = \frac{\text{unit price of labor}}{\text{unit price of capital}}.$$

This proportion is called the *Least-Cost Rule* (or *Equimarginal Rule*).

40. Least-Cost Rule Repeat Exercise 39 for the production function given by

$$f(x, y) = 100x^{0.6}y^{0.4}.$$

41. Production The production function for a company is given by

$$f(x, y) = 100x^{0.25}y^{0.75}$$

where x is the number of units of labor and y is the number of units of capital. Suppose that labor costs $48 per unit and capital costs $36 per unit. The total cost of labor and capital is limited to $100,000.

(a) Find the maximum production level for this manufacturer.

(b) Find the marginal productivity of money.

(c) Use the marginal productivity of money to find the maximum number of units that can be produced if $125,000 is available for labor and capital.

42. Production Repeat Exercise 41 for the production function given by

$$f(x, y) = 100x^{0.6}y^{0.4}.$$

43. Biology A microbiologist must prepare a culture medium in which to grow a certain type of bacteria. The percent of salt contained in this medium is given by

$$S = 12xyz$$

where x, y, and z are the nutrient solutions to be mixed in the medium. For the bacteria to grow, the medium must be 13% salt. Nutrient solutions x, y, and z cost $1, $2, and $3 per liter, respectively. How much of each nutrient solution should be used to minimize the cost of the culture medium?

44. Biology Repeat Exercise 43 for a salt-content model of

$$S = 0.01x^2y^2z^2.$$

45. Animal Shelter An animal shelter buys two different brands of dog food. The number of dogs that can be fed from x pounds of the first brand and y pounds of the second brand is given by the model

$$D(x, y) = -x^2 + 52x - y^2 + 44y + 256.$$

(a) The shelter orders 100 pounds of dog food. Use Lagrange multipliers to find the number of pounds of each brand of dog food that should be in the order so that the maximum number of dogs can be fed.

(b) What is the maximum number of dogs that can be fed?

46. Nutrition The number of grams of your favorite ice cream can be modeled by

$$G(x, y, z) = 0.05x^2 + 0.16xy + 0.25z^2$$

where x is the number of fat grams, y is the number of carbohydrate grams, and z is the number of protein grams. Use Lagrange multipliers to find the maximum number of grams of ice cream you can eat without consuming more than 400 calories. Assume that there are 9 calories per fat gram, 4 calories per carbohydrate gram, and 4 calories per protein gram.

47. Construction A rancher plans to use an existing stone wall and the side of a barn as a boundary for two adjacent rectangular corrals. Fencing for the perimeter costs $10 per foot. To separate the corrals, a fence that costs $4 per foot will divide the region. The total area of the two corrals is to be 6000 square feet.

(a) Use Lagrange multipliers to find the dimensions that will minimize the cost of the fencing.

(b) What is the minimum cost?

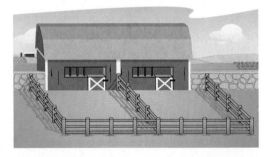

48. Office Space Partitions will be used in an office to form four equal work areas with a total area of 360 square feet (see figure). The partitions that are x feet long cost $100 per foot and the partitions that are y feet long cost $120 per foot.

(a) Use Lagrange multipliers to find the dimensions x and y that will minimize the cost of the partitions.

(b) What is the minimum cost?

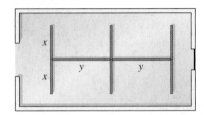

49. Investment Strategy An investor is considering three different stocks in which to invest $300,000. The average annual dividends for the stocks are

General Motors (G)	2.7%
PepsiCo, Inc. (P)	1.7%
Sara Lee (S)	2.4%.

The amount invested in PepsiCo, Inc. must follow the equation

$$3000(S) - 3000(G) + P^2 = 0.$$

How much should be invested in each stock to yield a maximum of dividends?

50. Investment Strategy An investor is considering three different stocks in which to invest $20,000. The average annual dividends for the stocks are

General Mills (G)	2.4%
Campbell Soup (C)	1.8%
Kellogg Co. (K)	1.9%.

The amount invested in Campbell Soup must follow the equation

$$1000(K) - 1000(G) + C^2 = 0.$$

How much should be invested in each stock to yield a maximum of dividends?

51. Advertising A private golf club is determining how to spend its $2700 advertising budget. The club knows from prior experience that the number of responses A is given by $A = 0.0001t^2pr^{1.5}$, where t is the number of cable television ads, p is the number of newspaper ads, and r is the number of radio ads. A cable television ad costs $30, a newspaper ad costs $12, and a radio ad costs $15.

(a) How much should be spent on each type of advertising to obtain the maximum number of responses? (Assume the golf club uses each type of advertising.)

(b) What is the maximum number of responses expected?

Section 13.7

Least Squares Regression Analysis

■ Find the sum of the squared errors for mathematical models.
■ Find the least squares regression lines for data.
■ Find the least squares regression quadratics for data.

Measuring the Accuracy of a Mathematical Model

When seeking a mathematical model to fit real-life data, you should try to find a model that is both as *simple* and as *accurate* as possible. For instance, a simple linear model for the points shown in Figure 13.37(a) is

$$f(x) = 1.8566x - 5.0246. \qquad \text{Linear model}$$

However, Figure 13.37(b) shows that by choosing a slightly more complicated quadratic model

$$g(x) = 0.1996x^2 - 0.7281x + 1.3749 \qquad \text{Quadratic model}$$

you can obtain significantly greater accuracy.

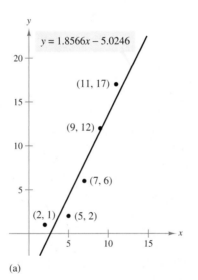

(a)

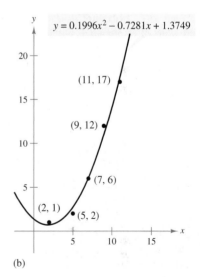

(b)

FIGURE 13.37

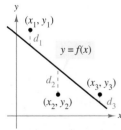

Sum of the squared errors:
$$S = d_1{}^2 + d_2{}^2 + d_3{}^2$$

FIGURE 13.38

To measure how well the model $y = f(x)$ fits a collection of points, sum the squares of the differences between the actual y-values and the model's y-values. This sum is called the **sum of the squared errors** and is denoted by S. Graphically, S can be interpreted as the sum of the squares of the vertical distances between the graph of f and the given points in the plane, as shown in Figure 13.38. If the model is a perfect fit, then $S = 0$. However, when a perfect fit is not feasible, you should use a model that minimizes S.

Definition of the Sum of the Squared Errors

The **sum of the squared errors** for the model $y = f(x)$ with respect to the points $(x_1, y_1), (x_2, y_2), \ldots, (x_n, y_n)$ is given by

$$S = [f(x_1) - y_1]^2 + [f(x_2) - y_2]^2 + \cdots + [f(x_n) - y_n]^2.$$

Example 1   Finding the Sum of the Squared Errors

Find the sum of the squared errors for the linear and quadratic models

$$f(x) = 1.8566x - 5.0246 \qquad \text{Linear model}$$
$$g(x) = 0.1996x^2 - 0.7281x + 1.3749 \qquad \text{Quadratic model}$$

(see Figure 13.37) with respect to the points

$$(2, 1), (5, 2), (7, 6), (9, 12), (11, 17).$$

SOLUTION   Begin by evaluating each model at the given x-values, as shown in the table.

x	2	5	7	9	11
Actual y-values	1	2	6	12	17
Linear model, $f(x)$	-1.3114	4.2584	7.9716	11.6848	15.3980
Quadratic model, $g(x)$	0.7171	2.7244	6.0586	10.9896	17.5174

For the linear model f, the sum of the squared errors is

$$S = (-1.3114 - 1)^2 + (4.2584 - 2)^2 + (7.9716 - 6)^2$$
$$+ (11.6848 - 12)^2 + (15.3980 - 17)^2$$
$$\approx 16.9959.$$

Similarly, the sum of the squared errors for the quadratic model g is

$$S = (0.7171 - 1)^2 + (2.7244 - 2)^2 + (6.0586 - 6)^2$$
$$+ (10.9896 - 12)^2 + (17.5174 - 17)^2$$
$$\approx 1.8968.$$

STUDY TIP

In Example 1, note that the sum of the squared errors for the quadratic model is less than the sum of the squared errors for the linear model, which confirms that the quadratic model is a better fit.

✓ **CHECKPOINT 1**

Find the sum of the squared errors for the linear and quadratic models

$$f(x) = 2.85x - 6.1$$
$$g(x) = 0.1964x^2 + 0.4929x - 0.6$$

with respect to the points $(2, 1), (4, 5), (6, 9), (8, 16), (10, 24)$. Then decide which model is a better fit. ■

Least Squares Regression Line

The sum of the squared errors can be used to determine which of several models is the best fit for a collection of data. In general, if the sum of the squared errors of f is less than the sum of the squared errors of g, then f is said to be a better fit for the data than g. In regression analysis, you consider all possible models of a certain type. The one that is defined to be the best-fitting model is the one with the least sum of the squared errors. Example 2 shows how to use the optimization techniques described in Section 13.5 to find the best-fitting linear model for a collection of data.

> **Algebra Review**
>
> For help in solving the system of equations in Example 2, see Example 2(b) in the *Chapter 13 Algebra Review*, on page 1014.

Example 2 **Finding the Best Linear Model**

Find the values of a and b such that the linear model

$$f(x) = ax + b$$

has a minimum sum of the squared errors for the points

$$(-3, 0), (-1, 1), (0, 2), (2, 3).$$

SOLUTION   The sum of the squared errors is

$$S = [f(x_1) - y_1]^2 + [f(x_2) - y_2]^2 + [f(x_3) - y_3]^2 + [f(x_4) - y_4]^2$$
$$= (-3a + b - 0)^2 + (-a + b - 1)^2 + (b - 2)^2 + (2a + b - 3)^2$$
$$= 14a^2 - 4ab + 4b^2 - 10a - 12b + 14.$$

To find the values of a and b for which S is a minimum, you can use the techniques described in Section 13.5. That is, find the partial derivatives of S.

$$\frac{\partial S}{\partial a} = 28a - 4b - 10 \qquad \text{Differentiate with respect to } a.$$

$$\frac{\partial S}{\partial b} = -4a + 8b - 12 \qquad \text{Differentiate with respect to } b.$$

Next, set each partial derivative equal to zero.

$$28a - 4b - 10 = 0 \qquad \text{Set } \partial S/\partial a \text{ equal to 0.}$$
$$-4a + 8b - 12 = 0 \qquad \text{Set } \partial S/\partial b \text{ equal to 0.}$$

The solution of this system of linear equations is

$$a = \frac{8}{13} \quad \text{and} \quad b = \frac{47}{26}.$$

So, the best-fitting linear model for the given points is

$$f(x) = \frac{8}{13}x + \frac{47}{26}.$$

The graph of this model is shown in Figure 13.39.

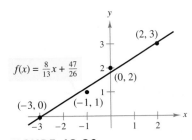

$f(x) = \frac{8}{13}x + \frac{47}{26}$

FIGURE 13.39

✓ CHECKPOINT 2

Find the values of a and b such that the linear model $f(x) = ax + b$ has a minimum sum of the squared errors for the points $(-2, 0)$, $(0, 2)$, $(2, 5)$, $(4, 7)$. ■

The line in Example 2 is called the **least squares regression line** for the given data. The solution shown in Example 2 can be generalized to find a formula for the least squares regression line, as shown below. Consider the linear model

$$f(x) = ax + b$$

and the points (x_1, y_1), (x_2, y_2), . . . , (x_n, y_n). The sum of the squared errors is

$$S = \sum_{i=1}^{n} [f(x_i) - y_i]^2 = \sum_{i=1}^{n} (ax_i + b - y_i)^2.$$

To minimize S, set the partial derivatives $\partial S / \partial a$ and $\partial S / \partial b$ equal to zero and solve for a and b. The results are summarized below.

The Least Squares Regression Line

The **least squares regression line** for the points

$$(x_1, y_1), (x_2, y_2), . . . , (x_n, y_n)$$

is $y = ax + b$, where

$$a = \frac{n \sum_{i=1}^{n} x_i y_i - \sum_{i=1}^{n} x_i \sum_{i=1}^{n} y_i}{n \sum_{i=1}^{n} x_i^2 - \left(\sum_{i=1}^{n} x_i \right)^2} \quad \text{and} \quad b = \frac{1}{n} \left(\sum_{i=1}^{n} y_i - a \sum_{i=1}^{n} x_i \right).$$

In the formula for the least squares regression line, note that if the x-values are symmetrically spaced about zero, then

$$\sum_{i=1}^{n} x_i = 0$$

and the formulas for a and b simplify to

$$a = \frac{n \sum_{i=1}^{n} x_i y_i}{n \sum_{i=1}^{n} x_i^2} \quad \text{and} \quad b = \frac{1}{n} \sum_{i=1}^{n} y_i.$$

Note also that only the *development* of the least squares regression line involves partial derivatives. The *application* of this formula is simply a matter of computing the values of a and b—a task that is performed much more simply on a calculator or a computer than by hand.

DISCOVERY

Graph the three points $(2, 2)$, $(2, 1)$, and $(2.1, 1.5)$ and visually estimate the least squares regression line for these data. Now use the formulas on this page or a graphing utility to show that the equation of the line is actually $y = 1.5$. In general, the least squares regression line for "nearly vertical" data can be quite unusual. Show that by interchanging the roles of x and y, you can obtain a better linear approximation.

Example 3 | **Modeling Hourly Wages**

The average hourly wages y (in dollars per hour) for production workers in manufacturing industries from 1998 through 2006 are shown in the table. Find the least squares regression line for the data and use the result to estimate the average hourly wage in 2010.   *(Source: U.S. Bureau of Labor Statistics)*

Year	1998	1999	2000	2001	2002	2003	2004	2005	2006
y	13.45	13.85	14.32	14.76	15.29	15.74	16.15	16.56	16.80

SOLUTION   Let t represent the year, with $t = 8$ corresponding to 1998. Then, you need to find the linear model that best fits the points

$(8, 13.45)$, $(9, 13.85)$, $(10, 14.32)$, $(11, 14.76)$, $(12, 15.29)$, $(13, 15.74)$, $(14, 16.15)$, $(15, 16.56)$, $(16, 16.80)$.

Using a calculator with a built-in least squares regression program, you can determine that the best-fitting line is $y = 9.98 + 0.436t$. With this model, you can estimate the 2010 average hourly wage, using $t = 20$, to be

$$y = 9.98 + 0.436(20) = \$18.70 \text{ per hour.}$$

This result is shown graphically in Figure 13.40.   ———

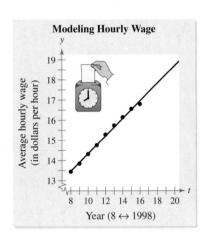

Modeling Hourly Wage

Average hourly wage (in dollars per hour)

Year (8 ↔ 1998)

FIGURE 13.40

✓ **CHECKPOINT 3**

The numbers of cellular phone subscribers y (in thousands) for the years 2001 through 2005 are shown in the table. Find the least squares regression line for the data and use the result to estimate the number of subscribers in 2010. Let t represent the year, with $t = 1$ corresponding to 2001.   *(Source: Cellular Telecommunications & Internet Association)*

Year	2001	2002	2003	2004	2005
y	128,375	140,767	158,722	182,140	207,896

TECHNOLOGY

Ⓣ Most graphing utilities and spreadsheet software programs have a built-in linear regression program. When you run such a program, the "r-value" gives a measure of how well the model fits the data. The closer the value of $|r|$ is to 1, the better the fit. For the data in Example 3, $r \approx 0.998$, which implies that the model is a very good fit. Use a graphing utility or a spreadsheet software program to find the least squares regression line and compare your results with those in Example 3. (Consult the user's manual of a graphing utility or a spreadsheet software program for specific instructions.)*

---

*Specific calculator keystroke instructions for operations in this and other technology boxes can be found at *college.hmco.com/info/larsonapplied*.

Least Squares Regression Quadratic

When using regression analysis to model data, remember that the least squares regression line provides only the best *linear* model for a set of data. It does not necessarily provide the best possible model. For instance, in Example 1, you saw that the quadratic model was a better fit than the linear model.

Regression analysis can be performed with many different types of models, such as exponential or logarithmic models. The following development shows how to find the best-fitting quadratic model for a collection of data points. Consider a quadratic model of the form

$$f(x) = ax^2 + bx + c.$$

The sum of the squared errors for this model is

$$S = \sum_{i=1}^{n} [f(x_i) - y_i]^2 = \sum_{i=1}^{n} (ax_i^2 + bx_i + c - y_i)^2.$$

To find the values of a, b, and c that minimize S, set the three partial derivatives, $\partial S / \partial a$, $\partial S / \partial b$, and $\partial S / \partial c$, equal to zero.

$$\frac{\partial S}{\partial a} = \sum_{i=1}^{n} 2x_i^2 (ax_i^2 + bx_i + c - y_i) = 0$$

$$\frac{\partial S}{\partial b} = \sum_{i=1}^{n} 2x_i (ax_i^2 + bx_i + c - y_i) = 0$$

$$\frac{\partial S}{\partial c} = \sum_{i=1}^{n} 2(ax_i^2 + bx_i + c - y_i) = 0$$

By expanding this system, you obtain the result given in the summary below.

Least Squares Regression Quadratic

The **least squares regression quadratic** for the points

$$(x_1, y_1), (x_2, y_2), \ldots, (x_n, y_n)$$

is $y = ax^2 + bx + c$, where a, b, and c are the solutions of the system of equations below.

$$a\sum_{i=1}^{n} x_i^4 + b\sum_{i=1}^{n} x_i^3 + c\sum_{i=1}^{n} x_i^2 = \sum_{i=1}^{n} x_i^2 y_i$$

$$a\sum_{i=1}^{n} x_i^3 + b\sum_{i=1}^{n} x_i^2 + c\sum_{i=1}^{n} x_i = \sum_{i=1}^{n} x_i y_i$$

$$a\sum_{i=1}^{n} x_i^2 + b\sum_{i=1}^{n} x_i + cn = \sum_{i=1}^{n} y_i$$

TECHNOLOGY

(T) Most graphing utilities have a built-in program for finding the least squares regression quadratic. This program works just like the program for the least squares line. You should use this program to verify your solutions to the exercises.

Example 4 | **Modeling Numbers of Newspapers**

The numbers y of daily morning newspapers in the United States from 1995 through 2005 are shown in the table. Find the least squares regression quadratic for the data and use the result to estimate the number of daily morning newspapers in 2010.   *(Source: Editor & Publisher Co.)*

Year	1995	1996	1997	1998	1999	2000	2001	2002	2003	2004	2005
y	656	686	705	721	736	766	776	776	787	813	817

Daily Morning Newspapers

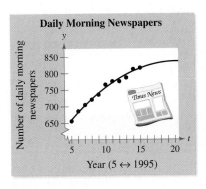

FIGURE 13.41

SOLUTION   Let t represent the year, with $t = 5$ corresponding to 1995. Then, you need to find the quadratic model that best fits the points

$(5, 656)$, $(6, 686)$, $(7, 705)$, $(8, 721)$, $(9, 736)$, $(10, 766)$, $(11, 776)$,
$(12, 776)$, $(13, 787)$, $(14, 813)$, $(15, 817)$.

Using a calculator with a built-in least squares regression program, you can determine that the best-fitting quadratic is $y = -0.76t^2 + 30.8t + 525$. With this model, you can estimate the number of daily morning newspapers in 2010, using $t = 20$, to be

$$y = -0.76(20)^2 + 30.8(20) + 525 = 837.$$

This result is shown graphically in Figure 13.41.

✓ **CHECKPOINT 4**

The per capita expenditures y for health services and supplies in dollars in the United States for selected years are listed in the table. Find the least squares regression quadratic for the data and use the result to estimate the per capita expenditure for health care in 2010. Let t represent the year, with $t = 9$ corresponding to 1999.   *(Source: U.S. Centers for Medicare and Medicaid Services)*

Year	1999	2000	2001	2002	2003	2004	2005
y	3818	4034	4340	4652	4966	5276	5598

CONCEPT CHECK

1. What are the two main goals when seeking a mathematical model to fit real-life data?
2. What does S, the sum of the squared errors, measure?
3. Describe how to find the least squares regression line for a given set of data.
4. Describe how to find the least squares regression quadratic for a given set of data.

Skills Review 13.7

The following warm-up exercises involve skills that were covered in earlier sections. You will use these skills in the exercise set for this section. For additional help, review Sections 0.2 and 13.4.

In Exercises 1 and 2, evaluate the expression.

1. $(2.5 - 1)^2 + (3.25 - 2)^2 + (4.1 - 3)^2$

2. $(1.1 - 1)^2 + (2.08 - 2)^2 + (2.95 - 3)^2$

In Exercises 3 and 4, find the partial derivatives of S.

3. $S = a^2 + 6b^2 - 4a - 8b - 4ab + 6$

4. $S = 4a^2 + 9b^2 - 6a - 4b - 2ab + 8$

In Exercises 5–10, evaluate the sum.

5. $\displaystyle\sum_{i=1}^{5} i$

6. $\displaystyle\sum_{i=1}^{6} 2i$

7. $\displaystyle\sum_{i=1}^{4} \frac{1}{i}$

8. $\displaystyle\sum_{i=1}^{3} i^2$

9. $\displaystyle\sum_{i=1}^{6} (2 - i)^2$

10. $\displaystyle\sum_{i=1}^{5} (30 - i^2)$

Exercises 13.7

See www.CalcChat.com for worked-out solutions to odd-numbered exercises.

In Exercises 1–4, (a) find the least squares regression line and (b) calculate S, the sum of the squared errors. Use the regression capabilities of a graphing utility or a spreadsheet to verify your results.

1.

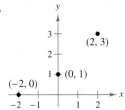

2.

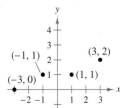

3.

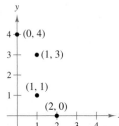

4.

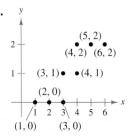

In Exercises 5–8, find the least squares regression line for the points. Use the regression capabilities of a graphing utility or a spreadsheet to verify your results. Then plot the points and graph the regression line.

5. $(-2, -1), (0, 0), (2, 3)$

6. $(-3, 0), (-1, 1), (1, 1), (3, 2)$

7. $(-2, 4), (-1, 1), (0, -1), (1, -3)$

8. $(-5, -3), (-4, -2), (-2, -1), (-1, 1)$

In Exercises 9–18, use the regression capabilities of a graphing utility or a spreadsheet to find the least squares regression line for the given points.

9. $(-2, 0), (-1, 1), (0, 1), (1, 2), (2, 3)$

10. $(-4, -1), (-2, 0), (2, 4), (4, 5)$

11. $(-2, 2), (2, 6), (3, 7)$

12. $(-5, 1), (1, 3), (2, 3), (2, 5)$

13. $(-3, 4), (-1, 2), (1, 1), (3, 0)$

14. $(-10, 10), (-5, 8), (3, 6), (7, 4), (5, 0)$

15. $(0, 0), (1, 1), (3, 4), (4, 2), (5, 5)$

16. $(1, 0), (3, 3), (5, 6)$

17. $(0, 6), (4, 3), (5, 0), (8, -4), (10, -5)$

18. $(6, 4), (1, 2), (3, 3), (8, 6), (11, 8), (13, 8)$

In Exercises 19–22, use the regression capabilities of a graphing utility or a spreadsheet to find the least squares regression quadratic for the given points. Then plot the points and graph the least squares regression quadratic.

19. $(-2, 0), (-1, 0), (0, 1), (1, 2), (2, 5)$

20. $(-4, 5), (-2, 6), (2, 6), (4, 2)$

21. $(0, 0), (2, 2), (3, 6), (4, 12)$

22. $(0, 10), (1, 9), (2, 6), (3, 0)$

Ⓣ Ⓢ In Exercises 23–26, use the regression capabilities of a graphing utility or a spreadsheet to find linear and quadratic models for the data. State which model best fits the data.

23. $(-4, 1), (-3, 2), (-2, 2), (-1, 4), (0, 6), (1, 8), (2, 9)$

24. $(-1, -4), (0, -3), (1, -3), (2, 0), (4, 5), (6, 9), (9, 3)$

25. $(0, 769), (1, 677), (2, 601), (3, 543), (4, 489), (5, 411)$

26. $(1, 10.3), (2, 14.2), (3, 18.9), (4, 23.7), (5, 29.1), (6, 35)$

27. Demand A store manager wants to know the demand y for an energy bar as a function of price x. The daily sales for three different prices of the product are listed in the table.

Price, x	$1.00	$1.25	$1.50
Demand, y	450	375	330

Ⓣ Ⓢ (a) Use the regression capabilities of a graphing utility or a spreadsheet to find the least squares regression line for the data.

(b) Estimate the demand when the price is $1.40.

(c) What price will create a demand of 500 energy bars?

28. Demand A hardware retailer wants to know the demand y for a tool as a function of price x. The monthly sales for four different prices of the tool are listed in the table.

Price, x	$25	$30	$35	$40
Demand, y	82	75	67	55

Ⓣ Ⓢ (a) Use the regression capabilities of a graphing utility or a spreadsheet to find the least squares regression line for the data.

(b) Estimate the demand when the price is $32.95.

(c) What price will create a demand of 83 tools?

29. Agriculture An agronomist used four test plots to determine the relationship between the wheat yield y (in bushels per acre) and the amount of fertilizer x (in hundreds of pounds per acre). The results are shown in the table.

Fertilizer, x	1.0	1.5	2.0	2.5
Yield, y	35	44	50	56

Ⓣ Ⓢ (a) Use the regression capabilities of a graphing utility or a spreadsheet to find the least squares regression line for the data.

(b) Estimate the yield for a fertilizer application of 160 pounds per acre.

Ⓣ Ⓢ **30. Finance: Median Income** In the table below are the median income levels for various age levels in the United States. Use the regression capabilities of a graphing utility or a spreadsheet to find the least squares regression quadratic for the data and use the resulting model to estimate the median income for someone who is 28 years old. (*Source: U.S. Census Bureau*)

Age level, x	20	30	40
Median income, y	28,800	47,400	58,100

Age level, x	50	60	70
Median income, y	62,400	52,300	26,000

Ⓣ Ⓢ **31. Infant Mortality** To study the numbers y of infant deaths per 1000 live births in the United States, a medical researcher obtains the data listed in the table. (*Source: U.S. National Center for Health Statistics*)

Year	1980	1985	1990	1995	2000	2005
Deaths, y	12.6	10.6	9.2	7.6	6.9	6.8

(a) Use the regression capabilities of a graphing utility or a spreadsheet to find the least squares regression line for the data and use this line to estimate the number of infant deaths in 2010. Let $t = 0$ represent 1980.

(b) Use the regression capabilities of a graphing utility or a spreadsheet to find the least squares regression quadratic for the data and use the model to estimate the number of infant deaths in 2010.

Ⓣ Ⓢ **32. Population Growth** The table gives the approximate world populations y (in billions) for six different years. (*Source: U.S. Census Bureau*)

Year	1800	1850	1900	1950	1990	2005
Time, t	-2	-1	0	1	1.8	2.1
Population, y	0.8	1.1	1.6	2.4	5.3	6.5

(a) During the 1800s, population growth was almost linear. Use the regression capabilities of a graphing utility or a spreadsheet to find a least squares regression line for those years and use the line to estimate the population in 1875.

(b) Use the regression capabilities of a graphing utility or a spreadsheet to find a least squares regression quadratic for the data from 1850 through 2005 and use the model to estimate the population in the year 2010.

(c) Even though the rate of growth of the population has begun to decline, most demographers believe the population size will pass the 8 billion mark sometime in the next 25 years. What do you think?

(T)(S) 33. *MAKE A DECISION: REVENUE* The revenues y (in millions of dollars) for Earthlink from 2000 through 2006 are shown in the table. *(Source: Earthlink, Inc.)*

Year	2000	2001	2002	2003
Revenue, y	986.6	1244.9	1357.4	1401.9

Year	2004	2005	2006
Revenue, y	1382.2	1290.1	1301.3

(a) Use a graphing utility or a spreadsheet to create a scatter plot of the data. Let $t = 0$ represent the year 2000.

(b) Use the regression capabilities of a graphing utility or a spreadsheet to find an appropriate model for the data.

(c) Explain why you chose the type of model that you created in part (b).

(T)(S) 34. *MAKE A DECISION: COMPUTERS AND INTERNET USERS* The global numbers of personal computers x (in millions) and Internet users y (in millions) from 1999 through 2005 are shown in the table. *(Source: International Telecommunication Union)*

Year	1999	2000	2001	2002
Personal computers, x	394.1	465.4	526.7	575.5
Internet users, y	275.5	390.3	489.9	618.4

Year	2003	2004	2005
Personal computers, x	636.6	776.6	808.7
Internet users, y	718.8	851.8	982.5

(a) Use a graphing utility or a spreadsheet to create a scatter plot of the data.

(b) Use the regression capabilities of a graphing utility or a spreadsheet to find an appropriate model for the data.

(c) Explain why you chose the type of model that you created in part (b).

(T)(S) In Exercises 35–38, use the regression capabilities of a graphing utility or a spreadsheet to find any model that best fits the data points.

35. $(1, 13)$, $(2, 16.5)$, $(4, 24)$, $(5, 28)$, $(8, 39)$, $(11, 50.25)$, $(17, 72)$, $(20, 85)$

36. $(1, 5.5)$, $(3, 7.75)$, $(6, 15.2)$, $(8, 23.5)$, $(11, 46)$, $(15, 110)$

37. $(1, 1.5)$, $(2.5, 8.5)$, $(5, 13.5)$, $(8, 16.7)$, $(9, 18)$, $(20, 22)$

38. $(0, 0.5)$, $(1, 7.6)$, $(3, 60)$, $(4.2, 117)$, $(5, 170)$, $(7.9, 380)$

In Exercises 39–44, plot the points and determine whether the data have positive, negative, or no linear correlation (see figures below). Then use a graphing utility to find the value of r and confirm your result. The number r is called the *correlation coefficient*. It is a measure of how well the model fits the data. Correlation coefficients vary between -1 and 1, and the closer $|r|$ is to 1, the better the model.

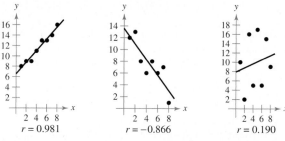

$r = 0.981$	$r = -0.866$	$r = 0.190$
Positive correlation	Negative correlation	No correlation

39. $(1, 4)$, $(2, 6)$, $(3, 8)$, $(4, 11)$, $(5, 13)$, $(6, 15)$

40. $(1, 7.5)$, $(2, 7)$, $(3, 7)$, $(4, 6)$, $(5, 5)$, $(6, 4.9)$

41. $(1, 3)$, $(2, 6)$, $(3, 2)$, $(4, 3)$, $(5, 9)$, $(6, 1)$

42. $(0.5, 2)$, $(0.75, 1.75)$, $(1, 3)$, $(1.5, 3.2)$, $(2, 3.7)$, $(2.6, 4)$

43. $(1, 36)$, $(2, 10)$, $(3, 0)$, $(4, 4)$, $(5, 16)$, $(6, 36)$

44. $(0.5, 9)$, $(1, 8.5)$, $(1.5, 7)$, $(2, 5.5)$, $(2.5, 5)$, $(3, 3.5)$

True or False? In Exercises 45–50, determine whether the statement is true or false. If it is false, explain why or give an example that shows it is false.

45. Data that are modeled by $y = 3.29x - 4.17$ have a negative correlation.

46. Data that are modeled by $y = -0.238x + 25$ have a negative correlation.

47. If the correlation coefficient is $r \approx -0.98781$, the model is a good fit.

48. A correlation coefficient of $r \approx 0.201$ implies that the data have no correlation.

49. A linear regression model with a positive correlation will have a slope that is greater than 0.

50. If the correlation coefficient for a linear regression model is close to -1, the regression line cannot be used to describe the data.

51. Extended Application To work an extended application analyzing the earnings per share, sales, and shareholder's equity of PepsiCo from 1999 through 2006, visit this text's website at *college.hmco.com*. *(Data Source: PepsiCo, Inc.)*

Section 13.8

Double Integrals and Area in the Plane

■ Evaluate double integrals.
■ Use double integrals to find the areas of regions.

Double Integrals

In Section 13.4, you learned that it is meaningful to differentiate functions of several variables by differentiating with respect to one variable at a time while holding the other variable(s) constant. It should not be surprising to learn that you can *integrate* functions of two or more variables using a similar procedure. For instance, if you are given the partial derivative

$$f_x(x, y) = 2xy \qquad \text{Partial with respect to } x$$

then, by holding y constant, you can integrate with respect to x to obtain

$$\int f_x(x, y) \, dx = f(x, y) + C(y)$$
$$= x^2y + C(y).$$

This procedure is called **partial integration with respect to x.** Note that the "constant of integration" $C(y)$ is assumed to be a function of y, because y is fixed during integration with respect to x. Similarly, if you are given the partial derivative

$$f_y(x, y) = x^2 + 2 \qquad \text{Partial with respect to } y$$

then, by holding x constant, you can integrate with respect to y to obtain

$$\int f_y(x, y) \, dy = f(x, y) + C(x)$$
$$= x^2y + 2y + C(x).$$

In this case, the "constant of integration" $C(x)$ is assumed to be a function of x, because x is fixed during integration with respect to y.

To evaluate a definite integral of a function of two or more variables, you can apply the Fundamental Theorem of Calculus to one variable while holding the other variable(s) constant, as shown.

$$\int_1^{2y} 2xy \, dx = x^2y \Big]_1^{2y} = (2y)^2y - (1)^2y$$

x is the variable of integration and y is fixed.

Replace x by the limits of integration.

$$= 4y^3 - y.$$

The result is a function of y.

Note that you omit the constant of integration, just as you do for a definite integral of a function of one variable.

Example 1 Finding Partial Integrals

Find each partial integral.

a. $\displaystyle\int_{1}^{x} (2x^2y^{-2} + 2y)\, dy$ **b.** $\displaystyle\int_{y}^{5y} \sqrt{x - y}\, dx$

SOLUTION

a. $\displaystyle\int_{1}^{x} (2x^2y^{-2} + 2y)\, dy = \left[\frac{-2x^2}{y} + y^2 \right]_{1}^{x}$ Hold x constant.

$$= \left(\frac{-2x^2}{x} + x^2 \right) - \left(\frac{-2x^2}{1} + 1 \right)$$

$$= 3x^2 - 2x - 1$$

b. $\displaystyle\int_{y}^{5y} \sqrt{x - y}\, dx = \left[\frac{2}{3}(x - y)^{3/2} \right]_{y}^{5y}$ Hold y constant.

$$= \frac{2}{3}[(5y - y)^{3/2} - (y - y)^{3/2}] = \frac{16}{3}y^{3/2}$$

✓ **CHECKPOINT 1**

Find each partial integral.

a. $\displaystyle\int_{1}^{x} (4xy + y^3)\, dy$

b. $\displaystyle\int_{y}^{y^2} \frac{1}{x + y}\, dx$ ▢

In Example 1(a), note that the definite integral defines a function of x and can *itself* be integrated. An "integral of an integral" is called a **double integral.** With a function of two variables, there are two types of double integrals.

$$\int_{a}^{b} \int_{g_1(x)}^{g_2(x)} f(x, y)\, dy\, dx = \int_{a}^{b} \left[\int_{g_1(x)}^{g_2(x)} f(x, y)\, dy \right] dx$$

$$\int_{a}^{b} \int_{g_1(y)}^{g_2(y)} f(x, y)\, dx\, dy = \int_{a}^{b} \left[\int_{g_1(y)}^{g_2(y)} f(x, y)\, dx \right] dy$$

STUDY TIP

Notice that the difference between the two types of double integrals is the order in which the integration is performed, *dy dx* or *dx dy*.

TECHNOLOGY

Ⓣ A symbolic integration utility can be used to evaluate double integrals. To do this, you need to enter the integrand, then integrate twice—once with respect to one of the variables and then with respect to the other variable. Use a symbolic integration utility to evaluate the double integral in Example 2.

Example 2 Evaluating a Double Integral

Evaluate $\displaystyle\int_{1}^{2} \int_{0}^{x} (2xy + 3)\, dy\, dx$.

SOLUTION

$$\int_{1}^{2} \int_{0}^{x} (2xy + 3)\, dy\, dx = \int_{1}^{2} \left[\int_{0}^{x} (2xy + 3)\, dy \right] dx$$

$$= \int_{1}^{2} \left[xy^2 + 3y \right]_{0}^{x} dx$$

$$= \int_{1}^{2} (x^3 + 3x)\, dx$$

$$= \left[\frac{x^4}{4} + \frac{3x^2}{2} \right]_{1}^{2}$$

$$= \left(\frac{2^4}{4} + \frac{3(2^2)}{2} \right) - \left(\frac{1^4}{4} + \frac{3(1^2)}{2} \right) = \frac{33}{4}$$

✓ **CHECKPOINT 2**

Evaluate the double integral.

$$\int_{1}^{2} \int_{0}^{x} (5x^2y - 2)\, dy\, dx$$ ▢

Finding Area with a Double Integral

One of the simplest applications of a double integral is finding the area of a plane region. For instance, consider the region R that is bounded by

$$a \le x \le b \quad \text{and} \quad g_1(x) \le y \le g_2(x).$$

Using the techniques described in Section 11.5, you know that the area of R is

$$\int_a^b [g_2(x) - g_1(x)]\, dx.$$

This same area is also given by the double integral

$$\int_a^b \int_{g_1(x)}^{g_2(x)} dy\, dx$$

because

$$\int_a^b \int_{g_1(x)}^{g_2(x)} dy\, dx = \int_a^b \left[y \right]_{g_1(x)}^{g_2(x)} dx = \int_a^b [g_2(x) - g_1(x)]\, dx.$$

Figure 13.42 shows the two basic types of plane regions whose areas can be determined by a double integral.

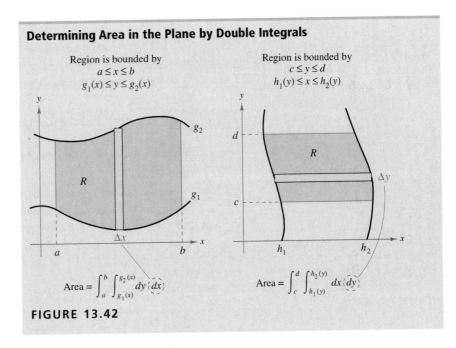

Determining Area in the Plane by Double Integrals

Region is bounded by
$a \le x \le b$
$g_1(x) \le y \le g_2(x)$

Region is bounded by
$c \le y \le d$
$h_1(y) \le x \le h_2(y)$

$$\text{Area} = \int_a^b \int_{g_1(x)}^{g_2(x)} dy\, dx$$

$$\text{Area} = \int_c^d \int_{h_1(y)}^{h_2(y)} dx\, dy$$

FIGURE 13.42

STUDY TIP

In Figure 13.42, note that the horizontal or vertical orientation of the narrow rectangle indicates the order of integration. The "outer" variable of integration always corresponds to the width of the rectangle. Notice also that the outer limits of integration for a double integral are constant, whereas the inner limits may be functions of the outer variable.

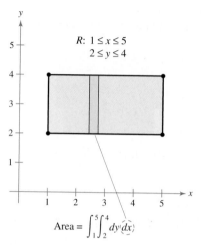

$R: 1 \le x \le 5$
$2 \le y \le 4$

$$\text{Area} = \int_1^5 \int_2^4 dy\,dx$$

FIGURE 13.43

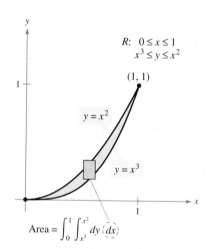

$R: \ 0 \le x \le 1$
$x^3 \le y \le x^2$

$(1, 1)$

$y = x^2$

$y = x^3$

$$\text{Area} = \int_0^1 \int_{x^3}^{x^2} dy\,dx$$

FIGURE 13.44

✓ **CHECKPOINT 4**

Use a double integral to find
the area of the region bounded by
the graphs of $y = 2x$ and $y = x^2$. ∎

Example 3 **Finding Area with a Double Integral**

Use a double integral to find the area of the rectangular region shown in Figure 13.43.

SOLUTION   The bounds for x are $1 \le x \le 5$ and the bounds for y are $2 \le y \le 4$. So, the area of the region is

$$\int_1^5 \int_2^4 dy\,dx = \int_1^5 \Big[y \Big]_2^4 dx \qquad \text{Integrate with respect to } y.$$

$$= \int_1^5 (4 - 2)\,dx \qquad \text{Apply Fundamental Theorem of Calculus.}$$

$$= \int_1^5 2\,dx \qquad \text{Simplify.}$$

$$= \Big[2x \Big]_1^5 \qquad \text{Integrate with respect to } x.$$

$$= 10 - 2 \qquad \text{Apply Fundamental Theorem of Calculus.}$$

$$= 8 \text{ square units.} \qquad \text{Simplify.}$$

You can confirm this by noting that the rectangle measures two units by four units.

✓ **CHECKPOINT 3**

Use a double integral to find the area of the rectangular region shown in Example 3 by integrating with respect to x and then with respect to y. ∎

Example 4 **Finding Area with a Double Integral**

Use a double integral to find the area of the region bounded by the graphs of $y = x^2$ and $y = x^3$.

SOLUTION   As shown in Figure 13.44, the two graphs intersect when $x = 0$ and $x = 1$. Choosing x to be the outer variable, the bounds for x are $0 \le x \le 1$ and the bounds for y are $x^3 \le y \le x^2$. This implies that the area of the region is

$$\int_0^1 \int_{x^3}^{x^2} dy\,dx = \int_0^1 \Big[y \Big]_{x^3}^{x^2} dx \qquad \text{Integrate with respect to } y.$$

$$= \int_0^1 (x^2 - x^3)\,dx \qquad \text{Apply Fundamental Theorem of Calculus.}$$

$$= \left[\frac{x^3}{3} - \frac{x^4}{4} \right]_0^1 \qquad \text{Integrate with respect to } x.$$

$$= \frac{1}{3} - \frac{1}{4} \qquad \text{Apply Fundamental Theorem of Calculus.}$$

$$= \frac{1}{12} \text{ square unit.} \qquad \text{Simplify.}$$

In setting up double integrals, the most difficult task is likely to be determining the correct limits of integration. This can be simplified by making a sketch of the region R and identifying the appropriate bounds for x and y.

Example 5   Changing the Order of Integration

For the double integral

$$\int_0^2 \int_{y^2}^4 dx \, dy$$

a. sketch the region R whose area is represented by the integral,

b. rewrite the integral so that x is the outer variable, and

c. show that both orders of integration yield the same value.

SOLUTION

a. From the limits of integration, you know that

$$y^2 \le x \le 4 \qquad \text{Variable bounds for } x$$

which means that the region R is bounded on the left by the parabola $x = y^2$ and on the right by the line $x = 4$. Furthermore, because

$$0 \le y \le 2 \qquad \text{Constant bounds for } y$$

you know that the region lies above the x-axis, as shown in Figure 13.45.

b. If you interchange the order of integration so that x is the outer variable, then x will have constant bounds of integration given by $0 \le x \le 4$. Solving for y in the equation $x = y^2$ implies that the bounds for y are $0 \le y \le \sqrt{x}$, as shown in Figure 13.46. So, with x as the outer variable, the integral can be written as

$$\int_0^4 \int_0^{\sqrt{x}} dy \, dx.$$

c. Both integrals yield the same value.

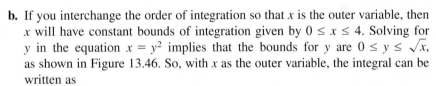

$$\int_0^2 \int_{y^2}^4 dx \, dy = \int_0^2 \left[x \right]_{y^2}^4 dy = \int_0^2 (4 - y^2) \, dy = \left[4y - \frac{y^3}{3} \right]_0^2 = \frac{16}{3}$$

$$\int_0^4 \int_0^{\sqrt{x}} dy \, dx = \int_0^4 \left[y \right]_0^{\sqrt{x}} dx = \int_0^4 \sqrt{x} \, dx = \left[\frac{2}{3} x^{3/2} \right]_0^4 = \frac{16}{3}$$

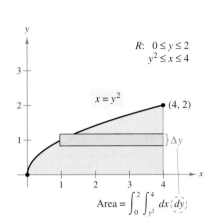

$$R: \ 0 \le y \le 2$$
$$y^2 \le x \le 4$$

$$x = y^2$$

$(4, 2)$

$\} \Delta y$

$$\text{Area} = \int_0^2 \int_{y^2}^4 dx \, \langle dy \rangle$$

FIGURE 13.45

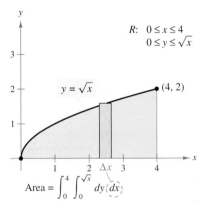

$$R: \ 0 \le x \le 4$$
$$0 \le y \le \sqrt{x}$$

$$y = \sqrt{x}$$

$(4, 2)$

$$\text{Area} = \int_0^4 \int_0^{\sqrt{x}} dy \, \langle dx \rangle$$

FIGURE 13.46

STUDY TIP

To designate a double integral or an area of a region without specifying a particular order of integration, you can use the symbol

$$\int_R \int dA$$

where $dA = dx \, dy$ or $dA = dy \, dx$.

✓ CHECKPOINT 5

For the double integral $\displaystyle\int_0^2 \int_{2y}^4 dx \, dy$,

a. sketch the region R whose area is represented by the integral,

b. rewrite the integral so that x is the outer variable, and

c. show that both orders of integration yield the same result. ■

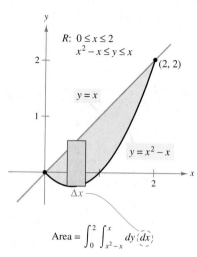

$R: 0 \le x \le 2$
$x^2 - x \le y \le x$

$(2, 2)$

$y = x$

$y = x^2 - x$

Δx

$\text{Area} = \int_0^2 \int_{x^2-x}^x dy\,(dx)$

FIGURE 13.47

| Example 6 | **Finding Area with a Double Integral** |

Use a double integral to calculate the area denoted by

$$\iint_R dA$$

where R is the region bounded by $y = x$ and $y = x^2 - x$.

SOLUTION   Begin by sketching the region R, as shown in Figure 13.47. From the sketch, you can see that vertical rectangles of width dx are more convenient than horizontal ones. So, x is the outer variable of integration and its constant bounds are $0 \le x \le 2$. This implies that the bounds for y are $x^2 - x \le y \le x$, and the area is given by

$$\iint_R dA = \int_0^2 \int_{x^2-x}^x dy\,dx \qquad \text{Substitute bounds for region.}$$

$$= \int_0^2 \Big[y \Big]_{x^2-x}^x dx \qquad \text{Integrate with respect to } y.$$

$$= \int_0^2 \big[x - (x^2 - x) \big]\, dx \qquad \text{Apply Fundamental Theorem of Calculus.}$$

$$= \int_0^2 (2x - x^2)\, dx \qquad \text{Simplify.}$$

$$= \Big[x^2 - \frac{x^3}{3} \Big]_0^2 \qquad \text{Integrate with respect to } x.$$

$$= 4 - \frac{8}{3} \qquad \text{Apply Fundamental Theorem of Calculus.}$$

$$= \frac{4}{3} \text{ square units.} \qquad \text{Simplify.}$$

✓ CHECKPOINT 6

Use a double integral to calculate the area denoted by $\int_R \int dA$ where R is the region bounded by $y = 2x + 3$ and $y = x^2$. ■

As you are working the exercises for this section, you should be aware that the primary uses of double integrals will be discussed in Section 13.9. Double integrals by way of areas in the plane have been introduced so that you can gain practice in finding the limits of integration. When setting up a double integral, remember that your first step should be to sketch the region R. After doing this, you have two choices of integration orders: $dx\,dy$ or $dy\,dx$.

CONCEPT CHECK

1. What is an "integral of an integral" called?
2. In the double integral $\int_0^2 \int_0^1 dy\,dx$, in what order is the integration performed? (Do not integrate.)
3. True or false: Changing the order of integration will sometimes change the value of a double integral.
4. To designate a double integral or an area of a region without specifying a particular order of integration, what symbol can you use?

| **Skills Review 13.8** | The following warm-up exercises involve skills that were covered in earlier sections. You will use these skills in the exercise set for this section. For additional help, review Sections 11.2–11.5. |

In Exercises 1–12, evaluate the definite integral.

1. $\displaystyle\int_0^1 dx$

2. $\displaystyle\int_0^2 3\,dy$

3. $\displaystyle\int_1^4 2x^2\,dx$

4. $\displaystyle\int_0^1 2x^3\,dx$

5. $\displaystyle\int_1^2 (x^3 - 2x + 4)\,dx$

6. $\displaystyle\int_0^2 (4 - y^2)\,dy$

7. $\displaystyle\int_1^2 \frac{2}{7x^2}\,dx$

8. $\displaystyle\int_1^4 \frac{2}{\sqrt{x}}\,dx$

9. $\displaystyle\int_0^2 \frac{2x}{x^2 + 1}\,dx$

10. $\displaystyle\int_2^e \frac{1}{y - 1}\,dy$

11. $\displaystyle\int_0^2 xe^{x^2 + 1}\,dx$

12. $\displaystyle\int_0^1 e^{-2y}\,dy$

In Exercises 13–16, sketch the region bounded by the graphs of the equations.

13. $y = x$, $y = 0$, $x = 3$

14. $y = x$, $y = 3$, $x = 0$

15. $y = 4 - x^2$, $y = 0$, $x = 0$

16. $y = x^2$, $y = 4x$

Exercises 13.8

See www.CalcChat.com for worked-out solutions to odd-numbered exercises.

In Exercises 1–10, evaluate the partial integral.

1. $\displaystyle\int_0^x (2x - y)\,dy$

2. $\displaystyle\int_x^{x^2} \frac{y}{x}\,dy$

3. $\displaystyle\int_1^{2y} \frac{y}{x}\,dx$

4. $\displaystyle\int_0^{e^y} y\,dx$

5. $\displaystyle\int_0^{\sqrt{4 - x^2}} x^2 y\,dy$

6. $\displaystyle\int_{x^2}^{\sqrt{x}} (x^2 + y^2)\,dy$

7. $\displaystyle\int_1^{e^y} \frac{y \ln x}{x}\,dx$

8. $\displaystyle\int_{-\sqrt{1 - y^2}}^{\sqrt{1 - y^2}} (x^2 + y^2)\,dx$

9. $\displaystyle\int_0^x ye^{xy}\,dy$

10. $\displaystyle\int_y^3 \frac{xy}{\sqrt{x^2 + 1}}\,dx$

In Exercises 11–24, evaluate the double integral.

11. $\displaystyle\int_0^1 \int_0^2 (x + y)\,dy\,dx$

12. $\displaystyle\int_0^2 \int_0^2 (6 - x^2)\,dy\,dx$

13. $\displaystyle\int_0^4 \int_0^3 xy\,dy\,dx$

14. $\displaystyle\int_0^1 \int_0^x \sqrt{1 - x^2}\,dy\,dx$

15. $\displaystyle\int_0^1 \int_0^y (x + y)\,dx\,dy$

16. $\displaystyle\int_0^2 \int_{3y^2 - 6y}^{2y - y^2} 3y\,dx\,dy$

17. $\displaystyle\int_1^2 \int_0^4 (3x^2 - 2y^2 + 1)\,dx\,dy$

18. $\displaystyle\int_0^1 \int_y^{2y} (1 + 2x^2 + 2y^2)\,dx\,dy$

19. $\displaystyle\int_0^2 \int_0^{\sqrt{1 - y^2}} -5xy\,dx\,dy$

20. $\displaystyle\int_0^4 \int_0^x \frac{2}{x^2 + 1}\,dy\,dx$

21. $\displaystyle\int_0^2 \int_0^{6x^2} x^3\,dy\,dx$

22. $\displaystyle\int_{-1}^1 \int_{-2}^2 (x^2 - y^2)\,dy\,dx$

23. $\displaystyle\int_0^\infty \int_0^\infty e^{-(x + y)/2}\,dy\,dx$

24. $\displaystyle\int_0^\infty \int_0^\infty xye^{-(x^2 + y^2)}\,dx\,dy$

In Exercises 25–32, sketch the region R whose area is given by the double integral. Then change the order of integration and show that both orders yield the same area.

25. $\displaystyle\int_0^1 \int_0^2 dy\, dx$

26. $\displaystyle\int_1^2 \int_2^4 dx\, dy$

27. $\displaystyle\int_0^1 \int_{2y}^2 dx\, dy$

28. $\displaystyle\int_0^4 \int_0^{\sqrt{x}} dy\, dx$

29. $\displaystyle\int_0^2 \int_{x/2}^1 dy\, dx$

30. $\displaystyle\int_0^4 \int_{\sqrt{x}}^2 dy\, dx$

31. $\displaystyle\int_0^1 \int_{y^2}^{\sqrt[3]{y}} dx\, dy$

32. $\displaystyle\int_{-2}^2 \int_0^{4-y^2} dx\, dy$

In Exercises 33 and 34, evaluate the double integral. Note that it is necessary to change the order of integration.

33. $\displaystyle\int_0^3 \int_y^3 e^{x^2} dx\, dy$

34. $\displaystyle\int_0^2 \int_x^2 e^{-y^2} dy\, dx$

In Exercises 35–40, use a double integral to find the area of the specified region.

35.
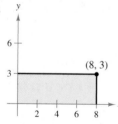

36.
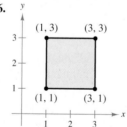

37.
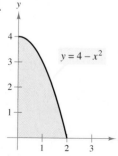

38.
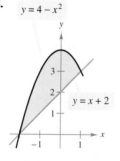

39.
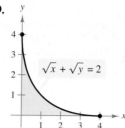

40.
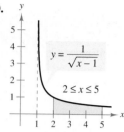

In Exercises 41–46, use a double integral to find the area of the region bounded by the graphs of the equations.

41. $y = 9 - x^2,\ y = 0$

42. $y = x^{3/2},\ y = x$

43. $2x - 3y = 0,\ x + y = 5,\ y = 0$

44. $xy = 9,\ y = x,\ y = 0,\ x = 9$

45. $y = x,\ y = 2x,\ x = 2$

46. $y = x^2 + 2x + 1,\ y = 3(x + 1)$

(T) In Exercises 47–54, use a symbolic integration utility to evaluate the double integral.

47. $\displaystyle\int_0^1 \int_0^2 e^{-x^2 - y^2} dx\, dy$

48. $\displaystyle\int_0^2 \int_{x^2}^{2x} (x^3 + 3y^2)\, dy\, dx$

49. $\displaystyle\int_1^2 \int_0^x e^{xy}\, dy\, dx$

50. $\displaystyle\int_1^2 \int_y^{2y} \ln(x + y)\, dx\, dy$

51. $\displaystyle\int_0^1 \int_x^1 \sqrt{1 - x^2}\, dy\, dx$

52. $\displaystyle\int_0^3 \int_0^{x^2} \sqrt{x}\sqrt{1 + x}\, dy\, dx$

53. $\displaystyle\int_0^2 \int_{\sqrt{4-x^2}}^{4-x^2/4} \frac{xy}{x^2 + y^2 + 1}\, dy\, dx$

54. $\displaystyle\int_0^4 \int_0^y \frac{2}{(x + 1)(y + 1)}\, dx\, dy$

True or False? In Exercises 55 and 56, determine whether the statement is true or false. If it is false, explain why or give an example that shows it is false.

55. $\displaystyle\int_{-1}^1 \int_{-2}^2 y\, dy\, dx = \int_{-1}^1 \int_{-2}^2 y\, dx\, dy$

56. $\displaystyle\int_2^5 \int_1^6 x\, dy\, dx = \int_1^6 \int_2^5 x\, dx\, dy$

Section 13.9
Applications of Double Integrals

■ Use double integrals to find the volumes of solids.
■ Use double integrals to find the average values of real-life models.

Volume of a Solid Region

In Section 13.8, you used double integrals as an alternative way to find the area of a plane region. In this section, you will study the primary uses of double integrals: to find the volume of a solid region and to find the average value of a function.

Consider a function $z = f(x, y)$ that is continuous and nonnegative over a region R. Let S be the solid region that lies between the xy-plane and the surface

$$z = f(x, y) \qquad \text{Surface lying above the } xy\text{-plane}$$

directly above the region R, as shown in Figure 13.48. You can find the volume of S by integrating $f(x, y)$ over the region R.

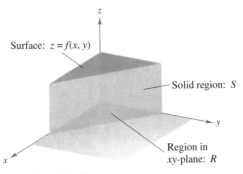

Surface: $z = f(x, y)$

Solid region: S

Region in xy-plane: R

FIGURE 13.48

Determining Volume with Double Integrals

If R is a bounded region in the xy-plane and f is continuous and nonnegative over R, then the **volume of the solid** region between the surface $z = f(x, y)$ and R is given by the double integral

$$\int_R \int f(x, y)\, dA$$

where $dA = dx\, dy$ or $dA = dy\, dx$.

| Example 1 | **Finding the Volume of a Solid** |

Find the volume of the solid region bounded in the first octant by the plane

$$z = 2 - x - 2y.$$

SOLUTION

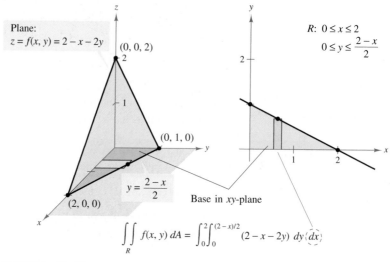

$$\iint\limits_{R} f(x, y)\, dA = \int_0^2 \int_0^{(2-x)/2} (2 - x - 2y)\, dy\, dx$$

FIGURE 13.49

To set up the double integral for the volume, it is helpful to sketch both the solid region and the plane region R in the xy-plane. In Figure 13.49, you can see that the region R is bounded by the lines $x = 0$, $y = 0$, and $y = \frac{1}{2}(2 - x)$. One way to set up the double integral is to choose x as the outer variable. With that choice, the constant bounds for x are $0 \leq x \leq 2$ and the variable bounds for y are $0 \leq y \leq \frac{1}{2}(2 - x)$. So, the volume of the solid region is

$$
\begin{aligned}
V &= \int_0^2 \int_0^{(2-x)/2} (2 - x - 2y)\, dy\, dx \\
&= \int_0^2 \left[(2 - x)y - y^2 \right]_0^{(2-x)/2} dx \\
&= \int_0^2 \left\{ (2 - x)\left(\frac{1}{2}\right)(2 - x) - \left[\frac{1}{2}(2 - x)\right]^2 \right\} dx \\
&= \frac{1}{4} \int_0^2 (2 - x)^2\, dx \\
&= \left[-\frac{1}{12}(2 - x)^3 \right]_0^2 \\
&= \frac{2}{3} \text{ cubic unit.}
\end{aligned}
$$

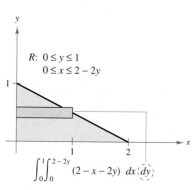

$$\int_0^1 \int_0^{2-2y} (2 - x - 2y)\, dx\, dy$$

FIGURE 13.50

✓CHECKPOINT 1

Find the volume of the solid region bounded in the first octant by the plane
$z = 4 - 2x - y.$ ■

In Example 1, the order of integration was arbitrary. Although the example used x as the outer variable, you could just as easily have used y as the outer variable. The next example describes a situation in which one order of integration is more convenient than the other.

Example 2   Comparing Different Orders of Integration

Find the volume under the surface $f(x, y) = e^{-x^2}$ bounded by the xz-plane and the planes $y = x$ and $x = 1$, as shown in Figure 13.51.

SOLUTION

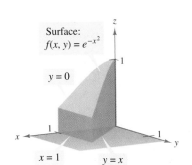

Surface:
$f(x, y) = e^{-x^2}$

$y = 0$

$x = 1$ $y = x$

FIGURE 13.51

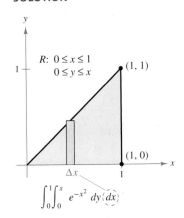

$R: 0 \le x \le 1$
$\quad\ 0 \le y \le x$

$(1, 1)$

$(1, 0)$

Δx

$$\int_0^1 \int_0^x e^{-x^2} \, dy \, dx$$

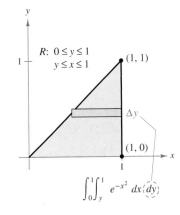

$R: 0 \le y \le 1$
$\quad\ y \le x \le 1$

$(1, 1)$

Δy

$(1, 0)$

$$\int_0^1 \int_y^1 e^{-x^2} \, dx \, dy$$

FIGURE 13.52

In the xy-plane, the bounds of region R are the lines $y = 0, x = 1$, and $y = x$. The two possible orders of integration are indicated in Figure 13.52. If you attempt to evaluate the two double integrals shown in the figure, you will discover that the one on the right involves finding the antiderivative of e^{-x^2}, which you know is not an elementary function. The integral on the left, however, can be evaluated more easily, as shown.

$$V = \int_0^1 \int_0^x e^{-x^2} \, dy \, dx$$

$$= \int_0^1 \left[e^{-x^2} y \right]_0^x dx$$

$$= \int_0^1 x e^{-x^2} \, dx$$

$$= \left[-\frac{1}{2} e^{-x^2} \right]_0^1$$

$$= -\frac{1}{2}\left(\frac{1}{e} - 1 \right) \approx 0.316 \text{ cubic unit}$$

TECHNOLOGY

(T) Use a symbolic integration utility to evaluate the double integral in Example 2.

✓ CHECKPOINT 2

Find the volume under the surface $f(x, y) = e^{x^2}$, bounded by the xz-plane and the planes $y = 2x$ and $x = 1$. ■

Guidelines for Finding the Volume of a Solid

1. Write the equation of the surface in the form $z = f(x, y)$ and sketch the solid region.

2. Sketch the region R in the xy-plane and determine the order and limits of integration.

3. Evaluate the double integral

$$\int_R \int f(x, y)\, dA$$

using the order and limits determined in the second step.

The first step above suggests that you sketch the three-dimensional solid region. This is a good suggestion, but it is not always feasible and is not as important as making a sketch of the two-dimensional region R.

Example 3   Finding the Volume of a Solid

Find the volume of the solid bounded above by the surface

$$f(x, y) = 6x^2 - 2xy$$

and below by the plane region R shown in Figure 13.53.

SOLUTION   Because the region R is bounded by the parabola $y = 3x - x^2$ and the line $y = x$, the limits for y are $x \le y \le 3x - x^2$. The limits for x are $0 \le x \le 2$, and the volume of the solid is

$$V = \int_0^2 \int_x^{3x-x^2} (6x^2 - 2xy)\, dy\, dx$$

$$= \int_0^2 \left[6x^2 y - xy^2 \right]_x^{3x-x^2} dx$$

$$= \int_0^2 \left[(18x^3 - 6x^4 - 9x^3 + 6x^4 - x^5) - (6x^3 - x^3) \right] dx$$

$$= \int_0^2 (4x^3 - x^5)\, dx$$

$$= \left[x^4 - \frac{x^6}{6} \right]_0^2$$

$$= \frac{16}{3} \text{ cubic units.}$$

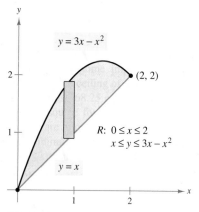

FIGURE 13.53

The figure shows the region with $y = 3x - x^2$, $y = x$, the point $(2, 2)$, and $R: 0 \le x \le 2$, $x \le y \le 3x - x^2$.

✓ CHECKPOINT 3

Find the volume of the solid bounded above by the surface $f(x, y) = 4x^2 + 2xy$ and below by the plane region bounded by $y = x^2$ and $y = 2x$. ∎

A *population density function* $p = f(x, y)$ is a model that describes density (in people per square unit) of a region. To find the population of a region R, evaluate the double integral

$$\int_R \int f(x, y) \, dA.$$

Example 4
MAKE A DECISION **Finding the Population of a Region**

The population density (in people per square mile) of the city shown in Figure 13.54 can be modeled by

$$f(x, y) = \frac{50,000}{x + |y| + 1}$$

where x and y are measured in miles. Approximate the city's population. Will the city's average population density be less than 10,000 people per square mile?

SOLUTION Because the model involves the absolute value of y, it follows that the population density is symmetrical about the x-axis. So, the population in the first quadrant is equal to the population in the fourth quadrant. This means that you can find the total population by doubling the population in the first quadrant.

$$\text{Population} = 2 \int_0^4 \int_0^5 \frac{50,000}{x + y + 1} \, dy \, dx$$

$$= 100,000 \int_0^4 \left[\ln(x + y + 1) \right]_0^5 dx$$

$$= 100,000 \int_0^4 \left[\ln(x + 6) - \ln(x + 1) \right] dx$$

$$= 100,000 \left[(x + 6) \ln(x + 6) - (x + 6) - \right.$$
$$\left. (x + 1) \ln(x + 1) + (x + 1) \right]_0^4$$

$$= 100,000 \left[(x + 6) \ln(x + 6) - (x + 1) \ln(x + 1) - 5 \right]_0^4$$

$$= 100,000 [10 \ln(10) - 5 \ln(5) - 5 - 6 \ln(6) + 5]$$

$$\approx 422,810 \text{ people}$$

So, the city's population is about 422,810. Because the city covers a region 4 miles wide and 10 miles long, its area is 40 square miles. So, the average population density is

$$\text{Average population density} = \frac{422,810}{40}$$

$$\approx 10,570 \text{ people per square mile.}$$

No, the city's average population density is not less than 10,000 people per square mile.

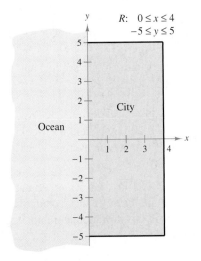

y

$R: \ 0 \le x \le 4$
$\quad -5 \le y \le 5$

City

Ocean

FIGURE 13.54

✓ **CHECKPOINT 4**

In Example 4, what integration technique was used to integrate

$$\int \left[\ln(x + 6) - \ln(x + 1) \right] dx?$$

Average Value of a Function over a Region

Average Value of a Function Over a Region

If f is integrable over the plane region R with area A, then its **average value** over R is

$$\text{Average value} = \frac{1}{A} \int_R \int f(x, y)\, dA.$$

Example 5 Finding Average Profit

A manufacturer determines that the profit for selling x units of one product and y units of a second product is modeled by

$$P = -(x - 200)^2 - (y - 100)^2 + 5000.$$

The weekly sales for product 1 vary between 150 and 200 units, and the weekly sales for product 2 vary between 80 and 100 units. Estimate the average weekly profit for the two products.

SOLUTION   Because $150 \le x \le 200$ and $80 \le y \le 100$, you can estimate the weekly profit to be the average of the profit function over the rectangular region shown in Figure 13.55. Because the area of this rectangular region is $(50)(20) = 1000$, it follows that the average profit V is

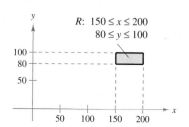

R: $150 \le x \le 200$
$80 \le y \le 100$

FIGURE 13.55

$$V = \frac{1}{1000} \int_{150}^{200} \int_{80}^{100} [-(x - 200)^2 - (y - 100)^2 + 5000]\, dy\, dx$$

$$= \frac{1}{1000} \int_{150}^{200} \left[-(x - 200)^2 y - \frac{(y - 100)^3}{3} + 5000y \right]_{80}^{100} dx$$

$$= \frac{1}{1000} \int_{150}^{200} \left[-20(x - 200)^2 - \frac{292{,}000}{3} \right] dx$$

$$= \frac{1}{3000} \left[-20(x - 200)^3 + 292{,}000x \right]_{150}^{200}$$

$$\approx \$4033.$$

✓**CHECKPOINT 5**

Find the average value of $f(x, y) = 4 - \frac{1}{2}x - \frac{1}{2}y$ over the region $0 \le x \le 2$ and $0 \le y \le 2$. ▪

CONCEPT CHECK

1. Complete the following: The double integral $\int_R \int f(x, y)\, dA$ gives the _____ of the solid region between the surface $z = f(x, y)$ and the bounded region in the xy-plane R.

2. Give the guidelines for finding the volume of a solid.

3. What does a population density function describe?

4. What is the average value of $f(x, y)$ over the plane region R?

The following warm-up exercises involve skills that were covered in earlier sections. You will use these skills in the exercise set for this section. For additional help, review Sections 11.4 and 13.8.

In Exercises 1–4, sketch the region that is described.

1. $0 \le x \le 2,\ 0 \le y \le 1$

2. $1 \le x \le 3,\ 2 \le y \le 3$

3. $0 \le x \le 4,\ 0 \le y \le 2x - 1$

4. $0 \le x \le 2,\ 0 \le y \le x^2$

In Exercises 5–10, evaluate the double integral.

5. $\displaystyle\int_0^1 \int_1^2 dy\,dx$

6. $\displaystyle\int_0^3 \int_1^3 dx\,dy$

7. $\displaystyle\int_0^1 \int_0^x x\,dy\,dx$

8. $\displaystyle\int_0^4 \int_1^y y\,dx\,dy$

9. $\displaystyle\int_1^3 \int_x^{x^2} 2\,dy\,dx$

10. $\displaystyle\int_0^1 \int_x^{-x^2+2} dy\,dx$

Exercises 13.9

In Exercises 1–8, sketch the region of integration and evaluate the double integral.

1. $\displaystyle\int_0^2 \int_0^1 (3x + 4y)\,dy\,dx$

2. $\displaystyle\int_0^3 \int_0^1 (2x + 6y)\,dy\,dx$

3. $\displaystyle\int_0^1 \int_y^{\sqrt{y}} x^2 y^2\,dx\,dy$

4. $\displaystyle\int_0^6 \int_{y/2}^3 (x + y)\,dx\,dy$

5. $\displaystyle\int_0^1 \int_0^{\sqrt{1-x^2}} y\,dy\,dx$

6. $\displaystyle\int_0^2 \int_0^{4-x^2} xy^2\,dy\,dx$

7. $\displaystyle\int_{-a}^a \int_{-\sqrt{a^2-x^2}}^{\sqrt{a^2-x^2}} dy\,dx$

8. $\displaystyle\int_0^a \int_0^{\sqrt{a^2-x^2}} dy\,dx$

In Exercises 9–12, set up the integral for both orders of integration and use the more convenient order to evaluate the integral over the region R.

9. $\displaystyle\int_R \int xy\,dA$

R: rectangle with vertices at $(0, 0)$, $(0, 5)$, $(3, 5)$, $(3, 0)$

10. $\displaystyle\int_R \int x\,dA$

R: semicircle bounded by $y = \sqrt{25 - x^2}$ and $y = 0$

11. $\displaystyle\int_R \int \frac{y}{x^2 + y^2}\,dA$

R: triangle bounded by $y = x$, $y = 2x$, $x = 2$

12. $\displaystyle\int_R \int \frac{y}{1 + x^2}\,dA$

R: region bounded by $y = 0$, $y = \sqrt{x}$, $x = 4$

In Exercises 13–22, use a double integral to find the volume of the specified solid.

13.

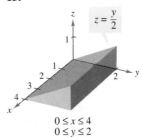

$z = \dfrac{y}{2}$

$0 \le x \le 4$
$0 \le y \le 2$

14.

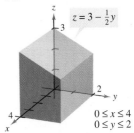

$z = 3 - \dfrac{1}{2}y$

$0 \le x \le 4$
$0 \le y \le 2$

15.

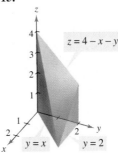

$z = 4 - x - y$

$y = x$ $y = 2$

16.

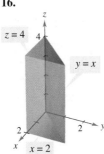

$z = 4$

$y = x$

$x = 2$

17.

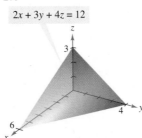

$2x + 3y + 4z = 12$

18.

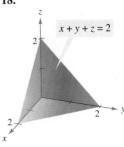

$x + y + z = 2$

19.

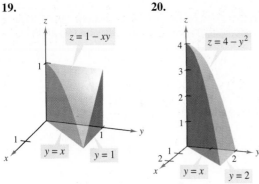

20.

21.

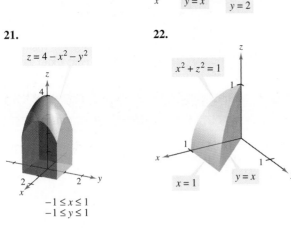

$-1 \le x \le 1$
$-1 \le y \le 1$

In Exercises 23–26, use a double integral to find the volume of the solid bounded by the graphs of the equations.

23. $z = xy$, $z = 0$, $y = 0$, $y = 4$, $x = 0$, $x = 1$

24. $z = x$, $z = 0$, $y = x$, $y = 0$, $x = 0$, $x = 4$

25. $z = x^2$, $z = 0$, $x = 0$, $x = 2$, $y = 0$, $y = 4$

26. $z = x + y$, $x^2 + y^2 = 4$ (first octant)

27. Population Density   The population density (in people per square mile) for a coastal town can be modeled by

$$f(x, y) = \frac{120{,}000}{(2 + x + y)^3}$$

where x and y are measured in miles. What is the population inside the rectangular area defined by the vertices $(0, 0)$, $(2, 0)$, $(0, 2)$, and $(2, 2)$?

28. Population Density   The population density (in people per square mile) for a coastal town on an island can be modeled by

$$f(x, y) = \frac{5000xe^y}{1 + 2x^2}$$

where x and y are measured in miles. What is the population inside the rectangular area defined by the vertices $(0, 0)$, $(4, 0)$, $(0, -2)$, and $(4, -2)$?

In Exercises 29–32, find the average value of $f(x, y)$ over the region R.

29. $f(x, y) = x$
   R: rectangle with vertices $(0, 0)$, $(4, 0)$, $(4, 2)$, $(0, 2)$

30. $f(x, y) = xy$
   R: rectangle with vertices $(0, 0)$, $(4, 0)$, $(4, 2)$, $(0, 2)$

31. $f(x, y) = x^2 + y^2$
   R: square with vertices $(0, 0)$, $(2, 0)$, $(2, 2)$, $(0, 2)$

32. $f(x, y) = e^{x+y}$
   R: triangle with vertices $(0, 0)$, $(0, 1)$, $(1, 1)$

33. Average Revenue   A company sells two products whose demand functions are given by

$$x_1 = 500 - 3p_1 \quad \text{and} \quad x_2 = 750 - 2.4p_2.$$

So, the total revenue is given by

$$R = x_1p_1 + x_2p_2.$$

Estimate the average revenue if the price p_1 varies between \$50 and \$75 and the price p_2 varies between \$100 and \$150.

34. Average Revenue   After 1 year, the company in Exercise 33 finds that the demand functions for its two products are given by

$$x_1 = 500 - 2.5p_1 \quad \text{and} \quad x_2 = 750 - 3p_2.$$

Repeat Exercise 33 using these demand functions.

35. Average Weekly Profit   A firm's weekly profit in marketing two products is given by

$$P = 192x_1 + 576x_2 - x_1^2 - 5x_2^2 - 2x_1x_2 - 5000$$

where x_1 and x_2 represent the numbers of units of each product sold weekly. Estimate the average weekly profit if x_1 varies between 40 and 50 units and x_2 varies between 45 and 50 units.

36. Average Weekly Profit   After a change in marketing, the weekly profit of the firm in Exercise 35 is given by

$$P = 200x_1 + 580x_2 - x_1^2 - 5x_2^2 - 2x_1x_2 - 7500.$$

Estimate the average weekly profit if x_1 varies between 55 and 65 units and x_2 varies between 50 and 60 units.

37. Average Production   The Cobb-Douglas production function for an automobile manufacturer is

$$f(x, y) = 100x^{0.6}y^{0.4}$$

where x is the number of units of labor and y is the number of units of capital. Estimate the average production level if the number of units of labor x varies between 200 and 250 and the number of units of capital y varies between 300 and 325.

38. Average Production   Repeat Exercise 37 for the production function given by

$$f(x, y) = x^{0.25}y^{0.75}.$$

Algebra Review

Solving Systems of Equations

Nonlinear System in Two Variables

$$\begin{cases} 4x + 3y = 6 \\ x^2 - y = 4 \end{cases}$$

Three of the sections in this chapter (13.5, 13.6, and 13.7) involve solutions of systems of equations. These systems can be linear or nonlinear, as shown at the left.

There are many techniques for solving a system of linear equations. Two of the more common ones are listed here.

Linear System in Three Variables

$$\begin{cases} -x + 2y + 4z = 2 \\ 2x - y + z = 0 \\ 6x + 2z = 3 \end{cases}$$

1. *Substitution*: Solve for one of the variables in one of the equations and substitute the value into another equation.

2. *Elimination*: Add multiples of one equation to a second equation to eliminate a variable in the second equation.

Example 1 Solving Systems of Equations

Solve each system of equations.

a. $\begin{cases} y - x^3 = 0 \\ x - y^3 = 0 \end{cases}$

b. $\begin{cases} -400p_1 + 300p_2 = -25 \\ 300p_1 - 360p_2 = -535 \end{cases}$

SOLUTION

a. Example 3, page 971

$\begin{cases} y - x^3 = 0 \\ x - y^3 = 0 \end{cases}$	Equation 1
	Equation 2
$y = x^3$	Solve for y in Equation 1.
$x - (x^3)^3 = 0$	Substitute x^3 for y in Equation 2.
$x - x^9 = 0$	$(x^m)^n = x^{mn}$
$x(x - 1)(x + 1)(x^2 + 1)(x^4 + 1) = 0$	Factor.
$x = 0$	Set factors equal to zero.
$x = 1$	Set factors equal to zero.
$x = -1$	Set factors equal to zero.

b. Example 4, page 972

$\begin{cases} -400p_1 + 300p_2 = -25 \\ 300p_1 - 360p_2 = -535 \end{cases}$	Equation 1
	Equation 2
$p_2 = \frac{1}{12}(16p_1 - 1)$	Solve for p_2 in Equation 1.
$300p_1 - 360\left(\frac{1}{12}\right)(16p_1 - 1) = -535$	Substitute for p_2 in Equation 2.
$300p_1 - 30(16p_1 - 1) = -535$	Multiply factors.
$-180p_1 = -565$	Combine like terms.
$p_1 = \frac{113}{36} \approx 3.14$	Divide each side by -180.
$p_2 = \frac{1}{12}\left[16\left(\frac{113}{36}\right) - 1\right]$	Find p_2 by substituting p_1.
$p_2 \approx 4.10$	Solve for p_2.

| **Example 2** | **Solving Systems of Equations** |

Solve each system of equations.

a. $\begin{cases} y(24 - 12x - 4y) = 0 \\ x(24 - 6x - 8y) = 0 \end{cases}$ b. $\begin{cases} 28a - 4b = 10 \\ -4a + 8b = 12 \end{cases}$

SOLUTION

a. Example 5, page 973

Before solving this system of equations, factor 4 out of the first equation and factor 2 out of the second equation.

$\begin{cases} y(24 - 12x - 4y) = 0 \\ x(24 - 6x - 8y) = 0 \end{cases}$ Original Equation 1
     Original Equation 2

$\begin{cases} y(4)(6 - 3x - y) = 0 \\ x(2)(12 - 3x - 4y) = 0 \end{cases}$ Factor 4 out of Equation 1.
     Factor 2 out of Equation 2.

$\begin{cases} y(6 - 3x - y) = 0 \\ x(12 - 3x - 4y) = 0 \end{cases}$ Equation 1
     Equation 2

In each equation, either factor can be 0, so you obtain four different linear systems. For the first system, substitute $y = 0$ into the second equation to obtain $x = 4$.

$\begin{cases} y = 0 \\ 12 - 3x - 4y = 0 \end{cases}$ $(4, 0)$ is a solution.

You can solve the second system by the method of elimination.

$\begin{cases} 6 - 3x - y = 0 \\ 12 - 3x - 4y = 0 \end{cases}$ $\left(\frac{4}{3}, 2\right)$ is a solution.

The third system is already solved.

$\begin{cases} y = 0 \\ x = 0 \end{cases}$ $(0, 0)$ is a solution.

You can solve the last system by substituting $x = 0$ into the first equation to obtain $y = 6$.

$\begin{cases} 6 - 3x - y = 0 \\ x = 0 \end{cases}$ $(0, 6)$ is a solution.

b. Example 2, page 989

$\begin{cases} 28a - 4b = 10 \\ -4a + 8b = 12 \end{cases}$ Equation 1
     Equation 2

$-2a + 4b = 6$ Divide Equation 2 by 2.

$26a = 16$ Add new equation to Equation 1.

$a = \frac{8}{13}$ Divide each side by 26.

$28\left(\frac{8}{13}\right) - 4b = 10$ Substitute for a in Equation 1.

$b = \frac{47}{26}$ Solve for b.

Chapter Summary and Study Strategies

After studying this chapter, you should have acquired the following skills.
The exercise numbers are keyed to the Review Exercises that begin on page 1017.
Answers to odd-numbered Review Exercises are given in the back of the text.

Section 13.1   Review Exercises

- Plot points in space.   *1, 2*
- Find the distance between two points in space.   *3, 4*
  $$d = \sqrt{(x_2 - x_1)^2 + (y_2 - y_1)^2 + (z_2 - z_1)^2}$$
- Find the midpoints of line segments in space.   *5, 6*
  $$\text{Midpoint} = \left(\frac{x_1 + x_2}{2}, \frac{y_1 + y_2}{2}, \frac{z_1 + z_2}{2}\right)$$
- Write the standard forms of the equations of spheres.   *7–10*
  $$(x - h)^2 + (y - k)^2 + (z - l)^2 = r^2$$
- Find the centers and radii of spheres.   *11, 12*
- Sketch the coordinate plane traces of spheres.   *13, 14*

Section 13.2

- Sketch planes in space.   *15–18*
- Classify quadric surfaces in space.   *19–26*

Section 13.3

- Evaluate functions of several variables.   *27, 28, 62*
- Find the domains and ranges of functions of several variables. *29, 30*
- Sketch the level curves of functions of two variables.   *31–34*
- Use functions of several variables to answer questions about real-life situations. *35–40*

Section 13.4

- Find the first partial derivatives of functions of several variables. *41–50*
  $$\frac{\partial z}{\partial x} = \lim_{\Delta x \to 0} \frac{f(x + \Delta x, y) - f(x, y)}{\Delta x} \qquad \frac{\partial z}{\partial y} = \lim_{\Delta y \to 0} \frac{f(x, y + \Delta y) - f(x, y)}{\Delta y}$$
- Find the slopes of surfaces in the x- and y-directions.   *51–54*
- Find the second partial derivatives of functions of several variables. *55–58*
- Use partial derivatives to answer questions about real-life situations. *59–61*

Section 13.5

- Find the relative extrema of functions of two variables.   *63–70*
- Use relative extrema to answer questions about real-life situations. *71, 72*

Section 13.6

<table>
<tr><td></td><td align="right">Review Exercises</td></tr>
</table>

	Review Exercises
■ Use Lagrange multipliers to find extrema of functions of several variables.	*73–78*
■ Use a spreadsheet to find the indicated extremum.	*79, 80*
■ Use Lagrange multipliers to answer questions about real-life situations.	*81, 82*

Section 13.7

■ Find the least squares regression lines, $y = ax + b$, for data and calculate the sum of the squared errors for data.	*83, 84*

$$a = \left[n\sum_{i=1}^{n} x_i y_i - \sum_{i=1}^{n} x_i \sum_{i=1}^{n} y_i\right] \bigg/ \left[n\sum_{i=1}^{n} x_i^2 - \left(\sum_{i=1}^{n} x_i\right)^2\right], \quad b = \frac{1}{n}\left(\sum_{i=1}^{n} y_i - a\sum_{i=1}^{n} x_i\right)$$

■ Use least squares regression lines to model real-life data.	*85, 86*
■ Find the least squares regression quadratics for data.	*87, 88*

Section 13.8

■ Evaluate double integrals.	*89–92*
■ Use double integrals to find the areas of regions.	*93–96*

Section 13.9

■ Use double integrals to find the volumes of solids.	*97, 98*

$$\text{Volume} = \int_R \int f(x, y)\, dA$$

■ Use double integrals to find the average values of real-life models.	*99, 100*

$$\text{Average value} = \frac{1}{A}\int_R \int f(x, y)\, dA$$

Study Strategies

■ **Comparing Two Dimensions with Three Dimensions**   Many of the formulas and techniques in this chapter are generalizations of formulas and techniques used in earlier chapters in the text. Here are several examples.

Two-Dimensional Coordinate System	Three-Dimensional Coordinate System
Distance Formula $d = \sqrt{(x_2 - x_1)^2 + (y_2 - y_1)^2}$	*Distance Formula* $d = \sqrt{(x_2 - x_1)^2 + (y_2 - y_1)^2 + (z_2 - z_1)^2}$
Midpoint Formula $\text{Midpoint} = \left(\dfrac{x_1 + x_2}{2}, \dfrac{y_1 + y_2}{2}\right)$	*Midpoint Formula* $\text{Midpoint} = \left(\dfrac{x_1 + x_2}{2}, \dfrac{y_1 + y_2}{2}, \dfrac{z_1 + z_2}{2}\right)$
Equation of Circle $(x - h)^2 + (y - k)^2 = r^2$	*Equation of Sphere* $(x - h)^2 + (y - k)^2 + (z - l)^2 = r^2$
Equation of Line $ax + by = c$	*Equation of Plane* $ax + by + cz = d$
Derivative of $y = f(x)$ $\dfrac{dy}{dx} = \lim\limits_{\Delta x \to 0} \dfrac{f(x + \Delta x) - f(x)}{\Delta x}$	*Partial Derivative of* $z = f(x, y)$ $\dfrac{\partial z}{\partial x} = \lim\limits_{\Delta x \to 0} \dfrac{f(x + \Delta x, y) - f(x, y)}{\Delta x}$
Area of Region $A = \displaystyle\int_a^b f(x)\, dx$	*Volume of Region* $V = \displaystyle\int_R \int f(x, y)\, dA$

Review Exercises

See www.CalcChat.com for worked-out solutions to odd-numbered exercises.

In Exercises 1 and 2, plot the points.

1. $(2, -1, 4), (-1, 3, -3)$

2. $(1, -2, -3), (-4, -3, 5)$

In Exercises 3 and 4, find the distance between the two points.

3. $(0, 0, 0), (2, 5, 9)$

4. $(-4, 1, 5), (1, 3, 7)$

In Exercises 5 and 6, find the midpoint of the line segment joining the two points.

5. $(2, 6, 4), (-4, 2, 8)$

6. $(5, 0, 7), (-1, -2, 9)$

In Exercises 7–10, find the standard form of the equation of the sphere.

7. Center: $(0, 1, 0)$; radius: 5

8. Center: $(4, -5, 3)$; radius: 10

9. Diameter endpoints: $(0, 0, 4), (4, 6, 0)$

10. Diameter endpoints: $(3, 4, 0), (5, 8, 2)$

In Exercises 11 and 12, find the center and radius of the sphere.

11. $x^2 + y^2 + z^2 + 4x - 2y - 8z + 5 = 0$

12. $x^2 + y^2 + z^2 + 4y - 10z - 7 = 0$

In Exercises 13 and 14, sketch the xy-trace of the sphere.

13. $(x + 2)^2 + (y - 1)^2 + (z - 3)^2 = 25$

14. $(x - 1)^2 + (y + 3)^2 + (z - 6)^2 = 72$

In Exercises 15–18, find the intercepts and sketch the graph of the plane.

15. $x + 2y + 3z = 6$

16. $2y + z = 4$

17. $3x - 6z = 12$

18. $4x - y + 2z = 8$

In Exercises 19–26, identify the surface.

19. $x^2 + y^2 + z^2 - 2x + 4y - 6z + 5 = 0$

20. $16x^2 + 16y^2 - 9z^2 = 0$

21. $x^2 + \dfrac{y^2}{16} + \dfrac{z^2}{9} = 1$

22. $x^2 - \dfrac{y^2}{16} - \dfrac{z^2}{9} = 1$

23. $z = \dfrac{x^2}{9} + y^2$

24. $-4x^2 + y^2 + z^2 = 4$

25. $z = \sqrt{x^2 + y^2}$

26. $z = 9x + 3y - 5$

In Exercises 27 and 28, find the function values.

27. $f(x, y) = xy^2$

(a) $f(2, 3)$

(b) $f(0, 1)$

(c) $f(-5, 7)$

(d) $f(-2, -4)$

28. $f(x, y) = \dfrac{x^2}{y}$

(a) $f(6, 9)$

(b) $f(8, 4)$

(c) $f(t, 2)$

(d) $f(r, r)$

In Exercises 29 and 30, describe the region R in the xy-plane that corresponds to the domain of the function. Then find the range of the function.

29. $f(x, y) = \sqrt{1 - x^2 - y^2}$

30. $f(x, y) = \dfrac{1}{x + y}$

In Exercises 31–34, describe the level curves of the function. Sketch the level curves for the given c-values.

31. $z = 10 - 2x - 5y, \ c = 0, 2, 4, 5, 10$

32. $z = \sqrt{9 - x^2 - y^2}, \ c = 0, 1, 2, 3$

33. $z = (xy)^2, \ c = 1, 4, 9, 12, 16$

34. $z = y - x^2, c = 0, \pm1, \pm2$

35. Meteorology The contour map shown below represents the average yearly precipitation for Iowa. *(Source: U.S. National Oceanic and Atmospheric Administration)*

(a) Discuss the use of color to represent the level curves.

(b) Which part of Iowa receives the most precipitation?

(c) Which part of Iowa receives the least precipitation?

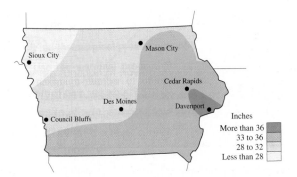

36. Population Density   The contour map below represents the population density of New York. *(Source: U.S. Bureau of Census)*

(a) Discuss the use of color to represent the level curves.

(b) Do the level curves correspond to equally spaced population densities?

(c) Describe how to obtain a more detailed contour map.

Persons per square mile
More than 250
101 to 250
50 to 100
Less than 50

37. Chemistry   The acidity of rainwater is measured in units called pH, and smaller pH values are increasingly acidic. The map shows the curves of equal pH and gives evidence that downwind of heavily industrialized areas, the acidity has been increasing. Using the level curves on the map, determine the direction of the prevailing winds in the northeastern United States.

38. Sales   The table gives the sales x (in billions of dollars), the shareholder's equity y (in billions of dollars), and the earnings per share z (in dollars) for Johnson & Johnson for the years 2000 through 2005. *(Source: Johnson & Johnson)*

Year	2000	2001	2002	2003	2004	2005
x	29.1	33.0	36.3	41.9	47.3	50.5
y	18.8	24.2	22.7	26.9	31.8	37.9
z	1.70	1.91	2.23	2.70	3.10	3.50

A model for these data is

$$z = f(x, y) = 0.078x + 0.008y - 0.767.$$

(T) (a) Use a graphing utility and the model to approximate z for the given values of x and y.

(b) Which of the two variables in this model has the greater influence on shareholder's equity?

(c) Simplify the expression for $f(x, 45)$ and interpret its meaning in the context of the problem.

39. Equation of Exchange   Economists use an equation of exchange to express the relation among money, prices, and business transactions. This equation can be written as

$$P = \frac{MV}{T}$$

where M is the money supply, V is the velocity of circulation, T is the total number of transactions, and P is the price level. Find P when $M = \$2500$, $V = 6$, and $T = 6000$.

40. Biomechanics   The Froude number F, defined as

$$F = \frac{v^2}{gl}$$

where v represents velocity, g represents gravitational acceleration, and l represents stride length, is an example of a "similarity criterion." Find the Froude number of a rabbit for which velocity is 2 meters per second, gravitational acceleration is 3 meters per second squared, and stride length is 0.75 meter.

In Exercises 41–50, find all first partial derivatives.

41. $f(x, y) = x^2y + 3xy + 2x - 5y$

42. $f(x, y) = 4xy + xy^2 - 3x^2y$

43. $z = \dfrac{x^2}{y^2}$

44. $z = (xy + 2x + 4y)^2$

45. $f(x, y) = \ln(2x + 3y)$

46. $f(x, y) = \ln\sqrt{2x + 3y}$

47. $f(x, y) = xe^y + ye^x$

48. $f(x, y) = x^2e^{-2y}$

49. $w = xyz^2$

50. $w = 3xy - 5xz + 2yz$

In Exercises 51–54, find the slope of the surface at the indicated point in (a) the x-direction and (b) the y-direction.

51. $z = 3x - 4y + 9$, $(3, 2, 10)$

52. $z = 4x^2 - y^2$, $(2, 4, 0)$

53. $z = 8 - x^2 - y^2$, $(1, 2, 3)$

54. $z = x^2 - y^2$, $(5, -4, 9)$

In Exercises 55–58, find all second partial derivatives.

55. $f(x, y) = 3x^2 - xy + 2y^3$ **56.** $f(x, y) = \dfrac{y}{x + y}$

57. $f(x, y) = \sqrt{1 + x + y}$ **58.** $f(x, y) = x^2e^{-y^2}$

59. Marginal Cost A company manufactures two models of skis: cross-country skis and downhill skis. The cost function for producing x pairs of cross-country skis and y pairs of downhill skis is given by

$$C = 15(xy)^{1/3} + 99x + 139y + 2293.$$

Find the marginal costs when $x = 500$ and $y = 250$.

60. Marginal Revenue At a baseball stadium, souvenir caps are sold at two locations. If x_1 and x_2 are the numbers of baseball caps sold at location 1 and location 2, respectively, then the total revenue for the caps is modeled by

$$R = 15x_1 + 16x_2 - \frac{1}{10}x_1^2 - \frac{1}{10}x_2^2 - \frac{1}{100}x_1 x_2.$$

Given that $x_1 = 50$ and $x_2 = 40$, find the marginal revenues at location 1 and at location 2.

61. Medical Science The surface area A of an average human body in square centimeters can be approximated by the model $A(w, h) = 101.4w^{0.425}h^{0.725}$, where w is the weight in pounds and h is the height in inches.

  (a) Determine the partial derivatives of A with respect to w and with respect to h.

  (b) Evaluate $\partial A / \partial w$ at $(180, 70)$. Explain your result.

62. Medicine In order to treat a certain bacterial infection, a combination of two drugs is being tested. Studies have shown that the duration D (in hours) of the infection in laboratory tests can be modeled by

$$D(x, y) = x^2 + 2y^2 - 18x - 24y + 2xy + 120$$

where x is the dosage in hundreds of milligrams of the first drug and y is the dosage in hundreds of milligrams of the second drug. Evaluate $D(5, 2.5)$ and $D(7.5, 8)$ and interpret your results.

In Exercises 63–70, find any critical points and relative extrema of the function.

63. $f(x, y) = x^2 + 2y^2$

64. $f(x, y) = x^3 - 3xy + y^2$

65. $f(x, y) = 1 - (x + 2)^2 + (y - 3)^2$

66. $f(x, y) = e^x - x + y^2$

67. $f(x, y) = x^3 + y^2 - xy$

68. $f(x, y) = y^2 + xy + 3y - 2x + 5$

69. $f(x, y) = x^3 + y^3 - 3x - 3y + 2$

70. $f(x, y) = -x^2 - y^2$

71. Revenue A company manufactures and sells two products. The demand functions for the products are given by

$$p_1 = 100 - x_1 \quad \text{and} \quad p_2 = 200 - 0.5x_2.$$

  (a) Find the total revenue function for x_1 and x_2.

  (b) Find x_1 and x_2 such that the revenue is maximized.

  (c) What is the maximum revenue?

72. Profit A company manufactures a product at two different locations. The costs of manufacturing x_1 units at plant 1 and x_2 units at plant 2 are modeled by $C_1 = 0.03x_1^2 + 4x_1 + 300$ and $C_2 = 0.05x_2^2 + 7x_2 + 175$, respectively. If the product sells for \$10 per unit, find x_1 and x_2 such that the profit, $P = 10(x_1 + x_2) - C_1 - C_2$, is maximized.

In Exercises 73–78, locate any extrema of the function by using Lagrange multipliers.

73. $f(x, y) = x^2 y$

  Constraint: $x + 2y = 2$

74. $f(x, y) = x^2 + y^2$

  Constraint: $x + y = 4$

75. $f(x, y, z) = xyz$

  Constraint: $x + 2y + z - 4 = 0$

76. $f(x, y, z) = x^2 z + yz$

  Constraint: $2x + y + z = 5$

77. $f(x, y, z) = x^2 + y^2 + z^2$

  Constraints: $x + z = 6, y + z = 8$

78. $f(x, y, z) = xyz$

  Constraints: $x + y + z = 32, x - y + z = 0$

Ⓢ In Exercises 79 and 80, use a spreadsheet to find the indicated extremum. In each case, assume that x, y, and z are nonnegative.

79. Maximize $f(x, y, z) = xy$

  Constraints: $x^2 + y^2 = 16, x - 2z = 0$

80. Minimize $f(x, y, z) = x^2 + y^2 + z^2$

  Constraints: $x - 2z = 4, x + y = 8$

81. Maximum Production Level The production function for a manufacturer is given by $f(x, y) = 4x + xy + 2y$. Assume that the total amount available for labor x and capital y is \$2000 and that units of labor and capital cost \$20 and \$4, respectively. Find the maximum production level for this manufacturer.

82. Minimum Cost A manufacturer has an order for 1000 units of wooden benches that can be produced at two locations. Let x_1 and x_2 be the numbers of units produced at the two locations. Find the number that should be produced at each location to meet the order and minimize cost if the cost function is given by

$$C = 0.25x_1^2 + 10x_1 + 0.15x_2^2 + 12x_2.$$

In Exercises 83 and 84, (a) use the method of least squares to find the least squares regression line and (b) calculate the sum of the squared errors.

83. $(-2, -3), (-1, -1), (1, 2), (3, 2)$

84. $(-3, -1), (-2, -1), (0, 0), (1, 1), (2, 1)$

85. Agriculture An agronomist used four test plots to determine the relationship between the wheat yield y (in bushels per acre) and the amount of fertilizer x (in hundreds of pounds per acre). The results are listed in the table.

Fertilizer, x	1.0	1.5	2.0	2.5
Yield, y	32	41	48	53

(T)(S) (a) Use the regression capabilities of a graphing utility or a spreadsheet to find the least squares regression line for the data.

(b) Estimate the yield for a fertilizer application of 20 pounds per acre.

86. Work Force The table gives the percents x and numbers y (in millions) of women in the work force for selected years. (*Source: U.S. Bureau of Labor Statistics*)

Year	1970	1975	1980	1985
Percent, x	43.3	46.3	51.5	54.5
Number, y	31.5	37.5	45.5	51.1

Year	1990	1995	2000	2005
Percent, x	57.5	58.9	59.9	59.3
Number, y	56.8	60.9	66.3	69.3

(T)(S) (a) Use the regression capabilities of a graphing utility or a spreadsheet to find the least squares regression line for the data.

(b) According to this model, approximately how many women enter the labor force for each one-point increase in the percent of women in the labor force?

(T)(S) In Exercises 87 and 88, use the regression capabilities of a graphing utility or a spreadsheet to find the least squares regression quadratic for the given points. Plot the points and graph the least squares regression quadratic.

87. $(-1, 9)$, $(0, 7)$, $(1, 5)$, $(2, 6)$, $(4, 23)$

88. $(0, 10)$, $(2, 9)$, $(3, 7)$, $(4, 4)$, $(5, 0)$

In Exercises 89–92, evaluate the double integral.

89. $\displaystyle\int_0^1 \int_0^{1+x} (4x - 2y)\, dy\, dx$

90. $\displaystyle\int_{-3}^3 \int_0^4 (x - y^2)\, dx\, dy$

91. $\displaystyle\int_1^2 \int_1^{2y} \frac{x}{y^2}\, dx\, dy$

92. $\displaystyle\int_0^4 \int_0^{\sqrt{16-x^2}} 2x\, dy\, dx$

In Exercises 93–96, use a double integral to find the area of the region.

93.
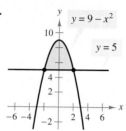

94.
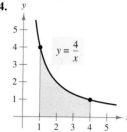

95.
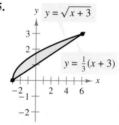

96.
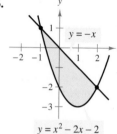

97. Find the volume of the solid bounded by the graphs of $z = (xy)^2$, $z = 0$, $y = 0$, $y = 4$, $x = 0$, and $x = 4$.

98. Find the volume of the solid bounded by the graphs of $z = x + y$, $z = 0$, $x = 0$, $x = 3$, $y = x$, and $y = 0$.

99. Average Elevation In a triangular coastal area, the elevation in miles above sea level at the point (x, y) is modeled by

$$f(x, y) = 0.25 - 0.025x - 0.01y$$

where x and y are measured in miles (see figure). Find the average elevation of the triangular area.

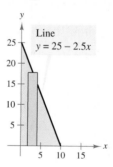

Figure for 99

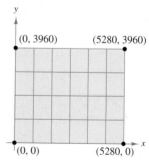

Figure for 100

100. Real Estate The value of real estate (in dollars per square foot) for a rectangular section of a city is given by

$$f(x, y) = 0.003x^{2/3}y^{3/4}$$

where x and y are measured in feet (see figure). Find the average value of real estate for this section.

Chapter Test

See www.CalcChat.com for worked-out solutions to odd-numbered exercises.

Take this test as you would take a test in class. When you are done, check your work against the answers given in the back of the book.

In Exercises 1–3, (a) plot the points on a three-dimensional coordinate system, (b) find the distance between the points, and (c) find the coordinates of the midpoint of the line segment joining the points.

1. $(1, -3, 0), (3, -1, 0)$ **2.** $(-2, 2, 3), (-4, 0, 2)$ **3.** $(3, -7, 2), (5, 11, -6)$

4. Find the center and radius of the sphere whose equation is

$$x^2 + y^2 + z^2 - 20x + 10y - 10z + 125 = 0.$$

In Exercise 5–7, identify the surface.

5. $3x - y - z = 0$ **6.** $36x^2 + 9y^2 - 4z^2 = 0$ **7.** $4x^2 - y^2 - 16z = 0$

In Exercises 8–10, find $f(3, 3)$ and $f(1, 1)$.

8. $f(x, y) = x^2 + xy + 1$ **9.** $f(x, y) = \dfrac{x + 2y}{3x - y}$ **10.** $f(x, y) = xy \ln \dfrac{x}{y}$

In Exercises 11 and 12, find f_x and f_y and evaluate each at the point $(10, -1)$.

11. $f(x, y) = 3x^2 + 9xy^2 - 2$ **12.** $f(x, y) = x\sqrt{x + y}$

In Exercises 13 and 14, find any critical points, relative extrema, and saddle points of the function.

13. $f(x, y) = 3x^2 + 4y^2 - 6x + 16y - 4$

14. $f(x, y) = 4xy - x^4 - y^4$

15. The production function for a manufacturer can be modeled by

$$f(x, y) = 60x^{0.7}y^{0.3}$$

where x is the number of units of labor and y is the number of units of capital. Each unit of labor costs $42 and each unit of capital costs $144. The total cost of labor and capital is limited to $240,000.

(a) Find the numbers of units of labor and capital needed to maximize production.

(b) Find the maximum production level for this manufacturer.

Exposure, x	Mortality, y
1.35	118.5
2.67	135.2
3.93	167.3
5.14	197.6
7.43	204.7

Table for 16

Ⓣ
Ⓢ

16. After contamination by a carcinogen, people in different geographic regions were assigned an exposure index to represent the degree of contamination. The table shows the exposure index x and the corresponding mortality y (per 100,000 people). Use the regression capabilities of a graphing utility or a spreadsheet to find the least squares regression quadratic for the data.

In Exercises 17 and 18, evaluate the double integral.

17. $\displaystyle\int_0^1 \int_x^1 (30x^2y - 1) \, dy \, dx$ **18.** $\displaystyle\int_0^{\sqrt{e-1}} \int_0^{2y} \dfrac{1}{y^2 + 1} \, dx \, dy$

19. Use a double integral to find the area of the region bounded by the graphs of $y = 3$ and $y = x^2 - 2x + 3$ (see figure).

20. Find the average value of $f(x, y) = x^2 + y$ over the region defined by a rectangle with vertices $(0, 0), (1, 0), (1, 3),$ and $(0, 3)$.

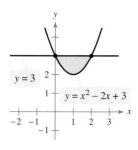

Figure for 19

Answers to Odd–Numbered Exercises and Tests

CHAPTER 0

SECTION 0.1   (page 8)

1. (a) Natural: $\{5\}$

   (b) Integer: $\{-9, 5\}$

   (c) Rational: $\left\{-9, -\frac{7}{2}, 5, \frac{2}{3}, 0.1\right\}$

   (d) Irrational: $\left\{\sqrt{2}\right\}$

3. (a) Natural: $\left\{12, 1, \sqrt{4}\right\}$ $\left(Note: \sqrt{4} = 2\right)$

   (b) Integer: $\left\{12, -13, 1, \sqrt{4}\right\}$

   (c) Rational: $\left\{12, -13, 1, \sqrt{4}, \frac{3}{2}\right\}$

   (d) Irrational: $\left\{\sqrt{6}\right\}$

5. (a) Natural: $\left\{\frac{8}{2}, 9\right\}$ $\left(Note: \frac{8}{2} = 4\right)$

   (b) Integer: $\left\{\frac{8}{2}, -4, 9\right\}$

   (c) Rational: $\left\{\frac{8}{2}, -\frac{8}{3}, -4, 9, 14.2\right\}$

   (d) Irrational: $\left\{\sqrt{10}\right\}$

7. $0.\overline{6}$ 9. $0.\overline{126}$ 11. $-1 < 2.5$

13. $\frac{3}{2} < 7$

15. $1 > -3.5$

17. $\frac{5}{6} > \frac{2}{3}$

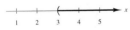

19. $\frac{204}{60}, \frac{31}{9}, 3.45, 2\sqrt{3}, \frac{7}{2}$ 21. $\frac{127}{90}, \frac{584}{413}, \frac{7071}{5000}, \sqrt{2}, \frac{47}{33}$

23. $x \le 4$ 25. $0 < x \le 3$

27. $x < 0$ denotes all negative real numbers.

29. $x \le 5$ denotes all real numbers less than or equal to 5.

31. $x > 3$ denotes all real numbers greater than 3.

33. $-2 < x < 2$ denotes all real numbers greater than -2 and less than 2.

35. $-1 \le x < 0$ denotes all real numbers greater than or equal to -1 and less than 0.

37. $x > 0$ 39. $5 < y \le 12$ 41. $A \ge 35$

43. $3.5\% \le r \le 6\%$ 45. 10   47. -6 49. -9

51. -1 53. $\pi - 3$ 55. $|-7| = |7|$

57. $|-3| > -|-3|$ 59. $-|-2| = -|2|$ 61. 4

63. $\frac{5}{2}$ 65. $\frac{7}{2}$ 67. 51   69. $\frac{128}{75}$ 71. $\left|z - \frac{3}{2}\right| > 1$

73. $|x + 10| \ge 6$ 75. $|y - 0| \ge 6 \Rightarrow |y| \ge 6$

77. $|x - m| > 5$ 79. 179 miles

81. $129°F$

		Passes Budget
$|a - b|$	$0.05b$	Variance Test
83. $\$876.55$	$\$1500$	No
85. $\$264.32$	$\$600$	Yes
87. $\$937.83$	$\$2040$	No

		Passes Quality
$|a - b|$	$0.0012b$	Control Test
89. 0.002	0.0168	Yes
91. 0.027	0.0192	No
93. 0.027	0.0216	No

95. (a) No. If $u > 0$ and $v < 0$ or if $u < 0$ and $v > 0$, then $|u + v| \ne |u| + |v|$.

   (b) Yes. If the signs of u and v are different, then $|u + v| < |u| + |v|$.

97. Answers will vary. Sample answer: The set of natural numbers includes only the integers greater than zero. The set of integers includes all numbers that have no fractional or decimal parts. The set of rational numbers includes all numbers that can be written as the quotient of two integers. Any real number that is not a rational number is in the set of irrational numbers.

SECTION 0.2   (page 18)

Skills Review   (page 18)

1. $-4 < -2$ 2. $0 > -3$ 3. $\sqrt{3} > 1.73$

4. $-\pi < -3$ 5. $|6 - 4| = 2$ 6. $|2 - (-2)| = 4$

7. $|0 - (-5)| = 5$ 8. $|3 - (-1)| = 4$

9. $|-7| + |7| = 7 + 7 = 14$

10. $-|8 - 10| = -|-2| = -2$

1. $7x, 4$ **3.** $x^2, -4x, 8$ **5.** $2x^2, -9x, 13$

7. -6 **9.** 6 **11.** (a) -10 (b) -6

13. (a) 14 (b) 2

15. (a) $\frac{1}{2}$ (b) Undefined. You cannot divide by zero.

17. 1 **19.** 35 **21.** $\frac{1}{4}$ **23.** Commutative (addition)

25. Inverse (addition)   **27.** Distributive Property

29. Inverse (multiplication)   **31.** Identity (addition)

33. Identity (multiplication)   **35.** Associative (addition)

37. $x(3y) = (x \cdot 3)y$ Associative (multiplication)
        $= (3x)y$ Commutative (multiplication)

39. $2^4 \cdot 3$ **41.** $2^4 \cdot 3 \cdot 5$ **43.** -14 **45.** $\frac{1}{24}$

47. $\frac{7}{20}$ **49.** $\frac{1}{12}$ **51.** -0.13 **53.** 1.56

55. 23.8 **57.** 46.25

59. (a) 21.2%

     (b) Health and Medicare: $\approx \$549$ billion

         Education and Veterans' Benefits: $\approx \$168$ billion

         Income Security: $\approx \$346$ billion

         Social Security: $\approx \$524$ billion

         Other: $\approx \$391$ billion

         National Defense: $\approx \$494$ billion

61. ≈ 1695 patients

63. (a) Scientific: 5 ⊗ ⦅ 18 ⊖ 2 $\boxed{y^x}$ 3 ⦆ ÷ 10 ⊜

       Graphing: 5 ⦅ 18 ⊖ 2 ⌃ 3 ⦆ ÷ 10 $\boxed{\text{ENTER}}$

     (b) Scientific: 6 $\boxed{x^2}$ $\boxed{+/-}$ ⊖ ⦅ 7 ⊕ ⦅ 2 $\boxed{+/-}$ ⦆

                $\boxed{y^x}$ 3 ⦆ ⊜

       Graphing: $\boxed{(-)}$ 6 $\boxed{x^2}$ ⊖ ⦅ 7 ⊕ ⦅ $\boxed{(-)}$ 2 ⦆ ⌃

                3 ⦆ $\boxed{\text{ENTER}}$

65. Food: 40%

     Vet care: 24%

     Supplies/OTC medicine: 24%

     Live animal purchases: 5%

     Grooming and boarding: 7%

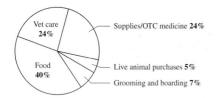

SECTION 0.3   *(page 27)*

Skills Review   *(page 27)*

1. 1 **2.** 5 **3.** 4 **4.** 4 **5.** $\frac{1}{4}$ **6.** 1

7. $\frac{3}{7}$ **8.** 0 **9.** $-\frac{1}{8}$ **10.** 1

---

1. 64 **3.** 8 **5.** 729 **7.** -81 **9.** $\frac{1}{2}$

11. 8 **13.** $-\frac{3}{10}$ **15.** 5184 **17.** $-\frac{3}{5}$ **19.** 1

21. 18 **23.** $\frac{7}{16}$ **25.** $-125z^3$ **27.** $16x^7$ **29.** $10x^4$

31. $-3z^7$ **33.** $\dfrac{5y^4}{2}$ **35.** $\dfrac{5184}{y^7}$ **37.** $\dfrac{5x}{3} + 5$

39. $\dfrac{7}{x}$ **41.** $\dfrac{1}{x}$ **43.** 27^n **45.** $1, x \neq 0$

47. $\dfrac{1}{(y+2)^3}$ **49.** $32y^2$ **51.** $\dfrac{10}{x}$ **53.** $\dfrac{125x^9}{y^{12}}$

55. 5.73×10^7 square miles   **57.** 9.46×10^{12} kilometers

59. 1×10^{-7} meter   **61.** $350{,}000{,}000$ air sacs

63. $0.00000000000000000001602$ coulomb

65. (a) 6.0×10^4 (b) 2.0×10^{11}

67. (a) 3.071×10^6 (b) 3.077×10^{10}

69. (a) 4.907×10^{17} (b) 1.479

71. (a) $(4.8 \times 10^{10})(2.5 \times 10^8) = 1.2 \times 10^{19}$

     (b) $\dfrac{1.2 \times 10^{-8}}{6.4 \times 10^{-6}} = 1.875 \times 10^{-3}$

73. (a) $\$22{,}477.40$ (b) $\$22{,}467.28$

     (c) $\$22{,}428.12$ (d) $\$22{,}327.67$

As the number of compoundings per year decreases, the balance in the account also decreases.

75. $\approx 5.19\%$ **77.** $\$210{,}048.59$

SECTION 0.4   *(page 36)*

Skills Review   *(page 36)*

1. $\frac{4}{27}$ **2.** 48 **3.** $-8x^3$ **4.** $6x^7$

5. $28x^6$ **6.** $\frac{1}{5}x^2$ **7.** $3z^4$ **8.** $\dfrac{25}{4x^2}$ **9.** 1

10. $(x+2)^{10}$

---

1. $9^{1/2} = 3$ **3.** $\sqrt[5]{32} = 2$ **5.** $\sqrt{196} = 14$

7. $(-216)^{1/3} = -6$ **9.** $81^{3/4} = 27$

11. $\sqrt[3]{125^2} = 25$ **13.** 3 **15.** 3 **17.** 2

19. -125 **21.** 4 **23.** 216 **25.** $\sqrt{6}$

27. $\dfrac{27}{8}$ **29.** -4 **31.** $2x\sqrt[3]{2x^2}$ **33.** $\dfrac{5\sqrt{3}\,y^2}{|x|}$

35. $\dfrac{2\sqrt[5]{2}}{y}$ **37.** 90   **39.** 45   **41.** $\dfrac{\sqrt{5}}{5}$

43. $4\sqrt[3]{4}$ **45.** $\dfrac{x(5+\sqrt{3})}{11}$ **47.** $3(\sqrt{6}-\sqrt{5})$

49. 25   **51.** $2^{1/2}$ **53.** $x^{3/2},\,x\neq 0$ **55.** 5

57. x^3 **59.** $8x^6y^3$ **61.** $2\sqrt[4]{2}$ **63.** $3^{1/2}=\sqrt{3}$

65. $\sqrt[3]{x}$ **67.** $2\sqrt{x}$ **69.** $31\sqrt{2}$ **71.** $-2\sqrt{y}$

73. 3.557   **75.** 2.006   **77.** 2.938   **79.** 0.382

81. ① 4 ⊖ ☑ 7 ① ① ÷ 3

83. $\sqrt{5}+\sqrt{3}>\sqrt{5+3}$ **85.** $5>\sqrt{3^2+2^2}$

87. $\sqrt{3}\cdot\sqrt[4]{3}>\sqrt[8]{3}$

89. 25 inches × 25 inches × 25 inches   **91.** $\approx 12.83\%$

93. No   **95.** ≈ 2.221 seconds   **97.** ≈ 0.021 inch

99. ≈ 494 vibrations per second

101. a; Higher notes have higher frequencies.

103. ≈ 40.2 miles per hour   **105.** $\approx 17.4°$F   **107.** 1

109. $a^0 = a^{n-n} = \dfrac{a^n}{a^n} = 1$

111. Radicals can be added together only if they have the same radicand and index.

MID-CHAPTER QUIZ   *(page 39)*

1. $-|-7|<|-7|$ **2.** $-(-3)=|-3|$

3. $x\geq 0$ **4.** $r\geq 95\%$

5. $-2\leq x<3$ denotes all numbers greater than or equal to -2 and less than 3.

6. $3x^2,\,-7x,\,2$ **7.** 3   **8.** -13 **9.** $\dfrac{5}{14}$ **10.** $\dfrac{11}{9}$

11. $-2x^7$ **12.** $\dfrac{1}{3}y^4$ **13.** $\dfrac{27x^6}{y^6}$ **14.** \$9527.79

15. -1 **16.** -64 **17.** 9   **18.** $-\sqrt[3]{3}$ **19.** $2\sqrt{3}$

20. 22 cm × 22 cm × 22 cm

SECTION 0.5   *(page 46)*

> **Skills Review**   *(page 46)*
>
> **1.** $42x^3$ **2.** $-20z^2$ **3.** $-27x^6$ **4.** $-3x^6$
>
> **5.** $\dfrac{9}{4}z^3,\,z\neq 0$ **6.** $4\sqrt{3}$ **7.** $\dfrac{9}{4x^2}$ **8.** 8
>
> **9.** $\sqrt{2}$ **10.** $-3x$

1. Degree: 2   **3.** Degree: 5
Leading coefficient: 2   Leading coefficient: 1

5. Degree: 5
Leading coefficient: 3

7. Polynomial, $-3x^3+2x+8$, degree 3

9. Not a polynomial

11. Polynomial, $-w^4+2w^3+w^2$, degree 4

13. (a) -3 (b) 1   (c) 5   (d) 17

15. (a) -10 (b) -1 (c) 4   (d) 5

17. $-2x-10$ **19.** $2x^3-4x-5$ **21.** $8x^3+29x^2+11$

23. $3x^3-6x^2+3x$ **25.** $4x^4-12x$

27. $9x^3-21x^2$ **29.** $x^2+7x+12$

31. $6x^2-7x-5$ **33.** x^2-25

35. $x^2+12x+36$ **37.** $4x^2-20xy+25y^2$

39. $x^2+2xy+y^2-6x-6y+9$

41. x^3+3x^2+3x+1 **43.** $8x^3-12x^2y+6xy^2-y^3$

45. $9y^4-1$ **47.** m^2-n^2-6m+9

49. $x-y$ **51.** x^4+x^2+1 **53.** $2x^2+2x$

55. The student omitted the middle term when squaring the binomial.
$$(x-3)^2 = x^2 - 2(x)(3) + 3^2 = x^2 - 6x + 9$$

57. $1000r^3+3000r^2+3000r+1000$ **59.** Yes; Yes; No

61. 66,988.76; 74,582.25; In the years 2004 and 2005, the total amounts of federal student aid disbursed were approximately \$66,988,760,000 and \$74,582,250,000.

63. $x=3$ inches: $V=4968$ cubic inches
$x=7$ inches: $V=7448$ cubic inches
$x=9$ inches: $V=7344$ cubic inches
$x=7$ inches produces the greatest volume.

65. $7x^2+14x+4$ **67.** Answers will vary.

SECTION 0.6   *(page 53)*

> **Skills Review**   *(page 53)*
>
> **1.** $15x^2-6x$ **2.** $-2y^2-2y$ **3.** $4x^2+12x+9$
>
> **4.** $9x^2-48x+64$ **5.** $2x^2+13x-24$
>
> **6.** $-5z^2-z+4$ **7.** $4y^2-1$ **8.** x^2-a^2
>
> **9.** $x^3+12x^2+48x+64$
>
> **10.** $8x^3-36x^2+54x-27$

1. $3(x+2)$ **3.** $3x(x^2-2)$ **5.** $(x-1)(x+5)$

7. $(x+6)(x-6)$ **9.** $(4x+3y)(4x-3y)$

11. $(x+1)(x-3)$ **13.** $(x-2)^2$ **15.** $(2y+3)^2$

17. $\left(y - \frac{1}{3}\right)^2$ **19.** $(x - 2)(x^2 + 2x + 4)$

21. $(y + 5)(y^2 - 5y + 25)$ **23.** $(2t - 1)(4t^2 + 2t + 1)$

25. $(x + 2)(x - 1)$ **27.** $(w - 2)(w - 3)$

29. $(y + 5)(y - 4)$ **31.** $(x - 20)(x - 10)$

33. $(3x - 2)(x - 1)$ **35.** $(3x + 1)(3x - 2)$

37. $(6x + 1)(x + 6)$ **39.** $(x - 1)(x^2 + 2)$

41. $(2x - 1)(x^2 - 3)$ **43.** $(2 - y^3)(3 + y)$

45. $4x(x - 2)$ **47.** $y(y - 3)(y + 3)$

49. $3(x + 4)(x - 4)$ **51.** $(x - 1)^2$ **53.** $(1 - 2x)^2$

55. $y(2y + 3)(y - 5)$ **57.** $2x(x - 2)(x + 1)$

59. $(3x + 1)(x^2 + 5)$ **61.** $x(x - 4)(x^2 + 1)$

63. $-x(x + 10)$ **65.** $(x + 1)^2(x - 1)^2$

67. $2(t - 2)(t^2 + 2t + 4)$

69.

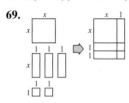

71.

73. $(2x - 3)$ feet

75. (a) $(x + 8)(x + 3)$; factoring by grouping

(b) $(3x - 5)(x + 4)$; factoring by trial and error

77. $c = \{7, 12, 15, 16\}$; Answers will vary.

79. Answers will vary. Sample answer:

(1) Find a combination of factors of 2 and -15 such that the outer and inner products add up to the middle term $-7x$.

$2x^2 - 7x - 15 = (2x + 3)(x - 5)$

(2) Rewrite -7 as the sum of two factors of the product $2(-15)$. Then factor by grouping.

$$2x^2 - 7x - 15 = 2x^2 - 10x + 3x - 15$$
$$= 2x(x - 5) + 3(x - 5)$$
$$= (x - 5)(2x + 3)$$

81. Box 1: $V = (a - b)a^2$

Box 2: $V = (a - b)ab$

Box 3: $V = (a - b)b^2$

Multiplying $(a - b)$ by each term of $(a^2 + ab + b^2)$ produces the volumes of the three boxes.

SECTION 0.7 *(page 60)*

Skills Review *(page 60)*

1. $5x^2(1 - 3x)$ **2.** $(4x + 3)(4x - 3)$

3. $(3x - 1)^2$ **4.** $(2y + 3)^2$ **5.** $(z + 3)(z + 1)$

6. $(x - 5)(x - 10)$ **7.** $(3 - x)(1 + 3x)$

8. $(3x - 1)(x - 15)$ **9.** $(s + 1)(s + 2)(s - 2)$

10. $(y + 4)(y^2 - 4y + 16)$

1. (a) No (b) Yes **3.** (a) Yes (b) Yes

5. All real numbers **7.** All real numbers except $x = 2$

9. All real numbers except $x = 0$ and $x = 4$

11. All real numbers greater than or equal to -1

13. $3x, \ x \neq 0$ **15.** $x - 2, \ x \neq 2, \ x \neq 0$

17. $x + 2, \ x \neq -2, \ x \neq 3$

19. $\dfrac{3x}{2}, \ x \neq 0$ **21.** $\dfrac{x}{2(x + 1)}, \ x \neq -1$

23. $-\dfrac{1}{2}, \ x \neq 5$ **25.** $-(x + 5), \ x \neq 5$

27. $\dfrac{x(x + 3)}{x - 2}, \ x \neq -2$ **29.** $\dfrac{y - 4}{y + 6}, \ y \neq 3$

31. $-1 - x^2, \ x \neq 2$ **33.** $z - 3$ **35.** $\dfrac{1}{5(x - 2)}, \ x \neq 1$

37. $-\dfrac{x(x + 7)}{x + 1}, \ x \neq 9$ **39.** $\dfrac{r + 1}{r}, \ r \neq 1$

41. $\dfrac{t - 3}{(t + 3)(t - 2)}, \ t \neq -2$ **43.** $\dfrac{x - 1}{x(x + 1)^2}, \ x \neq -2$

45. $\dfrac{3}{2}, \ x \neq -y$ **47.** $x(x + 1), \ x \neq -1, 0$

49. $x^2(x - 1)$ **51.** $(x + 5)(x - 7)$

53. $\dfrac{5x}{x - 2}$ **55.** $\dfrac{2x}{x - 4}$ **57.** $\dfrac{4x - 23}{x - 5}$

59. $\dfrac{x - 4}{(x + 2)(x - 2)(x - 1)}$ **61.** $\dfrac{2 - x}{x^2 + 1}, \ x \neq 0$

63. $\dfrac{1}{2}, \ x \neq 2$ **65.** $\dfrac{1}{x}, \ x \neq -1$ **67.** $\dfrac{2x - 1}{2x}, x > 0$

69. (a) 12% (b) $\dfrac{288(NM - P)}{N(12P + NM)}$; 12%

71. 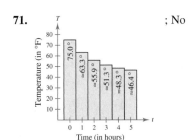 ; No

Temperature (in °F) vs Time (in hours). Bars: 75.0°, ≈63.3°, ≈55.9°, ≈51.3°, ≈48.3°, ≈46.4°

REVIEW EXERCISES   *(page 64)*

1. (a) Natural: $\{11\}$

(b) Integer: $\{11, -14\}$

(c) Rational: $\left\{11, -14, -\frac{8}{9}, \frac{5}{2}, 0.4\right\}$

(d) Irrational: $\left\{\sqrt{6}\right\}$

3. $-4 < -3$

Number line from -7 to -1, points at -4 and -3.

5. $x \le -6$ denotes all real numbers less than or equal to -6.

Number line from -8 to -4.

7. $x \ge 0$ **9.** $2 < x \le 5$ **11.** -14

13. $|-12| > -|12|$ **15.** 4   **17.** 10   **19.** $|x - 7| \ge 4$

21. $-2x^2, x, 4$ **23.** $3x^3, 7x, -4$ **25.** (a) 2   (b) 0

27. Distributive Property   **29.** Commutative (addition)

31. -1 **33.** $\frac{2}{3}$ **35.** 5   **37.** 0.10   **39.** -2

41. -5 **43.** $8x, \ x \ne 0$ **45.** $\frac{5}{x^4}$

47. 3.004×10^8 **49.** 864,400 miles

51. (a) 11,414.125   (b) 18,380.160

53.

Year	5	10	15
Balance	$2074.23	$2868.28	$3966.30

Year	20	25
Balance	$5484.67	$7584.30

55. $16^{1/2} = 4$ **57.** 13   **59.** $2x^2$ **61.** $2 + \sqrt{3}$

63. $-3\sqrt{x}$ **65.** $\sqrt{10}$ **67.** $\frac{1}{16}$ **69.** $\sqrt{5}$

71. 11.269   **73.** $-6x + 26$ **75.** $x^2 - 8x - 14$

77. $x^2 - x - 2$ **79.** $x^3 + 64$ **81.** $x^2 + 8x + 16$

83. 61.55

In 2005, the average sale price for a newly manufactured residential mobile home in the United States was $61,550.

85. $x = 0$: 106.6

$x = 5$: 202.5

There were 106,600,000 and 202,500,000 cell phone subscribers in the United States in 2000 and 2005, respectively.

87. $4(x + 3)(x - 3)$ **89.** $3x(x + 1)(x - 2)$

91. $x(x + 4)(x - 4)$ **93.** $(x + 3)(x - 3)(x - 2)$

95. All real numbers except $x = 3$

97. All real numbers   **99.** $3x$

101. $\frac{x - 2}{2}, \ x \ne -2$ **103.** $x - 5, \ x \ne -3$

105. $\frac{x + 3}{x - 1}, \ x \ne 0, 3$ **107.** $\frac{x - 1}{x - 3}, \ x \ne -1, \frac{1}{2}$

109. $\frac{3x^2 - 4x}{(x - 1)(x - 2)}$ **111.** $\frac{6}{x - 1}, \ x \ne -1$

113. $\frac{x + 1}{x - 1}, \ x \ne 0$ **115.** $\frac{y - x}{y + x}, \ xy \ne 0$

CHAPTER TEST   *(page 67)*

1. -12

2.

Year	5	10	15
Balance	$5813.18	$8448.26	$12,277.81

Year	20	25
Balance	$17,843.27	$25,931.52

The longer you leave the $4000 in the account, the more money you earn.

3. $-64x^6$ **4.** $-4\sqrt{x}$ **5.** 25   **6.** $4\left(\sqrt{3} - \sqrt{5}\right)$

7. $2x\sqrt{3x}$ **8.** $\frac{5 + \sqrt{7}}{9}$ **9.** $9x^2 + 42x + 49$

10. $-5x^2 + 29x$ **11.** $5(x + 4)(x - 4)$

12. $(2x + 3)^2$ **13.** $(x^2 - 3)(x - 6)$

14. $(x + 2)^2(x - 2)$ **15.** $\frac{1}{3}(x - 4), \ x \ne -4$

16. $\frac{x + 4}{3x + 5}, \ x \ne -3, \frac{5}{3}$ **17.** $\frac{4x^2 - 13x}{(x - 3)(x - 4)}$

18. $-\frac{x + 26}{(x + 5)(x - 2)}$

19. All real numbers greater than or equal to 2

20. All real numbers except $x = -1$

21. $\frac{2x^2 - 5x - 18}{5 + 5x - x^2}, \ x \ne 1, -2$

22. $x = 5$: 4.21, $x = 15$: 6.45

In 1995 and 2005, the average prices of a movie ticket in the United States were $4.21 and $6.45, respectively.

CHAPTER 1

SECTION 1.1   (page 76)

Skills Review   (page 76)

1. $-3x - 10$ **2.** $5x - 12$ **3.** x **4.** $x + 26$

5. $\dfrac{8x}{15}$ **6.** $\dfrac{3x}{4}$ **7.** $-\dfrac{1}{x(x+1)}$ **8.** $\dfrac{5}{x}$

9. $\dfrac{7x - 8}{x(x - 2)}$ **10.** $-\dfrac{2}{x^2 - 1}$

1. Identity   **3.** Conditional equation

5. Conditional equation

7. (a) No   (b) No   (c) Yes   (d) No

9. (a) Yes   (b) Yes   (c) No   (d) No

11. (a) Yes   (b) No   (c) No   (d) No

13. (a) No   (b) No   (c) No   (d) Yes

15. (a) Yes   (b) No   (c) No   (d) No

17. 5   **19.** -4 **21.** 3   **23.** 9   **25.** -26

27. -4 **29.** $-\frac{6}{5}$ **31.** 9   **33.** No solution

35. 10   **37.** 4   **39.** 3   **41.** 5   **43.** No solution

45. $\frac{11}{6}$ **47.** No solution   **49.** 0

51. All real numbers   **53.** No solution

55. Because substituting 2 for x in the equation produces division by zero, $x = 2$ cannot be a solution to the equation.

57. Extraneous solutions may arise when a fractional expression is multiplied by factors involving the variable.

59. Equivalent equations have the same solutions.

Example: $2x - 6 = 0$ and $x - 3 = 0$ both have the solution $x = 3$.

61. $x \approx 138.889$ **63.** $x \approx 62.372$ **65.** $x \approx 19.993$

67. Use the *table* feature in ASK mode or use the scientific calculator part of the graphing utility.

69. (a) 6.46 (b) 6.41; Yes   **71.** (a) 56.09 (b) 56.13; Yes

73. 2003 ($t \approx 12.81$)   **75.** 58.9 inches

77. 2003 ($t \approx 12.97$)   **79.** 2001 ($t \approx 10.98$)

SECTION 1.2   (page 87)

Skills Review   (page 87)

1. 14   **2.** 4   **3.** -3 **4.** 4   **5.** -2

6. 1   **7.** $\frac{2}{5}$ **8.** $\frac{10}{3}$ **9.** 6   **10.** $-\frac{11}{5}$

1. $x + (x + 1) = 2x + 1$ **3.** $50t$ **5.** $0.2x$ **7.** $6x$

9. $25x + 1200$ **11.** $5 + x = 8$ **13.** $\dfrac{r}{2} = 9$

15. $n + 2n = 15$ **17.** $525 = n + (n + 1)$; 262, 263

19. $5x - x = 148$; 37, 185

21. $n^2 - 5 = n(n + 1)$; $-5, -4$

23. Coworker's check: \$400

Your check: \$448

25. January: \$62,926.83; February: \$66,073.17

27. $\approx 37.03\%$ decrease   **29.** $\approx 39.42\%$ increase

31. $\approx 22.40\%$ increase   **33.** $\approx 128.57\%$

35. $\approx 71.43\%$ **37.** $\approx 54.17\%$ decrease

39. (a) \$37,800   (b) \$40,748.40   (c) \$44,578.75

41. (a) 719 million users   (b) ≈ 816.78 million users

(c) ≈ 1092.85 million users   (d) $\approx 118.57\%$

43. TV: 1564.2 hours

Radio, music: 1137.6 hours

Internet: 213.3 hours

Video games: 106.65 hours

Print media: 391.05 hours

Other: 142.2 hours

45. 15 feet $\times$ 22.5 feet   **47.** ≈ 5.7 years

49. 97 or greater   **51.** \$1411.76   **53.** $\approx 20.13\%$

55. \$18   **57.** \$361.25   **59.** $2\frac{1}{3}$ hours   **61.** $\frac{1}{3}$ hour

63. 1.28 seconds   **65.** 62.5 feet   **67.** \$781,080

69. \$10,500 at 6.5\% and \$4500 at 7.5\%

71. Stock A: \$2200

Stock B: \$2800

73. 11.43\%   **75.** 8571 units per month   **77.** ≈ 48 feet

79. ≈ 32.1 gallons   **81.** ≈ 12.31 miles per hour

83. $h = \dfrac{2A}{b}$ **85.** $l = \dfrac{V}{wh}$ **87.** $h = \dfrac{V}{\pi r^2}$

89. $C = \dfrac{S}{1 + R}$ **91.** $r = \dfrac{A - P}{Pt}$ **93.** $b = \dfrac{2A - ah}{h}$

95. $n = \dfrac{L + d - a}{d}$ **97.** $h = \dfrac{A}{2\pi r}$

99. $R_1 = \dfrac{R_2 f(n - 1)}{R_2 + f(n - 1)}$

101. Williams: \$18,700

Gonzalez: $\approx$ \$21,333

Walters: \$20,000

Gilbert: $\approx$ \$19,933

Hart: $\approx$ \$17,833

Team average: January: \$17,120, February: \$20,100, March: \$21,460

103. Williams: $\approx \$25,033$

Gonzalez: $\approx \$22,867$

Walters: $\$25,400$

Gilbert: $\approx \$27,467$

Hart: $\$28,100$

Reyes: $\approx \$24,967$

Sanders: $\approx \$13,633$

Team average: July: $\approx \$24,514$, August: $\approx \$25,157$,
September: $\$22,100$

105. "takes 30 minutes"; "from a depth of 150 feet"

SECTION 1.3   *(page 100)*

Skills Review   *(page 100)*

1. $\dfrac{\sqrt{14}}{10}$ **2.** $4\sqrt{2}$ **3.** 14 **4.** $\dfrac{\sqrt{10}}{4}$

5. $x(3x + 7)$ **6.** $(2x - 5)(2x + 5)$

7. $-(x - 7)(x - 15)$ **8.** $(x - 2)(x + 9)$

9. $(5x - 1)(2x + 3)$ **10.** $(6x - 1)(x - 12)$

---

1. $2x^2 + 5x - 3 = 0$ **3.** $x^2 - 25x = 0$

5. $x^2 - 6x + 7 = 0$ **7.** $2x^2 - 2x + 1 = 0$

9. $3x^2 - 60x - 10 = 0$ **11.** $4, -2$ **13.** $0, -\frac{1}{2}$

15. -5 **17.** $3, -\frac{1}{2}$ **19.** $2, -6$ **21.** $-2, -5$

23. ± 4 **25.** $\pm\sqrt{7} \approx \pm 2.65$ **27.** $\pm 2\sqrt{3} \approx \pm 3.46$

29. $12 + 3\sqrt{2} \approx 16.24$ **31.** $-2 + 2\sqrt{3} \approx 1.46$

$12 - 3\sqrt{2} \approx 7.76$ $\quad -2 - 2\sqrt{3} \approx -5.46$

33. ± 5 **35.** $\pm\sqrt{38} \approx \pm 6.16$ **37.** $\pm\dfrac{\sqrt{115}}{5} \approx \pm 2.14$

39. $\pm\dfrac{\sqrt{78}}{3} \approx \pm 2.94$ **41.** ± 8 **43.** 1 **45.** $\pm\dfrac{3}{4}$

47. $\frac{3}{2}$ **49.** $3, -11$ **51.** $\frac{3}{2}, -\frac{1}{2}$ **53.** $5, -\frac{10}{3}$

55. $9, 3$ **57.** $\frac{1}{5}, 1$ **59.** $-1, -5$ **61.** $-\frac{1}{2}$

63. Algebra argument:

$(x + 2)^2 = (x + 2)(x + 2)$ Definition of exponent

$\qquad\qquad = x^2 + 2x + 2x + 4$ FOIL

$\qquad\qquad = x^2 + 4x + 4$ Combine like terms.

So, $(x + 2)^2 \neq x^2 + 4$.

Graphing utility argument:

(1) Let $y_1 = (x + 2)^2$ and $y_2 = x^2 + 4$. Use the *table* feature with an arbitrary value of x (but not $x = 0$). The table will show that the values of y_1 are not the same as the values of y_2.

(2) Use the scientific calculator portion of the graphing utility to show that if $x = 5$, $(5 + 2)^2 = 49$ and $5^2 + 4 = 29$. So, $(x + 2)^2$ is not equal to $x^2 + 4$.

65. 34 feet $\times$ 48 feet

67. Base: $2\sqrt{2}$ feet

Height: $2\sqrt{2}$ feet

69. 5 feet   **71.** ≈ 3.54 seconds

73. ≈ 1.43 seconds   **75.** 42 seconds faster

77. ≈ 4.24 centimeters   **79.** 976 miles

81. ≈ 494.97 meters   **83.** $60,000$ units

85. 2012 $(t \approx 11.6)$

87. (a) 1987 $(t \approx 18.74)$

(b) Yes; the model is a good representation through 1890.

(c) Yes; the model is a good representation through 2006.

89. The model in Exercise 88 is *not* valid for the population in 2050 because it predicts 536,526,000 people (not 419,854,000).

91. 1 P.M. $(t \approx 12.96)$; No; the predicted temperature at 7 P.M. is about $145°$F, which is unreasonable.

93. 2003 $(t \approx 3.08)$

SECTION 1.4   *(page 110)*

Skills Review   *(page 110)*

1. $3\sqrt{17}$ **2.** $2\sqrt{3}$ **3.** $4\sqrt{6}$ **4.** $3\sqrt{73}$

5. $2, -1$ **6.** $\frac{3}{2}, -3$ **7.** $5, -1$ **8.** $\frac{1}{2}, -7$

9. $3, 2$ **10.** $4, -1$

---

1. One real solution   **3.** Two real solutions

5. No real solutions   **7.** Two real solutions

9. $\frac{1}{2}, -1$ **11.** $\frac{1}{4}, -\frac{3}{4}$ **13.** $1 \pm \sqrt{3}$

15. $-7 \pm \sqrt{5}$ **17.** $-4 \pm 2\sqrt{5}$

19. $\dfrac{2}{3} \pm \dfrac{\sqrt{7}}{3}$ **21.** $-\dfrac{1}{3} \pm \dfrac{\sqrt{11}}{6}$ **23.** $-\dfrac{1}{2} \pm \sqrt{2}$

25. $\dfrac{2}{7}$ **27.** $2 \pm \dfrac{\sqrt{6}}{2}$ **29.** $6 \pm \sqrt{11}$

31. $x \approx 0.976, -0.643$ **33.** $x \approx 0.561, 0.126$

35. No real solution   **37.** -11 **39.** $\pm\sqrt{10}$

41. $-\dfrac{3}{2} \pm \dfrac{\sqrt{5}}{2}$ **43.** $-2, 4$ **45.** $\dfrac{-1 \pm \sqrt{37}}{6}$

47. Real-life problems will vary; $50, 50$

49. Real-life problems will vary; $7, 8$ or $-8, -7$

51. 200 units   **53.** 653 units   **55.** 9 seats per row

57. 14 inches × 14 inches

59. (a) $s = -16t^2 + \frac{88}{3}t + 984$ (b) ≈ 845.33 feet

(c) ≈ 8.81 seconds

61. Moon: ≈ 14.9 seconds   **63.** Moon
Earth: ≈ 2.6 seconds

65. ≈ 259 miles; ≈ 541 miles

67. (a) 2003 ($t \approx 12.71$)   (b) 2005 ($t \approx 14.92$)

(c) No. The model's prediction of $8.79 billion is less than the expected sales.

69. 4:00 P.M. ($t \approx 3.9$)

71. Southbound: ≈ 550 miles per hour
Eastbound: ≈ 600 miles per hour

73. (a) $\approx 16.8°C$ (b) ≈ 2.5

75. 5279 units or 94,721 units   **77.** Answers will vary.

MID-CHAPTER QUIZ   *(page 114)*

1. $x = -6$ **2.** $x = 6$ **3.** $x = -2$ **4.** No solution

5. Use the *table* feature in ASK mode or the scientific calculator portion of the graphing utility.

6. 328.954   **7.** 431.398

8. $8.50x + 30,000 = 200,000$; 20,000 units

9. $300,000 = x(75 - 0.0002x)$; 4044 units or 370,956 units

10. $x = \frac{2}{3}, -5$ **11.** $x = \pm\sqrt{5}$; $x \approx \pm 2.24$

12. $x = -3 \pm \sqrt{17}$; $x \approx -7.12, 1.12$

13. $x = -1 \pm \sqrt{6}$; $x \approx -3.45, 1.45$

14. $x = \dfrac{-7 \pm \sqrt{73}}{6}$; $x \approx -2.59, 0.26$

15. $x \approx 1.568, -0.068$ **16.** No real solutions

17. One real solution

18. Answers will vary. *Sample answer:* Use the FOIL method $[(x + 3)^2 = (x + 3)(x + 3) = x^2 + 6x + 9]$, use the *table* feature of your graphing utility, or use the scientific calculator portion of your graphing utility to evaluate the solution.

19. ≈ 4.33 seconds   **20.** 8 inches × 8 inches × 6 inches

SECTION 1.5   *(page 123)*

> ### Skills Review   *(page 123)*
>
> **1.** 11   **2.** 20, -3 **3.** 5, -45 **4.** 0, $-\frac{1}{5}$
>
> **5.** $\frac{2}{3}, -2$ **6.** $\frac{11}{6}, -\frac{5}{2}$ **7.** 1, -5 **8.** $\frac{3}{2}, -\frac{5}{2}$
>
> **9.** $\dfrac{3 \pm \sqrt{5}}{2}$ **10.** $2 \pm \sqrt{2}$

1. $3, -1, 0$ **3.** $0, \pm\dfrac{3\sqrt{2}}{2}$ **5.** ± 3 **7.** $-3, 0$

9. $\pm 2, 7$ **11.** ± 1 **13.** $\pm\sqrt{11}, \pm 1$ **15.** ± 2

17. $\pm\frac{1}{2}, \pm 4$ **19.** $1, -2$ **21.** 50   **23.** 26

25. -16 **27.** $\frac{1}{4}$ **29.** 6, 5   **31.** 2, -5

33. 0   **35.** $-59, 69$ **37.** 1   **39.** $\pm\sqrt{69}$

41. $\dfrac{-3 \pm \sqrt{21}}{6}$ **43.** $4, -5$ **45.** -1 **47.** $1, -3$

49. $1, -3$ **51.** $3, -2$ **53.** $\sqrt{3}, -3$ **55.** $10, -1$

57. The quadratic equation was not written in general form before the values of a, b, and c were substituted in the Quadratic Formula. The general form for this equation is $3x^2 - 7x - 4 = 0$ ($a = 3$, $b = -7$, and $c = -4$), and the correct solution is

$$x = \frac{-(-7) \pm \sqrt{(-7)^2 - 4(3)(-4)}}{2(3)}.$$

59. $x \approx \pm 1.038$ **61.** $x \approx 16.756$ **63.** 34   **65.** 6%

67. $\approx 12.98\%$ **69.** 45,000 passengers   **71.** 63 years old

73. 67,760 units; It does not make sense for demand x or price p to be less than zero.

75. ≈ 12.12 feet

77. Least acceptable weight: 78.8 ounces
Greatest acceptable weight: 81.2 ounces

79. $8\frac{2}{11}$ hours

SECTION 1.6   *(page 134)*

> ### Skills Review   *(page 134)*
>
> **1.** $-\frac{1}{2}$ **2.** $-\frac{1}{6}$ **3.** -3 **4.** -6 **5.** $x \geq 0$
>
> **6.** $-3 < z < 10$ **7.** $P \leq 2$ **8.** $W \geq 200$
>
> **9.** 2, 7   **10.** 0, 1

1. $-1 \leq x \leq 5$; Bounded   **3.** $x > 11$; Unbounded

5. $x < -2$; Unbounded

7. c   **8.** h   **9.** f   **10.** e

11. g   **12.** a   **13.** b   **14.** d

15. (a) Yes   (b) No   (c) Yes   (d) No

17. (a) Yes   (b) No   (c) No   (d) Yes

19. (a) Yes   (b) Yes   (c) Yes   (d) No

21. (a) No   (b) Yes   (c) Yes   (d) Yes

23. If $2x > 6$, then $x > 3$.   **25.** If $2x \leq -8$, then $x \leq -4$.

27. If $2 - 4x > -10$, then $x < 3$.

29. If $-\frac{2}{3}x \geq -6$, then $x \leq 9$.

31. $x \geq 6$

33. $x > -4$

35. $x < 25$

37. $x > 2$

39. $x \leq -\frac{1}{3}$

41. $x < -18$

43. $x > \frac{2}{5}$

45. $2 \leq x < 4$

47. $-1 < x < 3$

49. $-\frac{9}{2} < x < \frac{15}{2}$

51. $-\frac{3}{4} < x < -\frac{1}{4}$

53. $-6 < x < 6$

55. $x < -6$ or $x > 6$

57. $-8 < x < 2$

59. $16 \leq x \leq 24$

61. $x < -\frac{1}{2}$ or $x > \frac{11}{2}$

63. $x \leq -7$ or $x \geq 13$

65. $4 < x < 5$

67. $x \leq -\frac{29}{2}$ or $x \geq -\frac{11}{2}$

69. No solution

71. $|x| \leq 2$ **73.** $|x - 9| \geq 3$ **75.** $|x - 12| \leq 10$

77. $|x + 3| > 5$ **79.** More than 250 miles

81. Greater than 8.67%   **83.** $33\frac{1}{3}$ weeks   **85.** 16 inches

87. (a)

x	10	20	30
R	$1399.50	$2799.00	$4198.50
C	$1820.00	$2790.00	$3760.00

x	40	50	60
R	$5598.00	$6997.50	$8397.00
C	$4730.00	$5700.00	$6670.00

(b) $x \geq 20$ units

89. Less than 29,687.5 miles   **91.** $x \geq 114.01$

93. 2008 $(t > 17.72)$ **95.** $[\approx 106.864, \approx 109.464]$

97. Undercharged or overcharged by as much as $0.25

99. $[65.8, 71.2]$

101. Minimum $= 20\%$; Maximum $= 80\%$

103. 2001 $(t > 11.32)$

105. (a) 1995 $(t > 4.86)$ (b) 2009 $(t > 19.33)$

SECTION 1.7   *(page 145)*

Skills Review   *(page 145)*

1. $y < -6$ **2.** $z > -\frac{9}{2}$ **3.** $-3 \leq x < 1$

4. $x \leq -5$ **5.** $-3 < x$ **6.** $5 < x < 7$

7. $-\frac{7}{2} \leq x \leq \frac{7}{2}$ **8.** $x < 2, x > 4$

9. $x < -6, x > -2$ **10.** $-2 \leq x \leq 6$

---

1. $(-\infty, -5), (-5, 5), (5, \infty)$

3. $(-\infty, -4), \left(-4, \frac{1}{2}\right), \left(\frac{1}{2}, \infty\right)$

5. $(-\infty, -1), (-1, 1), (1, \infty)$

7. $[-3, 3]$

9. $(-\infty, -2) \cup (2, \infty)$

11. $(-7, 3)$

13. $(-\infty, -5] \cup [1, \infty)$

15. $(-3, 2)$

17. $(-\infty, -1) \cup (1, \infty)$

19. $(-3, 1)$

21. $(-\infty, 0) \cup \left(0, \frac{3}{2}\right)$

23. $[-2, 0] \cup [2, \infty)$

25. $[-1, 1] \cup [2, \infty)$

27. $(-\infty, -1) \cup (0, 1)$

29. $(-\infty, -1) \cup (4, \infty)$

31. $(5, 15)$

33. $\left(-5, -\frac{3}{2}\right) \cup (-1, \infty)$

35. $\left(-\frac{3}{4}, 3\right) \cup [6, \infty)$

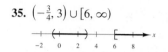

37. $(-\infty, -3] \cup [3, \infty)$ **39.** All real numbers

41. $\left[-\frac{9}{2}, \frac{9}{2}\right]$ **43.** $(-\infty, 2) \cup (5, \infty)$

45. All real numbers

47. The cube root of any real number is a real number.

49. $\left(\frac{5}{3}, \infty\right)$ **51.** $(-\infty, -3] \cup [0, 3]$

53. $[-2, \infty)$ **55.** $(-3.51, 3.51)$ **57.** $(-0.13, 25.13)$

59. $(2.26, 2.39)$ **61.** Between 2.5 and 10 seconds

63. Between about 13.8 meters and about 36.2 meters

65. (a) $90{,}000 \le x \le 100{,}000$ (b) $\$30 \le p \le \$32$

(c) about 185,967 units; The company should produce a maximum of about 185,967 units.

67. 14.5%   **69.** 2012 $(t > 22.45)$

71. 2008/2009 $(t > 18.94)$ **73.** $R_1 \ge 2$ ohms

REVIEW EXERCISES   (page 150)

1. Conditional equation

3. (a) No   (b) Yes   (c) Yes   (d) No

5. 13   **7.** $-\frac{1}{2}$ **9.** $-\frac{5}{3}$ **11.** $-\frac{2}{3}$ **13.** 377.778

15. 0.033   **17.** 12   **19.** $150 - x = 120$; 30 pounds

21. 29.5 feet × 59 feet   **23.** $20   **25.** $163.53

27. 2 hours   **29.** $751,664   **31.** $2\frac{2}{9}$ quarts

33. $-\frac{1}{2}, \frac{4}{3}$ **35.** 3, 8   **37.** $\pm\sqrt{11}, \approx \pm 3.32$

39. $-4 + 3\sqrt{2} \approx 0.24$

$-4 - 3\sqrt{2} \approx -8.24$

41. (1) Use the *table* feature in ASK mode with the variable equal to a solution.

(2) Use the scientific calculator portion of the graphing utility to evaluate the quadratic equation at a particular solution.

43. 15 feet × 27 feet   **45.** 200,000 units or 400,000 units

47. Two real solutions   **49.** $6 \pm \sqrt{6}$

51. $\dfrac{-19 \pm \sqrt{165}}{2}$ **53.** $-3 \pm 2\sqrt{3}$

55. 1.866, -0.283 **57.** 8.544, 0.162

59. Moon: ≈ 8.61 seconds

Earth: ≈ 3.54 seconds

61. $0, -1, 4$ **63.** $\pm 2, \pm 1$ **65.** $\frac{25}{4}$

67. No solution   **69.** $\pm 4\sqrt{2}$ **71.** $-3, \frac{7}{5}$

73. $2 \pm \sqrt{19}$ **75.** $900   **77.** $\approx 27.95\%$

79. $x < 11$

81. $-\frac{13}{2} < x < \frac{11}{2}$

83. $-12 < x < -8$

85. $x > 45$ units

87. $(-1, 3)$

89. $(-\infty, -3) \cup (0, 3)$

91. $\left(-\infty, \frac{6}{5}\right) \cup (4, \infty)$

93. $(-1.69, 1.69)$ **95.** $(1.65, 1.74)$ **97.** $[10, \infty)$

99. All real numbers   **101.** $(-\infty, 6] \cup [9, \infty)$

103. Between 3.65 and 4.72 seconds

105. (a) $-0.054x^2 + 1.43x < 8$

(b) $x < 8.03$ or $x > 18.45$

(c) No, 15 is not a solution of the inequality.

107. Between about 6.3 feet and about 23.7 feet.

109. Greater than 9.5%   **111.** $\$41.34 \le p \le \$58.66$

113. (a)

t	6	10	13	15
R	$1.58	$3.16	$5.18	$6.93

(b) Yes. The model predicts the revenue per share in 2007 to be $8.99.

(c) Yes. The model predicts that revenue per share will exceed $11.10 by 2009 $(t > 18.78)$.

CHAPTER TEST   (page 154)

1. $\frac{17}{23}$ **2.** (a) All real numbers   (b) $-3 \le x \le 3$

3. April: $325,786.00   **4.** $-\frac{5}{3}, \frac{1}{2}$ **5.** $4, -\frac{3}{2}$
May: $299,723.12

6. $\pm\sqrt{15}$ **7.** $\dfrac{-13 \pm \sqrt{69}}{2}$ **8.** $\dfrac{11 \pm \sqrt{145}}{6}$

9. 1.038, -0.446 **10.** $-\frac{7}{2}, \frac{13}{2}$

11. 4 (7 is extraneous.)   **12.** $-1, 1, -3, 3$ **13.** $-6, 6$

14. Selling either 341,421 units or 58,579 units will produce a revenue of $2,000,000.

15. $x < 3$

16. $x \le -4$ or $x \ge \frac{28}{5}$

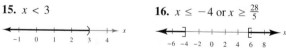

17. $(-11, -7)$

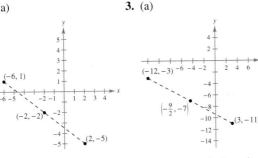

18. $(-\infty, -2] \cup [0, 2]$

19. Between 19,189 units and 143,311 units

20. 2008 $(t > 7.92)$

CUMULATIVE TEST: CHAPTERS 0–1
(page 155)

1. $-32x^6$ **2.** $3x^2\sqrt{2x}$ **3.** $\dfrac{3 + \sqrt{5}}{2}$

4. $\left(x + \sqrt{3}\right)\left(x - \sqrt{3}\right)(x - 6)$

5. $\dfrac{x + 4}{5}, x \neq 4$ **6.** $\dfrac{y - x}{x + y}, x \neq 0, y \neq 0$

7. (a) \$302.5 billion   (b) 2010 $(t \approx 9.83)$

8. $5, \frac{1}{2}$ **9.** $0.734, -1.022$ **10.** $\frac{8}{3}, -\frac{10}{3}$

11. $5 - 2\sqrt{2}$ **12.** $\pm 1, \pm 4$ **13.** $\pm 3\sqrt{2}$

14. $-3 < x < \frac{11}{3}$ **15.** $\left[-2\sqrt{2}, 0\right] \cup \left[2\sqrt{2}, \infty\right)$

16. $-\dfrac{16}{3} \leq x \leq \dfrac{26}{3}$

17. Between 10,263 units and 389,737 units

18. 2007 $(t > 7.48)$

CHAPTER 2

SECTION 2.1   *(page 167)*

Skills Review   *(page 167)*

1. 5   **2.** $3\sqrt{2}$ **3.** 1   **4.** -2

5. $3\left(\sqrt{2} + \sqrt{5}\right)$ **6.** $2\left(\sqrt{3} + \sqrt{11}\right)$ **7.** $-3, 11$

8. $9, 1$ **9.** $0, \pm 3$ **10.** ± 2

1. (a)   **3.** (a)

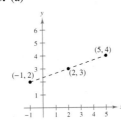

(b) 10   (c) $(-2, -2)$ (b) 17   (c) $\left(-\frac{9}{2}, -7\right)$

5. (a)

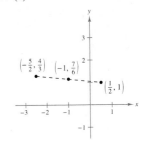

(b) $2\sqrt{10}$

(c) $(2, 3)$

7. (a)

(b) $\dfrac{\sqrt{82}}{3}$

(c) $\left(-1, \dfrac{7}{6}\right)$

9. (a)

(b) $\sqrt{47.65}$

(c) $(-0.35, 4.8)$

11. (a)

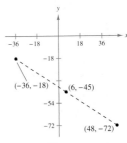

(b) $6\sqrt{277}$

(c) $(6, -45)$

13. 5   **15.** $\sqrt{109}$ **17.** $x = 15, -9$ **19.** $y = 9, -23$

21. (a) Yes   (b) No   **23.** (a) Yes   (b) Yes

25.

x	y
-2	-2.5
0	-1
1	-0.25
$\frac{4}{3}$	0
2	0.5

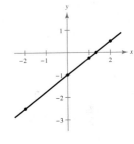

27. $\left(\frac{1}{2}, 0\right), (0, -1)$ **29.** $(-2, 0), (1, 0), (0, -2)$

31. $(0, 0), (-2, 0)$ **33.** $\left(\frac{4}{3}, 0\right), (0, 2)$

35. Every ordered pair on the x-axis has a y-coordinate of zero $[(x, 0)]$, so to find an x-intercept we let $y = 0$. Similarly, every ordered pair on the y-axis has an x-coordinate of zero $[(0, y)]$, so to find a y-intercept we let $x = 0$.

37. y-axis symmetry   **39.** x-axis symmetry

41. y-axis symmetry   **43.** Origin symmetry

45. Origin symmetry

47. x-axis, y-axis, and origin symmetry

49.

51.
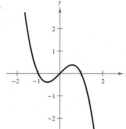

53. c   **54.** d   **55.** f   **56.** a   **57.** e   **58.** b

59.
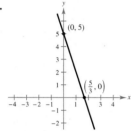

Intercepts:
$\left(\frac{5}{3}, 0\right)$, $(0, 5)$

Symmetry: none

61.
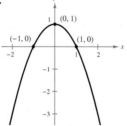

Intercepts:
$(-1, 0)$, $(1, 0)$, $(0, 1)$

Symmetry: y-axis

63.
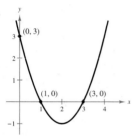

Intercepts:
$(3, 0)$, $(1, 0)$, $(0, 3)$

Symmetry: none

65.
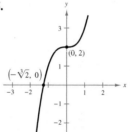

Intercepts:
$\left(-\sqrt[3]{2}, 0\right)$, $(0, 2)$

Symmetry: none

67.
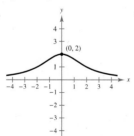

Intercept: $(0, 2)$

Symmetry: y-axis

69.
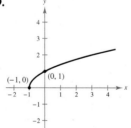

Intercepts: $(-1, 0)$, $(0, 1)$

Symmetry: none

71.
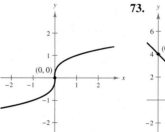

Intercept: $(0, 0)$

Symmetry: origin

73.
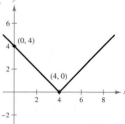

Intercepts: $(4, 0)$, $(0, 4)$

Symmetry: none

75.
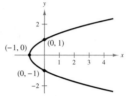

Intercepts: $(-1, 0)$, $(0, 1)$, $(0, -1)$

Symmetry: x-axis

77.
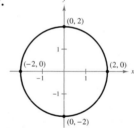

Intercepts: $(-2, 0)$, $(2, 0)$, $(0, 2)$, $(0, -2)$

Symmetry: x-axis, y-axis, origin

79. Radius: 2   **81.** Radius: $\sqrt{5}$

83. Center: $(3, 2)$; Radius: $\sqrt{5}$

85. $x^2 + y^2 = 9$ **87.** $(x + 4)^2 + (y - 1)^2 = 2$

89. $(x + 1)^2 + (y - 2)^2 = 5$

91. $(x - 1)^2 + (y - 1)^2 = 25$

93. $(x - 3)^2 + (y + 2)^2 = 16$

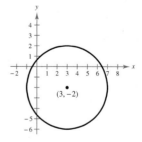

95. $(x - 2)^2 + (y + 3)^2 = 4$

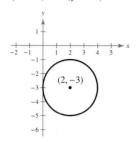

(2, -3)

97. $\left(x - \frac{1}{2}\right)^2 + \left(y - \frac{1}{2}\right)^2 = 2$

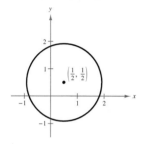

$\left(\frac{1}{2}, \frac{1}{2}\right)$

99. $\left(x + \frac{1}{2}\right)^2 + \left(y + \frac{5}{4}\right)^2 = \frac{9}{4}$

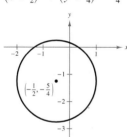

$\left(-\frac{1}{2}, -\frac{5}{4}\right)$

101. Center: $(3, -1)$; Radius: 5

$x^2 + y^2 - 6x + 2y - 15 = 0$

103. 610 dollars per fine ounce; 1980   **105.** $\approx 17\%$

107. (a) $(0, 377)$; It represents the population (in millions of people) of North America in 2000.

(b)

x	-20	-10	0	10	20
y	271	324	377	430	483

x	30	40	50
y	536	589	642

(c)

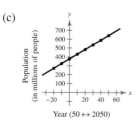

Year (50 ↔ 2050)

109. (a)

Year (6 ↔ 1996)

(b) 2006: $1.71; 2007: $1.76

The model's prediction for 2006 is close to Dollar Tree's prediction, but the model's prediction for 2007 is not close to Dollar Tree's prediction.

(c) The model does not support Dollar Tree's prediction. The model predicts an earnings per share of $1.80 in 2009. After 2009, the predicted values begin to decrease.

111. $x^2 + (y - 67.5)^2 = 4556.25$

SECTION 2.2   *(page 179)*

Skills Review   *(page 179)*

1. $-\frac{9}{2}$ **2.** $-\frac{13}{3}$ **3.** $-\frac{5}{4}$ **4.** $\frac{1}{2}$

5. $y = \frac{2}{3}x - \frac{5}{3}$ **6.** $y = -2x$

7. $y = 3x - 1$ **8.** $y = \frac{2}{3}x + 5$

9. $y = -2x + 7$ **10.** $y = x + 3$

1. 1   **3.** 0   **5.** -1 **7.** Positive

9.

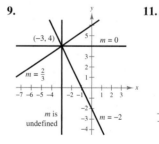

$m = 1$

11.

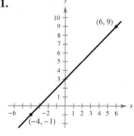

(6, 9)

$m = 1$

13.

$(-6, 4)$

$(-6, -1)$

m is undefined.

15.

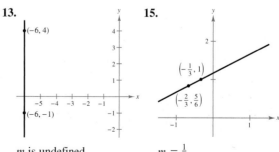

$\left(-\frac{1}{3}, 1\right)$

$\left(-\frac{2}{3}, \frac{5}{6}\right)$

$m = \frac{1}{2}$

17. Answers will vary.

Sample answer: $(3, -2), (-1, -2), (0, -2)$

19. Answers will vary.

Sample answer: $(2, -3), (2, -7), (2, 9)$

21. Answers will vary.

Sample answer: $(6, -5), (7, -4), (8, -3)$

23. Answers will vary.

Sample answer: $(2, 3), (-4, 0), (4, 4)$

25. $x - y - 7 = 0$ **27.** $4x + y + 8 = 0$

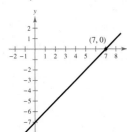   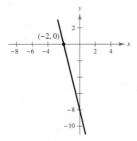

29. $2x + y = 0$ **31.** $x + 3y - 4 = 0$

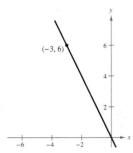   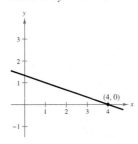

33. $x - 6 = 0$ **35.** $y + 7 = 0$

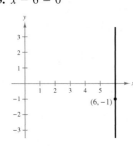   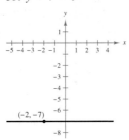

37. $8x - 6y - 17 = 0$ **39.** $m = 2, (0, -1)$

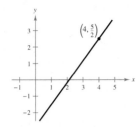   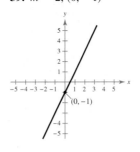

41. $m = 4, (0, -6)$ **43.** m is undefined; no y-intercept

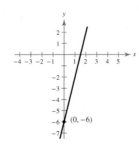   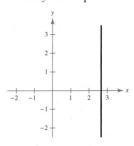

45. $m = -\frac{7}{6}, (0, 5)$ **47.** $m = 0, \left(0, \frac{7}{2}\right)$

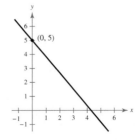   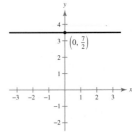

49. $3x - y - 1 = 0$ **51.** $x + 2y + 1 = 0$

53. $x + 9 = 0$ **55.** $y - 7 = 0$ **57.** $x + 2y - 3 = 0$

59. $2x - 5y + 1 = 0$

61. Answers will vary. Sample answer: You could graph a vertical line and pick two convenient points on the line to find the slope. Regardless of the points selected, the slope will have zero in the denominator. Division by zero is not possible, so the slope does not exist.

63. $4x - y - 4 = 0$ **65.** $x + y + 2 = 0$

67. $12x + 3y + 2 = 0$ **69.** Neither

71. Perpendicular   **73.** Parallel   **75.** Perpendicular

77. Parallel   **79.** Perpendicular   **81.** Neither

83. Parallel

85. (a) $2x - y - 10 = 0$ **87.** (a) $4x - 6y - 5 = 0$

(b) $x + 2y - 10 = 0$ (b) $36x + 24y + 7 = 0$

89. (a) $y = 0$

(b) $x + 1 = 0$

91. $F = \frac{9}{5}C + 32$ **93.** $A = 6.5t + 800$

95. Yes $\left(m = \frac{17}{180}\right)$ **97.** 4191 students

99. No; $179,000

101. $\approx$ $1968.4 million; No. If the actual yearly revenue followed a linear trend, then the yearly revenue in 2005 would be close to $1968.4 million.

103. $p = \frac{1}{33}d + 1$; $\frac{1}{33}$ atmosphere per foot

SECTION 2.3   *(page 189)*

Skills Review   *(page 189)*

1.
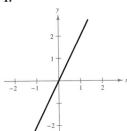

2.
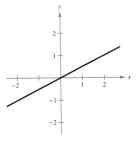

3.
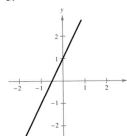

4.
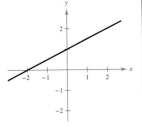

5. $y = x + 2$ **6.** $y = \frac{3}{2}x + 3$ **7.** $y = x + 2$

8. $y = \frac{6}{7}x + 4$ **9.** $5x + 40y - 213 = 0$

10. $29x + 60y - 448 = 0$

1.
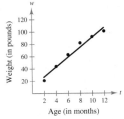

The model is a good fit for the actual data.

3.
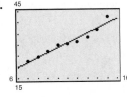

$y = 2.17x + 7.1$

The model is a good fit for the actual data.

5. $y = \dfrac{3}{8}x$ **7.** $y = 20x$ **9.** $y = \dfrac{3.2}{7}x$ or $y = \dfrac{16}{35}x$

11. $H = 3p$ **13.** $c = \frac{3}{5}d$ **15.** $I = 0.075P$

17. (a) $y = 0.0368x$ (b) $6808

19. (a) $C = \frac{33}{13}I$

(b)

Inches	5	10	20	25	30
Centimeters	12.69	25.38	50.77	63.46	76.15

21. $V = 125t + 1915, \quad 5 \le t \le 10$

23. $V = 30,400 - 2000t, \quad 5 \le t \le 10$

25. $V = 12,500t + 91,500, \quad 5 \le t \le 10$

27. (a) $h = 7000 - 20t$ (b) 2:13:50 P.M.

29. $V = 875 - 175t, \quad 0 \le t \le 5$ **31.** $S = 0.85L$

33. $W = 0.75x + 11.50$

35. (a) $P = 60t + 1300$ (b) 2020 deer

37. b; Slope $= -10$; The amount owed *decreases* by $10 per week.

39. a; Slope $= 0.48$; The amount received *increases* by $0.48 per mile driven.

41. No; Earning 10 points per coin would result in a positive slope.

43. Yes; Answers will vary.
Sample answer:

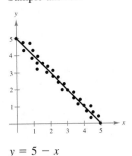

$y = 5 - x$

45. No

47. Yes; Answers will vary.
Sample answer:

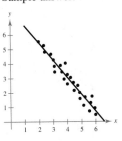

$y = -1.18x + 7.76$

49. (a)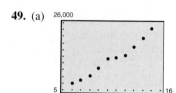

The data appear to be approximately linear.

(b) $A = 1779.3t - 3525$

(c) 1779.3; The amount spent on advertising increased by $1779.3 million each year.

(d) 2006: $24,943.8 million
2007: $26,723.1 million; Yes

51. (a)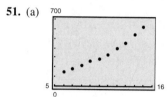

(b) Answers will vary.
Sample answer: $y = 52.88t - 203.9$

(c) $y = 50.49t - 187.2$

Using linear regression equation:

2006: $620.64 million
2007: $671.13 million

Using equation from part (b):

2006: $642.18 million
2007: $695.06 million

(d) The projections made by Sonic are higher than the predictions given by the models.

(e) No. Using the linear regression equation, the yearly revenue is expected to reach only about $873 million by 2011. Using the equation from part (b), the yearly revenue is expected to reach only about $907 million by 2011.

53. (a)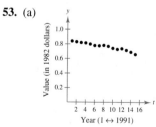

The data appear to be approximately linear.

(b) $y = -0.0115t + 0.840$

(c) 2007: 0.645, 2008: 0.633; Because the data followed a linear pattern from 1991 to 2005, you can assume that the estimates for 2007 and 2008 are reliable.

55. (a)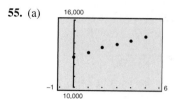

The data appear to be approximately linear.

(b) $E = 358.0t + 12,774$

(c) 2007: 15,280,000 employees
2009: 15,996,000 employees

(d)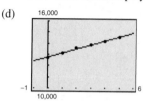

The predictions are most likely just about right because the model is a good fit for the actual data.

57. The model for population, because population tends to change at a consistent rate, whereas snowfall can be quite different year to year. You should use more than three data points to ensure the data can be represented accurately by a linear model.

SECTION 2.4   *(page 202)*

Skills Review   *(page 202)*

1. -73 **2.** 13   **3.** $2(x + 2)$ **4.** $-8(x - 2)$
5. $y = \frac{7}{5} - \frac{2}{5}x$ **6.** $y = \pm x$ **7.** $x \le -2, x \ge 2$
8. $-3 \le x \le 3$ **9.** All real numbers
10. $x \le 1, x \ge 2$

---

1. This is a function from A to B, because each element of A is matched with an element of B.

3. Not a function; The relationship does not match the element b of A with an element of B.

5. This is a function from A to B, because each element of A is matched with an element of B.

7. This is a function from A to B, because each element of A is matched with an element of B.

9. Not a function; The relationship assigns two elements of B to the element c of A.

11. Not a function from A to B; The relationship defines a function from B to A.

13. $\{(-2, 4), (-1, 1), (0, 0), (1, 1), (2, 4)\}$

15. $\{(-2, 0), (-1, 1), (0, \sqrt{2}), (1, \sqrt{3}), (2, 2)\}$

17. Not a function   **19.** Function   **21.** Function

23. Not a function   **25.** Function

27. (a) -6 (b) 34 (c) $6 - 4t$ (d) $2 - 4c$

29. (a) -1 (b) $\dfrac{1}{15}$ (c) $\dfrac{1}{t^2 - 2t}$ (d) $\dfrac{1}{t^2 - 1}$

31. (a) -1 (b) -9 (c) $2x - 5$ (d) $-\dfrac{5}{2}$

33. (a) 0 (b) 3 (c) $x^2 + 2x$ (d) -0.75

35. (a) 36π (b) 0 (c) $\dfrac{9\pi}{2}$ (d) $\dfrac{32\pi r^3}{3}$

37. (a) 1 (b) -7 (c) $3 - 2|x|$ (d) 2.5

39. (a) Undefined   (b) $-\dfrac{1}{16}$

   (c) $\dfrac{1}{y^2 + 4y - 12}, \quad y \neq -6, 2$

   (d) $\dfrac{1}{y^2 - 4y - 12}, \quad y \neq -2, 6$

41. (a) 1 (b) -1 (c) 1 (d) $\dfrac{|x - 1|}{x - 1}$

43. (a) -4 (b) 3 (c) -7 (d) 7

45. 5 **47.** ± 3 **49.** $0, 1, -1$

51. $\frac{10}{7}$ **53.** All real numbers x

55. All real numbers except $t = 0$

57. All real numbers y **59.** $-1 \leq x \leq 1$

61. All real numbers except $x = 0, -2$

63. All real numbers $x \geq -1$ except $x = 2$ **65.** $x > 0$

67. The domain of $f(x) = \sqrt{x - 2}$ is all real numbers $x \geq 2$, because an even root of a negative number is not a real number. The domain of $g(x) = \sqrt[3]{x - 2}$ is all real numbers. f and g have different domains because an odd root of a negative number is a real number, but an even root of a negative number is not a real number.

69. (a) $V = x(18 - 2x)^2$ (b) Domain: $0 < x < 9$

   (c) 400 cubic inches

71. (a) $C = 11.75x + 112,000$ (b) $R = 21.95x$

   (c) $10.2x - 112,000$

73. Yes $[y(30) = 6]$

75. 1995: \$37.55 billion
   2005: \$68.375 billion

77. (a)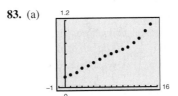

(b) Linear: $S = 135.24t - 923.3$

   Quadratic: $S = 12.843t^2 - 172.99t + 874.7$

(c)

Year	S (Actual)	S (Linear)	S (Quadratic)
1999	347.5	293.9	358.1
2000	438.3	429.1	429.1
2001	539.1	564.3	525.8
2002	652.0	699.6	648.2
2003	773.8	834.8	796.3
2004	969.2	970.1	970.1
2005	1177.6	1105.3	1169.5

The quadratic model is a better fit because its values for S are closest to the actual values of S.

79. (a) $C = 8000 + 2.95x$ (b) $\overline{C} = \dfrac{8000}{x} + 2.95$

(c)

x	100	1000	10,000	100,000
$\overline{C}$	82.95	10.95	3.75	3.03

   (d) Answers will vary. Sample answer: The average cost per unit decreases as x gets larger.

81. (a) $R = 12.00n - 0.05n^2, \quad n \geq 80$

(b)

n	90	100	110	120
R	\$675	\$700	\$715	\$720

n	130	140	150
R	\$715	\$700	\$675

   (c) Answers will vary. Sample answer: Maximum revenue of \$720 occurs when a group of 120 people charter a bus.

83. (a)

   (b) Linear: $D = 0.056t + 0.16$

   Quadratic: $D = 0.0015t^2 + 0.034t + 0.22$

(c) Linear   Quadratic

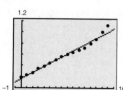   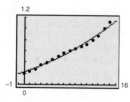

(d) The quadratic model is a better fit because its graph represents the actual values of D more closely than the linear model.

(e) 2006: $1.15, 2007: $1.23; The predictions given by the model are a little lower than the estimates given by Coca-Cola.

85. (a) Incorrect

(b) Correct

MID-CHAPTER QUIZ   *(page 207)*

1. (a)

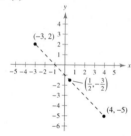

2. (a)

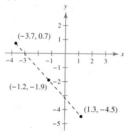

(b) $7\sqrt{2}$

(c) $\left(\frac{1}{2}, -\frac{3}{2}\right)$

(b) $\sqrt{52.04}$

(c) $(-1.2, -1.9)$

3. (a)

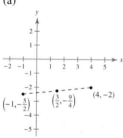

(b) $\dfrac{\sqrt{101}}{2}$

(c) $\left(\dfrac{3}{2}, -\dfrac{9}{4}\right)$

4. If the population follows a linear growth pattern, then the population will be 251,724 in 2009.

5. $2x - 3y + 9 = 0$

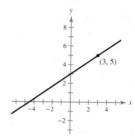

6. $y - 4 = 0$

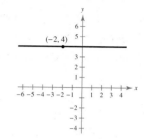

7. $x - 2 = 0$

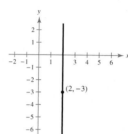

8. $2x + y + 9 = 0$

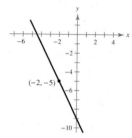

9.

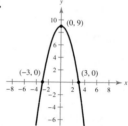

Intercepts: $(-3, 0), (3, 0), (0, 9)$

Symmetry: y-axis

10.

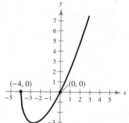

Intercepts: $(-4, 0), (0, 0)$

Symmetry: none

11.

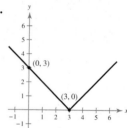

Intercepts: $(3, 0), (0, 3)$

Symmetry: none

12. $(x - 2)^2 + (y + 3)^2 = 16$

13. $x^2 + \left(y + \frac{1}{2}\right)^2 = 5$

14. $(x - 1)^2 + (y + 2)^2 = 9$

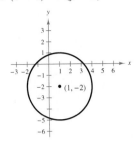

15. (a) 2   (b) -7 **16.** (a) 1   (b) -20

17. $x \geq 4$ **18.** All real numbers except $x = -2$

19.

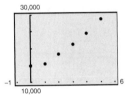

Linear: $C = 2808.4t + 13{,}542$

Quadratic: $C = 338.23t^2 + 1117.2t + 14{,}670$

20. Linear

2006: \$30,392,400,000

2007: \$33,200,800,000

Quadratic

2006: \$33,549,480,000

2007: \$39,063,670,000

21. $A = \dfrac{C^2}{4\pi}$

SECTION 2.5   *(page 215)*

Skills Review   *(page 215)*

1. 2   **2.** 0   **3.** $-\dfrac{3}{x}$ **4.** $x^2 + 3$ **5.** $0, \pm 4$

6. $\frac{1}{2}, 1$ **7.** All real numbers except $x = 4$

8. All real numbers except $x = 4, 5$ **9.** $t \leq \frac{5}{3}$

10. All real numbers

1. Domain: $[1, \infty)$; Range: $[0, \infty)$; 0

3. Domain: $(-\infty, \infty)$; Range: $(-\infty, 4]$; 4

5. Domain: $(-\infty, \infty)$; Range: $(-\infty, \infty)$; -1

7. Domain: $[-5, 5]$; Range: $[0, 5]$; 5

9. Function   **11.** Not a function

13. Increasing on $(-\infty, \infty)$; No change

15. Increasing on $(-\infty, 0)$ and $(2, \infty)$, decreasing on $(0, 2)$; behavior changes at $(0, 0)$ and $(2, -4)$.

17. Increasing on $(-1, 0)$ and $(1, \infty)$, decreasing on $(-\infty, -1)$ and $(0, 1)$; behavior changes at $(-1, -3)$, $(0, 0)$, and $(1, -3)$.

19. Increasing on $(-2, \infty)$, decreasing on $(-3, -2)$; behavior changes at $(-2, -2)$.

21.

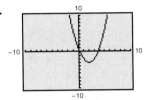

Minimum: $(2, -3)$

Increasing: $(2, \infty)$

Decreasing: $(-\infty, 2)$

23.

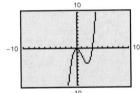

Relative maximum: $(0, 0)$

Relative minimum: $(2, -4)$

Increasing: $(-\infty, 0), (2, \infty)$

Decreasing: $(0, 2)$

25.

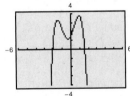

Relative maxima: $(-1.54, 3.29), (0.95, 3.77)$

Relative minimum: $(-0.34, 1.14)$

Increasing: $(-\infty, -1.54), (-0.34, 0.95)$

Decreasing: $(-1.54, -0.34), (0.95, \infty)$

27. Even   **29.** Odd   **31.** Odd

33. (a) 2   (b) 2   (c) -3 (d) -4

35. (a) 2   (b) 1   (c) -8 (d) -9

37.

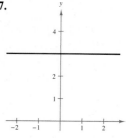

Even

39.

Neither even nor odd

41.

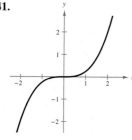

Odd

43.

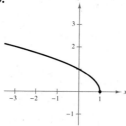

Neither even nor odd

45.

Neither even nor odd

47.

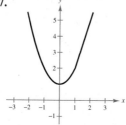

Neither even nor odd

49.

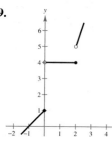

Neither even nor odd

51.

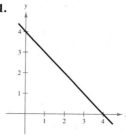

53.

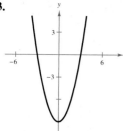

55.

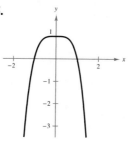

57.

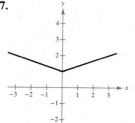

59.

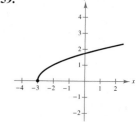

61.

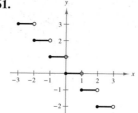

63.

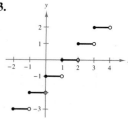

65. Maximum: $444.53 per ounce

Decreasing: 1995–2000

Increasing: 2000–2005

It is not realistic to assume that the price of gold will continue to follow this model, because the function decreases after the year 2005, and eventually yields negative values.

67.

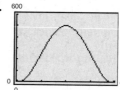

Increasing: 0 seconds to 2 seconds

Decreasing: 2 seconds to 4 seconds

Maximum change in volume: ≈ 501.9 milliliters

69. Maximum or minimum values may occur at endpoints.

71.

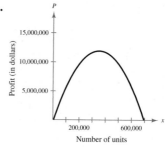

Approximately 350,000 units

73.

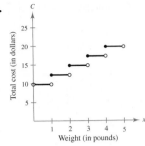

75.

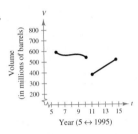

77. (a) $y = 25.215t^2 - 307.04t + 6263.8, \quad 5 \le t \le 11$

(b) $y = -301.450t^2 + 8763.99t - 56,490.3,$
$\quad 12 \le t \le 15$

(c)
$$y = \begin{cases} 25.215t^2 - 307.04t + 6263.8, & 5 \le t \le 11 \\ -301.450t^2 + 8763.99t - 56,490.3, & 12 \le t \le 15 \end{cases}$$

79. (a) Even; The graph is a reflection in the x-axis.

(b) Even; The graph is a reflection in the y-axis.

(c) Even; The graph is a vertical translation of f.

(d) Neither; The graph is a horizontal translation of f.

81. (a) $\left(\frac{5}{3}, -7\right)$

(b) $\left(\frac{5}{3}, 7\right)$

83. (a) $(-5, -1)$

(b) $(-5, 1)$

SECTION 2.6   *(page 225)*

Skills Review   *(page 225)*

1. 12   **2.** $\dfrac{-2x}{-x - 3}$ **3.** $0, \pm\sqrt{10}$ **4.** $\dfrac{4}{3}, -2$

5. $f(x) = -2$ **6.** $f(x) = -x$

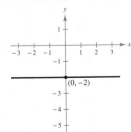

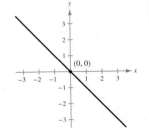

7. $f(x) = x + 5$

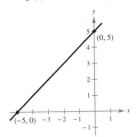

8. $f(x) = 2 - x$

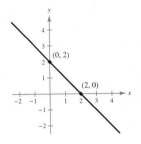

9. $f(x) = 3x - 4$

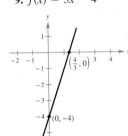

10. $f(x) = 9x + 10$

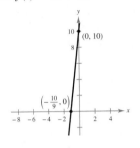

1. Shifted four units downward

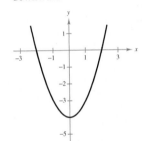

3. Shifted two units to the left

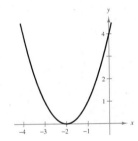

5. Shifted two units upward and two units to the right

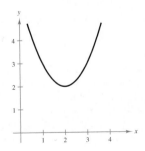

7. Reflected about the x-axis and shifted one unit upward

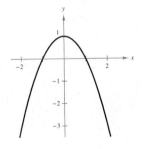

9. Shifted two units upward

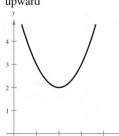

11. Shifted one unit to the right

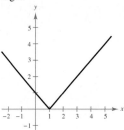

13. Reflected about the x-axis and shifted three units upward

15. Shifted one unit to the left and three units downward

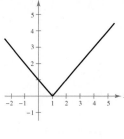

17. Shifted three units to the right

19. Shifted three units to the right and one unit upward

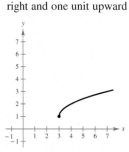

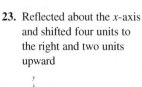

21. Vertically stretched by a factor of $\sqrt{2}$

23. Reflected about the x-axis and shifted four units to the right and two units upward

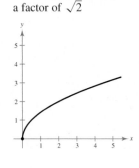

25. Reflected about the y-axis

27. Shifted one unit downward

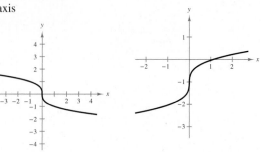

29. Reflected about the x-axis, shifted one unit to the left, and two units upward

31. Shifted one unit to the left and one unit downward

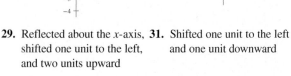

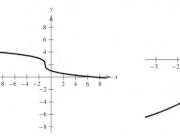

33. Vertically shrunk by a factor of $\frac{1}{2}$

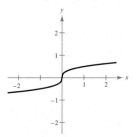

35. Common function: $y = x^3$

Transformation: shifted two units to the right

Equation: $y = (x - 2)^3$

37. Common function: $y = x^2$

Transformation: reflected about the x-axis

Equation: $y = -x^2$

39. Common function: $y = \sqrt{x}$

Transformation: reflected about the x-axis and shifted one unit upward

Equation: $y = -\sqrt{x} + 1$

41. (a) Vertical shift of two units

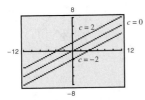

(b) Horizontal shift of two units

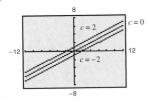

(c) Slope of the function changes

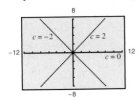

43. (a) $g(x) = (x - 1)^2 + 1$

(b) $g(x) = -(x + 1)^2$

45. (a)

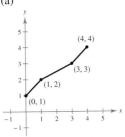

(b)

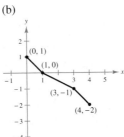

(c)

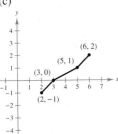

(d)

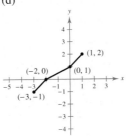

(e)

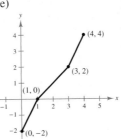

(f)

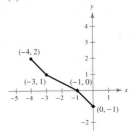

47. (a)

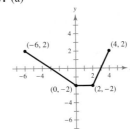

(b)

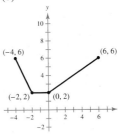

(c)

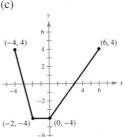

(d)

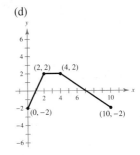

(e)

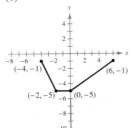

(f)

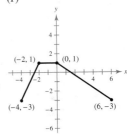

49. $y = x^3 - 2$ **51.** $y = 4x^3$

53. $g(x) = |x - 3| + 2$ **55.** $g(x) = -4|x|$

57. $h(x) = \sqrt{x - 4} - 3$ **59.** $h(x) = \frac{1}{2}\sqrt{x - 3}$

61. $g(x) = -x^3 + 3x^2 + 1$

63. Shifted one unit to the right and two units downward

$g(x) = (x - 1)^2 - 2$

65. (a)

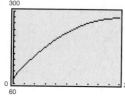

(b) $P(x) = 80 + 20x - 0.5x^2 - 25$

$P(x) = 55 + 20x - 0.5x^2, \quad 0 \le x \le 20$

Shifted 25 units down

(c) $P\left(\dfrac{x}{100}\right) = 80 + \dfrac{x}{5} - 0.00005x^2,$

$0 \le x \le 2000 \quad (x \text{ in dollars})$

Horizontal stretch

67.

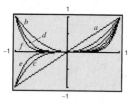

(a), (c), and (e) are odd functions; (b), (d), and (f) are even. Also, (a), (c), and (e) are increasing for all real numbers; (b), (d), and (f) are decreasing for all $x < 0$ and increasing for all $x > 0$.

SECTION 2.7   *(page 234)*

Skills Review   *(page 234)*

1. $\dfrac{1}{x(1-x)}$ **2.** $-\dfrac{12}{(x+3)(x-3)}$ **3.** $\dfrac{3x-2}{x(x-2)}$

4. $\dfrac{4x-5}{3(x-5)}$ **5.** $\dfrac{\sqrt{x^2-1}}{x+1}, \ x \ne 1$

6. $\dfrac{x+1}{x(x+2)}, \ x \ne 2$ **7.** $5(x-2), \ x \ne -2$

8. $\dfrac{x+1}{(x-2)(x+3)}, \ x \ne -5, -1, 0$

9. $\dfrac{1+5x}{3x-1}, \ x \ne 0$ **10.** $\dfrac{x+4}{4x}, \ x \ne 4$

1.

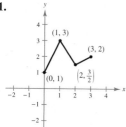

3.

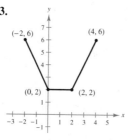

5. (a) $2x$ (b) 2 (c) $x^2 - 1$

    (d) $\dfrac{x+1}{x-1}$; Domain: $(-\infty, 1) \cup (1, \infty)$

7. (a) $x^2 - x + 1$ (b) $x^2 + x - 1$ (c) $x^2 - x^3$

    (d) $\dfrac{x^2}{1-x}$; Domain: $(-\infty, 1) \cup (1, \infty)$

9. (a) $x^2 + \sqrt{1-x} + 5$ (b) $x^2 - \sqrt{1-x} + 5$

    (c) $x^2\sqrt{1-x} + 5\sqrt{1-x}$

    (d) $\dfrac{x^2+5}{\sqrt{1-x}}$; Domain: $(-\infty, 1)$

11. (a) $\dfrac{x+1}{x^2}$ (b) $\dfrac{x-1}{x^2}$ (c) $\dfrac{1}{x^3}$

    (d) $x, x \ne 0$; Domain: $(-\infty, 0) \cup (0, \infty)$

13. 14 **15.** $-4t^2 + 4t + 3$ **17.** -6

19. $\frac{11}{23}$ **21.** 3 **23.** -6

25. (a) $6x + 15$ (b) $6x + 5$ (c) $9x$

27. (a) $9x^2 + 6x + 1$ (b) $3x^2 + 1$ (c) x^4

29. (a) $\sqrt{x^2 + 4}$ (b) $x + 4, x \ge -4$

31. (a) $x - \frac{8}{3}$ (b) $x - 8$ **33.** (a) $\sqrt[4]{x}$ (b) $\sqrt[4]{x}$

35. (a) $|x + 6|$ (b) $|x| + 6$

37. (a) All real numbers, or $(-\infty, \infty)$

    (b) $x \ge 0$, or $[0, \infty)$ (c) $x \ge 0$, or $[0, \infty)$

39. (a) All real numbers except $x = 0$, or $(-\infty, 0) \cup (0, \infty)$

    (b) All real numbers, or $(-\infty, \infty)$

    (c) All real numbers except $x = 2$, or $(-\infty, 2) \cup (2, \infty)$

41. (a) 3 (b) 0 **43.** (a) 0 (b) 4

45. Answers will vary.

    Sample answer: $f(x) = x^2, g(x) = 2x + 1$

47. Answers will vary.

    Sample answer: $f(x) = \sqrt[3]{x}, g(x) = x^2 - 4$

49. Answers will vary.

    Sample answer: $f(x) = \dfrac{1}{x}, g(x) = x + 2$

51. Answers will vary.

    Sample answer: $f(x) = x^2 + 2x, g(x) = x + 4$

53. $T = \frac{3}{4}x + \frac{1}{15}x^2$

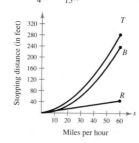

55. $(C \circ x)(t) = 1500t + 495$

$C \circ x$ represents the cost of producing x units in t hours.

57. $R_1 + R_2 = 917 - 6.7t$,
$t = 0, 1, 2, 3, 4, 5, 6, 7, 8$

Total sales were decreasing.

59. (a) $y_1 = 4.95t + 1.6$, $y_2 = 3.72t - 0.9$

(b)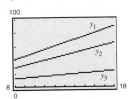

$y_3 = 1.23t + 2.5$;

y_3 represents the profits for 1998 through 2008;
$27,100

(c)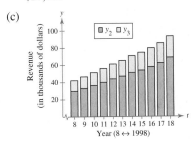

The heights of the bars represent the revenues for
1998 through 2008.

61. (a) $N(T(t)) = 100t^2 + 275$

(b) Approximately 2.18 hours

63. (a) $R(p) = p - 2000$ (b) $S(p) = 0.9p$

(c) $(R \circ S)(p) = 0.9p - 2000$; $(R \circ S)(p)$ represents the
factory rebate after the dealership discount.

$(S \circ R)(p) = 0.9(p - 2000)$; $(S \circ R)(p)$ represents the
dealership rebate after the factory discount.

(d) $(R \circ S)(20,500) = \$16,450$

$(S \circ R)(20,500) = \$16,650$

$16,450 is the lower cost because 10% of the price of
the car is larger than $2000.

65.

Year	2001	2002	2003	2004	2005
P/E	12.9	12.1	9.5	11.1	14.6

67. False. $(f \circ g)(x) = 6x + 1$ and $(g \circ f)(x) = 6x + 6$

69. Answers will vary.

SECTION 2.8   (page 245)

Skills Review   (page 245)

1. All real numbers   **2.** $[-1, \infty)$

3. All real numbers except $x = 0, 2$

4. All real numbers except $x = -\frac{5}{3}$ **5.** x

6. x **7.** x **8.** x **9.** $x = \frac{3}{2}y + 3$

10. $x = \dfrac{y^3}{2} + 2$

1. $f^{-1} = \{(4, 1), (5, 2), (6, 3), (7, 4)\}$

3. $f^{-1} = \{(1, -1), (2, -2), (3, -3), (4, -4)\}$

5. $f^{-1}(x) = \frac{1}{2}x$ **7.** $f^{-1}(x) = x + 5$

9. (a) $5\left(\dfrac{x - 1}{5}\right) + 1 = x$; $\dfrac{(5x + 1) - 1}{5} = x$

(b)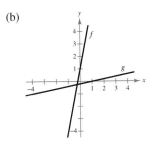

11. (a) $\left(\sqrt[3]{x}\right)^3 = x$; $\sqrt[3]{x^3} = x$

(b)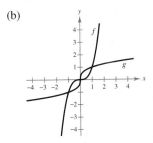

13. (a) $\sqrt{(x^2 + 4) - 4} = x$; $\left(\sqrt{x - 4}\right)^2 + 4 = x$

(b)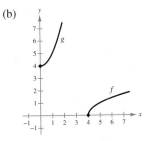

15. (a) $1 - \left(\sqrt[3]{1 - x}\right)^3 = x$; $\sqrt[3]{1 - (1 - x^3)} = x$

(b)

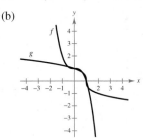

17.

x	0	1	2	3	4
$f^{-1}(x)$	-2	0	1	2	4

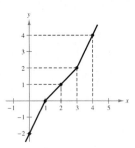

19.

x	-2	0	2	3
$f^{-1}(x)$	-4	-3	3	4

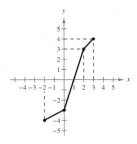

21. f doesn't have an inverse function.

23. $g^{-1}(x) = 8x$

25. p doesn't have an inverse function.

27. $f^{-1}(x) = \sqrt{x} - 3$, $x \geq 0$

29. $h^{-1}(x) = \dfrac{1}{x}$ **31.** $f^{-1}(x) = \dfrac{x^2 - 3}{2}$, $x \geq 0$

33. g doesn't have an inverse function.

35. $f^{-1}(x) = -\sqrt{25 - x}$, $x \leq 25$

37. Error: f^{-1} does not mean to take the reciprocal of $f(x)$.

39. $f^{-1}(x) = \dfrac{x + 3}{2}$ **41.** $f^{-1}(x) = \sqrt[5]{x}$

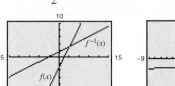
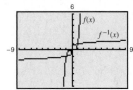

43. $f^{-1}(x) = x^2$, $x \geq 0$

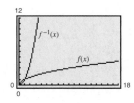

45. $f^{-1}(x) = \sqrt{16 - x^2}$, $0 \leq x \leq 4$

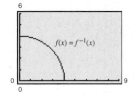

47. $f^{-1}(x) = x^3 - 2$

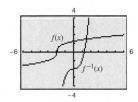

49. Because f is one-to-one, f has an inverse function.

51. f doesn't have an inverse function because two x-values share the same y-value.

53. g has an inverse function.

55. h doesn't have an inverse function.

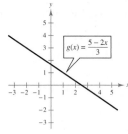
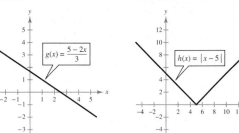

57. f doesn't have an inverse function.

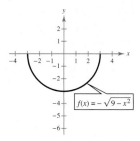

$f(x) = -\sqrt{9 - x^2}$

59. 32   **61.** 600

63. $(g^{-1} \circ f^{-1})(x) = \dfrac{x + 1}{2}$ **65.** $(f \circ g)^{-1}(x) = \dfrac{x + 1}{2}$

67. $C^{-1}(x) = \dfrac{x - 1500}{7.5}$; C^{-1} computes the number of T-shirts

that can be made for a cost of x.

Domain of C: $[0, \infty)$

Domain of C^{-1}: $[1500, \infty)$

69. (a)

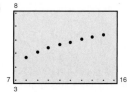

(b) $y = 0.235t + 2.95$

(c) $y^{-1} = \dfrac{t - 2.95}{0.235}$; y^{-1} represents the year in which the

average admission price is t dollars.

(d) 2012

71. After graphing $f(x) = x^2 + 1$, $x \geq 0$, and $f^{-1}(x) = \sqrt{x - 1}$,

it is observed that $f(x)$ and $f^{-1}(x)$ are reflections of each

other about the line $y = x$. Because of this reflection, inter-

changing the roles of x and y seems reasonable.

73. $f^{-1}(x) = \sqrt[3]{\dfrac{x^2 - 0.008}{0.0161}}$; $2.40

REVIEW EXERCISES   (page 250)

1. (a)

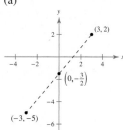

(b) $\sqrt{85} \approx 9.22$

(c) $\left(0, -\frac{3}{2}\right)$

3. (a)

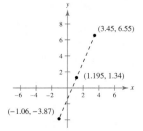

(b) $\sqrt{128.9165} \approx 11.35$

(c) $(1.195, 1.34)$

5. $x = -10$ or 30 **7.** (a) Yes   (b) Yes

9.

x	y
-2	3
0	2
2	1
3	$\frac{1}{2}$
4	0

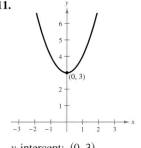

$y = -\frac{1}{2}x + 2$

11.

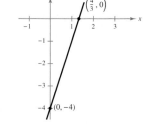

y-intercept: $(0, 3)$

Symmetry: y-axis

13.

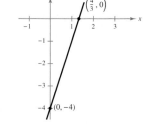

x-intercept: $\left(\frac{4}{3}, 0\right)$

y-intercept: $(0, -4)$

Symmetry: none

15.

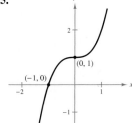

x-intercept: $(-1, 0)$

y-intercept: $(0, 1)$

Symmetry: none

17. $(x + 1)^2 + (y - 2)^2 = 36$

19. $(x - 2)^2 + (y + 3)^2 = 25$

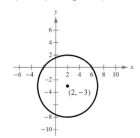

21.

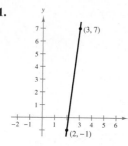

$m = 8$

23.

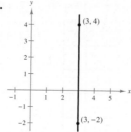

m is undefined.

25. $3x - 2y - 10 = 0$ **27.** $y = 6$

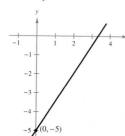

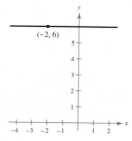

29. Slope: $\frac{5}{4}$;

y-intercept: $\left(0, \frac{11}{4}\right)$

31. Slope: undefined;

y-intercept: none

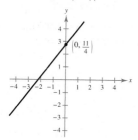

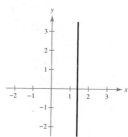

33. Parallel   **35.** Neither

37. (a) $5x - 4y - 23 = 0$ (b) $4x + 5y - 2 = 0$

39. (a) $y = -2$ (b) $x = -1$

41. $y = \frac{7}{3}x$ **43.** $y = 348x$ **45.** $A = 5r$

47. $a = \frac{3}{4}b$ **49.** $y = 0.0365x$; \$3723   **51.** \$336,000

53. $V = -12,950t + 135,000,\ 0 \le t \le 10$

55. y is a function of x.   **57.** y is a function of x.

59. Function; Every element of A is assigned to an element of B.

61. (a) -2 (b) -3 (c) -5 (d) $\sqrt{x + 7} - 5$

63. All real numbers x **65.** $x \ge -5$

67. $1 \le t < 4,\ t > 4$

69. The domain of $h(x)$ is all real numbers except $x = 0$, because division by zero is undefined. The domain of $k(x)$ is all real numbers except $x = -2$ and $x = 2$, because if $x = 2$ or $x = -2$, then $x^2 - 4$ equals zero, and division by zero is undefined. When a graphing utility and the *table* feature are used, $h(0)$ results in an error and $k(-2)$ or $k(2)$ also result in an error.

71. (a) $B = 6500\left(1 + \dfrac{0.0685}{4}\right)^{4t}$ (b) $t \ge 0$

73. (a) Domain: all real numbers

   Range: $[1, \infty)$

(b) Decreasing: $(-\infty, 0)$

   Increasing: $(0, \infty)$

(c) Even   (d) Minimum: $(0, 1)$

75. (a) Domain: all real numbers

   Range: all real numbers

(b) Decreasing: $\left(0, \frac{8}{3}\right)$

   Increasing: $(-\infty, 0) \cup \left(\frac{8}{3}, \infty\right)$

(c) Neither

(d) Relative minimum: $\left(\frac{8}{3}, -\frac{256}{27}\right)$

   Relative maximum: $(0, 0)$

77. y is a function of x.   **79.** y is not a function of x.

81. y is a function of x.

83.

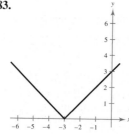

85.

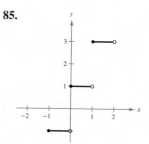

87.

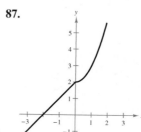

89.

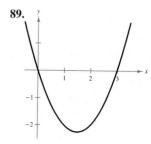

91.

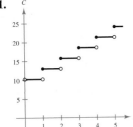

93. Reflected about the
x-axis and shifted two
units downward and
one unit to the right

95. Reflected about the
x-axis and shifted two
units to the right

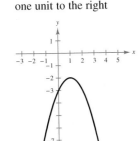

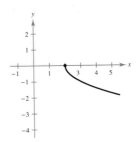

97. Shifted two units to the left

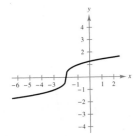

99. Common function: $y = \sqrt{x}$
Transformation: reflected about the y-axis and shifted
three units to the right
Equation: $y = \sqrt{3 - x}$ or $y = \sqrt{-(x - 3)}$

101. $(f + g)(x) = x^2 + 5x - 1$
$(f - g)(x) = -x^2 + x - 1$
$(fg)(x) = 3x^3 + 5x^2 - 2x$
$(f/g)(x) = \dfrac{3x - 1}{x^2 + 2x}$
Domain of f/g: $x < -2$, $-2 < x < 0$, $x > 0$, or
$(-\infty, -2) \cup (-2, 0) \cup (0, \infty)$

103. 9   **105.** 18

107. (a) $x^2 + 6x + 9$
Domain: All real numbers
(b) $x^2 + 3$
Domain: All real numbers

109. (a) $\dfrac{1}{3x + x^2}$
Domain: All real numbers except $x = 0$ and $x = -3$
(b) $\dfrac{3}{x} + \dfrac{1}{x^2}$, or $\dfrac{3x + 1}{x^2}$
Domain: All real numbers except $x = 0$

111. Answers will vary.
Sample answer: $f(x) = x^2$, $g(x) = 6x - 5$

113. Answers will vary.
Sample answer: $f(x) = \dfrac{1}{x^2}$, $g(x) = x - 1$

115. $R_1 + R_2 = 800.5 - 0.92t + 0.2t^2$, $t = 0, 1, 2, \ldots, 8$

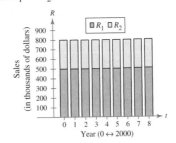

Total sales are increasing.

117. $g(f(x))$; The bonus is based on sales over \$500,000, and
is calculated by multiplying $x - 500,000$ by 0.03.

119. $f(g(x)) = 3\left(\dfrac{x - 5}{3}\right) + 5 = x$

$g(f(x)) = \dfrac{3x + 5 - 5}{3} = x$

$f(g(x)) = x = g(f(x))$, so f and g are inverse functions
of each other.

121. $f(x)$ does not have an inverse function.

123. $f^{-1}(x) = \dfrac{1}{x}$
(f is its own inverse function.)

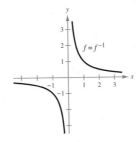

125. (a) $f^{-1}(x) = 2(x + 3) = 2x + 6$

(b)

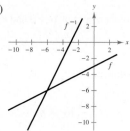

(c) $f^{-1}(f(x)) = 2\left(\frac{1}{2}x - 3\right) + 6$

$= x$

$f(f^{-1}(x)) = \frac{1}{2}(2x + 6) - 3$

$= x$

127. (a) $f^{-1}(x) = \sqrt{x}, x \geq 0$

(b)

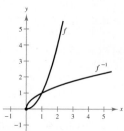

(c) $f^{-1}(f(x)) = \sqrt{x^2} = x, x \geq 0$

$f(f^{-1}(x)) = \left(\sqrt{x}\right)^2 = x, x \geq 0$

129. (a)

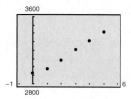

(b) $A = 103.5t + 2899$

(c) $A^{-1} = \dfrac{t - 2899}{103.5}$; 2007

CHAPTER TEST   *(page 255)*

1. Distance: $4\sqrt{5}$ **2.** Distance: ≈ 7.81

Midpoint: $(1, 0)$ Midpoint: $(0.44, 4.335)$

3. x-intercepts: $(-5, 0), (3, 0)$

y-intercept: $(0, -15)$

4. Symmetric with respect to the origin

5. $2x - 3y + 21 = 0$

6. $(x - 3)^2 + (y + 2)^2 = 16$

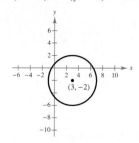

7. True. Each value of x corresponds to exactly one value of y.

8. False. The element -9 is not included in set B.

9. (a) Domain: All real numbers

Range: $(-\infty, 2]$

(b) Decreasing: $(0, \infty)$

Increasing: $(-\infty, 0)$

(c) Even

(d) Maximum: $(0, 2)$

10. (a) Domain: $(-\infty, -2] \cup [2, \infty)$

Range: $[0, \infty)$

(b) Decreasing: $(-\infty, -2)$

Increasing: $(2, \infty)$

(c) Even

(d) Minima: $(-2, 0), (2, 0)$

11.   **12.**

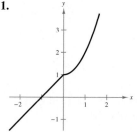 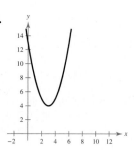

13. $(f - g)(x) = x^2 - 2x + 3$

14. $(fg)(x) = 2x^3 - x^2 + 4x - 2$

15. $(f \circ g)(x) = 4x^2 - 4x + 3$

16. $g^{-1}(x) = \frac{1}{2}x + \frac{1}{2}$

17. $V = 30,000 - 5200t$

18.

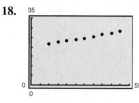

$P = 0.17t + 19.6$

CHAPTER 3

SECTION 3.1   *(page 265)*

Skills Review   *(page 265)*

1. $\frac{1}{2}, -6$ **2.** $-\frac{3}{5}, 3$ **3.** $\frac{3}{2}, -1$ **4.** -10

5. $3 \pm \sqrt{5}$ **6.** $-2 \pm \sqrt{3}$ **7.** $4 \pm \dfrac{\sqrt{14}}{2}$

8. $-5 \pm \dfrac{\sqrt{3}}{3}$ **9.** $-\dfrac{3}{2} \pm \dfrac{\sqrt{5}}{2}$ **10.** $-\dfrac{3}{2} \pm \dfrac{\sqrt{21}}{2}$

---

1. g   **2.** e   **3.** c   **4.** f

5. b   **6.** a   **7.** h   **8.** d

9. $y = -(x + 2)^2$ **11.** $y = (x - 3)^2 - 9$

13. $y = -2(x + 3)^2 + 3$

15. Compared with the graph of $y = x^2$, each output of $f(x) = 5x^2$ vertically stretches the graph by a factor of 5.

17. The graph of f is the graph of $y = x^2$ reflected in the x-axis, shifted to the left 1 unit and shifted upward 1 unit.

19. Intercept: $(0, 0)$ **21.** Intercepts: $(\pm 4, 0), (0, 16)$

Vertex: $(0, 0)$ Vertex: $(0, 16)$

 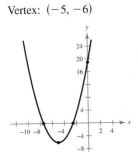

23. Intercepts: **25.** Intercepts: $(-1, 0), (0, 1)$

$\left(-5 \pm \sqrt{6}, 0\right), (0, 19)$ Vertex: $(-1, 0)$

Vertex: $(-5, -6)$

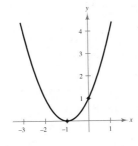

27. Intercepts: **29.** Intercept: $\left(0, \frac{5}{4}\right)$

$(1, 0), (-3, 0), (0, 3)$

Vertex: $(-1, 4)$ Vertex: $\left(\frac{1}{2}, 1\right)$

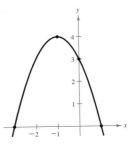 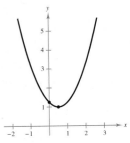

31. Intercepts: **33.** Intercept: $(0, 21)$

$\left(1 \pm \sqrt{6}, 0\right), (0, 5)$

Vertex: $(1, 6)$ Vertex: $\left(\frac{1}{2}, 20\right)$

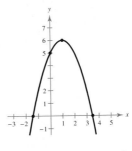 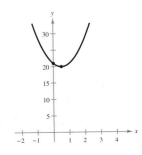

35. Intercepts: $\left(8 \pm 4\sqrt{2}, 0\right), (0, 8)$

Vertex: $(8, -8)$

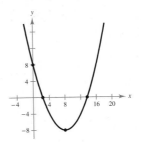

37. $y = -\frac{1}{2}(x - 2)^2 - 1$ **39.** $y = \frac{3}{4}(x - 5)^2 + 12$

41. Answers will vary.   **43.** Answers will vary.
Sample answer: Sample answer:

$f(x) = x^2 - x - 2$ $f(x) = x^2 - 10x$

$g(x) = -x^2 + x + 2$ $g(x) = -x^2 + 10x$

45. Answers will vary.
Sample answer:

$f(x) = 2x^2 + 7x + 3$

$g(x) = -2x^2 - 7x - 3$

47. $A = 100x - x^2$; $(50, 2500)$; The rectangle has the greatest area ($A = 2500$ square feet) when its width is 50 feet.

49. $x = 150$ feet, $y = 200$ feet; 300 feet × 200 feet

51. 25,000 units   **53.** 20 fixtures   **55.** 14 feet

57. (a)

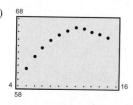

(b) $S = -0.132t^2 + 3.09t + 48.5$

(c)

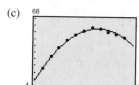

(d) 2002($t \approx 11.70$); No, according to the actual data, the year in which the number of basic cable subscribers was the greatest was 2001.

59. (a)

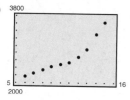

(b) $F = 19.53t^2 - 264.5t + 3149$

(c)

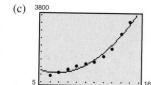

(d) \$4715.72

61. (a)

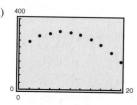

(b) $p(x) = -1.276x^2 + 21.63x + 236.5$

(c)

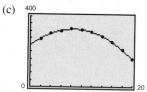

(d) $\approx (8.5, 328.2)$; Producing about 85,000 units yields the maximum profit, about \$32,820,000.

(e) Sample answer: Production costs may be growing faster than revenue, so profit decreases.

63. $f(x) = a\left(x + \dfrac{b}{2a}\right)^2 + \dfrac{4ac - b^2}{4a}$

SECTION 3.2   *(page 276)*

Skills Review   *(page 276)*

1. $(3x - 2)(4x + 5)$ **2.** $x(5x - 6)^2$

3. $z^2(12z + 5)(z + 1)$ **4.** $(y + 5)(y^2 - 5y + 25)$

5. $(x + 3)(x + 2)(x - 2)$ **6.** $(x + 2)(x^2 + 3)$

7. No real solution   **8.** $3 \pm \sqrt{5}$

9. $-\frac{1}{2} \pm \sqrt{3}$ **10.** ± 3

1. e   **2.** c   **3.** g   **4.** d   **5.** f   **6.** h

7. a   **8.** b

9.

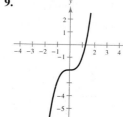

11.

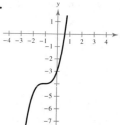

13.

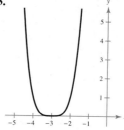

15.

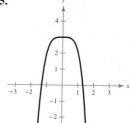

17. Rises to the left
Falls to the right

19. Rises to the left
Falls to the right

21. Rises to the left
Rises to the right

23. Rises to the left
Rises to the right

25. Falls to the left
Falls to the right

27. (a) 1   (b) 2   **29.** (a) 4   (b) 5

31. ± 3 **33.** -4 **35.** 1, -2 **37.** No real zeros

39. 2, 0   **41.** ± 1 **43.** $\pm\sqrt{5}$ **45.** 3

47.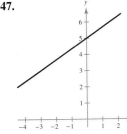

49.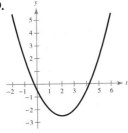

51.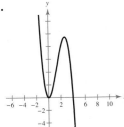

53.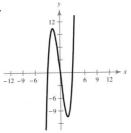

55.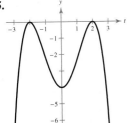

57.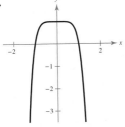

59. Answers will vary. Sample answers:

$a_n < 0$ $a_n > 0$

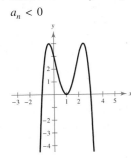   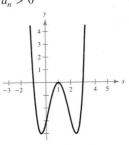

61. $f(x) = \frac{1}{2}(x^4 - 11x^3 + 28x^2)$

63. (a) 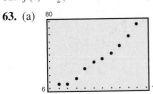   (b) Positive

(c) $C = 0.02025t^4 - 0.9103t^3 + 15.124t^2$
$- 106.31t + 309.0$

The model agrees with the prediction from part (b).

(d)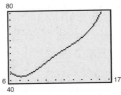

2007($t \approx 17.03$); The prediction does not seem reasonable. Through 2006, the number increased by at most 7 million in a year. To reach 92 million in 2007, the number would have had to increase by 14 million from 2006.

65. (200, 320)

67. Answers will vary; Domain: $0 < x < 6$

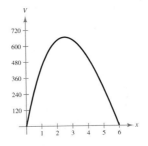

$x \approx 2.5$

69.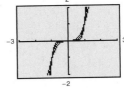

The functions have a common shape because their degrees are odd, but their graphs are not identical because they have different degrees.

SECTION 3.3   *(page 286)*

Skills Review   *(page 286)*

1. $x^3 - x^2 + 2x + 3$ **2.** $2x^3 + 4x^2 - 6x - 4$

3. $x^4 - 2x^3 + 4x^2 - 2x - 7$

4. $2x^4 + 12x^3 - 3x^2 - 18x - 5$

5. $(x - 3)(x - 1)$ **6.** $8(x + 2)(x - 5)$

7. $(3x + 5)(x - 1)$ **8.** $(3x - 4)^2$

9. $2x(x - 1)(2x - 3)$ **10.** $x(3x + 2)(2x + 1)$

1. $3x - 4$ **3.** $2x + 4$ **5.** $x + 3$

7. $x^3 + 3x^2 - 1$ **9.** $7 - \dfrac{25}{x + 4}$

11. $3x + 5 - \dfrac{2x - 3}{2x^2 + 1}$ **13.** $x + \dfrac{x - 27}{x^2 - 1}$

15. $x^2 - 6x + 17 - \dfrac{36}{x + 2}$

17. $2x^3 + 4x^2 - 2x - 8 - \dfrac{10x - 7}{x^2 - 2x + 1}$

19. $2x^2 - 3x + 5$ **21.** $4x^2 - 9$ **23.** $-x^2 + 10x - 25$

25. $x^3 - 7x^2 + 14x - 20 + \dfrac{84}{x + 3}$

27. $10x^3 + 10x^2 + 60x + 360 + \dfrac{1360}{x - 6}$

29. $2x^4 + 8x^3 + 2x^2 + 8x - 5 - \dfrac{7}{x - 4}$

31. $-3x^3 - 6x^2 - 12x - 24 - \dfrac{48}{x - 2}$

33. $-x^2 + 3x - 6 + \dfrac{11}{x + 1}$ **35.** $4x^2 + 14x - 30$

37. $f(x) = (x - 3)(x^2 + 4x) + 20;\ f(3) = 20$

39. $f(x) = \left(x - \tfrac{1}{3}\right)(3x^2 + 3x + 6);\ f\left(\tfrac{1}{3}\right) = 0$

41. $f(x) = \left(x - \sqrt{3}\right)\left[x^2 + \left(2 + \sqrt{3}\right)x + 2\sqrt{3}\right] - 6;$
$f\left(\sqrt{3}\right) = -6$

43. $f(x) = \left(x - 1 - \sqrt{3}\right)\left[2x^2 + \left(3 + 2\sqrt{3}\right)x\right.$
$\left. + \left(-5 + 5\sqrt{3}\right)\right];\ f\left(1 + \sqrt{3}\right) = 0$

45. (a) -69 (b) -2081 (c) -6 (d) 446

47. (a) 6 (b) -27 (c) -6.168 (d) 37

49. (a) 1 (b) -267 (c) $-\tfrac{11}{3}$ (d) -3.8

51. (a) Proof   (b) $x + 2$
(c) $f(x) = (x - 4)(x + 2)^2$ (d) $4, -2$
(e)

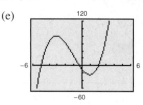

53. (a) Proof   (b) $x + 5$
(c) $f(x) = (3x + 1)(x - 2)(x + 5)$ (d) $-\tfrac{1}{3}, 2, -5$
(e)

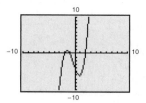

55. (a) Proof   (b) $x + \sqrt{3}$
(c) $f(x) = \left(x - \sqrt{3}\right)(x + 2)\left(x + \sqrt{3}\right)$
(d) $\sqrt{3}, -2, -\sqrt{3}$
(e)

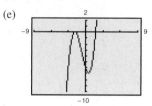

57. The second polynomial is a factor of the first polynomial.

59. e; $3, \dfrac{-1 \pm \sqrt{17}}{2}$ **61.** a; $-1, -2 \pm \sqrt{2}$

63. b; $-3, \pm\sqrt{5}$

65. Answers will vary. Sample answer:
$f(x) = 3x^3 - 13x^2 + 4x + 20$
$f(x) = -3x^3 + 13x^2 - 4x - 20$
Infinitely many polynomial functions

67. $x^2 - 7x + 10$ **69.** $3x^2 - x - 10$ **71.** $x^2 - 6x$

73. $x^2 + 4x + 3$ **75.** $x^2 + 10x + 24$ square feet

77. (a) \$199,978   (b) Proof

79. The remainder is 0.   **81.** $c = -210$

SECTION 3.4   *(page 298)*

Skills Review   *(page 298)*

1. $f(x) = 3x^3 - 8x^2 - 5x + 6$

2. $f(x) = 4x^4 - 3x^3 - 16x^2 + 12x$

3. $x^4 - 3x^3 + 5 + \dfrac{3}{x + 3}$

4. $3x^3 + 15x^2 - 9 - \dfrac{2}{x + (2/3)}$

5. $\tfrac{1}{2}, -3 \pm \sqrt{5}$ **6.** $10, -\tfrac{2}{3}, -\tfrac{3}{2}$ **7.** $-\tfrac{3}{4}, 2 \pm \sqrt{2}$

8. $\tfrac{2}{5}, -\tfrac{7}{2}, -2$ **9.** $\pm\sqrt{2}, \pm 1$ **10.** $\pm 2, \pm\sqrt{3}$

1. Possible: $\pm 1, \pm 2, \pm 4$
Actual: $-1, \pm 2$

3. $-3, \frac{1}{2}, 4$ **5.** $\pm\frac{1}{2}, \pm2$ **7.** $1, 2, 3$ **9.** $-1, -10$

11. $\frac{1}{2}, -1$ **13.** $\pm3, \pm\sqrt{2}$ **15.** $-1, 2$ **17.** $-6, \frac{1}{2}, 1$

19. $-2, 0, 1$

21. (a) $\pm1, \pm3, \pm\frac{1}{2}, \pm\frac{3}{2}, \pm\frac{1}{4}, \pm\frac{3}{4}, \pm\frac{1}{8}, \pm\frac{3}{8}, \pm\frac{1}{16}, \pm\frac{3}{16}, \pm\frac{1}{32}, \pm\frac{3}{32}$

(b) 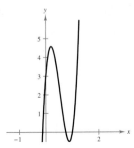   (c) $1, \frac{3}{4}, -\frac{1}{8}$

23. $f(1) = -2, f(2) = 7$ **25.** $f(2) = -6, f(3) = 44$

27. Real zero ≈ 0.7 **29.** Real zero ≈ 3.3

31. e; -1.769 **33.** d; 0.206 **35.** f; 2.769

37. $-1.164, 1.453$ **39.** $0.900, 1.100, 1.900$

41. $-1.453, 1.164$ **43.** $-2.177, 1.563$

45. d   **46.** a   **47.** b   **48.** c

49. (a) $V = x(18 - 2x)(15 - 2x)$

Domain: $0 < x < 7.5$

(b)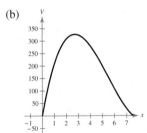

Approximate measurements:
2.72 inches $\times$ 12.56 inches $\times$ 9.56 inches

(c) $x \approx 0.448, 6, \approx 10.052$

A value of $x \approx 10.052$ inches is impossible because it would yield a negative length and width.

(d) $x = 6$

51. 18 inches $\times$ 18 inches $\times$ 36 inches

53. (a) $V = x^3 + 9x^2 + 26x + 24 = 120$

(b) 4 feet by 5 feet by 6 feet

55. 4.49 hours

57. (a)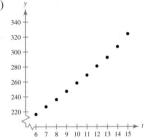

(b) Linear: $R = 2.255t - 15.27$

Quadratic: $R = 0.1295t^2 - 0.464t - 2.06$

Cubic: $R = 0.01105t^3 - 0.2184t^2 + 3.027t - 13.15$

Quartic: $R = 0.008724t^4 - 0.35535t^3 + 5.3735t^2$
$$- 33.612t + 73.59$$

(c) Linear   Quadratic

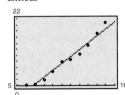   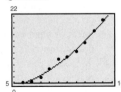

Cubic   Quartic

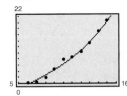   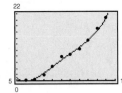

All four models fit the data well. The nonlinear models seem to fit a little better than the linear model because the curve of each graph tends to fit the points at the far left and far right better.

(d) Linear: 2013 $(t \approx 23.18)$

Quadratic: 2009 $(t \approx 19.25)$

Cubic: 2008 $(t \approx 18.33)$

Quartic: 2007 $(t \approx 16.92)$

Answers will vary. Sample answer: The higher the degree of the model, the faster the graph climbs after 2005, so the quartic model makes the earliest prediction of when the revenue per share will reach \$37. Based on the increasing trend of the last few data points, the year (2013) predicted by the linear model seems too late.

59. (a)

(b) Linear: $y = 11.65t + 143.5$

Quadratic: $y = 0.328t^2 + 4.76t + 177.0$

Cubic: $y = 0.0479t^3 - 1.180t^2 + 19.89t + 128.9$

Quartic: $y = 0.01219t^4 - 0.4643t^3 + 6.636t^2$
$$- 31.32t + 250.1$$

(c) Linear: 2012 ($t \approx 22.02$)

Quadratic: 2010 ($t \approx 19.81$)

Cubic: 2009 ($t \approx 18.64$)

Quartic: 2008 ($t \approx 17.79$)

Answers will vary. Sample answer: The long-term trend of the data points appears to be nearly linear, although there is an upward trend in the last three data points. If prices change according to the long-term trend, then the linear or quadratic model will give good predictions. If prices start rising more quickly as reflected in the last three data points, however, the cubic or quartic model may give better future predictions.

61. $\approx$ \$399,890 or \$744,400

63. (a)

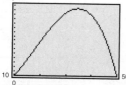

30 and ≈ 38.91

(b) You can solve $-x^3 + 54x^2 - 140x - 3000 = 14,400$ by rewriting the equation as $y = -x^3 + 54x^2 - 140x - 17,400$. Using the *table* feature of a graphing utility, you can approximate the solutions to be $x \approx -14.91$, $x = 30$, and $x \approx 38.91$. The company should charge \$38.91 to generate greater revenue.

65. No; setting $h = 64$ and solving the resulting equation yields imaginary roots.

67. No

69. (a) $x = -2, 1, 4$

(b) It touches the x-axis at $x = 1$ but does not cross the x-axis.

(c) f is at least fourth degree. The degree cannot be less than 3, because there are three zeros. The degree of f cannot be odd because its left-hand behavior matches its right-hand behavior. So, its degree cannot be 3.

(d) It is positive because f increases to the right and left.

(e) Answers will vary.
Sample answer: $f(x) = (x + 2)(x - 1)^2(x - 4)$

(f)

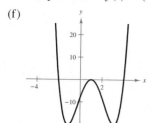

71. Answers will vary.

MID-CHAPTER QUIZ   (page 303)

1. Vertex: $(-1, -2)$

Intercepts: $\left(-1 \pm \sqrt{2}, 0\right), (0, -1)$

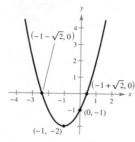

2. Vertex: $(0, 25)$

Intercepts: $(\pm 5, 0), (0, 25)$

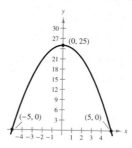

3. Rises to the left
Falls to the right

4. Rises to the left
Rises to the right

5. 293

6. $f(x) = (x - 1)(x^3 + x^2 - 4x - 4) + 0; f(1) = 0$

7. $f(x) = (x + 3)(x^2 + 2x - 8) + 0; f(-3) = 0$

8. $2x^2 + 5x - 12$ **9.** $\pm\sqrt{5}, -\frac{7}{2}$ **10.** $\pm 3, \pm\frac{1}{2}$

11. $1, -\frac{4}{3}$ **12.** $\frac{3}{2}$ **13.** $P = \$2,534,375; \$337,600$

14. (a)

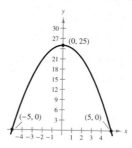

(b) Linear: $A(t) = 9.52t - 52.3$

Quadratic: $A(t) = -0.052t^2 + 10.67t - 58.1$

Cubic: $A(t) = 0.1014t^3 - 3.400t^2 + 45.69t - 173.3$

Quartic: $A(t) = -0.00495t^4 + 0.3194t^3 - 6.872t^2 + 69.34t - 231.2$

(c) Linear

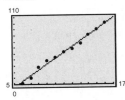

Quadratic

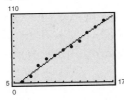

Cubic

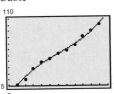

Quartic

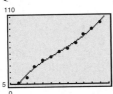

Each model could be considered a good fit, but the cubic and quartic models appear to fit the data a little better than the linear and quadratic models.

(d) Linear: 2011 ($t \approx 21.25$)

Quadratic: 2012 ($t \approx 21.82$)

Cubic: 2008 ($t \approx 18.49$)

Quartic: 2009 ($t \approx 18.94$)

Answers will vary. Sample answer: The linear and quadratic predictions are close to each other and the cubic and quartic predictions are close to each other. The cubic and quartic models give earlier predictions because their graphs rise faster in the years after 2006 than the graphs of the linear and quadratic models.

SECTION 3.5   *(page 312)*

Skills Review   *(page 312)*

1. $2\sqrt{3}$ **2.** $10\sqrt{5}$ **3.** $\sqrt{5}$ **4.** $-6\sqrt{3}$

5. 12   **6.** 48   **7.** $\dfrac{\sqrt{3}}{3}$ **8.** $\sqrt{2}$

9. $-\dfrac{1}{2} \pm \dfrac{\sqrt{5}}{2}$ **10.** $-1 \pm \sqrt{2}$

---

1. $i, -1, -i, 1, i, -1, -i, 1, i, -1, -i, 1, i, -1, -i, 1$
$i^{4n} = 1, i^{4n+1} = i, i^{4n+2} = -1, i^{4n+3} = -i,$
n is an integer.

3. $a = 7, b = 12$ **5.** $a = 4, b = -3$

7. $9 + 4i; 9 - 4i$ **9.** $-3 - 2\sqrt{3}i; -3 + 2\sqrt{3}i$

11. $-21; -21$ **13.** $-1 - 6i; -1 + 6i$

15. $-5i; 5i$ **17.** $-3; -3$ **19.** $2 + i$

21. $5 + 6i$ **23.** $3 - 3\sqrt{2}i$ **25.** $\dfrac{1}{6} + \dfrac{7}{6}i$

27. -14 **29.** $-2\sqrt{6}$ **31.** -10 **33.** $5 + i$

35. 25   **37.** $30 + 20i$ **39.** $-11 + 60i$ **41.** 8

43. $\left(16 + 4\sqrt{3}\right) + \left(-16\sqrt{2} + 2\sqrt{6}\right)i$ **45.** $\dfrac{4}{5} - \dfrac{3}{5}i$

47. $1 + \dfrac{1}{2}i$ **49.** $10 - 7i$ **51.** $\dfrac{1}{8}i$

53. $-\dfrac{44}{125} - \dfrac{8}{125}i$ **55.** $\dfrac{35}{29} + \dfrac{595}{29}i$

57. Error: $(3 - 2i)(3 + 2i) = 9 - 4i^2 = 9 + 4 = 13$
(not $9 - 4 = 5$)

59. $1 \pm i$ **61.** $-2 \pm \dfrac{1}{2}i$

63. $-\dfrac{3}{2}, -\dfrac{5}{2}$ **65.** $\dfrac{1}{8} \pm \dfrac{\sqrt{11}}{8}i$

67. $x = -\dfrac{1}{2} \pm \dfrac{\sqrt{7}}{2}i$

The graph has no x-intercepts, so the solutions must be complex conjugate pairs.

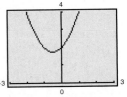

69. $x = -\dfrac{3}{2} \pm \dfrac{\sqrt{29}}{2}$

The x-intercepts of the graph correspond to the zeros of the function.

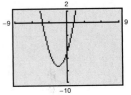

71.

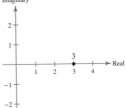

73.

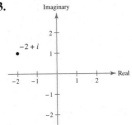

75.

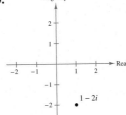

77. The complex number 0 is in the Mandelbrot Set because, for $c = 0$, the corresponding Mandelbrot sequence is $0, 0, 0, 0, 0, 0$, which is bounded.

79. The complex number 1 is not in the Mandelbrot Set because, for $c = 1$, the corresponding Mandelbrot sequence is 1, 2, 5, 26, 677, 458,330, . . . , which is unbounded.

81. The complex number $\frac{1}{2}i$ is in the Mandelbrot Set because, for $c = \frac{1}{2}i$, the corresponding Mandelbrot sequence is $\frac{1}{2}i$, $-\frac{1}{4} + \frac{1}{2}i$, $-\frac{3}{16} + \frac{1}{4}i$, $-\frac{7}{256} + \frac{13}{32}i$, $-\frac{10,767}{65,536} + \frac{1957}{4096}i$, $-\frac{864,513,055}{4,294,967,296} + \frac{46,037,845}{134,217,728}i$, which is bounded.

83. False. If the complex number is real, it equals its conjugate.

85. Answers will vary.

SECTION 3.6   *(page 320)*

Skills Review   *(page 320)*

1. $4 - \sqrt{29}i, 4 + \sqrt{29}i$ **2.** $-5 - 12i, -5 + 12i$

3. $-1 + 4\sqrt{2}i, -1 - 4\sqrt{2}i$ **4.** $6 + \frac{1}{2}i, 6 - \frac{1}{2}i$

5. $-13 + 9i$ **6.** $12 + 16i$ **7.** $26 + 22i$

8. 29   **9.** i **10.** $-9 + 46i$

---

1. 1   **3.** 3   **5.** 4   **7.** $\pm 4i; (x + 4i)(x - 4i)$

9. $\dfrac{5 \pm \sqrt{5}}{2}; \left(x - \dfrac{5 + \sqrt{5}}{2}\right)\left(x - \dfrac{5 - \sqrt{5}}{2}\right)$

11. $\pm 3, \pm 3i; (x - 3)(x + 3)(x - 3i)(x + 3i)$

13. $0, \pm \sqrt{5}i; x(x - \sqrt{5}i)(x + \sqrt{5}i)$

15. $-5, 8 \pm i; (x + 5)(x - 8 + i)(x - 8 - i)$

17. $2, 2 \pm i; (x - 2)(x - 2 + i)(x - 2 - i)$

19. $-5, 4 \pm 3i; (t + 5)(t - 4 + 3i)(t - 4 - 3i)$

21. $-10, -7 \pm 5i; (x + 10)(x + 7 - 5i)(x + 7 + 5i)$

23. $-5, -2 \pm \sqrt{3}i; (x + 5)(x + 2 - \sqrt{3}i)(x + 2 + \sqrt{3}i)$

25. $-\frac{3}{4}, 1 \pm \frac{1}{2}i; (4x + 3)(2x - 2 + i)(2x - 2 - i)$

27. $-\frac{1}{5}, 1 \pm \sqrt{5}i; (5x + 1)(x - 1 + \sqrt{5}i)(x - 1 - \sqrt{5}i)$

29. $2, \pm 2i; (x - 2)^2(x + 2i)(x - 2i)$

31. $\pm i, \pm 3i; (x + i)(x - i)(x + 3i)(x - 3i)$

33. $-4, 3, \pm i; (t + 4)^2(t - 3)(t + i)(t - i)$

35. Answers will vary.
Sample answer: $x^3 + 2x^2 + 9x + 18$

37. Answers will vary. Sample answer: $x^3 - 5x^2 + 9x - 5$

39. Answers will vary.
Sample answer: $x^5 + 4x^4 + 13x^3 + 52x^2 + 36x + 144$

41. Answers will vary.
Sample answer: $x^4 + 8x^3 + 9x^2 - 10x + 100$

43. Answers will vary.
Sample answer: $3x^4 - 17x^3 + 25x^2 + 23x - 22$

45. (a) $(x^2 - 8)(x^2 + 1)$

(b) $\left(x - 2\sqrt{2}\right)\left(x + 2\sqrt{2}\right)(x^2 + 1)$

(c) $\left(x - 2\sqrt{2}\right)\left(x + 2\sqrt{2}\right)(x - i)(x + i)$

47. (a) $(x^2 - 2x + 3)(x^2 - 3x - 5)$

(b) $(x^2 - 2x + 3)\left(x - \dfrac{3 + \sqrt{29}}{2}\right)\left(x - \dfrac{3 - \sqrt{29}}{2}\right)$

(c) $\left(x - 1 + \sqrt{2}i\right)\left(x - 1 - \sqrt{2}i\right)$
$\left(x - \dfrac{3 + \sqrt{29}}{2}\right)\left(x - \dfrac{3 - \sqrt{29}}{2}\right)$

49. $\pm 2i, \frac{7}{3}$ **51.** $\pm 6i, 1$ **53.** $-3 \pm i, \frac{1}{4}$

55. $1, 2, -3 \pm \sqrt{2}i$ **57.** $\frac{3}{4}, \frac{1}{2} \pm \frac{\sqrt{5}}{2}i$

59. $\pm 1, \pm 2$. The x-intercepts occur at the solutions of the equation.

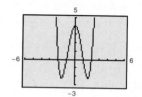

61. Answers will vary. Sample answer:

(a) $f(x) = x^4 - 7x^3 + 17x^2 - 17x + 6$
Zeros: 1, 1, 2, 3

(b) $g(x) = x^4 - 3x^3 + 3x^2 - 3x + 2$
Zeros: 1, 2, $\pm i$

(c) $h(x) = x^4 + 5x^2 + 4$
Zeros: $\pm i, \pm 2i$

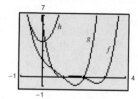

Similarities: f, g, and h all rise to the left and rise to the right.

Differences: f, g, and h have different numbers of x-intercepts.

63. There is no price p that would yield a profit of $9 million.

The graph of $y = P - 9,000,000$ has no intercepts, so there is no solution.

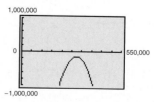

65. No. The conjugate pairs statement specifies polynomials with *real* coefficients. $f(x)$ has imaginary coefficients.

67. The imaginary zeros of f can only occur in conjugate pairs. f has only one unknown zero and no unpaired complex zeros, so the unknown zero must be real.

69. Imaginary zeros can only occur in conjugate pairs. Because f has three zeros, one or all of them must be real numbers.

71. Polynomials of odd degree eventually rise to one side and fall to the other side. So, f must cross the x-axis, which means f must have a real zero. You can show this by graphing several third-degree polynomials.

SECTION 3.7   (page 330)

Skills Review   (page 330)

1. $x(x - 4)$ **2.** $2x(x^2 - 3)$ **3.** $(x - 5)(x + 2)$

4. $(x - 5)(x - 2)$ **5.** $x(x + 1)(x + 3)$

6. $(x^2 - 2)(x - 4)$

7.    **8.**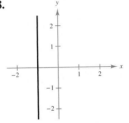

9. 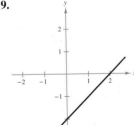   **10.**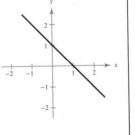

1. Domain: All $x \neq -1$; Horizontal asymptote: $y = 3$; Vertical asymptote: $x = -1$

3. Domain: All $x \neq 5$; Horizontal asymptote: $y = -1$; Vertical asymptote: $x = 5$

5. Domain: All reals; Horizontal asymptote: $y = 3$; Vertical asymptote: None

7. Domain: All $x \neq -4$; Horizontal asymptote: $y = 0$; Vertical asymptote: $x = -4$

9. (a) None
(b) None
(c) None

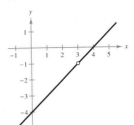

11. (a) $x = -1$
(b) None
(c) $y = x - 1$

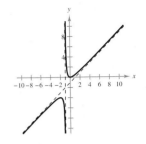

13. f   **14.** e   **15.** a   **16.** b   **17.** c   **18.** d

19. $g(x)$ shifts downward two units.

21. $g(x)$ is a reflection about the x-axis.

23. $g(x)$ shifts upward three units.

25. $g(x)$ is a reflection about the x-axis.

27. $g(x)$ shifts upward five units.

29. $g(x)$ is a reflection about the x-axis.

31. 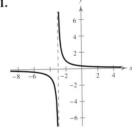   **33.**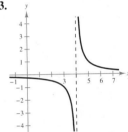

35. 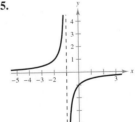   **37.**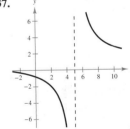

39. 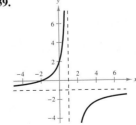   **41.**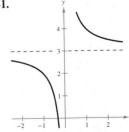

43.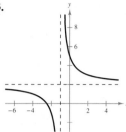

45.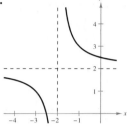

47.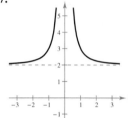

49.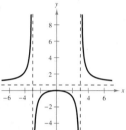

51.

53.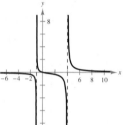

55.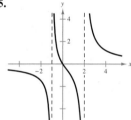

57. Answers will vary. Sample answer: $f(x) = \dfrac{2x^2}{x^2 + 1}$

59. Answers will vary. Sample answer: $f(x) = \dfrac{x^3}{x^2 + x - 2}$

61. Answers will vary. Sample answer: $f(x) = \dfrac{1}{x^2 + 1}$

63. Answers will vary. Sample answer: $f(x) = \dfrac{x}{x^2 - 2x - 3}$

65. No. Given

$$f(x) = \frac{a_n x^n + \cdots + a_0}{b_m x^m + \cdots + b_0}$$

if $n > m$, there is no horizontal asymptote and n must be greater than m for a slant asymptote to occur.

67. (a) \$176 million   (b) \$528 million   (c) \$1584 million

(d) No. The model has a vertical asymptote at $p = 100$.

69. (a) 318 deer; 500 deer; 900 deer   (b) 2500 deer

71. (a)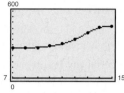

The model fits the data very well.

(b) \$366.8 billion; \$332.3 billion; \$319.1 billion; Answers will vary.

(c) $y \approx 292.7$; as time passes, the national defense outlays approach \$292.7 billion.

73. (a)

n	1	2	3	4	5
P	0.60	0.79	0.86	0.90	0.92

n	6	7	8	9	10
P	0.93	0.94	0.95	0.95	0.96

(b) 100%

75. (a)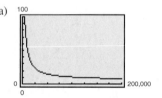

(b)

x	10,000	100,000	1,000,000	10,000,000
$\overline{C}$	\$51	\$10.50	\$6.45	\$6.05

Eventually, the average recycling cost per pound will approach the horizontal asymptote of \$6.

77. (a) $B = \dfrac{100.9708t + 6083.999}{3.0195t + 251.817}$, $5 \le t \le 15$

(b)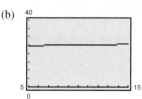

(c) ≈ 25.96 barrels per person

79. (a)

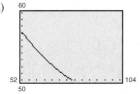

(b) 2008: 48.18 seconds; 2012: 48.43 seconds

(c) This model does not have a horizontal asymptote. A model with a horizontal asymptote would be reasonable for this type of data because the winning times should continue to improve, but it is not humanly possible for the winning times to decrease without bound.

REVIEW EXERCISES *(page 336)*

1.

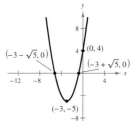

3.

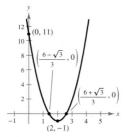

Vertex: $(-3, -5)$ Vertex: $(2, -1)$

Intercepts: Intercepts:

$(0, 4), \left(-3 \pm \sqrt{5}, 0\right)$ $(0, 11), \left(\dfrac{6 \pm \sqrt{3}}{3}, 0\right)$

5. $f(x) = \frac{7}{9}(x + 5)^2 - 1$

7. $A(x) = 250x - x^2$; 125 feet $\times$ 125 feet

9. (a)

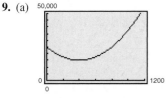

(b) About 385 units

(c) Write the equation of the quadratic function in standard form. The vertex is the minimum cost.

11. (a)

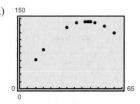

(b) $d(x) = -0.077x^2 + 6.59x + 2.4$

(c)

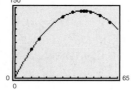

(d) $(42.8, 143.4)$; The vertex gives the angle $42.8°$ which results in the greatest distance of 143.4 meters.

13. Falls to the left   **15.** Falls to the left
Rises to the right   Falls to the right

17. ± 4 **19.** $0, 2, 5$

21. $x^2 - 3x + 1 - \dfrac{1}{2x + 1}$ **23.** $x^2 + 11x + 24$

25. $x^2 - 3x + 3$ **27.** (a) -10 (b) 11

29. (a) Proof   (b) $x - 2$

(c) $(x - 5)(x + 3)(x - 2)$ (d) $-3, 2, 5$

(e)

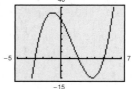

31. (a)

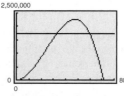

(b) Answers will vary.

$\approx \$325{,}167$

33. $\pm 1, \pm 3, \pm 5, \pm 15, \pm\frac{1}{2}, \pm\frac{3}{2}, \pm\frac{5}{2}, \pm\frac{15}{2}, \pm\frac{1}{4}, \pm\frac{3}{4}, \pm\frac{5}{4}, \pm\frac{15}{4}$

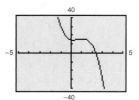

From the graph: $x \approx 2.357$, so the zero is not rational.

35. $-3, -1, 2$ **37.** $\pm 2, \pm\sqrt{5}$ **39.** $-\frac{1}{3}, \frac{3}{2}, 2$

41. $-1, 2$ **43.** $x \approx -2.3$ **45.** $-1.321, -0.283, 1.604$

47. (a)

(b) Linear: $R(t) = 3708.4t - 4690$

Quadratic: $R(t) = 157.78t^2 + 395.0t + 11,403$

Cubic: $R(t) = 2.427t^3 + 81.33t^2 + 1162.2t + 8967$

Quartic: $R(t) = 4.8829t^4 - 202.653t^3 + 3211.25t^2$
$$- 19,345.9t + 57,518$$

(c) Linear   Quadratic

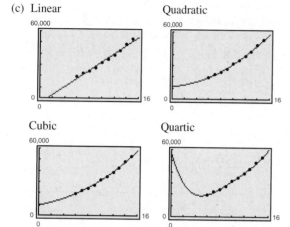

Cubic   Quartic

Each model could be considered a good fit for the data.

(d) Linear: 2009 ($t \approx 18.79$)

Quadratic: 2007 ($t \approx 17.22$)

Cubic: 2007 ($t \approx 17.14$)

Quartic: 2007 ($t \approx 16.59$)

Answers will vary. Sample answer: The quadratic, cubic, and quartic predictions are all around 2007. It appears that the curves of the graphs of these models fit the data points of the last few years better than the graph of the linear model. So, the linear prediction (2009) may not be as good.

49. $4\sqrt{2}i; -4\sqrt{2}i$ **51.** $-3 + 4i; -3 - 4i$

53. $5 + i$ **55.** $11 + 9\sqrt{3}i$ **57.** 89 **59.** $-10 - 8i$

61. $-7 + 24i$ **63.** 10 **65.** $3 - 2i$

67. $-3 - 4i$ **69.** $\dfrac{1 \pm \sqrt{23}i}{4}$ **71.** $\dfrac{-11 \pm \sqrt{73}}{8}$

73.

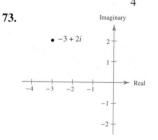

75. $\pm 3, \pm 3i; (x - 3)(x + 3)(x - 3i)(x + 3i)$

77. $-5, \pm\sqrt{3}i; (t + 5)(t - \sqrt{3}i)(t + \sqrt{3}i)$

79. $2, \pm\dfrac{3i}{2}; (x - 2)(2x - 3i)(2x + 3i)$

81. Answers will vary. Sample answer: $x^3 - x^2 + 9x - 9$

83. (a) $(x^2 + 8)(x^2 - 3)$

(b) $(x^2 + 8)(x - \sqrt{3})(x + \sqrt{3})$

(c) $(x + 2\sqrt{2}i)(x - 2\sqrt{2}i)(x - \sqrt{3})(x + \sqrt{3})$

85. $\pm 4i, \frac{1}{4}$ **87.** $-1 \pm 3i, -1, -4$

89. Domain: All $x \ne -2$

Vertical asymptote: $x = -2$

Horizontal asymptote: $y = 0$

91. Domain: All $x \ne \pm 3$

Vertical asymptotes: $x = -3, x = 3$

Horizontal asymptote: $y = 2$

93.   **95.**

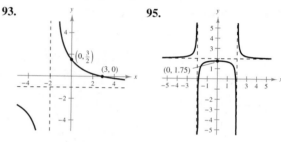

97. None

99. (a)

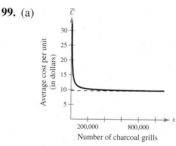

(b)

x	1000	10,000	100,000	1,000,000
$\overline{C}$	\$134.65	\$22.15	\$10.90	\$9.78

Eventually, the average cost per charcoal grill will approach the horizontal asymptote of \$9.65.

101. (a) 430,769 fish; 662,500 fish; 1,024,000 fish

(b) 1,666,667 fish

103. (a) \$35,000   (b) \$157,500   (c) \$10,395,000

(d) No. The model has a vertical asymptote at $p = 100$.

CHAPTER TEST   (page 340)

1.

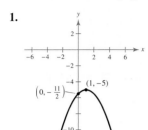

Vertex: $(1, -5)$

Intercept: $\left(0, -\frac{11}{2}\right)$

2. (a) Falls to the left   (b) Rises to the left
   Rises to the right   Rises to the right

3. $x^2 + 7x + 12$

4. $\pm 1, \pm 3, \pm 9, \pm 27, \pm\frac{1}{4}, \pm\frac{1}{2}, \pm\frac{3}{4}, \pm\frac{3}{2}, \pm\frac{9}{4}, \pm\frac{9}{2}, \pm\frac{27}{4}, \pm\frac{27}{2}$

$f\left(\frac{3}{2}\right) = 0$

$f\left(-\frac{3}{2}\right) = 0$

$f(x) = \left(x - \frac{3}{2}\right)\left(x + \frac{3}{2}\right)(x - 1)(x - 3)$

5. (a)

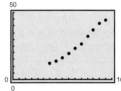

(b) Linear: $S = 3.836t - 13.79$

Quadratic: $S = 0.1415t^2 + 0.864t + 0.65$

Cubic: $S = -0.03857t^3 + 1.3564t^2 - 11.327t + 39.36$

(c) Linear   Quadratic

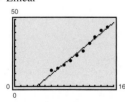   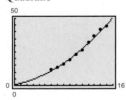

Cubic

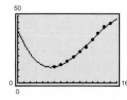

Each model could be considered a good fit for the data.

(d) Linear: 2007 $(t \approx 16.63)$

Quadratic: 2006 $(t \approx 15.87)$

Cubic: 2008 $(t \approx 17.91)$

Answers will vary. Sample answer: All three models give predictions that reflect the increasing data. Beyond 2008, however, the sales per share predicted by the cubic model begin to decrease, so the cubic model is probably not appropriate for future predictions. The linear and quadratic models show increases beyond 2008, but neither is certain to be a good predictor far into the future, because the data do not show a steady trend.

6. $16 - 3i$ **7.** $7 - 9i$ **8.** $27 - 4\sqrt{3}i$

9. $23 - 14i$ **10.** i

11. $x = \dfrac{-5 \pm \sqrt{3}i}{2}$ **12.** $x = \dfrac{5 \pm 3\sqrt{7}i}{4}$

13. $x^4 - 7x^3 + 19x^2 - 63x + 90$ **14.** $\pm\sqrt{5}i, -2$

15.

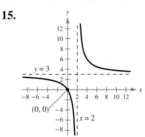

Domain: All $x \neq 2$

CHAPTER 4

SECTION 4.1   (page 350)

Skills Review   (page 350)

1. 5^x **2.** 3^{2x} **3.** 4^{3x} **4.** 10^x **5.** 4^{2x}

6. 4^{10x} **7.** $\left(\dfrac{3}{2}\right)^x$ **8.** 4^{3x} **9.** $\dfrac{1}{2^x}$ **10.** $\dfrac{3^x}{5^x}$

11. 2^x **12.** 3^x

---

1. 3.463   **3.** 95.946   **5.** 0.079   **7.** 54.598

9. 1.948   **11.** g   **12.** e   **13.** b

14. h   **15.** d   **16.** a   **17.** f   **18.** c

19.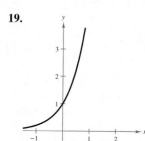

21.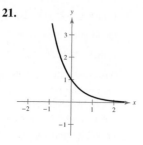

23.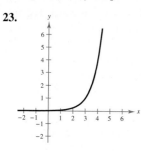

25.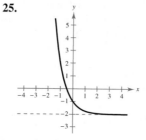

27.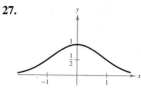

29.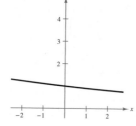

31.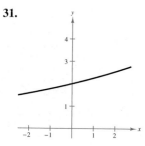

33.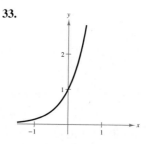

35.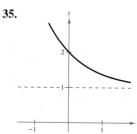

37.

n	1	2	4
A	\$7346.64	\$7401.22	\$7429.74

n	12	365	Continuous
A	\$7449.23	\$7458.80	\$7459.12

39.

n	1	2	4
A	\$24,115.73	\$25,714.29	\$26,602.23

n	12	365	Continuous
A	\$27,231.38	\$27,547.07	\$27,557.94

41.

t	1	10	20
P	\$91,393.12	\$40,656.97	\$16,529.89

t	30	40	50
P	\$6720.55	\$2732.37	\$1110.90

43.

t	1	10	20
P	\$90,521.24	\$36,940.70	\$13,646.15

t	30	40	50
P	\$5040.98	\$1862.17	\$687.90

45. The account paying 5% interest compounded quarterly earns more money. Even though the interest is compounded less frequently, the higher interest rate yields a higher return.

47. You should choose the account with the online access fee because it yields a higher return.

49. \$19,691.17   **51.** \$147,683.76

53. \$20,700.76   **55.** \$155,255.66

57. (a) \$182.91

    (b) \$29.58

    (c) 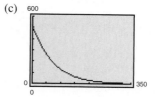   (d) \$117.19

59. (a) 100   (b) $\approx$ 110   (c) $\approx$ 121   (d) $\approx$ 158

61. \$7424.70   **63.** \$12,434.43

65. (a) 5907   (b) 6237   (c) 6767   (d) 7753

67. $\approx$ 30.42 kilograms

69. (a) 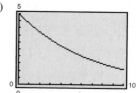   (b) $\approx$ 1.11 pounds

    (c) On the graph, when $P = 2.5$, $t \approx 4.6$ months.

71. (a) $410 million, $650 million, $2100 million

(b) $408.21 million, $649.49 million, $2073.90 million

73. (a) 22, 24, 25, 25.5   (b) 22.1, 24.0, 25.1, 25.4

75. Women tend to marry about 2 years younger than men do. The median ages of both have been rising, and the age difference is decreasing.

77. (a)

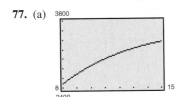

(b) 2.696 billion prescriptions, 3.127 billion prescriptions, 3.363 billion prescriptions

79. Answers will vary.

SECTION 4.2   *(page 361)*

Skills Review   *(page 361)*

1. 3   **2.** 0   **3.** −1   **4.** 1   **5.** 7.389

5. 7.389   **6.** 0.368

7. The graph of *g* is the graph of *f* shifted two units to the left.

8. The graph of *g* is the graph of *f* reflected about the *x*-axis.

9. The graph of *g* is the graph of *f* shifted downward one unit.

10. The graph of *g* is the graph of *f* reflected about the *y*-axis.

1. c   **2.** f   **3.** b   **4.** d   **5.** a   **6.** e

7. $\log_4 256 = 4$ **9.** $\log_{81} 3 = \frac{1}{4}$

11. $\log_6 \frac{1}{36} = -2$ **13.** $\ln e = 1$ **15.** $\ln 4 = x$

17. $4^2 = 16$ **19.** $2^{-1} = \frac{1}{2}$ **21.** $e^1 = e$

23. $5^{-1} = 0.2$ **25.** $27^{1/3} = 3$ **27.** 2

29. −4   **31.** $\frac{1}{3}$ **33.** 1   **35.** −4   **37.** 1

39. −4   **41.** 5   **43.** 2.538   **45.** −0.097

47. 0.452   **49.** 1.946   **51.** 2.913   **53.** 0.896

55.

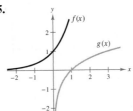

57.

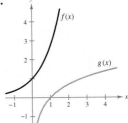

59. d   **60.** e   **61.** a   **62.** c   **63.** f   **64.** b

65. Domain: $(0, \infty)$

Asymptote: $x = 0$

x-intercept: $(1, 0)$

67. Domain: $(-4, \infty)$

Asymptote: $x = -4$

x-intercept: $(-3, 0)$

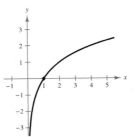

69. Domain: $(0, \infty)$

Asymptote: $x = 0$

x-intercept: $(1, 0)$

71. Domain: $(-\infty, 0)$

Asymptote: $x = 0$

x-intercept: $(-1, 0)$

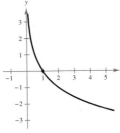

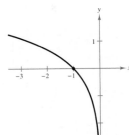

73. Domain: $(-1, \infty)$

Asymptote: $x = -1$

x-intercept: $(0, 0)$

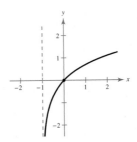

75. 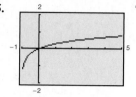   **77.**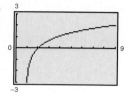

79.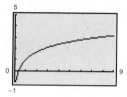

81. $t \approx 26.1$ years

83. (a) 78   (b) 68.2   (c) In 3 months $(t \approx 2.7)$

85. (a)

K	1	2	4	6	8	10	12
t	0	13.2	26.4	34.1	39.6	43.9	47.3

(b)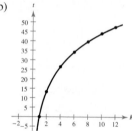

87.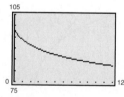

The domain $0 \le t \le 12$ covers the period of 12 months or 1 year. The scores range from the average score on the original exam to the average score after 12 months, so the range is $82.40 \le f(t) \le 98$

89.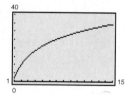

The domain $1 \le t \le 15$ represents the time period of 1 day to 15 days. The range $2 \le g(t) \le 34.5$ represents the possible units produced by the employees over the given time period.

91. (a) ≈ 29.4 years   (b) ≈ 33.3 years
    (c) ≈ 6 years   (d) ≈ 10.2 years
93. (a) False   (b) True   (c) True   (d) False

SECTION 4.3   (page 369)

Skills Review   (page 369)

1. 2   **2.** -5 **3.** -2 **4.** -3 **5.** e^5

6. $\dfrac{1}{e}$ **7.** e^6 **8.** 1   **9.** $y = x^{-2}$

10. $y = x^{1/2}$ **11.** $4^3 = 64$ **12.** $16^{1/2} = 4$

1. $\dfrac{\log_{10} 8}{\log_{10} 5}$ **3.** $\dfrac{\log_{10} 30}{\log_{10} e}$ **5.** $\dfrac{\log_{10} n}{\log_{10} 3}$ **7.** $\dfrac{\log_{10} x}{\log_{10} \frac{1}{5}}$

9. $\dfrac{\log_{10} \frac{3}{10}}{\log_{10} x}$ **11.** $\dfrac{\log_{10} x}{\log_{10} 2.6}$ **13.** $\dfrac{\ln 8}{\ln 5}$ **15.** $\dfrac{\ln 5}{\ln 10}$

17. $\dfrac{\ln n}{\ln 3}$ **19.** $\dfrac{\ln x}{\ln \frac{1}{5}}$ **21.** $\dfrac{\ln \frac{3}{10}}{\ln x}$ **23.** $\dfrac{\ln x}{\ln 2.6}$

25. 2.585   **27.** 1.079   **29.** 2.633   **31.** -0.683

33. -1.661 **35.** 2.322   **37.** 1.1833   **39.** -0.2084

41. 1.0686   **43.** 0.1781   **45.** 1.8957   **47.** -2.7124

49. 0.5708   **51.** $\frac{1}{3}$ **53.** $-\frac{1}{2}$ **55.** -3

57. $-1 - \log_9 2$ **59.** $\frac{1}{2} + \frac{1}{2} \log_7 10$

61. $-3 - \log_5 2$ **63.** $6 + \ln 5$ **65.** $6 + 5 \log_2 3$

67. $\log_3 4 + \log_3 n$ **69.** $\log_5 x - 2$ **71.** $4 \log_2 x$

73. $\frac{1}{2} \ln z$ **75.** $\ln x + \ln y + \ln z$ **77.** $\frac{1}{2} \ln(a - 1)$

79. $2 \ln(z - 1) - \ln z$ **81.** $\ln z - \frac{1}{3} \ln(z + 3)$

83. $\frac{1}{3}(\ln x - \ln y)$ **85.** $\frac{3}{4} \ln x + \frac{1}{4} \ln(x^2 + 3)$

87. $\log_3 5x$ **89.** $\log_4 \dfrac{8}{x}$ **91.** $\log_{10}(x + 4)^2$

93. $\ln \dfrac{1}{216x}$ **95.** $\ln \dfrac{\sqrt[3]{5x}}{x + 1}$ **97.** $\log_8 \dfrac{x - 2}{x + 2}$

99. $\ln \dfrac{9}{\sqrt{x^2 + 1}}$ **101.** $\ln \dfrac{x}{(x + 2)(x - 2)}$

103. $\ln y = \frac{1}{4} \ln x$ **105.** $\ln y = -\frac{1}{4} \ln x + \ln \frac{5}{2}$

107. $\ln y = \frac{2}{3} \ln x + \ln 0.070$ **109.** ≈ 26 decibels

111.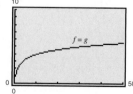

The two graphs are the same. The property is $\log_a(uv) = \log_a u + \log_a v$.

113. Choose a value for y and graph $\log_a x/\log_a y$ and $\log_a x - \log_a y$. Notice that the graphs are different. To demonstrate the correct property, graph $\log_a(x/y)$ and $\log_a x - \log_a y$, choosing a value for y.

115. Let $\log_a u = x$ and $\log_a v = y$.

$a^x = u$ and $a^y = v$

$u \cdot v = a^x \cdot a^y = a^{x+y}$

$\log_a a^{x+y} = x + y$

$\log_a uv = \log_a u + \log_a v$

117. Answers will vary.

MID-CHAPTER QUIZ   *(page 372)*

1.

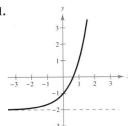

2.

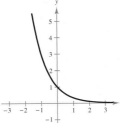

3.

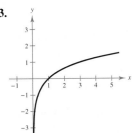

4.

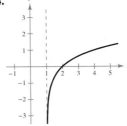

5. (a) Monthly: \$15,085.04

(b) Continuously: \$15,098.34

6. (a) 2000: 7.79 million   (b) 2009: 11.08 million
       2005: 9.47 million   2010: 11.53 million

7. (a) 100   (b) ≈ 364 (c) ≈ 1326

8. \$108.54   **9.** 2   **10.** 4   **11.** -2 **12.** 0

13.

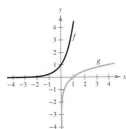

Domain of $f(x)$: $(-\infty, \infty)$

Domain of $g(x)$: $(0, \infty)$

The graphs are reflections of each other about the line $y = x$.

14. $\frac{3}{2}$ **15.** $\frac{6}{5}$ **16.** $\frac{1}{3}\log_{10} x + \frac{1}{3}\log_{10} y - \frac{1}{3}\log_{10} z$

17. $\ln(x^2 + 3) - 3\ln x$ **18.** $\ln \dfrac{xy}{3}$ **19.** $\log_{10} \dfrac{1}{64x^3}$

20. $\ln y = \frac{1}{3}\ln x$

SECTION 4.4   *(page 380)*

Skills Review   *(page 380)*

1. $\dfrac{\ln 3}{\ln 2}$ **2.** $1 + \dfrac{2}{\ln 4}$ **3.** $\dfrac{e}{2}$ **4.** $2e$

5. $2 \pm i$ **6.** $\frac{1}{2}, 1$ **7.** x **8.** $2x$

9. $2x$ **10.** $-x^2$

1. 3   **3.** -2 **5.** 2   **7.** 64   **9.** $\frac{1}{10}$ **11.** x^2

13. $x^2 + 1$ **15.** $x^3 - 7$ **17.** $x^3 - 8$ **19.** $x + 5$

21. x^2 **23.** $\ln 3 \approx 1.099$ **25.** $\log_3 8 \approx 1.893$

27. $\ln 28 \approx 3.332$ **29.** $\frac{1}{2}\log_3 80 \approx 1.994$ **31.** 2

33. $\log_3 28 + 1 \approx 4.033$ **35.** $3 - \log_2 565 \approx -6.142$

37. $\frac{1}{3}\log_{10} \frac{3}{2} \approx 0.059$ **39.** $\log_5 7 + 1 \approx 2.209$

41. $\frac{1}{3}\ln 12 \approx 0.828$ **43.** $\ln \frac{5}{3} \approx 0.511$

45. $\ln \frac{1}{2} \approx -0.693$ **47.** $\frac{1}{3} + \frac{1}{3}\log_2 \frac{8}{3} \approx 0.805$

49. $\ln 6 \approx 1.792, \ln 2 \approx 0.693$ **51.** $\ln 4 \approx 1.386$

53. $2\ln 75 \approx 8.635$ **55.** $\frac{1}{2}\ln 1498 \approx 3.656$

57. $\dfrac{\ln 4}{365 \ln\left(1 + \dfrac{0.065}{365}\right)} \approx 21.330$

59. $\dfrac{\ln 2}{12 \ln\left(1 + \dfrac{0.10}{12}\right)} \approx 6.960$

61. 10,000   **63.** $e^{-3} \approx 0.050$

65. $\dfrac{e^{2.4}}{2} \approx 5.512$ **67.** 5,000,000

69. $-1 + 3^{12/5} \approx 12.967$ **71.** $\frac{1}{5}e^{10/3} \approx 5.606$

73. $e^2 - 2 \approx 5.389$ **75.** $e^{-2/3} \approx 0.513$

77. No solution   **79.** $1 + \sqrt{1 + e} \approx 2.928$

81. No solution   **83.** 7   **85.** $\dfrac{-1 + \sqrt{17}}{2} \approx 1.562$

87. 2   **89.** $\dfrac{725 + 125\sqrt{33}}{8} \approx 180.384$

91. $y = 2x + 1$ **93.** $y = \dfrac{(x - 1)^2}{x + 2}$

95. 2.807   **97.** 20.086   **99.** ≈ 11.09 years

101. ≈ 15.15 years   **103.** 26 months

105. (a) ≈ 210 coin sets   (b) ≈ 588 coin sets

(c)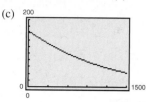

107. (a) ≈ 29.3 years   (b) ≈ 39.8 years

109. (a) 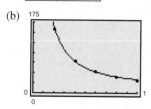   (b) and (c) 2001

111. (a)

x	y
0.2	162.6
0.4	78.5
0.6	52.5
0.8	40.5
1.0	33.9

(b)

The model is a good fit for the data.

(c) 1.197 meters

(d) No. To reduce the g's to fewer than 23 requires a crumple zone of more than 2.27 meters, a length that exceeds the front width of most cars.

113. $\log_b uv = \log_b u + \log_b v$

True by the Product Rule in Section 4.3.

115. $\log_b(u - v) = \log_b u - \log_b v$

False.

$1.95 \approx \log(100 - 10) \neq \log 100 - \log 10 = 1$

117. Yes. See Exercise 81.

SECTION 4.5   *(page 391)*

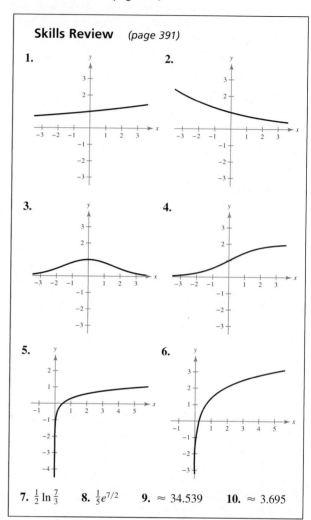

Skills Review   *(page 391)*

1.

2.

3.

4.

5.

6.

7. $\frac{1}{2} \ln \frac{7}{3}$ **8.** $\frac{1}{5}e^{7/2}$ **9.** ≈ 34.539   **10.** ≈ 3.695

Initial Investment	Annual % Rate	Time to Double	Amount After 10 Years
1. $5000	7%	9.90 years	$10,068.76
3. $500	6.93%	10 years	$1000.00
5. $1000	8.25%	8.40 years	$2281.88
7. $6392.79	11%	6.30 years	$19,205.00
9. $5000	8%	8.66 years	$11,127.70

Isotope	Half-Life (Years)	Initial Quantity	Amount After 1000 Years
11. ^{226}Ra	1599	4 g	2.59 g
13. ^{14}C	5715	3.95 g	3.5 g
15. ^{239}Pu	24,100	1.65 g	1.6 g

17. Exponential growth   **19.** Exponential decay

21. $C = 1, k = \frac{1}{4} \ln 10$ **23.** $C = 1, k = \frac{1}{4} \ln \frac{1}{4}$

25.

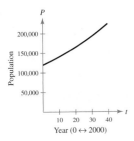

2025

27. ≈ 12.36 hours to double

≈ 19.59 hours to triple

29. 12,180 years   **31.** $\approx 9.92\%$

33. (a) $N = 40(1 - e^{-0.049t})$ (b) ≈ 42 days

35. (a)

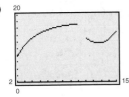

(b) S changed from a logarithmic function to a quadratic function.

37.

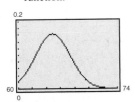

64.9 inches

39. (a)

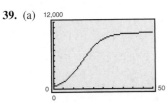

(b) ≈ 1252 fish

(c) ≈ 7.8 months

41. (a)

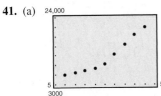

(b) $P = 3972.82e^{0.0328t}$

(c)

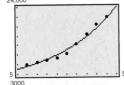

(d) 2022: $\approx 8,174,963$ people

2042: $\approx 15,753,714$ people

43. (a)

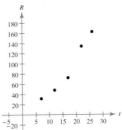

(b) $R = 16.58e^{0.0899t}$

(c)

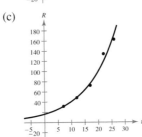

(d) 2009: $\approx \$224.8$ million

2010: $\approx \$246.0$ million

45. (a)

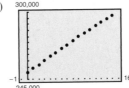

(b) $P = 251,372.02e^{0.0114t}$

(c) Linear model: $P = 3110.9t + 250,815$

Quadratic model:

$P = -20.55t^2 + 3419.2t + 250,096$

(d) Exponential model

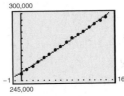

Linear model

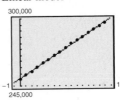

Quadratic model

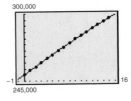

(e) Exponential model

  2008: 308,627,174 people

  2009: 312,165,655 people

  2010: 315,744,706 people

  Linear model

  2008: 306,811,200 people

  2009: 309,922,100 people

  2010: 313,033,000 people

  Quadratic model

  2008: 304,983,400 people

  2009: 307,642,250 people

  2010: 310,260,000 people

  Answers will vary. Sample answer: The predictions given by all three models are relatively close to each other and seem reasonable.

47. (a) ≈ 7.906 (b) ≈ 7.684

49. (a) 20 decibels   (b) 70 decibels

51. $\approx 1.585 \times 10^{-6}$ **53.** $\approx 31,623$ **55.** 3:00 A.M.

57.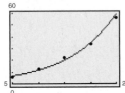

  The data fit an exponential model.

  Exponential model: $n = 3.9405e^{0.1086t}$

  A logistic growth model would be more appropriate for this data because after the initial rapid growth in productivity, the worker's production rate will eventually level off.

REVIEW EXERCISES   (page 398)

1. d   **2.** f   **3.** a   **4.** b

5. c   **6.** e   **7.** h   **8.** g

9. 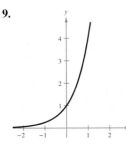   **11.**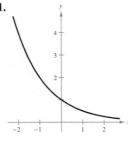

13. 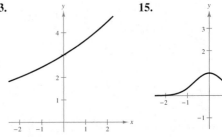   **15.**

17.

n	1	2	4
A	$13,681.11	$13,972.87	$14,127.43

n	12	365	Continuous
A	$14,233.93	$14,286.46	$14,288.26

19.

t	1	10	20
P	$185,548.70	$94,473.31	$44,626.03

t	30	40	50
P	$21,079.84	$9957.41	$4703.55

21. $8471.94   **23.** $\log_4 64 = 3$ **25.** $\ln 7.3890 \ldots = 2$

27. $3^4 = 81$ **29.** $e^0 = 1$ **31.** 5   **33.** 7   **35.** $-\frac{1}{2}$

37.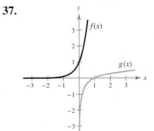

39. Domain: $(3, \infty)$

  Vertical asymptote: $x = 3$

  x-intercept: $(4, 0)$

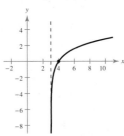

41. Domain: $(0, \infty)$

Vertical asymptote: $x = 0$

x-intercept: $(1, 0)$

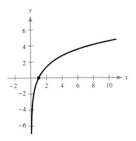

43. The average score decreased from 82 to about 68.

45. (a) 53.42 inches

(b)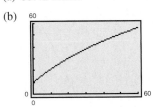

47. 2.096   **49.** 2.132   **51.** 0.9208   **53.** 0.2823

55. 2   **57.** 3.2   **59.** $\log_{10} x - \log_{10} y$

61. $\ln x + \frac{1}{2}\ln(x - 3)$ **63.** $4\log_5(y - 3)$ **65.** $\log_4 6$

67. $\ln\sqrt{x}$ **69.** $\ln\dfrac{x}{(x - 3)(x + 1)}$

71. $\ln y = \frac{4}{3}\ln x$ **73.** $\ln 8 \approx 2.079$ **75.** 1

77. $\ln 4 \approx 1.386$ **79.** $\frac{1}{3}e^{8.2} \approx 1213.650$

81. $\frac{1}{5}e^2 \approx 1.478$ **83.** $3e^2 \approx 22.167$ **85.** 7

87. (a) ≈ 197 desks   (b) ≈ 257 desks

89. 10.63 g

91. (a) 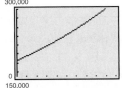   (b) 2016 ($t \approx 16.73$)

93. ≈ 9.93 hours to double

≈ 15.74 hours to triple

95. (a) $N = 50(1 - e^{-0.04838t})$ (b) 48 days

97. (a) $k \approx 0.2121$

(b) ≈ 196 deer; ≈ 294 deer; ≈ 383 deer

99. Yes

CHAPTER TEST   (page 402)

1. 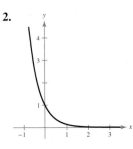   **2.**

3.   **4.**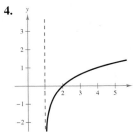

5. 87; ≈ 79.8; ≈ 76.5

6. After 12 months: ≈ 70.3

After 18 months: ≈ 67.8

Human memory diminishes slowly over time.

7. $2\ln x + 3\ln y - \ln z$

8. $\log_{10} 3 + \log_{10} x + \log_{10} y + 2\log_{10} z$

9. $\log_2 x + \frac{1}{3}\log_2(x - 2)$ **10.** $\frac{1}{5}\log_8(x^2 + 1)$

11. $\ln\dfrac{x^2 y^3}{z}$ **12.** $\log_{10}\sqrt[3]{x^2 y^2}$ **13.** $\dfrac{\log_2 21}{4}$

14. $\ln 6 \approx 1.792$

$\ln 2 \approx 0.693$

15. 127   **16.** $e^6 - 2 \approx 401.429$ **17.** ≈ 16.3 years

18. About 2015 ($t \approx 15.4$). Because the exponent on e is positive, the population of the city is growing.

19. ≈ 9.9 hours   **20.** ≈ 2.97 grams; ≈ 0.88 gram

21. Answers will vary. Sample answer:

The growth of the bear population will slow down as the population approaches the carrying capacity of the island. So, the population will grow logistically.

CUMULATIVE TEST: CHAPTERS 2–4
(page 403)

1. $x^2 + 3x - 4$ **2.** $x^2 - 3x + 6$

3. $3x^3 - 5x^2 + 3x - 5$ **4.** $\dfrac{x^2 + 1}{3x - 5}, x \neq \dfrac{5}{3}$

5. $9x^2 - 30x + 26$ **6.** $3x^2 - 2$

7.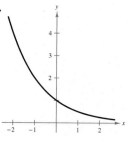

Domain: $(-\infty, \infty)$
Range: $[3, \infty)$

8.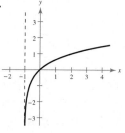

Domain: $(-\infty, 3) \cup (3, \infty)$
Range: $(-\infty, 0) \cup (0, \infty)$

9.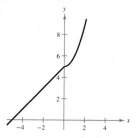

Domain: $(-\infty, \infty)$
Range: $(0, \infty)$

10.

Domain: $(-1, \infty)$
Range: $(-\infty, \infty)$

11.

Domain: $(-\infty, \infty)$
Range: $(-\infty, \infty)$

12. 75,000 units   **13.** $45 - 7i$ **14.** $-9 + 40i$

15. $-\dfrac{1}{29} - \dfrac{17}{29}i$ **16.** $\dfrac{5 \pm \sqrt{59}\,i}{6}$

17. $-4i, 4i, i, -i$. Because a given zero is $4i$ and $f(x)$ is a polynomial with real coefficients, $-4i$ is also a zero. Using these two zeros, you can form the factors $(x + 4i)$ and $(x - 4i)$. Multiplying these two factors produces $x^2 + 16$. Using long division to divide $x^2 + 16$ into f produces $x^2 + 1$. Then, factoring $x^2 + 1$ gives the zeros i and $-i$.

18. (a) $3x - 2 + \dfrac{2 - 3x}{2x^2 + 1}$

(b) $2x^3 - x^2 + 2x - 10 + \dfrac{25}{x + 2}$

19. $\ln 6 \approx 1.792$ **20.** $3 + e^{12} \approx 162{,}757.791$

21.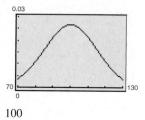

100

CHAPTER 5

SECTION 5.1   *(page 412)*

Skills Review   *(page 412)*

1. 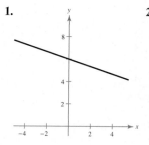   **2.**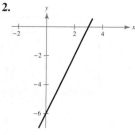

3. 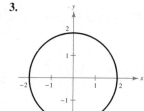   **4.**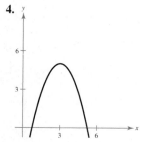

5. x **6.** $-37v$ **7.** $2x^2 + 9$ **8.** -1
9. $x = 6$ **10.** $y = 1$

1. (a) No   (b) Yes   **3.** (a) Yes   (b) No
5. (a) No   (b) No   **7.** $(2, -3)$ **9.** $(-1, 2), (2, 5)$
11. $(-2, -2), (0, 0), (2, 2)$ **13.** $(0, 4), \left(\frac{12}{5}, -\frac{16}{5}\right)$
15. $(-1, 0), (1, 0)$ **17.** $(-1, 1)$ **19.** $\left(\frac{1}{2}, 3\right)$
21. $(10, 3)$ **23.** $(1.5, 0.3)$ **25.** $\left(\frac{20}{3}, \frac{40}{3}\right)$
27. No solution
29. $\left(1 + \sqrt{2}, 2 + 2\sqrt{2}\right), \left(1 - \sqrt{2}, 2 - 2\sqrt{2}\right)$
31. $\left(\frac{29}{10}, \frac{21}{10}\right), (-2, 0)$ **33.** No solution   **35.** $(0, 1), (\pm 1, 0)$
37. $(2, 1)$ **39.** $(4, 3)$ **41.** $\left(\frac{5}{2}, \frac{3}{2}\right)$ **43.** $(2, 2), (4, 0)$
45. $(1, 4), (4, 7)$ **47.** $\left(4, -\frac{1}{2}\right)$ **49.** No solution

51. One solution   **53.** Two solutions   **55.** No solution

57. $\left(\frac{1}{2}, \frac{3}{4}\right), (-3, -1)$ **59.** $(1, 4), (4, 7)$ **61.** $(0, 1)$

63. No points of intersection   **65.** 192 units

67. 233,333 units   **69.** 1500 CDs   **71.** 1996 $(t \approx 5.70)$

73. Yes, at age 15.   **75.** \$15,000 at 8.5%, \$20,000 at 12%

77. \$150,000

79. According to the graphs, Federal Perkins Loan awards will exceed Federal Pell Grant awards. Both models eventually decrease and become negative. So, it is unlikely that these models will continue to be accurate.

SECTION 5.2   *(page 423)*

Skills Review   *(page 423)*

1.   **2.**

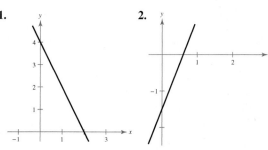

3. $x - y + 4 = 0$ **4.** $5x + 3y - 28 = 0$

5. $-\frac{1}{2}$ **6.** $\frac{7}{4}$ **7.** Perpendicular

8. Parallel   **9.** Neither parallel nor perpendicular

10. Perpendicular

---

1. $(2, 2)$ **3.** $(2, 0)$ **5.** Inconsistent

7. $(2a, 3a - 3)$, where a is any real number

9. $\left(-\frac{1}{3}, -\frac{2}{3}\right)$ **11.** $\left(2, \frac{1}{2}\right)$; consistent

13. $(-7, -13)$; consistent   **15.** $(4, -5)$; consistent

17. $(4, -1)$; consistent   **19.** $(40, 40)$; consistent

21. $\left(-\frac{6}{35}, \frac{43}{35}\right)$; consistent   **23.** No solution; inconsistent

25. $\left(\frac{18}{5}, \frac{3}{5}\right)$; consistent   **27.** $\left(\frac{19}{7}, -\frac{2}{7}\right)$; consistent

29. $\left(a, \frac{5}{6}a - \frac{1}{2}\right)$, where a is any real number; consistent

31. $\left(\frac{90}{31}, -\frac{67}{31}\right)$; consistent

33. No. The solution of the system is $(79,400, 398)$.

35. $\begin{cases} x + y = 13 \\ x - y = 3 \end{cases}$; $(8, 5)$ **37.** $\begin{cases} 2r + s = 8 \\ r - s = 7 \end{cases}$; $(5, -2)$

39. 550 miles per hour; 50 miles per hour

41. $6\frac{2}{3}$ gallons of 20%   **43.** \$10,000 at 9.5%
    $3\frac{1}{3}$ gallons of 50%   \$15,000 at 14%

45. Yes; Let the number of adult tickets $= a$ and the number of children's tickets $= c$. By solving the system of equations $a + c = 740$, $4688 = 8.5a + 4c$, you obtain $a = 384$, $c = 356$.

47. $x \approx 309,091$ units; $p \approx \$25.09$

49. $x = 2,000,000$ units; $p = \$100.00$

51. (a) Fast-food   Full-service

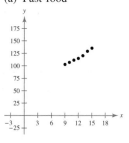   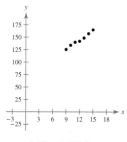

$y = 5.40x + 52.6$ $y = 6.19x + 70.2$

(b) No

53. 1380 units at \$810.60   **55.** $y = 0.97x + 2.1$

57. $y = 0.318x + 4.061$ **59.** $y = -2x + 4$

61. (a) $y = 161t + 605$

(b)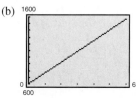

1571 cars

(c) You obtain the same model: $y = 161t + 605$.

63. Answers will vary. Sample answer:

$\begin{cases} 3x - y = 6 \\ 6x - 2y = 12 \end{cases}$; $(a, 3a - 6)$, a is any real number.

Using the method of elimination, you obtain the statement $0 = 0$, which is true for all values of the variables. So, the system has infinitely many solutions.

65. (a) Any value of $k \neq 3$ (b) $k = 3$

For the system to be inconsistent, the lines must have the same slope and different y-intercepts. So, any value of k except for $k = 3$ will produce an inconsistent system. For the system to be consistent (dependent), the lines must have the same slope and the same y-intercept. So, $k = 3$ will produce a consistent (dependent) system.

SECTION 5.3   *(page 435)*

> **Skills Review**   *(page 435)*
>
> **1.** $(15, 10)$ **2.** $\left(-2, -\frac{8}{3}\right)$ **3.** $(28, 4)$
>
> **4.** $(4, 3)$ **5.** Not a solution   **6.** Not a solution
>
> **7.** Solution   **8.** Solution   **9.** $5a + 2$
>
> **10.** $a + 13$

1. c   **2.** a   **3.** b   **4.** d

5. Yes. The system has a "stair-step" pattern with leading coefficients of 1.

7. No. The system has a "stair-step" pattern, but not all of its leading coefficients are 1.

9. $(4, -2, -2)$ **11.** $(2, -3, -2)$ **13.** $(-1, -6, 8)$

15. Inconsistent   **17.** $\left(1, -\frac{3}{2}, \frac{1}{2}\right)$

19. $(-3a + 10, 5a - 7, a)$ **21.** $\left(-4a + 13, -\frac{15}{2}a + \frac{45}{2}, a\right)$

23. Inconsistent   **25.** $(-3, 4, 2)$ **27.** $(3, -1, 2)$

29. $\left(\frac{3}{4}a, -2a, a\right)$ **31.** $(-5a + 3, -a - 5, a)$

33. Inconsistent   **35.** $(1, 1, 1, 1)$

37. Answers will vary. Sample answer:
$$\begin{cases} 2x - y + z = 9 \\ y + z = 1, \\ z = 2 \end{cases} \begin{cases} -x + 2y - 4z = -13 \\ x + y + z = 4 \\ x + z = 5 \end{cases}$$

39. Answers will vary. Sample answer:
$$\begin{cases} x - y + z = 3 \\ y - z = -2, \\ z = -3 \end{cases} \begin{cases} x + 3y + 4z = -26 \\ 4x - y - 5z = 24 \\ x + 2y = -9 \end{cases}$$

41. Answers will vary.
Sample answer:

$a = 3$: $(3, -2, 3)$

$a = 6$: $(6, 1, 5)$

$a = -3$: $(-3, -8, -1)$

43. Answers will vary.
Sample answer:

$a = 2$: $(1, 6, 5)$

$a = 4$: $(2, 12, 5)$

$a = 0$: $(0, 0, 5)$

45. $y = 2x^2 + 3x - 4$ **47.** $y = -4x^2 + 2x + 1$

49. $x^2 + y^2 - 4x = 0$ **51.** $x^2 + y^2 - 6x + 6y + 9 = 0$

53. $900,000 at 7%
$300,000 at 8%
$300,000 at 10%

55. 15,000 units of $15 candles
30,000 units of $10 candles
5000 units of $5 candles

57. 18 gallons of spray X
1 gallon of spray Y
6 gallons of spray Z

59. Invest $33,333.33 + 0.8a$ in certificates of deposit, $341,666.67 - 0.8a$ in municipal bonds, $125,000.00 - a$ in blue-chip stocks, and a in growth or speculative stocks, where $0 \le a \le 125,000$.

61. $y = 0.079x^2 + 0.63x + 2.9$

63. $y = -0.207x^2 - 0.89x + 5.1$

65. (a) $y = 0.421x^2 - 0.99x + 14.8$ (b) Same model

67. (a) $y = -0.371x^2 + 2.52x + 60.1$

(b) Same model   (c) $\approx 63.4\%$

69. (a) $y = 0.125x^2 - 2.55x + 18$

(b) $y = 0.125x^2 - 2.55x + 18$ (c) No

71. Yes. A system of linear equations can have three possible types of solutions: exactly one solution, infinitely many solutions, or no solution.

73. The solution $(-a, 2a - 1, a)$ is generated by solving for x and y in terms of z. The solution $(b, -2b - 1, -b)$ is generated by solving for y and z in terms of x. Both methods yield equivalent solutions.

75. Answers will vary.

MID-CHAPTER QUIZ   *(page 440)*

1. $(2, 5)$ **2.** No solution   **3.** $(1, 3)$

4. $\left(-\frac{2}{5} \pm \frac{2\sqrt{11}}{5}, \frac{1}{5} \pm \frac{4\sqrt{11}}{5}\right)$ **5.** 1500 units

6. 500,000 units   **7.** $(2, -1)$ **8.** $\left(1, \frac{3}{2}\right)$

9. $x = 5000$ units
$p = \$40$

10. $y = 0.62x + 40.0$ **11.** $(1, -2, 3)$

12. Answers will vary.

Sample answer: $(a + 6, a + 6, a)$, a is any real number.

13. Inconsistent   **14.** $y = -0.5829x^2 + 3.782x + 59.85$

SECTION 5.4   *(page 448)*

> **Skills Review**   *(page 448)*
>
> **1.** Line   **2.** Parabola   **3.** Circle   **4.** Parabola
>
> **5.** Line   **6.** Circle   **7.** $(1, 1)$ **8.** $(2, 0)$
>
> **9.** $(2, 1), \left(-\frac{5}{2}, -\frac{5}{4}\right)$ **10.** $(2, 3), (3, 2)$

1. d   **2.** b   **3.** a   **4.** c   **5.** f   **6.** e

7.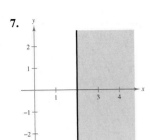

9.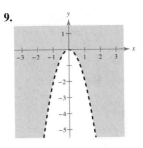

27.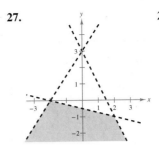

29.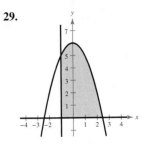

11.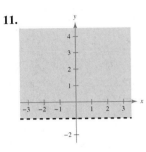

13.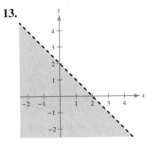

31.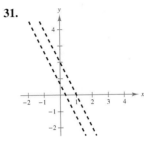

33.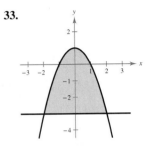

15.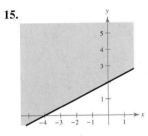

17.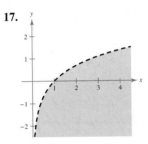

35.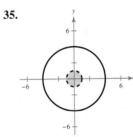

37.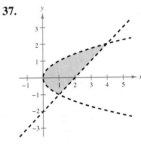

19.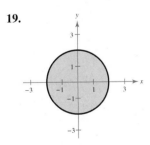

21.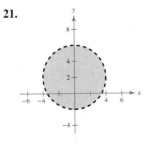

39.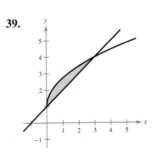

41.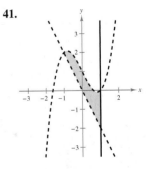

23.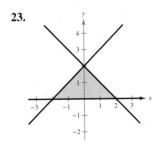

25.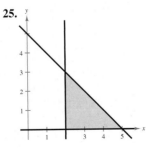

43.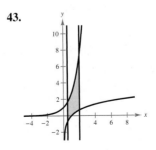

45. $0 \le x \le 8, 0 \le y \le 6$

47. $y \le \frac{4}{3}x, y \ge 0, y \le -4x + 16$

49. $x^2 + y^2 \le 16, x \ge 0, y \ge 0$

51. (a) $2x + \frac{3}{2}y \le 18, \frac{3}{2}x + \frac{3}{2}y \le 15, x \ge 0, y \ge 0$

(b)

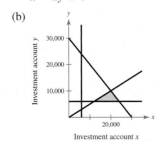

Number of tables
produced per day

53. Consumer surplus: \$4,777,001.41

Producer surplus: \$477,545.60

55. Consumer surplus: \$40,000,000

Producer surplus: \$20,000,000

57. The consumer surplus and producer surplus are equal when the slope of the demand equation is the negative of the slope of the supply equation.

59. (a) $x + y \le 30,000$

$x \ge 6000$

$y \ge 6000$

$x - 2y \ge 0$

(b)

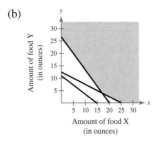

Investment account x

61. (a) $20x + 15y \ge 400$

$10x + 20y \ge 250$

$15x + 20y \ge 220$

$x \ge 0$

$y \ge 0$

(b)

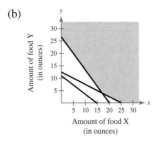

Amount of food X
(in ounces)

(c) Answers will vary. Sample answer: The nutritionist could give 10 ounces of food X and 15 ounces of food Y.

63. (a) $y \ge 0.5(220 - x)$

$y \le 0.75(220 - x)$

$x \ge 20$

$x \le 70$

(b)

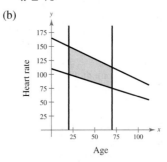

Age

(c) Answers will vary.

65. (a)

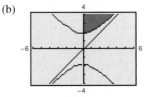

(b) \$308.35 billion

67. Answers will vary. Sample answer:

$$\begin{cases} y \le \frac{1}{3}x + 4 \\ y \le -\frac{1}{3}x + 4 \\ y \ge 0 \end{cases}$$

69. (a) $\begin{cases} \pi y^2 - \pi x^2 \ge 10 \\ \quad\quad\quad y > x \\ \quad\quad\quad x > 0 \end{cases}$

(b)

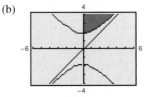

(c) The line is an asymptote to the boundary. The larger the circles, the closer the radii can be and the constraint will still be satisfied.

SECTION 5.5   (page 457)

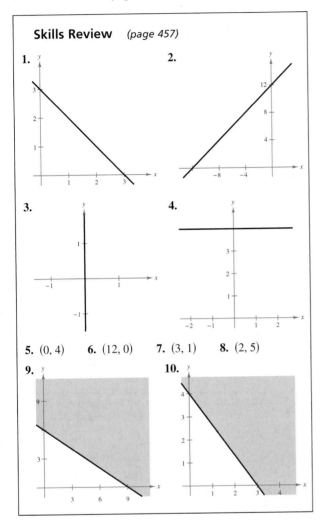

Skills Review   (page 457)

5. (0, 4)   **6.** (12, 0)   **7.** (3, 1)   **8.** (2, 5)

1. Minimum value at (0, 0): 0
   Maximum value at (6, 0): 36

3. Minimum value at (0, 0): 0
   Maximum value at (6, 0): 48

5. Minimum value at (0, 0): 0
   Maximum value at (3, 4): 17

7. Minimum value at (0, 0): 0
   Maximum value at (4, 0): 20

9. Minimum value at (0, 0): 0
   Maximum value at any point on the line segment
   connecting the points (5, 0) and (0, 3): 30

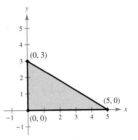

11. Minimum value at (0, 0): 0
   Maximum value at (5, 0): 45
   Same graph as in Answer 9

13. Minimum value at (5, 3): 35
   No maximum value

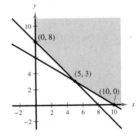

15. Minimum value at (10, 0): 20
   No maximum value
   Same graph as in Answer 13

17. Minimum value at (0, 0): 0
   Maximum value at any point on the line segment connecting
   the points (0, 20) and (10, 15): 40

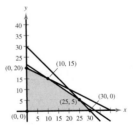

19. Minimum value at (0, 0): 0
   Maximum value at any point on the line segment connecting
   the points (25, 5) and (30, 0): 30
   Same graph as in Answer 17

21. Maximum value at $(3, 6)$: 12

23. Maximum value at any point on the line segment connecting the points $(0, 10)$ and $(3, 6)$: 30

25. Maximum value at $(4, 4)$: 28

27. Maximum value at $(7, 0)$: 84

29. Answers will vary. Sample answer: $z = 2x + 11y$

31. Answers will vary. Sample answer: $z = -x$

33. Answers will vary. Sample answer: $z = 2x + 5y$

35. Answers will vary. Sample answer: $z = 5x + 3y$

37. Crop A: 60 acres

    Crop B: 90 acres

    $33,150

39. Brand X: 3 bags

    Brand Y: 6 bags

    $240

41. Model A: 0 bicycles

    Model B: 1600 bicycles

    $120,000

43. 12 audits and 0 tax returns

45. Television: None

    Newspaper: $1,000,000

    250 million people

47. Type A: $62,500

    Type B: $187,500

    $26,875

49. Model A: 929 units

    Model B: 77 units

    $99,445

51. z is maximum at any point on the line segment connecting the vertices $(2, 0)$ and $\left(\frac{20}{19}, \frac{45}{19}\right)$.

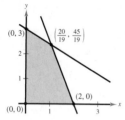

53. The constraint $x \le 10$ is extraneous. The maximum value of z occurs at $(0, 7)$.

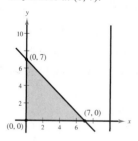

55. The constraint $2x + y \le 4$ is extraneous. The maximum value of z occurs at $(0, 1)$.

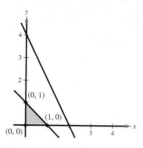

57. (a) Yes. The point $(1, 12)$ lies on the line segment connecting $(0, 14)$ and $(3, 8)$.

    (b) No. The point $(4, 6)$ lies outside the line segment connecting $(0, 14)$ and $(3, 8)$.

    (c) $(2, 10)$

59. Yes; The objective function also has the maximum value at any point on the line segment connecting the two vertices, so there are an infinite number of points that produce the maximum value.

REVIEW EXERCISES   *(page 462)*

1. $(-2, 4)$ **3.** $(8, -10)$ **5.** $(8, 6), (0, 10)$

7. $(1, 9), (1.5, 8.75)$

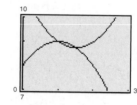

9. 800 plants

11. During the fourth month of the new format

13. $(3, -5)$

15. $\left(a, \frac{4}{3}a - \frac{10}{3}\right)$, where a is any real number

17. $\left(a, \frac{5}{8}a - \frac{7}{4}\right)$, where a is any real number

19. $(8, 9)$ **21.** The graph is a point. Solution: $(-1, 1)$

23. (a) $\begin{cases} 0.1x + 0.5y = 0.25(12) \\ x + y = 12 \end{cases}$

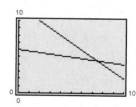

    (b) 7.5 gallons of 10% solution

    4.5 gallons of 50% solution

25. $x = 71,429$ units

$p = \$22.71$

27. $(3, 5, 2)$ **29.** $(2, -1, 3)$

31. $\left(\frac{1}{5}a + \frac{8}{5}, -\frac{6}{5}a + \frac{42}{5}, a\right)$, a is any real number

33. Inconsistent   **35.** $y = 2x^2 + x - 6$

37. $x^2 + y^2 - 4x + 2y - 4 = 0$

39. \$200,000 in certificates of deposit

    \$100,000 in municipal bonds

    \$15,000 in blue-chip stocks

    \$185,000 in growth or speculative stocks

41. $y = 1.01x + 1.54$

43. (a)

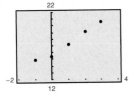

(b) $y = 1.49t + 15.9$

(c) $y = 0.150t^2 + 1.19t + 15.8$

(d) $y = 1.49t + 15.9$

    $y = 0.150t^2 + 1.19t + 15.8$

    They are the same as the regression models found in parts (b) and (c).

(e) Linear   Quadratic

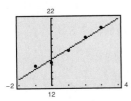 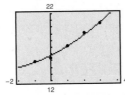

2006: \$21.86 billion   2006: \$22.96 billion

2007: \$23.35 billion   2007: \$25.50 billion

The predictions given by the quadratic model are greater than the predictions given by the linear model.

45.   **47.**

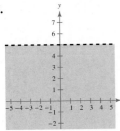 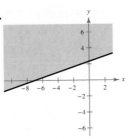

49.

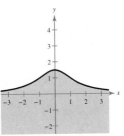

51.

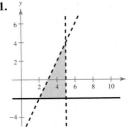

53.

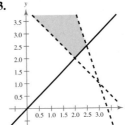

55.

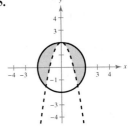

57.

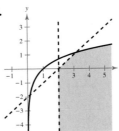

59. $\begin{cases} y \le 3x - 5 \\ y \le -x + 11 \\ y \ge \frac{1}{3}x + \frac{1}{3} \end{cases}$

61. Consumer surplus: \$4,000,000

    Producer surplus: \$6,000,000

63. $\begin{cases} 30x + 50y \ge 550,000 \\ x + y \ge 15,000 \\ x \ge 8000 \\ y \ge 4000 \end{cases}$

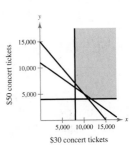

65. Minimum value at $(0, 0)$: 0

    Maximum value at $(3, 3)$: 81

67. Minimum value at $(0, 300)$: 18,000

    Maximum value at any point on the line segment connecting $(0, 500)$ and $(600, 0)$: 30,000

69. Minimum value at $(0, 0)$: 0
Maximum value at $(2, 5)$: 50

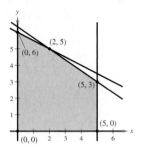

71. Minimum value at $(3, 0)$: 30
Maximum value at $(5, 4)$: 94

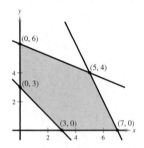

73. 700 units of the basic model
300 units of the deluxe model
$124,000

75. $270 model: 50 units
$455 model: 50 units
$3750

CHAPTER TEST   *(page 466)*

1. $(2, 4)$ **2.** $(-2, 5), (3, 0)$

3. $(1.68, 2.38), (-4.18, -26.88)$ **4.** $(3.36, -1.32)$

5. $(2, -1, 3)$ **6.** $(2, -3, 4)$

7. $60,000 at 9%   **8.** $(50,000, 34)$
$20,000 at 9.5%

9. $y = 0.450t^2 + 0.51t + 36.8$; 42.38 million

10.

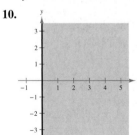

11.

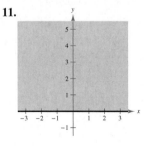

12.

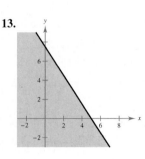

13.

14.

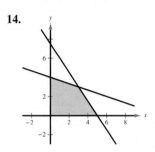

15. Minimum value at $(0, 0)$: 0
Maximum value at $(3, 3)$: 39

16. Model A: 297 units; model B: 570 units
$216,150; The profit model is $P = 200x + 275y$ with the constraints

$$\begin{cases} 3.5x + 8y \le 5600 \\ 2.5x + 2y \le 2000 \\ 1.3x + 0.9y \le 900 \\ x \ge 0 \\ y \ge 0 \end{cases}$$

and the maximum profit occurs at $(297, 570)$.

CHAPTER 6

SECTION 6.1   *(page 478)*

Skills Review   *(page 478)*

1. -3 **2.** 30   **3.** 6   **4.** $-\frac{1}{9}$ **5.** Solution

6. Not a solution   **7.** $(5, 2)$ **8.** $\left(\frac{12}{5}, -3\right)$

9. $(40, 14, 2)$ **10.** $\left(\frac{15}{2}, 4, 1\right)$

---

1. 2×3 **3.** 4×3 **5.** 4×2 **7.** 2×4

9. $\begin{bmatrix} 1 & 1 & 1 \\ 0 & -7 & -1 \end{bmatrix}$ **11.** $\begin{bmatrix} 1 & 0 & 14 & -11 \\ 0 & 1 & -2 & 2 \\ 0 & 0 & 1 & -7 \end{bmatrix}$

13. Add 5 times R_2 to R_1.

15. Interchange R_1 and R_2. Add 4 times R_1 to R_3.

17. Reduced row-echelon form

19. Not in row-echelon form   **21.** Not in row-echelon form

23. (a) $\begin{bmatrix} 1 & 1 & 2 \\ 0 & 1 & -9 \\ 2 & -1 & 1 \end{bmatrix}$ (b) $\begin{bmatrix} 1 & 1 & 2 \\ 0 & 1 & -9 \\ 0 & -3 & -3 \end{bmatrix}$

    (c) $\begin{bmatrix} 1 & 1 & 2 \\ 0 & 1 & -9 \\ 0 & 0 & -30 \end{bmatrix}$ (d) $\begin{bmatrix} 1 & 1 & 2 \\ 0 & 1 & -9 \\ 0 & 0 & 1 \end{bmatrix}$

25. $\begin{bmatrix} 1 & 2 & -1 & 5 \\ 0 & 1 & -1 & 1 \\ 0 & 0 & 1 & -2 \end{bmatrix}$ **27.** $\begin{bmatrix} 1 & -1 & -1 & 1 \\ 0 & 1 & 6 & 3 \\ 0 & 0 & 0 & 0 \end{bmatrix}$

29. $\begin{bmatrix} 1 & 0 & 0 \\ 0 & 1 & 0 \\ 0 & 0 & 1 \end{bmatrix}$ **31.** $\begin{bmatrix} 1 & 0 & 0 & 0 \\ 0 & 1 & 0 & 0 \\ 0 & 0 & 1 & 0 \\ 0 & 0 & 0 & 1 \end{bmatrix}$

33. $\begin{bmatrix} 1 & 0 \\ 0 & 1 \\ 0 & 0 \end{bmatrix}$

35. $\begin{cases} 2x + 4y = 6 \\ -x + 3y = -8 \end{cases}$ **37.** $\begin{cases} x \quad + 2z = -10 \\ \quad 3y - z = 5 \\ 4x + 2y \quad = 3 \end{cases}$

39. $\begin{bmatrix} 2 & -1 & \vdots & 3 \\ 5 & 7 & \vdots & 12 \end{bmatrix}$ **41.** $\begin{bmatrix} 1 & 10 & -3 & \vdots & 2 \\ 5 & -3 & 4 & \vdots & 0 \\ 2 & 4 & 0 & \vdots & 6 \end{bmatrix}$

43. $\begin{bmatrix} 9 & -3 & 20 & 1 & \vdots & 13 \\ 12 & 0 & -8 & 0 & \vdots & 5 \\ 1 & 2 & 3 & -4 & \vdots & -2 \\ -1 & -1 & 1 & 1 & \vdots & 1 \end{bmatrix}$

45. $\begin{cases} x - 5y = 6 \\ \quad y = -2 \end{cases}$ **47.** $\begin{cases} x + 3y - z = 15 \\ \quad y + 4z = -12 \\ \quad z = -5 \end{cases}$

    $(-4, -2)$ $(-14, 8, -5)$

49. $(-4, 6)$ **51.** $(-4, -8, 2)$

53. $(-2a - 4, -a + 6, a)$ **55.** $(3, 2)$ **57.** $(4, -2)$

59. $\left(\frac{1}{2}, -\frac{3}{4}\right)$ **61.** Inconsistent   **63.** $(-6, 8, 2)$

65. $\left(-\frac{3}{2}a + \frac{3}{2}, \frac{1}{3}a + \frac{1}{3}, a\right)$ **67.** Inconsistent

69. $(5a + 4, -3a + 2, a)$ **71.** $(0, 2 - 4a, a)$

73. $(-3b + 96a + 100, b, 52a + 54, a)$

75. $(0, 0)$ **77.** $(-2a, a, a)$ **79.** Yes; $(-1, 1, -3)$

81. No

83. \$1,200,000 was borrowed at 8%, \$200,000 was borrowed at 9%, and \$600,000 was borrowed at 12%.

85. Both are correct. Because there are infinitely many ordered triples that are solutions to this system, a solution can be written in many different ways. If $a = 3$, the ordered triple is $(3, -3, 5)$. You obtain the same triple when $b = 5$.

87. $y = 7.5t + 28$

133 new cases; Because the data values increased in a linear pattern, this estimate seems reasonable.

SECTION 6.2   *(page 492)*

Skills Review   *(page 492)*

1. -5 **2.** -7

3. Not in reduced row-echelon form

4. Not in reduced row-echelon form

5. $\begin{bmatrix} -5 & 10 & \vdots & 12 \\ 7 & -3 & \vdots & 0 \end{bmatrix}$

6. $\begin{bmatrix} 10 & 15 & -9 & \vdots & 42 \\ 6 & -5 & 0 & \vdots & 0 \end{bmatrix}$

7. $(0, 2)$ **8.** $(2 + a, 3 - a, a)$

9. $(1 - 2a, a, -1)$ **10.** $(2, -1, -1)$

1. $x = -3, y = 2$ **3.** $x = -2, y = 5$

5. (a) $\begin{bmatrix} 8 & -1 \\ 1 & 7 \end{bmatrix}$ (b) $\begin{bmatrix} 2 & -3 \\ 5 & -5 \end{bmatrix}$

    (c) $\begin{bmatrix} 15 & -6 \\ 9 & 3 \end{bmatrix}$ (d) $\begin{bmatrix} 9 & -8 \\ 13 & -9 \end{bmatrix}$

7. (a) $\begin{bmatrix} 7 & 3 \\ 1 & 9 \\ -2 & 15 \end{bmatrix}$ (b) $\begin{bmatrix} 5 & -5 \\ 3 & -1 \\ -4 & -5 \end{bmatrix}$

    (c) $\begin{bmatrix} 18 & -3 \\ 6 & 12 \\ -9 & 15 \end{bmatrix}$ (d) $\begin{bmatrix} 16 & -11 \\ 8 & 2 \\ -11 & -5 \end{bmatrix}$

9. (a) $\begin{bmatrix} 3 & 3 & -2 \\ -2 & 5 & 7 \\ 1 & -8 & 11 \end{bmatrix}$ (b) $\begin{bmatrix} 1 & 1 & 0 \\ 4 & -3 & -11 \\ 1 & 6 & -5 \end{bmatrix}$

    (c) $\begin{bmatrix} 6 & 6 & -3 \\ 3 & 3 & -6 \\ 3 & -3 & 9 \end{bmatrix}$ (d) $\begin{bmatrix} 4 & 4 & -1 \\ 9 & -5 & -24 \\ 3 & 11 & -7 \end{bmatrix}$

11. $\begin{bmatrix} -8 & -7 \\ 15 & -1 \end{bmatrix}$ **13.** $\begin{bmatrix} -24 & -4 & 12 \\ -12 & 32 & 12 \end{bmatrix}$

15. $\begin{bmatrix} 10 & 8 \\ -59 & 9 \end{bmatrix}$ **17.** $\begin{bmatrix} -17.143 & 2.143 \\ 11.571 & 10.286 \end{bmatrix}$

19. $\begin{bmatrix} -1.581 & -3.739 \\ -4.252 & -13.249 \\ 9.713 & -0.362 \end{bmatrix}$ **21.** $\begin{bmatrix} -6 & -9 \\ -1 & 0 \\ 17 & -10 \end{bmatrix}$

23. $\begin{bmatrix} 3 & 3 \\ -\frac{1}{2} & 0 \\ -\frac{13}{2} & \frac{11}{2} \end{bmatrix}$ **25.** $\begin{bmatrix} -7 & 10 & -12 & 17 \\ 6 & 21 & 7 & 15 \\ -3 & 3 & -5 & 6 \end{bmatrix}$

27. $\begin{bmatrix} -1 & 19 \\ 4 & -27 \\ 0 & 14 \end{bmatrix}$ **29.** $\begin{bmatrix} 1 & 0 & 0 \\ 0 & 1 & 0 \\ 0 & 0 & \frac{7}{2} \end{bmatrix}$

31. Not possible

33. $\begin{bmatrix} 41 & 7 & 7 \\ 42 & 5 & 25 \\ -10 & -25 & 45 \end{bmatrix}$ **35.** $\begin{bmatrix} 151 & 25 & 48 \\ 516 & 279 & 387 \\ 47 & -20 & 87 \end{bmatrix}$

37. Not possible

39. (a) $\begin{bmatrix} 0 & 15 \\ 6 & 12 \end{bmatrix}$ (b) $\begin{bmatrix} -2 & 2 \\ 31 & 14 \end{bmatrix}$ (c) $\begin{bmatrix} 9 & 6 \\ 12 & 12 \end{bmatrix}$

41. (a) $\begin{bmatrix} 3 & 5 \\ 1 & 6 \end{bmatrix}$ (b) $\begin{bmatrix} 11 & 5 & 0 \\ -7 & -4 & -1 \\ 14 & 8 & 2 \end{bmatrix}$

(c) Not possible

43. (a) $[11]$ (b) $\begin{bmatrix} -4 & 2 & 3 \\ 0 & 0 & 0 \\ -20 & 10 & 15 \end{bmatrix}$ (c) Not possible

45. (a) $\begin{bmatrix} -1 & 1 \\ -2 & 1 \end{bmatrix}\begin{bmatrix} x \\ y \end{bmatrix} = \begin{bmatrix} 4 \\ 0 \end{bmatrix}$ (b) $\begin{bmatrix} 4 \\ 8 \end{bmatrix}$

47. (a) $\begin{bmatrix} 1 & 2 \\ 3 & -1 \end{bmatrix}\begin{bmatrix} x \\ y \end{bmatrix} = \begin{bmatrix} 3 \\ 2 \end{bmatrix}$ (b) $\begin{bmatrix} 1 \\ 1 \end{bmatrix}$

49. (a) $\begin{bmatrix} 1 & -2 & 3 \\ -1 & 3 & -1 \\ 2 & -5 & 5 \end{bmatrix}\begin{bmatrix} x \\ y \\ z \end{bmatrix} = \begin{bmatrix} 9 \\ -6 \\ 17 \end{bmatrix}$ (b) $\begin{bmatrix} 1 \\ -1 \\ 2 \end{bmatrix}$

51. $\begin{bmatrix} 110 & 132 & 66 & 44 \\ 154 & 176 & 220 & 88 \end{bmatrix}$

53.

	Hotel x	Hotel y	Hotel z	
	97.75	105.80	126.50	Single
	115.00	138.00	149.50	Double
	126.50	149.50	161.00	Triple
	126.50	161.00	178.25	Quadruple

(Occupancy)

55. (a) $19,550 (b) $21,450

(c)

	Wholesale	Retail	
$ST =$	$15,850	$19,550	1
	$26,350	$30,975	2
	$21,450	$25,850	3

(Outlet)

ST represents the wholesale and retail prices of the computer inventories at the three outlets.

57. (a) $B = [2 \quad 0.5 \quad 3]$

(b)   Calories burned

120-lb   150-lb
person   person

$BA = [473.5 \quad 588.5]$

BA represents the total calories burned by each person.

59. Cannot perform operation.

61. Cannot perform operation. **63.** 2×2 **65.** 2×3

67. (a)

	X	Y	Z	
$D =$	3	1	2	Sacks
	0	2	3	Interceptions
	4	5	3	Key tackles

Each entry d_{ij} represents the number of each type of defensive play made by each player.

(b)   Sacks Interceptions Tackles

$B = [\$2000 \qquad \$1000 \qquad \$800]$

Each entry b_{ij} represents the bonus received for each type of play.

(c)   X   Y   Z

$BD = [\$9200 \quad \$8000 \quad \$9400]$

Each entry represents the bonus each player will receive.

(d) Player Z

69. $\begin{bmatrix} 0.40 & 0.15 & 0.15 \\ 0.28 & 0.53 & 0.17 \\ 0.32 & 0.32 & 0.68 \end{bmatrix}$

71. $\begin{bmatrix} -2 & 3 \\ 1 & -1 \end{bmatrix}$

Matrix is unique.

73. $AC = \begin{bmatrix} 12 & -6 & 9 \\ 16 & -8 & 12 \\ 4 & -2 & 3 \end{bmatrix} = BC$, but $A \neq B$.

75. True. $\begin{bmatrix} 3 & 2 \\ 1 & -4 \end{bmatrix}\begin{bmatrix} 1 & 0 \\ 0 & 1 \end{bmatrix} = \begin{bmatrix} 3 & 2 \\ 1 & -4 \end{bmatrix}$ and

$\begin{bmatrix} 1 & 0 \\ 0 & 1 \end{bmatrix}\begin{bmatrix} 3 & 2 \\ 1 & -4 \end{bmatrix} = \begin{bmatrix} 3 & 2 \\ 1 & -4 \end{bmatrix}$

77. (a) Gold subscribers: 28,750

Galaxy subscribers: 35,750

Nonsubscribers: 35,500

Multiply the original matrix by the 3×1 matrix

$\begin{bmatrix} 25,000 \\ 30,000 \\ 45,000 \end{bmatrix}$

which represents the current numbers of subscribers for each company and the number of nonsubscribers.

(b) Gold subscribers: 30,813

    Galaxy subscribers: 39,675

    Nonsubscribers: 29,513

    Multiply the original matrix by the 3×1 matrix

$$\begin{bmatrix} 28{,}750 \\ 35{,}750 \\ 35{,}500 \end{bmatrix}$$

    which represents the numbers of subscribers for each company and the number of nonsubscribers after 1 year.

(c) Gold subscribers: 31,947

    Galaxy subscribers: 42,330

    Nonsubscribers: 25,724

    Multiply the original matrix by the 3×1 matrix

$$\begin{bmatrix} 30{,}813 \\ 39{,}675 \\ 29{,}513 \end{bmatrix}$$

    which represents the numbers of subscribers for each company and the number of nonsubscribers after 2 years.

(d) The number of subscribers to each company is increasing each year. The number of nonsubscribers is decreasing each year.

SECTION 6.3   *(page 503)*

Skills Review   *(page 503)*

1. $\begin{bmatrix} 4 & 24 \\ 0 & -16 \\ 48 & 8 \end{bmatrix}$
  2. $\begin{bmatrix} \frac{11}{2} & 5 & 24 \\ \frac{1}{2} & 0 & 8 \\ 0 & 1 & 4 \end{bmatrix}$

3. $\begin{bmatrix} -5 & -2 & -13 \\ 4 & -13 & -2 \end{bmatrix}$
  4. $\begin{bmatrix} -13 & 11 \\ -19 & 21 \end{bmatrix}$

5. $\begin{bmatrix} 1 & 0 \\ 0 & 1 \end{bmatrix}$
  6. $\begin{bmatrix} 6 & 5 \\ 3 & -2 \end{bmatrix}$
  7. $\begin{bmatrix} 1 & 0 & 0 \\ 0 & 1 & 0 \\ 0 & 0 & 1 \end{bmatrix}$

8. $\begin{bmatrix} 1 & 0 & 0 \\ 0 & 1 & 0 \\ 0 & 0 & 1 \end{bmatrix}$
  9. $\left[\begin{array}{cc:cc} 1 & 0 & 3 & -2 \\ 0 & 1 & 4 & -3 \end{array}\right]$

10. $\left[\begin{array}{ccc:ccc} 1 & 0 & 0 & -6 & -4 & 3 \\ 0 & 1 & 0 & 11 & 6 & -5 \\ 0 & 0 & 1 & -2 & -1 & 1 \end{array}\right]$

1–9. $AB = I$ and $BA = I$
  11. $\begin{bmatrix} 7 & -2 \\ -3 & 1 \end{bmatrix}$

13. $\begin{bmatrix} 0 & -1 \\ 1 & 11 \end{bmatrix}$
  15. $\begin{bmatrix} \frac{1}{4} & \frac{1}{2} \\ -\frac{1}{4} & -1 \end{bmatrix}$
  17. $\begin{bmatrix} \frac{1}{2} & -\frac{1}{3} \\ \frac{1}{4} & 0 \end{bmatrix}$

19. Does not exist

21. $\begin{bmatrix} 1 & 1 & -1 \\ -3 & 2 & -1 \\ 3 & -3 & 2 \end{bmatrix}$
  23. $\dfrac{1}{2}\begin{bmatrix} -3 & 3 & 2 \\ 9 & -7 & -6 \\ -2 & 2 & 2 \end{bmatrix}$

25. $\begin{bmatrix} \frac{1}{3} & 0 & 0 \\ 0 & -\frac{1}{2} & 0 \\ 0 & 0 & \frac{1}{4} \end{bmatrix}$
  27. $\begin{bmatrix} 1 & 0 & 0 \\ -\frac{3}{4} & \frac{1}{4} & 0 \\ \frac{7}{20} & -\frac{1}{4} & \frac{1}{5} \end{bmatrix}$

29. Does not exist

31. $\begin{bmatrix} -175 & 37 & -13 \\ 95 & -20 & 7 \\ 14 & -3 & 1 \end{bmatrix}$
  33. $\dfrac{1}{11}\begin{bmatrix} 0 & -20 & 10 \\ -110 & 55 & 55 \\ 110 & -30 & -40 \end{bmatrix}$

35. Does not exist

37. $\begin{bmatrix} -24 & 7 & 1 & -2 \\ -10 & 3 & 0 & -1 \\ -29 & 7 & 3 & -2 \\ 12 & -3 & -1 & 1 \end{bmatrix}$
  39. $\dfrac{1}{19}\begin{bmatrix} 3 & 2 \\ -2 & 5 \end{bmatrix}$

41. Does not exist

43. $\dfrac{1}{59}\begin{bmatrix} 16 & 15 \\ -4 & 70 \end{bmatrix}$
  45. $(-2, 1)$
  47. $(4, 2)$

49. $(-2, 3)$
  51. $(2, 0)$
  53. $(3, 8, -11)$

55. $(2, 1, 0, 0)$
  57. $(2, -2)$
  59. Inconsistent

61. $(-4, -8)$
  63. $(-1, 3, 2)$
  65. $(5, 0, -2, 3)$

67. $\begin{cases} 2x + y + 3z = 16 \\ 4x - 2z = -2 \\ 3y + 2z = 1 \end{cases}$

69. AAA bonds: \$20,000
  71. AAA bonds: \$21,000

  A bonds: \$5000
         A bonds: \$5000

  B bonds: \$10,000
       B bonds: \$10,000

73. $I_1 = 4$ amperes, $I_2 = 1$ ampere, $I_3 = 5$ amperes

75. 100 bags of potting soil for seedlings

    100 bags of potting soil for general potting

    100 bags of potting soil for hardwood plants

77. 5 bags of potting soil for seedlings

    100 bags of potting soil for general potting

    120 bags of potting soil for hardwood plants

79. (a) $\begin{bmatrix} 99.28 & -17.66 & 0.76 \\ -17.66 & 3.17 & -0.14 \\ 0.76 & -0.14 & 0.006 \end{bmatrix}$

    $y = -0.08t^2 + 3.1t - 5$

  (b) 24.58 billion

  (c) The estimates are close and both seem reasonable.

81. $AB = \begin{bmatrix} 13 & 4 \\ 1 & 8 \end{bmatrix}$, $BA = \begin{bmatrix} 8 & 1 \\ 4 & 13 \end{bmatrix}$

Row 1 of AB is Row 2 of BA with reversed entries.
Row 2 of AB is Row 1 of BA with reversed entries.

83. Answers will vary. Sample answer:
If $k = 3$, then

$$\begin{bmatrix} 4 & 3 \\ -2 & 3 \end{bmatrix}^{-1} = \begin{bmatrix} \frac{1}{6} & -\frac{1}{6} \\ \frac{1}{9} & \frac{2}{9} \end{bmatrix}.$$

If $k = -\frac{3}{2}$, the matrix is singular.

85. True. The inverse of I_n is I_n.

87. Answers will vary.

MID-CHAPTER QUIZ   *(page 507)*

1. Any matrix with four rows and three columns

2. Any matrix with three rows and one column

3. $\begin{bmatrix} 3 & 2 & \vdots & -2 \\ 5 & -1 & \vdots & 19 \end{bmatrix}$ **4.** $\begin{bmatrix} 1 & 0 & 3 & \vdots & -5 \\ 1 & 2 & -1 & \vdots & 3 \\ 3 & 0 & 4 & \vdots & 0 \end{bmatrix}$

5. $(2.769, -5.154)$ **6.** $(4, -2, -3)$ **7.** $\begin{bmatrix} -2 & 2 \\ 5 & 15 \end{bmatrix}$

8. $\begin{bmatrix} 9 & -2 & 23 \\ 14 & -4 & 36 \end{bmatrix}$ **9.** $\begin{bmatrix} -1 & 10 \\ -11 & 3 \end{bmatrix}$

10. $\begin{bmatrix} -6 & -2 \\ 3 & -5 \end{bmatrix}$ **11.** $\begin{bmatrix} 3 & -2 \\ 1 & -\frac{1}{2} \end{bmatrix}$ **12.** Not possible

13. $\begin{bmatrix} -3 & 16 \\ -12 & 16 \end{bmatrix}$ **14.** $\begin{bmatrix} 2 & -10 \\ 7 & -11 \end{bmatrix}$

15. $22.80   **16.** $41.40   **17.** $59.60

18.

$$LW = \begin{bmatrix} \$22.80 & \$20.20 \\ \$47.80 & \$41.40 \\ \$66.50 & \$59.60 \end{bmatrix} \begin{matrix} A \\ B \\ C \end{matrix} \text{ Model}$$

with headers Plant 1, Plant 2.

LW represents the total labor costs for each model at each plant.

19. $(4, -2)$ **20.** $(4, 2, -3)$

SECTION 6.4   *(page 515)*

Skills Review   *(page 515)*

1. $\begin{bmatrix} 3 & 5 \\ 4 & 0 \end{bmatrix}$ **2.** $\begin{bmatrix} -2 & 8 \\ 2 & -4 \end{bmatrix}$

3. $\begin{bmatrix} 9 & -12 & 6 \\ 3 & 0 & -3 \\ 0 & 3 & -6 \end{bmatrix}$ **4.** $\begin{bmatrix} 0 & 8 & 12 \\ -4 & 8 & 12 \\ -8 & 4 & -8 \end{bmatrix}$

5. -22 **6.** 35 **7.** -15 **8.** $-\frac{1}{8}$

9. -45 **10.** -16

---

1. -5 **3.** 1 **5.** 3 **7.** 0 **9.** 5 **11.** 4

13. $\frac{11}{6}$ **15.** 0.14 **17.** -0.838 **19.** 248

21. (a) $M_{11} = -5, M_{12} = 2, M_{21} = 4, M_{22} = 3$
    (b) $C_{11} = -5, C_{12} = -2, C_{21} = -4, C_{22} = 3$

23. (a) $M_{11} = -4, M_{12} = -2, M_{21} = 1, M_{22} = 3$
    (b) $C_{11} = -4, C_{12} = 2, C_{21} = -1, C_{22} = 3$

25. (a) $M_{11} = 3, M_{12} = -4, M_{13} = 1, M_{21} = 2, M_{22} = 2,$
    $M_{23} = -4, M_{31} = -4, M_{32} = 10, M_{33} = 8$
    (b) $C_{11} = 3, C_{12} = 4, C_{13} = 1, C_{21} = -2, C_{22} = 2,$
    $C_{23} = 4, C_{31} = -4, C_{32} = -10, C_{33} = 8$

27. (a) $M_{11} = 30, M_{12} = 12, M_{13} = 11, M_{21} = -36,$
    $M_{22} = 26, M_{23} = 7, M_{31} = -4,$
    $M_{32} = -42, M_{33} = 12$
    (b) $C_{11} = 30, C_{12} = -12, C_{13} = 11, C_{21} = 36,$
    $C_{22} = 26, C_{23} = -7, C_{31} = -4,$
    $C_{32} = 42, C_{33} = 12$

29. (a) -99 (b) -99 **31.** (a) -145 (b) -145

33. (a) 170 (b) 170 **35.** -58 **37.** -30 **39.** 0

41. 0 **43.** -0.002 **45.** 0 **47.** -108 **49.** 0

51. 412 **53.** -126 **55.** 0 **57.** -336 **59.** 410

61. -6 **63.** -16 **65.** 120 **67.** 40 **69.** -168

71. -20 **73.** -18

75. (a) -3 (b) -2 (c) $\begin{bmatrix} -2 & 0 \\ 0 & -3 \end{bmatrix}$ (d) 6

77. (a) -8 (b) 0 (c) $\begin{bmatrix} -4 & 4 \\ 1 & -1 \end{bmatrix}$ (d) 0

79. (a) -21 (b) -19 (c) $\begin{bmatrix} 7 & 1 & 4 \\ -8 & 9 & -3 \\ 7 & -3 & 9 \end{bmatrix}$ (d) 399

81. (a) 2 (b) -6 (c) $\begin{bmatrix} 1 & 4 & 3 \\ -1 & 0 & 3 \\ 0 & 2 & 0 \end{bmatrix}$ (d) -12

83. Matrices will vary. The determinant of each matrix is the product of the entries on the main diagonal, which in this case equals -18.

85. Matrices will vary. The determinant of each matrix is the product of the entries on the main diagonal, which in this case equals 28.

87. Rows 2 and 4 are identical.

89. Row 4 is a multiple of Row 2.

91. True. If an entire row is zeros, then each cofactor in the expansion is multiplied by zero.

93–95. Answers will vary.

SECTION 6.5 *(page 524)*

Skills Review *(page 524)*

1. 1   **2.** 0   **3.** -8 **4.** x^2 **5.** 8   **6.** 60

7. $\begin{bmatrix} 7 & -3 \\ -2 & 1 \end{bmatrix}$ **8.** $\begin{bmatrix} 1 & 2 & -1 \\ -1 & -2 & 2 \\ 2 & 5 & 0 \end{bmatrix}$

9. $\begin{bmatrix} 0.14 \\ 0.31 \end{bmatrix}$ **10.** $\begin{bmatrix} 7 & 14 \end{bmatrix}$

---

1. 11   **3.** 28   **5.** $\frac{33}{8}$ **7.** $\frac{5}{2}$ **9.** 28

11. $y = \frac{16}{5}$ or $y = 0$ **13.** $y = -3$ or $y = -11$

15. 307.5 square miles   **17.** Collinear

19. Not collinear   **21.** Collinear   **23.** Not collinear

25. $y = -4$ **27.** $x = 3$ **29.** $x - 6y + 13 = 0$

31. $x + 3y - 5 = 0$ **33.** $x = -4$

35. $2x + 3y - 8 = 0$

37. $\begin{bmatrix} 3 & 15 \end{bmatrix}\begin{bmatrix} 13 & 5 \end{bmatrix}\begin{bmatrix} 0 & 8 \end{bmatrix}\begin{bmatrix} 15 & 13 \end{bmatrix}\begin{bmatrix} 5 & 0 \end{bmatrix}\begin{bmatrix} 19 & 15 \end{bmatrix}\begin{bmatrix} 15 & 14 \end{bmatrix}$
48, 81, 28, 51, 24, 40, 54, 95, 5, 10, 64, 113, 57, 100

39. $\begin{bmatrix} 3 & 1 & 12 \end{bmatrix}\begin{bmatrix} 12 & 0 & 13 \end{bmatrix}\begin{bmatrix} 5 & 0 & 20 \end{bmatrix}\begin{bmatrix} 15 & 13 & 15 \end{bmatrix}$
$\begin{bmatrix} 18 & 18 & 15 \end{bmatrix}\begin{bmatrix} 23 & 0 & 0 \end{bmatrix}$
$-68, 21, 35, -66, 14, 39, -115, 35, 60, -62, 15, 32,$
$-54, 12, 27, 23, -23, 0$

41. $1, -25, -65, 17, 15, -9, -12, -62, -119,$
$27, 51, 48, 43, 67, 48, 57, 111, 117$

43. $-5, -41, -87, 91, 207, 257, 11, -5, -41,$
$40, 80, 84, 76, 177, 227$

45. 34, 55, 43, 20, 35, 28, 19, 36, 33, 16, 24, 12, 56, 107, 111

47. HAPPY NEW YEAR   **49.** SOUND ALL CLEAR

51. SEND MORE MONEY

53. MEET ME TONIGHT RON

55. Because $\begin{bmatrix} 45 & -35 \end{bmatrix}\begin{bmatrix} w & x \\ y & z \end{bmatrix} = \begin{bmatrix} 10 & 15 \end{bmatrix}$

and $\begin{bmatrix} 38 & -30 \end{bmatrix}\begin{bmatrix} w & x \\ y & z \end{bmatrix} = \begin{bmatrix} 8 & 14 \end{bmatrix}$,

you can solve $\begin{bmatrix} 45 & -35 \\ 38 & -30 \end{bmatrix}\begin{bmatrix} w & x \\ y & z \end{bmatrix} = \begin{bmatrix} 10 & 15 \\ 8 & 14 \end{bmatrix}$;

JOHN RETURN TO BASE

57. Answers will vary.

REVIEW EXERCISES *(page 529)*

1. 2×4 **3.** $\begin{bmatrix} 1 & 3 & 0 & 2 \\ 0 & 1 & 1 & 2 \\ 0 & 0 & 1 & 2 \end{bmatrix}$ **5.** $\begin{bmatrix} 1 & 0 & 0 \\ 0 & 1 & 0 \\ 0 & 0 & 1 \end{bmatrix}$

7. $(3, -2)$ **9.** $(3, 2, -1)$ **11.** $(10 - 4a, a, a)$

13. Inconsistent

15. \$40,000 was borrowed at 8%, \$120,000 was borrowed at 10%, and \$40,000 was borrowed at 12%.

17. (a) $\begin{bmatrix} 3 & 7 \\ -4 & 4 \end{bmatrix}$ (b) $\begin{bmatrix} -5 & 3 \\ 8 & -2 \end{bmatrix}$

(c) $\begin{bmatrix} -4 & 20 \\ 8 & 4 \end{bmatrix}$ (d) $\begin{bmatrix} -16 & 14 \\ 26 & -5 \end{bmatrix}$

19. (a) $\begin{bmatrix} 3 & 4 & 2 & 1 \\ 3 & -5 & 6 & 0 \end{bmatrix}$ (b) $\begin{bmatrix} -1 & 2 & -6 & 11 \\ -3 & 7 & 0 & 4 \end{bmatrix}$

(c) $\begin{bmatrix} 4 & 12 & -8 & 24 \\ 0 & 4 & 12 & 8 \end{bmatrix}$ (d) $\begin{bmatrix} -2 & 9 & -20 & 39 \\ -9 & 22 & 3 & 14 \end{bmatrix}$

21. $\begin{bmatrix} 4 & -11 \\ -3 & 1 \\ -1 & -3 \end{bmatrix}$ **23.** $\frac{1}{2}\begin{bmatrix} 3 & -5 \\ 1 & 4 \\ 9 & 14 \end{bmatrix}$ **25.** $\begin{bmatrix} 8 \\ 5 \\ -6 \end{bmatrix}$

27. $\begin{bmatrix} 1 & 0 & 0 \\ 0 & 1 & 0 \\ 0 & 0 & 1 \end{bmatrix}$ **29.** Not possible

31. (a) $\begin{bmatrix} 4 \end{bmatrix}$ (b) $\begin{bmatrix} 2 & -6 & 8 \\ -2 & 6 & -8 \\ -1 & 3 & -4 \end{bmatrix}$ (c) Not possible

33. $\begin{bmatrix} 96 & 144 & 24 & 48 \\ 48 & 72 & 96 & 24 \\ 168 & 72 & 120 & 96 \end{bmatrix}$

35. (a) \$8325   (b) \$5200

(c)   Wholesale   Retail

$ST = \begin{bmatrix} \$5200 & \$8265 \\ \$5075 & \$7985 \\ \$5125 & \$8325 \end{bmatrix}\begin{matrix} 1 \\ 2 \\ 3 \end{matrix} \Big\} \text{Outlet}$

ST represents the wholesale and retail values of the car sound system inventory at each outlet.

37. $AB = I$ and $BA = I$

39. $\begin{bmatrix} -1 & 0 & 0 \\ 0 & \frac{1}{2} & 0 \\ 0 & 0 & \frac{1}{4} \end{bmatrix}$ **41.** $\begin{bmatrix} -5 & 3 \\ 2 & -1 \end{bmatrix}$ **43.** $(3, 4)$

45. $\left(2, \frac{1}{2}, 3\right)$ **47.** $(-6, -1)$ **49.** $(2, -1, -2)$

51. 10 units of fluid X

8 units of fluid Y

5 units of fluid Z

53. (a) $\begin{bmatrix} \frac{23}{5} & -\frac{33}{10} & \frac{1}{2} \\ -\frac{33}{10} & \frac{187}{70} & -\frac{3}{7} \\ \frac{1}{2} & -\frac{3}{7} & \frac{1}{14} \end{bmatrix}$

$y = 0.129t^2 - 1.29t + 4.4$

(b) 4.4%   (c) The estimate is too high.

55. 4   **57.** 0

59. (a) $M_{11} = 4, M_{12} = 7, M_{21} = -1, M_{22} = 2$

(b) $C_{11} = 4, C_{12} = -7, C_{21} = 1, C_{22} = 2$

61. (a) $M_{11} = 30, M_{12} = -12, M_{13} = -21,$

$M_{21} = 20, M_{22} = 19, M_{23} = 22,$

$M_{31} = 5, M_{32} = -2, M_{33} = 19$

(b) $C_{11} = 30, C_{12} = 12, C_{13} = -21,$

$C_{21} = -20, C_{22} = 19, C_{23} = -22,$

$C_{31} = 5, C_{32} = 2, C_{33} = 19$

63. 44; Answers will vary.   **65.** -12; Answers will vary.

67. -39; Answers will vary.   **69.** 10   **71.** 10

73. Not collinear   **75.** Collinear

77. $x + 15y - 38 = 0$ **79.** $x = 2$

81. $[20 \ 18][1 \ 14][19 \ 13][9 \ 20][0 \ 14][15 \ 23]$

94, 132, 44, 59, 77, 109, 78, 107, 42, 56, 99, 137

83. SEIZE THE DAY

85. (a) Because $[-57 \ \ -13]\begin{bmatrix} w & x \\ y & z \end{bmatrix} = [23 \ 5]$ and

$[91 \ \ 26]\begin{bmatrix} w & x \\ y & z \end{bmatrix} = [0 \ 13]$, you can solve

$\begin{bmatrix} -57 & -13 \\ 91 & 26 \end{bmatrix}\begin{bmatrix} w & x \\ y & z \end{bmatrix} = \begin{bmatrix} 23 & 5 \\ 0 & 13 \end{bmatrix}$.

(b) WE MISS YOU BIG BOB

CHAPTER TEST   *(page 533)*

1. $\begin{bmatrix} 2 & 1 & 4 & \vdots & 2 \\ 1 & 4 & -1 & \vdots & 0 \\ -1 & 3 & 3 & \vdots & -1 \end{bmatrix}$

2. $\begin{bmatrix} 3 & 4 & 2 & \vdots & 4 \\ 2 & 3 & 0 & \vdots & -2 \\ 0 & 2 & -3 & \vdots & -13 \end{bmatrix}$

3. $\left(\frac{7}{3}a - \frac{10}{3}, -\frac{8}{3}a + \frac{29}{3}, a\right)$ **4.** $(-12, -16, -6)$

5. $(2, -3, 1)$ **6.** $\begin{bmatrix} 2 & 4 \\ 7 & 13 \end{bmatrix}$ **7.** $\begin{bmatrix} -4 & -8 \\ 13 & 29 \end{bmatrix}$

8. $\begin{bmatrix} 1 \\ 11 \end{bmatrix}$ **9.** $\begin{bmatrix} 7 & 15 \\ 10 & 22 \end{bmatrix}$ **10.** $\frac{1}{5}\begin{bmatrix} 4 & 1 \\ 3 & 2 \end{bmatrix}$

11. $\begin{bmatrix} 1 & 0 \\ 0 & 1 \end{bmatrix}$ **12.** $\frac{1}{5}\begin{bmatrix} -9 & 16 & -6 \\ 6 & -9 & 4 \\ 4 & -6 & 1 \end{bmatrix}$ **13.** -17

14. -23 **15.** -20 **16.** $(2, -2, 3)$

17. Matrices will vary. Sample answer:

$\begin{bmatrix} 2 & -2 \\ -2 & 2 \end{bmatrix} \times \begin{bmatrix} 3 & 3 \\ 3 & 3 \end{bmatrix} = \begin{bmatrix} 0 & 0 \\ 0 & 0 \end{bmatrix}$

18. $\frac{21}{2}$ square units   **19.** Collinear

20. $-x + y + 3 = 0$

21. $BA = [\$384,000 \quad \$631,000]$

BA represents the total value of each product at each warehouse.

CHAPTER 7

SECTION 7.1   *(page 544)*

Skills Review   *(page 544)*

1. (a) 7   (b) $c^2 - 3c + 3$

(c) $x^2 + 2xh + h^2 - 3x - 3h + 3$

2. (a) -4 (b) 10   (c) $3t^2 + 4$ **3.** h **4.** 4

5. Domain: $(-\infty, 0) \cup (0, \infty)$

Range: $(-\infty, 0) \cup (0, \infty)$

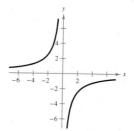

6. Domain: $[-6, 6]$ **7.** Domain: $(-\infty, \infty)$

Range: $[0, 6]$ Range: $[0, \infty)$

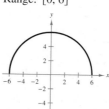
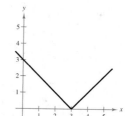

8. Domain: $(-\infty, 0) \cup (0, \infty)$

Range: $-\frac{1}{2}, \frac{1}{2}$

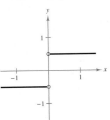

9. y is not a function of x.

10. y is a function of x.

9. (a) 1   (b) 3   **11.** (a) 1   (b) 3

13. (a) 12   (b) 27   (c) $\frac{1}{3}$

15. (a) 4   (b) 48   (c) 256

17. (a) 1   (b) 1   (c) 1

19. (a) 0   (b) 0   (c) 0

21. (a) 3   (b) -3 (c) Limit does not exist.

23. 4   **25.** -1 **27.** 0   **29.** 3   **31.** -2

33. $-\frac{3}{4}$ **35.** $\frac{35}{9}$ **37.** $\frac{1}{3}$ **39.** $-\frac{1}{20}$ **41.** 2

43. Limit does not exist.   **45.** Limit does not exist.

47. 12   **49.** Limit does not exist.   **51.** 2

53. -1 **55.** 2   **57.** $\dfrac{1}{2\sqrt{x+2}}$ **59.** $2t - 5$

1.

x	1.9	1.99	1.999	2
$f(x)$	8.8	8.98	8.998	?

x	2.001	2.01	2.1
$f(x)$	9.002	9.02	9.2

$\lim\limits_{x \to 2}(2x + 5) = 9$

3.

x	1.9	1.99	1.999	2
$f(x)$	0.2564	0.2506	0.2501	?

x	2.001	2.01	2.1
$f(x)$	0.2499	0.2494	0.2439

$\lim\limits_{x \to 2} \dfrac{x - 2}{x^2 - 4} = \dfrac{1}{4}$

5.

x	-0.1	-0.01	-0.001	0
$f(x)$	0.5132	0.5013	0.5001	?

x	0.001	0.01	0.1
$f(x)$	0.4999	0.4988	0.4881

$\lim\limits_{x \to 0} \dfrac{\sqrt{x + 1} - 1}{x} = 0.5$

7.

x	-0.5	-0.1	-0.01	-0.001	0
$f(x)$	-0.0714	-0.0641	-0.0627	-0.0625	?

$\lim\limits_{x \to 0^-} \dfrac{\dfrac{1}{x + 4} - \dfrac{1}{4}}{x} = -\dfrac{1}{16}$

61.

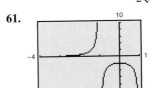

x	0	0.5	0.9	0.99
$f(x)$	-2	-2.67	-10.53	-100.5

x	0.999	0.9999	1
$f(x)$	-1000.5	$-10,000.5$	Undefined

$-\infty$

63.

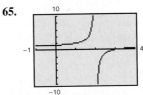

x	-3	-2.5	-2.1	-2.01
$f(x)$	-1	-2	-10	-100

x	-2.001	-2.0001	-2
$f(x)$	-1000	$-10,000$	Undefined

$-\infty$

65.

Limit does not exist.

67.

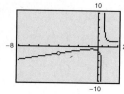

$-\frac{17}{9} \approx -1.8889$

69. (a) \$25,000   (b) 80%

(c) ∞; The cost function increases without bound as x approaches 100 from the left. Therefore, according to the model, it is not possible to remove 100% of the pollutants.

71. (a)

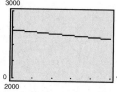

(b) For $x = 0.25$, $A \approx \$2685.06$.

For $x = \frac{1}{365}$, $A \approx \$2717.91$.

(c) $\lim\limits_{x \to 0^+} 1000(1 + 0.1x)^{10/x} = 1000e \approx \$2718.28$;

continuous compounding

SECTION 7.2   *(page 555)*

Skills Review   *(page 555)*

1. $\dfrac{x + 4}{x - 8}$ **2.** $\dfrac{x + 1}{x - 3}$ **3.** $\dfrac{x + 2}{2(x - 3)}$ **4.** $\dfrac{x - 4}{x - 2}$

5. $x = 0, -7$ **6.** $x = -5, 1$ **7.** $x = -\frac{2}{3}, -2$

8. $x = 0, 3, -8$ **9.** 13   **10.** -1

---

1. Continuous; The function is a polynomial.

3. Not continuous ($x \neq \pm 2$)

5. Continuous; The rational function's domain is the set of real numbers.

7. Not continuous ($x \neq 3$ and $x \neq 5$)

9. Not continuous ($x \neq \pm 2$)

11. $(-\infty, 0)$ and $(0, \infty)$; Explanations will vary. There is a discontinuity at $x = 0$, because $f(0)$ is not defined.

13. $(-\infty, -1)$ and $(-1, \infty)$; Explanations will vary. There is a discontinuity at $x = -1$, because $f(-1)$ is not defined.

15. $(-\infty, \infty)$; Explanations will vary.

17. $(-\infty, -1)$, $(-1, 1)$, and $(1, \infty)$; Explanations will vary. There are discontinuities at $x = \pm 1$, because $f(\pm 1)$ is not defined.

19. $(-\infty, \infty)$; Explanations will vary.

21. $(-\infty, 4)$, $(4, 5)$, and $(5, \infty)$; Explanations will vary. There are discontinuities at $x = 4$ and $x = 5$, because $f(4)$ and $f(5)$ are not defined.

23. Continuous on all intervals $\left(\dfrac{c}{2}, \dfrac{c}{2} + \dfrac{1}{2}\right)$, where c is an integer. Explanations will vary. There are discontinuities at $x = \dfrac{c}{2}$ where c is an integer, because $\lim\limits_{x \to c} f\left(\dfrac{c}{2}\right)$ does not exist.

25. $(-\infty, \infty)$; Explanations will vary.

27. $(-\infty, 2]$ and $(2, \infty)$; Explanations will vary. There is a discontinuity at $x = 2$, because $\lim\limits_{x \to 2} f(2)$ does not exist.

29. $(-\infty, -1)$ and $(-1, \infty)$; Explanations will vary. There is a discontinuity at $x = -1$, because $f(-1)$ is not defined.

31. Continuous on all intervals $(c, c + 1)$, where c is an integer. Explanations will vary. There are discontinuities at $x = c$ where c is an integer, because $\lim\limits_{x \to c} f(c)$ does not exist.

33. $(1, \infty)$; Explanations will vary.   **35.** Continuous

37. Nonremovable discontinuity at $x = 2$

39.

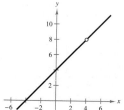

Continuous on $(-\infty, 4)$ and $(4, \infty)$

41.

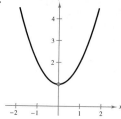

Continuous on $(-\infty, 0)$ and $(0, \infty)$

43.

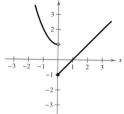

Continuous on $(-\infty, 0)$ and $(0, \infty)$

45. $a = 2$

47.

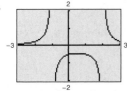

Not continuous at $x = 2$ and $x = -1$, because $f(-1)$ and $f(2)$ are not defined.

49.

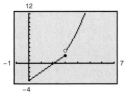

Not continuous at $x = 3$, because $\lim_{x \to 3} f(3)$ does not exist.

51.

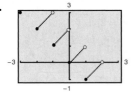

Not continuous at all integers c, because $\lim_{x \to c} f(c)$ does not exist.

53. $(-\infty, \infty)$

55. Continuous on all intervals $\left(\dfrac{c}{2}, \dfrac{c+1}{2} \right)$, where c is an integer.

57.

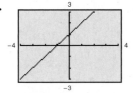

The graph of $f(x) = \dfrac{x^2 + x}{x}$ appears to be continuous on $[-4, 4]$, but f is not continuous at $x = 0$.

59. (a)

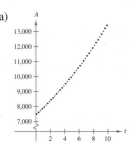

The graph has nonremovable discontinuities at $t = \frac{1}{4}, \frac{1}{2}, \frac{3}{4}, 1, \frac{5}{4}, \ldots$

(b) $11,379.17

61. $C = 12.80 - 2.50[\![1 - x]\!]$

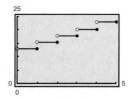

C is not continuous at $x = 1, 2, 3, \ldots$

63. (a)

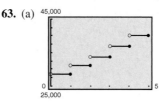

(b) $43,850.78

S is not continuous at $t = 1, 2, \ldots, 5$.

65. The model is continuous. The actual revenue probably would not be continuous, because the revenue is usually recorded over larger units of time (hourly, daily, or monthly). In these cases, the revenue may jump between different units of time.

SECTION 7.3   *(page 566)*

Skills Review   *(page 566)*

1. $x = 2$ **2.** $y = 2$ **3.** $y = -x + 2$

4. $2x$ **5.** $3x^2$ **6.** $\dfrac{1}{x^2}$ **7.** $2x$

8. $(-\infty, 1) \cup (1, \infty)$ **9.** $(-\infty, \infty)$

10. $(-\infty, 0) \cup (0, \infty)$

1.

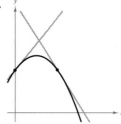

3.

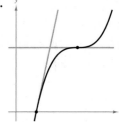

5. $m = 1$ **7.** $m = 0$ **9.** $m = -\frac{1}{3}$

11. 2002: $m \approx 200$ **13.** $t = 1$: $m \approx 65$
    2004: $m \approx 500$ $t = 8$: $m \approx 0$
                        $t = 12$: $m \approx -1000$

15. $f'(x) = -2$ **17.** $f'(x) = 0$
    $f'(2) = -2$ $f'(0) = 0$

19. $f'(x) = 2x$
$f'(2) = 4$

21. $f'(x) = 3x^2 - 1$
$f'(2) = 11$

23. $f'(x) = \dfrac{1}{\sqrt{x}}$

$f'(4) = \dfrac{1}{2}$

25. $f(x) = 3$
$f(x + \Delta x) = 3$
$f(x + \Delta x) - f(x) = 0$
$\dfrac{f(x + \Delta x) - f(x)}{\Delta x} = 0$
$\displaystyle\lim_{\Delta x \to 0} \dfrac{f(x + \Delta x) - f(x)}{\Delta x} = 0$

27. $f(x) = -5x$
$f(x + \Delta x) = -5x - 5\Delta x$
$f(x + \Delta x) - f(x) = -5\Delta x$
$\dfrac{f(x + \Delta x) - f(x)}{\Delta x} = -5$
$\displaystyle\lim_{\Delta x \to 0} \dfrac{f(x + \Delta x) - f(x)}{\Delta x} = -5$

29. $g(s) = \dfrac{1}{3}s + 2$

$g(s + \Delta s) = \dfrac{1}{3}s + \dfrac{1}{3}\Delta s + 2$

$g(s + \Delta s) - g(s) = \dfrac{1}{3}\Delta s$

$\dfrac{g(s + \Delta s) - g(s)}{\Delta s} = \dfrac{1}{3}$

$\displaystyle\lim_{\Delta s \to 0} \dfrac{g(s + \Delta s) - g(s)}{\Delta s} = \dfrac{1}{3}$

31. $f(x) = x^2 - 4$
$f(x + \Delta x) = x^2 + 2x\Delta x + (\Delta x)^2 - 4$
$f(x + \Delta x) - f(x) = 2x\Delta x + (\Delta x)^2$
$\dfrac{f(x + \Delta x) - f(x)}{\Delta x} = 2x + \Delta x$
$\displaystyle\lim_{\Delta x \to 0} \dfrac{f(x + \Delta x) - f(x)}{\Delta x} = 2x$

33. $h(t) = \sqrt{t - 1}$
$h(t + \Delta t) = \sqrt{t + \Delta t - 1}$
$h(t + \Delta t) - h(t) = \sqrt{t + \Delta t - 1} - \sqrt{t - 1}$
$\dfrac{h(t + \Delta t) - h(t)}{\Delta t} = \dfrac{1}{\sqrt{t + \Delta t - 1} + \sqrt{t - 1}}$
$\displaystyle\lim_{\Delta t \to 0} \dfrac{h(t + \Delta t) - h(t)}{\Delta t} = \dfrac{1}{2\sqrt{t - 1}}$

35. $f(t) = t^3 - 12t$
$f(t + \Delta t) = t^3 + 3t^2\Delta t + 3t(\Delta t)^2$
$\qquad\qquad + (\Delta t)^3 - 12t - 12\Delta t$
$f(t + \Delta t) - f(t) = 3t^2\Delta t + 3t(\Delta t)^2 + (\Delta t)^3 - 12\Delta t$
$\dfrac{f(t + \Delta t) - f(t)}{\Delta t} = 3t^2 + 3t\Delta t + (\Delta t)^2 - 12$
$\displaystyle\lim_{\Delta t \to 0} \dfrac{f(t + \Delta t) - f(t)}{\Delta t} = 3t^2 - 12$

37. $f(x) = \dfrac{1}{x + 2}$

$f(x + \Delta x) = \dfrac{1}{x + \Delta x + 2}$

$f(x + \Delta x) - f(x) = \dfrac{-\Delta x}{(x + \Delta x + 2)(x + 2)}$

$\dfrac{f(x + \Delta x) - f(x)}{\Delta x} = \dfrac{-1}{(x + \Delta x + 2)(x + 2)}$

$\displaystyle\lim_{\Delta x \to 0} \dfrac{f(x + \Delta x) - f(x)}{\Delta x} = -\dfrac{1}{(x + 2)^2}$

39. $y = 2x - 2$

41. $y = -6x - 3$

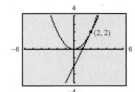

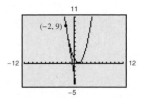

43. $y = \dfrac{x}{4} + 2$

45. $y = -x + 2$

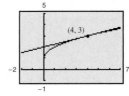

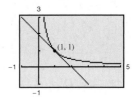

47. $y = -x + 1$

49. $y = -6x + 8$ and $y = -6x - 8$

51. $x \neq -3$ (node)   **53.** $x \neq 3$ (cusp)   **55.** $x > 1$

57. $x \neq 0$ (nonremovable discontinuity)

59. $x \neq 1$

61. $f(x) = -3x + 2$

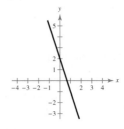

63.

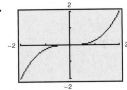

$f'(x) = \frac{3}{4}x^2$

x	-2	$-\frac{3}{2}$	-1	$-\frac{1}{2}$
$f(x)$	-2	-0.8438	-0.25	-0.0313
$f'(x)$	3	1.6875	0.75	0.1875

x	0	$\frac{1}{2}$	1	$\frac{3}{2}$	2
$f(x)$	0	0.0313	0.25	0.8438	2
$f'(x)$	0	0.1875	0.75	1.6875	3

65.

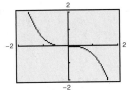

$f'(x) = -\frac{3}{2}x^2$

x	-2	$-\frac{3}{2}$	-1	$-\frac{1}{2}$
$f(x)$	4	1.6875	0.5	0.0625
$f'(x)$	-6	-3.375	-1.5	-0.375

x	0	$\frac{1}{2}$	1	$\frac{3}{2}$	2
$f(x)$	0	-0.0625	-0.5	-1.6875	-4
$f'(x)$	0	-0.375	-1.5	-3.375	-6

67. $f'(x) = 2x - 4$

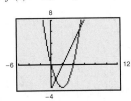

The x-intercept of the derivative indicates a point of horizontal tangency for f.

69. $f'(x) = 3x^2 - 3$

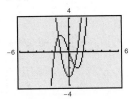

The x-intercepts of the derivative indicate points of horizontal tangency for f.

71. True   **73.** True

75.

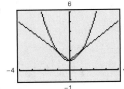

The graph of f is smooth at $(0, 1)$, but the graph of g has a sharp point at $(0, 1)$. The function g is not differentiable at $x = 0$.

SECTION 7.4   *(page 578)*

> ### Skills Review   *(page 578)*
>
> **1.** (a) 8 (b) 16 (c) $\frac{1}{2}$
> **2.** (a) $\frac{1}{36}$ (b) $\frac{1}{32}$ (c) $\frac{1}{64}$
> **3.** $4x(3x^2 + 1)$ **4.** $\frac{3}{2}x^{1/2}(x^{3/2} - 1)$ **5.** $\frac{1}{4x^{3/4}}$
> **6.** $x^2 - \frac{1}{x^{1/2}} + \frac{1}{3x^{2/3}}$ **7.** $0, -\frac{2}{3}$
> **8.** $0, \pm 1$ **9.** $-10, 2$ **10.** $-2, 12$

1. (a) 2 (b) $\frac{1}{2}$ **3.** (a) -1 (b) $-\frac{1}{3}$ **5.** 0
7. $4x^3$ **9.** 4 **11.** $2x + 5$ **13.** $-6t + 2$
15. $3t^2 - 2$ **17.** $\frac{16}{3}t^{1/3}$ **19.** $\frac{2}{\sqrt{x}}$ **21.** $-\frac{8}{x^3} + 4x$

23. Function: $y = \frac{1}{x^3}$

Rewrite: $y = x^{-3}$

Differentiate: $y' = -3x^{-4}$

Simplify: $y' = -\frac{3}{x^4}$

25. Function: $y = \frac{1}{(4x)^3}$

Rewrite: $y = \frac{1}{64}x^{-3}$

Differentiate: $y' = -\frac{3}{64}x^{-4}$

Simplify: $y' = -\frac{3}{64x^4}$

27. Function: $y = \frac{\sqrt{x}}{x}$

Rewrite: $y = x^{-1/2}$

Differentiate: $y' = -\frac{1}{2}x^{-3/2}$

Simplify: $y' = -\frac{1}{2x^{3/2}}$

29. -1 **31.** -2 **33.** 4 **35.** $2x + \frac{4}{x^2} + \frac{6}{x^3}$

37. $2x - 2 + \dfrac{8}{x^5}$ **39.** $3x^2 + 1$ **41.** $6x^2 + 16x - 1$

43. $\dfrac{2x^3 - 6}{x^3}$ **45.** $\dfrac{4x^3 - 2x - 10}{x^3}$ **47.** $\dfrac{4}{5x^{1/5}} + 1$

49. (a) $y = 2x - 2$ **51.** (a) $y = \dfrac{8}{15}x + \dfrac{22}{15}$

   (b) and (c)   (b) and (c)

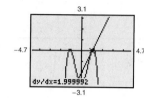   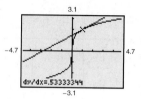

53. $(0, -1), \left(-\dfrac{\sqrt{6}}{2}, \dfrac{5}{4}\right), \left(\dfrac{\sqrt{6}}{2}, \dfrac{5}{4}\right)$ **55.** $(-5, -12.5)$

57. (a)

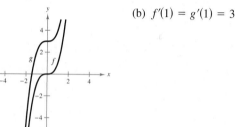

(b) $f'(1) = g'(1) = 3$

(c)

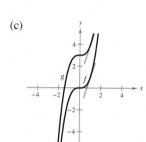

(d) $f' = g' = 3x^2$ for every value of x.

59. (a) 3   (b) 6   (c) -3 (d) 6

61. (a) 2001: 2.03

   2004: 249.01

   (b) The results are similar.

   (c) Millions of dollars/yr/yr

63. $P = 350x - 7000$

   $P' = 350$

65.

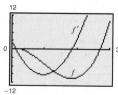

$(0.11, 0.14), (1.84, -10.49)$

67. False. Let $f(x) = x$ and $g(x) = x + 1$.

MID-CHAPTER QUIZ   (page 581)

1. 14   **2.** 2   **3.** Limit does not exist.

4. 7   **5.** $-\dfrac{1}{8}$ **6.** 0

7. $(-\infty, \infty)$; Explanations will vary.

8. $(-\infty, -2), (-2, \infty)$; Explanations will vary. There is a discontinuity at $x = -2$, because $f(-2)$ is not defined.

9. $(-\infty, 1), (1, \infty)$; Explanations will vary. There are discontinuities at $x = -3$ and $x = 1$, because $f(-3)$ and $f(1)$ are not defined.

10. $(-\infty, \infty)$; Explanations will vary.

11. $f(x) = -x + 2$

   $f(x + \Delta x) = -x - \Delta x + 2$

   $f(x + \Delta x) - f(x) = -\Delta x$

   $\dfrac{f(x + \Delta x) - f(x)}{\Delta x} = -1$

   $\displaystyle\lim_{\Delta x \to 0} \dfrac{f(x + \Delta x) - f(x)}{\Delta x} = -1$

   $f'(x) = -1$

   $f'(2) = -1$

12. $f(x) = \dfrac{4}{x}$

   $f(x + \Delta x) = \dfrac{4}{x + \Delta x}$

   $f(x + \Delta x) - f(x) = -\dfrac{4\Delta x}{x(x + \Delta x)}$

   $\dfrac{f(x + \Delta x) - f(x)}{\Delta x} = -\dfrac{4}{x(x + \Delta x)}$

   $\displaystyle\lim_{\Delta x \to 0} \dfrac{f(x + \Delta x) - f(x)}{\Delta x} = -\dfrac{4}{x^2}$

   $f'(x) = -\dfrac{4}{x^2}$

   $f'(1) = -4$

13. $f'(x) = 0$ **14.** $f'(x) = 19$ **15.** $f'(x) = -6x$

16. $f'(x) = \dfrac{3}{x^{3/4}}$ **17.** $f'(x) = -\dfrac{8}{x^3}$ **18.** $f'(x) = \dfrac{1}{\sqrt{x}}$

19. $y = -4x - 6$ **20.** $y = x$

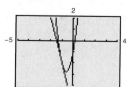   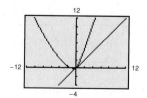

21. (a) $\dfrac{dS}{dt} = 0.5517t^2 - 1.6484t + 3.492$

(b) 2001: \$2.3953/yr
2004: \$5.7256/yr
2005: \$9.0425/yr

SECTION 7.5   *(page 593)*

Skills Review   *(page 593)*

1. 3   **2.** -7 **3.** $y' = 8x - 2$

4. $y' = -9t^2 + 4t$ **5.** $s' = -32t + 24$

6. $y' = -32x + 54$ **7.** $A' = -\frac{3}{5}r^2 + \frac{3}{5}r + \frac{1}{2}$

8. $y' = 2x^2 - 4x + 7$ **9.** $y' = 12 - \dfrac{x}{2500}$

10. $y' = 74 - \dfrac{3x^2}{10,000}$

---

1. (a) \$10.4 billion/yr   (b) \$7.4 billion/yr
(c) \$6.4 billion/yr   (d) \$16.6 billion/yr
(e) \$10.4 billion/yr   (f) \$11.4 billion/yr

3.
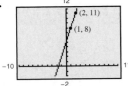

Average rate: 3
Instantaneous rates:
$f'(1) = f'(2) = 3$

5.
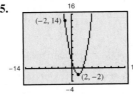

Average rate: -4
Instantaneous rates:
$h'(-2) = -8, h'(2) = 0$

7.
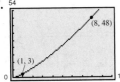

Average rate: $\frac{45}{7}$
Instantaneous rates:
$f'(1) = 4, f'(8) = 8$

9.
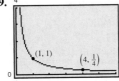

Average rate: $-\frac{1}{4}$
Instantaneous rates:
$f'(1) = -1, f'(4) = -\frac{1}{16}$

11.
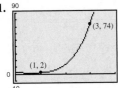

Average rate: 36
Instantaneous rates: $g'(1) = 2, g'(3) = 102$

13. (a) -500

The number of visitors to the park is decreasing at an average rate of 500 hundred thousand people per month from September to December.

(b) Answers will vary. The instantaneous rate of change at $t = 8$ is approximately 0.

15. (a) Average rate: $\frac{11}{27}$
Instantaneous rates: $E'(0) = \frac{1}{3}, E'(1) = \frac{4}{9}$

(b) Average rate: $\frac{11}{27}$
Instantaneous rates: $E'(1) = \frac{4}{9}, E'(2) = \frac{1}{3}$

(c) Average rate: $\frac{5}{27}$
Instantaneous rates: $E'(2) = \frac{1}{3}, E'(3) = 0$

(d) Average rate: $-\frac{7}{27}$
Instantaneous rates: $E'(3) = 0, E'(4) = -\frac{5}{9}$

17. (a) -80 ft/sec

(b) $s'(2) = -64$ ft/sec, $s'(3) = -96$ ft/sec

(c) $\dfrac{\sqrt{555}}{4} \approx 5.89$ sec   (d) $-8\sqrt{555} \approx -188.5$ ft/sec

19. 1.47 dollars   **21.** $470 - 0.5x$ dollars, $0 \le x \le 940$

23. $50 - x$ dollars   **25.** $-18x^2 + 16x + 200$ dollars

27. $-4x + 72$ dollars   **29.** $-0.0005x + 12.2$ dollars

31. (a) \$0.58   (b) \$0.60
(c) The results are nearly the same.

33. (a) \$4.95   (b) \$5.00
(c) The results are nearly the same.

35. (a)
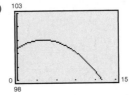

(b) For $t < 4$, positive; for $t > 4$, negative; shows when fever is going up and down.

(c) $T(0) = 100.4°F$
$T(4) = 101°F$
$T(8) = 100.4°F$
$T(12) = 98.6°F$

(d) $T'(t) = -0.075t + 0.3$
The rate of change of temperature

(e) $T'(0) = 0.3°F/hr$
$T'(4) = 0°F/hr$
$T'(8) = -0.3°F/hr$
$T'(12) = -0.6°F/hr$

37. (a) $R = 5x - 0.001x^2$

(b) $P = -0.001x^2 + 3.5x - 35$

(c)

x	600	1200	1800	2400	3000
dR/dx	3.8	2.6	1.4	0.2	-1
dP/dx	2.3	1.1	-0.1	-1.3	-2.5
P	1705	2725	3025	2605	1465

39. (a) $P = -0.0025x^2 + 2.65x - 25$

(b)

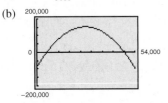

When $x = 300$, slope is positive.

When $x = 700$, slope is negative.

(c) $P'(300) = 1.15$; $P'(700) = -0.85$

41. (a) $P = -\frac{1}{3000}x^2 + 17.8x - 85{,}000$

(b)

When $x = 18{,}000$, slope is positive.

When $x = 36{,}000$, slope is negative.

(c) $P'(18{,}000) = 5.8$; $P'(36{,}000) = -6.2$

43. (a) \$0.33/unit   (b) \$0.13/unit

(c) \$0/unit   (d) $-\$0.08$/unit

$p'(2500) = 0$ indicates that $x = 2500$ is the optimal value

of x. So, $p = \dfrac{50}{\sqrt{x}} = \dfrac{50}{\sqrt{2500}} = \$1.00$.

45. $C = \dfrac{44{,}250}{x}$;

x	10	15	20	25
C	4425.00	2950.00	2212.50	1770.00
dC/dx	-442.5	-196.67	-110.63	-70.80

x	30	35	40
C	1475.00	1264.29	1106.25
dC/dx	-49.17	-36.12	-27.66

15 mi/gal; Explanations will vary.

47. (a) \$654.43   (b) \$1084.65   (c) \$1794.44

(d) Answers will vary.

SECTION 7.6   (page 605)

Skills Review   (page 605)

1. $2(3x^2 + 7x + 1)$ **2.** $4x^2(6 - 5x^2)$

3. $8x^2(x^2 + 2)^3 + (x^2 + 4)$

4. $(2x)(2x + 1)[2x + (2x + 1)^3]$

5. $\dfrac{23}{(2x + 7)^2}$ **6.** $-\dfrac{x^2 + 8x + 4}{(x^2 - 4)^2}$

7. $-\dfrac{2(x^2 + x - 1)}{(x^2 + 1)^2}$ **8.** $\dfrac{4(3x^4 - x^3 + 1)}{(1 - x^4)^2}$

9. $\dfrac{4x^3 - 3x^2 + 3}{x^2}$ **10.** $\dfrac{x^2 - 2x + 4}{(x - 1)^2}$

11. 11   **12.** 0   **13.** $-\frac{1}{4}$ **14.** $\frac{17}{4}$

1. $f'(2) = 15$; Product Rule

3. $f'(1) = 13$; Product Rule

5. $f'(0) = 0$; Constant Multiple Rule

7. $g'(4) = 11$; Product Rule

9. $h'(6) = -5$; Quotient Rule

11. $f'(3) = \frac{3}{4}$; Quotient Rule

13. $g'(6) = -11$; Quotient Rule

15. $f'(1) = \frac{2}{5}$; Quotient Rule

17. Function: $y = \dfrac{x^2 + 2x}{x}$

Rewrite: $y = x + 2$, $x \neq 0$

Differentiate: $y' = 1$, $x \neq 0$

Simplify: $y' = 1$, $x \neq 0$

19. Function: $y = \dfrac{7}{3x^3}$

Rewrite: $y = \dfrac{7}{3}x^{-3}$

Differentiate: $y' = -7x^{-4}$

Simplify: $y' = -\dfrac{7}{x^4}$

21. Function: $y = \dfrac{4x^2 - 3x}{8\sqrt{x}}$

Rewrite: $y = \dfrac{1}{2}x^{3/2} - \dfrac{3}{8}x^{1/2}$, $x \neq 0$

Differentiate: $y' = \dfrac{3}{4}x^{1/2} - \dfrac{3}{16}x^{-1/2}$

Simplify: $y' = \dfrac{3}{4}\sqrt{x} - \dfrac{3}{16\sqrt{x}}$

23. Function: $y = \dfrac{x^2 - 4x + 3}{x - 1}$

Rewrite: $y = x - 3, \, x \ne 1$

Differentiate: $y' = 1, \, x \ne 1$

Simplify: $y' = 1, \, x \ne 1$

25. $10x^4 + 12x^3 - 3x^2 - 18x - 15$; Product Rule

27. $12t^2(2t^3 - 1)$; Product Rule

29. $\dfrac{5}{6x^{1/6}} + \dfrac{1}{x^{2/3}}$; Product Rule

31. $-\dfrac{5}{(2x - 3)^2}$; Quotient Rule

33. $\dfrac{2}{(x + 1)^2}, \, x \ne 1$; Quotient Rule

35. $\dfrac{x^2 + 2x - 1}{(x + 1)^2}$; Quotient Rule

37. $\dfrac{3s^2 - 2s - 5}{2s^{3/2}}$; Quotient Rule

39. $\dfrac{2x^3 + 11x^2 - 8x - 17}{(x + 4)^2}$; Quotient Rule

41. $y = 5x - 2$ **43.** $y = \frac{3}{4}x - \frac{5}{4}$

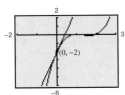 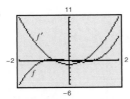

45. $y = -16x - 5$

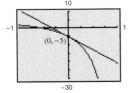

47. $(0, 0), (2, 4)$ **49.** $(0, 0), (\sqrt[3]{-4}, -2.117)$

51.   **53.**

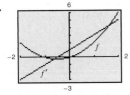

55. $-\$1.87$/unit

57. (a) -0.480/wk   (b) 0.120/wk   (c) 0.015/wk

59. 31.55 bacteria/hr

61. (a) $p = \dfrac{4000}{\sqrt{x}}$ (b) $C = 250x + 10{,}000$

(c) $P = 4000\sqrt{x} - 250x - 10{,}000$

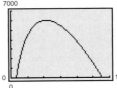

$\$500$/unit

63. (a)   (b)

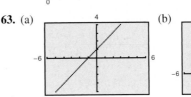

(c)

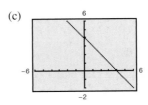

The graph of (c) would most likely represent a demand function. As the number of units increases, demand is likely to decrease, not increase as in (a) and (b).

65. (a) -38.125 (b) -10.37 (c) -3.80

Increasing the order size reduces the cost per item; Choices and explanations will vary.

67. $\dfrac{dP}{dt} = \dfrac{17{,}091 - 1773.4t + 39.5t^2}{(1000 - 128.2t + 4.34t^2)^2}$

$P'(8) = 0.0854$

$P'(10) = 0.1431$

$P'(12) = 0.2000$

$P'(14) = 0.0017$

The rate of change in price at year t

69. $f'(2) = 0$ **71.** $f'(2) = 14$ **73.** Answers will vary.

SECTION 7.7   *(page 615)*

Skills Review   *(page 615)*

1. $(1 - 5x)^{2/5}$ **2.** $(2x - 1)^{3/4}$

3. $(4x^2 + 1)^{-1/2}$ **4.** $(x - 6)^{-1/3}$

5. $x^{1/2}(1 - 2x)^{-1/3}$ **6.** $(2x)^{-1}(3 - 7x)^{3/2}$

7. $(x - 2)(3x^2 + 5)$ **8.** $(x - 1)(5\sqrt{x} - 1)$

9. $(x^2 + 1)^2(4 - x - x^3)$

10. $(3 - x^2)(x - 1)(x^2 + x + 1)$

$y = f(g(x))$	$u = g(x)$	$y = f(u)$

1. $y = (6x - 5)^4$ $u = 6x - 5$ $y = u^4$

3. $y = (4 - x^2)^{-1}$ $u = 4 - x^2$ $y = u^{-1}$

5. $y = \sqrt{5x - 2}$ $u = 5x - 2$ $y = \sqrt{u}$

7. $y = (3x + 1)^{-1}$ $u = 3x + 1$ $y = u^{-1}$

9. $\dfrac{dy}{du} = 2u$

$\dfrac{du}{dx} = 4$

$\dfrac{dy}{dx} = 32x + 56$

11. $\dfrac{dy}{du} = \dfrac{1}{2\sqrt{u}}$

$\dfrac{du}{dx} = -2x$

$\dfrac{dy}{dx} = -\dfrac{x}{\sqrt{3 - x^2}}$

13. $\dfrac{dy}{du} = \dfrac{2}{3u^{1/3}}$

$\dfrac{du}{dx} = 20x^3 - 2$

$\dfrac{dy}{dx} = \dfrac{40x^3 - 4}{3\sqrt[3]{5x^4 - 2x}}$

15. c   **17.** b   **19.** a   **21.** c   **23.** $6(2x - 7)^2$

25. $-6(4 - 2x)^2$ **27.** $6x(6 - x^2)(2 - x^2)$

29. $\dfrac{4x}{3(x^2 - 9)^{1/3}}$ **31.** $\dfrac{1}{2\sqrt{t + 1}}$ **33.** $\dfrac{4t + 5}{2\sqrt{2t^2 + 5t + 2}}$

35. $\dfrac{6x}{(9x^2 + 4)^{2/3}}$ **37.** $\dfrac{27}{4(2 - 9x)^{3/4}}$ **39.** $\dfrac{4x^2}{(4 - x^3)^{7/3}}$

41. $y = 216x - 378$ **43.** $y = \frac{8}{3}x - \frac{7}{3}$

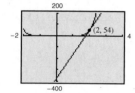

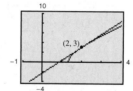

45. $y = x - 1$

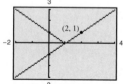

47. $f'(x) = \dfrac{1 - 3x^2 - 4x^{3/2}}{2\sqrt{x}(x^2 + 1)^2}$

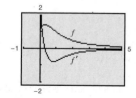

The zero of $f'(x)$ corresponds to the point on the graph of $f(x)$ where the tangent line is horizontal.

49. $f'(x) = -\dfrac{\sqrt{(x + 1)/x}}{2x(x + 1)}$

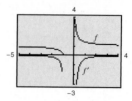

$f'(x)$ has no zeros.

In Exercises 51–65, the differentiation rule(s) used may vary. A sample answer is provided.

51. $-\dfrac{1}{(x - 2)^2}$; Chain Rule   **53.** $\dfrac{8}{(t + 2)^3}$; Chain Rule

55. $-\dfrac{2(2x - 3)}{(x^2 - 3x)^3}$; Chain Rule   **57.** $-\dfrac{2t}{(t^2 - 2)^2}$; Chain Rule

59. $27(x - 3)^2(4x - 3)$; Product Rule and Chain Rule

61. $\dfrac{3(x + 1)}{\sqrt{2x + 3}}$; Product Rule and Chain Rule

63. $\dfrac{t(5t - 8)}{2\sqrt{t - 2}}$; Product Rule and Chain Rule

65. $\dfrac{2(6 - 5x)(5x^2 - 12x + 5)}{(x^2 - 1)^3}$; Chain Rule and Quotient Rule

67. $y = \frac{8}{3}t + 4$

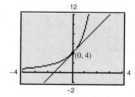

69. $y = -6t - 14$

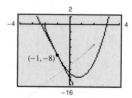

71. $y = -2x + 7$

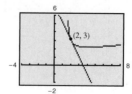

73. (a) $74.00 per 1%

   (b) $81.59 per 1%

   (c) $89.94 per 1%

75.

t	0	1	2	3	4
$\dfrac{dN}{dt}$	0	177.78	44.44	10.82	3.29

The rate of growth of N is decreasing.

77. (a) $V = \dfrac{10{,}000}{\sqrt[3]{t+1}}$

(b) $-\$1322.83/\text{yr}$

(c) $-\$524.97/\text{yr}$

79. False. $y' = \frac{1}{2}(1-x)^{-1/2}(-1) = -\frac{1}{2}(1-x)^{-1/2}$

81. (a) 15   (b) -10

REVIEW EXERCISES FOR CHAPTER 7
(page 621)

1. 7   **3.** 49   **5.** $\frac{10}{3}$ **7.** -2

9. $-\frac{1}{4}$ **11.** $-\infty$ **13.** Limit does not exist.

15. $-\frac{1}{16}$ **17.** $3x^2 - 1$ **19.** 0.5774

21. False, limit does not exist.

23. False, limit does not exist.

25. False, limit does not exist.

27. $(-\infty, -4)$ and $(-4, \infty)$; $f(-4)$ is undefined.

29. $(-\infty, -1)$ and $(-1, \infty)$; $f(-1)$ is undefined.

31. Continuous on all intervals $(c, c+1)$, where c is an integer; $\lim\limits_{x \to c} f(c)$ does not exist.

33. $(-\infty, 0)$ and $(0, \infty)$; $\lim\limits_{x \to 0} f(x)$ does not exist.

35. $a = 2$

37. (a)

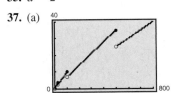

C is not continuous at $x = 25$, 100, and 500.

(b) $10

39. (a)

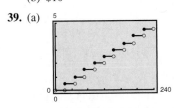

Continuous on all intervals $(24n, 24(n+1))$ where n is a whole number.

(b) $31.00

41. -2 **43.** 0

45. Answers will vary. Sample answer:

$t = 10$: slope $\approx \$7025$ million/yr; Sales were increasing by about $7025 million/yr in 2000.

$t = 13$: slope $\approx \$6750$ million/yr; Sales were increasing by about $6750 million/yr in 2003.

$t = 15$: slope $\approx \$10{,}250$ million/yr; Sales were increasing by about $10,250 million/yr in 2005.

47. $t = 0$: slope ≈ 180
$t = 4$: slope ≈ -70
$t = 6$: slope ≈ -900

49. -3; -3 **51.** $2x - 4$; -2

53. $\dfrac{1}{2\sqrt{x+9}}$; $\dfrac{1}{4}$ **55.** $-\dfrac{1}{(x-5)^2}$; -1

57. -3 **59.** 0   **61.** $\frac{1}{6}$ **63.** -5 **65.** 1   **67.** 0

69. $y = -\frac{4}{3}t + 2$ **71.** $y = 2x + 2$

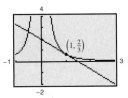
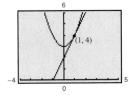

73. $y = -34x - 27$ **75.** $y = x - 1$

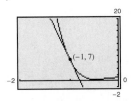
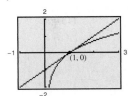

77. $y = -2x + 6$

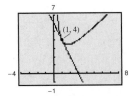

79. Average rate of change: 4

Instantaneous rate of change when $x = 0$: 3

Instantaneous rate of change when $x = 1$: 5

81. (a) $s(t) = -16t^2 + 276$ (b) -32 ft/sec

(c) $t = 2$: -64 ft/sec, $t = 3$: -96 ft/sec

(d) About 4.15 sec   (e) About -132.8 ft/sec

83. (a) About $7219 million/yr/yr

(b) 1999: about $8618 million/yr/yr
2005: about $10,279 million/yr/yr

(c) Sales were increasing in 1999 and 2005, and grew at a rate of about $7219 million over the period 1999–2005.

85. $R = 27.50x$

$C = 15x + 2500$

$P = 12.50x - 2500$

87. $\dfrac{dC}{dx} = 320$ **89.** $\dfrac{dC}{dx} = \dfrac{1.275}{\sqrt{x}}$

91. $\dfrac{dR}{dx} = 200 - \dfrac{2}{5}x$ **93.** $\dfrac{dR}{dx} = \dfrac{35(x - 4)}{2(x - 2)^{3/2}}$

95. $\dfrac{dP}{dx} = -0.0006x^2 + 12x - 1$

In Exercises 97–115, the differentiation rule(s) used may vary. A sample answer is provided.

97. $15x^2(1 - x^2)$; Power Rule

99. $16x^3 - 33x^2 + 12x$; Product Rule

101. $\dfrac{2(3 + 5x - 3x^2)}{(x^2 + 1)^2}$; Quotient Rule

103. $30x(5x^2 + 2)^2$; Chain Rule

105. $-\dfrac{1}{(x + 1)^{3/2}}$; Quotient Rule

107. $\dfrac{2x^2 + 1}{\sqrt{x^2 + 1}}$; Product Rule

109. $80x^4 - 24x^2 + 1$; Product Rule

111. $18x^5(x + 1)(2x + 3)^2$; Chain Rule

113. $x(x - 1)^4(7x - 2)$; Product Rule

115. $\dfrac{3(9t + 5)}{2\sqrt{3t + 1}(1 - 3t)^3}$; Quotient Rule

117. (a) $t = 1$: -6.63 $t = 3$: -6.5

   $t = 5$: -4.33 $t = 10$: -1.36

(b)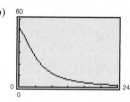

The rate of decrease is approaching zero.

CHAPTER TEST   *(page 625)*

1. -1 **2.** Limit does not exist.   **3.** 2 **4.** $\dfrac{1}{6}$

5. $(-\infty, \infty)$; Explanations will vary.

6. $(-\infty, 4)$ and $(4, \infty)$; Explanations will vary. There is a discontinuity at $x = 4$, because $f(4)$ is not defined.

7. $(-\infty, 5]$; Explanations will vary.

8. $(-\infty, \infty)$; Explanations will vary.

9. $f(x) = x^2 + 1$

$f(x + \Delta x) = x^2 + 2x\Delta x + \Delta x^2 + 1$

$f(x + \Delta x) - f(x) = 2x\Delta x + \Delta x^2$

$\dfrac{f(x + \Delta x) - f(x)}{\Delta x} = 2x + \Delta x$

$\displaystyle \lim_{\Delta x \to 0} \dfrac{f(x + \Delta x) - f(x)}{\Delta x} = 2x$

$f'(x) = 2x$

$f'(2) = 4$

10. $f(x) = \sqrt{x} - 2$

$f(x + \Delta x) = \sqrt{x + \Delta x} - 2$

$f(x + \Delta x) - f(x) = \sqrt{x + \Delta x} - \sqrt{x}$

$\dfrac{f(x + \Delta x) - f(x)}{\Delta x} = \dfrac{1}{\sqrt{x + \Delta x} + \sqrt{x}}$

$\displaystyle \lim_{\Delta x \to 0} \dfrac{f(x + \Delta x) - f(x)}{\Delta x} = \dfrac{1}{2\sqrt{x}}$

$f'(x) = \dfrac{1}{2\sqrt{x}}$

$f'(4) = \dfrac{1}{4}$

11. $f'(t) = 3t^2 + 2$ **12.** $f'(x) = 8x - 8$

13. $f'(x) = \dfrac{3\sqrt{x}}{2}$ **14.** $f'(x) = 2x$ **15.** $f'(x) = \dfrac{9}{x^4}$

16. $f'(x) = \dfrac{5 + x}{2\sqrt{x}} + \sqrt{x}$ **17.** $f'(x) = 36x^3 + 48x$

18. $f'(x) = -\dfrac{1}{\sqrt{1 - 2x}}$

19. $f'(x) = \dfrac{(10x + 1)(5x - 1)^2}{x^2} = 250x - 75 + \dfrac{1}{x^2}$

20. $y = 2x - 2$

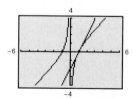

21. (a) \$169.80 million/yr

(b) 2001: \$68.84 million/yr
   2005: \$223.30 million/yr

(c) The annual sales of Bausch & Lomb from 2001 to 2005 increased on average by about \$169.80 million/yr, and the instantaneous rates of change for 2001 and 2005 are \$68.84 million/yr and \$223.30 million/yr, respectively.

22. $P = -0.016x^2 + 1460x - 715,000$

CHAPTER 8

SECTION 8.1   *(page 632)*

Skills Review   *(page 632)*

1. $t = 0, \frac{3}{2}$ **2.** $t = -2, 7$ **3.** $t = -2, 10$

4. $t = \dfrac{9 \pm 3\sqrt{10{,}249}}{32}$ **5.** $\dfrac{dy}{dx} = 6x^2 + 14x$

6. $\dfrac{dy}{dx} = 8x^3 + 18x^2 - 10x - 15$

7. $\dfrac{dy}{dx} = \dfrac{2x(x + 7)}{(2x + 7)^2}$ **8.** $\dfrac{dy}{dx} = -\dfrac{6x^2 + 10x + 15}{(2x^2 - 5)^2}$

9. Domain: $(-\infty, \infty)$ **10.** Domain: $[7, \infty)$

  Range: $[-4, \infty)$ Range: $[0, \infty)$

---

1. 0   **3.** 2   **5.** $2t - 8$ **7.** $\dfrac{9}{2t^4}$

9. $18(2 - x^2)(5x^2 - 2)$

11. $12(x^3 - 2x)^2(11x^4 - 16x^2 + 4)$ **13.** $\dfrac{4}{(x - 1)^3}$

15. $12x^2 + 24x + 16$ **17.** $60x^2 - 72x$

19. $120x + 360$ **21.** $-\dfrac{9}{2x^5}$ **23.** 260   **25.** $-\dfrac{1}{648}$

27. -126 **29.** $4x$ **31.** $\dfrac{1}{x^2}$ **33.** $12x^2 + 4$

35. $f''(x) = 6(x - 3) = 0$ when $x = 3$.

37. $f''(x) = 2(3x + 4) = 0$ when $x = -\frac{4}{3}$.

39. $f''(x) = \dfrac{x(2x^2 - 3)}{(x^2 - 1)^{3/2}} = 0$ when $x = \pm\dfrac{\sqrt{6}}{2}$.

41. $f''(x) = \dfrac{2x(x + 3)(x - 3)}{(x^2 + 3)^3}$

      $= 0$ when $x = 0$ or $x = \pm 3$.

43. (a) $s(t) = -16t^2 + 144t$

      $v(t) = -32t + 144$

      $a(t) = -32$

   (b) 4.5 sec; 324 ft

   (c) $v(9) = -144$ ft/sec, which is the same speed as the initial velocity

45.

t	0	10	20	30	40	50	60
$\dfrac{ds}{dt}$	0	45	60	67.5	72	75	77.1
$\dfrac{d^2s}{dt^2}$	9	2.25	1	0.56	0.36	0.25	0.18

As time increases, velocity increases and acceleration decreases.

47. $f(x) = x^2 - 6x + 6$

   $f'(x) = 2x - 6$

   $f''(x) = 2$

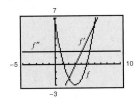

The degrees of the successive derivatives decrease by 1.

49.

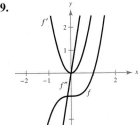

We know that the degrees of the successive derivatives decrease by 1.

51. (a) $y(t) = -0.2093t^3 + 1.637t^2 - 1.95t + 9.4$

   (b)

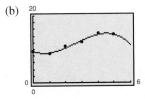

   The model fits the data well.

   (c) $y'(t) = -0.6279t^2 + 3.274t - 1.95$

      $y''(t) = -1.2558t + 3.274$

   (d) $y'(t) > 0$ on $[1, 4]$

   (e) 2002 $(t = 2.607)$

   (f) The first derivative is used to show that the retail value of motor homes is increasing in (d), and the retail value increased at the greatest rate at the zero of the second derivative as shown in (e).

53. False. The product rule is

   $[f(x)g(x)]' = f(x)g'(x) + g(x)f'(x)$.

55. True   **57.** $[xf(x)]^{(n)} = xf^{(n)}(x) + nf^{(n-1)}(x)$

SECTION 8.2   *(page 639)*

Skills Review   *(page 639)*

1. $y = x^2 - 2x$ **2.** $y = \dfrac{x - 3}{4}$

3. $y = 1, x \neq -6$ **4.** $y = -4, x \neq \pm\sqrt{3}$

5. $y = \pm\sqrt{5 - x^2}$ **6.** $y = \pm\sqrt{6 - x^2}$ **7.** $\dfrac{8}{3}$

8. $-\dfrac{1}{2}$ **9.** $\dfrac{5}{7}$ **10.** 1

1. $-\dfrac{y}{x}$ **3.** $-\dfrac{x}{y}$ **5.** $\dfrac{1 - xy^2}{x^2 y}$ **7.** $\dfrac{y}{8y - x}$

9. $-\dfrac{1}{10y - 2}$ **11.** $\dfrac{1}{2}$ **13.** $-\dfrac{x}{y}, 0$

15. $-\dfrac{y}{x + 1}, -\dfrac{1}{4}$ **17.** $\dfrac{y - 3x^2}{2y - x}, \dfrac{1}{2}$ **19.** $\dfrac{1 - 3x^2 y^3}{3x^3 y^2 - 1}, -1$

21. $-\sqrt{\dfrac{y}{x}}, -\dfrac{5}{4}$ **23.** $-\sqrt[3]{\dfrac{y}{x}}, -\dfrac{1}{2}$ **25.** 3

27. 0 **29.** $-\dfrac{\sqrt{5}}{3}$ **31.** $-\dfrac{x}{y}, \dfrac{4}{3}$ **33.** $\dfrac{1}{2y}, -\dfrac{1}{2}$

35. At $(8, 6)$: $y = -\dfrac{4}{3}x + \dfrac{50}{3}$

At $(-6, 8)$: $y = \dfrac{3}{4}x + \dfrac{25}{2}$

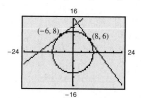

37. At $\left(1, \sqrt{5}\right)$: $15x - 2\sqrt{5}y - 5 = 0$

At $\left(1, -\sqrt{5}\right)$: $15x + 2\sqrt{5}y - 5 = 0$

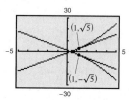

39. At $(0, 2)$: $y = 2$

At $(2, 0)$: $x = 2$

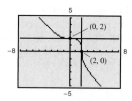

41. $-\dfrac{2}{p^2(0.00003x^2 + 0.1)}$ **43.** $-\dfrac{4xp}{2p^2 + 1}$

45. (a) -2

(b)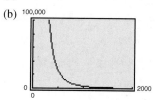

As more labor is used, less capital is available.

As more capital is used, less labor is available.

47. (a)

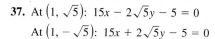

The numbers of cases of HIV/AIDS increases from 2001 to 2005.

(b) 2005

(c)

t	1	2	3	4	5
y	37.90	38.91	39.05	40.23	44.08
y'	2.130	0.251	0.347	2.288	5.565

2005

SECTION 8.3   *(page 647)*

Skills Review   *(page 647)*

1. $A = \pi r^2$ **2.** $V = \dfrac{4}{3}\pi r^3$ **3.** $S = 6s^2$

4. $V = s^3$ **5.** $V = \dfrac{1}{3}\pi r^2 h$ **6.** $A = \dfrac{1}{2}bh$

7. $-\dfrac{x}{y}$ **8.** $\dfrac{2x - 3y}{3x}$ **9.** $-\dfrac{2x + y}{x + 2}$

10. $-\dfrac{y^2 - y + 1}{2xy - 2y - x}$

1. (a) $\dfrac{3}{4}$ (b) 20   **3.** (a) $-\dfrac{5}{8}$ (b) $\dfrac{3}{2}$

5. (a) 36π in.^{2}/min

(b) 144π in.^{2}/min

7. If $\dfrac{dr}{dt}$ is constant, $\dfrac{dA}{dt} = 2\pi r \dfrac{dr}{dt}$ and so is proportional to r.

9. (a) $\dfrac{5}{2\pi}$ ft/min   (b) $\dfrac{5}{8\pi}$ ft/min

11. (a) 112.5 dollars/wk

(b) 7500 dollars/wk

(c) 7387.5 dollars/wk

13. (a) 9 cm³/sec

(b) 900 cm³/sec

15. (a) -12 cm/min

(b) 0 cm/min

(c) 4 cm/min

(d) 12 cm/min

17. (a) $-\frac{7}{12}$ ft/sec   (b) $-\frac{3}{2}$ ft/sec

(c) $-\frac{48}{7}$ ft/sec

19. (a) -750 mi/hr   (b) 20 min

21. -8.33 ft/sec   **23.** About 37.7 ft³/min

25. 4 units/wk

MID-CHAPTER QUIZ   *(page 649)*

1. $6x - 2$ **2.** $\dfrac{4}{9\sqrt[3]{(x-2)^7}}$ **3.** $6(x^2 + 1)(5x^2 + 1)$

4. $-\dfrac{60}{(2x+5)^3}$ **5.** $-\dfrac{1}{32}$ **6.** -120 **7.** -96

8. 864 ft; 48 ft/sec; -32 ft/sec²

9. $-\dfrac{2}{3}x + \dfrac{1}{3}$ **10.** $6x^2\sqrt{y}$ **11.** $-\dfrac{y-1}{x-1}$

12. $\dfrac{4xy}{3y^2 - 2x^2 + 1}$

13. $y = 2x + 1$

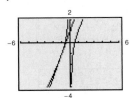

14. 2   **15.** $-\dfrac{2\sqrt{12}}{9}$

16. (a) \$190 per week

(b) \$20,000 per week

(c) \$19,810 per week

SECTION 8.4   *(page 657)*

> **Skills Review**   *(page 657)*
>
> **1.** $x = 0, x = 8$ **2.** $x = 0, x = 24$ **3.** $x = \pm 5$
>
> **4.** $x = 0$ **5.** $(-\infty, 3) \cup (3, \infty)$ **6.** $(-\infty, 1)$
>
> **7.** $(-\infty, -2) \cup (-2, 5) \cup (5, \infty)$ **8.** $\left(-\sqrt{3}, \sqrt{3}\right)$
>
> **9.** $x = -2$: -6 **10.** $x = -2$: 60
>
> $x = 0$: 2 $x = 0$: -4
>
> $x = 2$: -6 $x = 2$: 60
>
> **11.** $x = -2$: $-\frac{1}{3}$ **12.** $x = -2$: $\frac{1}{18}$
>
> $x = 0$: 1 $x = 0$: $-\frac{1}{8}$
>
> $x = 2$: 5 $x = 2$: $-\frac{3}{2}$

1. $f'(-1) = -\frac{8}{25}$ **3.** $f'(-3) = -\frac{2}{3}$

$f'(0) = 0$ $f'(-2)$ is undefined.

$f'(1) = \frac{8}{25}$ $f'(-1) = \frac{2}{3}$

5. Increasing on $(-\infty, -1)$

Decreasing on $(-1, \infty)$

7. Increasing on $(-1, 0)$ and $(1, \infty)$

Decreasing on $(-\infty, -1)$ and $(0, 1)$

9. No critical numbers

Increasing on $(-\infty, \infty)$

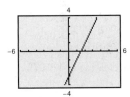

11. Critical number: $x = 1$ **13.** Critical number: $x = 3$

Increasing on $(-\infty, 1)$ Decreasing on $(-\infty, 3)$

Decreasing on $(1, \infty)$ Increasing on $(3, \infty)$

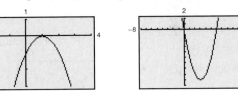

15. Critical numbers: $x = 0, x = 4$

Increasing on $(-\infty, 0)$ and $(4, \infty)$

Decreasing on $(0, 4)$

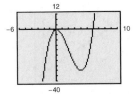

17. Critical numbers:

$x = -1, x = 1$

Decreasing on $(-\infty, -1)$

Increasing on $(1, \infty)$

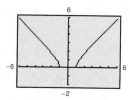

19. No critical numbers

Increasing on $(-\infty, \infty)$

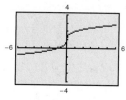

21. No critical numbers

Increasing on $(-\infty, \infty)$

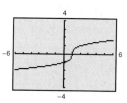

23. Critical number: $x = 1$

Increasing on $(-\infty, 1)$

Decreasing on $(1, \infty)$

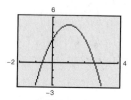

25. Critical numbers: $x = -1, x = -\frac{5}{3}$

Increasing on $\left(-\infty, -\frac{5}{3}\right)$ and $(-1, \infty)$

Decreasing on $\left(-\frac{5}{3}, -1\right)$

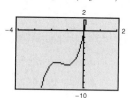

27. Critical numbers: $x = -1, x = -\frac{2}{3}$

Decreasing on $\left(-1, -\frac{2}{3}\right)$

Increasing on $\left(-\frac{2}{3}, \infty\right)$

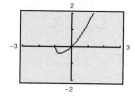

29. Critical numbers: $x = 0, x = \frac{3}{2}$

Decreasing on $\left(-\infty, \frac{3}{2}\right)$

Increasing on $\left(\frac{3}{2}, \infty\right)$

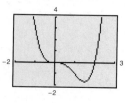

31. Critical numbers: $x = 2, x = -2$

Decreasing on $(-\infty, -2)$ and $(2, \infty)$

Increasing on $(-2, 2)$

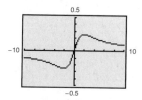

33. No critical numbers

Discontinuities: $x = \pm 4$

Increasing on $(-\infty, -4)$,

$(-4, 4)$, and $(4, \infty)$

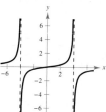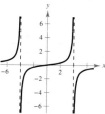

35. Critical number: $x = 0$

Discontinuity: $x = 0$

Increasing on $(-\infty, 0)$

Decreasing on $(0, \infty)$

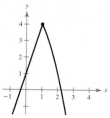

37. Critical number: $x = 1$

No discontinuity, but the function is not differentiable at $x = 1$.

Increasing on $(-\infty, 1)$

Decreasing on $(1, \infty)$

39. (a) Decreasing on $[1, 4.10)$

Increasing on $(4.10, \infty)$

(b)

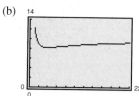

(c) $C = 9$ (or $900) when $x = 2$ and $x = 15$. Use an order size of $x = 4$, which will minimize the cost C.

41. (a)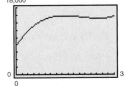

Increasing from 1970 to late 1986 and from late 1998 to 2004

Decreasing from late 1986 to late 1998

(b) $y' = 2.439t^2 - 111.4t + 1185.2$

Critical numbers: $t = 16.9$, $t = 28.8$

Therefore, the model is increasing from 1970 to late 1986 and from late 1998 to 2004 and decreasing from late 1986 to late 1998.

43. (a) $P = -\dfrac{1}{20,000}x^2 + 2.65x - 7500$

(b) Increasing on $[0, 26{,}500)$

Decreasing on $(26{,}500, 50{,}000]$

(c) The maximum profit occurs when the restaurant sells 26,500 hamburgers, the x-coordinate of the point at which the graph changes from increasing to decreasing.

SECTION 8.5　*(page 667)*

Skills Review　*(page 667)*

1. $0, \pm\frac{1}{2}$　　**2.** $-2, 5$　　**3.** 1　　**4.** $0, 125$

5. $-4 \pm \sqrt{17}$　　**6.** $1 \pm \sqrt{5}$

7. Negative　　**8.** Positive　　**9.** Positive

10. Negative　　**11.** Increasing　　**12.** Decreasing

1. Relative maximum: $(1, 5)$

3. Relative minimum: $(3, -9)$

5. Relative maximum: $\left(\frac{2}{3}, \frac{28}{9}\right)$

Relative minimum: $(1, 3)$

7. No relative extrema

9. Relative maximum: $(0, 15)$

Relative minimum: $(4, -17)$

11. Relative minima: $(-0.366, 0.75)$, $(1.37, 0.75)$

Relative maximum: $\left(\frac{1}{2}, \frac{21}{16}\right)$

13.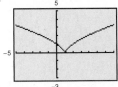

Relative minimum: $(1, 0)$

15.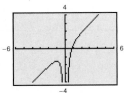

Relative maximum: $\left(-1, -\frac{3}{2}\right)$

17.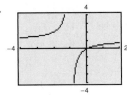

No relative extrema

19. Minimum: $(2, 2)$　　　**21.** Maximum: $(0, 5)$

Maximum: $(-1, 8)$　　　　　Minimum: $(3, -13)$

23. Minima: $(-1, -4), (2, -4)$　**25.** Maximum: $(2, 1)$

Maxima: $(0, 0), (3, 0)$　　　　Minimum: $\left(0, \frac{1}{3}\right)$

27. Maximum: $(-1, 5)$　　**29.** Maximum: $(-7, 4)$

Minimum: $(0, 0)$　　　　　Minimum: $(1, 0)$

31. 2, absolute maximum

33. Maximum: $(5, 7)$　　　**35.** Maximum: $(2, 2.\overline{6})$

Minimum: $(2.69, -5.55)$　　Minima: $(0, 0), (3, 0)$

37. Minimum: $(0, 0)$　　　**39.** Maximum: $\left(2, \frac{1}{2}\right)$

Maximum: $(1, 2)$　　　　　Minimum: $(0, 0)$

41. Maximum: $\left| f''\!\left(\sqrt[3]{-10 + \sqrt{108}}\right) \right| \approx 1.47$

43. Maximum: $\left| f^{(4)}(0) \right| = \dfrac{56}{81}$

45. Answers will vary. Example:

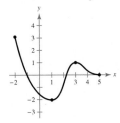

47. 82 units　　**49.** \$2.15

51. (a) Population tends to increase each year, so the minimum population occurred in 1790 and the maximum population occurred in 2000.

(b) Maximum population: 278.968 million

Minimum population: 3.775 million

(c) The minimum population was about 3.775 million in 1790 and the maximum population was about 278.968 million in 2000.

SECTION 8.6   *(page 676)*

Skills Review   *(page 676)*

1. $f''(x) = 48x^2 - 54x$

2. $g''(s) = 12s^2 - 18s + 2$

3. $g''(x) = 56x^6 + 120x^4 + 72x^2 + 8$

4. $f''(x) = \dfrac{4}{9(x-3)^{2/3}}$ 5. $h''(x) = \dfrac{190}{(5x-1)^3}$

6. $f''(x) = -\dfrac{42}{(3x+2)^3}$ 7. $x = \pm\dfrac{\sqrt{3}}{3}$

8. $x = 0, 3$ 9. $t = \pm 4$ 10. $x = 0, \pm 5$

---

1. Concave upward on $(-\infty, \infty)$

3. Concave upward on $\left(-\infty, -\frac{1}{2}\right)$
   Concave downward on $\left(-\frac{1}{2}, \infty\right)$

5. Concave upward on $(-\infty, -2)$ and $(2, \infty)$
   Concave downward on $(-2, 2)$

7. Concave upward on $(-\infty, 2)$
   Concave downward on $(2, \infty)$

9. Relative maximum: $(3, 9)$

11. Relative maximum: $(1, 3)$
    Relative minimum: $\left(\frac{7}{3}, \frac{49}{27}\right)$

13. Relative minimum: $(0, -3)$

15. Relative minimum: $(0, 1)$

17. Relative minima: $(-3, 0), (3, 0)$
    Relative maximum: $(0, 3)$

19. Relative maximum: $(0, 4)$

21. No relative extrema

23. Relative maximum: $(0, 0)$
    Relative minima: $(-0.5, -0.052), (1, -0.\overline{3})$

25. Relative maximum: $(2, 9)$
    Relative minimum: $(0, 5)$

27. Sign of $f'(x)$ on $(0, 2)$ is positive.
    Sign of $f''(x)$ on $(0, 2)$ is positive.

29. Sign of $f'(x)$ on $(0, 2)$ is negative.
    Sign of $f''(x)$ on $(0, 2)$ is negative.

31. $(3, 0)$ 33. $(1, 0), (3, -16)$

35. No inflection points   37. $\left(\frac{3}{2}, -\frac{1}{16}\right), (2, 0)$

39.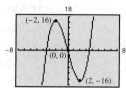

Relative maximum: $(-2, 16)$
Relative minimum: $(2, -16)$
Point of inflection: $(0, 0)$

41.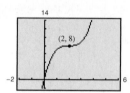

No relative extrema
Point of inflection: $(2, 8)$

43.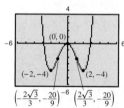

Relative maximum: $(0, 0)$
Relative minima: $(\pm 2, -4)$
Points of inflection:
$$\left(\pm\frac{2\sqrt{3}}{3}, -\frac{20}{9}\right)$$

45.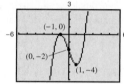

Relative maximum: $(-1, 0)$
Relative minimum: $(1, -4)$
Point of inflection: $(0, -2)$

47.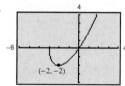

Relative minimum:
$(-2, -2)$
No inflection points

49.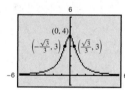

Relative maximum: $(0, 4)$
Points of inflection:
$$\left(\pm\frac{\sqrt{3}}{3}, 3\right)$$

51.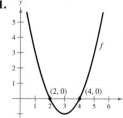

53.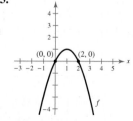

55.

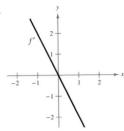

(a) f': Positive on $(-\infty, 0)$

　　f: Increasing on $(-\infty, 0)$

(b) f': Negative on $(0, \infty)$

　　f: Decreasing on $(0, \infty)$

(c) f': Not increasing

　　f: Not concave upward

(d) f': Decreasing on $(-\infty, \infty)$

　　f: Concave downward on $(-\infty, \infty)$

57. (a) f': Increasing on $(-\infty, \infty)$

(b) f: Concave upward on $(-\infty, \infty)$

(c) Relative minimum: $(-2.5, -6.25)$

　　No inflection points

(d)

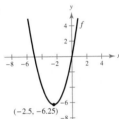

59. (a) f': Increasing on $(-\infty, 1)$

　　　Decreasing on $(1, \infty)$

(b) f: Concave upward on $(-\infty, 1)$

　　Concave downward on $(1, \infty)$

(c) No relative extrema

　　Point of inflection: $\left(1, -\frac{1}{3}\right)$

(d)

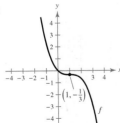

61. $(200, 320)$ **63.** 100 units   **65.** 8:30 P.M.

67. $\sqrt{3} \approx 1.732/\text{yr}$

69.

Relative minimum: $(0, -5)$

Relative maximum: $(3, 8.5)$

Point of inflection:

　$\left(\frac{2}{3}, -3.2963\right)$

When f' is positive, f is increasing. When f' is negative, f is decreasing. When f'' is positive, f is concave upward. When f'' is negative, f is concave downward.

71.

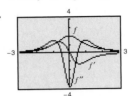

Relative maximum: $(0, 2)$

Points of inflection:

　$(0.58, 1.5), (-0.58, 1.5)$

When f' is positive, f is increasing. When f' is negative, f is decreasing. When f'' is positive, f is concave upward. When f'' is negative, f is concave downward.

73. 120 units

75. (a)

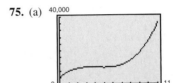

(b) November   (c) October   (d) October; April

77. (a) S' is increasing and $S'' > 0$.

(b) S' is increasing and positive and $S'' > 0$.

(c) S' is constant and $S'' = 0$.

(d) $S' = 0$ and $S'' = 0$.

(e) $S' < 0$ and $S'' > 0$.

(f) $S' > 0$ and there are no restrictions on S''.

79. Answers will vary.

REVIEW EXERCISES FOR CHAPTER 8
(page 683)

1. 6   **3.** $-\dfrac{120}{x^6}$ **5.** $\dfrac{35x^{3/2}}{2}$

7. $\dfrac{2}{x^{2/3}}$ **9.** 2   **11.** $\dfrac{512}{81}$

13. (a) $s(t) = -16t^2 + 5t + 30$ (b) About 1.534 sec

　　(c) About -44.09 ft/sec   (d) -32 ft/sec^2

15. $s'(t) = -\dfrac{2(t+1)}{(t^2 + 2t + 1)^2}$; $s''(t) = \dfrac{6}{(t^2 + 2t + 1)^2}$

17. $-\dfrac{2x + 3y}{3(x + y^2)}$ **19.** $\dfrac{2x - 8}{2y - 9}$

21. 5   **23.** 0   **25.** $y = \frac{1}{3}x + \frac{1}{3}$

27. $y = \frac{4}{3}x + \frac{2}{3}$ **29.** $\frac{1}{64}$ ft/min

31. (a) $P = -0.4x^3 + 3600x - 5200$

    (b) \$119,126.40

    (c) 5 units/wk

33. $x = 1$ **35.** $x = 0, x = 1$

37. Increasing on $\left(-\frac{1}{2}, \infty\right)$

    Decreasing on $\left(-\infty, -\frac{1}{2}\right)$

39. Increasing on $(-\infty, 3)$ and $(3, \infty)$

41. (a) $(1.38, 7.24)$ (b) $(1, 1.38), (7.24, 12)$

    (c) Normal monthly temperature is rising from early January to early July and decreasing from early July to early January.

    (d)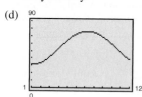

43. (a) Increasing on $(-\infty, -0.85)$

    Decreasing on $(-0.85, \infty)$

    (b) $-\infty$

    (c) Because $\dfrac{dN}{dt}$ is decreasing on $(-0.85, \infty)$, the value of N approaches $-\infty$ as t approaches ∞. This confirms the answer to part (b).

45. Relative maximum: $(0, -2)$

    Relative minimum: $(1, -4)$

47. Relative minimum: $(8, -52)$

49. Relative maxima: $(-1, 1), (1, 1)$

    Relative minimum: $(0, 0)$

51. Relative maximum: $(0, 6)$

53. Relative maximum: $(0, 0)$

    Relative minimum: $(4, 8)$

55. Maximum: $(0, 6)$ **57.** Maxima: $(-2, 17), (4, 17)$

    Minimum: $\left(-\frac{5}{2}, -\frac{1}{4}\right)$ Minima: $(-4, -15), (2, -15)$

59. Maximum: $(1, 3)$

    Minimum: $\left(3, 4\sqrt{3} - 9\right)$

61. Maximum: $\left(2, \dfrac{2\sqrt{5}}{5}\right)$ **63.** Maximum: $(1, 1)$

    Minimum: $(0, 0)$ Minimum: $(-1, -1)$

65. 1973

67. $x = \frac{137}{9} \approx 15.2$ yr

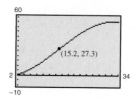

69. (a)

t	0	0.5	1	1.5	2	2.5	3
$C(t)$	0	0.06	0.11	0.15	0.17	0.18	0.17

    $t = 2.5$ hours

    (b)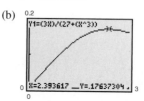

    $t \approx 2.39$ hours

    (c) $t \approx 2.38$ hours

71. Concave upward on $(2, \infty)$

    Concave downward on $(-\infty, 2)$

73. Concave upward on $\left(-\dfrac{2\sqrt{3}}{3}, \dfrac{2\sqrt{3}}{3}\right)$

    Concave downward on $\left(-\infty, -\dfrac{2\sqrt{3}}{3}\right)$ and $\left(\dfrac{2\sqrt{3}}{3}, \infty\right)$

75. $(0, 0), (4, -128)$

77. $(0, 0), (1.0652, 4.5244), (2.5348, 3.5246)$

79. Relative maximum: $\left(-\sqrt{3}, 6\sqrt{3}\right)$

    Relative minimum: $\left(\sqrt{3}, -6\sqrt{3}\right)$

81. Relative maxima: $\left(-\dfrac{\sqrt{2}}{2}, \dfrac{1}{2}\right), \left(\dfrac{\sqrt{2}}{2}, \dfrac{1}{2}\right)$

    Relative minimum: $(0, 0)$

83. (a)

    (b) Concave upward on $(-\infty, 2.48)$

    Concave downward on $(2.48, \infty)$

    (c) $(2.48, 1125.89)$

    (d) The concavity of the graph changes from upward to downward at the inflection point $(2.48, 1125.89)$.

CHAPTER TEST   *(page 687)*

1. 0 **2.** $-\dfrac{3}{8(3-x)^{5/2}}$ **3.** $-\dfrac{96}{(2x-1)^4}$

4. $\dfrac{dy}{dx} = -\dfrac{1+y}{x}$ **5.** $\dfrac{dy}{dx} = -\dfrac{1}{y-1}$ **6.** $\dfrac{dy}{dx} = \dfrac{x}{2y}$

7. Critical number: $x = 0$

Increasing on $(0, \infty)$

Decreasing on $(-\infty, 0)$

8. Critical numbers: $x = -2, x = 2$

Increasing on $(-\infty, -2)$ and $(2, \infty)$

Decreasing on $(-2, 2)$

9. Critical number: $x = 5$

Increasing on $(5, \infty)$

Decreasing on $(-\infty, 5)$

10.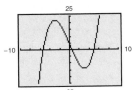

Relative minimum: $(3, -14)$

Relative maximum: $(-3, 22)$

11.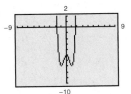

Relative minima: $(-1, -7)$ and $(1, -7)$

Relative maximum: $(0, -5)$

12.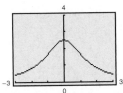

Relative maximum: $(0, 2.5)$

13. Minimum: $(-3, -1)$ **14.** Minimum: $(0, 0)$

Maximum: $(0, 8)$ Maximum: $(2.25, 9)$

15. Concave upward: $\left(\dfrac{\sqrt[3]{50}}{5}, \infty\right)$

Concave downward: $\left(-\infty, \dfrac{\sqrt[3]{50}}{5}\right)$

16. Concave upward: $\left(-\infty, -\dfrac{2\sqrt{2}}{3}\right)$ and $\left(\dfrac{2\sqrt{2}}{3}, \infty\right)$

Concave downward: $\left(-\dfrac{2\sqrt{2}}{3}, \dfrac{2\sqrt{2}}{3}\right)$

17. $(2, -2)$ **18.** $\left(\sqrt[3]{2}, -\dfrac{18\sqrt[3]{4}}{5}\right)$

19. Relative minimum: $(5.46, -135.14)$

Relative maximum: $(-1.46, 31.14)$

20. Relative minimum: $(3, -97.2)$

Relative maximum: $(-3, 97.2)$

21. (a) 3.75π cm³/min   (b) 15π cm³/min

22. (a) Late 1999; 2005

(b) Increasing from 1999 to late 1999.

Decreasing from late 1999 to 2005.

CHAPTER 9

SECTION 9.1   *(page 695)*

Skills Review   *(page 695)*

1. $x + \frac{1}{2}y = 12$ **2.** $2xy = 24$ **3.** $xy = 24$

4. $\sqrt{(x_2 - x_1)^2 + (y_2 - y_1)^2} = 10$

5. $x = -3$ **6.** $x = -\frac{2}{3}, 1$ **7.** $x = \pm 5$

8. $x = 4$ **9.** $x = \pm 1$ **10.** $x = \pm 3$

---

1. $60, 60$ **3.** $18, 9$ **5.** $\sqrt{192}, \sqrt{192}$

7. $l = w = 25$ m   **9.** $l = w = 8$ ft

11. $x = 25$ ft, $y = \frac{100}{3}$ ft

13. (a) Proof

(b) $V_1 = 99$ in.³

$V_2 = 125$ in.³

$V_3 = 117$ in.³

(c) 5 in. × 5 in. × 5 in.

15. $l = w = 2\sqrt[3]{5} \approx 3.42$ **17.** $x = 5$ m, $y = 3\frac{1}{3}$ m

$h = 4\sqrt[3]{5} \approx 6.84$

19. 1.056 ft³   **21.** 9 in. by 9 in.

23. Length: 3 units   **25.** Length: $5\sqrt{2}$ units

Width: 1.5 units   Width: $5\sqrt{2}/2$ units

27. Radius: about 1.51 in.

Height: about 3.02 in.

29. $(1, 1)$ **31.** $\left(3.5, \dfrac{\sqrt{14}}{2}\right)$

33. 18 in. × 18 in. × 36 in.

35. Radius: $\sqrt[3]{\dfrac{562.5}{\pi}} \approx 5.636$ ft

Height: about 22.545 ft

37. Side of square: $\dfrac{10\sqrt{3}}{9 + 4\sqrt{3}}$

Side of triangle: $\dfrac{30}{9 + 4\sqrt{3}}$

39. Width of rectangle: $\dfrac{100}{\pi} \approx 31.8$ m

Length of rectangle: 50 m

41. $w = 8\sqrt{3}$ in., $h = 8\sqrt{6}$ in.

43. (a) \$40,000 $(s = 40)$ (b) \$20,000 $(s = 20)$

SECTION 9.2 *(page 705)*

Skills Review *(page 705)*

1. 1   **2.** $\frac{6}{5}$ **3.** 2   **4.** $\frac{1}{2}$

5. $\dfrac{dC}{dx} = 1.2 + 0.006x$ **6.** $\dfrac{dP}{dx} = 0.02x + 11$

7. $\dfrac{dR}{dx} = 14 - \dfrac{x}{1000}$ **8.** $\dfrac{dR}{dx} = 3.4 - \dfrac{x}{750}$

9. $\dfrac{dP}{dx} = -1.4x + 7$ **10.** $\dfrac{dC}{dx} = 4.2 + 0.003x^2$

1. 2000 units   **3.** 200 units   **5.** 200 units

7. 50 units   **9.** \$60   **11.** \$67.50

13. 3 units

$$\overline{C}(3) = 17; \frac{dC}{dx} = 4x + 5; \text{ when } x = 3, \frac{dC}{dx} = 17$$

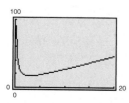

15. (a) \$70   (b) About \$40.63

17. The maximum profit occurs when $s = 10$ (or \$10,000).

The point of diminishing returns occurs at $s = \frac{35}{6}$ (or \$5833.33).

19. 200 players   **21.** \$50

23. C = cost under water + cost on land

$= 25(5280)\sqrt{x^2 + 0.25} + 18(5280)(6 - x)$

$= 132,000\sqrt{x^2 + 0.25} + 570,240 - 95,040x$

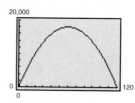

The line should run from the power station to a point across the river approximately 0.52 mile downstream.

$\left(\text{Exact: } \dfrac{9\sqrt{301}}{301} \text{ mi} \right)$

25. 60 mi/h

27. -3, elastic

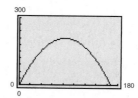

Elastic: $(0, 60)$

Inelastic: $(60, 120)$

29. $-\frac{2}{3}$, inelastic

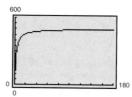

Elastic: $\left(0, 83\frac{1}{3}\right)$

Inelastic: $\left(83\frac{1}{3}, 166\frac{2}{3}\right)$

31. $-\frac{25}{23}$, elastic

Elastic: $(0, \infty)$

33. (a) $-\frac{11}{14}$ (b) $x = 500$ units, $p = \$10$

(c) Answers will vary.

35. 500 units $(x = 5)$

37. No; when $p = 5$, $x = 350$ and $\eta = -\frac{5}{7}$.

Because $|\eta| = \frac{5}{7} < 1$, demand is inelastic.

39. (a) 2006   (b) 2001

(c) 2006: $11.25 billion/yr

2001: $0.32 billion/yr

(d)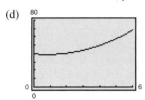

41. Demand function: a

Revenue function: c

Cost function: b

Profit function: d

43. Answers will vary.   **45.** Answers will vary.

SECTION 9.3   *(page 716)*

Skills Review   *(page 716)*

1. 3   **2.** 1   **3.** -11 **4.** 4   **5.** $-\frac{1}{4}$

6. -2 **7.** 0   **8.** 1

9. $\overline{C} = \dfrac{150}{x} + 3$ **10.** $\overline{C} = \dfrac{1900}{x} + 1.7 + 0.002x$

$\dfrac{dC}{dx} = 3$ $\dfrac{dC}{dx} = 1.7 + 0.004x$

11. $\overline{C} = 0.005x + 0.5 + \dfrac{1375}{x}$ **12.** $\overline{C} = \dfrac{760}{x} + 0.05$

$\dfrac{dC}{dx} = 0.01x + 0.5$ $\dfrac{dC}{dx} = 0.05$

---

1. Vertical asymptote: $x = 0$

Horizontal asymptote: $y = 1$

3. Vertical asymptotes: $x = -1, x = 2$

Horizontal asymptote: $y = 1$

5. Vertical asymptote: none

Horizontal asymptote: $y = \frac{3}{2}$

7. Vertical asymptotes: $x = \pm 2$

Horizontal asymptote: $y = \frac{1}{2}$

9. d   **10.** b   **11.** a   **12.** c

13. ∞ **15.** $-\infty$ **17.** $-\infty$ **19.** $-\infty$

21.

x	10^0	10^1	10^2	10^3
$f(x)$	2.000	0.348	0.101	0.032

x	10^4	10^5	10^6
$f(x)$	0.010	0.003	0.001

$\displaystyle\lim_{x\to\infty} \frac{x+1}{x\sqrt{x}} = 0$

23.

x	10^0	10^1	10^2	10^3
$f(x)$	0	49.5	49.995	49.99995

x	10^4	10^5	10^6
$f(x)$	50.0	50.0	50.0

$\displaystyle\lim_{x\to\infty} \frac{x^2-1}{0.02x^2} = 50$

25.

x	-10^6	-10^4	-10^2	10^0
$f(x)$	-2	-2	-1.9996	0.8944

x	10^2	10^4	10^6
$f(x)$	1.9996	2	2

$\displaystyle\lim_{x\to-\infty} \frac{2x}{\sqrt{x^2+4}} = -2, \ \lim_{x\to\infty} \frac{2x}{\sqrt{x^2+4}} = 2$

27. (a) ∞ (b) 5   (c) 0   **29.** (a) 0   (b) 1   (c) ∞

31. 2   **33.** 0   **35.** $-\infty$ **37.** ∞ **39.** 5

41. 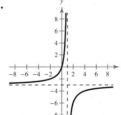   **43.**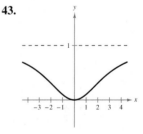

45. 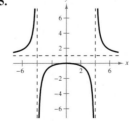   **47.**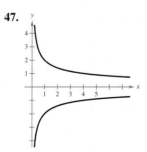

49.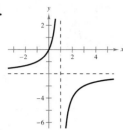

51.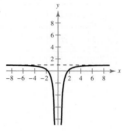

53.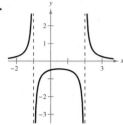

55.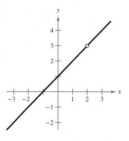

57.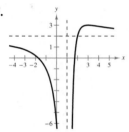

59. (a) $\overline{C} = 1.35 + \dfrac{4570}{x}$ (b) $47.05, $5.92   (c) $1.35

61. (a) $\overline{C} = 13.5 + \dfrac{45,750}{x}$

(b) $\overline{C}(100) = 471$; $\overline{C}(1000) = 59.25$

(c) $13.50; The cost approaches $13.50 as the number of PDAs produced increases.

63. (a) 25%: $176 million; 50%: $528 million;

75%: $1584 million

(b) ∞; The limit does not exist, which means the cost increases without bound as the government approaches 100% seizure of illegal drugs entering the country.

65. (a)

n	1	2	3	4	5
P	0.5	0.74	0.82	0.86	0.89

n	6	7	8	9	10
P	0.91	0.92	0.93	0.94	0.95

(b) 1

(c)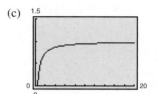

The percent of correct responses approaches 100% as the number of times the task is performed increases.

67. (a) $\overline{P} = 35.4 - \dfrac{15,000}{x}$

(b) $\overline{P}(1000) = 20.40; $\overline{P}(10,000) = 33.90;

$\overline{P}(100,000) = 35.25

(c) $35.40; Explanations will vary.

MID-CHAPTER QUIZ *(page 719)*

1. (a) 100 ft by 50 ft   (b) 5000 ft^2 **2.** $7\frac{1}{2}$ in. by 10 in.

3. 400 units   **4.** 70 units   **5.** $63   **6.** $.80

7. (a) 0.5   (b) Inelastic

(c)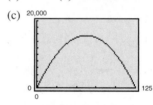

(d) Elastic: $(0, 62.5)$

Inelastic: $(62.5, 125)$

8. (a) 1   (b) Of unit elasticity

(c)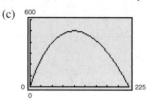

(d) Elastic: $(0, 100)$

Inelastic: $(100, 225)$

9. -1 **10.** $-\infty$ **11.** 10   **12.** $\frac{1}{3}$

13. -1 **14.** ∞

15. Vertical asymptote: $x = 1$

Horizontal asymptote: $y = 2$

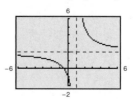

16. Vertical asymptotes: $x = 0$, $x = 2$

Horizontal asymptote: $y = 0$

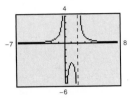

17. Vertical asymptote: $x = 3$

Horizontal asymptote: none

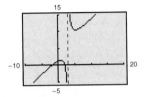

SECTION 9.4   (page 727)

Skills Review   (page 727)

1. Vertical asymptote: $x = 0$

Horizontal asymptote: $y = 0$

2. Vertical asymptote: $x = 2$

Horizontal asymptote: $y = 0$

3. Vertical asymptote: $x = -3$

Horizontal asymptote: $y = 40$

4. Vertical asymptotes: $x = 1$, $x = 3$

Horizontal asymptote: $y = 1$

5. Decreasing on $(-\infty, -2)$

Increasing on $(-2, \infty)$

6. Increasing on $(-\infty, -4)$

Decreasing on $(-4, \infty)$

7. Increasing on $(-\infty, -1)$ and $(1, \infty)$

Decreasing on $(-1, 1)$

8. Decreasing on $(-\infty, 0)$ and $\left(\sqrt[3]{2}, \infty\right)$

Increasing on $\left(0, \sqrt[3]{2}\right)$

9. Increasing on $(-\infty, 1)$ and $(1, \infty)$

10. Decreasing on $(-\infty, -3)$ and $\left(\frac{1}{3}, \infty\right)$

Increasing on $\left(-3, \frac{1}{3}\right)$

1.

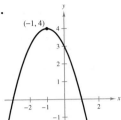

3.

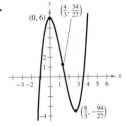

5.

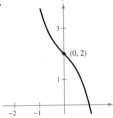

7.

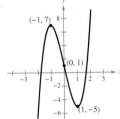

9.

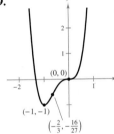

11.

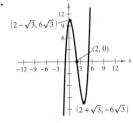

13.

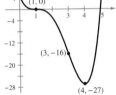

15.

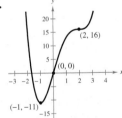

17.

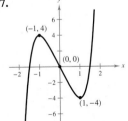

19.

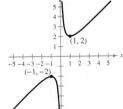

21.

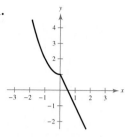

23.

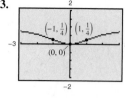

25.

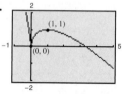

27.

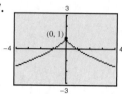

29.

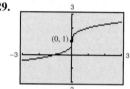

31.

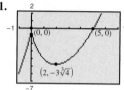

33.

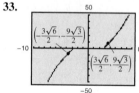

35.

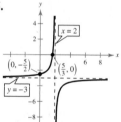

Domain: $(-\infty, 2) \cup (2, \infty)$

37.

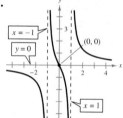

Domain: $(-\infty, -1) \cup (-1, 1) \cup (1, \infty)$

39.

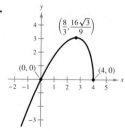

Domain: $(-\infty, 4]$

41.

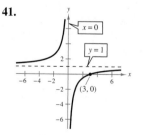

Domain: $(-\infty, 0) \cup (0, \infty)$

43.

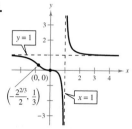

Domain: $(-\infty, 1) \cup (1, \infty)$

45. Answers will vary.

Sample answer: $f(x) = -x^3 + x^2 + x + 1$

47. Answers will vary.
Sample answer:

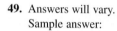

49. Answers will vary.
Sample answer:

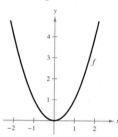

51. Answers will vary.
Sample answer:

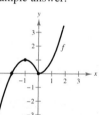

53. Answers will vary.
Sample answer:

$$y = \frac{1}{x - 5}$$

55. (a)

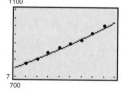

(b) $1099.31

The model fits the data well.

(c) No, because the benefits increase without bound as time approaches the year 2040 $(x = 50)$, and the benefits are negative for the years past 2040.

57. (a)

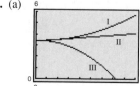

(b) Models I and II

(c) Model I; Model III; Model I; Explanations will vary.

59.

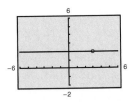

The rational function has the common factor $3 - x$ in the numerator and denominator. At $x = 3$, there is a hole in the graph, not a vertical asymptote.

SECTION 9.5 *(page 735)*

Skills Review *(page 735)*

1. $\dfrac{dC}{dx} = 0.18x$ **2.** $\dfrac{dC}{dx} = 0.15$

3. $\dfrac{dR}{dx} = 1.25 + 0.03\sqrt{x}$ **4.** $\dfrac{dR}{dx} = 15.5 - 3.1x$

5. $\dfrac{dP}{dx} = -\dfrac{0.01}{\sqrt[3]{x^2}} + 1.4$ **6.** $\dfrac{dP}{dx} = -0.04x + 25$

7. $\dfrac{dA}{dx} = \dfrac{\sqrt{3}}{2}x$ **8.** $\dfrac{dA}{dx} = 12x$ **9.** $\dfrac{dC}{dr} = 2\pi$

10. $\dfrac{dP}{dw} = 4$ **11.** $\dfrac{dS}{dr} = 8\pi r$ **12.** $\dfrac{dP}{dx} = 2 + \sqrt{2}$

13. $A = \pi r^2$ **14.** $A = x^2$

15. $V = x^3$ **16.** $V = \frac{4}{3}\pi r^3$

---

1. $dy = 6x\, dx$ **3.** $dy = 12(4x - 1)^2\, dx$

5. $dy = \dfrac{-x}{\sqrt{9 - x^2}}\, dx$ **7.** 0.1005   **9.** -0.013245

11. $dy = 0.6$ **13.** $dy = -0.04$

$\quad\ \Delta y = 0.6305$ $\quad\quad\ \Delta y \approx -0.0394$

15.

$dx = \Delta x$	dy	Δy	$\Delta y - dy$	$\dfrac{dy}{\Delta y}$
1.000	4.000	5.000	1.0000	0.8000
0.500	2.000	2.2500	0.2500	0.8889
0.100	0.400	0.4100	0.0100	0.9756
0.010	0.040	0.0401	0.0001	0.9975
0.001	0.004	0.0040	0.0000	1.0000

17.

$dx = \Delta x$	dy	Δy	$\Delta y - dy$	$\dfrac{dy}{\Delta y}$
1.000	-0.25000	-0.13889	0.11111	1.79999
0.500	-0.12500	-0.09000	0.03500	1.38889
0.100	-0.02500	-0.02324	0.00176	1.07573
0.010	-0.00250	-0.00248	0.00002	1.00806
0.001	-0.00025	-0.00025	0.00000	1.00000

19.

$dx = \Delta x$	dy	Δy	$\Delta y - dy$	$\dfrac{dy}{\Delta y}$
1.000	0.14865	0.12687	-0.02178	1.17167
0.500	0.07433	0.06823	-0.00610	1.08940
0.100	0.01487	0.01459	-0.00028	1.01919
0.010	0.00149	0.00148	-0.00001	1.00676
0.001	0.00015	0.00015	0.00000	1.00000

21. $y = 28x + 37$

For $\Delta x = -0.01$, $f(x + \Delta x) = -19.281302$ and $y(x + \Delta x) = -19.28$.

For $\Delta x = 0.01$, $f(x + \Delta x) = -18.721298$ and $y(x + \Delta x) = -18.72$.

23. $y = x$

For $\Delta x = -0.01$, $f(x + \Delta x) = -0.009999$ and $y(x + \Delta x) = -0.01$.

For $\Delta x = 0.01$, $f(x + \Delta x) = 0.009999$ and $y(x + \Delta x) = 0.01$.

25. $dP = 1160$

Percent change: about 2.7%

27. (a) $\Delta p = -0.25 = dp$ (b) $\Delta p = -0.25 = dp$

29. \$5.20   **31.** \$7.50

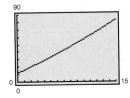

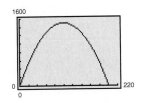

33. −$1250

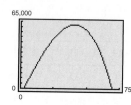

35. $R = -\frac{1}{3}x^2 + 100x$; $6

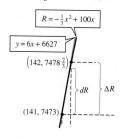

37. $P = -\dfrac{1}{2000}x^2 + 23x - 275,000$; −$5

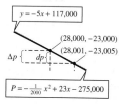

39. $\pm\dfrac{3}{8}$ in.², ±0.0026   **41.** $734.8 billion

43. $\dfrac{\sqrt{2}}{24} \approx 0.059$ m²   **45.** True

REVIEW EXERCISES FOR CHAPTER 9
(page 740)

1. 13, 13

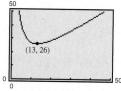

3. (a) 40 in. by 40 in. by 40 in.   (b) 64,000 in.³

5. (a) 59 trees   (b) 87,025 oranges

7. 9 feet from the shorter post

9. (a) 3 units   (b) 1 unit

11. 7.7 miles per hour   **13.** 125 units

15. Elastic: $(0, 75)$

Inelastic: $(75, 150)$

Demand is of unit
elasticity when $x = 75$.

17. Elastic: $(0, 150)$

Inelastic: $(150, 300)$

Demand is of unit
elasticity when $x = 150$.

19. $-\infty$ **21.** ∞ **23.** $\frac{2}{3}$ **25.** 0

27. Vertical asymptote: $x = 2$

Horizontal asymptote: $y = -2$

29. Vertical asymptote: $x = 0$

Horizontal asymptotes: $y = 2$ and $y = -2$

31. Vertical asymptote: $x = 4$

Horizontal asymptote: $y = 2$

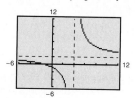

33. Vertical asymptotes: $x = -5, 2$

Horizontal asymptote: $y = 0$

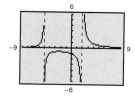

35. Horizontal asymptote: $y = 0$

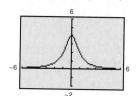

37. (a) 425; the temperature (in °F) of the oven

(b) 72; the temperature (in °F) of the room

39. (a) $\overline{C} = \dfrac{10,000 + 48.9x}{x}$ (b) 48.9

(c) $19.59 per unit; $19.595 per unit; $19.599 per unit

(d) $19.60 per unit

41.

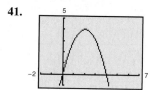

Intercepts: $(0, 0), (4, 0)$

Relative maximum: $(2, 4)$

Domain: $(-\infty, \infty)$

43.

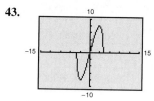

Intercepts: $(0, 0), (4, 0),$
$(-4, 0)$

Relative maximum: $(2\sqrt{2}, 8)$

Relative minimum:
$(-2\sqrt{2}, -8)$

Point of inflection: $(0, 0)$

Domain: $[-4, 4]$

45.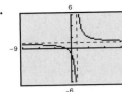

Intercepts: $(-1, 0), (0, -1)$
Horizontal asymptote: $y = 1$
Vertical asymptote: $x = 1$
Domain: $(-\infty, 1) \cup (1, \infty)$

47.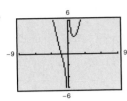

Intercept: $\left(-\sqrt[3]{2}, 0\right)$
Relative minimum: $(1, 3)$
Point of inflection: $\left(-\sqrt[3]{2}, 0\right)$
Vertical asymptote: $x = 0$
Domain: $(-\infty, 0) \cup (0, \infty)$

49. $dy = (1 - 2x)\, dx$ **51.** $dy = -\dfrac{x}{\sqrt{36 - x^2}}\, dx$

53. \$800   **55.** \$15.25

57. (a) $dA = 2x\Delta x, \quad \Delta A = 2x\Delta x + (\Delta x)^2$

(b) and (c)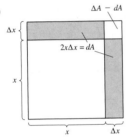

59. $\Delta p = -0.125$

$dp = -0.125$

The values are equal.

CHAPTER TEST   (page 743)

1. Vertical asymptote: $x = 1$
Horizontal asymptote: $y = 2$

2. Vertical asymptote: none
Horizontal asymptote: $y = 0$

3. Vertical asymptote: $x = 5$
Horizontal asymptote: $y = 3$

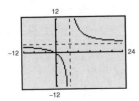

4. Vertical asymptotes: $x = 1$ and $x = -1$
Horizontal asymptote: $y = 1$

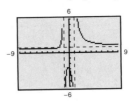

5. $-\infty$ **6.** ∞ **7.** ∞ **8.** 1   **9.** ∞ **10.** 3

11.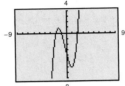

Intercepts: $(-2, 0), (-1, 0), (2, 0), (0, -4)$

Relative maximum: $\left(-\dfrac{\sqrt{13} + 1}{3}, \dfrac{26\sqrt{13} - 70}{27}\right)$

Relative minimum: $\left(\dfrac{\sqrt{13} - 1}{3}, -\dfrac{26\sqrt{13} + 70}{27}\right)$

Point of inflection: $\left(-\dfrac{1}{3}, -\dfrac{70}{27}\right)$

Domain: $(-\infty, \infty)$

12.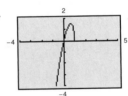

Intercepts: $(0, 0)$ and $(1, 0)$

Relative maximum: $\left(\dfrac{2}{3}, \dfrac{8\sqrt{3}}{9}\right)$

Domain: $(-\infty, 1]$

13. $dy = 10x\, dx$ **14.** $dy = \dfrac{-4}{(x + 3)^2}\, dx$

15. $dy = 3(x + 4)^2\, dx$ **16.** 125 m by 125 m

17. (a) 20 in. by 20 in. by 20 in.   (b) 2400 in.^{2}

18. $(312.5, 625)$

CHAPTER 10

SECTION 10.1   *(page 749)*

> ### Skills Review   *(page 749)*
>
> 1. Horizontal shift to the left two units
> 2. Reflection about the *x*-axis
> 3. Vertical shift down one unit
> 4. Reflection about the *y*-axis
> 5. Horizontal shift to the right one unit
> 6. Vertical shift up two units
> 7. Nonremovable discontinuity at $x = -4$
> 8. Continuous on $(-\infty, \infty)$
> 9. Discontinuous at $x = \pm 1$
> 10. Continuous on $(-\infty, \infty)$
> 11. 5   12. $\frac{4}{3}$ 13. $-9, 1$ 14. $2 \pm 2\sqrt{2}$
> 15. $1, -5$ 16. $\frac{1}{2}, 1$

1. (a) 625   (b) 9   (c) $16\sqrt{2}$
   (d) 9   (e) 125   (f) 4
3. (a) 3125   (b) $\frac{1}{5}$ (c) 625   (d) $\frac{1}{125}$
5. (a) $\frac{1}{5}$ (b) 27   (c) 5   (d) 4096
7. (a) 4   (b) $\frac{\sqrt{2}}{2} \approx 0.707$ (c) $\frac{1}{8}$ (d) $\frac{\sqrt{2}}{8} \approx 0.177$
9. (a) 0.907   (b) 348.912   (c) 1.796   (d) 1.308
11. 2 g   13. e   14. c   15. a
16. f   17. d   18. b

19. 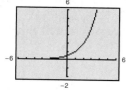   21.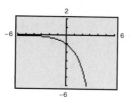

23. 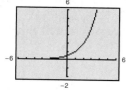   25.

27. 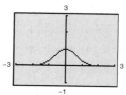   29.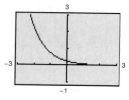

31. (a) $P(18) \approx 306.99$ million
    (b) $P(22) \approx 320.72$ million
33. (a) $V(5) \approx \$80{,}634.95$ (b) $V(20) \approx \$161{,}269.89$
35. $36.93
37. (a)

Year	1998	1999	2000	2001
Actual	152,500	161,000	169,000	175,200
Model	149,036	158,709	169,009	179,978

Year	2002	2003	2004	2005
Actual	187,600	195,000	221,000	240,900
Model	191,658	204,097	217,343	231,448

The model fits the data well. Explanations will vary.

(b)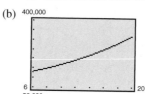

(c) 2009

SECTION 10.2   *(page 757)*

> ### Skills Review   *(page 757)*
>
> 1. Continuous on $(-\infty, \infty)$
> 2. Discontinuous for $x = \pm 2$
> 3. Discontinuous for $x = \pm\sqrt{3}$
> 4. Removable discontinuity at $x = 4$
> 5. 0   6. 0   7. 4   8. $\frac{1}{2}$ 9. $\frac{3}{2}$
> 10. 6   11. 0   12. 0

1. (a) e^7 (b) e^{12} (c) $\frac{1}{e^6}$ (d) 1
3. (a) e^5 (b) $e^{5/2}$ (c) e^6 (d) e^7
5. f   6. e   7. d   8. b   9. c   10. a

11.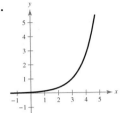

13.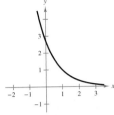

15.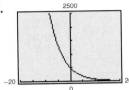

17.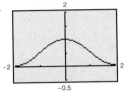

19.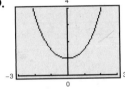

No horizontal asymptotes

Continuous on the entire
real number line

21.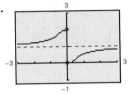

Horizontal asymptote: $y = 1$

Discontinuous at $x = 0$

23. (a)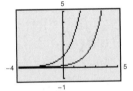

The graph of $g(x) = e^{x-2}$ is shifted horizontally two
units to the right.

(b)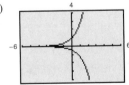

The graph of $h(x) = -\frac{1}{2}e^x$ decreases at a slower rate
than e^x increases.

(c)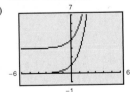

The graph of $q(x) = e^x + 3$ is shifted vertically three
units upward.

25.

n	1	2	4	12
A	1343.92	1346.86	1348.35	1349.35

n	365	Continuous compounding
A	1349.84	1349.86

27.

n	1	2	4	12
A	2191.12	2208.04	2216.72	2222.58

n	365	Continuous compounding
A	2225.44	2225.54

29.

t	1	10	20
P	96,078.94	67,032.00	44,932.90

t	30	40	50
P	30,119.42	20,189.65	13,533.53

31.

t	1	10	20
P	95,132.82	60,716.10	36,864.45

t	30	40	50
P	22,382.66	13,589.88	8251.24

33. $107,311.12

35. (a) 9%   (b) 9.20%   (c) 9.31%   (d) 9.38%

37. $12,500   **39.** $8751.92

41. (a) $849.53   (b) $421.12

$\lim\limits_{x \to \infty} p = 0$

43. (a) 0.1535   (b) 0.4866   (c) 0.8111

45. (a) The model fits the data well.

(b) $y = 421.60x + 1504.6$; The linear model fits the data
well, but the exponential model fits the data better.

(c) Exponential model: 2008

Linear model: 2010

47. (a)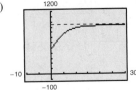

(b) Yes, $\lim\limits_{t\to\infty} \dfrac{925}{1 + e^{-0.3t}} = 925$

(c) $\lim\limits_{t\to\infty} \dfrac{1000}{1 + e^{-0.3t}} = 1000$

Models similar to this logistic growth model where $y = \dfrac{a}{1 + be^{-ct}}$ have a limit of a as $t \to \infty$.

49. (a) 0.731　(b) 11　(c) Yes, $\lim\limits_{n\to\infty} \dfrac{0.83}{1 + e^{-0.2n}} = 0.83$

51. Amount earned:

(a) $5267.71

(b) $5255.81

(c) $5243.23

You should choose the certificate of deposit in part (a) because it earns more money than the others.

SECTION 10.3　*(page 766)*

Skills Review　*(page 766)*

1. $\dfrac{1}{2}e^x(2x^2 - 1)$　**2.** $\dfrac{e^x(x + 1)}{x}$　**3.** $e^x(x - e^x)$

4. $e^{-x}(e^{2x} - x)$　**5.** $-\dfrac{6}{7x^3}$　**6.** $6x - \dfrac{1}{6}$

7. $6(2x^2 - x + 6)$　**8.** $\dfrac{t + 2}{2t^{3/2}}$

9. Relative maximum: $\left(-\dfrac{4\sqrt{3}}{3}, \dfrac{16\sqrt{3}}{9}\right)$

Relative minimum: $\left(\dfrac{4\sqrt{3}}{3}, -\dfrac{16\sqrt{3}}{9}\right)$

10. Relative maximum: $(0, 5)$

Relative minima: $(-1, 4), (1, 4)$

---

1. 3　**3.** -1　**5.** $5e^{5x}$　**7.** $-2xe^{-x^2}$

9. $\dfrac{2}{x^3}e^{-1/x^2}$　**11.** $e^{4x}(4x^2 + 2x + 4)$

13. $-\dfrac{6(e^x - e^{-x})}{(e^x + e^{-x})^4}$　**15.** $xe^x + e^x + 4e^{-x}$

17. $y = 2x - 3$　**19.** $y = \dfrac{4}{e^2}$　**21.** $y = 24x + 8$

23. $\dfrac{dy}{dx} = \dfrac{10 - e^y}{xe^y + 3}$　**25.** $\dfrac{dy}{dx} = \dfrac{e^{-x}(x^2 - 2x) + y}{4y - x}$

27. $6(3e^{3x} + 2e^{-2x})$　**29.** $5(e^{-x} - 10e^{-5x})$

31.

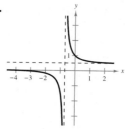

No relative extrema

No points of inflection

Horizontal asymptote to the right: $y = \dfrac{1}{2}$

Horizontal asymptote to the left: $y = 0$

Vertical asymptote: $x \approx -0.693$

33.

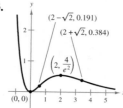

Relative minimum: $(0, 0)$

Relative maximum: $\left(2, \dfrac{4}{e^2}\right)$

Points of inflection: $\left(2 - \sqrt{2}, 0.191\right), \left(2 + \sqrt{2}, 0.384\right)$

Horizontal asymptote to the right: $y = 0$

35.

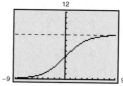

Horizontal asymptotes: $y = 0, y = 8$

37. $x = -\dfrac{1}{3}$　**39.** $x = 9$

41. (a)

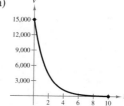

(b) $-$$5028.84/yr　(c) $-$$406.89/yr

(d) $v = -1497.2t + 15{,}000$

(e) In the exponential function, the initial rate of depreciation is greater than in the linear model. The linear model has a constant rate of depreciation.

43. (a) 1.66 words/min/week   (b) 2.30 words/min/week

(c) 1.74 words/min/week

45. $t = 1$: $-24.3\%/$week

$t = 3$: $-8.9\%/$week

47. (a)

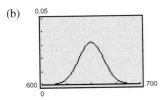

(b and c) 1996: 3.25 million people/yr

2000: 1.30 million people/yr

2005: 5.30 million people/yr

49. (a) $f(x) = \dfrac{1}{12.5\sqrt{2\pi}} e^{-(x-650)^2/312.5}$

(b)

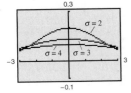

(c) $f'(x) = \dfrac{-4\sqrt{2}(x-650)e^{-2(x-650)^2/625}}{15{,}625\sqrt{\pi}}$

(d) Answers will vary.

51.

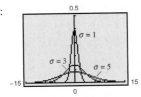

As σ increases, the graph becomes flatter.

53. Proof; maximum: $\left(0, \dfrac{1}{\sigma\sqrt{2\pi}}\right)$; answers will vary.

Sample answer:

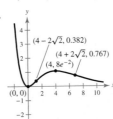

MID-CHAPTER QUIZ   (page 768)

1. 64   **2.** $\frac{8}{27}$ **3.** $3\sqrt[3]{3}$ **4.** $\frac{16}{81}$ **5.** 1024

6. 216   **7.** 27   **8.** $\sqrt{15}$ **9.** e^7 **10.** $e^{11/3}$

11. e^6 **12.** e^3

13.

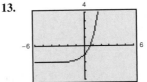

14.

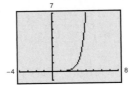

15.

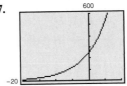

16.

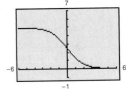

17.

18.

19. $23.22   **20.** (a) $3572.83   (b) $3573.74

21. $5e^{5x}$ **22.** e^{x-4} **23.** $5e^{x+2}$

24. $e^x(2-x)$ **25.** $y = -2x + 1$

26.

Relative maximum: $(4, 8e^{-2})$

Relative minimum: $(0, 0)$

Points of inflection: $\left(4 - 2\sqrt{2}, 0.382\right), \left(4 + 2\sqrt{2}, 0.767\right)$

Horizontal asymptote to the right: $y = 0$

SECTION 10.4   (page 775)

Skills Review   (page 775)

1. $\frac{1}{4}$ **2.** 64   **3.** 729   **4.** $\frac{8}{27}$ **5.** 1

6. $81e^4$ **7.** $\dfrac{e^3}{2}$ **8.** $\dfrac{125}{8e^3}$ **9.** $x > -4$

10. Any real number x **11.** $x < -1$ or $x > 1$

12. $x > 5$ **13.** $3462.03   **14.** $3374.65

1. $e^{0.6931\ldots} = 2$ **3.** $e^{-1.6094\ldots} = 0.2$ **5.** $\ln 1 = 0$

7. $\ln(0.0498\ldots) = -3$ **9.** c   **10.** d

11. b **12.** a

13. 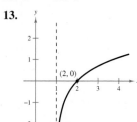 **15.**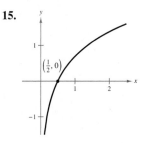

17.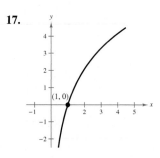

19. Answers will vary. **21.** Answers will vary.

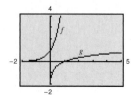 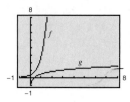

23. x^2 **25.** $5x + 2$ **27.** $2x - 1$

29. (a) 1.7917 (b) 0.4055 (c) 4.3944 (d) 0.5493

31. $\ln 2 - \ln 3$ **33.** $\ln 2 + \ln x + \ln y$

35. $\frac{1}{2}\ln(x^2 + 1)$ **37.** $\ln z + 2\ln(z - 1)$

39. $\ln 3 + \ln x + \ln(x + 1) - 2\ln(2x + 1)$

41. $\ln \dfrac{x - 2}{x + 2}$ **43.** $\ln \dfrac{x^3 y^2}{z^4}$ **45.** $\ln\left[\dfrac{x(x + 3)}{x + 4}\right]^3$

47. $\ln\left[\dfrac{x(x^2 + 1)}{x + 1}\right]^{3/2}$ **49.** $\ln \dfrac{(x + 1)^{1/3}}{(x - 1)^{2/3}}$

51. $x = 4$ **53.** $x = 1$

55. $x = \dfrac{e^{1.2}}{2} \approx 1.66$ **57.** $x = \dfrac{e^{8/3}}{5} \approx 2.88$

59. $x = \ln 4 - 1 \approx 0.3863$

61. $t = \dfrac{\ln 7 - \ln 3}{-0.2} \approx -4.2365$

63. $x = \frac{1}{2}\left(1 + \ln\frac{3}{2}\right) \approx 0.7027$

65. $x = -100 \ln\frac{3}{4} \approx 28.7682$

67. $x = \dfrac{\ln 15}{2 \ln 5} \approx 0.8413$ **69.** $t = \dfrac{\ln 2}{\ln 1.07} \approx 10.2448$

71. $t = \dfrac{\ln 3}{12 \ln[1 + (0.07/12)]} \approx 15.7402$

73. $t = \dfrac{\ln 30}{3 \ln[16 - (0.878/26)]} \approx 0.4092$

75. (a) 8.15 yr (b) 12.92 yr

77. (a) 14.21 yr (b) 13.89 yr
(c) 13.86 yr (d) 13.86 yr

79. (a) About 896 units (b) About 136 units

81. (a) $P(25) \approx 210{,}650$ (b) 2023

83. 9395 yr **85.** 12,484 yr

87. (a) 80 (b) 57.5 (c) 10 mo

89. (a)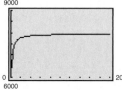

(b) $y = 7955.6$; This means that the orchard's yield approaches but does not reach 7955.6 pounds per acre as it increases in age.

(c) About 6.53 yr

91.

x	y	$\dfrac{\ln x}{\ln y}$	$\ln \dfrac{x}{y}$	$\ln x - \ln y$
1	2	0	-0.6931	-0.6931
3	4	0.7925	-0.2877	-0.2877
10	5	1.4307	0.6931	0.6931
4	0.5	-2	2.0794	2.0794

93.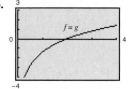

95. False. $f(x) = \ln x$ is undefined for $x \le 0$.

97. False. $f\left(\dfrac{x}{2}\right) = f(x) - f(2)$ **99.** False. $u = v^2$

101. 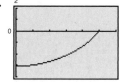 Answers will vary.

SECTION 10.5   *(page 784)*

Skills Review   *(page 784)*

1. $2\ln(x+1)$ **2.** $\ln x + \ln(x+1)$

3. $\ln x - \ln(x+1)$ **4.** $3[\ln x - \ln(x-3)]$

5. $\ln 4 + \ln x + \ln(x-7) - 2\ln x$

6. $3\ln x + \ln(x+1)$ **7.** $-\dfrac{y}{x+2y}$

8. $\dfrac{3 - 2xy + y^2}{x(x-2y)}$ **9.** $-12x + 2$ **10.** $-\dfrac{6}{x^4}$

1. 3   **3.** 2   **5.** $\dfrac{2}{x}$ **7.** $\dfrac{2x}{x^2+3}$ **9.** $\dfrac{1}{2(x-4)}$

11. $\dfrac{4}{x}(\ln x)^3$ **13.** $2\ln x + 2$ **15.** $\dfrac{2x^2-1}{x(x^2-1)}$

17. $\dfrac{1}{x(x+1)}$ **19.** $\dfrac{2}{3(x^2-1)}$ **21.** $-\dfrac{4}{x(4+x^2)}$

23. $e^{-x}\left(\dfrac{1}{x} - \ln x\right)$ **25.** $\dfrac{e^x - e^{-x}}{e^x + e^{-x}}$ **27.** $e^{x(\ln 2)}$

29. $\dfrac{1}{\ln 4}\ln x$ **31.** 1.404   **33.** 5.585   **35.** -0.631

37. -2.134 **39.** $(\ln 3)3^x$ **41.** $\dfrac{1}{x \ln 2}$

43. $(2\ln 4)4^{2x-3}$ **45.** $\dfrac{2x+6}{(x^2+6x)\ln 10}$

47. $2^x(1 + x\ln 2)$ **49.** $y = x - 1$

51. $y = \dfrac{1}{27\ln 3}x - \dfrac{1}{\ln 3} + 3$ **53.** $\dfrac{2xy}{3 - 2y^2}$

55. $\dfrac{y(1-6x^2)}{1+y}$ **57.** $y = x - 1$

59. $\dfrac{1}{2x}$ **61.** $\dfrac{1}{x}$ **63.** $(\ln 5)^2\, 5^x$

65. $\dfrac{d\beta}{dI} = \dfrac{10}{(\ln 10)I}$, so for $I = 10^{-4}$, the rate of change is about 43,429.4 db/w/cm^2.

67. 2, $y = 2x - 1$ **69.** $-\dfrac{8}{5}$, $y = -\dfrac{8}{5}x - 4$

71. $\dfrac{1}{\ln 2}$, $y = \dfrac{1}{\ln 2}x - \dfrac{1}{\ln 2}$

73.

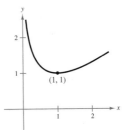

Relative minimum: $(1, 1)$

75.

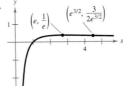

Relative maximum: $\left(e, \dfrac{1}{e}\right)$

Point of inflection: $\left(e^{3/2}, \dfrac{3}{2e^{3/2}}\right)$

77.

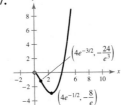

Relative minimum: $\left(4e^{-1/2}, \dfrac{-8}{e}\right)$

Point of inflection: $\left(4e^{-3/2}, \dfrac{-24}{e^3}\right)$

79. $-\dfrac{1}{p}, -\dfrac{1}{10}$

81. $p = 1000e^{-x}$

$\dfrac{dp}{dx} = -1000e^{-x}$

At $p = 10$, rate of change $= -10$.

$\dfrac{dp}{dx}$ and $\dfrac{dx}{dp}$ are reciprocals of each other.

83. (a) $\overline{C} = \dfrac{500 + 300x - 300\ln x}{x}$

(b) Minimum of 279.15 at $e^{8/3}$

85. (a)

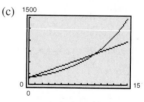

(b) $10.1625 billion/yr

87. (a) $I = 10^{8.3} \approx 199,526,231.5$

(b) $I = 10^{6.3} \approx 1,995,262.315$

(c) 10^R

(d) $\dfrac{dR}{dI} = \dfrac{1}{I \ln(10)}$

89. Answers will vary.

SECTION 10.6   (page 793)

Skills Review   (page 793)

1. $-\dfrac{1}{4} \ln 2$ **2.** $\dfrac{1}{5} \ln \dfrac{10}{3}$ **3.** $-\dfrac{\ln(25/16)}{0.01}$

4. $-\dfrac{\ln(11/16)}{0.02}$ **5.** $7.36e^{0.23t}$ **6.** $1.296e^{0.072t}$

7. $-33.6e^{-1.4t}$ **8.** $-0.025e^{-0.001t}$ **9.** 4

10. 12   **11.** $2x + 1$ **12.** $x^2 + 1$

1. $y = 2e^{0.1014t}$ **3.** $y = 4e^{-0.4159t}$

5. $y = 0.6687e^{0.4024t}$ **7.** $y = 10e^{2t}$, exponential growth

9. $y = 30e^{-4t}$, exponential decay

11. *Amount after 1000 years:* 6.48 g

Amount after 10,000 years: 0.13 g

13. *Initial quantity:* 6.73 g

Amount after 1000 years: 5.96 g

15. *Initial quantity:* 2.16 g

Amount after 10,000 years: 1.62 g

17. 68%   **19.** 15,642 yr

21. $k_1 = \dfrac{\ln 4}{12} \approx 0.1155$, so $y_1 = 5e^{0.1155t}$.

$k_2 = \dfrac{1}{6}$, so $y_2 = 5(2)^{t/6}$.

Explanations will vary.

23. (a) 1350   (b) $\dfrac{5 \ln 2}{\ln 3} \approx 3.15$ hr

(c) No. Answers will vary.

25. *Time to double:* 5.78 yr

Amount after 10 years: $3320.12

Amount after 25 years: $20,085.54

27. *Annual rate:* 8.66%

Amount after 10 years: $1783.04

Amount after 25 years: $6535.95

29. *Annual rate:* 9.50%

Time to double: 7.30 yr

Amount after 25 years: $5375.51

31. *Initial investment:* $6376.28

Time to double: 15.40 yr

Amount after 25 years: $19,640.33

33. $49,787.07   **35.** (a) Answers will vary.   (b) 6.17%

37.

Number of compoundings/yr	4	12
Effective yield	5.095%	5.116%

Number of compoundings/yr	365	Continuous
Effective yield	5.127%	5.127%

39. Answers will vary.

41. (a) $1486.1 million   (b) $964.4 million

(c)

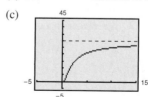

$t = 0$ corresponds to 1996. Answers will vary.

43. (a) $C = 30$

$k = \ln\left(\dfrac{1}{6}\right) \approx -1.7918$

(b) $30e^{-0.35836} = 20.9646$ or 20,965 units

(c)

45. About 36 days   **47.** $496,806

49. (a) $C = \dfrac{625}{64}$

$k = \dfrac{1}{100} \ln \dfrac{4}{5}$

(b) $x = 448$ units; $p = $3.59

51. 2046   **53.** Answers will vary.

REVIEW EXERCISES FOR CHAPTER 10
(page 800)

1. 8   **3.** 64   **5.** 1   **7.** $\frac{1}{6}$ **9.** e^{10}

11. e^3 **13.** $f(4) = 128$ **15.** $f(10) \approx 1.219$

17. (a) 1999: $P(9) \approx \$179.8$ million

2003: $P(13) \approx \$352.1$ million

2005: $P(15) \approx \$492.8$ million

(b) Answers will vary.

19.
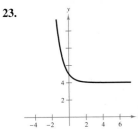

21.
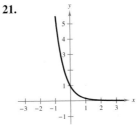

23.
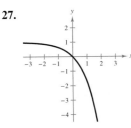

25.
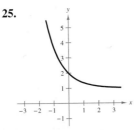

27.
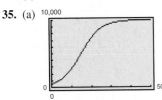

29. $7500
Explanations will vary.

31. (a) $5e \approx 13.59$ (b) $5e^{-1/2} \approx 3.03$

(c) $5e^9 \approx 40{,}515.42$

33. (a) $6e^{-3.4} \approx 0.2002$ (b) $6e^{-10} \approx 0.0003$

(c) $6e^{-20} \approx 1.2367 \times 10^{-8}$

35. (a)

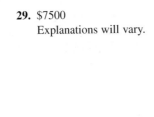

(b) $P \approx 1049$ fish

(c) Yes, P approaches 10,000 fish as t approaches ∞.

(d) The population is increasing most rapidly at the inflection point, which occurs around $t = 15$ months.

37.

n	1	2	4	12
A	$1216.65	$1218.99	$1220.19	$1221.00

n	365	Continuous compounding
A	$1221.39	$1221.40

39. b   **41.** (a) 6.14%   (b) 6.17%   **43.** $10,338.10

45. 1990: $P(0) = 29.7$ million

2000: $P(10) \approx 32.8$ million

2005: $P(15) \approx 34.5$ million

47. $8xe^{x^2}$ **49.** $\dfrac{1 - 2x}{e^{2x}}$ **51.** $4e^{2x}$ **53.** $-\dfrac{10e^{2x}}{(1 + e^{2x})^2}$

55.

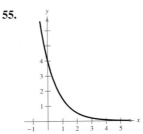

No relative extrema

No points of inflection

Horizontal asymptote: $y = 0$

57.

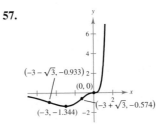

Relative minimum: $(-3, -1.344)$

Inflection points: $(0, 0)$, $\left(-3 + \sqrt{3}, -0.574\right)$, and $\left(-3 - \sqrt{3}, -0.933\right)$

Horizontal asymptote: $y = 0$

59.

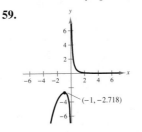

Relative maximum: $(-1, -2.718)$

Horizontal asymptote: $y = 0$

Vertical asymptote: $x = 0$

61.

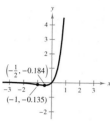

Relative minimum: $\left(-\frac{1}{2}, -0.184\right)$

Inflection point: $(-1, -0.135)$

Horizontal asymptote: $y = 0$

63. $e^{2.4849} \approx 12$ **65.** $\ln 4.4816 \approx 1.5$

67.

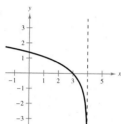

69.

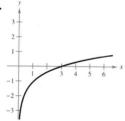

71. $\ln x + \frac{1}{2}\ln(x - 1)$ **73.** $2 \ln x - 3 \ln(x + 1)$

75. $3[\ln(1 - x) - \ln 3 - \ln x]$ **77.** 3

79. $e^{3e^{-1}} \approx 3.0151$ **81.** 1   **83.** $\frac{1}{2}(\ln 6 + 1) \approx 1.3959$

85. $\dfrac{3 + \sqrt{13}}{2} \approx 3.3028$ **87.** $-\dfrac{\ln(0.25)}{1.386} \approx 1.0002$

89. $\dfrac{\ln 1.1}{\ln 1.21} = 0.5$ **91.** $100 \ln\left(\dfrac{25}{4}\right) \approx 183.2581$

93. (a)

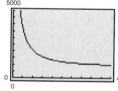

(b) A 30-year term has a smaller monthly payment, but the total amount paid is higher due to more interest.

95. $\dfrac{2}{x}$ **97.** $\dfrac{1}{x} + \dfrac{1}{x - 1} - \dfrac{1}{x - 2} = \dfrac{x^2 - 4x + 2}{x(x - 2)(x - 1)}$

99. 2   **101.** $\dfrac{1 - 3 \ln x}{x^4}$ **103.** $\dfrac{4x}{3(x^2 - 2)}$

105. $\dfrac{2}{x} + \dfrac{1}{2(x + 1)}$ **107.** $\dfrac{1}{1 + e^x}$

109.

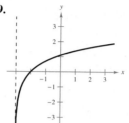

111.

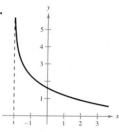

No relative extrema   No relative extrema
No points of inflection   No points of inflection

113. 2   **115.** 0   **117.** 1.594   **119.** 1.500

121. $\dfrac{2}{(2x - 1) \ln 3}$ **123.** $-\dfrac{2}{x \ln 2}$

125. (a)

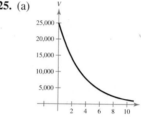

$t = 2$: \$14,062.50

(b) $t = 1$: $-$\$5394.04/yr   (c) $t \approx 5.6$ yr

$t = 4$: $-$\$2275.61/yr

127. $A = 500e^{-0.01277t}$ **129.** 27.9 yr

131. \$1048.2 million

CHAPTER TEST   *(page 804)*

1. 1   **2.** $\frac{1}{256}$ **3.** $e^{9/2}$ **4.** e^2

5.

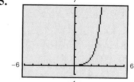

6.

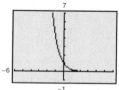

7.

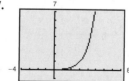

8.

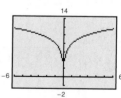

9.

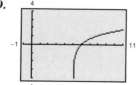

10.

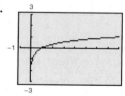

11. $\ln 3 - \ln 2$ **12.** $\frac{1}{2}\ln(x+y)$ **13.** $\ln(x+1) - \ln y$

14. $\ln[y(x+1)]$ **15.** $\ln\dfrac{8}{(x-1)^2}$ **16.** $\ln\dfrac{x^2 y}{z+4}$

17. $x \approx 3.197$ **18.** $x \approx 1.750$ **19.** $x \approx 58.371$

20. (a) 17.67 yr (b) 17.36 yr

(c) 17.33 yr (d) 17.33 yr

21. $-3e^{-3x}$ **22.** $7e^{x+2} + 2$

23. $\dfrac{2x}{3+x^2}$ **24.** $\dfrac{2}{x(x+2)}$

25. (a) \$828.58 million (b) \$24.95 million/yr

26. 59.4% **27.** 39.61 yr

CHAPTER 11

SECTION 11.1 *(page 814)*

Skills Review *(page 814)*

1. $x^{-1/2}$ **2.** $(2x)^{4/3}$ **3.** $5^{1/2}x^{3/2} + x^{5/2}$

4. $x^{-1/2} + x^{-2/3}$ **5.** $(x+1)^{5/2}$ **6.** $x^{1/6}$

7. -12 **8.** -10 **9.** 14 **10.** 14

1–7. Answers will vary. **9.** $6x + C$

11. $\dfrac{5}{3}t^3 + C$ **13.** $-\dfrac{5}{2x^2} + C$ **15.** $u + C$

17. $et + C$ **19.** $\dfrac{2}{5}y^{5/2} + C$

	Rewrite	Integrate	Simplify
21.	$\displaystyle\int x^{1/3}\,dx$	$\dfrac{x^{4/3}}{4/3} + C$	$\dfrac{3}{4}x^{4/3} + C$
23.	$\displaystyle\int x^{-3/2}\,dx$	$\dfrac{x^{-1/2}}{-1/2} + C$	$-\dfrac{2}{\sqrt{x}} + C$
25.	$\dfrac{1}{2}\displaystyle\int x^{-3}\,dx$	$\dfrac{1}{2}\left(\dfrac{x^{-2}}{-2}\right) + C$	$-\dfrac{1}{4x^2} + C$

27. $\dfrac{x^2}{2} + 3x + C$ **29.** $\dfrac{1}{4}x^4 + 2x + C$

31. $\dfrac{3}{4}x^{4/3} - \dfrac{3}{4}x^{2/3} + C$ **33.** $\dfrac{3}{5}x^{5/3} + C$

35. $-\dfrac{1}{3x^3} + C$ **37.** $2x - \dfrac{1}{2x^2} + C$

39. $\dfrac{3}{4}u^4 + \dfrac{1}{2}u^2 + C$ **41.** $x^3 + \dfrac{x^2}{2} - 2x + C$

43. $\dfrac{2}{7}y^{7/2} + C$

45.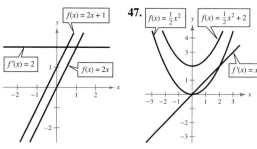

47.

49. $f(x) = 2x^2 + 6$ **51.** $f(x) = x^2 - 2x - 1$

53. $f(x) = -\dfrac{1}{x^2} + \dfrac{1}{x} + \dfrac{1}{2}$ **55.** $y = -\dfrac{5}{2}x^2 - 2x + 2$

57. $f(x) = x^2 - 6$ **59.** $f(x) = x^2 + x + 4$

61. $f(x) = \dfrac{9}{4}x^{4/3}$ **63.** $C = 85x + 5500$

65. $C = \dfrac{1}{10}\sqrt{x} + 4x + 750$

67. $R = 225x - \dfrac{3}{2}x^2,\ p = 225 - \dfrac{3}{2}x$

69. $P = -9x^2 + 1650x$ **71.** $P = -12x^2 + 805x + 68$

73. $s(t) = -16t^2 + 6000$; about 19.36 sec

75. (a) $C = x^2 - 12x + 125$ (b) \$2025

$\overline{C} = x - 12 + \dfrac{125}{x}$

(c) \$125 is fixed.

\$1900 is variable.

Examples will vary.

77. (a) $P(t) = 52.73t^2 + 2642.7t + 69{,}903.25$

(b) 273,912; Yes, this seems reasonable. Explanations will vary.

79. (a) $I(t) = -0.0625t^4 + 1.773t^3 - 9.67t^2$

$+ 21.03t - 0.212$ (in millions)

(b) 20.072 million; No, this does not seem reasonable. Explanations will vary. Sample answer: A sharp decline from 863 million users to about 20 million users from the year 2004 to the year 2012 does not seem to follow the trend over the past few years, which is always increasing.

SECTION 11.2 *(page 824)*

Skills Review *(page 824)*

1. $\dfrac{1}{2}x^4 + x + C$ **2.** $\dfrac{3}{2}x^2 + \dfrac{2}{3}x^{3/2} - 4x + C$

3. $-\dfrac{1}{x} + C$ **4.** $-\dfrac{1}{6t^2} + C$

5. $\dfrac{4}{7}t^{7/2} + \dfrac{2}{5}t^{5/2} + C$ **6.** $\dfrac{4}{5}x^{5/2} - \dfrac{2}{3}x^{3/2} + C$

7. $\dfrac{5x^3 - 4}{2x} + C$ **8.** $\dfrac{-6x^2 + 5}{3x^3} + C$

9. $\frac{1}{5}x^5 + \frac{2}{3}x^3 + x + C$

10. $\frac{1}{7}x^7 - \frac{4}{5}x^5 + \frac{1}{2}x^4 + \frac{4}{3}x^3 - 2x^2 + x + C$

11. $-\dfrac{5(x - 2)^4}{16}$ **12.** $-\dfrac{1}{12(x - 1)^2}$

13. $9(x^2 + 3)^{2/3}$ **14.** $-\dfrac{5}{(1 - x^3)^{1/2}}$

$\displaystyle\int u^n \dfrac{du}{dx}\, dx$	u	$\dfrac{du}{dx}$
1. $\displaystyle\int (5x^2 + 1)^2(10x)\, dx$	$5x^2 + 1$	$10x$
3. $\displaystyle\int \sqrt{1 - x^2}\,(-2x)\, dx$	$1 - x^2$	$-2x$

$\displaystyle\int u^n \dfrac{du}{dx}\, dx$	u	$\dfrac{du}{dx}$
5. $\displaystyle\int \left(4 + \dfrac{1}{x^2}\right)^5\left(\dfrac{-2}{x^3}\right) dx$	$4 + \dfrac{1}{x^2}$	$-\dfrac{2}{x^3}$
7. $\displaystyle\int (1 + \sqrt{x})^3\left(\dfrac{1}{2\sqrt{x}}\right) dx$	$1 + \sqrt{x}$	$\dfrac{1}{2\sqrt{x}}$

9. $\frac{1}{5}(1 + 2x)^5 + C$ **11.** $\frac{2}{3}(4x^2 - 5)^{3/2} + C$

13. $\frac{1}{5}(x - 1)^5 + C$ **15.** $\dfrac{(x^2 - 1)^8}{8} + C$

17. $-\dfrac{1}{3(1 + x^3)} + C$ **19.** $-\dfrac{1}{2(x^2 + 2x - 3)} + C$

21. $\sqrt{x^2 - 4x + 3} + C$ **23.** $-\frac{15}{8}(1 - u^2)^{4/3} + C$

25. $4\sqrt{1 + y^2} + C$ **27.** $-3\sqrt{2t + 3} + C$

29. $-\frac{1}{2}\sqrt{1 - x^4} + C$ **31.** $-\dfrac{1}{24}\left(1 + \dfrac{4}{t^2}\right)^3 + C$

33. $\dfrac{(x^3 + 3x + 9)^2}{6} + C$ **35.** $\frac{1}{4}(6x^2 - 1)^4 + C$

37. $-\frac{2}{45}(2 - 3x^3)^{5/2} + C$ **39.** $\sqrt{x^2 + 25} + C$

41. $\frac{2}{3}\sqrt{x^3 + 3x + 4} + C$

43. (a) $\frac{1}{3}x^3 - x^2 + x + C_1 = \frac{1}{3}(x - 1)^3 + C_2$

    (b) Answers differ by a constant: $C_1 = C_2 - \frac{1}{3}$

    (c) Answers will vary.

45. (a) $\dfrac{1}{6}x^6 - \dfrac{1}{2}x^4 + \dfrac{1}{2}x^2 + C_1 = \dfrac{(x^2 - 1)^3}{6} + C_2$

    (b) Answers differ by a constant: $C_1 = C_2 - \dfrac{1}{6}$

    (c) Answers will vary.

47. $f(x) = \frac{1}{3}[5 - (1 - x^2)^{3/2}]$

49. (a) $C = 8\sqrt{x + 1} + 18$

    (b)

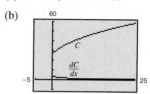

51. $x = \frac{1}{3}(p^2 - 25)^{3/2} + 24$ **53.** $x = \dfrac{6000}{\sqrt{p^2 - 16}} + 3000$

55. (a) $h = \sqrt{17.6t^2 + 1} + 5$ (b) 26 in.

57. (a) $Q = (x - 24{,}999)^{0.95} + 24{,}999$

    (b)

x	25,000	50,000	100,000	150,000
Q	25,000	40,067.14	67,786.18	94,512.29
$x - Q$	0	9932.86	32,213.82	55,487.71

    (c)

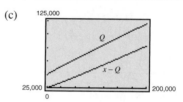

59. $-\frac{2}{3}x^{3/2} + \frac{2}{3}(x + 1)^{3/2} + C$

SECTION 11.3   (page 831)

Skills Review   (page 831)

1. $\left(\frac{5}{2}, \infty\right)$ **2.** $(-\infty, 2) \cup (3, \infty)$

3. $x + 2 - \dfrac{2}{x + 2}$ **4.** $x - 2 + \dfrac{1}{x - 4}$

5. $x + 8 + \dfrac{2x - 4}{x^2 - 4x}$ **6.** $x^2 - x - 4 + \dfrac{20x + 22}{x^2 + 5}$

7. $\dfrac{1}{4}x^4 - \dfrac{1}{x} + C$ **8.** $\dfrac{1}{2}x^2 + 2x + C$

9. $\dfrac{1}{2}x^2 - \dfrac{4}{x} + C$ **10.** $-\dfrac{1}{x} - \dfrac{3}{2x^2} + C$

1. $e^{2x} + C$ **3.** $\frac{1}{4}e^{4x} + C$ **5.** $-\frac{9}{2}e^{-x^2} + C$

7. $\frac{5}{3}e^{x^3} + C$ **9.** $\frac{1}{3}e^{x^3 + 3x^2 - 1} + C$ **11.** $-5e^{2-x} + C$

13. $\ln|x + 1| + C$ **15.** $-\frac{1}{2}\ln|3 - 2x| + C$

17. $\frac{2}{3}\ln|3x + 5| + C$ **19.** $\ln\sqrt{x^2 + 1} + C$

21. $\frac{1}{3}\ln|x^3 + 1| + C$ **23.** $\frac{1}{2}\ln|x^2 + 6x + 7| + C$

25. $\ln|\ln x| + C$ **27.** $\ln|1 - e^{-x}| + C$

29. $-\frac{1}{2}e^{2/x} + C$ **31.** $2e^{\sqrt{x}} + C$

33. $\frac{1}{2}e^{2x} - 4e^x + 4x + C$ **35.** $-\ln(1 + e^{-x}) + C$

37. $-2\ln|5 - e^{2x}| + C$

39. $e^x + 2x - e^{-x} + C$; Exponential Rule and General Power Rule

41. $-\frac{2}{3}(1 - e^x)^{3/2} + C$; Exponential Rule

43. $-\dfrac{1}{x - 1} + C$; General Power Rule

45. $2e^{2x-1} + C$; Exponential Rule

47. $\frac{1}{4}x^2 - 4\ln|x| + C$; General Power Rule and Logarithmic Rule

49. $2\ln(e^x + 1) + C$; Logarithmic Rule

51. $\frac{1}{2}x^2 + 3x + 8\ln|x - 1| + C$; General Power Rule and Logarithmic Rule

53. $\ln|e^x + x| + C$; Logarithmic Rule

55. $f(x) = \frac{1}{2}x^2 + 5x + 8\ln|x - 1| - 8$

57. (a) $P(t) = 1000[1 + \ln(1 + 0.25t)^{12}]$

(b) $P(3) \approx 7715$ bacteria   (c) $t \approx 6$ days

59. (a) $p = -50e^{-x/500} + 45.06$

(b)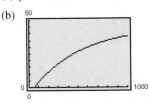

The price increases as the demand increases.

(c) 387

61. (a) $S = -7241.22e^{-t/4.2} + 42,721.88$ (in dollars)

(b) \$38,224.03

63. False. $\ln x^{1/2} = \frac{1}{2}\ln x$

MID-CHAPTER QUIZ   *(page 833)*

1. $3x + C$ **2.** $5x^2 + C$ **3.** $-\dfrac{1}{4x^4} + C$

4. $\dfrac{x^3}{3} - x^2 + 15x + C$ **5.** $\dfrac{x^3}{3} + 2x^2 + C$

6. $\dfrac{(6x + 1)^4}{4} + C$ **7.** $\dfrac{(x^2 - 5x)^2}{2} + C$

8. $-\dfrac{1}{2(x^3 + 3)^2} + C$ **9.** $\dfrac{2}{15}(5x + 2)^{3/2} + C$

10. $f(x) = 8x^2 + 1$ **11.** $f(x) = 3x^3 + 4x - 2$

12. (a) \$9.03   (b) \$509.03

13. $f(x) = \dfrac{2}{3}x^3 + x + 1$ **14.** $e^{5x+4} + C$

15. $\dfrac{x^2}{2} + e^{2x} + C$ **16.** $e^{x^3} + C$ **17.** $\ln|2x - 1| + C$

18. $-\ln|x^2 + 3| + C$ **19.** $3\ln|x^3 + 2x^2| + C$

20. (a) 1000 bolts   (b) About 8612 bolts

SECTION 11.4   *(page 843)*

Skills Review   *(page 843)*

1. $\frac{3}{2}x^2 + 7x + C$ **2.** $\frac{2}{5}x^{5/2} + \frac{4}{3}x^{3/2} + C$

3. $\dfrac{1}{5}\ln|x| + C$ **4.** $-\dfrac{1}{6e^{6x}} + C$ **5.** $-\dfrac{8}{5}$

6. $-\frac{62}{3}$ **7.** $C = 0.008x^{5/2} + 29,500x + C$

8. $R = x^2 + 9000x + C$

9. $P = 25,000x - 0.005x^2 + C$

10. $C = 0.01x^3 + 4600x + C$

1.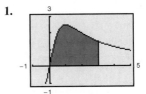

Positive

3.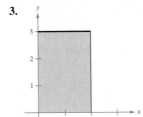

Area = 6

5.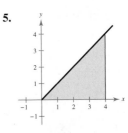

Area = 8

7.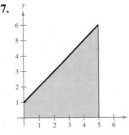

Area = $\frac{35}{2}$

9.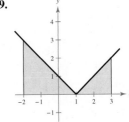

Area = $\frac{13}{2}$

11.

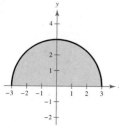

$$\text{Area} = \frac{9\pi}{2}$$

13. (a) 8   (b) 4   (c) -24 (d) 0

15. $\frac{1}{6}$ **17.** $\frac{1}{2}$ **19.** $6\left(1 - \frac{1}{e^2}\right)$ **21.** $8 \ln 2 + \frac{15}{2}$

23. 1   **25.** $-\frac{5}{2}$ **27.** $\frac{14}{3}$ **29.** $-\frac{15}{4}$ **31.** -4

33. $\frac{2}{3}$ **35.** $-\frac{27}{20}$ **37.** 2   **39.** $\frac{1}{2}(1 - e^{-2}) \approx 0.432$

41. $\frac{e^3 - e}{3} \approx 5.789$ **43.** $\frac{1}{3}\left[(e^2 + 1)^{3/2} - 2\sqrt{2}\right] \approx 7.157$

45. $\frac{1}{8} \ln 17 \approx 0.354$ **47.** 4   **49.** 4

51. $\frac{1}{2} \ln 5 - \frac{1}{2} \ln 8 \approx -0.235$

53. $2 \ln(2 + e^3) - 2 \ln 3 \approx 3.993$

55. Area $= 10$ **57.** Area $= \frac{1}{4}$

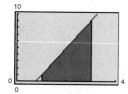   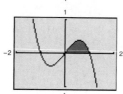

59. Area $= \ln 9$

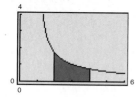

61. 10   **63.** $4 \ln 3 \approx 4.394$

65.

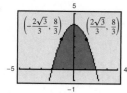

$$\text{Average} = \frac{8}{3}$$

$$x = \pm\frac{2\sqrt{3}}{3} \approx \pm1.155$$

67.

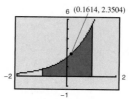

$$\text{Average} = e - e^{-1} \approx 2.3504$$

$$x = \ln\left(\frac{e - e^{-1}}{2}\right) \approx 0.1614$$

69.

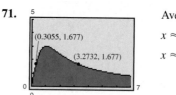

$$\text{Average} = \frac{4}{3}$$

$$x = \sqrt{2 + \frac{2\sqrt{5}}{3}} \approx 1.868$$

$$x = \sqrt{2 - \frac{2\sqrt{5}}{3}} \approx 0.714$$

71.

$$\text{Average} = \frac{3}{7} \ln 50 \approx 1.677$$

$$x \approx 0.3055$$

$$x \approx 3.2732$$

73. Even   **75.** Neither odd nor even

77. (a) $\frac{1}{3}$ (b) $\frac{2}{3}$ (c) $-\frac{1}{3}$

Explanations will vary.

79. \$6.75   **81.** \$22.50   **83.** \$3.97   **85.** \$1925.23

87. \$16,605.21   **89.** \$2500   **91.** \$4565.65

93. (a) \$137,000   (b) \$214,720.93   (c) \$338,393.53

95. \$2623.94   **97.** About 144.36 thousand kg

99. $\frac{2}{3}\sqrt{7} - \frac{1}{3}$ **101.** $\frac{39}{200}$

SECTION 11.5   (page 852)

Skills Review   (page 852)

1. $-x^2 + 3x + 2$ **2.** $-2x^2 + 4x + 4$

3. $-x^3 + 2x^2 + 4x - 5$ **4.** $x^3 - 6x - 1$

5. $(0, 4), (4, 4)$ **6.** $(1, -3), (2, -12)$

7. $(-3, 9), (2, 4)$ **8.** $(-2, -4), (0, 0), (2, 4)$

9. $(1, -2), (5, 10)$ **10.** $(1, e)$

1. 36   **3.** 9   **5.** $\frac{3}{2}$

7.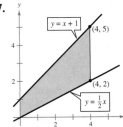

9.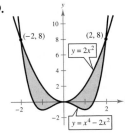

11.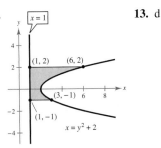

13. d

15.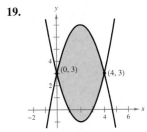

Area $= \frac{4}{5}$

17.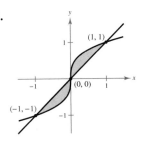

Area $= \frac{1}{2}$

19.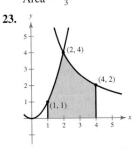

Area $= \frac{64}{3}$

21.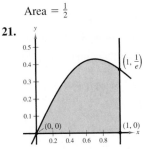

Area $= -\frac{1}{2}e^{-1} + \frac{1}{2}$

23.

Area $= \frac{7}{3} + 8 \ln 2$

25.

Area $= (2e + \ln 2) - 2e^{1/2}$

27.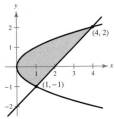

Area $= \frac{9}{2}$

29.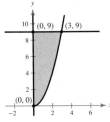

Area $= 18$

31.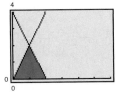

Area $= \int_0^1 2x \, dx + \int_1^2 (4 - 2x) \, dx$

33.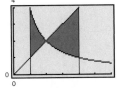

Area $= \int_1^2 \left(\frac{4}{x} - x\right) dx + \int_2^4 \left(x - \frac{4}{x}\right) dx$

35.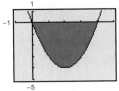

Area $= \frac{32}{3}$

37.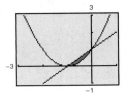

Area $= \frac{1}{6}$

39. 8

41. Consumer surplus $= 1600$

Producer surplus $= 400$

43. Consumer surplus $= 500$

Producer surplus $= 2000$

45. Offer 2 is better because the cumulative salary (area under the curve) is greater.

47. R_1, \$4.68 billion

49. \$300.6 million; Explanations will vary.

51. (a) 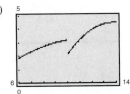   (b) 2.472 fewer pounds

53. Consumer surplus = $625,000   **55.** $337.33 million

Producer surplus = $1,375,000

57.

Quintile	Lowest	2nd	3rd	4th	Highest
Percent	2.81	6.98	14.57	27.01	45.73

59. Answers will vary.

SECTION 11.6   *(page 859)*

Skills Review   *(page 859)*

1. $\frac{1}{6}$ **2.** $\frac{3}{20}$ **3.** $\frac{7}{40}$ **4.** $\frac{13}{12}$ **5.** $\frac{61}{30}$ **6.** $\frac{53}{18}$

7. $\frac{2}{3}$ **8.** $\frac{4}{7}$ **9.** 0   **10.** 5

---

1. Midpoint Rule: 2   **3.** Midpoint Rule: 0.6730

Exact area: 2   Exact area: $\frac{2}{3} \approx 0.6667$

5. Midpoint Rule: 5.375   **7.** Midpoint Rule: 6.625

Exact area: $\frac{16}{3} \approx 5.333$ Exact area: $\frac{20}{3} \approx 6.667$

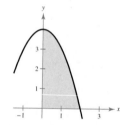   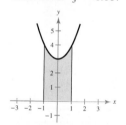

9. Midpoint Rule: 17.25   **11.** Midpoint Rule: 0.7578

Exact area: $\frac{52}{3} \approx 17.33$ Exact area: 0.75

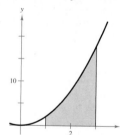   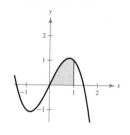

13. Midpoint Rule: 0.5703   **15.** Midpoint Rule: 6.9609

Exact area: $\frac{7}{12} \approx 0.5833$ Exact area: 6.75

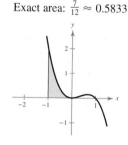

17. Area ≈ 54.6667,   **19.** Area ≈ 4.16,

$n = 31$ $n = 5$

21. Area ≈ 0.9163,   **23.** Midpoint Rule: 1.5

$n = 5$ Exact area: 1.5

25. Midpoint Rule: 25

Exact area: $\frac{76}{3} \approx 25.33$

27. Exact: 4

Trapezoidal Rule: 4.0625

Midpoint Rule: 3.9688

The Midpoint Rule is better in this example.

29. 1.1167   **31.** 1.55

33.

n	Midpoint Rule	Trapezoidal Rule
4	15.3965	15.6055
8	15.4480	15.5010
12	15.4578	15.4814
16	15.4613	15.4745
20	15.4628	15.4713

35. 4.8103

37. Answers will vary. Sample answers:

(a) 966 ft^2 (b) 966 ft^2

39. Midpoint Rule: 3.1468

Trapezoidal Rule: 3.1312

Graphing utility: 3.141593

REVIEW EXERCISES FOR CHAPTER 11
(page 865)

1. $16x + C$ **3.** $\frac{2}{3}x^3 + \frac{5}{2}x^2 + C$ **5.** $x^{2/3} + C$

7. $\frac{3}{7}x^{7/3} + \frac{3}{2}x^2 + C$ **9.** $\frac{4}{9}x^{9/2} - 2\sqrt{x} + C$

11. $f(x) = \frac{3}{2}x^2 + x - 2$ **13.** $f(x) = \frac{1}{6}x^4 - 8x + \frac{33}{2}$

15. (a) 2.5 sec   (b) 100 ft

(c) 1.25 sec   (d) 75 ft

17. $x + 5x^2 + \frac{25}{3}x^3 + C$ or $\frac{1}{15}(1 + 5x)^3 + C_1$

19. $\frac{2}{5}\sqrt{5x - 1} + C$ **21.** $\frac{1}{2}x^2 - x^4 + C$

23. $\frac{1}{4}(x^4 - 2x)^2 + C$

25. (a) 30.5 board-feet   (b) 125.2 board-feet

27. $-e^{-3x} + C$ **29.** $\frac{1}{2}e^{x^2 - 2x} + C$

31. $-\frac{1}{3}\ln|1 - x^3| + C$ **33.** $\frac{2}{3}x^{3/2} + 2x + 2x^{1/2} + C$

35.

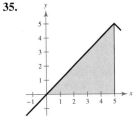

Area $= \frac{25}{2}$

37. $A = 4$ **39.** $A = \frac{32}{3}$ **41.** $A = \frac{8}{3}$ **43.** $A = 2 \ln 2$

45. (a) 13   (b) 7   (c) 11   (d) 50

47. 16   **49.** $\frac{422}{5}$ **51.** 0   **53.** 2

55. $\frac{1}{8}$ **57.** 3.899   **59.** 0

61.   **63.**

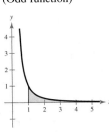   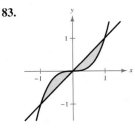

Area $= 6$ Area $= \frac{10}{3}$

65. Increases by \$700.25

67. Average value: $\frac{2}{5}$; $x = \frac{25}{4}$

69. Average value: $\frac{1}{3}(-1 + e^3) \approx 6.362$; $x \approx 3.150$

71. \$520.54; Explanations will vary.

73. (a) $B = -0.01955t^2 + 0.6108t - 1.818$

(b) According to the model, the price of beef per pound will never surpass \$3.25. The highest price is approximately \$2.95 per pound in 2005, and after that the prices decrease.

75. \$17,492.94

77. $\displaystyle\int_{-2}^{2} 6x^5 \, dx = 0$ **79.** $\displaystyle\int_{-2}^{-1} \frac{4}{x^2} \, dx = \int_{1}^{2} \frac{4}{x^2} \, dx = 2$

(Odd function)   (Symmetric about y-axis)

81.   **83.**

Area $= \frac{4}{5}$ Area $= \frac{1}{2}$

85.

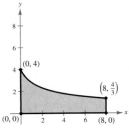

Area $= 16$

87.

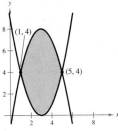

Area $= \frac{64}{3}$

89.

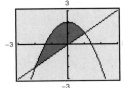

Area $= \frac{9}{2}$

91. Consumer surplus: 11,250

Producer surplus: 14,062.5

93. About \$1236.39 million less

95. About \$11,237.24 million more

97. $n = 4$: 13.3203   **99.** $n = 4$: 0.7867

$n = 20$: 13.7167 $n = 20$: 0.7855

101. Answers will vary. Sample answer: 381.6 mi^2

CHAPTER TEST   (page 869)

1. $3x^3 - 2x^2 + 13x + C$ **2.** $\dfrac{(x + 1)^3}{3} + C$

3. $\dfrac{2(x^4 - 7)^{3/2}}{3} + C$ **4.** $\dfrac{10x^{3/2}}{3} - 12x^{1/2} + C$

5. $5e^{3x} + C$ **6.** $\ln|x^3 - 11x| + C$

7. $f(x) = e^x + x$ **8.** $f(x) = \ln|x| + 2$

9. 8   **10.** 18   **11.** $\frac{2}{3}$

12. $2\sqrt{5} - 2\sqrt{2} \approx 1.644$ **13.** $\frac{1}{4}(e^{12} - 1) \approx 40,688.4$

14. $\ln 6 \approx 1.792$

15. (a) $S = \dfrac{15.7}{0.23}e^{0.23t} + 1679.49$ (b) \$2748.08 million

16.   **17.**

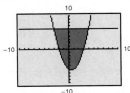   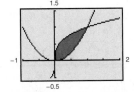

Area $= \frac{343}{6} \approx 57.167$ Area $= \frac{5}{12}$

18. Consumer surplus = 20 million

Producer surplus = 8 million

19. Midpoint Rule:

$\frac{63}{64} \approx 0.9844$

Exact area: 1

20. Midpoint Rule:

$\frac{21}{8} = 2.625$

Exact area: $\frac{8}{3} = 2.\overline{6}$

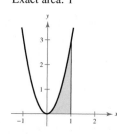

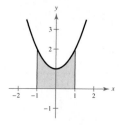

CHAPTER 12

SECTION 12.1   *(page 878)*

Skills Review   *(page 878)*

1. $\dfrac{1}{x+1}$ **2.** $\dfrac{2x}{x^2-1}$ **3.** $3x^2 e^{x^3}$

4. $-2xe^{-x^2}$ **5.** $e^x(x^2+2x)$ **6.** $e^{-2x}(1-2x)$

7. $\frac{64}{3}$ **8.** $\frac{4}{3}$ **9.** 36   **10.** 8

1. $u = x;\ dv = e^{3x}\,dx$ **3.** $u = \ln 2x;\ dv = x\,dx$

5. $\frac{1}{3}xe^{3x} - \frac{1}{9}e^{3x} + C$ **7.** $-x^2 e^{-x} - 2xe^{-x} - 2e^{-x} + C$

9. $x \ln 2x - x + C$ **11.** $\frac{1}{4}e^{4x} + C$

13. $\frac{1}{4}xe^{4x} - \frac{1}{16}e^{4x} + C$ **15.** $\frac{1}{2}e^{x^2} + C$

17. $-xe^{-x} - e^{-x} + C$ **19.** $2x^2 e^x - 4e^x x + 4e^x + C$

21. $\frac{1}{2}t^2 \ln(t+1) - \frac{1}{2}\ln|t+1| - \frac{1}{4}t^2 + \frac{1}{2}t + C$

23. $xe^x - 2e^x + C$ **25.** $-e^{1/t} + C$

27. $\frac{1}{2}x^2(\ln x)^2 - \frac{1}{2}x^2 \ln x + \frac{1}{4}x^2 + C$

29. $\frac{1}{3}(\ln x)^3 + C$ **31.** $-\frac{1}{x}(\ln x + 1) + C$

33. $\frac{2}{3}x(x-1)^{3/2} - \frac{4}{15}(x-1)^{5/2} + C$

35. $\frac{1}{4}x^4 + \frac{2}{3}x^3 + \frac{1}{2}x^2 + C$ **37.** $\dfrac{e^{2x}}{4(2x+1)} + C$

39. $e(2e-1) \approx 12.060$ **41.** $-12e^{-2} + 4 \approx 2.376$

43. $\frac{5}{36}e^6 + \frac{1}{36} \approx 56.060$ **45.** $2\ln 2 - 1 \approx 0.386$

47. Area $= 2e^2 + 6 \approx 20.778$

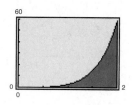

49. Area $= \frac{1}{9}(2e^3 + 1) \approx 4.575$

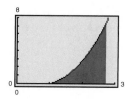

51. Proof   **53.** $\dfrac{e^{5x}}{125}(25x^2 - 10x + 2) + C$

55. $-\dfrac{1}{x}(1 + \ln x) + C$ **57.** $1 - 5e^{-4} \approx 0.908$

59. $\frac{1}{4}(e^2 + 1) \approx 2.097$ **61.** $\frac{3}{128} - \frac{379}{128}e^{-8} \approx 0.022$

63. $\dfrac{1,171,875}{256}\pi \approx 14,381.070$

65.

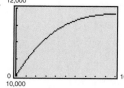

  (a) Increase   (b) 113,212 units   (c) 11,321 units/yr

67. (a) $3.2 \ln 2 - 0.2 \approx 2.018$

    (b) $12.8 \ln 4 - 7.2 \ln 3 - 1.8 \approx 8.035$

69. \$18,482.03   **71.** \$931,265.10   **73.** \$4103.07

75. (a) \$1,200,000   (b) \$1,094,142.27   **77.** \$45,957.78

79. (a) \$17,378.62   (b) \$3681.26   **81.** 4.254

SECTION 12.2   *(page 888)*

Skills Review   *(page 888)*

1. $(x-4)(x+4)$ **2.** $(x-5)(x+5)$

3. $(x-4)(x+3)$ **4.** $(x-2)(x+3)$

5. $x(x-2)(x+1)$ **6.** $x(x-2)^2$

7. $(x-2)(x-1)^2$ **8.** $(x-3)(x-1)^2$

9. $x + \dfrac{1}{x-2}$ **10.** $2x - 2 - \dfrac{1}{1-x}$

11. $x^2 - x - 2 - \dfrac{2}{x-2}$

12. $x^2 - x + 3 - \dfrac{4}{x+1}$

13. $x + 4 + \dfrac{6}{x-1},\quad x \neq -1$

14. $x + 3 + \dfrac{1}{x+1},\quad x \neq 1$

1. $\dfrac{5}{x-5} - \dfrac{3}{x+5}$ **3.** $\dfrac{9}{x-3} - \dfrac{1}{x}$

5. $\dfrac{1}{x-5} + \dfrac{3}{x+2}$ **7.** $\dfrac{3}{x} - \dfrac{5}{x^2}$

9. $\dfrac{1}{3(x-2)} + \dfrac{1}{(x-2)^2}$

11. $\dfrac{8}{x+1} - \dfrac{1}{(x+1)^2} + \dfrac{2}{(x+1)^3}$ **13.** $\dfrac{1}{2}\ln\left|\dfrac{x-1}{x+1}\right| + C$

15. $\dfrac{1}{4}\ln\left|\dfrac{x+4}{x-4}\right| + C$ **17.** $\ln\left|\dfrac{2x-1}{x}\right| + C$

19. $\ln\left|\dfrac{x-10}{x}\right| + C$ **21.** $\ln\left|\dfrac{x-1}{x+2}\right| + C$

23. $\dfrac{3}{2}\ln|2x-1| - 2\ln|x+1| + C$

25. $\ln\left|\dfrac{x(x+2)}{x-2}\right| + C$ **27.** $\dfrac{1}{2}(3\ln|x-4| - \ln|x|) + C$

29. $2\ln|x-1| + \dfrac{1}{x-1} + C$

31. $\ln|x| + 2\ln|x+1| + \dfrac{1}{x+1} + C$

33. $\dfrac{1}{6}\ln\dfrac{4}{7} \approx -0.093$ **35.** $-\dfrac{4}{5} + 2\ln\dfrac{5}{3} \approx 0.222$

37. $\dfrac{1}{2} - \ln 2 \approx -0.193$ **39.** $4\ln 2 + \dfrac{1}{2} \approx 3.273$

41. $12 - \dfrac{7}{2}\ln 7 \approx 5.189$ **43.** $5\ln 2 - \ln 5 \approx 1.856$

47. $\dfrac{1}{2a}\left(\dfrac{1}{a+x} + \dfrac{1}{a-x}\right)$ **49.** $\dfrac{1}{a}\left(\dfrac{1}{x} + \dfrac{1}{a-x}\right)$

51. Divide x^2 by $(x-5)$ because the degree of the numerator is greater than the degree of the denominator.

53. $y = \dfrac{1000}{1 + 9e^{-0.1656t}}$

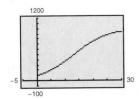

55. 1.077 thousand   **57.** $11,408$ million; 1426 million

59. The rate of growth is increasing on $[0, 3]$ for *P. aurelia* and on $[0, 2]$ for *P. caudatum*; the rate of growth is decreasing on $[3, \infty)$ for *P. aurelia* and on $[2, \infty)$ for *P. caudatum*; *P. aurelia* has a higher limiting population.

61. Answers will vary.

SECTION 12.3   (page 899)

> **Skills Review**   (page 899)
>
> **1.** $x^2 + 8x + 16$ **2.** $x^2 - 2x + 1$
>
> **3.** $x^2 + x + \dfrac{1}{4}$ **4.** $x^2 - \dfrac{2}{3}x + \dfrac{1}{9}$
>
> **5.** $\dfrac{2}{x} - \dfrac{2}{x+2}$ **6.** $-\dfrac{3}{4x} + \dfrac{3}{4(x-4)}$
>
> **7.** $\dfrac{3}{2(x-2)} - \dfrac{2}{x^2} - \dfrac{3}{2x}$ **8.** $\dfrac{4}{x} - \dfrac{3}{x+1} + \dfrac{2}{x-2}$
>
> **9.** $2e^x(x-1) + C$ **10.** $x^3\ln x - \dfrac{x^3}{3} + C$

1. $\dfrac{1}{9}\left(\dfrac{2}{2+3x} + \ln|2+3x|\right) + C$

3. $\dfrac{2(3x-4)}{27}\sqrt{2+3x} + C$ **5.** $\ln\left(x^2 + \sqrt{x^4-9}\right) + C$

7. $\dfrac{1}{2}(x^2 - 1)e^{x^2} + C$ **9.** $\ln\left|\dfrac{x}{1+x}\right| + C$

11. $-\dfrac{1}{3}\ln\left|\dfrac{3 + \sqrt{x^2+9}}{x}\right| + C$

13. $-\dfrac{1}{2}\ln\left|\dfrac{2 + \sqrt{4-x^2}}{x}\right| + C$

15. $\dfrac{1}{4}x^2(-1 + 2\ln x) + C$ **17.** $3x^2 - \ln(1 + e^{3x^2}) + C$

19. $\dfrac{1}{4}\left(x^2\sqrt{x^4-4} - 4\ln|x^2 + \sqrt{x^4-4}|\right) + C$

21. $\dfrac{1}{27}\left[\dfrac{4}{2+3t} - \dfrac{2}{(2+3t)^2} + \ln|2+3t|\right] + C$

23. $\dfrac{\sqrt{3}}{3}\ln\left|\dfrac{\sqrt{3+s} - \sqrt{3}}{\sqrt{3+s} + \sqrt{3}}\right| + C$

25. $-\dfrac{1}{2}x(2-x) + \ln|x+1| + C$

27. $\dfrac{1}{8}\left[\dfrac{-1}{2(3+2x)^2} + \dfrac{2}{(3+2x)^3} - \dfrac{9}{4(3+2x)^4}\right] + C$

29. $-\dfrac{\sqrt{1-x^2}}{x} + C$ **31.** $\dfrac{1}{9}x^3(-1 + 3\ln x) + C$

33. $\dfrac{1}{27}\left(3x - \dfrac{25}{3x-5} + 10\ln|3x-5|\right) + C$

35. $\dfrac{1}{9}(3\ln x - 4\ln|4 + 3\ln x|) + C$

37. Area $= \dfrac{40}{3}$

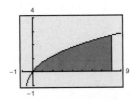

Area $= 13.\overline{3}$

39. Area $= \dfrac{1}{2}\left[4 + \ln\left(\dfrac{2}{1 + e^4}\right)\right]$

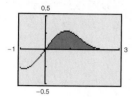

Area ≈ 0.3375

41. Area $= \dfrac{1}{4}\left[21\sqrt{5} - 8\ln\left(\sqrt{5} + 3\right) + 8\ln 2\right]$

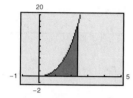

Area ≈ 9.8145

43. $\dfrac{-2\sqrt{2} + 4}{3} \approx 0.3905$ **45.** $-\dfrac{5}{9} + \ln\dfrac{9}{4} \approx 0.2554$

47. $12\left(2 + \ln\left|\dfrac{2}{1 + e^2}\right|\right) \approx 6.7946$

49. $-\dfrac{15}{4} + 8\ln 4 \approx 7.3404$ **51.** $(x^2 - 2x + 2)e^x + C$

53. $-\left(\dfrac{1}{x} + \ln\left|\dfrac{x}{x + 1}\right|\right) + C$ **55.** (a) 0.483 (b) 0.283

57.

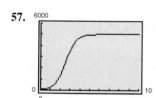

Average value: 401.40

59. $\$1138.43$ **61.** $\$0.50$ billion/yr

MID-CHAPTER QUIZ   *(page 901)*

1. $\dfrac{1}{5}xe^{5x} - \dfrac{1}{25}e^{5x} + C$ **2.** $3x\ln x - 3x + C$

3. $\dfrac{1}{2}x^2\ln x + x\ln x - \dfrac{1}{4}x^2 - x + C$

4. $\dfrac{2}{3}x(x + 3)^{3/2} - \dfrac{4}{15}(x + 3)^{5/2} + C$

5. $\dfrac{x^2}{4}\ln x - \dfrac{x^2}{8} + C$ **6.** $-\dfrac{1}{2}e^{-2x}\left(x^2 + x + \dfrac{1}{2}\right) + C$

7. Yes, $\$673,108.31 > \$650,000$.

8. $\ln\left|\dfrac{x - 5}{x + 5}\right| + C$ **9.** $3\ln|x + 4| - 2\ln|x - 2| + C$

10. $5\ln|x + 1| + \dfrac{6}{x + 1} + C$ **11.** $y = \dfrac{100,000}{1 + 3e^{-0.01186t}}$

12. $\dfrac{1}{4}(2x - \ln|1 + 2x|) + C$ **13.** $10\ln\left|\dfrac{x}{0.1 + 0.2x}\right| + C$

14. $\ln\left|x + \sqrt{x^2 - 16}\right| - \dfrac{\sqrt{x^2 - 16}}{x} + C$

15. $\dfrac{1}{2}\ln\left|\dfrac{\sqrt{4 + 9x} - 2}{\sqrt{4 + 9x} + 2}\right| + C$

16. $\dfrac{1}{4}[4x^2 - \ln(1 + e^{4x^2})] + C$ **17.** $x^2e^{x^2 + 1} + C$

18. About 515 stores   **19.** $\dfrac{8}{e} - 4 \approx -1.0570$

20. $e - 2 \approx 0.7183$ **21.** $\ln 4 + 2\ln 5 - 2\ln 2 \approx 3.2189$

22. $15(\ln 9 - \ln 5) \approx 8.8168$

23. $\dfrac{\sqrt{5}}{18} \approx 0.1242$ **24.** $\dfrac{1}{4}\left(\ln\dfrac{17}{19} - \ln\dfrac{7}{9}\right) \approx 0.0350$

SECTION 12.4   *(page 908)*

Skills Review   *(page 908)*

1. $\dfrac{2}{x^3}$ **2.** $-\dfrac{96}{(2x + 1)^4}$ **3.** $-\dfrac{12}{x^4}$ **4.** $6x - 4$

5. $16e^{2x}$ **6.** $e^{x^2}(4x^2 + 2)$ **7.** $(3, 18)$

8. $(1, 8)$ **9.** $n < -5\sqrt{10},\ n > 5\sqrt{10}$

10. $n < -5, n > 5$

	Exact Value	Trapezoidal Rule	Simpson's Rule
1.	2.6667	2.7500	2.6667
3.	8.4000	9.0625	8.4167
5.	4.0000	4.0625	4.0000
7.	0.6931	0.6941	0.6932
9.	5.3333	5.2650	5.3046
11.	12.6667	12.6640	12.6667
13.	0.6931	0.6970	0.6933

15. (a) 0.783 (b) 0.785 **17.** (a) 3.283 (b) 3.240

19. (a) 0.749 (b) 0.771 **21.** (a) 0.877 (b) 0.830

23. (a) 1.879 (b) 1.888 **25.** $\$21,831.20;\ \$21,836.98$

27. $\$678.36$ **29.** $0.3413 = 34.13\%$

31. $0.4999 = 49.99\%$ **33.** $89,500$ ft^2

35. (a) $|E| \le 0.5$ (b) $|E| = 0$

37. (a) $|E| \le \dfrac{5e}{64} \approx 0.212$ (b) $|E| \le \dfrac{13e}{1024} \approx 0.035$

39. (a) $n = 71$ (b) $n = 1$

41. (a) $n = 3280$ (b) $n = 60$ **43.** 19.5215

45. 3.6558 **47.** 23.375 **49.** 416.1 ft

51. (a) 17.171 billion board-feet/yr

(b) 17.082 billion board-feet/yr

(c) The results are approximately equal.

53. 58.912 mg   **55.** 1878 subscribers

SECTION 12.5   *(page 920)*

Skills Review   *(page 920)*

1. 9   **2.** 3   **3.** $-\frac{1}{8}$ **4.** Limit does not exist.

5. Limit does not exist.   **6.** -4

7. (a) $\frac{32}{3}b^3 - 16b^2 + 8b - \frac{4}{3}$ (b) $-\frac{4}{3}$

8. (a) $\dfrac{b^2 - b - 11}{(b - 2)^2(b - 5)}$ (b) $\dfrac{11}{20}$

9. (a) $\ln\left(\dfrac{5 - 3b^2}{b + 1}\right)$ (b) $\ln 5 \approx 1.609$

10. (a) $e^{-3b^2}(e^{6b^2} + 1)$ (b) 2

---

1. Improper; The integrand has an infinite discontinuity when $x = \frac{2}{3}$ and $0 \le \frac{2}{3} \le 1$.

3. Not improper; continuous on $[0, 1]$

5. Improper because the integrand has an infinite discontinuity when $x = 0$ and $0 \le 0 \le 4$; converges; 4

7. Improper because the integrand has an infinite discontinuity when $x = 1$ and $0 \le 1 \le 2$; converges; 6

9. Improper because the upper limit of integration is infinite; converges; 1

11. Converges; 1   **13.** Diverges   **15.** Diverges

17. Diverges   **19.** Diverges   **21.** Converges; 0

23. Diverges   **25.** Converges; 6   **27.** Diverges

29. Converges; 0   **31.** Converges; $\ln\left(\dfrac{4 + \sqrt{7}}{3}\right) \approx 0.7954$

33. 1

35.

x	1	10	25	50
xe^{-x}	0.3679	0.0005	0.0000	0.0000

37.

x	1	10	25	50
$x^2e^{-(1/2)x}$	0.6065	0.6738	0.0023	0.0000

39. 2   **41.** $\frac{1}{4}$

43. (a) 0.9495   (b) 0.0974   (c) 0.0027

45. $66,666.67   **47.** Yes, $360,000 < $400,000.

49. (a) $4,637,228   (b) $5,555,556

51. (a) $748,367.34   (b) $808,030.14   (c) $900,000.00

REVIEW EXERCISES FOR CHAPTER 12
(page 926)

1. $2\sqrt{x} \ln x - 4\sqrt{x} + C$ **3.** $xe^x + C$

5. $x^2e^{2x} - xe^{2x} + \frac{1}{2}e^{2x} + C$ **7.** $90,634.62

9. $865,958.50

11. (a) $8847.97, $7869.39, $7035.11   (b) $1,995,258.71

13. $90,237.67   **15.** $\dfrac{1}{5}\ln\left|\dfrac{x}{x + 5}\right| + C$

17. $6\ln|x + 2| - 5\ln|x - 3| + C$

19. $x - \frac{25}{8}\ln|x + 5| + \frac{9}{8}\ln|x - 3| + C$

21. (a) $y = \dfrac{10,000}{1 + 7e^{-0.106873t}}$

(b)

Time, t	0	3	6	12	24
Sales, y	1250	1645	2134	3400	6500

(c) $t \approx 28$ weeks

23. $\dfrac{1}{9}\left(\dfrac{2}{2 + 3x} + \ln|2 + 3x|\right) + C$

25. $\sqrt{x^2 + 25} - 5\ln\left|\dfrac{5 + \sqrt{x^2 + 25}}{x}\right| + C$

27. $\dfrac{1}{4}\ln\left|\dfrac{x - 2}{x + 2}\right| + C$ **29.** $\dfrac{8}{3}$

31. $2\sqrt{1 + x} + \ln\left|\dfrac{\sqrt{1 + x} - 1}{\sqrt{1 + x} + 1}\right| + C$

33. $(x - 5)^3e^{x-5} - 3(x - 5)^2e^{x-5} + 6(x - 6)e^{x-5} + C$

35. (a) 0.675   (b) 0.290   **37.** 0.705   **39.** 0.741

41. 0.376   **43.** 0.289   **45.** 9.0997   **47.** 0.017

49. Converges; 1   **51.** Diverges

53. Converges; 2   **55.** Converges; 2

57. (a) $989,050.57   (b) $1,666,666.67

59. (a) 0.441   (b) 0.119   (c) 0.015

CHAPTER TEST   *(page 929)*

1. $xe^{x+1} - e^{x+1} + C$ **2.** $3x^3 \ln x - x^3 + C$

3. $-3x^2e^{-x/3} - 18xe^{-x/3} - 54e^{-x/3} + C$

4. $1.95 per share   **5.** $\ln\left|\dfrac{x - 9}{x + 9}\right| + C$

6. $\dfrac{1}{3}\ln|3x + 1| + \dfrac{1}{3(3x + 1)} + C$

7. $2\ln|x| - \ln|x + 2| + C$

8. $\frac{1}{4}\left(\frac{7}{7+2x}+\ln|7+2x|\right)+C$

9. $x^3-\ln|1+e^{x^3}|+C$

10. $-\frac{2}{75}(2-5x^2)\sqrt{1+5x^2}+C$

11. $-1+\frac{3}{2}\ln 3\approx 0.6479$ **12.** $4\ln\left(\frac{48}{13}\right)\approx 5.2250$

13. $4\ln\left[3\left(\sqrt{17}-4\right)\right]+\sqrt{17}-5\approx -4.8613$

14. Trapezoid Rule: 0.2100; Exact: 0.2055

15. Simpson Rule: 41.3606; Exact: 41.1711

16. Converges; $\frac{1}{3}$ **17.** Converges; 12   **18.** Diverges

19. (a) \$498.75   (b) Plan B, because \$149 < \$498.75.

CHAPTER 13

SECTION 13.1   *(page 937)*

Skills Review   *(page 937)*

1. $2\sqrt{5}$ **2.** 5   **3.** 8   **4.** 8   **5.** $(4,7)$

6. $(1,0)$ **7.** $(0,3)$ **8.** $(-1,1)$

9. $(x-2)^2+(y-3)^2=4$

10. $(x-1)^2+(y-4)^2=25$

---

1. 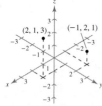   **3.**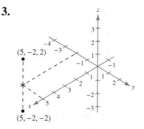

5. $A(2,3,4), B(-1,-2,2)$ **7.** $(-3,4,5)$

9. $(10,0,0)$ **11.** 0   **13.** $3\sqrt{2}$ **15.** $\sqrt{206}$

17. $(2,-5,3)$ **19.** $\left(\frac{1}{2},\frac{1}{2},-1\right)$ **21.** $(6,-3,5)$

23. $(1,2,1)$ **25.** $3, 3\sqrt{5}, 6$; right triangle

27. $2, 2\sqrt{5}, 2\sqrt{2}$; neither right nor isosceles

29. $(0,0,5), (2,2,6), (2,-4,9)$

31. $x^2+(y-2)^2+(z-2)^2=4$

33. $\left(x-\frac{3}{2}\right)^2+(y-2)^2+(z-1)^2=\frac{21}{4}$

35. $(x-1)^2+(y-1)^2+(z-5)^2=9$

37. $(x-1)^2+(y-3)^2+z^2=10$

39. $(x+2)^2+(y-1)^2+(z-1)^2=1$

41. Center: $\left(\frac{5}{2},0,0\right)$ **43.** Center: $(1,-3,-4)$

   Radius: $\frac{5}{2}$ Radius: 5

45. Center: $(1,3,2)$

   Radius: $\frac{5\sqrt{2}}{2}$

47. 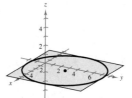   **49.**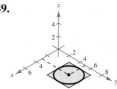

51. 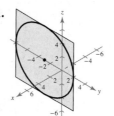   **53.**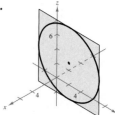

55. (a) 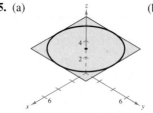   (b)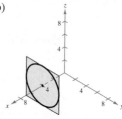

57. (a) 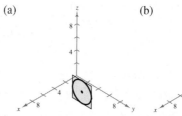   (b)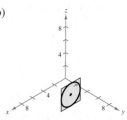

59. $(3,3,3)$ **61.** $x^2+y^2+z^2=6806.25$

SECTION 13.2   *(page 946)*

Skills Review   *(page 946)*

1. $(4,0), (0,3)$ **2.** $\left(-\frac{4}{3},0\right), (0,-8)$

3. $(1,0), (0,-2)$ **4.** $(-5,0), (0,-5)$

5. $(x-1)^2+(y-2)^2+(z-3)^2+1=0$

6. $(x-4)^2+(y+2)^2-(z+3)^2=0$

7. $(x+1)^2+(y-1)^2-z=0$

8. $(x-3)^2+(y+5)^2+(z+13)^2=1$

9. $x^2+y^2+z^2=\frac{1}{4}$ **10.** $x^2+y^2+z^2=4$

1.

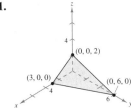

3.

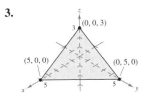

5.

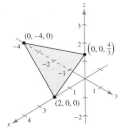

7.

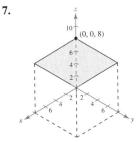

9.

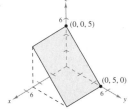

11.

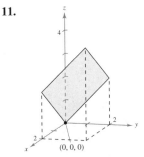

13. $\dfrac{6\sqrt{14}}{7}$ **15.** $\dfrac{8\sqrt{14}}{7}$ **17.** $\dfrac{13\sqrt{29}}{29}$ **19.** $\dfrac{28\sqrt{29}}{29}$

21. Perpendicular   **23.** Parallel   **25.** Parallel

27. Neither parallel nor perpendicular   **29.** Perpendicular

31. c   **32.** e   **33.** f   **34.** b   **35.** d   **36.** a

37. Trace in xy-plane $(z = 0)$: $y = x^2$ (parabola)

Trace in plane $y = 1$: $x^2 - z^2 = 1$ (hyperbola)

Trace in yz-plane $(x = 0)$: $y = -z^2$ (parabola)

39. Trace in xy-plane $(z = 0)$: $\dfrac{x^2}{4} + y^2 = 1$ (ellipse)

Trace in xz-plane $(y = 0)$: $\dfrac{x^2}{4} + z^2 = 1$ (ellipse)

Trace in yz-plane $(x = 0)$: $y^2 + z^2 = 1$ (circle)

41. Ellipsoid   **43.** Hyperboloid of one sheet

45. Elliptic paraboloid   **47.** Hyperbolic paraboloid

49. Hyperboloid of two sheets   **51.** Elliptic cone

53. Hyperbolic paraboloid

55. $(20, 0, 0)$ **57.** $(0, 0, 20)$

59. (a)

Year	1999	2000	2001
x	6.2	6.1	5.9
y	7.3	7.1	7.0
z (actual)	7.8	7.7	7.4
z (approximated)	7.8	7.7	7.5

Year	2002	2003	2004
x	5.8	5.6	5.5
y	7.0	6.9	6.9
z (actual)	7.3	7.2	6.9
z (approximated)	7.3	7.1	7.0

The approximated values of z are very close to the actual values.

(b) According to the model, increases in consumption of milk types y and z will correspond to an increase in consumption of milk type x.

SECTION 13.3   *(page 954)*

Skills Review   *(page 954)*

1. 11   **2.** -16 **3.** 7   **4.** 4   **5.** $(-\infty, \infty)$

6. $(-\infty, -3) \cup (-3, 0) \cup (0, \infty)$

7. $[5, \infty)$ **8.** $\left(-\infty, -\sqrt{5}\,\right] \cup \left[\sqrt{5}, \infty\right)$

9. 55.0104   **10.** 6.9165

1. (a) $\dfrac{3}{2}$ (b) $-\dfrac{1}{4}$ (c) 6   (d) $\dfrac{5}{y}$ (e) $\dfrac{x}{2}$ (f) $\dfrac{5}{t}$

3. (a) 5   (b) $3e^2$ (c) $2e^{-1}$ (d) $5e^y$ (e) xe^2 (f) te^t

5. (a) $\frac{2}{3}$ (b) 0   **7.** (a) 90π (b) 50π

9. (a) \$20,655.20   (b) \$1,397,672.67

11. (a) 0   (b) 6

13. (a) $x^2 + 2x\,\Delta x + (\Delta x)^2 - 2y$ (b) $-2, \Delta y \neq 0$

15. Domain: all points (x, y) inside and on the circle

$x^2 + y^2 = 16$

Range: $[0, 4]$

17. Domain: all points (x, y) such that $y \neq 0$

Range: $(0, \infty)$

19. All points inside and on the circle $x^2 + y^2 = 4$

21. All points (x, y)

23. All points (x, y) such that $x \neq 0$ and $y \neq 0$

25. All points (x, y) such that $y \geq 0$

27. The half-plane below the line $y = -x + 4$

29. b   **30.** d   **31.** a   **32.** c

33. The level curves are parallel lines.

35. The level curves are circles.

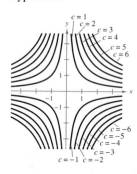

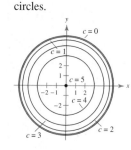

37. The level curves are hyperbolas.

39. The level curves are circles.

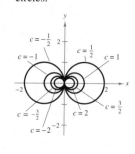

41. 135,540 units   **43.** (a) $15,250   (b) $18,425

45.

R \ I	0	0.03	0.05
0	$2593.74	$1929.99	$1592.33
0.28	$2004.23	$1491.34	$1230.42
0.35	$1877.14	$1396.77	$1152.40

47. (a) C   (b) A   (c) B

49. (a) $.663 earnings per share

(b) x; Explanations will vary. Sample answer: The x-variable has a greater influence on the earnings per share because the absolute value of its coefficient is larger than the absolute value of the coefficient of the y-term.

51. Answers will vary.

SECTION 13.4   (page 965)

Skills Review   (page 965)

1. $\dfrac{x}{\sqrt{x^2 + 3}}$ **2.** $-6x(3 - x^2)^2$ **3.** $e^{2t+1}(2t + 1)$

4. $\dfrac{e^{2x}(2 - 3e^{2x})}{\sqrt{1 - e^{2x}}}$ **5.** $-\dfrac{2}{3 - 2x}$ **6.** $\dfrac{3(t^2 - 2)}{2t(t^2 - 6)}$

7. $-\dfrac{10x}{(4x - 1)^3}$ **8.** $-\dfrac{(x + 2)^2(x^2 + 8x + 27)}{(x^2 - 9)^3}$

9. $f'(2) = 8$ **10.** $g'(2) = \dfrac{7}{2}$

1. $\dfrac{\partial z}{\partial x} = 3$; $\dfrac{\partial z}{\partial y} = 5$ **3.** $f_x(x, y) = 3$; $f_y(x, y) = -12y$

5. $f_x(x, y) = \dfrac{1}{y}$; $f_y(x, y) = -\dfrac{x}{y^2}$

7. $f_x(x, y) = \dfrac{x}{\sqrt{x^2 + y^2}}$; $f_y(x, y) = \dfrac{y}{\sqrt{x^2 + y^2}}$

9. $\dfrac{\partial z}{\partial x} = 2xe^{2y}$; $\dfrac{\partial z}{\partial y} = 2x^2e^{2y}$

11. $h_x(x, y) = -2xe^{-(x^2+y^2)}$; $h_y(x, y) = -2ye^{-(x^2+y^2)}$

13. $\dfrac{\partial z}{\partial x} = -\dfrac{2y}{x^2 - y^2}$, $\dfrac{\partial z}{\partial y} = \dfrac{2x}{x^2 - y^2}$

15. $f_x(x, y) = 3xye^{x-y}(2 + x)$

17. $g_x(x, y) = 3y^2e^{y-x}(1 - x)$ **19.** 9

21. $f_x(x, y) = 6x + y$, 13; $f_y(x, y) = x - 2y$, 0

23. $f_x(x, y) = 3ye^{3xy}$, 12; $f_y(x, y) = 3xe^{3xy}$, 0

25. $f_x(x, y) = -\dfrac{y^2}{(x - y)^2}$, $-\dfrac{1}{4}$; $f_y(x, y) = \dfrac{x^2}{(x - y)^2}$, $\dfrac{1}{4}$

27. $f_x(x, y) = \dfrac{2x}{x^2 + y^2}$, 2; $f_y(x, y) = \dfrac{2y}{x^2 + y^2}$, 0

29. $w_x = yz$

$w_y = xz$

$w_z = xy$

31. $w_x = -\dfrac{2z}{(x + y)^2}$

$w_y = -\dfrac{2z}{(x + y)^2}$

$w_z = \dfrac{2}{x + y}$

33. $w_x = \dfrac{x}{\sqrt{x^2 + y^2 + z^2}}$, $\dfrac{2}{3}$

$w_y = \dfrac{y}{\sqrt{x^2 + y^2 + z^2}}$, $-\dfrac{1}{3}$

$w_z = \dfrac{z}{\sqrt{x^2 + y^2 + z^2}}$, $\dfrac{2}{3}$

35. $w_x = \dfrac{x}{x^2 + y^2 + z^2}$, $\dfrac{3}{25}$

$w_y = \dfrac{y}{x^2 + y^2 + z^2}$, 0

$w_z = \dfrac{z}{x^2 + y^2 + z^2}$, $\dfrac{4}{25}$

37. $w_x = 2z^2 + 3yz,\ 2$

$w_y = 3xz - 12yz,\ 30$

$w_z = 4xz + 3xy - 6y^2,\ -1$

39. $(-6, 4)$ **41.** $(1, 1)$

43. (a) 2   (b) 1   **45.** (a) -2 (b) -2

47. $\dfrac{\partial^2 z}{\partial x^2} = 2$

$\dfrac{\partial^2 z}{\partial x \partial y} = \dfrac{\partial^2 z}{\partial y \partial x} = -2$

$\dfrac{\partial^2 z}{\partial y^2} = 6$

49. $\dfrac{\partial^2 z}{\partial x^2} = \dfrac{e^{2xy}(2x^2y^2 - 2xy + 1)}{2x^3}$

$\dfrac{\partial^2 z}{\partial x \partial y} = \dfrac{\partial^2 z}{\partial y \partial x} = ye^{2xy}$

$\dfrac{\partial^2 z}{\partial y^2} = xe^{2xy}$

51. $\dfrac{\partial^2 z}{\partial x^2} = 6x$

$\dfrac{\partial^2 z}{\partial y^2} = -8$

$\dfrac{\partial^2 z}{\partial y \partial x} = \dfrac{\partial^2 z}{\partial x \partial y} = 0$

53. $\dfrac{\partial^2 z}{\partial x^2} = \dfrac{2}{(x - y)^3}$

$\dfrac{\partial^2 z}{\partial x \partial y} = \dfrac{\partial^2 z}{\partial y \partial x} = \dfrac{-2}{(x - y)^3}$

$\dfrac{\partial^2 z}{\partial y^2} = \dfrac{2}{(x - y)^3}$

55. $f_{xx}(x, y) = 12x^2 - 6y^2,\ 12$

$f_{xy}(x, y) = -12xy,\ 0$

$f_{yy}(x, y) = -6x^2 + 2,\ -4$

$f_{yx}(x, y) = -12xy,\ 0$

57. $f_{xx}(x, y) = -\dfrac{1}{(x - y)^2},\ -1$

$f_{xy}(x, y) = \dfrac{1}{(x - y)^2},\ 1$

$f_{yy}(x, y) = -\dfrac{1}{(x - y)^2},\ -1$

$f_{yx}(x, y) = \dfrac{1}{(x - y)^2},\ 1$

59. (a) At $(120, 160)$, $\dfrac{\partial C}{\partial x} \approx 154.77$.

At $(120, 160)$, $\dfrac{\partial C}{\partial y} \approx 193.33$.

(b) Racing bikes; Explanations will vary. Sample answer: The y-variable has a greater influence on the cost because the absolute value of its coefficient is larger than the absolute value of the coefficient of the x-term.

61. (a) About 113.72   (b) About 97.47

63. Complementary

65. (a) $\dfrac{\partial z}{\partial x} = 1.25;\ \dfrac{\partial z}{\partial y} = -0.125$

(b) For every increase of 1.25 gallons of whole milk, there is an increase of one gallon of reduced-fat (1%) and skim milks. For every decrease of 0.125 gallon of whole milk, there is an increase of one gallon of reduced-fat (2%) milk.

67. $IQ_M(M, C) = \dfrac{100}{C}$, $IQ_M(12, 10) = 10$; For a child that has a current mental age of 12 years and chronological age of 10 years, the IQ is increasing at a rate of 10 IQ points for every increase of 1 year in the child's mental age.

$IQ_C(M, C) = \dfrac{-100M}{C^2}$, $IQ_C(12, 10) = -12$; For a child that has a current mental age of 12 years and chronological age of 10 years, the IQ is decreasing at a rate of 12 IQ points for every increase of 1 year in the child's chronological age.

69. An increase in either price will cause a decrease in the number of applicants.

71. Answers will vary.

SECTION 13.5   *(page 974)*

Skills Review   *(page 974)*

1. $(3, 2)$ **2.** $(11, 6)$ **3.** $(1, 4)$ **4.** $(4, 4)$

5. $(5, 2)$ **6.** $(3, -2)$ **7.** $(0, 0),\ (-1, 0)$

8. $(-2, 0),\ (2, -2)$

9. $\dfrac{\partial z}{\partial x} = 12x^2$ $\dfrac{\partial^2 z}{\partial y^2} = -6$

$\dfrac{\partial z}{\partial y} = -6y$ $\dfrac{\partial^2 z}{\partial x \partial y} = 0$

$\dfrac{\partial^2 z}{\partial x^2} = 24x$ $\dfrac{\partial^2 z}{\partial y \partial x} = 0$

10. $\dfrac{\partial z}{\partial x} = 10x^4$ $\dfrac{\partial^2 z}{\partial y^2} = -6y$

$\dfrac{\partial z}{\partial y} = -3y^2$ $\dfrac{\partial^2 z}{\partial x \partial y} = 0$

$\dfrac{\partial^2 z}{\partial x^2} = 40x^3$ $\dfrac{\partial^2 z}{\partial y \partial x} = 0$

11. $\dfrac{\partial z}{\partial x} = 4x^3 - \dfrac{\sqrt{xy}}{2x}$ $\dfrac{\partial^2 z}{\partial y^2} = \dfrac{\sqrt{xy}}{4y^2}$

$\dfrac{\partial z}{\partial y} = -\dfrac{\sqrt{xy}}{2y} + 2$ $\dfrac{\partial^2 z}{\partial x \partial y} = -\dfrac{\sqrt{xy}}{4xy}$

$\dfrac{\partial^2 z}{\partial x^2} = 12x^2 + \dfrac{\sqrt{xy}}{4x^2}$ $\dfrac{\partial^2 z}{\partial y \partial x} = -\dfrac{\sqrt{xy}}{4xy}$

12. $\dfrac{\partial z}{\partial x} = 4x - 3y$ $\dfrac{\partial^2 z}{\partial y^2} = 2$

$\dfrac{\partial z}{\partial y} = 2y - 3x$ $\dfrac{\partial^2 z}{\partial x \partial y} = -3$

$\dfrac{\partial^2 z}{\partial x^2} = 4$ $\dfrac{\partial^2 z}{\partial y \partial x} = -3$

13. $\dfrac{\partial z}{\partial x} = y^3 e^{xy^2}$ $\dfrac{\partial^2 z}{\partial y^2} = 4x^2 y^3 e^{xy^2} + 6xy e^{xy^2}$

$\dfrac{\partial z}{\partial y} = 2xy^2 e^{xy^2} + e^{xy^2}$ $\dfrac{\partial^2 z}{\partial x \partial y} = 2xy^4 e^{xy^2} + 3y^2 e^{xy^2}$

$\dfrac{\partial^2 z}{\partial x^2} = y^5 e^{xy^2}$ $\dfrac{\partial^2 z}{\partial y \partial x} = 2xy^4 e^{xy^2} + 3y^2 e^{xy^2}$

14. $\dfrac{\partial z}{\partial x} = e^{xy}(xy + 1)$ $\dfrac{\partial^2 z}{\partial y^2} = x^3 e^{xy}$

$\dfrac{\partial z}{\partial y} = x^2 e^{xy}$ $\dfrac{\partial^2 z}{\partial x \partial y} = xe^{xy}(xy + 2)$

$\dfrac{\partial^2 z}{\partial x^2} = ye^{xy}(xy + 2)$ $\dfrac{\partial^2 z}{\partial y \partial x} = xe^{xy}(xy + 2)$

1. Critical point: $(-2, -4)$

  No relative extrema

  $(-2, -4, 1)$ is a saddle point.

3. Critical point: $(0, 0)$

  Relative minimum: $(0, 0, 1)$

5. Relative minimum: $(1, 3, 0)$

7. Relative minimum: $(-1, 1, -4)$

9. Relative maximum: $(8, 16, 74)$

11. Relative minimum: $(2, 1, -7)$

13. Saddle point: $(-2, -2, -8)$

15. Saddle point: $(0, 0, 0)$

17. Relative maximum: $\left(\frac{1}{2}, \frac{1}{2}, e^{1/2}\right)$

  Relative minimum: $\left(-\frac{1}{2}, -\frac{1}{2}, -e^{1/2}\right)$

19. Saddle point: $(0, 0, 4)$

21. Insufficient information

23. $f(x_0, y_0)$ is a saddle point.

25. Relative minima: $(a, 0, 0)$, $(0, b, 0)$

  Second-Partials Test fails at $(a, 0)$ and $(0, b)$.

27. Saddle point: $(0, 0, 0)$

  Second-Partials Test fails at $(0, 0)$.

29. Relative minimum: $(0, 0, 0)$

  Second-Partials Test fails at $(0, 0)$.

31. Relative minimum: $(1, -3, 0)$

33. 10, 10, 10   **35.** 10, 10, 10   **37.** $x_1 = 3, x_2 = 6$

39. $p_1 = 2500, p_2 = 3000$ **41.** $x_1 \approx 94, x_2 \approx 157$

43. 32 in. $\times$ 16 in. $\times$ 16 in.

45. Base dimensions: 2 ft $\times$ 2 ft;

  Height: 1.5 ft; Minimum cost: \$1.80

47. Proof   **49.** $x = 1.25, y = 2.5$; \$4.625 million

51. True

MID-CHAPTER QUIZ   *(page 977)*

1. (a)   (b) 3   (c) $\left(0, \frac{5}{2}, 1\right)$

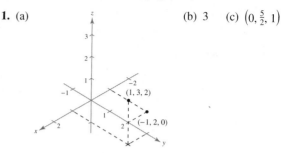

2. (a)

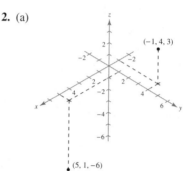

  (b) $3\sqrt{14}$ (c) $\left(2, \frac{5}{2}, -\frac{3}{2}\right)$

3. (a)

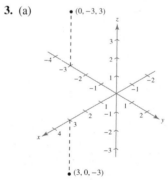

  (b) $3\sqrt{6}$ (c) $\left(\frac{3}{2}, -\frac{3}{2}, 0\right)$

4. $(x - 2)^2 + (y + 1)^2 + (z - 3)^2 = 16$

5. $(x - 1)^2 + (y - 4)^2 + (z + 2)^2 = 11$

6. Center: $(4, 1, 3)$; radius: 7

7.

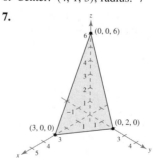

8.

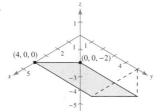

9.

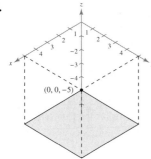

10. Ellipsoid   **11.** Hyperboloid of two sheets

12. Elliptic paraboloid

13. $f(1, 0) = 1$ **14.** $f(1, 0) = 2$

$\quad f(4, -1) = -5$ $f(4, -1) = 3\sqrt{7}$

15. $f(1, 0) = 0$

$\quad f(4, -1) = 0$

16. (a) Between 30° and 50°

    (b) Between 40° and 80°

    (c) Between 70° and 90°

17. $f_x = 2x - 3;\ f_x(-2, 3) = -7$

$\quad f_y = 4y - 1;\ f_y(-2, 3) = 11$

18. $f_x = \dfrac{y(3 + y)}{(x + y)^2};\ f_x(-2, 3) = 18$

$\quad f_y = \dfrac{-2xy - y^2 - 3x}{(x + y)^2};\ f_y(-2, 3) = 9$

19. Critical point: $(1, 0)$

    Relative minimum: $(1, 0, -3)$

20. Critical points: $(0, 0),\ \left(\frac{4}{3}, \frac{4}{3}\right)$

    Relative maximum: $\left(\frac{4}{3}, \frac{4}{3}, \frac{59}{27}\right)$

    Saddle point: $(0, 0, 1)$

21. $x = 80,\ y = 20;\ \$20,000$

22. $x^2 + y^2 + z^2 = 3963^2$

    Lines of longitude would be traces in planes passing through the z-axis. Each trace is a circle. Lines of latitude would be traces in planes parallel to the equator. They are circles.

SECTION 13.6   *(page 984)*

Skills Review *(page 984)*

1. $\left(\frac{7}{8}, \frac{1}{12}\right)$ **2.** $\left(-\frac{1}{24}, -\frac{7}{8}\right)$ **3.** $\left(\frac{55}{12}, -\frac{25}{12}\right)$

4. $\left(\frac{22}{23}, -\frac{3}{23}\right)$ **5.** $\left(\frac{5}{3}, \frac{1}{3}, 0\right)$ **6.** $\left(\frac{14}{19}, -\frac{10}{19}, -\frac{32}{57}\right)$

7. $f_x = 2xy + y^2$ **8.** $f_x = 50y^2(x + y)$

$\quad f_y = x^2 + 2xy$ $f_y = 50y(x + y)(x + 2y)$

9. $f_x = 3x^2 - 4xy + yz$ **10.** $f_x = yz + z^2$

$\quad f_y = -2x^2 + xz$ $f_y = xz + z^2$

$\quad f_z = xy$ $f_z = xy + 2xz + 2yz$

1. $f(5, 5) = 25$ **3.** $f(2, 2) = 8$ **5.** $f(\sqrt{2}, 1) = 1$

7. $f(25, 50) = 2600$ **9.** $f(1, 1) = 2$

11. $f(2, 2) = e^4$ **13.** $f(9, 6, 9) = 432$

15. $f\left(\frac{1}{3}, \frac{1}{3}, \frac{1}{3}\right) = \frac{1}{3}$ **17.** $f\left(\frac{\sqrt{3}}{3}, \frac{\sqrt{3}}{3}, \frac{\sqrt{3}}{3}\right) = \sqrt{3}$

19. $f(8, 16, 8) = 1024$

21. $f\left(\sqrt{\frac{10}{3}}, \frac{1}{2}\sqrt{\frac{10}{3}}, \sqrt{\frac{5}{3}}\right) = \dfrac{5\sqrt{15}}{9}$

23. $x = 4,\ y = \frac{2}{3},\ z = 2$ **25.** $40, 40, 40$ **27.** $\frac{S}{3}, \frac{S}{3}, \frac{S}{3}$

29. $3\sqrt{2}$ **31.** $\sqrt{3}$ **33.** 36 in. × 18 in. × 18 in.

35. Length = width = $\sqrt[3]{360} \approx 7.1$ ft

    Height = $\dfrac{480}{360^{2/3}} \approx 9.5$ ft

37. $x_1 = 752.5,\ x_2 = 1247.5$

    To minimize cost, let $x_1 = 753$ units and $x_2 = 1247$ units.

39. (a) $x = 50\sqrt{2} \approx 71$ (b) Answers will vary.

    $y = 200\sqrt{2} \approx 283$

41. (a) $f\left(\frac{3125}{6}, \frac{6250}{3}\right) \approx 147{,}314$ (b) 1.473

    (c) 184,142 units

43. $x = \sqrt[3]{0.065} \approx 0.402$ L

    $y = \frac{1}{2}\sqrt[3]{0.065} \approx 0.201$ L

    $z = \frac{1}{3}\sqrt[3]{0.065} \approx 0.134$ L

45. (a) $x = 52,\ y = 48$ (b) 64 dogs

47. (a) 50 ft × 120 ft   (b) \$2400

49. Stock G: \$157,791.67

    Stock P: \$8500.00

    Stock S: \$133,708.33

51. (a) Cable television: $1200

  Newspaper: $600

  Radio: $900

  (b) About 3718 responses

SECTION 13.7   *(page 994)*

Skills Review   *(page 994)*

1. 5.0225   **2.** 0.0189

3. $S_a = 2a - 4 - 4b$ **4.** $S_a = 8a - 6 - 2b$

   $S_b = 12b - 8 - 4a$ $S_b = 18b - 4 - 2a$

5. 15   **6.** 42   **7.** $\frac{25}{12}$

8. 14   **9.** 31   **10.** 95

---

1. (a) $y = \frac{3}{4}x + \frac{4}{3}$ (b) $\frac{1}{6}$

3. (a) $y = -2x + 4$ (b) 2

5. $y = x + \frac{2}{3}$ **7.** $y = -2.3x - 0.9$

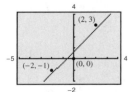 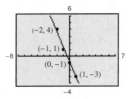

9. $y = 0.7x + 1.4$ **11.** $y = x + 4$

13. $y = -0.65x + 1.75$ **15.** $y = 0.8605x + 0.163$

17. $y = -1.1824x + 6.385$

19. $y = 0.4286x^2 + 1.2x + 0.74$ **21.** $y = x^2 - x$

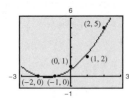 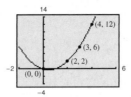

23. Linear: $y = 1.4x + 6$

  Quadratic: $y = 0.12x^2 + 1.7x + 6$

  The quadratic model is a better fit.

25. Linear: $y = -68.9x + 754$

  Quadratic: $y = 2.82x^2 - 83.0x + 763$

  The quadratic model is a better fit.

27. (a) $y = -240x + 685$ (b) 349   (c) $.77

29. (a) $y = 13.8x + 22.1$ (b) 44.18 bushels/acre

31. (a) $y = -0.238t + 11.93$;

  In 2010, $y \approx 4.8$ deaths per 1000 live births.

  (b) $y = 0.0088t^2 - 0.458t + 12.66$;

  In 2010, $y \approx 6.8$ deaths per 1000 live births.

33. (a)

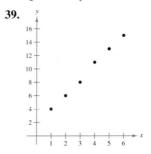

  (b) $y = -28.415t^2 + 208.33t + 1025.1$

  (c) Sample answer: The quadratic model has an "*r*-value" of about 0.95 ($r^2 \approx 0.91$) and the linear model has an "*r*-value" of about 0.58. Because $0.95 > 0.58$, the quadratic model is a better fit for the data.

35. Linear: $y = 3.757x + 9.03$

  Quadratic: $y = 0.006x^2 + 3.63x + 9.4$

  Either model is a good fit for the data.

37. Quadratic: $y = -0.087x^2 + 2.82x + 0.4$

39.

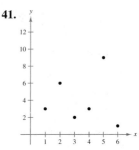

41.

Positive correlation,   No correlation, $r = 0$
$r \approx 0.9981$

43.

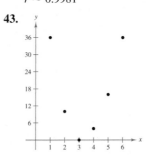

No correlation, $r \approx 0.0750$

45. False; The data modeled by $y = 3.29x - 4.17$ have a positive correlation.

47. True   **49.** True   **51.** Answers will vary.

SECTION 13.8   *(page 1003)*

Skills Review   *(page 1003)*

1. 1   **2.** 6   **3.** 42   **4.** $\frac{1}{2}$ **5.** $\frac{19}{4}$

6. $\frac{16}{3}$ **7.** $\frac{1}{7}$ **8.** 4   **9.** ln 5   **10.** $\ln(e - 1)$

11. $\frac{e}{2}(e^4 - 1)$ **12.** $\frac{1}{2}\left(1 - \frac{1}{e^2}\right)$

13. 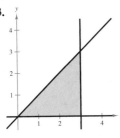   **14.**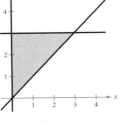

15. 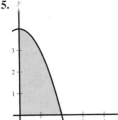   **16.**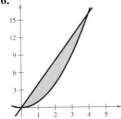

1. $\dfrac{3x^2}{2}$ **3.** $y \ln|2y|$ **5.** $x^2\left(2 - \frac{1}{2}x^2\right)$ **7.** $\dfrac{y^3}{2}$

9. $e^{x^2} - \dfrac{e^{x^2}}{x^2} + \dfrac{1}{x^2}$ **11.** 3   **13.** 36   **15.** $\dfrac{1}{2}$

17. $\frac{148}{3}$ **19.** 5   **21.** 64   **23.** 4

25.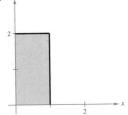

$$\int_0^1 \int_0^2 dy\, dx = \int_0^2 \int_0^1 dx\, dy = 2$$

27.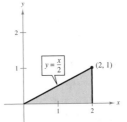

$$\int_0^1 \int_{2y}^2 dx\, dy = \int_0^2 \int_0^{x/2} dy\, dx = 1$$

29.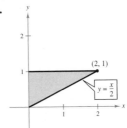

$$\int_0^2 \int_{x/2}^1 dy\, dx = \int_0^1 \int_0^{2y} dx\, dy = 1$$

31.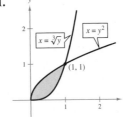

$$\int_0^1 \int_{y^2}^{\sqrt[3]{y}} dx\, dy = \int_0^1 \int_{x^3}^{\sqrt{x}} dy\, dx = \frac{5}{12}$$

33. $\frac{1}{2}(e^9 - 1) \approx 4051.042$ **35.** 24   **37.** $\frac{16}{3}$

39. $\frac{8}{3}$ **41.** 36   **43.** 5   **45.** 2   **47.** 0.6588

49. 8.1747   **51.** 0.4521   **53.** 1.1190   **55.** True

SECTION 13.9   *(page 1011)*

Skills Review   *(page 1011)*

1. 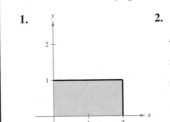   **2.**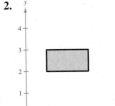

3.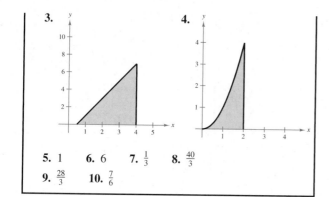

4.

5. 1   **6.** 6   **7.** $\frac{1}{3}$ **8.** $\frac{40}{3}$

9. $\frac{28}{3}$ **10.** $\frac{7}{6}$

1.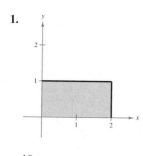

10

3.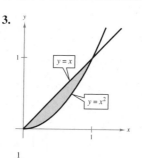

$\frac{1}{54}$

5.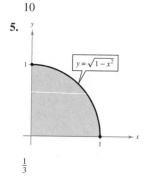

$\frac{1}{3}$

7.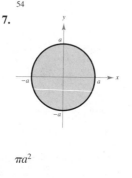

πa^2

9. $\displaystyle\int_0^3 \int_0^5 xy\, dy\, dx = \int_0^5 \int_0^3 xy\, dx\, dy = \frac{225}{4}$

11. $\displaystyle\int_0^2 \int_x^{2x} \frac{y}{x^2+y^2}\, dy\, dx = \int_0^2 \int_{y/2}^y \frac{y}{x^2+y^2}\, dx\, dy$

$\displaystyle + \int_2^4 \int_{y/2}^2 \frac{y}{x^2+y^2}\, dx\, dy = \ln\frac{5}{2}$

13. 4   **15.** 4   **17.** 12   **19.** $\frac{3}{8}$ **21.** $\frac{40}{3}$ **23.** 4

25. $\frac{32}{3}$ **27.** 10,000   **29.** 2   **31.** $\frac{8}{3}$ **33.** \$75,125

35. \$13,400   **37.** 25,645.24

REVIEW EXERCISES FOR CHAPTER 13
(page 1017)

1.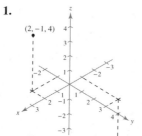

3. $\sqrt{110}$ **5.** $(-1, 4, 6)$

7. $x^2 + (y-1)^2 + z^2 = 25$

9. $(x-2)^2 + (y-3)^2 + (z-2)^2 = 17$

11. Center: $(-2, 1, 4)$; radius: 4

13. 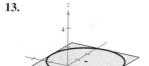   **15.**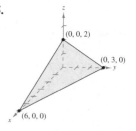

17.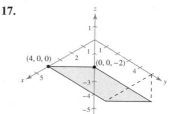

19. Sphere   **21.** Ellipsoid   **23.** Elliptic paraboloid

25. Top half of a circular cone

27. (a) 18   (b) 0   (c) -245 (d) -32

29. The domain is the set of all points inside or on the circle $x^2 + y^2 = 1$, and the range is $[0, 1]$.

31. The level curves are lines of slope $-\frac{2}{5}$.

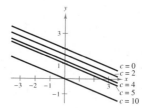

33. The level curves are hyperbolas.

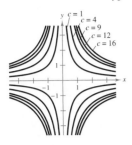

35. (a) As the color darkens from light green to dark green, the average yearly precipitation increases.

(b) The small eastern portion containing Davenport

(c) The northwestern portion containing Sioux City

37. Southwest to northeast   **39.** $2.50

41. $f_x = 2xy + 3y + 2$

$f_y = x^2 + 3x - 5$

43. $z_x = \dfrac{2x}{y^2}$

$z_y = \dfrac{-2x^2}{y^3}$

45. $f_x = \dfrac{2}{2x + 3y}$

$f_y = \dfrac{3}{2x + 3y}$

47. $f_x = ye^x + e^y$

$f_y = xe^y + e^x$

49. $w_x = yz^2$

$w_y = xz^2$

$w_z = 2xyz$

51. (a) $z_x = 3$ (b) $z_y = -4$

53. (a) $z_x = -2x$ (b) $z_y = -2y$

At $(1, 2, 3)$, $z_x = -2$.   At $(1, 2, 3)$, $z_y = -4$.

55. $f_{xx} = 6$

$f_{yy} = 12y$

$f_{xy} = f_{yx} = -1$

57. $f_{xx} = f_{yy} = f_{xy} = f_{yx} = \dfrac{-1}{4(1 + x + y)^{3/2}}$

59. $C_x(500, 250) = 99.50$

$C_y(500, 250) = 140$

61. (a) $A_w = 43.095w^{-0.575}h^{0.725}$

$A_h = 73.515w^{0.425}h^{-0.275}$

(b) ≈ 47.35;

The surface area of an average human body increases approximately 47.35 square centimeters per pound for a human who weighs 180 pounds and is 70 inches tall.

63. Critical point: $(0, 0)$

Relative minimum: $(0, 0, 0)$

65. Critical point: $(-2, 3)$

Saddle point: $(-2, 3, 1)$

67. Critical points: $(0, 0), \left(\frac{1}{6}, \frac{1}{12}\right)$

Relative minimum: $\left(\frac{1}{6}, \frac{1}{12}, -\frac{1}{432}\right)$

Saddle point: $(0, 0, 0)$

69. Critical points: $(1, 1), (-1, -1). (1, -1), (-1, 1)$

Relative minimum: $(1, 1, -2)$

Relative maximum: $(-1, -1, 6)$

Saddle points: $(1, -1, 2), (-1, 1, 2)$

71. (a) $R = -x_1^2 - 0.5x_2^2 + 100x_1 + 200x_2$

(b) $x_1 = 50, x_2 = 200$ (c) $22,500.00

73. At $\left(\frac{4}{3}, \frac{1}{3}\right)$, the relative maximum is $\frac{16}{27}$.

At $(0, 1)$, the relative minimum is 0.

75. At $\left(\frac{4}{3}, \frac{2}{3}, \frac{4}{3}\right)$, the relative maximum is $\frac{32}{27}$.

77. At $\left(\frac{4}{3}, \frac{10}{3}, \frac{14}{3}\right)$, the relative minimum is $\frac{104}{3}$.

79. At $\left(2\sqrt{2}, 2\sqrt{2}, \sqrt{2}\right)$, the relative maximum is 8.

81. $f(49.4, 253) \approx 13,202$

83. (a) $y = \frac{60}{59}x - \frac{15}{59}$ (b) 2.746

85. (a) $y = 14x + 19$ (b) 21.8 bushels/acre

87. $y = 1.71x^2 - 2.57x + 5.56$

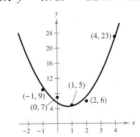

89. 1 **91.** $\frac{7}{4}$

93. $\displaystyle\int_{-2}^{2}\int_{5}^{9 - x^2} dy\, dx = \int_{5}^{9}\int_{-\sqrt{9 - y}}^{\sqrt{9 - y}} dx\, dy = \frac{32}{3}$

95. $\displaystyle\int_{-3}^{6}\int_{1/3(x + 3)}^{\sqrt{x + 3}} dy\, dx = \int_{0}^{3}\int_{3y - 3}^{y^2 - 3} dx\, dy = \frac{9}{2}$

97. $\frac{4096}{9}$ **99.** 0.0833 mi

CHAPTER TEST   *(page 1021)*

1. (a)

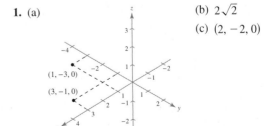

(b) $2\sqrt{2}$

(c) $(2, -2, 0)$

2. (a) 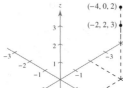   (b) 3

(−4, 0, 2)

(−2, 2, 3)

(c) $(−3, 1, 2.5)$

3. (a)   (b) $14\sqrt{2}$

(3, −7, 2)

(c) $(4, 2, −2)$

(5, 11, −6)

4. Center: $(10, −5, 5)$; radius: 5   **5.** Plane

6. Elliptic cone   **7.** Hyperbolic paraboloid

8. $f(3, 3) = 19$ **9.** $f(3, 3) = \frac{3}{2}$ **10.** $f(3, 3) = 0$

$f(1, 1) = 3$ $f(1, 1) = \frac{3}{2}$ $f(1, 1) = 0$

11. $f_x = 6x + 9y^2$; $f_x(10, −1) = 69$

$f_y = 18xy$; $f_y(10, −1) = −180$

12. $f_x = (x + y)^{1/2} + \dfrac{x}{2(x + y)^{1/2}}$; $f_x(10, −1) = \dfrac{14}{3}$

$f_y = \dfrac{x}{2(x + y)^{1/2}}$; $f_y(10, −1) = \dfrac{5}{3}$

13. Critical point: $(1, −2)$; Relative minimum: $(1, −2, −23)$

14. Critical points: $(0, 0), (1, 1), (−1, −1)$

Saddle point: $(0, 0, 0)$

Relative maxima: $(1, 1, 2), (−1, −1, 2)$

15. (a) $x = 4000$ units of labor, $y = 500$ units of capital

(b) About 128,613 units

16. $y = −1.839x^2 + 31.70x + 73.6$

17. $\frac{3}{2}$ **18.** 1   **19.** $\frac{4}{3}$ units^2 **20.** $\frac{11}{6}$

CHECKPOINTS

CHAPTER 0

SECTION 0.1   (page 5)

1. $x \geq 7$ denotes all real numbers greater than or equal to 7.

2. (a) $x \geq 5$ (b) $4 < y \leq 11$

3. 12   **4.** (a) $−|−6| = −|6|$ (b) $−|5| < |−5|$

5. 8   **6.** No

SECTION 0.2   (page 10)

1. (a) $8, −15x$ (b) $4x^2, −3y, −7$ **2.** 29   **3.** 10

4. (a) Multiplicative Identity Property

(b) Commutative Property of Addition

5. $\dfrac{11x}{12}$ **6.** $\dfrac{11}{12}$

7. Scientific calculator: $6 \;\boxed{\times}\; \boxed{(}\; 8 \;\boxed{y^x}\; 3 \;\boxed{-}\; 481 \;\boxed{)}\; \boxed{=}$

Graphing calculator: $6 \;\boxed{(}\; 8 \;\boxed{\wedge}\; 3 \;\boxed{-}\; 481 \;\boxed{)}\; \boxed{\text{ENTER}}$

8. 5.87

SECTION 0.3   (page 21)

1. 1024   **2.** $\dfrac{z^{12}}{27x^6}$ **3.** $\dfrac{7}{5}$ **4.** 3.45×10^{-3}

5. 428,000   **6.** 2000   **7.** 5   **8.** $7021.36

9. 3.8%

SECTION 0.4   (page 30)

1. 2   **2.** $3x^2\sqrt{2x}$ **3.** $3x\sqrt[3]{2x}$ **4.** $\dfrac{\sqrt[3]{2}}{2}$

5. $3 + \sqrt{3}$ **6.** 9   **7.** 4   **8.** $6\sqrt{x}$ **9.** 2.621

10. 0.388   **11.** No

SECTION 0.5   (page 40)

1. $−9x^2 + 3x + 7$; degree 2   **2.** Not a polynomial

3. $2x^2 + 5x + 4$ **4.** $3x^2 − 2x − 1$

5. $x^4 − 2x^3 + 2x^2 − 11x + 4$ **6.** $−x^2 + 9$

7. $x^2 − 8x + 16$ **8.** $x^3 − 9x^2 + 27x − 27$

9. $x^2 + 10x − y^2 + 25$ **10.** $12,282.98

11. Volume $= 4x^3 − 44x^2 + 120x$

$x = 2$ inches: $V = 96$ cubic inches

$x = 3$ inches: $V = 72$ cubic inches

SECTION 0.6   *(page 48)*

1. $(x + 1)(3x + 1)$ **2.** $x(x + 1)(x - 1)$

3. $4(5 + y)(5 - y)$ **4.** $(x - 6)^2$

5. $(y - 1)(y^2 + y + 1)$ **6.** $(x + 3)(x - 2)$

7. $(2x - 3)(x - 1)$ **8.** $(x + 1)(x^2 + 5)$

9. $(2x - 3)(x + 4)$

SECTION 0.7   *(page 55)*

1. All real numbers except $x = 5$ **2.** $\dfrac{2(x + 1)}{3}, x \neq 1$

3. $-\dfrac{3 + x}{2(x + 1)}, x \neq 1$ **4.** $\dfrac{1}{x + 1}, x \neq 2$

5. $\dfrac{8}{5}, x \neq -y$ **6.** $\dfrac{2(6 - x^2)}{3x}$ **7.** $\dfrac{9}{x}, x \neq 1$

8. $\dfrac{1}{3}, x \neq 3$

CHAPTER 1

SECTION 1.1   *(page 69)*

1. Identity   **2.** 2   **3.** $\frac{11}{3}$

4. Infinitely many solutions   **5.** No solution   **6.** 4

7. No solution   **8.** 6   **9.** 0.794   **10.** 2004 $(t = 4)$

SECTION 1.2   *(page 79)*

1. \$692.31   **2.** $S = 0.05(40,000) + 40,000$ **3.** 20%

4. 1.5%   **5.** 15 feet × 45 feet   **6.** 1.1 hours

7. 32 feet

8. \$200 was invested at 4% and \$800 was invested at 5%

9. About 1.27 feet

SECTION 1.3   *(page 93)*

1. $-3, 4$ **2.** 0, 2   **3.** -2 **4.** ± 2 **5.** $-3, 5$

6. 6 feet × 14 feet   **7.** 3 seconds   **8.** ≈ 8.5 feet

9. 2011 $(t \approx 10.82)$

SECTION 1.4   *(page 105)*

1. One real solution   **2.** $-1 \pm \sqrt{3}$ **3.** $\frac{1}{3}$

4. $-0.831, 1.511$ **5.** 2:00 P.M. $(t \approx 1.91)$

6. ≈ 5.1 seconds

SECTION 1.5   *(page 115)*

1. $0, \pm 1$ **2.** $1, \pm\sqrt{2}$ **3.** $\pm 2, \pm 1$ **4.** 12   **5.** 81

6. 1, 3   **7.** 1, 5   **8.** 20 ski club members

9. $\approx 4.5\%$ **10.** 4,010,025 copies

SECTION 1.6   *(page 127)*

1. (a) $2 \leq x < 7$; bounded

(b) $-\infty < x < 3$; unbounded

2. $x < 1$

3. $x \geq -6$ **4.** $-1 \leq x < 2$

5. $-9 \leq x \leq 5$ **6.** $x < -4$ or $x > 2$

7. More than 560 miles   **8.** At most $12\frac{1}{2}$ weeks

9. Undercharged by as much as \$0.08 or overcharged by as much as \$0.08

SECTION 1.7   *(page 139)*

1. $(-2, 1)$ **2.** $(-\infty, 1) \cup (2, \infty)$

3. (a) The solution set is empty.

(b) The solution set consists of all real numbers except 1.

4. $(-\infty, 2] \cup [3, \infty)$ **5.** At least \$35 and at most \$75

6. (a) $[-2, 2]$ (b) $(-\infty, \infty)$

CHAPTER 2

SECTION 2.1   *(page 158)*

1.

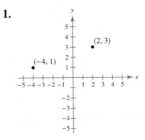

2. (a) 10   (b) $(2, -1)$ **3.** Yes

4.

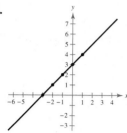

5.

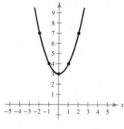

6. $\left(-\frac{2}{3}, 0\right), (0, 2)$ **7.** y-axis symmetry

8. $x^2 + (y - 1)^2 = 25$

9.

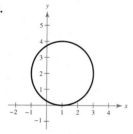

SECTION 2.2   *(page 172)*

1. $\frac{1}{2}$

2. (a) $y = -2x + 8$ (b) $y = \frac{3}{2}x + 9$

3. $y = 0.2x + 1$

4.

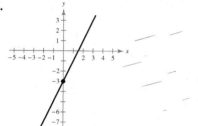

5. $y = 3x - 2$ **6.** $y = -4x + 4$

SECTION 2.3   *(page 182)*

1. The model approximates the weight of the puppy best for $t = 2$ months and worst for $t = 10$ months.

2. $y = 0.06x$ **3.** $y = 33.84615x$ **4.** 5 hours

5. 235,826 people   **6.** $V = -195t + 2300$

7. (a) $y = 17.11x + 518.2$

(b)

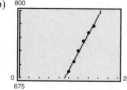

(c) $\approx 809{,}000$ employees

8.

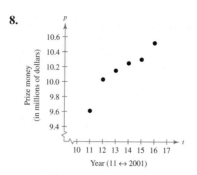

$p = 0.156t + 8.037$

SECTION 2.4   *(page 195)*

1. Yes   **2.** Yes   **3.** -3 **4.** 10, 2

5. All real numbers   **6.** $V = 4\pi h^3$ **7.** No

8. 534 cat cadavers

SECTION 2.5   *(page 208)*

1. Domain: $(-\infty, \infty)$
Range: $[-3, \infty)$

2. Yes

3. Decreasing on $\left(-\infty, -\frac{3}{2}\right)$ and increasing on $\left(-\frac{3}{2}, \infty\right)$

4. $(2, 2)$

5.

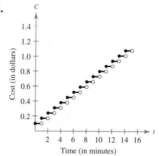

Less than 13 minutes

6. Neither

SECTION 2.6   *(page 220)*

1.

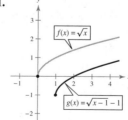

2. (a) The graph of *g* is a reflection of the graph of *f* in the *x*-axis.

(b) The graph of *h* is a reflection of the graph of *f* in the *y*-axis.

3. (a) 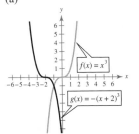   (b)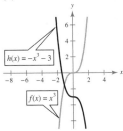

4. $h(x) = -(x - 2)^2 - 1$

5. (a) The graph of *g* is a vertical stretch of the graph of *f* by a factor of 4.

(b) The graph of *h* is a vertical shrink of the graph of *f* by a factor of $\frac{1}{4}$.

6.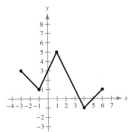

SECTION 2.7   *(page 228)*

1. $2x^2 + x - 1$ **2.** $-x - 7; -10$

3. Domain of $\frac{f}{h}$: All real numbers except $x = 3$

Domain of $\frac{h}{f}$: All real numbers except $x = 1$

4. $x^2 + 2x - 1$ **5.** $\sqrt{3 - x^2}; [-\sqrt{3}, \sqrt{3}]$

6. Answers will vary.

Sample answer: $f(x) = x^2 + 2, \ g(x) = x - 1$

$h(x) = (x - 1)^2 + 2 = f(x - 1) = f(g(x))$

7. *f* represents the number of Independent senators.

8. About 1 hour and 18 minutes

SECTION 2.8   *(page 238)*

1. $f^{-1}(x) = 6x$ **2.** $f^{-1}(x) = x - 10$

3. $(\sqrt[3]{x - 6})^3 + 6 = x; \ \sqrt[3]{(x^3 + 6)} - 6 = x$

4. $g(x)$ **5.** $f^{-1}(x) = \dfrac{x - 5}{4}$

6. 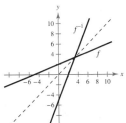   **7.**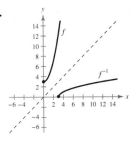

8. (a) *f* does not have an inverse function.

(b) *f* has an inverse function.

CHAPTER 3

SECTION 3.1   *(page 259)*

1.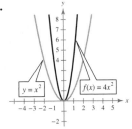

Compared with the graph of $y = x^2$, each output of $f(x) = 4x^2$ vertically stretches the graph by a factor of 4.

2. 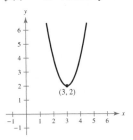   **3.**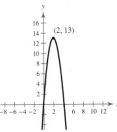

Vertex: $(3, 2)$ Vertex: $(2, 13)$

4. $f(x) = x^2 - 6x + 13$

5. ≈ 39.7 feet   **6.** ≈ 342 sparrows   **7.** 30 units

SECTION 3.2   *(page 270)*

1.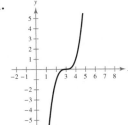

2. Falls to the left
Falls to the right

3. Rises to the left
Rises to the right

4. ± 2 **5.** $0, \pm 1$

6.

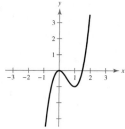

7.

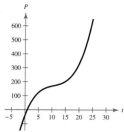

The median price was about $195,000 in 2004.

SECTION 3.3   *(page 279)*

1. $(x - 4)(x - 3)(x + 1)$ **2.** $x^2 - 2x + 4$

3. $5x^2 + 13 + \dfrac{2x + 13}{x^2 + 2x - 4}$ **4.** $2x^2 + 3x + 15 - \dfrac{5}{x - 5}$

5. 3

6. -2 | 1   6   7 -6 -8
            -2 -8 2   8

      1   4 -1 -4 0

-4 | 1   4 -1 -4
        -4 0   4

      1   0 -1 0

$f(x) = (x + 2)(x + 4)(x^2 - 1)$
$\quad\ = (x + 2)(x + 4)(x + 1)(x - 1)$

7. $\approx 10.8\%$

SECTION 3.4   *(page 289)*

1. No rational zeros   **2.** $-2, 1$ **3.** $-2, -1, \frac{1}{2}$

4. -2 **5.** The function has a zero between 1.3 and 1.4.

6. -1.290 **7.** $-0.247, 1.445, 2.802$ **8.** $x \approx 1.89$

9. $\approx \$289,000$

SECTION 3.5   *(page 305)*

1. (a) $5 + i$ (b) $-5i$

2. (a) $-8 + 12i$ (b) $16 - 30i$ **3.** $4 + i$

4. $4 + 3i$ **5.** $-\dfrac{3}{2} \pm \dfrac{\sqrt{7}}{2} i$

6.

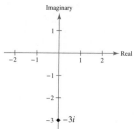

7. The complex number -3 is not in the Mandelbrot Set because for $c = -3$, the corresponding Mandelbrot sequence is $-3, 6, 33, 1086, 1, 179, 393, ...$, which is unbounded.

SECTION 3.6   *(page 315)*

1. Four zeros: $\pm \sqrt{6}, \pm \sqrt{6} i$

2. (a) $\pm 1, \pm 3i$; $f(x) = (x - 1)(x + 1)(x - 3i)(x + 3i)$

(b) $3, 3, -1, i, -i$;
$\quad g(x) = (x - 3)(x - 3)(x + 1)(x - i)(x + i)$

3. Answers will vary. Sample answer: $f(x) = x^4 - 5x^2 - 36$

4. (a) $(x^2 + 4)(x^2 - 3)$ (b) $(x^2 + 4)(x - \sqrt{3})(x + \sqrt{3})$

(c) $(x + 2i)(x - 2i)(x - \sqrt{3})(x + \sqrt{3})$

(d) Two irrational zeros and two imaginary zeros.

5. $\frac{5}{3}, \pm 4i$

SECTION 3.7   *(page 322)*

1. The domain of f is all real numbers except $x = 1$. As x approaches 1 from the left, $f(x)$ decreases without bound. As x approaches 1 from the right, $f(x)$ increases without bound.

2. Horizontal asymptote: $y = 1$
No vertical asymptotes.

3.

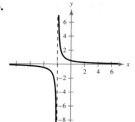

4.

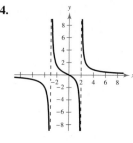

5. 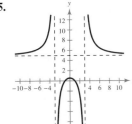 **6.**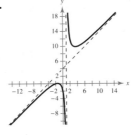

7. $1,066,667   **8.** ≈ 6.9 acres per person

CHAPTER 4

SECTION 4.1   *(page 342)*

1. 4.134

2. 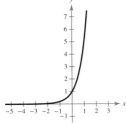   **3.**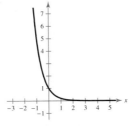

4. 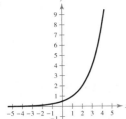   **5.** 403.4287935

6.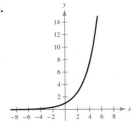

7. $7938.78   **8.** ≈ 9.971 pounds

SECTION 4.2   *(page 354)*

1. −2   **2.** 2.301   **3.** 0

4. 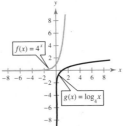   **5.**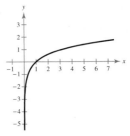

6. 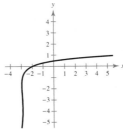   **7.** (a) 7   (b) 0

8. −2.303   **9.** $(-5, \infty)$

10. 103.0 in., 114.0 in., 128.2 in., 132.8 in., 141.7 in.

SECTION 4.3   *(page 364)*

1. 1.936   **2.** 1.936   **3.** $2 \log_{10} 5 - \log_{10} 3$

4. $-\ln \dfrac{2}{e} = -(\ln 2 - \ln e)$

$$= -\ln 2 + \ln e$$

$$= 1 - \ln 2$$

5. $\ln 2 + \ln m + 2 \ln n$ **6.** $\log_{10} \dfrac{(x+1)^2}{(x-1)^3}$

7. $\ln y = \frac{1}{2} \ln x$ **8.** 30 decibels

SECTION 4.4   *(page 373)*

1. (a) 4   (b) 216   **2.** $\log_6 84 \approx 2.473$

3. $\log_{10} 38 \approx 1.580$ **4.** $\frac{1}{2}(7 + \log_4 24) \approx 4.646$

5. $\ln 3 \approx 1.099, \ln 4 \approx 1.386$ **6.** $\frac{81}{2}$ **7.** 3

8. $e^4 \approx 54.598$ **9.** 4   **10.** ≈ 2.70 years

11. 2004 $(t \approx 14.03)$

SECTION 4.5   *(page 384)*

1. 2015 $(t \approx 24.83)$ **2.** $y = 3e^{0.19617x}$

3. ≈ 7681 years

4.

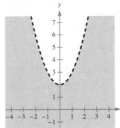

Average score: 503

5. 9 days   **6.** $\approx 2{,}511{,}886$

7. 1×10^{-5} mole of hydrogen per liter

8. Logistic growth model

CHAPTER 5

SECTION 5.1   *(page 406)*

1. $(5, 1)$

2. $11{,}500 is invested at 9% and $3500 is invested at 11%.

3. $(-4, -7), (2, 5)$ **4.** No solution   **5.** $(4, 0)$

6. ≈ 5455 pairs of shoes   **7.** Plan A

SECTION 5.2   *(page 415)*

1. $\left(2, -\frac{1}{3}\right)$ **2.** $(3, 2)$ **3.** $(-3, -3)$ **4.** $(1, -4)$

5. No solution   **6.** $(a, -a - 5)$

7. Speed of plane: ≈ 471.18 miles per hour
Speed of wind: ≈ 16.63 miles per hour

8. $(130{,}000, 22)$

SECTION 5.3   *(page 427)*

1. $(7, 2, 2)$ **2.** $\left(-\frac{1}{2}, \frac{1}{2}\right)$ **3.** $(1, 2, 3)$

4. No solution   **5.** $\left(-\frac{3}{5}a + \frac{3}{10}, -\frac{9}{5}a + \frac{2}{5}, a\right)$

6. $\left(\frac{1}{4}a, \frac{11}{4}a - 1, a\right)$

7. Answers will vary. Sample answer: $55,000 in certificates of deposit, $185,000 in municipal bonds, $105,000 in blue-chip stocks, and $15,000 in growth or speculative stocks

8. $y = 2x^2 - 2x + 3$

SECTION 5.4   *(page 441)*

1.

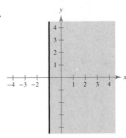

2.

3.

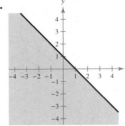

4.

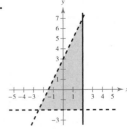

5.

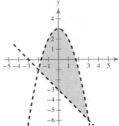

6. Consumer surplus: $845,000
Producer surplus: $845,000

7. No. The combination of 4 cups of dietary drink X and 1 cup of dietary drink Y does not meet all the minimum daily requirements.

SECTION 5.5   *(page 452)*

1. Maximum value at $(0, 3)$: 9

2. (a) Maximum value at $(27, 0)$: 135

(b) Minimum value at $(0, 0)$: 0

3. No maximum value

4. The maximum profit would be $2925, and it would occur at monthly production levels of 1050 units of product I and 150 units of product II.

5. The minimum cost would be $0.56 per day, and it would occur when 1 cup of drink X and 4 cups of drink Y were consumed each day.

CHAPTER 6

SECTION 6.1   *(page 468)*

1. 2×3 **2.** Multiply the second row by $\frac{1}{3}$.

3. $\begin{cases} x - 2y + 5z = 3 \\ \quad\quad y + 4z = -3 \\ \quad\quad\quad\quad z = 2 \end{cases}$

$(-29, -11, 2)$

4. Row-echelon form   **5.** $(-1, 0, -1, 3)$

6. No solution   **7.** $(6, 5, 5)$

8. It is not the same row-echelon form, but it does yield the same solution found in Example 8.

9. $(-9a - 10, -5a - 2, a)$

SECTION 6.2   *(page 482)*

1. $a_{11} = 5, a_{12} = 2, a_{21} = -1, a_{22} = 3$

2. $\begin{bmatrix} 6 & -2 \\ -2 & -3 \end{bmatrix}$

3. (a) $\begin{bmatrix} 4 & 8 & -2 \\ 0 & 2 & 6 \\ -6 & 4 & 10 \end{bmatrix}$ (b) $\begin{bmatrix} 4 & 2 & -5 \\ -7 & 6 & 5 \\ -8 & 4 & 12 \end{bmatrix}$

4. $\begin{bmatrix} 1 & 1 \\ 12 & -4 \end{bmatrix}$ **5.** $\begin{bmatrix} 5 & \frac{1}{2} \\ \frac{1}{2} & 3 \end{bmatrix}$ **6.** $\begin{bmatrix} 8 & -8 \\ 4 & -16 \\ -10 & 4 \end{bmatrix}$

7. $\begin{bmatrix} -3 & -22 \\ 3 & 10 \\ -5 & 10 \end{bmatrix}$ **8.** Not possible

9. $AB = [6], \ BA = \begin{bmatrix} 3 & -1 \\ -9 & 3 \end{bmatrix}$

10. $\begin{bmatrix} -2 & -3 \\ 6 & 1 \end{bmatrix}\begin{bmatrix} x_1 \\ x_2 \end{bmatrix} = \begin{bmatrix} -4 \\ -36 \end{bmatrix}$

$X = \begin{bmatrix} -7 \\ 6 \end{bmatrix}$

11. Company B

SECTION 6.3   *(page 497)*

1. $AB = I$ and $BA = I$

2. $\begin{bmatrix} 1 & 1 \\ -5 & -4 \end{bmatrix}$ **3.** $\begin{bmatrix} -4 & -2 & 5 \\ -2 & -1 & 2 \\ -1 & 0 & 1 \end{bmatrix}$

4. $\begin{bmatrix} \frac{1}{10} & \frac{3}{10} \\ \frac{2}{5} & \frac{1}{5} \end{bmatrix}$ **5.** $(-1, 2, 1)$

SECTION 6.4   *(page 509)*

1. -7

2. Minors: $M_{11} = -9, M_{12} = -10, M_{13} = 2, M_{21} = 5,$
$M_{22} = -2, M_{23} = -3, M_{31} = 13, M_{32} = 5,$
$M_{33} = -1$

Cofactors: $C_{11} = -9, C_{12} = 10, C_{13} = 2, C_{21} = -5,$
$C_{22} = -2, C_{23} = 3, C_{31} = 13, C_{32} = -5,$
$C_{33} = -1$

3. -32 **4.** -133 **5.** 27 **6.** -30

SECTION 6.5   *(page 518)*

1. 14 **2.** Not collinear   **3.** $x - y + 2 = 0$

4. $\begin{bmatrix} 15 & 23 & 12 \\ 14 & 15 & 3 \end{bmatrix}\begin{bmatrix} 19 & 0 & 1 \\ 20 & 21 & 18 \end{bmatrix}\begin{bmatrix} 18 & 5 & 0 \\ 14 & 1 & 12 \end{bmatrix}$

5. $110, -39, -59, 25, -21, -3, 23, -18, -5, 47, -20,$
$-24, 149, -56, -75, 87, -38, -37$

6. OWLS ARE NOCTURNAL

CHAPTER 7

SECTION 7.1

1. 6

2. (a) 4 (b) Does not exist   (c) 4

3. 5

4. 12

5. 7

6. $\frac{1}{4}$

7. (a) -1 (b) 1

8. 1

9. $\lim\limits_{x \to 1^-} f(x) = 12$ and $\lim\limits_{x \to 1^+} f(x) = 14$
$\lim\limits_{x \to 1^-} f(x) \neq \lim\limits_{x \to 1^+} f(x)$

10. Does not exist

SECTION 7.2

1. (a) f is continuous on the entire real line.
(b) f is continuous on the entire real line.

2. (a) f is continuous on $(-\infty, 1)$ and $(1, \infty)$.
(b) f is continuous on $(-\infty, 2)$ and $(2, \infty)$.
(c) f is continuous on the entire real line.

3. f is continuous on $[2, \infty)$.

4. f is continuous on $[-1, 5]$.

5.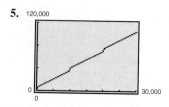

6. $A = 10,000(1 + 0.02)^{[[4t]]}$

SECTION 7.3

1. 3

2. For the months on the graph to the left of July, the tangent lines have positive slopes. For the months to the right of July, the tangent lines have negative slopes. The average daily temperature is increasing prior to July and decreasing after July.

3. 4

4. 2

5. $m = 8x$

At $(0, 1)$, $m = 0$.

At $(1, 5)$, $m = 8$.

6. $2x - 5$

7. $-\dfrac{4}{t^2}$

SECTION 7.4

1. (a) 0 (b) 0 (c) 0 (d) 0

2. (a) $4x^3$ (b) $-\dfrac{3}{x^4}$ (c) $2w$ (d) $-\dfrac{1}{t^2}$

3. $f'(x) = 3x^2$

$m = f'(-1) = 3$;

$m = f^{-1}(0) = 0$;

$m = f^{-1}(1) = 3$

4. (a) $8x$ (b) $\dfrac{8}{\sqrt{x}}$

5. (a) $\frac{1}{4}$ (b) $-\frac{2}{5}$

6. (a) $-\dfrac{9}{2x^3}$ (b) $-\dfrac{9}{8x^3}$

7. (a) $\dfrac{\sqrt{5}}{2\sqrt{x}}$ (b) $\dfrac{1}{3x^{2/3}}$

8. -1

9. $y = -x + 2$

10. $R'(13) \approx \$1.18/\text{yr}$

SECTION 7.5

1. (a) $0.5\overline{6}$ mg/ml/min

(b) 0 mg/ml/min

(c) -1.5 mg/ml/min

2. (a) -16 ft/sec (b) -48 ft/sec

(c) -80 ft/sec

3. When $t = 1.75$, $h'(1.75) = -56$ ft/sec.

When $t = 2$, $h'(2) = -64$ ft/sec.

4. $h = -16t^2 + 16t + 12$

$v = h' = -32t + 16$

5. When $x = 100$, $\dfrac{dP}{dx} = \$16/\text{unit}$.

Actual gain = \$16.06

6. $p = 11 - \dfrac{x}{2000}$

7. Revenue: $R = 2000x - 4x^2$

Marginal revenue: $\dfrac{dR}{dx} = 2000 - 8x$

8. $\dfrac{dP}{dx} = \$1.44/\text{unit}$

Actual increase in profit $\approx \$1.44$

SECTION 7.6

1. $-27x^2 + 12x + 24$

2. $\dfrac{2x^2 - 1}{x^2}$

3. (a) $18x^2 + 30x$ (b) $12x + 15$

4. $-\dfrac{22}{(5x - 2)^2}$

5. $y = \frac{8}{25}x - \frac{4}{5}$;

6. $\dfrac{-3x^2 + 4x + 8}{x^2(x + 4)^2}$

7. (a) $\frac{2}{5}x + \frac{4}{5}$ (b) $3x^3$

8. $\dfrac{2x^2 - 4x}{(x - 1)^2}$

9.

t	0	1	2	3	4	5	6	7
$\dfrac{dP}{dt}$	0	-50	-16	-6	-2.77	-1.48	-0.88	-0.56

As t increases, the rate at which the blood pressure drops decreases.

SECTION 7.7

1. (a) $u = g(x) = x + 1$

$y = f(u) = \dfrac{1}{\sqrt{u}}$

(b) $u = g(x) = x^2 + 2x + 5$

$y = f(u) = u^3$

2. $6x^2(x^3 + 1)$

3. $4(2x + 3)(x^2 + 3x)^3$

4. $y = \frac{1}{3}x + \frac{8}{3}$

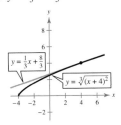

5. (a) $-\dfrac{8}{(2x + 1)^2}$ (b) $-\dfrac{6}{(x - 1)^4}$

6. $\dfrac{x(3x^2 + 2)}{\sqrt{x^2 + 1}}$

7. $-\dfrac{12(x + 1)}{(x - 5)^3}$

8. About $3.27/yr

CHAPTER 8

SECTION 8.1

1. $f'(x) = 18x^2 - 4x$, $f''(x) = 36x - 4$,

$f'''(x) = 36$, $f^{(4)}(x) = 0$

2. 18

3. $\dfrac{120}{x^6}$

4. $s(t) = -16t^2 + 64t + 80$

$v(t) = s'(t) = -32t + 64$

$a(t) = v'(t) = s''(t) = -32$

5. -9.8 m/sec^2

6.

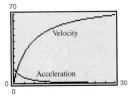

Acceleration approaches zero.

SECTION 8.2

1. $-\dfrac{2}{x^3}$

2. (a) $12x^2$ (b) $6y\dfrac{dy}{dx}$ (c) $1 + 5\dfrac{dy}{dx}$ (d) $y^3 + 3xy^2\dfrac{dy}{dx}$

3. $\frac{3}{4}$

4. $\dfrac{dy}{dx} = -\dfrac{x - 2}{y - 1}$

5. $\frac{5}{9}$

6. $\dfrac{dx}{dp} = -\dfrac{2}{p^2(0.002x + 1)}$

SECTION 8.3

1. 9

2. $12\pi \approx 37.7$ ft^2/sec

3. $72\pi \approx 226.2$ in.^{2}/min

4. $1500/day

5. $28,400/wk

SECTION 8.4

1. $f'(x) = 4x^3$

$f'(x) < 0$ if $x < 0$; therefore, f is decreasing on $(-\infty, 0)$.

$f'(x) > 0$ if $x > 0$; therefore, f is increasing on $(0, \infty)$.

2. $\dfrac{dW}{dt} = 0.116t + 0.19 > 0$ when $5 \le t \le 14$,

which implies that the consumption of bottled water was increasing from 1995 through 2004.

3. Increasing on $(-\infty, -2)$ and $(2, \infty)$

Decreasing on $(-2, 2)$

4. Increasing on $(0, \infty)$

Decreasing on $(-\infty, 0)$

5. Because $f'(x) = -3x^2 = 0$ when $x = 0$ and because f is decreasing on $(-\infty, 0) \cup (0, \infty)$, f is decreasing on $(-\infty, \infty)$.

6. $(0, 3000)$

SECTION 8.5

1. Relative maximum at $(-1, 5)$

   Relative minimum at $(1, -3)$

2. Relative minimum at $(3, -27)$

3. Relative maximum at $(1, 1)$

   Relative minimum at $(0, 0)$

4. Absolute maximum at $(0, 10)$

   Absolute minimum at $(4, -6)$

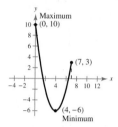

Checkpoint 5

x (units)	24,000	24,200	24,300	24,400
P (profit)	$24,760	$24,766	$24,767.50	$24,768

x (units)	24,500	24,600	24,800	25,000
P (profit)	$24,767.50	$24,766	$24,760	$24,750

SECTION 8.6

1. (a) $f'' = -4$; because $f''(x) < 0$ for all x, f is concave downward for all x.

   (b) $f''(x) = \dfrac{1}{2x^{3/2}}$; because $f''(x) > 0$ for all $x > 0$, f is concave upward for all $x > 0$.

2. Because $f''(x) > 0$ for $x < -\dfrac{2\sqrt{3}}{3}$ and $x > \dfrac{2\sqrt{3}}{3}$, f is concave upward on $\left(-\infty, -\dfrac{2\sqrt{3}}{3}\right)$ and $\left(\dfrac{2\sqrt{3}}{3}, \infty\right)$.

   Because $f''(x) < 0$ for $-\dfrac{2\sqrt{3}}{3} < x < \dfrac{2\sqrt{3}}{3}$, f is concave downward on $\left(-\dfrac{2\sqrt{3}}{3}, \dfrac{2\sqrt{3}}{3}\right)$.

3. f is concave upward on $(-\infty, 0)$ and $(1, \infty)$.

   f is concave downward on $(0, 1)$.

   Points of inflection: $(0, 1), (1, 0)$

4. Relative minimum: $(3, -26)$

5. Point of diminishing returns: $x = \$150$ thousand

CHAPTER 9

SECTION 9.1

1.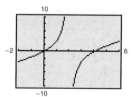

   Maximum volume $= 108$ in.^{3}

2. $x = 6, y = 12$

3. $\left(\sqrt{\dfrac{1}{2}}, \dfrac{7}{2}\right)$ and $\left(-\sqrt{\dfrac{1}{2}}, \dfrac{7}{2}\right)$

4. 8 in. by 12 in.

SECTION 9.2

1. 125 units yield a maximum revenue of $1,562,500.

2. 400 units

3. $6.25/unit

4. $4.00

5. Demand is elastic when $0 < x < 144$.

   Demand is inelastic when $144 < x < 324$.

   Demand is of unit elasticity when $x = 144$.

SECTION 9.3

1. (a) $\displaystyle\lim_{x \to 2^-} \dfrac{1}{x-2} = -\infty$; $\displaystyle\lim_{x \to 2^+} \dfrac{1}{x-2} = \infty$

   (b) $\displaystyle\lim_{x \to -3^-} \dfrac{-1}{x+3} = \infty$; $\displaystyle\lim_{x \to -3^+} \dfrac{-1}{x+3} = -\infty$

2. $x = 0, x = 4$

3. $x = 3$

4. $\displaystyle\lim_{x \to 2^-} \dfrac{x^2 - 4x}{x-2} = \infty$; $\displaystyle\lim_{x \to 2^+} \dfrac{x^2 - 4x}{x-2} = -\infty$

5. 2

6. (a) $y = 0$

(b) $y = \frac{1}{2}$

(c) No horizontal asymptote

7. $C = 0.75x + 25,000$

$\overline{C} = 0.75 + \dfrac{25,000}{x}$

$\lim\limits_{x \to \infty} \overline{C} = \$0.75/\text{unit}$

8. No, the cost function is not defined at $p = 100$, which implies that it is not possible to remove 100% of the pollutants.

SECTION 9.4

1.

	$f(x)$	$f'(x)$	$f''(x)$	Shape of graph
x in $(-\infty, -1)$		$-$	$+$	Decreasing, concave upward
$x = -1$	-32	0	$+$	Relative minimum
x in $(-1, 1)$		$+$	$+$	Increasing, concave upward
$x = 1$	-16	$+$	0	Point of inflection
x in $(1, 3)$		$+$	$-$	Increasing, concave downward
$x = 3$	0	0	$-$	Relative maximum
x in $(3, \infty)$		$-$	$-$	Decreasing, concave downward

2.

	$f(x)$	$f'(x)$	$f''(x)$	Shape of graph
x in $(-\infty, 0)$		$-$	$+$	Decreasing, concave upward
$x = 0$	5	0	0	Point of inflection
x in $(0, 2)$		$-$	$-$	Decreasing, concave downward
$x = 2$	-11	$-$	0	Point of inflection
x in $(2, 3)$		$-$	$+$	Decreasing, concave upward
$x = 3$	-22	0	$+$	Relative minimum
x in $(3, \infty)$		$+$	$+$	Increasing, concave upward

3.

	$f(x)$	$f'(x)$	$f''(x)$	Shape of graph
x in $(-\infty, 0)$		$+$	$-$	Increasing, concave downward
$x = 0$	0	0	$-$	Relative maximum
x in $(0, 1)$		$-$	$-$	Decreasing, concave downward
$x = 1$	Undef.	Undef.	Undef.	Vertical asymptote
x in $(1, 2)$		$-$	$+$	Decreasing, concave upward
$x = 2$	4	0	$+$	Relative minimum
x in $(2, \infty)$		$+$	$+$	Increasing, concave upward

4.

	$f(x)$	$f'(x)$	$f''(x)$	Shape of graph
x in $(-\infty, -1)$		$+$	$+$	Increasing, concave upward
$x = -1$	Undef.	Undef.	Undef.	Vertical asymptote
x in $(-1, 0)$		$+$	$-$	Increasing, concave downward
$x = 0$	-1	0	$-$	Relative maximum
x in $(0, 1)$		$-$	$-$	Decreasing, concave downward
$x = 1$	Undef.	Undef.	Undef.	Vertical asymptote
x in $(1, \infty)$		$-$	$+$	Decreasing, concave upward

5.

	$f(x)$	$f'(x)$	$f''(x)$	Shape of graph
x in $(0, 1)$		$-$	$+$	Decreasing, concave upward
$x = 1$	-4	0	$+$	Relative minimum
x in $(1, \infty)$		$+$	$+$	Increasing, concave upward

SECTION 9.5

1. $dy = 0.32$; $\Delta y = 0.32240801$

2. $dR = \$22$; $\Delta R = \$21$

3. $dP = \$10.96$; $\Delta P = \$10.98$

4. (a) $dy = 12x^2\,dx$ (b) $dy = \frac{2}{3}\,dx$

(c) $dy = (6x - 2)\,dx$ (d) $dy = -\frac{2}{x^3}\,dx$

5. $S = 1.96\pi$ in.$^2 \approx 6.1575$ in.^{2}

$dS = \pm 0.056\pi$ in.$^2 \approx \pm 0.1759$ in.^{2}

CHAPTER 10

SECTION 10.1

1. (a) 243   (b) 3   (c) 64

(d) 8   (e) $\frac{1}{2}$ (f) $\sqrt{10}$

2. (a) 5.453×10^{-13} (b) 1.621×10^{-13}

(c) 2.629×10^{-14}

3.

x	-2	-1	0	1	2
$f(x)$	$e^2 \approx 7.389$	$e \approx 2.718$	1	$\frac{1}{e} \approx 0.368$	$\frac{1}{e^2} \approx 0.135$

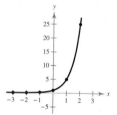

4.

x	-3	-2	-1	0	1	2	3
$f(x)$	9	5	3	2	$\frac{3}{2}$	$\frac{5}{4}$	$\frac{9}{8}$

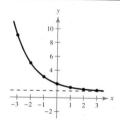

Horizontal asymptote: $y = 1$

SECTION 10.2

1.

x	-2	-1	0	1	2
$f(x)$	$e^2 \approx 7.389$	$e \approx 2.718$	1	$\frac{1}{e} \approx 0.368$	$\frac{1}{e^2} \approx 0.135$

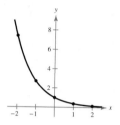

2. After 0 h, $y = 1.25$ g.

After 1 h, $y \approx 1.338$ g.

After 10 h, $y \approx 1.498$ g.

$\lim\limits_{t \to \infty} \dfrac{1.50}{1 + 0.2e^{-0.5t}} = 1.50$ g

3. (a) \$4870.38   (b) \$4902.71

(c) \$4918.66   (d) \$4919.21

All else being equal, the more often interest is compounded, the greater the balance.

4. (a) 7.12%   (b) 7.25%

5. \$16,712.90

SECTION 10.3

1. At $(0, 1)$, $y = x + 1$.

At $(1, e)$, $y = ex$.

2. (a) $3e^{3x}$ (b) $-\dfrac{6x^2}{e^{2x^3}}$ (c) $8xe^{x^2}$ (d) $-\dfrac{2}{e^{2x}}$

3. (a) $xe^x(x + 2)$ (b) $\frac{1}{2}(e^x - e^{-x})$

(c) $\dfrac{e^x(x - 2)}{x^3}$ (d) $e^x(x^2 + 2x - 1)$

4.

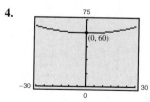

5. \$18.39/unit (80,000 units)

1.

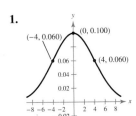

Points of inflection: $(-4, 0.060)$, $(4, 0.060)$

SECTION 10.4

1.

x	-1.5	-1	-0.5	0	0.5	1
$f(x)$	-0.693	0	0.405	0.693	0.916	1.099

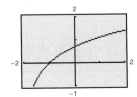

2. (a) 3 (b) $x + 1$

3. (a) $\ln 2 - \ln 5$ (b) $\frac{1}{3}\ln(x + 2)$

(c) $\ln x - \ln 5 - \ln y$ (d) $\ln x + 2\ln(x + 1)$

4. (a) $\ln x^4 y^3$ (b) $\ln \dfrac{x + 1}{(x + 3)^2}$

5. (a) $\ln 6$ (b) $5\ln 5$

6. (a) e^4 (b) e^3

7 7.9 yr

SECTION 10.5

1. $\dfrac{1}{x}$

2. (a) $\dfrac{2x}{x^2 - 4}$ (b) $x(1 + 2\ln x)$

(c) $\dfrac{2\ln x - 1}{x^3}$

3. $\dfrac{1}{3(x + 1)}$

4. $\dfrac{2}{x} + \dfrac{x}{x^2 + 1}$

5. Relative minimum: $(2, 2 - 2\ln 2) \approx (2, 0.6137)$

6. $\dfrac{dp}{dt} = -1.3\%/\text{mo}$

The average score would decrease at a greater rate than the model in Example 6.

7. (a) 4 (b) -2 (c) -5 (d) 3

8. (a) 2.322 (b) 2.631 (c) 3.161 (d) -0.5

9.

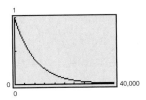

As time increases, the derivative approaches 0. The rate of change of the amount of carbon isotopes is proportional to the amount present.

SECTION 10.6

1. About 2113.7 yr

2. $y = 25e^{0.6931t}$

3. $r = \frac{1}{8}\ln 2 \approx 0.0866$ or 8.66%

4. About 12.42 mo

CHAPTER 1

SECTION 11.1

1. (a) $\displaystyle\int 3\,dx = 3x + C$

(b) $\displaystyle\int 2x\,dx = x^2 + C$

(c) $\displaystyle\int 9t^2\,dt = 3t^3 + C$

2. (a) $5x + C$ (b) $-r + C$ (c) $2t + C$

3. $\frac{5}{2}x^2 + C$

4. (a) $-\dfrac{1}{x} + C$ (b) $\dfrac{3}{4}x^{4/3} + C$

5. (a) $\frac{1}{2}x^2 + 4x + C$ (b) $x^4 - \frac{5}{2}x^2 + 2x + C$

6. $\frac{2}{3}x^{3/2} + 4x^{1/2} + C$

7. General solution: $F(x) = 2x^2 + 2x + C$

Particular solution: $F(x) = 2x^2 + 2x + 4$

8. $s(t) = -16t^2 + 32t + 48$. The ball hits the ground 3 seconds after it is thrown, with a velocity of -64 feet per second.

9. $C = -0.01x^2 + 28x + 12.01$

$C(200) = \$5212.01$

SECTION 11.2

1. (a) $\dfrac{(x^3 + 6x)^3}{3} + C$ (b) $\dfrac{2}{3}(x^2 - 2)^{3/2} + C$

2. $\dfrac{1}{36}(3x^4 + 1)^3 + C$

3. $2x^9 + \dfrac{12}{5}x^5 + 2x + C$

4. $\dfrac{5}{3}(x^2 + 1)^{3/2} + C$

5. $-\dfrac{1}{3}(1 - 2x)^{3/2} + C$

6. $\dfrac{1}{3}(x^2 + 4)^{3/2} + C$

7. About $32,068

SECTION 11.3

1. (a) $3e^x + C$ (b) $e^{5x} + C$

   (c) $e^x - \dfrac{x^2}{2} + C$

2. $\dfrac{1}{2}e^{2x+3} + C$

3. $2e^{x^2} + C$

4. (a) $2\ln|x| + C$ (b) $\ln|x^3| + C$ (c) $\ln|2x + 1| + C$

5. $\dfrac{1}{4}\ln|4x + 1| + C$

6. $\dfrac{3}{2}\ln(x^2 + 4) + C$

7. (a) $4x - 3\ln|x| - \dfrac{2}{x} + C$

   (b) $2\ln(1 + e^x) + C\,dx$

   (c) $\dfrac{x^2}{2} + x + 3\ln|x + 1| + C$

SECTION 11.4

1. $\dfrac{1}{2}(3)(12) = 18$

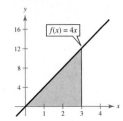

2. $\dfrac{22}{3}$ units²

3. 68

4. (a) $\dfrac{1}{4}(e^4 - 1) \approx 13.3995$

   (b) $-\ln 5 + \ln 2 \approx -0.9163$

5. $\dfrac{13}{2}$

6. (a) About $14.18 (b) $141.79

7. $13.70

8. (a) $\dfrac{2}{5}$ (b) 0

9. About $12,295.62

SECTION 11.5

1. $\dfrac{8}{3}$ units²

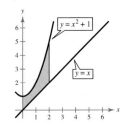

2. $\dfrac{32}{3}$ units²

3. $\dfrac{9}{2}$ units²

4. $\dfrac{253}{12}$ units²

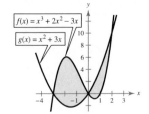

5. Consumer surplus: 40

   Producer surplus: 20

6. The company can save $39.36 million.

SECTION 11.6

1. $\dfrac{37}{8}$ units²

2. 0.436 unit²

3. 5.642 units²

4. About 1.463

CHAPTER 12

SECTION 12.1

1. $\dfrac{1}{2}xe^{2x} - \dfrac{1}{4}e^{2x} + C$

2. $\dfrac{x^2}{2}\ln x - \dfrac{1}{4}x^2 + C$

3. $\dfrac{d}{dx}[x\ln x - x + C] = x\left(\dfrac{1}{x}\right) + \ln x - 1$

   $\qquad\qquad\qquad\qquad = \ln x$

4. $e^x(x^3 - 3x^2 + 6x - 6) + C$

5. $e - 2$

6. $538,145

7. $721,632.08

SECTION 12.2

1. $\dfrac{5}{x+3} - \dfrac{4}{x+4}$

2. $\ln|x(x+2)^2| + \dfrac{1}{x+2} + C$

3. $\dfrac{1}{2}x^2 - 2x - \dfrac{1}{x} + 4\ln|x+1| + C$

4. $ky(1-y) = \dfrac{kbe - kt}{(1 + be^{-kt})^2}$

$$y = (1 + be^{-kt})^{-1}$$

$$\dfrac{dy}{dt} = \dfrac{kbe^{-kt}}{(1 + be^{-kt})^2}$$

Therefore, $\dfrac{dy}{dt} = ky(1-y)$

5. $y = 4$

6. $y = \dfrac{4000}{1 + 39e^{-0.31045t}}$

SECTION 12.3

1. $\frac{2}{3}(x-4)\sqrt{2+x} + C$ (Formula 19)

2. $\sqrt{x^2 + 16} - 4\ln\left|\dfrac{4 + \sqrt{x^2+16}}{x}\right| + C$
(Formula 23)

3. $\dfrac{1}{4}\ln\left|\dfrac{x-2}{x+2}\right| + C$ (Formula 29)

4. $\frac{1}{3}[1 - \ln(1+e) + \ln 2] \approx 0.12663$
(Formula 37)

5. $x(\ln x)^2 + 2x - 2x\ln x + C$ (Formula 42)

6. About 18.2%

SECTION 12.4

1. 3.2608

2. 3.1956

3. 1.154

SECTION 12.5

1. (a) Converges; $\frac{1}{2}$ (b) Diverges

2. 1

3. $\frac{1}{2}$

4. 2

5. Diverges

6. Diverges

7. 0.0038 or $\approx 0.4\%$

8. No, you do not have enough money to start the scholarship fund because you need \$125,000. (\$125,000 > \$120,000)

CHAPTER 13

SECTION 13.1

1.

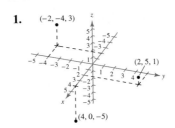

2. $2\sqrt{6}$

3. $\left(-\frac{5}{2}, 2, -2\right)$

4. $(x-4)^2 + (y-3)^2 + (z-2)^2 = 25$

5. $(x-1)^2 + (y-3)^2 + (z-2)^2 = 38$

6. Center: $(-3, 4, -1)$; radius: 6

7. $(x+1)^2 + (y-2)^2 = 16$

SECTION 13.2

1. x-intercept: $(4, 0, 0)$;

y-intercept: $(0, 2, 0)$;

z-intercept: $(0, 0, 8)$

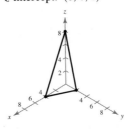

2. Hyperboloid of one sheet

xy-trace: circle, $x^2 + y^2 = 1$; yz-trace: hyperbola, $y^2 - z^2 = 1$; xz-trace: hyperbola, $x^2 - z^2 = 1$; $z = 3$ trace: circle, $x^2 + y^2 = 10$

3. (a) $\dfrac{x^2}{9} + \dfrac{y^2}{4} = z$; elliptic paraboloid

(b) $\dfrac{x^2}{4} + \dfrac{y^2}{9} - z^2 = 0$; elliptic cone

SECTION 13.3

1. (a) 0   (b) $\frac{9}{4}$

2. Domain: $x^2 + y^2 \le 9$

Range: $0 \le z \le 3$

3. Steep; nearly level

4. Alaska is mainly used for forest land. Alaska does not contain any manufacturing centers, but it does contain a mineral deposit of petroleum.

5. $f(1500, 1000) \approx 127{,}542$ units

$f(1000, 1500) \approx 117{,}608$ units

x, person-hours, has a greater effect on production.

6. (a) $M = \$733.76/\text{mo}$

(b) Total paid $= (30 \times 12) \times 733.76 = \$264{,}153.60$

SECTION 13.4

1. $\dfrac{\partial z}{\partial x} = 4x - 8xy^3$

$\dfrac{\partial z}{\partial y} = -12x^2y^2 + 4y^3$

2. $f_x(x, y) = 2xy^3;\ f_x(1, 2) = 16$

$f_y(x, y) = 3x^2y^2;\ f_y(1, 2) = 12$

3. In the x-direction: $f_x(1, -1, 49) = 8$

In the y-direction: $f_y(1, -1, 49) = -18$

4. Substitute product relationship

5. $\dfrac{\partial w}{\partial x} = xy + 2xy \ln(xz)$

$\dfrac{\partial w}{\partial y} = x^2 \ln xz$

$\dfrac{\partial w}{\partial z} = \dfrac{x^2y}{z}$

6. $f_{xx} = 8y^2$

$f_{yy} = 8x^2 + 8$

$f_{xy} = 16xy$

$f_{yx} = 16xy$

7. $f_{xx} = 0 \qquad f_{xy} = e^y \qquad f_{xz} = 2$

$f_{yx} = e^y \qquad f_{yy} = xe^y + 2 \qquad f_{yz} = 0$

$f_{zx} = 2 \qquad f_{zy} = 0 \qquad f_{zz} = 0$

SECTION 13.5

1. $f(-8, 2) = -64$: relative minimum

2. $f(0, 0) = 1$: relative maximum

3. $f(0, 0) = 0$: saddle point

4. $P(3.11, 3.81) = \$744.81$ maximum profit

5. $V\left(\frac{4}{3}, \frac{2}{3}, \frac{8}{3}\right) = \frac{64}{27}$ units^3

SECTION 13.6

1. $V\left(\frac{4}{3}, \frac{2}{3}, \frac{8}{3}\right) = \frac{64}{27}$ units^3

2. $f(187.5, 50) \approx 13{,}474$ units

3. About 26,740 units

4. $P(3.35, 4.26) = \$758.08$ maximum profit

5. $f(2, 0, 2) = 8$

SECTION 13.7

1. For $f(x)$, $S \approx 9.1$.

For $g(x)$, $S \approx 0.45715$.

The quadratic model is a better fit.

2. $f(x) = \frac{6}{5}x + \frac{23}{10}$

3. $y = 20{,}041.5t + 103{,}455.5$

In 2010, $y \approx 303{,}870.5$ subscribers

4. $y = 6.595t^2 + 143.50t + 1971.0$

In 2010, $y = \$7479$.

SECTION 13.8

1. (a) $\frac{1}{4}x^4 + 2x^3 - 2x - \frac{1}{4}$

(b) $\ln|y^2 + y| - \ln|2y|$

2. $\frac{25}{2}$

3. $\displaystyle\int_2^4 \int_1^5 dx\, dy = 8$

4. $\frac{4}{3}$

5. (a)

$R: 0 \le y \le 2$
$2y \le x \le 4$

(b) $\displaystyle\int_0^4 \int_0^{x/2} dy\, dx$

(c) $\displaystyle\int_0^2 \int_{2y}^4 dx\, dy = 4 = \int_0^4 \int_0^{x/2} dy\, dx$

6. $\displaystyle\int_{-1}^3 \int_{x^2}^{2x+3} dy\, dx = \dfrac{32}{3}$

SECTION 13.9

1. $\frac{16}{3}$

2. $e - 1$

3. $\frac{176}{15}$

4. Integration by parts

5. 3

Index